Single Variable Calculus

The Quadratic Formula

The roots of $ax^2 + bx + c = 0$ are

$$x = \frac{-b \pm \sqrt{b^2 - 4ac}}{2a}$$

Factorials

$$n! = n \times (n - 1) \times \ . \ . \ . \ \times 3 \times 2 \times 1, \qquad \binom{n}{k} = \frac{n!}{(k!(n - k)!}$$

The Binomial Theorem

$$(a + b)^n = a^n + \binom{n}{1} a^{n-1}b + \binom{n}{2} a^{n-2}b^2 + \ . \ . \ . + \binom{n}{k} a^{n-k}b^k + \ . \ . \ . + \binom{n}{n-1} ab^{n-1} + b^n$$

Trigonometric and Hyperbolic Identities

$$\cos^2 \theta + \sin^2 \theta = 1, \qquad 1 + \tan^2 \theta = \sec^2 \theta, \qquad \cosh^2 x - \sinh^2 x = 1$$

$$\cos^2 \theta = \frac{1 + \cos(2\theta)}{2}, \qquad \sin^2 \theta = \frac{1 - \cos(2\theta)}{2}$$

$$\cos^2 \frac{\theta}{2} = \frac{1 + \cos \theta}{2}, \qquad \sin^2 \frac{\theta}{2} = \frac{1 - \cos \theta}{2}$$

$$\sin(\theta + \phi) = \cos \theta \sin \phi + \sin \theta \cos \phi, \qquad \cos(\theta + \phi) = \cos \theta \cos \phi - \sin \theta \sin \phi$$

$$\sin 2\theta = 2 \cos \theta \sin \theta$$

$$\cos 2\theta = \cos^2 \theta - \sin^2 \theta = 2 \cos^2 \theta - 1 = 1 - 2 \sin^2 \theta$$

$$\sin(-\theta) = -\sin \theta \qquad \cos(-\theta) = \cos \theta$$

$$\sin\left(\frac{\pi}{2} - \theta\right) = \cos \theta \qquad \cos\left(\frac{\pi}{2} - \theta\right) = \sin \theta \qquad \tan\left(\frac{\pi}{2} - \theta\right) = \cot \theta$$

$$\sin(\pi - \theta) = \sin \theta \qquad \cos(\pi - \theta) = -\cos \theta \qquad \tan(\pi - \theta) = -\tan \theta$$

$$\sin(\pi + \theta) = -\sin \theta \qquad \cos(\pi + \theta) = -\cos \theta \qquad \tan(\pi + \theta) = \tan \theta$$

(Continued inside back cover)

Single Variable Calculus

Joe Repka

University of Toronto

WCB Wm. C. Brown Publishers
Dubuque, Iowa • Melbourne, Australia • Oxford, England

Book Team

Editor *Paula Christy-Heighton*
Developmental Editor *Theresa Grutz*
Publishing Services Specialist—Production *Julie Avery Kennedy*

Wm. C. Brown Publishers
A Division of Wm. C. Brown Communications, Inc.

Vice President and General Manager *Beverly Kolz*
Vice President, Director of Sales and Marketing *Virginia S. Moffat*
Marketing Manager *Julie Keck*
Advertising Manager *Janelle Keeffer*
Director of Production *Colleen A. Yonda*
Publishing Services Manager *Karen J. Slaght*

Wm. C. Brown Communications, Inc.

President and Chief Executive Officer *G. Franklin Lewis*
Corporate Vice President, President of WCB Manufacturing *Roger Meyer*
Vice President and Chief Financial Officer *Robert Chesterman*

Cover image © Tony Stone Worldwide/Stephen Johnson

The cover shows a computer-generated image of part of the Mandelbrot set. This set is defined in terms of a very simple formula, but it produces pictures of astounding variety and complexity. Each part of the picture reveals additional structure at higher and higher magnifications.

The Mandelbrot set is a key element in the study of Fractal Geometry, which is part of a larger field known as Chaos Theory. Many structures that occur in nature are too complicated to predict in detail but have overall behavior that can usefully be discussed in terms of Chaos Theory.

This approach is giving new insight into a wide range of problems, ranging from weather patterns to the mysteries of the brain.

Cover and interior design by York Production Services, Inc.

Copyediting, permissions, and production by York Production Services, Inc.

Illustrations by Debby Cohen

The credits section for this book begins on page 858 and is considered an extension of the copyright page.

A Times Mirror Company

Library of Congress Catalog Card Number: 92–74556

ISBN 0–697–15375–4

Printed in the United States of America by Wm. C. Brown Communications, Inc., 2460 Kerper Boulevard, Dubuque, IA 52001

10 9 8 7 6 5 4 3 2 1

. . . for Debby,

whose patience and help
and love make practically
anything *possible . . .*

Brief Contents

***Calculus with Analytic Geometry* presents all chapters.**
***Single Variable Calculus* presents Chapters 1 through 13.**

Detailed Contents

Calculus with Analytic Geometry **presents all chapters.**
Single Variable Calculus **presents Chapters 1 through 13.**

1 *Preliminaries 1*

2 *Limits and Continuity 62*

3 *Derivatives 128*

4 *Applications of Derivatives 227*

5 *Integration 346*

6 *Applications of Integration 413*

13 *Curves in the Plane* 796

14 *Vector Geometry* 824

15 *Curves and Surfaces in Space* 871

16 *Partial Derivatives 928*

17 *Multiple Integrals 1053*

18 *Vector Calculus 1141*

19 *Differential Equations 1211*

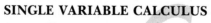

Preface

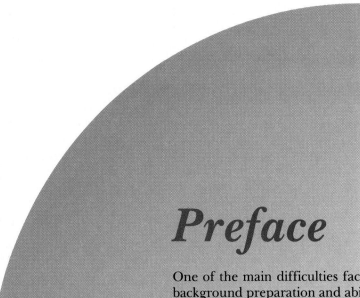

One of the main difficulties facing the teacher of calculus is the wide range of background preparation and abilities of the students, compounded by the diversity of their interests and goals. To make the subject comprehensible to the weaker students without boring the stronger ones is a difficult juggling act.

This text addresses the dilemma by presenting the material in an intuitive and reasonable way. The style is friendly without compromising accuracy and detail; the material is very accessible but has enough depth to challenge even the best students. Fine points and pitfalls are discussed in remarks in the text and are often elaborated in some of the later exercises at the end of a section; tricky computational steps and other minor difficulties are explained by notes in the margin without interrupting the exposition.

The presentation is enlivened by large numbers of examples drawn from many areas in the physical, life, and social sciences, and from business and finance. Most sections end with a "Point to Ponder," a discussion of some idea usually considered to be beyond the scope of an introductory text. Some involve technical subtleties, some involve extensions and generalizations, and others offer a glimpse into more advanced topics or other disciplines.

Many results are given complete proofs, but some proofs are omitted, especially when the level of complexity or difficulty limits their pedagogical value. In other cases, a proof is given for a special case, in order to convey the ideas without bogging down. Theoretical concepts are often discussed through examples and exercises rather than in dense technical language. Omissions are clearly acknowledged and proofs are not finessed.

Problem-solving strategies are discussed and conveniently summarized, but students are discouraged from memorizing cookbook approaches. Considerable attention is given to interpreting results, to warning the reader about complications and potential errors, and to considering shortcomings and limitations. An important part of the experience of learning calculus ought to be learning to analyze problems and to think about them in quantitative terms.

Problem sets are divided into a section of fairly routine and mechanical drill questions followed by a section of more varied exercises. These are not necessarily particularly difficult, but they are more interesting and usually require more thought. Students are encouraged not only to acquire a certain level of basic calculus skills, but also to develop simultaneously an ability to go beyond the rote and apply their skills effectively and critically.

A number of applications are given more sophisticated treatment than usual. For instance, Newton's method inspires a simple introduction to numerical analysis at a nontechnical level. Another example is a brief discussion of projectile motion using only single-variable techniques (simply by breaking the motion into horizontal and vertical components, without ever mentioning the word "vector"). This is done long before vector calculus and curves are available (though of course it makes a second appearance at that stage). In Section 7.5 there is a discussion of logistic growth, which may have more practical use than exponential growth, especially for students in the life and social sciences. This topic is not found in most other calculus texts.

Care has been taken wherever possible to avoid having later sections depend on earlier ones. Students do not usually read a calculus book through; they tend to "flip around," looking for what they need. As much as possible, the discussion in each section is self-contained. Though many of the examples in calculus relate to physics, there is an effort to keep physics intuition out of the mainstream of the exposition, to avoid alienating the students who have no physics background.

Graphics Calculators and Computers

One of the most exciting and challenging developments in mathematical education is the recent availability of graphics calculators and computers. Initially heralded as the breakthrough that would revolutionize the teaching of calculus, they have proved a little more difficult than expected to apply intelligently and effectively.

Some mathematics teachers have experienced considerable pressure to jump onto the technology bandwagon without adequate thought and preparation. Enthusiastic early claims have been pounced upon by university administrators who typically know very little about mathematics but are dazzled by anything that offers a solution to what they like to regard as "the calculus problem."

The situation is perhaps typified by "curve-sketching." Students in a traditional calculus course spend a lot of time sketching graphs. Computers and graphics calculators can now do this quickly and accurately, so why not just give the student a machine and save all that time and frustration?

The answer involves a question: Why do we have students sketch graphs? In fact, it is not particularly because we want to be able to look at the shape of the curve. Drawing graphs is an excellent way to practice most of the techniques and concepts of differential calculus. Working out a large number of examples consolidates these difficult ideas and helps the student acquire the intuition that is essential for the sophisticated use of calculus for max/min and other problems.

Eliminating this process is not necessarily a good thing. On the other hand, it is certainly possible to use the technology in concert with the traditional approach, in ways that enhance both.

A major downside to calculators and computers is the considerable amount of time required for students to learn how to use them, time *not* spent learning calculus. Another problem with in-course use of calculators and computers is that there are parts of the subject to which they are well suited and other parts where finding an appropriate use is much more problematical. But if a course is designed around expensive equipment that has been purchased by the students or by the institution, it is hard to justify letting it sit idle for substantial parts of the term. On the other hand, nothing turns students off faster than contrived examples that have been concocted just for the sake of having something.

Perhaps the single most important thing about calculators and computers is that they catch the students' attention and interest. *Anything* that does that can't be all bad. This text takes advantage of this effect by including graphics calculator or computer examples in most sections. These examples are designed to show how calculators can enhance our capabilities and in particular to illustrate some types of difficult problems that become tractable with a little help from our electronic friends.

Each example includes complete keystrokes for four of the most popular brands of graphics calculators, so that even a novice can quickly follow along and see how it works. Many include suggestions for modifications and related questions, to draw the reader into more creative use of the equipment. Even the instructor who has no computing experience can easily use the text examples to inject a little pizzazz into the classroom. The rest of the text does not depend on these examples; instructors can use them as much or as little as they want. This feature provides the perfect entry for the instructor who is interested in integrating calculators or computers into a course but who is not able to undertake the enormous task of structuring a course around them.

A few examples use a personal computer for tasks that are not possible on the graphics calculator.

Most sections end with a few exercises to be done with graphics calculators or computers. They are offered without much guidance, so in this form they are probably for the stronger students, preferably ones with some computing experience. Of course, an instructor can always include some classroom discussion or practice to help students approach these exercises.

The text will be accompanied by *The Calculator View,* by John Paulling, a more detailed supplement on the use of graphics calculators.

Calculator Example Keystrokes

The graphics calculator keystrokes are color-coded to match the keys on the calculators. In some cases light green is used to distinguish explanatory symbols from actual keystrokes. For instance, if the required key is a function key F5, and this appears with an on-screen menu item GRAPH, the keystroke F5 will be followed by (GRAPH) in parentheses and colored green. The hope is to make the code more comprehensible.

The keystrokes prepared on the Texas Instruments TI-85 will mostly work on the TI-81, either verbatim or with minor adjustments, but there are some exceptions.

Similar remarks apply to the SHARP EL-9300 and the EL-9200. In some places, the minus key − (i.e., "subtract") has to be replaced by (−) (i.e., "negative") at the beginning of expressions. Also for parametric and polar plots it may be necessary to rearrange the order of the nine RANGE parameters.

The CASIO fx-6300G is an inexpensive smaller version of the fx-7700GB, and they are quite similar, but there are some differences. A few keys have different colors on the two versions, so the printed color will not always be right for everybody. A few keys are actually different, and we have attempted to point this out. There are also a few things that the fx-7700GB can do that the fx-6300G cannot.

The Hewlett Packard HP48SX is in some ways the most sophisticated of the four types used here, but by the same token it is harder to learn and use. Programs that use the full power of the HP48SX can be nearly impenetrable to the novice, so I have opted for key sequences that are perhaps a little naive but easier to unwind. One of the main purposes of the graphics calculator examples is to make it easy for readers to get started with calculators, and I think it makes sense to sacrifice some efficiency and elegance in order to have keystrokes that can easily be modified to do related examples.

A final word of caution: All of these calculators have various settings that the user can change. If you get unexpected results or errors, it is quite likely because something has been changed from the default values. Clearing the previous example may solve the problem, and if not you should try resetting the calculator.

Organization of the Text

The organization of the text follows the standard calculus outline.

The trigonometric functions are differentiated in Chapter 3 and the derivatives of the logarithm and exponential are simply stated, so that these functions can be used in examples and exercises. Calculus courses increasingly introduce the transcendental functions early, finding that the difficulties are less serious than they used to be and the advantages are considerable.

For one thing, students are not particularly intimidated by these functions any more; they all recognize them from the buttons on their calculators. Beginning early allows the student to acquire a working familiarity with them. It also opens the door to many more interesting examples and applications, and makes various theoretical topics, such as the Chain Rule, much more interesting.

Chapters 4 and 6 give applications of differentiation and integration; there are a wide range of topics and innovative examples. Most instructors will likely include the first few sections and pick and choose among the more specialized topics. Section 4.12 covers l'Hôpital's rule, a topic that is often left until much later.

As much as possible, the chapters are written to be independent, to allow the greatest flexibility in the order of presentation. Chapter 7, on Logarithms and Exponential Functions, and Chapter 8, Techniques of Integration, are completely interchangeable, and Chapter 9 (Taylor's Formula) can be interchanged with most of Chapter 10 (Series). Section 10.8 (Taylor Series) brings the two streams together.

Chapter 7 contains an extremely brief discussion of differential equations, which are also the subject of Chapter 19.

Chapters 1 through 13 are available as *Single Variable Calculus,* while *Calculus with Analytic Geometry* contains Chapters 1 through 19.

The chapters on multivariable calculus do not assume a knowledge of linear algebra. They introduce the necessary vector techniques and determinants as they are needed.

Supplements

The Students' Solutions Manual, containing complete solutions to the odd-numbered exercises, was prepared by the author himself. The solutions are in the same style and use the same techniques as the text. They are prepared to a very high standard of accuracy and completeness.

Graphing calculator manuals, Derive® project manuals, and Mathematica® project manuals are also available for your students.

The Instructor's Solutions Manual and Instructor's Resource Manual have also been prepared by the author to follow the same approach and style found in the text.

Please contact your Wm. C. Brown Publishers representative for information on content and availability of the various supplements that support this text.

Reviewers and Acknowledgments

During its development this text has been class tested by many people; for their thoughtfulness and their suggestions, and the tact with which they have made them, I wish to thank: Stefan Bilaniuk, Trent University; Harriet Botta, York University; Julian Edward, M. P. Heble, Abe Igelfeld, Jerzy Jarosz, Val Jurdjevic, Anthony Lam, Philip Leah, Bong Lian, Dave Masson, Kunio Murasugi, D. W. Pravica, Felix Recio, Dave Roberts, Elizabeth Rowlinson, Andrew Schwartz, Luis Seco, Dipak Sen, Alexander Sobolev, Catherine Sulem, Ti Wang, all of the University of Toronto; and David Wehlau, Royal Military College.

Many others have made specific contributions or have added to my understanding of calculus and how to teach it. I want to thank: Ed Barbeau; Peter Botta; Ian Graham; George Giordano, Ryerson Polytechnical University; Barbara Japp, Clinical Physics Department, Princess Margaret Hospital; Emile Leblanc; Pierre Milman; Y. Park; Paul Selick; and John Vrolyk, George Brown College.

Among the many students who have pointed out errors in preliminary versions or made suggestions for improvements, I especially want to thank Lu Anne Burnham, Kong Wah Cora Chan, Nina Doan, David Hou, Tamarah Kagan, Shawn Langer, Wendy W. K. Lee, Dimitri Loukanidis, Ed MacDougall, Elena Poulos, Sharon M. L. Seow, Heng Sun, Erin Truscott, Cheryl Wein, and Stephanie Wiesenthal.

Various versions of the text have been thoroughly reviewed by an extensive team of experts. This is lonely work, but it is much appreciated. The reviewers were:

Dan Anderson
University of Iowa

Fred Brauer
University of Wisconsin–Madison

Mike Davidson
Cabrillo College

Kenneth Davis
Albion College

Edward Donley
Indiana University of Pennsylvania

Douglas Frank
Indian University–Penn

Dewey Furness
Ricks College

Stuart Goldenberg
California Polytechnic State University

Raymond Patrick Guzman
Pasadena City College

Gary S. Itzkowitz Ph.D.
Glassboro State College

Andrew Karantinos
The University of South Dakota

Judy Kasabian
El Camino College

John Krajewski
University of Wisconsin–Eau Claire

Robert F. Lax
Louisiana State University

Michael Mays
West Virginia University

Jim McKinney
California State Polytechnic University

Wesley J. Orser
Clark College

Dr. John Paulling
Nicholls State University

Susan Pfeifer
Butler County Community College

Claudia Pinter-Lucke
California State Polytechnic University

Benjamin F. Plybon
Miami University

Dr. Janice Rech
University of Nebraska at Omaha

Wayne Roberts
MacAlester College

William J. Roberts
Plymouth State College

Walter S. Sizer
Moorhead State University

Bruce Stephan
Webb Institute of Naval Architecture

J. R. Vanstone
University of Toronto

Gary L. Walls
University of Southern Mississippi

John L. Wulff
California State University at Sacramento

Anne Young
Loyola College

I owe an immeasurable debt to the magnificent staff in the University of Toronto Mathematics Department: Marie Bachtis, Pat Broughton, Ida Bulat, Nadia Cavaliere, Chibeck Graham, Beverley Leslie, Karin Smith, Haleh Vaez, and Yu Yuet-Wah.

The preparation of the graphics calculator examples in the text was made possible through the generosity of Texas Instruments, Hewlett Packard, CASIO, and the special help of Mike Brellisford at SHARP Electronics of Canada. I am indebted to "Uncle Dave" Roughley, who prepared the appropriate graphics output. I also want to thank "DesignCAD Keith" Campbell of American Small Business Computers for providing CAD images, and Giovanna Peel of the Ontario College of Art for all her help.

The people at York Graphic Services have done a wonderful job of putting together the manuscript with all its bells and whistles. I am especially grateful to Mary Jo Gregory, who patiently answered all my questions and more than once got a sore ear doing it.

It has been a pleasure for me to work with Wm. C. Brown Publishers; I want particularly to thank Earl McPeek, Theresa Grutz, Beth Kundert, the late Ed Jaffe, Shane Sheehan, and Wes Rafuse.

Finally I wish to thank Chandru Kriplani, without whom this book would never have been written.

Preliminaries

1

These diagrams give a visual proof of the Pythagorean Theorem. The pieces making up the two smaller squares can be reassembled into the large square.

The small figure at right shows what happens when the two smaller squares are of equal size.

INTRODUCTION

In this chapter we will discuss various concepts and formulas that will be needed later on. Many of them will be familiar, and it may not be necessary for you to spend much time reviewing them. On the other hand, it would probably be wise to read fairly carefully any sections that are new to you (a quick look at the summary at the end of the chapter should help you to identify unfamiliar topics). As you read later sections of the book, you will be able to use this chapter as a reference. Additional discussion and exercises are available in the Student's Solutions Manual.

Section 1.1 discusses the real numbers, which are fundamental to calculus and to much of the rest of mathematics. Section 1.2 is about inequalities, and Section 1.3 is about polynomials and rational functions (which simply means functions that are quotients of two polynomials). Section 1.4 discusses the concept of a function and the related idea of drawing a graph. Section 1.5 recalls some facts from geometry, principally the Pythagorean Theorem, facts about similar triangles, areas of simple regions, and volumes of simple geometric shapes. The remainder of the chapter is devoted to several specific types of graphs and functions. Section 1.6 is about conics, with particular emphasis on drawing their graphs, which is what will be most important for studying calculus. Section 1.7 is about trigonometric functions. One important item there is the idea of "radian measure" for angles; if you are not familiar with radians, you should be sure to read Section 1.7. Finally, Section 1.8 is about logarithms and exponentials.

1.1 The Real Numbers

RATIONAL NUMBERS
IRRATIONAL NUMBERS
SCIENTIFIC NOTATION

We are all familiar with the **natural numbers**, $\{1, 2, 3, 4, \ldots\}$. For many purposes it is necessary to consider negative numbers too, and this leads to the **integers** $\{\ldots, -4, -3, -2, -1, 0, 1, 2, 3, 4, \ldots\}$.

The most obvious drawback to the integers is that the quotient of two integers may not be an integer (e.g., $\frac{5}{9}$ is not an integer). This suggests considering all the numbers that are quotients of integers. These numbers are called **rational numbers**; the name is related to the word "ratio." Every rational number can be written in the form $r = \frac{a}{b}$, where a and b are integers. Of course the same rational number can be written as the quotient of many different pairs of integers; for instance,

$$\frac{-5}{9} = \frac{5}{-9} = \frac{-10}{18} = \frac{15}{-27}, \quad \text{etc.}$$

The rational number $r = \frac{a}{b}$ is said to be **in lowest terms** if a and b have no common divisors except 1 and -1. In the example above, $\frac{-5}{9}$ and $\frac{5}{-9}$ are in lowest terms (the only integers that divide ± 5 and ± 9 are 1 and -1), and the other two are not.

Rational numbers are very convenient and have the property that we can add, subtract, multiply, and divide them (except by zero) and the results will still be rational numbers.

On the other hand, the number π is not a rational number, though this is difficult to show, and as a result if the radius r of a circle is a rational number, then the area of the circle is *not* a rational number, since it equals πr^2 (see Exercise 5).

The square roots of some rational numbers are also rational numbers: $\sqrt{\frac{9}{4}} = \frac{3}{2}$, $\sqrt{\frac{1}{16}} = \frac{1}{4}$. However, there are many rational numbers whose square roots are not rational.

EXAMPLE 1 The number $\sqrt{2}$ is not a rational number.

PROOF Suppose there was some rational number $\frac{a}{b}$ so that $\left(\frac{a}{b}\right)^2 = 2$. We can assume that $\frac{a}{b}$ is in lowest terms because otherwise we could divide out the common factor in the numerator and denominator. Multiplying both sides of the equation by b^2, we find

$$a^2 = 2b^2.$$

Looking at this equation carefully, we see that the right side is an even integer, that is, divisible by 2. The same must be true of the left side, and the only way this could happen is for a to be divisible by 2 (after all if a is odd, then a^2 is odd too).

But if a is divisible by 2, then a^2 is divisible by 4, and this means that $2b^2$ is divisible by 4, so b^2 must be divisible by 2. The only way this can happen is for b to be divisible by 2.

We have just shown that a and b must both be divisible by 2, but we assumed at the beginning that they had no common divisors. This contradiction shows that our assumption that there was a rational number whose square is 2 is impossible, which proves that $\sqrt{2}$ is not rational.

This fact was first recognized by the ancient Greeks.

If we want to be able to find square roots, at least of nonnegative numbers, we have to expand the number system. The real numbers provide the solution to this and many related problems, and they are the most appropriate set of numbers for the study of calculus. The real numbers $\mathbb{R}$ are frequently represented by drawing a horizontal straight line, called the **real line**. We mark a point on it and label it 0 (zero); it is often in the middle of the line, but not always. We mark other points evenly spaced to the right and left of 0 and label them as the integers. To the right of 0 we place 1, 2, 3, and so on, in that order, and to the left of 0 we place -1, then -2, then -3, and so on. See Figure 1.1.1.

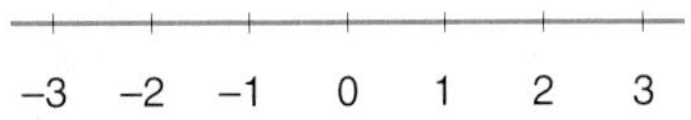

Figure 1.1.1

Between the integers we can mark any fractions we wish; for instance, $\frac{7}{2}$ is halfway between 3 and 4. It is less easy to mark an **irrational number** like π or $\sqrt{2}$; to do that, we use the decimal expansion of these numbers.

Every real number has a decimal expansion. For instance, $\pi = 3.141\,592\,65\ldots$, and $\sqrt{2} = 1.414\,213\,56\ldots$. The dots are very important. They tell us that the expansion does not end where we have stopped writing (in fact if the decimal expansion did stop, then the number would have to be a rational number).

We cannot write out the whole expansion of a number like π or $\sqrt{2}$, since it goes on forever, but when we write down some moderate number of digits (in the two examples above we have written nine digits), this gives us a pretty good *approximation* to the number in question.

Scientific Notation

For numbers that are very large or very small, it is often convenient to write them in **scientific notation**. This means writing a number as the product of a number between 1 and 10 and a power of 10, that is, as $a \times 10^n$, where $1 \leq a < 10$ and n is an integer.

The obvious advantage of this notation is that it eliminates having to write a large number of zeroes ($1,540,000,000$ can be written more neatly as 1.54×10^9, and certainly $0.000\,000\,000\,0037$ is harder to read than 3.7×10^{-12}).

The other advantage is that the number of significant digits is clear from the way the number is written. For instance, the meaning of a measurement written as 2000 is unclear, but 2×10^3, 2.0×10^3, and 2.00×10^3 (which in a sense all equal 2000) make it quite clear what is intended.

Exercises 1.1

1. Show that $\sqrt{8}$ is not a rational number.

*2. (i) Show that $\sqrt{3}$ is not a rational number.
(ii) Show that $\sqrt{6}$ is not a rational number.

*3. Explain the remark made in the text that a decimal expansion that terminates must represent a rational number.

*4. Show that the number whose decimal expansion is 19.680 314 314 314 314 . . . (with the three digits 314 repeated again and again indefinitely) is a rational number (by writing it as the quotient of two integers).

*5. If a is a rational number and b is an irrational number, show that (i) $a + b$ is irrational, (ii) ab is irrational unless $a = 0$.

1.2 Inequalities

ABSOLUTE VALUE
TRIANGLE INEQUALITY
INTERVAL
CLOSED/OPEN INTERVAL
UNION
INTERSECTION

One of the special properties of the real numbers is that there is a concept of "order," meaning that given two distinct real numbers, one of them is always less than the other. We use the conventional symbols: $a < b$ means a is less than b, $a > b$ means a is greater than b (which is the same thing as $b < a$), $a \leq b$ means a is less than or equal to b, and $a \geq b$ means a is greater than or equal to b. Each of these expressions is called an **inequality**. We sometimes refer to the ones with $<$ and $>$ as **strict inequalities**.

In terms of the picture of the real number line, to say that a number a is less than b means that a is to the left of b on the line (see Figure 1.2.1). We also say that a is "negative" if $a < 0$ and that a is "positive" if $a > 0$.

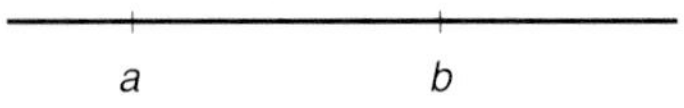

Figure 1.2.1

There are a number of properties of inequalities that we will be using.

Summary 1.2.1 Properties of Inequalities

(i) If $a < b$ and $b < c$, then $a < c$. (Transitivity Property)
(ii) If $a \leq b$ and $b \leq a$, then $a = b$.
(iii) If $a < b$ and $c \leq d$, then $a + c < b + d$.
(iv) If $a < b$ and $c > 0$, then $ac < bc$.
(v) If $a < b$ and $c < 0$, then $ac > bc$.
(vi) If $0 < a < b$, then $\frac{1}{a} > \frac{1}{b}$; if $a < b < 0$, then $\frac{1}{a} > \frac{1}{b}$.
(vii) $a^2 \geq 0$, for any $a \in \mathbb{R}$.

Properties (v) and (vi) say that an inequality is *reversed* if it is multiplied by a negative number or if both sides have the same sign and we take the reciprocal of both sides.

Property (vii) follows from Properties (iv) and (v).

Using Properties (iv) and (v) we also see that $ab > 0$ if a and b are both positive or both negative and $ab < 0$ if one of a and b is positive and the other is negative. In particular, the only way that the product ab can equal zero is for one (or both) of a and b to equal zero.

An important concept is the **absolute value** of a number a, written $|a|$. It is the "size" of the number and is *always nonnegative*. In symbols we write

$$|a| = \begin{cases} a, & \text{if } a \geq 0, \\ -a, & \text{if } a < 0. \end{cases}$$

> Note that if a is negative, then $|a| = -a$ (since $-a$ is positive . . .).

So $|3| = 3$, $|-3| = 3$, and $|0| = 0$. It is less clear what to do with $|\sqrt{12} - 4|$, but if we notice that $4^2 = 16 > 12$, we see that $4 > \sqrt{12}$. If we subtract 4 from both sides of this last inequality, we see that $0 > \sqrt{12} - 4$, so $\sqrt{12} - 4$ is negative and $|\sqrt{12} - 4| = -(\sqrt{12} - 4) = 4 - \sqrt{12}$.

We will often encounter inequalities involving absolute values. One of the most common types is $|x| \leq c$. If $x \geq 0$, we know that $|x| = x$, so the inequality means $x \leq c$. If $x \leq 0$, we know that $|x| = -x$, and the inequality means $-x \leq c$ or $x \geq -c$. Combining these two, we find that

$$|x| \leq c \quad \text{means exactly the same thing as} \quad -c \leq x \leq c.$$

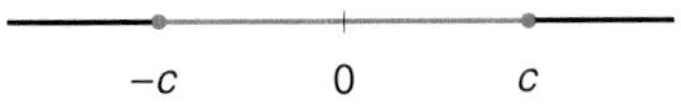

Figure 1.2.2

The set of all numbers x that satisfy this condition can be drawn on the line; they fill a segment between $-c$ and c. See Figure 1.2.2.

There is an extremely important inequality involving absolute values. It is called the triangle inequality.

THEOREM 1.2.2

> **The Triangle Inequality**
>
> If a and b are any real numbers, then
>
> $$|a + b| \leq |a| + |b|.$$

PROOF Since a equals either $|a|$ or $-|a|$, $-|a| \leq a \leq |a|$. Similarly, $-|b| \leq b \leq |b|$. Adding these two inequalities, we find that

$$-|a| - |b| \leq a + b \leq |a| + |b|$$

or

$$-(|a| + |b|) \leq a + b \leq |a| + |b|.$$

This is exactly the same as saying $|a + b| \leq |a| + |b|$, which is the result we wanted to show. ◆

> Recall $|x| \leq c$ means $-c \leq x \leq c$; here x is $a + b$ and c is $|a| + |b|$.

There is also a second version of the triangle inequality.

THEOREM 1.2.3

> **The Triangle Inequality (Second Version)**
>
> If a and b are real numbers, then
>
> $$|a - b| \geq |a| - |b|.$$

PROOF Applying the first version of the triangle inequality to $a = (a - b) + b$, we find $|(a - b) + b| \leq |a - b| + |b|$, or $|a| \leq |a - b| + |b|$. Rearranging gives the desired result. ◆

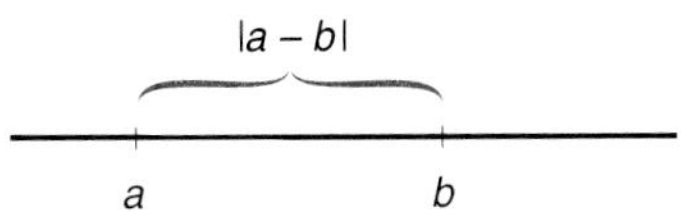

Figure 1.2.3

The absolute value can also describe distances. If a and b are real numbers, then $|a - b|$ is the **distance** between them. See Figure 1.2.3.

Figure 1.2.4

EXAMPLE 1 Consider the set described as

$$\{x : |x-2| \le 3\},$$

which we read as "the set of all x for which $|x-2| \le 3$."

Now $|x-2| \le 3$ means $-3 \le x-2 \le 3$; adding 2 to these inequalities gives $-1 \le x \le 5$, so the set consists of all x's that satisfy $-1 \le x \le 5$, that is, all x's lying between -1 and 5. See Figure 1.2.4.

Figure 1.2.5

Using the concept of distance frequently makes it easier to work with inequalities.

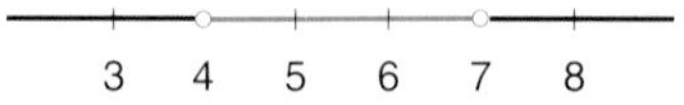

Figure 1.2.6

Intervals

If $a < b$, then the set of all points that lie between a and b is called an **interval**. The points a and b are called the **endpoints** of the interval, and it is important to specify whether or not the endpoints belong to the set.

An interval with its endpoints included is called a **closed interval**; one with the endpoints not included is called an **open interval**. We draw closed intervals with dots at the ends to indicate that the endpoints are included, and open intervals with "open" dots to show that those points are excluded. Figure 1.2.5 shows the closed interval from 1 to $2\frac{1}{2}$, and Figure 1.2.6 shows the open interval from 4 to 7.

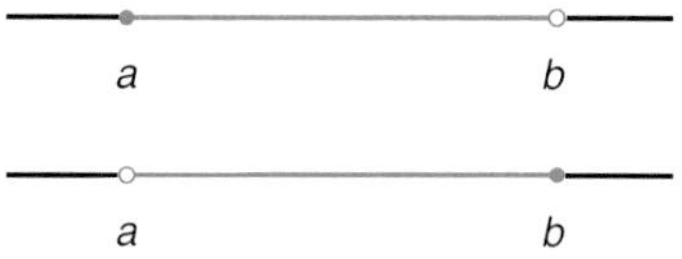

Figure 1.2.7

We write $[a, b]$ to represent the closed interval between a and b and (a, b) for the open interval between a and b. In symbols,

$$[a, b] = \{x : a \le x \le b\}, \qquad (a, b) = \{x : a < x < b\}.$$

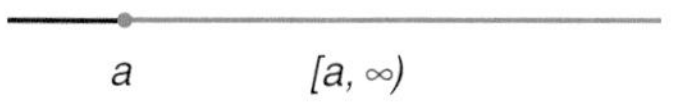

Figure 1.2.8

Occasionally it is necessary to consider **half-open intervals**, meaning ones that contain one endpoint but not the other. Adapting the above notation for closed and open intervals, we write

$$[a, b) = \{x : a \le x < b\}, \qquad (a, b] = \{x : a < x \le b\}.$$

These half-open intervals are shown in Figure 1.2.7.

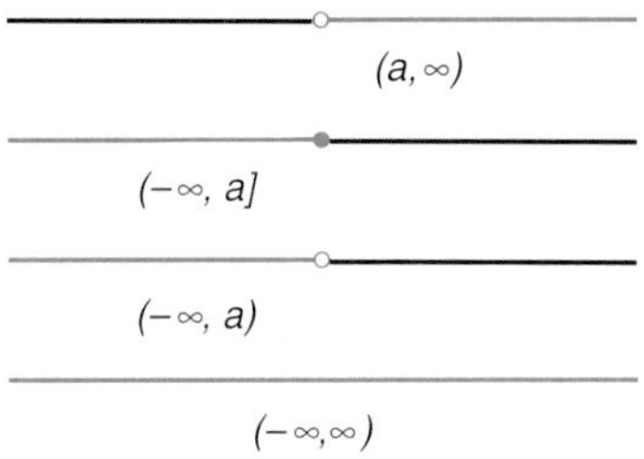

Figure 1.2.9

The other type of interval that is often used is called an **infinite interval**. We use the notation $[a, \infty)$ to refer to the "interval" shown in Figure 1.2.8, whose left endpoint is a and which extends out all the way along the real line. The symbol ∞ is *not* a real number. We call it "infinity," but it is extremely important to keep in mind that it is not a number. It is only a convenient way to summarize the idea of this type of set. We also use symbols like $(-\infty, 3)$, $(-\infty, -5]$ and $(-\infty, \infty)$ (which means the entire real line $\mathbb{R}$). See Figure 1.2.9.

$$[a, \infty) = \{x : a \le x\} \qquad (a, \infty) = \{x : a < x\}$$
$$(-\infty, a] = \{x : x \le a\} \qquad (-\infty, a) = \{x : x < a\}$$
$$(-\infty, \infty) = \text{the real line} = \mathbb{R}.$$

Figure 1.2.10

EXAMPLE 2 Consider the set

$$\{x : |x+2| > 2\}.$$

We write $|x+2| = |x-(-2)|$, which is the distance from x to the point -2.

The set in question is the set of all x's whose distance from -2 is greater than 2. From Figure 1.2.10 we see that the points whose distance from -2 is greater than 2 are all the points to the right of 0 and all the points to the left of -4.

In Example 1 we saw that $|x-2| \le 3$ is the same as $-3 \le x-2 \le 3$. But $|x+2| > 2$ is *not* the same as $-2 > x+2 > 2$. . . . If we want to write it out using inequalities, we have to say that $x+2 > 2$ *or* $x+2 < -2$.

We could write this out in symbols as follows:

$$\{x : |x + 2| > 2\} = (-\infty, -4) \cup (0, \infty).$$

We have rewritten the set as the union of two infinite intervals.

REMARK 1.2.4

We recall some useful notation. We write $A \cup B$ for the **union** of A and B, that is, the set of all points that belong to either or both of the sets. Similarly, we write $A \cap B$ for the **intersection** of A and B, that is, the set of all points that belong to both sets. In symbols,

$$A \cup B = \{x : x \in A \quad \text{or} \quad x \in B\}, \qquad A \cap B = \{x : x \in A \quad \text{and} \quad x \in B\}.$$

Exercises 1.2

Without using a calculator, write an expression for each of the following quantities without using any absolute value signs.

1. $|1 - \frac{5}{12} - \frac{3}{7} - \frac{1}{6}|$
2. $|1 + \frac{5}{12} - \sqrt{2}|$
3. $|\frac{5}{12} + \frac{3}{7} + \frac{1}{6} - 1|$
4. $|(1 + \sqrt{2})(1 - \sqrt{2})|$
5. $|\sqrt{3} - 2|$
6. $|3 - 2\sqrt{2}|$
7. $|3 - \sqrt{2} - \sqrt{3}|$
8. $|5 - \sqrt{5} - \sqrt{8}|$
9. $|x^2 + 3|$
10. $|x^2 + 2x + 2|$

Express each of the following sets as an interval or a union of intervals.

11. $\{x : |x - 1| \leq 4\}$
12. $\{x : |x + 3| < 1\}$
13. $\{x : |3x - 6| < 2\}$
14. $\{x : |4x + 1| \leq 2\}$
15. $\{x : |2x - 1| > 1\}$
16. $\{x : |3x + 5| \geq 4\}$
17. $\{x : |x - 3| \leq 4 \text{ and } |x - 1| \geq 1\}$
18. $\{x : |2x + 1| > 4 \text{ and } |x + 3| \geq 1\}$
19. $\{x : |1 - x^2| \geq 2\}$
20. $\{x : |x| \geq 3 \text{ and } |x + 3| \leq 2\}$

1.3 Polynomials and Rational Functions

COEFFICIENT
CONSTANT TERM
QUADRATIC/LINEAR POLYNOMIAL
COMPLETING THE SQUARE
QUADRATIC FORMULA
RATIONAL ROOT THEOREM
BINOMIAL THEOREM
FACTORIALS

DEFINITION 1.3.1

A **polynomial** is an expression $p(x)$ of the form

$$p(x) = a_n x^n + a_{n-1} x^{n-1} + a_{n-2} x^{n-2} + \ldots + a_2 x^2 + a_1 x + a_0,$$

where n is a nonnegative integer and each a_i is a real number. The numbers $a_n, a_{n-1}, \ldots, a_2, a_1, a_0$ are called the **coefficients** of the polynomial.

The highest power of x that occurs in the polynomial (which would be n in this polynomial, provided $a_n \neq 0$) is called the **degree** of the polynomial.

Notice that the **subscript** i on the coefficient a_i is just a label; it tells us which power of x is multiplied by a_i. We do not bother to write $a_0 x^0$, but just a_0 for the last term, which is called the **constant term**.

EXAMPLE 1 In the polynomial $3x^4 - 2x^3 + 5x^2 - 2x + 7$, the coefficients are $a_4 = 3$, $a_3 = -2$, $a_2 = 5$, $a_1 = -2$, and $a_0 = 7$. The degree of the polynomial is 4.

EXAMPLE 2 The formula $p(x) = 3x^2 - 5x + 2$ can be worked out for any specific value of x. For instance, if $x = 2$, then $p(2) = 3(2)^2 - 5(2) + 2 = 12 - 10 + 2 = 4$. On the other hand, if $x = 1$, then $p(1) = 3(1)^2 - 5(1) + 2 = 3 - 5 + 2 = 0$.

It is very easy to evaluate $p(0) = 3(0)^2 - 5(0) + 2 = 0 - 0 + 2 = 2$.

The same idea can be used with any polynomial. We call this process **evaluating** the polynomial $p(x)$ at the point x, and call the resulting number the **value of the polynomial** p at the point x.

Since x can *vary* and take on different values, we call it the **variable**.

Factoring

It is often important to be able to write a complicated polynomial as the product of two or more simpler polynomials. For instance, it is very easy to check what is known as the **difference of squares** formula:

$$a^2 - b^2 = (a + b)(a - b). \tag{1.3.2}$$

This formula makes sense when a and b are numbers (you can check it with $a = 5$ and $b = 3$, for instance), but it also works if a and b are polynomials.

EXAMPLE 3

(i) $49 - 9 = (7 + 3)(7 - 3) = (10)(4)$; (ii) $x^2 - 16 = (x + 4)(x - 4)$;
(iii) $(x + 1)^2 - 1 = (x + 2)x$.

The process of breaking up a polynomial $p(x)$ into a product of simpler factors is called **factoring** the polynomial. The simpler polynomials whose product is $p(x)$ are called the **factors** of $p(x)$. For instance, when we write $4x^2 - 25 = (2x + 5)(2x - 5)$, the polynomials $2x + 5$ and $2x - 5$ are the factors of $4x^2 - 25$.

Suppose we want to factor the polynomial $p(x) = x^2 - x - 6$. This is a **quadratic polynomial**, meaning one whose degree is 2. In this case it is possible to guess a factorization, namely, $x^2 - x - 6 = (x - 3)(x + 2)$.

This kind of guesswork approach can often be used.

EXAMPLE 4 Factor each of the following quadratic polynomials: (i) $x^2 + 2x + 1$; (ii) $x^2 + 2x - 8$; (iii) $2x^2 - 6x - 20$; (iv) $x^2 - 2$; (v) $x^2 + 1$.

Solution

(i) We just look for a factorization $(x - a)(x - b) = x^2 - (a + b) + ab$. The first thing to do is to try to find *integer* values for a and b. In this case the only possible pairs of integers a and b whose product is 1 are 1 and 1 or -1 and -1. Since $a + b = -2$, the obvious choice is $a = b = -1$, and the factorization is $x^2 + 2x + 1 = (x + 1)(x + 1) = (x + 1)^2$.

(ii) $x^2 + 2x - 8$. The only pairs of integers whose product is -8 are 1 and -8, or

-1 and 8, or 2 and -4, or -2 and 4. If the sum must also be -2, the only possibility is 2 and -4. The factorization is $x^2 + 2x - 8 = (x - 2)(x + 4)$.

(iii) We write $2x^2 - 6x - 20 = 2(x^2 - 3x - 10)$, and factor $x^2 - 3x - 10 = (x - 5)(x + 2)$, so $2x^2 - 6x - 20 = 2(x^2 - 3x - 10) = 2(x - 5)(x + 2)$.

(iv) $x^2 - 2$. Here we try looking for a factorization with integers but cannot find one.

Suppose that we knew a factorization $x^2 - 2 = (x - a)(x - b)$. If we let $x = a$, then the right side of this equation would be $(a - a)(a - b) = 0(a - b) = 0$. This suggests that we try to find a (or b) by looking for values of x that make $p(x) = 0$.

To solve $p(x) = 0$, we rewrite $x^2 - 2 = 0$ as $x^2 = 2$, which tells us that x should equal $\sqrt{2}$ (or $-\sqrt{2}$). We try letting a and b be $\sqrt{2}$ and $-\sqrt{2}$: $(x - \sqrt{2})(x + \sqrt{2}) = x^2 - 2$.

In hindsight, $x^2 - 2$ is a difference of squares, $x^2 - (\sqrt{2})^2$, so we might have guessed the factorization that way too.

(v) $x^2 + 1$. If it did factor as $(x - a)(x - b)$, then the polynomial $x^2 + 1$ would have to be zero when we let x equal a or b. But we know that $x^2 \geq 0$ for any value of x, so $x^2 + 1 \geq 1$ for any value of x. This means it can never be zero, so it is impossible to find a and b. The polynomial does not factor.

Some quadratic polynomials can be factored by just guessing integers a and b so that the polynomial equals $(x - a)(x - b)$. If the coefficient of x^2 in the original polynomial is not 1, then it is necessary to factor it out first.

Some polynomials cannot be factored with integer values for a and b. It may still be possible to guess a factorization, but if not, the next thing to do is look for values of x for which $p(x) = 0$. If we can find such a value, then it will serve as a in a factorization $p(x) = (x - a)(x - b)$.

We also saw in Example 4 that there are some polynomials that cannot be factored at all.

Completing the Square and the Quadratic Formula

We have already encountered some examples in which a quadratic polynomial factors as the square of a polynomial of degree 1 (polynomials of degree 1 are often called **linear polynomials**).

EXAMPLE 5 If the polynomial $x^2 - 4x + 12$ is a square, $(x - a)^2 = x^2 - 2a + a^2$, say, then its middle coefficient -4 would have to equal $-2a$, so $a = -2$. Since the constant term is *not* equal to $(-2)^2 = 4$, we see that the polynomial is not a square.

On the other hand, we can write it as $x^2 - 4x + 12 = (x^2 - 4x + 4) + 8$. Now the polynomial in parentheses is a square. We can write $x^2 - 4x + 12 = (x^2 - 4x + 4) + 8 = (x - 2)^2 + 8$. This process is called **completing the square**, for obvious reasons.

With the polynomial written as $(x - 2)^2 + 8$, we see immediately that it cannot be factored. After all, it is 8 plus a square, so for any $x \in \mathbb{R}$ it is always greater than or equal to 8, so it can *never* be zero. We can never find a and b so that the polynomial factors as $(x - a)(x - b)$.

Summary 1.3.3

Completing the Square

If $p(x) = x^2 + bx + c$, we can write $p(x)$ as the sum of the square of a linear polynomial and a number, as follows:

$$p(x) = x^2 + bx + c = x^2 + bx + \left(\frac{b}{2}\right)^2 - \left(\frac{b}{2}\right)^2 + c = \left(x + \frac{b}{2}\right)^2 + c - \left(\frac{b^2}{4}\right).$$

If $q(x) = ax^2 + bx + c$, we write $q(x) = a\left(x^2 + \frac{b}{a}x + \frac{c}{a}\right)$ and then complete the square of the polynomial inside the parentheses (for this to work, it is necessary that

$a \neq 0$). We find that

$$q(x) = a\left(x^2 + \frac{b}{a}x + \frac{c}{a}\right) = a\left(x^2 + \frac{b}{a}x + \left(\frac{b}{2a}\right)^2 - \left(\frac{b}{2a}\right)^2 + \frac{c}{a}\right)$$
$$= a\left(\left(x + \frac{b}{2a}\right)^2 + \frac{c}{a} - \left(\frac{b}{2a}\right)^2\right) = a\left(x + \frac{b}{2a}\right)^2 + c - \left(\frac{b^2}{4a}\right).$$

One advantage of completing the square in a quadratic polynomial is that it makes it easier to solve.

EXAMPLE 6 Find all solutions of $x^2 - 4x + 1 = 0$.

Solution We begin by completing the square on the left side, which gives us that $x^2 - 4x + 1 = (x^2 - 4x + 4) - 4 + 1 = (x - 2)^2 - 3$.

The original equation becomes $(x - 2)^2 - 3 = 0$, or $(x - 2)^2 = 3$. So $x - 2$ must be something whose square is 3; there are exactly two such numbers, $\sqrt{3}$ and $-\sqrt{3}$. The possibilities are $x - 2 = \sqrt{3}$ and $x - 2 = -\sqrt{3}$. The solutions of the equation are

$$x = 2 + \sqrt{3} \qquad \text{and} \qquad x = 2 - \sqrt{3}.$$

A common abbreviation for this is to write $x = 2 \pm \sqrt{3}$.

When we write the square root sign, it always means the **nonnegative square root**. This has the slightly surprising consequence that if $a < 0$ then $\sqrt{a^2} \neq a$; since $\sqrt{a^2} \geq 0$, we see that $\sqrt{a^2} = |a| = -a$.

The basic approach we have used in this example can be used to solve any quadratic equation. To find the solutions of $ax^2 + bx + c = 0$, we complete the square and find that the left side of the equation equals $a\left(x + \frac{b}{2a}\right)^2 + c - \left(\frac{b^2}{4a}\right)$. From this we can rewrite the original equation as $\left(x + \frac{b}{2a}\right)^2 = \frac{1}{a}\left(\left(\frac{b^2}{4a}\right) - c\right) = \frac{b^2}{4a^2} - \frac{c}{a}$.

To solve this, we should take square roots of both sides; this is possible *only if* the right side is nonnegative. In this case we find that $x + \frac{b}{2a} = \pm\sqrt{\frac{b^2}{4a^2} - \frac{c}{a}}$.

This amounts to $x = -\frac{b}{2a} \pm \sqrt{\frac{b^2}{4a^2} - \frac{c}{a}}$, and putting it over a common denominator, we find

$$x = \frac{-b \pm 2a\sqrt{\frac{b^2}{4a^2} - \frac{c}{a}}}{2a} = \frac{-b \pm \sqrt{4a^2\frac{b^2}{4a^2} - 4ac}}{2a} = \frac{-b \pm \sqrt{b^2 - 4ac}}{2a}.$$

This result is tremendously useful.

THEOREM 1.3.4

> **The Quadratic Formula**
>
> Consider the equation $ax^2 + bx + c = 0$, where $a \neq 0$. It has no solutions if $b^2 - 4ac < 0$, exactly one solution if $b^2 - 4ac = 0$, and two solutions if $b^2 - 4ac > 0$.
>
> If $b^2 - 4ac \geq 0$, the solutions are given by the formula
>
> $$x = \frac{-b \pm \sqrt{b^2 - 4ac}}{2a}.$$

The quantity $D = b^2 - 4ac$, which tells us whether or not there are solutions, is called the **discriminant** of the polynomial $ax^2 + bx + c$. Notice that when the discriminant is zero, the square root is zero, and this is why there is only one solution.

EXAMPLE 7(a) Find all solutions of the following equations: (i) $x^2 + x - 2 = 0$; (ii) $x^2 + x - 1 = 0$; (iii) $x^2 + x = -1$.

Solution

(i) The discriminant is $D = 1^2 - 4(1)(-2) = 9$ and the solutions are $x = \frac{-1 \pm \sqrt{1^2 - 4(1)(-2)}}{2(1)} = \frac{-1 \pm \sqrt{9}}{2} = \frac{-1 \pm 3}{2}$, that is, $\frac{2}{2}$ and $\frac{-4}{2}$, that is, 1 and -2. We might have guessed this if we tried factoring $x^2 + x - 2 = (x - 1)(x + 2)$.

(ii) Here $D = 1^2 - 4(1)(-1) = 5$, and the solutions are $x = \frac{-1 \pm \sqrt{5}}{2}$, that is, $\frac{-1 + \sqrt{5}}{2}$ and $\frac{-1 - \sqrt{5}}{2}$.

(iii) First we rewrite $x^2 + x = -1$ as $x^2 + x + 1 = 0$. The discriminant is $D = 1 - 4 = -3$, which is negative. This means that there are no solutions.

EXAMPLE 7(b) Describe (iv) $\{x : x^2 + 2x - 3 < 0\}$; (v) $\{x : x^2 - 4x + 2 \geq 0\}$.

Solution

(iv) We find $x^2 + 2x - 3 = x^2 + 2x + 1 - 1 - 3 = (x + 1)^2 - 4$. The condition $x^2 + 2x - 3 < 0$ amounts to $(x + 1)^2 - 4 < 0$, or $(x + 1)^2 < 4$. Taking square roots gives $|x + 1| < 2$, that is, $|x - (-1)| < 2$, so the set consists of all points whose distance from -1 is less than 2. It is the interval $(-3, 1)$.

(v) Here $x^2 - 4x + 2 = (x - 2)^2 - 2$, so the condition is $(x - 2)^2 - 2 \geq 0$, that is, $(x - 2)^2 \geq 2$. This amounts to $|x - 2| \geq \sqrt{2}$, which means $x \geq 2 + \sqrt{2}$ or $x \leq 2 - \sqrt{2}$. The set is $(-\infty, 2 - \sqrt{2}] \cup [2 + \sqrt{2}, \infty)$.

The Factor Theorem

When we studied quadratic polynomials, we found that if $p(a) = 0$, then $p(x)$ could be factored, one of the factors being $(x - a)$. From this it was easy to find the other factor. If $p(x)$ is any polynomial, not necessarily of degree 2, suppose we know that a is a root of $p(x)$, that is, that $p(a) = 0$. Our experience with quadratic polynomials suggests that $p(x)$ can be factored, with one of the factors being $(x - a)$.

THEOREM 1.3.5

The Factor Theorem

Suppose $p(x)$ is a polynomial, and suppose a is a **root** of $p(x)$, that is, a real number satisfying $p(a) = 0$. Then there is a polynomial $q(x)$ so that

$$p(x) = (x - a)q(x).$$

The other factor $q(x)$ can be found quite easily, using long division.

EXAMPLE 8 If $p(x) = x^4 - 2x^3 + 4x^2 - 5x - 6$, we see that $p(2) = 2^4 - 2(2^3) + 4(2^2) - 5(2) - 6 = 16 - 16 + 16 - 10 - 6 = 0$. The Factor Theorem says that we should be able to find a polynomial $q(x)$ so that $p(x) = (x - 2)q(x)$. To find $q(x)$, we divide both sides of this equation by $(x - 2)$ and find $q(x) = \frac{p(x)}{x-2}$. We have to divide, and this suggests long division. In fact, this is easier than it sounds.

At each stage, put in the term that makes the highest powers of x match up.

$$\begin{array}{r|l}
 & x^3 \qquad\quad + 4x + 3 \\ \hline
(x-2) & x^4 - 2x^3 + 4x^2 - 5x - 6 \\
 & x^4 - 2x^3 \\ \hline
 & \qquad 0 \; + 4x^2 - 5x - 6 \\
 & \qquad\quad 4x^2 - 8x \\ \hline
 & \qquad\qquad\quad 3x - 6 \\
 & \qquad\qquad\quad 3x - 6 \\ \hline
 & \qquad\qquad\qquad\quad 0
\end{array}$$

The quotient is $x^3 + 4x + 3$, and it is very easy to check that

$$(x - 2)(x^3 + 4x + 3) = x^4 - 2x^3 + 4x^2 - 5x - 6,$$

as desired.

EXAMPLE 9 Factor the polynomial $p(x) = x^2 + 6x + 6$.

Solution The obvious attempts to factor $p(x)$ by inspection do not work.

We can use the Quadratic Formula to find a root a, and the Factor Theorem assures us that $p(x)$ has $(x - a)$ as a factor.

The roots of $p(x)$ are $\frac{-6 \pm \sqrt{36-24}}{2} = -3 \pm \frac{\sqrt{12}}{2} = -3 \pm \frac{2\sqrt{3}}{2} = -3 \pm \sqrt{3}$, so $\left(x - (-3 + \sqrt{3})\right)$ and $\left(x - (-3 - \sqrt{3})\right)$ are both factors of $p(x)$, and their product $\left(x - (-3 + \sqrt{3})\right)\left(x - (-3 - \sqrt{3})\right) = (x + 3 - \sqrt{3})(x + 3 + \sqrt{3})$ must divide $p(x)$. We try multiplying out the product:

$$\begin{aligned}(x + 3 - \sqrt{3})(x + 3 + \sqrt{3}) &= x^2 + (3 - \sqrt{3} + 3 + \sqrt{3})x + (3 - \sqrt{3})(3 + \sqrt{3}) \\ &= x^2 + 6x + (9 - 3) = x^2 + 6x + 6.\end{aligned}$$

We see that the result is exactly equal to the original polynomial $p(x)$, so we have finished factoring $p(x)$:

$$p(x) = (x + 3 - \sqrt{3})(x + 3 + \sqrt{3}).$$

This procedure will always work for quadratic polynomials, provided there are roots. If $p(x) = ax^2 + bx + c$ has roots r_1 and r_2, then it factors as follows:

$$p(x) = ax^2 + bx + c = a(x - r_1)(x - r_2). \tag{1.3.6}$$

Although it can be extremely difficult to factor a polynomial, especially if its degree is large, there are two useful facts that can sometimes help us guess a root.

Suppose we know how to factor a polynomial $p(x) = k(x - r_1)(x - r_2) \times \ldots \times (x - r_n)$. Then the constant term of $p(x)$ is just $\pm k$ times the product of all the roots: $(-1)^n k r_1 \times \ldots \times r_n$. If k and all the roots happen to be integers, then this means that each root must divide the constant term of $p(x)$.

In fact, something similar is true about any integer root of $p(x)$, even if some of the other roots are not integers, provided every coefficient of $p(x)$ is an integer. There is also a related statement about rational roots.

THEOREM 1.3.7

The Rational Root Theorem

Suppose $p(x) = a_n x^n + a_{n-1} x^{n-1} + \ldots + a_1 x + a_0$ is a polynomial whose coefficients $a_0, a_1, \ldots, a_n$ are all *integers.*

(i) If r is a root of $p(x)$ that is an integer, then a_0 is an integer multiple of r, that is, $a_0 = rm$, for some integer m. If $r \neq 0$, this just means r divides a_0.

(ii) If $r = \frac{p}{q}$ is a root of $p(x)$ that is a rational number, then provided that $\frac{p}{q}$ is written in *lowest terms,* the constant term a_0 of $p(x)$ must be an integer multiple of the numerator p and the leading coefficient a_n must be an integer multiple of the denominator q.

The value of this theorem is that it limits the number of integers or rational numbers we have to consider when looking for a root.

EXAMPLE 10

(i) If $p(x) = x^3 - 5x^2 - 2x + 10$, we note that its coefficients are all integers, so any integer root must divide the constant term 10. This means that the only possible integer roots are $\pm 1, \pm 2, \pm 5, \pm 10$. We check and find that $x = 5$ is in fact a root. We saved time by not bothering to check $\pm 3, \pm 4$.

By long division we can easily find that $p(x) = x^3 - 5x^2 - 2x + 10 = (x - 5)(x^2 - 2)$. From this we see that the other two roots are $x = \pm\sqrt{2}$, which are not integers.

(ii) Consider the polynomial $q(x) = \left(x - \frac{1}{3}\right)(x - 3) = x^2 - \frac{10}{3}x + 1$. Notice that $x = 3$ is a root which is also an integer, but it does *not* divide the constant term 1. This is not a contradiction because the coefficients of $q(x)$ are not all integers (the middle one is $\frac{10}{3}$).

If the coefficients are not integers, the theorem simply does not apply.

(iii) Let $R(x) = x^4 + 2x^3 - x^2 - 6x - 6$. Since the coefficients of $R(x)$ are all integers, any integer root must divide -6. The only possibilities are $\pm 1, \pm 2, \pm 3, \pm 6$.

We evaluate $R(x)$ at each of these values of x and find that $R(x)$ is not zero at any of them. We conclude that $R(x)$ has no integer roots. (In fact, it can be written as $(x^2 - 3)(x^2 + 2x + 2)$; the second factor has no roots, so the roots of $R(x)$ are $x = \pm\sqrt{3}$.)

In part (iii) of this example we were able to check eight possible roots and then use the theorem to conclude that there are no integer roots. Without the theorem we might easily have spent our time checking larger and larger integers, reluctant to give up too soon.

The theorem allows us to save time by not checking unnecessary integers and tells us when there is no point in checking further.

Factorials and the Binomial Theorem

Frequently we have to take expressions and square them, cube them, or raise them to other powers. The Binomial Theorem gives a formula for the nth power of a sum $(a+b)$, where n is a positive integer. The quantities a and b can be numbers or polynomials or even more complicated expressions.

First we need some notation. We often need to use the product of the first n positive integers, $(1)(2)(3) \times \ldots \times (n-2)(n-1)(n)$. It is convenient to have a notation for this product; we write $(1)(2)(3) \times \ldots \times (n-2)(n-1)(n) = n!$ and read it "n **factorial**."

So, for instance, $3! = (1)(2)(3) = 6$, $5! = (1)(2)(3)(4)(5) = 120$, and so on. Factorials have the useful property that $n! = n(n-1)!$.

It would be convenient to be able to use this relation even when $n = 1$. To be able to do that, we need to say that $0! = 1$. This seems a bit peculiar at first, but in fact it is the sensible way to make the definition.

DEFINITION 1.3.8

Factorials

If n is a positive integer, we define $n!$ to be the product of the first n positive integers:

$$n! = (1)(2)(3) \times \ldots \times (n-2)(n-1)(n).$$

If $n = 0$, we define $0! = 1$.

The symbol $n!$ is read "***n* factorial.**"

Using factorials, we can define another very useful symbol. It can be interpreted as "the number of different ways of choosing k elements from a set of n elements." We will not be concerned with this interpretation here but simply make the definition.

DEFINITION 1.3.9

If n and k are integers with $0 \le k \le n$, then the symbol $\binom{n}{k}$ is defined as follows:

$$\binom{n}{k} = \frac{n!}{k!(n-k)!}.$$

We read $\binom{n}{k}$ as "n choose k."

EXAMPLE 11 Evaluate: (i) $\binom{3}{2}$; (ii) $\binom{4}{3}$; (iii) $\binom{5}{2}$; (iv) $\binom{6}{0}$.

Solution

(i) By the definition, $\binom{3}{2} = \frac{3!}{2!(3-2)!} = \frac{6}{2!1!} = \frac{6}{2} = 3$.

(ii) $\binom{4}{3} = \frac{4!}{3!(4-3)!} = \frac{24}{3!1!} = \frac{24}{6} = 4$.

(iii) $\binom{5}{2} = \frac{5!}{2!(5-2)!} = \frac{120}{2!3!} = \frac{120}{12} = 10.$

(iv) $\binom{6}{0} = \frac{6!}{0!(6-0)!} = \frac{720}{(1)6!} = \frac{720}{720} = 1.$

EXAMPLE 12 Suppose n is any positive integer. Then it is easy to evaluate

$$\binom{n}{0} = \frac{n!}{0!n!} = 1, \quad \binom{n}{1} = \frac{n!}{1!(n-1)!} = \frac{n!}{(n-1)!} = n,$$
$$\binom{n}{n-1} = \frac{n!}{(n-1)!(n-(n-1))!} = \frac{n!}{(n-1)!1!} = n, \quad \binom{n}{n} = \frac{n!}{n!0!} = 1.$$

With this notation it is now possible to state the Binomial Theorem.

THEOREM 1.3.10

The Binomial Theorem

If n is a positive integer, then

$$\begin{aligned}(a+b)^n &= a^n + \binom{n}{1}a^{n-1}b + \binom{n}{2}a^{n-2}b^2 + \ldots + \binom{n}{n-2}a^2b^{n-2} \\ &+\binom{n}{n-1}ab^{n-1} + b^n \\ &= a^n + \frac{n!}{1!(n-1)!}a^{n-1}b + \frac{n!}{2!(n-2)!}a^{n-2}b^2 + \ldots \\ &+\frac{n!}{(n-2)!2!}a^2b^{n-2} + \frac{n!}{(n-1)!1!}ab^{n-1} + b^n \\ &= a^n + na^{n-1}b + \frac{n!}{2!(n-2)!}a^{n-2}b^2 + \ldots + \frac{n!}{(n-2)!2!}a^2b^{n-2} \\ &+nab^{n-1} + b^n.\end{aligned}$$

(We have used Example 12 to simplify some of the expressions here).

Notice that each term in the sum contains the product of a power of a and a power of b; the two exponents always add up to n. Even the first and last terms can be seen this way if we remember that $b^0 = 1$ and $a^0 = 1$.

The coefficient of $a^{n-k}b^k$ is $\binom{n}{k}$.

EXAMPLE 13

(i) $(x+5)^2 = x^2 + \binom{2}{1}x(5) + 5^2 = x^2 + 2(5)x + 25 = x^2 + 10x + 25.$

(ii) $(x-7)^3 = x^3 + \binom{3}{1}x^2(-7) + \binom{3}{2}x(-7)^2 + (-7)^3$
$= x^3 + 3x^2(-7) + 3x(49) - 343 = x^3 - 21x^2 + 147x - 343.$

(iii) $(x^2-5x)^4 = (x^2)^4 + \binom{4}{1}(x^2)^3(-5x) + \binom{4}{2}(x^2)^2(-5x)^2 + \binom{4}{3}(x^2)(-5x)^3 +$
$(-5x)^4 = x^8 + 4x^6(-5x) + \frac{4!}{2!2!}x^4(25x^2) + 4x^2(-125x^3) + 625x^4 =$
$x^8 - 20x^7 + 150x^6 - 500x^5 + 625x^4.$

The Binomial Theorem with $n = 2, 3$ gives results about squares and cubes that we should remember:

$$(a+b)^2 = a^2 + 2ab + b^2, \qquad (a+b)^3 = a^3 + 3a^2b + 3ab^2 + b^3.$$

Rational Functions

Finally we discuss rational functions. Just as we constructed the rational numbers because we could not divide two integers, we can divide one polynomial by another and call the result a rational function.

So for instance $R(x) = \frac{x^2+x-5}{x-2}$ is a rational function. If we specify a value of x, then we can calculate what $R(x)$ is. There is one complication, however. The quotient does not make sense if the denominator is zero, because division by zero is not allowed. With this particular example, that means we cannot let $x = 2$.

DEFINITION 1.3.11

A **rational function** is an expression of the form $R(x) = \frac{p(x)}{q(x)}$, where $p(x)$ and $q(x)$ are both polynomials.

The rational function $R(x)$ is not defined at any x for which $q(x) = 0$, that is, where the denominator is zero.

EXAMPLE 14

(i) The rational function $R(x) = \frac{x^3+2x^2+3x+5}{x^2-x-6}$ is not defined when the denominator is zero. Since $x^2 - x - 6 = (x-3)(x+2)$, we see that the denominator is zero at $x = 3$ and $x = -2$, and nowhere else. The rational function $R(x)$ makes sense for all other values of x.

(ii) The rational function $Q(x) = \frac{-7}{x^2+4}$ is defined for *every* value of x because the denominator is never zero (it is 4 plus a square, so it can never be zero).

(iii) Consider the rational function $S(x) = \frac{x^2-9}{x-3}$. It does not make sense when $x = 3$ but can be evaluated at every other real number.

However, if we look at it carefully, we might recognize that the numerator is a difference of squares. If we write $\frac{x^2-9}{x-3} = \frac{(x-3)(x+3)}{x-3}$, it is very tempting to cancel out the $x - 3$ from top and bottom. This suggests that $S(x) = x + 3$.

This is true, *for every x except x* $= 3$. However, when $x = 3$, $S(x)$ is not defined and $x + 3$ is defined (it equals 6). The cancellation is correct when $x \neq 3$, but canceling amounts to *dividing* both numerator and denominator by $x - 3$. When $x = 3$, this would mean dividing by zero, which is not permissible.

When $x \neq 3$, we can cancel, and we find that for every such x, $S(x) = x + 3$. When $x = 3$, $S(x)$ does not make sense, but $x + 3$ does make sense. They are *not* the same.

Exercises 1.3

Factor the polynomial in each of the following questions.

1. $x^2 + x - 12$
2. $x^2 - x - 12$
3. $x^2 + 3x - 4$
4. $x^2 - 2x - 15$
5. $2x^2 + 4x + 2$
6. $4x^2 - 12x + 9$
7. $x^2 - 4$
8. $x^2 - 6$
9. $(x-7)^2 - 7$
10. $(x+2)^2 - 16$

Find the roots (if any) of each of the following polynomials.

11. $x^2 + 3x - 2$
12. $x^2 + 3x + 2$
13. $x^2 - 4x - 1$
14. $x^2 + 2x + 2$
15. $x^2 - x + 1$
16. $9x^2 + 12x + 4$
17. $x^2 - 6x$
18. $3x^2 + 4x + 1$
19. $5x^2 - 4x + \frac{1}{2}$
20. $4x^3 + 8x^2 + 3x$

Describe each of the following sets in terms of intervals.

21. $\{x : x^2 - 2x < 0\}$

22. $\{x : x^2 + 2x \leq 8\}$

23. $\{x : x^2 + 4x + 3 \geq 0\}$

24. $\{x : x^2 < 6x\}$

25. $\{x : x^2 \geq 16\}$

26. $\{x : x^2 - 4x + 5 \geq 0\}$

27. $\{x : 2x^2 - 8x + 4 \geq 0\}$

28. $\{x : x^2 + 6x + 10 \leq 0\}$

29. $\{x : x^2 - 8x + 19 \geq 0\}$

30. $\{x : 0 \leq x^2 + 4x \leq 12\}$

For each of the following polynomials $p(x)$, find a small integer a for which $p(a) = 0$, and then find $q(x)$ for which $p(x) = (x - a)q(x)$.

31. $p(x) = x^3 + 3x^2 + 4x + 2$

32. $p(x) = x^4 - 2x + 1$

33. $p(x) = x^3 - 3x^2 + 3x - 2$

34. $p(x) = 2x^4 + x^3 - 2x - 1$

35. $p(x) = x^3 - 1$

36. $p(x) = x^4 - 1$

37. $p(x) = x^3 - 27$

38. $p(x) = x^4 - 16$

39. $p(x) = x^4 - 5x^3 + 3x^2 - 11x$

40. $p(x) = x^8 - 1$

Expand each of the following.

41. $(x + 4)^3$

42. $(x - 3)^4$

43. $(x + 1)^5$

44. $(x - 1)^6$

45. $(x^2 - 2)^3$

46. $\left(x - \frac{1}{2}\right)^4$

47. $(1 - 2)^9$

48. $(3x - 2)^4$

49. $(1 - 4x)^3$

50. $(3\sqrt{2} - 2)^4$

1.4 Functions and Graphs

DOMAIN
RANGE
COORDINATES
AXES

In Section 1.3 we noted that a polynomial $p(x)$ could be "evaluated" at any particular value of x. For example, if $p(x) = x^3 - 2x + 7$, then $p(1) = 1^3 - 2(1) + 7 = 6$, $p(0) = 7$, and $p(-2) = (-2)^3 - 2(-2) + 7 = 3$. This is an example of a **function**.

The idea is that a function f takes a number a and gives us a result, which is often written $f(a)$. It will do this for various values of a. The result $f(a)$ is often called the **value of the function $f(x)$ at $x = a$**.

For instance if $f(x)$ is the polynomial $2x^2 + 3$, we can easily find its value at any point: $f(0) = 3$, $f(3) = 2(3^2) + 3 = 21$, and $f(-3) = 2(-3)^2 + 3 = 21$. These last two values are equal, and of course we realize that the same will be true of the values of this $f(x)$ at $x = a$ and $x = -a$, for any real number a.

Rational functions, as the name implies, are also functions. For instance, $R(x) = \frac{x^3+x-4}{x^2-1}$ can easily be evaluated. One difference, however, is that there are some values of x at which the function cannot be evaluated (in this case the function does not make any sense at $x = 1$ or at $x = -1$). This is not especially serious. A function $f(x)$ makes sense for some x's and not for others; this is frequently described by saying the function is *defined* for some x's and *not defined* for some others.

There are examples of functions that are not defined for large sets of x's. For instance, the square root function $f(x) = \sqrt{x}$ is defined only when $x \geq 0$.

The set of points at which a function is defined is called the **domain** of the function.

EXAMPLE 1

(i) The domain of any polynomial is the set of all real numbers (because the formula for a polynomial can be evaluated at any number $x = a$).

Very often the domain of a function will be clear from the context, but sometimes it will be less clear, and sometimes it will be necessary to specify the domain very carefully or deliberately change it.

For instance, we could define a function for $x \in [1,3) \cup (5,7)$ by $F(x) = \frac{x^5-2}{x-3}$. The rational function $\frac{x^5-2}{x-3}$ makes sense for every x except $x = 3$, but we have declared that the domain of $F(x)$ is exactly $[1,3) \cup (5,7)$.

We say that the domain of $F(x)$ is $[1,3) \cup (5,7)$, but the **natural domain** of the rational function $\frac{x^5-2}{x-3}$ is $(-\infty, 3) \cup (3, \infty)$. The natural domain is the set of numbers for which a formula makes sense, even if we specify some (smaller) set as the domain. This complication will not arise often.

(ii) The domain of a rational function $R(x) = \frac{p(x)}{q(x)}$ is the set of all x's for which $q(x) \neq 0$. It is convenient to write this in symbols as

$$\text{domain}\big(R(x)\big) = \{x : q(x) \neq 0\}.$$

(iii) The domain of the square root function is the set of all nonnegative numbers; if $f(x) = \sqrt{x}$, then $\text{domain}\big(f(x)\big) = \{x : x \geq 0\}$.

(iv) What is the domain of the function $g(x) = \sqrt{1-2x}$?

Solution We know that $\sqrt{1-2x}$ makes sense only if $1 - 2x \geq 0$, that is, if $1 \geq 2x$, that is, $\frac{1}{2} \geq x$. We see that $\text{domain}\big(g(x)\big) = \{x : x \leq \frac{1}{2}\}$.

(v) What is the domain of the function $h(x) = \sqrt{x/(x^2-1)}$?

Solution What we need to know is when the quantity inside the square root sign is nonnegative. First we notice that it does not even make sense if $x = \pm 1$, since the denominator is zero there. Second, a quotient is positive when both its numerator and denominator are positive or both are negative. The denominator is positive when $x^2 - 1 > 0$, that is, $x^2 > 1$, that is, $x > 1$ or $x < -1$. The denominator is negative when $-1 < x < 1$.

The numerator is positive when $x > 0$ and negative when $x < 0$, so the numerator and denominator are both positive when $x > 1$, and both negative when $-1 < x < 0$. The quotient inside the square root sign is positive when $x \in (-1, 0) \cup (1, \infty)$. It is zero when $x = 0$.

Combining these remarks, we see that the function $f(x)$ is defined when $x \in (-1, 0] \cup (1, \infty)$, so $\text{domain}\ \big(h(x)\big) = (-1, 0] \cup (1, \infty)$. This set is illustrated in Figure 1.4.1.

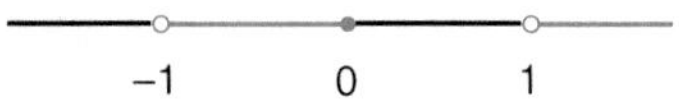

Figure 1.4.1

It is also necessary to discuss the set of values of a function, meaning the set of all numbers that are values $f(x)$ for some $x \in \text{domain}(f)$. This set is called the **range** of the function $f(x)$. In symbols we write

$$\text{range}(f) = \{f(x) : x \in \text{domain}(f)\} = \{y : y = f(x), \text{ for some } x \in \text{domain}(f)\}.$$

EXAMPLE 2

(i) What is the range of the polynomial $p(x) = x^2 + 2$?

Solution Since x^2 is always nonnegative, we see that $x^2 + 2 \geq 2$ for every x, and the range of $p(x)$ is contained in $[2, \infty)$.

The question is whether or not *every* element of $[2, \infty)$ is in the range. Suppose $r \geq 2$; let us try to find x for which $p(x) = r$. This means $x^2 + 2 = r$, or $x^2 = r - 2$. The solutions will be $x = \pm\sqrt{r-2}$, provided this makes sense, that is, provided that the quantity $r - 2$ inside the square root sign is nonnegative. This condition, $r - 2 \geq 0$, is the same as $r \geq 2$, which is exactly what we have assumed is true.

This shows that $\text{range}\big(p(x)\big) = [2, \infty)$.

(ii) What is the range of $q(x) = 3\sqrt{x} - 7$?

Solution We know that $\sqrt{x} \geq 0$ for every x in the domain of $\sqrt{x}$. This shows that $3\sqrt{x} \geq 0$, so $q(x) = 3\sqrt{x} - 7 \geq -7$, which makes us suspect that the range of $q(x)$ is $[-7, \infty)$.

We still have to show that every element of $[-7, \infty)$ is in the range; suppose $r \in [-7, \infty)$ and let us try to find x so that $q(x) = r$. In other words, we have to solve $3\sqrt{x} - 7 = r$, or $3\sqrt{x} = r + 7$, or $\sqrt{x} = \frac{r+7}{3}$. It is easy to find what x should be: $x = \left(\frac{r+7}{3}\right)^2$. The subtlety here is that with this value of x, the square root is $\sqrt{x} = \sqrt{\left((r+7)/3\right)^2} = \left|\frac{r+7}{3}\right|$, so we have to verify that this equals $\frac{r+7}{3}$. This amounts to checking that $\frac{r+7}{3} \geq 0$, and this is the same thing as $r + 7 \geq 0$, or $r \geq -7$. Once again, this is exactly what we have assumed (based on an intelligent guess at the answer).

We have shown that the range of $q(x)$ is $[-7, \infty)$. ◢

When we consider a polynomial, $p(x) = 5x^4 - 7x^3 + 2x^2 - 1$, for instance, it is reasonably easy to evaluate it at any point x. The same is true of rational functions. In a way this is misleading. Many functions cannot be described in a way that makes them easy to calculate.

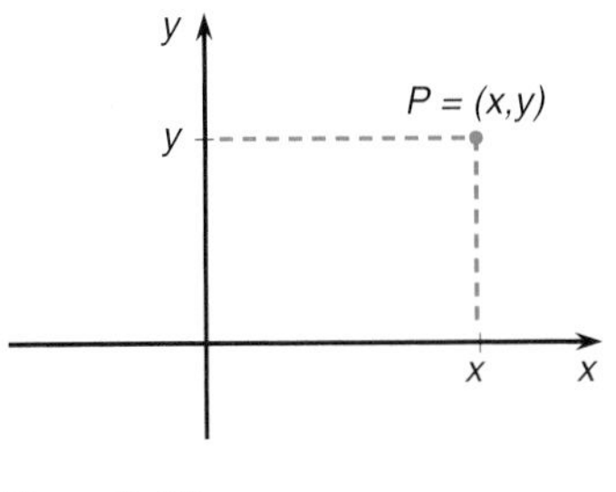

Figure 1.4.2

For one type of example, consider the polynomial $p(x) = x^7 + 3x^5 + 2x^3 + x + a$. It turns out that for any particular value of the constant a, there is exactly *one* value of x for which $p(x) = 0$, that is, for which $x^7 + 3x^5 + 2x^3 + x = -a$. This is not obvious (we will learn why it is true in Chapter 4), but accepting that it is true, let us define a function $f(x)$ as follows: For any real number a, $f(a)$ is that number x for which $x^7 + 3x^5 + 2x^3 + x = -a$.

This function $f(x)$ makes perfectly good sense, but it is not possible to write down a formula for it in terms of elementary functions (this is not to say that it is merely difficult, but rather that it cannot be done at all).

There are many other examples of functions that arise in quite natural ways but for which there is no nice formula. Part of the usefulness of calculus is that it provides means for dealing with some of these functions.

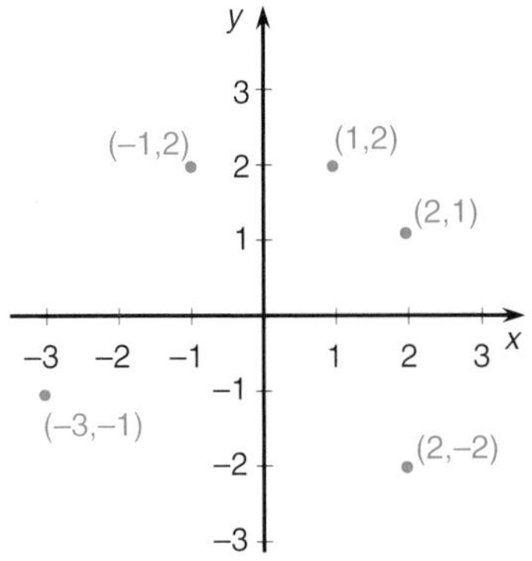

Figure 1.4.3

Another type of function we will need to use is one that is defined by one formula for some values of x and another for other values of x. For instance, consider the function

$$f(x) = \begin{cases} 2x + 7\sqrt{2}, & \text{if } x \leq 3, \\ \frac{1}{x^2-4}, & \text{if } x > 3. \end{cases}$$

This is a perfectly good function. The formula tells us quite clearly how to evaluate the function at every real number.

It is very important to become accustomed to the idea that not all functions are polynomials and rational functions.

Graphing Functions

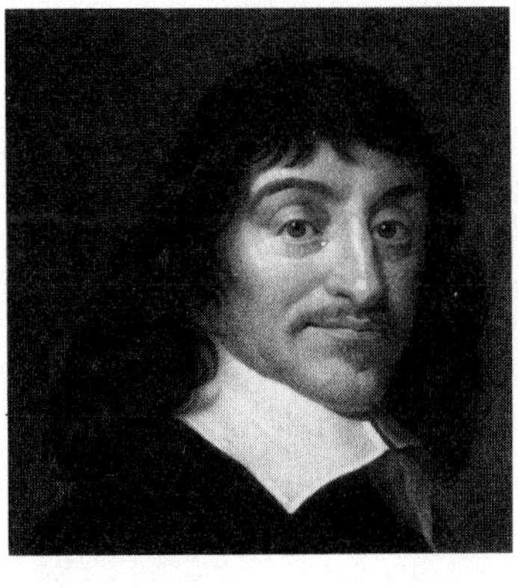

René Descartes

We recall how to describe points in the plane by their coordinates. We start with a horizontal line, which we call the **x-axis**, and a vertical line, which we call the **y-axis**. The point at which the axes cross is called the **origin**. The position of any point P in the plane can be described by giving an ordered pair of numbers (x, y). To find the corresponding point, we start at the origin, move x units along the x-axis, interpreting a positive number x to mean a movement to the right of the origin and a negative x to mean a movement to the left. Then we move y units in the vertical direction, interpreting positive values of y to refer to positions above the x-axis and negative values to refer to positions below the axis. See Figure 1.4.2.

Several points are shown in Figure 1.4.3, marked with their coordinates. The horizontal coordinate, usually denoted x, is called the **x-coordinate** or the **abscissa** of the point, and the vertical coordinate, usually denoted y, is called the **y-coordinate** or

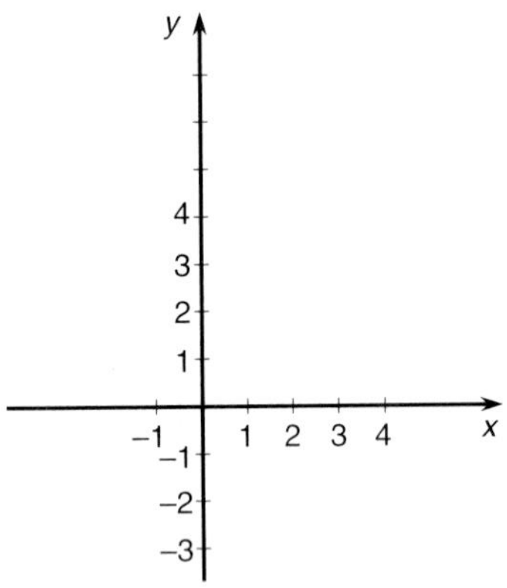

Figure 1.4.4

the **ordinate**. The ordered pair (x, y) are sometimes called the **Cartesian coordinates** of the point P, after the French philosopher and mathematician René Descartes (1596–1650).

Suppose $f(x)$ is a function. For each x in its domain, there is a value $f(x)$, and if we think of this value as the y-coordinate of a point, then we can draw the corresponding point on a graph. In other words, letting $y = f(x)$, we can plot the point (x, y).

Consider the function $f(x) = x^2 - 2x - 1$. Corresponding to each x we have the value $y = x^2 - 2x - 1$, and this is easy to evaluate for any particular x. For instance, if $x = -1$, we find that $y = (-1)^2 - 2(-1) - 1 = 1 + 2 - 1 = 2$, and we plot the point $(-1, 2)$. See Figure 1.4.4.

In a similar way we can find other points, such as $(0, -1)$, $(1, -2)$, $(2, -1)$, $(3, 2)$, $(4, 7)$, $(\frac{1}{2}, -\frac{7}{4})$, $(-1, 2)$, and so on. See Figure 1.4.5.

In Figure 1.4.6 we have plotted all the points with x moving from -1 to 4 in steps of $\frac{1}{2}$, and we begin to see a pattern emerging.

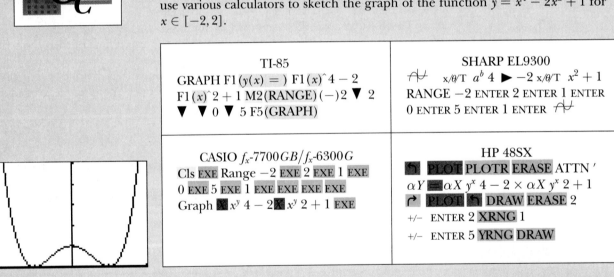

With the help of a graphics calculator it is possible to sketch the graph of a function quickly and accurately. For instance, the keystrokes below show how to use various calculators to sketch the graph of the function $y = x^4 - 2x^2 + 1$ for $x \in [-2, 2]$.

TI-85 GRAPH F1 (y(x) =) F1 (x)^4 − 2 F1 (x)^2 + 1 M2(RANGE) (−)2 ▼ 2 ▼ ▼ 0 ▼ 5 F5(GRAPH)	SHARP EL9300 ⩚ x/θ/T a^b 4 ▶ −2 x/θ/T x^2 + 1 RANGE −2 ENTER 2 ENTER 1 ENTER 0 ENTER 5 ENTER 1 ENTER ⩚
CASIO f_x-7700GB/f_x-6300G Cls EXE Range −2 EXE 2 EXE 1 EXE 0 EXE 5 EXE 1 EXE EXE EXE EXE Graph X x^y 4 − 2 X x^y 2 + 1 EXE	HP 48SX ↰ PLOT PLOTR ERASE ATTN ′ αY = αX y^x 4 − 2 × αX y^x 2 + 1 ↱ PLOT ↰ DRAW ERASE 2 +/− ENTER 2 XRNG 1 +/− ENTER 5 YRNG DRAW

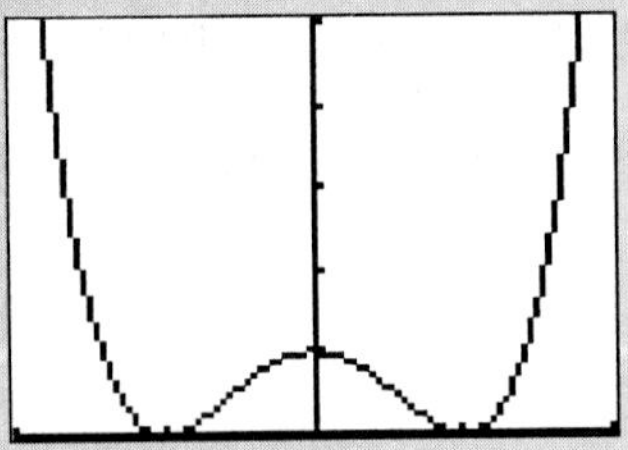

If you have a graphics calculator, you can use it to sketch the graphs of the functions we have been discussing. However, it is important to realize that it is not always easy to determine the overall shape of a graph. For instance, you could try editing any of the above sequences of keystrokes to graph the function $y = x^4 - 5x^3 + x$ for $x \in [-2, 2]$. You could then try the same function with $x \in [-3, 3]$ and with $-50 \leq y \leq 200$. Then you could try it with $x \in [-3, 6]$ and $-70 \leq y \leq 200$. The first two pictures will give seriously misleading impressions about the shape of the graph. . .

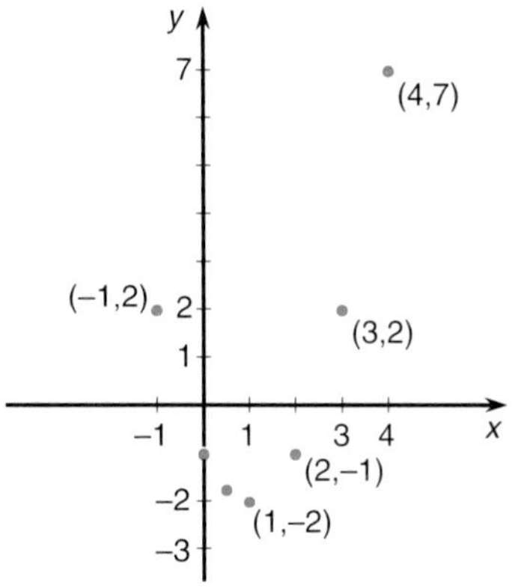

Figure 1.4.5

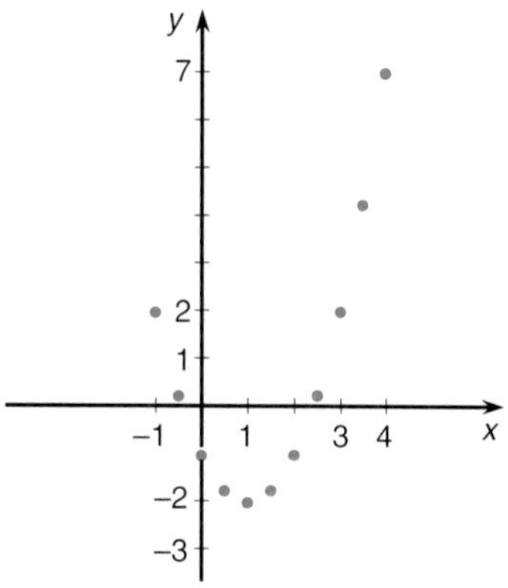

Figure 1.4.6

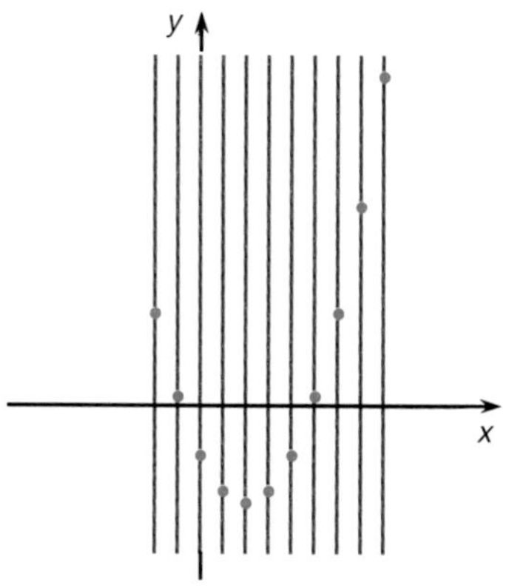

Figure 1.4.7

Since $f(x) = x^2 - 2x - 1$ is a polynomial, its domain is the set of all real numbers. There is exactly one point on the graph on each vertical line in the plane. See Figure 1.4.7.

If we could plot all the points $(x, f(x))$, we would obtain a curve that goes right across the plane from one side to the other. If we plot a reasonably large number of points, we can "fill in" the curve and get a fairly good picture of it (or at least of some part of it). This picture is called the **graph** of the function $f(x)$, or the graph of $y = f(x)$. In this particular case the curve is a "parabola," shown in Figure 1.4.8.

If we tried to graph a rational function like $R(x) = \frac{1}{x^2-1}$, for instance, we would realize that its domain is the set of all real numbers *except* ± 1. There can be no point on the graph with x-coordinate 1 or -1. On the other hand, for every other value of x, that is, for every x in the domain of f, there is exactly one point on the graph with that x as its x-coordinate. The graph of $y = \frac{1}{x^2-1}$ is shown in Figure 1.4.9.

The domain of $g(x) = \sqrt{x}$ is $\{x : x \geq 0\}$. There is a point $\big(x, g(x)\big)$ on the graph of $y = g(x)$ for each $x \geq 0$ (see Figure 1.4.10). Negative values of x are not in the domain of $g(x)$; in the picture there are no points on the graph lying to the left of the y-axis.

When we draw the graph of a function, there can never be more than one point on the graph on any vertical line. This observation shows that a circle cannot be the graph of a function, since there are vertical lines that intersect the circle in two points. See Figure 1.4.11.

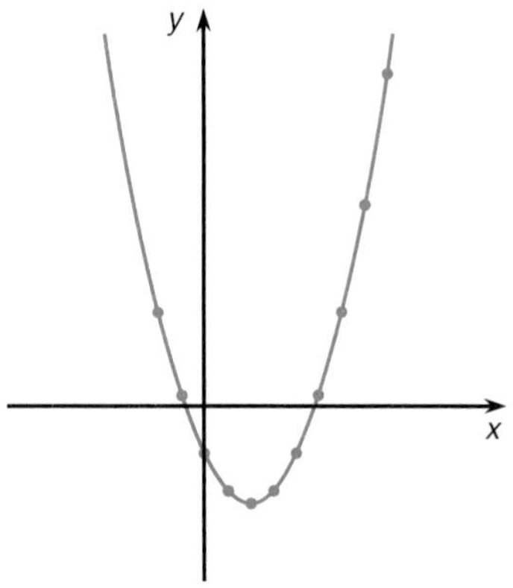

Figure 1.4.8

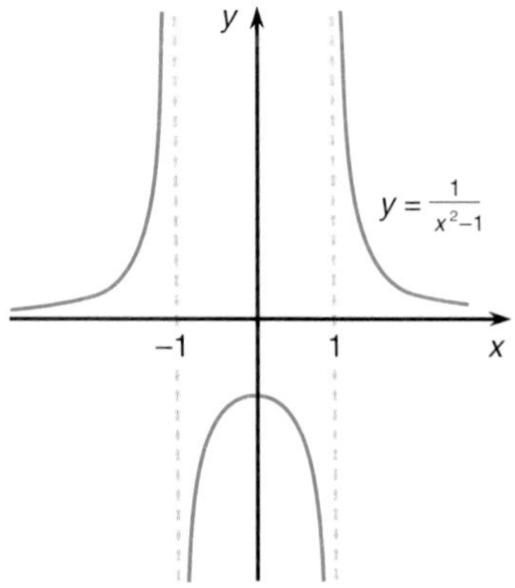

Figure 1.4.9

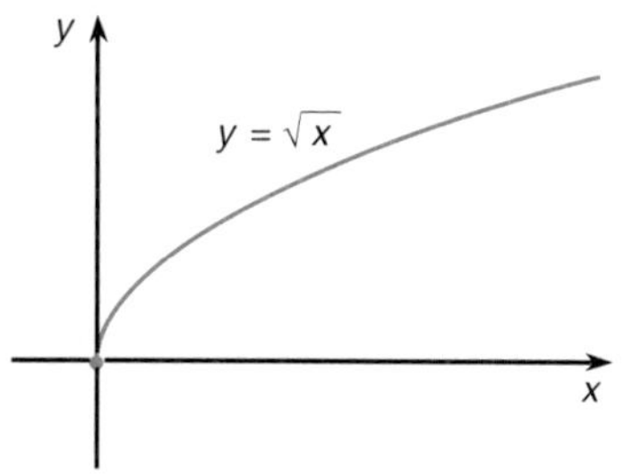

Figure 1.4.10

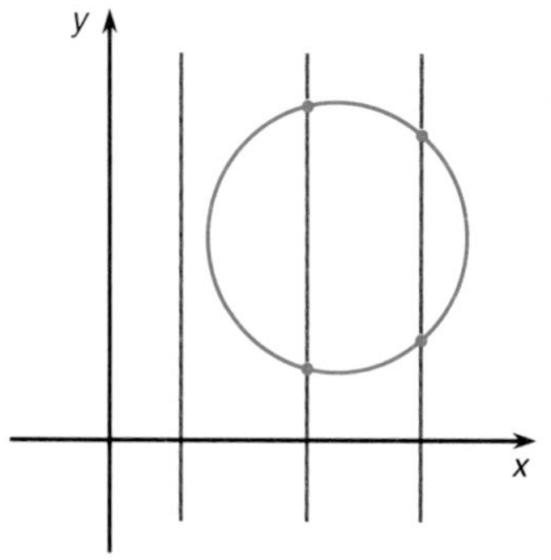

Figure 1.4.11

Exercises 1.4

Specify the domain of each of the following functions.

1. $f(x) = x^2 - 3x + 5$
2. $f(x) = x^5 - 4x^3 + 7x^2 - 4$
3. $f(x) = \frac{x^3-2x+8}{x+3}$
4. $f(x) = \frac{3x^4-2x+1}{x^2-16}$
5. $f(x) = 3x - 5\sqrt{x}$
6. $f(x) = 1 - 2\sqrt{3-4x}$
7. $f(x) = x^3 - \frac{1}{\sqrt{x+1}}$
8. $f(x) = \sqrt{x(x-4)}$
9. $f(x) = \frac{1}{1-\sqrt{x}}$
10. $f(x) = \frac{1}{1-\sqrt{x^2+4x+1}}$

Specify the range of each of the following functions.

11. $f(x) = x^2 + 5$
12. $f(x) = x^4 + 1$
13. $f(x) = (x-2)^2 + 3$
14. $f(x) = x^2 - 6x + 5$
15. $f(x) = 3x^2 - 5x + 2$
16. $f(x) = 4x^2 - 2x + 1$
17. $f(x) = \sqrt{x-6}$
18. $f(x) = 3 - 7\sqrt{2-x}$
19. $f(x) = \frac{1}{1+\sqrt{x}}$
20. $f(x) = x + 2\sqrt{x}$

Plot six points (or more) on the graph of each of the following functions, and then fill in the approximate shape of the rest of the graph.

21. $y = x^2 - 3$
22. $y = x^2 + 2$
23. $y = (x-1)^2 - 3$
24. $y = (x+2)^2 - 1$
25. $y = x^2 + 4x - 2$
26. $y = x^2 + 6x$
27. $y = 2x^2 + 1$
28. $y = 2x^2 + 4x + 3$
29. $y = 2x^2 + 3x - 8$
30. $y = \frac{1}{2}x^2 + 2x + 3$

1.5 Geometry

PYTHAGOREAN THEOREM
HYPOTENUSE
SIMILAR TRIANGLES
EQUATIONS OF LINES
INTERSECTING LINES
AREA

We begin with the Pythagorean Theorem, which is named after the ancient Greek mathematician Pythagoras (570–497 B.C.). Suppose $\triangle ABC$ is a **right-angled triangle**, with its right angle at B (i.e., $\angle ABC = 90°$). The side opposite the right angle (in this case the side AC) is called the **hypotenuse**. See Figure 1.5.1.

THEOREM 1.5.1

The Pythagorean Theorem

In a right-angled triangle the square of the length of the hypotenuse equals the sum of the squares of the lengths of the other two sides.

Consider a right-angled triangle $\triangle ABC$ with a right angle at B. If we write $|AB|$ for the length of the segment AB, and similarly for the other segments, the theorem can be stated as follows:

$$|AC|^2 = |AB|^2 + |BC|^2.$$

EXAMPLE 1

(i) Consider a right-angled triangle in which both sides adjacent to the right angle have length 1. Applying the Pythagorean Theorem, with $|AB| = |BC| = 1$, we find that $|AC|^2 = 1 + 1 = 2$, so $|AC| = \sqrt{2}$. The length of the hypotenuse is $\sqrt{2}$.

(ii) If the hypotenuse of a right-angled triangle has length 5 and one of the other sides has length 4, what is the length of the remaining side?

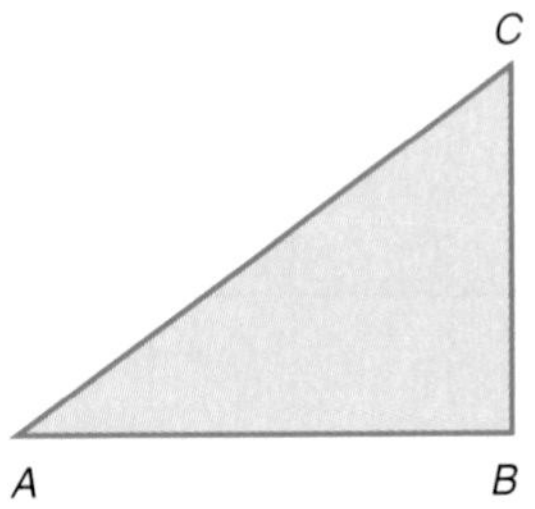

Figure 1.5.1

Solution If we assume that $|AC| = 5$ and $|AB| = 4$, then the theorem tells us that $5^2 = 4^2 + |BC|^2$, which means that $|BC|^2 = 25 - 16 = 9$, from which it follows that the length of the third side $|BC|$ is 3.

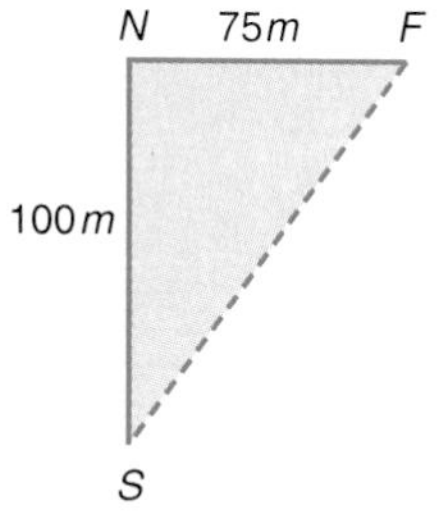

Figure 1.5.2

EXAMPLE 2 Suppose a person walks 100 meters due north and then turns and walks 75 m due east. How far will she be from her starting point?

Solution We begin by drawing a picture like the one in Figure 1.5.2, showing the starting point S, the point N which is 100 m due north of S, and the final point F which is 75 m due east of N. Then $\triangle SNF$ is a right-angled triangle with a right angle at N.

What we want to know is the distance between the starting point and the final point, which means the length of the segment SF. From the Pythagorean Theorem we see that

$$|SF|^2 = |SN|^2 + |NF|^2 = 100^2 + 75^2 = 15{,}625.$$

The length is the square root: $|SF| = \sqrt{15{,}625} = 125$. The person in the question is now 125 meters away from her starting point.

We can also use the Pythagorean Theorem to calculate distances in the plane.

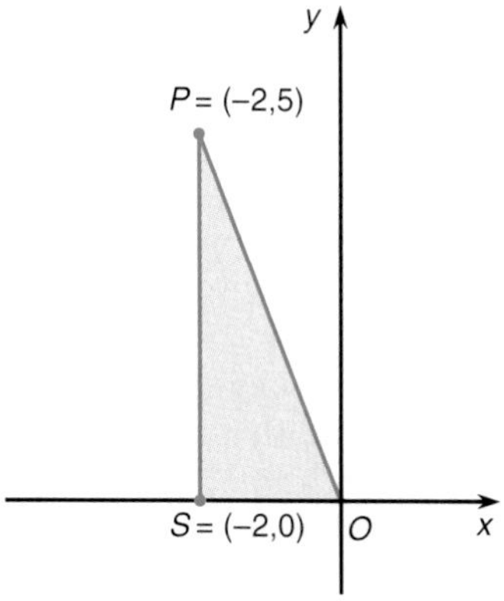

Figure 1.5.3

EXAMPLE 3 How far is the point $P = (-2, 5)$ from the origin $O = (0, 0)$?

Solution We begin by drawing a picture, showing the points P and O and a vertical line from P down to the x-axis. The point where this line meets the x-axis is the point $S = (-2, 0)$. See Figure 1.5.3.

The triangle PSO is a right-angled triangle with right angle at S. The distance $|OP|$ is the length of the hypotenuse.

The Pythagorean Theorem says that $|OP|^2 = |OS|^2 + |SP|^2 = 4 + 25 = 29$. The distance from P to O is $|OP| = \sqrt{29} \approx 5.385$.

The same procedure can be used to find the distance from any point to the origin, or the distance between any two points.

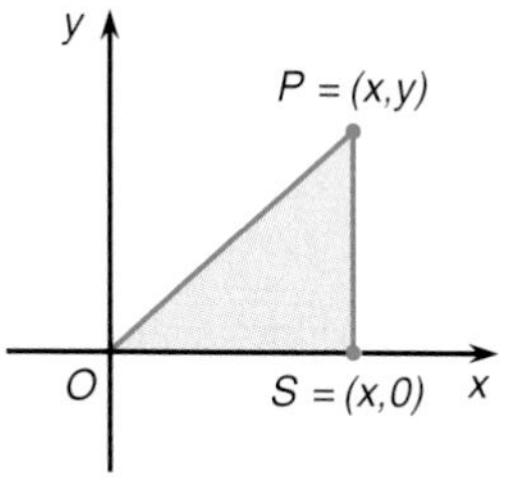

Figure 1.5.4

EXAMPLE 4

(i) Find the distance from $P = (x, y)$ to the origin $O = (0, 0)$.

(ii) Find the distance between $P = (x, y)$ and $Q = (a, b)$.

Solution

(i) As in the previous example, we draw the point S on the x-axis with the same x-coordinate as P, that is, $S = (x, 0)$. See Figure 1.5.4. Then $|OP|^2 = |OS|^2 + |SP|^2 = x^2 + y^2$, so $|OP| = \sqrt{x^2 + y^2}$.

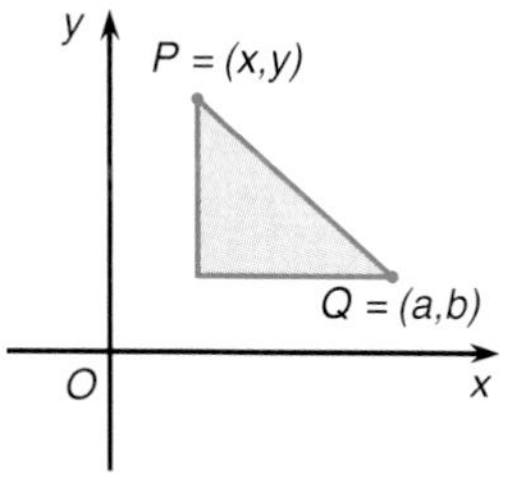

Figure 1.5.5

(ii) In Figure 1.5.5 we draw a right-angled triangle whose hypotenuse is PQ and with one vertical side and one horizontal side. The length of the horizontal side is the horizontal distance between P and Q, which is $|x - a|$. Similarly the length of the vertical side is $|y - b|$. The Pythagorean Theorem says that the distance between $P = (x, y)$ and $Q = (a, b)$ is

$$|PQ| = \sqrt{|PS|^2 + |SQ|^2} = \sqrt{(x - a)^2 + (y - b)^2}. \tag{1.5.2}$$

This formula for the distance between two points is extremely important and useful.

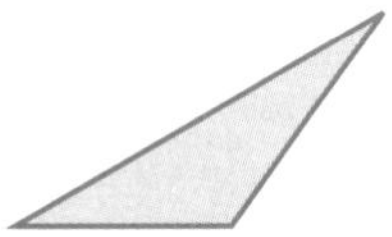

Figure 1.5.6

The distances between points in the plane satisfy the triangle inequality (see Theorem 1.2.2). What it says in this context is that the length of any side of a triangle is less than or equal to the sum of the lengths of the other two sides. See Figure 1.5.6.

THEOREM 1.5.3

> **The Triangle Inequality**
>
> If P, Q, and R are three points in the plane, then
>
> $$|PQ| \leq |PR| + |QR|.$$

Similar Triangles

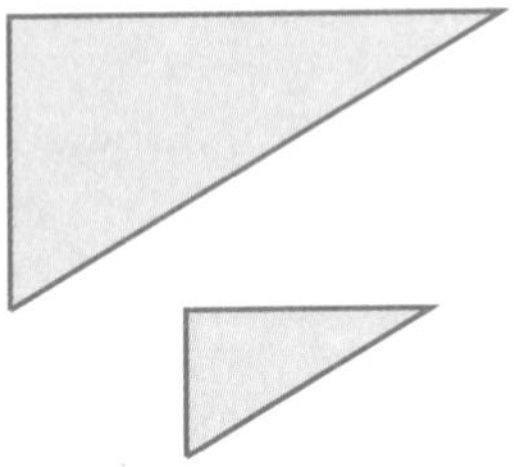

Figure 1.5.7

Two triangles $\triangle ABC$ and $\triangle DEF$ are **similar** if their respective angles are equal, that is, if $\angle ABC = \angle DEF$ and $\angle BCA = \angle EFD$ and $\angle CAB = \angle FDE$. In fact, since the angles of any triangle always add up to a total of $180°$, it is enough to check that any *two* pairs of angles are equal, since the third pair will then automatically have to be equal too.

In Figure 1.5.7 we see two right-angled triangles (notice that this means that one pair of angles, the right angles, are already equal). In fact the triangles are similar. (The angle at the extreme right end of each triangle is $30°$, so these two angles are equal. The remaining angle in each triangle is $180 - 90 - 30 = 60°$.)

The triangles are clearly not the same size; the hypotenuse of one is half as long as the hypotenuse of the other. Because the triangles are similar, that is, have the same angles, we can think of one as a "scaled-down" version of the other. In this particular case it is scaled down by a factor of $\frac{1}{2}$, and we expect that each side of the small triangle should be half as long as the corresponding side of the larger one.

The same idea will work for any similar triangles. The sides of one will all be the same multiple of the "corresponding" sides of the other. It is not immediately clear what is meant by "corresponding," but one good way to say it is to label sides by saying which angle they are *opposite*. So the hypotenuse of a right-angled triangle is the side opposite the right angle. In the example illustrated above, we can speak of the side opposite the $30°$ angle and the side opposite the $60°$ angle. By **corresponding sides** in two similar triangles we mean sides that are opposite equal angles.

THEOREM 1.5.4

> Suppose $\triangle ABC$ and $\triangle DEF$ are similar, with $\angle ABC = \angle DEF$ and $\angle BCA = \angle EFD$ (and so $\angle CAB = \angle FDE$). Then
>
> $$\frac{|AB|}{|DE|} = \frac{|AC|}{|DF|} = \frac{|BC|}{|EF|}.$$
>
> In words we say that the ratios of the lengths of each pair of corresponding sides are all the same.

EXAMPLE 5 Consider a person standing near a streetlight, as shown in Figure 1.5.8. If the person is 6 ft tall, standing 20 ft from the base of the streetlight, and the shadow is 8 ft long, how high is the streetlight above the ground?

Figure 1.5.8

Solution We draw the picture again, labeling the light A, the person's head C, the end of the shadow B, the person's feet D, and the base of the lamppost O. See Figure 1.5.9.

The large triangle $\triangle ABO$ is similar to the smaller triangle $\triangle CBD$. This is because they have the same angle at B and each has a right angle (at O and D, respectively). We know the person is 6 ft tall, so $|DC| = 6$, and we want to find $|OA|$, the height of the streetlight.

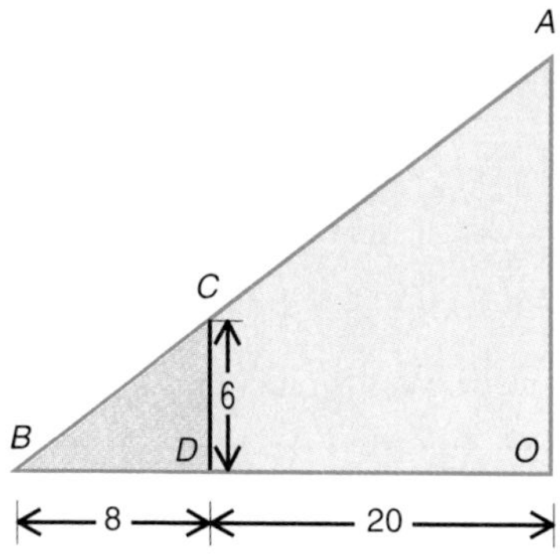

Figure 1.5.9

Now $\frac{|OA|}{|DC|} = \frac{|BO|}{|BD|}$, since these are two pairs of corresponding sides (OA and DC are both opposite to the angle at B, and BO and BD are opposite to angles $\angle BAO$ and $\angle BCD$ in their respective triangles).

We know that the length of the shadow is $|BD| = 8$, so the only thing left to find is $|BO| = |BD| + |DO| = 8 + 20 = 28$.

Putting all this together, we find that

$$\frac{|OA|}{6} = \frac{28}{8}, \qquad \text{so} \qquad |OA| = \frac{(6)(28)}{8} = 21.$$

The streetlight is 21 ft high.

Straight Lines

Consider three points P, Q, and R on a straight line ℓ. We draw some horizontal and vertical lines as shown in Figure 1.5.10 and label the points A, B, and C. There are three similar triangles in this diagram, $\triangle PQA$, $\triangle PRB$, and $\triangle QRC$. From 1.5.4 we conclude that $\frac{|QA|}{|RB|} = \frac{|PA|}{|PB|}$ and $\frac{|RB|}{|RC|} = \frac{|PB|}{|QC|}$, so

$$\frac{|QA|}{|PA|} = \frac{|RB|}{|PB|} = \frac{|RC|}{|QC|}.$$

Each of these ratios is obtained by taking two of the points on the line and taking the quotient of the vertical distance between them by the horizontal distance between them. Since the three points we started with could have been any three points on the line, we see that this ratio will be the same for any pair of points.

If the coordinates of P are (x, y) and the coordinates of Q are (u, v), then the vertical distance between them is $|y - v|$ and the horizontal distance is $|x - u|$. The ratio mentioned above is $\frac{|y-v|}{|x-u|}$, assuming that $x \neq u$.

If, for instance, this ratio equals 1, it means that the horizontal distance and vertical distance between P and Q are the same, which means that the line through the slopes at an angle of 45°. The line could slope up or down; because of the absolute value signs, the ratio would be the same.

This suggests that it might be better to consider the ratio $\frac{y-v}{x-u}$. If the line slopes upward, then the numerator and denominator will both have the same sign, while if the line slopes downward, they will have opposite signs (check this by looking at Figure 1.5.11).

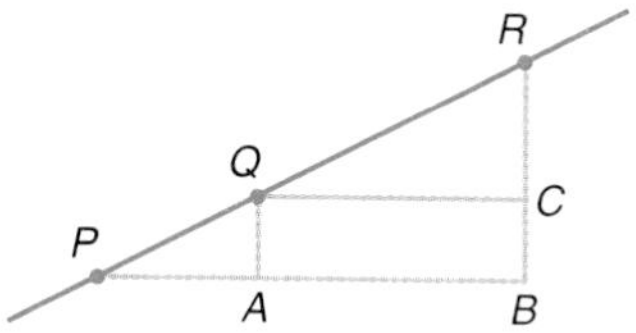

Figure 1.5.10

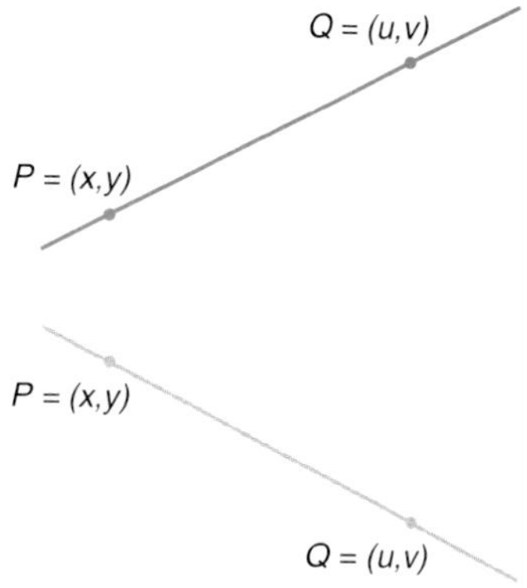

Figure 1.5.11

DEFINITION 1.5.5

If $P = (x, y)$ and $Q = (u, v)$ are two distinct points on the line ℓ, then provided $x \neq u$, the ratio

$$m = \frac{y - v}{x - u}$$

is called the **slope** of the line.

If $x = u$, the line ℓ must be a vertical line, and we say its slope is *undefined.*

REMARK 1.5.6

(i) If the slope of ℓ is $m = 0$, then $y = v$, that is, P and Q have the same y-coordinate, and ℓ is a horizontal line.

(ii) We have already noticed that $m > 0$ if ℓ slopes upward to the right, and $m < 0$ if ℓ slopes downward to the right. See Figure 1.5.12.

(iii) If $m = \frac{y-v}{x-u}$ is a large positive number, then $y - v$ is large in comparison to $x - u$, which means that the line slopes *steeply* upward. If m is a small positive number, it means that ℓ slopes gradually upward. If m is negative but close to zero, then ℓ slopes gradually downward, and if m is a negative number whose absolute value is large, then ℓ slopes steeply downward. See Figure 1.5.13.

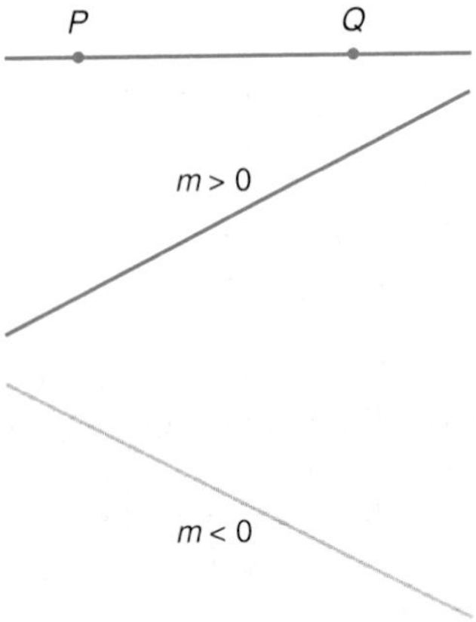

Figure 1.5.12

Now suppose ℓ is a line containing the point $Q = (x_0, y_0)$ and whose slope is m (in particular, this means the line is not vertical, since the slope is not defined for a vertical line). Let $P = (x, y)$ be any other point on the line ℓ (see Figure 1.5.14). Then we know that $\frac{y-y_0}{x-x_0} = m$. Multiplying by $x - x_0$, we find

$$y - y_0 = m(x - x_0), \qquad \text{or} \qquad y = mx + (y_0 - mx_0). \tag{1.5.7}$$

This is called the **point-slope form** of the equation of the line ℓ, because to write it, we need to know a point on the line and its slope. The points $P = (x, y)$ that satisfy it are exactly the points on the line.

If we let $b = y_0 - mx_0$, Equation 1.5.7 can be written as

$$y = mx + b. \tag{1.5.8}$$

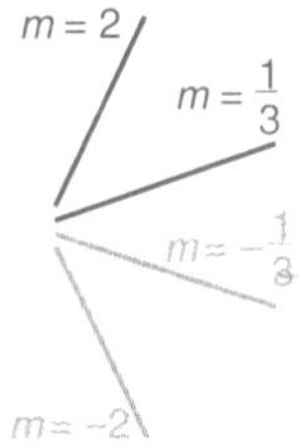

Figure 1.5.13

This is called the **slope-intercept form** of the equation of the line ℓ. It involves the slope m, but we need to discover the significance of the constant b. If we let $x = 0$, we find from Equation 1.5.8 that $y = b$. So the point $(0, b)$ is on the line; b is the y-coordinate of the point at which ℓ crosses the y-axis (see Figure 1.5.15). This y-coordinate b is called the **y-intercept** of ℓ, which explains the name.

EXAMPLE 6 Find the equation of each of the following lines: (i) the line containing the point $(3, 1)$ and having slope $m = -2$; (ii) the line containing the points $(1, 2)$ and $(5, -2)$; (iii) the line whose slope is $m = 3$ and whose y-intercept is $b = -2$.

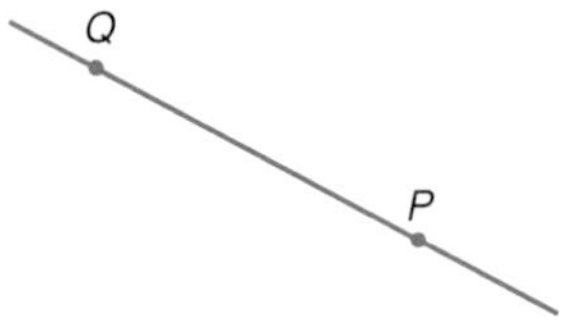

Figure 1.5.14

Solution

(i) The point-slope form is $y - 1 = -2(x - 3)$, that is, $y = 1 - 2x + 6$, or $y = 7 - 2x$. It can also be written as $y + 2x = 7$, or $y + 2x - 7 = 0$, or $2x + y - 7 = 0$, or $2x + y = 7$.

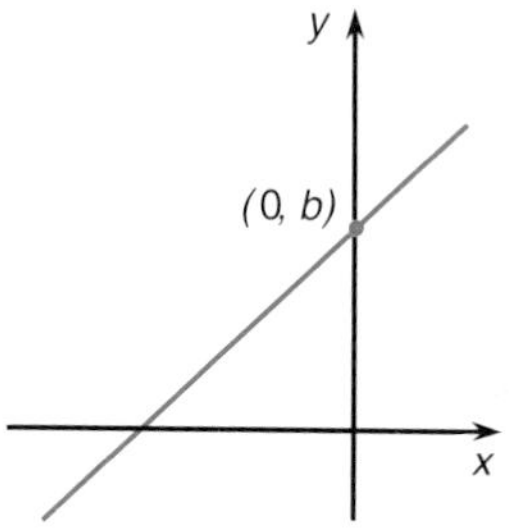

Figure 1.5.15

(ii) We need to find the slope. Using the two points we are given, we find that $m = \frac{2-(-2)}{1-5} = \frac{4}{-4} = -1$. Then from the point-slope form, the equation of the line is $y - 2 = -(x - 1)$, or $y + x - 3 = 0$.

(iii) The point-intercept form gives $y = 3x + (-2)$, or $y = 3x - 2$, which can also be written as $y - 3x + 2 = 0$.

We might observe that we have been able to write the equation of each of these lines in the form $ax + by + c = 0$, where a, b, and c are real numbers.

If a and b are not both zero, the equation

$$ax + by + c = 0 \tag{1.5.9}$$

is called the **general form of the equation of a line**. Any straight line has an equation in this form, and any equation in this form is the equation of a straight line.

REMARK 1.5.10

(i) The line $ax + by + c = 0$ is vertical if $b = 0$, and it is horizontal if $a = 0$. Notice that if $b \neq 0$, it is possible to rearrange and divide by b to obtain $y = -\frac{ax+c}{b}$. This expresses y as a function of x. If $b = 0$, the line is vertical, and it is not possible to express y as a function of x (recall that the graph of a function can intersect any vertical line in at most one point).

(ii) If $b \neq 0$, the line $ax + by + c = 0$ is not vertical. We have seen that the equation can be written as $y = -\frac{ax+c}{b}$, or $y = -\frac{a}{b}x - \frac{c}{b}$. This is the slope-intercept form of the equation; in particular, for the line defined by Equation 1.5.9, if $b \neq 0$,

$$y\text{-intercept} = -\frac{c}{b}, \qquad \text{slope} = -\frac{a}{b}. \tag{1.5.11}$$

While we are doing this, we might also like to find the **x-intercept**. A point (x, y) is on the x-axis if $y = 0$, so if the point $(x, 0)$ is on the line $ax + by + c = 0$, we see that $ax + b(0) + c = 0$, or $ax + c = 0$. If $a \neq 0$, this shows that the x-intercept is $x = -\frac{c}{a}$.

(iii) If $ax + by + c = 0$ is the equation of the line ℓ, then the equation $2ax + 2by + 2c = 0$ is *also* an equation of ℓ. So is the equation we get by multiplying by *any* nonzero number.

(iv) Suppose two lines ℓ_1 and ℓ_2 have the same slope. There are two possibilities: Either they are the same line or they never meet, that is, they have no points in common. In this latter case we say they are **parallel lines**. See Figure 1.5.16.

Figure 1.5.16

EXAMPLE 7 Find the point of intersection of the lines $3x + 2y + 1 = 0$ and $2x - y - 4 = 0$.

Solution If (x, y) is the point of intersection, it lies on both lines. This means it satisfies both equations:

$$3x + 2y + 1 = 0 \qquad \text{and} \qquad 2x - y - 4 = 0.$$

We use one of the equations to solve for one of the variables x and y in terms of the other. For instance, the first equation can be rearranged into $2y = -3x - 1$, so $y = -\frac{1}{2}(3x + 1)$. Then we "substitute" this expression for y into the second equation: $2x - y - 4 = 0$ becomes $2x + \frac{1}{2}(3x+1) - 4 = 0$, or $2x + \frac{3}{2}x + \frac{1}{2} - 4 = 0$. Combining terms, we find that $\frac{7}{2}x - \frac{7}{2} = 0$, or $\frac{7}{2}x = \frac{7}{2}$, or $x = 1$.

Next we take this value and substitute it into either equation to find the corresponding value of y. For instance, the first equation $3x + 2y + 1 = 0$ becomes $3(1) + 2y + 1 = 0$, or $2y + 4 = 0$, or $2y = -4$, so $y = -2$.

The point of intersection is $(1, -2)$.

You should check that substituting $x = 1$ into the second equation also gives $y = -2$.

The technique we used was to solve for one variable in terms of the other, using one of the equations, and then substitute the resulting expression into the other equation. This gives an equation with only one variable, so it is easily solved. This process is called "eliminating" one variable; in the solution above, we eliminated y to get an equation involving only x.

We could just as well have eliminated x and solved for y. Another thing to notice is that if we had used the second equation, $2x - y - 4 = 0$, we would have found $2x - 4 = y$, or $y = 2x - 4$. This is simpler because it does not involve any fractions. The same answer will result no matter which variable is eliminated, but sometimes one choice will give easier calculations than another. Specifically, if the coefficient of either x or y is 1 or -1 in either equation, then using that equation allows us to solve for that variable without dividing.

EXAMPLE 8 Find the point of intersection of the two lines $2x - 3y + 1 = 0$ and $-4x + 6y - 3 = 0$.

Solution We use the first equation to find $2x = 3y - 1$, so $x = \frac{1}{2}(3y-1)$. Substituting this into the second equation, we find

$$0 = -4x + 6y - 3 = -4\left(\frac{1}{2}\right)(3y-1) + 6y - 3$$
$$= -2(3y-1) + 6y - 3 = -6y + 2 + 6y - 3 = -1.$$

We have just found that $0 = -1$. This is clearly impossible. The reason is that in effect we *assumed* that there was a point (x, y) on both lines. These two lines are parallel (check their slopes), so there is no point of intersection. See Figure 1.5.17.

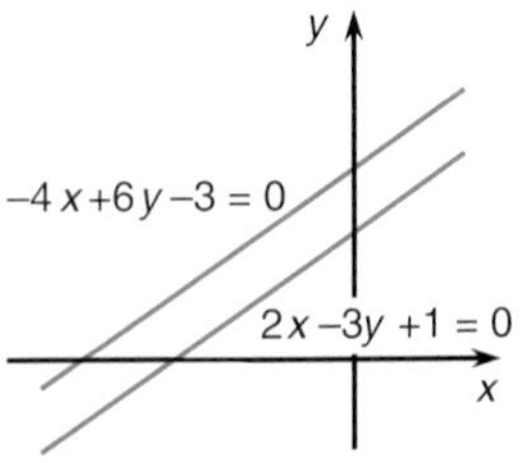

Figure 1.5.17

Finally consider two lines that are perpendicular. Let us assume that neither of them is horizontal (or vertical). In Figure 1.5.18 we have labeled the point of intersection I and have marked a point P on one of the lines. Also shown is a point Q on the other line, with the same y-coordinate as P. Finally we mark the point R between P and Q that is directly above (or below) I.

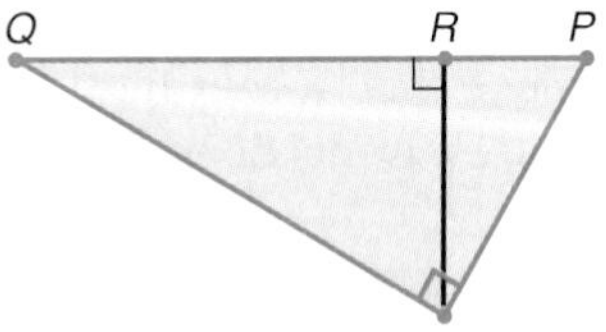

Figure 1.5.18

The right-angled triangles $\triangle QIP$ and $\triangle QRI$ share the angle $\angle PQI = \angle IQR$. Because of this, they are similar triangles.

In the same way, $\triangle QIP$ and $\triangle IRP$ are similar because they share the angle $\angle IPQ = \angle RPI$. Combining these facts, we see that $\triangle QRI$ is similar to $\triangle IRP$. Comparing corresponding sides, we get

$$\frac{|RP|}{|RI|} = \frac{|RI|}{|QR|}.$$

The right side of this equation is the absolute value of the slope of the line IQ. The left side is the reciprocal of $\frac{|RI|}{|RP|}$, which is the absolute value of the slope of IP. So the absolute values of the slopes of the two perpendicular lines are each other's reciprocals. Since one of the lines must slope up and the other down, we see that one must have positive slope and the other must have negative slope. This means that their slopes must be the *negatives* of each other's reciprocals.

Neither of these statements applies when one line is horizontal and the other is vertical, since the slope of a vertical line is not defined.

In fact the reverse is also true: If the slopes are negative reciprocals, then the lines are perpendicular.

THEOREM 1.5.12

> Suppose ℓ_1 and ℓ_2 are two lines with nonzero slopes m_1 and m_2, respectively. To say that the lines are perpendicular is equivalent to the condition
> $$m_1 = -\frac{1}{m_2}.$$

Graphing Inequalities

EXAMPLE 9 Describe each of the following sets of points: (i) $\{(x, y) : y + 2x \leq 4\}$; (ii) $\{(x, y) : 3x - y + 2 < 0\}$.

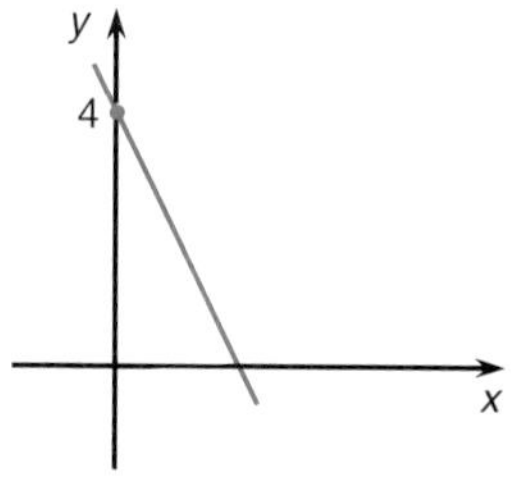

Figure 1.5.19

Solution

(i) We rewrite the condition $y + 2x \leq 4$ as $y \leq -2x + 4$. The corresponding equation $y = -2x + 4$ is the point-intercept form of the equation of a line, sketched in Figure 1.5.19.

The condition $y \leq -2x + 4$ means that for any particular value of x, we are asking that y should be less than or equal to the y-coordinate for the corresponding point on the line. The required points are exactly those lying on or below the line. See Figure 1.5.20.

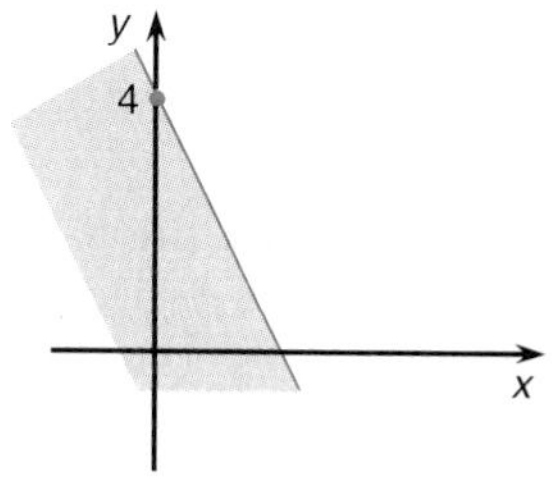

Figure 1.5.20

(ii) Again we rewrite the condition $3x - y + 2 < 0$ as $3x + 2 < y$, or $y > 3x + 2$. We sketch the corresponding line $y = 3x + 2$ and observe that $y > 3x + 2$ means that y is strictly greater than the y-coordinate of the point on that line. The set consists of all points that lie above the line $y = 3x + 2$ but not on it. See Figure 1.5.21.

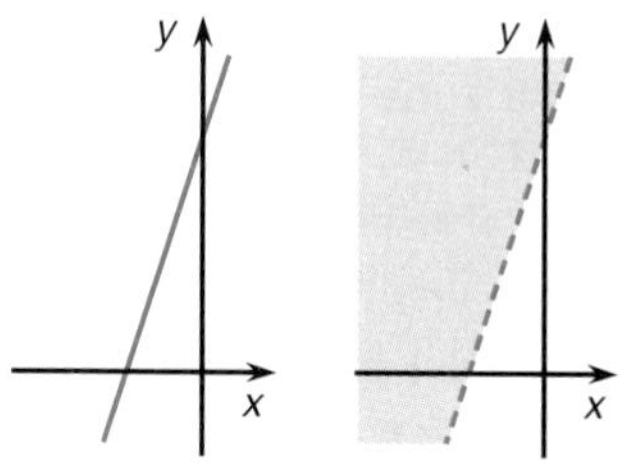

Figure 1.5.21

EXAMPLE 10 Find a formula to describe the set of all points that lie inside the circle of radius 2 centered at the point $(3, -2)$.

Solution The set consists of all points whose distance from the point $(3, -2)$ is less than 2. The distance from the point (x, y) to the point $(3, -2)$ is $\sqrt{(x-3)^2 + (y+2)^2}$, so we need $\sqrt{(x-3)^2 + (y+2)^2} < 2$. Squaring both sides, we get $(x-3)^2 + (y+2)^2 < 4$, or $x^2 - 6x + y^2 + 4y + 13 < 4$, or $x^2 - 6x + y^2 + 4y < -9$. See Figure 1.5.22.

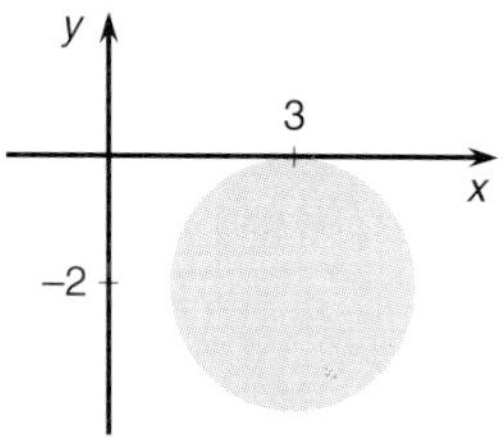

Figure 1.5.22

Notice that the condition for the points that lie *on* the circle is the *equation* $x^2 - 6x + y^2 + 4y = -9$.

EXAMPLE 11 Describe the set $\{(x, y) : x + y > 1 \text{ and } x^2 + y^2 \leq 9\}$.

Solution The first condition, $x + y > 1$, can be written as $y > -x + 1$, so it describes the points that lie above the line $y = -x + 1$, the region illustrated in Figure 1.5.23.

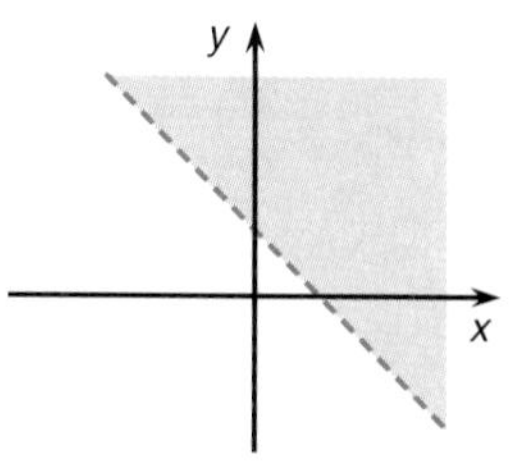

Figure 1.5.23

The second condition can be rewritten as $\sqrt{x^2 + y^2} \leq 3$. This inequality describes the points whose distance from the origin is less than or equal to 3. This is the set of all points lying inside or on the circle $x^2 + y^2 = 9$, the circle of radius 3 about the origin.

The required set is the set of all points that lie inside or on the circle of radius 3 about the origin and above the line $y = -x + 1$.

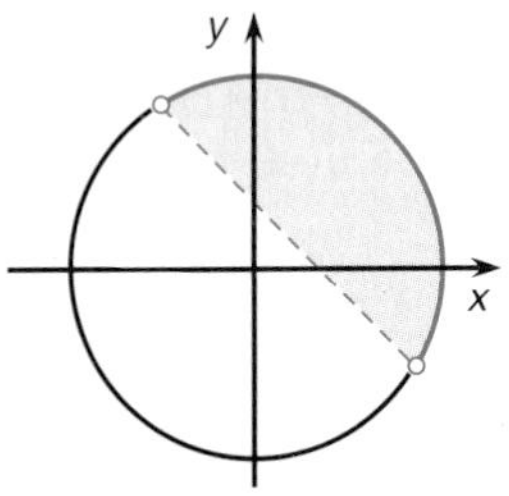

Figure 1.5.24

This set is illustrated in Figure 1.5.24. We use a solid line for the part of the circle that is included in the set and a dashed line for the line segment, which is not.

Areas

The **area of a rectangle** is the product of its length and width. The area of a circle of radius r is $A = \pi r^2$. See Figure 1.5.25.

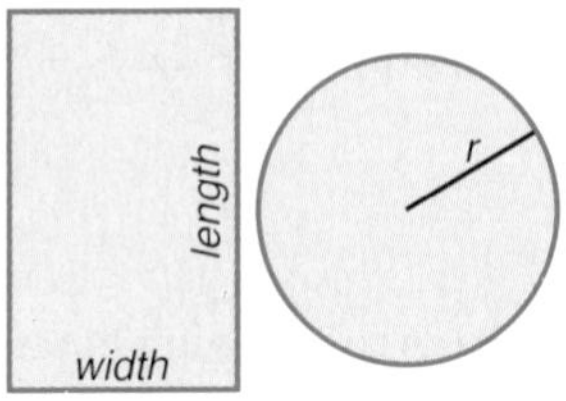

Figure 1.5.25

The **area of a triangle** is a little more difficult. We first consider the area of a parallelogram, as shown in Figure 1.5.26. The parallelogram is drawn with one side (the "base") horizontal. If we cut off the right side of this parallelogram and move that piece over to the left, we can reassemble it into a rectangle, as shown in Figure 1.5.27(i). This rectangle has the same base as the original parallelogram, and the height of the rectangle is what we might call the "height" of the parallelogram, the *perpendicular distance* from the base to the opposite side. See Figure 1.5.27(ii).

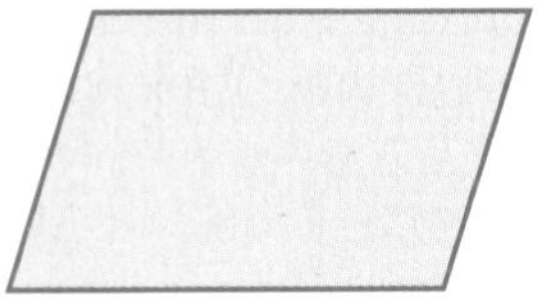

Figure 1.5.26

The **area of the parallelogram** equals the area of the rectangle we have just constructed, which equals the product of the base of the parallelogram and the height of the parallelogram.

Strictly speaking, this argument does not always work. If we consider the long thin parallelogram shown in Figure 1.5.28, we cannot simply cut off the right side and reassemble into a rectangle. However, even in this case the area does equal the base times the height.

We drew our parallelogram with its base horizontal, but of course not every parallelogram is in this position. If we want to find the area of a tilted parallelogram, we must decide to call one of the sides its base and interpret the height accordingly. In Figure 1.5.29 the "base" is vertical, and the "height" is horizontal. This is mildly surprising but causes no real difficulty.

As with parallelograms, the "base" and the "height" may not always refer to a horizontal side at the bottom and a vertical distance, but the height is always measured in a direction perpendicular to the base.

Now we return to the area of a triangle. We can add another triangle with the same area and end up with a parallelogram (see Figure 1.5.30). The area of the parallelogram is twice the area of the original triangle, and it is also the base times the height. Because of the way we constructed the parallelogram, its base is the base of the original triangle, and its height is the "height" of the triangle, meaning the perpendicular distance from the base to the opposite vertex. So

$$2 \times (\textit{Area of triangle}) = \textit{base} \times \textit{height},$$

so

$$\textit{Area of triangle} = \frac{1}{2}\textit{base} \times \textit{height}.$$

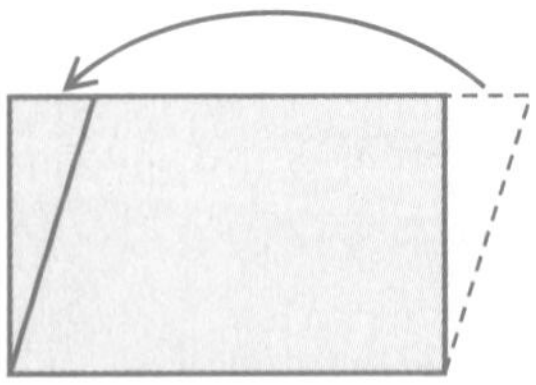

Figure 1.5.27(i)

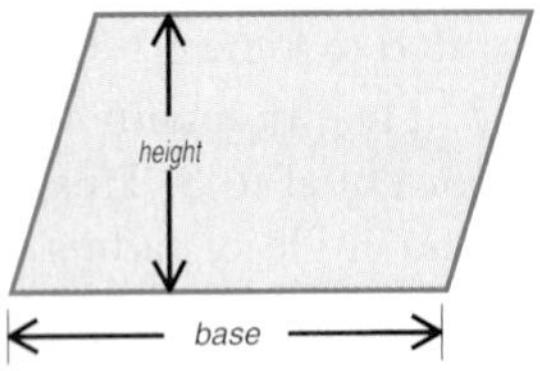

Figure 1.5.27(ii)

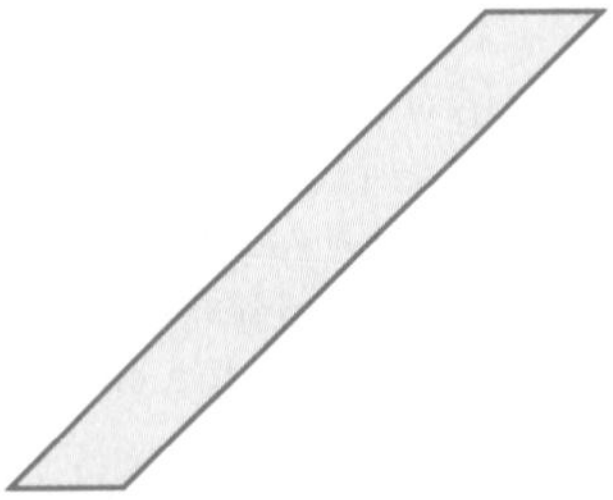

Figure 1.5.28

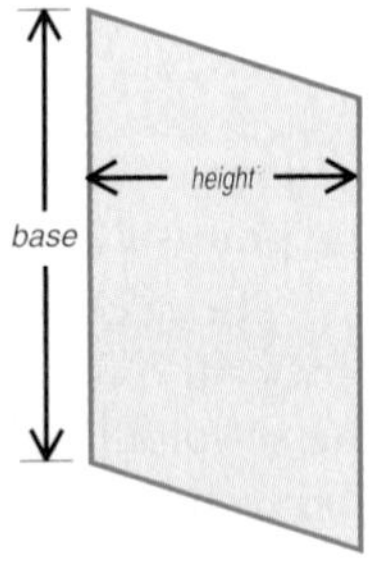

Figure 1.5.29

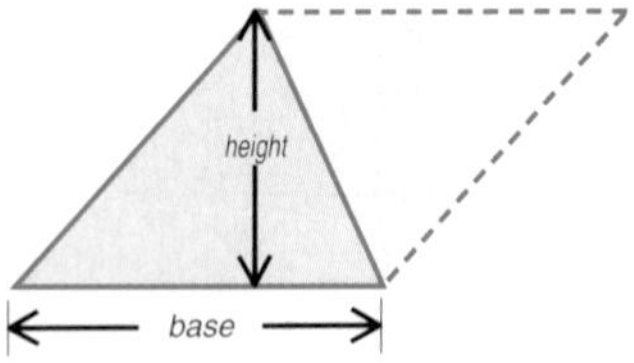

Figure 1.5.30

Summary 1.5.13 Areas

$$\begin{aligned} \textit{Area of rectangle} &= \textit{length} \times \textit{width} \\ \textit{Area of circle} &= \pi r^2, \qquad (r = \textit{radius}) \\ \big(\textit{Circumference of circle} &= 2\pi r, \qquad (r = \textit{radius})\big) \\ \textit{Area of parallelogram} &= \textit{base} \times \textit{height} \\ \textit{Area of triangle} &= \frac{1}{2}\textit{base} \times \textit{height} \end{aligned}$$

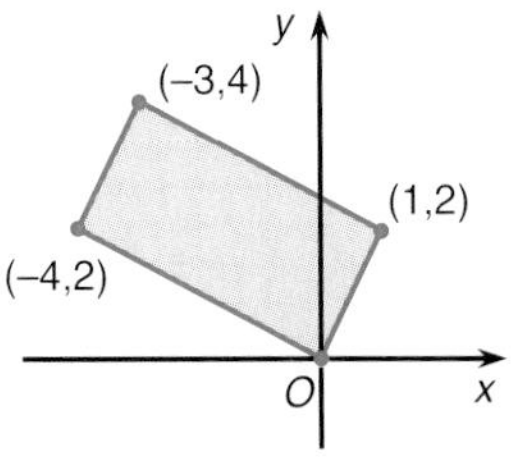

Figure 1.5.31

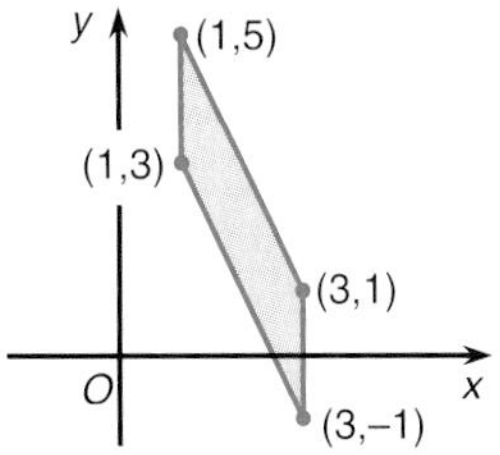

Figure 1.5.32

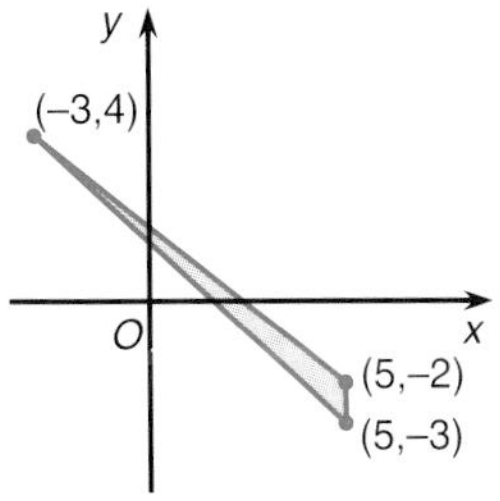

Figure 1.5.33

EXAMPLE 12 Find the area of each of the following figures: (i) the rectangle with vertices $(0, 0)$, $(1, 2)$, $(-4, 2)$, and $(-3, 4)$; (ii) the parallelogram with vertices $(1, 3)$, $(1, 5)$, $(3, -1)$ and $(3, 1)$; (iii) the triangle with vertices $(-3, 4)$, $(5, -3)$, and $(5, -2)$.

Solution

(i) Since we are told that the figure is a rectangle, all that is needed to find the area is to multiply the lengths of two adjacent sides. The length of the side joining $(0, 0)$ and $(1, 2)$ is $\sqrt{(1-0)^2 + (2-0)^2} = \sqrt{5}$. The length of the side joining $(0, 0)$ and $(-4, 2)$ is $\sqrt{(-4-0)^2 + (2-0)^2} = 2\sqrt{5}$. The area is the product of these two lengths, or $A = \sqrt{5} \times 2\sqrt{5} = 2(5) = 10$. See Figure 1.5.31.

(ii) The side joining $(1, 3)$ and $(1, 5)$ is vertical, so it will be convenient to use it as the base. Its length is 2. The "height" is the *horizontal* distance from this vertical side to the other side (through $(3, -1)$ and $(3, 1)$), which we see from Figure 1.5.32 is 2. The area is $A = \textit{base} \times \textit{height} = 2 \times 2 = 4$.

We chose the vertical side as the base to make the "height" easy to measure. If the base was a slanting side, it would be more difficult to measure the height, a distance that must be measured in the direction *perpendicular* to the base.

(iii) The side through $(5, -3)$ and $(5, -2)$ is vertical. We use it as the base; its length is 1 and the height is the distance from the vertical line $x = 5$ to the vertical line through the other vertex, that is, $x = -3$. The distance between these two lines is $5 - (-3) = 8$, and the area of the triangle is $\frac{1}{2}\textit{base} \times \textit{height} = \frac{1}{2}(1 \times 8) = 4$. See Figure 1.5.33.

We summarize some useful formulas about the volumes of various solid regions as follows.

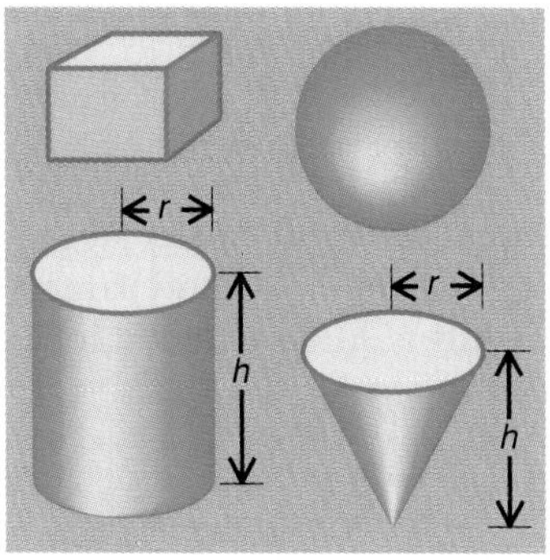

Figure 1.5.34

Summary 1.5.14 Volume Formulas (see Figure 1.5.34)

(i) The volume of a rectangular box is the product of its three dimensions: *length* × *width* × *height.*

(ii) The volume of a ball of radius r is $\frac{4}{3}\pi r^3$.

(iii) The volume of a cylinder of height h and with cross-sectional radius r is (*cross-sectional area* × *height*) $= \pi r^2 h$.

(iv) The volume of a cone of height h and radius r is $\frac{1}{3}\pi r^2 h$.

Exercises 1.5

Find each of the following.

1. The length of the hypotenuse of a right-angled triangle whose other two sides have lengths 2 and 24.
2. The length of the third side of a right-angled triangle whose hypotenuse has length 15 and another side of length 12.
3. The distance between the points $(1, -2)$ and $(4, 2)$.
4. The distance between the points $(5, 3)$ and $(-7, 4)$.
5. The distance from the point $(6, -8)$ to the origin.
6. The distance from the point $(3x - 1, 4 - x)$ to the origin.
7. The area of the triangle whose vertices are $(0, 0)$, $(0, 4)$, and $(2, 3)$.
8. The area of the triangle whose vertices are $(1, 3)$, $(1, 1)$, and $(-5, 4)$.
9. The area of the triangle whose vertices are $(-2, 1)$, $(2, 3)$, and $(6, 3)$.
10. The area of the triangle whose vertices are $(1, 1)$, $(2, 2)$, and $(5, 5)$.

Find the equation of each of the following straight lines.

11. The line through $(1, 2)$ and $(4, 5)$.
12. The line through $(-2, 4)$ and $(2, 2)$.
13. The line through $(3, 0)$ and $(-5, 4)$.
14. The line through $(-1, 3)$ and $(-1, 4)$.
15. The line through $(4, -6)$ with slope $m = 2$.
16. The line through $(-2, 7)$ with slope $m = -3$.
17. The line with slope $m = -1$, y-intercept $b = 3$.
18. The line with slope $m = 4$, y-intercept $b = -5$.
19. The line with slope $m = \frac{1}{3}$, x-intercept $d = -2$.
20. The line with slope $m = -\sqrt{2}$, x-intercept $d = 4$.

Find each of the following.

21. The slope of the line $3x + 2y - 4 = 0$.
22. The slope of the line $4x - y + 7 = 0$.
23. The equation of the line parallel to $x + 2y = 3$, through the point $(3, 4)$.
24. The equation of the line through $(-1, 3)$, parallel to $2x - y + 5 = 0$.
25. The y-intercept of the line $2x - 3y + 7 = 0$.
26. The y-intercept of the line $4x - 2y + 8 = 0$.
27. The x-intercept of the line $2x + y - 7 = 0$.
28. The x-intercept of the line $y - x = 3$.
29. The intersection of $x + 3y = 9$ and $3x - y - 7 = 0$.
30. The intersection of $2x - 4y - 6 = 0$ and $x - 2y = 5$.

1.6 Graphing Conics

PARABOLA
TRANSLATION OF AXES
ELLIPSE
HYPERBOLA
ASYMPTOTES

In this section we consider some important kinds of equations, with particular emphasis on drawing their graphs. The graphs include ellipses, hyperbolas, and parabolas. Each of them can be obtained by looking at the points of intersection of a flat plane with a double-ended cone (see Figure 1.6.1). For this reason these graphs are called **conics**.

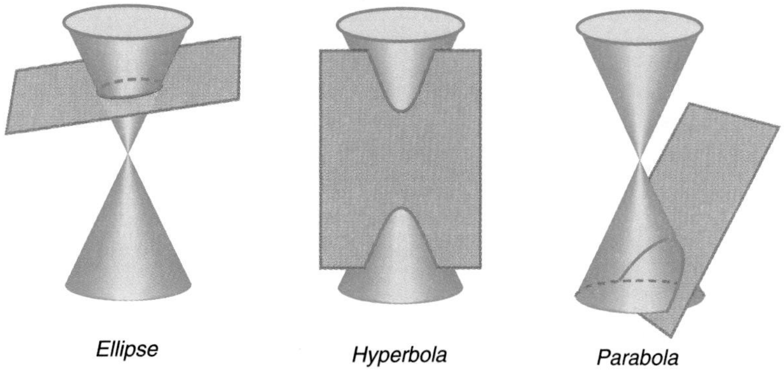

Figure 1.6.1

The equations we will be considering involve **polynomials in *two* variables**, x and y.

We will deal only with **quadratic polynomials**, which means polynomials whose degree is 2. Quadratic polynomials must contain a term of degree 2, that is, a term with x^2, y^2, or xy, and may also contain terms with x, y, and/or constants.

In fact, we will usually consider only quadratic polynomials that do not contain an xy-term, that is, polynomials of the form $ax^2 + by^2 + cx + dy + q$.

Parabolas

Consider the equation $y = x^2$. Every point on the graph of $y = x^2$ satisfies $y \geq 0$, so the graph never goes below the x-axis.

We also notice that the value of y will be the same for x and $-x$. We say that the graph will be "symmetrical" about the y-axis, meaning that it will be the same to the left of the y-axis as it is to the right of the axis.

We can plot a few points and begin to draw the graph. Its shape is the familiar **parabola** (see Figure 1.6.2).

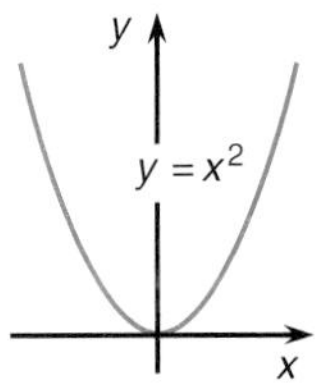

Figure 1.6.2

If we consider instead the equation $y = ax^2$, the graph will also be a parabola. If $a < 0$, then the parabola will open downward, while if $a > 0$, the parabola will open upward. The larger $|a|$ is, the more rapidly the points on the parabola will move away from the origin, that is, the longer and narrower the parabola will be.

Similar remarks apply to equations of the form $x = ay^2$. Here the graph will be a parabola that is symmetrical about the x-axis, opening to the right if $a > 0$ and to the left if $a < 0$. See Figure 1.6.3. In each case, the origin is called the **vertex** of the parabola.

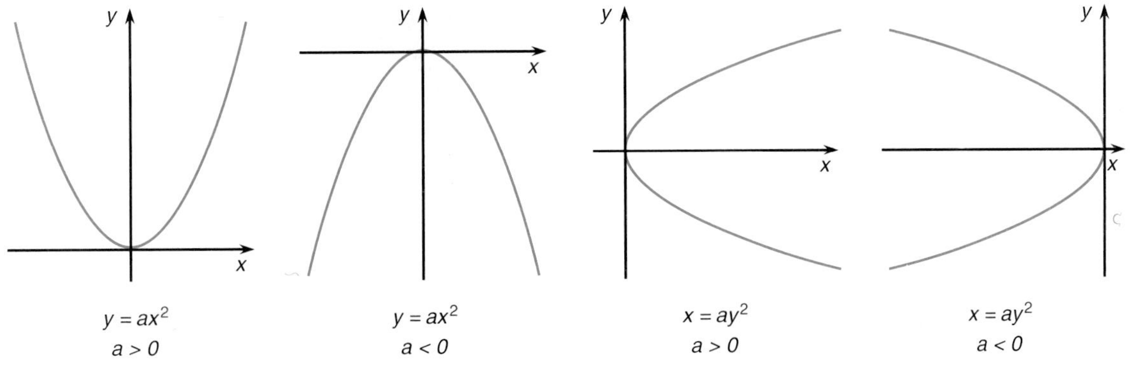

Figure 1.6.3

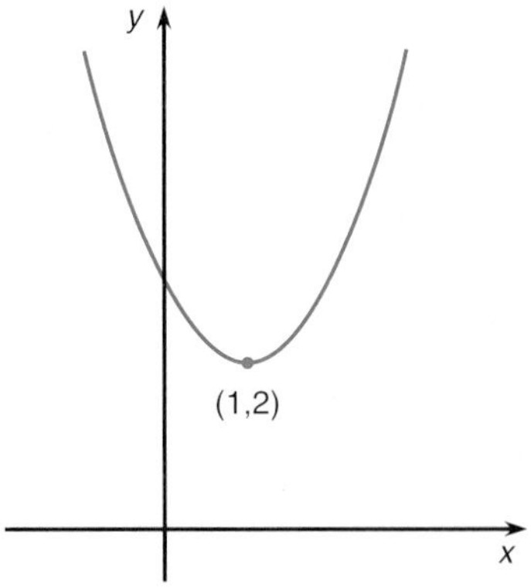

Figure 1.6.4

EXAMPLE 1 Now consider the equation $y = x^2 - 2x + 3$. To see what its graph is, we complete the square: $y = x^2 - 2x + 3 = (x-1)^2 + 2$. The quantity $(x - 1)$ measures the horizontal position starting at the value $x = 1$ instead of $x = 0$. This suggests that the equation $y = (x-1)^2$ is a parabola, except that its lowest point (its vertex) will be at $x = 1$.

The equation we are considering is obtained from this by adding 2. Adding 2 to the y-coordinate raises the whole graph up by two units. The graph of $y = x^2 - 2x + 3$ is a parabola that opens upward and whose vertex is the point $(1, 2)$. See Figure 1.6.4.

EXAMPLE 2 Consider the equation $y = ax^2 + bx + c$, with $a \neq 0$. By completing the square on the right side we can rewrite it as $y = a\left(x + \frac{b}{2a}\right)^2 + \left(c - \frac{b^2}{4a}\right)$, that is, $y - \left(c - \frac{b^2}{4a}\right) = a\left(x + \frac{b}{2a}\right)^2$. If we let $h = -\frac{b}{2a}$ and $k = c - \frac{b^2}{4a}$, then the equation becomes

$$y - k = a(x - h)^2.$$

We recognize this as the equation of a parabola. It opens upward if $a > 0$ and downward if $a < 0$. Its vertex is at the point (h, k). See Figure 1.6.5.

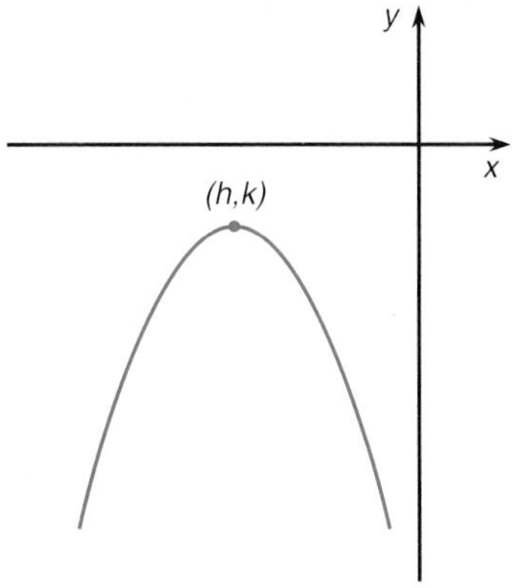

Figure 1.6.5

The approach used in this example suggests a useful idea. Suppose h and k are real numbers; we can plot the point $O' = (h, k)$. The quantities $x' = x - h$ and $y' = y - k$ measure the horizontal and vertical position of a point, *starting at the point* O' instead of starting at the origin.

We can draw "axes" through $O' = (h, k)$ and think of x' and y' as coordinates *relative to these axes.* The point O' is the origin for these axes and coordinates. See Figure 1.6.6.

As we saw in Example 1, an equation may be more easily recognized if it is expressed in terms of the coordinates x' and y'. We wrote $y = x^2 - 2x + 3$ as $y - 2 = x^2 - 2x + 1$. With $y' = y - 2$ and $x' = x - 1$ it becomes $y' = x'^2$, which is a parabola whose vertex is at $O' = (1, 2)$. In Figure 1.6.7 it is easy to draw this parabola if we mark the x'-axis and the y'-axis. After we have drawn the graph, we can ignore these axes and return to the original x-axis and y-axis.

This process is called "translating the axes."

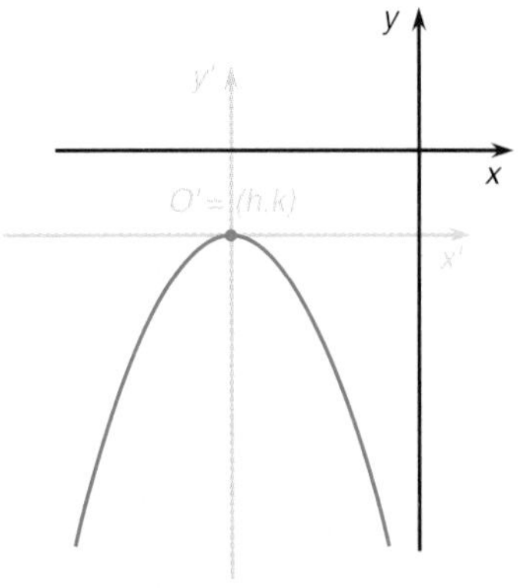

Figure 1.6.6

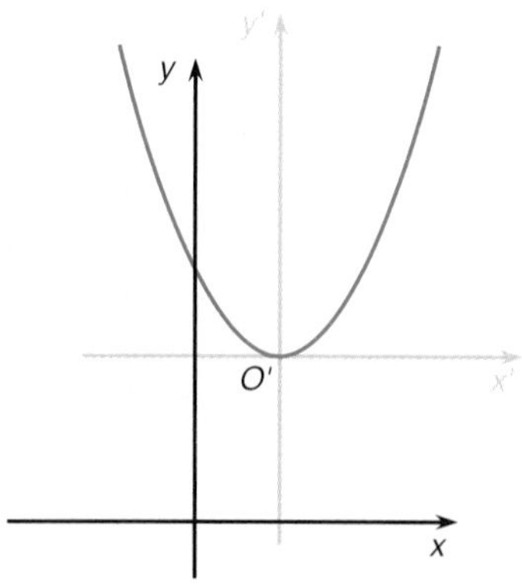

Figure 1.6.7

Summary 1.6.1

Translation of Axes

Suppose h and k are real numbers. Let $O' = (h, k)$, and let $x' = x - h$, $y' = y - k$. Draw the x'-axis and the y'-axis through O'.

To graph all the points (x, y) whose coordinates satisfy some equation $F(x, y) = 0$, it is equivalent to express $F(x, y)$ as some function of x' and y', that is,

$$F(x, y) = G(x', y') = G(x - h, y - k),$$

and graph all points (x', y') satisfying $G(x', y') = 0$; these points (x', y') must be plotted relative to the x'-axis and the y'-axis.

It is easy to find the function G. Observe that $x' = x - h$, so $x = x' + h$, and $y' = y - k$, so $y = y' + k$. Substitute these expressions into the equation above to find

$$G(x', y') = F(x, y) = F(x' + h, y' + k).$$

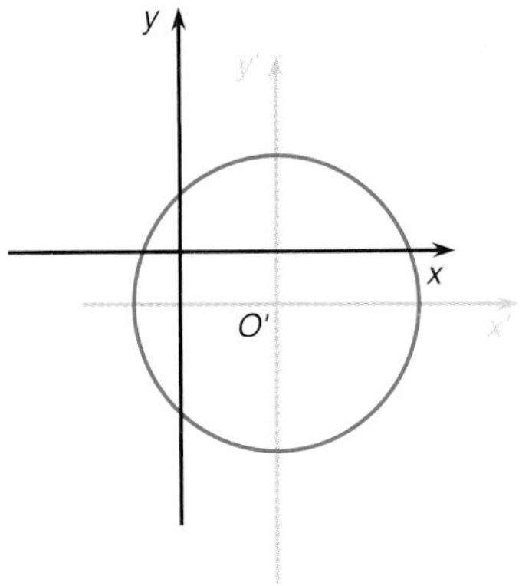

Figure 1.6.8

EXAMPLE 3 Consider the equation $x^2 - 4x + y^2 + 2y - 4 = 0$. We complete squares in both x and y, so $(x^2 - 4x) + (y^2 + 2y) - 4 = 0$ becomes $(x - 2)^2 - 4 + (y + 1)^2 - 1 - 4 = 0$, or $(x - 2)^2 + (y + 1)^2 - 9 = 0$.

This suggests that we try letting $x' = x - 2$ and $y' = y + 1$. Using the notation of the preceding summary, this means $h = 2$ and $k = -1$. It is easy to write the original equation in terms of x' and y': the "translated" equation is $x'^2 + y'^2 - 9 = 0$, or $x'^2 + y'^2 = 9$. The left-hand side is the square of the distance from (x', y') to the new origin O'. The equation $x'^2 + y'^2 = 9$ says that the distance from (x', y') to the origin O' is $\sqrt{9} = 3$. The set of all points satisfying this condition is the circle with center $O' = (2, -1)$ and radius 3. It is easily graphed (see Figure 1.6.8).

Translating the equation made it much easier to recognize that its graph was a circle.

REMARK 1.6.2

Whenever a quadratic polynomial contains both an x^2-term and an x-term, we can complete the square and translate coordinates to get a polynomial with an x'^2-term but no x'-term. Similarly, a polynomial with a y^2-term and a y-term can be converted into one with a y'^2-term but no y'-term by completing the square in y.

If the equation does not have an x^2-term, it is possible to combine the x-term and the constant term into x', resulting in an equation without a constant term. A similar procedure can be used to remove the constant term from an equation with no y^2-term.

In each case the effect of these changes in terms of the graph is to translate the origin, which can also be seen as moving the center of the conic.

In Example 3 we noticed that if $A > 0$, the graph of all points satisfying an equation $x^2 + y^2 = A$ is the graph of the circle centered at the origin with radius $\sqrt{A}$. We have already remarked that a circle cannot be the graph of a function, since there can be two values of y for the same x, but we can draw its graph, plotting all the points that satisfy the equation. The parabola $y = ax^2 + bx + c$ expresses y as a function of x, but for many conics, y will not be a function of x. In this situation we say that the equation gives a **relation** between x and y.

Ellipses

Consider the equation $x^2 + 4y^2 = 4$. It is not the equation of a circle (such as $x^2 + y^2 = R^2$), but it is similar. Let us rewrite it as $x^2 + (2y)^2 = 4$. If we let $y' = 2y$, the equation reads $x^2 + y'^2 = 4$. So (x, y') lies on the circle of radius $\sqrt{4} = 2$ with center at the origin; the problem is to interpret this in terms of the original variables x and y.

Changing from y to $y' = 2y$ amounts to changing the scale on the vertical axis, stretching everything by a factor of 2 in the vertical direction. We have a circle after stretching by a factor of 2 in the vertical direction, so before stretching, the figure must have been a "flattened out circle." It helps to get a more precise idea of the shape of the original figure by finding its x- and y-intercepts.

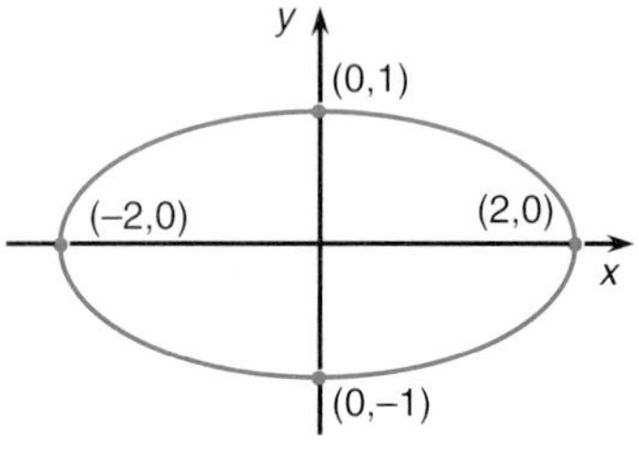

Figure 1.6.9

When $x = 0$, we find $0^2 + 4y^2 = 4$, so $y^2 = 1$ and $y = \pm 1$. When $y = 0$, we find $x^2 + 4(0^2) = 4$, so $x = \pm 2$. We plot $(0, \pm 1)$ and $(\pm 2, 0)$ and then fill in the rest of the graph, plotting additional points first if greater accuracy is required. See Figure 1.6.9.

The graph of $x^2 + 4y^2 = 4$ is an **ellipse**. The equations $2x^2 + 8y^2 = 10$, $x^2 + \frac{y^2}{9} = 16$, etc., are also ellipses. If we multiply one of these equations by a constant, the graph will not be changed, so there are many equations that represent the same ellipse. To remove this ambiguity, we frequently multiply or divide the equation by a constant so that the right-hand side becomes 1. It is customary to write the equation in the form

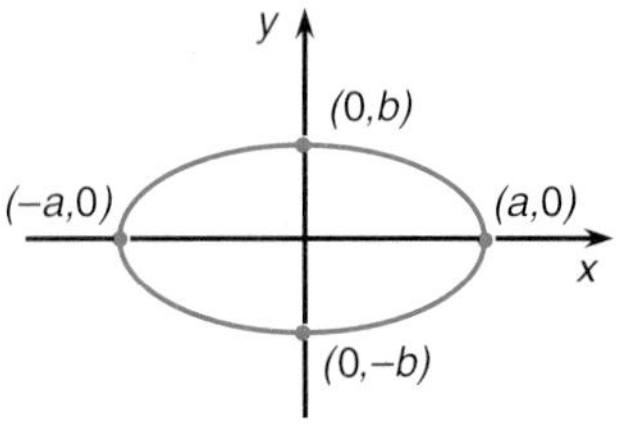

Figure 1.6.10(i)

$$\frac{x^2}{a^2} + \frac{y^2}{b^2} = 1 \quad \text{or} \quad \frac{x^2}{b^2} + \frac{y^2}{a^2} = 1, \tag{1.6.3}$$

with $a > b > 0$.

The graph of an equation of this form is an ellipse whose center is at the origin $(0, 0)$. For the first form, the intercepts are $(\pm a, 0)$ and $(0, \pm b)$. For the second form, the intercepts are $(\pm b, 0)$ and $(0, \pm a)$.

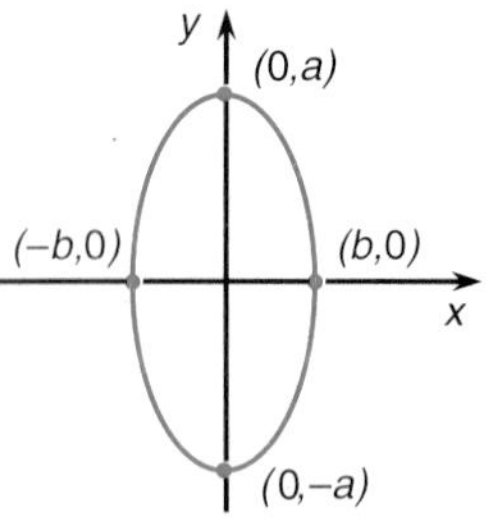

Figure 1.6.10(ii)

The distinction between the two forms involves whether the larger number a is in the denominator of the x^2-term or of the y^2-term. We see that for the first form, the x-intercepts $(\pm a, 0)$ are farther apart than the y-intercepts $(0, \pm b)$. This means that the corresponding ellipse is wider than it is tall (see Figure 1.6.10(i)). In the second type of equation the situation is reversed, and the ellipse is taller than it is wide (see Figure 1.6.10(ii)). It is to help distinguish these two cases that we always choose $a > b$.

EXAMPLE 4 Sketch the graph of the equation $4x^2 - 16x + 9y^2 + 18y - 11 = 0$.

Solution We begin by completing the square. The equation becomes $4(x^2 - 4x) + 9(y^2 + 2y) - 11 = 0$, or $4(x^2 - 4x + 4) - 16 + 9(y^2 + 2y + 1) - 9 - 11 = 0$, or $4(x - 2)^2 + 9(y + 1)^2 - 36 = 0$. We let $x' = x - 2$, $y' = y + 1$, and translate the axes to the new origin, $O' = (2, -1)$.

The equation now reads $4x'^2 + 9y'^2 = 36$. Following the model of Equa-

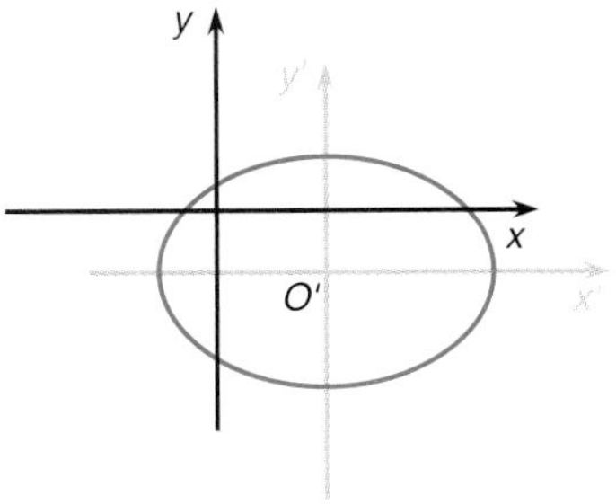

Figure 1.6.11

tions 1.6.3, we divide by 36 to obtain

$$\frac{x'^2}{9} + \frac{y'^2}{4} = 1.$$

This is the equation of an ellipse whose center is at the point O' and whose x'- and y'-intercepts are $(\pm 3, 0)$ and $(0, \pm 2)$, respectively.

It is easily graphed (see Figure 1.6.11). Note that the intercepts are described here in terms of x'- and y'-coordinates. Finally, we return to x and y coordinates and see that the graph is an ellipse whose center is at $(2, -1)$. The longest segment through the center from one side of the ellipse to the other goes from $(-1, -1)$ to $(5, -1)$ (i.e., the two x'-intercepts). The shortest such segment goes from $(2, -3)$ to $(2, 1)$.

These segments are called the **major axis** and **minor axis**, respectively. If the equation of an ellipse is written in either of the forms in Equations 1.6.3, then the major and minor axes go between the x- and y-intercepts. The major axis is the longer one (between $(a, 0)$ and $(-a, 0)$ for the first form of the equation and between $(0, -a)$ and $(0, a)$ for the second type).

The numbers a and b are half the lengths of the major and minor axes; they are sometimes called the **semi-major** and **semi-minor axes**, respectively.

If $a = b$, then the graph is a circle instead of an ellipse, and it does not make sense to talk about major and minor axes.

Hyperbolas

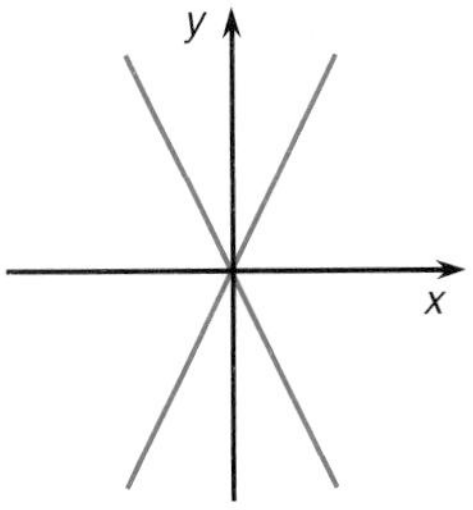

Figure 1.6.12

Consider the equation $4x^2 - y^2 = 4$. When $y = 0$, we see that $4x^2 = 4$, so $x = \pm 1$ and the x-intercepts are $(1, 0)$ and $(-1, 0)$. If we let $x = 0$, the equation becomes $-y^2 = 4$. This has no solutions, since y^2 is never negative, so there are no y-intercepts.

Next we ask what will happen if x is a large number. The equation can be written as $y^2 = 4x^2 - 4$, and if x is very large, the -4 on the right side will be so much smaller than the $4x^2$ that the equation is very nearly the same as $y^2 = 4x^2$.

> To illustrate this, suppose $x = 20$. Then $y^2 = 4x^2 - 4 = 4(20)^2 - 4 = 1596$. Taking the square root with a calculator, we see that $y \approx \pm 39.949\,97$. This is very close to what we get from $y^2 = 4x^2 = 1600$, from which $y = \pm 40$.

So as x moves away from 0, the graph lies very close to $y^2 = 4x^2$, or $y = \pm 2x$. The graphs of $y = 2x$ and $y = -2x$ are straight lines through the origin, with slopes 2 and -2, respectively. They are sketched in Figure 1.6.12.

The equation $y^2 = 4x^2 - 4$ has no solution unless the right side is nonnegative. This amounts to $4x^2 \geq 4$, or $x^2 \geq 1$, or $|x| \geq 1$. The graph has no points with $-1 < x < 1$.

Finally we remark that if a point (x, y) satisfies $4x^2 - y^2 = 4$, then so does $(x, -y)$, and so do $(-x, y)$ and $(-x, -y)$. This means that the graph is "symmetrical" about both axes, which means that "reflecting" it in either axis does not change the picture.

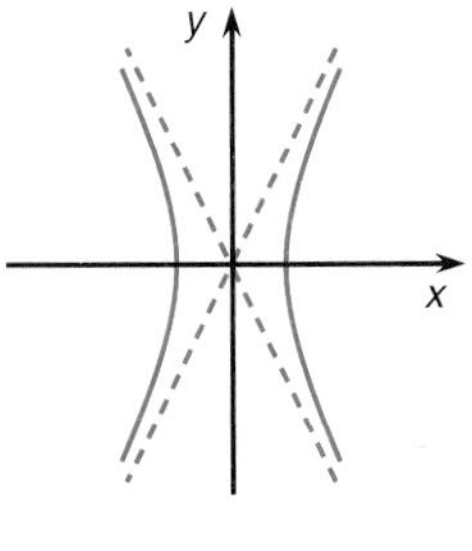

Figure 1.6.13

From all this we can guess the shape of the graph. We know its intercepts, we know there are no points with $-1 < x < 1$, and we know that as x becomes large, the graph approaches one or other of the lines $y = 2x$ and $y = -2x$. Combined with the symmetry, these facts suggest the shape shown in Figure 1.6.13. This conic is called a **hyperbola**. The lines $y = \pm 2x$ are called its **asymptotes**. The graph consists of two "pieces," which are called its **branches**.

As with the equation of an ellipse, it is convenient to divide the equation of a hyperbola by a suitable number. We choose the number so that the constant term of the equation is 1. So, for instance, the above equation $4x^2 - y^2 = 4$ becomes $x^2 - \frac{y^2}{4} = 1$.

Summary 1.6.4

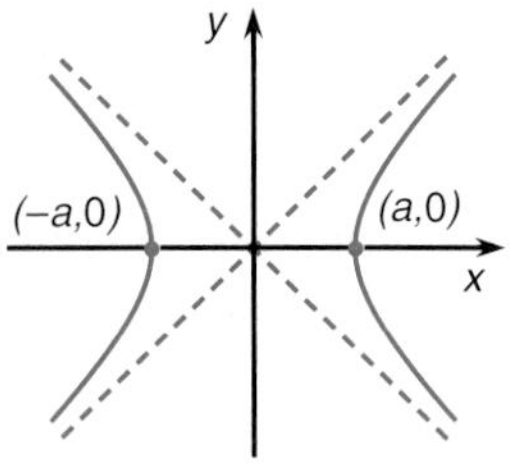

Figure 1.6.14(i)

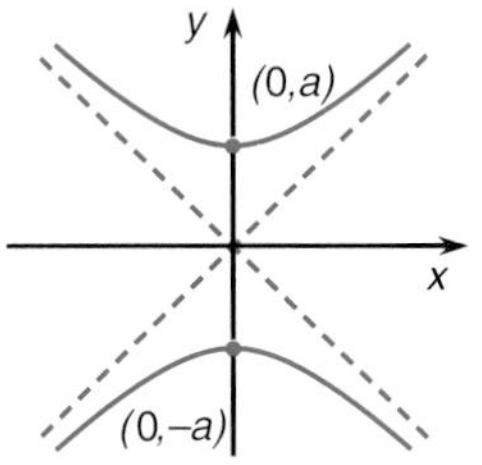

Figure 1.6.14(ii)

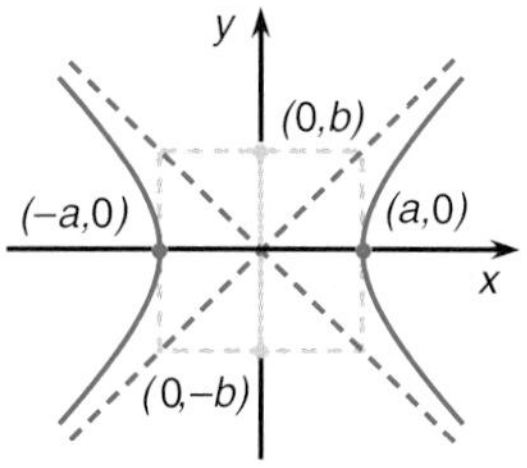

Figure 1.6.15

General Form of the Hyperbola

The equations

$$\frac{x^2}{a^2} - \frac{y^2}{b^2} = 1 \quad \text{and} \quad \frac{y^2}{a^2} - \frac{x^2}{b^2} = 1$$

are the equations of hyperbolas.

In the first case the x-intercepts are $(\pm a, 0)$, there are no points on the graph with $-a < x < a$, and the asymptotes are found by realizing that if x is large, then the equation is almost the same as $\frac{x^2}{a^2} = \frac{y^2}{b^2}$, or $\frac{x}{a} = \pm\frac{y}{b}$, or $y = \frac{b}{a}x$ and $y = -\frac{b}{a}x$. These are lines through the origin with slopes $\frac{b}{a}$ and $-\frac{b}{a}$. See Figure 1.6.14(i).

In the second case the roles of x and y are reversed. In this case there are no x-intercepts and no points on the graph with $-a < y < a$. The y-intercepts are $(0, \pm a)$, and the asymptotes are found by approximating the equation for large values of x by $\frac{y^2}{a^2} = \frac{x^2}{b^2}$, so they are the lines $\frac{y}{a} = \pm\frac{x}{b}$, or $y = \pm\frac{a}{b}x$. See Figure 1.6.14(ii).

Notice that we decide which denominator is a^2 and which is b^2 on the basis of the signs of the terms; a^2 is the denominator of the positive term (or, more precisely, the term with the same sign as the constant term on the right side). Unlike what happened with ellipses, it is *not* a question of the relative sizes of a and b.

In each case there are two intercepts (either two x-intercepts or two y-intercepts). The line segment joining them is called the **transverse axis**. Its length is $2a$. We also draw the segment that is perpendicular to the transverse axis, centered at the origin and of length $2b$; it is called the **conjugate axis**. (It is in order to be able to say that the transverse axis has length $2a$ that we changed around the order of the numbers a and b when we wrote the second form of the equation.) See Figure 1.6.15.

So for $\frac{x^2}{a^2} - \frac{y^2}{b^2} = 1$, the transverse axis joins $(-a, 0)$ and $(a, 0)$; the conjugate axis joins $(0, -b)$ and $(0, b)$. In the other case, $\frac{y^2}{a^2} - \frac{x^2}{b^2} = 1$, the transverse axis joins $(0, -a)$ and $(0, a)$, while the conjugate axis joins $(-b, 0)$ and $(b, 0)$.

If we fill in the rectangle passing through the ends of the transverse and conjugate axes (as in Figure 1.6.15), then its corners are on the asymptotes. This makes it easy to draw the asymptotes.

Rather than memorizing which formula represents which kind of hyperbola, it is preferable to realize that it is quite easy to distinguish the two cases simply by determining whether there are x-intercepts or y-intercepts. It is all a matter of whether the x^2-term or the y^2-term has the same sign as the constant term.

We are now able to sketch the graph of any quadratic equation that does not contain any xy-term. The idea is to inspect the equation to decide whether it is a parabola, an ellipse, a circle, or a hyperbola. We first complete squares if necessary and translate the axes accordingly.

Summary 1.6.5

Graphing Conics

Consider the quadratic equation $ax^2 + by^2 + cx + dy + q = 0$, which has no xy-term. To draw its graph, use the following steps.

(i) Complete squares if necessary; if the equation contains both an x^2-term and an x-term, that is, if $a \neq 0$ and $c \neq 0$, then complete the square in x. Similarly, if $b \neq 0$ and $d \neq 0$, complete the square in y.

(ii) Translate axes if necessary. If you found it necessary to complete squares in Step (i), and the result of completing the square in x is a term involving $x - h$, then let $x' = x - h$. If completing the square in y gives a term involving $y - k$, let $y' = y - k$.

If there is no x^2-term, then there must be a y^2-term. In this situation, combine the x-term and the constant term as $c(x - h)$, for some constants c and h, and let $x' = x - h$. Similarly, if there is no y^2-term, combine the y-term and the constant term as $d(y - k)$ and let $y' = y - k$.

(iii) If the axes have been translated, plot the "new origin" $O' = (h, k)$. (If you changed x to $x' = x - h$ but did not have to translate y, then $O' = (h, 0)$, that is, let $k = 0$. Similarly, if it was not necessary to translate x to x', then let $h = 0$.)

(iv) Express the equation in terms of x' and y' if translations were necessary. Inspect the equation to see whether it is the equation of a parabola, an ellipse, a circle, or a hyperbola. Then find the intercepts. For a hyperbola, find the asymptotes and mark them on the picture. Sketch the graph.

(v) Occasionally we encounter a quadratic equation that is not a parabola, an ellipse, a circle, or a hyperbola. This kind of graph is called a **degenerate conic**. There are several possibilities, but we describe just two of them. In an equation like $\frac{x^2}{a^2} + \frac{y^2}{b^2} = -1$, the left side is the sum of two squares, so it is never negative, and we see that no point (x, y) can satisfy the equation. There are *no points* on the graph.

Another possibility is an equation of the form $\frac{x^2}{a^2} - \frac{y^2}{b^2} = 0$. This is like a hyperbola except that the constant term is zero. Rearranging, we find that $\frac{x^2}{a^2} = \frac{y^2}{b^2}$, or $\frac{x}{a} = \pm\frac{y}{b}$. The graph is the two straight lines $y = \frac{b}{a}x$ and $y = -\frac{b}{a}x$. This is what results if a hyperbola "degenerates" into its two asymptotes.

EXAMPLE 5 Sketch the graph of the equation $9x^2 - 18x - 4y^2 + 32y = 19$.

Solution We complete the squares:

$$\begin{aligned}9x^2 - 18x - 4y^2 + 32y - 19 &= 9(x^2 - 2x) - 4(y^2 - 8y) - 19\\ &= 9(x^2 - 2x + 1) - 9 - 4(y^2 - 8y + 16) + 64 - 19\\ &= 9(x - 1)^2 - 4(y - 4)^2 + 36.\end{aligned}$$

If we let $h = 1$, $k = 4$, $x' = x - 1$, $y' = y - 4$, then the original equation becomes $9x'^2 - 4y'^2 = -36$, or

$$\frac{y'^2}{9} - \frac{x'^2}{4} = 1.$$

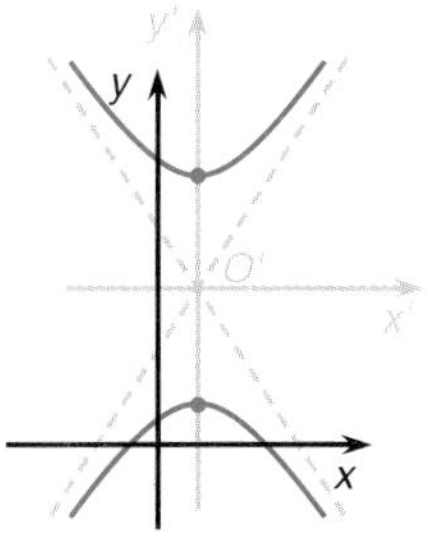

Figure 1.6.16

This is the equation of a hyperbola. Since the y'^2-term is positive, it is a hyperbola with transverse axis along the y'-axis and conjugate axis along the x'-axis. The center of the hyperbola is $O' = (h, k) = (1, 4)$. The intercepts on the y'-axis are $y' = \pm 3$. These points, where $x' = 0$ and $y' = \pm 3$, are the points where $x = x' + 1 = 0 + 1 = 1$ and $y = y' + 4 = 4 \pm 3 = 7, 1$, which are the points $(1, 1)$ and $(1, 7)$.

The conjugate axis has length $2\sqrt{4} = 4$; it runs between $(x', y') = (\pm 2, 0)$, that is, from $(x, y) = (-1, 4)$ to $(3, 4)$.

The asymptotes are the lines $\frac{y'^2}{9} = \frac{x'^2}{4}$, or $y' = \pm \frac{3}{2} x'$. They are easily sketched relative to the x'- and y'-axes.

Then it is comparatively easy to sketch in the shape of the hyperbola, making the graph symmetrical about both the x'- and y'-axes. See Figure 1.6.16.

Even without completing the squares, we might have noticed that the x^2-term and the y^2-term in the original equation had opposite signs, which suggests that the graph should be a hyperbola. We cannot be absolutely certain about this because it might have turned out to be a degenerate conic—a pair of straight lines. It is also important to observe that, even assuming that the graph is a hyperbola and just looking at the original equation, we could not easily tell which way the hyperbola opens (i.e., up/down or left/right). The constant term on the right side of the original equation has the same sign as the x^2-term, but after completing squares we have a constant with the same sign as the y'^2-term, and it is *this* that decides the type of the hyperbola.

Generally speaking, our approach to graphing conics has been to begin by completing squares and translating axes in order to simplify the equation. Then it is possible to recognize the type of conic from the simplified form of the equation. Rather than memorizing the various possibilities, we have concentrated on identifying the curve by finding intercepts, asymptotes, if any, and values of x or y for which there are no points on the curve. The calculations involved are simple and can frequently be done in your head. They use nothing more complicated than the fact that a sum of squares can never be negative.

We have observed that an ellipse is not the graph of a function, because some vertical lines intersect it twice. The same is true for hyperbolas and circles and some parabolas.

On the other hand, if we consider the ellipse $12x^2 + y^2 = 12$, it is tempting to write $y^2 = 12 - 12x^2$ and take the square root to solve for y:

$$y = \sqrt{12 - 12x^2}.$$

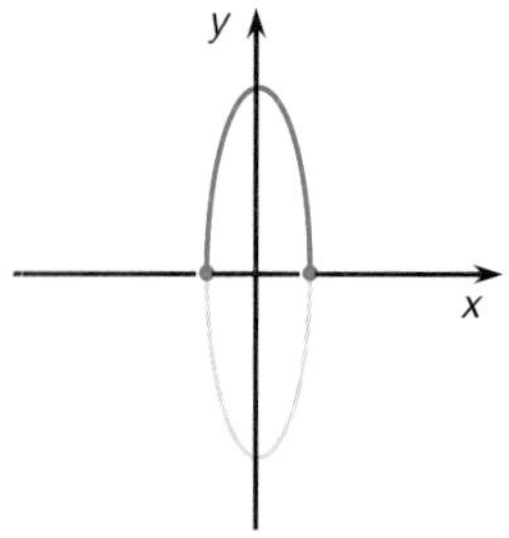

Figure 1.6.17

This formula does express y as a function of x. Since the square root is always nonnegative, the graph of this function $y = \sqrt{12 - 12x^2}$ is the *upper half* of the ellipse. See Figure 1.6.17.

We could also have tried letting $y = -\sqrt{12 - 12x^2}$, and the graph of this function is the lower half of the ellipse.

We could even have combined them in various ways, for instance,

$$y = f(x) = \begin{cases} \sqrt{12 - 12x^2}, & \text{if } x < 0, \\ -\sqrt{12 - 12x^2}, & \text{if } x \geq 0. \end{cases}$$

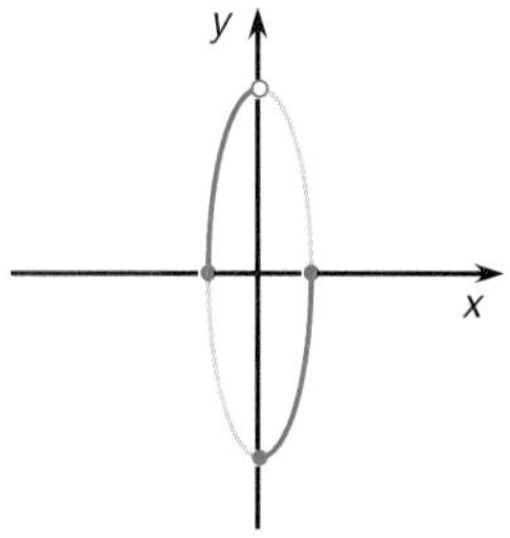

Figure 1.6.18

The graph of this function is the upper part of the ellipse some of the time and the lower part the rest of the time (see Figure 1.6.18). Clearly it is possible to do this sort of thing in many different and complicated ways.

Suppose we want to graph the ellipse $\frac{x^2}{4} + \frac{y^2}{9} = 1$ using a graphics calculator. Because the ellipse is not the graph of a function, we have to graph it in two parts.

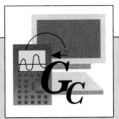

We begin by solving for y: $\frac{y^2}{9} = 1 - \frac{x^2}{4}$, so $y^2 = 9 - \frac{9x^2}{4}$. This means $y = \pm\sqrt{9 - \frac{9x^2}{4}} = \pm 3\sqrt{1 - \frac{x^2}{4}}$. We get the ellipse by graphing these *two* functions.

The other important thing to note is that the ellipse consists of points (x, y) with $-2 \leq x \leq 2$. Now we can graph the ellipse.

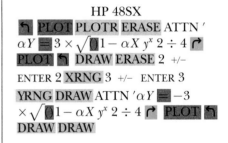

TI-85	SHARP EL9300
GRAPH F1 ($y(x)$ =) 3 √ (1− F1 (x) x^2 ÷ 4) ▼ (−) 3 √ (1− F1 (x) x^2 ÷ 4) M2 (RANGE) (−) 2 ▼ 2 ▼ ▼ (−) 3 ▼ 3 F5 (GRAPH)	[graph] 3 √ 1− x/θ/T x^2 ÷ 4 2ndF ▼ −3 √ 1− x/θ/T x^2 ÷ 4 RANGE −2 ENTER 2 ENTER 1 ENTER −3 ENTER 3 ENTER 1 ENTER [graph]
CASIO f_x-7700GB/f_x-6300G	**HP 48SX**
Cls EXE Range −2 EXE 2 EXE 1 EXE −3 EXE 3 EXE 1 EXE EXE EXE EXE Graph 3 √ (1 − X x^y 2 ÷ 4) EXE Graph −3 √ (1 − X x^y 2 ÷ 4) EXE	↰ PLOT PLOTR ERASE ATTN ' αY = 3 × √ () 1 − αX y^x 2 ÷ 4 ↱ PLOT ↰ DRAW ERASE 2 +/− ENTER 2 XRNG 3 +/− ENTER 3 YRNG DRAW ATTN ' αY = −3 × √ () 1 − αX y^x 2 ÷ 4 ↱ PLOT ↰ DRAW DRAW

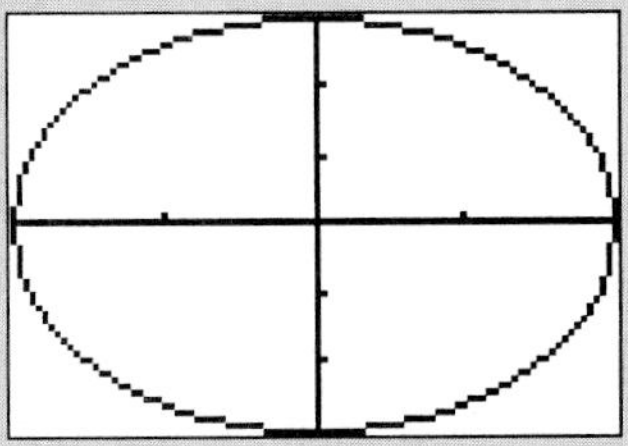

Be sure to take note of the vertical and horizontal scales when you identify the major and minor axes.

Now you might like to try graphing some other ellipses, such as $4x^2 + 3y^2 = 1$, and then some hyperbolas, such as $x^2 - y^2 = 1$, $4y^2 - 9x^2 = 16$. Try including asymptotes.

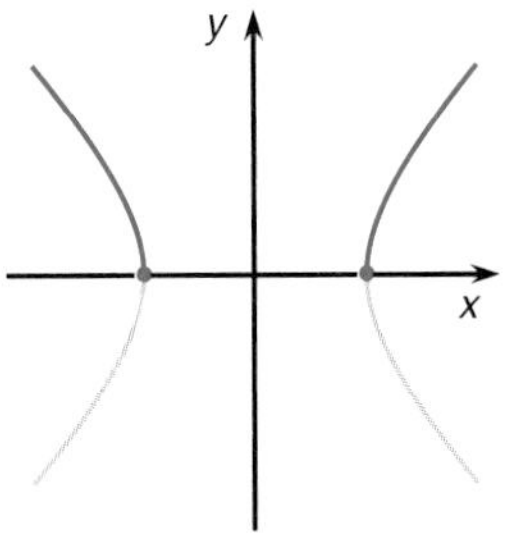

Figure 1.6.19

Many of the calculus techniques we will be learning are intended to be used on functions. There are times when we want to apply some of them to something like an ellipse but, to be able to do this it is first necessary to express part of the ellipse (say the upper or lower half) as the graph of a function and apply calculus to that function. Afterward, it may be possible to combine our conclusions about all the parts and draw some conclusion about the whole ellipse.

The same sort of thing will work for hyperbolas and parabolas. For instance, the hyperbola $\frac{x^2}{a^2} - \frac{y^2}{b^2} = 1$ can be divided into the graphs of two functions, $y = b\sqrt{x^2/a^2 - 1}$ and $y = -b\sqrt{x^2/a^2 - 1}$ (the top and bottom halves of the hyperbola; the top one is illustrated in Figure 1.6.19). Each of these functions has as its domain the set of all x satisfying $|x| \geq a$.

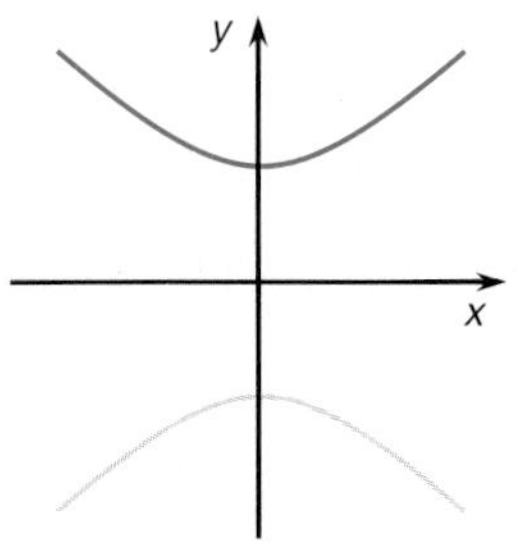

Figure 1.6.20

The hyperbola $\frac{x^2}{b^2} - \frac{y^2}{a^2} = -1$ can also be divided into the graphs of two functions, $y = a\sqrt{1 + x^2/b^2}$ and $y = -a\sqrt{1 + x^2/a^2}$ (the top and bottom branches of the hyperbola). Each of these functions has $\mathbb{R}$ as its domain. The top branch is illustrated in Figure 1.6.20.

Exercises 1.6

Identify each of the following conics, specifying its center, its intercepts, its axes, and its asymptotes, to the extent that they are applicable.

1. $16x^2 + 9y^2 = 144$
2. $4x^2 + 25y^2 = 100$
3. $x^2 - 4y^2 = 4$
4. $4y^2 - x^2 = 4$
5. $y^2 + 4x = 8$
6. $x^2 + y^2 = 4$
7. $x^2 + 4x + 9y^2 - 5 = 0$
8. $x^2 + y^2 - 6x + 2y + 9 = 0$
9. $x^2 + 2x - 4y - 5 = 0$
10. $x^2 + 4x + 4y^2 - 8y + 8 = 0$
11. $2x^2 + 8y^2 = 1$
12. $x^2 - y^2 = 1$
13. $12x^2 + 3y^2 = 27$
14. $\frac{4}{5}y^2 - 5x^2 = \frac{1}{5}$
15. $20x - 3y^2 = 2$
16. $5x^2 = 1 - 5y^2$
17. $1 - x^2 + 4x = 3y^2$
18. $9x^2 - 16y^2 = -1$
19. $16x + 25y^2 = 20$
20. $4x^2 - 24x - 9y^2 + 18y + 27 = 0$

Sketch the graph of each of the following conics, paying special attention to axes, intercepts, and asymptotes, whenever it makes sense to speak of them.

21. $4y + 16x^2 = 1$
22. $4x^2 + 25y^2 = 25$
23. $y^2 - x^2 - 1 = 0$
24. $y^2 = 1 - x^2$
25. $x^2 - 4y^2 = -2$
26. $x + 2y + y^2 + 3 = 0$
27. $4x^2 + 9y^2 - 7 = 0$
28. $x^2 = y - x$
29. $x^2 + y^2 + x + y - \frac{9}{2} = 0$
30. $4x^2 + 8x + 9y^2 + 5 = 0$

Find the equation of each of the following conics (the equations are all assumed to have no *xy*-term).

31. The circle with center $(-1, 3)$ and radius 2.
32. The ellipse with center $(0, 0)$, semi-minor axis $b = 3$, and horizontal major axis of length 8.
33. The hyperbola with asymptotes $y = \pm 2x$ and containing $(2, 0)$.
34. The parabola with vertex $(3, -5)$, opening downward, and containing $(4, -7)$.
35. The ellipse with center $(5, 2)$ containing $(3, 2)$ and $(5, 3)$.
36. The parabola opening to the right, with vertex $(-1, 2)$ and containing the origin.
37. The circle whose center is on the *x*-axis and that contains $(2, 3)$ and $(-6, 3)$.
38. The hyperbola whose center is $(-1, -2)$, whose asymptotes have slopes ± 3, and that contains $(-1, 0)$.
39. The ellipse with vertical major axis that is twice the length of the minor axis, which has center at $(2, -4)$ and contains the origin.
40. The hyperbola with transverse and conjugate axis both of length 2, transverse axis vertical, having its center at $(5, -2)$.

1.7 Trigonometric Functions

IDENTITY
RADIANS
LAW OF COSINES
ADDITION FORMULAS
POLAR COORDINATES

Consider the angle θ shown in Figure 1.7.1. It is drawn at the origin in the plane, and we have also drawn a circle of radius r, with its center at the origin. For convenience we have used the positive x-axis as one of the sides of the angle, and we label P, the point where the other side of the angle intersects the circle. We draw a vertical line through P and label Q, the point where this line intersects the x-axis. So the angle $\angle POQ$ equals θ.

The radius of the circle is $|OP| = r$. What we want to know is the lengths $|OQ|$ and $|PQ|$. If we draw the same sort of picture, with the same angle θ but with a larger circle, then r will be different, and so will the corresponding distances, $|OB|$ and $|AB|$ (see Figure 1.7.2).

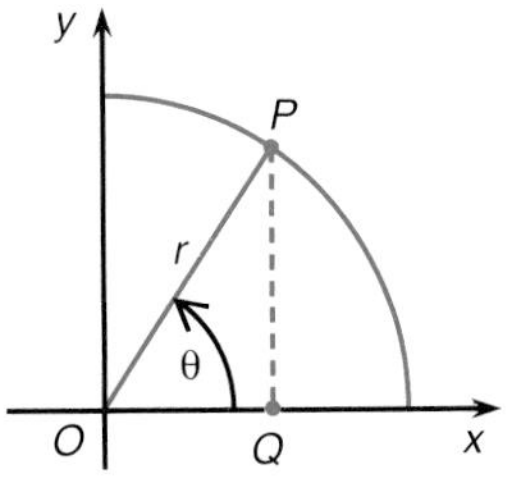

Figure 1.7.1

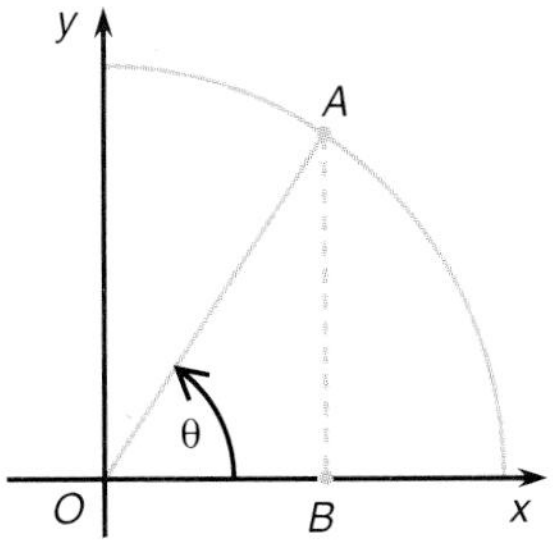

Figure 1.7.2

However, if we draw both pictures together, as in Figure 1.7.3, then we see that $\triangle AOB$ is similar to $\triangle POQ$, which means that the ratios of the lengths of corresponding sides are equal. In particular,

$$\frac{|OQ|}{|OP|} = \frac{|OB|}{|OA|}, \qquad \frac{|PQ|}{|OP|} = \frac{|AB|}{|OA|}, \qquad \frac{|PQ|}{|OQ|} = \frac{|AB|}{|OB|}.$$

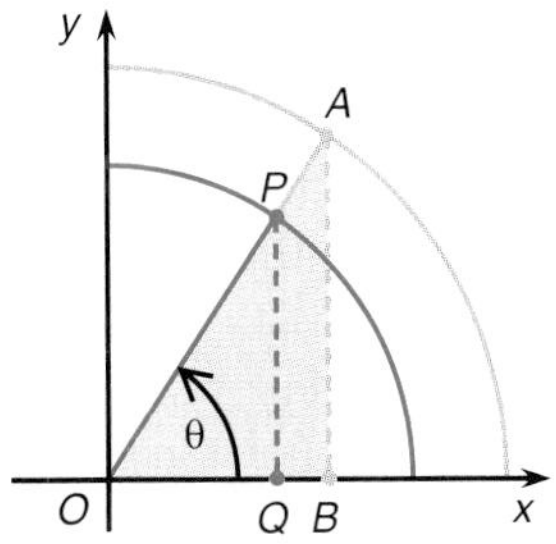

Figure 1.7.3

No matter how large or small the circle, if the angle at the center is θ, then these ratios will be the same. We need a way to express these ratios in terms of the angle θ.

The distance $|OP|$ is always r, the radius of the circle. If we write the coordinates of P as $P = (x, y)$, then $Q = (x, 0)$. In Figure 1.7.4, both x and y are positive, so $|OQ| = x$, and $|PQ| = y$. The ratios mentioned above can all be expressed in terms of x, y, and r.

These observations explain why the following definitions are made.

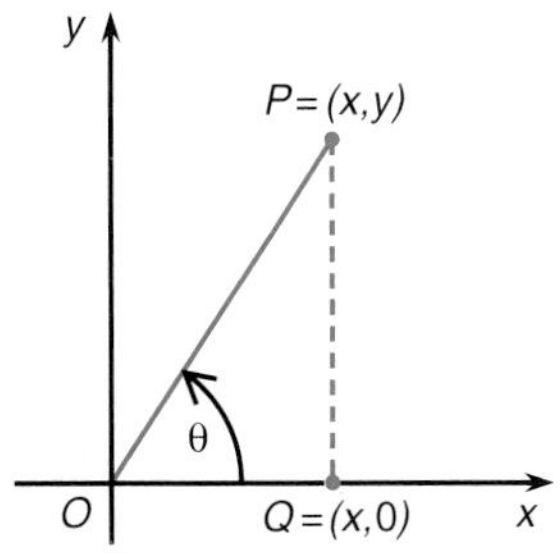

Figure 1.7.4

DEFINITION 1.7.1

For each angle θ we define the **trigonometric functions** $\cos\theta$, $\sin\theta$, $\tan\theta$, $\sec\theta$, $\csc\theta$, and $\cot\theta$ (abbreviations for *cosine, sine, tangent, secant, cosecant,* and *cotangent,* respectively).

Draw the positive x-axis, and draw another half-line starting at the origin that makes an angle of θ with the positive x-axis (see Figure 1.7.5). If θ is a positive angle, measure it counterclockwise from the positive x-axis; if it is negative, measure it clockwise from the positive x-axis. Now let $P = (x, y)$ be any point (other than the origin) on this second half-line. Let $r = |OP| = \sqrt{x^2 + y^2}$, and define

$$\cos\theta = \frac{x}{r}, \qquad \sin\theta = \frac{y}{r}.$$

If $x \neq 0$, we also define

$$\tan\theta = \frac{y}{x}, \qquad \sec\theta = \frac{r}{x}.$$

If $y \neq 0$, we also define

$$\cot\theta = \frac{x}{y}, \qquad \csc\theta = \frac{r}{y}.$$

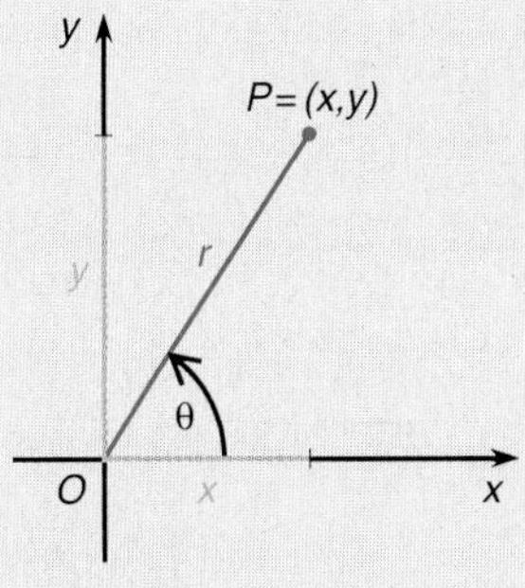

Figure 1.7.5

Each of these formulas defines a function of θ; they are known as the trigonometric functions.

In fact, all the trigonometric functions can be expressed in terms of the first two, $\sin\theta$ and $\cos\theta$. For instance,

$$\tan\theta = \frac{y}{x} = \frac{\frac{y}{r}}{\frac{x}{r}} = \frac{\sin\theta}{\cos\theta}, \qquad \cot\theta = \frac{\cos\theta}{\sin\theta},$$

$$\sec\theta = \frac{1}{\cos\theta}, \qquad \csc\theta = \frac{1}{\sin\theta}.$$

If we compare the angles θ and $\theta + 360°$, we see that when the line OP is drawn it will be the same for both angles. This means that $\sin(\theta + 360°) = \sin(\theta)$, and the same for all the other trigonometric functions (see Figure 1.7.6). The function $\sin\theta$ (or $\cos\theta$, etc.) "repeats itself after 360°"; this is frequently abbreviated by saying that $\sin\theta$ (and $\cos\theta$, etc.) are **periodic functions**.

In the diagrams we have drawn so far, the coordinates (x, y) of P are both positive. However, it is easy enough to find an angle for which x or y (or both) will be negative (see Figure 1.7.7). In this situation, some of the trigonometric functions will be negative.

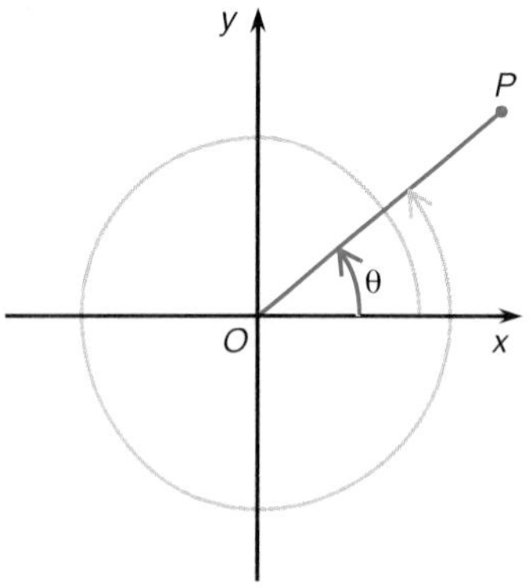

Figure 1.7.6

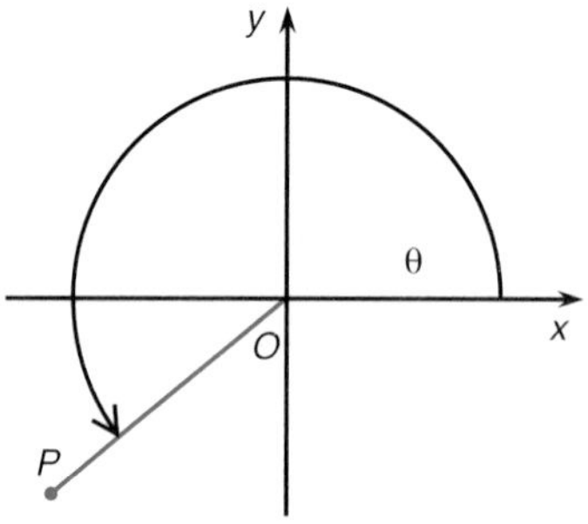

Figure 1.7.7

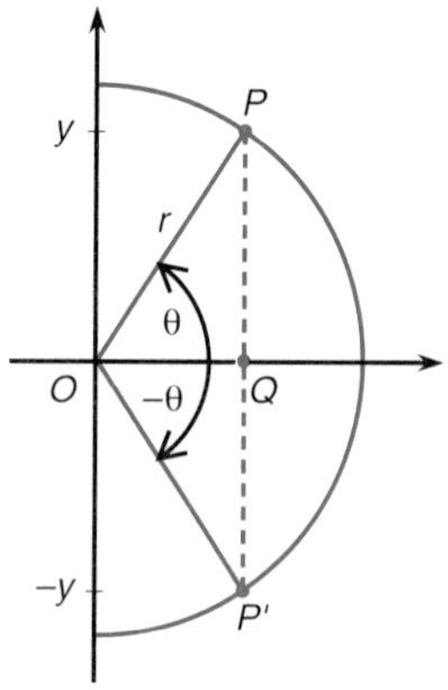

Figure 1.7.8

Figure 1.7.8 illustrates an angle θ and its negative $-\theta$, with the corresponding points P and P'. If $P = (x, y)$, then $P' = (x, -y)$, and the point $Q = (x, 0)$ is the same for both of them. The distances $|OP|$ and $|OP'|$ are the same, as are the distances $|PQ|$ and $|P'Q|$, so $\cos(-\theta) = \cos(\theta)$. However, $\sin(-\theta) = \frac{-y}{r} = -\frac{y}{r} = -\sin\theta$.

Notice that the domain of $\sin\theta$ or $\cos\theta$ is all real numbers θ. For the other functions it is necessary to exclude some points from the domain. For $\tan\theta$, for instance, it is necessary to exclude any angle for which $x = 0$. This means $\theta = 90°$, $\theta = -90°$, and any number that can be obtained by adding a multiple of 360° to either of these angles (such as $90 + 360 = 450°$ or $90 - 2 \times 360 = -630°$, etc).

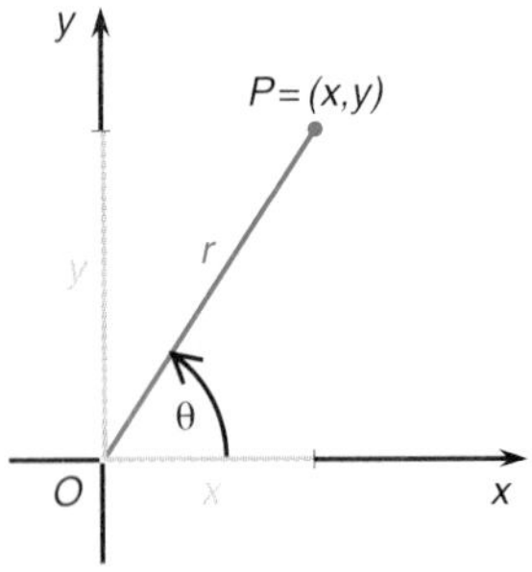

Figure 1.7.9

Figure 1.7.9 is like the diagram that was used to define $\cos\theta$ and $\sin\theta$. Recalling that with $P = (x, y)$ and $r = \sqrt{x^2 + y^2}$, we have $\cos\theta = \frac{x}{r}$ and $\sin\theta = \frac{y}{r}$, we can calculate

$$\cos^2\theta + \sin^2\theta = \left(\frac{x}{r}\right)^2 + \left(\frac{y}{r}\right)^2 = \frac{x^2 + y^2}{r^2} = \frac{x^2 + y^2}{x^2 + y^2} = 1.$$

This is a fundamental **identity** (the word "identity" means an equation that holds for every possible value of the variable).

THEOREM 1.7.2

> For every value of θ,
>
> $$\sin^2\theta + \cos^2\theta = 1.$$

If θ is any angle for which $\cos\theta \neq 0$, we can divide this identity by $\cos^2\theta$ and obtain

$$\frac{\sin^2\theta}{\cos^2\theta} + \frac{\cos^2\theta}{\cos^2\theta} = \frac{1}{\cos^2\theta},$$

that is,

$$\tan^2\theta + 1 = \sec^2\theta \qquad \text{or} \qquad \tan^2\theta = \sec^2\theta - 1. \tag{1.7.3}$$

Dividing the identity in Theorem 1.7.2 by $\sin^2\theta$ gives another identity, valid when $\sin\theta \neq 0$:

$$1 + \cot^2\theta = \csc^2\theta \qquad \text{or} \qquad \cot^2\theta = \csc^2\theta - 1. \tag{1.7.4}$$

Radian Measure

We are accustomed to measuring angles in degrees. This unit for angles can be traced back to an ancient Babylonian system of numbers. It is quite familiar to us, but apart from that there is very little reason to expect, for instance, that a right angle should happen to be 90°. To put it the other way around, why should $\frac{1}{90}$ of a right angle be chosen as the unit of measurement for angles?

It is possible to choose any units whatever to measure angles, but it turns out that there is one choice that is particularly convenient for doing calculus. The unfortunate thing is that it is fairly inconvenient for almost everything else. However, we will see in Chapter 3 that it makes calculus formulas very much simpler, and for this reason it is the choice we will adopt.

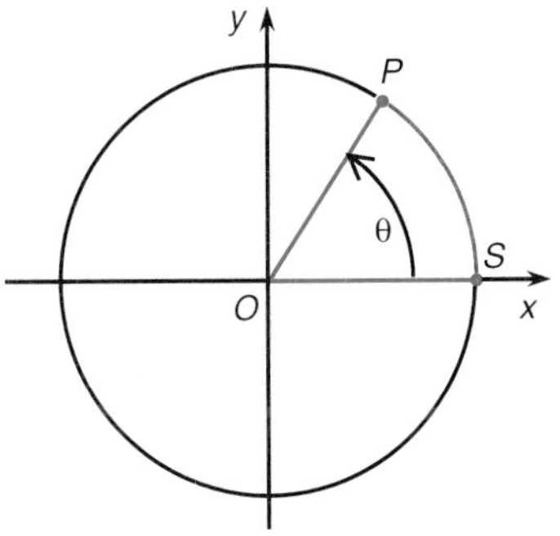

Figure 1.7.10

Consider the circle of radius $r = 1$, centered at the origin, as shown in Figure 1.7.10. This circle is called the **unit circle**; its circumference is $2\pi r = 2\pi$. Now if we are given an angle θ, let us measure the angle starting at the positive x-axis and rotating through an angle θ about the origin, rotating counterclockwise if the angle is positive, clockwise if it is negative. If $S = (1, 0)$ is the point where the circle intersects the positive x-axis, we can picture a point moving around the circle, starting at S, and moving through an angle θ about the origin. As before, let P be the point where it stops.

To measure the angle θ, we need some way of saying how far the point has moved in going from S to P. One fairly natural way to do this would be to measure the distance it has traveled along the circumference of the circle, with the understanding that a counterclockwise movement is measured with a positive distance and a clockwise movement with a negative number. This is **radian measure**.

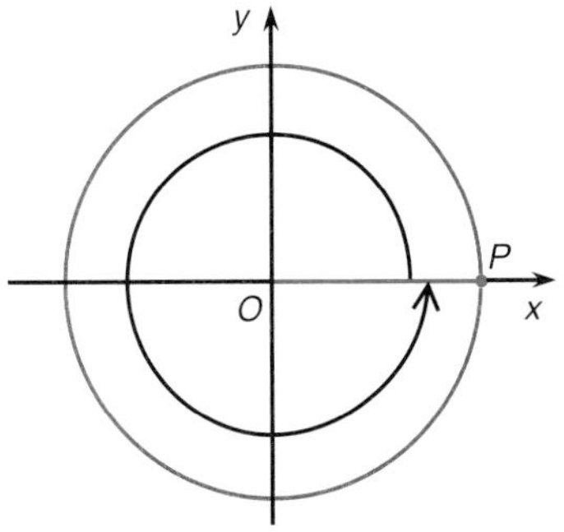

Figure 1.7.11

EXAMPLE 1

(i) An angle of 360° is the angle that goes all the way around the circle exactly once. The point on the circle goes around exactly once, so the distance it travels is the circumference of the circle, which is 2π (see Figure 1.7.11).

(ii) For an angle of $-45°$, the point moves *clockwise* $\frac{1}{8}$ of the way around the circle (since $\frac{45}{360} = \frac{1}{8}$), so the distance it travels is $\frac{1}{8}2\pi = \frac{\pi}{4}$. However, since the

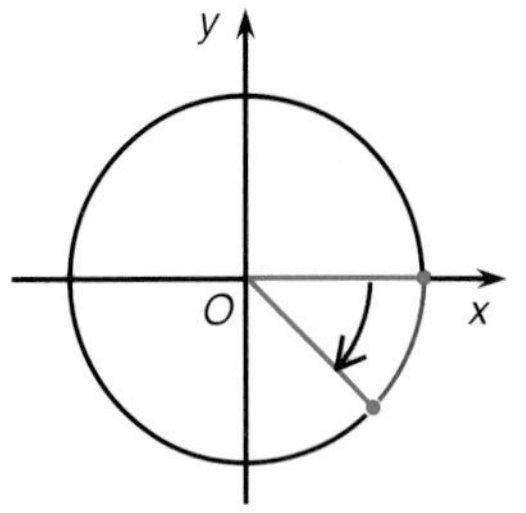

Figure 1.7.12

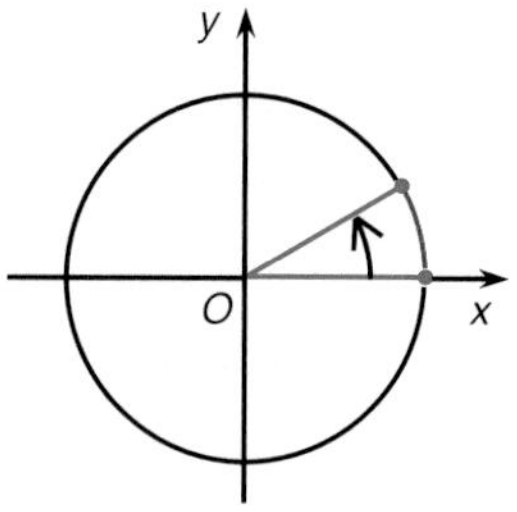

Figure 1.7.13

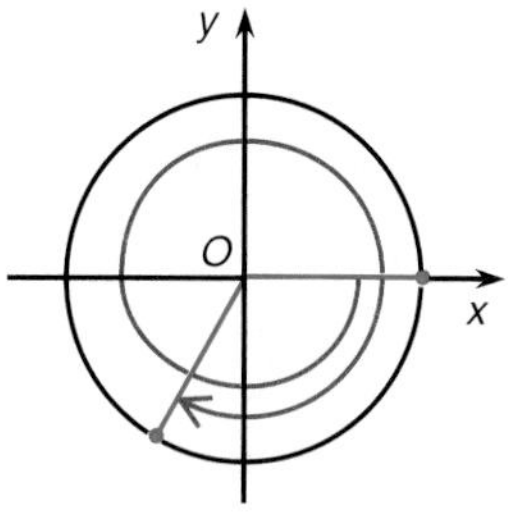

Figure 1.7.14

angle is in the negative (clockwise) direction, its radian measure is $-\frac{\pi}{4}$. See Figure 1.7.12.

(iii) An angle of $30°$ is $\frac{30}{360} = \frac{1}{12}$ of a circle, so its radian measure is $\frac{1}{12}(2\pi) = \frac{\pi}{6}$. See Figure 1.7.13.

(iv) For an angle of $-480°$ the radian measure is $\frac{-480}{360}(2\pi) = -\frac{8}{3}\pi = -\frac{8\pi}{3} \approx -8.377\,58$. See Figure 1.7.14.

Most calculators can operate either in "degree mode" or in "radian mode." It is important to develop the habit of putting your calculator into radian mode to avoid confusion.

It is very easy to convert degrees to radians and vice versa. Specifically, an angle of $\theta°$ is equal to $\frac{\theta}{360}(2\pi) = \frac{\theta}{180}\pi$ radians. In particular, an angle of $1°$ is $\frac{1}{360}(2\pi) = \frac{\pi}{180} \approx 0.017\,4533$ radians. The process can be reversed: An angle of t radians equals $\frac{t}{2\pi}\,360° = \frac{180t}{\pi}°$. In particular, 1 radian is $\frac{180(1)}{\pi} \approx 57.295\,78°$.

Summary 1.7.5

Radian/Degree Conversions

$$\theta° = \frac{\theta}{360}2\pi = \frac{\theta}{180}\pi \text{ radians}, \qquad t \text{ radians} = \frac{t}{2\pi}360° = \frac{180t}{\pi}°;$$

$$1° = \frac{\pi}{180} \approx 0.017\,4533 \text{ radians}, \qquad 1 \text{ radian} = \frac{180}{\pi}° \approx 57.295\,78°.$$

From now on, we will measure angles in radians unless otherwise specified.

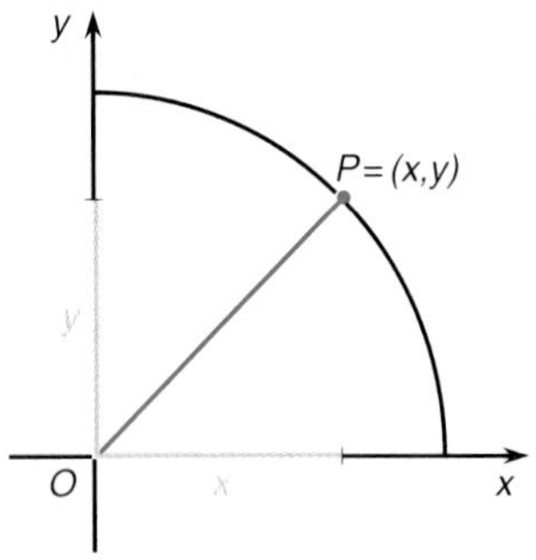

Figure 1.7.15

Special Values of Trigonometric Functions

There are some angles for which it is quite easy to find the sin and the cos, and we should become familiar with these values.

For instance, consider the angle $\frac{\pi}{4}$ (which we used to call $45°$). In Figure 1.7.15 we observe that for the point $P = (x, y)$ the numbers x and y are equal (P lies on the line through the origin that rises at an angle of $45°$, which is the line $y = x$). On the other hand, P lies on the "unit circle," the circle of radius 1 centered at the origin, whose equation is $x^2 + y^2 = 1$. If $x = y$ and $x^2 + y^2 = 1$, then $x^2 + x^2 = 1$, or $2x^2 = 1$, so $x^2 = \frac{1}{2}$ and $x = \pm\frac{1}{\sqrt{2}}$. From the picture it is clear that it is the positive

With a graphics calculator it is possible to sketch the graphs of any of the trigonometric functions. It is also possible to compare functions that can be constructed from them.

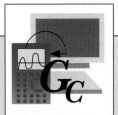

The keystrokes below show how to plot the functions $y = \sin x$ and $y = \sin(2x)$ for $x \in [-2\pi, 2\pi]$. The idea is that as x runs from -2π to 2π, $2x$ will run from -4π to 4π, so $\sin(2\pi)$ will go up and down twice as fast as $\sin x$. Be sure your calculator is in radian mode.

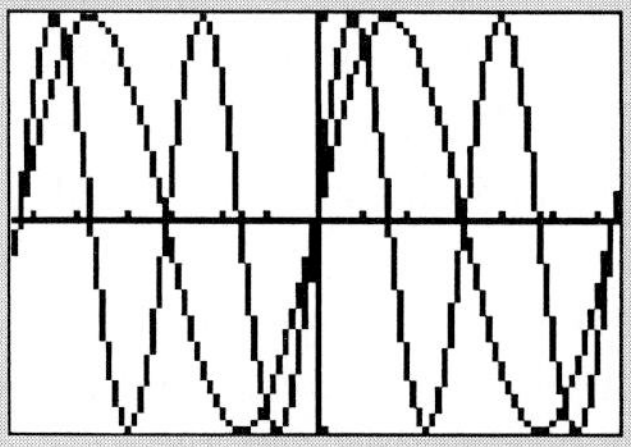

TI-85	SHARP EL9300
GRAPH F1 $(y(x) =)$ SIN F1(x) ▼ SIN (2 F1(x)) M2(RANGE) $(-)2\ \pi$ ▼ $2\ \pi$ ▼ ▼ $(-)1$ ▼ 1 F5(GRAPH)	sin x/θ/T 2ndF ▼ sin 2 x/θ/T RANGE $-2\ \pi$ ENTER $2\ \pi$ ENTER 1 ENTER -1.5 ENTER 1.5 ENTER 1 ENTER
CASIO f_x-7700GB/f_x-6300G	HP 48SX
Cls EXE Range $-2\ \pi$ EXE $2\ \pi$ EXE 1 EXE -1 EXE 1 EXE 1 EXE EXE EXE EXE Graph sin X EXE Graph sin(2X) EXE	PLOT PLOTR ERASE ATTN ' αY = SIN αX PLOT DRAW ERASE 2 π +/− × ENTER 2 π × XRNG 1 +/− ENTER 1 YRNG DRAW ATTN ' αY = SIN 2 × αX PLOT DRAW DRAW

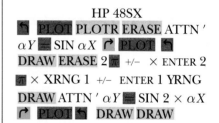

You might like to predict the behavior of $y = \sin(3x)$ before plotting it with $y = \sin x$. How will the graphs of $y = \cos x$ and $y = 2\cos(5x)$ compare? Plot $y = \sin(\frac{\pi}{2} - x)$; plot $y = \sin^2 x + \cos^2 x$. Try sketching $y = \frac{\sin 5x}{\sin(x/2)}$, for $x \in [-\pi, \pi]$.

value we want, so $x = y = \frac{1}{\sqrt{2}}$. From this we see that $\cos\frac{\pi}{4} = \frac{x}{r} = \left(\frac{1}{\sqrt{2}}\right)/1 = \frac{1}{\sqrt{2}}$ and $\sin\frac{\pi}{4} = \frac{y}{r} = \frac{x}{r} = \cos\frac{\pi}{4} = \frac{1}{\sqrt{2}}$.

For another example, let us consider an angle of $\frac{\pi}{3}$ (i.e., 60°). Since the angles in a triangle add up to 180°, a triangle whose sides are 1 unit long and whose angles are all equal must have angles of 60°. This means that our angle of $\frac{\pi}{3}$ forms one corner of the equilateral triangle shown in Figure 1.7.16. The sloping edge on the left side of the triangle is a radius of the unit circle, with length $r = 1$. The length $|OQ|$ is exactly half the bottom side of the triangle; since the sides of the triangle all have the same length, 1, the length $|OQ|$ is half as much, or $\frac{1}{2}$. We find that $\cos\frac{\pi}{3} = \frac{|OQ|}{r} = \frac{1}{2}$.

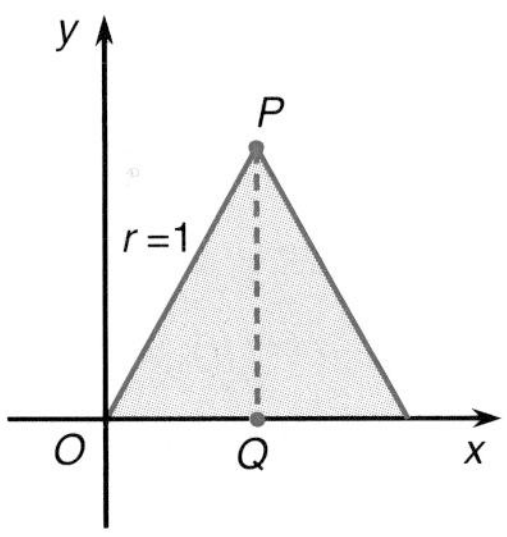

Figure 1.7.16

To find $\sin\frac{\pi}{3}$, we find the height $|QP|$, using the Pythagorean Theorem (Theorem 1.5.1). The right-angled triangle $\triangle POQ$ has base $\frac{1}{2}$ and hypotenuse of length 1, so the length of the vertical side is $\sqrt{1^2 - \left(\frac{1}{2}\right)^2} = \sqrt{1 - \frac{1}{4}} = \sqrt{\frac{3}{4}} = \frac{\sqrt{3}}{2}$. This shows that $\sin\frac{\pi}{3} = \frac{\sqrt{3}}{2}$.

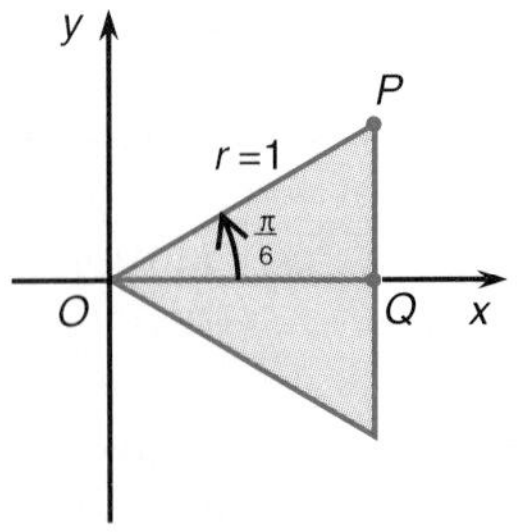

Figure 1.7.17

A similar argument can be applied to the angle $\frac{\pi}{6}$. This time the equilateral triangle has one side making an angle of $\frac{\pi}{6}$ with the positive x-axis and another side at an angle of $-\frac{\pi}{6}$ (see Figure 1.7.17). The horizontal segment OQ has length $\frac{\sqrt{3}}{2}$, and the vertical segment PQ has length $\frac{1}{2}$, so $\cos\frac{\pi}{6} = \frac{\sqrt{3}}{2}$ and $\sin\frac{\pi}{6} = \frac{1}{2}$.

Of course it is easy to see that $\cos(0) = 1$, $\sin(0) = 0$, $\cos\frac{\pi}{2} = 0$, $\sin\frac{\pi}{2} = 1$, $\cos\left(-\frac{\pi}{2}\right) = 0$, $\sin\left(-\frac{\pi}{2}\right) = -1$, $\cos\pi = -1$, and $\sin\pi = 0$.

Summary 1.7.6 Special Values of Trigonometric Functions

$$\begin{array}{llll}
\cos 0 = 1, & \sin 0 = 0, & \cos\pi = -1, & \sin\pi = 0, \\
\cos\frac{\pi}{2} = 0, & \sin\frac{\pi}{2} = 1, & \cos\frac{\pi}{4} = \frac{1}{\sqrt{2}}, & \sin\frac{\pi}{4} = \frac{1}{\sqrt{2}}, \\
\cos\frac{\pi}{3} = \frac{1}{2}, & \sin\frac{\pi}{3} = \frac{\sqrt{3}}{2}, & \cos\frac{\pi}{6} = \frac{\sqrt{3}}{2}, & \sin\frac{\pi}{6} = \frac{1}{2}.
\end{array}$$

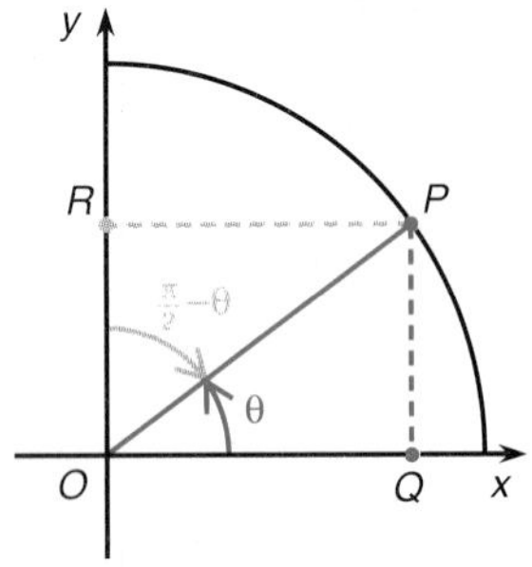

Figure 1.7.18

Now consider two angles whose sum is a right angle $\frac{\pi}{2}$. We can write them as θ and $\frac{\pi}{2} - \theta$. In Figure 1.7.18 the angle θ is drawn in the usual way, so that $\angle POQ = \theta$, and we see that $\angle POR = \frac{\pi}{2} - \theta$.

If $P = (x, y)$ is on the unit circle, then $r = 1$, so $\cos\theta = x$ and $\sin\theta = y$. Now let us consider the angle $\angle POR$. If we change our point of view and measure the angle starting from the positive y-axis and rotating *clockwise* to the line OP, then we can find $\cos\left(\frac{\pi}{2} - \theta\right)$ and $\sin\left(\frac{\pi}{2} - \theta\right)$ by looking at the triangle $\triangle POR$.

We started measuring the angle at the y-axis, so the cosine is the coordinate of P along the y-axis divided by the radius, that is, the cosine is $\frac{y}{r} = y$. The sine is the other coordinate divided by r, or $\frac{x}{r} = x$. We conclude the following:

THEOREM 1.7.7

For any angle θ we have

$$\cos\left(\frac{\pi}{2} - \theta\right) = \sin\theta \quad \text{and} \quad \sin\left(\frac{\pi}{2} - \theta\right) = \cos\theta.$$

If $P = (x, y)$ is the point on the unit circle determined by the angle θ as shown in Figure 1.7.19, then the corresponding point for the angle $-\theta$ is $P' = (x, -y)$. So $\cos(-\theta) = \frac{x}{r} = x = \cos(\theta)$ and $\sin(-\theta) = \frac{-y}{r} = -\frac{y}{r} = -y = -\sin(\theta)$.

From these formulas it is easy to see what happens to the other trigonometric functions if θ is replaced by $-\theta$.

THEOREM 1.7.8

The following formulas hold for every θ for which they make sense:

$$\begin{array}{lll}
\cos(-\theta) = \cos(\theta), & \sin(-\theta) = -\sin(\theta), & \tan(-\theta) = -\tan(\theta), \\
\sec(-\theta) = \sec(\theta), & \csc(-\theta) = -\csc(\theta), & \cot(-\theta) = -\cot(\theta).
\end{array}$$

These facts and the values of sin and cos that we found allow us to find the values of trigonometric functions at many angles.

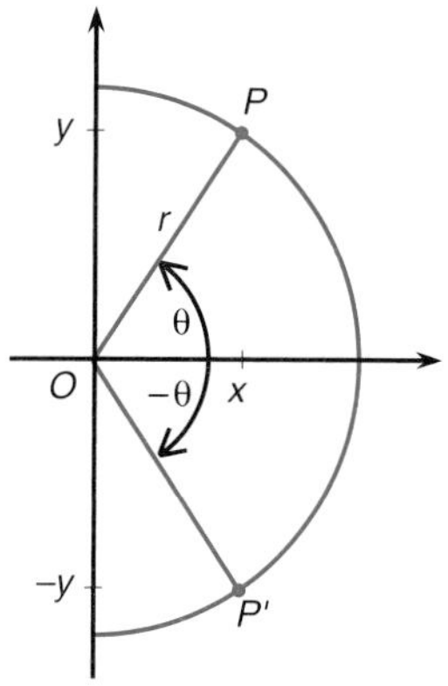

Figure 1.7.19

EXAMPLE 2 Find (i) $\sin\left(\frac{3\pi}{4}\right)$; (ii) $\cos\left(-\frac{2\pi}{3}\right)$; (iii) $\tan\left(\frac{7\pi}{6}\right)$.

Solution

(i) Since $\frac{3\pi}{4} + \left(-\frac{\pi}{4}\right) = \frac{\pi}{2}$, we know that $\sin\left(\frac{3\pi}{4}\right) = \cos\left(-\frac{\pi}{4}\right) = \cos\left(\frac{\pi}{4}\right) = \frac{1}{\sqrt{2}}$.

(ii) $\cos\left(-\frac{2\pi}{3}\right) = \cos\left(\frac{2\pi}{3}\right) = \cos\left(\frac{\pi}{2} - \left(-\frac{\pi}{6}\right)\right) = \sin\left(\left(-\frac{\pi}{6}\right)\right) = -\sin\left(\left(\frac{\pi}{6}\right)\right) = -\frac{1}{2}$.

(iii) To evaluate tan, it is simpler to begin with sin and cos: $\cos\frac{7\pi}{6} = \cos\left(\frac{\pi}{2} - \left(-\frac{2\pi}{3}\right)\right) = \sin\left(-\frac{2\pi}{3}\right) = -\sin\left(\frac{2\pi}{3}\right) = -\sin\left(\frac{\pi}{2} - \left(-\frac{\pi}{6}\right)\right) = -\cos\left(-\frac{\pi}{6}\right) = -\cos\left(\frac{\pi}{6}\right) = -\frac{\sqrt{3}}{2}$.

Similarly, $\sin\frac{7\pi}{6} = \sin\left(\frac{\pi}{2} - \left(-\frac{2\pi}{3}\right)\right) = \cos\left(-\frac{2\pi}{3}\right) = \cos\left(\frac{2\pi}{3}\right) = \cos\left(\frac{\pi}{2} - \left(-\frac{\pi}{6}\right)\right) = \sin\left(-\frac{\pi}{6}\right) = -\sin\left(\frac{\pi}{6}\right) = -\frac{1}{2}$.

From these two facts we see that

$$\tan\frac{7\pi}{6} = \frac{\sin(7\pi/6)}{\cos(7\pi/6)} = \frac{-1/2}{-\sqrt{3}/2} = \frac{1}{\sqrt{3}} \approx 0.577\,35.$$

Now that we can evaluate the trigonometric functions at many points, it is possible to plot a number of points and sketch their graphs, as in Figure 1.7.20.

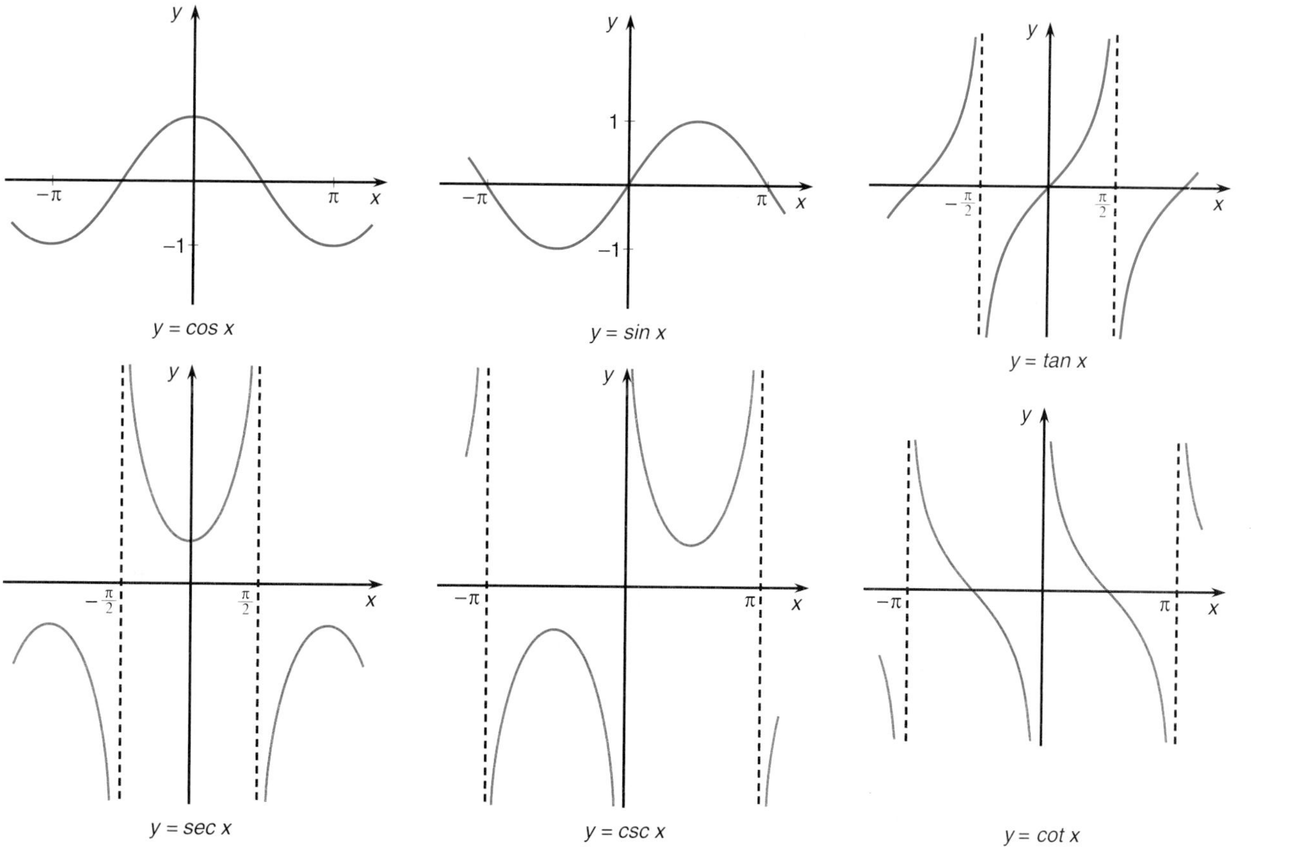

Figure 1.7.20

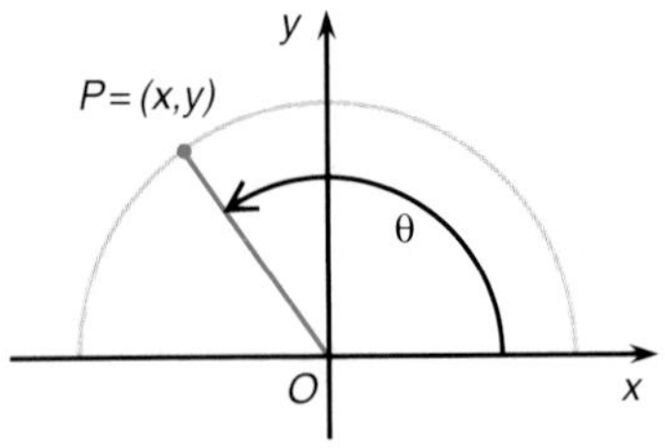

Figure 1.7.21

In Figure 1.7.21 we have drawn the point $P = (x, y)$, and labeled the angle θ between the positive x-axis and the line OP. If we let $r = \sqrt{x^2 + y^2}$, then P is on the circle of radius r centered at the origin. We know from Definition 1.7.1 that $\cos\theta = \frac{x}{r}$ and $\sin\theta = \frac{y}{r}$. These two equations can be rearranged to give

$$x = r\cos\theta \qquad \text{and} \qquad y = r\sin\theta. \tag{1.7.9}$$

We can find the coordinates of P if we know the length $|OP|$ and the angle θ.

Law of Cosines, Addition Formulas

The Pythagorean Theorem relates the squares of the lengths of the sides of a right-angled triangle. It is possible to find a relation among the lengths of the sides of any triangle, but it will depend on the angles as well as the lengths of the sides.

THEOREM 1.7.10

The Law of Cosines

Consider a triangle whose sides have lengths a, b, and c. Suppose the side of length c is opposite to an angle θ. Then

$$c^2 = a^2 + b^2 - 2ab\cos\theta.$$

Notice that if $\theta = \frac{\pi}{2}$, then the triangle has a right angle, and since $\cos\frac{\pi}{2} = 0$, the Law of Cosines reduces to the Pythagorean Theorem.

PROOF In Figure 1.7.22 we place the angle θ at the origin, the side of length a along the x-axis from O to $(a, 0)$, and the side of length b from the origin to the point $P = (x, y)$.

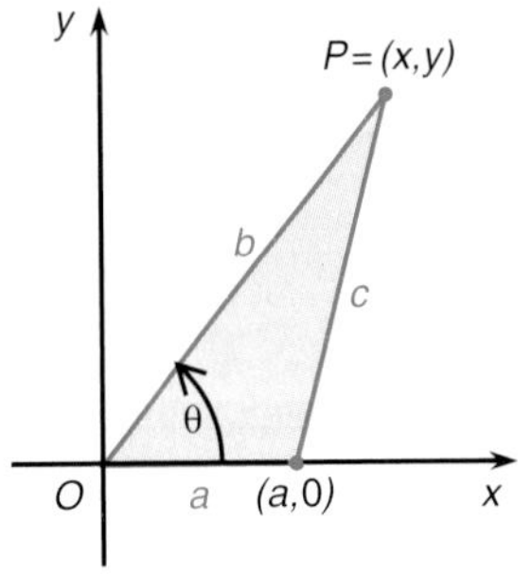

Figure 1.7.22

Then $b = \sqrt{x^2 + y^2}$, and by Equations 1.7.9 we see that $x = b\cos\theta$ and $y = b\sin\theta$. The side whose length is c lies between the points $P = (x, y)$ and $(a, 0)$, so $c = \sqrt{(x - a)^2 + (y - 0)^2}$.

From this we find $c^2 = (x - a)^2 + y^2 = x^2 - 2ax + a^2 + y^2 = (x^2 + y^2) + a^2 - 2ax$, and since $x^2 + y^2 = b^2$ and $x = b\cos\theta$, this becomes $c^2 = b^2 + a^2 - 2ab\cos\theta$, which is the desired formula. ◢

A natural and important question to ask is this: If we know the values of cos and sin for two angles θ and ϕ (ϕ is the Greek letter phi, which is sometimes pronounced "fee" and sometimes pronounced to rhyme with "eye"), how can we find $\cos(\theta + \phi)$?

It is actually slightly easier to find $\cos(\theta - \phi)$ first. Since $\cos(\theta - \phi) = \cos(\phi - \theta)$, we can assume $\theta > \phi$ (because otherwise we could interchange them, without

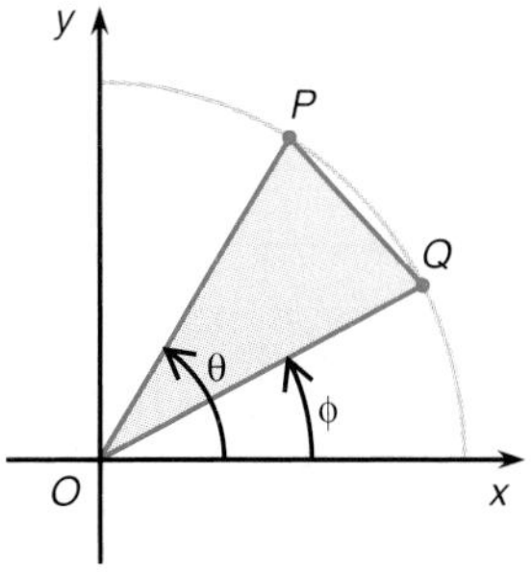

Figure 1.7.23

affecting the result). In Figure 1.7.23 we have drawn both angles θ and ϕ starting at the positive x-axis, and we have plotted the unit circle. Using Equations 1.7.9, we can find the coordinates of the points P and Q where the second arm of each angle intersects the circle. Since the radius of the circle is $r = 1$, we find that $P = (\cos\theta, \sin\theta)$ and $Q = (\cos\phi, \sin\phi)$.

Now we apply the Law of Cosines to the triangle $\triangle POQ$. The two sides OP and OQ both have length 1, and the length of the third side PQ satisfies

$$|PQ|^2 = c^2 = 1^2 + 1^2 - 2(1)(1)\cos(\theta - \phi) = 2 - 2\cos(\theta - \phi).$$

On the other hand, since $P = (\cos\theta, \sin\theta)$ and $Q = (\cos\phi, \sin\phi)$, we see that

$$\begin{aligned} |PQ|^2 &= (\cos\theta - \cos\phi)^2 + (\sin\theta - \sin\phi)^2 \\ &= \cos^2\theta - 2\cos\theta\cos\phi + \cos^2\phi + \sin^2\theta - 2\sin\theta\sin\phi + \sin^2\phi \\ &= \cos^2\theta + \sin^2\theta + \cos^2\phi + \sin^2\phi - 2(\cos\theta\cos\phi + \sin\theta\sin\phi) \\ &= 2 - 2(\cos\theta\cos\phi + \sin\theta\sin\phi). \end{aligned}$$

Equating these two expressions for $|PQ|^2$, we see that

$$\cos(\theta - \phi) = \cos\theta\cos\phi + \sin\theta\sin\phi.$$

If we replace ϕ in this equation with $-\phi$, then the left side becomes $\cos(\theta + \phi)$. Since $\cos(-\phi) = \cos\phi$ and $\sin(-\phi) = -\sin\phi$, the right side becomes $\cos\theta\cos\phi - \sin\theta\sin\phi$.

We could use a similar argument to find $\sin(\theta + \phi)$, but it is easier to reason as follows:

$$\begin{aligned} \sin(\theta + \phi) &= \cos\left(\frac{\pi}{2} - (\theta + \phi)\right) = \cos\left(\left(\frac{\pi}{2} - \theta\right) - \phi\right) \\ &= \cos\left(\frac{\pi}{2} - \theta\right)\cos\phi + \sin\left(\frac{\pi}{2} - \theta\right)\sin\phi \\ &= \sin\theta\cos\phi + \cos\theta\sin\phi. \end{aligned}$$

Again we can replace ϕ with $-\phi$ to find the formula for $\sin(\theta - \phi)$.

THEOREM 1.7.11

Addition Formulas

$$\begin{aligned} \cos(\theta + \phi) &= \cos\theta\cos\phi - \sin\theta\sin\phi, \\ \sin(\theta + \phi) &= \sin\theta\cos\phi + \cos\theta\sin\phi, \\ \cos(\theta - \phi) &= \cos\theta\cos\phi + \sin\theta\sin\phi, \\ \sin(\theta - \phi) &= \sin\theta\cos\phi - \cos\theta\sin\phi. \end{aligned}$$

If we let $\phi = \theta$ in these formulas, we can find formulas for $\cos(2\theta)$ and $\sin(2\theta)$.

THEOREM 1.7.12

Double-Angle Formulas

$$\cos(2\theta) = \cos^2\theta - \sin^2\theta, \qquad \sin(2\theta) = 2\cos\theta\sin\theta.$$

If we take the double-angle formula $\cos(2\theta) = \cos^2\theta - \sin^2\theta$ and substitute into it the identity $\sin^2\theta = 1 - \cos^2\theta$, we find $\cos(2\theta) = \cos^2\theta - (1 - \cos^2\theta) = 2\cos^2\theta - 1$. This can be rearranged as $2\cos^2\theta = 1 + \cos(2\theta)$, or $\cos^2\theta = \frac{1+\cos(2\theta)}{2}$.

We can use this to find that $\sin^2\theta = 1 - \cos^2\theta = 1 - \frac{1+\cos(2\theta)}{2} = \frac{1-\cos(2\theta)}{2}$.

These formulas can also be used to find $\cos^2\left(\frac{\theta}{2}\right)$ and $\sin^2\left(\frac{\theta}{2}\right)$ in terms of $\cos\theta$.

THEOREM 1.7.13

Half-Angle Formulas

$$\cos^2\theta = \frac{1+\cos(2\theta)}{2}, \qquad \sin^2\theta = \frac{1-\cos(2\theta)}{2}.$$

These formulas can be used to help us evaluate trigonometric functions at many additional points.

EXAMPLE 3 Find (i) $\sin\frac{\pi}{12}$; (ii) $\cos\frac{\pi}{8}$; (iii) $\cos\frac{11\pi}{12}$.

Solution

(i) From the half-angle formula we know that $\sin^2\frac{\pi}{12} = \frac{1}{2}\left(1 - \cos\left(2\left(\frac{\pi}{12}\right)\right)\right) = \frac{1}{2}\left(1 - \cos\frac{\pi}{6}\right) = \frac{1}{2}\left(1 - \frac{\sqrt{3}}{2}\right) = \frac{1}{2} - \frac{\sqrt{3}}{4} \approx 0.066\,987$. What we want is $\sin\frac{\pi}{12}$, which is either the square root of this number or the negative of the square root. Since $\frac{\pi}{12}$ is between 0 and the right angle $\frac{\pi}{2}$, we see that its sin must be positive, so $\sin\frac{\pi}{12} = \sqrt{\frac{1}{2} - \frac{\sqrt{3}}{4}} \approx 0.258\,819$.

(ii) The half-angle formula gives $\cos^2\frac{\pi}{8} = \frac{1}{2}\left(1 + \cos\left(2\left(\frac{\pi}{8}\right)\right)\right) = \frac{1}{2}\left(1 + \cos\frac{\pi}{4}\right) = \frac{1}{2}\left(1 + \frac{1}{\sqrt{2}}\right) = \frac{1}{2} + \frac{1}{2\sqrt{2}} \approx 0.853\,5534$. Since $\cos\frac{\pi}{8}$ is positive, $\cos\frac{\pi}{8} = \sqrt{\frac{1}{2} + \frac{1}{2\sqrt{2}}} \approx 0.923\,8795$.

(iii) Here $\cos^2\frac{11\pi}{12} = \frac{1}{2}\left(1 + \cos\frac{11\pi}{6}\right) = \frac{1}{2}\left(1 + \cos\left(2\pi - \frac{\pi}{6}\right)\right) = \frac{1}{2}\left(1 + \cos\left(-\frac{\pi}{6}\right)\right)$. We have used the fact that cos is periodic, having the same value at $2\pi + \theta$ as it has at θ.

Since $\cos\left(-\frac{\pi}{6}\right) = \cos\frac{\pi}{6} = \frac{\sqrt{3}}{2}$, we find $\cos^2\frac{11\pi}{12} = \frac{1}{2}\left(1 + \frac{\sqrt{3}}{2}\right) = \frac{1}{2} + \frac{\sqrt{3}}{4}$. Since $\frac{\pi}{2} < \frac{11\pi}{12} < \pi$, we see that if we start at the positive x-axis and rotate through an angle of $\frac{11\pi}{12}$, the resulting point will lie above the *negative* x-axis, so its x-coordinate will be negative. So $\cos\frac{11\pi}{12}$ is negative, and $\cos\frac{11\pi}{12} = -\sqrt{\frac{1}{2} + \sqrt{3}/4} \approx -0.965\,9258$.

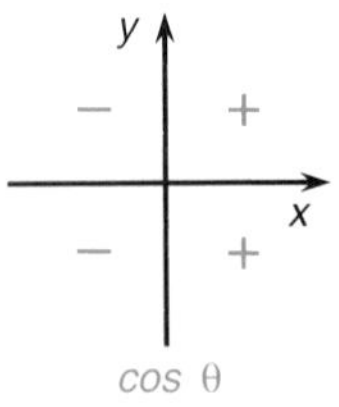

Figure 1.7.24(i)

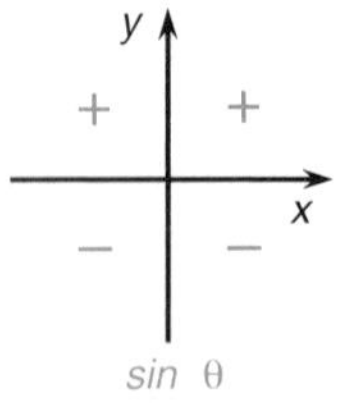

Figure 1.7.24(ii)

The half-angle formulas (Theorem 1.7.13) permit us to find $\cos^2\frac{\theta}{2}$ and $\sin^2\frac{\theta}{2}$. To find $\cos\frac{\theta}{2}$ or $\sin\frac{\theta}{2}$, we must take square roots, but it is also necessary to determine the correct *sign* of the answer. Figures 1.7.24(i) and (ii) show the regions in which cos and sin are positive and negative.

If we add the formulas for $\cos(\theta + \phi)$ and $\cos(\theta - \phi)$, we find that

$$\begin{aligned}\cos(\theta + \phi) + \cos(\theta - \phi) &= \cos\theta\cos\phi - \sin\theta\sin\phi + \cos\theta\cos\phi + \sin\theta\sin\phi \\ &= 2\cos\theta\cos\phi.\end{aligned}$$

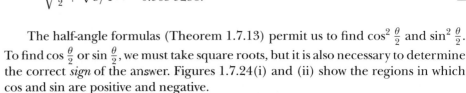

Turning this around, we find that $\cos\theta\cos\phi = \frac{1}{9}\left(\cos(\theta + \phi) + \cos(\theta - \phi)\right)$.

From the difference of the two addition formulas we find $\sin\theta\sin\phi = \frac{1}{2}\big(\cos(\theta-\phi) - \cos(\theta+\phi)\big)$. Working with the addition formulas for sin gives an additional identity.

THEOREM 1.7.14

$$\cos\theta\cos\phi = \frac{1}{2}\big(\cos(\theta+\phi) + \cos(\theta-\phi)\big),$$

$$\sin\theta\sin\phi = \frac{1}{2}\big(\cos(\theta-\phi) - \cos(\theta+\phi)\big),$$

$$\sin\theta\cos\phi = \frac{1}{2}\big(\sin(\theta+\phi) + \sin(\theta-\phi)\big).$$

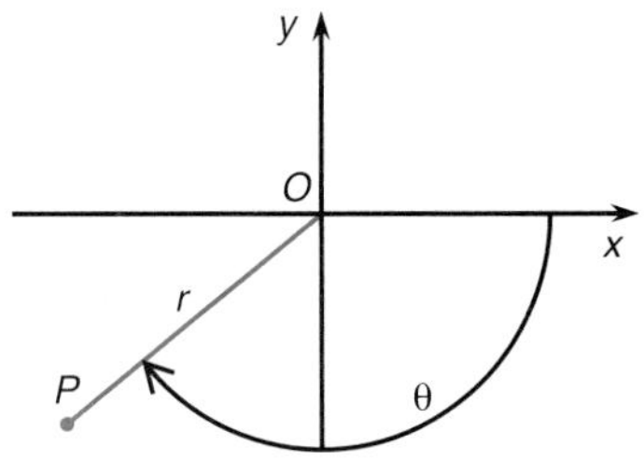

Figure 1.7.25

Polar Coordinates

Representing a point P in the plane by its coordinates (x, y) is quite familiar to us, but there is another way in which the same thing can be accomplished. Consider the point P shown in Figure 1.7.25, and join it to the origin O. The angle θ between the positive x-axis and the segment OP determines the "direction" in which the segment OP is "pointing." If we also know how far P is from the origin, then we will have determined the position of P exactly.

DEFINITION 1.7.15

Consider a point P in the plane. If the segment OP makes an angle θ with the positive x-axis, and if we let $r = |OP|$ be the distance from the origin to P, then the pair of numbers (r, θ) is called the **polar coordinates** of the point P.

The number r is sometimes called the **radius** or the **modulus** of P, and θ is called its **angle**, or its **polar angle**, or its **argument**.

There are several things to notice about this definition. One is that for any point P there are many possible choices for the angle θ. For instance, we could measure the angle θ in either direction from the positive x-axis. We could also measure an extra complete rotation about the origin, which would add 2π to θ.

So the angle θ is not completely determined. We can add 2π or 4π or -2π or *any* integer multiple of 2π to θ, and the resulting angle will still determine the same direction in the plane. See Figure 1.7.26.

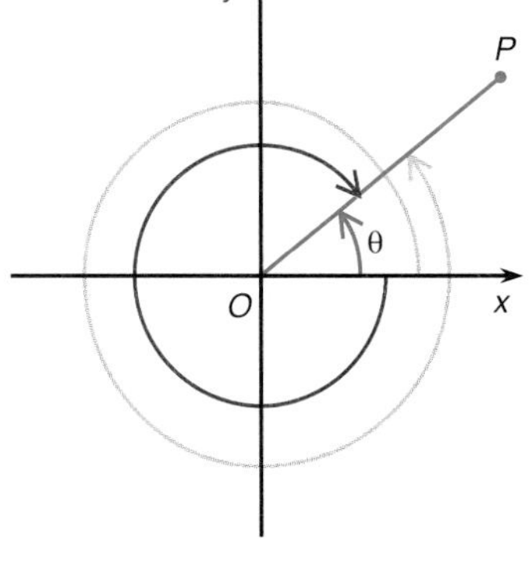

Figure 1.7.26

For instance, for the point $(0, 2)$ on the y-axis we could naturally choose the angle $\theta = \frac{\pi}{2}$. But $\frac{\pi}{2} + 2\pi = \frac{5\pi}{2}$ would be just as good. For that matter, so would $\frac{\pi}{2} - 2\pi = -\frac{3\pi}{2}$, or $\frac{9\pi}{2}$, etc. In fact if k is any integer, we could choose the angle to be $\frac{\pi}{2} + 2k\pi$.

In this situation we often write $\theta = \frac{\pi}{2} + 2k\pi$, to help remind ourselves that there are many possible values of θ.

Another complication arises with the point O, the origin. Its radius is $r = 0$, but *any* angle θ can be used. If we move a distance of zero in *any* direction, the point at which we will end up is the origin.

One obvious question is how can we go back and forth between polar coordinates (r, θ) and Cartesian coordinates (x, y). In other words, if we know the polar coordinates (r, θ) of a point P, how do we find its Cartesian (or "rectangular") coordinates (x, y), and vice versa?

In fact we have already answered one of these questions in Equations 1.7.9, where we saw that

$$x = r\cos\theta \qquad \text{and} \qquad y = r\sin\theta.$$

In the opposite direction it is a little more complicated. If P is the point whose rectangular coordinates are (x, y), then we know its modulus is $r = \sqrt{x^2 + y^2}$. If θ is its argument, then we know that $x = r\cos\theta$ and $y = r\sin\theta$. If $x \neq 0$, we divide these two equations and find that $\frac{y}{x} = \frac{\sin\theta}{\cos\theta} = \tan\theta$. This tells us that θ is an angle whose tangent equals $\frac{y}{x}$, which is often expressed by saying that $\theta = \arctan(\frac{y}{x})$.

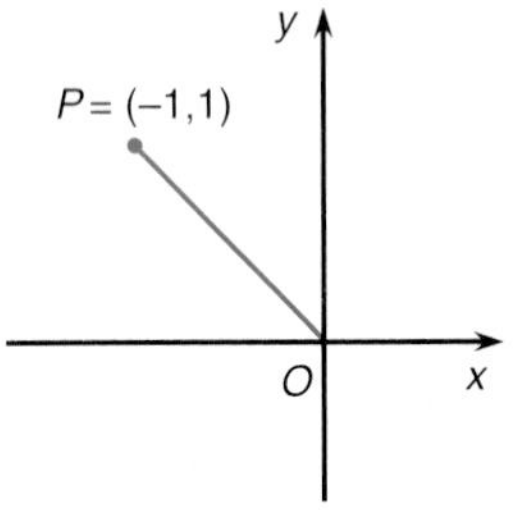

Figure 1.7.27

The complication is that this does not quite determine what θ is. For instance, if $P = (-1, 1)$, then $\frac{y}{x} = -1$. If we look for angles whose tangent equals -1, then there are many possibilities, but among them are $\theta = -\frac{\pi}{4}$ and $\theta = \frac{3\pi}{4}$. These two angles point in *opposite* directions from the origin.

This reflects the fact that the quotient of the y-coordinate by the x-coordinate is the same for $P = (x, y)$ and for $-P = (-x, -y)$; in both cases it equals $\frac{y}{x}$. However, by looking at the picture we can easily tell which is the right choice (in the case of $P = (-1, 1)$, one choice of the angle lies between 0 and π, so a suitable choice is $\frac{3\pi}{4}$); see Figure 1.7.27.

The polar coordinates of a point $P = (x, y)$ are (r, θ), where

$$r = \sqrt{x^2 + y^2} \qquad \text{and} \qquad \tan\theta = \frac{y}{x}.$$

If an angle θ can be found for which $\tan\theta = \frac{y}{x}$, then it is necessary to sketch a picture and decide whether the argument is this θ or $\theta + \pi$ (i.e., the opposite angle). If θ is one correct choice for the argument, then $\theta + 2k\pi$ is also a correct choice, for any integer k.

If $x = 0$, that is, P is on the vertical axis, then $\frac{y}{x}$ is undefined, but in this case, $\theta = \frac{\pi}{2} + 2k\pi$ if $y > 0$, and $\theta = -\frac{\pi}{2} + 2k\pi$ if $y < 0$.

EXAMPLE 4 (i) Find the rectangular coordinates of the point whose polar coordinates are $\left(2\sqrt{2}, -\frac{3\pi}{4}\right)$. Find the polar coordinates of the points (ii) $P = (-3, -\sqrt{3})$; (iii) $Q = (5, -2)$; (iv) $S = (0, -3)$.

Solution

(i) From Equations 1.7.9 we see that $x = r\cos\theta = 2\sqrt{2}\cos\left(-\frac{3\pi}{4}\right) = 2\sqrt{2}\cos\left(\frac{3\pi}{4}\right)$ $= 2\sqrt{2}\sin\left(\frac{\pi}{2} - \frac{3\pi}{4}\right) = 2\sqrt{2}\sin\left(-\frac{\pi}{4}\right) = -2\sqrt{2}\sin\frac{\pi}{4} = -2\sqrt{2}\,\frac{1}{\sqrt{2}} = -2.$

Similarly, $y = r\sin\theta = 2\sqrt{2}\sin\left(-\frac{3\pi}{4}\right) = -2\sqrt{2}\sin\left(\frac{3\pi}{4}\right) =$ $-2\sqrt{2}\cos\left(\frac{\pi}{2} - \frac{3\pi}{4}\right) = -2\sqrt{2}\cos\left(-\frac{\pi}{4}\right) = -2\sqrt{2}\cos\frac{\pi}{4} = -2\sqrt{2}\,\frac{1}{\sqrt{2}} = -2.$

The rectangular coordinates of the point in question are $(-2, -2)$.

(ii) We find that $r = \sqrt{(-3)^2 + (-\sqrt{3})^2} = \sqrt{9+3} = \sqrt{12} = 2\sqrt{3}$.

We also know that $\tan\theta = \frac{-\sqrt{3}}{-3} = \frac{1}{\sqrt{3}}$. If we think through the angles for which we know the values of the trigonometric functions, we know that $\sin\frac{\pi}{6} = \frac{1}{2}$ and $\cos\frac{\pi}{6} = \frac{\sqrt{3}}{2}$, so $\tan\frac{\pi}{6} = \frac{1}{2}/\frac{\sqrt{3}}{2} = \frac{1}{\sqrt{3}}$. We expect that either $\frac{\pi}{6}$ or $\frac{\pi}{6} + \pi = \frac{7\pi}{6}$ will be the correct argument.

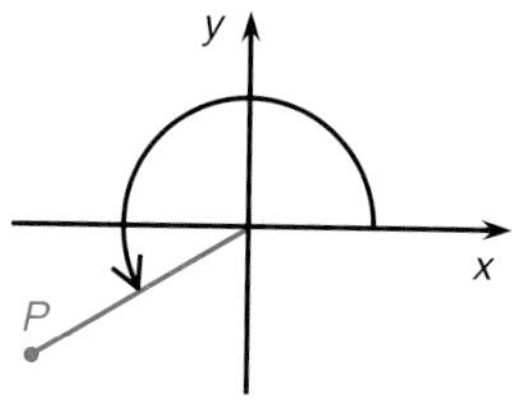

Figure 1.7.28

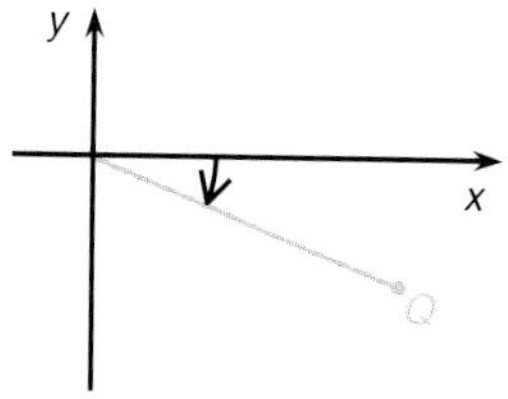

Figure 1.7.29

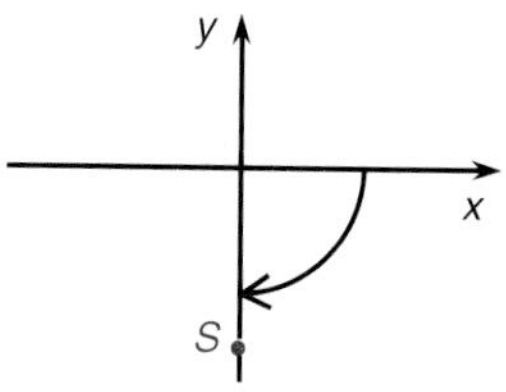

Figure 1.7.30

A quick look at Figure 1.7.28 will convince us that the angle is $\frac{7\pi}{6}$. The polar coordinates of the point $P = (-3, -\sqrt{3})$ are $\left(2\sqrt{3}, \frac{7\pi}{6} + 2k\pi\right)$, where k can be any integer.

(iii) For $Q = (5, -2)$ we find $r = \sqrt{5^2 + (-2)^2} = \sqrt{29}$ and $\tan\theta = -\frac{2}{5}$. None of the angles whose tangents we can work out easily has this as its tangent, so we resort to a calculator. Being careful to set it to *radian* mode, we find that the arctan (or "inverse tangent") of $-\frac{2}{5}$ is $\arctan\left(-\frac{2}{5}\right) \approx -0.380\,506$.

From Figure 1.7.29 we expect the argument to be a negative number between $-\pi$ and 0, which this is (recall that $\pi \approx 3.141\,59$). The angle we have found is the correct one. (If it were not, then we would simply add π to it.)

The polar coordinates of Q are $(\sqrt{29}, -0.380\,506 + 2k\pi)$, where k can be any integer and the argument is given as an approximate value.

(iv) For $S = (0, -3)$ the modulus is $r = 3$, but we cannot find $\frac{y}{x}$ because we would have to divide by zero. Fortunately, we see that S is on the vertical axis *below* the origin, so $\theta = -\frac{\pi}{2} + 2k\pi$. See Figure 1.7.30.

Exercises 1.7

Find the following values of trigonometric functions, without using a calculator except to check your work.

1. $\sin\left(-\frac{\pi}{2}\right)$
2. $\cos\left(\frac{9\pi}{2}\right)$
3. $\cos\left(\frac{3\pi}{4}\right)$
4. $\sin\left(-\frac{4\pi}{3}\right)$
5. $\tan\left(-\frac{5\pi}{3}\right)$
6. $\sec\left(\frac{2\pi}{3}\right)$
7. $\csc\left(\frac{5\pi}{4}\right)$
8. $\cot\left(\frac{5\pi}{6}\right)$
9. $\sin\left(-\frac{3\pi}{4}\right)$
10. $\tan\left(\frac{7\pi}{2}\right)$
11. $\sin\frac{\pi}{8}$
12. $\cos\frac{\pi}{12}$
13. $\sin\frac{5\pi}{12}$
14. $\cos\frac{7\pi}{8}$
15. $\cos\frac{11\pi}{12}$
16. $\sin\frac{17\pi}{8}$
17. $\tan\frac{11\pi}{8}$
18. $\sec\frac{13\pi}{12}$
19. $\tan\frac{15\pi}{8}$
20. $\csc\frac{3\pi}{12}$

Find the rectangular coordinates of each of the points whose polar coordinates are given below.

21. $(3, \pi)$
22. $\left(\frac{1}{2}, 4\pi\right)$
23. $\left(2, \frac{\pi}{2}\right)$
24. $\left(5, -\frac{\pi}{2}\right)$
25. $\left(1, -\frac{\pi}{3}\right)$
26. $\left(2, \frac{7\pi}{2}\right)$
27. $\left(4, -\frac{5\pi}{4}\right)$
28. $\left(8, \frac{11\pi}{4}\right)$
29. $\left(2, \frac{5\pi}{6}\right)$
30. $\left(2, \frac{3\pi}{8}\right)$

Find the polar coordinates of each of the points whose Cartesian coordinates are given below.

31. $(3, 0)$
32. $(-2, 0)$
33. $(0, -2)$
34. $\left(0, \frac{1}{2}\right)$
35. $(2, -2)$
36. $(4, 4\sqrt{3})$
37. $(-4, -4)$
38. $(0, 0)$
39. $\left(\frac{1}{4}, -\frac{1}{4\sqrt{3}}\right)$
40. $(\cos x, \sin x)$

1.8 Logarithms and Exponentials

EXPONENT
LOGARITHM
BASE
COMMON LOGARITHMS
NATURAL LOGARITHMS

Taking powers is a familiar concept: If n is any positive integer and x is any real number, x^n means the product of x with itself n times, that is, $x \times x \times \ldots \times x$, with a total of n factors. The number n is called the **exponent**.

The product $x^m x^n$ is the product of two products, one containing m factors all equal to x and the other containing n such factors. So $x^m x^n$ is the product of $m + n$ factors, all equal to x, so it is x^{m+n}. If x is any real number and m and n are positive integers,

$$x^m x^n = x^{m+n}. \tag{1.8.1}$$

Another useful property of powers is illustrated by looking at $(x^2)^3$, the product of three copies of x^2. Since each of them is the product of two copies of x, the whole thing is the product of six copies of x, or x^6. In exactly the same way, $(x^m)^n$ is the product of mn copies of x. In other words, for any real x and any positive integers m and n,

$$(x^m)^n = x^{mn}. \tag{1.8.2}$$

Property 1.8.1 suggests a way to extend the definition to integer exponents. If we want all integer powers to satisfy the same property, then we should expect that $x^0 x^n = x^{0+n} = x^n$. For this to be true for all values of x, x^0 must equal 1.

Moreover, $x^{-n}x^n = x^{-n+n} = x^0 = 1$, so $x^{-n} = \frac{1}{x^n}$. This allows us to define x^n for any integer n and any real number x, except that when $x = 0$, it only makes sense to consider x^n for $n > 0$.

$$x^0 = 1 \qquad \text{and} \qquad x^{-n} = \frac{1}{x^n}. \tag{1.8.3}$$

It is also possible to define what is meant by x^s even if s is not a rational number, provided that $x > 0$. For instance, knowing that $\sqrt{2}$ is not a rational number, we could ask what is meant by $x^{\sqrt{2}}$. Roughly speaking, the idea is to find rational numbers r that are closer and closer to $\sqrt{2}$ and hope that x^r will get close to some number w as r gets close to $\sqrt{2}$. It turns out that this is exactly what happens, so we can let $x^{\sqrt{2}} = w$. For any real number s we can find rational numbers r that get closer and closer to s; the numbers x^r will then get closer and closer to some number, which is what we call x^s. The details of this approach will be discussed more carefully in Chapter 7.

Next it is possible to consider powers x^r, where the exponent r is a rational number. We expect Property 1.8.1 to hold for them too.

For instance, with $r = \frac{1}{2}$ we expect that $x^r x^r = x^{2r}$, that is, $x^{1/2}x^{1/2} = x^1$, which means $(x^{1/2})^2 = x$. This says that $x^{1/2}$ is something whose square equals x. This is not possible if $x < 0$, and if $x \geq 0$, it means that $x^{1/2} = \sqrt{x}$. Similarly, $x^{1/3}$ is the cube root of x, $x^{1/4}$ is the fourth root of x, and so on.

If Property 1.8.2 holds for rational exponents, then $x^{m/n} = x^{(1/n)m} = (x^{1/n})^m$, the mth power of the nth root of x. This makes sense whenever $x > 0$ and $\frac{m}{n}$ is a rational number (if $\frac{m}{n} < 0$, we can assume that $n > 0$ and $m < 0$, so it makes sense to consider $x^{1/n}$ and $(x^{1/n})^m$).

Logarithms

Suppose $a > 0$ and we want to find a real number s so that $10^s = a$. For instance, if $a = 1000$, then $s = 3$; if $a = \frac{1}{10}$, then $s = -1$; if $a = \sqrt{10}$, then $s = \frac{1}{2}$. This number s is called the **logarithm** of a, and it is often written $s = \log a$. It is determined by the property that $10^{\log a} = a$. Another way of putting it is that $\log(10^s) = s$; the logarithm of a positive number is the exponent to which we have to raise 10 in order to get that number:

$$10^{\log a} = a \qquad \text{and} \qquad \log(10^s) = s. \tag{1.8.4}$$

Property 1.8.1 tells us that

$$10^{\log(ac)} = ac = (10^{\log a})(10^{\log c}) = 10^{\log a + \log c}.$$

Matching exponents at the left and right ends of this equation, we find that $\log(ac) = \log a + \log c$, an extremely important property of logarithms. In much the same way,

$$10^{\log(a^s)} = a^s = (a)^s = (10^{\log a})^s = 10^{s \log a},$$

so $\log(a^s) = s \log a$. In particular, with $s = -1$ we see that $\log(a^{-1}) = -\log a$, that is, $\log \frac{1}{a} = -\log a$. From this we find that $\log \frac{a}{c} = \log\left(a \times \frac{1}{c}\right) = \log a + \log \frac{1}{c} = \log a - \log c$. These properties of log are summarized below. For any $a, c > 0$,

$$\begin{aligned} \log(ac) &= \log a + \log c, & \log(a^s) &= s \log a, \\ \log \tfrac{1}{a} &= -\log a, & \log \tfrac{a}{c} &= \log a - \log c. \end{aligned} \tag{1.8.5}$$

There is no particular reason why we should choose powers of 10 when we define logarithms. It would have been just as good to use 2, say, and look for a number s so that $2^s = a$. This number s is called the logarithm to the base 2 of a. It is the exponent to which 2 must be raised to get the number a.

We could have used any positive number $b \neq 1$ in the same way; if $b^s = a$, we say that s is the logarithm to the base b of a, and we write $s = \log_b(a)$. There are properties analogous to Equations 1.8.4:

$$b^{\log_b a} = a \qquad \text{and} \qquad \log_b(b^s) = s. \tag{1.8.6}$$

There are also properties analogous to Equations 1.8.5 for logarithms to the base b:

$$\begin{aligned} \log_b(ac) &= \log_b a + \log_b c, & \log_b(a^s) &= s \log_b a, \\ \log_b \tfrac{1}{a} &= -\log_b a, & \log_b \tfrac{a}{c} &= \log_b a - \log_b c. \end{aligned} \tag{1.8.7}$$

In particular, when we write $\log a$, we mean $\log_{10} a$. Because logarithms to the base 10 are used so often, they are often called **common logarithms**.

EXAMPLE 1 Find (i) $\log \frac{100}{\sqrt{10}}$; (ii) $\log_2(16)$; (iii) $\log_3 \frac{1}{27}$.

Solution

(i) By Properties 1.8.5, $\log \frac{100}{\sqrt{10}} = \log 100 - \log(\sqrt{10}) = 2 - \log(10^{1/2}) = 2 - \frac{1}{2} = \frac{3}{2}$.

(ii) $\log_2(16) = \log_2(2^4) = 4$.

(iii) $\log_3 \frac{1}{27} = -\log_3 27 = -\log_3(3^3) = -3$.

There is one other base that is used quite frequently for logarithms. It is like radian measure for trigonometric functions in the sense that it is particularly well suited for use in calculus, even though it seems fairly inconvenient for many other purposes. Because logarithms to this base are so well adapted for calculus, they are called **natural logarithms**, and we write $\ln x$ for the natural logarithm of x. The base for natural logarithms is the number $e \approx 2.718\,281\,83$, that is, $\ln x = \log_e x$. So $\ln a$ is defined for any $a > 0$, and

$$e^{\ln a} = a, \qquad \ln(e^s) = s, \qquad e^{-\ln a} = \frac{1}{a}. \tag{1.8.8}$$

Because natural logarithms are logarithms to the base e, Properties 1.8.7 hold for them too:

$$\ln(ac) = \ln a + \ln c, \qquad \ln(a^s) = s \ln a,$$
$$\ln \tfrac{1}{a} = -\ln a, \qquad \ln \tfrac{a}{c} = \ln a - \ln c. \tag{1.8.9}$$

The domain of the function $y = \ln x$ is the set of all positive real numbers x; its graph is sketched in Figure 1.8.1. Notice that $\ln 1 = 0$ and that $\ln x > 0$ when $x > 1$ and $\ln x < 0$ when $0 < x < 1$; $\ln x$ does not make sense if $x \le 0$.

The domain of $y = e^x$ is the set of all real numbers x; the graph of $y = e^x$ is sketched in Figure 1.8.2. Notice that $e^x > 0$ for every $x \in \mathbb{R}$. The function $y = e^x$ is often called the **exponential function**. It grows very rapidly as x increases. For instance, $e^1 = e \approx 2.718\,281\,83$, $e^{10} \approx 22,026.4658$, $e^{20} \approx 485,165,195.4$, $e^{30} \approx 1.068\,647 \times 10^{13}$, $e^{100} \approx 2.688 \times 10^{43}$.

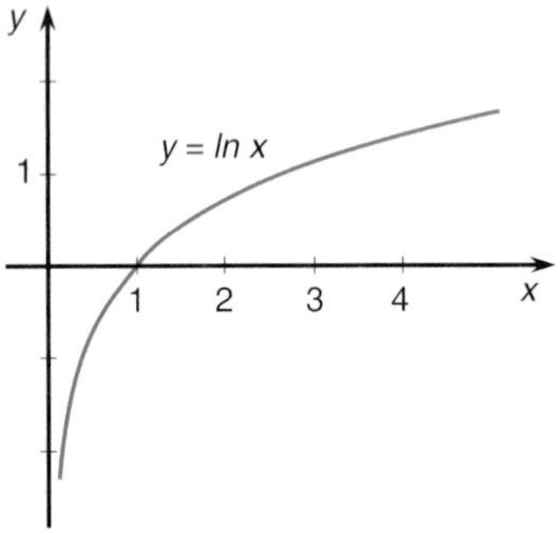

Figure 1.8.1

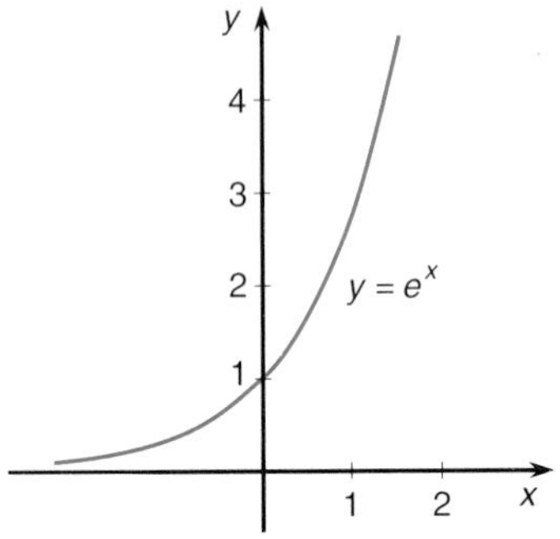

Figure 1.8.2

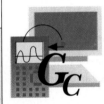

The fact that $\ln(2x) = \ln(2) + \ln(x)$ suggests that the graph of $y = \ln(2x)$ should be the graph of $y = \ln x$ "shifted up" by the constant $\ln(2)$.

We can use a graphics calculator to help visualize this fact.

TI-85	SHARP EL9300
GRAPH F1 ($y(x) =$) LN F1(x) ▼ LN (2 F1(x)) M2(RANGE) 0.1 ▼ 2 ▼ ▼ (−)3 ▼ 1.5 F5(GRAPH)	ln x/θ/T 2ndF ▼ ln 2 x/θ/T RANGE .1 ENTER 2 ENTER 1 ENTER −3 ENTER 1.5 ENTER 1 ENTER
CASIO f_x-7700GB/f_x-6300G	**HP 48SX**
Cls EXE Range .1 EXE 2 EXE 1 EXE −3 EXE 1.5 EXE 1 EXE EXE EXE EXE Graph ln X EXE Graph ln(2X) EXE	↰ PLOT PLOTR ERASE ATTN ′ αY = LN αX ↱ PLOT ↰ DRAW ERASE .1 ENTER 2 XRNG 3 +/− ENTER 1.5 YRNG DRAW ATTN ′ αY = LN 2 × αX ↱ PLOT ↰ DRAW DRAW

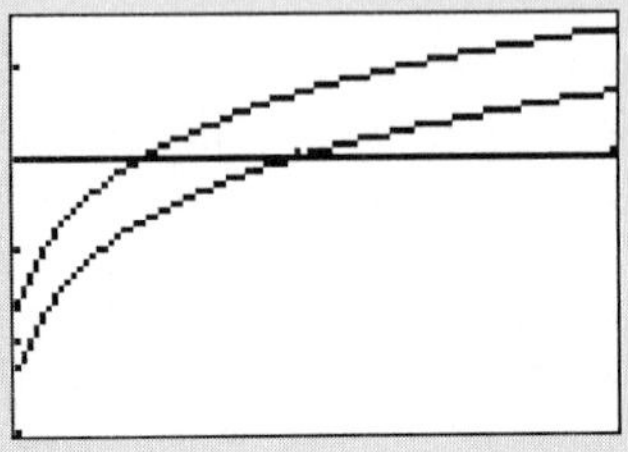

Try graphing $y = \ln(x)$ and $y = \ln(2x) - \ln(2)$ together; also try $y = \ln(x)$ and $y = \ln(x^3)$.

Exercises 1.8

Suppose a and b are two positive real numbers, and let $A = \log a$, $B = \log b$. Express each of the following in terms of A and B.

1. $\log ab$
2. $\log a^2 b^3$
3. $\log \frac{a^2}{b}$
4. $\log a^{-7} b$
5. $\log \sqrt{a}$
6. $\log \sqrt{ab}$
7. $\log a^{1/3}$
8. $\log b^{\pi}$
9. $\log a^b$
10. $\log b^{a+3}$

Evaluate each of the following.

11. $\log_2(32)$
12. $\log_5\left(\frac{1}{125}\right)$
13. $\log_4(2)$
14. $\log_9\left(\frac{1}{3}\right)$
15. $\log_{27}\left(\frac{1}{3}\right)$
16. $\log_3(81)$
17. $\log_{1/2}(8)$
18. $\log_{1/3}(27)$
19. $\log_{1/4}\left(\frac{1}{2}\right)$
20. $\log_{e^{-2}}(e)$

Chapter Summary

§1.1
natural numbers
rational numbers in lowest terms
real line
irrational numbers
scientific notation

§1.2
inequality
strict inequality
properties of inequalities
absolute value
triangle inequality (1.2.2, 1.2.3)
distance
interval
endpoint
closed interval
open interval
half-open interval
infinite interval
union, $A \cup B$
intersection, $A \cap B$

§1.3
polynomial
coefficient
degree
subscript
constant term
evaluating
value of a polynomial
variable
factoring
factor
difference of squares
quadratic polynomial
linear polynomial
completing the square
discriminant
Quadratic Formula (1.3.4)
nonnegative square root
root
Factor Theorem (1.3.5)
Rational Root Theorem (1.3.7)
factorials
Binomial Theorem (1.3.10)
rational function

§1.4
function
value of a function
domain
natural domain
range
origin
x-axis
y-axis
x-coordinate (abscissa)
y-coordinate (ordinate)
Cartesian coordinates
graph

§1.5
right-angled triangle
hypotenuse
Pythagorean Theorem (1.5.1)
similar triangles (1.5.4)
corresponding sides
slope
point–slope form (1.5.7)
slope–intercept form (1.5.8)
general form of the equation of a line (1.5.9)
x-intercept
y-intercept
parallel lines
area of rectangle
area of circle
area of parallelogram
area of triangle

§1.6
conics
polynomial in two variables
quadratic polynomials
parabola
vertex (of a parabola)
translation of axes
ellipse
relation
semi-major and semi-minor axes
major and minor axes
intercepts
hyperbola
asymptotes
transverse axis
conjugate axis
branches

§1.7
trigonometric functions
unit circle
periodic functions

radian measure
radian/degree conversions
identity
$\cos^2\theta + \sin^2\theta = 1$ (1.7.2)
Law of Cosines (1.7.10)
addition formulas (1.7.11)
double-angle formulas (1.7.12)
half-angle formulas (1.7.13)
polar coordinates (1.7.15)
radius
polar angle
modulus
argument
§1.8
exponent
logarithm
base
common logarithms
natural logarithms
$e \approx 2.718\,281\,83$
$\log_b(ac) = \log_b a + \log_b c$
exponential function

Review Exercises

Find an equivalent way of expressing each of the following without using absolute value signs.

1. $0 < |x+3| \le 2$
2. $1 \le |x-1| < 2$
3. $1 < \left|\frac{1}{x}\right| < 4$
4. $3 \le \frac{2}{|x+2|}$
5. $\{x : 1 < |x+6| < 8\}$
6. $\{x : |x+1| \ge 3\}$
7. $\{x : |1-x^2| \le \frac{1}{2}\}$
8. $\{x : 1 < |x^2-4| < 2\}$
9. $\{x : |x^3-4x^2+x+1| \le -2\}$
10. $\{x : \frac{1}{2} \le \frac{1}{|x+7|} \le \frac{1}{4}\}$

Factor each of the following polynomials as much as possible.

11. $x^2+7x-60$
12. $x^2+7x+12$
13. x^2+x-1
14. $x^2+3x-40$
15. x^3-2x^2-35x
16. x^4+3x^2+2
17. x^4-2x^2-15
18. x^3+x^2+x+1
19. x^5-x^3
20. x^6+4x^4

Find the coefficient of x^3 in each of the following expressions.

21. $(x+3)^4$
22. $(x-2)^5$
23. $x^2(x+6)^2$
24. $(x+4)^{10}$
25. $x^2(x-4)(x+4)$
26. $\frac{1}{x}\left(32-(x+2)^5\right)$
27. $x^2\left((1+x)^2-(1-x)^2\right)$
28. $x(x^2+1)^3$
29. $(x^3-2)^4$
30. $(x^2-7)^{11}$

Find each of the following.

31. The domain of $\frac{1}{x^2-3}$.
32. The domain of $\frac{x}{x^2-9}$.
33. The domain of $\sqrt{\frac{1}{1-x}}$.
34. The domain of $x \ln x$.
35. The range of x^2-6x+2.
36. The range of $1-x+2x^2$.
37. The distance between the points $(2,3)$ and $(1,-4)$.
38. The distance between the points $(1,-5)$ and $(2,0)$.
39. The distance from the y-axis to the point $(-4,6)$.
40. The distance from the x-axis to the point $(3,-5)$.
41. The length of the hypotenuse in a right-angled triangle whose other sides have lengths 3 and 7.
42. The length of the third side in a right-angled triangle with hypotenuse of length 4 and one other side of length 3.
43. The area of the triangle whose vertices are $(0,0)$, $(4,2)$, and $(3,0)$.
44. The area of the triangle whose vertices are $(1,1)$, $(3,-3)$, and $(3,2)$.
45. The area of the triangle whose vertices are $(-1,3)$, $(2,5)$, and $(3,3)$.
46. The area of the parallelogram whose vertices are $(1,4)$, $(-1,4)$, $(7,-1)$, and $(5,-1)$.
47. The equation of the straight line containing $(2,5)$ and $(3,4)$.
48. The equation of the line through $(1,-3)$ and with slope -3.
49. The equation of the line through $(0,-3)$ and with x-intercept -2.
50. The equation of the line through $(0,-3)$ and perpendicular to $x+3y=7$.
51. The equation of the ellipse whose center is $(0,0)$ and that contains $(0,2)$ and $(-3,0)$.
52. The equation of the ellipse whose center is $(2,-1)$ and that has vertical major axis of length 4 and minor axis of length 2.
53. The equation of the circle whose center is $(5,4)$ and whose radius is 3.
54. The equation of the parabola that opens upward, whose vertex is $(2,3)$, and that contains the point $(3,4)$.

55. The equation of the hyperbola with asymptotes $y = \pm 2x$ that contains $(0, 3)$.

56. The equation of the hyperbola with asymptotes $2x = \pm(3y + 6)$ and containing $(1, -2)$.

57. The points of intersection of the line $y = x + 3$ and the ellipse $x^2 + 4y^2 = 8$.

58. The points of intersection of the circle $x^2 + y^2 = 4$ and the ellipse $x^2 + 2y^2 = 4$.

59. The center of the ellipse $x^2 + 4x + 9y^2 - 18y + 5 = 0$.

60. The semi-major axis of the ellipse $3x^2 - 6x + 4y^2 + 3y = 0$.

Identify each of the following conics (state what type it is, its center, and, where applicable, find major and minor axes, in which direction it opens, and asymptotes).

61. $x^2 - 4y = 8$

62. $x^2 + 4x + y^2 - 6y = 0$

63. $x^2 - y^2 = -4$

64. $4x^2 - 9y^2 = -36$

65. $x^2 + 4y^2 - 12y = 8$

66. $2x^2 - 4x + 3y^2 + 6y = 4$

67. $16x^2 - 9y^2 = -1$

68. $4x - 3y^2 = -1$

69. $x^2 + 4x + 3y^2 + 3 = 0$

70. $x^2 - y^2 + 2y = 1$

Evaluate each of the following without using a calculator.

71. $\cos\left(-\frac{5\pi}{4}\right)$

72. $\sin \frac{7\pi}{6}$

73. $\cos \frac{5\pi}{8}$

74. $\sec\left(-\frac{5\pi}{12}\right)$

75. $\tan \frac{5\pi}{3}$

76. $\sec^2 \frac{5\pi}{3}$

77. $\sin\left(\frac{\pi}{4} + \frac{2\pi}{3}\right)$

78. $\cos 3\pi$

79. $\cot \frac{4\pi}{3}$

80. $2 \cos \frac{\pi}{8} \sin \frac{\pi}{8}$

Convert each of the following rectangular coordinates (x, y) into polar coordinates (r, θ) or vice versa, using a calculator if necessary to get approximate values.

81. $(x, y) = (-4, -4)$

82. $(x, y) = (5, -2)$

83. $(x, y) = (3, -6)$

84. $(x, y) = (-2, 2\sqrt{3})$

85. $(x, y) = (0, -8)$

86. $(x, y) = (-5, 0)$

87. $(r, \theta) = (1, -\pi)$

88. $(r, \theta) = \left(2, \frac{\pi}{4}\right)$

89. $(r, \theta) = \left(7, \frac{5\pi}{4}\right)$

90. $(r, \theta) = \left(\frac{1}{2}, -\frac{7\pi}{6}\right)$

Assume $a > 0$, $b > 0$. Express each of the following expressions in terms of a, b, $\ln a$, $\ln b$, and any necessary constants (for constants such as ln 10, for instance, do not evaluate them on your calculator, but simply write them down in this form. For example, for a question that asks you to rewrite $\ln(10^s)$, write $\ln(10^s) = s \ln 10$).

91. $\ln(10a^2 b)$

92. $\ln(a^{2s})$

93. $\log a$

94. $\log_2(a^4)$

95. $\ln \frac{a}{2b}$

96. $\ln(e^5)$

97. $e^{-3 \ln a}$

98. $\ln(a^b)$

99. $\ln(2a \times 3^b)$

100. $\ln\left(\frac{ab}{ba}\right)$

Limits and Continuity

2

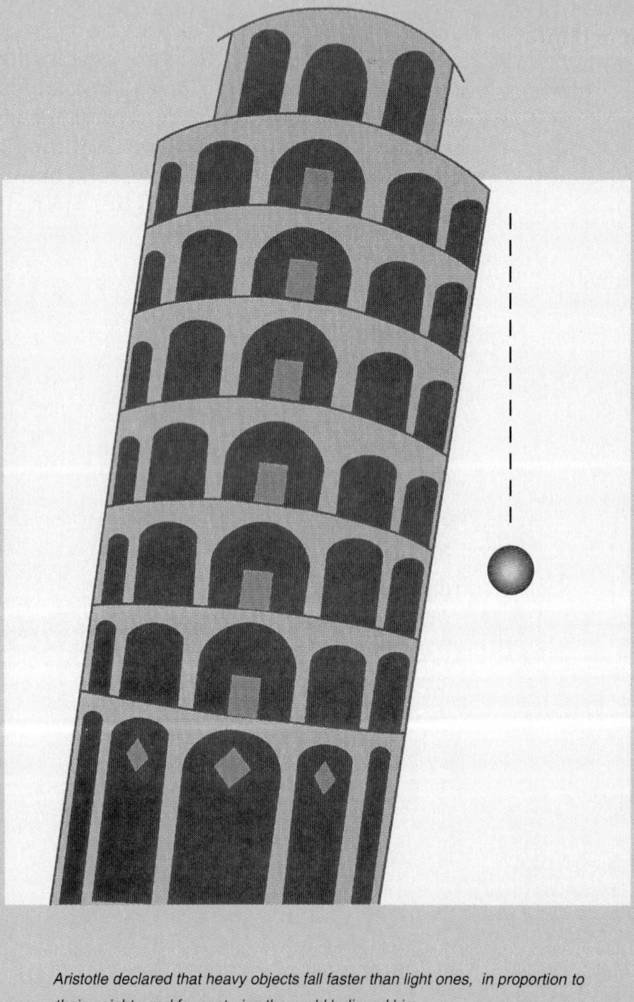

Aristotle declared that heavy objects fall faster than light ones, in proportion to their weight, and for centuries the world believed him. Finally Galileo showed otherwise, reputedly by dropping a cannonball and a musket ball from the tower of Pisa. He is often regarded as the father of experimental science (Aristotle purported to have conducted similar experiments, but presumably never did). Galileo's work lead to Newton's study of gravitation, which in turn led to the development of calculus.

INTRODUCTION

This chapter is devoted to the ideas of limits and continuity. These are subtle concepts that are at the foundation of the study of calculus.

Section 2.1 serves to introduce the notion of limits by considering how it arises in various contexts. Section 2.2 shows how to work with limits of polynomials and rational functions; if you are not sufficiently familiar with polynomials and rational functions, you might like to review Section 1.3 before proceeding. In Section 2.3 we learn techniques for dealing with many different types of limits and the rules that govern their use. Section 2.4 considers some specialized types of limits: one-sided limits, infinite limits, and limits at infinity. Having learned to work with limits, in Section 2.5 we use them to discuss the idea of continuity. In Section 2.6 we study limits as they apply to the trigonometric functions; for this it is very important to be familiar with the concept of radian measure, which is discussed in Section 1.7. Finally, in Section 2.7 there is a discussion of the mathematically precise definition of limits. This is a difficult topic, which is why we have concentrated on using and understanding limits before considering it.

2.1 Introduction to Limits

TANGENT LINE
SLOPE
AREA

The fundamental idea on which calculus is based is the concept of a limit. We are drawn to this concept in the course of attempting to solve several different types of problems, some of which will be described in this section.

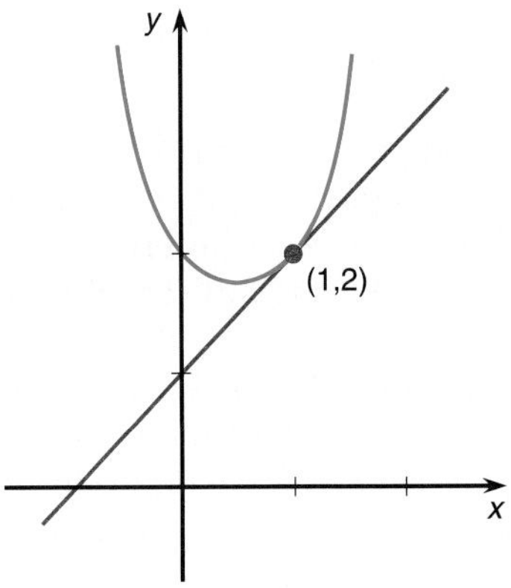

Figure 2.1.1

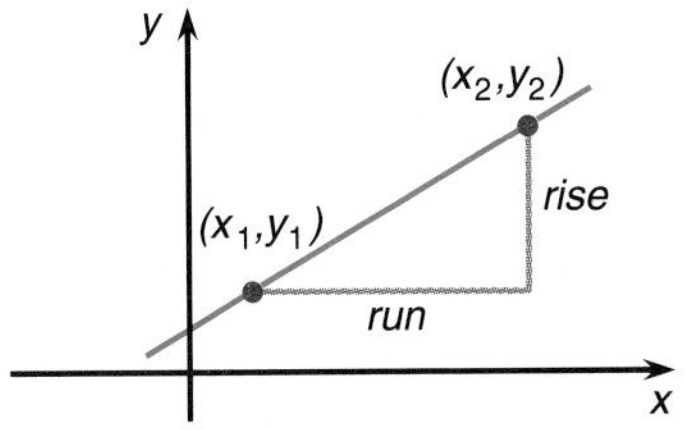

Figure 2.1.2

EXAMPLE 1 Suppose we look at a sketch of the parabola $y = x^2 - x + 2$ and note that the point $(1, 2)$ is on the graph (see Figure 2.1.1). We might want to find the equation of the tangent line to the parabola at that point. Obviously the tangent goes through the point $(1, 2)$, but to find its equation, we need to know either another point that it goes through or else its slope.

The way we usually find the slope of a line is to find two points on it and use the formula

$$Slope = \frac{rise}{run} = \frac{y_2 - y_1}{x_2 - x_1},$$

where (x_1, y_1) and (x_2, y_2) are the two points (see Figure 2.1.2). The difficulty with a tangent line is that we usually know only one point on it, namely, the point of tangency.

We could try to get around this problem by picking a second point on the graph and drawing the secant—the line through this point and $(1, 2)$. If the second point were chosen to be fairly close to $(1, 2)$, then the secant should be fairly close to the tangent line. Figure 2.1.3 illustrates a few such secants.

As the second point gets closer and closer to $(1, 2)$, without actually reaching it, the secant gets closer and closer to the tangent. More important, its slope gets closer and closer to the slope of the tangent, which is what we want to find.

Suppose the second point is (x, y). Because we are choosing it to be a point on the graph, it must satisfy the equation of the parabola. In other words, $y = x^2 - x + 2$, and we can write the point (x, y) as $(x, x^2 - x + 2)$ (see Figure 2.1.4).

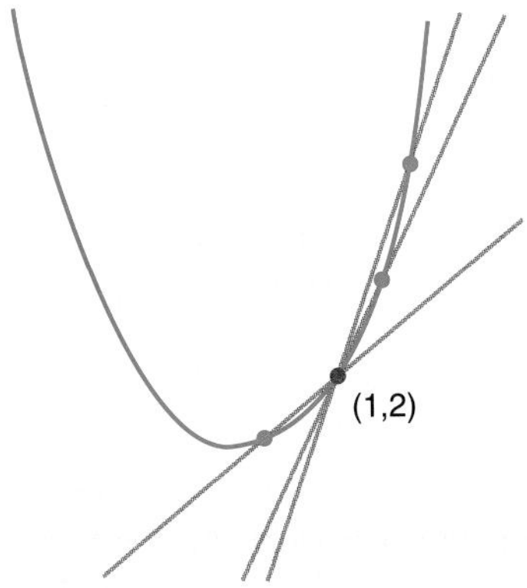

Figure 2.1.3

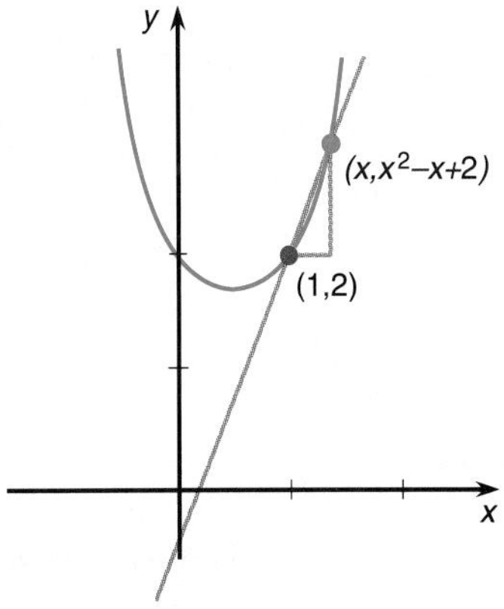

Figure 2.1.4

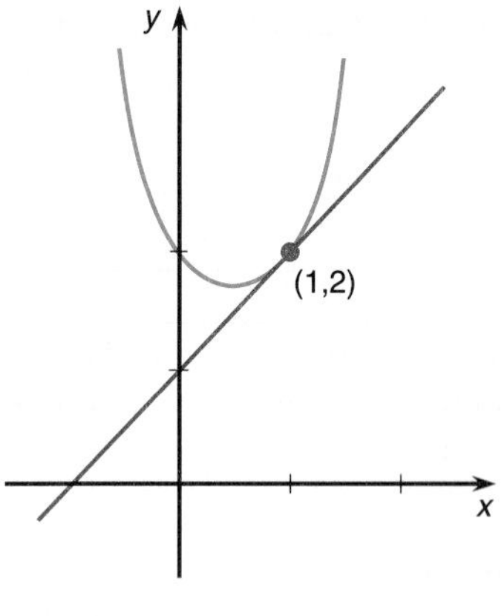

Figure 2.1.5

We can calculate the slope of the secant using the formula given above:

$$Slope = \frac{y_2 - y_1}{x_2 - x_1} = \frac{(x^2 - x + 2) - 2}{x - 1} = \frac{x^2 - x}{x - 1} = \frac{x(x - 1)}{x - 1} = x.$$

For this last step we divided by $x - 1$, so we have to know that $x - 1$ is not equal to zero, but it is all right because we are choosing the second point to be different from $(1, 2)$, so $x \neq 1$. As the second point approaches $(1, 2)$, the value of x must approach 1, so the slope of the secant, which is x, also approaches 1. This suggests that the slope of the tangent must be 1. See Figure 2.1.5.

Knowing this and the fact that the tangent passes through $(1, 2)$, we can find its equation using the point-slope form of the equation of a line (cf. Equations 1.5.7):

$$y - 2 = (1)(x - 1), \qquad \text{or} \qquad y - 2 = x - 1,$$

or $x - y + 1 = 0$.

Finding the slope of a tangent line is one of the most important ideas in calculus, and while the example we have just done is misleadingly simple in some respects, it does illustrate the general approach. In particular, it shows how we can approximate the tangent by secants and let them get closer and closer to the tangent. The slopes of the secants should then approach the slope of the tangent.

Unfortunately we have been vague about what we mean by letting secants "approach" the tangent and by their slopes "approaching" the slope of the tangent. The value they approach is called the limit, and the process of approaching a limit is the subject of this chapter. This introductory section will use several examples to explore some of the ways in which limits arise and how they are used.

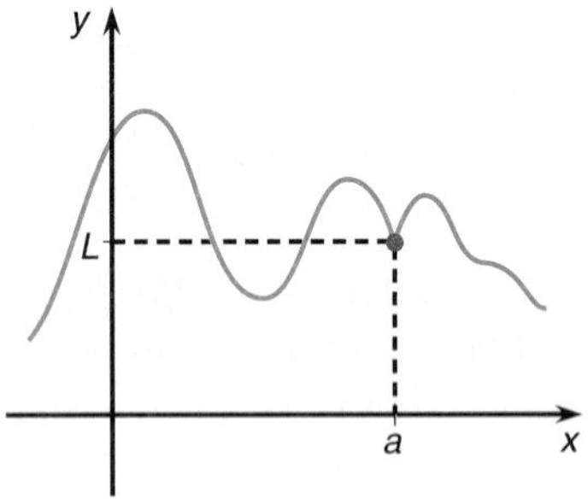

Figure 2.1.6

If $f(x)$ is a function, we can consider what happens to the value of $f(x)$ as x approaches some fixed value a. If the values of $f(x)$ approach some value L, as shown in the example in Figure 2.1.6, we call L the limit of $f(x)$ as x approaches a, and we write

$$L = \lim_{x \to a} f(x).$$

EXAMPLE 2 Find the equation of the line that is tangent to the curve $y = x^3 + 2x^2 - 2x + 1$ at the point $(-1, 4)$.

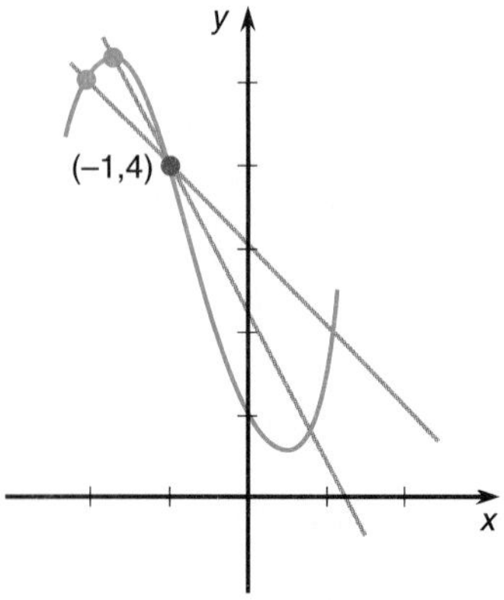

Figure 2.1.7

Solution If (x, y) is a point on the curve, we know that it satisfies the equation, so we can write the point as $(x, x^3 + 2x^2 - 2x + 1)$ (see Figure 2.1.7). We calculate the slope of the secant through this point and the point $(-1, 4)$, using the formula given above, and find it equals

$$\frac{(x^3 + 2x^2 - 2x + 1) - 4}{x - (-1)} = \frac{x^3 + 2x^2 - 2x - 3}{x + 1}.$$

It is easy to see that the numerator factors: we find that $x^3 + 2x^2 - 2x - 3 = (x + 1)(x^2 + x - 3)$, (verify this), so the slope is

$$\frac{(x + 1)(x^2 + x - 3)}{x + 1} = x^2 + x - 3.$$

Once again we must be careful not to divide by zero, so this formula for the slope of the secant makes sense only when $x \neq -1$. What we want is to let the point $(x, y) = (x, x^3 + 2x^2 - 2x + 1)$ approach $(-1, 4)$ and see what happens to the slope.

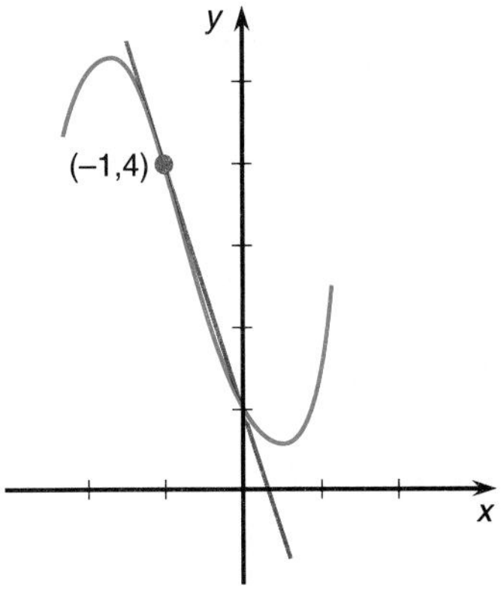

Figure 2.1.8

In other words, we want to consider

$$\lim_{x \to -1} (x^2 + x - 3).$$

Now as x approaches -1, it seems reasonable to expect that $x^2 + x - 3$ will approach $(-1)^2 + (-1) - 3 = -3$. In other words,

$$\lim_{x \to -1} (x^2 + x - 3) = -3.$$

This limit of the slopes of the secants will be the slope of the tangent line. See Figure 2.1.8.

Now we can find the equation of the tangent line, since we know that it passes through $(-1, 4)$ and has slope equal to -3. The equation is

$$y - 4 = (-3)\big(x - (-1)\big) \qquad \text{or} \qquad y - 4 = -3(x + 1) \qquad \text{or}$$
$$3x + y - 1 = 0.$$

If a vehicle travels at a constant speed for 15 minutes, say, and covers a distance of 10 miles, then we can easily find the speed by dividing the distance by the time. In this case we find that the speed is $\frac{10\,miles}{\frac{1}{4}\,hour} = 40$ miles/hour. If the vehicle is not traveling at a constant speed, then this same procedure can be used to find the *average* speed. But there is the question of what we mean by the speed of the vehicle at a particular moment, which is sometimes called the "instantaneous speed." This is a very familiar notion; after all, it is exactly what the speedometer in a car tells us.

If a car is traveling at 40 mph, we cannot conclude that it will have to travel 40 miles in the next hour. We could say that if the car continued at exactly the same speed for an hour, then it would go 40 miles; but in practice this is both unlikely and awkward to use.

Slightly better would be to say that if it continued for another 15 minutes at the same speed, then it would go 10 miles. But carrying this idea further, we could say that in one minute it would go $\frac{2}{3}$ mile or that in one second it would go $\frac{2}{3}\frac{1}{60} = \frac{1}{90}$ mile.

The advantage is that for a car moving at 40 mph, there isn't much need to worry about the speed changing in 1 second. And if it does, or if we want better accuracy, we could consider $\frac{1}{10}$ second, $\frac{1}{100}$ second, etc.

What we are doing is finding the average speed over a brief period of time. The shorter the period of time, the better we will be able to approximate the instantaneous speed. To eliminate the question of just how short the time should be, we could take the limit as it approaches zero. This limit is exactly what we mean by the instantaneous speed. In fact, it was to deal with this question that calculus was originally invented.

Suppose an object held at a height A feet above the ground is released and allowed to fall, as shown in Figure 2.1.9. The height of the object t seconds after its release will be approximately

$$h(t) = A - 16t^2,$$

where this height is measured in feet. Notice that when $t = 0$, this formula says that the height is A feet, which means that the object has not begun to fall, and that as t increases, the height will decrease. Of course the object will eventually hit the ground and stop moving, and from that point on, this formula will not apply.

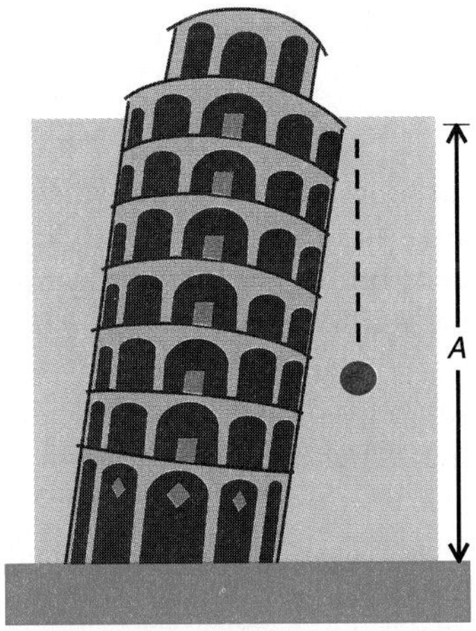

Figure 2.1.9

EXAMPLE 3 Suppose an object falls from a height of 64 ft above the ground. How fast will it be falling one second after its release? When will it hit the ground? What will be its speed of impact?

Notice that since the object is falling, we expect $h(1) > h(t)$, so we have taken $h(1) - h(t)$ in order to get a positive distance.

Solution We use the above formula with $A = 64$. If t is slightly greater than 1, then between 1 sec and t sec the object will fall a distance of $h(1) - h(t) = \big(64 - 16(1)^2\big) - (64 - 16t^2) = 64 - 16 - 64 + 16t^2 = 16(t^2 - 1)$ ft.

The average speed of the object between 1 sec and t sec will be

$$\frac{distance}{time} = \frac{16(t^2-1)}{t-1} = \frac{16(t+1)(t-1)}{t-1} = 16(t+1) \text{ ft/sec.}$$

This calculation involves dividing by $t - 1$, so it cannot be done when $t = 1$, but we can find the instantaneous speed at the moment when $t = 1$ by taking a limit. The speed of the falling object 1 sec after its release will be the limit as t approaches 1 of the average speed over the brief interval from 1 to t:

$$\textit{Speed after 1 sec} = \lim_{t\to 1} 16(t+1) = 16(1+1) = 32 \text{ ft/sec.}$$

To find when the object hits the ground we remark that what this means is just that the height above the ground should be zero, that is, that $h(t) = 0$. This says $0 = h(t) = 64 - 16t^2$, or $64 = 16t^2$, so $t^2 = \frac{64}{16} = 4$ and $t = \sqrt{4} = 2$. The object will hit the ground 2 seconds after its release.

To find the speed with which it hits the ground, we can pretend that it just continues falling after 2 seconds and do the same sort of calculation that we just completed.

The average speed between 2 sec and t sec is $\frac{h(2)-h(t)}{t-2} = \frac{(64-16(2)^2)-(64-16t^2)}{t-2} = \frac{(64-64)-(64-16t^2)}{t-2} = \frac{16t^2-64}{t-2}$. We can factor the numerator as a difference of squares to get $\frac{(4t+8)(4t-8)}{t-2} = \frac{(4t+8)4(t-2)}{t-2} = 4(4t+8) = 16t + 32$.

The instantaneous speed with which the falling object hits the ground is

$$\lim_{t\to 2}(16t+32) = 16(2) + 32 = 64 \text{ ft/sec.}$$

The use of limits to find instantaneous speed is actually quite similar to the way we used them to find the slope of a tangent line to a curve, even though the two problems look somewhat different. We now discuss an entirely different way in which limits arise.

EXAMPLE 4 Suppose we want to figure out the area inside a circle. For simplicity, assume the radius of the circle is 1. The familiar formula that the area of a circle is πr^2 if the radius is r tells us that in this case the area is π, but how would we figure this out starting from scratch, or for that matter, how would we figure out what π is?

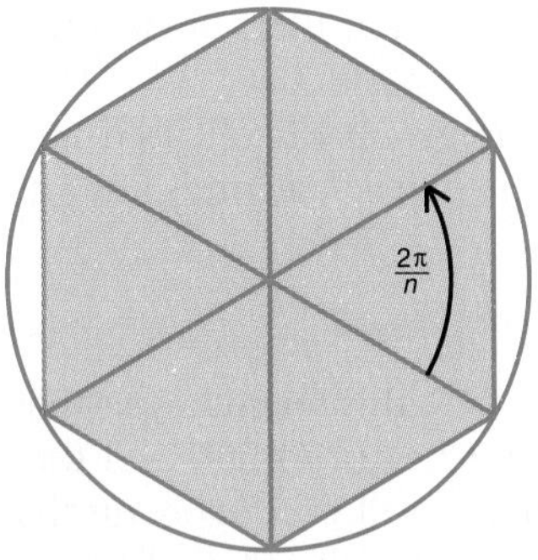

Figure 2.1.10(i)

One way to go about it would be to approximate the region inside the circle by regions whose areas are easier to calculate. Figure 2.1.10(i) shows how the region inside the circle can be approximated by n congruent triangles. To figure out the angle in each of these triangles at the vertex which is the center of the circle, we note that n of them exactly make up the angle that goes all around the circle. Since this angle measures 2π radians, each of the n equal pieces must measure $\frac{2\pi}{n}$ radians.

The area inside the circle is approximated by the total area of the n triangles, which is n times the area of one of them. Consider the triangle that lies along the

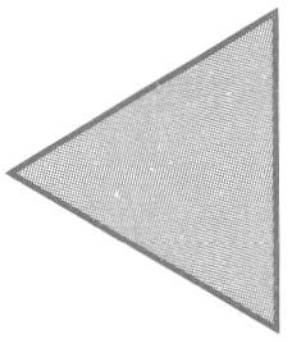

Figure 2.1.10(ii)

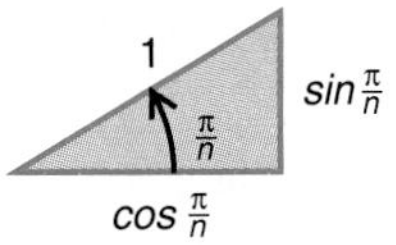

Figure 2.1.10(iii)

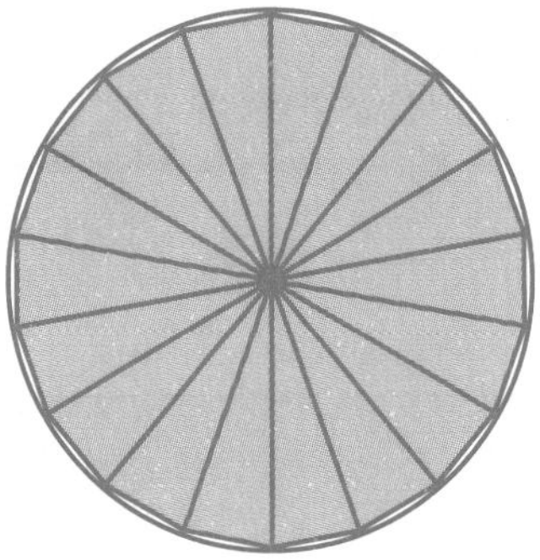

Figure 2.1.11

positive x-axis, as shown in Figure 2.1.10(ii), and look at the upper half of it, the smaller triangle illustrated in Figure 2.1.10(iii). To find its area, we need to know its base and its height. The angle inside this little triangle at the center of the circle is half the angle in the bigger triangle, or $\frac{2\pi}{2n} = \frac{\pi}{n}$. By the definitions of the trigonometric functions (cf. Definition 1.7.1), the base of the triangle is $\cos\frac{\pi}{n}$ and its height is $\sin\frac{\pi}{n}$. The area of the little triangle is therefore $\frac{1}{2}\mathit{base} \times \mathit{height} = \frac{1}{2}\cos\left(\frac{\pi}{n}\right)\sin\left(\frac{\pi}{n}\right)$.

The bigger triangle then has area twice as large, or $\cos\left(\frac{\pi}{n}\right)\sin\left(\frac{\pi}{n}\right)$. The n congruent triangles that approximate the region inside the circle have total area $n\cos\left(\frac{\pi}{n}\right)\sin\left(\frac{\pi}{n}\right)$. This approximates the area of the circle.

To get better and better approximations, we should let n get larger and larger (see Figure 2.1.11). We can describe this by saying that n should "approach infinity," which we write as $n \longrightarrow \infty$. Then we can consider better and better approximations and take their limit as $n \longrightarrow \infty$; we find that the area inside the circle is

$$\mathit{Area} = \lim_{n\longrightarrow\infty} n\cos\left(\frac{\pi}{n}\right)\sin\left(\frac{\pi}{n}\right).$$

Again we have expressed something in terms of a limit. This limit is not easy to evaluate, but the point is that we have been able to use it to give a precise meaning to the difficult idea of the area of a circle.

The last example brought us face to face with the possibility of encountering perfectly reasonable limits that are difficult to evaluate. Let us look at another example.

EXAMPLE 5

Consider the following limit:

$$\lim_{x\longrightarrow 0} \frac{\sin x}{x}.$$

As usual, we are measuring x in radians. There is no obvious way to go about evaluating this limit, but we could try to get an idea of what is going on by working out what we get for some values of x that are close to 0.

Using a calculator, being sure to set it to measure angles in radians, we can evaluate $\frac{\sin x}{x}$ for some small values of x:

x	$\frac{\sin x}{x}$
1.0	0.84147099
0.5	0.95885108
−0.5	0.95885108
0.2	0.99334665
−0.2	0.99334665
0.1	0.99833417
−0.1	0.99833417
0.01	0.99998333
−0.01	0.99998333
0.001	0.99999983
−0.001	0.99999983
0.0001	0.999999998

These values certainly make it appear as though $\frac{\sin x}{x}$ approaches the limit 1 as $x \to 0$. This is true, but it is difficult to prove. In passing, it is worth noticing that it does not work if angles are measured in degrees.

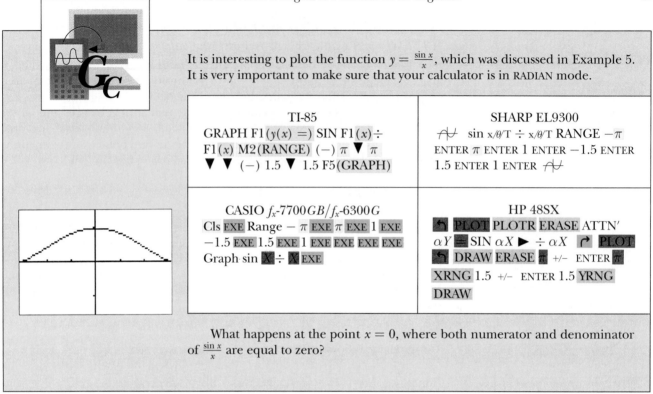

It is interesting to plot the function $y = \frac{\sin x}{x}$, which was discussed in Example 5. It is very important to make sure that your calculator is in RADIAN mode.

TI-85	SHARP EL9300
GRAPH F1 ($y(x)$ =) SIN F1 (x) ÷ F1 (x) M2 (RANGE) (−) π ▼ π ▼ ▼ (−) 1.5 ▼ 1.5 F5 (GRAPH)	sin x/θ/T ÷ x/θ/T RANGE $-\pi$ ENTER π ENTER 1 ENTER −1.5 ENTER 1.5 ENTER 1 ENTER
CASIO f_x-7700GB/f_x-6300G	HP 48SX
Cls EXE Range − π EXE π EXE 1 EXE −1.5 EXE 1.5 EXE 1 EXE EXE EXE EXE Graph sin X ÷ X EXE	↰ PLOT PLOTR ERASE ATTN′ αY = SIN αX ▶ ÷ αX ↱ PLOT ↰ DRAW ERASE π +/− ENTER π XRNG 1.5 +/− ENTER 1.5 YRNG DRAW

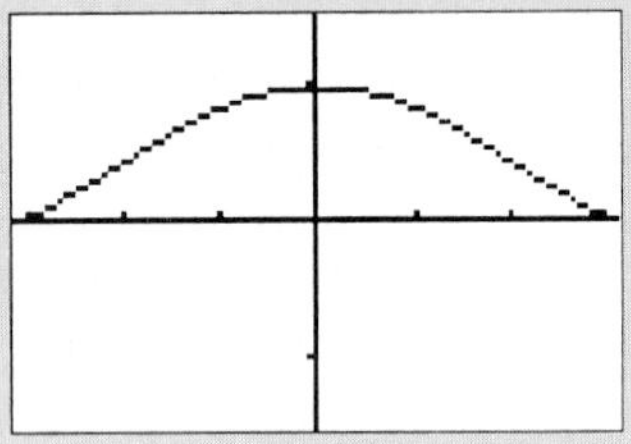

What happens at the point $x = 0$, where both numerator and denominator of $\frac{\sin x}{x}$ are equal to zero?

Finally, we look at two examples to show how things can go wrong with limits.

EXAMPLE 6 What happens to the function $f(x) = \frac{x}{|x|}$ as $x \to 0$?

Solution Notice that the function $f(x)$ is not defined when $x = 0$. When $x > 0$, we know that $|x| = x$, so $f(x) = \frac{x}{|x|} = \frac{x}{x} = 1$.

On the other hand, when $x < 0$, it is easy to see that $|x| = -x$; after all, if $x < 0$, then $-x > 0$. (For example, when $x = -3$, we see that $|x| = |-3| = 3$, and this is the same thing as $-(-3)$.) So when $x < 0, f(x) = \frac{x}{|x|} = \frac{x}{-x} = -1$. The graph of $f(x)$ is sketched in Figure 2.1.12.

Now as $x \to 0$, it depends whether x is negative or positive. If x approaches 0 from the left, that is, through negative values of x, then $f(x)$ will equal -1 all the time and so certainly approaches -1 as a limit. On the other hand, if x approaches 0 from the right, through positive values of x, then $f(x)$ can only approach the value 1 as its limit.

There is no single value that we can call the limit in this case, so we have to say that the limit does not exist. Actually, in this particular example we could talk about the "limit from the left side" or the "limit from the right side," both of which do exist, but it is possible to make up examples with even worse behavior in which even the limits from the left and right do not exist.

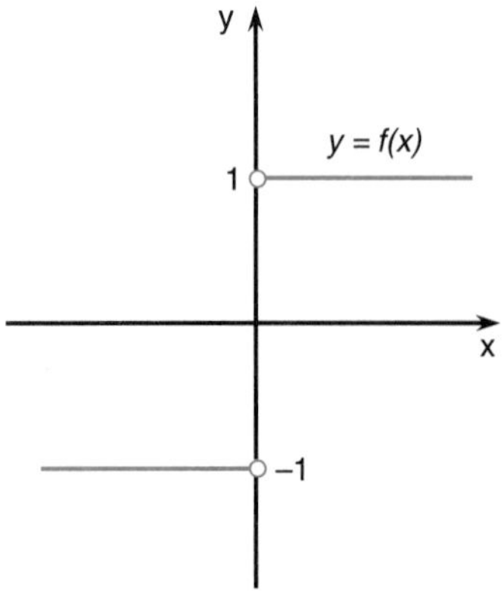

Figure 2.1.12

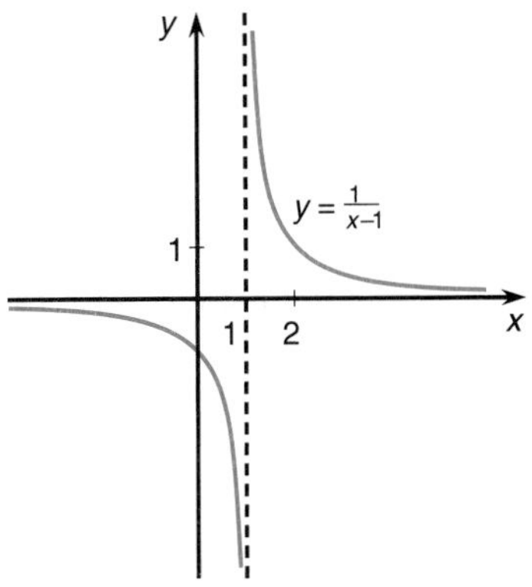

Figure 2.1.13

The fact that it may be difficult to tell whether or not a limit exists is one of the things that make them difficult to work with.

EXAMPLE 7 What happens to the function $g(x) = \frac{1}{x-1}$ as $x \to 1$?

Solution The function $g(x)$ is not defined when $x = 1$. If x is a number slightly bigger than 1, then $x - 1$ is a small positive number, and $g(x) = \frac{1}{x-1}$ is a large positive number. The closer x gets to 1, the smaller $x - 1$ gets, and the larger $g(x) = \frac{1}{x-1}$ gets.

In fact, $g(x)$ doesn't have any limit L as $x \to 1$, because $g(x)$ gets bigger than any L when x gets sufficiently close to 1. To make matters worse, if x approaches 1 from the left, $x - 1$ is negative, and so is $g(x) = \frac{1}{x-1}$, and as x gets close to 1, $g(x)$ becomes larger and larger in the negative direction. See Figure 2.1.13.

The next section will be devoted to learning to work with limits.

Exercises 2.1

I

Find the slope of the tangent line to the graph of each of the given functions $y = f(x)$ at the specified point.

1. $f(x) = x^2$, at the point $(2, 4)$
2. $f(x) = x^2 + 2x + 1$, at the point $(-2, 1)$
3. $f(x) = 3x^2 - 4x + 1$, at the point $(1, 0)$
4. $f(x) = 2x^2 - 12x + 6$, at the point $(3, -12)$
5. $f(x) = x^3 + 2x - 3$, at the point $(-1, -6)$
6. $f(x) = x^3 - 2x^2 - 3x - 3$, at the point $(0, -3)$
7. $f(x) = x^3 - 8$, at the point $(2, 0)$
8. $f(x) = x^4 + x^2 + 1$, at the point $(-1, 3)$
9. $f(x) = x^4$, at the point $(0, 0)$
10. $f(x) = 3x + 1$, at the point $(1, 4)$

Find the equation of the tangent line to the graph of each of the given functions $y = f(x)$ at the specified point.

11. $f(x) = x^2 - 2$, at the point $(1, -1)$
12. $f(x) = x^2 - 2x + 1$, at the point $(1, 0)$
13. $f(x) = x^2 + 3x + 5$, at the point $(-1, 3)$
14. $f(x) = 3x^2 + 5x - 6$, at the point $(-2, -4)$
15. $f(x) = 5x^2 + x$, at the point $(-2, 18)$
16. $f(x) = x^3$, at the point $(2, 8)$
17. $f(x) = x^3 - x - 2$, at the point $(-2, -8)$
18. $f(x) = x^3 + x^2 + x + 1$, at the point $(0, 1)$
19. $f(x) = x^4$, at the point $(-2, 16)$
20. $f(x) = x^4 + 2x^2 + 4$, at the point $(1, 7)$

II

21. Find the equation of the tangent line to the parabola $y = x^2$ at the point (a, a^2).
22. If an object is released from a height of 144 ft above the ground, how long will it take for it to reach the ground, and what will be its speed of impact?
23. If an object is released and allowed to fall to the ground, from what height was it dropped if it takes 4 sec to reach the earth?
24. Suppose an object falls from a height A ft above the ground. How long will it take to hit the ground?

*25. Suppose an object falls from a height A ft above the ground. At what speed will it be falling at the moment t sec after its release?

*26. If an object is released and allowed to fall to the ground, from what height was it dropped if it hits the earth with an impact speed of 100 ft/sec?

27. Suppose an apple falls from the top of a tree and lands on Newton's head at a speed of 16 ft/sec. If Newton is sitting with his head 4 ft above ground level, how high is the tree?

*28. Suppose an object is dropped to the ground from a height of A ft. If its speed of impact is S ft/sec, find A, the height from which it was dropped.

*29. Suppose an object is thrown straight up into the air. It can be shown that its height in feet above the ground after t sec will be given by an expression of the form

$$h(t) = -16t^2 + vt + A.$$

Obviously, A ft is the height from which the object is thrown (since $h(0) = A$).

(i) Show that $|v|$ is the initial speed with which the object is thrown, that is, the instantaneous speed when $t = 0$.

(ii) What does it mean when v is positive, negative, or zero?

(iii) At what moment (i.e., at what value of t) will the object reach the highest point in its flight, and what is the greatest height it will reach? (*Hint:* Complete the square.)

(iv) If a ball is thrown upward from a height of 4 ft above the ground at a speed of 20 ft/sec, when will it hit the ground?

(v) If someone on a balcony throws a ball upward from a height of 28 ft above the ground and finds that it hits the ground $1\frac{3}{4}$ sec later, how fast was it thrown?

***30.** In this question we discuss an alternative method for solving part (iii) of the previous question. Knowing that the height above the ground of the object is $h(t) = -16t^2 + vt + A$, find its instantaneous speed at the moment when $t = u$, where u is some fixed but unspecified number. Then to find when the object reaches the highest point in its flight, observe that this will happen at the moment when the speed is exactly equal to 0, when the object has stopped rising but not yet begun to fall.

III

31. For a circle of radius 1 we found that the area can be expressed as the limit

$$\lim_{n\to\infty} n \cos\left(\frac{\pi}{n}\right) \sin\left(\frac{\pi}{n}\right).$$

Use a calculator or computer to evaluate this function for various values of n (e.g., $n = 4, 8, 20, 50, 100, \ldots$). What is the first value of n for which the answer is correct to four decimal places? What is the first value of n for which your calculator gives the same answer with n and $n + 1$? How large does n have to be so that your calculator gives the same answer as its built-in value of π, which equals the area of the circle? (*Hint:* Be sure you are using *radians.*)

32. Use a calculator or computer to estimate the area of a circle of radius 3. How large does n have to be to get an answer that is correct to four decimal places?

33. Estimate to four decimal places the area of the upper semicircle of radius 4. How large does n have to be to get an answer that is correct to four decimal places?

34. Estimate to within four decimal places the area of the part of the circle of radius 4 that lies below the line $y = x$ and above the line $y = -x$. How large does n have to be to get an answer that is correct to four decimal places?

35. Use a graphics calculator or computer to graph the function $y = \frac{\sin x}{x}$ for $-2 \le x \le 2$. Then graph it for $-0.1 \le x \le 0.1$ and for $-0.01 \le x \le 0.01$. Compare with Example 5.

36. How do the graphs in the previous question change if you set your machine to measure angles in degrees?

37. (i) In Example 3 we discussed a falling object whose height is $h(t) = 64 - 16t^2$. Graph this function and by magnifying the graph estimate how far the object falls in the last 1 sec, the last 0.1 sec, and the last 0.01 sec.

(ii) Graph the function $d(t)$ that tells how far the object falls between t and $t + 0.1$. Graph the function $D(t)$ that tells how far the object falls between t and $t + 0.01$. Compare these graphs with the result of Example 3.

2.2 Limits of Polynomials and Rational Functions

LIMITS OF SUMS
LIMITS OF PRODUCTS
LIMITS OF QUOTIENTS

To define the notion of limit with precision is surprisingly tricky. We won't attempt it until Section 2.7. Fortunately, in many situations it is possible to work with limits using only our intuitive sense of what they mean.

In this section we explore the basic rules that apply to limit calculations. In particular, we learn to evaluate limits of polynomials and rational functions (cf. Definitions 1.3.1 and 1.3.11).

Suppose $f(x)$ is a function whose domain may be the entire real line $\mathbb{R}$ or perhaps only some subset of it. Suppose $a \in \mathbb{R}$ is fixed. We can consider what happens to $f(x)$ when x "approaches" a. This can make sense even if a is not in the domain of f, but we should assume $f(x)$ is defined at every *other* point in an interval (r, s) containing a. This means $f(x)$ is defined on $(r, a) \cup (a, s)$, which is sometimes called a "punctured interval." Of course $f(x)$ might also be defined at $x = a$.

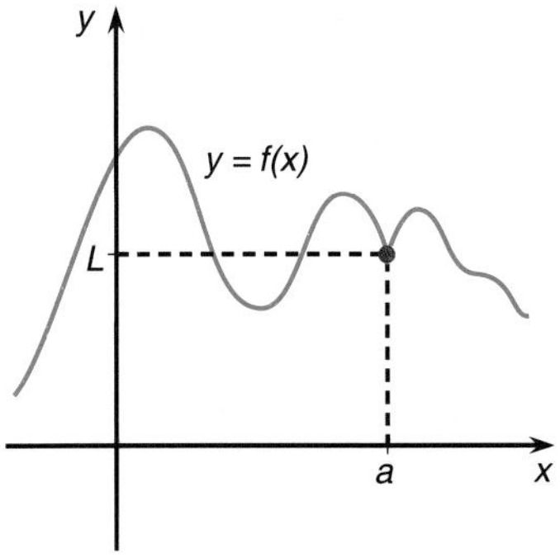

Figure 2.2.1

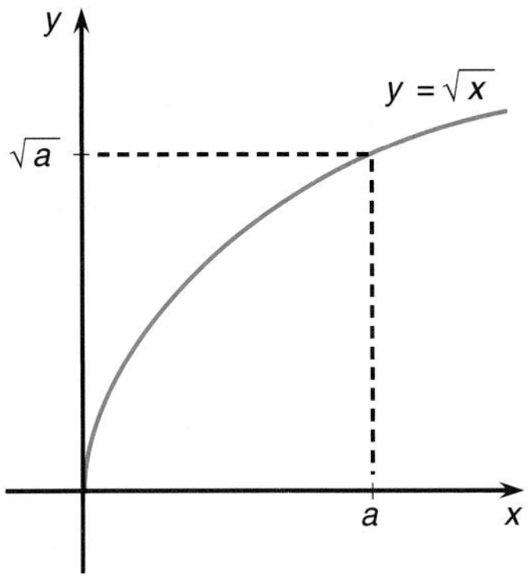

Figure 2.2.2

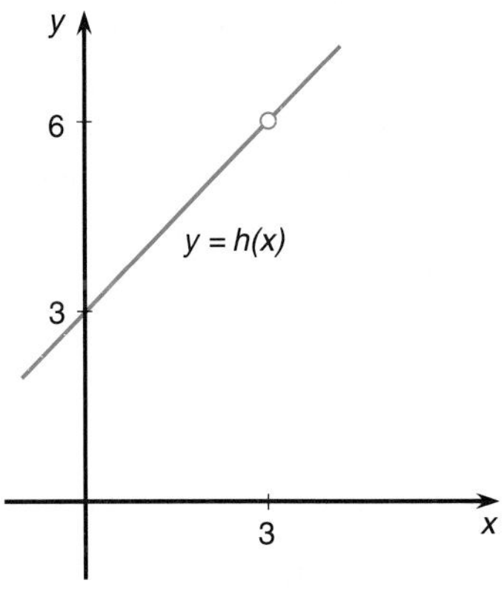

Figure 2.2.3

If there is some number L so that $f(x)$ approaches L when x approaches a (see Figure 2.2.1), we say that L is the **limit** of $f(x)$ as x approaches a or that $f(x)$ tends to L as $x \longrightarrow a$, and we write

$$\lim_{x \longrightarrow a} f(x) = L.$$

We have seen, in Examples 2.1.6 and 2.1.7, that there are cases in which no limit exists. When we write $\lim_{x \longrightarrow a} f(x) = L$, we mean both that the limit exists and that it equals L.

EXAMPLE 1

(i) Let $g(x) = \sqrt{x}$ and consider the behavior of $g(x)$ as x approaches a, where a is some positive number. From Figure 2.2.2 it certainly looks as if $y = g(x)$ approaches $\sqrt{a}$ when x approaches a, and if you were to try with your calculator, you would find that for values of x near a, the value of $g(x) = \sqrt{x}$ would be close to $\sqrt{a}$. This leads us to suspect that

$$\lim_{x \longrightarrow a} g(x) = \lim_{x \longrightarrow a} \sqrt{x} = \sqrt{a},$$

which is in fact correct.

(ii) Next we consider the behavior of $h(x) = \frac{x^2-9}{x-3}$ as x approaches 3. Notice that the number 3 is not in the domain of h, because $h(x) = \frac{x^2-9}{x-3}$ makes sense only when $x \neq 3$. So we consider what happens to $h(x)$ when x approaches 3 without ever equaling 3.

Now $h(x) = \frac{x^2-9}{x-3} = \frac{(x+3)(x-3)}{x-3} = x + 3$, whenever $x \neq 3$. So $h(x)$ is the function whose domain is all real numbers except 3 and that equals $x + 3$ for all such numbers x. We sketch the graph of $h(x)$ in Figure 2.2.3 and leave a little empty circle at the point $(3, 6)$ to indicate that this point is left out. It certainly does appear that as x approaches 3, the point $\big(x, h(x)\big)$ on the graph approaches the missing point or, to put it another way, that

$$\lim_{x \longrightarrow 3} h(x) = 6.$$

There are two types of limits that are particularly easy to calculate. One we have already encountered in Example 2.1.1; there we noticed that

$$\lim_{x \longrightarrow 1} x = 1.$$

This simply says that as x approaches 1, the value of x approaches 1, which is difficult to dispute. There is no reason to restrict ourselves to the number 1; 2 or $-\sqrt{3}$ would do just as well, or in fact any fixed number a.

PROPOSITION 2.2.1

> If a is any constant, then
>
> $$\lim_{x \longrightarrow a} x = a.$$

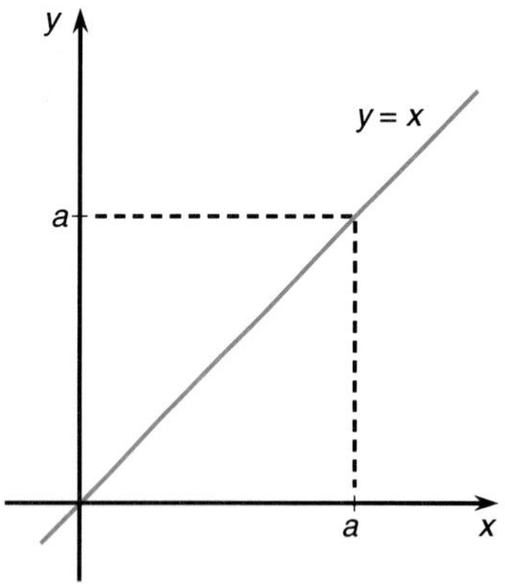

Figure 2.2.4

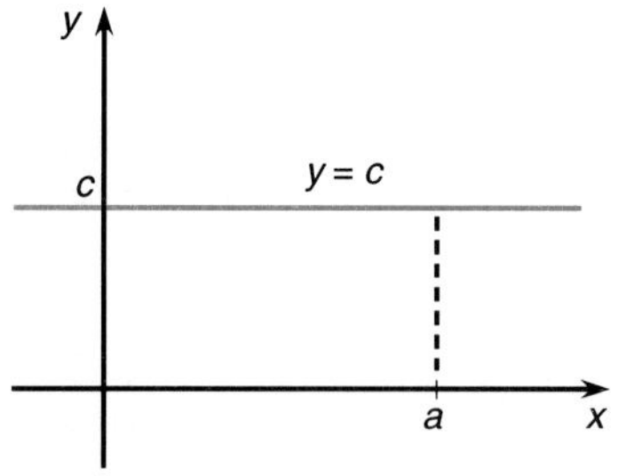

Figure 2.2.5

This result is illustrated in Figure 2.2.4, which shows the graph of $y = x$.

The other very easy limit is even easier. If $f(x)$ is the constant function $f(x) = c$, for some fixed number c, then no matter what x is, $f(x)$ always equals c (see Figure 2.2.5). So if x approaches a, $f(x)$ "approaches" c—in fact it equals c all the time. In symbols this says the following.

PROPOSITION 2.2.2

If a and c are any constants, then

$$\lim_{x \to a} c = c.$$

Now suppose we want to find $\lim_{x \to a} x^2$. If we write $x^2 = x \times x$, and realize that as $x \to a$ the value of x approaches the limit a, then we should expect that the product $x \times x$ should approach $a \times a$, the product of the limits.

A similar line of thought should convince us that as $x \to a$, the function $p(x) = x^2 + x$ should approach $a^2 + a$, the sum of the limits of x^2 and x.

This sort of thing ought to work not just for these particular functions x and x^2, but for any functions that have limits as $x \to a$. If $f(x)$ and $g(x)$ are two such functions, then their sum $f(x) + g(x)$ ought to have a limit, and the limit ought to be the sum of the limits of $f(x)$ and $g(x)$. Also the product $f(x)g(x)$ ought to have a limit, and the limit should be the product of the limits of $f(x)$ and $g(x)$. If one of the functions is a constant function, say $g(x) = c$, then Proposition 2.2.2 tells us its limit is just c, so the limit of $cf(x)$ is just c times the limit of $f(x)$.

We record these important observations as a theorem.

THEOREM 2.2.3

Suppose c is a constant, $f(x)$ and $g(x)$ are functions, and a is a number so that

$$\lim_{x \to a} f(x) = L \quad \text{and} \quad \lim_{x \to a} g(x) = M.$$

(Recall that this means both limits exist and they have the specified values.)

Then

(i) $\lim_{x \to a} \big(f(x) + g(x)\big) = L + M$,

(ii) $\lim_{x \to a} f(x)g(x) = LM$,

(iii) $\lim_{x \to a} cf(x) = cL$.

(i.e., the limits exist and have the given values).

Combining this theorem with the previous two propositions allows us to calculate many limits.

EXAMPLE 2

(i) Evaluate $\lim_{x\to 2}(x^2+3x)$.

Solution (i) We regard x^2+3x as the sum of two functions, x^2 and $3x$. Theorem 2.2.3 says that its limit will be the sum of the limits of these two functions, and it also tells us how to evaluate those limits. After all, $x^2=(x)(x)$, and by Proposition 2.2.1, x approaches the limit 2 as $x\to 2$. So by part (ii) of Theorem 2.2.3, $\lim_{x\to 2}x^2=(\lim_{x\to 2}x)(\lim_{x\to 2}x)=(2)(2)=4$.

We can also use part (iii) of Theorem 2.2.3 to find the limit of $3x$, by writing it as the product of the constant function 3 and the function x, so $\lim_{x\to 2}3x=3(\lim_{x\to 2}x)=3(2)=6$.

Finally we use these results and part (i) of Theorem 2.2.3 to see that $\lim_{x\to 2}x^2+3x=(\lim_{x\to 2}x^2)+(\lim_{x\to 2}3x)=(4)+(6)=10$. Here we used Propositions 2.2.1 and 2.2.2 to evaluate the limits of x and 3, respectively.

(ii) Evaluate $\lim_{x\to -4}(2x^3-5x^2+2x+7)$.

Solution By repeated use of part (i) of Theorem 2.2.3 we can express this limit as the sum of the limits of $2x^3$, $-5x^2$, $2x$, and the constant function 7. By part (i) and repeated use of part (ii) we can express each of these as a product of the limits of some constant function and the function x, the appropriate number of times. So, for instance, we can write $2x^3=2(x)(x)(x)$ and find that its limit as $x\to -4$ is $2(-4)(-4)(-4)=2(-4)^3=-128$.

Doing this for the other three and combining, we find that

$$\begin{aligned}\lim_{x\to -4}(2x^3-5x^2+2x+7) &= \left(\lim_{x\to -4}2x^3\right)+\left(\lim_{x\to -4}-5x^2\right)+\left(\lim_{x\to -4}2x\right)\\ &\quad+\left(\lim_{x\to -4}7\right)\\ &= -128+(-5)(-4)(-4)+2(-4)+(7)\\ &= -128-80-8+7=-209.\end{aligned}$$

REMARK 2.2.4 In doing the last example, we realized that Theorem 2.2.3 applies not just to sums and products of two functions, but to sums and products of any (finite) number of functions.

In Example 2 we found the limit of $2x^3$ as $x\to -4$. In fact, we could do the same thing for any power of x, as x approaches any number a, and of course we would find that

$$\lim_{x\to a}x^n=a^n.$$

Now suppose we consider a polynomial such as $p(x)=3x^7-2x^4+9x+2$, and let $x\to a$. Then each of the powers of x in $p(x)$ tends to the corresponding power of a, and because of Theorem 2.2.3 the whole polynomial $3x^7-2x^4+9x+2$ tends to $3a^7-2a^4+9a+2$, which equals $p(a)$. In other words,

$$\lim_{x\to a}p(x)=p(a),$$

or the limit of the polynomial $p(x)$ as $x\to a$ is just the value $p(a)$.

This calculation was done for a specific polynomial, $3x^7 - 2x^4 + 9x + 2$, but exactly the same argument works for any polynomial. If $p(x) = c_n x^n + c_{n-1}x^{n-1} + \ldots + c_2 x^2 + c_1 x + c_0$ and we let $x \to a$, then each power of x will approach the corresponding power of a, and $p(x)$ will approach $c_n a^n + c_{n-1}a^{n-1} + \ldots + c_2 a^2 + c_1 a + c_0 = p(a)$. We have proved a very important and useful result.

THEOREM 2.2.5

(i) If n is any positive integer, then

$$\lim_{x \to a} x^n = a^n.$$

(ii) If $p(x)$ is any polynomial, then

$$\lim_{x \to a} p(x) = p(a).$$

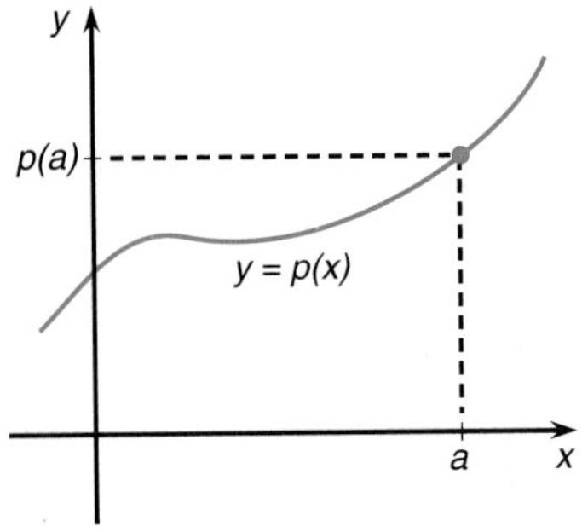

Figure 2.2.6

Notice that what this result says is that as $x \to a$, the polynomial $p(x)$ approaches the "correct" value, the value that $p(x)$ has when x actually equals a. See Figure 2.2.6.

EXAMPLE 3 Evaluate the following limits: (i) $\lim_{x \to -1} (x^3 - 2x^2 + 4x + 8)$, (ii) $\lim_{x \to 4} (x^4 - 4x^3 + x - 4)$, (iii) $\lim_{x \to \sqrt{2}} (x^4 - 2x^3 + 6x^2 + x - 5)$.

Solution

(i) We merely apply the theorem to the polynomial $p(x) = x^3 - 2x^2 + 4x + 8$ and find that $\lim_{x \to -1} (x^3 - 2x^2 + 4x + 8) = p(-1) = (-1)^3 - 2(-1)^2 + 4(-1) + 8 = -1 - 2 - 4 + 8 = 1$.

(ii) $\lim_{x \to 4} (x^4 - 4x^3 + x - 4) = 4^4 - 4(4^3) + 4 - 4 = 256 - 4(64) + 4 - 4 = 256 - 256 + 4 - 4 = 0$.

(iii) $\lim_{x \to \sqrt{2}} (x^4 - 2x^3 + 6x^2 + x - 5) = (\sqrt{2})^4 - 2(\sqrt{2})^3 + 6(\sqrt{2})^2 + \sqrt{2} - 5 = 2^2 - 2(2\sqrt{2}) + 6(2) + \sqrt{2} - 5 = 4 - 4(\sqrt{2}) + 12 + \sqrt{2} - 5 = 11 - 3\sqrt{2} \approx 6.757$.

In simplifying this last limit, wherever possible we combined square roots of 2 (using the fact that $\sqrt{2}\sqrt{2} = 2$) and then collected first the terms that no longer contained any square roots and second the terms that did contain $\sqrt{2}$. Only then did we use a calculator to approximate the final answer.

Since Theorem 2.2.3 says that limits "preserve" sums and products, it is reasonable to ask what happens with quotients. The same sort of reasoning that led us to expect the limit of a sum to be the sum of the limits and the limit of a product to be the product of the limits should also apply to quotients, but there is a complication because we have to avoid dividing by zero.

If $\lim_{x \to a} g(x) = M$, then as $x \to a$, $\frac{1}{g(x)}$ ought to approach $\frac{1}{M}$, provided $M \neq 0$. If we write

$$\frac{f(x)}{g(x)} = f(x)\frac{1}{g(x)},$$

then we can apply Theorem 2.2.3(ii) about limits of products to see that the quotient of $f(x)$ and $g(x)$ should have as its limit the quotient of the limits of $f(x)$ and $g(x)$.

THEOREM 2.2.6

> Suppose $f(x)$ and $g(x)$ are functions and a is a number satisfying
>
> $$\lim_{x\to a} f(x) = L \quad \text{and} \quad \lim_{x\to a} g(x) = M.$$
>
> Suppose too that $M \neq 0$. Then
>
> (i) $\lim_{x\to a} \frac{1}{g(x)} = \frac{1}{M}$,
>
> (ii) $\lim_{x\to a} \frac{f(x)}{g(x)} = \frac{L}{M}$.
>
> (iii) In particular, if $f(x) = \frac{p(x)}{q(x)}$ is a rational function, then
>
> $$\lim_{x\to a} f(x) = \frac{p(a)}{q(a)} = f(a),$$
>
> whenever $f(a)$ is defined (i.e., whenever $q(a) \neq 0$).

EXAMPLE 4 Find (i) $\lim_{x\to 3} \frac{x^2+1}{x+1}$; (ii) $\lim_{x\to -1} \frac{x+1}{x^2+1}$.

Solution

(i) As $x \to 3$, we know that the numerator approaches $3^2 + 1 = 10$ and the denominator approaches $3 + 1 = 4$, so the quotient approaches $\frac{10}{4} = \frac{5}{2}$.

(ii) The denominator approaches the limit $(-1)^2 + 1 = 2$, and the numerator approaches the limit $(-1) + 1 = 0$, so $\lim_{x\to -1} \frac{x+1}{x^2+1} = \frac{0}{2} = 0$.

Example 4 illustrates that the limit of the numerator does not have to be nonzero for the theorem to work; the only condition is that the limit of the denominator is nonzero.

We should also consider what goes wrong when the limit of the denominator does equal zero. In fact, we have already seen one such problem in Example 2.1.7, where we considered the behavior as $x \to 1$ of $g(x) = \frac{1}{x-1}$. There we realized that as x approaches 1 from the right side, $g(x)$ becomes very large and does not approach any limit. As x approaches 1 from the left side, $g(x)$ becomes very large in the negative direction and does not approach any limit.

This illustrates that when the limit of the denominator is zero, it may be that the quotient does not have a limit. It can also happen that the quotient does have a limit, as the following example demonstrates.

EXAMPLE 5 Find (i) $\lim_{x\to -3} \frac{2x^2+6x}{x+3}$; (ii) $\lim_{x\to 5} \frac{x^2-10x+25}{x-5}$.

Solution

(i) $\lim_{x\to -3} \frac{2x^2+6x}{x+3} = \lim_{x\to -3} \frac{2x(x+3)}{x+3} = \lim_{x\to -3} 2x = -6$.

(ii) $\lim_{x\to 5} \frac{x^2-10x+25}{x-5} = \lim_{x\to 5} \frac{(x-5)^2}{x-5} = \lim_{x\to 5} (x-5) = 0$.

We see that when the limit of the denominator is zero, the limit of the quotient is unpredictable. It might not exist, as in Example 2.1.7, but it also might exist, as in the above example, and may even equal zero.

With a graphics calculator it is possible to sketch the graph of the function $y = \frac{1}{x-1}$ from Example 2.1.7, to see why it has no limit as $x \to 1$.

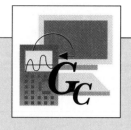

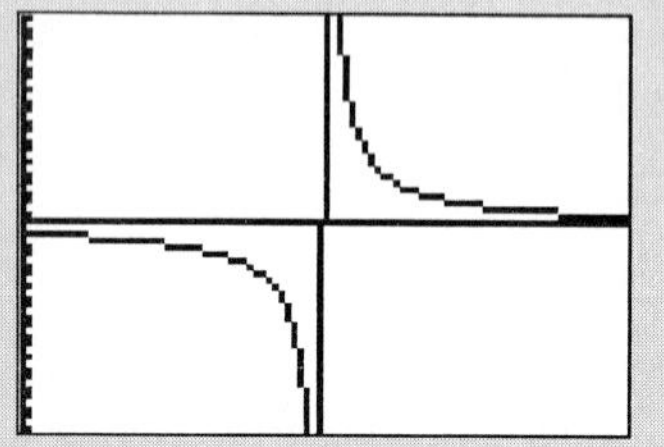

TI-85 GRAPH F1 ($y(x)$ =) 1 ÷ (F1 (x) −1) M2(RANGE) 0 ▼ 2 ▼ ▼ (−) 20 ▼ 20 F5(GRAPH)	SHARP EL9300 1 ÷ (x/θ/T − 1) RANGE 0 ENTER 2 ENTER 1 ENTER −20 ENTER 20 ENTER 1 ENTER
CASIO f_x-7700GB/f_x-6300G Cls EXE Range 0 EXE 2 EXE 1 EXE −20 EXE 20 EXE 1 EXE EXE EXE EXE Graph 1 ÷ (X − 1 EXE	HP 48SX PLOT PLOTR ERASE ATTN′ αY = 1 ÷ () αX − 1 PLOT DRAW ERASE 0 ENTER 2 XRNG 20 +/− ENTER 20 YRNG DRAW

You might like to try varying the vertical scale in your plot.

Summary 2.2.7

Limits of Rational Functions

To evaluate a limit of a rational function,

$$\lim_{x \to a} \frac{p(x)}{q(x)},$$

first evaluate the limits of the numerator and the denominator. If $q(a)$, the limit of the denominator, is not zero, then Theorem 2.2.6 can be used without trouble, and

$$\lim_{x \to a} \frac{p(x)}{q(x)} = \frac{p(a)}{q(a)}.$$

If the limit of the denominator is zero, that is, $q(a) = 0$, and the limit of the numerator is nonzero, that is, $p(a) \neq 0$, then, as in Example 2.1.7, the rational function will have no limit as $x \to a$.

If the numerator and denominator both tend to 0 as $x \to a$, that is, $p(a) = q(a) = 0$, then the Factor Theorem (Theorem 1.3.5) (p. 12) says that both $p(x)$ and $q(x)$ are divisible by $x - a$. Perform the long division (cf. Example 1.3.8, p. 12), and write $p(x) = (x - a)P(x)$, $q(x) = (x - a)Q(x)$. Then, provided $x \neq a$, $\frac{p(x)}{q(x)} = \frac{(x-a)P(x)}{(x-a)Q(x)} = \frac{P(x)}{Q(x)}$, so

$$\lim_{x \to a} \frac{p(x)}{q(x)} = \lim_{x \to a} \frac{P(x)}{Q(x)}.$$

Now we can start again, trying to evaluate this limit. With luck, at least one of the numerator and denominator will have a nonzero value at a, and then we can deal with the limit. If both are zero at a, then we may have to divide again, but this process will eventually stop.

EXAMPLE 6 Find $\lim_{x\to 2} \frac{x^3-3x^2+4}{x^3-6x^2+12x-8}$.

Solution Both numerator and denominator are zero when $x = 2$. So both can be divided by $(x - 2)$. We find that $x^3 - 3x^2 + 4 = (x - 2)(x^2 - x - 2)$ and that $x^3 - 6x^2 + 12x - 8 = (x - 2)(x^2 - 4x + 4)$. Accordingly,

$$\lim_{x\to 2} \frac{x^3 - 3x^2 + 4}{x^3 - 6x^2 + 12x - 8} = \lim_{x\to 2} \frac{(x-2)(x^2-x-2)}{(x-2)(x^2-4x+4)} = \lim_{x\to 2} \frac{x^2-x-2}{x^2-4x+4}.$$

Again both numerator and denominator are zero at $x = 2$, so we divide again:

$$\lim_{x\to 2} \frac{x^2-x-2}{x^2-4x+4} = \lim_{x\to 2} \frac{(x-2)(x+1)}{(x-2)^2} = \lim_{x\to 2} \frac{(x+1)}{(x-2)}.$$

Here the numerator tends to 3 as $x \to 2$, and the denominator tends to 0. So we know that the limit of the quotient does not exist, and the limit that we started with does not exist.

Theorem 2.2.5 and our work with polynomials, which have the "expected" limit as x approaches any a, might lead us to think that the limit as $x \to a$ of $f(x)$ will always be related to the value $f(a)$ of $f(x)$ at a. But in the above examples we have discussed limits of functions that are not even defined at $x = a$. In fact, though it is very convenient when a limit does equal the value of the function, the limit itself does not necessarily equal that value, even if it is defined.

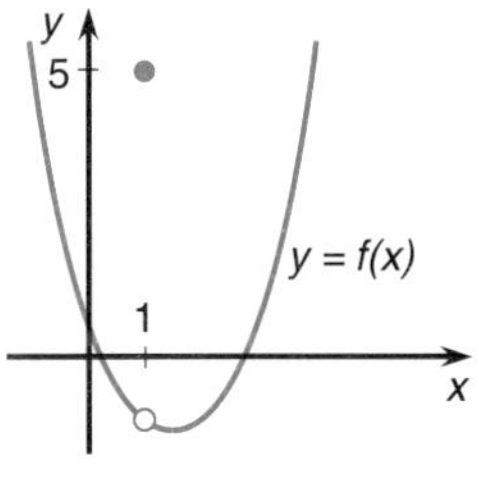

Figure 2.2.7

EXAMPLE 7 Let $f(x)$ be the function defined as follows:

$$f(x) = \begin{cases} x^2 - 3x + 1, & \text{if } x \neq 1, \\ 5, & \text{if } x = 1. \end{cases}$$

In Figure 2.2.7 we leave an empty circle at the point $(1, -1)$ to indicate that that point on the parabola is not included in the graph of $f(x)$. Then as $x \to 1$, as usual we are concerned only with x "approaching" 1 but not actually equaling 1. So the value of $f(x)$ right at $x = 1$ is irrelevant. As far as the limit process is concerned, the limit as $x \to 1$ of $f(x)$ is exactly the same as the limit of the polynomial $x^2 - 3x + 1$, since they are equal at all points other than $x = 1$.

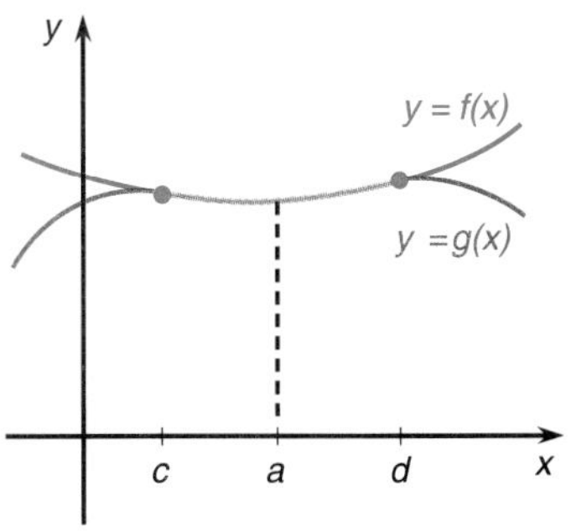

Figure 2.2.8

This example shows that the limit of a function as $x \to a$ may exist and be different from the value of the function at $x = a$. Of course this example was specially made up to demonstrate this phenomenon, and it may seem rather contrived, but the point it illustrates is important.

Just as the limit $\lim_{x\to a} f(x)$ does not depend on $f(a)$, nor even on whether $f(x)$ is defined when $x = a$, the limit also does not depend on the values of $f(x)$ for x far away from a. Using Figure 2.2.8, we see that if there is some interval (c, d) containing a so that $f(x) = g(x)$ for every $x \in (c, d)$, with the possible exception of a itself, then $f(x)$ and $g(x)$ have the same limit as $x \to a$. We record this useful result.

PROPOSITION 2.2.8

Suppose $f(x)$ and $g(x)$ are functions and a a number. Suppose there is an interval (c, d) containing a (i.e., $c < a < d$) so that for every $x \in (c, d)$ except possibly $x = a$, we have

$$f(x) = g(x).$$

Then

$$\lim_{x \to a} f(x) = \lim_{x \to a} g(x),$$

in the sense that if either of these limits exists, then the other also exists and they are equal.

EXAMPLE 8

(i) Notice that we can use Proposition 2.2.8 to establish the limit discussed in Example 7. After all, with $f(x)$ as defined in that example and $g(x) = x^2 - 3x + 1$, we see that these functions are equal everywhere except at $x = 1$. So the proposition applies, using any interval containing 1, for example, $(0, 2)$. We conclude that $\lim_{x\to 1} f(x) = \lim_{x\to 1} g(x) = \lim_{x\to 1} (x^2 - 3x + 1) = -1$. See Figure 2.2.9.

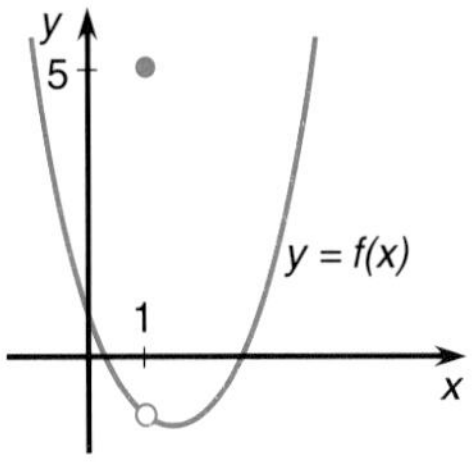

Figure 2.2.9

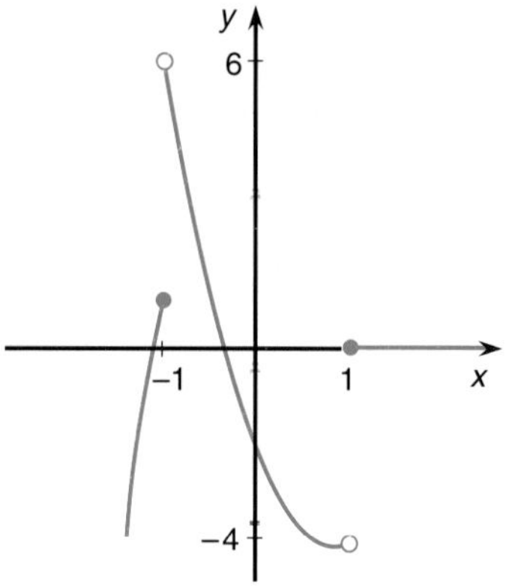

Figure 2.2.10

(ii) Define a function as follows:

$$f(x) = \begin{cases} 3x^2 - 5x - 2, & \text{if } -1 < x < 1, \\ 2x^3 - 3x^2 + 6, & \text{if } x \leq -1, \\ 0, & \text{if } x \geq 1. \end{cases}$$

Then for any a with $-1 < a < 1$, $f(x)$ behaves like $3x^2 - 5x - 2$ as $x \to a$. For example, $\lim_{x\to 0} f(x) = \lim_{x\to 0} (3x^2 - 5x - 2) = -2$. Similarly, $\lim_{x\to 1/2} f(x) = -\frac{15}{4}$. See Figure 2.2.10.

The function $f(x)$ does not have a limit as $x \to 1$ or $x \to -1$.

For any a with $a < -1$, $\lim_{x\to a} f(x)$ will be the same as $\lim_{x\to a} (2x^3 - 3x^2 + 6)$, which equals $2a^3 - 3a^2 + 6$. For any a with $a > 1$, $\lim_{x\to a} f(x)$ will equal $\lim_{x\to a} 0 = 0$.

Another observation we should make is that although in the examples we have done so far we have always used x as the variable, there is really no reason to insist on this choice. Any other letter would do just as well, as long as we are consistent. So for example, we could write

$$\lim_{t\to 7}(t^2-3t+1)=29, \qquad \lim_{u\to -1}\frac{3u+7}{u^2+1}=\frac{4}{2}=2,$$

$$\lim_{s\to 0}\frac{s^3-3s^2+5s}{s}=\lim_{s\to 0}(s^2-3s+5)=5.$$

Exercises 2.2

I

Find each of the following limits.

1. $\lim_{x\to 2}(x^2-2x+1)$
2. $\lim_{x\to 4}(x^2-10)$
3. $\lim_{x\to 1}(3x^2-x-5)$
4. $\lim_{x\to -3}(x^3+20)$
5. $\lim_{x\to 0}(x^4+17x-1)$
6. $\lim_{x\to 3}(x^4-1)$
7. $\lim_{t\to -1}(t^3+3t^2-1)$
8. $\lim_{x\to\sqrt{3}}(x^4+2x^3-4x-1)$
9. $\lim_{t\to\sqrt{5}}(t^5-3t^3+t-1)$
10. $\lim_{u\to\sqrt{2}}(u^7+2u^6-4u^3+2u-1)$
11. $\lim_{x\to 2}\frac{x^2+x-1}{x^3-x^2+x-5}$
12. $\lim_{t\to -1}\frac{t^2+2t+2}{t^3+3t^2+4t-1}$
13. $\lim_{s\to 0}\frac{s^9+s^4+5}{s^5-3s^4+s^2-s+5}$
14. $\lim_{r\to 4}\frac{r^2-2r-6}{r^2+2r-11}$
15. $\lim_{x\to -1}\frac{x^5+2x^4-1}{x^3+2x^2+3x-1}$
16. $\lim_{u\to 0}\frac{u^3-2u^2+u-1}{u^2-u+20}$
17. $\lim_{t\to\sqrt{3}}\frac{t^4-6t^2+9}{t^2-3}$
18. $\lim_{y\to 2^{1/3}}\frac{y^3-2}{y^6-4}$
19. $\lim_{x\to(1+\sqrt{2})}\frac{(x-1)^4-4}{(x-1)^2-2}$
20. $\lim_{u\to(1+\sqrt{5})}\frac{u^2-2u-4}{(u-1)^2-5}$

In each of the following cases, decide whether or not the limit exists. If it does, calculate it. If it does not, explain why.

21. $\lim_{x\to 1}\frac{x^2-1}{x-1}$
22. $\lim_{t\to -1}\frac{t^2+2t+1}{t^2+3t+2}$
23. $\lim_{r\to 3}\frac{r^2-r-6}{r-3}$
24. $\lim_{u\to 0}\frac{2u^3-3u^2+7u}{u^3-3u^2}$
25. $\lim_{y\to -2}\frac{y^2+4y+4}{y^2-4}$
26. $\lim_{x\to -1}\frac{2x^2+4x+2}{x^3+3x^2+3x+1}$
27. $\lim_{t\to -1}\frac{7t^3+14t^2+7t}{t^2+2t+1}$
28. $\lim_{s\to 0}\frac{3s^9-s^5}{s^7-2s^6+3s^5}$
29. $\lim_{x\to 1}\frac{x^3-x^2+x-1}{x^4-x^3+x^2-x+1}$
30. $\lim_{u\to 2}\left(\frac{u^2-4u+4}{u^2-4u+3}\right)^3$
31. $\lim_{u\to 3}(u^2-5u+1)$
32. $\lim_{t\to 1}\frac{t^2-3t+1}{2t^3-t^2+t+2}$
33. $\lim_{x\to -2}\frac{x^2+2x}{x^2+1}$
34. $\lim_{r\to 2}\frac{r^3+r^2-5r-2}{r^3-2r^2-r+2}$
35. $\lim_{y\to 0}\frac{2y^3-6y^2+13y}{y^3-2y^2}$
36. $\lim_{x\to 1}\frac{2x^3+x-3}{3x^3-3x^2-3x+3}$
37. $\lim_{u\to 2}\frac{5u^2-4u-10}{u^3+u-6}$
38. $\lim_{r\to -1}\frac{2r^2-2}{3r^2-3}$
39. $\lim_{x\to 1}\frac{x^2-2x+1}{2(x^2-1)^2}$
40. $\lim_{t\to -1}\left(\frac{3t+3}{t^3+1}\right)^4$

For each of the following functions $f(x)$, find all values a so that $f(x)$ does not have a limit as $x\to a$. Find the value of the limit for all other values of a (i.e., your answer will depend on a).

41. $f(x)=\frac{x^2-9}{x-3}$
42. $f(x)=\frac{x-3}{x^2-9}$
43. $f(x)=\frac{(x-1)(x-2)(x-5)}{(x+1)(x+2)}$

44. $f(x) = \frac{(x+3)^2(x+1)}{x^2-9}$

45. $f(x) = \frac{x(x+1)(x-2)^2}{(x+1)(x-1)(x-2)}$

46. $f(x) = \frac{x^3-2x^2+x}{x^2-x}$

47. $f(x) = \frac{\sqrt{(x-1)^2}}{x-1}$

48. $f(x) = \frac{\sqrt{(x-4)^4}}{(x-4)^{3/2}}$

49. $f(x) = \frac{x^2-c^2}{(x-c)^2}$

50. $f(x) = \frac{(x-c)^2}{(x-c)^3}$

II

***51.** Suppose $f(x)$ and $g(x)$ are polynomials that can be written in the form $f(x) = (x-a)^m F(x)$, $g(x) = (x-a)^n G(x)$, where F and G are polynomials satisfying $F(a) \neq 0$, $G(a) \neq 0$. Evaluate the following limit, when it exists:

$$\lim_{x \to a} \frac{f(x)}{g(x)}.$$

(Your answer will depend on the numbers m and n, and should include a precise statement about when the limit exists.)

III

52. Consider the function $f(x) = \frac{x^3-3x^2+4}{x^3-6x^2+12x-8}$, which we encountered in Example 6. Use a graphics calculator or a computer to sketch the graph. Begin with the graph for $-5 \leq x \leq 5$ and then zoom in to higher magnification around $x = 2$. Notice in particular how the graph behaves to the left and to the right of $x = 2$.

Find the limit, if it exists, of each of the following functions at the points where the denominator equals 0. Graph each function to confirm your answer.

53. $\frac{x^2+3x+2}{x^2-1}$

54. $\frac{x^3-3x^2}{x^2-9}$

55. $\frac{x+1}{x-1}$

56. $\frac{x-1}{x+1}$

57. $\frac{x^3+4x^2+3x}{x+3}$

58. $\frac{x^4-8x^2+16}{x+2}$

2.3 Working with Limits

POWERS
COMPOSITE FUNCTIONS
THE SQUEEZE THEOREM

We have seen that limits of polynomials and rational functions are very easy to find, but of course it would be desirable to be able to find limits of other, more complicated functions. In this section we learn how to find limits of powers of x with nonintegral exponents (e.g., $x^{2/3}$), and also how to deal with limits of functions like $\sqrt{x+1}$, which are "composite" functions, meaning that one first takes the function $f(x) = x + 1$ and then "composes" it with the square root function, to get $\sqrt{f(x)} = \sqrt{x+1}$.

The section ends with a more technical result, the Squeeze Theorem, concerning functions whose graphs lie between the graphs of two other functions whose limits are known. This result is of great theoretical importance and is often used to find new limit rules, but for us its practical applicability will be limited.

Limits of Noninteger Powers of x

EXAMPLE 1 If $a > 0$, find $\lim_{x \to a} \sqrt{x}$.

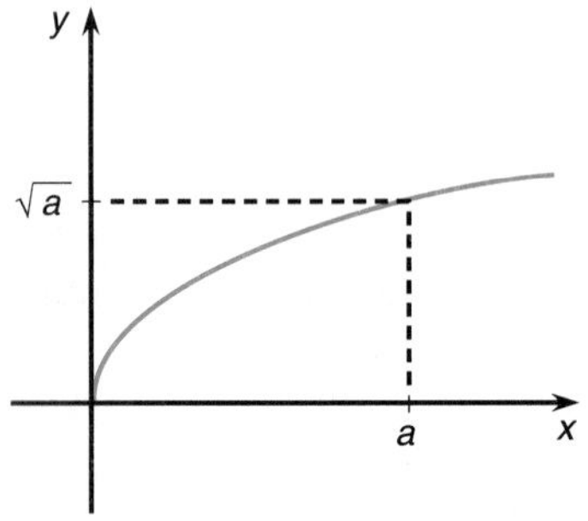

Figure 2.3.1

Solution The graph of $y = \sqrt{x}$ is shown in Figure 2.3.1. It seems reasonable to guess that $\lim_{x \to a} \sqrt{x} = \sqrt{a}$, which is equivalent to showing that

$$\lim_{x \to a} (\sqrt{x} - \sqrt{a}) = 0.$$

It is this that we shall show. Since $\sqrt{x}$ always means the nonnegative square root, we know that $0 \leq \sqrt{x}$.

Write

$$\sqrt{x} - \sqrt{a} = (\sqrt{x} - \sqrt{a}) \frac{\sqrt{x} + \sqrt{a}}{\sqrt{x} + \sqrt{a}}.$$

Multiplying out the difference of squares in the numerator, we find this equals

$$\frac{x-a}{\sqrt{x}+\sqrt{a}}.$$

Now $0 \leq \sqrt{x}$, so

When the denominator gets smaller, the quotient gets larger. . . .

$$\frac{1}{\sqrt{x}+\sqrt{a}} \leq \frac{1}{0+\sqrt{a}} = \frac{1}{\sqrt{a}}.$$

So when $x - a > 0$,

The right-hand inequality comes from multiplying the previous one by $x - a$ . . .

$$0 < \sqrt{x} - \sqrt{a} = \frac{x-a}{\sqrt{x}+\sqrt{a}} \leq \frac{x-a}{\sqrt{a}},$$

and when $x - a < 0$,

Multiplying by the negative number $x - a$ reverses the inequality. . . .

$$0 > \sqrt{x} - \sqrt{a} \geq \frac{x-a}{\sqrt{a}}.$$

The function

$$\frac{x-a}{\sqrt{a}}$$

is just $x - a$ times the constant $\frac{1}{\sqrt{a}}$. By Theorem 2.2.3(iii) its limit as $x \longrightarrow a$ is 0.

We have shown that $\sqrt{x} - \sqrt{a}$ lies between the constant function 0 and $\frac{x-a}{\sqrt{a}}$. These two functions both approach 0 as $x \longrightarrow a$. The situation is illustrated in Figure 2.3.2.

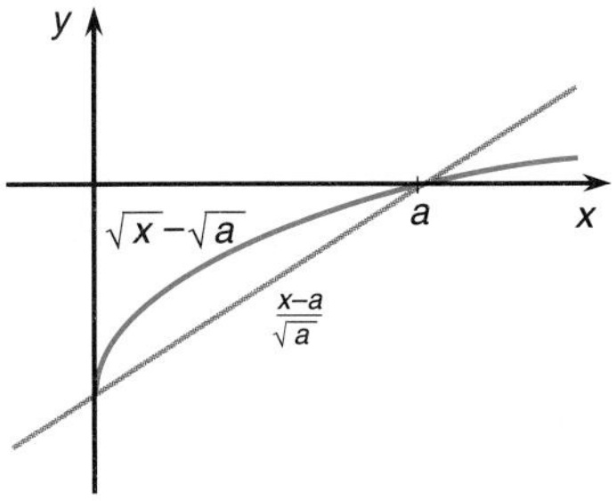

Figure 2.3.2

If two functions both approach 0 as $x \longrightarrow a$, it certainly seems that anything that lies between them should be forced to approach 0 also. This is a particular instance of what we shall call the Squeeze Theorem (Theorem 2.3.5), which says that something that is "squeezed" between two functions whose limits are known to be the same must itself approach the same limit.

Accepting this reasoning, we have proved what we wanted to know, that

$$\lim_{x \longrightarrow a} (\sqrt{x} - \sqrt{a}) = 0,$$

or

$$\lim_{x \longrightarrow a} \sqrt{x} = \sqrt{a}.$$

Strictly speaking, the argument we have given only works when $a > 0$, but something similar is true even when $a = 0$. A comparable result holds not just for square roots, but for roots of any order, or indeed for any power of x, to the extent that it makes sense.

THEOREM 2.3.1

Let r be any real number. Then

$$\lim_{x \longrightarrow a} x^r = a^r,$$

provided it makes sense.

Specifically, if $r \leq 0$, then it is necessary to require that $a \neq 0$. Moreover, it does not make sense to consider nonpositive values of a except when r is a rational number, $r = \frac{p}{q}$, with q an odd number.

EXAMPLE 2

(i) We know that $x^{1/2} = \sqrt{x}$ makes sense only for $x \geq 0$, and we have just seen that for $a > 0$, $\lim_{x \to a} \sqrt{x} = \sqrt{a}$.

(ii) The cube root of x makes sense for any $x \in \mathbb{R}$ (e.g., $(-8)^{1/3} = -2$). So we can write $\lim_{x \to -8} x^{1/3} = (-8)^{1/3} = -2$, or for any $a \in \mathbb{R}$, $\lim_{x \to a} x^{1/3} = a^{1/3}$.

(iii) The function $x^{-8/5}$, which can be written as

$$(x^{-1/5})^8 = \frac{1}{(x^{1/5})^8},$$

makes sense for all x except $x = 0$ (because in that case the denominator would be zero). So for $a \neq 0$ we know that $\lim_{x \to a} x^{-8/5} = a^{-8/5}$. For example, if $a = -32$, we get $\lim_{x \to -32} x^{-8/5} = (-32)^{-8/5} = \left((-32)^{-1/5}\right)^8 = \left(-\frac{1}{2}\right)^8 = \frac{1}{256}$.

Now that we know how to deal with limits of powers of x and with sums, products, and quotients of limits, we can evaluate a great many complicated limits. The idea is to evaluate the simple parts of a complicated limit and then see how they combine.

EXAMPLE 3

(i) Evaluate

$$\lim_{x \to 3} \frac{x^2 - 9 - \sqrt{x}}{x^{3/2}}.$$

Solution We can easily find that the limit of the numerator is $3^2 - 9 - \sqrt{3} = 9 - 9 - \sqrt{3} = -\sqrt{3}$ and that the limit of the denominator is $(3)^{3/2} = 3\sqrt{3}$. Applying Theorem 2.2.6(ii), we find that

$$\lim_{x \to 3} \frac{x^2 - 9 - \sqrt{x}}{x^{3/2}} = \frac{-\sqrt{3}}{3\sqrt{3}} = -\frac{1}{3}.$$

(ii) Find

$$\lim_{x \to -1} \frac{x^{1/5} + 3}{x + 1}.$$

Solution The limit of the numerator is $(-1)^{1/5} + 3 = -1 + 3 = 2$, and the limit of the denominator is 0. This means that as $x \to -1$, the numerator approaches 2 and the denominator approaches 0, so the quotient is something near 2 divided by something very small. This means it will be very large, either very large positive or very large negative, and it will continue to get larger. So the limit does not exist.

(iii) Find

$$\lim_{x \to 0} \frac{x^{1/3} + x^{7/3}}{x^3 - 2x}.$$

Solution This time we find that the numerator and denominator both tend to 0, so the theorem on quotients does not apply. In a situation like this, something else is needed, and the first thing to try is factoring the numerator and denominator. We can write $x^{1/3} + x^{7/3} = x^{1/3}(1 + x^2)$, so

$$\frac{x^{1/3} + x^{7/3}}{x^3 - 2x} = \frac{x^{1/3}(1 + x^2)}{x(x^2 - 2)} = \frac{1 + x^2}{x^{2/3}(x^2 - 2)}.$$

Now we can try again to evaluate the limit. This time we find that the numerator has 1 as its limit and the denominator tends to 0, so as above the quotient does not have a limit.

> This example illustrates the basic approach to finding limits of complicated expressions. Try evaluating the limit of each simple piece separately, and then put them together. If you encounter a quotient in which the denominator tends to 0 and the numerator tends to a limit other than 0, then the limit of the quotient does not exist. If you encounter a quotient in which both numerator and denominator approach 0, then try to rearrange the expression to make it easier to work with. A good first step is to attempt to factor either of them and cancel any common factors.

EXAMPLE 4 Find

$$\lim_{t \to 0} \frac{|t|^2 + |t|^{6/5}}{|t|^{1/5} + |t|^{1/10}}.$$

Solution Note that both numerator and denominator tend to 0. The thing to look for is something we can divide out of both the numerator and the denominator, and the most obvious choice is a power of $|t|$.

Dividing by a power of $|t|$ amounts to changing the exponents on all the powers of $|t|$ occurring in both numerator and denominator. After doing that we still need to find the limits of numerator and denominator, so we should avoid having negative powers of $|t|$, which do not tend to any limit as $t \longrightarrow 0$. The natural thing to do is divide by the lowest power of $|t|$ that occurs, namely, $|t|^{1/10}$. We find that

$$\lim_{t \to 0} \frac{|t|^2 + |t|^{6/5}}{|t|^{1/5} + |t|^{1/10}} = \lim_{t \to 0} \frac{|t|^{19/10} + |t|^{11/10}}{|t|^{1/10} + 1}.$$

Now the numerator still tends to 0, but the denominator tends to 1 as $t \longrightarrow 0$, so

$$\lim_{t \to 0} \frac{|t|^2 + |t|^{6/5}}{|t|^{1/5} + |t|^{1/10}} = \frac{0}{1} = 0.$$

EXAMPLE 5 Find $\lim_{u \to 8} \frac{u-8}{u^{1/3}-2}$.

Solution Both numerator and denominator tend to 0 as u tends to 8, but this time we cannot divide out by a power of u, since both numerator and denominator contain constant terms. Instead, we should try factoring them both.

On the face of it, the numerator cannot be factored, but suppose we take a hint from the fact that the denominator contains $u^{1/3}$ and write $x = u^{1/3}$. Then the

As $u \to 8$, we have $x = u^{1/3} \to 2$. This suggests evaluating $x^3 - 8$ at $x = 2$; we get 0, so $(x - 2)$ divides $x^3 - 8$.

numerator is $x^3 - 8$, which does factor: $x^3 - 8 = (x-2)(x^2 + 2x + 4)$. Returning to the variable u, we find the following:

$$\frac{u-8}{u^{1/3}-2} = \frac{(u^{1/3}-2)(u^{2/3}+2u^{1/3}+4)}{u^{1/3}-2} = u^{2/3} + 2u^{1/3} + 4,$$

so

$$\lim_{u\to 8} \frac{u-8}{u^{1/3}-2} = \lim_{u\to 8} (u^{2/3} + 2u^{1/3} + 4) = 8^{2/3} + 2(8)^{1/3} + 4$$
$$= 4 + 4 + 4 = 12.$$

Limits of Composite Functions

EXAMPLE 6 Suppose we want to find

$$\lim_{x\to 2} \sqrt{x+1}.$$

We would like to say that if $x \to 2$, then $x + 1$ approaches 3, and as t approaches 3, we know that $\sqrt{t}$ approaches $\sqrt{3}$. Letting $t = x + 1$, we expect $\sqrt{x+1}$ to approach $\sqrt{3}$.

This is true, and it is a particular case of the following result.

THEOREM 2.3.2

Suppose f and g are functions, with $\lim_{x\to a} f(x) = L$, $\lim_{t\to L} g(t) = M$, and $g(L) = M$. Then

$$\lim_{x\to a} g\big(f(x)\big) = M.$$

This can also be written as

$$\lim_{x\to a} g\big(f(x)\big) = \lim_{t\to L} g(t) = M.$$

The condition $g(L) = M$ says that not only should $g(t)$ have a limit as $t \to L$, but it should have the "right value" at $t = L$. In the exercises we shall see the theorem fails without this condition.

In Example 6 we have $f(x) = x + 1$, $g(t) = \sqrt{t}$, $a = 2$, $L = 3$, and $M = \sqrt{3}$. The theorem tells us how to evaluate the limit of $g\big(f(x)\big)$, which is the function obtained by first applying to x the function f and then taking the answer you get and applying g to that. In the example this means $\sqrt{x+1}$. In the statement of the theorem we use t as the variable for $g(t)$ in an attempt to avoid confusion, since we intend to substitute $f(x)$ for t later on.

Also notice that the statement of the theorem considers the limit L of $f(x)$ as $x \to a$, and then considers the limit of $g(t)$ as t approaches *this limit* L. Of course this is reasonable, since the idea is to evaluate the limit of $g\big(f(x)\big)$, and $f(x)$ will be approaching L.

EXAMPLE 7 Find the following limits: (i) $\lim_{x\to 4} \sqrt{x^2 - 1}$; (ii) $\lim_{x\to 1} (x+2)^5$; (iii) $\lim_{x\to 0} (1 + \sqrt{x+4})^3$; (iv) $\lim_{x\to 0} \frac{2-\sqrt{4-x}}{x}$; (v) Find $\lim_{t\to 0} t^{1/3}$ without using Theorem 2.3.1, *assuming* the limit exists.

Solution To solve any of these questions, the idea is to figure out first what functions f and g we need to use in order to apply the theorem.

(i) We can write $\sqrt{x^2-1} = \sqrt{f(x)}$, with $f(x) = x^2 - 1$. Letting $g(t) = \sqrt{t}$, we are ready for the theorem. Since $\lim_{x\to 4} f(x) = \lim_{x\to 4}(x^2-1) = 15$ and $\lim_{t\to 15} g(t) = \lim_{t\to 15}\sqrt{t} = \sqrt{15} = g(15)$, the theorem tells us that $\lim_{x\to 4}\sqrt{x^2-1} = \sqrt{15}$.

(ii) Here we let $f(x) = x+2, g(t) = t^5$. So $g\big(f(x)\big) = g(x+2) = (x+2)^5$. As $x \to 1$, $f(x)$ tends to 3, and $g\big(f(x)\big)$ tends to $g(3) = 3^5 = 243$. So $\lim_{x\to 1}(x+2)^5 = 243$.

(iii) For this part we need to evaluate the limit of a function that is the cube of something. As in part (ii), the limit will be the cube of the limit of the something. So we must begin by finding that "inner" limit, that is, $\lim_{x\to 0}(1+\sqrt{x+4})$. But as $x \to 0$, we know that $x + 4 \to 4$, so $\sqrt{x+4} \to \sqrt{4} = 2$.

This means $\lim_{x\to 0}(1+\sqrt{x+4}) = 1 + 2 = 3$, so

$$\lim_{x\to 0}(1+\sqrt{x+4})^3 = 3^3 = 27.$$

(iv) The denominator obviously tends to 0. For the numerator, notice that as $x \to 0$, $4 - x$ tends to 4, so $\sqrt{4-x}$ tends to $\sqrt{4} = 2$, and $2 - \sqrt{4-x}$ tends to 0.

Here there is no obvious way to divide out any common factor from the numerator and denominator. Instead, we should remember Example 5 and try to get rid of the square root by multiplying both numerator and denominator by $2 + \sqrt{4-x}$. We find

$$\begin{aligned}\lim_{x\to 0}\frac{2-\sqrt{4-x}}{x} &= \lim_{x\to 0}\frac{2-\sqrt{4-x}}{x}\times\frac{2+\sqrt{4-x}}{2+\sqrt{4-x}} = \lim_{x\to 0}\frac{4-(4-x)}{x(2+\sqrt{4-x})}\\ &= \lim_{x\to 0}\frac{x}{x(2+\sqrt{4-x})} = \lim_{x\to 0}\frac{1}{2+\sqrt{4-x}} = \frac{1}{2+\sqrt{4}}\\ &= \frac{1}{4}.\end{aligned}$$

(v) We can use the second formula in Theorem 2.3.2. Letting $t = x^3$, we observe that $\lim_{x\to 0} t = \lim_{x\to 0} x^3 = 0$. So with $g(t) = t^{1/3}$, $f(x) = x^3$, we find $\lim_{x\to 0} f(x) = 0$, and

$$\lim_{t\to 0} t^{1/3} = \lim_{t\to 0} g(t) = \lim_{x\to 0} g(x^3) = \lim_{x\to 0} x = 0.$$

Notice the way we used the theorem in part (v) of Example 7. In effect we turned it around: Here it is easy to find the limit of $g\big(f(x)\big) = x$, and we use this to find the limit of $g(t)$. This approach cannot be used to show the limit exists, but it does allow us to find the value, *assuming* the limit exists.

Another fact we need to record involves the sign of a limit. If f is always positive, say, then it seems likely that the same should be true of its limits. In fact, this is not quite true.

EXAMPLE 8 Let

$$f(u) = \begin{cases} |u|, & \text{if } u \neq 0\\ 6, & \text{if } u = 0.\end{cases}$$

Its graph is shown in Figure 2.3.3. By Proposition 2.2.8, $f(u)$ has the same limit as $|u|$, so

$$\lim_{u \to 0} f(u) = \lim_{u \to 0} |u| = 0.$$

Even though the function was always positive, the limit is zero.

On the other hand, it could never be negative, and this is the result we want.

THEOREM 2.3.3

> If $f(x)$ is a function, a is a number so that $\lim_{x \to a} f(x)$ exists, and (c, d) is an interval containing a (i.e., $c < a < d$) so that $f(x) \geq 0$ for all $x \in (c, d)$, except possibly $x = a$, then
>
> $$\lim_{x \to a} f(x) \geq 0.$$

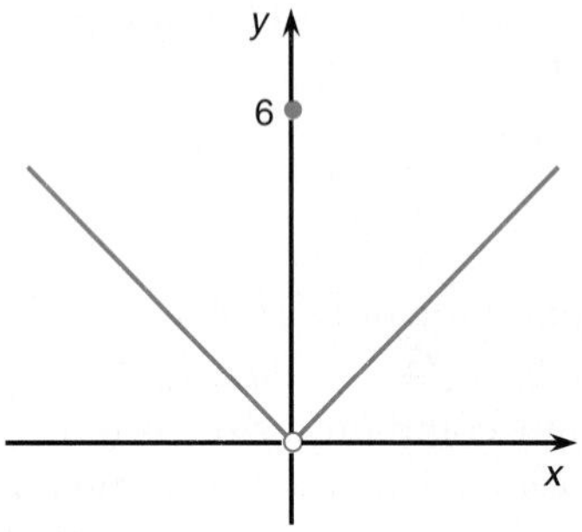

Figure 2.3.3

EXAMPLE 9

(i) Suppose, in the notation of Theorem 2.3.3, that $f(x) \leq 0$ for all $x \in (c, d)$, except possibly $x = a$. Then $-f(x) \geq 0$, and the theorem tells us that $\lim_{x \to a} \big(-f(x)\big) \geq 0$. But by Theorem 2.2.3(iii), $\lim_{x \to a} \big(-f(x)\big) = -\lim_{x \to a} f(x)$, so

$$\lim_{x \to a} f(x) \leq 0.$$

(ii) Suppose K is a constant and $f(x) \geq K$ for all $x \in (c, d)$. Then $f(x) - K \geq 0$, so $\lim_{x \to a} \big(f(x) - K\big) \geq 0$, but this says that $\big(\lim_{x \to a} f(x)\big) - (\lim_{x \to a} K) \geq 0$, or

$$\lim_{x \to a} f(x) \geq K.$$

A similar result can be found if $f(x) \leq K$.

One final fact we need to record is a complement to Theorem 2.3.3. That theorem states that if $f(x) \geq 0$ on some interval around a, and $\lim_{x \to a} f(x)$ exists, then $\lim_{x \to a} f(x) \geq 0$. The following theorem is more or less the reverse.

THEOREM 2.3.4

> Suppose $f(x)$ is a function and $a \in \mathbb{R}$ so that $\lim_{x \to a} f(x)$ exists.
>
> If $\lim_{x \to a} f(x) > 0$, then there is some interval (c, d) containing a (i.e., $c < a < d$) so that for every $x \in (c, d)$, except possibly $x = a$ itself,
>
> $$f(x) > 0.$$

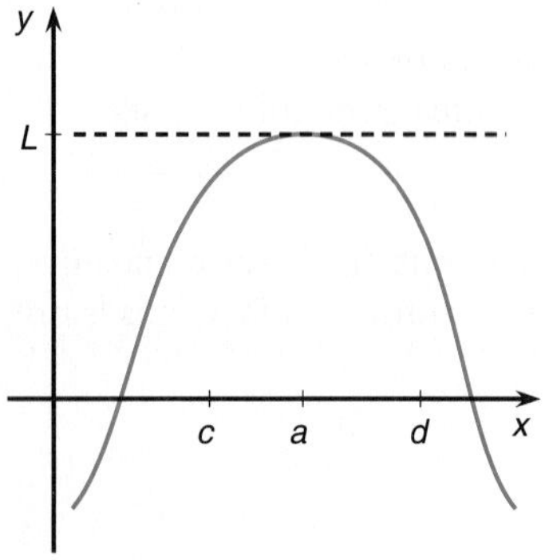

Figure 2.3.4

The idea is very simple. If $f(x)$ approaches some positive limit L as $x \to a$, then close to a, $f(x)$ must get so close to L that it is bigger than 0. This result will be used in the next chapter. See Figure 2.3.4.

The Squeeze Theorem

There are many limits that we can now evaluate, but it would be desirable to extend our techniques for finding limits to other types of functions. One useful way to do this involves comparing a difficult limit with easier ones that are already known (cf. Example 1).

Specifically, suppose we have two functions that are both known to have a limit as $x \longrightarrow a$, and also suppose they have the same limit as $x \longrightarrow a$. The situation is illustrated in Figure 2.3.5. The most obvious case would be the one in which both functions are polynomials or rational functions, since we have no trouble evaluating their limits.

Now suppose there is another function, possibly quite complicated, whose graph lies between those two graphs. Then as $x \longrightarrow a$, this function is "squeezed" between the other two and is forced to approach the same limit as they do. See Figure 2.3.6.

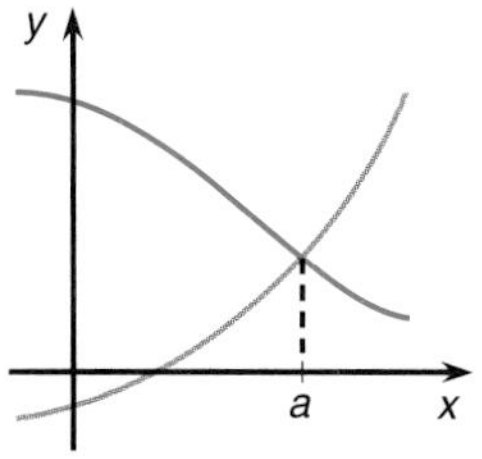

Figure 2.3.5

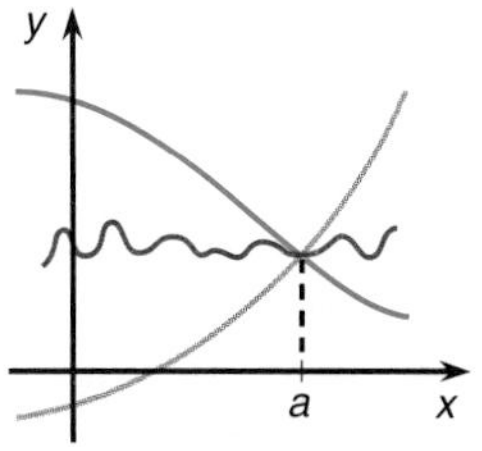

Figure 2.3.6

EXAMPLE 10 Find $\lim_{x\longrightarrow 0} x\sin\left(\frac{1}{x}\right)$.

Solution The behavior of $\sin\left(\frac{1}{x}\right)$ as $x \longrightarrow 0$ is very complicated. After all, if x approaches 0 from the right, then it is small and positive, so $\frac{1}{x}$ is large and positive and gets larger and larger. Now as u gets larger and larger, $\sin(u)$ goes up and down infinitely many times. So as x approaches 0 from the right, $\sin\left(\frac{1}{x}\right)$ goes up and down infinitely often between $y = 1$ and $y = -1$. Its behavior to the left of 0 is similar. In particular, it does not approach any limit. See Figure 2.3.7.

To get an idea of the function $x\sin\left(\frac{1}{x}\right)$, we first draw the two lines $y = x$ and $y = -x$ (see Figure 2.3.8). Then think of the function $x\sin\left(\frac{1}{x}\right)$ as x multiplied by the function $\sin\left(\frac{1}{x}\right)$, which goes back and forth between 1 and -1. So the resulting product function goes back and forth between x and $-x$, and we can visualize it as what you get by squeezing $\sin\left(\frac{1}{x}\right)$ between the two lines $y = x$ and $y = -x$. See Figure 2.3.9(i). Since $\sin\left(\frac{1}{x}\right)$, being a sine, is never bigger than 1 or smaller than -1, we see that when $x > 0$, the product $x\sin\left(\frac{1}{x}\right)$ never gets above the line $y = x$ or below $y = -x$. When $x < 0$, it never lies below $y = x$ or above $y = -x$. See Figure 2.3.9(ii).

Now we are also set up to find the limit. Both the straight lines $y = x$ and $y = -x$ go through the origin or, in terms of limits,

$$\lim_{x\longrightarrow 0} x = \lim_{x\longrightarrow 0} (-x) = 0.$$

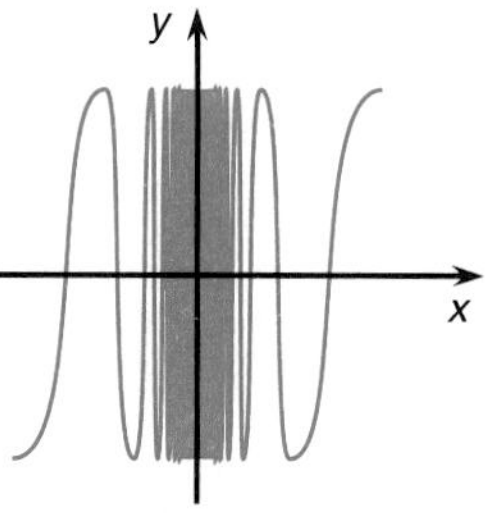

Figure 2.3.7

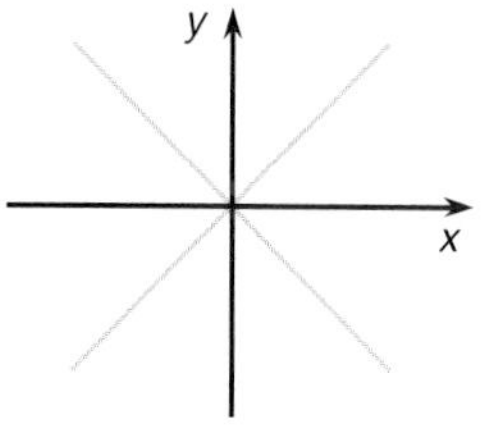

Figure 2.3.8

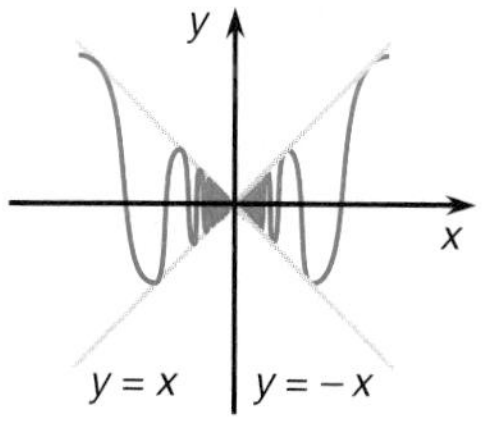

Figure 2.3.9(i)

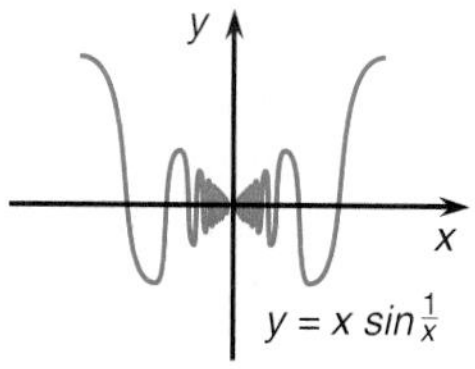

Figure 2.3.9(ii)

The function $x\sin\left(\frac{1}{x}\right)$ is "squeezed" between these two lines, so it must have the same limit:

$$\lim_{x\longrightarrow 0}\left(x\sin\left(\frac{1}{x}\right)\right) = 0.$$

This example and Example 1 illustrate the use of the following result.

THEOREM 2.3.5

The Squeeze Theorem

Suppose $g(x)$ and $h(x)$ are functions and a is some number so that

$$\lim_{x\to a} g(x) = \lim_{x\to a} h(x) = L.$$

Suppose $f(x)$ is another function and (c, d) is an interval containing a (i.e., $c < a < d$). Suppose that on the interval (c, d), except possibly at $x = a$, the graph of the function $f(x)$ lies between the graphs of $g(x)$ and $h(x)$, that is, either

$$g(x) \le f(x) \le h(x)$$

or

$$h(x) \le f(x) \le g(x).$$

Then

$$\lim_{x\to a} f(x) = L.$$

See Figure 2.3.10.

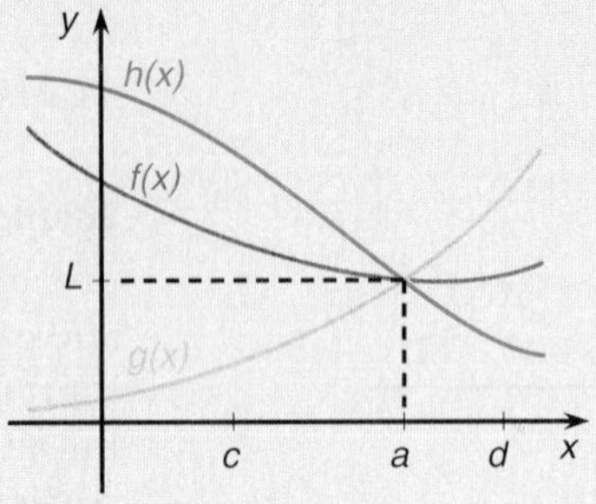

Figure 2.3.10

EXAMPLE 11 Suppose $F = F(x)$ is a function that is defined for every $x \in (0, 2)$. Find $\lim_{x\to 1} \frac{x^2+(F(x))^2}{1+(F(x))^2}$.

Solution We begin by writing the function in the form

$$\frac{x^2 + \left(F(x)\right)^2}{1 + \left(F(x)\right)^2} = \frac{x^2 - 1 + 1 + \left(F(x)\right)^2}{1 + \left(F(x)\right)^2} = \frac{x^2 - 1}{1 + \left(F(x)\right)^2} + \frac{1 + \left(F(x)\right)^2}{1 + \left(F(x)\right)^2}$$

$$= \frac{x^2 - 1}{1 + \left(F(x)\right)^2} + 1.$$

Since the constant function 1 tends to the limit 1, we should concentrate on finding $\lim_{x\to 1} \frac{x^2-1}{1+(F(x))^2}$.

Consider the interval $(0, 2)$, which contains 1. On the left half of this interval, $(0, 1)$, the numerator $x^2 - 1$ is always negative, and on the right half, $(1, 2)$, the numerator is always positive. The denominator is always positive, being a sum of squares, so the quotient is also negative on $(0, 1)$ and positive on $(1, 2)$.

We also observe that $1 + \left(F(x)\right)^2 \ge 1$, since the square is nonnegative, and taking reciprocals, we find that $\frac{1}{1+(F(x))^2} \le \frac{1}{1} = 1$. (Notice that the inequality is reversed by taking reciprocals.) On the interval $(1, 2)$, where $x^2 - 1$ is positive, we multiply this last inequality by $x^2 - 1$ to find that $\frac{x^2-1}{1+(F(x))^2} \le \frac{x^2-1}{1} = x^2 - 1$.

On the interval $(0, 1)$, where $x^2 - 1$ is negative, we multiply the inequality by $x^2 - 1$, being careful to reverse the inequality, and find that $\frac{x^2-1}{1+(F(x))^2} \ge \frac{x^2-1}{1} = x^2 - 1$.

Combining all this information, we see that on $(1, 2)$,

$$0 \le \frac{x^2 - 1}{1 + \left(F(x)\right)^2} \le x^2 - 1,$$

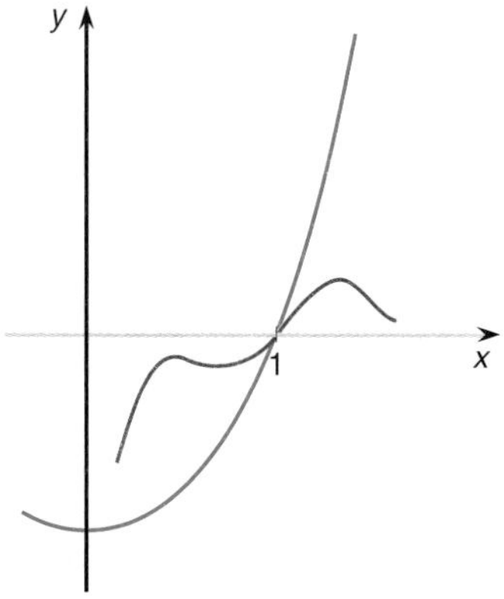

Figure 2.3.11

and on $(0, 1)$,

$$0 \geq \frac{x^2 - 1}{1 + \left(F(x)\right)^2} \geq x^2 - 1.$$

In both cases, the function $\frac{x^2-1}{1+(F(x))^2}$ lies between the functions 0 and $x^2 - 1$ (see Figure 2.3.11). Since $\lim_{x \to 1} (x^2 - 1) = 0 = \lim_{x \to 1} 0$, the Squeeze Theorem applies, and we can conclude that $\lim_{x \to 1} \frac{x^2-1}{1+(F(x))^2} = 0$.

From this we find that the original limit is

$$\lim_{x \to 1} \left(\frac{x^2 - 1}{1 + \left(F(x)\right)^2} + 1 \right) = \lim_{x \to 1} \frac{x^2 - 1}{1 + \left(F(x)\right)^2} + \lim_{x \to 1} 1 = 0 + 1 = 1.$$

Notice that it was possible to find this limit without knowing anything at all about F. Notice too that the function $\frac{x^2-1}{1+(F(x))^2}$ is above 0 and below $x^2 - 1$ when $x > 1$ and the other way around when $x < 1$. What is important is that it is *between* them all the time.

Theorem 2.3.5 requires that $f(x)$ lie between $g(x)$ and $h(x)$ for x in some interval containing a. This is very much in the spirit of Proposition 2.2.8, which says that the behavior of $f(x)$ far away from a is irrelevant to the limit as $x \to a$. It is useful

With a graphics calculator we can try out Example 11 with various functions $F(x)$. For instance, we try $F(x) = \frac{1}{2} \cos x = 0.5 \cos x$.

TI-85 GRAPH F1($y(x)$ =) (F1(x) x^2 + (.5 COS F1(x))x^2 ÷ (1 + (.5 COS F1(x)) x^2) M2(RANGE) 0.1 ▼ 1.9 ▼ 1 ▼ (−) 1 ▼ 3 ▼ 1 F5(GRAPH)	SHARP EL9300 (x/θ/T x^2+ (.5 cos x/θ/T) x^2) ÷ (1 + (.5 cos x/θ/T) x^2) RANGE .1 ENTER 1.9 ENTER 1 ENTER −1 ENTER 3 ENTER 1 ENTER
CASIO f_x-7700GB/f_x-6300G Cls EXE Range .1 EXE 1.9 EXE 1 EXE −1 EXE 3 EXE 1 EXE EXE EXE EXE Graph (X x^2 + (.5 cos X) x^2) ÷(1 + (.5 cos X) x^2) EXE	HP 48SX ↰ PLOT PLOTR ERASE ATTN′ αY = () αX y^x2 + () .5 × COS αX ► ► y^x 2 ► ÷ () 1 + () .5 × COS αX ► ► y^x 2 ↱ PLOT ↰ DRAW ERASE .1 ENTER 1.9 XRNG 1 +/− ENTER 3 YRNG DRAW

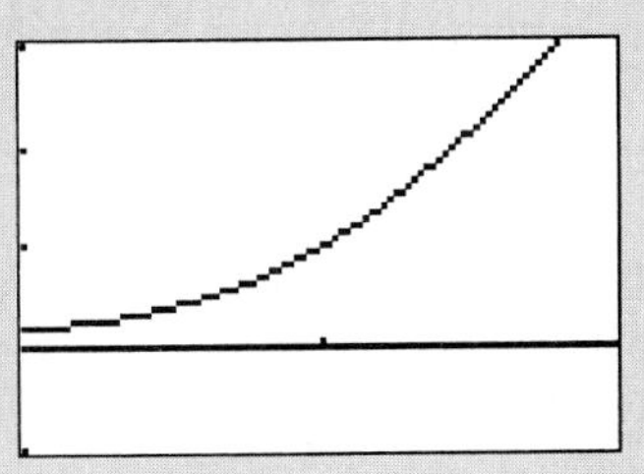

Locate the position of $x = 1$ in the picture. You could also draw the horizontal line $y = 1$ to make it easier to see the limit.

Now you can edit these keystroke sequences and try other functions, such as $F(x) = \sin \pi x$, $F(x) = x^3 - 2$, etc.

because it is often much easier to find suitable g and h on such an interval than on the whole line. See Figure 2.3.12.

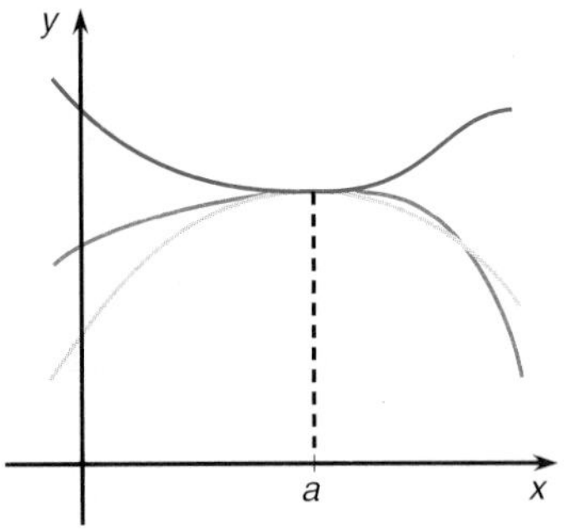

Figure 2.3.12

> The hard thing about using Theorem 2.3.5 is finding the functions $g(x)$ and $h(x)$. Usually we are just given $f(x)$, and we have to think of $g(x)$ and $h(x)$ for ourselves. The idea is to replace difficult components of $f(x)$, like $\frac{1}{1+(F(x))^2}$ in the above example, with simpler things that are bigger or smaller (in the example, 0 is smaller than $\frac{1}{1+\left(F(x)\right)^2}$, and 1 is bigger). Similarly, in Example 10 we replaced the messy function $\sin\left(\frac{1}{x}\right)$ with the smaller function -1 and the larger function 1.

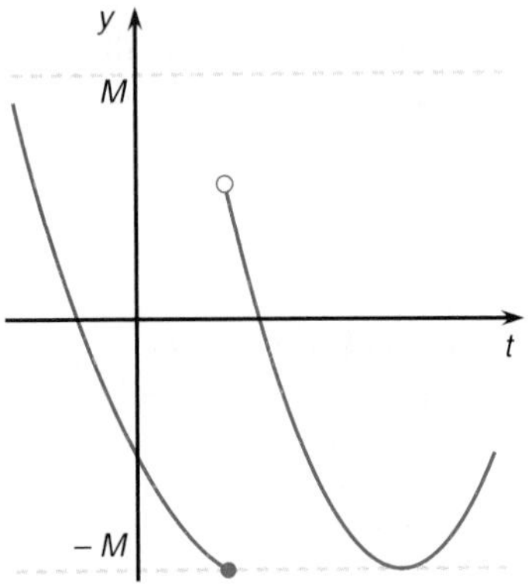

Figure 2.3.13

EXAMPLE 12 Let $f(t)$ be a function such that

$$|f(t)| \leq M, \qquad \text{for all } t,$$

where M is some positive constant (we say that f is "bounded" by M). Show that

$$\lim_{t\to 0} tf(t) = 0.$$

Solution To say that $|f(t)| \leq M$ is to say that $-M \leq f(t) \leq M$, as shown in Figure 2.3.13, which implies that

$$|tf(t)| \leq M|t|,$$

which means that the graph of $tf(t)$ lies between the graphs of $y = Mt$ and $y = -Mt$. See Figure 2.3.14.

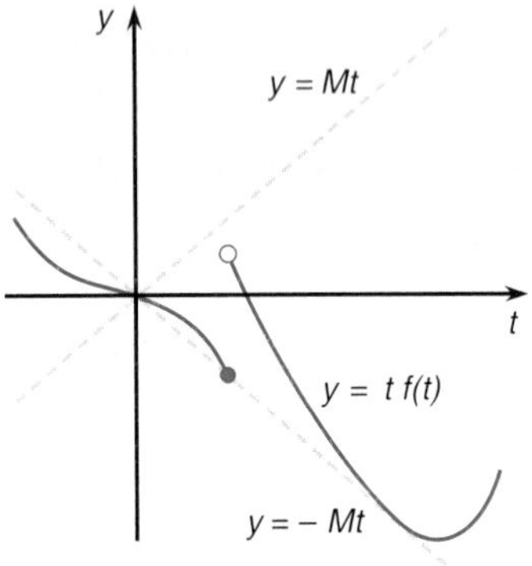

Figure 2.3.14

Since $\lim_{t\to 0}(-Mt) = \lim_{t\to 0} Mt = 0$, the Squeeze Theorem applies, and we conclude that

$$\lim_{t\to 0} tf(t) = 0.$$

The Squeeze Theorem is extremely useful for establishing some limits that we don't know, but as the examples suggest, it is often difficult to use. The good news is that we will not use it very often, and when we do, it will usually be to see how to evaluate some new type of limit, as we did for the square root function in Example 1.

The Meaning of Limits

We end this section with a brief discussion of the fundamental problem of what a limit is. We have repeatedly used phrases like "approaches" and "tends to," but some of our examples show that these words are not really adequate to describe precisely what we mean.

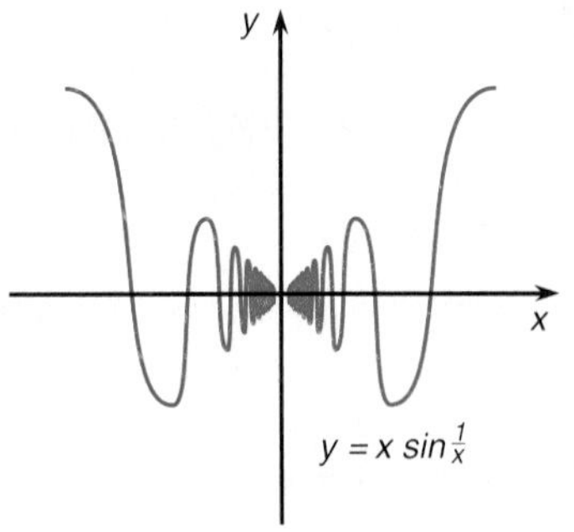

Figure 2.3.15

For instance, the function $f(x) = x\sin\left(\frac{1}{x}\right)$, discussed in Example 10, has limit equal to 0 as x tends to 0, but from Figure 2.3.15 we see that it does not always "approach" 0. On the contrary, it approaches 0 (and actually reaches it) and then moves away again. In fact, it does this infinitely many times, so we cannot justify saying that it keeps approaching 0.

What is true is that it does get close to 0, and while it may move away again, it does not move away too far. By making x very small, we can make sure that $f(x)$ is as close as we like to 0. This idea can be made into a precise definition of limit, but since it is fairly complicated, it will be postponed until Section 2.7.

Exercises 2.3

I

Exercises 1–10 use Theorem 2.3.1, about limits of nonintegral powers of x; Exercises 11–20 use Theorem 2.3.2, about limits of composite functions; Exercises 21–30 use Theorem 2.3.5, the Squeeze Theorem.

Find each of the following limits, explaining your reasoning.

1. $\lim_{x\to 0} \frac{x^{1/3}-2x^{5/3}}{2x^{1/3}-3x^2}$

2. $\lim_{t\to 0} \frac{4(-t)^{1/5}+t^{3/5}}{(-t)^{1/7}}$

3. $\lim_{y\to 0} \frac{3y^{2/3}-5y}{(2y-y^{1/3})^2}$

4. $\lim_{r\to 1} \frac{r^3-3r^{5/3}+\sqrt{r}}{r^2+r^{7/2}}$

5. $\lim_{x\to 9} \frac{3-\sqrt{x}}{x-9}$

6. $\lim_{x\to -1} \frac{x+1}{x^{1/3}+1}$

7. $\lim_{x\to 16} \frac{x-16}{2-x^{1/4}}$

8. $\lim_{x\to 0} \frac{|x|^{5/2}+x^2}{x\sqrt{|x|}}$

9. $\lim_{x\to 3} \frac{x-3}{3^{1/2}-x^{1/2}}$

10. $\lim_{x\to 0} \frac{(x-4)^2-16}{\sqrt{|x|}}$

11. $\lim_{x\to 5} \sqrt{x+4}$

12. $\lim_{t\to 3} \sqrt{t^3-2}$

13. $\lim_{y\to 1} (y^2-3y+1)^{1/3}$

14. $\lim_{x\to 2} \frac{(x^2-1)^{3/2}}{\sqrt{x+1}}$

15. $\lim_{x\to 1} (3x-\sqrt{x+1})^{5/2}$

16. $\lim_{y\to -1} \frac{(y^2+2)^{1/2}}{(10+y)^{1/4}}$

17. $\lim_{t\to 3} \frac{2-\sqrt{t+1}}{t-3}$

18. $\lim_{s\to 0} \frac{s^3-2s^2}{3-\sqrt{9-s}}$

19. $\lim_{r\to -1} \frac{r^{1/3}+1}{1-r^{2/5}}$

20. $\lim_{x\to 0} \frac{(8-x)^{1/3}-2}{x}$

21. $\lim_{x\to 0} x^2 \sin\left(\frac{1}{x}\right)$

22. $\lim_{x\to 0} x^3 \cos\left(\frac{1}{x}\right)$

23. $\lim_{x\to 0} x \sin\left(\frac{1}{x^2-2x+13}\right)$

24. $\lim_{x\to 1} (x-1)\cos\left(\frac{1}{x^2}\right)$

25. $\lim_{x\to 1} (x^2-1)\sin\left(\frac{1}{x-1}\right)$

26. $\lim_{x\to 0} (x^2-2x)\cos\left(\frac{x^2-x+1}{x^2+x}\right)$

27. $\lim_{x\to 0} \frac{x}{3-\cos(\frac{1}{x})}$

28. $\lim_{t\to 0} \frac{3t^2-2t^3\cos(t^{-1})}{t^3-t^2}$

29. $\lim_{u\to 0} u^2 \sin^2\left(\frac{1}{u^3}\right)$

30. $\lim_{s\to 1} \frac{(s-1)^2\sin((s-1)^{-1})}{s^2-1}$

Find whether each of the following limits exists. Evaluate those that do exist and explain why the others do not.

31. $\lim_{x\to 0} \frac{2x^{1/3}-4x^2}{x^{2/3}-7x^3}$

32. $\lim_{t\to 1} \frac{(t-1)-8(t-1)^{4/5}}{t-1+2(t-1)^{2/3}}$

33. $\lim_{u\to 1} \frac{u^{3/2}-1}{u-1}$

34. $\lim_{x\to 0} \frac{\sqrt{9-x}-2}{x}$

35. $\lim_{x\to 0} \frac{1}{\sin(1/x)}$

36. $\lim_{x\to 0} \left(\cos^2\left(\frac{1}{x}\right)+\sin^2\left(\frac{1}{x}\right)\right)$

37. $\lim_{u\to 9} \frac{u^{1/2}-3}{u^2-81}$

38. $\lim_{x\to 16} \frac{4-\sqrt{x}}{2-x^{1/4}}$

39. $\lim_{s\to 0} \frac{3-\sqrt{9-s}}{s^3+2s^2}$

40. $\lim_{t\to 2} \frac{t^2-4}{t^{3/2}-2t^{1/2}}$

II

41. Suppose $r, s, t, u \in \mathbb{R}$ satisfy $r < s$ and $t < u$. Consider the following limit:

$$\lim_{x\to 0} \frac{|x|^r+|x|^s}{|x|^t+|x|^u}.$$

(i) Find conditions on r, s, t, u so that the limit does or does not exist.

(ii) Evaluate the limit when it does exist.

(iii) Do parts (i) and (ii) again with the hypotheses changed to $r \le s$ and $t \le u$.

42. Let c be a positive real number. Evaluate

(i) $\lim_{x\to c^2} \frac{x^{1/2}-c}{x-c^2}$,

(ii) $\lim_{x\to c^{1/2}} \frac{x-c^{1/2}}{x^{1/2}-c^{1/4}}$,

(iii) $\lim_{x\to 0} \frac{x^{1/3}-c}{x}$.

***43.** Several times we have used the fact that

$$\lim_{x\to a} |x| = |a|,$$

which is true, but we have never explained why. Using the theorems of this section, give an argument to explain why it is true. (*Hint:* You might want to deal separately with the cases where $a = 0$ or $a \neq 0$.)

***44.** The hypotheses of Theorem 2.3.2 include the requirement that $g(L) = M$ in addition to the statement about the limit of $g(t)$. Find an example to show that this is really necessary. In other words, find an example that satisfies all the conditions of the theorem except that $g(L) \neq M$, and for which the conclusion of the theorem is false. (*Hint:* Try letting $f(x) = x$.)

***45.** Suppose $f(x)$ has a limit as $x \to a$ for every $a \in \mathbb{R}$.
(i) Express $\lim_{x\to a} f(x + c)$ as a limit of $f(x)$.
(ii) Express $\lim_{x\to a} f(c - x)$ as a limit of $f(x)$.
(iii) Can you say when your results for parts (i) and (ii) remain true if $f(x)$ has a limit as $x \to a$ only for certain values of a?

***46.** (i) Suppose $f(x)$ and $g(x)$ are functions so that $|f(x)| \le M$ for all x (i.e., $f(x)$ is "bounded" by M) and $\lim_{x\to a} g(x) = 0$. Show that $\lim_{x\to a} f(x)g(x) = 0$.
(ii) Suppose $\lim_{x\to a} g(x) = 0$ and that $-4 \le f(x) \le 3$ for all x. Show that

$$\lim_{x\to a} f(x)g(x) = 0.$$

(Be careful; it is possible that $g(x) < 0$.)

III

47. In Example 11 (p. 88) we discussed a limit involving a function $F(x)$ without knowing what $F(x)$ is. Use a graphics calculator or computer to sketch the graph of $y = \frac{x^2+(F(x))^2}{1+(F(x))^2}$ for $x \in (0, 2)$ if $F(x) = \sin x$. Check that the limit as $x \to 1$ does equal 1.

48. Use a graphics calculator or computer to sketch the graph of $y = \frac{x^2+(F(x))^2}{1+(F(x))^2}$ for $x \in (0, 2)$, if (i) $F(x) = \ln x$, (ii) $F(x) = \tan\frac{\pi x}{4}$. Check that in both cases the limit as $x \to 1$ does equal 1.

49. Graph the function occurring in the limit of Example 4 for $x \in [-1, 1]$. Then magnify the graph by using $[-0.1, 0.1]$ or $[-0.01, 0.01]$ to convince yourself of the limit.

50. Plot the function $y = \frac{|x|^{3/8}-2|x|^{1/3}}{|x|^{1/4}(4|x|^{5/3}+3|x|^{1/12})}$ and, by looking at the graph, try to determine the limit as $x \to 0$. Evaluate the limit and confirm your answer.

51. We considered the function $y = \sin\frac{1}{x}$ in Example 10 (p. 87). Draw its graph for $x \in [-1, 1]$, $x \in [-\frac{1}{10}, \frac{1}{10}]$, $x \in [-\frac{1}{100}, \frac{1}{100}]$, $x \in [-\frac{1}{1000}, \frac{1}{1000}]$.

52. Draw the graph of the function $y = x \sin\frac{1}{x}$ for $x \in [-1, 1]$, $x \in [-\frac{1}{10}, \frac{1}{10}]$, $x \in [-\frac{1}{100}, \frac{1}{100}]$, $x \in [-\frac{1}{1000}, \frac{1}{1000}]$.

Find the limit, if it exists, of each of the following functions as $x \to 1$. Graph each function to confirm your answer.

53. $\frac{x-1}{\sqrt{x}-1}$

54. $\frac{x-1}{x^{1/3}-1}$

2.4 One-Sided and Infinite Limits and Limits at Infinity

LIMITS FROM ABOVE/BELOW
INFINITE LIMITS
LIMITS AT INFINITY
LIMITS OF RATIONAL FUNCTIONS

We have already encountered situations in which a function $f(x)$ approaches one value if x approaches a from the right side and a different value if x approaches a from the left (cf. Example 2.1.6, in which the function $f(x) = \frac{x}{|x|}$ was seen to approach 1 if x approaches 0 from the right and -1 if x approaches 0 from the left; see Figure 2.4.1). This phenomenon is sufficiently common that there is special terminology for it.

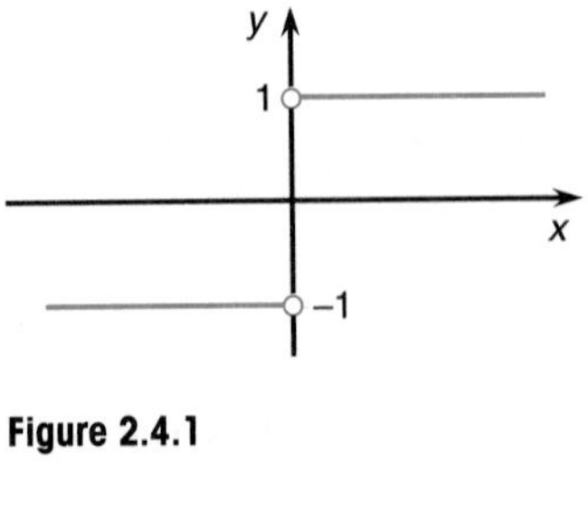

Figure 2.4.1

DEFINITION 2.4.1

One-Sided Limits

A function $f(x)$ is said to have a limit **from the right** or **from above** if $f(x)$ approaches some value L as x approaches a through values to the right of a, that is, with $x > a$. (Strictly speaking, we need to assume that $f(x)$ is defined for every x in an interval (a, s), for some $s > a$, and possibly at $x = a$.) In this case we write

$$\lim_{x \to a^+} f(x) = L.$$

Similarly, we can speak of limits **from the left** or **from below** (assuming that $f(x)$ is defined on (r, a), for some $r < a$, and possibly at $x = a$) and write

$$\lim_{x \to a^-} f(x) = M.$$

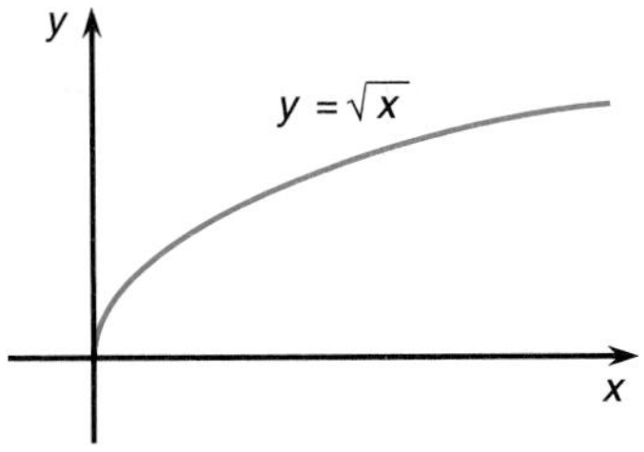

Figure 2.4.2

One-sided limits sometimes arise with functions that are only defined for x on one side of a. An example of this is

$$\lim_{x \to 0^+} \sqrt{x} = 0,$$

since $\sqrt{x}$ only makes sense when $x \geq 0$. See Figure 2.4.2.

But as we have already seen with $f(x) = \frac{x}{|x|}$, they can also be found with functions that are defined on both sides of the point $x = a$.

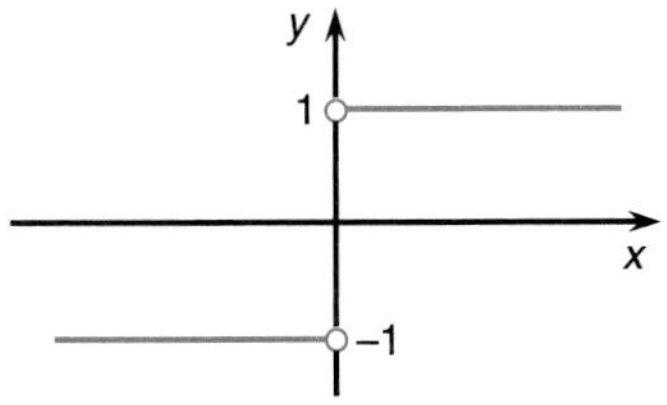

Figure 2.4.3

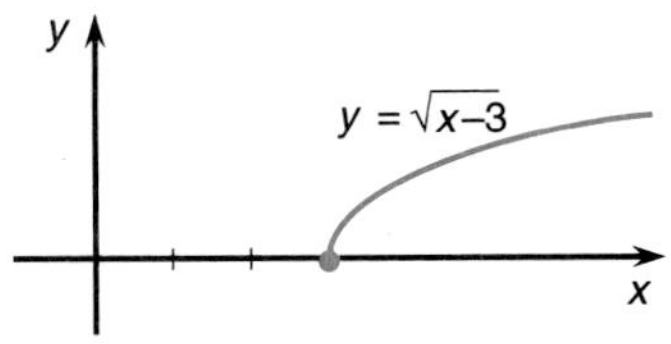

Figure 2.4.4

EXAMPLE 1

(i) The function $f(x) = \frac{x}{|x|}$, graphed in Figure 2.4.3, satisfies

$$\lim_{x \to 0^+} f(x) = 1, \qquad \lim_{x \to 0^-} f(x) = -1.$$

(ii) The function $g(t) = \sqrt{t-3}$ is only defined when $t \geq 3$, since the square root only makes sense for positive numbers (see Figure 2.4.4). If $t \to 3^+$, that is, $t > 3$ and t approaches 3, then $t - 3$ approaches 0, so its square root should approach 0 too. In other words, $g(t)$ approaches 0, or

$$\lim_{t \to 3^+} \sqrt{t-3} = 0.$$

(iii) Let $h(u) = \frac{\sqrt{(u-1)^2}}{u^2-u}$. Recall that the square root sign always means the nonnegative square root, so $\sqrt{(u-1)^2} \geq 0$. If $u > 1$, then $u - 1 > 0$, and $\sqrt{(u-1)^2} = u - 1$. But if $u < 1$, then $u - 1 < 0$, and the nonnegative square root of $(u-1)^2$ is not $u - 1$ but rather $-(u-1) = 1 - u$.

So if $u > 1$, then

$$\frac{\sqrt{(u-1)^2}}{u^2 - u} = \frac{u-1}{u(u-1)} = \frac{1}{u},$$

and

$$\lim_{u \to 1^+} \frac{\sqrt{(u-1)^2}}{u^2 - u} = \lim_{u \to 1^+} \frac{1}{u} = 1.$$

Similarly, if $u < 1$, then

$$\frac{\sqrt{(u-1)^2}}{u^2-u} = \frac{1-u}{u(u-1)} = -\frac{1}{u},$$

and

$$\lim_{u\to 1^-} \frac{\sqrt{(u-1)^2}}{u^2-u} = \lim_{u\to 1^-} -\frac{1}{u} = -1.$$

The nice thing about one-sided limits is that almost all the techniques we have for dealing with limits work just the same way for one-sided limits. Theorems 2.2.3 and 2.2.6 about limits of sums, products, and quotients can be stated for one-sided limits without any changes. The one-sided limits of polynomials and rational functions are both equal to the two-sided limits (cf. Theorems 2.2.5 and 2.2.6 and the discussion after Example 2.2.5).

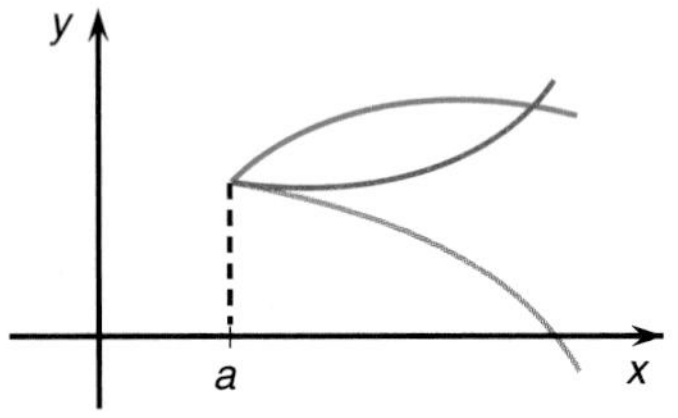

Figure 2.4.5

Proposition 2.2.8, which says that two functions f and g have the same limit as $x \to a$ if $f(x) = g(x)$ for all $x \neq a$ in some interval containing a, has a one-sided analog. It says that the one-sided limits are equal if f and g are equal on some interval on the appropriate side of a.

The Squeeze Theorem (Theorem 2.3.5) works for the one-sided limit of a function that lies between two other functions in an interval on one side of a. See Figure 2.4.5.

Theorem 2.3.3, concerning the nonnegativity of the limit of a nonnegative function, carries over to one-sided limits.

The one-sided limits of a power of x with noninteger exponent equal the two-sided limit (Theorem 2.3.1). We can also fill in the gap we left about limits of these powers at $x = 0$. Indeed,

$$\lim_{x\to 0^+} x^r = 0, \qquad \text{whenever } r > 0,$$

$$\lim_{x\to 0^-} x^r = 0, \qquad \text{whenever } r > 0 \text{ and}$$

$r = \frac{a}{b}$ is a rational number with b an odd number (i.e., whenever it makes sense to consider negative x).

The one slight complication arises with Theorem 2.3.2, the result about the limit of a composite function $g\big(f(x)\big)$. To get a one-sided version of it, we need to know that the one-sided limit of $f(x)$ is L and that the *two-sided* limit of $g(t)$ as $t \to L$ equals $g(L)$. It is not enough to know that $g(t)$ has a one-sided limit.

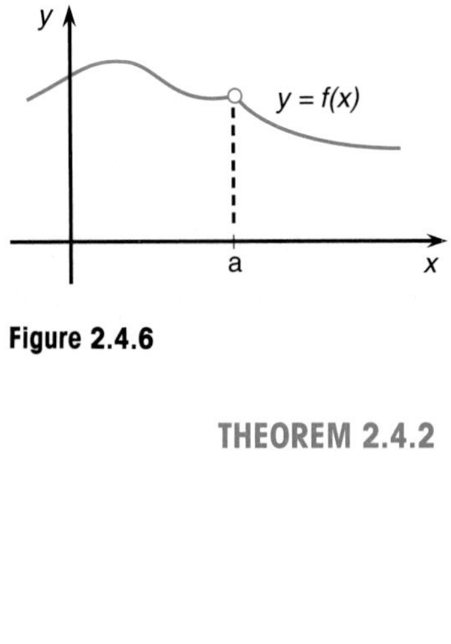

Figure 2.4.6

But apart from this minor subtlety, we can work with one-sided limits just as we do with two-sided limits. The main usefulness of one-sided limits comes from their relation to two-sided limits. If $\lim_{x\to a} f(x) = L$, then both one-sided limits exist, and they must also equal L. (If $f(x)$ approaches L as $x \to a$, then it must certainly approach L as $x \to a^+$ or $x \to a^-$.) The converse is also true: If both one-sided limits exist and have the same value, then the two-sided limit exists and equals the one-sided limits. See Figure 2.4.6. We record these facts for later use.

THEOREM 2.4.2

If

$$\lim_{x\to a} f(x) = L,$$

then $\lim_{x\to a^+} f(x)$ and $\lim_{x\to a^-} f(x)$ both exist, and

$$\lim_{x\to a^+} f(x) = \lim_{x\to a^-} f(x) = L.$$

Conversely, if the one-sided limits exist and are equal, that is,

$$\lim_{x \to a^+} f(x) = \lim_{x \to a^-} f(x) = L,$$

then the two-sided limit exists and equals the common value:

$$\lim_{x \to a} f(x) = L.$$

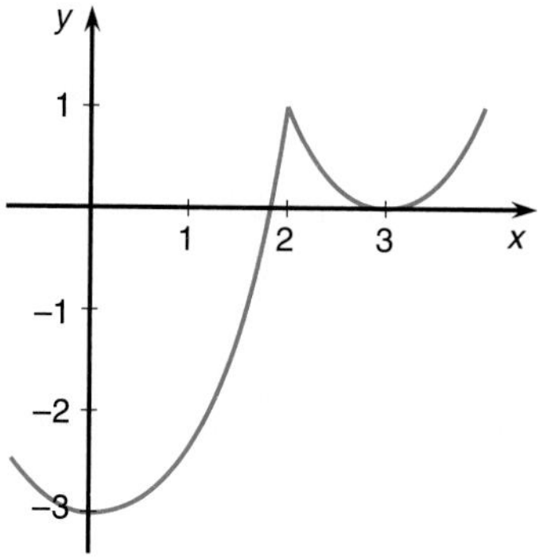

Figure 2.4.7

EXAMPLE 2

(i) Let $f(x)$ be defined as follows:

$$f(x) = \begin{cases} x^2 - 3, & \text{if } x \le 2, \\ x^2 - 6x + 9, & \text{if } x > 2. \end{cases}$$

We consider the one-sided limits of $f(x)$ as x approaches 2 from either side. The function is graphed in Figure 2.4.7.

First, when $x < 2$, $f(x)$ equals the polynomial $x^2 - 3$. By the one-sided version of Proposition 2.2.8, we know that $f(x)$ and $x^2 - 3$ will have the same limit from the left. Theorem 2.2.5(ii) tells us how to find the two-sided limit of the polynomial $x^2 - 3$, which will then also be the one-sided limit, by Theorem 2.4.2. All in all,

$$\lim_{x \to 2^-} f(x) = \lim_{x \to 2^-} (x^2 - 3) = \lim_{x \to 2} (x^2 - 3) = 1.$$

Similarly,

$$\lim_{x \to 2^+} f(x) = \lim_{x \to 2^+} (x^2 - 6x + 9) = \lim_{x \to 2} (x^2 - 6x + 9) = 1.$$

Using the above theorem again, we see that since the one-sided limits exist and are equal, the two-sided limit exists and equals them, or

$$\lim_{x \to 2} f(x) = 1.$$

(ii) If

$$g(x) = \begin{cases} -1, & \text{if } x < -1, \\ x^2, & \text{if } -1 < x < 1, \\ 2x - 1, & \text{if } x > 1, \end{cases}$$

then, reasoning as above we find that

$$\lim_{x \to -1^-} g(x) = \lim_{x \to -1^-} (-1) = -1, \qquad \lim_{x \to -1^+} g(x) = \lim_{x \to -1^+} (x^2) = 1,$$
$$\lim_{x \to 1^-} g(x) = \lim_{x \to 1^-} (x^2) = 1, \qquad \lim_{x \to 1^+} g(x) = \lim_{x \to 1^+} (2x - 1) = 1.$$

Since the one-sided limits at -1 are different, the two-sided limit there does not exist, but at $x = 1$, where the one-sided limits are both the same, the two-sided limit also exists, and

$$\lim_{x \to 1} g(x) = 1.$$

See Figure 2.4.8.

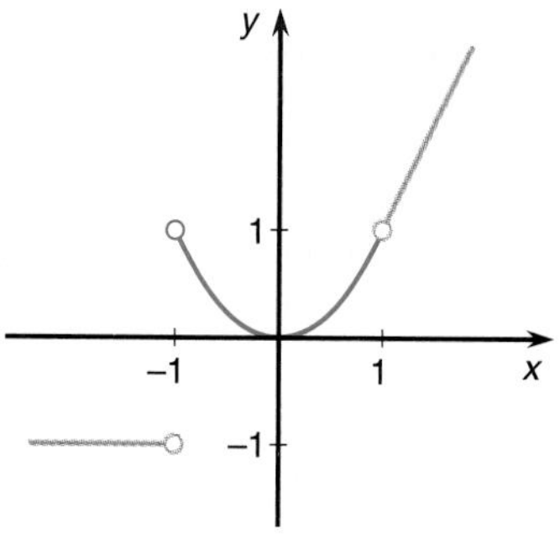

Figure 2.4.8

There are different ways to get a graphics calculator to plot functions that are defined by different formulas on different intervals. Here is one:

Consider the function

$$f(x) = \begin{cases} x^3 + 1, & \text{if } x < 0, \\ x^2, & \text{if } x > 0. \end{cases}$$

We begin by noting that $\frac{|x|}{x}$ equals 1 for $x > 0$ and -1 for $x < 0$. So $1 + \frac{|x|}{x}$ equals 2 and 0, respectively, and $\left(1 + \frac{|x|}{x}\right)/2$ is 1 when $x > 0$ and 0 when $x < 0$. Similarly, $\left(1 - \frac{|x|}{x}\right)/2$ is 0 when $x > 0$ and 1 when $x < 0$.

The function $f(x)$ given above can be expressed as

$$f(x) = (x^3 + 1)\frac{\left(1 - \frac{|x|}{x}\right)}{2} + x^2\frac{\left(1 + \frac{|x|}{x}\right)}{2}.$$

We plot the graph.

TI-85	SHARP EL9300
GRAPH F1(y(x) =) (F1(x)^3 + 1) ×(1− MATH-F1-F5(abs) EXIT (F1(x))÷ F1(x)) ÷ 2+ F1(x)x^2 ×(1+ MATH-F1-F5(abs) EXIT (F1(x))÷ F1(x)) ÷ 2 M2(RANGE) (−)1.5 ▼ 1.5 ▼ ▼ (−)3 ▼ 2 F5(GRAPH)	(x/θ/T a^b3 ▶ +1) × (1− MATH 1 x/θ/T ▶ ÷ x/θ/T) ÷ 2 + x/θ/T x^2 × (1+ MATH 1 x/θ/T ▶ ÷ x/θ/T) ÷ 2 RANGE −1.5 ENTER 1.5 ENTER 1 ENTER −3 ENTER 2 ENTER 1 ENTER
CASIO f_x-7700GB/f_x-6300G	**HP 48SX**
Cls EXE Range −1.5 EXE 1.5 EXE 1 EXE −3 EXE 2 EXE 1 EXE EXE EXE EXE Graph (X x^y 3 + 1) × (1− *Abs* (X) ÷ X) ÷ 2 + X x^2 × (1 + *Abs* (X) ÷ X) ÷ 2 EXE	↰ PLOT PLOTR ERASE ATTN′ αY = () αX y^x 3 + 1 ▶ × () 1− MTH PARTS ABS αX ▶ ÷ αX ▶ ÷ 2 + αX y^x 2 × () 1+ MTH PARTS ABS αX ▶ ÷ αX ▶ ÷ 2 ↱ PLOT ↰ DRAW ERASE 1.5 +/− ENTER 1.5 XRNG 3 +/− ENTER 2 YRNG DRAW

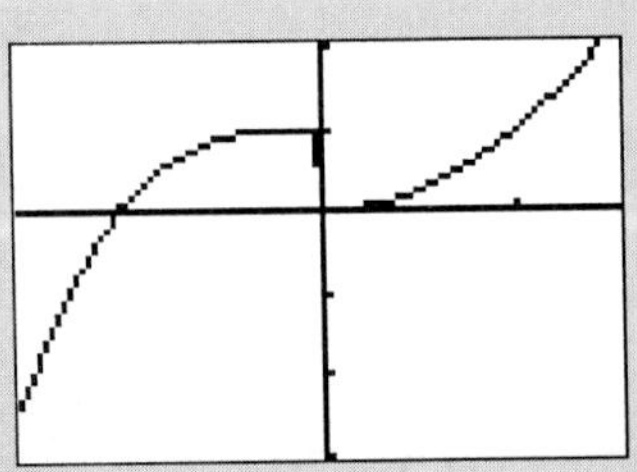

How would you adapt this approach to plot the function

$$f(x) = \begin{cases} \cos x, & \text{if } x < 1, \\ x^2 - 2, & \text{if } x > 1? \end{cases}$$

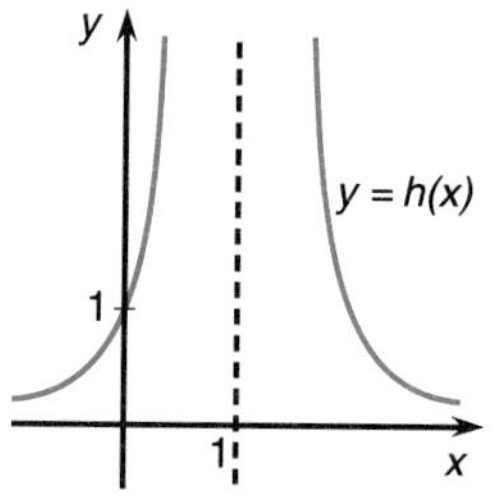

Figure 2.4.9

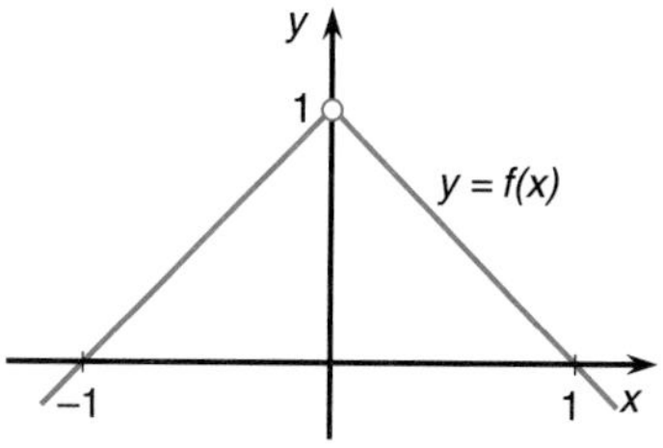

Figure 2.4.10

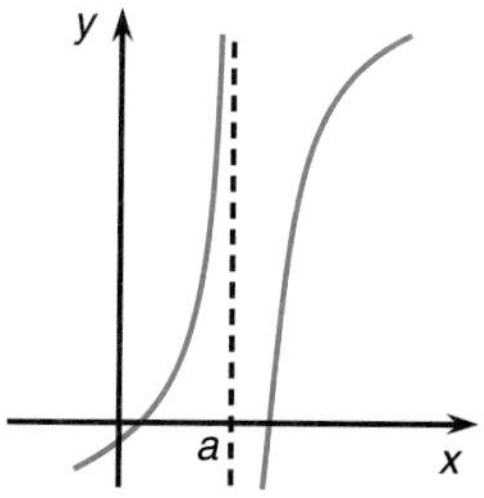

Figure 2.4.11

Infinite Limits

We have considered functions like $h(x) = \frac{1}{(x-1)^2}$, which does not have a limit as x approaches 1. This is because the denominator tends to 0, so the quotient becomes larger and larger, eventually getting bigger than any possible limit L. See Figure 2.4.9.

In a situation like this we say that $h(x)$ becomes infinitely large, or "tends to infinity," and we write $h(x) \to \infty$.

There are several things to be careful about. One is that it is not enough to say merely that a function keeps getting larger. For instance, the function $f(x) = 1 - |x|$, illustrated in Figure 2.4.10, keeps getting larger as x approaches 0 but certainly does not tend to infinity. To tend to infinity, a function must get larger and stay larger than any fixed number L, provided that x is sufficiently close to a (cf. the discussion of ordinary limits at the end of Section 2.3).

Another difficulty is that we must be careful about the sign of the function. If it is getting large in the *negative* direction, then it will tend to *negative infinity*. It frequently happens that a function will tend to infinity as x approaches a from one side and tend to negative infinity as x approaches a from the other side, and then we must talk about one-sided infinite limits. See Figure 2.4.11.

Finally, while we talk about infinite limits and will shortly begin using the usual sort of limit symbols for them, it is very important to bear in mind that infinite limits are *not* limits in the ordinary sense. When we described limits right at the beginning of Section 2.2, we said that there had to be a *number* L so that $f(x)$ approached L. Infinity and negative infinity are *not* numbers. So when we say that $f(x)$ tends to infinity, this means in particular that it does *not* have a limit in the ordinary sense.

This apparent confusion is not as bad as it first appears, and is justified by the usefulness of being able to talk about infinite limits. We just have to bear in mind that infinite limits, though perfectly legitimate, are different from ordinary limits.

DEFINITION 2.4.3

Infinite Limits

If $f(x)$ tends to infinity as $x \to a$, we say the "limit of $f(x)$ as x tends to a is **infinity**," and we write

$$\lim_{x \to a} f(x) = \infty.$$

Similarly we can talk about the limit being **negative infinity**, and we write

$$\lim_{x \to a} f(x) = -\infty.$$

One-sided limits can also be infinite, and the obvious meaning will be conveyed by

$$\lim_{x \to a^+} f(x) = \infty, \quad \lim_{x \to a^-} f(x) = -\infty, \text{ etc.}$$

In each of these situations it is also appropriate to say that $f(x)$ *does not* have a limit as $x \to a$ (or $x \to a^+$, etc.).

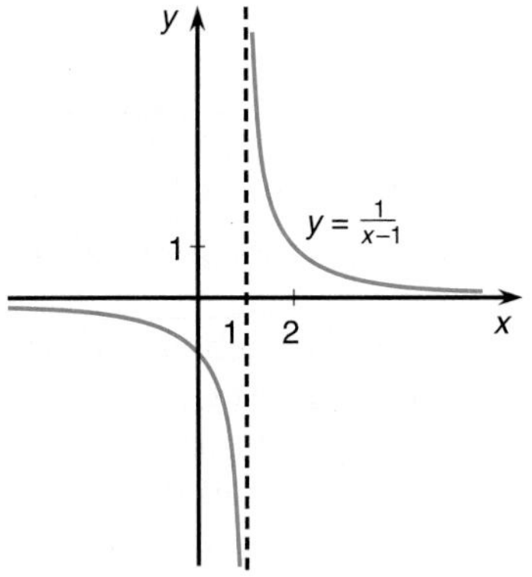

Figure 2.4.12

EXAMPLE 3

(i) Consider the function $f(x) = \frac{1}{x-1}$, shown in Figure 2.4.12. As $x \to 1$, we know that $(x-1) \to 0$. The complication is that if $x > 1$, then $x - 1$ is positive, so $\frac{1}{x-1}$ is positive and gets very large, but when $x < 1$, the function is negative and gets very large in the negative direction. In other words,

$$\lim_{x\to 1^+} \frac{1}{x-1} = \infty, \qquad \lim_{x\to 1^-} \frac{1}{x-1} = -\infty.$$

None of $\lim_{x\to 1} f(x)$, $\lim_{x\to 1^+} f(x)$, or $\lim_{x\to 1^-} f(x)$ exists (i.e., the *ordinary* limits do not exist, even though some of the corresponding infinite ones do).

(ii) If $g(x) = \frac{x-1}{(x-2)^3(x-3)^2}$, then we know that $g(x)$ has (ordinary) limits as x approaches any number except 2 or 3. What happens at these points?

Suppose x is very close to 2. We may as well assume that it is bigger than 1 and less than 3, so $x - 1 > 0$, $x - 3 < 0$, and $(x-3)^2 > 0$. Now $(x-2)^3$ will be positive when $x > 2$ and negative when $x < 2$, and we can use all this to find the sign of $g(x)$. Indeed,

$$g(x) = \frac{x-1}{(x-2)^3(x-3)^2} = \frac{1}{(x-2)^3}\,\frac{x-1}{(x-3)^2}.$$

We have just seen that for x close to 2, the last quotient is always positive, and the previous one depends on the sign of $x - 2$. So for x near 2, $g(x)$ is positive when $x > 2$ and negative when $x < 2$.

Since the denominator of $g(x)$ approaches zero as x approaches 2 and the numerator approaches something other than zero (namely, 1), $g(x)$ will tend to infinity or negative infinity depending from which side x approaches 2:

$$\lim_{x\to 2^+} g(x) = \infty, \qquad \lim_{x\to 2^-} g(x) = -\infty.$$

Similarly, if x is near 3 (specifically if $x > 2$), then $\frac{x-1}{(x-2)^3} > 0$, and since $(x-3)^2 \geq 0$ for any x but tends to 0 as $x \to 3$, we write $g(x) = \frac{x-1}{(x-2)^3}\frac{1}{(x-3)^2}$ and see that

$$\lim_{x\to 3^+} g(x) = \infty, \qquad \lim_{x\to 3^-} g(x) = \infty.$$

Since the function tends to infinity as x approaches 3 from either side,

$$\lim_{x\to 3} g(x) = \infty.$$

In the discussion after Example 2.2.5, we saw that a rational function $\frac{p(x)}{q(x)}$ has a limit as $x \to a$ if $q(a) \neq 0$. If $q(a) = 0$, the situation is more complicated. If both $q(a)$ and $p(a)$ are zero, then $q(x)$ and $p(x)$ can both be divided by $x - a$, and the common factor can be canceled.

Repeating this process as many times as necessary, we may rewrite the function so that at most one of $q(a)$ and $p(a)$ is zero. If $q(a) \neq 0$, then the function has a limit, which we already know how to find.

The remaining case occurs if $q(a) = 0$ and $p(a) \neq 0$. Since $q(a) = 0$, we know by the Factor Theorem (Theorem 1.3.5) that $q(x)$ is divisible by $x - a$. We divide and write $q(x) = (x-a)R(x)$ for some polynomial $R(x)$. If $R(a) = 0$, we divide again, and so on until we can write $q(x) = (x-a)^n Q(x)$, with some positive integer n and some polynomial $Q(x)$ that satisfies $Q(a) \neq 0$.

We have divided out the largest possible power of $(x-a)$ from $q(x)$.

Then

$$\frac{p(x)}{q(x)} = \frac{1}{(x-a)^n}\frac{p(x)}{Q(x)}.$$

The last quotient, $\frac{p(x)}{Q(x)}$, is one in which neither numerator nor denominator is zero at $x = a$, so it has a limit, namely, $\frac{p(a)}{Q(a)}$, which is not zero.

The behavior of $\frac{1}{(x-a)^n}$ as x approaches a depends on whether n is even or odd. If n is even, then $(x-a)^n$ is a square and is always nonnegative, so $\frac{1}{(x-a)^n} \to \infty$. If n is odd, then $(x-a)^n$ will have the same sign as $x - a$, so

$$\lim_{x\to a^-} \frac{1}{(x-a)^n} = -\infty, \qquad \lim_{x\to a^+} \frac{1}{(x-a)^n} = \infty.$$

Finally, the original function $\frac{p(x)}{q(x)}$ is the product of $\frac{1}{(x-a)^n}$ with something that tends to the nonzero number $\frac{p(a)}{Q(a)}$. If this number is positive, then $\frac{p(x)}{q(x)}$ will tend to ∞ or $-\infty$ exactly as $\frac{1}{(x-a)^n}$ does. If the number $\frac{p(a)}{Q(a)}$ is negative, then $\frac{p(x)}{q(x)}$ will tend to $-\infty$ when $\frac{1}{(x-a)^n} \to \infty$, and to $+\infty$ when $\frac{1}{(x-a)^n} \to -\infty$.

Summary 2.4.4

Limits of Rational Functions

1. If $q(a) \neq 0$, then $\lim_{x\to a} \frac{p(x)}{q(x)} = \frac{p(a)}{q(a)}$.
2. If $p(a) = q(a) = 0$, then divide $p(x)$ and $q(x)$ by $(x - a)$. Repeat until $p(a) \neq 0$ and/or $q(a) \neq 0$.
3. If $\frac{p(x)}{q(x)}$ is a rational function and $q(a) = 0, p(a) \neq 0$, write

$$q(x) = (x-a)^n Q(x), \qquad \text{with } Q(a) \neq 0.$$

Suppose $\frac{p(a)}{Q(a)} > 0$.

(i) If n is even, then $\lim_{x\to a} \frac{p(x)}{q(x)} = \infty$.

(ii) If n is odd, then $\lim_{x\to a^-} \frac{p(x)}{q(x)} = -\infty, \lim_{x\to a^+} \frac{p(x)}{q(x)} = +\infty$.

If $\frac{p(a)}{Q(a)} < 0$, then the signs are all reversed, that is, $-\infty$ is changed to ∞ and vice versa.

These rules may seem complicated, but they are really quite easy when you bear in mind the behavior as $x \to a$ of $\frac{1}{(x-a)^n}$. Do not memorize rules like this; it is much better to figure out how they work so that memorizing them becomes unnecessary.

EXAMPLE 4 Discuss the limiting behavior of $\frac{x^3-4x^2+5x+1}{x^3-x^2-x+1}$ as x approaches 1 and -1.

Solution Note that the numerator is not zero when $x = 1$ or $x = -1$ and that the denominator is zero at both these points.

Dividing the denominator by $(x+1)$, we get $x^3 - x^2 - x + 1 = (x+1)(x^2 - 2x + 1) = (x+1)(x-1)^2$. To discuss the limiting behavior at $x = -1$, we write

$$\frac{x^3 - 4x^2 + 5x + 1}{x^3 - x^2 - x + 1} = \frac{1}{x+1} \frac{x^3 - 4x^2 + 5x + 1}{(x-1)^2}.$$

Evaluating the last quotient at $x = -1$, we find that it equals $\frac{-1-4-5+1}{4} = -\frac{9}{4}$. In particular, it is negative. The exponent n on $(x+1)$ is $n = 1$. As above, we conclude that

$$\lim_{x \to -1^-} \frac{p(x)}{q(x)} = \infty, \qquad \lim_{x \to -1^+} \frac{p(x)}{q(x)} = -\infty.$$

To discuss the behavior at $x = 1$, we write

$$\frac{x^3 - 4x^2 + 5x + 1}{x^3 - x^2 - x + 1} = \frac{1}{(x-1)^2} \frac{x^3 - 4x^2 + 5x + 1}{x+1}.$$

Evaluating the last quotient at $x = 1$, we find that it equals $\frac{1-4+5+1}{1+1} = \frac{3}{2} > 0$. In this case the exponent n is 2, so

$$\lim_{x \to 1} \frac{p(x)}{q(x)} = \infty.$$

Limits at Infinity

We have considered what happens to $f(x)$ as x approaches some value a, and have found that $f(x)$ sometimes approaches infinity. Another possibility is to consider what happens to $f(x)$ as x tends to infinity.

EXAMPLE 5 Discuss the behavior as x tends to infinity of (i) $f(x) = \frac{1}{1+x^2}$; (ii) $g(x) = \frac{2x^3 - 2x + 3}{x^3 + 1}$.

Solution

(i) As $x \to \infty$, $1 + x^2$ gets very large, so $\frac{1}{1+x^2}$ gets very small and approaches the limit 0. We write

$$\lim_{x \to \infty} \frac{1}{1+x^2} = 0.$$

The same argument shows that

$$\lim_{x \to -\infty} \frac{1}{1+x^2} = 0.$$

(ii) Notice that as $x \to \infty$, since x gets very large, $\frac{1}{x} \to 0$, so $\frac{1}{x^2} \to 0$, $\frac{1}{x^3} \to 0$, etc. Dividing both numerator and denominator of $g(x)$ by x^3, we get

$$g(x) = \frac{2x^3 - 2x + 3}{x^3 + 1} = \frac{2 - \frac{2}{x^2} + \frac{3}{x^3}}{1 + \frac{1}{x^3}}.$$

By the remarks above, the numerator tends to $2 - 0 + 0 = 2$ and the denominator tends to $1 + 0 = 1$ as $x \to \infty$, so the quotient tends to $\frac{2}{1} = 2$. In other words,

$$\lim_{x\to\infty} g(x) = \lim_{x\to\infty} \frac{2x^3 - 2x + 3}{x^3 + 1} = 2.$$

This example illustrates the technique to use when finding the limit at infinity (or at negative infinity) of any rational function. Factor out the highest power of x that occurs in the denominator. In the simplest case, when this power is the same for both numerator and denominator, we can cancel them out. In general, cancel as much as possible and then evaluate the limit, remembering that $\frac{1}{x}$ or any positive power of $\frac{1}{x}$ must tend to 0 as $x \to \infty$ or $x \to -\infty$.

EXAMPLE 6 Find the following (possibly infinite) limits:
(i) $\lim_{x\to\infty} \frac{x^4-3x^3+2x-1}{2x^4+2x^2+2}$; (ii) $\lim_{x\to-\infty} \frac{3x^3-x-2}{2x^3+4}$; (iii) $\lim_{x\to-\infty} \frac{2x^3+x^2-1}{x^4+1}$;
(iv) $\lim_{x\to\infty} \frac{x^3-x+2}{x^2-1}$; (v) $\lim_{x\to-\infty} (3x^3 + 2x^2 - x + 1)$.

Solution

(i) The highest power of x that occurs in the denominator is x^4, so we divide by it and write

$$\lim_{x\to\infty} \frac{x^4 - 3x^3 + 2x - 1}{2x^4 + 2x^2 + 2} = \lim_{x\to\infty} \frac{1 - \frac{3}{x} + \frac{2}{x^3} - \frac{1}{x^4}}{2 + \frac{2}{x^2} + \frac{2}{x^4}} = \frac{1 - 0 + 0 - 0}{2 + 0 + 0} = \frac{1}{2}.$$

(ii) The highest power is x^3, so

$$\lim_{x\to-\infty} \frac{3x^3 - x - 2}{2x^3 + 4} = \lim_{x\to-\infty} \frac{3 - \frac{1}{x^2} - \frac{2}{x^3}}{2 + \frac{4}{x^3}} = \frac{3}{2}.$$

(iii) The highest power is x^4, and

$$\lim_{x\to-\infty} \frac{2x^3 + x^2 - 1}{x^4 + 1} = \lim_{x\to-\infty} \frac{\frac{2}{x} + \frac{1}{x^2} - \frac{1}{x^4}}{1 + \frac{1}{x^4}} = \frac{0 + 0 - 0}{1 + 0} = 0.$$

(iv) Dividing by x^2, we find

$$\lim_{x\to\infty} \frac{x^3 - x + 2}{x^2 - 1} = \lim_{x\to\infty} \frac{x - \frac{1}{x} + \frac{2}{x^2}}{1 - \frac{1}{x^2}}.$$

The x in the numerator tends to ∞ and $-\frac{1}{x} + \frac{2}{x^2} \to 0$, so the numerator tends to ∞. The denominator tends to 1, so the rational function is something that tends to ∞ over something that tends to 1. Accordingly,

$$\lim_{x\to\infty} \frac{x^3 - x + 2}{x^2 - 1} = \infty.$$

Similar considerations allow us to see that

$$\lim_{x\to-\infty} \frac{x^3 - x + 2}{x^2 - 1} = -\infty.$$

(v) In this case the function $3x^3 + 2x^2 - x + 1$ is a polynomial; there is no denominator (or, if you prefer, the denominator is the constant 1). The highest power of x occurring in the numerator is x^3, so we can write $3x^3 + 2x^2 - x + 1 = x^3\left(3 + \frac{2}{x} - \frac{1}{x^2} + \frac{1}{x^3}\right)$. As $x \to -\infty$, the quantity in parentheses tends to $(3 + 0 - 0 + 0) = 3$; the factor x^3 tends to $-\infty$, so the product tends to $-\infty$.

Summary 2.4.5

Limits of Rational Functions at $\pm\infty$

To find the limits at $\pm\infty$ of the rational function

$$\frac{p(x)}{q(x)} = \frac{a_m x^m + a_{m-1}x^{m-1} + \ldots + a_1 x + a_0}{b_n x^n + b_{n-1}x^{n-1} + \ldots + b_1 x + b_0},$$

we factor out x^n, the highest power of x occurring in the denominator. We can write

$$\frac{p(x)}{q(x)} = \frac{x^m\left(a_m + \frac{a_{m-1}}{x} + \ldots + \frac{a_1}{x^{m-1}} + \frac{a_0}{x^m}\right)}{x^n\left(b_n + \frac{b_{n-1}}{x} + \ldots + \frac{b_1}{x^{n-1}} + \frac{b_0}{x^n}\right)}$$

$$= \frac{x^{m-n}\left(a_m + \frac{a_{m-1}}{x} + \ldots + \frac{a_1}{x^{m-1}} + \frac{a_0}{x^m}\right)}{b_n + \frac{b_{n-1}}{x} + \ldots + \frac{b_1}{x^{n-1}} + \frac{b_0}{x^n}}.$$

First, if $m = n$, then $x^{m-n} = x^0 = 1$ and

$$\lim_{x\to\infty} \frac{p(x)}{q(x)} = \lim_{x\to-\infty} \frac{p(x)}{q(x)} = \frac{a_m}{b_m}.$$

Second, if $m < n$, then $m - n < 0$ and $x^{m-n} \to 0$. Since the rest of the formula tends to $\frac{a_m}{b_n}$ and the original rational function is the product of these two factors,

$$\lim_{x\to\infty} \frac{p(x)}{q(x)} = \lim_{x\to-\infty} \frac{p(x)}{q(x)} = 0.$$

Third, if $m > n$, x^{m-n} will get large and the other factors will still tend to $\frac{a_m}{b_n}$. Specifically, if $m - n$ is even, then $x^{m-n} \to \infty$. If $m - n$ is odd, then $x^{m-n} \to \infty$ as $x \to \infty$, and $x^{m-n} \to -\infty$ as $x \to -\infty$. So if $\frac{a_m}{b_n} > 0$, then

$$\lim_{x\to\infty} \frac{p(x)}{q(x)} = \infty, \qquad \lim_{x\to-\infty} \frac{p(x)}{q(x)} = \begin{cases} \infty, & \text{if } m-n \text{ is even,} \\ -\infty, & \text{if } m-n \text{ is odd.} \end{cases}$$

If $\frac{a_m}{b_n} < 0$, then

$$\lim_{x\to\infty} \frac{p(x)}{q(x)} = -\infty, \qquad \lim_{x\to-\infty} \frac{p(x)}{q(x)} = \begin{cases} -\infty, & \text{if } m-n \text{ is even,} \\ \infty, & \text{if } m-n \text{ is odd.} \end{cases}$$

Once again, these rules are too complicated to memorize. After doing a few exercises you will get the idea and be able to figure out what you need without having to memorize the details.

Limits at infinity behave pretty much like one-sided limits (after all, x approaches ∞ only from one side, namely, from below, and approaches $-\infty$ only from above). The results we have recognized for one-sided limits (see the discussion after Example 1) carry over to limits at infinity without much change.

We have seen how to evaluate limits of rational functions, and in a sense this includes polynomials (rational functions whose denominator happens to be 1). Of course the limit of any nonconstant polynomial as $x \to \infty$ or $x \to -\infty$ is infinite (i.e., either ∞ or $-\infty$).

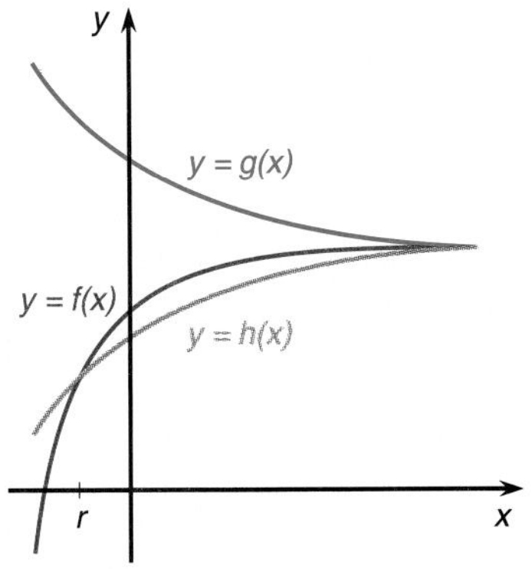

Figure 2.4.13

In keeping with the remark about the one-sided version of Theorem 2.3.2, about composite functions, there is a version in which $x \to a$ is replaced by $x \to \infty$ or $x \to -\infty$, that is, $L = \infty$ or $-\infty$. However, M must be an ordinary finite limit.

Also, to make sense of versions at infinity of Proposition 2.2.8, about two functions being equal on an interval having the same limit, and the Squeeze Theorem, (Theorem 2.3.5), we have to find a way of expressing the idea of an interval on one side of ∞ (or $-\infty$). In fact, this is easy to do: An interval to the left of ∞ will be an interval of the form (r, ∞), meaning $\{x : r < x\}$. An interval to the right of $-\infty$ will be an interval of the form $(-\infty, s)$, meaning $\{x : x < s\}$. Replacing the intervals in Theorems 2.2.8 and 2.3.5 with this kind of "intervals," we obtain versions of them for limits at infinity. See Figure 2.4.13.

The rules for infinite limits, that is, limits in which the value of the limit is ∞ or $-\infty$, are less straightforward. Since they are *not* limits in the ordinary sense, we should not expect them to obey the ordinary rules. In particular, we cannot perform ordinary arithmetic with infinity, so even Theorems 2.2.3 and 2.2.6 about sums, products, and quotients of limits don't necessarily make sense when infinite limits are involved.

EXAMPLE 7 Let

$$f(x) = \frac{2}{(x-1)^2}, \qquad g(x) = \frac{-1}{(x-1)^2},$$
$$h(x) = \frac{-3}{(x-1)^2}, \qquad k(x) = 2x - \frac{2}{(x-1)^2}.$$

We see immediately that

$$\lim_{x\to 1} f(x) = \infty, \qquad \lim_{x\to 1} g(x) = -\infty,$$
$$\lim_{x\to 1} h(x) = -\infty, \qquad \lim_{x\to 1} k(x) = -\infty.$$

Also

$$\lim_{x\to 1} \big(f(x) + g(x)\big) = \lim_{x\to 1} \frac{1}{(x-1)^2} = \infty,$$
$$\lim_{x\to 1} \big(f(x) + h(x)\big) = \lim_{x\to 1} \frac{-1}{(x-1)^2} = -\infty,$$
$$\lim_{x\to 1} \big(f(x) + k(x)\big) = \lim_{x\to 1} 2x = 2.$$

This illustrates that adding functions with infinite limits gives unpredictable results: Adding two functions whose limits are ∞ and $-\infty$ can result in a function whose limit is ∞, $-\infty$ or even a finite limit.

There are some things we can say with assurance, however.

THEOREM 2.4.6

Suppose $f(x)$ and $g(x)$ are two functions. If $\lim_{x\to a} f(x) = \infty$ and $\lim_{x\to a} g(x)$ is either finite or ∞ (but not $-\infty$), then

$$\lim_{x\to a} \big(f(x) + g(x)\big) = \infty.$$

A similar result holds if one limit is $-\infty$ and the other is finite or $-\infty$; the limit of the sum is then $-\infty$.

If one limit is ∞ and the other is $-\infty$, then the limit of the sum is unpredictable and depends on the specific functions.

If $\lim_{x\to a} f(x) = \infty$ or $-\infty$, and $\lim_{x\to a} g(x) \neq 0$, (i.e., the limit of $g(x)$ can be finite or infinite but not zero), then the limit of the product $f(x)g(x)$ is either ∞ or $-\infty$. To decide which, we merely consider the signs of the two limits, as we would for ordinary products:

$$\lim_{x\to a} f(x)g(x) = \begin{cases} \infty, & \text{if the signs of the limits are the same,} \\ -\infty, & \text{if the signs of the limits are different.} \end{cases}$$

(Here we understand that ∞ has a positive sign and $-\infty$ has a negative sign.)

If one of the limits is zero and the other is ∞ or $-\infty$, then the limit of the product is unpredictable: It could be ∞, $-\infty$, 0, or any nonzero finite limit. It also might not exist at all.

All the above remarks also apply to one-sided limits and to limits at infinity (i.e., when $a = \infty$ or $a = -\infty$).

Exercises 2.4

I

Evaluate each of the following finite or infinite limits.

1. $\lim_{x\to 2^+} \frac{x^3+3}{x-2}$
2. $\lim_{t\to 2^-} \frac{t^3+3}{t-2}$
3. $\lim_{x\to 0^+} \frac{x}{x}$
4. $\lim_{x\to 0^-} \frac{x}{x}$
5. $\lim_{u\to\infty} \frac{u^3-u^2+3}{2u^3-u+1}$
6. $\lim_{x\to-\infty} \frac{x^2-1}{x^3+1}$
7. $\lim_{x\to 0^+} \frac{x^3-3x^2+x}{x^3-2x^2}$
8. $\lim_{x\to 2^+} \frac{x^2-2x+3}{x^2-4}$
9. $\lim_{x\to\infty} (x^3 - 3x^2 + x - 1)$
10. $\lim_{x\to-\infty} \frac{x^3-2x+1}{2x^2+2x+3}$

Decide whether each of the following limits exists, and evaluate those that do, explaining your reasons. For those limits that do not exist, distinguish those that tend to ∞ or $-\infty$ and those that do neither.

11. $\lim_{r\to 3} \frac{r+6}{r^2-9}$
12. $\lim_{u\to\infty} \sqrt{u^2+1}$
13. $\lim_{x\to 0^+} \frac{x^2+1}{x}$
14. $\lim_{x\to 0^-} \frac{x^2+1}{x}$
15. $\lim_{x\to 1^+} \frac{x^2+3x+1}{x^2-1}$
16. $\lim_{x\to 1^-} \frac{x^3+1}{x^2-1}$
17. $\lim_{t\to 1} \frac{1}{\sqrt{(t-1)^2}}$
18. $\lim_{y\to 0^+} y^{-1/3}$
19. $\lim_{x\to 0^+} \frac{1}{x^2} \sin(1/x)$
20. $\lim_{x\to 0^-} \frac{1}{x} \sin^2(1/x)$

II

*21. Find $\lim_{x\to 0^+} \sqrt{x}$.

*22. Some of the functions in Exercises 31–40 of Exercises 2.2 do not have limits at the specified points. Find the one-sided and/or two-sided infinite limits at these points if they exist.

23. Find $\lim_{x\to\infty} (2x^4 - 3x^3 + 7x - 1)$.

24. Find $\lim_{x\to-\infty} (4 - 5x + 2x^2 - 5x^3)$.

*25. Assume that $f(x)$ has one-sided limits at every value of x.
   (i) Evaluate $\lim_{x\to a^+} g(x)$, $\lim_{x\to a^-} g(x)$, if $g(x) = f(x+c)$.
   (ii) Evaluate $\lim_{x\to a^+} h(x)$, $\lim_{x\to a^-} h(x)$, if $h(x) = f(c-x)$.

*26. The functions in Exercises 41–50 in Exercises 2.2 do not have limits at certain points. Find the one-sided and/or two-sided infinite limits at these points.

27. Let $f(x)$ be the function defined in Example 2.2.8(ii). Find (i) $\lim_{x\to -1^-} f(x)$, (ii) $\lim_{x\to -1^+} f(x)$, (iii) $\lim_{x\to 1^-} f(x)$, (iv) $\lim_{x\to 1^+} f(x)$.

*28. Assume that $p(x)$ and $q(x)$ are polynomials, and define

$$f(x) = \begin{cases} p(x), & \text{if } x < a, \\ q(x), & \text{if } x > a, \\ A, & \text{if } x = a. \end{cases}$$

Evaluate (i) $\lim_{x\to a^+} f(x)$, (ii) $\lim_{x\to a^-} f(x)$, (iii) $\lim_{x\to b} f(x)$ for any $b \neq a$.

*29. Find examples of functions with the following behavior at some point a.

(i) $\lim_{x\to a^+} f(x) = +\infty, \lim_{x\to a^-} f(x) = -\infty$.

(ii) $\lim_{x\to a} |f(x)| = L$, but $f(x)$ has no limit as $x \to a$.

(iii) $\lim_{x\to a} |f(x)| = \infty$, but $f(x)$ has no limit, including ∞ or $-\infty$, as $x \to a$.

*30. Find examples of functions $f(x)$ and $g(x)$ and a number a so that

$$\lim_{x\to a} f(x) = 0, \qquad \lim_{x\to a} g(x) = \infty,$$

but for which $\lim_{x\to a} f(x)g(x)$ equals (i) 0, (ii) ∞, (iii) $-\infty$, (iv) 3, (v) -2.

III

For each of the following functions, describe the one-sided limits from both sides at the specified point(s) $x = a$ (including possible infinite one-sided limits). Use a graphics calculator or a computer to sketch the graph of the function and confirm your description.

31. $f(x) = \frac{x}{|x|}, \quad a = 0$

32. $f(x) = \frac{1}{x-1}, \quad a = 1$

33. $f(x) = \frac{x}{(x-1)^2}, \quad a = 1$

34. $f(x) = \frac{1}{x^3(x-1)^2}, \quad a = 1$

35. $f(x) = \frac{x}{|x-1|}, \quad a = 1$

36. $f(x) = \frac{2x-1}{x^2-1}, \quad a = \pm 1$

37. Let $f(x) = \frac{x^3-3x+2}{(x^2-1)x^2}$. At each point where $f(x)$ is not defined, predict the one-sided limits (including possible infinite one-sided limits). Use a graphics calculator or a computer to sketch the graph of the function and confirm your results.

38. Let $g(x) = \frac{x^3+x^2-6x}{(x^2-4)^2(x-2)^2}$. At each point where $f(x)$ is not defined, predict the one-sided limits (including possible infinite one-sided limits). Use a graphics calculator or a computer to sketch the graph of the function and confirm your results.

For each of the following two functions it is not easy to solve for the points at which the denominator is zero. (i) Use your computer or graphics calculator to find approximate values for the zeroes of the denominator. Zoom to higher magnification to find the values to three decimal places. Also note the sign of the denominator on either side of each zero. (ii) Predict the one-sided behavior of each function near each of the points you found in (i). Graph the function and confirm your predictions.

39. $F(x) = \frac{x^2-2x+3}{x^3-3x^2+2x-1}$

40. $G(x) = \frac{3x-5}{x^4-4x^2+x+1}$

2.5 Continuity

- POLYNOMIALS
- RATIONAL FUNCTIONS
- JUMP DISCONTINUITY
- REMOVABLE DISCONTINUITY
- INFINITE DISCONTINUITY
- INTERMEDIATE VALUE THEOREM
- CLOSED INTERVALS

In this section we discuss what it means for a function to be "continuous." In terms of the picture, it means that the graph of the function can be drawn "continuously," that is, without lifting the pencil from the page.

A function that "jumps" or has a "hole" in it is not continuous (see Figure 2.5.1(i)).

Unfortunately, this simple notion is not easy to express in a mathematically precise way. It turns out that the best way to do it uses limits. When a function is continuous, it should have a limit as $x \to a$ for any value of a in the domain. Moreover, to avoid "jumps," the value of the function at $x = a$ should equal the limit. (It should be clear from Figure 2.5.1(ii) that if the value $f(a)$ does not equal the limit $\lim_{x\to a} f(x)$, then the function is not continuous at $x = a$.)

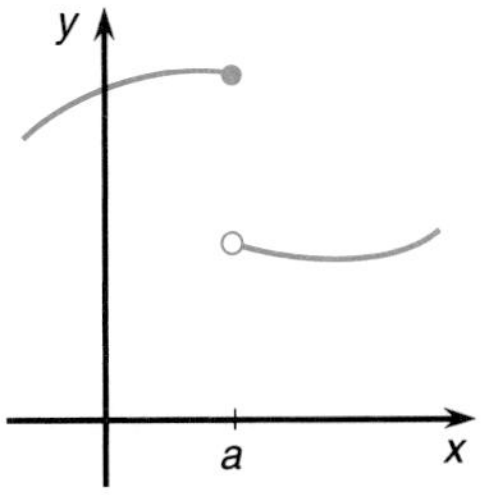

Figure 2.5.1(i)

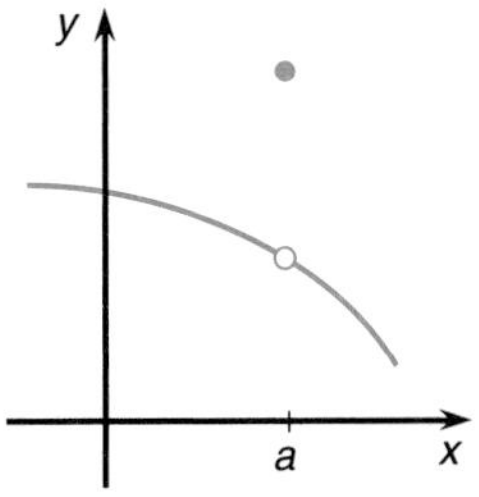

Figure 2.5.1(ii)

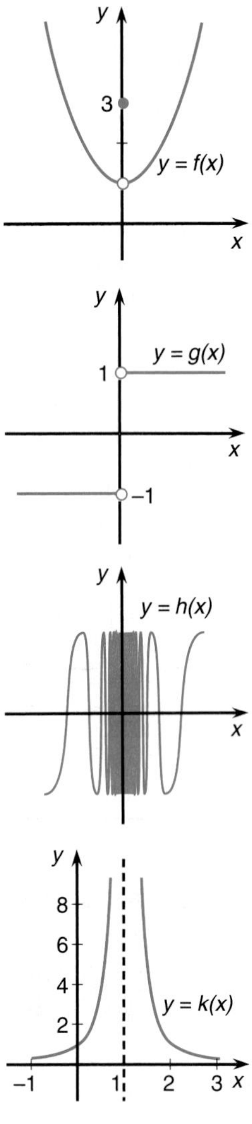

Figure 2.5.2

In Theorem 2.2.5 we saw that polynomials $f(x)$ have the "right" limit as x approaches a, namely, $f(a)$. Rational functions $f(x)$ also have the "right" limit, at least when $f(a)$ makes sense (cf. Theorem 2.2.6). But it is easy to think of functions that do not.

EXAMPLE 1 If

$$f(x) = \begin{cases} x^2 + 1, & \text{whenever } x \neq 0, \\ 3, & \text{when } x = 0, \end{cases}$$

$$g(x) = \frac{x}{|x|},$$

$$h(x) = \sin(1/x),$$

$$k(x) = \frac{1}{(x-1)^2},$$

then we see immediately that

$$\lim_{x \to 0} f(x) = 1 \neq f(0),$$

$$\lim_{x \to 0^+} g(x) = 1 \neq -1 = \lim_{x \to 0^-} g(x),$$

and $h(x) = \sin(1/x)$ does not have a limit, nor even one-sided limits, as $x \to 0$. The function $k(x)$ does not have a (finite) limit as $x \to 1$, but

$$\lim_{x \to 1} k(x) = \infty.$$

See Figure 2.5.2.

The function $f(x)$ has a limit at $x = 0$, but f was deliberately changed to make it have the "wrong" value. The function $g(x)$ has different one-sided limits and jumps from one to the other, and $h(x)$ has no limit at all. Each of these types of behavior is unlike what we know happens for polynomials and rational functions. The function $k(x)$ is a rational function, but $x = 1$ is a point where its denominator is zero, and $k(x)$ does not have a (finite) limit as $x \to 1$.

DEFINITION 2.5.1

A function $f(x)$ is **continuous at** a if $f(x)$ has a limit as x approaches a and

$$\lim_{x \to a} f(x) = f(a).$$

If $a < b$, we say "$f(x)$ is continuous on the open interval $(a, b) = \{x : a < x < b\}$" to mean that $f(x)$ is continuous at every point of the interval.

If $a < b$, we say "$f(x)$ is continuous on the closed interval $[a, b] = \{x : a \leq x \leq b\}$" to mean that $f(x)$ is continuous at every point of the open interval $(a, b) = \{x : a < x < b\}$ and "continuous from the right" at $x = a$ and "continuous from the left" at $x = b$, that is,

$$f(a) = \lim_{x \to a^+} f(x), \qquad f(b) = \lim_{x \to b^-} f(x).$$

A function $f(x)$ is "continuous" if it is continuous at every $a \in \mathbb{R}$.

EXAMPLE 2

(i) Every polynomial is continuous (by Theorem 2.2.5).

(ii) Every rational function is continuous except at the points where its denominator vanishes (by Theorem 2.2.6).

(iii) $F(x) = |x|$ is continuous.

(iv) The function $\frac{x}{|x|}$ is not continuous at $x = 0$ but is continuous everywhere else.

(v) $\sin(1/x)$ is not continuous at $x = 0$.

DEFINITION 2.5.2

(i) The function $f(x)$ has a **removable discontinuity** at $x = a$ if $\lim_{x\to a} f(x)$ exists and $f(a)$ either does not equal the limit or is not defined. The discontinuity can be "removed" by changing $f(x)$ at $x = a$ so that $f(a) = \lim_{x\to a} f(x)$.

(ii) The function $f(x)$ has a **jump discontinuity** at $x = a$ if both one-sided limits exist but are unequal, that is, $\lim_{x\to a^+} f(x) \neq \lim_{x\to a^-} f(x)$.

(iii) The function $f(x)$ has an **infinite discontinuity** at $x = a$ if at least one of the one-sided limits $\lim_{x\to a^+} f(x)$, $\lim_{x\to a^-} f(x)$ is ∞ or $-\infty$.

Notice that a function can fail to be continuous at $x = a$ without having any of these three types of discontinuity.

EXAMPLE 3 In Example 1 the function f has a removable discontinuity at $x = 0$, the function $g(x) = \frac{x}{|x|}$ has a jump discontinuity there, and the function $k(x) = \frac{1}{(x-1)^2}$ has an infinite discontinuity at $x = 1$. The function $h(x) = \sin\left(\frac{1}{x}\right)$ is not continuous at $x = 0$, but it does not have a removable or a jump or an infinite discontinuity. See Figure 2.5.2.

THEOREM 2.5.3

Suppose $f(x)$ and $g(x)$ are continuous at $x = a$. Then so are

(i) $f(x) + g(x)$,

(ii) $cf(x)$ for any $c \in \mathbb{R}$,

(iii) $f(x)g(x)$,

(iv) $\frac{f(x)}{g(x)}$, provided $g(a) \neq 0$.

PROOF These are immediate consequences of the corresponding facts about limits (Theorems 2.2.3 and 2.2.6).

THEOREM 2.5.4

Suppose $f(x)$ is continuous at $x = a$ and $g(x)$ is continuous at $x = f(a)$. Define the composite function $h(x)$ by

$$h(x) = g\big(f(x)\big).$$

Then $h(x)$ is continuous at $x = a$.

PROOF This is also an immediate consequence of the corresponding theorem about limits of composite functions, Theorem 2.3.2.

EXAMPLE 4 Find the points where each of the following functions is continuous: (i) $h(x) = \sqrt{x^2 + x}$; (ii) $k(x) = \left(\frac{x^2-16}{x+2}\right)^{-1/3}$.

Solution

(i) We apply the theorem with $f(x) = x^2 + x$ and $g(x) = \sqrt{x}$. Since $f(x)$, being a polynomial, is continuous everywhere and $g(x) = \sqrt{x}$ is continuous whenever it is defined, that is, whenever $x \geq 0$, we see that the composite function is continuous whenever it is defined. And it is defined whenever the quantity inside the square root sign is nonnegative, that is, when $x^2 + x \geq 0$.

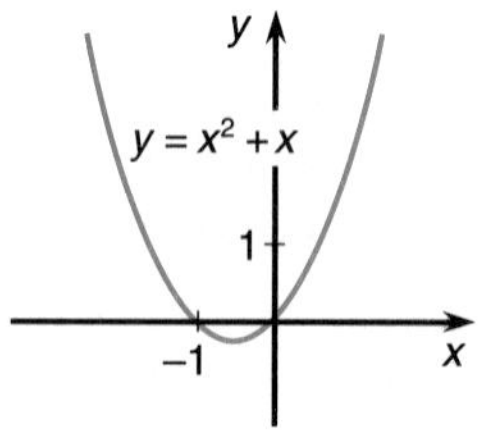

Figure 2.5.3

Drawing a sketch of the parabola $y = x^2 + x$ in Figure 2.5.3, we see that it is negative precisely when $-1 < x < 0$, which means that the composite function $h(x)$ is defined everywhere else. So $h(x) = \sqrt{x^2 + x}$ is defined and continuous when $x \leq -1$ and when $x \geq 0$. Notice that the continuity at $x = -1$ and $x = 0$ is only one-sided; $h(x)$ is continuous "from the left" at $x = -1$ and "from the right" at $x = 0$.

(ii) Here we write $k(x) = g\big(f(x)\big)$ with $f(x) = \frac{x^2-16}{x+2}$ and $g(x) = x^{-1/3}$. Now $g(x)$ is defined for every x except $x = 0$, and $f(x)$ is defined for every x except $x = -2$. So $k(x) = g\big(f(x)\big)$ will be defined for every value of x except $x = -2$ and any values of x where $f(x) = 0$. It is easy to see that $f(x) = 0$ precisely when $x = 4, -4$, so $k(x)$ is defined except at $x = -2, 4, -4$.

Moreover, $f(x)$ is continuous at every point except $x = -2$, and $g(x)$ is continuous everywhere except $x = 0$, so the composite function will be continuous at every x except $x = -2, 4, -4$.

Up to now, we have had several results that dealt with open intervals (a, b), that is, the set of all points between a and b but not including a or b. Now we must also consider closed intervals, that is, intervals that include their endpoints. We write

$$[a, b] = \{x : a \leq x \leq b\}.$$

There are three important and profound facts about the behavior of continuous functions on closed intervals. To prove them requires a subtle understanding of the properties of the real numbers. We shall content ourselves instead with stating the results, illustrating them with examples, and trying to explore their significance.

THEOREM 2.5.5

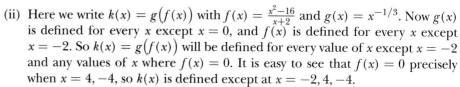

Intermediate Value Theorem

Suppose $a < b$ and $f(x)$ is continuous on the closed interval $[a, b]$. Suppose k is a number that lies between the values of $f(x)$ at $x = a$ and $x = b$ (i.e., $f(a) < k < f(b)$ or $f(a) > k > f(b)$). Then there is a number c between a and b (i.e., $a < c < b$) such that

$$f(c) = k.$$

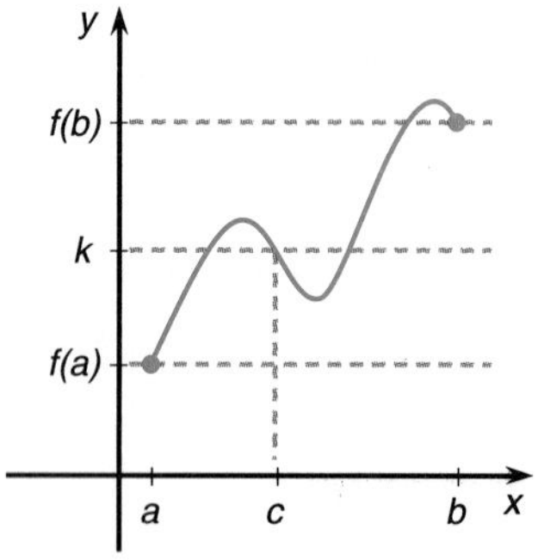

Figure 2.5.4

The theorem is illustrated in Figure 2.5.4. The number k lies between the values of f at a and b (so k is an "intermediate value"). The graph "can't get from $f(a)$ to $f(b)$ without passing through k." The figure also illustrates that there may be more than one possible choice of c.

THEOREM 2.5.6

> Suppose $a < b$ and $f(x)$ is continuous on $[a, b]$. Then $f(x)$ is "bounded" on $[a, b]$, that is, there is a constant M so that $|f(x)| \leq M$, for all $x \in [a, b]$.

Figure 2.5.5(i) illustrates the theorem, showing that the graph of the function $f(x)$ between a and b doesn't get bigger than M (nor smaller than $-M$). We also see that the result is false if the interval is not closed: The function $y = \frac{1}{x}$ is continuous on the open interval $(0, 1)$ but is not bounded (as $x \to 0^+, \frac{1}{x} \to \infty$). See Figure 2.5.5(ii). The idea of the theorem is that if $f(x)$ is continuous on $[a, b]$, it is "tied down" at both ends and can't get too big in between.

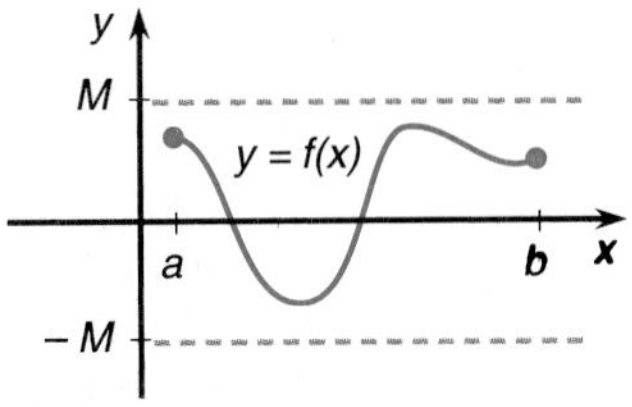

Figure 2.5.5(i)

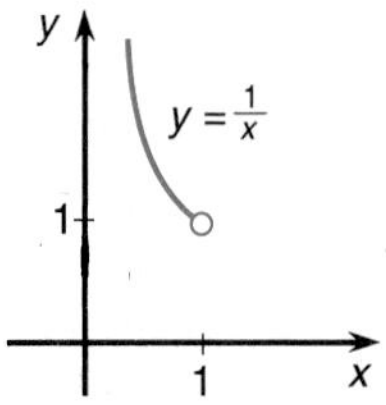

Figure 2.5.5(ii)

THEOREM 2.5.7

> **Extreme Value Theorem**
>
> Suppose $a < b$ and $f(x)$ is continuous on $[a, b]$. Then there is a $c \in [a, b]$ so that $f(c) \geq f(x)$, for all $x \in [a, b]$. There is also a $d \in [a, b]$ so that $f(d) \leq f(x)$, for all $x \in [a, b]$.

The theorem says that $f(c)$ is the "maximum value" of $f(x)$ on $[a, b]$, and $f(d)$ is the "minimum." The theorem can be abbreviated by saying "$f(x)$ achieves its maximum (minimum) value at some point of $[a, b]$."

For the function $f(x)$ illustrated in Figure 2.5.6(i), the minimum occurs at $d = a$, and there are two possible choices for c (corresponding to the two "humps" in the graph). Figure 2.5.6(ii) shows that for this theorem too the interval must be closed: The function $y = x^2$ is continuous on $(0, 1)$ but has neither a maximum nor a minimum there (the maximum and minimum are at the "missing" endpoints $x = 1$ and $x = 0$).

Theorems 2.5.5 and 2.5.7 both assert the existence of a value of x at which $f(x)$ has a special value. This can be extremely useful, but the shortcoming of these results is that neither gives us any idea at all about how to find the x in question. In fact, these are both very difficult problems.

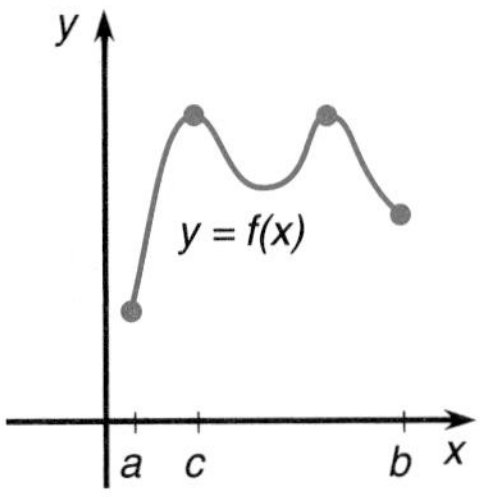

Figure 2.5.6(i)

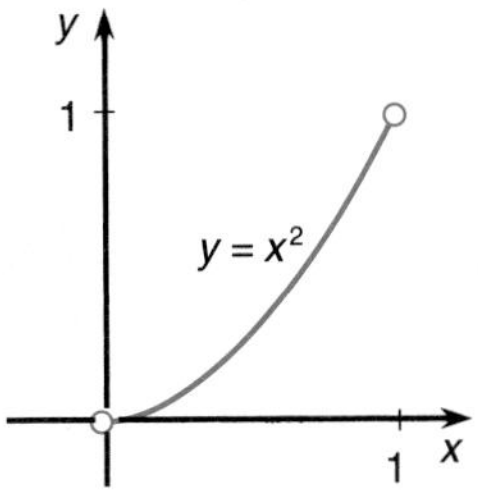

Figure 2.5.6(ii)

EXAMPLE 5

(i) Show that $f(x) = x^4 + 2x^3 - 2x^2 + x + 2$ has a zero between $x = -3$ and $x = -2$.

(ii) Find consecutive integers n and $n+1$ so that $g(x) = x - \sqrt{x} - 10$ has a zero between n and $n+1$.

Solution

(i) We evaluate $f(-3) = (-3)^4 + 2(-3)^3 - 2(-3)^2 - 3 + 2 = 81 - 54 - 18 - 3 + 2 = 8$ and $f(-2) = (-2)^4 + 2(-2)^3 - 2(-2)^2 - 2 + 2 = 16 - 16 - 8 - 2 + 2 = -8$. Since $f(x)$ is positive at $x = -3$ and negative at $x = -2$, the Intermediate Value Theorem says that it must have a zero between these points.

(ii) We evaluate $g(x)$ at various integers x: $g(0) = -10$, $g(1) = 1 - 1 - 10 = -10$, $g(2) = 2 - \sqrt{2} - 10 = -\sqrt{2} - 8$. Each of these numbers is negative.

We can continue in this way, but let us observe that if we write $g(x) = -\sqrt{x} - (10 - x)$, then it will certainly be negative if $10 - x$ is positive, that is, if $x < 10$. Since we are looking for an integer at which g is positive, we may as well start with $x = 10$.

Now $g(10) = -\sqrt{10} + 10 - 10 = -\sqrt{10} < 0$, $g(11) = -\sqrt{11} + 11 - 10 = 1 - \sqrt{11}$. This is also negative, since $\sqrt{11} > 1$. With $x = 12$ we find that $g(12) = 2 - \sqrt{12}$. Since $\sqrt{12} > \sqrt{9} = 3$, we see that $g(12) < 0$; similarly, $g(13) = 3 - \sqrt{13}$, and this is also negative, since $\sqrt{13} > 3$. The next value is $g(14) = 4 - \sqrt{14}$, and this is positive, since $\sqrt{14} < \sqrt{16} = 4$.

By the Intermediate Value Theorem we conclude that there is a zero between $x = 13$ and $x = 14$.

On the facing page, we use a graphics calculator for a similar type of problem.

Techniques for finding the points at which a function achieves its maximum and minimum values are an important part of calculus, and we shall explore them quite fully later on.

For the present we must content ourselves with the observation that it is sometimes not too difficult to find explicit bounds for a function on a closed interval, as promised by Theorem 2.5.6.

EXAMPLE 6 Find a bound for each of the following functions on the specified interval.

(i) $f(x) = x^3 - 2x^2 + 4x - 1$, on $[-3, 2]$,

(ii) $g(x) = \frac{3x^2-2}{x^2+4}$, on $[-1, 1]$.

Solution

(i) The existence of a bound is guaranteed by Theorem 2.5.6. We note that if $x \in [-3, 2]$, then $|x| \le 3$, and applying the triangle law (cf. Theorem 1.2.2), we find that $|f(x)| = |x^3 - 2x^2 + 4x - 1| \le |x^3| + |2x^2| + |4x| + |1| \le 3^3 + 2(3^2) + 4(3) + 1 = 27 + 18 + 12 + 1 = 58$. So $|f(x)| \le 58$ for $x \in [-3, 2]$. This does not say that we have found the best possible bound for $f(x)$. It might be possible to find smaller numbers that are also bounds for $f(x)$ on $[-3, 2]$,

The polynomial $x^4 - 3x^2 + 2x + 1$ is negative at $x = -1$ and positive at $x = 0$. We know there must be a zero in $[-1, 0]$. We can plot the graph using a graphics calculator and then zoom in or use the equation-solving feature to find the zero. On some calculators it is not even necessary to plot the graph first.

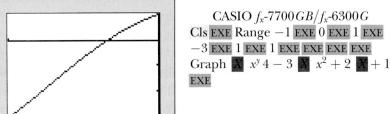

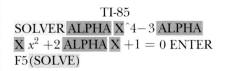

TI-85 SOLVER ALPHA X ^4 − 3 ALPHA X x^2 +2 ALPHA X +1 = 0 ENTER F5(SOLVE)	SHARP EL9300 SOLVER EQTN CL x/θ/T a^b 4 ▶ − 3 x/θ/T x^2 + 2 x/θ/T +1 = 0 ENTER −1 ENTER ENTER ENTER
CASIO f_x-7700GB/f_x-6300G Cls EXE Range −1 EXE 0 EXE 1 EXE −3 EXE 1 EXE 1 EXE EXE EXE EXE Graph X x^y 4 − 3 X x^2 + 2 X + 1 EXE	HP 48SX ↰ PLOT PLOTR ERASE ATTN′ αY = αX y^x 4 − 3 × αX y^x 2 + 2 ×αX + 1 ↱ PLOT ↰ DRAW ERASE 1 +/− ENTER 0 XRNG 3 +/− ENTER 1 YRNG DRAW FCN ROOT

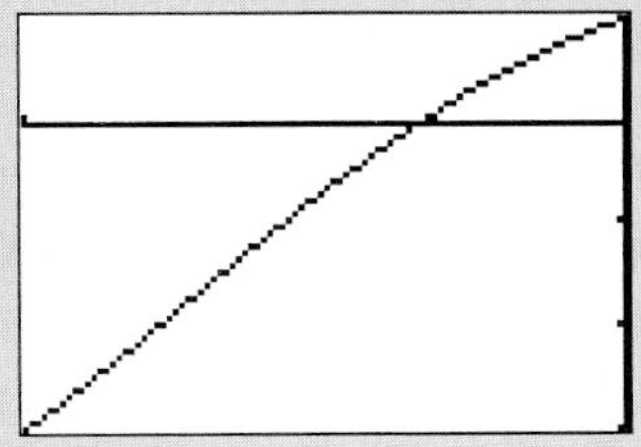

We find that the zero is at approximately $x \approx -0.336532274$ (the exact result may vary on different machines).

but at least we have found one bound. In this case, in fact, we happen to have found the best possible bound, since $f(-3) = -58$.

(ii) This part is trickier. To show that a quotient is smaller than something, we need to show that the numerator is smaller than something and the denominator is *bigger* than something.

On $[-1, 1]$, $|x| \le 1$, so the numerator satisfies $|3x^2 - 2| \le |3x^2| + |2| \le 3 + 2 = 5$. For the denominator, on the other hand, remembering that we want an inequality going the other way, we use the other form of the triangle law (cf. Theorem 1.2.3) (p. 5), and find that $|x^2 + 4| \ge |4| - |x^2| \ge 4 - 1 = 3$.

So on the specified interval,

$$|g(x)| \le \frac{5}{x^2 + 4} \le \frac{5}{3}.$$

In fact, we could also have noticed that since $x^2 \ge 0$, it follows that $x^2 + 4 \ge 4$, so $|g(x)| \le \frac{5}{4}$, which is smaller.

This example illustrates two ideas. One is the use of the triangle law to find simple bounds for polynomials. The other is that if we are asked to find a bound for a function on some interval, it is only necessary to find some bound, and not to worry about finding the smallest possible bound. Of course that might be an interesting question in its own right, but if all we need is some bound, any bound, that tends to be much easier to do.

Exercises 2.5

I

Find all points at which each of the following functions (a) is continuous, (b) has a removable discontinuity, (c) has a jump discontinuity, (d) has an infinite discontinuity, or (e) none of the above.

1. $f(x) = x^7 - 2x^2 + x$

2. $f(x) = \frac{x}{|x|}$

3. $f(x) = \frac{x^3-2x+1}{x(x^2+1)}$

4. $f(x) = \frac{x^2-9}{x+3}$

5. $f(x) = \frac{1}{x^2-3}\frac{x}{|x|}$

6. $f(x) = \frac{x^2}{x^2-x}$

7. $f(x) = \frac{x}{|x|}(x^2+3)$

8. $f(x) = \frac{x}{|x|}(x^3-4x)$

9. $f(x) = (3x+1)\left(\frac{x}{|x|}\right)^2$

10. $f(x) = \frac{x-1}{|x-1|} + \frac{x-2}{|x-3|}$

Decide which of the following functions are bounded on the stated domain. For those that are, specify a bound M.

11. $f(x) = x^2 - 1$, on the interval $[-1, 3]$

12. $f(x) = x^2 - 1$, on the interval $(-1, 3)$

13. $f(u) = \frac{1}{u-1}$, on the interval $[0, 2]$

14. $f(u) = \frac{1}{u-1}$, on the interval $[-1, 0]$

15. $f(x) = \sin x$, on the entire real axis

16. $f(x) = -x^2 + 3x + 1$, on the entire real axis

17. $f(x) = \sin\frac{1}{x}$, on the interval $(0, 1)$

18. $f(x) = 3x^4 - 2x^3 + 5x^2 - 2x + 4$, on the interval $[-2, 1]$.

19. $f(x) = \frac{2x^2+x-1}{x^3+3}$, on the interval $[-1, 1]$.

20. $f(x) = \frac{3x^2-2x+4}{x^4+2}$, on the interval $[0, 1]$.

Decide which of the following achieve their maximum and/or minimum values on the stated domain. For those that do, specify points at which the maximum and/or minimum is achieved. (If you're stuck, draw a sketch.)

21. $f(x) = x^2 - 2x + 3$, on the interval $[-1, 3]$

22. $f(x) = x^2 - 2x + 3$, on the interval $(-1, 3)$

23. $f(x) = \frac{1}{x}\sin\frac{1}{x}$, on the interval $(0, 1)$

24. $f(x) = \sin x$, on the interval $\left(-\frac{\pi}{4}, 3\pi\right)$

25. $f(t) = \frac{1}{t+2}$, on the interval $[0, 1]$

26. $f(s) = \frac{s^2-1}{s-1}$, on the interval $[0, 1)$

27. $f(u) = |u - 4|$, on the interval $[-2, 5]$

28. $f(x) = 3x + 5$, on the interval $[2, 9]$

29. $f(t) = \frac{1}{t^2-4t+2}$, on the interval $[0, 5]$

30. $f(u) = u^3 - 3u + 2$, on the interval $(-2, 2)$ (*Hint:* Consider $u = -2, -1, 0, 1, 2$.)

II

***31.** Show that the square root function $f(x) = \sqrt{x}$ is continuous on $[0, \infty)$:

(i) Show that $f(x)$ is "continuous from the right" at $x = 0$, that is, $f(0) = \lim_{x\to 0^+} f(x)$.

(ii) Show that $f(x)$ is continuous at every $a > 0$.

***32.** Show that the power function $g(x) = x^r$ is continuous wherever it makes sense. You should deal separately with the cases in which r is positive or negative and with the points where $x > 0$, $x = 0$, or $x < 0$.

33. Find all points where each of the following functions is continuous:

(i) $f(x) = (x+4)^{1/4}$

(ii) $g(t) = \frac{t+1}{|t+1|}$

(iii) $h(u) = \begin{cases} \frac{u(u-1)^2}{|u(u-1)^2|}, & \text{if } u \neq 0, 1 \\ 1, & \text{if } u = 0 \text{ or } 1 \end{cases}$

***34.** Show that any polynomial is bounded on any open interval (a, b).

35. Show that $f(x) = 3x^4 - 8x^3 - 18x^2 + 6$ has a zero between $x = 0$ and $x = 1$, and another zero between $x = -1$ and $x = 0$.

36. Find consecutive integers so that $g(x) = x - 20 - x^{1/3}$ has a zero between them.

***37.** Give an example to show that the conclusion of Theorem 2.5.4 may not be true if $g(x)$ has a removable discontinuity at $x = f(a)$.

***38.** Find examples to show that each of Theorems 2.5.5, 2.5.6, and 2.5.7 fails if we do not require the function to be continuous.

***39.** Which (if any) of Theorems 2.5.5, 2.5.6, and 2.5.7 fail and which (if any) remain true for functions that are continuous on $[a, b]$ except for one removable discontinuity at a point in (a, b)? (Give examples of any that fail.)

***40.** (i) Show that any polynomial whose degree is odd must have at least one root.

(ii) Find examples of nonconstant polynomials without any (real) roots.

41. In Example 5(i) (p. 110) we showed that $f(x) = x^4 + 2x^3 - 2x^2 + x + 2$ has a zero between $x = -3$ and $x = -2$. By evaluating $f\left(-\frac{5}{2}\right)$, decide whether the zero is between -3 and $-\frac{5}{2}$ or between $-\frac{5}{2}$ and -2.

42. Give examples of functions with (i) a removable discontinuity at $x = 2$, (ii) removable discontinuities at $x = -2$ and $x = 3$, (iii) a jump discontinuity at $x = -1$ and a removable discontinuity at $x = 2$.

III

43. In Example 5(i) we saw that $f(x) = x^4 + 2x^3 - 2x^2 + x + 2$ has a zero between $x = -3$ and $x = -2$. Use a graphics calculator or computer to find this point, correct to four decimal places.

44. In Example 5(ii) we saw that $g(x) = x - \sqrt{x} - 10$ has a zero between $x = 13$ and $x = 14$. Use a graphics calculator or computer to find this point, correct to four decimal places.

45. Use a graphics calculator or computer to locate the zeroes of $x^5 - 2x^4 + 1$. Estimate each to four decimal places. How many are there?

46. Use a graphics calculator or computer to locate the zeroes of $x^5 - x^4 + 1$. Estimate each to four decimal places. How many are there?

47. Sketch the graph of $y = x^3 - 2x^2 + 4x - 1$ for $x \in [-3, 2]$. Confirm the bound that we found in Example 6(i).

48. Sketch the graph of $y = \frac{3x^2-2}{x^2+4}$ for $x \in [-1, 1]$. Confirm the bound that we found in Example 6(ii). Can you use your graph to find a better bound?

POINT TO PONDER

You might be wondering at this point why we bother with the concept of continuity. After all, the most familiar kinds of functions, polynomials and rational functions, are continuous except where rational functions are not defined. In the next section we shall learn, among other things, that the same is true of the trigonometric functions $\sin x$, $\cos x$, etc. Moreover, the examples we have considered of functions that are not continuous seem rather contrived. They have been deliberately concocted to display unusual behavior.

In a sense, it is precisely because the "nice" functions to which we are most accustomed are continuous that this property is useful. It becomes one of the ways in which we can say that some function is "nice," that is, that at least in this respect it is like polynomials and rational functions. See Figure 2.5.7.

The other importance of continuity is that in doing calculus we are from time to time forced to deal with functions that are not continuous. In Example 2.1.1 we encountered the function $\frac{x^2-x}{x-1}$, which is not even defined at $x = 1$. Yet it was precisely its behavior at this point that interested us.

To study calculus, we need to confront discontinuous functions, and we need to appreciate the significance of continuity to help understand them.

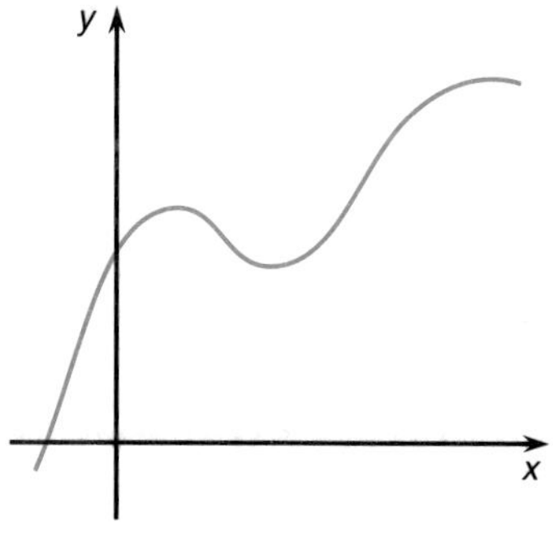

Figure 2.5.7

2.6 Limits of Trigonometric Functions

CONTINUITY
$\lim_{x\to 0}\frac{\sin x}{x}=1$
$\lim_{x\to 0}\frac{\cos x-1}{x}=0$

In this brief section we show that the trigonometric functions $\sin x$ and $\cos x$ are continuous, from which it follows that the other trigonometric functions, $\tan x$, $\cot x$, $\sec x$, and $\csc x$, are continuous whenever they are defined.

We also establish two limits involving trigonometric functions that will be needed later on.

It should be emphasized that the angles are always measured in radians. This might appear to be a nuisance, but the work of this section would actually be considerably more difficult if we were to use degrees. If you are not comfortable with radian measure, it would make sense to review it now, before continuing (see Section 1.7).

If we draw the graphs of $\cos x$ and $\sin x$, they appear to be continuous functions (see Figure 2.6.1), but it takes a little work to prove it. We first prove that they are continuous at $x = 0$ and use this to prove continuity everywhere.

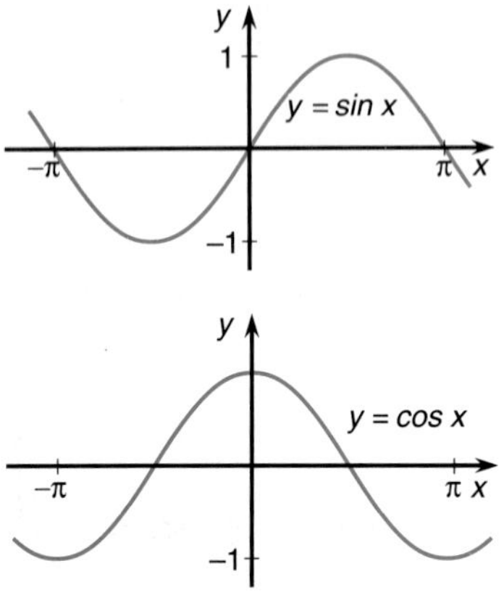

Figure 2.6.1

> **THEOREM 2.6.1**
>
> (i) $\lim_{x\to 0}\sin x = 0$,
> (ii) $\lim_{x\to 0}\cos x = 1$,
> (iii) $\lim_{x\to a}\sin x = \sin a$, for every $a \in \mathbb{R}$,
> (iv) $\lim_{x\to a}\cos x = \cos a$, for every $a \in \mathbb{R}$.

PROOF

(i) Suppose that the angle AOC in Figure 2.6.2 has radian measure x, and that OA and OB are each 1 unit long. Then AC has length $\sin x$, OC has length $\cos x$, and by the definition of radian measure the arc of the circle from A to B has length x.

Now ABC is a right triangle, so its hypotenuse AB is longer than either of its other sides. But its hypotenuse, the straight line AB, is definitely shorter than the circular arc from A to B. So: length of $AC <$ length $AB <$ length of arc from A to B, that is,

$$\sin x < \text{ length } AB < x.$$

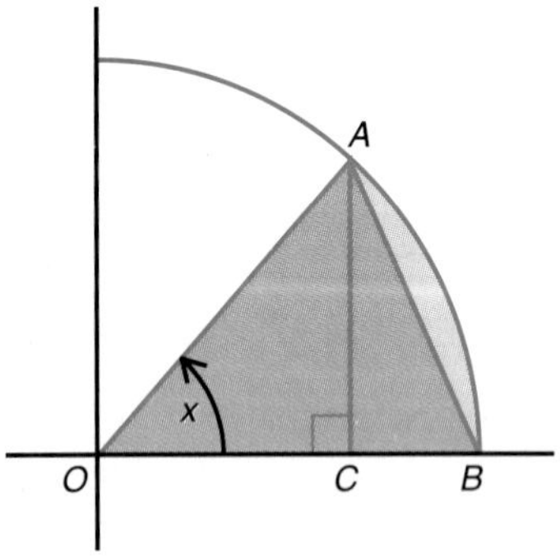

Figure 2.6.2

If we draw the same sort of figure with a negative angle x, then both $\sin x$ and x are negative and $x < \sin x < 0$. In both cases, $|\sin x| < |x|$, that is, $\sin x$ lies between x and $-x$.

Now apply the Squeeze Theorem (Theorem 2.3.5). We have just seen that $\sin x$ lies between the functions x and $-x$, and we know that both of them approach 0 as $x \to 0$. From this, part (i) is immediate.

(ii) This proof is similar, since CB is also shorter than AB. Since the length of OC is $\cos x$ and the length of OB is 1, the length of CB is $1 - \cos x$. The length of CB is less than the length of the hypotenuse AB, which is smaller than x, so $|1 - \cos x| < |x|$, and (ii) follows from the Squeeze Theorem, as above.

(iii), (iv) We notice that instead of $x \longrightarrow a$, we can write $x = a + h$ and let $h \longrightarrow 0$. So $\lim_{x\longrightarrow a} \sin x = \lim_{h\longrightarrow 0} \sin(a+h) = \lim_{h\longrightarrow 0} \big(\sin(a)\cos(h) + \cos(a)\sin(h)\big)$, using the addition formula for sin (cf. Theorem 1.7.11). This then equals

$$\begin{aligned}\lim_{h\longrightarrow 0} \sin(a)\cos(h) + \lim_{h\longrightarrow 0} \cos(a)\sin(h) \\ &= \sin(a)\lim_{h\longrightarrow 0}\cos(h) + \cos(a)\lim_{h\longrightarrow 0}\sin(h) \\ &= \sin(a)(1) + \cos(a)(0) = \sin a,\end{aligned}$$

so (iii) follows.

Similarly,

$$\begin{aligned}\lim_{x\longrightarrow a}\cos x &= \lim_{h\longrightarrow 0}\cos(a+h) = \lim_{h\longrightarrow 0}\big(\cos(a)\cos(h) - \sin(a)\sin(h)\big) \\ &= \lim_{h\longrightarrow 0}\cos(a)\cos(h) - \lim_{h\longrightarrow 0}\sin(a)\sin(h) \\ &= \cos(a)\lim_{h\longrightarrow 0}\cos(h) - \sin(a)\lim_{h\longrightarrow 0}\sin(h) \\ &= \cos(a)(1) - \sin(a)(0) = \cos a,\end{aligned}$$

as desired.

COROLLARY 2.6.2

(i) $\sin x$ and $\cos x$ are continuous.

(ii) $\tan x$, $\cot x$, $\sec x$, and $\csc x$ are continuous wherever they are defined.

PROOF (i) is just a restatement of parts (iii) and (iv) of the theorem while (ii) follows from it by Theorem 2.5.3(iv).

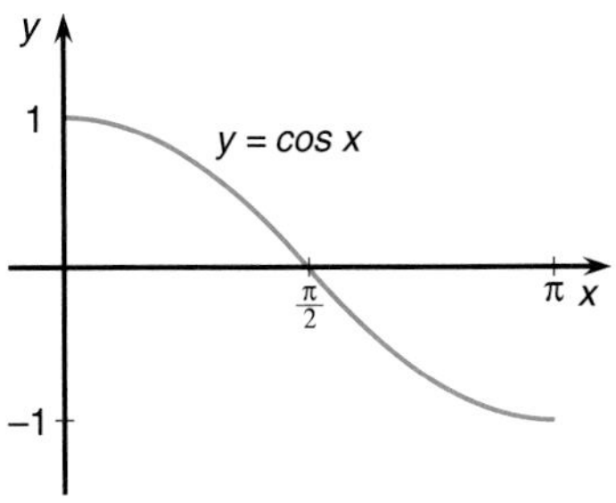

Figure 2.6.3

EXAMPLE 1 Evaluate (i) $\lim_{x\longrightarrow\pi/2} \frac{\sin x - \cos(2x)}{3x}$; (ii) $\lim_{x\longrightarrow(\frac{\pi}{2})^+} \tan x$.

Solution

(i) $\lim_{x\longrightarrow\pi/2} 3x = \frac{3\pi}{2} \neq 0$, so

$$\lim_{x\longrightarrow\pi/2} \frac{\sin x - \cos(2x)}{3x} = \frac{\sin(\pi/2) - \cos(\pi)}{3\pi/2} = \frac{1-(-1)}{3\pi/2} = \frac{4}{3\pi}.$$

(ii) Write $\tan x = \frac{\sin x}{\cos x}$. As $x \longrightarrow \frac{\pi}{2}^+$, we know that $\sin x \longrightarrow \sin\frac{\pi}{2} = 1$ and also that $\cos x \longrightarrow \cos\frac{\pi}{2} = 0$. So the answer is that the limit does not exist.

However, we might well ask about the possibility that there is an infinite limit. Since the numerator tends to 1, the question amounts to asking whether or not the denominator either stays positive (in which case the quotient will approach ∞) or stays negative (in which case the quotient will approach $-\infty$). As $x \longrightarrow \frac{\pi}{2}^+$, we see from Figure 2.6.3 that $\cos x < 0$, at least for values of x reasonably near $\frac{\pi}{2}$, so the quotient is negative, and we see that

$$\lim_{x\longrightarrow\frac{\pi}{2}^+} \tan x = -\infty.$$

(If we were to consider the limit as $x \longrightarrow \frac{\pi}{2}^-$, we would find that it is $+\infty$, for similar reasons, which certainly is consistent with the graph of $y = \tan x$ shown in Figure 2.6.4. This shows, incidentally, that there is no two-sided limit, not even an infinite one.)

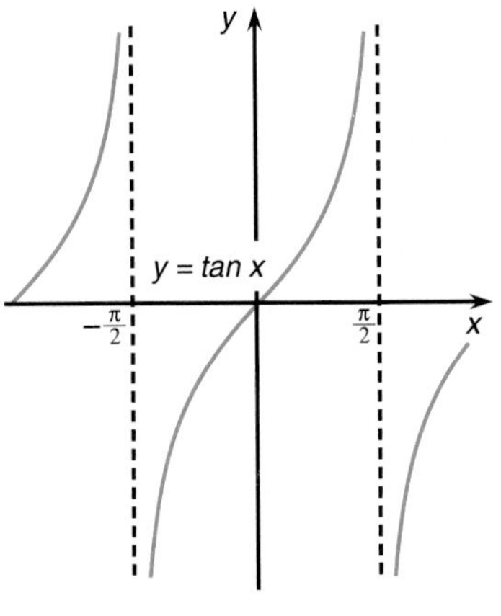

Figure 2.6.4

THEOREM 2.6.3

> (i) $\lim_{x\to 0} \frac{\sin x}{x} = 1$,
>
> (ii) $\lim_{x\to 0} \frac{\cos x - 1}{x} = 0$.

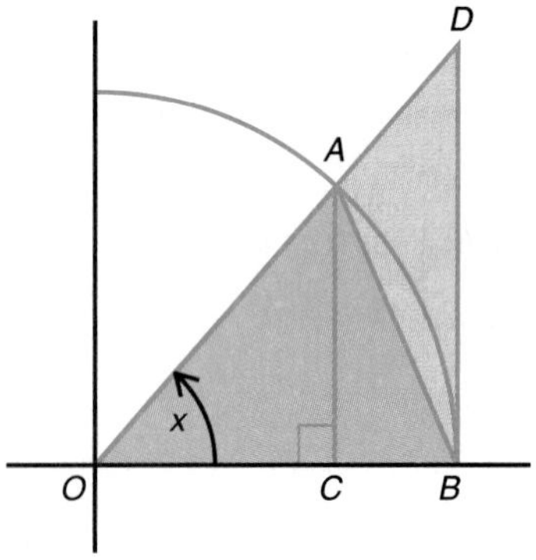

Figure 2.6.5

PROOF We consider first the case in which $x > 0$; refer to Figure 2.6.5. Since the circle has radius $r = 1$, the whole circle has area $\pi r^2 = \pi$. The area of a circular sector like AOB depends on the angle AOB. If the sector was the full circle, the angle would be 2π radians, and the area would be π. If the sector was half a circle, the angle would be π, and the area would be $\frac{\pi}{2}$. In general, the area is proportional to the angle, and we find that the area in square units is one-half the radian measure of the angle. In our case the angle is x, so the area of the circular sector AOB is $\frac{x}{2}$.

From the diagram we see that the triangle AOB is contained in the circular sector AOB, which is itself contained in triangle DOB. The area of triangle AOB is $\frac{1}{2}|OB||AC| = \frac{1}{2}(1)\sin x$.

To find the length $|BD|$, we note that $\frac{|BD|}{|OB|} = \tan x$. Since $|OB| = 1$, we see that $|BD| = \tan x$.

Now the area of triangle DOB is $\frac{1}{2}|OB||DB| = \frac{1}{2}(1)\tan x = \frac{1}{2}\frac{\sin x}{\cos x}$. Comparing these with $\frac{x}{2}$, the area of the circular sector AOB, we get

$$\frac{1}{2}\sin x < \frac{x}{2} < \frac{1}{2}\frac{\sin x}{\cos x}.$$

Dividing by $\frac{1}{2}\sin x$, we get

$$1 < \frac{x}{\sin x} < \frac{1}{\cos x}.$$

Taking reciprocals, we reverse the inequalities and get

$$1 > \frac{\sin x}{x} > \cos x.$$

We showed this under the assumption that $x > 0$. A nearly identical argument can be made if $x < 0$, but it is quicker to notice that $\sin(-x) = -\sin x$ (cf. Theorem 1.7.8, p. 48). Now if $x < 0$, write it as $x = -t$ for some positive t. Then

$$\frac{\sin x}{x} = \frac{\sin(-t)}{-t} = \frac{-\sin t}{-t} = \frac{\sin t}{t}.$$

But we found that this last function lies between 1 and $\cos t = \cos(-x) = \cos x$. Since $\lim_{x\to 0} 1 = \lim_{x\to 0}\cos x = 1$, the Squeeze Theorem applies and proves part (i).

For part (ii), write

$$\begin{aligned}
\lim_{x\to 0}\frac{\cos x - 1}{x} &= \lim_{x\to 0}\frac{\cos x - 1}{x}\,\frac{\cos x + 1}{\cos x + 1} \\
&= \lim_{x\to 0}\frac{\cos^2 x - 1}{x(\cos x + 1)} = \lim_{x\to 0}\frac{-\sin^2 x}{x(\cos x + 1)} \\
&= -\lim_{x\to 0}\frac{\sin x}{x}\,\frac{\sin x}{\cos x + 1} = -\left(\lim_{x\to 0}\frac{\sin x}{x}\right)\left(\lim_{x\to 0}\frac{\sin x}{\cos x + 1}\right) \\
&= -1\left(\frac{0}{2}\right) = 0,
\end{aligned}$$

as required. ◢

> We use the identity $\cos^2 x - 1 = -\sin^2 x$.

> Notice how we split off $\frac{\sin x}{x}$, whose limit we know . . .

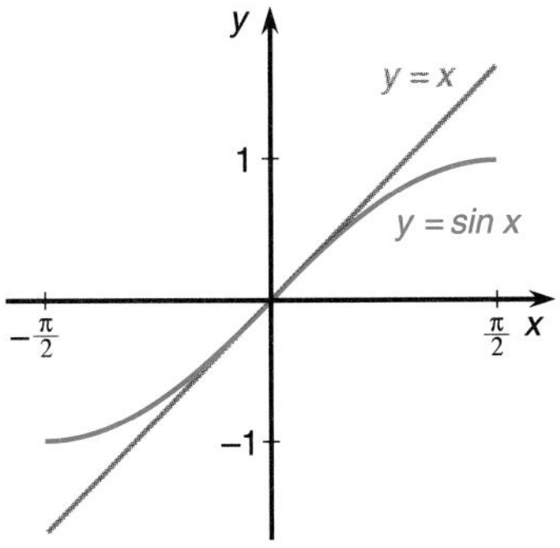

Figure 2.6.6

Part (i) of Theorem 2.6.3 says that as x approaches 0, the value of $\sin x$ gets very close to x. In Figure 2.6.6 the graph of $y = \sin x$ is plotted together with the graph of $y = x$. The idea is that near $x = 0$ the two graphs are very close together, which is certainly apparent from the picture.

EXAMPLE 2 Evaluate $\lim_{x \to 0} \frac{2x+\sin(5x)}{\sin(3x)}$.

Solution Write

$$\begin{aligned}\frac{2x+\sin(5x)}{\sin(3x)} &= \frac{2x}{\sin(3x)} + \frac{\sin(5x)}{\sin(3x)} = \frac{2}{3}\frac{3x}{\sin(3x)} + \frac{\sin(5x)}{5x}\frac{5x}{3x}\frac{3x}{\sin(3x)} \\ &= \frac{2}{3}\frac{3x}{\sin(3x)} + \frac{5}{3}\frac{\sin(5x)}{5x}\frac{3x}{\sin(3x)}.\end{aligned}$$

Now, as $x \to 0$, we have $5x \to 0$. Writing $u = 5x$, we see that $\lim_{x \to 0} \frac{\sin(5x)}{5x} = \lim_{u \to 0} \frac{\sin u}{u} = 1$.

To work this example, we try to find expressions of the form $\frac{\sin u}{u}$ or $\frac{u}{\sin u}$, whose limits we know as $u \to 0$.

Similarly, $3x \to 0$ and $\frac{3x}{\sin(3x)} \to 1$. Combining everything, we find that

$$\lim_{x \to 0} \frac{2x+\sin(5x)}{\sin(3x)} = \frac{2}{3} + \frac{5}{3} = \frac{7}{3}.$$

We can use the same approach as in Example 2 to evaluate $\lim_{x \to 0} \frac{\sin 5x}{\sin(x/2)}$. This is an interesting function to graph.

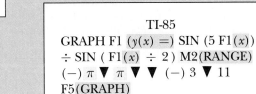

TI-85	SHARP EL9300
GRAPH F1 ($y(x) =$) SIN (5 F1(x)) ÷ SIN (F1(x) ÷ 2) M2(RANGE) (−) π ▼ π ▼ ▼ (−) 3 ▼ 11 F5(GRAPH)	sin (5 x/θ/T) ÷ sin (x/θ/T ÷ 2) RANGE $-\pi$ ENTER π ENTER 1 ENTER −3 ENTER 11 ENTER 1 ENTER
CASIO f_x-7700GB/f_x-6300G	**HP 48SX**
Cls EXE Range $-\pi$ EXE π EXE 1 EXE −3 EXE 11 EXE 1 EXE EXE EXE EXE Graph sin (5 X) ÷ sin (X ÷ 2) EXE	↰ PLOT PLOTR ERASE ATTN′ αY = SIN 5 × αX ▶ ÷ SIN αX÷ 2 ↱ PLOT ↰ DRAW ERASE π +/− ENTER π XRNG 3 +/− ENTER 11 YRNG DRAW

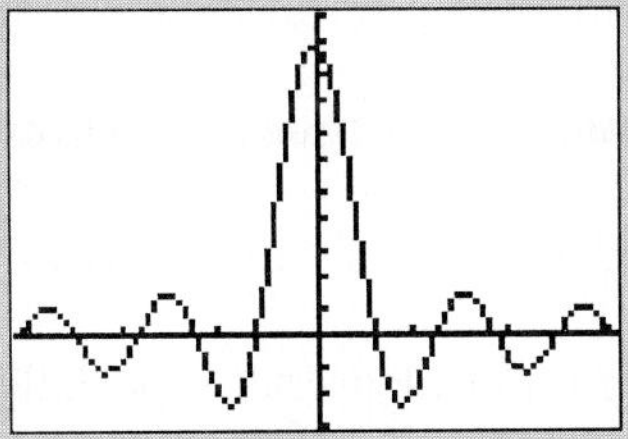

Check that the picture agrees with your calculation of the limit as $x \to 0$. Then find $\lim_{x \to 0} \frac{\sin kx}{\sin(x/2)}$ for different values of k and try plotting the corresponding functions.

Exercises 2.6

I

Evaluate the following finite or infinite limits.

1. $\lim_{x\to 0} \frac{3x^2+x-1+\sin(x+\pi/2)}{\cos(2x)}$
2. $\lim_{x\to \pi/2} \frac{x^2+\cos x}{\sin x}$
3. $\lim_{x\to \pi} \frac{\cos^2 x+\sin(x/2)}{\cos x}$
4. $\lim_{x\to \pi/4} \tan x$
5. $\lim_{x\to \pi/2} \csc^3 x$
6. $\lim_{x\to \pi} x + \sec x$
7. $\lim_{x\to 0} \frac{\sin x}{3x}$
8. $\lim_{x\to 0} \frac{\cos^2 x-1}{x}$
9. $\lim_{x\to \pi/2} \tan^2 x$
10. $\lim_{x\to \pi} \left(\cos^2(7x) + \sin^2(7x)\right)$

Decide whether or not each of the following finite or infinite limits exists, and evaluate those that do.

11. $\lim_{x\to 0} \frac{\cos x}{x}$
12. $\lim_{x\to 0} \frac{\sin(7x)}{\sin(4x)}$
13. $\lim_{x\to 0} \frac{\sin x}{\cos x-1}$
14. $\lim_{x\to 0+} \frac{\sin 2x}{\cos x-1}$
15. $\lim_{x\to 0} \frac{\cos(2x)-2}{x}$
16. $\lim_{x\to 0} \frac{\sin x}{2x^2+x}$
17. $\lim_{x\to 2} \sin(x^2 - 3x + 2)$
18. $\lim_{x\to \frac{1}{2}} \tan(\pi x)$
19. $\lim_{x\to 0-} \tan x$
20. $\lim_{x\to -1} \csc^2(\pi x)$

Evaluate the following finite or infinite limits.

21. $\lim_{x\to 2} \sin\left(\pi(x^2 - 4)\right)$
22. $\lim_{x\to -1} \cos\left(\pi(1 + x^2)\right)$
23. $\lim_{t\to 3} \frac{\sin(9-t^2)}{9-t^2}$
24. $\lim_{t\to 3} \frac{\sin(9-t^2)}{3-t}$
25. $\lim_{u\to \frac{\pi}{2}} \frac{\sin(u)-1}{u-\frac{\pi}{2}}$
26. $\lim_{r\to -\frac{\pi}{2}} \frac{\cos r}{r+\frac{\pi}{2}}$
27. $\lim_{t\to \frac{1}{2}} \frac{\cos(\pi t)}{\sin(2\pi t)}$
28. $\lim_{x\to 2} \frac{\sin(\pi x)}{\cos(\frac{\pi}{4} x)}$
29. $\lim_{x\to -1} \frac{\tan(\pi x)}{\tan(2\pi x)}$
30. $\lim_{x\to 0} \frac{\cos^2(2x)-1+\sin^2(2x)}{x}$

II

31. (i) Find all values of a at which $\tan x$ does not have a limit, that is, at which $\lim_{x\to a} \tan x$ does not exist.
    (ii) For the values you found in part (i), decide whether or not $\tan x$ has one-sided limits, infinite limits, one-sided infinite limits, or none of the above.
32. Repeat Exercise 31 for the function $\sec x$.
33. Repeat Exercise 31 for the functions $\cot x$ and $\tan^2 x$.

*34. In Example 2.1.4 (p. 66) we expressed the area of a circle of radius $r = 1$ as a limit. Evaluate the limit.

III

35. Use a graphics calculator or computer to graph the function $y = \frac{\cos x-1}{x}$ for $x \in [-1, 1]$, for $x \in [-0.1, 0.1]$, and for $x \in [-0.01, 0.01]$. Confirm the limit in Theorem 2.6.3(ii).
36. Graph the function $y = \frac{\sin 3x}{x}$ for $x \in [-1, 1]$, for $x \in [-0.1, 0.1]$, and for $x \in [-0.01, 0.01]$. Confirm that the limit is as expected.

POINT TO PONDER

In this section we have studied various limits involving trigonometric functions. The two fundamental results were Theorem 2.6.1, which just says that $\sin x$ and $\cos x$ are continuous (i.e., that $\lim_{x\to a} \sin x = \sin a$ and $\lim_{x\to a} \cos x = \cos a$), and Theorem 2.6.3, which found that $\lim_{x\to 0} \frac{\sin x}{x} = 1$ and $\lim_{x\to 0} \frac{\cos x-1}{x} = 0$.

The arguments we gave for all these limits were quite geometrical. For Theorem 2.6.1 we had to talk about the length of a piece of the arc of a circle, and in Theorem 2.6.3 we needed to talk about the area of a sector of a circle.

These arguments are quite plausible, especially if you look at the pictures, and they are correct. However, we have not really defined what is meant by the length of an arc of a circle, and we do not really know how to find the area of a circular sector. So the arguments we have used for these limits rely on concepts that we must take for granted.

In fact, both these concepts will be discussed at considerable length later on. Calculating areas is one of the main uses of the technique of integration, which is the subject of Chapters 5 and 6, and the lengths of curves will be discussed in Chapter 6 and again in Chapter 8.

None of this should really trouble anybody too much. We have not really given careful proofs of many of our results and have chosen instead to concentrate on developing our skills in working with them. On the other hand, it is interesting to acknowledge our shortcomings.

2.7 The Precise Definition of Limits

EPSILON (ϵ)
DELTA (δ)

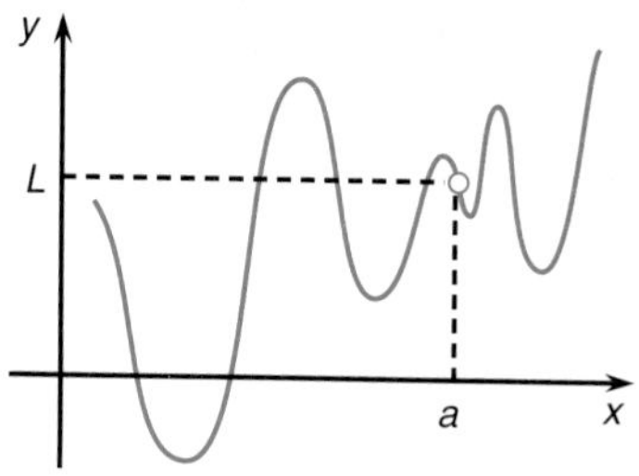

Figure 2.7.1

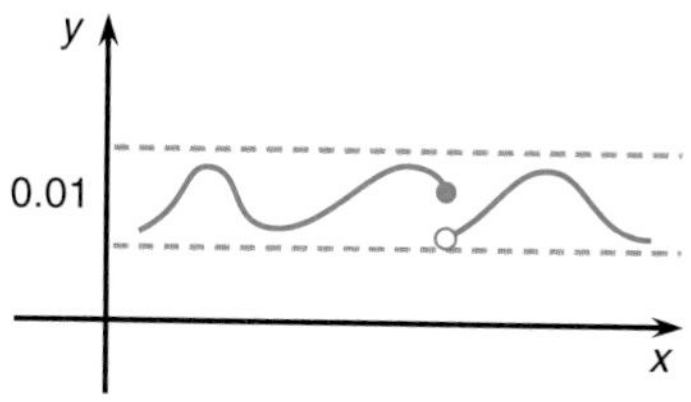

Figure 2.7.2

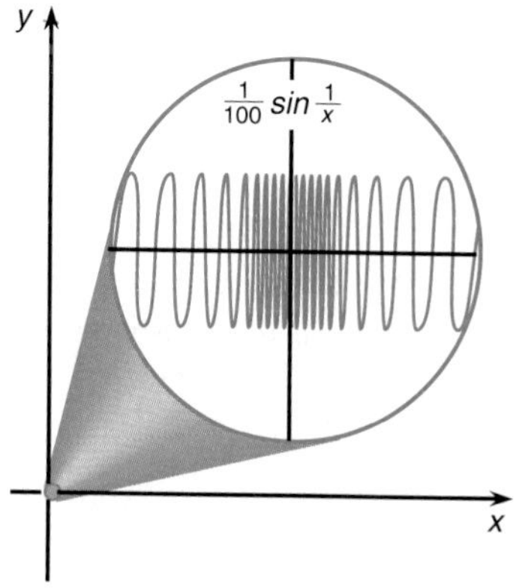

Figure 2.7.3

In this section we discuss how limits can be defined with mathematical precision. We have contented ourselves up to now with a vague description and found that it was enough for our purposes, enabling us to develop a good working knowledge of limits and a feel for what they mean. We have also hinted that the precise definition is difficult to understand and use.

The problem is to state exactly what we mean when we say that $f(x)$ "approaches" L when x "approaches" a. We have already mentioned (at the end of Section 2.3) that it is not simply a matter of saying that $f(x)$ keeps getting closer to L. There we mentioned that $f(x)$ can approach and perhaps even *equal* L and then move away again; what is more important is that it not move away too far. See Figure 2.7.1.

So we should begin by saying that $f(x)$ gets close to L and stays close, but this leaves the question of "how close?" (see Figure 2.7.2). It would not be good enough to say within 0.01 or even within 0.00000001 because it would always be possible to find a function that did stay that close but did not have a limit (after all, think of $\frac{1}{100}\sin\frac{1}{x}$ as $x \to 0\ldots$). See Figure 2.7.3.

In fact, no specified degree of closeness will work. What is needed is to say that $f(x)$ will eventually get as close as we like to L. In other words, it will eventually get within 0.01 of L, and eventually within 0.0001, and eventually within 0.00000001, and so on. This is not to say that it will satisfy all these conditions simultaneously, but rather that it will eventually satisfy any particular one of them.

So to say that $\lim_{x\to a} f(x) = L$ is to say that if we are given any particular degree of closeness, we can be sure that $f(x)$ will eventually be that close to L. By "eventually" we mean when x is sufficiently close to a.

To specify what we have been calling a "degree of closeness," we should give a number and say "not farther than this much away."

So we need to say that given any positive number $\epsilon > 0$, we can be sure that $f(x)$ will "eventually" be a distance less than ϵ away from L. (The Greek letter $\boldsymbol{\epsilon}$ **(epsilon)** is traditionally used in this context.)

By "eventually" we mean that we can be sure it will be true once x is within some small distance of a. To specify this small distance, we can say there must be a positive number $\delta > 0$ so that $f(x)$ will be within ϵ of L whenever x is within a distance δ of a. (The Greek letter $\boldsymbol{\delta}$ **(delta)** is traditionally used here.)

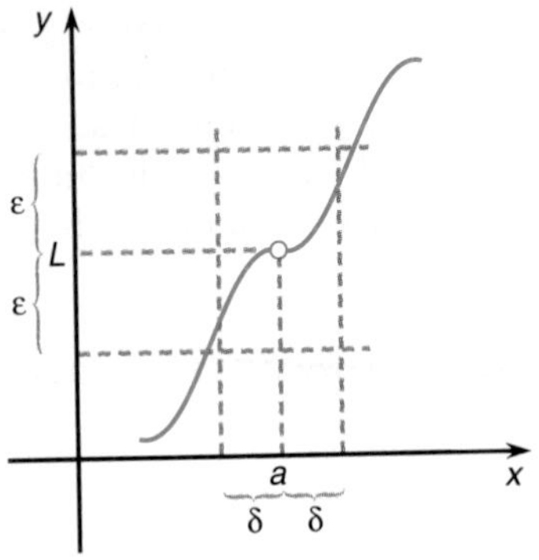

Figure 2.7.4

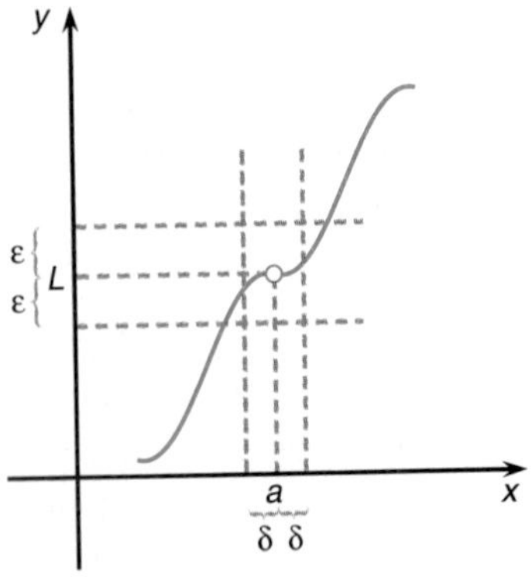

Figure 2.7.5

In Figure 2.7.4 it is possible to see how δ is used as a measure of the horizontal closeness of x to a and ϵ is used as a measure of the (vertical) closeness of $f(x)$ to L. We see that when x is within a distance of δ from a, the value of $f(x)$ is within ϵ of L.

The important thing to bear in mind is that it must work for any choice of $\epsilon > 0$. That is, given any $\epsilon > 0$, there must be a $\delta > 0$ so that $|f(x) - L| < \epsilon$ whenever x is within δ of a. We should expect that δ will depend on ϵ, that is, that for different choices of ϵ it will be necessary to use different choices of δ. See Figure 2.7.5.

The above discussion is an attempt to explain and justify the following definition.

DEFINITION 2.7.1

Suppose the interval (c, d) contains the point a, and suppose the function $f = f(x)$ is defined at every point in (c, d) except possibly the point a. A number L is said to be the **limit** of $f(x)$ as x approaches a, and we write

$$\lim_{x \to a} f(x) = L,$$

provided that for every $\epsilon > 0$ there exists a $\delta > 0$ so that $|f(x) - L| < \epsilon$ whenever $0 < |x - a| < \delta$.

Notice that $f(x)$ has to be defined at every point "near $x = a$" but not necessarily at a itself.

EXAMPLE 1 In Section 2.1 and again in Section 2.2 we agreed that $\lim_{x\to 1} x = 1$. Let us see how the definition applies in this case. Notice that $f(x) = x$, $a = 1$, and $L = 1$.

What is needed is to make x close to 1 or to make $|x - 1|$ small. Suppose we are given $\epsilon > 0$ and want $|x - 1| < \epsilon$. To do this, we are allowed to insist that x be close to 1, that is, that $|x - 1|$ be less than some well-chosen δ.

The answer is almost too easy. To make $|f(x) - L| < \epsilon$, that is, $|x - 1| < \epsilon$, all we need is to require that $|x - 1| < \delta$, with $\delta = \epsilon$.

So for any $\epsilon > 0$ we certainly know how to pick a $\delta > 0$: Just let $\delta = \epsilon$. If $0 < |x - 1| < \delta$, then $|f(x) - L| = |x - 1|$ is indeed less than ϵ. This proves that $\lim_{x\to 1} x = 1$.

On the other hand, it was a little too simple.

EXAMPLE 2 Let $f(x) = 2x$ and show that $\lim_{x\to -3} f(x) = -6$.

Solution $|f(x) - (-6)| = |2x + 6| = 2|x + 3| = 2|x - (-3)|$. Given any $\epsilon > 0$, we want to make $|f(x) - (-6)| < \epsilon$, that is, $2|x - (-3)| < \epsilon$, which is obviously true if $|x - (-3)| < \epsilon/2$.

Letting $\delta = \epsilon/2$, we see that if $|x - (-3)| < \delta = \epsilon/2$, then $2|x - (-3)| < \epsilon$, as needed. Notice how δ is different for different values of ϵ.

EXAMPLE 3 Let $f(x) = x^2 - x + 3$. What is the limit of $f(x)$ as x approaches 2?

Solution We know that the limit is $f(2) = 5$. And $|f(x) - 5| = |x^2 - x - 2| = |(x-2)(x+1)| = |x-2||x+1|$. To make this less than ϵ, we deal with each factor separately.

First, we notice that if x is close to 2, this restricts the size of $|x+1|$. Indeed, $|x+1| = |x-2+3| \leq |x-2| + 3$. We are going to assume that $|x-2| < \delta$, so $|x+1| < \delta + 3$. If $\delta \leq 1$, this shows that $|x+1| < 1 + 3 = 4$. Knowing this, we could

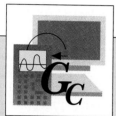

A graphics calculator can be used to visualize how a particular δ works with a particular ϵ. The drawback is that it is usually difficult to do anything that shows anything about *every* ϵ.

We plot the function from Example 3. Suppose we want to consider $\epsilon = 0.01$. The example suggests choosing $\delta = 0.0025$. This means considering $x \in [2 - 0.0025, 2 + 0.0025]$, and letting the vertical range be between $5 - \epsilon$ and $5 + \epsilon$, that is, $4.99 \leq y \leq 5.01$.

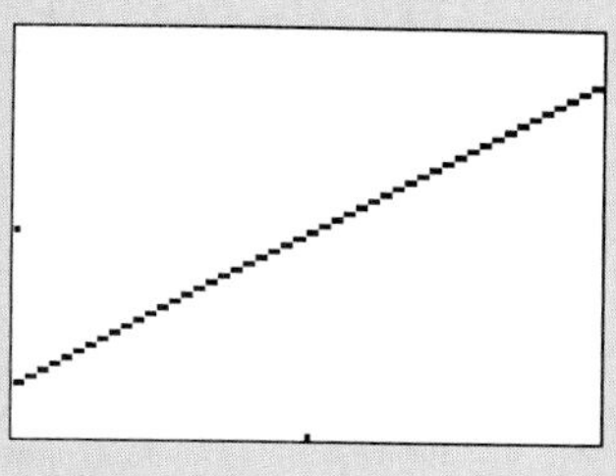

TI-85 GRAPH F1 ($y(x) =$) F1 (x) x^2 – F1 (x) + 3 M2 (RANGE) 1.9975 ▼ 2.0025 ▼ ▼ 4.99 ▼ 5.01 F5 (GRAPH)	SHARP EL9300 x/θ/T x^2 – x/θ/T + 3 RANGE 1.9975 ENTER 2.0025 ENTER .001 ENTER 4.99 ENTER 5.01 ENTER .01 ENTER
CASIO f_x-7700GB/f_x-6300G Cls EXE Range 1.9975 EXE 2.0025 EXE .001 EXE 4.99 EXE 5.01 EXE .01 EXE EXE EXE EXE Graph X x^2 – X + 3 EXE	HP 48SX ↰ PLOT PLOTR ERASE ATTN' αY = αX y^x 2 – αX + 3 ↱ PLOT ↰ DRAW ERASE 1.9975 ENTER 2.0025 XRNG 4.99 ENTER 5.01 YRNG DRAW

For each $x \in [5 - \delta, 5 + \delta]$, the function lies in the specified range $5 - \epsilon \leq y \leq 5 + \epsilon$, that is, $4.99 \leq y \leq 5.01$. This means that the δ we have chosen is sufficiently small. If the graph ran off the top or the bottom of the picture, we would have to choose a smaller δ. In fact, from the picture it appears that we could have chosen a slightly bigger δ . . .

also ask that $|x-2| < \epsilon/4$, so that $|f(x)-5| = |x-2||x+1| < (\epsilon/4)(4) = \epsilon$, as required. A value of δ that will achieve all this is therefore the minimum of 1 and $\epsilon/4$, that is, $\delta = \min(1, \epsilon/4)$.

If $\delta = \min(1, \epsilon/4)$ and $0 < |x-2| < \delta$, then $|f(x)-5| < \epsilon$, and we have found that $\lim_{x\to 2} f(x) = 5$.

This is not the only δ that will work; obviously, any smaller δ will be just as good. Fortunately, all that is required is to find one suitable value.

EXAMPLE 4 Let $f(x) = -3$ be the constant function. What is the limit of $f(x)$ as $x \to 10$?

Solution We know that the limit is -3. And $|f(x)-(-3)| = |-3-(-3)| = 0$, which is less than ϵ for any $\epsilon > 0$. So δ could be anything at all; for example, $\delta = 1$ works for any $\epsilon > 0$. If $0 < |x-10| < 1$, then $|f(x)-(-3)| = |-3+3| = 0 < \epsilon$, as needed. Of course, the same argument works for any constant function.

EXAMPLE 5 Show that the function $f(x) = \frac{x}{|x|}$ does not have a limit as $x \to 0$.

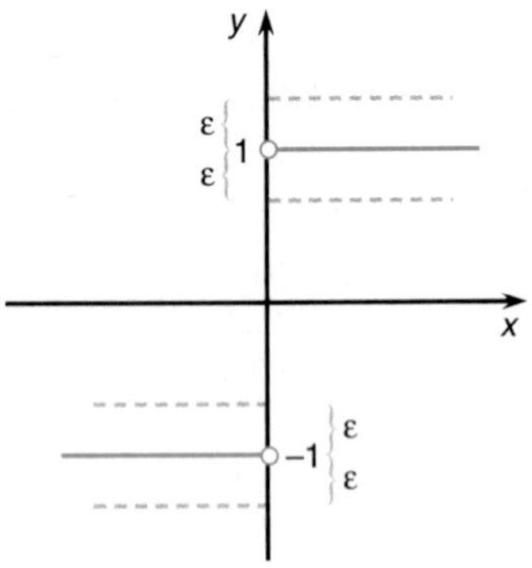

Figure 2.7.6

Solution Suppose it did have a limit L, and suppose $\epsilon < 1$. Then there would be a $\delta > 0$ so that $|f(x)-L| < \epsilon$ whenever $0 < |x-0| < \delta$, that is, $0 < |x| < \delta$. If $-\delta < x < 0, f(x) = -1$, so $|-1-L| < \epsilon < 1$, so $L < 0$. See Figure 2.7.6.

On the other hand, if $0 < x < \delta, f(x) = 1$, and $|1-L| < \epsilon < 1$, so $L > 0$. But L cannot be greater than 0 *and* less than 0, and this contradiction proves that there is no limit.

Notice that for a limit to exist, it must be possible to find a δ for every ϵ, but to show that it doesn't exist, all that is needed is one value of ϵ for which no δ can be found.

EXAMPLE 6 Evaluate $\lim_{x\to 2} \frac{1}{3x+1}$.

Solution We know what the limit should be: $\lim_{x\to 2} \frac{1}{3x+1} = \frac{1}{3(2)+1} = \frac{1}{7}$.

Suppose $\epsilon > 0$ is given. We must consider the difference $\frac{1}{3x+1} - \frac{1}{7}$ and make it small. Rewriting

$$\frac{1}{3x+1} - \frac{1}{7} = \frac{7-(3x+1)}{7(3x+1)} = \frac{6-3x}{21x+7},$$

we have made the difference into a rational function whose numerator is $6-3x = -3(x-2)$. As $x \to 2$, this numerator approaches zero, so to make the quotient small, we only have to make sure the denominator does not also become small.

We are going to consider a condition that $0 < |x-2| < \delta$. The possible trouble with the denominator would occur when it gets small, which happens when x gets close to $-\frac{1}{3}$. If, for instance, we assume that $\delta \le 1$, then $|x-2| < 1$, and this means that $x > 1$. If $x > 1$, then $21x > 21$ and $21x+7 > 28$.

So, with this understanding that $\delta \le 1$, we see that

$$\left|\frac{6-3x}{21x+7}\right| \le \frac{|6-3x|}{28} = \frac{3}{28}|x-2|.$$

This is what we have to make smaller than ϵ, and that will certainly be true if $|x-2| < \frac{28}{3}\epsilon$. In other words, we can choose δ to be less than or equal to $\frac{28}{3}\epsilon$.

Bearing in mind our earlier assumption about δ, we let

$$\delta = \min\left(1, \frac{28}{3}\epsilon\right).$$

If $0 < |x-2| < \delta$, then

$$\left|\frac{1}{3x+1} - \frac{1}{7}\right| = \left|\frac{3(2-x)}{21x+7}\right| \le \frac{3}{28}|x-2| < \frac{3}{28}\frac{28}{3}\epsilon = \epsilon.$$

For any ϵ we have shown how to find δ, which proves that

$$\lim_{x\to 2} \frac{1}{3x+1} = \frac{1}{7},$$

as desired.

The examples that we have just done demonstrate that Definition 2.7.1 is not easy to use. Even relatively simple limits can be quite difficult to prove.

It can help us to understand and appreciate why certain things work the way they do, but as a practical matter, knowing the definition will not even be much use when it comes to evaluating any of the limits we have encountered so far.

This is why we have postponed this section to the end of the chapter. It was not really necessary to learn it to work out the limits we need, and it is undeniably difficult. On the other hand, it can be very interesting, and it is very important for theoretical work with limits.

It is clearly too cumbersome to have to use the definition directly every time we evaluate a limit. The idea is that all the theorems about limits can be proved rigorously by starting with the definition. Then, of course, it is possible to use them, as we have already been doing, without going back to ϵ and δ all the time. We have chosen to bypass the time-consuming proofs of the theorems and concentrate instead on learning to use them.

However, we illustrate the approach by giving the proof of Theorem 2.2.3(i), which says that the limit of the sum of two functions with limits is the sum of their limits.

THEOREM 2.2.3

Suppose $f(x)$ and $g(x)$ are functions and a is a number so that

$$\lim_{x\to a} f(x) = L \quad \text{and} \quad \lim_{x\to a} g(x) = M.$$

(Recall that this means that both limits exist and they have the specified values.)

Then

$$\lim_{x\to a} \big(f(x) + g(x)\big) = L + M.$$

PROOF Given $\epsilon > 0$, we must show that there is a $\delta > 0$ so that

$$\left|\big(f(x)+g(x)\big)-(L+M)\right|<\epsilon,$$

whenever $0 < |x-a| < \delta$.

Notice that $\left|\big(f(x)+g(x)\big)-(L+M)\right| = \left|\big(f(x)-L\big)+\big(g(x)-M\big)\right| \le |f(x)-L| + |g(x)-M|$. The last two terms are the things we know can be made small because $f(x) \to L$ and $g(x) \to M$.

To make their sum less than ϵ, it would be enough to make each of them less than $\frac{\epsilon}{2}$. Since

$$\lim_{x\to a} f(x) = L,$$

given the number $\frac{\epsilon}{2}$, we know there exists a $\delta_1 > 0$ so that

$$|f(x)-L| < \frac{\epsilon}{2} \qquad \text{whenever} \qquad 0 < |x-a| < \delta_1.$$

Similarly, there is a $\delta_2 > 0$ so that

$$|g(x)-M| < \frac{\epsilon}{2} \qquad \text{whenever} \qquad 0 < |x-a| < \delta_2.$$

Now let

$$\delta = \min(\delta_1, \delta_2).$$

Whenever $0 < |x-a| < \delta$, then $0 < |x-a| < \delta_1$ and $0 < |x-a| < \delta_2$. So

$$|f(x)-L| < \frac{\epsilon}{2} \qquad \text{and} \qquad |g(x)-M| < \frac{\epsilon}{2}.$$

From this we find that

$$\left|\big(f(x)+g(x)\big)-(L+M)\right| \le |f(x)-L| + |g(x)-M| < \frac{\epsilon}{2}+\frac{\epsilon}{2} = \epsilon.$$

We have shown that, given any $\epsilon > 0$, it is possible to find a $\delta > 0$ so that $\left|\big(f(x)+g(x)\big)-(L+M)\right| < \epsilon$ whenever $0 < |x-a| < \delta$, and the proof is complete.

Exercises 2.7

I

In each of the following cases, find $\lim_{x\to a} f(x)$, and then use Definition 2.7.1 to prove that your answer is correct.

1. $f(x) = x - 4$, $a = -1$
2. $f(x) = 3x + 2$, $a = \frac{1}{3}$
3. $f(x) = x^2 + 3x - 1$, $a = 2$
4. $f(x) = 2x^2 - x + 1$, $a = -2$
5. $f(x) = x^3 + 9$, $a = -2$
6. $f(x) = x^3 - x^2 + 2x - 1$, $a = 1$
7. $f(x) = 2x + 1$, $a = -\frac{1}{2}$
8. $f(x) = x^{10}$, $a = 1$
9. $f(x) = x(x+1)$, $a = 2$
10. $f(x) = (x+1)(x-1)$, $a = 1$

In each of the following cases, use Definition 2.7.1 to determine whether or not $f(x)$ has a limit as x approaches a, and determine the limit if it exists.

11. $f(x) = \frac{2}{x+1}$, $a = -1$
12. $f(x) = \frac{x^2-1}{x+1}$, $a = 1$
13. $f(x) = \frac{x^2+1}{x-1}$, $a = 1$
14. $f(x) = |x|$, $a = 0$
15. $f(x) = \frac{x^2}{|x|}$, $a = 0$
16. $f(x) = \frac{|x|}{x^2}$, $a = 0$

17. $f(x) = 2 - x^2 \sin(1/x)$, $a = 0$

18. $f(x) = \frac{x^3+2x^2+x-4}{x-1}$, $a = 1$

19. $f(x) = \frac{x^2+x\sin(1/x)}{x}$, $a = 0$

20. $f(x) = \frac{x-x^2\sin(1/x)}{x}$, $a = 0$

Decide whether or not each of the following functions has a limit at the point $a = 0$ and prove that your answer is correct, using Definition 2.7.1.

21. $f(x) = \cos(1/x)$

22. $g(t) = 1/t$

23. $h(x) = \cos^2(1/x)$

24. $k(x) = \cos^2(1/x) + \sin^2(1/x)$

II

25. Use Definition 2.7.1 to show that

$$\lim_{x \to a} x = a.$$

***26.** Use Definition 2.7.1 to show that

(i) $\lim_{x\to a}(cx + d) = ca + d$, assuming $c \neq 0$.

(ii) $\lim_{x\to a}(Ax^2 + Bx + C) = Aa^2 + Ba + C$.

***27.** (i) Mimic Definition 2.7.1 of a limit to formulate definitions in terms of ϵ and δ for one-sided limits

$$\lim_{x\to a^+} f(x) \qquad \text{and} \qquad \lim_{x\to a^-} f(x).$$

(ii) Use your definition to find the one-sided limits at $x = 0$ of $f(x) = \frac{x}{|x|}$.

(iii) Using your definitions, prove Theorem 2.4.2, that $\lim_{x\to a} f(x)$ exists and equals L if and only if both $\lim_{x\to a^+} f(x)$ and $\lim_{x\to a^-} f(x)$ exist and both equal L.

***28.** Use your definition from Exercise 27 to prove that $\lim_{x\to 0^+}\sqrt{x} = 0$. (This problem was unexpectedly difficult earlier on, but if we use ϵ's and δ's, it is actually not too bad.)

III

***29.** Let $f(x) = x^{1/3} + x^{2/5}$. Then $\lim_{x\to 0} f(x) = 0$. Use a graphics calculator or computer to find a value of δ that corresponds to each of the following values of ϵ: $\epsilon = 0.1$, $\epsilon = 0.01$, $\epsilon = 0.001$, $\epsilon = 0.0001$.

***30.** With $f(x)$ as in the previous question, we see that $\lim_{x\to 1} f(x) = 2$. Use a graphics calculator or computer to find a value of δ that corresponds to each of the following values of ϵ: $\epsilon = 0.1$, $\epsilon = 0.01$, $\epsilon = 0.001$, $\epsilon = 0.0001$.

POINT TO PONDER

At first glance, many people find the definition of a limit (Definition 2.7.1) confusing and complicated. Try to find a better one, bearing in mind that it should give the same answers as Definition 2.7.1 for each of the examples and exercises we have tried. (This is a problem for which there is probably no solution. If you think about it for a while, you will see that any other definition is just as complicated, or even worse.)

Chapter Summary

§2.1
tangent line
area

§2.2
limit
limits of sums
limits of products
limits of quotients
limits of polynomials

§2.3
limits of powers x^r
limits of composite functions
The Squeeze Theorem

§2.4
one-sided limits
limits from above/below
limits from the right/left
infinite limits
limits of rational functions
infinity
negative infinity
limits at infinity

§2.5
continuous function
continuity of polynomials
continuity of rational functions
removable discontinuity

jump discontinuity
infinite discontinuity
continuity of composite functions
Intermediate Value Theorem
boundedness of continuous functions on $[a, b]$
continuous functions achieve maximum/minimum values on $[a, b]$
Extreme Value Theorem

§2.6
limits of trigonometric functions
continuity of trigonometric functions
$\lim_{x\to 0} \frac{\sin x}{x} = 1$
$\lim_{x\to 0} \frac{\cos x - 1}{x} = 0$

§2.7
limit
precise definition of limits
epsilon (ϵ)
delta (δ)

Review Exercises

I

Calculate each of the following, provided that it exists.

1. $\lim_{x\to 0} (x^3 - 4x^2 + 2x - 1)$
2. $\lim_{x\to -1} (x^2 + 2x + 1)$
3. $\lim_{x\to 2} \frac{x+1}{x+2}$
4. $\lim_{x\to -1} \frac{1+3x-x^2}{x^2+x+3}$
5. $\lim_{x\to 1} \frac{x^2-1}{x-1}$
6. $\lim_{x\to -1} \frac{x^3+x^2+x+1}{x+1}$
7. $\lim_{x\to 4} \frac{x^2-3x-4}{x-3}$
8. $\lim_{x\to -3} \frac{x^2+2x-1}{x^2-9}$
9. $\lim_{x\to 0} \frac{2x^3-3x^2+5x}{x^2-2x}$
10. $\lim_{x\to -1} \frac{(x+1)^5-4}{(x+1)^2-x+2}$
11. $\lim_{x\to 1} \sqrt{x^2+3}$
12. $\lim_{x\to -1} (x^2 - 2x + 1)^{-3/2}$
13. $\lim_{x\to\sqrt{5}} \frac{x^2-5}{x-\sqrt{5}}$
14. $\lim_{x\to -1} \frac{4-(x^2-6x+1)^{2/3}}{2-(x^2-6x+1)^{1/3}}$
15. $\lim_{x\to 8} \frac{x^{1/3}-2}{x-8}$
16. $\lim_{x\to -27} \frac{x^{1/3}+3}{x+27}$
17. $\lim_{x\to 81} \frac{x^{3/4}-27}{9-\sqrt{x}}$
18. $\lim_{x\to 2} \frac{1}{3-\sqrt{2x^2+1}}$
19. $\lim_{x\to -8} \left(\frac{64-x^2}{2x+16}\right)^{1/3}$
20. $\lim_{x\to\sqrt{3}} \frac{x^4-8}{x^2-2}$
21. $\lim_{x\to 1} \sin(\pi x)$
22. $\lim_{x\to -1} \tan(\pi x) + 2$
23. $\lim_{x\to 1/2} \sin(\pi x)$
24. $\lim_{x\to 1} \cot\left(\frac{\pi}{4} x\right)$
25. $\lim_{x\to 0} \frac{1}{x} \sin(\pi x)$
26. $\lim_{x\to 1} \cos\left(\frac{\pi}{2} x\right)\left(\frac{1}{x-1}\right)$
27. $\lim_{x\to 0} \frac{\sin(\pi x)}{\sin(2x)}$
28. $\lim_{x\to 0} \frac{1-\cos(\pi x)}{x}$
29. $\lim_{x\to 0} 3x \cot x$
30. $\lim_{x\to 0} \frac{1}{x}(3x - \cos^2 x + \cos x)$
31. $\lim_{x\to 0^+} \sqrt{x}$
32. $\lim_{x\to 0^-} \frac{|x|^2}{x^2}$
33. $\lim_{x\to 0^+} \frac{x}{|1-x|}$
34. $\lim_{x\to 1^-} \frac{\sin x}{|x|}$
35. $\lim_{x\to 0^-} \frac{2-x}{|4x|}$
36. $\lim_{x\to 1^+} \frac{\sin(\pi x)}{|x-1|}$
37. $\lim_{x\to 0^-} x|\cot 2x|$
38. $\lim_{x\to 3^-} \frac{\tan(\pi x)}{|9-x^2|}$
39. $\lim_{x\to\pi^-} \frac{\sin(2x)}{\sin|x|}$
40. $\lim_{x\to 0^-} \frac{\sin^2(\pi x)}{x}$

Evaluate each of the following finite or infinite limits, provided that it exists.

41. $\lim_{x\to\infty} \frac{x^3-4x+2}{2x^3+5}$
42. $\lim_{x\to -\infty} \frac{x^3+2x^2-x-4}{3x-2}$
43. $\lim_{x\to 0^+} \cot(-3x)$
44. $\lim_{x\to -\infty} \frac{\sin x}{3x}$

45. $\lim_{x\to 1^+} \frac{1}{x^2-1}$

46. $\lim_{x\to -\infty} \frac{x^3+x+2}{2x^2-3x}$

47. $\lim_{x\to 2} \frac{1}{x^3-8}$

48. $\lim_{x\to \infty} \frac{x+2}{2x\sqrt{x}-2x+1}$

49. $\lim_{x\to 2^-} \frac{1}{\sin(\pi x)}$

50. $\lim_{x\to 0^-} \frac{\cos^2 x+\sin^2 x}{x}$

II

51. Suppose a is a positive number. Evaluate

$$\lim_{x\to a^2} \frac{x-a^2}{a-\sqrt{x}}.$$

52. Suppose a is any real number. Evaluate

$$\lim_{x\to a^3} \frac{x-a^3}{a-x^{1/3}}.$$

53. To evaluate the limit

$$\lim_{x\to 0} \frac{\sin(2x)\cos(2x)}{3x},$$

it is possible to proceed in two different ways. One would be to break the numerator up as a product and deal with each part separately, and the other would be to recognize that the numerator can be expressed in terms of $\sin(4x)$. Evaluate the limit both ways and compare your answers.

***54.** Suppose it is known that for every $x \in (1,3)$, $|k(x)| \leq \sin^2 x$. Find

$$\lim_{x\to 2} \frac{x^2-4}{2-k(x)}.$$

***55.** Find examples of functions $f(x)$ and $g(x)$ and a number a so that

$$\lim_{x\to a} f(x) = 0, \qquad \lim_{x\to a} g(x) = \infty,$$

and for which $\lim_{x\to a^-} f(x)g(x) = -1$ and $\lim_{x\to a^+} f(x)g(x) = 1$.

***56.** Find examples of functions $f(x)$ and $g(x)$ and a number a so that

$$\lim_{x\to a} f(x) = 0, \qquad \lim_{x\to a} g(x) = \infty,$$

and for which $\lim_{x\to a^-} f(x)g(x)$ and $\lim_{x\to a^+} f(x)g(x)$ do not exist, even as infinite limits.

***57.** At each of the points where the function in Example 2.5.4(ii) is not continuous, describe the type of discontinuity.

***58.** Find (i) $\lim_{x\to 0^+} \frac{\sin x}{\sqrt{x}}$, (ii) $\lim_{x\to 0^+} \frac{\sin(\sqrt{x})}{\sqrt{x}}$, (iii) $\lim_{x\to 0} \frac{\sin x}{x^{2/3}}$.

III

59. Use a graphics calculator or computer to draw the graph of the function $y = x^2 \sin \frac{1}{x}$ for $x \in [-1,1]$, $x \in [-\frac{1}{10}, \frac{1}{10}]$, $x \in [-\frac{1}{100}, \frac{1}{100}]$, $x \in [-\frac{1}{1000}, \frac{1}{1000}]$.

60. In Section 2.1 we noted that the height of an object dropped from 128 ft at time $t = 0$ is $h(t) = 128 - 16t^2$. (i) Graph $A(t)$, the average speed of the object from 0 to t. (ii) Graph $S(t)$, the average speed of the object between t and $t + 0.1$. Repeat with $T(t)$, the average speed between t and $t + 0.01$.

For each of the following functions it is not easy to solve for the points at which the denominator is zero. (i) Use your computer or graphics calculator to find approximate values for the zeroes of the denominator. Zoom to higher magnification to find the values to three decimal places. Also note the sign of the denominator on either side of each zero. (ii) Predict the one-sided behavior of each function near each of the points you found in (i). Then graph the function and confirm your predictions.

61. $\frac{3x-2}{x^3-4x+1}$

62. $\frac{x^3-3x}{x^5+3x^3+2}$

63. $\frac{x^3-3x^2+3x-1}{x^3+2x^2-x-2}$

64. $\frac{x^2-x+3}{x^4-2x^3+5x^2-4x+4}$

Derivatives 3

Newton studied gravity, reputedly after watching an apple fall from a tree. Combining this work with his recently developed techniques of calculus, he was able to show how the laws of gravity explained the motions of the planets. This had been a great mystery, and his solution established both Newton's reputation and the importance of calculus.

INTRODUCTION

In this chapter we introduce and study the concept of the derivative. It is the first of two basic ideas in calculus (the other one is the integral, which we will study in Chapter 5). In Section 3.1 we discuss several different examples, which help to show where the idea arises and how it may be useful. In Section 3.2, derivatives are defined, and we learn the basic rules for working with them. In particular, we learn how to differentiate (i.e, find the derivatives of) polynomials and rational functions.

In Section 3.3 we see how derivatives can be used to solve what are known as *maximum/minimum problems,* that is, finding the value of x at which a particular function $f(x)$ has its maximum or minimum value. Solving this type of problem is an extremely important application of calculus. In Section 3.4 we learn how to find the derivatives of trigonometric functions. In Section 3.5 we learn the derivatives of logarithms and exponentials, and we also discuss what are called *hyperbolic functions* and their derivatives.

Section 3.6 is devoted to what is known as the *Chain Rule,* which is a formula for finding the derivative of composite functions. This idea is used in Section 3.7 to develop a technique called "implicit differentiation," which often saves a great deal of work in doing calculations. Derivatives of complicated functions are often found more easily by beginning with their logarithms, using a process known as *logarithmic differentiation.* Section 3.8 studies what are called the *inverse trigonometric functions* and the *inverse hyperbolic functions* and discusses finding their derivatives.

Finally, in Section 3.9 there is a discussion of second and higher-order derivatives. The idea is that if we begin with a function and find its derivative, the derivative will be a function in its own right. The derivative of this function is called the *second derivative* of the original function (since it is obtained by performing twice the operation of finding the derivative).

3.1 Introduction to Differentiation

TANGENT LINE
SLOPE
DIFFERENCE QUOTIENT
DISPLACEMENT
VELOCITY
RATE OF CHANGE

In Section 2.1 we discussed the question of finding the slope of the tangent line to a curve $y = f(x)$ at some specified point on the curve. Ordinarily, the way to find the slope of a line is to start with two points on the line, (x_1, y_1) and (x_2, y_2), say, and to use the formula

$$slope = \frac{rise}{run} = \frac{y_2 - y_1}{x_2 - x_1}.$$

See Figure 3.1.1. The difficulty with applying this approach to tangents is that while we know the point of tangency, call it $\big(a, f(a)\big)$, there is often no apparent way to find a second point on the tangent line. See Figure 3.1.2.

One way around this problem is to consider a second point on the curve, $\big(x, f(x)\big)$, and consider the line through both it and the first point $\big(a, f(a)\big)$. This line is called a secant, and if $\big(x, f(x)\big)$ is chosen to be close to $\big(a, f(a)\big)$, it will likely be close to the tangent line, as Figure 3.1.3 shows. It also seems reasonable to expect

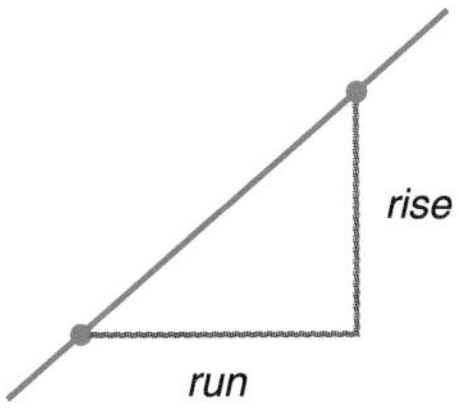

Figure 3.1.1

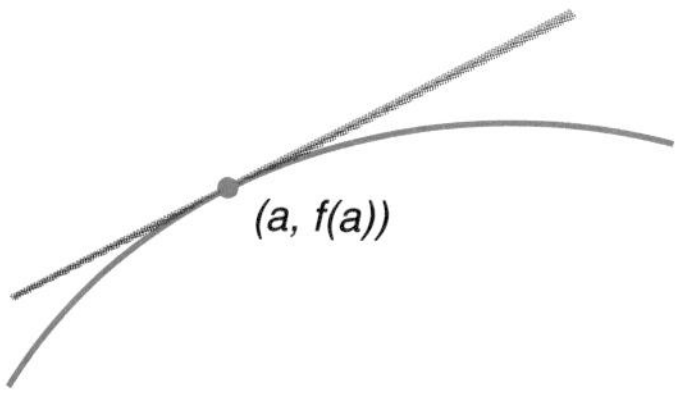

Figure 3.1.2

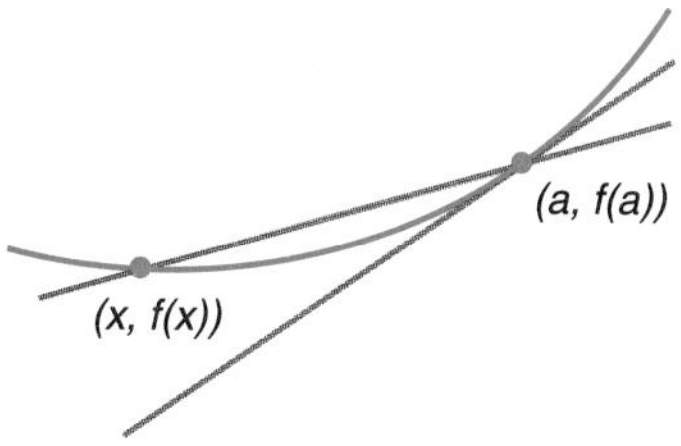

Figure 3.1.3

that its slope will be close to the slope of the tangent line, which is what we want to find. Rather than worry about the awkward question of "how close?," we can let x approach a and find the limit of the slopes of the corresponding secants.

EXAMPLE 1 (i) Find the slope of the line that is tangent to the curve $y = x^3 - 2x + 3$ at the point $(1, 2)$.

If (x, y) is on the curve, then $y = x^3 - 2x + 3$.

Solution First observe that the point $(1, 2)$ is indeed on the curve. Then consider a second point, $(x, x^3 - 2x + 3)$, and the secant through that point and $(1, 2)$. See Figure 3.1.4.

Using the formula we can find the slope of the secant; it is

$$slope = \frac{rise}{run} = \frac{(x^3 - 2x + 3) - 2}{x - 1} = \frac{x^3 - 2x + 1}{x - 1}.$$

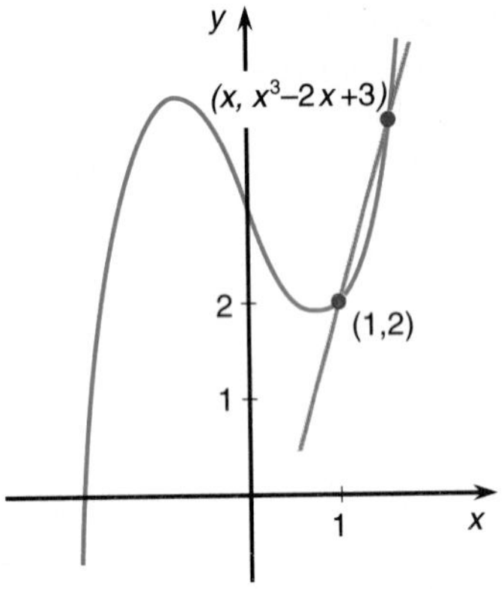

Figure 3.1.4

The slope of the tangent should be the limit of the slope of the secant as $x \to 1$, so we need to evaluate $\lim_{x \to 1} \frac{x^3 - 2x + 1}{x - 1}$. Applying our techniques for limits of rational functions (see Summary 2.2.7), we notice that both numerator and denominator tend to zero as $x \to 1$. Dividing the numerator by $x - 1$ (see Theorem 1.3.5, the Factor Theorem, p. 12), we find

$$\frac{x^3 - 2x + 1}{x - 1} = \frac{(x - 1)(x^2 + x - 1)}{x - 1} = x^2 + x - 1.$$

Taking the limit is easy now:

$$\lim_{x \to 1} \frac{x^3 - 2x + 1}{x - 1} = \lim_{x \to 1} (x^2 + x - 1) = 1.$$

The slope of the tangent at $(1, 2)$ should be the limit of the slopes of the secants as $x \to 1$, which is 1 (see Figure 3.1.5).

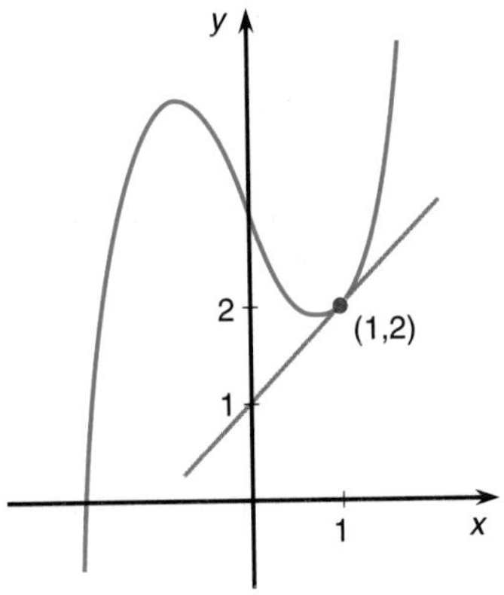

Figure 3.1.5

(ii) Find the slope of the tangent to the parabola $y = x^2$ at the point (a, a^2).

Solution Consider the secant through (a, a^2) and (x, x^2). Its slope is $\frac{x^2 - a^2}{x - a}$. The slope of the tangent should be

$$\lim_{x \to a} \frac{x^2 - a^2}{x - a} = \lim_{x \to a} \frac{(x - a)(x + a)}{x - a} = \lim_{x \to a} (x + a) = 2a.$$

Notice that the slope of the tangent will be different at different points. For large values of a, that is, at points far out along the parabola, the slope will be large, meaning that the tangent rises steeply up to the right. When $a = 0$, that is, at the origin, the slope of the tangent is zero, meaning that it is a horizontal line. When $a < 0$, the slope is negative, meaning that the tangent line slopes down to the right (see Figure 3.1.6).

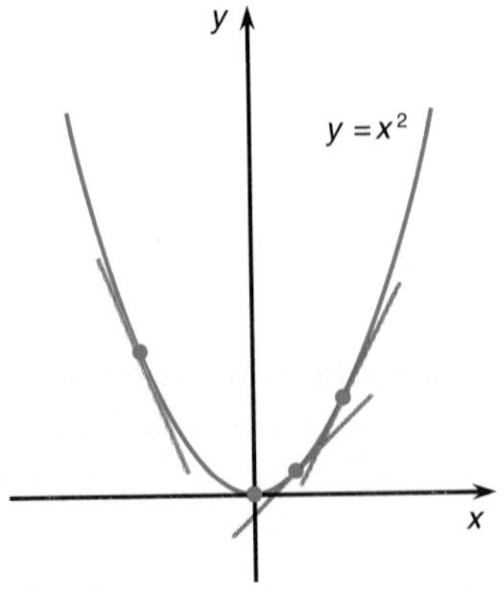

Figure 3.1.6

With a graphics calculator we can plot the function $y = x^2$ and some of its tangent lines. For instance, by Example 1, the tangent line at $(1, 1)$ has slope $2(1) = 2$. Since it passes through $(1, 1)$, its equation is $\frac{y-1}{x-1} = 2$, that is, $y - 1 = 2(x - 1)$, that is, $y = 2x - 1$. Similarly, the tangent line at $(-1, 1)$ has slope -2 and equation $\frac{y-1}{x+1} = -2$, that is, $y = 1 - 2(x + 1) = -2x - 1$.

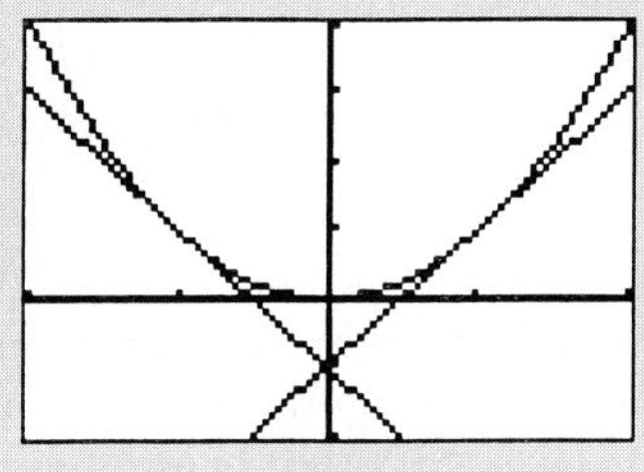

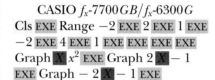

TI-85	SHARP EL9300
GRAPH F1 ($y(x)$ =) F1 (x) x^2 ▼ 2 F1 (x) − 1 ▼ (−)2 F1 (x) − 1 M2(RANGE) (−)2 ▼ 2 ▼ ▼ (−)2 ▼ 4 F5(GRAPH)	[graph key] x/θ/T x^2 2ndF ▼ 2 x/θ/T − 1 2ndF ▼ − 2 x/θ/T − 1 RANGE −2 ENTER 2 ENTER 1 ENTER −2 ENTER 4 ENTER 1 ENTER [graph key]
CASIO f_x-7700GB/f_x-6300G	**HP 48SX**
Cls EXE Range −2 EXE 2 EXE 1 EXE −2 EXE 4 EXE 1 EXE EXE EXE EXE Graph X x^2 EXE Graph 2 X − 1 EXE Graph − 2 X − 1 EXE	↰ PLOT PLOTR ERASE ATTN ′ αY = αX y^x 2 ↱ PLOT ↰ DRAW ERASE 2 +/− ENTER 2 XRNG 2 +/− ENTER 4 YRNG DRAW ATTN ′ αY = 2 × αX − 1 ↱ PLOT ↰ DRAW DRAW ATTN ′ αY = 2 +/− × αX − 1 ↱ PLOT ↰ DRAW DRAW

Upon reflection it is clear that we have used the notion of a tangent line without ever really saying what it meant, and then figured out what its slope ought to be. On the other hand, now that we know what its slope ought to be, we are in a position to say what a tangent line is.

DEFINITION 3.1.1

Consider the function $y = f(x)$, and suppose a is a point in the domain of f, so $\big(a, f(a)\big)$ is a point on the graph of $y = f(x)$. The **tangent line** to $y = f(x)$ at the point $\big(a, f(a)\big)$ is the straight line through $\big(a, f(a)\big)$ whose slope is

$$m = \lim_{x \to a} \frac{f(x) - f(a)}{x - a},$$

provided the limit exists. In particular, if the limit m does exist, the equation of the line is

$$y - f(a) = m(x - a), \qquad \text{or, equivalently,} \qquad y = f(a) + m(x - a).$$

If the limit does not exist, then neither does the tangent line.

For obvious reasons, the quantity $\frac{f(x)-f(a)}{x-a}$ is called a **difference quotient**. It equals the slope of the secant through $(a, f(a))$ and $(x, f(x))$.

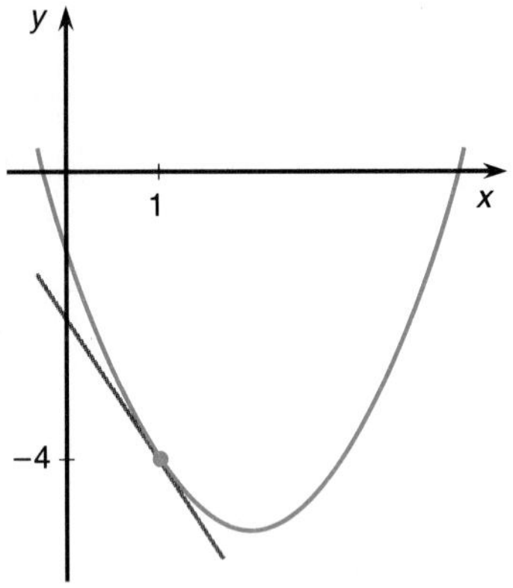

Figure 3.1.7

EXAMPLE 2 (i) Find the equation of the tangent line to the curve $y = f(x)$, where $f(x) = x^2 - 4x - 1$, at the point $(1, -4)$.

Solution The slope of the tangent line is

$$m = \lim_{x\to1} \frac{f(x) - f(1)}{x - 1} = \lim_{x\to1} \frac{(x^2 - 4x - 1) - \left(1^2 - 4(1) - 1\right)}{x - 1}$$

$$= \lim_{x\to1} \frac{x^2 - 4x + 3}{x - 1} = \lim_{x\to1} \frac{(x-1)(x-3)}{x - 1} = \lim_{x\to1} (x - 3) = -2.$$

Since the line passes through $(1, -4)$, its equation is

$$y - (-4) = -2(x - 1), \qquad \text{or} \qquad y = -2x - 2.$$

See Figure 3.1.7.

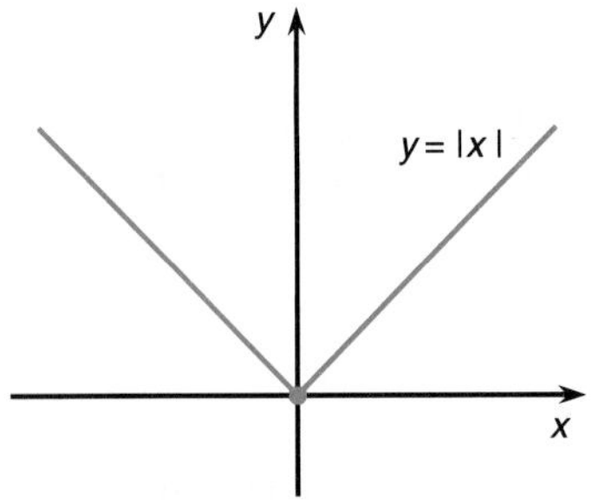

Figure 3.1.8

(ii) If we consider the graph of $y = f(x)$, where $f(x) = |x|$, and ask about the tangent line at the point $(0, 0)$, we have to evaluate the limit

$$\lim_{x\to0} \frac{f(x) - f(0)}{x - 0} = \lim_{x\to0} \frac{|x| - 0}{x} = \lim_{x\to0} \frac{|x|}{x}.$$

As we saw in Example 2.4.1 (i), the function $\frac{|x|}{x}$ approaches different limits from above and below ($+1$ as $x \to 0^+$ and -1 as $x \to 0^-$). Accordingly, the (two-sided) limit does not exist (by Theorem 2.4.2), and therefore the tangent line does not exist (see Figure 3.1.8). The idea is that there is no single line that is an obvious candidate to be the tangent line; if we approach 0 from the left, it appears that the slope of the tangent should be -1, but if we approach from the right, the slope appears to be $+1$.

Figure 3.1.9

Another idea that we discussed in Section 2.1 was the concept of instantaneous speed. Suppose an object is moving in a straight line, and suppose some point on the line has been marked as a reference for making measurements (following the usual convention, we label this point O and call it the origin). See Figure 3.1.9.

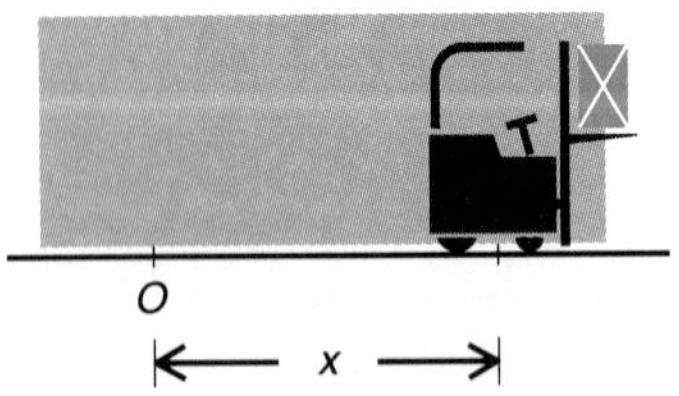

Figure 3.1.10

Suppose we know where the moving object is at any particular moment. In effect this means that we know its position relative to the origin at any time t, or more specifically that we know a function $f(t)$ that tells us its position at time t. See Figure 3.1.10. Once again we follow the convention and measure positions on the line in such a way that when $f(t) > 0$, it represents a point to the right of the origin, and when $f(t) < 0$, it represents a point to the left of the origin. The absolute value $|f(t)|$ indicates the distance from the origin to the object. For example, if $f(3) = -2$, we know that at time $t = 3$ the object is 2 units to the left of the origin. Similarly, if $f(1.5) = \sqrt{2}$, then at time $t = 1.5$ the object is $\sqrt{2} \approx 1.41421$ units to the right of the origin. We say that $f(t)$ is the **displacement** of the object from the origin at time t. Displacement measures the distance, but its sign also tells us the direction from the origin. In contrast, when we speak of distance, we always mean a nonnegative number, and no direction is implied.

The average speed of a moving object is simply the distance it travels divided by the time it takes. To find the average speed of the object between times $t = 2$ and

$t = 4$, for example, we could consider the quotient

$$\frac{f(4) - f(2)}{4 - 2} = \frac{f(4) - f(2)}{2}.$$

There is a catch, however. The numerator is the difference between the *displacements* of the object at the two times $t = 2$ and $t = 4$, which is almost the same as the distance between the corresponding positions, except that it might be negative. The difference quotient we have written down is the change in the displacement divided by the length of time. If the quotient is negative, it represents the average speed of a motion to the left, and if it is positive, it is the average speed of a motion to the right. Strictly speaking, between $t = 2$ and $t = 4$ the object might easily move to the right some of the time and to the left some of the time, but the sign of the quotient tells us the direction of the *net* motion.

To distinguish this from the average speed, which we understand always to be nonnegative, it is called the average **velocity**. The absolute value of the velocity is the speed of a motion, and the sign of the velocity indicates the direction (positive means to the right and negative means to the left).

The difference quotient we considered above, the change in the displacement divided by the time, represents the average velocity of the object between $t = 2$ and $t = 4$. If we want to know its velocity at the "instant" when $t = 2$, we cannot simply take a quotient. Instead, we should take difference quotients corresponding to very brief intervals starting at $t = 2$ and evaluate the limit as the length of the intervals approaches zero.

Specifically, if we consider an interval of time of length h, say, what we mean is the interval from $t = 2$ to $t = 2 + h$. The change in the displacement is $f(2 + h) - f(2)$, and the length of time elapsed is $(2 + h) - 2 = h$. The average velocity is

$$\frac{\textit{change in displacement}}{\textit{time}} = \frac{f(2 + h) - f(2)}{h}.$$

To find the instantaneous velocity, we should let $h \to 0$ and evaluate the limit. In other words, the instantaneous velocity at $t = 2$ is

$$\lim_{h \to 0} \frac{f(2 + h) - f(2)}{h}.$$

This limit is very similar to the one we used in Definition 3.1.1 for the slope of the tangent line. It arises so often that there is a special name and notation for it.

DEFINITION 3.1.2

The **derivative** of $f(x)$ at $x = a$, written $f'(a)$, equals the following limit, provided it exists:

$$f'(a) = \lim_{x \to a} \frac{f(x) - f(a)}{x - a} = \lim_{h \to 0} \frac{f(a + h) - f(a)}{h}.$$

(Notice that the two apparently different limits are really the same, since letting $h \to 0$ makes $(a + h) \to a$, so the first expression is obtained from the second by replacing $a + h$ with x.)

Both $\frac{f(x)-f(a)}{x-a}$ and $\frac{f(a+h)-f(a)}{h}$ are called **difference quotients**.

EXAMPLE 3 (i) Suppose the displacement of a moving object at time t sec is given, in feet, by the function $f(t) = t^2 - 8t - 4$. What is its (instantaneous) velocity at time $t = 3$?

Solution The instantaneous velocity at $t = 3$ is

$$f'(3) = \lim_{h\to 0} \frac{f(3+h) - f(3)}{h}.$$

Simplifying the difference quotient, we find that

$$\begin{aligned}\frac{f(3+h) - f(3)}{h} &= \frac{\left((3+h)^2 - 8(3+h) - 4\right) - \left(3^2 - 8(3) - 4\right)}{h} \\ &= \frac{9 + 6h + h^2 - 24 - 8h - 4 - 9 + 24 + 4}{h} = \frac{-2h + h^2}{h} \\ &= -2 + h \text{ ft/sec.}\end{aligned}$$

So the instantaneous velocity at $t = 3$ is

$$f'(3) = \lim_{h\to 0}(-2 + h) = -2 \text{ ft/sec.}$$

This means that at the instant $t = 3$ sec the object is moving to the left at a speed of 2 ft/sec.

The difference quotient $\frac{f(3+h)-f(3)}{h}$ is precisely the average velocity of the moving object between the times $t = 3$ and $t = 3 + h$.

(ii) At what time t will the above moving object be exactly at rest, that is, have velocity equal to zero?

Solution The velocity at $t = a$ is

$$\begin{aligned}f'(a) &= \lim_{h\to 0} \frac{f(a+h) - f(a)}{h} \\ &= \lim_{h\to 0} \frac{\left((a+h)^2 - 8(a+h) - 4\right) - (a^2 - 8a - 4)}{h} \\ &= \lim_{h\to 0} \frac{a^2 + 2ah + h^2 - 8a - 8h - 4 - a^2 + 8a + 4}{h} \\ &= \lim_{h\to 0} \frac{2ah + h^2 - 8h}{h} \\ &= \lim_{h\to 0}(2a + h - 8) = 2a - 8.\end{aligned}$$

This equals zero precisely when $a = 4$.

Notice that when $a < 4$, the velocity at $t = a$ will be negative, and when $a > 4$, the velocity will be positive. So the object will be moving to the left until $t = 4$, and after $t = 4$ it will be moving to the right. The instant when $t = 4$ is the instant at which the object comes to rest as it "turns around."

Now that we have the notation for derivatives, it is easier to deal with tangent lines.

EXAMPLE 4 (i) Find the slope of the tangent at each point of $y = x^2 - 2x + 2$.

Solution Let $f(x) = x^2 - 2x + 2$. Any point on the graph can be written in the form $(a, a^2 - 2a + 2)$, and the slope of the tangent at that point will just be

$$\begin{aligned} f'(a) &= \lim_{h \to 0} \frac{f(a+h) - f(a)}{h} \\ &= \lim_{h \to 0} \frac{(a+h)^2 - 2(a+h) + 2 - (a^2 - 2a + 2)}{h} \\ &= \lim_{h \to 0} \frac{a^2 + 2ah + h^2 - 2a - 2h + 2 - a^2 + 2a - 2}{h} \\ &= \lim_{h \to 0} \frac{2ah + h^2 - 2h}{h} = \lim_{h \to 0} (2a + h - 2) \\ &= 2a - 2. \end{aligned}$$

(ii) Find the equation(s) of the tangent lines to the curve $y = x^2 - 2x + 2$ that pass through the point $(2, 0)$.

Solution Observe that the point $(2, 0)$ does not lie on the curve (it does not satisfy the equation). So what the question asks us is to find lines that are tangent to the curve at some point on the curve and that also happen to pass through $(2, 0)$.

Consider the tangent at the point $(a, a^2 - 2a + 2)$. Its equation is $y = f(a) + f'(a)(x - a)$. (Notice that we have written $f'(a)$ instead of the awkward limit m that appeared in Definition 3.1.1.) The question we must answer is for what values of a does this line pass through $(2, 0)$?

Substituting into this equation of the tangent the fact that $f(a) = a^2 - 2a + 2$ and the result of part (i) that $f'(a) = 2a - 2$, we find that the tangent is

$$y = a^2 - 2a + 2 + (2a - 2)(x - a) = a^2 - 2a + 2 + (2a - 2)x - 2a^2 + 2a,$$

or

$$y = (2a - 2)x - a^2 + 2.$$

For $(2, 0)$ to be on this line means $0 = (2a - 2)(2) - a^2 + 2 = 4a - 4 - a^2 + 2$, or $a^2 - 4a + 2 = 0$.

Solving by using the Quadratic Formula (Theorem 1.3.4), we find

$$a = \frac{4 \pm \sqrt{16 - 8}}{2} = 2 \pm \frac{\sqrt{8}}{2} = 2 \pm \sqrt{2}.$$

We have found that the tangent lines to $y = f(x)$ that pass through $(2, 0)$ are the ones whose points of tangency are at $x = 2 + \sqrt{2}$ and $x = 2 - \sqrt{2}$.

We check this result: When $a = 2 + \sqrt{2} \approx 3.4142$, the tangent line, as we found above, is

$$y = (2a - 2)x - a^2 + 2 = (4 + 2\sqrt{2} - 2)x - (4 + 4\sqrt{2} + 2) + 2,$$

or

$$y = (2 + 2\sqrt{2})x - 4 - 4\sqrt{2}.$$

This line certainly passes through $(2, 0)$.

A similar calculation shows that when $a = 2 - \sqrt{2} \approx 0.5858$, the tangent line is

$$y = (2 - 2\sqrt{2})x - 4 + 4\sqrt{2},$$

which also contains $(2, 0)$. See Figure 3.1.11.

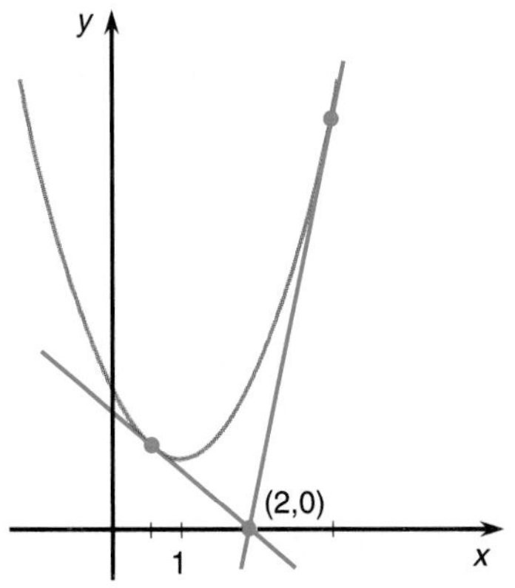

Figure 3.1.11

EXAMPLE 5 **Ideal Gas Law** Suppose a quantity of some kind of gas is confined in a container at a fixed temperature. The **Ideal Gas Law** of physics says that the volume of the gas, V, and the pressure, P, are related by the following equation:

$$V = \frac{k}{P},$$

where k is some constant.

In other words, the volume is *inversely proportional* to the pressure, which is a precise way of stating the obvious fact that the harder you squeeze the gas, the smaller will be the volume it occupies.

If the pressure is P, then the volume will be $V = \frac{k}{P}$. If we change the pressure slightly, the volume will change in response. For instance, if we change the pressure from P to $P + h$, then the volume will change to $V = \frac{k}{P+h}$.

Thinking of V as a function of P, we can find its derivative. For fixed P,

$$\begin{aligned} V'(P) &= \lim_{h \to 0} \frac{V(P+h) - V(P)}{h} = \lim_{h \to 0} \frac{\frac{k}{P+h} - \frac{k}{P}}{h} = \lim_{h \to 0} \frac{\frac{kP - k(P+h)}{(P+h)P}}{h} \\ &= \lim_{h \to 0} \frac{\frac{-kh}{(P+h)P}}{h} = \lim_{h \to 0} \frac{-k}{(P+h)P} \\ &= \frac{-k}{P^2}. \end{aligned}$$

In doing this calculation we think of P as a *constant*, since it is some *fixed* value of the pressure from which we start. So, for instance, in the evaluation of the last limit we regard it as a rational function in the variable h, and the other two symbols, k and P, are constants.

The difference quotient whose limit we took was the ratio of the change in V to the change in P. We think of the change in volume V as the result of the change in pressure P. If the limit, that is, the derivative, is large, it suggests that a small change in pressure will result in quite a large change in volume; if the derivative is small, a change in the pressure will cause only a small change in the volume. The fact that the derivative is negative reflects the fact that an increase in the pressure results in a *decrease* in the volume and vice versa. If, for example, $V'(P) = -2$, it means that an increase in the pressure by 1 unit, starting from P, should be expected to cause a decrease in the volume of about 2 units.

These considerations lead us to think of the derivative $V'(P)$ as the *rate of change* of the volume with respect to the pressure.

In the earlier discussion of velocity we could also have thought of the velocity as the rate of change of the displacement with respect to time. The bigger the velocity is, the faster the displacement will change as time passes.

> In general, if $y = f(x)$ is some function of the variable x, the derivative $f'(a)$ can be thought of as the **rate of change** of y with respect to x at $x = a$. If we imagine ourselves making a small change in the value of x near a, the derivative tells us what kind of change we should expect in the value of $y = f(x)$. The sign of the derivative tells us whether to expect an increase or a decrease.

Now consider another type of example. Suppose a factory produces a single product, blenders, for instance. Let N be the number of blenders produced in a day, the "level of production." The profit P earned by the factory will depend on N, probably in a very complicated way (it will be best when the factory is operating at its proper capacity; operating too far below or above capacity will likely reduce profits).

Now suppose somebody is able to describe the way profits depend on production level by giving us a function $P = f(N)$. In other words, for any given level of production N, the value $P = f(N)$ will be the resulting profit. In actual situations, such a function is usually impossible to find exactly, though it is not so difficult to make approximations.

If the factory is producing N_0 blenders per day, it would be reasonable to want to know whether this is a good level of production. One way to accomplish that would be to ask what would be the effect on the profit of increasing or decreasing production. An answer to this question would certainly help in deciding how best to run the factory.

To answer the question, we could ask what happens to P if we increase production by some small amount h. The new level of production would be $N_0 + h$, and the corresponding profit would be $P = f(N_0 + h)$. The change in the profit would be $f(N_0 + h) - f(N_0)$, and to some extent the question is answered by knowing this change. Specifically, if it is positive, then increased production increases profit, and if it is negative, then increased production decreases profit. However, this simple observation does not take into account the size of h. It is clearly important to know how much of a change in P will result from a particular change in N.

The difference quotient $\frac{f(N_0+h)-f(N_0)}{h}$ represents the change in profit *per unit change in production*. If this quotient equals \$5, for example, it means that increasing production by h blenders will result in a \$$5h$ increase in profit, or \$5 per additional blender produced.

In keeping with the spirit of the previous examples we can let $h \to 0$ and consider the limit

$$\lim_{h \to 0} \frac{f(N_0 + h) - f(N_0)}{h},$$

which we can think of as the *rate of change of profit per unit change in production*. What it tells us is roughly how much profit will increase if we increase production by one unit. In economics this limit is called the **marginal increase in profit**.

EXAMPLE 6 For a factory producing N blenders per day, suppose the profit, in dollars, is given by $P = P(N) = 6{,}000N - N^2 - 8{,}985{,}000$.

We find that the derivative $P'(N)$ equals

$$\lim_{h \to 0} \frac{6{,}000(N + h) - (N + h)^2 - 8{,}985{,}000 - (6{,}000N - N^2 - 8{,}985{,}000)}{h}$$

$$= \lim_{h \to 0} \frac{6{,}000N + 6{,}000h - N^2 - 2hN - h^2 - 8{,}985{,}000 - 6{,}000N + N^2 + 8{,}985{,}000)}{h}$$

$$= \lim_{h \to 0} \frac{6{,}000h - 2hN - h^2}{h}$$

$$= \lim_{h \to 0} (6{,}000 - 2N - h)$$

$$= 6{,}000 - 2N.$$

The interesting thing about this derivative is that it is easy to see when it is negative and when it is positive. In fact, it is positive when $N < 3{,}000$ and negative when $N > 3{,}000$. This suggests that if $N < 3{,}000$, a small increase in production would result in an increase in profits but that if $N > 3{,}000$, a small increase in production would result in lower profits.

From this it seems reasonable to expect that the most profit will be obtained with $N = 3{,}000$, that is, that the most profitable way to operate the factory is to produce 3,000 blenders per day.

We can check this to some extent by direct calculation. If $N = 3{,}000$, we find that $P = P(3{,}000) = 6{,}000(3{,}000) - 3{,}000^2 - 8{,}985{,}000 = 18{,}000{,}000 - 9{,}000{,}000 - 8{,}985{,}000 = 15{,}000$, or a profit of \$5.00 for each of the 3,000 blenders produced. If, on the other hand, we were to produce one more: $N = 3,001$, then $P = 6{,}000(3,001) - 3,001^2 - 8,985{,}000 = 18{,}006{,}000 - 9{,}006,001 - 80{,}985{,}000 = 14{,}999$. In other words, total profits would decrease by \$1.00. You can check that if $N = 2{,}999$, the profit would also be less than 15,000.

The above calculation encourages us to think that our reasoning may be correct, but it certainly does not prove conclusively that it is impossible to get higher profits by using some other value of N.

On the other hand, if we take the formula $P = 6{,}000N - N^2 - 8{,}985{,}000$ and complete the square, we find

$$\begin{aligned} P &= -(N^2 - 6{,}000N + 8{,}985{,}000) = -(N^2 - 6{,}000N + 9{,}000{,}000 - 15{,}000) \\ &= -(N^2 - 6{,}000N + 9{,}000{,}000) + 15{,}000 = 15{,}000 - (N - 3{,}000)^2. \end{aligned}$$

This last expression is 15,000 minus a square, and since any square must always be nonnegative, the expression must always be less than or equal to 15,000. Moreover, the only way for it to equal 15,000 is for the square to be zero, which can happen only if $N = 3{,}000$.

So in fact the highest possible profit is 15,000, which will be obtained precisely when $N = 3{,}000$. ◢

It is important to mention once again that while this is fine as far as it goes, we assumed at the beginning that we knew how to express the profit P as a function of production N, but in the real world it is unlikely that we could do that.

Exercises 3.1

I

For each of the following functions $f(x)$, calculate the limit of the difference quotient to find the derivative $f'(a)$ at the specified point a.

1. $f(x) = x^2 + 1, a = 2$

2. $f(x) = x^2 - 3, a = -1$

3. $f(x) = x^2 - x, a = 1$

4. $f(x) = x^2 - 3x, a = 0$

5. $f(x) = 2x^2 - x + 1, a = 2$

6. $f(x) = 3x^2 - 2x + 1, a = -3$

7. $f(x) = -3x^2 + 2x - 2, a = 0$

8. $f(x) = -x^2 + 4x, a = 7$

9. $f(x) = x^3, a = 1$

10. $f(x) = x^3 + 4x - 1, a = -2$

11.– 20. For each of the functions $f(x)$ in Exercises 1 through 10, find the equation of the line that is tangent to the graph of $y = f(x)$ at the point $(a, f(a))$.

II

21. Find all values a for which the tangent line to $y = f(x) = x^2 - 2x + 3$ at $(a, f(a))$ is horizontal.

22. Find all values a for which the tangent line to $y = f(x) = 2x^2 + 2x - 5$ at $(a, f(a))$ is horizontal.

23. (i) Find all tangent lines to the graph of $y = x^2$ that pass through the point $(0, -1)$.
(ii) Find all tangent lines to the graph of $y = x^2$ that pass through the point $(0, 1)$. (Draw a picture.)

24. (i) If the height of a falling object t sec after its release is $H(t) = 128 - 16t^2$ ft, find its velocity $H'(t)$ at time t.
(ii) What is the significance of the sign of the velocity?

25. Suppose a ball is thrown upward in such a way that its height above the ground after t sec in flight is $H(t) = 6 + 32t - 16t^2$ ft.
(i) Find its velocity after t sec in flight.
(ii) When is it rising and when is it falling?
(iii) What is the greatest height it will reach?

26. Suppose a train starts at the origin and moves along a straight track. Suppose after t minutes it has gone $1000t - 50t^2$ ft.
(i) What is its velocity after t min in feet per minute?
(ii) What is the significance of the sign of the velocity?
(iii) When will the train come to rest?
(iv) How far will it have traveled at the moment when it comes to rest?
(v) What will be the greatest speed it reaches before coming to rest?
(vi) Express the answer to part (v) in miles per hour.
(vii) Where will the train be 20 min after it starts?

27. Suppose a balloon contains 0.4 quart of air at a pressure of 15 pounds/sq in (psi).
(i) In the terminology of Example 5, find the constant k for this situation.
(ii) Suppose the pressure is increased to 30 psi. What will be the volume?
(iii) Compare the rate of change of volume with respect to pressure in these two situations (i.e., when $P = 15$ and $P = 30$). In which case will increasing the pressure by 1 psi cause the greater change in the volume?
(iv) Use your answer to part (iii) when $P = 15$ psi to estimate what the volume will be if the pressure is increased by 1 psi to 16 psi.
(v) Calculate the exact volume when $P = 16$ psi, using the formula of Example 5. (The discrepancy arises because the rate of change we found when $P = 15$ psi will not be the rate of change when $P = 16$ psi or 15.5 psi or other points in between. It will be reasonably close, so this approach can be used for estimating, but it will not give exact answers.)

28. Suppose the profit of a factory producing N electric blankets per week is given by $P = 400N - N^2 - 38{,}000$ dollars.
(i) What would you recommend the manager do to increase profits if the level of production was $N = 100$?
(ii) What would you suggest if $N = 500$?
(iii) Can you suggest a strategy that will describe what to do for any value of N?
(iv) What is the greatest possible profit this factory can make under the present circumstances, and how can they make this level of profit?

***29.** (i) Use Definition 3.1.2 (p. 133) to find the derivative $f'(1)$ if $f(x) = Ax^2 + Bx + C$, where A, B, C are all constants.
(ii) Use Definition 3.1.2 to find the derivative $f'(a)$, where $f(x)$ is as above and a is some fixed but unspecified number.

***30.** (i) Use Definition 3.1.2 to find the derivative at $x = 2$ of $f(x) = x^3$.
(ii) Use Definition 3.1.2 to find the derivative $f'(a)$, where $f(x) = x^3$.

***31.** Let $f(x) = |x|$. Specify the points a at which the derivative $f'(a)$ does not exist, and find the derivative at all the points where it does exist.

***32.** (i) Show that the derivative of a straight line $y = mx + b$ equals the slope of the line.
(ii) Notice that the straight line $y = mx + b$ is never a vertical line. What goes wrong with part (i) above if we try to do the same thing for a vertical line?

III

33. In Example 1 (p. 129) we discussed the tangent line to $y = x^3 - 2x + 3$ at the point $(1, 2)$. Use a graphics calculator or computer to draw this curve and the secant lines through (i) $(1, 2)$ and $(2, 7)$, (ii) $(1, 2)$ and $\left(\frac{3}{2}, \frac{27}{8}\right)$, (iii) $(1, 2)$ and $(1.1, 2.131)$.

34. In Example 2 (p. 132) we discussed the tangent line to $y = f(x) = x^2 - 4x - 1$ at the point $(1, -4)$. Graph the curve and the secant through $(1, -4)$ and $\big(1 + h, f(1 + h)\big)$, where (i) $h = 1$, (ii) $h = \frac{1}{2}$, (iii) $h = 0.1$, (iv) $h = 0.01$.

35. Consider the moving object of Example 3 (p. 134). At $t = 3$ its displacement is $f(3) = 3^2 - 8(3) - 4 = -19$. Let $D(h)$ be the net displacement away from this point at time $t = 3 + h$, that is, $D(h) = f(3 + h) - f(3)$.
(i) Graph $D(h)$.
(ii) Let $V(h)$ be the object's average velocity between $t = 3$ and $t = 3 + h$. Graph $V(h)$, and use the

picture to confirm the result we found for the instantaneous velocity at $t = 3$.

36. Use your graphics calculator or computer as in Exercise 35 to find the instantaneous velocity of the same moving object at $t = 6$.

37. Graph the profit function $P(N) = 6{,}000N - N^2 - 8{,}985{,}000$ of Example 6. We found that the highest profit is $P = 15{,}000$, obtained when $N = 3{,}000$. Use the picture to find all the values of N for which the profit is at least 10,000.

3.2 Differentiation Rules; Derivatives of Polynomials and Rational Functions

DERIVATIVES OF POLYNOMIALS
PRODUCT RULE
QUOTIENT RULE
DERIVATIVES OF RATIONAL FUNCTIONS
EQUATIONS OF TANGENT LINES

We have defined the derivative of a function $f(x)$ at a point a to be the limit of the difference quotient:

$$f'(a) = \lim_{h \to 0} \frac{f(a+h) - f(a)}{h},$$

provided the limit exists. We saw in Example 3.1.2(ii) that the derivative does not always exist (in that example we saw that $f(x) = |x|$ does not have a derivative at $x = 0$). If the derivative $f'(a)$ exists (i.e., the above limit exists), we say the function $f(x)$ is **differentiable** at $x = a$. To say that $f(x)$ is "differentiable on an interval (c, d)" means that $f(x)$ is differentiable at every $a \in (c, d)$. If $f(x)$ is differentiable at every point in its domain, we simply say that $f(x)$ is differentiable.

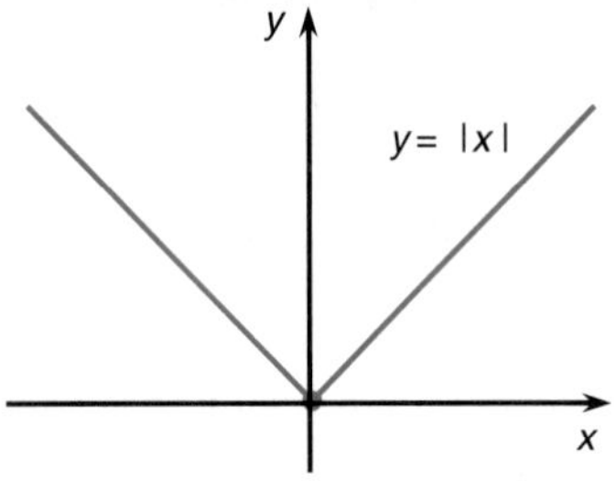

Figure 3.2.1

Roughly speaking, the derivative $f'(a)$ is the slope of the tangent line to the graph of $y = f(x)$ at the point $(a, f(a))$. The reason why $f(x) = |x|$ is not differentiable at $x = 0$ is that it does not have a tangent line there, and that happens because the graph has a "corner" in it (see Figure 3.2.1).

If the function $f(x)$ has a tangent at $x = a$, the graph of $y = f(x)$ must lie "close" to the graph of the tangent line near the point $(a, f(a))$. This suggests that $f(x)$ ought to be continuous at $x = a$. Certainly, if $f(x)$ had a removable or jump discontinuity at $x = a$, it is hard to imagine that it could have a tangent there. See Figure 3.2.2.

THEOREM 3.2.1

Suppose $f(x)$ is differentiable at $x = a$. Then $f(x)$ must be **continuous** at $x = a$.

PROOF To show that $\lim_{x \to a} f(x) = f(a)$ amounts to the same thing as showing that $\lim_{x \to a}\big(f(x) - f(a)\big) = 0$. And

> Note that dividing by $x - a$ is not allowed at $x = a$, but to find the limit, we need to worry only about the points where $x \neq a \ldots$

$$\begin{aligned}\lim_{x \to a}\big(f(x) - f(a)\big) &= \lim_{x \to a} \frac{f(x) - f(a)}{x - a}(x - a)\\ &= \left(\lim_{x \to a} \frac{f(x) - f(a)}{x - a}\right)\left(\lim_{x \to a}(x - a)\right)\\ &= \big(f'(a)\big)(0) = 0.\end{aligned}$$

So $\lim_{x \to a} f(x) = f(a)$, which says that $f(x)$ is continuous at $x = a$, as required.

Figure 3.2.2

The theorem says that if a function is differentiable, then it will also be continuous. However, being continuous is not enough to guarantee that a function will be differentiable, as the function $f(x) = |x|$ shows. After all, $f(x)$ is continuous everywhere, but it is not differentiable at $x = 0$.

As with limits, there are some derivatives that are very easy to evaluate.

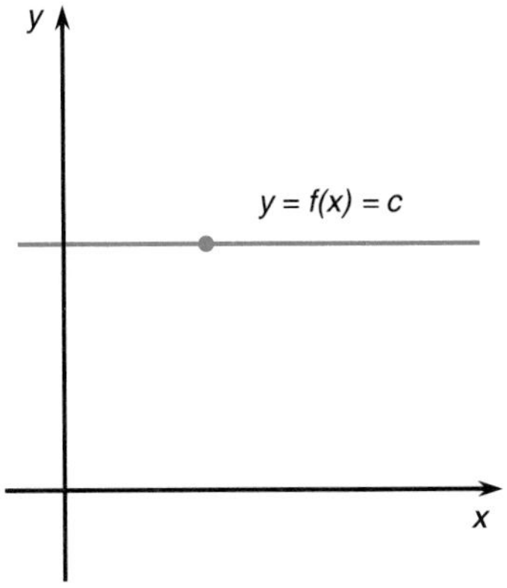

Figure 3.2.3

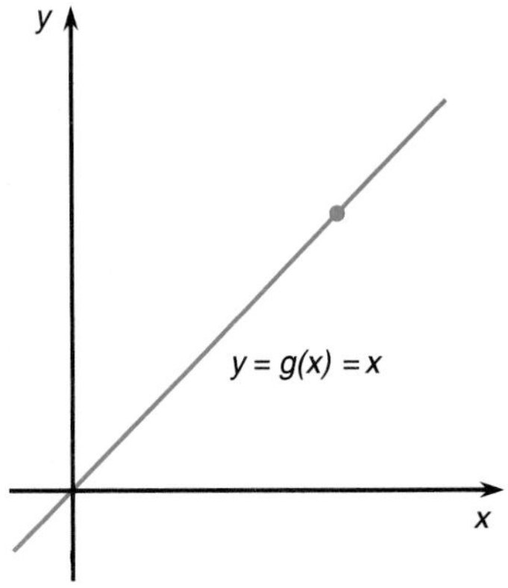

Figure 3.2.4

EXAMPLE 1 Let $f(x) = c$ be a constant function; let $g(x) = x$. Find $f'(x)$ and $g'(x)$ for all values of x.

Solution We simply write down the difference quotients and evaluate the limits:

$$f'(x) = \lim_{h\to 0} \frac{f(x+h) - f(x)}{h} = \lim_{h\to 0} \frac{c-c}{h} = \lim_{h\to 0} \frac{0}{h} = 0.$$

Similarly,

$$g'(x) = \lim_{h\to 0} \frac{g(x+h) - g(x)}{h} = \lim_{h\to 0} \frac{(x+h)-x}{h} = \lim_{h\to 0} \frac{h}{h} = 1.$$

In other words, the derivative of any constant function is zero, and the derivative of $g(x) = x$ is the constant function 1. In particular, constant functions and the function $g(x) = x$ are differentiable at every point. See Figure 3.2.3, which shows that the tangent line to any constant function is a horizontal line, so its slope is zero, and Figure 3.2.4, which shows that the tangent line to $y = g(x) = x$ is the line $y = x$, whose slope is 1.

There is another notation for derivatives with which we should become familiar. When we defined $f'(a)$, the number a was fixed, but we have already seen that it is often possible to work out a formula for $f'(a)$ that tells what it will be for any choice of a. In this light the derivative itself becomes a function. We write $f'(x)$ for the *function* whose value is $f'(a)$ at each point $x = a$ where $f(x)$ is differentiable. The process of finding the derivative $f'(x)$ from $f(x)$ is called differentiation.

The disadvantage of this notation is that the function usually needs to have a name (like $f(x)$ or $g(x)$, etc.). The alternate notation for the derivative of $f(x)$ is

$$\frac{df}{dx} = \frac{d}{dx}(f).$$

These symbols are read "$d\,f$ by $d\,x$" and "d by $d\,x$ of f." The notation includes the idea that $\frac{df}{dx}$ is a function (whose value at x is $\frac{df}{dx}(x)$), but it also allows us to write, for instance,

$$\frac{d}{dx}(x^3 - 3x^2 + 5x - 2)$$

for the function that is the derivative of $x^3 - 3x^2 + 5x - 2$, without having to give a name to that polynomial. This may not seem important, but it can be very useful.

For instance, the results of Example 1 can be written as

$$\frac{d}{dx}(c) = 0, \qquad \frac{d}{dx}(x) = 1.$$

When we write $\frac{df}{dx}$ or $\frac{d}{dx}(f)$, it is not a fraction. The symbol makes sense only as a whole and cannot be broken into its parts. This is emphasized by the way we read it, which is different from the way we would read it if it were a quotient.

If we were considering functions for which the variable was t, say, or u, we would write $\frac{d}{dt}(f)$ or $\frac{d}{du}(u^3 - 2u + 7)$, etc.

Suppose $f(x)$ and $g(x)$ are two functions that are both differentiable at $x = a$. What can we say about their sum, $f(x) + g(x)$? We could attempt to find the derivative of the sum at $x = a$ by evaluating the limit of the difference quotient:

$$\frac{d(f+g)}{dx}(a) = \lim_{h\to 0} \frac{\big(f(a+h) + g(a+h)\big) - \big(f(a) + g(a)\big)}{h}$$

$$= \lim_{h\to 0} \frac{\big(f(a+h)-f(a)\big)+\big(g(a+h)-g(a)\big)}{h}$$
$$= \lim_{h\to 0} \left(\frac{\big(f(a+h)-f(a)\big)}{h} + \frac{\big(g(a+h)-g(a)\big)}{h}\right).$$

Here we have the limit of the sum of two functions. We know that by Theorem 2.2.3(i) it equals the sum of the two limits. Also we know what the two limits are; the first difference quotient approaches $f'(a)$, and the second approaches $g'(a)$. So the derivative of the sum $f+g$ equals the sum of the derivatives of f and g.

A similar but easier calculation shows that the derivative of cf, with c a constant, is just $cf'(a)$, which follows from the corresponding fact about limits (Theorem 2.2.3(iii)).

We have just given the proof of the following extremely important theorem. The statement is written in both derivative notations to help us become accustomed to using them both.

THEOREM 3.2.2

Suppose c is a constant and $f(x)$ and $g(x)$ are two functions that are both differentiable at $x=a$. Then the functions $f+g$ and cf are both differentiable at $x=a$, and their derivatives are given by the following:

$$\frac{d(f+g)}{dx}(a) = \frac{df}{dx}(a) + \frac{dg}{dx}(a),$$
$$\frac{d(cf)}{dx} = c\frac{df}{dx}.$$

Alternatively, these formulas can be written as

$$(f+g)'(a) = f'(a) + g'(a),$$
$$(cf)'(a) = cf'(a),$$

and, if the point a is not important, as

$$\frac{d(f+g)}{dx} = \frac{df}{dx} + \frac{dg}{dx} \quad \text{or} \quad (f+g)' = f' + g',$$
$$\frac{d(cf)}{dx} = c\frac{df}{dx} \quad \text{or} \quad (cf)' = cf'.$$

Of course the result about derivatives of sums applies to sums of three or more functions too.

EXAMPLE 2

(i) Suppose $a, b \in \mathbb{R}$ are constants. Find the derivative of $y = ax + b$.

Solution Differentiating, we have

$$\frac{dy}{dx} = \frac{d}{dx}(ax+b) = \frac{d}{dx}(ax) + \frac{d}{dx}(b)$$
$$= a\frac{d}{dx}(x) + \frac{d}{dx}(b) = a(1) + 0 = a.$$

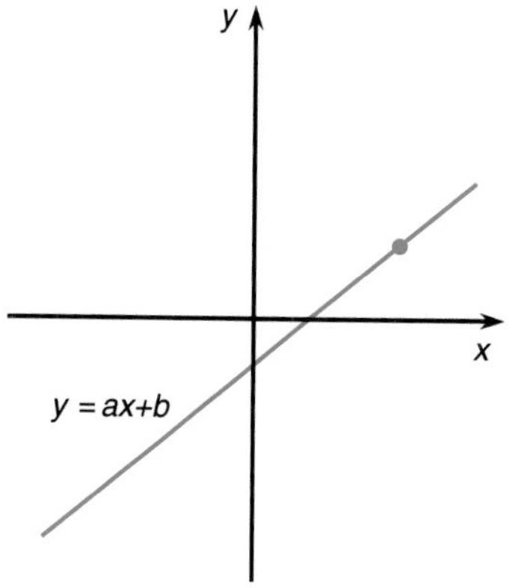

Figure 3.2.5

Here we used Theorem 3.2.2 for the second and third equalities and Example 1 for the fourth.

(ii) Find the tangent line to the graph $y = ax + b$ at any point on the graph.

Solution The result of part (i) says that the derivative of a straight line $y = ax + b$ equals the slope of the line (which is just a). If we combine this observation with the definition of the tangent line (Definition 3.1.1), we find that the tangent line to $y = ax + b$ at any point on the graph is a line with the same slope as $y = ax + b$ and passing through some point on $y = ax + b$; the only way this can happen is for the tangent line to be $y = ax + b$ also. In other words, the tangent line to a straight line is the line itself, which is hardly surprising. See Figure 3.2.5.

Suppose we want to differentiate a polynomial $p(x) = a_m x^m + a_{m-1}x^{m-1} + \ldots + a_1 x + a_0$.

Using Theorem 3.2.2, we find that

$$\frac{dp}{dx} = a_m \frac{d}{dx}(x^m) + a_{m-1}\frac{d}{dx}(x^{m-1}) + \ldots + a_1 \frac{d}{dx}(x) + a_0 \frac{d}{dx}(1),$$

so to find the derivative, all we really need is to be able to find the derivative of x^n for $n = 1, 2, 3, \ldots, m$. Of course we already know the derivative of $x = x^1$; by Example 1 it is the constant function 1.

To find the derivative of x^n, for $n = 2, 3, 4, \ldots$, we must consider the difference quotient $\frac{(x+h)^n - x^n}{h}$. First we recall the Binomial Theorem (Theorem 1.3.10, p. 15), which says that

$$(x+h)^n = x^n + \binom{n}{n-1}x^{n-1}h + \binom{n}{n-2}x^{n-2}h^2 + \ldots + \binom{n}{1}xh^{n-1} + \binom{n}{0}h^n.$$

Recalling that $\binom{n}{n-1} = \frac{n!}{(n-1)!1!} = n$, we find that

$$(x+h)^n - x^n = nx^{n-1}h + \binom{n}{n-2}x^{n-2}h^2 + \ldots + \binom{n}{1}xh^{n-1} + \binom{n}{0}h^n.$$

Dividing by h to get the difference quotient, we find

$$\frac{(x+h)^n - x^n}{h} = nx^{n-1} + \binom{n}{n-2}x^{n-2}h + \ldots + \binom{n}{1}xh^{n-2} + \binom{n}{0}h^{n-1}.$$

To find $\frac{d}{dx}(x^n)$, we let $h \to 0$ and take the limit. In the above expression, each term on the right side after the first contains some positive power of h, so as $h \to 0$, each term except the first tends to 0. In other words,

$$\begin{aligned}\frac{d}{dx}(x^n) &= \lim_{h\to 0}\frac{(x+h)^n - x^n}{h}\\ &= \lim_{h\to 0}\left(nx^{n-1} + \binom{n}{n-2}x^{n-2}h + \ldots + \binom{n}{1}xh^{n-2} + \binom{n}{0}h^{n-1}\right)\\ &= nx^{n-1}.\end{aligned}$$

We have just proved that the derivative of x^n is nx^{n-1}. To find the derivative of a power x^n, we reduce the exponent from n to $n-1$ and multiply by the original exponent n, to get nx^{n-1}. Notice that when $x \neq 0$, this is consistent with our result that $\frac{d}{dx}(x) = 1$, since $x = x^1$, and $1(x^0) = 1(1) = 1$. Moreover, if we think of $1 = x^0$, then the derivative of 1 should be $0(x^{-1})$, which is more or less correct. This last formula equals 0 whenever it is defined, and the derivative of 1 is 0. Unfortunately, $0(x^{-1})$ is not defined when $x = 0$, so we should not really use this formula. Similarly, x^0 is not defined when $x = 0$, so the formula should not be used for the derivative of $x = x^1$ either. On the other hand, it is sometimes helpful as a mnemonic device.

We record these important facts as a theorem.

THEOREM 3.2.3 Let n be any positive integer. Then x^n is differentiable at every point x, and for $n \geq 2$,

$$\frac{d}{dx}(x^n) = nx^{n-1}.$$

When $n = 1$ we know that

$$\frac{d}{dx}(x) = 1.$$

Also the constant function 1 is differentiable at every x, and

$$\frac{d}{dx}(1) = 0.$$

EXAMPLE 3 Differentiate the following functions: (i) x^3; (ii) x^5; (iii) $4x^3 - 2x^2 + 5x - 1$.

Solution We simply apply Theorem 3.2.3 to find that (i) $\frac{d}{dx}(x^3) = 3x^2$, (ii) $\frac{d}{dx}(x^5) = 5x^4$.

For (iii) we must use Theorem 3.2.2 to break the polynomial into pieces:

$$\begin{aligned}\frac{d}{dx}(4x^3 - 2x^2 + 5x - 1) &= \frac{d}{dx}(4x^3) + \frac{d}{dx}(-2x^2) + \frac{d}{dx}(5x) + \frac{d}{dx}(-1)\\ &= 4\frac{d}{dx}(x^3) - 2\frac{d}{dx}(x^2) + 5\frac{d}{dx}(x) - 1\frac{d}{dx}(1).\end{aligned}$$

Now we can take the derivative of each power of x and find that the above expression equals

$$4(3x^2) - 2(2x^1) + 5(1) - 1(0) = 12x^2 - 4x + 5.$$

The procedure we just used to find the derivative of $4x^3 - 2x^2 + 5x - 1$ will work for any polynomial. It is not necessary to write out all the steps; with a little practice, most of them can be done in your head. Theorem 3.2.2 says that we can work with each term of the polynomial separately. It also says that the derivative of the term $a_n x^n$ will be the product of a_n with the derivative of x^n.

So we work our way through the polynomial. For each term $a_n x^n$ we can write down its derivative: $a_n(nx^{n-1})$. When we have done this for each term, we will have found the derivative of the original polynomial.

EXAMPLE 4 Find the derivative of each of the following polynomials: (i) $3x^3 - 7x^2 - x + 4$; (ii) $2x^{18} - 3x^4 + x + 2$.

Solution

(i) We write down the polynomial and then find its derivative term by term:

$$\frac{d}{dx}(3x^3 - 7x^2 - x + 4) = 9x^2 - 14x - 1.$$

Notice that we did not bother to write down the derivative of the constant term, since it is 0 anyway, and that we did not write the derivative of $-x$ as $-1(1)$ but simply combined them all at one step.

(ii) Again,

$$\frac{d}{dx}(2x^{18} - 3x^4 + x + 2) = 36x^{17} - 12x^3 + 1.$$

We have established the following result.

THEOREM 3.2.4

Consider a polynomial

$$p(x) = a_m x^m + a_{m-1}x^{m-1} + \ldots + a_1 x + a_0.$$

It is differentiable at every point x, and its derivative is given by

$$\frac{dp}{dx} = ma_m x^{m-1} + (m-1)a_{m-1}x^{m-2} + \ldots + a_1.$$

This makes it very much easier to do problems involving derivatives.

EXAMPLE 5 Find the tangent line to the curve $y = 4x^5 - 5x^3 + 3x^2 + 2x - 2$ at the point $(1, 2)$.

See the Graphic Calculator box on the next page for a plot of this example.

Solution $\frac{dy}{dx} = 20x^4 - 15x^2 + 6x + 2$. So the slope of the tangent, which is just the value of the derivative at the point $x = 1$, is $20 - 15 + 6 + 2 = 13$. Accordingly, the equation of the tangent line can be found by using Definition 3.1.1. It is

$$y = 2 + 13(x - 1) \qquad \text{or} \qquad y = 13x - 11.$$

(Notice that this line does indeed go through the point $(1, 2)$. This is an easy way to check that your answer is reasonable.)

EXAMPLE 6 Find all the points at which the tangent line to the curve $y = 2x^3 - 9x^2 + 12x - 6$ is horizontal.

Solution For the tangent line to be horizontal, its slope must equal 0. The slope of the tangent is just the derivative, which is

$$6x^2 - 18x + 12.$$

To find when this is zero, we factor it:

$$6x^2 - 18x + 12 = 6(x^2 - 3x + 2) = 6(x - 2)(x - 1),$$

which is zero precisely when $x = 1$ or $x = 2$.

So there are exactly two points where the tangent line is horizontal. They are $(1, -1)$ and $(2, -2)$. See Figure 3.2.6.

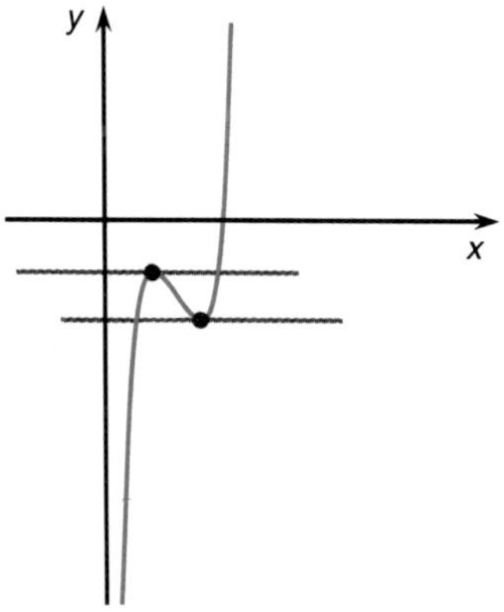

Figure 3.2.6

With a graphics calculator it is possible to plot both the function and the tangent line from Example 5.

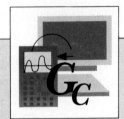

TI-85	SHARP EL9300
GRAPH F1 (y(x) =) 4 F1 (x) ^5 − 5 F1 (x) ^3 + 3 F1 (x) x^2 + 2 F1 (x) − 2 ▼ M2(RANGE) (−) 2 ▼ 2 ▼ ▼ (−) 10 ▼ 10 F5(GRAPH)	4 x/θ/T a^b 5 ▶ − 5 x/θ/T a^b 3 ▶ + 3 x/θ/T x^2 + 2 x/θ/T − 2 2ndF ▼ 13 x/θ/T −11 RANGE −2 ENTER 2 ENTER 1 ENTER −10 ENTER 10 ENTER 1 ENTER
CASIO f_x-7700GB/f_x-6300G	**HP 48SX**
Cls EXE Range −2 EXE 2 EXE 1 EXE −10 EXE 10 EXE 1 EXE EXE EXE EXE Graph 4 X x^y 5 − 5 X x^y 3 + 3 X x^2 + 2 X − 2 EXE Graph 13 X − 11 EXE	PLOT PLOTR ERASE ATTN ′ αY = 4 × αX y^x 5 − 5 × αX y^x 3 + 3 × αX y^x 2 + 2 × αX − 2 PLOT DRAW ERASE 2 +/− ENTER 2 XRNG 10 +/− ENTER 10 YRNG DRAW ATTN ′ αY = 13 × αX− 11 PLOT DRAW DRAW

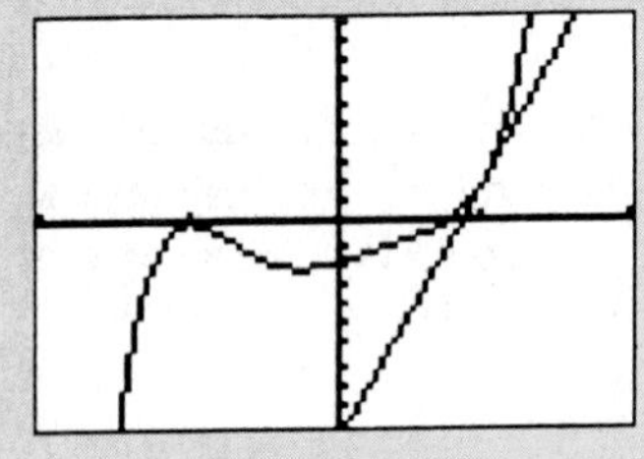

EXAMPLE 7 **Weather Balloon** A weather balloon is released at time $t = 0$ and allowed to rise. Suppose its height above the ground at time t sec is given by $H(t) = \frac{1}{3}t^3 - 5t^2 + 23t$ meters.

(i) Find the velocity of the balloon t sec after its release.

(ii) Find when the balloon is rising and when (if ever) it is falling.

(iii) Because of unforeseen problems, the balloon bursts after 11 sec of flight. How fast is it rising at the moment it bursts?

Solution

(iv) The velocity is simply the derivative $H'(t) = 3\left(\frac{1}{3}t^2\right) - 2(5t) + 23 = t^2 - 10t + 23$ m/sec.

(v) The derivative is the rate of change of the height, so when $H'(t) > 0$, it means that the height is increasing, or that the balloon is rising. Similarly, $H'(t) < 0$ means the balloon is falling. We must study the sign of $H'(t)$.

The easiest way to determine the sign of a quadratic expression like $H'(t) = t^2 - 10t + 23$ is to complete the square:

$$t^2 - 10t + 23 = (t^2 - 10t + 25) - 2 = (t - 5)^2 - 2.$$

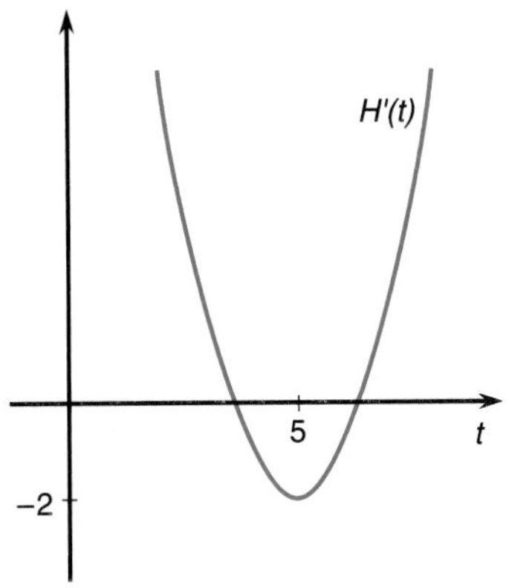

Figure 3.2.7

This is obviously positive when $(t-5)^2 > 2$ and negative when $(t-5)^2 < 2$. Now $(t-5)^2 < 2$ precisely when $|t-5| < \sqrt{2}$, which means $-\sqrt{2} < t-5 < \sqrt{2}$, which is the same as $5-\sqrt{2} < t < 5+\sqrt{2}$.

So the derivative is negative, which means the balloon is falling, when t is between $5-\sqrt{2} \approx 3.586$ sec and $5+\sqrt{2} \approx 6.414$ sec. The derivative is zero, which means the balloon is neither rising nor falling, when $t = 5 \pm \sqrt{2} \approx 3.586$ or 6.414 sec. The rest of the time the derivative is positive, which means the balloon is rising when $t < 5-\sqrt{2}$ and when $t > 5+\sqrt{2}$. See the illustration of the graph of $H'(t)$ in Figure 3.2.7.

(vi) To find how fast the balloon is rising 11 sec into its flight, we simply find $H'(11) = (11)^2 - 10(11) + 23 = 121 - 110 + 23 = 34$ m/sec. The balloon is rising at a speed of 34 m/sec.

The solution to part (ii) of Example 7 was not particularly hard, but a little thought should convince you that to do it without knowing about derivatives would be quite difficult.

We used Theorem 2.2.3, which says that limits "preserve" sums and that constants can be "pulled outside" of limits, to prove Theorem 3.2.2, which says that differentiation "preserves" sums and that constants can be "pulled outside" of derivatives.

Since the rules for the limit of a product or a quotient are also quite simple, we might ask about the derivatives of products and quotients. Unfortunately, the rules for them are not quite so simple.

THEOREM 3.2.5

The Product Rule

If $f(x)$ and $g(x)$ are both differentiable at $x = a$, then the product $H(x) = f(x)g(x)$ is also differentiable at $x = a$, and its derivative is given as follows:

$$H'(a) = f(a)g'(a) + f'(a)g(a).$$

This may also be written as

$$\frac{d}{dx}\big(f(x)g(x)\big)(a) = f(a)\frac{dg}{dx}(a) + \frac{df}{dx}(a)g(a).$$

If it is not particularly important to specify the point a, the **Product Rule** can be written more briefly as

$$(fg)' = fg' + f'g \qquad \text{or} \qquad \frac{d}{dx}(fg) = f\frac{dg}{dx} + \frac{df}{dx}g.$$

PROOF The difference quotient for $H(x)$ is

$$\frac{H(a+h) - H(a)}{h} = \frac{f(a+h)g(a+h) - f(a)g(a)}{h}$$
$$= \frac{f(a+h)g(a+h) - f(a+h)g(a) + f(a+h)g(a) - f(a)g(a)}{h}.$$

We have added and subtracted the term $f(a+h)g(a)$ in the numerator, which of course changes nothing. But when we regroup the terms, the difference quotient becomes easier to work with:

$$\frac{f(a+h)\big(g(a+h)-g(a)\big)+\big(f(a+h)-f(a)\big)g(a)}{h}$$
$$=f(a+h)\frac{g(a+h)-g(a)}{h}+\frac{f(a+h)-f(a)}{h}g(a).$$

The second term is the difference quotient for $f(x)$ multiplied by $g(a)$, so as $h \to 0$, it will tend to $f'(a)g(a)$. The first term is $f(a+h)$ times the difference quotient for $g(x)$. As $h \to 0$, the difference quotient will tend to $g'(a)$, and $f(a+h)$ will tend to $f(a)$, since $f(x)$ is continuous at $x=a$, by Theorem 3.2.1. In symbols,

$$\begin{aligned}
H'(a) &= \lim_{h\to 0}\frac{H(a+h)-H(a)}{h}\\
&= \lim_{h\to 0}\left(f(a+h)\frac{g(a+h)-g(a)}{h}+\frac{f(a+h)-f(a)}{h}g(a)\right)\\
&= \lim_{h\to 0}\left(f(a+h)\frac{g(a+h)-g(a)}{h}\right)+\lim_{h\to 0}\left(\frac{f(a+h)-f(a)}{h}g(a)\right)\\
&= \left(\lim_{h\to 0}f(a+h)\right)\left(\lim_{h\to 0}\frac{g(a+h)-g(a)}{h}\right)+\left(\lim_{h\to 0}\frac{f(a+h)-f(a)}{h}\right)g(a)\\
&= f(a)g'(a)+f'(a)g(a).
\end{aligned}$$

The idea of this proof was to rearrange the difference quotient into pieces, in each of which only one function (f or g) changed. In each piece we found a difference quotient for f or g times some other factor, so we were able to take the limit. The last ingredient was realizing that $\lim_{h\to 0}f(a+h)=f(a)$, that is, that $f(x)$ is continuous at $x=a$, which is a consequence of Theorem 3.2.1.

Since the Product Rule makes sense wherever both $f(x)$ and $g(x)$ are differentiable, we can also write it in the form

$$(fg)'(x)=f(x)g'(x)+f'(x)g(x),$$

with the understanding that it makes sense wherever the right side of the formula makes sense. Since polynomials, for example, are differentiable everywhere, this form of the Product Rule is particularly convenient for them.

EXAMPLE 8 Differentiate the following products: (i) $(x+3)(x^3-2x-4)$; (ii) $(t^2+3t+4)^2$; (iii) $(x^2+1)(x^3-2)(2x^2+x)$.

Solution

(i) We simply apply the Product Rule, with $f(x)=x+3$ and $g(x)=x^3-2x-4$. We find that

$$\begin{aligned}
\frac{d}{dx}\big((x+3)(x^3-2x-4)\big) &= (x+3)(3x^2-2)+(1)(x^3-2x-4)\\
&= 3x^3+9x^2-2x-6+x^3-2x-4\\
&= 4x^3+9x^2-4x-10.
\end{aligned}$$

We could have multiplied out the product $f(x)g(x)$ first and then just differentiated the resulting polynomial. You should try it and check that the answer you get is the same.

(ii) This time we let $f(t) = g(t) = t^2 + 3t + 4$:

$$\frac{d}{dt}\left((t^2 + 3t + 4)^2\right) = (t^2 + 3t + 4)(2t + 3) + (2t + 3)(t^2 + 3t + 4).$$

Noticing that these two terms are really the same, we combine them and find that the derivative is

$$2(t^2 + 3t + 4)(2t + 3) = 2(2t^3 + 9t^2 + 17t + 12) = 4t^3 + 18t^2 + 34t + 24.$$

(iii) We think of the product as

$$(x^2 + 1)(x^3 - 2)(2x^2 + x) = (x^2 + 1)\left((x^3 - 2)(2x^2 + x)\right).$$

We apply the Product Rule with $f(x) = x^2 + 1, g(x) = (x^3 - 2)(2x^2 + x)$. To do this, we need to know $g'(x)$.

To find $g'(x)$, we apply the Product Rule and find

$$g'(x) = (x^3 - 2)(4x + 1) + (3x^2)(2x^2 + x)$$
$$= 4x^4 + x^3 - 8x - 2 + 6x^4 + 3x^3 = 10x^4 + 4x^3 - 8x - 2.$$

Now the original derivative is

$$\frac{d}{dx}\Big((x^2 + 1)\left((x^3 - 2)(2x^2 + x)\right)\Big) = f(x)g'(x) + f'(x)g(x)$$
$$= (x^2 + 1)(10x^4 + 4x^3 - 8x - 2) + (2x)(x^3 - 2)(2x^2 + x)$$
$$= 10x^6 + 4x^5 + 10x^4 - 4x^3 - 2x^2 - 8x - 2 + (2x^4 - 4x)(2x^2 + x)$$
$$= 10x^6 + 4x^5 + 10x^4 - 4x^3 - 2x^2 - 8x - 2 + 4x^6 + 2x^5 - 8x^3 - 4x^2$$
$$= 14x^6 + 6x^5 + 10x^4 - 12x^3 - 6x^2 - 8x - 2.$$

The diligent student can compare this result with what is obtained by multiplying the three factors first and then differentiating.

Next we should find how to differentiate a quotient $\frac{f(x)}{g(x)}$ at $x = a$, assuming that f and g are both differentiable at $x = a$ and that $g(a) \neq 0$. Notice that if $g(x)$ is differentiable at $x = a$, then Theorem 3.2.1 says it is continuous at $x = a$. If in addition $g(a) \neq 0$, then by Theorem 2.3.4, for x sufficiently near a, $g(x)$ cannot be zero either (close to a, $g(x)$ must be closer to $g(a)$ than 0 is and therefore cannot be 0). The importance of this is that for small values of h we know that $g(a + h) \neq 0$, which we need to know, since $g(a + h)$ will appear in the denominator of the difference quotient.

If we write $Q(x) = \frac{f(x)}{g(x)}$, then the difference quotient is

$$\frac{Q(a + h) - Q(a)}{h} = \frac{1}{h}\left(Q(a + h) - Q(a)\right) = \frac{1}{h}\left(\frac{f(a + h)}{g(a + h)} - \frac{f(a)}{g(a)}\right)$$
$$= \frac{1}{h}\frac{f(a + h)g(a) - f(a)g(a + h)}{g(a + h)g(a)}$$
$$= \frac{1}{g(a + h)g(a)} \times \frac{1}{h} \times \left(f(a + h)g(a) - f(a)g(a + h)\right).$$

Again we can use the trick of adding and subtracting the same quantity in the numerator, to allow breaking the numerator into pieces in each of which only one function changes. The last expression equals

We added and subtracted $f(a)g(a)$.

$$\frac{1}{g(a+h)g(a)} \frac{f(a+h)g(a) - f(a)g(a) + f(a)g(a) - f(a)g(a+h)}{h}$$
$$= \frac{1}{g(a+h)g(a)} \frac{\big(f(a+h) - f(a)\big)g(a) + f(a)\big(g(a) - g(a+h)\big)}{h}$$
$$= \frac{1}{g(a+h)g(a)} \left(\frac{f(a+h) - f(a)}{h} g(a) - f(a) \frac{g(a+h) - g(a)}{h}\right).$$

Taking the limit as $h \to 0$ is now easy. The first quotient tends to $\frac{1}{g(a)^2}$, the second to $f'(a)$, and the third to $g'(a)$.

We have just shown that the derivative of $\frac{f(x)}{g(x)}$ at $x = a$ is $\frac{f'(a)g(a) - f(a)g'(a)}{g(a)^2}$. This result is known as the Quotient Rule.

THEOREM 3.2.6

The Quotient Rule

Suppose $f(x)$ and $g(x)$ are differentiable at $x = a$ and $g(a) \neq 0$. Then $\frac{f(x)}{g(x)}$ is differentiable at $x = a$, and its derivative there is

$$\left(\frac{f}{g}\right)'(a) = \frac{f'(a)g(a) - f(a)g'(a)}{g(a)^2}.$$

If it is not necessary to specify the point a, this can be written as

$$\left(\frac{f}{g}\right)' = \frac{f'g - fg'}{g^2}.$$

EXAMPLE 9 Differentiate (i) $Q(x) = \frac{x+1}{x-1}$; (ii) $R = \frac{1}{x^2+2}$; (iii) $S = \frac{x^3+3x^2-x-2}{2x-1}$.

Solution

(i) $Q'(x) = \frac{(1)(x-1)-(x+1)(1)}{(x-1)^2} = \frac{-2}{(x-1)^2}$.

(ii) $R'(x) = \frac{(0)(x^2+2)-(1)(2x)}{(x^2+2)^2} = -\frac{2x}{(x^2+2)^2} = -2x(x^2+2)^{-2}$.

(iii) $S'(x) = \frac{(3x^2+6x-1)(2x-1)-(x^3+3x^2-x-2)(2)}{(2x-1)^2} = \frac{(6x^3+9x^2-8x+1)-(2x^3+6x^2-2x-4)}{(2x-1)^2}$
$= \frac{4x^3+3x^2-6x+5}{(2x-1)^2}$.

The use of the product and quotient rules can be complicated and sometimes even tedious, but there is nothing essentially difficult about them. It is, however, absolutely essential to memorize the rules themselves.

Incidentally, since we already know how to differentiate any polynomial, the Quotient Rule tells us how to differentiate any rational function, as the above example illustrates.

THEOREM 3.2.7

> If $R(x) = \frac{f(x)}{g(x)}$ is a rational function, then $R(x)$ is differentiable at every point a for which $g(a) \neq 0$.
>
> Moreover, its derivative is given by the Quotient Rule:
>
> $$R'(x) = \frac{f'(x)g(x) - f(x)g'(x)}{g(x)^2},$$
>
> assuming $g(x) \neq 0$.

EXAMPLE 10 Let $f(x) = \frac{x}{1+x^2}$. Find all values of x at which $f'(x) = 0$.

Solution Notice that the denominator is never zero, so the function is differentiable everywhere. Using the Quotient Rule, we find that its derivative is

$$\frac{(1)(1+x^2) - x(2x)}{(1+x^2)^2} = \frac{1+x^2-2x^2}{(1+x^2)^2} = \frac{1-x^2}{(1+x^2)^2}.$$

The question asks when this is zero. Since the denominator is never zero, this is equivalent to asking when the numerator is zero, which is easy: $1 - x^2 = 0$ precisely when $x = \pm 1$.

EXAMPLE 11 Find the derivative of (i) $\frac{1}{x}$, (ii) $\frac{1}{x^n}$, where n is a positive integer.

Solution

(i) Apply the Quotient Rule, with $f(x) = 1$ and $g(x) = x$. We find that

$$\frac{d}{dx}\left(\frac{1}{x}\right) = \frac{0(x) - 1(1)}{x^2} = -\frac{1}{x^2}.$$

(ii) Similarly,

$$\frac{d}{dx}\left(\frac{1}{x^n}\right) = \frac{0(x) - 1(nx^{n-1})}{(x^n)^2} = -\frac{nx^{n-1}}{x^{2n}} = -nx^{-n-1}.$$

Notice that the formula established in Example 11(ii) is consistent with the formula for the derivative of x^n when $n > 0$, as stated in Theorem 3.2.3. We record this result below.

THEOREM 3.2.8

> Let n be any integer. If $n > 0$, then x^n is differentiable at every x. If $n < 0$, then x^n is differentiable except at $x = 0$. The constant function 1 is differentiable at every x.
>
> The derivatives of x and 1 are given as follows:
>
> $$\frac{d}{dx}(x) = 1, \qquad \frac{d}{dx}(1) = 0.$$
>
> For $n \neq 0, 1$, whenever x^n is differentiable, its derivative is given as follows:
>
> $$\frac{d}{dx}(x^n) = nx^{n-1}.$$

Exercises 3.2

I

Find the derivative of each of the following functions.

1. $x^3 - 2x^2 + x - 1$
2. $x^4 + 2x^3 - 5x + 7$
3. $2x^3 + 3x^2 + 6x - 4$
4. $3x^{67}$
5. $(3x^2 - 5)(2x + 3)$
6. $(2x + 1)(x - 1)$
7. $(4x + 3)^2$
8. $(x^2 - 2x)^2$
9. $\frac{2x+1}{x^2+1}$
10. $\frac{1}{1+x^4}$

Find the derivatives of the following functions wherever they are differentiable. Specify all values of x at which they are not differentiable.

11. $\frac{2x-1}{x-1}$
12. $\frac{5x+2}{x}$
13. $\frac{x^2}{x^2-4}$
14. $\frac{x}{x^2+3x+2}$
15. $\frac{x^2}{3x^4-2x^3}$
16. $\frac{x-1}{(x-1)^2}$
17. $\frac{3x^2+x-1}{x^2+3x-4}$
18. $\frac{x^3+x^2+x+1}{x+3}$
19. $\frac{x^4}{1+x^3}$
20. $\frac{1+x^3}{x^4}$

II

21. Differentiate the following functions, specifying any points where they are not differentiable:
    (i) $(x + 1)(x^2 + x)(3x - 2)$
    (ii) $(4x^3 - 1)(3x + 3)x^2$
    (iii) $(x + 2)\frac{x}{x^2-x-2}$
22. Find the derivative of each of the following:
    (i) $(2x^2 + x - 1)^2$
    (ii) $(2x^2 + x - 1)^4$
    (iii) $(2x^2 + x - 1)^8$
    (iv) $(2x^2 + x - 1)^{16}$
23. Find the tangent line to each of the following curves at the specified point:
    (i) $y = x^3 - 2x^2 + 2x + 1, (1, 2)$
    (ii) $y = x^4 + x^2 + 2x, (-1, 0)$
24. Find the tangent line to each of the following curves at the specified point:
    (i) $y = x^5 - 3x^4 + 4x^2 - 5, (2, -5)$
    (ii) $y = x^3 - 2x + 4, (a, a^3 - 2a + 4)$
25. Find all tangent lines to the curve $y = x^3$ that pass through the point $(0, 16)$.
26. (i) Find all tangent lines to the curve $y = x^4$ that pass through the point $(0, 1)$.
    (ii) Find all tangent lines to the curve $y = x^4$ that pass through the point $(0, -3)$.
27. (i) Calculate the derivative of $H(x) = x^2$ in two ways. First do it by simply applying the formula. Second, write $H(x) = f(x)g(x)$ with $f(x) = g(x) = x$ and apply the Product Rule.
    (ii) With $f(x)$, $g(x)$ as above, calculate the product of the derivatives, $f'(x)g'(x)$, and remark that it is definitely different from the derivative of $H = fg$.

*28. (i) Find a Product Rule for the derivative of the product of three functions: $P(x) = f(x)g(x)h(x)$.
    (ii) A Product Rule for $P = fgh$ can be found by writing $P = fgh = (fg)h$ and then applying the Product Rule for two functions twice. But it is also possible to write $P = f(gh)$ and then apply the Product Rule. Check that these two calculations give the same answer for the derivative P'.

29. If you are familiar with the idea of proof by induction, use it to prove the formula

$$\frac{d}{dx}(x^n) = nx^{n-1}, \quad \text{for} \quad n \geq 1.$$

*30. (i) Find the derivative of $p(x)^3$ if $p(x)$ is a polynomial. (Express your answer in terms of $p(x)$ and $p'(x)$.)
    (ii) Find the derivative of $f(x)^3$ at $x = a$ if $f(x)$ is a function that is differentiable at $x = a$.
    (iii) Repeat part (ii) with $f(x)^4$ in place of $f(x)^3$.
    (iv) Repeat part (ii) with $f(x)^n$ in place of $f(x)^3$, where n is a positive integer. (*Hint:* Use induction.)

*31. Suppose the height of a helicopter above the ground at time t sec is given by $H(t) = t^3 - 3t^2 + 6$ ft.
    (i) When is the helicopter rising and when is it falling?
    (ii) What is the lowest point in its flight? (Think about when it is rising and falling, according to your answer to part (i).)
    (iii) Will it touch the ground?

*32. (i) If $f(x)$ is differentiable at $x = a$ and $f(a) \neq 0$, show that the derivative of $\frac{1}{f(x)}$ at $x = a$ equals $-\frac{f'(a)}{f(a)^2}$.
    (ii) Find the derivative of $f(x)^{-2}$ at $x = a$.
    (iii) Find the derivative of $f(x)^{-n}$ at $x = a$ for any positive integer n.

*33. Suppose $g(x)$ is a function that is differentiable on an interval (c, d) and such that $g(x) > 0$ for all $x \in (c, d)$. Let $f(x) = \sqrt{g(x)}$.

(i) Express $f'(x)$ in terms of $g(x)$ and $g'(x)$ for $x \in (c, d)$. (*Hint:* Write $g(x) = f(x)f(x)$.)

(ii) Use part (i) to find the derivative of $f(x) = \sqrt{1+x^2}$.

(iii) Find $\frac{d}{dx}\left((x^2+2x+2)^{1/2}\right)$.

(iv) If $g(x)$ is differentiable on (c, d), figure out the derivative of $f(x) = \left(g(x)\right)^{1/3}$, when it exists, and specify when it does and does not exist.

*34. Find all points x at which $f'(x) = 0$ if (i) $f(x) = \frac{x}{x^2+2}$, (ii) $f(x) = \frac{x^3}{x^2+2}$.

*35. Let $f(x) = x^4 + 2x^3 - x^2 + 3$, and let $g(x) = f'(x)$.

(i) Calculate $g(x)$.

(ii) Calculate $g'(x)$. This function, the derivative of the derivative of $f(x)$, is called the second derivative of $f(x)$.

(iii) What is the second derivative of $g(x)$?

*36. (i) Suppose $f(x)$ is differentiable at $x = a$. Suppose $g(x)$ is a function and suppose there is an interval (c, d) containing a (i.e., $c < a < d$) so that $f(x) = g(x)$ for every $x \in (c, d)$. Show that $g(x)$ is differentiable at $x = a$ and that $g'(a) = f'(a)$.

(ii) Use part (i) to find the derivative of $f(x) = \frac{x-1}{x-1}$.

(iii) Use part (i) to find the derivative of $f(x) = \frac{x^2-9}{x+3}$.

*37. Suppose that neither of the polynomials $p(x)$ and $q(x)$ is the zero polynomial, and suppose that the derivative of their product $p(x)q(x)$ equals the product of their derivatives. Show that $p(x)$ and $q(x)$ must both be constant polynomials.

*38. Use the result of Exercise 32(i) and the Product Rule to find the Quotient Rule.

III

39. Use a graphics calculator or computer to sketch the curve $y = x^3 - 3x^2 + 1$, and on the same picture draw the tangent lines at $(0, 1)$, $(1, -1)$, $(2, -3)$.

40. Sketch the curve $y = \frac{x}{x-1}$, and on the same picture draw the tangent lines at $(0, 0)$, $(2, 2)$, $\left(3, \frac{3}{2}\right)$.

41. Sketch the curve $y = 3x^4 - 2x^3 + 2x^2 + 2$, and on the same picture draw the tangent lines at $(0, 2)$, $(-1, 9)$, $(1, 5)$.

42. We have not yet discussed the derivative of the exponential function $e^x = \exp(x)$, but we can use computer graphics to estimate its derivative. Graph the difference quotient $Q(h) = \frac{\exp(0+h)-\exp(0)}{h}$ and use the graph to estimate the derivative $\exp'(0)$. Using your estimate, draw the tangent line to $y = \exp(x)$ at $(0, 1)$. Graph $y = \exp(x)$ on the same picture, and see whether your line does appear to be a tangent.

43. Use the idea of Exercise 42 to estimate $\exp'(1)$ correct to four decimal places.

44. Use the idea of Exercise 42 to estimate the derivative of $f(x) = \sin x$ at $x = 0$ (with x in *radians*).

45. Use the idea of Exercise 42 to estimate the derivative of $f(x) = \sin x$ at $x = 1$ correct to four decimal places.

POINT TO PONDER

Now that we have learned the Product Rule and become familiar with its use, let us remember that our first guess would probably have been that the derivative of a product $f(t)g(t)$ ought to be the product of the derivatives of f and g. We have seen otherwise, and the proof of the Product Rule is not difficult, but on the other hand it is not completely clear where the correct formula comes from.

Suppose $f(t)$ and $g(t)$ are two functions that are both differentiable at $t = a$. Let $A(t) = f(t)g(t)$ be the product, which we can think of as the area of a rectangle whose sides have lengths $f(t)$ and $g(t)$ at time t. We should think of the rectangle changing size as time passes.

The derivative of $A(t)$, if it exists, should be the rate of change of the area of this rectangle. The numerator of the difference quotient, $A(a+h) - A(a) = f(a+h)g(a+h) - f(a)g(a)$, is the difference between the areas of the rectangles at time $t = a+h$ and $t = a$. In Figure 3.2.8 we see the rectangle at time $t = a$ and then the rectangle at time $t = a+h$. The illustration

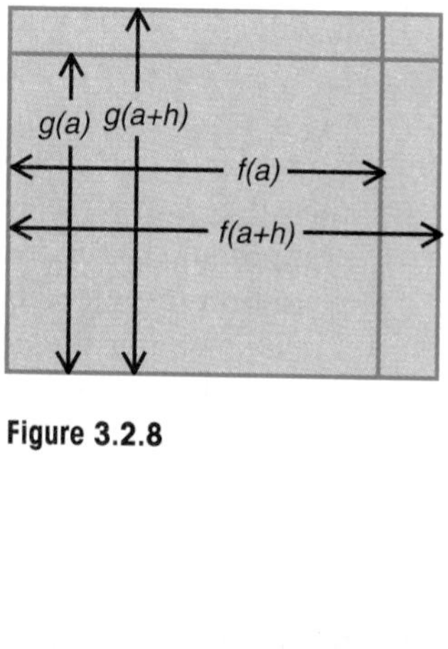

Figure 3.2.8

shows the rectangle at $t = a + h$ being slightly larger in both dimensions than the one at $t = a$, though there are other possibilities.

The numerator of the difference quotient is the change in the area of the rectangle, which is the area of the extra little strips along two sides of the rectangle. Let us ignore the tiny rectangle in the corner and concentrate on the two long thin strips (see Figure 3.2.9). Their long sides have lengths $f(a)$ and $g(a)$, respectively. The height of the horizontal strip at the top is just the change in $g(t)$ between $t = a$ and $t = a + h$, or in other words, $g(a + h) - g(a)$. The width of the vertical strip at the right side is $f(a + h) - f(a)$, the change in $f(t)$ between $t = a$ and $t = a + h$.

So, ignoring that little rectangle in the corner, the numerator of the difference quotient is $f(a)\big(g(a + h) - g(a)\big) + g(a)\big(f(a + h) - f(a)\big)$. If we form the difference quotient, it is

$$\frac{f(a)\big(g(a+h) - g(a)\big) + g(a)\big(f(a+h) - f(a)\big)}{h}$$
$$= f(a)\frac{g(a+h) - g(a)}{h} + g(a)\frac{f(a+h) - f(a)}{h}.$$

Taking the limit, we know that the first quotient tends to $g'(a)$ and the second to $f'(a)$, which leads us to expect that $A'(t) = f(a)g'(a) + g(a)f'(a)$.

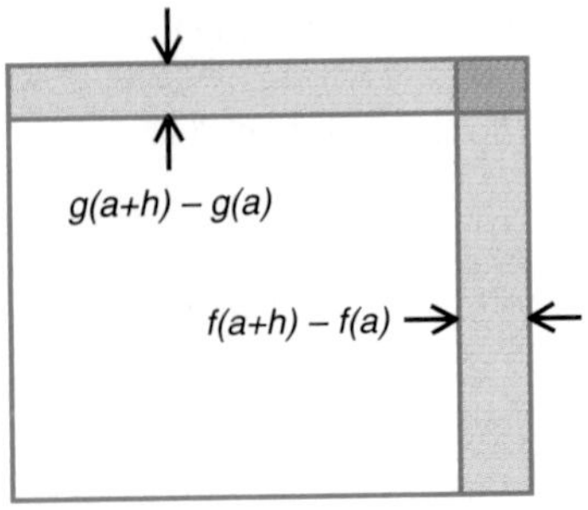

Figure 3.2.9

Roughly speaking, the two terms in this formula come from the areas of those two strips. If we think of the strips having their long side constant and their short side changing, their areas will be the constant length times the changing thickness, and the rate of change of their areas will be the constant length times the rate of change of the thickness, or $f(a)g'(a)$ in one case and $g(a)f'(a)$ in the other.

This explains to some extent why the rate of change of the whole area (i.e., of the product $f(t)g(t)$) ought to be given by the Product Rule $(fg)' = fg' + f'g$.

The argument we have just given is not entirely airtight, since we have simply ignored the area of the little rectangle in the corner. In fact, as we have already seen, that area tends to zero in the limit .

3.3 Maximum/Minimum Problems

CLOSED INTERVAL
ENDPOINTS

In this section we begin to see how useful derivatives can be, by learning how to use them to solve a very important type of problem. There are a great many problems that can be described as follows: For some function $f(x)$, at what point x will $f(x)$ have its maximum (or minimum) value? These are known as **maximum/minimum problems.**

This sort of problem arises in a huge variety of situations. The function $f(x)$ might represent the profit for a factory as a function of production, and the problem would be to find the level of production that would result in the highest profits. On the other hand, the function $f(x)$ could represent the cost of operating a piece of equipment as a function of the speed at which it runs; in this case the problem would be to find the speed that minimizes the cost.

In a chemical reaction we might be able to express the concentration of one of the products as a function $f(t)$ of time, and it could be extremely useful to know when the concentration is highest (if it is a desirable product) or lowest (if it is undesirable).

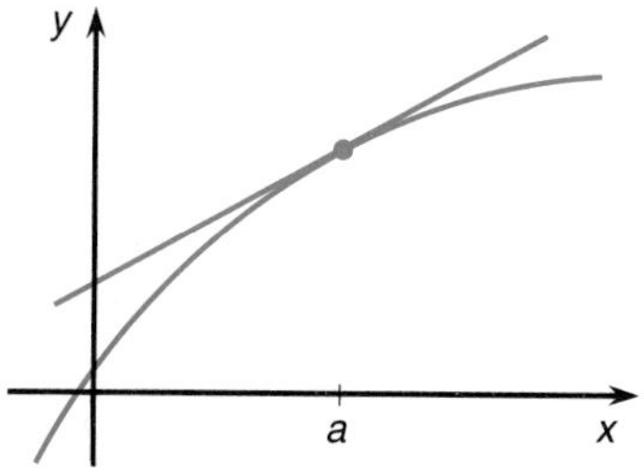

Figure 3.3.1

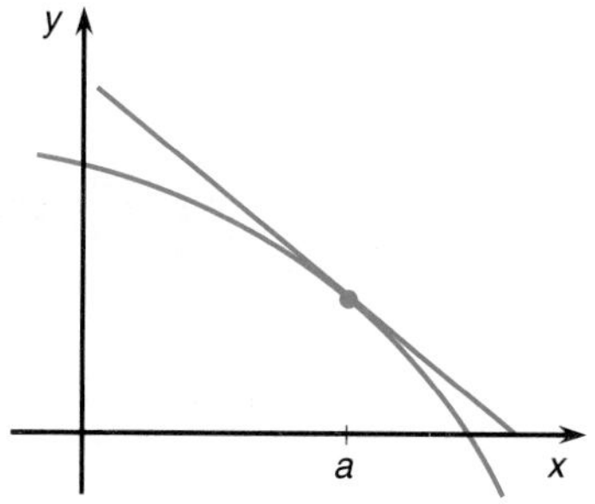

Figure 3.3.2

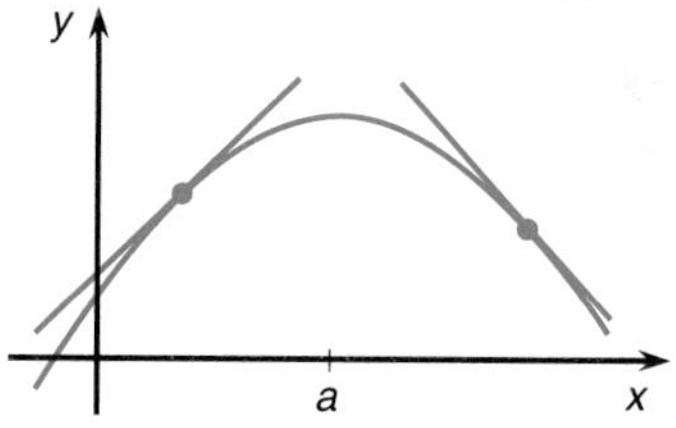

Figure 3.3.3

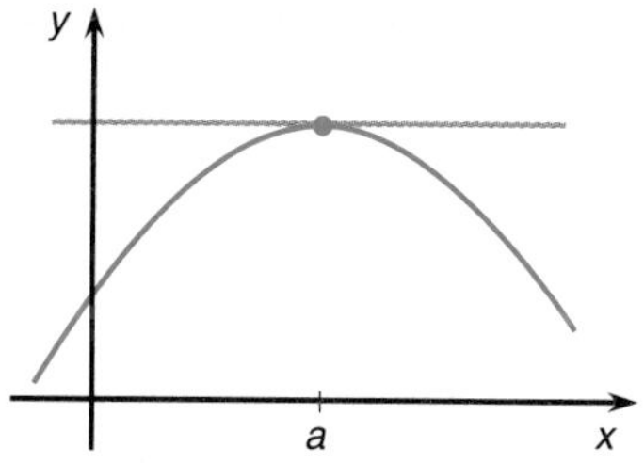

Figure 3.3.4

Foresters might like to know when to cut a stand of trees in order to get the greatest amount of usable timber; farmers might want to know when to harvest crops in order to get the highest yield or the lowest waste; engineers might want to know what shape of steel beam gives the greatest strength; pedestrians might want to know the shortest route from one place to another.

Suppose we are looking for the maximum or minimum of a differentiable function $f(x)$. Recall that the derivative $f'(a)$ is the slope of the tangent at $x = a$. If the slope of the tangent is positive, it means that the tangent slopes upward to the right, as in Figure 3.3.1. In this situation we say that the function $f(x)$ is increasing because for values of x slightly to the right of $x = a$ the value $f(x)$ will be higher than $f(a)$, that is, $f(x)$ gets bigger as x gets bigger. Similarly, if $f'(a) < 0$, the tangent line at $x = a$ slopes down to the right, and we say the function $f(x)$ is decreasing. See Figure 3.3.2.

If $f(x)$ has its maximum value when $x = a$, then as Figure 3.3.3 shows, it should be increasing to the left of $x = a$ and decreasing to the right of $x = a$. This is just another way of saying that the function comes up to its maximum as $x \to a^-$ and then goes down as x moves away on the right side of a. The derivative will be positive to the left of $x = a$ and negative to the right of $x = a$.

This strongly suggests that right at $x = a$ the derivative $f'(a)$ should equal zero. On the graph this means that the tangent line at $x = a$ should be horizontal (i.e., have slope equal to zero), which is also consistent with the idea that the graph has a sort of "hump" at $x = a$. See Figure 3.3.4.

Now working backward, if we do not know where $f(x)$ has its maximum value, we expect it to be at a point where $f'(a) = 0$. So we could start by finding all the points where $f'(a) = 0$. With luck there will not be too many of them, and then we should easily be able to decide at which of them $f(x)$ has the biggest value. Minimum values of $f(x)$ could occur in "valleys," so the same procedure will simultaneously find the minimum value.

There is a complication, however.

EXAMPLE 1 (i) Find the maximum and minimum values of $y = f(x) = x^2$.

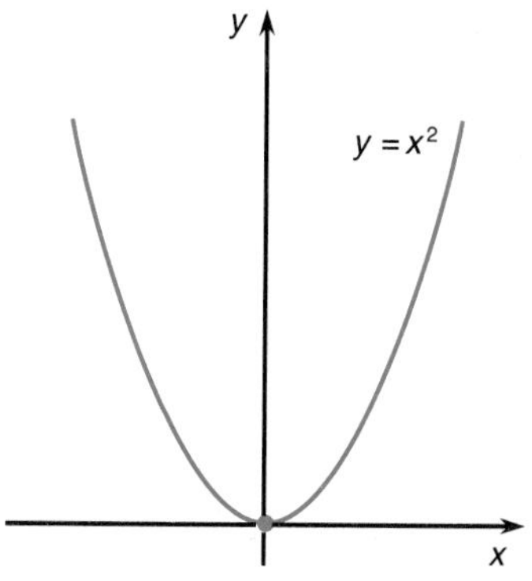

Figure 3.3.5

Solution The derivative is $\frac{dy}{dx} = 2x$, which is zero when $x = 0$ and nowhere else. This reflects the familiar fact that the tangent line at $x = 0$ (which is just the x-axis) has slope equal to 0. The function obviously has a minimum at $x = 0$ (after all, it is 0 there and positive everywhere else). Finding where $f'(x) = 0$ helped us to locate where $f(x)$ has its minimum value. See Figure 3.3.5.

But it does not seem to help find the maximum value. The reason is fairly obvious from the picture: There is no maximum value. In fact, $\lim_{x\to-\infty} f(x) = \infty$ and $\lim_{x\to\infty} f(x) = \infty$, which means that $f(x)$ cannot have any maximum because it gets larger than any number we might choose.

So the answer to the problem is that the minimum value of $f(x) = x^2$ is 0, which occurs when $x = 0$, and there is no maximum value.

(ii) Find the maximum and minimum values of $f(x) = x^3$.

Solution We could begin by noticing that $f'(x) = 3x^2$, which is zero when $x = 0$, but if we remark that $\lim_{x\to-\infty} f(x) = -\infty$ and $\lim_{x\to\infty} f(x) = \infty$, we see that $f(x)$ gets very large as $x \to \infty$—bigger than any possible maximum—and very small as

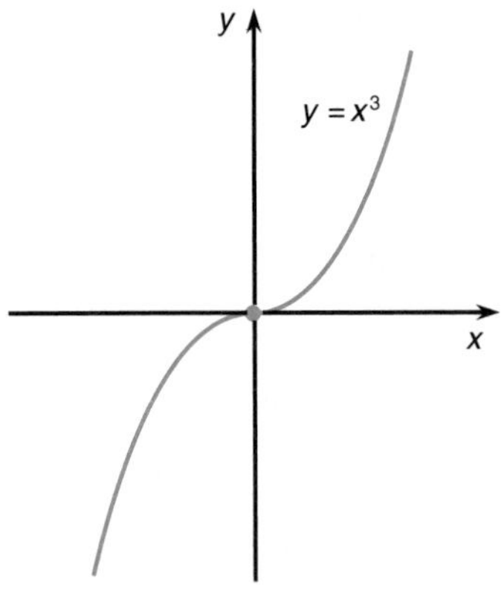

Figure 3.3.6

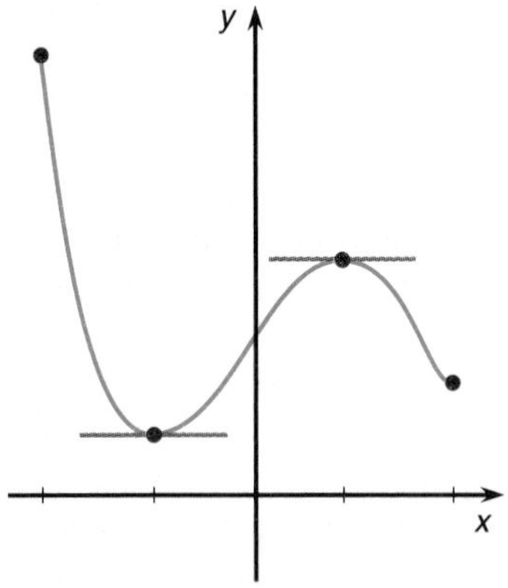

Figure 3.3.7

$x \to -\infty$—smaller than any possible minimum. As Figure 3.3.6 suggests, there is no maximum or minimum.

One way around this difficulty is to appeal to Theorems 2.5.6 and 2.5.7. They say that any continuous function on a closed interval must be **bounded,** that is, it cannot go off to infinity, and that it actually does have a maximum and a minimum value.

So given a differentiable function, such as a polynomial, for instance, on a closed interval, we know it must have a maximum and a minimum there. To find them, we could start by finding all points where the derivative is zero. These will be all the points where the tangent is horizontal, so they will include all the "humps" and "valleys." However, as Figure 3.3.7 shows, there is also the possibility that the maximum or minimum might occur at one of the endpoints of the interval (in the illustration the maximum occurs at the left endpoint). This can happen without the derivative being zero there because the maximum is in a sense artificial. It is there only because the interval comes to an end, not because the function goes up and then comes down again.

So we must revise our approach. First, we should find out where the derivative is zero. Then we should evaluate the function at each of those points *and* at the endpoints of the interval. The biggest value will be the maximum, and the smallest will be the minimum, and we will also know at which point(s) each of them occurs.

EXAMPLE 2 Find the maximum and minimum values of the function $y = f(x) = -3x^4 + 4x^3 + 12x^2 - 2$ on the interval $[-2, 3]$.

Solution The derivative is $f'(x) = -12x^3 + 12x^2 + 24x$. To find where this equals zero, we should factor it:

$$-12x^3 + 12x^2 + 24x = -12x(x^2 - x - 2) = -12x(x-2)(x+1).$$

So $f'(x) = 0$ when $x = -1, 0, 2$.

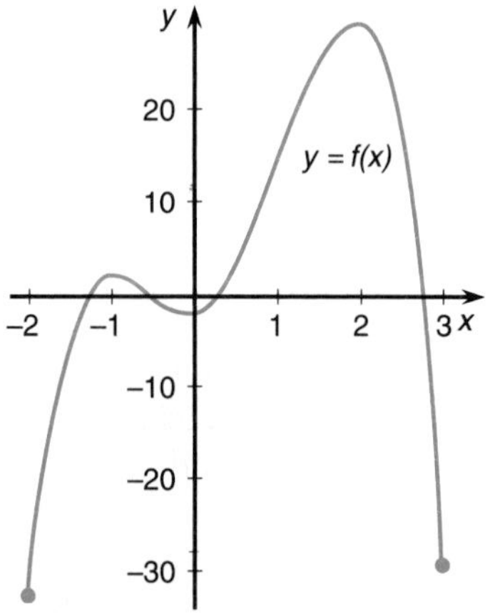

Figure 3.3.8

We can evaluate $f(x)$ at each of these points and at the endpoints $x = -2$ and $x = 3$. We find $f(-2) = -34$, $f(-1) = 3$, $f(0) = -2$, $f(2) = 30$, and $f(3) = -29$. The biggest value is 30, which is the value of $f(x)$ when $x = 2$, and the smallest is -34, which is the value of $f(x)$ when $x = -2$. Figure 3.3.8 is drawn with different scales on the two axes in order to fit the whole graph in.

This technique is extremely useful, so we summarize it for future reference.

Summary 3.3.1 Maximum/Minimum Problems on Closed Intervals

Suppose $f(x)$ is continuous on the closed interval $[c, d]$ and differentiable on the corresponding open interval (c, d), and we want to find the maximum or minimum values of $f(x)$ on $[c, d]$.

1. If it is not stated directly in the problem, identify the function $f(x)$ whose maximum or minimum values we need to find, and the interval $[c, d]$ on which it is defined.
2. Find the derivative $f'(x)$, checking that it is defined on the whole interval (c, d).
3. Solve the equation $f'(x) = 0$ to find all the points $a \in (c, d)$ at which $f'(a) = 0$. The points at which $f(x)$ achieves its maximum and minimum values must be among these points and the endpoints c and d.
4. Evaluate $f(x)$ at $x = c$, $x = d$, and all points where $f'(x) = 0$; the largest value will be the maximum, and the smallest will be the minimum. Note that the maximum and/or the minimum value may occur at more than one point.

Notice that the hypothesis that $f(x)$ be continuous on $[c, d]$ and differentiable on (c, d) is satisfied for any polynomial $f(x)$ by Definition 2.5.2(i) and Theorem 3.2.4. It also holds for a rational function, provided that the denominator is never zero on $[c, d]$, by Definition 2.5.2(ii) and Theorem 3.2.6.

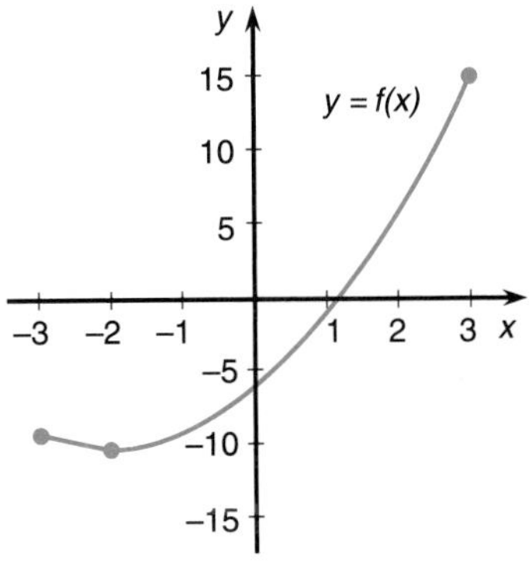

Figure 3.3.9

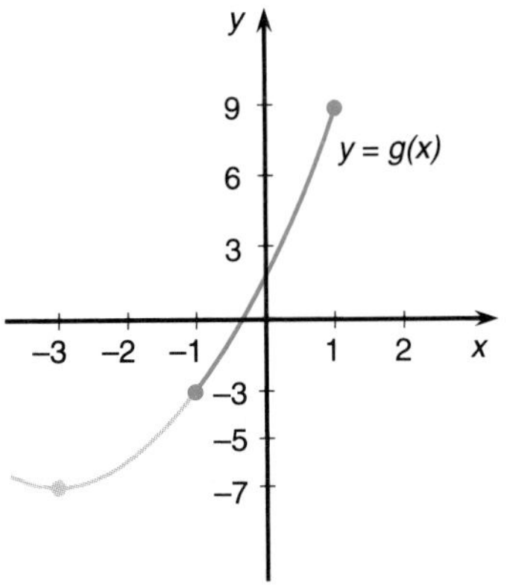

Figure 3.3.10

EXAMPLE 3 Find the maximum and minimum values of each of the following functions on the specified interval: (i) $f(x) = x^2 + 4x - 6$, on $[-3, 3]$; (ii) $g(x) = x^2 + 6x + 2$, on $[-1, 1]$; (iii) $h(x) = x^3 - 3x + 1$, on $[-2, 2]$; (iv) $k(t) = t^4 - 2t^2 + 1$, on $[-1, 2]$.

Solution

(i) We find that $f'(x) = 2x + 4 = 2(x + 2)$, which is zero when $x = -2$. So we have to evaluate $f(x)$ at $x = -2$ and at the endpoints, $x = -3, 3$.

Now $f(-2) = -10, f(-3) = -9$, and $f(3) = 15$. Clearly the maximum is 15, which occurs at the right endpoint $x = 3$. The minimum is -10, which occurs at $x = -2$. The graph of $y = f(x)$ is of course a parabola; we could have found where the minimum was by completing the square, but finding where $f'(x) = 0$ is actually easier. See Figure 3.3.9.

(ii) Again the graph will be a parabola. The derivative is $g'(x) = 2x + 6$, which is zero when $x = -3$, so that is the minimum point on the graph of the parabola. Unfortunately, the problem asked us to find the maximum and minimum on the interval $[-1, 1]$. The point $x = -3$ where the derivative is zero is outside this interval.

The procedure we are following is to find all points in the interval at which the derivative is zero and evaluate the function at them and at the endpoints. In this case, since there are no points in the interval at which the derivative is zero, this just means evaluating at the endpoints.

We find that $g(-1) = -3, g(1) = 9$, so the minimum is -3, which occurs at $x = -1$, and the maximum is 9, which occurs at $x = 1$.

If we look at Figure 3.3.10, we realize that the entire interval $[-1, 1]$ is to the right of the minimum point and the graph keeps increasing all along the

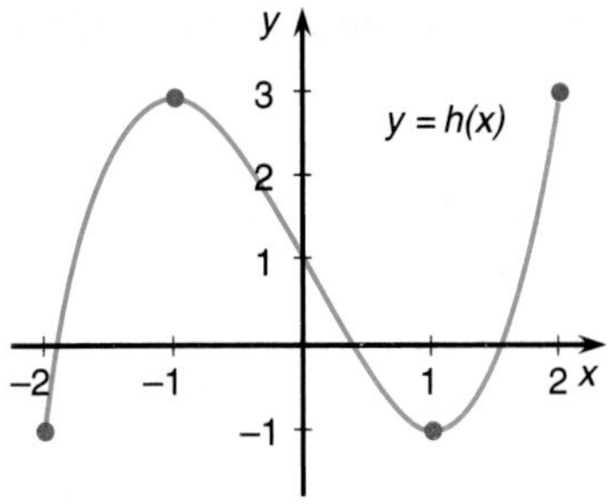

Figure 3.3.11

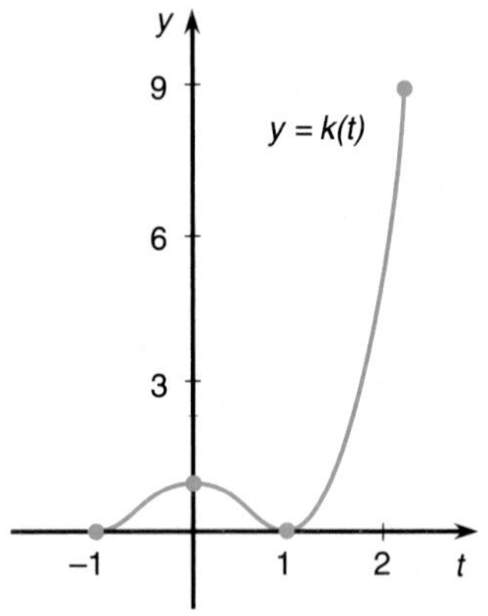

Figure 3.3.12

interval, so it is not surprising that the minimum will be at the left endpoint and the maximum at the right endpoint.

(iii) Here we calculate $h'(x) = 3x^2 - 3 = 3(x^2 - 1)$, which can be factored as $h'(x) = 3(x+1)(x-1)$, so we see that $h'(x) = 0$ exactly at $x = \pm 1$.

We evaluate $h(x)$ at $x = -2, -1, 1, 2$: $h(-2) = -1$, $h(-1) = 3$, $h(1) = -1$, and $h(2) = 3$. The maximum value is 3, and it occurs at both $x = -1$ and $x = 2$. The minimum value is -1, and it occurs at both $x = -2$ and $x = 1$. See Figure 3.3.11.

(iv) The derivative is $k'(t) = 4t^3 - 4t = 4t(t^2 - 1) = 4t(t+1)(t-1)$, which is zero at $t = 0, -1, 1$. Notice that $t = -1$ is a point at which the derivative is zero and is also one of the endpoints of the interval.

Evaluating $k(-1) = 0$, $k(0) = 1$, $k(1) = 0$, $k(2) = 9$, we see that the maximum is 9, which occurs at $t = 2$. The minimum is 0, which occurs at $t = -1$ and at $t = 1$. See Figure 3.3.12.

All the functions in Example 3 are polynomials, but it has already been remarked that the technique outlined in Summary 3.3.1 for finding maximum and minimum values will work for rational functions too, provided that the denominator is never zero on the interval in question.

EXAMPLE 4 Find the maximum and minimum values of each of the following functions on the specified interval: (i) $R(u) = \frac{1}{u^2+1}$ on $[-1, 2]$; (ii) $S(x) = \frac{x+1}{x-1}$ on $[-2, 2]$; (iii) $T(x) = \frac{x+1}{x-1}$ on $[-2, 0]$.

Solution

(i) Applying the quotient rule (Theorem 3.2.6), we can calculate the derivative $R'(u) = \frac{(0)(u^2+1)-1(2u)}{(u^2+1)^2} = -\frac{2u}{(u^2+1)^2}$. It is easier than it looks to find when this is zero, because we can forget about the denominator and simply ask when the numerator is zero, which is obviously when $u = 0$.

Evaluating at this point and at the endpoints, we find that $R(-1) = \frac{1}{2}$, $R(0) = 1$, $R(2) = \frac{1}{5}$. The minimum value is $\frac{1}{5}$, which occurs at $u = 2$, and the maximum is 1, which occurs at $u = 0$.

(ii) We notice that the denominator is zero at $x = 1$, which lies *inside* the interval $[-2, 2]$, so our technique is not applicable. (In fact, there is neither a maximum nor a minimum on this interval, since it is easy to see that $\lim_{x\to 1^-} S(x) = -\infty$, $\lim_{x\to 1^+} S(x) = \infty$. . . .)

The important thing is to remember to check that the denominator of a rational function is not zero anywhere in the interval before you try to apply Summary 3.3.1.

(iii) This time the denominator is zero at $x = 1$, but this point is *outside* the interval $[-2, 0]$. We differentiate: $T'(x) = \frac{(1)(x-1)-(x+1)(1)}{(x-1)^2} = -\frac{2}{(x-1)^2}$, which is never zero on the specified interval.

Evaluating at the endpoints, we find $T(-2) = \frac{1}{3}$, $T(0) = -1$. The minimum, -1, occurs at the right endpoint $x = 0$, and the maximum, $\frac{1}{3}$, occurs at the left endpoint $x = -2$.

With a graphics calculator it is possible to plot the functions from Example 4. It is best to plot them *after* you have found the maximum and minimum values as a way of checking your work.

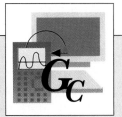

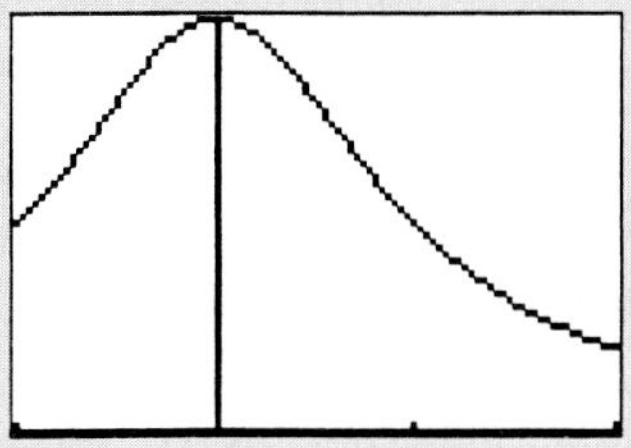

TI-85 GRAPH F1 ($y(x)$ =) 1 ÷ (F1 (x) x^2+ 1) M2(RANGE) (−) 1 ▼ 2 ▼ ▼ 0 ▼ 1 F5(GRAPH)	SHARP EL9300 1 ÷ (x/θ/T x^2 + 1) RANGE − 1 ENTER 2 ENTER 1 ENTER 0 ENTER 1 ENTER 1 ENTER
CASIO f_x-7700GB/f_x-6300G Cls EXE Range −1 EXE 2 EXE 1 EXE 0 EXE 1 EXE 1 EXE EXE EXE EXE Graph 1 ÷ (X x^2 + 1) EXE	HP 48SX PLOT PLOTR ERASE ATTN ′ αY = 1 ÷ () αX y^x 2 + 1 PLOT DRAW ERASE 1 +/− ENTER 2 XRNG 0 ENTER 1 YRNG DRAW

Try graphing the other functions from the example, and compare the results of the example with the pictures.

We can use these ideas to solve many interesting problems.

Helicopter bringing hikers to mountain summit in the Alsek Range in southeast Alaska.

EXAMPLE 5 **Helicopter** Suppose the height above the ground of a helicopter is given by $H(t) = 60t^2 - 6t^3$ ft, where t is the time, in minutes, after the takeoff.

(i) What is the greatest height the helicopter will reach in the first ten minutes of its flight?

(ii) At what moment will it be climbing the fastest?

(iii) Notice that $H(10) = 0$, which means that after flying for 10 minutes the helicopter will land again. If a helicopter touches down while descending at a rate in excess of 5 ft/sec, it will likely sustain damage. Calculate the rate of descent and decide whether or not this helicopter will land safely.

Solution Part (i) asks us to find the maximum value of $H(t)$ for t in the interval $[0, 10]$, that is, for $0 \le t \le 10$. We find the derivative is $H'(t) = 120t - 18t^2 = 6t(20 - 3t)$. This is zero when $t = 0$ or $t = \frac{20}{3}$. Both these times are in the interval $[0, 10]$, so the maximum could occur at $t = 0, \frac{20}{3}$, or 10.

We evaluate and find that $H(0) = 0, H\left(\frac{20}{3}\right) = \frac{8000}{9} \approx 888.9, H(10) = 0$. The maximum height is approximately 888.9 ft, which is reached $\frac{20}{3}$ minutes after takeoff, or 6 min, 40 sec after takeoff.

Part (ii) asks us to find the maximum value of the rate at which the helicopter is climbing. As we know, the rate at which the helicopter is climbing is exactly the rate of change of the height, which is $H'(t)$. We have just calculated that $H'(t) = 120t - 18t^2$, and to find the maximum of *this* function, we have to find *its* derivative.

The derivative of $H'(t)$, that is, the derivative of the derivative of $H(t)$, is called the second derivative of $H(t)$ and is written $H''(t)$. It is easily calculated:

$$H''(t) = 120 - 36t.$$

This is zero when $t = \frac{120}{36} = \frac{10}{3}$.

Evaluating $H'(0) = 0$, $H'(\frac{10}{3}) = 200$, $H'(10) = -600$, we see that the helicopter will be climbing fastest at the moment $\frac{10}{3} = 3\frac{1}{3}$ min after takeoff, which means 3 min, 20 sec after takeoff. At this moment the rate of climb will be $H'(\frac{10}{3}) = 200$ ft/min, or $\frac{200}{60} = 3\frac{1}{3}$ ft/sec.

For part (iii) we evaluate $H'(10) = -600$. The minus sign reflects the fact that the helicopter is descending. Its rate of descent is 600 ft/min, which we must convert to $\frac{600}{60}$ ft/sec $= 10$ ft/sec.

The helicopter will land too fast and will likely be damaged.

In Example 5 the function $H(t)$ was given to us as part of the question, so the solution amounted to finding various maximum values of specified functions. A more interesting but more difficult type of problem arises when we are given a situation and a question to answer but have to figure out the function for ourselves.

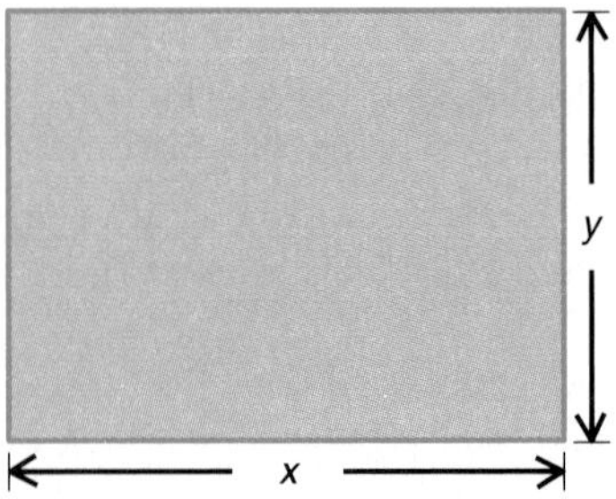

Figure 3.3.13

EXAMPLE 6 Suppose we wish to enclose a rectangular field in a fence. If we are given 600 running feet of fencing material, what is the field of largest area that we can enclose?

Solution First we draw a diagram. Label the lengths of two adjacent sides of the field by x and y (see Figure 3.3.13). The area of the field is then $A = xy$.

Since the perimeter of the field is $2x + 2y$, this must equal the total length of fencing, that is,

$$2x + 2y = 600.$$

We can solve for y. We know $2y = 600 - 2x$, so $y = 300 - x$. From this we see that $A = xy = x(300 - x) = 300x - x^2$.

We have expressed the area A as a function of x, the length of one side of the field: $A(x) = 300x - x^2$. The problem is to find the maximum possible value of $A(x)$. Since x is a length, it must be positive, and it cannot be bigger than 300; if it were, then the two sides of length x would use up more than the total fencing available. In other words, what we need is the maximum value of $A(x)$ for $x \in [0, 300]$. This is a problem we know how to solve.

The derivative is $A'(x) = 300 - 2x$, which is zero when $x = 150$. Evaluating here and at the endpoints, we find $A(0) = 0$, $A(150) = 22{,}500$, $A(300) = 0$, so the maximum occurs when $x = 150$.

Since $y = 300 - x$, we see that when $x = 150$, the value of y is also 150. In other words, the field of largest possible area is obtained by making a square field with all four sides 150 ft long. Its area is 22,500 sq ft.

Summary 3.3.2

To do Example 6, it was necessary to decide what it was that was to be maximized (or minimized). In this case the area of the field was to be maximized. Then this quantity had to be expressed as a function. The complication is that there may be more than one variable (in this case x and y, the lengths of the two sides). To avoid this difficulty, it is necessary to use the other information in the problem to *eliminate* all but one of them. Typically, this is done by solving for some variables in terms of the others until only one is left (in Example 6 we solved for y and found $y = 300 - x$). Substituting back into the function results in a function with only one variable. Information in the problem or just common sense should also allow us to restrict our attention to values of the variable in some closed interval, and then it should be possible to solve the problem by using the technique outlined in Summary 3.3.1.

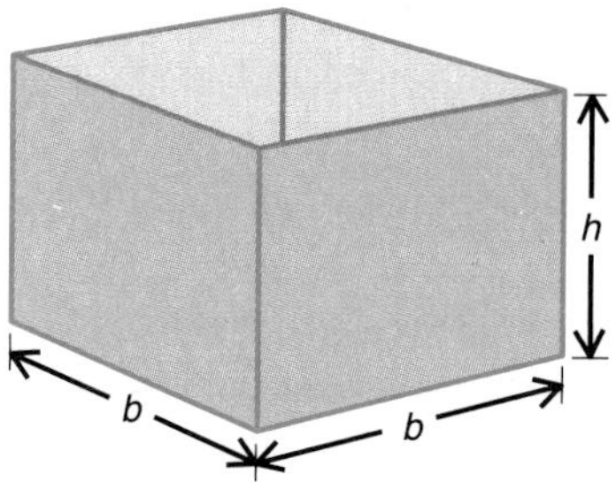

Figure 3.3.14

EXAMPLE 7 **Volume of Box** An open rectangular box is made so that its bottom is a square and the total surface area (i.e., the outside surface area) is 12 square units. What is the maximum possible volume of the box? (By an open rectangular box, we mean a box without a lid, that is, with a bottom but no top.)

Solution Let b be the length of each side of the base, and let h be the height of the box (see Figure 3.3.14). The area of the bottom is b^2 and the area of each side is bh, so the total surface area is $b^2 + 4bh$. The volume is $V = b^2h$.

At this point we have the function representing the volume, which we want to maximize, but there are two variables, b and h. We need to eliminate one of them.

Knowing that $b^2 + 4bh = 12$, we could try to solve for either b or h. If we try to solve for b, it will be messy because we have a quadratic equation in the variable b and the solution will involve square roots. On the other hand, it is easy to solve for h.

From $4bh = 12 - b^2$ we get $h = \frac{12-b^2}{4b}$.

Substituting this back into the formula for V, we find

$$V = b^2h = b^2\frac{12 - b^2}{4b} = \frac{1}{4}(12b - b^3).$$

We have to use some initiative to find the appropriate interval . . .

This is a function of only one variable, which is what we want. Once again, a length should not be negative, so $b \geq 0$. On the other hand, the total area is 12 and the area of the base is b^2, so we must have $b \leq \sqrt{12}$. The problem is to find the maximum value of $V = \frac{1}{4}(12b - b^3)$ for $b \in [0, \sqrt{12}]$.

The derivative is $V'(b) = \frac{1}{4}(12 - 3b^2)$, which is zero when $12 = 3b^2$, or $b^2 = 4$. This happens when $b = \pm 2$, but of these two possibilities, only $2 \in [0, \sqrt{12}]$. Evaluating at this point and the endpoints, we find $V(0) = 0$, $V(2) = \frac{1}{4}(24 - 8) = 4$, $V(\sqrt{12}) = \frac{1}{4}\left(12\sqrt{12} - (\sqrt{12})^3\right) = \frac{1}{4}\left(12\sqrt{12} - (\sqrt{12})^2\sqrt{12}\right) = 0$. The maximum value is $V = 4$, which occurs when $b = 2$.

The largest possible volume is 4 cubic units. A box of this volume can be made by making the sides of the base 2 units long, in which case the height of the box is $h = \frac{12-b^2}{4b} = \frac{12-4}{8} = 1$ unit.

Notice that we have used letters (b, h, V for "base," "height," "Volume") that help us remember what they stand for.

Example 7 illustrates that the possible choices of which variable to eliminate may not all be equally easy to work with. It is *possible* to solve this problem by eliminating b and working with h, but it is definitely harder and cannot easily be done with our present knowledge of derivatives.

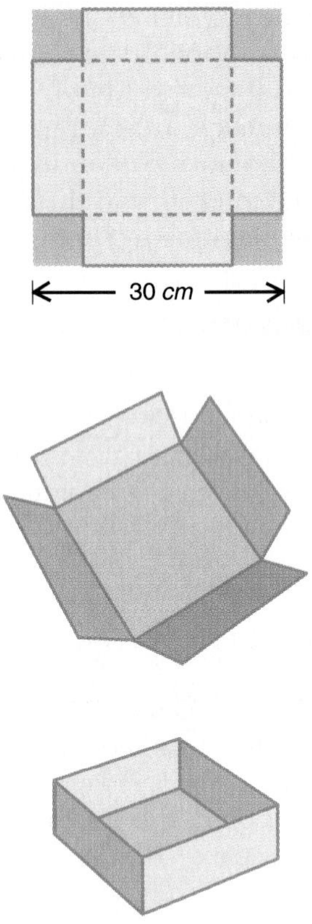

Figure 3.3.15

EXAMPLE 8 **Folded Box** Starting with a square piece of sheet metal whose sides are each 30 cm long, an open rectangular box is made by cutting out four equal squares from the corners and folding up the sides, as shown in Figure 3.3.15. What is the box of largest volume that can be made in this way?

Solution It is the volume that is to be maximized, but we have to decide what to use as our variable. Let us try letting x be the length of the sides of the little squares cut out of the corners. When the box is folded up, this becomes the height of the box. The base of the box is a square, outlined by dotted lines in Figure 3.3.15.

The length of the side of this "inner" square is $30 - 2x$ (the length of the side of the big square minus two corners). So the area of the base is $(30 - 2x)^2$, and the volume of the box is $V = V(x) = x(30 - 2x)^2 = x(900 - 120x + 4x^2) = 900x - 120x^2 + 4x^3$. Certainly, x cannot be less than 0 nor greater than 15 (since that would mean that the four corners used up all of the big square).

The derivative is $V'(x) = 900 - 240x + 12x^2$. To find when it is zero, we use the Quadratic Formula. The two points are

$$\begin{aligned} x &= \frac{240 \pm \sqrt{(-240)^2 - 4(12)900}}{24} \\ &= \frac{240 \pm \sqrt{57,600 - 43,200}}{24} = \frac{240 \pm \sqrt{14,400}}{24} \\ &= \frac{240 \pm 120}{24} = 5, 15. \end{aligned}$$

Evaluating at these points and the endpoints gives $V(0) = 0$, $V(5) = 2,000$, and $V(15) = 0$, so the maximum occurs when $x = 5$.

The largest possible box is made by cutting out corners that measure 5 cm on each side. The base will be 20 cm on each side, and the volume will be 2,000 cu cm.

> It is interesting to notice that the derivative is zero at $x = 15$. This corresponds to removing from each corner a square with 15 cm sides, which would leave no bottom at all. It is not uncommon to get some ridiculous possible answers like this. We call them **extraneous solutions,** and they will be eliminated when we evaluate the function.

EXAMPLE 9 **Fruit Tree Yield** A certain species of fruit tree yields 20 bushels of fruit if it is grown with the trees evenly spaced so that there are 100 trees to the acre. A farmer realizes that putting more trees to the acre will increase the total yield, even though crowding will reduce the yield per tree.

Experience has taught the farmer that each additional tree per acre will reduce the yield per tree by $\frac{1}{10}$ bushel. He also knows that more than 250 trees per acre will result in poor quality fruit. What number of trees per acre will give the highest total yield?

Solution Let N be the number of trees per acre. First we can calculate the yield per tree. It is 20 when $N = 100$ and reduces by $\frac{1}{10}$ for every tree over $N = 100$. The number of trees over 100 is of course $N - 100$, so the reduction in yield per tree is $\frac{1}{10}(N - 100)$. From this we see that the yield per tree is $20 - \frac{1}{10}(N - 100)$.

The total yield from N trees is therefore

$$Y = N\left(20 - \frac{1}{10}(N - 100)\right) = N\left(20 - \frac{1}{10}N + 10\right) = N\left(30 - \frac{1}{10}N\right)$$
$$= 30N - \frac{1}{10}N^2.$$

We calculate the derivative $\frac{dY}{dN} = 30 - \frac{1}{5}N$. It is zero when $N = 150$. Evaluating the yield at this point and at the endpoints 100, 250 shows that $Y(100) = 2{,}000$, $Y(150) = 2{,}250$, $Y(250) = 1{,}250$.

The highest yield will be obtained by planting 150 trees per acre.

> Example 9 is typical of many problems in which two competing factors affect the function. In this case, increasing the number of trees both reduced the yield per tree and increased the number of trees yielding fruit. It is not easy to guess the best answer with this kind of problem.

EXAMPLE 10 **Closest Point on a Curve** Find the point on the curve $y = x^2 + 2x - 1$ that is closest to the point $(2, -2)$.

Solution The distance from $(2, -2)$ to (x, y) is

$$D = \sqrt{(x-2)^2 + \big(y - (-2)\big)^2}.$$

If (x, y) is on the curve, we have $y = x^2 + 2x - 1$, so

$$D = \sqrt{(x-2)^2 + (x^2 + 2x - 1 + 2)^2} = \sqrt{(x-2)^2 + (x^2 + 2x + 1)^2}$$
$$= \sqrt{x^2 - 4x + 4 + x^4 + 4x^3 + 6x^2 + 4x + 1}$$
$$= \sqrt{x^4 + 4x^3 + 7x^2 + 5}.$$

We want to find the minimum of D, but the square root is awkward. It will be much easier to minimize $D^2 = x^4 + 4x^3 + 7x^2 + 5$, and a point at which D^2 has a minimum value will also be a point at which D has a minimum value and vice versa. (Here we are relying on the fact that $D \geq 0$.)

The derivative of D^2 is

$$(D^2)'(x) = 4x^3 + 12x^2 + 14x = x(4x^2 + 12x + 14).$$

We can find when the quadratic factor $4x^2 + 12x + 14$ is zero by applying the Quadratic Formula. We find that the discriminant is $b^2 - 4ac = (12)^2 - 4(4)14 = 144 - 224 = -80$, so there are no roots. This says that the only place where the derivative is zero is $x = 0$.

The corresponding point on the curve is $(0, -1)$, and the value of D^2 when $x = 0$ is 5. This means that $D = \sqrt{5}$, which means that the point $(0, -1)$ is a distance of $\sqrt{5}$ from $(2, -2)$.

The other ingredient we need is an interval on which to find the minimum. We already have one point, $(0, -1)$, which is $\sqrt{5}$ units away from $(2, -2)$, so there is no need to look at any points that are farther away than that. In particular, the minimum cannot occur at any point (x, y) for which x is more than $\sqrt{5}$ units from 2, or, to make it easier to work with, when x is more than 3 units from 2. So the minimum cannot occur where $x > 5$ or $x < -1$, so it must occur for some $x \in [-1, 5]$.

We found that $D^2 = 5$ when $x = 2$. Also $D^2 = (x-2)^2 + (y+2)^2 \geq (x-2)^2$. Since $(x-2)^2 \geq 3^3 = 9$ outside $[-1, 5]$, we see that $D^2 \geq 9$ outside $[-1, 5]$, so the minimum must occur *inside* $[-1, 5]$.

We have seen that the derivative of D^2 is zero when $x = 0$, so we evaluate D^2 at this point and at the endpoints: $D^2(-1) = 9, D^2(0) = 5, D^2(5) = 1305$. The minimum is certainly at $x = 0$.

The point on the curve $y = x^2 + 2x - 1$ that is closest to $(2, -2)$ is $(0, -1)$. The distance between these two points is $\sqrt{5}$.

One trick we learned in this example was to minimize (or maximize) a square root by minimizing (or maximizing) its square. Apart from that, the hardest part of the solution was figuring out what interval would be appropriate.

Exercises 3.3

I

Find the maximum and minimum values of each of the following functions on the specified interval (identify *all* the points where the maximum and minimum values occur).

1. $f(x) = x^2 + 4x + 3$, on $[-1, 1]$
2. $g(x) = x^2 + 4x + 3$, on $[-3, -1]$
3. $h(t) = -3t^2 + 3t - 2$, on $[0, 1]$
4. $k(u) = u^3 + 3$, on $[-2, 4]$
5. $F(s) = s^4 - 4s^2 + 1$, on $[-2, 1]$
6. $G(x) = 3x^4 - 8x^3 + 6x^2$, on $[-2, 2]$
7. $H(x) = x^4 - 4x^3 + 4x^2 + 3$, on $[0, 3]$
8. $K(t) = t^5 - 15t^3 + 7$, on $[0, 5]$
9. $L(u) = 2u^7 - 14u + 2$, on $[-2, 2]$
10. $R(x) = x(x-2)(x-4)$, on $[0, 2]$
11. $f(x) = \frac{1}{2+x^2}$, on $[-4, 4]$
12. $f(x) = \frac{x+2}{3-x}$, on $[-1, 1]$
13. $f(x) = -\frac{x+2}{3-x}$, on $[-1, 1]$
14. $f(x) = \frac{2}{x}$, on $[1, 4]$
15. $f(x) = \frac{x^2}{x+3}$, on $[-2, 2]$
16. $f(x) = \frac{x+3}{x^2}$, on $[-8, -2]$
17. $f(x) = \frac{x^3}{x-1}$, on $[-1, 0]$
18. $f(x) = \frac{x-1}{x^3}$, on $[1, 2]$
19. $f(x) = \frac{x^2-1}{x+1}$, on $[0, 2]$
20. $f(x) = \frac{x-1}{x^2-1}$, on $[2, 4]$

Practice your differentiation skills by finding the derivative of each of the following functions.

21. $f(x) = 3x^7 - 2x^4 + x - 1$
22. $f(x) = 12x^{12} + 4x^4 + x$
23. $f(x) = \frac{x-1}{x+1}$
24. $f(x) = \frac{1}{x^2+1}$
25. $f(x) = \frac{1}{2x+3}$
26. $f(x) = \frac{x^2+2}{x^2+1}$
27. $f(x) = x^{-3}(2x+1)$
28. $f(x) = x^{-k}(x^k + 1)$
29. $f(x) = (x^4 + 2x^2 + 4)^{-1}$
30. $f(x) = (x^2 + 1)^{-2}$

II

31. Suppose we wish to enclose a rectangular field of area 2,500 square meters. What lengths of the sides will permit us to do this with a minimum of fencing material?
32. What is the largest possible volume of a closed rectangular box (i.e., including a lid) whose base is square and whose (outside) surface area is 54 square inches?
33. Suppose an open box is made from a rectangular sheet of metal whose dimensions are 20 cm by 30 cm by cutting equal squares from the corners and folding up the sides. What is the largest possible volume for such a box?

34. A rectangular pasture is to be made beside a river by fencing three sides and letting the straight riverbank be the fourth side (see Figure 3.3.16). How should this be done in order to make the pasture's area as large as possible if 300 yd of fencing material is available?

Figure 3.3.16

35. An open rectangular box must be made with a square base and a volume of 100 cubic inches. Find how to make such a box with a minimum of material (ignoring the thickness of the material and any waste in the cutting).

36. What are the dimensions of the rectangle whose perimeter is P and that has the greatest possible area?

37. (i) What point on the line $y = 2x + 1$ is closest to the point $(0, 0)$?

(ii) What point on the line $y = 3 - x$ is closest to the point $(2, 2)$?

38. (i) What point on the curve $y = x^2$ is closest to the point $(0, -3)$?

(ii) What point on the curve $y = x^2 - 4x + 4$ is closest to the point $(-1, 0)$?

***39.** A rectangle whose perimeter is 50 cm is rotated around one of its sides to form a right circular cylinder (see Figure 3.3.17). Find the dimensions of the rectangle that results in the cylinder with the largest possible volume.

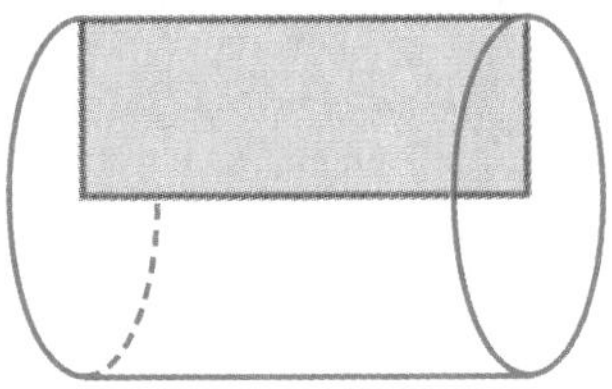

Figure 3.3.17

***40.** A farmer has 30 plants in a plot, and each yields an average of 7 kg of vegetables. For each additional 10 plants that are put in the plot, crowding causes the average yield to decrease by 1 kg per plant. How many plants should she have to maximize the total yield?

***41.** A piece of wire 1 m long is cut into two pieces. One piece is bent into the shape of a square, the other into a circle. What should the lengths of the pieces be in order that the sum of the areas of the circle and the square should be (i) a minimum, (ii) a maximum?

***42.** A piece of wire 16 ft long is cut into two pieces. Each piece is bent into a square. What should the lengths of the pieces be in order that the sum of the areas of the two squares should be (i) a minimum, (ii) a maximum?

***43.** Market research has shown that if a certain store sells a kitchen gadget for \$10, they will be able to sell 800 units. Moreover, for each additional dollar added to the price, the number the store will sell decreases by 100.

(i) At what price should the store sell the gadgets in order to maximize the total revenue?

(ii) If the gadgets cost the store \$5 each, what price will maximize the profits?

***44.** To set up the equipment for a certain manufacturing process costs \$1,000, and once it is set up it costs \$10 for each unit produced. If 1,000 units are produced, they will sell for \$15 each, and experience has shown that for each additional 1,000 units produced, the unit price will have to be lowered by \$1 in order to sell them all. How many units should be produced to maximize profits, and at what price should they sell?

III

For many functions $f(x)$ it is difficult to find exactly where $f'(x) = 0$, but it is possible to use computer graphics to get good estimates.

For the following two functions $f(x)$, use a graphics calculator or computer to graph $y = f'(x)$ at various magnifications and estimate each zero of $f'(x)$, accurate to three decimal places. Use this information and the graph of $y = f(x)$ to estimate the maximum and minimum values of $f(x)$ on the specified interval.

45. $f(x) = x^5 - x^3 + 2x^2 - x + 1$, for $x \in [-1, 1]$

46. $f(x) = x^4 - 2x^3 + x$, for $x \in [-0.7, 0.7]$

Use a graphics calculator or computer to estimate the maximum and minimum values of each of the following functions on the specified interval, correct to four decimal places.

47. $f(x) = x^4 - 4x^2 + 3x - 2, \quad [-2, 2]$

48. $f(x) = x^6 - 2x^4 + x, \quad [-1.5, 1]$

49. $f(x) = x \sin x, \quad [0, \pi]$

50. $f(x) = x^2 \cos(\pi x), \quad [-1, 1]$

3.4 Derivatives of Trigonometric Functions

$\sin'(x) = \cos x$
$\cos'(x) = -\sin x$

In this section we learn how to differentiate the trigonometric functions. To calculate the derivatives, we must evaluate some limits involving trigonometric functions, so it will be necessary to refer to results about limits established in Chapter 2, in particular Theorem 2.6.3. You may wish to refresh your memory about that result before proceeding. Use will also be made of the addition formulas (Theorem 1.7.11 p. 51) for sin and cos and of various trigonometric identities (e.g., Theorem 1.7.2), so you may wish to review them too.

Throughout this section, angles will be measured in radians unless it is explicitly stated otherwise. By the end of the section it will be reasonably clear that this choice is a good one and that using degrees would make differentiation much more complicated.

To find the derivative of sin x, we must evaluate

$$\lim_{h\to 0}\frac{\sin(x+h)-\sin(x)}{h}.$$

To simplify the numerator, we appeal to the addition formula for sin (Theorem 1.7.11), which says

$$\sin(x+h) = \sin(x)\cos(h) + \cos(x)\sin(h).$$

Substituting this into the limit gives

$$\begin{aligned}\frac{d}{dx}(\sin x) &= \lim_{h\to 0}\frac{\sin(x+h)-\sin(x)}{h}\\ &= \lim_{h\to 0}\frac{\sin(x)\cos(h)+\cos(x)\sin(h)-\sin(x)}{h}\\ &= \lim_{h\to 0}\frac{\sin(x)\cos(h)-\sin(x)+\cos(x)\sin(h)}{h}\\ &= \lim_{h\to 0}\left(\frac{\sin(x)\big(\cos(h)-1\big)}{h}+\frac{\cos(x)\sin(h)}{h}\right)\\ &= \lim_{h\to 0}\frac{\sin(x)\big(\cos(h)-1\big)}{h}+\lim_{h\to 0}\frac{\cos(x)\sin(h)}{h}\\ &= \sin(x)\lim_{h\to 0}\frac{\cos(h)-1}{h}+\cos(x)\lim_{h\to 0}\frac{\sin(h)}{h}.\end{aligned}$$

At this point we apply Theorem 2.6.3 to evaluate the two limits. It tells us that

$$\lim_{h\to 0}\frac{\cos h-1}{h}=0 \quad \text{and} \quad \lim_{h\to 0}\frac{\sin h}{h}=1.$$

Substituting these limits into the above formula, we find that

$$\frac{d}{dx}(\sin x) = \sin(x)(0)+\cos(x)(1) = \cos x.$$

In other words, the derivative of sin x is cos x.

A similar computation can be used to find the derivative of cos x. Before you read the next few lines, which tell how to do it, remind yourself of the addition formula for cos x (Theorem 1.7.11), and try to work out the derivative for yourself.

The addition formula for $\cos x$ is
$\cos(x+h) = \cos(x)\cos(h) - \sin(x)\sin(h)$.

Using this, we can calculate

$$\begin{aligned}\frac{d}{dx}(\cos x) &= \lim_{h\to 0}\frac{\cos(x+h)-\cos(x)}{h}\\ &= \lim_{h\to 0}\frac{\cos(x)\cos(h)-\sin(x)\sin(h)-\cos(x)}{h}\\ &= \lim_{h\to 0}\frac{\cos(x)\cos(h)-\cos(x)-\sin(x)\sin(h)}{h}\\ &= \lim_{h\to 0}\frac{\cos(x)\big(\cos(h)-1\big)-\sin(x)\sin(h)}{h}\\ &= \lim_{h\to 0}\frac{\cos(x)\big(\cos(h)-1\big)}{h}-\lim_{h\to 0}\frac{\sin(x)\sin(h)}{h}\\ &= \cos(x)\lim_{h\to 0}\frac{\cos(h)-1}{h}-\sin(x)\lim_{h\to 0}\frac{\sin(h)}{h}.\end{aligned}$$

Again substituting for these two limits using Theorem 2.6.3, we find that

$$\frac{d}{dx}(\cos x) = \cos(x)(0) - \sin(x)(1) = -\sin x.$$

In other words, the derivative of $\cos x$ is $-\sin x$. We have just proved the following result.

THEOREM 3.4.1

The functions $\sin x$ and $\cos x$ are differentiable at every x, and

$$\frac{d}{dx}(\sin x) = \cos x, \qquad \frac{d}{dx}(\cos x) = -\sin x.$$

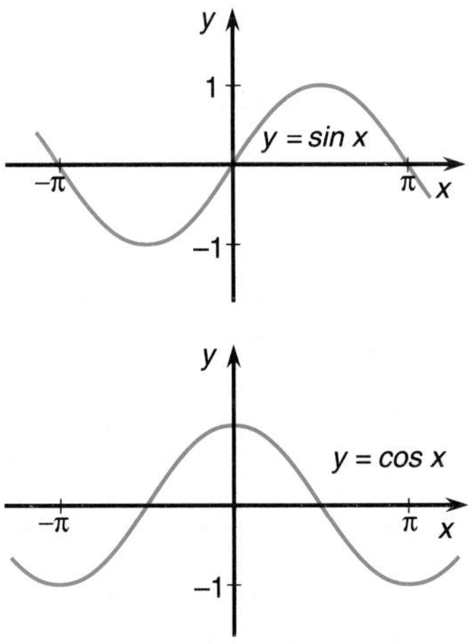

Figure 3.4.1

These two derivatives are extremely important and should be memorized. If you are having trouble remembering which one has the minus sign, consider Figure 3.4.1. It shows the graphs of $\sin x$ and $\cos x$. At $x = 0$ the graph of $\sin x$ passes through the origin. The tangent at that point clearly slopes upward to the right, which means that its slope is positive.

In fact, the slope of its tangent is $\sin'(0)$, which we have just seen is $\cos(0) = 1$. This confirms that the derivative of $\sin x$ should be $\cos x$ and cannot have a minus sign.

It is interesting to look at the two graphs in Figure 3.4.1 and observe how the value of $\cos x$ changes to reflect the slope of the tangent to $y = \sin x$; $\cos x$ is positive when $\sin x$ is increasing and negative when $\sin x$ is decreasing. Similar observations can be made about the value of $-\sin x$, the derivative of $\cos x$.

EXAMPLE 1 Differentiate the following functions: (i) $x^2 \sin x$; (ii) $\sin^2 x$; (iii) $\frac{1}{\cos x}$; (iv) $\tan x$.

Solution

(i) Writing $x^2 \sin x$ as the product of x^2 and $\sin x$ and using the Product Rule and Theorem 3.4.1, we find that its derivative is $2x \sin x + x^2 \cos x$.

(ii) Again we use the Product Rule, thinking of $\sin^2 x$ as the product of $\sin x$ with itself. We find that its derivative is

$$\cos x \sin x + \sin x \cos x = 2 \sin x \cos x.$$

(iii) For this question we use the Quotient Rule, and find that, provided $\cos x \neq 0$,

$$\frac{d}{dx}\left(\frac{1}{\cos x}\right) = \frac{0(\cos x) - 1(-\sin x)}{\cos^2 x} = \frac{\sin x}{\cos^2 x} = \frac{\sin x}{\cos x}\frac{1}{\cos x}$$
$$= \tan x \sec x.$$

Since the original function $\frac{1}{\cos x}$ is the same thing as $\sec x$, we can write this as

$$\frac{d}{dx}(\sec x) = \sec x \tan x,$$

which makes sense whenever $\cos x \neq 0$, which means x cannot be $\pm\frac{\pi}{2}$, $\pm\frac{3\pi}{2}$, etc. In other words, x cannot be an *odd* multiple of $\frac{\pi}{2}$. In symbols this can be written as $x \neq (2n+1)\frac{\pi}{2}$, for any integer n.

(iv) To use Theorem 3.4.1, we first express $\tan x$ in terms of $\sin x$ and $\cos x$, which is easy, since $\tan x = \frac{\sin x}{\cos x}$. Then we can apply the quotient rule to obtain that, provided $\cos x \neq 0$,

$$\frac{d}{dx}(\tan x) = \frac{d}{dx}\left(\frac{\sin x}{\cos x}\right) = \frac{\cos x \cos x - \sin x(-\sin x)}{\cos^2 x}$$
$$= \frac{\cos^2 x + \sin^2 x}{\cos^2 x} = \frac{1}{\cos^2 x}$$
$$= \sec^2 x.$$

Here we used the identity $\cos^2 x + \sin^2 x = 1$ (cf. Theorem 1.7.2).

This is true whenever $x \neq (2n+1)\frac{\pi}{2}$, for any integer n.

Example 1 suggests that we can use the quotient rule in combination with Theorem 3.4.1 to find the derivatives of the other trigonometric functions. Parts (iii) and (iv) of Example 1 asked us to find the derivatives of $\sec x$ and $\tan x$, respectively. Similar calculations give the derivatives of $\csc x$ and $\cot x$.

THEOREM 3.4.2

The trigonometric functions $\sec x$, $\tan x$, $\csc x$, and $\cot x$ are differentiable wherever they are defined. Moreover, their derivatives are as follows:

$$\tfrac{d}{dx}(\sec x) = \sec x \tan x, \qquad \tfrac{d}{dx}(\csc x) = -\csc x \cot x,$$
$$\tfrac{d}{dx}(\tan x) = \sec^2 x, \qquad \tfrac{d}{dx}(\cot x) = -\csc^2 x.$$

You might find it helpful to remember when differentiating the trigonometric functions sin, cos, sec, tan, csc, and cot that three of the six derivatives contain a minus sign, and they are exactly the derivatives of the "co-" functions: cos (cosine), csc (cosecant), and cot (cotangent). If you learn the derivatives of sec and tan, you can get the derivatives of the corresponding "co" functions csc and cot by changing the functions in the derivative to the corresponding "co" functions and putting in a minus sign.

PROOF Parts (iii) and (iv) of Example 1 were $\sec x$ and $\tan x$. The others are very similar:

$$\begin{aligned}\frac{d}{dx}(\csc x) &= \frac{d}{dx}\left(\frac{1}{\sin x}\right) = \frac{0(\sin x) - 1(\cos x)}{\sin^2 x} \\ &= -\frac{\cos x}{\sin^2 x} = -\frac{\cos x}{\sin x}\frac{1}{\sin x} = -\csc x \cot x. \\ \frac{d}{dx}(\cot x) &= \frac{d}{dx}\left(\frac{\cos x}{\sin x}\right) = \frac{-\sin x \sin x - \cos x \cos x}{\sin^2 x} \\ &= -\frac{\sin^2 x + \cos^2 x}{\sin^2 x} = -\frac{1}{\sin^2 x} = -\csc^2 x.\end{aligned}$$

These results hold whenever the functions involved are defined. For $\sec x$ and $\tan x$ this is whenever $\cos x \neq 0$, that is, when $x \neq (2n+1)\frac{\pi}{2}$, for any integer n. For $\csc x$ and $\cot x$ it is whenever $\sin x \neq 0$, that is, when $x \neq n\pi$, for any integer n. ◢

It is interesting to use a graphics calculator to plot a function together with its derivative. For instance, we plot $y = \cot x$ and its derivative $y = -\csc^2 x = -\frac{1}{\sin^2 x}$.

TI-85	SHARP EL9300
GRAPH F1 $(y(x) =)$ 1÷ TAN F1 (x) ▼ (−) 1 ÷ (SIN F1 (x)) x^2 M2(RANGE) 0 ▼ π ▼ ▼ (−) 10 ▼ 10 F5(GRAPH)	1 ÷ tan x/θ/T 2ndF ▼ −1 ÷ (sin x/θ/T) x^2 RANGE 0 ENTER π ENTER 1 ENTER −10 ENTER 10 ENTER 1 ENTER
CASIO f_x-7700GB/f_x-6300G	HP 48SX
Cls EXE Range 0 EXE π EXE 1 EXE −10 EXE 10 EXE 1 EXE EXE EXE EXE Graph 1 ÷ tan X EXE Graph − 1 ÷ (sin X) x^2 EXE	↰ PLOT PLOTR ERASE ATTN ' αY = 1÷ TAN αX ↱ PLOT ↰ DRAW ERASE 0 ENTER π XRNG 10 +/− ENTER 10 YRNG DRAW ATTN ' αY = 1 +/− ÷ (SIN αX ▶ y^x 2 ↱ PLOT ↰ DRAW DRAW

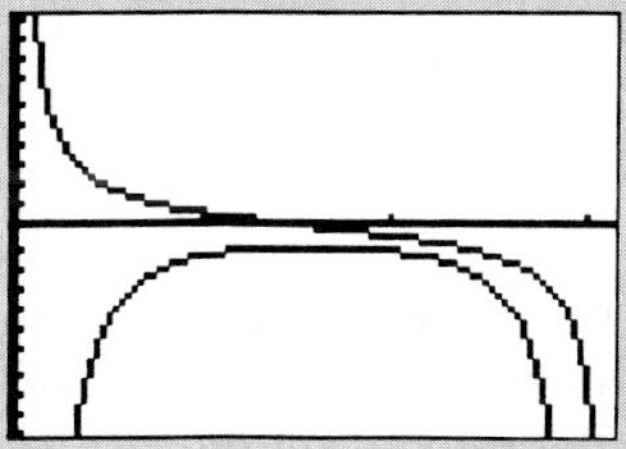

Identify which graph is which in the picture. Notice how the function $y = \cot x$ decreases more rapidly when its derivative is more negative.

EXAMPLE 2 Find all points at which the derivative of each of the following functions is zero: (i) $\tan x$; (ii) $\sec x$; (iii) $\cos(2x)$.

Solution

(i) The derivative of $\tan x$ is $\sec^2 x = \frac{1}{\cos^2 x}$. It is defined only when $\cos x \neq 0$, and it is never zero (the numerator is always 1).

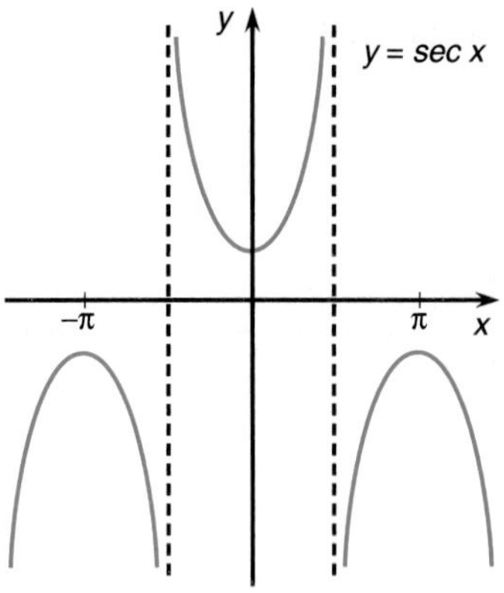

Figure 3.4.2

(ii) The derivative of $\sec x$ is $\sec x \tan x = \frac{\sin x}{\cos^2 x}$. This will be zero whenever $\sin x = 0$, which happens whenever x is an integer multiple of π, that is, when $x = n\pi$, for any integer n. We see from Figure 3.4.2 that these are exactly the points at which the tangent line is horizontal.

(iii) To do this, we observe that $\cos(2x) = \cos(x + x)$. By the addition formula (Theorem 1.7.11) this is $\cos(2x) = \cos x \cos x - \sin x \sin x = \cos^2 x - \sin^2 x$. So its derivative is

$$\begin{aligned}\frac{d}{dx}\big(\cos(2x)\big) &= \frac{d}{dx}(\cos^2 x - \sin^2 x) = \frac{d}{dx}(\cos x \cos x - \sin x \sin x)\\ &= -\sin x \cos x + \cos x(-\sin x) - \cos x \sin x - \sin x \cos x\\ &= -4 \cos x \sin x.\end{aligned}$$

This is zero whenever $\cos x$ or $\sin x$ is zero, that is, whenever $x = n\frac{\pi}{2}$, for any integer n. This is also very reasonable when we look at the graph of $y = \cos(2x)$ in Figure 3.4.3.

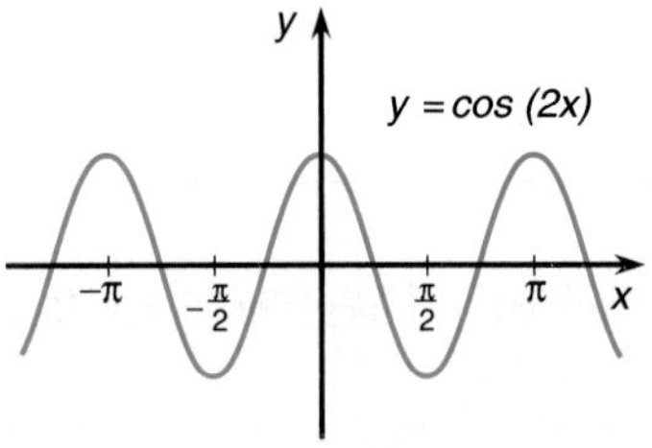

Figure 3.4.3

EXAMPLE 3 Let c be any nonzero real number, and find the derivative of $f(x) = \sin(cx)$.

Solution We know that

$$\begin{aligned}f'(x) &= \lim_{h\to 0} \frac{f(x+h) - f(x)}{h} = \lim_{h\to 0} \frac{\sin\big(c(x+h)\big) - \sin(cx)}{h}\\ &= \lim_{h\to 0} \frac{\sin(cx + ch) - \sin(cx)}{h}\\ &= \left(\lim_{h\to 0} \frac{\sin(cx + ch) - \sin(cx)}{ch}\right) c.\end{aligned}$$

Now as $h \to 0$, the product ch will also tend to 0, so if we write $u = ch$, the above limit becomes $\lim_{u\to 0} \frac{\sin(cx+u) - \sin(cx)}{u}$, which is just the derivative of sin evaluated at cx, which is $\cos(cx)$.

Including the constant c at the end of the line, we find that

$$\frac{d}{dx}\big(\sin(cx)\big) = \cos(cx)\, c.$$

The argument we just gave is quite similar to the argument used in Example 2.6.2. A few moments' thought should convince you that there is nothing particularly special about the function $\sin x$. We could have started with any differentiable function $g(x)$ and let $f(x) = g(cx)$. Then we would have found that $f(x)$ is differentiable at every x for which g is differentiable at cx. Moreover, the derivative of $f(x)$ is $g'(cx)\, c$.

We record this result for future reference.

THEOREM 3.4.3

Let $g(x)$ be a function, let c be a nonzero real number, and define $f(x) = g(cx)$. If $g(x)$ is differentiable at $x = a$, then $f(x)$ is differentiable at $x = \frac{a}{c}$, and, whenever f is differentiable,

$$f'(x) = g'(cx)\, c.$$

EXAMPLE 4 Find the derivative of (i) $\cos(3x)$; (ii) $\tan(-5x)$; (iii) $\sin(2x)$.

Solution

(i) By Theorem 3.4.3,

$$\frac{d}{dx}\big(\cos(3x)\big) = \big(-\sin(3x)\big)(3) = -3\sin(3x).$$

(ii) Similarly,

$$\frac{d}{dx}\big(\tan(-5x)\big) = \sec^2(-5x)(-5) = -5\sec^2(-5x).$$

(iii) The same sort of reasoning tells us the derivative of $\sin(2x)$ is $\cos(2x)(2) = 2\cos(2x)$. However, it is interesting to do this calculation another way. Using the addition formula for sin (Theorem 1.7.11), we write $\sin(2x) = \sin(x + x) = \cos x \sin x + \sin x \cos x = 2\cos x \sin x$. Now we can use the Product Rule to differentiate, and we find

$$\begin{aligned}\frac{d}{dx}\big(\sin(2x)\big) &= \frac{d}{dx}(2\cos x \sin x) = 2(-\sin x \sin x + \cos x \cos x)\\ &= 2(\cos^2 x - \sin^2 x).\end{aligned}$$

On the face of it we have two different answers for the derivative of $\sin(2x)$. It turns out that they are really the same, because the addition formula (Theorem 1.7.11) for cos tells us that $\cos(2x) = \cos x \cos x - \sin x \sin x = \cos^2 x - \sin^2 x$.

The last example illustrates an important fact. Finding a derivative in two different ways sometimes gives two apparently different answers. When this happens, there is always the possibility that we have made an error somewhere, but it is also possible that the two formulas are actually equal and merely appear to be different. This is especially likely to happen with trigonometric functions, because the various trigonometric identities say that apparently different formulas may actually be equal.

EXAMPLE 5 Find the maximum and minimum values of $f(t) = \sin t \cos t$ for $t \in \left[0, \frac{\pi}{2}\right]$.

Solution The derivative is $f'(t) = \cos t \cos t + \sin t(-\sin t) = \cos^2 t - \sin^2 t$. To find the maximum and minimum values, we must find when the derivative is zero, that is, when $\cos^2 t = \sin^2 t$.

Since $\cos^2 t + \sin^2 t = 1$, we have $\sin^2 t = 1 - \cos^2 t$, and substituting, we see that we are asking when does $\cos^2 t = 1 - \cos^2 t$, or $2\cos^2 t = 1$, or $\cos^2 t = \frac{1}{2}$. This happens when $\cos t = \pm\frac{1}{\sqrt{2}}$, which happens when $t = \pm\frac{\pi}{4}, \pm\frac{3\pi}{4}, \ldots$.

The only one of these points that lies in the specified interval $\left[0, \frac{\pi}{2}\right]$ is $\frac{\pi}{4}$. Evaluating there and at the endpoints, we find that $f(0) = 0, f\left(\frac{\pi}{4}\right) = \sin\frac{\pi}{4}\cos\frac{\pi}{4} = \frac{1}{\sqrt{2}}\frac{1}{\sqrt{2}} = \frac{1}{2}, f\left(\frac{\pi}{2}\right) = \sin\frac{\pi}{2}\cos\frac{\pi}{2} = 1(0) = 0$.

The maximum value of $f(t)$ on $\left[0, \frac{\pi}{2}\right]$ is $\frac{1}{2}$, which occurs when $t = \frac{\pi}{4}$. The minimum value is 0, which occurs when $t = 0$ and when $t = \frac{\pi}{2}$.

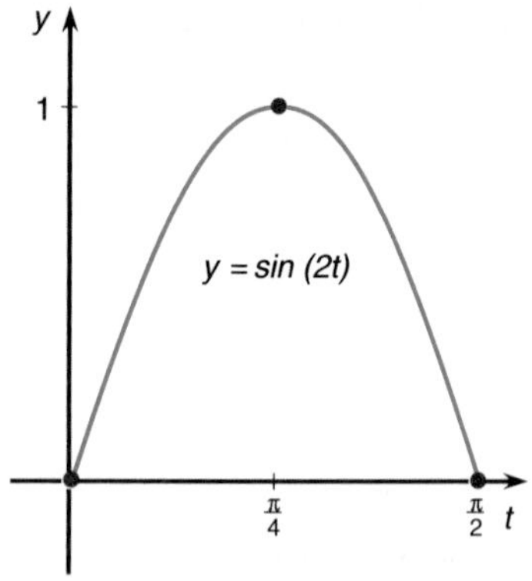

Figure 3.4.4

If we realize that $\sin t \cos t = \frac{1}{2}\sin(2t)$, the answer makes good sense. After all, on the interval $[0, \pi]$ the function $\sin t$ has one "hump" from 0 up to 1. The function $\sin(2t)$ has a corresponding "hump" on the interval $\left[0, \frac{\pi}{2}\right]$, with maximum in the middle at $t = \frac{\pi}{4}$. See Figure 3.4.4.

EXAMPLE 6 **Degrees Versus Radians** In this example we explore what happens if we try to measure angles in degrees instead of radians. Recall that an angle of $180°$ is the same thing as π radians. This means that $1°$ must equal $\frac{\pi}{180}$ radians.

For the purposes of this example only, we write $\text{sindeg}(x)$ for the function sin, with x measured in *degrees*. So, for example, $\text{sindeg}(90) = 1$, $\text{sindeg}(180) = 0$, $\text{sindeg}(270) = -1$, $\text{sindeg}(45) = \frac{1}{\sqrt{2}}$, etc.

As we have just noted, x degrees equals $\frac{\pi x}{180}$ radians. This allows us to write

$$\text{sindeg}(x) = \sin\left(\frac{\pi x}{180}\right),$$

which just expresses the conversion from degrees into radians.

What we want to do is to differentiate $\text{sindeg}(x)$. We are able to do it by applying Theorem 3.4.3 to the above conversion formula. After all, the formula says that $\text{sindeg}(x)$ equals $\sin(cx)$ with $c = \frac{\pi}{180}$, so Theorem 3.4.3 tells us that

$$\begin{aligned}\frac{d}{dx}\big(\text{sindeg}(x)\big) &= \sin'(cx)\,(c) = c\cos(cx)\\ &= \frac{\pi}{180}\cos\left(\frac{\pi x}{180}\right).\end{aligned}$$

If we write $\text{cosdeg}(x)$ for the cos function with the angle x measured in *degrees*, this can be written

$$\frac{d}{dx}\big(\text{sindeg}(x)\big) = \frac{\pi}{180}\text{cosdeg}(x).$$

What this says is that if we measure angles in degrees, then the derivative of sin is not cos, but cos multiplied by that awkward constant $\frac{\pi}{180}$. When we first use radians, it seems strange to have π turning up all the time, but now we see that by doing that we make the derivative of sin very simple. Moreover, if we refuse to use radians and insist on degrees, then the π turns up in the differentiation formula anyway.

In a sense the use of radians is a choice to have the unpleasant constant in the angles so that everything else will work out well. There is no way to avoid having something awkward somewhere, and this choice reduces the confusion in many calculations.

Exercises 3.4

I

Find the derivative of each of the following functions, stating where the derivative does and does not exist.

1. $f(x) = \cos^2 x$
2. $g(t) = t\cos t$
3. $h(u) = \frac{1}{2-\cos u}$
4. $k(r) = \frac{\cos r}{1+r^2}$
5. $F(w) = (w^2 - 2w - 1)\sin(2w)$
6. $G(x) = \tan(x/2)$
7. $H(t) = \frac{\sin(2t)}{\sin(3t)}$
8. $K(u) = \frac{\sin u}{u}$
9. $L(r) = r^3\tan r - \sec r$

10. $T(x) = \sin^2(2x) + \cos^2(2x)$

11. $f(x) = \sec^2 x$

12. $f(x) = \csc^2 x$

13. $f(x) = \sin x \csc x$

14. $f(x) = \sin x \sec x$

15. $f(x) = \cos x \csc x$

16. $f(x) = \cot(3x)$

17. $f(x) = \cot x \csc x$

18. $f(x) = \frac{1}{\csc x}$

19. $f(x) = \frac{1}{\cot x}$

20. $f(x) = 1 + \sec^2 x$

Find all points at which the derivative of each of the following functions is zero.

21. $f(x) = \csc x$

22. $g(t) = \cos t \sin t$

23. $h(u) = \sin^2 u$

24. $k(s) = \cot s$

25. $f(v) = \sin(2v)$

26. $g(x) = \sin(3x)$

27. $h(r) = r - \tan r$

28. $k(t) = 2t - \tan(2t)$

29. $f(u) = 3u - \cot(7u)$

30. $g(s) = 7s + \cos s + 2 \sin s$

Practice differentiation by finding the derivative of each of the following functions.

31. $f(x) = (1 + 2x^2) \sin x$

32. $f(x) = \frac{x + \sin x}{3x^2 + 4}$

33. $f(x) = \frac{1}{1 + \sin^2 x}$

34. $f(x) = \frac{1 - \cos^2 x}{1 - \cos x}$

35. $f(x) = \frac{\cos x}{x^2 + 2x + 5}$

36. $f(x) = \frac{\sin^2 x}{1 + \cos x}$

37. $f(x) = (x^3 - 8)(\sin x + 2 \cos x)$

38. $f(x) = (\sin^2 x + 3)^{-2}$

39. $f(x) = x^2 + 2x \tan x + \tan^2 x$

40. $f(x) = (3x + \sec x)^3$

II

41. Find all points at which each of the following functions has a horizontal tangent line: (i) $g(\theta) = \tan\theta + 2\sec\theta$, (ii) $f(\theta) = \tan\theta - \sec\theta$.

42. Find all points at which the tangent line to $y = \cot x$ has slope equal to -2.

43. Make a rough sketch of the graph of $y = \sin x$ for $x \in [-\pi, \pi]$, and indicate the parts of the interval $[-\pi, \pi]$ where the derivative is positive, the parts where it is negative, and the places where it is zero.

44. Make a rough sketch of the graph of $y = \sec x$ for $x \in [-\pi, \pi]$, and indicate the parts of the interval $[-\pi, \pi]$ where the derivative is positive, the parts where it is negative, and the places where it is zero. Confirm your geometrical impressions by calculating the derivative.

***45.** Let $g(x) = x \sin x$.

(i) Evaluate $g'(x)$ and $g'(0)$.

(ii) Show that there must be at least one value of x between $\frac{\pi}{2}$ and $\frac{3\pi}{2}$ at which the derivative g' is zero, that is, $g'(x) = 0$. (*Hint:* Apply the Intermediate Value Theorem (Theorem 2.5.5) to g'.)

***46.** (i) Let $f(x) = \sin\left(x + \frac{\pi}{2}\right)$. Show that the derivative is $f'(x) = \cos\left(x + \frac{\pi}{2}\right)$.

(ii) Let c be any fixed real number and define $g(x) = \sin(x + c)$. Find the derivative $g'(x)$.

(iii) What is the derivative of $h(x) = \sin(3x + 5)$?

(iv) If c and d are constants, what is the derivative of $k(x) = \sin(cx + d)$?

***47.** Find all points x at which the tangent line to $y = \sec x$ has slope equal to $\sqrt{2}$.

***48.** Are there any values of x at which the tangents to $y = \cos x$ and $y = \sin x$ have the same slope?

III

49. Use a graphics calculator or computer to plot both $\sin x$ and its derivative on the same graph. Confirm that the derivative is zero where the function has a horizontal tangent, and that where the derivative is positive the function slopes up to the right.

50. Plot $y = \cos x$ and its derivative on the same graph, checking the behavior of the function where the derivative is zero, positive, or negative.

51. Plot $y = \tan x$ and its derivative on the same graph. Notice how the function rises (or falls) faster when the derivative is large (or is negative with large absolute value).

52. Plot each of $y = \sec x$ and $y = \csc x$ along with its derivative.

***53.** Plot the difference quotient $\frac{f(x+h) - f(x)}{h}$ for an appropriate (small) value of h to get an approximation to the graph of $f'(x)$. Use this technique to verify Theorem 3.4.3 (p. 170) for (i) $g(x) = \sin x$, with $c = 2, -1, 3, \frac{1}{5}, 0$; (ii) $g(x) = x^2$, $c = 2, -2, 3, -\frac{1}{10}, 0$.

3.5 Derivatives of Logarithms, Exponentials, and Hyperbolic Functions

COMMON LOGARITHMS
NATURAL LOGARITHMS
EXPONENTIAL FUNCTION
HYPERBOLIC FUNCTIONS AND THEIR DERIVATIVES

In this brief section we learn the derivatives of several important functions. We have not yet learned much about the functions themselves; that will have to wait until Chapter 7. However, the differentiation rules are easily stated, so we can begin to work with them now.

We begin with a brief review of what we do know about logarithms (Section 1.8). Much as trigonometric functions can be defined using either degrees or radians, there are two standard ways of defining logarithms. The familiar way is to say that the logarithm of x is the power to which you have to raise the number 10 to get x, that is,

$$x = 10^{\log x},$$

or

$$\log(10^s) = s. \tag{3.5.1}$$

So, for instance, $\log 10 = 1, \log(1{,}000) = 3, \log(1) = 0, \log\left(\frac{1}{10}\right) = -1$, etc. Strictly speaking, $\log x$ only makes sense if $x > 0$, since 10 raised to any power has to be positive.

Much of the usefulness of logarithms arises from the fact that

$$\log(xy) = \log x + \log y, \tag{3.5.2}$$

which is just another way of saying that $10^{a+b} = 10^a 10^b$, since

$$10^{\log x + \log y} = 10^{\log x} 10^{\log y} = xy.$$

Logarithms of this kind are called **common logarithms**, or *logarithms to the base* 10. However, we could have used some other number instead of 10 when we defined logarithms. It is not unusual to consider logarithms to the base 2, sometimes written $\log_2$. They are defined by the relation

$$x = 2^{\log_2(x)}.$$

So, for example, $\log_2(8) = 3, \log_2\left(\frac{1}{16}\right) = -4, \log_2(1) = 0$, etc.

Just as trigonometric functions were defined using radians, an apparently peculiar choice of angle measurement, there is an unexpected choice that is often made for the base of logarithms. It is a number called e, and it is approximately 2.718 2818. Logarithms to the base e are so important that they are given a special name. They are called **natural logarithms**, and for any positive number x we write $\ln x$ for the natural logarithm of x. It satisfies relations analogous to Equation 3.5.1:

$$x = e^{\ln x},$$

or

$$\ln(e^s) = s.$$

So $\ln(e) = 1, \ln(e^2) = 2, \ln(1) = 0, \ln(10) \approx 2.302\,585$, etc. Natural logarithms have the disadvantage that they have complicated values for "nice" numbers like 10.

The reason for using them is that they make derivatives much easier to work with, just as radian measure does for trigonometric functions.

The natural logarithm of a product is the sum of the natural logarithms, just as for common logarithms:

$$\ln(xy) = \ln x + \ln y. \tag{3.5.3}$$

Recall that for any numbers r, s, b with $b > 0$, we know that $(b^r)^s = b^{rs}$. If we write this out with e in place of b and $r = \ln(10)$, we find

$$(e^{\ln(10)})^s = e^{\ln(10)(s)}.$$

Since $e^{\ln(10)} = 10$, this says that

$$10^s = e^{\ln(10)(s)}.$$

Now suppose $s = \log x = \log_{10}(x)$. Then $10^s = 10^{\log(x)} = x$. The above equation becomes

$$x = e^{\ln(10)\log x}.$$

Applying ln to both sides "gets rid" of the e on the right side and gives

$$\ln x = \ln(10)\log x,$$

from which we find

It is easy to remember which way this relation goes by letting $x = 10$, so $\log x = 1$ . . .

$$\log x = \frac{1}{\ln 10}\ln x. \tag{3.5.4}$$

The graph of $y = \ln x$ is illustrated in Figure 3.5.1. Notice that it crosses the x-axis at $x = 1$, since $\ln(1) = 0$. Notice too that $\ln x > 0$ when $x > 1$ and $\ln x < 0$ when $0 < x < 1$. The function $\ln x$ is not defined when $x \le 0$.

The derivative of $\ln x$ is easy to remember, though we will not be able to prove it until Chapter 7.

THEOREM 3.5.5

The function $\ln x$ is differentiable for all $x > 0$, and

$$\frac{d}{dx}(\ln x) = \frac{1}{x}.$$

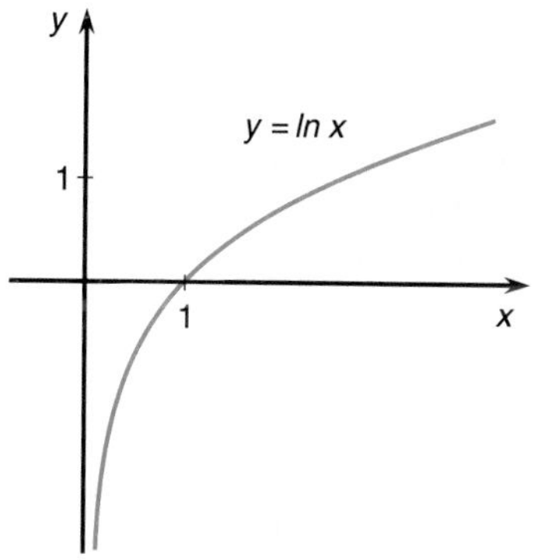

Figure 3.5.1

Refer to the graph of $y = \ln x$ and observe that the derivative is always positive, as the function is always increasing. It is also clear from the picture that the function increases more steeply for small values of x and less steeply for large x. This is reflected in the fact that the derivative $\frac{1}{x}$ is large when x is small, meaning that the slope of the tangent is large then, and smaller when x is larger, meaning that the slope of the tangent is smaller there.

EXAMPLE 1 Differentiate the following functions: (i) $f(x) = x^2 \ln x$; (ii) $g(t) = \ln^2 t$; (iii) $h(u) = \ln(4u^3)$; (iv) $k(s) = \ln\left(\frac{1}{s}\right)$; (v) $F(x) = x\ln x - x$.

Solution

(i) Applying the Product Rule and Theorem 3.5.5, we find $f'(x) = 2x\ln x + x^2\frac{d}{dx}(\ln x) = 2x\ln x + x^2\frac{1}{x} = 2x\ln x + x$.

(ii) Writing $\ln^2 t = \ln t \ln t$ and applying the Product Rule, we find $g'(t) = \frac{1}{t}\ln t + (\ln t)\frac{1}{t} = \frac{2}{t}\ln t$.

(iii) For this part we use the way logarithms work with products (Equation 3.5.3). Realizing that $\ln(4u^3) = \ln 4 + \ln u^3 = \ln 4 + \ln\big((u)(u)(u)\big) = \ln 4 + \ln u + \ln u + \ln u = \ln 4 + 3 \ln u$, we can differentiate. Notice that ln 4 is a constant ($\ln 4 \approx 1.38629$), so its derivative is 0.

We have found $h'(u) = 0 + 3\frac{1}{u} = \frac{3}{u}$.

(iv) The rule about logarithms of products can be used to figure out the logarithm of a quotient. Notice that $x = \frac{x}{y}y$, so $\ln x = \ln\big(\frac{x}{y}y\big) = \ln\frac{x}{y} + \ln y$. Rearranging, we find

$$\ln\frac{x}{y} = \ln x - \ln y,$$

so the logarithm of a quotient is the difference of the logarithms of the numerator and the denominator.

Applying this to our problem, we find that $\ln\frac{1}{s} = \ln 1 - \ln s = 0 - \ln s = -\ln s$, so $k'(s) = -\frac{1}{s}$.

(v) $F'(x) = 1(\ln x) + x\frac{1}{x} - 1 = \ln x + 1 - 1 = \ln x$.

EXAMPLE 2 **Derivative of the Common Logarithm** Let $f(x) = \log x = \log_{10}(x)$, and find its derivative $f'(x)$.

Solution We know from Equation 3.5.4 that $f(x) = \log x = \frac{1}{\ln 10}\ln x$, so

$$f'(x) = \frac{1}{\ln 10}\frac{d}{dx}(\ln x) = \frac{1}{\ln 10}\frac{1}{x} \approx 0.4342945\frac{1}{x}.$$

Example 2 shows why it is awkward to use common (base 10) logarithms when differentiating (just as it is awkward to measure angles in degrees for trigonometric functions). It is because the derivative of ln x is so nice that this kind of logarithm is called "natural."

EXAMPLE 3 Find all points where the derivative of $f(x) = x \ln x$ is zero.

Solution The function $f(x)$ is defined and differentiable whenever $x > 0$. Its derivative is $f'(x) = 1(\ln x) + x\frac{1}{x} = \ln x + 1$. For $f'(x) = 0$ we must have $\ln x = -1$, which means $x = e^{\ln x} = e^{-1} = \frac{1}{e}$. So $f'(x) = 0$ at $x = \frac{1}{e}$ and nowhere else.

EXAMPLE 4 Find the maximum and minimum values of $g(t) = t^2 \ln t - \frac{3}{2}t^2$ on the interval $[1, 10]$.

Solution Notice that $g(t)$ is defined for all $t > 0$, and is differentiable for every such t. Since a differentiable function is continuous, by Theorem 3.2.1, we can apply the technique of Summary 3.3.1 to find its maximum and minimum values.

The derivative is $g'(t) = 2t \ln t + t^2\frac{1}{t} - 3t = 2t \ln t + t - 3t = 2t \ln t - 2t = 2t(\ln t - 1)$. This will be zero only when $\ln t - 1 = 0$, that is, $\ln t = 1$. (Notice that we do not consider $t = 0$, first because it is not in the interval $[1, 10]$ and second because $g(t)$ is not defined at $t = 0$.) If $\ln t = 1$, then $t = e^{\ln t} = e^1 = e$. We evaluate the function $g(t)$ at this point $t = e$ and at the endpoints $t = 1$ and $t = 10$. We find that $g(e) = e^2 \ln(e) - \frac{3}{2}e^2 = e^2\big(\ln(e) - \frac{3}{2}\big) = e^2\big(1 - \frac{3}{2}\big) = -\frac{1}{2}e^2 \approx -3.694\,528$. Also $g(1) = 1(\ln 1) - \frac{3}{2}(1)^2 = 1(0) - \frac{3}{2} = -\frac{3}{2}$, and $g(10) = 10^2 \ln(10) - \frac{3}{2}(10)^2 \approx 100(2.302\,585 - 1.5) = 80.2585$.

The smallest of these values is $-3.694\,528$, which occurs when $t = e$, and the largest is 80.2585, which occurs when $t = 10$. The minimum value of $g(t)$ on $[1, 10]$ is $-\frac{1}{2}e^2 \approx -3.694\,528$, which occurs when $t = e$. The maximum value is $100\left(\ln(10) - \frac{3}{2}\right)$, which occurs when $t = 10$.

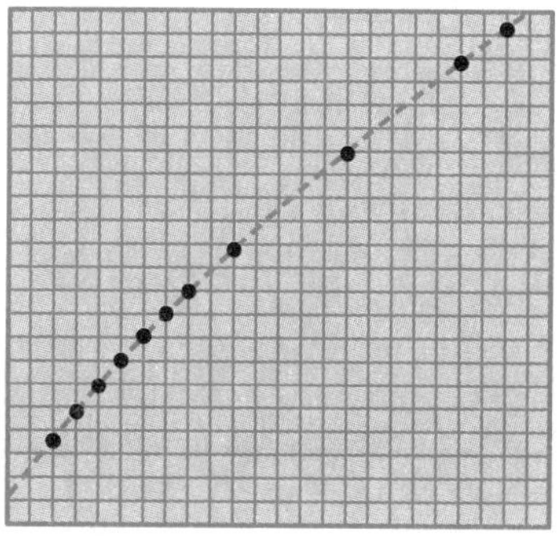

Figure 3.5.2(i)

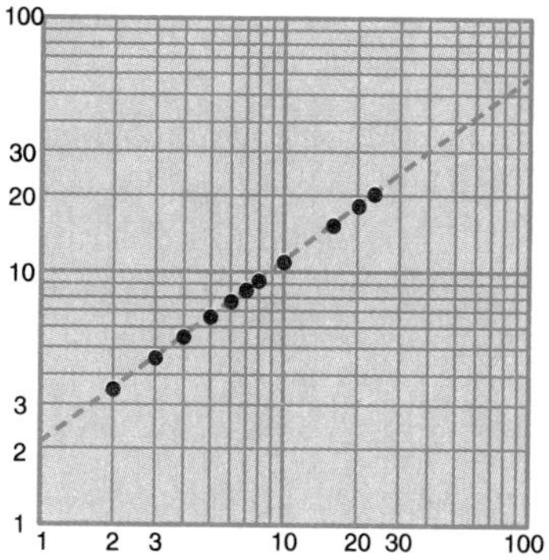

Figure 3.5.2(ii)

Log–Log Graph Paper

It is quite common for two quantities x and y to be related by a "power function" like $y = cx^k$. This situation could arise for the length and the circumference of a growing bone, two quantities measured in a chemical reaction, the intensity of a stimulus and the resulting response of a nerve cell, the length and surface area of an amoeba, the dose of some medication and the measured effect, and many other examples.

If a number of measurements can be taken, they can be plotted on a graph and used to estimate the constants c and k. However, it is not easy to guess the exponent k from the shape of the curve.

If we take logarithms of both sides of the equation $y = cx^k$, it becomes

$$\log y = \log c + k \log x.$$

Writing $Y = \log y$, $C = \log c$, and $X = \log x$, we see that the points (X, Y) lie on the straight line $Y = C + kX$. If we plot this line, its slope will be the constant k, and its Y-intercept will be C.

This can be done by finding and plotting the logarithms of all the coordinates, but it is most easily accomplished by using log–log graph paper.

The scales on the paper are "logarithmic," so when we plot (x, y) the horizontal and vertical distances from the origin are $\log x$ and $\log y$. In effect we have plotted (X, Y). In Figure 3.5.2(i) we have plotted a number of points on ordinary graph paper and estimated the shape of a curve joining them. It is difficult to guess the exponent.

However, in Figure 3.5.2(ii) we have plotted the same points on log–log paper, and the slope is easily measured to be about $\frac{3}{4}$. The value of y at the vertical intercept is about 2.1, so the original curve is approximately $y = 2.1x^{3/4}$. (Notice that on log–log paper the vertical intercept occurs where $X = 0$, that is, where $x = 1$.)

The Exponential Function

Next we consider the exponential function e^x, which is sometimes written $\exp(x)$. It is related to the natural logarithm function by the relation

$$\ln(e^x) = x,$$

or

$$\exp(\ln x) = x. \qquad (3.5.6)$$

(The first of these equations is true for all x, and the second makes sense only when $x > 0$.) Notice that if x is an integer, then e^x is just the ordinary power of the number e:

$$\exp(n) = e^n \approx (2.7182818)^n.$$

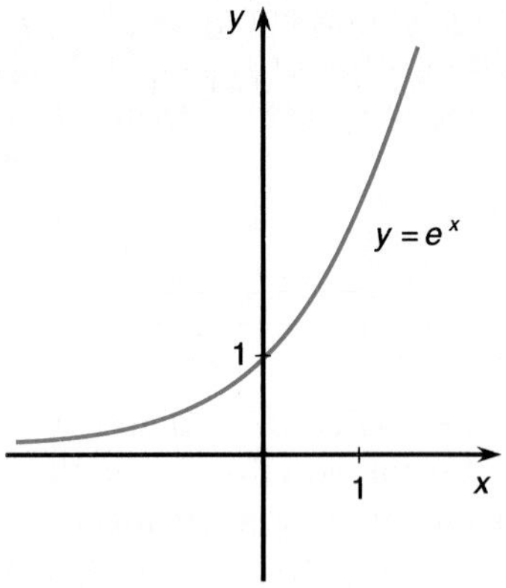

Figure 3.5.3

The graph of $y = e^x$ is illustrated in Figure 3.5.3. Notice that it is defined for all values of x. We see from the graph that $e^0 = 1$, as we should expect. Also notice that $e^x > 1$ when $x > 0$ and $e^x < 1$ when $x < 0$, but that $e^x > 0$ for all values of x. The function e^x increases very fast for large values of x.

Because of the rule about the logarithm of a product (Equation 3.5.3), we observe that

$$\ln(e^{a+b}) = a + b = \ln(e^a) + \ln(e^b) = \ln(e^a e^b).$$

Applying the exponential function to the beginning and end of this equality, we find that

$$\exp\big(\ln(e^{a+b})\big) = \exp\big(\ln(e^a e^b)\big),$$

or, using Equation 3.5.6,

$$e^{a+b} = e^a e^b. \tag{3.5.7}$$

THEOREM 3.5.8

> The function $f(x) = \exp(x) = e^x$ is defined and differentiable for all x. Its derivative is given by
>
> $$\frac{d}{dx}(e^x) = e^x.$$
>
> In words, the exponential function is its own derivative.

It is instructive to reexamine the graph of $y = e^x$ in light of this result. When x is negative, the graph is nearly flat, so its derivative is nearly zero, and the function itself is very small, that is, nearly zero. At $x = 0$ the graph is rising and its derivative is $e^0 = 1$. When $x > 0$, the derivative is bigger than 1 and the function is increasing very fast.

EXAMPLE 5 Find the derivative of each of the following functions: (i) $f(x) = e^{1+x}$; (ii) $g(t) = t^2 e^t$; (iii) $h(u) = e^{-u}$.

Solution

(i) By Equation 3.5.7, $e^{1+x} = e^1 e^x = ee^x$, so $f'(x) = e\frac{d}{dx}(e^x) = ee^x = e^{1+x}$.

(ii) $g'(t) = 2te^t + t^2 e^t = (2t + t^2)e^t$.

(iii) There are two ways to approach this. One is to observe that $e^{-u}e^u = e^{-u+u} = e^0 = 1$, using Equation 3.5.7, so

$$e^{-u} = \frac{1}{e^u}. \tag{3.5.9}$$

From this the derivative can be found by using the Quotient Rule:

$$h'(u) = \frac{0(e^u) - 1(e^u)}{(e^u)^2} = -\frac{e^u}{(e^u)^2} = -\frac{1}{e^u} = -e^{-u}.$$

In other words, $h'(u) = -h(u)$.

Another way to do this question is to appeal to Theorem 3.4.3, letting $c = -1$. So $h(u) = F(-u) = F(cu)$, with $F(x) = e^x$ and $c = -1$. Theorem 3.4.3 then tells us that $h'(u) = F'(cu)c = e^{cu}(-1) = -e^{-u} = -h(u)$. ◢

If we think about ordinary powers with integer exponents, we recall that $b^{mn} = (b^m)^n = (b^n)^m$. We can use this formula to define powers in which the exponent is *not* necessarily an integer. If $b > 0$, then $b = e^{\ln b}$, and we can define $b^r = e^{r \ln b}$.

DEFINITION 3.5.10

If $b > 0$ and $r \in \mathbb{R}$, define the power b^r by

$$b^r = e^{r \ln b} = \exp(r \ln b).$$

EXAMPLE 6 Let $a > 0$ and define $f(x) = a^x$. Find the derivative $f'(x)$.

Solution Notice that $f(x) = a^x = \exp(x \ln a)$. Now $\ln a$ is a constant, so we can write $f(x) = e^{cx}$, with $c = \ln a$, and Theorem 3.4.3 tells us that $f'(x) = e^{cx}c = \exp(x \ln a) \ln a = a^x \ln a$. ◢

Hyperbolic Functions

There are some important functions that are built up from the exponential function. In various ways they are similar or at least analogous to the trigonometric functions, and their names reflect this. They are called hyperbolic sine, hyperbolic cosine, hyperbolic tangent, etc. The symbols used for these functions are the symbols used for trigonometric functions with the letter "h" (for hyperbolic) added at the end.

DEFINITION 3.5.11

The **hyperbolic functions** are defined as follows:

$$\sinh x = \frac{e^x - e^{-x}}{2}, \qquad \cosh x = \frac{e^x + e^{-x}}{2},$$
$$\tanh x = \frac{\sinh x}{\cosh x} = \frac{e^x - e^{-x}}{e^x + e^{-x}}, \qquad \coth x = \frac{\cosh x}{\sinh x} = \frac{e^x + e^{-x}}{e^x - e^{-x}},$$
$$\operatorname{sech} x = \frac{1}{\cosh x} = \frac{2}{e^x + e^{-x}}, \qquad \operatorname{csch} x = \frac{1}{\sinh x} = \frac{2}{e^x - e^{-x}}.$$

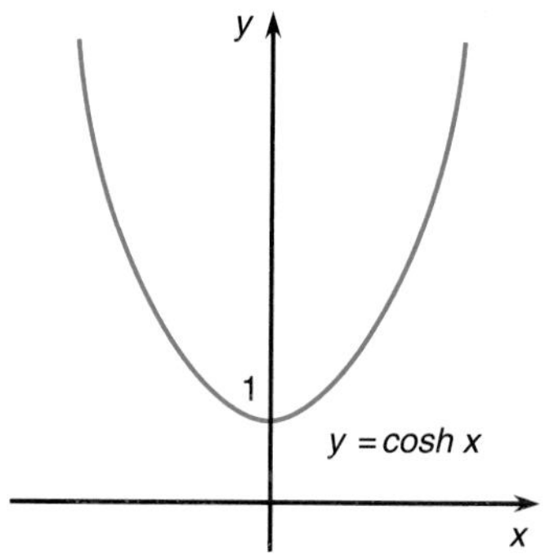

Figure 3.5.4

They can be read as hyperbolic sine, hyperbolic cos, etc., but often they are read more or less phonetically, with $\cosh x$ pronounced "cawsh x," $\sinh x$ pronounced "sinch x," and $\tanh x$ pronounced "tansh x." There seems to be no consensus about how to pronounce sech x, coth x, or csch x, but fortunately they are not used as frequently.

The graphs of $\cosh x$ and $\sinh x$ are illustrated in Figures 3.5.4 and 3.5.5. Notice that, unlike the trigonometric functions they are not bounded (i.e., they get arbitrarily large), and they are not periodic (i.e., they do not repeat after 2π).

Despite these differences, we have said that in many ways the hyperbolic functions behave like trigonometric functions. For instance,

$$\begin{aligned}
\cosh^2 x - \sinh^2 x &= \left(\frac{e^x + e^{-x}}{2}\right)^2 - \left(\frac{e^x - e^{-x}}{2}\right)^2 \\
&= \frac{e^{2x} + 2e^x e^{-x} + e^{-2x}}{4} - \frac{e^{2x} - 2e^x e^{-x} + e^{-2x}}{4} \\
&= \frac{e^{2x} + 2(1) + e^{-2x} - e^{2x} + 2(1) - e^{-2x}}{4} = \frac{4}{4} \\
&= 1.
\end{aligned}$$

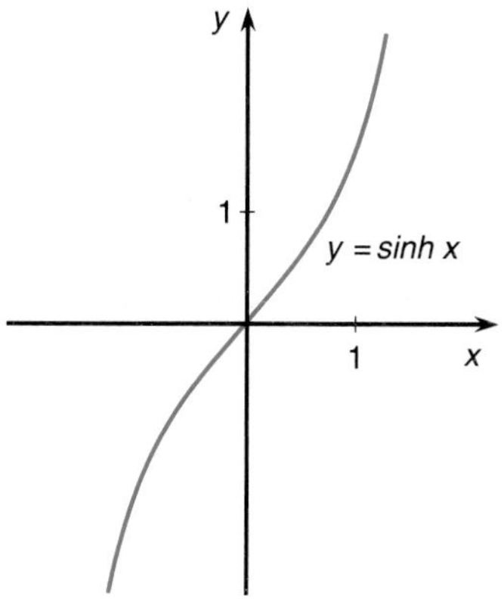

Figure 3.5.5

In other words, we have established the hyperbolic version of the fundamental trigonometric identity ($\cos^2\theta + \sin^2\theta = 1$), namely,

$$\cosh^2 x - \sinh^2 x = 1. \tag{3.5.12}$$

Notice that there is a minus sign in the hyperbolic identity but that it is remarkably similar to the corresponding trigonometric identity.

EXAMPLE 7 Find the derivatives of the functions sinh x, cosh x, and tanh x.

Solution $\frac{d}{dx}(\sinh x) = \frac{d}{dx}\left(\frac{1}{2}(e^x - e^{-x})\right) = \frac{1}{2}\left(e^x - e^{-x}(-1)\right) = \frac{e^x+e^{-x}}{2} = \cosh x$. Similarly, $\frac{d}{dx}(\cosh x) = \frac{d}{dx}\left(\frac{1}{2}(e^x + e^{-x})\right) = \frac{1}{2}\left(e^x + e^{-x}(-1)\right) = \frac{e^x-e^{-x}}{2} = \sinh x$. Using the above results and the Quotient Rule, we find that $\frac{d}{dx}(\tanh x) = \frac{d}{dx}\left(\frac{\sinh x}{\cosh x}\right) = \frac{\cosh x \cosh x - \sinh x \sinh x}{\cosh^2 x}$. Applying Equation 3.5.12 to the numerator, we see that this equals $\frac{1}{\cosh^2 x} = \text{sech}^2 x$.

We record the derivatives of all the hyperbolic functions as a theorem. We have actually just worked out the first three of them; the other three, which are similar, will be done as an exercise at the end of the section.

THEOREM 3.5.13

Derivatives of Hyperbolic Functions

The hyperbolic functions are differentiable at every point at which they are defined, and their derivatives are as follows:

$$\frac{d}{dx}(\sinh x) = \cosh x, \qquad \frac{d}{dx}(\cosh x) = \sinh x,$$
$$\frac{d}{dx}(\tanh x) = \text{sech}^2 x, \qquad \frac{d}{dx}(\coth x) = -\text{csch}^2 x,$$
$$\frac{d}{dx}(\text{sech } x) = -\text{sech } x \tanh x, \qquad \frac{d}{dx}(\text{csch } x) = -\text{csch } x \coth x.$$

These formulas are strikingly similar to the formulas for the derivatives of the trigonometric functions (Theorems 3.4.1, 3.4.2). In fact, they are direct translations (adding an "h" to each trigonometric function), *except* for the fact that some of the signs change. For the two basic ones, cosh x and sinh x, there is no minus sign in their derivatives, which makes them easier to remember than the trigonometric ones.

EXAMPLE 8 Find all points where there is a horizontal tangent line to the graph of (i) the hyperbolic sine, (ii) the hyperbolic cosine.

Solution

(i) The derivative of sinh x is cosh x. For the tangent to be horizontal we need the derivative to be zero, that is, $\cosh x = 0$. This amounts to $\frac{1}{2}(e^x + e^{-x}) = 0$, or $e^x + e^{-x} = 0$, which is impossible, since e^x and e^{-x} are both positive. In other words, the tangent to $y = \sinh x$ is never horizontal, which is consistent with the illustration of the graph.

(ii) The derivative of cosh x is sinh x, which is zero when $e^x - e^{-x} = 0$, or $e^x = e^{-x}$. Multiplying by e^x, we find that this occurs when $e^x e^x = e^x e^{-x} = 1$, which means $e^x = \pm 1$. But e^x is always positive, so this means $e^x = 1$ and $x = 0$. The only place where the tangent is horizontal is $x = 0$.

It is interesting to use a graphics calculator to plot each hyperbolic function together with its derivative. For instance, we plot $y = \tanh x$ and its derivative $y = \operatorname{sech}^2 x = \frac{1}{\cosh^2 x}$.

TI-85 GRAPH F1 ($y(x)$ =) MATH F4 F3 (tanh) EXIT F1 (x)) ▼ 1 ÷ (MATH F4 F2 (cosh) EXIT F1 (x)) x^2 M2 (RANGE) (−) 3 ▼ 3 ▼ ▼ (−) 1 ▼ 1 F5 (GRAPH)	SHARP EL9300 MATH ▼ ▶ ▼ ▼ ENTER x/θ/T 2ndF ▼ (MATH ▲ ENTER x/θ/T) a^b (−) 2 RANGE − 3 ENTER 3 ENTER 1 ENTER −1 ENTER 1 ENTER 1 ENTER
CASIO f_x-7700GB/f_x-6300G Cls EXE Range − 3 EXE 3 EXE 1 EXE − 1 EXE 1 EXE 1 EXE EXE EXE EXE Graph hyp tan X EXE Graph (hyp cos X) x^y − 2 EXE (Use MATH HYP on fx7700GB)	HP 48SX ↰ PLOT PLOTR ERASE ATTN ′ αY = MTH HYP TANH αX ↱ PLOT ↰ DRAW ERASE 3 +/− ENTER 3 XRNG 1 +/− ENTER 1 YRNG DRAW ATTN ′ αY = MTH HYP COSH αX ▶ y^x 2 +/− ↱ PLOT ↰ DRAW DRAW

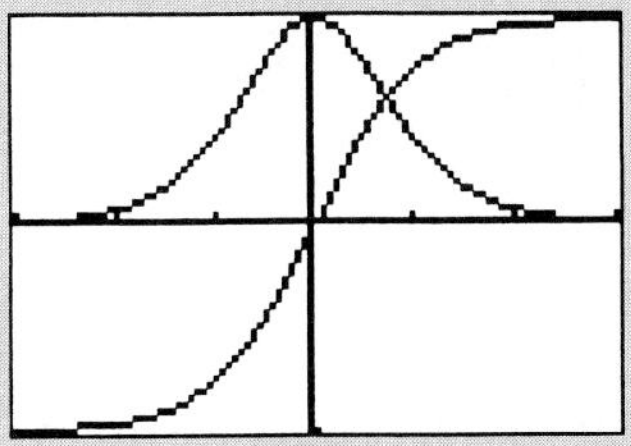

Identify which graph is which in the picture. Notice how the function $y = \tanh x$ increases more rapidly when its derivative is larger. Try plotting the other hyperbolic functions with their derivatives.

EXAMPLE 9 Find the minimum value of (i) $f(t) = t + \frac{1}{t}$ for $t > 0$; (ii) $\cosh x$.

Solution

(i) We cannot apply the technique for finding maximum and minimum values of a function on a closed interval, since the function is defined for all positive values of t. However, we could still ask where $f'(t) = 0$. Since $f'(t) = 1 + \frac{-1}{t^2}$, we see that $f'(t) = 0$ means $\frac{1}{t^2} = 1$, or $t^2 = 1$, or $t = \pm 1$. Since we are assuming $t > 0$, the only place where the derivative is zero is $t = 1$. The value of the function at this point is $f(1) = 2$.

Now an interesting thing about the function f is that $f(t) = t + \frac{1}{t} > t$. So for example, if $t > 2$, then $f(t) > 2$. Also $f(t) = t + \frac{1}{t} > \frac{1}{t}$. So if $t < \frac{1}{2}$, then taking reciprocals, $\frac{1}{t} > 2$, and $f(t) > \frac{1}{t} > 2$. We have found that $f(1) = 2$, and that $f(t) > 2$ when $t > 2$ and when $t < \frac{1}{2}$. This means that in looking for the minimum value there is no need to consider values of t that are greater than 2 or less than $\frac{1}{2}$.

In other words, we need only look for the minimum value of $f(t)$ for t in the closed interval $\left[\frac{1}{2}, 2\right]$, and this we know how to do. In fact we have already found that $t = 1$ is the only point where the derivative is zero, so we need only evaluate $f(t)$ at this point and the endpoints: $f(1) = 2, f\left(\frac{1}{2}\right) = \frac{1}{2} + 2 = \frac{5}{2}, f(2) = 2 + \frac{1}{2} = \frac{5}{2}$. The minimum value is $f(1) = 2$.

(ii) Writing $\cosh x = \frac{1}{2}(e^x + e^{-x}) = \frac{1}{2}\left(e^x + \frac{1}{e^x}\right) = \frac{1}{2}f(e^x)$, with $f(t) = t + \frac{1}{t}$, as in part (i), we can use the result of part (i) to help solve part (ii).

Indeed, we know that e^x is always positive, so part (i) says that the smallest possible value of $f(e^x)$ will be 2 and that it will occur when $e^x = 1$, that is, when $x = 0$.

The minimum value of $\cosh x$ is $\frac{1}{2}(2) = 1$, which occurs when $x = 0$.

Notice that

$$\tanh x = \frac{e^x - e^{-x}}{e^x + e^{-x}} < \frac{e^x + e^{-x}}{e^x + e^{-x}} = 1,$$

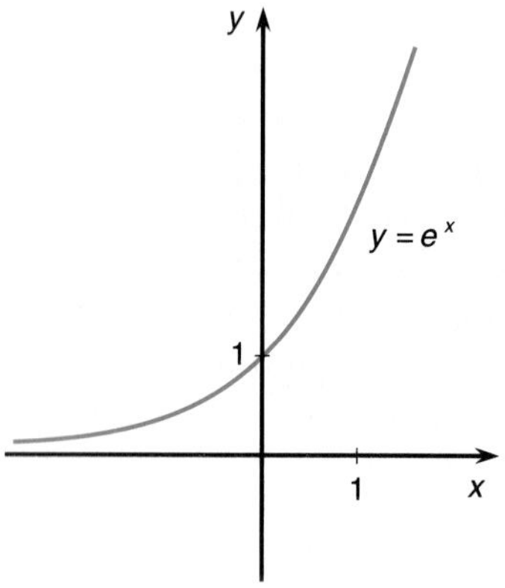

Figure 3.5.6

so $\tanh x$ is always less than 1. On the other hand,

$$\tanh x = \frac{e^x - e^{-x}}{e^x + e^{-x}} = \frac{1 - e^{-2x}}{1 + e^{-2x}}.$$

If x is very large, then from the graph of e^x in Figure 3.5.6 it seems reasonable to think that e^x is also very large, so $e^{-2x} = \frac{1}{e^{2x}}$ will be very small and the above quotient will be close to 1. In fact this is true, and

$$\lim_{x \to \infty} \tanh x = 1.$$

A similar argument shows that $\lim_{x \to -\infty} \tanh x = -1$. The graph of $y = \tanh x$ is sketched in Figure 3.5.7. Notice that the function is always increasing but that it is bounded, that is, it never gets bigger than 1 (nor smaller than −1).

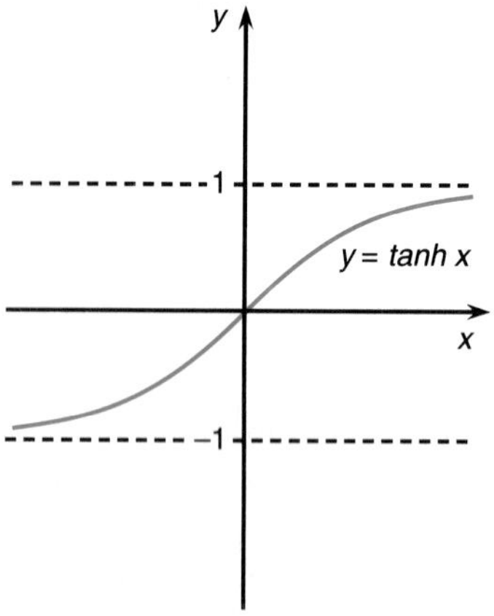

Figure 3.5.7

EXAMPLE 10 **Bacterial Growth** Consider a culture of bacteria growing in a glass jar. Suppose the number of bacteria present after t hours have elapsed is $F(t) = 10^5 + 10^6 \tanh\left(\frac{t}{10}\right)$. From the remarks just made about tanh always being less than 1, we see that $F(t)$ will always be less than $10^5 + 10^6 = 1.1 \times 10^6$. On the other hand, as time passes and $t \to \infty$, the number of bacteria $F(t)$ will approach 1.1×10^6.

The biological significance is that the jar cannot support more than 1.1×10^6 bacteria, perhaps because of lack of space, perhaps because of limited food or oxygen, or perhaps because of a number of such considerations. At any rate, the population cannot exceed 1.1×10^6, which is called the carrying capacity of the jar environment.

On the other hand, the bacteria will tend to grow and fill up the jar as much as possible. As they approach the maximum, their growth will slow down (because of lack of space, food, etc.). So the graph of $\tanh t$ has roughly the correct form to describe this kind of growth.

Of course, if the bacteria are left alone, they will eventually die and the population will decrease, but until that happens, a function like $F(t)$ will describe their growth fairly well.

The question we wish to answer is how long does it take for the culture to reach a population of 7×10^5, that is, for what t does $F(t)$ equal $7 \times 10^5 = 700{,}000$?

Solution What we have to do is solve

$$10^5 + 10^6 \tanh\left(\frac{t}{10}\right) = 7 \times 10^5,$$

or

$$10^6 \tanh\left(\frac{t}{10}\right) = 6 \times 10^5,$$

or

$$\tanh\left(\frac{t}{10}\right) = 0.6.$$

To do this, let us write $u = \exp\left(\frac{t}{10}\right)$ so that $\tanh\left(\frac{t}{10}\right) = \frac{u-u^{-1}}{u+u^{-1}}$. Then we need to solve $\frac{u-u^{-1}}{u+u^{-1}} = 0.6$, or $10(u - u^{-1}) = 6(u + u^{-1})$, or $(10 - 6)u = (10 + 6)u^{-1}$, or $4u = 16u^{-1}$, or $4u^2 = 16$, $u^2 = \frac{16}{4} = 4$.

From this we conclude that $u = \pm 2$. But since u was defined to be an exponential, it is always positive, so $u = 2$, or to return to the original variable,

$$e^{t/10} = 2.$$

Taking natural logarithms of both sides, we find that $\frac{t}{10} = \ln 2 \approx 0.693$, so $t \approx 6.93$.

The answer to our question is that it will take about 6.93 hours for the culture to number 7×10^5.

Exercises 3.5

I

Find the derivative of each of the following functions, specifying any points where it is not defined.

1. $f(t) = t^3 \ln t$
2. $g(u) = \ln^3 u$
3. $h(x) = x \ln x^2$
4. $k(s) = s \ln^2 s$
5. $H(r) = e^r \ln r$
6. $L(z) = (z^3 - 1)e^z$
7. $F(x) = \frac{x}{e^{2x}}$
8. $G(t) = te^{t-6}$
9. $K(u) = e^{\ln u}$
10. $T(w) = \ln(e^w)$
11. $f(x) = \ln(3x)$
12. $g(x) = e^{-7x}$
13. $h(t) = \ln(te^t)$
14. $k(u) = e^{2u+\ln u}$
15. $F(t) = \frac{1}{\ln t}$
16. $G(x) = \frac{1}{e^{\pi x-1}}$
17. $H(x) = \ln(x^2)$
18. $K(x) = \ln x^n$
19. $R(x) = e^{2x+4+2\ln x}$
20. $T(x) = e^{\sin^2 x+\cos^2 x}$
21. $f(x) = \cosh x \sinh x$
22. $g(x) = \cosh^2 x$
23. $h(x) = x \tanh 2x$
24. $k(x) = \frac{1}{\coth x}$
25. $U(x) = e^x \sinh x$
26. $V(x) = \sinh(\ln x)$
27. $L(x) = \cosh^2 x - \sinh^2 x$
28. $F(x) = \operatorname{sech}^2 x$
29. $M(x) = \sinh 3x + \cosh 4x$
30. $F(x) = \cosh x \tanh x$

Practice your skill at differentiation by finding the derivatives of the following functions.

31. $f(x) = \frac{\ln x}{2x^3-4}$
32. $f(x) = \cos x + 3e^{-x} \tan x$
33. $f(x) = \frac{x^2+6}{e^x+\sin x+2}$
34. $f(x) = \frac{1}{(x-3)^2+1} \tan x$
35. $f(x) = (2 + x)e^x \cos x$
36. $f(x) = \frac{e^{-x}}{1+e^{-2x}}$
37. $f(x) = \frac{x^2+3}{\tan^2 x+2+\sin x}$
38. $f(x) = \sin^2 x + \sin x + 1$
39. $f(x) = \sin^3 x$
40. $f(x) = \frac{\sin^3 x-1}{\sin x-1}$

II

41. Verify the differentiation formulas (Theorem 3.5.13) for sech x, csch x, and coth x.
42. (i) Find an expression for $\log_2 x$ in terms of $\ln x$, in analogy with Equation 3.5.4.

(ii) Find an expression for $\log_2 x$ in terms of $\log_{10} x$.

(iii) If a, b are both greater than 1, say what we should mean by $\log_a x$, $\log_b x$, and find an expression for $\log_a x$ in terms of $\log_b x$.

43. Find the derivative of the function $f(x) = \log_2 x$.

44. Find all numbers a for which the tangent line to the graph of $y = \ln x$ at $x = a$ passes through the origin.

45. Find all numbers a for which the tangent line to the graph of $y = \exp x$ at $x = a$ passes through the origin.

46. (i) Find the maximum and minimum values of $\cosh x$ on $[-1, 1]$.

(ii) Find the maximum and minimum values of $\operatorname{sech} x$ on $[-1, 1]$.

(iii) Find the maximum and minimum values of $\tanh x$ on $[-2, 3]$.

***47.** Let $c > 0$. Find the maximum and minimum values on the interval $[-c, c]$ of each of the following functions: (i) $\cosh x$, (ii) $\operatorname{sech} x$, (iii) $\sinh x$.

***48.** Find the addition formula for $\sinh x$, that is, a formula for $\sinh(x + y)$.

***49.** (i) Find the addition formula for $\cosh x$, that is, a formula for $\cosh(x + y)$.

(ii) Use the answer to part (i) to give a formula for $\cosh(2x)$.

50. Recall the trigonometric identity $1 + \tan^2 x = \sec^2 x$. Find an analogous identity for hyperbolic functions.

***51.** Find a function $f(x)$ whose derivative is $f'(x) = \ln^2 x$. (*Hint:* Look at Example 1(v), p. 176.)

***52.** (i) Let $f(x) = e^x$ and pretend that we do not know its derivative except that we know $f'(0) = 1$. Use this and other properties of e^x to show that $f'(x) = f(x)$ for all x.

(ii) Let $g(x) = \ln x$ and assume that we only know that $g'(1) = 1$. Use this and other properties of $\ln x$ to show that $\frac{d}{dx}(\ln x) = \frac{1}{x}$ for all $x > 0$.

***53.** Suppose a culture of bacteria is growing in a jar in such a way that the number of bacteria after t minutes is $F(t) = 10^5 \times \left(3 + \tanh\left(\frac{t}{100}\right)\right)$.

(i) What is the carrying capacity of the jar?

(ii) How long will it take until the culture reaches 90% of the carrying capacity?

***54.** Suppose bacteria are known to grow so that the number of milligrams of bacteria after t hours is given by $M(t) = 2 + 17\tanh(kt)$, where k is some constant. Suppose we measure and find that when $t = 2$, the value of M is $M(2) = 10$ (i.e., there are 10 mg after 2 hours). Find the value of the constant k.

III

55. Use a graphics calculator or computer to draw the graphs of $y = \ln x$ and $y = \ln(2x)$ on the same picture, and confirm that $\ln(2x) = \ln x + \ln 2$.

56. Draw the graphs of $y = e^x$ and $y = 2e^x$ on the same picture, and confirm that $2e^x$ equals $e^{x+\ln 2}$, whose graph is the graph of e^x shifted to the left by a distance of $\ln 2$.

57. Draw the graph of $y = \ln x$. Using successive magnifications, find the value of e to four decimal places by noting where the graph crosses the horizontal line $y = 1$.

58. Use the graph of $y = e^x$ to estimate $\ln 10$ to four decimal places.

59. Sketch the graphs of $y = e^x$, $y = 2^x$, $y = 3^x$, and $y = \left(\frac{1}{2}\right)^x$ on the same graph. How many times do any two of them cross?

60. Compare the graphs of $y = \sinh x$ and the difference quotient $Q(x) = \frac{\cosh(x+h) - \cosh(x)}{h}$, with (i) $h = 1$, (ii) $h = 0.5$, (iii) $h = 0.1$.

61. Use the graph of $y = \sinh x$ to find the value of x for which $\sinh x = 1000$. Can you solve for this value without using the graph?

62. We found that the bacterial culture in Example 10 will number 7×10^5 after 6.93 hours. Graph the function $F(t)$ and use the picture to find (i) how many there will be after another 6.93 hours, (ii) when there will be 3.5×10^5.

POINT TO PONDER

The trigonometric functions $\sin t$ and $\cos t$ were first introduced for dealing with various problems involving angles and circles. They satisfy the fundamental identity $\cos^2 t + \sin^2 t = 1$, which can be interpreted to mean that the point $(\cos t, \sin t)$ is on the circle of radius 1 whose center is the origin (after all, the equation of this circle is $x^2 + y^2 = 1$, which is satisfied by $x = \cos t$, $y = \sin t$).

The fundamental identity for the hyperbolic functions is $\cosh^2 t - \sinh^2 t = 1$, which means that the point $(\cosh t, \sinh t)$ satisfies the equation $x^2 - y^2 = 1$. Of course this is the equation of a hyperbola, and we see that the points $(\cosh t, \sinh t)$ all lie on the hyperbola. In fact, since $\cosh t = \frac{1}{2}(e^t + e^{-t})$ is always positive, these points are all on the right branch of the hyperbola. As t takes on every possible real value, the point $(\cosh t, \sinh t)$ sweeps out the whole right branch of the hyperbola. See Figure 3.5.8.

Of course this is why these functions are called "hyperbolic"; in some old-fashioned books the trigonometric functions are referred to as "circular functions."

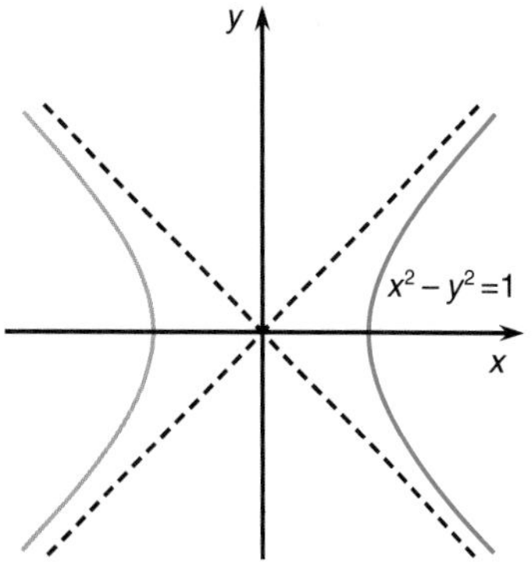

Figure 3.5.8

3.6 The Chain Rule

COMPOSITE FUNCTIONS
DERIVATIVES OF POWERS

In this section we learn to differentiate composite functions. This will allow us to work with a great many new functions built up from the ones we already know about. Before proceeding, you should be familiar and comfortable with the derivatives of polynomials, rational functions, trigonometric functions, logarithms, and exponentials.

In Theorem 2.3.2 we discussed the limit of a composite function $h(t) = g\big(f(t)\big)$. The *Chain Rule* is a technique for evaluating the derivative of such a composite function. (The name suggests that the functions fit together to form a chain.)

To get some idea of what to expect, let us think of the derivative as a rate of change. If $f'(a) = 2$, for example, it means that for t near a, $f(t)$ changes approximately twice as fast as t does. To put it another way, a small change in t will result in a change of approximately twice as much in $f(t)$. So if t changes from $t = a$ to $t = a + h$, say, then we expect $f(t)$ to change from $f(a)$ by *about* $2h$, to something close to $f(a) + 2h$.

To consider the composite function $h(t) = g\big(f(t)\big)$, assume $g(u)$ is differentiable at $u = f(a)$. For the sake of our example, suppose $g'\big(f(a)\big) = 3$. As above, this means that the value of $g(u)$ changes approximately 3 times as fast as the value of u, provided that u is near $u = f(a)$.

Now suppose we allow t to change slightly near $t = a$ and ask what happens to $h(t) = g\big(f(t)\big)$. As t changes, we know that $f(t)$ changes by approximately twice as much. Then $g\big(f(t)\big)$ should change by approximately 3 times as much as that, or $(3)(2) = 6$ times as much as the change in t.

This suggests that the rate of change of $h(t) = g\big(f(t)\big)$ should be 6, or, to put it another way, that $h'(a) = 6$. Of course, the values we chose for $f'(a)$ and $g'\big(f(a)\big)$ were not important; what is important is that the derivative of $h(t) = g\big(f(t)\big)$ is the product of the derivatives $f'(a)$ and $g'\big(f(a)\big)$. This is what is known as the Chain Rule.

Notice that when we form the composite function $h(t) = g\big(f(t)\big)$ and consider its behavior near $t = a$, we must of course know what $f(t)$ does for t near $t = a$. But as for $g(u)$, what matters is its behavior near $u = f(a)$. In fact, a moment's reflection will convince you that what $g(u)$ does near $u = a$ is irrelevant. It is possible that $g(u)$ might not even be defined at $u = a$. This is why the derivative of g that enters into the Chain Rule for $h(t)$ is $g'\big(f(a)\big)$.

EXAMPLE 1 Let $h(t) = (t^2 - 5)^3$. If we want to find its derivative $h'(t)$, we can express $h(t)$ as the composition of two functions: $h(t) = g\big(f(t)\big)$, with $g(u) = u^3$ and $f(t) = t^2 - 5$.

Of course, we know that $f'(t) = 2t$ and $g'(u) = 3u^2$, so the Chain Rule says that

$$\begin{aligned} h'(t) &= f'(t)g'\big(f(t)\big) = (2t)3\big(f(t)\big)^2 = 6t(t^2 - 5)^2 = 6t(t^4 - 10t^2 + 25) \\ &= 6t^5 - 60t^3 + 150t. \end{aligned}$$

To check this, since after all we do not really know the Chain Rule yet, we could instead just expand $h(t)$ and differentiate. Doing this, we find

$$h(t) = (t^2 - 5)^3 = t^6 - 15t^4 + 75t^2 - 125.$$

Differentiating, we get

$$h'(t) = 6t^5 - 60t^3 + 150t,$$

and we do get the same result both ways.

Depending on how quickly we could expand $(t^2 - 5)^3$, the two methods are about equally easy in this example, but there are many examples in which the Chain Rule is much quicker, and there are many more in which there is no way to differentiate without the Chain Rule.

EXAMPLE 2 Differentiate $h(x) = \cos(x^2 - 3x + 7)$.

Solution If we write $h(x) = g\big(f(x)\big)$, with $g(u) = \cos u$ and $f(x) = x^2 - 3x + 7$, then we expect that the derivative of $h(x)$ will be

$$\begin{aligned} h'(x) &= f'(x)g'\big(f(x)\big) = (2x - 3)\big(-\sin(x^2 - 3x + 7)\big) \\ &= (3 - 2x)\sin(x^2 - 3x + 7). \end{aligned}$$

Notice that in this case there is no other way to find the derivative.

Now that we have begun to see how the Chain Rule is used, it is time to state it precisely.

THEOREM 3.6.1

The Chain Rule

Suppose $f(x)$ is differentiable at $x = a$ and $g(u)$ is differentiable at $u = f(a)$. Then $h(x) = g\big(f(x)\big)$ is differentiable at $x = a$, and its derivative is

$$h'(a) = g'\big(f(a)\big)f'(a).$$

As a function, the derivative is

$$\frac{d}{dx}\big(h(x)\big) = \frac{d}{dx}\Big(g\big(f(x)\big)\Big) = g'\big(f(x)\big)f'(x).$$

PROOF The proof of the Chain Rule is actually quite delicate. You may wish to skip over it the first time through and proceed to see how it is used in the subsequent examples.

To evaluate $h'(x) = \lim_{k \to 0} \frac{h(x+k)-h(x)}{k}$, the natural thing to try first is writing

$$\begin{aligned}\frac{h(x+k)-h(x)}{k} &= \frac{g\big(f(x+k)\big)-g\big(f(x)\big)}{k} \\ &= \frac{g\big(f(x+k)\big)-g\big(f(x)\big)}{f(x+k)-f(x)} \times \frac{f(x+k)-f(x)}{k},\end{aligned}$$

hoping that as $k \to 0$, the first factor will tend to $g'\big(f(a)\big)$ and the second will tend to $f'(a)$. This is more or less the right idea, but the difficulty is that division by $f(x+k)-f(x)$ may not be permissible because $f(x+k)-f(x)$ may be zero even when $k \neq 0$.

Instead we fix the value of x and define a useful function $Q(r)$ for small values of r by

$$Q(r) = \begin{cases} \frac{g(f(x)+r)-g(f(x))}{r}, & \text{if } r \neq 0, \\ g'\big(f(x)\big), & \text{if } r = 0. \end{cases}$$

Since $\lim_{r \to 0} Q(r) = \lim_{r \to 0} \frac{g(f(x)+r)-g(f(x))}{r} = g'\big(f(x)\big) = Q(0)$, we see that $Q(r)$ is continuous at $r = 0$.

Observe that as $k \to 0$, $f(x+k) \to f(x)$, so $f(x+k)-f(x) \to 0$. By Theorem 2.3.2, about the limit of a composite function, we see that as $k \to 0$, we have $Q\big(f(x+k)-f(x)\big) \to Q(0) = g'\big(f(x)\big)$.

Now consider the product

$$Q\big(f(x+k)-f(x)\big)\frac{f(x+k)-f(x)}{k}.$$

If $f(x+k) \neq f(x)$, then the product equals

$$\begin{aligned}&\frac{g\big(f(x)+f(x+k)-f(x)\big)-g\big(f(x)\big)}{f(x+k)-f(x)} \times \frac{f(x+k)-f(x)}{k} \\ &\qquad = \frac{g\big(f(x+k)\big)-g\big(f(x)\big)}{k}.\end{aligned}$$

If, on the other hand, for some nonzero k, $f(x+k) = f(x)$, that is, $f(x+k) - f(x) = 0$, then the first factor in the product, $Q\big(f(x+k)-f(x)\big)$, is equal to $Q(0) = g'\big(f(x)\big)$, and the second is obviously zero. So the product is 0, which equals the same difference quotient $\frac{g(f(x+k))-g(f(x))}{k}$ in this case too. So

$$\begin{aligned}\lim_{k \to 0} \frac{g\big(f(x+k)\big)-g\big(f(x)\big)}{k} &= \lim_{k \to 0}\left(Q\big(f(x+k)-f(x)\big)\frac{f(x+k)-f(x)}{k}\right) \\ &= \left(\lim_{k \to 0} Q\big(f(x+k)-f(x)\big)\right)\left(\lim_{k \to 0}\frac{f(x+k)-f(x)}{k}\right) \\ &= Q(0)f'(x) = g'\big(f(x)\big)f'(x),\end{aligned}$$

as desired. ◾

The formula

$$\frac{d}{dx}\Big(g\big(f(x)\big)\Big) = g'\big(f(x)\big)f'(x)$$

is called the Chain Rule.

EXAMPLE 3 Find the derivative of each of the following functions: (i) $h(x) = \sin(x^2+1)$; (ii) $h(x) = (x^2-5x+2)^7$; (iii) $h(x) = e^{2\tan x}$; (iv) $h(x) = \sin^4(x^3-2)$.

To find $h(x) = g\big(f(x)\big)$, we first evaluate $f(x)$ and then apply g to the result. In part (i) we get $\sin(x^2+1)$ by evaluating $f(x) = x^2+1$ and then taking sin of that. In part (ii) we evaluate $f(x) = x^2-5x+2$ and find its seventh power.

Solution Each of these functions $h(x)$ can be expressed as a composite function $h(x) = g\big(f(x)\big)$.

(i) Here we let $g(u) = \sin u$, $f(x) = x^2+1$. The Chain Rule tells us that $h'(x) = g'\big(f(x)\big)f'(x) = \cos(x^2+1)(2x) = 2x\cos(x^2+1)$.

(ii) Here $g(u) = u^7$ and $f(x) = x^2-5x+2$. So $\frac{d}{dx}\big((x^2-5x+2)^7\big) = 7(x^2-5x+2)^6(2x-5)$. Notice that in this case to expand the power first and then differentiate would be an unpleasant task.

(iii) Here $g(u) = e^u$ and $f(x) = 2\tan x$. So $h'(x) = \exp(2\tan x)(2\sec^2 x)$ (since $g'(u) = g(u) = e^u$). This formula only makes sense when $\tan x$ is differentiable, which means when $x \neq (2n+1)\frac{\pi}{2}$, for any integer n.

(iv) This function $h(x) = \sin^4(x^3-2)$ is the fourth power of the function $\sin(x^3-2)$, so if we let $g(u) = u^4$ and $f(x) = \sin(x^3-2)$, then we know that

$$h'(x) = 4\big(f(x)\big)^3 f'(x) = 4\sin^3(x^3-2)f'(x).$$

However, to find $f'(x)$, we have to use the Chain Rule a second time. Writing $f(x) = \sin(x^3-2) = G\big(F(x)\big)$, with $G(u) = \sin u$ and $F(x) = x^3-2$, we see that

$$f'(x) = G'\big(F(x)\big)F'(x) = \cos(x^3-2)(3x^2).$$

Substituting this into the formula for $h'(x)$, we find that

$$h'(x) = 4\sin^3(x^3-2)\cos(x^3-2)(3x^2) = 12x^2\sin^3(x^3-2)\cos(x^3-2).$$

The last part of Example 3 illustrates the possibility of using the Chain Rule repeatedly to differentiate a function that is the composition of three or more functions.

With a little practice, the Chain Rule becomes easy to use, and problems like the ones above can be done mostly in your head without having to write out carefully which functions you are combining to form the composite function. If you are having difficulty figuring out which function is which, consider that in a composite function $h(x) = g\big(f(x)\big)$, what happens first is that x is changed into $f(x)$, and *then* the result is changed into $g\big(f(x)\big)$.

So, for instance, to construct the function $(x^9-3x^5+2)^{11}$, we first transform x to x^9-3x^5+2 and *then* take the eleventh power. Written as a composite function, $(x^9-3x^5+2)^{11} = g\big(f(x)\big)$, with $f(x) = x^9-3x^5+2$ and $g(u) = u^{11}$.

With a graphics calculator it is possible to sketch the graph of the function $h(x) = \sin(x^2+1)$ of part (i) of Example 3 together with the graph of its derivative.

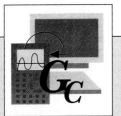

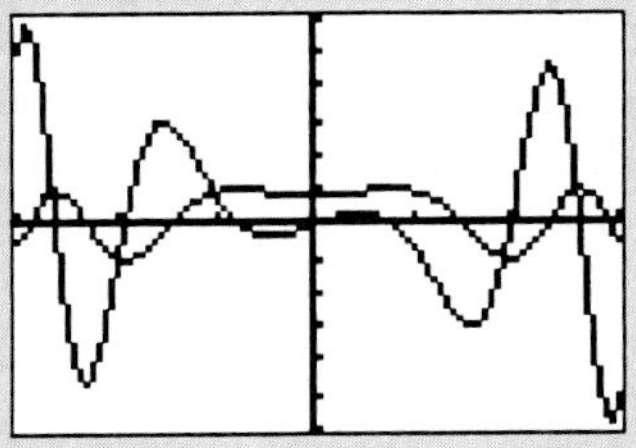

TI-85 GRAPH F1 $(y(x) =)$ SIN(F1 (x) x^2+ 1) ▼ 2 F1 (x) COS (F1 (x) x^2 + 1) M2(RANGE) (−) 3 ▼ 3 ▼ ▼ (−) 6 ▼ 6 F5(GRAPH)	SHARP EL9300 ⎍ sin (x/θ/T x^2 + 1) 2ndF ▼ 2 x/θ/T cos (x/θ/T x^2 + 1) RANGE −3 ENTER 3 ENTER 1 ENTER − 6 ENTER 6 ENTER 1 ENTER ⎍
CASIO f_x-7700GB/f_x-6300G Cls EXE Range −3 EXE 3 EXE 1 EXE −6 EXE 6 EXE 1 EXE EXE EXE EXE Graph sin (X x^2 + 1) EXE Graph 2 X cos (X x^2 + 1) EXE	HP 48SX ↰ PLOT PLOTR ERASE ATTN ′ αY = SIN αX y^x 2+ 1 ↱ PLOT ↰ DRAW ERASE 3 +/− ENTER 3 XRNG 6 +/− ENTER 6 YRNG DRAW ATTN ′ αY = 2 × αX × COS αX y^x 2 + 1 ↱ PLOT ↰ DRAW DRAW

Be sure to identify which curve is $h(x)$ and which is the derivative $h'(x)$. Notice how the absolute value of the derivative is larger when the function is changing more rapidly. You might like to try plotting these curves on different intervals, or plotting the other functions from the example.

In describing composite functions we have always been careful to write $g(u)$ and $f(x)$, using different variables for the two functions to avoid confusion. In making the composite function we let $u = f(x)$, so $h(x) = g(u) = g\big(f(x)\big)$. The Chain Rule involves two derivatives; the first is $g'(u)$, which can be written $\frac{d}{du}\big(g(u)\big)$ or $\frac{dg}{du}$. The second is $f'(x) = \frac{d}{dx}\big(f(x)\big)$, and with the understanding that $u = f(x)$, this can be written $\frac{du}{dx}$.

Now the Chain Rule says that $h'(x) = g'\big(f(x)\big)f'(x) = g'(u)f'(x)$, or

$$\frac{dh}{dx} = \frac{dg}{du}\frac{du}{dx}. \tag{3.6.2}$$

This version of the Chain Rule can be very useful, but it is important to interpret it correctly. Specifically, the right side of Equation 3.6.2 should not be interpreted as the product of two fractions (cf. the discussion after Example 3.2.1). If the derivative symbols are read as fractions, then there is a great temptation to cancel the du from denominator and numerator. In a way this is helpful for remembering the Chain Rule, but it is not a correct calculation.

The version of the Chain Rule given in Equation 3.6.2 is sometimes helpful for finding derivatives. We return to the example mentioned above.

EXAMPLE 4 Differentiate (i) $(x^9 - 3x^5 + 2)^{11}$; (ii) $\cos(\ln x)$.

Solution

(i) Writing $(x^9 - 3x^5 + 2)^{11} = f(x)^{11}$, to differentiate, we use Equation 3.6.2 and take the derivative of the eleventh power first. The idea is that the derivative of the eleventh power of some function starts, as we expect, with 11 times the tenth power of the function, but the Chain Rule (Equation 3.6.2) says that we have to multiply that by the derivative of the function:

$$\frac{d}{dx}\big((x^9 - 3x^5 + 2)^{11}\big) = 11(x^9 - 3x^5 + 2)^{10}(9x^8 - 15x^4).$$

(ii) To differentiate $\cos(\ln x)$, we notice that the derivative of cos of some function is $-$ sin of the function multiplied by the derivative of that function. So

$$\frac{d}{dx}\big(\cos(\ln x)\big) = -\sin(\ln x)\frac{1}{x}.$$

EXAMPLE 5 In Theorem 3.4.3 we discussed the derivative of a function of the form $g(cx)$, where c was some nonzero constant. We now can see that this function is the composite function $g\big(f(x)\big)$, with $f(x) = cx$. Since $f'(x) = c$, the result of Theorem 3.4.3 is an immediate consequence of the Chain Rule:

$$\frac{d}{dx}\big(g(cx)\big) = g'(cx)\,c.$$

EXAMPLE 6 Let $r \in \mathbb{R}$ and define $f(x) = x^r$. Find the derivative $f'(x)$ and state where it exists.

Solution Using Definition 3.5.10, the definition of a power, if $x > 0$ we write

$$f(x) = x^r = e^{r \ln x}.$$

The Chain Rule tells us that

$$f'(x) = e^{r \ln x}\frac{d}{dx}(r \ln x) = x^r\left(r\frac{1}{x}\right) = rx^{r-1}.$$

This will certainly make sense whenever $x > 0$.

However, there are some powers that make sense even for negative values of x, such as $x^{1/3}$. In fact, as we noticed in Theorem 2.3.1, this will happen for x^r when r is a rational number, $r = \frac{p}{q}$, with the denominator q an *odd* number.

We can also use the Chain Rule to evaluate the derivative of this type of function when $x < 0$. We start with $F(x) = x^{1/q}$, where q is an odd integer. Write $g(u) = u^q$, and notice that $h(x) = g\big(F(x)\big) = F(x)^q = (x^{1/q})^q = x$. So of course $h'(x) = 1$, but it is also possible to calculate $h'(x)$ using the Chain Rule.

We find that $h'(x) = g'\big(F(x)\big)F'(x) = qF(x)^{q-1}F'(x)$. But $h'(x) = 1$, so

$$1 = q(x^{1/q})^{q-1}\,F'(x),$$

and

$$F'(x) = \frac{1}{qx^{\frac{q-1}{q}}} = \frac{1}{q}x^{-(1-1/q)} = \frac{1}{q}x^{1/q-1}.$$

Now we consider the function $f(x) = x^{p/q} = (x^{1/q})^p = F(x)^p$, and differentiate using the Chain Rule and the above derivative of $x^{1/q}$. We find that

$$\begin{aligned}\frac{d}{dx}(x^{p/q}) &= \frac{d}{dx}\left((x^{1/q})^p\right) = p(x^{1/q})^{p-1}\left(\frac{1}{q}x^{1/q-1}\right)\\ &= \frac{p}{q}x^{\frac{p-1}{q}+\frac{1}{q}-1} = \frac{p}{q}x^{p/q-1}.\end{aligned}$$

Strictly speaking, to apply the Chain Rule, we have to know that the function $x^{1/q}$ is differentiable. It is, but we cannot really show why, since we do not actually know very much about the power function. Later on it will be possible to discuss this point with greater precision, but for now the result will be very useful.

In summary, then, we have shown that the derivative of $f(x) = x^r$ is $f'(x) = rx^{r-1}$, whenever it makes sense. This is consistent with Theorems 3.2.3 and 3.2.8, which say that $\frac{d}{dx}(x^n) = nx^{n-1}$, for any integer n, provided that the formula makes sense. We record this important result.

THEOREM 3.6.3

The derivative of the rth power function $f(x) = x^r$ is

$$\frac{d}{dx}(x^r) = rx^{r-1},$$

whenever it makes sense.

Specifically, the function x^r is differentiable whenever $x > 0$. If $r = \frac{p}{q}$ is a rational number whose denominator q is odd, then x^r is differentiable at every x if $r \geq 1$ and at every x except $x = 0$ if $r < 1$.

The most important case of Theorem 3.6.3 is $r = \frac{1}{2}$, since $x^{1/2} = \sqrt{x}$.

COROLLARY 3.6.4

The function $f(x) = \sqrt{x} = x^{1/2}$ is differentiable for every $x > 0$, and its derivative is

$$\frac{d}{dx}(x^{1/2}) = \frac{1}{2}x^{-1/2} = \frac{1}{2x^{1/2}},$$

or

$$\frac{d}{dx}(\sqrt{x}) = \frac{1}{2\sqrt{x}}.$$

We often apply the Chain Rule to a power $f(x)^n$ of a function $f(x)$.

COROLLARY 3.6.5

The Generalized Power Rule

$$\frac{d}{dx}\left(f(x)^r\right) = rf(x)^{r-1}f'(x),$$

whenever it makes sense.

EXAMPLE 7 Find the derivative of (i) $h(x) = \sqrt{x^2 + 2x + 2}$; (ii) $\sqrt{1 + \sin\theta}$.

Solution

(i) $h(x) = g\big(f(x)\big)$, with $f(x) = x^2 + 2x + 2$ and $g(u) = \sqrt{u}$. So $g'(u) = \frac{1}{2}u^{-1/2}$, and

$$h'(x) = \frac{1}{2}(x^2 + 2x + 2)^{-1/2}(2x + 2) = \frac{x + 1}{\sqrt{x^2 + 2x + 2}}.$$

For this formula to make sense we have to know that $x^2 + 2x + 2$ is not zero, but completing the square shows that $x^2 + 2x + 2 = x^2 + 2x + 1 + 1 = (x + 1)^2 + 1 > 1$, so it is never zero and there is no difficulty.

(ii) Using Equation 3.6.2, we see that

$$\begin{aligned}\frac{d}{d\theta}(\sqrt{1 + \sin\theta}) &= \frac{1}{2\sqrt{1 + \sin\theta}}\frac{d}{d\theta}(1 + \sin\theta) = \frac{1}{2\sqrt{1 + \sin\theta}}\cos\theta \\ &= \frac{\cos\theta}{2\sqrt{1 + \sin\theta}}.\end{aligned}$$

This will be true whenever $1 + \sin\theta > 0$, which is the same thing as $\sin\theta > -1$. Since $\sin\theta \geq -1$ for every θ, it will be true except when $\sin\theta = -1$. The function will be differentiable and will have the derivative stated above except when $\theta = -\frac{\pi}{2} + 2n\pi$, for some integer n.

Notice that apart from anything else we have to avoid these points because the formula for the derivative will have a zero in the denominator there.

The Chain Rule can also be used to find the derivatives of functions that arise in other problems. For instance, there are many maximum/minimum problems in which the function to be maximized or minimized must be differentiated with the Chain Rule.

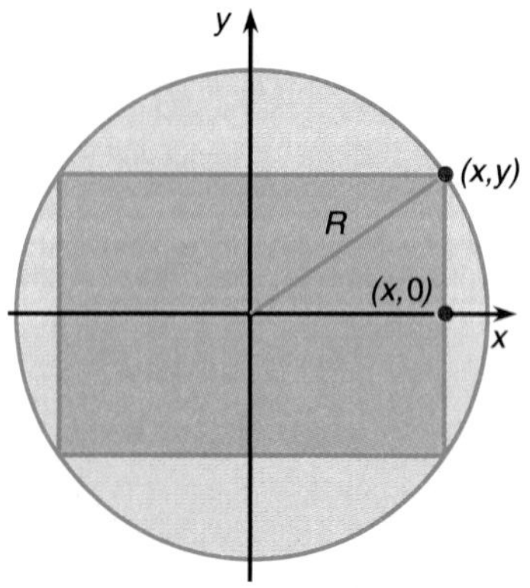

Figure 3.6.1

EXAMPLE 8 **Largest Beam in a Log** A rectangular beam is cut from a cylindrical log of radius R. What is the largest possible cross-sectional area of the beam?

Solution Suppose we place the origin at the center of the circular cross-section of the log (see Figure 3.6.1). Since the radius is R, the circle at the outside of the log is $x^2 + y^2 = R^2$. If the point where the edge of the beam crosses the positive x-axis is $(x, 0)$, as shown, then the width of the beam is $2x$. Since the upper right corner of the beam lies on the circle, we can calculate that its coordinates must be (x, y) with $x^2 + y^2 = R^2$, or $y = \sqrt{R^2 - x^2}$. The height of the beam is then $2y = 2\sqrt{R^2 - x^2}$.

The cross-sectional area of the beam is the product of the width and the height, or $A = (2x)(2y) = 4xy = 4x\sqrt{R^2 - x^2}$. We must find the maximum value of this function, and we know that $x \geq 0$ and $x \leq R$, so we must find the maximum on $[0, R]$.

The derivative is

$$\begin{aligned}A'(x) &= \frac{d}{dx}(4x\sqrt{R^2 - x^2}) = 4\sqrt{R^2 - x^2} + 4x\frac{1}{2}(R^2 - x^2)^{-1/2}(-2x) \\ &= 4\sqrt{R^2 - x^2} - \frac{4x^2}{\sqrt{R^2 - x^2}}.\end{aligned}$$

The derivative is zero when

$$4\sqrt{R^2 - x^2} = \frac{4x^2}{\sqrt{R^2 - x^2}}.$$

Multiplying by $\sqrt{R^2 - x^2}$, we see that this is the same as

$$4(R^2 - x^2) = 4x^2,$$

or $4R^2 = 8x^2$,

or $x^2 = \frac{1}{2}R^2$,

or $x = \frac{1}{\sqrt{2}}R$.

Without even evaluating, we realize that this must give the maximum area, since the area is zero when $x = 0$ or $x = R$ (just look at the picture). So the answer is that the maximum area occurs when $x = \frac{1}{\sqrt{2}}R$, in which case $y = \sqrt{R^2 - x^2} = \sqrt{R^2 - \frac{1}{2}R^2} = \sqrt{\frac{1}{2}R^2} = \frac{1}{\sqrt{2}}R$. The beam with the largest cross-sectional area has a square cross-section, and the area is $2R^2$.

Exercises 3.6

I

Find the derivative of each of the following functions.

1. $h(x) = (3x^4 - 2x^2 - 1)^3$
2. $h(x) = (x^5 - 1)^{24}$
3. $h(x) = (x^2 + 4x + 5)^{-2}$
4. $h(x) = \sin^{-2} x$
5. $h(x) = \cos(2x + 1)$
6. $h(x) = \sin(x^2 + 3)$
7. $h(x) = \ln(x^2 + 4)$
8. $h(x) = e^{x^2 - x}$
9. $h(x) = \cosh(\sin x)$
10. $h(x) = \ln\big(\exp(x)\big)$

Find the derivative of each of the following functions, stating where it exists.

11. $f(t) = \sqrt{t^2 - 2t + 1}$
12. $k(x) = \sqrt{\sin x}$
13. $g(x) = x^{1/3}$
14. $h(x) = (x^3 + 1)^{2/3}$
15. $F(r) = \ln(1 + \cos r)$
16. $K(x) = \sin(\sqrt{x})$
17. $G(x) = e^{\ln x}$
18. $H(x) = \sin^2(\sqrt{x}) + \cos^2(\sqrt{x})$
19. $f(x) = \cosh(\sinh x)$
20. $g(x) = \sqrt{\tanh x}$

Find the derivative of each of the following functions.

21. $f(x) = \ln^3(x^2 + 7)$
22. $g(x) = \frac{1}{2+\cos x^2}$
23. $h(x) = \sin\big(\ln(x^4 + 2)\big)$
24. $k(x) = \ln\big(\cosh(x^2)\big)$
25. $P(x) = 10^{x^2}$
26. $Q(x) = 15^{x+3}$
27. $F(x) = \sqrt{\cosh x}$
28. $G(x) = \frac{1}{\exp(\cos x)}$
29. $K(x) = \cosh^2\big(\ln(e^{2x})\big) - \sinh^2\big(\ln(e^{2x})\big)$
30. $H(x) = \exp\Big(\ln\big(5 + \cos(2x)\big)\Big)$
31. $f(x) = \frac{1}{(x^2+2)^2}$
32. $g(x) = \frac{1}{x^2+3}$
33. $h(x) = \sqrt{4 - \cos x}$
34. $k(x) = (x^2 + 4x + 7)^{7/2}$
35. $W(x) = \left(\frac{x+3}{x^2+1}\right)^4$
36. $P(x) = \exp\big(\sin(x^2 + 1)\big)$
37. $F(x) = \sin\big(\exp(x^2 + 1)\big)$
38. $G(x) = \frac{1}{\sqrt{1+\cos^2 x}}$
39. $K(x) = (x^5)^{1/7}$
40. $H(x) = (1 + \sin^2 x)^{1/2}$

II

41. Find a Chain Rule for $F(x) = h\Big(g\big(f(x)\big)\Big)$, assuming that f, g, h are differentiable at the appropriate points.

42. Let $f(t) = t^2$, $g(s) = 2s - 1$, and calculate the derivatives of $h(x) = g\big(f(x)\big)$ and $k(x) = f\big(g(x)\big)$. Observe that they are different, which shows that the Chain Rule is not simply a matter of multiplying the two derivatives together.

***43.** In Example 3.3.10 we had to maximize a distance that was expressed as a square root $D = \sqrt{(x-2)^2 + (x^2+2x+1)^2}$. In that example we solved the problem by working not with D itself but with its square, D^2. Now that we know how to differentiate a square root, try to work that example again, this time working directly with D. (Of course, you should get the same answer, and you will also see that working with D^2 is actually easier, since the derivative of the square root is more complicated to work with.)

44. Find all points at which the graph of each of the following functions has a horizontal tangent:
(i) $f(x) = x\sqrt{x^2+1}$ (ii) $f(x) = x^2\sqrt{x^2+1}$
(iii) $f(x) = x\exp(-x^2)$ (iv) $f(x) = \exp(\sin x)$

45. Find the maximum and minimum value of each of the following functions on the interval $[-1, 1]$:
(i) $f(x) = x^2\sqrt{x^2+1}$ (ii) $g(x) = x\exp(-x^2)$

46. Suppose a rectangular beam is cut from a cylindrical log of radius R, as in Example 8. For what dimensions of the beam will the perimeter of the rectangular cross-section be a maximum?

***47.** The strength of a rectangular beam is proportional to the product of its width and the square of the height of its cross-section. If we want to cut the strongest possible beam from a cylindrical log whose radius is R, what should be the dimensions of the beam?

***48.** In a semicircle of radius R, a rectangle is inscribed with one side along the semicircle's diameter and the two opposite corners on the semicircle (see Figure 3.6.2). What is the largest possible area for such a rectangle?

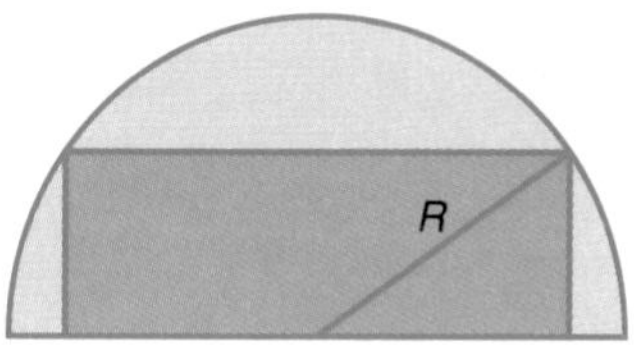

Figure 3.6.2

***49.** Suppose $f(t)$ is a function that satisfies $f(\tan x) = x$, for every $x \in \left(-\frac{\pi}{2}, \frac{\pi}{2}\right)$. (So $f(t)$ is a function that "undoes" what $\tan x$ does, much as e^t "undoes" what $\ln x$ does. A function like this is called an *inverse function;* so e^t is the inverse of $\ln x$, and $f(x)$ is the inverse of $\tan x$.)

(i) Apply the Chain Rule to the equation $f(\tan x) = x$ to find an expression for the derivative $f'(\tan x)$.

(ii) The formula found in part (i) expresses $f'(\tan x)$ in terms of trigonometric functions of x. Use a trigonometric identity to express $f'(t)$ in terms of t.

***50.** Repeat Exercise 49 with $\sin x$ for $x \in \left(-\frac{\pi}{2}, \frac{\pi}{2}\right)$ in place of $\tan x$.

III

51. Use a graphics calculator or computer to sketch the graphs of $y = x^{1/2}$ and its derivative $y' = \frac{1}{2}x^{-1/2}$ on the same picture. Notice how the value of the derivative gives the slope of the original curve.

52. Sketch the graphs of $y = x^{1/3}$ and its derivative $y' = \frac{1}{3}x^{-2/3}$ on the same picture.

53. Sketch the graphs of $y = x^{3/2}$ and its derivative $y' = \frac{3}{2}x^{1/2}$ on the same picture.

54. In Example 8 we found the formula $A = 4x\sqrt{R^2 - x^2}$ for the area of the beam with one side x. Sketch the graph of the area with $R = 2$ and $R = 5$ and confirm our result for the maximum in these cases.

55. Observe that if $y = e^{f(x)}$, then $y' = e^{f(x)}f'(x)$, which is zero precisely when $f'(x) = 0$. So $e^{f(x)}$ has a horizontal tangent where $f(x)$ does. Confirm this by sketching $f(x)$ and $e^{f(x)}$ on the same picture, with (i) $f(x) = x^2$, (ii) $f(x) = -x^3$, (iii) $f(x) = x^3 - x^2$.

POINT TO PONDER

In defining the derivative we considered the difference quotient

$$\frac{f(x)-f(a)}{x-a}=\frac{f(a+h)-f(a)}{h}.$$

The numerator is the difference between the values of f at two values of the variable x, and the denominator is the difference between these two values of x.

In many books a difference quotient like this is written as $\frac{\Delta f}{\Delta x}$. The symbol Δ is the capital Greek letter delta; it corresponds to our letter d, and for this reason it reminds us of the word "difference." So $\frac{\Delta f}{\Delta x}$ means "the difference between the values of f divided by the difference between the values of x."

The derivative of f at a is the limit of this quotient as $x \to a$, which is the same thing as $\Delta x \to 0$. In symbols,

$$f'(a)=\lim_{x\to a}\frac{f(x)-f(a)}{x-a}=\lim_{h\to 0}\frac{f(a+h)-f(a)}{h}$$
$$=\lim_{\Delta x\to 0}\frac{\Delta f}{\Delta x}.$$

The idea of this notation is that as we take the limit, the Greek letter Δ changes into the letter d, and the derivative is

$$\lim_{\Delta x\to 0}\frac{\Delta f}{\Delta x}=\frac{df}{dx}.$$

We have not used this notation (and will not), but we mention it here because you might encounter it elsewhere. It is really just a way of referring to difference quotients.

3.7 Implicit Differentiation

IMPLICIT FUNCTION
MAXIMUM/MINIMUM PROBLEMS
LOGARITHMIC DIFFERENTIATION

We frequently encounter equations relating x and y that cannot readily be solved explicitly for y as a function of x or for x as a function of y. Such an equation may nonetheless determine y as a function of x or vice versa; such a function is called an **implicit function**.

The equation of the circle $x^2+y^2=1$ can be solved for y in terms of x. In this case we can solve explicitly $y=\sqrt{1-x^2}$ or $y=-\sqrt{1-x^2}$ (see the graphs of these two functions in Figure 3.7.1). Notice that there is no single function that will give all the points on the graph of the circle, since each value of x between -1 and 1 corresponds to *two* points on the circle. On the other hand, the graphs of these two functions together give all the points on the circle. However, let us pretend for a moment that we do not know how to find these solutions (either we do not know about square roots or perhaps we simply failed to notice how to simplify).

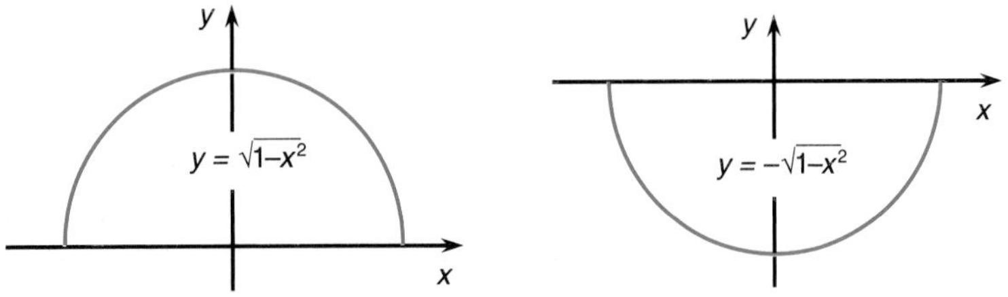

Figure 3.7.1

Suppose we have some point (a, b) that is on the curve (i.e., that satisfies the equation). Then if we change x a little bit near a, it seems reasonable to expect the point to move a little bit along the curve and y to change a little bit near b. We could ask what is $\frac{dy}{dx}$, the rate of change of y with respect to x. This could be interpreted as the slope of the tangent through (a, b), which is exactly what it would be if we knew how to write y as a function of x explicitly. See Figure 3.7.2.

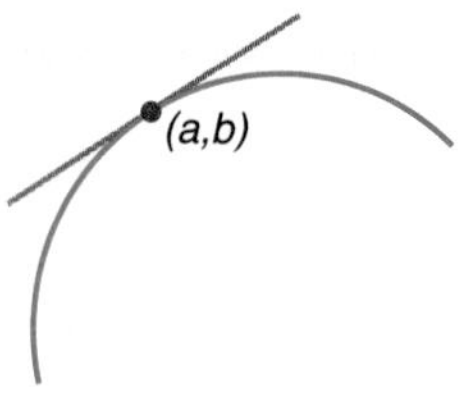

Figure 3.7.2

If we simply *assume* that y is a function of x, even though we have no formula for it, we can differentiate the original equation and find a formula for the derivative $y' = \frac{dy}{dx}$.

EXAMPLE 1 Find $\frac{dy}{dx}$ if x and y satisfy $x^2 + y^2 = 1$.

Solution Differentiating $x^2 + y^2 = 1$, we get

$$2x + \frac{d}{dx}(y^2) = 0,$$

and using the Chain Rule, we have $\frac{d}{dx}(y^2) = 2y\frac{dy}{dx} = 2yy'$. Substituting this into the previous equation, we find that

> If y is a function of x, the Generalized Power Rule (Corollary 3.6.5) gives $\frac{d}{dx}y^n = ny^{n-1}y'$; here we have the case with $n = 2$.

$$2x + 2yy' = 0,$$

which we can solve: $yy' = -x$, so $y' = -\frac{x}{y}$, provided $y \neq 0$.

We can check that this agrees with ordinary differentiation if $y = -\sqrt{1 - x^2}$, say. After all, in this case,

$$y' = \frac{d}{dx}\left(-(1 - x^2)^{1/2}\right) = -\frac{1}{2}(1 - x^2)^{-1/2}(-2x) = x(1 - x^2)^{-1/2}$$
$$= \frac{-x}{-(1 - x^2)^{-1/2}} = -\frac{x}{y}.$$

A similar calculation works for the other function $y = \sqrt{1 - x^2}$.

We found the derivative without expressing y as a function of x, simply by differentiating the equation $x^2 + y^2 = 1$ and using the Chain Rule. This process is called **implicit differentiation**, since we assume that y is implicitly determined as a function of x, even though we have not written y explicitly as a function. In this particular case it was possible after all to write y as a function of x, and we checked that the derivative of this function was the same as what we found by the process of implicit differentiation.

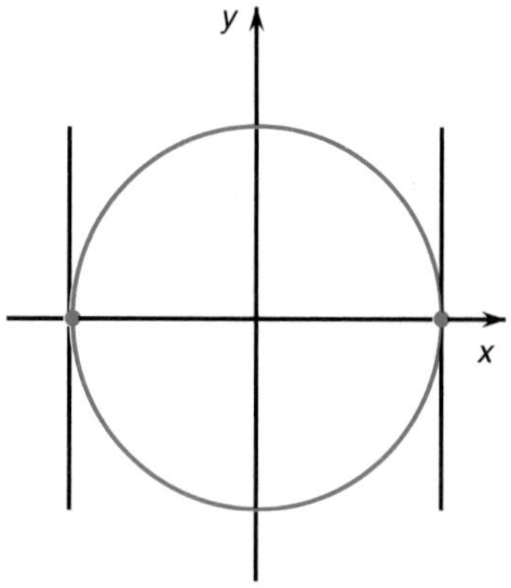

Figure 3.7.3

Several things need to be mentioned here. One is that the result for y' may include y instead of just x in the formula. This is perhaps inconvenient but not disastrous. Second, the derivative may not always exist. In the above example it does not exist when $y = 0$ (i.e., at $(1, 0)$ and $(-1, 0)$, where the tangents to the circle are vertical; see Figure 3.7.3). Third, we could just as easily have assumed that x was a function of y and found $\frac{dx}{dy}$. Fourth, when we assume that y is a function of x or vice versa, we can probably assume that this is true for some values of x, but we should not expect that there will be a function that works for all values of x, as the circle example shows. Finally, although it seems reasonable, our expectation that a small change in x will result in a small change in y is not always valid. For example, if we start at the point $(0, 0)$, which satisfies the equation $x^2 + y^2 = 0$, and change x a little bit, then there will be no y that satisfies the equation (since $(0,0)$ is the only solution of $x^2 + y^2 = 0$).

Nonetheless, the assumption that y is a function of x or vice versa, at least near some starting point (a, b), is reasonable and usually justified. We shall not concern ourselves with it, but simply use the process of implicit differentiation to find derivatives.

Summary 3.7.1

Implicit Differentiation

Suppose x and y satisfy some relation, say, $f(x, y) = 0$. Assuming that y is a function of x, at least for certain values of x, differentiate the relation, using the Chain Rule.

The result will be an equation involving x, y, and y', which can be solved for y'. Finding y' in this way is called implicit differentiation.

EXAMPLE 2 Find $y' = \frac{dy}{dx}$ if (x, y) satisfies $xy^3 - 2x^2y + x^3 - 1 = 0$.

Solution No obvious solution for y in terms of x springs to mind, but using the Product Rule and the Chain Rule, we can differentiate both sides of the equation with respect to x:

$$y^3 + x(3y^2)y' - 4xy - 2x^2y' + 3x^2 = 0,$$

that is,

$$(3xy^2 - 2x^2)y' = 4xy - 3x^2 - y^3.$$

So

$$y' = \frac{4xy - 3x^2 - y^3}{3xy^2 - 2x^2}.$$

This is of limited use unless we know a specific point on the curve, but noting that $(1,0)$ satisfies the equation, we can substitute these values for x and y and find that the derivative $\frac{dy}{dx}$ at the point $(1,0)$ has the value $\frac{dy}{dx} = \frac{-3}{-2} = \frac{3}{2}$.

EXAMPLE 3

(i) Find the equation of the tangent line to the circle $x^2 + y^2 = 2$ at the point $(1,1)$.

Solution Differentiating implicitly, we find $2x + 2yy' = 0$, so $yy' = -x$ and $y' = -\frac{x}{y}$. At the point in question, $(1,1)$, the slope of the tangent is therefore $y' = -1$. By the point-slope form, the equation of the tangent line is $y - 1 = (-1)(x - 1)$, or $x + y = 2$.

Of course, it is possible to solve for y as a function of x in this case and differentiate to find $\frac{dy}{dx}$. However, if you try it, you will see that the calculations are more complicated if you do it that way.

(ii) Find the equation of the tangent line to the curve $x^3 + xy^2 - y^3 + 3 = 0$ at the point $(1, 2)$.

Solution This time we find $3x^2 + y^2 + 2xyy' - 3y^2y' = 0$, or $(3x^2 + y^2) + (2xy - 3y^2)y' = 0$, so

$$y' = -\frac{3x^2 + y^2}{2xy - 3y^2}.$$

At the specified point, this equals $\frac{7}{8}$, so the slope of the tangent is $\frac{7}{8}$, and the equation of the tangent line is $y - 2 = \frac{7}{8}(x - 1)$, or

$$7x - 8y + 9 = 0.$$

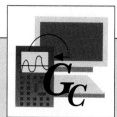

With a graphics calculator it is possible to sketch the circle and the tangent line of part (i) of Example 3. For the circle we write $y^2 = 2 - x^2$, so $y = \pm\sqrt{2 - x^2}$, and we graph both semicircles.

TI-85	SHARP EL9300
GRAPH F1 ($y(x)$ =) √(2− F1 (x) x^2) ▼ (−) √(2− F1 (x) x^2) ▼ 2 − F1 (x) M2 (RANGE) (−) 3 ▼ 3 ▼ ▼ (−) 2 ▼ 2 F5 (GRAPH)	√ 2 − x/θ/T x^2 2ndF ▼ − √ 2 − x/θ/T x^2 2ndF ▼ 2 − x/θ/T RANGE −3 ENTER 3 ENTER 1 ENTER − 2 ENTER 2 ENTER 1 ENTER
CASIO f_x-7700GB/f_x-6300G	**HP 48SX**
Cls EXE Range −3 EXE 3 EXE 1 EXE −2 EXE 2 EXE 1 EXE EXE EXE EXE Graph √ (2 − X x^2) EXE Graph − √ (2 − X x^2) EXE Graph 2 − X EXE	↰ PLOT PLOTR ERASE ATTN ′ αY = √x ( ) 2 − αX y^x 2 ↱ PLOT ↰ DRAW ERASE 3 +/− ENTER 3 XRNG 2 +/− ENTER 2 YRNG DRAW ATTN ′ αY = −√x ( ) 2 − αX y^x 2 ↱ PLOT ↰ DRAW DRAW ATTN ′ αY = 2 − αX ↱ PLOT ↰ DRAW DRAW

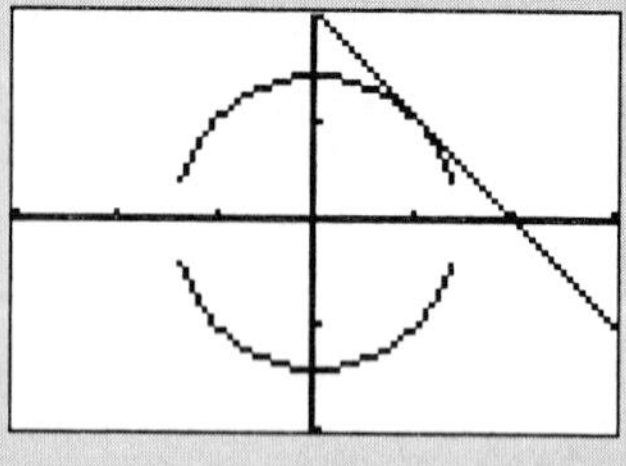

Differences in the horizontal and vertical scales may make the circle appear elliptical.

Implicit Differentiation and Maximum/Minimum Problems

An important use of implicit differentiation arises in maximum/minimum problems. In Example 3.3.6 we considered how to maximize the area A of a rectangular field whose perimeter is 600 ft. We approached the problem by letting the sides of the field be x and y, so $A = xy$ and the condition about the perimeter is $2x + 2y = 600$. It was easy to solve this last equation for y in terms of x and substitute the answer into the formula for A. However, sometimes it is difficult to solve for either variable. In such a case we can "assume" that y is a function of x and use implicit differentiation.

The derivative of the function we wish to maximize or minimize (A' in the above example) will contain y'. But implicit differentiation of the other equation relating x and y will allow us to solve for y', and it may be relatively easy to substitute this into the formula for A' and find where $A' = 0$.

EXAMPLE 4 Find the maximum value of $f(x, y) = x + y$ on the ellipse $2x^2 + 3xy + 3y^2 = 8$.

Solution Differentiating both equations implicitly, we find that $\frac{df}{dx} = 1 + y'$ and that $4x + 3y + 3xy' + 6yy' = 0$. We need to find where $\frac{df}{dx} = 0$, that is, $1 + y' = 0$, that is, $y' = -1$.

Substituting $y' = -1$ into the other equation gives

$$4x + 3y - 3x - 6y = 0,$$

or $\quad x - 3y = 0,$

or $\quad x = 3y.$

Looking for points on the ellipse that satisfy $x = 3y$, we substitute $x = 3y$ into $2x^2 + 3xy + 3y^2 = 8$ and get

$$18y^2 + 9y^2 + 3y^2 = 8,$$

that is, $\quad 30y^2 = 8,$

so $\quad y^2 = \frac{4}{15}, \qquad y = \pm\frac{2}{\sqrt{15}},$

and $\quad x = 3y = \pm\frac{6}{\sqrt{15}} = \pm 2\sqrt{\frac{3}{5}}.$

The two points on the ellipse at which $\frac{df}{dx} = 0$ are $\left(2\sqrt{\frac{3}{5}}, \frac{2}{\sqrt{15}}\right)$ and $\left(-2\sqrt{\frac{3}{5}}, -\frac{2}{\sqrt{15}}\right)$. It is clear that $f(x, y) = x + y$ is larger at the first point and smaller at the second one. If we knew there had to be a maximum and a minimum, they would have to be at these points.

We can see that this is in fact true, as follows. The equation of the ellipse is a quadratic equation. Without actually writing it all out, we know we can solve for y as either of two functions of x, one the "upper half" of the ellipse, and the other the "lower half," corresponding to taking the plus sign or the minus sign in the Quadratic Formula. Each of these functions y is a continuous function on a closed interval, and consequently so is $x + y$. By Theorem 2.5.7, $x + y$ has a maximum and a minimum on each piece of the ellipse and hence on all of it.

The maximum value of $f(x, y) = x + y$ occurs at the point $\left(2\sqrt{\frac{3}{5}}, \frac{2}{\sqrt{15}}\right)$. The value of f at this point is $2\sqrt{\frac{3}{5}} + \frac{2}{\sqrt{15}} = \frac{8}{\sqrt{15}} \approx 2.06559$.

This example illustrates the principal drawback in using implicit differentiation for maximum/minimum problems: While it often makes it easy to find points at which the derivative is zero, it may still be very difficult to decide whether the function has a maximum or minimum value there, and the technique offers no help about whether a maximum or minimum value exists at all.

Logarithmic Differentiation

When we have to differentiate a complicated expression, the work required may sometimes be reduced by differentiating not the expression itself but rather its logarithm. This is especially true with functions involving powers or complicated products or quotients. The idea is that the logarithm of such a function may be simpler to work with than the function itself.

EXAMPLE 5 Differentiate the function $f(x) = \frac{\sqrt{x^4+2}}{(3+\cos x)^{1/5}}$.

Solution To differentiate directly would require an elaborate application of the Quotient and Chain Rules. Instead, we let $y = f(x)$ and consider

$$\ln y = \ln\big(f(x)\big) = \frac{1}{2}\ln(x^4+2) - \frac{1}{5}\ln(3+\cos x).$$

Its derivative is

$$\frac{d}{dx}\Big(\ln\big(f(x)\big)\Big) = \frac{1}{2} \times \frac{4x^3}{x^4+2} - \frac{1}{5} \times \frac{(-\sin x)}{3+\cos x}.$$

We have just found an expression for $\frac{d}{dx}(\ln y) = \frac{1}{y}\frac{dy}{dx}$, so we can multiply by $y = f(x)$ to find

$$\frac{dy}{dx} = y\frac{d}{dx}(\ln y) = f(x)\frac{d}{dx}\Big(\ln\big(f(x)\big)\Big)$$
$$= \frac{\sqrt{x^4+2}}{(3+\cos x)^{1/5}}\left(\frac{2x^3}{x^4+2} + \frac{\sin x}{15+5\cos x}\right).$$

The calculation we have just done is probably easier than differentiating the original expression using the Quotient Rule. The advantage of *logarithmic differentiation* is that taking the logarithm "pulls apart" quotients and products and allows us to differentiate each factor separately.

In Example 5 we neglected to use absolute values, but a quick look will convince you that the functions involved there are always positive ...

Strictly speaking, where $f(x)$ is negative, we should really consider $\ln|f(x)|$, but, interestingly, it turns out that the formula we get will be the same. For instance, suppose $f(x) < 0$ for x in some interval. Then for these values of x, $|f(x)| = -f(x)$ and $|f(x)|' = \frac{d}{dx}\big(|f(x)|\big) = \frac{d}{dx}(-f) = -f'$, so $\frac{d}{dx}\big(\ln|f(x)|\big) = \frac{|f(x)|'}{|f(x)|} = \frac{-f'(x)}{-f(x)} = \frac{f'(x)}{f(x)}$. The formula $\frac{d}{dx}\big(\ln|f(x)|\big) = \frac{f'(x)}{f(x)}$ is not really affected by the absolute value.

This means that we can work without worrying about signs. It is often quite difficult to decide when a function is positive and when it is negative, so it is convenient to avoid this complication.

We now summarize the procedure.

Summary 3.7.2

Logarithmic Differentiation

To differentiate $y = f(x)$, we follow these steps:

1. Write out $\ln|f(x)|$, using the properties of the natural logarithm to "pull apart" products, quotients, and powers. Express $\ln|f(x)|$ as a sum of terms $\ln|g(x)|$ for various functions $g(x)$.

2. Differentiate to find an expression for each term: $\frac{d}{dx}\left(\ln|g(x)|\right) = \frac{d}{dx}(\ln|g|) = \frac{1}{g}\frac{dg}{dx} = \frac{g'}{g}$. To do this, it is not really necessary to consider absolute values.
3. Multiply by the original function to find the derivative

$$\frac{df}{dx} = \frac{dy}{dx} = y\frac{d}{dx}(\ln|y|) = f(x)\frac{d}{dx}\left(\ln|f(x)|\right).$$

EXAMPLE 6 Differentiate $f(x) = \sqrt{x(x^2-2)(x^3+4x+1)}$.

Solution Letting $y = f(x)$, we write

$$\ln|y| = \frac{1}{2}(\ln|x| + \ln|x^2-2| + \ln|x^3+4x+1|).$$

Differentiating, we find

$$\frac{d}{dx}(\ln|y|) = \frac{1}{2}\left(\frac{1}{x} + \frac{2x}{x^2-2} + \frac{3x^2+4}{x^3+4x+1}\right),$$

so

$$\begin{aligned}\frac{dy}{dx} &= y\frac{d}{dx}(\ln|y|)\\ &= \frac{1}{2}\sqrt{x(x^2-2)(x^3+4x+1)}\left(\frac{1}{x} + \frac{2x}{x^2-2} + \frac{3x^2+4}{x^3+4x+1}\right).\end{aligned}$$

Exercises 3.7

I

Find $\frac{dy}{dx}$ on each of the following curves.

1. $x^2 + xy - y^2 = 3$
2. $xy^2 - 3x^2y + y^3 + 1 = 0$
3. $\frac{xy+y^3}{1+x} = 1$
4. $xy - y^2 \sin x = \pi$
5. $3x^2 - 9y^3 + xy + 5 = 0$
6. $xe^{xy} = 2$
7. $x = \sin y$
8. $x^2 = \sin(y^2)$
9. $\cos x \sin y = \frac{1}{2}$
10. $x = e^y$

11.–20. Find $\frac{dx}{dy}$ for each of the curves in Exercises 1 through 10.

Differentiate each of the following.

21. $f(x) = (x^4 + x^2)(x^3 - 8)(x^2 + x + 1)$
22. $f(x) = \sin x \tan x \sec^2 x$
23. $f(x) = x(x^3 + 4x)^{1/3}$
24. $f(x) = \sqrt{(1+x^2)(1+x^4)(1+x^6)}$
25. $f(x) = \sqrt{(x+1)(x+2)(x+3)}$
26. $f(x) = \left(x(x+1)(x-2)\right)^{3/4}$
27. $f(x) = \frac{\sqrt{x^3+1}}{(x^2+2)^4}$
28. $f(x) = \frac{(x+1)^{8/9}}{x^{3/4}}$
29. $f(x) = \left(\frac{(x+1)(x-2)}{(x+3)(x^2-9)}\right)^{3/7}$
30. $f(x) = \sqrt{x(x+2)\ln(x)}$

II

31. Use implicit differentiation to do Example 3.3.6 (p. 160) again.
32. Use implicit differentiation to do Example 3.3.7 (p. 161) again.
33. Use implicit differentiation to do Example 3.3.8 (p. 162) again.

***34.** A rectangle is drawn inside the ellipse $9x^2 + 4y^2 = 36$ so that two of its sides are horizontal and two are vertical. Find the rectangle with the largest possible area.

35. Find the equation of the tangent line to each of the following curves at the specified point: (i) $x^2 - 3xy + y^2 + 1 = 0$ at $(1, 1)$; (ii) $x^3 + 3xy^2 - xy = 8$ at $(2, 0)$.

36. Find the equation of the tangent line to each of the following curves at the specified point: (i) $x^2 + 4y^2 = 1$ at $(-1, 0)$; (ii) $xy^3 - 2x^2y^2 - 4xy + 8 = 0$ at $(1, 2)$.

37. Find the equation of the tangent line to each of the following curves at the specified point: (i) $x \sin y + y \cos x = \pi$ at $(0, \pi)$; (ii) $xe^y + x^2y - xy^3 = 3$ at $(3, 0)$.

***38.** If (x, y) lies on the circle $x^2 + y^2 = 1$, then we have seen that $\frac{dy}{dx}$ does not exist at the points $(-1, 0)$ and $(1, 0)$.

(i) Find $\frac{dx}{dy}$ at these two points.

(ii) Use your answer to part (i) to find the equations of the tangent lines at those two points. (Be careful; when you interpret $\frac{dx}{dy}$ as a slope, it is the slope of the tangent to the graph of x as a function of y.)

***39.** Assuming that $f(x, y) = 3x - 2y$ has a maximum and a minimum value on the curve $3x^2 + xy + y^2 = 1$, find them.

***40.** Assuming that $f(x, y) = 2x - 3y$ has a maximum and a minimum value on the curve $x^2 - xy + y^2 - 2x = 1$, find them.

***41.** Assuming that $f(x, y) = xy$ has a maximum and a minimum value on the curve $x^2 + xy + y^2 = 12$, find them.

***42.** Show that $f(x, y) = 2y - x$ does not have a maximum value nor a minimum value on the curve $y^2 = x^2 + 2x + 2$, even though implicit differentiation does find two points at which $\frac{dy}{dx} = 0$. Draw a picture.

43. Solve Examples 5 (p. 200) and 6 (p. 201) without using logarithmic differentiation. This should convince you of the value of using it.

***44.** Let $f(x) = (x^2 - 3)^5(x^3 - 2x)^{-1/7}$.

(i) Find the values of x for which $f(x) > 0$ and those for which $f(x) < 0$.

(ii) Find $\frac{df}{dx}$ using logarithmic differentiation, using part (i) to find a formula for $|f(x)|$ on each interval where the sign stays the same.

(iii) Find $\frac{df}{dx}$ using logarithmic differentiation but without worrying about signs. Confirm the remarks made before Summary 3.7.2 (p. 200) to the effect that the answers will be the same.

III

45. Use a graphics calculator or computer and the result of Example 1 (p. 196) to sketch the circle $x^2 + y^2 = 1$ and its tangent line at (i) $\left(\frac{1}{\sqrt{2}}; -\frac{1}{\sqrt{2}}\right)$; (ii) $\left(-\frac{1}{2}, \frac{\sqrt{3}}{2}\right)$; (iii) $\left(\frac{3}{5}, \frac{4}{5}\right)$. (*Hint:* To sketch the circle, work with two semicircles.)

46. Sketch the ellipse $x^2 + 4y^2 = 4$. (Work with two halves.) Find the equation of the tangent at the point $\left(-1, \frac{\sqrt{3}}{2}\right)$, and sketch it on the same picture.

***47.** In Exercise 39 we considered the curve $3x^2 + xy + y^2 = 1$. Solve for y (think of the equation as a quadratic equation in y). You will get two solutions; graph both of them and identify the curve.

48. Graph the curve $x^2 - xy + y^2 - 2x = 1$ from Exercise 40.

49. Graph the curve $x^2 + xy + y^2 = 12$ from Exercise 41.

50. Graph the curve $y^2 = x^2 + 2x + 2$ from Exercise 42.

51. Sketch the graph of the function $f(x) = (x^2 - 3)^5(x^3 - 2x)^{-1/7}$ from Exercise 44. Compare the picture with your answer to part (i) of that exercise.

3.8 Derivatives of Inverse Trigonometric and Inverse Hyperbolic Functions

INVERSE FUNCTION
ONE-TO-ONE
arcsin x
arctan x
GRAPHS OF INVERSE FUNCTIONS
INVERSE TRIGONOMETRIC AND INVERSE HYPERBOLIC FUNCTIONS AND THEIR DERIVATIVES

This section presents the difficult concept of inverse functions and applies it in particular to trigonometric and hyperbolic functions. The most important result is Theorem 3.8.5, which lists the derivatives of the inverse trigonometric functions. We will need to know the trigonometric identities (Theorem 1.7.2, Equation 1.7.3, Theorem 1.7.11), the derivatives of trigonometric functions (Theorems 3.4.1 and 3.4.2), and the Chain Rule. The last part of the section deals with inverse hyperbolic functions and their derivatives. These topics are of less importance, but to study them, it is necessary to know the derivatives of hyperbolic functions (Theorem 3.5.13) and the hyperbolic function identity (Equation 3.5.12).

EXAMPLE 1 If $f(x) = 2x + 4$ and $g(u) = \frac{u}{2} - 2$, we notice that their composition is $g(f(x)) = \frac{f(x)}{2} - 2 = \frac{2x+4}{2} - 2 = x + 2 - 2 = x$. This says that whatever the function f does to x, the function g "undoes" it and takes us back to the original value of x. For example, if $x = 3$, then $f(x) = f(3) = 2(3) + 4 = 10$, and $g(10) = \frac{10}{2} - 2 = 5 - 2 = 3$, which is the value of x with which we started.

In this situation the function g is called the inverse of f and is often written $g(u) = f^{-1}(u)$. It is very important not to confuse this with the reciprocal $\frac{1}{f(u)}$. (In the above example, $\frac{1}{f(u)} = \frac{1}{2u+4}$, which is quite different from the inverse function $f^{-1}(u) = g(u) = \frac{u}{2} - 2$.)

DEFINITION 3.8.1

If $f(x)$ is a function and $g(u)$ is another function so that

$$g(f(x)) = x$$

for all x in the domain of f, then $g(u)$ is called the **inverse** of $f(x)$, and we write $g(u) = f^{-1}(u)$.

The domain of the inverse function $f^{-1}(u)$ is the set of all values of the function $f(x)$, that is, the domain of $f^{-1}(u)$ is $\{u : u = f(x)$, for some $x \in$ domain $(f(x))\}$.

EXAMPLE 2

(i) The function $\ln u$ is the inverse of e^x, since $\ln(e^x) = x$ for every $x \in \mathbb{R}$.

(ii) The function $\ln x$, defined for all $x > 0$, has as its inverse function $g(u) = e^u$, since

$$e^{\ln x} = x,$$

for all $x > 0$. Notice that it does not make sense to ask what happens for values of x that are not in the domain of f.

EXAMPLE 3 Consider the functions $f(x) = x^2$ and $g(u) = \sqrt{u}$. If $x \geq 0$, then $g(f(x)) = \sqrt{x^2} = x$. But if $x = -3$, for example, then $g(f(-3)) = g((-3)^2) = g(9) = \sqrt{9} = 3 \neq -3$ (recall that $\sqrt{a}$ is always nonnegative). So $g(f(x)) = x$ if $x \geq 0$ but not if $x < 0$, so we can say that $g = f^{-1}$ if the domain of $f(x)$ is $[0, \infty)$.

What complicates this example is that $f(x) = x^2$ may have the same value at different points x. For instance, $f(3) = f(-3)$, and indeed $f(-a) = f(a)$ for any a. So if we try to make a function $g = f^{-1}$ that "undoes" what $f(x)$ does, we have to make a choice (the square root function is conventionally taken to be nonnegative). We could make a different choice: $h(u) = -\sqrt{u}$ is $f^{-1}(u)$ if the domain of f is $(-\infty, 0]$.

In any case, what is essential is that the domain of $f(x)$ not contain two points at which f takes the same value.

DEFINITION 3.8.2

A function $f(x)$ is **one-to-one** on its domain D if $f(s) \neq f(t)$ for every pair of distinct points $s, t \in D$.

The terminology suggests that the function takes only *one* x to any *one* y.

To talk about the inverse of a function, we must consider the original function on a domain where it is one-to-one. If $f(x)$ is one-to-one, then f^{-1} is automatically defined (just by "undoing" whatever f does).

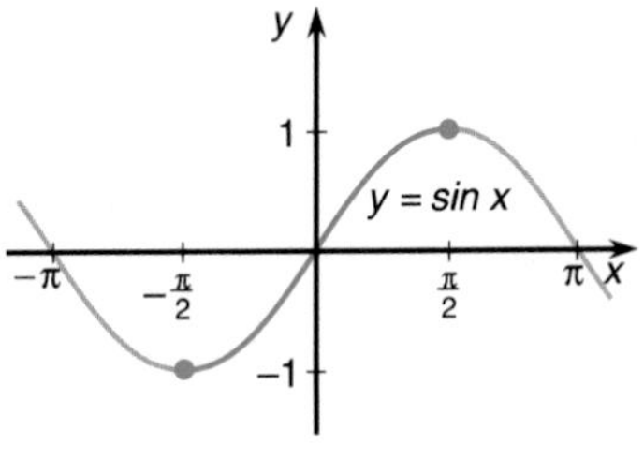

Figure 3.8.1

EXAMPLE 4 Consider the function $y = \sin x$, illustrated in Figure 3.8.1. It is certainly not one-to-one. However, if we look at $f(x) = \sin x$ on the interval $\left[-\frac{\pi}{2}, \frac{\pi}{2}\right]$, it *is* one-to-one there (because it is increasing, so it cannot "double back" and take the same value twice). So it has an inverse function, $f^{-1}(u)$, defined for $u \in [-1, 1]$ and taking values in the interval $\left[-\frac{\pi}{2}, \frac{\pi}{2}\right]$. Because we so frequently write trigonometric functions with exponents, the notation $\sin^{-1} u$ is especially prone to misinterpretation. Instead, we call this particular inverse function arcsin u.

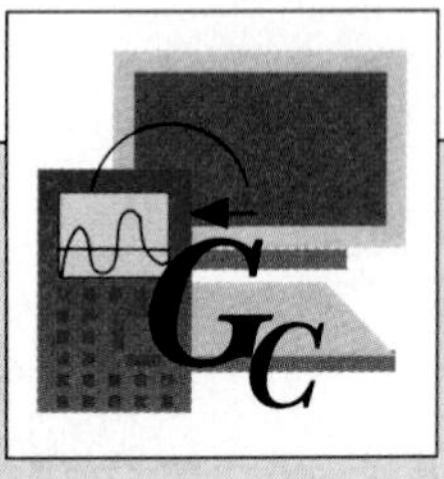

It is usually not easy to evaluate inverse functions, but a graphics calculator can be used to get approximate evaluations.

To find arcsin $\frac{1}{3}$, that is, the number x so that $\sin x = \frac{1}{3}$, we plot $y = \sin x$ and $y = \frac{1}{3}$.

TI-85 GRAPH F1($y(x) =$) SIN F1(x) ▼ 1 ÷ 3 M2(RANGE) 0 ▼ $\pi \div 2$ ▼ ▼ 0 ▼ 1 F5(GRAPH)	SHARP EL9300 ⩘ SIN x/θ/T 2ndF ▼ 1 ÷ 3 RANGE 0 ENTER π ÷ 2 ENTER 1 ENTER 0 ENTER 1 ENTER 1 ENTER ⩘
CASIO f_x-7700GB/f_x-6300G Cls EXE Range 0 EXE π ÷ 2 EXE 1 EXE 0 EXE 1 EXE 1 EXE EXE EXE EXE Graph sin ■ EXE Graph 1 ÷ 3 EXE	HP 48SX ■ PLOT PLOTR ERASE ATTN ′ αY ■ SIN αX ↱ PLOT ■ DRAW ERASE ■ ENTER 2 ÷ 0 SWAP XRNG 0 ENTER 1 YRNG DRAW ATTN ′ αY ■ 1 ÷ 3 ↱ PLOT ■ DRAW DRAW

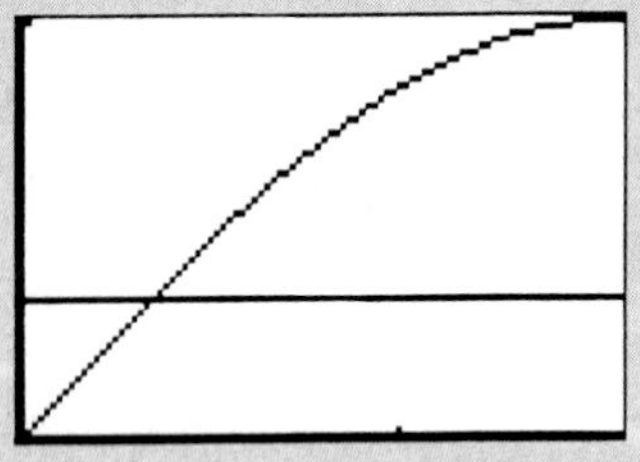

Then by zooming in it is possible to find an approximate value for arcsin $\frac{1}{3}$, the value of x where the curve crosses the line. Depending on which calculator you are using, it is about arcsin $\frac{1}{3} \approx 0.339836909$. It is also possible to use the SOLVE feature of your calculator and find where $\frac{1}{3} - \sin x = 0\ldots$ Most calculators have an inverse tangent key, so in this particular case, that is the easiest of all.

EXAMPLE 5 Suppose we look at the graph of $y = \tan x$, or more specifically the part of the graph with $-\frac{\pi}{2} < x < \frac{\pi}{2}$. It seems from Figure 3.8.2 that $\lim_{x \to -\frac{\pi}{2}^+} \tan x = -\infty$ and $\lim_{x \to \frac{\pi}{2}^-} \tan x = \infty$. By the Intermediate Value Theorem (Theorem 2.5.5) this implies that the function $y = \tan x$ takes on every real value y for some x in the interval $\left(-\frac{\pi}{2}, \frac{\pi}{2}\right)$. Moreover, the function is always increasing, so it cannot take on

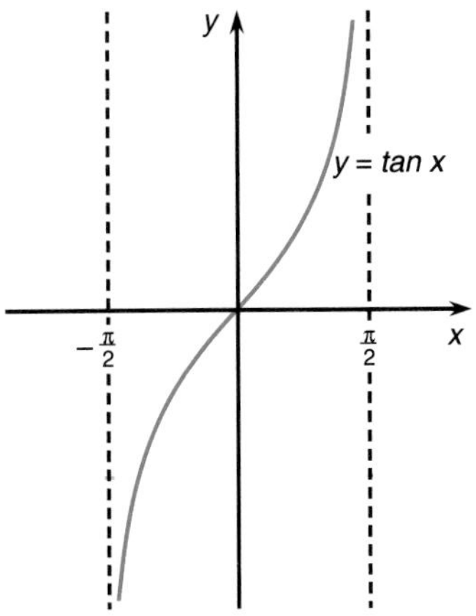

Figure 3.8.2

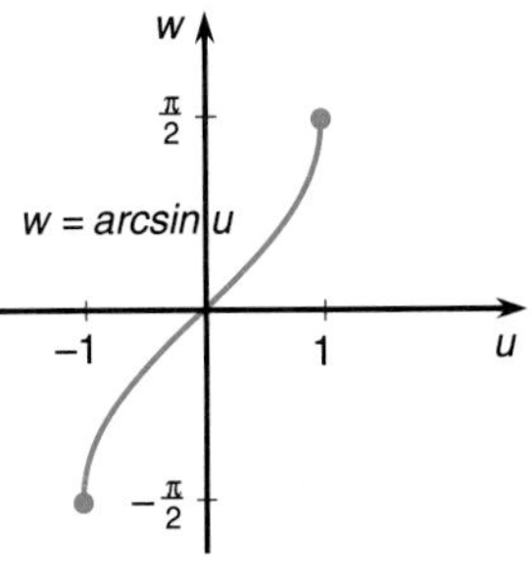

Figure 3.8.3

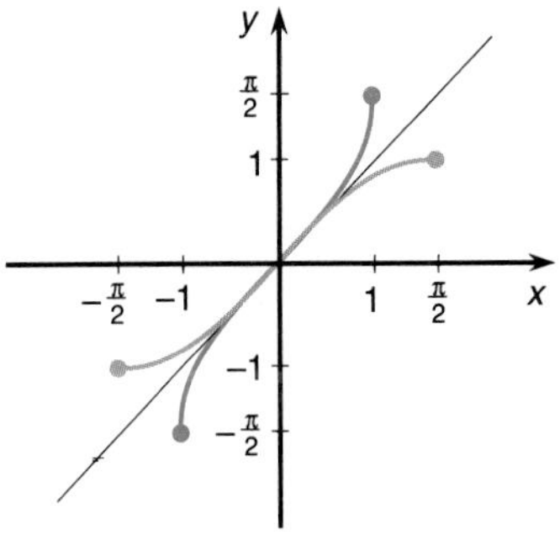

Figure 3.8.4

the same value twice and must be one-to-one. In short, the function $y = \tan x$ on the interval $\left(-\frac{\pi}{2}, \frac{\pi}{2}\right)$ takes on every real value exactly once.

So although it might not be easy to do in practice, it is possible in principle to find what value of x will result in some particular value of y. For instance, if $y = 1$, we can find an x so that $\tan x = 1$, namely, $x = \frac{\pi}{4}$. Of course, that one was deliberately chosen to be easy. It would be harder to specify the value of x for which $\tan x = 3\sqrt{2}$, say. Still, we do know that there must be such an x.

In effect, this means there is an inverse function $g(t)$ so that for every $t \in \mathbb{R}$,

$$\tan\big(g(t)\big) = t.$$

This function g is called arctan t.

If we graph the function arcsin u, which was defined in Example 4, we notice that $w = \arcsin u$ means $u = \sin w$. Because of this, the graph of $w = \arcsin u$ relative to the u-axis and the w-axis is related to the graph of $u = \sin w$ on the w-axis and the u-axis. Unfortunately, the order of the two axes is reversed, but if you take the graph of $w = \arcsin u$ and turn it 90° clockwise, you get the graph of $u = \sin w$ upside-down (i.e., the graph of $u = -\sin w$). Try it with Figure 3.8.3.

Another way of saying this is that the graph of $y = f^{-1}(x)$ is the graph of $y = f(x)$ *reflected* in the line $y = x$, as shown in Figure 3.8.4 (see the box below). The tricky thing about this is that the same letter, x, is used for the variable of $f(x)$ and for the variable of $f^{-1}(x)$, whereas if we start with $y = f(x)$, then f^{-1} should "undo" what f does and take y back to x, suggesting y as the variable for f^{-1}.

In dealing with inverse functions it is very important to keep clearly in mind what is the independent variable for each function. As with composite functions, it often helps to avoid confusion if you use a different letter altogether, for example, $f(x)$ and $f^{-1}(u)$.

Graphs of Inverse Functions

Suppose $f(x)$ is a function that has an inverse $f^{-1}(x)$. If (a, b) is a point on the graph of $y = f(x)$, that is, $b = f(a)$, then $a = f^{-1}(b)$, so (b, a) is on the graph of $y = f^{-1}(x)$. The graph of $y = f^{-1}(x)$ is obtained by reversing the coordinates of the points on the graph of $y = f(x)$, changing (a, b) to (b, a).

On the other hand, what is the effect of reflecting a graph in the line $y = x$? From Figure 3.8.4 we realize that reflecting interchanges the positive x-axis and the positive y-axis. A point (a, b), which is found by moving a units along the x-axis and b units along the y-axis, will be reflected into the point (b, a), which is b units along the x-axis and a units along the y-axis.

This shows that the graph of $y = f^{-1}(x)$ is obtained by reflecting the graph of $y = f(x)$ in the line $y = x$.

To return to Example 4, it is possible to choose other domains on which $\sin x$ is one-to-one (e.g., $\left[\frac{\pi}{2}, \pi\right]$, $\left[\frac{\pi}{2}, \frac{3\pi}{2}\right]$, $\left[0, \frac{\pi}{2}\right] \cup \left(\pi, \frac{3\pi}{2}\right]$, etc.), and $\sin x$ has an inverse on each of them. However, most often we use the interval $\left[\frac{-\pi}{2}, \frac{\pi}{2}\right]$, and then arcsin u is the corresponding inverse function.

It is also easy to verify that $\cos x$ is decreasing on $[0, \pi]$, so it is one-to-one there. Also, $\tan x$ is increasing on $\left(-\frac{\pi}{2}, \frac{\pi}{2}\right)$, $\cot x$ is decreasing on $(0, \pi)$, $\sec x$ is positive and increasing on $\left[0, \frac{\pi}{2}\right)$ and negative and decreasing on $\left[\pi, \frac{3\pi}{2}\right)$, and $\csc x$ is negative and increasing on $\left(-\pi, -\frac{\pi}{2}\right]$ and positive and decreasing on $\left(0, \frac{\pi}{2}\right]$. So each of these functions is one-to-one on the specified domain and has an inverse there (see Figure 3.8.5).

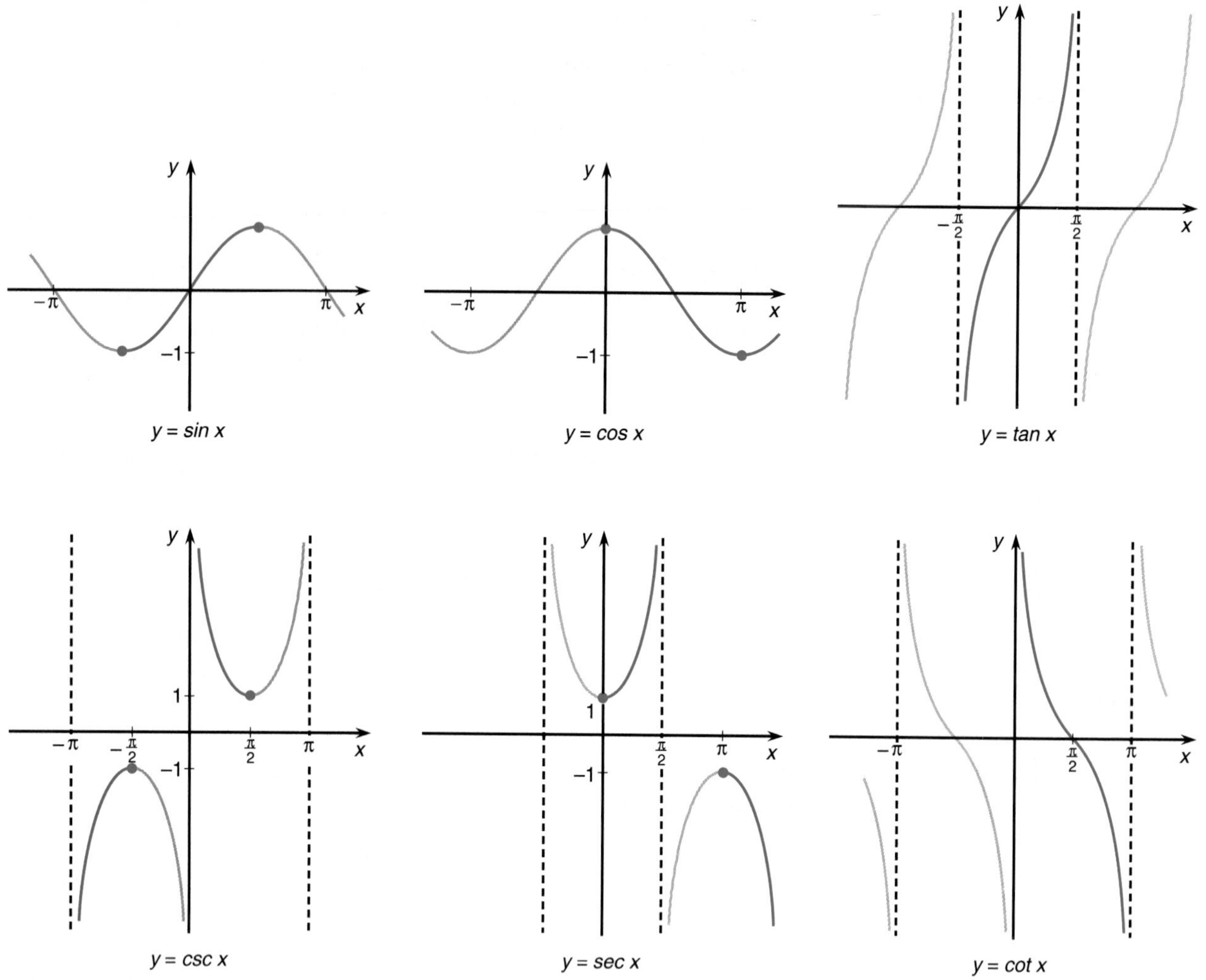

Figure 3.8.5

DEFINITION 3.8.3

The **standard inverse functions of the trigonometric functions** are defined as follows.

For $u \in [-1, 1]$,

$$\arcsin u \in \left[-\frac{\pi}{2}, \frac{\pi}{2}\right] \quad \text{and} \quad \arccos u \in [0, \pi].$$

For any $u \in \mathbb{R}$,

$$\arctan u \in \left(-\frac{\pi}{2}, \frac{\pi}{2}\right) \quad \text{and} \quad \operatorname{arccot} u \in (0, \pi).$$

For u with $|u| \geq 1$,

$$\text{arcsec } u \in \left[0, \frac{\pi}{2}\right) \cup \left[\pi, \frac{3\pi}{2}\right) \quad \text{and} \quad \text{arccsc } u \in \left(-\pi, -\frac{\pi}{2}\right] \cup \left(0, \frac{\pi}{2}\right].$$

Moreover, if $w = \arcsin u$, then $u = \sin w$; if $w = \arccos u$, then $u = \cos w$; if $w = \arctan u$, then $u = \tan w$; if $w = \text{arccot } u$, then $u = \cot w$; if $w = \text{arcsec } u$, then $u = \sec w$; if $w = \text{arccsc } u$, then $u = \csc w$.

EXAMPLE 6 The graphs of $\tan x$, $\arctan u$, $\sec x$, and $\text{arcsec}\, u$ are sketched in Figure 3.8.6. Note how the graphs of $\tan x$ and $\arctan u$ are each other's reflections in the line $y = x$ and similarly for $\sec x$ and $\text{arcsec } u$.

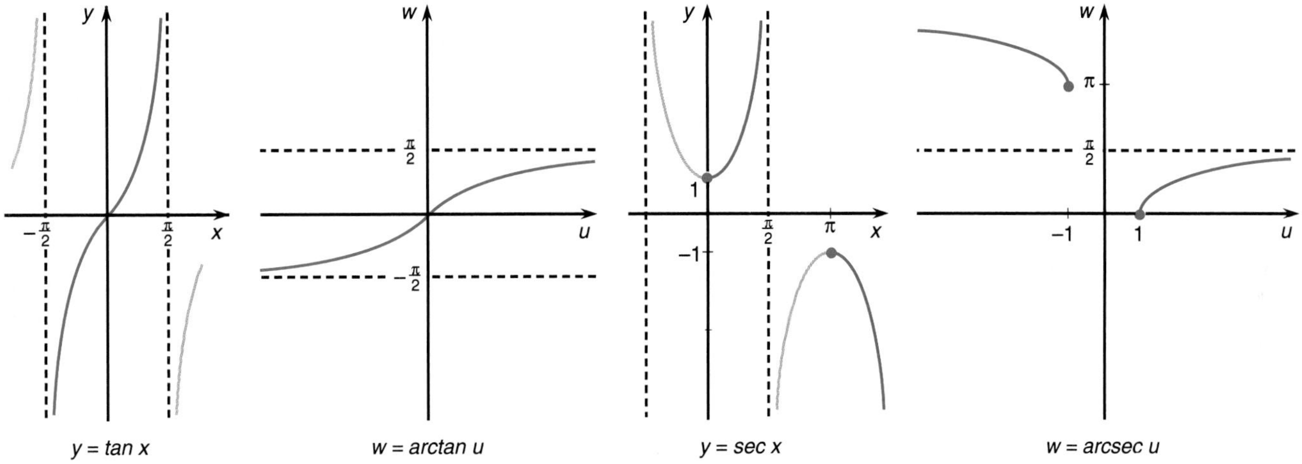

Figure 3.8.6

Next we turn to the question of derivatives of inverse functions. If $f(x)$ is a function that has an inverse, say, $f^{-1}(u)$, then we know that $f^{-1}\big(f(x)\big) = x$, for all x in the domain of f. Differentiating this equation, we see that

$$\frac{d}{dx}\Big(f^{-1}\big(f(x)\big)\Big) = \frac{dx}{dx} = 1.$$

On the other hand, applying the Chain Rule to the left side suggests that

$$(f^{-1})'\big(f(x)\big)f'(x) = 1,$$

so

$$(f^{-1})'\big(f(x)\big) = \frac{1}{f'(x)},$$

or

$$(f^{-1})'(u) = \frac{1}{f'\big(f^{-1}(u)\big)}.$$

The difficulty is that to be able to apply the Chain Rule, we have to know that f^{-1} is differentiable, and this is not always true. For instance, the graph of $w = \text{arcsec } u$

above is not even continuous, so it is certainly not always differentiable. Also if $f'(x) = 0$, then the above formula does not make sense, and in fact f^{-1} will not be differentiable at the corresponding point $u = f(x)$.

If we think of the graph of f^{-1} as the reflection in the line $y = x$ of the graph of f, then a point at which $f'(x) = 0$ is a point at which the tangent line to $y = f(x)$ is horizontal. This point will be reflected into a point on the graph of f^{-1} at which the tangent line is *vertical*, which is why the derivative does not exist.

On the other hand, this idea also suggests that if $y = f(x)$ has a tangent line that is not horizontal, then at the corresponding point on the graph of $w = f^{-1}(u)$ there will be a tangent line that is not vertical. In particular, f^{-1} should be differentiable there.

EXAMPLE 7

(i) Recall the functions $f(x) = 2x + 4$ and $g(u) = \frac{u}{2} - 2$, which were defined in Example 1. We find that $g'(u) = \frac{1}{2}$, $f'(x) = 2$, so in this case it certainly is true that $(f^{-1})'(u) = g'(u) = \frac{1}{2} = \frac{1}{f'(f^{-1}(u))}$.

(ii) If $F(x) = e^x$, then from Example 2 we know that $F^{-1}(u) = \ln u$. In this case, $(F^{-1})'(u) = \frac{d}{du}(\ln u) = \frac{1}{u}$. On the other hand, $F'(x) = e^x = F(x)$, so $F'\big(F^{-1}(u)\big) = F\big(F^{-1}(u)\big) = \exp(\ln u) = u$, and we see that here too $(F^{-1})'(u) = \frac{1}{u} = \frac{1}{F'(F^{-1}(u))}$.

(iii) Considering the functions $f(x) = x^2$ and $g(u) = \sqrt{u}$, as defined in Example 3, we notice that $g = f^{-1}$ and $(f^{-1})'(u) = g'(u) = \frac{1}{2\sqrt{u}}$. On the other hand, $f'\big(f^{-1}(u)\big) = 2f^{-1}(u) = 2g(u) = 2\sqrt{u}$. So once again $(f^{-1})'(u) = \frac{1}{2\sqrt{u}} = \frac{1}{f'\big(f^{-1}(u)\big)}$, provided $u > 0$.

Example 7 illustrates that the formula suggested by the Chain Rule for the derivative of an inverse function does work, although we do have to be careful about the derivative of the original function being nonzero.

THEOREM 3.8.4

Suppose a function $f(x)$, defined on some domain D, has an inverse function. Then $f^{-1}(u)$ is differentiable at $u = a$ if $f(x)$ is differentiable at $x = f^{-1}(a)$ *provided* $f'\big(f^{-1}(a)\big) \neq 0$. In this case,

$$(f^{-1})'(a) = \frac{1}{f'\big(f^{-1}(a)\big)}.$$

In Example 7 we verified that this formula works for various examples of inverse functions. But its usefulness derives from the fact that it can be used even if we do not know much about the inverse function.

EXAMPLE 8 Let $f(x) = x^5 + 2x + 1$. Although it is not obvious, this function is one-to-one on the whole real line, so it has an inverse f^{-1}. Assuming this fact, we can use Theorem 3.8.4 to find the derivative $(f^{-1})'(u)$.

Since $f'(x) = 5x^4 + 2$, we see that $f'(x) \geq 2$, so $f'(x) \neq 0$ for any x, so f^{-1} is differentiable at every point in its domain, and the formula says that

$$(f^{-1})'(u) = \frac{1}{f'\big(f^{-1}(u)\big)} = \frac{1}{5\big(f^{-1}(u)\big)^4 + 2}.$$

If we evaluate $f(2) = 32 + 4 + 1 = 37$, for instance, we see that $(f^{-1})'(37) = \frac{1}{f'(f^{-1}(37))} = \frac{1}{f'(2)} = \frac{1}{82}$. We could also write

$$(f^{-1})'(u) = \frac{1}{5x^4 + 2}.$$

Notice that the formula is inevitably awkward. Either the derivative $(f^{-1})'(u)$ is expressed in terms of $f^{-1}(u)$ or else it can be expressed more simply in terms of x, but then it is the value of $(f^{-1})'$ at $u = f(x)$.

EXAMPLE 9 Letting $L(x) = \ln x$, for $x > 0$, we know from Example 2(ii) that $L^{-1}(u) = e^u$. So

$$\frac{d}{du}(e^u) = (L^{-1})'(u) = \frac{1}{L'\big(L^{-1}(u)\big)} = \frac{1}{L'(e^u)} = \frac{1}{\frac{1}{e^u}} = e^u.$$

This result is not new, but we see that it follows from knowing the derivative of $\ln x$. The interesting thing is that the derivative of e^u is expressed as a function of u itself, rather than as a function of $L^{-1}(u)$.

Notice that we used the fact that $\ln'(w) = \frac{1}{w}$, but that we had to evaluate $\frac{1}{w}$ at $w = L^{-1}(u) = e^u$.

In Example 9 our knowledge of the functions involved allowed us to write a formula for the derivative that was very simple. Most notably, it avoids the difficulty of expressing the derivative of f^{-1} at u in terms of $f^{-1}(u)$. The occurrence of f^{-1} in the formula is concealed.

There are other examples of this phenomenon for familiar functions. One of them we have already seen.

EXAMPLE 10 Let n be a positive integer, and define $R(x) = x^{1/n}$ for $x > 0$. Then $R(x) = f^{-1}(x)$, where $f(x) = x^n$.

We can calculate the derivative of R:

$$R'(x) = \frac{1}{f'\big(R(x)\big)} = \frac{1}{n\big(R(x)\big)^{n-1}} = \frac{1}{n(x^{1/n})^{n-1}}$$
$$= \frac{1}{nx^{\frac{n-1}{n}}} = \frac{1}{n}x^{\frac{1-n}{n}} = \frac{1}{n}x^{1/n-1}.$$

Once again the derivative $R'(x)$ is expressed in terms of x. This was possible because we have an explicit formula for $R(x)$.

Next we want to find the derivatives of the inverse trigonometric functions.

EXAMPLE 11 Find the derivative of $\arctan x$.

Solution By Theorem 3.8.4, we can differentiate

$$\frac{d}{dx}(\arctan x) = \frac{1}{\tan'(\arctan x)} = \frac{1}{\sec^2(\arctan x)}.$$

In a sense this is the answer, but we can do better. Recall the trigonometric identity $\sec^2 t = 1 + \tan^2 t$ (Equation 1.7.3). So the denominator of the derivative above is

Note that $\tan(\arctan x) = x$, so $\tan^2(\arctan x) = x^2$.

$\sec^2(\arctan x) = 1 + \tan^2(\arctan x) = 1 + x^2$. In other words,

$$\frac{d}{dx}(\arctan x) = \frac{1}{1+x^2}.$$

The other inverse trigonometric functions can also be differentiated.

THEOREM 3.8.5

Derivatives of Inverse Trigonometric Functions

The standard inverse trigonometric functions (cf. Definition 3.8.3) are differentiable except at the points that are excluded by Theorem 3.8.4 (i.e., f^{-1} is *not* differentiable at u if $f'\big(f^{-1}(u)\big) = 0$).

Moreover, their derivatives are as follows:

$$\arcsin'(u) = \tfrac{1}{\sqrt{1-u^2}}, \qquad \arccos'(u) = -\tfrac{1}{\sqrt{1-u^2}},$$

$$\arctan'(u) = \tfrac{1}{1+u^2}, \qquad \operatorname{arccot}'(u) = -\tfrac{1}{1+u^2},$$

$$\operatorname{arcsec}'(u) = \tfrac{1}{u\sqrt{u^2-1}}, \qquad \operatorname{arccsc}'(u) = -\tfrac{1}{u\sqrt{u^2-1}}.$$

PROOF The differentiability of the inverse trigonometric functions as stated above is a direct consequence of Theorem 3.8.4. We have already established the formula for the derivative of arctan u in Example 11.

Now consider the function $\arcsin(u)$. By Theorem 3.8.4 its derivative is

$$\arcsin'(u) = \frac{1}{\cos(\arcsin u)},$$

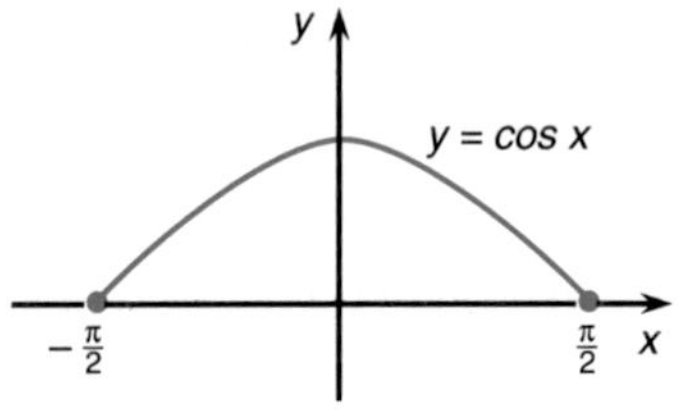

Figure 3.8.7

whenever this formula makes sense. From Definition 3.8.3 we know that the domain of arcsin u is $[-1, 1]$ and that $\arcsin u \in \left[-\frac{\pi}{2}, \frac{\pi}{2}\right]$. Looking at the graph of $\cos x$ for $x \in \left[-\frac{\pi}{2}, \frac{\pi}{2}\right]$, as shown in Figure 3.8.7, we notice that $\cos x \geq 0$ for all x in this interval.

We also know (from Theorem 1.7.2) that $\cos^2 x = 1 - \sin^2 x$, and the non-negativity of $\cos(\arcsin u)$ implies that

$$\cos(\arcsin u) = \sqrt{1 - \sin^2(\arcsin u)} = \sqrt{1 - u^2}.$$

Substituting into the formula for the derivative, we find that

$$\arcsin'(u) = \frac{1}{\sqrt{1-u^2}},$$

as desired.

Now consider the function arccsc u. Its derivative is

$$\operatorname{arccsc}'(u) = \frac{1}{\csc'\big(\operatorname{arccsc}(u)\big)}.$$

Now $\csc'(x) = -\csc(x)\cot(x)$, so this time the problem is to express

$$-\csc\big(\operatorname{arccsc}(u)\big)\cot\big(\operatorname{arccsc}(u)\big)$$

in terms of u. The first factor, $-\csc\big(\operatorname{arccsc}(u)\big)$, is of course just $-u$. For the second, $\cot\big(\operatorname{arccsc}(u)\big)$, we need to use a trigonometric identity.

If you do not happen to remember any identity involving $\cot x$, the thing to do is write the standard identity $\cos^2 x + \sin^2 x = 1$ and recognize that to get something involving $\cot x = \frac{\cos x}{\sin x}$, we should divide by $\sin^2 x$. What we find is

$$\cot^2 x + 1 = \frac{1}{\sin^2 x} = \csc^2 x,$$

so

$$\cot^2 x = \csc^2 x - 1.$$

Again we consult Definition 3.8.3 and find that $\operatorname{arccsc}(u) \in \left(-\pi, -\frac{\pi}{2}\right] \cup \left(0, \frac{\pi}{2}\right]$. The important thing about these intervals is that $\cot x$ is nonnegative on both of them. In particular, this means that $\cot x = \sqrt{\csc^2 x - 1}$, so

$$\cot\big(\operatorname{arccsc}(u)\big) = \sqrt{\csc^2\big(\operatorname{arccsc}(u)\big) - 1} = \sqrt{u^2 - 1}.$$

Substituting this into the formula for the derivative, we find that

$$\operatorname{arccsc}'(u) = -\frac{1}{u\sqrt{u^2 - 1}},$$

as desired.

The formulas for the derivatives of the remaining inverse trigonometric functions are derived in a similar way, being careful about signs, and are left for the exercises.

> The calculation we have just done for the derivative of arccsc x relied on the fact that $\cot x \geq 0$ whenever x is in the range of arccsc. When we defined arccsc and arcsec in Definition 3.8.3, we made what might have seemed to be slightly unexpected choices of intervals, but it was done to make these calculations work out well. It is possible to use other choices, but then the derivatives of arccsc and arcsec will be more complicated.

EXAMPLE 12 Find the derivative of $f(x) = \arctan(2x^2 - x)$.

Solution Using the Chain Rule, we find that

$$f'(x) = \arctan'(2x^2 - x)(4x - 1) = \frac{1}{1 + (2x^2 - x)^2}(4x - 1) = \frac{4x - 1}{1 + (2x^2 - x)^2}$$
$$= \frac{4x - 1}{4x^4 - 4x^3 + x^2 + 1}.$$

EXAMPLE 13 Find all points at which the tangent line to $y = \arcsin\left(\frac{1}{2+x^2}\right)$ is horizontal.

Solution The derivative is

$$\frac{dy}{dx} = \arcsin'\left(\frac{1}{2 + x^2}\right)\frac{d}{dx}\left(\frac{1}{2 + x^2}\right) = \frac{1}{\sqrt{1 - \left(\frac{1}{2+x^2}\right)^2}}\frac{0(2 + x^2) - 1(2x)}{(2 + x^2)^2}$$
$$= \frac{1}{\sqrt{1 - \left(\frac{1}{2+x^2}\right)^2}}\frac{-2x}{(2 + x^2)^2}.$$

The first factor is never zero, and the second is zero precisely when $x = 0$. So the answer to the problem is that the only horizontal tangent is at $x = 0$.

EXAMPLE 14 Suppose an observer is watching an aircraft as it approaches. The aircraft is flying directly toward the observer at an altitude of 2 miles. Its speed is 250 mph. As the aircraft approaches, its angle of elevation (the angle between the ground and the observer's line of sight to the aircraft) will change.

(i) How fast is this angle changing when the aircraft is a horizontal distance of 4 miles from the observer?

(ii) Suppose that at the moment when the aircraft is a horizontal distance of 2 miles from the observer, the angle of elevation is increasing at a rate of 1 degree per second. How fast is the aircraft flying (assuming that its altitude is still 2 miles)?

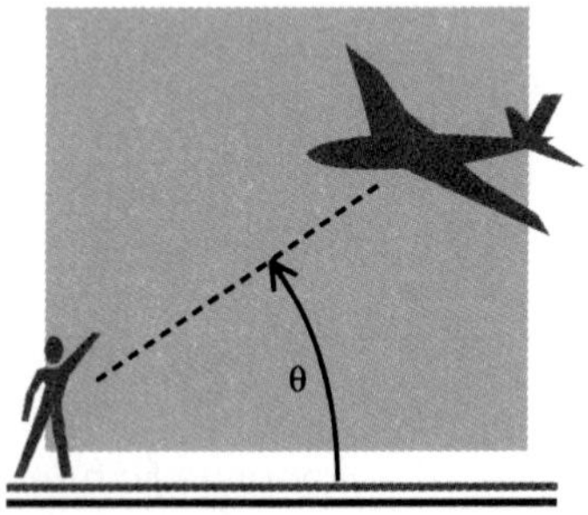

Figure 3.8.8(i)

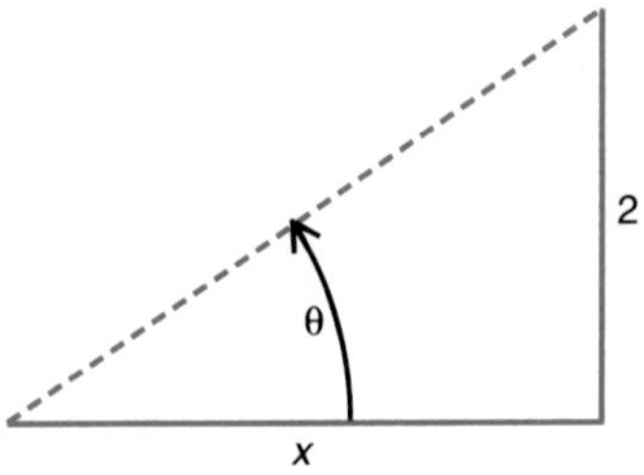

Figure 3.8.8(ii)

Solution

(i) We begin by drawing a diagram in Figure 3.8.8(i), marking the angle of elevation as θ. If we let x represent the (changing) horizontal distance from the observer to the aircraft, then from Figure 3.8.8(ii) we find that $\tan\theta = \frac{2}{x}$. This amounts to the same thing as saying that $\theta = \arctan\frac{2}{x}$.

What we want to know is the rate of change of θ. It is easily calculated:

$$\frac{d}{dt}(\theta) = \frac{d}{dt}\left(\arctan\frac{2}{x}\right) = \frac{1}{1+\frac{4}{x^2}}\left(-\frac{2}{x^2}\right)\frac{dx}{dt} = -\frac{2}{x^2+4}\frac{dx}{dt}.$$

At the moment in question, $x = 4$, and since the aircraft is *approaching* the observer, $\frac{dx}{dt} = -250$. From this we find that $\frac{d\theta}{dt} = -\frac{2}{4^2+4}(-250) = 25$.

Since the speed was measured in miles per *hour*, this rate of change is measured in radians per hour. It would be more convenient to express it in degrees per second. We find that 25 radians per hour equals $25 \times \frac{180}{\pi} \approx 1432.4$ degrees per hour, which is $25 \times \frac{180}{\pi} \times \frac{1}{3600} \approx 0.3979°/\text{sec}$.

(ii) As before, we find that $\frac{d}{dt}(\theta) = -\frac{2}{x^2+4}\frac{dx}{dt}$, with x measured in miles and $\frac{d\theta}{dt}$ measured in radians per hour. Converting this to degrees per second, we see that the rate of change of θ is $-\left(\frac{2}{x^2+4}\frac{dx}{dt} \times \frac{180}{\pi} \times \frac{1}{3600}\right)^{\circ}/\text{sec}$.

At the specified moment, $x = 2$, and this rate of change equals $1°/\text{sec}$; that is,

$$1 = -\frac{2}{2^2+4}\frac{dx}{dt}\frac{180}{\pi}\frac{1}{3600} = -\frac{360}{8\pi(3600)}\frac{dx}{dt}.$$

From this we find that $\frac{dx}{dt} = (-1)\frac{8\pi(3600)}{360} \approx -251.3$.

The aircraft is approaching the observer at a speed of about 251.3 mph.

Existence of Inverse Functions

One problem that we have largely avoided is deciding whether a function $f(x)$ has an inverse function, that is, whether it is one-to-one. Generally speaking, this is difficult to do, but there is a useful criterion that applies to many of the situations that interest us.

The idea is something we used in Examples 4 and 5. There we noted that if a function was increasing on an interval $[a, b]$, then it could not "double back" and take the same value twice, so it must be one-to-one on $[a, b]$. We are familiar enough with the graphs of the trigonometric functions to be able to see where they are increasing, so we were able to find intervals on which they have inverses. Of course it also works for a function that is always decreasing on some interval; in order *not* to be one-to-one on an interval a function must both increase and decrease there.

To apply this idea to functions that are less familiar, we need a way to decide when they are increasing or decreasing. In the discussion before Example 3.3.1, we noticed that the derivative $f'(a)$, being the slope of the tangent to $y = f(x)$ at $x = a$, gives an indication of the behavior of the function. We expect that when $f'(a) > 0$, the function $f(x)$ should be increasing, at least for values of x very close to $x = a$ (see Figure 3.8.9), and when $f'(a) < 0$, then $f(x)$ should be decreasing near $x = a$ (see Figure 3.8.10). In particular, if $f'(x) > 0$ for all x in some interval, then $f(x)$ should

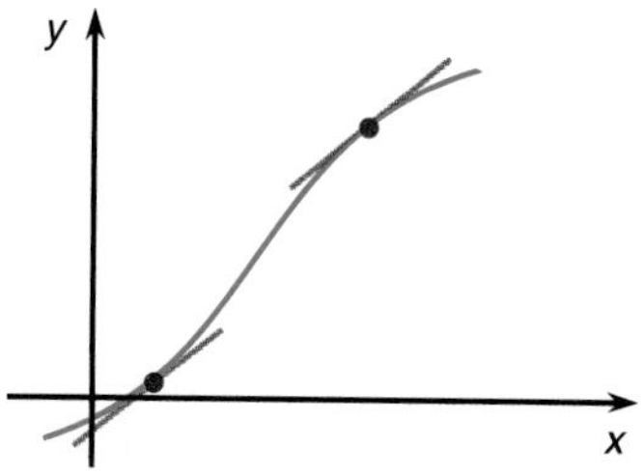

Figure 3.8.9

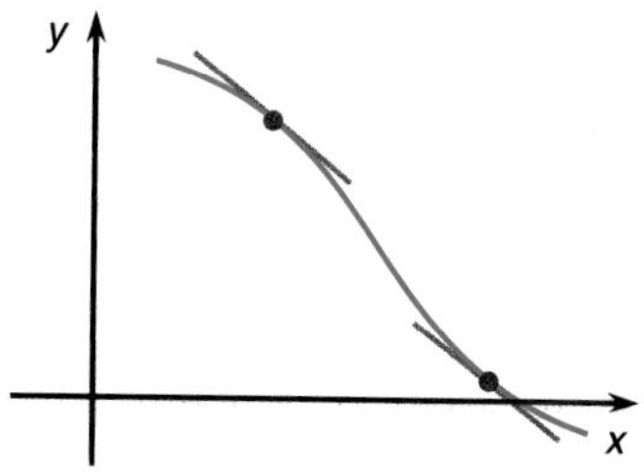

Figure 3.8.10

be increasing on that interval. Similarly, if $f'(x) < 0$ for every x in some interval, then we expect that the function $f(x)$ is decreasing on that interval.

Now suppose that $f'(x) > 0$ for every $x \in (a, b)$. Then $f(x)$ should be increasing on the interval (a, b) and must therefore be one-to-one on (a, b). Accordingly, it has an inverse function f^{-1}.

The value of this criterion is that it gives us something to calculate, namely, the derivative, and if we are able to show the derivative is always positive, then it tells us there must be an inverse function. It does not depend on our knowledge of the function, apart from being able to find and work with its derivative.

Strictly speaking, we have not really yet developed the techniques for discussing when a function is increasing or decreasing (this will be done in the next chapter), but the ideas suggested in the discussion above are in fact correct. We summarize the result as a theorem.

THEOREM 3.8.6

If $f(x)$ is continuous on $[a, b]$, differentiable on (a, b), and $f'(x) > 0$ for every $x \in (a, b)$, then $f(x)$ is one-to-one on $[a, b]$. The inverse function $f^{-1}(u)$ is defined for $u \in \left[f(a), f(b)\right]$, and is differentiable on $\left(f(a), f(b)\right)$, and

$$(f^{-1})'\left(f(x)\right) = \frac{1}{f'(x)},$$

that is,

$$(f^{-1})'(u) = \frac{1}{f'\left(f^{-1}(u)\right)}.$$

The theorem is also true under the condition that $f'(x) < 0$ for every $x \in (a, b)$, except that then $f(a) > f(b)$, so f^{-1} is defined on $\left[f(b), f(a)\right]$ and differentiable on $\left(f(b), f(a)\right)$. The formula for $(f^{-1})'$ remains the same.

There is also a version of the theorem for open intervals, replacing square brackets with round brackets (i.e., with $f(x)$ continuous and differentiable on (a, b) and $f'(x) > 0$ there), except that we cannot then speak of $f(a)$ and $f(b)$. In this case the theorem says simply that there is an inverse function, f^{-1} and that it is defined and differentiable on the set of all values taken by f on (a, b), that is, on $\{y\colon y = f(x)$ for some $x \in (a, b)\}$.

Comparable remarks apply to half-open intervals (e.g., if $f(x)$ is continuous on $[a, b)$ and differentiable on (a, b) or continuous on $(a, b]$ and differentiable on (a, b) and with $f'(x) > 0$ for all $x \in (a, b)$, then there is an inverse function defined on the set of values taken on by $f(x)$ for $x \in [a, b)$ or $x \in (a, b]$, respectively, and differentiable on $\{y : y = f(x)$ for some $x \in (a, b)\})$.

The theorem is also true with $a = -\infty$ and/or $b = \infty$, though of course then the interval must be open at that end.

Finally, all of these versions still hold true if $f'(x) < 0$ on the interval (a, b), and for all of them the formula for the derivative of f^{-1} is the same.

EXAMPLE 15

(i) The function $\tan x$ has as its derivative $\sec^2 x$, which of course is positive whenever it is defined. In particular, on the interval $\left(-\frac{\pi}{2}, \frac{\pi}{2}\right)$ the derivative is always positive, and the open interval version of the above theorem (Theorem 3.8.6) asserts that the function arctan exists.

(ii) The derivative of $\sin x$ on $\left[-\frac{\pi}{2}, \frac{\pi}{2}\right]$ is $\cos x$, which is positive on the corresponding open interval $\left(-\frac{\pi}{2}, \frac{\pi}{2}\right)$. This implies the existence of the function arcsin.

(iii) The half-open interval versions of Theorem 3.8.6 show that sec is one-to-one on $\left[0, \frac{\pi}{2}\right)$ (since the derivative, $\sec x \tan x$, is positive on $\left(0, \frac{\pi}{2}\right)$) and similarly that it is one-to-one on $\left[\pi, \frac{3\pi}{2}\right)$ (since the derivative is negative there).

To show that it is one-to-one on the union of these two intervals $\left[0, \frac{\pi}{2}\right) \cup \left[\pi, \frac{3\pi}{2}\right)$, we remark that the value of $\sec x$ is positive when $x \in \left[0, \frac{\pi}{2}\right)$ and negative when $x \in \left[\pi, \frac{3\pi}{2}\right)$. Because of this, there is no possibility that $\sec x$ could have the same value for some $x \in \left[0, \frac{\pi}{2}\right)$ as it does for some $x \in \left[\pi, \frac{3\pi}{2}\right)$, so it must be one-to-one on the union of the two intervals.

So Theorem 3.8.6 asserts the existence of the inverse function arcsec u on each of the two intervals, and we can combine them to define arcsec u on the union of the two intervals.

(iv) In Example 8 we discussed the function $f(x) = x^5 + 2x + 1$. Its derivative, $f'(x) = 5x^4 + 2$, is clearly positive on the whole real line, so Theorem 3.8.6 says that $f(x)$ is one-to-one and has an inverse function. When we did Example 8, we had to assume that this was true.

Inverse Hyperbolic Functions

It is also possible to find inverses for the hyperbolic functions (Definition 3.5.11), though, as with the trigonometric functions, it is necessary to restrict the domains of some of them. To begin with, we consider the graphs of the hyperbolic functions in Figure 3.8.11.

From the pictures we see that $\cosh x$ is increasing on $[0, \infty)$ and that $\sinh x$ and $\tanh x$ are increasing on the whole real line. So $\cosh x$ is one-to-one on the positive half-line $[0, \infty)$, and $\sinh x$ and $\tanh x$ are both one-to-one on the entire real line.

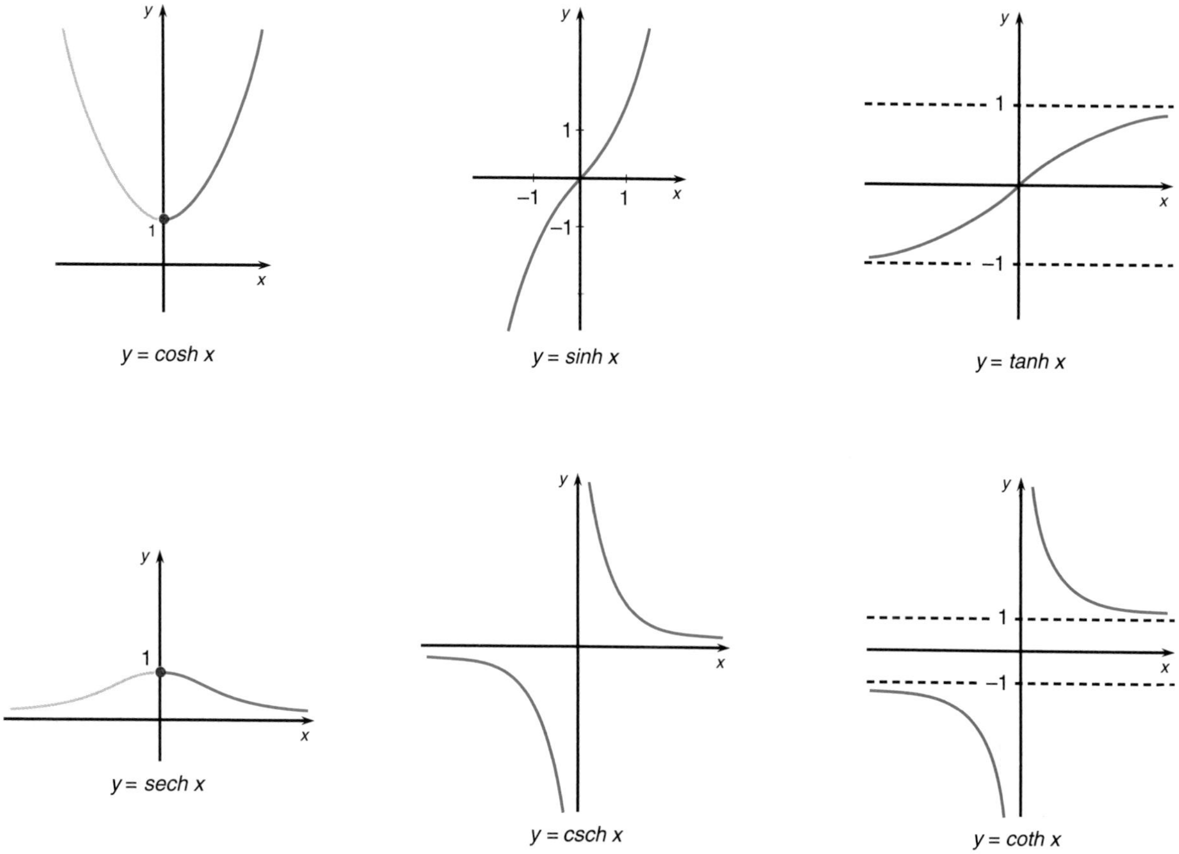

Figure 3.8.11

Also sech x is decreasing on $[0, \infty)$ and one-to-one there, while csch x and coth x are both decreasing on $(-\infty, 0)$ and on $(0, \infty)$ and are one-to-one on the union $(-\infty, 0) \cup (0, \infty)$.

It is possible to check these observations by using the criterion given in Theorem 3.8.6, and this allows us to define the inverse hyperbolic functions.

DEFINITION 3.8.7

The **standard inverse functions of the hyperbolic functions** are defined as follows:

For $u \in [1, \infty)$, arccosh $u \in [0, \infty)$.

For $u \in \mathbb{R}$, arcsinh $u \in \mathbb{R}$, that is, takes all values.

For u with $|u| < 1$, arctanh $u \in \mathbb{R}$, that is, takes all values.

For u with $|u| > 1$, arccoth $u \in (-\infty, 0) \cup (0, \infty)$.

For $u \in (0, 1]$, arcsech $u \in [0, \infty)$.

For $u \in (-\infty, 0) \cup (0, \infty)$, arccsch $u \in (-\infty, 0) \cup (0, \infty)$.

Moreover, if $w =$ arcsinh u, then $u = \sinh w$; if $w =$ arccosh u, then $u = \cosh w$; if $w =$ arctanh u, then $u = \tanh w$; if $w =$ arccoth u, then $u = \coth w$; if $w =$ arcsech u, then $u =$ sech w; if $w =$ arccsch u, then $u =$ csch w.

EXAMPLE 16 The graphs of $\cosh x$, $\operatorname{arccosh} u$, $\tanh x$, and $\operatorname{arctanh} u$ are sketched in Figure 3.8.12.

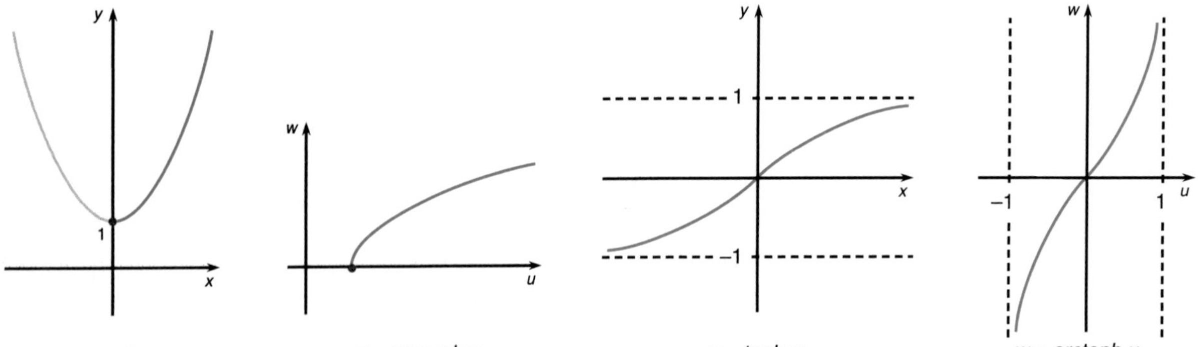

$y = \cosh x$ $w = \operatorname{arccosh} u$ $y = \tanh x$ $w = \operatorname{arctanh} u$

Figure 3.8.12

The derivatives of the inverse hyperbolic functions can be calculated in much the way we did for the inverse trigonometric functions.

EXAMPLE 17 Find the derivative of $\operatorname{arctanh} u$.

Solution From Theorem 3.5.13 we know that $\tanh'(u) = \operatorname{sech}^2 u$. Noting that $\operatorname{sech} u \neq 0$ for every u, we apply Theorem 3.8.4 to see that $\operatorname{arctanh}'(u) = \frac{1}{\operatorname{sech}^2(\operatorname{arctanh}(u))}$.

Recalling the identity $\cosh^2 t - \sinh^2 t = 1$ from Equation 3.5.12, we divide both sides by $\cosh^2 t$ to get that $1 - \tanh^2 t = \operatorname{sech}^2 t$. So

$$\begin{aligned}\operatorname{arctanh}'(u) &= \frac{1}{\operatorname{sech}^2\big(\operatorname{arctanh}(u)\big)} \\ &= \frac{1}{1 - \tanh^2\big(\operatorname{arctanh}(u)\big)} = \frac{1}{1 - u^2}.\end{aligned}$$

The other inverse hyperbolic functions can also all be differentiated.

THEOREM 3.8.8

Derivatives of Inverse Hyperbolic Functions

The inverse hyperbolic functions defined in Definition 3.8.7 are differentiable except at the points that are excluded by Theorem 3.8.4. In particular, $\operatorname{arccosh} u$ is not differentiable at $u = 1$, and $\operatorname{arcsech} u$ is not differentiable at $u = 1$. Apart from these exceptions, the inverse hyperbolic functions are differentiable everywhere they are defined.

Moreover, their derivatives are as follows:

$$\operatorname{arcsinh}'(u) = \frac{1}{\sqrt{1 + u^2}},$$

$$\operatorname{arccosh}'(u) = \frac{1}{\sqrt{u^2 - 1}}, \quad u > 1,$$

$$\operatorname{arctanh}'(u) = \frac{1}{1-u^2}, \quad |u| < 1,$$

$$\operatorname{arccoth}'(u) = \frac{1}{1-u^2}, \quad |u| > 1,$$

$$\operatorname{arcsech}'(u) = -\frac{1}{u\sqrt{1-u^2}}, \quad 0 < u < 1,$$

$$\operatorname{arccsch}'(u) = -\frac{1}{|u|\sqrt{1+u^2}}, \quad u \neq 0.$$

PROOF The differentiability follows from Theorem 3.8.4. We have already derived the formula for the derivative of arctanh$u$ in Example 17. The other formulas are established in a similar way (cf. the proof of Theorem 3.8.5) and are left for the exercises. ◢

In truth the inverse hyperbolic functions and their derivatives are only rarely used. When they are, it is usually cosh, sinh, tanh, and their inverse functions that appear; the other three are almost never seen. The formulas for their derivatives appear here for the sake of completeness and as examples. It is worthwhile to try to work some of them out for yourself (see the exercises), but they are not worth committing to memory.

EXAMPLE 18 Evaluate the derivative of $\operatorname{arcsinh}(t^2+1)$.

Solution Using the formula from Theorem 3.8.8 and the Chain Rule, we see that

$$\frac{d}{dt}\left(\operatorname{arcsinh}(t^2+1)\right) = \frac{1}{\sqrt{1+(t^2+1)^2}}(2t) = \frac{2t}{\sqrt{t^4+2t^2+2}}.$$ ◢

Exercises 3.8

I

Find the derivative of each of the following functions.

1. $\arctan(2x+4)$
2. $\arcsin\left(\frac{1}{2}\sin x\right)$
3. $\arccos\left(\frac{1}{1+x^2}\right)$
4. $\arctan(\tan x)$
5. $\operatorname{arcsec}(4+x^2)$
6. $\operatorname{arccot}(\sin x)$
7. $\operatorname{arccsc}(5+e^x)$
8. $\ln(2+\arctan^2 x)$
9. $\exp\left(\operatorname{arccot}(2x)\right)$
10. $\cos(\arctan x)$
11. $\operatorname{arccosh}(\sec^2 x)$
12. $\operatorname{arcsinh}(\tan x)$
13. $\operatorname{arccoth}(x^2+2x+3)$
14. $\operatorname{arctanh}\left(\frac{1}{2}\sin x\right)$
15. $\ln\left(\operatorname{arccosh}(x)\right)$
16. $\operatorname{arctanh}^2\left(\frac{1}{x^2+3}\right)$
17. $\arcsin(\tanh x)$
18. $\operatorname{arcsech}\left(1-\tanh^2(x)\right)$
19. $\operatorname{arctanh}\left(\operatorname{sech}(x)\right)$
20. $\operatorname{arcsinh}(\tanh x)$

II

For each of the following functions defined on the specified interval, show that the function has an inverse, and calculate the derivative of the inverse function.

21. $g(x) = 5x - 1$, on $[1, 9]$
22. $f(x) = x^2 - 2x + 3$, on $[0, 1]$
23. $h(x) = -2x^2 + 4x - 1$, on $(-\infty, 1]$
24. $F(x) = x^3 - 12x + 1$, on $[-2, 2]$
25. $H(x) = x^3 + 6x^2 + 4$, on $[-2, -1]$
26. $k(x) = \ln(1+x^2)$, on $[0, \infty)$

Find the formula for the derivative of each of the following functions.

*27. arccot u

*28. (i) arccos u, (ii) arcsec u

***29.** (i) arccoth u, (ii) arcsinh u

***30.** (i) arccosh u, (ii) arcsech u, (iii) arccsch u

***31.** (i) What is the inverse of arcsin x? What is its domain?
(ii) What is the inverse of $\arctan(2x+1)$?

***32.** (i) If we know that $\frac{d}{dx}(\ln x) = \frac{1}{x}$ and that e^u is the inverse of $\ln x$, show how to use these facts to conclude that $\frac{d}{du}(e^u) = e^u$.
(ii) If we know that $\frac{d}{dx}(e^x) = e^x$ and that $\ln u$ is the inverse of e^x, show how to use these facts to conclude that $\frac{d}{du}(\ln u) = \frac{1}{u}$.

***33.** (i) Suppose $f(x)$ and $g(t)$ are both one-to-one. If the composite function $g\big(f(x)\big)$ is defined, show that it is also one-to-one.
(ii) In the situation of part (i), find a formula for the inverse function of $g\big(f(x)\big)$ in terms of the inverses of f and g.

***34.** Find examples of functions $f(x)$ and $g(t)$ for which
(i) $g\big(f(x)\big)$ is one-to-one even though $g(t)$ is not.
(ii) g is one-to-one but $g\big(f(x)\big)$ is not.
(iii) f is one-to-one but $g\big(f(x)\big)$ is not.

***35.** Suppose $-\frac{3\pi}{2} < x < -\frac{\pi}{2}$. What is the function $F(x) = \arctan(\tan x)$? (*Hint:* It does not equal the function $f(x) = x$ on $\left(-\frac{3\pi}{2}, -\frac{\pi}{2}\right)$.)

***36.** Suppose the aircraft of Example 14 flies past again at an altitude of 1 mile. If the angle of elevation is increasing at 1 degree per second when the aircraft is a horizontal distance of 3 miles from the observer, how fast is the aircraft flying?

37. After the aircraft of Example 14 has gone 3 miles past the observer, its angle of elevation is decreasing at the rate of $\frac{4}{5}^\circ$/sec. How fast is it flying?

***38.** Verify that $\csc x$ is one-to-one on $\left[-\frac{\pi}{2}, 0\right) \cup \left(0, \frac{\pi}{2}\right]$. For this exercise *only*, let Arccsc x be the corresponding inverse function. Show that its derivative is $\frac{d}{dx}(\text{Arccsc } x) = -\frac{1}{|x|\sqrt{x^2-1}}$.

39. If we weigh an object in pounds, we might want to convert its weight to kilograms or vice versa. Let $K(x)$ be the function which converts a weight x lb to the corresponding value in kilograms.
Let $P(x)$ be the function that converts a weight x kg to the corresponding value in pounds.
(i) Find formulas for $K(x)$ and $P(x)$. (Note that 1 kg equals 2.2 lb.)
(ii) Show that $P(x)$ and $K(x)$ are inverse functions.
(iii) Find the derivatives $P'(x)$ and $K'(x)$, and confirm that they satisfy the expected relation for derivatives of inverse functions.

40. If we measure a temperature on the Fahrenheit scale, we might want to convert it to the Celsius scale or vice versa. Let $F(t)$ be the function that converts a temperature t°C to the corresponding value on the Fahrenheit scale.
Let $C(t)$ be the function which converts a temperature t°F to the corresponding value on the Celsius scale.
(i) Find formulas for $F(t)$ and $C(t)$. (Use the facts that the freezing point of water is 0°C and 32°F and that the boiling point of water is 100°C and 212°F; figure out how many Fahrenheit degrees correspond to one Celsius degree. . .)
(ii) Show that $F(t)$ and $C(t)$ are inverse functions.
(iii) Find the derivatives $F'(t)$ and $C'(t)$, and confirm that they satisfy the expected relation for derivatives of inverse functions.

III

41. Use a graphics calculator or computer to graph $y = \arctan x$ and $y = \frac{1}{\tan x} = \tan^{-1} x = \cot x$ on the same picture, and convince yourself once and for all that these are very different functions.

42. Graph $y = \arcsin x$ and $y = \frac{1}{\sin x} = \sin^{-1} x = \csc x$ on the same picture.

43. Graph $y = \ln x$ and $y = \frac{1}{e^x} = e^{-x}$ on the same picture.

44. Graph $y = \tan x$ at various magnifications and find, to four decimal places, the value of arctan 2, that is, the point where $\tan x$ crosses the line $y = 2$.

45. Use the graph of $y = \sin x$ to find $\arcsin \frac{1}{2}$ to four decimal places.

46. Graph the function $y = x^5 + 2x + 1$ of Example 8 (p. 208) and confirm from the picture that it is one-to-one (cf. Example 15(iv)).

3.9 Second and Higher-Order Derivatives

DERIVATIVE OF ORDER k
$f^{(k)}(x)$
ACCELERATION

If we start with a differentiable function $f(x)$, then its derivative is a function $f'(x)$. We might want to differentiate this function; its derivative is conventionally denoted $f''(x)$ and is called the **second derivative** of $f(x)$.

EXAMPLE 1

(i) If $f(x) = x^5 + 3x^4 - 2x^3 + x^2 - 7$, then its derivative is $f'(x) = 5x^4 + 12x^3 - 6x^2 + 2x$. The second derivative is $f''(x) = 20x^3 + 36x^2 - 12x + 2$.

(ii) If $g(t) = \ln(t^2 + 4)$, then

$$g'(t) = \frac{1}{t^2+4}(2t) = \frac{2t}{t^2+4}.$$

Also

$$g''(t) = \frac{d}{dt}\left(\frac{2t}{t^2+4}\right) = \frac{2(t^2+4) - (2t)(2t)}{(t^2+4)^2} = \frac{8-2t^2}{(t^2+4)^2}.$$

If we want to use the other notation for derivatives, we can write

$$f''(x) = \frac{d}{dx}\left(\frac{df}{dx}\right),$$

but a more convenient notation is the following:

$$f''(x) = \frac{d^2f}{dx^2} = \frac{d^2}{dx^2}\big(f(x)\big).$$

The expression in the middle of the above equation is read "d two f by d x squared," *not* "d squared f over d x squared." The expression on the right side is read "d two by d x squared of $f(x)$."

Of course, we might continue taking derivatives. We could write $f'''(x)$ for the third derivative, and so on, but before long this will get confusing. Instead, we usually write $f^{(n)}(x)$ for the nth derivative of $f(x)$, where n could be 3, 4, 5, etc., and occasionally we even use this notation for $n = 2$ or $n = 1$ (i.e., $f^{(2)}(x) = f''(x)$, and $f^{(1)}(x) = f'(x)$). The alternative notation for the **derivative of order n** is

$$f^{(n)}(x) = \frac{d^nf}{dx^n} = \frac{d^n}{dx^n}\big(f(x)\big).$$

EXAMPLE 2

(i) If we return to Example 1(i), where $f(x) = x^5 + 3x^4 - 2x^3 + x^2 - 7$, $f'(x) = 5x^4 + 12x^3 - 6x^2 + 2x$, and $f''(x) = 20x^3 + 36x^2 - 12x + 2$, then we can continue differentiating and find $f^{(3)}(x) = \frac{d}{dx}\big(f''(x)\big) = 60x^2 + 72x - 12$.

Similarly, $f^{(4)}(x) = 120x + 72$, $f^{(5)}(x) = 120$, and $f^{(6)}(x) = 0$.

Subsequent derivatives are all zero, that is, $f^{(n)}(x) = 0$ for all $n \geq 6$.

(ii) Find $G^{(4)}(t)$ if $G(t) = \ln t$.

Solution $G'(t) = \frac{d}{dt}(\ln t) = \frac{1}{t}$. Writing this as t^{-1} (by which we mean the -1 power, *not* the inverse function), we see that its derivative is $G''(t) = -t^{-2}$. Differentiating again, we find that $G^{(3)}(t) = -(-2)t^{-3} = 2t^{-3}$. Finally, $G^{(4)}(t) = 2(-3)t^{-4} = -6t^{-4}$.

Part of the usefulness of second derivatives comes from their physical interpretation. In Sections 2.1 and 3.1 we discussed the idea of velocity. Recall that if an object is moving along a straight line in such a way that its displacement from the origin at time t is given by some function $s(t)$, then the derivative $s'(t)$ is the rate of change of $s(t)$, which is what we call the velocity.

Now consider the second derivative $s''(t)$. It is the rate of change of the velocity $s'(t)$, and this is what is known in physics as the acceleration. After all, if a car is moving along at a constant velocity, we say it is not accelerating, but if the velocity is increasing, we say the car is accelerating.

In the discussion before Example 2.1.3, it was mentioned that if an object held A ft above the ground is released and allowed to fall, its height above the ground t sec after release will be $h(t) = A - 16t^2$. Differentiating, we find that its velocity is $h'(t) = -32t$ ft/sec. This means that the velocity is changing; it starts, when $t = 0$, with the velocity $h'(0) = -32(0) = 0$ (i.e., at the moment of release it is not moving), but after 1 second it is falling at the rate of 32 ft/sec, etc. Notice that the velocity is negative, reflecting the fact that the object is *falling*, that is, its height is decreasing.

Since the velocity is changing, we expect the acceleration to be nonzero. In fact, the acceleration is $h''(t) = \frac{d}{dt}\big(h'(t)\big) = \frac{d}{dt}(-32t) = -32$ ft/sec^2.

The acceleration is constant. This means that each second the velocity changes by the same amount, in this case -32 ft/sec. So from $h'(0) = 0$ ft/sec, it changes after 1 sec to $h'(1) = -32$ ft/sec, and after another second to $h'(2) = -64$ ft/sec, etc. The fact that the acceleration is negative simply means that the velocity is decreasing (becoming more negative), since the acceleration is the rate of change of the velocity.

This represents a fundamental fact in practical physics. Near the surface of the earth a falling object will accelerate due to the force of gravity, and its acceleration will be approximately

$$-\,32 \text{ ft/sec}^2$$

(which means 32 ft/sec^2 downward). This is the same thing as

$$-\,9.81 \approx -9.8 \text{ m/sec}^2.$$

Of course it is not quite so simple as that. Air resistance will have the effect of slowing down the object's fall, especially for very light objects (a feather does not accelerate nearly as fast as -32 ft/sec^2 . . .). Still, it is quite accurate for reasonably heavy objects or, more important, for objects whose surface area is relatively small in comparison with their weight, in which case the air resistance will be insignificant relative to the force of gravity.

Notice that the units in which acceleration is measured are ft/sec^2 or m/sec^2, which can also be written as ft/sec/sec and m/sec/sec, respectively. This is because the acceleration is the rate of change of the velocity, so it tells us, for instance, by how many ft/sec the velocity is changing in one second. To say the acceleration is -32 ft/sec^2 means that with each passing second the velocity changes by -32 ft/sec. The rate of change of the velocity is -32 ft/sec *per second*, or -32 ft/sec/sec, or -32 ft/sec^2.

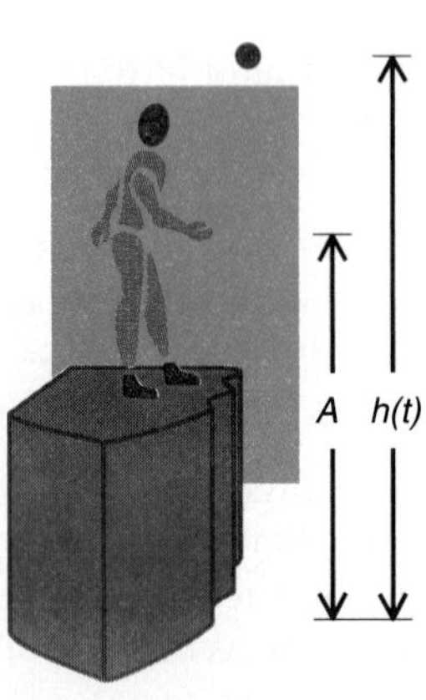

Figure 3.9.1

EXAMPLE 3 Suppose an object is thrown vertically so that it starts at a height A ft above the ground. Suppose moreover that it starts with a velocity B ft/sec. See Figure 3.9.1.

Let $h(t)$ represent the height of the object above the ground at a moment t sec after it is thrown. We know that $h(0) = A$ and that $h'(0) = B$. From the preceding discussion we know that $h''(x) = -32$.

From this we would like to find a formula for $h(t)$. However, rather than try to do that right away, let us ask first for a formula for $h'(t)$. What we need is a function whose derivative is the constant function -32.

One function that obviously satisfies that requirement is $-32t$; its derivative is certainly -32. We want $h'(t)$ to be a function whose derivative is -32 and whose value at $t = 0$ is B. To achieve this, we could let $h'(t) = -32t + B$, which clearly satisfies both conditions. Notice that adding the constant B does not change the derivative.

In fact, as we shall see later, the only functions whose derivative equals -32 are the functions $-32t + c$, where c is some constant. Taking this for granted, the only one of these functions whose value at $t = 0$ is B is

$$h'(t) = -32t + B.$$

The next step is to ask what are the possibilities for $h(t)$? After all, we want $h'(t) = -32t + B$. There is one function that obviously satisfies that condition, namely, $-16t^2 + Bt$. On the other hand, as we have already remarked, adding a constant would not change the derivative, so we should really consider functions of the form $-16t^2 + Bt + C$, where C is any constant.

Now, taking into account the additional condition that $h(0) = A$, we find that if $h(t) = -16t^2 + Bt + C$, then $h(0) = C$, so C must be A. In other words, we have found that the height above the ground at time t must be given by

$$h(t) = -16t^2 + Bt + A.$$

In summary, the height above the ground of an object that has been thrown into the air is

$$h(t) = -16t^2 + V_0t + H_0,$$

where $V_0 = h'(0)$ is the initial velocity and $H_0 = h(0)$ is the initial height. Notice that if $V_0 > 0$, it means the object was thrown upward; if $V_0 < 0$, it means the object was thrown downward; and if $V_0 = 0$, it means the object started at rest, that is, it was *dropped* rather than thrown.

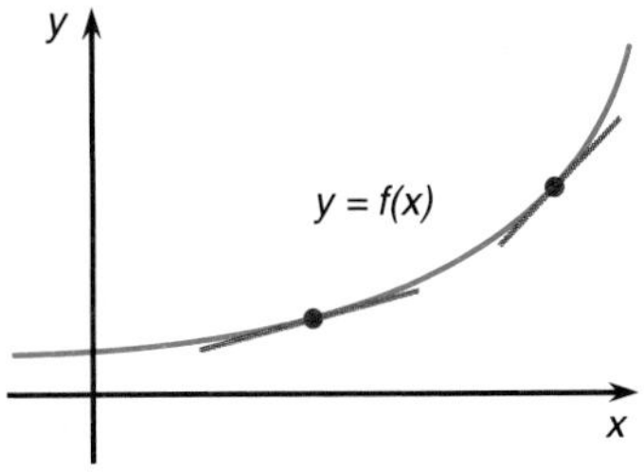

Figure 3.9.2

The other main use of the second derivative arises from its application to drawing graphs. After all, as the rate of change of the derivative of a function $f(x)$, it tells us how the derivative is changing, and specifically it tells us whether the derivative is increasing or decreasing.

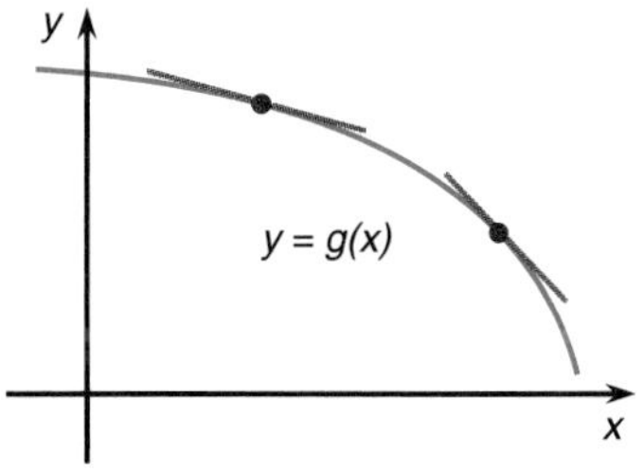

Figure 3.9.3

For the curve $y = f(x)$, illustrated in Figure 3.9.2, the derivative is always positive, that is, the tangent line has positive slope, so it always slopes upward to the right. But we can say more than this. As we move from left to right along the curve, the tangent line slopes upward more steeply. This means that the slope of the tangent is larger and larger as the point of tangency moves from left to right. In terms of the derivative it means that as x increases, the derivative $f'(x)$ increases.

Now to say that $f'(x)$ is an increasing function suggests that its derivative is positive, that is, that $f''(t) > 0$, or at least that $f''(t) \geq 0$.

In Figure 3.9.3 the graph is decreasing, that is, $g'(x) \leq 0$. But as x increases, the graph slopes more steeply downward, that is, $g'(x)$ becomes more negative. This suggests that $g''(x) \leq 0$.

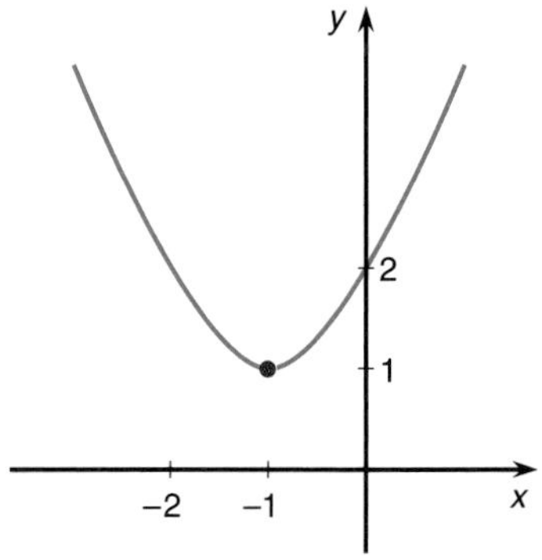

Figure 3.9.4

EXAMPLE 4 Consider the function $y = x^2 + 2x + 2$, whose graph is illustrated in Figure 3.9.4.

Its derivative is $\frac{dy}{dx} = 2x + 2$. We have already discussed in Section 3.3 that to look for maximum or minimum values of y, we should look for places where $y' = 0$. In this case the only such point is $x = -1$. Now consider the second derivative, $\frac{d^2y}{dx^2} = \frac{d}{dx}(2x + 2) = 2$.

The fact that the second derivative is positive means that the first derivative is increasing. Since it is zero when $x = -1$, this suggests that it should be negative to the left of $x = -1$ and positive to the right of $x = -1$. This in turn suggests that the graph is decreasing for points to the left of $x = -1$ and increasing at points to the right of $x = -1$. The only way this can happen is for $f(x)$ to have a *minimum* value at $x = -1$.

Of course, in this case it is obvious from the picture that this is so, but in fact it is often possible to use this sort of reasoning to decide whether $f(x)$ has a maximum or a minimum value at some particular point where $f'(x) = 0$.

EXAMPLE 5 Let $f(x) = x^3 - 3x + 2$ for $x \in [-2, 3]$. We find that $f'(x) = 3x^2 - 3 = 3(x^2 - 1)$, which is zero when $x = \pm 1$. Calculating that $f''(x) = 6x$, we see that $f''(-1) = -6$ and $f''(1) = 6$.

Near $x = 1$, this suggests that $f'(x)$ is increasing, and since it equals zero at $x = 1$, this means it must be negative to the left of $x = 1$ and positive to the right, so the graph decreases down to $x = 1$ and then increases again. This means that $f(x)$ should have a minimum at $x = 1$.

Similarly, at $x = -1$ the derivative is zero, but the second derivative is negative, so the derivative is decreasing. This suggests that the derivative is positive to the left of $x = -1$ and negative to the right, so the function increases up to its value at $x = -1$ and then decreases again. The function $f(x)$ should have a maximum value at $x = -1$.

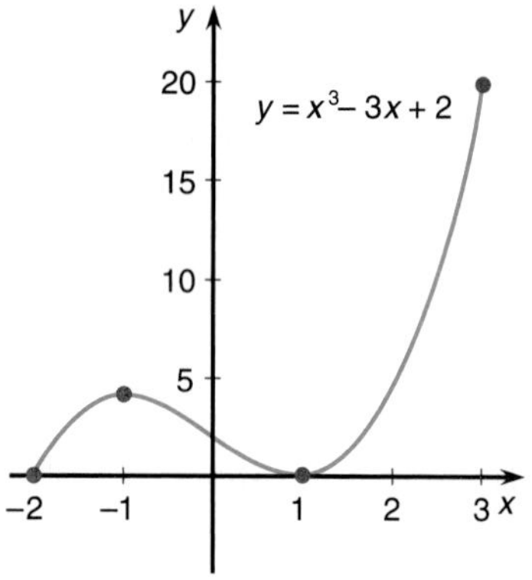

Figure 3.9.5

The graph of $y = f(x)$ is illustrated in Figure 3.9.5. It is clear that $f(x)$ has a minimum at $x = 1$, but there is also an endpoint minimum at $x = -2$. More distressing is that the maximum value occurs at the endpoint $x = 3$, rather than where we expected it, at $x = -1$. The argument we gave suggested that the value of $f(x)$ at $x = -1$ should be bigger than its values at nearby points, and that is true, but it said nothing about what the function might do somewhere else. In this case it did indeed get larger at the right endpoint.

The use of second derivatives to find maximum and minimum values is fairly tricky. Specifically, it can only be used to describe what happens for points near the one in question. There is another difficulty in that it might also happen that $f''(x) = 0$, in which case the sort of argument we have been using simply does not apply.

Nonetheless, the second derivative is extremely useful as a tool for finding maximum and minimum values, and in Section 4.4 we will learn in greater detail how to use it. For now, these examples have been presented to give some indication of our purpose in studying second derivatives.

Exercises 3.9

I

Evaluate the following.

1. $\frac{d^2}{dx^2}(x^4 - 3x^3 + 2x^2 - x + 3)$
2. $\frac{d^2}{dx^2}(\tan x)$
3. $\frac{d^2}{dx^2}(e^{x^2+1})$
4. $\frac{d^2}{dx^2}(\cosh x)$
5. $\frac{d^3}{dx^3}(e^{7x+1})$
6. $\frac{d^8}{dx^8}(\sinh x)$
7. $\frac{d^3}{dx^3}(\sin x)$
8. $\frac{d^9}{dx^9}(\cos x)$
9. $\frac{d^7}{dx^7}(13x^5 + 7x^4 - 21x^3 + \sqrt{2}x^2 + 9x - 1)$
10. $\frac{d^3}{dx^3}(\sqrt{x})$
11. $\frac{d^2}{dx^2}\left(\frac{1}{x+1}\right)$
12. $\frac{d^2}{dx^2}\left(\frac{1}{x^2+1}\right)$

13. $\frac{d^3}{dt^3}\left(\frac{1}{3t+2}\right)$

14. $\frac{d^4}{dx^4}\left(\cos(2x)\right)$

15. $\frac{d^3}{dx^3}\left(\frac{1}{e^x+1}\right)$

16. $\frac{d^3}{dx^3}(\arctan x)$

17. $\frac{d^2}{dx^2}(\arcsin x)$

18. $\frac{d^3}{dx^3}\left(\ln(x^2+4)\right)$

19. $\frac{d^3}{dx^3}\left(\exp(\sin x)\right)$

20. $\frac{d^3}{dx^3}\left(\exp\left(\ln(x^2+4)\right)\right)$

II

21. If n and k are positive integers, find a formula for the kth derivative of x^n. (Your answer will likely distinguish the case in which $k \leq n$ from the case in which $k > n$. You might also find it helpful to use the factorial notation (Definition 1.3.8); recall that $m! = m(m-1)(m-2) \times \ldots \times (3)(2)(1)$.)

If n and k are positive integers, find formulas for the following.

22. $\frac{d^k}{dx^k}(x^{-n})$

23. $\frac{d^k}{dx^k}(\ln x)$

24. Show that $f(x) = x^2 - 2x - 6$ cannot have a maximum value, by finding all the points where $f'(x) = 0$ and evaluating the second derivative.

***25.** Let $h(x) = 2x^6 - 15x^4 + 24x^2 + 3$.

(i) Find all the points at which $h'(x) = 0$. (*Hint:* You should be able to factor $h'(x)$ just by looking at it for a while.)

(ii) Evaluate $h''(x)$ at each of the points you found in part (i). What does this tell you about the behavior of $h(x)$ near each of those points? (It may tell you more for some of them than it does for others.)

***26.** Let $g(t) = \sec t$ for $t \in \left(-\frac{\pi}{2}, \frac{\pi}{2}\right)$.

(i) Find the first and second derivatives of $g(t)$.

(ii) By considering the first and second derivatives at $t = 0$, show that $g(t)$ has a minimum value at $t = 0$.

27. An object is thrown straight up from ground level at a speed of 96 ft/sec.

(i) How long after it is thrown will it land again?

(ii) What is the maximum height it will reach?

28. (i) An object is dropped from a height of 100 ft. How long will it take to reach the ground?

(ii) If instead the object was thrown down, with a downward speed of 20 ft/sec, how long would it take to fall?

(iii) Compare the difference between the two times that are the answers to parts (i) and (ii) with the time it would take an object falling at the *constant* speed 20 ft/sec to fall a distance of 100 ft.

29. At what speed must an object be thrown straight up in order to reach a maximum height of 27 m?

***30.** On the planet Vruggles an object falling near the planet's surface experiences an acceleration due to gravity of 6 boingles per snoom per snoom. (The boingle is the Vrugglian unit of distance; the snoom is the unit of time.)

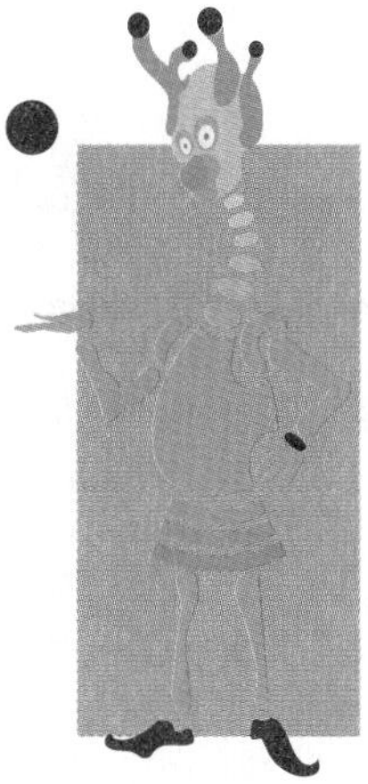

(i) How many snooms will it take an object to fall 40 boingles if it is dropped from rest?

(ii) How far will an object dropped from rest fall in 5 snooms?

(iii) If a ball is thrown straight up from the surface of Vruggles at a speed of 14 boingles per snoom, what is the maximum height it will reach, and how long will it take to fall back to the surface?

***31.** Suppose $f(x)$ is a polynomial of degree n.

(i) Show that $f^{(k)}(x)$ is the zero polynomial if $k > n$.

(ii) What can you say about $f^{(n)}(x)$?

***32.** Find a formula for $\frac{d^n}{dx^n}(\sin x)$. (*Hint:* n can be written in exactly one of the following forms: $n = 4k+1$, $n = 4k+2$, $n = 4k+3$, or $n = 4k$.)

Find a formula for the following.

***33.** $\frac{d^n}{dx^n}(\cos x)$

***34.** $\frac{d^n}{dx^n}(e^{2x})$

***35.** $\frac{d^n}{dx^n}(e^{2x+1})$

If a and b are any constants, find a formula for the following.

***36.** $\frac{d^n}{dx^n}(e^{ax+b})$

***37.** $\frac{d^n}{dx^n}\left(\sin(ax+b)\right)$

***38.** $\frac{d^n}{dx^n}\left(\ln(ax+b)\right)$

III

39. Consider an object thrown upward with a velocity of 50 ft/sec from ground level. Use a graphics calculator or computer to estimate the maximum height it reaches, and compare with the answer you can calculate.

40. Consider an object thrown upward with a velocity of 14 m/sec from 2 m above ground level. Use a graphics calculator or computer to estimate the maximum height it reaches.

41. Sketch the function $y = x^2 + 2x + 2$ and its derivative on the same picture. Compare with the discussion in Example 4 (p. 221).

42. On one picture, graph $f(x) = x^3 - 3x + 2$ and $f'(x)$ and $f''(x)$ for $x \in [-2, 3]$ (cf. Example 5, p. 222).

POINT TO PONDER

Consider an object moving with velocity $v(t)$ and acceleration $a(t)$. Suppose $a(t) > 0$. This means that the velocity $v(t)$ is increasing, but if v happens to be negative, the *speed* $|v(t)|$ will be *decreasing*. This is because the acceleration is in a direction opposite to the motion, so the object slows down.

You can imagine a spaceship flying along; to speed it up, the crew fire the rockets in one direction; to slow it down, they fire the rockets in the opposite direction.

The situation can be summarized by saying that the speed increases if $v(t)$ and $a(t)$ have the same sign, and it decreases if they have opposite signs. Positive acceleration does not necessarily mean that the object moves faster.

Chapter Summary

§3.1
tangent line
slope of tangent line
difference quotient
derivative
displacement
velocity
rate of change
Ideal Gas Law
marginal increase in profit

§3.2
differentiable function
continuity of differentiable functions
$\frac{d}{dx}(constant) = 0$
$\frac{d}{dx}(x^n) = nx^{n-1}$
derivatives of polynomials
Product Rule
finding equations of tangent lines
Quotient Rule
derivatives of rational functions

§3.3
maximum/minimum problems on closed intervals
endpoints
boundedness
extraneous solutions

§3.4
derivatives of trigonometric functions

§3.5
$10^{\log x} = x$
$\log(10^s) = s$
$\log(xy) = \log x + \log y$
common logarithms
$e^{\ln x} = x$
$\ln(e^s) = s$
$\ln(xy) = \ln x + \ln y$
natural logarithms
$\frac{d}{dx}(\ln x) = \frac{1}{x}$, for $x > 0$
$\frac{d}{dx}(e^x) = e^x$
$b^r = e^{r \ln b}$, for $b > 0$
hyperbolic functions
$\cosh^2 x - \sinh^2 x = 1$
derivatives of hyperbolic functions

§3.6
the Chain Rule
the Generalized Power Rule: $\frac{d}{dx}y^r = ry^{r-1}y'$
$\frac{d}{dx}(x^r) = rx^{r-1}$

§3.7
implicit function
implicit differentiation
logarithmic differentiation

§3.8
one-to-one
inverse function
arcsin
arctan
graphs of inverse functions
standard inverse functions of the trigonometric functions
derivatives of inverse functions
derivatives of inverse trigonometric functions
existence of inverse functions
standard inverse functions of the hyperbolic functions
derivatives of inverse hyperbolic functions

§3.9
second derivative
$f^{(n)}(x)$
derivative of order n
$\frac{d^n}{dx^n}\big(f(x)\big)$

Review Exercises

I

Calculate the derivative of each of the following functions, specifying all points at which it is not differentiable.

1. $f(x) = 4x^5 - 3x^2 + 12$
2. $g(x) = 17 - 2x + 8x^2$
3. $h(x) = 2x^7 - 4x^4 - 3$
4. $k(x) = 1 + x^{3000}$
5. $F(x) = 2x - \sqrt{x}$
6. $G(x) = x^2 - 5x^{1/3}$
7. $H(x) = x^{3/5} + x^{-1/2}$
8. $K(x) = x^3 - 5x^{28/4}$
9. $R(x) = \frac{x^2-4}{x^2+1}$
10. $S(x) = \frac{x^3+x^2+x+1}{x+1}$
11. $f(x) = 3\cos x \sin x$
12. $g(x) = \sin(-7x)$
13. $h(x) = (1 + 3x^2)\tan x$
14. $k(x) = \frac{\sin x}{1+x}$
15. $F(x) = \sqrt{x^4 + 2}$
16. $G(x) = (x^3 - 8)^{1/3}$
17. $H(x) = (x^4 + x^2)^{-1/2}$
18. $K(x) = (1 - x)^{8/6}$
19. $L(x) = e^{2x-5}\cos x$
20. $M(x) = \ln(2 + x^2)\cot x$

Calculate the derivative of each of the following functions.

21. $f(x) = \sinh(x^2 + 3x - 2)$
22. $g(x) = \cosh(2x - 4)$
23. $h(x) = \ln(\cosh x)$
24. $k(x) = (\sinh x)^{-2/3}$
25. $F(x) = \tanh(3x^4 - 2x)$
26. $G(x) = \operatorname{sech}^{-2}(3x^3 - 2)$
27. $H(x) = x\ln x$
28. $K(x) = \frac{x}{\sinh x}$
29. $T(x) = \sin^3 x \cosh^5 x$
30. $U(x) = \cot x \tan x$

Find the maximum and minimum values of each of the following functions on the specified interval.

31. $f(x) = x^2 + 6x - 2$, on $[-5, 0]$
32. $f(x) = x^2 - 2x + 5$, on $[3, 6]$
33. $f(x) = x^3 - 12x + 4$, on $[0, 3]$
34. $f(x) = x^3 - 6x + 1$, on $[-1, 3]$
35. $f(x) = \frac{1}{1+x^2}$, on $[-2, 3]$
36. $f(x) = \frac{3}{4-x}$, on $[-2, 2]$
37. $f(x) = \ln x$, on $[1, 8]$
38. $f(x) = \sin x$, on $\left[0, \frac{3}{4}\pi\right]$
39. $f(x) = \tan x$, on $\left[-\frac{\pi}{4}, 0\right]$
40. $f(x) = \sin^2 x + \cos^2 x$, on $\left[-\pi, \frac{4}{5}\pi\right]$

In each of the following situations, find the derivative $\frac{dy}{dx}$.

41. $x^2 + 3xy - y^2 = 3$
42. $xy^3 + x^3y = 16$
43. $\sin(xy) = \frac{1}{2}$
44. $x^2e^y + 2y = 4$
45. $\frac{1}{y} = \frac{1}{x}$
46. $\sin x + \cos y = 0$
47. $y\ln x = x^2$
48. $\sqrt{xy^3 + y} = 5$
49. $\frac{1}{x+y} = e^{xy}$
50. $xy^5 - 3x^2y^2 - 3x = 4y - 2$

Find the derivatives of each the following functions.

51. $\arctan(x + 3)$
52. $\arcsin(1 - x^2)$
53. $\arccos(\sin x)$
54. $\arctan(e^{1-x^2})$
55. $\operatorname{arccosh}(1 + 2x^2)$
56. $\operatorname{arctanh}(\sin x)$
57. $\exp\big(\arctan(2x)\big)$
58. $\arctan(\ln x)$
59. $\ln(\arcsin x)$
60. $\arctan(2\tan x)$

II

***61.** (i) Find a formula for arccosh u as follows. Write out the equation $\cosh x = u$ by expressing $\cosh x$ in terms of e^x and e^{-x}. Multiply through by e^x, and the resulting equation will be a quadratic equation in e^x (i.e., an equation involving e^x and its square). Solve this equation to get an expression for e^x. Finally, use this equation to solve for $x = \operatorname{arccosh} u$.

(ii) Differentiate the formula that you obtained in part (i), and compare the derivative with what we already know.

***62.** (i) Find a formula for arcsinh u, following the idea used in Exercise 61.

(ii) Compare the derivative obtained from your answer to part (i) with the derivative as we already know it.

***63.** Suppose x and y satisfy $2x\sin^2 y - 3\cos(2x) = 4 - 2x\cos^2 y$. Explain why it is not possible to use this relation to find $\frac{dy}{dx}$.

***64.** Verify that $\sec x$ is one-to-one on $\left[0, \frac{\pi}{2}\right) \cup \left(\frac{\pi}{2}, \pi\right]$. For this exercise *only*, let Arcsec x be the corresponding inverse function. Find its derivative $\frac{d}{dx}(\text{Arcsec } x)$.

65. If $f(x)$ and $g(x)$ are differentiable functions that are never zero, verify the Product Rule for $(fg)'$ using logarithmic differentiation.

66. If $f(x)$ and $g(x)$ are differentiable functions that are never zero, verify the Quotient Rule for $\left(\frac{f}{g}\right)'$ using logarithmic differentiation.

***67.** (i) The function $|x|$ is not differentiable at $x = 0$, but verify that at every other x, $\frac{d}{dx}(|x|) = \frac{x}{|x|}$.

(ii) Use part (i) and the Chain Rule to find $\frac{d}{dx}(|\sin x|)$.

68. Find (i) $\frac{d}{dx}(|e^x \cos x|)$, (ii) $\frac{d}{d\theta}(|\tan\theta|)$.

III

69. Use a graphics calculator or computer to compare the graphs of $y = \operatorname{arcsinh} x$ and $y = \ln(x + \sqrt{x^2+1})$.

70. Use a graphics calculator or computer to compare the graphs of $y = \operatorname{arcsec} x$ and $y = \operatorname{arccsc} x$.

Suppose $f(x)$ is some differentiable function. If h is small, we expect the difference quotient $Q_h(x) = \frac{f(x+h)-f(x)}{h}$ to be close to $f'(x)$. For each of the following functions $f(x)$ and values of h, sketch $f'(x)$ and the approximation $Q_h(x)$ on the same picture. With sufficient magnification you should be able to see that they are not exactly the same.

71. $f(x) = x^2$, $h = 0.5$

72. $f(x) = x^2$, $h = 0.1$

73. $f(x) = x^3 - 2x^2$, $h = 0.2$

74. $f(x) = \sin x$, $h = 0.1$

75. $f(x) = \frac{1}{x-1}$, $h = 0.01$

76. $f(x) = \ln x$, $h = 0.05$

77. $f(x) = \arctan x$, $h = 0.1$

78. $f(x) = \tanh x$, $h = 0.5$

79. $f(x) = e^{\sin x}$, $h = 0.1$

80. $f(x) = \ln(e^x)$, $h = 1$

***81.** Graph $|\tan x|$ and its derivative on the same picture (cf. Exercise 68(ii)).

Applications of Derivatives 4

INTRODUCTION

In Chapter 3 we learned about differentiation. We saw how the idea came from looking for the slope of the tangent to a curve and how the same concept arose when we tried to describe the notion of instantaneous velocity or rates of change in general.

We learned how to differentiate many types of functions, and we have become reasonably comfortable with these calculations. In the process we saw that derivatives could be used for various purposes. In addition to finding the slopes of tangent lines and working with velocity and acceleration, we found that they could be used to help find maximum and minimum values of functions.

In this chapter we will learn to use derivatives to solve a great many different types of problems. We will return to maximum/minimum problems and study them more systematically. We will study the use of derivatives and second derivatives in drawing graphs. The interpretation of velocity and acceleration as derivatives will be discussed at greater length. But we will also learn about a number of other types of problems that can successfully be approached using derivatives.

We begin in Section 4.1 with a type of problem known as *related rates;* this refers to a situation in which two different quantities are changing at rates that are somehow related. In Section 4.2 we discuss an important result called the *Mean Value Theorem.* It is the technical tool that allows us to refine the observation we have already made that a function is increasing if its derivative is positive and decreasing if the derivative is negative. We also learn the *First Derivative Test,* which is used to identify maximum and minimum values. In Section 4.3 we introduce the idea of critical points, which is fundamental to the study of maximum/minimum problems, and in Section 4.4 we learn the *Second Derivative Test,* which helps to distinguish maximum points from minimum points.

In Section 4.5 these ideas are applied to *curve sketching,* that is, drawing graphs, and in Section 4.6, two additional refinements are added to our techniques for curve sketching, namely, *concavity* and *inflection points.* In Section 4.7 we discuss maximum/minimum problems at greater length; you are encouraged to do a large number of exercises from this section.

In Section 4.8 we consider the problem of finding an *antiderivative,* which means a function whose derivative equals some specified function. This concept will be considered in much greater depth in Chapters 5, 6, and 8, but it is of immediate use in Section 4.9 for the discussion of velocity and acceleration of moving objects.

Section 4.10 deals with *linear approximation and differentials,* which are techniques that use derivatives to help calculate approximate values of functions that cannot be evaluated exactly. The subject of Section 4.11 is *Newton's Method,* another technique of the same sort; it is used to find approximate values for the roots of equations.

In Section 4.12 we consider *l'Hôpital's Rule,* a technique that uses derivatives to evaluate limits of quotients in situations in which both the numerator and denominator tend to zero. Finally, Section 4.13 discusses *marginal cost, revenue, and profit,* some of the most important applications of the derivative to business and economics.

You may not want to study all these various topics. Depending on your interests, you may choose to skip some of them, but you should probably read Sections 4.1–4.8 at least and add whichever of the remaining sections appeal to you.

Many of the practical problems that we discuss can be broken into two parts. The second part is to apply the techniques of calculus to obtain some information about the behavior of some function. But before it is possible to do that, it is necessary to have the function. Frequently, the first part of the solution is to express the practical question in terms of some function, and very often this is the most difficult aspect of the problem.

4.1 Related Rates

IMPLICIT DIFFERENTIATION
RATE OF CHANGE

For this section it is necessary to be able to differentiate quickly and reliably. It is also important to know about implicit differentiation (cf. Section 3.7), and we need to use both the Pythagorean Theorem (Theorem 1.5.1) and facts about similar triangles (Theorem 1.5.4).

We have already encountered (in Section 3.1 and later) the idea that $\frac{df}{dt}$ represents the rate of change of $f(t)$. There are many situations in which there are two or more quantities, each of them changing with time, in such a way that their rates of change are related in some way. The problem addressed in this section is finding the derivative of one function in terms of the derivative of some other function.

EXAMPLE 1 **Expanding Balloon** Suppose air is being pumped into a spherical balloon in such a way that its volume is increasing by 40 cu in/sec. How fast is the radius of the balloon increasing at the moment when the radius equals 5 in?

Solution Letting the radius of the balloon be r, we know that its volume, the volume of a sphere of radius r, is $V = \frac{4}{3}\pi r^3$. We are also told that $\frac{dV}{dt} = 40$.

At this point we could take the equation $V = \frac{4}{3}\pi r^3$ and solve it for r. It would then be possible to calculate $\frac{dr}{dt}$, which is what we want. But in fact it is easier to perform an implicit differentiation. We find that

$$\frac{dV}{dt} = \frac{d}{dt}\left(\frac{4}{3}\pi r^3\right) = \frac{4}{3}\pi(3r^2)\frac{dr}{dt} = 4\pi r^2\frac{dr}{dt}.$$

Knowing that the left side of the above equation, $\frac{dV}{dt}$, equals 40, we can now solve for $\frac{dr}{dt}$. We find that

$$\frac{dr}{dt} = \frac{1}{4\pi r^2}\frac{dV}{dt} = \frac{10}{\pi r^2}.$$

When $r = 5$, this tells us that $\frac{dr}{dt} = \frac{10}{25\pi} = \frac{2}{5\pi} \approx 0.1273$. So when the radius of the balloon is 5 in, it is increasing at the rate of $\frac{2}{5\pi}$ in/sec.

As was remarked earlier, this problem could also have been done by solving for r as a function of V and then differentiating. In fact, the calculation is somewhat more complicated that way, and using implicit differentiation actually saves work.

In the course of doing the above calculation we ignored the units, since we expected the answer to come out in inches per second. It is often a good idea to carry the units along through the computation, partly just to be sure of the units for the final answer and partly as a way of checking that the calculation is correct.

EXAMPLE 2 **Leaking Balloon** Suppose a spherical balloon is leaking air so that its surface area is decreasing at the rate of 1 sq in/sec. At the moment when the radius is 6 in, how fast is the volume decreasing?

Solution We need to calculate $\frac{dV}{dt} = \frac{d}{dt}\left(\frac{4}{3}\pi r^3\right) = 4\pi r^2\frac{dr}{dt}$. What we know is the rate of change of the surface area.

The formula for the surface area of a sphere of radius r is $A = 4\pi r^2$; we will learn how to derive this formula in Chapter 6, but for now we must just accept it, as we do the formula for the volume.

Again we differentiate implicitly and find that

$$\frac{dA}{dt} = \frac{d}{dt}(4\pi r^2) = 8\pi r\frac{dr}{dt}.$$

From this we find that $\frac{dr}{dt} = \frac{1}{8\pi r}\frac{dA}{dt}$. Substituting this into the formula for the derivative of V, we find that

$$\frac{dV}{dt} = 4\pi r^2\frac{dr}{dt} = 4\pi r^2\frac{1}{8\pi r}\frac{dA}{dt} = \frac{r}{2}\frac{dA}{dt}.$$

The question asked us to find $\frac{dV}{dt}$ at the moment when $r = 6$ in, assuming that $\frac{dA}{dt} = -1\ \text{in}^2/\text{sec}$. We see that

$$\frac{dV}{dt} = \frac{r}{2}\frac{dA}{dt} = \frac{6\ \text{in}}{2}(-1\ \text{in}^2/\ \text{sec}) = -3\ \text{in}^3/\text{sec}.$$

At the specified moment the volume of the balloon is decreasing at the rate of 3 cu in/sec.

There are several things to notice here. One is that the sign of the derivative being negative indicates that the volume is decreasing. This is of course only common sense in this example, but sometimes it is not easy to guess whether a quantity is increasing or decreasing, and close attention to signs will tell us. Another is that after we found a formula relating the two derivatives, we put in numerical values, with units included, and multiplied and divided units just as we would variables. The answer ended up being a number with units. Here again there were no surprises, though occasionally it is not entirely clear what units will be used for the answer. More important is that if we make a mistake in the calculation, there is a reasonable chance that the units will be affected, and this will help us to detect the error. Finally observe that in working with units we wrote cubic inches as in^3 and square inches as in^2 so as to be able to multiply and divide them easily. This is consistent with the way we wrote the units for acceleration as ft/sec^2 or m/sec^2, etc.

There are many types of problems in which two distances are related in some geometrical way. It is often possible to find an equation expressing this relationship by using the Pythagorean Theorem (Theorem 1.5.1) or results about similar triangles. Implicit differentiation then gives a relationship between their rates of change.

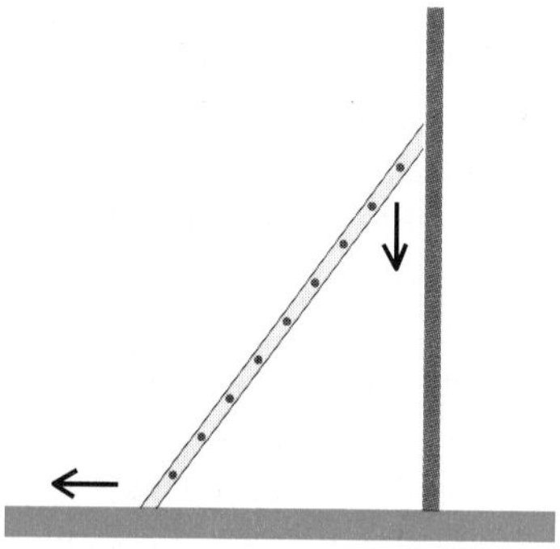

Figure 4.1.1

EXAMPLE 3 **Sliding Ladder** Consider a ladder leaning against a wall. Suppose the foot of the ladder begins to slip away from the wall so that the top of the ladder begins to slide down (see Figure 4.1.1).

If the ladder is 10 ft long and the foot of the ladder is moving away from the wall at a rate of 4 in/sec, how fast is the top descending at the moment when the top is 8 ft above the floor level?

Solution The ladder, the wall, and the floor make a right-angled triangle as shown in Figure 4.1.2. If we let x be the horizontal distance from the wall to the foot of the ladder, in feet, and y the vertical distance from the floor to the top of the ladder, also in feet, then the Pythagorean Theorem says that $x^2 + y^2 = 10^2 = 100$.

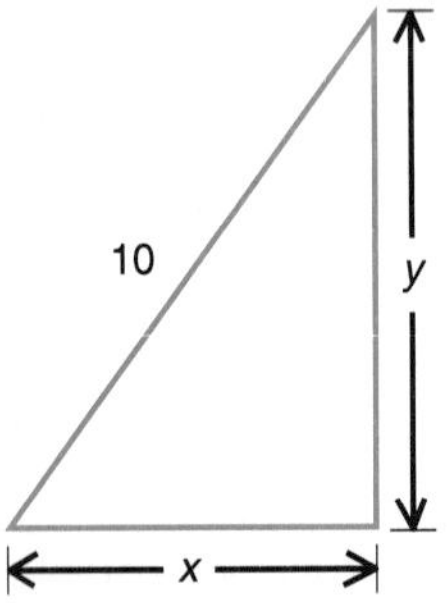

Figure 4.1.2

Differentiating this equation and recognizing that both x and y are changing quantities, so they should be regarded as functions of t, we find that

$$2x\frac{dx}{dt} + 2y\frac{dy}{dt} = 0,$$

that is,

$$2y\frac{dy}{dt} = -2x\frac{dx}{dt},$$

so

$$\frac{dy}{dt} = -\frac{x}{y}\frac{dx}{dt}.$$

The question asks us to find the rate of descent of the top of the ladder, which is just $\frac{dy}{dt}$. At the moment in question, $y = 8$, and since the Pythagorean Theorem says that $x^2 + y^2 = 100$, we solve and find that $x^2 = 100 - 8^2 = 100 - 64 = 36$. Since x, being a distance, is positive, we must have $x = 6$.

So the rate at which the top of the ladder is descending is

$$\frac{dy}{dt} = -\frac{x}{y}\frac{dx}{dt} = -\frac{6 \text{ ft}}{8 \text{ ft}}(4 \text{ in/sec}) = -\frac{3}{4}(4 \text{ in/sec}) = -3 \text{ in/sec}.$$

At the specified moment the top of the ladder is sliding down the wall at the rate of 3 in/sec.

Notice that the negative sign means that y is decreasing, which confirms our intuition that the top of the ladder is descending. Notice also that although part of the calculation was in feet and part in inches, it all worked out. It would have been possible to do this problem by converting 4 in/sec to $\frac{1}{3}$ ft/sec and proceeding from there, but in fact this was not necessary. Carrying the units along in the computation showed that leaving them as they were was all right; however, to do a calculation using both inches and feet simultaneously without keeping track of units would be unthinkable.

Finally, it is important to notice that there are some quantities in the example that change with time, such as x and y, and others that are constant, such as the length of the ladder. Quantities that change with time must be regarded as functions of t.

Summary 4.1.1 Related Rates

It is difficult to give hard and fast rules that will apply exactly to all related rates problems, but the general idea is roughly the same.

There will be one quantity whose rate of change is known and another whose rate of change we wish to find. If a relation can be found between these two quantities, then implicit differentiation will give an equation that can be solved to express the derivative of one of them in terms of the derivative of the other. The solution may involve various other quantities, so their values at the moment in question will have to be found.

It may be difficult to find a relation between the two quantities but easier to find relations between them and other quantities. In this situation we can express the rate of change of one quantity in terms of the rate of change of a second, that rate of change in terms of a third, and so on until we get to one we know. This situation appeared in Example 2, in which we expressed the rate of change of the volume in terms of the rate of change of the radius and the rate of change of the radius in terms of the rate of change of the area. Together these expressions allowed us to find the rate of change of the volume in terms of the rate of change of the area.

There is no generally applicable method for finding a relation between two quantities. If in doubt, it is usually an excellent first step to draw a diagram. The relation being sought may be a consequence of simple geometry.

A vital step in solving one of these problems is identifying the quantity whose rate of change we know and the one whose rate of change we want. Specifying exactly what they are will often help to uncover the relation between them.

Notice that the question frequently asks for the rate of change of some quantity "at the moment when some other quantity equals some fixed number." It is important to recognize that this other quantity is also varying, so it must be regarded as a function of time. It would be a mistake just to let it equal the given fixed number. Only after differentiating and solving for the required derivative should we substitute in the values of this and other quantities at the specified moment. On the other hand, if a quantity does not vary with time, there is no need to label it with a variable. Its constant value can be substituted before differentiating.

It is always important to keep careful track of units and signs.

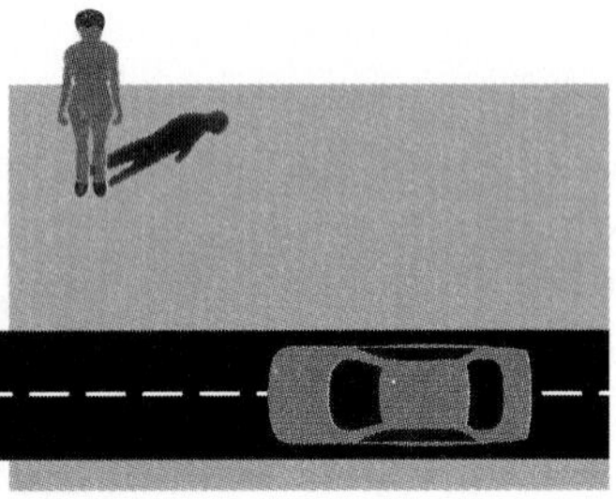

Figure 4.1.3

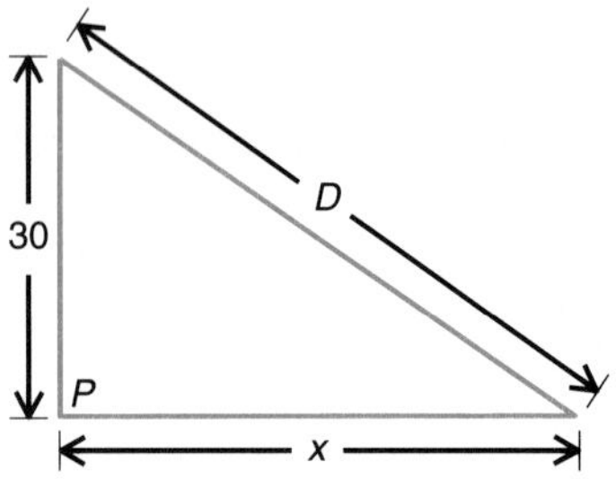

Figure 4.1.4

EXAMPLE 4 A car is traveling along a straight road at a speed of 30 m/sec. An observer is located at a point 30 m off to the side of the road. How fast is the distance between the car and the observer increasing at the moment when the car is 40 m past the point on the road that is closest to the observer?

Solution First we draw a diagram of the situation, shown in Figure 4.1.3. A second, more idealized diagram shows the distance of 30 m between the observer and P, the point on the road that is closest to the observer (see Figure 4.1.4). Notice that the triangle determined by P, the observer, and the car has a right angle at P.

If we let x be the distance from P to the car and let D be the distance from the observer to the car, then $D^2 = 30^2 + x^2 = 900 + x^2$, and what we want to find is the rate of change of D, that is, $\frac{dD}{dt}$. The other piece of information we know is that x is changing at a rate of 30 m/sec, that is, $\frac{dx}{dt} = 30$.

Implicitly differentiating the equation $D^2 = 900 + x^2$, we get

$$2D\frac{dD}{dt} = 2x\frac{dx}{dt},$$

so

$$\frac{dD}{dt} = \frac{x}{D}\frac{dx}{dt}.$$

At the moment in question, $x = 40$ m and $D = \sqrt{30^2 + 40^2}$ m $= 50$ m, so

$$\frac{dD}{dt} = \frac{40 \text{ m}}{50 \text{ m}} 30 \text{ m/sec} = 24 \text{ m/sec}.$$

The distance between the observer and the car is increasing at a rate of 24 m/sec.

Notice that while we could have figured out how to write x and D as functions of t, it was not really necessary, since we were able to use implicit differentiation. In many problems it is essential to use this approach, since expressing the various quantities as functions of t may be difficult or impossible.

Also notice that we could have written D as a square root and differentiated that, but as we noticed on an earlier occasion (Example 3.3.10, p. 163), differentiating the square root is an unnecessary complication.

Figure 4.1.5

EXAMPLE 5 **Lamppost and Shadow** A girl who is 5 ft tall is approaching a post that holds a lamp 15 ft above the ground (see Figure 4.1.5). If she is walking at a speed of 4 ft/sec, how fast is the end of her shadow moving when she is 17 ft away from the base of the lamppost?

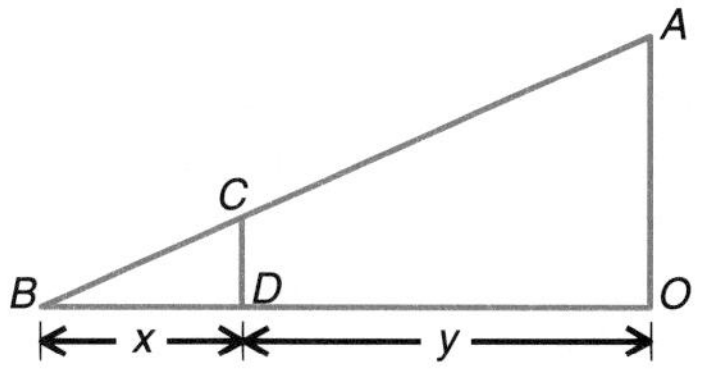

Figure 4.1.6

Solution We make a diagram, labeling the base of the lamppost O, the light A, the end of the shadow B, the girl's feet D, and her head C (see Figure 4.1.6). Notice that the triangles BCD and BAO are similar (cf. Theorem 1.5.4, p. 24).

If we let x and y be the lengths BD and DO, respectively, then the distance from the end of the shadow to the base of the post is $x + y$, and we need to find $\frac{d}{dt}(x + y)$. By similar triangles we know that

$$\frac{x+y}{15} = \frac{x}{5},$$

so

$$5x + 5y = 15x, \qquad \text{or} \qquad 5y = 10x,$$

so

$$y = 2x, \qquad \text{and} \qquad y' = 2x'.$$

Note that $x = x(t)$ and $y = y(t)$ are both functions of t, the time.

But y' is the velocity at which the girl approaches the lamppost, so, noting that y is *decreasing*, we see that $y' = -4$ ft/sec, $x' = -2$ ft/sec, and $x' + y' = -4 - 2 = -6$ ft/sec.

The end of the shadow is approaching the lamppost at a speed of 6 ft/sec, which is faster than the girl is walking. Incidentally, the length of the shadow is x, and $x' = -2$ ft/sec, so the length of her shadow is decreasing by 2 ft/sec.

Notice that the distance from the lamppost (17 ft) is irrelevant in this problem.

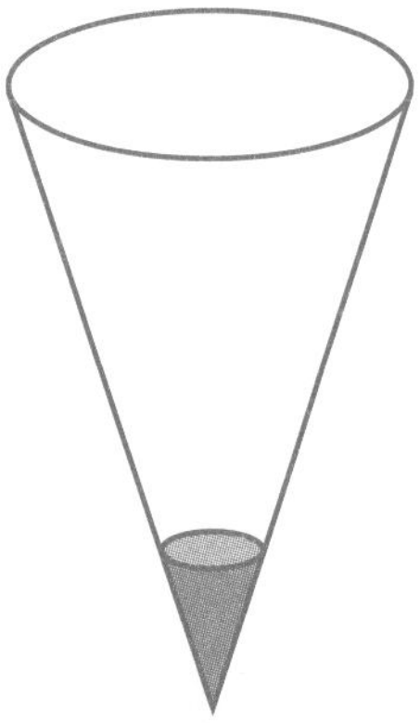

Figure 4.1.7

EXAMPLE 6 **Conical Water Tank** A water tank has the shape of a cone, with the vertex pointing straight down (see Figure 4.1.7). The tank is 12 ft high, and its radius (at the top) is 4 ft. Water flows out at the bottom. When the water is 3 ft deep at its deepest point, the water is flowing out at a rate of 30 cu in/sec. Find the rate at which the level of the surface of the water is falling at that moment.

Solution Recall that a cone of radius r and height h has volume $V = \frac{1}{3}\pi r^2 h$ (see Summary 1.5.14, p. 31). Let h be the depth of the water and r be the radius of the surface of the water, as shown in Figure 4.1.8.

Similar triangles tell us that $\frac{12}{4} = \frac{h}{r}$, or $h = 3r$. Consequently,

$$V = \frac{1}{3}\pi r^2 h = \frac{1}{3}\pi r^2(3r) = \pi r^3.$$

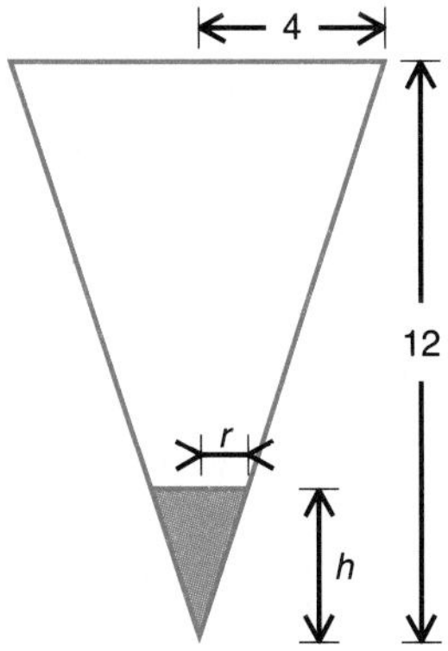

Figure 4.1.8

Differentiating with respect to time gives $V' = 3\pi r^2 r'$, so $r' = \frac{V'}{3\pi r^2}$. At the moment in question, $h = 3$ ft, so $r = \frac{h}{3} = 1$ ft $= 12$ in, and $V' = -30$ in^3/sec. Accordingly,

$$r' = \frac{V'}{3\pi r^2} = \frac{-30}{3\pi(12)^2} = -\frac{5}{72\pi} \text{ in/sec.}$$

Since $h = 3r$, we see that $h' = 3r'$, so

$$h' = -3\frac{5}{72\pi} = -\frac{5}{24\pi} \text{ in/sec.}$$

The water level is falling at the rate of $\frac{5}{24\pi} \approx 0.0663$ in/sec.

Notice that in this question we had to be careful to use consistent units. The distances were measured in feet, but the flow rate was measured in cu in/sec, so we had to convert one or the other (ft to in or in^3/sec to ft^3/sec). Notice also that there is not enough information to determine what happens at other times (presumably, the flow rate will vary as the depth of the water changes), but that is not necessary for this solution.

Figure 4.1.9

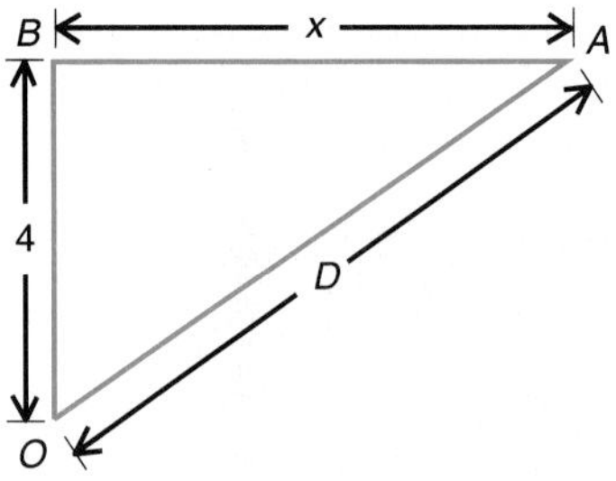

Figure 4.1.10

EXAMPLE 7 **Distance to Airplane** An airplane is flying horizontally in a straight line at 600 mph. One minute from now, it will be directly over an observer on the ground at an altitude of 4 miles (see Figure 4.1.9). How fast is the distance from the plane to the observer changing at the present moment?

Solution We begin by drawing a diagram (see Figure 4.1.10). The observer is on the ground at point O, the airplane is at point A, and the flight path of the airplane passes through point B, which is directly over O at an altitude of 4 mi.

The distance AB is the distance the airplane will travel in 1 min $= \frac{1}{60}$ hr, which is $\frac{1}{60}600 = 10$ miles. We let x be the (horizontal) distance from B to the plane, so $\frac{dx}{dt} = -600$ mph. We let D be the distance from the plane to the observer, so what we are asked to find is $\frac{dD}{dt}$.

Since triangle ABO has a right angle at B, the Pythagorean Theorem says that

$$D^2 = 4^2 + x^2,$$

and implicit differentiation gives

$$2D\frac{dD}{dt} = 2x\frac{dx}{dt},$$

so

$$\frac{dD}{dt} = \frac{x}{D}\frac{dx}{dt}.$$

At the specified moment, $x = 10$, so $D = \sqrt{4^2 + 10^2} = \sqrt{116} = 2\sqrt{29}$. Substituting in the formula for $\frac{dD}{dt}$, we find that

$$\frac{dD}{dt} = \frac{x}{D}\frac{dx}{dt} = \frac{10 \text{ mi}}{2\sqrt{29} \text{ mi}}(-600 \text{ mph}) = -\frac{3000}{\sqrt{29}} \text{ mph} \approx -557 \text{ mph}.$$

We conclude that the distance between the plane and the observer is decreasing at the rate of 557 mph.

Exercises 4.1

1. How fast would the volume of a spherical balloon have to be increasing in order for the radius to be increasing at a rate of $\frac{1}{4}$ in/sec at the moment when the radius is 6 in?

2. How fast is the surface area of a spherical balloon increasing if the radius is 4 in and the volume is increasing at the rate of 10 cu in/sec?

3. A 16-ft ladder is leaning against a vertical wall, and the bottom of the ladder begins to slide away from the wall. How fast is the top of the ladder moving when the foot is 10 ft from the wall, sliding at 6 in/sec?

4. A car approaches a man who is standing 100 ft in front of a brick wall so that the car's headlights shine on the wall, leaving a shadow of the man. If the head-

lights are 2 ft above ground level and the man is 6 ft tall, how fast is the top of the shadow moving when the car is 50 ft from the man and moving at 10 ft/sec?

5. In the situation of Example 6 (p. 233), how fast is the area of the top surface of the water decreasing at the specified moment?

6. Two cars pass a crossroads. One is traveling due north at 60 km/hr. The other passes the crossroads one-half hour after the first, traveling due west at 90 km/hr. How fast is the distance between them changing one hour after the second car passes the crossroads?

7. A weather balloon is rising vertically at the rate of 10 m/sec. A cyclist passes directly underneath it, traveling in a straight line at 3 m/sec, at a moment when the balloon is 40 m above the cyclist. If they both maintain the same speed and direction, how fast will they be moving apart 2 sec later?

***8.** Vandals turn the conical tank of Example 6 upside down and insert a hose to fill it up. If water is flowing in at the rate of 30 cu in/sec, how fast is the level rising when the water is 6 ft deep?

***9.** For the inverted tank of Exercise 8, how fast must the water be flowing in to make the area of the water surface decrease at the rate of 1 sq ft/min when the water is 6 ft deep?

***10.** A water trough is 8 ft long and has as its cross-section an equilateral triangle with a horizontal top. If water is flowing in at the rate of 2 cu ft/min, how fast is the water level rising when the water is 6 in deep?

***11.** A family of pigs is steadily drinking water from the water trough described in Exercise 10. When the water in the trough is 3 in deep, the water level is dropping at the rate of $\frac{1}{2}$ in/min. If the pigs continue to drink water at the same rate, how long will it be until the trough is empty? (*Hint:* The answer is not 6 min.)

***12.** A motorist who had been charged with speeding on the basis of evidence from a police radar trap claimed that she should be acquitted. She argued as follows. The police car with the radar unit was parked 50 ft off to the side of the road. The radar measured not the speed with which the car was traveling along the road, but the rate of change of the distance from the car to the radar. When the car was 200 ft up the road from the point on the road that is closest to the radar, the radar registered that this speed was 80 mph. Since the law says that it is illegal to travel at 80 mph along the road but says nothing about the rate of change of the distance to a point 50 ft to one side, the motorist argued that the charge ought to be dismissed. The judge had never studied calculus, and let the motorist off with a warning not to do it again. How fast was the motorist in fact traveling along the road? (Assume the road was perfectly straight, and be careful about units.)

***13.** A rope 15 m long passes through a pulley 8 m above the ground. On one end of the rope is a weight; the other end is held 1.5 m above the ground by a man who is walking away from the weight at a speed of 2 m/sec. How fast is the weight moving when the man is 7 m from the spot directly under the weight?

***14.** A light plane passes over point A at an altitude of 3 miles, traveling due west at 200 mph. One and a half minutes later, an observer 4 miles due south of point A catches sight of the plane. How fast is the distance from the observer to the plane increasing at that moment?

***15.** A woman is aiming a camera at a moving train, pivoting so as to follow its motion. The train is traveling due south on a straight track at 60 mph. The woman is standing 100 ft due west of the track, on the same level as the headlight on the engine; she holds the headlight at the center of her viewfinder. See Figure 4.1.11.

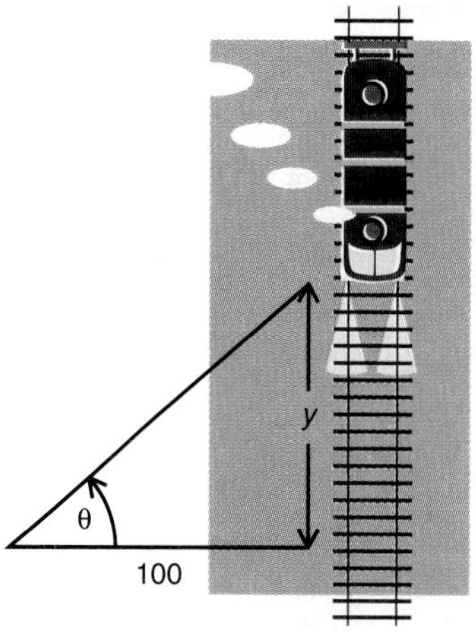

Figure 4.1.11

(i) At the moment when the headlight is due east of the woman, that is, when her line of vision points east, how fast is she turning? (Express your answer in radians per second and in revolutions per minute.)

(ii) At the moment when the headlight is exactly northeast of the woman, that is, when her line of vision points northeast, how fast is she turning? (Express your answer in radians per second and in revolutions per minute.)

(iii) At the moment specified in part (ii), how fast is the distance between the woman and the headlight decreasing?

(iv) Suppose a second train passes by, and at the moment when the woman's line of vision to its headlight points northeast she is revolving at 4 rpm (revolutions per minute) to keep aligned with the headlight. How fast is the train traveling?

4.2 Mean Value Theorem, Increasing and Decreasing Functions

LOCAL AND ABSOLUTE MAXIMUM/MINIMUM
ROLLE'S THEOREM
MEAN VALUE THEOREM
AVERAGE SLOPE
INCREASING/DECREASING FUNCTION
FIRST DERIVATIVE TEST

When we first explored the concept of the derivative, we thought of it as the slope of the tangent line to a curve. Later on, in Section 3.3 we realized that if the derivative was positive, the tangent sloped upward to the right, so it seemed reasonable to think that the function itself must "slope" upward also, at least near the point of tangency. We described this behavior by saying the function was increasing. Similarly, a negative derivative suggested that the curve was decreasing.

This idea was very useful. It allowed us to look for maximum and minimum values of a function by looking for points at which the derivative is zero. In Section 3.8 it gave us a criterion for a function to have an inverse (Theorem 3.8.6) because a function that is always increasing (or always decreasing) must be one-to-one.

However, although it seems very reasonable, we never really established why a function with a positive derivative must be increasing. This turns out to be quite subtle and difficult, and it is the purpose of this section to develop the techniques that will enable us to do it.

First we need some terminology.

DEFINITION 4.2.1

A function $f(x)$ has its **maximum value** at $x = a$ if $f(a) \geq f(x)$ for all x in the domain of f. In other words, f has its largest value at $x = a$. In this case we sometimes say that $x = a$ is a **maximum** for $f(x)$. The plural of maximum is *maxima.*

A function $f(x)$ has its **minimum value** at $x = a$ if $f(a) \leq f(x)$ for all x in the domain of f. In other words, f has its smallest value at $x = a$. In this case we sometimes say that $x = a$ is a **minimum** for $f(x)$. The plural of minimum is *minima.*

Saying that $x = a$ is a maximum or a minimum for $f(x)$ can seem a little confusing at first, but it is very convenient. A maximum is the point $x = a$ at which $f(x)$ has its largest value. That value, $f(a)$, is referred to as the maximum value. Similar remarks apply to minima and minimum values.

EXAMPLE 1

(i) $f(x) = x^2 + 3$ has its minimum value at $x = 0$ (the minimum value is $f(0) = 3$). It has no maximum value. (See Figure 4.2.1(i).)

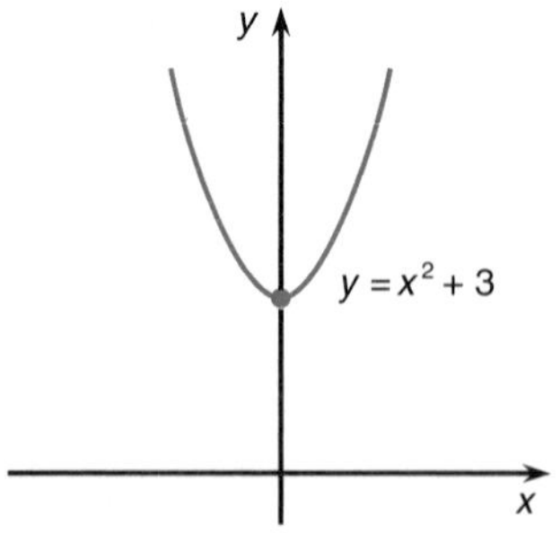

Figure 4.2.1(i)

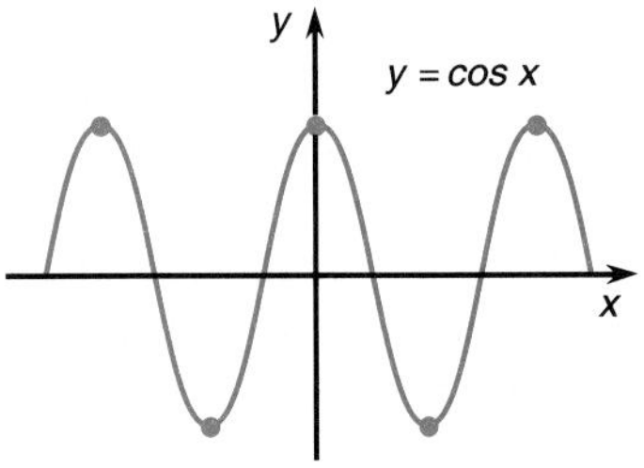

Figure 4.2.1(ii)

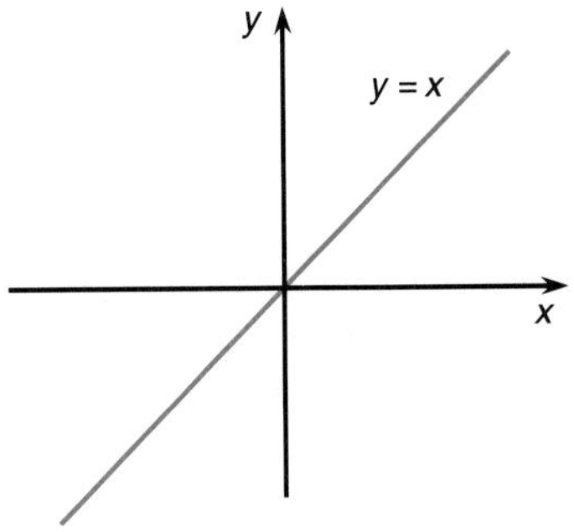

Figure 4.2.1(iii)

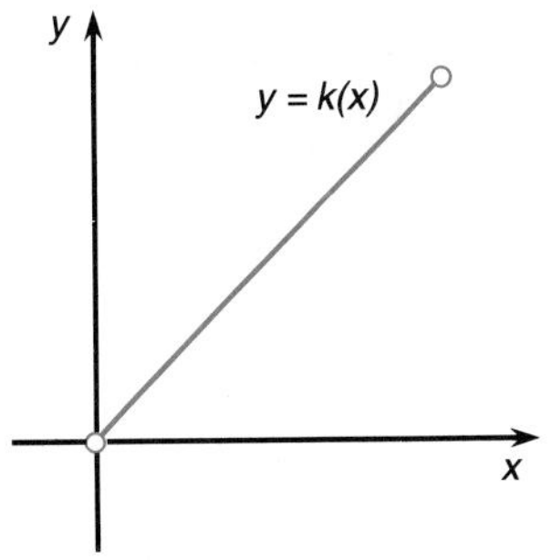

Figure 4.2.1(iv)

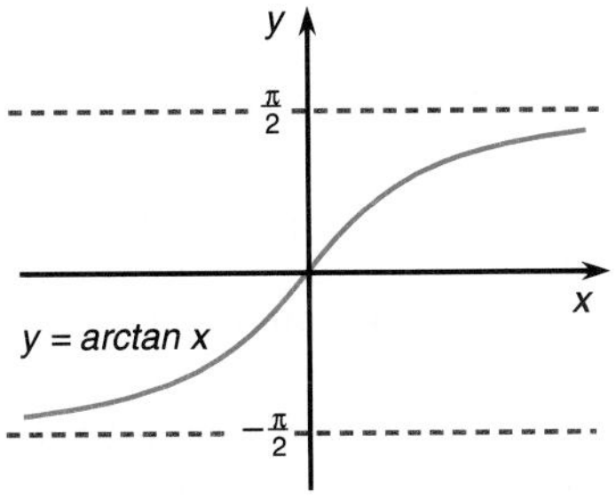

Figure 4.2.1(v)

(ii) The function $g(x) = \cos x$ has maxima at $x = 0, \pm 2\pi, \pm 4\pi, \ldots$. It has its minima at $x = \pm\pi, \pm 3\pi, \pm 5\pi, \ldots$. The maximum value is 1, and the minimum value is -1. (See Figure 4.2.1(ii).) Notice that the maximum and/or minimum value of a function can occur at more than one point.

(iii) The function $h(x) = x$ has neither a maximum nor a minimum value. It gets arbitrarily large and arbitrarily small. (See Figure 4.2.1(iii).)

(iv) The function $k(x) = x$, for $x \in (0, 1)$, is bounded: It never gets bigger than 1 nor smaller than 0. Still it has no maximum and no minimum. The maximum and minimum values should be "at" the missing endpoints 1 and 0. (See Figure 4.2.1(iv).)

(v) The function $p(x) = \arctan x$ is bounded but has no maximum or minimum. (See Figure 4.2.1(v).) The function never quite reaches what might appear to be its maximum and minimum values, namely, $\frac{\pi}{2}$ and $-\frac{\pi}{2}$, respectively.

EXAMPLE 2 Consider the function $f(x) = x^3 - 3x$, for $x \in [-2, 3]$. We encountered a similar function in Example 3.3.3(iii) (p. 158). The graph is sketched in Figure 4.2.2.

We can calculate $f'(x) = 3x^2 - 3 = 3(x^2 - 1)$, which is zero at $x = \pm 1$, negative when $x \in (-1, 1)$, and positive when $x < -1$ or $x > 1$. This leads us to suspect that $f(x)$ will have a minimum at $x = 1$ and a maximum at $x = -1$. The trouble is that $f(3) = 18$, which is greater than $f(-1) = 2$, so there is definitely not a maximum at $x = -1$. It happens that there is a minimum at $x = 1$, but there is also an endpoint minimum at $x = -2$.

On the other hand, from Figure 4.2.2 we see that the behavior of $f(x)$ for x *near* -1 is the sort of thing we expect near a maximum. The function does go up and then come down, so $f(-1)$ is bigger than $f(x)$ for every x near $x = -1$. It turns out that $x = -1$ is not a maximum because of something $f(x)$ does somewhere else, at $x = 3$, which is relatively far away.

Example 2 illustrates a common phenomenon. It frequently happens that a function $f(x)$ has what appears to be a maximum (or a minimum) at $x = a$, provided that we consider only x's that are near a. In this situation we say that $f(x)$ has a local maximum (or local minimum) at $x = a$.

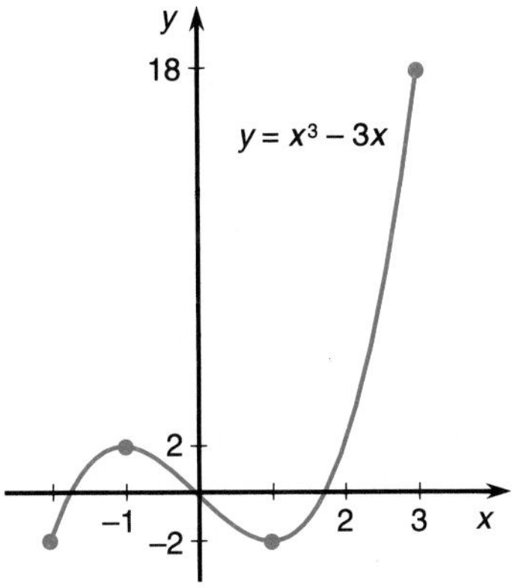

Figure 4.2.2

DEFINITION 4.2.2

A function $f(x)$ has a **local maximum** at the point $x = a$ if there is an interval (c, d) containing a so that $f(a) \geq f(x)$ for every $x \in (c, d)$ that is in the domain of f. We sometimes say that "$x = a$ is a local maximum for $f(x)$."

A function $f(x)$ has a **local minimum** at the point $x = a$ if there is an interval (c, d) containing a so that $f(a) \leq f(x)$ for every $x \in (c, d)$ that is in the domain of f. We sometimes say that "$x = a$ is a local minimum for $f(x)$."

If we want to emphasize that a maximum is not a local maximum, we sometimes refer to it as an **absolute maximum**. Similarly, the term **absolute minimum** means exactly the same thing as "minimum," but is used when it is important to distinguish it from a local minimum. Another term for local maximum or minimum is **relative maximum** or **relative minimum.**

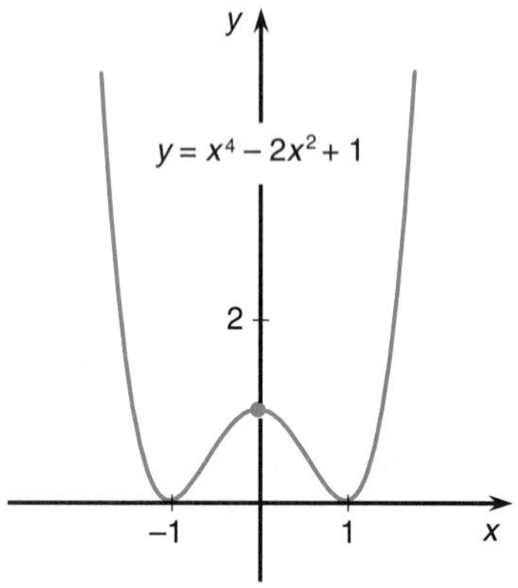

Figure 4.2.3

EXAMPLE 3

(i) If $x = a$ is a maximum or minimum for $f(x)$, then it is also a local maximum or a local minimum, respectively. The converse is not true, as Example 2 demonstrates. There, as we saw in Figure 4.2.2, the point $x = -1$ is a local maximum but not an absolute maximum.

(ii) For the function $f(x)$ of Example 2, $x = -2$ and $x = 1$ are both absolute minima (and hence also local minima). There is only one absolute maximum, namely, $x = 3$, but there is also a local maximum at $x = -1$.

(iii) Let $h(x) = (x^2 - 1)^2 = x^4 - 2x^2 + 1$. The graph of $y = h(x)$ is sketched in Figure 4.2.3. Since $h(1) = h(-1) = 0$ and $h(x)$, being a square, is always greater than or equal to zero, we see that $x = \pm 1$ are both absolute minima for h. From Figure 4.2.3 it appears that $x = 0$ is a local maximum. We can confirm this as follows. When $x \in (-1, 1)$, $0 \leq x^2 < 1$, so $-1 \leq x^2 - 1 < 0$, and $h(x) = (x^2 - 1)^2 \leq 1$. Since $h(0) = 1$, this shows that $h(0) \geq h(x)$ for all $x \in (-1, 1)$, so $x = 0$ is a local maximum for h. However, it is not an absolute maximum, since, for instance, $h(3) = 64 > h(0)$.

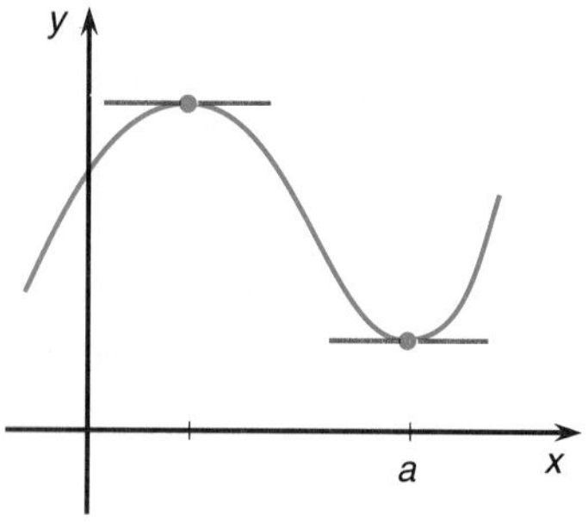

Figure 4.2.4

In Section 3.3 we suggested that if $f(x)$ has a maximum or minimum at $x = a$, then we should expect that the graph should have a sort of "hump" or "valley" at $x = a$ (see Figure 4.2.4). It seems reasonable to expect that the tangent line should be horizontal, that is, $f'(a) = 0$. This is true, provided that $f(x)$ is differentiable, and in fact it is true even at *local* maxima and minima.

THEOREM 4.2.3

Suppose the function $f(x)$ is differentiable at $x = a$ and has a local maximum or a local minimum at $x = a$. Then $f'(a) = 0$.

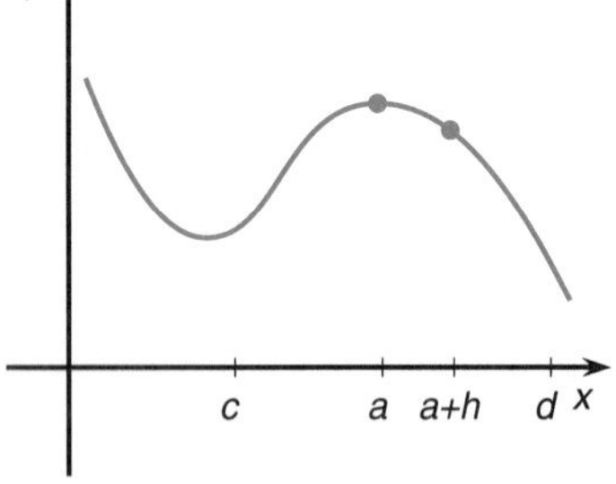

Figure 4.2.5

PROOF Remark that because $f(x)$ is differentiable at $x = a$, the domain of f contains an open interval around $x = a$.

First suppose $f(x)$ has a local maximum at $x = a$. So there is an interval (c, d) containing a so that $f(a) \geq f(x)$ for all $x \in (c, d)$. (By the above remark we can assume that (c, d) is contained in the domain of f.) This means that $f(a + h) \leq f(a)$ whenever h is small enough that $a + h \in (c, d)$. See Figure 4.2.5.

Consider the one-sided limit $\lim_{h \to 0^+} \frac{f(a+h) - f(a)}{h}$. Since $f(a + h) \leq f(a)$ if $a + h \in (c, d)$ and $h > 0$, the difference quotient satisfies $\frac{f(a+h) - f(a)}{h} \leq 0$ for all $h > 0$ small enough that $a + h \in (c, d)$.

The one-sided version of Theorem 2.3.3 implies that

$$\lim_{h\to 0^+} \frac{f(a+h)-f(a)}{h} \le 0.$$

Similarly, if $h < 0$ is small enough that $a + h \in (c, d)$, then $f(a+h) - f(a) \le 0$. Because $h < 0$, the difference quotient satisfies $\frac{f(a+h)-f(a)}{h} \ge 0$. The one-sided version of Theorem 2.3.3 says that

$$\lim_{h\to 0^-} \frac{f(a+h)-f(a)}{h} \ge 0.$$

Since $f'(a)$ equals both one-sided limits, we have $f'(a) \le 0$ and $f'(a) \ge 0$. We conclude $f'(a) = 0$.

But the differentiability of $f(x)$ at $x = a$ implies that these two one-sided limits are equal and both equal to $f'(a)$ (by Theorem 2.4.2). The only way this can happen is for them both to be zero, that is, $f'(a) = 0$, as desired.

The case in which $f(x)$ has a local minimum at $x = a$ is similar.

EXAMPLE 4

(i) The function $g(x) = x^2$ has a minimum at $x = 0$ and $g'(0) = 0$. See Figure 4.2.6(i).

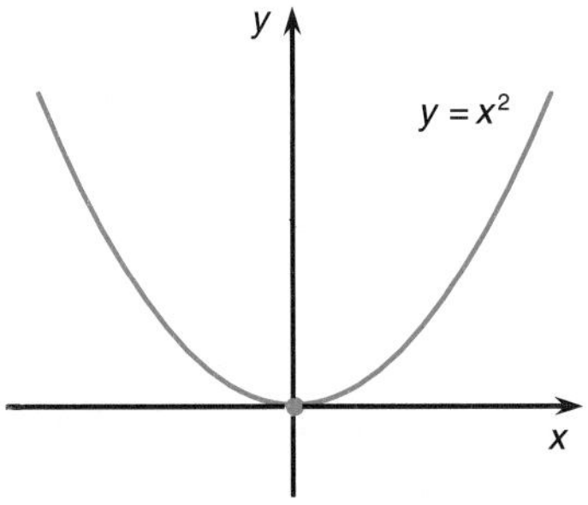

Figure 4.2.6(i)

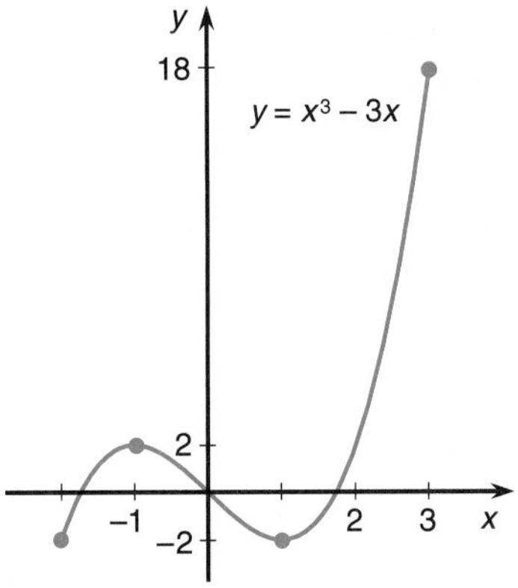

Figure 4.2.6(ii)

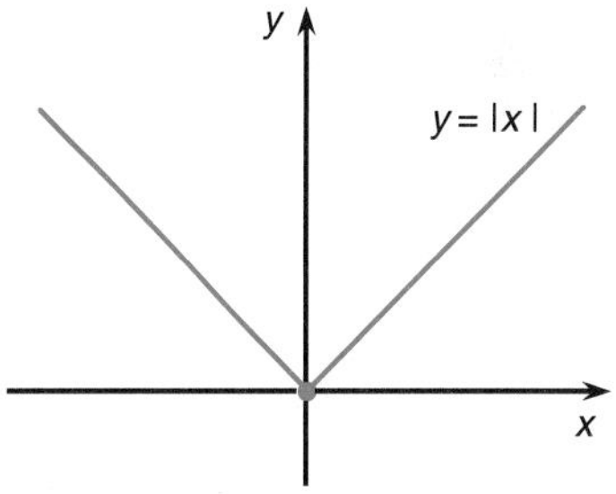

Figure 4.2.6(iii)

(ii) On $[-2, 3]$ the function $f(x) = x^3 - 3x$ of Example 2 has a local maximum at $x = -1$ and $f'(-1) = 0$. It has a minimum at $x = 1$ and $f'(1) = 0$. See Figure 4.2.6(ii).

(iii) On the other hand, notice that the same $f(x)$ has a minimum at $x = -2$ and a maximum at $x = 3$ but $f'(x)$ is not zero at either of these points. This does not contradict the theorem because $f(x)$ is not differentiable at these points. (It is not possible to take the limit *from both sides* at either of these points, since f is not defined on an open interval containing either of them.) This is typical of what happens at endpoint maxima and minima.

(iv) The function $A(t) = |t|$ has its minimum at $t = 0$, but $A(t)$ is not differentiable at $t = 0$, so Theorem 4.2.3 does not apply. See Figure 4.2.6(iii).

> The immediate use of Theorem 4.2.3 is for finding the local maxima and minima of a function $f(x)$. The theorem implies that all local maxima and minima must occur at points at which f' equals zero or f is not differentiable.

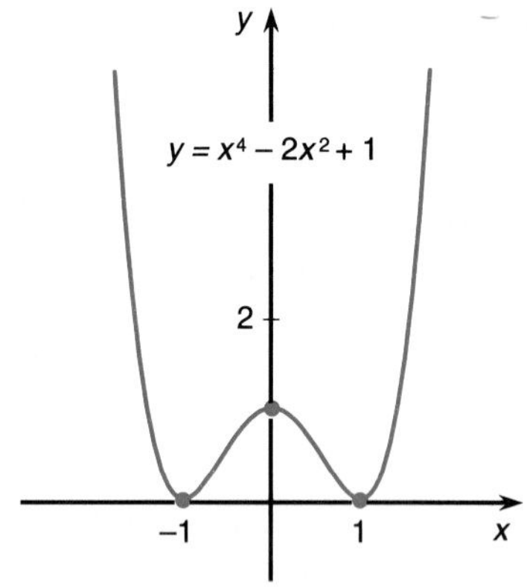

Figure 4.2.7(i)

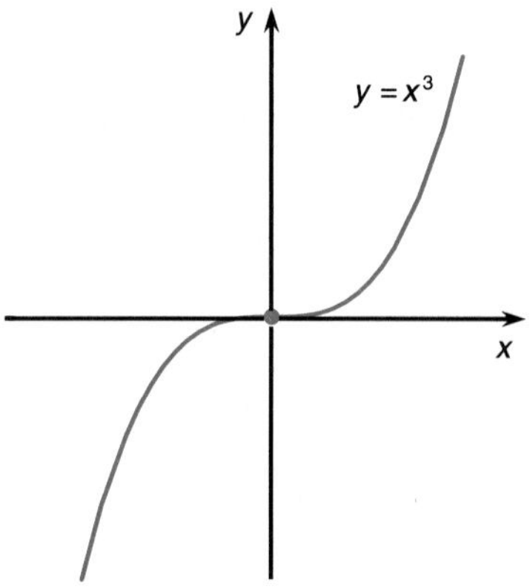

Figure 4.2.7(ii)

EXAMPLE 5

(i) Recall the function $h(x) = (x^2 - 1)^2 = x^4 - 2x^2 + 1$ of Example 3(iii). Suppose we want to locate all its local maxima and minima. Theorem 4.2.3 tells us to look where $h'(x) = 0$.

We calculate that $h'(x) = 4x^3 - 4x = 4x(x^2 - 1)$, which is zero at $x = 0$ and at $x = \pm 1$ and nowhere else. Since $h(x)$ is a polynomial, it is differentiable at every x. We know from the discussion in Example 3(iii) that $h(x)$ has absolute minima at $x = \pm 1$ and a local maximum at $x = 0$. See Figure 4.2.7(i).

Since the theorem tells us that all local maxima and minima occur at points where $h(x)$ is not differentiable or at points where $h'(x) = 0$, we know that there cannot be any other local (or absolute) maxima or minima; we have found them all.

(ii) Let $K(s) = s^3$. Then $K'(s) = 3s^2$, which is zero only at $s = 0$. Since $K(s) = s^3$ is positive when $s > 0$ and negative when $s < 0$, the point $s = 0$ cannot be a local maximum or a local minimum. Since $K(x)$ is differentiable everywhere, it cannot have any local maxima or minima. This is consistent with the illustration in Figure 4.2.7(ii).

Example 5 illustrates that the derivative may be zero without the function having a local maximum or local minimum. Still, finding where the derivative is not defined or equals zero does narrow down the search to a relatively small number of points. We are at least assured that there can be no other points that are local maxima or local minima.

This approach to finding local maxima and minima is very useful, and we intend to explore it at greater length in the next section.

Rolle's Theorem and the Mean Value Theorem

Consider a function $f(x)$ on a closed interval $[a, b]$ that takes the same value at both endpoints, that is, $f(a) = f(b)$. If the graph moves up, say, as x moves to the right of a, then it must come down again to get back to the original level at $x = b$. It seems reasonable to expect that there will be a point at which it stops increasing and starts to decrease, and at that point the tangent line should be horizontal. See Figure 4.2.8.

Similar remarks apply if the graph decreases to the right of a, and in both cases we expect that there should be some point at which the derivative is zero. Of course this will only make sense if we assume the function is continuous on $[a, b]$ and differentiable everywhere on (a, b). This important result is named after the French mathematician Michel Rolle (1652–1719).

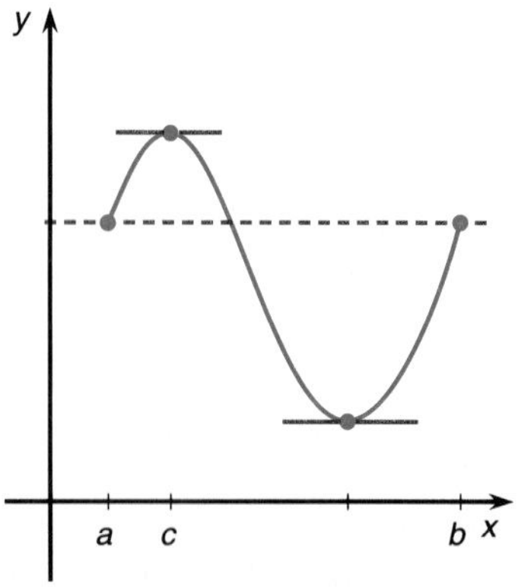

Figure 4.2.8

THEOREM 4.2.4

Rolle's Theorem

Suppose the function $f(x)$ is continuous on the closed interval $[a, b]$ and differentiable on the corresponding open interval (a, b) (i.e., $f'(x)$ exists for every x with $a < x < b$). Suppose $f(a) = f(b)$.

Then there is at least one $c \in (a, b)$ for which $f'(c) = 0$.

Theorem 4.2.4 is illustrated in Figure 4.2.8, which shows two possible points at which the derivative is zero $\big(f'(c) = 0\big)$.

PROOF If $f(x)$ is a constant function, then we know $f'(x) = 0$ for all x (by Example 3.2.1) and the conclusion holds. Now suppose $f(x)$ is not constant. It is a continuous function on the closed interval $[a, b]$, so by Theorem 2.5.7 it must achieve its maximum and minimum values.

Moreover, since $f(x)$ is not constant, the maximum and minimum cannot both equal $f(a)$, so at least one of them does not occur at $x = a$ or at $x = b$. This means that there is some $c \in (a, b)$ so that $f(x)$ has a maximum or a minimum at $x = c$. Theorem 4.2.3 implies $f'(c) = 0$. ◢

In fact we need another result that is closely related to Rolle's Theorem but slightly more complicated.

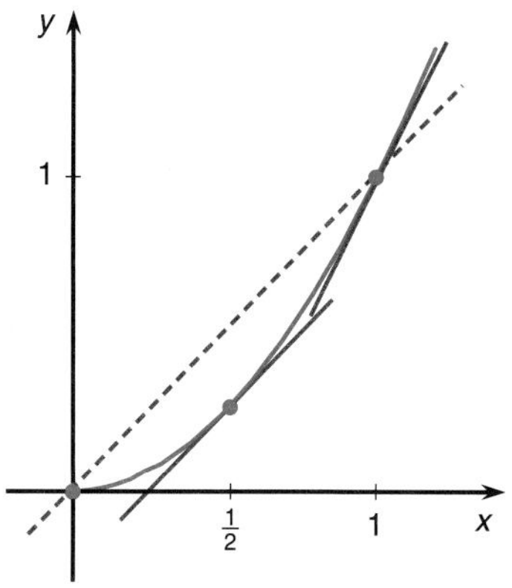

Figure 4.2.9

EXAMPLE 6 Consider the curve $y = x^2$ for x between 0 and 1, which is illustrated in Figure 4.2.9. The dashed line through the endpoints has slope equal to 1. Various tangent lines are illustrated too. The tangent at $x = 0$ has slope $m = 0$, and as x increases, the slope of the tangent increases until at $x = 1$ the tangent has slope $m = 2$. As the slope of the tangent increases from 0 to 2, it seems reasonable that at some point it will have to equal 1, the slope of the line through the endpoints.

We can think of the slope of the dashed line as the **average slope** between the two endpoints, and the above discussion leads us to expect that there should be some point at which the slope of the tangent actually equals the average slope. (In this particular case, $f(x) = x^2$, so $f'(x) = 2x$ and $f'\left(\frac{1}{2}\right) = 1$, that is, we see that the slope of the tangent at the point $x = \frac{1}{2}$ is equal to the average slope.) ◢

In this example, what made it work was that there were two points ($x = 0$ and $x = 1$) at which the slopes of the tangent lines were less than and greater than the average slope, respectively, so somewhere in between there should be a point where the tangent has the average slope. This sort of thing should always happen, because it seems unlikely that the slope of the tangent could, for instance, always be greater than the average slope. After all, that would seem to mean that the graph was always rising faster than average. However, this discussion does not constitute a proof.

THEOREM 4.2.5

The Mean Value Theorem

Suppose the function $f(x)$ is continuous on the closed interval $[a, b]$ and differentiable on the corresponding open interval (a, b). Then there is at least one $c \in (a, b)$ at which

$$f'(c) = \frac{f(b) - f(a)}{b - a}.$$

Geometrically, this means that the slope of the tangent at $x = c$ equals the slope of the line through the curve's endpoints, which may be regarded as the average or "mean" slope.

PROOF Let $F(x) = f(x) - \frac{x-a}{b-a}\big(f(b) - f(a)\big)$. Notice that

$$F(x) - f(x) = -\frac{x - a}{b - a}\big(f(b) - f(a)\big) = -\frac{f(b) - f(a)}{b - a}x + \frac{f(b) - f(a)}{b - a}a,$$

which is a polynomial in x, so it is continuous on $[a, b]$ and differentiable on (a, b). Then $F(x)$ is the sum of $f(x)$ and this polynomial, two functions that are both continuous on $[a, b]$ and differentiable on (a, b), so $F(x)$ itself is also continuous on $[a, b]$ and differentiable on (a, b). Moreover, $F(a) = f(a) - 0 = f(a)$, and $F(b) = f(b) - \frac{b-a}{b-a}\left(f(b) - f(a)\right) = f(b) - f(b) + f(a) = f(a) = F(a)$; we see that $F(x)$ has the same value at $x = a$ and $x = b$.

So Rolle's Theorem applies to $F(x)$ and says there is a $c \in (a, b)$ at which $F'(c) = 0$. But

$$F'(x) = f'(x) - \frac{1}{b-a}\left(f(b) - f(a)\right),$$

so $F'(c) = 0$ implies that

$$f'(c) - \frac{f(b) - f(a)}{b-a} = 0,$$

so

$$f'(c) = \frac{f(b) - f(a)}{b-a},$$

as desired.

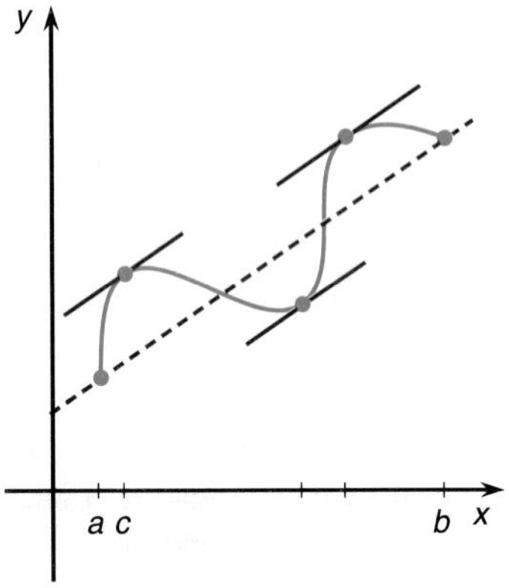

Figure 4.2.10

REMARK 4.2.6

Figure 4.2.10 also illustrates the Mean Value Theorem. Three points are marked at which the slope of the tangent equals the average slope, that is, the slope of the dashed line, which is $m = \frac{f(b)-f(a)}{b-a}$. It should be clear that this is just a "tilted" version of Rolle's Theorem, and indeed the proof just given amounted to "tilting" the graph of $y = f(x)$ until the dotted line was horizontal and then applying Rolle's Theorem.

EXAMPLE 7 Let $f(x) = 2x^3 - 6x^2 + 6x - 3$, for $0 \le x \le 2$. Find the value or values of x at which $f'(x)$ equals the average slope between $x = 0$ and $x = 2$.

Solution $f(0) = -3$, $f(2) = 1$, so the average slope, that is, the slope of the line through $(0, -3)$ and $(2, 1)$, is $\frac{1-(-3)}{2-0} = \frac{4}{2} = 2$. Also $f'(x) = 6x^2 - 12x + 6$, so we need to solve $6x^2 - 12x + 6 = 2$, or $6x^2 - 12x + 4 = 0$, or $3x^2 - 6x + 2 = 0$. By the Quadratic Formula (1.3.4, p. 11), the solutions are

$$x = \frac{6 \pm \sqrt{36-24}}{6} = \frac{6 \pm \sqrt{12}}{6} = 1 \pm \frac{\sqrt{3}}{3}.$$

Since $\sqrt{3} < 2$, these solutions both lie between 0 and 2; in fact they are approximately 0.4226 and 1.5774. The graph of f is illustrated in Figure 4.2.11.

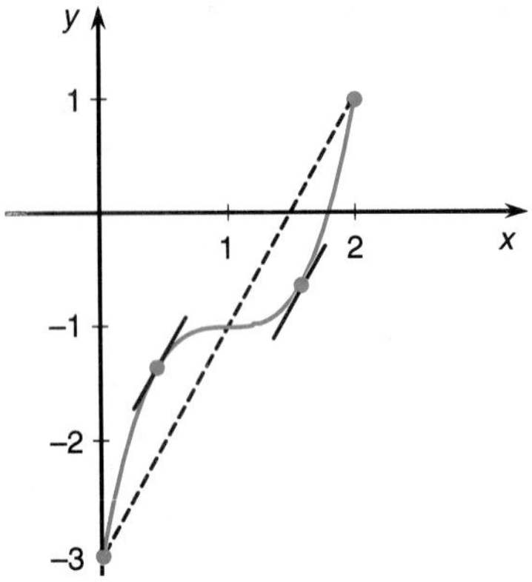

Figure 4.2.11

The main importance of the Mean Value Theorem is that it allows us to describe increasing and decreasing functions in terms of the signs of their derivatives.

DEFINITION 4.2.7

> A function $f(x)$ is said to be **increasing** (resp. **decreasing**) on an interval if whenever u and v are in the interval with $u < v$, we have $f(u) \le f(v)$ (resp. $f(u) \ge f(v)$).
>
> We say $f(x)$ is increasing (resp. decreasing) if it is increasing (resp. decreasing) on every interval in its domain.

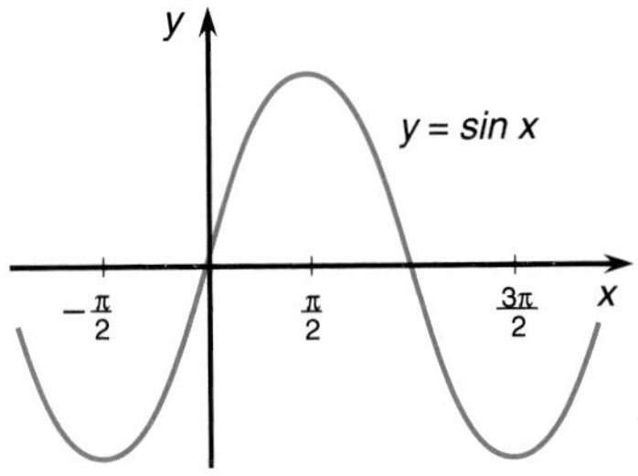

Figure 4.2.12(i)

EXAMPLE 8

(i) The function $f(x) = \sin x$ is increasing on the interval $\left[-\frac{\pi}{2}, \frac{\pi}{2}\right]$ and decreasing on $\left[\frac{\pi}{2}, \frac{3\pi}{2}\right]$. See Figure 4.2.12(i).

(ii) The function $g(x) = x^2$ is decreasing on $(-\infty, 0]$ and increasing on $[0, \infty)$. See Figure 4.2.12(ii).

(iii) The function $h(x) = x$ is increasing. See Figure 4.2.12(iii).

(iv) Any constant function is both increasing and decreasing (!). See Figure 4.2.12(iv).

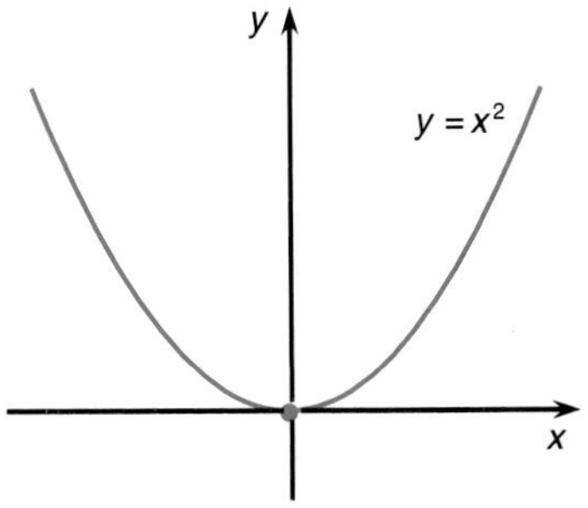

Figure 4.2.12(ii)

We are finally able to prove what we have long suspected: that a function with a positive derivative is increasing and one with a negative derivative is decreasing.

> **THEOREM 4.2.8**
>
> If $f(x)$ is differentiable on (a, b) and $f'(x) \ge 0$ for all $x \in (a, b)$, then $f(x)$ is increasing on (a, b). If $f'(x) \le 0$ for all $x \in (a, b)$, then $f(x)$ is decreasing on (a, b).

PROOF Suppose $f'(x) \ge 0$ for all $x \in (a, b)$ but $f(x)$ is not increasing. Then we can find $u, v \in (a, b)$ with $u < v$ and $f(u) > f(v)$. The Mean Value Theorem (Theorem 4.2.5) applied to $f(x)$ says that there is a c between u and v with $f'(c) = \frac{f(v)-f(u)}{v-u}$. Since the numerator is negative and the denominator positive, this implies $f'(c) < 0$, which is impossible because we assumed $f'(x) \ge 0$ for all $x \in (a, b)$.

The case with $f'(x) \le 0$ for all $x \in (a, b)$ is similar.

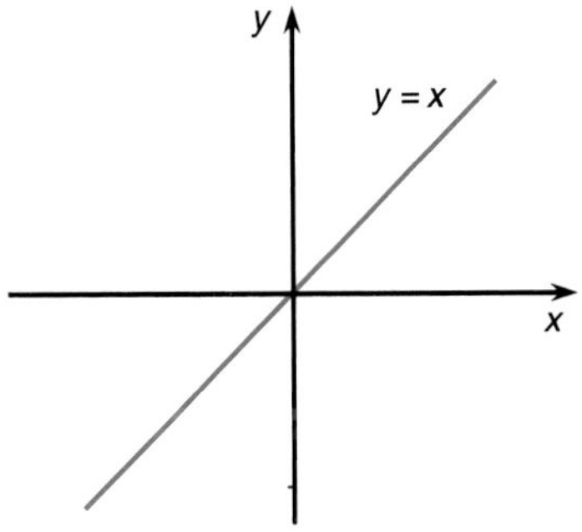

Figure 4.2.12(iii)

This result was stated for open intervals because it is easier to talk about functions being differentiable on open intervals, but a similar result holds for closed intervals.

> **COROLLARY 4.2.9**
>
> Suppose $f(x)$ is continuous on $[a, b]$ and differentiable on (a, b). If $f'(x) \ge 0$ (resp. $f'(x) \le 0$) for all $x \in (a, b)$, then $f(x)$ is increasing (resp. decreasing) on $[a, b]$.
>
> Similar statements apply if $f(x)$ is continuous on $(a, b]$ or on $[a, b)$ and differentiable on (a, b).

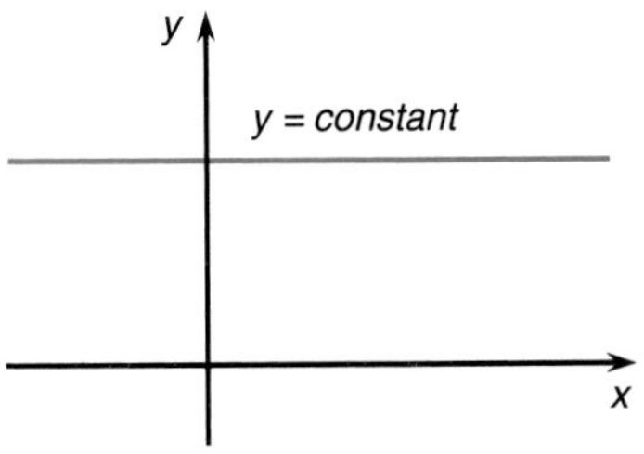

Figure 4.2.12(iv)

PROOF The only questions not covered by the theorem involve the endpoints. Suppose $f'(x) \ge 0$ on (a, b) but $f(b) < f(c)$ for some $c \in (a, b)$. Since $f(x)$ is increasing on (a, b), we know that $f(c) \le f(x)$ for all x with $c \le x < b$. By the one-sided version of Example 2.3.9(ii), $\lim_{x \to b^-} f(x) \ge f(c)$. Since the left side of this inequality equals $f(b)$, by the continuity of $f(x)$, we see that $f(b) \geq f(c)$. But we assumed $f(b) < f(c)$; this contradiction shows that $f(x)$ must in fact be increasing on $(a, b]$. A similar proof shows that $f(x)$ is increasing on $[a, b)$ and hence on $[a, b]$. The case in which $f'(x) \leq 0$ can be proved by applying this first part of the result to $-f(x)$.

Knowing when a function is increasing and decreasing allows us to check when it has a maximum or minimum. The following result is obvious but very useful.

THEOREM 4.2.10

The First Derivative Test

Suppose $a \in (c, d)$, and $f(x)$ is a function that is continuous on (c, d) and differentiable at every point in (c, d) except *possibly* $x = a$.

If $f'(x) > 0$ on (c, a) and $f'(x) < 0$ on (a, d), that is, $f(x)$ is increasing to the left of $x = a$ and decreasing to the right, then $f(x)$ has a local maximum at $x = a$.

If $f'(x) < 0$ on (c, a) and $f'(x) > 0$ on (a, d), that is, $f(x)$ is decreasing to the left of $x = a$ and increasing to the right, then $f(x)$ has a local minimum at $x = a$.

If $f'(x)$ has the same sign on (c, a) and (a, d) (either both positive or both negative), then there is no local extremum at $x = a$.

Suppose $f(x)$ is a function whose domain is an interval (possibly infinite) containing a. Suppose $f(x)$ is differentiable at every point in the interval except possibly $x = a$ and the endpoints. If $f'(x) \geq 0$ for every $x < a$ and $f'(x) \leq 0$ for every $x > a$, then there is an absolute maximum at $x = a$. If $f'(x) \leq 0$ for every $x < a$ and $f'(x) \geq 0$ for every $x > a$, then there is an absolute minimum at $x = a$.

EXAMPLE 9 We reconsider Example 8 with the help of Theorem 4.2.8 and Theorem 4.2.10.

(i) For $f(x) = \sin x$ we know that $f'(x) = \cos x$, so $f'(x) \geq 0$ for $x \in \left[-\frac{\pi}{2}, \frac{\pi}{2}\right]$ and $f'(x) \leq 0$ for $x \in \left[\frac{\pi}{2}, \frac{3\pi}{2}\right]$. By the First Derivative Test there is an (absolute) maximum at $x = \frac{\pi}{2}$.

(ii) For $g(x) = x^2$ we know $g'(x) = 2x$, so $g'(x) \leq 0$ for $x \in (-\infty, 0]$ and $g'(x) \geq 0$ for $x \in [0, \infty)$. By the First Derivative Test there is an (absolute) minimum at $x = 0$.

(iii) For $h(x) = x$ the derivative is $h'(x) = 1$, which is always positive. There are no local maxima or minima.

(iv) The derivative of a constant function is always zero.

The statements we made in Example 8 can now all be proved by applying Theorem 4.2.8 and Corollary 4.2.9. Strictly speaking, although they all seemed perfectly reasonable, some of these assertions would have been quite difficult to prove before. The functions are all graphed in Figure 4.2.13.

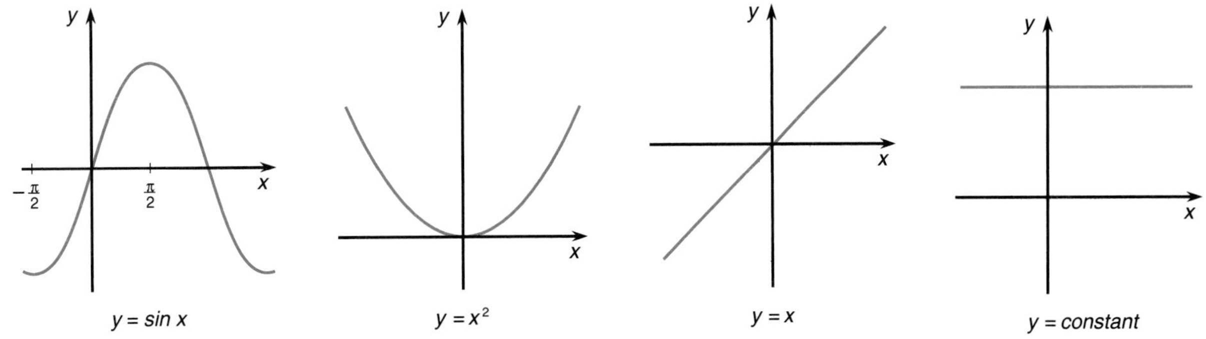

Figure 4.2.13

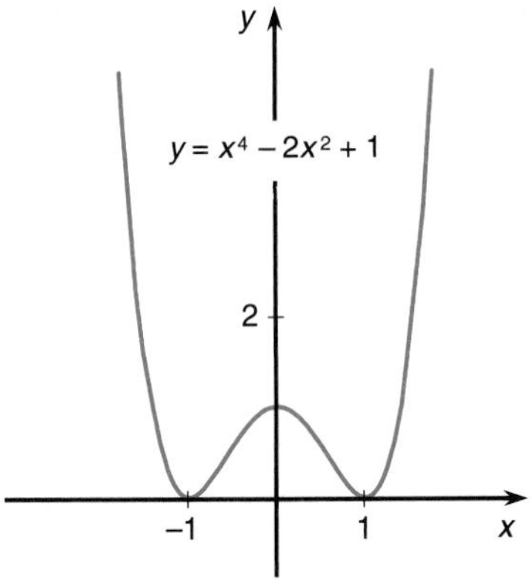

Figure 4.2.14

EXAMPLE 10 Describe the regions where the function $f(x) = x^4 - 2x^2 + 1$ is increasing or decreasing.

Solution $f'(x) = 4x^3 - 4x = 4x(x^2 - 1)$. The sign of $f'(x)$ depends on the sign of the two factors, $4x$ and $x^2 - 1$. The first, $4x$, is negative when x is negative and positive when x is positive. The second, $x^2 - 1$, is positive when $x < -1$ or $x > 1$ and negative when $-1 < x < 1$. Consequently, their product $f'(x)$ is negative on $(-\infty, -1)$ and $(0, 1)$ and positive on $(-1, 0)$ and $(1, \infty)$. The function $f(x)$ is decreasing on $(-\infty, -1]$ and $[0, 1]$ and increasing on $[-1, 0]$ and $[1, \infty)$.

Notice that we can include the endpoints in the intervals on which $f(x)$ is increasing or decreasing. See Figure 4.2.14.

We have observed that the derivative of a constant function is zero. We are about to show that the converse is true too: that a function whose derivative is zero must be constant. This seems plausible, but we need the Mean Value Theorem to prove it. It is a fact that will be very useful.

THEOREM 4.2.11

(i) If $f(x)$ is differentiable on (a, b) and $f'(x) = 0$ for all $x \in (a, b)$, then $f(x)$ is a constant function.

(ii) If $f(x)$ and $g(x)$ are both differentiable on (a, b) and $f'(x) = g'(x)$ for all $x \in (a, b)$, then $f(x) - g(x)$ is a constant, that is, $f(x) = g(x) + C$, for some constant C.

PROOF Part (ii) follows by applying part (i) to $f(x) - g(x)$. To prove part (i), suppose that $f(x)$ is not constant. Then there are $u, v \in (a, b)$ with $u < v$ and $f(u) \neq f(v)$. The Mean Value Theorem says that there is a c between u and v with $f'(c) = \frac{f(v)-f(u)}{v-u} \neq 0$, contradicting the hypothesis. ◢

EXAMPLE 11 If $f(x) = \sin^2 x$ and $g(x) = -\cos^2 x$, then by the Product Rule,

$$f'(x) = \sin x \cos x + \cos x \sin x = 2 \sin x \cos x,$$

and

$$g'(x) = -\cos x(-\sin x) - (-\sin x)\cos x = 2\sin x \cos x$$
$$= f'(x).$$

So $f(x) - g(x)$ is a constant; in fact we already know this, because $f(x) - g(x) = \sin^2 x - (-\cos^2 x) = \sin^2 x + \cos^2 x = 1$.

EXAMPLE 12 Suppose $f(x)$ is a differentiable function with $f'(x) = 6x^2 - 4x + 3$ and $f(0) = 7$. What is $f(x)$?

Solution It is not difficult to think of a function whose derivative is $6x^2 - 4x + 3$. After all, we can work with each term separately, and a function whose derivative is 3 is $3x$, a function whose derivative is $-4x$ is $-2x^2$, and a moment's thought shows that a function whose derivative is $6x^2$ is $2x^3$. Adding these together, we find that $g(x) = 2x^3 - 2x^2 + 3x$ has the desired derivative, that is, $g'(x) = 6x^2 - 4x + 3$. By part (ii) of Theorem 4.2.11, $f(x)$ must be $g(x) + C$, for some constant C.

To find C, we notice that $f(0) = 7$, so $7 = g(0) + C = 0 - 0 + 0 + C = C$. The answer is that $f(x) = 2x^3 - 3x^2 + 3x + 7$.

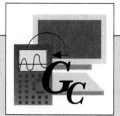

With a graphics calculator we can plot both the function $f(x)$ and the derivative $f'(x)$ from Example 12.

TI-85 GRAPH F1 ($y(x)$ =) 2 F1 (x)^3 − 2 F1 (x)^2 + 3 F1 (x) + 7 ▼ 6 F1 (x) x^2 − 4 F1 (x) + 3 M2(RANGE) (−)2 ▼ 2 ▼ ▼ (−) 10 ▼ 20 F5(GRAPH)	SHARP EL9300 2 x/θ/T a^b 3 ▶ − 2 x/θ/T x^2 + 3 x/θ/T + 7 2ndF ▼ 6 x/θ/T x^2 − 4 x/θ/T + 3 RANGE −2 ENTER 2 ENTER 1 ENTER −10 ENTER 20 ENTER 5 ENTER
CASIO f_x-7700GB/f_x-6300G Cls EXE Range −2 EXE 2 EXE 1 EXE −10 EXE 20 EXE 1 EXE EXE EXE EXE Graph 2 X x^y 3 − 2 X x^2 + 3 X + 7 EXE Graph 6 X x^2 − 4 X + 3 EXE	HP 48SX ↰ PLOT PLOTR ERASE ATTN ′ αY = 2 × αX y^x 3 − 2 × αX y^x 2 + 3 × αX + 7 ↱ PLOT ↰ DRAW ERASE 2 +/− ENTER 2 XRNG 10 +/− ENTER 20 YRNG DRAW ATTN ′ αY = 6 × αX y^x 2 − 4 × αX + 3 ↱ PLOT ↰ DRAW DRAW

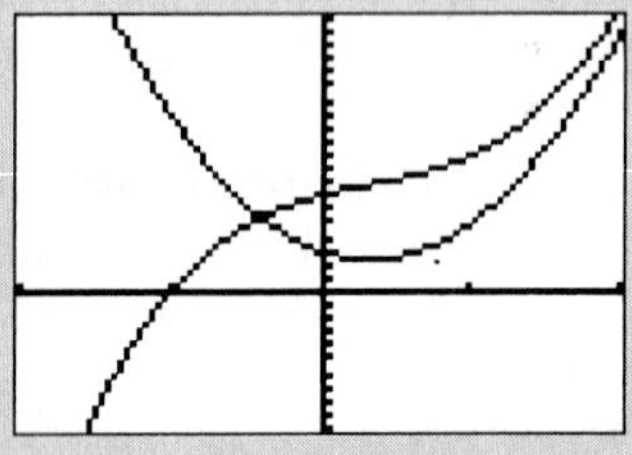

Notice that $f(0) = 7$, and consider how $f'(x)$ gives the rate of increase of $f(x)$ (i.e., the slope of the tangent).

EXAMPLE 13 (i) Find all functions whose derivative equals $x^4 - 2x + 1$.

Solution As above, one function with the specified derivative is easy to find. We can let $g(x) = \frac{1}{5}x^5 - x^2 + x$ and observe that $g'(x) = \frac{1}{5}5x^4 - 2x + 1 = x^4 - 2x + 1$.

By part (ii) of Theorem 4.2.11, any function with the same derivative must be of the form $g(x) + C$, so it must be $\frac{1}{5}x^5 - x^2 + x + C$, for some constant C.

(ii) Find all functions whose derivative equals $2x + \sin x$.

Solution One function with the specified derivative is $x^2 - \cos x$ (we simply look for something whose derivative is $2x$, and x^2 comes to mind, and for something whose derivative is $\sin x$, and $-\cos x$ comes to mind).

Again by part (ii) of Theorem 4.2.11, any function with the same derivative must be this one plus a constant. The answer is that all functions whose derivative is $2x + \sin x$ are of the form $x^2 - \cos x + C$.

In doing these problems the first step was to guess something with the specified derivative. To do this, we could work with each term of a sum separately, since derivatives work that way. Having found one function with the desired derivative, we applied the theorem to see that every such function is obtained by adding a constant to the one we know.

Given a function $h(x)$, the process of finding a function whose derivative is $h(x)$ is called finding an *antiderivative* of $h(x)$. It is an extremely important question, which we shall consider more fully in Section 4.8 and in Chapters 5 and 8.

Exercises 4.2

I

Find where each of the following functions has local and/or absolute maxima and/or minima.

1. $f(x) = \sin x$
2. $f(x) = x^6$
3. $f(x) = 2 - \cos x$
4. $f(x) = 1 - x^4$
5. $f(x) = 2 - |x|$
6. $f(x) = |2 - |x||$
7. $f(x) = \sqrt{|x|}$
8. $f(x) = x^3 - 3x$
9. $f(x) = \cosh x$
10. $f(x) = \sec x$

In each of the following situations, find a point c so that $f'(c)$ equals the average slope of $f(x)$ on the specified interval.

11. $f(x) = x^2$, $[-1, 0]$
12. $f(x) = x^2$, $[1, 2]$
13. $f(x) = x^3$, $[0, 1]$
14. $f(x) = x^3 + 1$, $[-1, 1]$
15. $f(x) = x^6$, $[-3, 3]$
16. $f(x) = x^6 - 2x^4 + x^2 - 3$, $[-4, 4]$
17. $f(x) = \sin x$, $[0, 2\pi]$
18. $f(x) = \cos x$, $[-\frac{\pi}{7}, \frac{\pi}{7}]$
19. $f(x) = x^3 - 2x^2 + x + 1$, $[0, 1]$
20. $f(x) = x^4 - x^3$, $[0, 1]$

Find the regions where each of the following functions is increasing or decreasing.

21. $f(x) = 2x + 5$
22. $g(x) = x^2 + 3$
23. $h(x) = x^3 + 6x - 2$
24. $k(t) = t^3 - 2t^2 + t + 3$
25. $F(x) = x^2 - 4x + 2$
26. $G(u) = \frac{|u|}{u}$
27. $H(\theta) = \tan \theta$
28. $K(t) = \arctan t$
29. $R(x) = e^x$
30. $T(x) = x^4 - 4x^3 + 4x^2 + 3$

II

31. Let $f(x) = x^2$, $b > a$, and find $c \in [a, b]$ so that $f'(c)$ equals the average slope of $f(x)$ on the interval $[a, b]$.

32. Let $f(x) = x^2 + 4x + 3$, and find c so that $f'(c)$ equals the average slope of $f(x)$ on the interval $[a, b]$.

33. Let $g(x) = x^8 - 3x^6 + 4x^4 - x^2 - 4$. For any $a > 0$, consider the interval $[-a, a]$ and find a number $c \in [-a, a]$ so that $g'(c) = \frac{g(a) - g(-a)}{2a}$.

***34.** Let $P(t)$ be any polynomial and let $f(x) = P(x^2)$. For any $a > 0$, find $c \in [-a, a]$ so that $f'(c) = \frac{f(a) - f(-a)}{2a}$ (cf. Exercise 33).

35. Find a function whose derivative is (i) $4x^3 - 4x - 3$, (ii) $x^4 - 3x^3 + 2x - 1$.

36. Find a function whose derivative is (i) $x^3 + \cos x$, (ii) e^{2x}.

37. In each part, find a function that satisfies the given conditions: (i) $f'(x) = 8x^3 - 6x^2 + 4x - 3$ and $f(1) = -1$, (ii) $f'(x) = \cos x$ and $f(0) = 5$.

38. In each part, find a function that satisfies the given conditions: (i) $f'(x) = 0$ and $f(3) = 4$, (ii) $f'(x) = \frac{1}{1+x^2}$ and $f(0) = 1$.

***39.** For $x \geq 0$, let $h(x) = \sqrt{x}$. Let $a > 0$ and find a point between 0 and a at which the slope of the tangent equals the average slope of the graph of $y = h(x)$ between $x = 0$ and $x = a$.

***40.** Find an example in which the conclusion of Rolle's Theorem (Theorem 4.2.4, p. 240) fails for a function that is not differentiable at exactly one point.

***41.** (i) Suppose $f(x)$ has a local maximum at $x = a$. Show that $g(x) = 3 - f(x)$ has a local minimum at $x = a$.

(ii) Suppose $f(x)$ has a local minimum at $x = a$. What can you say about $g(x) = -f(x)$?

(iii) What happens in parts (i) and (ii) if we replace the word "local" with the word "absolute"?

***42.** (i) Suppose $f'(x) = 2$, for every x. Show that the graph of $y = f(x)$ is a straight line.

(ii) What can you say if $f'(x) = C$, for some constant C?

(iii) What can you say if $f'(x) = 2$ for every $x \neq 3$ but $f(x)$ is not differentiable at $x = 3$?

***43.** If $f(x)$ is differentiable at every x and c is some fixed point, decide whether it is always possible to find $a \neq b$ so that

$$f'(c) = \frac{f(b) - f(a)}{b - a}.$$

***44.** Suppose we know the following facts about the exponential function: $\frac{d}{dx}(e^x) = e^x$, $e^{a+b} = e^a e^b$, $e = e^1 > 2$, and $e^x > 0$ for all x. Using these facts, prove that $\lim_{x \to \infty} e^x = \infty$ and that $\lim_{x \to -\infty} e^x = 0$. (*Hint:* For $n > 0$, $e^n = e^1 \times e^1 \times \ldots \times e^1 > 2^n$.)

***45.** Show that the graph of $y = x^2$ lies *above* any of its tangent lines.

***46.** Show that the graph of $y = \ln x$ lies *below* any of its tangent lines.

***47.** (i) Show that if $a > 0$, then the tangent line to the graph of $y = x^3$ at the point (a, a^3) passes below the origin (i.e., its y-intercept is negative).

(ii) What happens if $a < 0$?

***48.** Suppose $r < s$ and $f(x)$ is a twice-differentiable function on (r, s), with $f''(x) > 0$ for all $x \in (r, s)$. Suppose $a \in (r, s)$ and show that on the interval (r, s) the graph of $y = f(x)$ lies above the tangent line to $y = f(x)$ at $x = a$.

III

Use a graphics calculator or computer to sketch each of the following functions on the specified interval. Locate the local maxima and minima, and use greater magnification to identify them correct to three decimal places.

49. $x^4 - 3x^3 + 2x$, $[-2, 3]$

50. $\sin(x^2)$, $[-3, 3]$

51. $x^3 \sin x$, $[0, 2\pi]$

52. $2 \sin x - x$, $[0, \pi]$

***53.** Let $f(x) = x^4 - 3x^3 + x^2 + x$, for $x \in [0, 2]$. Estimate (to four decimal places) the location of two points $c \in [0, 2]$ for which $f'(c)$ equals the average slope.

4.3 Critical Points and Extrema

STATIONARY POINT
LOCAL/ABSOLUTE EXTREMUM

This brief section introduces some terminology and presents some examples.

From Theorem 4.2.3 (p. 238) we know that if a function $f(x)$ is differentiable at $x = a$ and $f(x)$ has a local or absolute maximum or minimum at $x = a$, then $f'(a) = 0$. From this we realized that to find the local maxima and minima, we should first locate all points at which the derivative does not exist or at which it does exist and equals zero.

Such points play such an important role that there is a special terminology for them.

DEFINITION 4.3.1

> Suppose $f(x)$ is a function. A point a in the domain of $f(x)$ is called a **critical point** of $f(x)$ if $f'(a) = 0$ *or* if $f(x)$ is not differentiable at $x = a$. Sometimes the critical points at which the derivative does exist (and so equals zero) are called **stationary points**.

Very often we have to find maximum and/or minimum values of a function, but frequently we do not know what to expect. It is irritating to keep having to refer to "maxima or minima," and it is much more convenient to have a single term to cover both possibilities.

DEFINITION 4.3.2

> A point $x = a$ is called an **extremum** for $f(x)$ if it is either a maximum or a minimum for $f(x)$. It is a **local extremum** if it is a local maximum or a local minimum, and one can speak of an **absolute extremum** to distinguish it from a local extremum. The plural of extremum is extrema, and an alternate term for extremum is *extreme point.* The value $f(a)$ of the function $f(x)$ at the extremum $x = a$ is an *extreme value.*

The process of finding the extrema of a function $f(x)$ has two steps. The first is to locate the critical points of $f(x)$. The next is to try to decide which of them are local or absolute maxima or minima and which of them are not extrema at all. Sometimes it is possible to do this by relatively simple methods, requiring only a certain amount of ingenuity, but there are occasions when it can be extremely difficult.

EXAMPLE 1 Find all critical points of $f(x) = \frac{1}{1+x^2}$. Decide which of them are maxima or minima.

Solution By the Quotient Rule, $f'(x) = \frac{-2x}{(1+x^2)^2}$. For this to equal zero, the numerator must equal zero, so $x = 0$. Since the derivative exists at every x, this is the only critical point.

Since $1 + x^2 \geq 1$, we see that for any x, $f(x) = \frac{1}{1+x^2} \leq \frac{1}{1} = 1 = f(0)$, so $x = 0$ is an (absolute) maximum for $f(x)$.

Taking reciprocals reverses inequalities.

EXAMPLE 2 Find all critical points of $f(x) = \frac{x}{1+x^2}$, and find which are maxima or minima.

Solution The derivative is

$$f'(x) = \frac{1(1+x^2) - x(2x)}{(1+x^2)^2} = \frac{1-x^2}{(1+x^2)^2},$$

so $f'(x) = 0$ precisely when $1 - x^2 = 0$, that is, when $x = \pm 1$. Since the derivative exists at every x, these are the only critical points.

Notice that $f(-1) = -\frac{1}{2}$, $f(1) = \frac{1}{2}$, and $\lim_{x\to\infty} f(x) = \lim_{x\to-\infty} f(x) = 0$. The derivative $f'(x)$ is negative when $x < -1$ or $x > 1$ and positive when $-1 < x < 1$.

Accordingly, $f(x)$ is decreasing on $(-\infty, -1]$ and on $[1, \infty)$ and increasing on $[-1, 1]$. Since $f(x)$ decreases on $[1, \infty)$ but is never negative there, $f(x)$ must have an absolute minimum at $x = -1$ and similarly an absolute maximum at $x = 1$. The graph of $y = f(x)$ is illustrated in Figure 4.3.1.

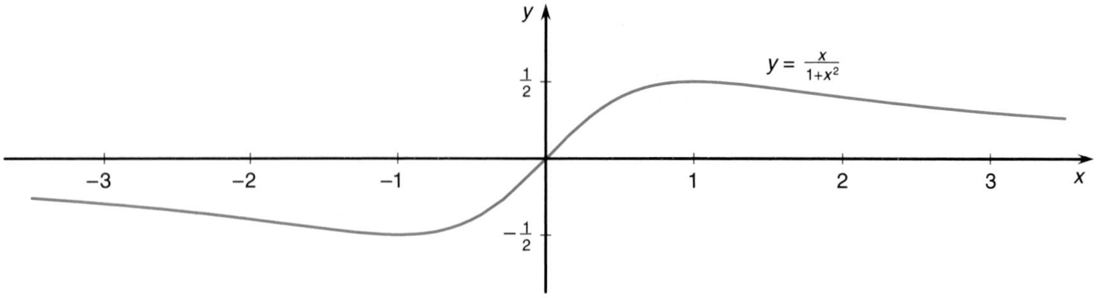

Figure 4.3.1

EXAMPLE 3 Find all local extrema of $g(t) = |t|$.

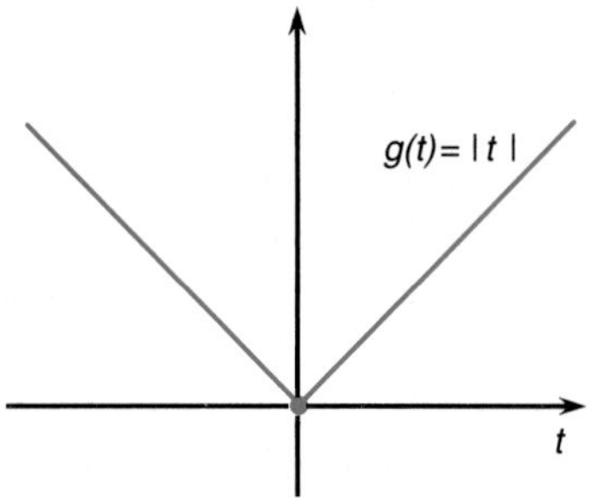

Figure 4.3.2

Solution When $t > 0$, $g(t) = t$, so $g'(t) = 1$. (Here we notice that the limit that defines $g'(t)$ only depends on the behavior of the function near t, by Proposition 2.2.8; the derivative of $g(t)$ will be the same as the derivative of t for $t > 0$.) Similarly, for $t < 0$, $g(t) = -t$ and $g'(t) = -1$. As we have already remarked, the derivative does not exist at $t = 0$.

So the only critical point is $t = 0$. Of course, it is an absolute minimum for $g(t)$, since for all t, $g(t) = |t| \geq 0 = g(0)$. See Figure 4.3.2.

EXAMPLE 4 Find and classify the critical points of $f(x) = (x-1)e^x$.

See the Graphics Calculator Example on the facing page for a plot of this example.

Solution We find that $f'(x) = e^x + (x-1)e^x = (1 + x - 1)e^x = xe^x$. Since e^x is never zero, this derivative is zero precisely when $x = 0$. Since the derivative exists for all values of x, the only critical point is $x = 0$.

When the question asks us to *classify* the critical points, it means that we should decide which are local or absolute maxima or minima. In this case we can easily determine the sign of $f'(x)$. After all, e^x is always positive, so the sign of xe^x is the same as the sign of x.

In particular, when $x < 0$, the derivative is negative and $f(x)$ is decreasing; when $x > 0$, the derivative is positive and $f(x)$ is increasing. By the First Derivativé Test the critical point $x = 0$ is a minimum.

EXAMPLE 5 Find and classify the critical points of $h(x) = x^5 + 5x^2$.

Solution We differentiate: $h'(x) = 5x^4 + 10x = 5x(x^3 + 2)$. So $h'(x)$ exists for every x and equals zero when $x = 0$ or when $x^3 + 2 = 0$, that is, when $x^3 = -2$, which means $x = -2^{1/3}$. These two points are the only critical points.

We see from the formula for $h'(x)$ that when $x > 0$, we have $h'(x) > 0$. Also notice that the factor $x^3 + 2$ is positive when $x > -2^{1/3}$ and negative when $x < -2^{1/3}$. So when $x < -2^{1/3}$, both factors, $5x$ and $x^3 + 2$, are negative and $h'(x) > 0$. When $-2^{1/3} < x < 0$, then the factor $5x$ is negative and the factor $x^3 + 2$ is positive. Their product, $h'(x)$, must be negative.

With a graphics calculator we can plot both the function $f(x)$ and the derivative $f'(x)$ from Example 4.

TI-85	SHARP EL9300
GRAPH F1 $(y(x) =)$ (F1 (x) − 1) e^x F1 (x) ▼ F1 (x) e^x F1 (x) M2(RANGE) (−) 2 ▼ 2 ▼ ▼ (−) 1 ▼ 10 F5(GRAPH) CLEAR	(x/θ/T − 1) e^x x/θ/T 2ndF ▼ x/θ/T e^x x/θ/T RANGE −2 ENTER 2 ENTER 1 ENTER −1 ENTER 10 ENTER 1 ENTER
CASIO f_x-7700GB/f_x-6300G	HP 48SX
Cls EXE Range −2 EXE 2 EXE 1 EXE −1 EXE 10 EXE 1 EXE EXE EXE EXE Graph (X − 1) e^x X EXE Graph X e^x X EXE	↰ PLOT PLOTR ERASE ATTN ' αY = (αX − 1 ▶ × e^x αX ↱ PLOT ↰ DRAW ERASE 2 +/− ENTER 2 XRNG 3 +/− ENTER 10 YRNG DRAW ATTN ' αY = αX × e^x αX ↱ PLOT ↰ DRAW DRAW

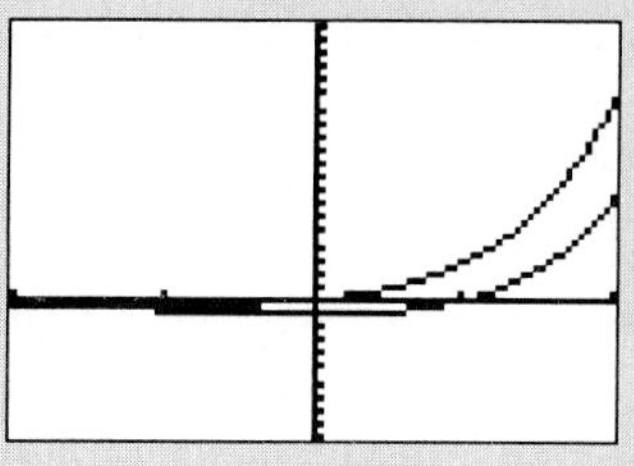

We see the critical point and note that the derivative is zero there. Observe that the calculator picture alone is not really enough to tell us what happens as $x \to \pm\infty$.

$-2^{1/3}$ 0

+ + + − − − − − − + + +

Figure 4.3.3

We sketch the real line in Figure 4.3.3, marking the critical points. We also indicate the sign of the derivative $h'(x)$ (where the derivative is positive, we place a +, and where it is negative, we place a −). Notice that the derivative $h'(x)$ can only change sign at a critical point, in other words that it has the same sign on any interval in the domain that does not contain any critical points.

Using Figure 4.3.3, we find that the function $h(x)$ is increasing for $x < -2^{1/3}$, decreasing between $-2^{1/3}$ and 0, and increasing for $x > 0$. This means that $x = -2^{1/3}$ is a local maximum and $x = 0$ is a local minimum.

Finally, note that $\lim_{x\to\infty} h(x) = \infty$ and $\lim_{x\to-\infty} h(x) = -\infty$, which implies that $h(x)$ has neither an absolute maximum nor an absolute minimum, since it gets arbitrarily large and also arbitrarily large in the negative direction. So neither critical point is an absolute extremum.

EXAMPLE 6 Find and classify the critical points of $R(t) = t^3e^t$.

Solution The derivative is $R'(t) = 3t^2e^t + t^3e^t = (3t^2 + t^3)e^t = t^2(3 + t)e^t$. It is zero at $t = 0$ and $t = -3$. These are the only critical points.

−3   0

− − − + + + + + + + + +

Figure 4.3.4

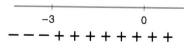

As above, we sketch the real line in Figure 4.3.4, marking the critical points, and we indicate the sign of the derivative on each part of the domain, that is, on each interval between critical points (or between a critical point and the end of the line). Notice that when $t > 0$, each factor in the derivative is positive, so $R'(t) > 0$. When $t < -3$, we have $t^2 > 0$, $3 + t < 0$ and $e^t > 0$, so their product, $R'(t)$, is negative. When $-3 < t < 0$, the factors are all positive, so $R'(t) > 0$.

This means that the function is decreasing to the left of $t = -3$ and increasing to the right of $t = -3$, which implies that $t = -3$ is a minimum for $R(t)$. Notice that the critical point $t = 0$ is neither a local maximum nor a local minimum. Also notice that the derivative does not change sign at $t = 0$, even though it is a critical point.

In Example 6 it was possible to figure out the sign of the derivative on each interval by working out the sign of each factor. On the other hand, as we observed in Example 5, the sign of the derivative can only change at a critical point, so to find what the sign is on each interval, we need only evaluate the derivative at one point in each interval.

For instance, in Example 6 the derivative is $R'(t) = 3t^2e^t + t^3e^t$, which we can easily evaluate at, say, $x = 1$, $x = -1$, and $x = -4$. We find that $R'(1) = 3e^1 + e^1 = 4e$, $R'(-1) = 3e^{-1} - e^{-1} = 2e^{-1} = \frac{2}{e}$, and $R'(-4) = 3(16e^{-4}) - 64e^{-4} = -16e^{-4}$. This last number is negative and the other two are positive, which shows that the derivative $R'(x)$ is positive at one point in $(0, \infty)$, positive at one point in $(-3, 0)$, and negative at one point in $(-\infty, -3)$. Since the sign of the derivative does not change on each of these intervals, we see that $R'(x)$ is positive for $x > 0$, positive for $-3 < x < 0$, and negative for $x < -3$, which agrees with what we found earlier.

To do this calculation, we had to choose a point in each interval in the domain between critical points. We chose $x = 1, -1, -4$ because they would be relatively easy to work with (better than fractions, for example), but it is possible to do it with any other convenient choice of one point in each interval between critical points.

In this case the two methods of finding the sign of the derivative $R'(x)$ are more or less equally easy, but there are cases in which evaluating at selected points is the only practical way to do it. In many other cases it is the easiest thing to try.

EXAMPLE 7 Find and classify all critical points of the function $k(t) = t^3(t^2 - 1)^2$, defined for $t \in [-2, 2]$.

Solution We find that the derivative is $k'(t) = 3t^2(t^2 - 1)^2 + 2t^3(t^2 - 1)(2t) = t^2(t^2 - 1)(3t^2 - 3 + 4t^2) = t^2(t^2 - 1)(7t^2 - 3)$. It is defined for all $t \in (-2, 2)$ but strictly speaking not at $t = \pm 2$. Even though the formula makes sense at $t = \pm 2$, the function $k(t)$ is not differentiable there because it is not defined on an open interval containing $t = -2$ or $t = 2$.

The derivative is zero at $t = 0$, $t = \pm 1$, $t = \pm\sqrt{\frac{3}{7}}$. The critical points of the function $k(t)$ are $t = 0$, $t = \pm 1$, $t = \pm\sqrt{\frac{3}{7}}$, and $t = \pm 2$. (The derivative is zero at $t = 0, \pm 1, \pm\sqrt{\frac{3}{7}}$ and does not exist at $t = \pm 2$.)

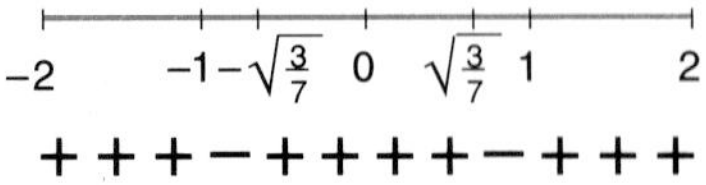

Figure 4.3.5(i)

As before, we draw the real axis in Figure 4.3.5(i), marking the critical points. We could work out the sign of each factor of $k'(t)$ on each interval, but it is probably easier to evaluate. For a point between 0 and $\sqrt{\frac{3}{7}}$ we could try $t = \sqrt{\frac{2}{7}}$; between $\sqrt{\frac{3}{7}}$ and 1 we could try $t = \sqrt{\frac{4}{7}}$. Between 1 and 2 we can use $t = \frac{3}{2}$. Then

$$k'\left(\sqrt{\frac{2}{7}}\right) = \left(\sqrt{\frac{2}{7}}\right)^2\left(\left(\sqrt{\frac{2}{7}}\right)^2 - 1\right)\left(7\left(\sqrt{\frac{2}{7}}\right)^2 - 3\right)$$

$$= \frac{2}{7}\left(\frac{2}{7} - 1\right)\left(7\left(\frac{2}{7}\right) - 3\right) = \frac{2}{7}\left(-\frac{5}{7}\right)(-1) = \frac{10}{49} > 0,$$

$$k'\left(\sqrt{\frac{4}{7}}\right) = \left(\sqrt{\frac{4}{7}}\right)^2\left(\left(\sqrt{\frac{4}{7}}\right)^2 - 1\right)\left(7\left(\sqrt{\frac{4}{7}}\right)^2 - 3\right)$$
$$= \frac{4}{7}\left(\frac{4}{7} - 1\right)\left(7\left(\frac{4}{7}\right) - 3\right)$$
$$= \frac{4}{7}\left(-\frac{3}{7}\right)(1) = -\frac{12}{49} < 0,$$
$$k'\left(\frac{3}{2}\right) = \left(\frac{3}{2}\right)^2\left(\left(\frac{3}{2}\right)^2 - 1\right)\left(7\left(\frac{3}{2}\right)^2 - 3\right)$$
$$= \frac{9}{4}\left(\frac{9}{4} - 1\right)\left(7\left(\frac{9}{4}\right) - 3\right)$$
$$= \frac{9}{4}\left(\frac{5}{4}\right)\left(\frac{63}{4} - \frac{12}{4}\right) > 0.$$

So $k'(t)$ is positive between $t = 0$ and $t = \sqrt{\frac{3}{7}}$, negative between $t = \sqrt{\frac{3}{7}}$ and $t = 1$, and positive between $t = 1$ and $t = 2$.

We could work it out for negative values of t, but since $k'(-t) = k'(t)$, it is not necessary to do any work. We mark the signs of the derivative on Figure 4.3.5(i).

Using the First Derivative Test, we see that there will be local maxima at $t = \sqrt{\frac{3}{7}}$, $t = 2$, and $t = -1$ and local minima at $t = -\sqrt{\frac{3}{7}}$, $t = -2$, and $t = 1$. There is no local maximum or minimum at $t = 0$.

To determine absolute maxima and minima, we evaluate:

$$k(0) = 0, \qquad k(\pm 1) = 0,$$
$$k\left(\sqrt{\frac{3}{7}}\right) = \left(\frac{3}{7}\right)^{3/2}\left(\frac{3}{7} - 1\right)^2 = \left(\frac{3}{7}\right)^{3/2}\left(\frac{16}{49}\right) \approx 0.091613,$$
$$k\left(-\sqrt{\frac{3}{7}}\right) = -\left(\frac{3}{7}\right)^{3/2}\left(\frac{3}{7} - 1\right)^2 = -\left(\frac{3}{7}\right)^{3/2}\left(\frac{16}{49}\right) \approx -0.091613,$$
$$k(2) = 8(4-1)^2 = 72, \qquad k(-2) = 8(4-1)^2 = -72.$$

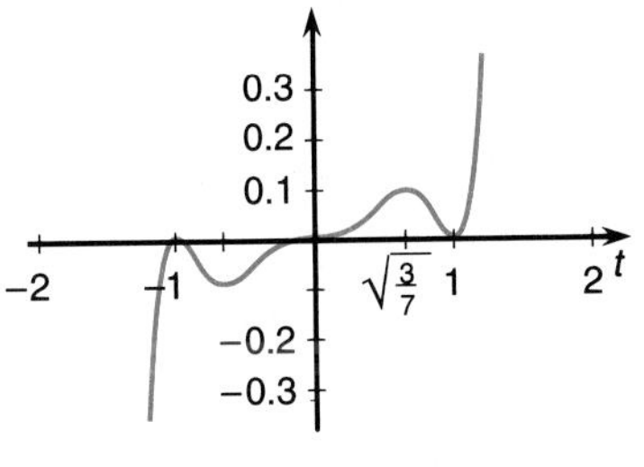

Figure 4.3.5(ii)

There is an absolute maximum at $t = 2$, an absolute minimum at $t = -2$, local maxima at $t = \sqrt{\frac{3}{7}}$ and $t = -1$, and local minima at $t = -\sqrt{\frac{3}{7}}$ and $t = 1$.

We sketch part of the graph in Figure 4.3.5(ii).

Notice that we must include the endpoints in our list of critical points, since $k(t)$ is not defined on open intervals containing them and so cannot possibly be differentiable at either of them. This may seem confusing because the function $f(t) = t^3(t^2 - 1)^2$ *is* differentiable at $t = \pm 2$; it is very important to distinguish this function from $k(t)$, which is only defined for $t \in [-2, 2]$. These two functions are given by the same formula when they are both defined, but they are defined on different domains, so they are *different functions.* In particular they do not have the same critical points.

If we had not made this distinction, we would not have found the absolute maximum at the endpoint $t = 2$.

Exercises 4.3

I

Find all critical points of the following functions.

1. $f(x) = x^3 - 12x$
2. $f(x) = x^4 - 18x^2 + 3$
3. $f(x) = \cos x$
4. $f(x) = \tan x$
5. $f(x) = \exp(2x - 4)$
6. $f(x) = \sin \frac{1}{x}$
7. $f(x) = \exp(\sin x)$
8. $f(x) = \sin(e^x)$
9. $f(x) = |x^2 - 1|$
10. $f(x) = \exp|x|$

Find all critical points of the following functions, and classify which are local or absolute maxima or minima.

11. $f(x) = x^5$
12. $f(x) = 3x^5 - 5x^3 + 2$
13. $f(x) = x^5 + 2x^3$
14. $f(x) = |x^2 - 4x + 3|$
15. $f(x) = (x^2 - 4)^2$
16. $f(x) = (x^2 - 9)^3$
17. $f(x) = \tan(2x)$
18. $f(x) = \csc x$
19. $f(x) = \exp(x^2)$
20. $f(x) = \arctan(x^2)$

For each of the following functions, find and classify all the critical points.

21. $f(x) = x^2 - 2, x \in [-1, 2]$
22. $f(x) = x^2 - 2, x \in (-1, 2)$
23. $f(x) = x^3 + 1, x \in [-3, 3]$
24. $f(x) = -3x^3, x \in (-1, 5)$
25. $f(x) = x^3 - 12x, x \in [-3, 3)$
26. $f(x) = \ln x - x, x > 0$
27. $f(x) = \sin x, x \in \left[0, \frac{3\pi}{4}\right]$
28. $f(x) = e^x \sin x, x \in [0, 2\pi]$
29. $f(x) = \exp(x^2 + 2x), x \in [0, 3]$
30. $f(x) = 1 + \tan^2 x - \sec^2 x, x \in (-1, 1)$

II

31. Suppose $a \neq b$ and let $f(x) = (x - a)(x - b)$. Find and classify the critical points of $f(x)$. (Draw a sketch.)
32. (i) Find and classify the critical points of $F(x) = \exp(x^2 - 2x - 3)$.
    (ii) Find and classify the critical points of $G(x) = \exp(-x^2 + 2x + 3)$.

*33. (i) Suppose $f(x)$ is a function that is defined and differentiable on the whole real line. Let $H(x) = e^{f(x)}$, and show that the critical points of $H(x)$ are exactly the same as the critical points of $f(x)$.
    (ii) Show that the critical points of $H(x)$ are of the same "type" for $H(x)$ as they are for $f(x)$, that is, a local maximum for $f(x)$ is a local maximum for $H(x)$, etc.
    (iii) If $K(x) = e^{-f(x)}$, what statements can you make about the critical points of $K(x)$ in analogy with parts (i) and (ii)?

*34. Suppose $f(x)$ can be written in the form $f(x) = (x - a)^k g(x)$, where $k \geq 2$ is an integer and $g(x)$ is a function that is differentiable at $x = a$ and that satisfies $g(a) \neq 0$.
    (i) Show that $x = a$ is a critical point of $f(x)$.
    (ii) Give conditions (in terms of k and $g(a)$) for this critical point to be a local maximum, a local minimum, or neither.
    (iii) Think of examples of functions $f(x)$ that illustrate each type of critical point discussed in part (ii).

*35. (i) Show that every polynomial of even degree has at least one critical point.
    (ii) Find examples of polynomials without any critical points (find examples with at least two different degrees, more if possible).

*36. Find and classify all critical points of $f(x) = |x - 1|^3$.

III

Use a graphics calculator or computer to graph the following functions.

37. $f(x) = \frac{1}{1+x^2}$ (Example 1, p. 249)
38. $f(x) = \frac{x}{1+x^2}$ (Example 2, p. 249)
39. $f(x) = (x - 1)e^x$ (Example 4, p. 250)
40. $h(x) = x^5 + 5x^2$ (Example 5, p. 250)
41. $R(t) = t^3 e^t$ (Example 6, p. 251)
42. $k(t) = (t - 1)\sqrt{t}$ (Example 7, p. 252)

4.4 The Second Derivative Test and Classifying Extrema

LOCAL MAXIMUM
LOCAL MINIMUM

Sections 4.2 and 4.3 showed how derivatives could be used to determine when a differentiable function was increasing or decreasing and then used this idea to see that local extrema occur at critical points. We observed that it is possible to have a critical point that is not a local extremum. This is what happens with $f(x) = x^3$; the

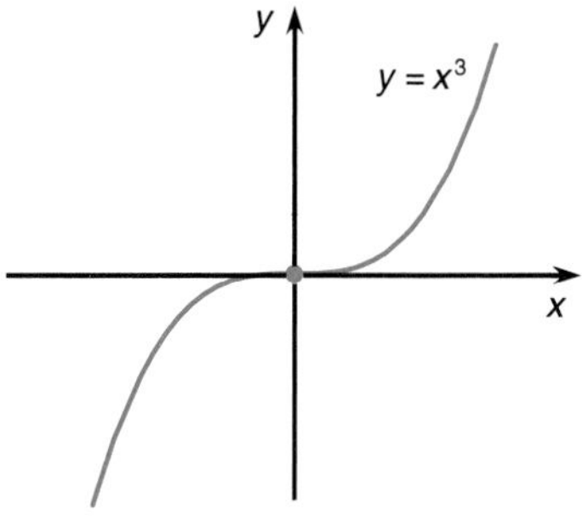

Figure 4.4.1

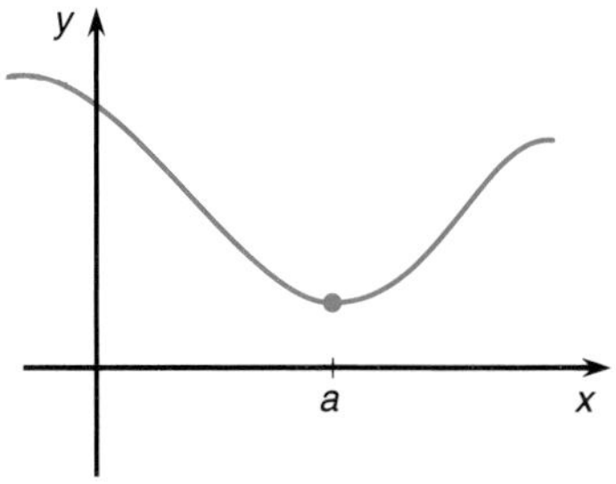

Figure 4.4.2

function has exactly one critical point, at $x = 0$, but it is increasing on the entire real line and has no local extrema at all, let alone absolute extrema. See Figure 4.4.1.

If we want to find the extrema of some function $f(x)$, we can start by finding all the critical points, but then it is necessary to figure out which of them, if any, are in fact extrema and which kind they are (local or absolute maxima or minima). In several examples in Section 4.3 it was possible to do this by drawing a picture and remembering various facts (especially the fact that a square is always nonnegative), but a lot depended on our familiarity with the particular example. In this section we show how the second derivative can frequently be used to identify local maxima and local minima. The method is not always applicable, but when it does work, it is very useful and relatively easy.

Suppose the function $f(x)$ has a local minimum at $x = a$. Then from Figure 4.4.2 we expect that $f(x)$ will be decreasing to the left of $x = a$ and increasing to the right of $x = a$, at least if we stay nearby. If $f(x)$ is differentiable, this means that the derivative $f'(x)$ should be negative for $x < a$ and positive for $x > a$ (and of course zero at the minimum $x = a$).

This in turn leads us to expect that the derivative $f'(x)$ should be increasing near $x = a$, from something negative to something positive. To say that the derivative is increasing suggests that *its* derivative, the second derivative $f''(x)$, should be positive near $x = a$.

This is not foolproof because we know that it is possible for a function to be increasing but to have its derivative equal zero at some point (as with x^3 at $x = 0$), but at the very least we expect that if $x = a$ is a local minimum for $f(x)$, then $f''(a) \geq 0$. Similarly, the second derivative should be less than or equal to zero at a local maximum.

EXAMPLE 1

(i) For $f(x) = x^2$, which has an absolute minimum at $x = 0$, we calculate and find that $f'(x) = 2x$, $f''(x) = 2$, which is certainly greater than 0. See Figure 4.4.3(i).

(ii) For $g(x) = \sin x$, which has an absolute maximum at $x = \frac{\pi}{2}$, we calculate $g'(x) = \cos x$ and $g''(x) = -\sin x$ and observe that $g''\left(\frac{\pi}{2}\right) = -\sin\frac{\pi}{2} = -1$, which is negative. See Figure 4.4.3(ii).

(iii) For $h(t) = t^4$, which has an absolute minimum at $t = 0$, we find that $h'(t) = 4t^3$ and $h''(t) = 12t^2$, so $h''(0) = 0$. This illustrates that it is definitely possible to have a minimum at which the second derivative is not positive but actually equal to zero. See Figure 4.4.3(iii).

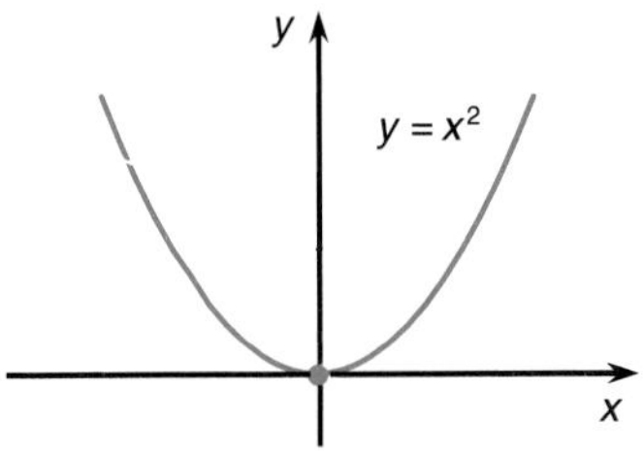

Figure 4.4.3(i)

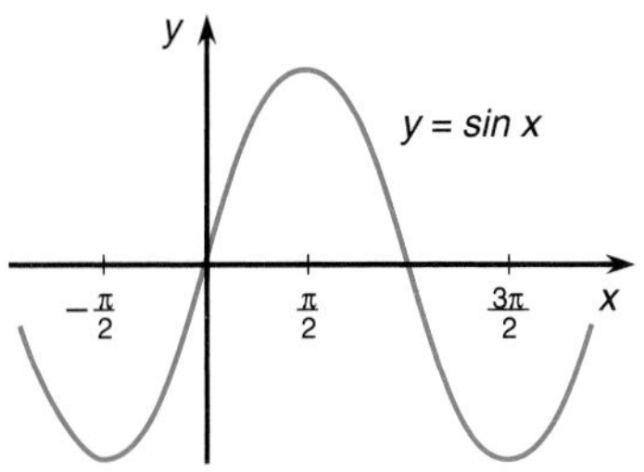

Figure 4.4.3(ii)

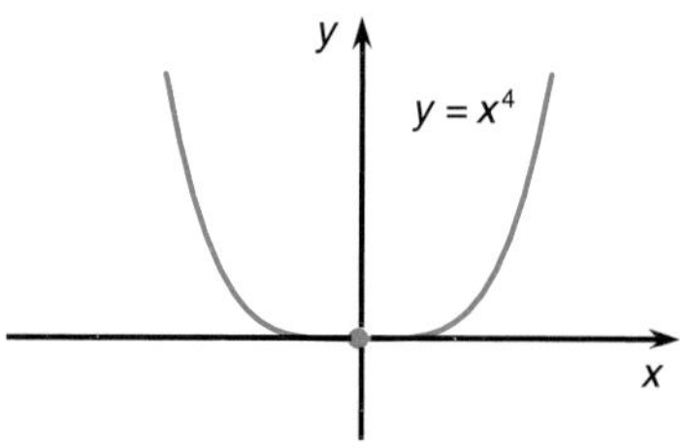

Figure 4.4.3(iii)

The importance of this idea comes from using it the other way around. If we know that $x = a$ is a critical point of $f(x)$ and that $f''(a)$ exists and is positive, then we can conclude that $x = a$ is a local minimum of $f(x)$; if $f''(a) < 0$, then $x = a$ is a local maximum.

THEOREM 4.4.1

The Second Derivative Test

Suppose $f(x)$ is differentiable on an interval containing $x = a$ and $f'(x)$ is itself differentiable at $x = a$. Suppose too that $f'(a) = 0$, that is, that $x = a$ is a critical point.

If $f''(a) < 0$, then $x = a$ is a local maximum.

If $f''(a) > 0$, then $x = a$ is a local minimum.

If $f''(a) = 0$, the test is inconclusive.

PROOF Suppose $f''(a) < 0$. Since f'' is the derivative of f', this says that

$$0 > f''(a) = \lim_{x \to a} \frac{f'(x) - f'(a)}{x - a} = \lim_{x \to a} \frac{f'(x)}{x - a}.$$

Note that $f'(a) = 0$.

Applying Theorem 2.3.4, we see that when x is sufficiently close to a, $\frac{f'(x)}{x-a}$ is negative. Since the denominator $x - a$ is negative when $x < a$ and positive when $x > a$, the numerator $f'(x)$ must be positive when $x < a$ and x is close to a, and it must be negative when $x > a$ and close to a.

Applying Corollary 4.2.9 (p. 243) to $f(x)$, we see that, close to $x = a$, $f(x)$ is increasing for $x \le a$ and decreasing for $x \ge a$. This implies that $x = a$ is a local maximum.

If $f''(a) > 0$, we may use a similar proof, or simply consider the function $-f(x)$, whose second derivative is $-f''(x)$, which is negative at $x = a$. The argument just completed implies that $-f(x)$ has a local maximum at $x = a$, so $f(x)$ has a local minimum there. ◀

EXAMPLE 2 Find and classify the local extrema of the function $f(x) = x^4 - 2x^2 + 1$, which we have already considered in Example 4.2.3(iii) (p. 238).

Solution The derivative is $f'(x) = 4x^3 - 4x = 4x(x^2 - 1)$, so the critical points are $x = -1, 0, 1$. Now $f''(x) = 12x^2 - 4$, so $f''(-1) = 12 - 4 = 8$, $f''(0) = -4$, and $f''(1) = 8$, so by the Second Derivative Test (Theorem 4.4.1) the points $x = \pm 1$ are local minima and $x = 0$ is a local maximum. See Figure 4.4.4. ◀

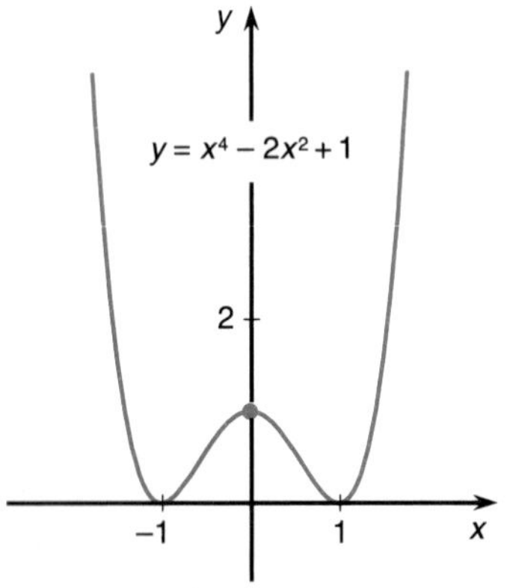

Figure 4.4.4

In fact, $f(x)$ has *absolute* minima at $x = \pm 1$. The Second Derivative Test does not distinguish between local and absolute extrema.

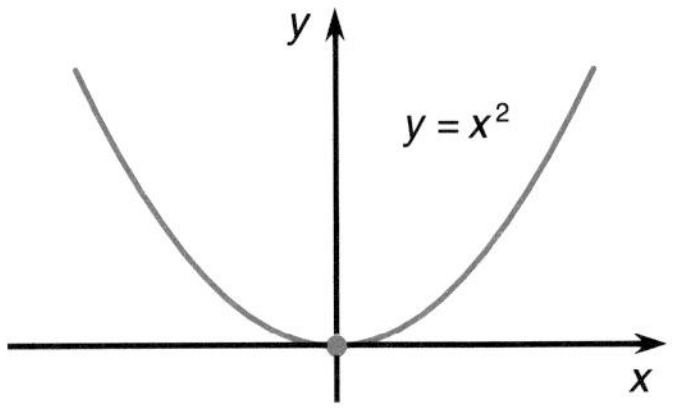

Figure 4.4.5(i)

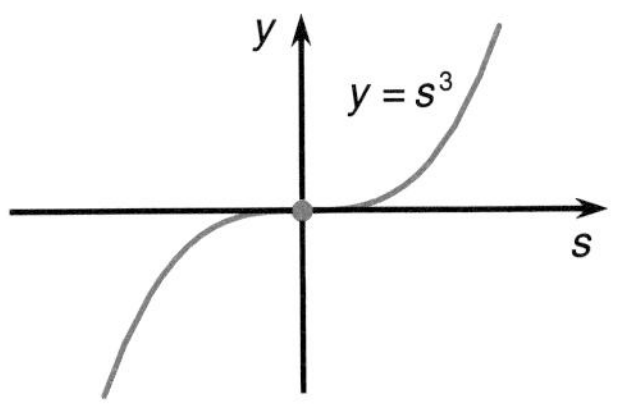

Figure 4.4.5(ii)

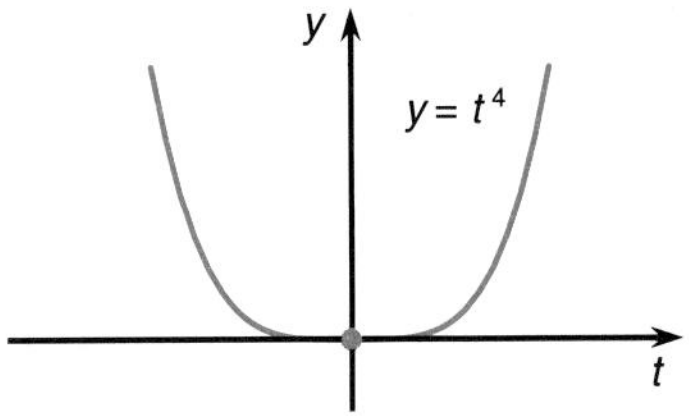

Figure 4.4.5(iii)

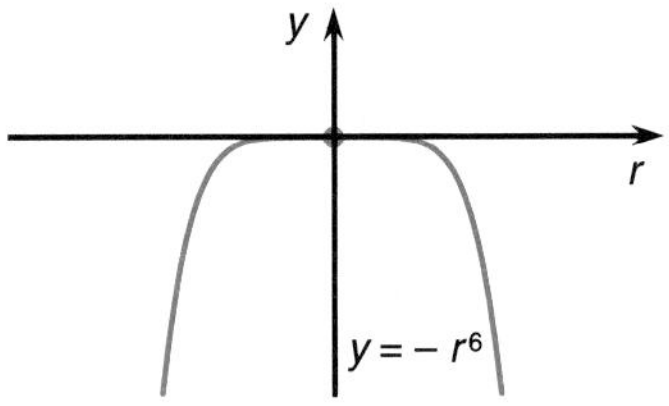

Figure 4.4.5(iv)

EXAMPLE 3 Find and classify the local extrema of (i) $f(x) = x^2$, (ii) $K(s) = s^3$, (iii) $h(t) = t^4$, and (iv) $g(r) = -r^6$. Their graphs are illustrated in Figures 4.4.5(i), (ii), (iii), and (iv).

Solution

(i) $f'(x) = 2x$ and $f''(x) = 2$. The only critical point is $x = 0$, and since $f''(0) = 2 > 0$, the Second Derivative Test tells us that $x = 0$ is a local minimum. In fact we know it is an absolute minimum, since for any $x \neq 0$ we know that $f(x) = x^2 > 0 = f(0)$.

(ii) $K'(s) = 3s^2$ and $K''(s) = 6s$. The only critical point is $s = 0$, and since $K''(0) = 0$, the Second Derivative Test does not apply, and we are unable to use it to decide whether the critical point $s = 0$ is a local extremum. In fact it is not because, as we observed in Example 4.2.5(ii) (p. 240), the function $K(s)$ has no local extrema.

(iii) $h'(t) = 4t^3$ and $h''(t) = 12t^2$. The only critical point is $t = 0$, and $h''(0) = 0$, so the Second Derivative Test is not applicable. In this case we can easily see that $t = 0$ is in fact an absolute minimum, since $h(t) = t^4 \geq 0 = h(0)$, since $t^4 = (t^2)^2$, being a square, is always nonnegative.

(iv) $g'(r) = -6r^5$ and $g''(r) = -30r^4$, so the only critical point is at $r = 0$. Since $g''(0) = 0$, the Second Derivative Test does not apply. In this particular case it is easy to see that $r = 0$ is in fact an absolute maximum, since $g(r) = -r^6 = -(r^3)^2$, being the negative of a square, is never bigger than $0 = g(0)$.

Example 3 illustrates not only how we can use the Second Derivative Test, but also how it cannot be used when the second derivative is zero at a particular critical point. Parts (ii), (iii), and (iv) of Example 3 demonstrate that in this situation it is possible to have maxima or minima or points that are neither.

Despite these shortcomings, the Second Derivative Test is extremely useful. When it can be applied, that is, when the second derivative is nonzero, it is the easiest way to identify local extrema.

The way the Second Derivative Test works is to describe the behavior of a function $f(x)$ for values of x very near to some critical point $x = a$. It can tell us whether the function has a local maximum or local minimum at $x = a$. On the other hand, it is unable to tell us whether or not any of these local extrema are absolute extrema. That depends on the behavior of the function on its entire domain, not just very near $x = a$.

To identify absolute extrema, it is necessary to use our knowledge of the function. There is no simple test that can give that sort of information.

In addition to the Second Derivative Test and using the first derivative to find regions of increase and decrease, it is often helpful to find limits and one-sided limits of $f(x)$ as x approaches $\pm\infty$ or points of discontinuity or points that are not in the domain of $f(x)$.

EXAMPLE 4 (i) Find the extrema of the function $f(x) = x^3 + 3x^2 + 2$ on the interval $[-4, 1]$.

Solution Since $f'(x) = 3x^2 + 6x = 3x(x + 2)$, the critical points are $x = 0$, $x = -2$, and the endpoints $x = -4$ and $x = 1$. Since $f(x)$ is continuous on a closed interval, it must have extrema, and they must be among these four points.

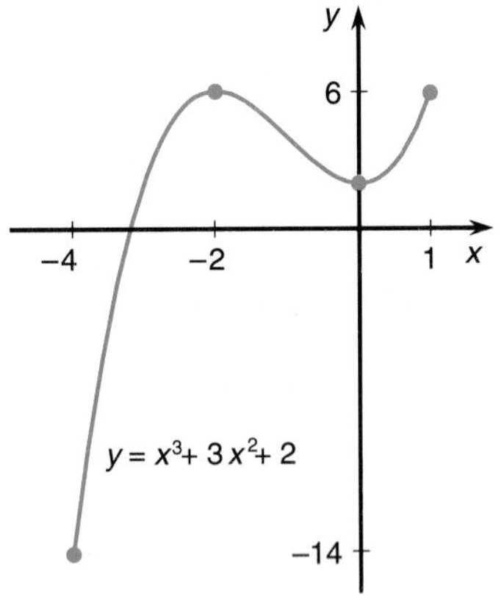

Figure 4.4.6

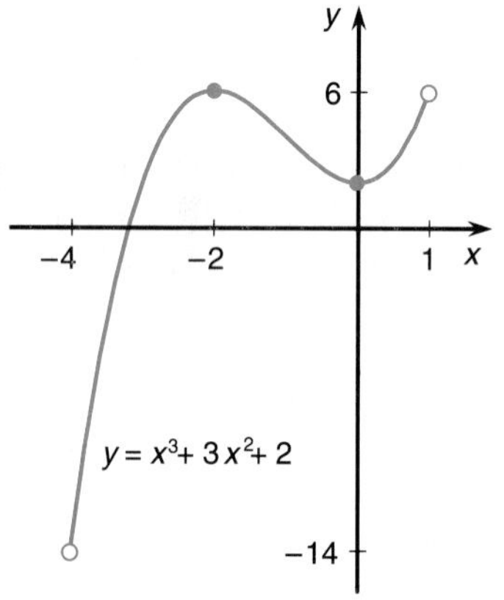

Figure 4.4.7

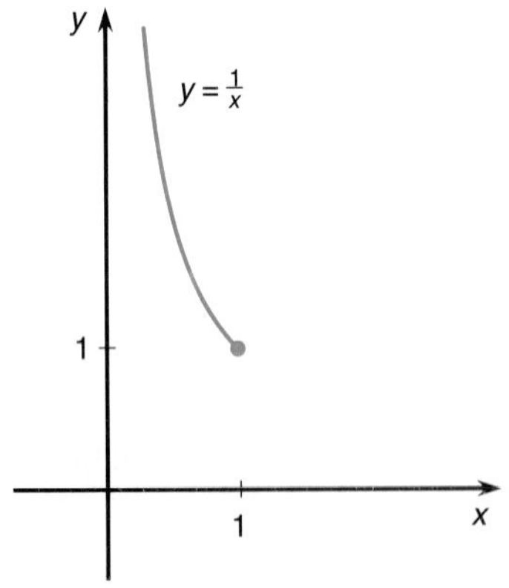

Figure 4.4.8

Evaluating, we find $f(-4) = -14$, $f(-2) = 6$, $f(0) = 2$, and $f(1) = 6$. The minimum value is -14, occurring at $x = -4$, and the maximum value is 6, occurring at $x = -2$ and $x = 1$.

The second derivative is $f''(x) = 6x + 6$, so $f''(0) = 6$, which is positive, so the Second Derivative Test says that there is a local minimum at $x = 0$.

The graph of $y = f(x)$ is sketched in Figure 4.4.6.

(ii) Find the extrema of the function $F(x) = x^3 + 3x^2 + 2$ on the interval $(-4, 1)$.

Solution This time the endpoints are not included in the domain. Since by the definition of a critical point (Definition 4.3.1, p. 249), a critical point of $F(x)$ must be in the domain of $F(x)$, the only critical points of $F(x)$ are $x = -2$ and $x = 0$.

We know that $F(-2) = 6$ and $F(0) = 2$ and also that

$$\lim_{x\to -4^+} F(x) = -14 \quad \text{and} \quad \lim_{x\to 1^-} F(x) = 6.$$

The first limit says that values of $F(x)$ approach -14, which tells us that the value 2 at $x = 0$ is not an absolute minimum (even though the Second Derivative Test tells us it is a local minimum, since $F''(0) = 6 > 0$). So there is no absolute minimum, the absolute maximum occurs at $x = -2$, and there is a local minimum at $x = 0$.

The graph of $y = F(x)$ is illustrated in Figure 4.4.7.

EXAMPLE 5 Find the extrema of $g(x) = \frac{1}{x}$ on the interval $(0, 1]$.

Solution Since $g'(x) = -\frac{1}{x^2}$, which is never zero, the only critical point is the endpoint $x = 1$. Notice that $x = 0$ is not in the domain of $g(x)$.

Since $g'(x) < 0$ for all $x \in (0, 1]$, we can apply Corollary 4.2.9 (p. 243) to see that $g(x)$ is decreasing on $(0, 1]$. This implies that the absolute minimum occurs at the endpoint $x = 1$.

Since $\lim_{x\to 0^+} g(x) = \infty$, there is no maximum. The graph of $y = g(x)$ is sketched in Figure 4.4.8.

EXAMPLE 6 Find the extrema of $h(x) = 3x^4 + 4x^3 - 24x^2 - 48x + 7$.

Solution $h'(x) = 12x^3 + 12x^2 - 48x - 48 = 12(x^3 + x^2 - 4x - 4)$. We want to find when this derivative is zero. In a situation like this, the easiest thing to do is to look for a rational root using the Rational Root Theorem (Theorem 1.3.7, p. 13). We see that $h'(0) = -48 \neq 0$, $h'(1) = -72 \neq 0$, $h'(-1) = 0$, so $x = -1$ is a critical point.

Now since $x = -1$ is a root of $h'(x)$, the Factor Theorem (Theorem 1.3.5, p. 12) tells us that $h'(x)$ can be divided by $x - (-1) = x + 1$. We perform a long division (cf. Example 1.3.8, p. 12), and find that

$$h'(x) = 12(x^3 + x^2 - 4x - 4) = 12(x + 1)(x^2 - 4) = 12(x + 1)(x - 2)(x + 2).$$

The critical points are $x = -1$, $x = 2$, and $x = -2$.

The second derivative is $h''(x) = 36x^2 + 24x - 48$, so we see that $h''(-2) = 48$, $h''(-1) = -36$, and $h''(2) = 144$; $x = -1$ is a local maximum, and $x = \pm 2$ are local minima. Since $h(-2) = 23$, $h(2) = -105$, there is an absolute minimum at $x = 2$ and a local minimum at $x = -2$. Since $\lim_{x\to\pm\infty} h(x) = \infty$, there is no absolute maximum.

It is interesting to use a graphics calculator to plot the function $f(x)$ from Example 4 on the same picture with its derivative $f'(x)$ and its second derivative.

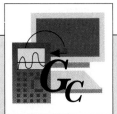

TI-85 GRAPH F1 $(y(x) =)$ F1 (x) ^3 + 3 F1 (x) x^2 + 2 ▼ 3 F1 (x) x^2 + 6 F1 (x) ▼ 6 F1 (x) + 6 M2(RANGE) (−) 4 ▼ 1 ▼ ▼ (−) 20 ▼ 10 F5(GRAPH)	SHARP EL9300 x/θ/T a^b 3 ▶ + 3 x/θ/T x^2 + 2 2ndF ▼ 3 x/θ/T x^2 + 6 x/θ/T 2ndF ▼ 6 x/θ/T + 6 RANGE − 4 ENTER 1 ENTER 1 ENTER −20 ENTER 10 ENTER 5 ENTER
CASIO f_x-7700GB/f_x-6300G Cls EXE Range −4 EXE 1 EXE 1 EXE −20 EXE 10 EXE 5 EXE EXE EXE EXE Graph X x^y 3 + 3 X x^2 + 2 EXE Graph 3 X x^2 + 6 X EXE Graph 6 X + 6 EXE	HP 48SX ↰ PLOT PLOTR ERASE ATTN ′ αY = αX y^x 3 + 3 × αX y^x 2 + 2 ↱ PLOT ↰ DRAW ERASE 4 +/− ENTER 1 XRNG 20 +/− ENTER 10 YRNG DRAW ATTN ′ αY = 3 × αX y^x 2 + 6 × αX ↱ PLOT ↰ DRAW DRAW ATTN ′ αY = 6 × αX + 6 ↱ PLOT ↰ DRAW DRAW

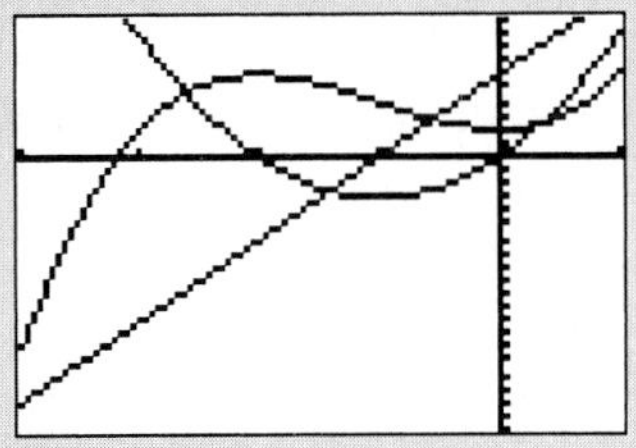

Observe the sign of the second derivative at each critical point, and note the corresponding behavior of the first derivative and the original function.

EXAMPLE 7 Find the extrema of

$$f(x) = \frac{x^2 - 1}{x^2 + 2}.$$

Solution The derivative is

$$f'(x) = \frac{2x(x^2+2) - 2x(x^2-1)}{(x^2+2)^2} = \frac{2x^3 + 4x - 2x^3 + 2x}{(x^2+2)^2}$$

$$= \frac{6x}{(x^2+2)^2}.$$

So the only critical point is $x = 0$. (Since the denominator is never zero, the function $f(x)$ is always differentiable.)

Notice that $\lim_{x\to\infty} f(x) = \lim_{x\to-\infty} f(x) = 1$ and $f(0) = -\frac{1}{2}$. From the formula for $f'(x)$ we see that $f'(x) < 0$ when $x < 0$ and $f'(x) > 0$ when $x > 0$, so $f(x)$ is decreasing for $x < 0$ and increasing for $x > 0$. By the First Derivative Test we conclude that $f(x)$ has a minimum at $x = 0$ (with less trouble than it would have been to calculate the second derivative) and has no maximum.

In this example we see that analyzing the regions of increase and decrease of $f(x)$ may be easier than applying the Second Derivative Test, especially if the second derivative is complicated to calculate. Of course, there is also the advantage that this approach is always applicable, unlike the Second Derivative Test, which tells us nothing when the second derivative is not defined or equals zero at a critical point.

EXAMPLE 8 Find all local and absolute extrema of $g(t) = \frac{t^3-3t}{1+t^2}$.

Solution The derivative is

$$g'(t) = \frac{(3t^2-3)(1+t^2) - 2t(t^3-3t)}{(1+t^2)^2} = \frac{3t^4 - 3 - 2t^4 + 6t^2}{(1+t^2)^2} = \frac{t^4+6t^2-3}{(1+t^2)^2}.$$

To find the critical points, we must solve $t^4 + 6t^2 - 3 = 0$.

To do this, we solve for t^2 first because the equation is quadratic in t^2; that is, letting $u = t^2$, the equation becomes $u^2 + 6u - 3 = 0$, and so

$$u = \frac{-6 \pm \sqrt{36+12}}{2} = -3 \pm 2\sqrt{3}.$$

Since $\sqrt{3} > 1.5$, we know $2\sqrt{3} > 3$, so $-3 + 2\sqrt{3} > 0$, and we can let $t = \pm\sqrt{-3+2\sqrt{3}} \approx \pm 0.68125$. The other value $u = -3 - 2\sqrt{3}$ is negative and cannot equal t^2 for any real number t.

To work out the second derivative would be straightforward but messy. Instead, it is relatively easy to calculate $g'(-1) = 1$, $g'(0) = -3$, and $g'(1) = 1$, so $g(t)$ is increasing for $t < -\sqrt{-3+2\sqrt{3}}$ and for $t > \sqrt{-3+2\sqrt{3}}$, and decreasing for $-\sqrt{-3+2\sqrt{3}} < t < \sqrt{-3+2\sqrt{3}}$. This shows that $-\sqrt{-3+2\sqrt{3}}$ is a local maximum and $\sqrt{-3+2\sqrt{3}}$ is a local minimum. Since $\lim_{t\to\infty} g(t) = \infty$ and $\lim_{t\to-\infty} g(t) = -\infty$, there are no absolute extrema.

EXAMPLE 9 Consider the function $F(x)$ that is graphed in Figure 4.4.9. At which of the indicated points is $F(x)$ not differentiable, and at which of them is $F'(x) < 0$,

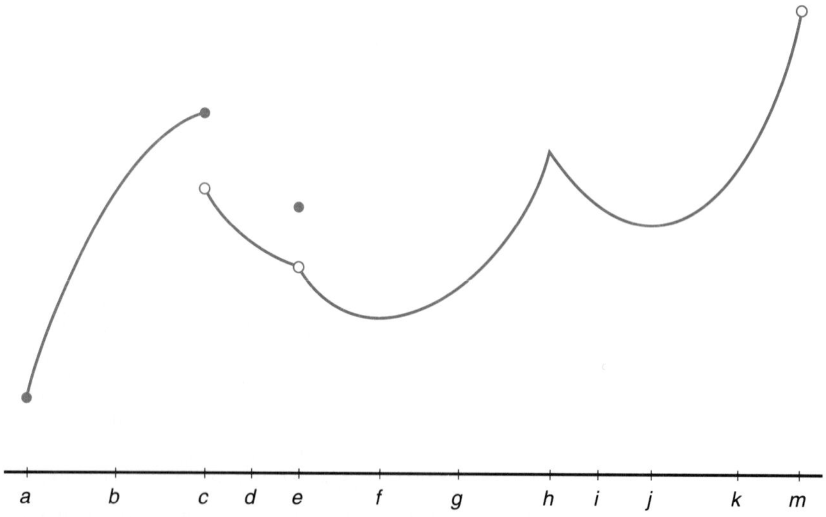

Figure 4.4.9

$F'(x) > 0, F'(x) = 0$? At which of the points does $F(x)$ have a jump discontinuity or a removable discontinuity? Describe all local and absolute extrema of $F(x)$.

Solution $F(x)$ is not differentiable at $x = a$ (the endpoint), $x = c$, $x = e$, $x = h$, and $x = m$. At $x = b$ the tangent line slopes upward to the right, so $F'(b) > 0$; also $F'(d) < 0, F'(f) = 0, F'(g) > 0, F'(i) < 0, F'(j) = 0$, and $F'(k) > 0$. There is a jump discontinuity at $x = c$ and a removable discontinuity at $x = e$. This certainly implies that $F(x)$ is not differentiable at those points. It is continuous at $x = h$ but not differentiable there.

There are local minima at $x = a, f$, and j; local maxima at $x = c, e$, and h; an absolute minimum at $x = a$; and no absolute maxima.

EXAMPLE 10 Find the extrema of $h(\theta) = \sec\theta$ for $-\frac{\pi}{2} < \theta < \frac{\pi}{2}$.

Solution The derivative is $h'(\theta) = \sec(\theta)\tan(\theta)$, and $\sec\theta = \frac{1}{\cos\theta}$ is never zero for $\theta \in \left(-\frac{\pi}{2}, \frac{\pi}{2}\right)$. Also the only $\theta \in \left(-\frac{\pi}{2}, \frac{\pi}{2}\right)$ for which $\tan\theta = 0$ is $\theta = 0$, so $\theta = 0$ is the only critical point.

Since $h''(\theta) = \sec\theta\tan\theta\tan\theta + \sec\theta\sec^2\theta = \sec\theta\tan^2\theta + \sec^3\theta$, we find $h''(0) = 0 + 1 = 1$, so $\theta = 0$ is a local minimum. Since $\cos\theta > 0$ for $\theta \in \left(-\frac{\pi}{2}, \frac{\pi}{2}\right)$, we see that $\sec\theta > 0$ also, and $\lim_{\theta\to-\frac{\pi}{2}^+}\sec\theta = \lim_{\theta\to\frac{\pi}{2}^-}\sec\theta = \infty$, so $\theta = 0$ is the absolute minimum, and there are no maxima.

Exercises 4.4

I

Find all local and absolute extrema for each of the following functions. (If the domain is not specified, assume that it is the whole real line.)

1. $f(x) = x^2 - 2x - 2$
2. $f(x) = x^2 + 4x - 6$
3. $f(x) = x^3 - 6x + 3$
4. $f(x) = x^3 - 6x + 3, x \in [-2, 2]$
5. $f(x) = x^3 - 6x + 3, x \in [-1, 1]$
6. $f(x) = x^3 - 6x + 3, x \in [-1, 2]$
7. $f(x) = x^3 - 6x + 3, x \in (-2, 2)$
8. $f(x) = x^3 - 6x + 3, x \in [-2, 1)$
9. $f(x) = x^4 - 8x^2 + 16$
10. $f(x) = 2x^4 + x - 1, x \in [-2, 2]$
11. $g(x) = x^4 - 2x^3 + 3$
12. $h(t) = \frac{t+1}{t^2+1}$
13. $S(\theta) = \sin^2\theta$
14. $T(\phi) = \tan\phi, \phi \in \left[-\frac{\pi}{4}, \frac{\pi}{4}\right]$
15. $f(\theta) = \cos^2\theta$
16. $G(t) = \frac{1}{t+1}, t \geq 0$
17. $H(\theta) = \csc\theta, \theta \in (-\pi, 0)$
18. $k(x) = \frac{3x}{x^2+4}$
19. $K(t) = (t - 1)e^t$
20. $F(x) = \cos\left(x - \frac{\pi}{4}\right)$

II

21. (i) Find and classify all critical points of the constant function $K(x) = 3$.
    (ii) Find and classify all critical points of the function $A(t) = \frac{|t|}{t}$.
22. Find and classify all critical points of the quadratic function $q(x) = ax^2 + bx + c$. (Assume $a \neq 0$.)

*23. (i) Let $C(t) = t^3 + at^2 + bt + c$. Find conditions on a, b, and c so that $C(t)$ will have no critical points, exactly one critical point, or two critical points.
    (ii) Classify the critical points in each of the above cases.
    (iii) Find examples of specific numbers a, b, and c that illustrate the various possibilities.

*24. If $Q(x) = x^4 + ax^3 + bx^2 + cx + d$, what are the possible answers to the questions: (i) How many critical points does $Q(x)$ have? (ii) Of what types can the critical points of $Q(x)$ be (local minimum, absolute maximum, etc.)? Give examples.

25. Consider the function $f(x)$ that is graphed in Figure 4.4.10. At which of the indicated points is $f(x)$ not differentiable, and at which of them does it have a removable or jump discontinuity? At which points is the derivative $f'(x)$ positive, negative, or zero? Describe all local and absolute extrema.

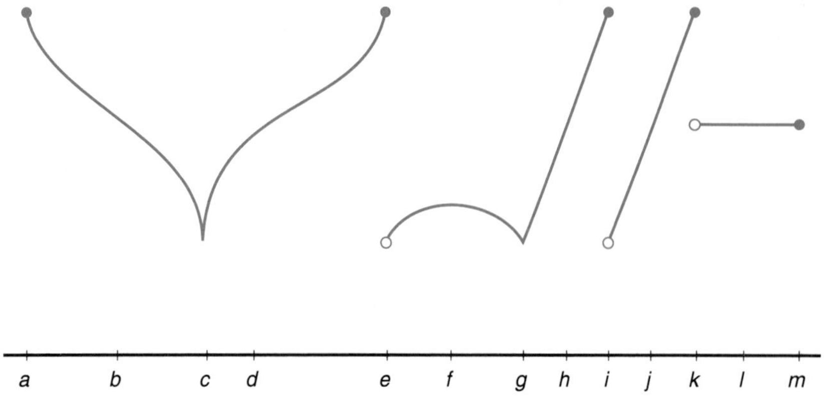

Figure 4.4.10

26. Repeat Exercise 25 for the function $g(x)$ graphed in Figure 4.4.11.

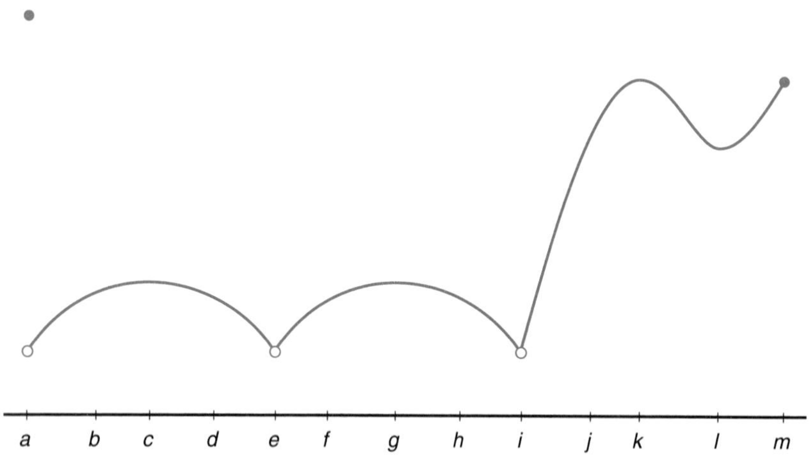

Figure 4.4.11

***27.** Let k be a positive integer and define $f(x) = x^k$. Find and classify the critical points of $f(x)$. (Your answer should depend on k.)

***28.** (i) Find and classify all critical points of $f(x) = \sinh x$. (You may need to use the fact that $\lim_{x\to\infty} e^x = \infty$.)

(ii) Find and classify all critical points of $g(x) = \cosh x$.

III

Use a graphics calculator or computer to graph the following functions.

29. $h(x) = 3x^4 + 4x^3 - 24x^2 - 48x + 7$ (Example 6, p. 258)

30. $f(x) = \frac{x^2-1}{x^2+2}$ (Example 7, p. 259)

31. $g(t) = \frac{t^3-3t}{1+t^2}$ (Example 8, p. 260)

32. $h(\theta) = \sec\theta$ (Example 10, p. 261)

4.5 Curve Sketching and Asymptotes

EVEN/ODD FUNCTIONS
ASYMPTOTE
VERTICAL ASYMPTOTE
RATIONAL FUNCTIONS

The material of the last few sections gives a great deal of information about the behavior of the graph of a function $y = f(x)$. We know how to find when the graph is increasing or decreasing, and we know how to find local and absolute extrema. All of this can be used to sketch the graph and obtain a reasonably accurate picture of what it looks like. This can be very useful in itself, but the process of sketching the graph will also help us to understand the function $f(x)$.

Computers and graphics calculators can sketch graphs quite quickly and accurately, but it is recommended that you attempt the examples and exercises of this section without these aids and use them only as a means of checking your work afterward, if at all. The finished sketch of a graph may be of considerable use and interest, but learning how to produce it is also an excellent way to exercise your understanding of the concepts we have learned.

In this section we learn how to sketch the graphs of functions, concentrating on polynomials and rational functions. These ideas will be refined and extended to other types of functions in the next section.

EXAMPLE 1 Sketch the graph of $y = f(x) = x^2 + 4x + 2$.

Solution First we calculate the derivative $f'(x) = 2x + 4$. The only critical point is $x = -2$. Since $f''(x) = 2 > 0$, this critical point is a local minimum.

The derivative $f'(x) = 2x + 4$ is positive when $2x + 4 > 0$, that is, $2x > -4$, that is, when $x > -2$, and $f'(x) < 0$ when $x < -2$. From this we conclude that $f(x)$ is decreasing when $x < -2$ and increasing when $x > -2$ and in particular that $x = -2$ is an absolute minimum. We sketch the real axis in Figure 4.5.1, indicating the sign of the derivative $f'(x)$.

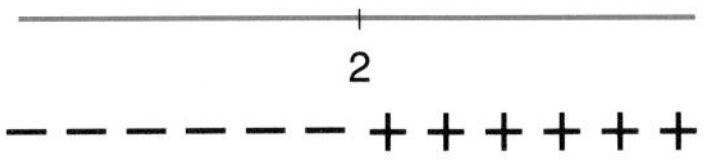

Figure 4.5.1

It is easy to see that $\lim_{x \to -\infty} f(x) = \lim_{x \to \infty} f(x) = \infty$, which shows that there is no absolute maximum, and we expect the graph to rise "up to infinity" at either end.

We also get some additional idea of the behavior of the graph by noticing that not only is $f'(x) = 2x + 4 = 2\big(x - (-2)\big)$ positive when $x > -2$, but it gets bigger and bigger the farther x gets to the right of -2. Similarly, it gets more and more negative the farther x is to the left of -2. This means that the tangents to $y = f(x)$ are steeper and steeper the farther the point of tangency is from $x = -2$. The graph rises more and more steeply as we move farther and farther to the right of $x = -2$ and falls more and more steeply as we move farther and farther to the left of $x = -2$.

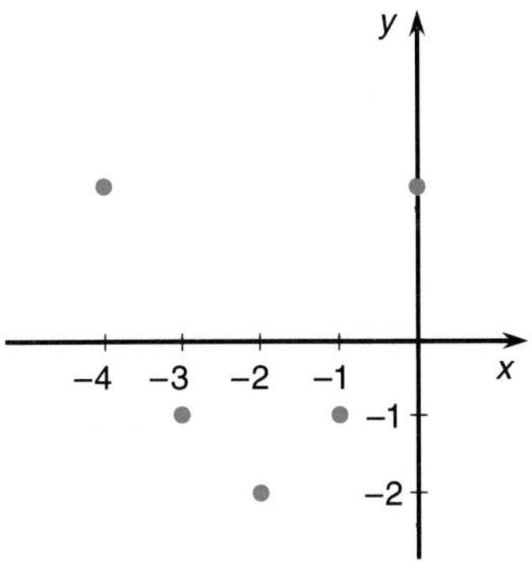

Figure 4.5.2

Finally, we plot a few points on the graph to give us a rough idea of where it lies. The point on the graph corresponding to the critical point $x = -2$ is $\big(-2, f(-2)\big) = (-2, -2)$. We can also plot a few nearby points, such as $\big(-1, f(-1)\big) = (-1, -1)$ and $(-3, -1)$, $(0, 2)$, and $(-4, 2)$ (see Figure 4.5.2). While we are at it, we should attempt to find the x-intercepts, the points at which the curve crosses the x-axis. To do this, we simply solve for $y = 0$, that is, for $x^2 + 4x + 2 = 0$. We can use the Quadratic Formula or simply complete the square: $x^2 + 4x + 2 = x^2 + 4x + 4 - 2 = (x + 2)^2 - 2$. This is zero when $x + 2 = \pm\sqrt{2}$, that is, when $x = -2 \pm \sqrt{2}$. So these are the x-intercepts, $x = -2 \pm \sqrt{2} \approx -0.586, -3.414$.

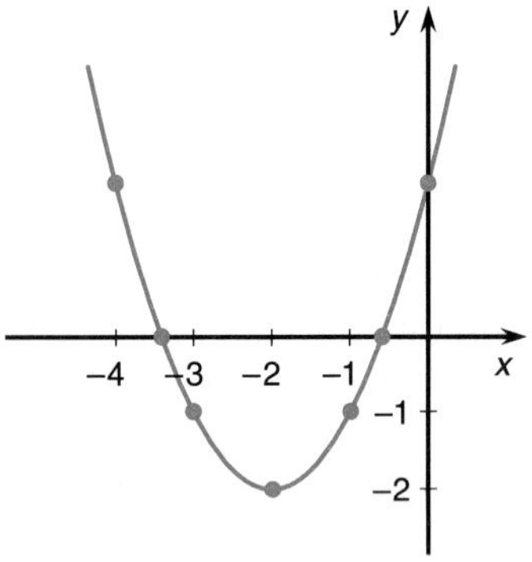

Figure 4.5.3

Finally, we attempt to "fill in" the curve, joining the points we have plotted, bearing in mind what we know about critical points and regions of increase and decrease. In this particular instance we can hardly be surprised to find that the graph is a parabola. See Figure 4.5.3.

Example 1 illustrates the essential ideas we have at our disposal for sketching graphs. We find the derivative and use it to locate the critical points. The sign of the derivative between critical points tells us where the graph is increasing or decreasing. Using that information and/or the second derivative, we classify the critical points and find local extrema.

Limits of the function at infinity or as we approach points not in the domain are also useful for describing the behavior of the graph of $y = f(x)$ and may in particular be helpful for deciding whether or not there are absolute extrema.

To this largely qualitative information we can add exact information by plotting some points on the graph. The more points we plot, the more detailed and precise will be our information. It makes sense to plot points whose coordinates are easy to calculate. In the above example we tried letting x equal several small integers, which tend to be convenient for computations. Notice that $x = 0$ is likely to be particularly easy to work with, especially if $f(x)$ is a polynomial. It amounts to finding the y-intercept, and if possible we should also find the x-intercepts, that is, the points at which $f(x) = 0$.

EXAMPLE 2 Sketch the graph of $y = f(x)$, where $f(x) = \frac{x}{1+x^2}$.

Solution We have already considered this function in Example 4.3.2 (p. 249). There we found that its derivative was

$$f'(x) = \frac{1(1+x^2) - x(2x)}{(1+x^2)^2} = \frac{1-x^2}{(1+x^2)^2},$$

so the critical points are $x = \pm 1$.

Remember that it is enough to evaluate f' at one point in each interval.

Drawing the real line and marking the critical points, we can easily find the sign of the derivative on each of the three intervals. Indeed, $f'(0) = 1, f'(-2) = -\frac{3}{5}$, and $f'(2) = -\frac{3}{5}$, so the derivative is positive between $x = -1$ and $x = 1$ and negative on $(-\infty, -1)$ and $(1, \infty)$. See Figure 4.5.4.

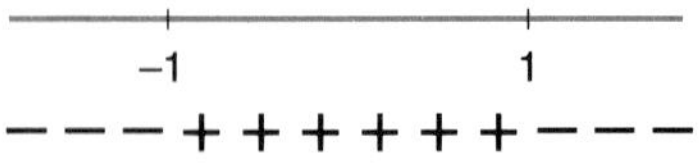

Figure 4.5.4

We also find that $\lim_{x\to-\infty} f(x) = \lim_{x\to\infty} f(x) = 0$, and notice that since the denominator of $f(x)$ is always positive, the function $f(x)$ itself is positive when $x > 0$ and negative when $x < 0$. The only x-intercept is at $x = 0$.

It is easy to plot a number of points, such as $(0, 0)$, $\left(1, \frac{1}{2}\right)$, $\left(-1, -\frac{1}{2}\right)$, $\left(2, \frac{2}{5}\right)$, $\left(-2, -\frac{2}{5}\right)$, $\left(3, \frac{3}{10}\right)$, and $\left(-3, -\frac{3}{10}\right)$. From these points we can fill in the rest of the curve, bearing in mind all the qualitative information we have discovered. See Figure 4.5.5.

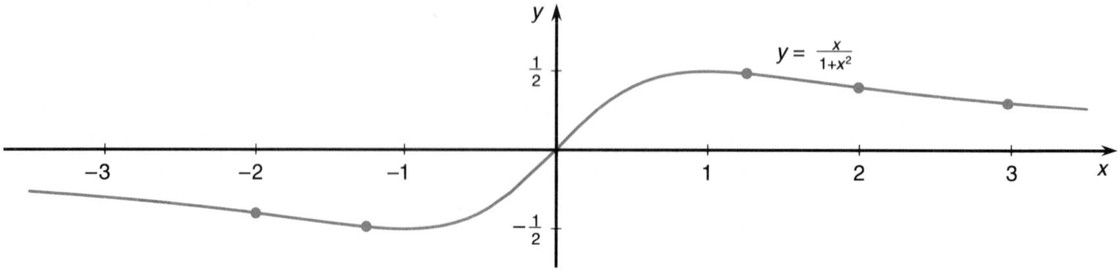

Figure 4.5.5

EXAMPLE 3 Sketch the graph of $y = g(x)$ if $g(x) = x^3 - 3x^2 + 2$.

Solution We calculate the derivative $g'(x) = 3x^2 - 6x = 3x(x-2)$, so the critical points are $x = 0$ and $x = 2$. It is very easy to compute $g'(-1) = 9$, $g'(1) = -3$, and $g'(3) = 9$, from which we realize that $g'(x)$ is positive on $(-\infty, 0)$ and $(2, \infty)$ and negative on $(0, 2)$. From this we are able to draw the real line and mark the sign of the derivative on each of the three intervals determined by the critical points. See Figure 4.5.6.

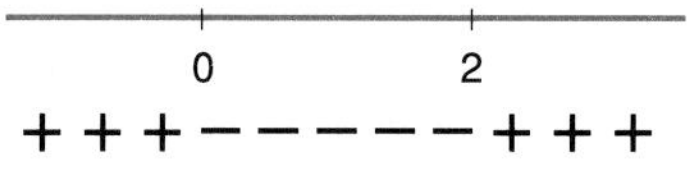

Figure 4.5.6

Without even using the Second Derivative Test, we see that $x = 0$ is a local maximum and $x = 2$ is a local minimum. Since

$$\lim_{x\to-\infty} g(x) = -\infty \qquad \text{and} \qquad \lim_{x\to\infty} g(x) = \infty,$$

we see that there are no absolute extrema.

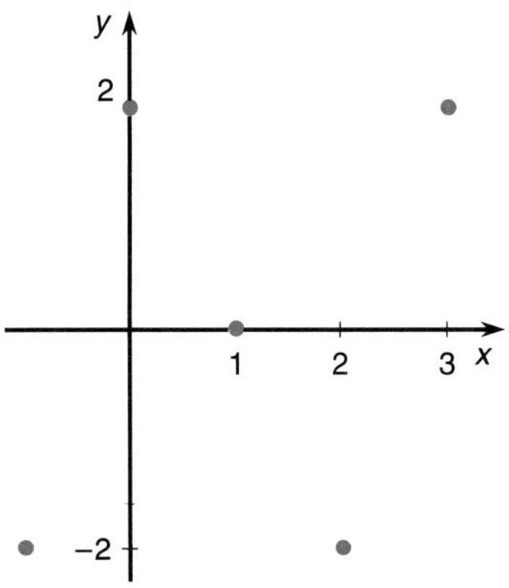

Figure 4.5.7

We plot some points, including the local extrema at the critical points: $(0, 2)$, $(2, -2)$, $(1, 0)$, $(3, 2)$, and $(-1, -2)$ (see Figure 4.5.7). By accident we have found one x-intercept, at $x = 1$. The Factor Theorem (Theorem 1.3.5, p. 12) says that we can divide $g(x)$ by $x - 1$. Carrying out the long division, we find $g(x) = x^3 - 3x^2 + 2 = (x-1)(x^2 - 2x - 2)$, so to find the other points at which $g(x) = 0$, we need to solve $x^2 - 2x - 2 = 0$. From the Quadratic Formula we find that the other x-intercepts are $1 \pm \sqrt{3} \approx -0.7321, 2.7321$. From this it is possible to fill in the shape of the curve (see Figure 4.5.8).

EXAMPLE 4 Sketch the graph of $y = h(x)$, where $h(x) = \frac{1}{x^2+2}$.

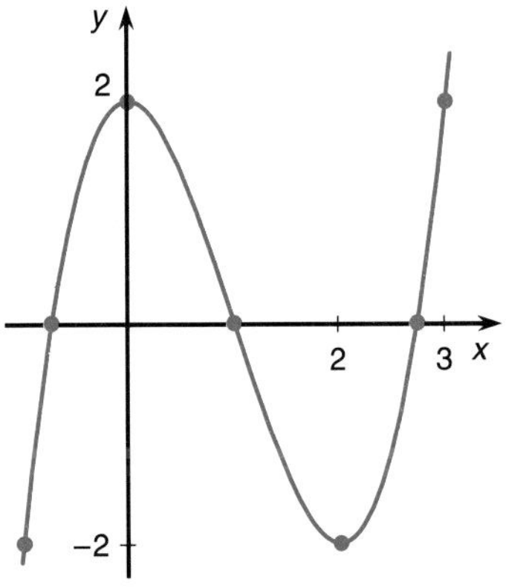

Figure 4.5.8

Solution Writing $h(x) = (x^2 + 2)^{-1}$ and applying the Chain Rule, we find that the derivative is

$$h'(x) = -\frac{1}{(x^2+2)^2}(2x) = -\frac{2x}{(x^2+2)^2}.$$

The only critical point is $x = 0$, and we see that $h'(x) > 0$ when $x < 0$ and $h'(x) < 0$ when $x > 0$, so $x = 0$ is an absolute maximum.

The function $h(x)$ is always positive, and $\lim_{x\to\infty} h(x) = \lim_{x\to-\infty} h(x) = 0$. From this we see that $h(x)$ never equals zero, but it gets arbitrarily close to zero as x gets large, so there is no absolute minimum.

We sketch in a few points, such as $\left(0, \frac{1}{2}\right)$, $\left(1, \frac{1}{3}\right)$, $\left(-1, \frac{1}{3}\right)$, $\left(2, \frac{1}{6}\right)$, and $\left(-2, \frac{1}{6}\right)$, and remark that $h(x)$ is never zero, so there are no x-intercepts. At this stage we notice that $h(-x) = h(x)$ for any x, which can save us trouble when we start to plot points (i.e., when we work out a point, we get another one free). It also tells us that when we plot the curve, it will be "symmetric" about the y-axis, that is, the part of the curve to the left of the axis will be the reflection of the part to the right.

Keeping this in mind and remembering what we know about the regions of increase and decrease, we can plot the curve. See Figure 4.5.9.

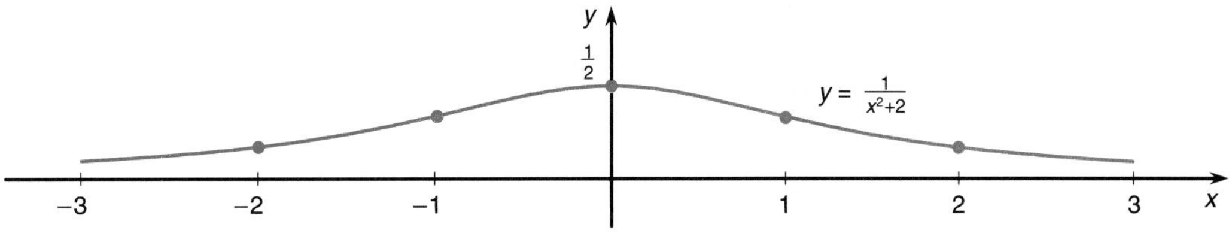

Figure 4.5.9

Using the symmetry of the graph is a very important idea. Not only can it save us a lot of time and effort, but it is an important geometrical property of a curve and should not be missed. The function of Example 4 satisfied $h(x) = h(-x)$. This had the effect of making the graph symmetrical about the y-axis, that is, the part of the graph to the right of the origin is the reflection of the part to the left. Equivalently, it means that reflecting in the y-axis leaves it unchanged.

Another possibility, illustrated by the function $f(x) = \frac{x}{1+x^2}$ of Example 2, is that a function may satisfy $f(-x) = -f(x)$. In terms of the picture, this means that the part of the graph to the right of the origin is obtained by reflecting the part to the left of the origin in the y-axis and then taking its negative. Another way to describe it is that we can draw a line from each point on the graph through the origin and out the same distance on the other side of the origin, and it will end in another point on the graph. Two such points are said to be each other's reflections in the origin, so we can say that reflecting the graph in the origin leaves it unchanged. See Figure 4.5.10.

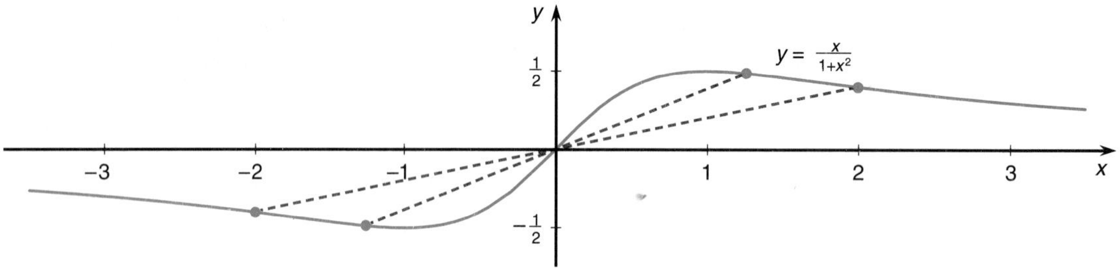

Figure 4.5.10

DEFINITION 4.5.1

> A function $f(x)$ is **even** if $f(-x) = f(x)$ for every x in the domain of f. In particular, if x is in the domain of f, then $-x$ must also be in the domain.
>
> A function $f(x)$ is **odd** if $f(-x) = -f(x)$ for every x in the domain of f. In particular, if x is in the domain of f, then $-x$ must also be in the domain.

Notice that some functions are even, some are odd, but many are neither even nor odd.

EXAMPLE 5 The function $f(x) = \frac{x}{1+x^2}$ of Example 2 is odd; the function $h(x) = \frac{1}{x^2+2}$ of Example 4 is even; the functions $x^2 + 4x + 2$ of Example 1 and $x^3 - 3x^2 + 2$ of Example 3 are neither even nor odd.

> We have just seen that the significance of even and odd functions is that the graph of an even function is symmetric about the y-axis and the graph of an odd function is left unchanged by reflection through the origin.

EXAMPLE 6 Sketch the graph of $F(x) = \frac{|x|}{x}\left(\sqrt{|x|} - |x|\right)$.

Solution Notice that the function is not defined when $x = 0$ (because of the x in the denominator). In dealing with formulas involving absolute values we have already seen that the thing to do is consider separately what happens when $x > 0$ and when $x < 0$.

First consider the case in which $x > 0$. In this case, $F(x) = \frac{x}{x}(\sqrt{x} - x) = \sqrt{x} - x = x^{1/2} - x$. The derivative is

$$F'(x) = \frac{1}{2}x^{-1/2} - 1 = \frac{1}{2\sqrt{x}} - 1.$$

To find critical points, we see that this will be zero precisely when $2\sqrt{x} = 1$, that is, $\sqrt{x} = \frac{1}{2}$, that is, $x = \frac{1}{4}$.

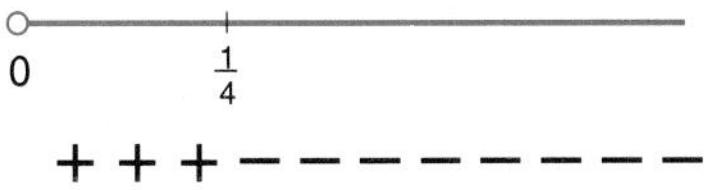

Figure 4.5.11

We notice that when $x > \frac{1}{4}$, then $\sqrt{x} > \frac{1}{2}$, $\frac{1}{2\sqrt{x}} < 1$, so $F'(x) < 0$. Similarly, when $0 < x < \frac{1}{4}$, $F'(x) > 0$. In Figure 4.5.11 we draw the positive real axis and mark on it the critical point and this information about the sign of the derivative. In particular we see that the critical point $x = \frac{1}{4}$ is an absolute maximum (at least on the *positive* axis).

To look for x-intercepts on the positive axis, we must find when $\sqrt{x} - x = 0$, that is, when $\sqrt{x} = x$. The only positive x that satisfies this is $x = 1$, so it is the x-intercept.

Next we should find out what happens as $x \to 0$ and as $x \to \infty$. We see that

$$\lim_{x\to 0^+} F(x) = \left(\lim_{x\to 0^+} \sqrt{x}\right) - \left(\lim_{x\to 0^+} x\right) = 0 - 0 = 0.$$

However, the behavior at ∞ is more difficult to describe.

We expect that as $x \to \infty$, $\sqrt{x}$ will tend to infinity. Indeed, $\sqrt{x}$ gets bigger and bigger. After all, if A is any large number, $\sqrt{x} > A$ whenever $x > A^2$. This says that $\sqrt{x}$ eventually gets bigger than any fixed number A, which is exactly what it means to say $\lim_{x\to\infty} \sqrt{x} = \infty$. As $x \to \infty$, $\sqrt{x}$ and x both tend to infinity, so we do not know what will happen to their difference. It seems reasonable to think that $\sqrt{x}$ is smaller than x (at least for $x > 1$), so $\sqrt{x}$ will go to infinity more slowly than x itself, so perhaps we should expect that $\sqrt{x} - x$ will tend to $-\infty$.

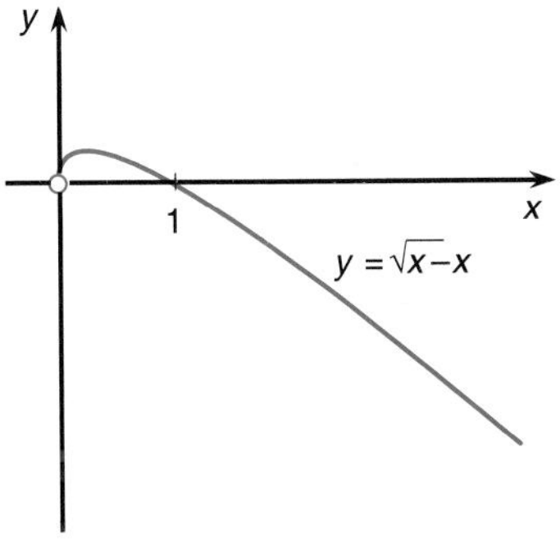

Figure 4.5.12

The easiest way to show this would be to show that $F(x)$ is less than some function that we know tends to $-\infty$, at least for large values of x. Suppose $x > 4$. Then $x = \sqrt{x}\sqrt{x} > 2\sqrt{x}$ and $-x < -2\sqrt{x}$. So for $x > 4$, we see that $F(x) = \sqrt{x} - x < \sqrt{x} - 2\sqrt{x} = -\sqrt{x}$. Since we have just seen that $\sqrt{x} \to \infty$, we know that $-\sqrt{x} \to -\infty$, so $\lim_{x\to\infty} F(x) = -\infty$.

Using all the information we have collected, we can draw a sketch of the part of the graph corresponding to positive values of x. See Figure 4.5.12.

The next step is to consider what happens when $x < 0$. It is possible to analyze this case in much the same way as we have just done when $x > 0$, but it is much easier to notice that $F(x)$ is an odd function. (After all, the parts of the formula involving $|x|$ do not change when x is replaced by $-x$, and the quotient $\frac{|x|}{x}$ will change to its negative.)

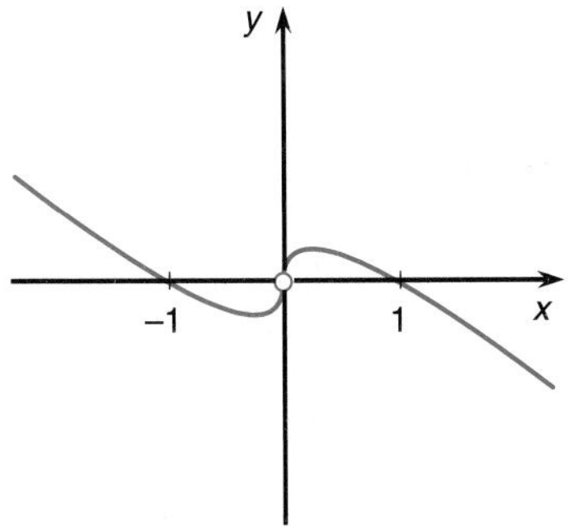

Figure 4.5.13

From this we realize that the part of the graph to the left of the origin is just the reflection through the origin of the part to the right of the origin, which we have already sketched. We can just fill in the remainder of the sketch without further trouble. See Figure 4.5.13.

In drawing the graph we noticed that the function is not defined at $x = 0$, so we have left a little open circle in the picture. On the other hand, both one-sided limits exist there, and in fact they are both equal to 0. This means that there is a removable discontinuity at $x = 0$. We can also remark that there are no absolute extrema, since the function tends to ∞ at one end and to $-\infty$ at the other. ◢

Example 6 demonstrates that recognizing even or odd functions can save a great deal of work.

EXAMPLE 7 Sketch the graph of $G(x) = \frac{x^2}{x^2+2}$.

Solution We could begin by finding the derivative of $G(x)$ and looking for critical points, but in this case there is an easy way to do the problem. We notice that

$$G(x) = \frac{x^2}{x^2+2} = \frac{x^2+2}{x^2+2} - \frac{2}{x^2+2} = 1 - \frac{2}{x^2+2}.$$

We added and subtracted $\frac{2}{x^2+2}$.

In Example 4, we sketched the graph of $h(x) = \frac{1}{x^2+2}$. The graph of $y = \frac{2}{x^2+2}$ will be the graph of $h(x)$ made twice as high, that is, stretched by a factor of 2 in the vertical direction. The graph of $y = -\frac{2}{x^2+2}$ will be *that* graph turned upside down. See Figure 4.5.14.

Since $G(x) = 1 - \frac{2}{x^2+2}$, its graph will be the graph of $y = -\frac{2}{x^2+2}$ shifted up by one unit (after all, this is the effect of adding 1 to a function). See Figure 4.5.15.

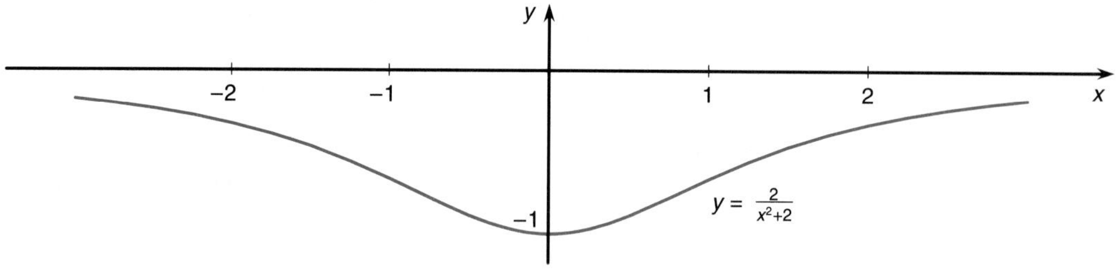

Figure 4.5.14

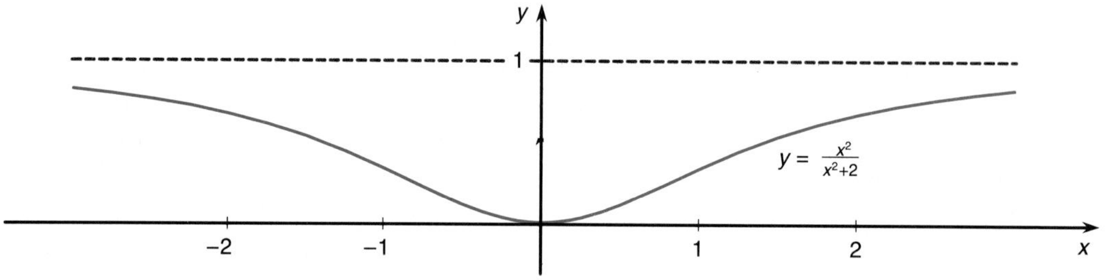

Figure 4.5.15

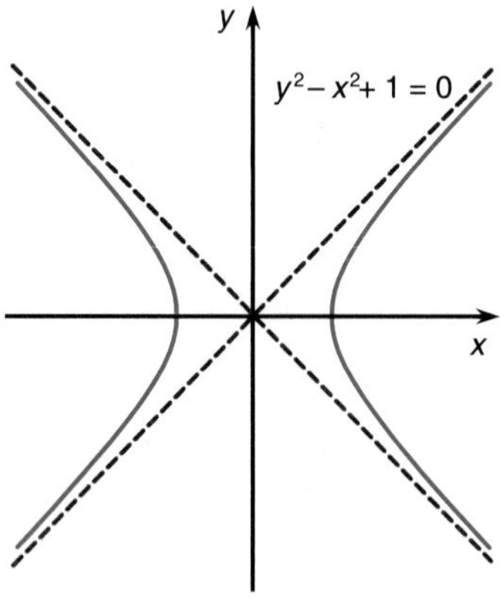

Figure 4.5.16

Asymptotes

One feature of the graph we have just drawn is that the function approaches the limit 1 as $x \to \pm\infty$. In the sketch we see how the curve gets closer and closer to the horizontal line $y = 1$. The functions $f(x) = \frac{x}{1+x^2}$, sketched in Example 2, and $h(x) = \frac{1}{x^2+2}$, sketched in Example 4, also have the property that as $x \to \infty$ and as $x \to -\infty$, the function approaches a certain line (in both these cases it is the line $y = 0$).

Other examples of similar behavior can be found in the graph of a hyperbola (cf. Summary 1.6.4, p. 38). For instance, in Figure 4.5.16 we see the hyperbola $y^2 - x^2 + 1 = 0$. As it stands, this equation does not determine y as a function of x (since there are two values of y for each x). But of course we can choose part of the hyperbola that does define y as a function of x. The most obvious thing to do is to let $y = T(x) = \sqrt{x^2 - 1}$, which defines a function whose graph is the "top half" of

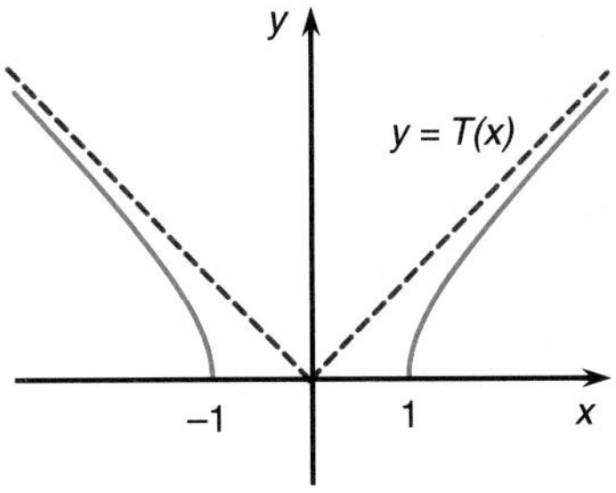

Figure 4.5.17

the hyperbola. We know that as $x \to \pm\infty$ the hyperbola gets close to one of the two lines we call its asymptotes, getting closer the farther it goes from the origin.

This can be seen from the formula that defines $T(x)$, as follows. Suppose x is a very large number. Then $x^2 - 1$ is close to x^2, and $T(x) = \sqrt{x^2 - 1}$ will be close to $\sqrt{x^2} = x$. This suggests that the graph of $y = T(x)$ will be close to the graph of $y = x$ when x is very large, although it is not clear how close. A similar argument can be used to suggest that when x is a very large negative number, the graph of $y = T(x)$ is close to the graph of the line $y = |x| = -x$. See Figure 4.5.17. The behavior of the bottom half of the hyperbola is similar.

Another similar kind of behavior is found in the function $y = R(x) = \frac{1}{x}$. As $x \to \pm\infty$, the function $R(x)$ tends to zero, which means that the graph of $y = \frac{1}{x}$ gets closer and closer to the line $y = 0$.

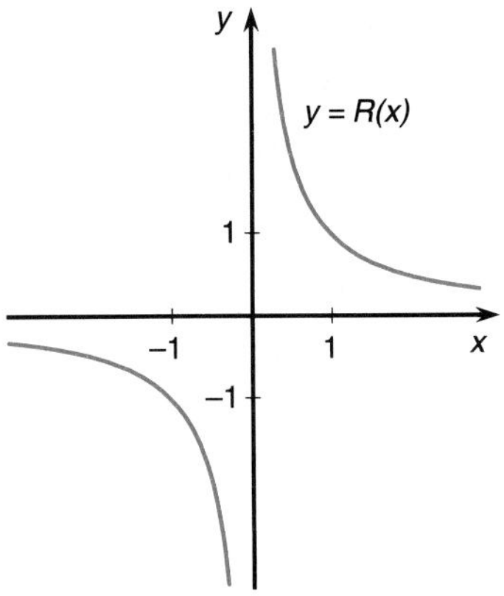

Figure 4.5.18

But we should also consider what happens near $x = 0$. As x approaches 0 from above, that is, $x \to 0^+$, we know that $R(x) = \frac{1}{x}$ approaches ∞. Similarly, as $x \to 0^-$, x is negative, and $R(x) = \frac{1}{x}$ approaches $-\infty$. In Figure 4.5.18 we see that the graph of $y = R(x)$ gets closer and closer to the y-axis, moving farther and farther away from the origin.

What all of these situations have in common is that the graph of a function approaches some straight line. This situation occurs quite frequently, and there is special terminology for it. Notice that any straight line that is not vertical can be written in the form $y = ax + b$.

DEFINITION 4.5.2

A nonvertical line $y = ax + b$ is an **asymptote** for the function $f(x)$ if the graph of $y = f(x)$ approaches the line as $x \to \infty$ or as $x \to -\infty$. Symbolically, this means

$$\lim_{x\to\infty} \big(f(x) - (ax + b)\big) = 0$$

or

$$\lim_{x\to-\infty} \big(f(x) - (ax + b)\big) = 0.$$

In this situation we also say that $f(x)$ approaches the line $y = ax + b$ **asymptotically** as $x \to \infty$ (or as $x \to -\infty$, as the case may be).

The vertical line $x = a$ is a **vertical asymptote** for the function $f(x)$ if $\lim_{x\to a^+} f(x) = \pm\infty$ or if $\lim_{x\to a^-} f(x) = \pm\infty$. In this situation we also say that $f(x)$ approaches the line $x = a$ asymptotically as $x \to a^+$ (or as $x \to a^-$, as the case may be).

An asymptote that is neither vertical nor horizontal is called an **oblique asymptote**.

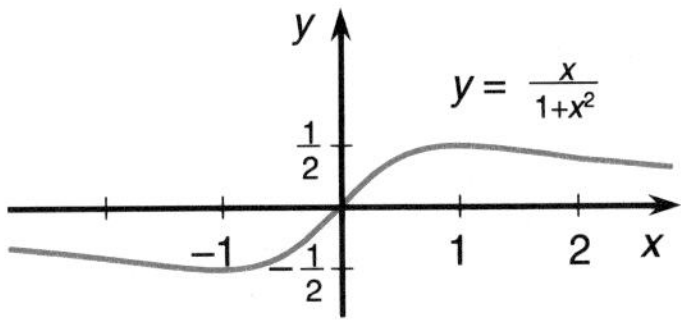

Figure 4.5.19(i)

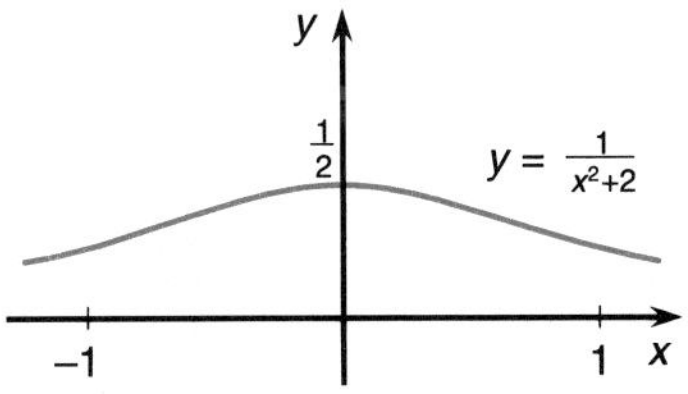

Figure 4.5.19(ii)

EXAMPLE 8

(i) The functions $f(x) = \frac{x}{1+x^2}$ of Example 2 and $h(x) = \frac{1}{x^2+2}$ of Example 4 both approach the line $y = 0$ asymptotically as $x \to \pm\infty$. See Figures 4.5.19(i) and 4.5.19(ii).

(ii) The function $G(x) = \frac{x^2}{x^2+2}$ of Example 7 has the line $y = 1$ as an asymptote as $x \to \pm\infty$. See Figure 4.5.19(iii).

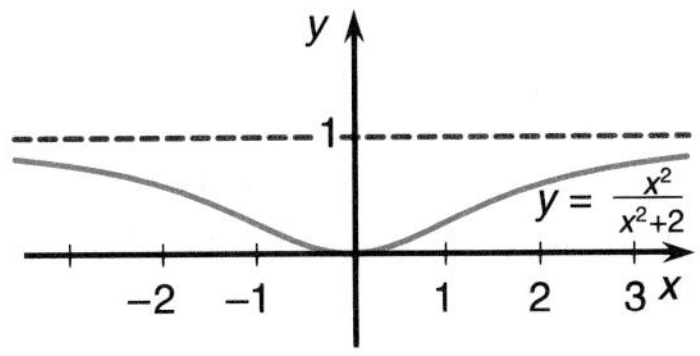

Figure 4.5.19(iii)

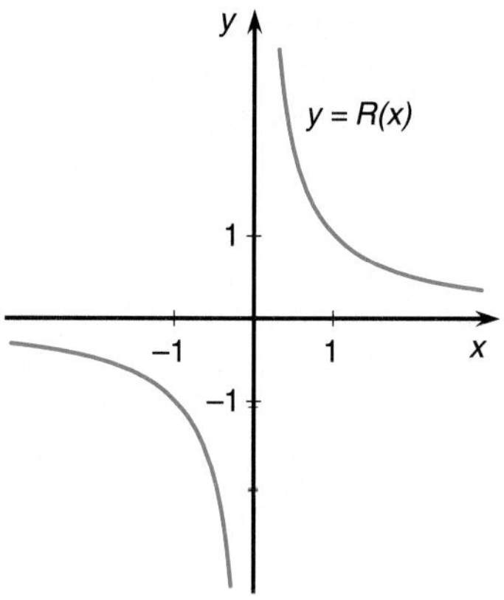

Figure 4.5.19(iv)

(iii) The function $R(x) = \frac{1}{x}$ discussed above has the y-axis $x = 0$ as an asymptote as $x \to 0$. It also approaches the x-axis $y = 0$ asymptotically as $x \to \pm\infty$. See Figure 4.5.19(iv).

(iv) As we have already suggested, the function $T(x) = \sqrt{x^2 - 1}$ approaches the line $y = x$ asymptotically as $x \to \infty$ and approaches the line $y = -x$ asymptotically as $x \to -\infty$. Strictly speaking, we should verify this fact.

What we need to check is that $\lim_{x\to\infty} \big(T(x) - x\big) = 0$. Assuming $x > 0$, we write

$$T(x) - x = \sqrt{x^2 - 1} - x.$$

Now we need a way to deal with the square root; the usual trick in this situation is to multiply and divide by the *sum* of the two terms. What results is

$$T(x) - x = \big(\sqrt{x^2 - 1} - x\big)\frac{\sqrt{x^2 - 1} + x}{\sqrt{x^2 - 1} + x} = \big((x^2 - 1) - x^2\big)\frac{1}{\sqrt{x^2 - 1} + x}$$

$$= \frac{-1}{\sqrt{x^2 - 1} + x}.$$

Since both terms in the denominator tend to ∞ as $x \to \infty$, this last quotient tends to zero, as desired.

With a graphics calculator we can plot the function $y = T(x)$ of Example 8(iv) on the same picture with the asymptotes $y = \pm x$.

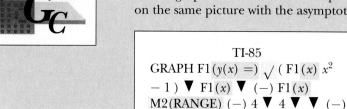

TI-85 GRAPH F1 (y(x) =) √ (F1 (x) x^2 − 1) ▼ F1 (x) ▼ (−) F1 (x) M2 (RANGE) (−) 4 ▼ 4 ▼ ▼ (−) 4 ▼ 4 F5 (GRAPH)	SHARP EL9300 [graph key] √ x/θ/T x^2 − 1 2ndF ▼ x/θ/T 2ndF ▼ − x/θ/T RANGE −4 ENTER 4 ENTER 1 ENTER −4 ENTER 4 ENTER 1 ENTER [graph key]
CASIO f_x-7700GB/f_x-6300G Cls EXE Range −4 EXE 4 EXE 1 EXE −4 EXE 4 EXE 1 EXE EXE EXE EXE Graph √ (X x^2 − 1) EXE Graph X EXE Graph − X EXE	HP 48SX ↰ PLOT PLOTR ERASE ATTN ′ αY = √x () αX y^x 2 − 1 ↱ PLOT ↰ DRAW ERASE 4 +/− ENTER 4 XRNG 4 +/− ENTER 4 YRNG DRAW ATTN ′ αY = αX ↱ PLOT ↰ DRAW DRAW ATTN ′ αY = − αX ↱ PLOT ↰ DRAW DRAW

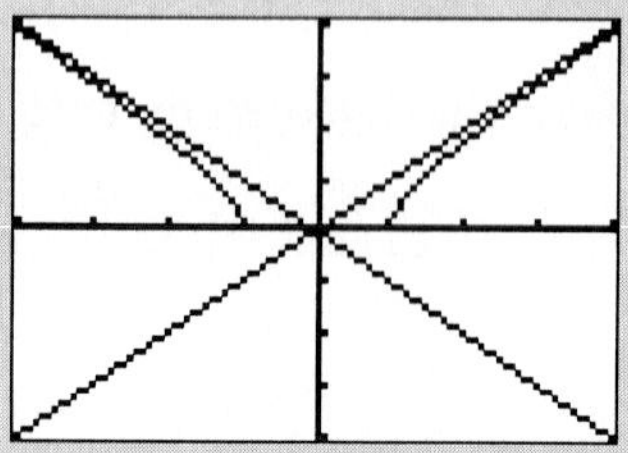

You might like to try plotting different parts of the graph (i.e., for different values of x); you might also like to include the bottom branch of the hyperbola. How close is the hyperbola to its asymptote when $x = 4$? when $x = 10$?

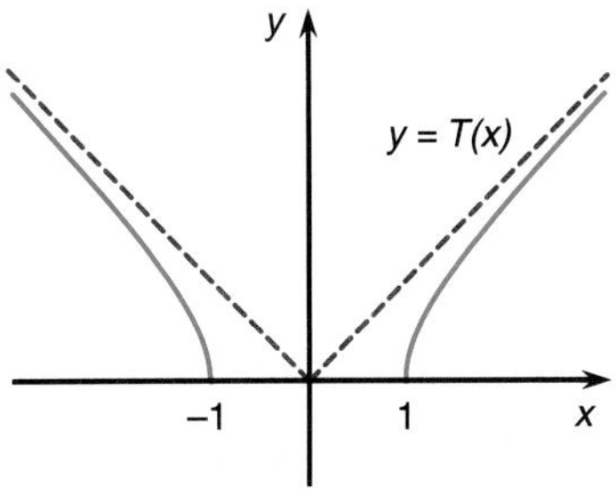

Figure 4.5.19(v)

This shows that $\lim_{x\to\infty}\big(T(x)-x\big)=0$, which shows that $T(x)$ tends asymptotically to the line $y=x$ as $x\to\infty$. A similar argument shows that $T(x)$ tends asymptotically to the line $y=-x$ as $x\to-\infty$. See Figure 4.5.19(v).

EXAMPLE 9 Sketch the graph of $y=Q(x)$, where $Q(x)=\frac{x+1}{x-1}$.

Solution The function is not defined at $x=1$. Elsewhere its derivative is $Q'(x)=\frac{1(x-1)-(x+1)1}{(x-1)^2}=\frac{-2}{(x-1)^2}$. The derivative is never zero, and since the denominator is a square, $Q'(x)$ is always negative. In other words, the function is decreasing on each interval in its domain.

The limits at $\pm\infty$ both exist and are easy to evaluate. In fact, we find that $\lim_{x\to\infty}Q(x)=\lim_{x\to-\infty}Q(x)=1$. The horizontal line $y=1$ is an asymptote for $Q(x)$ (at both ∞ and $-\infty$).

As $x\to1^+$, the numerator is positive and the denominator is positive and tends to zero. Accordingly, the quotient tends to $+\infty$. As $x\to1^-$, the numerator tends to 2 and the denominator tends to 0 through negative values. Accordingly, the quotient tends to $-\infty$. So $Q(x)$ tends asymptotically to the vertical line $x=1$ as $x\to1$ from either side.

The x-intercept is $x=-1$, and it is easy to plot a few points on the graph. For instance, we evaluate and find $(-1,0)$, $(0,-1)$, $\left(-2,\frac{1}{3}\right)$, $\left(-3,\frac{1}{2}\right)$, $\left(\frac{1}{2},-3\right)$, $\left(\frac{3}{2},5\right)$, $(2,3)$, and $(3,2)$.

From all this it is relatively easy to sketch in the graph (see Figure 4.5.20). Notice how we draw the curve to show the asymptotic behavior at $\pm\infty$ and at $x=1$.

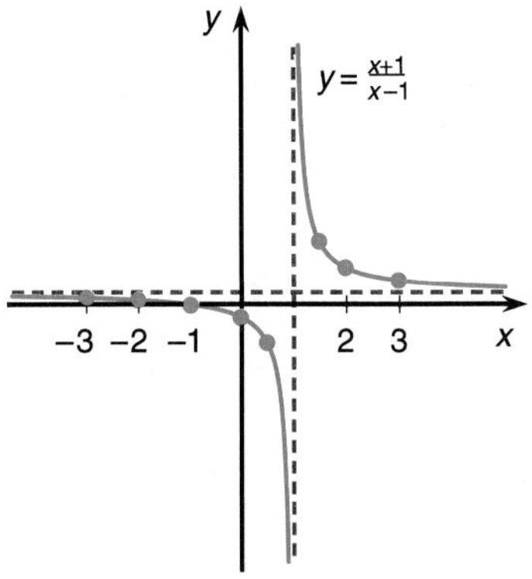

Figure 4.5.20

Another thing to notice about Example 9 is that the function is decreasing on $(-\infty,1)$ and on $(1,\infty)$, but it is not decreasing on the whole line (after all, it is bigger to the right of $x=1$ than it is to the left of $x=1$). The key point is that since the function is not defined at $x=1$, its behavior on one side of $x=1$ is not well related to its behavior on the other side. It decreases on the left side, but it "jumps up" at $x=1$ before decreasing on the right side. This is why we were careful to say the function is decreasing "on each interval in its domain."

Example 9 also illustrates the way vertical asymptotes can arise with rational functions. If $R(x)=\frac{p(x)}{q(x)}$ is a rational function, suppose $x=a$ is a point for which $q(a)=0$ but $p(a)\neq0$. Then depending on the sign of $p(a)$ and on the sign of $q(x)$ as $x\to a^+$ and as $x\to a^-$, the function $R(x)$ will tend to ∞ or to $-\infty$ as $x\to a^+$ and as $x\to a^-$. The vertical line $x=a$ will be an asymptote for the function $R(x)$.

EXAMPLE 10 Sketch the graph of $y=f(x)=\frac{x+1}{x^3-2x^2+x}$.

Solution First we notice that the denominator factors: $x^3-2x^2+x=x(x^2-2x+1)=x(x-1)^2$, so the function is undefined when $x=0$ or $x=1$. Next we calculate the derivative:

$$\begin{aligned}f'(x)&=\frac{1(x^3-2x^2+x)-(x+1)(3x^2-4x+1)}{(x^3-2x^2+x)^2}\\&=\frac{x^3-2x^2+x-3x^3+x^2+3x-1}{(x^3-2x^2+x)^2}\\&=\frac{-2x^3-x^2+4x-1}{(x^3-2x^2+x)^2}.\end{aligned}$$

We try some small values of x in the numerator and find that it is zero when $x = 1$, so the numerator is divisible by $x - 1$. We find $-2x^3 - x^2 + 4x - 1 = (x-1)(-2x^2 - 3x + 1)$.

So the derivative is

$$f'(x) = \frac{-2x^3 - x^2 + 4x - 1}{(x^3 - 2x^2 + x)^2} = \frac{(x-1)(-2x^2 - 3x + 1)}{\left(x(x-1)^2\right)^2} = \frac{-2x^2 - 3x + 1}{x^2(x-1)^3}.$$

Notice that if we had not factored out the $(x - 1)$ from the numerator and denominator, we might have thought that $f'(x) = 0$ when $x = 1$. This would be incorrect, since the function is not even defined when $x = 1$.

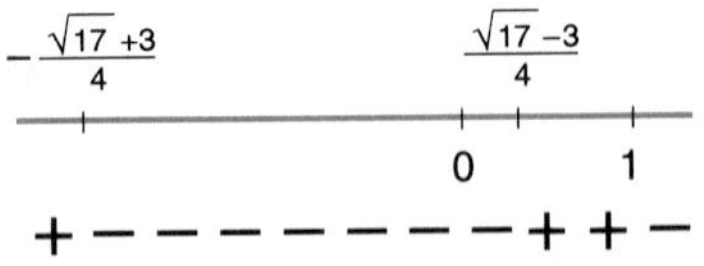

Figure 4.5.21

The critical points are the roots of $-2x^2 - 3x + 1$, which are $\frac{3 \pm \sqrt{17}}{-4} \approx -1.781, 0.281$. We draw the real line in Figure 4.5.21, marking these two points and $x = 0, 1$. By evaluating at a point in each interval (e.g., $x = -2$, $x = -1$, $x = \frac{1}{10}$, $x = \frac{1}{2}$, and $x = 2$), we see that $f'(x)$ is positive on $\left(-\infty, -\frac{3+\sqrt{17}}{4}\right)$ and on $\left(-\frac{3-\sqrt{17}}{4}, 1\right)$, and negative on the other three intervals.

As $x \to \pm\infty$, $f(x) \to 0$, so $y = 0$ is an asymptote at both ends of the graph. As $x \to 0^+$, the denominator, which equals $x(x-1)^2$, is positive and approaches zero. The numerator approaches 1, so the quotient $f(x)$ approaches $+\infty$. As $x \to 0^-$, the denominator is negative, the numerator approaches 1, and $f(x) \to -\infty$. As $x \to 1^+$, the denominator is positive and approaches 0. The numerator approaches 2, so $f(x) \to \infty$. As $x \to 1^-$, both numerator and denominator are positive and $f(x) \to \infty$. We see that $x = 0$ and $x = 1$ are both asymptotes for the graph of $y = f(x)$.

From this information we see that there are no absolute extrema. There is a local maximum at $x = -\frac{3+\sqrt{17}}{4}$, and a local minimum at $x = \frac{\sqrt{17}-3}{4}$. The only x-intercept is at $x = -1$. We should plot the local extrema on our graph; for this purpose, approximate values are perfectly adequate. We find $f(-1.781) \approx 0.0567$, $f(0.281) \approx 8.818$.

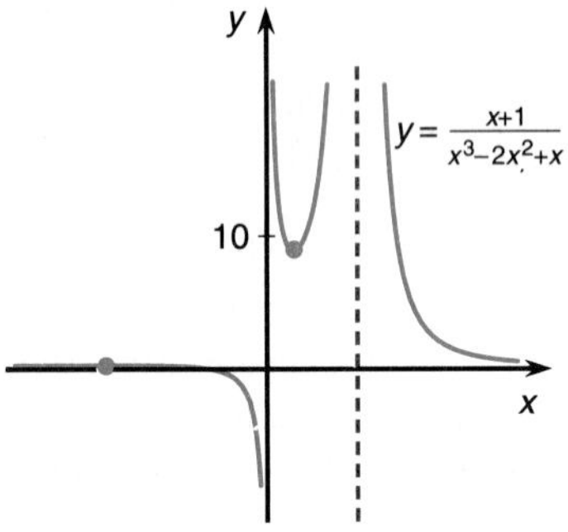

Figure 4.5.22

With the information we now have, it is possible to sketch the graph of $y = f(x)$ reasonably accurately. See Figure 4.5.22.

There is one other type of situation in which asymptotes are readily seen to play a role. Suppose $R(x) = \frac{p(x)}{q(x)}$ is a rational function in which the degree of $p(x)$ is 1 larger than the degree of $q(x)$. In this situation we know that $R(x)$ will tend to ∞ or $-\infty$ as $x \to \infty$ and as $x \to -\infty$.

EXAMPLE 11 Sketch the graph of $y = \frac{x^2-1}{2x+4}$.

Solution For convenience we write $g(x) = \frac{x^2-1}{2x+4}$. Then $g(x)$ is not defined when $x = -2$. At other points,

$$g'(x) = \frac{2x(2x+4) - (x^2-1)2}{(2x+4)^2} = \frac{4x^2 + 8x - 2x^2 + 2}{(2x+4)^2}$$

$$= \frac{2x^2 + 8x + 2}{4(x+2)^2} = \frac{x^2 + 4x + 1}{2(x+2)^2}.$$

The critical points are $x = \frac{-4 \pm \sqrt{12}}{2} = -2 \pm \sqrt{3} \approx -0.268, -3.732$.

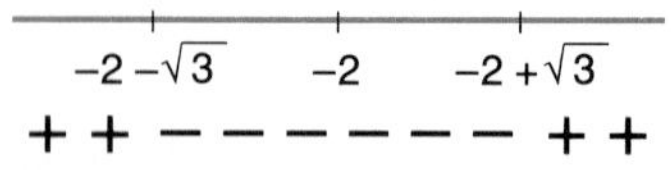

Figure 4.5.23

By evaluating at $x = -4$, $x = -3$, $x = -1$, and $x = 0$ we find that $g'(x)$ is negative on $(-2 - \sqrt{3}, -2)$ and on $(-2, -2 + \sqrt{3})$ and positive on $(-\infty, -2 - \sqrt{3})$ and on $(-2 + \sqrt{3}, \infty)$ (see Figure 4.5.23).

As $x \to -2^+$, the numerator of $g(x)$ tends to 3 and the denominator, which is obviously positive, tends to zero. We see that $\lim_{x\to -2^+} g(x) = \infty$. Similarly, as $x \to -2^-$, the numerator tends to 3 and the denominator is negative and tends to zero, so $\lim_{x\to -2^-} g(x) = -\infty$. From this it is apparent that there are no absolute extrema and that there is a local maximum at $x = -2 - \sqrt{3}$ and a local minimum at $x = -2 + \sqrt{3}$.

Finally, we should consider what happens as $x \to \pm\infty$. To do this, we perform a long division and write $g(x) = \frac{x^2-1}{2x+4}$ as a polynomial plus a remainder term. We find

$$g(x) = \frac{x^2 - 1}{2x + 4} = \frac{1}{2}x - 1 + \frac{3}{2x + 4}.$$

Now as $x \to \pm\infty$, the last term $\frac{3}{2x+4}$ tends to zero. This says that as $x \to \pm\infty$, the graph of the function $g(x)$ gets very close to the function $\frac{1}{2}x - 1$. This last function is of course a straight line, and we have just seen that $g(x)$ tends asymptotically to it as $x \to \pm\infty$.

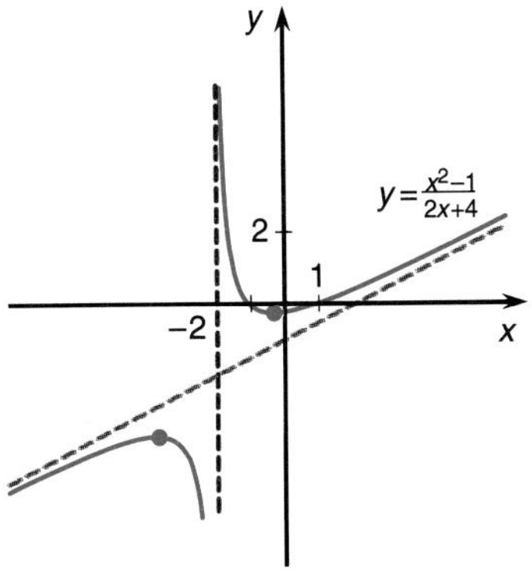

Figure 4.5.24

Keeping in mind everything we have learned, we sketch the graph of $y = g(x)$ in Figure 4.5.24. The asymptotes $x = -2$ and $y = \frac{1}{2}x - 1$ are marked as dashed lines. Notice that the remainder term $\frac{3}{2x+4}$, which tells us the difference between $g(x)$ and the straight line, is positive when $x > -2$ and negative when $x < -2$, which says that the graph of $y = g(x)$ is above the asymptote when $x > -2$ and below the asymptote when $x < -2$. ◢

We began by considering a rational function $R(x) = \frac{p(x)}{q(x)}$ in which the degree of $p(x)$ was 1 larger than the degree of $q(x)$. If we do a long division in this situation, the first term in the answer will be a nonzero constant times x, and the second will be a constant (possibly zero). We get $R(x) = ax + b + \frac{r(x)}{q(x)}$.

In the remainder term the degree of the numerator is smaller than the degree of the denominator, so as $x \to \pm\infty$, $\frac{r(x)}{q(x)}$ will tend to zero.

THEOREM 4.5.3

For a rational function $R(x) = \frac{p(x)}{q(x)}$ in which the degree of $p(x)$ is 1 greater than the degree of $q(x)$, we can write

$$R(x) = ax + b + \frac{r(x)}{q(x)},$$

with $a \neq 0$ and the degree of $r(x)$ less than the degree of $q(x)$. Since $\lim_{x\to\pm\infty} \frac{r(x)}{q(x)} = 0$, we see that $R(x)$ tends asymptotically to the line $y = ax + b$ as $x \to \pm\infty$.

If the degrees of $p(x)$ and $q(x)$ are equal, we can write

$$R(x) = a + \frac{r(x)}{q(x)},$$

with the degree of $r(x)$ less than the degree of $q(x)$ and $a \neq 0$. In this case, $R(x)$ tends asymptotically to the horizontal line $y = a$ as $x \to \pm\infty$.

PROOF The degree of the remainder term $r(x)$ is less than the degree of $q(x)$ because otherwise we could continue the long division. ◢

EXAMPLE 12 Sketch the graph of $y = h(x) = \frac{x^3}{x^2+1}$.

Solution The denominator is never zero, so the function is defined and differentiable everywhere. Its derivative is

$$h'(x) = \frac{3x^2(x^2+1) - x^3(2x)}{(x^2+1)^2} = \frac{x^4+3x^2}{(x^2+1)^2} = \frac{x^2(x^2+3)}{(x^2+1)^2}.$$

The only critical point is $x = 0$, and since everything in the derivative is a square, the derivative is positive everywhere except at $x = 0$. It is also clear that $\lim_{x\to-\infty} h(x) = -\infty$ and $\lim_{x\to\infty} h(x) = \infty$, and there are no local maxima or minima. The only x-intercept is $x = 0$. We observe that $h(x)$ is an odd function.

Performing a long division, we find that

$$h(x) = \frac{x^3}{x^2+1} = x - \frac{x}{x^2+1}.$$

As $x \to \pm\infty$, the graph of $y = h(x)$ tends asymptotically to the line $y = x$.

We plot a few points, for example, $(0, 0)$, $\left(1, \frac{1}{2}\right)$, $\left(-1, -\frac{1}{2}\right)$, $\left(2, \frac{8}{5}\right)$, and $\left(-2, -\frac{8}{5}\right)$. We are now able to sketch the graph of $y = h(x)$. See Figure 4.5.25.

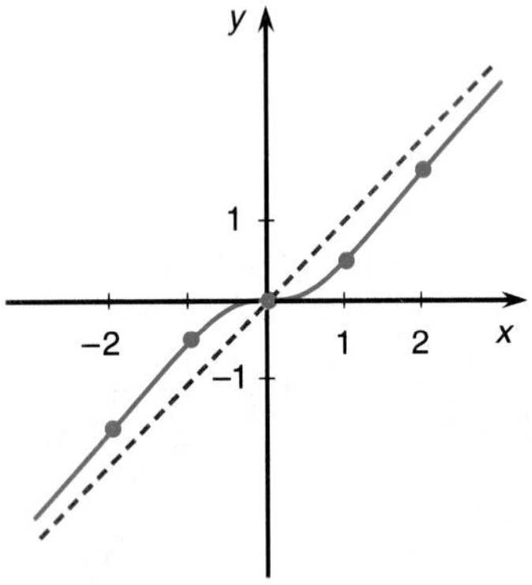

Figure 4.5.25

After we performed the long division in Example 12, we might have observed that since $h(x) = x - \frac{x}{x^2+1}$, we could think of its graph as being obtained from the graph of the straight line $y = x$ by shifting down by the graph of $y = \frac{x}{x^2+1}$. We graphed this last function in Example 2, and if you tilt this page 45° clockwise, you can see that the graph of $h(x)$ does appear to be the negative of $y = \frac{x}{x^2+1}$ added to the line $y = x$. (In fact it is stretched out because of the tilting, but the shape is still roughly the same.)

Finally, we summarize the steps we have been using to sketch the graphs of rational functions.

Summary 4.5.4 Sketching Rational Functions

1. Find where the function is undefined, by finding where the denominator is zero. If you cannot factor the denominator, you can use the Rational Root Theorem (Theorem 1.3.7, p. 13) in hopes of finding a small integer value at which the denominator is zero. If the denominator is zero at $x = a$, be sure to check whether the numerator is also zero at $x = a$; if it is, it is possible to factor out $(x - a)$ from both numerator and denominator. Check whether the function happens to be either an even function or an odd function.
2. Find the derivative. Find all critical points. By evaluating at one point in each interval in the domain between critical points, find the sign of the derivative on each interval. Draw a picture of the domain, marking all this information. Find the local extrema.
3. For each point $x = a$ at which the denominator is zero, study the behavior of the rational function as $x \to a^+$ and as $x \to a^-$. This involves the signs of numerator and denominator. Also determine the behavior as $x \to \pm\infty$.

4. If the degree of the numerator is greater than the degree of the denominator by 1, then perform a long division and find the asymptotes at $\pm\infty$. At this stage it should be possible to identify any absolute extrema.
5. Plot a few points, including local and absolute extrema and all x-intercepts that can be found. Sketch the curve.

You might like to review the examples of this section, many of which are rational functions, in light of this summary.

Exercises 4.5

I

Find all asymptotes for each of the following functions.

1. $f(x) = \frac{x^2}{x-1}$
2. $f(x) = \frac{x^2}{x+2}$
3. $f(x) = \frac{x^2+2}{x^2-3}$
4. $f(x) = \frac{x^2+x-1}{x+2}$
5. $f(x) = \frac{x^3}{x^2-4}$
6. $f(x) = \frac{2x^3-x^2+1}{3x^2-1}$
7. $f(x) = \frac{x^3}{x^2+x}$
8. $f(x) = \frac{x^4-x}{2x^2-3x+1}$
9. $f(x) = \frac{x^7}{(x-3)(x+2)x^2}$
10. $f(x) = \frac{x^2-1}{x+1}$

Sketch the graphs of the following functions, showing all asymptotes.

11. $f(x) = \frac{x}{x-1}$
12. $f(x) = \frac{x^2}{x-1}$
13. $f(x) = \frac{x^3}{x^2-1}$
14. $f(x) = \frac{x^3}{x^2-4}$
15. $f(x) = \frac{x^3-3x^2+3x-1}{x^2-2x}$
16. $f(x) = \frac{x^3}{x^2+4}$
17. $f(x) = \frac{1}{x}$
18. $f(x) = \frac{x^3+x^2+4}{x^2+4}$
19. $f(x) = \frac{x^3+x+1}{x^2+1}$
20. $f(x) = \frac{x^4-2x^2-3}{x^2+1}$

II

21. Sketch the graph of $y = f(x) = \frac{x^2-9}{2x+4}$.

22. Sketch the graph of $y = g(x) = \frac{4x^2-1}{3x-6}$.

*23. Consider a rational function $R(x) = \frac{p(x)}{q(x)}$, where $q(x) = (x-a)^k g(x)$ and $g(a) \neq 0$, $p(a) \neq 0$, $k \geq 0$. Under what circumstances will $\lim_{x\to a^+} R(x) = \lim_{x\to a^-} R(x)$?

24. (i) Show that if $f(x)$ is even and $g(x)$ is odd, then $f(x)g(x)$ is odd.
    (ii) Show that if $f(x)$ and $g(x)$ are both even or both odd, then their product is even.
    (iii) Show that the sum of two even functions is even and the sum of two odd functions is odd.

*25. Find conditions on a polynomial $P(x) = a_n x^x + a_{n-1}x^{n-1} + \ldots + a_1 x + a_0$ that guarantee that $P(x)$ is (i) even, (ii) odd, (iii) neither even nor odd.

**26. Let $F(x)$ be a function whose domain is the whole real line. Show that it is possible to write $F(x) = F_{even}(x) + F_{odd}(x)$, where $F_{even}(x)$ is an even function and $F_{odd}(x)$ is an odd function.

*27. (i) Let $f(x) = 3x^4 - 2x^3 - x^2 + 7x - 1$. Write $f(x)$ as the sum of an even function and an odd function.
    (ii) If $P(x) = a_n x^n + a_{n-1}x^{n-1} + \ldots + a_1 x + a_0$, write $P(x)$ as the sum of an even and an odd function.

*28. Let $g(x)$ be both even and odd. What can you say about $g(x)$?

*29. If $H(x) = H_e(x) + H_o(x)$, where H_e and H_o are even and odd, respectively, show that this can be done in only one way, that is, show that if $H(x) = f(x) + g(x)$ with f even and g odd, then $f(x) = H_e(x)$ and $g(x) = H_o(x)$.

*30. Find an example of a function $G(x)$ that cannot be written as the sum of an even function and an odd function. (*Hint:* Think about the domain of $G(x)$.)

*31. (i) Suppose $f(x)$ is an odd function that is differentiable at every x. Show that $f'(x)$ is an even function.
    (ii) Suppose $f(x)$ is an even function that is differentiable at every x. What can you say about $f'(x)$?
    (iii) What happens in parts (i) and (ii) if we remove the assumption that $f(x)$ is differentiable everywhere and just let $f'(x)$ be the derivative when it does exist and be undefined when it does not?

*32. Write each of the following functions as the sum of an even function and an odd function: (i) $\cos\left(x + \frac{\pi}{4}\right)$, (ii) e^x, (iii) $(x-3)^4$.

III

Use a graphics calculator or computer to sketch the following functions considered in the examples in the text.

33. $f(x)$ (Example 2)

34. $g(x)$ (Example 3)

35. $h(x)$ (Example 4)

36. $G(x)$ (Example 7)

37. $Q(x)$ (Example 9)

38. $f(x)$ (Example 10)

39. $g(x)$ (Example 11)

40. $h(x)$ (Example 12)

41. Sketch the function $G(x)$ of Example 7 (p. 268) on the same graph with $y = -\frac{2}{x^2+2}$.

42. Use a graphics calculator or computer to find how large x has to be before $T(x) = \sqrt{x^2 - 1}$ stays within 0.001 of x (cf. Example 8(iv), p. 270).

POINT TO PONDER

In the examples of asymptotes we have considered so far, the graph of a function has usually approached the asymptotic line from above or below (cf. Figures 4.5.26(i) and 4.5.26(ii)). However, it is perfectly possible for the function to cross the asymptote.

As an example, consider the graph of $y = f(x) = \frac{x}{2} + \frac{1}{x}\sin(5x)$, shown in Figure 4.5.26(iii). As $x \to \pm\infty$, we see that $\frac{1}{x}\sin(5x) \to 0$, so $f(x)$ approaches the line $y = \frac{x}{2}$, which is the asymptote. However, the function $f(x)$ crosses and recrosses this line as $x \to \pm\infty$.

Notice that $\lim_{x\to 0} f(x) = \lim_{x\to 0} \frac{\sin(5x)}{x} = 5$.

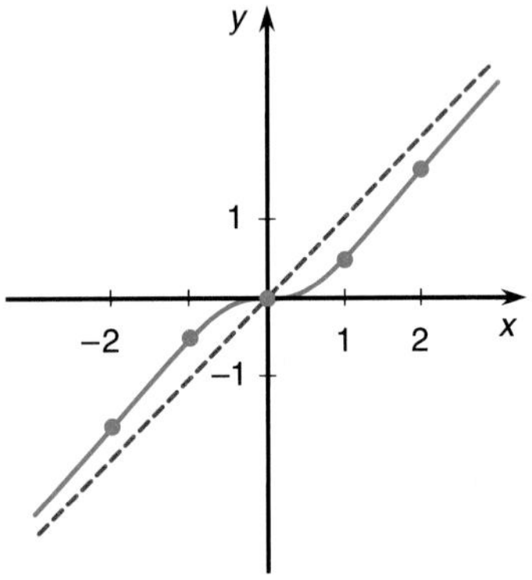

Figure 4.5.26(i)

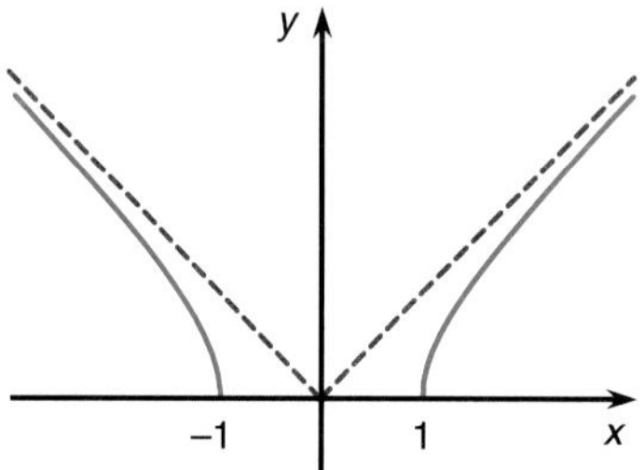

Figure 4.5.26(ii)

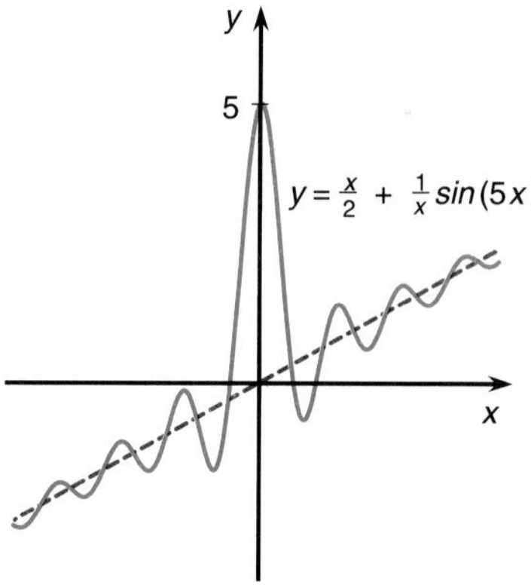

Figure 4.5.26(iii)

4.6 Concavity and Inflection Points; Applications to Curve Sketching

CONCAVE UP
CONCAVE DOWN
SECOND DERIVATIVE AND CONCAVITY
INFLECTION POINT
CUSP

In this section we introduce the notion of concavity and discuss how to test for it and how to apply it to curve sketching problems. In the process we learn about *inflection points*, which are points where the concavity changes.

In Figure 4.6.1 there are several short curves. The first one is increasing, the second is decreasing, and the third is decreasing at first and then increasing. What they all have in common is the way in which they are bending.

This may be clearer by comparison with Figure 4.6.2, in which there are also three curves. One is increasing, one is decreasing, and one is increasing at first and then decreasing.

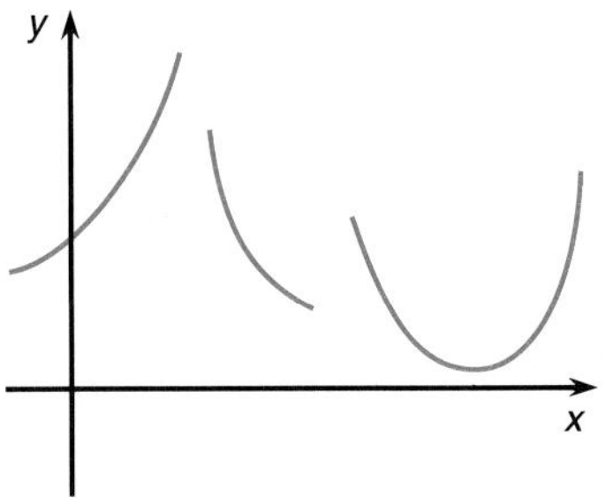

Figure 4.6.1

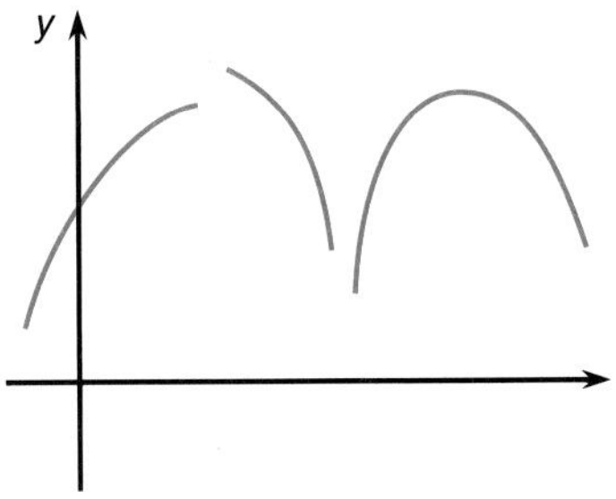

Figure 4.6.2

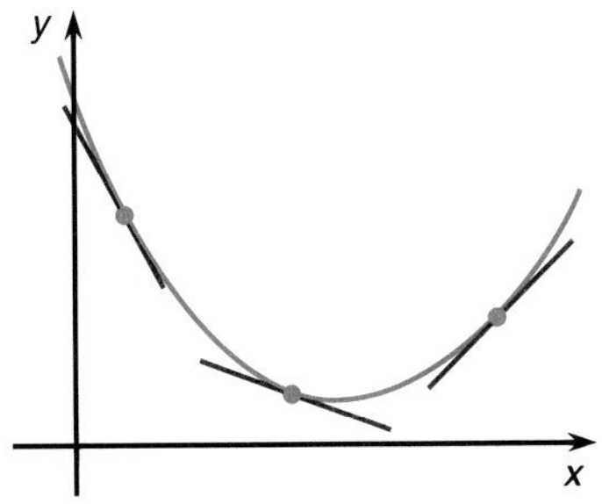

Figure 4.6.3

When a curve bends, we can think of it having a concave side and a convex side. The three curves in Figure 4.6.1 all have their concave side facing up, while the ones in Figure 4.6.2 all have their concave side facing down. Notice that this is independent of whether a curve is increasing or decreasing, in the sense that the concave side can face up for increasing or decreasing curves and can likewise face down for increasing or decreasing curves.

The first thing we need to do is make a precise definition of what we mean by concavity. Once that is accomplished, we should look for ways to find out which side of a particular curve is concave. We will only consider the question at a point where the function is differentiable.

One great advantage of this restriction is that it means that we can draw the tangent line. In Figure 4.6.3 we have drawn a curve that is "concave up," that is, whose concave side faces upward. The tangent lines have been drawn in at several points, and from the picture we see that the curve lies *above* each tangent line.

Figure 4.6.4 shows a function that is "concave down," and the curve lies below all its tangent lines. This certainly suggests a way of defining what we mean by "concave up" and "concave down," but we have to be careful, as the next example shows.

EXAMPLE 1 Consider the function $y = g(x) = x^3 - 3x^2 + 2$, whose graph we sketched in Example 4.5.3 (p. 265) (see Figure 4.6.5). The part of the graph near $x = 0$, which looks like a hump, is concave down. However, the part of the curve near $x = 2$, which looks like a valley, is concave up. In the first place, this suggests that it is perfectly possible for a function to be concave up at some points and concave down at other points.

Now if we try the idea of looking at the tangent lines, we find that the part of the curve that is the hump does in fact lie below the tangent line at $x = 0$ (which of course is horizontal). However, the curve bends up again and crosses this tangent line and eventually rises above it. Similarly, the part of the curve that is the valley does lie above the tangent line at $x = 2$, but part of the curve farther to the left bends down and actually crosses below the tangent.

We have chosen the two horizontal tangents because they are easy to draw, but similar remarks apply to other tangent lines. The curve may well lie above or below the tangent near the point of tangency, but farther away it may loop back and cross the line.

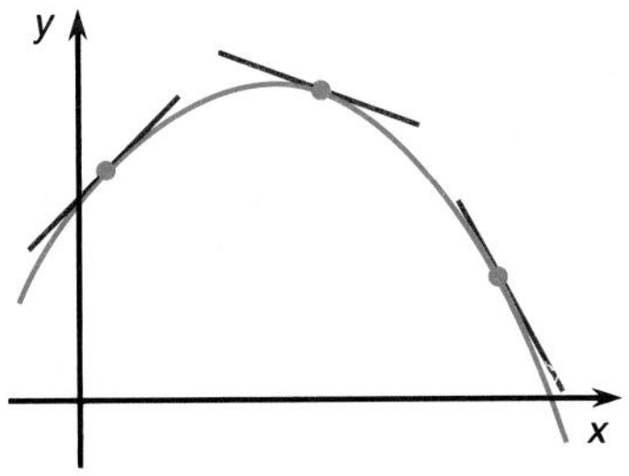

Figure 4.6.4

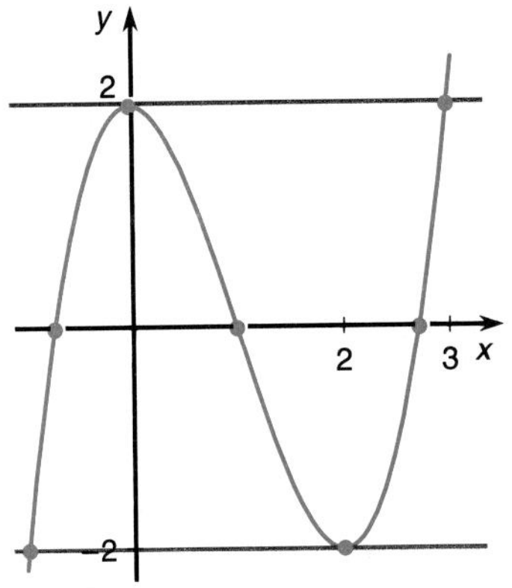

Figure 4.6.5

The answer to this difficulty is to realize that the property of a curve being concave up or concave down at some point $x = a$ is something that has to do only with its behavior near that point. So the definition should be that the curve lies above or below the tangent line *near* $x = a$, that is, on some open interval containing $x = a$.

Recall that the tangent line to $y = f(x)$ at the point $x = a$ has the equation $y = f(a) + f'(a)(x - a)$ (cf. Definition 3.1.1).

DEFINITION 4.6.1

A function $f(x)$ that is differentiable at $x = a$ is **concave up** at $x = a$ if there is an interval (c, d) containing a so that the graph $y = f(x)$ lies above the tangent line to $f(x)$ at $x = a$ at every point in the interval (c, d) except $x = a$. It is **concave down** at $x = a$ if there is an interval (c, d) containing a so that the graph $y = f(x)$ lies below the tangent line to $f(x)$ at $x = a$ at every point in the interval (c, d) except $x = a$.

In symbols this means that $f(x)$ is concave up at $x = a$ if

$$f(x) > f(a) + f'(a)(x - a), \qquad \text{for every } x \in (c, d) \text{ except } x = a$$

and concave down if

$$f(x) < f(a) + f'(a)(x - a) \qquad \text{for every } x \in (c, d) \text{ except } x = a.$$

A function is said to be concave up (respectively concave down) on an interval if it is concave up (respectively concave down) at every point in that interval. It is said to be concave up (respectively concave down) if it has this property at every point in its domain.

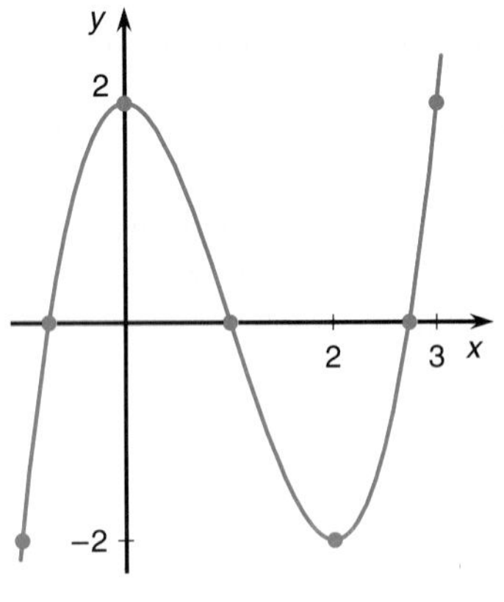

Figure 4.6.6

EXAMPLE 2

(i) The function $g(x) = x^3 - 3x^2 + 2$ discussed in Example 1 is concave down on $(-\infty, 1)$ and concave up on $(1, \infty)$ (see Figure 4.6.6). Notice that although the curve does eventually rise above the horizontal tangent at $x = 0$, there is an interval (e.g., $(-1, 1)$ will do in this case) on which the curve is always below the tangent. It is possible to find such an interval for the tangent at any point $x = a$ with $a < 1$ (although the interval might not continue very far to the right of a if a is very close to 1). If $a > 1$, an interval can be found on which the curve is above the tangent at $x = a$.

(ii) The parabola $y = x^2$ is concave up (i.e., concave up at every point). See Figure 4.6.7.

(iii) The function $y = x^3$, sketched in Figure 4.6.8, is concave down on $(-\infty, 0)$ and concave up on $(0, \infty)$.

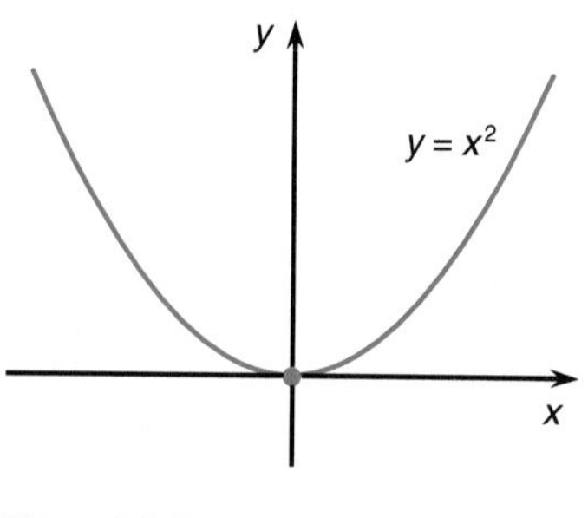

Figure 4.6.7

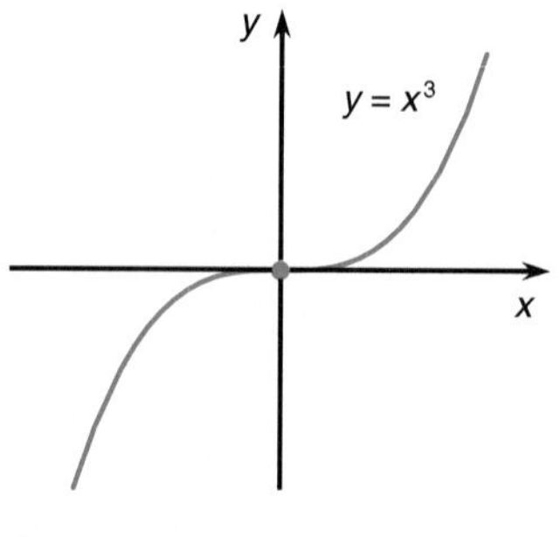

Figure 4.6.8

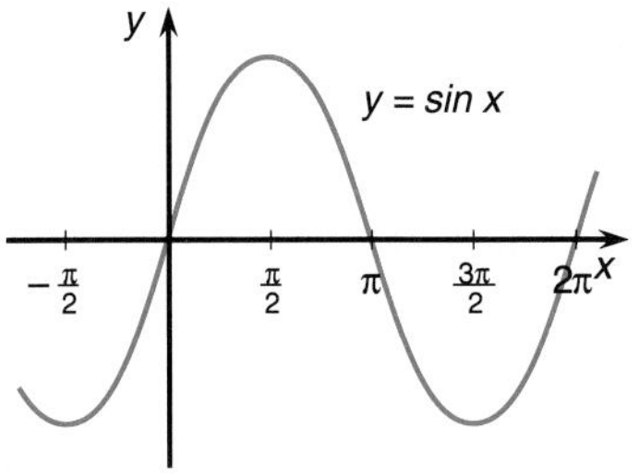

Figure 4.6.9

(iv) The function $\sin x$ is concave down on $(0, \pi)$ and concave up on $(\pi, 2\pi)$. See Figure 4.6.9.

(v) The function $\sqrt{x}$ is concave down on $(0, \infty)$. Notice that it is only defined for $x \geq 0$, and it does not make sense to ask about its concavity at $x = 0$, since it is not differentiable there. See Figure 4.6.10.

Knowing the concavity of $f(x)$ is a very good way to improve the accuracy of our graphs. What we want is to be able to say where the function is concave up and where it is concave down, just as we began our sketching efforts by looking for where it was increasing or decreasing. Regions of increase and decrease were identified by using the derivative $f'(x)$, and it turns out that the concavity can be determined from the second derivative $f''(x)$.

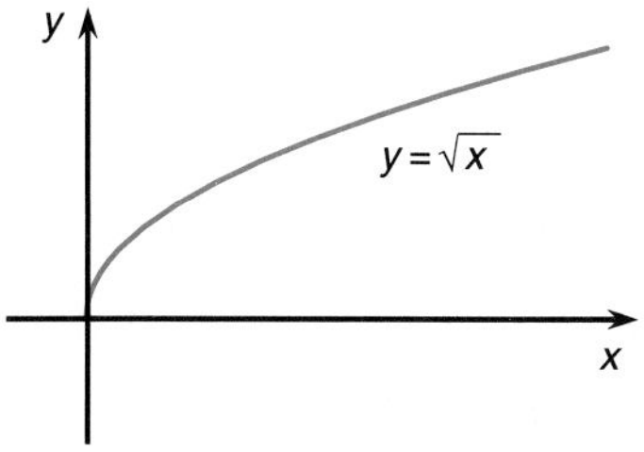

Figure 4.6.10

In Figure 4.6.11 we see a curve that is concave up. At the left end, the curve is decreasing, so its derivative is negative. As we move to the right, the tangents slope less steeply, so the derivative becomes closer to zero. At the point $x = a$ the tangent is horizontal and the slope is zero. Farther to the right, the slope becomes positive, and the tangents slope more and more steeply upward, meaning that the slope becomes larger and larger.

In short, the derivative is increasing all the time. If we look at the graph of a function that is concave down, we see in a similar fashion that its derivative is always decreasing. See Figure 4.6.12.

If the derivative is increasing, it means that the tangent lines slope more and more steeply up (or less and less steeply down), which suggests that the curve is bending toward the region above the curve, which in turn suggests that it will be concave up. A decreasing derivative suggests a curve that is concave down. We can detect when the derivative is increasing or decreasing by looking at *its* derivative, the second derivative.

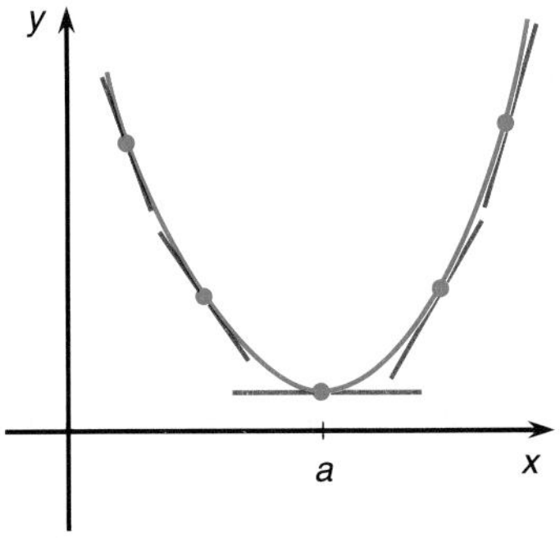

Figure 4.6.11

THEOREM 4.6.2

The Second Derivative and Concavity

Suppose $f''(x)$ exists at a point $x = a$.

If $f''(a) > 0$, then $f(x)$ is concave up at $x = a$.

If $f''(a) < 0$, then $f(x)$ is concave down at $x = a$.

The proof of Theorem 4.6.2 is surprisingly tricky. Instead of reading it right away, you might prefer to skip ahead to Example 3 to get some idea of how the theorem is used. Using the result is actually quite easy.

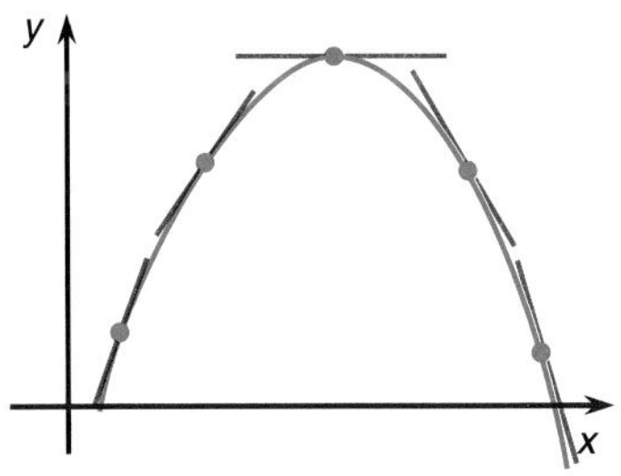

Figure 4.6.12

PROOF Suppose $f''(a) > 0$. In particular, this implies that $f'(x)$ must exist on some open interval containing $x = a$. Moreover, since $f''(a) = \lim_{x \to a} \frac{f'(x) - f'(a)}{x - a}$ is *positive*, we can apply Theorem 2.3.4 to conclude that there is an open interval (r, s) containing a so that

$$\frac{f'(x) - f'(a)}{x - a} > 0, \qquad \text{whenever } x \in (r, s) \qquad (\text{except} \quad x = a).$$

We will show that on this interval (r, s), the curve $y = f(x)$ lies above the tangent line to the curve at $x = a$. The vertical distance between the curve $y = f(x)$ and the tangent line $y = f(a) + f'(a)(x - a)$ at any particular value of x is

$$D = f(x) - \big(f(a) + f'(a)(x - a)\big)$$

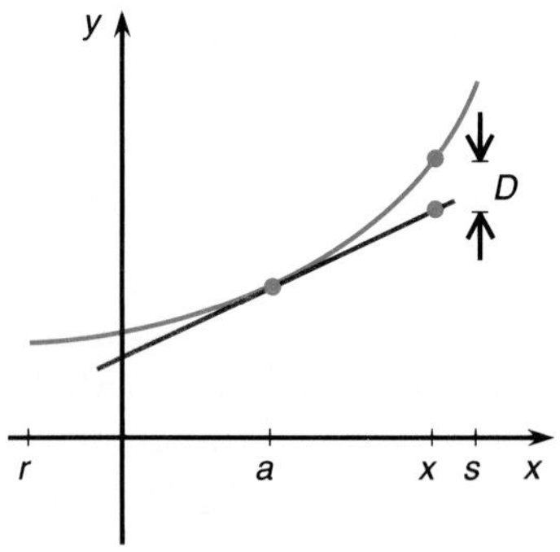

Figure 4.6.13

(see Figure 4.6.13). We have to show that $D > 0$ whenever $x \in (r, s)$, except for $x = a$.

Now

$$D = f(x) - \big(f(a) + f'(a)(x-a)\big) = f(x) - f(a) - f'(a)(x-a)$$
$$= \left(\frac{f(x) - f(a)}{x - a} - f'(a)\right)(x - a).$$

The Mean Value Theorem (Theorem 4.2.5) shows that the average slope $\frac{f(x)-f(a)}{x-a}$ equals $f'(c)$ for some c between a and x. In particular, $c \in (r, s)$. So

$$D = \big(f'(c) - f'(a)\big)(x-a)$$
$$= \frac{f'(c) - f'(a)}{c - a}(c-a)(x-a).$$

The first factor is positive, since $c \in (r, s)$. Since c and x are on the same side of a, the factors $(c - a)$ and $(x - a)$ have the same sign, so the product $(c - a)(x - a)$ is also positive. This shows that $D > 0$ whenever $x \in (r, s)$, as desired.

The case in which $f''(a) < 0$ can be proved similarly, or by considering the function $-f(x)$.

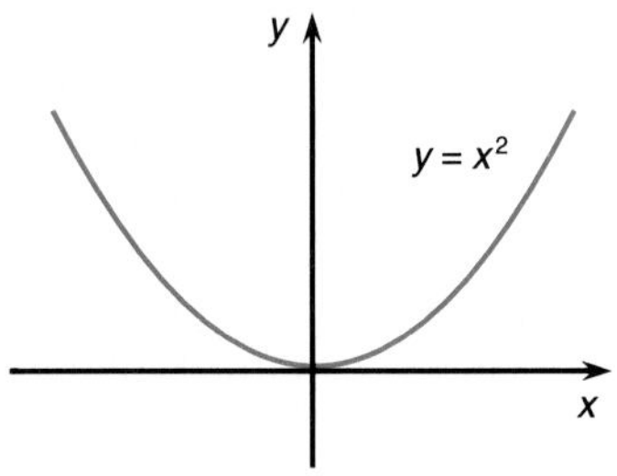

Figure 4.6.14(i)

EXAMPLE 3

(i) The function $f(x) = x^2$ satisfies $f'(x) = 2x$, $f''(x) = 2$. Since the second derivative is always positive, the function is concave up at every point. See Figure 4.6.14(i).

(ii) The function $g(x) = x^3$ satisfies $g'(x) = 3x^2$, $g''(x) = 6x$. The function $g''(x) = 6x$ is positive when $x > 0$ and negative when $x < 0$, so Theorem 4.6.2 tells us that $g(x)$ is concave up when $x > 0$ and concave down when $x < 0$. See Figure 4.6.14(ii).

(iii) The function $h(x) = e^x$ satisfies $h'(x) = h''(x) = e^x$. Since e^x is always positive, the function $h(x)$ is concave up at every point. See Figure 4.6.14(iii).

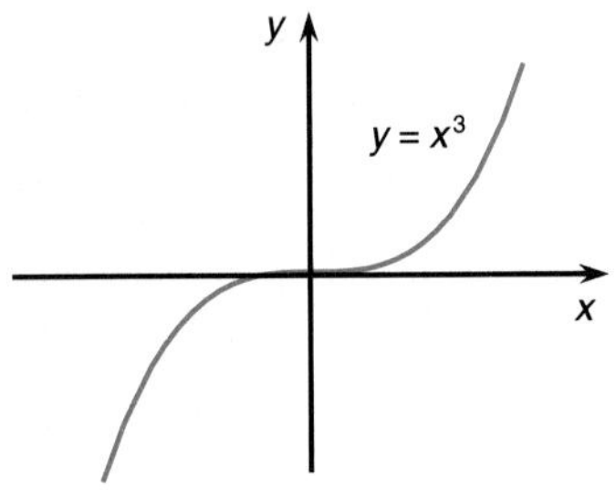

Figure 4.6.14(ii)

Now when we sketch the graph of a function, we can determine not only the regions in which it is increasing or decreasing, but also the regions in which it is concave up or concave down. Just as we drew a picture of the real line and marked on it the sign of the derivative, we can mark the type of concavity. One way to do this is simply to draw little U-shaped curves in the regions where the function is concave up and inverted U's where it is concave down (see Figure 4.6.15). This information can easily be found from the second derivative and will be useful when the sketch of the curve is drawn.

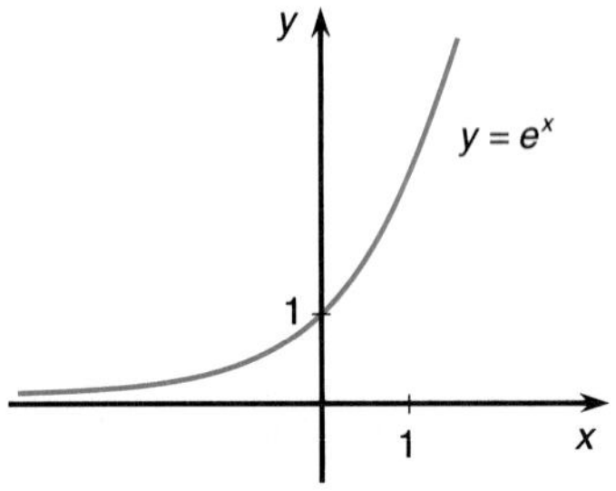

Figure 4.6.14(iii)

EXAMPLE 4 Recall the function $h(x) = \frac{1}{x^2+2}$, which we considered in Example 4.5.4. There we calculated the derivative

$$h'(x) = -\frac{2x}{(x^2+2)^2}.$$

Differentiating this, we find the second derivative:

$$h''(x) = -\frac{2(x^2+2)^2 - (2x)2(x^2+2)2x}{(x^2+2)^4} = -\frac{2x^2 + 4 - 8x^2}{(x^2+2)^3}$$
$$= \frac{6x^2 - 4}{(x^2+2)^3}.$$

Figure 4.6.15

The denominator is always positive. The numerator is positive when $x < -\sqrt{\frac{2}{3}}$ and when $x > \sqrt{\frac{2}{3}}$.

We conclude that the second derivative is positive on $\left(-\infty, -\sqrt{\frac{2}{3}}\right)$ and on $\left(\sqrt{\frac{2}{3}}, \infty\right)$ and similarly that it is negative on $\left(-\sqrt{\frac{2}{3}}, \sqrt{\frac{2}{3}}\right)$.

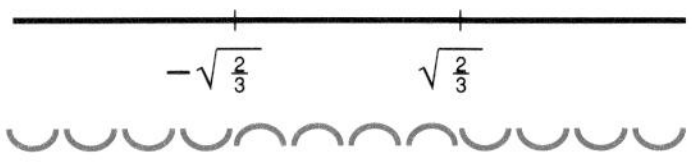

Figure 4.6.16

We draw the real line in Figure 4.6.16, marking these intervals and indicating the concavity of $h(x)$, marking concave up regions (where $h''(x) > 0$) with little U-shaped curves and concave down regions with little inverted U-shaped curves.

When we sketch the graph, we can use this information to give the graph the correct shape. See Figure 4.6.17.

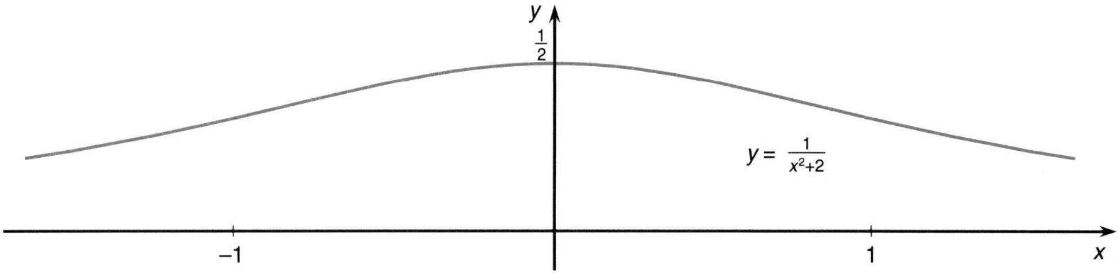

Figure 4.6.17

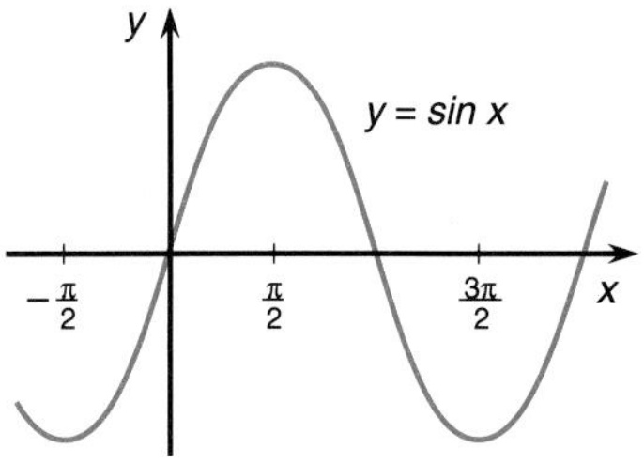

Figure 4.6.18

EXAMPLE 5 Sketch the graph of $y = S(\theta) = \sec\theta$.

Solution Note that $S(\theta) = \frac{1}{\cos\theta}$ is not defined when $\cos\theta = 0$, that is, when $\theta = \pm\frac{\pi}{2}, \pm\frac{3\pi}{2}, \pm\frac{5\pi}{2}, \ldots$. Recall that $S'(\theta) = \sec\theta\tan\theta = \frac{\sin\theta}{\cos^2\theta}$. So $S'(\theta) = 0$ for $\theta = n\pi$, with $n = 0, \pm 1, \pm 2, \ldots$.

Let us consider the graph between $\theta = 0$ and $\theta = 2\pi$. For $\theta \in \left(0, \frac{\pi}{2}\right)$, $\sin\theta$ is positive (see Figure 4.6.18), and since $\cos^2\theta > 0$, we see that $S'(\theta) > 0$ for $\theta \in \left(0, \frac{\pi}{2}\right)$. Similarly, $\sin\theta > 0$ for $\theta \in \left(\frac{\pi}{2}, \pi\right)$, so $S'(\theta) > 0$ there, and $S'(\theta) < 0$ for $\theta \in \left(\pi, \frac{3\pi}{2}\right)$ and for $\theta \in \left(\frac{3\pi}{2}, 2\pi\right)$.

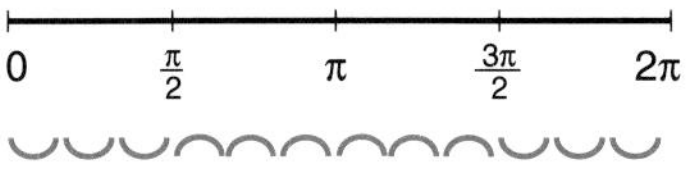

Figure 4.6.19

By the Product Rule, $S''(\theta) = \sec\theta\tan^2\theta + \sec^3\theta = \sec\theta(\tan^2\theta + \sec^2\theta)$. Since the sum of squares is never negative, $S''(\theta)$ is positive for $\theta \in \left(0, \frac{\pi}{2}\right)$ and $\theta \in \left(\frac{3\pi}{2}, 2\pi\right)$. It is negative for $\theta \in \left(\frac{\pi}{2}, \frac{3\pi}{2}\right)$. We sketch the part of the real line from 0 to 2π in Figure 4.6.19 and mark the concavity of $S(\theta)$ in the usual way.

As $\theta \to \frac{\pi}{2}^-$, $\cos\theta$ approaches zero from above, that is, $\cos\theta > 0$, so $S(\theta) = \sec\theta = \frac{1}{\cos\theta} \to \infty$, that is, $\lim_{\theta\to\pi/2^-} S(\theta) = \infty$. As $\theta \to \frac{\pi}{2}^+$, $\cos\theta < 0$ and $\lim_{\theta\to\pi/2^+} S(\theta) = -\infty$.

Similarly,

$$\lim_{\theta\to 3\pi/2^-} S(\theta) = -\infty \quad \text{and} \quad \lim_{\theta\to 3\pi/2^+} S(\theta) = \infty.$$

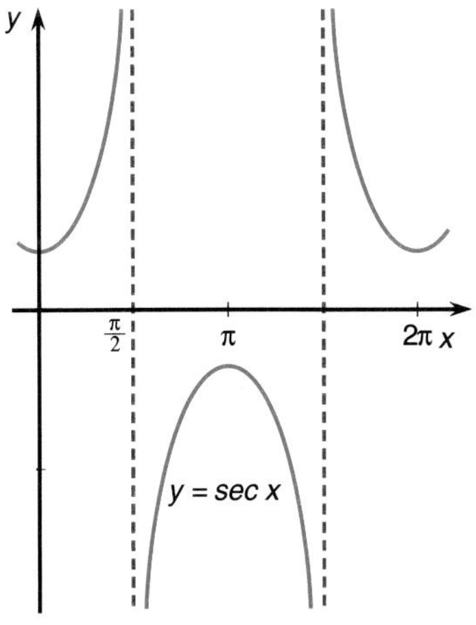

Figure 4.6.20

Putting all this together, we can sketch the graph of $S(\theta)$ for $\theta \in [0, 2\pi]$. Since $\cos\theta$ is periodic, that is, repeats itself after 2π, its reciprocal $\sec\theta$ is also periodic, and we can sketch as much of the graph as we wish. See Figure 4.6.20.

Inflection Points

When we sketch the graph of $y = x^3$, we find that the second derivative $\frac{d^2y}{dx^2}$ equals $6x$, which is positive when $x > 0$ and negative when $x < 0$. It is interesting to look

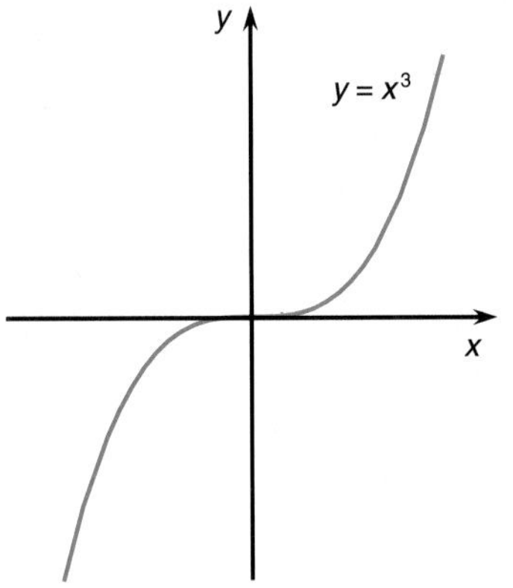

Figure 4.6.21

more closely at the point $(0, 0)$ at which the concavity changes from concave down to concave up (see Figure 4.6.21).

There is a special term for this type of point.

DEFINITION 4.6.3
A point on the graph of f is an **inflection point** for $f(x)$ if the concavity changes there. In other words, $(a, f(a))$ is an inflection point if $f(x)$ is continuous at $x = a$ and is concave up on some interval (c, a) on the left side of a and concave down on some interval (a, d) on the right side of a, or vice versa. We say that "$f(x)$ has an inflection point at $x = a$."

Notice that the function $f(x)$ does not have to be differentiable at $x = a$, but it does have to be defined there and continuous.

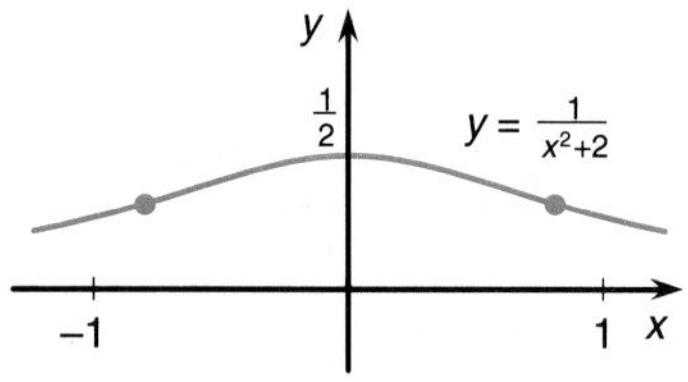

Figure 4.6.22

EXAMPLE 6

(i) For the function $h(x) = \frac{1}{x^2+2}$, which we graphed in Example 4, we found that the function was concave up for $x < -\sqrt{\frac{2}{3}}$ and for $x > \sqrt{\frac{2}{3}}$ and concave down for $-\sqrt{\frac{2}{3}} < x < \sqrt{\frac{2}{3}}$. It has inflection points at $x = \sqrt{\frac{2}{3}}, -\sqrt{\frac{2}{3}}$. Since the second derivative does not change sign at any other points, they are the only ones. See Figure 4.6.22.

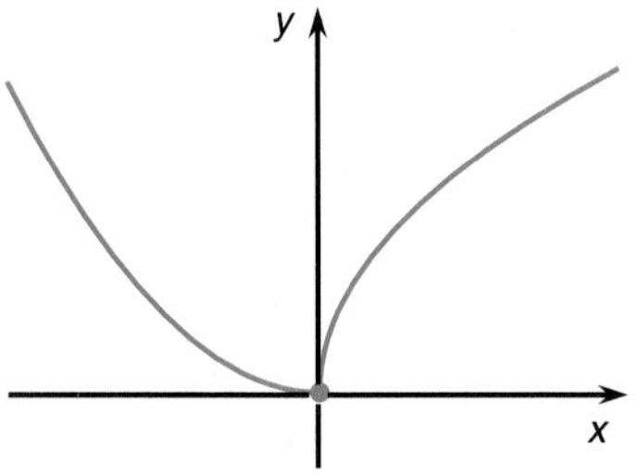

Figure 4.6.23

(ii) The function $C(x) = x^3$ has second derivative $6x$, which changes from negative to positive at $x = 0$, so there is an inflection point at $x = 0$, and it is the only one (cf. Figure 4.6.21).

(iii) Let

$$F(x) = \begin{cases} x^2, & \text{when } x \le 0, \\ \sqrt{x}, & \text{when } x \ge 0. \end{cases}$$

The graph of $F(x)$ is easily sketched (after all, it is just the familiar parabola $y = x^2$ to the left of the y-axis and the familiar square root to the right). See Figure 4.6.23.

When $x < 0$, $F'(x) = \frac{d}{dx}(x^2) = 2x$ and $F''(x) = 2 > 0$. When $x > 0$, $F'(x) = \frac{d}{dx}(\sqrt{x}) = \frac{1}{2}x^{-1/2}$ and $F''(x) = -\frac{1}{4}x^{-3/2} < 0$. This means that $F(x)$ is concave up when $x < 0$ and concave down when $x > 0$, so by definition it has an inflection point at $x = 0$. There are no others.

Notice that $F(x)$ is not differentiable at $x = 0$. This is because the limit from the left of the difference quotient is 0, the derivative of x^2 at $x = 0$, and the limit from the right is ∞. This reflects the fact that the tangent line to $y = x^2$ at $x = 0$ is horizontal (i.e., has slope 0) and the tangent lines to $y = \sqrt{x}$ near $x = 0$ approach the vertical. It is possible to have an inflection point even when the function is not differentiable. It frequently happens at a point where the graph has a "corner," as in this example. This inflection point also happens to be a minimum.

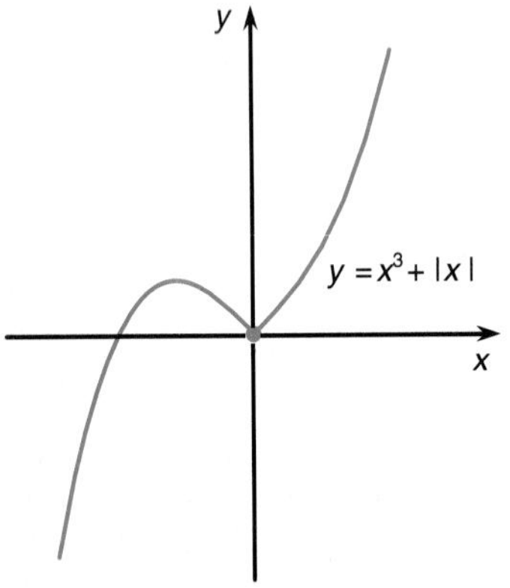

Figure 4.6.24

(iv) Let $H(x) = x^3 + |x|$. When $x < 0$, $H(x) = x^3 - x$, $H'(x) = 3x^2 - 1$, $H''(x) = 6x$. When $x > 0$, $H(x) = x^3 + x$, $H'(x) = 3x^2 + 1$, $H''(x) = 6x$. The origin is an inflection point; see Figure 4.6.24.

(v) Let $g(x) = x^4 - 6x^2 + 8x - 2$. Then $g'(x) = 4x^3 - 12x + 8$ and $g''(x) = 12x^2 - 12$. Trying a few small values of x, we find that $g'(1) = 0$, so we can divide by $(x - 1)$. We find that $g'(x) = 4x^3 - 12x + 8 = 4(x^3 - 3x + 2) = 4(x-1)(x^2 + x - 2) = 4(x-1)(x+2)(x-1) = 4(x-1)^2(x+2)$. Evaluating at $x = -3$, $x = 0$, and $x = 2$, we find that $g'(x) > 0$ on $(-2, 1)$ and $(1, \infty)$, and $g'(x) < 0$ on $(-\infty, -2)$. At the critical point $x = -2$, the second derivative is $g''(-2) = 36 > 0$, so $x = -2$ is a local minimum. At $x = 1$ the second derivative is zero, but since the function is increasing on both sides of this point, it is neither a local maximum nor a local minimum.

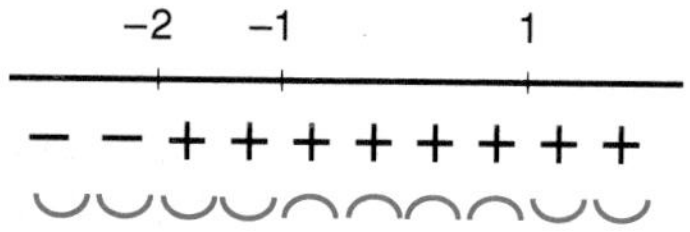

Figure 4.6.25

Observing that the second derivative is continuous everywhere, we see that the only way to have an inflection point, a point at which the second derivative changes from negative to positive or vice versa, is to have $g''(a) = 0$. So we look for points at which $g''(x) = 0$ and find that the only possibilities are ± 1. Also $g''(x) = 12x^2 - 12$ is positive on $(-\infty, -1)$ and on $(1, \infty)$ and negative on $(-1, 1)$. This shows that $g(x)$ has inflection points at both $x = 1$ and $x = -1$. We draw the real axis in Figure 4.6.25, marking the critical points and the inflection points and indicating the sign of the first derivative and the type of concavity.

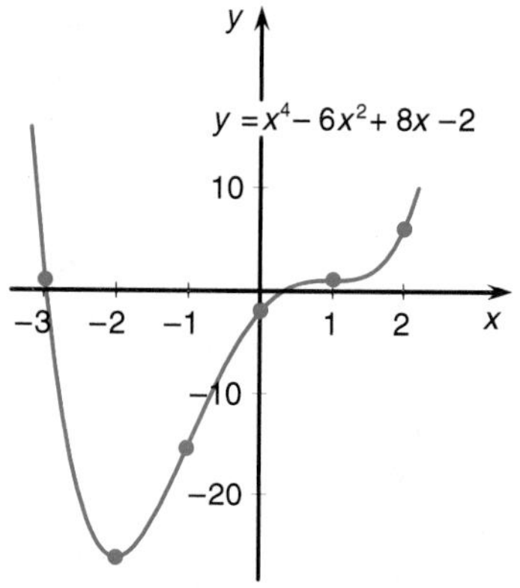

Figure 4.6.26

Plotting a few points, for example, $(0, -2)$, $(1, 1)$, $(2, 6)$, $(-2, -26)$, and $(-1, -17)$, we are able to fill in the curve, being careful to draw the change of concavity at the inflection points. See Figure 4.6.26.

> Example 6 illustrates two important ideas. One is that the function does not have to be differentiable at an inflection point. The other is that if the second derivative exists and is continuous at an inflection point, then the second derivative must be zero there.

So to find the inflection points of a function $f(x)$, we should consider any points at which the second derivative does not exist and any points at which the second derivative equals zero (i.e., at the critical points of $f'(x)$ *rather than* those of $f(x)$). To decide whether or not such a point is an inflection point, it is necessary to decide whether or not the second derivative has different signs on either side.

EXAMPLE 7 Find all the inflection points of (i) $f(x) = x^4 - 24x^2 + 64x + 1$, (ii) $g(x) = x^4 - 3x + 7$, (iii) $h(x) = \sin x$, (iv) $k(x) = \tan x$, (v) $r(x) = x^{1/3}$.

Solution

(i) We differentiate and find that $f'(x) = 4x^3 - 48x + 64$, $f''(x) = 12x^2 - 48$. This is continuous everywhere and zero at $x = \pm 2$. Since $f''(x) = 12x^2 - 48$ is positive when $x < -2$ and when $x > 2$, and negative when $-2 < x < 2$, there are inflection points at $x = -2$, $x = 2$, and they are the only ones.

(ii) $g'(x) = 4x^3 - 3$, $g''(x) = 12x^2$. The only point at which $g''(x) = 0$ is $x = 0$. Since $g''(x) > 0$ at every other x, there are no inflection points.

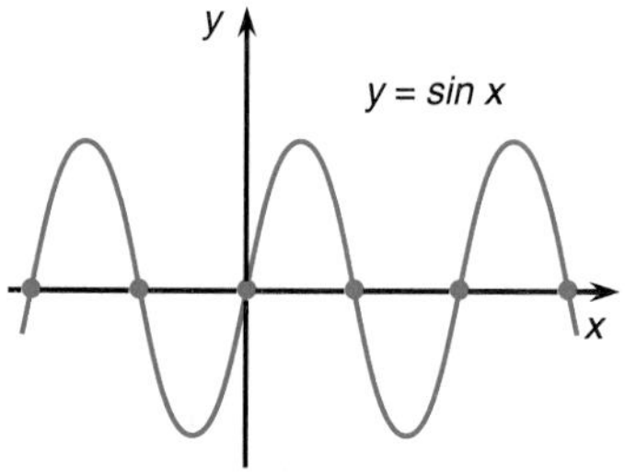

Figure 4.6.27

(iii) $h'(x) = \cos x$, $h''(x) = -\sin x$. The second derivative is zero at $x = n\pi$, for every integer n. Since the graph of $h''(x) = -\sin x$ crosses the axis at each of these points, it has different signs on either side of each of them and so has an inflection point at each of them. There are no others. See Figure 4.6.27.

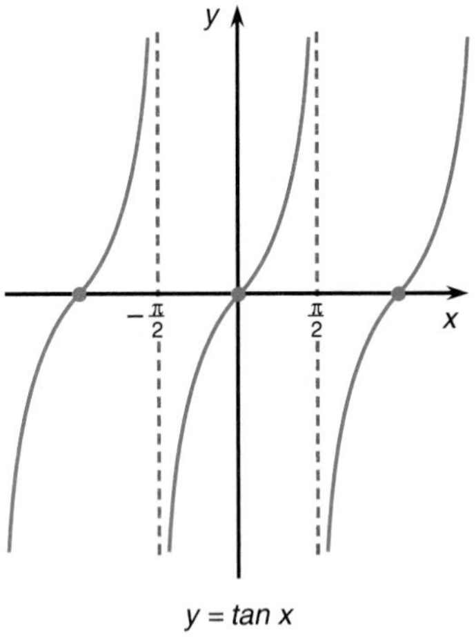

Figure 4.6.28

(iv) $k'(x) = \sec^2 x$, $k''(x) = 2\sec x \sec x \tan x = 2\sec^2 x \tan x$. This function is not defined whenever its denominator ($\cos^3 x$) is zero, that is, at $x = \frac{\pi}{2} + n\pi$, for any integer n. However, the original function $k(x) = \tan x$ is not defined there either, so there cannot be an inflection point at any of these points.

The second derivative is zero whenever $\tan x = 0$, which means $x = n\pi$, for any integer n. Since $\tan x$ is periodic and repeats itself after π, that is, $\tan(x + \pi) = \tan x$ for every x, it will be enough to consider the point $x = 0$. We consider the values of $k''(x)$ near $x = 0$.

If x is a small positive number, $\cos x$ is close to 1 and $\sin x$ is near 0 and positive. So $\tan x$ is positive. On the other hand, if x is near zero and negative, then $\cos x$ is near 1 and $\sin x$ is small and negative. In this case, $\tan x < 0$. From these observations we see that the second derivative does change sign at $x = 0$, so there is an inflection point. Because of the periodic nature of $\tan x$, the same is true of $x = n\pi$ for every integer n. See Figure 4.6.28.

(v) $r'(x) = \frac{1}{3}x^{-2/3}$, which exists whenever $x \neq 0$. Differentiating again, we find that $r''(x) = -\frac{2}{9}x^{-5/3}$, which also exists for all $x \neq 0$ and is never zero. The only possible inflection point is at $x = 0$, where the second derivative does not exist.

We observe that $x^{-5/3}$ has the same sign as $x^{1/3}$, which has the same sign as x. So $r''(x) = -\frac{2}{9}x^{-5/3}$ is positive when $x < 0$ and negative when $x > 0$. The graph is concave up for $x < 0$ and concave down for $x > 0$, so there is an inflection point at $x = 0$.

If we realize that $r(x) = x^{1/3}$ is the inverse function of $C(x) = x^3$, then its graph is the reflection in the line $y = x$ of the graph of $y = C(x) = x^3$, so we can draw it easily (see Figure 4.6.29). The tangent line to $y = x^3$ at the origin is horizontal; it is reflected into the tangent line to $y = r(x) = x^{1/3}$, which is vertical. This can also be seen by considering

$$\lim_{x \to 0} r'(x) = \lim_{x \to 0} \left(\frac{1}{3}x^{-2/3} \right) = \infty$$

and recognizing that a slope that tends to infinity corresponds to a vertical tangent.

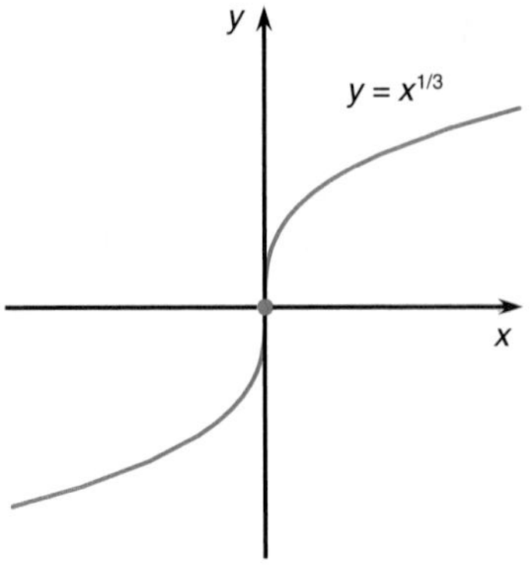

Figure 4.6.29

Summary 4.6.4

Inflection Points

The important things to remember about inflection points are the following:

1. If the second derivative exists at an inflection point, it must equal zero.
2. The second derivative may not exist at an inflection point; even the first derivative may not exist at an inflection point, but the function itself must be defined and continuous there.
3. To find the inflection points of a function $f(x)$, find all points where $f''(x) = 0$ and all points where $f''(x)$ does not exist but $f(x)$ is defined. To check whether each of these candidates is an inflection point, determine whether the sign of the second derivative $f''(x)$ changes there. As we learned when finding the sign of the first derivative of a function, it is enough to evaluate at one convenient point in each interval in the domain of $f''(x)$ between points where $f''(x) = 0$ or where f'' does not exist.

With a graphics calculator we can locate inflection points and discuss the concavity of a curve. We plot a function $y = f(x)$ on the same picture with the second derivative $f''(x)$. For instance, consider $f(x) = x^4 - 6x^2 + 3$ (so $f''(x) = 12x^2 - 12$).

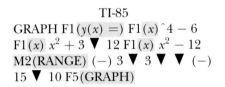

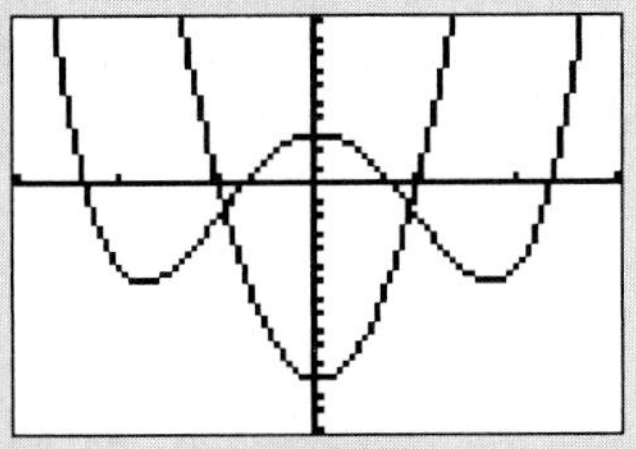

TI-85 GRAPH F1 (y(x) =) F1 (x) ^4 − 6 F1 (x) x^2 + 3 ▼ 12 F1 (x) x^2 − 12 M2(RANGE) (−) 3 ▼ 3 ▼ ▼ (−) 15 ▼ 10 F5(GRAPH)	SHARP EL9300 [graph key] x/θ/T a^b 4 ▶ − 6 x/θ/T x^2 + 3 2ndF ▼ 12 x/θ/T x^2 − 12 RANGE −3 ENTER 3 ENTER 1 ENTER −15 ENTER 10 ENTER 5 ENTER [graph key]
CASIO f_x-7700GB/f_x-6300G Cls EXE Range − 3 EXE 3 EXE 1 EXE − 15 EXE 10 EXE 5 EXE EXE EXE EXE Graph X x^y 4 − 6 X x^2 + 3 EXE Graph 12 X x^2 − 12 EXE	HP 48SX ↰ PLOT PLOTR ERASE ATTN ′ αY = αX y^x 4 − 6 × αX y^x 2 + 3 ↱ PLOT ↰ DRAW ERASE 3 +/− ENTER 3 XRNG 15 +/− ENTER 10 YRNG DRAW ATTN ′ αY = 12 × αX y^x 2 − 12 ↱ PLOT ↰ DRAW DRAW

We note how the second derivative changes sign at two points and observe how the function $f(x)$ changes concavity at these points, which are inflection points. You might also like to plot the first derivative on the same picture.

Cusps

Besides the behavior exhibited in Example 7(v) of a function with an inflection point and a vertical tangent line, there is another phenomenon associated with vertical tangents.

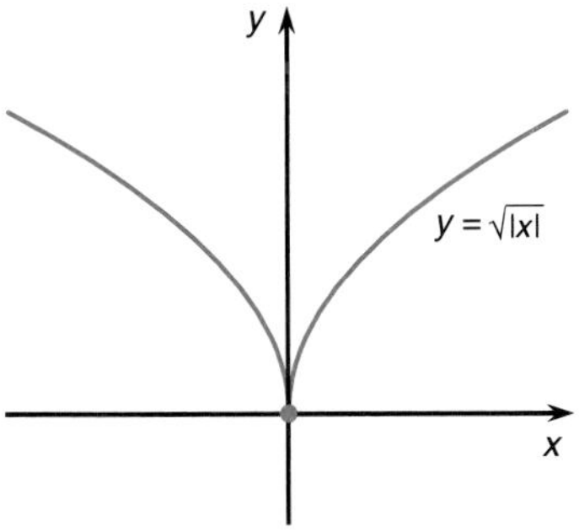

Figure 4.6.30

EXAMPLE 8 Let $f(x) = (x^2)^{1/4}$. When $x > 0$, this is of course the same thing as $x^{2(1/4)} = x^{1/2} = \sqrt{x}$. When $x < 0$, we notice that $x^2 = |x|^2$, so $f(x) = (x^2)^{1/4} = (|x|^2)^{1/4} = (|x|)^{2(1/4)} = |x|^{1/2} = \sqrt{|x|}$.

We can easily sketch the graph of $f(x)$ from this information; it is simply the graph of $y = \sqrt{x}$ for $x > 0$ and the graph of $y = \sqrt{|x|}$ for $x < 0$ (see Figure 4.6.30). We may ask what happens to the derivative at $x = 0$, and find that for $x > 0$,

$$f'(x) = \frac{d}{dx}(\sqrt{x}) = \frac{1}{2}x^{-1/2} = \frac{1}{2\sqrt{x}}.$$

The function is even, so its graph is symmetric about the y-axis. When $x < 0$, $f(x) = \sqrt{|x|} = \sqrt{-x}$ (since $x < 0$, we know that $-x$ is positive . . .). So when $x < 0$,

$$f'(x) = \frac{d}{dx}(\sqrt{-x}) = \frac{1}{2}(-x)^{-1/2}(-1) = -\frac{1}{2\sqrt{-x}}.$$

In each case the denominator is positive and tends to 0 as x approaches 0. This means that

$$\lim_{x\to 0^+} f(x) = \infty \qquad \text{and} \qquad \lim_{x\to 0^-} f(x) = -\infty.$$

The slope of the tangent approaches ∞ from one side and $-\infty$ from the other. The curve appears to come to a point. Notice that the concavity is in the same direction on both sides of the point.

DEFINITION 4.6.5

> A point $(a, f(a))$ is called a **cusp** for $f(x)$ if $f(x)$ is continuous at $x = a$ and $f'(x)$ tends to ∞ as x approaches a from one side and tends to $-\infty$ as x approaches a from the other side. We say that "$f(x)$ has a cusp at $x = a$."
>
> A cusp is never an inflection point because $f(x)$ will be concave in the same direction on both sides of $x = a$. The function $f(x)$ always has a local maximum or a local minimum at any point where it has a cusp.

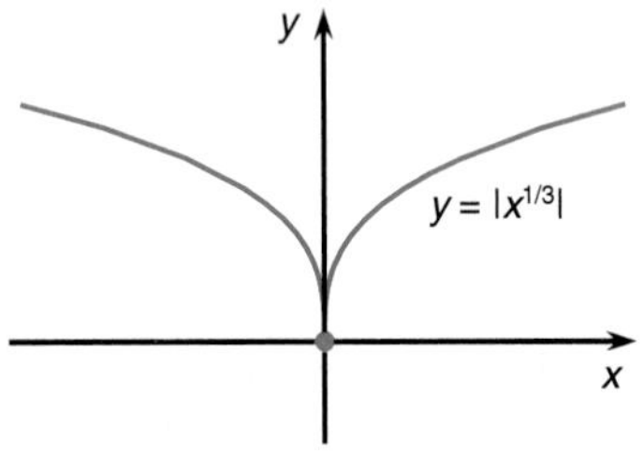

Figure 4.6.31

EXAMPLE 9

(i) Let $g(x) = |x^{1/3}|$. We sketched $x^{1/3}$ in Example 7(v), so it is very easy to sketch $y = g(x)$ (see Figure 4.6.31). If $x > 0$, we find that $g(x) = x^{1/3}$, so $g'(x) = \frac{1}{3}x^{-2/3}$, which tends to ∞ as $x \to 0^+$. On the other hand, if $x < 0$, $g(x) = -x^{1/3}$ and $g'(x) = -\frac{1}{3}x^{-2/3}$, which is negative, and $g'(x) \to -\infty$ as $x \to 0^-$.

We see that $g(x)$ has a cusp at $x = 0$.

(ii) Let $h(x) = \frac{1}{x^2}$. Then $h'(x) = \frac{d}{dx}(x^{-2}) = -2x^{-3}$. Since x^{-3} has the same sign as x, we find that

$$\lim_{x\to 0^+} h'(x) = -\infty \qquad \text{and} \qquad \lim_{x\to 0^-} h'(x) = \infty.$$

However, $h(x)$ does not have a cusp at $x = 0$ because the function $h(x)$ itself is not continuous there (it goes to infinity). See Figure 4.6.32.

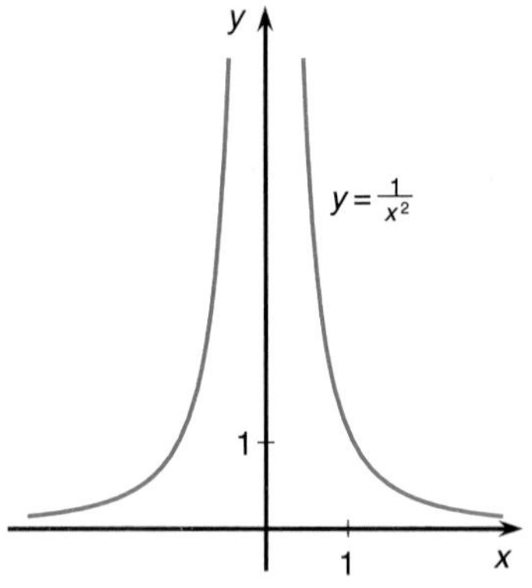

Figure 4.6.32

Inflection points are an important piece of information about a function and should always be marked on the graph. The same is true of cusps, but they are relatively rare.

EXAMPLE 10 In the graph pictured in Figure 4.6.33, mark all the regions in which the function is increasing and the regions in which it is decreasing. Mark the regions

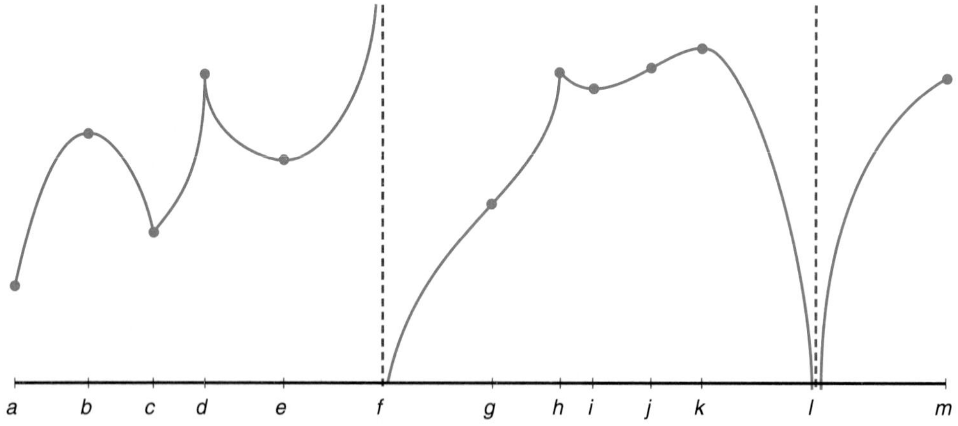

Figure 4.6.33

where it is concave up and the regions where it is concave down. Identify all extrema, inflection points, and cusps. (The dashed lines are asymptotes.)

Solution The graph is increasing on (a, b), (c, d), (e, f), (f, h), (i, k), and (ℓ, m). It is decreasing on (b, c), (d, e), (h, i), and (k, ℓ). It is concave up on (c, d), (d, f), (g, h), and (h, j) and concave down on (a, c), (f, g), (j, ℓ), and (ℓ, m).

There are no absolute extrema. There are local maxima at b, d, h, k, and m and local minima at a, c, e, and i. There are inflection points at c, g, and j and a cusp at d. Notice that h is neither an inflection point nor a cusp.

We draw the axis and indicate the regions of increase and decrease by marking the sign of the derivative. We indicate the concavity with little U-shaped curves. Notice that the concavity changes at f, but it is not an inflection point, since the function is not defined there. See Figure 4.6.34.

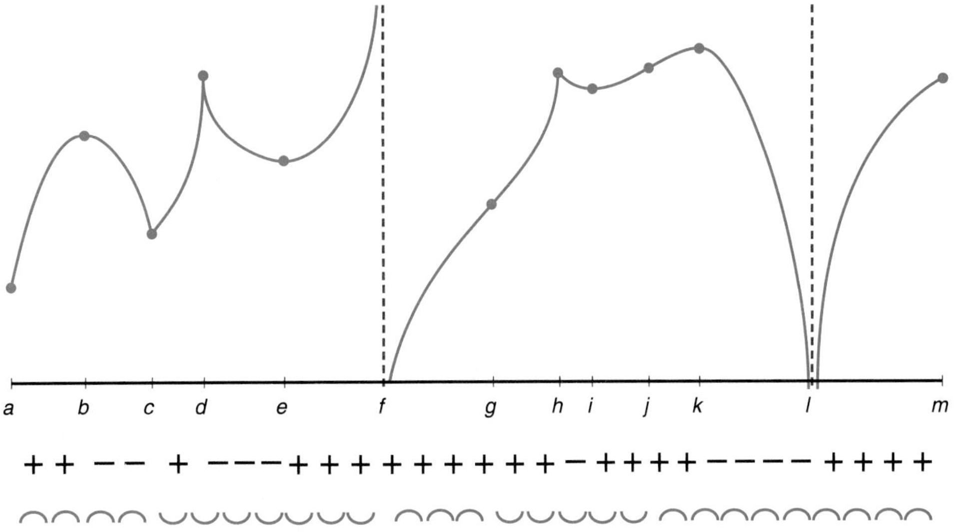

Figure 4.6.34

EXAMPLE 11 Sketch the graph of the function $f(x) = \frac{x}{1+x^2}$, which we have already considered in Example 4.5.2 (p. 264).

Solution In Example 4.5.2 we found that

$$f'(x) = \frac{1 - x^2}{(1 + x^2)^2},$$

so we can differentiate again to find the second derivative:

$$\begin{aligned} f''(x) &= \frac{-2x(1 + x^2)^2 - (1 - x^2)(2)(1 + x^2)2x}{(1 + x^2)^4} \\ &= \frac{-2x(1 + x^2) - (1 - x^2)(2)(2x)}{(1 + x^2)^3} \\ &= \frac{-2x - 2x^3 - 4x + 4x^3}{(1 + x^2)^3} = \frac{2x^3 - 6x}{(1 + x^2)^3} \\ &= \frac{2x(x^2 - 3)}{(1 + x^2)^3}. \end{aligned}$$

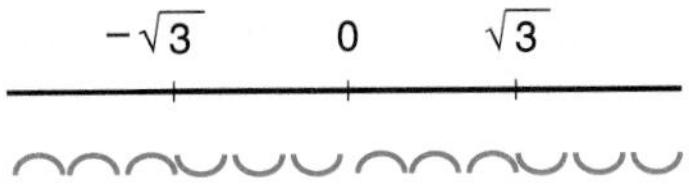

Figure 4.6.35

The denominator is always positive, and the numerator is positive on $(-\sqrt{3}, 0)$ and $(\sqrt{3}, \infty)$ and negative on $(-\infty, -\sqrt{3})$ and $(0, \sqrt{3})$. The function is concave up on the first two intervals and concave down on the last two. There are inflection points at $x = -\sqrt{3}, 0$, and $\sqrt{3}$.

We mark all this information on the real line in Figure 4.6.35, and then we sketch the graph in Figure 4.6.36, incorporating all we know about the curve.

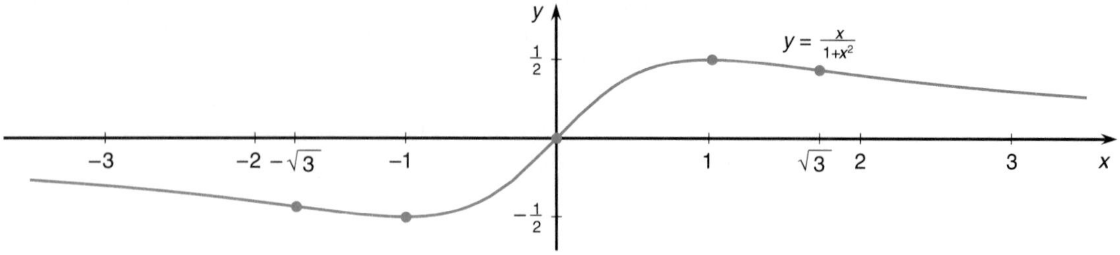

Figure 4.6.36

In the examples we have done so far, the functions were simple enough that it was always easy to find exactly where the second derivative was zero. As the next example shows, this is certainly not always the case.

EXAMPLE 12 Sketch the graph of $G(x) = \frac{e^x}{1+x^2}$.

Solution The derivative is

$$G'(x) = \frac{e^x(1+x^2) - e^x(2x)}{(1+x^2)^2} = \frac{e^x(x^2 - 2x + 1)}{(1+x^2)^2} = \frac{e^x(x-1)^2}{(1+x^2)^2}.$$

It is positive except at $x = 1$, where it is zero, so the function is always increasing. In Section 4.12 we will learn how to show that $\lim_{x\to-\infty} G(x) = 0$ and $\lim_{x\to\infty} G(x) = \infty$.

The second derivative is found by differentiating $G'(x)$:

$$\begin{aligned} G''(x) &= \frac{\left(e^x(x-1)^2 + e^x(2)(x-1)\right)(1+x^2)^2 - e^x(x-1)^2(2)(1+x^2)(2x)}{(1+x^2)^4} \\ &= \frac{e^x(1+x^2)(x-1)\left((x-1)(x^2+1) + 2(x^2+1) - 4x(x-1)\right)}{(1+x^2)^4} \\ &= e^x(x-1)\frac{x^3 - x^2 + x - 1 + 2x^2 + 2 - 4x^2 + 4x}{(1+x^2)^3} \\ &= e^x(x-1)\frac{x^3 - 3x^2 + 5x + 1}{(1+x^2)^3}. \end{aligned}$$

So $G''(x) = 0$ when $x = 1$, but to find any other points at which $G''(x) = 0$, we must solve $A(x) = 0$, where for convenience we write $A(x) = x^3 - 3x^2 + 5x + 1$. We first look for integer roots of $A(x)$, but the Rational Root Theorem (Theorem 1.3.7, p. 13) tells us we only need to check ± 1, and neither of them works. Next we calculate the derivative $A'(x) = 3x^2 - 6x + 5 = 3(x^2 - 2x + 1) + 2 = 3(x-1)^2 + 2$, and having completed the square, we see that $A'(x) > 0$ for every x. This means that $A(x)$ is always increasing, so it has only one zero.

There is a complicated formula for the zero of a cubic equation, but instead we illustrate the possibility of finding the numerical value approximately.

Noticing that $A(-1) = -8 < 0$ and $A(0) = 1 > 0$, we see that the zero must lie between these two points $x = -1$ and $x = 0$ (by the Intermediate Value Theorem (Theorem 2.5.5)). We could narrow down its location more precisely; for example, $A(-\frac{1}{2}) = -2\frac{3}{8}$, and since this is negative, the zero must lie between it and $x = 0$. We could then evaluate at $x = -\frac{1}{4}$ and find whether the zero is to the left or right of it, and then evaluate at $x = -\frac{3}{8}$ or $x = -\frac{1}{8}$, whichever is closer, and so on.

Continuing this process, we could find that the zero is approximately $x \approx -0.18$. Knowing that $A(x)$ is negative to the left of this point and positive to the right of it, we see from the formula for $G''(x)$ that $G''(x)$ is positive to the left of this point and to the right of $x = 1$ and negative between them.

So $G''(x)$ changes sign at $x = 1$ and at the point we have approximated, so there are inflection points at both of them. We can plot the curve now (see Figure 4.6.37). Notice that two significant digits is amply accurate for purposes of plotting the graph.

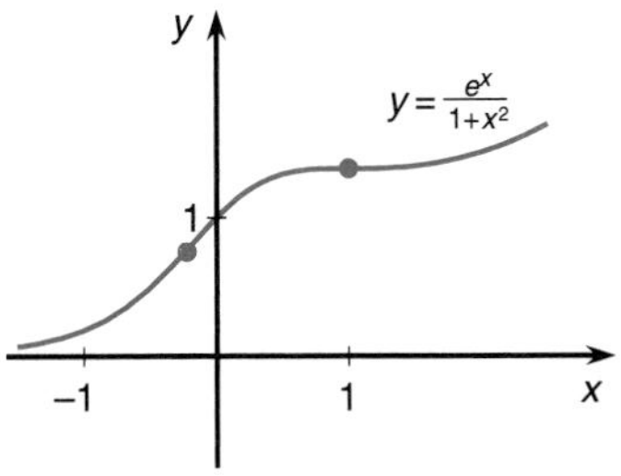

Figure 4.6.37

There are other ways to calculate approximate solutions of equations, and they will be the subject of Sections 4.10 and 4.11.

Exercises 4.6

I

For each of the following functions, find the regions on which it is concave up and the regions on which it is concave down.

1. $f(x) = x^2 - 3x + 11$
2. $f(x) = x^3 + 1$
3. $f(x) = x^3 + 2x - 3$
4. $f(x) = x^3 - 2x^2 - x + 7$
5. $f(x) = \frac{1}{x^2+4}$
6. $f(x) = \frac{x^2}{x^2+4}$
7. $f(x) = \cos(2x)$
8. $f(x) = xe^x$
9. $f(x) = e^{-x^2}$
10. $f(x) = x \ln x$

Find all the inflection points of each of the following functions.

11. $f(x) = 2x^3 + 7x - 1$
12. $f(x) = x^4 - 2x^3 - 12x^2 + 5x + 7$
13. $f(x) = x^5 - 10x^2 - x + 2$
14. $f(x) = 3x^5 - 10x^3 - 7x + 13$
15. $f(x) = \cos x$
16. $f(x) = \cot(2x)$
17. $f(x) = \cos x \sin x$
18. $f(x) = \exp(-x^2)$
19. $f(x) = \sinh x$
20. $f(x) = \frac{x^3}{|x|}$

II

Sketch the graph of each of the following functions, indicating all extrema and inflection points.

21. $y = f(x) = x^3 - 6x + 7$
22. $y = g(x) = x^3 + 6x^2 + 12x + 1$
23. $y = h(x) = x^3 + 6x - 2$
24. $y = k(x) = (x-2)^3$
25. $y = F(x) = x^4 - 2x^2 + 1$
26. $y = G(x) = 3x^4 - 8x^3 + 6x^2 - 3$
27. $y = H(x) = \frac{x^2}{x-1}$
28. $y = K(x) = \frac{x}{x^2-1}$
29. $y = L(x) = \frac{x+1}{x^2-1}$
30. $y = M(x) = \frac{x^2+2x+2}{x+2}$
31. $y = N(x) = e^{-x^2}$
32. $y = P(x) = xe^x$
33. $y = Q(x) = \arctan x$
34. $y = R(x) = x \arctan x$
35. $y = S(x) = x \ln x$, for $x > 0$
36. $y = T(x) = \ln(1 + x^2)$

***37.** Let $k \geq 3$ be an integer and define $f(x) = x^k$. For which values of k does $f(x)$ have an inflection point at $x = 0$?

***38.** Suppose $F(x)$ is a function for which we know the concavity, that is, for which we know all the regions on which it is concave up or concave down and all the inflection points. Let $G(x) = F(x) + 3x - 2$.

(i) Show that $F(x)$ and $G(x)$ have inflection points and upward and downward concavity at exactly the same places.

(ii) Show that the same is true of $H(x) = F(x) + ax + b$ for any constants a and b.

39. Determine which of the following functions has an inflection point at $x = 0$ and which of them has a cusp:
(i) $f(x) = x^{1/5}$, (ii) $g(x) = x^{3/7}$, (iii) $h(x) = x^{4/5}$, (iv) $k(x) = x^{4/3}$.

***40.** Let p, q be positive integers with q odd and $p < q$. Let $f(x) = x^{p/q}$. Specify conditions on p and q that determine whether $f(x)$ has an inflection point (with a vertical tangent) or a cusp at $x = 0$.

***41.** Carry out the approximation method of Example 12 (p. 288) sufficiently many times to find the third significant digit in the approximation of the location of the inflection point of $G(x)$.

****42.** Sketch the graph of $k(x) = \arccos(1 - |x|)$, for $x \in [-1, 1]$.

III

43. (i) Use a graphics calculator or computer to sketch the function $g(x) = x^3 - 3x^2 + 2$ of Example 2(i) (p. 278). (ii) On the same graph, draw the tangent line at $a = \frac{3}{4}$. (iii) Find an interval around a on which the curve lies below the tangent line. (iv) Repeat parts (ii) and (iii) with $a = 0.99$.

44. Predict the regions of upward and downward concavity for $R(\theta) = \csc\theta$ (cf. Example 5, p. 281). Use a graphics calculator or computer to sketch the graph and verify your answer.

45. Sketch the function $g(x) = x^4 - 6x^2 + 8x - 2$ of Example 6(v) (p. 283), along with $g'(x)$ and $g''(x)$ on the same picture.

46. Sketch the function $f(x) = x^4 - 24x^2 + 64x + 1$ of Example 7(i) (p. 283), along with $f'(x)$ and $f''(x)$ on the same picture.

47. Sketch the function $g(x) = x^4 - 3x + 7$ of Example 7(ii), along with $g'(x)$ and $g''(x)$ on the same picture.

48. Sketch the graphs of $G(x) = \frac{e^x}{1+x^2}$ and $A(x) = x^3 - 3x^2 + 5x + 1$ on the same picture (see Example 12, p. 288). Find the root of $A(x)$ correct to six decimal places.

4.7 Applied Maximum/Minimum Problems

EXTRANEOUS SOLUTIONS
INTEGER SOLUTIONS

In Section 3.3 we saw how derivatives could be used to find maximum and minimum values of functions on closed intervals, and we learned how to use this technique to solve a variety of problems. We began to run into difficulties with some problems when the functions we found turned out not to be defined on closed intervals.

Now that our knowledge of extrema is much more detailed and our skill at differentiating more fully developed, we can deal with a wider range of problems.

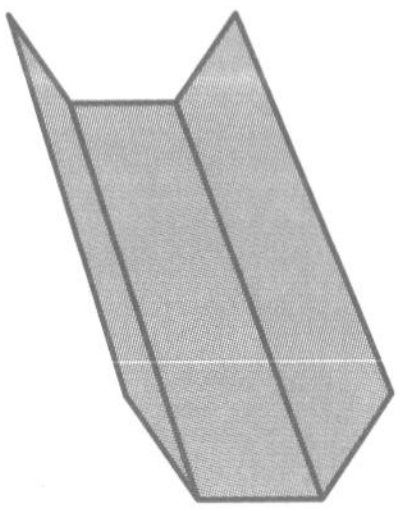

Figure 4.7.1

EXAMPLE 1 **Drinking Trough** A drinking trough is made by taking a long piece of sheet metal and folding up two sloping sides (see Figure 4.7.1). Suppose the metal is 75 cm wide and the folds divide it into three equal widths (of 25 cm each). Suppose that the two sides are folded up at the same angle. At what angle should they be folded up in order to make a trough of the largest possible volume?

Figure 4.7.2

Solution The volume of the trough is just the length times the cross-sectional area, so to solve the problem, we must find out how to maximize the cross-sectional area. We draw a diagram of the cross-section in Figure 4.7.2, and we mark the angle θ as shown, the angle through which the sides are bent up.

Figure 4.7.3

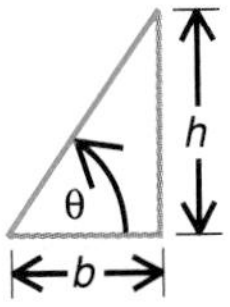

Figure 4.7.4

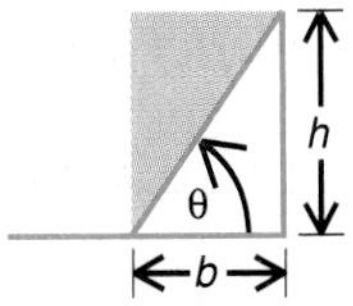

Figure 4.7.5

Figure 4.7.6

Next we must express the area of the cross-section as a function of θ. To do this, we divide it into a rectangle and two right-angled triangles, as shown in Figure 4.7.3. The base of the rectangle is 25 cm, and to find its height, we redraw in Figure 4.7.4 the part of the diagram that involves the right side of the trough. In the resulting right-angled triangle, whose height h we want to find, the hypotenuse has length 25 cm, so the height is $h = 25 \sin \theta$. While we are looking at that triangle, we observe that its base is $b = 25 \cos \theta$.

Returning to the rectangle in the middle, we have just found that its height, in centimeters, is $25 \sin \theta$. Since its base is 25, its area is the product $(25)(25 \sin \theta) = 625 \sin \theta$.

Now we turn our attention to the two triangles at the sides of the trough. They both have the same area, so let us concentrate on the one at the right side, shown in Figure 4.7.5. Its height is the same as the height of the rectangle that we just calculated, and its width is the same as the base of the bottom triangle, which as we observed above is $25 \cos \theta$. The area of the triangle is $\frac{1}{2} \textit{width} \times \textit{height} = \frac{1}{2} 25 \cos \theta \times 25 \sin \theta = \frac{625}{2} \cos \theta \sin \theta$. The combined area of the triangles at the left and right sides is twice as much, or $625 \cos \theta \sin \theta$.

The total area of the cross-section is $A(\theta) = 625 \sin \theta + 625 \cos \theta \sin \theta$. To maximize it, we calculate its derivative:

$$A'(\theta) = 625 \cos \theta + 625(-\sin^2 \theta + \cos^2 \theta).$$

This will equal zero when $\cos \theta + \cos^2 \theta - \sin^2 \theta = 0$, but this appears difficult to solve.

We begin by writing $\sin^2 \theta = 1 - \cos^2 \theta$, so the equation becomes

$$\cos \theta + \cos^2 \theta - 1 + \cos^2 \theta = 0$$

or

$$2 \cos^2 \theta + \cos \theta - 1 = 0.$$

What we want is to solve this for θ, but it will be easier to start by solving it for $\cos \theta$. In fact it is a quadratic equation in $\cos \theta$.

Letting $u = \cos \theta$, we can write the equation as

$$2u^2 + u - 1 = 0.$$

We can use the quadratic formula or just factor $2u^2 + u - 1 = (2u - 1)(u + 1)$, so the solutions are $u = \frac{1}{2}, -1$. The answer we want is $u = \frac{1}{2}$. After all, since $u = \cos \theta$, to say that $u = -1$ is to say that $\cos \theta = -1$ or that $\theta = \pi$. This would correspond to folding the side all the way over onto the bottom of the trough (see Figure 4.7.6), which would give cross-sectional area equal to 0; this point is an endpoint minimum.

To say that $u = \cos \theta = \frac{1}{2}$ is to say that $\theta = \frac{\pi}{3}$. The trough with the maximum volume is made by bending up the sides at an angle of $\frac{\pi}{3}$ radians, that is, $60°$. ◢

Example 1 illustrates that it is perfectly possible to solve maximum/minimum problems involving trigonometric functions. One thing to observe is that we had to use a trigonometric identity to convert one stage of the problem into something we could solve. A second is that we solved not for the angle θ directly, but for $\cos \theta$, from which we were able to find θ.

The most important remark, however, is that somebody had to come up with the idea to use the angle θ as the variable. In Section 3.3 we realized that choosing the right quantity to be the variable is a vital and frequently difficult part of solving maximum/minimum problems. If you attempt Example 1 using some other variable, you will find that it is not easy.

With a graphics calculator we can plot the function $y = A(\theta)$ of Example 1. In fact, we plot $y = 625 \sin x + 625 \cos x \sin x$. While we are at it, we can plot the derivative $625 \cos x + 625(\cos^2 x - \sin^2 x)$ too.

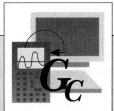

TI-85	SHARP EL9300
GRAPH F1 ($y(x)$ =) 625 SIN F1 (x) + 625 SIN F1 (x) COS F1 (x) ▼ 625 COS F1 (x) + 625 ((COS F1 (x)) x^2 −(SIN F1 (x)) x^2) M2(RANGE) 0 ▼ π ▼ ▼ (−) 800 ▼ 1200 F5(GRAPH)	625 sin x/θ/T + 625 sin x/θ/T cos x/θ/T 2ndF ▼ 625 cos x/θ/T + 625 ((cos x/θ/T) x^2 − (sin x/θ/T) x^2) RANGE 0 ENTER π ENTER 1 ENTER − 800 ENTER 1200 ENTER 200 ENTER
CASIO f_x-7700GB/f_x-6300G	HP 48SX
Cls EXE Range 0 EXE π EXE 1 EXE −800 EXE 1200 EXE 200 EXE EXE EXE EXE Graph 625 sin X + 625 sin X cos X EXE Graph 625 cos X + 625 ((cos X) x^2 − (sin X) x^2) EXE	PLOT PLOTR ERASE ATTN ′ αY = 625 × SIN αX ▶ + 625 × SIN αX ▶ × COS αX PLOT DRAW ERASE 0 ENTER π XRNG 800 +/− ENTER 1200 YRNG DRAW ATTN ′ αY = 625 × COS αX ▶ + 625 × ( ) COS αX ▶ y^x 2 − SIN αX ▶ y^x 2 PLOT DRAW DRAW

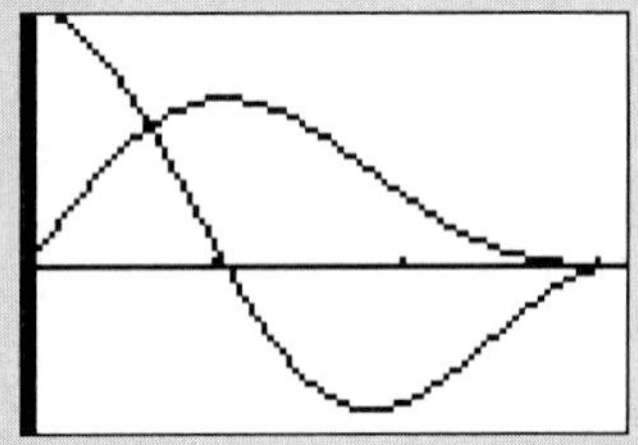

We see the maximum value and note the sign of the derivative in various regions.

EXAMPLE 2 **Window** A window is made so that it is 2 ft wide, in the shape of a rectangle with an isosceles triangle on top (see Figure 4.7.7). A lazy architect has 12 ft of weatherstripping available, and rather than go to the hardware store for more, he decides to design the window to have a perimeter of exactly 12 ft and the largest possible area. What will the dimensions of the window be?

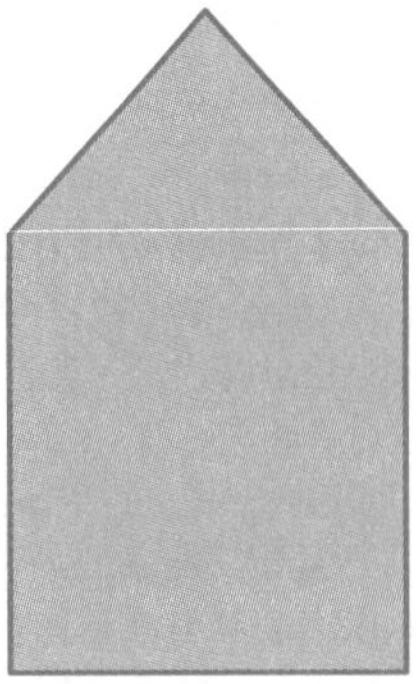

Figure 4.7.7

Solution This time we choose our variable to be the angle θ between the top edge of the window and the horizontal, as shown in Figure 4.7.8. We know that the width of the window is 2 ft, but we do not know the height h of the rectangular part of the window.

On the other hand, we do know that the perimeter is 12 ft. It is also possible to figure out the perimeter in terms of h and θ. If we draw the little triangle made by half of the top triangle, we find a right-angled triangle whose base is 1 ft long and with an angle θ as shown in Figure 4.7.9. So $\cos\theta$ equals 1 over the length of the hypotenuse, and the length of the hypotenuse is $\frac{1}{\cos\theta} = \sec\theta$. Accordingly, the perimeter of the entire window is $2 + 2h + 2\sec\theta$.

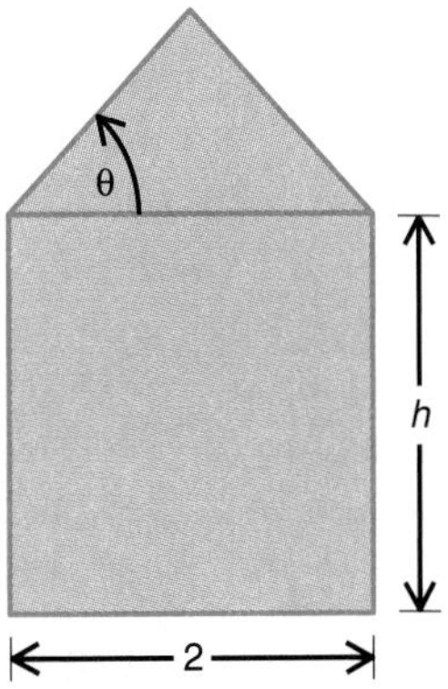

Figure 4.7.8

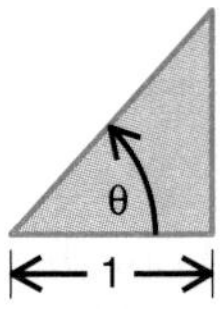

Figure 4.7.9

From the equation $12 = 2 + 2h + 2\sec\theta$, we find that $10 = 2h + 2\sec\theta$, $5 = h + \sec\theta$, so

$$h = 5 - \sec\theta.$$

To find the area of the triangle at the top of the window, we need to observe that the height of this triangle is the same as the height of the little triangle we considered above. That height is $\tan\theta$. The base of the big triangle is 2, so its area is $\frac{1}{2}\,\textit{base} \times \textit{height} = \frac{1}{2}2\tan\theta = \tan\theta$.

In the little triangle, $\frac{\textit{height}}{\textit{base}} = \tan\theta$, and the base has length 1, so $\textit{height} = \tan\theta$

The total area of the window is

$$\begin{aligned} A = A(\theta) &= 2h + \tan\theta \\ &= 10 - 2\sec\theta + \tan\theta. \end{aligned}$$

Here we used $h = 5 - \sec\theta$.

Its derivative is

$$\begin{aligned} A'(\theta) &= -2\sec\theta\tan\theta + \sec^2\theta \\ &= \sec\theta(-2\tan\theta + \sec\theta). \end{aligned}$$

For this to equal zero means that

$$\tan\theta = \frac{1}{2}\sec\theta.$$

Multiplying this equation by $\cos\theta$, we get

$$\cos\theta\tan\theta = \frac{1}{2}\cos\theta\sec\theta,$$

or

$$\sin\theta = \frac{1}{2}.$$

Since the angle in an isosceles triangle must be between 0 and $\frac{\pi}{2}$, this means that $\theta = \frac{\pi}{6}$.

What remains is to convince ourselves that this really is the angle that results in the maximum area.

One way to do this is to rewrite the derivative:

$$A'(\theta) = \sec\theta(\sec\theta - 2\tan\theta)$$
$$= \sec^2\theta(1 - 2\sin\theta),$$

(since $\tan\theta = \frac{\sin\theta}{\cos\theta} = \sin\theta\sec\theta$).

In this form we see that the derivative is positive when $\sin\theta < \frac{1}{2}$ and negative when $\sin\theta > \frac{1}{2}$, which is to say $A'(\theta) > 0$ when $\theta < \frac{\pi}{6}$ and $A'(\theta) < 0$ when $\theta > \frac{\pi}{6}$. This means that $A(\theta)$ is increasing on $\left[0, \frac{\pi}{6}\right)$ and decreasing on $\left(\frac{\pi}{6}, \frac{\pi}{2}\right)$, so $\theta = \frac{\pi}{6}$ is indeed a maximum.

The dimensions of the window of greatest possible area are as follows. The width is of course 2 ft. The height of the rectangular part is $h = 5 - \sec\theta = 5 - \sec\frac{\pi}{6} = 5 - \frac{2}{\sqrt{3}} \approx 3.845$ ft. The length of each slanted side at the top is $\sec\theta = \frac{2}{\sqrt{3}} \approx 1.155$ ft.

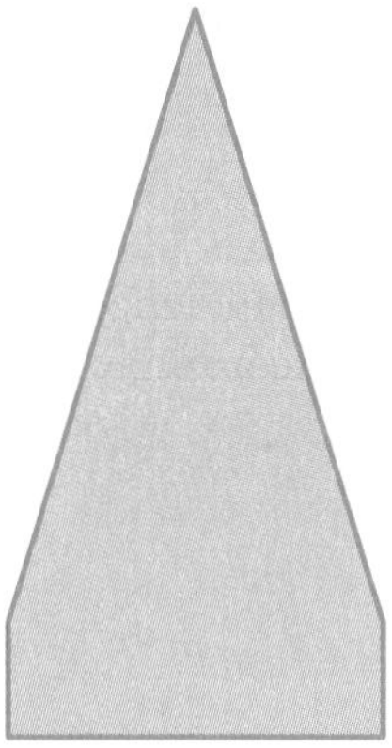
Figure 4.7.10

We have been using our intuition about the behavior of the graph of $A(\theta)$ to decide whether or not the point in question was a maximum. Finding where the derivative was zero was the easy part of the problem. The hard parts were choosing the variable and setting up the function in the first place and deciding whether or not the point we found was really a maximum.

Also observe that once the function $A(\theta)$ was constructed, we thought of it as a function on $\left[0, \frac{\pi}{2}\right)$. In fact this doesn't make sense in terms of the original problem, since if θ is nearly $\frac{\pi}{2}$, that is, nearly a right angle, then the triangle will be so high that the perimeter will have to be greater than 12 (see Figure 4.7.10). If we calculate $A(\theta)$ for values of θ very near $\frac{\pi}{2}$, we will find that it is negative, so it does not make sense to think of it as an area.

Still it does make sense to look for the maximum of this function, and if it happens to be at a point at which the problem does make sense, then we have solved the problem. This is in fact exactly what we did.

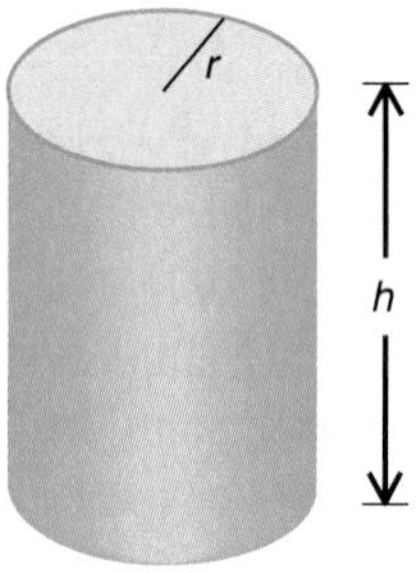

Figure 4.7.11

EXAMPLE 3 **Cylindrical Drum** A cylindrical storage drum is to be made with a volume of 16 cu ft. How can it be made so that the surface area is a minimum, that is, so that it will take the least possible amount of material to construct it?

Solution Suppose the radius and height of the cylinder are r and h, respectively (see Figure 4.7.11). We know that the volume is the height times the cross-sectional area, that is,

$$V = \pi r^2 h.$$

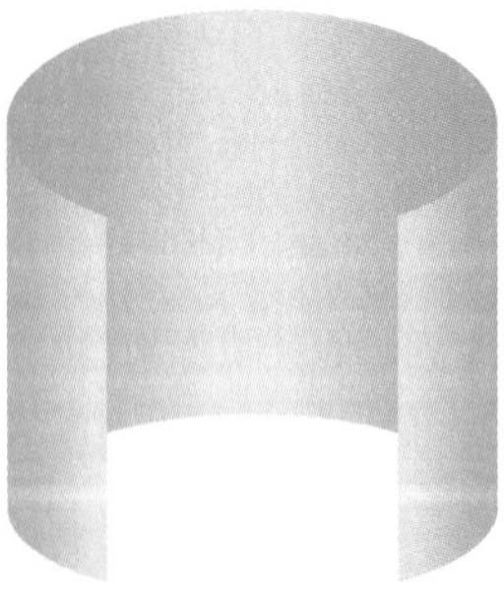
Figure 4.7.12

The top and the bottom each have area πr^2. To find the area of the curved sides, cut off the ends, slit the open cylinder, and unroll it into a rectangle as shown in Figures 4.7.12 and 4.7.13. The rectangle will have one dimension equal to h, the height of the cylinder, and the other equal to the circumference, or $2\pi r$. Its area will be $2\pi rh$. The total surface area of the cylinder is $A = 2\pi r^2 + 2\pi rh$.

Knowing that $V = \pi r^2 h = 16$, we can solve and find that

$$h = \frac{16}{\pi r^2}.$$

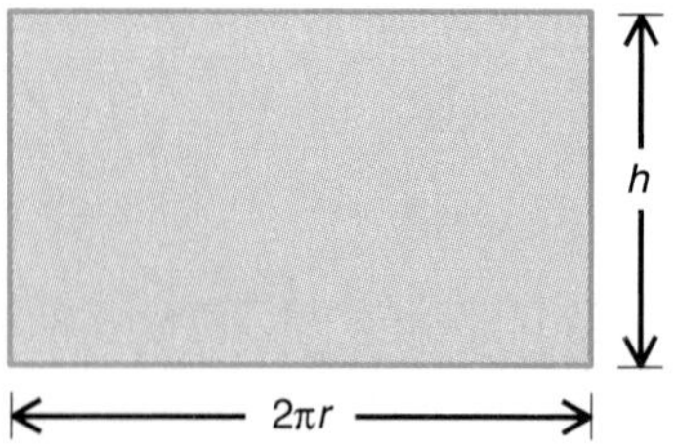

Figure 4.7.13

Substituting this into the area formula, we find that

$$A = 2\pi r^2 + 2\pi rh = 2\pi r^2 + 2\pi r\frac{16}{\pi r^2} = 2\pi r^2 + \frac{32}{r}.$$

The derivative is

$$A' = 4\pi r - \frac{32}{r^2}.$$

This is zero when $4\pi r = \frac{32}{r^2}$, that is, when $4\pi r^3 = 32$, that is, when $r^3 = \frac{8}{\pi}$, or

$$r = \frac{2}{(\pi)^{1/3}} \approx 1.366.$$

The function $A = A(r) = 2\pi r^2 + \frac{32}{r}$ is defined on $(0, \infty)$. It is easy to find its limiting behavior at both ends of its domain: $\lim_{r\to 0^+} A(r) = \infty$ and $\lim_{r\to\infty} A(r) = \infty$. Since the function is differentiable on the whole interval and has only one critical point, that critical point must be a minimum.

The drum of minimum area has radius approximately equal to 1.366 ft and height approximately equal to 2.731 ft.

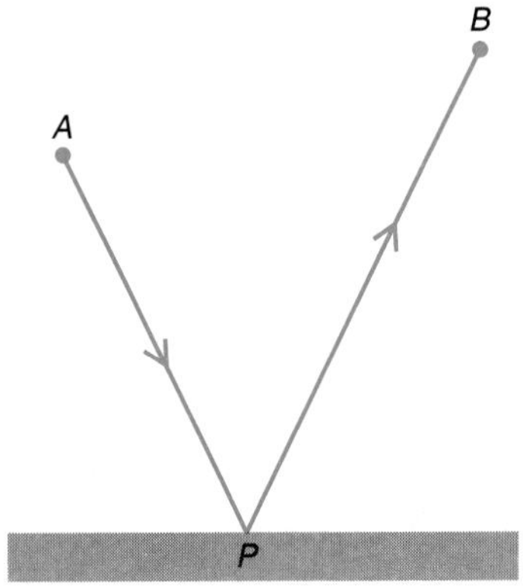

Figure 4.7.14

EXAMPLE 4 **Angle of Reflection** Consider a mirror lying horizontally as shown in Figure 4.7.14. Light travels from a source at point A, is reflected from the mirror at P, and travels on to point B. It is a fundamental idea in optics, known as *Fermat's Principle,* that light travels in such a way as to take the minimum amount of time. Since this amounts to the same thing as traveling the minimum possible distance, we will be able to find the point on the mirror from which the ray of light reflects. Which point of reflection P will require the light to travel the shortest possible total distance?

Solution The problem does not specify any units of distance, so we must choose our own. We let the height of A above the mirror be h and the height of B above the mirror be k. The horizontal distance between A and B (i.e., measured along the surface of the mirror) we call d. As our variable we try the horizontal distance x from the point directly beneath A to the point of reflection P, as shown in Figure 4.7.15.

It is now easy to find the total distance traveled by the light in going from A to P and from P to B. It is

$$D = \sqrt{h^2 + x^2} + \sqrt{k^2 + (d - x)^2},$$

using the Pythagorean Theorem (Theorem 1.5.1, p. 22).

To find the minimum possible D, we calculate

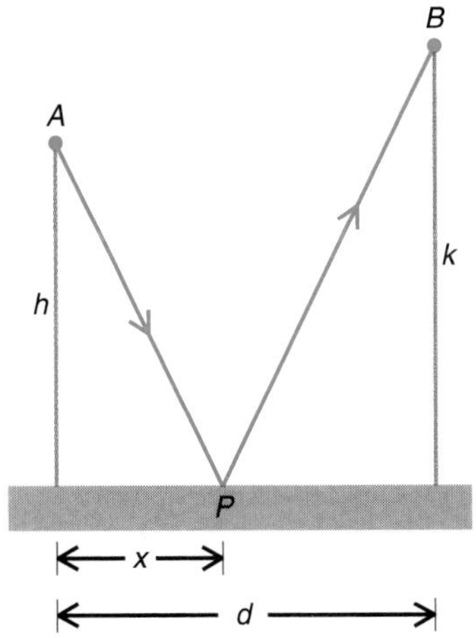

Figure 4.7.15

$$\begin{aligned} D'(x) &= \frac{1}{2}\frac{1}{\sqrt{h^2 + x^2}}(2x) + \frac{1}{2}\frac{1}{\sqrt{k^2 + (d - x)^2}}\big(2(d - x)(-1)\big) \\ &= \frac{x}{\sqrt{h^2 + x^2}} - \frac{d - x}{\sqrt{k^2 + (d - x)^2}}. \end{aligned}$$

This is zero when

$$\frac{x}{\sqrt{h^2 + x^2}} = \frac{d - x}{\sqrt{k^2 + (d - x)^2}}.$$

The denominators of the two sides are just the distances traveled by the ray of light between A and P and between P and B, respectively.

If we call these distances d_1 and d_2, respectively, then the condition becomes

$$\frac{x}{d_1} = \frac{d - x}{d_2}.$$

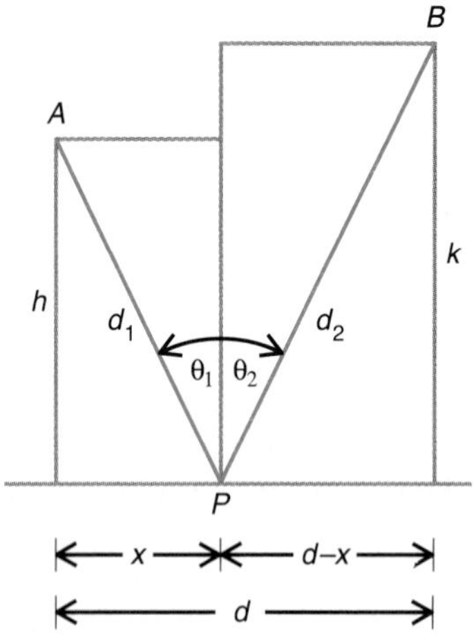

Figure 4.7.16

From Figure 4.7.16 we see that these quotients are the sines of the angles between the two rays and the vertical. If we call these angles θ_1 and θ_2, as shown in the picture, then the condition is that $\sin\theta_1 = \sin\theta_2$.

Since both angles are between 0 and $\frac{\pi}{2}$, this implies that the angles themselves are equal, that is, $\theta_1 = \theta_2$.

The ray of light from A to P, the ray that hits the mirror, is called the *incident ray*. The ray from P to B is called the *reflected ray*, the angle θ_1 is the *angle of incidence*, and the angle θ_2 the *angle of reflection*. What we have shown is that when light is reflected from a mirror, *the angle of incidence equals the angle of reflection*, a law that is well known in physics.

EXAMPLE 5 **Trip from Island** A woman on an island wishes to go to a small town on the mainland. The island is 1 mile from shore. From the point on the coast closest to the island it is 4 miles along the straight shoreline to the town. See Figure 4.7.17.

If the woman can row her boat at a speed of 2 mph and walk at a speed of 4 mph along the beach, where should she land the boat in order to get to town as quickly as possible?

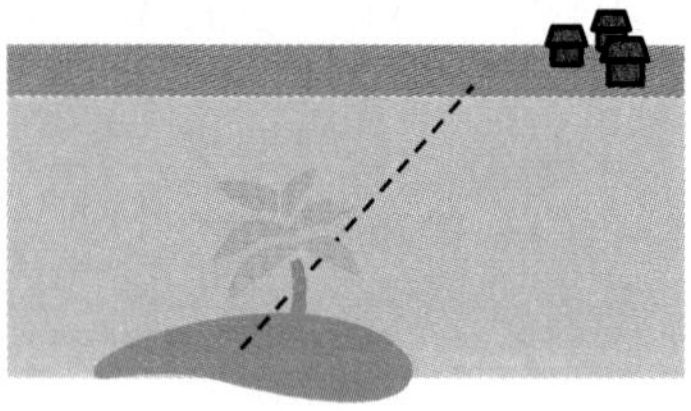

Figure 4.7.17

Solution We call the point on the shore that is closest to the island P. Let x be the distance from P to the point at which the boat lands. See Figure 4.7.18.

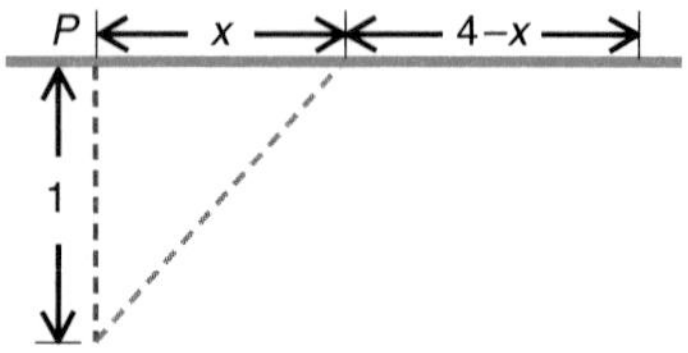

Figure 4.7.18

The woman must walk $4 - x$ miles along the beach at 4 mph, so the time it takes her to walk is $\frac{4-x}{4}$ hr. The quickest way to row to the specified point on shore will of course be to row in a straight line, as illustrated. The Pythagorean Theorem (Theorem 1.5.1) tells us that the distance traveled by the boat is $\sqrt{1+x^2}$ mi, so the time it takes is $\frac{\sqrt{1+x^2}}{2}$ hr.

$time = \frac{distance}{speed}$.

The total time for the trip is the sum of these two times. The total time, in hours, is

$$T = \frac{4-x}{4} + \frac{\sqrt{1+x^2}}{2}.$$

The problem is to minimize T. To do this, we find the derivative

$$T'(x) = -\frac{1}{4} + \frac{1}{2}\frac{1}{2\sqrt{1+x^2}}(2x) = -\frac{1}{4} + \frac{x}{2\sqrt{1+x^2}}.$$

To find critical points, we ask when this equals zero, which is when

$$\frac{x}{2\sqrt{1+x^2}} = \frac{1}{4}$$

or

$$2x = \sqrt{1+x^2}$$

or

$$4x^2 = 1 + x^2$$

or

$$3x^2 = 1$$

or

$$x = \pm\frac{1}{\sqrt{3}}.$$

In fact the negative value we have found for x is not a critical point, as is easily seen by evaluating $T'\left(-\frac{1}{\sqrt{3}}\right)$. (It arose because we squared both sides of the second equation above and in doing so lost track of the sign.) The function $T(x)$ is differentiable at every x and has exactly one critical point, at $x = \frac{1}{\sqrt{3}}$.

Notice from the formula for $T'(x)$ that $T'(x) < 0$ whenever $x < 0$. This means that $T(x)$ is decreasing on $(-\infty, 0)$, and since the sign of the derivative can change only at a critical point, $T(x)$ must be decreasing on $\left(-\infty, \frac{1}{\sqrt{3}}\right)$. Next we consider the sign of the derivative to the right of $x = \frac{1}{\sqrt{3}}$. By the usual sort of argument it is enough to calculate $T'(x)$ at a single convenient point x. Since $1 > \frac{1}{\sqrt{3}}$, we evaluate $T'(1) = -\frac{1}{4} + \frac{1}{2\sqrt{2}}$. Since $\sqrt{2} < 2$, $2\sqrt{2} < 4$, and $\frac{1}{2\sqrt{2}} > \frac{1}{4}$, so $T'(1) > 0$, which means that $T'(x) > 0$ to the right of $x = \frac{1}{\sqrt{3}}$ and the function $T(x)$ is increasing there.

Putting these facts together, we find that the minimum of $T(x)$ must occur at the critical point $\frac{1}{\sqrt{3}}$. The woman will minimize the time it takes to get to town by rowing to the point $\frac{1}{\sqrt{3}} \approx 0.577$ mi along the beach from the point closest to the island. Using this route, her trip will take about 1.433 hr.

In doing this example, techniques that were developed for purposes of sketching graphs were used to simplify the problem.

In hindsight, it is certainly ridiculous to expect that the best route would be to land farther from town than the closest point to the island, so we could have deduced that it is only necessary to consider positive values of x. Similarly, landing beyond the town would not be sensible, so we can restrict attention to $x \in [0, 4]$. By these remarks it is possible to reduce the problem to one on a closed interval, which can be approached by the techniques of Summary 3.3.1 (p. 157).

To do the problem this way would require evaluations at the endpoints $x = 0$ and $x = 4$. This approach is not more difficult, but we have shown how graphical considerations can be used to reduce the complexity of a problem. There are instances in which major simplifications result from this idea.

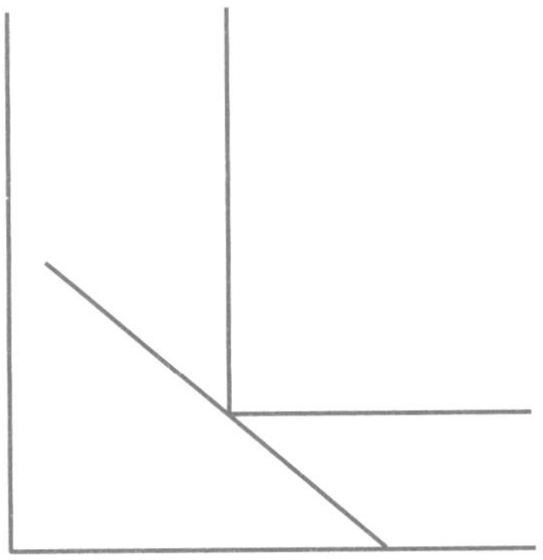

Figure 4.7.19

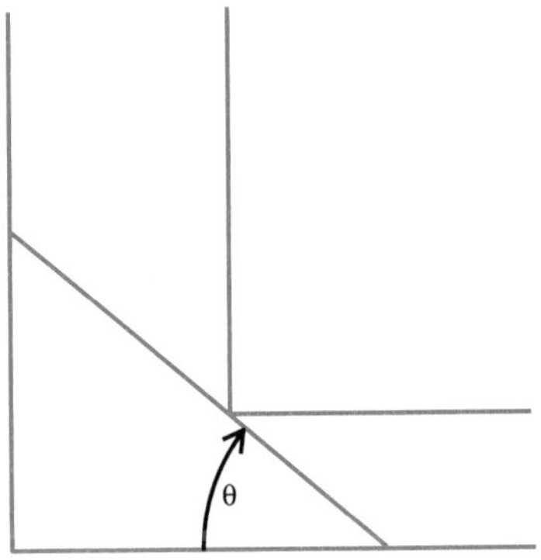

Figure 4.7.20

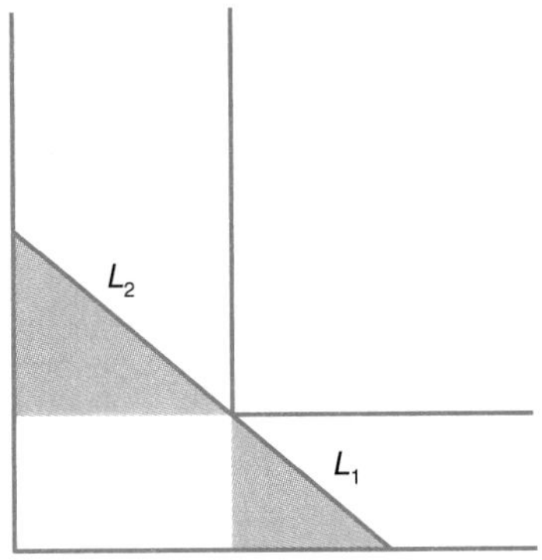

Figure 4.7.21

EXAMPLE 6 **Carrying a Tube Around a Corner** Movers wish to carry a long tube around a corner in a corridor, as shown in Figure 4.7.19. The tube cannot be tilted because it is filled with delicate electronic equipment. If the corridors are 2 m wide and 1 m wide and the tube is 4.2 m long, will the movers be able to get it around the corner?

Solution The question we shall answer is slightly different: What is the longest object that can be carried around the corner?

We consider a long, thin rod placed in the corridor as shown in Figure 4.7.20. Letting θ be the angle marked, we think of the length of the rod as a function L of θ. If we find the **minimum** of this function, we will have found the shortest space available for a rod going around the corner, so we will know the longest rod that can go around the corner.

To find the length of the rod as a function of θ, we consider two small right-angled triangles as shown in Figure 4.7.21. The length of the rod is the sum of the lengths of the hypotenuses of the two triangles. We find the length L_2 of the hypotenuse of the upper left triangle by observing that the base of that triangle has length 2 (the width of the corridor), so $\cos\theta = \frac{2}{L_2}$ and $L_2 = \frac{2}{\cos\theta} = 2\sec\theta$.

To find the length of the hypotenuse of the other triangle, we observe that $\frac{1}{L_1} = \sin\theta$, so $L_1 = \frac{1}{\sin\theta} = \csc\theta$.

Adding together the lengths of the hypotenuses, we find the length of the rod is

$$L = L_1 + L_2 = \csc\theta + 2\sec\theta.$$

To minimize L, we find its derivative:

$$L'(\theta) = -\csc\theta\cot\theta + 2\sec\theta\tan\theta = -\frac{\cos\theta}{\sin^2\theta} + 2\frac{\sin\theta}{\cos^2\theta}.$$

To find when the derivative is zero, we cross-multiply and find

$$\begin{aligned} 2\sin^3\theta &= \cos^3\theta \\ \frac{\sin^3\theta}{\cos^3\theta} &= \frac{1}{2} \\ \tan\theta &= 2^{-1/3} \\ \theta &= \arctan(2^{-1/3}). \end{aligned}$$

We can think of $L(\theta)$ as a function defined for $\theta \in \left(0, \frac{\pi}{2}\right)$ (in terms of the picture of the rod in the corridor it makes no sense for θ to be greater than $\frac{\pi}{2}$). We have found the only place the derivative is zero. Since $\lim_{\theta\to 0^+} L(\theta) = \infty$ and $\lim_{\theta\to\pi/2^-} L(\theta) = \infty$, the point we have found is a minimum.

We conclude that the longest rod that can be carried around the corner has length $L\left(\arctan(2^{-1/3})\right) = \csc\left(\arctan(2^{-1/3})\right) + 2\sec\left(\arctan(2^{-1/3})\right) \approx 4.1619$ m. In particular the tube whose length is 4.2 m will not fit around the corner.

Notice that we did not consider the thickness of the tube. We showed that an object 4.2 m long will not fit around the corner, no matter how thin it is. If the answer had been that the thin rod of a certain length would fit, then it would have been necessary to consider what thickness it could have and still fit, and this actually turns out to be more difficult.

EXAMPLE 7 **An Inventory Problem** A store sells an average of one television set per day, every day of the year. It costs \$70 to store a television set for a year. It costs \$100 to place an order for more sets.

The problem is to decide how often to place an order so as to minimize the overall costs. Ordering often will reduce storage costs but increase ordering charges; ordering infrequently will reduce ordering charges but increase storage costs. What is the best strategy?

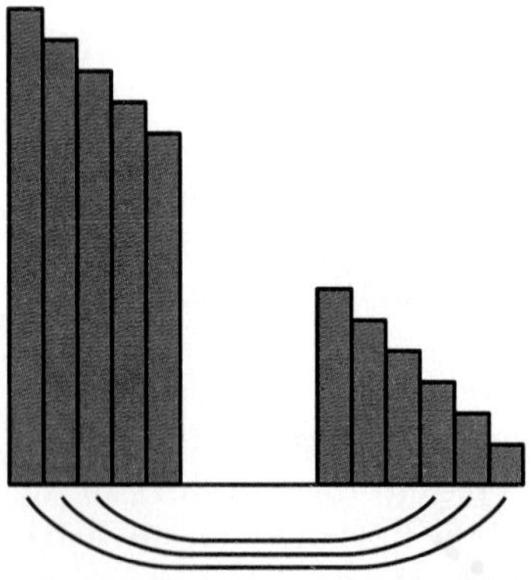

Figure 4.7.22

Solution The question assumes that one television is sold each day. In fact this might not be a very good assumption, since it ignores seasonal variations and random fluctuations. However, under this assumption, suppose an order is placed every x days.

Then the number of orders per year will be $\frac{365}{x}$, and the cost of ordering will be $100\left(\frac{365}{x}\right)$ dollars. Next we have to find the storage costs.

If an order is placed every x days, it will have to be for exactly x sets (one to be sold each day until the next order). What we need to know is the average number of sets in storage.

On the day a new order arrives, there will be x sets in storage; on the last day before the next order, there will be just one set left. Figure 4.7.22 shows a bar graph of the number of sets in storage on each day between consecutive orders, and we can make the following observation. The total of the numbers for the first day and the last day is $x + 1$, so the average for these two days is $\frac{x+1}{2}$. Then if we consider the second day and pair it with the second-last day, we see that the second day has one fewer set than the first day but the second-last has one more set than the last day, so the total for these two days remains the same, $x + 1$, and the average is $\frac{x+1}{2}$.

Similarly, the average for the third and third-last days is $\frac{x+1}{2}$, and so on. What we find is that all the days can be grouped in pairs for which the average number of sets in storage is $\frac{x+1}{2}$, *except* that if x is odd, there will be one day right in the middle that will not be paired with any other day. But if x is odd, then the day in the middle is the day on which the number of sets in storage is $\frac{x+1}{2}$.

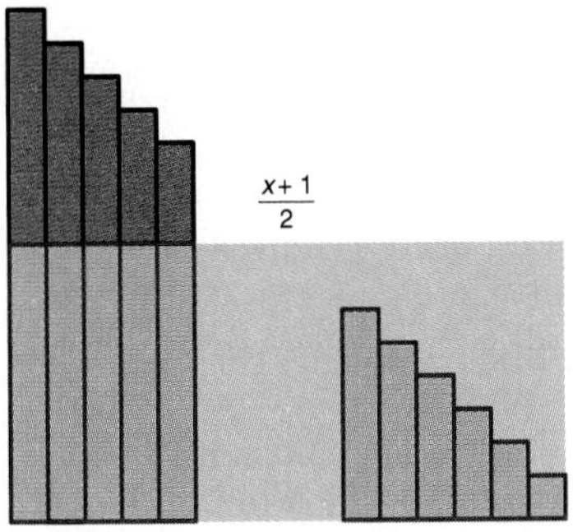

Figure 4.7.23

So in any case the overall average number of sets in storage is $\frac{x+1}{2}$. In terms of Figure 4.7.22 this amounts to "leveling off" the bar graph until all the bars have the same height and observing that this height will then be $\frac{x+1}{2}$ (see Figure 4.7.23). Knowing that it costs \$70 to store a set for a year, we conclude that to store an average of $\frac{x+1}{2}$ sets will cost $\$70\frac{x+1}{2} = \$35(x + 1)$ per year. After all, over the entire year there are, on the average, $\frac{x+1}{2}$ sets in storage.

We want to minimize the sum of the cost of ordering and the cost of storing the sets. This total "inventory cost" is

$$C(x) = \frac{365}{x}(100) + 35(x + 1).$$

Its derivative is

$$C'(x) = -\frac{36{,}500}{x^2} + 35,$$

which is zero when

$$35 = \frac{36{,}500}{x^2}, \qquad \text{or} \qquad x^2 = \frac{36{,}500}{35}$$

or

$$x = \sqrt{\frac{36{,}500}{35}} \approx 32.29.$$

Since $C(x)$ tends to ∞ as $x \to 0^+$ and as $x \to \infty$, this point must be the absolute minimum. The most efficient scheme would be to order every 32.29 days.

Unfortunately, it is not practical to order once every 32.29 days. In a situation like this it is necessary to find the best *integer* solution. A quick look at the formula for $C'(x)$ will convince you that the function $C(x)$ is decreasing to the left of the critical point and increasing to the right of it, so we only have to consider the two nearest integers, 32 and 33. It is easy to evaluate $C(x)$ for $x = 32, 33$, and what we find is that $C(32) \approx \$2295.63$ and $C(33) \approx \$2296.06$. Ordering every 32 days will be slightly better than ordering every 33 days and better than any other scheme.

Problems in textbooks frequently work out to very "nice" answers, just to prevent bogging down in the calculations, but it is very important to recognize that in practice the answer to a problem is unlikely to work out to a nice round number. Sometimes this is unimportant, but very often it is necessary to give some thought, as we did in this example, to finding an answer that is an integer, even though it might not be the absolute minimum.

Exercises 4.7

I

1. Suppose a trough was made, as in Example 1 (p. 290), by bending up the sides of a sheet of metal. This time, assume that the metal is 100 cm wide and that two sides are folded up at the same angle, with each side 25 cm wide (so the bottom is 50 cm wide).
   (i) Find the cosine of the angle at which the sides should be bent up in order to make the trough with the maximum possible volume.
   (ii) Use a calculator to find an approximate value for the corresponding angle.
2. A rectangular page is to have margins of 2 in at the top and the bottom and 1 in at each side and is to contain a rectangular printed space with area 50 sq in in the middle. What dimensions for the page will minimize the total amount of paper?
3. Suppose the height above the ground of a helicopter is given by $H(t) = \frac{5{,}000t^2}{t^2+16}$ ft, where t is the time, in minutes, after the takeoff.
   (i) What is the greatest height it will reach in the first 10 minutes of its flight?
   (ii) At what moment will it be climbing the fastest?
4. Suppose the cylindrical storage drum of Example 3 (p. 294) is to be made without a top.
   (i) If the idea is to make a drum of the same volume (16 cu ft) with a minimum of material, that is, minimum surface area (*not* including the top), how do you expect the best possible dimensions will compare with those found in Example 3 with the top included? Will the radius be smaller and the height bigger or vice versa?
   (ii) Find the exact answer, that is, find the radius and height of the topless drum of volume 16 cu ft for which the total area of sides and bottom is a minimum.
5. Two towns on opposite sides of a river want to be joined by a telephone line. The river is perfectly straight and 1 km wide. One town is exactly 3 km upstream from the other. Suppose laying underwater telephone cable costs $100,000 per kilometer but laying cable on dry land costs only $50,000 per kilometer. What will be the least expensive route for the cable?
6. Suppose a second cable company makes a last-minute bid for the telephone cable joining the two towns of Exercise 5. The second company offers to lay cable on dry land for only $40,000 per kilometer, but expect that laying underwater cable will cost $120,000 per kilometer.
   (i) First try to guess how the best route for the second company to take will differ from the best route for the first company.
   (ii) With each company following the route that is best for it, which company will be able to do the job for less?
   (iii) The town suggests that the companies share the contract, one doing the underwater work and the other doing the dry land part. If the town takes the lower bid for each section of cable, what will be the best route?
   (iv) Will this shared approach cost the town less than either contractor alone?
7. A company wishes to manufacture a tin can, closed both at the top and the bottom, with a volume of 48 cu in. If the material for the top and bottom costs twice as much per square inch as the material for the rounded sides, what are the dimensions of the cheapest can that fills the requirements?
8. (i) If the can in Exercise 7 is replaced with one that has no top, do you expect the radius of the cheapest can to be smaller or larger than the radius found in Exercise 7?
   (ii) Find the cheapest way to make a can with a bottom and no top, whose volume is 48 cu in, assuming that the material for the bottom is twice as expensive as the material for the sides.
9. (i) If Exercise 8 is changed by having the material for the bottom cost half as much as the material for the sides, what effect do you expect this to have on the answer?
   (ii) Find the cheapest way to make an open can (bottom but no top) with volume 48 cu in, if the bottom is made from material that is half as expensive per square inch as the sides.

*10. A cube and a sphere are to be made so that the sum of their surface areas is 12. What are the maximum possible and minimum possible total volume of the two objects? (A sphere of radius r has volume $\frac{4}{3}\pi r^3$ and surface area $4\pi r^2$.)

11. Suppose A is a point 2 units above a horizontal mirror and B is another point 4 units above the mirror and 1 unit to the right of A. (i) Find the point of reflection that results in the shortest total distance for a light ray traveling from A to the mirror and then to B. (ii) Compare the total distance using this point of reflection with the distance obtained by reflecting in the point on the mirror $\frac{1}{2}$ unit to the right of A, that is, halfway between A and B in the horizontal direction.

12. Billiard balls bounce according to the same law that governs the reflection of light: that the angle of incidence equals the angle of reflection. Use this fact to find where a pool shark should aim a ball if he wants to bounce it off the side of the table and hit another ball. Specifically, suppose the first ball is 10 in from the side and the ball at which it is aimed is 25 in from the side. Suppose that the distance between the two balls in the direction parallel to the side of the table is 28 in, as shown in Figure 4.7.24. Find the point on the side of the table at which the first ball should be aimed in order to bounce and hit the second ball.

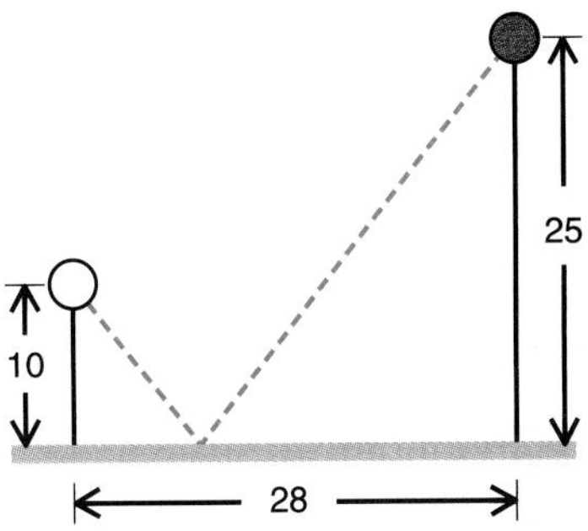

Figure 4.7.24

***13.** A more advanced player wants to make a shot in which a ball bounces twice before hitting another ball, as shown in Figure 4.7.25. For the sake of simplicity, let us suppose that the corner of the table is at the origin.

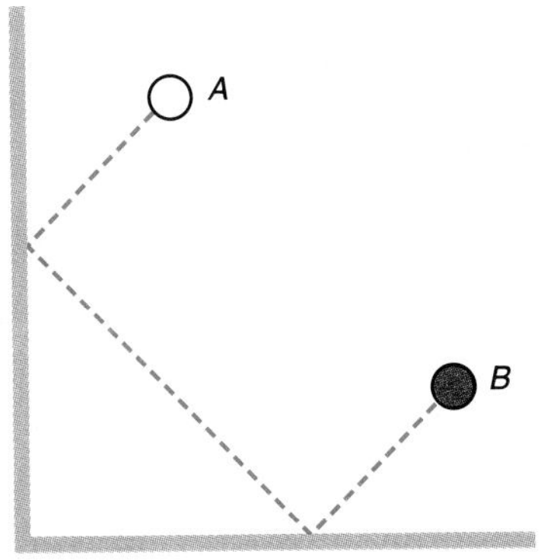

Figure 4.7.25

(i) Consider a ball whose position is point A, whose coordinates are $(3, 9)$ (i.e., it is 3 in to the right and 9 in up from the origin). Suppose the idea is to make this ball bounce off the left wall (the y-axis), then off the bottom wall (the x-axis), and then roll to point B, whose coordinates are $(9, 3)$. Find the angles at which the ball hits the two walls. (*Hint:* Call the angles θ and ϕ, as shown in Figure 4.7.26; notice that there is a relationship between them. Try solving for $\tan\theta$.)

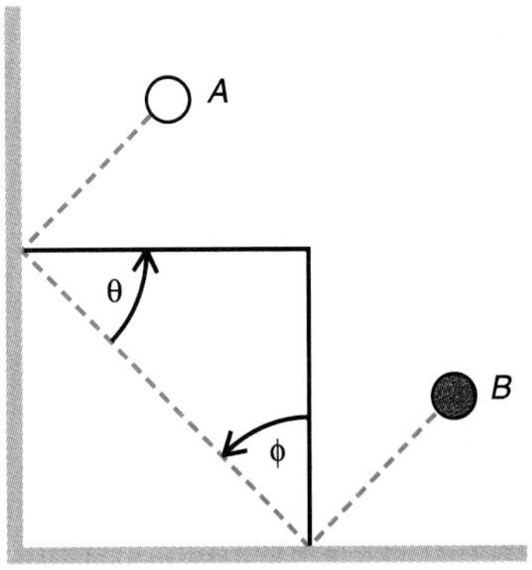

Figure 4.7.26

(ii) If A is the point $(4, 12)$ and B is the point $(8, 4)$, find the angles for the two "reflections." (*Hint:* Again solve for the tangent of one of the angles, and then use your calculator to find the actual angle.)

***14.** In finding the law for light reflecting from a mirror, we mentioned that light travels in such a way as to minimize the time of travel and observed that this is the same thing as minimizing the distance. This observation is based on the assumption that the speed of light is always the same, but it is no longer true if the speed can vary. In fact, the speed of light depends on the medium in which it is traveling. The speed of light that is usually quoted in physics texts is the speed in a vacuum, but light travels slightly less fast in air and slower still in glass or water. This fact is behind the everyday observation that an object partially submerged in water appears to bend at the point where it enters the water.

(i) Consider a ray of light traveling from air into water as shown in Figure 4.7.27. If the speed of

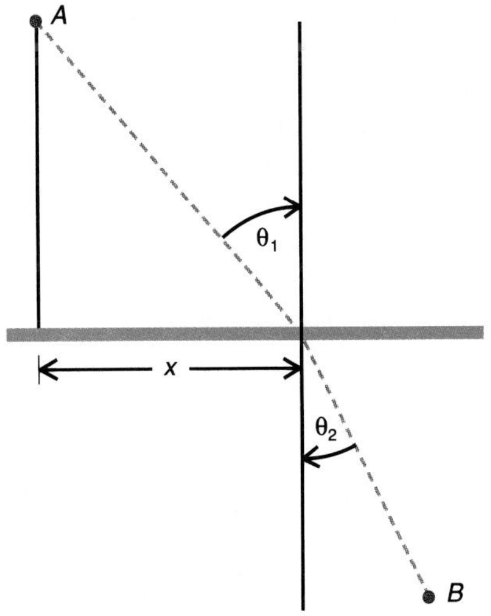

Figure 4.7.27

light in air is v_1 and the speed of light in water is v_2, express the total time for the light to travel from A to B as a function of the variable x (cf. Example 4, p. 295).

(ii) If we label the angles θ_1 and θ_2 as shown, show that the minimum time occurs when x is chosen in such a way that

$$\frac{\sin \theta_1}{\sin \theta_2} = \frac{v_1}{v_2}.$$

The quotient $\frac{v_1}{v_2}$ is called the *index of refraction* of water relative to air and measures how much an object immersed in water appears to bend. The above formula relating the angle of incidence and the angle of refraction to the index of refraction is called **Snell's Law**.

(iii) Assuming that the speed of light in air is known (it is very close to the speed of light in a vacuum, or about 3×10^{10} cm/sec), describe an experiment to determine the speed of light in water.

(iv) Carry out the experiment.

15. Find the maximum length of a thin rod that can be carried without tilting around the corner made by two corridors at right angles that have widths 2 yd and 3 yd. (When the question speaks of a "thin" rod, it means that the thickness is so small that it can be ignored.)

16. What is the maximum length of a thin rod that can be carried without tilting around the corner made by two corridors at right angles that have widths a and b?

***17.** Solve the inventory problem of Example 7 (p. 298) again, this time assuming that the prudent store manager wants to keep at least five sets in stock at all times. How does this affect the strategy for when to order and the total cost?

18. (i) If the storage costs remain the same but the cost of ordering new sets rises, what sort of change do you expect there will be to the best strategy for ordering?

(ii) If the cost of ordering rises to \$120 and the storage costs remain the same, what would be the best interval between successive orders?

(iii) Some time after the ordering cost rose to \$120, news arrived that the storage costs had gone up too. The manager did a calculation and found that the best length of time between orders should be 28 days. What are the new storage charges?

II

The preceding problems were more or less in the same order as the corresponding examples in the text and were sufficiently similar that it was usually clear which sort of technique to apply to each of them. Generally speaking, figuring out which technique to use, and specifically choosing what should be the variable, is one of the most difficult parts of a maximum/minimum problem.

With this in mind, the following problems are in no particular order, and a big part of the question is deciding how to approach them. Of course, in many cases several different approaches are possible, but even then some may be easier to work with than others.

19. A cylinder is placed inside a right circular cone as shown in Figure 4.7.28. Suppose the radius and height of the cone are r and h, respectively. What is the largest possible volume of the cylinder, and what are the dimensions of the cylinder of largest possible volume?

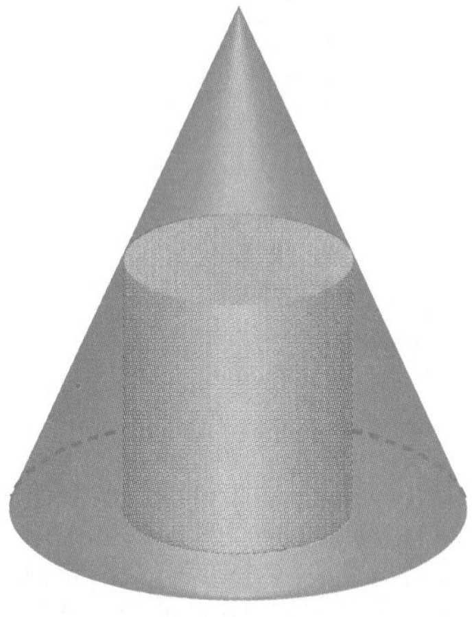

Figure 4.7.28

20. A sphere has radius 1. What is the largest possible volume of a cylinder that can be placed inside the sphere?

21. The intensity of illumination from a small source of light is inversely proportional to the square of the distance from the source, that is, proportional to $\frac{1}{r^2}$, if r is the distance from the source. This is one of many examples in physics of what is known as an *inverse square law*.

Two light bulbs are 100 m apart. One is four times as bright as the other. What point on the straight line between them is darkest, that is, at what point on the line is the total illumination at a minimum?

22. Fences are to be used to enclose a rectangular field and divide it into two pieces by a divider running parallel to two of the sides. If 1200 ft of fencing material are available, what is the maximum possible total area of the enclosure?

23. If a right circular cone has height h and radius r, the distance from the vertex to the circle at the opposite end, measured along the surface of the cone, is $S = \sqrt{h^2 + r^2}$. This quantity is sometimes called the *slant height* of the cone (see Figure 4.7.29). What is the largest possible volume of a cone whose slant height is 40?

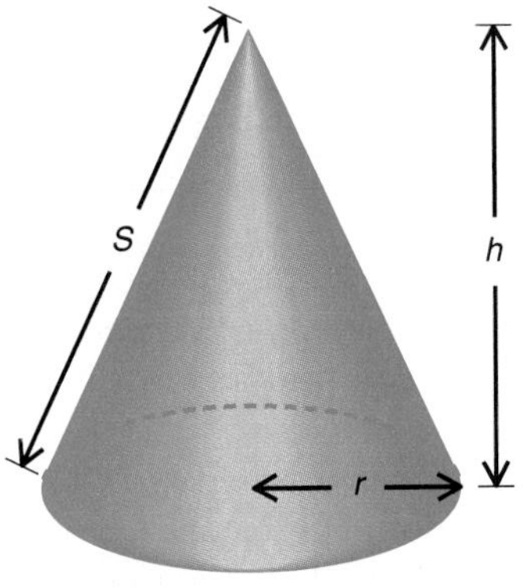

Figure 4.7.29

24. A container is made in the shape of a cylinder with a flat base and a hemispherical cap, as shown in Figure 4.7.30. If the total surface area of the container is to be 80 sq in, what is the maximum possible volume?

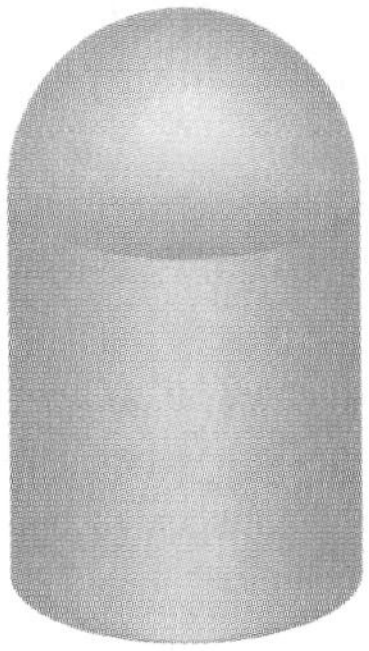

Figure 4.7.30

25. A silo is made in the shape of a cylinder with a hemispherical cap, the shape illustrated for Exercise 24. The material available to cover the silo, that is, to cover the cap and the curved sides but not the bottom, will cover 2000 sq ft.

(i) What dimensions will result in a silo with the maximum possible volume?

(ii) For many purposes only the cylindrical part of the silo can be used for storage (i.e., the cap remains empty). What dimensions will result in a silo with the maximum possible storage capacity?

26. A tank for compressed gas is made in the shape of a cylinder with a hemispherical cap at each end (see Figure 4.7.31). What is the largest possible volume for such a tank if the surface area is 1 sq m?

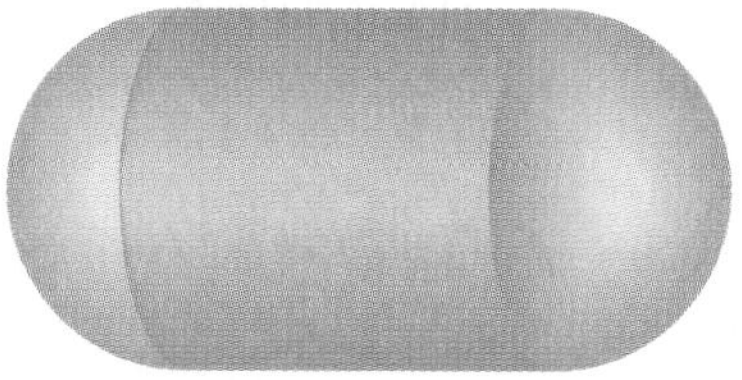

Figure 4.7.31

27. A family having a house built must choose the amount of insulation to install. They have been told that they can have as many layers of insulation as they please, and that each layer will add \$100 to their annual mortgage payment. On the other hand, each layer will reduce the heating bill by 20% (so, for example, one layer would reduce heating costs to 80% of the costs for an uninsulated house, and a second layer would reduce costs to 80% of *that*, or 64% of the costs for an uninsulated house . . .).

Thermograph showing the distribution of heat on the external surface of a house. Different levels of heat are indicated by different colors.

(i) Assuming that the only consideration is to minimize the combined cost of mortgage payments and heating bills, how many layers of insulation would be best if the annual heating costs for an uninsulated house are \$1100?

(ii) What would you recommend if the annual heating costs for an uninsulated house are \$1910?

(iii) If the heating costs are very low for a house in a warm climate, it will not be worthwhile (from the purely financial point of view) to insulate the house. Find the maximum annual heating cost for an uninsulated house for which it is economically unfeasible to insulate at all.

***28.** A rectangular painting 5 ft high is hung on a gallery wall with its bottom edge 10 ft above the floor. A person whose eye level is 5 ft above floor level is looking at it. How far from the wall holding the painting must the person stand to maximize the angle between her lines of sight to the bottom and top of the painting?

III

GC

29. Use a graphics calculator or computer to sketch the function $A(\theta) = 10 - 2\sec\theta + \tan\theta$ of Example 2 (p. 292), for $\theta \in \left[0, \frac{\pi}{2}\right)$. Estimate to four decimal places the value of θ at which $A(\theta)$ crosses the axis to become negative. Can you explain in terms of the original problem why this happens?

30. Sketch the function $L(\theta) = \csc\theta + 2\sec\theta$ of Example 6 (p. 297), and use the graph to verify the answer found in the example.

4.8 Antiderivatives

FUNCTIONS WITH THE SAME DERIVATIVE
ADDING A CONSTANT
COMMON ANTIDERIVATIVES

In Section 4.2 we touched on the idea that two functions with the same derivative must differ by a constant. To put it another way, given a function $f(x)$, if we can guess some function $F(x)$ whose derivative is $f(x)$, then we can find all such functions simply by adding constants to $F(x)$.

It is not terribly difficult to imagine situations in which it is necessary to find a function whose derivative equals some particular function. For instance, if we know that an object moves in such a way that its velocity at time t is $f(t)$, then its position at time t (or more properly its displacement) will be a function $F(t)$ whose derivative is $f(t)$.

More generally, if the rate of change of some quantity is known to be $f(t)$, then the quantity itself will be a function whose derivative is $f(t)$. We often do know the velocity or rate of change of something; if we know the velocity of an airplane, it can be quite important to know where the airplane will be after 3 hours have elapsed, for instance. We might know the rate at which a chemical reaction is proceeding and want to know how much of the product will have been made after a specified period of time. Knowing the rate of population growth, we might like to predict the population ten years from now.

This type of problem is extremely important and will be discussed at greater length in the next four chapters, but in this brief section we make some preliminary remarks that will be sufficient for the purposes of the next section on velocity and acceleration.

We already observed that if $p(x)$ is a polynomial, it is quite easy to find a polynomial whose derivative is $p(x)$.

EXAMPLE 1 Find all functions $F(x)$ defined for all $x \in \mathbb{R}$ and for which $F'(x) = x^3 - 2x^2 + 8x - 3$.

Solution We begin by looking for just one such function $F(x)$, and to do that, we work with each term of the polynomial separately. A function whose derivative is x^3 is $\frac{1}{4}x^4$, a function whose derivative is $-2x^2$ is $-\frac{2}{3}x^3$, a function whose derivative is $8x$ is $4x^2$, and a function whose derivative is -3 is $-3x$.

Because of the way derivatives add (Theorem 3.2.2), we can put these functions together to find a function whose derivative is the sum of x^3, $-2x^2$, $8x$, and -3. Let $G(x) = \frac{1}{4}x^4 - \frac{2}{3}x^3 + 4x^2 - 3x$. It is obvious that $G'(x) = x^3 - 2x^2 + 8x - 3$.

The question asked for *all* such functions, but this is easier than it sounds. Theorem 4.2.11(ii) says that any such function must differ from $G(x)$ by a constant, which means that it must equal $G(x) + C$, for some constant C. In other words, every function $F(x)$ that satisfies $F'(x) = x^3 - 2x^2 + 8x - 3$ must be of the form

$$F(x) = \frac{1}{4}x^4 - \frac{2}{3}x^3 + 4x^2 - 3x + C,$$

for some constant C. We can find as many particular examples as we like by choosing different values for C (e.g., $C = 0, 1, \sqrt{2}, -\frac{7}{3}, \ldots$).

Finding a function with a specified derivative is so common a problem that it is convenient to have terminology for it. Thinking of this process as the reverse of taking the derivative suggests a name.

DEFINITION 4.8.1

> Let $f(x)$ be a function. A function $F(x)$ is called an **antiderivative** of $f(x)$ if
>
> $$F'(x) = f(x),$$
>
> wherever $f(x)$ is defined.

EXAMPLE 2 Find antiderivatives of (i) $f(x) = \cos x$, (ii) $g(x) = x^{-10}$, (iii) $h(x) = e^x$, (iv) $k(x) = \sqrt{x}$, (v) $r(x) = \frac{1}{1+x^2}$.

Solution

(i) By observation, one antiderivative of $f(x) = \cos x$ is $F(x) = \sin x$. Any antiderivative of $f(x)$ is of the form $\sin x + C$, for some constant C.

(ii) Recalling that differentiating a power of x has the effect of lowering the exponent by 1 and multiplying by the original exponent, we see that to have derivative equal to $g(x) = x^{-10}$, a function would have to involve a power of x whose exponent is higher by 1, that is, x^{-9}. But the derivative of x^{-9} is $-9x^{-10}$, so to get rid of the -9 we must let $G(x) = \frac{1}{-9}x^{-9}$. Then $G'(x) = g(x) = x^{-10}$, as required. The antiderivatives of $g(x)$ are of the form $\frac{1}{-9}x^{-9} + C$, for some constant C.

Notice that $g(x)$ is not defined at $x = 0$ and neither is any of its antiderivatives, so strictly speaking, it is possible to have two different constants C, one when $x < 0$ and another when $x > 0$.

(iii) It is easy to guess an antiderivative for $h(x)$. Any antiderivative for $h(x)$ must be of the form $H(x) = e^x + C$, for some constant C.

(iv) Rewriting $k(x) = \sqrt{x} = x^{1/2}$, we realize that an antiderivative for $k(x)$ is $\frac{2}{3}x^{3/2}$.

If that was too quick, notice that it should be some constant times x raised to the power $\frac{1}{2} + 1 = \frac{3}{2}$. The constant you can figure out by calculating that $\frac{d}{dx}(x^{3/2}) = \frac{3}{2}x^{1/2}$ and then putting in the constant $\frac{2}{3}$ to cancel the $\frac{3}{2}$.

Every antiderivative of $k(x)$ is of the form $K(x) = \frac{2}{3}x^{3/2} + C$, for some constant C.

(v) Here we have to remember our formulas for derivatives and recall that $r(x) = \frac{1}{1+x^2}$ is the derivative of $\arctan x$. From this we see that every antiderivative of $r(x)$ must be of the form $R(x) = \arctan x + C$, for some constant C. ◢

> In the last example we went to great pains to include the constant C with each antiderivative. It is important to get into the habit of doing this; most of the time it seems to be unnecessary, but there are situations in which it is vital to remember this step. If it is not something you do automatically, then you are liable to forget it in one of these situations, with disastrous consequences.

We record as a theorem what we have already observed about adding constants to antiderivatives.

THEOREM 4.8.2

> If $F(x)$ is an antiderivative of $f(x)$ on an open interval, then so is $G(x) = F(x) + C$, for any constant C. Moreover, every antiderivative of $f(x)$ on the interval is of this form for some constant C.

PROOF The theorem is really just a restatement of Theorem 4.2.11(ii).

EXAMPLE 3 Let $f(x) = 3x^2$ for every $x \neq 0$, and let $f(x)$ not be defined for $x = 0$. Find every antiderivative of $f(x)$.

Solution Of course, we know that x^3 is one antiderivative of $3x^2$ and $x^3 + C$ is an antiderivative for any constant C. But what makes this example more subtle is that the domain of $f(x)$ is not the whole real line $\mathbb{R}$ but is made up of two intervals $(-\infty, 0)$ and $(0, \infty)$.

The theorem says that on each of these intervals, any antiderivative of $f(x)$ must equal $x^3 + C$, for some constant C. What it does *not* say is that it has to be the same constant for both intervals. We can get an antiderivative by letting

$$F(x) = \begin{cases} x^3 + C, & \text{if } x < 0, \\ x^3 + D, & \text{if } x > 0, \end{cases}$$

where C and D are any constants.

In fact these are all the antiderivatives of $f(x)$, since any antiderivative must equal the sum of x^3 and some constant on each of the two intervals in the domain.

> The reason for the above example is to illustrate the importance of the domain for questions involving antiderivatives and to explain the need for the open interval in Theorem 4.8.2.

It is quite easy to find antiderivatives for x^r for any real exponent r, *except* $r = -1$.

EXAMPLE 4 Find the antiderivatives of $f(x) = \frac{1}{x}$.

Solution When $x > 0$, we know that one antiderivative is $\ln x$, so any antiderivative must be of the form $\ln x + C$.

When $x < 0$, the situation is a bit more complicated. Suppose we let $u = -x$ and consider the function $G(x) = \ln u = \ln(-x)$. By the Chain Rule, $G'(x) = \frac{d}{dx}(\ln u) = \frac{1}{u}\frac{du}{dx} = \frac{1}{-x}(-1) = \frac{1}{x}$. In other words, $G(x) = \ln(-x)$ is an antiderivative of $\frac{1}{x}$ when $x < 0$. (Notice that $G(x)$ is not defined when $x > 0$, since $\ln t$ makes sense only when $t > 0$.)

When $x < 0$, we see that $-x = |x|$, so we can combine the cases in which $x > 0$ and $x < 0$ by letting $H(x) = \ln|x|$. This is an antiderivative of $f(x) = \frac{1}{x}$, and it is defined whenever $x \neq 0$, that is, whenever $f(x)$ is defined.

As in the previous example, any antiderivative of $f(x)$ must differ from $H(x)$ by a constant on each of the two intervals in the domain of $f(x)$, that is, on $(-\infty, 0)$ and $(0, \infty)$. So any antiderivative of $f(x) = \frac{1}{x}$ must be of the form

$$F(x) = \begin{cases} \ln|x| + C, & \text{when } x < 0, \\ \ln|x| + D, & \text{when } x > 0, \end{cases}$$

where C and D are constants.

This is a bit cumbersome, so we often sacrifice some accuracy and say that an antiderivative of $\frac{1}{x}$ is $\ln|x| + C$. This is true, but it neglects the fact that there are other antiderivatives. Similar remarks can be made about antiderivatives of x^r if $r < 0$, since this function is also not defined at $x = 0$ (so its domain is $(-\infty, 0) \cup (0, \infty)$).

Lists of derivatives can now be read the other way around to give lists of antiderivatives. We record a few of the most familiar.

Summary 4.8.3

Common Antiderivatives

For each of the following functions $f(x)$, its antiderivatives are of the form $F(x)$, with $F(x)$ as given in the list below:

$f(x) = x^r, \quad r \neq 0, -1$	$F(x) = \frac{1}{r+1}x^{r+1} + C$
$f(x) = 1$	$F(x) = x + C$
$f(x) = \frac{1}{x}$	$F(x) = \ln|x| + C$
$f(x) = \sin x$	$F(x) = -\cos x + C$
$f(x) = \cos x$	$F(x) = \sin x + C$
$f(x) = e^x$	$F(x) = e^x + C$

For $f(x) = \frac{1}{x}$ and $f(x) = x^r$ with $r < 0$, it is possible to have different constants for $x < 0$ and for $x > 0$ (cf. Example 4).

We can guess antiderivatives for many functions because they are easily recognized as derivatives of known functions. This process is often made easier by breaking up a function as a sum of simpler ones and dealing with each summand separately.

A more complicated problem is to recognize a derivative that is obtained by using the Chain Rule.

EXAMPLE 5 Find the antiderivatives of each of the following functions: (i) $f(x) = \sin 3x$; (ii) $g(x) = (x+4)^7$; (iii) $h(x) = (2x-3)^4$; (iv) $k(x) = e^{-5x}$; (v) $r(x) = xe^{x^2}$.

Solution

(i) If we let $u = 3x$, then $\sin 3x = \sin u$, which makes us think of the derivative of $-\cos u$. We could guess that the antiderivative of $\sin 3x$ should be something involving $-\cos 3x$.

The Chain Rule says that taking the derivative of $-\cos 3x = -\cos u$ results in the product of $\sin 3x$ with the derivative of $u = 3x$. This last derivative is just 3, so we have to "fix up" the function by dividing by 3. The antiderivatives of $f(x) = \sin 3x$ are all of the form $F(x) = -\frac{1}{3}\cos 3x + C$.

(ii) Similarly, $g(x) = (x+4)^7$ reminds us of the derivative of $\frac{1}{8}u^8$, with $u = x + 4$, except for the factor of $\frac{du}{dx}$. This time we are lucky and $\frac{du}{dx} = 1$, so the antiderivatives of $g(x) = (x+4)^7$ are $G(x) = \frac{1}{8}(x+4)^8 + C$.

(iii) If we write $h(x) = (2x-3)^4 = \frac{1}{2}(2x-3)^4 \times 2$, we recognize the last two factors $(2x-3)^4 \times 2$ as the derivative of $\frac{1}{5}(2x-3)^5$. We conclude that the antiderivatives of $h(x)$ are all of the form $H(x) = \frac{1}{2}\left(\frac{1}{5}\right)(2x-3)^5 + C = \frac{1}{10}(2x-3)^5 + C$.

(iv) Noting that $\frac{d}{dx}(e^{-5x}) = e^{-5x}(-5)$, we might write $k(x) = e^{-5x} = -\frac{1}{5}e^{-5x}(-5)$, from which it is easy to see that its antiderivatives are $K(x) = -\frac{1}{5}e^{-5x} + C$.

(v) This part is more complicated. The function $r(x) = xe^{x^2}$ makes us think of e^u with $u = x^2$. If we differentiate $e^u = e^{x^2}$, we obtain $e^u\frac{du}{dx}$. Since $\frac{du}{dx} = 2x$, this derivative is very nearly equal to $r(x)$. We find that the antiderivatives of $r(x) = xe^{x^2}$ are all of the form $R(x) = \frac{1}{2}e^{x^2} + C$.

Let us examine more closely the last part of Example 5. The reason it worked is that $r(x) = xe^{x^2}$ is the product of two things. One of them, x, is, apart from a constant, the derivative of the function $u = x^2$ occurring in the other, and the other one, e^u, is itself a derivative: $e^u = \frac{d}{du}(e^u)$.

This is exactly what is needed to recognize something as the result of a Chain Rule differentiation. The Chain Rule says that

$$\frac{d}{dx}\Big(G\big(u(x)\big)\Big) = G'(u)\frac{du}{dx},$$

so we must look for some familiar derivative $\big(G'(u)\big)$ evaluated at u, multiplied by the derivative of u. Just to complicate things, we can expect to have to multiply by constants.

EXAMPLE 6 Find antiderivatives for each of the following: (i) $\frac{x}{(x^2+3)^2}$; (ii) $\frac{x}{\sqrt{x^2+1}}$; (iii) $x^2\cos(x^3+4)$.

Solution

(i) We begin by looking for a suitable u, whose derivative appears in the function too. In this case we notice the function x^2+3, whose derivative $2x$ is essentially the numerator (apart from a factor of 2). We can write

$$\frac{x}{(x^2+3)^2} = \frac{1}{2}\frac{1}{(x^2+3)^2}(2x),$$

and let $u = x^2+3$. We see that

$$\frac{x}{(x^2+3)^2} = \frac{1}{2}u^{-2}\frac{du}{dx}.$$

Since u^{-2} is the derivative $\frac{d}{du}(-u^{-1})$ (by Summary 4.8.3), we can see that the antiderivative ought to be $\frac{1}{2}(-u^{-1}) + C = -\frac{1}{2}\frac{1}{x^2+3} + C$.

(ii) Here again we recognize that the x in the numerator is $\frac{1}{2}$ the derivative of the quantity inside the square root. If we let $u = x^2+1$, the function becomes $\frac{x}{\sqrt{u}} = xu^{-1/2} = u^{-1/2}\left(\frac{1}{2}\right)\frac{du}{dx}$. Since $\frac{1}{2}u^{-1/2}$ equals $\frac{d}{du}(u^{1/2})$, the antiderivatives of the original function are $u^{1/2} + C = \sqrt{u} + C = \sqrt{x^2+1} + C$.

(iii) Here the factor x^2 looks a lot like the derivative of x^3+4. If we let $u = x^3+4$, the function is $x^2\cos(x^3+4) = \frac{1}{3}3x^2\cos(x^3+4) = \frac{1}{3}\cos u\frac{du}{dx}$. From this we see that the antiderivatives are all of the form $\frac{1}{3}\sin(x^3+4) + C$.

Notice that it is always easy to check these answers; just differentiate.

EXAMPLE 7 Suppose $f(x)$ is differentiable on the whole real line and $f(x) \neq 0$ for any x. Find the antiderivatives of the function $\frac{f'(x)}{f(x)}$.

Solution The presence of the derivative $f'(x)$ in the numerator suggests that this might be the result of a differentiation using the Chain Rule. If we write it in the slightly different form $\frac{1}{f(x)}f'(x)$, we see immediately that it equals the derivative of $\ln|f(x)|$.

The fact that $f(x)$ is differentiable at every x and never zero implies that $\ln|f(x)|$ is differentiable at every x. The domain is the whole real line, and since $f(x)$ is continuous and never zero, it must be always positive or always negative. (By the Intermediate Value Theorem it cannot change sign.) From this we see that every antiderivative of $\frac{f'(x)}{f(x)}$ must equal $\ln|f(x)| + C$, for some constant C, by Theorem 4.8.2.

Another type of problem we can solve is finding an antiderivative of some function $f(x)$ that also satisfies some additional condition. This has the effect of determining the constant that we usually add to antiderivatives, removing the ambiguity.

EXAMPLE 8 (i) Find the antiderivative $F(x)$ of $f(x) = x^2 - 2x + 1$ that satisfies $F(0) = 3$.

Solution We know that any antiderivative must be of the form $F(x) = \frac{1}{3}x^3 - x^2 + x + C$, for some constant C. On the other hand, if we substitute $x = 0$ in this formula, we find that $F(0) = \frac{1}{3}0^3 - 0^2 + 0 + C = C$. The constant C must equal 3.

The solution to the problem is the function $F(x) = \frac{1}{3}x^3 - x^2 + x + 3$.

(ii) Find the antiderivative $G(x)$ of $g(x) = e^{2x}$ that satisfies $G(0) = -1$.

Solution Any antiderivative $G(x)$ must be of the form $G(x) = \frac{1}{2}e^{2x} + C$, for some constant C. The condition says that $-1 = G(0) = \frac{1}{2}e^0 + C = \frac{1}{2} + C$. This implies that $C = -\frac{3}{2}$.

The required function is $G(x) = \frac{1}{2}e^{2x} - \frac{3}{2}$.

(iii) Find an antiderivative $H(x)$ of $h(x) = \frac{1}{x-1}$ that satisfies $H(0) = 4$ and $H(2) = 5$.

Solution We know from Example 4 that $H(x)$ must be obtained from the function $\ln|x-1|$ by adding one constant when $x - 1 > 0$ and another constant when $x - 1 < 0$.

When $x = 2$, we see that $\ln|x-1| = \ln|2-1| = \ln 1 = 0$, so the constant that is added to $\ln|x-1|$ when $x > 1$ must be $H(2) = 5$.

Similarly, when $x = 0$, $\ln|x-1| = \ln 1 = 0$, so the constant added to $\ln|x-1|$ when $x < 1$ must be $H(0) = 4$. We see that

$$H(x) = \begin{cases} \ln|x-1| + 5, & \text{if } x > 1, \\ \ln|x-1| + 4, & \text{if } x < 1. \end{cases}$$

(iv) Find an antiderivative $K(x)$ of $k(x) = \frac{1}{x}$ that satisfies $K(1) = -2$ and $K(2) = 5$.

Solution The antiderivative $K(x)$ must be of the form

$$K(x) = \begin{cases} \ln|x| + C, & \text{if } x > 0, \\ \ln|x| + D, & \text{if } x < 0, \end{cases}$$

for some constants C and D.

From this we see that $K(1) = \ln|1| + C = 0 + C = C$. Since $K(1) = -2$, we see that $C = -2$.

Also since $2 > 0$, we see that $K(2) = \ln|2| + C = \ln 2 + (-2) \approx -1.307 \neq 5$.

This says that an antiderivative of $\frac{1}{x}$ that equals -2 at $x = 1$ cannot equal 5 at $x = 2$. It is not possible to find any function satisfying the requirements of the question.

With a graphics calculator we can plot the solution to part (i) of Example 8. We also plot $F(x) = \frac{1}{3}x^3 - x^2 + x + C$, with $C = 0, 4$.

TI-85	SHARP EL9300
GRAPH F1 (y(x) =) F1 (x) ^3 ÷ 3 − F1 (x) x^2 + F1 (x) + 3 ▼ F1 (x) ^3 ÷ 3 − F1 (x) x^2 + F1 (x) + 0 ▼ F1 (x) ^3 ÷ 3 − F1 (x) x^2 + F1 (x) + 4 M2 (RANGE) (−) 2 ▼ 3 ▼ ▼ (−) 8 ▼ 7 F5 (GRAPH)	[graph key] x/θ/T a^b 3 ▶ ÷ 3 − x/θ/T x^2 + x/θ/T + 3 2ndF ▼ x/θ/T a^b 3 ▶ ÷ 3 − x/θ/T x^2 + x/θ/T 2ndF ▼ x/θ/T a^b 3 ▶ ÷ 3 − x/θ/T x^2 + x/θ/T + 4 RANGE −2 ENTER 3 ENTER 1 ENTER −8 ENTER 7 ENTER 1 ENTER [graph key]
CASIO f_x**-7700GB/**f_x**-6300G**	**HP 48SX**
Cls EXE Range −2 EXE 3 EXE 1 EXE −8 EXE 7 EXE 1 EXE EXE EXE EXE Graph X x^y 3 ÷ 3 − X x^2 + X + 3 EXE Graph X x^y 3 ÷ 3 − X x^2 + X EXE Graph X x^y 3 ÷ 3 − X x^2 + X + 4 EXE	↰ PLOT PLOTR ERASE ATTN ′ αY = αX y^x 3 ÷ 3 − αX y^x 2 + αX + 3 ↱ PLOT ↰ DRAW ERASE 2 +/− ENTER 3 XRNG 8 +/− ENTER 7 YRNG DRAW ATTN ′ αY = αX y^x 3 ÷ 3 − αX y^x 2 + αX ↱ PLOT ↰ DRAW DRAW ATTN ′ αY = αX y^x 3 ÷ 3 − αX y^x 2 + αX + 4 ↱ PLOT ↰ DRAW DRAW

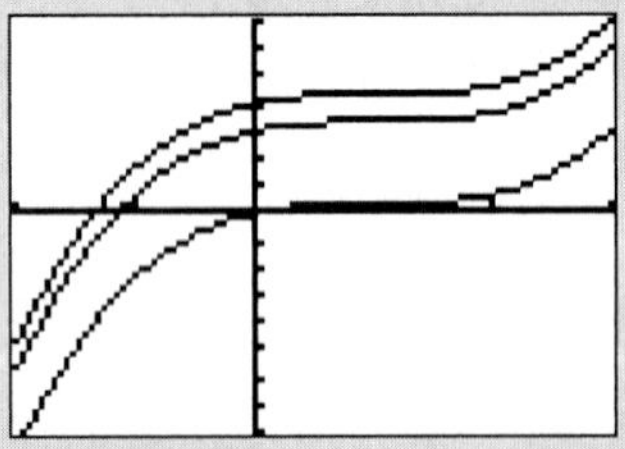

Notice that these functions are all the same function shifted up or down by the addition of a constant, so their derivatives are the same. Exactly one of them has the required value 3 at $x = 0$.

EXAMPLE 9 **Bacterial Growth** Bacteria and other one-celled organisms reproduce by dividing. One cell will divide into two. The rate at which they divide will depend on various factors, such as the temperature, the amount of nutrients available, and so on.

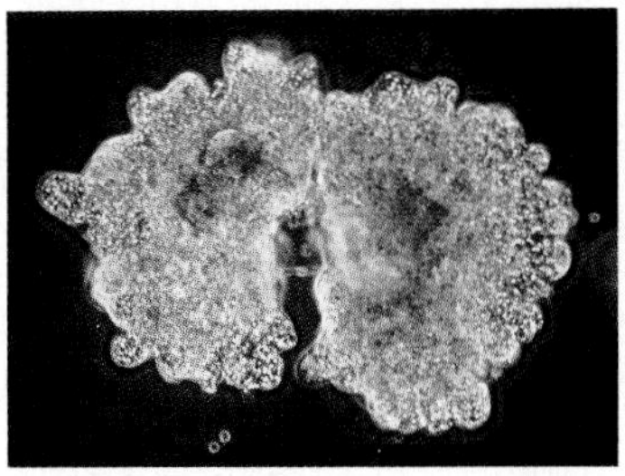

This microscopic view shows the Amoeba proteus *undergoing binary fission, "dividing in two."*

If all these extraneous factors remain the same, then the rate of reproduction (the number of additional cells created per unit of time) will be proportional to the number of cells already present. This simply means that the more cells there are, the more divisions there will be: If one culture contains twice as many cells as another, then the larger one will undergo twice as many divisions as the smaller one, so it will grow twice as fast.

If we write $P(t)$ for the "population," the number of cells in a particular culture at time t, this amounts to saying that

$$\frac{dP}{dt} = kP,$$

where k is some constant, depending on various environmental factors as well as the units in which time t is measured.

Writing $P'(t)$ for $\frac{dP}{dt}$, we find that the above equation is $P'(t) = kP(t)$. Assuming that the population is never zero (a reasonable assumption), we can divide by P and find

$$\frac{P'(t)}{P(t)} = k.$$

By Example 7 we know that an antiderivative of the left side of this equation is $\ln|P(t)|$. This equals $\ln P(t)$, since $P(t)$, being a population, is never negative. An antiderivative of the right side is kt, and these two antiderivatives must differ by a constant. In other words, $\ln P(t) = kt + C$, for some constant C.

To recover a formula for $P(t)$ itself, we take the exponential of each side and find that

$$\begin{aligned} P(t) &= e^{kt+C} = e^{kt}e^{C} \\ &= C'e^{kt}. \end{aligned}$$

Here C' is some constant (it equals e^C, the exponential of the other constant C). The notation does *not* mean that it is a derivative; it merely indicates that it is a different constant. Notice that the constant C was the sort of constant that is added to an antiderivative, but since we exponentiated everything, it was converted into a constant that *multiplies* the answer. If we forgot to add the constant C to the antiderivative, then we would not have obtained the constant C', and our answer would be ridiculous. It would say, for instance, that any bacterial culture starts with exactly one organism at the beginning, since $P(0) = e^{k\times 0} = e^0 = 1$.

Since the constant C came up in the middle of the calculation but it is the final answer that is of the most interest, we often rename the constants and use C to refer to C'. This way we can write

$$P(t) = Ce^{kt}.$$

We say the population of bacteria grows *exponentially*.

This kind of growth is extremely important in a huge variety of contexts. It will arise whenever the rate of growth (or decrease) of some quantity is proportional to that quantity. We will discuss it at greater length in Chapter 7, but mention in passing that it is relevant to such diverse problems as the spread of epidemics, radioactive decay, compound interest, and making yogurt, to name only a few. In each of these situations the rate of change may be proportional to the amount currently present.

EXAMPLE 10 **Bacterial Culture** Suppose a bacterial culture weighs 10 mg (milligrams). Suppose that 1 hour later it weighs 20 mg. What will it weigh 30 min after that?

Solution From Example 9 we know that the weight of the culture (which is just a constant times the number of organisms) will be given by a formula of the form

$$W(t) = Ce^{kt},$$

where $W(t)$ is the weight of the culture at time t, in milligrams, and C and k are constants. We may choose to measure time in the most convenient units, which in this case appear to be hours. Next we should find C and k.

If $t = 0$ at the starting time, then the problem tells us that $W(0) = 10$. On the other hand, $W(0) = Ce^{k\times 0} = Ce^0 = C$. This shows that $C = 10$.

The other piece of information we have is that $W(1) = 20$. From the formula we see that $W(1) = Ce^k = 10e^k$, so

$$10e^k = 20,$$
$$e^k = 2,$$

so

$$k = \ln 2.$$

Having calculated both C and k, we know that

$$W(t) = 10e^{t\ln 2}.$$

The question asked us to find the weight of the culture $1\frac{1}{2}$ hours after the starting point. This is $W\left(\frac{3}{2}\right) = 10e^{3/2\ln 2} = 10(e^{\ln 2})^{3/2} = 10(2^{3/2}) = 10(2\sqrt{2}) = 20\sqrt{2} \approx 28.3$.

After one and one-half hours the culture will weigh about 28.3 mg.

In the course of doing this example we have noticed that in a formula for exponential growth

$$P(t) = Ce^{kt},$$

the constant C is exactly equal to $P(0)$, the *initial population*. The constant k, which depends on the units used, has to do with how fast the population grows. It is often called the *growth constant*.

Exercises 4.8

I

Find all antiderivatives of each of the following functions.

1. $f(x) = 6x^2 - 4x + 3$
2. $g(x) = x^3 - 4x^2 + 3x - 2$
3. $h(x) = \sin x$
4. $j(x) = \csc^2 x$
5. $r(x) = \frac{1}{\sqrt{1-x^2}}$
6. $k(x) = x^2 - 2x + \cos x$
7. $s(x) = \sqrt{x} + x^4$
8. $t(x) = x^{7/4}$
9. $v(x) = \frac{3}{\sqrt{x}}$
10. $w(x) = \frac{x^2}{1+x^2}$
11. $f(x) = x\sin(2x^2 - 3)$
12. $g(x) = x^2\cos(x^3 + 2)$
13. $h(x) = (x + 1)\exp(x^2 + 2x - 5)$

14. $k(x) = (x^2 - 1)\sin(x^3 - 3x + 7)$

15. $q(x) = \frac{x}{x^2+1}$

16. $r(x) = \frac{(x-1)}{x^2-2x+3}$

17. $s(x) = x^3\sqrt{x^4 + 4}$

18. $t(x) = \frac{1}{1+(x-2)^2}$

19. $v(x) = \frac{1}{e^{-x}+1}$

20. $w(x) = \cos x \sin^2 x$

II

21. Find an antiderivative $F(x)$ of $f(x) = x^3 - 2x + 3$ that satisfies $F(0) = 5$.

22. Find an antiderivative $G(x)$ of $g(x) = x - \cos x$ that satisfies $G(0) = -2$.

23. Find an antiderivative $H(x)$ of $h(x) = 3x^2 - 4x + 1$ that satisfies $H(1) = 1$.

***24.** Is it possible to find an antiderivative $K(x)$ of $k(x) = x^2 + 3x - 2$ that satisfies $K'(3) = 3$? If so, find it.

Find *all* antiderivatives of each of the following functions.

***25.** $f(x) = \frac{|x|}{x}$

***26.** $g(x) = \frac{3}{x}$

***27.** $h(x) = \frac{1}{x-3}$

***28.** $k(x) = \frac{1}{(x-2)^3}$

***29.** $f(x) = \sec^2 x$

***30.** $g(x) = \frac{\cos x}{\sin^2 x}$

***31.** $h(x) = \tan x$

***32.** $h(x) = \cot x$

33. Suppose a bacterial culture containing 2000 organisms doubles in 40 min.

(i) How many bacteria will there be 1 hour after the beginning?

(ii) How long will it take until there are 10,000 organisms?

34. A technician forgot to weigh a bacterial culture when it began growing. He weighed it after 3 hours and found that it weighed 16 mg and weighed it again two hours later, when it weighed 32 mg.

(i) How much did it weigh at the beginning?

(ii) What will it weigh 11 hr after the beginning?

***35.** Find all antiderivatives $K(x)$ of $k(x) = \sec^2 x$ that satisfy $K\left(-\frac{\pi}{4}\right) = 3$ and $K\left(\frac{\pi}{4}\right) = 5$.

***36.** Show that any antiderivative $F(x)$ of $\sec x \tan x$ must satisfy $F\left(\frac{\pi}{6}\right) = F\left(-\frac{\pi}{6}\right)$.

***37.** Find as many functions $f(x)$ as you can that satisfy $f''(x) - f'(x) - f(x) = 0$. (*Hint:* Try letting $f(x) = e^{cx}$ and solve for c.)

***38.** Find as many functions $f(x)$ as you can that satisfy $f''(x) + 2f'(x) - 3f(x) = 0$. (*Hint:* Try letting $f(x) = e^{cx}$ and solve for c.)

***39.** Find as many functions $f(x)$ as you can that satisfy $f''(x) + f'(x) - 6f(x) = e^{4x}$.

****40.** Suppose $f(x)$ is a function that is defined on the whole real number line and $F(x)$ is an antiderivative of $f(x)$.

(i) If $f(x)$ is odd, show that $F(x)$ must be even.

(ii) Show by example that the corresponding result with "odd" and "even" interchanged is not necessarily true.

(iii) Show by example that if $f(x)$ is not defined on the whole real line, then even the result of part (i) need not be true.

III

41. (i) Use a graphics calculator or computer to sketch $f(x) = \frac{x}{1+x^2}$. By noting where $f(x)$ is positive or negative, predict the shape of the graph of an antiderivative $F(x)$ of $f(x)$.

(ii) Find the antiderivative $F(x)$ that satisfies $F(0) = 0$ and graph it on the same picture, comparing the result with your predictions.

42. (i) Sketch $g(x) = x^3 - 3x^2$ and predict the shape of the graph of an antiderivative $G(x)$.

(ii) Find the antiderivative $G(x)$ with $G(0) = 4$ and sketch it on the same picture.

43. (i) Sketch the graph of the exponential growth function $f(t) = 7e^{3t}$. Use the picture to find the value of t for which $f(t) = 2f(0)$.

(ii) Calculate the answer and compare with what you found graphically.

44. (i) Sketch the growth function $g(t) = 2e^{t/4}$. Find to four decimal places how long it takes for $g(t)$ to double starting at $t = 0$.

(ii) Find how long it takes to double starting at $t = 2$.

(iii) Find how long it takes to double starting at $t = 0.3$.

4.9 Velocity and Acceleration

CONSTANT ACCELERATION
VERTICAL AND HORIZONTAL COMPONENTS

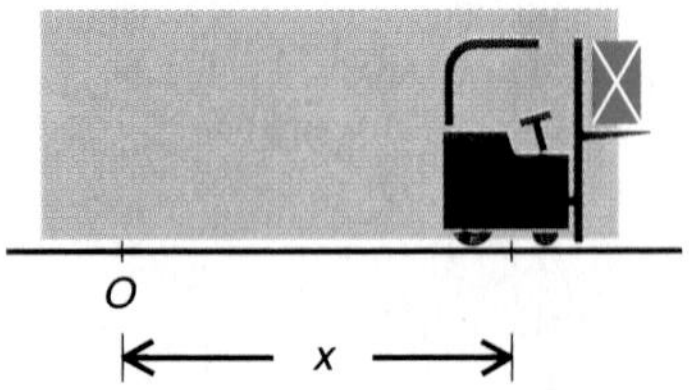

Figure 4.9.1

We recall the discussion in Section 3.1 of the derivative of a distance or a displacement with respect to time and its interpretation as the speed or velocity with which the distance is changing.

Suppose an object is moving along a straight line. Fixing some point on the line, we can call it the origin and measure x, the distance from the object to that point (see Figure 4.9.1). However, if the object moves in both directions, we may need to speak of negative as well as positive values of x, and the word "distance" is usually used only for a positive measurement. To get around this, we use the word "displacement," and the displacement, x, of the moving object from the origin can be any real number. Positive displacements will be in one direction—on one side of the origin—and negative displacements will be in the other direction; it is important to be clear which is which.

The derivative $\frac{dx}{dt}$ of displacement with respect to time is the rate of change of the displacement, which is related to what we call the "speed." But to incorporate the possibility of different signs, we call it the **velocity** of the moving object.

EXAMPLE 1 Suppose an object is moving along the x-axis so that after t seconds it is at the point $x = t^2 - 2t - 1$. Then its velocity is $\frac{dx}{dt} = 2t - 2$. Notice that when $t < 1$, the velocity is negative and the object is moving to the left; when $t > 1$, we have $\frac{dx}{dt} > 0$, and the motion is to the right. At $t = 1$, $\frac{dx}{dt} = 0$, and the object is (momentarily) stationary.

Frequently the word "speed" is used to refer to the absolute value of the velocity. So at times $t = 0$ and $t = 2$ in the above example the object is moving at velocities of -2 units per second and $+2$ units per second, respectively, but at both times its speed is 2 units per second. This is analogous to the distinction between distance and displacement: At times $t = 1$ and $t = 3$ the object is at displacements of $x = -2$ and $x = +2$, respectively, but at both these times it is at a distance of 2 units from the origin.

EXAMPLE 2 Suppose a moving object has velocity $v(t) = t^3 - t^2 + 3t + 2$ at time t. When $t = 3$, how far will it be from where it was when $t = 0$?

Solution We know that the velocity, $v = t^3 - t^2 + 3t + 2$, is $\frac{dx}{dt}$, so the displacement x, an antiderivative of this function, must be $\frac{1}{4}t^4 - \frac{1}{3}t^3 + \frac{3}{2}t^2 + 2t + c$, for some constant c. The problem doesn't give us enough information to evaluate c (we would have to know where the object is at some particular time), but fortunately this doesn't matter. When $t = 3$, $x = \frac{1}{4}3^4 - \frac{1}{3}3^3 + \frac{3}{2}3^2 + 2(3) + c = \frac{81}{4} - 9 + \frac{27}{2} + 6 + c = \frac{123}{4} + c$, and when $t = 0$, $x = 0 - 0 + 0 + 0 + c = c$, so the distance between these two points is $\left(\frac{123}{4} + c\right) - c = \frac{123}{4} = 30\frac{3}{4}$.

The derivative of the velocity is called the **acceleration**, and of course it is the second derivative of the displacement. The units of acceleration are of the form *(length)/(time)*^{2}, for example, ft/sec^2 or cm/sec^2. If, for example, the velocity is decreasing by 3 ft/sec every second, the acceleration is -3 ft/sec/sec, which we write as -3 ft/sec^2.

With a graphics calculator we can plot the antiderivative we found in Example 2. It is necessary to use x as the variable. It is also necessary to choose a particular value of c; the simplest choice is $c = 0$.

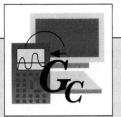

TI-85 GRAPH F1 ($y(x)$ =) F1 (x) ^4 ÷ 4 − F1 (x) ^3 ÷ 3 + 3 F1 (x) c^2 ÷ 2 + 2 F1 (x) M2 (RANGE) 0 ▼ 3 ▼ ▼ 0 ▼ 35 F5 (GRAPH)	SHARP EL9300 x/θ/T a^b 4 ▶ ÷ 4 − x/θ/T a^b 3 ▶ ÷ 3 + 3 x/θ/T x^2 ÷ 2 + 2 x/θ/T RANGE 0 ENTER 3 ENTER 1 ENTER 0 ENTER 35 ENTER 5 ENTER
CASIO f_x-7700GB/f_x-6300G Cls EXE Range 0 EXE 3 EXE 1 EXE 0 EXE 35 EXE 5 EXE EXE EXE EXE Graph X x^y 4 ÷ 4 − X x^y 3 ÷ 3 + 3 X x^2 ÷ 2 + 2 X EXE	HP 48SX PLOT PLOTR ERASE ATTN ' αY = αX y^x 4 ÷ 4 − αX y^x 3 ÷ 3 + 3 × αX y^x 2 ÷ 2 + 2 × αX PLOT DRAW ERASE 0 ENTER 3 XRNG 0 ENTER 35 YRNG DRAW

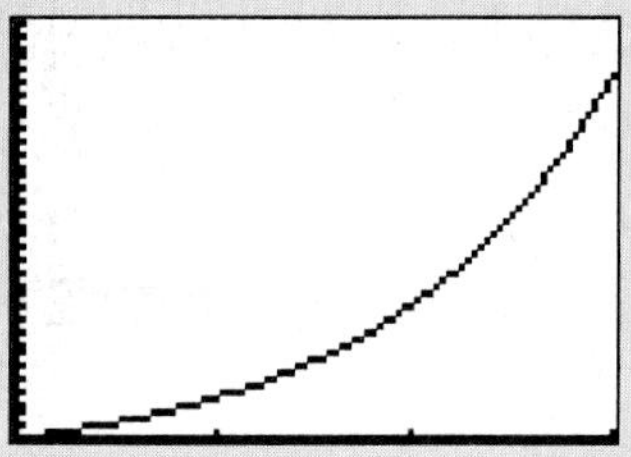

From the picture we see quickly that the object moves about 30 units between $x = 0$ and $x = 3$. We could get a more accurate value by zooming in.

A falling object near the earth's surface moves with a constant acceleration, called g (to remind us of gravity). The value of g is about 32.16 ft/sec^2 (often approximated by 32 ft/sec^2) or 9.8 m/sec^2 (or 980 cm/sec^2).

EXAMPLE 3 **Ball Thrown Straight Up** A girl throws a ball straight up in the air. It leaves her hand, 5 ft above the ground, traveling at a speed of 50 ft/sec. When will it hit the ground?

Solution Write $h(t)$ for the height of the ball above the ground, in feet, at time t. We know that $h(0) = 5$, $h'(0) = 50$, and that the acceleration $h''(t)$ is constant, 32 ft/sec^2 *downward*. This means that $h''(t) = -32$.

Now if $h''(t) = -32$, we must have $h'(t) = -32t + c$, for some constant c. But $h'(0) = 50$, so $50 = -32(0) + c = 0 + c = c$, that is, $c = 50$. So $h'(t) = -32t + 50$.

$h'(t)$ is an antiderivative of $h''(t)$.

But then the antiderivative $h(t)$ must be $-16t^2 + 50t + d$ for some constant d. But $h(0) = 5$, so $5 = h(0) = 16(0^2) + 50(0) + d = d$, so $d = 5$ and $h(t) = -16t^2 + 50t + 5$. To solve the problem, we must find when $h(t) = 0$, which happens when

$$t = \frac{-50 \pm \sqrt{50^2 - 4(-16)5}}{2(-16)} = \frac{50 \pm \sqrt{2820}}{32} \approx \frac{50 \pm 53.1}{32}$$
$$\approx -0.096875 \quad \text{or} \quad 3.22.$$

The first of these is negative, representing a time before the ball was thrown, so it is an extraneous solution. The second is the right answer: $t \approx \frac{103.1}{32} \approx 3.22$ sec. The ball will hit the ground 3.22 seconds after it is released.

REMARK 4.9.1

Motion with Constant Acceleration

Example 3 illustrates a general phenomenon. Suppose an object is moving with constant acceleration a. Writing x for its displacement, we find that $x' = at + c$ and

$$x = \frac{1}{2}at^2 + ct + d.$$

When $t = 0, x = d$ and $x' = c$, so c is the *initial velocity* and d is the *initial displacement.* In particular, with $a = -32$ ft/sec^2 for a falling object, the height above the ground is

$$h(t) = -16t^2 + ct + d.$$

EXAMPLE 4 **Constant Acceleration** (i) If a car moves with constant acceleration 10 ft/sec^2, beginning with velocity 20 ft/sec, how far will it travel in 10 sec?

Solution The question tells us that the acceleration (in ft/sec^2) is 10 and that the initial velocity (in ft/sec) is 20. The question does not mention an initial displacement. In effect this means that we can place the origin wherever we please, and a natural place to put it is the starting point, which amounts to saying that the initial displacement is 0.

From Remark 4.9.1 we see that the displacement at time t (time measured in *seconds*) is

$$x(t) = \frac{1}{2}10t^2 + 20t + 0 = 5t^2 + 20t.$$

The displacement of the car after 10 sec will be $x(10) = 5(10)^2 + 20(10) = 700$. After 10 sec, the car will have traveled 700 ft.

El Capitan, in Yosemite National Park in California, is about 3600 feet high. How long would it take to fall off?

(ii) The driver of a car moving at a speed of 40 m/sec slams on the brakes. The car is subjected to a constant deceleration of 5 m/sec^2. How far does it travel before coming to a stop?

Solution First we must find the time at which the car comes to rest, that is, the time at which the velocity is zero. The fact that the car is decelerating means that the acceleration is negative (the acceleration is the rate of change of the velocity, which is decreasing). So the acceleration is -5, and the initial velocity is 40.

The displacement is $x(t) = \frac{1}{2}(-5)t^2 + 40t + d$, where d is the initial displacement, so the velocity is $x'(t) = -5t + 40$. This is zero when $t = 8$. In other words, the car comes to a standstill after 8 sec.

If we let the origin be the starting point, that is, let $d = 0$, then the displacement at the moment when the car stops will be

$$x(8) = \frac{1}{2}(-5)8^2 + 40(8) + 0 = 160.$$

The car will travel 160 m before stopping.

(iii) A man seated on the edge of a high cliff throws a small heavy object straight up at a speed of 32 ft/sec. It hits the ground at the bottom of the cliff exactly 5 sec later. How high is the cliff?

Solution We could measure height either from the ground level at the bottom of the cliff or from the top of the cliff (or anywhere else, for that matter). Let us measure from the bottom of the cliff and let d be the height of the cliff, in feet.

Then the initial height (displacement) of the object is d (it is d ft above ground level). Since the initial velocity is 32 ft/sec and the acceleration is -32 ft/sec^2, we see that the height at time t is

$$h(t) = -16t^2 + 32t + d.$$

We know that the object hits the ground when $t = 5$, which means that $h(5) = 0$. In terms of the formula, this means that

$$0 = -16(5)^2 + 32(5) + d = -400 + 160 + d = -240 + d.$$

This is only possible if $d = 240$. We have just found that the height of the cliff is 240 ft.

You might like to rework Example 4(iii), making the measurement of height from the top of the cliff. In this case the initial displacement (i.e., initial height) will be zero. Of course, the answer should be the same.

REMARK 4.9.2

If somebody throws a rock, it will travel both vertically and horizontally. The surprising thing is that **the vertical and horizontal components of the motion** can be considered independently. That is, the vertical motion (height above the ground) will be described by the formula for the height of a falling object (see Remark 4.9.1), and its horizontal velocity will be constant (ignoring air resistance, which would actually slow it down slightly).

Figure 4.9.2(i)

EXAMPLE 5 A rock is catapulted into the air from ground level so that its path makes an angle of 30° above the ground (see Figure 4.9.2(i)). If its speed is 25 m/sec, how far from its starting point will it fall?

This miniature shows an "engine of war," a catapult from the fifteenth century.

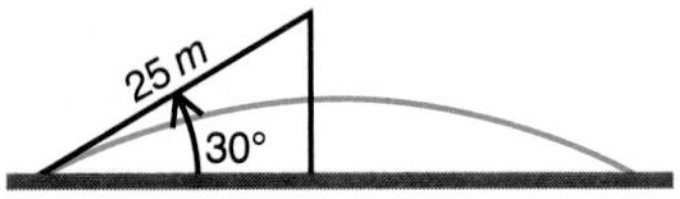

Figure 4.9.2(ii)

Solution The rock is moving at 25 m/sec in a direction 30° above the horizontal. If it continued in a straight line for 1 second, it would travel 25 m along the hypotenuse of the triangle shown in Figure 4.9.2(ii). The horizontal distance it would travel is $25\cos(30°)$ m $= 25\frac{\sqrt{3}}{2}$ m ≈ 21.65 m, and the vertical distance (the vertical side of the triangle) is $25\sin(30°)$ m $= 12.5$ m. This tells us that the initial horizontal velocity is 21.65 m/sec and the initial vertical velocity is 12.5 m/sec.

By Remark 4.9.1, the height is given by $h(t) = \frac{1}{2}at^2 + ct + d$, where the acceleration $a = g = -9.8$ m/sec^2, and the initial height d is zero, since the rock is catapulted from ground level. The initial vertical velocity c is 12.5 m/sec. So $h(t) = -4.9t^2 + 12.5t = t(-4.9t + 12.5)$, and $h(t) = 0$ when $t = 0$ or when $t = \frac{12.5}{4.9} \approx 2.55$ sec. (The solution $t = 0$ tells us what we already know, that the rock starts at ground level initially; the other answer is the one we want.)

Meanwhile, by Remark 4.9.2, the horizontal motion has constant velocity $x' = 21.65$, so $x(t) = 21.65t + c$. If we measure horizontal displacement x from the starting point, $x(0) = 0$, so $c = 0$ and $x(t) = 21.65t$. When the rock hits the ground, $t = 2.55$ sec and $x = 21.65(2.55) \approx 55.2$ m. The rock lands 55.2 m away from its starting point.

Exercises 4.9

I

Evaluate the following.

1. The acceleration at time t of a moving object whose displacement at time t is $x = 5t^3 - 3t^2 + 7t - 12$.
2. The displacement at time t of a moving object whose acceleration at time t is $6t - 2$, whose initial displacement is 6, and whose initial velocity is 5.
3. The displacement at time t of a moving object whose acceleration is given by $12t - 4$ and whose displacement is 2 at $t = 1$ and 12 at $t = 3$.
4. The time or times at which an object is at rest if its acceleration is 12 ft/sec^2 and its initial velocity and displacement are -6 ft/sec and -3 ft, respectively.
5. The maximum height reached by a ball thrown from a point 5 ft above ground level at a speed of 64 ft/sec straight up.
6. The height of a building from which it takes a penny exactly 2 sec to fall to the ground.
7. The time when an object returns to its starting point if its acceleration is -14 m/sec^2 and its initial velocity is 20 m/sec.
8. The greatest distance from its starting point which the object in Exercise 7 reaches before returning to its starting point.
9. The length of time it takes a penny to fall to the ground from a building 144 ft tall.
10. The displacement at $t = 3$ of an object moving with constant acceleration whose displacements at times $t = 0$, $t = 1$, and $t = 2$ are $x = -1$, $x = 2$, and $x = 4$, respectively.

II

11. A ball is projected straight upward from ground level with an initial speed of 60 ft/sec.
    (i) When will it hit the ground?
    (ii) What is the maximum height it will reach?
12. A car traveling at 60 mph drives off the edge of a cliff.
    (i) If the drop is 100 ft to the lake below, how long will it be before the car hits the water?
    (ii) How far from the base of the vertical cliff will the point of impact be?
13. How fast must a ball be thrown upward if it is to reach the top of a building that is 15 m high? (Assume that the ball leaves the thrower's hand at a point 1.5 m above ground level.)
14. Applying a car's brakes causes a constant deceleration of 45 ft/sec^2.
    (i) How long (in seconds) does it take the car to stop after the brakes are applied with the car moving at 50 mph?
    (ii) How far (in feet) does the car in part (i) travel before stopping?
    (iii) Repeat parts (i) and (ii) with initial speeds of 60 mph and 70 mph.
15. A driver sees a child in the road and slams on her brakes. Suppose she is traveling at 40 mph, her

brakes cause a deceleration of 30 ft/sec^2, and her reaction time (between seeing the child and applying the brakes) is 0.8 sec.

(i) If she first sees the child at a distance of 105 ft, is there time to stop?

(ii) How far away must she see the child if her speed is 60 mph?

16. (i) If a daredevil in a barrel takes 3 seconds to descend to the foot of a waterfall, how high is the waterfall?

(ii) If the waterfall is 160 ft high, how long will the daredevil fall?

***17.** A football is thrown at an angle of 45° above the horizontal. It leaves the quarterback's hand at a point 6 ft above the ground, and the highest point in its flight is 30 ft above the ground.

(i) How far will it fly if it is caught when it is again 6 ft above the ground?

(ii) How far will it fly if it is allowed to hit the ground?

***18.** Repeat Exercise 17 with an angle of 26°.

***19.** A projectile is catapulted into the air so that its path makes an angle θ with respect to the ground. Find a formula (involving θ and the initial velocity v_0) for the distance it travels before hitting the ground again. Assume that the ground is perfectly flat and ignore air resistance.

***20.** Using the result of Exercise 19, find the angle at which the projectile should be fired in order to maximize the distance traveled. What is the maximum possible distance?

21. A spaceship is at rest in space. It fires rockets for 10 seconds, during which time it undergoes a constant acceleration of 15 m/sec^2. It coasts for 20 seconds, then fires the rockets for an additional 10 seconds. After a further 20 seconds' coasting, how far has the spaceship traveled from its starting position?

22. A small, heavy object is dropped from an airplane.

(i) If the airplane is traveling at 300 mph and at an altitude of 20,000 ft, how long will the object fall, and what will be the horizontal distance it travels from the drop point to the point of impact?

(ii) Repeat part (i) for an aircraft traveling at 600 mph.

(iii) At what angle will the objects in parts (i) and (ii) strike the ground?

23. Experiment has shown that at large distances from the earth, an object will fall toward the earth with an acceleration due to gravity equal to $\frac{k}{R^2}$, where k is some constant and R is the distance from the object to the center of the earth. The earth's radius is 4000 miles, and we know that at the surface of the earth the acceleration due to gravity is 32 ft/sec^2. What is the value of the constant k?

24. A car decelerates at 10 m/sec^2. (i) Find its stopping distance, starting at a speed of 50 km/hr. (ii) Do you expect the stopping distance from 100 km/hr to be exactly twice as far or less or more than twice as far? (iii) Find the stopping distance from 100 km/hr.

III

25. Suppose an object moves so that its acceleration at time t is $x''(t) = 20t^3 - 24t$ and its initial velocity $x'(0)$ is zero. (i) Find a formula for the displacement $x(t)$. (ii) Use a graphics calculator or computer to find the smallest time $t > 0$ at which the object is 7 units from its initial position.

26. Two objects move in a straight line, both beginning with displacement 0 and velocity 0. One has constant acceleration 1, and the other has acceleration $x''(t) = 2\cos t$. Use a graphics calculator or computer to find when they collide. How far will they have traveled before colliding?

4.10 Linear Approximation and Differentials

ERROR
ERROR PROPAGATION
PERCENTAGE ERROR

In this section we discuss an application of the derivative to calculating approximate values of functions. It can be extremely useful in situations in which precise calculations are difficult or impossible.

Many functions are actually very difficult to calculate explicitly, except perhaps for a few special values. Even something relatively simple like $\sqrt{63}$ can easily be found by using a calculator, but it is less clear what to do if you want more than eight or ten digits' accuracy. Other functions such as $\sin x$ or $\ln x$ can be *approximated* with a calculator, but there is no way to give an exact evaluation except for a few familiar points.

This is usually not a matter of great concern, since the accuracy that is available on a calculator is generally sufficient for most practical purposes, but a more subtle problem would be to evaluate a function that is described as an inverse function. For instance, the function $g(x) = x^3 + 2x - 5$ has as its derivative $g'(x) = 3x^2 + 2$. This derivative is always positive, so the function $g(x)$ is always increasing, which implies that it has an inverse function, call it $f(x)$. How would we begin to find, say, $f(3)$?

We could try guessing and then try refining our guesses, using the Intermediate Value Theorem to narrow down where the value lies. In this case, for instance, $g(1) = -2$ and $g(2) = 7$, so $g(x)$ must equal 3 for some x between 1 and 2. We followed this approach in Example 4.6.12 (p. 288), realizing that there had to be a zero between two points where a function had opposite signs, but it was fairly cumbersome. There is another approach involving the derivative $f'(x)$.

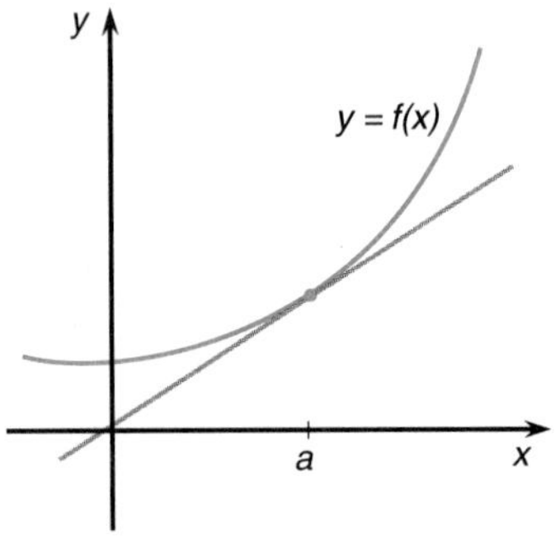

Figure 4.10.1

If $f(x)$ is differentiable at $x = a$, then we can draw the tangent to the graph $y = f(x)$ at the point $(a, f(a))$. Near that point, the tangent line lies very close to the graph of $y = f(x)$. For values of x farther away from a, the tangent can easily be quite far from the graph, but for x near a the tangent line is a good approximation to the curve $y = f(x)$. See Figure 4.10.1.

This fact can be very useful. It is very easy to calculate precisely any point on a straight line, so we can find good approximations to $f(x)$ provided that x is close to a.

If $y = f(x)$, we found the derivative $f'(a)$ by taking the limit, as $h \to 0$, of $\frac{f(a+h)-f(a)}{h}$. This idea can be used in reverse: If we already know $f(a)$ and $f'(a)$, then we can estimate nearby values of $f(x)$.

Since $f'(a) = \lim_{h\to 0} \frac{f(a+h)-f(a)}{h}$, we see that for small values of h,

$$\frac{f(a+h) - f(a)}{h} \approx f'(a),$$

that is,

$$f(a+h) - f(a) \approx f'(a)h.$$

This can be written in the form

$$f(a+h) \approx f(a) + f'(a)h \tag{4.10.1}$$

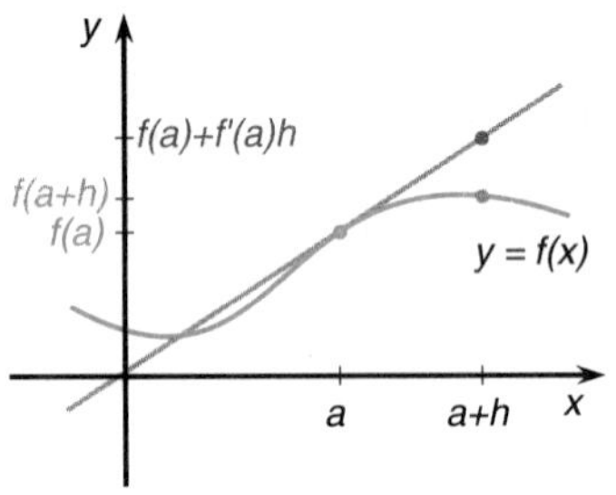

Figure 4.10.2

Figure 4.10.2 shows the tangent line to $y = f(x)$ at $x = a$. Starting from the point $(a, f(a))$, we move along the tangent through a horizontal distance of h to find the point $(a + h, f(a) + f'(a)h)$. The y-coordinate of this point is given by the right side of Formula 4.10.1 and is an approximation to the actual value of $f(a + h)$. The actual value of $f(a + h)$ is the value of y for the point *on the curve* $y = f(x)$; the approximation comes from the corresponding point *on the tangent line*. In Figure 4.10.2 the actual value is slightly lower than the approximation.

This way of approximating $f(a + h)$ is called **linear approximation** because, instead of finding the exact point on the curve, we went along the tangent *line*. Notice that the closer $a + h$ is to a, that is, the smaller h is, the better the approximation will be.

EXAMPLE 1 Use linear approximation to estimate $\sqrt{63}$.

Solution Let $y = f(x) = \sqrt{x} = x^{1/2}$. We know that $f(64) = 8$, and $f'(x) = \frac{1}{2}x^{-1/2}$, so $f'(64) = \frac{1}{16}$. We want to apply Formula 4.10.1 with $a = 64$, $a + h = 63$, so $h = -1$.

We find $f(63) = f(64-1) \approx f(64) + f'(64)(-1) = 8 + \frac{1}{16}(-1) = 8 - \frac{1}{16} = 7\frac{15}{16} = 7.9375$. In fact, $\sqrt{63} \approx 7.937254\ldots$, so our approximation is reasonably good.

Just how good the approximation is depends on various things, particularly on how small h is. Of course, approximating square roots is not very useful in the age of pocket calculators, but understanding the technique will lead to other applications.

To make this approach work, what is needed is a point $x = a$ for which the values of $f(a)$ and the derivative $f'(a)$ are known. We want an approximate value for f at some nearby point. We write this point as $x = a + h$, so that h is the (small) amount by which it is necessary to move to get from a to $a + h$. We can think of h as the *change* in x from the known point a to the unknown point $a + h$.

EXAMPLE 2 Estimate $\sin(46°)$.

Solution We want to apply Formula 4.10.1 with $f(\theta) = \sin\theta$. It seems reasonable to use 45° as our starting point a, so then $h = 1°$. The catch is that we are interpreting $f'(\theta) = \cos\theta$ as the rate of change, and it is the rate of change of $f(\theta) = \sin\theta$ with respect to θ, with θ measured in *radians*.

So we have to convert 45° to radians: $a = 45 \times \frac{\pi}{180} = \frac{\pi}{4}$ and $h = 1\left(\frac{\pi}{180}\right)$, so

$$\sin(46°) = \sin\left(\frac{\pi}{4} + \frac{\pi}{180}\right) \approx \sin\left(\frac{\pi}{4}\right) + \cos\left(\frac{\pi}{4}\right)\frac{\pi}{180} = \frac{1}{\sqrt{2}} + \frac{1}{\sqrt{2}}\frac{\pi}{180}$$
$$\approx 0.7194.$$

You can check how close this approximation is by comparing with your calculator.

With a graphics calculator we can plot the function $y = \sin x$ and the tangent line to it at $x = \frac{\pi}{4}$. Since $f'(\frac{\pi}{4}) = \cos\frac{\pi}{4} = \frac{1}{\sqrt{2}}$, the tangent line is $y = \frac{1}{\sqrt{2}} + \frac{1}{\sqrt{2}}(x - \frac{\pi}{4})$.

TI-85	SHARP EL9300
GRAPH F1(y(x) =) SIN F1(x) ▼ 1 ÷ √2 + (F1(x) − π ÷ 4) ÷ √2 M2(RANGE) 0 ▼ π ÷ 2 ▼ ▼ 0 ▼ 2 F5(GRAPH)	[graph key] sin x/θ/T 2ndF ▼ 1 ÷ √2 ▶ + (x/θ/T − π ÷ 4) ÷ √2 RANGE 0 ENTER π ÷ 2 ENTER 1 ENTER 0 ENTER 2 ENTER 1 ENTER [graph key]
CASIO f_x-7700GB/f_x-6300G	**HP 48SX**
Cls EXE Range 0 EXE π ÷ 2 EXE 1 EXE 0 EXE 2 EXE 1 EXE EXE EXE EXE Graph sin X EXE Graph 1 ÷ √2 + (X − π ÷ 4) ÷ √2 EXE	↰ PLOT PLOTR ERASE ATTN ′ αY = SIN αX ↱ PLOT ↰ DRAW ERASE 0 ENTER π 2 ÷ XRNG 0 ENTER 2 YRNG DRAW ATTN ′ αY = 1 ÷ √x 2 + () αX − π ÷ 4 ▶ ÷ √x 2 ↱ PLOT ↰ DRAW DRAW

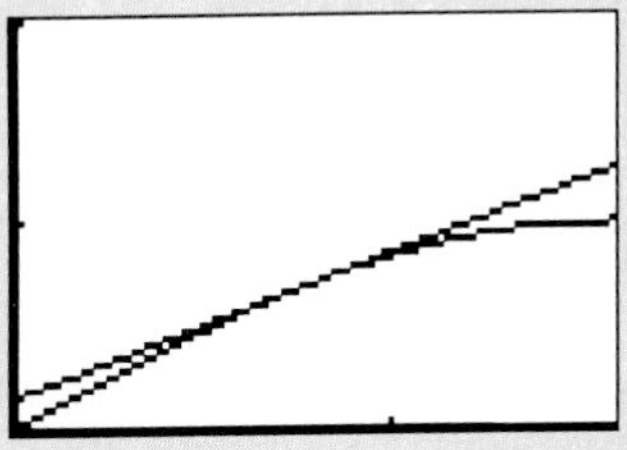

Now try graphing the same functions for $x \in \left[\frac{\pi}{4}, \frac{\pi}{4} + \frac{\pi}{180}\right]$, (with $0.7 \le y \le 0.72$). The tangent line is extremely close to the curve...

Differentials

If $y = f(x)$ is differentiable, we can take the equation $\frac{dy}{dx} = f'(x)$ and, pretending that dx is a number, just "multiply" by it, to get

$$dy = f'(x)\,dx. \tag{4.10.2}$$

The symbols dx and dy are called **differentials.** Equation 4.10.2 is just another way of writing $\frac{dy}{dx} = f'(x)$, and it means exactly the same thing, no more and no less.

EXAMPLE 3 Find the differential dy if (i) $y = 3x^3 - \cos x$, (ii) $xy^2 - y^3 + 2x^2y^2 - xy = 1$.

Solution

(i) $\frac{dy}{dx} = 9x^2 + \sin x$, so $dy = (9x^2 + \sin x)\,dx$.

(ii) Here we differentiate implicitly: $y^2 + 2xyy' - 3y^2y' + 4xy^2 + 4x^2yy' - y - xy' = 0$, so $(2xy - 3y^2 + 4x^2y - x)y' = -y^2 - 4xy^2 + y$, and $\frac{dy}{dx} = y' = \frac{-y^2-4xy^2+y}{2xy-3y^2+4x^2y-x}$. Therefore $dy = \frac{-y^2-4xy^2+y}{2xy-3y^2+4x^2y-x}\,dx$.

Although we were careful to observe that $dy = f'(x)\,dx$ is just another way of saying $\frac{df}{dx} = f'(x)$, the formula with differentials can help us to remember Formula 4.10.1 for linear approximation. After all, Formula 4.10.1 says that $f(a+h) - f(a) \approx f'(a)h$. This is quite similar to Equation 4.10.2, the differences being that there is a dx in place of h on the right side and dy in place of $f(a+h) - f(a)$ on the left and that an equals sign has replaced the "approximately equals" sign $\approx$.

If we think of h as the change in x between a and $a + h$, and if we replaced it with dx, we should interpret dx as a small change in x (the letter d might remind us of "difference"). The natural thing would be to let dy represent the corresponding change in y. Now when x changes from a to $a + h$, y will change from $f(a)$ to $f(a+h)$. The change in y will be $f(a+h) - f(a)$, which is exactly the left side of Formula 4.10.1.

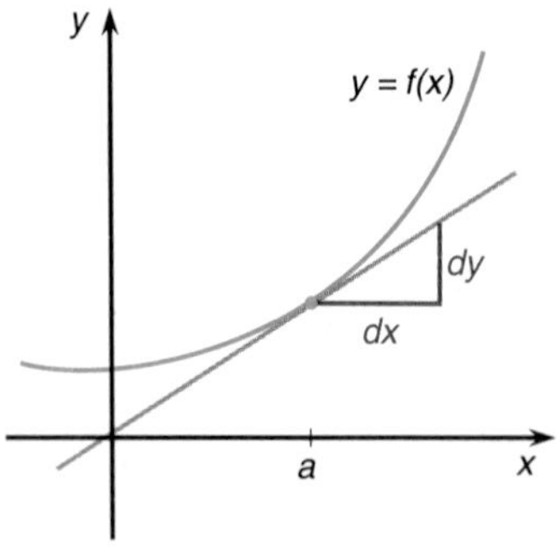

Figure 4.10.3

So Equation 4.10.2, which says $dy = f'(x)\,dx$, can be interpreted as meaning that the change in y is $f'(x)$ times the change in x. Since $f'(x)$ is the slope of the tangent line, this is the familiar statement that the slope is the rise divided by the run (see Figure 4.10.3). The reason that Equation 4.10.2 is an equality rather than only an approximation is that it refers to changes along the tangent line.

Differentials tell us how a change dx in x is converted into a change dy in y *along the tangent line.* Since the whole idea of linear approximation was to use the tangent line to approximate the curve, differentials tell us *approximately* how the function f changes.

EXAMPLE 4 Notice that the point $(1, 1)$ satisfies the equation in Example 3(ii). Find the approximate value of y for a point (x, y) satisfying the same equation (i.e., "on the curve") with $x = \frac{11}{10}$.

Solution By Example 3(ii), $dy = \frac{-y^2-4xy^2+y}{2xy-3y^2+4x^2y-x}\,dx$. This suggests that if x changes by dx, the corresponding change in y will be approximately

$$dy = \frac{-y^2 - 4x^2y + y}{2xy - 3y^3 + 4x^2y - x}\,dx.$$

When $x = 1, y = 1$, this equals $\frac{-4}{2}dx = -2dx$. So if x increases from $x = 1$ to $x = \frac{11}{10}$, that is, $dx = \frac{1}{10}$, then the change in y will be approximately $dy = -2dx = -\frac{2}{10} = -\frac{1}{5}$. Accordingly, the desired value of y is approximately obtained by adding this change to the original value, which is $y = 1$. The desired approximation is $1 - \frac{1}{5} = \frac{4}{5}$.The point "on the curve" is approximately $\left(\frac{11}{10}, \frac{4}{5}\right)$.

Notice that there is no obvious way to find approximate solutions to this sort of problem on a calculator, except perhaps by trial and error. The method of linear approximation allows us to find approximate solutions to equations we don't know how to solve exactly, and this can be very useful.

In Example 4 the differentials were used as a convenient way to summarize the calculations needed to do the linear approximation. They have other uses involving functions with several variables, but these will not be discussed in this text. There is one additional type of example to be considered here.

It concerns what is sometimes called *error propagation.* Most measurements are made subject to possible errors. Here the word "error" is used in the technical sense of "inaccuracy" rather than in the sense of "mistake."

If slightly inaccurate measurements are used to do calculations, the answers that result will themselves be inaccurate. It is very important to know how large an error in the answer can result from a specified error in the data. Frequently, people make the mistake of assuming that the error in the answer will be no worse than the error in the original data, but this is false.

EXAMPLE 5 **Measurement Error** Suppose somebody measures a square piece of metal and finds that its side is $x = 30$ cm long, with a possible error of ± 3 mm $= \pm 0.3$ cm. This error could come from sloppiness on the part of the person making the measurement, but it might also reflect irregularities in the shape of the square (one side might be slightly longer than another, for example).

Using this measurement, it is easy to see that the area of the square is $A = x^2 = 900$ sq cm. The question is to find the possible error in this value.

Solution We think of this as an approximation problem. Knowing the value of $A = x^2$ when $x = 30$, we ask by how much A might vary if x varies by up to 0.3.

Letting $dx = 0.3$, we see that $dA = 2x\,dx = 2(30)(0.3) = 18$. This says that the change in A resulting from a change of 0.3 in x is *approximately* 18. It means that a possible error of 3 mm in the measurement of the side of the square will result in a possible error of 18 sq cm in the area.

Of course, at first glance this sounds much bigger than the original error of 0.3 cm, but it is natural for the skeptical to think that although it is a bigger number, it only really makes sense to consider it relative to the number 900. This suggests that we consider the *percentage error.*

If a measurement M is made with an error of E, then the **percentage error** is $\frac{E}{M} \times 100\%$. It expresses the amount of the error as a percentage of the measured value M.

EXAMPLE 6 **Percentage Error** What happens to the percentage error in Example 5?

Solution The percentage error in the original measurement is $\frac{0.3}{30} \times 100\% = 1\%$. The percentage error in the calculated area is $\frac{18}{900} \times 100\% = 2\%$. Calculating the area *doubled* the percentage error.

Doing calculations has the unfortunate effect of piling error upon error. The result is that errors may compound, much the way that compound interest builds interest upon interest. This tendency of errors to progress through the calculation is what is known as error propagation. The effect is often that the errors grow and grow.

In an elaborate calculation, which may involve several measurements, each with its possible errors, the errors may become so large as to make the final answer meaningless. After all, if the possible error is as large as the answer, then the answer is quite unreliable.

This is a serious problem. Errors do sometimes grow alarmingly fast. It can be very important to be able to keep track of just how large they may be.

Exercises 4.10

I

Use linear approximation to estimate each of the following, and compare your answer with what you get with a calculator.

1. $\sqrt{65}$
2. $\sqrt{64.1}$
3. $\sqrt{63.999}$
4. $(126)^{1/3}$
5. $(31)^{3/5}$
6. $(32.01)^{3/5}$
7. $\cos(44°)$
8. $\sin(45.01°)$
9. $\frac{1}{\tan(136°)}$
10. $\cot(-89°)$
11. $\frac{1}{101}$
12. $\frac{1}{(100.01)^3}$

Find the differential dy in each of the following situations.

13. $y = \frac{1}{x+1}$
14. $y = x^3 - 4x + 2$
15. $y = \frac{x}{\sin x}$
16. $y = \tan\theta$
17. $x^2y - y^2x + 2 = 0$
18. $6x^2 - 8y^2 = 48$
19. $x + xy\sin x + y^2\frac{\cos x}{x} = 1$
20. $\cos(x)\sin(y) = \frac{1}{2}$

II

21. (i) Noting that $(1, 2)$ satisfies the equation $x^2y - xy^2 + 2xy = 2$, find an approximate value of y so that $(1.01, y)$ is also on that curve (i.e., satisfies the same equation).
    (ii) Substitute in the equation to see how nearly the point you found does satisfy it.
22. Repeat Exercise 21, this time approximating a point on the same curve that has the form $(x, 1.98)$. Check how close it comes to satisfying the equation.
23. A rough block of wood in the shape of a cube is measured and found to have sides 40 in long, with a possible error of 1 in.
    (i) What is the possible error in the volume of the block calculated from this information?
    (ii) Compare the percentage errors in the original measurement and the volume.
24. (i) What will be the possible error in the volume of a sphere calculated from a measurement of the radius that may be wrong by as much as 1%?
    (ii) What will be the possible error in the surface area of the sphere mentioned in part (i)? (Recall that the surface area of a sphere of radius r is $4\pi r^2$.)
25. An artisan makes her own soap, which she sells in cylindrical cakes, each of which has its height exactly equal to its diameter. She feels that, because of irregularities in the shape, it is not possible to measure these dimensions with better than 1% accuracy. She wants to be able to assert with confidence that each cake contains at least 6 cu in of soap. What size (i.e., what height and diameter) cakes should she make?

*26. In studying the effect of calculations upon errors we used differentials, which is another way of saying that we used linear approximation. This is accurate for small errors but becomes less so as the size of errors increases.
    (i) Suppose a quantity x is measured to within an accuracy of 5%. Use differentials to estimate the resulting possible error in the quantity $y = x^{20}$.
    (ii) For the sake of argument, suppose the measured value of x is exactly $x = 1$. If the error is the maximum possible, that is, 5%, it could be that $x = 1.05$. Calculate $y = x^{20}$ and see how much the error actually is.

III

27. Let $f(x) = x^3 + x^2 + 2x + 5$ and note that $f(-1) = 3$ and $f(-2) = -3$, so $f(x)$ has a zero somewhere in $[-2, -1]$. (i) Use linear approximation to estimate the zero. Do the calculation starting at $a = -2$ and at $a = -1$; compare the values of $f(x)$ at the two answers you find. (ii) Use a graphics calculator or computer to graph $y = f(x)$. Try to decide from the picture which estimate is likely to be better. (It may help to draw the two tangent lines.) (iii) Use the graph to find the zero to four decimal places.

***28.** Let $g(x) = x^4 - 20x^2 - x + 97$. Noting that $g(-3) = 1$ and $g'(-3) = 11$, use linear approximation to estimate where $g(x) = 0$. Use a graphics calculator or computer to find to four decimal places where $g(x) = 0$. Use the picture to explain what went wrong.

POINT TO PONDER

Suppose we want to approximate the value of $\sin(\pi e^{0.01})$. We might begin by trying to approximate $t = e^{0.01}$, using the function $f(x) = e^x$, and then try to approximate the value of $\sin(\pi t)$, using the function $g(u) = \sin(\pi u)$. This would involve two approximations. It would also be necessary to evaluate, or at least approximate, the values $g(t)$ and $g'(t)$ in order to perform the second approximation.

On the other hand, we could let $F(x) = \sin(\pi e^x)$ and observe that when $x = 0$, we know that $F(0) = \sin(\pi) = 0$. Since $F'(x) = \cos(\pi e^x)(\pi e^x)$, we see that $F'(0) = \cos(\pi)(\pi) = -\pi$. The linear approximation to F gives the approximation $F(0.01) \approx -\pi(0.01) \approx -0.031\,4159$.

It is easier to do all the approximating at once, rather than doing it step by step. In some cases it can even be much more difficult to do some of the steps than it is to do the single calculation.

4.11 Newton's Method

LINEAR APPROXIMATION
ITERATION

There are many situations in which we want to know the solution of some equation but are unable to think of a way to solve it. Sometimes it will be sufficient for our purposes to be able to find an approximate value for the solution. Newton's Method is a technique for finding approximate solutions of equations. It uses the idea of linear approximation, but having made one linear approximation, it goes on to make a second, a third, and so on, hoping to improve the approximation each time.

Suppose $y = f(x)$ is a differentiable function, and suppose we want to find a zero of $f(x)$, that is, a value of x for which $f(x) = 0$. Starting at some point $x = a$, consider the tangent line at $(a, f(a))$. With luck, it will cross the x-axis near a point at which the curve $y = f(x)$ crosses the x-axis (as shown in Figure 4.11.1). It is easy to find the point at which the tangent line crosses the x-axis, and we can hope this will give a value of x that is closer to a zero of $f(x)$ than $x = a$ was. What we have done is use the idea of linear approximation from Section 4.10 to find an x that is close to a zero of f. Needless to say, this is likely to work better in cases in which $x = a$ is already close to a zero of $f(x)$.

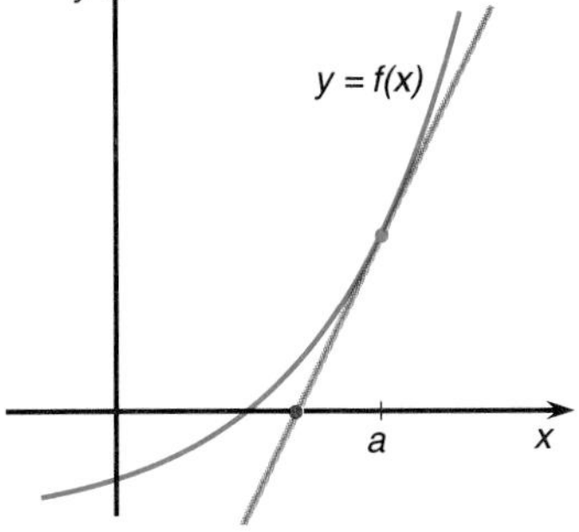

Figure 4.11.1

Since the tangent line to $y = f(x)$ at $(a, f(a))$ is $y = f(a) + f'(a)(x - a)$, to find the point where it crosses the x-axis, we need to let $y = 0$, that is, $0 = f(a) +$

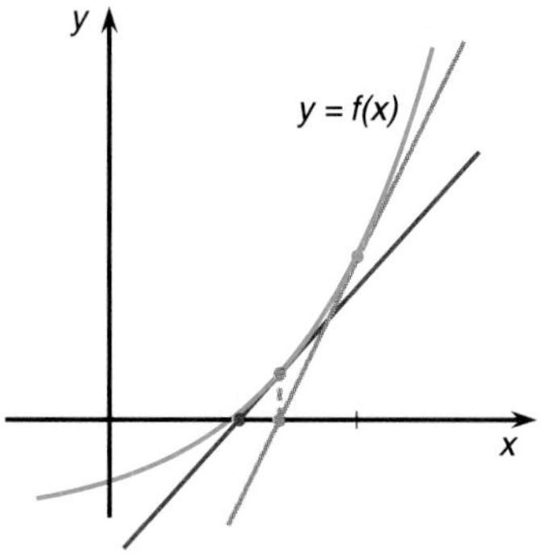

Figure 4.11.2

$f'(a)(x - a)$, so $x - a = -\frac{f(a)}{f'(a)}$, or

$$x = a - \frac{f(a)}{f'(a)}.$$

Having found this x, we hope that we have moved closer to a zero of f. On the other hand, it is unlikely that x will be exactly a zero, and it might well not even be terribly close to one. The idea that is known as Newton's Method is to apply the process again, this time using the tangent line at the new point $(x, f(x))$ (see Figure 4.11.2). With luck, this will move even closer to a zero. After that, we apply it again and again. The result will be a sequence of numbers $x_1, x_2, x_3, \ldots$, which ought to get closer and closer to a zero of f. In particular, if everything works as we hope, by applying the process enough times we should be able to find an approximate value for a zero of f to as many digits' accuracy as we may want.

Summary 4.11.1 Newton's Method

Given a differentiable function $f(x)$, Newton's Method is a procedure for producing a sequence of values of x, labeled $x_1, x_2, x_3, \ldots$, etc. Each successive x_n is obtained from the previous one as follows:

$$x_n = x_{n-1} - \frac{f(x_{n-1})}{f'(x_{n-1})}.$$

The first value, x_1, can be chosen at will but if possible should be chosen near a zero of $f(x)$. Often, though not always, the values $x_1, x_2, x_3, \ldots$, etc. will be better and better approximations of a zero of $f(x)$.

EXAMPLE 1 Use Newton's Method to approximate $\sqrt{2}$.

Solution The square root of 2 is a zero of the polynomial $f(x) = x^2 - 2$. By the Intermediate Value Theorem (Theorem 2.5.5), $f(x)$ will have a root between $x = 1$ and $x = 2$. We try Newton's Method with $x_1 = 2$. Then

$$x_2 = x_1 - \frac{f(x_1)}{f'(x_1)} = 2 - \frac{2^2 - 2}{2(2)} = 2 - \frac{2}{4} = 1.5,$$

$$x_3 = x_2 - \frac{f(x_2)}{f'(x_2)} = 1.5 - \frac{(1.5)^2 - 2}{2(1.5)} \approx 1.416667,$$

$$x_4 = x_3 - \frac{f(x_3)}{f'(x_3)} \approx 1.416667 - \frac{(1.416667)^2 - 2}{2(1.416667)} \approx 1.414216.$$

Similarly, $x_5 \approx 1.414\,214$, $x_6 \approx 1.414\,214$. Here only seven significant digits are given, but in fact x_6 will be correct to ten significant digits if you keep track of them all.

Example 1 shows the power of Newton's Method. In a relatively few steps, this method gives surprisingly accurate approximations. The square root key on many pocket calculators uses this method.

Sir Isaac Newton (1642–1727)

EXAMPLE 2 Find an approximate value for a zero of $f(x) = x^5 - 3x^3 + 2x - 1$.

Solution Noting that $f(0) = -1, f(1) = -1$, and $f(2) = 11$, we see that there must be a zero between $x = 1$ and $x = 2$, by the Intermediate Value Theorem (Theorem 2.5.5). We try $x_1 = 1$. Then $x_2 = x_1 - \frac{f(x_1)}{f'(x_1)} = 1 - \frac{f(1)}{f'(1)} = 1 - \frac{-1}{-2} = 0.5$; $x_3 = x_2 - \frac{f(x_2)}{f'(x_2)} = 0.5 - \frac{f(0.5)}{f'(0.5)} = 6$. Continuing in this way, we find $x_4 \approx 4.8407\ldots$, $x_5 \approx 3.9236\ldots$, $x_6 \approx 3.2029\ldots$, $x_7 \approx 2.6434\ldots$, $x_8 \approx 2.2180\ldots$, $x_9 \approx 1.9084\ldots$, $x_{10} \approx 1.7036\ldots$, $x_{11} \approx 1.5967\ldots$, $x_{12} \approx 1.5656\ldots$, $x_{13} \approx 1.5631\ldots$, $x_{14} \approx 1.563\,0959$, where except for the last value we have written down only four decimal places. Continuing the process further does not change the answer on a calculator with eight significant digits, and substituting x_{14} into $f(x)$, we find $f(x_{14}) \approx 0.000\,000\,004$, and x_{14} is indeed very nearly a zero of $f(x)$.

There are several things to notice about Example 2. One is that the first few approximations jumped around: $1, \frac{1}{2}, 6$. Second, though we were looking for a root between $x = 1$ and $x = 2$, the first approximation took us in the wrong direction (from $x_1 = 1$ to $x_2 = \frac{1}{2}$), and the second approximation took us much too far in the opposite direction (to $x_3 = 6$). Nonetheless, the process did eventually give a good approximation to a zero. It might easily happen that the process could find a zero other than where we expect it. (In this case we were looking between $x = 1$ and $x = 2$, but having gone to $x_2 = \frac{1}{2}$, it might have found a root to the left of $x = 1$, etc.)

Another observation is that there does not seem to be any way to find an exact solution. Even with relatively simple polynomials like this, it may be absolutely impossible to find a root exactly. There are very many situations in which finding a root is necessary, and Newton's Method is often the best thing to use.

EXAMPLE 3 Find a value of x for which $x \sin x = \frac{1}{2}$.

Solution There is no apparent way to solve this exactly, but we can approximate an answer. Letting $f(x) = x \sin x - \frac{1}{2}$, we note that $f(0) = -\frac{1}{2}$, $f(\pi/2) = \pi/2 - \frac{1}{2} > 0$, so there is a zero between $x = 0$ and $x = \pi/2$, by the Intermediate Value Theorem.

We apply Newton's Method, trying $x_1 = 0$. Then $x_2 = x_1 - \frac{f(x_1)}{f'(x_1)}$. Calculating $f'(x) = \sin x + x \cos x$, we find $f'(0) = 0$, so the formula makes no sense. In this situation the thing to do is to try a different x_1. How about $x_1 = \frac{\pi}{2}$? Then $x_2 = \frac{\pi}{2} - \frac{\pi/2 - 1/2}{1} = \frac{1}{2}$. Continuing, with the aid of a calculator, we find

$$x_3 \approx 0.783\,470\,336,$$
$$x_4 \approx 0.741\,490\,738,$$
$$x_5 \approx 0.740\,841\,123,$$
$$x_6 \approx 0.740\,840\,955,$$
$$x_7 \approx 0.740\,840\,955.$$

Further cycles of the procedure (often called **iterations**) will not change the answer on a calculator that displays nine digits. Be sure to have your calculator evaluate $\sin x$ in radian mode, that is, with x measured in radians.

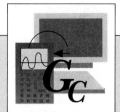

If you have a programmable calculator or a computer, you can program it to perform one iteration, that is, to find x_{n+1} from x_n. With each cycle you will see a new approximation (some intermediate results may also be displayed); it is entertaining to watch as they approach the answer.

```
           .5
 .7834703364
 .7414907384
 .7408411235
 .7408409551
 .7408409551
 .7408409551
```

TI-85	SHARP EL9300
PRGM F2(EDIT) NEWT ENTER X − (X SIN X −.5) ÷ (SIN X + X COS X) ENTER ANS STO X ENTER EXIT $\pi \div 2$ STO X ENTER PRGM F1 F1 (NEWT) ENTER ENTER ENTER . . .	▼ ▼ (NEW) ENTER ENTER NEWT ENTER X = X − (X sin X − .5) ÷ (sin X + X cos X) ENTER COMMAND A 1 (Print) X ENTER COMMAND A 6 (End) ENTER $\pi \div 2$ STO X (RUN) ENTER (NEWT) ENTER ENTER ENTER ENTER ENTER ENTER
CASIO f_x-6300G	**HP 48SX**
Cls EXE MODE 2 EXE Ans → X : X − (X sin X − .5) ÷ (sin X + X cos X) : Ans → X : MODE 1 $\pi \div 2$ EXE Prog 0 EXE EXE EXE EXE EXE EXE (on the f_x-7700GB, : is PRGM F6, and Prog is PRGM F3.)	ATTN CLR ≪ ≫ → αX ′ αX − ( ) αX × SIN αX ▶ − .5 ▶ ÷ ( ) SIN αX ▶ + αX × COS αX ▶ ENTER ′ NEWTON STO π 2 ÷ VAR NEWTON VAR NEWTON VAR NEWTON VAR NEWTON VAR NEWTON VAR NEWTON

Now edit the program to work for other starting points (e.g., $x = 1$) or for some of the other functions we have discussed.

However, it is important to recognize that Newton's Method is not foolproof.

EXAMPLE 4 Apply Newton's Method to $f(x) = x^3 - 5x$, starting with $x_1 = 1$.

Solution $x_2 = x_1 - \frac{f(x_1)}{f'(x_1)} = 1 - \frac{1-5}{3-5} = -1$, and $x_3 = x_2 - \frac{f(x_2)}{f'(x_2)} = -1 - \frac{-1+5}{3-5} = 1$.

Continued iterations give us alternately 1 and -1, and they never approach a zero. Figure 4.11.3 shows what is happening. We know there is a zero at $x = 0$, but Newton's Method keeps jumping back and forth from one side to the other.

The tangent line at $x = 1$ crosses the axis at $x = -1$, and vice versa.

If we chose an x_1 between -1 and 1, Newton's Method would give approximations to the zero at $x = 0$. Values of x_1 outside $[-1, 1]$ would result in approximations to one or other of the other two zeroes. (Factoring the polynomial, $f(x) = x^3 - 5x = x(x^2 - 5)$, we see that the other zeroes are $\pm\sqrt{5}$.)

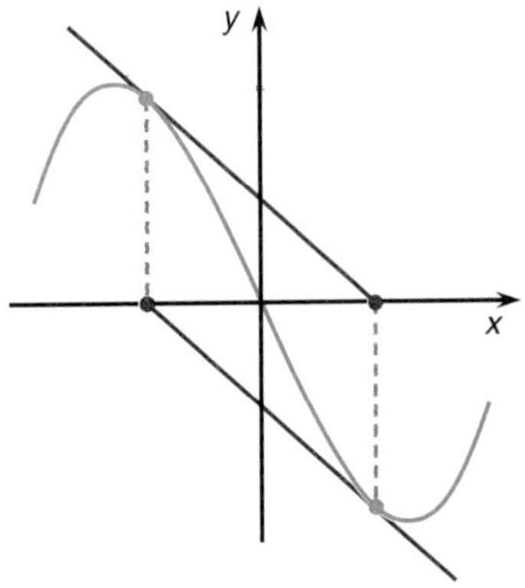

Figure 4.11.3

It is also possible to have examples in which successive values of x_n go to infinity. However, if a particular choice of x_1 does not work, just try a different one; most often there are only a few x_1's that lead to trouble, and the procedure usually works well with most choices of x_1. It is also true that Newton's Method will always approach a root provided that the first choice x_1 is sufficiently close to the root.

Finally, it is worth remarking that in our earlier work on maximum/minimum problems and curve sketching we always worked with polynomials or other functions for which we could find the zeroes. In practice, most polynomials are difficult or impossible to solve, but Newton's Method allows us to find approximate solutions. For a practical maximum/minimum problem or for purposes of sketching a curve, approximate answers are usually perfectly adequate. We leave some examples to the exercises.

Exercises 4.11

In the following exercises, if you have a calculator, try to get as much accuracy as your calculator will permit; if you're doing it by hand, try for three or four significant digits.

I

Find approximate values of the following.

1. $\sqrt{19}$

2. $(103)^{1/3}$

3. $\sqrt{53}$

4. $(71)^{-3/4}$

5. A root of $f(x) = x^3 + 3x^2 + 5x - 2$

6. A root of $g(x) = x^4 - x^3 - 2x^2 + 1$

7. A root of $h(t) = t^7 + 2t - 1$

8. A value of x so that $x^3 - 3x^2 + 4x = 7$

9. A value of t so that $t^4 - 2t^3 + 3t = 4$

10. A value of x so that $x \sin x = 1$

Remembering to use radians, approximate each of the following.

11. $\arcsin(3/4)$

12. $\arccos(0.12)$

13. $\arctan(-1.7)$

14. $\operatorname{arcsec}(31)$

15. A solution of $\cos x + \sin x = 1/3$

16. A solution of $\tan x + 2 \sin x = 3$

17. A value of x so that $3x^2 - \sin x = 5$

18. A value of x satisfying $\frac{\sin x}{x} = \frac{1}{2}$

II

19. Apply Newton's Method to Exercise 4.10.1 (p. 324) to get better approximations.

20. Apply Newton's Method to Exercises 4.10.2 through 4.10.6.

***21.** Apply Newton's Method to Exercise 4.10.21.

***22.** The function $f(x) = x^2 + 5$ has no zeroes. What happens if you apply Newton's Method to it? Draw a picture.

***23.** Work out an exact formula for x_n in terms of x_1 for Newton's Method applied to the function $f(x) = x^2$ (whose only zero is at $x = 0$).

***24.** Apply Newton's Method to the function $f(x) = \frac{x}{1+x^2}$, starting with $x_1 = 2$. Perform five iterations, and then sketch the graph and use it to explain your results.

25. Use Newton's Method to find an approximate value for the minimum of $f(x) = x^4 - 2x^3 + x^2 - x + 1$. Assume that we know there is only one critical point.

26. Make an approximate sketch of $g(x) = 2x^4 + 5x^3 - 7x^2 + 8x + 1$.

***27.** The function $F(x) = x \sin x$ can be thought of as "sin x squeezed inside the envelope of $y = x$ and $y = -x$." In particular, it has a local maximum between 0 and π, a local minimum between π and 2π, a local maximum between 2π and 3π, etc.

(i) Before using Newton's Method to find the approximate locations of these local extrema, try to answer two questions: (a) Will the local maximum between 0 and π be to the left or to the right of $\pi/2$, the local maximum of sin x? (b) Consider the local maximum between $2n\pi$ and $(2n + 1)\pi$. As n increases, will the critical point be shifted more, less, or the same amount away from the midpoint $2n\pi + \pi/2$?

(ii) Use Newton's Method to approximate a few of the critical points, and see whether they confirm your answers to questions (a) and (b) in part (i).

***28.** Repeat Exercise 27 for $G(x) = x^2 \sin x$, adding one more preliminary question: (c) Where are the critical points of G in relation to the critical points of $F(x) = x \sin x$—in the same places, slightly to the left, or slightly to the right?

***29.** Suppose we want to sketch the graph of $F(x) = x \sin x$ for $x \in [-\pi, \pi]$. The derivative is $F'(x) = \sin x + x \cos x$, so it is easy to see there is a critical point at $x = 0$. The second derivative is $F''(x) = \cos x + \cos x - x \sin x = 2\cos x - x \sin x$.

(i) Argue that there is exactly one zero of $F'(x)$ in $(0, \pi)$ and that it is in $\left(\frac{\pi}{2}, \pi\right)$.

(ii) Use Newton's Method to find approximate values for *all* solution(s) of $F'(x) = 0$ with $x \in [-\pi, \pi]$. (Save some trouble by noting that $F'(x)$ is an odd function.)

(iii) Argue that there is exactly one zero of $F''(x)$ in $(0, \pi)$ and that it is in $\left(0, \frac{\pi}{2}\right)$.

(iv) Use Newton's Method to find approximate values for *all* solution(s) of $F''(x) = 0$ with $x \in [-\pi, \pi]$.

(v) Sketch the graph of $y = F(x)$ for $x \in [-\pi, \pi]$.

***30.** When we use linear approximation to approximate $\sqrt{26}$, we consider the function $f(x) = \sqrt{x} = x^{1/2}$ and work with the tangent line to the graph of $y = f(x)$ at the point $(25, 5)$. If we use Newton's Method for the same purpose, we consider the function $g(x) = x^2 - 26$ and look for a zero of this function. We should try starting with a point that we know is reasonably close to a zero, such as $(5, -1)$.

(i) Perform the linear approximation to $f(x)$, starting at the point $(25, 5)$.

(ii) Perform one single iteration of Newton's Method for the function $g(u)$, starting at the point $x = 5$, that is, let $x_1 = 5$ and find x_2.

(iii) Compare the approximations to $\sqrt{26}$ you get in parts (i) and (ii). Explain the comparison.

III

31. If you have a programmable calculator or a computer, write a program to find the next iteration of Newton's Method for the function $f(x) = x^5 - 3x^3 + 2x - 1$ of Example 2 (p. 327). Use it to confirm the calculations of that example, and then try it with some different choices for x_1.

32. Modify your program of the previous problem to work for the function $f(x) = x^3 - 5x$ of Example 4 (p. 328), and try some different values of x_1 to confirm the remarks at the end of Example 4. (Check that the answers you get for several choices of $|x_1| > 1$ are approximately $\pm\sqrt{5}$, by squaring them.)

33. Sketch the graph of the function in Exercise 9, and use the picture to decide which starting values x_1 will cause Newton's Method to converge to which zero.

34. In Exercise 17 we found a root of $f(x) = 3x^2 - \sin x - 5$. Note that $f(0) = -5$, $f(-2) = 7 - \sin(-2) = 7 + \sin(2) \geq 6$. The Intermediate Value Theorem says that there is a zero between $x = 0$ and $x = -2$. Since $f(-1) = 3 - 5 - \sin(1) = -2 - \sin(1) < 0$, it is between $x = -1$ and $x = -2$. Try to find it with Newton's Method.

POINT TO PONDER

It is relatively easy to program a computer or calculator to do an iteration of Newton's Method for some function $f(x)$. Much harder is to decide how many iterations it should perform before stopping. Suppose we are approximating $\sqrt{a}$ but don't know the size of a in advance. We could stop after there is an iteration that does not change the answer, but that is risky: It could happen that because of round-off errors the answer keeps changing back and forth by $+1$ and -1 in the very last digit and never actually stops. Another possibility would be to stop after some fixed number of iterations, say, ten. This would be fine for moderately sized a's but might cause trouble for very large ones. Another possibility would be to stop if the last change is small relative to the current answer, that is, if $\frac{x_{n+1} - x_n}{x_n}$ is small (say, less than or equal to 10^{-8}).

Another problem is choosing x_1. We could let $x_1 = a$. This is good for small a's but wasteful for big values. We could let $x_1 = a/2$, but this is worse for small values (e.g., if $a \le 1$). We could always let $x_1 = 1$, no matter what the value of a.

Try out some of these strategies for various values of a (including very large and very small ones), and see whether you can find a good system for choosing x_1 and deciding when to stop.

This sort of question belongs to the beginnings of a subject called Numerical Analysis.

4.12 L'Hôpital's Rule

ONE-SIDED LIMITS
INFINITE LIMITS
INDETERMINATE FORMS
GENERALIZED MEAN VALUE THEOREM

The Marquis de l'Hôpital (1661–1704)

Johann Bernoulli (1667–1748)

L'Hôpital's Rule gives a way to evaluate limits of the form $\lim_{x\to a} \frac{f(x)}{g(x)}$ if the numerator and denominator both tend to zero, that is, $\lim_{x\to a} f(x) = \lim_{x\to a} g(x) = 0$. To try to see what to do in this situation, suppose first that both $f(x)$ and $g(x)$ are differentiable at $x = a$. In particular this implies that they are both continuous there, and because we know what the limits are, we see that $f(a) = g(a) = 0$.

This allows us to write

$$\frac{f(x)}{g(x)} = \frac{f(x) - f(a)}{g(x) - g(a)},$$

which makes us begin to think of difference quotients. Writing

$$\frac{f(x)}{g(x)} = \frac{f(x) - f(a)}{g(x) - g(a)} = \frac{f(x) - f(a)}{g(x) - g(a)} \times \frac{x-a}{x-a} = \frac{f(x) - f(a)}{x-a} \times \frac{x-a}{g(x) - g(a)}$$
$$= \frac{f(x) - f(a)}{x-a} \times \frac{1}{\frac{g(x)-g(a)}{x-a}},$$

we can let $x \to a$, and the first factor will tend to $f'(a)$ and the second will tend to $\frac{1}{g'(a)}$. This suggests that

$$\lim_{x\to a} \frac{f(x)}{g(x)} = \frac{f'(a)}{g'(a)}.$$

There are several difficulties with the argument we just gave, among them the possibility that $g'(a)$ or $g(x)$ might equal zero, so various quotients we have written down might not be defined. Nonetheless, it certainly indicates that the limit of $\frac{f}{g}$ has something to do with $\frac{f'}{g'}$. The precise statement is what is known as l'Hôpital's Rule, in honor of the French nobleman the Marquis de l'Hôpital (1661–1704), who was the patron of the Swiss mathematician John Bernoulli, who actually invented the rule.

The rule says that if the numerator and denominator both tend to 0, then the limit of the quotient $\frac{f(x)}{g(x)}$ equals the limit of the quotient of the derivatives $\frac{f'(x)}{g'(x)}$.

EXAMPLE 1 For example, to evaluate $\lim_{x\to 0} \frac{\sin x}{x}$, we note that both numerator and denominator tend to 0, but $\sin' x = \cos x$, $\frac{d}{dx}(x) = 1$, and $\lim_{x\to 0} \frac{\cos x}{1} = 1$, which as we know does equal $\lim_{x\to 0} \frac{\sin x}{x}$.

Unfortunately, a proof of l'Hôpital's Rule is not as simple as the rule itself. It hinges on a generalization of the Mean Value Theorem, Theorem 4.12.5. We postpone the proof until the end of the section and concentrate instead on the use of the rule.

THEOREM 4.12.1

L'Hôpital's Rule

Suppose there is an open interval (r, s) containing $x = a$ so that $f(x)$ and $g(x)$ are differentiable at every $x \in (r, s)$ except possibly $x = a$, and that $g'(x) \neq 0$ for every $x \in (r, s)$ except possibly $x = a$. Suppose too that $\lim_{x\to a} f(x) = \lim_{x\to a} g(x) = 0$ and that $\lim_{x\to a} \frac{f'(x)}{g'(x)}$ exists. Then

$$\lim_{x\to a} \frac{f(x)}{g(x)} = \lim_{x\to a} \frac{f'(x)}{g'(x)}.$$

In addition, if $\lim_{x\to a} \frac{f'(x)}{g'(x)}$ is $+\infty$ or $-\infty$, then the equality still holds.

COROLLARY 4.12.2

(i) **L'Hôpital's Rule for one-sided limits:** Suppose f and g are differentiable on (a, s), $g'(x) \neq 0$ for all $x \in (a, s)$, $\lim_{x\to a^+} f(x) = 0$, $\lim_{x\to a^+} g(x) = 0$, and $\lim_{x\to a^+} \frac{f'(x)}{g'(x)}$ exists. Then

$$\lim_{x\to a^+} \frac{f(x)}{g(x)} = \lim_{x\to a^+} \frac{f'(x)}{g'(x)},$$

where infinite limits are included. A similar statement holds for limits from below, using an interval of the form (r, a).

(ii) **L'Hôpital's Rule for limits at $\pm\infty$:** Suppose f and g are differentiable on (r, ∞), $g'(x) \neq 0$ for all $x \in (r, \infty)$, $\lim_{x\to\infty} f(x) = 0$, $\lim_{x\to\infty} g(x) = 0$, and $\lim_{x\to\infty} \frac{f'(x)}{g'(x)}$ exists. Then

$$\lim_{x\to\infty} \frac{f(x)}{g(x)} = \lim_{x\to\infty} \frac{f'(x)}{g'(x)},$$

where infinite limits are included. A similar statement holds for limits as $x \to -\infty$, using an interval of the form $(-\infty, r)$.

EXAMPLE 2 Evaluate $\lim_{x\to 1} \frac{\ln(x)}{\sin(\pi x)}$.

Solution Since $\lim_{x\to 1} \ln(x) = \ln(1) = 0$ and $\lim_{x\to 1} \sin(\pi x) = \sin \pi = 0$, we can apply l'Hôpital's Rule. Now $\frac{d}{dx}(\ln x) = \frac{1}{x}$ and $\frac{d}{dx}\big(\sin(\pi x)\big) = \pi \cos(\pi x)$, so $\lim_{x\to 1} \frac{\ln(x)}{\sin(\pi x)} = \lim_{x\to 1} \frac{1/x}{\pi \cos(\pi x)} = \frac{1/1}{\pi \cos \pi} = \frac{1}{-\pi} = -\frac{1}{\pi}$.

EXAMPLE 3 Find $\lim_{x\to \pi/2^-} \frac{\cos x}{1-\sin x}$.

Solution First we check that numerator and denominator do both tend to zero as $x \to \frac{\pi}{2}^-$. They are both continuous functions that equal zero at $x = \frac{\pi}{2}$, so their limits are both zero. This means we can apply l'Hôpital's Rule, so $\lim_{x\to \pi/2^-} \frac{\cos x}{1-\sin x} =$

With a graphics calculator we can plot the original function $y = \frac{\ln x}{\sin(\pi x)}$ from Example 2 and also the quotient of the derivatives $\frac{1/x}{\pi\cos(\pi x)}$.

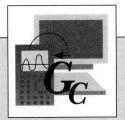

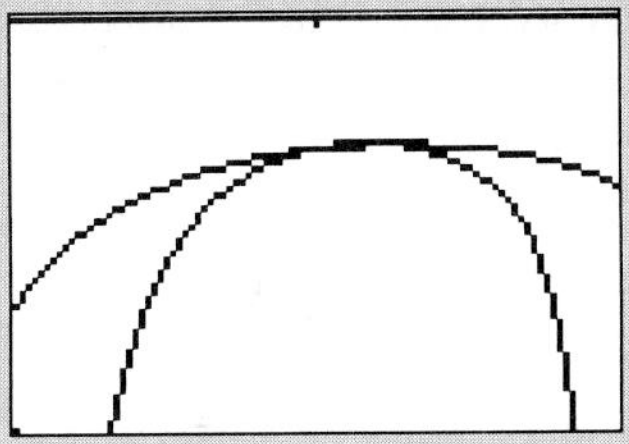

TI-85	SHARP EL9300
GRAPH F1 $(y(x) =)$ LN F1 (x) ÷ SIN (π F1(x)) ▼ F1(x) x^{-1} ÷ (π COS (π F1(x))) M2(RANGE) .5 ▼ 1.5 ▼ ▼ (−) 1 ▼ 0 F5(GRAPH)	ln x/θ/T ÷ (sin (π x/θ/T)) 2ndF ▼ 1 ÷ x/θ/T ÷ (π cos (π x/θ/T)) RANGE .5 ENTER 1.5 ENTER 1 ENTER −1 ENTER 0 ENTER 1 ENTER
CASIO f_x-7700GB/f_x-6300G	**HP 48SX**
Cls EXE Range .5 EXE 1.5 EXE 1 EXE −1 EXE 0 EXE 1 EXE EXE EXE EXE Graph ln X ÷ (sin (π X)) EXE Graph 1 ÷ X ÷ (π cos (π X)) EXE	↰ PLOT PLOTR ERASE ATTN ′ αY = LN αX ▶ ÷ () SIN π × αX ↱ PLOT ↰ DRAW ERASE .5 ENTER 1.5 XRNG 1 +/− ENTER 0 YRNG DRAW ATTN ′ αY = 1 ÷ αX ÷ () π × COS π × αX ↱ PLOT ↰ DRAW DRAW

The two functions are different, but we see that the limits are the same as $x \to 1$. Explain why the limits are not the same at other points . . .

$\lim_{x\to\pi/2^-} \frac{-\sin x}{-\cos x} = \lim_{x\to\pi/2^-} \tan x$. Now $\lim_{x\to\pi/2^-} \tan x = \infty$ (since $\sin x \to 1$ and $\cos x \to 0$ and both are positive for $x \in (0, \pi/2)$). So

$$\lim_{x\to\pi/2^-} \frac{\cos x}{1-\sin x} = \lim_{x\to\pi/2^-} \frac{-\sin x}{-\cos x} = \infty.$$

EXAMPLE 4 Evaluate $\lim_{x\to 0} \frac{\sin(x^3)}{\sin(x^2)}$.

Solution $\lim_{x\to 0} \frac{\sin(x^3)}{\sin(x^2)} = \lim_{x\to 0} \frac{\cos(x^3)(3x^2)}{\cos(x^2)(2x)}$. Here we used l'Hôpital's Rule, only to find that it resulted in another quotient in which both numerator and denominator tend to 0.

But dividing out the common factor of x gives $\lim_{x\to 0} \frac{\cos(x^3)(3x)}{\cos(x^2)(2)}$. As $x \to 0$, the numerator tends to $\cos(0)(3)(0) = (1)(3)(0) = 0$, but the denominator tends to $\cos(0)(2) = 2$. The quotient tends to $\frac{0}{2} = 0$, and we have found that $\lim_{x\to 0} \frac{\sin(x^3)}{\sin(x^2)} = 0$.

It is often necessary to use l'Hôpital's Rule more than once.

EXAMPLE 5 Evaluate $\lim_{x \to \pi} \frac{\sin\left((x-\pi)^2\right)}{\sin^2 x}$.

Solution Both numerator and denominator tend to 0 as $x \to \pi$, so we may apply l'Hôpital's Rule and find $\lim_{x \to \pi} \frac{\sin\left((x-\pi)^2\right)}{\sin^2 x} = \lim_{x \to \pi} \frac{\cos\left((x-\pi)^2\right) 2(x-\pi)}{2\sin(x)\cos(x)}$. The numerator tends to $\cos(0)(2)(0) = 0$, and the denominator tends to $2\sin(\pi)\cos(\pi) = (2)(0)(-1) = 0$. Applying l'Hôpital's Rule again, we get $\lim_{x \to \pi} \frac{-\sin\left((x-\pi)^2\right) \times 2(x-\pi) \times 2(x-\pi) + \cos\left((x-\pi)^2\right)(2)}{2\cos^2 x - 2\sin^2 x}$. This time the numerator tends to $\sin(0)(2)(0)(2)(0) + \cos(0)(2) = (0)(0)(0) + (1)(2) = 2$, and the denominator tends to $2\cos^2(\pi) - 2\sin^2(\pi) = 2(1)^2 - 2(0)^2 = 2$, so $\lim_{x \to \pi} \frac{\sin\left((x-\pi)^2\right)}{\sin^2 x} = \frac{2}{2} = 1$.

> There is a word of warning that must be given here. After you use l'Hôpital's Rule a few times, especially if you use it repeatedly as in Example 5, there is a temptation to be sloppy about verifying the hypotheses. The important one is that both numerator and denominator must tend to zero, and using l'Hôpital's Rule in situations in which this is not true leads to serious errors.
>
> As a simple example, consider $\lim_{x \to 1} \frac{x-1}{x}$. In fact this quotient tends to $\frac{0}{1} = 0$ as $x \to 1$, but if we neglected to notice that the denominator does not tend to zero and mistakenly applied l'Hôpital's Rule, we would obtain the incorrect answer that the limit is 1.

Finally, there is a version of l'Hôpital's Rule for the situation in which $f(x)$ and $g(x)$ both have infinite limits. The proof is quite difficult, so we merely state the result.

THEOREM 4.12.3

L'Hôpital's Rule for Infinite Limits

Suppose $\lim_{x \to a} f(x) = \infty$ or $-\infty$ and $\lim_{x \to a} g(x) = \infty$ or $-\infty$, and $f(x)$ and $g(x)$ are both differentiable on (r, a) and (a, s) for some r, s with $r < a < s$. Suppose that $\lim_{x \to a} \frac{f'(x)}{g'(x)}$ exists. Then

$$\lim_{x \to a} \frac{f(x)}{g(x)} = \lim_{x \to a} \frac{f'(x)}{g'(x)},$$

where infinite limits are included. One-sided versions of this result also hold, as do versions for limits as $x \to \infty$ or $x \to -\infty$.

EXAMPLE 6 Find $\lim_{x \to 1^+} \frac{\exp\left(\frac{1}{x-1}\right)}{\tan(\pi x/2)}$.

Solution $\lim_{x \to 1^+} \tan(\pi x/2) = \lim_{x \to \pi/2^+} \tan x = -\infty$ and $\lim_{x \to 1^+} \frac{1}{x-1} = +\infty$, so $\lim_{x \to 1^+} \exp\left(\frac{1}{x-1}\right) = \infty$ (since $\lim_{t \to \infty} e^t = \infty$). Applying Theorem 4.12.3, we consider

$$\lim_{x\to 1^+} \frac{\exp\left(\frac{1}{x-1}\right)\frac{-1}{(x-1)^2}}{\sec^2(\pi x/2)(\pi/2)} = -\frac{2}{\pi}\lim_{x\to 1^+} \exp\left(\frac{1}{x-1}\right)\frac{1/(x-1)^2}{1/\cos^2(\pi x/2)}$$
$$= -\frac{2}{\pi}\lim_{x\to 1^+} \exp\left(\frac{1}{x-1}\right)\frac{\cos^2(\pi x/2)}{(x-1)^2}.$$

Now $\lim_{x\to 1^+} \frac{\cos(\pi x/2)}{x-1} = \lim_{x\to 1^+} \frac{-\sin(\pi x/2)(\pi/2)}{1} = -\sin(\pi/2)(\pi/2) = -\frac{\pi}{2}$, so $\lim_{x\to 1^+} \frac{\cos^2(\pi x/2)}{(x-1)^2} = \left(-\frac{\pi}{2}\right)^2$.

The limit in question is $-\frac{2}{\pi}\left(\frac{\pi^2}{4}\right)\lim_{x\to 1^+}\exp\left(\frac{1}{x-1}\right) = -\infty$.

There are some other types of limits for which l'Hôpital's Rule can be used.

EXAMPLE 7 Find $\lim_{x\to 0^+} x\ln x$.

Solution Here we have a *product* of two functions, x and $\ln x$, instead of a quotient. One of them, x, goes to 0 and the other, $\ln x$, goes to $-\infty$. To be able to use l'Hôpital's Rule, we need to convert the product into a quotient. One way to accomplish this is to write

$$x\ln x = \frac{\ln x}{\frac{1}{x}}.$$

As $x \to 0^+$, the numerator of this quotient tends to $-\infty$ and the denominator to ∞. We apply l'Hôpital's Rule:

$$\lim_{x\to 0^+} x\ln x = \lim_{x\to 0^+} \frac{\ln x}{\frac{1}{x}} = \lim_{x\to 0^+} \frac{\frac{1}{x}}{-\frac{1}{x^2}} = \lim_{x\to 0^+}(-x) = 0.$$

EXAMPLE 8 Find (i) $\lim_{x\to\infty}(x-\ln x)$; (ii) $\lim_{x\to 0^+}(\csc x - \cot x)$.

Solution

(i) Since $x \to \infty$ and $\ln x \to \infty$, the behavior of their difference is unpredictable. However, we can write

$$x - \ln x = x\left(1 - \frac{\ln x}{x}\right).$$

Observing that $\lim_{x\to\infty} \frac{\ln x}{x} = \lim_{x\to\infty} \frac{1/x}{1} = 0$, so $\left(1 - \frac{\ln x}{x}\right) \to 1$, we find that

$$\lim_{x\to\infty} x\left(1 - \frac{\ln x}{x}\right) = \left(\lim_{x\to\infty} x\right)\left(\lim_{x\to\infty}\left(1 - \frac{\ln x}{x}\right)\right) = \left(\lim_{x\to\infty} x\right)(1) = \infty.$$

(ii) Since $\csc x$ and $\cot x$ both tend to ∞ as $x \to 0^+$, we have to rearrange the function. Writing

$$\csc x - \cot x = \frac{1}{\sin x} - \frac{\cos x}{\sin x} = \frac{1-\cos x}{\sin x},$$

we now have the quotient of two functions that both tend to 0. L'Hôpital's Rule applies:

$$\lim_{x\to 0+} (\csc x - \cot x) = \lim_{x\to 0+} \frac{1-\cos x}{\sin x} = \lim_{x\to 0+} \frac{\sin x}{\cos x} = 0.$$

Summary 4.12.4 Indeterminate Forms

1. L'Hôpital's Rule is used to find the limit of a quotient $\frac{f(x)}{g(x)}$, where $f(x)$ and $g(x)$ both tend to 0 or both tend to $\pm\infty$. We refer to these types of limits as indeterminate forms of type $\frac{0}{0}$ and $\frac{\infty}{\infty}$. These symbols indicate the limiting behavior of the numerator and denominator.
2. For a product $f(x)g(x)$ in which $f(x) \to 0$ and $g(x) \to \pm\infty$ it is possible to write

$$f(x)g(x) = \frac{f(x)}{\frac{1}{g(x)}}.$$

In the quotient the numerator and denominator both tend to 0, so l'Hôpital's Rule can be applied. Sometimes it may be easier to write

$$f(x)g(x) = \frac{g(x)}{\frac{1}{f(x)}}.$$

If $\frac{1}{f(x)}$ tends to ∞ or $-\infty$, then l'Hôpital's Rule may be applied (cf. Example 7).

We refer to the limit of $f(x)g(x)$ as an indeterminate form of type $0 \times \infty$.

3. For a difference $f(x) - g(x)$ in which $f(x)$ and $g(x)$ both tend to ∞ (or $-\infty$), it is often possible to perform algebraic manipulations that convert the function into a quotient (cf. Example 8). In particular, if $f(x)$ and $g(x)$ are quotients, it is often helpful to write $f(x) - g(x)$ over a common denominator (cf. Example 8(ii)).

The limit of the function $f(x) - g(x)$ is called an indeterminate form of type $\infty - \infty$.

Proof of l'Hôpital's Rule

Now that we have learned how to use l'Hôpital's Rule, we might want to see how it is proved. The rough argument given at the beginning of this section provides a reasonably good idea of why a result like this ought to be true, but a correct proof is surprisingly delicate and tricky. Understanding the proof will not likely offer much help in using l'Hôpital's Rule, which is probably more important.

The proof of l'Hôpital's Rule rests on the following theorem, which is a generalization of the Mean Value Theorem (Theorem 4.2.5, p. 241).

THEOREM 4.12.5

Suppose $f(x)$ and $g(x)$ are continuous functions on $[a, b]$ that are differentiable on (a, b). Then there is a $c \in (a, b)$ so that

$$\big(f(b) - f(a)\big)g'(c) = \big(g(b) - g(a)\big)f'(c).$$

PROOF Let $H(x) = \big(f(b) - f(a)\big)g(x) - \big(g(b) - g(a)\big)f(x)$. Then $H(x)$ is also continuous on $[a, b]$ and differentiable on (a, b).

Note that

$$\begin{aligned} H(a) &= \big(f(b) - f(a)\big)g(a) - \big(g(b) - g(a)\big)f(a) \\ &= f(b)g(a) - f(a)g(a) - g(b)f(a) + g(a)f(a) \\ &= f(b)g(a) - g(b)f(a). \end{aligned}$$

Also

$$\begin{aligned} H(b) &= \big(f(b) - f(a)\big)g(b) - \big(g(b) - g(a)\big)f(b) \\ &= f(b)g(b) - f(a)g(b) - g(b)f(b) + g(a)f(b) \\ &= g(a)f(b) - f(a)g(b). \end{aligned}$$

In particular, $H(b) = H(a)$. So Rolle's Theorem (Theorem 4.2.4, p. 240) applies to $H(x)$ and says that there is a $c \in (a, b)$ so that $0 = H'(c) = \big(f(b) - f(a)\big)g'(c) - \big(g(b) - g(a)\big)f'(c)$. But this is exactly what we wanted to prove. ∎

We said that Theorem 4.12.5 is a generalization of the Mean Value Theorem. Notice that if $g(x) = x$, then the statement of Theorem 4.12.5 is exactly the statement of the Mean Value Theorem.

With this theorem available it is now possible to prove l'Hôpital's Rule.

PROOF OF L'HÔPITAL'S RULE Since none of the hypotheses depends on the values of $f(x)$ and $g(x)$ at $x = a$, we can change them, if necessary, so that $f(a) = 0, g(a) = 0$, making $f(x)$ and $g(x)$ continuous at $x = a$. Then, being differentiable at the other points, they are continuous on all of (r, s).

Fix $x \in (r, s)$ with $x > a$. We consider the interval $[a, x]$ and apply Theorem 4.12.5. So there is a $c \in (a, x)$ so that $\big(f(x) - f(a)\big)g'(c) = \big(g(x) - g(a)\big)f'(c)$.

Notice that if $g(x) = g(a)$, Rolle's Theorem would say that g' has a zero between a and x, contradicting the hypotheses, so $g(x) - g(a) \neq 0$, and we may divide by it and by $g'(c)$ to obtain

$$\frac{f(x) - f(a)}{g(x) - g(a)} = \frac{f'(c)}{g'(c)}.$$

Since $f(a) = g(a) = 0$, this says that $\frac{f(x)}{g(x)} = \frac{f'(c)}{g'(c)}$, with $c \in (a, x)$.

A similar argument shows that if $x < a$, then $\frac{f(x)}{g(x)} = \frac{f'(c)}{g'(c)}$, for some c with $c \in (x, a)$. In particular, since c is always closer to a than x is, c must approach a whenever x approaches a.

Consequently, $\lim_{x\to a}\frac{f(x)}{g(x)} = \lim_{c\to a}\frac{f'(c)}{g'(c)}$, since we are assuming that the right-hand limit exists (or could be ∞ or $-\infty$).

This completes the proof of l'Hôpital's Rule (Theorem 4.12.1).

In fact, since we worked with $x > a$ and then with $x < a$ separately, the above proof is also a proof of the one-sided version of l'Hôpital's Rule, Corollary 4.12.2(i).

Exercises 4.12

I

Evaluate the following limits.

1. $\lim_{x\to 0}\frac{x^2+2x}{\sin x}$
2. $\lim_{x\to 2}\frac{x^2-3x+2}{x^2-4x+3}$
3. $\lim_{x\to 3}\frac{x^4-2x^3-4x^2+2x+3}{x^3-4x^2+4x-3}$
4. $\lim_{x\to -1}\frac{x+1}{x^2-1}$
5. $\lim_{x\to 0}\frac{1-\cos x}{x\sin x}$
6. $\lim_{x\to \pi/2}\frac{\cos^2 x}{1-\sin x}$
7. $\lim_{x\to 0}\frac{\arctan x}{\sin x}$
8. $\lim_{x\to \pi/2}\frac{1-\sin x}{\operatorname{arccot}(x)}$
9. $\lim_{x\to 0}\frac{\sin^2 x}{\sin(x^2)}$
10. $\lim_{x\to 1}\frac{x^2-1}{x^2+1}$
11. $\lim_{x\to 0^+}\frac{\sin x}{1-\cos x}$
12. $\lim_{x\to 0^-}\frac{\sin x}{1-\cos x}$
13. $\lim_{x\to 1^-}\frac{\arccos x}{\ln x}$
14. $\lim_{x\to 0}\frac{x-\sin x}{x^3}$
15. $\lim_{x\to 1^+}\frac{(\ln x)^2}{(x-1)^3}$
16. $\lim_{x\to 3^-}\frac{\sin \pi x}{x^3-2x^2-3x}$
17. $\lim_{x\to 1^+}\frac{(\ln x)^3}{(x-1)^2}$
18. $\lim_{x\to e^-}\frac{1-\ln x}{(x-e)^2}$
19. $\lim_{x\to 4^+}\frac{\sin \pi x}{x^2-3x-2}$
20. $\lim_{x\to 0}\frac{\sin^7 x}{x^7}$
21. $\lim_{x\to 0^+} x^2\ln x$
22. $\lim_{x\to 0} x\cot(2x)$
23. $\lim_{x\to 0}\left(\frac{1}{x}-\csc x\right)$
24. $\lim_{x\to 0^+}\sqrt{x}\ln x$
25. $\lim_{x\to \infty} xe^{-x}$
26. $\lim_{x\to \infty}(x-x^2)$
27. $\lim_{x\to 1^+}\ln(x)\tan\frac{\pi x}{2}$
28. $\lim_{x\to 2}\sin(\pi x^2)\cot(\pi x)$
29. $\lim_{x\to 0}\frac{1}{\ln(x)\arctan(x)}$
30. $\lim_{x\to -\infty}\left(x+\sqrt{x^2+1}\right)$

II

*31. (i) Evaluate $\lim_{x\to \pi/2}\frac{\cos^3 x}{(2x-\pi)^3}$. (ii) If k is a positive integer, evaluate $\lim_{x\to 0}\frac{\sin^k x}{x^k}$.

*32. (i) Evaluate $\lim_{x\to 3}\frac{(x-3)^3}{\sin^3(\pi x)}$. (ii) If k is a positive integer, evaluate $\lim_{x\to 0}\frac{x^k}{\sin^k(\pi x)}$. (iii) If k is a positive integer, evaluate $\lim_{x\to 0}\frac{(1-\cos x)^k}{x^{2k}}$.

*33. Let $r > 0$; find $\lim_{x\to 0^+} x^r\ln x$.

*34. Evaluate $\lim_{x\to \pi/2^-}(2x-\pi)\tan x$.

**35. If $f(x)$ is differentiable and $f'(x)$ is continuous, prove that $\lim_{h\to 0}\frac{f(x+h)-f(x-h)}{2h} = f'(x)$.

**36. Suppose the first and second derivatives of $f(x)$ exist at every x. Evaluate

$$\lim_{h\to 0}\frac{f(x+h)+f(x-h)-2f(x)}{h^2}.$$

**37. Prove the version of l'Hôpital's Rule for limits as $x\to\infty$ with both numerator and denominator tending to zero. (*Hint:* Let $x = \frac{1}{t}$, as $t\to 0^+$.)

**38. Prove a weaker form of the version of l'Hôpital's Rule that applies for limits as $x\to a$ in which $\lim_{x\to a} f(x) = \infty = \lim_{x\to a} g(x)$. Assume that *both* $\lim_{x\to a}\frac{f(x)}{g(x)}$ and $\lim_{x\to a}\frac{f'(x)}{g'(x)}$ exist and are nonzero, and under this assumption prove the formula. (*Hint:* $\frac{f}{g} = \frac{1/g}{1/f}$.)

III

39. Let $f(x) = \sin x$ and $g(x) = x$. Use a graphics calculator or computer to graph $\frac{f(x)}{g(x)}$ and $\frac{f'(x)}{g'(x)}$ on the same picture. Compare their limits as $x\to 0^+$.

40. With $f(x) = \tan^2 x$ and $g(x) = x^2$, sketch $\frac{f}{g}$, $\frac{f'}{g'}$, and $\frac{f''}{g''}$ on the same picture. Compare their limiting behavior near $x = 0$.

4.13 Marginal Cost, Revenue, and Profit

DEMAND FUNCTION
PRICE FUNCTION

In Section 3.1 we saw how various questions coming from the world of business and economics lead quite naturally to the use of derivatives. In this section we will make a more systematic study of these applications.

Suppose a factory produces television sets and the cost of producing x sets is $C(x)$. If the amount for which x sets will sell—the revenue—is $R(x)$, then the profit to be made by producing x sets is $P(x) = R(x) - C(x)$.

Consider the derivatives of these functions. Since $C'(x) = \lim_{h\to 0} \frac{C(x+h)-C(x)}{h}$, if x is large, we will likely get a good approximation to $C'(x)$ by taking $h = 1$ and considering $\frac{C(x+1)-C(x)}{1} = C(x+1) - C(x)$. This is the cost of producing one more set if the factory is already producing x of them. The derivative $C'(x)$ is called the **marginal cost function.**

Similarly, $R'(x)$ is the **marginal revenue function** and $P'(x)$ is the **marginal profit function.** They represent, approximately, the revenue and profit, respectively, to be gained by producing one additional set if current production is x.

EXAMPLE 1 **Television Factory** Studies have shown that for a certain television factory, $C(x) = -\frac{1}{10}x^2 + 500x + 30{,}000$ dollars, at least for values of x up to $x = 1000$. Market research indicates that $R(x) = -\frac{1}{2}x^2 + 800x - 10{,}000$ dollars. So $P(x) = R(x) - C(x) = -\frac{2}{5}x^2 + 300x - 40{,}000$ dollars.

The marginal cost is $C'(x) = -\frac{1}{5}x + 500$, the marginal revenue is $R'(x) = -x + 800$, and the marginal profit is $P'(x) = -\frac{4}{5}x + 300$. Let us try to interpret these functions.

The marginal cost $C'(x)$, the approximate cost of producing one additional set, is $500 - \frac{1}{5}x$. This means it costs about \$500 to produce each additional set when production is small, but as volume increases, the unit cost falls, presumably because of increased efficiency and economies of scale. The marginal revenue is $R'(x) = 800 - x$, which means that sets bring in about \$800 each for the first few, but the amount falls as the number increases. This could, for example, mean that as the market becomes saturated, it becomes increasingly difficult, and hence increasingly expensive, to sell additional sets. It could also mean that to sell more, they must offer discounts.

The marginal profit is $P'(x) = -\frac{4}{5}x + 300$. It is about \$300 per set at first but drops off. This says that producing additional sets brings smaller and smaller added profits.

The above example is rather simple, but it illustrates the essential ideas. Presuming that the factory's owners wish to maximize profits, we look for a maximum of $P(x)$.

EXAMPLE 2 In the situation of Example 1, what level of production will give maximum profit?

Solution We want to maximize $P(x) = -\frac{2}{5}x^2 + 300x - 40{,}000$.

Now $P'(x) = -\frac{4}{5}x + 300$, so the only critical point is $x = 375$. Since $P''(x) = -\frac{4}{5} < 0$, it must be an absolute maximum. The maximum profits will be obtained by producing 375 television sets, and the profit will be $P(375) = \$16{,}250$.

It is interesting to calculate $P(376) = \$16{,}249.60$. This shows that producing one set more than the optimal 375 will actually reduce profits by 40 cents, because costs will be increasing faster than revenues.

Example 2 illustrates a fundamental result in economics.

PRINCIPLE 4.13.1

> The maximum profit will occur at a point where the marginal profit is zero, that is, where marginal cost and marginal revenue are equal.

Strictly speaking, we should assume that $P(x)$ is differentiable, and we should worry about endpoint maxima (e.g., $x = 0$), etc., so this "principle" is a little vague. However, in a situation like this, in which we have simplified and idealized the problem considerably, the general idea is what is important, rather than the technical details.

In a sense, this "principle" is nothing more than Theorem 4.2.3 (p. 238) (that extrema occur at critical points), but it is more than that. What we have described is a simplified special case of a principle with much wider applicability. We should be pleased that our knowledge of calculus makes it seem so natural and easy.

In practice it may be easier to find the marginal cost function $C'(x)$ rather than $C(x)$ itself. Frequently, the assumption is made that $C'(x) = ax + b$ for some constants a and b. Then $C(x)$ must be of the form $C(x) = \frac{a}{2}x^2 + bx + c$ for some constant c. This constant c can be interpreted as the fixed cost, the cost that will be incurred even if no units are produced. It is usually the result of overhead costs or setup costs for a production run.

In practice it might also be easier to determine how many items will sell if they are offered at a given price p, rather than the revenue function $R(x)$. If we assume that the number that will sell is $x = n - up$, then this represents a plausible situation in which higher prices result in fewer sales, or smaller demand. For this reason, when we solve for p, the resulting equation

$$p = p(x) = \frac{n}{u} - \frac{1}{u}x$$

is called the **demand function** or simply the **price function.**

Of course, the total revenue is simply $R(x) = xp(x)$, the unit price times the number of units sold.

EXAMPLE 3 **Kettle Production** Market studies have shown that if a certain model of kettle is sold for \$20, it will be possible to sell 1600 units, but that at \$22 only 1200 will sell. Assume that the number x that will sell at price $\$p$ is a linear function, that is, $x = n - up$, for some constants u and n. If the marginal cost is also assumed to be linear, that is, $C'(x) = ax + b$ for constants a and b, and the cost of producing one additional kettle is \$12 if 1000 are already being produced and \$10 if 1400 are being produced, what production level will maximize profits?

Solution We are assuming that $x = n - up$, and we know that when $p = 20$, x is 1600 and when $p = 22$, x is 1200. This says that $1600 = n - 20u$ and $1200 = n - 22u$. Subtracting these two equations gives $400 = 2u$, so $u = 200$. Substituting this into either equation gives $n = 5600$. So $x = 5600 - 200p$, and $200p = 5600 - x$, so $p = 28 - (1/200)x$. The total revenue is $R(x) = xp = 28x - \frac{1}{200}x^2$.

Similarly, $C'(x) = ax + b$, and, approximately, $C'(1000) = 12$, $C'(1400) = 10$, that is, $12 = 1000a + b$ and $10 = 1400a + b$. Subtracting, we get $2 = -400a$, $a = -\frac{1}{200}$, and substituting, $b = 17$, so $C'(x) = -\frac{1}{200}x + 17$.

The marginal profit is $P'(x) = R'(x) - C'(x) = 28 - \frac{1}{100}x - (-\frac{1}{200}x + 17) = 11 - \frac{1}{200}x$, which is zero when $x = 2200$. Since $P''(x) = -\frac{1}{200} < 0$, this is a maximum. So producing 2200 kettles will maximize profits.

Notice that to solve this problem, we didn't actually have to know the cost function $C(x)$ but just the marginal cost function $C'(x)$. If in addition we knew that the fixed costs were \$10,000, that is, $C(0) = 10{,}000$, then from $C'(x) = -\frac{1}{200}x + 17$ we would find that $C(x) = -\frac{1}{400}x^2 + 17x + 10{,}000$. So at a production level of 2200 kettles the total cost is $C(2200) = \$35{,}300$. Also $R(2200) = \$37{,}400$, so the total profit is $P(2200) = R(2200) - C(2200) = \$2100$.

The average cost, or cost per unit, is $\frac{C(x)}{x}$, which, for $x = 2200$, works out to about \$16.05. The average revenue is $\frac{R(x)}{x}$, and for $x = 2200$ this is \$17.00. The average or per unit profit is $\frac{P(x)}{x} \approx \$0.95$.

Exercises 4.13

1. Find the production level x that will maximize profits if the cost function is $C(x) = 300 + 20x - \frac{1}{100}x^2$ and the revenue function is $R(x) = 38x - \frac{1}{10}x^2$.

2. Find the optimal production level for a company whose cost function for producing x units per week is $C(x) = 200 + 30x - \frac{1}{50}x^2$ and whose sales will be $x = 875 - 25p$ if the unit selling price is p.

3. A manufacturer needs to choose one of two products to manufacture. For one, a type of electric motor, the cost function is $C_M(x) = 400 + 15x - \frac{1}{80}x^2$, and the revenue function is $R_M(x) = 20x - \frac{1}{40}x^2$. For the other, a kind of generator, the marginal cost is $C_G'(x) = 18 - \frac{1}{60}x$, and the sales will be $x = 960 - 40p$ if the unit price is p. Supposing that the fixed costs for generators are $C_G(0) = 1000$, which product can yield higher profits?

4. Suppose that for a certain process the revenue function is $R(x) = -\frac{1}{60}x^2 + 13x - 50$ and the marginal cost is $C'(x) = 10 - \frac{1}{50}x$. (i) If $C(0) = 100$, find the production level that maximizes profits. (ii) In the above situation, suppose it is practical to run the process only in batches of 100. How many batches would maximize profits?

5. Assume that the marginal cost $C'(x)$ is a linear function of x. Assume that the number x of units that will sell at price \$$p$ is a linear function of p. Studies indicate that $C'(45) = 24$ and $C'(180) = 21$. Moreover, they show that if $p = 20$, x will be 450, and if $p = 26$, x will be 225.
   (i) Find the optimal production level.
   (ii) If fixed costs are \$500, what will be the profit per unit at the optimal production level (as found in part (i))?
   (iii) If fixed costs are \$500, what will be the maximum profit per unit?
   (iv) What conclusion should you draw if fixed costs are \$1000?

*6. Show that when the average cost (i.e., unit cost) is a minimum, the marginal cost equals the average cost.

*7. What conclusions would you draw if marginal cost and marginal revenue are both constant? In particular, how would you decide what level of production to recommend?

*8. In a certain jurisdiction a person whose annual taxable income is \$$x$ must pay taxes of $T(x) = \frac{1}{10}x + \frac{1}{1{,}000{,}000}x^2$ (for $x \le 400{,}000$).
   (i) Explain why the marginal tax $T'(x)$ is (approximately) the amount of additional tax paid by a taxpayer with taxable income \$$x$ if he earns an additional \$1.
   (ii) If a taxpayer's taxable income is \$30,000, how much will he be able to keep of an additional dollar of income after taxes are paid?
   (iii) How much of each dollar earned does the above taxpayer keep?
   (iv) Repeat parts (ii) and (iii) for a taxpayer whose taxable income is \$300,000.
   (v) If the taxpayer in part (ii) and his wife (whose taxable income is \$60,000) have an opportunity to make an investment that will earn an extra dollar, explain why it makes sense for this investment income to be in his name rather than hers. How much difference does it make?

Chapter Summary

§4.1
related rates
implicit differentiation

§4.2
maximum value
minimum value
maximum
minimum
local maximum
local minimum
absolute maximum
absolute minimum
relative maximum
relative minimum
Rolle's Theorem (Theorem 4.2.4, p. 240)
average slope
Mean Value Theorem (Theorem 4.2.5, p. 241)
increasing/decreasing function
First Derivative Test (Theorem 4.2.10, p. 244)
If f' is always zero, then f is constant.

§4.3
critical point
stationary point
extremum
local extremum
absolute extremum

§4.4
Second Derivative Test (Theorem 4.4.1, p. 256)

§4.5
curve sketching
even function
odd function
asymptote
vertical asymptote
oblique asymptote
sketching rational functions

§4.6
concave up
concave down
second derivative and concavity
inflection point
cusp

§4.7
maximum/minimum problems
Snell's Law

§4.8
antiderivatives

§4.9
velocity
acceleration
motion with constant acceleration
vertical and horizontal components of motion

§4.10
linear approximation
differentials
percentage error
error propagation

§4.11
Newton's Method (Summary 4.11.1, p. 326)
$x_n = x_{n-1} - \frac{f(x_{n-1})}{f'(x_{n-1})}$

§4.12
l'Hôpital's Rule (Theorem 4.12.1, p. 332)
l'Hôpital's Rule for one-sided limits
l'Hôpital's Rule for limits at infinity
l'Hôpital's Rule for infinite limits
indeterminate forms

§4.13
marginal cost
marginal revenue
marginal profit
demand function
price function

Review Exercises

I

1. If the volume of a spherical balloon is increasing at the rate of 50 cm^3/sec, how fast is its radius increasing at the moment when the radius is 10 cm?

2. If a spherical balloon is deflating, how fast is its surface area decreasing at the moment when its radius is 4 in and its volume is decreasing by 1 in^3/sec?

3. A 20-ft ladder has slipped down a wall, so Fred is pushing it back. How fast is the top of the ladder rising up the wall at the moment when the bottom of the ladder is 12 ft from the base of the wall, moving toward the wall at a speed of 1.5 ft/sec?

4. Diane is standing on a 15-ft ladder when the bottom starts to slide away from the wall at a speed of 1 in/sec. She jumps free when she notices that the

top of the ladder is sliding down at the rate of $\frac{1}{2}$ in/sec. How far is it from the top of the ladder to the ground at this moment?

5. Consider a conical tank with its vertex pointing downward; suppose the radius at the top is 2 m and the tank is 3 m high. If water is leaking out at the rate of 0.002 m^3/hr, how fast is the area of the surface of the water in the tank decreasing when the water is 2 m deep?

6. In the situation of Exercise 5, how fast is the area decreasing when the water is 1 m deep?

7. A ship is approaching a lighthouse from the east at a speed of 10 knots. A sailboat is sailing due south at 7 knots on a course that will take it past the lighthouse at a distance of 4 nautical miles to the east (see figure below). How fast are the vessels approaching each other when the ship is 6 nautical miles due east of the lighthouse and the sailboat is 1 nautical mile north of the imaginary line between the ship and the lighthouse?

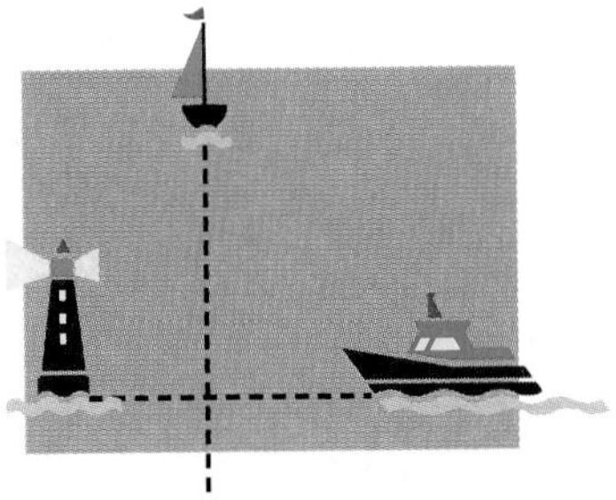

8. A mouse runs in a straight line across the floor, at a speed of 1 ft/sec. It does not see the stalking cat and passes the cat at a distance of 8 ft at the closest point. How fast is the mouse moving away from the watchful cat exactly ten seconds later?

9. A retired scientist is hoping to perfect a cylindrical balloon, which has the property that no matter how much it is inflated, it still maintains the same basic shape, that of a cylinder whose height is four times its radius. How fast would the volume of this balloon be increasing at the moment when its radius is 3 in and the radius is increasing at the rate of $\frac{1}{8}$ in/sec?

10. How fast would the surface area of the cylindrical balloon in Exercise 9 be increasing if the volume was increasing at the rate of 5 in^3/sec and the radius was 4 in?

Find the regions where each of the following functions is increasing or decreasing.

11. $f(x) = x^2 - 4x + 9$
12. $f(x) = -x^2 - 7x + 1$
13. $f(x) = x^3 - 9x + 2$
14. $f(x) = -x^3 - 2x - 4$
15. $f(x) = xe^x$
16. $f(x) = \cosh x$
17. $f(x) = \operatorname{sech} x$
18. $f(x) = \sec x$
19. $f(x) = x \ln x$
20. $f(x) = \cos^2 x - \sin^2 x$

Find all critical points of the following functions, and classify which are local or absolute maxima or minima.

21. $f(x) = x^7$
22. $f(x) = x^8$
23. $f(x) = x^4 + 2x^2 + 3$
24. $f(x) = x^2 e^x$
25. $f(x) = |x^2 - 2x - 3|$
26. $f(x) = |x^3 - 1|$
27. $f(x) = \cot \frac{x}{2}$
28. $f(x) = \frac{1}{1+x}$
29. $f(x) = \frac{x^4}{1+x^4}$
30. $f(x) = \operatorname{sech} x$

Find all local and absolute extrema for the following functions. (If the domain is not specified, assume that it is $\mathbb{R}$.)

31. $f(x) = x^3 - x^2 + x + 2$
32. $f(x) = x^3 - x^2 - x + 3$
33. $f(x) = x^3 - 6x,\ x \in [-1, 1]$
34. $f(x) = x^5 - 5x + 3$
35. $f(x) = |x^2 - 3x|,\ x \in [0, 2]$
36. $f(x) = \frac{1}{2x-3},\ x \in [0, 1]$
37. $f(x) = x + |x|,\ x \in [-3, 1]$
38. $f(x) = \cosh x \sinh x,\ x \in [-1, 1]$
39. $f(x) = (x^2 + x)e^{2x}$
40. $f(x) = \ln(x^2 + 1),\ x \in [-2, 3]$

Sketch the graphs of the following functions.

41. $f(x) = \frac{x}{x^2+4}$
42. $f(x) = \frac{x^2+4}{x}$
43. $f(x) = \frac{x^2}{x+3}$
44. $f(x) = x^2 e^{2x}$
45. $f(x) = \cos^2 x$
46. $f(x) = \frac{x}{2-3x}$
47. $f(x) = \frac{x^2-1}{x}$
48. $f(x) = \frac{x}{x^2-2x}$
49. $f(x) = \frac{x}{x^2-9}$
50. $f(x) = \frac{1}{x} + \frac{1}{x^2}$

Find all antiderivatives of each of the following functions.

51. $f(x) = x^4 - 3x^2 + 4x - 1$
52. $f(x) = e^{3x-1}$
53. $f(x) = 3x^{9/4} + 3x - 4$
54. $f(x) = 3x + \sin x$
55. $f(x) = \sqrt{x+7}$
56. $f(x) = \cos(-x)$
57. $f(x) = \sin(5x)$
58. $f(x) = 2e^x - \cos(3x)$
59. $f(x) = x^{-4} + 2x$
60. $f(x) = 6x^{-1/3}$

Find each of the following.

61. The length of time a ball will be in the air if it is thrown straight up from ground level at a speed of 256 ft/sec.

62. The speed with which a ball must be thrown straight up from ground level in order for it to reach a maximum height of 60 m.

63. The height of a building if a brick takes 2.6 sec to fall from its roof.

64. The speed with which a ball must be thrown upward in order for it to take 3 sec to return to the point from which it was launched.

65. The total distance traveled by a car as it decelerates from 50 mph at a rate of 20 ft/sec^2.

66. The acceleration of a car that starts at rest and accelerates at a constant rate, reaching 100 km/hr in 10 sec.

67. The maximum height reached by a ball that is thrown straight up from ground level at 15 m/sec.

68. The constant deceleration necessary for a "soft landing" of a lunar probe that is 100 m above the moon's surface and falling at a rate of 20 m/sec.

69. The constant rate of deceleration needed for a car traveling at 70 mph to stop in a distance of 300 ft.

70. The average speed of the car in Exercise 69 over the period of its deceleration.

Estimate each of the following using linear approximation.

71. $\sqrt{80.5}$

72. $(26.99)^{1/3}$

73. $\frac{1}{40.01}$

74. $(24.99)^{-1/2}$

75. $(63.999)^{5/3}$

76. $\left(1+(1.01)^2\right)^{-1}$

77. $\sin(30.2°)$

78. $\tan(59.85°)$

79. $\sec(44.97°)$

80. $\csc(-29.95°)$

Use Newton's Method to find approximate values of the following.

81. $\sqrt{41}$

82. $\sqrt{55}$

83. $(40)^{1/3}$

84. $(12)^{-1/5}$

85. $\arctan 5$

86. $\arcsin\left(-\frac{2}{3}\right)$

87. A root of $x^3 - 4x + 2$

88. A root of $x^5 - 3x^3 + x - 1$

89. A root of $x^4 - 4x^2 + 2$

90. A solution of $x \ln x = 6$

Evaluate the following limits, provided that they exist.

91. $\lim_{x\to 0} \frac{x^2-3x}{x^2-2}$

92. $\lim_{x\to 1} \frac{x^2-x}{x^2+x-2}$

93. $\lim_{x\to 0} \frac{x^2-3x}{\sin(4x)}$

94. $\lim_{x\to -2} \frac{x^2+2x}{\sin(\pi x)}$

95. $\lim_{x\to\infty} \frac{x^3-2x^2}{e^{2x}}$

96. $\lim_{x\to 1^+} (x-1)\ln(x^2-1)$

97. $\lim_{x\to -\infty} xe^{-x}$

98. $\lim_{x\to 3} \frac{x^2-2x+1}{\cos(\pi \frac{x}{2})}$

99. $\lim_{x\to 0} (1-e^x)x^{-1/2}$

100. $\lim_{x\to 3} \frac{\sin x}{\sin(\sqrt{x})}$

II

***101.** The light on the roof of a police car rotates once every second, flashing along a fence that is 50 ft from the light at the nearest point.

(i) How fast is the spot of light moving along the fence as it passes the point that is closest to the car?

(ii) How fast is the spot of light moving as it passes the point on the fence that is 40 ft away from the point that is closest to the car (assuming that the fence is straight)?

***102.** Two planks, one 10 ft long and the other $7\frac{1}{2}$ ft long, are hinged together at one end. A boy is lifting the hinged joint straight up over his head, allowing the other ends of the planks to drag along the ground (see figure below). How fast is the unhinged end of the 10-ft plank moving toward the boy at the moment when the hinge is 6 ft above the ground if the unhinged end of the short plank is moving toward him at the rate of $\frac{1}{2}$ ft/sec?

103. A rectangular sheet of cardboard measuring $8\frac{1}{2} \times 11$ inches is used to make an open box by cutting equal squares from the four corners and folding up the sides. What is the largest possible volume of a box that can be made in this way?

104. Find the point(s) on the parabola $y = 3x^2 + x - 1$ that are closest to $(2, -3)$.

***105.** Sketch the graph of $f(x) = \frac{\sin x}{x}$, for $x \in \left(0, \frac{\pi}{2}\right]$.

***106.** Let $g(\theta) = |\sin\theta|$.

(i) Sketch the graph of $g(\theta)$ for $\theta \in [-\pi, \pi]$.

(ii) Does g have a cusp at $\theta = 0$?

107. Find all antiderivatives of $\frac{3x^2-2x}{x}$.

***108.** Let $f(x)$ be the function defined for every nonzero real number x by $f(x) = |x|$; $f(x)$ is undefined for $x = 0$. Find all antiderivatives of $f(x)$.

***109.** Suppose a building has n stories (ignoring the basement), each of them 10 ft high (including the thickness of the floors). Find a formula (depending on n) for the length of time it will take a small heavy object to fall down a stairwell from the top floor to the ground floor. A lazy person proposes to drop a ball down a stairwell, time its fall, and use the formula instead of going outside and counting the number of stories. Why would it be less reliable to use this approach to find the number of stories in a building with over ten stories than for one with under three stories?

***110.** A toy rocket accelerates at the rate of 100 ft/sec^2 for 2 sec until its fuel is burned and then coasts.

(i) What height will it reach, assuming that it flies straight up?

(ii) How long will its flight last?

111. Find the differential dy if (i) $y = x\sin x$, (ii) $x^2y^3 - 4x^3y = 5$, (iii) $e^x\sin(x+y) = \frac{1}{2}$.

112. Find the differential dy if (i) $x^2 - 2y^4 = 2 - 7x$, (ii) $\sin(xy^2) = x - y$, (iii) $\tan(x + y^3) = 6$.

***113.** Use linear approximation to estimate $10^{1.99}$.

114. (i) What is the possible percentage error in the calculation of the volume of a cylinder if its radius is exactly 3 units and its height is measured to within 2%?

(ii) What is the possible percentage error in the calculation of the volume of the cylinder if its height is exactly 2 units and its radius is measured to within 2%?

115. What production level should be used by a company wishing to maximize its profits if the cost function is $C(x) = 500 + 30x - \frac{1}{80}x^2$ and the revenue function is $R(x) = -1000 + 50x - \frac{1}{40}x^2$ when the production level is x units per day?

116. Find the optimal price and level of production for a gadget if it costs $C(x) = 20{,}000 + 170x - x^2$ dollars to produce x units per month and the manufacturer can sell $x = 180 - \frac{9}{10}p$ units when the unit price is \$$p$.

***117.** (i) Suppose $f(x)$ is an odd function defined on all of $\mathbb{R}$. Show that $f^{(k)}(0) = 0$ for every even integer $k \geq 0$ for which $f^{(k)}(0)$ exists.

(ii) Suppose $g(x)$ is an even function defined on all of $\mathbb{R}$. Show that $g^{(k)}(0) = 0$ for every odd integer $k \geq 0$ for which $g^{(k)}(0)$ exists.

118. Sketch the graph of the function $f(x) = x\sqrt{1-x^2}$ on $[-1, 1]$.

***119.** Show that if $a \neq 0$, then $f(x) = x^3 + ax^2$ always has a local extremum. What about absolute extrema?

***120.** Discuss the extrema of $g(x) = x^4 + ax^3$.

***121.** A 15-ft ladder that is leaning against a wall begins to slide down. How fast is the angle between the ground and the ladder changing at the moment when the top of the ladder is 12 ft above the ground and descending at a rate of 6 in/sec?

122. (i) Suppose $r > s$ and $f(x)$ is differentiable on (r, s). Fix $a \in (r, s)$ and let $P_a^0(x)$ be the polynomial of degree 0, that is, a constant polynomial, that has the same value as $f(x)$ at the point $x = a$. Show that if $\frac{f(x)-P_a^0(x)}{x-a} \geq 0$ for every $a \neq x \in (r, s)$, then $f'(a) \geq 0$ for every $a \in (r, s)$.

(ii) Suppose $f(x)$ is twice differentiable on (r, s). Fix $a, b \in (r, s)$ and let $P_{a,b}^1(x)$ be the polynomial of degree 1 that has the same value as $f(x)$ at the points $x = a$ and $x = b$. Show that if $\frac{f(x)-P_{a,b}^1(x)}{(x-a)(x-b)} > 0$ for every $a, b \in (r, s)$ and every $x \in (r, s)$ except $x = a, b$, then $f''(a) \geq 0$ for every $a \in (r, s)$.

(iii) Suppose $f(x)$ is three times differentiable on (r, s). Fix $a, b, c \in (r, s)$ and let $P_{a,b,c}^2(x)$ be the polynomial of degree 2 that has the same value as $f(x)$ at the points $x = a$, $x = b$, and $x = c$. Show that if $\frac{f(x)-P_{a,b,c}^2(x)}{(x-a)(x-b)(x-c)} > 0$ for every $a, b, c \in (r, s)$ and every $x \in (r, s)$ except $x = a, b, c$, then $f^{(3)}(a) \geq 0$ for every $a \in (r, s)$.

(iv) Can you guess a generalization of this result to the fourth derivative, $f^{(4)}(x)$?

(v) Can you guess a generalization of this result to the nth derivative, $f^{(n)}(x)$?

***123.** Suppose we want to sketch the graph of $F(x) = e^{\sin x}$ for $x \in [-\pi, \pi]$. The derivative is $F'(x) = e^{\sin x}\cos x$, so it is easy to find the critical points.

But if we compute the second derivative, $F''(x) = e^{\sin x}\cos^2 x + e^{\sin x}(-\sin x) = e^{\sin x}(\cos^2 x - \sin x)$, we see that to find the inflection points, we must know where $\cos^2 x - \sin x = 0$, that is, when $\cos^2 x = \sin x$.

(i) Use Newton's Method to find approximate values for *all* solution(s) of $F''(x) = 0$ with $x \in [-\pi, \pi]$.

(ii) Sketch the graph of $y = F(x)$ for $x \in [-\pi, \pi]$.

Integration 5

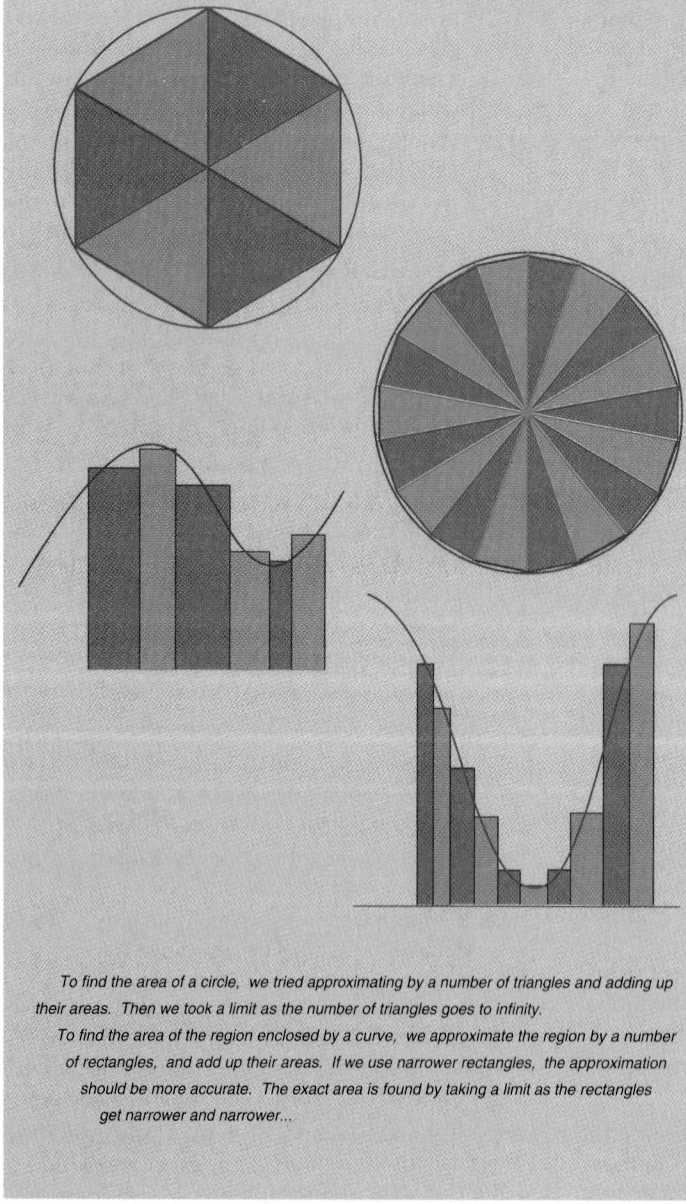

To find the area of a circle, we tried approximating by a number of triangles and adding up their areas. Then we took a limit as the number of triangles goes to infinity.

To find the area of the region enclosed by a curve, we approximate the region by a number of rectangles, and add up their areas. If we use narrower rectangles, the approximation should be more accurate. The exact area is found by taking a limit as the rectangles get narrower and narrower...

INTRODUCTION

The preceding two chapters have developed the theory of derivatives, which is known as differential calculus. This chapter and Chapters 6 and 8 develop the other major branch of the subject, which is known as *integral calculus.*

Integration is a technique, much the way differentiation is. It arises when we try to find the area of a region with curved boundaries. For instance, it will eventually allow us to explain why the area of a circle is πr^2.

The notion of differentiation also arose in answer to a geometrical problem, trying to find the equation of a tangent line. When we learned how to answer that problem, we found that the ideas we had explored could also be used to answer other questions. Something similar happens with integration.

In addition to enabling us to find areas, it can be used for many other things. One of the more unexpected ones is finding antiderivatives. In Section 4.8 we discussed the idea of an antiderivative of $f(x)$, meaning a function $F(x)$ satisfying $F'(x) = f(x)$. We found that this was a useful notion that had applications to questions involving velocity or other rates of change, but we realized that actually finding antiderivatives could sometimes be quite difficult.

It turns out that the technique of integration can be used to find antiderivatives. In fact, we will see that integration more or less amounts to the same thing as finding antiderivatives. It may still be a difficult problem, but integration will help us to approach it systematically.

Sections 5.1 through 5.4 discuss integration in terms of the problem of finding areas. We begin in Section 5.1 by trying to approximate the area under a graph. Section 5.2 is a technical section on summation notation, which is needed for Sections 5.3 and 5.4, which use Riemann sums to define integrals. Section 5.5 establishes the link between the problem of finding areas and the problem of finding antiderivatives, which is known as the Fundamental Theorem of Calculus. Then we begin to learn how to work with integrals, learning to evaluate definite integrals in Section 5.6 and exploring the technique of substitution in Section 5.7.

5.1 The Area Under a Curve

RECTANGLES APPROXIMATION

Suppose $f(x)$ is a continuous function with $f(x) \geq 0$ for all $x \in [a, b]$. The graph of $y = f(x)$ lies above the x-axis, as shown in Figure 5.1.1, and we might ask: What is the area of the shaded region—the region below the graph of $y = f(x)$? Strictly speaking, it is the region whose boundaries are the graph of $y = f(x)$, the x-axis, and the two vertical lines $x = a$ and $x = b$.

There is one situation in which it is easy to find the area—when $y = f(x)$ is a horizontal straight line and the region is a rectangle, as seen in Figure 5.1.2.

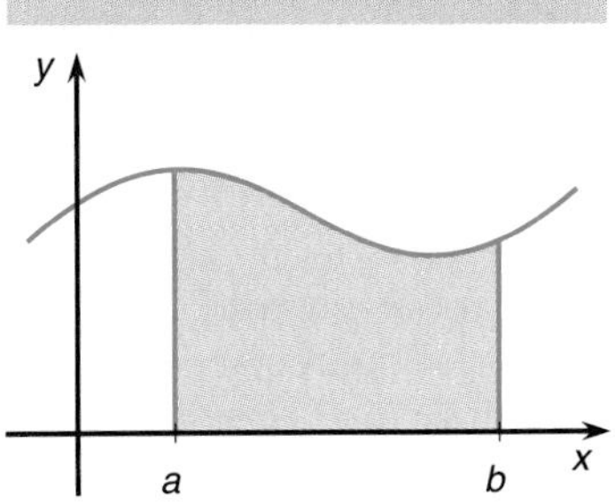

Figure 5.1.1

EXAMPLE 1 If $f(x) = c$ is a constant function, with $c \geq 0$, then the region under the graph is a rectangle. Its height is c, and its width is $b - a$, the horizontal distance from $x = a$ to $x = b$.

Its area is

$$\textit{height} \times \textit{width} = c(b - a).$$

Notice that if $c = 0$, the "rectangle" is really just a line segment, so it is not surprising that the formula says that its area is zero.

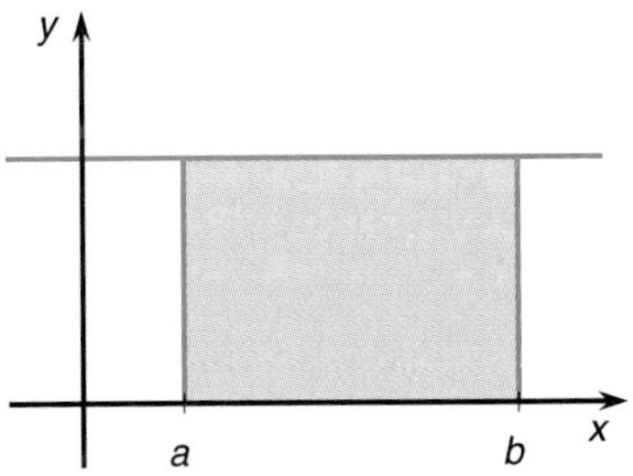

Figure 5.1.2

For a curve that is not a straight line the problem is more difficult. Rather than calculating the area directly, we can try approximating it by filling the region with rectangles, as shown in Figure 5.1.3. We know how to calculate the total area of the rectangles, and this should be approximately equal to the area under the curve.

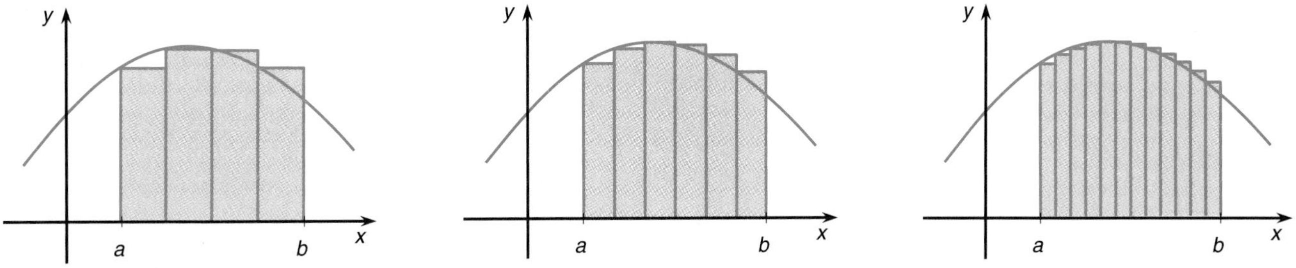

Figure 5.1.3

If we use a larger number of rectangles, as in the second and third figures, the approximation should be better. The total area of the rectangles will be very close to the area of the region under the curve. (The error, that is, the difference between the actual area of the region and the sum of the areas of the rectangles, involves the areas of the little regions between the curve and the tops of the rectangles, which are certainly smaller in the second figure than in the first, and smaller yet in the third).

This suggests trying the approximation with a large number of rectangles, say, n of them, and then taking the limit as $n \to \infty$. This is a good idea but difficult to carry out in practice. To do a simple example, we need to know the following fact.

THEOREM 5.1.1

If n is a positive integer, then

$$1^2 + 2^2 + 3^2 + \ldots + (n-1)^2 + n^2 = \frac{n(n+1)(2n+1)}{6}.$$

This formula for the sum of the squares of the integers from 1 to n will be proved in Example 5.2.5 but first let us use it to find an area.

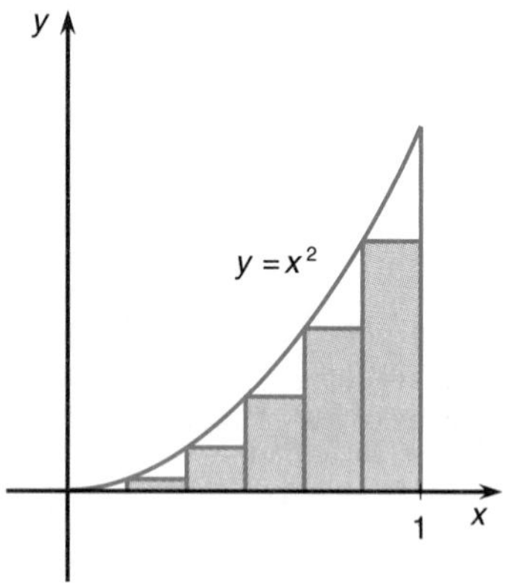

Figure 5.1.4

If we divide $[0, 1]$ into n equal segments, then the width of each segment is $\frac{1}{n}$.

EXAMPLE 2 Find the area under the curve $y = f(x) = x^2$ between $x = 0$ and $x = 1$.

Solution We divide the interval $[0, 1]$ into n equal segments, and construct a rectangle on each of them, as shown in Figure 5.1.4 (in the illustration, $n = 6$). We let the height of each rectangle be the height of the function at the left endpoint of its segment, so the curve passes through the upper left corner of the rectangle.

The first rectangle lies over the interval $\left[0, \frac{1}{n}\right]$. Its height is the height of $f(x)$ at $x = 0$, so this "rectangle" has area equal to 0.

The second rectangle lies over the interval $\left[\frac{1}{n}, \frac{2}{n}\right]$, so its height is $f\left(\frac{1}{n}\right) = \frac{1}{n^2}$, and its width is $\frac{1}{n}$. Its area is $\frac{1}{n^2} \times \frac{1}{n} = \frac{1}{n^3}$.

The third rectangle lies over the interval $\left[\frac{2}{n}, \frac{3}{n}\right]$, so its height is $f\left(\frac{2}{n}\right) = \frac{2^2}{n^2}$, and its width is $\frac{1}{n}$. Its area is $\frac{2^2}{n^2} \times \frac{1}{n} = \frac{2^2}{n^3}$.

The kth rectangle lies over the interval $\left[\frac{k-1}{n}, \frac{k}{n}\right]$, so its height is $f\left(\frac{k-1}{n}\right) = \left(\frac{k-1}{n}\right)^2$, and its width is $\frac{1}{n}$. Its area is $\left(\frac{k-1}{n}\right)^2 \frac{1}{n} = \frac{(k-1)^2}{n^3}$.

The last rectangle lies over the interval $\left[\frac{n-1}{n}, \frac{n}{n}\right] = \left[\frac{n-1}{n}, 1\right]$, so its height is $f\left(\frac{n-1}{n}\right) = \left(\frac{n-1}{n}\right)^2$, and its area is $\frac{(n-1)^2}{n^3}$.

The total area of all the rectangles is then

$$0 + \frac{1}{n^3} + \frac{2^2}{n^3} + \ldots + \frac{(k-1)^2}{n^3} + \ldots + \frac{(n-1)^2}{n^3}.$$

There is a common denominator of n^3. Factoring it out, we find the simpler expression

$$\frac{1}{n^3}\left(0 + 1 + 2^2 + 3^2 + \ldots + (n-1)^2\right) = \frac{1}{n^3}\left(1^2 + 2^2 + 3^2 + \ldots + (n-1)^2\right).$$

To simplify this, we need Theorem 5.1.1. Using the formula with n replaced by

$n-1$, we see that the sum is

$$1^2+2^2+\ldots+(n-1)^2=\frac{(n-1)\big((n-1)+1\big)\big(2(n-1)+1\big)}{6}$$
$$=\frac{(n-1)(n)(2n-1)}{6},$$

so the total area of the rectangles is

$$\frac{1}{n^3}\left(1^2+2^2+\ldots+(n-1)^2\right)=\frac{1}{n^3}\left(\frac{(n-1)n(2n-1)}{6}\right)$$
$$=\frac{(n-1)(2n-1)}{6n^2}=\frac{2n^2-3n+1}{6n^2}.$$

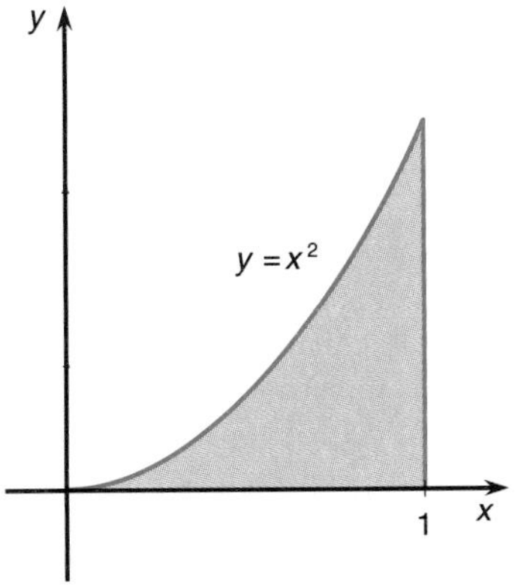

Figure 5.1.5

The last step is to take the limit as $n\to\infty$. This amounts to considering more and more rectangles, each of which is narrower and narrower; the total area approaches the area under the curve as shown in Figure 5.1.5.

The area of the region under the curve is

$$\lim_{n\to\infty}\left(\frac{2n^2-3n+1}{6n^2}\right)=\frac{2}{6}=\frac{1}{3}.$$

The idea of approximating the area by large numbers of rectangles and then taking limits does seem to work, but it is awfully complicated in practice. After all, $y=x^2$ is just about the simplest curve we can think of other than a straight line, and still it took a lot of work to find the area. The rest of this chapter is devoted to developing a more practical technique for finding areas.

Despite the difficulties, Example 2 does illustrate the method of approximating areas by filling a region with rectangles.

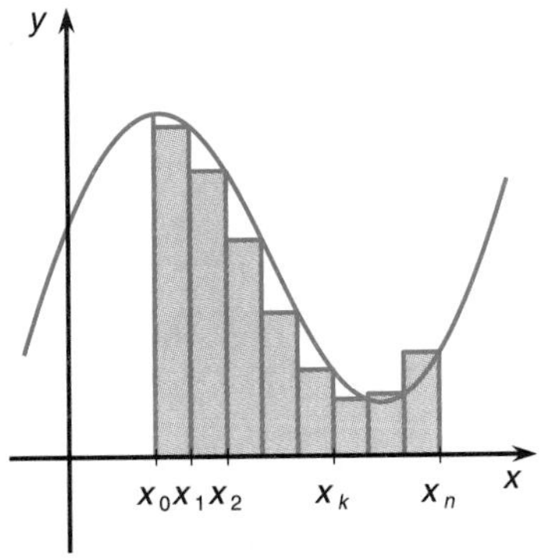

Figure 5.1.6

EXAMPLE 3 If $f(x)\geq 0$ for all $x\in[a,b]$, approximate the area under the curve $y=f(x)$ between $x=a$ and $x=b$, by dividing the interval into n equal subintervals and using rectangles whose height equals the height of the curve at the *right* endpoint of each of subinterval (see Figure 5.1.6).

> The kth point x_k is found by moving k subintervals from $x=a$. Since each subinterval has width $\frac{b-a}{n}$, $x_k=a+k\frac{b-a}{n}$.

Solution The length of the interval $[a,b]$ is $b-a$, so the length of each subinterval will be $\frac{b-a}{n}$. Suppose we label the endpoints of the subintervals as x_0, x_1, $x_2,\ldots,$ x_n, so that $x_0=a$, $x_1=a+\frac{b-a}{n}$, $x_2=a+2\frac{b-a}{n}$, $x_3=a+3\frac{b-a}{n},\ldots,$ $x_k=a+k\frac{b-a}{n},\ldots,x_{n-1}=a+(n-1)\frac{b-a}{n}$, $x_n=a+n\frac{b-a}{n}=b$.

The kth subinterval will be $[x_{k-1},x_k]=\left[a+(k-1)\frac{b-a}{n},a+k\frac{b-a}{n}\right]$. Its right endpoint is $x_k=a+k\frac{b-a}{n}$, so the rectangle over this subinterval has height $f(x_k)$. Its area is $f(x_k)\frac{b-a}{n}$.

The total area of all the rectangles is

$$f(x_1)\frac{b-a}{n}+f(x_2)\frac{b-a}{n}+f(x_3)\frac{b-a}{n}+\ldots+f(x_n)\frac{b-a}{n}$$
$$=\big(f(x_1)+f(x_2)+f(x_3)+\ldots+f(x_n)\big)\frac{b-a}{n}.$$

In Example 3 we chose the height of our rectangles by taking the value of the function at the right endpoint of each subinterval; in Example 2 we used the value

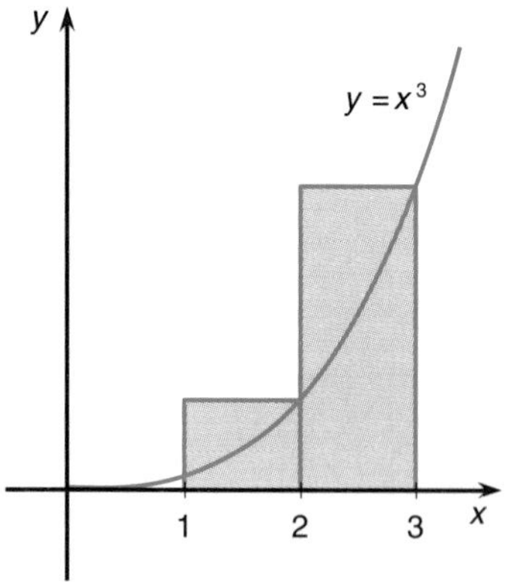

Figure 5.1.7

of the function at the left endpoints. We could also have used the midpoints of the subintervals, but the calculations would be slightly more complicated.

Different choices will give different values when we approximate the area under a curve. Our ultimate goal is to take a limit and get the *exact* value of the area, and when we do this, the differences should disappear.

EXAMPLE 4 Use the formula from Example 3 to approximate the area under the curve $y = f(x) = x^3$ between $x = 1$ and $x = 3$, using the values $n = 2, 4, 6$.

Solution With $n = 2$, as shown in Figure 5.1.7, we have $x_0 = 1, x_1 = 2, x_2 = 3$. The width of each rectangle is $\frac{b-a}{n} = \frac{3-1}{2} = 1$, so the total area of the rectangles is

$$\big(f(x_1) + f(x_2)\big)\frac{b-a}{n} = \big(f(2) + f(3)\big) \times 1$$

$$= f(2) + f(3) = 2^3 + 3^3 = 8 + 27 = 35.$$

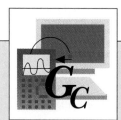

A programmable calculator or a computer can be programmed to evaluate the sum in Example 4. The advantage is that the machine can easily do the calculation with larger values of n. The width of each rectangle is $\frac{b-a}{n} = \frac{2}{n}$.

TI-85 PRGM F2(EDIT) RIEMANN ENTER F3 F1(Input) N EXIT ENTER 0 STO I ENTER 0 STO S ENTER Lbl B ENTER I+1 STO I ENTER S + (1 + I × 2 ÷ N) ^3 STO S ENTER F4 F1(If) (I TEST F2(<) N) ENTER Goto B ENTER S × 2 ÷ N ENTER EXIT EXIT EXIT PRGM F1(NAME) RIEMANN ENTER 6 ENTER . . .	SHARP EL9300 ▼ ▼ (NEW) ENTER ENTER RIEMANN ENTER COMMAND A 3 (Input) *N* ENTER *I* = 0 ENTER *S* = 0 ENTER COMMAND B 1 (Label) 1 ENTER *I* = *I* + 1 ENTER *S* = *S* + (1 + *I* × 2 ÷ *N*) a^b 3 ENTER COMMAND 3 (If) *I* COMMAND C 2 (<) *N* COMMAND B 2 (Goto) 1 ENTER *S* = *S* × 2 ÷ *N* ENTER COMMAND A 1 (Print) *S* ENTER COMMAND A 6 (End) ENTER (RUN) ENTER (RIEMANN) ENTER 6 ENTER
CASIO f_x-6300*G* (f_x-7700*GB*) MODE 2 EXE ? → *N* : 0 → *I* : 0 → *S* : Lbl 1 : *I* + 1 → *I* : *S* + (1 + *I* × 2 ÷ *N*) x^y 3 → *S* : *I* < *N* ⇒ Goto 1: *S* × 2 ÷ *N* ◢ MODE 1 Prog 0 EXE 6 EXE (on the f_x-7700*GB*, : is PRGM F6, and Prog is PRGM F3.)	HP 48SX ATTN ↱ CLR ≪≫ "'' KEYIN N ▶ SPC PROMPT SPC ′ *N* ▶ STO 0 SPC 1 SPC *N* SPC FOR *I* SPC *I* SPC 2 × *N* ÷ 1 + 3 y^x + NEXT ▶ 2 × *N* ÷ ▶ ENTER ′ RIEMANN STO VAR RIEMANN 6 ENTER ↰ CONT

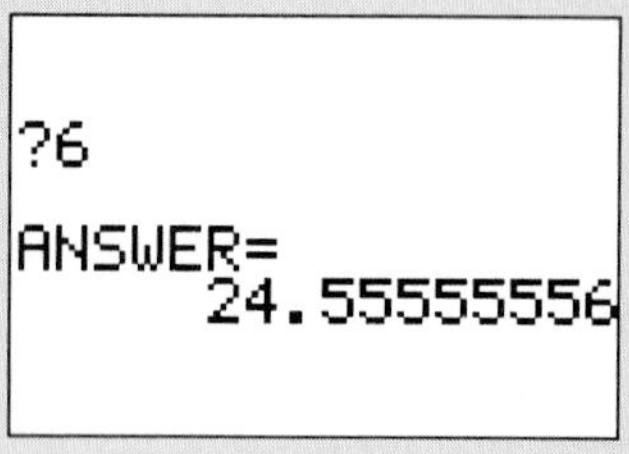

Now use the program with other values of n or for some other functions.

With $n = 4$ we have $x_0 = 1,\ x_1 = \frac{3}{2},\ x_2 = 2,\ x_3 = \frac{5}{2},\ x_4 = 3$. The width of each rectangle is $\frac{b-a}{n} = \frac{3-1}{4} = \frac{1}{2}$, so the total area of the rectangles is

$$\begin{aligned}\big(f(x_1) + f(x_2) + f(x_3) + f(x_4)\big)\frac{b-a}{n} &= \left(f\left(\frac{3}{2}\right) + f(2) + f\left(\frac{5}{2}\right) + f(3)\right) \times \frac{1}{2}\\ &= \left(\left(\frac{3}{2}\right)^3 + 2^3 + \left(\frac{5}{2}\right)^3 + 3^3\right) \times \frac{1}{2}\\ &= \left(\frac{27}{8} + 8 + \frac{125}{8} + 27\right) \times \frac{1}{2}\\ &= 27.\end{aligned}$$

With $n = 6$ we have $x_0 = 1,\ x_1 = \frac{4}{3},\ x_2 = \frac{5}{3},\ x_3 = 2,\ x_4 = \frac{7}{3},\ x_5 = \frac{8}{3},\ x_6 = 3$. The width of each rectangles is $\frac{b-a}{n} = \frac{3-1}{6} = \frac{1}{3}$, so the total area of the rectangles is

$$\begin{aligned}&\big(f(x_1) + f(x_2) + f(x_3) + f(x_4) + f(x_5) + f(x_6)\big)\frac{b-a}{n}\\ &= \left(\left(\frac{4}{3}\right)^3 + \left(\frac{5}{3}\right)^3 + 2^3 + \left(\frac{7}{3}\right)^3 + \left(\frac{8}{3}\right)^3 + 3^3\right) \times \frac{1}{3}\\ &= \left(\frac{64}{27} + \frac{125}{27} + 8 + \frac{343}{27} + \frac{512}{27} + 27\right) \times \frac{1}{3}\\ &= 24\frac{5}{9}.\end{aligned}$$

In principle this formula can be used with larger and larger values of n, giving better and better approximations to the area under the curve.

However, this process can only approximate the area, and we want to find a way to calculate it exactly. That is the purpose of the rest of this chapter.

Exercises 5.1

I

Use the formula from Example 3 to approximate the area under each of the following curves $f(x)$ on the given interval, dividing the interval into n equal subintervals with the specified value of n.

1. $f(x) = x^2$ on $[0, 1]$; $n = 3$
2. $f(x) = x^3$ on $[0, 1]$; $n = 4$
3. $f(x) = x^2$ on $[0, 1]$; $n = 5$
4. $f(x) = x^2$ on $[0, 4]$; $n = 4$
5. $f(x) = 3x^2$ on $[0, 1]$; $n = 5$
6. $f(x) = x^2$ on $[-4, 0]$; $n = 4$
7. $f(x) = 2x + 1$ on $[0, 3]$; $n = 6$
8. $f(x) = x^2$ on $[-4, 4]$; $n = 8$
9. $f(x) = x^2 + 3$ on $[0, 1]$; $n = 6$
10. $f(x) = x^3$ on $[0, 3]$; $n = 3$
11. $f(x) = \sin x$ on $[0, \pi]$; $n = 2$
12. $f(x) = \sin x$ on $[0, \pi]$; $n = 4$
13. $f(x) = \cos x$ on $\left[-\frac{\pi}{2}, \frac{\pi}{2}\right]$; $n = 2$
14. $f(x) = \cos x$ on $\left[0, \frac{\pi}{2}\right]$; $n = 2$
15. $f(x) = \sqrt{x}$ on $[0, 4]$; $n = 4$
16. $f(x) = \frac{1}{1+x^2}$ on $[-1, 1]$; $n = 4$
17. $f(x) = \sin^2 x$ on $[0, 4\pi]$; $n = 4$
18. $f(x) = x^4 + 2x + 1$ on $[0, 3]$; $n = 3$
19. $f(x) = \sin^2 x$ on $[0, 4\pi]$; $n = 8$
20. $f(x) = \frac{1}{x}$ on $[1, 6]$; $n = 5$

II

21. Find the area under the curve $y = x^2$ between $x = -1$ and $x = 0$.

22. Find the area under the curve $y = x^2$ between $x = 0$ and $x = 2$.

23. Find the area under the curve $y = x^2$ between $x = 1$ and $x = 2$.

24. Find the area under the curve $y = 5x^2$ between $x = 0$ and $x = 1$.

***25.** Consider the straight line $y = f(x) = 2x + 3$, which lies above the x-axis on $[0, 2]$.

(i) Use the formula from Example 3 to approximate the area under the line between $x = 0$ and $x = 2$, dividing the interval $[0, 2]$ into four subintervals of equal length.

(ii) If we use rectangles whose height is the value of $f(x)$ at the *left* endpoint of each subinterval, do you expect the approximation to be greater than, less than or the same as the answer to part (i)?

(iii) Use rectangles whose height is the value of $f(x)$ at the *left* endpoint of each subinterval to approximate the area of the region we considered in part (i).

(iv) Compare the approximations of parts (i) and (ii) and compare them both with the actual area under the graph.

***26.** Consider the constant function $f(x) = -3$ on the interval $[0, 1]$. Suppose we apply the formula from Example 3 with $n = 2$, even though the function $f(x)$ is *negative* on this interval. Compare the number that results from this calculation with the area enclosed between the graph of $y = f(x)$, the x-axis, and the lines $x = 0$ and $x = 1$. Explain what has happened.

***27.** If $g(x) = x^2$, and we use the formula from Example 3 to approximate the area under the graph of $y = g(x)$ between $x = -4$ and $x = -1$, will the approximation be greater than the actual area or less than the actual area? (*Hint:* Draw a picture.)

***28.** If $h(x) = \sqrt{x}$ and $0 \leq a < b$, suppose we use the formula from Example 3 to approximate the area under the graph of $y = h(x)$ between $x = a$ and $x = b$. Will the approximation be greater than the actual area or less than the actual area? (*Hint:* Draw a picture.)

III

29. For this exercise you need a personal computer or programmable calculator set to evaluate the formula from Example 3 for $f(x) = x^2$ between $x = 0$ and $x = 2$, using n equal subintervals. Evaluate the approximate area under $y = x^2$ between 0 and 2 with $n = 2, 4, 10, 100$.

30. Use a personal computer or programmable calculator to approximate the area under $y = x^3$ between $x = 1$ and $x = 4$. Use the formula from Example 3 with $n = 4, 16, 64, 256$.

31. Use a personal computer or programmable calculator to approximate the area under $y = \sin x$ between $x = 0$ and $x = \frac{\pi}{2}$. Use the formula from Example 3 with $n = 2, 8, 50, 200$. Can you guess the actual area?

POINT TO PONDER

In this section we have talked about the notion of the area of a region in the plane, but we never really specified what it means. This is partly because it is quite difficult to be precise about areas.

There are some regions whose areas we can calculate explicitly. In principle we know how to find the area of any triangle, and from that we can find the area of any region that can be built up from triangles. From Figure 5.1.8 we see how even a complicated polygon (region with a finite number of straight line segments as its boundary) can be broken into triangles, so we could find its area.

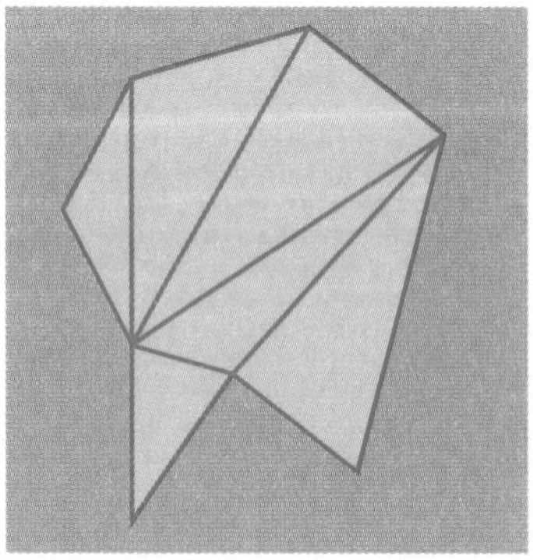

Figure 5.1.8

In fact these are the only regions we really know how to work with. We know that the area of a circle of radius r is πr^2, but this is a fact we have learned, rather than something we can explain. (For instance, it is not easy to show that the constant must be π; try to convince yourself that it cannot be $\frac{22}{7}$, which is a number close to π ...).

One thing about area is that we have a good deal of

intuition about what it ought to be. So a possible avenue through which to approach it is to state all the properties we can think of that area should satisfy. For instance, in discussing polygons, we implicitly assumed that areas "add," that is, that the area of a region composed of two nonoverlapping regions should be the sum of their areas. Try to think of other properties that area should satisfy.

There is a surprising fact about areas. Our intuition suggests that it ought to be possible to find the area of any region, but in fact it can be shown that if we specify some reasonable properties for area, then it is impossible to find the area of some regions. To put it another way, there are some regions that are so weird that the idea of area does not make sense for them. However, this is a very difficult subject, and it is extraordinarily hard to find any examples of such a region.

5.2 Summation Notation

INDEX OF SUMMATION
SUM OF INTEGERS
SUM OF SQUARES

In Section 5.1 we developed the idea of approximating the area under a curve by filling the region with rectangles. In Example 5.1.3 we divided the interval into n equal subintervals and on each subinterval constructed the rectangle whose height equals the height of the curve at the right endpoint. Then we calculated the total area of these rectangles.

This provides a formula for calculating approximate areas, but it involves a fairly cumbersome sum with a large number of terms. To deal more efficiently with such sums, we devote this section to introducing a notation for them.

Suppose we want to add up the first n positive integers,

$$1 + 2 + 3 + \ldots + (n - 1) + n,$$

or the squares of the first n positive integers,

$$1^2 + 2^2 + 3^2 + \ldots + (n - 1)^2 + n^2.$$

Both of these sums can be expressed in the form

$$a_1 + a_2 + a_3 + \ldots + a_{n-1} + a_n,$$

where each a_k is some number determined by k. In the first sum, the sum of the first n positive integers, we can let $a_k = k$, so $a_1 = 1$, $a_2 = 2$, $a_n = n$, etc. For the second we can let $a_k = k^2$, so $a_1 = 1^2 = 1$, $a_2 = 2^2 = 4$, $a_3 = 3^2 = 9$, $a_n = n^2$, etc.

A great many interesting sums can be written as

$$a_1 + a_2 + a_3 + \ldots + a_{n-1} + a_n,$$

for some numbers a_k. This is the sum of the a_k's, as k runs from $k = 1$ to $k = n$. We think of k starting out at $k = 1$ and the sum starting out with a_1. Then k increases to $k = 2$, and we add a_2. Next k increases to $k = 3$, and we add a_3. This continues until k reaches its final value $k = n$, at which point we add a_n and then stop.

There is no particular reason to insist that the sum begin at $k = 1$; we can consider sums beginning and ending with any specified integers:

$$a_m + a_{m+1} + a_{m+2} + \ldots + a_{n-1} + a_n,$$

where m and n are any integers satisfying $m \leq n$.

This process is so common, so useful, and so important that there is special notation for it.

DEFINITION 5.2.1

Suppose m and n are integers with $m \leq n$. Suppose that for each k between m and n (i.e., $m \leq k \leq n$), a_k is a real number.

The **summation notation**

$$\sum_{k=m}^{n} a_k$$

is an abbreviation for the sum of the a_k's as k ranges from $k = m$ to $k = n$. Symbolically,

$$\sum_{k=m}^{n} a_k = a_m + a_{m+1} + a_{m+2} + \ldots + a_{n-1} + a_n.$$

The left side of the above equation is read "the sum as k runs from m to n of a_k" or "the sum of a_k as k runs from m to n."

The summation sign $\sum$ is a stylized capital sigma. Sigma is the letter in the Greek alphabet that corresponds to our letter S, so it reminds us of "sum."

EXAMPLE 1

(i) For the sum of the first n positive integers we can write

$$\sum_{k=1}^{n} k = 1 + 2 + 3 + \ldots + (n-1) + n.$$

(ii) For the sum of the squares of the first n positive integers we write

$$\sum_{k=1}^{n} k^2 = 1 + 4 + 9 + \ldots + (n-1)^2 + n^2.$$

By Theorem 5.1.1 we know what this sum equals. It is

$$\sum_{k=1}^{n} k^2 = \frac{n(n+1)(2n+1)}{6}.$$

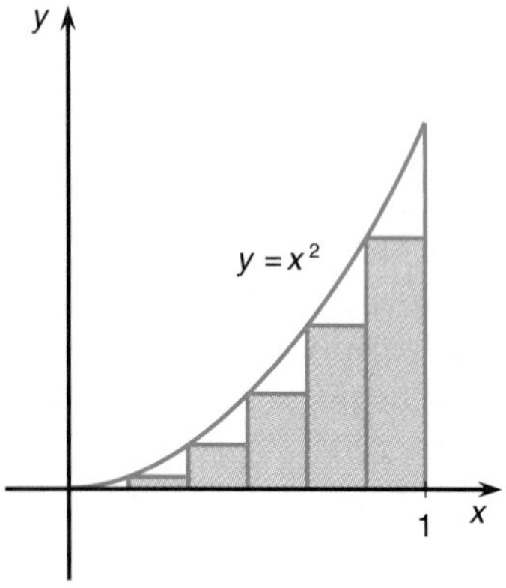

Figure 5.2.1

EXAMPLE 2 We reconsider Example 5.1.2 (p. 348), in which we calculated the approximation to the area under the curve $y = f(x) = x^2$ between $x = 0$ and $x = 1$ by n rectangles.

In that example we found that the kth rectangle lies above the interval $\left[\frac{k-1}{n}, \frac{k}{n}\right]$. Its width is $\frac{1}{n}$, and we let its height be the height of the curve at the left endpoint, that is, $f\left(\frac{k-1}{n}\right)$ (see Figure 5.2.1).

The area of the kth rectangle is therefore $f\left(\frac{k-1}{n}\right) \times \frac{1}{n}$, and the total area of all the rectangles is the sum

$$\sum_{k=1}^{n} f\left(\frac{k-1}{n}\right) \times \frac{1}{n} = \sum_{k=1}^{n} \frac{(k-1)^2}{n^2} \times \frac{1}{n} = \sum_{k=1}^{n} \frac{(k-1)^2}{n^3} = \frac{1}{n^3} \sum_{k=1}^{n} (k-1)^2.$$

We observe that while this sum is written in terms of k, it is really $k - 1$ that is most important. While k runs from $k = 1$ to $k = n$, we see that $k - 1$ runs from 0

A programmable calculator or a computer can be programmed to evaluate the sum in part (ii) of Example 1. We can use the machine to test the formula for fairly large values of n.

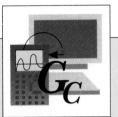

TI-85	SHARP EL9300
PRGM F2(EDIT) SUM ENTER F3 F1(Input) N EXIT ENTER 0 STO K ENTER 0 STO S ENTER Lbl B ENTER K + 1 STO K ENTER S + K^2 STO S ENTER F4 F1(If) (K TEST F2($<$) N) ENTER Goto B ENTER S ENTER EXIT EXIT EXIT PRGM F1(NAME) SUM ENTER 20 ENTER	[icon] ▼ ▼ (NEW) ENTER ENTER SUM ENTER COMMAND A 3 (Input) N ENTER K = 0 ENTER S = 0 ENTER COMMAND B 1 (Label) 1 ENTER K = K + 1 ENTER S = S + K x^2 ENTER COMMAND 3 (If) K COMMAND C 2 ($<$) N COMMAND B 2 (Goto) 1 ENTER N ENTER COMMAND A 1 (Print) S ENTER COMMAND A 6 (End) ENTER [icon] [icon] (RUN) ENTER (SUM) ENTER 20 ENTER
CASIO f_x-6300G (f_x-7700GB)	**HP 48SX**
MODE 2 EXE ? → N : 0 → K : 0 → S : Lbl 1 : K + 1 → K : S + K x^2 → S : K $<$ N ⇒ Goto 1 : S ◢ MODE 1 Prog 0 EXE 20 EXE (on the f_x-7700GB, : is PRGM F6, and Prog is PRGM F3.)	ATTN ↱ CLR «» "" KEYIN N ▶ SPC PROMPT SPC ′ N ▶ STO 0 SPC 1 SPC N SPC FOR K SPC K SPC x^2 + NEXT ▼ ENTER ′ SUM STO VAR SUM 20 ENTER ↰ CONT

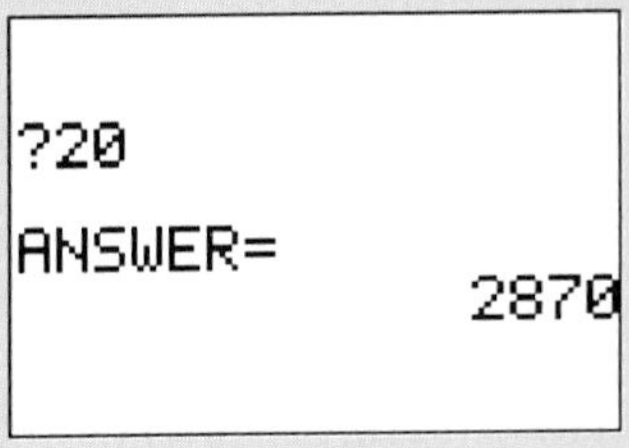

Now use the program with other values of n. Modify your program to find $\sum_{k=1}^{n} k^3$. Can you guess a formula for this sum (in terms of n)?

to $n-1$. In an attempt to avoid confusion, let us write $\ell = k - 1$, so the last sum is exactly the same thing as

$$\sum_{\ell=0}^{n-1} \ell^2.$$

Moreover, the first term in this sum is $0^2 = 0$, which we can safely ignore. In other words, the sum will be the same if we allow ℓ to run from $\ell = 1$ to $\ell = n - 1$ instead of starting at $\ell = 0$. This means that the sum equals

$$\sum_{\ell=1}^{n-1} \ell^2.$$

This can be evaluated by part (ii) of Example 1.

We find that the total area of the n rectangles is

$$\frac{\sum_{k=1}^{n}(k-1)^2}{n^3} = \frac{1}{n^3}\sum_{\ell=1}^{n-1}\ell^2 = \frac{1}{n^3}\left(\frac{(n-1)n(2n-1)}{6}\right)$$
$$= \frac{(n-1)(2n-1)}{6n^2} = \frac{2n^2-3n+1}{6n^2}.$$

This is exactly the same as the calculation we did in Example 5.1.2, except that there we had to write out each sum in detail. The summation notation allows us to write the computation in a neater, more efficient way.

The final observation in Example 2 of Section 5.1 was that we could take the limit as $n \to \infty$ of this approximation by n rectangles, and we find that the area under the graph of $y = f(x) = x^2$ between $x = 0$ and $x = 1$ is

$$\lim_{n\to\infty} \frac{2n^2-3n+1}{6n^2} = \frac{1}{3}.$$

Summation notation can be used to restate the result of Example 5.1.3 more neatly.

EXAMPLE 3 Approximation by Rectangles

Suppose $f(x)$ is a continuous function satisfying $f(x) \geq 0$ for all $x \in [a, b]$. Then the area under the graph of the curve $y = f(x)$ between $x = a$ and $x = b$ can be approximated by

$$\frac{b-a}{n}\sum_{k=1}^{n} f(x_k),$$

where $x_k = a + k\frac{b-a}{n}$.

This is the total area of the rectangles constructed by dividing the interval $[a, b]$ into n equal subintervals and constructing on each subinterval a rectangle whose height equals the height of the curve at the right endpoint.

There is essentially no difference between this formula and the one in the original example (Example 5.1.3). The summation notation used in this version makes it more compact and may make some calculations easier to follow, as we saw in Example 2.

When we perform calculations with summation notation, we need to know how it can be manipulated. First is the elementary observation that the distributive law is satisfied.

PROPOSITION 5.2.2

If C is any real number, m and n are integers with $m \leq n$, and a_k is a real number for each k with $m \leq k \leq n$, then the distributive law holds:

$$\sum_{k=m}^{n} Ca_k = C\sum_{k=m}^{n} a_k.$$

Next is the equally obvious remark that it is permissible to combine sums.

PROPOSITION 5.2.3

Suppose that m and n are integers with $m \leq n$ and a_k and b_k are real numbers for each k with $m \leq k \leq n$. Then

$$\sum_{k=m}^{n} a_k + \sum_{k=m}^{n} b_k = \sum_{k=m}^{n} (a_k + b_k).$$

Next we recall that in the calculation of Example 2 we encountered the sum

$$\sum_{k=1}^{n} (k-1)^2.$$

Recognizing that the terms of the sum were more simply expressed as functions of $k-1$ rather than k, we let $\ell = k-1$. As k runs from 1 to n, we see that $\ell = k-1$ runs from 0 to $n-1$.

Rather than using k to label the terms in the sum, we can use ℓ, and this means we can rewrite the sum as

$$\sum_{\ell=0}^{n-1} \ell^2.$$

This equals the previous sum, and this technique is very useful. It amounts to changing the label k, sometimes called the **index of summation,** to transform the sum into something simpler or better suited to our purpose.

The important thing to realize here is that when we write a sum

$$\sum_{k=m}^{n} a_k,$$

what matters is the numbers a_k that are being summed, not the indices by which they are labeled. The index k can be shifted without changing the value of the sum.

PROPOSITION 5.2.4

Suppose $m \leq n$ are integers and that for each k with $m \leq k \leq n$, a_k is a real number. Then the sum

$$\sum_{k=m}^{n} a_k$$

is the same as the sum obtained by *shifting* the index k by any fixed integer s. Letting $\ell = k+s$, we have

$$\sum_{k=m}^{n} a_k = \sum_{\ell=m+s}^{n+s} a_{\ell-s}.$$

(Notice how this was done: When $k=m$, we see that $\ell = k+s$ must equal $m+s$, which tells us where the sum over ℓ should begin. Similarly, when $k=n$, we see that $\ell = n+s$, which tells us where the sum should end. Finally, since $\ell = k+s$, we can solve and find $k = \ell - s$, so the terms a_k in the sum are replaced with $a_{\ell-s}$.)

EXAMPLE 4 The sum

$$\sum_{m=5}^{12} (m-4)^2$$

involves terms of the form $(m-4)^2$. It seems reasonable to try writing them using $m-4$ as the index, rather than m.

Letting $h = m - 4$, we find that when $m = 5$, we have $h = m - 4 = 5 - 4 = 1$. Also when $m = 12$, we have $h = m - 4 = 8$. This allows us to write

$$\sum_{m=5}^{12} (m-4)^2 = \sum_{h=1}^{8} h^2.$$

This last sum is the sum of the squares of the first eight positive integers, which we can evaluate using Theorem 5.1.1, which was repeated in Example 1 (ii). The formula says that the sum of the squares of the first n positive integers is $\frac{n(n+1)(2n+1)}{6}$. In this example we are interested in the case in which $n = 8$, so the sum is $\frac{(8)(9)(17)}{6} = 204$.

We have just evaluated the original sum:

$$\sum_{m=5}^{12} (m-4)^2 = 204.$$

Rearranging the sum turned it into another sum that we recognized, which made it easy to evaluate. Without this trick, the only way to evaluate would have been to add up all the terms individually. Of course, in this case that would not have been tremendously difficult, but it is easy to imagine more complicated examples in which adding up all the terms would be extremely tedious.

We conclude this section with a proof of the formula for the sum of the squares of the first n positive integers (cf. Example 1 (ii)).

EXAMPLE 5 Prove Theorem 5.1.1:

$$\sum_{k=1}^{n} k^2 = \frac{n(n+1)(2n+1)}{6}.$$

Solution We begin by finding a formula for the sum

$$S = 1 + 2 + 3 + \ldots + (n-1) + n$$

of the first n positive integers. To do this, we consider the sum

$$T = \sum_{k=0}^{n} \left((k+1)^2 - k^2\right).$$

By Proposition 5.2.3 this sum equals

$$\sum_{k=0}^{n} (k+1)^2 - \sum_{k=0}^{n} k^2.$$

It is possible to adjust the second sum by noticing that when $k = 0$, the term being added is $k^2 = 0^2 = 0$, so the sum will not change if we omit that term. In other words, the second sum can be replaced by $\sum_{k=1}^{n} k^2$.

Now we apply Proposition 5.2.4 to the first sum to shift the index. Adding up $(k+1)^2$ as k runs from 0 to n is exactly the same as adding k^2 with k running from 1 to $n+1$, so the first sum equals $\sum_{k=1}^{n+1} k^2$. By writing the last term (with $k = n+1$) separately we can rewrite this as $(n+1)^2 + \sum_{k=1}^{n} k^2$.

Combining, we find that

$$T = \sum_{k=0}^{n} \left((k+1)^2 - k^2\right) = \sum_{k=0}^{n} (k+1)^2 - \sum_{k=0}^{n} k^2 = \sum_{k=1}^{n+1} k^2 - \sum_{k=1}^{n} k^2$$
$$= (n+1)^2 + \sum_{k=1}^{n} k^2 - \sum_{k=1}^{n} k^2 = (n+1)^2.$$

On the other hand, we can rewrite T using Proposition 5.2.3:

$$T = \sum_{k=0}^{n} \left((k+1)^2 - k^2\right) = \sum_{k=0}^{n} (k^2 + 2k + 1 - k^2) = \sum_{k=0}^{n} (2k+1).$$

By Proposition 5.2.2 we see that

$$T = \sum_{k=0}^{n} 2k + \sum_{k=0}^{n} 1 = 2\sum_{k=0}^{n} k + \sum_{k=0}^{n} 1.$$

The last sum amounts to adding 1 to itself $n+1$ times (once for $k=0$, once for $k=1$, etc., up to $k=n$), so it equals $n+1$. The second-last sum is unchanged by omitting the term with $k=0$, so we find that

Karl Friedrich Gauss (1777–1855)

$$T = 2\left(\sum_{k=1}^{n} k\right) + (n+1).$$

On the other hand, we already know that $T = (n+1)^2$, so, writing $S = \sum_{k=1}^{n} k$, we have

$$(n+1)^2 = 2S + (n+1)$$

or

$$(n+1)^2 - (n+1) = 2S$$

or

$$(n+1)(n+1-1) = 2S$$

or

$$(n+1)n = 2S.$$

From this we find that the sum S of the first n positive integers is given by

$$S = \sum_{k=1}^{n} k = \frac{n(n+1)}{2}. \qquad (5.2.5)$$

One of the great mathematicians of all time was Karl Friedrich Gauss (1777–1855). His elementary school teacher asked the class to add up the numbers from 1 to 100, expecting to keep them busy for a long time. Young Gauss found Formula 5.2.5 and instantly wrote down the correct answer.

To prove the formula for the sum of the squares of the first n positive integers, we proceed in a similar way. Consider the sum

$$U = \sum_{k=0}^{n} \left((k+1)^3 - k^3\right) = \sum_{k=0}^{n} (k+1)^3 - \sum_{k=0}^{n} k^3 = \sum_{k=1}^{n+1} k^3 - \sum_{k=1}^{n} k^3$$
$$= (n+1)^3 + \sum_{k=1}^{n} k^3 - \sum_{k=1}^{n} k^3 = (n+1)^3.$$

Here, just as in the previous argument, we broke the sum into two parts, omitted the irrelevant term with $k = 0$ from the second part, shifted the index of the first part, and split off the term with $k = n + 1$.

On the other hand, we can evaluate U another way:

$$\begin{aligned} U &= \sum_{k=0}^{n} \left((k+1)^3 - k^3\right) = \sum_{k=0}^{n} \left(k^3 + 3k^2 + 3k + 1 - k^3\right) \\ &= \sum_{k=0}^{n} (3k^2 + 3k + 1) = 3\sum_{k=0}^{n} k^2 + 3\sum_{k=0}^{n} k + \sum_{k=0}^{n} 1 \\ &= 3\sum_{k=1}^{n} k^2 + 3\sum_{k=1}^{n} k + (n+1) = 3\sum_{k=1}^{n} k^2 + 3\frac{n(n+1)}{2} + (n+1). \end{aligned}$$

Here we used the formula $\sum_{k=1}^{n} k = \frac{n(n+1)}{2}$.

Comparing this with our other evaluation of U, we see that

$$(n+1)^3 = 3\sum_{k=1}^{n} k^2 + 3\frac{n(n+1)}{2} + (n+1),$$

so

$$\begin{aligned} 3\sum_{k=1}^{n} k^2 &= (n+1)^3 - 3\frac{n(n+1)}{2} - (n+1) \\ &= (n+1)\left((n+1)^2 - \frac{3n}{2} - 1\right) \\ &= (n+1)\left(n^2 + 2n + 1 - \frac{3n}{2} - 1\right) \\ &= (n+1)\left(n^2 + \frac{n}{2}\right) \\ &= \frac{1}{2}(n+1)(2n^2 + n). \end{aligned}$$

Dividing by 3, we find

$$\sum_{k=1}^{n} k^2 = \frac{n(n+1)(2n+1)}{6},$$

which is Theorem 5.1.1, as desired. ◢

A similar procedure can be followed to find the sum of the cubes of the first n positive integers, or the fourth powers, etc. The value of using the summation notation in these arguments should be apparent. They would be much more tedious if each sum were written out in full.

Theorem 5.1.1 can also be proved by using the technique known as mathematical induction, which is discussed in the appendix. However, the proof just given does not require a knowledge of induction, and it also has the advantage that it discovers the answer for us. To use induction, it is necessary to know beforehand what the formula is supposed to be. Since the formula was announced in Theorem 5.1.1, we knew what to expect, but without that it would not have been too easy to guess the answer. If you don't believe that, try to guess a formula for the sum of the *cubes* of the first n positive integers. . . .

Exercises 5.2

I

Evaluate each of the following sums.

1. $\sum_{k=1}^{5} k^3$
2. $\sum_{k=2}^{4} k^4$
3. $\sum_{k=0}^{6} (k^2 - 3)$
4. $\sum_{k=-2}^{2} k^5$
5. $\sum_{k=2}^{4} (4k + 1)$
6. $\sum_{\ell=-1}^{4} (\ell^2 - \ell)$
7. $\sum_{h=1}^{6} (h^2 + 5)$
8. $\sum_{h=-6}^{-1} (h^2 + 5)$
9. $\sum_{i=1}^{12} i^2$
10. $\sum_{j=3}^{3} j^4$
11. $\sum_{j=-3}^{3} (2j + 1)$
12. $\sum_{k=0}^{n} k^2 - \sum_{i=1}^{n-1} i^2$
13. $\sum_{j=1}^{m} (j - 1)^2$
14. $\sum_{r=0}^{n} (r + 1)^2$
15. $\sum_{i=1}^{m} (i + 1)^2$
16. $\sum_{r=0}^{n-1} (r + 1)^2 - \sum_{u=2}^{n+1} (u - 1)^2$
17. $\sum_{j=-m}^{m} (j^2 + 2j + 1)$
18. $\sum_{k=1}^{n} (n - k)$
19. $\sum_{i=2}^{9} 1$
20. $\sum_{k=1}^{n} \left(k - \frac{n+1}{2}\right)$

II

21. Suppose m and n are positive integers satisfying $m \leq n$. Find a formula for

$$\sum_{k=m}^{n} k.$$

What is the meaning of your formula when $m = n$?

22. Suppose m and n are positive integers satisfying $m \leq n$. Find a formula for

$$\sum_{k=m}^{n} k^2.$$

What is the meaning of your formula when $m = n$?

*23. Suppose m and n are integers, *not necessarily positive*, satisfying $m \leq n$. Find a formula for

$$\sum_{k=m}^{n} k.$$

What is the meaning of your formula when $m = n$?

*24. Suppose m and n are integers, *not necessarily positive*, satisfying $m \leq n$. Find a formula for

$$\sum_{k=m}^{n} k^2.$$

What is the meaning of your formula when $m = n$?

25. (i) Evaluate $\sum_{i=1}^{5} (-1)^i i$.
(ii) Evaluate $\sum_{i=1}^{8} (-1)^i i$.
(iii) If n is a positive integer, find a formula for $\sum_{i=1}^{n} (-1)^i i$. (*Hint:* You may want to distinguish between the cases in which n is even and odd.)

*26. (i) Evaluate $\sum_{j=1}^{5} (-1)^j j^2$.
(ii) Evaluate $\sum_{j=1}^{8} (-1)^j j^2$.
(iii) If n is a positive integer, find a formula for $\sum_{j=1}^{n} (-1)^j j^2$. (*Hint:* You may want to distinguish between the cases in which n is even and odd.)

*27. Let n be a positive integer and evaluate $\sum_{k=0}^{n} \cos(k\pi)$.

*28. Let n be a positive integer and evaluate $\sum_{k=0}^{n} \cos\left(k\frac{\pi}{2}\right)$.

29. Use summation notation to write the Binomial Theorem more conveniently. In other words, write the expression given by the Binomial Theorem (Theorem 1.3.10) for $(a + b)^n$ as a summation.

*30. Find a formula for the sum of the cubes of the first n positive integers, that is, for

$$\sum_{\ell=1}^{n} \ell^3.$$

*31. Evaluate (i) $\sum_{k=1}^{5}\left(\sum_{j=1}^{3} jk\right)$, (ii) $\sum_{j=1}^{3}\left(\sum_{k=1}^{5} jk\right)$.

*32. Evaluate

$$\sum_{k=1}^{n}\left(\sum_{j=1}^{k} j\right).$$

III

33. Use a personal computer or programmable calculator to evaluate $\sum_{n=1}^{100} n$; compare with the expected answer.

34. Evaluate $\sum_{k=1}^{50} k^2$ and compare with the expected answer.

35. Evaluate $\sum_{m=0}^{45} 2^{-m}$.

36. Evaluate $\sum_{n=0}^{70} n3^{-n}$.

Use the result of Example 3(p. 356) to approximate the area under the graph of each of the following functions, with the specified n.

37. $x^2 - x^3$, $x \in [0, 1]$, $n = 100$
38. $\cos x$, $x \in \left[0, \frac{\pi}{2}\right]$, $n = 40$
39. $\ln x$, $x \in [1, e]$, $n = 50$
40. e^{-x^2}, $x \in [0, 5]$, $n = 50$

5.3 Riemann Sums

PARTITION
SUBINTERVAL
RIEMANN SUM
INTEGRAL
INTEGRABLE
LIMITS OF INTEGRATION
SIGNED AREA

In Section 5.1 we approximated areas with rectangles (see Figure 5.3.1). There we used subintervals of equal width and rectangles whose height was equal to the height of the curve at the left or right endpoint. To develop ways of calculating areas exactly, we need a different kind of approximation. Ironically, this approach is usually more cumbersome for making approximations, but it is better adapted to the theoretical development which leads to exact evaluations.

As before, we consider the area under a nonnegative function $f(x)$ on an interval $[a, b]$. We will divide the interval $[a, b]$ up into a number of **subintervals**, as shown in Figure 5.3.2. If the endpoints of these subintervals (in increasing order) are $a = x_0, x_1, x_2, \ldots, x_n = b$, then the subintervals are $[x_0, x_1], [x_1, x_2], \ldots, [x_{n-1}, x_n]$; there are n subintervals, and the kth one is $[x_{k-1}, x_k]$.

Notice that the subintervals are not all required to have the same length. We shall write $\Delta x_k = x_k - x_{k-1}$ for the length of the kth subinterval.

DEFINITION 5.3.1

If $a = x_0 < x_1 < x_2 < \ldots < x_n = b$, then the set

$$\mathcal{P} = \{x_0, x_1, x_2, \ldots, x_n\}$$

is called a **partition** of the interval $[a, b]$. The length of the largest subinterval,

$$\max_{1\le k\le n} \Delta x_k = \max_{1\le k\le n} (x_k - x_{k-1}),$$

is called the **norm** of $\mathcal{P}$ (or the **mesh** of $\mathcal{P}$), and is denoted $\|\mathcal{P}\|$. (The symbol Δ is the capital Greek letter delta. It corresponds to our letter D, which may remind us that Δx_k is the *difference* $x_k - x_{k-1}$ between adjacent points in the partition.)

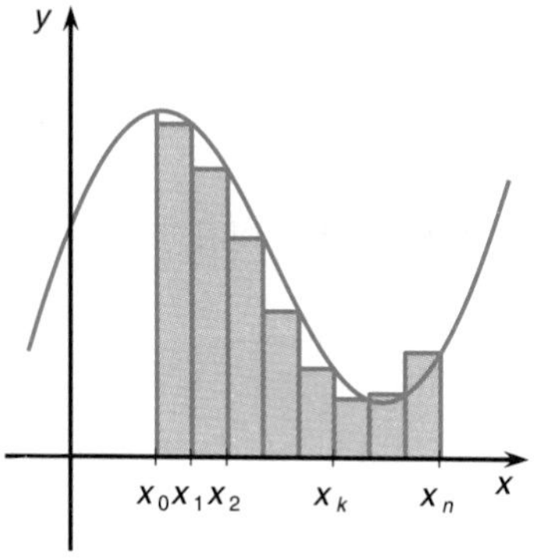

Figure 5.3.1

We shall approximate the area under $y = f(x)$ using partitions $\mathcal{P}$ and then take a sort of limit as $\|\mathcal{P}\| \to 0$, that is, as the subintervals all get smaller and smaller.

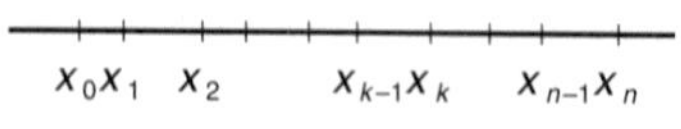

Figure 5.3.2

EXAMPLE 1 In the interval $[0, 1]$, consider the partitions $\mathcal{P}_1 = \{0, \frac{1}{4}, \frac{1}{2}, \frac{3}{4}, 1\}$, $\mathcal{P}_2 = \{0, \frac{1}{3}, \frac{2}{3}, 1\}$. Note that $\|\mathcal{P}_1\| = \frac{1}{4}$ and $\|\mathcal{P}_2\| = \frac{1}{3}$.

Neither of $\mathcal{P}_1$ and $\mathcal{P}_2$ contains the other. One partition that contains both of them is $\mathcal{P}_3 = \{0, \frac{1}{4}, \frac{1}{3}, \frac{1}{2}, \frac{2}{3}, \frac{3}{4}, 1\}$. Another is $\mathcal{P}_4$, which is obtained by taking all fractions between 0 and 1 that can be written with denominator 12, so

$$\begin{aligned}\mathcal{P}_4 &= \left\{0, \frac{1}{12}, \frac{2}{12}, \frac{3}{12}, \frac{4}{12}, \frac{5}{12}, \frac{6}{12}, \frac{7}{12}, \frac{8}{12}, \frac{9}{12}, \frac{10}{12}, \frac{11}{12}, \frac{12}{12} = 1\right\} \\ &= \left\{0, \frac{1}{12}, \frac{1}{6}, \frac{1}{4}, \frac{1}{3}, \frac{5}{12}, \frac{1}{2}, \frac{7}{12}, \frac{2}{3}, \frac{3}{4}, \frac{5}{6}, \frac{11}{12}, 1\right\}.\end{aligned}$$

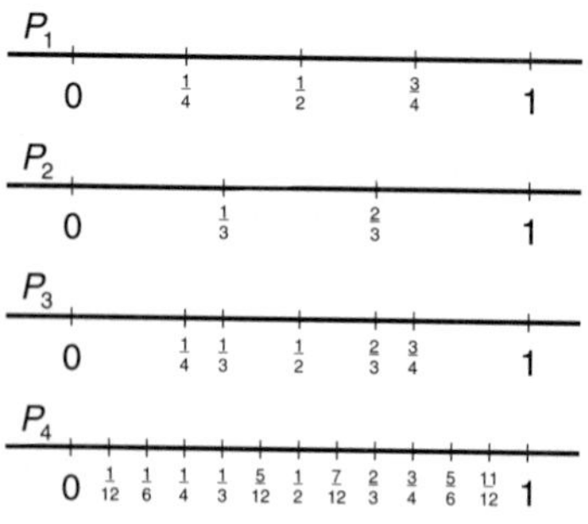

Figure 5.3.3

Each of the partitions $\mathcal{P}_1, \mathcal{P}_2$, and $\mathcal{P}_4$ has the property that the subintervals it determines are all of equal length; $\mathcal{P}_3$ does not have this property. See Figure 5.3.3.

Suppose we are given a partition $\mathcal{P} = \{x_0, \ldots, x_n\}$ of $[a, b]$, and for each $k = 1, 2, \ldots, n$ a point $x_k^* \in [x_{k-1}, x_k]$. There are of course many ways of choosing the points $\{x_k^*\}$.

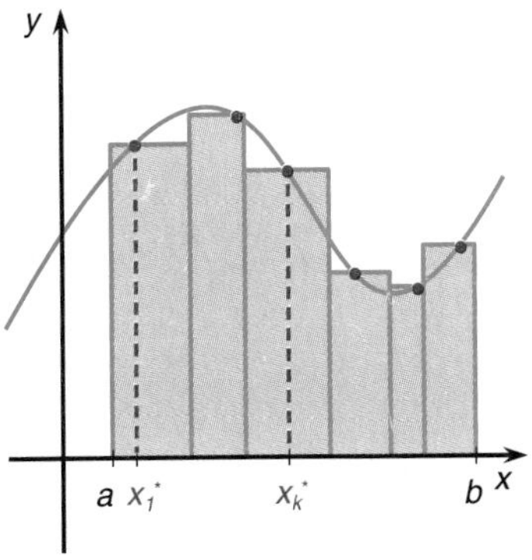

Figure 5.3.4

If $f(x)$ is a continuous nonnegative function on $[a, b]$, we can evaluate $f(x_k^*)$ for each k; it is the height of the graph above the x-axis at the point x_k^*. We could try approximating the area under the graph of $y = f(x)$ by adding up the areas of the rectangles whose heights are $f(x_k^*)$ and whose bases are the intervals $[x_{k-1}, x_k]$. See Figure 5.3.4.

Different choices of $\{x_k^*\}$ will result in different sums, but after all they are only approximations.

> **DEFINITION 5.3.2**
>
> The **Riemann sum** for $f(x)$ associated to the partition
>
> $\mathcal{P} = \{x_0, \ldots, x_n\}$ of $[a, b]$
>
> and to the choices $x_k^* \in [x_{k-1}, x_k]$ is
>
> $$S(f, \mathcal{P}, \{x_1^*, x_2^*, \ldots, x_n^*\}) = \sum_{k=1}^{n} f(x_k^*)\Delta x_k = \sum_{k=1}^{n} f(x_k^*)(x_k - x_{k-1}).$$

Georg F. B. Riemann (1826–1866)

If $f(x) \geq 0$ for all $x \in [a, b]$, this sum is the total area of the rectangles with bases $[x_{k-1}, x_k]$ and heights $f(x_k^*)$ as illustrated in Figure 5.3.4. But the sum makes sense even if $f(x)$ is negative at some values of x. If $f(x_k^*) < 0$, then $f(x_k^*)\Delta x_k$ is the *negative* of the area of the rectangle. In fact this is what we want; we shall call it the **signed area** of the rectangle, meaning the negative of the area if the rectangle is below the x-axis and the actual area if it is above the x-axis. In Figure 5.3.5 the rectangles with positive signed area are shaded blue; those with negative signed area are shaded purple.

The Riemann sum $S(f, \mathcal{P}, \{x_1^*, \ldots, x_n^*\})$ is the sum of the signed areas of the rectangles with bases $[x_{k-1}, x_k]$ and "heights" $f(x_k^*)$.

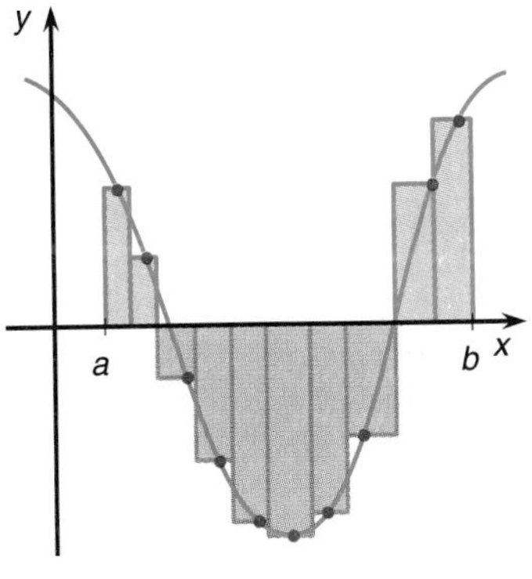

Figure 5.3.5

> Note that $x_0 = 1$, $x_1 = 2$, $x_2 = 3$, $x_3 = 4$. Also $\Delta x_1 = x_1 - x_0 = 2 - 1$, $\Delta x_2 = x_2 - x_1 = 3 - 2$, etc.

EXAMPLE 2

(i) With $f(x) = 4x^2 - 1$ on $[1, 4]$ and the partition $\mathcal{P} = \{1, 2, 3, 4\}$ we choose the midpoints $x_1^* = \frac{3}{2} \in [1, 2]$, $x_2^* = \frac{5}{2} \in [2, 3]$, $x_3^* = \frac{7}{2} \in [3, 4]$, and the Riemann sum is

$$\begin{aligned} S(f, \mathcal{P}, \{x_k^*\}) &= \sum_{k=1}^{3} f(x_k^*)\Delta x_k \\ &= f\left(\frac{3}{2}\right)(2-1) + f\left(\frac{5}{2}\right)(3-2) + f\left(\frac{7}{2}\right)(4-3) \\ &= \left(4\left(\frac{3}{2}\right)^2 - 1\right) + \left(4\left(\frac{5}{2}\right)^2 - 1\right) + \left(4\left(\frac{7}{2}\right)^2 - 1\right) \\ &= \left(4\left(\frac{9}{4}\right) - 1\right) + \left(4\left(\frac{25}{4}\right) - 1\right) + \left(4\left(\frac{49}{4}\right) - 1\right) \\ &= 8 + 24 + 48 = 80. \end{aligned}$$

(ii) With $f(x) = \sin x$ on $[0, \pi]$, and the partition $\mathcal{P} = \left\{0, \frac{\pi}{4}, \frac{\pi}{2}, \frac{3\pi}{4}, \pi\right\}$, we choose the left endpoints of the subintervals

$$x_1^* = 0, \qquad x_2^* = \frac{\pi}{4}, \qquad x_3^* = \frac{\pi}{2}, \qquad x_4^* = \frac{3\pi}{4}.$$

The corresponding Riemann sum is

$$\begin{aligned}
S(f,\mathcal{P},\{x_k^*\}) &= \sum_{k=1}^{4} f(x_k^*)\Delta x_k \\
&= f(0)\left(\frac{\pi}{4}-0\right)+f\left(\frac{\pi}{4}\right)\left(\frac{\pi}{2}-\frac{\pi}{4}\right)+f\left(\frac{\pi}{2}\right)\left(\frac{3\pi}{4}-\frac{\pi}{2}\right) \\
&\quad +f\left(\frac{3\pi}{4}\right)\left(\pi-\frac{3\pi}{4}\right) \\
&= \sin(0)\left(\frac{\pi}{4}\right)+\sin\left(\frac{\pi}{4}\right)\left(\frac{\pi}{4}\right)+\sin\left(\frac{\pi}{2}\right)\left(\frac{\pi}{4}\right) \\
&\quad +\sin\left(\frac{3\pi}{4}\right)\left(\frac{\pi}{4}\right) \\
&= 0+\frac{1}{\sqrt{2}}\left(\frac{\pi}{4}\right)+(1)\left(\frac{\pi}{4}\right)+\frac{1}{\sqrt{2}}\left(\frac{\pi}{4}\right)=(1+\sqrt{2})\frac{\pi}{4}.
\end{aligned}$$

(iii) Suppose $0 \le a < b$ and let $\mathcal{P}_n$ be the partition that divides the interval $[a, b]$ into n equal subintervals. Let $f(x) = x^m$, with m a fixed positive integer.

If $x_k^* = x_{k-1}$ is the left endpoint of the kth subinterval $[x_{k-1}, x_k]$, let us calculate

$$S(f,\mathcal{P}_n,\{x_1^*,\ldots,x_n^*\}).$$

> The interval $[a, b]$ has length $b - a$, so each little subinterval has length $\frac{b-a}{n}$. The point $x_k = a + k\frac{b-a}{n}$ is k subintervals to the right of $x = a$.

Since x_k is the point that is k subintervals to the right of a, we see that $x_k = a + k\frac{b-a}{n}$. So $\mathcal{P}_n = \left\{a, a+\frac{b-a}{n}, a+2\frac{b-a}{n}, a+3\frac{b-a}{n}, \ldots, a+(n-1)\frac{b-a}{n}, b\right\}$. Since $\Delta x_k = \frac{b-a}{n}$, we find that

$$\begin{aligned}
S(f,\mathcal{P}_n,\{x_1^*,\ldots,x_n^*\}) &= \sum_{k=1}^{n} f(x_k^*)\Delta x_k = \sum_{k=1}^{n} f(x_{k-1})\frac{b-a}{n} \\
&= \frac{b-a}{n}\sum_{k=1}^{n} f\left(a+(k-1)\frac{b-a}{n}\right) \\
&= \frac{b-a}{n}\sum_{k=1}^{n}\left(a+(k-1)\frac{b-a}{n}\right)^m.
\end{aligned}$$

(iv) With $f(x) = x^m$ and $\mathcal{P}_n$ as in part (iii), we calculate the Riemann sum associated to the choice where $x_k^{**} = x_k$ is the right endpoint of the interval $[x_{k-1}, x_k]$.

Here $x_k^{**} = x_k = a + k\frac{b-a}{n}$, so

$$S(f,\mathcal{P}_n,\{x_1^{**},\ldots,x_n^{**}\}) = \sum_{k=1}^{n} f(x_k^{**})\Delta x_k = \frac{b-a}{n}\sum_{k=1}^{n}\left(a+k\frac{b-a}{n}\right)^m.$$

We expect that the sums in parts (iii) and (iv) of Example 2 should approximate the area under the graph of $y = f(x) = x^m$ between $x = a$ and $x = b$. Consider their difference,

$$\begin{aligned}
&S(f,\mathcal{P}_n,\{x_1^{**},\ldots,x_n^{**}\}) - S(f,\mathcal{P}_n,\{x_1^*,\ldots,x_n^*\}) \\
&= \frac{b-a}{n}\sum_{k=1}^{n}\left(a+k\frac{b-a}{n}\right)^m - \frac{b-a}{n}\sum_{k=1}^{n}\left(a+(k-1)\frac{b-a}{n}\right)^m
\end{aligned}$$

A programmable calculator or a computer can be programmed to evaluate the sum in part (iv) of Example 2 for specific values of a and b. For example, we try it with $a = 0$ and $b = 2$.

TI-85

PRGM F2 (EDIT) RSUM ENTER F3 F1 (Input) M ENTER M3 F1 (Input) N EXIT ENTER 0 STO K ENTER 0 STO S ENTER Lbl B ENTER K+1 STO K ENTER S+(K × 2 ÷ N)^M STO S ENTER F4 F1 (If) (K TEST F2(<) N) ENTER Goto B ENTER S × 2 ÷ N ENTER EXIT EXIT EXIT PRGM F1 (NAME) RSUM ENTER 3 ENTER 10 ENTER

SHARP EL9300

▼ ▼ (NEW) ENTER ENTER RSUM ENTER COMMAND A 3 (Input) *M* ENTER COMMAND A 3 (Input) *N* ENTER *K* = 0 ENTER *S* = 0 ENTER COMMAND B 1 (Label) 1 ENTER *K* = *K* + 1 ENTER *S* = *S* + (*K* × 2 ÷ *N*) a^b *M* ENTER COMMAND 3 (If) *K* COMMAND C 2 (<) *N* COMMAND B 2 (Goto) 1 ENTER *S* = *S* × 2 ÷ *N* ENTER COMMAND A 1 (Print) *S* ENTER COMMAND A 6 (End) ENTER (RUN) ENTER (RSUM) ENTER 3 ENTER 10 ENTER

CASIO f_x-6300G (f_x-7700GB)

MODE 2 EXE ? → M : ? → N : 0 → K : 0 → S : Lbl 1 : K + 1 → K : S + (K × 2 ÷ N) x^y M → S : K < N ⇒ Goto 1 : S × 2 ÷ N ◢ MODE 1 Prog 0 EXE 3 EXE 10 EXE (on the f_x-7700GB, : is PRGM F6, and Prog is PRGM F3.)

HP 48SX

ATTN ↱ CLR ≪ ≫ "" KEYIN M AND N ▶ SPC PROMPT SPC ′ *N* ▶ STO ′ *M* ▶ STO 0 SPC 1 SPC *N* SPC FOR *K* SPC *K* SPC 2 × *N* ÷ *M* y^x + NEXT ▶ 2 × *N* ÷ ▶ ENTER ′ RSUM STO VAR RSUM 3 ENTER 10 ENTER ↰ CONT

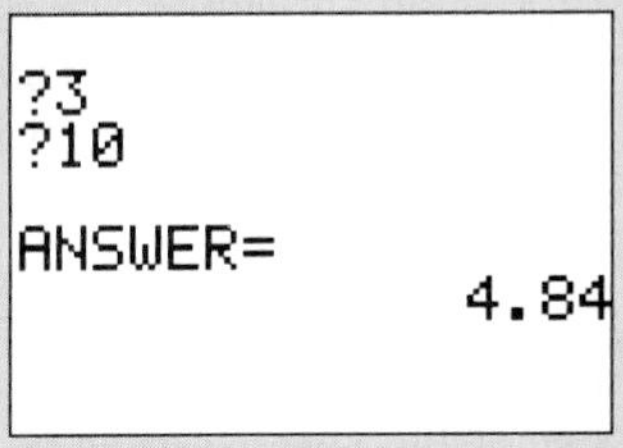

Now use the program with various other values of m and n. Can you guess a formula for $\int_0^2 x^m dx$?

$$= \frac{b-a}{n}\left(\sum_{k=1}^{n}\left(a + k\frac{b-a}{n}\right)^m - \sum_{k=1}^{n}\left(a + (k-1)\frac{b-a}{n}\right)^m\right).$$

Most of the terms in these last two sums cancel each other. We can see this by changing the second sum, using Proposition 5.2.4 (p. 357). Letting $\ell = k - 1$, we have

$$\sum_{k=1}^{n}\left(a + (k-1)\frac{b-a}{n}\right)^m = \sum_{\ell=0}^{n-1}\left(a + \ell\frac{b-a}{n}\right)^m.$$

With this in mind we can write the previous difference as

$$S\big(f, \mathcal{P}_n, \{x_1^{**}, \ldots, x_n^{**}\}\big) - S\big(f, \mathcal{P}_n, \{x_1^{*}, \ldots, x_n^{*}\}\big)$$

$$= \frac{b-a}{n}\left(\sum_{k=1}^{n}\left(a+k\frac{b-a}{n}\right)^m - \sum_{\ell=0}^{n-1}\left(a+\ell\frac{b-a}{n}\right)^m\right)$$

$$= \frac{b-a}{n}\left(\sum_{k=1}^{n-1}\left(a+k\frac{b-a}{n}\right)^m + \left(a+n\frac{b-a}{n}\right)^m\right.$$

$$\left. - \sum_{\ell=1}^{n-1}\left(a+\ell\frac{b-a}{n}\right)^m - \left(a+0\times\frac{b-a}{n}\right)^m\right).$$

We split off the term with $k = n$ from the first sum and the term with $\ell = 0$ from the second.

Since the two summations in this last formula are the same, they cancel, and we are left with

$$S(f, \mathcal{P}_n, \{x_1^{**}, \ldots, x_n^{**}\}) - S(f, \mathcal{P}_n, \{x_1^*, \ldots, x_n^*\})$$

$$= \frac{b-a}{n}\left(\left(a+n\frac{b-a}{n}\right)^m - \left(a+0\frac{b-a}{n}\right)^m\right)$$

$$= \frac{b-a}{n}\left((a+b-a)^m - a^m\right)$$

$$= \frac{b-a}{n}(b^m - a^m).$$

In particular, if we let $n \to \infty$, this difference tends to zero, which says that the two different types of sums (left endpoints or right endpoints) get closer and closer together. This is what we expect, since we are hoping that each type of sum will tend to some limit, and both limits will be the same.

We want to say that the value of this common limit is the area under the graph, but the difficulty is that we have considered only two possible choices, $x_k^* = x_{k-1}$ and $x_k^{**} = x_k$ (left and right endpoints of each subinterval). For that matter, we have considered only the special partitions $\mathcal{P}_n$ that divide $[a, b]$ into equal subintervals. Different choices of partition or of points in each subinterval could conceivably lead to different limits (or to no limit at all).

What is needed is a way to consider every partition $\mathcal{P}$ and all possible choices $\{x_k^*\}$ of points in the intervals determined by $\mathcal{P}$ and to take a limit as the partitions become finer and finer, that is, as $\|\mathcal{P}\| \to 0$. This sort of limit is different from the ones we considered earlier. Roughly speaking, what it means is that for any partition composed of very small subintervals and any choice of points $\{x_k^*\}$ in those subintervals, the Riemann sum should be very close to the value of the limit.

As we did with ordinary limits in Chapter 2, we will not stop to discuss the technical details of this type of limit. Instead, we will proceed to see how it is used and present the precise definition at the end of this section.

DEFINITION 5.3.3

Suppose $f(x)$ is a function on the interval $[a, b]$. Consider all possible partitions $\mathcal{P}$ of $[a, b]$, and for each $\mathcal{P}$ consider all possible choices $\{x_k^*\}$ of a point in each subinterval determined by $\mathcal{P}$. For such a partition $\mathcal{P}$ and choices $\{x_k^*\}$ we form the Riemann sum

$$S(f, \mathcal{P}, \{x_k^*\}).$$

The function $f(x)$ is said to be **integrable** on $[a, b]$ if the following limit exists:

$$\lim_{\|\mathcal{P}\|\to 0} S(f, \mathcal{P}, \{x_k^*\}).$$

In this case the value of the limit is called the **integral** of $f(x)$ from a to b and is written

$$\int_a^b f(x)\,dx.$$

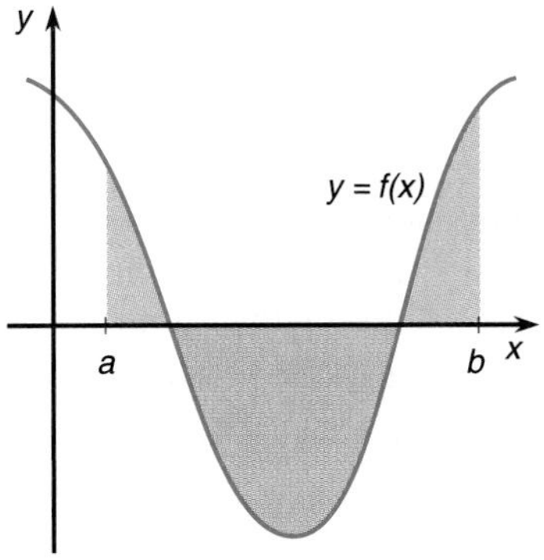

Figure 5.3.6

The way we have defined the integral leads us to expect that $\int_a^b f(x)\,dx$ represents the signed area under the graph of $y = f(x)$ between $x = a$ and $x = b$. See Figure 5.3.6.

In this notation the number a is called the **lower limit of integration** and b is the **upper limit of integration**. They indicate the interval over which the integral is considered, and the word "limit" here means "endpoint" rather than limit in the sense of the previous definition or in the sense in which it was used in Chapter 2.

The integral sign $\int$ is a stylized capital S; the idea is that as the Riemann *sums* $\sum_{k=1}^{n} f(x_k^*)\Delta x_k$ tend to a limit, the stylized Greek letter $\sum$ (sigma) changes into the corresponding Roman letter S. Similarly, the notation Δx_k changes into dx. The function $f(x)$ is called the **integrand**.

EXAMPLE 3 Suppose $f(x) = c$ is a constant function. Show that $f(x)$ is integrable on any interval $[a, b]$, and find $\int_a^b f(x)\,dx$.

Solution If $\mathcal{P} = \{x_0, x_1, \ldots, x_n\}$ is any partition and $x_k^* \in [x_{k-1}, x_k]$, then the corresponding Riemann sum is

What makes this easy is that $f(x_k^*) = c$ for *any* point x_k^* . . .

$$\sum_{k=1}^{n} f(x_k^*)\Delta x_k = \sum_{k=1}^{n} c\Delta x_k = c\sum_{k=1}^{n} \Delta x_k.$$

Since $\sum_{k=1}^{n} \Delta x_k$ is the sum of the lengths of all the subintervals, it equals the total length of the interval $[a, b]$, or $b - a$.

This shows that every Riemann sum equals $c(b - a)$, so they certainly "tend" to this limit. In other words, $f(x) = c$ is integrable on $[a, b]$, and

$$\int_a^b c\,dx = c(b - a).$$

This is what we expect from our interpretation of the integral as a signed area. (See Figure 5.3.7, which illustrates what happens when $c > 0$ and when $c < 0$. In each case the area of the shaded region is $|c|(b - a)$, and the signed area is $c(b - a)$.)

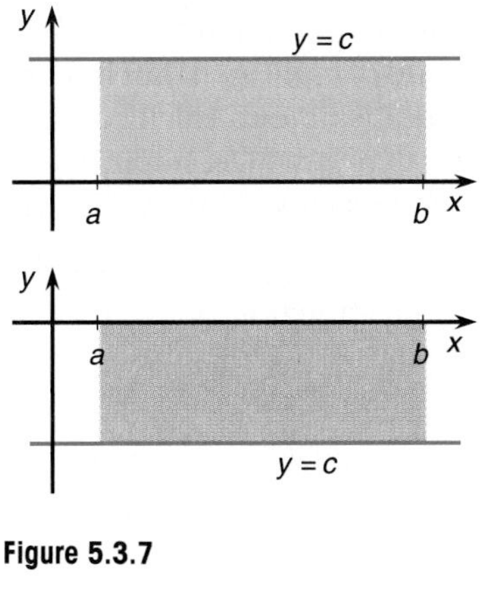

Figure 5.3.7

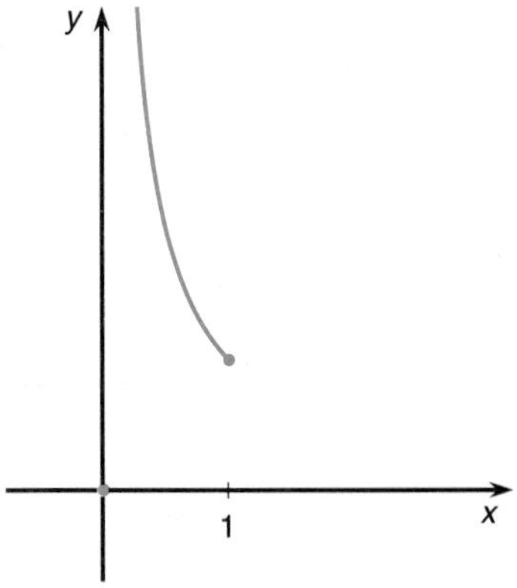

Figure 5.3.8

On the other hand, there are functions that are not integrable.

EXAMPLE 4 Define the function $g(x)$ on the interval $[0, 1]$ by

$$g(x) = \begin{cases} \frac{1}{x}, & \text{if } x > 0, \\ 0, & \text{if } x = 0 \end{cases}$$

(see Figure 5.3.8).

We will show that $g(x)$ is not integrable on $[0, 1]$. For each positive integer n, let $\mathcal{P}_n$ be the partition that divides $[0, 1]$ into n equal subintervals. The first subinterval determined by $\mathcal{P}_n$ is $\left[0, \frac{1}{n}\right]$. Define $\{x_k^*\}$, a choice of points in the subintervals determined by $\mathcal{P}_n$, by $x_1^* = \frac{1}{n^2}$, and for each $k > 1$, $x_k^* = \frac{k}{n}$. For each $k > 1$, x_k^* is the right endpoint of the kth subinterval. Notice that the point $x_1^* = \frac{1}{n^2}$ is in the first subinterval $\left[0, \frac{1}{n}\right]$.

The corresponding Riemann sum is

$$\begin{aligned} S\big(g, \mathcal{P}_n, \{x_k^*\}\big) &= \sum_{k=1}^{n} g(x_k^*)\Delta x_k = g(x_1^*)\Delta x_1 + \sum_{k=2}^{n} g\left(\frac{k}{n}\right)\Delta x_k \\ &= \frac{1}{\frac{1}{n^2}}\frac{1}{n} + \sum_{k=2}^{n} \frac{n}{k}\frac{1}{n} \\ &= n + \sum_{k=2}^{n} \frac{1}{k}. \end{aligned}$$

Without actually evaluating it, we see that the final sum must be positive. This shows that the Riemann sum is greater than n. Now if $g(x)$ were integrable, we would know that as $n \to \infty$, $\|\mathcal{P}_n\| = \frac{1}{n} \to 0$, so the Riemann sums $S\big(g, \mathcal{P}_n, \{x_k^*\}\big)$ would have to approach a limit (the value of the integral). But since the nth sum is bigger than n, they cannot approach any (finite) limit. This shows that the function $g(x)$ is not integrable on $[0, 1]$.

The essential fact about $g(x)$ that prevented it from being integrable is that it is not bounded on $[0, 1]$. In fact it is also possible to find bounded functions that are not integrable.

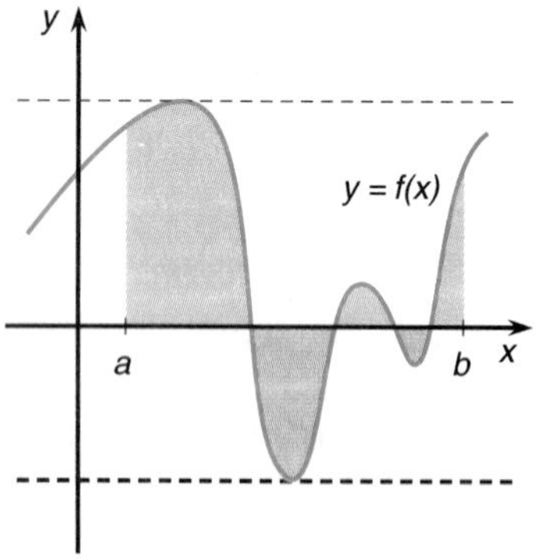

Figure 5.3.9

Example 4 shows that there are functions that are not integrable. The important thing is that most "reasonable" functions are in fact integrable, which means that it makes sense to speak of the "area" under the graph of the function.

In particular, if $f(x)$ is continuous, then by Theorem 2.5.6 it is *bounded*, and our intuition suggests that the graph encloses a geometrical region (see Figure 5.3.9). The region may be quite complicated, but it seems reasonable to expect that its area is something we should be able to calculate. Calculating the (signed) area can be overwhelmingly difficult in practice, but it is possible in principle, in the sense that a continuous function on a closed interval is always integrable.

THEOREM 5.3.4 If $f(x)$ is a continuous function on a closed interval $[a, b]$, then $f(x)$ is integrable on $[a, b]$.

This theorem is of great importance, but it is extremely difficult to prove. Notice that it does not tell us anything about how to calculate the value of the integral

$$\int_a^b f(x)\,dx.$$

Nonetheless, it does give us enough information that we can do the calculations for a few special cases.

EXAMPLE 5 Suppose $a < b$.

(i) Find $\int_a^b x\,dx$, that is, find the integral from a to b of the function $g(x) = x$.

(ii) Find $\int_a^b x^2 dx$, that is, find the integral from a to b of the function $f(x) = x^2$.

Solution

(i) Theorem 5.3.4 tells us that $g(x) = x$ is integrable over any interval $[a, b]$. What this means is that Riemann sums $S(g, \mathcal{P}, \{x_k^*\})$ tend to a limit provided that $\|\mathcal{P}\| \to 0$, and the limit equals $\int_a^b x\,dx$.

So to calculate the integral, we have only to find this limit for some convenient partitions and choices of points $\{x_k^*\}$. In other words, knowing that the Riemann sums converge to the same limit, no matter what the partitions, we do not have to find the limit for every possible partition. It will be enough to find it for some intelligently chosen partitions.

Not surprisingly, we choose the partitions that divide $[a, b]$ into subintervals of equal length. Let $\mathcal{P}_n$ divide $[a, b]$ into n equal subintervals. For the points in the subintervals we choose the right endpoints, that is,

$$x_k^* = a + k\frac{b-a}{n}.$$

The corresponding Riemann sum is

$$\begin{aligned} S(g, \mathcal{P}_n, \{x_k^*\}) &= \sum_{k=1}^{n} g(x_k^*)\Delta x_k = \sum_{k=1}^{n}\left(a + k\frac{b-a}{n}\right)\frac{b-a}{n} \\ &= \frac{b-a}{n}\sum_{k=1}^{n}\left(a + k\frac{b-a}{n}\right) = \frac{b-a}{n}\left(a\sum_{k=1}^{n}1 + \frac{b-a}{n}\sum_{k=1}^{n}k\right) \\ &= \frac{b-a}{n}\left(an + \frac{b-a}{n}\frac{n(n+1)}{2}\right) \\ &= (b-a)a + (b-a)^2\frac{n+1}{2n}. \end{aligned}$$

Note: $\sum_{k=1}^{n} 1 = n$ and $\sum_{k=1}^{n} k = \frac{n(n+1)}{2}$.

Now we let $n \to \infty$ in this last formula. Since $\frac{n+1}{n} \to 1$, we see that the Riemann sums tend to the limit

$$\begin{aligned} (b-a)a + (b-a)^2\frac{1}{2} &= ba - a^2 + \frac{1}{2}(b^2 - 2ab + a^2) \\ &= \frac{1}{2}(b^2 - a^2). \end{aligned}$$

What we have just shown is that

$$\int_a^b x\,dx = \frac{1}{2}(b^2 - a^2).$$

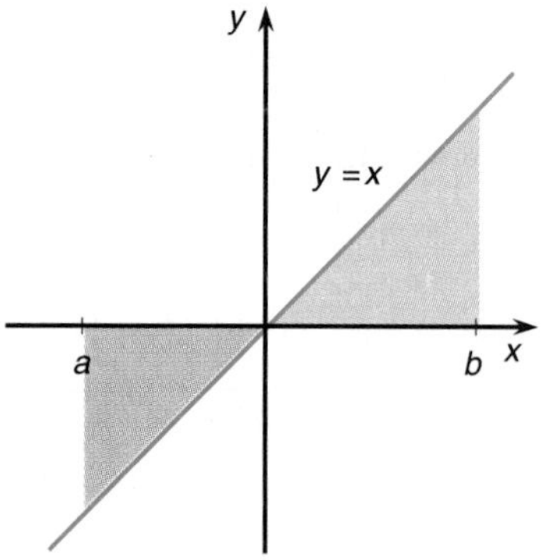

Figure 5.3.10

$\sum_{k=1}^{n} 1$ is obtained by adding n 1's, so it equals n. We have also used Formula 5.2.5 and the result of Example 5.2.5.

You might like to compare this with the signed area under the graph of $y = x$ between $x = a$ and $x = b$ (see Figure 5.3.10).

(ii) Following a similar procedure, using the same partition $\mathcal{P}_n$ and the same points $\{x_k^*\}$, we calculate the Riemann sum

$$\begin{aligned}
S(f, \mathcal{P}_n, \{x_k^*\}) &= \sum_{k=1}^{n} f(x_k^*)\Delta x_k = \sum_{k=1}^{n}\left(a + k\frac{b-a}{n}\right)^2 \frac{b-a}{n} \\
&= \frac{b-a}{n}\sum_{k=1}^{n}\left(a^2 + 2ak\frac{b-a}{n} + k^2\left(\frac{b-a}{n}\right)^2\right) \\
&= \frac{b-a}{n}\left(a^2\sum_{k=1}^{n} 1 + 2a\frac{b-a}{n}\sum_{k=1}^{n} k + \left(\frac{b-a}{n}\right)^2\sum_{k=1}^{n} k^2\right) \\
&= \frac{b-a}{n}\left(a^2 n + 2a\frac{b-a}{n}\frac{n(n+1)}{2} \right.\\
&\qquad \left. + \left(\frac{b-a}{n}\right)^2 \frac{n(n+1)(2n+1)}{6}\right) \\
&= (b-a)a^2 + a(b-a)^2\frac{n+1}{n} \\
&\qquad + \frac{(b-a)^3}{6}\frac{(n+1)(2n+1)}{n^2}.
\end{aligned}$$

Now we let $n \to \infty$ in this last formula. Since $\frac{n+1}{n} \to 1$ and $\frac{(n+1)(2n+1)}{n^2} = \frac{n+1}{n}\frac{2n+1}{n} \to (1)(2) = 2$, we see that the Riemann sums tend to the limit

$$\begin{aligned}
&(b-a)a^2 + a(b-a)^2 + \frac{1}{3}(b-a)^3 \\
&\quad = a^2b - a^3 + ab^2 - 2a^2b + a^3 + \frac{1}{3}(b^3 - 3ab^2 + 3a^2b - a^3) \\
&\quad = \frac{1}{3}(b^3 - a^3).
\end{aligned}$$

What we have just shown is that

$$\int_a^b x^2\,dx = \frac{1}{3}(b^3 - a^3).$$

See Figure 5.3.11.

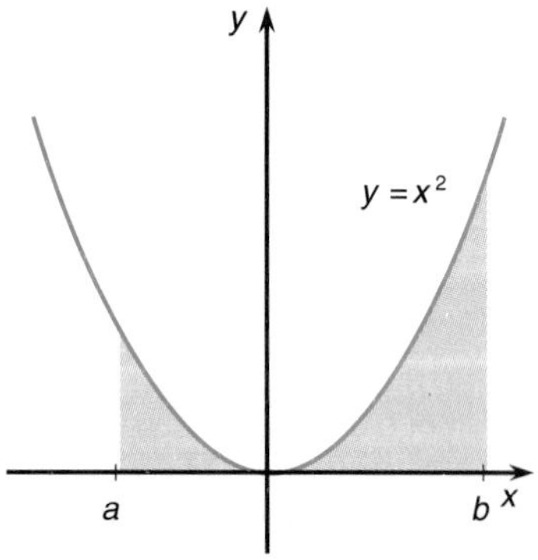

Figure 5.3.11

Unfortunately, this technique is too cumbersome to be very practical. We will have to find better ways to calculate integrals if they are to be of much use to us.

The Precise Definition of the Integral

Definition 5.3.3 of the integral of $f(x)$ over $[a, b]$ used a special kind of limit,

$$\lim_{\|\mathcal{P}\|\to 0} S(f, \mathcal{P}, \{x_k^*\}).$$

The idea was to consider all possible partitions $\mathcal{P}$ of $[a, b]$ and all possible choices $\{x_k^*\}$ of a point in each of the corresponding subintervals and to take a limit as the mesh $\|\mathcal{P}\|$ of the partitions tends to zero.

This seems a reasonable thing to do, but it is certainly different from the limits we considered previously, and we never specified precisely what it means.

DEFINITION 5.3.5

If $f(x)$ is a function on the interval $[a, b]$, then to say

$$\lim_{\|\mathcal{P}\|\to 0} S(f, \mathcal{P}, \{x_k^*\}) = L$$

means the following:

Given any $\epsilon > 0$, it is possible to find a $\delta > 0$ so that for every partition $\mathcal{P}$ of $[a, b]$ with $\|\mathcal{P}\| < \delta$ and for every choice of $x_k^* \in [x_{k-1}, x_k]$, the kth subinterval corresponding to $\mathcal{P}$, we have

$$|S(f, \mathcal{P}, \{x_k^*\}) - L| < \epsilon.$$

In this case, $f(k)$ is integrable on $[a, b]$ and $\int_a^b f(x)\,dx = L$.

The idea is that no matter how close we want the Riemann sum to be to the limit L, as measured by ϵ, it is possible to find a measure δ of the mesh of partitions so that any Riemann sum corresponding to a partition whose mesh is smaller than δ must be close enough to L. In other words, it is possible to ensure that Riemann sums are as close as we want to L by requiring that the mesh of the partition is sufficiently small.

There is a close analogy between this definition and the precise definition of ordinary limits (Definition 2.7.1, p. 120). But as with ordinary limits, the precise definition is not needed for us to be able to work with integrals.

One reason for this, as we saw in Example 5, is that once we know the limit exists, it is not necessary to consider all partitions and choices $\{x_k^*\}$ to calculate it. And for continuous functions we already know that the limit exists, by Theorem 5.3.4.

In fact, we will soon learn techniques for calculating integrals that do not require Riemann sums at all.

Exercises 5.3

I

In each of the following situations, let $\mathcal{P}_n$ be the partition that divides the specified interval into n subintervals of equal length, where n is the given number. Let x_k^* be the left endpoint of the kth subinterval determined by $\mathcal{P}_n$, let $f(x)$ be the given function, and calculate the Riemann sum $S(f, \mathcal{P}_n, \{x_k^*\})$.

1. $f(x) = x$ on $[0, 1]$; $n = 4$

2. $f(x) = x$ on $[-1, 0]$; $n = 3$

3. $f(x) = x^2$ on $[0, 1]$; $n = 4$

4. $f(x) = x$ on $[-1, 2]$; $n = 3$

5. $f(x) = x + 3$ on $[0, 4]$; $n = 4$

6. $f(x) = 2x$ on $[-1, 2]$; $n = 3$

7. $f(x) = x^3 - 2$ on $[-2, 3]$; $n = 5$

8. $f(x) = 7$ on $[-5, 9]$; $n = 13$

9. $f(x) = -x$ on $[0, 1]$; $n = 4$

10. $f(x) = \frac{1}{x}$ on $[1, 6]$; $n = 5$

In each of the following questions, let $\mathcal{P}_n$ be the partition that divides the specified interval into n subintervals of equal length, where n is the given number. Let x_k^{**} be the right endpoint of the kth subinterval determined by $\mathcal{P}_n$, let $f(x)$ be the given function, and calculate the Riemann sum $S(f, \mathcal{P}_n, \{x_k^{**}\})$.

11. $f(x) = \sin x$ on $[0, \pi]$; $n = 2$

12. $f(x) = \cos(\pi x)$ on $[-1, 1]$; $n = 4$

13. $f(x) = \frac{1}{1+x^2}$ on $[0, 3]$; $n = 3$

14. $f(x) = \frac{1}{1+x^2}$ on $[-3, 0]$; $n = 3$

15. $f(x) = 0$ on $[-3, 7]$; $n = 5$

16. $f(x) = x^2 - 3x + 1$ on $[-2, 2]$; $n = 4$

17. $f(x) = x + \sin x$ on $[-\pi, \pi]$; $n = 4$

18. $f(x) = \frac{1}{x}$ on $[-2, 2]$; $n = 3$

19. $f(x) = e^x$ on $[-1, 2]$; $n = 3$

20. $f(x) = \sin^2 x$ on $[-\pi, \pi]$; $n = 2$

II

21. Use Theorem 5.3.4 (p. 368) to show that the function $g(x) = \frac{1}{x}$ is integrable on $\left[\frac{1}{2}, 1\right]$.

*__22.__ Show that any rational function $R(x) = \frac{p(x)}{q(x)}$ is integrable on any interval $[a, b]$ on which $q(x)$ is never zero.

23. Draw a picture and use your knowledge of geometry (triangles and rectangles) to calculate the signed area between the straight line $y = x$ and the x-axis between $x = a$ and $x = b$. (Distinguish three cases: $0 \le a \le b$, $a \le 0 \le b$, and $a \le b \le 0$.) Compare your answer with the result of Example 5 (p. 369).

*__24.__ Let $f(x) = cx + d$ be the equation of a straight line; assume $c \neq 0$. Let $a < b$, and consider the interval $[a, b]$.

(i) If $\mathcal{P}_n$ is the partition that divides $[a, b]$ into n subintervals of equal length, let $x_k^* = a + \frac{k-1}{n}(b - a)$ be the left endpoint of the kth subinterval and x_k^{**} be the right endpoint. Calculate the Riemann sums corresponding to these two choices, namely,

$$S\big(f, \mathcal{P}_n, \{x_k^*\}\big) \quad \text{and} \quad S\big(f, \mathcal{P}_n, \{x_k^{**}\}\big).$$

(ii) Find the limit as $n \to \infty$ of each of these two Riemann sums.

(iii) Suppose $\{x_k^{***}\}$ is *any* choice of points in the subintervals determined by $\mathcal{P}_n$. Show that the corresponding Riemann sum lies between

$$S\big(f, \mathcal{P}_n, \{x_k^*\}\big) \quad \text{and} \quad S\big(f, \mathcal{P}_n, \{x_k^{**}\}\big).$$

(*Hint:* First suppose $c > 0$, and notice that this means that $f(x)$ is an increasing function. Draw a picture.)

(iv) Show that no matter what points $\{x_k^{***}\}$ we choose in the subintervals determined by $\mathcal{P}_n$, the Riemann sums

$$S\big(f, \mathcal{P}_n, \{x_k^{***}\}\big)$$

converge to a limit. Calculate the limit.

(v) Notice that to show that $f(x)$ is integrable on $[a, b]$, we would have to show that the Riemann sums associated to *every* possible partition converge to a limit, not just the ones associated to the special partitions $\mathcal{P}_n$. However, since Theorem 5.3.4 allows us to conclude that $f(x)$, being continuous, is integrable on $[a, b]$, we know that they must converge, and from part (iv) we know the value of the limit. What is the value of

$$\int_a^b f(x)\,dx,$$

and how does it compare with the signed area under the graph? (*Hint:* To calculate the signed area geometrically, consider separately the parts of the graph above and below the x-axis.)

25. Let $f(x) = x$, $g(x) = x^3 - x$, $h(x) = x^5 - 2x^3$. Each of these functions is an odd function. We wish to work out various Riemann sums for these functions. For $f(x)$ we will consider the interval $[-1, 1]$; for $g(x)$ we will use the interval $[-2, 2]$; and for $h(x)$ we will use the interval $[-4, 4]$. In each case, let $\mathcal{P}_n$ be the partition that divides the specified interval into n equal subintervals, and let $\{x_k^*\}$ be the *midpoint* of the kth subinterval.

(i) Calculate the Riemann sums for $f(x)$, $g(x)$, and $h(x)$ on $[-1, 1]$, $[-2, 2]$, and $[-4, 4]$, respectively, using the partition $\mathcal{P}_n$ and the points $\{x_k^*\}$, with $n = 2$.

(ii) Repeat part (i) with $n = 4$.

(iii) Repeat part (i) with $n = 2$ but using the left endpoints of the subintervals rather than the midpoints as the choices $\{x_k^*\}$.

**__26.__ (i) After doing the previous exercise, show that if $f(x)$ is any odd function that is integrable on the interval $[-1, 1]$, then

$$\int_{-1}^{1} f(x)\,dx = 0.$$

(ii) If a is any positive number and $f(x)$ is any odd function that is integrable on the interval $[-a, a]$, show that

$$\int_{-a}^{a} f(x)\,dx = 0.$$

27. Find $\|\mathcal{P}_3\|$ and $\|\mathcal{P}_4\|$, where $\mathcal{P}_3$ and $\mathcal{P}_4$ are as in Example 1 (p. 362).

**__28.__ Let $f(x)$ be the function that is zero except at $x = 1$, where it equals 1 (see Figure 5.3.12). In symbols,

$$f(x) = \begin{cases} 0, & \text{if } x \neq 1, \\ 1, & \text{if } x = 1. \end{cases}$$

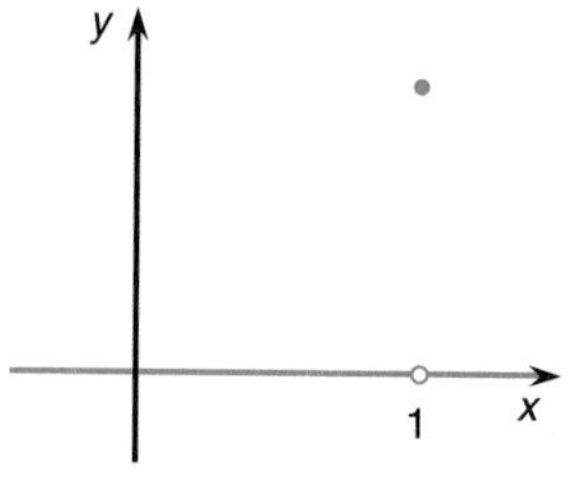

Figure 5.3.12

Show that $f(x)$ is integrable on any interval $[a, b]$. (*Hint:* It is easy if $1 \notin [a, b]$. If $1 \in [a, b]$, notice that in any Riemann sum, all the terms are zero except possibly one or two.)

****29.** Let $g(x)$ be the function defined as follows:

$$g(x) = \begin{cases} 0, & \text{if } x \leq 0, \\ 1, & \text{if } x > 0. \end{cases}$$

Show that $g(x)$ is integrable on any interval $[a, b]$ (see Figure 5.3.13).

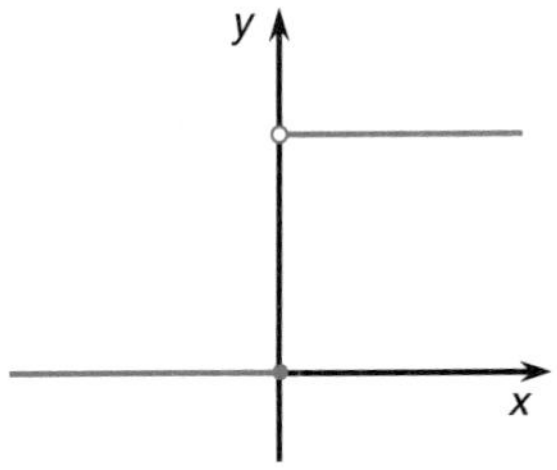

Figure 5.3.13

***30.** Assuming that the function $g(x)$ of the previous question is integrable, evaluate

$$\int_{-1}^{1} g(x)\,dx.$$

III

If you have a programmable calculator or can program a computer, make up a program that computes the Riemann sum for some specified function $f(x)$ on an interval $[a, b]$ relative to the partition $\mathcal{P}_n$ that divides $[a, b]$ into n equal subintervals, using the right endpoints as the chosen points $\{x_k^*\}$. Use your program (or one from a software package) to evaluate the Riemann sum $S\big(f, \mathcal{P}_n, \{x_k^*\}\big)$ in each of the following cases (x_k^* always refers to right endpoints).

31. $f(x) = x^4 - 3x$ on $[0, 2]$; $n = 20$

32. $f(x) = e^{2x}$ on $[-1, 1]$; $n = 10$

33. $f(x) = \tan x$ on $\left[0, \frac{\pi}{4}\right]$; $n = 15$

34. $f(x) = \frac{1}{x}$ on $[1, 2]$; $n = 25$.

***35.** If you have suitable software, evaluate some Riemann sums for the functions $f(x) = x^m$ on the interval $[0, 1]$, where m is a positive integer and x_k^* is the right endpoint of the kth subinterval.

(i) Evaluate the Riemann sums $S\big(f, \mathcal{P}_n, \{x_k^*\}\big)$ for $m = 1, 2, 3, 4, 5$, with $n = 4, 10, 50$.

(ii) Using the information from part (i), try to see a pattern and try to guess the limits. In other words, try to guess the value of

$$\int_0^1 x^m\,dx.$$

(*Hint:* It might be easier to guess the reciprocal, $\dfrac{1}{\int_0^1 x^m\,dx}$.)

5.4 Properties of Integrals

INTEGRAL MEAN VALUE THEOREM
JUMP/REMOVABLE/INFINITE DISCONTINUITIES

We can establish some basic properties of integrals immediately.

THEOREM 5.4.1

Suppose $f(x)$ and $g(x)$ are integrable on $[a, b]$. Then

(i) for any constant K the function $Kf(x)$ is also integrable on $[a, b]$, and

$$\int_a^b Kf(x)\,dx = K\int_a^b f(x)\,dx.$$

(ii) $f(x) + g(x)$ is integrable on $[a, b]$, and

$$\int_a^b \big(f(x) + g(x)\big)\,dx = \int_a^b f(x)\,dx + \int_a^b g(x)\,dx.$$

PROOF For part (i), notice that any Riemann sum $\sum_{k=1}^{n} Kf(x_k^*)\Delta x_k$ for $Kf(x)$ can be rewritten as $K\big(\sum_{k=1}^{n} f(x_k^*)\Delta x_k\big)$, which is K times a Riemann sum for $f(x)$. Since the Riemann sums for $f(x)$ tend to the limit $\int_a^b f(x)\,dx$ as $\|\mathcal{P}\| \to 0$, we see that the Riemann sums for $Kf(x)$ tend to $K\int_a^b f(x)\,dx$. This shows both that $Kf(x)$ is integrable and that its integral has the right value.

For part (ii), note that any Riemann sum for $f(x) + g(x)$ is

$$\sum_{k=1}^{n} \big(f(x_k^*) + g(x_k^*)\big)\Delta x_k = \sum_{k=1}^{n} f(x_k^*)\Delta x_k + \sum_{k=1}^{n} g(x_k^*)\Delta x_k,$$

which is the sum of a Riemann sum for $f(x)$ and one for $g(x)$ (with the same partition). Since Riemann sums for $f(x)$ and for $g(x)$ tend to the limits $\int_a^b f(x)\,dx$ and $\int_a^b g(x)\,dx$, respectively, as $\|\mathcal{P}\| \to 0$, their sum tends to $\int_a^b f(x)\,dx + \int_a^b g(x)\,dx$. So Riemann sums for $f(x) + g(x)$ converge to this limit as $\|\mathcal{P}\| \to 0$, proving (ii).

THEOREM 5.4.2

Suppose $a < b$ and $f(x)$ is a nonnegative function that is integrable on $[a, b]$. Then its integral is also nonnegative:

$$\int_a^b f(x)\,dx \geq 0.$$

PROOF For any partition $\mathcal{P} = \{x_0, x_1, x_2, \ldots, x_n\}$ and any choice $x_k^* \in [x_{k-1}, x_k]$, consider the corresponding Riemann sum

$$\sum_{k=1}^{n} f(x_k^*)\Delta x_k.$$

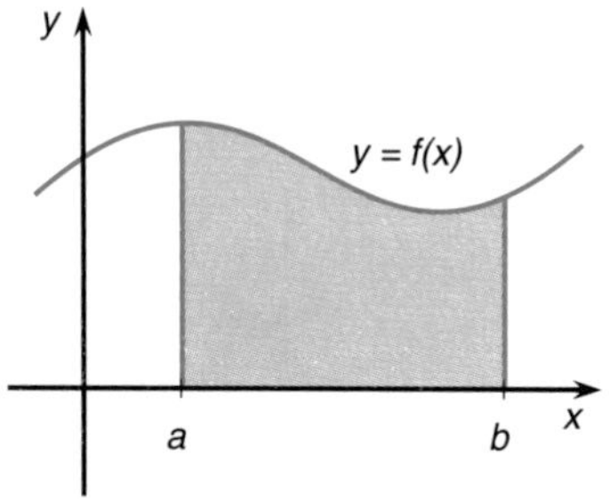

Figure 5.4.1

Each term in the sum is nonnegative; after all, $f(x) \geq 0$ for every x, so $f(x_k^*) \geq 0$, and Δx_k, the length of the kth subinterval, is positive.

Every Riemann sum is nonnegative, so the integral, which is a limit of Riemann sums, must be nonnegative.

What this theorem says seems geometrically obvious: If $f(x)$ lies above the x-axis on $[a, b]$, then the corresponding signed area is nonnegative (see Figure 5.4.1). Perhaps it is better to say that the theorem confirms that this property of integrals is *consistent* with familiar properties of areas.

THEOREM 5.4.3

Suppose $f(x)$ is integrable on $[a, b]$ and M is a number.

(i) Suppose $f(x) \leq M$ for every $x \in [a, b]$. Then

$$\int_a^b f(x)\,dx \leq M(b - a).$$

(ii) Suppose $M \leq f(x)$ for every $x \in [a, b]$. Then

$$M(b - a) \leq \int_a^b f(x)\,dx.$$

(iii) Suppose $|f(x)| \leq M$ for every $x \in [a, b]$. Then

$$\left|\int_a^b f(x)\,dx\right| \leq M(b - a).$$

PROOF It is possible to prove these results using Riemann sums directly, but it is easier to use the preceding result.

To prove part (i), observe that since $f(x) \leq M$, we can subtract $f(x)$ from both sides to find $0 \leq M - f(x)$, that is, $M - f(x) \geq 0$ for all $x \in [a, b]$. By Theorem 5.4.2,

$$\int_a^b \big(M - f(x)\big)\,dx \geq 0.$$

But $\int_a^b \big(M - f(x)\big)\,dx = \int_a^b M\,dx - \int_a^b f(x)\,dx$, by Theorem 5.4.1(ii), and $\int_a^b M\,dx = M(b - a)$, by Example 5.3.3 (p. 367). So

$$\begin{aligned} 0 \leq \int_a^b \big(M - f(x)\big)\,dx &= \int_a^b M\,dx - \int_a^b f(x)\,dx \\ &= M(b-a) - \int_a^b f(x)\,dx. \end{aligned}$$

Adding $\int_a^b f(x)\,dx$ to both sides, we find

$$\int_a^b f(x)\,dx \leq M(b - a),$$

as needed.

Similarly, for part (ii) we know that $M \leq f(x)$, so $f(x) - M \geq 0$ for all $x \in [a, b]$. Accordingly,

$$\begin{aligned} 0 \leq \int_a^b \big(f(x) - M\big)\,dx &= \int_a^b f(x)\,dx - \int_a^b M\,dx \\ &= \int_a^b f(x)\,dx - M(b-a). \end{aligned}$$

Adding $M(b - a)$ to both sides gives

$$M(b - a) \leq \int_a^b f(x)\,dx.$$

(We could also have proved (ii) by applying part (i) to $-f(x)$ and $-M$.)

For part (iii) we notice that $|f(x)| \leq M$ is exactly the same as $-M \leq f(x) \leq M$. Using (i) and (ii), we conclude that

$$-M(b - a) \leq \int_a^b f(x)\,dx \leq M(b - a).$$

Then we notice that this is exactly equivalent to the conclusion of the theorem:

$$\left|\int_a^b f(x)\,dx\right| \leq M(b - a).$$

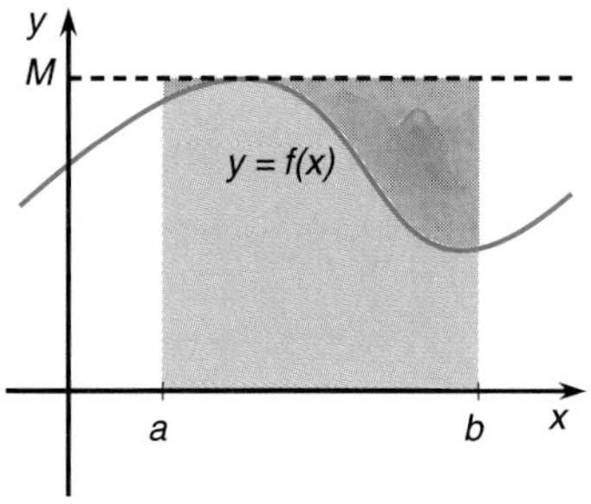

Figure 5.4.2(i)

If $f(x) \geq 0$ for all $x \in [a, b]$, we can think in terms of areas, and part (i) of Theorem 5.4.3 makes the very reasonable statement that the area under the graph of $y = f(x)$ cannot be more than the area of the larger region under the horizontal line $y = M$. After all, the area under this line between $x = a$ and $x = b$ is precisely $M(b - a)$ (see Figure 5.4.2(i)).

In the case in which $f(x)$ may not always be nonnegative, part (iii) says that the signed area cannot be bigger than $M(b - a)$ nor smaller than $-M(b - a)$. If the signed area is positive, it cannot be greater than the area under the horizontal line

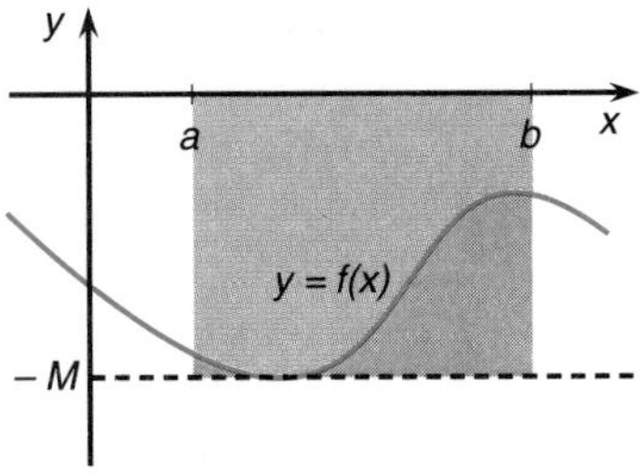

Figure 5.4.2(ii)

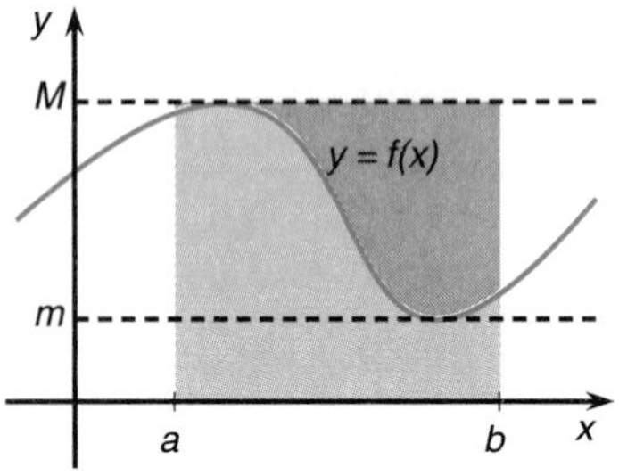

Figure 5.4.3

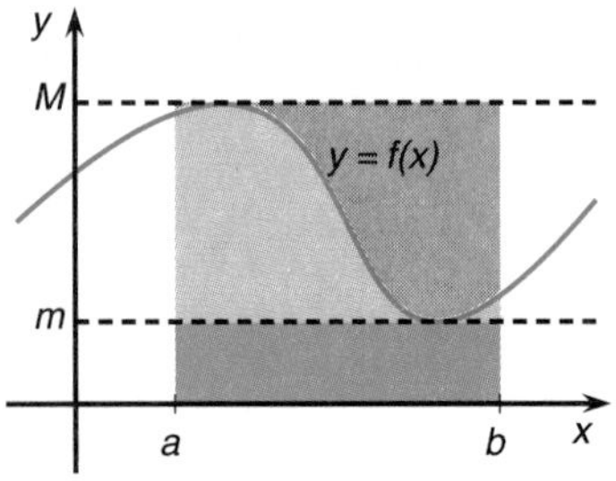

Figure 5.4.4

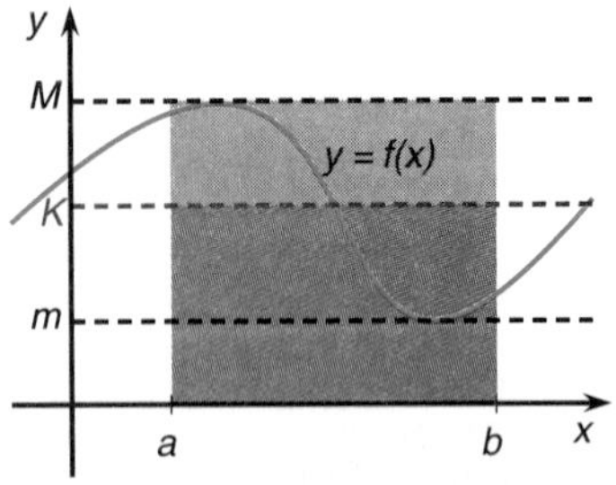

Figure 5.4.5

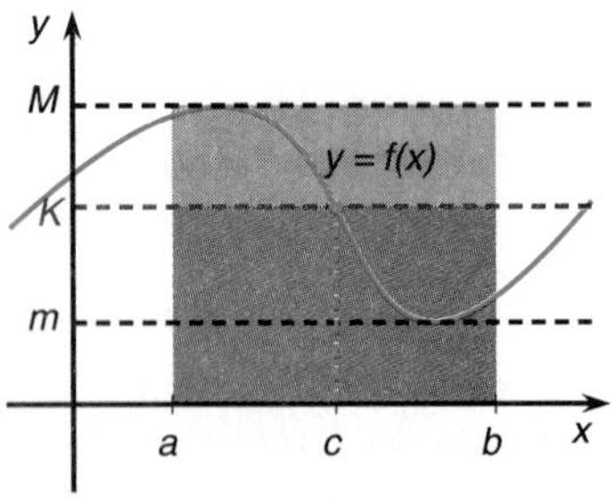

Figure 5.4.6

$y = M$. If it is negative, it cannot be less than the signed area above the line $y = -M$ (which is negative). See Figure 5.4.2(ii).

There will be times when we are unable to evaluate some integral exactly, but knowing a restriction on its magnitude will still be useful. The importance of Theorem 5.4.3 is that it restricts the size of the integral without our having to evaluate it.

Next we are going to prove a Mean Value Theorem for integrals. Suppose $f(x)$ is a continuous function on $[a, b]$. Then by Theorem 2.5.7 (p. 109) it has maximum and minimum values on $[a, b]$. Let M be the maximum value and m the minimum value of $f(x)$ on $[a, b]$, as shown in Figure 5.4.3.

The preceding theorem says that $\int_a^b f(x)\,dx$ lies between $m(b-a)$ and $M(b-a)$, which are just the signed areas determined by the horizontal lines $y = m$ and $y = M$. In the figure we have drawn m and M both positive, but this is not necessary.

The integral of $f(x)$ is some number lying between the signed areas of the lower rectangle and the upper rectangle (see Figure 5.4.4). We can think of it as the signed area of some intermediate rectangle, as shown in Figure 5.4.5. We can let K be the height of this intermediate rectangle and observe that of course K lies between m and M, the maximum and minimum values of $f(x)$.

It makes sense to think of K as the average or mean value of $f(x)$. Because $f(x)$ takes on the values m and M somewhere in $[a, b]$, the Intermediate Value Theorem (Theorem 2.5.5, p. 108) says that there is some $c \in [a, b]$ so that $f(c) = K$. In other words, there is some point in the interval $[a, b]$ at which $f(x)$ takes on its average value (see Figure 5.4.6).

This is what is known as the Integral Mean Value Theorem. The argument we have just given suggests where the idea comes from, but it relied on our geometrical intuition to find the "intermediate rectangle." Below we give a proof that is less subjective, though it is perhaps harder to understand.

THEOREM 5.4.4

The Integral Mean Value Theorem

Suppose $f(x)$ is continuous on $[a, b]$. Then there is some $c \in [a, b]$ so that

$$\int_a^b f(x)\,dx = f(c)(b-a).$$

PROOF Let M and m be the maximum and minimum values, respectively, of $f(x)$ on $[a, b]$. We know that they exist by Theorem 2.5.7. In particular, $m \le f(x) \le M$ for all $x \in [a, b]$, so by Theorem 5.4.3,

$$m(b-a) \le \int_a^b f(x)\,dx \le M(b-a).$$

Dividing by the positive number $b - a$ does not change the inequalities, so we see that

$$m \le \frac{1}{b-a}\int_a^b f(x)\,dx \le M.$$

This shows that the number $K = \frac{1}{b-a}\int_a^b f(x)\,dx$ lies between m and M, that is, $m \le K \le M$. Since there are points in the interval $[a, b]$ at which $f(x)$ equals m and M, the Intermediate Value Theorem (Theorem 2.5.5) tells us that there is a number $c \in [a, b]$ so that $K = f(c)$.

This means that $\frac{1}{b-a}\int_a^b f(x)\,dx = f(c)$, which is the same thing as

$$\int_a^b f(x)\,dx = f(c)(b-a),$$

which is the desired result.

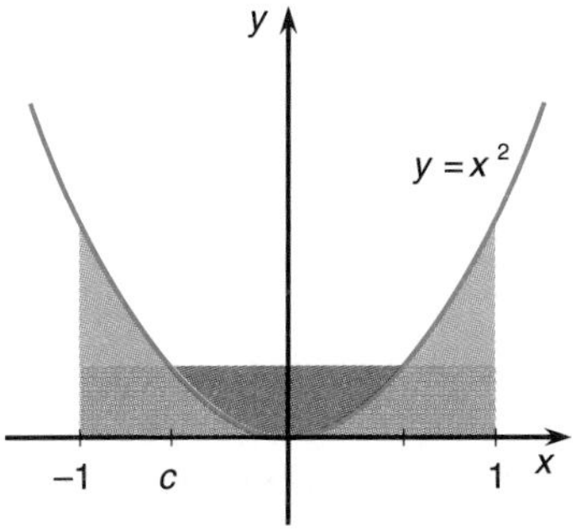

Figure 5.4.7

EXAMPLE 1 For the function $f(x) = x^2$ on the interval $[-1, 1]$ we can evaluate the integral using Example 5.3.5 (p. 369). We calculate $\int_{-1}^{1} x^2 dx = \frac{1}{3}\left(1^3 - (-1)^3\right) = \frac{2}{3}$. The length of the interval is 2, and the theorem says that there must be a point c so that $f(c)(b-a) = \int_{-1}^{1} x^2 dx$, that is, so that $f(c)(2) = \frac{2}{3}$, that is, $f(c) = \frac{1}{3}$.

Since $f(x) = x^2$, it is obvious in this case that c can be either $\frac{1}{\sqrt{3}}$ or $-\frac{1}{\sqrt{3}}$. Each of these points satisfies the condition given in Theorem 5.4.4, and both of them are in the interval $[-1, 1]$, so we have verified the theorem in this easy case. See Figure 5.4.7.

In a sense this example is misleading because it was relatively simple to find the necessary point. The theorem is more important in a situation in which it is not easy to find the point because it assures us that there is such a point, even though it may be extremely difficult to find explicitly.

Before long we will be able to find that $\int_{-1}^{2}(x^3 - 2x)\,dx = \frac{3}{4}$. The Integral Mean Value Theorem tells us there is a $c \in [-1, 2]$ so that $f(c)(3) = \frac{3}{4}$, that is, $f(c) = \frac{1}{4}$. (Here $f(x)$ is the integrand, $f(x) = x^3 - 2x$.) Using a graphics calculator, we sketch the graphs of $y = f(x)$ and the constant function $y = \frac{1}{4}$.

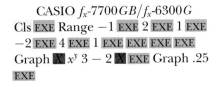

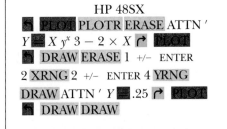

TI-85 GRAPH F1 (y(x) =) F1 (x) ^ 3 − 2 F1 (x) ▼ .25 M2(RANGE) (−) 1 ▼ 2 ▼ ▼ (−) 2 ▼ 4 F5(GRAPH)	SHARP EL9300 ⁄⁄ x/θ/T a^b 3 ▶ − 2 x/θ/T 2ndF ▼ .25 RANGE −1 ENTER 2 ENTER 1 ENTER −2 ENTER 4 ENTER 1 ENTER ⁄⁄
CASIO f_x-7700GB/f_x-6300G Cls EXE Range −1 EXE 2 EXE 1 EXE −2 EXE 4 EXE 1 EXE EXE EXE EXE Graph X x^y 3 − 2 X EXE Graph .25 EXE	HP 48SX ◣ PLOT PLOTR ERASE ATTN ′ Y ■ X y^x 3 − 2 × X ↱ PLOT ◣ DRAW ERASE 1 +/− ENTER 2 XRNG 2 +/− ENTER 4 YRNG DRAW ATTN ′ Y ■ .25 ↱ PLOT ◣ DRAW DRAW

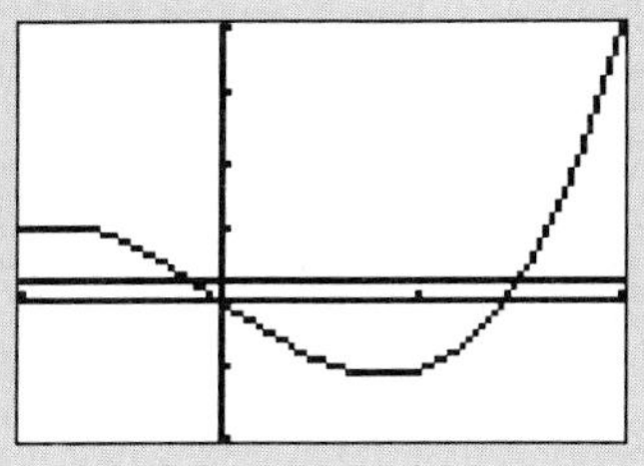

By zooming in (or using the SOLVE feature), we find that $c \approx -0.126\,000\,193$ or $c \approx 1.472\,997\,601$. (Note that there is also a solution to the left of $x = -1$, but it is not in the required interval. . .).

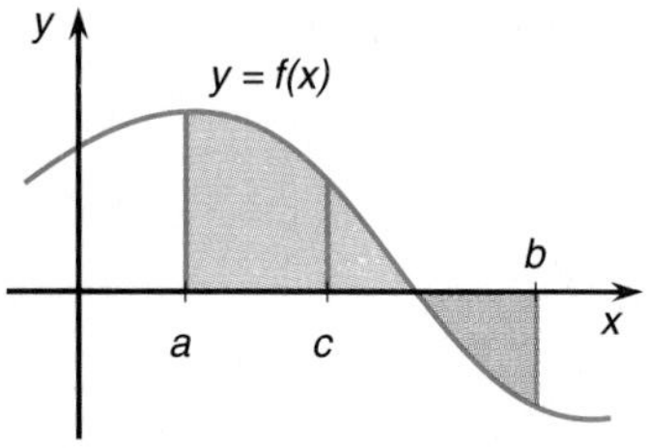

Figure 5.4.8

Suppose $f(x)$ is a function on $[a, b]$ and c is some point between a and b, that is, $a < c < b$. Then we can think of $f(x)$ as a function on the interval $[a, c]$ or on the interval $[c, b]$.

Now suppose $f(x)$ is integrable on $[a, c]$ and on $[c, b]$. It seems reasonable to expect that the total of the signed areas determined by the graph of $y = f(x)$ between $x = a$ and $x = b$ will equal the sum of the signed areas between a and c and the signed areas between c and b (see Figure 5.4.8). This is in fact true. In particular, the function $f(x)$ is integrable on $[a, b]$, and its integral equals the sum of the two smaller integrals.

THEOREM 5.4.5

If $a < c < b$ and $f(x)$ is integrable on $[a, c]$ and on $[c, b]$, then $f(x)$ is integrable on $[a, b]$, and

$$\int_a^b f(x)\,dx = \int_a^c f(x)\,dx + \int_c^b f(x)\,dx.$$

EXAMPLE 2

(i) The continuous function $f(x) = x^2$ is integrable on any closed interval. For instance, it is integrable on both $[-1, 1]$ and $[1, 3]$. Using Example 5.3.5 we can evaluate the corresponding integrals:

$$\int_{-1}^{1} x^2\,dx = \frac{1}{3}\left(1^3 - (-1)^3\right) = \frac{2}{3},$$

$$\int_{1}^{3} x^2\,dx = \frac{1}{3}\left(3^3 - 1^3\right) = \frac{26}{3}.$$

On the other hand, we can also evaluate

$$\int_{-1}^{3} x^2\,dx = \frac{1}{3}\left(3^3 - (-1)^3\right) = \frac{28}{3}.$$

It is comforting to see that Theorem 5.4.5 is verified in this case:

$$\int_{-1}^{1} x^2\,dx + \int_{1}^{3} x^2\,dx = \frac{2}{3} + \frac{26}{3} = \frac{28}{3} = \int_{-1}^{3} x^2\,dx.$$

(ii) In fact, if $a < c < b$, we see that

$$\begin{aligned}\int_a^c x^2\,dx + \int_c^b x^2\,dx &= \frac{1}{3}(c^3 - a^3) + \frac{1}{3}(b^3 - c^3) \\ &= \frac{1}{3}(c^3 - a^3 + b^3 - c^3) = \frac{1}{3}(b^3 - a^3) \\ &= \int_a^b x^2\,dx.\end{aligned}$$

This confirms Theorem 5.4.5 for any $a < c < b$ for the specific function $f(x) = x^2$. We were able to do this because we knew how to evaluate the integral of x^2 by Example 5.3.5.

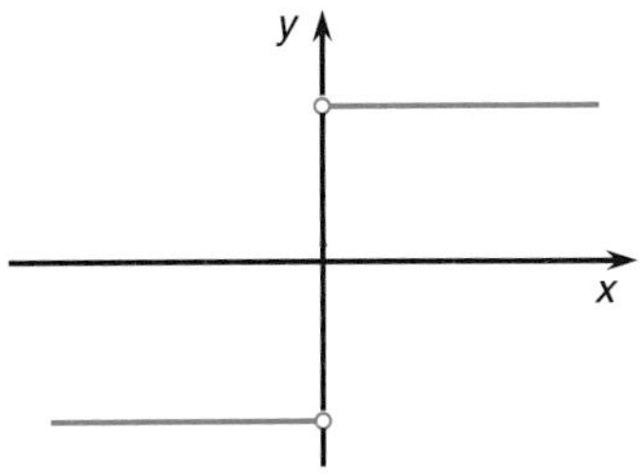

Figure 5.4.9

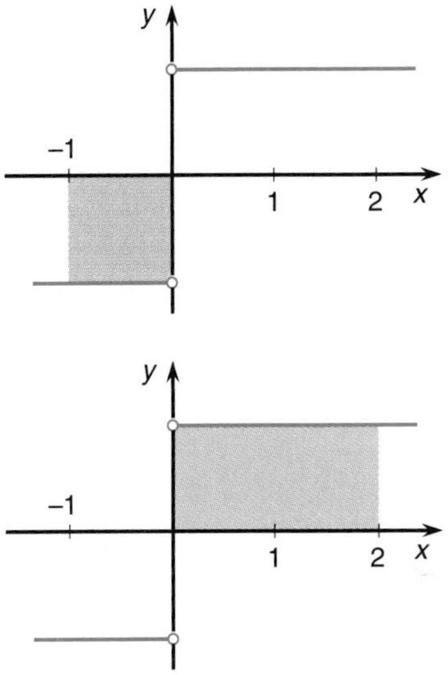

Figure 5.4.10

Integrating Functions with Removable or Jump Discontinuities

EXAMPLE 3 Let $f(x) = \frac{|x|}{x}$, the function whose graph is illustrated in Figure 5.4.9, and calculate $\int_{-1}^{2} f(x)\,dx$.

Solution The difficulty is that since $f(x)$ is not continuous, we cannot use Theorem 5.3.4 (p. 368) to conclude that it is integrable. On the other hand, Theorem 5.4.5 suggests that the way to evaluate the integral would be to evaluate the integral from $x = -1$ to $x = 0$ and the integral from $x = 0$ to $x = 2$, and add them.

Between $x = -1$ and $x = 0$, we know that $f(x) = -1$, that is, $f(x)$ is a constant function, so it is integrable. Strictly speaking, $f(x)$ is not defined at $x = 0$, so we cannot really use Theorem 5.3.4 to say that it is integrable on $[-1, 0]$. However, $f(x)$ is continuous on $[-1, 0)$ with a removable discontinuity at $x = 0$. In a situation like this we will allow ourselves to remove the discontinuity, making the function integrable on $[-1, 0]$.

Similarly, $f(x)$ is continuous on $(0, 2]$, with a removable discontinuity at $x = 0$. Again we can make the function integrable by removing the discontinuity.

The integral can now be evaluated by doing it in pieces: See Figure 5.4.10.

$$\begin{aligned}
\int_{-1}^{2} \frac{|x|}{x}\,dx &= \int_{-1}^{0} \frac{|x|}{x}\,dx + \int_{0}^{2} \frac{|x|}{x}\,dx \\
&= \int_{-1}^{0} (-1)\,dx + \int_{0}^{2} 1\,dx \\
&= (-1)\big(0 - (-1)\big) + 1(2 - 0) \\
&= 1.
\end{aligned}$$

The answer we obtain is certainly consistent with our interpretation of the integral as the total signed area.

In this example we used the idea of Theorem 5.4.5 to integrate $f(x)$ over the interval $[-1, 2]$, even though $f(x)$ is not defined at the point $x = 0$. It has a jump discontinuity at $x = 0$, so there is no way to change $f(x)$ to make it continuous there. On the other hand, if we look at the part of $f(x)$ on the interval $[-1, 0]$, it has a **removable discontinuity** at $x = 0$. Removing it, we can integrate.

Similarly, we can integrate $f(x)$ from $x = 0$ to $x = 2$ after changing the function at $x = 0$. Curiously, the change we make to $f(x)$ in order to integrate it over $[-1, 0]$ is not the same as the change that is necessary to integrate it over $[0, 2]$. (For $[-1, 0]$ we define $f(0) = -1$, and for $[0, 2]$ we define $f(0) = 1$.) Having integrated over both small intervals, we can add the results and call the sum the integral of $f(x)$ over the large interval $[-1, 2]$.

Strictly speaking, the Definition 5.3.3 of integrability was only made for functions that were defined at every point in the interval. On the other hand, the approach we have just used seems very reasonable, and it gives results that fit our geometrical intuition. We want to be able to integrate functions of this type, that is, functions with a jump or removable discontinuity at some point or, for that matter, at several points $d_1, d_2, \ldots, d_r$.

The first observation is that if $f(x)$ is defined at every point of $[a, b]$ and is continuous there except for a removable or jump discontinuity at one point $c \in [a, b]$, then it is integrable on $[a, b]$. This result is not very difficult to prove, but we leave the proof to the exercises.

THEOREM 5.4.6

Suppose $f(x)$ is defined at every point in $[a, b]$ and is continuous except for a removable discontinuity or a jump discontinuity at $x = c$. Then $f(x)$ is integrable on $[a, b]$.

This also works if there are removable or jump discontinuities at several points $d_1, d_2, \ldots, d_r$.

Another common situation is that $f(x)$ might not even be defined at certain points $d_1, d_2, \ldots, d_r$. Theorem 5.4.6 suggests that its actual values at these points are not important, so we can try to choose values for $f(d_1), f(d_2), \ldots, f(d_r)$ in such a way that $f(x)$ has at worst a removable or a jump discontinuity at each of them. If this is possible, the resulting "extended" function is integrable, by the theorem, and we will say that the original function is integrable and that its integral equals the integral of the extended function.

DEFINITION 5.4.7

Suppose $f(x)$ is defined on $[a, b]$, except possibly at the points $d_1, d_2, \ldots, d_r$. If $f(x)$ has at worst a removable or a jump discontinuity at each of them, then we say that $f(x)$ is integrable on $[a, b]$, and

$$\int_a^b f(x)\,dx = \int_a^{d_1} f(x)\,dx + \int_{d_1}^{d_2} f(x)\,dx + \ldots + \int_{d_{r-1}}^{d_r} f(x)\,dx + \int_{d_r}^{b} f(x)\,dx.$$

If $f(x)$ has a removable discontinuity, we evaluate the integral of $f(x)$ by "removing" the discontinuity and integrating the resulting function. If the function has jump discontinuities, it is necessary to evaluate the integral of each continuous piece separately and add the results together, using Theorem 5.4.5, just as we did in Example 3.

To integrate over the subintervals, it might be necessary to remove some removable discontinuities at the endpoints.

EXAMPLE 4 Evaluate (i) $\int_0^4 \frac{x^2-4}{x-2}\,dx$; (ii) $\int_{-2}^{2} \frac{|t-1|}{t-1}(t-2)\,dt$.

Solution

(i) The integrand equals $x + 2$ at every point except $x = 2$, where it is not defined. So by Definition 5.4.7 we have

$$\begin{aligned}\int_0^4 \frac{x^2-4}{x-2}\,dx &= \int_0^2 (x+2)\,dx + \int_2^4 (x+2)\,dx \\ &= \int_0^2 x\,dx + \int_0^2 2\,dx + \int_2^4 x\,dx + \int_2^4 2\,dx \\ &= \frac{1}{2}(2^2 - 0^2) + 2(2-0) + \frac{1}{2}(4^2 - 2^2) + 2(4-2) \\ &= 16.\end{aligned}$$

Note that this equals $\int_0^4 (x+2)\,dx$, the integral we get by removing the discontinuity . . .

(ii) The integrand is continuous except at $t = 1$. Accordingly, we proceed to break the integral into two pieces:

$$\begin{aligned}\int_{-2}^{2} \frac{|t-1|}{t-1}(t-2)\,dt &= \int_{-2}^{1} \frac{|t-1|}{t-1}(t-2)\,dt + \int_{1}^{2} \frac{|t-1|}{t-1}(t-2)\,dt \\ &= \int_{-2}^{1}(-1)(t-2)\,dt + \int_{1}^{2}(1)(t-2)\,dt \\ &= -\int_{-2}^{1} t\,dt + \int_{-2}^{1} 2\,dt + \int_{1}^{2} t\,dt - \int_{1}^{2} 2\,dt \\ &= -\frac{1}{2}\big(1^2-(-2)^2\big) + 2\big(1-(-2)\big) + \frac{1}{2}(2^2-1^2) - 2(2-1) \\ &= 7.\end{aligned}$$

Note that $\frac{|t-1|}{t-1}$ equals -1 when $t < 1$ and 1 when $t > 1$.

Exercises 5.4

I

For the following exercises, recall that there are several integrals that we know how to calculate. Specifically, we know that $\int_a^b C\,dx = C(b-a)$ for any constant C, by Example 5.3.3 (p. 367). Also, by Example 5.3.5 (p. 369), we know that $\int_a^b x\,dx = \frac{1}{2}(b^2-a^2)$ and that $\int_a^b x^2 dx = \frac{1}{3}(b^3-a^3)$. Using these facts and Theorem 5.4.1 (p. 373), evaluate the following integrals.

1. $\int_0^1 (x+1)\,dx$
2. $\int_{-1}^2 2x^2\,dx$
3. $\int_0^3 (x^2-1)\,dx$
4. $\int_{-2}^2 5x\,dx$
5. $\int_1^4 (x^2+x-1)\,dx$
6. $\int_{-3}^4 (3x^2+2)\,dx$
7. $\int_2^3 (x^2+2x+5)\,dx$
8. $\int_3^5 (6x^2+4x-3)\,dx$
9. $\int_{-1}^1 (6x^2+17x+2)\,dx$
10. $\int_{-\pi}^{\pi} (\sin^2 x + \cos^2 x)\,dx$

Suppose $f(x)$, $g(x)$, and $h(x)$ are continuous functions on $[1,-1]$, and suppose we know the following facts:

$$\int_{-1}^{0} f(x)\,dx = -2,\quad \int_{-1}^{0} g(x)\,dx = 3,\quad \int_{-1}^{1} h(x)\,dx = 5,$$

$$\int_{0}^{1} f(x)\,dx = 1,\quad \int_{-1}^{1} g(x)\,dx = 7,\quad \int_{0}^{1} h(x)\,dx = 11.$$

Use these facts to evaluate the following integrals.

11. $\int_{-1}^1 f(x)\,dx$
12. $\int_{-1}^1 \big(f(x)+g(x)\big)\,dx$
13. $\int_0^1 g(x)\,dx$
14. $\int_0^1 \big(f(x)+3h(x)\big)\,dx$
15. $\int_{-1}^0 h(x)\,dx$
16. $\int_{-1}^0 \big(2g(x)-f(x)\big)\,dx$
17. $\int_0^1 \big(4g(x)-2x^2+3\big)\,dx$
18. $\int_{-1}^1 \big(2h(x)-f(x)+4x^2-6x\big)\,dx$
19. $\int_{-1}^0 \big(g(x)-f(x)+3h(x)-3x^2+2x\big)\,dx$
20. $\int_0^1 \big(h(x)+3g(x)+f(x)-6x^2-2x\big)\,dx$

II

21. Verify Theorem 5.4.5 (p. 378) for the function $f(x) = x$ on the intervals $[-2,1]$ and $[1,2]$.
22. Verify Theorem 5.4.5 for the function $g(x) = x^2 - 2x + 4$ on the intervals $[-3,-1]$ and $[-1,4]$.

*23. Let $a < c < b$. Verify Theorem 5.4.5 for the function $f(x) = x$ on the intervals $[a,c]$ and $[c,b]$. (*Hint:* Use Example 5.3.5, p. 369).

*24. Let $a < c < b$. Verify Theorem 5.4.5 for the function $f(x) = 3x^2 - 4x + 7$ on the intervals $[a,c]$ and $[c,b]$. (*Hint:* Use Example 5.3.5).

Verify the Integral Mean Value Theorem (Theorem 5.4.4) for each of the following functions.

25. $f(x) = x$ on the interval $[-2, 3]$

***26.** $f(x) = x$ on the interval $[a, b]$

27. $f(x) = x^2 - 2x + 3$ on the interval $[0, 4]$

***28.** $f(x) = x^2$ on the interval $[a, b]$

29. Find constants c and d so that the function $f(x) = cx + d$ satisfies

$$\int_0^1 f(x)\,dx = 4, \qquad \int_1^3 f(x)\,dx = 14.$$

30. Find constants c and d so that the function $f(x) = cx + d$ satisfies

$$\int_{-1}^2 f(x)\,dx = 3, \qquad \int_2^3 f(x)\,dx = 9.$$

31. Find constants b, c, and d so that the function $g(x) = bx^2 + cx + d$ satisfies

$$\int_{-1}^0 g(x)\,dx = -1, \int_0^1 g(x)\,dx = 1, \int_1^2 g(x)\,dx = 9.$$

32. Find constants b, c, and d so that the function $h(x) = bx^2 + cx + d$ satisfies

$$\int_0^2 h(x)\,dx = 10, \int_{-2}^2 h(x)\,dx = 20, h(1) = 4.$$

***33.** Use Theorem 5.4.3 (p. 374) to show that

$$0 \le \int_0^1 x^3\,dx \le 1.$$

***34.** Suppose $f(x)$ and $g(x)$ are integrable on $[a, b]$. Show that we cannot expect that $\int_a^b f(x)g(x)\,dx$ will equal the product of the integrals of $f(x)$ and $g(x)$, by finding an example in which they are different.

***35.** Suppose $f(x)$ and $g(x)$ are integrable on $[a, b]$, and

$$f(x) \le g(x)$$

for every $x \in [a, b]$. Show that

$$\int_a^b f(x)\,dx \le \int_a^b g(x)\,dx.$$

***36.** (i) Show that

$$1 \le \int_1^2 x^4\,dx \le 16.$$

(ii) By writing $\int_1^2 x^4\,dx = \int_1^{3/2} x^4\,dx + \int_{3/2}^2 x^4\,dx$, find better inequalities, that is, show that $C \le \int_1^2 x^4\,dx \le D$ for some constants C and D. These inequalities will be "better" than those in part (i) if $C > 1$ and $D < 16$, because they will then narrow down more closely what the value of the integral might be.

***37.** Prove Theorem 5.4.6 (p. 380).

***38.** Apply Theorem 5.4.3 to find a number K so that $\left|\int_{-1}^2 (x^2 + 1)\sin(2x)\,dx\right| \le K$.

III

39. It turns out that $\int_1^4 \sqrt{x}\,dx = \frac{14}{3}$. (i) Use a calculator to find (to four decimal places) the corresponding number $c \in [1, 4]$, which is promised by the Integral Mean Value Theorem (5.4.4). (ii) Use a graphics calculator or computer to sketch $\sqrt{x}$ and the corresponding rectangle on the same graph.

40. Sketch the graphs of $y = x^2$ and $y = \frac{1}{3}$ for $x \in [-1, 1]$ on the same graph. Compare with Example 1 (p. 377).

In each of the following situations the value of an integral is given. Use a graphics calculator or computer to find the number c promised by Theorem 5.4.4, correct to five decimal places.

41. $\int_0^\pi \sin x\,dx = 2$

42. $\int_{-1}^1 x^5\,dx = 0$

43. $\int_1^e \frac{1}{x}\,dx = 1$

44. $\int_0^1 (x^3 - 2x)\,dx = -\frac{3}{4}$

POINT TO PONDER

We have used various facts about limits: In the proof of Theorem 5.4.1 we needed to know that the limit of a sum is the sum of the limits and that multiplying by a constant and taking a limit gives exactly the same result as taking the limit and then multiplying by the constant. For the proof of Theorem 5.4.2 we used the fact that the limit of nonnegative numbers is nonnegative.

Strictly speaking, these are facts we learned in Chapter 2 about conventional limits. Here we have been using analogous facts about the special kind of limits we have been discussing in connection with Riemann sums (limits as the mesh of the partition goes to zero). Convince yourself that the results carry over to this new situation.

5.5 Indefinite Integrals and the Fundamental Theorem of Calculus

DEFINITE/INDEFINITE INTEGRALS
VARIABLE OF INTEGRATION
DUMMY VARIABLE
ANTIDERIVATIVE

In this section we explore the ideas behind the Fundamental Theorem of Calculus, which establishes the link between differentiation and integration. We will find that integration is closely related to finding antiderivatives, that is, to "undoing" the process of differentiation.

Suppose $f(x)$ is a continuous function on $[a, b]$. By Theorem 5.3.4 (p. 368), $f(x)$ is integrable on $[a, b]$. For that matter, for any c between a and b, $f(x)$ is continuous on $[a, c]$, so it is integrable on $[a, c]$. The integral

$$\int_a^c f(x)\,dx$$

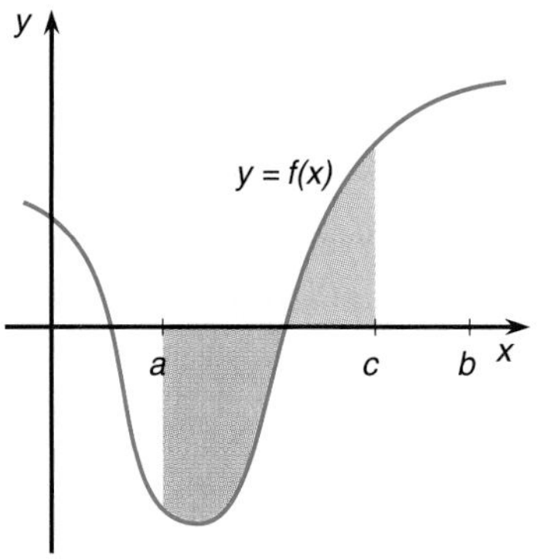

Figure 5.5.1

will depend on c, taking on different values as c changes (see Figure 5.5.1). We can think of c as a variable and of the integral as a function of c.

To emphasize that c is a variable, we can replace it by the letter x and let

$$F(x) = \int_a^x f(t)\,dt, \qquad \text{for any } x \in [a, b].$$

Notice that the variable inside the integral has been changed to t to avoid confusion. It makes no difference what letter is used there as long as it is different from any other symbol being used.

Similarly, there is no particular reason that the lower limit of integration should always be a; for any fixed $d \in [a, b]$ we could consider

$$\int_d^x f(t)\,dt.$$

This leads to trouble in the situation in which $x \leq d$. We avoid the problem by making a definition.

DEFINITION 5.5.1

(i) If $f(x)$ is any function defined at $x = a$, define

$$\int_a^a f(x)\,dx = 0.$$

(ii) If $a < b$ and $f(x)$ is integrable on $[a, b]$, define

$$\int_b^a f(x)\,dx = -\int_a^b f(x)\,dx.$$

REMARK 5.5.2

Part (i) of Definition 5.5.1 has the effect of making the formula in Theorem 5.4.5 (p. 378) remain valid if $c = a$ or $c = b$. It can also be interpreted to say there is zero area between $x = a$ and $x = a$. Part (ii) makes the formula hold even if $c < a$ or $c > b$, provided that the function $f(x)$ is integrable on the larger interval.

DEFINITION 5.5.3

> Suppose $f(x)$ is integrable on $[a, b]$ and $d \in [a, b]$. For $x \in [a, b]$ we can define
>
> $$F(x) = \int_d^x f(t)\,dt.$$
>
> The function $F(x)$ is called an **indefinite integral** of $f(x)$. It is a function on $[a, b]$. The name suggests that the integral can have different values as x changes.

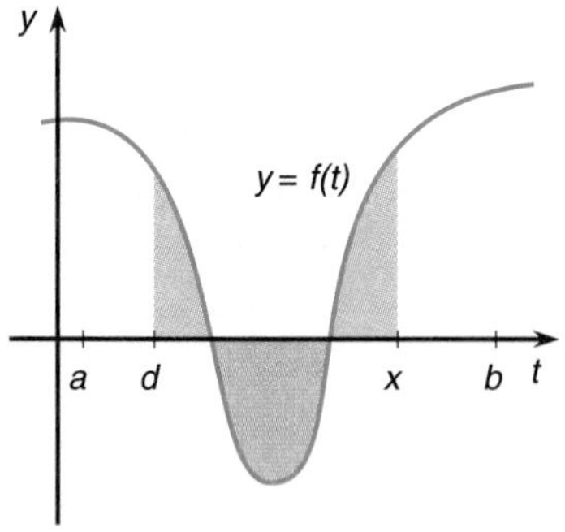

Figure 5.5.2

If $d < x$, the indefinite integral $F(x)$ represents the signed area of the region between $y = f(t)$ and the horizontal axis, between $t = d$ and $t = x$ (see Figure 5.5.2). As x changes, the region changes, and the indefinite integral represents the signed area of different regions.

EXAMPLE 1

(i) If $f(x) = x$, we notice that $f(x)$ is integrable on any interval (since it is always continuous). If d is any fixed number, we can define the indefinite integral

$$F(x) = \int_d^x f(t)\,dt = \int_d^x t\,dt.$$

This last integral can be evaluated, using Example 5.3.5(p. 369). It equals $\frac{1}{2}(x^2 - d^2)$.

So we have found an indefinite integral of $f(x) = x$. It is $\frac{1}{2}(x^2 - d^2)$, where d is any fixed number. See Figure 5.5.3(i).

Notice the way t was used as the variable in the integral. Since x was used as the upper limit of integration, it would have led to confusion if we had used it again inside the integral; this way it is clear that t is a variable that can take on any value between $t = d$ and $t = x$.

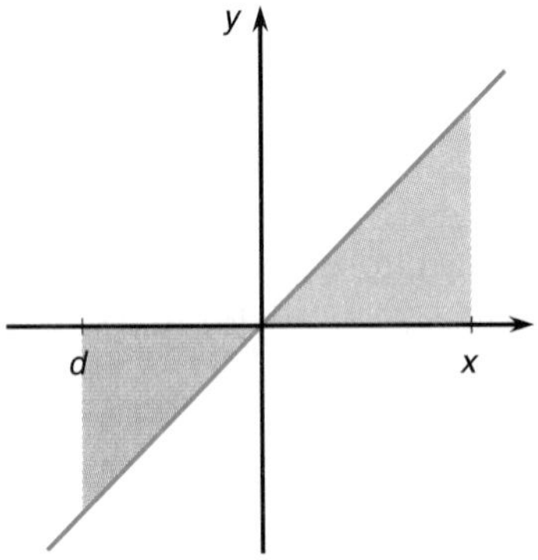

Figure 5.5.3(i)

(ii) Similarly, for $g(x) = c$, where c is a constant, we can define an indefinite integral

$$G(x) = \int_d^x g(t)\,dt = \int_d^x c\,dt.$$

Using Example 5.3.3 (p. 367), we can evaluate this integral; we find that

$$G(x) = c(x - d).$$

See Figure 5.5.3(ii).

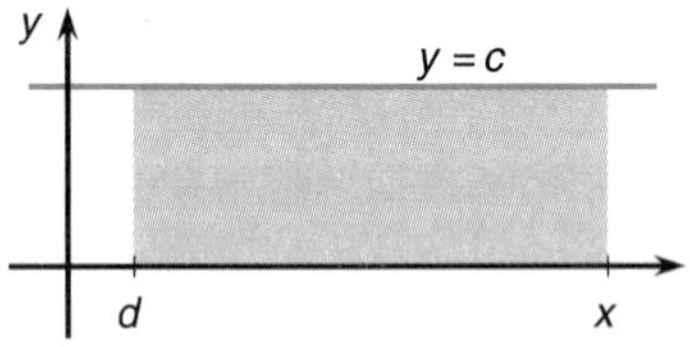

Figure 5.5.3(ii)

(iii) For $h(x) = x^2$ we can define the indefinite integral

$$H(x) = \int_d^x h(t)\,dt = \int_d^x t^2\,dt.$$

Using Example 5.3.5, we evaluate and find that

$$H(x) = \frac{1}{3}(x^3 - d^3).$$

See Figure 5.5.3(iii).

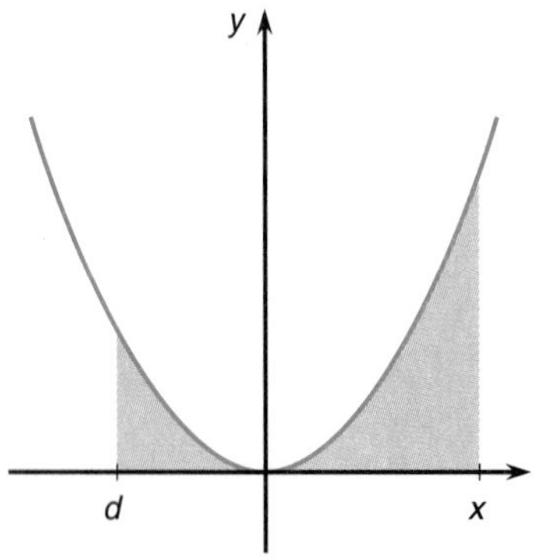

Figure 5.5.3(iii)

EXAMPLE 2 The indefinite integral $F(x)$ of Definition 5.5.3 depends on the choice of $d \in [a, b]$. Consider two possible choices $d_1, d_2 \in [a, b]$ and define

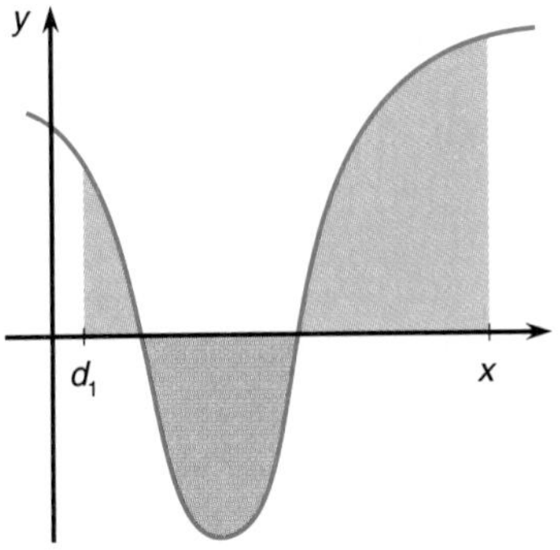

Figure 5.5.4(i)

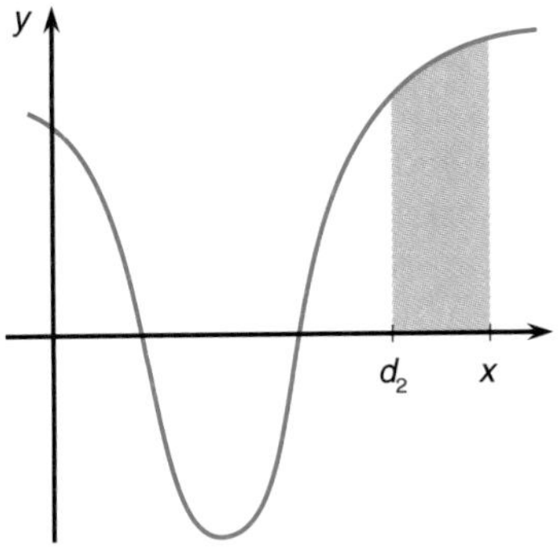

Figure 5.5.4(ii)

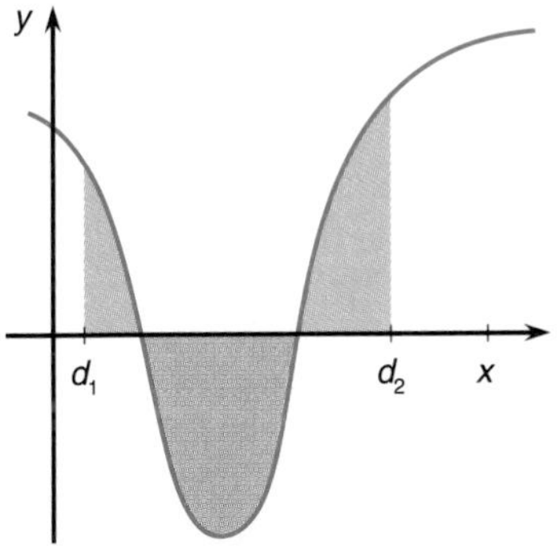

Figure 5.5.4(iii)

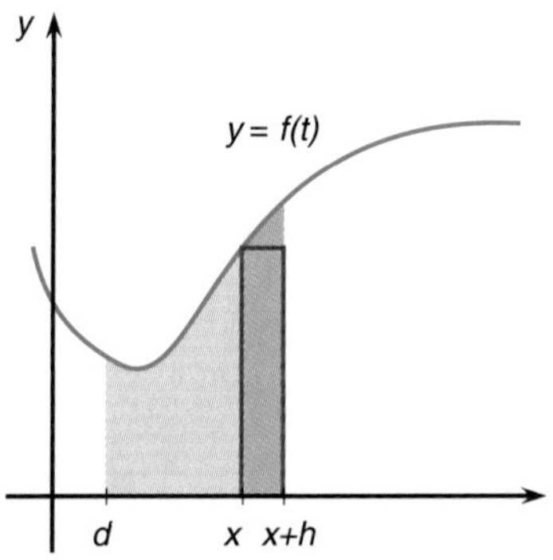

Figure 5.5.5

$$F_1(x) = \int_{d_1}^{x} f(t)\,dt \qquad \text{and} \qquad F_2(x) = \int_{d_2}^{x} f(t)\,dt.$$

Then

$$\begin{aligned} F_1(x) - F_2(x) &= \int_{d_1}^{x} f(t)\,dt - \int_{d_2}^{x} f(t)\,dt \\ &= \int_{d_1}^{x} f(t)\,dt + \int_{x}^{d_2} f(t)\,dt, \end{aligned}$$

by Definition 5.5.1(ii), and by Remark 5.5.2 this equals

$$\int_{d_1}^{d_2} f(t)\,dt.$$

Notice that the above formula for $F_1(x) - F_2(x)$ is independent of x, that is, it is a constant function. We have just showed that different choices d_1 and d_2 result in indefinite integrals $F_1(x)$ and $F_2(x)$ that differ by a constant. Figures 5.5.4(i), (ii), (iii) show how the difference between $F_1(x)$ and $F_2(x)$ is the signed area between d_1 and d_2.

In particular, since two differentiable functions that differ by a constant have the same derivative, if we knew that $F_1(x)$ and $F_2(x)$ were differentiable, we would know that $F_1'(x) = F_2'(x)$. In fact, not only are they differentiable, but in a sense the derivative is the most important thing about them, as we will soon see.

Derivative of an Indefinite Integral

Suppose $f(x)$ is a continuous function on $[a, b]$. Fix $d \in [a, b]$, and consider the indefinite integral

$$F(x) = \int_{d}^{x} f(t)\,dt.$$

The derivative of $F(x)$ is defined by

$$F'(x) = \lim_{h \to 0} \frac{F(x+h) - F(x)}{h}.$$

To discuss the derivative, we must consider the difference quotient:

$$\begin{aligned} \frac{F(x+h) - F(x)}{h} &= \frac{1}{h}\left(\int_{d}^{x+h} f(t)\,dt - \int_{d}^{x} f(t)\,dt\right) \\ &= \frac{1}{h}\left(\int_{d}^{x+h} f(t)\,dt + \int_{x}^{d} f(t)\,dt\right). \end{aligned}$$

For the last equality we used Definition 5.5.1(ii). Using Theorem 5.4.5 and Remark 5.5.2, we can combine the two integrals in the last line to find that the difference quotient equals

$$\frac{1}{h}\left(\int_{x}^{x+h} f(t)\,dt\right).$$

Suppose for the moment that h is a small positive number. The situation is illustrated in Figure 5.5.5.

The numerator of the difference quotient is the integral

$$\int_x^{x+h} f(t)\,dt,$$

which represents the signed area of the shaded region. This is approximately the same as the area of the tall rectangle shown, whose height is $f(x)$. (Note that x is the point at the left side of the rectangle.) Since the width of the rectangle is h, its (signed) area is $hf(x)$.

The difference quotient is then approximately this area divided by h, or $\frac{hf(x)}{h} = f(x)$. As $h \to 0$, the difference quotient is close to $f(x)$. This suggests the remarkable fact that

$$F'(x) = f(x),$$

which is known as the Fundamental Theorem of Calculus (what makes it fundamental, at least in part, is that it relates derivatives and integrals, the two main ingredients of calculus). In words, what it says is that the derivative of an indefinite integral equals the integrand. Another way to say it is that if we start with a continuous function $f(x)$, construct its indefinite integral $F(x)$, and then take the derivative of that, what we end up with is the original function $f(x)$. So differentiation has the effect of "undoing" the process of forming the indefinite integral.

In making this approximate discussion we assumed that h was positive, and in the illustration we implicitly assumed $f(x) > 0$. You might like to check that it still works if either or both of h and $f(x)$ are negative.

The argument we just gave relied upon noticing that the area of the region under the graph of $y = f(t)$ between $t = x$ and $t = x + h$ was "approximately" equal to the area of a certain rectangle, which does seem reasonable when h is small. It was then necessary to take the limit as $h \to 0$ and to assume that we could ignore the word "approximately." In fact it is possible to show that this all works, but it is not easy. The proof given below uses the Integral Mean Value Theorem (Theorem 5.4.4). Even this proof is not very easy. Since after all it is the result itself that interests us most, you may prefer to jump over the proof and look at the examples that follow it to see how the result can be used.

THEOREM 5.5.4

The Fundamental Theorem of Calculus

Suppose $f(x)$ is continuous on $[a, b]$ and $d \in [a, b]$. Then the indefinite integral

$$F(x) = \int_d^x f(t)\,dt$$

is differentiable on (a, b), and

$$F'(x) = f(x),$$

that is,

$$\frac{d}{dx}\left(\int_d^x f(t)\,dt\right) = f(x).$$

PROOF The difference quotient is

$$\frac{F(x+h) - F(x)}{h} = \frac{1}{h}\left(\int_d^{x+h} f(x)\,dx - \int_d^x f(x)\,dx\right)$$

$$= \frac{1}{h}\left(\int_d^{x+h} f(x)\,dx + \int_x^d f(x)\,dx\right)$$
$$= \frac{1}{h}\left(\int_x^{x+h} f(x)\,dx\right).$$

Assume for the moment that $h > 0$ and apply the Integral Mean Value Theorem (Theorem 5.4.4) to this last integral. We find that it can be written as

$$\int_x^{x+h} f(x)\,dx = f(c)(x+h-x) = f(c)h,$$

for some $c \in [x, x+h]$. Dividing this by h, we find that the difference quotient equals $f(c)$.

As $h \to 0^+$, the point c will be squeezed between x and $x+h$, which are both approaching x. Accordingly, we see that $c \to x$, and by the continuity of $f(x)$ we know that $f(c) \to f(x)$. Since $f(c)$ equals the difference quotient, we have just shown that the difference quotient approaches $f(x)$ as $h \to 0^+$.

An entirely similar argument applies in the case in which $h < 0$, and we see that

$$\lim_{h\to 0} \frac{F(x+h) - F(x)}{h} = f(x),$$

in the sense that the limit exists and equals $f(x)$. Theorem 5.5.4 is proved. ◾

> Observe that in this proof, x does not change. It plays the role of a constant rather than a variable. Notice that the proof actually works at the endpoints too, showing that $F(x)$ has a one-sided derivative at each endpoint.

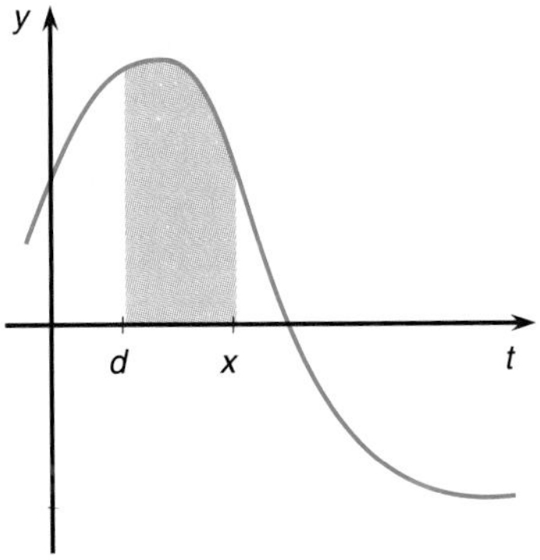

Figure 5.5.6(i)

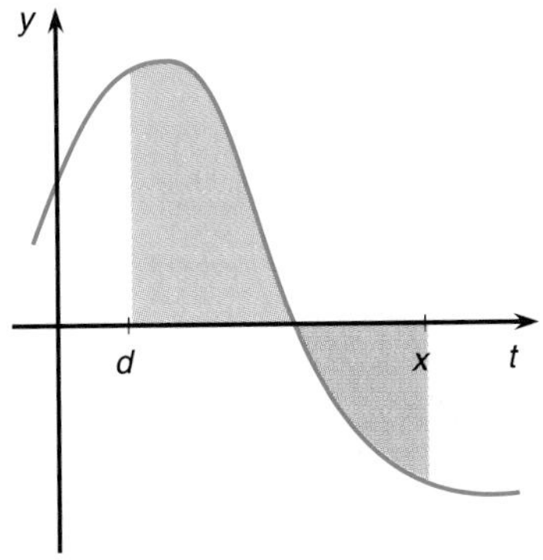

Figure 5.5.6(ii)

It might help to visualize Theorem 5.5.4 in the following way. The derivative $F'(x)$ is the rate at which $F(x)$ changes as x increases. Imagine x moving along the horizontal axis, sweeping out larger and larger regions as it goes. As x moves, the region changes, and the signed area changes too. How can we visualize the rate of change of the signed area?

If $f(x)$ is a positive number, then as x moves to the right, the shaded region is increasing, and the part being added on is above the axis, so it has positive signed area. The bigger $f(x)$ is, the larger will be the area of the region being added on, and the larger will be the rate at which the total signed area is increasing. See Figure 5.5.6(i).

Similarly, if $f(x) < 0$, the region being added on has negative signed area, so the total signed area is decreasing as x moves to the right (see Figure 5.5.6(ii)). Moreover, the rate at which $F(x)$ is decreasing will depend on how far below the axis $f(x)$ is. This is consistent with the fact that the derivative of $F(x)$ equals $f(x)$.

EXAMPLE 3 We can easily check the theorem for the indefinite integrals we found in Example 1. For $f(x) = x$, we found the indefinite integral

$$F(x) = \frac{1}{2}(x^2 - d^2).$$

The derivative of $F(x)$ is $F'(x) = \frac{1}{2}(2x) = x$. Since this equals the original function $f(x)$, we have shown that in this case, $F'(x) = f(x)$, as required.

For $g(x) = c$ we have the indefinite integral $G(x) = c(x - d)$. Its derivative is $G'(x) = c$, which again equals the original function $g(x)$. Once again Theorem 5.5.4 is verified.

For $h(x) = x^2$ we found the indefinite integral

$$H(x) = \frac{1}{3}(x^3 - d^3),$$

whose derivative is $H'(x) = \frac{1}{3}3x^2 = x^2 = h(x)$. Theorem 5.5.4 is confirmed here too.

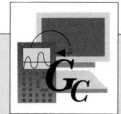

With a graphics calculator it is possible to plot both the function $h(x) = x^2$ and the indefinite integral $H(x) = \frac{1}{3}(x^3 - d^3)$. For simplicity we let $d = 0$ and plot $H(x) = \frac{x^3}{3}$.

TI-85 GRAPH F1 (y(x) =) F1 (x) ^2 ▼ F1 (x) ^3 ÷ 3 M2 (RANGE) (−) 2 ▼ 2 ▼ ▼ (−) 4 ▼ 4 F5 (GRAPH)	SHARP EL9300 [graph key] x/θ/T x^2 2ndF ▼ x/θ/T a^b 3 ► ÷ 3 RANGE − 2 ENTER 2 ENTER 1 ENTER −4 ENTER 4 ENTER 1 ENTER [graph key]
CASIO f_x-7700GB/f_x-6300G Cls EXE Range − 2 EXE 2 EXE 1 EXE − 4 EXE 4 EXE 1 EXE EXE EXE EXE Graph X x^2 EXE Graph X x^y 3 ÷ 3 EXE	HP 48SX ↰ PLOT PLOTR ERASE ATTN ′ Y = X y^x 2 ↱ PLOT ↰ DRAW ERASE 2 +/− ENTER 2 XRNG 4 +/− ENTER 4 YRNG DRAW ATTN ′ Y = X y^x 3 ÷ 3 ↱ PLOT ↰ DRAW DRAW

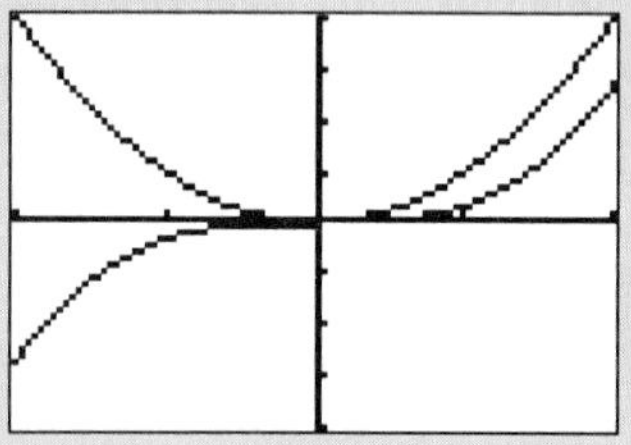

Identify $h(x)$ and $H(x)$; notice how the indefinite integral increases faster when the graph of $y = h(x)$ is higher above the x-axis.

EXAMPLE 4 Find the critical points of

$$F(x) = \int_1^x (t^2 - 1)\,dt.$$

Solution The integrand $t^2 - 1$ is continuous on the whole real line, so the Fundamental Theorem tells us that any of its indefinite integrals $F(x)$ is differentiable at every point. Moreover, it tells us that the derivative is

$$F'(x) = x^2 - 1.$$

This function is zero precisely at $x = \pm 1$, and these are the critical points of $F(x)$.

Notice the way the variables x and t were used in Example 4. The variable x is the variable used for the indefinite integral $F(x)$. It serves as the upper limit of integration in

$$F(x) = \int_1^x (t^2 - 1)\,dt.$$

Because x was used this way, we used a different letter, t, as the variable inside the integral.

The tricky thing is that when we find the derivative $F'(x)$, it should of course be a function of x. The Fundamental Theorem says that this derivative equals the integrand, but we have to be careful. In the integral itself the integrand is a function of t. Here, when it equals the derivative of $F(x)$, it must be expressed as a function of the variable x.

The integrand is $t^2 - 1$; the derivative of $F(x)$ is $x^2 - 1$.

It does not really matter what letter we use inside the integral, as long as it is not used anywhere else. The variable inside the integral is sometimes called the **integration variable** or **variable of integration**. It is also sometimes referred to as a **dummy variable** to emphasize the fact that it can be changed without affecting the integral. (This is somewhat analogous to the situation we encountered in Section 5.2, where the index of summation could be changed without affecting the value of a sum.)

The choice of integration variable is not important. In using the Fundamental Theorem it is important to express the *derivative* in terms of the appropriate variable.

EXAMPLE 5 Find the regions where $G(x)$ is increasing or decreasing if

$$G(x) = \int_{-1}^x (u+1)^3 e^u\,du.$$

Solution $G'(x) = (x+1)^3 e^x$, and we have to discuss the sign of $G'(x)$. Now $(x+1)^3$ is negative when $x + 1 < 0$, that is, when $x < -1$. Also $(x+1)^3 > 0$ when $x + 1 > 0$, that is, when $x > -1$, and $(x+1)^3 = 0$ when $x = -1$.

Since $e^x > 0$ for all x, the same remarks apply to $G'(x) = (x+1)^3 e^x$. This shows that $G(x)$ is increasing for $x > -1$, is decreasing for $x < -1$, and has a minimum at $x = -1$.

EXAMPLE 6 Find the derivative of (i) $H(x) = \int_0^{x^2} \sin\theta\,d\theta$; (ii) $K(x) = \int_x^3 \frac{e^t}{t^2+1}\,dt$.

Solution For part (i) we have to use the Chain Rule (Theorem 3.6.1, p. 186). If $F(x) = \int_0^x \sin\theta\,d\theta$, then $H(x) = F(x^2)$, so

$$H'(x) = F'(x^2)(2x) = \sin(x^2)(2x).$$

(Here we have evaluated F' using the Fundamental Theorem.)

For part (ii), write

$$K(x) = \int_x^3 \frac{e^t}{t^2+1}\,dt = -\int_3^x \frac{e^t}{t^2+1}\,dt, \qquad \text{so} \qquad K'(x) = -\frac{e^x}{x^2+1}.$$

What the Fundamental Theorem of Calculus says is that $\int_d^x f(x)\,dx$ is an **antiderivative** of $f(x)$ (using the terminology of Section 4.8). It will turn out to be very useful to discuss antiderivatives in terms of indefinite integrals.

In fact, we never really had a good notation for antiderivatives, so we can take advantage of the connection with indefinite integrals. We will write $\int f(x)\,dx$, an integral without limits of integration, to mean an antiderivative of $f(x)$. As we mentioned in Section 4.8, it is very important to write antiderivatives with a constant added at the end.

DEFINITION 5.5.5

Suppose $f(x)$ is a continuous function on some interval. Then we write

$$\int f(x)\,dx$$

to mean its general antiderivative. The word "general" means that rather than just one specific antiderivative, we are referring to all possible antiderivatives, which are obtained by adding a constant to any one of them.

EXAMPLE 7 Evaluate the following: (i) $\int x^2\,dx$; (ii) $\int \sin\theta\,d\theta$; (iii) $\int \frac{1}{t^2+1}\,dt$.

Solution

(i) We can easily guess one antiderivative for x^2, namely, $\frac{1}{3}x^3$. The general antiderivative is obtained by adding any constant c to this, so

$$\int x^2\,dx = \frac{1}{3}x^3 + c.$$

(ii) Similarly, an antiderivative of $\sin\theta$ is $-\cos\theta$, so

$$\int \sin\theta\,d\theta = -\cos\theta + c.$$

(iii) We know an antiderivative of $\frac{1}{t^2+1}$ is $\arctan t$, so the general antiderivative is

$$\int \frac{1}{t^2+1}\,dt = \arctan t + c.$$

Exercises 5.5

I

Find the derivative of each of the following functions.

1. $F(x) = \int_0^x (t^2 - 1)\,dt$
2. $F(x) = \int_a^x e^u\,du$
3. $g(x) = \int_{-2}^x \sin u\,du$
4. $h(x) = \int_1^x \frac{t-1}{t^2+1}\,dt$
5. $G(t) = \int_1^t e^u\,du$
6. $H(u) = \int_a^u \frac{1-x}{e^x+1}\,dx$
7. $K(x) = \int_1^x (e^u - u^2)\,du$
8. $k(x) = \int_2^x e^{-u^2}\,du$
9. $L(x) = \int_{-1}^x 7\,du$
10. $R(x) = \int_0^x 0\,dt$
11. $f(x) = \int_0^{x^2} \cos u\,du$
12. $g(x) = \int_0^{1+x^2} e^t\,dt$

13. $h(x) = \int_{-x}^{2} e^{t+1}\,dt$

14. $k(t) = \int_{t}^{t^2} e^x\,dx$

15. $F(u) = \int_{1-u}^{0} \sin x\,dx$

16. $G(t) = \int_{t}^{t+1} \ln x\,dx$

17. $H(r) = \int_{\cos r}^{\sin r} (t^2 - 1)\,dt$

18. $K(x) = \int_{1}^{\exp(x)} \ln u\,du$

19. $R(t) = \int_{-t}^{t} (u^3 - u)\,du$

20. $T(x) = \int_{-x}^{x} \sin u\,du$

II

21. Find the regions of increase and decrease of $F(x) = \int_1^x (t^3 - 4t)\,dt$.

22. Find the maximum value of

$$F(x) = \int_0^x (1 - t^3)\,dt.$$

(Express your answer as an integral, but do not evaluate it.)

23. Find and classify the local extrema of

$$G(x) = \int_{-x}^{x} (t^3 + 4t^2 - t - 1)\,dt.$$

24. Suppose $f(t)$ is continuous on $[-a, a]$, and define

$$H(x) = \int_{-x}^{x} f(t)\,dt,$$

for all $x \in [0, a]$. Find a formula for $H'(x)$.

***25.** Find an antiderivative for the continuous function $f(x) = |x|$.

***26.** Let $F(x) = \int_0^x |t|\,dt$. Calculate $F'(x)$ and $F''(x)$, stating at which points $F'(x)$ exists and at which points $F''(x)$ exists.

***27.** In discussing the Integral Mean Value Theorem (Theorem 5.4.4, p. 376) we mentioned that the quantity

$$\frac{1}{b-a}\int_a^b g(x)\,dx$$

could be interpreted as the average value of $g(x)$ on the interval $[a, b]$. Show that if $f(x)$ is differentiable on $[a, b]$ and $f'(x)$ is continuous on $[a, b]$, then the quantity

$$\frac{f(b) - f(a)}{b - a}$$

is in this sense equal to the average value of $f'(x)$ on $[a, b]$.

This says that the average rate of change $\frac{f(b)-f(a)}{b-a}$ equals the average value of the rate of change $f'(x)$. This is not very surprising, but it shows that our notion of the average value of a function is a reasonable one.

***28.** Show that every continuous function $f(x)$ on a closed interval $[a, b]$ has an antiderivative. In other words, show that there is a function $F(x)$ on $[a, b]$ so that $F'(x) = f(x)$ for every $x \in (a, b)$ and so that the one-sided derivative of $F(x)$ from above exists at $x = a$ and equals $f(a)$ and the one-sided derivative of $F(x)$ from below exists at $x = b$ and equals $f(b)$.

***29.** Suppose $f(x)$ is a continuous function on $[a, b]$ satisfying

$$F(x) = \int_a^x f(t)\,dt = 0, \qquad \text{for all} \quad x \in [a, b].$$

Show that $f(x) = 0$, for every $x \in [a, b]$.

***30.** (i) Suppose $f(x)$ is a continuous function on $[a, b]$ satisfying

$$F(x) = \int_a^x f(t)\,dt = x - a, \text{ for all } x \in [a, b].$$

Show that $f(x) = 1$, for every $x \in [a, b]$.

(ii) Suppose $f(x)$ and $g(x)$ are continuous functions on $[a, b]$ satisfying

$$\int_a^x f(t)\,dt = \int_a^x g(t)\,dt, \text{ for all } x \in [a, b].$$

Show that $f(x) = g(x)$, for every $x \in [a, b]$.

***31.** Suppose $f(x)$ is an odd function defined on the whole real line, that is, $f(-x) = -f(x)$ for all x. Let $F(x) = \int_{-x}^{x} f(t)\,dt$. (i) Evaluate $F'(x)$. (ii) Evaluate $F(0)$. (iii) What can you say about the function $F(x)$? (*Hint:* You should be able to say something quite forceful; if you are having trouble, draw a picture and interpret $F(x)$ as a signed area.)

***32.** Suppose $F(x)$ is a continuous function whose derivative is not defined at $x = 0$ but so that

$$F'(x) = \begin{cases} 0, & \text{if } x < 0, \\ 2, & \text{if } x > 0. \end{cases}$$

What can you say about $F(x)$?

III

33. Use your programmable calculator or personal computer to give numerical confirmation of the Fundamental Theorem of Calculus. Let $F(x) = \int_1^x e^t\,dt$, and consider $F'(1)$, which should equal $e^1 = e \approx 2.718282$. With $h > 0$, consider the difference quotient $\frac{1}{h}\left(\int_1^{1+h} e^t\,dt\right)$. (i) Use the result of Example 5.2.3 (p. 356) with $n = 10$ to estimate this integral for various values of h. Estimate the difference quotient for $h = 1$, $h = 0.1$, $h = 0.01$, and $h = 0.001$. Do they approach the correct value? (ii) Try the same calculations with $n = 20$.

34. Repeat part (i) of Exercise 33 with $h = -1, -0.1, -0.01, -0.001$.

Repeat Exercise 33(i) to estimate the derivative of each of the following indefinite integrals at the specified value of x.

35. $F(x) = \int_0^x (t^5 - 2t^2)\,dt$; $x = 1$

36. $F(x) = \int_0^x \sin u\,du$; $x = \frac{\pi}{4}$

37. $F(x) = \int_{-1}^x \arctan t\,dt$; $x = 0$

38. $F(x) = \int_1^x \frac{1}{t}\,dt$; $x = 1$

5.6 Definite Integrals

FUNDAMENTAL THEOREM OF CALCULUS (SECOND VERSION)

EVALUATING DEFINITE INTEGRALS

The greatest value of the Fundamental Theorem of Calculus is that it allows us to calculate integrals easily. We have discussed indefinite integrals, by which we mean functions of the form

$$F(x) = \int_d^x f(t)\,dt.$$

As x varies, we are considering the signed area of a changing region, and it is in this sense that the integral is "indefinite."

We also want to be able to calculate integrals of the form $\int_a^b f(t)\,dt$, with specific numbers a and b as the limits of integration. Such an integral is called a **definite integral**.

EXAMPLE 1 Calculate $\int_1^2 (2x + 6x^2)\,dx$.

Solution Consider the indefinite integral

$$F(x) = \int_1^x (2t + 6t^2)\,dt.$$

The integral we want to evaluate is just $F(2)$. By the Fundamental Theorem of Calculus (Theorem 5.5.4) we know that

$$F'(x) = 2x + 6x^2.$$

So $F(x)$ is an antiderivative of $2x + 6x^2$, so it must be of the form $F(x) = x^2 + 2x^3 + c$, for some constant c. (Note that the derivative of $x^2 + 2x^3$ equals $2x + 6x^2 = F'(x)$, and recall that two functions with the same derivative must differ by a constant, by Theorem 4.2.11 (p. 245). Here this means that $F(x) - (x^2 + 2x^3)$ is a constant, that is, $F(x) = x^2 + 2x^3 + c$ for some $c \in \mathbb{R}$.)

On the other hand, $F(1) = \int_1^1 (2x + 6x^2)\,dx = 0$, that is, $0 = F(1) = 1^2 + 2(1^3) + c = 3 + c$, so $c = -3$. This shows that $F(x) = x^2 + 2x^3 - 3$.

In particular,

$$\int_1^2 (2x + 6x^2)\,dx = F(2) = 2^2 + 2(2^3) - 3 = 17.$$

In Example 1 we could have figured out the integral by using the formulas of Example 5.3.5 for the integrals of x and x^2, but it is easy to think of examples for which we have no such formulas available.

EXAMPLE 2 Evaluate the definite integral

$$\int_{-\pi/2}^{\pi/2} \sin(u)\,du.$$

Solution If $F(x) = \int_{-\pi/2}^{x} \sin(u)\,du$, then $F'(x) = \sin x$, so $F(x) = -\cos x + c$, for some constant c.

Since $0 = \int_{-\pi/2}^{-\pi/2} \sin(u)\,du = F\left(-\frac{\pi}{2}\right) = -\cos\left(-\frac{\pi}{2}\right) + c = 0 + c$, we find that $c = 0$, so $F(x) = -\cos x$.

In particular,

$$\int_{-\pi/2}^{\pi/2} \sin(u)\,du = F\left(\frac{\pi}{2}\right) = -\cos\left(\frac{\pi}{2}\right) = 0.$$

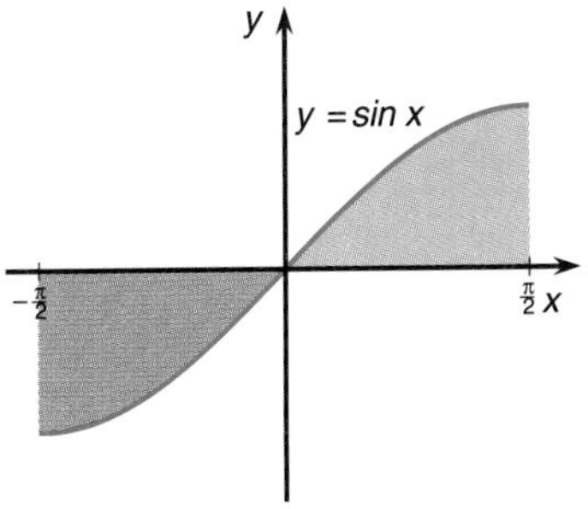

Figure 5.6.1

Figure 5.6.1 shows that the integral is the sum of two signed areas. The region between $x = -\frac{\pi}{2}$ and $x = 0$ has negative signed area, and the region between $x = 0$ and $x = \frac{\pi}{2}$ has positive signed area. Since the two regions have the same (unsigned) area, the signed areas cancel, resulting in 0.

The preceding examples demonstrate a technique for calculating the definite integral $\int_a^b f(x)\,dx$ that is vastly easier than finding the limit of Riemann sums. However, to make it work, we have to be able to guess an antiderivative $F(x)$ of the integrand $f(x)$. Sometimes this can be difficult or even impossible, but frequently it is reasonably easy, as in the above examples.

Suppose we want to calculate $\int_a^b f(x)\,dx$, and suppose we are able to find an antiderivative $F(x)$ of $f(x)$, that is, a function satisfying $F'(x) = f(x)$.

If $a < d < b$ and we compare $F(x)$ with the indefinite integral $\int_d^x f(t)\,dt$, we see that their derivatives are the same (both equaling $f(x)$), so they must differ by a constant, that is,

$$\int_d^x f(t)\,dt = F(x) + c,$$

for some constant c.

Now

$$\int_a^b f(x)\,dx = \int_a^d f(x)\,dx + \int_d^b f(x)\,dx,$$

by Theorem 5.4.5. The right side of this equation equals

$$-\int_d^a f(x)\,dx + \int_d^b f(x)\,dx,$$

by Definition 5.5.1, and this equals $-\big(F(a) + c\big) + \big(F(b) + c\big) = F(b) - F(a)$. We have just proved the following result.

THEOREM 5.6.1

The Fundamental Theorem of Calculus (Second Version)

If $f(x)$ is continuous on $[a, b]$ and $F'(x) = f(x)$ on $[a, b]$, then

$$\int_a^b f(x)\,dx = F(b) - F(a).$$

If we know an antiderivative $F(x)$ of $f(x)$, Theorem 5.6.1 says that $\int_a^b f(x)\,dx$ equals the difference of the values of $F(x)$ at $x = b$ and $x = a$. Since this will arise so often, we have special notation for it. We write

$$F(x)\Big|_a^b = F(b) - F(a).$$

The symbol on the left side is read "$F(x)$, evaluated from a to b" or "$F(x)$, evaluated between a and b."

The formula in Theorem 5.6.1 can now be written as

$$\int_a^b f(x)\,dx = F(x)\Big|_a^b.$$

EXAMPLE 3 Evaluate the following definite integrals: (i) $\int_0^1 x\,dx$; (ii) $\int_1^3 3x^2\,dx$; (iii) $\int_1^2 x^3\,dx$; (iv) $\int_{-1}^2 (3x^2 - 2x + 5)\,dx$.

Solution For each integral we need to find an antiderivative. In part (i), for an antiderivative of x we can take $F(x) = \frac{1}{2}x^2$. So

$$\int_0^1 x\,dx = F(x)\Big|_0^1 = \frac{1}{2}x^2\Big|_0^1 = \frac{1}{2}(1)^2 - \frac{1}{2}(0)^2$$
$$= \frac{1}{2}.$$

For part (ii) we can let $F(x) = x^3$ be the antiderivative, and

$$\int_1^3 3x^2\,dx = F(x)\Big|_1^3 = x^3\Big|_1^3 = 3^3 - 1^3$$
$$= 26.$$

For part (iii), take $F(x) = \frac{1}{4}x^4$, and

$$\int_1^2 x^3\,dx = \frac{1}{4}x^4\Big|_1^2 = \frac{1}{4}(2)^4 - \frac{1}{4}(1)^4$$
$$= \frac{15}{4}.$$

For part (iv) we can let $F(x) = x^3 - x^2 + 5x$, so

$$\int_{-1}^2 (3x^2 - 2x + 5)\,dx = (x^3 - x^2 + 5x)\Big|_{-1}^2$$
$$= \left(2^3 - 2^2 + 5(2)\right) - \left((-1)^3 - (-1)^2 + 5(-1)\right)$$
$$= (8 - 4 + 10) - (-1 - 1 - 5)$$
$$= 21.$$

Parts (i), (ii), and (iv) of this example could have been done by using the formulas of Example 5.3.5, but part (iii) could not.

With a graphics calculator we plot the integrand $3x^2 - 2x + 5$ from part (iv) of Example 3, for $x \in [-1, 2]$.

TI-85 GRAPH F1 $(y(x) =)$ 3 F1 (x) ^ 2 − 2 F1 (x) + 5 M2(RANGE) (−) 1 ▼ 2 ▼ ▼ 0 ▼ 9 F5(GRAPH)	SHARP EL9300 3 x/θ/T x^2 − 2 x/θ/T + 5 RANGE −1 ENTER 2 ENTER 1 ENTER 0 ENTER 9 ENTER 1 ENTER
CASIO f_x-7700GB/f_x-6300G Cls EXE Range −1 EXE 2 EXE 1 EXE 0 EXE 9 EXE 1 EXE EXE EXE EXE Graph 3 X x^2 − 2 X + 5 EXE	HP 48SX ↰ PLOT PLOTR ERASE ATTN ' Y = 3 × X y^x 2 − 2 × X + 5 ↱ PLOT ↰ DRAW ERASE 1 +/− ENTER 2 XRNG 0 ENTER 9 YRNG DRAW

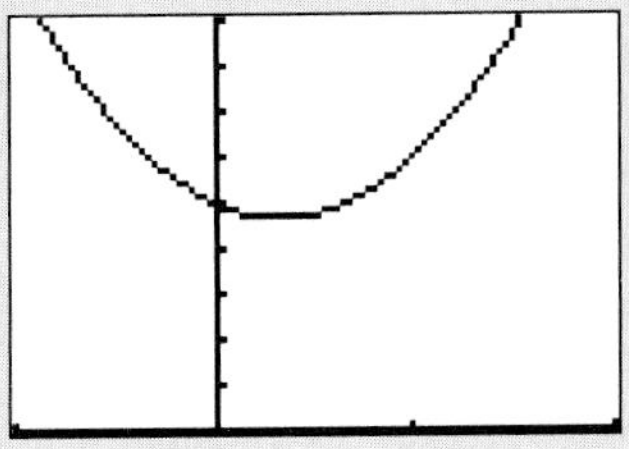

We can try to estimate the area under the graph by looking at the picture. There is a rectangular region 3 units wide and slightly more than 4 units high lying between the x-axis and the lowest point on the graph. This has an area slightly over 12. The remaining part of the region (under the curved ends of the graph) appears to have a total area that is less than the area of this rectangle, but more than half as big.

This suggests the actual area is between $12 + 6 = 18$ and $12 + 12 = 24$, which we see is correct.

EXAMPLE 4 In using this technique for evaluating definite integrals we have simply guessed an antiderivative $F(x)$ of $f(x)$. This may or may not be possible, but even if it can be done, there will be many possible choices for $F(x)$. After all, if $F(x)$ is an antiderivative of $f(x)$, then so is $F(x) + c$ for any constant c. Happily, if we use $F(x) + c$ to evaluate a definite integral of $f(x)$, we get

$$\int_a^b f(x)\,dx = \big(F(x) + c\big)\Big|_a^b = \big(F(b) + c\big) - \big(F(a) + c\big)$$
$$= F(b) + c - F(a) - c = F(b) - F(a)$$
$$= F(x)\Big|_a^b .$$

In other words, the c's cancel, and choosing different antiderivatives does not affect the answer. It will be enough to choose any convenient antiderivative, such as $\frac{1}{3}x^3$ for x^2, $-\cos x$ for $\sin x$, etc.

EXAMPLE 5 Evaluate

(i) $\int_0^{\pi/2} \sin\theta\, d\theta$; (ii) $\int_0^{\pi/4} \sec^2 u\, du$; (iii) $\int_a^b x^n\, dx$, $(n \neq -1)$;

(iv) $\int_0^1 e^{-5x} dx$; (v) $\int_{-1}^0 u^2 \cos(u^3+4)\, du$; (vi) $\int_0^1 \frac{1}{1+x^2} dx$.

Solution

(i) $\int_0^{\pi/2} \sin\theta\, d\theta = -\cos\theta\Big|_0^{\pi/2} = -\cos\left(\frac{\pi}{2}\right) - \big(-\cos(0)\big) = -0 - (-1) = 1.$

(ii) For this one we have to remember that $\frac{d}{du}\tan u = \sec^2 u$, and since $\sec^2 u$ is continuous on $\left[0, \frac{\pi}{4}\right]$,

$$\int_0^{\pi/4} \sec^2 u\, du = \tan u\Big|_0^{\pi/4} = \tan\left(\frac{\pi}{4}\right) - \tan(0) = 1 - 0$$
$$= 1.$$

(iii) An antiderivative of x^n is obviously $\frac{1}{n+1}x^{n+1}$, so

$$\int_a^b x^n\, dx = \frac{1}{n+1}x^{n+1}\Big|_a^b = \frac{1}{n+1}b^{n+1} - \frac{1}{n+1}a^{n+1}$$
$$= \frac{1}{n+1}\left(b^{n+1} - a^{n+1}\right).$$

(iv) If we think for a moment, it is possible to guess an antiderivative for e^{-5x}, namely, $F(x) = -\frac{1}{5}e^{-5x}$ (cf. Example 4.8.5(iv), p. 308).

$$\int_0^1 e^{-5x} dx = -\frac{1}{5}e^{-5x}\Big|_0^1 = -\frac{1}{5}e^{-5} - \left(-\frac{1}{5}e^0\right)$$
$$= \frac{1}{5}(1 - e^{-5}).$$

(v) To guess an antiderivative for $u^2\cos(u^3+4)$, we should think of the Chain Rule. Writing $u^2\cos(u^3+4) = \left(\frac{1}{3}\right)3u^2\cos(u^3+4)$, we see that an antiderivative is $\frac{1}{3}\sin(u^3+4)$ (cf. Example 4.8.6(iii), p. 308). From this we get that

$$\int_{-1}^0 u^2\cos(u^3+4)\, du = \frac{1}{3}\sin(u^3+4)\Big|_{-1}^0$$
$$= \frac{1}{3}\sin(4) - \frac{1}{3}\sin(3) = \frac{1}{3}\big(\sin(4) - \sin(3)\big)$$
$$\approx -0.8979.$$

(vi) Recall that $\frac{1}{1+x^2}$ is the derivative of $\arctan x$, so

$$\int_0^1 \frac{1}{1+x^2} dx = \arctan x\Big|_0^1 = \arctan(1) - \arctan(0) = \frac{\pi}{4} - 0$$
$$= \frac{\pi}{4}.$$

In doing these problems we needed to have an antiderivative $\int f(x)\,dx$. Previously, we emphasized that the general antiderivative should always include a constant added at the end. Example 4 tells us that *for this purpose,* the evaluation of definite integrals, that constant is irrelevant. So we allow ourselves to omit it and write, for instance,

$$\int_{-\pi}^{\pi} \cos\theta\, d\theta = \sin\theta\Big|_{-\pi}^{\pi} = \sin(\pi) - \sin(-\pi)$$
$$= 0.$$

When we can guess an antiderivative of the integrand, it is easy to evaluate definite integrals. For some examples, guessing an antiderivative may be *extremely* difficult. Chapter 8 will be devoted to developing techniques for finding indefinite integrals (i.e., antiderivatives) of many different types of functions.

Exercises 5.6

I

Evaluate the following integrals.

1. $\int_0^2 (x^4 - 2x)\,dx$
2. $\int_{-2}^1 x^7\,dx$
3. $\int_1^3 (4x^3 - 6x^2 - 4x)\,dx$
4. $\int_{-2}^2 (x^7 - 3x^3)\,dx$
5. $\int_0^\pi \sin u\,du$
6. $\int_{-\pi}^0 \sin t\,dt$
7. $\int_{-1}^1 \sin x\,dx$
8. $\int_{-1}^1 \cos x\,dx$
9. $\int_0^1 e^x\,dx$
10. $\int_1^e \frac{1}{x}\,dx$
11. $\int_0^1 e^{2x}\,dx$
12. $\int_{-1}^0 e^{-x}\,dx$
13. $\int_0^\pi \sin(2t)\,dt$
14. $\int_0^\pi \sin(5t)\,dt$
15. $\int_0^3 \sin(\pi u)\,du$
16. $\int_0^2 \sec^2\left(\frac{\pi}{8}x\right)dx$
17. $\int_0^4 \sqrt{x}\,dx$
18. $\int_0^1 \cosh t\,dt$
19. $\int_0^1 x(x^2+3)^{17}\,dx$
20. $\int_0^1 \frac{x}{x^2+1}\,dx$
21. $\int_0^3 |x-2|\,dx$
22. $\int_1^4 |x-2|\,dx$
23. $\int_0^3 |x^2-4|\,dx$
24. $\int_{-4}^4 |x^2-4|\,dx$
25. $\int_{-4}^4 |x^2-x-6|\,dx$
26. $\int_0^3 |t-1||t-2|\,dt$
27. $\int_{-2}^3 |x-2||x^2-1|\,dx$
28. $\int_{-3}^2 |x^3-x|\,dx$
29. $\int_0^2 |x-1|^2\,dx$
30. $\int_{-3}^3 |x^2-x-2|\,dx$

II

Evaluate the following.

31. (i) $\int_0^{\pi/2} \cos(2x)\,dx$; (ii) $\int_{-1}^1 e^{x/2}\,dx$; (iii) $\int_{-1}^0 \sin(\pi x)\,dx$
32. (i) $\int_{-\pi/2}^0 \cos^2 t \sin t\,dt$; (ii) $\int_0^1 \frac{3}{u^2+1}\,du$; (iii) $\int_0^1 \frac{u^2}{u^2+1}\,du$

If you have studied the hyperbolic functions in Section 3.5, evaluate the following definite integrals.

33. (i) $\int_0^1 \cosh t\,dt$; (ii) $\int_{-3}^3 \sinh u\,du$; (iii) $\int_0^2 \frac{1}{\cosh^2 x}\,dx$
34. (i) $\int_0^1 \cosh t\,dt - \int_{-1}^0 \cosh t\,dt$; (ii) $\int_{-1}^0 \sinh 4t\,dt$

The next four questions concern the functions $f(x)$ and $g(x)$ defined below. Let

$$f(x) = \begin{cases} x, & \text{if } x \le -1, \\ x^2 - 2, & \text{if } -1 \le x \le 2, \\ x^3 - 1, & \text{if } 2 < x; \end{cases}$$

$$g(x) = \begin{cases} x^2, & \text{if } x \le 0, \\ 0, & \text{if } 0 \le x < 1, \\ 3, & \text{if } 1 \le x. \end{cases}$$

35. Evaluate (i) $\int_{-2}^3 f(x)\,dx$, (ii) $\int_{-1}^3 f(x)\,dx$, (iii) $\int_{-3}^3 f(x)\,dx$.
36. With $f(x)$ and $g(x)$ as above, evaluate (i) $\int_0^2 g(x)\,dx$, (ii) $\int_{-1}^1 g(x)\,dx$, (iii) $\int_{-1}^2 g(x)\,dx$.
37. With $f(x)$ and $g(x)$ as above, evaluate (i) $\int_0^2 f(x)g(x)\,dx$, (ii) $\int_{-1}^3 \big(f(x) - 2g(x)\big)\,dx$, (iii) $\int_{-2}^2 \sqrt{g(x)}\,dx$.
38. With $f(x)$ and $g(x)$ as above, evaluate (i) $\int_0^1 f(x)g(x)\,dx$, (ii) $\int_{-2}^{-1} \frac{1}{g(x)}\,dx$, (iii) $\int_{-2}^0 f(x)^2 g(x)\,dx$.

*39. (i) Suppose $a > 0$ and evaluate $\int_{-a}^a x\,dx$.
(ii) Suppose $a > 0$ and evaluate $\int_{-a}^a x^3\,dx$.
(iii) Suppose $a > 0$ and evaluate $\int_{-a}^a x^k\,dx$, where k is any odd positive integer.

***40.** Suppose $a > 0$ and $f(x)$ is an odd function. What can you say about $\int_{-a}^{a} f(x)\,dx$?

***41.** (i) Suppose $a > 0$ and evaluate $\int_0^a x^2 dx$, $\int_{-a}^0 x^2 dx$.

(ii) Suppose $a > 0$ and evaluate $\int_0^a x^4 dx$, $\int_{-a}^0 x^4 dx$.

(iii) Suppose $a > 0$ and evaluate $\int_0^a x^k dx$, $\int_{-a}^0 x^k dx$, where k is any positive even integer.

***42.** Suppose $a > 0$ and $f(x)$ is an even function. What can you say about $\int_{-a}^0 f(x)\,dx - \int_0^a f(x)\,dx$?

***43.** Evaluate

$$\int_0^1 \left(\int_1^x 2t\,dt \right) dx.$$

***44.** Evaluate

$$\int_{-1}^2 \left(\int_1^2 xt\,dt \right) dx.$$

***45.** The second version of the Fundamental Theorem of Calculus (Theorem 5.6.1) was proved from the first version of the Fundamental Theorem of Calculus (Theorem 5.5.4). Show that it can be done the other way around, that is, that if we assume Theorem 5.6.1, we can use it to prove Theorem 5.5.4, provided that we know that the integrand $f(x)$ has an antiderivative.

***46.** The function $f(x) = x^2$ is always nonnegative, so its integral should always be nonnegative, by Theorem 5.4.2. However,

$$\int_0^{-1} f(x)\,dx = \frac{1}{3}\left((-1)^3 - 0^3\right) = -\frac{1}{3}.$$

Explain why this is not really a contradiction.

III

For each of the following integrals, evaluate the Riemann sum approximation from Example 5.2.3 (p. 356) with $n = 100$ and compare with the exact value.

47. $\int_0^4 x^3 dx$

48. $\int_0^\pi \sin\theta\,d\theta$

49. $\int_{-1}^1 \frac{dt}{1+t^2}$

50. $\int_0^{\pi/4} \frac{1}{\cos^2\theta}\,d\theta$

POINT TO PONDER

Now that we have finally developed a convenient way to integrate, it might be interesting to remind ourselves that the chapter began with a question about areas. Try to trace the connection between (signed) areas under curves and expressions of the form $F(b) - F(a)$.

In particular, convince yourself that the two coincide for polygonal regions, that is, regions with straight sides (i.e., integrating functions whose graphs are made up of pieces of straight lines).

For more complicated regions there is no a priori concept of area (cf. the Point to Ponder in Section 5.1), so integrals can provide a way of defining what areas mean. We should at least be sure that they satisfy the properties we expect of areas. If you thought of some of these properties in Section 5.1, try to check that they hold for areas defined by integrals.

5.7 Integration by Substitution

CHAIN RULE
DEFINITE INTEGRALS

In Section 5.6 we learned that evaluating definite integrals is quite easy *provided* that we know an antiderivative of the integrand, that is, an indefinite integral. When we discussed finding antiderivatives in Section 4.8, we found that there were many instances in which it was possible to do it by guessing, but others were more complicated. Now that we have seen the importance of being able to find indefinite integrals, we will have to learn techniques for it. This will be the purpose of all of Chapter 8, but in this section we will begin with the first of these techniques.

When we first encountered antiderivatives in Section 4.8, one kind of function whose antiderivative was less easy to guess was the kind that resulted from applying the Chain Rule to a composite function. We developed an approach to help with this type of problem, and it is the purpose of this section to make that approach more systematic. Being able to speak in terms of indefinite integrals instead of antiderivatives makes it easier to do than when we first attempted the problem.

We will need to use the concept of differentials from Section 4.10.

EXAMPLE 1 Find an antiderivative of xe^{x^2}, that is, find $\int xe^{x^2}\,dx$.

Solution From Section 4.8 we recall how to approach this sort of question. Writing $xe^{x^2} = \frac{1}{2}2xe^{x^2}$, we recognize the factor $2x$ as the derivative of the function x^2, so $2xe^{x^2}$ is the derivative of $e^{x^2} = e^u$, with $u = x^2$. So an antiderivative of xe^{x^2} is $\frac{1}{2}e^{x^2} + c$.

This solves the problem, but let us pursue a little further what happens if we try to write it in terms of indefinite integrals.

The approach we just used suggests letting $u = x^2$. So $e^{x^2} = e^u$, and we could try substituting this into the integral:

$$\int xe^{x^2}\,dx = \int xe^u\,dx.$$

The difficulty here is that there are two different variables, x and u, appearing in the same integral. Since $u = x^2$, we could consider trying to write x as a function of u (i.e., solving for x as a function of u). Doing this could help us get rid of the factor x at the beginning of the integrand, but it would leave us with the apparently more difficult problem of what to do about the dx at the end.

Fortunately, the solution is not so difficult. We deal with du and dx by thinking of them as differentials, and since $u = x^2$, we know that

$du = \frac{du}{dx}\,du$ and $\frac{du}{dx} = \frac{d}{dx}(x^2) = 2x$.

$$du = 2x\,dx.$$

Dividing by 2, we get $\frac{1}{2}du = x\,dx$. This allows us to rewrite the integral:

$$\int xe^{x^2}\,dx = \int e^{x^2}x\,dx = \int e^u\frac{1}{2}\,du$$
$$= \frac{1}{2}\int e^u\,du.$$

Notice that the factor x in the original integrand combined with the dx to give $\frac{1}{2}du$.

The point of all this is that it is very easy to evaluate the last integral: $\frac{1}{2}\int e^u\,du = \frac{1}{2}e^u + c$. In a sense this is the answer, but the original problem was stated in terms of the variable x, so we ought to convert our answer back into a function of x. Since $u = x^2$, nothing could be easier:

$$\frac{1}{2}e^u + c = \frac{1}{2}e^{x^2} + c.$$

This seems like a lot of work to do something that we already knew how to do with less trouble, but in fact this approach will be extremely useful.

EXAMPLE 2 Find $\int x^2(x^3-1)^{11}\,dx$.

Solution We begin by observing that the x^2 factor looks almost like the derivative of the quantity inside parentheses. So we let that quantity equal u, that is, $u = x^3 - 1$.
Then we work with differentials:

$$du = \frac{du}{dx}dx = 3x^2\,dx.$$

Since the original integral contained $x^2\,dx$, we can divide the differentials by 3 to get $\frac{1}{3}du = x^2\,dx$.
Now the original integral can be written as

$$\begin{aligned}\int x^2(x^3-1)^{11}\,dx &= \int (x^3-1)^{11}x^2\,dx = \int u^{11}\frac{1}{3}du \\ &= \frac{1}{3}\int u^{11}\,du = \frac{1}{3}\frac{1}{12}u^{12} + c \\ &= \frac{1}{36}(x^3-1)^{12} + c.\end{aligned}$$

There are several things to notice here. One is that because u is defined as a function of x, it is relatively easy to find the differential du, so that is what we did. However, the way we used this formula was really the other way around, allowing us to change dx into something involving du.

In fact it was not dx but rather $x^2\,dx$ that we expressed in terms of du. The key to the whole calculation was recognizing that that factor x^2 looked like the derivative of $x^3 - 1$, so if we let $u = x^3 - 1$ then $du = 3x^2\,dx$. Because of this we could express the integral in terms of u, and the $x^2\,dx$ would turn into a constant times du, leaving the simple integrand u^{11}. In making this recognition we were able to ignore the constant 3, knowing that it would be taken care of in the computations.

The second thing to notice is the very last step, in which the answer was converted back from a function of u to a function of x. This is easy but important.

Finally, as always, we should not forget the "$+c$" at the end of the indefinite integral.

The technique we have illustrated with these two examples is called *integration by substitution* because we substitute the variable u for some function of x and remove the variable x. The idea and the calculations are relatively easy; what can sometimes make it difficult is the absence of a hard and fast rule. All we can do is guess a substitution, try it out, and see if it works. We will agree that it works if the original integral is converted into one that is easily integrated. Otherwise, it is probably not of much use.

It is important to recognize that there is a process of guessing, of trial and error. Nobody expects to see the correct substitution immediately and be certain that it will work. It is necessary to guess and then to try out your guess and, if it does not work, to guess again.

One vital requirement for any substitution to succeed is that the original integral (in the variable x, say) should be converted into one in terms of u, say. There should be no x's in the integral after the substitution. If some occurrences of x do remain, we should try to express them in terms of u. If this is not possible, the attempt must be abandoned and a new substitution found.

Summary 5.7.1

Integration by Substitution

Consider the indefinite integral $\int f(x)\,dx$.

1. Look at the integrand; try to find in it a factor $u'(x)$, so that the remaining part of the integrand can be expressed as a (relatively) simple function of u. At this stage, do not worry too much about constants.
2. Make the substitution. Letting u be the function suggested by the observation just made, calculate the differential $du = u'(x)\,dx$, and rewrite the integral in terms of u. If there are any x's left in the integral, try to express them in terms of u; if this is impossible, go back to Step 1 and look for another substitution.
3. Evaluate the indefinite integral in terms of u, being sure to include the constant c at the end.
4. Substitute back for u in the resulting function to write it as a function of the original variable x.

If this procedure is unsuccessful, something else may be necessary. One trick that is frequently helpful is to express the integral as a sum of two or more integrals and then work with each of them separately. Sometimes different substitutions will work for different parts of the integrand. Far in the back of your mind you should know that there are some integrals that cannot be evaluated by the technique of substitution, but you should not let that knowledge discourage you from trying.

The best thing about indefinite integrals is that it is very easy to check your work. Once you have finished, all that is needed to verify the answer is to differentiate it. The resulting derivative should equal the original integrand.

EXAMPLE 3 Evaluate the following indefinite integrals:

(i) $\int \cos^2 x \sin x\,dx$, (iv) $\int e^{2-3u}du$,

(ii) $\int (x+1)\sqrt{x^2+2x+4}\,dx$, (v) $\int x(x+1)^6 dx$,

(iii) $\int (3t-1)^3 dt$, (vi) $\int \tan\theta\,d\theta$.

Solution

(i) Here we notice that, up to a minus sign, $\sin x$ is the derivative of $\cos x$. Letting $u = \cos x$, we find $du = -\sin x\,dx$, so

$$\int \cos^2 x \sin x\,dx = \int u^2(-1)\,du = -\int u^2 du = -\frac{1}{3}u^3 + c$$
$$= -\frac{1}{3}\cos^3 x + c.$$

(ii) $\int (x+1)\sqrt{x^2+2x+4}\,dx$. Here we can guess that the factor $x+1$ is (approximately) the derivative of the quantity inside the square root sign. Letting u equal that quantity, $u = x^2+2x+4$, we find $du = (2x+2)\,dx = 2(x+1)\,dx$.

So

$$\int (x+1)\sqrt{x^2+2x+4}\,dx = \int \sqrt{x^2+2x+4}(x+1)\,dx = \int \sqrt{u}\frac{1}{2}du$$
$$= \frac{1}{2}\int u^{1/2}du = \frac{1}{2}\frac{1}{3/2}u^{3/2} + c$$
$$= \frac{1}{3}(x^2+2x+4)^{3/2} + c.$$

This result is easily checked by differentiation.

(iii) $\int (3t-1)^3 dt$. At first glance there does not seem to be anything that is the derivative of anything else here. However, a nice choice for the function u would be $3t-1$, the function inside parentheses. Its derivative is just 3, and with our usual understanding about constants we should try this substitution.

If $u = 3t-1$, then $du = 3\,dt$, and

$$\int (3t-1)^3 dt = \int u^3 \frac{1}{3} du = \frac{1}{3}\int u^3 du = \frac{1}{3} \times \frac{1}{4}u^4 + c$$
$$= \frac{1}{12}(3t-1)^4 + c.$$

Notice that in this case the original integral was expressed in terms of the dummy variable t instead of x, but that had no effect on the procedure we used.

(iv) $\int e^{2-3u}du$. Here again, there is no apparent derivative until we recognize that the exponent $2-3u$ has as its derivative the constant -3. Since in this case the original integral is in terms of the variable u, it would be confusing to use the same letter for the substitution.

Letting $v = 2-3u$, we find $dv = (-3)\,du$, and

$$\int e^{2-3u}du = \int e^v\left(-\frac{1}{3}\right)dv = -\frac{1}{3}\int e^v dv = -\frac{1}{3}e^v + c$$
$$= -\frac{1}{3}e^{2-3u} + c.$$

(v) $\int x(x+1)^6 dx$. Our first guess here might be to let $u = x+1$. In this case, $du = dx$, which looks fine, but when we substitute, we get $\int x(x+1)^6 dx = \int xu^6 du$, and there is still that factor of x in the integral. We could despair, and simply multiply out $(x+1)^6$ and integrate each term. However, first we should try to express that x in terms of u. After all, if $u = x+1$, then $x = u-1$. So we are able to perform the integration:

$$\int x(x+1)^6 dx = \int xu^6 du = \int (u-1)u^6 du$$
$$= \int (u^7 - u^6)\,du = \frac{1}{8}u^8 - \frac{1}{7}u^7 + c$$
$$= \frac{1}{8}(x+1)^8 - \frac{1}{7}(x+1)^7 + c.$$

This was definitely easier than expanding $(x+1)^6$ and integrating. It has the additional advantage that the answer is expressed in a much more compact form.

(vi) $\int \tan\theta\, d\theta$. At first glance there seems to be nothing we can do with $\tan\theta$, but as always with trigonometric functions, the thing to do when you are stumped is to write everything in terms of $\cos x$ and $\sin x$.

$$\int \tan\theta\, d\theta = \int \frac{\sin\theta}{\cos\theta}\, d\theta.$$

Now things look better. Each of $\sin\theta$ and $\cos\theta$ is the other's derivative (up to the usual sign). If we tried $u = \sin\theta$, we would find $du = \cos\theta\, d\theta$. The difficulty with this is that it is not what appears in the integral. There we have $\frac{1}{\cos\theta}\, d\theta$, which is quite different.

Not discouraged, we try the other one. Letting $u = \cos\theta$, we have $du = -\sin\theta\, d\theta$, so

$$\begin{aligned}\int \tan\theta\, d\theta &= \int \frac{\sin\theta}{\cos\theta}\, d\theta = \int \frac{1}{u}(-1)\, du = -\int \frac{1}{u}\, du \\ &= -\ln|u| + c = -\ln|\cos\theta| + c \\ &= \ln|\sec\theta| + c.\end{aligned}$$

Part (vi) of Example 3 is interesting in that neither the integrand nor the integral is defined for all real numbers. Depending on the use to which we intend to put it, it will be important to be careful about how it is interpreted.

Notice too that we have allowed ourselves to write

$$\int \frac{1}{u}\, du = \ln|u| + c,$$

instead of the more complicated antiderivative discussed in Example 4.8.4 (p. 306). This is pure laziness, but it will not cause much trouble. It is our intention to use this technique for evaluating definite integrals, and in light of Example 5.6.4 (p. 395), omitting the extra constant there will not matter.

Now that we have begun to get used to the technique of substitution, it is desirable to have a more efficient way of writing down what we are doing. A convenient system is to write down the substitution in a little box beside the integral. Since it is necessary to find the differential of the function being substituted, we write that in the box too.

EXAMPLE 4 Find the following indefinite integrals:

(i) $\int \frac{\sin(\sqrt{t})}{\sqrt{t}}\, dt$, (iii) $\int x^3\sqrt{x^2+2}\, dx$,

(ii) $\int \tan^3 x \sec^2 x\, dx$, (iv) $\int \frac{1}{4+x^2}\, dx$.

Solution

(i) $\int \frac{\sin(\sqrt{t})}{\sqrt{t}}\, dt$. It is often helpful to rewrite square roots as powers with exponent $\frac{1}{2}$. Doing this, we find

$$\begin{aligned}\int \frac{\sin(\sqrt{t})}{\sqrt{t}}\, dt &= \int \frac{\sin(t^{1/2})}{t^{1/2}}\, dt \\ &= \int \sin(t^{1/2})\, t^{-1/2}\, dt.\end{aligned}$$

Since, apart from a constant, $t^{-1/2}$ is the derivative of $t^{1/2}$, this suggests what substitution we should try:

$$\int \frac{\sin(\sqrt{t})}{\sqrt{t}}\,dt = \int \sin(t^{1/2})t^{-1/2}\,dt$$

$u = t^{1/2}$
$du = \frac{1}{2}t^{-1/2}dt$

$$= \int (\sin u)2\,du = 2\int \sin u\,du$$
$$= -2\cos u + c = -2\cos(t^{1/2}) + c$$
$$= -2\cos(\sqrt{t}) + c.$$

(ii) $\int \tan^3 x \sec^2 x\,dx$. For this one we ought to stare at it for a while, and with luck we will see what to do.

$$\int \tan^3 x \sec^2 x\,dx = \int u^3\,du$$

$u = \tan x$
$du = \sec^2 x\,dx$

$$= \frac{1}{4}u^4 + c$$
$$= \frac{1}{4}\tan^4 x + c.$$

(iii) $\int x^3\sqrt{x^2+2}\,dx$. Here it is not immediately clear what to do. If we try $u = x^3$, we find its derivative disguised in the x^2, but that term is inside the square root, where it does not easily combine with the dx to make the differential du.

A second attempt would be to try the function inside the square root:

$$\int x^3\sqrt{x^2+2}\,dx = \int x^2\sqrt{x^2+2}\,x\,dx$$

$u = x^2 + 2$
$du = 2x\,dx$

Since $u = x^2 + 2$, we find $x^2 = u - 2$.

$$= \int x^2\sqrt{u}\frac{1}{2}du = \frac{1}{2}\int (u-2)u^{1/2}du$$
$$= \frac{1}{2}\int (u^{3/2} - 2u^{1/2})\,du = \frac{1}{2}\left(\frac{2}{5}u^{5/2} - 2\times\frac{2}{3}u^{3/2}\right) + c$$
$$= \frac{1}{5}(x^2+2)^{5/2} - \frac{2}{3}(x^2+2)^{3/2} + c.$$

(iv) $\int \frac{1}{4+x^2}\,dx$. Here there does not seem to be any obvious substitution available, but it will help to rewrite the integrand:

$$\frac{1}{4+x^2} = \frac{1}{4}\left(\frac{1}{1+\frac{x^2}{4}}\right) = \frac{1}{4}\left(\frac{1}{1+\left(\frac{x}{2}\right)^2}\right).$$

This suggests the substitution $u = \frac{x}{2}$:

$$\int \frac{1}{4+x^2}\,dx = \frac{1}{4}\int \frac{1}{1+\left(\frac{x}{2}\right)^2}\,dx$$

$u = \frac{x}{2}$
$du = \frac{1}{2}dx$

$$= \frac{1}{4}\int \frac{1}{1+u^2}(2)\,du = \frac{1}{2}\int \frac{1}{1+u^2}\,du$$
$$= \frac{1}{2}\arctan u + c$$
$$= \frac{1}{2}\arctan\left(\frac{x}{2}\right) + c.$$

We include two integrals that will be needed in the next chapter.

EXAMPLE 5 Find the following indefinite integrals:

(i) $\int \cos^2 x\,dx$, (ii) $\int \sin^2 x\,dx$.

Solution

(i) Since no obvious substitution presents itself, it is necessary to resort to a trick. Recalling the trigonometric identity from Theorem 1.7.12 (p. 51),

$$\cos 2x = \cos^2 x - \sin^2 x,$$

we write $\sin^2 x = 1 - \cos^2 x$ and get

$$\cos 2x = \cos^2 x - (1 - \cos^2 x) = 2\cos^2 x - 1.$$

Adding 1 to both sides gives $\cos(2x) + 1 = 2\cos^2 x$, or $\cos^2 x = \frac{\cos(2x)+1}{2}$. So

$$\int \cos^2 x\,dx = \int \left(\frac{\cos(2x)+1}{2}\right) dx = \frac{1}{2}\int \cos(2x)\,dx + \frac{1}{2}\int 1\,dx$$

$u = 2x$
$du = 2\,dx$

$$= \frac{1}{2}\left(\frac{1}{2}\right)\int \cos u\,du + \frac{1}{2}\int 1 dx$$

$$= \frac{1}{4}\sin u + \frac{x}{2} + c$$

$$= \frac{1}{4}\sin(2x) + \frac{x}{2} + c.$$

(ii) We can integrate $\int \sin^2 x\,dx$ by a similar trick, or by using the fact that $\sin^2 x = 1 - \cos^2 x$, so

$$\int \sin^2 x\,dx = \int (1 - \cos^2 x)\,dx = \int 1\,dx - \int \cos^2 x\,dx$$

$$= x - \left(\frac{1}{4}\sin(2x) + \frac{x}{2}\right) + c$$

$$= \frac{x}{2} - \frac{1}{4}\sin(2x) + c.$$

Notice that when we did the integrations in part (ii), we wrote $\frac{1}{4}\sin(2x) + \frac{x}{2}$ for $\int \cos^2 x\,dx$, without the added constant, and added a constant c after all the integrations. This way we end up with the usual "$+c$" at the end. If we included the constant in the integral of $\cos^2 x$, then our final formula would have ended with "$-c$." This complication is unnecessary.

Essentially all that matters is that there must be a "$+c$" with any indefinite integral. As long as it is there when we finish, we do not have to worry about where it came from. In particular, if we have a formula that involves the sum of several integrals, it is not necessary to add a constant to each of them; a single constant is enough. In the above calculation, where we integrated 1 and $\cos^2 x$, all we needed was *one* indefinite integral for each of them. We chose x and $\frac{1}{4}\sin(2x) + \frac{x}{2}$, respectively, and then added the required constant at the end.

Definite Integrals and Substitution

One of the main purposes of finding indefinite integrals is to use them to calculate definite integrals. If we have a definite integral to evaluate, we may find it necessary to use substitution to find the indefinite integral, after which the definite integral can be found in the usual way, using Theorem 5.6.1.

EXAMPLE 6 Evaluate

$$\int_0^{\pi/4} \sec^2 t \sin t\, dt.$$

Solution First we find the indefinite integral $\int \sec^2 t \sin t\, dt$. The presence of the factor $\sec^2 t$ suggests the $\tan t$ substitution, but that does not seem to get us anywhere. Instead, we write $\sec^2 t = \frac{1}{\cos^2 t}$ and try

$$\int \sec^2 t \sin t\, dt = \int \frac{1}{\cos^2 t} \sin t\, dt$$

$u = \cos t$
$du = -\sin t\, dt$

$$= \int u^{-2}(-1)\, du = -\int u^{-2} du$$

$$= -\frac{1}{-1} u^{-1} + c = \frac{1}{\cos t} + c$$

$$= \sec t + c.$$

Now it is easy to evaluate the definite integral:

$$\int_0^{\pi/4} \sec^2 t \sin t\, dt = \sec t \Big|_0^{\pi/4} = \sqrt{2} - 1 \approx 0.41421.$$

This is a perfectly reasonable way to evaluate integrals. It is also possible to do it another way, in which the substitution is performed on the definite integral rather than the indefinite integral. The idea is to transform the definite integral in the original variable into a definite integral in the substituted variable. To do this, it is necessary to change the limits of integration in accordance with the substitution.

EXAMPLE 7 Find

$$\int_2^3 \frac{t}{t-1} dt.$$

Solution After some thought, we realize that a good substitution to make would be $u = t - 1$. In this case we find that $du = dt$ and the numerator of the integrand is $t = u + 1$.

It is easy to write down the new form of the integral, but we have to consider what happens to the limits of integration as we make the substitution.

When $t = 2$, we see that $u = t - 1 = 1$, and when $t = 3$, $u = t - 1 = 2$. So we can write the integral as follows:

$$\int_2^3 \frac{t}{t-1} dt = \int_1^2 \frac{u+1}{u} du = \int_1^2 \left(\frac{u}{u} + \frac{1}{u}\right) du$$

$u = t - 1$
$du = dt$

$$= \int_1^2 \left(1 + \frac{1}{u}\right) du = (u + \ln|u|)\Big|_1^2$$
$$= (2 + \ln 2) - (1 + \ln 1) = 1 + \ln 2$$
$$\approx 1.693.$$

Either of these methods will work perfectly well, and it is largely a matter of personal taste which one you wish to use. Finding the indefinite integral and then evaluating the definite integral gives the appearance of involving a separate step, but the second method involves figuring out the new limits of integration.

There is one additional subtlety that occasionally causes some trouble. Fortunately, it does not usually matter, but for the second approach to work, it is necessary that the substituted function have a continuous derivative on the interval of integration and that the integrand be continuous and have an antiderivative that is defined at every point of the interval.

EXAMPLE 8 Evaluate

$$\int_{-\pi/2}^{\pi/2} \frac{\cos x}{\sin^2 x}\, dx.$$

Solution This is quite similar to Example 6. We are tempted to try $u = \sin x$, in which case $du = \cos x\, dx$. When $x = -\frac{\pi}{2}$, $u = \sin x = \sin\left(-\frac{\pi}{2}\right) = -1$, and when $x = \frac{\pi}{2}$, $u = \sin\left(\frac{\pi}{2}\right) = 1$.

This seems to suggest the following calculation:

$$\int_{-\pi/2}^{\pi/2} \frac{\cos x}{\sin^2 x}\, dx = \int_{-1}^{1} u^{-2}\, du$$

$u = \sin x$
$du = \cos x\, dx$

$$= \frac{1}{-1} u^{-1}\Big|_{-1}^{1} = -\left(1 - (-1)\right)$$
$$= -2.$$

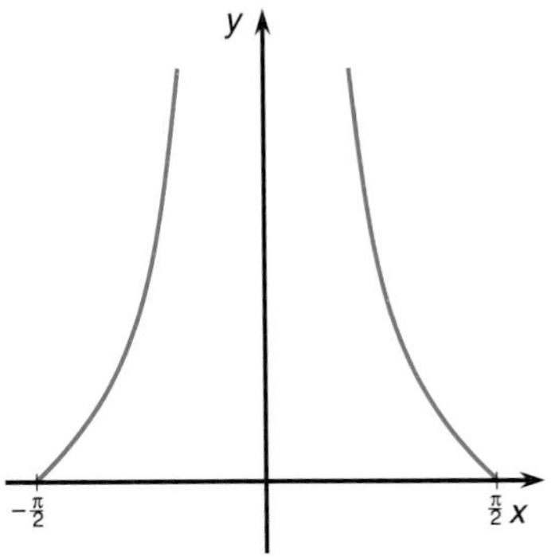

Figure 5.7.1

This calculation is *wrong!* For one thing, for x between $-\frac{\pi}{2}$ and $\frac{\pi}{2}$, $\cos x$ is always nonnegative, so the whole integrand is nonnegative. Theorem 5.4.2 says that the integral of a nonnegative integrand should be nonnegative, so it certainly could not equal -2.

The reason it is not permissible to do the calculation suggested above is that the integrand is not defined at $x = 0$, which lies in the interval of integration $\left[-\frac{\pi}{2}, \frac{\pi}{2}\right]$ (see Figure 5.7.1).

In fact the integrand is not integrable over this interval (because it is not bounded there), so any calculation that purports to integrate it will give a meaningless result. If an integrand goes to infinity somewhere in an interval, we cannot integrate it over that interval.

We conclude with a careful statement of the theorem that explains when we can perform a substitution and change the limits of integration.

THEOREM 5.7.2

Suppose $g(x)$ is a function whose derivative $g'(x)$ exists and is continuous on an interval $[a, b]$.

Suppose $f(x)$ is a function that is continuous and has an antiderivative $F(x)$ on an interval that contains all the values of $g(x)$ for $x \in [a, b]$. Then

$$\int_a^b f\big(g(x)\big)g'(x)\,dx = \int_{g(a)}^{g(b)} f(u)\,du = F(u)\Big|_{g(a)}^{g(b)}.$$

PROOF Consider the function $h(x) = F\big(g(x)\big)$. It is defined on $[a, b]$, and it is differentiable there. Moreover, by the Chain Rule,

$$h'(x) = F'\big(g(x)\big)g'(x) = f\big(g(x)\big)g'(x).$$

In other words, $h(x)$ is an indefinite integral for $f\big(g(x)\big)g'(x)$. By the second version of the Fundamental Theorem (Theorem 5.6.1) we can evaluate the definite integral using $h(x)$:

$$\int_a^b f\big(g(x)\big)g'(x)\,dx = h(x)\Big|_a^b = h(b) - h(a)$$
$$= F\big(g(b)\big) - F\big(g(a)\big).$$

But since $F(x)$ is an indefinite integral of $f(x)$, we know that

$$F\big(g(b)\big) - F\big(g(a)\big) = \int_{g(a)}^{g(b)} f(u)\,du.$$

This is the desired result. In this last formula, u is just a dummy variable; we could have used x (or t, or anything else), but there is less danger of confusion with u. ◢

This theorem is the precise statement behind the technique of integration by substitution. If we can find a function $u = g(x)$ so that the integrand can be written in the form $f\big(g(x)\big)g'(x) = f(u)\,u'$, then the original integral $\int f\big(g(x)\big)g'(x)\,dx$ can be transformed into $\int f(u)\,du$. It also tells us what happens to the limits of integration.

The technique of substitution that we have learned in this section is extremely useful. It is not easy to master, especially since it involves considerable guesswork. Nothing can take the place of practice and experience, which can help suggest what will and what will not work.

It is very important to do a large number of exercises to consolidate your skill at doing substitutions.

Exercises 5.7

I

Evaluate the following definite or indefinite integrals.

1. $\int e^{4x-2}dx$
2. $\int x\cos(x^2)\,dx$
3. $\int t^2(t^3-2)^7\,dt$
4. $\int (x+2)(x^2+4x-3)^5\,dx$
5. $\int_0^3 (t-2)^4\,dt$
6. $\int_1^2 \frac{1}{(x+1)^2}\,dx$
7. $\int_{-1}^2 \sin \pi x\,dx$
8. $\int_0^1 x^2e^{x^3}\,dx$
9. $\int \tan(\pi x)\sec^3(\pi x)\,dx$
10. $\int x^2\cos(5x^3+7)\,dx$
11. $\int_0^{\pi/4} \sin^2 x\,dx$
12. $\int_0^{\pi} \sin^2 x\,dx$
13. $\int_0^{2\pi} \cos^2\theta\,d\theta$
14. $\int_{-\pi}^{\pi} \sin^2 t\,dt$
15. $\int \cos^2(3x)\,dx$
16. $\int \sin^2(-5x)\,dx$
17. $\int_0^1 \cos^2(\pi x)\,dx$
18. $\int_0^1 \frac{1}{1+x}\,dx$
19. $\int_2^3 \frac{x^2}{x^3-1}\,dx$
20. $\int \frac{x^k}{1+x^{k+1}}\,dx,\ k \neq -1$
21. $\int \frac{x}{\sqrt{1+x^2}}\,dx$
22. $\int_0^1 x^2\sqrt{1-x^3}\,dx$
23. $\int \frac{1}{\sqrt{x}(\sqrt{x}+2)^3}\,dx$
24. $\int_0^1 s\sqrt{s+1}\,ds$
25. $\int_1^3 t(3-t)^{1/3}\,dt$
26. $\int e^{ax+b}\,dx$

27. $\int \cos(at)\,dt$

28. $\int_a^b \sin(\pi t)\,dt$

29. $\int (x^2+x)\sqrt{2x^3+3x^2+2}\,dx$

30. $\int_0^a \cos\left(\frac{2\pi x}{a}\right)dx$

II

31. Find $\int \frac{1}{9+t^2}\,dt$.

32. Let $a>0$. Find $\int \frac{1}{a^2+t^2}\,dt$.

33. Let $a>0$ and let $f(x)$ be a function that is integrable on $[0,a]$. Show that $\int_0^a f(a-x)\,dx = \int_0^a f(x)\,dx$. (Draw a picture.)

34. Let $a<b$ and let $f(x)$ be a function that is integrable on $[a,b]$. Show that

$$\int_a^b f(a+b-x)\,dx = \int_a^b f(x)\,dx.$$

(Draw a picture.)

35. For any number a, evaluate

(i) $\displaystyle\int_a^{a+2\pi} \sin x\,dx$, (ii) $\displaystyle\int_a^{a+2\pi} \sin^2 x\,dx$.

(*Hint:* The answers to parts (i) and (ii) do not depend on a.)

***36.** Differentiate $g(x) = x\ln x$, and use your answer to help you guess the indefinite integral

$$\int \ln x\,dx.$$

***37.** Suppose k and ℓ are nonnegative integers. Consider

$$\int \cos^k x \sin^{2\ell+1} x\,dx,$$

that is, the integral of the product of any power of $\cos x$ and an odd power of $\sin x$.

(i) Show that this integral can always be transformed into an integral that in principle we know how to integrate, by making the substitution $u=\cos x$ and writing $\sin^2 x = 1-\cos^2 x$. (Do not actually integrate the transformed integral.) Evaluate

(ii) $\int \cos^3 x \sin^3 x\,dx$,

(iii) $\int \sin^3 x\,dx$,

(iv) $\int \cos^4 x \sin^3 x\,dx$,

(v) $\int \cos^3 x \sin^2 x\,dx$.

(vi) Show that a similar trick will work for $\int \cos^k x \sin^\ell x\,dx$, provided that at least one of k and ℓ is odd.

***38.** (i) Integrate $\int \cos^4 x\,dx$. (*Hint:* Look at Example 5, p. 405).

(ii) Integrate $\int \sin^4 x\,dx$.

***39.** If we consider the integral $\int_a^b f(-x)\,dx$ and apply the substitution $u=-x$, we appear to have a minus sign in the resulting integral. But if the function $f(x)$ is always positive, for instance, then so is the function $f(-x)$, and its integral should also be positive. We appear to have found an equation saying that the (positive) integral of $f(-x)$ equals minus a (positive) integral of $f(x)$! Explain why there is really no difficulty here.

***40.** Consider the integral $\int_0^1 x^2(4-x^3)^6\,dx$. It is clear that the integrand is nonnegative on $[0,1]$, but after the substitution $u=4-x^3$ we obtain an integral of $-\frac{1}{3}u^6$, which is always *nonpositive.* Explain why this is not contradictory.

III

For each of the following integrals, evaluate the Riemann sum approximation from Example 5.2.3 with $n=100$ and compare with the exact value.

41. $\int_0^1 \sin(\pi x)\,dx$

42. $\int_0^1 xe^{x^2}\,dx$

43. $\int_0^1 \frac{1}{4+t^2}\,dt$

44. $\int_0^\pi \cos^2(2x)\,dx$

POINT TO PONDER

Try to see what the process of substitution has to do with signed areas. First consider the integral

$$\int_a^b f(x+1)\,dx$$

and the substitution $u = x+1$. Draw a picture.

Second try to figure out what happens in more complicated examples such as

$$\int_1^2 f(6-2x)\,dx \quad \text{or} \quad \int_0^1 3x^2(x^3+1)^3\,dx.$$

Chapter Summary

§5.1
area under a curve
rectangle
approximating the area

§5.2
summation notation
index of summation
distributive law
$\sum_{k=1}^{n} k = \frac{n(n+1)}{2}$
$\sum_{k=1}^{n} k^2 = \frac{n(n+1)(2n+1)}{6}$

§5.3
partition
mesh, norm
subinterval
Riemann sum
signed area
integrable
integral $\int_a^b f(x)\,dx$
integrand
lower limit of integration
upper limit of integration
limits of integration
precise definition of integral (Definition 5.3.5)
limit as $\|\mathcal{P}\| \to 0$

§5.4
Integral Mean Value Theorem (Theorem 5.4.4)
integrating functions with jump or removable discontinuities

§5.5
indefinite integral
Fundamental Theorem of Calculus (Theorem 5.5.4)
variable of integration
dummy variable
antiderivative
$\int f(x)\,dx$

§5.6
definite integral
Fundamental Theorem of Calculus (second version) (Theorem 5.6.1)
evaluating definite integrals

§5.7
integration by substitution
substitution and definite integrals

Review Exercises

I

Use the result of Example 5.2.3 (p. 356) to approximate the area under the graph of $y = f(x)$ on the specified interval, dividing the interval into n subintervals of equal length, with the specified value of n.

1. $f(x) = x^2$ on $[0, 2]$; $n = 4$
2. $f(x) = x^2$ on $[0, 2]$; $n = 6$
3. $f(x) = x^2$ on $[-3, 3]$; $n = 4$
4. $f(x) = |x^3|$ on $[-3, 3]$; $n = 6$
5. $f(x) = \frac{1}{x}$ on $[1, 2]$; $n = 3$
6. $f(x) = -\frac{1}{x}$ on $[-2, -1]$; $n = 3$
7. $f(x) = 7$ on $[-3, 4]$; $n = 9$
8. $f(x) = \sin x$ on $\left[0, \frac{\pi}{2}\right]$; $n = 2$
9. $f(x) = -x$ on $[-5, -1]$; $n = 8$
10. $f(x) = \cos x$ on $\left[0, \frac{\pi}{2}\right]$; $n = 2$

Evaluate the following (assume $m, n \geq 1$).

11. $\sum_{k=0}^{5} \ln(k + 1)$
12. $\sum_{k=1}^{n} \ln k$
13. $\sum_{k=0}^{4} 2^{-k}$
14. $\sum_{i=1}^{7} (i - 4)$
15. $\sum_{j=0}^{3} xj^3$
16. $\sum_{k=1}^{4} (k^3 - 2k^2 + k + 2)$
17. $\sum_{k=1}^{m} (k^2 - 2k + 1)$
18. $\sum_{\ell=m-4}^{m} (\ell + 2)$
19. $\sum_{i=-n}^{n} i$
20. $\sum_{j=-m}^{m} j^2$

In each of the following situations, let $\mathcal{P}_n$ be the partition that divides the interval $[0, 4]$ into n subintervals of equal length. Let x_k^ℓ, x_k^r, and x_k^m be the left endpoint, the right endpoint, and the midpoint, respectively, of the kth subinterval. Calculate the following Riemann sums.

21. $S\left(x^2+1, \mathcal{P}_4, \{x_k^r\}\right)$

22. $S\left(x+2, \mathcal{P}_4, \{x_k^m\}\right)$

23. $S\left(3-6x, \mathcal{P}_3, \{x_k^\ell\}\right)$

24. $S\left(x^4, \mathcal{P}_4, \{x_k^m\}\right)$

25. $S\left(\sin(\pi x), \mathcal{P}_4, \{x_k^r\}\right)$

26. $S\left(\sin(\pi x), \mathcal{P}_4, \{x_k^m\}\right)$

27. $S\left(\cos(\pi x), \mathcal{P}_4, \{x_k^\ell\}\right)$

28. $S\left(\cos(\pi x), \mathcal{P}_4, \{x_k^m\}\right)$

29. $S\left(\frac{1}{x}, \mathcal{P}_4, \{x_k^r\}\right)$

30. $S\left(\frac{1}{x+1}, \mathcal{P}_4, \{x_k^\ell\}\right)$

Differentiate the following functions of x.

31. $\int_0^x (t^3 - 2t + 1)\,dt$

32. $\int_1^{x^2} \cos u\,du$

33. $\int_{-x}^x (t^7 - 5t^5 + 11t^3 + 6t - 1)\,dt$

34. $\int_{\cos x}^{\sin x} t\,dt$

35. $x\left(\int_1^x (t^2 - 1)\,dt\right)$

36. $\int_1^{\ln x} e^t\,dt$

37. $\int_1^{\exp x} \ln t\,dt$

38. $\left(\int_0^x \sin t\,dt\right)\left(\int_{-x}^0 \cos t\,dt\right)$

39. $\left(\int_1^x t\,dt\right)\left(\int_0^x \frac{2t}{(t^2-1)^2}\,dt\right)$

40. $\int_0^3 (x + 3t)\,dt$

Integrate each of the following.

41. $\int_0^5 (x^3 - 3x^2 + 4)\,dx$

42. $\int_1^3 e^t\,dt$

43. $\int_1^2 (4 - 2x^3)\,dx$

44. $\int_0^x e^t\,dt$

45. $\int_0^\pi \sin x\,dx$

46. $\int_1^x \frac{1}{t}\,dt$

47. $\int_0^{\pi/4} \sec^2 x\,dx$

48. $\int_0^1 \sqrt{x}\,dx$

49. $\int_0^8 t^{1/3}(t-2)\,dt$

50. $\int_1^4 x^{-1/2}\,dx$

51. $\int x\sqrt{1+2x^2}\,dx$

52. $\int_0^1 x\sqrt{2-x^2}\,dx$

53. $\int_1^2 t(t^3 + 2t - 3)\,dt$

54. $\int 5x\sin(x^2 - 1)\,dx$

55. $\int \frac{x-1}{\sqrt{x+1}}\,dx$

56. $\int e^{\ln x}\,dx$

57. $\int \exp(\sqrt{t})\,t^{-1/2}\,dt$

58. $\int_1^8 t^{-2/3}\sin\left(\frac{\pi}{2}(t^{1/3})\right)dt$

59. $\int \frac{x-1}{x+1}\,dx$

60. $\int_0^\pi (\sin^2 t + \cos^2 t)\,dt$

II

61. What is the average of the integers from 1 to n?

62. What is the average of the squares of the integers from 1 to n?

***63.** Suppose $r \neq 1$. We want to evaluate $S = \sum_{k=0}^n r^k$. Consider $A = (1-r)S$. Expand A by rearranging the sum; you should be able to express it as a simple expression without summations. Then divide by $1-r$ to find an expression for S. A sum of this sort (consecutive powers of r) is called a *geometric sum*.

***64.** Let r be any nonzero number. Find an expression without a summation sign for

$$\sum_{k=m}^{n} r^k.$$

***65.** Evaluate the integral $\int_{-1}^1 \sqrt{1-x^2}\,dx$ by thinking of it as the area under the graph of a function.

***66.** Suppose $f(x)$ is the function whose graph is illustrated below. Sketch the graph of its indefinite integral $F(x) = \int_0^x f(t)\,dt$.

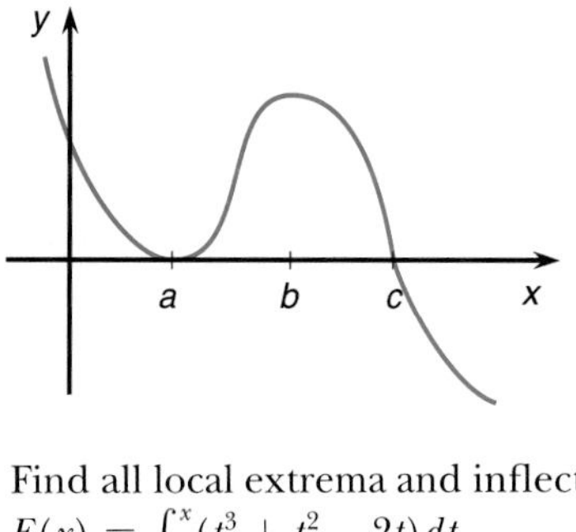

67. Find all local extrema and inflection points of $E(x) = \int_1^x (t^3 + t^2 - 2t)\,dt$.

68. Find all local extrema and inflection points of $G(x) = \int_1^x (t^2 - 4t + 3)e^t\,dt$.

***69.** Find all local extrema and inflection points of $H(x) = \int_{-x}^{1} (t \ln t - 2t)\, dt$.

***70.** Find all local extrema and inflection points of $K(x) = \int_{2}^{\exp(x)} (t^2 - 1)\, dt$.

***71.** Suppose $f(x)$ is a continuous function and $F(x) = \int_d^x f(t)\, dt$. If $F(x)$ has a local maximum at $x = a$, what can we say about the behavior of $f(x)$ at $x = a$?

***72.** Suppose $f(x)$ is a differentiable function and $F(x) = \int_d^x f(t)\, dt$. If $f(x)$ has a local maximum at $x = a$, what can we say about the behavior of $F(x)$ at $x = a$?

***73.** Suppose we are told that $\int_0^{\pi/2} x \cos x\, dx = \frac{\pi}{2} - 1$. Assuming this fact, calculate $\int_0^{\pi/4} x \cos(2x)\, dx$.

***74.** (i) Differentiate $x \sin x$ and guess an antiderivative for $x \cos x$.
(ii) Guess an antiderivative for $x \sin x$.

75. Evaluate the following: (i) $\int_{-2}^{2} |x^2 - x|\, dx$, (ii) $\int_0^{\pi} \cos t \sin t\, dt$, (iii) $\int_0^{4} |(x-2)(x-3)|\, dx$.

***76.** Evaluate: (i) $\int_0^{2} |x(x^2-1)^3|\, dx$, (ii) $\int_0^{\pi} \cos^3 t \sin t\, dt$.

***77.** Find an example of a function $f(x)$ that is differentiable at every $x \in \mathbb{R}$ but so that $f'(x)$ is *not* differentiable at $x = 0$.

***78.** Find an example of a function $f(x)$ so that $f(x)$ and $f'(x)$ are differentiable at every $x \in \mathbb{R}$ but so that $f''(x)$ is *not* differentiable at $x = 0$.

***79.** If $f(x) = \frac{1}{1+x^2}$, note that $F(x) = \frac{\pi}{3} + \arctan x$ is an antiderivative of $f(x)$. Find a number d so that $F(x) = \int_d^x f(t)\, dt$.

****80.** If $f(x) = \cos x$, then $F(x) = 5 + \sin x$ is an antiderivative. (i) Show that for any x, $4 \le F(x) \le 6$. (ii) Show that the indefinite integral

$$G(x) = \int_d^x f(t)\, dt = \int_d^x \cos t\, dt$$

always satisfies $-2 \le G(x) \le 2$, for *any* choice of the starting point d. (*Hint:* Draw a picture.) So although $F(x)$ is an antiderivative it is *never* the signed area under $y = f(x)$. (iii) Explain this.

III

For each of the following functions $f(x)$, find an antiderivative $F(x)$ and use a graphics calculator or computer to graph $f(x)$ and $F(x)$ on the same picture. Interpret $F(x)$ in terms of signed areas.

81. $f(x) = x$

82. $f(x) = x^2$

83. $f(x) = x^3$

84. $f(x) = x^3 - 3x$

85. $f(x) = e^x$

86. $f(x) = \frac{1}{x}$, $x > 0$

87. $f(x) = \sec^2 x$, $|x| < \frac{\pi}{2}$

88. $f(x) = \cos^2 x$

Applications of Integration

6

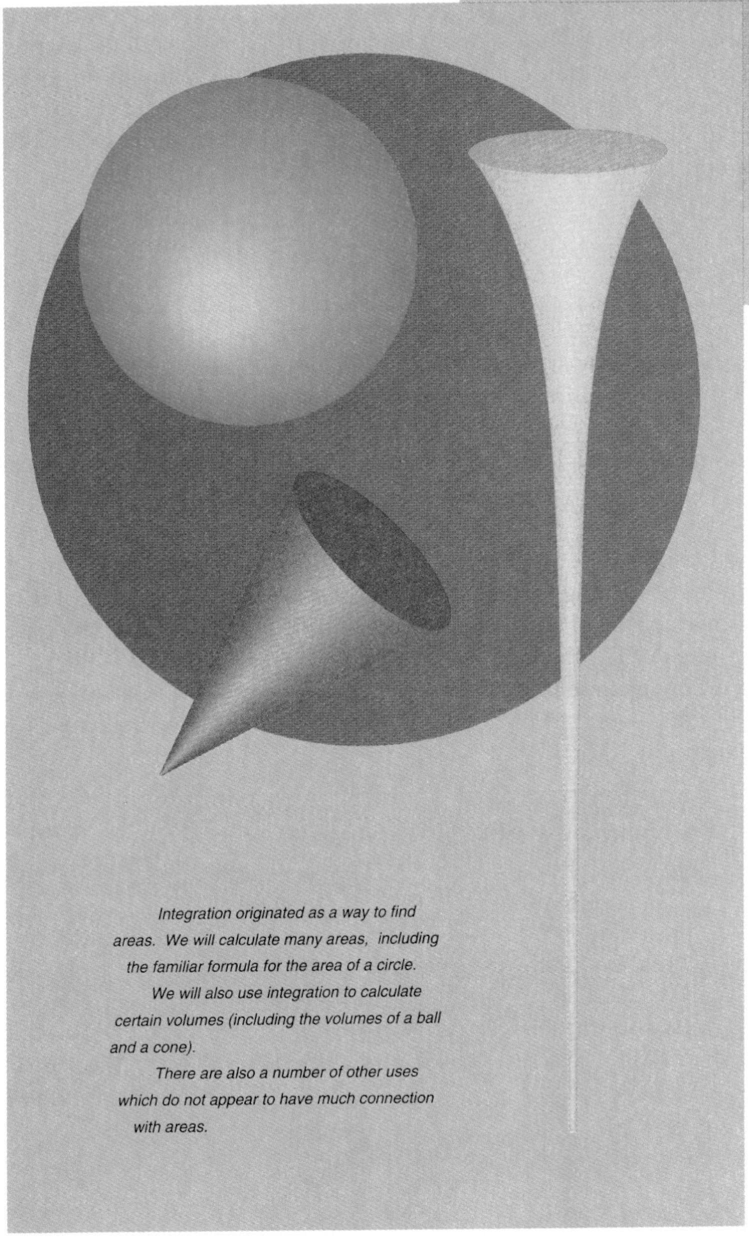

Integration originated as a way to find areas. We will calculate many areas, including the familiar formula for the area of a circle.

We will also use integration to calculate certain volumes (including the volumes of a ball and a cone).

There are also a number of other uses which do not appear to have much connection with areas.

INTRODUCTION

This chapter will explore some of the uses of integration for practical purposes. We have already found that being able to integrate is essentially the same thing as being able to find antiderivatives, and in Section 4.9 we saw the importance of antiderivatives for questions involving velocity and acceleration. If we know, for instance, the velocity of a moving object, then we may want to know its position (i.e., displacement), which is an antiderivative of the velocity. Similarly, the velocity itself is an antiderivative of the acceleration. A similar idea applies in any situation in which the rate of change of some quantity is known; the actual value of that quantity is an indefinite integral of the rate of change.

We have also hinted that integrals have something to do with averages. This arose in the Integral Mean Value Theorem (Theorem 5.4.4), where the number $\frac{1}{b-a}\int_a^b f(x)\,dx$ was regarded as the average value of the function $f(x)$ over the interval $[a, b]$.

Both these ideas will be developed more fully in this chapter. However, we begin with something more basic.

After all, the idea of integration arose in the first place as a means of finding areas. Now that the theory is available to us, it should be possible to use it to find areas of various geometrical regions. Not only will we do this, but we will also be able to use it to find volumes of solid regions, lengths of curves, and areas of curved surfaces.

Section 6.1 concerns areas, and the related Section 6.2 discusses the techniques needed for these calculations. Sections 6.3 and 6.4 concern volume computations, and Section 6.5 discusses the average value of a function. The next four sections are about more specialized topics, and you may wish to omit some of them. Section 6.6 is about work (in the sense in which the term is used in physics), and Section 6.7 discusses the calculation of moments and centroids, a very important topic in physics having to do with various things including rotational motion. Section 6.8, which deals with pressure and forces in a fluid, makes some use of the results of Section 6.7, and Section 6.9 discusses the lengths of curves and the areas of surfaces (such as the surface area of a sphere, etc.). Finally, Section 6.10 is about "numerical integration," which means finding numerical approximations to the value of an integral when none of the methods we know gives us the exact answer. In practice there are many instances in which this is necessary, and numerical integration, especially with the aid of computers or sophisticated calculators, is extremely useful.

6.1 Areas

SIGNED AREA
REGION BETWEEN TWO CURVES

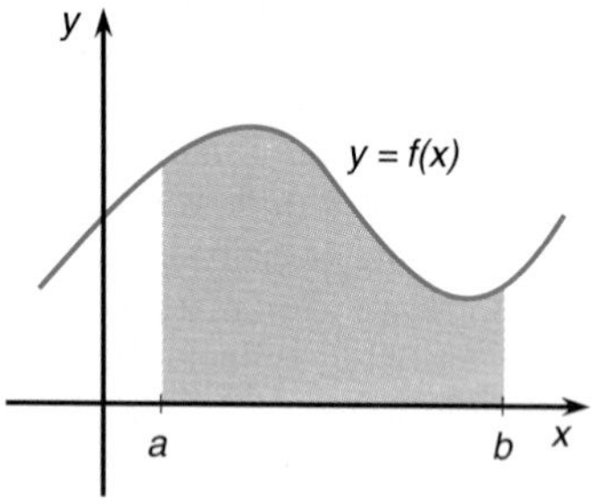

Figure 6.1.1

The integral originally arose in connection with the area under a curve, and it is very useful for calculating areas. Strictly speaking, we do not really know what is meant by the area of a peculiarly shaped region. However, if we consider the region illustrated in Figure 6.1.1, the region under the graph of a nonnegative continuous function $f(x)$ between $x = a$ and $x = b$, it is reasonable to regard $\int_a^b f(x)\,dx$ as the area.

Indeed, the integral was defined as a limit, obtained by approximating the region by collections of rectangles and summing their areas. So if the area of the region is to mean anything at all and be consistent with the areas of rectangles, we can use the integral to define what area means.

If we do not assume that $f(x)$ is nonnegative between $x = a$ and $x = b$, then of course the integral will give the signed area; we can obtain the actual (unsigned) area by integrating $|f(x)|$.

DEFINITION 6.1.1

If $f(x)$ is a nonnegative continuous function on $[a, b]$, consider the region bounded by the graph $y = f(x)$, the x-axis, and the lines $x = a$ and $x = b$. Then the **area** of this region is defined to be the integral

$$\int_a^b f(x)\,dx.$$

If $f(x)$ is a continuous function on $[a, b]$, the area of the region bounded by the graph $y = f(x)$, the x-axis, and the lines $x = a$ and $x = b$ is defined to be the integral

$$\int_a^b |f(x)|\,dx.$$

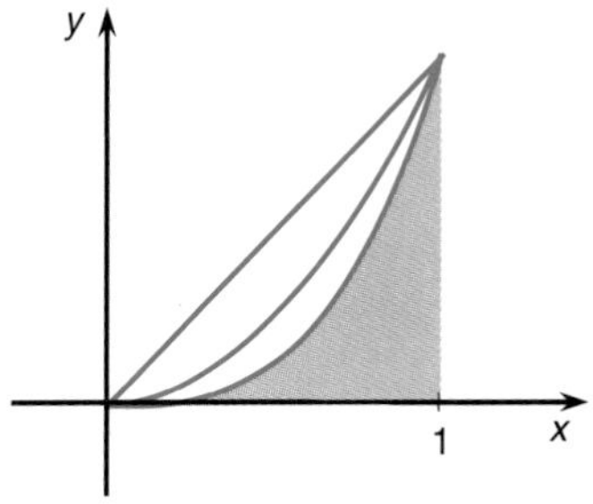

Figure 6.1.2

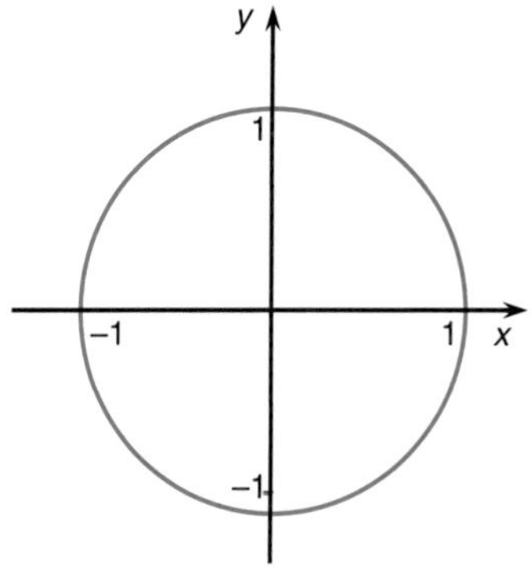

Figure 6.1.3

EXAMPLE 1 If $f(x) = x^n$, then the area under the curve $y = f(x) = x^n$ between $x = 0$ and $x = 1$ is

$$\int_0^1 x^n\,dx = \frac{1}{n+1}x^{n+1}\bigg|_0^1 = \frac{1}{n+1}.$$

As n increases, the curves $y = x^n$ are lower and lower on the interval $[0, 1]$, so their integrals get smaller and smaller (Figure 6.1.2 shows that $y = x, y = x^2, y = x^3$.)

One area we would like to calculate is the area of a circle (strictly speaking, this means the area of the region inside the circle; see Figure 6.1.3). In fact it is easier to calculate the area of a semicircle. If we consider the circle of radius 1 centered at the origin, its equation is $x^2 + y^2 = 1$, and we know that the top half of the circle is the graph of $y = \sqrt{1 - x^2}$. To find the area, we will have to know how to integrate this function. It can be done by using a technique we will learn in Section 8.3, but for now we simply state the result.

EXAMPLE 2 Check the indefinite integral

$$\int \sqrt{1 - x^2}\,dx = \frac{1}{2}\left(x\sqrt{1 - x^2} + \arcsin x\right) + c.$$

Solution All that is needed is to differentiate the right side:

$$\begin{aligned}
&\frac{d}{dx}\left(\frac{1}{2}\left(x\sqrt{1 - x^2} + \arcsin x\right) + c\right)\\
&= \frac{1}{2}\left(\sqrt{1 - x^2} + \frac{1}{2}x\frac{1}{\sqrt{1 - x^2}}(-2x) + \frac{1}{\sqrt{1 - x^2}}\right)\\
&= \frac{1}{2}\left(\frac{1 - x^2}{\sqrt{1 - x^2}} - \frac{x^2}{\sqrt{1 - x^2}} + \frac{1}{\sqrt{1 - x^2}}\right)\\
&= \sqrt{1 - x^2}.
\end{aligned}$$

In the following box we describe a trick for finding this integral.

Consider the indefinite integral $\int \sqrt{1 - x^2}\,dx$ of Example 2.

There is no obvious way to integrate by substitution. Instead, we look at the function $x\sqrt{1 - x^2}$. It is not the answer we want, but it may help. Differentiating, we find:

$$\begin{aligned}
\frac{d}{dx}\left(x\sqrt{1 - x^2}\right) &= \sqrt{1 - x^2} + x\frac{1}{2}(1 - x^2)^{-1/2}(-2x)\\
&= \sqrt{1 - x^2} - x^2(1 - x^2)^{-1/2} = \frac{1 - x^2}{\sqrt{1 - x^2}} - \frac{x^2}{\sqrt{1 - x^2}}\\
&= \frac{1 - 2x^2}{\sqrt{1 - x^2}}.
\end{aligned}$$

This does not appear to be much good, but suppose for a moment that the 1 in the numerator was a 2. Then we would have

$$\frac{2-2x^2}{\sqrt{1-x^2}} = 2\frac{1-x^2}{\sqrt{1-x^2}} = 2\sqrt{1-x^2},$$

which is practically what we need.

The function $x\sqrt{1-x^2}$ is not the antiderivative we want, but if we could add to it something that would add $\frac{1}{\sqrt{1-x^2}}$ to its derivative, we would be close. And that is easy if we remember that we know a function whose derivative equals $\frac{1}{\sqrt{1-x^2}}$, namely, arcsin x.

So consider $x\sqrt{1-x^2} + \arcsin x$, whose derivative is

$$\frac{1-2x^2}{\sqrt{1-x^2}} + \frac{1}{\sqrt{1-x^2}} = \frac{2-2x^2}{\sqrt{1-x^2}} = 2\sqrt{1-x^2},$$

which means

$$\int \sqrt{1-x^2}\,dx = \frac{1}{2}(x\sqrt{1-x^2} + \arcsin x) + c.$$

EXAMPLE 3 **Area of Circle** Find the area of the circle of radius r.

Solution We may as well let the center of the circle be at the origin, so the equation of the circle is $x^2 + y^2 = r^2$. The upper semicircle is the graph of the function $y = \sqrt{r^2 - x^2}$ illustrated in Figure 6.1.4. It will be easier to calculate the area under this semicircle and then double it to get the area of the circle.

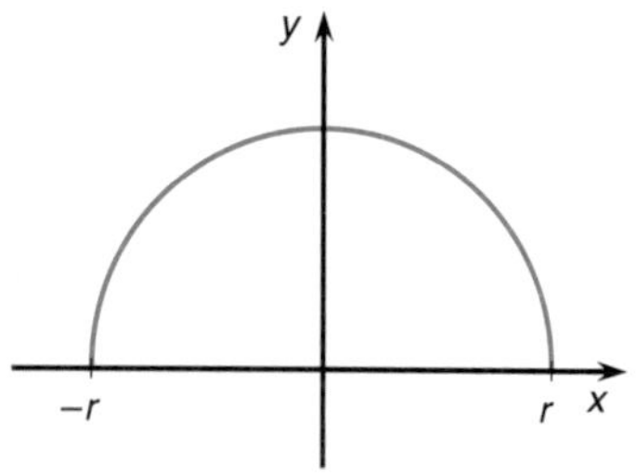

Figure 6.1.4

The semicircular region is the region under the function $y = \sqrt{r^2 - x^2}$ between $x = -r$ and $x = r$. So the area of the region under the semicircle is

$$\int_{-r}^{r} \sqrt{r^2 - x^2}\,dx.$$

This is not quite the same as the integral we discussed in the preceding example, because of the r^2 inside the square root. To deal with that, we write the integrand as $\sqrt{r^2 - x^2} = \sqrt{r^2\left(1 - \left(\frac{x}{r}\right)^2\right)} = r\sqrt{1 - \left(\frac{x}{r}\right)^2}$.

This suggests the substitution $u = \frac{x}{r}$. We notice that as x goes from $-r$ to r, the substituted variable $u = \frac{x}{r}$ goes from -1 to 1, so the integral becomes

$$\int_{-r}^{r} \sqrt{r^2 - x^2}\,dx = \int_{-r}^{r} \sqrt{r^2\left(1 - \left(\frac{x}{r}\right)^2\right)}\,dx = \int_{-r}^{r} r\sqrt{1 - \left(\frac{x}{r}\right)^2}\,dx$$

$u = \frac{x}{r}$
$du = \frac{1}{r}dx$

$$= r\int_{-1}^{1} \sqrt{1-u^2}\,r\,du = r^2 \int_{-1}^{1} \sqrt{1-u^2}\,du$$

$$= r^2\frac{1}{2}\left(u\sqrt{1-u^2} + \arcsin u\right)\Big|_{-1}^{1}$$

$$= \frac{r^2}{2}\left(1\sqrt{0} + \arcsin(1) - (-1)\sqrt{0} - \arcsin(-1)\right)$$

$$= \frac{r^2}{2}\left(0 + \frac{\pi}{2} - 0 - \left(-\frac{\pi}{2}\right)\right)$$

$$= \frac{1}{2}\pi r^2.$$

This is the area of the semicircle; the area of the circle is twice as much, or πr^2. Of course, we are not surprised, but we should be impressed, at least slightly. After all, this is something we have been told for a long time; now we finally see why it is so.

Example 3 illustrates several interesting points. First, it is important to have a function whose graph encloses the region whose area we want to find. Second, it is frequently convenient to work with only part of the region (in this case the upper semicircle) and determine the remaining areas separately or, as here, by symmetry. Third, it is necessary to find the limits of integration. Sometimes they will be made obvious by drawing a picture, but other times they will have to be calculated.

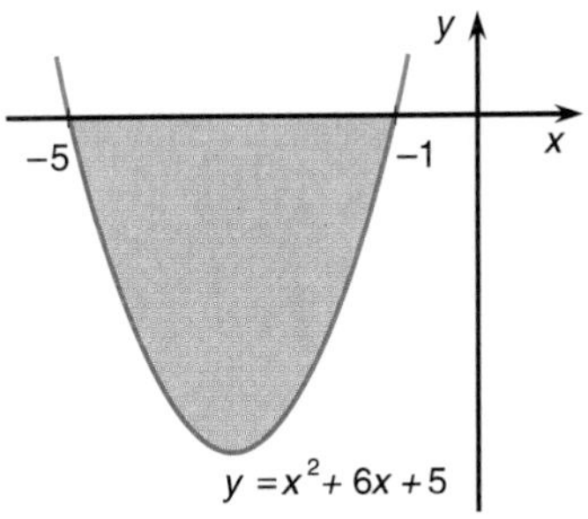

Figure 6.1.5

EXAMPLE 4 Find the area of the region enclosed between the x-axis and the curve $y = x^2 + 6x + 5$.

Solution We begin with a sketch in Figure 6.1.5 of the curve $y = x^2 + 6x + 5$, in which the specified region is shaded. To find the limits of integration, it is necessary to find the x-intercepts—the points where $x^2 + 6x + 5 = 0$. This is easily done by using the Quadratic Formula (Theorem 1.3.4) or by factoring:

$$0 = x^2 + 6x + 5 = (x+5)(x+1),$$

so

$$x = -5, \ -1.$$

The picture suggests that $f(x)$ is negative for $x \in (-5, -1)$. We could confirm this by completing the square: $x^2 + 6x + 5 = (x+3)^2 - 4$, which is negative when $(x+3)^2 < 4$, that is, when $|x+3| < 2$, that is, when $x \in (-5, -1)$. Since $f(x) < 0$, we see that $|f(x)| = -f(x)$. The corresponding integral is easy to evaluate:

$$\int_{-5}^{-1} |f(x)|\,dx = \int_{-5}^{-1} \left(-(x^2 + 6x + 5)\right)dx = -\left(\frac{1}{3}x^3 + 3x^2 + 5x\right)\Bigg|_{-5}^{-1}$$

$$= -\left(\frac{1}{3}(-1) + 3(1) + 5(-1)\right) + \left(\frac{1}{3}(-125) + 3(25) + 5(-5)\right)$$

$$= \frac{32}{3}.$$

The desired area is $\frac{32}{3} = 10\frac{2}{3}$.

If we did not realize that the graph was below the axis, that is, that $f(x)$ was negative, we would have integrated $f(x)$ and obtained the incorrect answer $-\frac{32}{3}$.

EXAMPLE 5 Find the area of the region enclosed between the x-axis and the curve $y = x^3 - 3x$.

To find where $y = x^3 - 3x$ crosses the x-axis, we need to find where $y = 0$, so we solve $x^3 - 3x = 0$.

Solution The curve is sketched in Figure 6.1.6. To find where it crosses the x-axis, we solve $0 = x^3 - 3x = x(x^2 - 3) = x(x + \sqrt{3})(x - \sqrt{3})$. The zeroes are at $x = -\sqrt{3},\ 0,\ \sqrt{3}$. The region in question consists of two parts, the part between $x = -\sqrt{3}$ and $x = 0$, and the part between $x = 0$ and $x = \sqrt{3}$.

The area of the first of these is

$$\begin{aligned}\int_{-\sqrt{3}}^{0} (x^3 - 3x)\,dx &= \left(\frac{1}{4}x^4 - \frac{3}{2}x^2\right)\Bigg|_{-\sqrt{3}}^{0} \\ &= \left(\frac{1}{4}(0) - \frac{3}{2}(0)\right) - \left(\frac{1}{4}(-\sqrt{3})^4 - \frac{3}{2}(-\sqrt{3})^2\right) \\ &= 0 - \left(\frac{9}{4} - \frac{9}{2}\right) \\ &= \frac{9}{4}.\end{aligned}$$

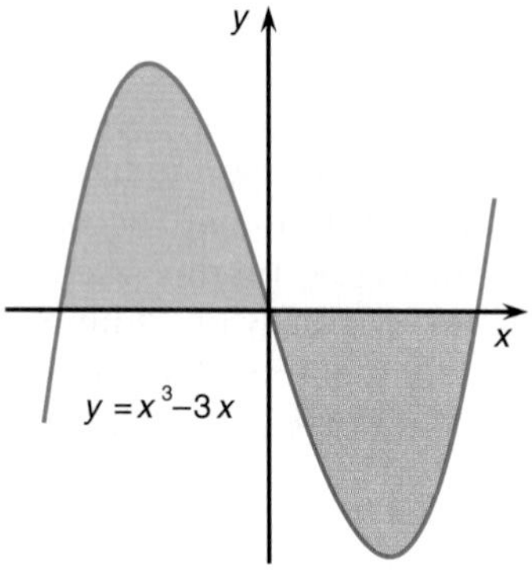

Figure 6.1.6

Since the second part of the region lies below the x-axis, its area is the integral of $|x^3 - 3x| = -(x^3 - 3x)$:

$$\begin{aligned}\int_{0}^{\sqrt{3}} \left(-(x^3 - 3x)\right)dx &= -\left(\frac{1}{4}x^4 - \frac{3}{2}x^2\right)\Bigg|_{0}^{\sqrt{3}} \\ &= -\left(\frac{1}{4}(\sqrt{3})^4 - \frac{3}{2}(\sqrt{3})^2\right) + \left(\frac{1}{4}(0) - \frac{3}{2}(0)\right) \\ &= -\left(\frac{9}{4} - \frac{9}{2}\right) + 0 = \frac{9}{4}.\end{aligned}$$

The total area is the sum of the areas of the two parts, or $\frac{9}{4} + \frac{9}{4} = \frac{9}{2}$.

It was essential to divide the region into parts according to whether the curve was above or below the axis. If we did not do that but simply integrated all at once, we would evaluate:

$$\begin{aligned}\int_{-\sqrt{3}}^{\sqrt{3}} (x^3 - 3x)\,dx &= \left(\frac{1}{4}x^4 - \frac{3}{2}x^2\right)\Bigg|_{-\sqrt{3}}^{\sqrt{3}} \\ &= \left(\frac{1}{4}(\sqrt{3})^4 - \frac{3}{2}(\sqrt{3})^2\right) - \left(\frac{1}{4}(-\sqrt{3})^4 - \frac{3}{2}(-\sqrt{3})^2\right) \\ &= \left(\frac{9}{4} - \frac{9}{2}\right) - \left(\frac{9}{4} - \frac{9}{2}\right) \\ &= 0.\end{aligned}$$

In this case the regions above and below the axis have the same area, so the signed areas cancel. This illustrates the importance of figuring out where the integrand is positive and where it is negative. In practice the integral is usually evaluated by dividing the region into pieces according to the sign of the integrand and integrating each piece separately.

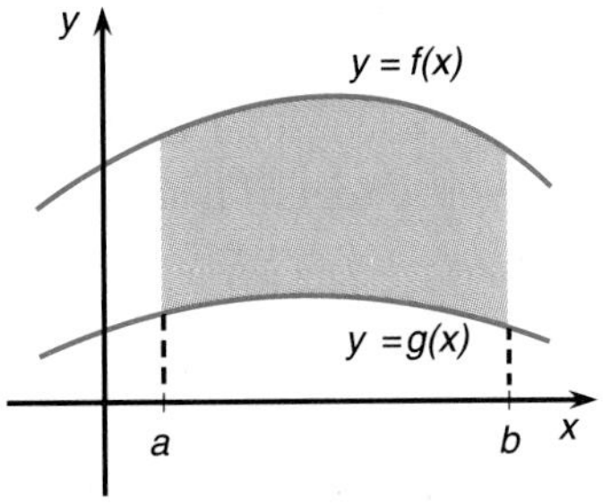

Figure 6.1.7(i)

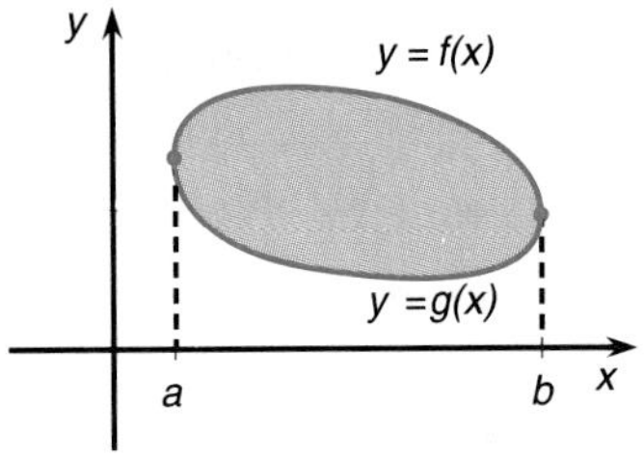

Figure 6.1.7(ii)

Another problem that frequently arises involves regions lying between two curves. In Figure 6.1.7 there are two examples. In each case the shaded region lies between $x = a$ and $x = b$ and between $y = f(x)$ and $y = g(x)$. Moreover, in each case $f(x) \geq g(x)$ for all $x \in [a, b]$.

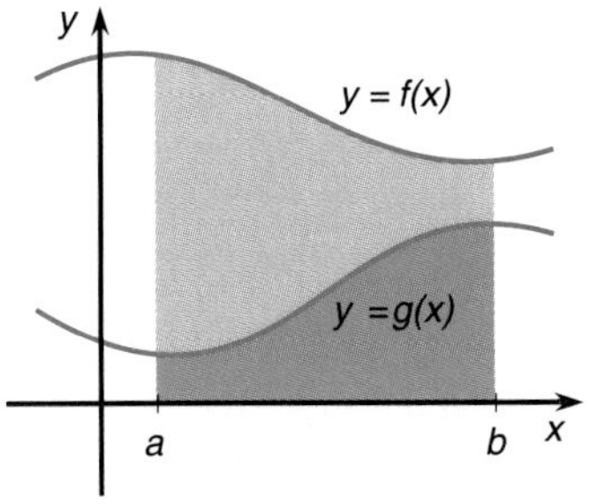

Figure 6.1.8

To discuss the area of such a region, we refer to Figure 6.1.8, in which there are two continuous functions $f(x)$ and $g(x)$ with $f(x) \geq g(x)$ for all $x \in [a, b]$. Now $\int_a^b f(x)\,dx$ is the total area of both shaded regions, and $\int_a^b g(x)\,dx$ is the area of the lower shaded region. So the area of the upper shaded region—the region between $y = f(x)$ and $y = g(x)$—is their difference:

$$\int_a^b f(x)\,dx - \int_a^b g(x)\,dx = \int_a^b \big(f(x) - g(x)\big)\,dx.$$

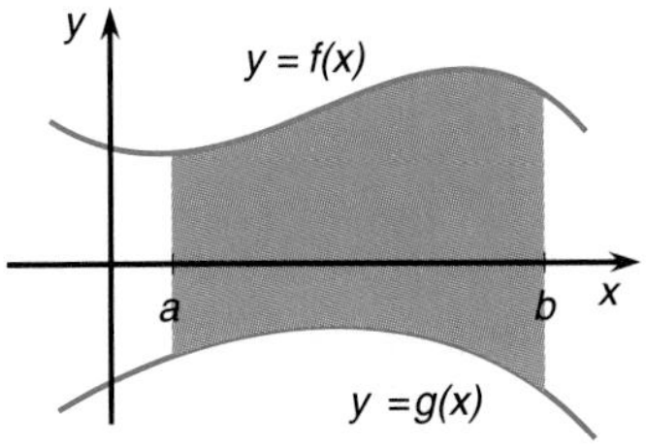

Figure 6.1.9

This is still reasonable if either $f(x)$ or $g(x)$ is negative, provided $f(x) \geq g(x)$. In the case illustrated in Figure 6.1.9 we see that $f(x) \geq 0 \geq g(x)$, so the area of the upper shaded region is $\int_a^b f(x)\,dx$, the area of the lower shaded region is $-\int_a^b g(x)\,dx$, and the area of the whole region is their sum, $\int_a^b f(x)\,dx + \left(-\int_a^b g(x)\,dx\right) = \int_a^b \big(f(x) - g(x)\big)\,dx$.

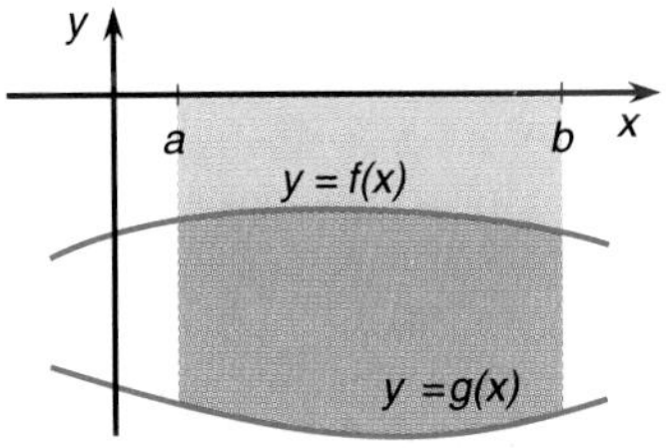

Figure 6.1.10

In the case shown in Figure 6.1.10, with $0 \geq f(x) \geq g(x)$, the total area of both parts is $-\int_a^b g(x)\,dx$, the area of the upper part is $-\int_a^b f(x)\,dx$, and the area of the region between the curves is $-\int_a^b g(x)\,dx - \left(-\int_a^b f(x)\,dx\right) = \int_a^b \big(f(x) - g(x)\big)\,dx$.

DEFINITION 6.1.2

Area of the Region Between Two Curves

If $f(x)$ and $g(x)$ are continuous functions satisfying $f(x) \geq g(x)$ for all $x \in [a, b]$, consider the region between $y = f(x)$ and $y = g(x)$, that is, $\{(x, y) : a \leq x \leq b,\ f(x) \geq y \geq g(x)\}$. The area of this region is defined to be

$$\int_a^b \big(f(x) - g(x)\big)\,dx.$$

If $f(x)$ and $g(x)$ are continuous functions on $[a, b]$, the area of the region between $y = f(x)$ and $y = g(x)$ and between $x = a$ and $x = b$ is defined to be

$$\int_a^b |f(x) - g(x)|\,dx.$$

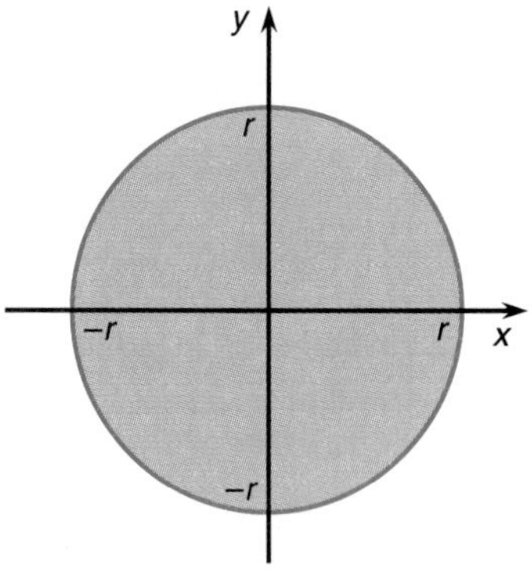

Figure 6.1.11

EXAMPLE 6 Find the area of a circle of radius r (again).

Solution As in Example 3, we consider the circle $x^2 + y^2 = r^2$ (see Figure 6.1.11). The region in question is the region between the upper semicircle $y = \sqrt{r^2 - x^2}$ and the lower semicircle $y = -\sqrt{r^2 - x^2}$. Its area is

$$\int_{-r}^{r} \left(\sqrt{r^2 - x^2} - (-\sqrt{r^2 - x^2})\right) dx = 2\int_{-r}^{r} \sqrt{r^2 - x^2}\,dx.$$

We have already evaluated this integral in Example 3; there we found that the integral equals $\frac{1}{2}\pi r^2$, so the area of the circle is πr^2.

EXAMPLE 7 Find the area of the region enclosed between the curves $y = x^3$ and $y = x$.

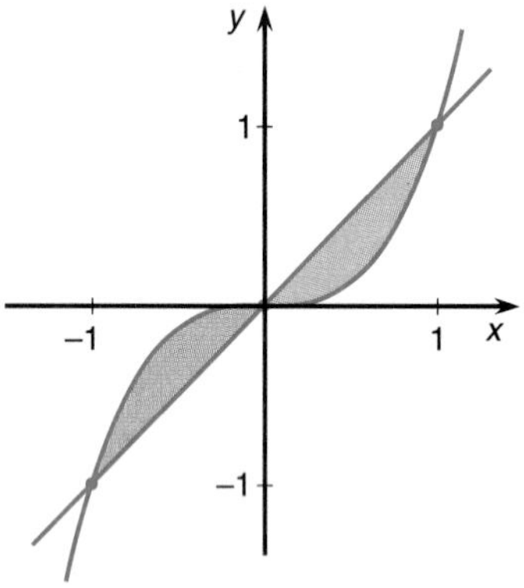

Figure 6.1.12

Solution For questions about the region enclosed between two curves, the first step is to find the points where the two curves intersect. In this case, that will happen when $x^3 = x$, that is, when $x^3 - x = 0$, or $0 = x(x^2 - 1) = x(x + 1)(x - 1)$, which means $x = -1,\ 0,\ 1$. The points of intersection are $(-1, -1)$, $(0, 0)$, and $(1, 1)$, and we sketch them in Figure 6.1.12.

The part of the region lying between $x = -1$ and $x = 0$ lies below $y = x^3$ and above $y = x$, so its area is

$$\begin{aligned}\int_{-1}^{0} (x^3 - x)\,dx &= \left(\frac{1}{4}x^4 - \frac{1}{2}x^2\right)\Big|_{-1}^{0}\\ &= \left(\frac{1}{4}(0) - \frac{1}{2}(0)\right) - \left(\frac{1}{4}(-1)^4 - \frac{1}{2}(-1)^2\right)\\ &= \frac{1}{4}.\end{aligned}$$

Between $x = 0$ and $x = 1$, the region lies below $y = x$ and above $y = x^3$, so the area of this part of the region is

$$\begin{aligned}\int_{0}^{1} (x - x^3)\,dx &= \left(\frac{1}{2}x^2 - \frac{1}{4}x^4\right)\Big|_{0}^{1}\\ &= \left(\frac{1}{2}(1^2) - \frac{1}{4}(1^4)\right) - \left(\frac{1}{2}(0) - \frac{1}{4}(0)\right)\\ &= \frac{1}{4}.\end{aligned}$$

The total area of the region is the sum of these two parts, or $\frac{1}{4} + \frac{1}{4} = \frac{1}{2}$.

Notice that it was necessary to break the region into parts depending upon which function was greater. This is analogous to what we did earlier in dealing with the region between the x-axis and a single function, where it was necessary to decide where the function was positive (i.e., above the axis) and where it was negative (below the axis).

Also notice that whenever we speak of a region "enclosed" between two curves without mentioning an interval $[a, b]$, we are referring to the part that is *completely* enclosed. For instance in Example 7 we ignore the region between the curves to the right of $x = 1$ and the region between the curves to the left of $x = -1$.

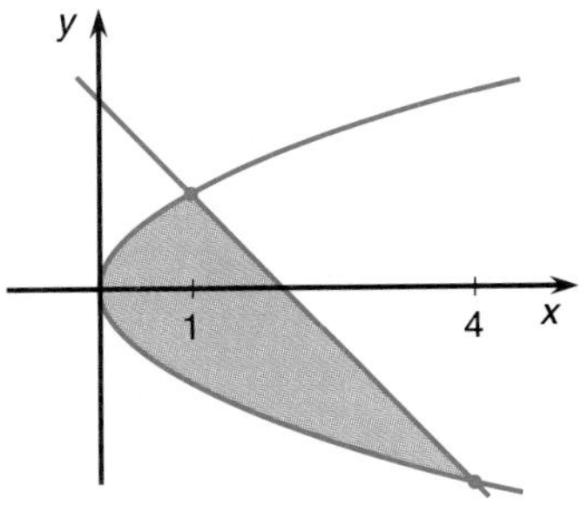

Figure 6.1.13

EXAMPLE 8 Find the area of the region enclosed by the parabola $y^2 = x$ and the line $x + y = 2$.

Solution The region is illustrated in Figure 6.1.13. We find the points of intersection of the line and the parabola by solving for points that satisfy *both* equations. From $x + y = 2$ we get $x = 2 - y$, so $y^2 = x = 2 - y$, or $y^2 + y - 2 = 0$. This we factor as $(y - 1)(y + 2) = 0$ to find $y = 1, \ -2$. The points of intersection are $(1, 1)$ and $(4, -2)$.

We divide the region into two parts as shown in Figure 6.1.14. The part between $x = 0$ and $x = 1$ is the region between the curves $y = \sqrt{x}$ and $y = -\sqrt{x}$, so its area is

$$\int_0^1 \left(\sqrt{x} - (-\sqrt{x})\right) dx = 2\int_0^1 \sqrt{x}\, dx = 2\left(\frac{2}{3}x^{3/2}\right)\Big|_0^1$$
$$= \frac{4}{3}.$$

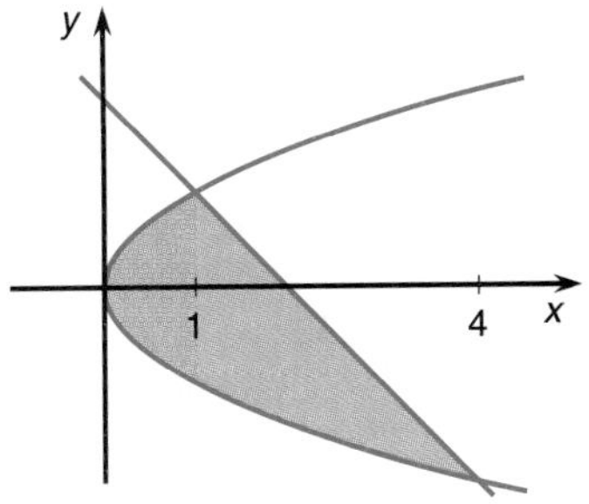

Figure 6.1.14

The part of the region between $x = 1$ and $x = 4$ is the region between $y = 2 - x$ and $y = -\sqrt{x}$, so its area is

$$\int_1^4 \left(2 - x - (-\sqrt{x})\right) dx = \left(2x - \frac{1}{2}x^2 + \frac{2}{3}x^{3/2}\right)\Big|_1^4$$
$$= \left(8 - 8 + \frac{2}{3}(8)\right) - \left(2 - \frac{1}{2} + \frac{2}{3}\right) = \frac{19}{6}.$$

The total area of the original region is $\frac{4}{3} + \frac{19}{6} = \frac{9}{2}$.

Notice that different parts of the region lie between different pairs of curves. To integrate, it is necessary to consider only one such part at a time. Also notice that while the curves were given to us as $y^2 = x$ and $x + y = 2$, it was necessary to express them with y as a function of x in order to be able to integrate. In the case of $y^2 = x$ there were two possible functions of this sort, $y = \sqrt{x}$ and $y = -\sqrt{x}$. Each of them figured in the calculation.

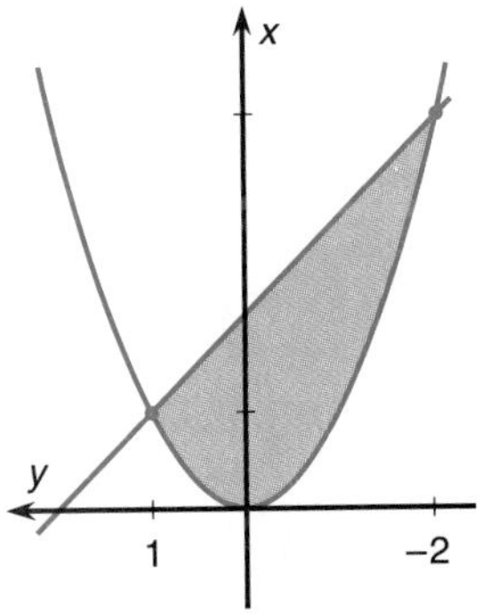

Figure 6.1.15

EXAMPLE 9 Example 8 can also be done another way. Instead of thinking of $y^2 = x$ and $x + y = 2$ as functions $y = \sqrt{x}$, $y = -\sqrt{x}$, and $y = 2 - x$ in which x is the variable, we could just as well think of $x = y^2$ and $x = 2 - y$ as functions of y (see Figure 6.1.15, in which the axes have been rotated). The function $x = 2 - y$ is greater than $x = y^2$ for y between -2 and 1, so the area of the region is

$$\int_{-2}^1 (2 - y - y^2)\, dy = \left(2y - \frac{1}{2}y^2 - \frac{1}{3}y^3\right)\Big|_{-2}^1$$
$$= \left(2 - \frac{1}{2} - \frac{1}{3}\right) - \left(-4 - \frac{1}{2}(-2)^2 - \frac{1}{3}(-2)^3\right) = \frac{9}{2}.$$

When we rotate the axes, as in Figure 6.1.15, the positive y-axis points to the *left*.

The answer we obtain is exactly the same, but this way the calculation was less work. Specifically, using y it was possible to do the integration for the entire region in one piece instead of having to divide it up. This example illustrates the useful fact that considerable simplifications will sometimes result from using y as the variable of integration for a region, or for some part of it.

EXAMPLE 10 **Metal Tulips** Decorative tulips are made of painted sheet metal as shown in Figure 6.1.16. Suppose the scale is in inches, and suppose a 1″ square of sheet metal weighs $\frac{3}{4}$ oz. By finding the weight of each tulip, decide whether or not

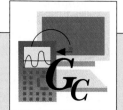

A graphics calculator can be used to find the area enclosed between two curves when it is not easy to solve explicitly for the points where they intersect.

For instance, to find the area between $y = x^4 - 3x + 1$ and $y = x$, we plot the graphs, solve for the points a and b where $x^4 - 3x + 1 = x$, i.e., $x^4 - 4x + 1 = 0$, and then evaluate the area $\int_a^b \left(x - (x^4 - 3x + 1)\right) dx = \left(-\frac{x^5}{5} + 2x^2 - x\right)\Big|_a^b$.

TI-85

GRAPH F1 ($y(x) =$) F1 (x) ^ 4 − 3 F1 (x) + 1 ▼ F1 (x) M2 (RANGE) (−) 1 ▼ 2 ▼ ▼ (−) 1 ▼ 3 F5 (GRAPH) SOLVER X ^ 4 − 4 X + 1 = 0 ENTER 0 F5 (SOLVE) EXIT X STO A ENTER SOLVER ENTER 1 F5 (SOLVE) EXIT X STO B ENTER (−) B ^ 5 ÷ 5 + 2 B ^ 2 − B + A ^ 5 ÷ 5 − 2 A ^ 2 + A ENTER

SHARP EL9300

[graph key] x/θ/T a^b 4 ► − 3 x/θ/T + 1 2ndF ▼ x/θ/T RANGE −1 ENTER 2 ENTER 1 ENTER −1 ENTER 3 ENTER 1 ENTER [graph key] SOLVER x/θ/T a^b 4 ► − 3 x/θ/T + 1 = x/θ/T ENTER (−) 1 ENTER ENTER ENTER ENTER ENTER [calc key] X STO A SOLVER ENTER ENTER ENTER 2 ENTER ENTER ENTER [calc key] X STO B − B a^b 5 ► ÷ 5 + 2 B x^2 − B + A a^b 5 ► ÷ 5 − 2 A x^2 + A ENTER

CASIO f_x-6300G (f_x-7700GB)

Range −1 EXE 2 EXE 1 EXE −1 EXE 3 EXE 1 EXE EXE EXE EXE Graph X x^y 4 − 3 X + 1 EXE Graph X EXE Trace ► (move cursor to intersection) X → A EXE Graph X EXE Trace ► (move cursor to intersection) X → B EXE (−) B x^y 5 ÷ 5 + 2 B x^2 − B + A x^y 5 ÷ 5 − 2 A x^2 + A EXE (Use Zoom for more accuracy)

HP 48SX

↰ EQUATION X ↰ = X y^x 4 ► − 3 × X + 1 ENTER ↰ SOLVE NEW αE αQ αN ENTER 30 +/− ↱ CST NXT SF ↰ PLOT PTYPE FUNC PLOTR 1 +/− ENTER 2 XRNG 1 +/− ENTER 3 YRNG DRAW ◄ (Move cursor to intersection) FCN ISECT ATTN ATTN X ENTER ′ A STO DRAW ◄ (Move cursor to intersection) FCN ISECT ATTN ATTN X ENTER ′ B STO B +/− 5 y^x 5 ÷ B x^2 2 × + B − A ENTER 5 y^x 5 ÷ + A ENTER x^2 2 × − A +

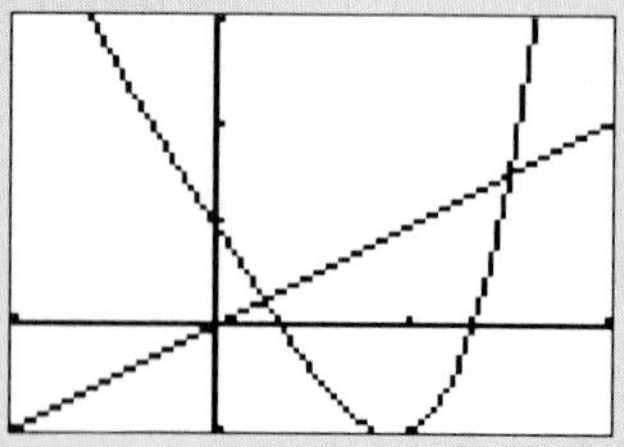

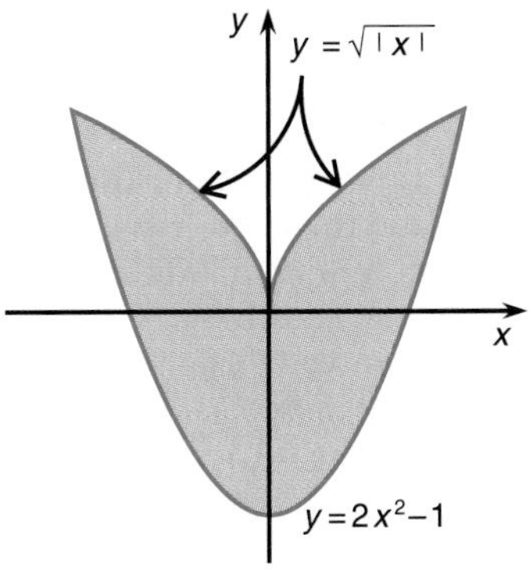

Figure 6.1.16

a "bouquet" of a dozen tulips can be suspended by a nylon thread that can support a load of one pound.

Solution The weight of the tulip will be proportional to its area, and we are told that the metal weighs $\frac{3}{4}$ oz per square inch. To find the area, we divide the tulip into left and right halves; they are symmetrical, so the total area is twice the area of the right half. The upper right corner is the point $(1, 1)$, so the area of the right half is

$$\int_0^1 \left(\sqrt{x} - (2x^2 - 1)\right) dx = \left(\frac{2}{3}x^{3/2} - \frac{2}{3}x^3 + x\right)\Big|_0^1 = 1 \text{ sq in.}$$

The total area of a tulip is 2 sq in, and the weight of the tulip is $2 \times \frac{3}{4} = \frac{3}{2}$ oz.

A dozen tulips weigh 18 oz, so they will break the thread. A stronger support will be needed.

Example 10 illustrates the useful fact that for a thin object made of uniform material, the weight is proportional to the area. Note that when we found the "total area" of a tulip it was the area of one side, not both the front and the back.

Summary 6.1.3

Areas of Regions in the Plane

1. Find curves that enclose the region whose area we want, expressing y as a function of x.
2. Distinguish intervals on which one curve is always above the other curve. This will involve solving for points of intersection of two curves (or of one curve with the x-axis).
3. Integrate over each interval and add the answers.
4. Consider writing x as a function of y to simplify some or all of the integrals.

Exercises 6.1

I

In each of the following situations, find the area of the region between the given function $y = f(x)$ and the x-axis over the given interval $[a, b]$, that is, between $x = a$ and $x = b$.

1. $f(x) = x^2$ over $[-1, 1]$

2. $f(x) = x^2$ over $[0, 4]$

3. $f(x) = x^3$ over $[-2, 1]$

4. $f(x) = x^5$ over $[-4, 4]$

5. $f(x) = \sin x$ over $\left[0, \frac{\pi}{2}\right]$

6. $f(x) = \sin x$ over $[0, 2\pi]$

7. $f(x) = \cos x$ over $\left[0, \frac{\pi}{2}\right]$

8. $f(x) = \sin^2 x$ over $\left[0, \frac{\pi}{2}\right]$

9. $f(x) = \frac{1}{x}$ over $[1, 2]$

10. $f(x) = 1/(1 + x^2)$ over $[-1, 1]$

In each of the following situations, find the area of the region enclosed between the given function $y = f(x)$ and the x-axis.

11. $f(x) = 1 - x^2$

12. $f(x) = 4 - x^2$

13. $f(x) = 2 - x^2$

14. $f(x) = x^2 - 2x - 3$

15. $f(x) = x^3 - 4x$

16. $f(x) = x^3 - 6x$

17. $f(x) = x^4 - 16$

18. $f(x) = 1 - \frac{2}{1+x^2}$

19. $f(x) = |x| - 3$

20. $f(x) = x\sqrt{|4 - x^2|}$

In each of the following situations, find the area of the region enclosed between the given functions $y = f(x)$ and $y = g(x)$.

21. $f(x) = x^3,\ g(x) = 4x$

22. $f(x) = 2x,\ g(x) = x^3$

23. $f(x) = x\ ,\ g(x) = x^5$

24. $f(x) = x^5,\ g(x) = 16x$

25. $f(x) = x^2\ ,\ g(x) = 2x^2 - 1$

26. $f(x) = x^2 - 1,\ g(x) = 11 - 2x^2$

27. $f(x) = x^2 - 2x + 1,\ g(x) = x + 1$

28. $f(x) = 2 - x^2,\ g(x) = |x|$

29. $f(x) = 3 - x,\ g(x) = x^2 + 2x + 3$

30. $f(x) = x^4 - 1,\ g(x) = 2x^2 - 2$

II

31. (i) Find the area of the region enclosed between $y = \sin x$ and the x-axis between $x = -\frac{\pi}{2}$ and $x = 0$.

(ii) Find the area of the region enclosed between $y = \sin(2x)$ and the x-axis between $x = -\frac{\pi}{2}$ and $x = 0$.

32. (i) Find the area of the region enclosed between $y = \cos x$ and the x-axis between $x = -\frac{\pi}{2}$ and $x = 0$.

(ii) Find the area of the region enclosed between $y = \cos(3x)$ and the x-axis between $x = -\frac{\pi}{2}$ and $x = 0$.

33. Find the area of the region enclosed between $y = f(x) = x^3 - 3x + 2$ and the x-axis. (*Hint:* $f(1) = 0$.)

34. Find the area of the region enclosed between $y = f(x) = (x^2 - 1)^2$ and the x-axis (cf. Example 4.4.2, p. 256).

***35.** Find the area of the region enclosed between $y = |3x|$ and $y = x^2$.

***36.** Find the area of the region enclosed between $y = x^2$ and $y = 19 - |5x - 5|$.

***37.** Suppose t is some number between -2 and 2, and consider the region between the x-axis and the curve $y = x^3 - 4x$ between $x = -2$ and $x = t$. (i) Find the area of this region. (ii) For what value(s) of t is the area of the specified region equal to $5\frac{3}{4}$?

***38.** Consider the region A enclosed between $y = x - 1$ and $y = 1 - x^2$. Also consider the region B enclosed between the line $3x + 2y = 0$ and the curve $y = x^2 - 1$. (i) Find the area of region A. (ii) Find the area of region B. (iii) Let C be the region that is the intersection of A and B. Find the area of C.

***39.** Sheet metal is cut into pieces shaped like the region enclosed between $y = x^2$ and $y = \sqrt{x}$ (with all measurements in centimeters). See Figure 6.1.17. Eight of them are glued to a plastic disk to make a "daisy." If the sheet metal weighs 3 grams per square centimeter, what is the weight of the daisy, ignoring the weight of the disk?

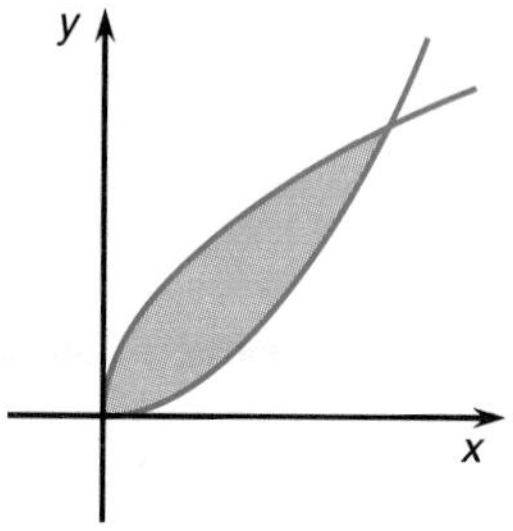

Figure 6.1.17

****40.** Consider the region that lies inside the circle $x^2 + y^2 = 1$ and above the line $y = \frac{1}{2}$. Find the area of this region. (You will need to use your calculator to obtain a numerical value.)

***41.** Suppose $y = f(x)$ intersects the x-axis at only finitely many points. Show that the region enclosed between the x-axis and the function $y = f(x)$ has the same area as the region enclosed between the x-axis and the function $y = |f(x)|$.

***42.** Find an example to show that the region between $f(x)$ and $g(x)$ over an interval $[a, b]$ may not have the same area as the region between $|f(x)|$ and $|g(x)|$ over $[a, b]$.

III

43. Use a graphics calculator or computer to estimate the points x where $\sin x = \frac{x}{2}$ and use this information to find (approximately) the area between the graphs of $y = \frac{x}{2}$ and $y = \sin x$, with $x \geq 0$.

44. Find (approximately) the area lying above the line $y = \frac{x}{3}$ and below $y = \cos x$, with $x \geq 0$.

POINT TO PONDER

As we have already seen in dealing with derivatives of trigonometric functions in Section 3.4, the number π is not accidental. It is a fundamental and profound ingredient of any mathematics involving circles, among other things.

Figure 6.1.18 shows the circle of radius 1. Its area equals $2\int_{-1}^{1}\sqrt{1-x^2}\,dx$, and this integral does not appear to have anything to do with π. Where does the π in the answer "come from"?

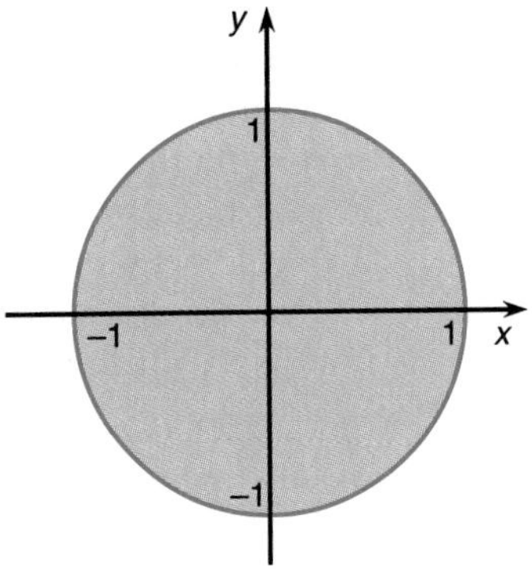

Figure 6.1.18

6.2 Setting Up Integrals; Limits of Integration

STRIP OF WIDTH *dx*
SPECIFIED RATE OF CHANGE

In Section 6.1 we saw how integration could be used to calculate the areas of geometric regions. The idea was simple enough, but in practice there are two steps that may be difficult. The first was finding the function to integrate, and the second was finding the limits of integration. Usually, the function to be integrated either was given to us explicitly or could be found with relative ease from a picture, but sometimes more effort was required. It frequently took some work to specify the interval over which the function should be integrated.

To some extent these problems must be solved on a case-by-case basis, but in this brief section we discuss some ideas that may make them slightly more manageable.

When we defined integrals, we added the areas of little rectangles of height $f(x_i^*)$ and width Δx_i, as illustrated in Figure 6.2.1, and approximated the integral by a sum

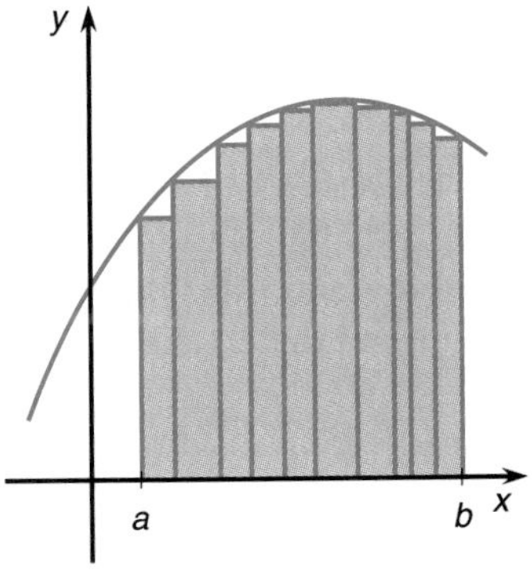

Figure 6.2.1

$$\sum_{i=1}^{n} f(x_i^*)\,\Delta x_i.$$

After taking a limit this became the integral

$$\int_a^b f(x)\,dx.$$

This suggests viewing the integral as a sort of infinite sum of the "areas" of little rectangles of height $f(x)$ and "width" dx. This does not really make sense, but it is often a very helpful way to set up integrals.

EXAMPLE 1 **Area of Ellipse** Assuming $a > 0$ and $b > 0$, find the area of the ellipse

$$\frac{x^2}{a^2} + \frac{y^2}{b^2} = 1.$$

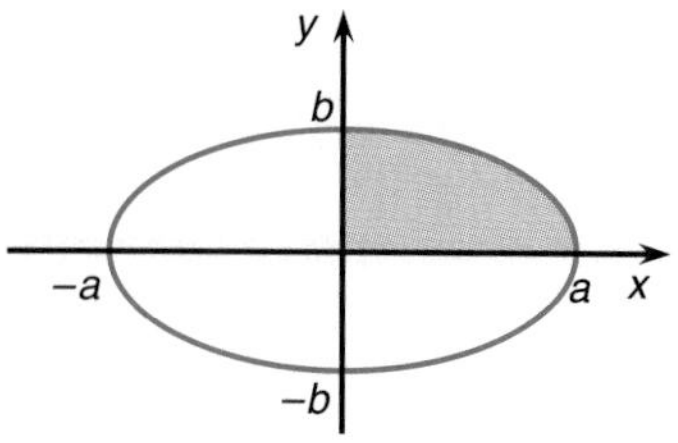

Figure 6.2.2

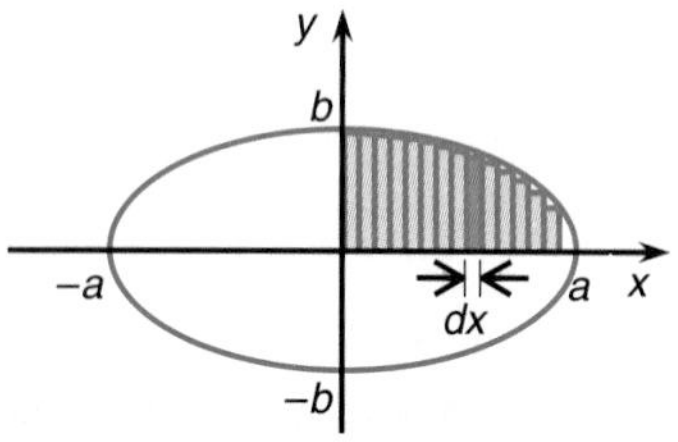

Figure 6.2.3

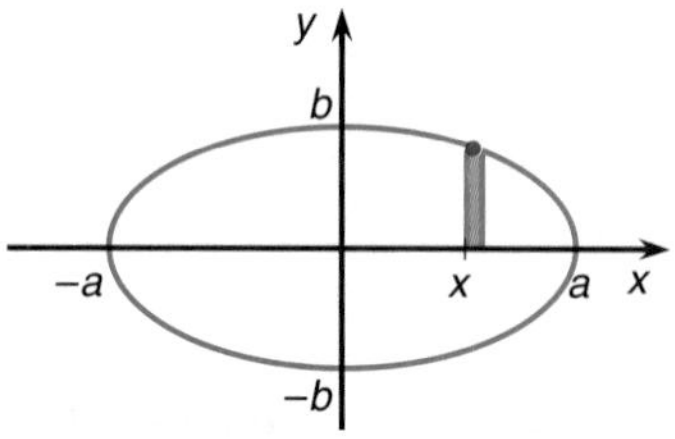

Figure 6.2.4

Solution The area will be 4 times the area of the part of the region that lies in the first quadrant, the shaded part of the region in Figure 6.2.2.

Suppose we think of this region as being covered with thin little vertical rectangles, as shown in Figure 6.2.3. Each rectangle has width dx, which we regard as a very very small thickness, and we want to figure out the height. Of course, the height is different for different rectangles, so we need to find the height of the rectangle depending on its position. Specifically, if we suppose that the base of the rectangle is at the point x on the x-axis, we should find its height as a function of x. Since the width of the rectangle is so small, we allow ourselves not to worry about the difference between its left side and right side.

Now think about the point at the top of the rectangle (see Figure 6.2.4). We know its x-coordinate is x, and if we let y be its y-coordinate, then we know that the point (x, y) lies on the ellipse, so its coordinates satisfy the equation $\frac{x^2}{a^2} + \frac{y^2}{b^2} = 1$. We can solve this to express y in terms of x. Indeed,

$$\frac{x^2}{a^2} + \frac{y^2}{b^2} = 1$$

so

$$\frac{y^2}{b^2} = 1 - \frac{x^2}{a^2}$$

or

$$y^2 = b^2\left(1 - \frac{x^2}{a^2}\right)$$

and

$$y = b\sqrt{1 - \frac{x^2}{a^2}}.$$

This is the height of the rectangle at x, and its "width" is dx, so its "area" is $b\sqrt{1 - x^2/a^2}\,dx$. To get the area of the whole region, the upper right quarter of the ellipse, we "add up" these rectangles starting with $x = 0$ and going out to the end of the ellipse, which is $x = a$. "Adding up" means integrating, so the area is

$$\int_0^a b\sqrt{1 - \frac{x^2}{a^2}}\,dx = b\int_0^a \sqrt{1 - \frac{x^2}{a^2}}\,dx$$

$u = \frac{x}{a}$ $du = \frac{1}{a}dx$

$$= b\int_0^1 \sqrt{1 - u^2}\,a\,du = ab\int_0^1 \sqrt{1 - u^2}\,du$$

$$= (ab)\frac{1}{2}\left(u\sqrt{1 - u^2} + \arcsin u\right)\Big|_0^1$$

$$= \frac{ab}{2}\left(0 + \arcsin(1) - 0 - \arcsin(0)\right)$$

$$= \frac{\pi ab}{4}.$$

The area of the whole ellipse is four times as much, or πab.

Notice that if $a = b = r$, then the "ellipse" is just a circle of radius r, whose area is πr^2.

Also notice that this approach using little rectangles of width dx is not ultimately any different than what we did before. It amounts to noticing that the region we wanted is the region lying below the curve $y = b\sqrt{1 - x^2/a^2}$, between $x = 0$ and $x = a$, so its area is the corresponding integral. The advantage is that the rectangles are more geometrical, and that may make it easier to visualize what is being done.

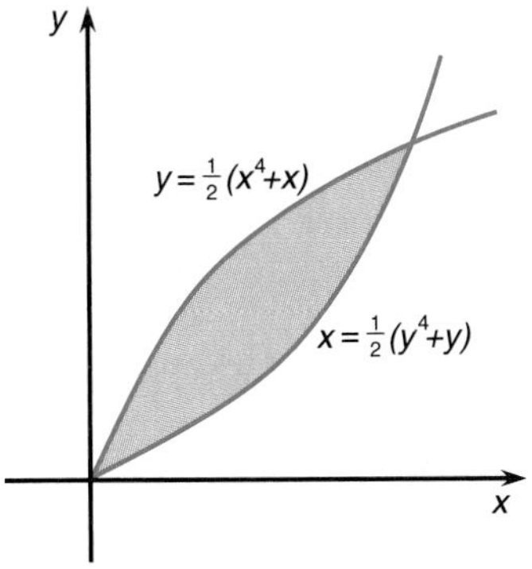

Figure 6.2.5

EXAMPLE 2 Find the area of the region enclosed by the curves $y = \frac{1}{2}(x^4 + x)$ and $x = \frac{1}{2}(y^4 + y)$, as shown in Figure 6.2.5.

Solution The first thought is to regard this as the region between two curves $y = f(x)$ and $y = g(x)$. The lower curve is $y = g(x) = \frac{1}{2}(x^4 + x)$, but it is not easy to see how to write the upper curve $x = \frac{1}{2}(y^4 + y)$ in the form $y = f(x)$. On the other hand, it is clearly already written with x as a function of y.

The trick is to divide the region into two parts, separated by the line $y = x$ (see Figure 6.2.6). The region below the line can be filled with little rectangles of width dx whose height is the distance from the line down to the bottom curve, that is, $x - \frac{1}{2}(x^4 + x) = \frac{1}{2}(x - x^4)$. The area of the bottom piece is then the "sum" of these little areas, or

$$\int_0^1 \frac{1}{2}(x - x^4)\,dx = \frac{1}{2}\left(\frac{x^2}{2} - \frac{x^5}{5}\right)\Bigg|_0^1 = \frac{3}{20}.$$

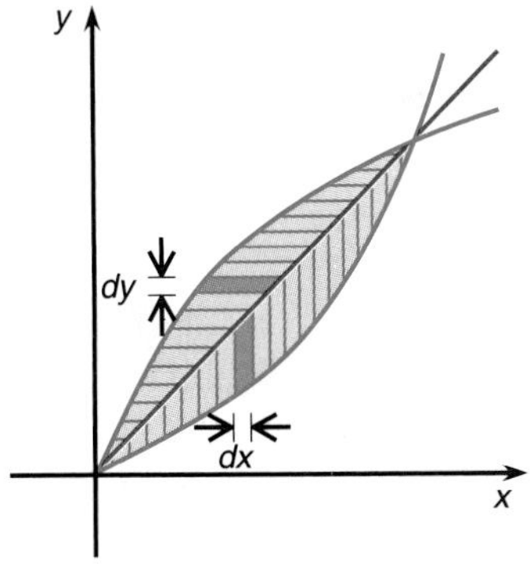

Figure 6.2.6

For the top piece we use horizontal rectangles of depth dy and length $y - \frac{1}{2}(y^4 + y) = \frac{1}{2}(y - y^4)$. The area of such a rectangle is $\frac{1}{2}(y - y^4)\,dy$, so the area of the region is

$$\int_0^1 \frac{1}{2}(y - y^4)\,dy = \frac{1}{2}\left(\frac{y^2}{2} - \frac{y^5}{5}\right)\Bigg|_0^1 = \frac{3}{20}.$$

The original region has area $\frac{3}{20} + \frac{3}{20} = \frac{3}{10}$.

To do the integral for the upper piece, we needed to express the upper curve and the line with x as a function of y, so that the length of the little rectangle could be written in terms of y. The (narrow) width of the rectangle is dy, and we integrated in terms of y, that is, using y as the variable of integration. The limits of integration were apparent from the picture.

What makes this example work is that the line $y = x$ can be used either way (as a function of x for the lower piece and as a function of y for the upper region). In hindsight we notice that the upper and lower regions are symmetrical, so we only had to evaluate one integral and then double it.

Finally, we remark again that the little rectangles were not really needed. The lower region was after all the region between two curves, $y = x$ and $y = \frac{1}{2}(x + x^4)$, so we know how to find its area by integrating their difference. Similarly, for the upper region we have to think of y as the variable, but once we have made this choice, the region is again enclosed between two curves, and all that is needed is to integrate their difference. Using rectangles of width dx (or dy) may help us to keep track of what we are doing. It does not change the calculation or give us different answers, but if it makes the situation clearer, then it is useful.

A graphics calculator can help find the area enclosed between two curves when it is not easy to solve explicitly for the points where they intersect.

For instance, to find the area between $y = \cos x$ and $y = x^2$, we plot the graphs, solve for the points a and b where $\cos x = x^2$, i.e., $\cos x - x^2 = 0$, and then evaluate the area $\int_a^b (\cos x - x^2)\,dx = \left(\sin x - \frac{x^3}{3}\right)\Big|_a^b$. We save some trouble by noting that $a = -b$.

TI-85

GRAPH F1 $(y(x) =)$ COS F1 (x) ▼ F1 (x) ^ 2 M2(RANGE) (−) 1 ▼ 1 ▼ ▼ (−) 1 ▼ 2 F5(GRAPH) SOLVER COS X − X ^ 2 = 0 ENTER 0 F5(SOLVE) EXIT X STO B ENTER (−) X STO A ENTER SIN B − B ^3 ÷ 3− SIN A + A ^3 ÷ 3 ENTER

SHARP EL9300

(graph key) cos x/θ/T 2ndF ▼ x/θ/T x^2 RANGE −1 ENTER 1 ENTER 1 ENTER −1 ENTER 2 ENTER 1 ENTER (graph key) SOLVER cos x/θ/T = x/θ/T x^2 ENTER (−) 1 ENTER ENTER ENTER ENTER ENTER (calc key) X STO A (−) X STO B sin B − B a^b 3 ▶ ÷ 3 − sin A + A a^b 3 ▶ ÷ 3 ENTER

CASIO f_x-6300G (f_x-7700GB)

Range −1 EXE 1 EXE 1 EXE −1 EXE 2 EXE 1 EXE EXE EXE EXE Graph cos X EXE Graph X x^2 EXE Trace ▶ (move cursor to intersection) X → A EXE (−) X → B EXE sin B − B x^y 3 ÷ 3 − sin A + A x^y 3 ÷ 3 EXE (Use Zoom for more accuracy)

HP 48SX

↰ EQUATION X y^x 2 ▶ ↰ = COS X ENTER ↰ SOLVE NEW αE αQ αN ENTER 30 +/− ↱ CST NXT SF ↰ PLOT PTYPE FUNC PLOTR 1 +/− ENTER 1 XRNG 1 +/− ENTER 2 YRNG DRAW ◀ (Move cursor to intersection) FCN ISECT ATTN ATTN X ENTER ′ A STO A +/− ′ B STO B SIN B ENTER 3 y^x 3 ÷ − A SIN − A ENTER 3 y^x 3 ÷ +

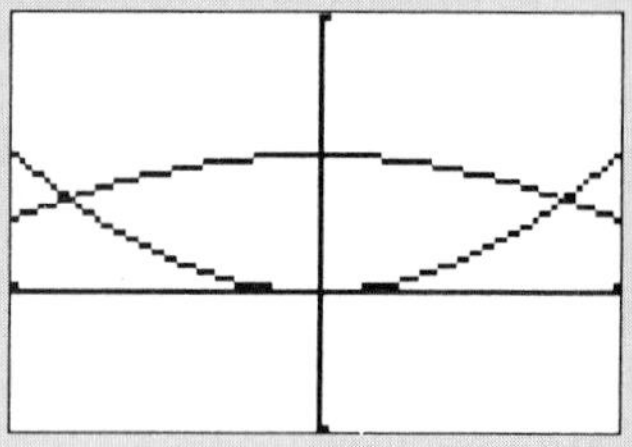

In this example we did not really need the graphs to know which curve is higher and how often they cross. . .

Summary 6.2.1

Setting Up Area Integrals

1. Draw a picture, indicating points where the curves cross each other.
2. Draw little rectangles of width dx, whose height can be expressed as a function of x. It may be necessary to break the region into pieces and find different formulas for the different subregions. It may also be necessary to use y as the variable for some of the subregions (and dy as the "width" of the corresponding rectangles).

3. Figure out the limits of integration from the picture. They will be the values of x (or y) where each subregion begins and ends. Set up the integral(s), being careful that the integrands are never negative.
4. Integrate and add up the areas of the subregions.

Functions with Specified Rate of Change

Suppose $F(t)$ is some changing quantity. Possible examples are almost endless, but they include such things as the temperature or concentration in a chemical reaction, the displacement of a moving object, the weight of a bacterial culture, the cost of a commodity, the success rate for somebody learning a skill, the population of a country, the volume of a deflating balloon, and so on. If the variable t represents time, then $F(t)$ represents the value of the quantity at time t and $F'(t)$ is its rate of change.

Now suppose the rate of change is known, that is, it is specified that at time t the rate of change of $F(t)$ is some continuous function $f(t)$. This just means $f(t) = F'(t)$, or that $F(t)$ is an antiderivative of $f(t)$. The second version of the Fundamental Theorem of Calculus (Theorem 5.6.1) says that $\int_a^b f(t)\,dt = F(b) - F(a)$, or $\int_a^b F'(t)\,dt = F(b) - F(a)$, which can be written

$$F(b) = F(a) + \int_a^b F'(t)\,dt. \tag{6.2.2}$$

In other words, if we know the rate of change $F'(t)$ and we know the value $F(a)$ of $F(t)$ at some time $t = a$, then we can find the value of $F(t)$ at any other time $t = b$.

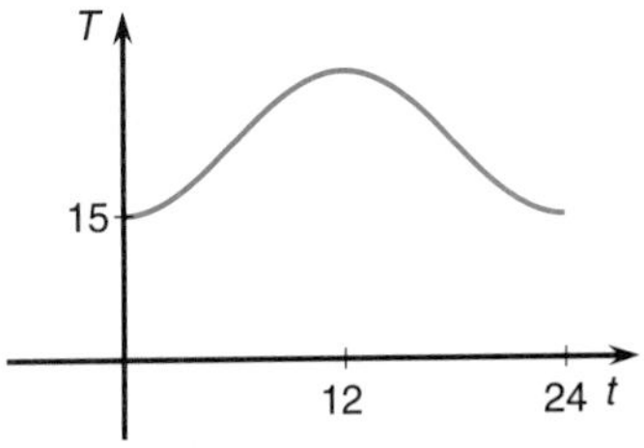

Figure 6.2.7(i)

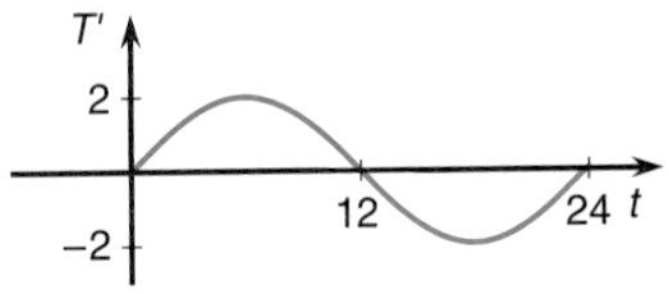

Figure 6.2.7(ii)

EXAMPLE 3 **Atmospheric Temperature** Let t represent time in hours, $t = 0$ being midnight. Suppose that the outdoor temperature on a certain day is 15°C at midnight, and its rate of change at time t is given by

$$2\sin\left(\frac{\pi t}{12}\right) \text{degrees/hour.}$$

What is the temperature at 3 P.M.?

Solution Let $T(t)$ represent the temperature (in degrees Celsius) at time t. The question tells us that $T(0) = 15$ and $T'(t) = 2\sin\left(\frac{\pi t}{12}\right)$. See Figures 6.2.7(i) and (ii).

In the diagrams we see that $T'(t)$ is positive when $0 < t < 12$. This means that the temperature increases from midnight until noon, and since $T'(t) < 0$ when $12 < t < 24$, it decreases from noon until midnight. The temperature is highest at noon and lowest at midnight.

$$u = \frac{\pi t}{12}$$
$$du = \frac{\pi}{12}dt$$

What we want to find is the temperature at 3 P.M., which is $T(15)$. (Notice that noon is at $t = 12$, so 3 P.M. is 3 hours later, or $t = 15$.) By Equation 6.2.2, we have that

$$\begin{aligned}
T(15) &= T(0) + \int_0^{15} T'(t)\,dt = 15 + \int_0^{15} 2\sin\left(\frac{\pi t}{12}\right)dt \\
&= 15 + 2\int_0^{5\pi/4} \sin(u)\left(\frac{12}{\pi}\right)du = 15 + \frac{24}{\pi}\int_0^{5\pi/4} \sin u\,du \\
&= 15 + \frac{24}{\pi}(-\cos u)\Big|_0^{5\pi/4} = 15 + \frac{24}{\pi}\left(\frac{1}{\sqrt{2}} - (-1)\right) = 15 + \frac{24}{\pi}\left(1 + \frac{1}{\sqrt{2}}\right) \\
&\approx 28.0°\text{C} \approx 82°\text{F}.
\end{aligned}$$

EXAMPLE 4 **Volume of a Balloon** Suppose air is being released from a balloon so that at time t (measured in seconds) the volume is decreasing at an instantaneous rate of $\left(10 - \frac{1}{2}t\right)$ cu in/sec.

If the volume at $t = 0$ is 100 cu in, what is the volume 5 sec later?

Solution If $V(t)$ represents the volume of the balloon at time t, in cu in, we know that $V(0) = 100$ and that $V'(t) = -\left(10 - \frac{1}{2}t\right)$ (notice the initial minus sign, which is there because the volume is decreasing). From Equation 6.2.2 we see that the volume after 5 sec is

$$\begin{aligned} V(5) &= V(0) + \int_0^5 V'(t)\,dt = 100 - \int_0^5 \left(10 - \frac{1}{2}t\right) dt \\ &= 100 - \left(10t - \frac{1}{4}t^2\right)\Bigg|_0^5 \\ &= 56\frac{1}{4}\text{ cu in.} \end{aligned}$$

It is also possible to consider the idea of rates of change in terms of little changes over a tiny period of time dt.

EXAMPLE 5 Consider an object moving so that its velocity is $(16 - t^2)$ m/sec at time t. How far will it move before coming to rest?

Solution Notice that the velocity will equal 0 when $t = 4$, which means that the object will come to rest after 4 sec.

At any particular moment the velocity is $16 - t^2$, which means that in a very short interval of time, say, of length I, the object will travel a distance of approximately $(16 - t^2) \times I$. If we consider what happens during an extremely short interval of time, say, of length dt, we can allow ourselves to ignore the approximation and say that it will travel a distance of $(16 - t^2)\,dt$. The total distance traveled should be the integral of these tiny distances, or

$$\int_0^4 (16 - t^2)\,dt = \left(16t - \frac{1}{3}t^3\right)\Bigg|_0^4 = 42\frac{2}{3}\text{ m.}$$

Once again there is nothing here that we could not have done without ever mentioning dt, but this approach may make it easier to understand.

It is really a matter of taste. If you like this approach, you should use it; if you do not like it, you should probably avoid it.

Exercises 6.2

I

Find the area of the region enclosed by the graphs of the given curves.

1. $y = x^2$, $x = y^2$
2. $y = x^3$, $x = y^3$
3. $y^2 + 2y = x - 2$, $x = 5$
4. $y^3 = x + 1$, $x = 0$, $y = 0$
5. $x = |y^3|$, $x = 8$
6. $y = x^2 - 2x$, $y^2 + 2y + 2 = x$
7. $x = ye^{y^2}$, $x = 0$, $y = 1$
8. $x = |y^3 + y|$, $x = 2$
9. $x = y(y - 1)(y - 2)$, $x = 0$
10. $xy = 1$, $x = 2$, $y = 3$

In each of the following situations we are given the derivative $F'(t)$ of some function $F(t)$ and one value $F(a)$, for some a. Find the required value $F(b)$.

11. $F'(t) = 3t^2 - 2, F(0) = 2$; find $F(3)$.

12. $F'(t) = \sin t, F(0) = 0$; find $F(\pi)$.

13. $F'(t) = 1, F(2) = -5$; find $F(4)$.

14. $F'(t) = t, F(3) = 1$; find $F(\sqrt{2})$.

15. $F'(t) = t^2 - 3, F(5) = 2$; find $F(0)$.

16. $F'(t) = |t|, F(2) = 4$; find $F(-2)$.

17. $F'(t) = 1/(t^2+1), F(-1) = 0$; find $F(1)$.

18. $F'(t) = 1/t, F(1) = 6$; find $F(2)$.

19. $F'(t) = e^{2t+1}, F(0) = 0$; find $F\left(\frac{1}{2}\right)$.

20. $F'(t) = \cos^2(3t) + \sin^2(3t), F(1) = -2$; find $F(5)$.

II

21. Find the area of the region enclosed between the curve $x = y^2$ and the right semicircle $x^2 + y^2 = 2$, $x \geq 0$.

22. Find the area of the region enclosed by the graphs of $x = y^3 - 3y$ and $y = x$.

***23.** Find the area of the region consisting of all points lying between the graphs of $y = \sin x$ and $y = 1 - \sin x$, with $0 \leq x \leq \pi$.

***24.** Find the area of the region enclosed by the graphs of $x = 0$, $y = 3$, and $y = x^3 + 2x$.

****25.** Let m and n be positive integers, $m \neq n$.

(i) Describe the region enclosed between the graphs of $y = x^n$ and $y = x^m$.

(ii) Find the area of this region.

****26.** Let m and n be positive integers, $m \neq n$.

(i) Describe the region enclosed between the graphs of $y = x^{1/n}$ and $x = y^{1/m}$.

(ii) Find the area of this region.

Find the areas of the triangles whose vertices are specified below.

27. $(0,0)$, $(3,0)$, and $(1,2)$

28. $(0,0)$, $(1,1)$, and $(1,2)$

***29.** $(0,0)$, $(2,1)$, and $(1,2)$

***30.** $(-1,0)$, $(2,1)$, and $(1,-2)$

***31.** Find the area of the triangle whose vertices are the points $(0,0)$, (a,b), and (c,d), where $0 \leq a \leq c$.

***32.** Use integration to verify the formula for the area A of a triangle, $A = \frac{1}{2}bh$, where b is the length of the base and h is the height.

33. (i) In the situation of Example 3 (p. 429), what will be the temperature at 9 A.M.?

(ii) What will be the temperature at 5 P.M.?

***34.** Find a formula for the temperature at time t in the situation of Example 3. (If necessary, simplify so that your formula does not contain an integral sign.)

35. In the situation of Example 4, what will be the volume of the balloon 10 sec after it begins deflating?

***36.** (i) In the situation of Example 4, at what time will the volume be a minimum?

(ii) At the time mentioned in part (i), what will the volume of the balloon be?

37. In the situation of Example 5, how far will the object travel between $t = 1$ and $t = 3$, that is, between the moment 1 sec after starting and the moment 3 sec after starting?

****38.** Suppose an object moves so that its velocity at time t is $t^3 - 6t^2 + 5t + 6$ ft/sec.

(i) Verify that at time $t = 4$ it will have returned to its starting point, that is, the position it was at when $t = 0$.

(ii) What is the total distance it will have traveled between $t = 0$ and $t = 4$? (*Hint:* The answer is not 0.)

***39.** Consider water flowing out of a tank. As the water level falls, the pressure drops and the flow is slower. Suppose the water is flowing out so that at time t the rate of flow is $25e^{-t}$ cu ft/min, with t measured in minutes.

(i) If there are 25 cu ft of water in the tank when $t = 0$, how much will there be 2 min later?

(ii) How long will it be until there is only 1 cu ft left?

***40.** If money in a bank account grows with *continuous compounding*, it earns interest on its interest and accrues more rapidly than it would with simple interest. If the original amount is A_0 and the rate of interest is $r\%$, then the rate at which the balance is increasing at time t is $A_0 \frac{r}{100} e^{rt/100}$, with t measured in years.

(i) If \$1,000 is placed in an account at an interest rate of 10% with continuous compounding, how much will be there after 1 year?

(ii) By what percentage has the balance increased in the 1-year period? (This is known as the effective annual interest rate, meaning the rate of simple interest that will give the same increase over 1 year as the compound interest does.)

(iii) What is the effective annual interest rate for continuously compounded interest at a rate of 12%?

(iv) What is the effective annual interest rate for continuously compounded interest at a rate of 18%?

(v) How much money would have to be invested at 15%, compounded continuously, so that after 6 *months* there would be \$5,000?

*41. Consider the parallelogram shown in Figure 6.2.8, whose base is b units long and whose height is h units. Find its area by integration.

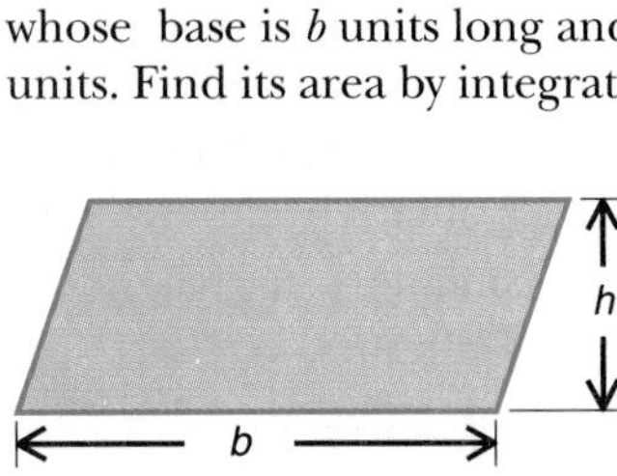

Figure 6.2.8

*42. Notice that the previous question mentioned the height and the length of the base but no other information. Clearly, the area of the parallelogram depends on the height and the length of the base but *not* on the amount by which the parallelogram "tilts." (i) Convince yourself that this is so by a geometrical argument showing that the parallelogram has the same area as the rectangle with base b and height h. (ii) If the other side has length a and the angle between two adjacent sides is θ, find a formula for the area in terms of a, b, and θ. (iii) What happens if you use the *other* angle between two adjacent sides?

III

Use a graphics calculator or computer to help you to find the approximate area of the region enclosed between each of the following pairs of curves.

43. $y = x^3 - 2x - 1$, $y = x$

44. $y = \sin x$, $y = x^3$

45. $y = \cos x$, $y = x^4$

46. $y = e^x$, $y = 4 + \ln x$, $x > 0$

47. $y = x^3 + x$, $x = y^2 - 1$

48. $y = \sec^2 x$, $y = 2 - x^2$, $|x| \le 1$

6.3 Volumes of Solids of Revolution by Slices

SOLID OF REVOLUTION
CYLINDRICAL SLICES

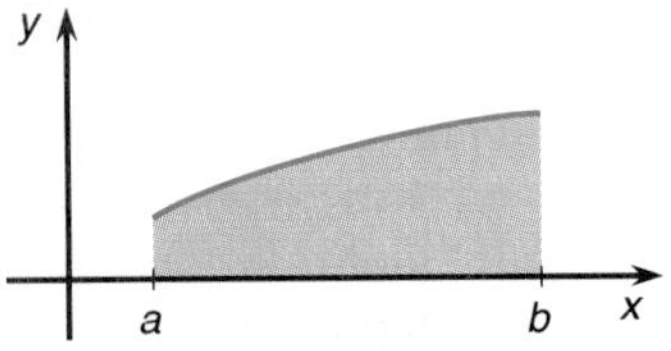

Figure 6.3.1

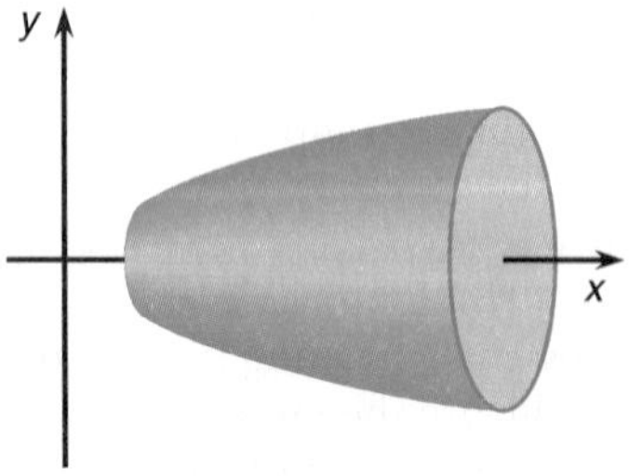

Figure 6.3.2

Integrals can be used to calculate volumes of solid regions in three-dimensional space. We concentrate on what are called *solids of revolution.* If $f(x)$ is a nonnegative function on $[a, b]$, draw its graph $y = f(x)$ as shown in Figure 6.3.1. Then take the xy-plane, containing the graph, and rotate it in three-dimensional space around the x-axis. As the graph revolves, the region under it (shaded in Figure 6.3.1) "sweeps out" a solid three-dimensional region, as shown in Figure 6.3.2. This is a **solid of revolution**.

To find its volume, consider a tiny piece of the interval $[a, b]$ of width dx. Look at the narrow strip of the region under the graph of $y = f(x)$ whose base is this piece of the interval (see Figure 6.3.3). The height of the strip is $f(x)$.

Consider the part of the solid swept out by this strip, as illustrated in Figure 6.3.4. It is a "slice," which is approximately a thin disk. For our purposes a more useful way of seeing it is as a cylinder on its side with radius $f(x)$ and "height" dx. The volume of a cylinder with radius r and height h is $\pi r^2 h$ (cf. Summary 1.5.14(iii)), so the volume of this cylinder is $\pi f(x)^2 dx$.

This suggests that to get the volume of the whole solid, we ought to add up the volumes of all these little slices (cylinders), as shown in Figure 6.3.5, that is,

$$\textit{Volume} = \int_a^b \pi f(x)^2 dx.$$

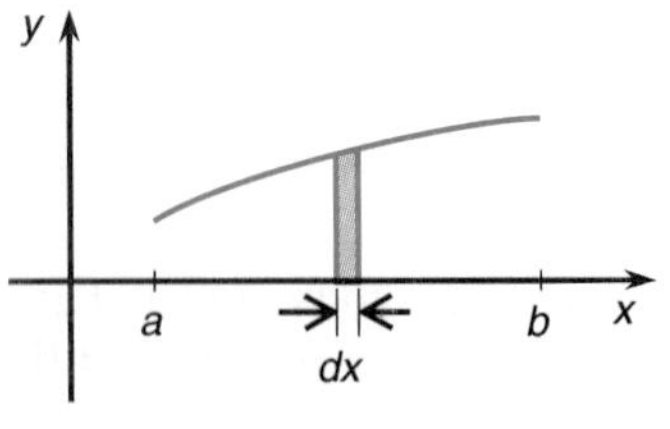

Figure 6.3.3

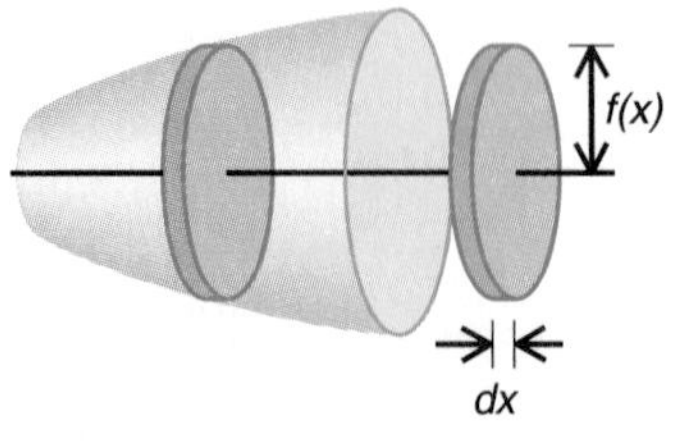

Figure 6.3.4

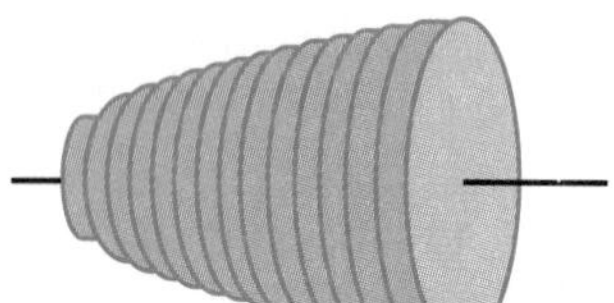

Figure 6.3.5

It is possible to make this argument more precise, by partitioning $[a, b]$, approximating the volume by the sum of the volumes of cylinders of height Δx_k, and taking a limit. The same integral results, so it seems reasonable to make a definition.

DEFINITION 6.3.1

> If $f(x) \geq 0$ for all $x \in [a, b]$, then the volume of the solid of revolution obtained by rotating the graph $y = f(x)$ about the x-axis is
>
> $$\textit{Volume} = \int_a^b \pi f(x)^2 \, dx.$$

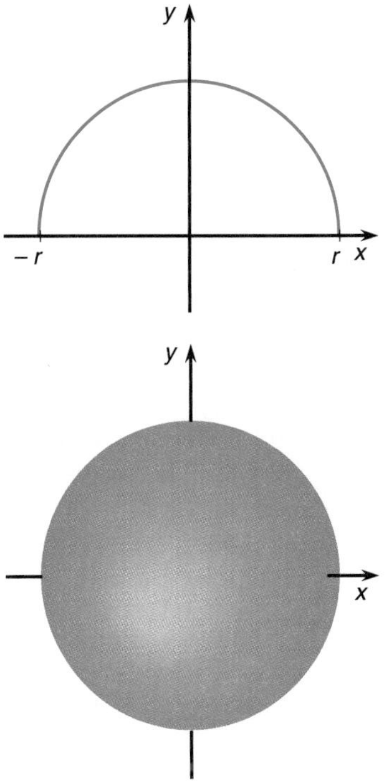

Figure 6.3.6

EXAMPLE 1 **Volume of a Ball** Find the volume of a ball of radius r.

Solution We may as well work with the ball whose center is at the origin. It is obtained by revolving the upper semicircle $y = \sqrt{r^2 - x^2}$ around the x-axis, as shown in Figure 6.3.6. So its volume is

$$\int_{-r}^{r} \pi f(x)^2 dx = \pi \int_{-r}^{r} (r^2 - x^2)\, dx = \pi \left(r^2 x - \frac{1}{3} x^3 \right) \Big|_{-r}^{r}$$

$$= \pi \left(r^3 - \frac{1}{3} r^3 - \left(-r^3 + \frac{1}{3} r^3 \right) \right)$$

$$= \frac{4}{3} \pi r^3.$$

EXAMPLE 2 **Volume of a Headlight** Find the volume of the headlight obtained by revolving about the x-axis the parabola $y = \sqrt{x}$ for $0 \leq x \leq 1$. This solid is called a *paraboloid* (see Figure 6.3.7).

Solution The volume equals

$$\int_0^1 \pi (\sqrt{x})^2 dx = \pi \int_0^1 x\, dx = \frac{\pi}{2} x^2 \Big|_0^1 = \frac{\pi}{2}.$$

EXAMPLE 3 **Volume of Cone** Find the formula for the volume of a right circular cone of radius r and height h (see Figure 6.3.8).

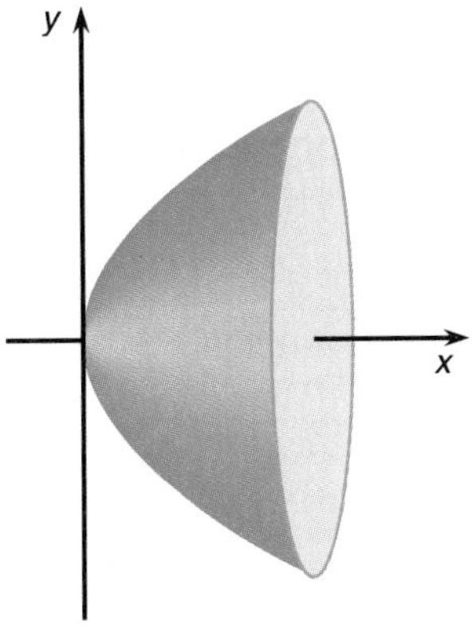

Figure 6.3.7

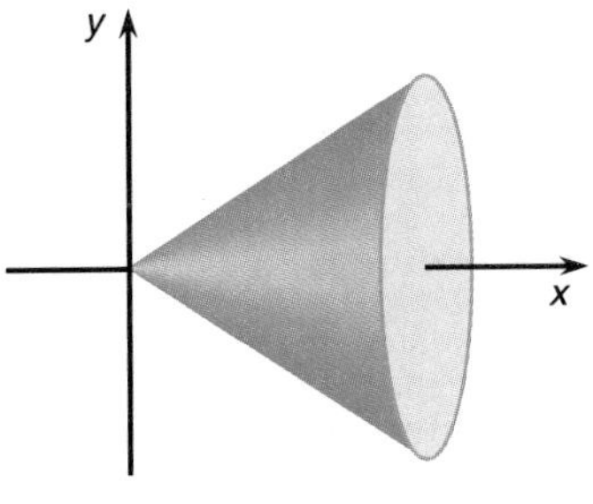

Figure 6.3.8

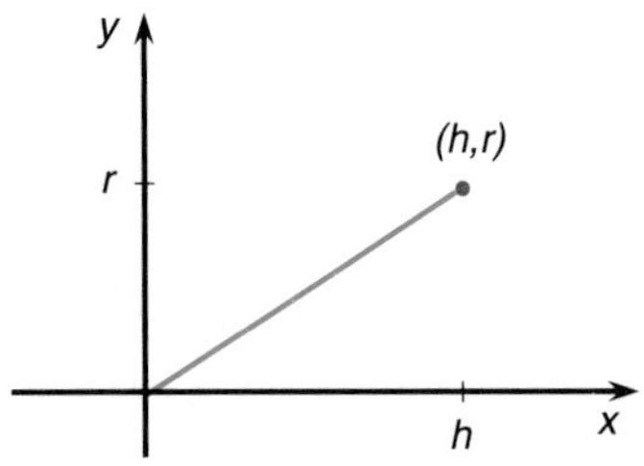

Figure 6.3.9

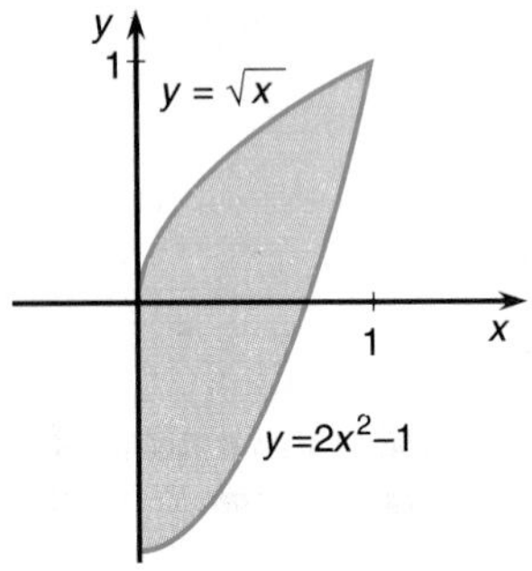

Figure 6.3.10

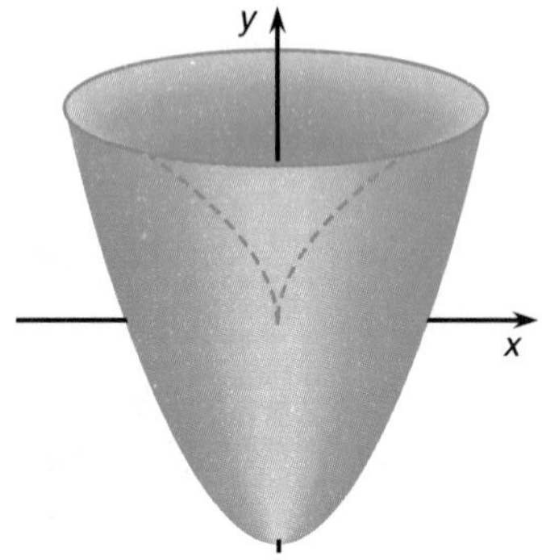

Figure 6.3.11

Since we are integrating with respect to y, the radius is the horizontal distance out from the y-axis. This is why we write $x = f(y)$ and integrate $\pi f(y)^2$.

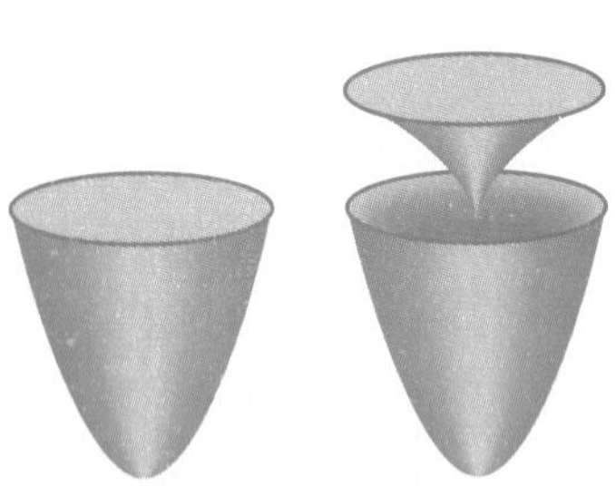
Figure 6.3.12

Solution We need to express the cone as a solid of revolution. It is obtained by revolving about the x-axis a straight line through the origin. The catch is to find the equation of the line.

To do this, we need two points on the line. One is the origin; what are the coordinates of the point at the upper right corner? As illustrated in Figure 6.3.9, the point is (h, r). This is slightly confusing because the cone is lying on its side, so the "height" is horizontal, but we see from Figure 6.3.8 that it is this dimension that we call the height of the cone.

We find the equation of the line passing through $(0, 0)$ and (h, r). It is $\frac{y-0}{x-0} = \frac{r-0}{h-0}$, or

$$y = \frac{r}{h}x.$$

We are interested in the function $f(x) = \frac{r}{h}x$ with x running from $x = 0$ to $x = h$.

The corresponding volume is

$$\int_0^h \pi\left(\frac{r}{h}x\right)^2 dx = \pi\frac{r^2}{h^2}\int_0^h x^2\,dx = \frac{1}{3}\pi\frac{r^2}{h^2}x^3\Big|_0^h = \frac{1}{3}\pi r^2 h.$$

EXAMPLE 4 Consider the region in the plane bounded by the y-axis and the curves $y = \sqrt{x}$ and $y = 2x^2 - 1$, shown in Figure 6.3.10. Find the volume of the solid of revolution obtained by revolving this region about the y-axis (see Figure 6.3.11).

Solution Since the region is rotated about the y-axis, we must integrate with respect to y. It is easiest to obtain this solid by taking the paraboloid formed by rotating $y = 2x^2 - 1$ and removing the solid obtained by rotating $y = \sqrt{x}$. The volume we want will be the difference of these volumes. See Figure 6.3.12.

To find the volume of the paraboloid, we must express the parabola $y = 2x^2 - 1$ with x as a function of y, $x = f(y)$. Since $y = 2x^2 - 1$, we see that $y + 1 = 2x^2$, $x^2 = \frac{1}{2}(y+1)$, $x = \sqrt{\frac{1}{2}(y+1)} = f(y)$, and the volume of the paraboloid is

$$\int_{-1}^{1} \pi\big(f(y)\big)^2 dy = \pi\int_{-1}^{1}\left(\sqrt{\frac{1}{2}(y+1)}\right)^2 dy$$

$$= \pi\int_{-1}^{1}\frac{1}{2}(y+1)\,dy = \frac{\pi}{2}\left(\frac{1}{2}y^2 + y\right)\Big|_{-1}^{1}$$

$$= \pi.$$

The other curve $y = \sqrt{x}$ can be written as $x = g(y) = y^2$, and the corresponding solid has volume

$$\int_0^1 \pi\big(g(y)\big)^2 dy = \int_0^1 \pi(y^2)^2 dy = \pi\frac{1}{5}y^5\Big|_0^1 = \frac{\pi}{5}.$$

The original solid has volume $\pi - \frac{1}{5}\pi = \frac{4\pi}{5}$.

EXAMPLE 5 Find the volume of the solid obtained by revolving about the x-axis the region between the line $y = \frac{x}{2}$ and the parabola $x = y^2$, between $x = 0$ and $x = 1$ (see Figure 6.3.13).

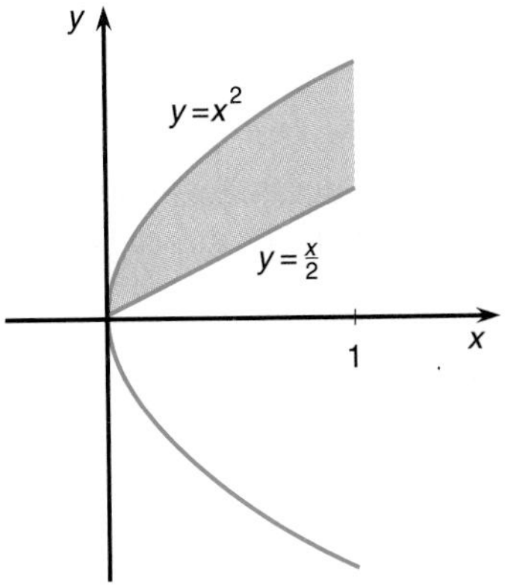

Figure 6.3.13

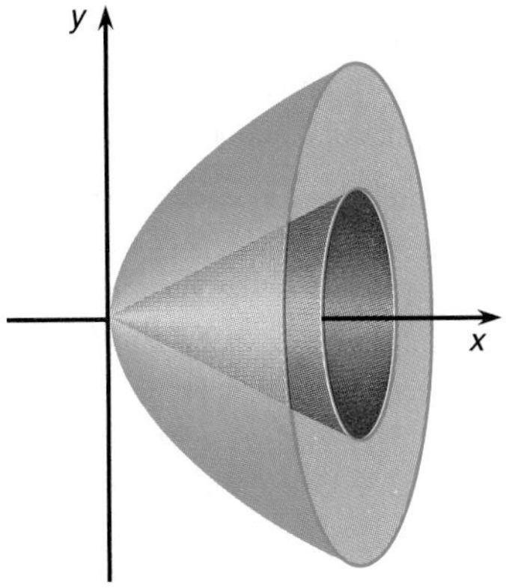

Figure 6.3.14

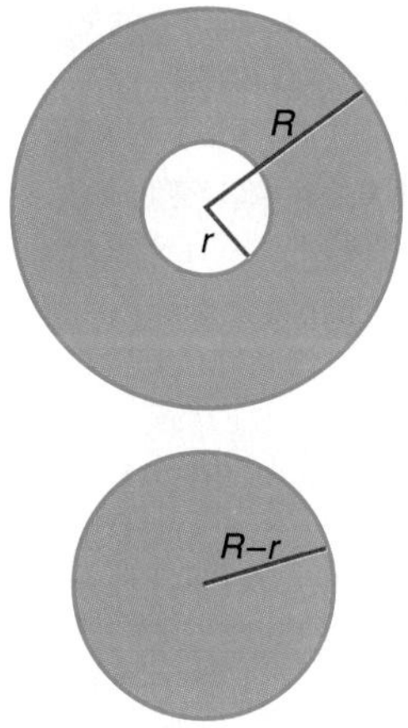

Figure 6.3.15

Solution Again we find the volumes of two regions and subtract. The volume we want is the volume of the paraboloid minus the volume of the cone (see Figure 6.3.14).

For the paraboloid we are summing up the volumes of slices of thickness dx and radius $y = \sqrt{x}$. The volume is $\int_0^1 \pi(\sqrt{x})^2 dx = \pi \int_0^1 x\,dx = \frac{\pi}{2}$. For the cone we could use the formula from Example 3 or simply calculate $\int_0^1 \pi\left(\frac{x}{2}\right)^2 dx = \frac{\pi}{12}$. The volume of the solid is $\frac{\pi}{2} - \frac{\pi}{12} = \frac{5\pi}{12}$.

There is a temptation to try a shortcut here. In analogy with the procedure for finding the area between two curves, we might try subtracting the lower curve $y = \frac{x}{2}$ from the upper curve $y = \sqrt{x}$ and integrating $\int_0^1 \pi\left(\sqrt{x} - \frac{x}{2}\right)^2 dx$. It is easy to check this gives $\frac{11\pi}{60}$, the *wrong answer.* The reason is that $\left(\sqrt{x} - \frac{x}{2}\right)^2$ does not equal $(\sqrt{x})^2 - \left(\frac{x}{2}\right)^2$.

Geometrically, the reason is that the area between circles of radius R and r is bigger than the area of a circle of radius $R - r$, as shown in Figure 6.3.15. This means that the thin slice between two cylinders has greater volume than the thin cylinder of radius $R - r$. If you are not convinced by the picture, try working it out with $R = 2$ and $r = 1$.

If $f(x) \geq g(x) \geq 0$ on $[a, b]$, then the volume obtained by revolving the region between $y = f(x)$ and $y = g(x)$ about the x-axis is

$$\int_a^b \pi\left(f(x)^2 - g(x)^2\right) dx.$$

EXAMPLE 6 Find the volume of the solid obtained by revolving about the x-axis the region between $f(x) = x^2$ and $g(x) = 1$ for x between $x = -2$ and $x = 2$.

Solution We begin by drawing a sketch of the situation in Figure 6.3.16. The thing that complicates this problem is that some of the time $f(x)$ is greater than $g(x)$, and

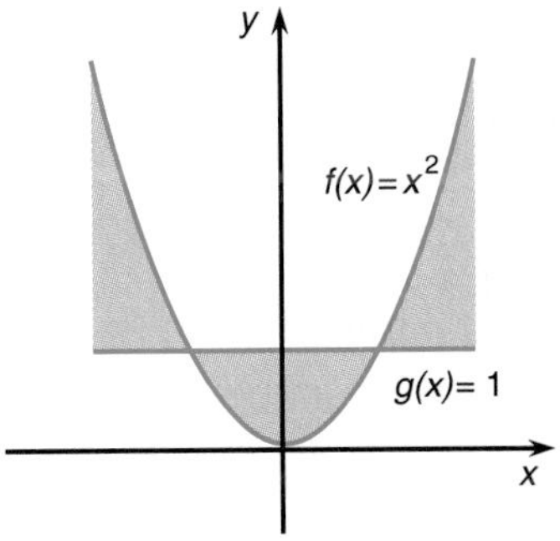

Figure 6.3.16

some of the time it is the other way around. What is required is to find the volume of the solid obtained by revolving the bigger function and subtract from it the volume of the solid obtained from the smaller function. The only way to do this is to work on smaller intervals, on each of which $f(x) \geq g(x)$ all the time or $f(x) \leq g(x)$ all the time.

From the picture we see that $f(x) \leq g(x)$ on $[-1, 1]$ and $f(x) \geq g(x)$ on $[-2, -1]$ and on $[1, 2]$. We work with each of these intervals separately.

If it is not easy to solve explicitly for the points where two curves intersect, a graphics calculator can help find the volume obtained by revolving the region enclosed between the curves.

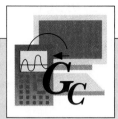

For instance, to find the volume obtained by revolving the region enclosed between $y = x^4$ and $y = x^3 + 1$, we plot the graphs, solve for the points a and b where $x^4 = x^3 + 1$, i.e., $x^4 - x^3 - 1 = 0$, and then evaluate the volume

$$\pi \int_a^b \left((x^3+1)^2 - (x^4)^2\right) dx = \pi\left(\frac{x^7}{7} + \frac{x^4}{2} + x - \frac{x^9}{9}\right)\Big|_a^b.$$

TI-85	SHARP EL9300
GRAPH F1(y(x) =) F1(x) ^ 4 ▼ F1(x) ^3 + 1 M2(RANGE) (−) 2 ▼ 2 ▼ ▼ (−) 1 ▼ 5 F5(GRAPH) SOLVER X^ 4 − X^ 3 − 1 = 0 ENTER 0 F5(SOLVE) EXIT X STO A ENTER SOLVER ENTER 1 F5(SOLVE) EXIT X STO B ENTER π × (B^ 7 ÷ 7 + B^ 4 ÷ 2 + B − B^ 9 ÷ 9 − A^ 7 ÷ 7 − A^ 4 ÷ 2 − A + A^ 9 ÷ 9) ENTER	x/θ/T a^b 3 ▶ + 1 2ndF ▼ x/θ/T a^b 4 RANGE −2 ENTER 2 ENTER 1 ENTER −1 ENTER 5 ENTER 1 ENTER SOLVER x/θ/T a^b 3 ▶ + 1 = x/θ/T a^b 4 ENTER (−) 1 ENTER ENTER ENTER ENTER ENTER X STO A SOLVER ENTER 1 ENTER ENTER ENTER ENTER ENTER X STO B π × (B a^b 7 ▶ ÷ 7 + B a^b 4 ▶ ÷ 2 + B − B a^b 9 ▶ ÷ 9 − A a^b 7 ▶ ÷ 7 − A a^b 4 ▶ ÷ 2 − A + A a^b 9 ▶ ÷ 9) ENTER
CASIO f_x-6300G (f_x-7700GB)	**HP 48SX**
Range −2 EXE 2 EXE 1 EXE −1 EXE 5 EXE 1 EXE EXE EXE EXE Graph X x^y 3 + 1 EXE Graph X x^y 4 EXE Trace ▶ (move cursor to intersection) X → A EXE Graph X x^y 4 EXE Trace ▶ (move cursor to intersection) X → B EXE π × (B x^y 7 ÷ 7 + B x^y 4 ÷ 2 + B − B x^y 9 ÷ 9 − A x^y 7 ÷ 7 − A x^y 4 ÷ 2 − A + A x^y 9 ÷ 9) EXE (Use Zoom for more accuracy)	↰ EQUATION X y^x 3 ▶ + 1 ↰ = X y^x 4 ENTER ↰ SOLVE NEW αE αQ αN ENTER 30 +/− ↱ CST NXT SF ↰ PLOT PTYPE FUNC PLOTR 2 +/− ENTER 2 XRNG 1 +/− ENTER 5 YRNG DRAW ◀ (Move cursor to intersection) FCN ISECT ATTN ATTN X ENTER ′ A STO DRAW ◀ (Move cursor to intersection) FCN ISECT ATTN ATTN X ENTER ′ B STO B ENTER 7 y^x 7 ÷ B ENTER 4 y^x 2 ÷ + B ENTER + B ENTER 9 y^x 9 ÷ − A ENTER 7 y^x 7 ÷ − A ENTER 4 y^x 2 ÷ − A ENTER − A ENTER 9 y^x 9 ÷ + π ×

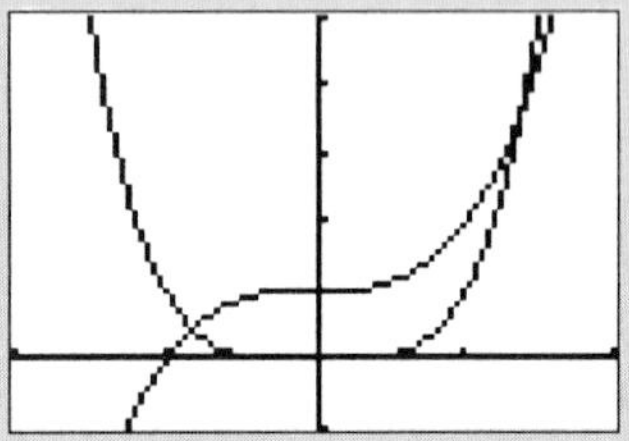

On $[-2,-1]$, the volume is

$$\int_{-2}^{-1} \pi\big(f(x)^2 - g(x)^2\big)\,dx = \pi\int_{-2}^{-1}\big((x^2)^2 - 1^2\big)\,dx = \pi\int_{-2}^{-1}\big(x^4 - 1\big)\,dx$$
$$= \pi\left(\frac{x^5}{5} - x\right)\Big|_{-2}^{-1} = \pi\left(\frac{-1}{5} + 1 + \frac{32}{5} - 2\right) = \frac{26}{5}\pi.$$

On $[-1,1]$, what we need is

$$\int_{-1}^{1} \pi\big(g(x)^2 - f(x)^2\big)\,dx = \pi\int_{-1}^{1}(1 - x^4)\,dx = \pi\left(x - \frac{x^5}{5}\right)\Big|_{-1}^{1} = \pi\left(2 - \frac{2}{5}\right)$$
$$= \frac{8}{5}\pi.$$

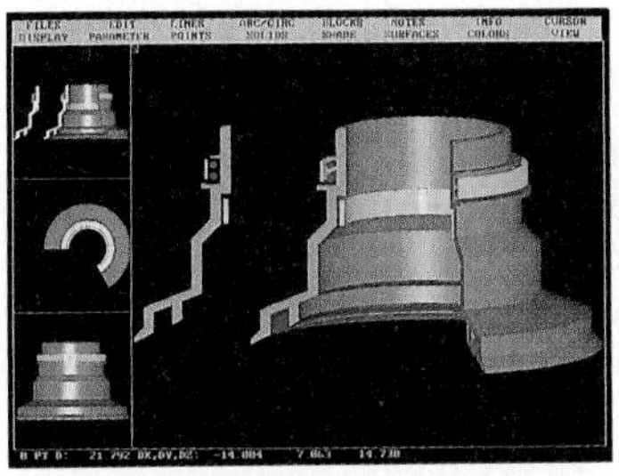

This flange is the region contained inside one solid of revolution and outside another solid of revolution. The image was generated by a computer using DesignCAD 3D *software made by* American Small Business Computers. *Note the blue plastic seal and the ease with which the computer can produce different views of the object.*

On $[1,2]$ the calculation is $\pi\int_1^2 (x^4 - 1)\,dx = \pi\left(\frac{1}{5}x^5 - x\right)\Big|_1^2 = \frac{26}{5}\pi$.

The total volume we want is the sum of the volumes for the three intervals, $\frac{26}{5}\pi + \frac{8}{5}\pi + \frac{26}{5}\pi = 12\pi$.

It is very important to recognize that incorrect answers will result from finding the volumes resulting from both functions over the entire interval $[-2,2]$ and then subtracting. In this example,

$$\int_{-2}^{2} \pi f(x)^2\,dx = \pi\frac{1}{5}x^5\Big|_{-2}^{2} = \frac{64}{5}\pi \quad \text{and} \quad \int_{-2}^{2} \pi g(x)^2\,dx = \pi x\Big|_{-2}^{2} = 4\pi.$$

The difference of these two is not the correct answer.

> We see that solids obtained by revolving the region between two curves are more complicated. It is essential to work on smaller intervals on which one function is always bigger and the other is always smaller.

Summary 6.3.2

Volumes of Revolution by Slices

1. Identify the axis about which the solid is revolved. The corresponding variable will be the variable of integration (i.e., integrate with respect to x for solids of revolution about the x-axis and with respect to y for solids of revolution about the y-axis). We visualize the thin "slice" moving along the axis as x (or y) moves from the lower limit of integration to the upper limit of integration.
2. Express the function f whose graph is being revolved in terms of the variable of integration.
3. Find the limits of integration, drawing a picture if necessary. Set up the integral, $\int_a^b \pi f(x)^2 dx$ or $\int_a^b \pi f(y)^2 dy$ as the case may be.
4. Integrate.
5. For a solid obtained by revolving about an axis the region between two curves, remember to break the interval into smaller intervals so that on each of them

one of the functions is always bigger than the other. If $f(x) \geq g(x) \geq 0$ on $[a, b]$, then the volume obtained by revolving the region between $y = f(x)$ and $y = g(x)$ around the x-axis is

$$\int_a^b \pi\left(f(x)^2 - g(x)^2\right) dx.$$

An analogous formula works for revolution about the y-axis. Finally, combine the results from all the intervals.

Exercises 6.3

I

In each of the following situations, find the volume of the solid obtained by revolving the given function $f(x)$, defined on the specified interval $[a, b]$, about the x-axis. In each case, make a sketch of the solid.

1. $f(x) = 3x$ on $[0, 2]$

2. $f(x) = x^2$ on $[1, 2]$

3. $f(x) = 2 - x$ on $[0, 2]$

4. $f(x) = 2 + x^2$ on $[0, 1]$

5. $f(x) = e^x$ on $[0, 1]$

6. $f(x) = \sin x$ on $[0, \pi]$

7. $f(x) = x^2 - x$ on $[0, 1]$

8. $f(x) = 1 - x^2$ on $[-1, 1]$

9. $f(x) = 2$ on $[2, 5]$

10. $f(x) = x^2$ on $[-1, 1]$

In each of the following situations, find the volume of the solid obtained by revolving about the x-axis the region between the given functions $f(x)$ and $g(x)$ over the specified interval $[a, b]$. In each case, make a sketch of the solid.

11. $f(x) = 2x, g(x) = x$ over $[0, 1]$

12. $f(x) = x^2, g(x) = x$ over $[0, 1]$

13. $f(x) = x, g(x) = \sin x$ over $[0, \pi]$

14. $f(x) = x^3, g(x) = x^2$ over $[1, 2]$

15. $f(x) = 2 - x^2, g(x) = 1$ over $[-1, 1]$

16. $f(x) = |x|, g(x) = x^2$ over $[-2, 2]$

17. $f(x) = 2, g(x) = 1$ over $[0, h]$

18. $f(x) = x, g(x) = 3$ over $[0, 2]$

19. $f(x) = M, g(x) = N$ over $[0, L]$

20. $f(x) = 1 + \sin x, g(x) = 2 + \sin x$ over $\left[0, \frac{\pi}{2}\right]$

In each of the following situations, find the volume of the solid obtained by revolving about the x-axis the region enclosed between the given functions $f(x)$ and $g(x)$. In each case, make a sketch of the solid.

21. $f(x) = x^3$, $g(x) = x^2$

22. $f(x) = x^2$, $g(x) = 1 - x^2$

23. $f(x) = x^2$, $g(x) = 3$

24. $f(x) = x^4$, $g(x) = 16$

25. $f(x) = x^4$, $g(x) = x^6$

26. $f(x) = \sqrt{|x|}$, $g(x) = x^4$

27. $f(x) = 2 - |x - 1|$, $g(x) = 1$

28. $f(x) = |x^2 - x|$, $g(x) = 0$

29. $f(x) = 1$, $g(x) = \sqrt{2 - x^2}$

30. $f(x) = \frac{1}{\sqrt{2}}x^2$, $g(x) = \left(1 + x^2\right)^{-1/2}$

II

***31.** Find the volume of the solid obtained by revolving about the x-axis the region between the graphs of the functions $f(x) = x$ and $g(x) = x^3$ for x between $x = 0$ and $x = 2$. Draw a picture.

***32.** Find the volume of the solid obtained by revolving about the x-axis the region between the graphs of the functions $f(x) = x$ and $g(x) = x + \sin x$ for x between $x = 0$ and $x = 2\pi$. Draw a picture.

***33.** Consider the functions $f(x) = \sec x$ and $g(x) = 2$ for $x \in \left(-\frac{\pi}{2}, \frac{\pi}{2}\right)$. Rotate the region lying above $f(x)$ and below $g(x)$ about the x-axis, and compute the volume of the resulting solid.

***34.** The curves $y = \cos x$ and $y = 1 - \cos x$ cross exactly once for $0 \leq x \leq \frac{\pi}{2}$. Consider the region lying between these curves, to the right of the y-axis and to the left of the point where the curves cross. Find the volume of the solid obtained when this region is rotated about the x-axis.

***35.** Consider the circle whose radius is 1 and whose center is at the point $(0, 2)$.

(i) Describe the solid obtained by revolving the region inside this circle about the x-axis.

(ii) Find the volume of this solid.

***36.** Suppose $R > r$ and consider the circle whose radius is r and whose center is at the point $(0, R)$. Find the volume of the solid obtained by revolving the region inside this circle about the x-axis.

***37.** Suppose $a, b > 0$. Find the volume of the "ellipsoid" obtained by revolving about the x-axis the upper half of the ellipse $\frac{x^2}{a^2} + \frac{y^2}{b^2} = 1$. What happens if $a = b$?

***38.** Suppose $a, b > 0$. Find the volume of the ellipsoid obtained by revolving the ellipse $\frac{x^2}{a^2} + \frac{y^2}{b^2} = 1$ about the y-axis.

***39.** When we talked about solids of revolution and calculating their volumes, we considered a *nonnegative* function $f(x)$. Discuss what happens if $f(x)$ is allowed to have negative values too. In particular, what is the geometrical meaning of $\int_{-1}^{1} \pi(x)^2 dx$, that is, $f(x) = x$? Make a picture.

***40.** Find the volume of the slice of width 2 units cut from the center of a ball of radius 2 units.

III

In each of the following situations, use a graphics calculator or computer to find the points of intersection of the graphs of $y = f(x)$ and $y = g(x)$. Find the volume of the solid obtained by revolving about the x-axis the region enclosed between the graphs.

41. $f(x) = 3x + 5, \quad g(x) = x^3 + 4$

42. $f(x) = \cos x, \quad g(x) = x^2$

43. $f(x) = 2 \sin x, \quad g(x) = x^3 + x$

44. $f(x) = \frac{1}{4}e^x, \quad g(x) = \sqrt{x}$

6.4 Volumes by Cylindrical Shells

SLICES vs SHELLS

There is another way of calculating volumes of solids of revolution that sometimes results in simpler calculations. In the previous section we built up a solid out of thin slices, as shown in Figure 6.4.1. We thought of the thin slice moving along the axis of rotation, "sweeping out" the solid.

Instead of this, we could build it up out of **cylindrical shells**, as shown in Figure 6.4.2. We can imagine the shells starting as a narrow tube around the axis of rotation and expanding to the outside of the solid, sweeping out the whole volume as they go. "Adding up," that is, integrating, their volumes will give us a way to find the volume of the solid.

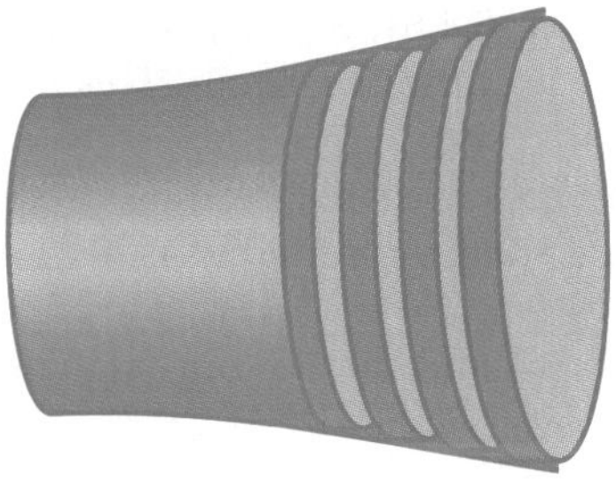

Figure 6.4.1

EXAMPLE 1 Use cylindrical shells to find the volume of the paraboloid found by rotating $y = \sqrt{x}$, $0 \le x \le 4$, about the x-axis.

Solution As x runs from 0 to 4, $y = \sqrt{x}$ runs from 0 to 2 (see Figure 6.4.3). For each value of y we obtain a cylinder as shown in Figure 6.4.4. Its radius is y and its

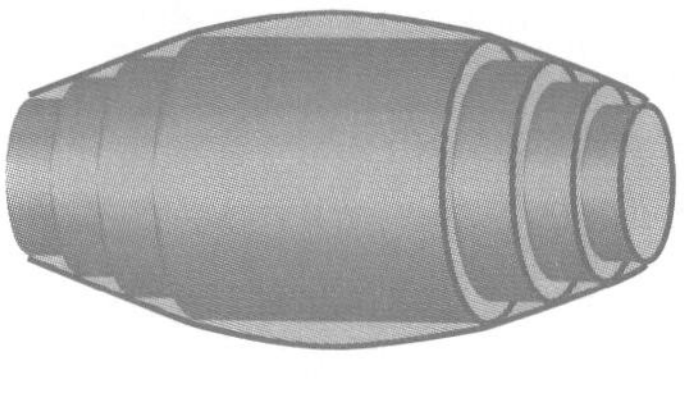

Figure 6.4.2

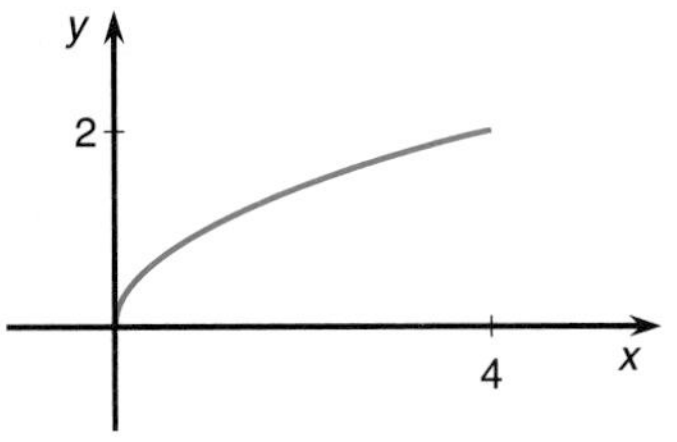

Figure 6.4.3

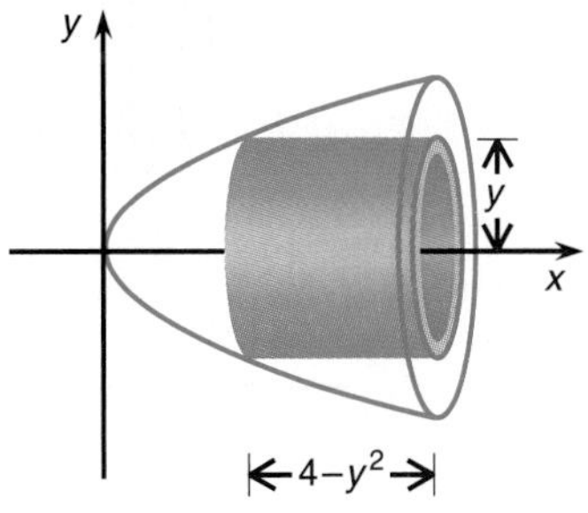

Figure 6.4.4

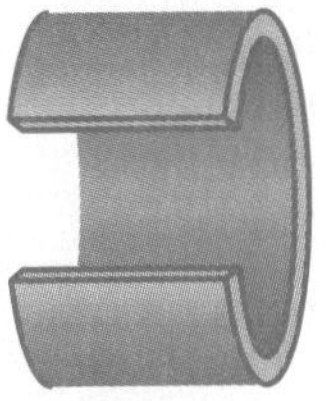
Figure 6.4.5

length is the horizontal distance from the curve to the right end of the paraboloid. The curve is $y = \sqrt{x}$ or $x = y^2$, and the right end of the paraboloid is $x = 4$. The horizontal distance between the curve and this vertical line is $4 - y^2$.

To find the volume of the cylindrical shell, we slit it and flatten it, as shown in Figure 6.4.5. What results is the flat rectangle shown in Figure 6.4.6, whose thickness is dy and with one side $4 - y^2$. The length of the other side used to be the circumference of a circle with radius y, so it is $2\pi y$. The volume of the shell equals the volume of this rectangle, $2\pi y(4 - y^2)\,dy$. We obtain the volume of the whole paraboloid by integrating:

$$\begin{aligned} \textit{Volume} &= \int_0^2 2\pi y(4 - y^2)\,dy = 2\pi \int_0^2 (4y - y^3)\,dy = 2\pi\left(2y^2 - \frac{1}{4}y^4\right)\Bigg|_0^2 \\ &= 8\pi. \end{aligned}$$

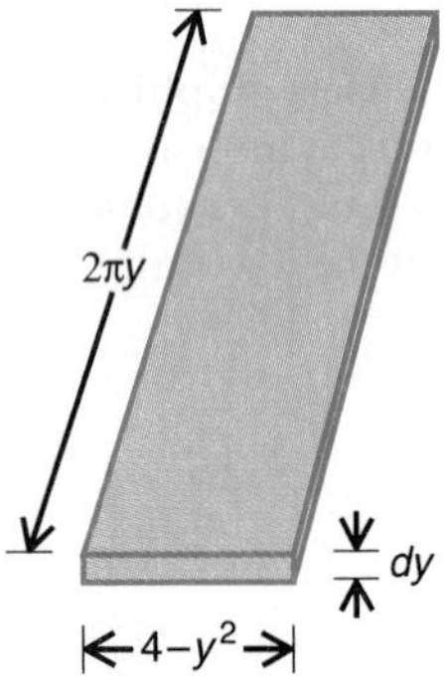

Figure 6.4.6

The example we have just finished suggests that the volume of a cylindrical shell is (*length*) × (*circumference*) × (*thickness*) = (*length*) × 2π(*radius*) × (*thickness*), and this can be used to set up integrals.

In Example 1 the cylindrical shells calculation was no better (perhaps slightly worse) than what we had to do with the method of slices, but there are times when it is clearly easier.

EXAMPLE 2 Consider the curve $y = x^3 - x^2 + x$ between $x = 1$ and $x = 2$, and the region between it and the y-axis, shown in Figure 6.4.7. Find the volume of the solid obtained by rotating this region about the y-axis.

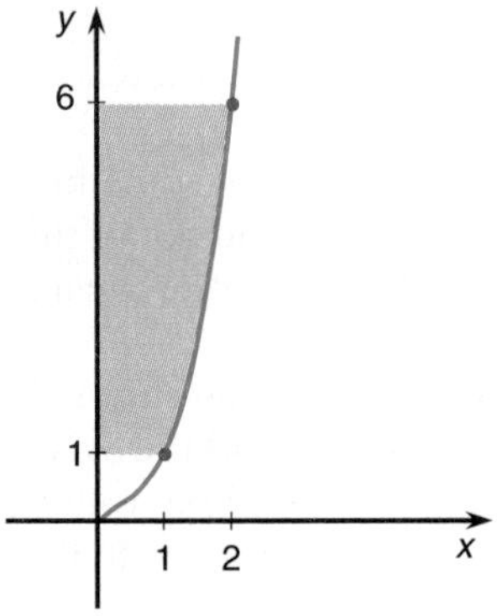

Figure 6.4.7

Solution The derivative $\frac{dy}{dx} = 3x^2 - 2x + 1$ is a quadratic polynomial with no zeroes, so it is always positive, and the graph of $y = x^3 - x^2 + x$ is always increasing.

To apply the method of slices, we would have to find the radius of the slice. This means expressing x as a function of y (see Figure 6.4.8). Since y is a cubic polynomial in x, this will not be easy. Instead, we can use cylindrical shells. This will require integrating with respect to x.

As x goes from 1 to 2, $y = x^3 - x^2 + x$ goes from 1 to 6. We will let x move from 1 to 2 and for each value of x consider the cylindrical shell whose radius is x. The height of the cylindrical shell shown in Figure 6.4.9 is $6 - y = 6 - x^3 + x^2 - x$; after all this is just the difference of the heights at the top (6) and the bottom (y). The thickness of the shell is dx.

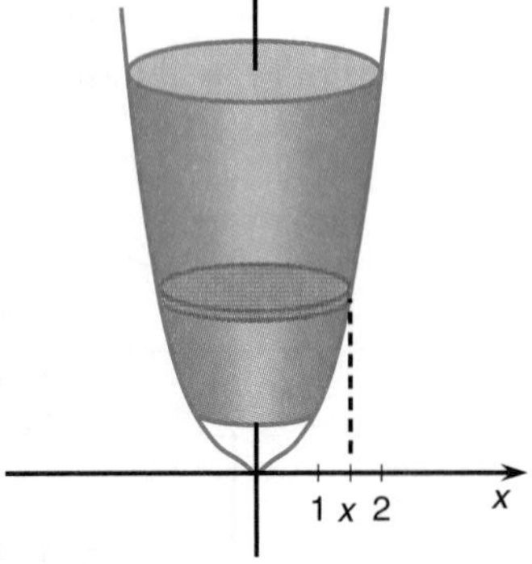

Figure 6.4.8

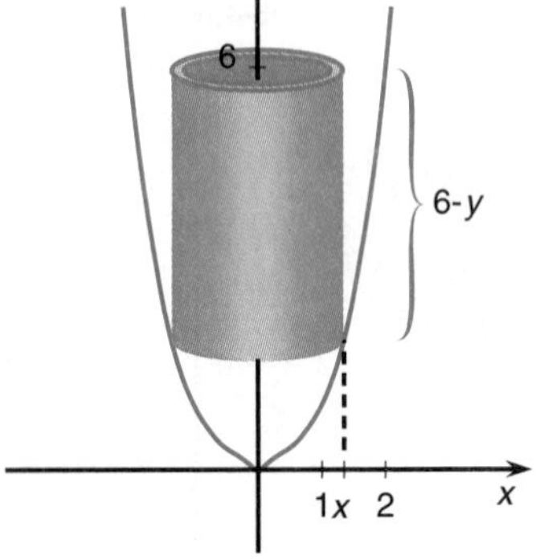

Figure 6.4.9

A graphics calculator can be helpful when it is not easy to solve explicitly for the points of intersection.

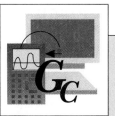

For instance, to find the volume obtained by revolving about the x-axis the region enclosed between $x = y^4 + 1$ and $x = 5y - y^3$, we plot the graphs and solve for the points a and b where $y^4 + 1 = 5y - y^3$, i.e., $y^4 + y^3 - 5y + 1 = 0$. Since we plot the graphs with "y" as the variable, the positive x-axis points up and the positive y-axis points to the right on the calculator screen. For the actual situation, it is necessary to reflect the picture, as in the second illustration. Then we evaluate the volume $2\pi \int_a^b y(5y - y^3 - (y^4+1))\,dy = 2\pi\left(5\frac{y^3}{3} - \frac{y^5}{5} - \frac{y^6}{6} - \frac{y^2}{2}\right)\Big|_a^b$.

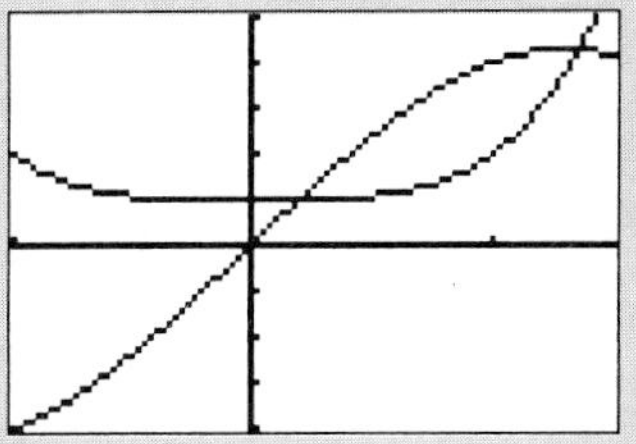

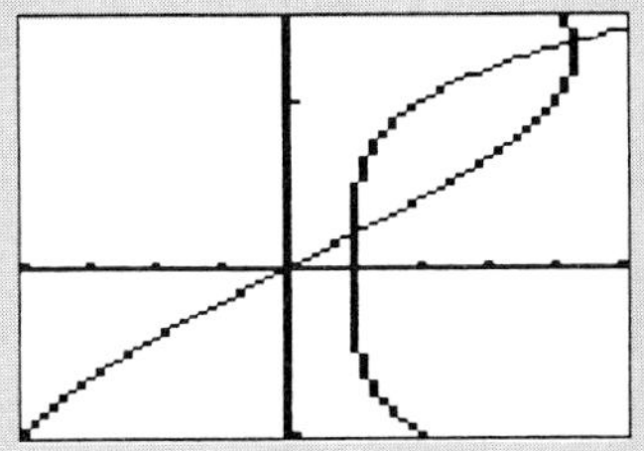

TI-85

GRAPH F1 ($y(x)$ =) 5 F1 (x) − F1 (x) ^3 ▼ F1 (x) ^4 + 1 M2 (RANGE) (−) 1 ▼ 1.5 ▼ ▼ (−) 4 ▼ 5 F5 (GRAPH) SOLVER X ^ 4 + X ^ 3 − 5 X + 1 = 0 ENTER 0 F5 (SOLVE) EXIT X STO A ENTER SOLVER ENTER 1 F5 (SOLVE) EXIT X STO B ENTER 2 π × (5 B ^ 3 ÷ 3 − B ^ 5 ÷ 5 − B ^ 6 ÷ 6 − B ^ 2 ÷ 2 − 5 A ^ 3 ÷ 3 + A ^ 5 ÷ 5 + A ^ 6 ÷ 6 + A ^ 2 ÷ 2) ENTER

SHARP EL9300

[graph key] x/θ/T a^b 4 ▶ + 1 2ndF ▼ 5 x/θ/T − x/θ/T a^b 3 RANGE −1 ENTER 1.5 ENTER 1 ENTER −4 ENTER 5 ENTER 1 ENTER [graph key] SOLVER x/θ/T a^b 4 ▶ + 1 = 5 x/θ/T − x/θ/T a^b 3 ENTER (−) 1 ENTER ENTER ENTER ENTER ENTER [calc key] X STO A SOLVER ENTER 1 ENTER ENTER ENTER ENTER ENTER [calc key] X STO B 2 π × (5 B a^b 3 ▶ ÷ 3 − B a^b 5 ▶ ÷ 5 − B a^b 6 ▶ ÷ 6 − B x^2 ÷ 2 − 5 A a^b 3 ▶ ÷ 3 + A a^b 5 ▶ ÷ 5 + A a^b 6 ▶ ÷ 6 + A x^2 ÷ 2) ENTER

CASIO f_x-6300G (f_x-7700GB)

Range −1 EXE 1.5 EXE 1 EXE −4 EXE 5 EXE 1 EXE EXE EXE EXE Graph X x^y 4 + 1 EXE Graph 5 X − X x^y 3 EXE Trace ▶ (move cursor to intersection) X → A EXE Graph 5 X − X x^y 3 EXE Trace ▶ (move cursor to intersection) X → B EXE 2 π × (5 B x^y 3 ÷ 3 − B x^y 5 ÷ 5 − B x^y 6 ÷ 6 − B x^y 2 ÷ 2 − 5 A x^y 3 ÷ 3 + A x^y 5 ÷ 5 + A x^y 6 ÷ 6 + A x^y 2 ÷ 2) EXE (Use Zoom for more accuracy)

HP 48SX

↰ EQUATION X y^x 4 ▶ + 1 ↰ = 5 × X − X y^x 3 ENTER ↰ SOLVE NEW αE αQ αN ENTER 30 +/−

↱ CST NXT SF ↰ PLOT PTYPE FUNC PLOTR 1 +/− ENTER 1.5 XRNG 4 +/− ENTER 5 YRNG DRAW

◀ (Move cursor to intersection) FCN ISECT ATTN ATTN X ENTER ′ A STO DRAW ◀ (Move cursor to intersection) FCN ISECT ATTN ATTN X ENTER ′ B STO B ENTER 3 y^x 3 ÷ 5 × B ENTER 5 y^x 5 ÷ − B ENTER 6 y^x 6 ÷ − B ENTER 2 y^x 2 ÷ − A ENTER 3 y^x 3 ÷ 5 × − A ENTER 5 y^x 5 ÷ + A ENTER 6 y^x 6 ÷ + A ENTER 2 y^x 2 ÷ + π × 2 ×

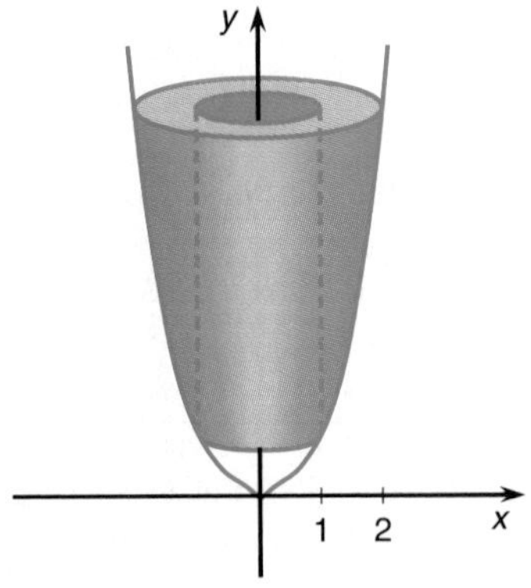

Figure 6.4.10

The volume of the shell is $2\pi x(6-y)\,dx = 2\pi x(6-x^3+x^2-x)\,dx = 2\pi(-x^4+x^3-x^2+6x)\,dx$. The volume is obtained by integrating:

$$2\pi\int_1^2(-x^4+x^3-x^2+6x)\,dx = 2\pi\left(-\frac{1}{5}x^5+\frac{1}{4}x^4-\frac{1}{3}x^3+3x^2\right)\Bigg|_1^2$$
$$= \frac{253\pi}{30}.$$

This is the volume of the solid obtained by revolving the part of the region between $x=1$ and $x=2$; this is the solid without its central "core" (see Figure 6.4.10).

The core is a cylinder of radius 1 and height $6-1=5$, so its volume is 5π. The total volume is $\frac{253\pi}{30}+5\pi = \frac{403\pi}{30} \approx 42.202$.

If we attempt to write down a formula for the volume of a solid of revolution using the technique of cylindrical shells, either the result will be too restricted to be used in many situations or it will be unnecessarily complicated. Rather than memorizing formulas, it is far better to figure out what is needed in individual cases.

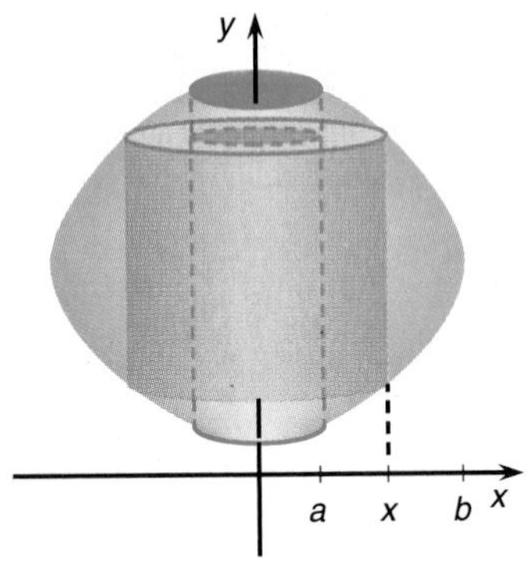

Some people find it helpful to visualize the method of cylindrical shells in the following way. Consider a solid obtained by revolving a region about the y-axis, as shown in the illustration. To find its volume, we integrate with respect to x, with x running from the inner radius a to the outer radius b. At each particular value of x we consider the cylinder whose radius is x and whose height is the vertical cross-section of the solid at radius x. The total volume equals the integral

$$\int_a^b 2\pi x(\textit{height})dx.$$

Observe that the integrand $2\pi x(\textit{height})$ is exactly the surface area of the cylinder of radius x. We can think of the integral as the integral as the radius goes from $x=a$ to $x=b$ of the surface area of the cylinder of radius x inside the solid.

Summary 6.4.1 Volumes of Revolution by Cylindrical Shells

1. Identify the axis about which the solid is revolved. The *other* variable will be the variable of integration (i.e., integrate with respect to y for solids of revolution about the x-axis and with respect to x for solids of revolution about the y-axis). We visualize the thin "shell" expanding out from the axis as the radius y (or x) moves from the smallest value (at the inside of the solid) to its largest value (at the outside of the solid).
2. Express the function f whose graph is being revolved in terms of the variable of integration.

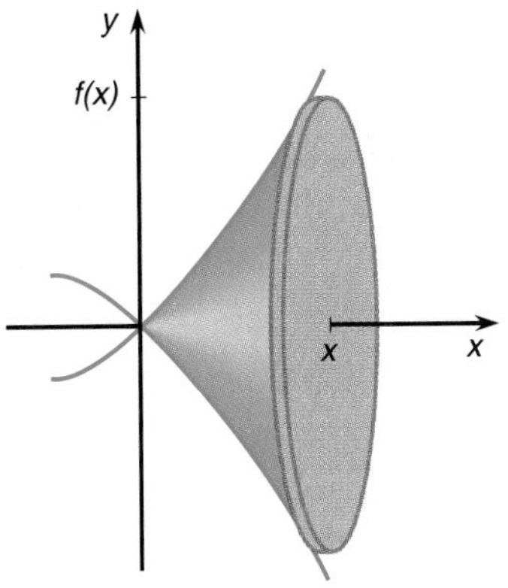

Figure 6.4.11

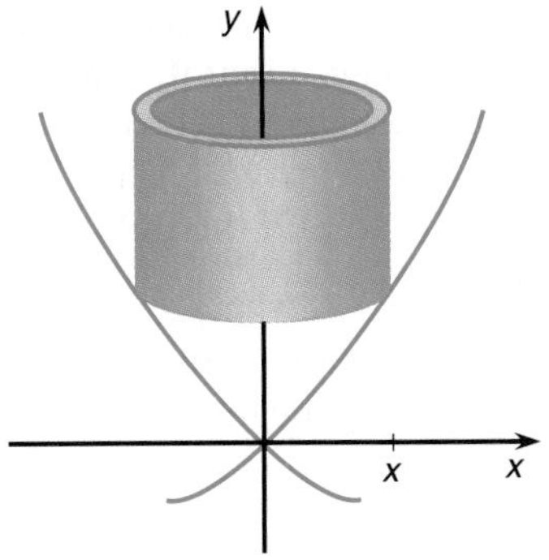

Figure 6.4.12

3. Find the limits of integration, drawing a picture if necessary. The integration will go from the smallest radius of the solid up to the largest radius. Set up the integral, $2\pi \int_a^b x(\textit{height})\,dx$ or $2\pi \int_a^b y(\textit{height})\,dy$ as the case may be, where "height" refers to the height of the shell in each case. (Note that the "height" may be horizontal.)
4. Integrate.

For a solid obtained by revolving the region between two curves around an axis, it may be necessary to break the interval into smaller intervals so that on each of them one of the functions is always greater than the other. Then find the volumes of the solid constructed from each function and subtract. It may be necessary to use shells for some parts and slices for others. Finally, combine the results from all the intervals.

Slices Vs. Shells

If we rotate a curve $y = f(x)$ about the x-axis, it is usually easy to use the method of slices, because the radius of the slice is $f(x)$, if x is the point where the slice crosses the x-axis (see Figure 6.4.11).

But if we rotate the same curve about the y-axis, then the radius of a slice is x, and we have to express it as a function of y. If $f(x)$ is a quadratic polynomial, this is not too difficult, as we saw in Examples 6.3.4 and 6.3.5, but in general it is complicated and likely to result in a difficult integral. In such a case the method of cylindrical shells is more likely to be suitable. There we would be integrating with respect to x, the radius of the shell would be x, and the length of the shell would be something involving the function $f(x)$. See Figure 6.4.12.

Similar remarks apply to rotating $x = g(y)$ about the y-axis (for which slices are likely best) or about the x-axis (for which shells are likely better).

It is true, however, that there are occasionally instances in which these general guidelines fail and what appears to be the sensible approach gives an integral that is hard to evaluate. In such a case it is sometimes possible to get the answer using the other approach.

So far we have only considered solids formed by revolving about one of the axes, but it is possible to use any straight line.

EXAMPLE 3 Find the volume of the region obtained by revolving the region enclosed between $y = x$ and $y = x^2$ about the line $x = 2$. See Figure 6.4.13.

Solution We begin by sketching the situation in Figure 6.4.14. If we try the method of slices, we will have to do two calculations, one for the outer region between $y = x$

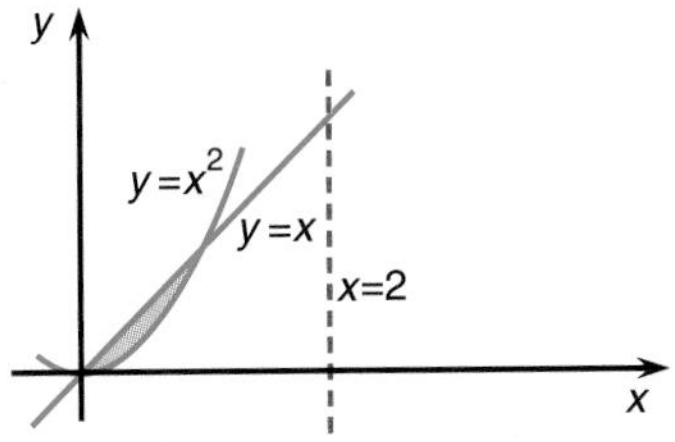

Figure 6.4.13

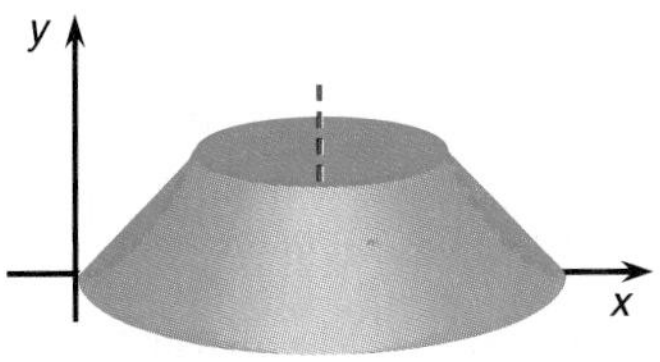

Figure 6.4.14

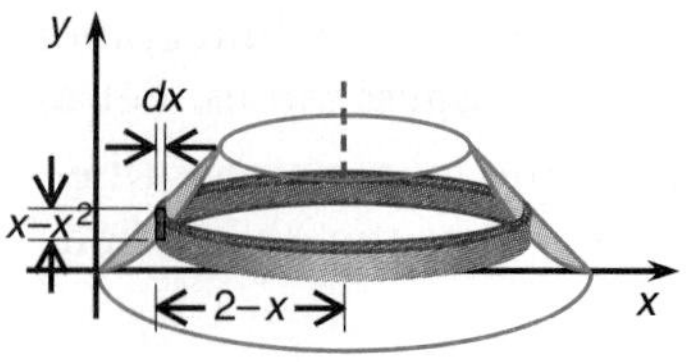

Figure 6.4.15

and $x = 2$, with $0 \leq y \leq 1$, and one for the inner region between $y = x^2$ and $x = 2$, with $0 \leq y \leq 1$, and then subtract the resulting volumes.

Instead, we try cylindrical shells. From Figure 6.4.15 we see that we should consider values of x between $x = 0$ and $x = 1$. The narrow strip with width dx going from $y = x^2$ to $y = x$ has height $x - x^2$. If we revolve this strip about the line $x = 2$, we obtain a cylindrical shell. Its radius is the distance to the line $x = 2$, or $2 - x$. The volume of the shell is $2\pi\,(radius)\,(height)\,dx = 2\pi\,(2 - x)\,(x - x^2)\,dx = 2\pi\,(x^3 - 3x^2 + 2x)\,dx$.

We integrate to find the volume of the whole solid:

$$2\pi \int_0^1 (x^3 - 3x^2 + 2x)\,dx = 2\pi\left(\frac{1}{4}x^4 - x^3 + x^2\right)\Bigg|_0^1 = 2\pi\left(\frac{1}{4}\right)$$
$$= \frac{\pi}{2}.$$

Rather than trying to memorize formulas for all the possible ways of calculating volumes of revolution about various lines, it is better to realize that it is fairly easy to construct the integral as we have just done. The idea is to rely on the picture and the idea of slices or cylindrical shells, as the case may be. The one new ingredient if the line is not one of the axes is having to find the radius of the slice or the shell, which is just the distance to the line.

Finally, we remark that volumes are related to mass or weight, especially for objects made of material that has uniform density. This simply means that every cubic inch of the material weighs the same amount, which means that the total weight is just that constant times the volume.

EXAMPLE 4 **Flashlight Reflector** The glass reflector for a large flashlight is made in the shape of a solid of revolution. It is obtained by revolving about the x-axis the region between the curves $x = \frac{1}{2}y^2$ and $x = \frac{1}{2}y^2 + \frac{1}{8}$, for $0 \leq x \leq 2$. All measurements are made using inches as units. If the glass used has a density of 1.50 oz/cu in, how much does the reflector weigh?

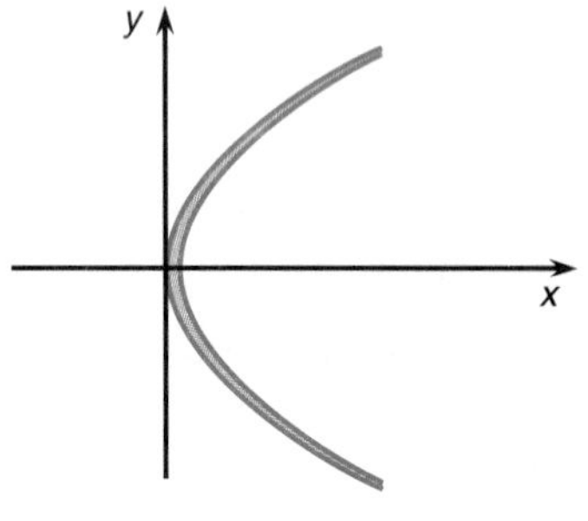

Figure 6.4.16

Solution We find the volume of the glass reflector. This is the situation of the volume between two solids of revolution. The outer solid is obtained by revolving about the x-axis the curve $x = \frac{1}{2}y^2$ for $0 \leq x \leq 2$. We use slices and note that $y^2 = 2x$, so its volume is

$$\pi \int_0^2 2x\,dx = \pi x^2\Big|_0^2 = 4\pi.$$

The inner solid is obtained by rotating the curve $x = \frac{1}{2}y^2 + \frac{1}{8}$ about the x-axis. Notice that for this curve, x is never less than $\frac{1}{8}$ (see Figure 6.4.16). We find that x should run from $\frac{1}{8}$ to $x = 2$, and noting that $y^2 = 2x - \frac{1}{4}$, we find that the volume is

$$\pi \int_{1/8}^2 \left(2x - \frac{1}{4}\right) dx = \pi\left(x^2 - \frac{1}{4}x\right)\Bigg|_{1/8}^2 = 3\frac{33}{64}\pi.$$

The volume of the reflector is the difference of these two volumes, or $4\pi - 3\frac{33}{64}\pi = \frac{31}{64}\pi \approx 1.52$ cu in.

Knowing that the density of the glass is 1.50 oz/cu in, we see that the reflector weighs $1.52 \times 1.50 \approx 2.28$ oz.

Exercises 6.4

I

In each of the following, revolve the region under the graph of the specified function over the given interval about the x-axis and find the volume of the resulting solid.

1. $x = y^5 + 2y$, for $y \in [0, 1]$
2. $x = y^3 + 3y^2 + 1$, for $y \in [0, 2]$
3. $x = y^3 + 3y$, for $x \in [0, 4]$
4. $x = y^4 + y^2 + 2$, for $x \in [2, 4]$
5. $x = y^2 + 1$, for $y \in [0, 2]$
6. $x = y^4 - 1$, for $x \in [0, 15]$
7. $x = y^3 + y$, for $y \in [0, 1]$
8. $x = y^3 + 2y^2$, for $y \in [0, 1]$
9. $x = y^{3/2}$, for $y \in [1, 4]$
10. $x = y^{4/3}$, for $y \in [1, 2]$
11. $x = y + y^{2/3}$, for $y \in [1, 8]$
12. $x = \frac{\sin y}{y}$, for $y \in \left[\frac{\pi}{4}, \frac{\pi}{2}\right]$
13. $y = \sqrt{1 - x^2}$, for $x \in [-1, 1]$
14. $y = \sqrt{R^2 - x^2}$, for $x \in [0, R]$
15. $y = \sqrt{1 - x^2}$, for $x \in \left[\frac{1}{2}, 1\right]$
16. $y = \sqrt{2 - x^2}$, for $x \in [0, 1]$
17. $x = y^{1/3} + y^{2/5}$, for $y \in [1, 3]$
18. $x = \frac{1}{1+y^2}$, for $y \in [2, 4]$
19. $x^2 = 1 + y^2$, for $y \in [0, 1]$, $x > 0$
20. $x^3 = \frac{1}{1+y^2}$, for $y \in [0, 2]$

In each of the following situations there are two or more curves (and/or lines). Revolve the region bounded by the given curves about the specified line and find the volume of the resulting solid.

21. $y = x^3 + x$, the x-axis, $x = 1$, revolved about the y-axis
22. $y = x^5 + 2x$, the x-axis, $x = 2$, revolved about the y-axis
23. $x = y$, $x + y = 2$, the x-axis, revolved about the x-axis
24. $x = y$, $x + y = 2$, the x-axis, revolved about the y-axis
25. $x = y^3$, $y = x^2$, revolved about the x-axis
26. $y = \sqrt{4 - x^2}$, $-2 \le x \le 2$, the x-axis, revolved about the x-axis
27. $x = y^2$, $x = 4$, revolved about the y-axis
28. $x = y^2$, $x = 4$, revolved about the x-axis
29. $y^4 + y = x$, the y-axis, $y = 1$, revolved about the x-axis
30. $x = y^3 + 4y$, the x-axis, $x = 5$, revolved about the x-axis
31. $y = x^3$, $y = x$, $0 \le x \le 1$, revolved about the y-axis
32. $y = x^5$, $y = x^2$, revolved about the line $x = 3$
33. $y = x^2$, $x = y^2$, revolved about the line $x = 2$
34. $y = \exp(x^2)$, the y-axis, $y = e$, revolved about the y-axis
35. $y = x^3$, $y = 4x$, between the x-axis and $y = 8$, revolved about the y-axis
36. $x^2 + y^2 = 1$, revolved about the line $x = 1$

II

***37.** Find the volume of the region obtained by revolving about the y-axis the region under the graph of $y = 3x^2 - 12x + 14$, for $x \in [0, 4]$.

***38.** Find the volume of the region obtained by revolving about the y-axis the region under the graph of $y = \arctan(x^2)$, for $x \in [0, 1]$.

***39.** Find the volume of the region obtained by revolving the region enclosed between $y = x$ and $y = x^2$ about the line $x = 4$.

***40.** Find the volume of the region obtained by revolving the region enclosed between $y = x$ and $y = x^2$ about the line $y = 1$.

****41.** Find the volume of the region obtained by revolving the region enclosed between $y = x$, the x-axis, and $x = 1$ about the line $y = x$.

****42.** Find the volume of the region obtained by revolving the region enclosed between $y = x$ and $y = x^2$ about the line $y = x$.

***43.** Find the volume that remains if a cylindrical hole of radius 1 in is bored exactly through the center of a sphere of radius 3 in.

***44.** Suppose $R > r$ and find the volume that remains if a cylindrical hole of radius r is bored exactly through the center of a sphere of radius R (see Figure 6.4.17).

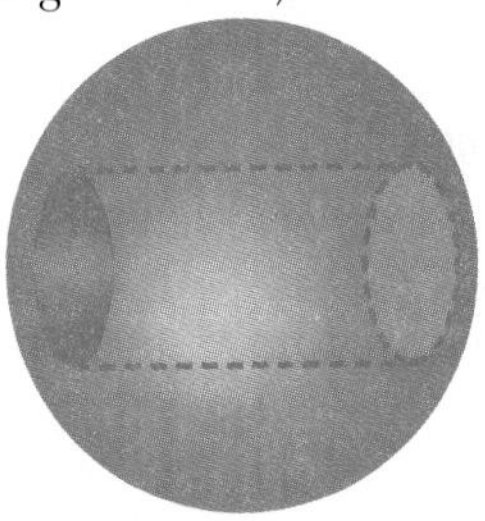

Figure 6.4.17

***45.** Let m be a positive integer and consider the solid formed by revolving about the x-axis the region under the curve $y = x^m$, for $0 \le x \le a$. Calculate the volume of this solid in two different ways, first using the technique of slices and second using cylindrical shells. Of course, the answers should be the same. (If you are having trouble, try it first with $m = 1$ or $m = 2$.)

**46. A cylindrical hole is bored exactly through the center of a sphere in such a way that the length of the hole in the part that remains is 6 cm, as shown in Figure 6.4.18. Find the volume of the part of the sphere that is left. (*Hint:* The question does not mention the radius of the original sphere, which appears to be necessary, but in fact it is not needed.)

Figure 6.4.18

47. A bowl is made in the shape of the solid of revolution obtained by revolving about the y-axis the region between the curves $4y = 3x^2 + 4$ and $4y = x^4$. Suppose all measurements are made in inches, and suppose the clay from which the bowl is made weighs 2 oz/cu in. How much does the bowl weigh?

48. A piston head is made in the shape of a cylinder with a smaller cylindrical hole bored in its end, as shown in Figure 6.4.19. Suppose the large cylinder has radius 3 cm and height 4 cm and the cylindrical hole has radius 1 cm and height 1 cm. If the material from which the piston head is constructed has a density of 7.6 g/cm^3, will the piston head meet the design specification that it must weigh no more than 850 grams?

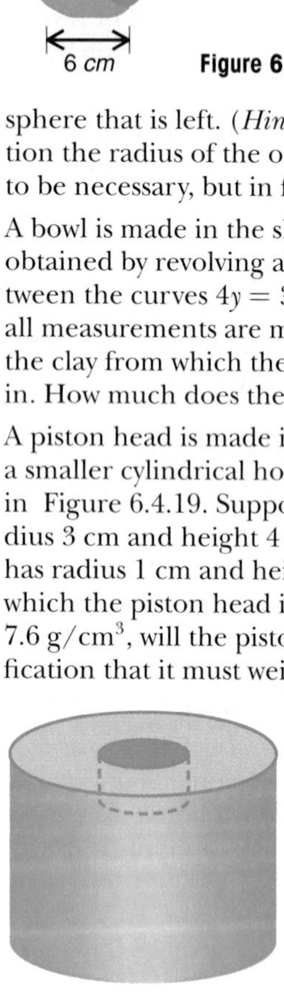
Figure 6.4.19

*__49.__ In the situation of Example 4 (p. 444), calculate the volume of the reflector using cylindrical shells. (Be careful near the top.) Of course, the answer should be the same as when it was done with slices.

50. The horn of a musical instrument is made in the shape obtained by revolving the curve $y = e^{x/2}$, for $x \in [-3, 2]$, about the x-axis. If the units are inches, find the volume enclosed in the horn (i.e., if the ends were covered with flat sheets of plastic, how much water would it hold?).

51. A child's top is made by revolving the curve $x = y - 2 + \sqrt{4 - 2y}$, for $0 \le y \le 2$, about the y-axis (see Figure 6.4.20).

(i) Assuming that the units are inches, find the volume of the top.

(ii) If the top is made of wood whose density is 0.277 oz/cu in, how much does it weigh?

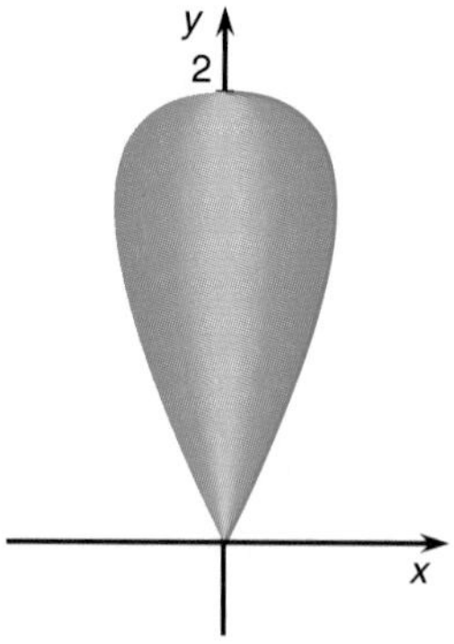

Figure 6.4.20

*__52.__ Consider the curve $x = y^3 + 1$, for $x \in [0, 2]$. Find the volume of the solid obtained by revolving around the x-axis the region between the curve and the x-axis.

III

Use a graphics calculator or computer to find the volume of the solid obtained by revolving about the x-axis the region enclosed between each of the following pairs of functions.

53. $x = 3y - y^3, \quad x = y^4 + 0.5$

54. $x = \frac{2}{1+y^2}, \quad x = y^2 - 3y + 2$

55. $x = \cos(y^2), \quad x = (y - 1)^2$

56. $x = \frac{1}{2+y^2}, \quad x = \frac{2y^2}{y^4+1}$

6.5 Average Values

AVERAGE/MEAN
VELOCITY/SPEED

Meteorologists frequently speak of average temperatures. If we take the temperature reading at many different times during the course of the day, these values can easily be averaged. Suppose we measure t in hours, running from $t = 0$ at midnight to $t = 24$ the following midnight, and suppose the temperature at time t is $T(t)$. If we take n evenly spaced temperature readings, at times $0 = t_1^*, t_2^*, t_3^*, \ldots, t_n^* = 24$, then the approximate average temperature is $\frac{1}{n}\left(T(t_1^*) + \ldots + T(t_n^*)\right)$.

This looks a lot like a Riemann sum; indeed, it equals

$$\frac{1}{24}\left(T(t_1^*)\frac{24}{n} + \ldots + T(t_n^*)\frac{24}{n}\right),$$

and the sum in the parentheses is a Riemann sum. (Note that partitioning $[0, 24]$ into n equal subintervals gives each subinterval a length of $\frac{24}{n}$.) As n increases, these averages tend to a limit, which is

$$\frac{1}{24}\int_0^{24} T(t)\,dt.$$

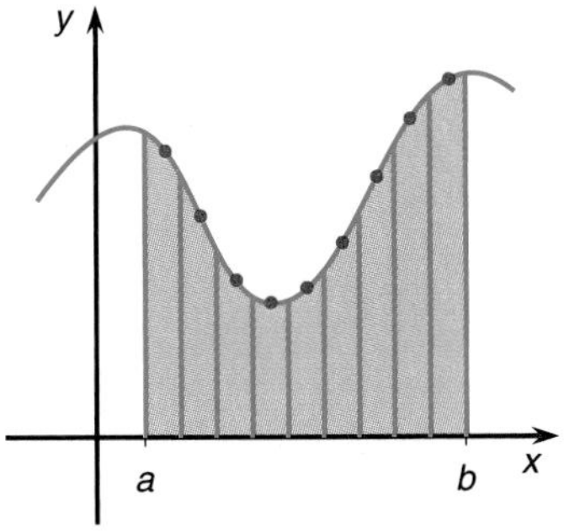

Figure 6.5.1

The same argument applies to any function $f(x)$ defined on $[a, b]$. Average the values of $f(x)$ at n evenly spaced points (see Figure 6.5.1), and let $n \to \infty$. The averages will tend to the limit $\frac{1}{b-a}\int_a^b f(x)\,dx$. This suggests the following definition.

DEFINITION 6.5.1

If $f(x)$ is integrable on $[a, b]$, then the **average value** of $f(x)$ on $[a, b]$ is

$$\frac{1}{b-a}\int_a^b f(x)\,dx.$$

Roughly speaking, in a case in which the variable x represents time, say, what this means is to "add up" all the values of $f(x)$, in the sense of integrating them, and then divide by the total time.

EXAMPLE 1 **Average Temperature** Suppose the outdoor temperature on a particular day is $T(t) = 60 + 16\sin^2\left(\frac{\pi t}{24}\right)^\circ$F, with time measured in hours and $t = 0$ at midnight. What is the average temperature for this 24-hour period?

Solution By the definition above, the average temperature is

$$\frac{1}{24}\int_0^{24}\left(60 + 16\sin^2\left(\frac{\pi t}{24}\right)\right)dt = \frac{1}{24}\left(\int_0^{24} 60\,dt + 16\int_0^{24}\sin^2\left(\frac{\pi t}{24}\right)dt\right)$$

$u = \frac{\pi t}{24}$
$du = \frac{\pi}{24}dt$

$$= \frac{60}{24}t\Big|_0^{24} + \frac{16}{24}\int_0^{\pi}\sin^2(u)\frac{24}{\pi}du$$

$$= \frac{60}{24}24 + \frac{16}{24}\frac{24}{\pi}\int_0^{\pi}\sin^2(u)\,du$$

$$= 60 + \frac{16}{\pi}\left(\frac{1}{2}u - \frac{1}{4}\sin(2u)\right)\Big|_0^{\pi}$$

$$= 60 + \frac{16}{\pi}\frac{\pi}{2}$$

$$= 68^\circ\text{F}.$$

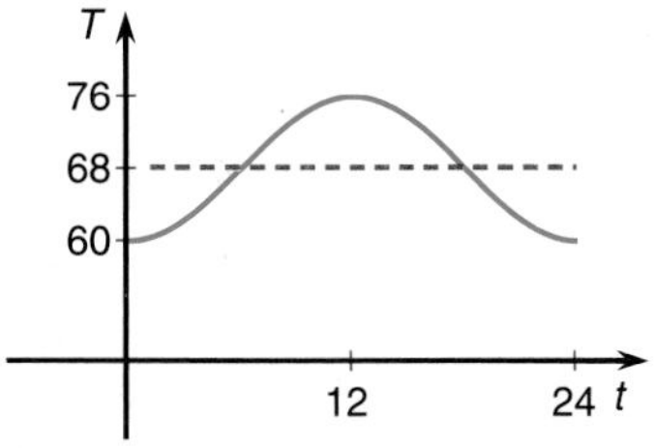

Figure 6.5.2

The graph of $T(t)$ is sketched in Figure 6.5.2, and from the picture it seems very reasonable to expect that the average temperature is 68°F.

In a case like this, in which the function is so symmetrical, it is not difficult to guess the average value, but for more complicated functions that will not be possible.

EXAMPLE 2 **Average Velocity** Suppose an object is moving in a straight line so that at time t its velocity is $v(t) = 2t - t^2$.

(i) Find the average velocity between $t = 0$ and $t = 1$.

(ii) Find the average velocity between $t = 0$ and $t = 3$.

(iii) Find the average velocity between $t = 2$ and $t = 4$.

Solution

(i) The average velocity between $t = 0$ and $t = 1$ is $\frac{1}{1-0}\int_0^1 v(t)\,dt = \left(t^2 - \frac{1}{3}t^3\right)\Big|_0^1 = \frac{2}{3}$.

(ii) The average velocity between $t = 0$ and $t = 3$ is $\frac{1}{3-0}\int_0^3 v(t)\,dt = \frac{1}{3}\left(t^2 - \frac{1}{3}t^3\right)\Big|_0^3 = 0$.

This reflects the fact that at $t = 3$ the object is back where it started when $t = 0$. It does not mean that it did not move, but that its motion in the positive direction was canceled by its return motion in the negative direction. You might like to sketch the graph of its displacement $s(t) = s(0) + \int_0^t v(u)\,du = s(0) + t^2 - \frac{1}{3}t^3$.

(iii) The average velocity between $t = 2$ and $t = 4$ is $\frac{1}{4-2}\int_2^4 v(t)\,dt = \frac{1}{2}\left(t^2 - \frac{1}{3}t^3\right)\Big|_2^4 = \frac{1}{2}\left(16 - \frac{1}{3}64 - 4 + \frac{1}{3}8\right) = -\frac{10}{3}$. The negative sign means that the net displacement and therefore the average velocity are both in the negative direction.

In this particular example we can check our calculations against the familiar notion of average velocity, which after all is just the net displacement divided by the time.

We can find the displacement $s(t)$ by integrating $v(t)$, that is, $s(b) = s(a) + \int_a^b v(t)\,dt$.

As t runs from a to b, say, the object moves a *net* displacement of $s(b) - s(a) = \int_a^b v(t)\,dt$. Its average velocity over this time is

$$\frac{\textit{Displacement}}{\textit{Time}} = \frac{s(b) - s(a)}{b - a} = \frac{1}{b-a}\int_a^b v(t)\,dt.$$

This confirms the calculations that we have done and reassures us that the definition of average value (Definition 6.5.1) is a good definition.

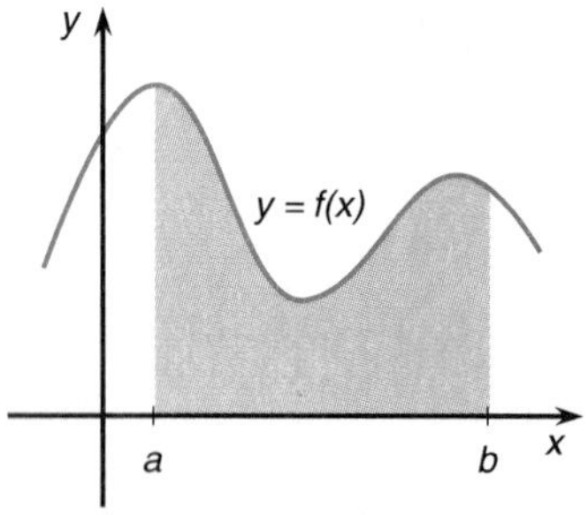

Figure 6.5.3

EXAMPLE 3 **Sandbox** Figure 6.5.3 is the graph of a continuous nonnegative function $f(x)$ on $[a, b]$, with the region under the graph shaded. Think of the graph as the top of the sand in a sandbox, and try to "level it off," that is, push the sand from the "peaks" into the "valleys" to get a flat top on the sand, as illustrated by the shaded region in Figure 6.5.4.

With a graphics calculator we can plot the velocity from Example 2, and plot the various average values, too.

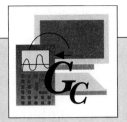

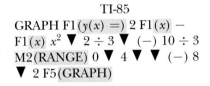

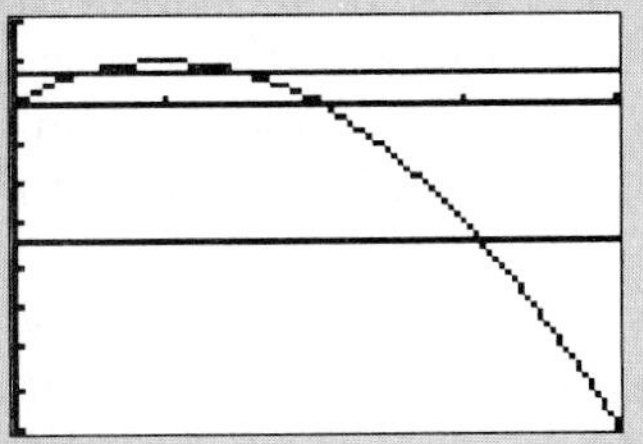

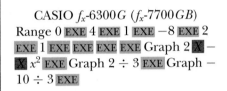

TI-85 GRAPH F1 (y(x) =) 2 F1 (x) − F1 (x) x^2 ▼ 2 ÷ 3 ▼ (−) 10 ÷ 3 M2(RANGE) 0 ▼ 4 ▼ ▼ (−) 8 ▼ 2 F5(GRAPH)	SHARP EL9300 2 x/θ/T − x/θ/T x^2 2ndF ▼ 2 ÷ 3 2ndF ▼ − 10 ÷ 3 RANGE 0 ENTER 4 ENTER 1 ENTER −8 ENTER 2 ENTER 1 ENTER
CASIO f_x-6300G (f_x-7700GB) Range 0 EXE 4 EXE 1 EXE −8 EXE 2 EXE 1 EXE EXE EXE EXE Graph 2 X − X x^2 EXE Graph 2 ÷ 3 EXE Graph − 10 ÷ 3 EXE	HP 48SX PLOT PLOTR ERASE ATTN ' Y = 2 × X − X y^x 2 PLOT DRAW ERASE 0 ENTER 4 XRNG 8 +/− ENTER 2 YRNG DRAW ATTN ' Y = 2 ÷ 3 PLOT DRAW DRAW ATTN ' Y = 10 ÷ 3 +/− PLOT DRAW DRAW

You might like to adapt this example to plot the velocity for $t \in [0, 1]$ with the corresponding average, and similarly for $t \in [0, 3]$ or for $t \in [2, 4]$ with the corresponding averages.

As described in the example, you could also plot the displacement $s(t)$; to do this you have to choose a value for $s(0)$ (one obvious choice is $s(0) = 0 \ldots$).

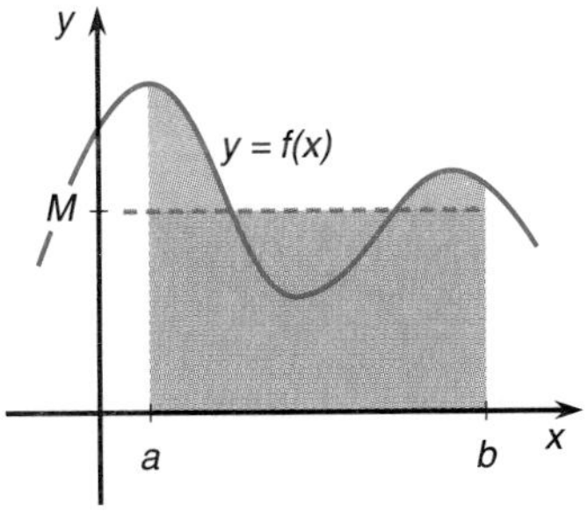

Figure 6.5.4

We think of the height M of the sand *after leveling* as the "average height" of the original curve. The area of the original shaded region is $\int_a^b f(x)\,dx$, and the same sand fills up the shaded rectangle whose height is M and whose base is $b - a$. The area of the rectangle is $M(b - a)$, so $M(b - a) = \int_a^b f(x)\,dx$, and we see that

$$M = \frac{1}{b - a} \int_a^b f(x)\,dx.$$

This is consistent with our definition (Definition 6.5.1) of the average.

We encountered the idea of the average or mean value of a function in Section 5.4, in connection with the Integral Mean Value Theorem (Theorem 5.4.4). The argument given there before the statement of the theorem was based essentially on this picture of the "sandbox."

EXAMPLE 4 **Average Speed of a Car** Suppose a car moves so that its velocity at time t (in hours) is $v(t) = 15(4t - 3t^2)$ mph. Find the displacement of the car 2 hours after it starts, and find its average speed over those 2 hours.

Solution We can assume the displacement of the car is $s = 0$ at $t = 0$. Then to find its displacement at $t = 2$, we integrate:

$$s(2) = \int_0^2 s'(t)\,dt = \int_0^2 v(t)\,dt = 15\int_0^2 (4t - 3t^2)\,dt = 15(2t^2 - t^3)\Big|_0^2$$
$$= 0.$$

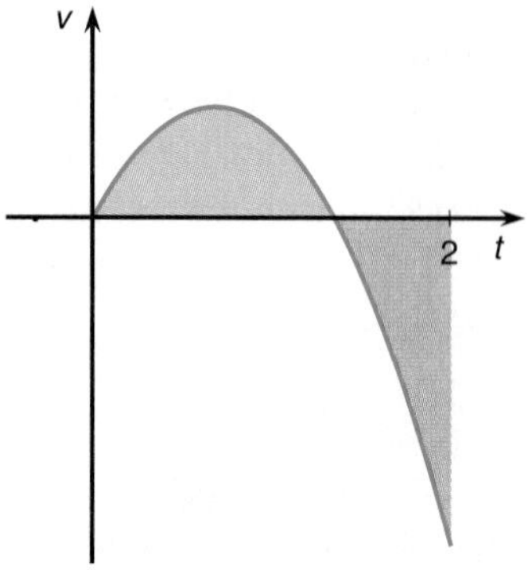

Figure 6.5.5

This means that the car has returned to its starting point after 2 hours. It also means that the average velocity is 0 (cf. Example 2). However, this is not the same as the average speed.

After all, the speed is the absolute value of the velocity. The reason the average velocity is 0 is that some of the time the velocity is positive and some of the time it is negative, and the integral of the positive part and the integral of the negative part cancel (see Figure 6.5.5).

On the other hand, the speed is always positive (being an absolute value), so that sort of cancellation cannot occur. To find the average speed, we have to find the absolute value of $v(t)$, and to do that, we must know where the graph of $v(t)$ crosses the horizontal axis. We find that $v(t) = 0$ means $15(4t - 3t^2) = 0$, or $0 = 4t - 3t^2 = t(4 - 3t)$, which happens exactly when $t = 0$ or when $t = \frac{4}{3}$. From the sketch of the graph (or by evaluating $v(1)$ and $v(2)$, for example) we see that $v(t)$ is positive for $0 < t < \frac{4}{3}$ and negative for $t > \frac{4}{3}$.

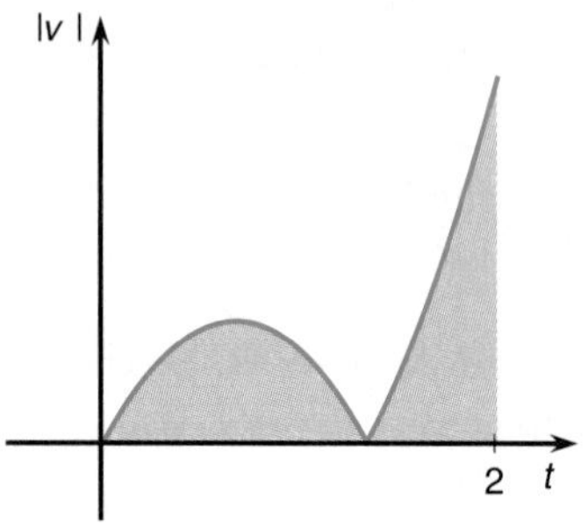

Figure 6.5.6

So to find the average speed, we must integrate the speed $|v(t)|$, which equals $v(t)$ for $0 < t < \frac{4}{3}$ and equals $-v(t)$ for $t > \frac{4}{3}$ (see Figure 6.5.6). The average speed equals

$$\frac{1}{2-0}\int_0^2 |v(t)|\,dt = \frac{1}{2}\left(\int_0^{4/3} v(t)\,dt + \int_{4/3}^2 \big(-v(t)\big)\,dt\right)$$
$$= \frac{15}{2}\left(\int_0^{4/3} (4t - 3t^2)\,dt + \int_{4/3}^2 (3t^2 - 4t)\,dt\right)$$
$$= \frac{15}{2}\left((2t^2 - t^3)\Big|_0^{4/3} + (t^3 - 2t^2)\Big|_{4/3}^2\right)$$
$$= \frac{15}{2}\left(\frac{32}{9} - \frac{64}{27} + 8 - 8 - \frac{64}{27} + \frac{32}{9}\right)$$
$$= 17\frac{7}{9} \quad \text{mph.}$$

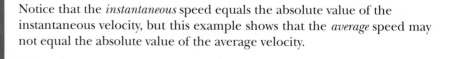

Notice that the *instantaneous* speed equals the absolute value of the instantaneous velocity, but this example shows that the *average* speed may not equal the absolute value of the average velocity.

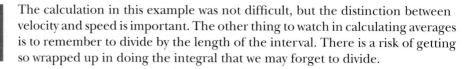

The calculation in this example was not difficult, but the distinction between velocity and speed is important. The other thing to watch in calculating averages is to remember to divide by the length of the interval. There is a risk of getting so wrapped up in doing the integral that we may forget to divide.

Most of the averages considered so far have been averages *with respect to time.* It is also possible to consider averages with respect to other quantities.

EXAMPLE 5 **Average Radius of Cone** Consider a long conical rod, 2 m long, whose diameter at one end is 2 cm and whose diameter at the other end is 6 cm. As we move from one end to the other, find the average radius and the average cross-sectional area.

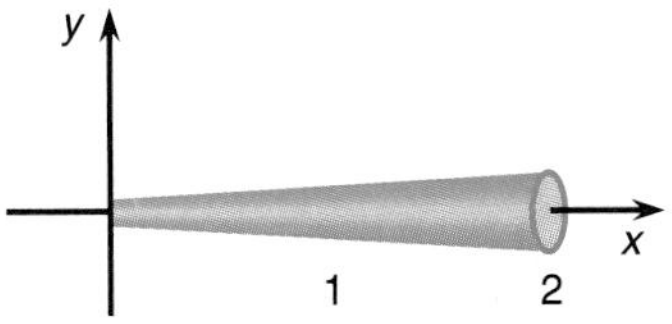

Figure 6.5.7

This is the line that is the top of the cone.

Solution

(i) For convenience we place the rod along the x-axis with its narrow end at the origin (see Figure 6.5.7). Measuring in meters, we see that the wider end will be at $x = 2$. The conical shape is obtained by revolving a straight line about the x-axis; since the radius at the small end is 0.01 and the radius at the large end is 0.03, the line passes through $(0, 0.01)$ and $(2, 0.03)$. The equation of the line through these points is easily found. It is

$$y = 0.01 + 0.01x.$$

As x moves from $x = 0$ to $x = 2$, the corresponding radius of the cone is just the height of this line above the axis, which is just $y = 0.01 + 0.01x$. The average radius is easily found. It is

$$\frac{1}{2-0}\int_0^2 y\,dx = \frac{1}{2}\int_0^2 (0.01 + 0.01x)\,dx = \frac{1}{2}\left(0.01x + 0.005x^2\right)\Big|_0^2$$
$$= 0.02 \text{ m} = 2 \text{ cm}.$$

In a sense this is hardly surprising, since the radius goes from 1 cm to 3 cm in a straight line.

(ii) Since the radius is $y = 0.01 + 0.01x$, the cross-sectional area is $\pi r^2 = \pi y^2 = \pi(0.01 + 0.01x)^2 = \frac{\pi}{10{,}000}(1+x)^2 = \frac{\pi}{10{,}000}(1 + 2x + x^2)$. The average cross-sectional area is

$$\frac{1}{2-0}\int_0^2 \frac{\pi}{10{,}000}(1 + 2x + x^2)\,dx = \frac{\pi}{20{,}000}\int_0^2 (1 + 2x + x^2)\,dx$$
$$= \frac{\pi}{20{,}000}\left(x + x^2 + \frac{1}{3}x^3\right)\Big|_0^2 = \frac{13\pi}{30{,}000} \text{ m}^2$$
$$= \frac{13\pi}{3} \text{ cm}^2 \approx 13.61 \text{ cm}^2.$$

Notice that this is not just halfway between the areas at the two ends.

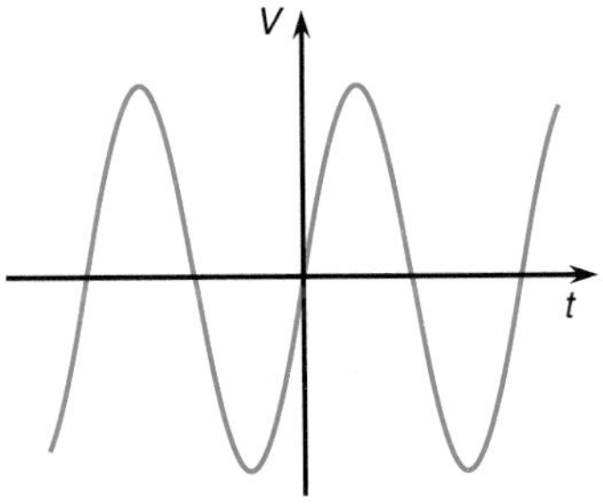

Figure 6.5.8

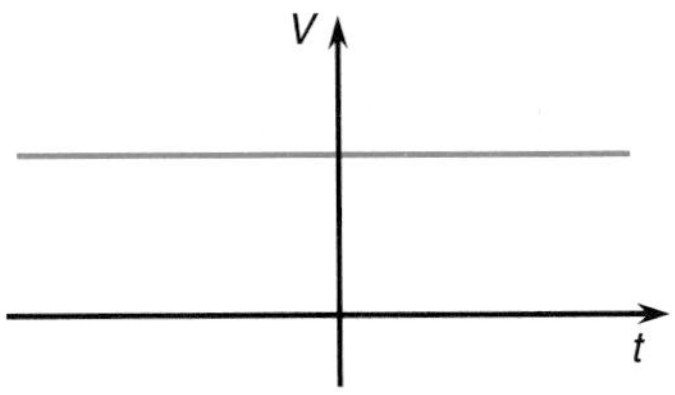

Figure 6.5.9

EXAMPLE 6 **Electricity and RMS Averages** Electricity comes from various sources, but in everyday life there are two main types. Ordinary household current is what is called "Alternating Current," or AC; this means that the voltage alternates from positive to negative. The graph of the voltage is a sine curve, as shown in Figure 6.5.8; it changes at the rate of 60 cycles per second, so one complete cycle takes $\frac{1}{60}$ sec. Figure 6.5.9 illustrates "Direct Current" (DC), such as the current that comes from a battery. Its voltage stays constant.

It is easy to measure a DC voltage, but for AC it is less clear what to do. One approach would be to measure the maximum voltage (the height of the peak), but in a way this is misleading. Most of the time the voltage is less than the peak value.

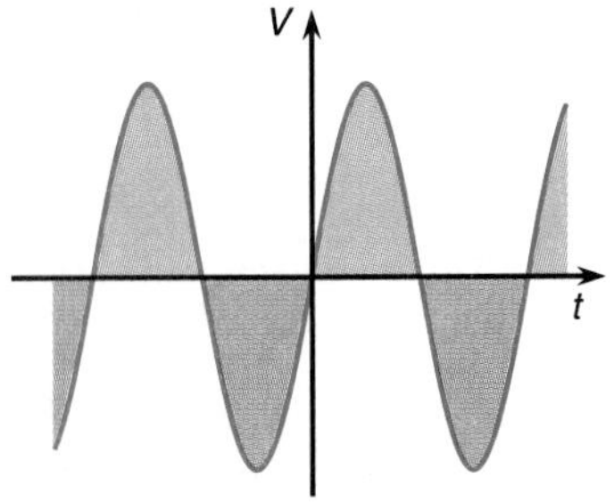

Figure 6.5.10

Another idea would be to find the average voltage. The difficulty here is that since the voltage is negative half the time and positive half the time, the average is zero (see Figure 6.5.10).

We could try averaging the absolute value, but it turns out that it is better to find the average of the *square* of the voltage and then take the square root of that average. This is called the *Root Mean Square* average, and it is extremely important in science and engineering.

For instance, with 60-cycle household current, the voltage can be described as a function of time by $V(t) = A\sin(120\pi t)$. When t increases by $\frac{1}{60}$, we see that $120\pi t$ increases by 2π, so the voltage goes through one complete cycle.

Let us average over one cycle. We find

$$\frac{1}{\frac{1}{60}}\int_0^{1/60} V^2(t)\,dt = 60\int_0^{1/60} A^2\sin^2(120\pi t)\,dt$$

$u = 120\pi t$
$du = 120\pi\ dt$

$$= 60A^2\int_0^{2\pi}\sin^2(u)\frac{1}{120\pi}du = \frac{A^2}{2\pi}\left(\frac{u}{2} - \frac{1}{4}\sin(2u)\right)\Big|_0^{2\pi}$$

$$= \frac{A^2}{2\pi}\pi = \frac{A^2}{2}.$$

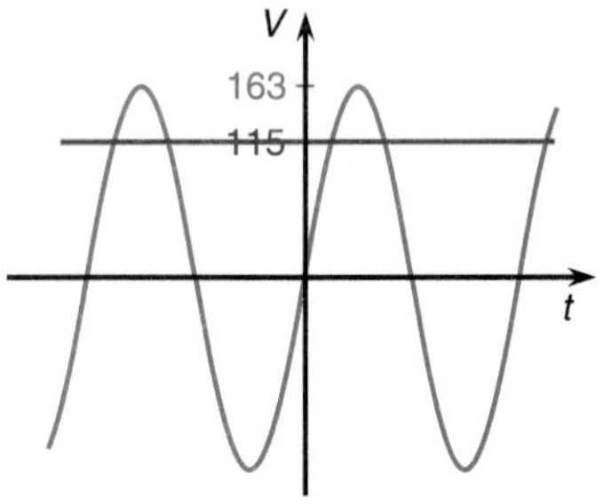

Figure 6.5.11

This is the average of V^2; its square root is $\frac{A}{\sqrt{2}}$. The root mean square (RMS) average voltage is $\frac{A}{\sqrt{2}}$. When we read that household electricity comes at 115 volts, this means that the RMS average is 115. In particular, $A = 115\sqrt{2}$. Since the maximum value of $V(t) = A\sin(120\pi t)$ equals A, we see that the peak voltage is $115\sqrt{2} \approx 162.6$ volts (see Figure 6.5.11).

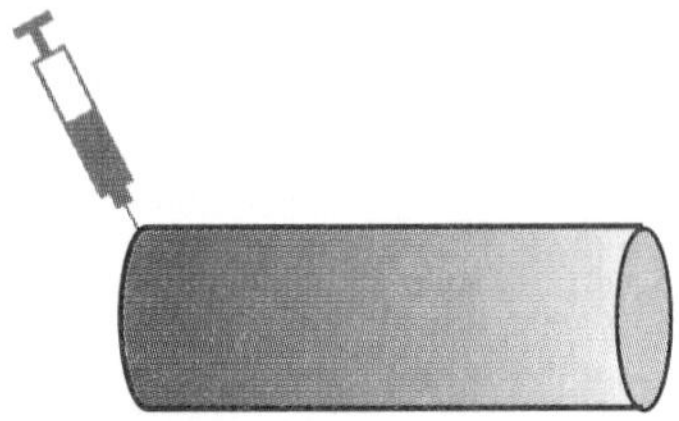
Figure 6.5.12

EXAMPLE 7 **Drug Concentration** A certain blood vessel has the approximate shape of a cylinder, 200 mm long and with radius 2 mm. A drug is injected at one end (see Figure 6.5.12) and rapidly spreads along the blood vessel.

Let x measure distance along the cylinder from the point of injection, in millimeters. One second after the injection, the concentration of the drug at a distance x along the blood vessel is

$$C(x) = 80e^{-x/40}\%.$$

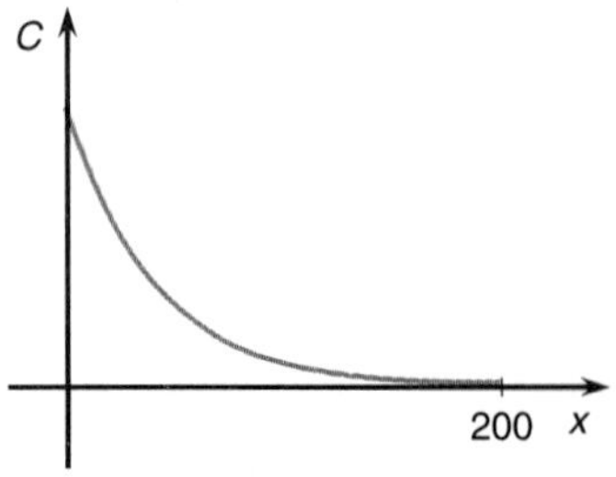

Figure 6.5.13

Figure 6.5.13 graphs the concentration as a function of x, showing that it is highest at the point of injection and trails off to lower and lower concentrations farther away.

We want to find the average concentration for $x \in [0, 200]$, which is

$$\frac{1}{200}\int_0^{200} C(x)\,dx = \frac{1}{200}\int_0^{200} 80e^{-x/40}dx = \frac{2}{5}\int_0^{200} e^{-x/40}dx$$

$u = \frac{x}{40}$
$du = \frac{1}{40}dx$

$$= \frac{2}{5}\int_0^{5} e^{-u}(40)\,du = 16\int_0^5 e^{-u}du = -16e^{-u}\Big|_0^5$$

$$= -16(e^{-5} - 1) = 16(1 - e^{-5}) \approx 15.89\%.$$

One way of interpreting this result is that if the drug were evenly distributed along the whole length of the tube, its concentration would be 15.89% (see

Figure 6.5.14). Since the total volume of the cylinder is $\pi r^2 h = \pi 2^2(200) = 800\pi \approx 2513.3\ \text{mm}^3 = 2.5133\ \text{cm}^3$, we can also find that the total amount of the drug present is 15.89% of this, or

$$0.1589 \times 2.5133 \approx 0.3994\ \text{cm}^3.$$

It appears that the original injection was about 0.4 cc (cubic centimeters) of the drug.

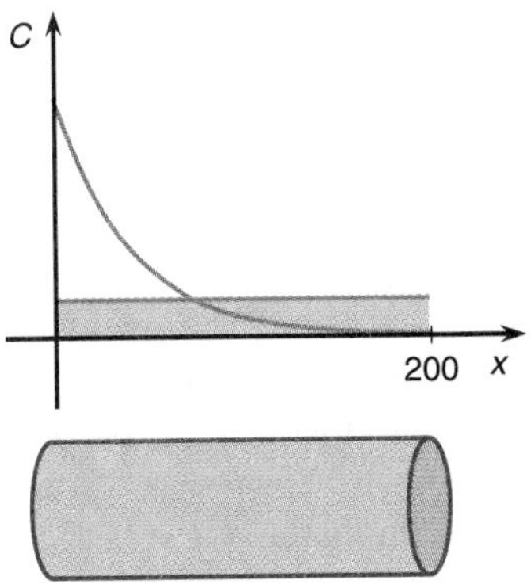

Figure 6.5.14

Example 7 ignores irregularities in the shape of the blood vessel and the pumping motion of the blood, which tends to carry the drug along the vessel, but it shows how averages can be used.

EXAMPLE 8 **Blood Flow and Poiseuille's Law** The walls of a blood vessel exert friction on the blood as it flows past. The effect is that the blood flows more slowly near the walls than it does in the center. Poiseuille's Law says that in a tube of radius R the fluid velocity at a distance r from the center is

$$v(r) = k(R^2 - r^2),$$

where k is some constant (see Figure 6.5.15). When $r = R$ (i.e., right at the walls), the velocity is zero, and the velocity is largest in the center (where $r = 0$ and $v(0) = kR^2$).

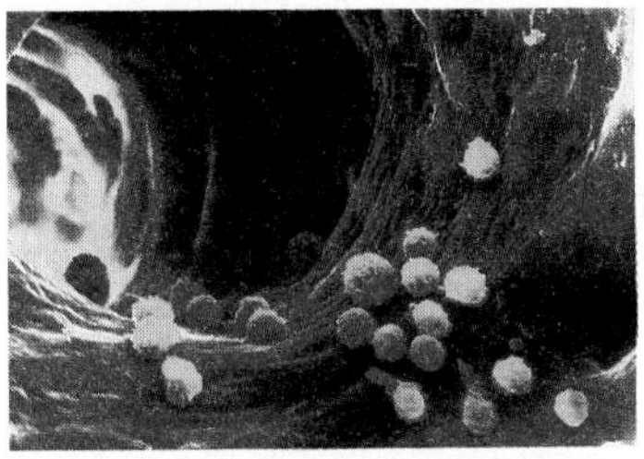

False-color scanning electron micrograph of the interior of an artery in the human liver, showing lymphocytes (white blood cells) adhering to the inner surface. Magnification: 800×.

In Figure 6.5.16 we imagine the blood in "layers," each a fixed distance from the center. The blood in each thin layer travels at the same velocity, but the inner layers travel faster than the outer layers.

Now consider the average velocity. We could find the average of $v(r)$ for $r \in [0, R]$. It is

$$\frac{1}{R}\int_0^R v(r)\,dr = \frac{k}{R}\int_0^R (R^2 - r^2)\,dr = \frac{k}{R}\left(R^2 r - \frac{r^3}{3}\right)\Bigg|_0^R = \frac{k}{R}\left(R^3 - \frac{R^3}{3}\right)$$
$$= \frac{2}{3}kR^2.$$

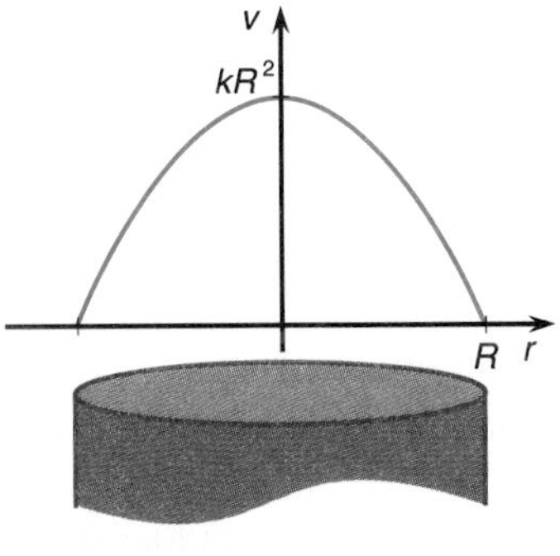

Figure 6.5.15

However, this average may be misleading. From the picture we see that the slow outer layers are bigger than the fast inner layers. This suggests that most of the blood is moving slowly, so our average, which counts each speed equally, is probably too high.

Another kind of average can be found by asking how much blood passes by in 1 second. To answer this, we consider the blood in a thin layer at a distance r from the center. The velocity of this layer is $v(r)$, so the length that passes a fixed point in 1 sec is $v(r)(1) = v(r)$. Since the circumference is $2\pi r$, the surface area of this part of the layer is $2\pi r v(r)$, and if the thickness of the layer is dr, the volume passing by in this layer is $2\pi r v(r)\,dr$ (see Figure 6.5.17).

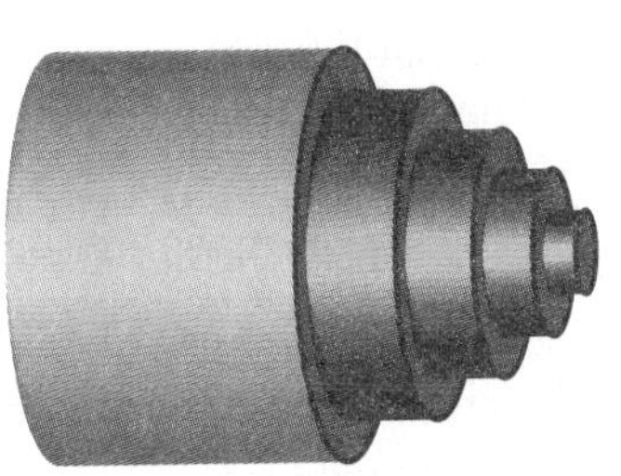
Figure 6.5.16

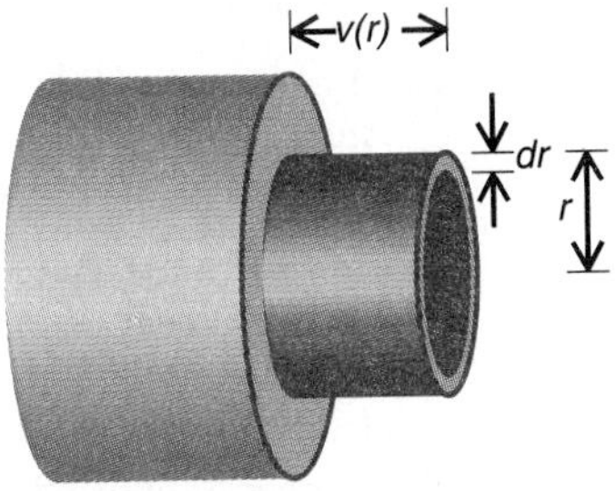

Figure 6.5.17

The total flow in 1 sec is the "sum" of the flows in each layer, that is,

$$\int_0^R 2\pi r v(r)\,dr = 2\pi k\int_0^R (rR^2 - r^3)\,dr = 2\pi k\left(R^2\frac{r^2}{2} - \frac{r^4}{4}\right)\Bigg|_0^R = 2\pi k\frac{R^4}{4}$$
$$= \frac{\pi k R^4}{2}.$$

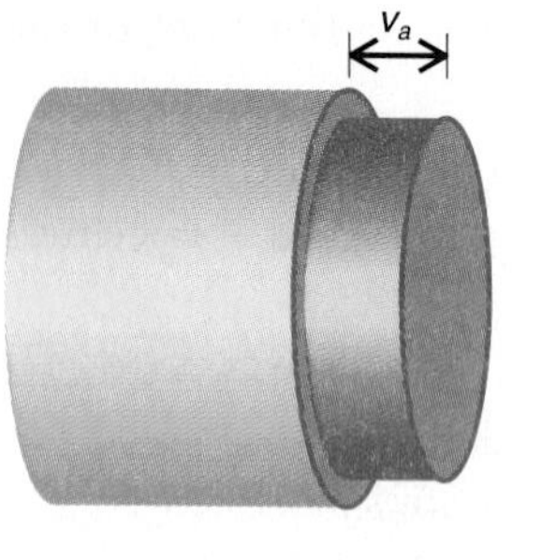

Figure 6.5.18

To say that the average velocity is v_a should mean that the total flow would be the same if *all* the blood moved at velocity v_a. If the blood all flowed at v_a, then in 1 sec the amount flowing past would be a cylinder of length $v_a(1) = v_a$, as shown in Figure 6.5.18. Since the cross-sectional area is πR^2, the volume passing in 1 sec is $\pi R^2 v_a$.

If this equals the total flow $\frac{\pi k R^4}{2}$, we find that $v_a = \frac{kR^2}{2}$. As expected, this is less than the value $\frac{2kR^2}{3}$ that we found earlier. The smaller average is more meaningful because it reflects more accurately the way the blood flows.

This example shows that deciding what an average actually means can be quite subtle. Care must be used in interpreting averages.

Exercises 6.5

I

In each of the following situations, find the average value of the given function over the specified interval.

1. $f(x) = x^2 + 1$ over $[-1, 1]$
2. $g(x) = x^3 + 3$ over $[-1, 1]$
3. $h(x) = e^x$ over $[0, 1]$
4. $k(x) = \sin x$ over $[0, \pi]$
5. $r(u) = \frac{1}{u}$ over $[1, 5]$
6. $p(t) = \sin t$ over $\left[0, \frac{\pi}{2}\right]$
7. $q(v) = \sin v$ over $[-\pi, \pi]$
8. $s(x) = \sin^2 x$ over $[0, 2\pi]$
9. $w(t) = \sec^2 t$ over $\left[0, \frac{\pi}{4}\right]$
10. $\ell(x) = x\sin(\pi x^2)$ over $[0, 1]$
11. $f(x) = \frac{1}{x^2+1}$ over $[-1, 1]$
12. $f(x) = \frac{x}{x^2+1}$ over $[-1, 1]$
13. $f(x) = \frac{x}{\sqrt{x^2+1}}$ over $[-1, 1]$
14. $f(x) = x\exp(x^2)$ over $[0, 2]$
15. $f(x) = \sin(5\pi x)$ over $[0, 3]$
16. $f(x) = \sin x\cos^3 x$ over $\left[0, \frac{\pi}{2}\right]$
17. $f(x) = x^3\sin(\pi x^4)$ over $[-1, 1]$
18. $f(x) = \sec^2 x\tan^2 x$ over $\left[0, \frac{\pi}{4}\right]$
19. $f(x) = \sin x e^{\cos x}$ over $\left[0, \frac{\pi}{2}\right]$
20. $f(x) = \sin(x)\ln(\cos^2 x + \sin^2 x)$ over $[0, \pi]$

II

21. (i) Find the average value of $f(x) = 3$ over the interval $[1, 2]$.
    (ii) Find the average value of $f(x) = -7$ over the interval $[-1, 4]$.
    (iii) Find the average value of the constant function $f(x) = C$ over the interval $[a, b]$.
22. Define $f(x)$ as follows:
$$f(x) = \begin{cases} 0, & \text{if } x \le 2, \\ 1, & \text{if } x > 2. \end{cases}$$
    (i) Find the average value of $f(x)$ over the interval $[0, 4]$.
    (ii) Find the average value of $f(x)$ over the interval $[1, 5]$.

*23. (i) Find the average radius of the horn in Exercise 50 of Section 6.4.
    (ii) Find the average cross-sectional area of the horn.

*24. A car travels 50 km at 50 km/hr and then travels the next 50 km at 100 km/hr. What is its average speed? (*Hint:* The answer is not 75 km/hr.)

*25. (i) Find the average radius of the cross-sections of a sphere of radius 1 as the cross-sections move from one side of the sphere to the other.
    (ii) Find the average cross-sectional area of the sphere.

*26. Repeat Exercise 25 for a sphere of radius r.

*27. Consider a bacterial culture in which the weight of the bacteria present at time t is $P(t) = 2e^{t/20}$, where

t is measured in hours and the weight is measured in grams.

(i) Find the average weight of the culture for the 24-hour period beginning at $t = 0$.

(ii) Find the average weight of the culture for the 6-hour period beginning at $t = 4$.

***28.** Suppose the population of a city is given by $P(t) = 2{,}000{,}000e^{(t-1980)/15}$, with t being the date, measured in years.

(i) What was the population in 1990?

(ii) What was the average population between 1980 and 1990?

(iii) What was the average population between 1980 and 1985?

***29.** (i) If the velocity of a moving object is $v(t) = t^2 - 4t + 3$, what is its average velocity between $t = 0$ and $t = 4$?

(ii) What is its average speed for the same period?

***30.** (i) Suppose the displacement of a moving object, in meters, is $s(t) = t^3 - 6t^2 + 9t + 3$. What is its average velocity between $t = 0$ and $t = 4$?

(ii) What is its average speed for the same period?

***31.** A motorist knows that on a certain freeway his time is monitored to detect speeding. Two tollbooths are 10 miles apart. He sets out from one and travels 5 miles at an average speed of 80 mph. Suddenly, he realizes that he must slow down to avoid a ticket. At what average speed must he travel the second 5 miles in order to have an average speed of 55 mph on the whole 10-mile stretch?

***32.** Suppose $f(x)$ is an integrable function on $[a, b]$ and that M and m are numbers for which $m \le f(x) \le M$ for all $x \in [a, b]$. Show that the average value of $f(x)$ on $[a, b]$ lies between m and M.

***33.** (i) If $f(x)$ and $g(x)$ are integrable functions on $[a, b]$, show that the average of $f(x) + g(x)$ over $[a, b]$ equals the sum of the averages of $f(x)$ and $g(x)$.

(ii) Express the average of $3f(x) - 5g(x)$ in terms of the averages of $f(x)$ and $g(x)$.

***34.** If $f(x)$ and $g(x)$ are integrable functions on $[a, b]$, show by example that the average of $f(x)g(x)$ over $[a, b]$ does not necessarily equal the product of the averages of $f(x)$ and $g(x)$.

35. (i) Find the average value of x^3 over the interval $[-1, 1]$.

(ii) Find the average value of x^7 over the interval $[-3, 3]$.

(iii) Find the average value of $x^5 + 4x^3 - 2x$ over the interval $[-2, 2]$.

***36.** (i) Suppose $f(x)$ is an odd function and $a > 0$. Find the average value of $f(x)$ on the interval $[-a, a]$.

(ii) If $g(x)$ is an even function, compare the average values of $g(x)$ on $[-a, a]$ and $[0, a]$.

***37.** For the voltage $V(t) = A\sin(120\pi t)$ of Example 6 (p. 451), find the RMS average over the intervals (i) $\left[0, \frac{1}{120}\right]$, (ii) $\left[\frac{1}{240}, \frac{1}{120}\right]$, (iii) $\left[\frac{1}{160}, \frac{1}{96}\right]$.

***38.** If electricity with voltage V flows through a resistance R, such as a toaster or a light bulb, the power consumed is $P = \frac{V^2}{R}$. (i) Consider the AC voltage $V(t)$ of Example 6 passing through a fixed resistance R, and find the average power for one cycle (i.e., between $t = 0$ and $t = \frac{1}{60}$). (ii) Suppose a (constant) DC voltage through the same resistance consumes the same average power. Show that this constant DC voltage equals the RMS average voltage corresponding to the AC voltage $V(t)$. Show that it is different from the average of the absolute value $|V(t)|$.

This shows why RMS is a good kind of average; it gives the correct value for the electrical power, which is very often the most important quantity to measure.

39. The lower section of a tree trunk is approximately a cylinder of length 10 ft and radius 3 in. Somebody has carelessly spilled a toxic chemical near the roots, and it has entered the tree. If h measures the height above ground, in feet, the concentration of the chemical in the tree at height h is $C(h) = 1.7e^{-h}$ ppm (parts per million). (i) What is the average concentration of the chemical in the lowest 10 ft of the trunk? (ii) What is the average concentration in the lowest 5 ft of the trunk? (iii) Experience has shown that long-term concentrations above 0.5 ppm will kill a tree. Assuming that the chemical disperses evenly through the 10-ft section of trunk and no more enters from the roots, will the tree survive? (iv) How much chemical (in cubic inches) is in this 10-ft section of trunk?

***40.** In Example 8 (p. 453) we decided that a sort of weighted average was appropriate. (i) Find the average velocity, in this sense, for the blood in the center of the tube, the part with $0 \le r \le \frac{R}{2}$. (ii) Find the average velocity for the part of the tube with $\frac{R}{2} \le r \le R$. (iii) Without checking, do you expect the average for the whole tube to be exactly halfway between these averages, or smaller, or greater? (iv) Check and explain.

III

41. (i) Use a personal computer or programmable calculator to approximate the average temperature in Example 1 (p. 447) by averaging the temperature at every hour on the hour from midnight to 11 P.M. (ii) What happens if you take the temperature every half hour? Every 5 minutes?

42. Approximate the average velocity and average speed of the car in Example 4 (p. 449), by averaging the values measured every 5 minutes.

43. Approximate the average value of $\frac{1}{1+x^3}$ over $[0, 4]$.

44. Approximate the average value of $\arcsin x$ over $[0, 1]$.

POINT TO PONDER

In Example 5 (p. 451) we were not surprised to find that the average radius of the cone was halfway between the radii at the two ends. However, a quick check will show that the same is not true of the average cross-sectional area. Another way to say this is that if we take the average radius r_a and compute the corresponding cross-sectional area πr_a^2, what we get will *not* equal the average cross-sectional area.

Explain why this happens and why we should not be surprised.

6.6 Work

FORCE
KINETIC ENERGY
POTENTIAL ENERGY
HOOKE'S LAW
GRAVITATIONAL FORCE AND WORK

In this section we discuss the notion of work. When a force acts on an object and the object moves, the force performs **work** on the object. This is another way of saying that the force changes the total energy of the object, for example, by making it move faster, which increases its kinetic energy.

There is a simple formula that says that the work done is the product of the force and the distance through which the object moves. This is fine if the force is constant, but many forces vary in intensity, and the simple formula does not apply to them. It turns out that the way to deal with this case is to use integration.

When an object moves a distance s in a straight line under the influence of a force F, the force performs work on the object, and the amount of work done is

$$W = Fs.$$

Performing work means changing the energy of the object. This can happen because of changing its **potential energy** (by changing its position, e.g., lifting an object up increases its potential energy), changing its **kinetic energy** (by changing its speed; the faster an object is moving, the greater is its kinetic energy), or a combination of the two.

Perhaps the most familiar force is gravity.

EXAMPLE 1 **Work and the Force of Gravity** Near the earth's surface, an object experiences a force due to gravity, which is called its weight. If an object of weight m at rest at ground level is raised to a height h above the ground, then it travels a distance h. The force required to lift it against gravity is just m, so the work performed is $W = mh$. Since the object starts and ends at rest, there is no kinetic energy involved, and we see that the potential energy has changed by mh.

If instead the object is allowed to fall to the ground from a height h, it moves a distance h under the gravitational force m, so its potential energy decreases by mh. This potential energy is converted into kinetic energy, since the object is moving when it hits the ground. When the object hits the ground, its kinetic energy is converted into various types of energy, including the energy in the sound of the impact, kinetic energy in the object if it bounces, kinetic energy in any little pebbles or dust it may kick up, and so on.

Figure 6.6.1

At first it may seem puzzling that we insist that the object must move in order for work to be done on it. Imagine yourself pushing against a brick wall (see Figure 6.6.1). You may push quite hard and even tire yourself out, but nothing happens to the wall. After you stop, the wall will certainly not be moving, so it has no kinetic energy, and since you have not moved it, you cannot possibly have changed its potential energy. No work has been done to the wall.

The difficulty really arises from the way the word is used. In physics, doing work on an object means changing its energy. In everyday life we feel that if we push hard on something, we have done work, even if it did not move.

Strictly speaking, the force and the motion must point along the same line. If you move an object horizontally, that is, without changing its height, the force of gravity does no work on it. Note, however, that pointing along the same line does allow for the possibility that the force and the motion may point in opposite directions along the same line. Work *is* done by a force that slows a moving object by pushing in the direction opposite to its motion.

A more difficult problem arises when an object is moving under the influence of a force that is not constant. We break the motion up into a succession of small motions, over each of which the force is essentially constant. For each little motion we calculate the work done and then add them up.

Suppose the object moves along the horizontal axis and that its displacement changes from $s = a$ to $s = b$. Suppose that the force, being different at different points along the object's path, can be expressed as a function $F(s)$, that is, a function depending on the displacement s.

We partition the interval $[a, b]$ into n subintervals, each of which is sufficiently short that the value of the force does not change appreciably on the subinterval. If the length of the ith subinterval is Δs_i and s_i^* is any point in the ith subinterval, then the work done as the object moves along the ith subinterval is approximately $F(s_i^*)\Delta s_i$, the product of the force and the displacement. The total work done as the object moves from $s = a$ to $s = b$ is approximately

$$\sum_{i=1}^{n} F(s_i^*)\Delta s_i.$$

To improve the approximation, we must take finer and finer partitions, and the approximations to the work approach the limit $\int_a^b f(s)\,ds$. This suggests the following.

DEFINITION 6.6.1

The work done by a force $F(s)$ on an object that moves from $s = a$ to $s = b$ along the same line as the direction of the force is

$$W = \int_a^b F(s)\,ds.$$

Just as with area and volume calculations it is possible to think in terms of very small displacements ds to help setting up the integral.

EXAMPLE 2 **Springs and Hooke's Law** The force exerted by a spring is proportional to the displacement of the end of the spring from its rest position, and in

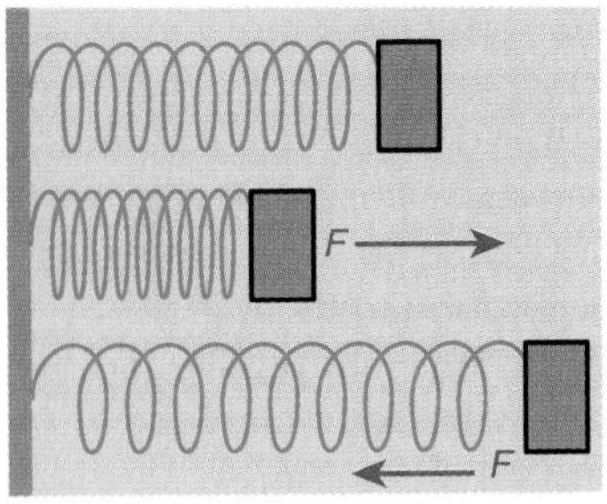

Figure 6.6.2

the direction opposite to the displacement. This is Hooke's Law, named after the English physicist Robert Hooke (1635–1703) and illustrated in Figure 6.6.2.

More informally, what it says is, roughly, that the force of the spring tries to push it back to its rest position (i.e., if you push the spring to the right, it pushes to the left and vice versa) and that the more it is displaced (compressed or extended), the harder it will push back. The top picture shows the spring in its rest position, the second shows it compressed, and the third shows it extended (i.e., stretched).

Hooke's Law says that the force exerted by the spring is proportional to the displacement, that is, when the spring is displaced an amount s from its rest position, the force it exerts is

$$F = -ks \tag{6.6.2}$$

for some positive constant k; the minus sign reflects the fact that the force is in the direction opposite to the displacement. Calculate the work done by the spring when the displacement of the end of the spring is changed from $s = a$ to $s = b$.

Solution The displacement changes from $s = a$ to $s = b$. At any particular displacement s the force exerted by the spring is $F(s) = -ks$, and the work done in making a small change in displacement ds will be $-ks\,ds$. Adding up the work done in all the little displacements (i.e., integrating), we find that the total work done is

$$W = \int_a^b (-ks)\,ds = -k\int_a^b s\,ds = -k\frac{s^2}{2}\bigg|_a^b$$
$$= -\frac{k}{2}(b^2 - a^2).$$

> Notice that a and b are the displacements of the end of the spring *from the rest position,* not the length of the spring.

At this point we must discuss the units of force and work.

> If force is measured in pounds and displacement in feet, then work is measured in **foot-pounds (ft-lb)**. In the metric system, force is measured in **newtons (N)**, so work is measured in **newton-meters (N-m)**, which are more often called **joules (J)**. An object whose mass is 1 kg weighs 9.80 N at the earth's surface (i.e., gravity exerts a force of 9.80 N on it). An object whose mass is 1 lb weighs 1 lb at the earth's surface. This apparently confusing statement simply means that the units of mass (pounds) correspond to the units of force (pounds) and that unlike what happens in the metric system, there is no conversion factor. It is sometimes necessary to measure mass in a different unit, the "slug," but we will use pounds.

EXAMPLE 3 Find the work done by a person stretching a spring starting from its rest position until it is 1 ft longer than at rest if it exerts a 1 lb force when stretched 1 in.

Solution If we measure displacement s in feet, we know that when $s = \frac{1}{12}$ (i.e., one inch), $F = -1$ lb, that is, in the direction opposite to the stretching.

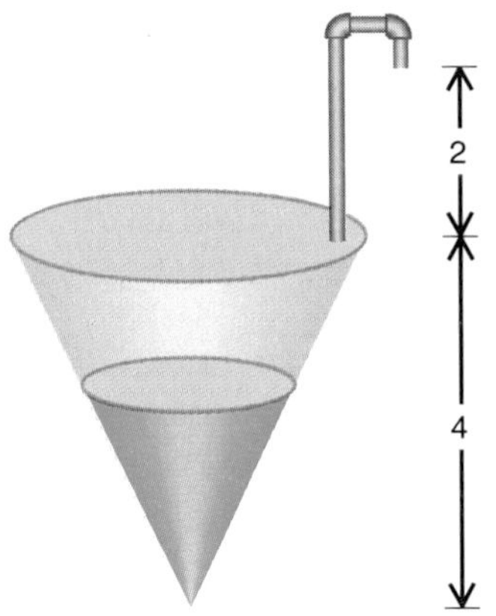

Figure 6.6.3

Since $F = -ks$, we see that $k = 12$. By Example 2, with $a = 0$, $b = 1$, we find that the work done *by the spring* is

$$W = -\frac{k}{2}(b^2 - a^2) = -6 \text{ ft-lb}.$$

The negative sign occurs because the spring pulls in the direction opposite to the motion. The person doing the stretching pulls in the same direction as the motion and so does work of 6 ft-lb.

EXAMPLE 4 **Conical Water Tank** Find the work done in pumping all the water out of a conical tank up to a point 2 m above the top of the tank, as shown in Figure 6.6.3. The tank has height 4 m and radius 2 m at the top.

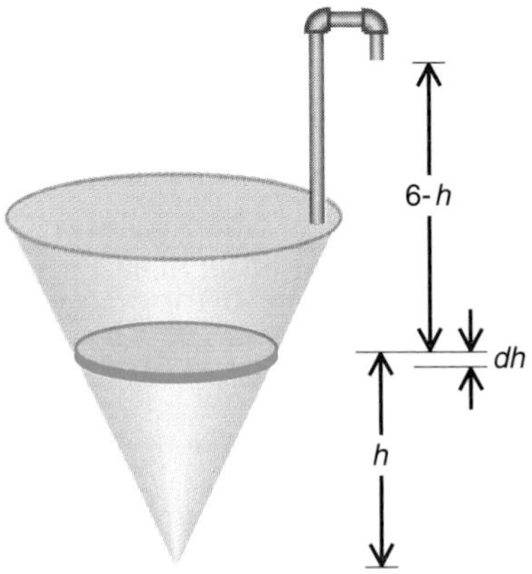

Figure 6.6.4

Solution Let h be a variable measuring the height above the base of the tank. Consider a thin horizontal slice of water at height h with thickness dh. It will be lifted through a height of $6 - h$ (see Figure 6.6.4). The volume of the slice is (*Area of circle*) $\times$ (*thickness*) $= \pi r^2 dh$, where r is the radius of the circle at height h. By similar triangles in Figure 6.6.5 we see that $\frac{r}{h} = \frac{2}{4}$, that is, $r = \frac{1}{2}h$.

So the slice has volume $\pi\left(\frac{1}{2}h\right)^2 dh = \frac{1}{4}\pi h^2 dh$, in cubic meters. Since 1 cubic meter contains 1000 liters, and since the mass of 1 liter of water is 1 kg, we know that the mass of 1 m^3 of water is 1000 kg. The force of gravity on 1 cubic meter of water is $1000(9.80)$ N $= 9800$ N. Therefore the force of gravity on the slice of water we are considering is $\frac{1}{4}\pi h^2 dh(9800) = 2450\pi h^2 dh$ N, and the work done in pumping it up is $(6 - h)2450\pi h^2 dh$ J.

Adding up the work done for all the slices, that is, for h running from $h = 0$ to $h = 4$, we find that the total work done is

$$\begin{aligned} W &= \int_0^4 (6 - h)2450\pi h^2 dh = 2450\pi \int_0^4 (6h^2 - h^3)\, dh \\ &= 2450\pi\left(2h^3 - \frac{1}{4}h^4\right)\Big|_0^4 \\ &\approx 493{,}000 \text{ J}. \end{aligned}$$

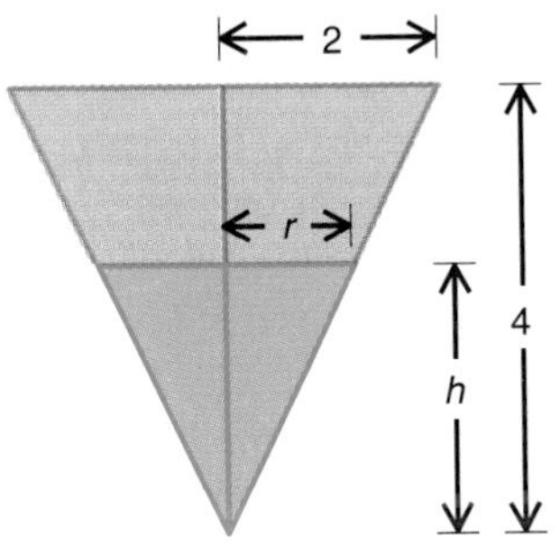

Figure 6.6.5

Gravitational Force and Work

Near the surface of the earth, the force of gravity is related to the mass of an object in a simple way. The force exerted by gravity on a mass of 1 kg is 9.80 N; the force exerted by gravity on a mass of 1 lb is 1 lb. In other words, the force of gravity is a constant times the mass.

For an object that is not necessarily very close to the earth's surface, the force exerted by gravity is $\frac{Km}{r^2}$, where r is the distance from the center of the earth, m is the mass of the object, and K is some constant (see Figure 6.6.6). To find K, we compare this formula with what we know at the earth's surface.

For example, if r is measured in miles and m in pounds, we should have that $\frac{Km}{r^2} = m$ when the object is at the earth's surface (where its weight is m lb). At the earth's surface, r is the earth's radius, or about 4000 miles. So $K \approx 1.6 \times 10^7$ mi^2.

If r is measured in kilometers and m in kilograms, then at the earth's surface, $r \approx 6400$ and $\frac{Km}{(6400)^2} = 9.8m$ Newtons. So $K \approx 9.8(6400)^2$ N km^2 kg $\approx 4.0 \times 10^8$ N km^2/kg.

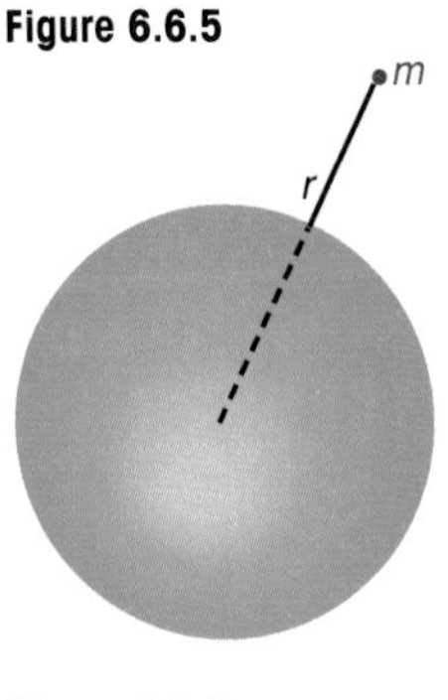

Figure 6.6.6

EXAMPLE 5 **Satellite Launch** Find the energy required to lift a 100-lb satellite to a point 1000 mi above the earth's surface, ignoring the energy needed to lift the rocket and its fuel.

Solution The satellite rises from the earth's surface (where $r = 4000$ mi) to its destination at $r = 5000$ mi. At any particular r, the force of gravity is $F(r) = \frac{Km}{r^2} = \frac{1.6\times 10^7\times 100}{r^2}$ lb, and the work done is

$$\begin{aligned}\int_{4000}^{5000} F(r)\,dr &= \int_{4000}^{5000} 1.6\times 10^9 r^{-2}\,dr \\ &= 1.6\times 10^9\left(-\frac{1}{r}\right)\Bigg|_{4000}^{5000} = 1.6\times 10^9\left(-\frac{1}{5000}+\frac{1}{4000}\right) \\ &= 8.0\times 10^4.\end{aligned}$$

Since the force was measured in pounds and distances were in miles, the work just calculated is in "mile-pounds." Since 1 mile = 5280 ft, 1 mile-pound = 5280 ft-lb, and the work done is $8.0 \times 10^4 \times 5280 \approx 4.2 \times 10^8 = 420{,}000{,}000$ ft-lb.

Since an actual launch involves lifting the booster rocket and its fuel, which weigh many times as much as the payload, and also usually imparts kinetic energy to the satellite, it actually requires far more energy than what we have calculated to launch a satellite.

Exercises 6.6

I

In each of the following situations, calculate the work done by the force $F(s)$ in moving an object the specified displacement.

1. $F(s) = 3s^2$ lb, moving from $s = 0$ ft to $s = 3$ ft.
2. $F(s) = 12\left(1 - \frac{s}{2}\right)$ lb, moving from $s = 0$ ft to $s = 2$ ft.
3. $F(s) = 3\sin(\pi s)$ lb, moving from $s = 0$ ft to $s = 3$ ft.
4. $F(s) = s\sin(\pi s^2)$ N, moving from $s = 0$ m to $s = 2$ m.
5. $F(s) = \sin(\pi s)\cos(\pi s)$ N, moving from $s = -2$ m to $s = 2$ m.
6. $F(s) = \sin^3 s\cos^4 s$ lb, moving from $s = -4$ ft to $s = 4$ ft.
7. $F(s) = 6$ lb, moving from $s = 1$ ft to $s = 6$ ft.
8. $F(s) = -7$ N, moving from $s = 2$ m to $s = -3$ m.
9. $F(s) = 4s^2$ lb, moving from $s = 2$ ft to $s = -4$ ft.
10. $F(s) = 5s^3$ N, moving from $s = -1$ m to $s = -5$ m.

The following questions concern a spring. The first information tells how far some particular force will stretch the spring from its rest position. Use this to calculate the work done by a person moving the spring between the specified displacements. (All displacements are measured relative to the rest position.)

11. Stretched 1 ft with a force of 2 lb; moved from $s = -1$ ft to $s = 2$ ft.
12. Stretched 1 in with a force of 1 lb; moved from $s = 0$ ft to $s = 3$ ft.
13. Stretched 0.1 m with a force of 1 N; moved from $s = -0.2$ m to $s = 0.3$ m.
14. Stretched 20 cm with a force of 3 N; moved from $s = 0$ cm to $s = 45$ cm.
15. Stretched 1 ft with a force of 30 lb; moved from $s = 1$ ft to $s = -2$ ft.
16. Stretched 1 mm with a force of 0.2 N; moved from $s = -1$ cm to $s = 1$ cm.
17. Stretched 0.1 in with a force of 3 oz; moved from $s = -0.4$ in to $s = -0.5$ in.
18. Stretched 1 in with a force of 30 lb; moved from $s = 0$ ft to $s = 0.1$ ft.
19. Stretched 1 ft with a force of 240 lb; moved from $s = -2$ in to $s = 3$ in.
20. Stretched 1 cm with a force of 2 N; moved from $s = -1$ in to $s = 2$ in.

II

21. A spherical tank of radius 5 ft is resting with its base at ground level. How much work is done in pumping

it full of water if the water comes from a reservoir at ground level? (One cubic foot of water weighs 62.4 lb.)

22. Calculate the work done in pumping all the water out of a full spherical tank of radius 2 m and up to a point 3 m above the top of the tank.

23. A conical tank (with the vertex at the bottom) has height 6 ft, and the radius at the top is 2 ft. The tank has water in it to a depth of 2 ft. Calculate the work done in filling it to a depth of 5 ft, pumping the water up from a river that is 3 ft below the bottom of the tank.

***24.** Suppose a spherical tank of radius R ft is full of water. Show that the work done in pumping all the water out of the tank to an outlet valve located right at the top of the tank is exactly the same as the work done in raising the same amount of water (i.e., the amount that would fill the tank) a height of R ft.

***25.** Suppose a spring exerts a force of 4 lb when it is extended 1 inch. How far would it be extended if, starting from the rest position, 12 ft-lb of work was done in extending it?

***26.** (i) Suppose a spring starts at the rest position, and by doing 1 ft-lb of work it is possible to extend it 1 ft. How much farther would it extend if somebody performed an additional 1 ft-lb of work in extending it?

(ii) How much work would it take to stretch it from an extension of 1 ft to an extension of 2 ft?

***27.** Suppose it takes 8 ft-lb of work to extend a spring 1 ft from its rest position.

(i) What force will the spring exert when it is extended 1 ft?

(ii) How much work will it take to extend it an additional foot (i.e., to an extension of 2 ft)?

***28.** A spring is compressed 3 in from its rest position. When it is released, it does 6 ft-lb of work pushing against a door until it reaches its rest position.

(i) How much force will this spring exert when extended 4 in?

(ii) How much work will it take to move the spring from its first position, compressed 3 in, to the position where it is 4 in extended?

***29.** (i) Figure out how much work it would take to lift a 1-lb weight to a height h ft above the surface of the earth, using the formula $F = \frac{Km}{r^2}$ for the force due to gravity.

(ii) To send a spaceship far off into space, it is necessary to lift it free of the earth's gravitational field. This amounts to giving it enough energy so that it will not eventually fall back to earth. Find out how much work is needed to lift a 1-lb object free of the earth's gravity by letting $h \to \infty$ in your answer to part (i).

***30.** Consider a spring that exerts a force of 10 lb when it is extended 1 ft from the rest position.

(i) Suppose the spring is extended 1 ft from the rest position, and a person moves it until it is *compressed* 1 ft from the rest position. Show that the work done by the spring in the course of this movement equals zero.

(ii) Explain why no work is done in this situation.

***31.** We have two different formulas for the force due to gravity near the earth's surface. On the one hand, the simple approximate version is that an object whose mass is m lb experiences a force of m lb due to gravity. On the other hand, we have seen that the force on an object of mass m is actually $\frac{Km}{r^2}$, if r is the distance from the object to the center of the earth. These two formulas are different, but for objects near the earth's surface they give very similar results.

(i) If an object weighs 1 lb at the earth's surface, how much does it weigh 1 mile above the earth's surface?

(ii) If an object weighs 1 lb at the earth's surface, how much does it weigh 1 ft above the earth's surface?

***32.** In this question we compare the calculation of the work done in lifting a payload depending on whether we use the actual formula for the force due to gravity or the approximate formula (cf. Exercise 31). (i) Calculate the work done in lifting a load weighing 1 lb at the earth's surface to a height of 1 mile above the earth's surface. Do the calculation using both the correct formula for the force due to gravity and the approximation, and compare the answers.

Repeat part (i), this time lifting the load to a height of (ii) 10 miles, (iii) 100 miles, (iv) 10,000 miles.

***33.** If an object weighing m lb is moving at a speed of v ft/sec, then its kinetic energy *in foot-pounds* is $\frac{1}{64}mv^2$. (If you have studied kinetic energy in physics, you have probably learned a formula for kinetic energy that looks different. The reason is that there is a conversion factor because we have measured mass in pounds and energy in ft-lb, a unit that is well suited to work performed by a force but poorly adapted to kinetic energy.)

(i) If a force of 3 lb acts on an object weighing 4 lb and pushes it 6 ft, how much work does it perform?

(ii) If the object in part (i) began at rest and all the work is converted into kinetic energy, how fast will the object be moving when the force stops?

***34.** (i) Suppose an object weighing m lb is lifted to a height of h ft above ground level. What is its potential energy (i.e., the work done to it in the process of lifting)? (Use the approximate formula for the force due to gravity.)

(ii) Suppose the object in part (i) drops to the ground. At the moment of impact, all the potential energy it used to have has been converted into kinetic energy. Use this fact and the information in Exercise 33 to calculate the speed of impact.

(iii) From your answer to part (ii), conclude that Galileo was right: The speed of impact does not depend on the weight of the object, that is, heavy objects fall no faster than light ones.

(iv) Check that your answer to part (ii) agrees with the answer that could be obtained by using Remark 4.9.1 (p. 316).

*35. A rope is 10 feet long and weighs 1 lb. It is lying coiled on the ground, and somebody comes and lifts one end straight up (see Figure 6.6.7). Find the work done in lifting the end of the rope to a height of (i) 5 ft, (ii) 15 ft.

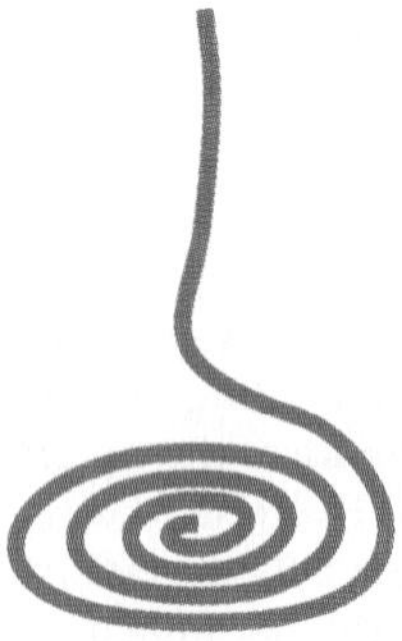

Figure 6.6.7

*36. Water leaks out of a bucket at a constant rate of $\frac{1}{5}$ liter/sec. Starting at ground level, the bucket is lifted at a constant rate of 1 m every 3 sec. The bucket contains 6 liters of water at the beginning, and the bucket itself weighs 1 kg. Find the work that is performed in lifting the bucket to a height of (i) 5 m, (ii) 10 m, (iii) 15 m.

*37. The bucket from Exercise 36 starts with 5 liters of water and is lifted at a rate of 1 m every 4 sec. The lifting force performs work of 150 J. How far has the bucket risen?

*38. A force pushing a vehicle performs work. Suppose the force is doubled; presumably, the speed will increase. For the sake of example, let us assume that the speed doubles too. Then in 1 second the object will travel twice as far and experience twice as much force, so the work done on it will be *quadrupled.* Since it has gone twice as far, the energy required to make it go a fixed distance has *doubled.* This rough idea is behind the familiar notion that it is less fuel-efficient to drive a car very fast.

If the propelling force doubles, it is actually unlikely that the speed will double too. Suppose that the speed is proportional to the cube root of the force. What will be the effect on the energy required to drive a fixed distance?

III

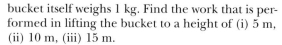

39. Water is pumped into an empty spherical tank of radius R from the level of the bottom of the tank.

(i) Find a formula for the work done in pumping in water to a depth d above the bottom.

(ii) Suppose the radius is 2 m and the pumping has used 200,000 J of energy. Use a graphics calculator or computer to find the depth to which the tank has been filled.

6.7 Moments and Centroids

FULCRUM
CENTER OF MASS
UNIFORM DENSITY
SYMMETRY

If you take a thin disk of metal or cardboard, it is possible to balance it on a sharp point held exactly at its center, as jugglers sometimes do with plates (see Figure 6.7.1).

If you take a thin flat object of a more irregular shape, as shown in Figure 6.7.2, it should still be possible to do this, but it will be harder to find the point at which

Figure 6.7.1

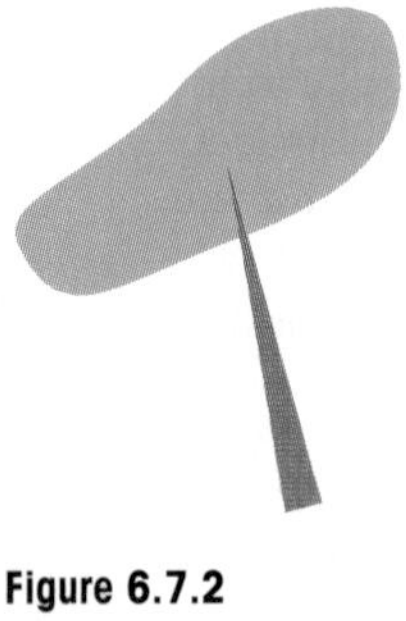

Figure 6.7.2

it will balance, the so-called *centroid* or *center of mass*, or *center of gravity*. It may be possible to find the centroid by trial and error, but the purpose of this section is to learn how to calculate it.

The results we find here will be used in the later parts of the next section, Section 6.8, on fluid pressure.

Perhaps the simplest example of balancing is the seesaw. We are all familiar with the phenomenon of people of different weights being able to balance by sitting at different distances from the support or **fulcrum**.

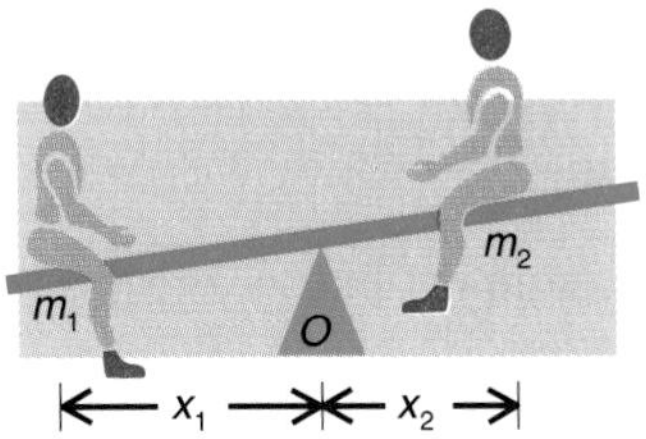

Figure 6.7.3

Let x_1 and x_2 be the displacements of two people from the fulcrum O and write m_1 and m_2 for their respective weights, as in Figure 6.7.3.

A weight on the seesaw will tend to make the seesaw rotate about O. The heavier the weight m or the greater its displacement x from O, the greater will be the tendency to turn. This suggests that the tendency to turn depends on the product mx.

The quantity $m_i x_i$ represents the tendency of the ith person to make the seesaw rotate clockwise about O. (If the quantity is negative, it represents a tendency to rotate counterclockwise.) The total tendency to rotate caused by both persons is $m_1x_1 + m_2x_2$.

The seesaw will balance when there is no net tendency to turn, that is, when

$$m_1x_1 + m_2x_2 = 0.$$

Notice that since the people are on opposite sides of the fulcrum, one displacement is positive and the other is negative.

This formula reflects the familiar fact that a heavy person should sit closer to the fulcrum than a light person in order to balance. (For example, if the second person is twice as heavy as the first, then $m_2 = 2m_1$, so to balance, $0 = m_1x_1 + m_2x_2 = m_1x_1 + 2m_1x_2 = m_1(x_1 + 2x_2)$. This means that $x_2 = -\frac{1}{2}x_1$; the heavy person sits half as far from O as the light person.

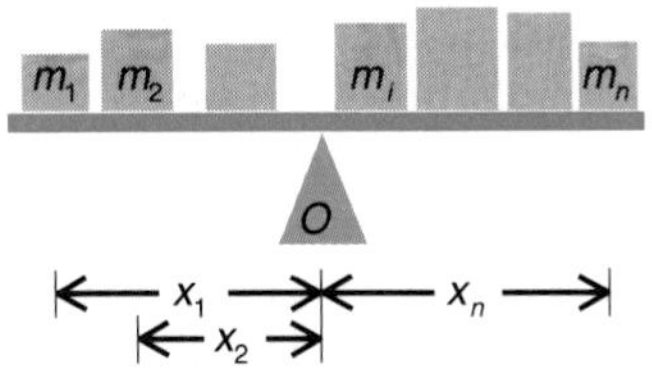

Figure 6.7.4(i)

DEFINITION 6.7.1

If an object of weight m is at a horizontal displacement x from O, the quantity

$$mx$$

is called the **moment** of the object **about the point** O.

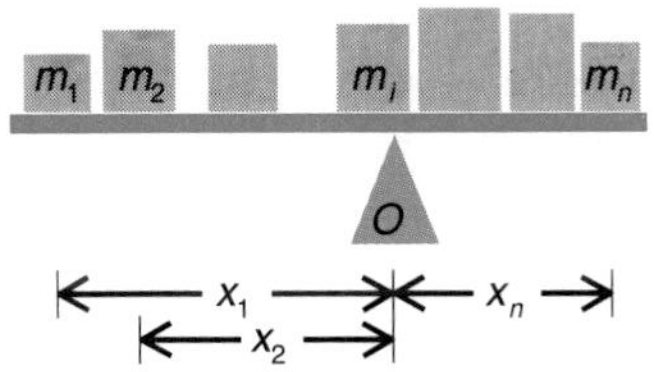

Figure 6.7.4(ii)

Suppose objects weighing $m_1, m_2, \ldots, m_n$ are placed on a seesaw supported at O, and let x_i be the displacement (positive or negative) of the ith object from O (see Figure 6.7.4(i)). The seesaw will balance when the moments cancel, that is, when the total moment is zero (see Figure 6.7.4(ii)):

$$\sum_{i=1}^{n} m_i x_i = 0.$$

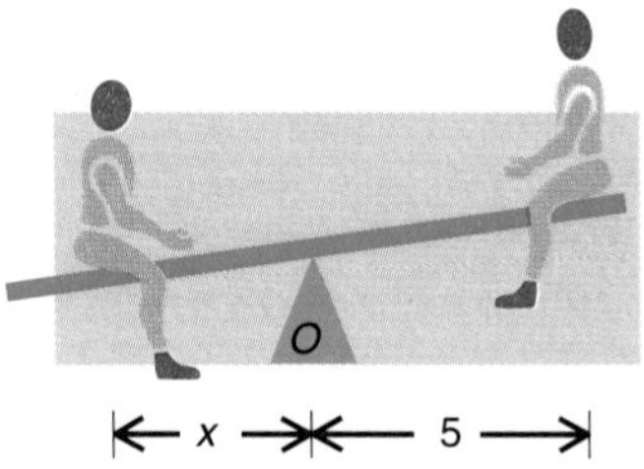

Figure 6.7.5

EXAMPLE 1 **Seesaw** If one person weighing 40 lb sits 5 ft from the fulcrum of a seesaw, where will a person weighing 60 lb have to sit for them to balance?

Solution The situation is illustrated in Figure 6.7.5, where we have written x for the displacement from the 60-lb person to the fulcrum. The 40-lb person has a displacement of 5 ft from the fulcrum. The total moment is

$$40(5) + 60x = 200 + 60x.$$

The seesaw will balance when this is zero, that is, when $x = -\frac{200}{60} = -\frac{10}{3}$. The negative sign indicates that this displacement is in the direction opposite to the displacement of the 40-lb person. To balance the seesaw, the 60-lb person must sit $3\frac{1}{3}$ ft from the fulcrum, on the side opposite the other person.

Notice that we ignored the weight of the seesaw itself, presuming that it balances when nobody is on it, so that it contributes nothing to the total moment about the fulcrum.

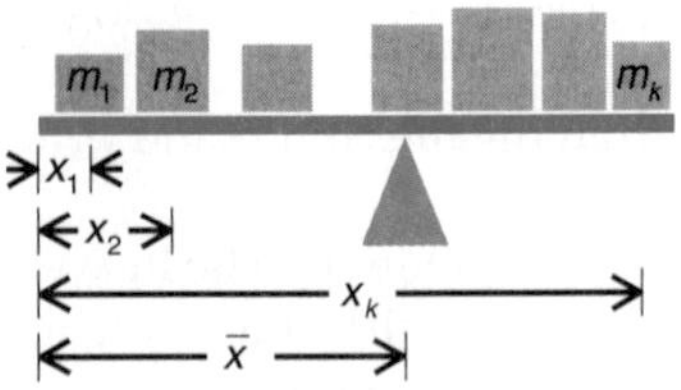

Figure 6.7.6

Now suppose objects with weights $m_1, m_2, \ldots, m_k$ are positioned at displacements $x_1, x_2, \ldots, x_k$ along a thin rod. Let us try to find the point $\bar{x}$ at which the rod can be balanced (see Figure 6.7.6).

If the rod is placed on a pointed support at $\bar{x}$, the support exerts an *upward* force equal to the total weight $M = m_1 + \ldots + m_k$. For the rod to balance, the corresponding moment must cancel the sum of the moments due to the little weights. Since the force of the support is in the direction opposite to the weights, it is $-M$, and we find

$$(-M)\bar{x} + m_1x_1 + \ldots + m_kx_k = 0,$$

or

$$M\bar{x} = m_1x_1 + \ldots + m_kx_k,$$

so

$$\bar{x} = \frac{m_1x_1 + \ldots + m_kx_k}{M} = \frac{m_1x_1 + \ldots + m_kx_k}{m_1 + \ldots + m_k}.$$

Figure 6.7.7

Now suppose the rod with k weights is replaced by a more realistic thin rod whose weight is distributed continuously along its length (see Figure 6.7.7). We could think of this in terms of letting k, the number of weights, become larger and larger.

If we use the variable x to describe displacement along the rod, with $x \in [a, b]$, then the distribution of weight is conveniently described by a density function, usually denoted $\rho(x)$ (ρ is the Greek letter "rho"). To say that $\rho(x)$ is the density function means that a small piece of the rod centered at x and having length dx has weight $\rho(x)\,dx$ (see Figure 6.7.8). Another way of thinking of it is that the density $\rho(x)$ tells the weight per unit *length* at the point x.

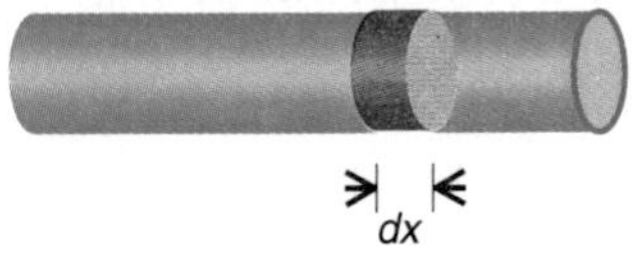

Figure 6.7.8

> You may be accustomed to using the word "density" in physics to mean the weight per unit *volume*. For a thin rod we measure only one dimension, the length, so we express the density as weight per unit length. In everyday speech we might say that some electrical wire weighs 3 oz per foot or that kite string weights 1 oz per hundred feet. . .

If the weight of a little piece of length dx is $\rho(x)\,dx$, then the total weight M of the rod is found by "adding up" all these little weights; it is $M = \int_a^b \rho(x)\,dx$. The little piece of length dx centered at x has weight $\rho(x)\,dx$, so its moment about the origin is equal to $x\rho(x)\,dx$. The total moment about the origin of the whole rod is the sum of these moments, or $\int_a^b x\rho(x)\,dx$.

For the rod to balance at a point $\bar{x}$, the moment of the rod must equal the moment obtained by placing the total weight M at $\bar{x}$; that is, $\int_a^b x\rho(x)\,dx = M\bar{x} = \bar{x}\int_a^b \rho(x)\,dx$. Dividing, we find

$$\bar{x} = \frac{1}{M}\int_a^b x\rho(x)\,dx = \frac{\int_a^b x\rho(x)\,dx}{\int_a^b \rho(x)\,dx}. \tag{6.7.2}$$

This formula gives the location of the point at which the object balances, which is called the **center of mass**. By far the most common situation is the one in which the density is constant, that is, the rod has **uniform density**. In this case the center of mass is often called the **centroid**.

EXAMPLE 2 Consider a rod of uniform density whose length is L meters. Find its centroid.

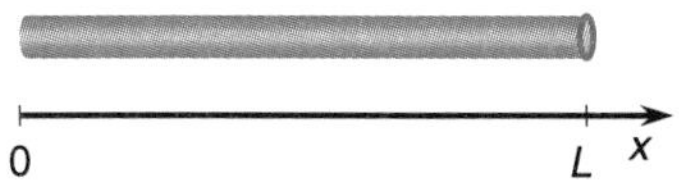

Figure 6.7.9

Solution Let us place the rod with one end at the origin and the rest of it lying along the positive x-axis. The variable x will run from $x = 0$ to $x = L$, and the density function is a constant, $\rho(x) = c$, for some number c. See Figure 6.7.9.

To find the centroid, we need to know the total weight $M = \int_0^L \rho(x)\,dx = \int_0^L c\,dx = c(L-0) = cL$. The total moment about the origin is $\int_0^L x\rho(x)\,dx = \int_0^L cx\,dx = \frac{c}{2}x^2\Big|_0^L = \frac{cL^2}{2}$.

Using the formula above, we find that the centroid is

$$\bar{x} = \frac{\int_0^L x\rho(x)\,dx}{\int_0^L \rho(x)\,dx} = \frac{\frac{cL^2}{2}}{cL} = \frac{L}{2}.$$

The centroid of the rod is $\frac{L}{2}$ meters from the end, which is the same thing as saying the centroid is precisely in the middle, which is hardly surprising.

Notice that the constant c does not appear in the answer. As long as the density is uniform, its exact value is irrelevant to the location of the centroid.

EXAMPLE 3 Consider a rod that is 3 ft long whose density is proportional to the distance from one of the ends. Find its center of mass.

Solution Again we place the rod on the x-axis with one end at the origin and the other on the positive axis. The variable x ranges from $x = 0$ to $x = 3$.

The density is proportional to the distance from the origin, say, which means that it equals a constant times x. We write $\rho(x) = cx$, for some constant c (whose value we do not know). The total weight is $\int_0^3 \rho(x)\,dx = \int_0^3 cx\,dx = \frac{c}{2}x^2\Big|_0^3 = \frac{9c}{2}$.

The total moment about the origin is $\int_0^3 x\rho(x)\,dx = \int_0^3 cx^2\,dx = \frac{c}{3}x^3\Big|_0^3 = 9c$. Using the formula for the center of mass, we find that

$$\bar{x} = \frac{\int_0^3 x\rho(x)\,dx}{\int_0^3 \rho(x)\,dx} = \frac{9c}{\frac{9c}{2}} = 2.$$

The center of mass is 2 ft from the end at the origin, that is, the end at which the density is lower.

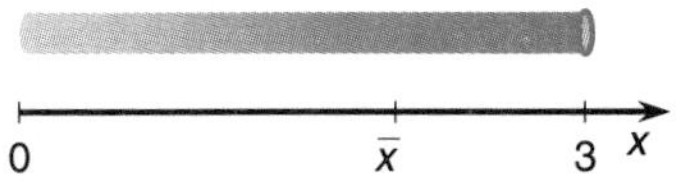

Figure 6.7.10

It is worth noticing that the center of mass is two-thirds of the way from the light end to the heavy end. It makes sense that it ought to be close to where the mass is concentrated, that is, closer to the heavy end. See Figure 6.7.10.

Also remark that the constant c again did not appear in the answer. This is not surprising if we consider what will happen to the quotient

$$\bar{x} = \frac{\int_a^b x\rho(x)\,dx}{\int_a^b \rho(x)\,dx}$$

if we multiply the density $\rho(x)$ by some constant k. Both the numerator and the denominator will be multiplied by k, so the quotient will remain the same. In other words, multiplying the density by a constant does not affect the location of the center of mass. In particular, we can assume that the constant c used in either of the preceding examples equals 1 and we will still get the correct answer. This observation is worth recording.

REMARK 6.7.3

If we are considering a rod of uniform density, we may as well assume that the density is the constant function $\rho(x) = 1$. In this case the centroid is

$$\bar{x} = \frac{\int_a^b x\,dx}{\int_a^b 1\,dx} = \frac{\int_a^b x\,dx}{b-a} = \frac{1}{b-a}\int_a^b x\,dx.$$

This is exactly the same thing as the *average value* of the function x over the interval $[a, b]$ (cf. Definition 6.5.1, p. 447), so it is not surprising that it turns out to be the value of x at the middle of the rod.

If the density $\rho(x)$ is not a constant, we can still think of the center of mass as a sort of "weighted average" of the function x; we count more heavily the values of x where the rod is heavier and divide by the total weighting. So $\bar{x}$ is a sort of average value of x over the whole rod, taking its density into account.

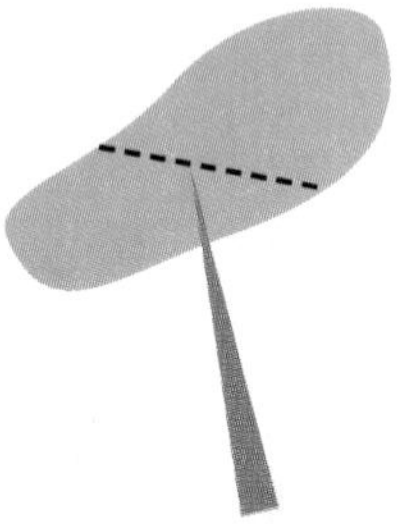

Figure 6.7.11

Next we want to consider the center of mass of a region in the plane. If we attempt to balance a flat object on a point but make a poor choice of point, what will happen is that the object will start to tilt and eventually fall off. When it is tilting, what it is really doing is *rotating* about some line (see Figure 6.7.11). Since moments have to do with an object's tendency to rotate, we should look for a kind of moment having to do with rotation about a line.

Consider the thin flat plate in Figure 6.7.12, which can rotate freely about the y-axis. A small object of weight m is placed on the plate. What will be the tendency of this weight to make the plate rotate about the axis?

It seems reasonable to consider the product of the weight m and some displacement. Rather than the displacement from the origin, it makes more sense to consider the displacement from the axis (after all, moving the object up or down parallel to the y-axis should not affect its tendency to rotate the plate). If the object is at the point (x, y) in the plane, then its displacement from the y-axis is simply x. This suggests that the object's moment about the y-axis ought to be mx. Similar considerations apply to moments about the x-axis.

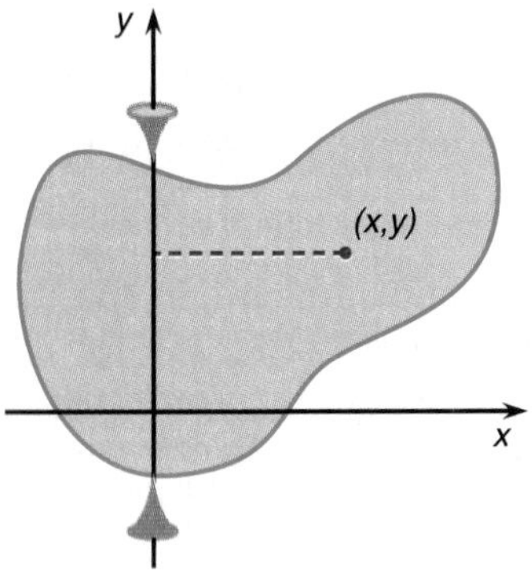

Figure 6.7.12

DEFINITION 6.7.4

An object of weight m positioned at the point (x, y) has its moment about the x-axis defined by

$$M_x = my$$

and its moment about the y-axis defined by

$$M_y = mx.$$

Notice that M_x, the moment about the x-axis, involves y, and M_y, the moment about the y-axis, involves x. This is because the moments involve the displacement away from the axis rather than a displacement *along* it.

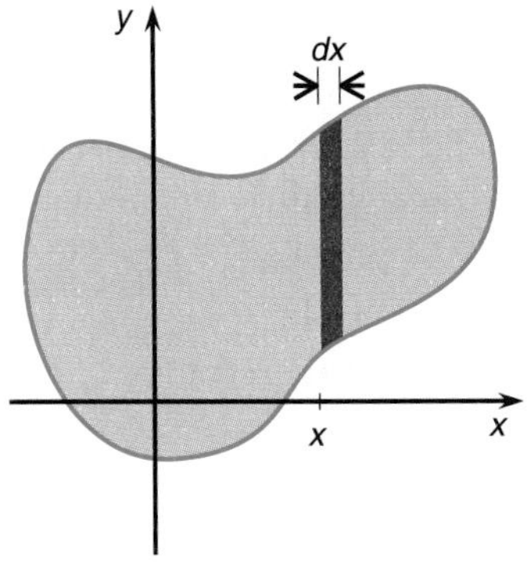

Figure 6.7.13

Now suppose we have a flat object in the shape of some region in the plane, with some specified density. Then, as before, we want to find the total moment of the object about each of the axes and use them to find the center of mass. The most obvious case is that of a flat plate of uniform density.

To find the total moment M_y about the y-axis, we consider the narrow strip in the region of width dx and at a displacement x from the y-axis. Its area is dx times its height, and its weight is $(area) \times (density)$. (Recall that we are assuming that the density equals the constant ρ.) See Figure 6.7.13.

The moment of this strip is $x \times (weight) = x(area)\rho = x(height)\rho\, dx$, and we integrate to get the total moment M_y. The total weight M will be the integral of the weights of all these strips, that is, the integral of $(height)\rho\, dx$.

In particular, if the region is the region lying between two curves $y = f(x)$ and $y = g(x)$, say, for $x \in [a, b]$, and $f(x) \geq g(x)$ for all $x \in [a, b]$, then the height of the strip at a displacement x from the y-axis is $f(x) - g(x)$. See Figure 6.7.14.

Similar remarks apply to the moment about the x-axis. See Figure 6.7.15.

DEFINITION 6.7.5

Suppose $f(x) \geq g(x)$, for all $x \in [a, b]$. Consider the region lying between the curves $y = f(x)$ and $y = g(x)$ and between the lines $x = a$ and $x = b$, with uniform density ρ.

The moment M_y of this region about the y-axis and the total weight M are given as follows:

$$M_y = \int_a^b x\big(f(x) - g(x)\big)\rho\, dx, \qquad M = \int_a^b \big(f(x) - g(x)\big)\rho\, dx.$$

Similarly, for a region lying between $y = c$ and $y = d$ and between the curves $x = h(y)$ and $x = \ell(y)$ with $h(y) \geq \ell(y)$ for all $y \in [c, d]$, if the density is ρ, then

$$M_x = \int_c^d y\big(h(y) - \ell(y)\big)\rho\, dy, \qquad M = \int_c^d \big(h(y) - \ell(y)\big)\rho\, dy.$$

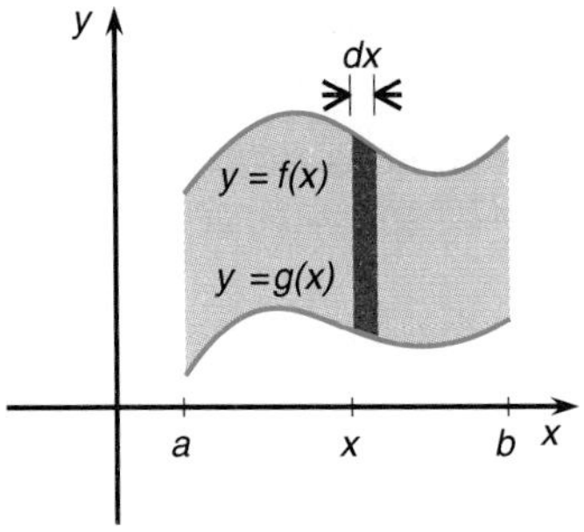

Figure 6.7.14

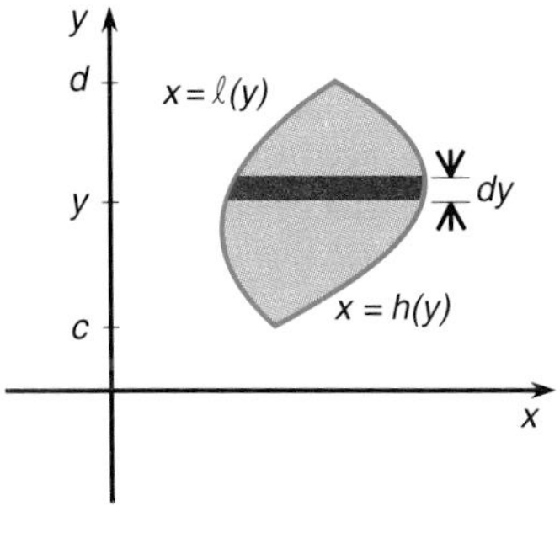

Figure 6.7.15

The centroid will have the property that the total weight concentrated at the centroid will have the same moments about the axes as the original object. If we write $(\bar{x}, \bar{y})$ for the centroid, this means that

$$M_y = M\bar{x}, \qquad M_x = M\bar{y}.$$

These equations can be solved to find the centroid:

$$\bar{x} = \frac{M_y}{M}, \qquad \bar{y} = \frac{M_x}{M}. \tag{6.7.6}$$

EXAMPLE 4 **Centroid of Rectangle** Find the centroid of the rectangle of uniform density that lies between $(0, 0)$, $(0, 2)$, $(5, 0)$, and $(5, 2)$.

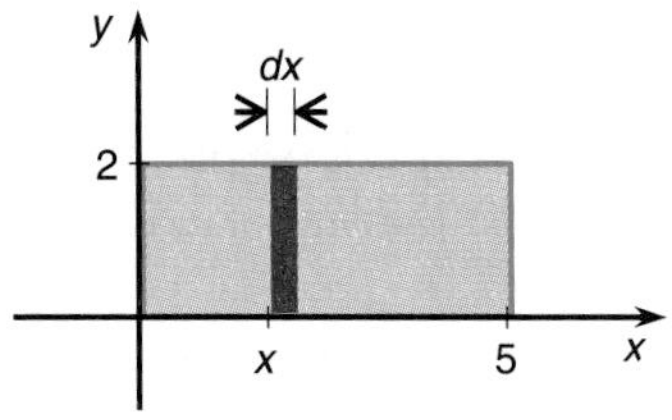

Figure 6.7.16

Solution First we observe that by analogy with Remark 6.7.3 we can assume that the density is the constant function 1. The rectangle lies between $y = f(x) = 2$ and $y = g(x) = 0$ and between the vertical lines $x = h(y) = 5$ and $x = \ell(y) = 0$. The height of the vertical strip of width dx that lies at a displacement x from the y-axis is $f(x) - g(x) = 2$ for each $x \in [0, 5]$, and $M_y = \int_0^5 x(height)\rho\, dx = \int_0^5 x(2)(1)\, dx = \int_0^5 2x\, dx = x^2\big|_0^5 = 25$. See Figure 6.7.16.

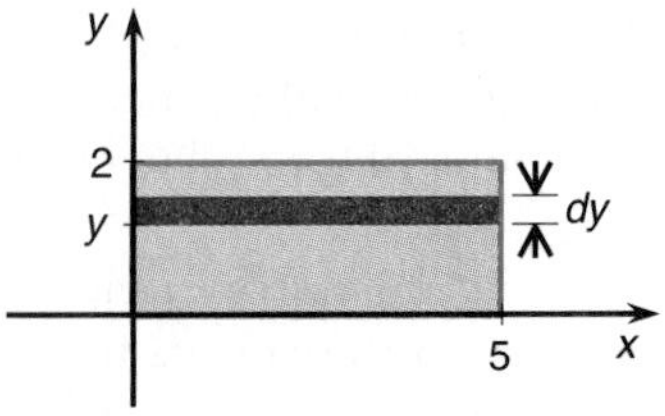

Figure 6.7.17

The total weight is $M = \int_0^5 (height)\rho\, dx = \int_0^5 2\, dx = 10$, so the x-coordinate $\bar{x}$ of the centroid is $\bar{x} = \frac{M_y}{M} = \frac{25}{10} = \frac{5}{2}$.

Similarly, to find the y-coordinate $\bar{y}$ of the centroid, we consider narrow horizontal strips of height dy (see Figure 6.7.17). Each such strip has width 5, and y ranges from $y = 0$ to $y = 2$, so $M_x = \int_0^2 y\big(h(y) - \ell(y)\big)\rho\, dy = \int_0^2 5y\, dy = \frac{5}{2}y^2\big|_0^2 = 10$. Since we have already found that $M = 10$, we find that $\bar{y} = \frac{M_x}{M} = \frac{10}{10} = 1$.

The centroid is $(\bar{x}, \bar{y}) = \left(\frac{5}{2}, 1\right)$, which is the exact center of the rectangle.

■ In fact it is easy to see that the centroid of any rectangle is its center.

For long thin objects the density represented the weight per unit length. Now we are considering thin flat objects, and it makes sense for "density" to refer to weight per unit area. In ordinary situations we might say that some new roofing material weighs 15 kg per square meter or that some sails are being made from special cloth that weighs only 2 lb per hundred square feet.

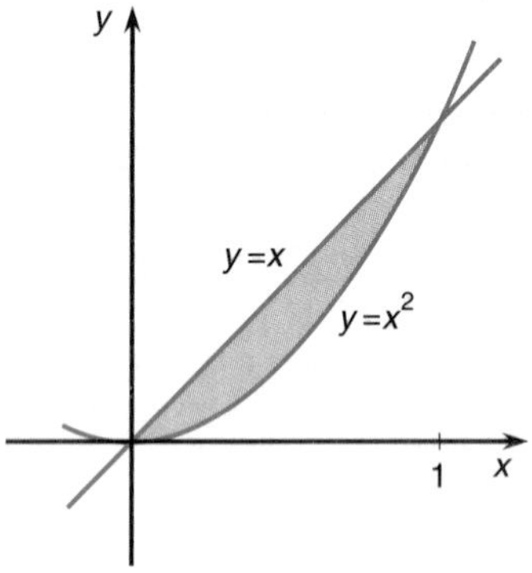

Figure 6.7.18

EXAMPLE 5 Find the centroid of the region with uniform density enclosed between the curves $y = x$ and $y = x^2$, as shown in Figure 6.7.18.

Solution The region lies between the graphs of the functions $y = f(x) = x$ and $y = g(x) = x^2$, with $f(x) \geq g(x)$ for all $x \in [0, 1]$. Letting $\rho = 1$, we find

$$M_y = \int_0^1 x\big(x - x^2\big)\rho\, dx = \int_0^1 (x^2 - x^3)\, dx = \left(\frac{1}{3}x^3 - \frac{1}{4}x^4\right)\bigg|_0^1$$
$$= \frac{1}{12}.$$

Also the total weight is $M = \int_0^1 (x - x^2)\rho\, dx = \left(\frac{1}{2}x^2 - \frac{1}{3}x^3\right)\big|_0^1 = \frac{1}{6}$.

Looking at the region the other way, with y as the variable, we see that it lies between $x = h(y) = \sqrt{y}$ and $x = \ell(y) = y$, with $h(y) \geq \ell(y)$ for all $y \in [0, 1]$. Consequently,

$$M_x = \int_0^1 y(\sqrt{y} - y)\rho\, dy = \int_0^1 (y^{3/2} - y^2)\, dy = \left(\frac{2}{5}y^{5/2} - \frac{1}{3}y^3\right)\bigg|_0^1$$
$$= \frac{1}{15}.$$

From all this we find $\bar{x} = \frac{M_y}{M} = \frac{1}{12}\Big/\frac{1}{6} = \frac{1}{2}$ and $\bar{y} = \frac{M_x}{M} = \frac{1}{15}\Big/\frac{1}{6} = \frac{2}{5}$. The centroid is the point $(\bar{x}, \bar{y}) = \left(\frac{1}{2}, \frac{2}{5}\right)$.

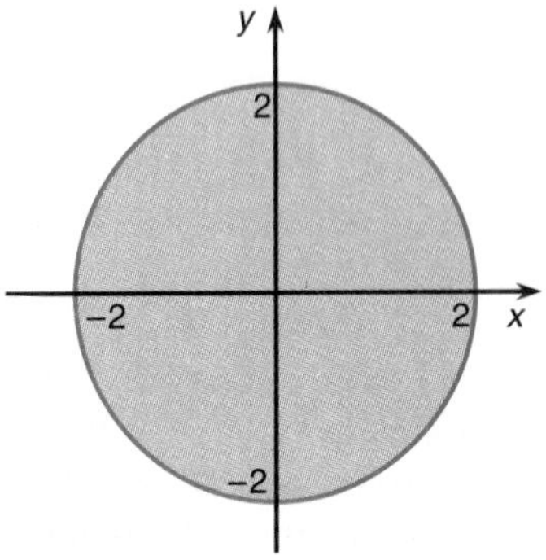

Figure 6.7.19

EXAMPLE 6 **Centroid of Disk** Find the centroid of a disk of radius 2 with uniform density.

Solution We may as well assume that the center of the disk is at the origin, so the circle enclosing the disk is $x^2 + y^2 = 4$ (see Figure 6.7.19). The disk is the region lying between $y = f(x) = \sqrt{4 - x^2}$ and $y = g(x) = -\sqrt{4 - x^2}$.

To find M_y, we consider a vertical strip of width dx at a displacement x from the y-axis. The height of the strip is $f(x) - g(x) = 2\sqrt{4 - x^2}$, and if the density is $\rho = 1$,

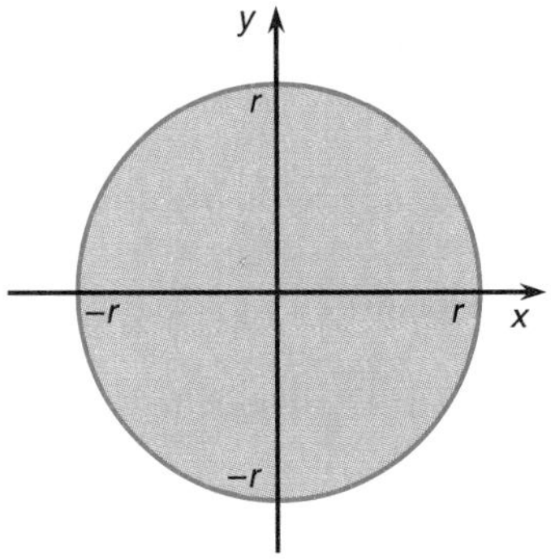

Figure 6.7.20

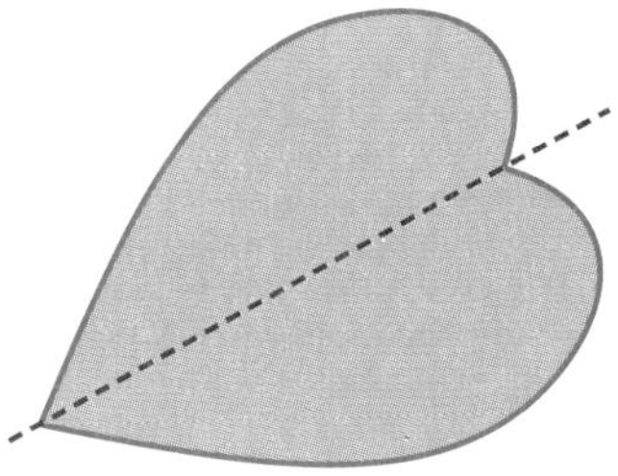

Figure 6.7.21

then

$$M_y = \int_{-2}^{2} x(1)\big(f(x) - g(x)\big)\,dx = \int_{-2}^{2} 2x\sqrt{4 - x^2}\,dx$$

$u = 4 - x^2$
$du = -2x\,dx$

$$= \int_{0}^{0} \sqrt{u}(-1)\,du = -\frac{2}{3}u^{3/2}\Big|_{0}^{0}$$
$$= 0.$$

Of course, the calculation for M_x is almost identical, and we find that $M_x = 0$. Without even calculating the total weight M we see that the centroid is $(\bar{x}, \bar{y}) = (0, 0)$. In other words, the centroid of the disk is its center.

Virtually the same calculation works for a disk of any radius, and we see that the centroid of any disk is its center.

In a way, if we think of the moments M_y and M_x as the *average values* of x and y, respectively, it seems obvious from Figure 6.7.20 that they should both equal zero. This is because the part of the disk to the left of the y-axis and the part to the right are symmetrical, so they contribute the same amount to the average, except that they have opposite signs (i.e., x is positive on the right side and negative on the left side). The two parts cancel, and the total moment is zero.

The same idea can be applied in other situations. If an object is symmetrical about some line, as in Figure 6.7.21, then it has no tendency to rotate one way or the other about that line, so its centroid, the point upon which it balances, must be on the line.

THEOREM 6.7.7

Suppose a thin plate with uniform density is symmetrical about some line L. Then the centroid lies on the line L.

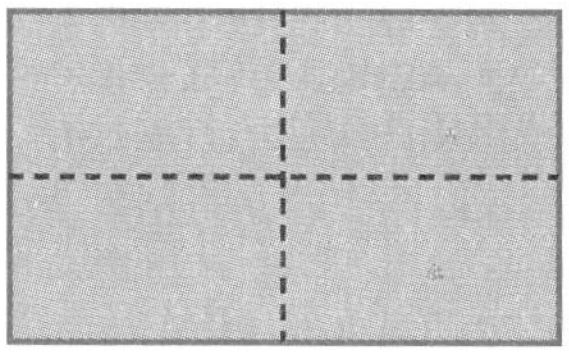

Figure 6.7.22

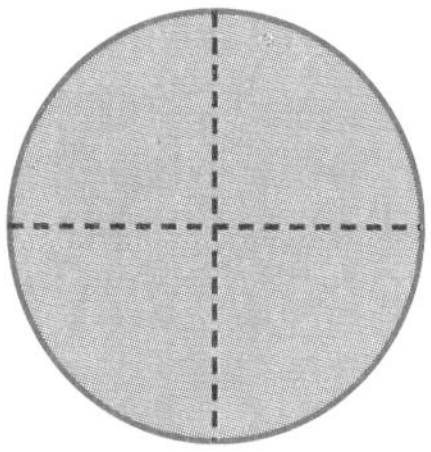

Figure 6.7.23

EXAMPLE 7 This observation immediately explains and justifies the remarks that the centroid of a rectangle is its center and that the centroid of a circle is its center. For example, a rectangle whose sides are parallel to the axes is symmetrical about the horizontal line through its center and about the vertical line through its center, so the centroid lies on both these lines (see Figure 6.7.22). This means that it must be the point of intersection of these lines, which is the center.

Similarly, a disk is symmetrical about *every* line through its center, so the centroid lies on all these lines. The only point on all these lines is the center (see Figure 6.7.23).

We can also see that the centroid of the region inside an ellipse is the center of the ellipse, since the ellipse is symmetrical about the major axis and the minor axis, which intersect at the center. See Figure 6.7.24.

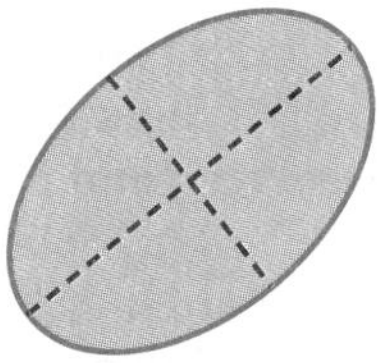

Figure 6.7.24

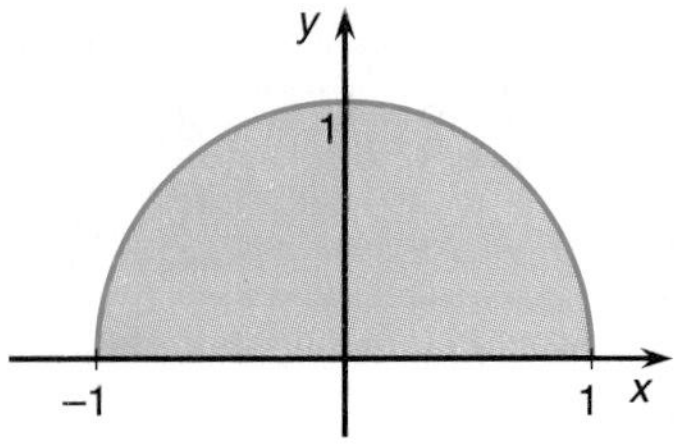

Figure 6.7.25

EXAMPLE 8 Centroid of Semidisk Find the centroid of the region inside the upper semicircle $y = \sqrt{1 - x^2}$, shown in Figure 6.7.25.

Solution When we are discussing the *centroid*, we assume that the region has uniform density, so we may as well let $\rho = 1$.

The region is certainly symmetric about the y-axis, so we know that the centroid lies on the y-axis, which means $\bar{x} = 0$.

To find $\bar{y}$, we think of the region as the region between the two sides of the semicircle, that is, the curves $x = h(y) = \sqrt{1 - y^2}$ and $x = \ell(y) = -\sqrt{1 - y^2}$, for $0 \leq y \leq 1$. So

$$M_x = \int_0^1 y\big(h(y) - \ell(y)\big)\rho\, dy = \int_0^1 2y\sqrt{1 - y^2}\, dy$$

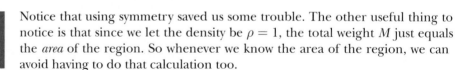

$u = 1 - y^2$
$du = -2y\, dy$

$$= -\int_1^0 \sqrt{u}\, du = \int_0^1 u^{1/2} du = \frac{2}{3}u^{3/2}\Big|_0^1$$

$$= \frac{2}{3}.$$

We also need to evaluate the total weight

$$M = \int_0^1 \big(h(y) - \ell(y)\big)\rho\, dy = \int_0^1 2\sqrt{1 - y^2}\, dy.$$

We could evaluate this integral by using Example 6.1.2, but we could also realize that it is precisely the area of the semicircle, so it equals $\frac{\pi}{2}$.

Now we can find $\bar{y} = \frac{M_x}{M} = \frac{2/3}{\pi/2} = \frac{4}{3\pi} \approx 0.4244$. The centroid of the semicircle is $(\bar{x}, \bar{y}) = \left(0, \frac{4}{3\pi}\right) \approx (0, 0.4244)$. See Figure 6.7.26.

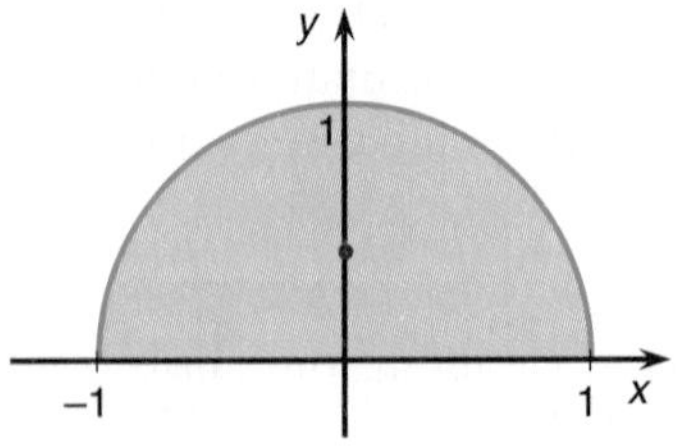

Figure 6.7.26

Notice that using symmetry saved us some trouble. The other useful thing to notice is that since we let the density be $\rho = 1$, the total weight M just equals the *area* of the region. So whenever we know the area of the region, we can avoid having to do that calculation too.

Having found the centroids of rectangles, circles, semicircles, and ellipses, we should ask about the centroid of a triangle.

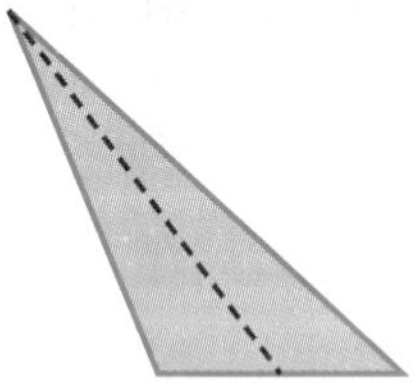

Figure 6.7.27

Consider the triangle shown in Figure 6.7.27. The dotted line is one of the *medians*, meaning the lines from one of the vertices to the middle of the opposite side. If we break up the triangle into narrow horizontal strips of height dy, as in Figure 6.7.28, then the median cuts each strip in half. The parts of the strip on either side of the median have moments about the median that cancel because they extend the same distance from the line but in opposite directions. This means that the moment of the strip about the median is zero. Each of these strips has no tendency to rotate about the line because it balances.

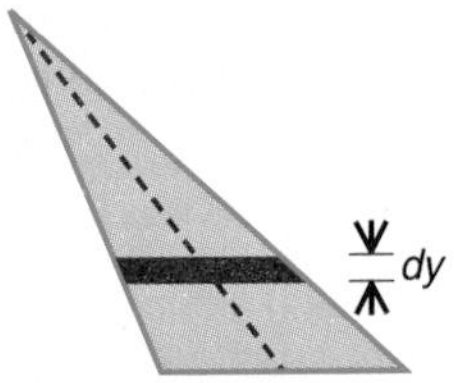

Figure 6.7.28

The total moment of the triangle about the median is the "sum" (i.e., the integral) of the moments of these little strips. Adding up zero many times gives zero, and we see that the total moment about the median is zero. (This argument is not exactly airtight, but it is essentially correct, and it is not terribly difficult to make it precise. One of the exercises suggests how to do it.)

As we observed above, just before Theorem 6.7.7, to say that a region has moment zero about some line means that the centroid must lie on that line. What we have just said indicates that the centroid of a triangle lies on the median. In fact, the same argument works for the other two medians, so the centroid is on each median.

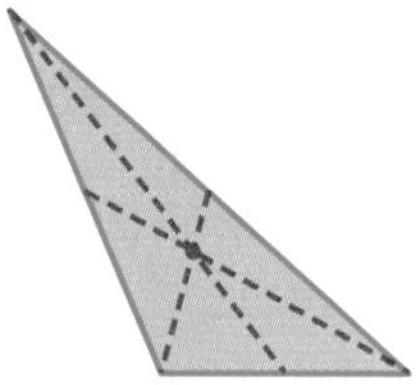

Figure 6.7.29

This means that the centroid of the triangle is the point where the medians intersect, shown in Figure 6.7.29.

Happily, this point is extremely easy to find. If the vertices of the triangle are (x_1, y_1), (x_2, y_2), and (x_3, y_3), then the centroid is obtained by "averaging" the coordinates of the vertices:

$$\bar{x} = \frac{x_1 + x_2 + x_3}{3}, \qquad \bar{y} = \frac{y_1 + y_2 + y_3}{3}. \tag{6.7.8}$$

EXAMPLE 9 **Centroid of Triangle** Find the centroid of the triangle whose vertices are $(1, 2)$, $(3, 5)$, and $(2, -4)$.

Solution We simply apply Equations 6.7.8:

$$\bar{x} = \frac{1 + 3 + 2}{3} = 2, \qquad \bar{y} = \frac{2 + 5 - 4}{3} = 1,$$

so the centroid is the point $(2, 1)$.

Frequently, a region is composed of simpler parts, and it is possible to find the centroid of the whole region from knowledge of the centroids and weights of the parts.

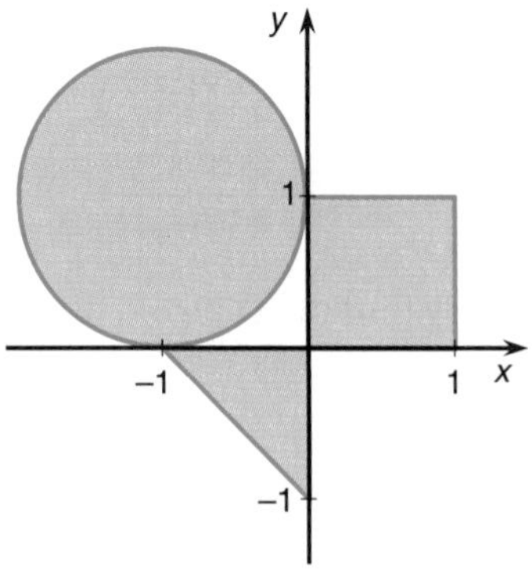

Figure 6.7.30

EXAMPLE 10 Consider a region that is composed of simpler regions. An example is illustrated in Figure 6.7.30. We know the centroids of the disk, the triangle, and the square without doing any work. We want to use this information to find the centroid of the whole shaded region.

Instead of the centroid, it is better to begin with the moments about the axes. First look at the square. Its centroid is, of course, its center, which is $\left(\frac{1}{2}, \frac{1}{2}\right)$. Letting the density be $\rho = 1$, we know that the weight of the square equals its area, which is 1. Since $\bar{x} = \frac{M_y}{M}$, we find that $M_y = M\bar{x} = 1\left(\frac{1}{2}\right) = \frac{1}{2}$. Similarly, the other moment of the square is $M_x = M\bar{y} = \frac{1}{2}$.

It is equally easy to find the moments of the triangle about the axes. By Equations 6.7.8, the centroid is $(\bar{x}, \bar{y}) = \left(-\frac{1}{3}, -\frac{1}{3}\right)$. Since M equals the area, which is $\frac{1}{2}$, we find that $M_x = M\bar{y} = \frac{1}{2}\left(-\frac{1}{3}\right) = -\frac{1}{6}$ and $M_y = M\bar{x} = -\frac{1}{6}$.

Finally, the radius of the circle is 1, so the area is $\pi r^2 = \pi$, so its moments are $M_x = M\bar{y} = \pi(1) = \pi$ and $M_y = M\bar{x} = \pi(-1) = -\pi$.

The moment of the whole region is the sum of the moments of the parts, that is, $M_x = \frac{1}{2} - \frac{1}{6} + \pi = \frac{1}{3} + \pi$ and $M_y = \frac{1}{2} - \frac{1}{6} - \pi = \frac{1}{3} - \pi$. The total weight of the whole region is the sum of the weights of the parts, so $M = 1 + \frac{1}{2} + \pi = \frac{3}{2} + \pi$.

Combining all this, we find the centroid of the whole region.

$$\bar{x} = \frac{M_y}{M} = \frac{\frac{1}{3} - \pi}{\frac{3}{2} + \pi} = \frac{2 - 6\pi}{9 + 6\pi}, \qquad \bar{y} = \frac{M_x}{M} = \frac{\frac{1}{3} + \pi}{\frac{3}{2} + \pi} = \frac{2 + 6\pi}{9 + 6\pi},$$

and the centroid is $\left(\frac{2-6\pi}{9+6\pi}, \frac{2+6\pi}{9+6\pi}\right) \approx (-0.605, 0.749)$.

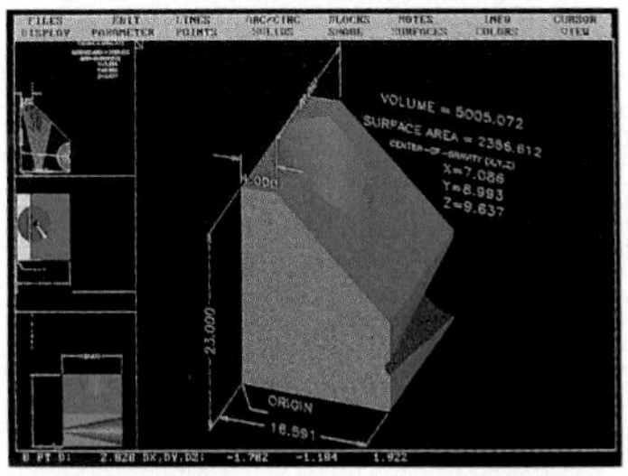

DesignCAD 3D *software was used to produce this image of a machine component. A designer can adjust the shape of the object and see the results on the screen. The computer calculates the center of gravity, which may affect the operation of the component.*

> Notice that we add up the moments M_x for all the parts of the region to find the corresponding moment for the whole region. It is *not* correct simply to average the coordinates of the centroids of the subregions.

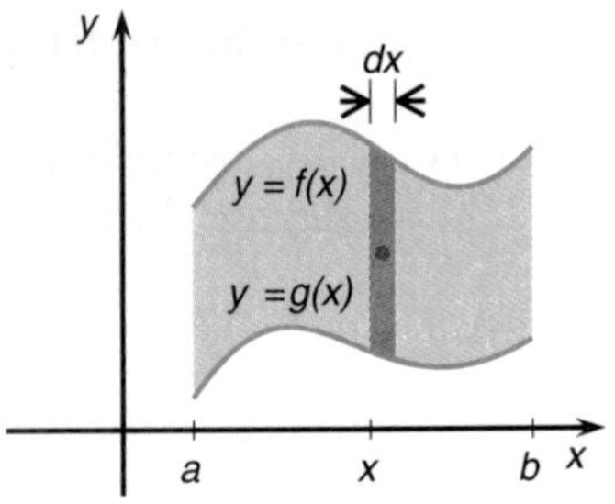

Figure 6.7.31

Returning to the question of calculating moments, we consider again the region between $y = f(x)$ and $y = g(x)$, shown in Figure 6.7.31. The centroid of a thin vertical strip will be at its center, that is, halfway up. Since the top of the strip is at a height $f(x)$ and the bottom at $g(x)$, the midpoint will be at height $\frac{1}{2}\big(f(x) + g(x)\big)$. The moment of the strip about the x-axis will be $\frac{1}{2}\big(f(x) + g(x)\big)\,(area)\rho$.

The area of the strip is $\big(f(x) - g(x)\big)\,dx$, so to find the total moment M_x about the x-axis, we "add up" the moments of these strips. In other words, we integrate: $M_x = \int_a^b \frac{1}{2}\big(f(x) + g(x)\big)\big(f(x) - g(x)\big)\rho\,dx = \frac{1}{2}\int_a^b \big(f(x)^2 - g(x)^2\big)\rho\,dx$. We summarize this result below.

THEOREM 6.7.9

Suppose $f(x) \geq g(x)$, for all $x \in [a, b]$. Consider the region lying between the curves $y = f(x)$ and $y = g(x)$ and between the lines $x = a$ and $x = b$, with uniform density ρ.

The moment M_x of this region about the x-axis is given as follows:

$$M_x = \frac{1}{2}\int_a^b \big(f(x)^2 - g(x)^2\big)\rho\,dx.$$

Similar remarks apply to the moment M_y of a region between two curves $x = h(y)$ and $x = \ell(y)$ (cf. Definition 6.7.5).

See the Graphics Calculator Example on the facing page.

EXAMPLE 11 **Centroid of Semidisk Revisited** We can now find the centroid of the upper semicircle more easily (cf. Example 8). The semicircle lies between $f(x) = \sqrt{1 - x^2}$ and $g(x) = 0$, so the formula gives

$$M_x = \frac{1}{2}\int_{-1}^{1} (1 - x^2 - 0)\,dx = \frac{1}{2}\left(x - \frac{x^3}{3}\right)\bigg|_{-1}^{1} = \frac{2}{3}.$$

This is certainly easier than the calculation we did in Example 8.

Suppose $f(x) \geq g(x) \geq 0$ on $[a, b]$, with $0 \leq a < b$, and let R be the region between their graphs. Consider the solid formed by revolving R around the y-axis, and let us find its volume. Using cylindrical shells, the volume is

$$Volume = \int_a^b 2\pi x\big(f(x) - g(x)\big)\,dx = 2\pi\int_a^b x\big(f(x) - g(x)\big)\,dx.$$

This last integral is the moment M_y of the region R about the y-axis, so we can say that the volume is $2\pi M_y$. (In talking about the moment, we are assuming that the density is $\rho = 1$.) Now if the centroid of R is $(\bar{x}, \bar{y})$, then $\bar{x} = \frac{M_y}{M}$, so the volume can be written as

$$Volume = 2\pi M_y = 2\pi M\bar{x}.$$

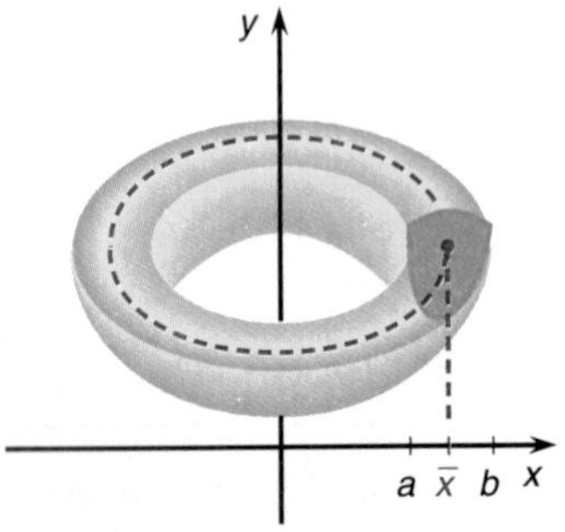

Figure 6.7.32

It is interesting to interpret this result geometrically. Since we have let the density be $\rho = 1$, the mass M is really just the area of R. And $2\pi\bar{x}$ is the circumference of a circle of radius $\bar{x}$. In Figure 6.7.32 we see the region R with its centroid marked, and we see how R is revolved around the y-axis. We also see how the centroid moves around in a circle whose radius is $\bar{x}$.

So $2\pi\bar{x}$ is the distance traveled by the centroid when we revolve R. We have just shown that for a region R lying between two graphs, the volume of the corresponding solid of revolution can be expressed in terms of the area of R and its centroid. This

With a graphics calculator we can plot the semicircle from Example 11, and plot the centroid

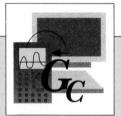

$$(\bar{x}, \bar{y}) = \left(\frac{M_y}{M}, \frac{M_x}{M}\right) = \left(0, \frac{2/3}{\pi/2}\right) = \left(0, \frac{4}{3\pi}\right).$$

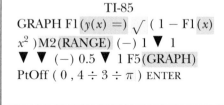

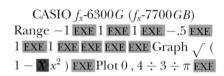

TI-85	SHARP EL9300
GRAPH F1 (y(x) =) √ (1 − F1 (x) x^2) M2 (RANGE) (−) 1 ▼ 1 ▼ ▼ (−) 0.5 ▼ 1 F5 (GRAPH) PtOff (0 , 4 ÷ 3 ÷ π) ENTER	√ 1 − x/θ/T x^2 RANGE −1 ENTER 1 ENTER 1 ENTER −.5 ENTER 1 ENTER 1 ENTER 2ndF PLOT ▼ ▶ ▼ ENTER 0 ENTER 4 ÷ 3 ÷ π ENTER
CASIO f_x-6300G (f_x-7700GB)	**HP48SX**
Range −1 EXE 1 EXE 1 EXE −.5 EXE 1 EXE 1 EXE EXE EXE EXE Graph √ (1 − X x^2) EXE Plot 0 , 4 ÷ 3 ÷ π EXE	↰ PLOT PLOTR ERASE ATTN ′ Y = $\sqrt{x}$ () 1 − X y^x 2 ↱ PLOT ↰ DRAW ERASE 1 +/− ENTER 1 XRNG .5 +/− ENTER 1 YRNG ATTN 0 ENTER 4 ENTER 3 ÷ π ÷ PRG OBJ NXT R → C PRG DSPL NXT C → PX PIXOF ↰ PLOT PLOTR DRAW

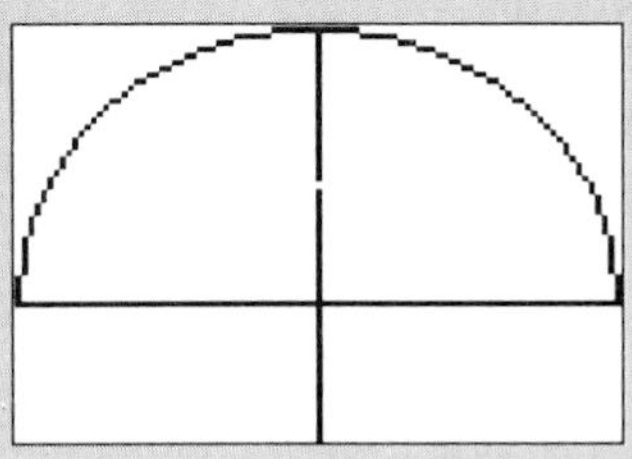

As usual, the semicircle may be somewhat distorted by the unequal horizontal and vertical scales. Can you adjust the scales to remove the distortion?

famous result is named after the Greek mathematician Pappus, who lived in the fourth century A.D. In fact the result is true for more complicated regions too.

THEOREM 6.7.10

Pappus' Theorem

Suppose R is a region in the plane with centroid $(\bar{x}, \bar{y})$. If R is revolved around an axis, the volume of the resulting solid is the product of the area A of R and the distance traveled by its centroid.

In particular, if R is revolved around the y-axis, the resulting solid has volume $2\pi\bar{x}A$, and if R is revolved around the x-axis, the resulting solid has volume $2\pi\bar{y}A$.

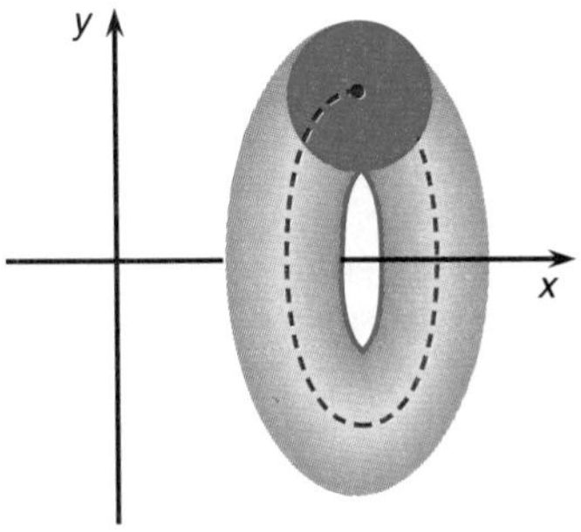

Figure 6.7.33

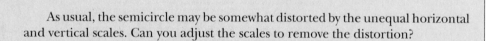

EXAMPLE 12 **Volume of a Doughnut** Find the volume of the "doughnut" shape swept out when the disk of radius 1 centered at (3, 2) is revolved around the x-axis (see Figure 6.7.33).

Solution To do this by integration is not extremely difficult, but it is quite messy. But since we know that the centroid of a disk is its center, we know that in this case, $\bar{y} = 2$. The area is $A = \pi$, so by Pappus' Theorem the volume of the doughnut is

$$2\pi\bar{y}A = 2\pi(2)\pi = 4\pi^2.$$

Exercises 6.7

I

Find the center of mass of each of the following objects.

1. A thin rod lying along the interval $[5, 9]$ and having uniform density.
2. A thin rod lying along the x-axis between $x = -2$ and $x = 6$ with density $\rho(x) = x + 2$.
3. A thin rod lying along the x-axis between $x = -3$ and $x = 3$ with density $\rho(x) = 3 + x^2$.
4. A thin rod of uniform density lying along the x-axis between $x = a$ and $x = a + L$.
5. A thin rod lying along the x-axis between $x = 1$ and $x = 2$ with density $\rho(x) = x^2 - 1$.
6. A thin rod lying along the x-axis between $x = 1$ and $x = 4$ with density $\rho(x) = \frac{1}{x}$.
7. A thin rod lying along the x-axis between $x = 0$ and $x = 1$ with density $\rho(x) = \frac{x}{1+x^2}$.
8. A thin rod lying along the x-axis between $x = 0$ and $x = a$ with density $\rho(x) = x^r$, where $a > 0$ and $r > 0$.
9. A thin cone lying along the x-axis between $x = 2$ and $x = 4$, with vertex at $x = 2$ and radius 0.2 at $x = 4$, whose density is proportional to its cross-sectional area.
10. An object composed of two thin rods of the same uniform density, one lying between $x = -4$ and $x = -1$ and the other lying between $x = 1$ and $x = 5$.

Find the centroid of each of the regions described below, assuming that they are all of uniform density.

11. The right semidisk $\{(x, y) : 0 \le x \le \sqrt{1 - y^2}, -1 \le y \le 1\}$.
12. The part of the region inside the circle $x^2 + y^2 = 9$ that lies in the first quadrant (i.e., with $x \ge 0, y \ge 0$).
13. The region inside the bottom half of the ellipse $x^2 + 2y^2 = 4$ (i.e., the intersection of the region inside the ellipse and the region below the x-axis).
14. The intersection of the region inside the ellipse $3x^2 + 4y^2 = 25$ and the third quadrant, that is, the region where $x, y \le 0$.
15. The region inside the triangle with vertices $(-2, 1)$, $(-5, 2)$, and $(1, 3)$.
16. The region inside the triangle with vertices $(2, 3)$, $(-1, 2)$, and $(2, 1)$.
17. The region enclosed between the curve $y = x^2$ and the line $y = 4$.
18. The region enclosed between the curve $y = x^2$ and the line $y = x + 2$.
19. The region enclosed between the curves $y = \sqrt{x}$ and $y = x^2$.
20. The region between the curves $y = x^r$ and $y = x^{1/r}$, for $0 \le x \le 1$, where $r > 0$.

II

21. Suppose the rod of Example 3 (p. 465) is turned around so that its density is proportional to the distance between x and 3 (for $x \in [0, 3]$). Find its center of mass.
22. If a person weighing 40 kg sits on a seesaw 1 meter from the fulcrum and a person weighing 30 kg sits on the other side at a distance of 1.5 meters from the fulcrum, where must a dog weighing 10 kg sit to make the seesaw balance?
23. Find the center of mass of a rod of length L whose density is proportional to the square of the distance from one of the ends.
24. (i) Find the center of mass of a rod of length L whose density is proportional to the cube of the distance from one of the ends.
    (ii) Find the center of mass of a rod of length L whose density is proportional to the sixth power of the distance from one of the ends.

*25. From Example 3 we know that if the density of a thin rod between $x = 0$ and $x = 3$ is $\rho(x) = x$, then the center of mass is $\bar{x} = 2$.
    (i) Suppose the density is changed from x to $\rho(x) = c + x$, where c is a positive constant. Do you expect the center of mass to stay in the same place, to move to the left (i.e., $\bar{x} < 2$), or to move to the right (i.e., $\bar{x} > 2$)?
    (ii) Find $\bar{x}$ when $c = 3$. Is your answer consistent with your answer to part (i)?
    (iii) If the constant c in part (i) is allowed to in-

crease, what effect do you expect it to have on the position of the center of mass?

(iv) Find $\bar{x}$ when $c = 4$ and when $c = 6$. Is your answer consistent with your answer to part (iii)?

***26.** Verify Formula 6.7.8 (p. 471) for the centroid of a triangle. (*Hint:* Place the triangle with one vertex at the origin and its base along the x-axis. Calculate M_x and M_y by integrating.)

***27.** Find the centroid of the region illustrated in Figure 6.7.34.

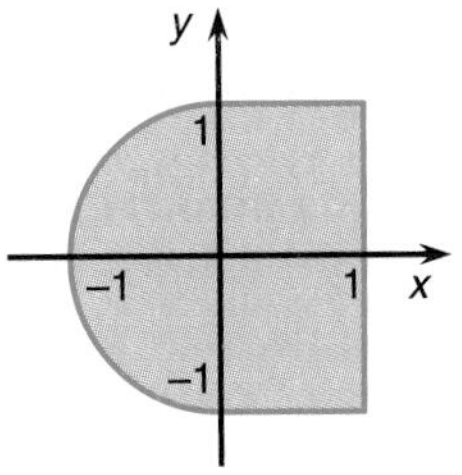

Figure 6.7.34

***28.** Find the centroid of the region illustrated in Figure 6.7.35.

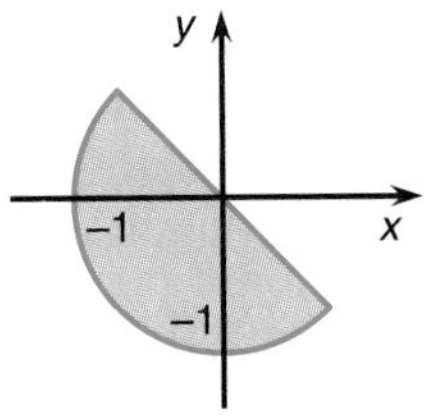

Figure 6.7.35

***29.** Find the centroid of the region that consists of all points lying above the x-axis, below the semicircle $y = \sqrt{4 - x^2}$, and above the semicircle $y = \sqrt{1 - x^2}$.

***30.** Find the centroid of the region that consists of the region inside the circle of radius 2 with center at the origin *and* the region inside the circle of radius 1 with center at the point $(5, 5)$.

***31.** Find the centroid of the region that consists of the union of the region inside the circle of radius 1 with center at the origin, the region inside the circle of radius 2 with center at the point $(0, 8)$, and the rectangular region between the lines $y = 0$, $y = 6$, $x = 1$, and $x = -1$.

32. (i) Find the centroid of the region enclosed between the curve $y = x^3 - 3x$ and the x-axis.

(ii) Find the centroid of the region enclosed between the curve $y = x^3 - 3x$ and the x-axis and to the right of the y-axis.

***33.** Find the centroid of the region that consists of all points lying inside the circle $x^2 + y^2 = 9$ and outside the square whose corners are $(1, 1)$, $(1, -1)$, $(-1, 1)$, and $(-1, -1)$.

***34.** Find the centroid of the region that consists of all points lying inside the circle $x^2 + y^2 = 9$ and outside the square whose corners are $(1, 1)$, $(1, 0)$, $(0, 1)$, and $(0, 0)$.

***35.** Find the centroid of the parallelogram whose vertices are $(0, 0)$, $(3, 0)$, $(1, 2)$, and $(4, 2)$.

***36.** Find the centroid of the parallelogram whose vertices are $(0, 0)$, $(a, 0)$, (b, c), and $(a + b, c)$.

37. Find the centroid of the region enclosed between the curve $y = x^3$ and the line $y = x$.

38. Find the centroid of the region enclosed between the curves $y = x^2$ and $y = -x^2 + 2x + 4$.

***39.** Find the centroid of the region consisting of two disks, one of radius 1 and the other of radius 2, that touch at exactly one point.

***40.** Find the centroid of the region consisting of two disks, one of radius 1 and the other of radius r, that touch at exactly one point.

***41.** Find the center of mass of the region consisting of two disks touching at a single point, if both disks have the same radius but the density is $\rho = 1$ on one of them and $\rho = 2$ on the other.

***42.** Consider a disk of radius 2 with uniform density and place on top of it a disk of radius 1 with the same density so that the smaller disk just touches the outside edge of the larger one at a single point. Find the center of mass of the resulting object. (It might help to visualize two disks made of sheet metal, one welded onto the other.)

***43.** Consider the square enclosed between the axes and the lines $x = 2$ and $y = 2$. Let the density function ρ equal r on the smaller square enclosed between the axes and the lines $x = 1$ and $y = 1$, and let ρ equal 1 on the rest of the big square. Find the center of mass of this region if $r = 2$. See Figure 6.7.36.

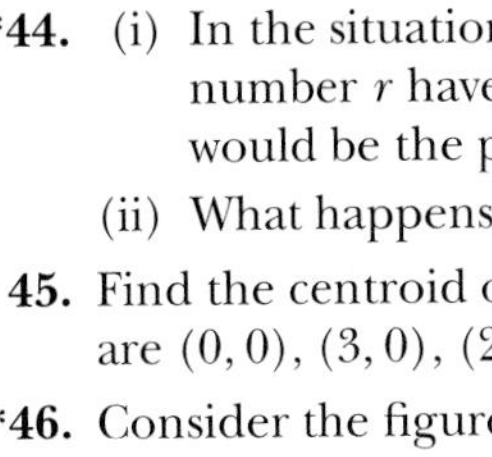

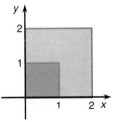

Figure 6.7.36

***44.** (i) In the situation of Exercise 43, what would the number r have to be so that the center of mass would be the point $\left(\frac{3}{4}, \frac{3}{4}\right)$?

(ii) What happens to the center of mass as $r \to \infty$?

45. Find the centroid of the quadrilateral whose vertices are $(0, 0)$, $(3, 0)$, $(2, 1)$, and $(1, 1)$.

***46.** Consider the figure obtained by taking the three points $(-2, 1)$, $(0, 2)$, and $(4, 4)$, joining each pair of these points with a line segment, and looking at the set of all points either lying on one of the segments or enclosed by them.

(i) Find the centroid of the region described above.

(ii) Find the area of the region described above.

47. Find the volume of the solid obtained by revolving the disk of Example 12 around the y-axis.

48. Find the volume of the doughnut obtained by revolving the disk $(x-a)^2+(y-b)^2 \leq R^2$ around the x-axis. (Assume $0 < R < b$.)

49. Find the volume of the solid obtained by revolving the upper half of the disk in Example 12 around (i) the y-axis, (ii) the x-axis.

50. Find the volume of the solid obtained by revolving around the x-axis the triangle with vertices $(1, 2)$, $(2, 1)$, and $(3, 3)$.

6.8 Fluid Pressure and Force

DENSITY OF A FLUID
PASCAL'S PRINCIPLE
FLUID FORCE AND CENTROIDS
BUOYANCY

This section is devoted to a discussion of the pressure exerted by a fluid. The fundamental idea is to use integration to "add up" the forces exerted by the fluid at various points on the surface of a submerged object. Later on, it will become clear that many of the calculations we do can be simplified by using the notion of centroid, which was introduced in Section 6.7.

If we imagine an object submerged in a liquid, it seems reasonable to expect that the pressure will depend on the depth; the deeper the object is, the greater will be the force exerted on it by the liquid.

Now suppose we want to find the total force exerted on the object. The tricky thing is that the pressure is not constant. It will be greater at the bottom of the object than at the top, since the bottom is deeper in the liquid. To find the total force, we have to "add up" the force at each point. Since the pressure varies, what is involved here is really an integral rather than a sum.

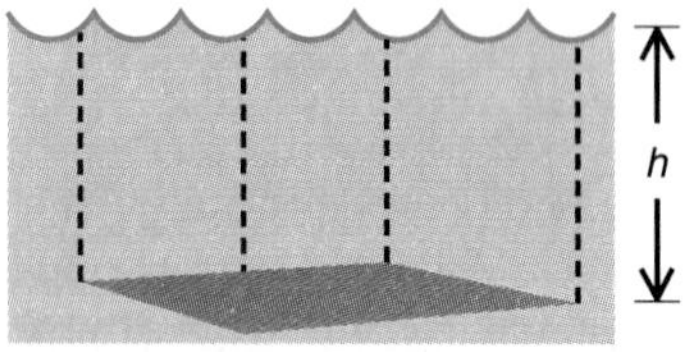

Figure 6.8.1

Consider a square flat plate whose area is 1 sq ft submerged in water at a depth h ft, lying horizontally, as shown in Figure 6.8.1. What is the force with which the water presses on the top surface of the plate? One reasonable way to approach this question would be to consider the water lying directly above the plate (see Figure 6.8.2). It fills a rectangular solid region of height h and of horizontal cross-sectional area 1 sq ft. The volume of this solid is $h \times 1 \times 1 = h$ ft^3. The density of water is 62.4 lb/ft^3, so the weight of the water over the plate is $62.4h$ lb.

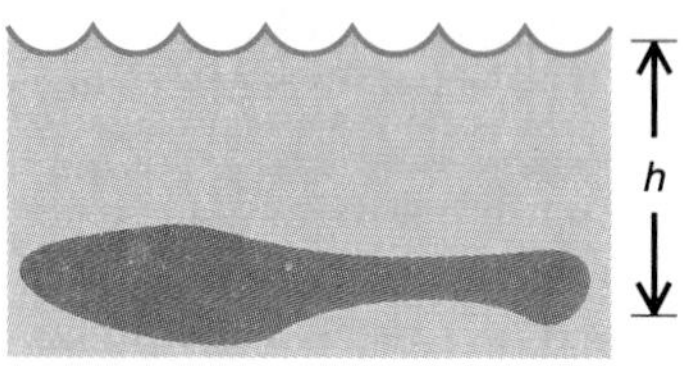

Figure 6.8.2

Whatever the fluid, it is conventional to write ρ, the Greek letter "rho," for its density (i.e., how much the fluid weighs per unit volume). So the weight of the fluid lying directly over the plate is $h\rho$ (h and ρ must be in compatible units, such as ft and lb/ft^3 as above, kg and kg/m^3, in and lb/in^3, etc., and we have assumed that the area of the top of the plate is one square unit). What this says is that the force exerted on the top of a horizontal flat plate of surface area 1 square unit is $h\rho$, where h is its depth and ρ is the density of the fluid.

In fact, there is nothing about this argument that makes it necessary to consider a square plate, or one whose area is 1 square unit. If a flat plate of area A sq ft is lying horizontally in water at a depth of h ft, the same argument shows that the water presses on it with a force of $62.4Ah$ lb (see Figure 6.8.3).

Figure 6.8.3

Another way to say this is that the water presses on the plate with a **pressure** of $62.4h$ lb/sq ft. This is what pressure means, the force per unit area. The advantage of speaking in terms of pressure is that it is not necessary to mention the area of the plate.

Pressures are also frequently measured in *pounds per square inch*, abbreviated *psi*. Since there are $12 \times 12 = 144$ square inches in 1 square foot, we see that the pressure exerted by water on a horizontal flat plate at a depth of h ft is $\frac{62.4h}{144} \approx 0.433h$ psi.

In the metric system we find that the water lying over a horizontal flat plate of area 1 m^2 at a depth of h m has volume $1 \times h = h$ m^3. Since a cubic meter of water weighs 1000 kg, the pressure on a horizontal plate at a depth of h m is $1000h$ kg/m^2 = $9800h$ N/m^2.

> Regardless of the units being used, as long as they are compatible, the pressure on a horizontal plate at depth h in a fluid of density ρ is $h\rho$.

EXAMPLE 1 (i) Find the pressure exerted by water at a depth of 100 ft. (ii) If the top surface area of a large flounder is 2 sq ft, what is the total force exerted on the top of the flounder by the water at a depth of 100 ft?

Solution

(i) In units of pounds per square foot the pressure of the water at a depth of $h = 100$ ft is $62.4h = 6240$ lb/sq ft.

(ii) If the water pressure is 6240 lb/sq ft, then on an area of 2 sq ft the water exerts a force of $2 \times 6240 = 12{,}480$ lb, or a little over 6 tons.

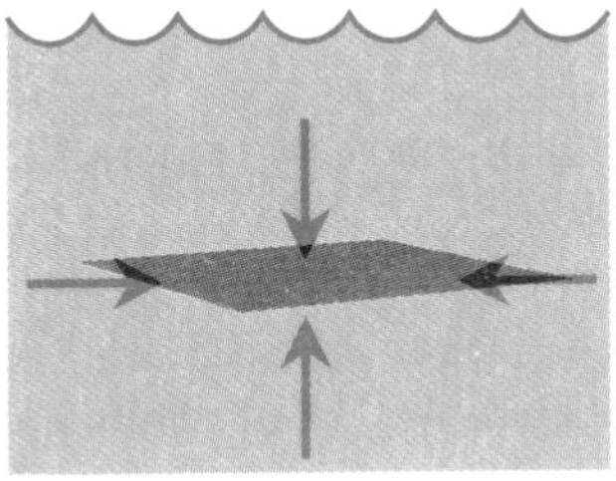

Figure 6.8.4

So far we have concentrated on flat horizontal objects (even going so far as to consider a flat fish like the flounder). The remarkable thing about pressure in a fluid is that it is the same in all directions. That is, at a depth of 100 ft in water, as in Example 1, the water exerts downward pressure of 6240 lb/ft^2 on a horizontal plate. If we ask what the pressure is on the underside of the plate, we realize that it must be *upward* pressure, and it is not clear why the water should be pressing up at all.

However, suppose the pressure on the bottom of the plate was, say, 5000 lb/ft^2. Then a plate of area 1 ft^2 would experience downward force of 6240 lb and upward force of 5000 lb. The net force would be 1240 lb downward, which would push the plate downward very hard (and very fast). The fact is that water pressure does not push a plate down or up (or sideways). The only way this can occur is for the pressure to be the same in all directions (up, down, north, south, etc.). See Figure 6.8.4. This fundamental fact is known as Pascal's Principle, after the French mathematician and physicist Blaise Pascal (1623–1662).

THEOREM 6.8.1

> **Pascal's Principle**
>
> Fluid pressure is the same in all directions. At a depth of h in a fluid of density ρ the pressure equals $h\rho$.

This will allow us to calculate the force exerted by water pressure on objects that are not necessarily flat and horizontal. The new subtlety that arises here is that since the pressure depends on the depth, the pressure on an object is not uniform but will be different at different places on its surface.

EXAMPLE 2 Consider a square flat plate, each side 2 ft long, suspended vertically in water. If the top edge is horizontal and at a depth of 20 ft, what is the total force exerted on one side of the plate by the water?

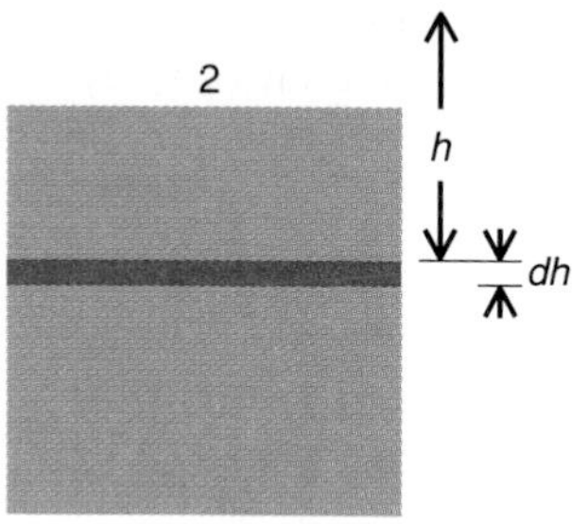

Figure 6.8.5

Solution We know that the pressure at a depth h is $62.4h$ lb/ft^2. The plate lies between the depths of 20 ft at its top and 22 ft at its lower edge. Consider the depth h as a variable, ranging from $h = 20$ to $h = 22$.

Now at any particular depth h, imagine a narrow strip on the surface of the plate of height dh as shown in Figure 6.8.5. Its width is the width of the plate, or 2 ft. The area of this strip is $2\,dh$, and the pressure on it (at depth h) is $62.4h$ lb/ft^2. Accordingly, the force on it is *(pressure)* $\times$ *(area)* $= 62.4h(2)\,dh = 124.8h\,dh$.

To find the total force, it is necessary to add up the forces on all such little strips as h ranges from $h = 20$ to $h = 22$. As usual, "adding up" means integrating. The total force on the plate is

$$\int_{20}^{22} 124.8h\,dh = 124.8\int_{20}^{22} h\,dh = 124.8\left(\frac{1}{2}h^2\right)\Bigg|_{20}^{22} = 62.4\left(22^2 - 20^2\right)$$
$$\approx 5240 \text{ lb.}$$

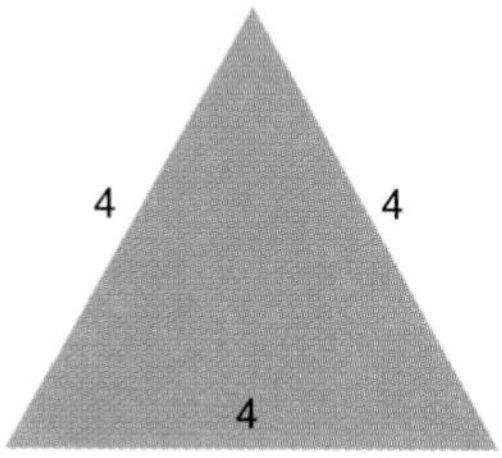

Figure 6.8.6

EXAMPLE 3 Consider a triangular plate suspended vertically in water as shown in Figure 6.8.6. The triangle is equilateral with edges 4 ft long. Its bottom edge is horizontal and submerged to a depth of 30 ft. Find the total force exerted on one side of the plate by the water.

Solution Let h be a variable representing the depth of water. First we should figure out the range of values for h on the plate. To find the height of the plate, that is, the vertical distance from the base to the top, we use the Pythagorean Theorem (Theorem 1.5.1). In one of the little right-angled triangles that make up half the large equilateral triangle, the hypotenuse is 4 ft, the length of one of the sides (see Figure 6.8.7). The horizontal side at the bottom is 2 ft, so the length of the vertical side is $\sqrt{4^2 - 2^2} = \sqrt{12} = 2\sqrt{3}$.

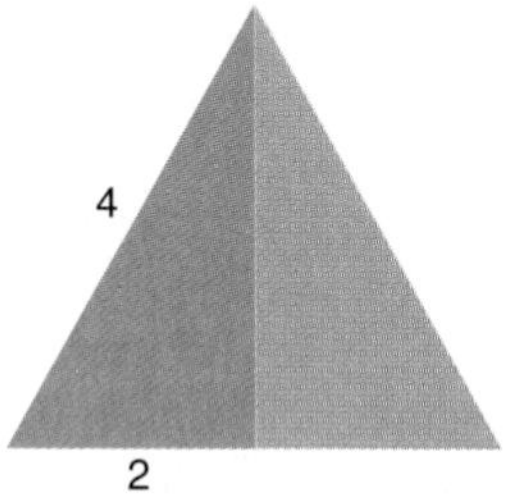

Figure 6.8.7

Since the bottom of the triangle is at a depth of 30 ft, the top is at a depth of $30 - 2\sqrt{3}$ ft. The variable h, measuring depth in feet, ranges from $h = 30 - 2\sqrt{3}$ to $h = 30$.

Next consider a small horizontal strip on the triangular plate, of height dh. To find its area, we must find its width, which varies as the strip moves down the height of the triangle.

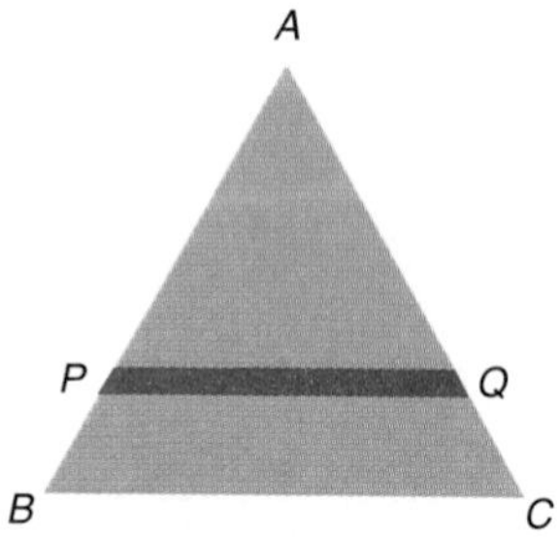

Figure 6.8.8

We draw the triangle in Figure 6.8.8, labeling its vertices A, B, and C and the ends of the strip P and Q. Then the triangle APQ is similar to the triangle ABC. The height of ABC is $2\sqrt{3}$, and if h measures the depth at which the strip PQ lies, then the height of APQ is $h - (30 - 2\sqrt{3})$, since the top of the triangle, A, is at a depth of $30 - 2\sqrt{3}$. Since the length of BC is 4, we can use similar triangles to conclude that the length of PQ is $\frac{(\textit{height of } APQ)}{(\textit{height of } ABC)} \times 4 = \frac{h-(30-2\sqrt{3})}{2\sqrt{3}}(4) = 2\frac{h-(30-2\sqrt{3})}{\sqrt{3}}$.

The area of the strip is therefore $(\textit{length } PQ)\,dh = 2\frac{h-(30-2\sqrt{3})}{\sqrt{3}}\,dh$. Since the pressure at a depth of h ft is $h\rho = 62.4h$ lb/ft^2, the force on the strip is $(\textit{area}) \times (\textit{pressure}) = 2\frac{h-(30-2\sqrt{3})}{\sqrt{3}}\,dh \times 62.4h = 124.8h\frac{h-(30-2\sqrt{3})}{\sqrt{3}}\,dh$.

The total force on the triangle is obtained by integrating. It is

$$\int_{30-2\sqrt{3}}^{30} 124.8h\frac{h-(30-2\sqrt{3})}{\sqrt{3}}\,dh = \frac{124.8}{\sqrt{3}}\int_{30-2\sqrt{3}}^{30}\left(h^2 - (30-2\sqrt{3})h\right)dh$$
$$= \frac{124.8}{\sqrt{3}}\left(\frac{h^3}{3} - (30-2\sqrt{3})\frac{h^2}{2}\right)\Bigg|_{30-2\sqrt{3}}^{30}$$
$$\approx 12{,}500 \text{ lb.}$$

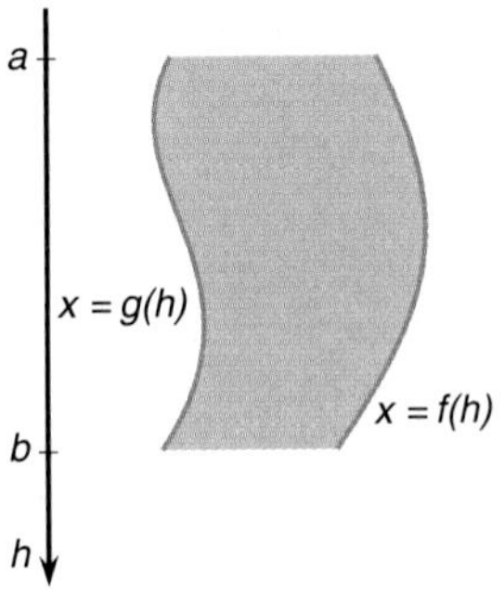

Figure 6.8.9

Before proceeding, we should pause and analyze the calculations we are doing. The variable in the vertical direction is h. It is "upside-down," in the sense that large values of the depth h refer to points far *down* in the fluid. Let us continue to use x for the variable in the horizontal direction.

Consider a plate suspended vertically in the fluid. Suppose it can be described as lying between two curves $x = f(h)$ and $x = g(h)$, for values of h with $a \le h \le b$. Also suppose that $f(h) \ge g(h)$ for all $h \in [a, b]$. See Figure 6.8.9.

Then the width of the narrow horizontal strip at depth h is $f(h) - g(h)$, so the area of the strip is $\big(f(h) - g(h)\big)\,dh$. The force on the strip is (*pressure*) (*area*) $=$ $h\big(f(h) - g(h)\big)\rho\,dh$, and we integrate to find the total force $\int_a^b h\big(f(h) - g(h)\big)\rho\,dh$. See Figure 6.8.10.

This last integral is *exactly* the same as the integral needed to find the moment M_x of the plate about the x-axis, assuming that the plate has uniform density ρ (cf. Definition 6.7.5, p. 467).

On the other hand, the moment equals $M_x = M\bar{h}$, the product of the total "weight" and the vertical coordinate of the centroid. Moreover, the total "weight" is $M = \int_a^b \big(f(h) - g(h)\big)\rho\,dh = \rho\int_a^b \big(f(h) - g(h)\big)\,dh$. This is the product of ρ and the area of the plate.

The force $\int h\big(f(h) - g(h)\big)\rho\,dh$ equals $M_x = M\bar{h} = A\bar{h}\rho = A(\bar{h}\rho)$, and $\bar{h}\rho$ is the pressure at $\bar{h}$.

Putting all the pieces together, we see that the total force on the plate equals $M_x = M\bar{h} = A\rho\bar{h}$, where A is the area of the plate. Since $\bar{h}\rho$ is just the pressure at depth $\bar{h}$, by Theorem 6.8.1, and multiplying pressure by area gives the total force, what this says is that the total force on the plate is exactly the same as the force $A\bar{h}\rho$ that would be exerted on the plate if it were lying horizontally at a depth of $\bar{h}$.

We have just established the following extremely useful result.

THEOREM 6.8.2

The force exerted by a fluid on a flat plate equals the force that would be exerted on the plate if it were suspended horizontally at the depth where its centroid now lies.

If the plate has area A and is suspended with its centroid at depth $\bar{h}$ in a fluid whose density is ρ, then the total force on the plate is $\bar{h}\rho A$.

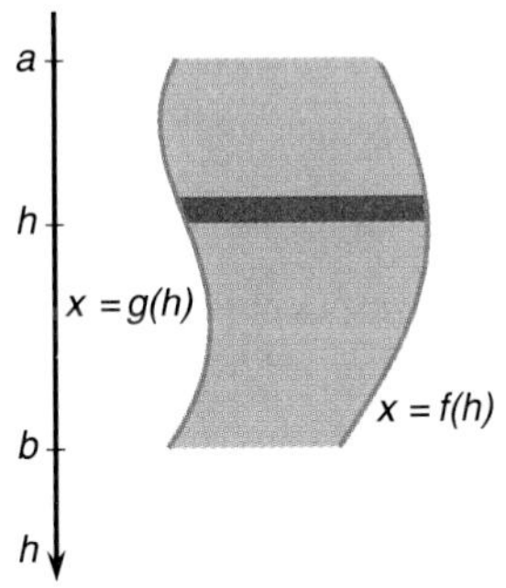

Figure 6.8.10

EXAMPLE 4 Consider a circular plate of radius 6 in suspended vertically in water with its center at a depth of 4 ft. See Figure 6.8.11. What is the force exerted by the water on one side of the plate?

Solution The centroid of a disk is its center, so the depth of the centroid is $\bar{h} = 4$ ft, the depth of the center of the disk. The radius of the disk is 6 in, but in order to have all the measurements in compatible units we convert this to $r = \frac{1}{2}$ ft.

The area of the disk is $\pi r^2 = \frac{\pi}{4}$. The pressure at a depth of $\bar{h}$ is $\bar{h}\rho = 62.4\bar{h} =$ $62.4(4) = 249.6\ \text{lb/ft}^2$, so the total force on the disk equals the force exerted on an area of $\frac{\pi}{4}\ \text{ft}^2$ by a pressure of $249.6\ \text{lb/ft}^2$. The total force is $249.6\left(\frac{\pi}{4}\right) \approx 196$ lb.

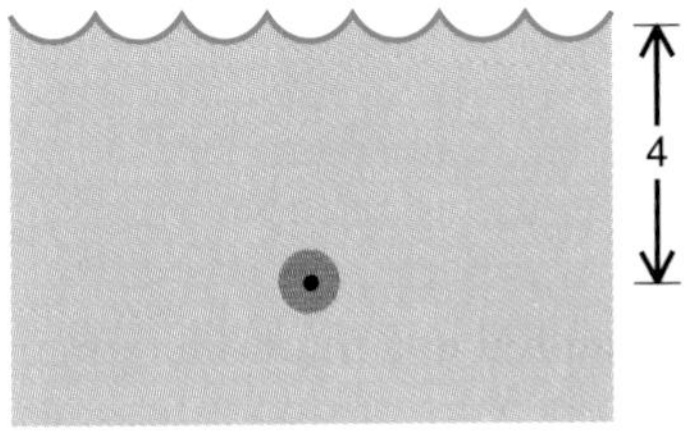

Figure 6.8.11

Using Theorem 6.8.2 simplified this example considerably. For purposes of comparison you might like to find the total force by integration, as in Examples 2 and 3.

EXAMPLE 5 **Underwater Observation Tank** Find the total force exerted on an underwater observation tank built in the shape of a cube whose sides are all 2 meters long if the bottom of the tank is at a depth of 25 meters. Express the answer in newtons.

Solution The bottom of the cube is a square of area 4 m^2 at a depth of 25 m, so the force on it is $(area)\,(pressure) = 4(25)9800 = 980{,}000$ N. Similarly, the top of the cube is at a depth of 23 m, so the force on it is $4(23)9800 = 901{,}600$ N.

The other sides can be considered separately. Each of them is a square with depth ranging from $h = 23$ to $h = 25$. The centroid, being the center of the square, is at a depth of $\bar{h} = 24$ m. Since the area of the side is 4 m^2, the total force on it can be found by using Theorem 6.8.2. It is $\bar{h}\rho A = 24(9800)4 = 940{,}800$ N.

This is the force on each of the four side walls, so the total force on all six walls is $980{,}000 + 901{,}600 + 4 \times 940{,}800 = 5{,}644{,}800$ N.

The examples we have considered so far have all involved flat surfaces that are either horizontal or vertical. Naturally, it may be important to deal with curved or sloping surfaces. In fact, curved surfaces will be the subject of the next section, but we can do examples of sloping flat surfaces.

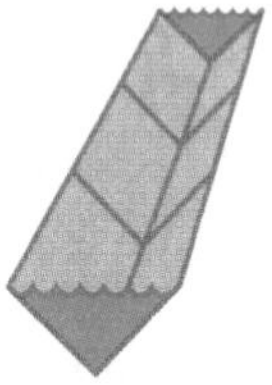
Figure 6.8.12

EXAMPLE 6 **Force in Drainage Ditch** Consider a drainage trench built so that its cross-section has the shape of a V. Suppose the sides of the trench are at an angle of 45° above the horizontal and that they are made of prefabricated concrete slabs, each 10 ft square. See Figure 6.8.12.

Find the force exerted on one of these slabs when the trench is full of water.

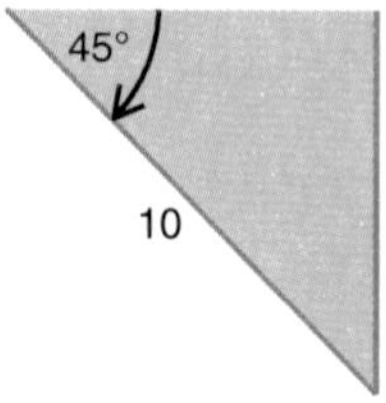

Figure 6.8.13

Solution First we find the depth of the water. The surface of the slab forms the hypotenuse of a right-angled triangle, in which the other angles are both 45°. Accordingly, the depth of the water, which is the vertical side of the triangle, is $10\sin(45°) = \frac{10}{\sqrt{2}} \approx 7.07$ ft (see Figure 6.8.13).

The centroid, which is just the center of the rectangular slab, is halfway down, so it is at a depth of $\bar{h} = \frac{1}{2}\frac{10}{\sqrt{2}} = \frac{5}{\sqrt{2}} \approx 3.54$ ft. The pressure on the slab is $\bar{h}\rho A = \frac{5}{\sqrt{2}}(62.4)100 \approx 22{,}060$ lb, or slightly over 11 tons.

It is also possible to solve this problem directly by integrating, as we now demonstrate, partly to show how much easier it is to use centroids and partly to illustrate the approach needed for objects that are tilted.

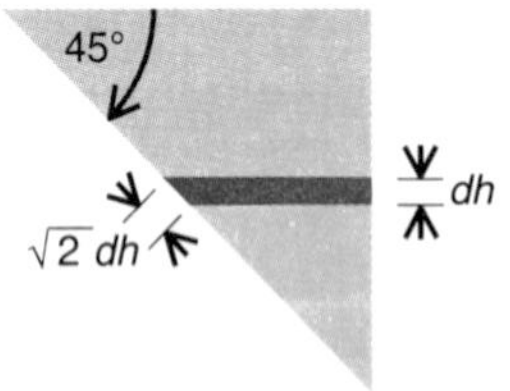

Figure 6.8.14

We let h be a variable representing the depth. It runs from $h = 0$ to $h = \frac{10}{\sqrt{2}}$. Now consider what happens as h changes by a small amount dh. There is a corresponding little strip on the sloping wall of the trench. However, the sloping height of the strip on the wall does not equal dh, but rather $\frac{dh}{\sin(45°)} = \sqrt{2}\,dh$. See Figure 6.8.14.

Since the slab of concrete is a square, the width of the strip always equals 10 ft, so the area of the strip is $10\sqrt{2}\,dh$ ft^2. The pressure at depth h is $62.4h$ lb/ft^2, so the total force on the slab is

$$\int_0^{10/\sqrt{2}} 62.4h \times 10\sqrt{2}\,dh = 624\sqrt{2}\int_0^{10/\sqrt{2}} h\,dh = 624\sqrt{2}\frac{h^2}{2}\bigg|_0^{10/\sqrt{2}}$$
$$\approx 22{,}060 \text{ lb.}$$

Buoyancy

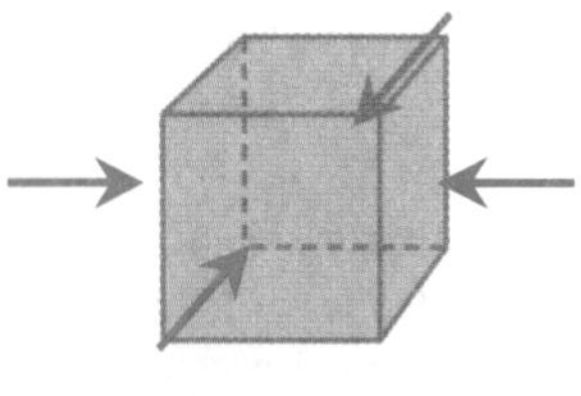
Figure 6.8.15

Consider a cube, whose sides are each k units long, submerged in a fluid. The force on each of the vertical sides is the same, so the forces on pairs of opposite sides cancel, and there is no *net* horizontal force. See Figure 6.8.15.

However, the top and the bottom are at different depths, so the forces on them are not the same. In fact, suppose the top is at a depth h; the bottom is then at depth $h + k$, since each side has length k. The pressure on the top is $h\rho$, and the pressure

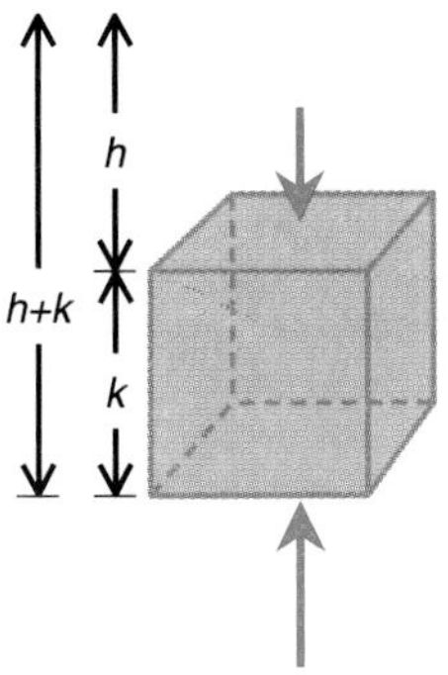

Figure 6.8.16

on the bottom is $(h+k)\rho$. Since each has area k^2, the total force on the top is $hk^2\rho$ *downward*, and the total force on the bottom is $(h+k)k^2\rho$ *upward*. See Figure 6.8.16.

The difference between the force on the bottom and the force on the top is $(h+k)k^2\rho - hk^2\rho = k^3\rho$. Since k^3 is exactly the volume of the cube and ρ is the density of the fluid, the quantity $k^3\rho$ is just the weight of the fluid it would take to fill the cube.

The cube experiences a net upward force equal to the weight of the fluid it would take to fill it up, that is, the amount of fluid "displaced" by the cube.

In fact, the same is true of any submerged object, and the upward force is known as its **buoyancy**. If the object weighs less than the fluid it displaces, then the buoyancy is greater than the weight, and the object floats. If it weighs more than the displaced fluid, it sinks.

This is the famous discovery about which Archimedes is reputed to have shouted "Eureka!" as he jumped out of his bathtub.

EXAMPLE 7 Floating Block of Wood Suppose a rectangular block of pine measuring 1 ft $\times$ 1 ft $\times$ 2 ft is floating in water with its long side vertical. If the density of the wood is 30 lb/ft^3, how much of the block will be above the surface?

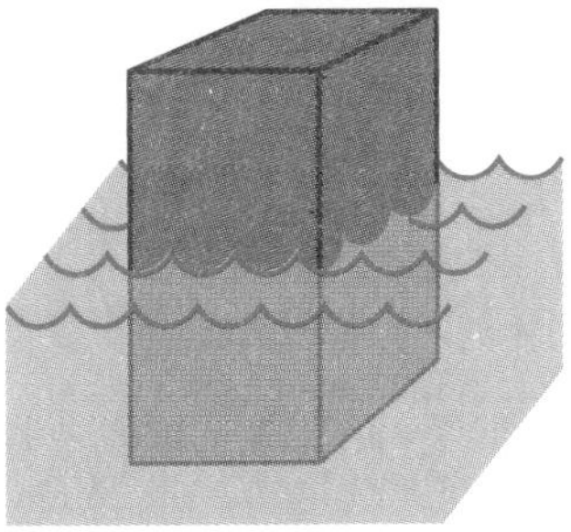

Figure 6.8.17

Solution If the block is floating, the buoyant force due to the displaced water must exactly equal the weight of the block. Since the volume of the block is 2 ft^3 and its density is 30 lb/ft^3, the block weighs $2 \times 30 = 60$ lb.

The amount of water displaced must weigh 60 lb to balance this weight. Now if the height of the block above the surface is t ft, then the height of the submerged part is $2-t$, and the volume of the submerged part is $1 \times 1 \times (2-t) = 2 - t$ ft^3. For this to weigh 60 lb we must have $60 = 62.4 \times (2-t) = 124.8 - 62.4t$, or $62.4t = 124.8 - 60 = 64.8$, so $t = \frac{64.8}{62.4} \approx 1.04$ ft.

The height to which the block protrudes above the surface is approximately 1.04 ft. See Figure 6.8.17.

Exercises 6.8

I

Each of the flat plates described below is submerged in water. Find the total force exerted by the water on one side of the plate.

1. A square plate, 3 ft on each side, lying horizontally at a depth of 11 ft.
2. A square plate, 2 m on each side, lying horizontally at a depth of 3 m.
3. A square plate, 5 in on each side, lying horizontally at a depth of 6 ft.
4. A square plate, 1 ft on each side, lying horizontally at a depth of 4 m.
5. A square plate, 5 ft on each side, lying vertically with its bottom edge (horizontal) at a depth of 20 ft.
6. A square plate, 3 m on each side, lying vertically with its bottom edge at a depth of 10 m.
7. A square plate, 15 in on each side, lying vertically with its bottom edge at a depth of 8 ft.
8. A square plate, 25 in on each side, lying vertically with its bottom edge at a depth of 1 ft.
9. A round plate, with radius 3 ft, lying vertically with its center at a depth of 10 ft.
10. A round plate, with radius 0.5 m, lying vertically with its center at a depth of 50 m.

Each of the flat plates described below is submerged in water. Find the total force exerted by the water on one side of the plate.

11. A round plate, with radius 30 in, lying vertically with its center at a depth of 20 ft.
12. A round plate, with radius 30 cm, lying vertically with its center at a depth of 4 m.

13. A round plate, with radius 1 ft, lying horizontally at a depth of 26 ft.

14. An elliptical plate, with semiaxes 3 ft and 4 ft, lying with major axis vertical and the lowest point at a depth of 24 ft.

15. An equilateral triangle, with sides 8 ft, lying vertically with its bottom edge horizontal at a depth of 100 ft.

16. An equilateral triangle, with sides 3 ft, lying vertically with its top edge horizontal at a depth of 40 ft.

17. An equilateral triangle, with sides 2 m, lying horizontally at a depth of 14 ft.

18. An equilateral triangle, with sides 20 in, lying vertically with one edge vertical and the opposite vertex at a depth of 40 ft.

19. A right-angled triangle, with base 4 ft long, lying at a depth of 10 ft, and vertical side 3 ft high.

20. A right-angled triangle, with base 6 ft, at a depth of 30 ft and vertical side 6 ft high.

Each of the flat plates described below is submerged in water. Find the total force exerted by the water on one side of the plate.

21. A square plate, 3 ft on each side, lying at a 45° angle with its bottom edge at a depth of 10 ft.

22. A square plate, 2 ft on each side, lying at a 30° angle above the horizontal with its bottom edge at a depth of 10 ft.

23. A round plate, with radius 3 ft, lying at an angle of 45° above the horizontal with its center at a depth of 10 ft.

24. A round plate, with radius 4 m, lying at an angle of 60° above the horizontal with its lowest point at a depth of 30 m.

25. A round plate, with radius 6 ft, lying at an angle of 45° above the horizontal with its center at the surface of the water.

26. A square plate, with sides 4 ft long, lying at an angle of 30° above the horizontal with its bottom edge at a depth of 1 ft.

27. An equilateral triangle, with sides 2 ft, lying at an angle of 45° above the horizontal with its bottom edge at a depth of 100 ft.

28. An equilateral triangle, with sides 5 m, lying at an angle of 30° above the horizontal with its bottom edge at a depth of 10 m.

29. An equilateral triangle, with sides 3 ft, lying at an angle of 60° above the horizontal with its top edge at a depth of 15 ft.

30. A right-angled triangle, with base 4 ft long, lying at a depth of 20 ft, and another side 4 ft long, lying at a 60° angle above the horizontal.

II

31. Find the pressure exerted by the water on one side of a square plate whose sides are all 4 ft long. The plate is lying vertically with each of its sides at an angle of 45° above the horizontal so that the lowest point on the plate is at a depth of 40 ft. See Figure 6.8.18.

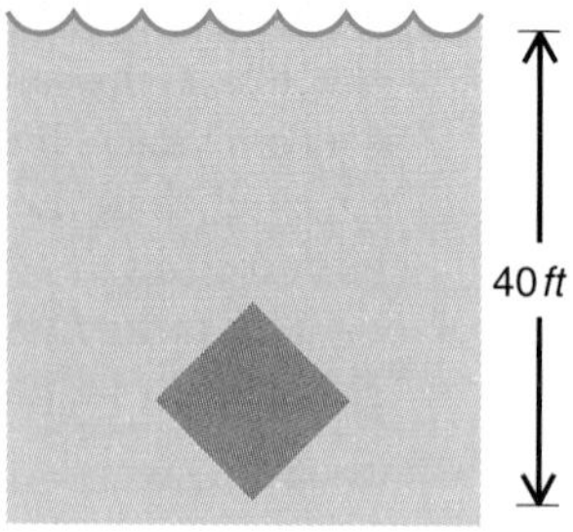

Figure 6.8.18

32. What is the total force on one side of a square plate of area 1 sq ft if the plate is suspended vertically in a liquid whose density is 50 lb/ft^3, in such a way that its bottom edge lies horizontally at a depth of 3 ft?

***33.** Suppose the slab of Example 6 (p. 480) was at an angle of 30° above the horizontal. What would be the total force exerted on it by water that just reaches up to the top of the slab?

***34.** Suppose the slab of Example 6 was at an angle of 60° above the horizontal and submerged so that its lower edge was at a depth of 50 ft. What would be the total force exerted on one side of it by the water?

***35.** In Example 4 (p. 479) we found the total force by locating the centroid. Do this example directly by evaluating an integral, without using centroids at all.

***36.** In Example 5 (p. 479) we found the total force by locating the centroid. Do this example directly by integrating, without using centroids at all.

37. (i) A barge is built in the shape of a rectangular box, 20 ft wide, 40 ft long, and 8 ft high. If the barge weighs 60 tons, how much water will it draw (i.e., to what depth will it sink)?

(ii) If the barge is loaded with a cargo weighing 100 tons, how much water will it draw?

(iii) How heavy a cargo can the barge hold without sinking?

***38.** Suppose a block in the shape of a cube is floating in water with its bottom horizontal. If the block is made of material whose density is ρ, figure out a formula for the fraction of the block that is below the surface.

***39.** People swimming in seawater find that they float higher in the water, that is, with less of their bodies submerged, than when they swim in fresh water. What does this say about the densities of fresh water and salt water?

People swimming in the Dead Sea float even higher in the water. What does this say about the density of Dead Sea water?

*40. If the density of seawater is 64.3 lb/ft^3 and the density of ice is 57.5 lb/ft^3, what fraction of an iceberg protrudes above the surface of the ocean?

*41. Consider a trough that is 10 ft long, 2 ft high, and 1 ft wide at the base and 3 ft wide at the top, as shown in Figure 6.8.19. The ends of the trough are made of flat sheet metal; they are vertical.

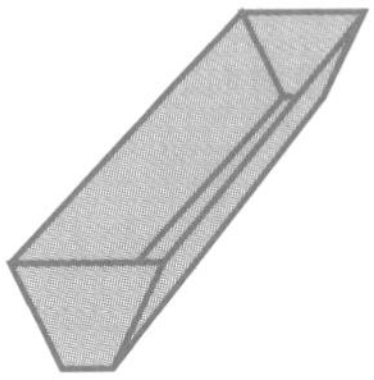

Figure 6.8.19

(i) Suppose the trough is full of water. What is the force exerted by the water on the end of the trough?

(ii) What is the force on the end of the trough if there is water in the trough to a depth of 1 ft?

**42. This problem should be attempted only if you have read Section 6.6 on work.

A hollow cylinder with open ends and radius 3 in stands vertically with its bottom under water and its top above the surface. A piston fits inside the cylinder, moving freely but making a watertight seal (see Figure 6.8.20). How much work is done in pushing the piston down to a depth of 4 ft? (*Hint:* The upward force on the piston comes from the water pressure on the piston head.)

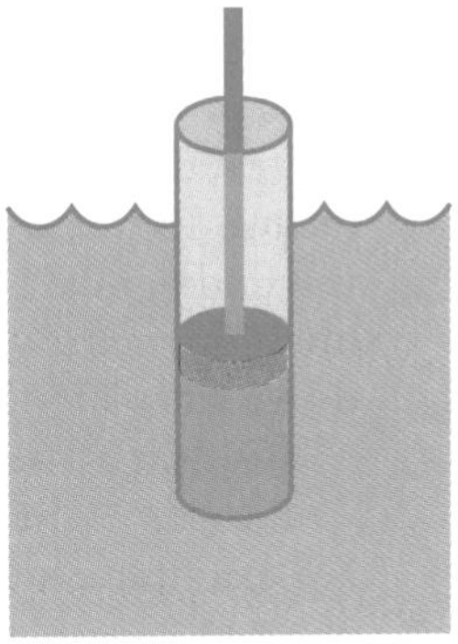

Figure 6.8.20

*43. Consider the cube of Example 5. Show that the magnitude of the net upward buoyant force exactly equals the force due to gravity on the amount of water that it would take to fill the cube.

*44. Find the total force exerted by the water on one side of a round plate, with radius 30 in, lying vertically with its center at a depth of 2 ft.

POINT TO PONDER

When we figured out the pressure at a depth h, we took a flat plate lying horizontally and considered the fluid lying directly over it. This gave us the correct answer, but if we think about it a little more, we may begin to have doubts.

For instance, it is possible that fluid not directly above the plate might still have some effect on it. More troubling is the possibility that there might not be that much fluid lying over the plate, owing to the shape of the container.

As shown in Figure 6.8.21(i), the container could have a narrow neck, in which case there is only a thin column of fluid over the plate. It could even be (as in Figure 6.8.21(ii)) that the plate is lying in a part of the container that is covered so it does not even reach to the surface level. Could it be that the pressure is different in these situations?

Figure 6.8.21(i)

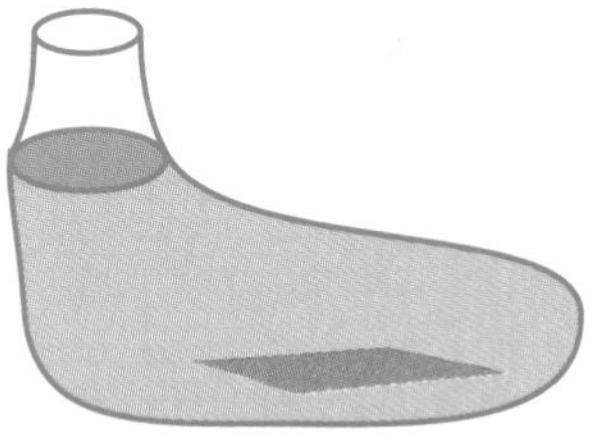

Figure 6.8.21(ii)

In fact, our original argument does give the correct value, despite these objections. Describe how each of the apparent objections can be answered.

6.9 Arclength; Areas of Surfaces of Revolution

SURFACE OF REVOLUTION
SURFACE AREA

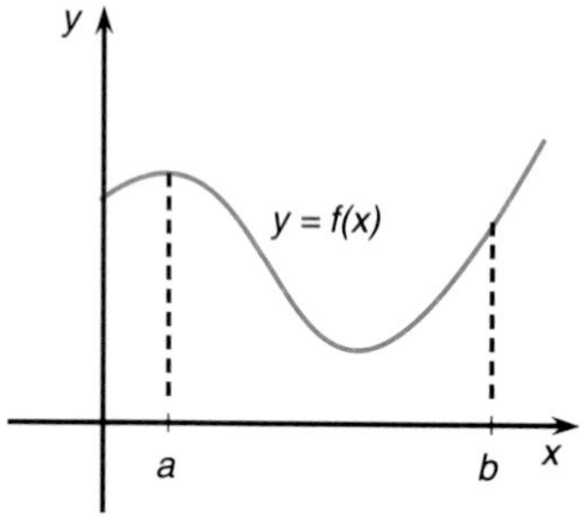

Figure 6.9.1

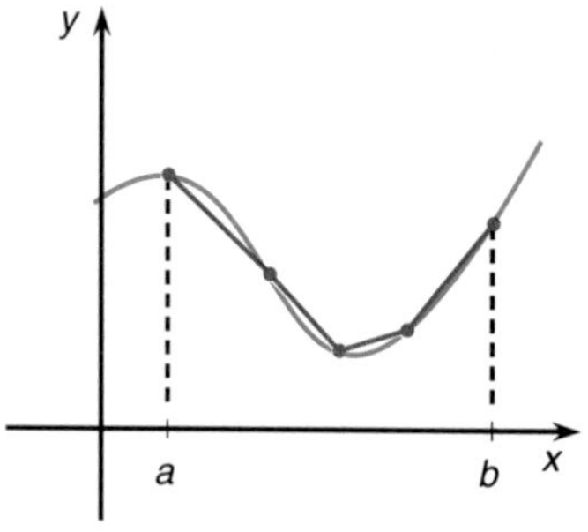

Figure 6.9.2

If we consider a curve drawn in the plane, we might want to find its length. In this section we will discuss this question, beginning by asking what we mean by the length of a curve. We are accustomed to the idea that a circle of radius r has circumference (i.e., length) equal to $2\pi r$; we have used this fact in discussing trigonometric functions in Section 2.6 and for finding volumes of solids of revolution in Section 6.4. However, if we think about it carefully, it becomes less clear what this familiar fact means. In this section we learn how to define and calculate the length of a curve using integrals. Having done this, we will use what we have learned to ask a related question, namely, what do we mean by the surface area of a curved surface, and how do we calculate it?

Consider the graph of a function $y = f(x)$, shown in Figure 6.9.1. We might want to know the length of the part of the curve that lies between $x = a$ and $x = b$. For example, we might like to know the circumference of a circle (strictly speaking, it would be easier to deal with a semicircle $y = \sqrt{r^2 - x^2}$). It could also be that the curve represents the path of a moving object and what we want to know is the total distance it travels.

Based on our experience with similar questions, a reasonable approach to try would be to *approximate* the curve with a number of short straight line segments as in Figure 6.9.2. The sum of their lengths ought to approximate what we should mean by the length of the curve. If we use more and more segments, the sums of their lengths might approach a limit, which should be the length of the curve.

Suppose $\mathcal{P} = \{x_0, \ldots, x_n\}$ is a partition of $[a, b]$. The corresponding points on the curve are $\big(x_0, f(x_0)\big), \big(x_1, f(x_1)\big), \ldots, \big(x_n, f(x_n)\big)$. The kth segment joining consecutive points is the segment from $\big(x_{k-1}, f(x_{k-1})\big)$ to $\big(x_k, f(x_k)\big)$. Its length is

$$\sqrt{(x_k - x_{k-1})^2 + \big(f(x_k) - f(x_{k-1})\big)^2} = \sqrt{(\Delta x_k)^2 + \big(f(x_k) - f(x_{k-1})\big)^2}.$$

The Mean Value Theorem (Theorem 4.2.5, p. 241), applied to the function $f(x)$ on the interval $[x_{k-1}, x_k]$, says that there is some $x_k^* \in [x_{k-1}, x_k]$ so that

$$f(x_k) - f(x_{k-1}) = f'(x_k^*)(x_k - x_{k-1}) = f'(x_k^*)\Delta x_k.$$

Using this, we find that the length of the kth segment is

$$\sqrt{(\Delta x_k)^2 + \big(f'(x_k^*)\Delta x_k\big)^2} = \sqrt{\big(1 + f'(x_k^*)^2\big)\Delta x_k^2} = \sqrt{1 + f'(x_k^*)^2}\,\Delta x_k.$$

We approximate the length of the curve by the sum of the lengths of all the segments, that is,

$$\textit{length} \approx \sum_{k=1}^{n} \sqrt{1 + f'(x_k^*)^2}\,\Delta x_k.$$

As $n \to \infty$, we are considering more and more segments, and if $\|\mathcal{P}\| \to 0$, then it seems reasonable to expect that the segments will all become shorter and shorter and the sum of their lengths will approach the length of the curve. As $n \to \infty$, the sum approaches the integral $\int_a^b \sqrt{1 + f'(x)^2}\,dx$.

Rather than worrying any more about what is meant by the "length of a curve," we can simply use this integral to define the length, knowing that the argument we have just made makes it seem reasonable.

DEFINITION 6.9.1

Suppose $f(x)$ is a differentiable function on the interval $[a, b]$ and $f'(x)$ is continuous on $[a, b]$. The length of the curve $y = f(x)$ for $x \in [a, b]$, sometimes called the **arclength**, is defined to be

$$s = \int_a^b \sqrt{1 + f'(x)^2}\,dx.$$

It is traditional to use the letter s to represent arclength, and it is also common to encapsulate this definition by writing

$$ds = \sqrt{(dx)^2 + (dy)^2} = \sqrt{1 + \left(\frac{dy}{dx}\right)^2}\,dx.$$

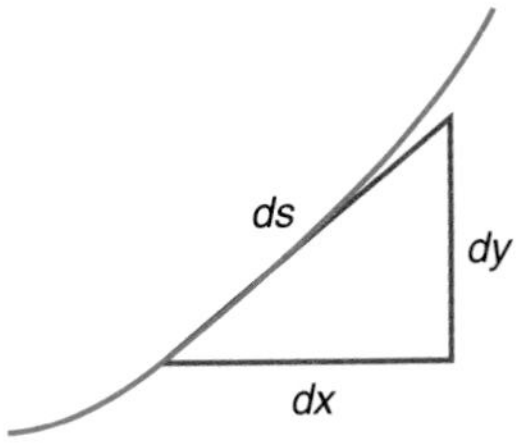

Figure 6.9.3

It may help to remember this last formula if we consider Figure 6.9.3, in which ds, the differential of the arclength, is the hypotenuse of the triangle whose base is dx and whose height is dy.

EXAMPLE 1 Find the length of the curve $y = x^{3/2}$ for x between $x = 1$ and $x = 2$.

Solution Applying the definition with $f(x) = x^{3/2}$, we find $f'(x) = \frac{3}{2}x^{1/2}$, and the arclength of the curve equals

$$s = \int_1^2 \sqrt{1 + \left(\frac{3}{2}x^{1/2}\right)^2}\,dx = \int_1^2 \sqrt{1 + \frac{9}{4}x}\,dx$$

$$\boxed{\begin{aligned} u &= 1 + \frac{9}{4}x \\ du &= \frac{9}{4}\,dx \end{aligned}}$$

$$= \int_{13/4}^{11/2} u^{1/2}\frac{4}{9}\,du$$

$$= \frac{4}{9}\frac{2}{3}u^{3/2}\Big|_{13/4}^{11/2}$$

$$\approx 2.086.$$

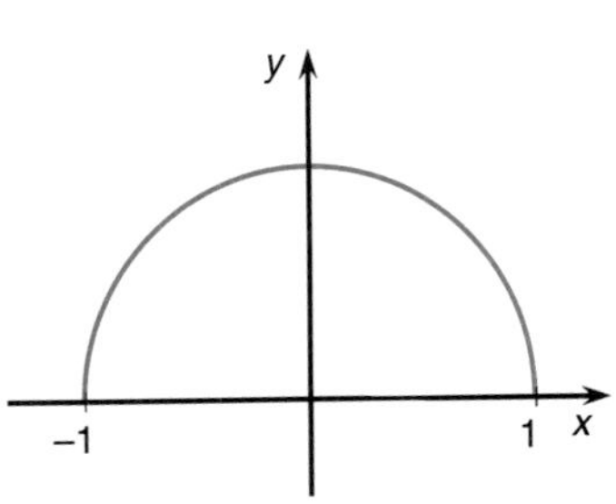

Figure 6.9.4

EXAMPLE 2 **Circumference of a Circle** Find the circumference of a circle of radius 1.

Solution It is easier to find the arclength of the upper semicircle and double it. We consider the function $f(x) = \sqrt{1 - x^2}$, for $x \in [-1, 1]$, whose graph is the upper semicircle (see Figure 6.9.4). We find $f'(x) = \frac{1}{2}(1 - x^2)^{-1/2}(-2x) = -\frac{x}{\sqrt{1-x^2}}$.

The length of the semicircle is

$$s = \int_{-1}^1 \sqrt{1 + f'(x)^2}\,dx = \int_{-1}^1 \sqrt{1 + \left(\frac{x^2}{1 - x^2}\right)}\,dx = \int_{-1}^1 \sqrt{\frac{(1 - x^2) + x^2}{1 - x^2}}\,dx$$

$$= \int_{-1}^1 \sqrt{\frac{1}{1 - x^2}}\,dx = \int_{-1}^1 \frac{1}{\sqrt{1 - x^2}}\,dx = \arcsin x\Big|_{-1}^1 = \frac{\pi}{2} - \left(-\frac{\pi}{2}\right) = \pi.$$

The length of the semicircle is π, so the length of the circle of radius 1 is 2π.

With a programmable calculator we could try approximating the arclength by dividing the interval $[-1, 1]$ into n equal subintervals and adding up the distances between adjacent pairs of points...

TI-85

PRGM F2(EDIT) ARCL ENTER F3 F1(Input) N EXIT ENTER 0 STO K ENTER 0 STO S ENTER Lbl B ENTER K + 1 STO K ENTER S + √ ((2 ÷ N) ^ 2 + (√ (1 − ((−) 1 + 2 K ÷ N) x^2) − √ (1 − ((−) 1 + 2(K −1) ÷ N) x^2)) x^2) STO S ENTER F4 F1(If) (K TEST F2(<) N) ENTER Goto B ENTER S ENTER EXIT EXIT EXIT PRGM F1(NAME) ARCL ENTER 20 ENTER

SHARP EL9300

▼ ▼ (NEW) ENTER ENTER ARCL ENTER COMMAND A 3 (Input) N ENTER K = 0 ENTER S = 0 ENTER COMMAND B 1 (Label) 1 ENTER K = K + 1 ENTER A = √ 1 − ((−) 1 + 2 K ÷ N) x^2 ENTER B = √ 1 − ((−) 1 + 2 (K − 1) ÷ N) x^2 ENTER S = S + √ (2 ÷ N) x^2 + (A − B) x^2) ENTER COMMAND 3 (If) K COMMAND C 2 (<) N COMMAND B 2 (Goto) 1 ENTER COMMAND A 1 (Print) S ENTER COMMAND A 6 (End) ENTER (RUN) ENTER (ARCL) ENTER 20 ENTER

CASIO f_x-6300G (f_x-7700GB)

MODE 2 EXE ? → N : 0 → K : 0 → S : Lbl 1 : K + 1 → K : S + √ ((2 ÷ N) x^2 + (√ (1 − ((−) 1 + 2 K ÷ N) x^2) − √ (1 − ((−) 1 + 2 (K − 1) ÷ N) x^2)) x^2) → S : K < N ⇒ Goto 1 : S ◢ MODE 1 Prog 0 EXE 20 EXE (on the f_x-7700GB, : is PRGM F6, and Prog is PRGM F3.)

HP 48SX

ATTN ↱ CLR ≪≫ "" KEY IN N ▶ SPC PROMPT SPC ′ N ▶ STO 0 SPC 1 SPC N SPC FOR K SPC 2 SPC N ÷ x^2 1 +/− SPC 2 SPC K × N ÷ + x^2 +/− 1 + $\sqrt{x}$ 1 +/− SPC 2 SPC K SPC 1 − × N ÷ + x^2 +/− 1 + $\sqrt{x}$ − x^2 + $\sqrt{x}$ + NEXT ▶ ▶ ENTER ′ ARCL STO VAR ARCL 20 ENTER ↰ CONT

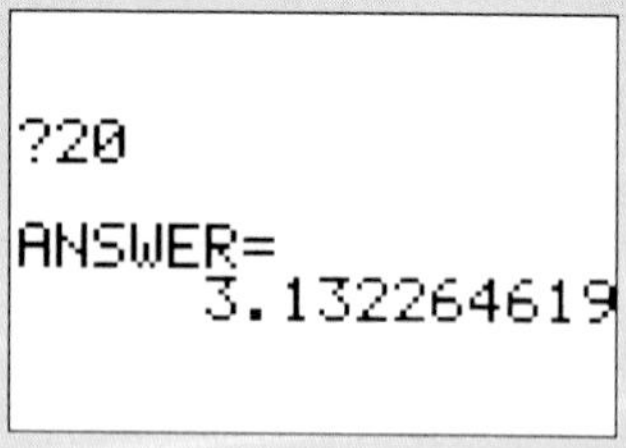

It is interesting to see how close the approximation is to π for various values of n.

The unpleasant thing about calculating arclength is that the integrals that result are often difficult to evaluate. For instance, if we try to find the length of some piece of the parabola $y = x^2$, we have to evaluate the integral $\int_a^b \sqrt{1 + 4x^2}\, dx$, which we do not know how to do.

After we have learned some additional techniques for finding integrals, which is the subject of Chapter 8, we will be able to evaluate the arclengths of more curves.

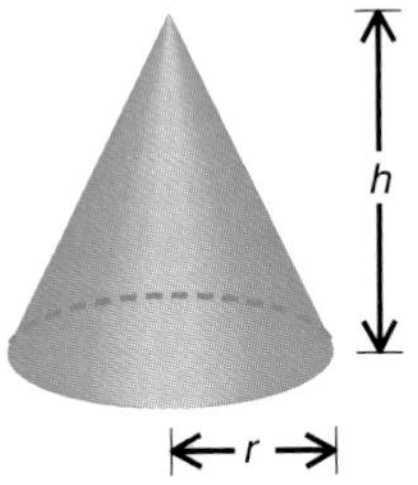

Figure 6.9.5(i)

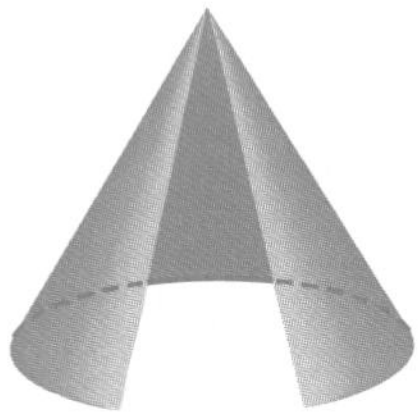

Figure 6.9.5(ii)

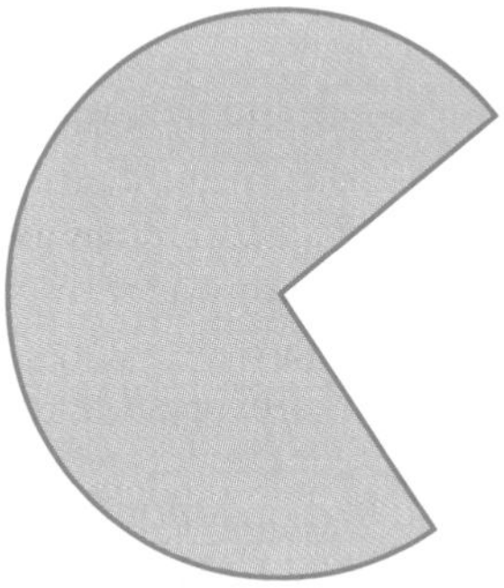

Figure 6.9.5(iii)

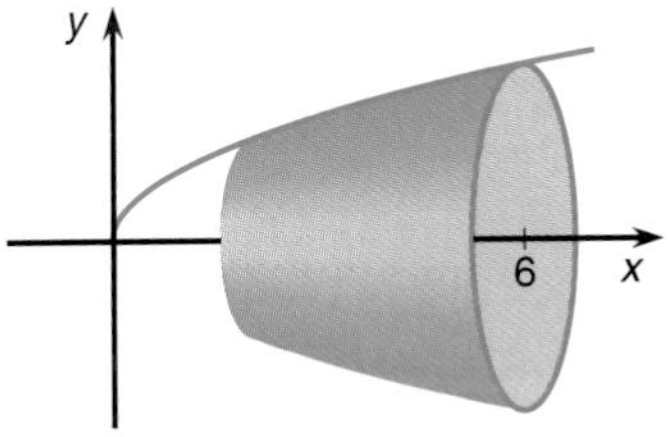

Figure 6.9.6

Surface Areas of Surfaces of Revolution

In many instances it is desirable to find the surface area of some region in space. For example, we might need to know the surface area of a sphere in order to figure out the amount of paint needed to cover a water tank; the surface area of an ellipsoid might be needed to estimate the rate at which a one-celled organism can absorb oxygen through its outer membrane; the surface area of a cone would tell us how much sheet metal was needed to make a funnel. There are many other examples, but it is interesting to notice that the ones mentioned so far are all **surfaces of revolution**, that is, the type of surface obtained by revolving a curve about a line (or, to put it another way, the outside surface of a solid of revolution). It turns out that areas are reasonably easy to find for this type of surface.

EXAMPLE 3 **Area of a Cone** Consider the cone shown in Figure 6.9.5(i), whose height is h and whose radius at the bottom is r. Find its surface area.

Solution To find the area of the cone, slit it open and spread it out flat, as shown in Figures 6.9.5(ii) and 6.9.5(iii). What results will be a sector of a disk. The radius of the disk will be the *slant height* of the cone, that is, the distance from top to bottom *along the sloping side*. By the Pythagorean Theorem (Theorem 1.5.1) this is $R = \sqrt{r^2 + h^2}$.

The circumference of the bottom of the cone is $2\pi r$; the circumference of the flattened disk is $2\pi R = 2\pi\sqrt{r^2 + h^2}$. The sector filled by the flattened cone covers a distance of $2\pi r$ at the outside edge, so the fraction of the whole disk that it covers is $\frac{2\pi r}{2\pi R} = \frac{r}{R} = \frac{r}{\sqrt{r^2+h^2}}$.

Since the area of the whole disk of radius R is πR^2, the area of the sector is $\frac{r}{R}\pi R^2 = \pi r R = \pi r\sqrt{r^2 + h^2}$.

Example 3 was done by using a trick and some simple geometry, but we cannot rely on such methods to work for more complicated curves. After all, the cone is obtained as a surface of revolution by revolving the simplest of all curves, a straight line. For more general surfaces it will be necessary to use integration.

EXAMPLE 4 Consider the surface of revolution obtained by revolving the curve $y = f(x) = \sqrt{x}$ about the x-axis. Find the surface area of the part of this surface that lies between $x = 2$ and $x = 6$ (see Figure 6.9.6).

Solution We consider a short interval of length dx on the x-axis and look at the corresponding short piece of the curve $y = f(x) = \sqrt{x}$. When it is revolved about the axis, it will sweep out a small band in the surface, as shown in Figure 6.9.7(i).

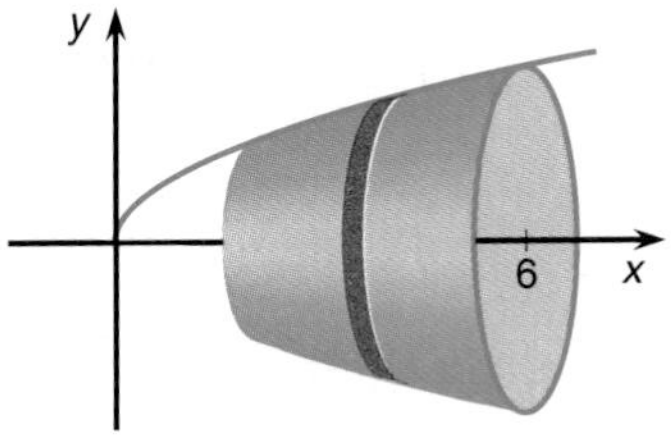

Figure 6.9.7(i)

Figure 6.9.7(ii)

To find the area of this narrow band, slit it and unwind it into a flat strip, as shown in Figure 6.9.7(ii). The width of the strip is just the length of the little piece of curve that we revolved, which is $ds = \sqrt{1+f'(x)^2}dx$. The length of the strip is the circumference of the band, which is 2π times its radius. Since the radius of the band is just the height of the curve above the x-axis, it is $f(x) = \sqrt{x}$, and the circumference is $2\pi\sqrt{x}$.

This suggests that the area of the band is $2\pi f(x)\,ds = 2\pi\sqrt{x}\sqrt{1+f'(x)^2}dx$. The surface area S of the whole surface of revolution is obtained by integrating:

$$S = \int 2\pi f(x)\,ds = 2\pi\int_2^6 \sqrt{x}\sqrt{1+f'(x)^2}dx = 2\pi\int_2^6 \sqrt{x}\sqrt{1+\left(\frac{1}{2}x^{-1/2}\right)^2}dx$$

$u = x + \frac{1}{4}$
$du = dx$

$$= 2\pi\int_2^6 \sqrt{x}\sqrt{1+\frac{1}{4x}}dx = 2\pi\int_2^6 \sqrt{x+\frac{1}{4}}dx = 2\pi\int_{9/4}^{25/4}\sqrt{u}\,du$$

$$= 2\pi\frac{2}{3}u^{3/2}\Big|_{9/4}^{25/4} = \frac{4}{3}\pi\left(\left(\frac{25}{4}\right)^{3/2} - \left(\frac{9}{4}\right)^{3/2}\right) = \frac{4}{3}\pi\left(\left(\frac{5}{2}\right)^3 - \left(\frac{3}{2}\right)^3\right)$$

$$= \frac{49}{3}\pi \approx 51.3.$$

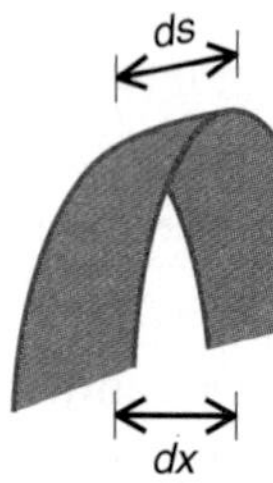

Figure 6.9.8

There are several things to notice here. One is that it is important not to confuse the arclength s with the surface area, which is traditionally denoted S. Second, even though we allowed x to change by the small amount dx, the piece of curve we revolved changed by ds. The idea is that the piece of curve is likely not horizontal, so it will be longer than the horizontal piece dx, and ds is the corresponding length on the slant, as shown in Figure 6.9.8. The effect of this is that the revolved band has width ds and area $2\pi f(x)\,ds$.

To find the surface area, we integrate: $S = \int 2\pi f(x)\,ds$. Notice that at this point it is not possible to put in the limits of integration because the limits are expressed in terms of the variable x and the variable of integration here is s.

The final step is to rewrite the integral in terms of x and then evaluate.

The technique illustrated in the preceding example can be used for any surface of revolution, and in fact the resulting integral can be taken as the definition of surface area.

DEFINITION 6.9.2

Suppose $f(x)$ is a differentiable function for $x \in [a, b]$, with $f(x) \geq 0$ for all $x \in [a, b]$, and $f'(x)$ continuous on $[a, b]$. If the graph of $y = f(x)$ is revolved about the x-axis, the resulting surface of revolution has **surface area** given by the following formula:

$$S = \int 2\pi f(x)\,ds = 2\pi\int_a^b f(x)\sqrt{1+f'(x)^2}\,dx.$$

If $f(x)$ is not necessarily nonnegative on $[a, b]$, the formula must be modified to

$$S = \int 2\pi|f(x)|\,ds = 2\pi\int_a^b |f(x)|\sqrt{1+f'(x)^2}\,dx.$$

EXAMPLE 5 **Area of a Sphere** Find the surface area of a sphere of radius r.

Solution We think of the sphere as the surface obtained by revolving the curve $y = f(x) = \sqrt{r^2 - x^2}$, for $-r \le x \le r$, about the x-axis. The difficulty is that the function $f(x)$ is not differentiable at the endpoints $x = \pm r$. Because of this, we cannot find the surface area of the whole sphere, which would involve integrating from $x = -r$ to $x = r$.

Instead, we consider the part of the sphere between $x = -a$ and $x = a$, for some a with $0 < a < r$ (see Figure 6.9.9). This is the surface obtained by removing equal "caps" from both sides. It is a surface of revolution, and the function $f(x)$ is differentiable on $[-a, a]$, so Definition 6.9.2 applies, and we can find the surface area.

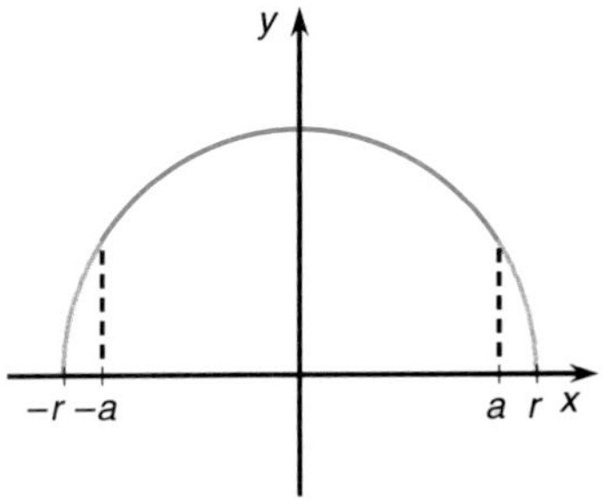

Figure 6.9.9

The formula tells us that the surface area is

$$\begin{aligned} S &= \int 2\pi f(x)\,ds = 2\pi \int_{-a}^{a} f(x)\sqrt{1 + f'(x)^2}\,dx \\ &= 2\pi \int_{-a}^{a} \sqrt{r^2 - x^2}\sqrt{1 + \left(\frac{-x}{\sqrt{r^2 - x^2}}\right)^2}\,dx \\ &= 2\pi \int_{-a}^{a} \sqrt{r^2 - x^2}\sqrt{1 + \left(\frac{x^2}{r^2 - x^2}\right)}\,dx \\ &= 2\pi \int_{-a}^{a} \sqrt{r^2 - x^2}\sqrt{\frac{(r^2 - x^2) + x^2}{r^2 - x^2}}\,dx = 2\pi \int_{-a}^{a} \sqrt{r^2}\,dx \\ &= 2\pi\sqrt{r^2} \int_{-a}^{a} dx = 2\pi r x\Big|_{-a}^{a} \\ &= 4\pi r a. \end{aligned}$$

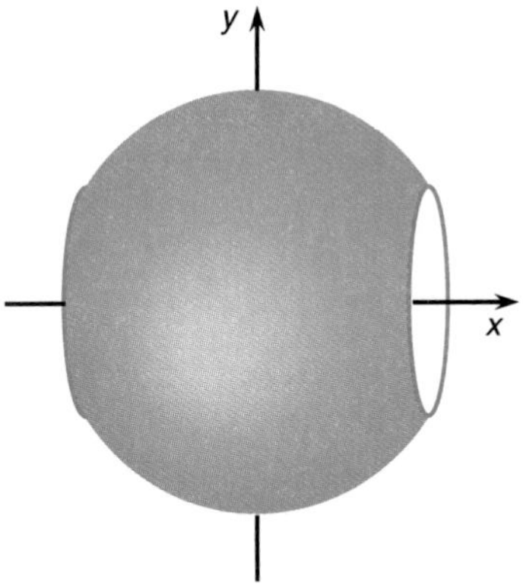

Figure 6.9.10

This is the area of the surface obtained by removing two equal caps from the sphere (see Figure 6.9.10). To find the area of the whole sphere, we let a tend to r and take the limit. This amounts to letting the caps shrink so that the surface that is left approaches the whole sphere. We find that the surface area of the sphere equals

$$S = \lim_{a \to r} 4\pi r a = 4\pi r^2.$$

As a mnemonic device, if you remember that the volume of the sphere is $\frac{4}{3}\pi r^3$, then its surface area $4\pi r^2$ is the derivative of $\frac{4}{3}\pi r^3$.

This formula was used in Section 4.1, even though we did not know how to derive it at that time.

It is also possible to form surfaces by revolving curves about the y-axis.

EXAMPLE 6 Find the area of the surface obtained by revolving about the y-axis the curve $y = x^2$, for $1 \le x \le 2$.

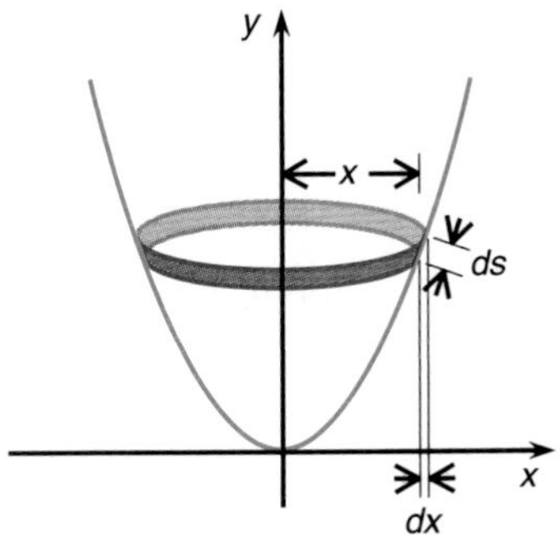

Figure 6.9.11

Solution There are two ways to approach this problem. One is to use y as the variable, write $x = \sqrt{y}$, and proceed as we did in Example 4.

On the other hand, we could also let x be the variable. In this case, as x passes through a small interval of width dx, the corresponding part of the curve will sweep out a band in the surface, as shown in Figure 6.9.11. The radius of the band is x, so its circumference is $2\pi x$. Its width is $ds = \sqrt{1 + \left(\frac{dy}{dx}\right)^2}\,dx$, so the area is

$2\pi x\sqrt{1+(y')^2}\,dx$. The total area of the surface is

$$S=\int_1^2 2\pi x\sqrt{1+\left(\frac{dy}{dx}\right)^2}\,dx=2\pi\int_1^2 x\sqrt{1+(2x)^2}\,dx=2\pi\int_1^2 x\sqrt{1+4x^2}\,dx$$

$u=1+4x^2$
$du=8x\,dx$

$$=2\pi\int_5^{17}\sqrt{u}\frac{1}{8}\,du=\frac{\pi}{4}\frac{2}{3}u^{3/2}\Big|_5^{17}$$

$$=\frac{\pi}{6}(17^{3/2}-5^{3/2})=\frac{\pi}{6}(17\sqrt{17}-5\sqrt{5})$$

$$\approx 30.85.$$

This illustrates a second way of finding surface areas. In the example, y was expressed as a function of x and the graph was revolved about the y-axis, but of course a similar formula can be given if x is expressed as a function of y and the graph is revolved about the x-axis.

All told, there are four possible formulas. Frequently, two of them will apply to the same situation. We list all four below, noting that the first one is the formula of Definition 6.9.2. We have used absolute value signs, but in practice it is often unnecessary, since the functions concerned are usually nonnegative anyway.

As in Definition 6.9.2, we are unable to include the limits of integration on the integrals with respect to s; it is usually only after we express them in terms of x or y that we can find the limits.

THEOREM 6.9.3

Surface Area for Surfaces of Revolution

Suppose $f(x)$ is differentiable and $f'(x)$ is continuous on $[a,b]$. The surface obtained by revolving the graph of $y=f(x)$ about the x-axis has area

$$S=\int 2\pi|f(x)|\,ds=2\pi\int_a^b |f(x)|\sqrt{1+f'(x)^2}\,dx.$$

The surface obtained by revolving the graph of $y=f(x)$ about the y-axis has area

$$S=\int 2\pi|x|\,ds=2\pi\int_a^b |x|\sqrt{1+f'(x)^2}\,dx.$$

Suppose $g(y)$ is differentiable and $g'(y)$ is continuous on $[c,d]$. The surface obtained by revolving the graph of $x=g(y)$ about the y-axis has area

$$S=\int 2\pi|g(y)|\,ds=2\pi\int_c^d |g(y)|\sqrt{1+g'(y)^2}\,dy.$$

The surface obtained by revolving the graph of $x=g(y)$ about the x-axis has area

$$S=\int 2\pi|y|\,ds=2\pi\int_c^d |y|\sqrt{1+g'(y)^2}\,dy.$$

Since many graphs can be expressed as functions of x or as functions of y, it often happens that two of these formulas are applicable to a particular problem. In this case it is really just a matter of which approach gives the easier integral.

Figure 6.9.12

EXAMPLE 7 **Areas of Cell Membranes** Consider two one-celled organisms. One is spherical, and the other has the elongated shape of the solid of revolution obtained by revolving the ellipse $x^2 + 2y^2 = 1$ around the x-axis (see Figure 6.9.12). If the two microorganisms have the same volume, it is reasonable to assume that their metabolisms are similar and in particular that they have similar requirements for nutrients.

Nutrients are absorbed through the cell membrane, so the amount that can be absorbed is proportional to the surface area. It has been suggested that the shapes of organisms may be determined by their relative efficiencies. Let us compare the two surface areas.

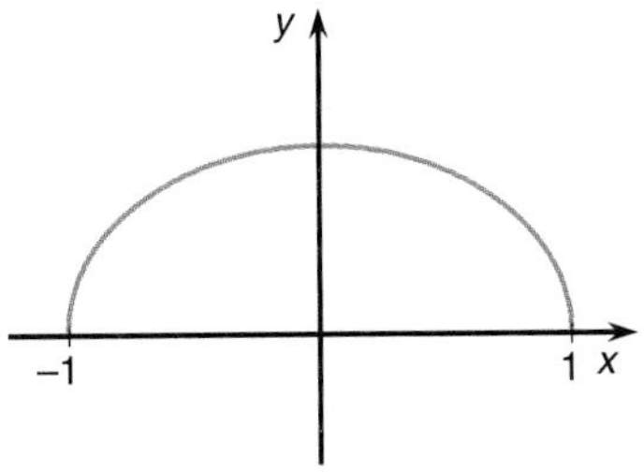

Figure 6.9.13

We begin by finding the volume of the elongated cell. Using the function $y = \frac{1}{\sqrt{2}}\sqrt{1 - x^2}$ illustrated in Figure 6.9.13, we find that the volume of the solid of revolution is

$$V = \int_{-1}^{1} \pi y^2\,dx = \frac{\pi}{2}\int_{-1}^{1}(1 - x^2)\,dx = \frac{\pi}{2}\left(x - \frac{x^3}{3}\right)\Bigg|_{-1}^{1} = \frac{2\pi}{3}.$$

Since the volume of the spherical cell is $\frac{4}{3}\pi r^3$ and it must equal this value $\frac{2\pi}{3}$, we find that the radius satisfies $r^3 = \frac{1}{2}$, so $r = 2^{-1/3}$. By Example 5 the surface area of the spherical cell is $4\pi r^2 = 4\pi 2^{-2/3} = 2^2\pi 2^{-2/3} = 2^{4/3}\pi \approx 7.916$.

Now we find the surface area of the elongated cell. It is

$$\begin{aligned} S &= \int 2\pi y\,ds = \pi\sqrt{2}\int_{-1}^{1}\sqrt{1 - x^2}\sqrt{1 + \frac{1}{2}\frac{x^2}{1 - x^2}}\,dx \\ &= \pi\sqrt{2}\int_{-1}^{1}\sqrt{1 - x^2 + \frac{x^2}{2}}\,dx = \pi\sqrt{2}\int_{-1}^{1}\sqrt{1 - \frac{x^2}{2}}\,dx. \end{aligned}$$

We know how to integrate $\sqrt{1 - x^2}$ from Example 6.1.2; the last integral can be changed by making the substitution $u = \frac{x}{\sqrt{2}}$, $du = \frac{1}{\sqrt{2}}dx$, so

$$\begin{aligned} S &= \pi\sqrt{2}\int_{-1}^{1}\sqrt{1 - \frac{x^2}{2}}\,dx = \pi\sqrt{2}\int_{-1/\sqrt{2}}^{1/\sqrt{2}}\sqrt{1 - u^2}\sqrt{2}\,du \\ &= 2\pi\int_{-1/\sqrt{2}}^{1/\sqrt{2}}\sqrt{1 - u^2}\,du = \pi\left(u\sqrt{1 - u^2} + \arcsin u\right)\Bigg|_{-1/\sqrt{2}}^{1/\sqrt{2}} \\ &= \pi\left(\frac{1}{\sqrt{2}}\frac{1}{\sqrt{2}} + \frac{\pi}{4} + \frac{1}{\sqrt{2}}\frac{1}{\sqrt{2}} + \frac{\pi}{4}\right) = \pi\left(1 + \frac{\pi}{2}\right) \\ &= \pi + \frac{\pi^2}{2} \approx 8.076. \end{aligned}$$

We find that the area of the elongated cell is only about 2% larger than the area of the spherical cell of the same volume, and this does not seem large enough to be likely to make a decisive difference in the structure of the organisms. ◢

Exercises 6.9

I

Find the arclength of each of the following curves.

1. $y = x^{3/2}$ for $x \in [2, 4]$
2. $y = 2x^{3/2}$ for $x \in [a, b]$
3. $y = x + 2$ for $x \in [-1, 3]$
4. $x = 2y - 3$ for $y \in [2, 3]$
5. $y = x^{3/2} + 7$ for $x \in [2, 4]$
6. $y = 4x^{3/2} - 12$ for $y \in [-8, 20]$
7. $y = (x + 2)^{3/2}$ for $x \in [-1, 2]$
8. $x = (y - 6)^{3/2} + 4$ for $y \in [7, 10]$
9. $x = y^{2/3}$ for $y \in [1, 8]$
10. $y = x^{2/3}$ for $x \in [8, 27]$

Find the surface area of the surface obtained by revolving each of the following curves about the specified axis.

11. $y = 3\sqrt{x}$, $x \in [1, 5]$; x-axis
12. $y = \sqrt{5x}$, $x \in [2, 3]$; x-axis
13. $y = 3x + 1$, $x \in [0, 2]$; x-axis
14. $y = x$, $x \in [-2, 2]$; x-axis
15. $y = 3x + 1$, $x \in [0, 2]$; y-axis
16. $y = |x|$, $x \in [-2, 2]$; x-axis
17. $y = 3$, $x \in [0, 7]$; x-axis
18. $y = x^3$, $x \in [0, 3]$; x-axis
19. $x = r$, $y \in [0, h]$; y-axis
20. $y = x$, $x \in [0, 3]$; y-axis

II

21. Find the arclength of a circle of radius r.
22. Set up (but do not attempt to evaluate) the integral that expresses the length of the parabola $y = x^2$ for x between $x = 0$ and $x = 1$.
23. Verify the formula for the area of a cone found in Example 3 (p. 487) by using Definition 6.9.2 (p. 488) and integrating.

*24. (i) Suppose the function $f(x) = cx + d$ is defined for all $x \in [a, b]$ and is positive there. Find the area of the surface obtained by revolving it about the x-axis.
(ii) Suppose the function $f(x) = cx + d$ is negative for all $x \in [a, b]$. Find the area of the surface obtained by revolving it about the x-axis.
(iii) What happens if the function $f(x) = cx + d$ is positive for some $x \in [a, b]$ and negative for some $x \in [a, b]$? Find the area of the surface obtained by revolving it about the x-axis.

25. Consider a cone of height h with its vertex pointing down and radius r at the top. Suppose $0 < \ell < h$ and consider the part of the cone that lies above the height ℓ. Find the area of this part of the surface.

*26. Consider Figure 6.9.14. The left-hand curve is $y = \sqrt{x}$, and the right-hand curve is (part of) the circle $x^2 + y^2 = 1$.

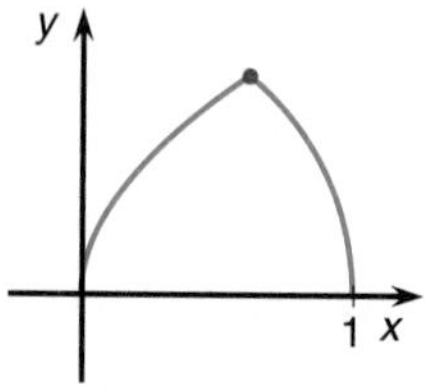

Figure 6.9.14

(i) Find the point of intersection of the two curves.
(ii) Find the area of the surface obtained by revolving the two curves about the x-axis.

*27. Consider the sphere of radius 4 with center at the origin, and find the surface area of the "cap," the part of the sphere that lies above $y = 3$.

*28. Consider the sphere of radius R with center at the origin, and find the surface area of the "cap," the part of the sphere that lies above $y = h$, where $0 < h < R$.

*29. (i) Find the area of the outer surface of the region obtained by drilling a hole of radius $\frac{1}{2}$ precisely through the center of a sphere of radius 3.
(ii) Find the total surface area of the solid described in part (i), that is, including the area of the surface inside the hole.

*30. (i) Assume that the earth is a perfect sphere of radius 4000 miles, and find a formula for the surface area of the part of the earth that lies north of latitude $\ell°$.
(ii) What latitude $\ell°$ has the property that the surface to the north of it has area equal to exactly one-quarter of the surface area of the whole planet?

*31. Suppose $0 < a < b$, and find the area of the surface obtained by revolving $y = \sqrt{x}$ for $a \leq x \leq b$ about the x-axis.

*32. Noticing that $y = \sqrt{x}$ is not differentiable at $x = 0$, find the area of the surface obtained by revolving $y = \sqrt{x}$ for $0 \leq x \leq b$ about the x-axis.

33. Find the total surface area of an ice cream cone consisting of a cone of height 4 in and radius 1 in at the

top, covered with ice cream in the shape of a hemisphere.

*34. Find the area of the surface obtained by revolving the curve $y = 3 - \sqrt{x}$ for $1 \le x \le 4$ about the line $y = 3$.

*35. Compare the surface areas of the solid obtained by revolving around the x-axis the curve $x^2 + 4y^2 = 1$ and the sphere of the same volume.

**36. Let $a > 1$ and revolve $x^2 + a^2y^2 = 1$ around the x-axis. (i) Find the surface area of the resulting solid. (ii) Compare with the surface area of the sphere with the same volume when $a = 3, 5, 10$.

37. Use a personal computer or programmable calculator to approximate the arclength of the parabola $y = x^2$ between $(0, 0)$ and $(1, 1)$, using the result of Example 5.2.3 (p. 356) with $n = 10$.

*38. (i) Do you expect the arclength of $y = x^2$ for $x \in [1, 2]$ to be greater or less than the arclength of $y = x^2$ for $x \in [0, 1]$, and by approximately what factor? (You might like to draw the graphs.)
(ii) Use Example 5.2.3 to check your guess.

Use Example 5.2.3 to estimate the surface area of the solid obtained by revolving each of the following around the x-axis.

39. $y = x^3$, $x \in [1, 3]$

40. $y = 1 + x^2$, $x \in [-1, 1]$

41. $y = e^x$, $x \in [0, 1]$

42. $y = \frac{1}{1+x^2}$, $x \in [0, 1]$

43. $y = \sin x$, $x \in [0, \pi]$

44. $y = \cos x$, $x \in [0, \pi]$

6.10 Numerical Integration; The Trapezoidal Rule and Simpson's Rule

TRAPEZOIDAL RULE
SIMPSON'S RULE
ERROR ESTIMATES

We have learned how to evaluate many integrals, and in Chapter 8 we will learn techniques that will permit us to evaluate many more, but there have also been disconcerting suggestions that there are some integrals that arise in fairly ordinary ways that we will never be able to do. Not only are there some that are extraordinarily difficult to find, but there are also some that cannot be done, in the sense that it can be proved that the indefinite integral cannot be expressed in terms of the functions we know, that is, rational functions, powers, roots, trigonometric functions, inverse trigonometric functions, logarithms and exponentials, and combinations of these.

In any case there are times when it is necessary to find the numerical value of a definite integral, or at least a good approximation to it, without being able to find the indefinite integral. In a sense this is how we defined integrals in the first place, by approximating areas under graphs with collections of rectangles.

It is possible to evaluate Riemann sums to approximate the value of an integral, as we did in Example 5.1.3 (p. 349). In this section we also learn two other methods that usually give better approximations. They are called the Trapezoidal Rule and Simpson's Rule.

There are two basic problems. One is that it would be desirable to perform these calculations efficiently. They are repetitive (adding up many similar terms) and are well suited to computers and sophisticated calculators, but with all such calculations efficiency is important. In addition to speed, it is a question of accuracy, since round-off errors accumulate and become more significant as the calculation becomes longer. This question is important and interesting, but it will not be discussed here.

The second problem is that it is all very well to have an accurate approximation, but its usefulness is very limited unless we know just how accurate it is. To put it the other way around, we know that the Riemann sums associated to the partitions dividing an interval into n subintervals of equal length will eventually converge to the integral. But suppose we need to know the value of some integral *correct to two decimal places.* How do we choose how big n must be to ensure this level of accuracy?

Suppose we want to estimate $\int_a^b f(x)\,dx$ using Riemann sums. To simplify things, let us decide to use the partition $\mathcal{P}_n$ that divides $[a, b]$ into n equal subintervals and choose x_k^* to be the midpoint of the kth subinterval. The corresponding Riemann sum $R_n = \sum_{k=1}^n f(x_k^*)\Delta x_k = \frac{b-a}{n}\sum_{k=1}^n f(x_k^*)$ is an approximation to the integral.

Assuming that $f(x)$ is differentiable on $[a, b]$, let B_1 be a "bound" for $f'(x)$, that is, a number satisfying $|f'(x)| \leq B_1$ for all $x \in [a, b]$. Then the accuracy with which the Riemann sum R_n approximates the integral can be expressed as follows:

$$\left|\int_a^b f(x)\,dx - R_n\right| \leq \frac{B_1(b-a)^2}{2n}. \qquad (6.10.1)$$

The approximation by a Riemann sum is accurate to at least this amount. In particular, we can use this result to decide how big n needs to be to ensure an approximation of specified accuracy.

EXAMPLE 1 Consider the integral $\int_0^1 x^4\,dx$. How large does n have to be to ensure that the Riemann sum R_n will be accurate to two decimal places?

Solution To be accurate to two decimal places means to be accurate to within 0.005. In this case the derivative of the integrand is $\frac{d}{dx}(x^4) = 4x^3$, which is always less than or equal to $B_1 = 4$ on $[0, 1]$.

By Inequality 6.10.1 the Riemann sum associated to the midpoints of $\mathcal{P}_n$ is accurate to within $B_1\left(\frac{1^2}{2n}\right) = \frac{B_1}{2n} = \frac{2}{n}$. To make this less than or equal to 0.005, we need $\frac{2}{n} \leq 0.005$, or $n \geq 400$.

The difficulty here is that accuracy to two decimal places is not terribly good, and evaluating a Riemann sum with 400 terms in it is quite tedious. We have done this example to illustrate the principle, but it is usually not very practical to use straight Riemann sums to estimate integrals. Fortunately, other methods are available.

The reason that Inequality 6.10.1 involves the size of the derivative is roughly speaking that the error in approximating the integral by the sum arises because the values of $f(x)$ on a particular subinterval will differ from the value of $f(x)$ at the midpoint. The amount by which the values of $f(x)$ can change on a subinterval depends on the derivative $f'(x)$, the rate of change of $f(x)$.

We will prove Inequality 6.10.1 at the end of this section, in Example 5, but for now we will proceed to other approximation techniques.

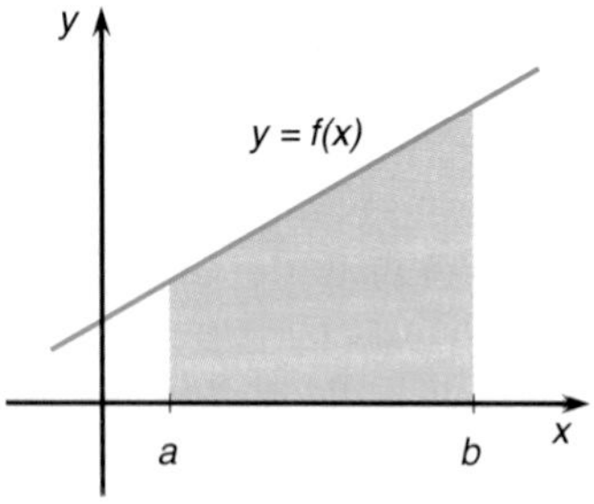

Figure 6.10.1

The Trapezoidal Rule

Suppose $y = f(x)$ is a straight line, as illustrated in Figure 6.10.1. We would like to find the area of the shaded region, which is called a **trapezoid**.

EXAMPLE 2 Find the area of the trapezoid illustrated in Figure 6.10.1.

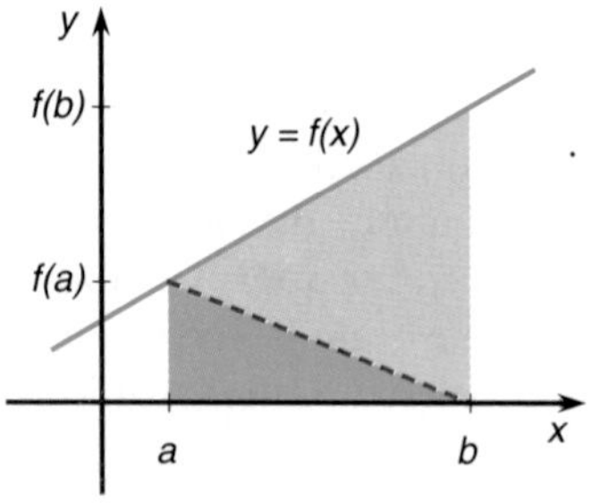

Figure 6.10.2

Solution In Figure 6.10.2 a dashed line has been drawn on the trapezoid, dividing it into two triangles. A simple way to find the area of the trapezoid is to find the area of each triangle and add them. To find the area of the upper triangle, we think of its vertical right side as the "base," which has length $f(b)$, and the "height" will be the distance from this vertical line to the opposite corner, which is just $b-a$ (the length of the interval $[a, b]$). So the area is $\frac{1}{2}$*base* $\times$ *height* $= \frac{1}{2}f(b)(b-a)$.

Similarly, the area of the lower triangle is $\frac{1}{2}f(a)(b-a)$, and the total area of the trapezoid is

$$\frac{f(a)+f(b)}{2}(b-a).$$

We can think of $\frac{f(a)+f(b)}{2}$ as the average height of the line $y=f(x)$, and the formula then says the area of the trapezoid is the length of the interval $[a, b]$ multiplied by the average height of the curve. This idea is illustrated in Figure 6.10.3.

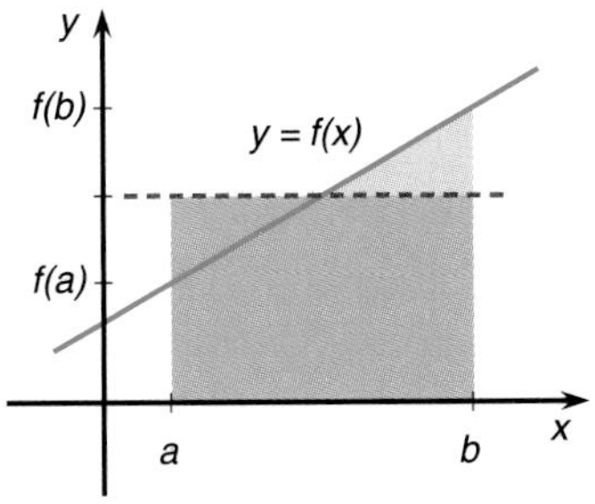

Figure 6.10.3

To approximate the area of the region under a graph, we can try filling the region with trapezoids, as shown in Figure 6.10.4. We can easily calculate the total area of the trapezoids, which should be approximately equal to the area under the curve.

If we use a larger number of trapezoids, the approximation should be better. This suggests trying the approximation with a large number of trapezoids, say, n of them.

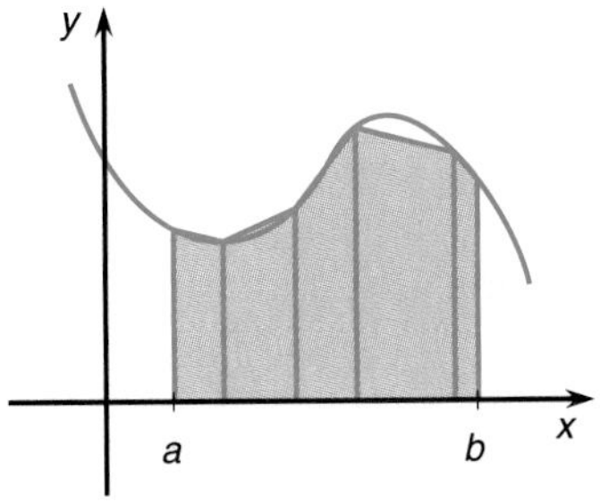

Figure 6.10.4

EXAMPLE 3 Estimate the area under the curve $y=f(x)=x^2$ between $x=0$ and $x=1$, using n trapezoids of equal width.

Solution We divide the interval $[0,1]$ into n equal segments and construct a trapezoid on each of them, as shown in Figure 6.10.5 (in the illustration, $n=4$). The first trapezoid lies over the interval $[0, \frac{1}{n}]$. The heights of its sides (at $x=0$ and $x=\frac{1}{n}$) are $f(0)=0^2=0$ and $f\left(\frac{1}{n}\right)=\left(\frac{1}{n}\right)^2=\frac{1}{n^2}$. So as we have seen in Example 1, the area of the trapezoid is $\frac{1}{2}\left(f(0)+f\left(\frac{1}{n}\right)\right)\left(\frac{1}{n}-0\right)=\frac{1}{2}\left(0+\frac{1}{n^2}\right)\left(\frac{1}{n}\right)$. Similarly, the area of the second trapezoid, which lies over $\left[\frac{1}{n}, \frac{2}{n}\right]$, is $\frac{1}{2}\left(f\left(\frac{1}{n}\right)+f\left(\frac{2}{n}\right)\right)\left(\frac{2}{n}-\frac{1}{n}\right)=\frac{1}{2}\left(\frac{1}{n^2}+\frac{4}{n^2}\right)\left(\frac{1}{n}\right)$.

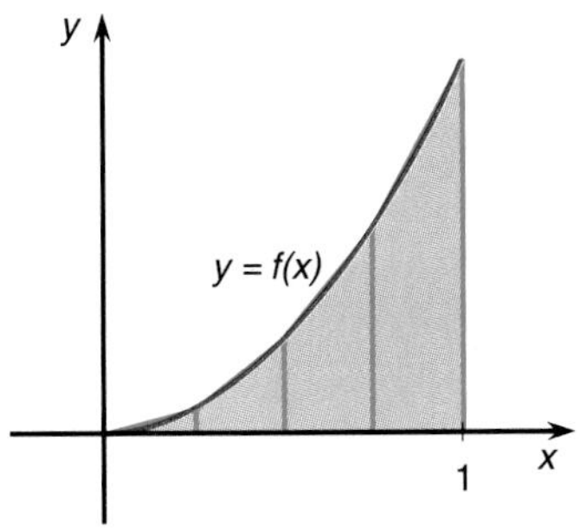

Figure 6.10.5

The kth trapezoid lies over the interval $\left[\frac{k-1}{n}, \frac{k}{n}\right]$, and its area is

$$\frac{1}{2}\left(f\left(\frac{k-1}{n}\right)+f\left(\frac{k}{n}\right)\right)\left(\frac{k}{n}-\frac{k-1}{n}\right)=\frac{1}{2}\left(\frac{(k-1)^2}{n^2}+\frac{k^2}{n^2}\right)\left(\frac{1}{n}\right).$$

The total area of all the trapezoids is then

$$\begin{aligned}\frac{1}{2}\sum_{k=1}^{n}\left(\frac{(k-1)^2}{n^2}+\frac{k^2}{n^2}\right)\frac{1}{n} &= \frac{1}{2n^3}\sum_{k=1}^{n}\left((k-1)^2+k^2\right)\\ &= \frac{1}{2n^3}\left(\sum_{k=1}^{n}(k-1)^2+\sum_{k=1}^{n}k^2\right)\\ &= \frac{1}{2n^3}\left(\sum_{j=1}^{n-1}j^2+\sum_{k=1}^{n}k^2\right).\end{aligned}$$

For the second-last sum we shifted to $j=k-1$ and then dropped off the term with $j=0$ because it equals zero.

Using Theorem 5.1.1 (p. 348), we see that the total area of the trapezoids is

$$\frac{1}{2n^3}\left(\frac{(n-1)n(2n-1)}{6}+\frac{n(n+1)(2n+1)}{6}\right)$$

For the first sum we use Theorem 5.1.1 with n replaced by $n-1$.

$$= \frac{1}{12n^2}\big((n-1)(2n-1)+(n+1)(2n+1)\big)$$
$$= \frac{1}{12n^2}\big(2n^2-3n+1+2n^2+3n+1\big)$$
$$= \frac{1}{12n^2}(4n^2+2) = \frac{2}{12n^2}(2n^2+1)$$
$$= \frac{1}{6}\left(\frac{2n^2+1}{n^2}\right).$$

As we did in Section 5.1 with rectangles, we could take the limit as $n \to \infty$. This amounts to considering more and more trapezoids, each of which is narrower and narrower; the total area approaches the area under the curve.

The area of the region under the curve is $\lim_{n\to\infty} \frac{1}{6}\left(\frac{2n^2+1}{n^2}\right) = \frac{2}{6} = \frac{1}{3}$.

The same approach can be used to approximate the area under other curves.

EXAMPLE 4 If $f(x) \geq 0$ for all $x \in [a, b]$, approximate the area under the curve $y = f(x)$ between $x = a$ and $x = b$ by dividing the interval into n equal subintervals and using trapezoids.

Solution The length of each subinterval will be $\frac{b-a}{n}$. We label the endpoints of the subintervals as $x_0, x_1, x_2, \ldots, x_n$, so $x_0 = a$, $x_1 = a + \frac{b-a}{n}$, $x_2 = a + 2\frac{b-a}{n}$, $x_3 = a + 3\frac{b-a}{n}, \ldots, x_k = a + k\frac{b-a}{n}, \ldots, x_{n-1} = a + (n-1)\frac{b-a}{n}$, $x_n = a + n\frac{b-a}{n} = b$.

The kth point x_k is found by moving k subintervals from $x = a$. Since each subinterval has width $\frac{b-a}{n}$, $x_k = a + k\frac{b-a}{n}$.

The kth subinterval will be $[x_{k-1}, x_k] = \left[a + (k-1)\frac{b-a}{n}, a + k\frac{b-a}{n}\right]$. The trapezoid over this subinterval has sides whose heights are $f(x_{k-1})$ and $f(x_k)$, so its area is $\frac{f(x_{k-1})+f(x_k)}{2}\frac{(b-a)}{n}$.

The total area of all the trapezoids is

$$\sum_{k=1}^{n} \frac{f(x_{k-1})+f(x_k)}{2}\frac{(b-a)}{n} = \frac{(b-a)}{2n}\sum_{k=1}^{n}\big(f(x_{k-1})+f(x_k)\big)$$
$$= \frac{(b-a)}{2n}\left(\sum_{k=1}^{n} f(x_{k-1}) + \sum_{k=1}^{n} f(x_k)\right)$$
$$= \frac{(b-a)}{2n}\left(f(x_0) + \left(\sum_{j=1}^{n-1} f(x_j)\right) + \left(\sum_{k=1}^{n-1} f(x_k)\right) + f(x_n)\right)$$
$$= \frac{b-a}{2n}\big(f(x_0)+f(x_n)\big) + \frac{b-a}{n}\sum_{k=1}^{n-1} f(x_k).$$

The two summations in the second-last line are equal; we combine them and cancel the 2 in the original denominator $2n$.

Note that $f(a) = f(x_0)$, $f(b) = f(x_n)$.

The formula that we obtained in Example 4 can be used even if $f(x)$ is not nonnegative. In this case it approximates the *signed* area. The formula is important enough to have a name.

THEOREM 6.10.2

The Trapezoidal Rule

The signed area enclosed by the graph of $y = f(x)$ between $x = a$ and $x = b$ can be approximated by

$$T_n = \big(f(a)+f(b)\big)\frac{b-a}{2n} + \big(f(x_1)+f(x_2)+\ldots+f(x_{n-1})\big)\frac{b-a}{n}$$

$$= \big(f(x_0) + 2f(x_1) + 2f(x_2) + \ldots + 2f(x_{n-2}) + 2f(x_{n-1}) + f(x_n)\big)\frac{(b-a)}{2n}$$

$$= \frac{b-a}{2n}\big(f(x_0) + f(x_n)\big) + \frac{b-a}{n}\sum_{k=1}^{n-1} f(x_k),$$

where $x_k = a + k\frac{b-a}{n}$, for $k = 0, 1, \ldots, n$.

T_n is the total area of the trapezoids constructed by dividing the interval $[a, b]$ into n equal subintervals.

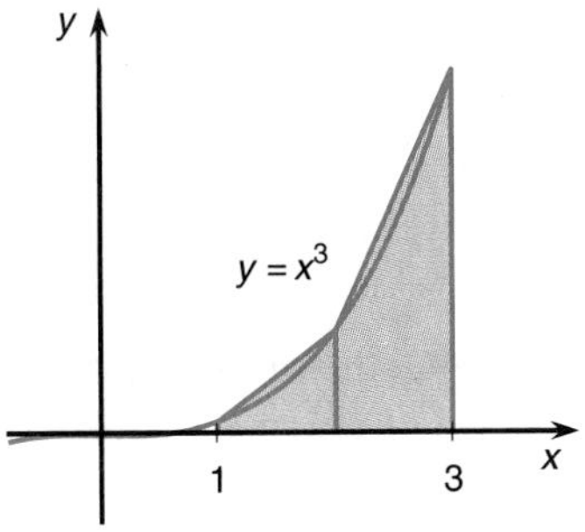

Figure 6.10.6

EXAMPLE 5 Use the Trapezoidal Rule to approximate the area under the curve $y = x^3$ between $x = 1$ and $x = 3$, using the values $n = 2,\ 4,\ 6$.

Solution With $n = 2$, as shown in Figure 6.10.6, we have $x_0 = 1$, $x_1 = 2$, $x_2 = 3$, and

$$T_2 = \big(f(1) + f(3)\big)\frac{2}{4} + \big(f(2)\big)\frac{2}{2} = (1 + 27)\frac{1}{2} + 8(1)$$
$$= 14 + 8 = 22.$$

With $n = 4$ we have $x_0 = 1,\ x_1 = \frac{3}{2},\ x_2 = 2,\ x_3 = \frac{5}{2},\ x_4 = 3$, and

$$T_4 = \big(f(1) + f(3)\big)\frac{2}{8} + \left(f\left(\frac{3}{2}\right) + f(2) + f\left(\frac{5}{2}\right)\right)\frac{2}{4}$$
$$= (1 + 27)\frac{1}{4} + \left(\frac{27}{8} + 8 + \frac{125}{8}\right)\frac{2}{4} = 7 + \left(\frac{152}{8} + 8\right)\frac{2}{4} = 7 + (27)\frac{1}{2}$$
$$= 20\frac{1}{2}.$$

With $n = 6$ we have $x_0 = 1$, $x_1 = \frac{4}{3}$, $x_2 = \frac{5}{3}$, $x_3 = 2$, $x_4 = \frac{7}{3}$, $x_5 = \frac{8}{3}$, $x_6 = 3$, and

$$T_6 = \big(f(1) + f(3)\big)\frac{2}{12} + \left(f\left(\frac{4}{3}\right) + f\left(\frac{5}{3}\right) + f(2) + f\left(\frac{7}{3}\right) + f\left(\frac{8}{3}\right)\right)\frac{2}{6}$$
$$= (1 + 27)\frac{1}{6} + \left(\frac{64}{27} + \frac{125}{27} + 8 + \frac{343}{27} + \frac{512}{27}\right)\frac{2}{6}$$
$$= \frac{14}{3} + \left(\frac{1044}{27} + 8\right)\frac{1}{3} = \frac{14}{3} + \left(38\frac{2}{3} + 8\right)\frac{1}{3}$$
$$= 20\frac{2}{9}.$$

It is interesting to compare this example with Example 5.1.4 (p. 350), which approximated the same area with Riemann sums. Since we now know that the actual area is $\int_1^3 x^3\,dx = \frac{x^4}{4}\Big|_1^3 = 20$, we find that the Trapezoidal Rule gives much better approximations.

Suppose we want to estimate $\int_a^b f(x)\,dx$ using the Trapezoidal Rule. It turns out that the accuracy we can expect will depend on the size of the second derivative. Assume that $f(x)$ is twice-differentiable on $[a, b]$, and suppose that B_2 is a bound for $f''(x)$ on $[a, b]$, that is, assume $|f''(x)| \leq B_2$ for all $x \in [a, b]$. Let T_n be the Trapezoidal Rule approximation to $\int_a^b f(x)\,dx$ using n subintervals. Then

$$\left|\int_a^b f(x)\,dx - T_n\right| \leq \frac{B_2(b-a)^3}{12n^2}. \tag{6.10.3}$$

For comparison we consider the same integral that we approximated by Riemann sums in Example 1.

EXAMPLE 6 Consider the integral $\int_0^1 x^4\,dx$. How large does n have to be to ensure that the Trapezoidal Rule approximation T_n will be accurate to two decimal places?

Solution In this case the second derivative of the integrand is $\frac{d^2}{dx^2}(x^4) = 12x^2$, so for a bound we can choose $B_2 = 12$.

By Inequality 6.10.3 the trapezoidal approximation T_n will be accurate to within $B_2\left(1^3/(12n^2)\right) = \frac{1}{n^2}$. To make this less than or equal to 0.005, we need $\frac{1}{n^2} \leq 0.005$, or $n^2 \geq 200$. This amounts to $n \geq \sqrt{200} \approx 14.14$. Since n must be an integer, we should use $n = 15$.

This is a *vast* improvement over the approximation R_n by Riemann sums. Fifteen terms is a sufficiently small sum that it is reasonable to evaluate it on a calculator. Doing this and writing down only five decimal places, we find the points at which we evaluate are $x_0 = 0$, $x_1 = \frac{1}{15}, \ldots, x_{15} = 1$, so

$$\begin{aligned}
T_{15} &= \frac{b-a}{2n}\left(f(x_0) + f(x_{15}) + 2\sum_{k=1}^{14} f(x_k)\right) \\
&= \frac{1}{30}\left(0 + 1 + 2\left(\left(\frac{1}{15}\right)^4 + \left(\frac{2}{15}\right)^4 + \ldots + \left(\frac{14}{15}\right)^4\right)\right) \\
&\approx \frac{1}{30}\big(1 + 2(0.00002 + 0.00032 + 0.0016 + 0.00506 + 0.01235 + 0.0256 \\
&\quad +0.04743 + 0.08091 + 0.1296 + 0.19753 + 0.28920 + 0.4096 \\
&\quad +0.56417 + 0.75883)\big) \\
&\approx \frac{1}{30} 6.04444 \approx 0.20148.
\end{aligned}$$

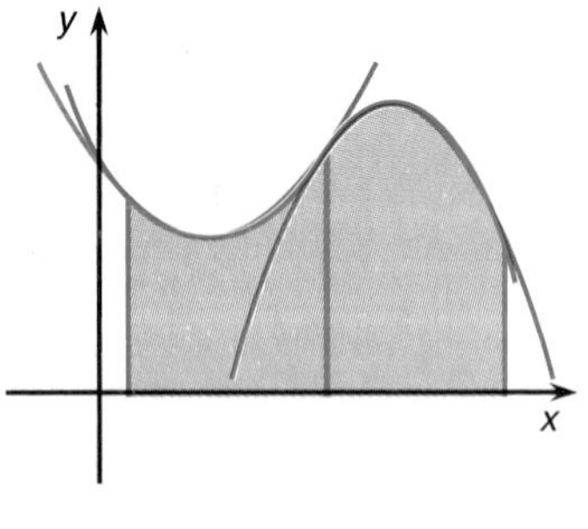

Figure 6.10.7

Simpson's Rule

There is yet another formula for approximating the value of an integral. It is called **Simpson's Rule** after the English mathematician Thomas Simpson (1710–1761), even though he did not invent the rule.

The Riemann sum approximated the area under a graph using rectangles, and the Trapezoidal Rule used trapezoids. To try to get a better approximation, it seems desirable to use some fairly simple region that will fit the shape of the curve more closely. Specifically, it is helpful to use a region that is curved (see Figure 6.10.7).

Probably the simplest of all curved regions is the region under a parabola. We could try to approximate the region under a curve with many pieces, each of which is the region under a suitable parabola. This procedure leads to Simpson's Rule.

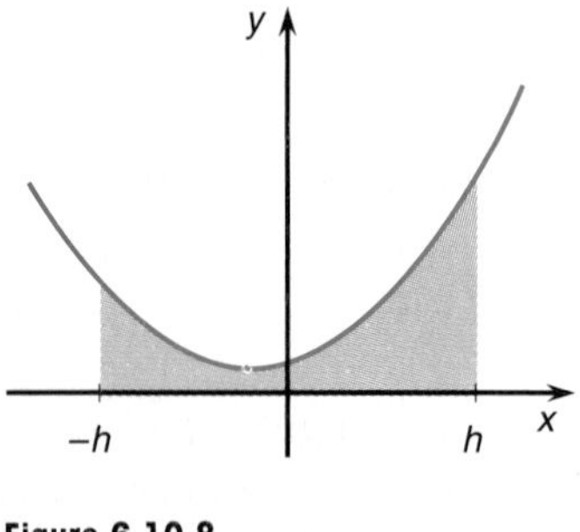

Figure 6.10.8

To begin with, consider the parabola $y = p(x) = ax^2 + bx + c$. We will look at the part of the parabola lying over the interval $[-h, h]$, shown in Figure 6.10.8, and compute its integral $\int_{-h}^{h} p(x)\,dx = \int_{-h}^{h}(ax^2 + bx + c)\,dx = \left(\frac{a}{3}x^3 + \frac{b}{2}x^2 + cx\right)\Big|_{-h}^{h} = \frac{2}{3}ah^3 + 2ch$.

The values of $p(x)$ at the left and right endpoints and at the midpoint of the interval are

$$Y_\ell = p(-h) = ah^2 - bh + c \qquad Y_m = p(0) = c$$
$$Y_r = p(h) = ah^2 + bh + c.$$

With a little algebra it is easy to express the integral we just calculated in terms of these three values. It turns out that

$$\int_{-h}^{h} p(x)\,dx = \frac{2}{3}ah^3 + 2ch = \frac{h}{3}\big(Y_\ell + 4Y_m + Y_r\big). \tag{6.10.4}$$

This formula expresses the integral of the parabola over the interval $[-h, h]$ in terms of the values of the function at the endpoints and the midpoint. It makes no mention of where the interval is. If we consider any interval of length $2h$ and integrate $p(x)$ over it, the integral will be given by the same formula in terms of the values of $p(x)$ at the endpoints and midpoint of the interval. This amounts to shifting the parabola and the interval $[-h, h]$ to the left or to the right, which does not change the integral.

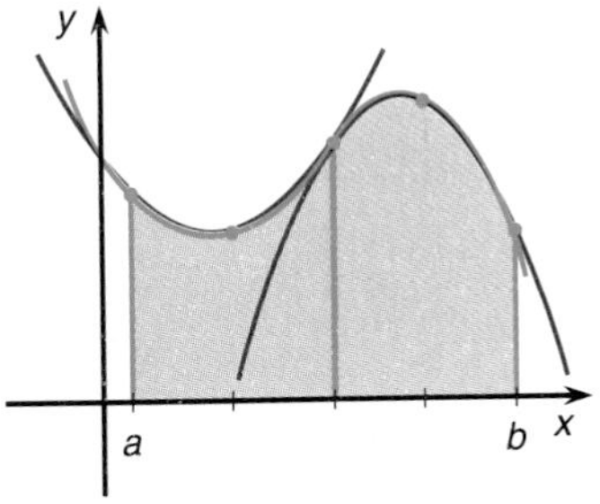

Figure 6.10.9

Now we turn to the problem of approximating the integral $\int_a^b f(x)\,dx$. Suppose n is an even integer, and let $\mathcal{P}_n = \{x_0, \ldots, x_n\}$ be the partition that divides the interval $[a, b]$ into n subintervals of equal length. On the first two subintervals we approximate the curve $y = f(x)$ by a parabola. On the next two subintervals we approximate it by another parabola, and so on. On pairs of adjacent subintervals we approximate the integrand $f(x)$ by parabolas. See Figure 6.10.9.

It is not necessary to figure out the equations of the parabolas. We simply choose the parabola that goes through the same points as the graph of $y = f(x)$ at the endpoints and midpoint of the double subintervals. Having done this, we approximate the integral of $f(x)$ with the sum of the integrals of the little parabolas, which we can find by using Equation 6.10.4.

For instance, the width of each little subinterval is $h = \frac{b-a}{n}$, and for the first two subintervals the value of the function at the left endpoint is $Y_\ell = f(x_0)$, the value at the midpoint is $Y_m = f(x_1)$, and the value at the right endpoint is $Y_r = f(x_2)$. The integral of the parabola is found by Equation 6.10.4. It is $\frac{h}{3}\big(Y_\ell + 4Y_m + Y_r\big) = \frac{b-a}{3n}\big(f(x_0) + 4f(x_1) + f(x_2)\big)$.

Similarly, we can find the integral of the parabolic approximation over the third and fourth subintervals. It is $\frac{h}{3}\big(Y_\ell + 4Y_m + Y_r\big) = \frac{b-a}{3n}\big(f(x_2) + 4f(x_3) + f(x_4)\big)$.

We do this for the fifth and sixth subintervals, the seventh and eighth, and so on. Finally, we add up the approximation for each of these pairs of subintervals to get the approximation to the whole integral.

The values of $f(x)$ at $x = x_0$ and $x = x_n$ are multiplied by 1. The values at the odd-numbered points $x = x_1$, $x = x_3, \ldots$ are multiplied by 4, while the values at the even–numbered points $x = x_2$, $x = x_4, \ldots$ are multiplied by 2.

THEOREM 6.10.5

Simpson's Rule

If n is an even integer, the integral $\int_a^b f(x)\,dx$ can be approximated by

$$S_n = \frac{b-a}{3n}\big(f(x_0) + 4f(x_1) + 2f(x_2) + 4f(x_3) + 2f(x_4) + \ldots + 4f(x_{n-1}) + f(x_n)\big),$$

where $x_k = a + k\frac{b-a}{n}$.

Moreover, if the fourth derivative $f^{(4)}(x)$ exists on $[a, b]$ and B_4 is a number satisfying $|f^{(4)}(x)| \le B_4$ for all $x \in [a, b]$, then

$$\left| \int_a^b f(x)\,dx - S_n \right| \le \frac{B_4(b-a)^5}{180n^4}.$$

EXAMPLE 7 Consider the integral $\int_0^1 x^4\,dx$. Find a Simpson's Rule approximation S_n that is accurate to two decimal places.

Solution In this case the fourth derivative of the integrand is $\frac{d^4}{dx^4}(x^4) = 24$, so we can take $B_4 = 24$, and we need n to be large enough that $B_4(1^5)/(180)n^4 \le 0.005$, that is, so that $24/(180n^4) \le 0.005$, or $n^4 \ge \frac{80}{3} = 26\frac{2}{3}$. This amounts to $n \ge 2.27$, approximately, but since n must be an *even* integer, we must take $n = 4$.

The corresponding approximation by Simpson's Rule is

$$\begin{aligned} S_4 &= \frac{1}{3n}\left(f(0) + 4f\left(\frac{1}{4}\right) + 2f\left(\frac{2}{4}\right) + 4f\left(\frac{3}{4}\right) + f(1)\right) \\ &\approx \frac{1}{12}(0 + 4 \times 0.00391 + 2 \times 0.0625 + 4 \times 0.31641 + 1) \\ &\approx \frac{1}{12}(2.40625) \\ &\approx 0.20052. \end{aligned}$$

The work involved in finding the Simpson's Rule approximation S_n is hardly any more than what is needed for the Trapezoidal Rule *with the same n*, but the accuracy is usually much better. So to get some particular accuracy, Simpson's Rule usually requires a smaller n and consequently is *less* work. Notice that the approximation we obtained in Example 7 is quite good. The integral equals 0.2, and the approximation differs from this by 0.00052, which is much better than the 0.005 we asked for. It is in fact typical of Simpson's Rule that very often the approximation is considerably more accurate than expected. Of course, this is an additional advantage.

The examples we have done so far have all been estimating the integral $\int_0^1 x^4\,dx$, which we can easily evaluate exactly. This was convenient for purposes of comparison, but of course the real interest is in estimating integrals that cannot be evaluated exactly.

EXAMPLE 8 Estimate $\int_0^2 \sqrt{1+x^2}\,dx$ to within an error of less than 0.001.

Solution Writing $f(x) = \sqrt{1+x^2}$, we find that

$$\begin{aligned} f'(x) &= x(1+x^2)^{-1/2}, \\ f''(x) &= (1+x^2)^{-1/2} - \frac{x}{2}(1+x^2)^{-3/2}(2x) = (1+x^2-x^2)(1+x^2)^{-3/2} \\ &= (1+x^2)^{-3/2}, \\ f^{(3)}(x) &= -\frac{3}{2}(1+x^2)^{-5/2}(2x) = -3x(1+x^2)^{-5/2}, \end{aligned}$$

$$f^{(4)}(x) = -3(1+x^2)^{-5/2} + \frac{5}{2}(3x)(1+x^2)^{-7/2}(2x)$$
$$= \left(-3(1+x^2) + 15x^2\right)(1+x^2)^{-7/2} = (12x^2-3)(1+x^2)^{-7/2}.$$

We need a bound B_4 for this fourth derivative on $[0,2]$. There are many possible ways of doing this. First, we can observe that $1+x^2 \geq 1$, so $(1+x^2)^{-7/2} \leq 1$. Since $x \in [0,2]$, we know that $x^2 \leq 4$, so $|12x^2 - 3| \leq |12x^2| + 3 \leq 48 + 3 = 51$. Combining these two observations, we find that

$$|f^{(4)}(x)| \leq \frac{51}{1} = 51,$$

so we can let $B_4 = 51$.

We could also be more sophisticated and recognize that since $0 \leq 12x^2 \leq 48$, we have $-3 \leq 12x^2 - 3 \leq 48 - 3 = 45$, and then let $B_4 = 45$.

So the Simpson's Rule approximation with n subintervals is accurate to within $B_4(b-a)^5/(180n^4) = 51(2-0)^5/(180n^4) = 408/(45n^4)$. To make this less than 0.001, we need to have $408/(45n^4) \leq 0.001$, or $n^4 \geq \frac{27{,}200}{3}$. This amounts to $n \geq 9.76$, approximately, so we should take $n = 10$, the next higher even integer.

It is frequently useful to summarize a Simpson's Rule calculation in the form of a table, as shown in Table 6.10.1. Each half of each horizontal row corresponds to one of the points of the partition, which is listed in the second column. The next column lists the value of the function $f(x)$ at that point, correct to five decimal places, and the last column shows the number (1, 2, or 4) by which that value is multiplied in the sum in Simpson's Rule.

Table 6.10.1

k	x_k	$f(x_k)$	**Multiplier**	k	x_k	$f(x_k)$	**Multiplier**
0	0	1	1	6	1.2	1.56205	2
1	0.2	1.01980	4	7	1.4	1.72047	4
2	0.4	1.07703	2	8	1.6	1.88680	2
3	0.6	1.16619	4	9	1.8	2.05913	4
4	0.8	1.28062	2	10	2	2.23607	1
5	1	1.41421	4				

Now it is reasonably easy to calculate S_n. We multiply each of the values $f(x_k)$ in Table 6.10.1 by the corresponding multiplier (1, 2, or 4) and add the resulting numbers. In this case the total is approximately 44.36827. To find S_n, we multiply this by $\frac{b-a}{3n} = \frac{2}{3(10)} = \frac{1}{15}$. In Table 6.10.1 we wrote down only five decimal places, but if we use a calculator or computer with more accuracy, we find that $S_{10} \approx 2.9578848$.

In Chapter 8 we will learn techniques by which this integral can be evaluated, and its actual value is $\sqrt{5} + \frac{1}{2}\ln|2+\sqrt{5}| \approx 2.9578857$. Simpson's Rule gives us an approximation that is far more accurate than what we expected.

See the Graphics Calculator Example on the next page.

There is another remark to be made here. The calculations for Simpson's Rule are slightly tedious, but they are not difficult. The hard part was estimating the size of $f^{(4)}(x)$ in order to find B_4. We went to some lengths to find a slightly better estimate (45 instead of 51), but in fact it was not worthwhile. If you work it out using $B_4 = 45$, you will find that what is needed to get the desired accuracy is $n \geq 9.46$, approximately, so we would still end up using $n = 10$. The point is that slight improvements in the value of B_4 tend to disappear when you take the fourth root.

As a result of this, it is probably not worth the effort to be too careful about finding the smallest possible bound.

At worst, choosing the wrong bound will result in having to evaluate Simpson's Rule with a few more terms in it, which will almost certainly be less trouble than worrying about the bound.

A programmable calculator or a computer can be used to evaluate Simpson's Rule. For instance, we try the integral from Example 8.

TI-85

PRGM F2 (EDIT) SIMPSON ENTER F3 F1 (Input) N EXIT ENTER 1 STO I ENTER $\sqrt{}$ (1 + 0 ^ 2) STO S ENTER S + $\sqrt{}$ (1 + 2 ^ 2) + 4 × $\sqrt{}$ (1 + (2 ÷ N) ^ 2) STO S ENTER Lbl B ENTER S + 2 × $\sqrt{}$ (1 + (I × 4 ÷ N) ^ 2) STO S ENTER S + 4 × $\sqrt{}$ (1 + (2 ÷ N + I × 4 ÷ N) ^ 2) STO S ENTER I + 1 STO I ENTER F4 F1 (If) (I TEST F2 (<) (N ÷ 2)) ENTER Goto B ENTER S × 2 ÷ (3 × N) ENTER EXIT EXIT EXIT PRGM F1 (NAME) SIMPSON ENTER 10 ENTER

SHARP EL9300

▼ ▼ (NEW) ENTER ENTER SIMP ENTER COMMAND A 3 (Input) N ENTER I = 1 ENTER S = $\sqrt{}$ 1 + 0 x^2 ENTER S = S + $\sqrt{}$ (1 + 2 x^2 ▶ + 4 × $\sqrt{}$ 1 + (2 ÷ N) x^2 ▶ ENTER COMMAND B 1 (Label) 1 ENTER S = S + 2 × $\sqrt{}$ (1 + (I × 4 ÷ N) x^2) ENTER S = S + 4 × $\sqrt{}$ (1 + (2 ÷ N + I × 4 ÷ N) x^2) ENTER I = I + 1 ENTER COMMAND 3 (If) I COMMAND C 2 (<) N ÷ 2 COMMAND B 2 (Goto) 1 ENTER S = S × 2 ÷ 3 ÷ N ENTER COMMAND A 1 (Print) S ENTER COMMAND A 6 (End) ENTER (RUN) ENTER (SIMP) ENTER 10 ENTER

CASIO f_x-6300G (f_x-7700GB)

MODE 2 EXE ? → N : 1 → I : $\sqrt{}$ (1 + 0 x^2) → S : S + $\sqrt{}$ (1 + 2 x^2) + 4 × $\sqrt{}$ (1 + (2 ÷ N) x^2) → S : Lbl 1 : S + 2 × $\sqrt{}$ (1 + (I × 4 ÷ N) x^2) → S : S + 4 × $\sqrt{}$ (1 + (2 ÷ N + I × 4 ÷ N) x^2) → S : I + 1 → I : I < N ÷ 2 ⇒ Goto 1 : S × 2 ÷ 3 ÷ N ◢ MODE 1 Prog 0 EXE 10 EXE (on the f_x-7700GB, : is PRGM F6, and Prog is PRGM F3.)

HP 48SX

ATTN ↱ CLR ≪ ≫ " " KEY IN N ▶ SPC PROMPT SPC ′ N ▶ STO 1 SPC 0 x^2 + $\sqrt{x}$ 1 SPC 2 x^2 + $\sqrt{x}$ + 2 SPC N ÷ x^2 1 + $\sqrt{x}$ 4 × + 1 SPC N SPC 2 ÷ 1 − FOR SPC I SPC I SPC 4 × N ÷ x^2 1 + $\sqrt{x}$ 2 × + I SPC 4 × N ÷ 2 SPC N ÷ + x^2 1 + $\sqrt{x}$ 4 × + NEXT ▶ 2 × N ÷ 3 ÷ ▶ ENTER ′ SIMP STO VAR SIMP 10 ENTER ↰ CONT

```
?10
ANSWER=
 2.957884775
```

How large an n do you have to use to get the answer 2.957 885 715 (which is $\sqrt{5} + \frac{1}{2}\ln(2+\sqrt{5})$ to nine decimal places)?

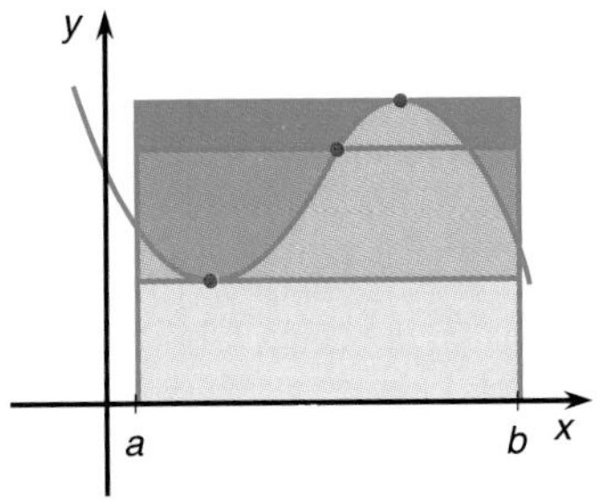

Figure 6.10.10

EXAMPLE 9 In this example we prove Inequality 6.10.1 about the accuracy of approximation by Riemann sums.

In Figure 6.10.10 there is one subinterval. The part of the integral corresponding to this subinterval is the shaded area. The corresponding part of the Riemann sum is the area of the middle rectangle, which may be greater or less than the integral or perhaps even equal to it.

Also shown are the big rectangle whose height is the maximum value of $f(x)$ on the subinterval, and the small rectangle whose height is the minimum value of $f(x)$ on the subinterval. We see that the actual integral lies between the areas of these two rectangles and that the area of the rectangle belonging to the Riemann sum with x_k^* equal to the midpoint is also between them.

In particular, the midpoint Riemann sum approximation R_n is no farther from the value of the integral than it is from at least one of the Riemann sums associated to the maximum value or the minimum value on each subinterval. So we can get an inequality describing the accuracy of the Riemann sum approximation by estimating the size of these differences.

Suppose x_k^M and x_k^m are points in the kth subinterval at which $f(x)$ achieves its maximum and minimum values, respectively. We want to estimate $f(x_k^M) - f(x_k^*)$ and $f(x_k^m) - f(x_k^*)$. Assuming that $f(x)$ is differentiable, we know that $f(x_k^M) - f(x_k^*) = \int_{x_k^*}^{x_k^M} f'(x)\,dx$. Assume that B_1 is a positive number chosen so that $|f'(x)| \le B_1$, for all $x \in [a, b]$; we can use Inequality 5.4.3(iii) to conclude that

$$f(x_k^M) - f(x_k^*) = \int_{x_k^*}^{x_k^M} f'(x)\,dx \le B_1 \left| x_k^M - x_k^* \right|.$$

Since x_k^M cannot be farther than $\frac{b-a}{2n}$ from the midpoint x_k^*, we see that $f(x_k^M) - f(x_k^*) \le B_1 \frac{b-a}{2n}$. This is true for each subinterval, so the difference between the "maximum Riemann sum" and the "midpoint Riemann sum" R_n is

$$\begin{aligned}\sum_{k=1}^{n} f(x_k^M)\Delta x_k - \sum_{k=1}^{n} f(x_k^*)\Delta x_k &= \frac{b-a}{n}\sum_{k=1}^{n}\left(f(x_k^M) - f(x_k^*)\right)\\ &\le \frac{b-a}{n}\sum_{k=1}^{n}\left(B_1\frac{b-a}{2n}\right)\\ &= B_1\left(\frac{(b-a)^2}{2n^2}\right)\sum_{k=1}^{n} 1 = B_1\frac{(b-a)^2}{2n}.\end{aligned}$$

A similar argument shows that the difference between R_n and the "minimum Riemann sum" is also less than or equal to $B_1(b-a)^2/(2n)$, and from this we conclude that the difference between R_n and the actual integral is also less than or equal to the same number. This is Inequality 6.10.1.

This argument is given here to illustrate the types of techniques used to prove the accuracy estimates. The estimates for the Trapezoidal Rule and for Simpson's Rule are established by more complicated but similar arguments. For our purposes, what is most important is using them.

Exercises 6.10

I

For each of the following integrals, find the Trapezoidal Rule approximation T_n and the Simpson's Rule approximation S_n, for the specified n. Compare these values with the actual value of the integral.

1. $\int_0^4 x\,dx,\ n = 4$
2. $\int_1^9 x^4\,dx,\ n = 8$
3. $\int_1^4 \sqrt{x}\,dx,\ n = 6$
4. $\int_{-1}^3 (x^3 - 2x)\,dx,\ n = 4$
5. $\int_0^\pi \sin x\,dx,\ n = 4$
6. $\int_0^\pi \sin^2 x\,dx,\ n = 4$
7. $\int_1^5 \frac{1}{x}\,dx,\ n = 8$
8. $\int_1^3 \frac{1}{x},\,dx,\ n = 6$
9. $\int_0^2 \frac{x}{1+x^2}\,dx,\ n = 4$
10. $\int_{-3}^3 (x^5 - 7x^3 + 2x)\,dx,\ n = 6$

For each of the following integrals, find a value of n that will ensure that the Simpson's Rule approximation S_n will be correct with an error of less than 0.001. Calculate the corresponding S_n and compare it with the exact value of the integral.

11. $\int_0^4 x^4\,dx$
12. $\int_0^2 x^5\,dx$
13. $\int_0^{\pi/2} \sin x\,dx$
14. $\int_0^1 e^x\,dx$
15. $\int_{-1}^0 \frac{1}{x-1}\,dx$
16. $\int_0^4 e^{x/4}\,dx$
17. $\int_0^1 x\sqrt{1+x^2}\,dx$
18. $\int_{-1}^1 \cos(\pi x)\,dx$
19. $\int_{-\pi}^\pi \tan(x/4)\,dx$
20. $\int_1^3 \frac{1}{x}\ln(x)\,dx$

For each of the following integrals, find a value of n that will ensure that the Simpson's Rule approximation S_n will be correct with an error of less than 0.001. Calculate the corresponding S_n.

21. $\int_1^2 \ln x\,dx$
22. $\int_1^2 e^x\,dx$
23. $\int_{-1}^0 e^x\,dx$
24. $\int_0^1 \sin(\pi x)\,dx$
25. $\int_1^2 \cos(\pi x)\,dx$
26. $\int_2^4 \cos\left(\frac{\pi}{2}x\right)\,dx$
27. $\int_2^3 x^{5/3}\,dx$
28. $\int_5^7 x^{4/3}\,dx$
29. $\int_0^1 x\sin x\,dx$
30. $\int_{-1}^0 x^2\sin(\pi x)\,dx$
31. $\int_0^1 \sqrt{1+x^2}\,dx$
32. $\int_0^{1/2} \sqrt{1+x^2}\,dx$
33. $\int_0^1 \frac{1}{\sqrt{1+x^2}}\,dx$
34. $\int_0^{1/2} \frac{1}{\sqrt{1+x^2}}\,dx$
35. $\int_0^1 \sqrt{1+2x^2}\,dx$
36. $\int_0^2 \sqrt{1+x^2/2}\,dx$
37. $\int_0^1 \sqrt{1+x^3}\,dx$
38. $\int_0^1 \sqrt{1+2x^3}\,dx$
39. $\int_0^1 \frac{1}{1+x^3}\,dx$
40. $\int_0^1 \frac{1}{1+x^4}\,dx$

II

*41. (i) Consider the part of the parabola $y = x^2$ between $x = 0$ and $x = 1$. For what value of n will Simpson's Rule give an approximation S_n to the arclength of this curve that is accurate to within 0.001?

(ii) Find an approximation to the arclength of the curve described in part (i), accurate to within 0.001.

*42. (i) Consider the part of the parabola $y = \frac{1}{4}x^2$ between $x = 0$ and $x = 4$. For what value of n will Simpson's Rule give an approximation S_n to the arclength of this curve that is accurate to within 0.001?

(ii) Find an approximation to the arclength of the curve described in part (i), accurate to within 0.001.

**43. With what n can we estimate the arclength of the curve $y = x^3$ between $x = -1$ and $x = 1$ to within 0.001 by Simpson's Rule S_n? (Do not actually evaluate S_n unless you have a computer or programmable calculator.)

**44. With what n can we estimate the arclength of the curve $y = x^4$ between $x = 1$ and $x = 2$ to within 0.001 by Simpson's Rule S_n? (Do not actually evaluate S_n unless you have a computer or programmable calculator.)

*45. Use Simpson's Rule to estimate to within 0.001 the arclength of the curve $y = \sin x$ between $x = 0$ and $x = \pi$.

*46. Use Simpson's Rule to estimate to within 0.001 the arclength of the curve $y = e^x$ between $x = 0$ and $x = 1$.

47. Consider the "multipliers" in the formula of Simpson's Rule, and show that the multipliers in the formula for S_n add up to $3n$. (This means that $\frac{1}{3n}$ times the sum is really a kind of *weighted average*, i.e., the average of $3n$ values of $f(x)$, some of which have been repeated; S_n, which is just this average times $b - a$,

approximates the integral, which equals $b - a$ times the (integral) average value of $f(x)$.)

***48.** (i) Show that a parabola of the form $y = ax^2 + bx + c$ is determined by any three points on it (i.e., given three points, there is only one such parabola containing them, i.e., two parabolas containing the three points must actually be the same).

(ii) Can you think of three points that do not lie in such a parabola?

***49.** Show that the Simpson's Rule approximation S_n gives an exact evaluation of the integral of any polynomial $p(x)$ whose degree is 3 or less, using *any* positive even integer n.

***50.** (i) Suppose that $f(x)$ is an odd function, $a > 0$, and show that for any positive even n, S_n will exactly equal the integral $\int_{-a}^{a} f(x)\,dx$.

(ii) Evaluate the Simpson's Rule approximation S_{20} for $\int_{-6}^{6} (3 + x\cos x)\,dx$.

***51.** Use Simpson's Rule to approximate ln 2 to within 0.001. (*Hint:* $\ln 2 = \int_1^2 \frac{1}{x}\,dx$.)

***52.** Use Simpson's Rule to approximate ln 10 to within 0.1.

***53.** Suppose a and b are two quantities to which we have approximations A and B, respectively. Suppose we know that $|A - a| \le 0.001$ and $|B - b| \le 0.1$ and that $A = 4.7$ and $B = 32.5$.

(i) How good an approximation to the product ab is AB (i.e., estimate the size of the error $|AB - ab|$)?

(ii) How good an approximation to $\frac{a}{b}$ is $\frac{A}{B}$?

***54.** (i) Combine the ideas of the previous three problems to estimate the value of $\log_{10}(2)$.

(ii) What degree of accuracy can you guarantee for your estimate?

(iii) Compare your estimate with the actual value.

***55.** Recall that $\int_0^1 1/(1 + x^2)\,dx = \arctan x\Big|_0^1 = \pi/4$. We can use Simpson's Rule to approximate $\pi/4$ (and hence π itself).

(i) What value of n is needed to ensure that S_n is within 0.001 of $\pi/4$?

(ii) What value of n is needed to ensure that S_n is within 0.0001 of $\pi/4$?

(iii) What value of n is needed to ensure that S_n is within 0.000001 of $\pi/4$?

(iv) Estimate π to within 0.001, and compare your estimate with the actual value.

****56.** Estimate the coordinates of the centroid of the region lying underneath the curve $y = (1 + x^2)^{-1/2}$ between $x = 0$ and $x = 1$. Several integrals are involved here. For each one, first try to evaluate it exactly. Failing that, use Simpson's Rule to estimate it to within 0.001. What accuracy can you guarantee for your answer?

III

****57.** If you have a computer or programmable calculator, program it to evaluate the Simpson's Rule approximation S_n for a specified integrand $f(x)$. Use your program to check your calculations for the exercises in this section.

58. Use Simpson's Rule and a personal computer or programmable calculator to estimate π to within 0.000001 (cf. Exercise 55).

59. Estimate ln 2 to within 0.000001.

***60.** Use Simpson's Rule to approximate ln 10 to within 0.001.

Use Simpson's Rule S_n to estimate the surface areas of the surfaces obtained by revolving the following functions around the x-axis. Instead of estimating how large n must be, try it with various values of n until the answer seems to have stopped changing in the first five significant digits.

61. $y = \frac{1}{x^2}$, $x \in [1, 2]$

62. $y = x^4 + 1$, $x \in [-2, 2]$

63. $y = \sin^2 x$, $x \in [0, \pi]$

64. $y = \ln x$, $x \in [1, e]$

65. $y = \sec x$, $x \in \left[-\frac{\pi}{4}, \frac{\pi}{4}\right]$

66. $y = \arctan x$, $x \in [0, 1]$

Chapter Summary

§6.1
area
signed area
region between two curves

§6.2
small change dx
strip of width dx
function with specified rate of change

§6.3
solid of revolution
volume by slices (Definition 6.3.1, p. 433)

§6.4
cylindrical shells

§6.5
average value (Definition 6.5.1, p. 447)
velocity/speed

§6.6
work
force
potential energy
kinetic energy
Hooke's Law (Equation 6.6.2, p. 458)
foot-pound (ft-lb)
newton (N)
newton-meter (N-m)
joule (J)

§6.7
centroid
fulcrum
moment about a point
center of mass
uniform density
moment about a line (Definition 6.7.5, p. 467)
symmetry
Pappus' Theorem (Theorem 6.7.10)

§6.8
pressure
density of a fluid
Pascal's Principle (Theorem 6.8.1, p. 477)
fluid force and centroids
buoyancy

§6.9
arclength
$ds = \sqrt{(dx)^2 + (dy)^2} = \sqrt{1 + (y')^2}\, dx$
surface of revolution
surface area
$dS = 2\pi f(x)\, ds$

§6.10
Trapezoidal Rule (Theorem 6.10.2, p. 496)
Simpson's Rule (Theorem 6.10.5, p. 499)
error estimates

Review Exercises

I

In each of the following situations, find the area of the region enclosed between the functions $f(x)$ and $g(x)$.

1. $f(x) = x^2,\ g(x) = 2 - x^2$
2. $f(x) = x^2,\ g(x) = 9$
3. $f(x) = 4 - x^2,\ g(x) = -2x$
4. $f(x) = x^3 - 3x,\ g(x) = 2x^2$
5. $f(x) = |x^3|,\ g(x) = |x|$
6. $f(x) = x^2 - 4,\ g(x) = |2x - 1|$
7. $f(x) = \sqrt{8 - x^2},\ g(x) = 2$
8. $f(x) = x^4 - 7x^2 + 10,\ g(x) = x^2 - 6$
9. $f(x) = x^2,\ g(x) = \sqrt{2 - x^2}$
10. $f(x) = x,\ g(x) = \sqrt{x}$

Find the areas of the following regions.

11. Inside $x^2 + y^2 = 1$ and above $y = x$.
12. The intersection of the interiors of $x^2 + y^2 = 1$ and $5x^2 + y^2 = 2$.
13. The intersection of a disk of radius 4 and a square with the same center and sides of length 6.
14. The part of the square in Exercise 13 that is outside the disk.
15. The intersection of two disks of radius 5 with centers 40 units apart.
16. The intersection of two disks of radius r with centers c units apart. (Assume $c < r$.)
17. The intersection of a disk of radius 3 with an ellipse with the same center and semi-minor axis 2 and semi-major axis 4.
18. The intersection of a disk of radius r with an ellipse with the same center and semi-minor axis b and semi-major axis a, with $b < r < a$.
19. The region above the parabola $y = x^2 - 2$ and inside the disk of radius 2 centered at the origin.
20. The region consisting of all points above the parabola $y = x^2$ lying at a distance of 1 unit or less from the point $(1, 0)$.

Find the volume of the solid of revolution formed by revolving the given function $y = f(x)$ on the specified interval about the x-axis. Sketch the solid.

21. $f(x) = \sin x$ on $\left[0, \frac{\pi}{2}\right]$
22. $f(x) = x - 2$ on $[3, 4]$
23. $f(x) = |x|$ on $[-2, 4]$
24. $f(x) = |x^3|$ on $[-1, 1]$
25. $f(x) = \cos^2 x - \sin^2 x$ on $\left[0, \frac{\pi}{4}\right]$
26. $f(x) = \cos x \sin x$ on $\left[0, \frac{\pi}{2}\right]$
27. $f(x) = (1 + x^2)^{-1/2}$ on $[0, 1]$
28. $f(x) = x(x - 1)$ on $[0, 1]$
29. $f(x) = \sec \frac{\pi x}{4}$ on $[-1, 1]$
30. $f(x) = e^x$ on $[-1, 0]$

Find the average value of each of the following functions over the specified interval.

31. $f(x) = x^2 + 3$ on $[0, 2]$
32. $f(x) = x^3$ on $[-2, 2]$
33. $f(x) = \frac{1}{x}$ on $[1, 2]$
34. $f(x) = 3$ on $[-7, 11]$
35. $f(x) = \sin \pi x$ on $[3, 4]$
36. $f(x) = \sin \pi x$ on $[3, 6]$
37. $f(x) = \cos \pi x$ on $[3, 4]$
38. $f(x) = \sin x$ on $\left[0, \frac{\pi}{2}\right]$
39. $f(x) = |x^2 - 1|$ on $[-2, 2]$
40. $f(x) = 1/(1 + x^2)$ on $[0, \sqrt{3}]$

Find the work done by each of the following.

41. Lifting a 3-lb weight through a height of 5 ft.
42. Lifting a 10-kg weight through a height of 3.7 m.
43. Lifting a 30-g weight through a height of 60 cm.
44. Lifting a 3-oz weight through a height of 14 in.
45. Compressing by 1 ft a spring that can be compressed 1 in from rest by a force of 3 lb.
46. A spring relaxing from a stretch of 3 ft to a stretch of 2.5 ft, if it takes 13 lb to compress the spring 1 ft from rest.
47. Gravity on a 10-lb bag of potatoes as it falls 3 ft from the counter to the floor.
48. Gravity on a 5-lb steel ball as it falls 8 ft and then bounces up 7 ft.
49. Pumping all the water from a conical tank of height 10 ft and radius 4 ft at the top and raising it to a point 5 ft above the top of the tank.
50. Pumping the water from a spherical tank of radius 10 ft that is half full and raising it to the outlet valve at the top of the tank.

Find the centroid of each of the following regions.

51. The triangle with vertices $(1, 2)$, $(2, 4)$, $(-3, 0)$.
52. The triangle with vertices $(0, -1)$, $(20, 4)$, $(-3, -1)$.
53. The parallelogram with vertices $(0, 0)$, $(2, 0)$, $(1, 3)$, $(3, 3)$.
54. The parallelogram with vertices $(1, 1)$, $(4, 4)$, $(2, 3)$, $(5, 6)$.
55. The union of the triangle with vertices $(0, 0)$, $(1, 2)$, $(3, 2)$ and the disk of radius 1 centered at $(1, -1)$.
56. The union of the disk of radius 2 centered at $(1, 1)$ and the disk of radius 4 centered at $(10, 0)$.
57. The union of the disk of radius 1 centered at $(0, 0)$ and the disk of radius 1 centered at $(1, 1)$.
58. The union of the disk of radius 1 centered at $(0, 0)$ and the square with vertices $(3, 3)$, $(3, 5)$, $(5, 3)$, $(5, 5)$.
59. The union of the disk of radius 1 centered at $(0, 0)$ and the square with vertices $(0, 0)$, $(1, 0)$, $(0, 1)$, $(1, 1)$.
60. The union of the disk of radius 4 centered at $(0, 0)$ and the square with vertices $(0, 0)$, $(1, 0)$, $(0, 1)$, $(1, 1)$, with density $\rho = 2$ on the square and density $\rho = 1$ on the rest of the disk.

Find the total force on one side of each of the following.

61. A flat plate of area 3 sq ft lying horizontally at a depth of 40 ft in water.
62. A flat plate of area 5 m^2 lying horizontally at a depth of 2 m in water.
63. A square plate with sides 2 ft long, standing vertically in water with its bottom edge at a depth of 20 ft.
64. An equilateral triangle with sides 5 ft long, standing vertically in water with its bottom edge horizontal at a depth of 10 ft.
65. A circular plate of radius 3 m, lying at an angle of 45° above the horizontal, with its lowest point at a depth of 6 m.
66. A circular plate of radius 6 ft, lying at an angle of 30° above the horizontal with its highest point at a depth of 40 ft.
67. A semicircular plate of radius 3 ft standing vertically in water with its bottom straight edge horizontal at a depth of 10 ft.
68. A semicircular plate of radius 2 m standing vertically in water with its straight edge uppermost at a depth of 5 m.
69. A circular plate of radius 4 ft, standing vertically with its lowest point resting on the bottom of a stream that is 6 ft deep.
70. A semicircular plate of radius 1 ft lying at a 60° angle above the horizontal with its straight edge flat on the bottom at a depth of 4 ft.

Find the surface area of the surface obtained by revolving the specified function about the x-axis.

71. $f(x) = x$, for $x \in [3, 6]$
72. $f(x) = 3x$, for $x \in [4, 10]$
73. $f(x) = \sqrt{x}$, for $x \in [2, 4]$
74. $f(x) = (3x)^{2/3}$, for $x \in \left[0, \frac{1}{3}\right]$
75. $f(x) = \sqrt{9 - x^2}$, for $x \in [0, 1]$
76. $f(x) = \sqrt{1 - x^2/4}$, for $x \in [-1, 0]$

77. $f(x) = |x - 3|$, for $x \in [0, 6]$

78. $f(x) = x^3$, for $x \in [1, 8]$

79. $f(x) = (x - 1)^3$, for $x \in [2, 9]$

80. $f(x) = \sqrt{2 - x^2}$, for $x \in [0, 1]$

In each of the following cases, find how large n must be so that the Simpson's Rule approximation S_n should be within 0.001 of the given integral. Evaluate S_n for this value of n.

81. $\int_0^1 (3x^5 - 2x + 1)\, dx$

82. $\int_1^3 x^6\, dx$

83. $\int_0^{\pi/2} \cos x\, dx$

84. $\int_0^{\pi} \sin x\, dx$

85. $\int_{-1}^{1} \sin x\, dx$

86. $\int_0^{\pi} \sin^2 x\, dx$

87. $\int_1^2 \sqrt{2x^2 - 1}\, dx$

88. $\int_3^4 \sqrt{1 + x^2}\, dx$

89. $\int_{-2}^{2} x\sqrt{x^4 + x^2 + 2}\, dx$

90. $\int_1^4 \frac{1}{x}\, dx$

II

***91.** Find the volume of the intersection of two balls of radius 5 whose centers are 8 units apart.

***92.** Find the volume of the intersection of two balls of radius r whose centers are d units apart, with $d < 2r$.

***93.** Find the centroid of the region lying between the x-axis and the curve $y = \sin x$, for $0 \leq x \leq 2\pi$.

***94.** Find the centroid of the region lying between the x-axis and the curve $y = \sin x$, for $0 \leq x \leq \pi$.

Do each of the following examples, using Theorem 6.7.9 (p. 472) to find M_x.

***95.** Example 6.7.4 (p. 467)

***96.** Example 6.7.5 (p. 468)

***97.** Example 6.7.6 (p. 468)

***98.** Consider a sphere of radius R and two horizontal planes that are d units apart. Find the surface area of the part of the sphere lying between the two planes. (Assume that both planes intersect the sphere.) Notice that the problem does not say where the planes are; there appears to be insufficient information to solve the problem. In fact, what you will find is that the area of the part of the sphere in question does not depend on where the planes are, but only on how far apart they are. Use this fact to give a snappy answer to Exercise 30(ii) of Section 6.9.

***99.** Verify Formula 6.10.4 (p. 499) for the integral of a quadratic function over the interval $[-h, h]$.

***100.** Two mathematicians on a fishing trip are having an argument. One says that ln 7 is greater than 2, and the other says that it is less than 2. They are many miles from the nearest calculator or textbook, but there is a nice sandy beach on which they can do calculations with a stick. Since $\ln 7 = \int_1^7 \frac{1}{x}\, dx$, the question can be solved by approximating this integral by Simpson's Rule. The tricky thing is to approximate it with sufficient accuracy.

(i) Find the Simpson's Rule approximation S_2, and calculate the degree of accuracy we can guarantee for S_2. Based on these two pieces of information, can you tell whether or not $\ln 7 > 2$?

(ii) Is $\ln 7 > 2$ or not? (No fair using a calculator.)

***101.** Suppose the graph of $y = f(x)$ is a straight line. Find the average value of $f(x)$ over the interval $[a, b]$, and show that it equals $f(c)$, where c is the midpoint of the interval.

***102.** Find an example of a nonnegative function $f(x)$ and an interval $[a, b]$ so that the average value of $f(x)$ on $[a, b]$ does not equal the RMS average, that is, the square root of the average of $f(x)^2$.

***103.** Consider the solid obtained by revolving around the x-axis the square whose vertices are $(1, 1)$, $(1, 2)$, $(2, 2)$, $(2, 1)$. Find its volume using (i) slices, (ii) cylindrical shells, (iii) Pappus' Theorem.

***104.** Consider the solid obtained by revolving around the x-axis the square whose vertices are $(1, 1)$, $(2, 0)$, $(2, 2)$, $(3, 1)$. Find its volume using (i) slices, (ii) cylindrical shells, (iii) Pappus' Theorem.

***105.** Compare the surface areas of the solid obtained by revolving around the x-axis the curve $y = \sqrt{1 - |x|}$, $-1 \leq x \leq 1$, and the sphere of the same volume.

****106.** Let $a > 0$ and revolve $y = a\sqrt{1 - |x|}$, $-1 \leq x \leq 1$, around the x-axis. (i) Find the surface area of the resulting solid. (ii) Compare with the surface area of the sphere with the same volume when $a = 2, 5, 10$.

III

***107.** How do you expect the arclengths of $y = x^2$, for $x \in [0, 1]$, $y = \frac{x^2}{4}$, for $x \in [0, 2]$, and $y = \frac{x^2}{2}$, for $x \in [0, 2]$ to compare? (Sketch all three graphs.) Use Simpson's Rule to estimate each to five decimal places. What are the ratios of the answers you get? Explain.

Logarithms and Exponential Functions

7

Legend says that an ancient king was so pleased with the new game of chess that he offered a huge reward to its inventor. The wise man asked the king to put one grain of wheat on the first square of the chessboard, two on the second, four on the third, and so on, doubling each time.

Calculate the amount of wheat on the sixty-fourth square, assuming a grain of wheat weighs 50 mg and bearing in mind that the annual worldwide wheat harvest is about 500 million metric tonnes.

Incidentally, the phrase "check mate" is apparently a corruption of "shah matt", which in old Persian means "the king is dead".

INTRODUCTION

We have been working with the exponential and natural logarithm functions for some time, but we have not really studied them very carefully in their own right. In this chapter we learn more about these functions and their practical applications.

The exponential function turns out to be extremely useful. It describes a great many processes of growth or change. Among the examples of exponential growth are population growth in an unrestricted environment (such as a bacterial culture or a population of animals or humans), the concentration of various ingredients or products in many chemical reactions, the amount remaining of an element undergoing radioactive decay, the temperature of an object that is being warmed or cooled by contact with another object at a different temperature, and the amount of money in an account growing under continuously compounded interest.

We begin by studying logarithms in Section 7.1, move from there to exponentials in Section 7.2, and then discuss various applications, including exponential growth and decay in Sections 7.3 and 7.4 and logistic growth in Section 7.5. The applications in Sections 7.3 and 7.4 can be studied any time after you have covered Chapter 6, and Section 7.5 can be covered after Chapter 5.

The final section, Section 7.6, concerns *differential equations*, which are equations involving the *derivatives* of a function. This section is optional.

The relevance of differential equations to this chapter comes from two different observations. The first is that the fundamental fact about all the processes mentioned above is that each is described by a function whose rate of change is related to how big the function is (e.g., the more bacteria there are, the greater the rate at which the population grows). This amounts to a differential equation relating the *derivative* of a function to the *values* of the function. The second thing is that exponential functions play a key role in the solution of several important types of differential equations.

7.1 Logarithms

NATURAL LOGARITHM
$e = 2.71828\ldots$

When we studied the derivatives of trigonometric functions in Section 3.4, we were able to figure out what the derivatives should be by finding some moderately difficult limits and performing some clever calculations. On the other hand, when we got to logarithms and exponentials in Section 3.5, we were able to work with the derivatives, but the actual statement of what the derivatives are was something we just had to learn and take on trust, with no real explanation. The reason for the difference is that we have a definite picture of where the trigonometric functions come from. They have to do with various ratios related to an angle. There is no comparable starting point for logarithms or exponentials.

On the other hand, it is clear in hindsight that we need to have logarithms to be able to evaluate certain common integrals. The function $\ln x$ is an antiderivative of $\frac{1}{x}$.

In light of what we know about indefinite integrals, $\ln x$ is completely determined, as a function on the positive real numbers, by this fact and its value at one point, for example, $\ln 1 = 0$. So if we want to give a precise definition of $\ln x$, for $x > 0$, we can do it this way.

DEFINITION 7.1.1

The **natural logarithm** $\ln x$ is defined for all $x > 0$ by

$$\ln x = \int_1^x \frac{1}{t}\,dt.$$

We notice immediately from the definition that $\ln(1) = 0$ and by the Fundamental Theorem of Calculus that the derivative is $\frac{d}{dx}\left(\ln x\right) = \frac{1}{x}$.

The other basic thing about logarithms is the way they behave with respect to products. It is interesting to see that this is also a relatively simple consequence of the definition.

THEOREM 7.1.2

If a and b are positive numbers, then

$$\ln(ab) = \ln a + \ln b.$$

PROOF We can write

$$\ln(ab) = \int_1^{ab} \frac{1}{t}\,dt = \int_1^{a} \frac{1}{t}\,dt + \int_a^{ab} \frac{1}{t}\,dt = \ln a + \int_a^{ab} \frac{1}{t}\,dt.$$

Notice that ab might, for instance, be less than 1, in which case the first integral would be negative, and similar remarks apply to the other integrals too.

It remains to show that the last integral equals $\ln b$. To do this, we make the substitution $u = \frac{t}{a}$, $du = \frac{1}{a}dt$. The integral becomes

$$\int_a^{ab} \frac{1}{t}\,dt = \int_1^{(ab)/a} \frac{1}{au}(a)\,du = \int_1^{b} \frac{1}{u}\,du = \ln b.$$

This completes the proof. ■

Repeating this result n times, we see that $\ln(a^n) = n \ln a$, for any positive integer n.

Combining Theorem 7.1.2 with the fact that $\ln(1) = 0$, we see that if $a > 0$, then $0 = \ln 1 = \ln\left(a\frac{1}{a}\right) = \ln a + \ln \frac{1}{a}$, or, after rearranging, $\ln \frac{1}{a} = -\ln a$. From this we also see that if n is a negative integer, $\ln(a^n) = \ln\left(\frac{1}{a^{-n}}\right) = -\ln(a^{-n}) = -(-n)\ln a = n \ln a$, so this formula is valid for any integer n and any $a > 0$.

We make these rearrangements so that we can work with a positive exponent $-n$...

Since the integrand $\frac{1}{t}$ is always positive, it is also apparent that $\ln x = \int_1^x \frac{1}{t}dt$ is positive if $x > 1$. On the other hand, if $0 < x < 1$, then the limits of integration are "the wrong way around," and $\ln x = \int_1^x \frac{1}{t}dt = -\int_x^1 \frac{1}{t}dt$, which is negative. In other words, $\ln x > 0$ if $x > 1$ and $\ln x < 0$ if $0 < x < 1$.

Since the derivative of $\ln x$ is $\frac{1}{x}$, the derivative is positive for all $x > 0$, and the function is increasing. The second derivative is $\frac{d}{dx}\left(\frac{1}{x}\right) = -\frac{1}{x^2}$, which is negative, so the graph of $y = \ln x$ is concave down.

Now consider $\ln(2^n) = n \ln 2$. We know that $y = \ln x$ is increasing, so $\ln 2 > \ln 1 = 0$. So $\ln 2$ is a positive number, and as $n \to \infty$, $n \ln 2 \to \infty$ also. This means that $\ln(2^n) \to \infty$ as $n \to \infty$, so $\ln(2^n)$ becomes larger than any fixed number if n is large enough. Since $\ln x$ is increasing, this shows that $\lim_{x\to\infty} \ln x = \infty$.

We could use a similar argument to find $\lim_{x\to 0^+} \ln x$, or we could simply remark that as $x \to 0^+$, we must have $\frac{1}{x} \to \infty$, so

$$\lim_{x\to 0^+} \ln x = \lim_{1/x\to\infty} \ln x = \lim_{1/x\to\infty} \left(-\ln \frac{1}{x}\right) = -\lim_{u\to\infty} \ln u = -\infty.$$

Notice how we substituted $u = \frac{1}{x}$.

We collect all these facts together.

Summary 7.1.3 Properties of ln *x*

1. For any $x > 0$, $\ln x$ is defined by

$$\ln x = \int_1^x \frac{1}{t}\,dt.$$

2. For any positive numbers a and b and any integer n,

$$\ln(ab) = \ln a + \ln b, \qquad \ln\left(\frac{a}{b}\right) = \ln a - \ln b, \qquad \ln(a^n) = n \ln a.$$

3. $\ln(1) = 0$.

4. $\frac{d}{dx}(\ln x) = \frac{1}{x}$; the function $\ln x$ is increasing and concave downward.
5. If $x > 1$, then $\ln x > 0$; and if $0 < x < 1$, then $\ln x < 0$.
6. $\lim_{x\to\infty} \ln x = \infty$ and $\lim_{x\to 0^+} \ln x = -\infty$.

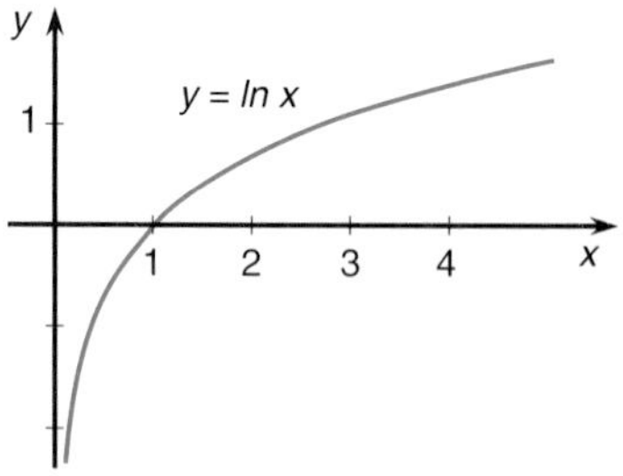

Figure 7.1.1

From Properties 3, 4, 5, and 6 it is possible to sketch the graph of $y = \ln x$.

On the one hand, we know that $\lim_{x\to\infty} \ln x = \infty$, but on the other hand, the derivative is $\frac{1}{x}$, which becomes smaller and smaller, tending to 0 as $x \to \infty$. In terms of the graph, this means that the graph is less and less steep, closer and closer to being horizontal. See Figure 7.1.1.

It seems contradictory that the graph could tend to infinity and yet have tangents that tend to the horizontal, but this is what happens. The idea is that while $\ln x$ does tend to infinity, it does so very *slowly*.

Certainly it goes to infinity more slowly than the straight line $y = x$, as shown in Figure 7.1.2, but the same is true of $y = \sqrt{x}$, for example, the graph of which is shown in Figure 7.1.3. It is not very clear from the pictures whether $\ln x$ or $\sqrt{x}$ goes

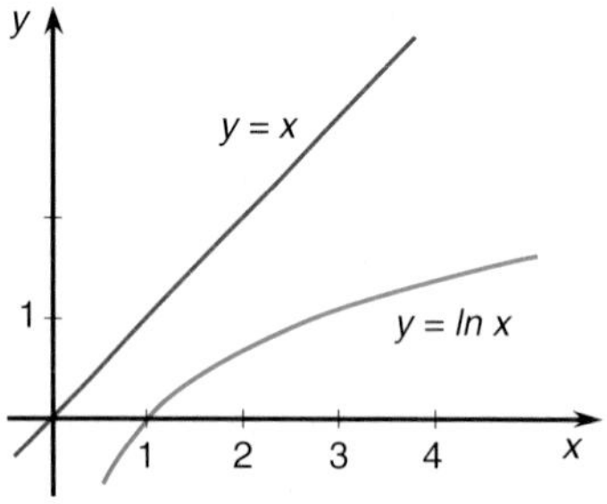

Figure 7.1.2

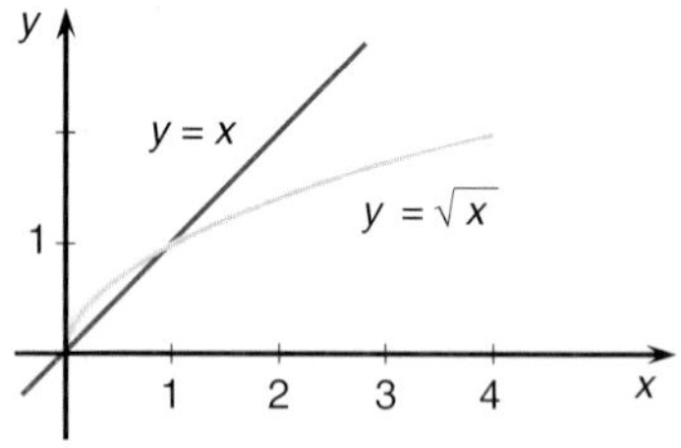

Figure 7.1.3

to infinity more slowly. For that matter we could also consider other powers x^r. If r is any positive number, $\lim_{x\to\infty} x^r = \infty$, but if r is small (i.e., $0 < r < 1$), the graph of $y = x^r$ grows more and more slowly as x increases.

It would be interesting to compare $y = \ln x$ and $y = x^r$ for various positive values of r to see which of them grows most slowly. The answer is contained in the following theorem.

THEOREM 7.1.4

If r is any positive number, then

$$\lim_{x\to\infty} \frac{\ln x}{x^r} = 0.$$

In words, we say that "$\ln x$ grows more slowly than any power x^r."

PROOF We use the version of l'Hôpital's Rule that applies to infinite limits at infinity. It tells us that

$$\lim_{x\to\infty} \frac{\ln x}{x^r} = \lim_{x\to\infty} \frac{\frac{1}{x}}{rx^{r-1}} = \lim_{x\to\infty} \frac{1}{rx^r} = 0.$$

With a graphics calculator we can plot $y = \frac{\ln x}{x^r}$ for various values of r, to see the decrease predicted by the theorem. For instance, with $r = 0.75$,

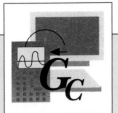

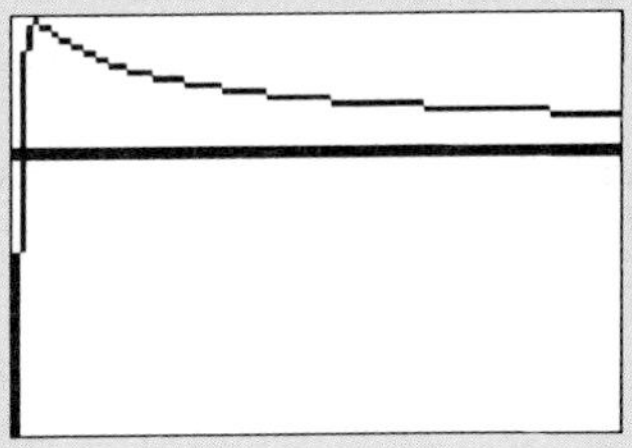

TI-85	SHARP EL9300
GRAPH F1 ($y(x)$ =) LN F1 (x) ÷ F1 (x) ^ .75 M2 (RANGE) .5 ▼ 100 ▼ ▼ (−) 1 ▼ .5 F5 (GRAPH)	[graph key] ln x/θ/T ÷ (x/θ/T a^b .75 RANGE .5 ENTER 100 ENTER 10 ENTER −1 ENTER .5 ENTER .5 ENTER [graph key]
CASIO f_x-6300G (f_x-7700GB)	**HP 48SX**
Range .5 EXE 100 EXE 10 EXE −1 EXE .5 EXE .5 EXE EXE EXE EXE Graph ln X ÷ (X x^y .75) EXE	↰ PLOT PLOTR ERASE ATTN ′ Y = LN X ▶ ÷ X y^x .75 ↱ PLOT ↰ DRAW ERASE .5 ENTER 100 XRNG 1 +/− ENTER .5 YRNG DRAW

You could modify these commands to try it with different values of r. You may have to plot the curve for a wider range of x in order to see the decreasing behavior. For instance, try it with $r = 0.1$ or $r = 0.01$ . . .

Leonhard Euler (1707–1783)

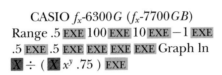

In anticipation of the next section we remark that since $\ln x$ equals zero when $x = 1$ and tends to infinity as $x \to \infty$, the graph must at some point cross the horizontal line $y = 1$ (see Figure 7.1.4). In other words, there is some number x whose natural logarithm equals 1. This number is called e, and it is approximately equal to 2.718 281 83. The symbol e, which stands for "exponential," was first introduced by the Swiss mathematician Leonhard Euler (1707–1783).

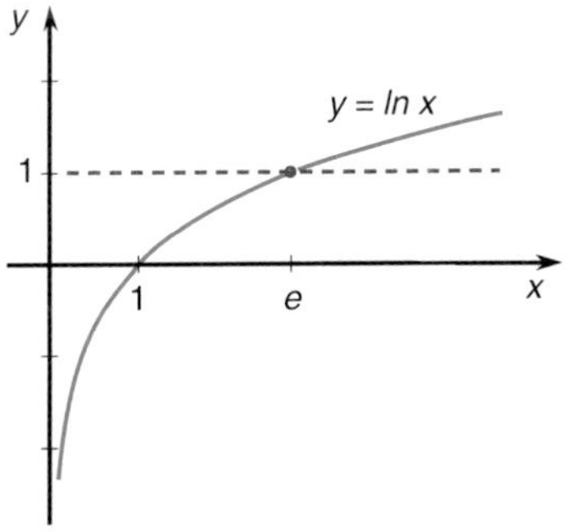

Figure 7.1.4

Exercises 7.1

I

Suppose A and B are positive real numbers, and write $a = \ln A$ and $b = \ln B$. Suppose n and m are integers. Write each of the following quantities in terms of a, b, m, and n.

1. $\ln(AB)$

2. $\ln(A^4 B)$

3. $\ln\left(\frac{A}{B^2}\right)$

4. $\ln \frac{1}{A^2}$

5. $\ln(A^m B^n)$

6. $\ln\big((1)(A)\big)$

7. $\ln(A^{-m})$

8. $\ln(A^m / B^n)$

9. $\ln(\sqrt{A})$

10. $\ln(\sqrt{AB})$

Evaluate the following limits.

11. $\lim_{x\to\infty} \frac{\ln x}{x^3}$

12. $\lim_{x\to\infty} \frac{\ln x}{x+1}$

13. $\lim_{x\to\infty} \frac{\ln(x^5)}{x^{1/3}}$

14. $\lim_{x\to\infty} \frac{x \ln x}{x+4}$

15. $\lim_{x\to 0^+} x \ln x$

16. $\lim_{x\to 0^+} \sqrt{x} \ln x$

17. $\lim_{x\to 2^+} (x-2)^{2/13} \ln(x-2)$

18. $\lim_{x\to 0^+} \sqrt{x} \ln(\sqrt{x})$

19. $\lim_{x\to 1^+} \frac{\ln x}{x-1}$

20. $\lim_{x\to e^+} \frac{x-e}{1-\ln x}$

II

21. Use an argument like the one in Theorem 7.1.2 (p. 510) (making a substitution in an integral) to show directly that for $a > 0$, $\ln \frac{1}{a} = -\ln a$.

22. Suppose $r > 0$ and use an argument like the one in Theorem 7.1.4 (p. 512) to evaluate

$$\lim_{x\to 0^+} (x^r \ln x).$$

23. The formula $\ln(a^n) = n \ln a$ (Summary 7.1.3, Property 2, p. 511) was established in the case in which n is an integer. Give an argument to show that it is still true for $n = \frac{1}{2}$.

***24.** (i) Show that for any $a > 0$ and any positive integer n, $\ln(a^{1/n}) = \frac{1}{n} \ln a$. (Notice that $a^{1/n}$ means the positive nth root of a.)

(ii) Show that for any $a > 0$ and any positive rational number r, $\ln(a^r) = r \ln a$.

(iii) Show that for any $a > 0$ and any rational number r, $\ln(a^r) = r \ln a$.

***25.** Evaluate $\lim_{x\to\infty} (x - \ln x)$.

26. (i) What is the domain of the function $f(x) = \ln(x^2)$?

(ii) Make a quick sketch of the graphs of $y = f(x) = \ln(x^2)$ and of $y = 2 \ln x$.

***27.** Imagine the graph of $y = \ln x$, with the units measured in inches. (i) If you go out along the positive x-axis for a distance of 1 mile (= 5280 ft) from the origin, what will be the value of $\ln x$? (First find x, recalling that there are 12 inches in one foot.) (ii) What will be the value of $\ln x$ if you travel the distance to the moon (about 240,000 miles)? (iii) What about the distance to the sun (about 93,000,000 miles)?

***28.** Now imagine the graph of $y = \ln x$ with units measured in centimeters. How far is it necessary to go from the origin before $\ln x = 100$? (Express your answer in light-years: one light-year is the distance light travels in one year, going at 300,000 km/sec.)

29. In Theorem 7.1.2 (p. 510) we used the integral definition to prove that $\ln(ab) = \ln a + \ln b$. Prove this result another way, by comparing the derivatives of $f(x) = \ln x$ and $g(x) = \ln(ax)$.

30. If n is a positive integer, prove that $\ln(x^n) = n \ln x$, for $x > 0$, by considering the derivative of $h(x) = \ln(x^n)$.

III

31. Use a graphics calculator or computer to sketch the graph of $y = \ln x$, and use the picture to find as many digits as possible in the expression for e. (Recall that $\ln e = 1$.)

32. Use the graph of $y = e^x$ to find as many digits as possible in the expression for $\ln(10)$.

7.2 Exponentials and Powers

EXPONENTIAL FUNCTION
POWERS
LOGARITHM
BASE
COMMON LOGARITHM

In Section 7.1 we studied the natural logarithm. We defined it as the integral $\ln x = \int_1^x \frac{1}{t}\,dt$ and were able to discover most of the familiar properties of the logarithm as simple consequences of the definition.

One of those properties is the fact that $\ln x$ is an increasing function because the derivative $\frac{d}{dx}(\ln x) = \frac{1}{x}$ is positive. We realized in Section 3.8 that an increasing function could not take the same value twice (i.e., at two different values of x); see Figure 7.2.1. In Definition 3.8.2 (p. 203) we called a function with this property *one-to-one* and observed that a function that is one-to-one always has an inverse function.

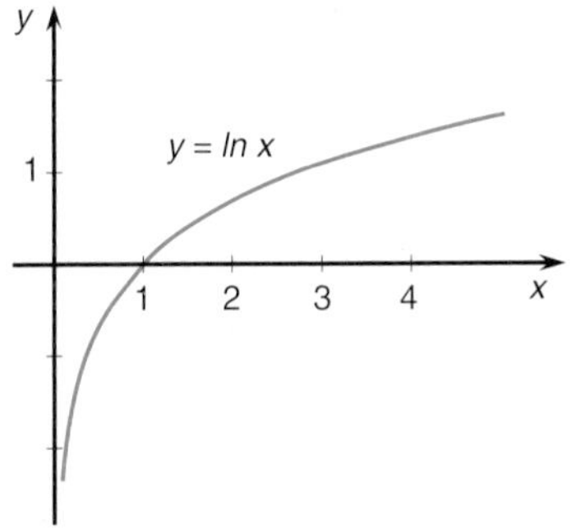

Figure 7.2.1

We define the inverse function of $\ln x$ to be the **exponential function** e^x, sometimes written $\exp(x)$. In other words,

$$e^{\ln x} = x, \quad \text{if } x > 0 \qquad \text{and} \qquad \ln(e^x) = x, \quad \text{for any } x. \tag{7.2.1}$$

Notice that since $\ln x$ is only defined for $x > 0$, its inverse, the exponential function, only has positive values, that is, $e^x > 0$, for every x.

Many properties of e^x can easily be deduced from the corresponding properties of $\ln x$, using Equations 7.2.1. For instance,

$$e^{a+b} = \exp(a+b) = \exp\big(\ln(e^a) + \ln(e^b)\big) = \exp\big(\ln(e^a e^b)\big) = e^a e^b.$$

Repeating this several times shows that $e^{nx} = (e^x)^n$, for any positive integer n. (If you know about proofs by induction, then you will recognize this as a result that should be proved that way.)

Since $\ln(1) = 0$, we know from Equations 7.2.1 that $e^0 = 1$, and from this and the above formula about products it follows easily that $e^{-x} = 1/e^x$.

In Theorem 3.8.4 we used the Chain Rule to calculate the derivative of an inverse function. In the present case, if we let $y = e^x$, then $x = \ln(y)$, and the derivative is

$$\frac{dy}{dx} = \frac{d}{dx}(e^x) = \frac{1}{\frac{d}{dy}(\ln y)} = \frac{1}{1/y} = y = e^x.$$

This explains the formula $\frac{d}{dx}(e^x) = e^x$, which we have been using since Chapter 3.

It also shows that the derivative of e^x is always positive, so the function is increasing. In particular, since $e^0 = 1$, we see that $e^1 > 1$, so $e^n = e^{n(1)} = (e^1)^n$; this is the nth power of a number bigger than 1, so it grows larger and larger and tends to infinity. Together with the fact that e^x is increasing, this shows that $\lim_{x\to\infty} e^x = \infty$. An argument like the one used to show $\lim_{x\to 0^+} \ln x = -\infty$ shows that $\lim_{x\to-\infty} e^x = 0$.

With these facts and the trivial observation that the second derivative is also positive, it is easy to sketch the graph of $y = e^x$. We summarize.

Summary 7.2.2

Properties of e^x

1. e^x is the inverse function of $\ln x$. This means that

$$e^{\ln x} = x, \quad \text{if } x > 0 \qquad \text{and} \qquad \ln(e^x) = x, \quad \text{for any } x.$$

2. $e^{a+b} = e^a e^b$, and for any integer n, $e^{na} = (e^a)^n$.
3. $e^0 = 1$.
4. $\frac{d}{dx}(e^x) = e^x$; $y = e^x$ is increasing and concave up.
5. $e^x > 0$ for any x; $0 < e^x < 1$ if $x < 0$, and $e^x > 1$ if $x > 0$.
6. $\lim_{x\to\infty} e^x = \infty$ and $\lim_{x\to-\infty} e^x = 0$.
7. The function $f(x) = e^x$ is uniquely determined by the facts that $\frac{d}{dx}(f) = f$ and $f(0) = 1$; that is, it is the only function defined for every $x \in \mathbb{R}$ and satisfying $f' = f$ and $f(0) = 1$.

Since $2e^x = e^{x+\ln 2}$, we see that the graph of $y = 2e^x$ is the graph of $y = e^x$ *shifted to the left* by a distance of ln 2.

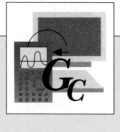

With a graphics calculator we can plot $y = e^x$, $y = 2e^x$, $y = 4e^x$, and $y = 8e^x$ and see how these curves are all horizontal shifts of each other.

TI-85	SHARP EL9300
GRAPH F1$(y(x) =)$ e^x F1(x) ▼ 2 e^x F1(x) ▼ 4 e^x F1(x) ▼ 8 e^x F1(x) M2(RANGE) $(-)$ 1 ▼ 3 ▼ ▼ 0 ▼ 20 F5(GRAPH)	⩙ e^x x/θ/T 2ndF ▼ 2 e^x x/θ/T 2ndF ▼ 4 e^x x/θ/T 2ndF ▼ 8 e^x x/θ/T RANGE -1 ENTER 3 ENTER 1 ENTER 0 ENTER 20 ENTER 5 ENTER ⩙
CASIO f_x-6300G (f_x-7700GB)	**HP 48SX**
Range -1 EXE 3 EXE 1 EXE 0 EXE 20 EXE 5 EXE EXE EXE EXE Graph e^x X EXE Graph 2 e^x X EXE Graph 4 e^x X EXE Graph 8 e^x X EXE	↰ PLOT PLOTR ERASE ATTN ′ Y = e^x X ↱ PLOT ↰ DRAW ERASE 1 ENTER +/− 3 XRNG 1 +/− ENTER 20 YRNG DRAW ATTN ′ Y = 2 × e^x X ↱ PLOT ↰ DRAW DRAW ATTN ′ Y = 4 × e^x X ↱ PLOT ↰ DRAW DRAW ATTN ′ Y = 8 × e^x X ↱ PLOT ↰ DRAW DRAW

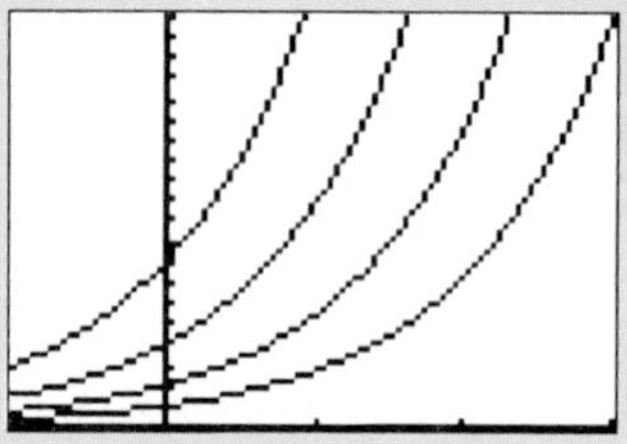

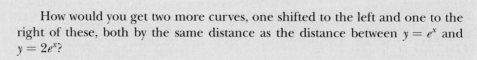

How would you get two more curves, one shifted to the left and one to the right of these, both by the same distance as the distance between $y = e^x$ and $y = 2e^x$?

The graph of $y = e^x$ is shown in Figure 7.2.2. As always happens with inverse functions, we can think of it as shown in Figure 7.2.3 as the result of reflecting the graph of $y = \ln x$ in the line $y = x$ (cf. the discussion after Example 3.8.5, p. 204).

Powers

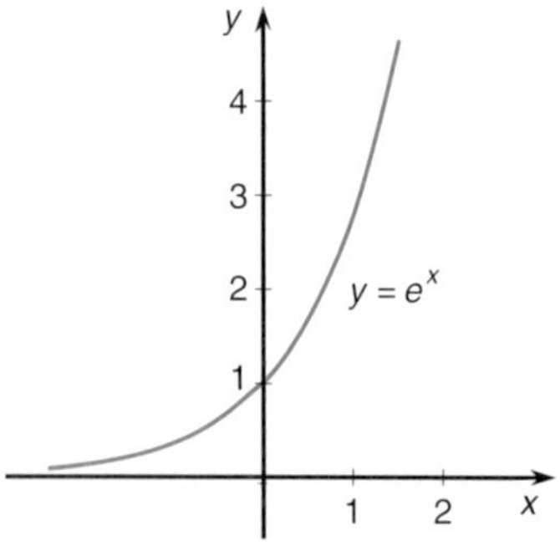

Figure 7.2.2

Now we can use the exponential function to define powers. What makes this a subtle problem is that we already know what we mean by a^n if n is an integer. Even if r is a rational number, a^r has a definite meaning, at least if $a > 0$. So if we want to define a^x for *any* real number x, it will be necessary to check that our definition agrees with the original meaning in case x is a rational number and $a > 0$.

If n is an integer, the power a^n simply refers to the product of n factors of a, if n is positive, and to a product of factors of $\frac{1}{a}$, if $n < 0$. Fractional powers refer to roots ($a^{m/n}$ means the mth power of the nth root of a or, equivalently, the nth root of the mth power of a).

We know that for any $a > 0$ and any integer n, $a^n = (e^{\ln a})^n = e^{n \ln a}$. Next consider the mth root $a^{1/m}$, for any positive integer m. It is the (unique) posi-

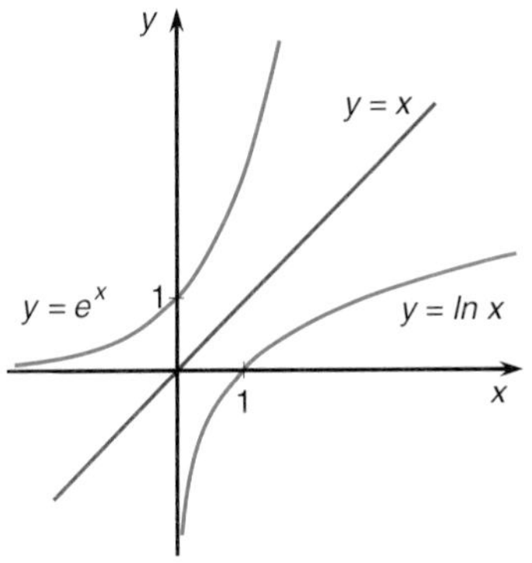

Figure 7.2.3

tive number whose mth power equals a, that is, $a^{1/m}a^{1/m}a^{1/m}\ldots a^{1/m} = a$, where the product on the left side consists of m factors of $a^{1/m}$. Applying the natural logarithm function to both sides of this equation, we find $\ln(a^{1/m}) + \ln(a^{1/m}) + \ln(a^{1/m}) + \ldots + \ln(a^{1/m}) = \ln(a)$, and since there are m terms on the left, this says that $m\ln(a^{1/m}) = \ln a$, or $\ln(a^{1/m}) = \frac{\ln a}{m}$. Exponentiating both sides of the equation, we find $a^{1/m} = \exp\left(\frac{\ln a}{m}\right)$. The left side means the mth root of a, and the right side means the value of $\exp x$ at $x = \frac{\ln a}{m}$.

Now consider the $\frac{n}{m}$th power, for any integers $m \neq 0$ and n. We already know that $\exp\left(\frac{\ln a}{m}\right) = a^{1/m}$, and by the formula about powers in Property 2 of Summary 7.2.2 we can conclude that $\exp\left(n\frac{\ln a}{m}\right) = (a^{1/m})^n$, or $\exp\left(\frac{n}{m}\ln a\right) = a^{n/m}$.

To put it another way, for any $a > 0$ and any rational number r we can express the rational power by using the simple exponential formula

$$a^r = e^{r\ln a}.$$

This formula makes sense for any $r \in \mathbb{R}$, so we can use it to define *any* power.

DEFINITION 7.2.3

Suppose $a > 0$ and x is any real number. Then the **power** a^x is defined by

$$a^x = e^{x\ln a}.$$

The remarks preceding the definition show that when x is a rational number, this definition agrees with the usual meaning of the power a^x.

THEOREM 7.2.4

If $a > 0$, and u, v are any real numbers, then

(i) $a^u a^v = a^{u+v}$,

(ii) $(a^u)^v = a^{uv}$.

PROOF

(i) We simply evaluate

$$a^u a^v = e^{u\ln a}e^{v\ln a} = e^{(u\ln a + v\ln a)} = e^{(u+v)\ln a} = a^{(u+v)}.$$

(ii) Similarly,

$$\begin{aligned}(a^u)^v &= \exp\big(v\ln(a^u)\big) = \exp\big(v\ln(e^{u\ln a})\big)\\ &= \exp\big((v)u\ln a\big) = \exp\big((uv)\ln a\big) = a^{uv}.\end{aligned}$$

It is interesting to regard x^r as a function of x, for $x > 0$.

EXAMPLE 1 Let $r \in \mathbb{R}$ be fixed. Find the derivative of the function $y = x^r$, for $x > 0$.

Solution Since $y = x^r = e^{r\ln x}$, we see that

$$\frac{dy}{dx} = \frac{d}{dx}(e^{r\ln x}) = e^{r\ln x}\frac{d}{dx}(r\ln x) = e^{r\ln x}\left(\frac{r}{x}\right) = x^r\frac{r}{x} = rx^{r-1}.$$

In other words, the familiar formula for the derivative of x^r is valid for any power r. Previously, we knew it only in the case in which r is rational.

The graph of $y = e^x$ rises very fast. Certainly, it goes up faster than the straight line $y = x$, but it is less clear what happens when e^x is compared with other powers x^t. In analogy with the result of Theorem 7.1.4 comparing the growth of x^r with that of $\ln x$, we consider $\lim_{x\to\infty} (x^r/e^x)$.

THEOREM 7.2.5

For any real number r,

$$\lim_{x\to\infty} \left(\frac{x^r}{e^x}\right) = 0.$$

In words, we say that "e^x grows faster than any power of x."

PROOF Consider

$$\ln\left(\frac{x^r}{e^x}\right) = r\ln x - x = \left(r\frac{\ln x}{x} - 1\right)x.$$

In Theorem 7.1.4 we saw that $\lim_{x\to\infty} (\ln x)/x = 0$. So $\left(\frac{r\ln x}{x} - 1\right) \to -1$. Since $x \to \infty$, we see that $\ln\big((x^r)/(e^x)\big) = \left(\frac{r\ln x}{x} - 1\right)x \to -\infty$.

If the logarithm of some quantity tends to $-\infty$, then the quantity itself must tend to 0, since $\lim_{x\to-\infty} e^x = 0$. This completes the proof, showing that $(x^r)/(e^x) \to 0$.

In addition to the rth power function $f(x) = x^r$, it is also possible to consider the function $g(x) = a^x$, where a is any fixed positive number. This is a function that is defined for all values of x, and it is possible to consider its derivative.

EXAMPLE 2 Let $a > 0$ be fixed. Find $\frac{d}{dx}(a^x)$.

Solution We write $a^x = e^{x\ln a}$, so

$$\frac{d}{dx}(a^x) = \frac{d}{dx}(e^{x\ln a}) = e^{x\ln a}\frac{d}{dx}(x\ln a) = a^x \ln a.$$

Note how we wrote $e^{x\ln a}$ as a^x at the end.

Limits of Powers

In evaluating a limit of the form $\lim_{x\to a}\big(f(x)^{g(x)}\big)$, where the base $f(x)$ is a positive function of x, it is frequently helpful to write out the power as

$$f(x)^{g(x)} = e^{g(x)\ln(f(x))}.$$

If we can evaluate the limit of this exponent, say, $\lim_{x\to a} g(x)\ln\big(f(x)\big) = L$, for instance, then we can conclude that

$$\lim_{x\to a}\big(f(x)^{g(x)}\big) = \lim_{x\to a}\big(e^{g(x)\ln(f(x))}\big) = e^L.$$

Frequently, in questions of this kind we find that in the product $g(x)\ln(f(x))$ one factor will tend to ∞ and the other to 0. In such a case it is often helpful to rearrange it into a quotient and use l'Hôpital's Rule. We write

$$g(x)\ln\big(f(x)\big) = \frac{\ln\big(f(x)\big)}{\frac{1}{g(x)}} = \frac{g(x)}{\frac{1}{\ln(f(x))}},$$

and apply l'Hôpital's Rule to either of these quotients. Offhand, it is probably simpler to use the one in which both numerator and denominator tend to zero rather than the other, in which they both tend to ∞.

EXAMPLE 3 Find $\lim_{x\to\infty}\left(1+\frac{1}{x}\right)^x$.

Solution We write $\left(1+\frac{1}{x}\right)^x = \exp\left(x\ln\left(1+\frac{1}{x}\right)\right)$. As $x\to\infty$, $1+\frac{1}{x}\to 1$, so $\ln\left(1+\frac{1}{x}\right)\to 0$. Following the suggestion made in the discussion above, we write $x\ln\left(1+\frac{1}{x}\right) = \frac{\ln(1+1/x)}{1/x}$ and apply l'Hôpital's Rule. We find

$$\lim_{x\to\infty} x\ln\left(1+\frac{1}{x}\right) = \lim_{x\to\infty}\frac{\ln\left(1+\frac{1}{x}\right)}{\frac{1}{x}} = \lim_{x\to\infty}\frac{\frac{1}{1+\frac{1}{x}}\left(-\frac{1}{x^2}\right)}{-\left(\frac{1}{x^2}\right)}$$

$$= \lim_{x\to\infty}\frac{1}{1+\frac{1}{x}} = 1.$$

This tells us that the natural logarithm of $\left(1+\frac{1}{x}\right)^x$ tends to 1, so $\left(1+\frac{1}{x}\right)^x$ itself tends to $e^1 = e$. We have just found that

$$\lim_{x\to\infty}\left(1+\frac{1}{x}\right)^x = e. \tag{7.2.6}$$

This result is very useful.

EXAMPLE 4 Suppose a is any fixed real number. Find $\lim_{x\to\infty}\left(1+\frac{a}{x}\right)^x$.

Solution As above, we find the limit of the logarithm $x\ln\left(1+\frac{a}{x}\right)$, using l'Hôpital's Rule:

$$\lim_{x\to\infty}\left(x\ln\left(1+\frac{a}{x}\right)\right) = \lim_{x\to\infty}\frac{\ln\left(1+\frac{a}{x}\right)}{\frac{1}{x}} = \lim_{x\to\infty}\frac{\frac{1}{1+\frac{a}{x}}\left(-\frac{a}{x^2}\right)}{-\frac{1}{x^2}}$$

$$= \lim_{x\to\infty}\frac{1}{1+\frac{a}{x}}\frac{a}{1} = a.$$

Taking exponentials, we see that

$$\lim_{x\to\infty}\left(1+\frac{a}{x}\right)^x = e^a. \tag{7.2.7}$$

Of course, Example 4 includes the previous example, by letting $a = 1$. They will be used in Section 7.4 for the study of compound interest.

Logarithms to Other Bases

The natural logarithm has the property that

$$a = e^{\ln a},$$

for any positive number a.

We could try to replace the number e in this formula with some other positive number and look for an analogous formula. Suppose we fix some positive number

$b \neq 1$ and consider the function $y = b^x$. Given any $a > 0$, we could look for the value of x for which $b^x = a$.

In fact it is easy to find this value of x. After all, if we write $b^x = e^{x \ln b}$, then we see that for this to equal a we must have $x \ln b = \ln a$, or $x = (\ln a)/(\ln b)$.

DEFINITION 7.2.8

Suppose b is a fixed positive number, $b \neq 1$, and consider the function $y = b^x$. We refer to b as the **base** for this function.

Moreover, define the **logarithm to the base *b***, written $\log_b$, by

$$a = b^{\log_b(a)},$$

for every $a > 0$.

We saw above that

$$\log_b(a) = \frac{\ln a}{\ln b}.$$

REMARK 7.2.9

From the preceding formula we see that $\log_b$ satisfies many of the same properties as the natural logarithm. Specifically, for $b > 0$, with $b \neq 1$,

(i) $\log_b(xy) = \log_b(x) + \log_b(y)$, for any $x,\ y > 0$.

(ii) $\log_b(1) = 0$.

(iii) $\log_b(x^r) = r \log_b(x)$, for any $x > 0$ and any power r.

(iv) We can easily find $\frac{d}{dx}\big(\log_b(x)\big)$. In fact, using the formula for $\log_b x$, we see that

$$\frac{d}{dx}\big(\log_b(x)\big) = \frac{d}{dx}\left(\frac{\ln(x)}{\ln(b)}\right) = \frac{1}{\ln(b)}\frac{d}{dx}(\ln x) = \frac{1}{\ln(b)}\frac{1}{x}.$$

The most frequently used bases are $b = e$, $b = 10$, and $b = 2$, though in principle it is possible to use any positive number $b \neq 1$ as the base for logarithms. The base $b = 10$ in particular is used so often that logarithms to the base 10 are often called **common logarithms**.

EXAMPLE 5

(i) $\log_{10}(e) = \ln(e)/\ln(10) = \frac{1}{\ln(10)} \approx 0.434294$.

(ii) $\log_2(\sqrt{2}) = \frac{1}{2}$. This we should realize immediately without calculating anything, since after all, $\log_2(\sqrt{2})$ ought to be the power to which we have to raise 2 to get $\sqrt{2}$, so it must equal $\frac{1}{2}$.

(iii) Find $\log_3(81)$. This we might be able to guess, but instead we can calculate $\log_3(81) = \big(\ln(81)\big)/\big(\ln(3)\big)$. This will not help much unless we use a calculator or unless we remember that $81 = (9)(9) = 3^4$. With this in mind we find $\ln(81) = \ln(3^4) = 4\ln(3)$, so $\log_3(81) = \big(\ln(81)\big)/\big(\ln(3)\big) = \big(4\ln(3)\big)/\big(\ln(3)\big) = 4$.

In a sense we ought to have guessed this because all it says is that 81 equals the number 3 raised to the fourth power.

Exercises 7.2

I

Evaluate each of the following. If any limit does not exist, check to see whether it equals $\pm\infty$.

1. $\lim_{x\to\infty} e^{-x}$
2. $\lim_{x\to\infty} xe^{-x}$
3. $\lim_{x\to\infty} \left(\frac{x}{\pi^x}\right)$
4. $\lim_{x\to-\infty} 2^{-x}$
5. $\lim_{x\to-\infty} (\sqrt{2})^x$
6. $\lim_{x\to 0^-} \left(\frac{1}{2}\right)^{1/x}$
7. $\frac{d}{dx}(x^{\sqrt{5}})$
8. $\frac{d}{dx}\left((\sqrt{5})^x\right)$
9. $\lim_{x\to 0+} (x^{-2}e^{-1/x})$
10. $\lim_{x\to\infty} \left(\frac{\pi^x}{x^\pi}\right)$

Evaluate each of the following.

11. $\lim_{x\to\infty} \frac{e^{x/2}}{x^3+3x-2}$
12. $\lim_{x\to-\infty} \left((2x^4+3x+1)e^x\right)$
13. $\int x^{\sqrt{2}}\,dx$
14. $\int x^{2\pi}\,dx$
15. $\int 2^x\,dx$
16. $\int 5^{2x}\,dx$
17. $\int x4^{x^2}\,dx$
18. $\int x^2\pi^{x^3}\,dx$
19. $\int x^{-1/2}10^{\sqrt{x}}\,dx$
20. $\int 2^x e^{7x}\,dx$

Evaluate each of the following.

21. $\log_{10}(100)$
22. $\log_{1/2}(16)$
23. $\log_\pi(\pi^2)$
24. $\log_3(9^{-4/5})$
25. $\frac{d}{dx}\left(\log_{10}(x)\right)$
26. $\frac{d}{dx}\left(\log_5(x^2+3)\right)$
27. $\lim_{x\to\infty} \log_{1/4}(x)$
28. $\lim_{x\to 0+} \left(\log_{10}(3x)\right)/\left(\log_2(x)\right)$
29. $\int \frac{\log_{10}(x)}{\log_3(x)}\,dx$
30. $\int \frac{\log_2(x)}{x}\,dx$

Evaluate each of the following finite or infinite limits.

31. $\lim_{x\to\infty} \left(1+\frac{3}{x}\right)^x$
32. $\lim_{x\to\infty} \left(1+\frac{1}{2x}\right)^x$
33. $\lim_{x\to\infty} \left(1+\frac{1}{x}\right)^{-x}$
34. $\lim_{x\to-\infty} \left(1+\frac{1}{x}\right)^x$
35. $\lim_{x\to\infty} \left(1-\frac{2}{x}\right)^{2x}$
36. $\lim_{x\to\infty} \left(1-\frac{1}{\pi x}\right)^{x/2}$
37. $\lim_{x\to\infty} \left(1-\frac{1}{x}\right)^x/\left(1+\frac{1}{x}\right)^x$
38. $\lim_{x\to\infty} \left(1+\frac{1}{x^2}\right)^{1/x}$
39. $\lim_{x\to\infty} \left(1+\frac{1}{x^2}\right)^x$
40. $\lim_{x\to\infty} \left(1+\frac{1}{x}\right)^{x^2}$

II

41. Calculate the derivative of $F(x) = (e^2)^x$. Verify that it equals the derivative of $f(x) = e^{2x}$.
42. Suppose $a > 0$ and $b > 0$ and show that for any x, $a^x b^x = (ab)^x$.
43. What is the minimum value of $f(x) = 6^x + 6^{-x}$?
44. What is the minimum value of $f(x) = 2^x + 4^{-x}$?
45. Show that for any $a > 1$ the function $y = a^x$ goes to infinity faster than any power of x in the sense that $\lim_{x\to\infty} (x^r)/(a^x) = 0$.
46. Show that for any $b > 1$ the function $y = \log_b(x)$ goes to infinity more slowly than any positive power of x, in the sense that $\lim_{x\to\infty} \left(\log_b(x)\right)/(x^r) = 0$, for any $r > 0$.

*47. Show that for any $a > 1$ the function $y = a^x$ goes to infinity faster than any polynomial $p(x)$, in the sense that $\lim_{x\to\infty} \left(p(x)\right)/(a^x) = 0$.

*48. Show that for any $b > 1$ the function $y = \log_b(x)$ goes to infinity more slowly than any polynomial $p(x)$, in the sense that $\lim_{x\to\infty} \left(\log_b(x)\right)/\left(p(x)\right) = 0$. What happens if $0 < b < 1$?

49. Find $\lim_{x\to\infty} \left(\frac{1}{6}\right)^x$.

*50. Suppose $a > 0$ and $b > 0$ and let $f(x) = a^x + b^{-x}$. Find the value(s) of x at which $f(x)$ has an absolute minimum. Your answer should contain a condition on a and b, without which there is no absolute minimum.

51. Let $f(x) = 5^x$, defined for every real number x. Find the regions where $f(x)$ is increasing and decreasing.

*52. Suppose $a > 0$ and let $f(x) = a^x$. Find the regions where $f(x)$ is increasing and decreasing. (*Hint:* Your answer will depend on the value of a.)

*53. For $x > 0$, define $f(x) = x^x$. (i) Find the derivative $f'(x)$. (ii) Find $\lim_{x\to 0+} f(x)$. (iii) Classify the extrema of $f(x)$. (iv) Sketch the graph of $y = f(x)$.

54. If a, b, and c are positive numbers with $a \neq 1$ and $b \neq 1$, show that

$$\left(\log_a(b)\right)\left(\log_b(c)\right) = \log_a(c).$$

***55.** Prove the statement made in Property 7 of Summary 7.2.2. (*Hint:* Let $f(x) = e^x$, and suppose $g(x)$ is another function satisfying $g' = g$ and $g(0) = 1$. Since $f(x)$ is never zero, it is possible to consider the function $\frac{g(x)}{f(x)}$; find its derivative . . .)

56. Find the derivative of x^{x^x}, for $x > 0$.

III

57. Use a graphics calculator or computer to sketch the graphs of $y = \ln x$, $y = \log_{10}(x)$, and $y = \frac{\ln(x)}{\ln(10)}$ on the same picture.

58. Use a graphics calculator or computer to sketch the graphs of $y = e^{\sin x}$ and $y = \sin x$ on the same picture.

POINT TO PONDER

In Definition 7.2.3 we defined powers a^r. In Example 1 we found how to differentiate the power x^r; the result was the unsurprising formula $\frac{d}{dx}(x^r) = rx^{r-1}$, for any $r \in \mathbb{R}$.

Using this, we were able to apply l'Hôpital's Rule to find in Theorem 7.2.5 that

$$\lim_{x \to \infty} \frac{x^r}{e^x} = 0,$$

for any $r \in \mathbb{R}$.

In fact we used a similar argument in Theorem 7.1.4 (p. 512) to show that

$$\lim_{x \to \infty} \frac{\ln x}{x^r} = 0,$$

for any $r \in \mathbb{R}$. However, at that time we did not really know what x^r means unless r is a rational number, and we certainly did not know how to differentiate it if r is irrational.

So, strictly speaking, the result of Theorem 7.1.4 was really proved only for rational exponents r, but now, with the benefit of Example 1, we can prove it for every $r \in \mathbb{R}$. The proof is exactly the same.

7.3 Exponential Growth and Decay

GROWTH CONSTANT
DECAY CONSTANT
INITIAL VALUE
HALF-LIFE

There are many practical situations in which the rate of change of some quantity is proportional to the size of the quantity. The first example is population growth, for which the rate of change, that is, the number of new offspring per unit time, is proportional to the population. After all, the larger the population, the more offspring there will be.

This idea applies to many different types of population, including a culture of bacteria, the human population of some region, an animal population, or even some kinds of plant growth. In each of these situations it seems reasonable to assume that the rate of growth is proportional to the population.

In fact it is usually more subtle than that. Imagine a bacterial culture in a jar. The bacteria multiply, and the population grows. It grows faster and faster, but eventually there is a problem because it begins to run out of room. Except in science fiction it is not realistic to think that the culture will keep on growing until it bursts out of the jar. Before the jar is completely full, the growth will slow down.

This will happen for perfectly sensible reasons. For one thing the culture may begin to use up the nutrients in the jar. With less food available the bacteria cannot reproduce as quickly. Even simple overcrowding may reduce the efficiency of the bacterial metabolism.

In any case, the growth will slow down before exceeding the "limits" imposed on its growth by the nature and size of the environment. Similar observations can be made about more complicated populations. Human populations grow at slower rates under the effects of drastic overcrowding. The population of deer in a forest grows dramatically for a time but then falls off as the available food supply becomes insufficient to support all the animals.

Another consideration that we have not even mentioned is that in any natural population there are deaths as well as reproduction. This will of course affect the *net* population growth. In fact in many situations this particular question does not really change the behavior of a population in any qualitative way, so to ignore it may actually not be too serious.

The factors that limit growth and the mechanisms by which they affect the population are numerous and extremely complicated. We do not intend to study them here, other than to remark that such factors do exist and do exert a strong influence upon the growth of various populations. We intend to consider situations in which the rate of growth is proportional to the population. Typically, we can expect this to happen when the population is low in comparison to the levels that the environment can support because in this case the limitations are not likely to have much effect.

Consider a function $f(t)$. We let the variable t represent time, so we may think of $f(t)$ as a changing quantity. Its derivative $\frac{df}{dt}$ is the rate of change of $f(t)$. For convenience, let us assume the values of the function $f(t)$ are always positive, that is, $f(t) > 0$ for all t. An example that you may wish to keep in mind is to let $f(t)$ be the number of bacteria in a jar at time t.

Now suppose that we know that the rate of change of $f(t)$ equals a constant times the function $f(t)$ itself. Symbolically,

$$f'(t) = kf(t),$$

where k is some constant.

From this we see that $\frac{f'(t)}{f(t)} = k$. The left side of this equation is the derivative of $\ln(f(t))$, so it says that $\frac{d}{dt}\Big(\ln(f(t))\Big) = k$. This means that $\ln(f(t))$ is an antiderivative of the constant function k, and that means that it must equal $kt + c'$, for some constant c'.

If $\ln(f(t)) = kt + c'$, then $f(t) = e^{\ln(f(t))} = e^{kt+c'}$. By the properties of the exponential function this can be rewritten as $e^{kt+c'} = e^{kt}e^{c'} = ce^{kt}$, where $c = e^{c'}$ is a constant. What we have found is that

$$f(t) = ce^{kt}. \tag{7.3.1}$$

> This sort of exponential formula describes the behavior of any function whose derivative equals a constant times the original function. For this reason a process like this is often called **exponential growth**. The constant k is called the **growth constant**. If $k > 0$, then $f(t)$ is an increasing function of t, and if $k < 0$, it is decreasing (in which case it is sometimes called **exponential decay** and k is called the **decay constant**).

Notice that if we let $t = 0$ in Equation 7.3.1, we find $f(0) = ce^0 = c$. This means that the constant c equals the value of $f(t)$ at $t = 0$; for obvious reasons it is sometimes called the **initial value**. The behavior of an exponential growth process is completely determined once we know the initial value c and the growth constant k.

EXAMPLE 1 Suppose the function $f(t)$ grows exponentially. If $f(0) = 4$ and $f(2) = 20$, find $f(5)$, that is, the value of $f(t)$ when $t = 5$.

Solution We are assuming that $f(t)$ is of the form $f(t) = ce^{kt}$, for some constants c and k. The fact that $f(0) = 4$ tells us that $c = 4$, so $f(t) = 4e^{kt}$.

The other piece of information we have is that $f(2) = 20$. But we know that $f(2) = 4e^{k(2)}$, so $4e^{2k} = 20$ and $e^{2k} = \frac{20}{4} = 5$. Taking logarithms of both sides, we find that $2k = \ln(5)$, or $k = \frac{1}{2}\ln(5)$.

Now we know both constants, so it is possible to write the function precisely: $f(t) = 4e^{t\ln(5)/2}$. In particular, when $t = 5$,

$$f(5) = 4e^{5\ln(5)/2} = 4(e^{\ln(5)})^{5/2} = 4(5)^{5/2} = 100\sqrt{5} \approx 223.6068.$$

The idea was to use the values of $f(t)$ at the two times $t = 0$ and $t = 2$, where the values are known, to calculate the constants c and k. Once these have been found, we know the formula for $f(t)$ at any time t, and calculating it for any particular t is easy.

EXAMPLE 2 Suppose $g(t)$ is a function whose derivative equals a constant times the function itself. If $g(1) = 9$ and $g(4) = 3$, at what time t will $g(t) = 1$?

Solution Once again we write the function in the form of Equation 7.3.1, $g(t) = ce^{kt}$. What we know is that $g(1) = 9$ and $g(4) = 3$. In symbols this means

$$ce^k = 9 \qquad \text{and} \qquad ce^{4k} = 3.$$

Dividing these two equations, we find that

$$\frac{ce^{4k}}{ce^k} = \frac{3}{9} = \frac{1}{3}, \quad \text{or} \quad e^{3k} = \frac{1}{3}.$$

Taking logarithms, we find that $3k = \ln\frac{1}{3} = -\ln(3)$, so $k = -\frac{1}{3}\ln(3)$.

Next we have to find c. Since we are not told the value of $g(t)$ at $t = 0$, it is necessary to use the information we do have, including the value we have just found for k.

We know that $9 = g(1) = ce^k = ce^{-\ln(3)/3} = c(e^{\ln(3)})^{-1/3} = c(3)^{-1/3}$. From this we find that $c = 9/(3)^{-1/3} = 9(3^{1/3}) = 3^{7/3}$. So $g(t) = ce^{kt} = 3^{7/3}e^{-t\ln(3)/3}$.

To find when this equals 1, we write

$$3^{7/3}e^{-t\ln(3)/3} = 1$$

$$e^{-t\ln(3)/3} = \frac{1}{3^{7/3}}$$

$$-t\ln(3)/3 = -\ln(3^{7/3})$$

So

$$t = 3\frac{\ln(3^{7/3})}{\ln(3)} = 3\frac{\frac{7}{3}\ln(3)}{\ln(3)}$$
$$= 7.$$

The function $g(t)$ equals 1 when $t = 7$.

There are several things to notice about this example. One is that k is negative. This means that the function is "decaying," that is, decreasing; in hindsight this was obvious from the fact that $g(4) < g(1)$.

Second, even though $c = g(0)$ was not given to us in the question, we were able to use the values of $g(t)$ at two times t to find what k is. Knowing this, we were able to find c from either of the known values of $g(t)$.

Finally, it is interesting to observe that in the three units of time between $t = 1$ and $t = 4$ the function decreased by a factor of $\frac{1}{3}$ (from $g(1) = 9$ to $g(4) = 3$). For it to decrease to 1, a further decrease by a factor of $\frac{1}{3}$, it took an additional 3 units of time (from $t = 4$ to $t = 7$). This is in fact characteristic of exponential growth. In a particular length of time, say, I units long (I for "interval"), the function will always change by the same factor, no matter where it starts. In Example 2 the function decreases by a factor of $\frac{1}{3}$ whenever the time increases by 3 units.

The reason is actually very simple. Suppose $f(t) = ce^{kt}$, with $c \neq 0$. If we compare the values of the function at times t and $t + I$, their quotient is

$$\frac{ce^{kt}}{ce^{k(t+I)}} = e^{kt-k(t+I)} = e^{-kI}.$$

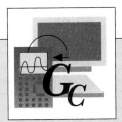

With a graphics calculator we can plot the function $y = g(t)$ from Example 2; we also plot the constant function $y = 1$.

TI-85	SHARP EL9300
3 ^ (7 ÷ 3) STO C ENTER (−) LN (3) ÷ 3 STO K ENTER GRAPH F1($y(x)$ =) C × e^x (K × F1(x)) ▼ 1 M2(RANGE) 1 ▼ 10 ▼ ▼ (−) 1 ▼ 9 F5(GRAPH)	3 a^b 7 ÷ 3 ▶ × e^x (−) x/θ/T × ln 3 ÷ 3 2ndF ▼ 1 RANGE 1 ENTER 10 ENTER 1 ENTER −1 ENTER 9 ENTER 1 ENTER
CASIO f_x-6300G (f_x-7700GB)	**HP 48SX**
Range 1 EXE 10 EXE 1 EXE −1 EXE 9 EXE 1 EXE EXE EXE EXE Graph 3 x^y (7 ÷ 3) × e^x ((−) X × ln (3) ÷ 3) EXE Graph 1 EXE	↰ PLOT PLOTR ERASE ATTN ′ Y = 3 y^x () 7 ÷ 3 ▶ × e^x X × LN 3 ▶ ÷ 3 +/− ↱ PLOT ↰ DRAW ERASE 1 ENTER 10 XRNG 1 +/− ENTER 9 YRNG DRAW ATTN ′ Y = 1 ↱ PLOT ↰ DRAW DRAW

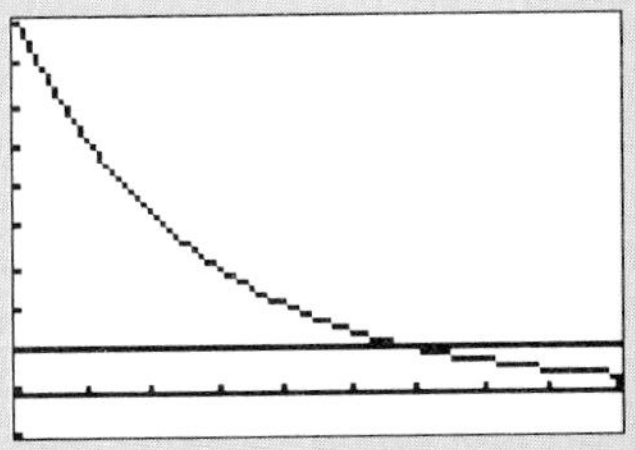

Notice where $y = g(t)$ crosses $y = 1$ . . .

This number e^{-kI} is independent of t. In other words, the ratio of $f(t)$ to $f(t+I)$ is the same for any t. This is true for any function of exponential growth (or decay).

This remark suggests that it might be interesting to ask how long it will take a particular exponential function to change by some factor R. The most familiar example of this idea occurs in the study of radioactive decay, which we will discuss in Section 7.4.

A substance undergoing radioactive decay is transformed into other substances, and the amount of it that remains at time t is a decaying exponential function $f(t) = ce^{kt}$, with $k < 0$. As usual, the constant k determines how quickly it decays, but a convenient way to describe it is the **half-life**, the length of time it takes for the substance to diminish by one-half. For example, suppose the half-life of a substance was 1 year. Then in one year half of it would decay, and in another year half of what was left would decay, leaving one-quarter of the original amount. In a period of 5 years, a sample of the substance would reduce by half, by half again, by half again, by half again, and by half once more, leaving only $\left(\frac{1}{2}\right)\left(\frac{1}{2}\right)\left(\frac{1}{2}\right)\left(\frac{1}{2}\right)\left(\frac{1}{2}\right) = \frac{1}{32}$ of the original amount.

The half-life is a handy way to describe the rate of decay. It is fairly easy to visualize, so it is used more often than the decay constant k.

Of course, the half-life and the decay constant are closely related.

EXAMPLE 3 Find the half-life of the decay process that is described by the function $f(t) = ce^{-kt}$. (Notice that to emphasize that the process is a decay, we have used the constant $-k$, with $k > 0$.)

Solution We know that $f(0) = c$; let us find the value of t for which $f(t) = \frac{c}{2}$. This amounts to

$$ce^{-kt} = \frac{c}{2},$$

or

$$e^{-kt} = \frac{1}{2},$$

or

$$-kt = \ln\left(\frac{1}{2}\right) = -\ln(2),$$

or

$$t = \frac{-\ln(2)}{-k} = \frac{\ln(2)}{k}.$$

The half-life is $\ln(2)/k$.

EXAMPLE 4 **Half-Life of a Radioactive Decay** Suppose some radioactive substance decays in such a way that after 100 years the substance has diminished to 98.79% of its original amount. What is the half-life?

Solution If we write the amount remaining at time t as $f(t) = ce^{-kt}$, then what we know is that $f(100) = 0.9879 f(0)$. This means that

$$ce^{-100k} = 0.9879c,$$

or

$$e^{-100k} = 0.9879,$$

or

$$-100k = \ln(0.9879),$$

so

$$k = -\frac{\ln(0.9879)}{100}$$
$$\approx 1.2174 \times 10^{-4}.$$

Combining this with the result of Example 3, we find that the half-life is $\ln(2)/k \approx 5700$ years.

This happens to be the approximate half-life of the isotope carbon 14, which is of great importance because of its use for estimating the date of archaeological finds. This is also something we will discuss in Section 7.4.

EXAMPLE 5 **Yogurt** Yogurt is made by placing a small amount of yogurt in some warm milk and letting it sit. The bacteria in the yogurt thrive in the warm milk and multiply. In the process, the milk is turned into yogurt.

Suppose we start with 10 grams of yogurt (about 2 teaspoons). Suppose too that the bacteria divide in such a way that the amount of yogurt doubles every 30 minutes. Let us measure time t in hours; the culture doubles every half hour.

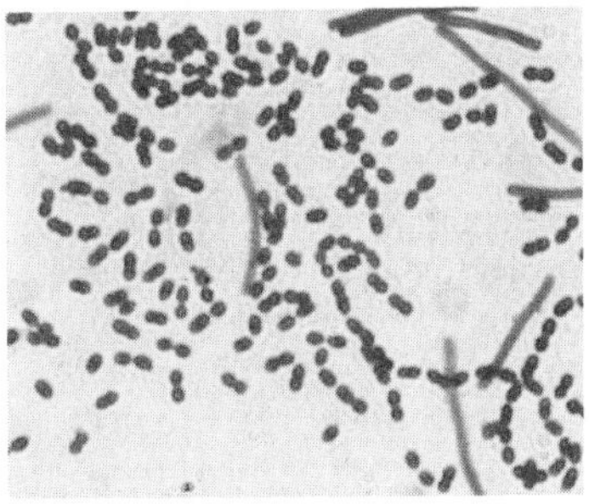

This 1000× magnification of a smear shows spherical Streptococcus lactis *and rod-shaped* Lactobacillus bulgaris *bacteria in yogurt. The* Lactobacillus bulgaris *bacteria are found in fermenting animal and plant products and produce lactic acid from the fermentation of carbohydrates. They also live in the alimentary tract and vagina of humans and other animals, where they protect against more dangerous bacteria.*

If the amount of yogurt present at time t is $Y(t) = ce^{kt}$, measured in grams, then $c = 10$ and $Y\left(\frac{1}{2}\right) = 20$, so $e^{k/2} = 2$ and $\frac{k}{2} = \ln(2)$, $k = 2\ln(2)$. The number of grams of yogurt present at time t is $Y(t) = 10e^{2t\ln(2)}$.

After 3 hours there will be about $Y(3) = 10e^{2(3)\ln(2)} = 10(e^{\ln(2)})^6 = 10(2^6) = 640$ g of yogurt, or a little under 3 cups.

If we consider what will happen in 24 hours, we find that $Y(24) = 10e^{2(24)\ln(2)} = 10(2)^{48} \approx 2.8 \times 10^{15}$ g. To get some idea of the size of this quantity, let us convert it to 2.8×10^{12} kg, which is 2.8×10^9 tonnes. (The "tonne," the metric ton, is 1000 kg, or about 2200 lb, not much different from the ordinary ton, which is 2000 lb.)

Assuming that the density of yogurt is the same as that of water, 1 tonne occupies 1 cubic meter, so after growing for 24 hours, the yogurt will occupy about 2.8×10^9 m^3. If all this yogurt were gathered into a cube-shaped tank, each side of the tank would be $(2.8 \times 10^9)^{1/3} \approx 1400$ m long. This is slightly under one mile.

That is a lot of yogurt.

> The point of this example is not to tell us anything about yogurt, but to demonstrate the pitfalls that lie in assuming that something will continue indefinitely to grow exponentially. The growth of yogurt will be approximately exponential for a while, but of course it will slow down when it has used up most of the milk available to it. The interesting thing about this example is to see just how fast exponential growth increases. The growth rate suggested here for yogurt is fairly reasonable at the outset, and yet within 24 hours it has become completely ridiculous.

Exercises 7.3

I

In each of the following cases, assume that the function $f(t)$ is of the form of Equation 7.3.1.

1. Find $f(4)$, if $f(0) = 3, f(2) = 6$.

2. Find $f(6)$, if $f(0) = 2, f(2) = 1$.

3. Find $f(3)$, if $f(0) = 1, f(2) = 4$.

4. Find $f(10)$, if $f(0) = 2, f(3) = 20$.

5. Find $f(7)$, if $f(1) = 2, f(3) = 8$.

6. Find $f(4)$, if $f(2) = 5, f(6) = 10$.

7. Find $f(5)$, if $f(2) = 3, f(4) = \frac{1}{3}$.

8. Find $f(0)$, if $f(4) = 8, f(7) = 2$.

9. Find $f(-1)$, if $f(4) = 1, f(6) = 4$.

10. Find $f(3)$, if $f(-3) = 6, f(0) = 18$.

Find the half-life of each of the decay processes described below. Time t is measured in years.

11. $f(0) = 10, f(1) = 5$

12. $f(0) = 2, f(1) = \sqrt{2}$

13. $f(2) = 1, f(5) = 0.75$

14. $f(2) = 6, f(3) = 4$

15. $f(1) = 8, f(2) = 2$

16. $f(-1) = 4, f(1) = 3$

17. $f(6) = 16, f(7) = 1$

18. $f(0) = 1, f(1) = 0.99$

19. $f(-3) = 12, f(0) = 9$

20. $f(-4) = 7, f(-1) = 2$

II

21. If a bacterial culture in an unrestricted environment doubles every half hour, how long will it take to reach ten times its original size?

22. If a bacterial culture in an unrestricted environment increases by 10% each hour, how long will it take to double in size? We can think of this time as a sort of average time for the bacteria to divide in two, in the sense that this is how long it takes for the total number to double.

23. If the half-life of a decay process is 3 years, how long will it take for a sample to reduce to 90% of its original size?

24. If the half-life of a decay process is 8 years, how long will it take for a sample to reduce to 10% of its original size?

25. If it takes 30 years for a sample to reduce to 40% of its original size, what is the half-life of the process?

26. If it takes 30,000 years for a sample to reduce to 1% of its original size, what is the half-life of the process?

27. A bacterial culture is growing exponentially. One hour after it begins growing, it weighs 0.5 g; four hours after *that*, it weighs 3.7 g. How much did it weigh at the beginning?

***28.** Grandfather put some money in a bank account 40 years ago. It has been increasing by 6% each year ever since. Assume that the amount grows exponentially (which is correct for continuously compounded interest, as we shall see in the next section). If there is $3085.72 in the account now, how much did Grandfather deposit 40 years ago?

29. Suppose the yogurt of Example 5 (p. 527) were allowed to grow exponentially for 48 hours. Figure out how much yogurt would be produced, and figure out how this compares with the volume of the earth.

***30.** Suppose a population of small organisms has a rate of reproduction that is proportional to the population. Suppose the organisms die at a rate that is also proportional to the population. (Roughly speaking, this means that a certain percentage of them die in a given length of time, so the more of them there are, the more will die.) Show that the function $P(t)$, which expresses the total population at time t, is of the form of Equation 7.3.1. What is the significance of the sign of the constant k?

III

31. Use a graphics calculator or computer to sketch the graph of $y = f(t) = e^{t\ln(2)}$. Check the graph at several points to confirm that the value of $f(t)$ doubles whenever t increases by 1. (Be sure to use some values of t that are not integers.)

POINT TO PONDER

When we developed Formula 7.3.1 for a function whose rate of change is proportional to the function itself, we found that $\frac{f'(t)}{f(t)} = k$. Integrating, we saw that $\ln\big(f(t)\big) = kt + c'$.

In this formula the constant c' is just a constant of integration. However, notice the way in which it is changed when we exponentiate:

$$f(t) = e^{\ln(f(t))} = e^{c'}e^{kt}.$$

The constant that was *added* to kt has been transformed into one by which e^{kt} is *multiplied.* Notice the importance of having added the constant of integration. If we had not done so, we would have found that $f(t) = e^{kt}$. This is one of the possible answers, but it is certainly not the only one, and forgetting to add the constant would have caused us to miss all the others.

Notice that we cannot just add a constant wherever we like. Check that if d is a constant, the function $g(t) = ce^{kt} + d$ does *not* have a rate of change that is proportional to the function $g(t)$ (unless $d = 0$).

Finally notice that when we found the function $f(t) = ce^{kt}$, the constant $c = e^{c'}$ is an exponential, so strictly speaking, it is *positive.* There is no reason it should not be positive, but on the other hand a function with a negative constant will still have the basic property that its derivative is a constant times the function itself. Explain why our argument misses the functions with negative constants c.

7.4 Examples of Exponential Growth and Decay

BACTERIAL GROWTH
HUMAN POPULATIONS
RADIOACTIVE DECAY
CARBON DATING
COMPOUND INTEREST
PRESENT VALUE
NEWTON'S LAW OF COOLING
CONCENTRATION OF IMPURITIES

In Section 7.4 we discussed the behavior of a function whose rate of change is proportional to the function itself, that is, $f'(t) = kf(t)$, for some (nonzero) constant k. We found that such a function must be of the form $f(t) = ce^{kt}$, with some constant c, the initial value. The function $f(t)$ is increasing (growing) if $k > 0$ and decreasing (decaying) if $k < 0$.

In this section we discuss a number of practical problems that fit this description. In the back of our minds we ought to remember Example 7.3.5 (p. 527), which showed that applying the mathematical consequences of exponential growth to situations in which they are not justified can easily give ridiculous results. It is particularly dangerous to assume that the trend will continue for very large or very small values of t.

EXAMPLE 1 Bacterial Growth We have already considered the example of a culture of bacteria. Bacteria are organisms that usually consist of a single cell; the way they reproduce is for each cell to divide into two.

How frequently they do this depends on many factors, including such things as the temperature and the amounts and types of nutrients available. If the environment is reasonably stable, we can assume that all these factors remain unchanged for a period of time. Under these conditions it makes sense to expect that the cells will reproduce at a regular rate.

In other words, if T minutes, say, is the average time between divisions of a particular cell, then every time T minutes elapse, we should expect that, on the

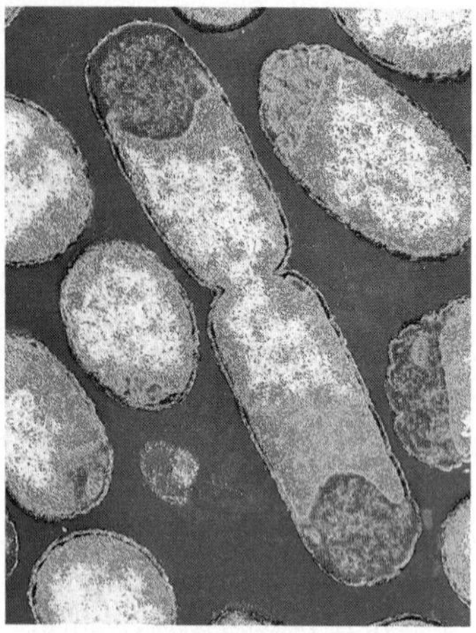

False-color transmission electron micrograph of the bacterium Escherichia coli, *a normal inhabitant of the human intestine. The pair in the center are in the act of separating following binary fission, the process by which a bacterium divides into two. Under certain conditions,* E. coli *bacteria may become harmful and cause infection in the intestinal or urinogenital tracts.* E. coli *is the organism most commonly used in genetic engineering experiments. Magnification: 16,500×.*

average, each cell will have divided once. The effect of this is that the number of cells doubles every T minutes.

For some types of bacteria, under ideal conditions, the average time T between divisions can be less than 15 minutes.

Let $N(t)$ be the number of bacteria present at time t. We should consider what happens in some brief period of length h. Starting at some fixed time t, when the population is $N(t)$, we know that between t and $t+T$ the population will double. In this interval, in effect, each of the original cells will divide exactly once. In a shorter interval, some fraction of them will double.

If we choose an interval of length $\frac{T}{2}$, it seems reasonable to expect half of them to divide. During an interval of length $\frac{T}{6}$, one-sixth of the original cells should divide. During an interval of length h the proportion of the cells that divide ought to be $\frac{h}{T}$.

If h is small, this means that the number of cells that divide will be $\frac{h}{T}N(t)$. The result will be that this many new cells will be created; symbolically, this can be expressed as

$$N(t+h) - N(t) = \frac{h}{T}N(t).$$

To find the derivative of $N(t)$, we form the difference quotient and take the limit:

$$\frac{d}{dt}\big(N(t)\big) = \lim_{h\to 0}\frac{N(t+h)-N(t)}{h} = \lim_{h\to 0}\frac{1}{h}\frac{h}{T}N(t) = \frac{N(t)}{T}.$$

We have just shown that $N'(t) = \frac{1}{T}N(t)$, that is, that $N'(t)$ is proportional to $N(t)$.

From this it follows from the results of the previous section that the population at time t is an exponential function: $N(t) = ce^{kt}$, for some constants c and k. ◢

We can summarize the above argument as follows. The number of cell divisions in a fixed brief period of time is proportional to the number of cells present. To put it the other way around, the *proportion* of the cells that divide in a given short length of time is a constant, independent of the number of cells present.

This means that the rate of change of the population is proportional to the population, and from this we can conclude the growth is exponential.

The tricky thing about this argument is the assumption that the environment is stable. Eventually, the bacteria themselves will destabilize it if they are allowed to grow unhindered. They will begin to use up the available space or some of the nutrients, and their growth rate will diminish.

Other factors can easily enter the picture too, such as the presence of an organism that eats the bacteria that we are considering.

In spite of all the difficulties, it is still true that bacterial cultures do conform to the rules of exponential growth while the environment remains stable.

EXAMPLE 2 If a bacterial culture increases by 10% in one hour, how long will it take to increase to five times its original size?

Solution Let $N(t)$ be the population at time t, with t measured in hours. We know that $N(t) = ce^{kt}$, for some constants c and k.

We also know that $N(1) = 1.10 \times N(0)$, that is, $ce^k = 1.10c$. Dividing by c, we find that $e^k = 1.10$, so $k = \ln(1.10) \approx 0.09531$.

The question asks us to find the value of t for which $N(t) = 5N(0)$. This amounts to $ce^{kt} = 5c$, and dividing by c, we find $e^{kt} = 5$, or $kt = \ln(5)$. Knowing the value of k from above, we see that the time we want is

$$t = \frac{\ln(5)}{k} = \frac{\ln(5)}{\ln(1.10)} \approx 16.89 \text{ hr.}$$

One interesting feature of Example 2 is that there is not enough information to determine the value of the constant c. However, it is still possible to solve the problem. Notice that in the calculations the constant c just "divides out."

EXAMPLE 3 **Human Populations** A population of people grows under the influence of a remarkably complex environment. So many factors affect the growth rate that it is extremely difficult to make accurate predictions.

For one thing, people migrate from region to region, so population growth or decrease due to births and deaths may be masked by immigration and emigration. Diseases and wars may dramatically increase the death rate; medical advances may reduce it. Social and religious factors may have a huge effect on the birth rate; perhaps the most obvious instance is the difference between cultures that do and do not accept contraception.

Another complication is that birth rates depend not so much on the total population as on the population of women of childbearing age.

With all these subtleties to consider, making precise statements about human population growth is hazardous. Even the experts are not terribly reliable, which is hardly surprising, considering the nature of the problem.

Nonetheless, if we assume that the birth and death rates are constant and ignore the effects of migration, the population will grow (or diminish) exponentially, for exactly the same reasons that applied to the bacterial culture discussed in Example 1. We will make these assumptions and see what simple conclusions can be drawn as a result. We will not make any serious attempt to discuss the validity of the assumptions.

EXAMPLE 4 In a country with no immigration or emigration, the population is 10,000,000. One month later, 15,000 births and 6,000 deaths have occurred. Assuming that the birth and death rates are constant, what will the population be 5 years after the starting point?

Solution From the assumptions we can conclude that the population is described by an exponential function $P(t) = ce^{kt}$. Suppose the time is measured in years, with $t = 0$ at the time when the population is 10,000,000.

We know that the initial value is $c = P(0) = 10{,}000{,}000$. We know that one month after that, that is, when $t = \frac{1}{12}$, the population is $10{,}000{,}000 + 15{,}000 - 6{,}000 = 10{,}009{,}000$. This means that $10{,}000{,}000e^{k/12} = 10{,}009{,}000$, so $\frac{k}{12} = \ln\left(\frac{10{,}009{,}000}{10{,}000{,}000}\right) = \ln(1.0009)$.

We find that $P(t) = 10{,}000{,}000e^{12t\ln(1.0009)}$. What we want to evaluate is $P(5) = 10{,}000{,}000e^{12(5)\ln(1.0009)} \approx 10{,}555{,}000$.

Of course, the same sort of calculations can be made for populations of animals, subject to the usual assumptions. The situation may be complicated by predators. If the population of rabbits, for instance, begins to grow, then the population of foxes will likely increase. It might well happen that the foxes will grow so fast that, in combination with shortages of food, they will begin to reduce the rabbit population. Before long, the rabbit shortage will devastate the fox population. Because of the resulting reduced predation, the rabbits will rebound.

What happens is a cycle in which the populations grow and fall in a fairly regular pattern. External factors can disrupt the cycle, but there have been wilderness studies that appear to confirm this phenomenon. Folklore almost always seems to suggest that the natural cycles are seven years long.

EXAMPLE 5 **Radioactive Decay** Radioactive decay is a particularly attractive example because there is no need to make questionable assumptions.

Certain atomic nuclei are unstable. This means that they have a tendency to break apart into pieces, and this "fission" is accompanied by the release of a large amount of energy. Roughly speaking, the energy that is released is what used to be holding the unstable nucleus together.

Consider a sample of some type of unstable nuclei. Let h be a length of time. During any period of length h, the same fixed fraction of the nuclei will decay, no matter how many there were at the beginning of the period.

By an argument similar to the one in Example 1 we conclude that the number of unstable nuclei remaining is an exponentially decaying function of time t. The length of time it takes for half the unstable nuclei of a certain type to decay is called the **half-life** of that type of nuclei. It varies from tiny fractions of a second for some nuclei to billions of years for others, and it is a convenient way to measure how fast the radioactive decay occurs.

The nucleus of an atom is made up of protons and neutrons. The protons carry a positive electric charge, and the neutrons are electrically neutral. Electrons are a type of particle of much lighter weight that carry a negative charge. They are attracted to the positive charge of the nucleus, and it is the combination of the nucleus surrounded by electrons that constitutes an atom.

There is a tendency for the number of electrons to be the same as the number of protons; this has the effect that the positive and negative charges balance, giving the atom no net charge.

Broadly speaking, chemistry is the study of what happens when different atoms interact, which they do essentially through their electrons. One way this happens, for example, is for two nearby atoms to "share" some of their electrons. The resulting pair of atoms, bound together by their electrons, is a simple example of a *molecule.* More complicated molecules may have hundreds or even thousands of atoms. The essential thing is that the chemical behavior is determined by the electrons, and the number of electrons is the same as the number of protons in the nucleus. This number is called the *atomic number* of the nucleus.

For example, the atomic number of carbon is 6, meaning that the carbon nucleus contains 6 protons. But there is also the question of how many neutrons the nucleus contains.

For carbon the answer is usually 6. This means that the nucleus contains 6 protons and 6 neutrons, or a total of 12 particles. Such a nucleus is denoted ^{12}C, with the C standing for carbon and the mass number 12 giving the total number of nuclear particles.

However, it is occasionally possible to form a nucleus with 6 protons and only 5 neutrons. This nucleus, ^{11}C, is unstable. The combination of 6 protons and 5 neutrons does not hold together nearly as well as ^{12}C, and it tends to break apart. The half-life of ^{11}C is about $\frac{1}{3}$ of an hour.

From the chemical point of view, however, ^{11}C is nearly indistinguishable from ^{12}C. It can combine with two oxygen atoms to make carbon dioxide, for example, just as ^{12}C can. We say that ^{11}C is an **isotope** of carbon (meaning that it has atomic number 6).

In addition to the usual stable isotope ^{12}C and the unstable isotope ^{11}C, there is another isotope ^{14}C, with 8 neutrons. It is unstable, but much less so than ^{11}C. Its half-life is about 5730 years.

EXAMPLE 6 The half-life of uranium ^{238}U is 4.5 billion years. If the quantity of this isotope in a rock sample has diminished to 90% of its original amount, how old is the rock sample?

Solution We know that the amount of ^{238}U remaining t years after the formation of the rock is $Q(t) = ce^{-kt}$, for some constants c and k. Notice that we use a negative sign to help us remember that it is a process of decay.

Since the half-life is 4.5 billion years, we find from Example 7.3.4 that $k = \frac{\ln 2}{4.5\times 10^9}$.

What we need is the time t for which $Q(t) = 0.90Q(0)$. This amounts to $e^{-kt} = 0.90$, or $-kt = \ln(0.90)$, which means

$$t = \frac{\ln(0.90)}{-k} = -\frac{\ln(0.90)\,(4.5\times 10^9)}{\ln 2} \approx 6.84\times 10^8.$$

The rock is about 684 million years old.

In practice, the method is a little more complicated than this example appears to indicate. The ^{238}U isotope decays (through several intermediate steps) into the

If we want to plot the function of Example 6 we have to fix some value of c. For simplicity we let $c = 1$ and plot $n = Q(t) = e^{-t\ln(2)/(4.5\times 10^9)}$.

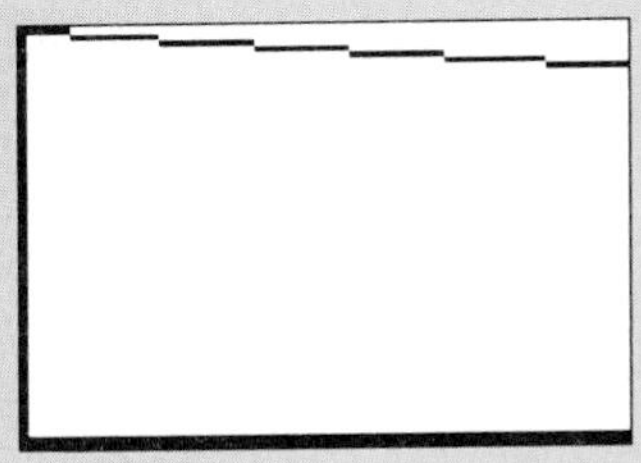

TI-85	SHARP EL9300
GRAPH F1(y(x) =) e^x ((−) F1(x) × LN (2) ÷ 4500000000) M2(RANGE) 0 ▼ 684000000 ▼ ▼ 0 ▼ 1 F5(GRAPH)	e^x (−) x/θ/T × ln 2 ÷ 4500000000 RANGE 0 ENTER 684000000 ENTER 100000000 ENTER 0 ENTER 1 ENTER 1 ENTER
CASIO f_x-6300G (f_x-7700GB)	**HP 48SX**
Range 0 EXE 684000000 EXE 100000000 EXE 0 EXE 1 EXE 1 EXE EXE EXE EXE Graph e^x ((−) X × ln (2) ÷ 4500000000 EXE	↰ PLOT PLOTR ERASE ATTN ' Y = e^x X × LN 2 ► ÷ 4500000000 +/- ↱ PLOT ↰ DRAW ERASE 0 ENTER 684000000 XRNG 0 ENTER 1 YRNG DRAW

We see how the function decreases to 90% of the original amount. Notice that the part of the curve we have plotted is nearly a straight line...

lead isotope ^{206}Pb. (The Latin name for lead is *plumbum*, which we recognize in our word "plumber," someone who works with lead pipes.) It is possible to measure how much radioactive decay has occurred by comparing the amounts of ^{238}U and ^{206}Pb. The greater the proportion of ^{206}Pb in a rock, the longer it has been in existence. The rock might contain some lead that did not result from radioactive decay, but happily it turns out that this will be a different isotope, ^{204}Pb.

Other combinations of isotopes can also be used to estimate the age of rocks. The first requirement is to find an isotope with a half-life that is roughly comparable to the likely age of the rock. (It is no use trying to date a rock that is millions of years old with an isotope whose half-life is 20 minutes.) Second, the original isotope and/or the products of the decay must be distinguishable from isotopes that may arise in other ways.

The best-known isotope dating is the use of ^{14}C to date archaeological sites.

EXAMPLE 7 **Carbon Dating** It has already been mentioned that the isotope ^{14}C has a half-life of 5730 years. This is a convenient unit for archaeological materials, whose ages are likely to range from 100 years to perhaps 10,000 or 20,000 years.

Living plants take carbon dioxide from the air and by photosynthesis convert it into products they can use. These processes are *chemical* in nature, so they do not distinguish between carbon dioxide containing ^{12}C and carbon dioxide containing ^{14}C. The effect is that the carbon atoms in any living plant will consist of ^{12}C and ^{14}C in approximately the same proportions as the carbon atoms in the atmosphere.

Liquid scintillation counter used to measure the amount of the carbon-14 isotope in a sample being dated.

However, once the plant dies, it no longer absorbs carbon dioxide from the air. Since the ^{14}C isotope undergoes radioactive decay, the amount of it begins to diminish. By measuring the relative amounts of the two isotopes it is possible to measure the length of time the plant has been dead.

The process can be used to date pieces of wood from excavated buildings, for instance, and since animals consume plant material, it can also be used to date animal tissues.

The entire idea relies on the assumption that the proportion of the two isotopes in the atmosphere has remained the same for thousands of years. The ^{14}C isotope is created in nature by the bombardment of nitrogen nuclei in the atmosphere by radiation from space, and the amount of it that is produced appears to have remained unchanged. Comparisons of ^{14}C dates for sites whose age is corroborated by historical data bear out the reliability of the technique. Atmospheric nuclear testing has resulted in elevated ^{14}C levels, so it will not be possible to use carbon dating for organic material deposited in the recent past.

EXAMPLE 8 If a sample of charred wood from an excavated campsite is found to have 63% of the original amount of ^{14}C, how old is the site?

Solution As usual, we know that the number of ^{14}C nuclei can be expressed as a function ce^{-kt}. We know that the half-life is 5730 years, which means that $k = \frac{\ln 2}{5730}$.

If t is the age of the site, we know that $e^{-kt} = 0.63$, or $-kt = \ln(0.63)$, so

$$t = -\frac{\ln(0.63)}{k} = \frac{-5730 \ln(0.63)}{\ln 2} \approx 3820.$$

The site is approximately 3820 years old.

Notice that we have been writing exponentially decaying functions in the form ce^{-kt}, with a minus sign in the exponent. This has the effect of making the constant k positive.

It is just as good to omit the minus sign; what matters is to be consistent. Either put it in throughout a particular example or leave it out all the time. It is a question of taste.

EXAMPLE 9 **Compound Interest** Consider money in a bank account earning 12% annual interest. This means that each year the amount increases by 12%, in effect multiplying the original amount by $\left(1+\frac{12}{100}\right)$. We say the balance increases by a **factor** of $\left(1+\frac{12}{100}\right)$; the word "factor" suggests that we *multiply* by this number.

Suppose the interest is compounded semiannually. This means that 6% interest is paid after 6 months (multiplying the original amount by $\left(1+\frac{6}{100}\right)$), and at the end of the year another 6% interest is paid. This has the effect of multiplying by $\left(1+\frac{6}{100}\right)$ again, and the overall effect is to multiply the original amount by $\left(1+\frac{6}{100}\right)^2$.

In one year the amount will increase by a factor of $\left(1+\frac{6}{100}\right)^2 = 1.1236$. This is slightly more than the factor $1+\frac{12}{100} = 1.12$ that arises from a single payment of 12% interest. The slight difference comes from paying 6% interest for the second half of the year on the first installment of interest.

The interest could also be compounded quarterly, that is, every three months. The overall effect would be to increase the original amount by a factor

$$\left(1+\frac{3}{100}\right)\left(1+\frac{3}{100}\right)\left(1+\frac{3}{100}\right)\left(1+\frac{3}{100}\right) = \left(1+\frac{3}{100}\right)^4 \approx 1.1255.$$

Many banks compound interest monthly. With 12% annual interest, in one year this increases the original amount by $\left(1+\frac{1}{100}\right)^{12} \approx 1.126825$.

The 12% interest rate is called the **nominal interest rate**. We have just seen that with a nominal rate of 12% and monthly compounding, the annual increase is actually about 12.6825%. This is called the **effective annual interest rate**.

EXAMPLE 10

(i) What is the effective annual interest rate with quarterly compounding and a nominal rate of 10%?

(ii) How does this compare with monthly compounding at a nominal rate of 9.9%?

Solution

(i) Quarterly compounding at a nominal 10% annual rate means paying 2.5% each quarter. Over one year this has the effect of increasing the original deposit by a factor of $\left(1+\frac{2.5}{100}\right)^4 \approx 1.1038$, so the effective annual interest rate is about 10.38%.

(ii) Monthly compounding means paying $\frac{1}{12}$ of the nominal rate each month. In this case that means increasing the deposit each month by a factor of $\left(1+\frac{1}{12}\frac{9.9}{100}\right)$. Over a year, the deposit grows by $\left(1+\frac{1}{12}\frac{9.9}{100}\right)^{12} \approx 1.1036$, and the effective annual interest rate is about 10.36%, nearly as much as the rate in part (i).

If the nominal interest rate is $p\%$, let $r = \frac{p}{100}$ so that for simple interest (i.e., not compounded) the effect in one year is to increase the amount by a factor of $(1 + r)$.

Compounding semiannually has the effect of increasing in one year by $\left(1 + \frac{r}{2}\right)^2$; for quarterly compounding, the increase is $\left(1 + \frac{r}{4}\right)^4$. If the interest is compounded n times in one year, the balance increases by the factor

$$\left(1 + \frac{r}{n}\right)^n. \tag{7.4.1}$$

For example, to compound the interest daily, that is, $n = 365$ times a year, we multiply by $\left(1 + \frac{r}{365}\right)^{365}$.

The more often the interest is compounded, the higher will be the effective annual interest. Instead of compounding monthly or daily, we could compound hourly, or every minute, or every second. Compounding more often raises the effective rate, so it is to the advantage of the owner of the account to have interest compounded as often as possible.

If we try compounding every second, say, then in one year the original amount will increase by $\left(1 + \frac{r}{n}\right)^n$, where n is the number of seconds in a year, or about $(365)(24)(60)(60) = 31{,}536{,}000$. We could try compounding every one one-thousandth of a second in hopes of raising the effective rate. This would mean letting $n = 31{,}536{,}000{,}000$. We could try letting n, the number of compoundings, get larger and larger, but $\left(1 + \frac{r}{n}\right)^n$ does not get arbitrarily large. In fact, we saw in Example 7.2.4 (p. 519) that

$$\lim_{n\to\infty} \left(1 + \frac{r}{n}\right)^n = e^r.$$

No matter how often the interest is compounded, the annual increase will never exceed e^r. Rather than take a very large n, corresponding to compounding very many times, we can take this limit and regard it as the effect of **compounding continuously**.

EXAMPLE 11 What is the effective annual rate if the nominal rate is 10% and interest is compounded continuously?

Solution The factor by which the original amount is increased in one year is $\lim_{n\to\infty} \left(1 + \frac{10}{100n}\right)^n = e^{1/10} \approx 1.10517$. The effective annual interest rate is approximately 10.517%.

It is interesting to compare this with the effective rate resulting from daily compounding. In that situation the amount increases by $\left(1 + \frac{1}{365}\frac{10}{100}\right)^{365} \approx 1.1051558$, and the effective annual rate is approximately 10.51558%.

The difference between continuous compounding and daily compounding is remarkably small.

It should come as no surprise that in an account with continuously compounded interest, the total amount of money is an exponentially growing function of time t. Essentially, this is because the rate at which the total amount increases is proportional to the amount. (Notice that the *rate* at which the total amount increases is *not* the same thing as the interest *rate;* the interest rate is constant.)

EXAMPLE 12 Mary deposits \$1000 in an account with continuously compounded interest at a nominal rate of 8%. One hundred and ten days later, she withdraws \$700 from the account. How much will be in the account 40 days after this withdrawal?

Solution Let t represent time, measured in days, with $t = 0$ at the time of the deposit. If $A(t)$ is the amount in the account at time t, in dollars, we know that it grows exponentially, so it must be of the form $A(t) = ce^{kt}$.

The complication is that while it grows according to a function of this form for t between 0 and 110, we know that it changes at $t = 110$ because of the withdrawal. From that time on, it continues to grow exponentially (at least until the next deposit or withdrawal).

First we should find the constant k. If 8% interest is compounded n times annually, the resulting effective increase is $\left(1 + \frac{1}{n}\frac{8}{100}\right)^n$. For continuous compounding we let $n \to \infty$ and take the limit

$$\lim_{n\to\infty}\left(1 + \frac{8}{100n}\right)^n = e^{8/100}.$$

(To evaluate this limit, we used Equation 7.2.7, p. 519).

This means that $A(365) = e^{8/100}A(0)$, or $ce^{365k} = e^{8/100}c$. Dividing by c, we find that $e^{365k} = e^{8/100}$, so $365k = \frac{8}{100}$ and $k = \frac{8}{36{,}500}$.

First, consider the part of the function between $t = 0$ and $t = 110$. Let us give the name $a(t)$ to this part of the function. Its initial value is 1000, so $a(t) = 1000e^{8t/36500}$, for $0 \le t \le 110$. In particular, just before Mary's withdrawal is made, the amount in the account is $a(110) = 1000e^{880/36500} \approx \$1024.40$.

Now consider separately the amount of money in the account for t between 110 and 150. It will be growing with the same rate of interest, so it will be an exponential function with the *same* growth constant k. Let us use the name $b(t)$ for this function. Its value at $t = 110$ is the amount left after Mary makes her withdrawal, which is $1000e^{880/36500} - 700 \approx 324.40$. So for $110 \le t \le 150$ the amount is $b(t) = c'e^{kt}$, with $k = \frac{8}{36{,}500}$ as above. To find c', we note that $b(110) = 324.40$, so $c'e^{880/36500} = 324.40$, and $c' = 324.40/e^{880/36500} \approx 316.67$. Using this, we are finally able to evaluate $A(150) = b(150) = 316.67e^{150k} = 316.67e^{1200/36500} \approx 327.25$.

At the specified time the amount in the account is \$327.25.

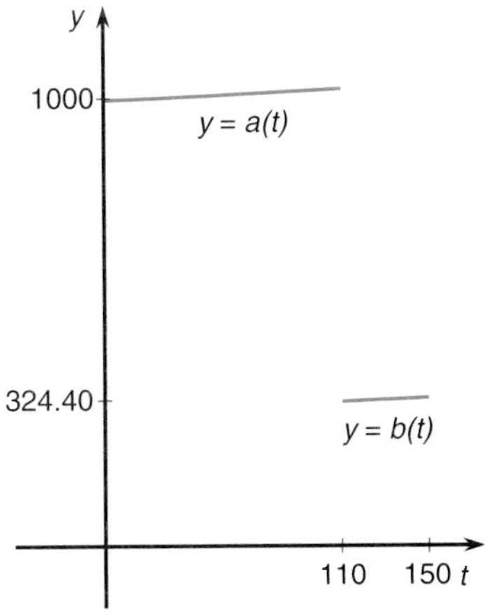

Figure 7.4.1

In Figure 7.4.1 we see that the graph of $y = A(t)$ has a discontinuity at $t = 110$, corresponding to Mary's withdrawal. In the above calculation we first considered the initial part of the curve and used it to find the amount in the account after 110 days. Then we found a formula for the part of the function after $t = 110$, and we evaluated it at $t = 110$. This evaluation was made after the withdrawal, so it was $b(110) = 1024.40 - 700 = 324.40$.

In working with $b(t) = c'e^{kt}$ we had to do a small calculation to find c'. This is because the "initial value" we have for $b(t)$ is not at $t = 0$ but rather at $t = 110$. An alternate way to deal with this is to let $b(t) = c''e^{k(t-110)}$. The advantage of this is that when $t = 110$, the right side of this formula is c'', which we know without any work is $c'' = b(110) = 324.40$. It is easy to verify that this form of the function $b(t)$ is exactly the same as the one used above. (You should verify that you get the same value for $b(150)$.)

Either way of writing $b(t)$ is perfectly acceptable.

In Example 12 we addressed the following problem. If the nominal interest rate is $p\%$, what is the effective rate with continuous compounding? We know that with compounding n times per year the effective rate is $\left(1 + \frac{p}{100n}\right)^n$, and to find the effective rate for continuous compounding, we let $n \to \infty$ and evaluate the limit. Using Formula 7.2.7, we find that in one year an amount of money will increase by the factor

$$\lim_{n\to\infty}\left(1 + \frac{p}{100n}\right)^n = e^{p/100}. \tag{7.4.2}$$

To find the effective annual interest rate *as a percentage*, we take $(e^{p/100} - 1) \times 100\%$.

Incidentally, this reasoning also shows that the growth constant k for the total amount ce^{kt} is $k = \frac{p}{100}$.

EXAMPLE 13 **Present Value** Suppose a construction company does some work on Albert's house, and Albert agrees to pay the company \$5000 two years from now. He says that he has no money but that he will be able to pay later, and the company was anxious to have the business, since it came during the slow season.

In fact, Albert has plenty of money, but he is just being sneaky. He figures he might as well pay later and earn interest on the money in the interim. However, Albert likes to play the horses, and he is afraid that he might lose all his extra cash, so he decides it would be prudent to put enough money in the bank to cover his obligation.

He finds a bank that will pay him 7% annual interest, compounded continuously, and decides to deposit his money there. Certainly, he does not have to deposit \$5000, because the money will earn interest and grow over the two-year period, so he has to figure out exactly how much he must deposit in order to have \$5000 in two years.

Let us write \$$D$ for the amount of the deposit. What we need is to find how much this deposit will be worth after two years. We know that in one year the deposit will grow by a factor of $e^{7/100} \approx 1.0725$. In two years it will do this twice, which means that it will increase by a factor of $\left(e^{7/100}\right)^2 = e^{14/100} \approx 1.1503$.

What we want is that $e^{14/100}D = 5000$, which means $D = 5000e^{-14/100} \approx 4346.79$. The answer to Albert's question is that he should deposit \$4346.79 in the bank. With interest at a nominal 7%, compounded continuously, this will grow in two years to \$5000, and he will be able to use it to pay off his debt when it is due.

In a sense, then, an obligation to pay \$5000 in two years is an obligation that right now amounts to \$4346.79. We say that the **present value** of the obligation is \$4346.79.

From the point of view of the construction company, things look slightly different. The company was happy to have the job, but not having been paid, it now has to borrow money. The best deal the company can find is from a bank that will lend it money at 12% interest, compounded semiannually. On the strength of Albert's promise to pay \$5000 in two years, the company can borrow the amount of money that will grow to \$5000 in two years with 12% interest and semiannual compounding.

An amount $\$L$ will grow in one year to $L(1.06)^2$ and in two years to $L(1.06)^4 \approx 1.2625L$. For this increased amount to equal 5000 we must have $L = 5000/(1.06)^4 \approx 3960.47$. The amount the bank will agree to lend is \$3960.47.

So to the company, the present value of Albert's obligation is \$3960.47. The difference arises because of the different interest rates. It is very important to remember that calculations of present value depend on the interest rate used. In practical situations it is often impossible to predict the interest rates over the period of an investment, so calculations of present value are necessarily contingent upon the assumptions being made. Further complications are introduced by inflation, but we will not consider them here.

EXAMPLE 14 **Lottery Ticket** What is the present value of a "million-dollar" winning lottery ticket that promises to pay \$100,000 right away and a further \$100,000 each year for a total of ten payments? (Assume 8% interest, compounded quarterly.)

Solution First let us figure out that in one year an initial amount will increase by a factor of $(1.02)^4 \approx 1.0824$.

Now let us consider each payment separately. The amount paid right now has present value \$100,000. The present value of the payment to be made in one year is the amount that will grow to 100,000 in one year, or $100{,}000/(1.02)^4 \approx 92{,}384.54$. Similarly, the present value of the payment to be made in two years is $100{,}000/(1.02)^8 \approx 85{,}349.04$. We continue like this, finding that the present value of the payment to be made in k years is $100{,}000/(1.02)^{4k}$. For instance, the last payment, which will be paid in 9 years (not 10), has present value $100{,}000/(1.02)^{36} \approx 49{,}022.32$.

The present value of the ticket is the sum of the present values of all the payments, which is

$$
\begin{aligned}
100{,}000 + \frac{100{,}000}{(1.02)^4} + \frac{100{,}000}{(1.02)^8} + \cdots + \frac{100{,}000}{(1.02)^{36}} \\
&\approx 100{,}000 + 92{,}384.54 + 85{,}349.04 + 78{,}849.32 + 72{,}844.58 \\
&\quad + 67{,}297.13 + 62{,}172.15 + 57{,}437.46 + 53{,}063.33 + 49{,}022.32 \\
&= \$718{,}419.86.
\end{aligned}
$$

The present value of the lottery ticket is only \$718,419.86, less than three-quarters of a million dollars.

It is also possible to find the present value of an amount paid some time ago or the present value of a combination of various amounts at various times. As with the lottery ticket in Example 14, the idea is to work with each amount separately and add them up at the end.

EXAMPLE 15 **Christmas Trees** Five years ago, Alice planted 2000 Christmas tree seedlings at a cost of \$2.00 each for materials and labor. One year later and each succeeding year, she had them sprayed, at a cost of \$700 each time.

Five and a half years from now, she will harvest the trees. The harvesting and shipping will cost \$4.00 per tree, and she will be able to sell them for \$22.00 per tree.

Find the present value of her operation, assuming 9% annual interest, compounded continuously.

Solution Note that with this interest rate an amount of money increases each year by a factor of $e^{0.09} \approx 1.09417$.

It may help to draw a line, representing time in years, and mark the various events on it as in Figure 7.4.2. The dot at the left end represents the planting 5 years ago (i.e., at $t = -5$). The evenly spaced dots (at one-year intervals) represent the sprayings. The large dot represents the present time. The dot at the right end, one-half year after the last spraying, represents the harvesting.

The initial expense was $2000 \times \$2 = \$4000$ for planting. In subsequent years there is an annual expense of \$700, and at the time of harvesting there is an expense of $2000 \times \$4 = \$8000$. We draw the line again in Figure 7.4.3, marking these amounts on it.

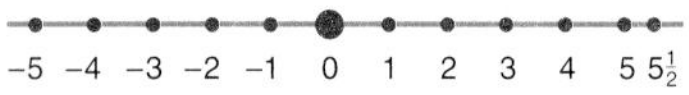

Figure 7.4.2

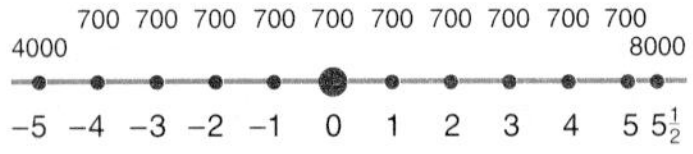

Figure 7.4.3

The next step is to find the present value of each of these amounts. The easy one is the \$700 expense at the present time, whose present value is, of course, \$700. The present value of the \$700 expense one year from now is $700/e^{0.09} = \$639.75$. Similarly, we can find the present value of the expense two years from now, which is $700/(e^{0.09})^2 = \$584.69$, and so on. The present values of the spraying expenses for three, four, and five years from now are $700/(e^{0.09})^3 = \$534.37$, $700/(e^{0.09})^4 = \$488.37$, and $700/(e^{0.09})^5 = \$446.34$.

The present value of the \$700 expense one year ago means how much \$700 deposited one year ago would be worth today, which is $700 \times (e^{0.09}) = \$765.92$. Similarly, the expense from two years ago would now be worth $700 \times (e^{0.09})^2 = \$838.05$. The expenses from three and four years ago have present values $700 \times (e^{0.09})^3 = \$916.98$ and $700 \times (e^{0.09})^4 = \$1003.33$. The present value of the planting expense is $4000 \times (e^{0.09})^5 = \$6273.25$.

The harvesting will occur in five and a half years, that is, at time $t = 5\frac{1}{2}$. The factor by which an amount will increase in $5\frac{1}{2}$ years is $(e^{0.09})^{5.5} = e^{5.5\times 0.09} = e^{0.495}$, so the present value of the harvesting cost is $8000/e^{0.495} = \$4876.57$.

We add up all these present values to find the total present value of all the expenses. It comes out to \$18,067.62.

The one remaining thing is the money that Alice will earn by selling the trees. It will be $2000 \times 22 = \$44{,}000$ and will be obtained in $5\frac{1}{2}$ years. Its present value is $44{,}000/e^{0.495} = \$26{,}821.12$.

The net present value of the operation is the difference between the present value of the assets (\$26,821.12) and the expenses (\$18,067.62), or \$8,753.50. Because this amount is positive, Alice can expect to make a profit, provided interest rates do not change and there are no unforeseen complications (such as forest fires).

EXAMPLE 16 **Newton's Law of Cooling** If you hold your hand in a refrigerator, it will begin to cool off. If you hold your hand in the freezer, it will cool off more quickly. This is a simple example to illustrate an important principle, which is that an object that is warmer than its surroundings will cool off and moreover that it will do so more quickly the cooler the surroundings are.

Let us consider a reasonably small object at temperature T placed in surroundings where the temperature is A. (We use the letter A to remind us of the **ambient temperature**, which simply means the temperature of the surroundings.) Let us make the further assumption that A is constant. (The idea here is that the warmth from the small object is not enough to have any significant warming effect on the surroundings.) **Newton's Law of Cooling** says that the rate of change of T is proportional to $T - A$; a warm object will cool down at a rate proportional to the difference between its temperature and the ambient temperature. Symbolically,

$$\frac{dT}{dt} = -k(T - A) \tag{7.4.3}$$

for some positive constant k. The negative sign is there to ensure that if $T > A$, then $\frac{dT}{dt}$ is negative, meaning that the small warm object will cool down. Of course, it also works the other way around: A small cold object will warm up, and in either case the rate of change depends on the difference between the two temperatures.

Since A is assumed to be constant, Equation 7.4.3 can be written as $\frac{d}{dt}(T - A) = -k(T - A)$. The rate of change of the function $T - A$ is proportional to the function itself. As usual, this has the consequence that $T - A$ is an exponential function of t. We can write $(T - A)(t) = ce^{-kt}$, that is, $T(t) - A = ce^{-kt}$, or $T(t) = A + ce^{-kt}$, where the constant c is the initial value of $T - A$.

EXAMPLE 17 A metal spatula is being used to stir-fry some food; its blade is at a temperature of 400°F. The blade is plunged into a sink full of tepid water whose temperature is 110°F. After 2 sec in the water the spatula blade has cooled to 270°F. How long will it take it to cool to 150°F so that the cook can safely handle it?

Solution We assume that there is enough water in the sink so that its temperature will not change appreciably. This means that the ambient temperature is $A = 110$. From Equation 7.4.3 we know that $T(t) = A + ce^{-kt} = 110 + ce^{-kt}$. The constant c is the initial value of $T - A$, which is $400 - 110 = 290$.

To find k, we use the fact that $T(2) = 270$, which means that $110 + 290e^{-2k} = 270$, or $290e^{-2k} = 160$, so $e^{-2k} = \frac{160}{290}$. We take logarithms and see that $-2k = \ln\left(\frac{160}{290}\right)$ and $k = -\frac{1}{2}\ln\left(\frac{16}{29}\right)$.

We need to know when $T(t) = 150$, which means $150 = 110 + 290e^{-kt} = 110 + 290e^{t\ln(16/29)/2}$, or $290e^{t\ln(16/29)/2} = 40$. Dividing by 290 and taking logarithms gives $\frac{t}{2}\ln\left(\frac{16}{29}\right) = \ln\left(\frac{40}{290}\right)$, so $t = 2\ln\left(\frac{40}{290}\right)\frac{1}{\ln(16/29)} \approx 6.66$.

The spatula will be cool enough after 6.66 sec.

> The thing to remember about questions of this type is that it is not the temperature $T(t)$ that is an exponential function of t, but rather the function $T(t) - A$, the difference between the two temperatures.

EXAMPLE 18 **Concentration of Impurities** Imagine a polluted lake. Clean water flows into the lake at one end, and dirty water flows out at the other. Suppose that the original source of the pollution has been cleaned up, so the flow of water

Discharge of production waste into a river in Copperhill, Tennessee.

will gradually reduce the concentration of the contaminants in the lake. It is a very important question to estimate the amount of pollution remaining and/or the length of time it will take to flush it out.

Unfortunately, the pollution will never disappear completely, but the concentration will diminish. If we can assume that the fresh water flowing into the lake mixes perfectly with the water already there, which is a fairly reasonable assumption in most cases, then it turns out that the concentration of contaminants will decrease as an exponential function of time.

Consider a lake containing 100,000,000 cubic meters of water. Suppose the rivers flow in and out at the rate of 100,000 m^3 per day. Also suppose that the concentration of the contaminant is 0.1%.

Let t be the time in days, and let $P(t)$ be the number of cubic meters of contaminant at time t. In one day, 100,000 m^3 of polluted water flow out of the lake. This is $\frac{100{,}000}{100{,}000{,}000} = \frac{1}{1000}$ of the water in the lake. So the amount of contaminant that flows out in this water will be $\frac{1}{1000}$ of all the contaminant present, or $\frac{P(t)}{1000}$.

This says that $P(t)$ decreases at the rate of $\frac{P(t)}{1000}$ m^3 per day. In other words, $\frac{dP}{dt} = -\frac{1}{1000}P(t)$.

From this we see that $P(t) = ce^{-t/1000}$. The constant c, the initial value of $P(t)$, is easily calculated to be 0.1% of the volume of the lake, or 100,000. So the amount of contaminant in the lake is $P(t) = 100{,}000e^{-t/1000}$ m^3, with t measured in days.

For example, after 1 year the amount of contaminant left will be

$$P(365) = 100{,}000e^{-365/1000} \approx 69{,}420\ m^3.$$

The concentration will have diminished from 0.1% to $\frac{69{,}420}{100{,}000{,}000} \times 100\% = 0.06942\%$.

We might ask how long it will take for it to reduce to $\frac{1}{10}$ of its original level, or 0.01%. This means that we need to find t for which $e^{-t/1000} = \frac{1}{10}$, or $-\frac{t}{1000} = \ln\left(\frac{1}{10}\right)$, or $t = -1000\ln(0.1) \approx 2300$. It will take approximately 2300 days, or about 6.3 years for the pollution to reduce to $\frac{1}{10}$ of its original level.

In practice there might be several different rivers flowing into and out of a lake, and some of the ones flowing in might bring in new contaminants. In this kind of situation it is necessary to figure out the *net* inflow and outflow of contaminants, but then the same sort of analysis can be applied. The result will be that the difference between the current level of pollution and the level flowing in will decrease exponentially. We can expect that as $t \to \infty$, the concentration of contaminants in the lake will approach as a limit the concentration of contaminants flowing in. We cannot expect the lake to become cleaner than the water flowing into it.

Exercises 7.4

I

Find the half-life of each of the following decay processes.

1. The original amount reduces to 20% in 3 hr.
2. The original amount reduces to 44% in 9 min.
3. The original amount reduces by 15% in 1 day.
4. The original amount reduces by 89% in 28 sec.
5. After 4 hr the amount is 10 oz; after 6 hr it is 8 oz.

6. After 1 yr the amount is 65 g; after 3 yr it is 30 g.

Find the decay constants for the processes with the following half-lives.

7. 2800 yr

8. 30 days

9. 34 min

10. 2.1 sec

II

11. In one year the population of a small Balkan country grows from 3,200,000 to 3,220,000.

(i) Assuming that the conditions for exponential growth are satisfied, how long will it be before the population reaches 4,000,000?

(ii) How long is it since the population was 3,000,000?

12. If the annual birth rate in a certain region is 1.4% and the annual death rate is 0.6%, how long will it take to have a 10% increase in population, assuming the growth is exponential?

13. Suppose a substance decays radioactively with half-life h. What is the decay constant k in the formula $Q(t) = ce^{-kt}$ for the amount remaining at time t?

14. If a substance grows according to the formula $Q(t) = ce^{kt}$, how long will it take to double?

15. Suppose a substance decays radioactively with half-life h. Suppose the initial amount of the substance is I. Find a formula for the amount remaining at time t.

***16.** (i) Suppose a radioactive substance decays with half-life h. If the amount of the substance has reduced to r times as much as there was at the beginning of the decay (with $0 < r < 1$), how long has the substance been decaying?

(ii) Check your answer by applying this formula to Example 8 (p. 534).

***17.** (i) The half-life of the isotope polonium 210 is 140 days. Make a quick estimate in your head of how long it will take for a sample to decay to 1% of its original size.

(ii) Calculate the exact answer to the question posed in part (i). How good was your estimate?

18. If the half-life of a radioactive substance is h, how long will it take to diminish to p% of its original size?

19. Compare the effective annual interest resulting from daily compounding at a nominal annual rate of 9.9% with the effective rate for quarterly compounding at a nominal 10%.

***20.** In the text it was stated that the effective interest rate increases as the number of times interest is compounded increases. Prove that this is correct, assuming that the nominal rate is fixed.

21. If the nominal annual interest rate is 10%, what is the effective rate with continuous compounding?

22. What nominal rate of annual interest will give an effective rate of 10% with continuous compounding?

23. If the nominal annual interest rate is p%, what is the effective rate with continuous compounding? Express your answer in terms of $r = \frac{p}{100}$.

24. What nominal rate of annual interest will give an effective rate of p% with continuous compounding? Express your answer in terms of $r = \frac{p}{100}$.

25. A bank account containing \$1500 earns interest at the rate of 6%, compounded continuously. No withdrawals or deposits are made for 6 weeks, after which \$1000 is deposited. Two weeks after that, \$300 is withdrawn. Calculate the amount in the account 12 weeks after this withdrawal.

26. Repeat Exercise 25 with the nominal interest rate of 8%.

***27.** Three thousand dollars are deposited in an account bearing interest at the nominal rate of 7.5%, compounded continuously. After 90 days the interest rate rises to 8%. What is the total in the account 1 year after the initial deposit?

28. An account is opened with a deposit of \$100, and an additional \$100 is deposited each month from then on. If the nominal interest is 9%, compounded continuously, how much money will be in the account after 18 months? (Include the nineteenth deposit.)

29. Compare the effective annual interest rates obtained with a nominal rate of 8% if the compounding is (i) quarterly, (ii) monthly, (iii) daily, (iv) hourly, (v) continuous.

30. Compare the effective annual interest rates obtained with a nominal rate of 11% if the compounding is (i) quarterly, (ii) monthly, (iii) daily, (iv) hourly, (v) continuous.

***31.** Work out the effective annual interest rate for daily compounding at 6% nominal interest. Do it two different ways, using a calculator: (i) by multiplying out $\left(1 + \frac{1}{365}\frac{6}{100}\right)^{365}$ (i.e., 364 multiplications) and (ii) using logarithms. Any slight difference in the answers shows the effect of round-off errors.

***32.** If we repeat Exercise 31 with 12% nominal interest, do you expect the effective rate to be twice what resulted from 6% nominal interest, or more, or less? Do the calculation and check your intuition.

***33.** (i) With interest at a nominal rate of 7% per annum, compare the effective rates obtained by compounding daily in an ordinary year (with 365 days) and in a leap year (with 366 days).

(ii) How much money would there have to be in an account for the difference in the amounts of annual interest to equal 1 cent?

***34.** A car loan is paid off in 36 equal monthly installments (each payment is at the end of the month). If the interest is compounded continuously and the nominal rate is 12% per annum, what will the size of the monthly payment be for an original loan of $8500?

35. What is the present value of an annuity that will pay its owner $20,000 per year for the next 15 years, assuming 8% annual interest, compounded continuously? (Assume that the first payment is today and there are a total of 15 payments.)

36. Uncle Fred makes a deposit of $500 each year for each of his nieces and nephews, beginning on their first birthday and ending on their twentieth.

(i) Assuming an effective annual interest rate of 8%, what is the present value of the deposit on the tenth birthday of one of the nephews or nieces?

(ii) What is the present value of the deposit on the twentieth birthday of one of the nephews or nieces?

(iii) What is the present value of the deposit on the day one of the nephews or nieces is born?

37. For a period of time in the early 1980s the number of reported cases of AIDS was doubling every six months. Ignoring the discrepancies between the number of actual cases and the number of reported cases and supposing that this trend continued, how long after the first case would it take for the entire world population of approximately 5,000,000,000 to be infected?

***38.** Discuss the assumptions underlying Exercise 37. How reasonable do you think they are? What significance can be attached to the answer you obtained?

39. A lake contains 500,000,000 ft^3 of water contaminated with chemicals whose concentration is 300 ppm (parts per million). If clean water flows in at the rate of 5,000,000 ft^3 per day, how long will it be until the chemicals are below the minimum standard for drinking water of 10 ppm? (Assume that the clean water mixes perfectly with the lake water and that the outflow from the lake is the same as the inflow.)

40. Suppose a lake just like the one in Exercise 39 (same volume, same water flow) has never been polluted and contains pure water. Suppose that a paper mill opens 3 miles upriver from this lake with the result that the water flowing in is heavily polluted, with the concentration of the contaminants equal to 1.3%.

(i) How long will it be before the lake water is contaminated at the level of 10 ppm?

(ii) Show that in the long run the concentration of pollutants in the lake will approach the level in the water flowing in.

41. A lake is round, with a radius of 1 mile and an average depth of 50 ft. A river flows into it; at a point where the river is 40 ft wide and has an average depth of 10 ft, the river flows at a rate of 6 inches per second. This is the only inflow to the lake; a river flows out at the opposite side. The water flowing in is perfectly clean. Suppose the water in the lake is polluted with contaminants at a concentration of 0.07%. Assuming that the water flowing in mixes perfectly with the lake water, how long will it take for the concentration of the pollutant to reach 0.02%?

****42.** Two chemists are cleaning flasks after an experiment. Each flask has a capacity of 1 liter and contains 10 g of salt dissolved in 1 liter of water. It is important to remove as much salt as possible because even minute amounts may invalidate the results of the next experiment. Sara empties her flask, leaving 10 milliliters behind. She fills the flask with clean water from the tap, empties it again (except for 10 ml), and so on. In all, she empties the flask six times. Dave, on the other hand, places his flask beneath the tap and lets the water run. Water runs into the flask and mixes with the solution inside, and some of the solution runs over the rim of the flask and down the drain. He leaves it long enough for 5 liters of water to run into the flask and then empties it except for 10 ml. Assume that in both cases the solutions always mix perfectly. The question we want to answer is which method does a better job of cleaning out the salt. Notice that they use almost exactly the same amount of water.

(i) Try to decide which method leaves less salt behind, using your powers of reasoning but not making any calculations.

(ii) Calculate how much salt remains in each flask at the end of the cleaning process. By what factor is one method preferable over the other?

43. The rate at which a chemical reaction takes place depends on many things, but if most of them are held fixed (e.g., constant temperature, constant volume, a consistent amount of agitation of the mixture, constant illumination, etc.), the main factor will be the concentration of the reactants. Consider a simple reaction in which a single compound decomposes, in the presence of heat, into two products. Let us assume that the temperature, the volume, and the illumination are all held constant. Let us also assume that the amount of the reactant remaining at any given time is an exponentially decreasing function of time (a fairly reasonable assumption under these conditions).

(i) Suppose the reaction begins with 1.3 g of the

reactant and that after 25 min there is 0.9 g remaining. How much will there be after 2 hours and 5 minutes?

(ii) How long will it be until there is 0.5 g remaining?

*44. Suppose a different chemical reaction proceeds in such a way that the rate of change of the concentration of a certain reactant is proportional to the square of the concentration of that reactant (a reaction of this type is said to be "of second order"). Find a formula for the concentration of the reactant in question as a function of time. Interpret any constants in your formula (e.g., as "initial amounts," etc.).

III

45. Use a programmable calculator or computer to find the effective rate for nominal 6% interest compounded every *hour* by multiplying out the product (cf. Exercise 31).

46. Use a programmable calculator or computer to find the effective rate for nominal 12% interest compounded every *hour* by multiplying out the product (cf. Exercise 32).

POINT TO PONDER

In Example 7.3.5 (p. 527) we saw the effect of continuing calculations of exponential growth for too long a time, beyond the point at which the assumptions are reasonable. In discussing radioactive decay we said that it satisfies the assumptions for exponential decay very well, and it is true that the calculations for this type of problem are extremely accurate. On the other hand, there is a limit to their reliability too.

Consider a sample of radon. It is a chemically inert but radioactive gas, formed as a product of the radioactive decay of radium. Radon decays with a half-life of 3.82 days. A sample of radon weighing 1 gram contains approximately 2.7×10^{21} atoms.

Calculate how long it would take until there are only 1000 atoms left. Then calculate how long it would take until there is only 1 atom left.

At some point the last atom will decay, and there will be no radon left at all. It is not possible to predict exactly when it (or any other single atom) will decay, but it is possible to calculate the probability that it will have decayed by a certain time. After it does decay, the exponential function, which is never zero, will be (slightly) incorrect in that it will suggest that there is still some radon present.

There is a fundamental difficulty here that arises from the fact that the exponential function is a continuous function, whereas matter is composed of individual atoms. The number of atoms of radon remaining is not a continuous function; it has a jump discontinuity every time an atom decays.

As long as the number of atoms is quite large, the jumps will be relatively small, and the continuous function will be a very accurate approximation to the actual (discontinuous) function. When there are only a handful of atoms, then the randomness in their behavior will become apparent, and the exponential function will be less accurate.

The difficulty is more theoretical than practical because usually for us to be able to measure the number of atoms, they have to be fairly numerous. The situation in which there are few enough for their individual behavior to be noticeable is usually beyond our ability to perceive.

7.5 Logistic Growth

LOGISTIC EQUATION
CARRYING CAPACITY

In Sections 7.3 and 7.4 we discussed exponential growth. We saw how that concept could be applied to the growth of organisms from bacteria to human beings, but we also gained some appreciation of the shortcomings of the approach.

The fundamental difficulty is that while the assumptions from which we can conclude that the growth will be exponential may well hold for a while, they will cease to be valid as the population grows. This can occur in various ways, but usually the growth rate will slow down as the population becomes sufficiently large that the existing population interferes with further unrestricted growth.

In this section we discuss a formula for population growth that attempts to take these ideas into consideration. It is a fairly simple idea, but it turns out to give a surprisingly accurate formula in many situations. Unfortunately, the actual calculations involved are a little less simple than the corresponding ones for exponential growth. They are not tremendously difficult, but they are more complicated.

This section has no bearing on anything appearing in later sections and so may be omitted. However, it should be of special interest to anybody studying life sciences, since the ideas introduced here are widely used in those areas.

The basic assumption underlying the theory of exponential growth is that $\frac{dP}{dt}$, the rate of change of the population, is proportional to the population P. This is equivalent to assuming that the birth rate is *constant.* (The birth rate means the number of births per member of the population per unit of time.) This assumption is frequently quite reasonable for a while, but as the population grows, it will be unable to continue to grow as rapidly. To make a more realistic mathematical model, we will have to incorporate the idea that the birth rate will decrease if the population grows very large.

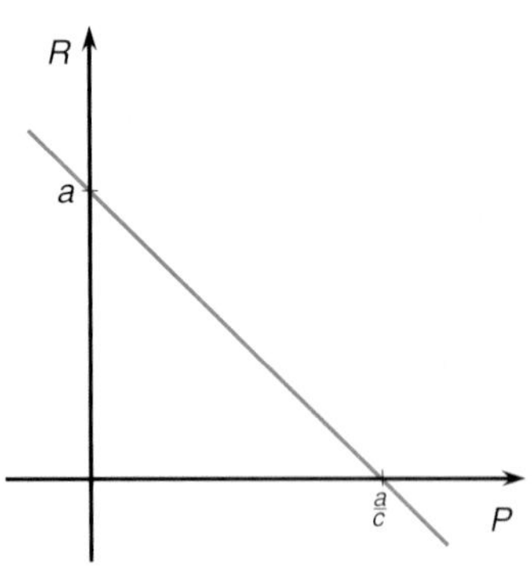

Figure 7.5.1

To simplify the discussion, let us consider a population that does not die, such as bacteria. Instead of assuming that the birth rate is a constant, let us try the simplest decreasing function, a straight line with negative slope, as shown in Figure 7.5.1. This means that we should assume that the birth rate is a function of the form $a - cP$, where a and c are constants, with $c > 0$.

If the birth rate is $a - cP$, then the number of births per unit time will be $(a - cP)P$, and this means that $\frac{dP}{dt} = (a - cP)P$. Letting $L = \frac{a}{c}$, we can rewrite this equation as

$$\frac{dP}{dt} = cP(L - P). \tag{7.5.1}$$

This is known as the **logistic equation**.

To solve it, we rearrange it in the form

$$\frac{dP}{P(L-P)} = c\,dt,$$

and try to integrate.

To integrate the left side, it will be necessary to use a trick. We notice that $\frac{1}{P(L-P)} = \frac{1}{L}\left(\frac{1}{P} + \frac{1}{L-P}\right)$.

This permits us to integrate the left side of the equation above:

$$\begin{aligned}\int \frac{dP}{P(L-P)} &= \frac{1}{L}\int\left(\frac{1}{P} + \frac{1}{L-P}\right)dP\\ &= \frac{1}{L}\big(\ln P - \ln(L-P)\big) + d = \frac{1}{L}\ln\frac{P}{L-P} + d.\end{aligned}$$

Since the integral of the right side of the equation is ct plus a constant, we get $\frac{1}{L}\ln\frac{P}{L-P} = ct + d'$. Multiplying by L and exponentiating, we see that $\frac{P}{L-P} = Ce^{cLt}$, where the constant C can be determined by evaluating at $t = 0$. If we write P_0 for the population at $t = 0$, then we find that $C = P_0/(L - P_0)$, so

$$\frac{P}{L-P} = \frac{P_0 e^{cLt}}{L-P_0}.$$

Cross-multiplying, we find that $P(L - P_0) = P_0 e^{cLt}(L - P)$.

Collecting the terms containing P, we have $P(L - P_0 + P_0 e^{cLt}) = LP_0 e^{cLt}$, and dividing gives

$$P = P(t) = \frac{LP_0 e^{cLt}}{L - P_0(1 - e^{cLt})} = \frac{P_0 L}{P_0 + (L - P_0)e^{-cLt}}. \tag{7.5.2}$$

In deriving this formula we have implicitly assumed that $L > P$ because we said $\int \frac{dP}{L-P} = -\ln(L-P) + constant.$

Several things are immediately apparent from this formula. It is easy to evaluate the limit of $P(t)$ as $t \to \infty$. Since we know that c and L are positive, we know that $\lim_{t\to\infty}(e^{-cLt}) = 0$. From this we see that

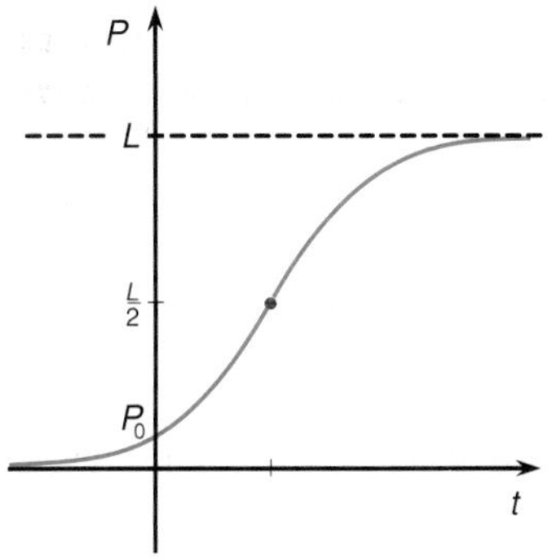

Figure 7.5.2

$$\lim_{t\to\infty} P(t) = \frac{P_0 L}{P_0 + 0} = L.$$

The population will approach the limit L as t tends to infinity. As $t \to -\infty$, we know that $e^{-cLt} \to \infty$, so the denominator tends to infinity, and $P(t) \to 0$. The shape of the graph of $y = P(t)$ is shown in Figure 7.5.2. It is straightforward but messy to calculate the second derivative $P''(t)$. From this calculation it is possible to see that there is exactly one inflection point. It occurs at the value of t that satisfies $e^{-cLt} = P_0/(L - P_0)$. The value of $P(t)$ at this particular value of t is $\frac{L}{2}$. In terms of the picture, this means that the graph is concave upward until it reaches halfway up from 0 to its limiting value L, and after that it is concave downward.

The limiting population L is often called the **carrying capacity** of the environment. If we think of bacteria in a jar, L represents the maximum number of bacteria that can survive in the jar. It would have a similar interpretation if we were talking about a population of rabbits, deer, or humans.

The logistic formula has proven to be remarkably accurate in many situations. It has had striking successes in predicting human population growth in countries. (It has also failed sometimes but usually because the situation changed in some way, such as a change in the law permitting a large immigration or a medical advance that saved many lives.)

It was mentioned above that we have implicitly assumed that $P < L$, that is, that the population stays below the carrying capacity. If we place a population that is too large in the environment, we should expect it to decrease toward the carrying capacity, and in fact this is reflected in the behavior of any solution of Equation 7.5.1 that begins with a population that is "too large."

Unfortunately, it is more difficult to figure out the constants associated to a logistic curve than it is for exponential curves. In particular, since there are three of them (the initial population P_0, the carrying capacity L, and the constant c in the exponent, which is a sort of growth constant), it is necessary to know the value of $P(t)$ for three values of t.

In terms of the calculation it turns out to be very much easier to work out the constants if we happen to know the values of $P(t)$ for three values of t that are evenly spaced.

EXAMPLE 1 Suppose the population of a small town follows a logistic curve. If the population was 6,000 in 1970, 10,000 in 1980 and 15,000 in 1990, what will it be in the year 2000?

Solution Let us write the population in the form of Equation 7.5.2, $P(t) = \frac{P_0 L}{P_0+(L-P_0)\exp(-cLt)}$. It will be more convenient to measure population in thousands, and it will also be convenient to let t be the time in years, with $t = 0$ at the year 1980. We know that P_0 is the *initial population*, meaning the population when $t = 0$, which is $P_0 = 10$ (meaning that in 1980 the population was 10 thousand).

This means that

$$P(t) = \frac{10L}{10 + (L - 10)e^{-cLt}}.$$

We also know that $P(-10) = 6$ and $P(10) = 15$. Writing these out, we have

$$\frac{10L}{10 + (L - 10)e^{10cL}} = 6,$$
$$\frac{10L}{10 + (L - 10)e^{-10cL}} = 15.$$

Cross-multiplying in each of these equations, we obtain

$$10L = 60 + 6(L - 10)e^{10cL},$$
$$10L = 150 + 15(L - 10)e^{-10cL}$$

or

$$10(L - 6) = 6(L - 10)e^{10cL},$$
$$10(L - 15) = 15(L - 10)e^{-10cL}.$$

What makes these equations difficult to solve is the presence of the exponential factors on the right side. But if we multiply the right sides of the last two equations, we see that the two exponential factors cancel. Multiplying the two left sides and the two right sides, we obtain

$$\begin{aligned}100(L - 6)(L - 15) &= 6(15)(L - 10)^2,\\ 100(L^2 - 21L + 90) &= 90(L^2 - 20L + 100),\\ 100L^2 - 2100L + 9000 &= 90L^2 - 1800L + 9000,\\ 10L^2 - 300L &= 0,\\ 10L(L - 30) &= 0.\end{aligned}$$

The two possible solutions are $L = 0$ and $L = 30$; the first is an extraneous solution. The carrying capacity is $L = 30$ (i.e., a population of 30,000).

Now it is relatively easy to solve for the constant c. For instance, we have already found that $10(L - 6) = 6(L - 10)e^{10cL}$; putting in the value $L = 30$, we find that

$$\begin{aligned}10(24) &= 6(20)e^{300c},\\ 240 &= 120e^{300c},\\ 300c &= \ln 2,\\ c &= \frac{\ln 2}{300}.\end{aligned}$$

We are now able to write the logistic population formula:

$$P(t) = \frac{300}{10 + 20e^{-t\ln(2)/10}}.$$

From this it is easy to answer the question about the population in the year 2000. Since $t = 0$ in the year 1980, the year 2000 occurs when $t = 20$, so the population then is

$$P(20) = \frac{300}{10 + 20e^{-2\ln(2)}} = \frac{300}{10 + 20(2)^{-2}} = \frac{300}{15} = 20.$$

According to our convention, the population in the year 2000 will be 20,000.

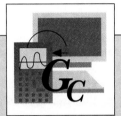

With a graphics calculator we can plot the function of the previous example.

TI-85	SHARP EL9300
GRAPH F1 ($y(x) =$) 300 ÷ (10 + 20 e^x ((−) F1 (x) × LN (2) ÷ 10)) M2(RANGE) (−) 60 ▼ 60 ▼ 10 ▼ (−) 10 ▼ 30 ▼ 10 F5(GRAPH)	∿ 300 ÷ (10 + 20 e^x (−) x/θ/T × ln 2 ÷ 10 ▶) RANGE −60 ENTER 60 ENTER 10 ENTER − 10 ENTER 30 ENTER 10 ENTER ∿
CASIO f_x-6300G (f_x-7700GB)	**HP 48SX**
Range −60 EXE 60 EXE 10 EXE − 10 EXE 30 EXE 10 EXE EXE EXE EXE Graph 300 ÷ (10 + 20 e^x ((−) X × ln (2) ÷ 10)) EXE	↰ PLOT PLOTR ERASE ATTN ' Y = 300 ÷ () 10 + 20 × e^x X × LN 2 ▶ ÷ 10 +/− ↱ PLOT ↰ DRAW ERASE 60 ENTER +/− 60 XRNG 10 +/− ENTER 30 YRNG DRAW

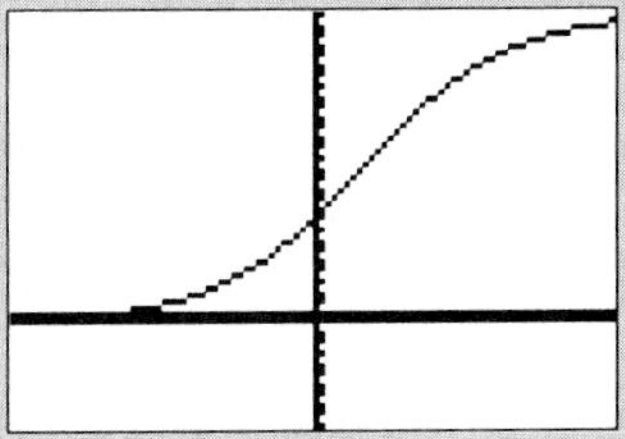

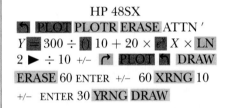

We see the characteristic property that the curve approaches a limit as time grows very large.

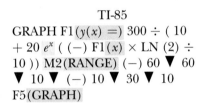

Incidentally, in the course of solving this problem we have found that the carrying capacity is 30,000, so we expect that the population of the town will level off as it approaches 30,000. This kind of unlimited long-term prediction is notoriously unreliable and should be taken with a large grain of salt in any realistic application.

Notice that it was because we chose times that were evenly spaced that we were able to use the trick of multiplying to get rid of the exponential factors. It is possible to find the constants c, L, and P_0 from the values of the population at any three points, but it is much more difficult. We will not consider the general case.

Exercises 7.5

I

In each of the following situations, find the constants in Formula 7.5.2, assuming that the function is logistic.

1. $P(-1) = 1, P(0) = 2, P(1) = 3$

2. $P(2) = 5, P(4) = 7, P(6) = 9$

3. $P(0) = 8, P(1) = 10, P(2) = 11$

4. $P(4) = 1, P(5) = 9, P(6) = 12$

5. $P(1950) = 15{,}000, P(1960) = 20{,}000, P(1970) = 25{,}000$

6. $P(1962) = 20{,}000, P(1966) = 22{,}000, P(1970) = 23{,}000$

7. $P(10) = 1, P(11) = 44, P(12) = 45$

8. $P(0) = 3, P(1) = 5, P(2) = 6$

9. $P(-3) = 4, P(-2) = 5, P(-1) = 6$

10. $P(-h) = a, P(0) = 2a, P(h) = 3a$

II

11. A culture of bacteria weighs 2 mg after growing for 1 day, 4 mg after two more days, and 6 mg two days after that.

(i) What is the carrying capacity of their container, assuming logistic growth?

(ii) How long will it be after growth begins until the culture reaches 90% of the carrying capacity?

12. A shipwreck releases rabbits on a small island. After 5 years there are 500 rabbits; after 2 years more there are 800 rabbits; 2 years after that there are 1000 rabbits.

(i) Assuming that the rabbit population follows a logistic curve, what is the carrying capacity of the island for rabbits?

(ii) How long after the shipwreck will the rabbits have reached half the carrying capacity?

***13.** Confirm the assertions made in the text about the location of the inflection point of $y = P(t)$ (i.e., that there is exactly one inflection point, that it occurs when $e^{-cLt} = P_0/(L - P_0)$, and that the value of $P(t)$ at the inflection point is $\frac{L}{2}$). You may find it helpful to use the logistic equation (Equation 7.5.1) instead of calculating the derivative $P'(t)$ directly.

***14.** Show that the logistic curve is symmetrical about its inflection point. (This means that if you rotate the graph 180° around the inflection point, you get the same graph back again.)

***15.** What happens if $L < P_0$? (Give a qualitative description of the type of solution we should expect, and sketch the graph. Discuss how you would interpret your answer in a real situation, such as a population of rabbits on an island.)

***16.** We remarked that it can be quite difficult to find the constants for a logistic curve if we know the values at three unevenly spaced points. Modify the idea demonstrated in Example 1 (p. 548) to deal with the following case: Find the logistic function $P(t)$ that satisfies $P(-1) = 1$, $P(0) = 2$, and $P(2) = 6$. (The carrying capacity L does not work out to a nice round number.)

17. Suppose a bacterial culture is weighed at different times. The first recorded weight was 5 mg; two days later the weight was 7 mg, and two days after that the weight was 9 mg.

(i) Find the logistic curve that fits these data.

(ii) How long before the first measurement did the culture weigh 0.1 mg?

(iii) Suppose we have lost the third measurement, and consider what happens if we use the first two (5 mg and, two days later, 7 mg) and assume that the growth follows an *exponential* curve. Use this approach to answer part (ii), and compare the two answers.

(iv) Use the logistic curve and the exponential curve of part (iii) to find when the culture will weigh 13 mg. Compare the two answers.

***18.** Show that the logistic function (Equation 7.5.2) can also be written in the form

$$P(t) = \frac{L}{1 + e^{-cL(t-t_0)}}$$

if t_0 is the point where $P(t)$ has its inflection point.

***19.** Consider the logistic curve

$$Q(t) = \frac{k}{1 + e^{d-at}}.$$

(i) What is the carrying capacity?

(ii) At what time t does the inflection point occur?

(iii) If we write the function $Q(t)$ in the form of Equation 7.5.2, what are the constants P_0 and c?

***20.** Show that it is not possible to find a logistic curve for which $P(0) = 9$, $P(1) = 10$, and $P(2) = 12$.

***21.** Suppose twin lakes are separated by a barrier. Suppose fish grow in one lake according to the logistic equation

$$P(t) = \frac{L}{1 + e^{d-at}},$$

with t in years. Suppose that at some time t_0 the barrier is removed. If the second lake had the same number of fish, growing according to the same logistic equation, it seems reasonable to expect that the equation for the *total* population of the two joined lakes will now be $T(t) = 2P(t) = \frac{2L}{1+\exp(d-at)}$.

But suppose the barrier is removed at time t_0 and the second lake has no fish in it. What will be the formula for $T(t)$ in this situation? The carrying capacity will have doubled, but what will happen to the constants a and d? (*Hint:* Try redistributing the fish in the obvious way: half in each lake. At this point it may be easier to think about what happens in just one of the lakes and combine them later.)

***22.** Suppose that if they are left unmolested, the fish in a lake grow according to the logistic equation

$$P(t) = \frac{L}{1 + e^{d-at}},$$

where t is the time in years. Suppose too that once a year there is a brief fishing season, during which the population is decreased, but that for the rest of the year no fish are caught. The problem is to optimize our fishing in the long run. If we take too many fish this year, the remaining population will be so small that it will grow slowly, and there will not be very many fish next year. If we take too few, in the first place we lose out on our catch, and in the second place the population will be large and the logistic curve will limit their growth; not many will grow to replace the ones we took.

(i) Find the best strategy, that is, the one that gives us the biggest overall catch in the long run. (*Hint:* When we catch some fish at a certain time t_0 and reduce the population at t_0 to A, this amounts to shifting the logistic curve to the right until its value at t_0 is A. In terms of the formula given above, we adjust the constant d but leave a and L unchanged.)

(ii) Discuss *briefly* the best strategy if we want to maximize our catch only for 1, 2, or 3 years.

III

23. Use a graphics calculator or computer to discuss the effects on a logistic curve of changing the constants P_0, L, and c. For instance, what is the effect of changing P_0 if L and c remain the same? (Sketch several curves.) What is the effect of doubling L and P_0, leaving c the same? What happens if L is doubled but P_0 and c remain unchanged? What is the effect of changing c and leaving L and P_0 the same? What will be the shape of the curve if c is negative?

7.6 Differential Equations

SEPARABLE EQUATION
GENERAL SOLUTION
PARTICULAR SOLUTION
ORDER
CHARACTERISTIC POLYNOMIAL
AUXILIARY EQUATION

In this brief section we begin to explore the subject of differential equations, which arose in the discussion of exponential growth in Section 7.3 and in the discussion of logistic growth in Section 7.5.

A differential equation expresses a relation between a function and one or more of its derivatives. We will use the letter t as the variable; this is because in many of the applications the equation deals with the rate of change of some quantity, so the natural variable is *time*. Functions of t will sometimes be written in the usual way as $f(t)$, $g(t)$, and so on, but sometimes we will also use the letter x. The understanding is that x is not a variable but a *function* of t. To emphasize this, we sometimes write $x = x(t)$.

Separable Equations

A differential equation is called **separable** if it can be written in the form

$$F(x)\frac{dx}{dt} = G(t)$$

for some functions F and G. This can be rearranged to

$$F(x)\,dx = G(t)\,dt,$$

with all the x's on the left side and all the t's on the right side. In this form it is not too useful, but we can integrate to obtain

$$\int F(x)\,dx = \int G(t)\,dt + c.$$

This formula expresses a relation between x and t that does not involve derivatives. In particular cases we may be able to solve it to write x as a function of t explicitly.

Strictly speaking, the constant c is understood when we write indefinite integrals. However, this is one instance in which it is vitally important to remember the constant, so we write it out to help remind ourselves.

EXAMPLE 1 Solve the differential equation $\frac{dx}{dt} = xt$.

Solution To solve a differential equation means to find a function that satisfies the equation. In this case we are looking for the function $x = x(t)$.

As suggested above, we rearrange the equation to $\frac{dx}{x} = t\,dt$. Integrating, we find $\int \frac{1}{x} dx = \int t\,dt$, or $\ln|x| = \frac{1}{2}t^2 + c'$.

To find a formula for the function $x = x(t)$, we exponentiate the last equation and find $|x| = e^{c'} e^{t^2/2}$. In rearranging the equation we divided by x, so in effect we have assumed that x is never zero. This means that it cannot change sign. We simplify notation by writing $x = ce^{t^2/2}$, with $c = e^{c'}$ if $x > 0$ and $c = -e^{c'}$ if $x < 0$. We have found the solution

$$x = x(t) = ce^{t^2/2}.$$

The constant $e^{c'}$, being an exponential, is always positive. So $c = \pm e^{c'}$ is never zero. However, the constant function $x(t) = 0$ is also a solution of the original equation; it can be seen as the result of letting $c = 0$ in the solution we have just found. It often happens that we find a solution containing an unspecified constant, with some restrictions on the constant. In this situation we may find that we still get solutions by letting the constant have some of the forbidden values. However, when we proceed this way, it is important to check that the final answer does actually satisfy the original equation.

In solving a separable equation we will encounter a constant of integration. The solution containing this unspecified constant is called the **general solution**. If we know the value of the solution at any particular value of t, then this information can be used to find the constant, and it is then possible to write down the precise solution (i.e., without any undetermined constants). In this case we call it a **particular solution**.

EXAMPLE 2 Find the solution of $\frac{dx}{dt} = 2x - 14$ that satisfies $x(0) = 1$.

Solution We rearrange the equation in the form $\frac{dx}{x-7} = 2\,dt$. This is in the standard form for a separable equation, and we integrate it to obtain $\ln|x - 7| = 2t + c'$.

To find an expression for the function $x(t)$ itself, we exponentiate: $|x - 7| = e^{c'} e^{2t}$. Letting $c = \pm e^{c'}$, we write $x - 7 = ce^{2t}$, and rearranging, we have

$$x = x(t) = 7 + ce^{2t}.$$

To find the particular solution that satisfies $x(0) = 1$, we must find the correct value for the constant c. Knowing that $x(t)$ is given by the formula above, we see that $1 = x(0) = 7 + ce^{(2)(0)} = 7 + c$, so c must equal -6.

The required particular solution is $x = x(t) = 7 - 6e^{2t}$.

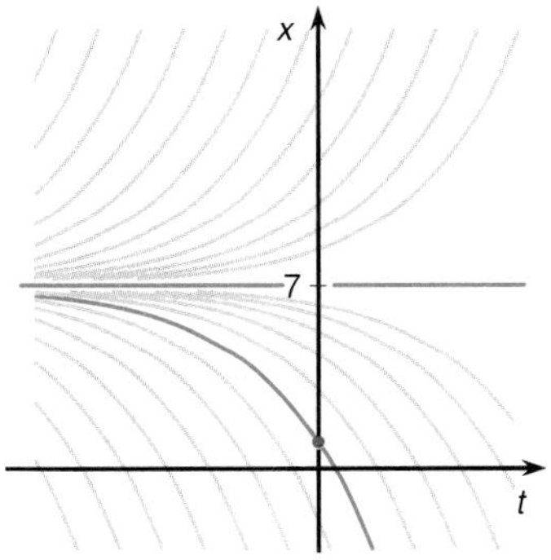

Figure 7.6.1

Notice how the condition $x(0) = 1$ forced the constant c to be negative. Also notice how we used the constant c' when we integrated so that we could use the simpler constant c in the final formula. This is a matter of style and has no other significance.

When we obtain the general solution as above, there is a constant that can have any fixed value. Each different value of the constant gives a function that is a particular solution of the differential equation. There is a whole "family" of solutions. See Figure 7.6.1.

The value of the function $x(t)$ at a particular point is known as an **initial value**. Roughly speaking, the idea is that once we know the value of the function at some "starting time," then the differential equation, which specifies how the function "changes," will determine its values at all other points.

Linear Differential Equations with Constant Coefficients

Consider the differential equation $\frac{df}{dt} = kf$, where k is a constant. We already know its solutions, but it is interesting to realize that it is an example of a separable equation. We write $\frac{df}{f} = k\,dt$, so $\ln|f| = kt + c'$. Exponentiating, we find $f = f(t) = ce^{kt}$, with c a constant.

In fact this is exactly how we solved this type of equation in Section 7.3.

If we write the equation just considered in the form $f' = kf$, or $f' - kf = 0$, we see that it is constructed by multiplying each of f and f' by a constant, adding them up, and setting the resulting formula equal to zero.

Now consider the same sort of thing with f, f', and f''.

EXAMPLE 3 Solve the equation $f'' + f' - 2f = 0$.

Solution The function $f(t)$ must be one whose derivatives f' and f'' are very closely related to f and each other. Taking our cue from the equation $f' - kf = 0$, we make a guess and try letting $f(t) = e^{kt}$, for some constant k.

This means that $f' = ke^{kt} = kf$ and $f'' = k^2e^{kt} = k^2f$. Substituting these derivatives into the original equation, we find

$$0 = f'' + f' - 2f = k^2f + kf - 2f = (k^2 + k - 2)f.$$

The exponential function $f(t) = e^{kt}$ is never zero, so this equation implies that $k^2 + k - 2 = 0$.

This is just an ordinary quadratic equation for the number k. We solve it by the quadratic formula, or in this case simply by factoring: $0 = k^2 + k - 2 = (k + 2)(k - 1)$, so $k = -2$ or $k = 1$.

We have just found two solutions of the equation: $f_1(t) = e^{-2t}$ and $f_2(t) = e^t$. It is easy to check that they do both satisfy the equation.

It is also easy to check that if a and b are any constants, then $f(t) = af_1(t) + bf_2(t)$ is also a solution. After all, $f' = af_1' + bf_2'$ and $f'' = af_1'' + bf_2''$, so

$$\begin{aligned} f'' + f' - 2f &= (af_1'' + bf_2'') + (af_1' + bf_2') - 2(af_1 + bf_2) \\ &= a(f_1'' + f_1' - 2f_1) + b(f_2'' + f_2' - 2f_2) \\ &= a(0) + b(0) = 0. \end{aligned}$$

We group f_1 and its derivatives together, and f_2 and its derivatives together. Each group equals zero because f_1 and f_2 satisfy the original equation.

It is less obvious but also true that this is the general solution, that is, that it gives us all the solutions.

With a graphics calculator we can plot some of the solutions of the previous example. For instance, we try $f(t) = af_1(t) + bf_2(t)$, with $a = 1$, $b = 0$; $a = 0$, $b = 1$; $a = b = 1$; $a = 1$, $b = -2$.

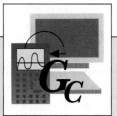

TI-85	SHARP EL9300
GRAPH F1 $(y(x) =)$ e^x ((−) 2 F1 (x)) ▼ e^x F1 (x) ▼ e^x ((−) 2 F1 (x)) + e^x F1 (x) ▼ e^x ((−) 2 F1 (x)) − 2 e^x F1 (x) M2(RANGE) (−)1 ▼ 1 ▼ ▼ (−)5 ▼ 10 F5(GRAPH)	e^x (−) 2 x/θ/T 2ndF ▼ e^x x/θ/T 2ndF ▼ e^x (−) 2 x/θ/T ▶ + e^x x/θ/T 2ndF ▼ e^x (−) 2 x/θ/T ▶ − 2 e^x x/θ/T RANGE −1 ENTER 1 ENTER 1 ENTER −5 ENTER 10 ENTER 5 ENTER
CASIO f_x-6300G (f_x-7700GB)	HP 48SX
Range −1 EXE 1 EXE 1 EXE −5 EXE 10 EXE 5 EXE EXE EXE EXE Graph e^x ((−) 2 X) EXE Graph e^x X EXE Graph e^x ((−) 2 X) + e^x X EXE Graph e^x ((−) 2 X) − 2 e^x X EXE	↰ PLOT PLOTR ERASE ATTN ' Y = e^x 2 +/− × X ↱ PLOT ↰ DRAW ERASE 1 ENTER +/− 1 XRNG 5 +/− ENTER 10 YRNG DRAW ATTN ' Y = e^x X ↱ PLOT ↰ DRAW DRAW ATTN ' Y = e^x 2 +/− × X ▶ + e^x X ↱ PLOT ↰ DRAW DRAW ATTN ' Y = e^x 2 +/− × X ▶ − 2 × e^x X ↱ PLOT ↰ DRAW DRAW

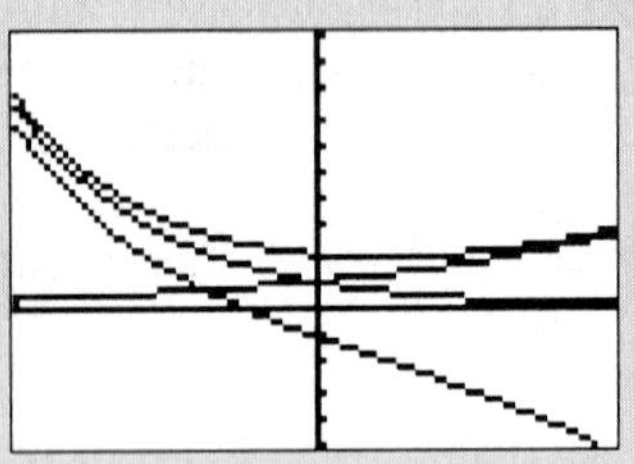

The solutions behave in quite different ways. Identify which is which.

EXAMPLE 4 Find the general solution of $f''' - f'' - 6f' = 0$.

Solution Again we try letting $f(t) = e^{kt}$ and find that $f' = kf$, $f'' = k^2 f$ and $f''' = k^3 f$. The differential equation becomes $0 = f''' - f'' - 6f' = \left(k^3 - k^2 - 6k\right)f = 0$.

Since $f(t)$ is never zero, this means that $k^3 - k^2 - 6k = 0$, and we can factor: $0 = k(k^2 - k - 6) = k(k-3)(k+2)$, so $k = 0,\ 3$, or -2.

We find three solutions, $f_1(t) = e^{0(t)} = 1$, $f_2(t) = e^{3t}$, and $f_3(t) = e^{-2t}$. As above, *every* solution is of the form $c_1 f_1 + c_2 f_2 + c_3 f_3$, for some constants c_1, c_2, and c_3, and any choice of these constants does give a solution.

A pattern is beginning to emerge from these examples. Consider the differential equation

$$a_n f^{(n)} + a_{n-1} f^{(n-1)} + \ldots + a_2 f'' + a_1 f' + a_0 f = 0.$$

We try letting $f(t) = e^{kt}$, for some constant k, so $f' = kf$, $f'' = k^2 f, \ldots, f^{(n)} = k^n f$, and the equation becomes

$$(a_n k^n + a_{n-1} k^{n-1} + \ldots + a_2 k^2 + a_1 k + a_0) f = 0.$$

Since f is never zero, we have to solve

$$a_n k^n + a_{n-1}k^{n-1} + \ldots + a_2 k^2 + a_1 k + a_0 = 0.$$

This is called the **auxiliary equation**, and its left side is called the **characteristic polynomial** of the original differential equation.

The number n, the order of the highest derivative that occurs in the equation, is called the **order** of the differential equation.

Find the roots $\{k_i\}$ of the characteristic polynomial. For each root k_i the function $f(t) = e^{k_i t}$ is a solution of the differential equation, and if there are n distinct roots, then every solution is of the form

$$c_1 e^{k_1 t} + c_2 e^{k_2 t} + \ldots + c_n e^{k_n t}.$$

This expression is the **general solution** of the differential equation. This approach is exactly what we used to solve the examples above.

There are two possible complications. One is that some roots of the characteristic polynomial may not be real numbers (e.g., the polynomial $k^2 + 1$ has no real roots).

A second difficulty is that some of the roots may be "repeated."

EXAMPLE 5 Solve $f'' + 4f' + 4f = 0$.

Solution The characteristic polynomial is $k^2 + 4k + 4$, which we can factor: $k^2 + 4k + 4 = (k+2)^2$. The only root is $k = -2$, but it is "repeated."

The function $f_1(t) = e^{-2t}$ is a solution of the differential equation, but from the earlier examples we expect to find a second solution. It is not immediately clear why it works, but it is easily checked that $f_2(t) = te^{-2t}$ is also a solution.

It turns out that the general solution is $c_1 f_1 + c_2 f_2 = c_1 e^{-2t} + c_2 te^{-2t}$. This means that every solution is of this form, for some constants c_1 and c_2.

Summary 7.6.1

The powers of t go up to one less than m, the number of times the root r is repeated.

For each root k_i of the characteristic polynomial we get solutions in this way. If a root r is repeated m times, that is, the characteristic polynomial is divisible by the factor $(k - r)^m$, then $f_1(t) = e^{rt}, f_2(t) = te^{rt}, f_3(t) = t^2 e^{rt}, \ldots, f_m(t) = t^{m-1}e^{rt}$ are all solutions of the differential equation. In this way we find solutions corresponding to each root of the characteristic polynomial. We multiply each of them by a constant and add them up to get a more complicated solution, including solutions corresponding to all the roots.

If the total number of roots, *including repetitions*, equals n, the order of the original equation, then this procedure will result in the general solution; that is, it will give all the solutions of the equation.

EXAMPLE 6 Find the general solution of the differential equation $f^{(6)} - f^{(5)} - 2f^{(4)} + 2f^{(3)} + f'' - f' = 0$.

Solution The characteristic polynomial is $k^6 - k^5 - 2k^4 + 2k^3 + k^2 - k$. We can easily see that it is divisible by k, and it equals $k(k^5 - k^4 - 2k^3 + 2k^2 + k - 1)$. The next problem is to factor this fifth-degree polynomial.

To do this, we try substituting small integer values of k in hopes of finding a root, using the rational root theorem (Theorem 1.3.7) to help choose possible values of k. Our very first attempt, with $k = 1$, gives 0, meaning that the polynomial is divisible by $k - 1$. Dividing, we get $k^5 - k^4 - 2k^3 + 2k^2 + k - 1 = (k-1)(k^4 - 2k^2 + 1)$.

We could try the same approach again, but if we are lucky, we might notice that the last factor is a square, the square of $(k^2 - 1) = (k-1)(k+1)$.

We are now able to factor the characteristic polynomial completely. We have that $k^6 - k^5 - 2k^4 + 2k^3 + k^2 - k = k(k^5 - k^4 - 2k^3 + 2k^2 + k - 1) = k(k-1)(k^4 - 2k^2 + 1) = k(k-1)(k^2-1)^2 = k(k-1)(k-1)^2(k+1)^2 = k(k-1)^3(k+1)^2$.

From the above discussion we see that the root $k = 0$ occurs once; the corresponding solution is $e^{0(t)} = 1$. The root $k = 1$ occurs three times; the corresponding solutions are e^t, te^t, and t^2e^t. The root $k = -1$ occurs twice; the corresponding solutions are e^{-t} and te^{-t}.

Multiplying each of these solutions by a constant and adding, we find the solution

$$c_1 + c_2e^t + c_3te^t + c_4t^2e^t + c_5e^{-t} + c_6te^{-t}.$$

The root $k = 0$ occurs once, the root $k = 1$ occurs three times, and the root $k = -1$ occurs twice. The total number of roots, *counting repetitions*, is 6. Since this equals the order of the original equation, the formula above is the general solution, that is, every solution must be of that form.

Finally, there is the question of finding a particular solution for a differential equation.

For instance, consider the equation $f'' + 4f' + 4f = 0$, which we solved in Example 5. The solution contained two constants, c_1 and c_2. In order for us to be able to solve for them, we will need *two* pieces of information about the solution. For an equation of order n we expect the general solution will have n constants in it, and to determine them all, it will take n pieces of information.

So instead of an initial value, the value of the solution at some point, we need the values of the function at n points, or the values of $f, f', \ldots, f^{(n-1)}$ at some point, or some other combination of n pieces of information.

EXAMPLE 7 Find the solution of $f'' + 4f' + 4f = 0$ that satisfies $f(0) = 1$ and $f'(0) = 4$.

Solution We know that the solution is of the form $f(t) = c_1e^{-2t} + c_2te^{-2t}$, for some constants c_1 and c_2. The value of this function at $t = 0$ is $f(0) = c_1$. The derivative is $f'(t) = -2c_1e^{-2t} - 2c_2te^{-2t} + c_2e^{-2t}$, so $f'(0) = -2c_1 + c_2$.

The condition specified in the question means $c_1 = 1$ and $-2c_1 + c_2 = 4$. Substituting $c_1 = 1$ into the second equation gives $c_2 = 6$.

The required particular solution is $f(t) = e^{-2t} + 6te^{-2t}$.

Exercises 7.6

I

Solve the following differential equations.

1. $\frac{dx}{dt} = xt^2$
2. $\frac{dx}{dt} = x^2t$
3. $\frac{dx}{dt} = \frac{t}{x}$
4. $\frac{dx}{dt} = tx^{-2}$
5. $\frac{dx}{dt} = \frac{x}{t}$
6. $\frac{dx}{dt} = x\sin t$
7. $\frac{dx}{dt} = (x-2)t^2$
8. $\frac{dx}{dt} = x^2\sin t$
9. $\frac{dx}{dt} = e^{x+t}$
10. $\frac{dx}{dt} = t(x^2+4)$

Find the solution of each of the following equations with the given initial value.

11. $\frac{dx}{dt} = \frac{x}{2}, \quad x(0) = 2$

12. $\frac{dx}{dt} = 3xt, \quad x(0) = 4$

13. $\frac{dx}{dt} = xt^2, \quad x(1) = 3$

14. $\frac{dx}{dt} = x^2 \cos t, \quad x(\pi) = 1$

15. $\frac{dx}{dt} = tx \ln x, \quad x(1) = 1$

16. $\frac{dx}{dt} = x \ln x, \quad x(0) = \frac{1}{2}$

17. $\frac{dx}{dt} = \sec x, \quad x(0) = \frac{\pi}{4}$

18. $\frac{dx}{dt} = (x^2 + 1)t, \quad x(0) = 1$

19. $\frac{dx}{dt} = t\sqrt{x}, \quad x(4) = 5$

20. $\frac{dx}{dt} = \cos^2 x \sin^2 t, \quad x(0) = \frac{\pi}{4}$

Find the general solution of each of the following differential equations.

21. $f'' - f' - 12f = 0$

22. $f'' + 2f' - 8f = 0$

23. $f''' + 3f'' + 2f' = 0$

24. $f''' - f' = 0$

25. $f'' + 6f' + 9f = 0$

26. $f''' + 3f'' + 3f' + f = 0$

27. $f''' + 2f'' + f' = 0$

28. $f''' + f'' - f' - f = 0$

29. $f^{(4)} - 3f^{(3)} + 3f'' - f' = 0$

30. $f^{(6)} + 6f^{(5)} + 12f^{(4)} + 8f^{(3)} = 0$

Find a solution of each of the following differential equations that satisfies the given initial conditions.

31. $f'' - f = 0, \quad f(0) = f'(0) = 1$

32. $f'' - 4f = 0, \quad f(0) = f(1) = 4$

33. $f'' + 3f' - 4f = 0, \quad f(0) = f'(0) = 0$

34. $f'' - 6f' + 9f = 0, \quad f(0) = 0, f'(0) = 1$

35. $f'' = 0, \quad f(1) = 0, f(3) = 2$

36. $f''' = 0, \quad f(0) = 0 = f'(0), f''(0) = 2$

37. $f''' = 0, \quad f(1) = f'(1) = f''(1) = 0$

38. $f''' = 0, \quad f(1) = 0 = f'(1), f''(1) = 4$

39. $f''' - 2f'' - 15f' = 0, \quad f(0) = f'(0) = 0, f''(0) = 1$

40. $f' = f, \quad f(1) = 0 = f'(1), f'(0) = 0$

II

For each of the following equations, sketch several solutions, that is, find the general solution and then sketch the curve you get by picking several different values of the constant(s). Be careful to label the axes correctly.

41. $\frac{dx}{dt} = x$

42. $\frac{dx}{dt} = x^2$

43. $f'' - f' = 0$

44. $f'' = 0$

45. $f''' = 0$

46. $f'' = 2$

47. Suppose m is a positive integer and consider the differential equation $f^{(m)} = 0$.

(i) What is its general solution?

(ii) Find the particular solution that satisfies $f(0) = f'(0) = \ldots = f^{(m-1)}(0) = 0$.

48. Find the particular solution of $f^{(m)} = 0$ that satisfies $f(1) = f'(1) = \ldots = f^{(m-1)}(1) = 0$.

III

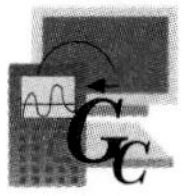

49. Find the general solution $f(t) = ae^{ct} + be^{dt}$ of $f'' + 2f' - 3f = 0$, with $c < d$. Use a graphics calculator or computer to sketch on one picture the solutions with $a = b = 1$; $a = b = 2$; $a = b = -1$; $a = 0, b = 1$; $a = 1, b = 0$; $a = -1, b = 1$; $a = 1, b = -1$.

50. The general solution of $f''' - 2f'' - 8f' = 0$ is $f(t) = a + be^{4t} + ce^{-2t}$. Sketch the graph of $2 + 5e^{-2t}$ (i.e., $a = 2, b = 0, c = 5$). Now keeping $a = 2$ and $c = 5$ fixed, sketch the graph of $2 + be^{4t} + 5e^{-2t}$ for several values of b. By trial and error, using the graphs, try to find the value of b for which $f(0) = f(1)$. Can you calculate the correct value of b?

51. Find and sketch the solution of $f''' - f'' - f' + f = 0$ satisfying $f(0) = f(1) = 1, f'(0) = -e$.

52. Find and sketch the solution of $f''' - f'' - f' + f = 0$ satisfying $f(0) = 0, f(1) = 1, f(-1) = 0$.

Find the general solution of each of the following differential equations.

53. $f''' - 4f' + f = 0$

54. $f^{(4)} - 5f^{(2)} + f^{(1)} + f = 0$

55. $f^{(5)} - 5f^{(3)} + 2f^{(2)} = 0$

56. $f^{(4)} - 5f^{(2)} + f^{(1)} = 0$

Chapter Summary

§7.1
natural logarithm
$e \approx 2.71828$

§7.2
exponential function
powers: x^a (Definition 7.2.3, p. 517)
base
logarithm to the base b
$\log_b(x)$
common logarithm

§7.3
exponential growth
growth constant
exponential decay
decay constant
initial value
half-life

§7.4
isotope
factor
compounding
nominal interest rate
effective annual interest rate
continuous compounding
present value
ambient temperature
Newton's Law of Cooling (Example 16, p. 540)

§7.5
logistic equation (Equation 7.5.1, p. 546)
logistic growth
carrying capacity

§7.6
separable equation
general solution
particular solution
auxiliary equation
order
characteristic polynomial

Review Exercises

I

Evaluate each of the following.

1. $\frac{d}{dx}\left(x^{2x}\right)$
2. $\frac{d}{dx}\left((2x)^x\right)$
3. $\frac{d}{dx}\left(x^{3x+5}\right)$
4. $\frac{d}{dx}\left(x^{x^2}\right)$
5. $\frac{d}{dx}\left(x^{\log_2(x)}\right)$
6. $\frac{d}{dx}\left((1+x^2)^{\sin x}\right)$
7. $\lim_{x\to 0^+}\left(x^{-\ln x}\right)$
8. $\lim_{x\to\infty}\left(x^{1/\ln x}\right)$
9. $\lim_{x\to 1^+}(\ln x)^{\ln x}$
10. $\lim_{x\to 0^+}(-\ln x)^x$

Find each of the following.

11. The half-life of a substance that diminishes by 5% in 10 minutes.
12. How long it takes a substance to reduce by 20% if its half-life is 3 years.
13. The size after 10 years of an exponentially growing population that begins at 10,000 and grows by 80 in one month.
14. The initial size of an exponentially growing population that is 10,000 after 8 years and 11,000 after 11 years.
15. The temperature after 10 minutes of an object that cools from 80°C to 70°C in 1 min with an ambient temperature of 20°C.
16. The initial temperature of an object that is at 50°F after 3 min and 48°F after 4 min with an ambient temperature of 40°F.
17. The length of time for a bank account to double with 8% interest compounded continuously.
18. The value after 100 months of an initial deposit of \$1000 with 6% annual interest compounded continuously.
19. The present value of a contract to make eight payments of \$400, one today and one every three months after that, assuming 9.5% interest compounded continuously.
20. The present value of a contract that will pay Josie \$200 each month for one year, starting two months from today, in return for her promise to pay back \$3000 three years from today, assuming 12% interest, compounded monthly.

Find the general solution of each of the following differential equations.

21. $f'' + f' - f = 0$
22. $f'' - 8f' + 16f = 0$
23. $f''' + 3f'' = 0$
24. $f^{(4)} = 9f^{(2)}$
25. $\frac{dx}{dt} = x^4 \sec^2 t$
26. $\frac{dx}{dt} = -7x$

27. $t^4(t+1)^2\frac{du}{dt} = t^4 + t^2 + 2t + 1$

28. $\frac{d^2v}{dt^2} = \frac{dv}{dt} + 6v$

29. $\frac{dy}{dx} = -(x+1)y$

30. $\frac{dP}{dt} = P - P^2$

In each of the following situations, find a solution of the differential equation that satisfies the given initial conditions.

31. $f'' + f' - 6f = 0,\ f(0) = 1,\ f(1) = -1$

32. $2f'' + 5f' + 2f = 0,\ f(1) = f'(1) = 4$

33. $f''' - 16f' = 0,\ f(0) = f(1) = f(2) = 0$

34. $\frac{f}{f'} = 2,\ f(1) = 4$

35. $f'' = 25f,\ f(0) = 1,\ f'(0) = 5$

36. $f'' - f' - 2f = 0,\ f(0) = f'(0),\ f(1) = 4$

37. $f^{(4)} = f^{(2)},\ f(0) = 1,\ f'(0) = f''(0) = f'''(0) = 0$

38. $f''' - 2f'' + f' = 0,\ f(0) = f'(0) = f''(0) = 1$

39. $f'' = \pi^2 f,\ f(0) = 1,\ f'(0) = -1$

40. $f''' + 8f = 0,\ f(0) = 1$

II

41. If $f(x)$ is a nonzero differentiable function, verify the formula for $\frac{d}{dx}(f^n)$ using logarithmic differentiation.

42. If $f(x)$ is a differentiable function, verify the formula for $\frac{d}{dx}(e^{f(x)})$ using logarithmic differentiation.

*43. Which is bigger: $(10^{10})^{10}$ or $10^{(10^{10})}$?

*44. (i) If $a > 1$, which of $(a^a)^a$ and $a^{(a^a)}$ is bigger and why?
(ii) What happens if $0 < a < 1$?

*45. A population growing according to a logistic curve begins at exactly half the carrying capacity. One year later it has increased by 2%. How long will it take to reach 75% of the carrying capacity?

*46. A small warm object following Newton's law of cooling begins at 150°F. One minute later it is at 125°F, and one minute after that it is at 115°F. What is the ambient temperature?

*47. (Spread of Rumors) Social scientists study how rumors spread (and similar analyses may apply to the spread of cultural changes and some epidemics).

Consider a population of P people, and let $f(t)$ be the number who have heard the rumor at time t. The more people who have heard the rumor, the more opportunities they will have to pass it on. However, these opportunities also depend on the number of people who have not heard it yet, which is $P - f(t)$. So it seems reasonable that the rate at which new people hear the rumor should be proportional to the product $f(t)\left(P - f(t)\right)$, that is, $f'(t) = Kf(t)\left(P - f(t)\right)$, for some constant K.

Consider a small town with population $P = 10{,}000$. One hundred people hear about a scandal at work. One day later 300 people know about it. How long does our theory predict it will take for half the town to know about it? How long for 90% of the town to know?

Techniques of Integration 8

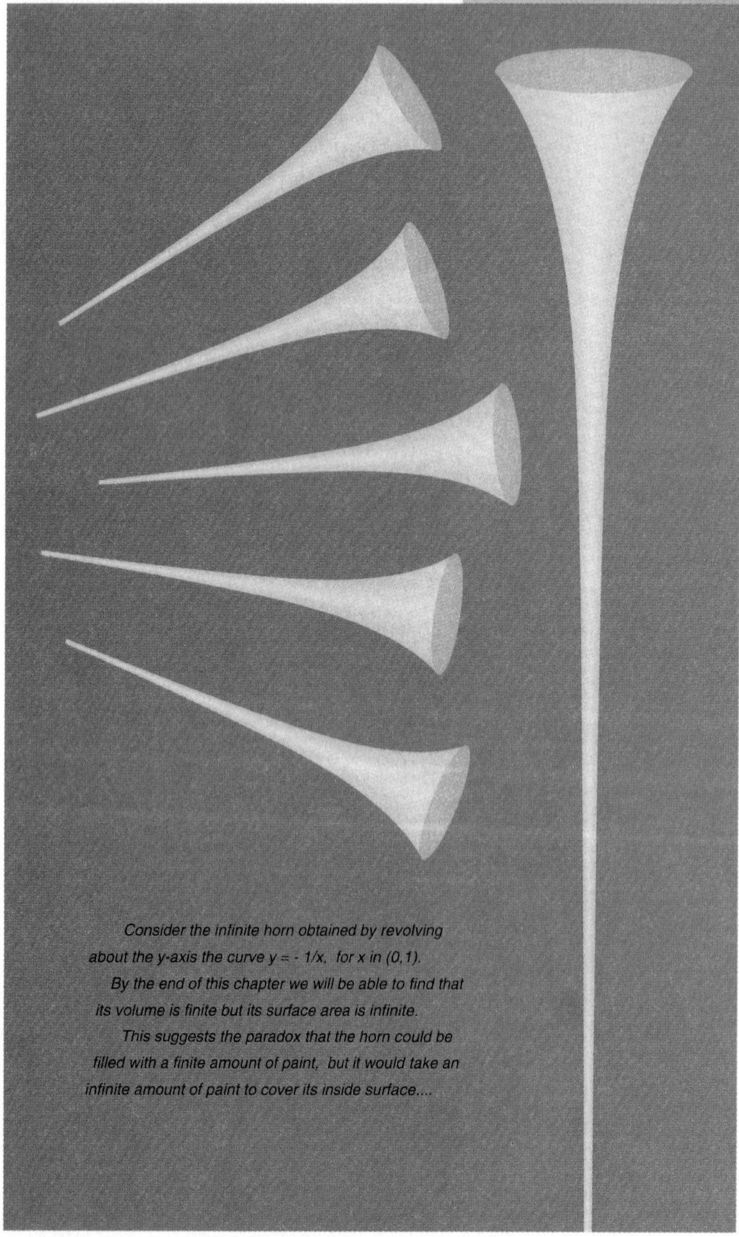

Consider the infinite horn obtained by revolving about the y-axis the curve $y = -1/x$, for x in $(0,1)$.
By the end of this chapter we will be able to find that its volume is finite but its surface area is infinite.
This suggests the paradox that the horn could be filled with a finite amount of paint, but it would take an infinite amount of paint to cover its inside surface....

INTRODUCTION

We have seen that the Fundamental Theorem of Calculus (Theorem 5.6.1) gives a good way to evaluate definite integrals:

$$\int_a^b f(x)\,dx = F(x)\Big|_a^b = F(b) - F(a),$$

where $F(x)$ is an antiderivative of the integrand $f(x)$, that is, $F'(x) = f(x)$. In Section 5.7 we discussed the technique of substitution, which allowed us to extend considerably the list of functions we could integrate.

This chapter is devoted to learning a number of techniques that will allow us to integrate many additional types of functions.

It is important to mention that tables of integrals are available. A short list appears inside the covers of this book, and more comprehensive ones can be found in your library. But no table can contain all the integrals you will ever encounter. The idea is to use the table in conjunction with the technique of substitution and the other techniques we shall learn; in this way you will greatly expand your capacity to find indefinite integrals.

The techniques to be learned in this chapter take time and practice to become familiar, and the only way to master them is to do a great many examples and exercises. Probably more so than in any other chapter of this book, it is absolutely *vital* that you work out a large number of problems for yourself. Ask for help, discuss them with your friends, but ultimately you will learn how to integrate by doing problems on your own.

Remember that when you look up an answer in the back of the book or compare with somebody else's work it is possible to have two apparently different answers that are really the same, or the same apart from the addition of a constant. So be careful not to discard your work immediately if it does not match exactly. The good news is that it is always possible to check an integral easily by differentiating.

The exercises in each section are largely ones that use the techniques learned in that section. However, each exercise set ends with a Drill, which is a list of integrals that may require any of the methods learned up to that point. The idea, of course, is that part of the problem of integrating is recognizing which technique to use, and this drill will help you to practice.

The various sections of this chapter are heavily interrelated. Section 8.1, "Integration by Parts," discusses the integral version of the product rule and its use for simplifying integrals. Section 8.2 discusses certain integrals of trigonometric functions and uses several results from Section 8.1. Section 8.3 describes a technique for converting other types of integrals into integrals of trigonometric functions, for which techniques from Section 8.2 are frequently required. The ideas developed in Section 8.4 convert another class of integrals into the kind that are discussed in Section 8.3. Section 8.5 describes how to integrate rational functions. This can always be done, assuming that we know how to factor the denominator of the integrand. In some cases it is necessary to make use of results from Sections 8.3 and 8.4. Section 8.6 describes a trick that converts integrals of trigonometric functions into integrals of rational functions, which can be evaluated as in Section 8.5. The trick is complicated, but it may be necessary in cases in which Section 8.2 does not apply. Section 8.7 discusses some applications, mostly ones we could have done in Chapter 6 except for being unable to evaluate the integrals, something we are now able to do. Finally, Section 8.8 discusses improper integrals, integrals in which the integrand or limits of integration are allowed to become infinite.

Before beginning, we list a number of integrals we already know. Most of them are obtained from familiar rules for differentiation. Some of the formulas contain a constant a; it is often easier to recognize the formula in the case with $a = 1$ (the general case then follows by a simple substitution).

Ignore the last four formulas if you have not studied hyperbolic functions.

$\int x^r\,dx = \frac{x^{r+1}}{r+1} + c, \text{ if } r \neq -1$	$\int \frac{1}{x}\,dx = \ln\|x\| + c$
$\int a\,dx = ax + c$	$\int e^{ax}\,dx = \frac{1}{a}e^{ax} + c$
$\int \sin ax\,dx = -\frac{1}{a}\cos ax + c$	$\int \cos ax\,dx = \frac{1}{a}\sin ax + c$
$\int \sec^2 ax\,dx = \frac{1}{a}\tan ax + c$	$\int \sec ax \tan ax\,dx = \frac{1}{a}\sec ax + c$
$\int \csc^2 ax\,dx = -\frac{1}{a}\cot ax + c$	$\int \csc ax \cot ax\,dx = -\frac{1}{a}\csc ax + c$
$\int \tan ax\,dx = -\frac{1}{a}\ln\|\cos ax\| + c$	$\int \frac{1}{a^2+x^2}\,dx = \frac{1}{a}\arctan\frac{x}{a} + c$
$= \frac{1}{a}\ln\|\sec ax\| + c$	$\int \frac{1}{x\sqrt{x^2-a^2}}\,dx = \frac{1}{a}\operatorname{arcsec}\frac{x}{a} + c$
$\int \cot ax\,dx = \frac{1}{a}\ln\|\sin ax\| + c$	$\int \frac{1}{\sqrt{a^2-x^2}}\,dx = \arcsin\frac{x}{a} + c$
$\int \cos^2 ax\,dx = \frac{x}{2} + \frac{1}{4a}\sin 2ax + c$	$\int \sin^2 ax\,dx = \frac{x}{2} - \frac{1}{4a}\sin 2ax + c$
$\int \sinh ax\,dx = \frac{1}{a}\cosh ax + c$	$\int \cosh ax\,dx = \frac{1}{a}\sinh ax + c$
$\int \operatorname{sech}^2 ax\,dx = \frac{1}{a}\tanh ax + c$	$\int \operatorname{sech} ax \tanh ax\,dx = -\frac{1}{a}\operatorname{sech} ax + c$

8.1 Integration by Parts

DEFINITE INTEGRALS BY PARTS
REDUCTION FORMULAS

The technique of substitution (Section 5.7) is in effect the integral version of the Chain Rule. We now consider the integral version of the Product Rule. The integral $\int f(x)g(x)\,dx$ of a product of two functions does not in general equal the product of the integrals of $f(x)$ and $g(x)$ (e.g., try it with $f(x) = x$ and $g(x) = x \ldots$).

The Product Rule (Theorem 3.2.5) says

$$(uv)' = u'v + uv'.$$

Rearranging, we have $uv' = (uv)' - vu'$, and if we integrate, we find

$$\int uv'\,dx = \int (uv)'\,dx - \int vu'\,dx.$$

In this formula we note that the second integral means an antiderivative of the derivative of uv, so we can just replace it with uv. The formula becomes

$$\int uv'\,dx = uv - \int vu'\,dx. \tag{8.1.1}$$

Strictly speaking, we ought to have added a constant c when we did that indefinite integration, but since the formula still contains two other indefinite integrals, we can just absorb the constant into them.

Formula 8.1.1 is the basis of the technique known as *integration by parts.*

EXAMPLE 1 Find $\int x \sin x\,dx$.

Solution We use Formula 8.1.1, letting $u = x$ and $v' = \sin x$. If $v' = \sin x$, then $v = \int \sin x\,dx = -\cos x + c$. Once again we can omit the constant c until the last step, so we take $v = -\cos x$. Then

$$\begin{aligned}\int x \sin x\,dx &= \int uv'\,dx = uv - \int vu'\,dx \\ &= x(-\cos x) - \int (-\cos x)(1)\,dx = -x\cos x + \int \cos x\,dx \\ &= -x\cos x + \sin x + c.\end{aligned}$$

It is easy to check by differentiating that this is indeed an antiderivative of $x \sin x$. Notice that we did add the constant c at the end of the calculation.

It is possible to write Formula 8.1.1 for integration by parts slightly differently using differentials. Noting that $u'(x)\,dx = du$ and $v'(x)\,dx = dv$, we can write the formula as

$$\int u\,dv = uv - \int v\,du.$$

EXAMPLE 2 Find $\int x^2 e^x\,dx$.

Solution Let $u = x^2$, $dv = e^x\,dx$. Then $du = 2x\,dx$, and $v = \int dv = \int e^x\,dx = e^x$

(again we omit constants until the end), and

$$\int x^2 e^x dx = \int u\,dv = uv - \int v\,du$$
$$= x^2 e^x - \int e^x (2x)\,dx = x^2 e^x - 2\int xe^x dx.$$

We must apply integration by parts again to evaluate the last integral. Letting $U = x$, $dV = e^x dx$, we find $dU = dx$, $V = \int dV = \int e^x dx = e^x$, so

$$\int xe^x dx = UV - \int V\,dU = xe^x - \int e^x (1)\,dx = xe^x - e^x.$$

From this we find that the original integral equals

$$\int x^2 e^x dx = x^2 e^x - 2\int xe^x dx = x^2 e^x - 2(xe^x - e^x) + c$$
$$= x^2 e^x - 2xe^x + 2e^x + c.$$

Notice the constant c in the final answer. In this example it was necessary to use integration by parts twice to get the final answer.

THEOREM 8.1.2

Integration by Parts

$$\int uv'\,dx = uv - \int vu'\,dx$$
$$\int u(x)v'(x)\,dx = u(x)v(x) - \int v(x)u'(x)\,dx$$
$$\int u\,dv = uv - \int v\,du.$$

These formulas hold on any interval where u and v are differentiable with continuous derivatives.

To apply integration by parts to an indefinite integral $\int f(x)\,dx$, it is necessary to break $f(x)$ into two "parts," $f = uv'$. This has to be done in such a way that v' can be integrated, that is, it must be possible to find $v = \int v'\,dx$. Moreover, the integral $\int vu'\,dx = \int v\,du$ must be easier to integrate than $\int uv'\,dx = \int u\,dv$. As with substitutions, it takes practice and experience to be able to see a suitable u and v. We have also seen that it may be necessary to apply the process more than once.

EXAMPLE 3 (i) $\int \cos^2 x\,dx$.

Solution We have already seen (Example 5.7.5, p. 405) how to integrate using the half-angle formula, but we try integrating by parts, letting $u = \cos x$, $v' = \cos x$. So $v = \sin x$, and $\int \cos^2 dx = \int u\,dv = uv - \int v\,du = \cos x \sin x - \int \sin x(-\sin x)\,dx = \cos x \sin x + \int \sin^2 x\,dx$.

This does not seem to be much of an improvement, but we write $\sin^2 x = 1 - \cos^2 x$, so

$$\int \cos^2 x\,dx = \cos x \sin x + \int (1-\cos^2 x)\,dx$$
$$= \cos x \sin x + \int 1\,dx - \int \cos^2 x\,dx$$
$$= \cos x \sin x + x - \int \cos^2 x\,dx.$$

Moving the last integral from the right side to the left, we find

$$2\int \cos^2 x\,dx = \cos x \sin x + x + c', \quad \text{or} \quad \int \cos^2 x\,dx = \frac{1}{2}\cos x \sin x + \frac{x}{2} + c.$$

(Notice how we added c' when we combined the two integrals and then let $c = \frac{1}{2}c'$.)

(ii) $\int e^x \cos x\,dx$.

Solution Letting $u = e^x$, $v' = \cos x$, so $u' = e^x$, $v = \sin x$, we find

$$\int e^x \cos x\,dx = uv - \int vu'\,dx = e^x \sin x - \int e^x \sin x\,dx.$$

This does not seem too useful, but let us proceed and do it again, this time with $U = e^x$, $V' = \sin x$, $U' = e^x$, $V = -\cos x$. So

$$\int e^x \sin x\,dx = UV - \int VU'\,dx = -e^x \cos x + \int e^x \cos x\,dx.$$

So

$$\int e^x \cos x\,dx = e^x \sin x - \int e^x \sin x\,dx$$
$$= e^x \sin x - \left(-e^x \cos x + \int e^x \cos x\,dx\right)$$
$$= e^x \sin x + e^x \cos x - \int e^x \cos x\,dx.$$

Moving the last integral to the left side gives

$$2\int e^x \cos x\,dx = e^x \sin x + e^x \cos x + c',$$

so

$$\int e^x \cos x\,dx = \frac{1}{2}e^x(\sin x + \cos x) + c,$$

being careful, as usual, to add the constant.

This example illustrates an idea that can often be used with integrals involving $\cos x$ or $\sin x$. Performing an integration by parts can change a $\cos x$ into a $\sin x$ or vice versa (perhaps with a minus sign). Doing it twice or applying a trigonometric identity may change it back again. But it often happens that the two integrals do not cancel. (In both parts of the above example they had opposite signs on opposite sides of the equation, so they added up to twice the original integral.)

EXAMPLE 4 Supposing n is a positive integer, find $\int \cos^n x\,dx$.

Solution Write the integrand as $\cos^n x = (\cos^{n-1} x)(\cos x)$, and integrate by parts, with $u = \cos^{n-1} x$ and $dv = \cos x\,dx$; so $v = \sin x$. We find that

$$\begin{aligned}\int \cos^n x\,dx &= \int \cos^{n-1} x \cos x\,dx = \int u\,dv = uv - \int v\,du \\ &= \cos^{n-1} x \sin x - \int (\sin x)(n-1)(\cos^{n-2} x)(-\sin x)\,dx \\ &= \cos^{n-1} x \sin x + (n-1)\int \cos^{n-2} x \sin^2 x\,dx \\ &= \cos^{n-1} x \sin x + (n-1)\int \cos^{n-2} x(1-\cos^2 x)\,dx \\ &= \cos^{n-1} x \sin x + (n-1)\int \cos^{n-2} x\,dx - (n-1)\int \cos^n x\,dx.\end{aligned}$$

Moving the last integral to the left side, we find

$$\int \cos^n x\,dx + (n-1)\int \cos^n x\,dx = \cos^{n-1} x \sin x + (n-1)\int \cos^{n-2} x\,dx,$$

or

$$n\int \cos^n x\,dx = \cos^{n-1} x \sin x + (n-1)\int \cos^{n-2} x\,dx,$$

so

$$\int \cos^n x\,dx = \frac{1}{n}\cos^{n-1} x \sin x + \frac{n-1}{n}\int \cos^{n-2} x\,dx.$$

This does not tell us how to integrate $\cos^n x$ because the formula still contains an unevaluated integral. However, that integral is $\int \cos^{n-2} x\,dx$, with the exponent of $\cos x$ reduced by 2. So for instance, if $n = 3$ or $n = 4$, the unevaluated integral would be the integral of $\cos x$ or $\cos^2 x$, respectively, either of which we know how to do.

Whatever n we start with, we can apply the formula to reduce the integral to something involving the integral of $\cos^{n-2} x$. If necessary, we can do it again and reduce to the integral of $\cos^{n-4} x$. Continuing as many times as needed, we eventually get to the integral of $\cos x$ or the integral of $\cos^2 x$, either of which we can do.

So this formula, which *reduces* the integral to a simpler one, provides a means for evaluating the integral of any power $\cos^n x$, even though it does not provide an actual formula for every n. A formula of this type is called a **reduction formula**. An exactly parallel computation establishes a similar formula for the integral of $\sin^n x$.

THEOREM 8.1.3

Reduction Formulas

Suppose n is a positive integer. Then

$$\int \cos^n x\,dx = \frac{1}{n}\cos^{n-1} x \sin x + \frac{n-1}{n}\int \cos^{n-2} x\,dx,$$

$$\int \sin^n x\,dx = -\frac{1}{n}\sin^{n-1} x \cos x + \frac{n-1}{n}\int \sin^{n-2} x\,dx.$$

EXAMPLE 5 (i) Evaluate $\int \cos^5 x\,dx$, (ii) $\int \sin^4 x\,dx$.

Solution

(i) The reduction formula for $\cos^n x$ shows that $\int \cos^5 x\,dx = \frac{1}{5}\cos^4 x \sin x + \frac{4}{5}\int \cos^3 x\,dx$. Now it is necessary to perform another reduction for the integral of $\cos^3 x$.

We find $\int \cos^3 x\,dx = \frac{1}{3}\cos^2 x \sin x + \frac{2}{3}\int \cos x\,dx = \frac{1}{3}\cos^2 x \sin x + \frac{2}{3}\sin x$. Substituting this into the previous expression, we find that

$$\begin{aligned}\int \cos^5 x\,dx &= \frac{1}{5}\cos^4 x \sin x + \frac{4}{5}\int \cos^3 x\,dx \\ &= \frac{1}{5}\cos^4 x \sin x + \frac{4}{5}\left(\frac{1}{3}\cos^2 x \sin x + \frac{2}{3}\sin x\right) + c \\ &= \frac{1}{5}\cos^4 x \sin x + \frac{4}{15}\cos^2 x \sin x + \frac{8}{15}\sin x + c.\end{aligned}$$

(ii) Similarly, $\int \sin^4 x\,dx = -\frac{1}{4}\sin^3 x \cos x + \frac{3}{4}\int \sin^2 x\,dx$. We could evaluate this last integral using Example 5.7.5, but we can also just apply the reduction formula again: $\int \sin^2 x\,dx = -\frac{1}{2}\sin x \cos x + \frac{1}{2}\int \sin^0 x\,dx = -\frac{1}{2}\sin x \cos x + \frac{1}{2}\int 1\,dx = -\frac{1}{2}\sin x \cos x + \frac{x}{2}$.

Combining, we find

$$\begin{aligned}\int \sin^4 x\,dx &= -\frac{1}{4}\sin^3 x \cos x + \frac{3}{4}\int \sin^2 x\,dx \\ &= -\frac{1}{4}\sin^3 x \cos x - \frac{3}{8}\sin x \cos x + \frac{3x}{8} + c.\end{aligned}$$

It is also possible to use integration by parts for definite integrals. After all, to evaluate $\int_a^b uv'\,dx$, we need an antiderivative of uv', and the formula tells us that we can use $F(x) = uv - \int vu'\,dx$.

THEOREM 8.1.4

Integration by Parts for Definite Integrals

$$\int_a^b u(x)v'(x)\,dx = u(x)v(x)\Big|_a^b - \int_a^b v(x)u'(x)\,dx.$$

This formula holds whenever u and v are differentiable on $[a, b]$ and u' and v' are continuous on $[a, b]$.

EXAMPLE 6 Evaluate $\int_1^4 x \ln x\,dx$.

Solution A first guess might be to let $u = x$, since du would then be easy to find, but in that case we would have $v' = \ln x$, which we do not know how to integrate to find v. A second guess, which seems a little peculiar, is to let $u = \ln x$ and $v' = x$.

We summarize this information in a box beside the calculation:

$u = \ln x, \quad dv = x\,dx$
$du = \frac{1}{x}dx, \quad v = \frac{1}{2}x^2$

$$\begin{aligned}\int_1^4 x\ln x\,dx &= \frac{1}{2}x^2\ln x\Big|_1^4 - \int_1^4 \frac{1}{2}x^2\left(\frac{1}{x}\right)dx\\ &= \frac{1}{2}(4^2)\ln(4) - \frac{1}{2}(1^2)\ln(1) - \frac{1}{2}\int_1^4 x\,dx\\ &= 8\ln(4) - \frac{1}{2}(0) - \frac{1}{4}x^2\Big|_1^4\\ &= 8\ln(4) - \left(4 - \frac{1}{4}\right)\\ &\approx 7.340.\end{aligned}$$

In Example 6 we found that it is sometimes necessary to make unexpected choices for u and v' in order to integrate by parts.

EXAMPLE 7 $\int \ln x\,dx$.

Solution We do not know how to integrate $\ln x$, so we certainly cannot use it for v'. We might try letting it be u, but then what is dv? The answer is almost too easy: Let $dv = dx$. Strangely enough, it works:

$u = \ln x, \quad dv = dx$
$du = \frac{1}{x}\,dx, \quad v = x$

$$\begin{aligned}\int \ln x\,dx &= x\ln x - \int x\left(\frac{1}{x}\right)dx\\ &= x\ln x - \int 1\,dx\\ &= x\ln x - x + c.\end{aligned}$$

This trick will not always work, but if we are faced with an unknown integrand, we can try letting it be u and letting $dv = dx$. The result will be a formula involving the integral of xu', which may not be an improvement, but there are occasions when it is. It is particularly useful for the integrals of inverse trigonometric functions.

EXAMPLE 8 $\int_0^1 \arctan x\,dx$.

Solution Following the suggestion made above,

$u = \arctan x, \quad v' = 1$
$u' = \frac{1}{1+x^2}, \quad v = x$

$$\begin{aligned}\int_0^1 \arctan x\,dx &= uv\Big|_0^1 - \int_0^1 vu'\,dx\\ &= x\arctan x\Big|_0^1 - \int_0^1 x\frac{1}{1+x^2}\,dx\\ &= 1\left(\frac{\pi}{4}\right) - 0(0) - \int_0^1 \frac{x}{1+x^2}\,dx.\end{aligned}$$

For this last integral we make the substitution $U = 1 + x^2$.

$$\int_0^1 \arctan x\,dx = \frac{\pi}{4} - \int_0^1 \frac{x}{1+x^2}\,dx$$

$U = 1 + x^2$
$dU = 2x\,dx$

$$= \frac{\pi}{4} - \int_1^2 \frac{1}{U}\frac{1}{2}\,dU$$

$$= \frac{\pi}{4} - \frac{1}{2}\ln U\Big|_1^2 = \frac{\pi}{4} - \frac{1}{2}\ln(2)$$

$$\approx 0.439.$$

Integration by parts can often be used for an integral of the form $\int xf(x)\,dx$, letting $u = x$, $v' = f(x)$. Similarly, $\int x^2 f(x)\,dx$ suggests letting $u = x^2$, $v' = f(x)$, and for $\int x^k f(x)\,dx$ we can try $u = x^k$, $v' = f(x)$. These approaches reduce the power of x but depend on being able to integrate $f(x)$. It is important not to forget that integrals like these sometimes require integration by substitution. We integrate $\int x\cos(x^2)\,dx$ by letting $u = x^2$, $du = 2x\,dx$, so $\int x\cos(x^2)\,dx = \frac{1}{2}\int \cos u\,du = \frac{1}{2}\sin u + c = \frac{1}{2}\sin(x^2) + c$; integration by parts will not work.

Exercises 8.1

I

Evaluate each of the following indefinite integrals.

1. $\int xe^{-x}dx$
2. $\int x\sin(2x)\,dx$
3. $\int x^2e^{3x}dx$
4. $\int \cos^4 t\,dt$
5. $\int \sin^5(2u)\,du$
6. $\int x^2\cos(3x)\,dx$
7. $\int x\sec^2 x\,dx$
8. $\int e^x \sin x\,dx$
9. $\int (x^2+3x-2)e^x dx$
10. $\int \ln(1+4x)\,dx$
11. $\int \arcsin x\,dx$
12. $\int x\ln x\,dx$
13. $\int 4x/(1+x^2)\,dx$
14. $\int x\ln(x^4)\,dx$
15. $\int x^2\ln(x^3)\,dx$
16. $\int e^{ax}\cos(bx)\,dx$
17. $\int x^2\ln x\,dx$
18. $\int (\ln x)^2 dx$
19. $\int e^{2x}\big(3-\sin(4x)\big)\,dx$
20. $\int 2x\arctan x\,dx$

Evaluate each of the following definite integrals.

21. $\int_{-1}^1 x^2e^{-2x}dx$
22. $\int_1^3 x^3\ln x\,dx$
23. $\int_0^1 e^{2x}\cos \pi x\,dx$
24. $\int_0^{2\pi} \sin x\cos(2x)\,dx$
25. $\int_{-\pi}^{\pi} \sin(2x)\sin(3x)\,dx$
26. $\int_1^4 \ln x\,dx$
27. $\int_0^2 t^2\sin\big(\pi(1-2t)\big)\,dt$
28. $\int_1^4 (\ln u)^3 du$
29. $\int_1^2 \sqrt{x}\arctan(\sqrt{x})\,dx$
30. $\int_0^2 x^2\cos^4(\pi x^3)\,dx$

II

31. Establish the reduction formula for $\int \sin^n x\,dx$ given in Theorem 8.1.3 (p. 565).

*32. (i) Let n be any integer except -1. Find $\int x^n\ln x\,dx$.
(ii) What happens if $n = -1$? Evaluate $\int \frac{1}{x}\ln x\,dx$.

*33. Let n be any positive integer. Find a reduction formula for $\int (\ln x)^n dx$. Can you use it to get a final formula for $\int (\ln x)^n dx$?

*34. Suppose n is an integer with $n \geq 3$. Find a reduction formula for $\int \sec^n x\,dx$. You may wish to use the identity $\tan^2 x = \sec^2 x - 1$ (cf. Equation 1.7.3).

*35. Establish the following reduction formula:

$$\int x^n\cos x\,dx = x^n\sin x - n\int x^{n-1}\sin x\,dx.$$

*36. Integrate $\int x^4\cos x\,dx$.

*37. Establish a reduction formula for $\int x^n\sin x\,dx$.

*38. Integrate $\int t^5\sin t\,dt$.

*39. Establish the following reduction formula:

$$\int x^n e^{-x^2}dx = -\frac{1}{2}x^{n-1}e^{-x^2} + \frac{n-1}{2}\int x^{n-2}e^{-x^2}dx.$$

*40. Establish a reduction formula for $\int x^n e^x dx$.

**41. (i) Suppose n and m are positive integers with $m \neq n$. Evaluate

$$\int_0^{2\pi} \cos(m\theta)\cos(n\theta)\,d\theta.$$

(ii) What happens if we evaluate the integral with $m = -n$?

**42. Suppose n and m are integers. Evaluate $\int_0^{2\pi} \cos(m\theta)\sin(n\theta)\,d\theta$.

Drill

43. $\int \sin^2 x \cos^2 x\,dx$

44. $\int \sin x \cos x\,dx$

45. $\int x \sin \frac{x}{2}\,dx$

46. $\int (x+3)e^{2x}dx$

47. $\int x^2 \sin(x^3+4)\,dx$

48. $\int xe^{x^2}\,dx$

49. $\int \frac{x}{e^{5x}}\,dx$

50. $\int \frac{x}{1+x}\,dx$

51. $\int \cos x \sin(2x)\,dx$

52. $\int (\cos^2 x - \sin^2 x)\,dx$

8.2 Trigonometric Integrals

INTEGRATING $\cos^m x \sin^n x$

INTEGRATING $\tan^m x \sec^n x$

We have already learned how to evaluate many integrals of trigonometric functions, including powers of $\sin x$ and $\cos x$, certain products of powers of $\sin x$ and $\cos x$, and various other kinds. The first technique we used was substitution, which covers many cases, especially if it is supplemented with clever use of trigonometric identities (cf. Examples 5.7.3(i), (vi) (p. 401); 5.7.4(ii); 5.7.5, and 5.7.6).

In Section 8.1 we saw how integration by parts can be used to establish reduction formulas that enable us to do many more integrals.

In this section we try to organize these results and learn a few additional tricks. Apart from their intrinsic interest, these methods are important to us because in the next section we will learn techniques for converting many other types of integrals into integrals of trigonometric functions. This will be helpful only to the extent that we are then able to evaluate the resulting trigonometric integrals.

Before beginning, we recall the fundamental trigonometric identity $\cos^2 x + \sin^2 x = 1$ (Theorem 1.7.2). Dividing it by $\cos^2 x$, we find $1 + \tan^2 x = \sec^2 x$ (Equation 1.7.3). We record these for convenience:

$$\cos^2 x + \sin^2 x = 1, \tag{8.2.1}$$

$$1 + \tan^2 x = \sec^2 x. \tag{8.2.2}$$

Now suppose m and n are nonnegative integers and consider the integral

$$\int \cos^m x \sin^n x\,dx.$$

If either m or n is odd, it is possible to evaluate using substitution.

EXAMPLE 1 $\int \cos^4 x \sin^3 x\,dx$.

Solution Write the integrand as $\cos^4 x \sin^2 x \sin x$. The idea is to use that last $\sin x$ as the derivative (apart from a sign) of $\cos x$ and perform a substitution. To make this work, it will be necessary to express the $\sin^2 x$ in terms of $\cos x$, but nothing could be easier. We simply apply Equation 8.2.1:

$$\int \cos^4 x \sin^3 x\,dx = \int \cos^4 x(1-\cos^2 x)\sin x\,dx = \int u^4(1-u^2)(-1)\,du$$

$$\boxed{\begin{array}{l} u = \cos x \\ du = -\sin x\,dx \end{array}} \quad = \int (u^6 - u^4)\,du = \frac{1}{7}u^7 - \frac{1}{5}u^5 + c$$
$$= \frac{1}{7}\cos^7 x - \frac{1}{5}\cos^5 x + c.$$

Of course, if m is odd, similar techniques will work, using the substitution $u = \sin x$.

This leaves the more difficult situation in which m and n are both even numbers. If either m or n is zero, then we are integrating a power of $\sin x$ or $\cos x$, and the reduction formulas of the previous section will apply.

If m and n are both even and positive, then two different approaches are possible. One is to use the half-angle identities $\cos^2 x = \frac{1}{2}\big(1 + \cos(2x)\big)$ and $\sin^2 x = \frac{1}{2}\big(1 - \cos(2x)\big)$ (1.7.13). Using them it is possible to express $\int \cos^m x \sin^n x\,dx$ as a combination of integrals involving lower powers of $\cos(2x)$. These can all be done by using the reduction formulas (Theorem 8.1.3, p. 565).

Another possibility is to develop a reduction formula especially designed for this situation.

EXAMPLE 2 We integrate by parts. Making a substitution

$$\boxed{\begin{array}{ll} u = \cos^{m-1} x \sin^n x, & dv = \cos x\,dx, \\ du = \big(-(m-1)\cos^{m-2} x \sin^{n+1} x + n\cos^m x \sin^{n-1} x\big)\,dx, & v = \sin x, \end{array}}$$

we find

$$\int \cos^m x \sin^n x\,dx = \cos^{m-1} x \sin^{n+1} x + (m-1)\int \cos^{m-2} x \sin^{n+2} x\,dx$$
$$- n\int \cos^m x \sin^n x\,dx$$
$$= \cos^{m-1} x \sin^{n+1} x + (m-1)\int \cos^{m-2} x \sin^n x(1 - \cos^2 x)\,dx$$
$$- n\int \cos^m x \sin^n x\,dx$$
$$= \cos^{m-1} x \sin^{n+1} x + (m-1)\int \cos^{m-2} x \sin^n x\,dx$$
$$- (m-1)\int \cos^m x \sin^n x\,dx - n\int \cos^m x \sin^n x\,dx.$$

The last two terms are constants times the original integral.

Moving the integrals of $\cos^m x \sin^n x$ to the left side, we find that

$$(m+n)\int \cos^m x \sin^n x\,dx = \cos^{m-1} x \sin^{n+1} x + (m-1)\int \cos^{m-2} x \sin^n x\,dx,$$

so

$$\int \cos^m x \sin^n x\,dx = \frac{1}{m+n}\cos^{m-1} x \sin^{n+1} x + \frac{m-1}{m+n}\int \cos^{m-2} x \sin^n x\,dx.$$

This is the desired reduction formula. A similar calculation finds a formula that reduces the power of $\sin x$ instead of the power of $\cos x$.

THEOREM 8.2.3

Reduction Formulas

(i) $\int \cos^m x \sin^n x\,dx = \frac{1}{m+n}\cos^{m-1} x \sin^{n+1} x + \frac{m-1}{m+n}\int \cos^{m-2} x \sin^n x\,dx,$

(ii) $\int \cos^m x \sin^n x\,dx = -\frac{1}{m+n}\cos^{m+1} x \sin^{n-1} x + \frac{n-1}{m+n}\int \cos^m x \sin^{n-2} x\,dx.$

EXAMPLE 3 $\int \cos^2 x \sin^2 x\,dx.$

Solution Using Theorem 8.2.3(i), we have

$$\begin{aligned}\int \cos^2 x \sin^2 x\,dx &= \frac{1}{4}\cos x \sin^3 x + \frac{1}{4}\int \sin^2 x\,dx\\ &= \frac{1}{4}\cos x \sin^3 x + \frac{1}{4}\int \frac{1}{2}\big(1-\cos(2x)\big)\,dx\\ &= \frac{1}{4}\cos x \sin^3 x + \frac{1}{8}\left(x - \frac{1}{2}\sin(2x)\right) + c\\ &= \frac{1}{4}\cos x \sin^3 x + \frac{x}{8} - \frac{1}{8}\cos x \sin x + c.\end{aligned}$$

An alternative approach to this question is to use the identity $\sin 2x = 2\cos x \sin x$, so $\cos^2 x \sin^2 x = \frac{1}{4}\sin^2 2x$.

We summarize the approach to integrals of products of powers of $\cos x$ and $\sin x$.

Summary 8.2.4

$\int \cos^m x \sin^n x\,dx$

Suppose m and n are nonnegative integers, and consider $\int \cos^m x \sin^n x\,dx$.

1. If m or n equals zero, then the integrand is a power of $\sin x$ or $\cos x$, and the reduction formulas (Theorem 8.1.3) from the previous section can be used:

$$\int \cos^m x\,dx = \frac{1}{m}\cos^{m-1} x \sin x + \frac{m-1}{m}\int \cos^{m-2} x\,dx, \qquad (8.1.3(\text{i}))$$

$$\int \sin^n x\,dx = -\frac{1}{n}\sin^{n-1} x \cos x + \frac{n-1}{n}\int \sin^{n-2} x\,dx. \qquad (8.1.3(\text{ii}))$$

2. If m is odd, it is possible to use the substitution $u = \sin x$, $du = \cos x\,dx$. It is necessary to convert the remaining factor of $\cos^{m-1} x$ into an expression involving $\sin x$, using the identity $\cos^2 x = 1 - \sin^2 x$. (Note that $m - 1$ is even.)

   Similarly, if n is odd, it is possible to use the substitution $u = \cos x$, $du = -\sin x\,dx$. It is necessary to convert the remaining factor of $\sin^{n-1} x$ into an expression involving $\cos x$, using the identity $\sin^2 x = 1 - \cos^2 x$. (Note that $n - 1$ is even.)

3. If both m and n are even, several approaches are possible. Since $\cos^m x = (\cos^2 x)^{m/2} = (1 - \sin^2 x)^{m/2}$, the integral can be written as a combination of integrals of powers of $\sin x$. It could also be converted into powers of $\cos x$; in either case it can be integrated as in part 1 above. It is also possible to use the half-angle formulas to transform the integrand into an expression involving

powers of $\cos 2x$, which can then be integrated as in part 1. Finally, if you can remember them, it is possible to use the reduction formulas 8.2.3:

$$\text{(i)} \int \cos^m x \sin^n x \, dx = \frac{1}{m+n} \cos^{m-1} x \sin^{n+1} x + \frac{m-1}{m+n} \int \cos^{m-2} x \sin^n x \, dx,$$

$$\text{(ii)} \int \cos^m x \sin^n x \, dx = -\frac{1}{m+n} \cos^{m+1} x \sin^{n-1} x + \frac{n-1}{m+n} \int \cos^m x \sin^{n-2} x \, dx.$$

(8.2.3)

At least one of these methods will work for any integral of the form $\int \cos^m x \sin^n x \, dx$, with m and n nonnegative integers. On the other hand, there are instances in which the same integral can be done in different ways (e.g., if m and n are both odd, then it is possible to substitute $u = \sin x$ or $u = \cos x$). In a situation like this it is not unusual to get results that appear to be different, even though they are in fact the same *up to the addition of a constant.*

EXAMPLE 4 $\int \cos^3 x \sin^3 x \, dx$.

Solution One way, using $u = \sin x$, $du = \cos x \, dx$, we find

$$\begin{aligned} \int \cos^3 x \sin^3 x \, dx &= \int \cos^2 x \sin^3 x \cos x \, dx = \int (1 - \sin^2 x) \sin^3 x \cos x \, dx \\ &= \int (1 - u^2) u^3 \, du = \frac{1}{4} u^4 - \frac{1}{6} u^6 + c \\ &= \frac{1}{4} \sin^4 x - \frac{1}{6} \sin^6 x + c. \end{aligned}$$

On the other hand, a similar calculation with $u = \cos x$, $du = -\sin x \, dx$ gives us

$$\begin{aligned} \int \cos^3 x \sin^3 x \, dx &= \int \cos^3 x \sin^2 x \sin x \, dx = \int \cos^3 x (1 - \cos^2 x) \sin x \, dx \\ &= -\int u^3 (1 - u^2) \, du = -\frac{1}{4} u^4 + \frac{1}{6} u^6 + c' \\ &= -\frac{1}{4} \cos^4 x + \frac{1}{6} \cos^6 x + c'. \end{aligned}$$

Notice that we used a different constant c' for this formula.

These two answers do not appear to be the same. It is possible to show that the difference of the two formulas is a constant. In other words, if the constants c and c' are chosen in a suitable way, the formulas are in fact the same.

Indeed, it will suffice to consider the derivative of the difference $\frac{1}{6}(\cos^6 x + \sin^6 x) - \frac{1}{4}(\cos^4 x + \sin^4 x) + c' - c$. The derivative equals

$$\begin{aligned} &\frac{1}{6}\left(6 \cos^5 x(-\sin x) + 6 \sin^5 x \cos x\right) - \frac{1}{4}\left(4 \cos^3 x(-\sin x) + 4 \sin^3 x \cos x\right) \\ &\quad = \cos x \sin x (\sin^4 x - \cos^4 x + \cos^2 x - \sin^2 x) \\ &\quad = \cos x \sin x \left((\sin^2 x + \cos^2 x)(\sin^2 x - \cos^2 x) + \cos^2 x - \sin^2 x\right) \end{aligned}$$

Knowing that the difference is a constant, we can evaluate at any convenient point to find out what it is. Using $x = 0$, we find that the difference is $\frac{1}{6} - \frac{1}{4} = -\frac{1}{12}$.

$$= \cos x \sin x(\sin^2 x - \cos^2 x)(\sin^2 x + \cos^2 x - 1)$$
$$= 0.$$

The derivative is zero, so the two answers differ by a constant.

Integrals Involving Powers of tan *x* and sec *x*

Next we consider integrals of the form $\int \tan^m x \sec^n x\, dx$, where m and n are non-negative integers.

The following example provides an important and useful result. It uses a very clever trick, which is difficult to explain other than by saying that it just happens to work.

EXAMPLE 5 $\int \sec x\, dx$.

Solution The trick is to multiply the integrand by

$$\frac{\sec x + \tan x}{\sec x + \tan x}.$$

So

$$\int \sec x\, dx = \int \sec x \frac{\sec x + \tan x}{\sec x + \tan x}\, dx = \int \frac{\sec^2 x + \sec x \tan x}{\sec x + \tan x}\, dx.$$

The numerator of this integrand is the derivative of the denominator. So letting $u = \sec x + \tan x$, we find

$$\int \sec x\, dx = \int \frac{\sec^2 x + \sec x \tan x}{\sec x + \tan x}\, dx = \int \frac{du}{u} = \ln|u| + c$$
$$= \ln|\sec x + \tan x| + c.$$

We already know how to integrate tan x; in Example 5.7.3(vi) we used the substitution $u = \cos x$, $du = -\sin x\, dx$ to find

$$\int \tan x\, dx = \int \frac{\sin x}{\cos x}\, dx = \int \frac{1}{u}(-1)\, du = -\ln|u| + c = \ln|\sec x| + c.$$

These two results can be combined with reduction formulas to integrate $\int \tan^m x\, dx$ or $\int \sec^n x\, dx$. For the mth power of the tangent, with $m \geq 2$, we write

$$\int \tan^m x\, dx = \int \tan^{m-2} x \tan^2 x\, dx = \int \tan^{m-2} x(\sec^2 x - 1)\, dx$$
$$= \int \tan^{m-2} x \sec^2 x\, dx - \int \tan^{m-2} x\, dx.$$

The first of these two integrals is easily evaluated using the substitution $u = \tan x$, $du = \sec^2 x$, and we find $\int \tan^{m-2} x \sec^2 x\, dx = \int u^{m-2} du = \frac{1}{m-1} u^{m-1} + c = \frac{1}{m-1} \tan^{m-1} x + c$. Absorbing the constant, we have $\int \tan^m x\, dx = \frac{1}{m-1} \tan^{m-1} x - \int \tan^{m-2} x\, dx$.

For powers of sec x there is a reduction formula, which can be proved by integration by parts, much as Theorem 8.1.3 was done. We record these two reduction formulas below.

THEOREM 8.2.5

Reduction Formulas

Suppose m is an integer with $m \geq 2$ and n is an integer with $n \geq 2$. Then

(i) $\int \tan^m x\, dx = \frac{1}{m-1} \tan^{m-1} x - \int \tan^{m-2} x\, dx,$

(ii) $\int \sec^n x\, dx = \frac{1}{n-1} \sec^{n-2} x \tan x + \frac{n-2}{n-1} \int \sec^{n-2} x\, dx.$

EXAMPLE 6 (i) $\int \tan^3 x\, dx$, (ii) $\int \sec^4 x\, dx$.

Solution

(i) By Theorem 8.2.5(i), with $m = 3$, we find

$$\int \tan^3 x\, dx = \frac{1}{2} \tan^2 x - \int \tan x\, dx = \frac{1}{2} \tan^2 x - \ln|\sec x| + c.$$

(ii) By Theorem 8.2.5(ii), with $n = 4$, we find

$$\begin{aligned} \int \sec^4 x\, dx &= \frac{1}{3} \sec^2 x \tan x + \frac{2}{3} \int \sec^2 x\, dx \\ &= \frac{1}{3} \sec^2 x \tan x + \frac{2}{3} \tan x + c. \end{aligned}$$

By applying the reduction formulas, repeatedly if necessary, it is possible to reduce the integral of $\tan^m x$ to an integral of $\tan x$, as above, or an integral of $\tan^0 x = 1$. Similarly, the integral of any positive power $\sec^n x$ can be reduced to $\int \sec^2 x\, dx = \tan x + c$, as above, or $\int \sec x\, dx = \ln|\sec x + \tan x| + c$. In each case it is straightforward to evaluate the integral. Of course, if m or n is large, the reduction process will become quite complicated.

Finally, we turn to $\int \tan^m x \sec^n x\, dx$. If m or n is zero, the remarks above tell us what to do. Otherwise, we should look for a substitution. If n is *even*, we can write $\sec^n x = \sec^{n-2} x \sec^2 x$ and use the factor $\sec^2 x\, dx$ as du for the substitution $u = \tan x$. To make this work, it is necessary to convert the remaining $\sec^{n-2} x$ into an expression involving powers of $\tan x$, using $\sec^2 x = 1 + \tan^2 x$ (Equation 8.2.2). (Notice that $n - 2$ is even.)

If m is *odd*, we write $\tan^m x \sec^n x = \tan^{m-1} x \sec^{n-1} x \tan x \sec x$. Since $m - 1$ is even, we can use Equation 8.2.2 to convert $\tan^{m-1} x$ into an expression in powers of $\sec x$ and then evaluate the integral using the substitution $u = \sec x$, $du = \tan x \sec x\, dx$.

These two approaches will deal with every case except one in which m is even and n is odd. In this case, use Equation 8.2.2 to convert $\tan^m x$ into a sum of powers of $\sec x$, so all the resulting integrands will be powers of $\sec x$, which we know how to do.

EXAMPLE 7 (i) $\int \tan^3 x \sec^4 x\, dx$, (ii) $\int \tan^3 x \sec^3 x\, dx$, (iii) $\int \tan^2 x \sec^3 x\, dx$.

Solution

(i) One way to do this is the $u = \tan x$ substitution. In this case, $du = \sec^2 x$, and the remaining factor of $\sec^2 x$ in the integrand must be converted to an expression in tan x, using Equation 8.2.2:

$$\begin{aligned}\int \tan^3 x \sec^4 x\, dx &= \int \tan^3 x \sec^2 x \sec^2 x\, dx \\ &= \int \tan^3 x(1 + \tan^2 x) \sec^2 x\, dx \\ &= \int u^3(1 + u^2)\, du = \frac{1}{4}u^4 + \frac{1}{6}u^6 + c \\ &= \frac{1}{4}\tan^4 x + \frac{1}{6}\tan^6 x + c.\end{aligned}$$

(ii) $\int \tan^3 x \sec^3 x\, dx$. Here we can try the substitution $u = \sec x$, $du = \sec x \tan x$. This time it will be necessary to convert the remaining factor $\tan^2 x$ into an expression involving $\sec x$:

$$\begin{aligned}\int \tan^3 x \sec^3 x\, dx &= \int \tan^2 x \sec^2 x \sec x \tan x\, dx \\ &= \int (\sec^2 x - 1) \sec^2 x \sec x \tan x\, dx \\ &= \int (u^2 - 1) u^2\, du = \frac{1}{5}u^5 - \frac{1}{3}u^3 + c \\ &= \frac{1}{5}\sec^5 x - \frac{1}{3}\sec^3 x + c.\end{aligned}$$

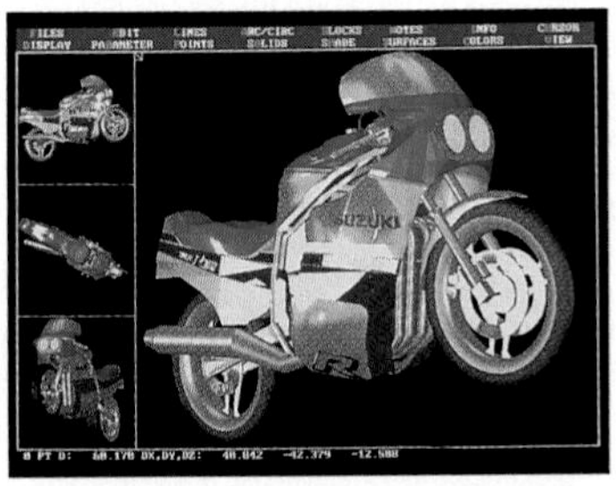

DesignCAD 3D *software was used to produce this image of a motorcycle. Computer graphics help designers visualize new products quickly and accurately, and are especially useful for trying out modifications. (Suzuki is a registered trademark of Suzuki Motor Corporation, Japan.)*

(iii) $\int \tan^2 x \sec^3 x\, dx$. Here we must change everything into $\sec x$:

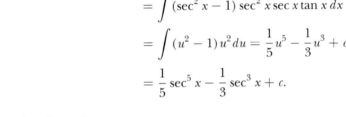

$$\int \tan^2 x \sec^3 x\, dx = \int (\sec^2 x - 1) \sec^3 x\, dx = \int \sec^5 x\, dx - \int \sec^3 x\, dx.$$

These two integrals can be evaluated by using the reduction formula of Theorem 8.2.5(ii). We find that

$$\int \sec^5 x\, dx = \frac{1}{4}\sec^3 x \tan x + \frac{3}{4}\int \sec^3 x\, dx,$$

so

$$\begin{aligned}\int \tan^2 x \sec^3 x\, dx &= \int \sec^5 x\, dx - \int \sec^3 x\, dx \\ &= \frac{1}{4}\sec^3 x \tan x - \frac{1}{4}\int \sec^3 x\, dx.\end{aligned}$$

This last integral can be evaluated by using the reduction formula again, which gives

$$\begin{aligned}\int \sec^3 x\, dx &= \frac{1}{2}\sec x \tan x + \frac{1}{2}\int \sec x\, dx \\ &= \frac{1}{2}\sec x \tan x + \frac{1}{2}\ln|\sec x + \tan x| + c.\end{aligned}$$

Finally, we can combine and find the original integral:

$$\int \tan^2 x \sec^3 x\, dx$$

$$= \frac{1}{4}\sec^3 x \tan x - \frac{1}{4}\left(\frac{1}{2}\sec x \tan x + \frac{1}{2}\ln|\sec x + \tan x| + c\right)$$

$$= \frac{1}{4}\sec^3 x \tan x - \frac{1}{8}\sec x \tan x - \frac{1}{8}\ln|\sec x + \tan x| + c'.$$

Notice that we added the constant c' in the last line. We had the constant c, but it was multiplied by $-\frac{1}{4}$, and it is simpler to write c' than $-\frac{1}{4}c$.

Summary 8.2.6

$\int \tan^m x \sec^n x\, dx$

Let m and n be nonnegative integers and consider $\int \tan^m x \sec^n x\, dx$.

1. If $m = 0$, then we are integrating a power of $\sec x$. If $n = 0$, the integrand is a power of $\tan x$. In either case it is possible to apply a reduction formula.

   Suppose m is an integer with $m \geq 2$ and n is an integer with $n \geq 2$. Then

   (i) $$\int \tan^m x\, dx = \frac{1}{m-1}\tan^{m-1} x - \int \tan^{m-2} x\, dx,$$

   (ii) $$\int \sec^n x\, dx = \frac{1}{n-1}\sec^{n-2} x \tan x + \frac{n-2}{n-1}\int \sec^{n-2} x\, dx. \qquad (8.2.5)$$

   By using these it is possible to express the answer in terms of $\int \sec^2 x\, dx = \tan x + c$ or $\int \sec x\, dx = \ln|\sec x + \tan x| + c$ (cf. Example 5) or in terms of $\int \tan x\, dx = \ln|\sec x| + c$ or $\int 1\, dx = x + c$.
2. If n is nonzero and even, it is possible to use the substitution $u = \tan x$, $du = \sec^2 x\, dx$. It is necessary to convert the remaining factor $\sec^{n-2} x$ into an expression in $\tan x$, using Equation 8.2.2.
3. If m is odd and $n \neq 0$, it is possible to use the substitution $u = \sec x$, $du = \sec x \tan x\, dx$. It is necessary to convert the remaining factor $\tan^{m-1} x$ into an expression in $\sec x$, using Equation 8.2.2.
4. If m is even and n is odd, it is possible to use Equation 8.2.2 to convert $\tan^m x$ into an expression involving $\sec x$. The integral will be a sum of terms, each of which is a constant times the integral of some power of $\sec x$. These can be done as in part 1.

We have seen how to evaluate many types of trigonometric integrals. Of course, there are others that do not fit into any of the categories we have discussed. If you encounter such an integral, you should probably first look to see whether any obvious substitution occurs to you. Failing that, it may help to write everything out in terms of $\sin x$ and $\cos x$. You should also consider integrating by parts.

Exercises 8.2

I

Integrate each of the following.

1. $\int \sin^3 x\,dx$
2. $\int \cos^4 x\,dx$
3. $\int \sin^4 x\,dx$
4. $\int \cos^5 x\,dx$
5. $\int \sec^3 x\,dx$
6. $\int \sec^4 x\,dx$
7. $\int \tan^3 x\,dx$
8. $\int \tan^4 x\,dx$
9. $\int \sin^3 2x\,dx$
10. $\int \cos^3 \frac{x}{3}\,dx$
11. $\int \cos x \sin^3 x\,dx$
12. $\int \cos^3 x \sin x\,dx$
13. $\int \cos^2 x \sin^3 x\,dx$
14. $\int \cos^4 x \sin^2 x\,dx$
15. $\int \tan^3 x \sec^2 x\,dx$
16. $\int \tan^2 x \sec^4 x\,dx$
17. $\int \tan^3 x \sec x\,dx$
18. $\int \tan^2 x \sec x\,dx$
19. $\int \cos^3 2x \sin^3 2x\,dx$
20. $\int \cos^2 x \sin^4 x\,dx$
21. $\int \cos^3 x \sin^7 x\,dx$
22. $\int \cos^4 x \sin^4 x\,dx$
23. $\int \cos^2 3x \sin^2 3x\,dx$
24. $\int (1 - \cos^2 x) \sin^4 x\,dx$
25. $\int \tan^5(3x) \sec^2(3x)\,dx$
26. $\int \tan^5 \frac{x}{2} \sec^3 \frac{x}{2}\,dx$
27. $\int \cos x \sin 2x\,dx$
28. $\int \sec^7 x \sin x\,dx$
29. $\int \tan^4 x \sin x\,dx$
30. $\int (\sin^4 x + 2\cos^2 x \sin^2 x + \cos^4 x)\,dx$

Evaluate each of the following.

31. $\int_0^{\pi} \sin^2 x\,dx$
32. $\int_0^{\pi} \sin^2 x \cos x\,dx$
33. $\int_0^{\pi/2} \sin^2 x \cos^3 x\,dx$
34. $\int_0^{\pi/2} \sin^2 x \cos x\,dx$
35. $\int_{-\pi/2}^{\pi/2} \sin^3 x \cos^2 x\,dx$
36. $\int_{-\pi}^{\pi} \sin^5 x \cos^6 x\,dx$
37. $\int_0^1 \sin^2(\pi x) \cos^2(\pi x)\,dx$
38. $\int_{-\pi/2}^{0} \sin^3 x \cos^3 x\,dx$
39. $\int_0^{2\pi} \sin^2 x \cos^3 x\,dx$
40. $\int_{-\pi}^{0} \sin^2(2x) \cos^2(2x)\,dx$

II

41. Find the integral in Example 3 (p. 571) using the half-angle formula. Compare the two answers, making sure they are the same up to the addition of a constant.

*42. Do the integral $\int \tan^3 x\,dx$ of Example 6(i) (p. 574) a different way, by writing $\tan x$ out in terms of $\cos x$ and $\sin x$ and making a substitution. Reconcile the answer you get with the answer given for Example 6(i).

*43. Use the trigonometric identity (Equation 8.2.1) to show directly that the two answers to Example 4 (p. 572) agree up to a constant. What is the constant?

**44. (i) Consider the function $\sin x + \cos x$, and show that it can be written in the form $K\cos(x + s)$ for some numbers K and s. Find K and s. (*Hint:* Think about the formula $\cos(x + s) = \cos x \cos s - \sin x \sin s$.)
(ii) Integrate $\int \frac{1}{\sin x + \cos x}\,dx$.
(iii) Assume that a and b are real numbers that are not both zero. Integrate $\int \frac{1}{a \sin x + b \cos x}\,dx$.

*45. (i) Consider the integral $\int_0^{2\pi} \cos^6 x \sin^3 x\,dx$. Argue that it must equal zero, without actually evaluating it.
(ii) Find the integral $\int_0^{2\pi} \cos^5 x \sin^8 x\,dx$.

*46. (i) Suppose m and n are nonnegative integers. Consider the integral

$$\int_{-\pi}^{\pi} \cos^m x \sin^n x\,dx.$$

Show that it equals zero if either of m and n is odd.
(ii) Argue that if both m and n are positive even integers, then $\int_0^{2\pi} \cos^m x \sin^n x\,dx > 0$.

*47. Find the arclength of the curve $y = \frac{1}{2}\ln(\cos^2 x)$ for x between $x = 0$ and $x = \frac{\pi}{4}$.

*48. Find $\int \csc^2 x \cot^3 x\,dx$.

**49. (i) Find $\int \csc x\,dx$. (ii) Find a reduction formula for $\int \csc^m x\,dx$.

*50. Describe how you would set about integrating $\int \cot^m x \csc^n x\,dx$ if m and n are nonnegative integers. (Describe different approaches for different cases.)

*51. (i) $\int \csc^2 x \cot^2 x\,dx$, (ii) $\int \csc^3 x \cot^3 x\,dx$.

*52. (i) $\int \csc^4 x \cot^3 x\,dx$, (ii) $\int \csc^3 x \cot^4 x\,dx$.

53. Find $\int \frac{1}{\cos^5 x}\,dx$.

54. Prove the reduction formula of Theorem 8.2.5(ii) (p. 574).

Drill

55. $\int \sin x \cos^3 x\,dx$

56. $\int x \sin 2x\,dx$

57. $\int x\sqrt{x^2+1}\,dx$

58. $\int x/\sqrt{x^2+1}\,dx$

59. $\int x(x^2+1)^{3/2}\,dx$

60. $\int e^{2x} \cos x\,dx$

61. $\int x \cos(x^2-3)\,dx$

62. $\int (x+5) \cos x \sin x\,dx$

63. $\int \frac{\sin^4 x}{\cos^2 x}\,dx$

64. $\int \frac{1}{1+\sin\theta}\,d\theta$

8.3 Trigonometric Substitutions

TRIGONOMETRIC IDENTITIES
RANGE OF VARIABLES

In Section 8.2 we learned how to do many trigonometric integrals, that is, integrals of trigonometric functions. That is an interesting question in its own right, but the main reason for studying it there is that we will need it for the work of this section. Here we discover ways of changing many different types of integrals into trigonometric integrals. The hope is that the resulting integrals can be done by using the techniques of the last section.

We begin with some examples.

EXAMPLE 1 Find (i) $\int \sqrt{1-x^2}\,dx$, (ii) $\int_0^3 \sqrt{9-x^2}\,dx$.

Solution

(i) The problem is that we cannot "get at" the expression inside the square root sign. Things would be better if $1-x^2$ were equal to the square of something so that the square root could be removed.

At this point, a little bell should be ringing in our memories, reminding us that there is something very familiar about 1 minus the square of something equaling the square of something else. It is the trigonometric identity $\cos^2\theta + \sin^2\theta = 1$, which can be rewritten as $1-\sin^2\theta = \cos^2\theta$.

This suggests that we might approach the integral $\int \sqrt{1-x^2}\,dx$ by letting $x = \sin\theta$. If we do this, the integrand becomes $\sqrt{1-x^2} = \sqrt{1-\sin^2\theta} = \sqrt{\cos^2\theta} = \cos\theta$. (Actually, we should be more careful; the square root is always nonnegative and $\cos\theta$ may be negative, but let us worry about that later.)

The other thing to do with a substitution is to consider the dx. If $x = \sin\theta$, then $dx = \cos\theta\,d\theta$. We can write the integral as

$$\int \sqrt{1-x^2}\,dx = \int \sqrt{1-\sin^2\theta}\cos\theta\,d\theta = \int \cos^2\theta\,d\theta$$
$$= \frac{\theta}{2} + \frac{1}{2}\cos\theta\sin\theta + c.$$

Recall that $\int \cos^2\theta\,d\theta = \frac{1}{2}(\theta + \cos\theta\sin\theta) + c$ (cf. Example 8.1.3, p. 563).

The last step is to return to the original variable x. This is slightly more tricky than what we encountered with other substitutions. We know that $x = \sin\theta$. This means that $\theta = \arcsin x$. With this we are able to convert most of the above formula back into a function of x, but the one remaining difficulty is the function $\cos\theta$. To deal with it, we must write $\cos^2\theta = 1 - \sin^2\theta = 1 - x^2$, so $\cos\theta = \sqrt{1-x^2}$. Putting all the pieces together, we find the result we mentioned in Example 6.1.2:

$$\int \sqrt{1-x^2}\,dx = \frac{1}{2}(\arcsin x + x\sqrt{1-x^2}) + c.$$

It might be instructive to verify this result by differentiating it. The derivative equals

$$\begin{aligned}\frac{1}{2}\left(\frac{1}{\sqrt{1-x^2}} + \sqrt{1-x^2} - \frac{x^2}{\sqrt{1-x^2}}\right) &= \frac{1}{2}\left(\frac{1-x^2}{\sqrt{1-x^2}} + \sqrt{1-x^2}\right)\\ &= \frac{1}{2}(\sqrt{1-x^2} + \sqrt{1-x^2})\\ &= \sqrt{1-x^2}.\end{aligned}$$

This confirms our integration. Notice that to integrate, we had to use $\sqrt{1-\sin^2\theta} = \cos\theta$, which means that we have assumed that $\cos\theta \geq 0$.

(ii) Now consider $\int_0^3 \sqrt{9-x^2}\,dx$. This time it is not 1 minus a square, but 9 minus a square. Since there is no convenient trigonometric identity involving 9, it is necessary to adjust the integrand:

$$\int_0^3 \sqrt{9-x^2}\,dx = \int_0^3 3\sqrt{1-\left(\frac{x}{3}\right)^2}\,dx = 3\int_0^3 \sqrt{1-\left(\frac{x}{3}\right)^2}\,dx.$$

The quantity $1-\left(\frac{x}{3}\right)^2$ is reminiscent of the left side of the identity $1-\sin^2\theta = \cos^2\theta$. We take advantage of this observation by letting $\frac{x}{3} = \sin\theta$. As x goes from $x=0$ to $x=3$, the quantity $\frac{x}{3}$ goes from 0 to 1. In order for $\sin\theta$ to go from 0 to 1 we let θ go from $\theta = 0$ to $\theta = \frac{\pi}{2}$.

Rewriting the substitution as $x = 3\sin\theta$, we see that $dx = 3\cos\theta\,d\theta$, and $\sqrt{1-\left(\frac{x}{3}\right)^2} = \sqrt{1-\sin^2\theta} = \sqrt{\cos^2\theta} = \cos\theta$. Notice that this time, since we know $0 \leq \theta \leq \frac{\pi}{2}$, we know that $\cos\theta \geq 0$, and it was correct to take the square root as we did.

Now we can integrate:

$$\begin{aligned}\int_0^3 \sqrt{9-x^2}\,dx = 3\int_0^3 \sqrt{1-\left(\frac{x}{3}\right)^2}\,dx &= 3\int_0^{\pi/2} \sqrt{1-\sin^2\theta}\,3\cos\theta\,d\theta\\ &= 9\int_0^{\pi/2} \cos^2\theta\,d\theta = 9\left(\frac{\theta}{2} + \frac{1}{2}\cos\theta\sin\theta\right)\Big|_0^{\pi/2}\\ &= \frac{9}{2}\left(\frac{\pi}{2} - 0\right) - \frac{9}{2}(0-0) = \frac{9\pi}{4}.\end{aligned}$$

These examples illustrate the technique of trigonometric substitution. It is often useful for integrals involving an expression of one of the following forms:

$$\sqrt{a^2-x^2}, \qquad \sqrt{a^2+x^2}, \qquad \sqrt{x^2-a^2}.$$

The quantities inside the square root signs remind us of the left sides of the trigonometric identities

$$1-\sin^2\theta=\cos^2\theta, \qquad 1+\tan^2\theta=\sec^2\theta, \qquad \sec^2\theta-1=\tan^2\theta,$$

respectively. Accordingly, we try the following substitutions.

Summary 8.3.1 Trigonometric Substitutions

Expression in integrand	Substitution	Range
$\sqrt{a^2-x^2}$, $a>0$	$x=a\sin\theta$	$-\frac{\pi}{2}\le\theta\le\frac{\pi}{2}$
$\sqrt{a^2+x^2}$, $a>0$	$x=a\tan\theta$	$-\frac{\pi}{2}<\theta<\frac{\pi}{2}$
$\sqrt{x^2-a^2}$, $a>0$	$x=a\sec\theta$	$0\le\theta<\frac{\pi}{2}$ or $\pi\le\theta<\frac{3\pi}{2}$

There are two reasons for making the restrictions on the values of θ. First there is the question of the sign of the square root. For instance, $\sqrt{a^2-x^2}=\sqrt{a^2(1-\sin^2\theta)}=a\sqrt{\cos^2\theta}$, and when $-\frac{\pi}{2}\le\theta\le\frac{\pi}{2}$, we have $\cos\theta\ge 0$, so $\sqrt{\cos^2\theta}=\cos\theta$, and $\sqrt{a^2-x^2}=a\cos\theta$.

Similarly, when $-\frac{\pi}{2}<\theta<\frac{\pi}{2}$, we have $\sec\theta>0$, so $\sqrt{a^2+a^2\tan^2\theta}=a\sec\theta$, and when $0\le\theta<\frac{\pi}{2}$ or $\pi\le\theta<\frac{3\pi}{2}$, we know that $\tan\theta>0$, and $\sqrt{a^2\sec^2\theta-a^2}=a\tan\theta$.

Second, the results of evaluating the new integral after one of these substitutions may involve θ. If $x=a\sin\theta$, that is, $\frac{x}{a}=\sin\theta$, then $\theta=\arcsin\left(\frac{x}{a}\right)$. The substitutions $x=a\tan\theta$ and $x=a\sec\theta$ amount to $\theta=\arctan\left(\frac{x}{a}\right)$ and $\theta=\operatorname{arcsec}\left(\frac{x}{a}\right)$, respectively. The restrictions on the range of θ are chosen in a way that is consistent with the standard definitions of the inverse trigonometric functions (cf. Definition 3.8.3, p. 206).

EXAMPLE 2 Find $\int\frac{1}{x\sqrt{x^2-2}}\,dx$.

Solution Following the above suggestions, we try $x=\sqrt{2}\sec\theta$. So $\sqrt{x^2-2}=\sqrt{2\sec^2\theta-2}=\sqrt{2}\tan\theta$, $dx=\sqrt{2}\sec\theta\tan\theta\,d\theta$, and

$$\int\frac{dx}{x\sqrt{x^2-2}}=\frac{1}{\sqrt{2}}\int\frac{\sqrt{2}\sec\theta\tan\theta}{\sqrt{2}\sec\theta\tan\theta}d\theta=\frac{1}{\sqrt{2}}\int 1\,d\theta=\frac{1}{\sqrt{2}}\theta+c.$$

Now $\frac{x}{\sqrt{2}}=\sec\theta$, so $\theta=\operatorname{arcsec}\left(\frac{x}{\sqrt{2}}\right)$, and

$$\int\frac{dx}{x\sqrt{x^2-2}}=\frac{1}{\sqrt{2}}\operatorname{arcsec}\left(\frac{x}{\sqrt{2}}\right)+c.$$

Had we recognized that the integrand $\frac{1}{x\sqrt{x^2-2}}$ is very similar to $\frac{1}{x\sqrt{x^2-1}}$, which is the derivative of arcsec x (cf. Theorem 3.8.5, p. 210), we might have done this example more easily with the substitution $u = \operatorname{arcsec}\left(\frac{x}{\sqrt{2}}\right)$. This remark serves to emphasize the value of knowing the derivatives of as many functions as possible, but meanwhile the example illustrates trigonometric substitution.

Here the answer was simply a constant times θ, which is an inverse trigonometric function of the original variable x. Frequently, the integral will involve trigonometric functions of θ.

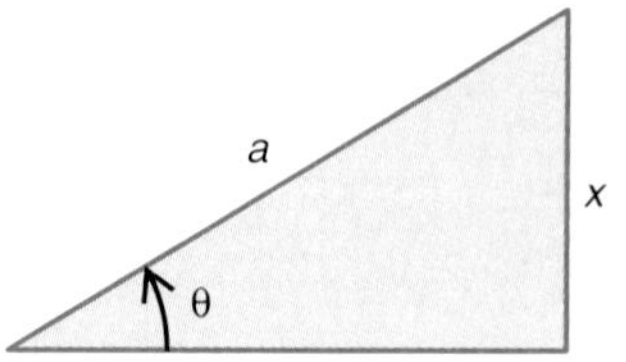

Figure 8.3.1(i)

EXAMPLE 3 $\int \frac{dx}{x^2\sqrt{a^2-x^2}}$.

Solution Try the substitution $x = a\sin\theta$, so $dx = a\cos\theta\, d\theta$, and

$$\int \frac{dx}{x^2\sqrt{a^2-x^2}} = \int \frac{a\cos\theta}{(a^2\sin^2\theta)(a\cos\theta)}\, d\theta = \frac{1}{a^2}\int \csc^2\theta\, d\theta$$

$$= -\frac{1}{a^2}\cot\theta + c.$$

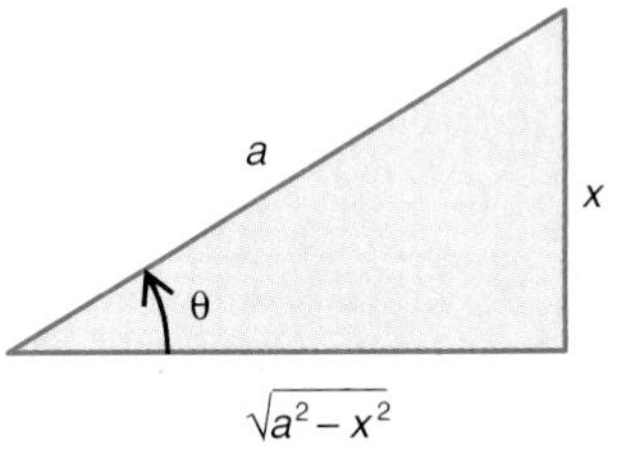

Figure 8.3.1(ii)

Since $\sin\theta$ equals the vertical side over the hypotenuse and $\sin\theta = \frac{x}{a}$, we let the hypotenuse be a and the vertical side be x.

To express $\cot\theta$ in terms of the original variable x, we draw in Figure 8.3.1(i) a right-angled triangle containing the angle θ. Since our substitution is $\sin\theta = \frac{x}{a}$, we fill in the lengths of the hypotenuse and the vertical side. Finally we use the Pythagorean Theorem (Theorem 1.5.1) to fill in the third side as shown in Figure 8.3.1(ii).

From the figure we can read off that $\cot\theta = \frac{\sqrt{a^2-x^2}}{x}$, so

$$\int \frac{dx}{x^2\sqrt{a^2-x^2}} = -\frac{1}{a^2}\cot\theta + c = -\frac{1}{a^2}\frac{\sqrt{a^2-x^2}}{x} + c.$$

The same approach can be used for the other trigonometric substitutions.

EXAMPLE 4 $\int \frac{\sqrt{x^2-9}}{x}\, dx$.

Solution $x = 3\sec\theta$, $dx = 3\sec\theta\tan\theta\, d\theta$, so

$$\int \frac{\sqrt{x^2-9}}{x}\, dx = \int \frac{3\sqrt{\tan^2\theta}}{3\sec\theta} 3\sec\theta\tan\theta\, d\theta = 3\int \tan^2\theta\, d\theta$$

$$= 3\int (\sec^2\theta - 1)\, d\theta$$

$$= 3\tan\theta - 3\theta + c.$$

Here we use the identity $\tan^2\theta = \sec^2\theta - 1$.

In Figure 8.3.2(i) we illustrate the substitution $\frac{x}{3} = \sec\theta$, or $\cos\theta = \frac{3}{x}$. In Figure 8.3.2(ii) the Pythagorean Theorem is used to fill in the length of the third side.

When we draw these pictures, it is convenient to draw θ as a positive angle and x as a positive number, but once we find the antiderivative, it will work whenever it is defined.

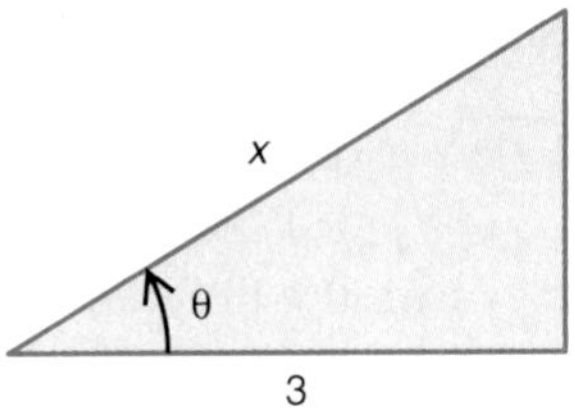

Figure 8.3.2(i)

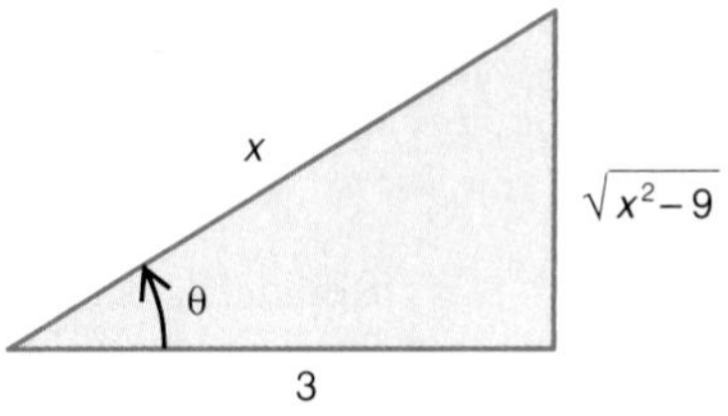

Figure 8.3.2(ii)

We see that $\tan\theta = \frac{\sqrt{x^2-9}}{3}$, so

$$\int \frac{\sqrt{x^2-9}}{x}\,dx = 3\frac{\sqrt{x^2-9}}{3} - 3\,\mathrm{arcsec}\left(\frac{x}{3}\right) + c$$
$$= \sqrt{x^2-9} - 3\,\mathrm{arcsec}\left(\frac{x}{3}\right) + c.$$

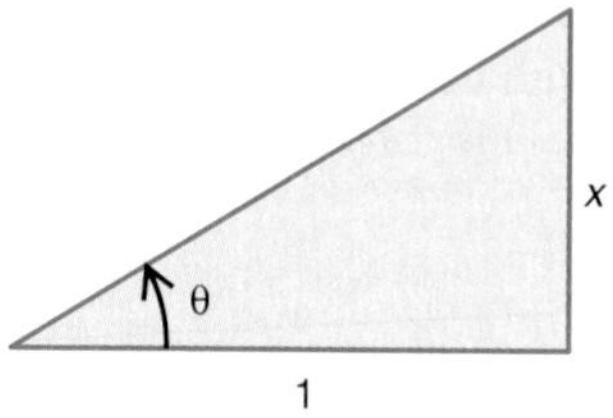

Figure 8.3.3(i)

EXAMPLE 5 $\int \frac{dx}{x^2\sqrt{1+x^2}}$.

Solution Let $x = \tan\theta$, so $dx = \sec^2\theta\,d\theta$, and

$$\int \frac{dx}{x^2\sqrt{1+x^2}} = \int \frac{\sec^2\theta\,d\theta}{\tan^2\theta\sqrt{1+\tan^2\theta}} = \int \frac{\sec^2\theta\,d\theta}{\tan^2\theta\sec\theta}$$
$$= \int \frac{\sec\theta}{\tan^2\theta}\,d\theta = \int \frac{1}{\cos\theta\frac{\sin^2\theta}{\cos^2\theta}}\,d\theta$$
$$= \int \frac{\cos\theta}{\sin^2\theta}\,d\theta.$$

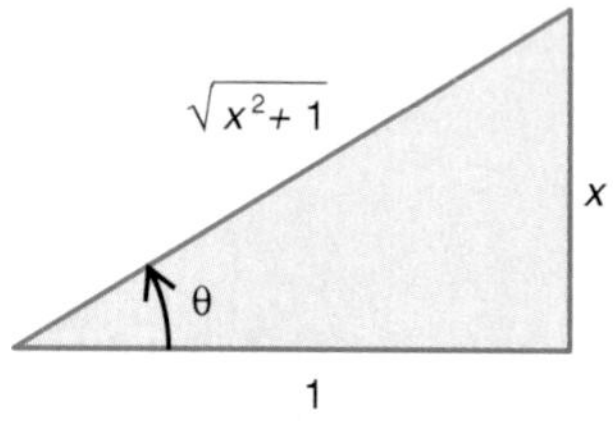

Figure 8.3.3(ii)

After the substitution $u = \sin\theta$, $du = \cos\theta\,d\theta$, the integral becomes $\int \frac{du}{u^2} = -\frac{1}{u} + c = -\frac{1}{\sin\theta} + c$.

The triangle illustrating $x = \tan\theta$ is shown in Figure 8.3.3. Filling in the hypotenuse by the Pythagorean Theorem, we see that $\sin\theta = \frac{x}{\sqrt{1+x^2}}$, so

$$\int \frac{dx}{x^2\sqrt{1+x^2}} = -\frac{1}{\sin\theta} + c = -\frac{\sqrt{1+x^2}}{x} + c.$$

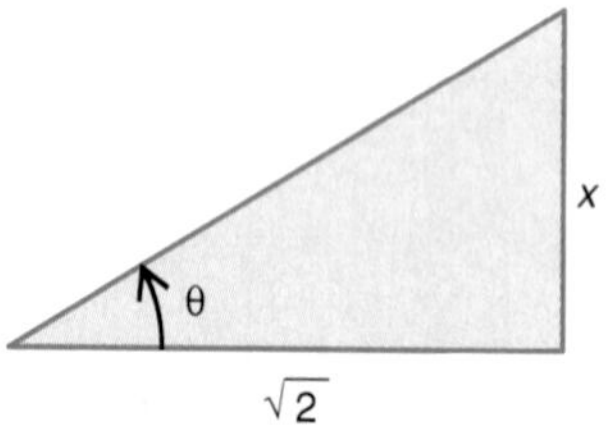

Figure 8.3.4(i)

EXAMPLE 6 $\int \frac{dx}{\sqrt{2+x^2}}$.

Solution We want the quantity inside the square root to be $2(1+\tan^2\theta)$, which means $x^2 = 2\tan^2\theta$, so we let $x = \sqrt{2}\tan\theta$, and $dx = \sqrt{2}\sec^2\theta\,d\theta$. With this substitution we integrate:

> The $\sqrt{2}$ in the numerator cancels the $\sqrt{2}$ in the denominator.

$$\int \frac{dx}{\sqrt{2+x^2}} = \int \frac{\sec^2\theta}{\sqrt{1+\tan^2\theta}}\,d\theta = \int \frac{\sec^2\theta}{\sec\theta}\,d\theta = \int \sec\theta\,d\theta$$
$$= \ln|\sec\theta + \tan\theta| + c,$$

using Example 8.2.5 (p. 573).

Using Figures 8.3.4(i) and 8.3.4(ii), we find that $\sec\theta = \frac{1}{\cos\theta} = \frac{\sqrt{2+x^2}}{\sqrt{2}}$, so

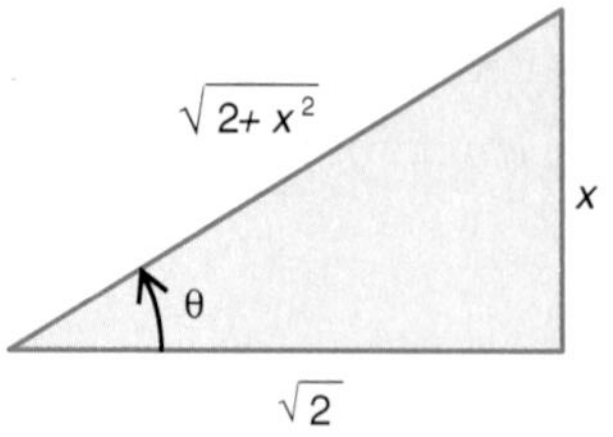

Figure 8.3.4(ii)

$$\int \frac{dx}{\sqrt{2+x^2}} = \ln\left|\frac{\sqrt{2+x^2}}{\sqrt{2}} + \frac{x}{\sqrt{2}}\right| + c = \ln\left(\frac{1}{\sqrt{2}}|\sqrt{2+x^2}+x|\right) + c$$
$$= \ln|\sqrt{2+x^2}+x| + \ln\frac{1}{\sqrt{2}} + c.$$

Replacing the constant c with the different constant $c' = \ln\frac{1}{\sqrt{2}} + c$, we have

$$\int \frac{dx}{\sqrt{2+x^2}} = \ln|\sqrt{2+x^2}+x| + c'.$$

In Section 6.9 we noted that we could not evaluate the integrals that arose from trying to find the arclength of a parabolic arc. In Example 6.10.8 (p. 500) we estimated such an integral. Now we can calculate it exactly.

EXAMPLE 7 $\int_0^2 \sqrt{1+x^2}\,dx$.

Solution We try $x = \tan\theta$, so $dx = \sec^2\theta d\theta$. As x goes from $x = 0$ to $x = 2$, the variable θ will go from $\theta = 0$ to $\theta = \arctan 2$, and

$$\int_0^2 \sqrt{1+x^2}dx = \int_0^{\arctan 2} \sqrt{\sec^2\theta}\sec^2\theta d\theta = \int_0^{\arctan 2} \sec^3\theta d\theta.$$

This is the type of integral that should be done by using the reduction formula of Theorem 8.2.5(ii). We find that $\int \sec^3\theta d\theta = \frac{1}{2}\sec\theta\tan\theta + \frac{1}{2}\int\sec\theta d\theta = \frac{1}{2}\sec\theta\tan\theta + \frac{1}{2}\ln|\sec\theta+\tan\theta| + c$, so

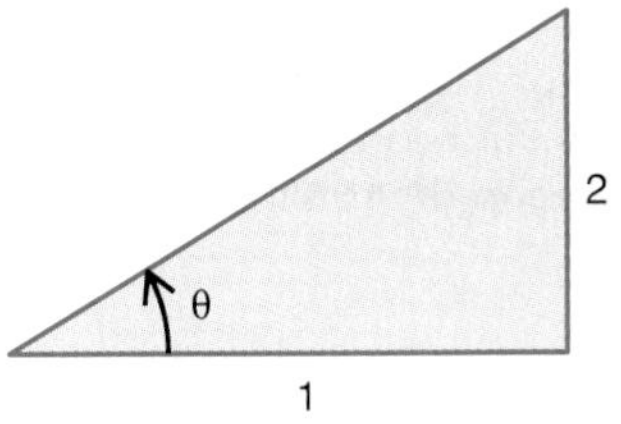

Figure 8.3.5(i)

$$\begin{aligned}\int_0^2 \sqrt{1+x^2}\,dx &= \int_0^{\arctan 2} \sec^3\theta d\theta \\ &= \left(\frac{1}{2}\sec\theta\tan\theta + \frac{1}{2}\ln|\sec\theta+\tan\theta|\right)\Big|_0^{\arctan 2} \\ &= \left(\frac{1}{2}\sec(\arctan 2)\tan(\arctan 2) + \frac{1}{2}\ln|\sec(\arctan 2)+\tan(\arctan 2)|\right).\end{aligned}$$

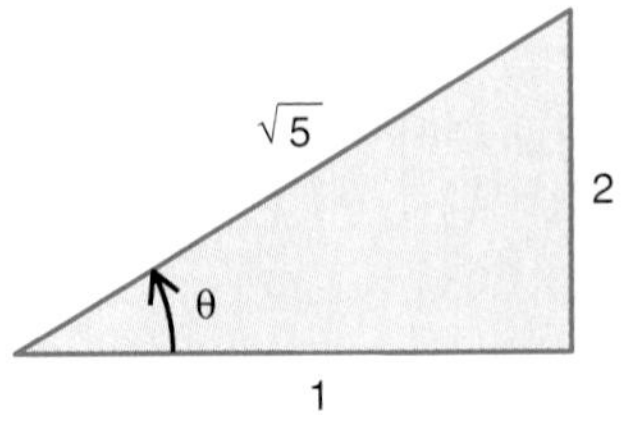

Figure 8.3.5(ii)

From the triangle shown in Figures 8.3.5(i) and (ii) we see that if $\tan\theta = 2$, then $\sec\theta = \frac{1}{\cos\theta} = \frac{1}{1/\sqrt{5}}$, that is, $\sec(\arctan 2) = \sqrt{5}$. Using this, it is now possible to evaluate the integral:

$$\begin{aligned}&\int_0^2 \sqrt{1+x^2}\,dx \\ &= \left(\frac{1}{2}\sec(\arctan 2)\tan(\arctan 2) + \frac{1}{2}\ln|\sec(\arctan 2)+\tan(\arctan 2)|\right) \\ &= \left(\frac{1}{2}\sqrt{5}(2) + \frac{1}{2}\ln|\sqrt{5}+2|\right) = \sqrt{5} + \frac{1}{2}\ln|2+\sqrt{5}|.\end{aligned}$$

The result of this example was mentioned in Example 6.10.8, in which a numerical approximation to this integral was found.

Exercises 8.3

I

Evaluate the following integrals.

1. $\int \sqrt{4-x^2}\,dx$
2. $\int \sqrt{3-x^2}\,dx$
3. $\int \frac{1}{x\sqrt{x^2-1}}\,dx$
4. $\int \frac{1}{x\sqrt{3x^2-1}}\,dx$
5. $\int \frac{1}{x^2\sqrt{1-x^2}}\,dx$
6. $\int \frac{1}{x^2\sqrt{4-x^2}}\,dx$
7. $\int \frac{1}{x}\sqrt{x^2-4}\,dx$
8. $\int \frac{1}{x}\sqrt{x^2-a^2}\,dx$
9. $\int \frac{1}{x^2\sqrt{x^2+4}}\,dx$
10. $\int \frac{1}{\sqrt{x^2+a^2}}\,dx$
11. $\int (1-x^2)^{3/2}dx$
12. $\int (4-x^2)^{-1/2}dx$
13. $\int (9-4x^2)^{1/2}dx$

14. $\int (4+9x^2)^{-1/2}dx$
15. $\int (1+5x^2)^{3/2}dx$
16. $\int (4+9x^2)^{-3/2}dx$
17. $\int x(8+x^2)^{3/2}dx$
18. $\int x^2(5-x^2)^{-1/2}dx$
19. $\int x(x^2-1)^{1/2}dx$
20. $\int x^5(4-x^2)^{-1/2}dx$
21. $\int (1+x^2)^{-2}dx$
22. $\int x^3(9-x^2)^{-1/2}dx$
23. $\int (5-x^2)^{-3/2}dx$
24. $\int \frac{1}{9+6x^2+x^4}\,dx$
25. $\int 2x\sqrt{1-x^4}\,dx$
26. $\int 2x\sqrt{1+x^4}\,dx$
27. $\int x^{-3}(1+x^4)^{-1/2}dx$
28. $\int x^3\sqrt{1+x^4}\,dx$
29. $\int (x+3)\sqrt{4-x^2}\,dx$
30. $\int x^7\sqrt{1+x^8}\,dx$

Evaluate the following definite integrals.

31. $\int_0^1 \sqrt{1-x^2}\,dx$
32. $\int_{-1}^1 \sqrt{2-x^2}\,dx$
33. $\int_1^2 x^3\sqrt{x^2-1}\,dx$
34. $\int_2^4 x\sqrt{x^2-1}\,dx$
35. $\int_0^1 \frac{1}{(2-x^2)^2}\,dx$
36. $\int_2^4 \frac{\sqrt{x^2-1}}{x}\,dx$
37. $\int_0^2 \frac{1}{4+x^2}\,dx$
38. $\int_{-1}^1 x^5\sqrt{4-x^2}\,dx$
39. $\int_0^3 \frac{x}{1+x^2}\,dx$
40. $\int_0^1 x(2-x^2)^{-9/2}dx$

II

41. Find the arclength of the parabola $y=x^2$ from $x=0$ to $x=1$.
42. Find the arclength of the parabola $y=\frac{1}{2}x^2$ from $x=1$ to $x=4$.
43. Find the arclength of the parabola $y=x^2$ from $x=a$ to $x=b$.

*44. Evaluate $\int \frac{1}{1+x^2}\,dx$ using the substitution $x=\tan\theta$.

45. (i) Evaluate $\int 2x\sqrt{1-x^2}\,dx$ by the trigonometric substitution $x=\sin\theta$.
(ii) Compare your answer with what you get by making the more conventional substitution $u=1-x^2$.

46. Integrate $\int \sqrt{1-x^2}\,dx$, using the substitution $x=\cos\theta$, and check that your answer is consistent with what happens using the $\sin\theta$ substitution.

*47. Recall the hyperbolic identity $\cosh^2 t-\sinh^2 t=1$ from Equation 3.5.12 (p 180). Use it and a suitable substitution to evaluate $\int \frac{1}{\sqrt{1+x^2}}\,dx$.

**48. Evaluate the integral of the previous question, using a trigonometric substitution, and confirm that the two answers are the same (up to the usual constant).

Drill

49. $\int e^x/(e^{2x}+1)\,dx$
50. $\int \cos x\sin x\sqrt{1+\sin^2 x}\,dx$
51. $\int x\sqrt{4+x^2}\,dx$
52. $\int x^2\sqrt{4-x^2}\,dx$
53. $\int x^2\sqrt{9+x^2}\,dx$
54. $\int xe^{x^2+1}dx$
55. $\int \cosh x\sinh x\,dx$
56. $\int x\ln(1+x^2)\,dx$
57. $\int x\sin(x^2)\sin(2x^2)\,dx$
58. $\int x\ln(x^4)\,dx$

POINT TO PONDER

When we first encountered substitutions in Section 5.7, the situation was something like $\int 2x\cos(1-x^2)\,dx$. The idea there was to make a substitution $u=1-x^2$. In this section we encounter an integral like $\int \sqrt{1-x^2}\,dx$, and we make the substitution $x=\sin\theta$. Notice that the substitution is written quite differently. In the first case we wrote the substituted variable u as a function of the original variable x; in the second it is the other way around, and we have written the original variable x as a function of the new variable θ. In this second situation it is quite straightforward to find dx because we have a formula for x as a function of θ. This is actually easier than what we are used to doing with substitutions of the first type.

Of course, it is possible to write the trigonometric substitution in a form that is consistent with the other kind. If we take $x=\sin\theta$ and write it as $\theta=\arcsin x$, then it looks more like a conventional substitution. However, there is the drawback that $d\theta$ will be expressed with a function of x. In this case, $d\theta=1/\sqrt{1-x^2}\,dx$, and in order to be able to integrate we would have to convert that square root back into a function of θ. This would be an unnecessary complication.

Converting the result of our integration back into an expression using the original variable may be more difficult for trigonometric substitutions.

So trigonometric substitutions are in a sense the same kind of thing as the substitutions we already know, but there are some differences.

8.4 Completing the Square in Quadratic Expressions in Integrands

COMPLETING SQUARES
TRIGONOMETRIC SUBSTITUTION

The subject of this section is best described by considering an example.

EXAMPLE 1 $\int \frac{1}{x^2+2x+2}\,dx$.

Solution It is easy to complete the square (cf. Summary 1.3.3, p. 9): $x^2 + 2x + 2 = x^2 + 2x + 1 + 1 = (x+1)^2 + 1$. So the integral can be written $\int \frac{1}{x^2+2x+2}\,dx = \int \frac{1}{(x+1)^2+1}\,dx$, which suggests the substitution $u = x + 1$, $du = dx$. We find

$$\int \frac{1}{x^2+2x+2}\,dx = \int \frac{1}{(x+1)^2+1}\,dx = \int \frac{1}{u^2+1}\,du$$
$$= \arctan u + c = \arctan(x+1) + c.$$

The idea is simple enough. Starting with a quadratic expression $Ax^2 + Bx + C$, we can complete the square and make a substitution to obtain an expression of the form $Au^2 \pm a^2$. The advantage is that this last expression is the kind we may be able to approach using trigonometric substitutions as in Section 8.3. Even if that does not work, it is simpler, and something else might occur to us.

EXAMPLE 2 $\int \sqrt{-2x^2 + 12x - 10}\,dx$.

Solution We begin by pulling out the common factor of 2 from inside the square root: $\int \sqrt{-2x^2+12x-10}\,dx = \sqrt{2}\int \sqrt{-x^2+6x-5}\,dx$. It is easy to complete the square: $-x^2+6x-5 = -(x^2-6x+5) = -(x^2-6x+9-4) = -\left((x-3)^2-4\right) = 4-(x-3)^2$. Making the substitution $u = x-3$, $du = dx$, we find

$$\int \sqrt{-2x^2+12x-10}\,dx = \sqrt{2}\int \sqrt{4-(x-3)^2}\,dx = \sqrt{2}\int \sqrt{4-u^2}\,du.$$

This is the type of integral we know how to evaluate with a trigonometric substitution. Letting $u = 2\sin\theta$, so $du = 2\cos\theta\, d\theta$, we find

$$\sqrt{2}\int \sqrt{4-u^2}\,du = \sqrt{2}\int \sqrt{4-4\sin^2\theta}\,2\cos\theta\, d\theta$$
$$= 2\sqrt{2}\int \sqrt{4(1-\sin^2\theta)}\cos\theta\, d\theta = 4\sqrt{2}\int \cos^2\theta\, d\theta$$
$$= 2\sqrt{2}(\theta + \cos\theta\sin\theta) + c.$$

Recall that $\int \cos^2\theta\, d\theta = \frac{1}{2}(\theta + \cos\theta\sin\theta) + c$ (cf. Example 8.1.3, p. 563).

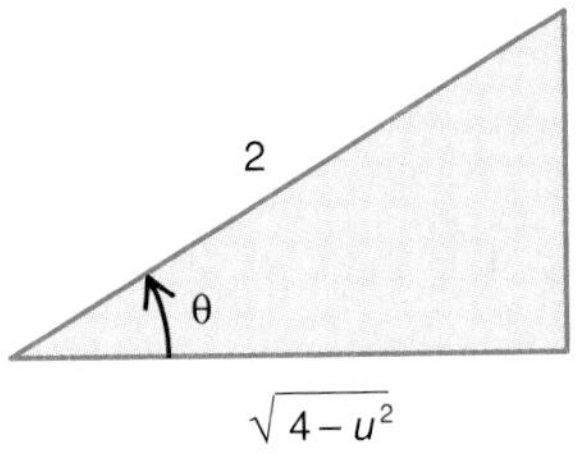

Figure 8.4.1

Considering the triangle shown in Figure 8.4.1, we see that if $u = 2\sin\theta$, that is, $\frac{u}{2} = \sin\theta$, then $\cos\theta = \frac{1}{2}\sqrt{4-u^2}$. We find that

$$\int \sqrt{-2x^2+12x-10}\,dx = \sqrt{2}\int \sqrt{4-u^2}\,du = 2\sqrt{2}(\theta + \cos\theta\sin\theta) + c$$
$$= 2\sqrt{2}\left(\arcsin\left(\frac{u}{2}\right) + \frac{1}{4}u\sqrt{4-u^2}\right) + c$$

$$= 2\sqrt{2}\arcsin\left(\frac{x-3}{2}\right) + \frac{1}{\sqrt{2}}(x-3)\sqrt{4-(x-3)^2} + c.$$

EXAMPLE 3 $\displaystyle\int \frac{x^2-2x+3}{x^2+4x+8}\,dx.$

Solution What is confusing about this integrand is that there are *two* quadratic expressions. We can complete the square in each one without difficulty. The problem is that then we must choose a substitution. The substitution that makes the numerator into something simple (u^2+2) makes the denominator into something less good $(u^2+6u+13)$, and vice versa. The question is which approach will give us a more manageable integral?

The answer is that it is more important to simplify the denominator. We complete the square: $x^2+4x+8 = x^2+4x+4+4 = (x+2)^2+4$, and this suggests the substitution $u = x+2$, $du = dx$. As usual, it is necessary to express the numerator in terms of u.

Since $u = x+2$, we know that $x = u-2$, so the numerator is $x^2-2x+3 = (u-2)^2 - 2(u-2) + 3 = u^2 - 4u + 4 - 2u + 4 + 3 = u^2 - 6u + 11$. The integral becomes

$$\int \frac{x^2-2x+3}{x^2+4x+8}\,dx = \int \frac{x^2-2x+3}{(x+2)^2+4}\,dx = \int \frac{u^2-6u+11}{u^2+4}\,du.$$

Now what? We know how to integrate things like $\frac{u}{u^2+4}$ and things like $\frac{1}{u^2+4}$. The difficulty is that u^2 term in the numerator. The solution is to do a long division. The integrand can be expressed as the sum of a polynomial (i.e., with no denominator) *plus* a rational function in which the numerator has no powers of u higher than the first power.

In this particular case this is easily accomplished without long division just by writing

$$\frac{u^2-6u+11}{u^2+4} = \frac{u^2+4-6u+7}{u^2+4} = \frac{u^2+4}{u^2+4} + \frac{-6u+7}{u^2+4} = 1 + \frac{-6u+7}{u^2+4}.$$

Now we can integrate:

$$\begin{aligned}\int \frac{x^2-2x+3}{x^2+4x+8}\,dx &= \int \frac{u^2-6u+11}{u^2+4}\,du = \int \left(1 + \frac{-6u+7}{u^2+4}\right)du \\ &= \int 1\,du - \int \frac{6u}{u^2+4}\,du + \int \frac{7}{u^2+4}\,du.\end{aligned}$$

The first integral is trivial, the second can easily be done by substituting $v = u^2+4$, so $dv = 2u\,du$, and the third can be done by substituting $w = \frac{u}{2}$, so $dw = \frac{1}{2}\,du$. We find

$$\begin{aligned}\int \frac{x^2-2x+3}{x^2+4x+8}\,dx &= \int 1\,du - \int \frac{6u}{u^2+4}\,du + \int \frac{7}{u^2+4}\,du \\ &= u - 3\int \frac{1}{v}\,dv + 7\int \frac{2}{4w^2+4}\,dw \\ &= u - 3\ln|v| + \frac{7}{2}\arctan w + c\end{aligned}$$

Since $u^2 + 4 > 0$, we can write $\ln(u^2 + 4)$ without the absolute value.

$$= u - 3\ln(u^2 + 4) + \frac{7}{2}\arctan\left(\frac{u}{2}\right) + c$$

$$= x + 2 - 3\ln\big((x + 2)^2 + 4\big) + \frac{7}{2}\arctan\left(\frac{x + 2}{2}\right) + c$$

$$= x - 3\ln(x^2 + 4x + 8) + \frac{7}{2}\arctan\left(\frac{x}{2} + 1\right) + c'.$$

Notice how the constants c and $+2$ were combined into one constant c'. Also notice that the integral obtained by completing the square in the denominator and then substituting had to be broken up into three separate pieces, which could be integrated in fairly simple but different ways.

> In this case we completed the square in the denominator. Generally speaking, if there are two or more quadratic expressions, it is probably best to complete the square in the one that is *least accessible.* For instance, if one is inside a square root sign, it would be a better choice than one outside the square root. However, there are no overall rules, and it is really a question of trial and error.

EXAMPLE 4 $\int \frac{x}{(x^2-2x+2)^2}\,dx$.

Solution The square is easily completed in the denominator: $x^2 - 2x + 2 = x^2 - 2x + 1 + 1 = (x - 1)^2 + 1$. This suggests the substitution $u = x - 1$, $du = dx$, and we find that the numerator becomes $x = u + 1$:

$$\int \frac{x}{(x^2 - 2x + 2)^2}\,dx = \int \frac{u + 1}{(u^2 + 1)^2}\,du = \int \frac{u}{(u^2 + 1)^2}\,du + \int \frac{1}{(u^2 + 1)^2}\,du.$$

The first of these integrals can be evaluated using the substitution $v = u^2 + 1$, $dv = 2u\,du$. We find $\int u/(u^2 + 1)^2 du = \frac{1}{2}\int v^{-2} dv = -\frac{1}{2}\frac{1}{v} + c = -\frac{1}{2}(u^2 + 1)^{-1} + c = -\frac{1}{2}(x^2 - 2x + 2)^{-1} + c$.

For the second integral the work of the previous section suggests letting $u = \tan\theta$, so $du = \sec^2\theta\,d\theta$. We find

$$\int \frac{1}{(u^2 + 1)^2}\,du = \int \frac{1}{(\tan^2\theta + 1)^2}\sec^2\theta\,d\theta$$

$$= \int \frac{\sec^2\theta}{(\sec^2\theta)^2}\,d\theta$$

$$= \int \sec^{-2}\theta\,d\theta = \int \cos^2\theta\,d\theta$$

$$= \frac{1}{2}(\theta + \cos\theta\sin\theta) + c'.$$

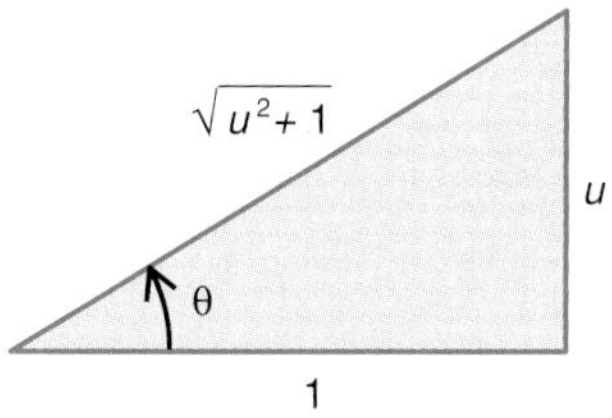

Figure 8.4.2

Since $u = \tan\theta$, we see from Figure 8.4.2 that $\cos\theta = 1/\sqrt{1 + u^2}$. Also $\sin\theta = u/\sqrt{1 + u^2}$ and $\theta = \arctan u$. So

$$\int \frac{1}{(u^2 + 1)^2}\,du = \frac{1}{2}(\theta + \cos\theta\sin\theta) + c'$$

$$= \frac{1}{2}\left(\arctan u + \frac{1}{\sqrt{1+u^2}}\frac{u}{\sqrt{1+u^2}}\right) + c'$$
$$= \frac{1}{2}\left(\arctan u + \frac{u}{1+u^2}\right) + c'$$
$$= \frac{1}{2}\left(\arctan(x-1) + \frac{x-1}{x^2-2x+2}\right) + c'.$$

Combining, we find

$$\int \frac{x}{(x^2-2x+2)^2}\,dx = \int \frac{u}{(u^2+1)^2}\,du + \int \frac{1}{(u^2+1)^2}\,du$$
$$= -\frac{1}{2}(x^2-2x+2)^{-1}$$
$$+\frac{1}{2}\left(\arctan(x-1) + \frac{x-1}{x^2-2x+2}\right) + c$$
$$= \frac{1}{2}\left(\arctan(x-1) + \frac{x-2}{x^2-2x+2}\right) + c.$$

REMARK 8.4.1

The technique of Example 4 can be used to evaluate any integral of the form $\int \frac{rx+s}{(Ax^2+Bx+C)^n}\,dx$. The idea is to complete the square in the denominator and make the appropriate substitution so that the denominator will become something of the form $(\pm u^2 \pm a^2)^n$.

Writing the numerator in terms of u, we can break the integral into two parts, one a constant times $\int u/(\pm u^2 \pm a^2)^n du$ and the other a constant times $\int 1/(\pm u^2 \pm a^2)^n du$. The first of these can be evaluated using the substitution $v = \pm u^2 \pm a^2$, and the second can be done using the appropriate trigonometric substitution (cf. Summary 8.3.1, p. 580).

Exercises 8.4

I

Evaluate the following integrals.

1. $\int \frac{1}{x^2-4x+5}\,dx$
2. $\int \frac{1}{x^2-2x+5}\,dx$
3. $\int \frac{1}{x^2+6x+14}\,dx$
4. $\int \frac{x}{x^2+2x+2}\,dx$
5. $\int \frac{x+1}{x^2-2x+4}\,dx$
6. $\int \frac{1-3x}{x^2+4x+3}\,dx$
7. $\int \frac{x^2+3}{x^2-2x+2}\,dx$
8. $\int \frac{2x^2}{x^2+4x+4}\,dx$
9. $\int \sqrt{3-2x-x^2}\,dx$
10. $\int \sqrt{7+6x-x^2}\,dx$
11. $\int 1/\sqrt{16-6x-x^2}\,dx$
12. $\int 1/\sqrt{8+2x-x^2}\,dx$
13. $\int \frac{1}{4x-x^2-8}\,dx$
14. $\int \frac{x}{8x-20-x^2}\,dx$
15. $\int \frac{1}{2x^2-4x+4}\,dx$
16. $\int \frac{1}{2x^2+6x+5}\,dx$
17. $\int \frac{x}{\sqrt{x^2+2x+2}}\,dx$
18. $\int (x+2)(x^2-4x+8)^{-1/2}dx$
19. $\int (x+1)(x^2-4x+5)^{-3/2}\,dx$
20. $\int x\sqrt{8+2x-x^2}\,dx$
21. $\int (x+1)/\sqrt{3+2x-x^2}\,dx$
22. $\int x(x^2+6x+10)^{-5/2}dx$
23. $\int (x+1)(x^2+2x+17)^{-17/2}dx$
24. $\int (1-x^2)^{-1/2}dx$
25. $\int (5-4x-x^2)^{-3/2}dx$
26. $\int (6-2x-3x^2)^{-1/2}dx$

27. $\int \frac{x+2}{(x^2+4x+13)^5}\,dx$

28. $\int x^2\sqrt{1-x^2}\,dx$

29. $\int x^3\sqrt{1-x^2}\,dx$

30. $\int x^2\sqrt{2-2x-x^2}\,dx$

II

*31. Evaluate $\int \frac{x}{(x^2-2x+2)^3}\,dx$.

*32. (i) Let $q(x) = ax^2 + bx + c$ be a quadratic polynomial. Consider a quotient $\frac{p(x)}{q(x)^3}$, where $p(x)$ is a polynomial of degree less than 6. Show that $\frac{p(x)}{q(x)^3}$ can always be written in the form

$$\frac{p(x)}{q(x)^3} = \frac{Ax+B}{q(x)^3} + \frac{Cx+D}{q(x)^2} + \frac{Ex+F}{q(x)}$$

for suitable constants A, B, C, D, E, F. (*Hint:* Write $p(x) = rx^5 + sx^4 + tx^3 + ux^2 + vx + w$. What must E equal?)

(ii) If $p(x)$ is a polynomial of *any* degree, show that $\frac{p(x)}{q(x)^3}$ can be written as the sum of a polynomial $Q(x)$ and an expression like the one in part (i). What is the degree of $Q(x)$? (*Hint:* Start with long division, resulting in a quotient plus a remainder; consider separately the cases in which the degree of $p(x)$ is or is not less than 6.)

Drill

33. $\int \frac{x+1}{x^2+2x+10}\,dx$

34. $\int x\sqrt{x^2+4x+8}\,dx$

35. $\int x\sqrt{x-1}\,dx$

36. $\int \sqrt{2x-x^2}\,dx$

37. $\int (x-2)^3\sqrt{5+4x-x^2}\,dx$

38. $\int e^x\sqrt{e^x-1}\,dx$

39. $\int e^{x/2}\sqrt{e^x-4}\,dx$

40. $\int \frac{\sin t}{\cos^2 t-2\cos t+2}\,dt$

41. $\int \frac{1}{x}\sqrt{\ln x-\ln^2 x}\,dx$

42. $\int \frac{\sqrt{2+\tan^2\theta}}{\cos^2\theta}\,d\theta$

8.5 Partial Fractions

RATIONAL FUNCTION
PROPER RATIONAL FUNCTION
IRREDUCIBLE QUADRATIC FACTOR
DISCRIMINANT

In this section we learn a technique for integrating rational functions (functions that are quotients of polynomials). The idea hinges on being able to factor the denominator. Assuming that this is possible, there is a way to express the rational function as a sum of rational functions of several relatively simple types, each of which we can integrate using techniques from the previous sections.

Of course, it is not always possible to see how to factor a polynomial, so in a sense the method is not as complete as it sounds. However, it is extremely useful, and we will always assume that we do know how to factor the denominator.

EXAMPLE 1 If we consider $\frac{1}{1+x} + \frac{1}{1-x}$, we can write it over a common denominator $1-x^2$, and $\frac{1}{1+x} + \frac{1}{1-x} = \frac{(1-x)+(1+x)}{1-x^2} = \frac{2}{1-x^2}$. Integrating, we find

$$\int \frac{2}{1-x^2}\,dx = \int \frac{1}{1+x}\,dx + \int \frac{1}{1-x}\,dx$$
$$= \ln|1+x| - \ln|1-x| + c = \ln\left|\frac{1+x}{1-x}\right| + c.$$

Here we used substitutions $u = 1+x$ and $v = 1-x$ in the two integrals and simplified the answer using properties of the natural logarithm.

This example took a sum of two simple functions, whose integrals we can find very easily, and expressed it as a single rational function over a common denominator. This is really the reverse of what we want; the useful thing would be to take a more complicated rational function and express it as a sum of simple ones. The example suggests the following. Suppose we know how to factor the denominator. Then it might be possible to express the original integrand as a sum of terms, each of which is equal to a constant divided by one of the factors of the denominator. This is not always quite correct, but sometimes it is, and it is a very good idea.

EXAMPLE 2 $\int \frac{2x+6}{x^3+x^2-6x}\, dx$.

Solution Following the suggestion made above, we factor the denominator. Since there is no constant term, it is divisible by x, so $x^3 + x^2 - 6x = x(x^2 + x - 6) = x(x+3)(x-2)$.

We want to try to write

$$\frac{2x+6}{x^3+x^2-6x} = \frac{A}{x} + \frac{B}{x+3} + \frac{C}{x-2}$$

for suitable constants A, B, C. Clearing denominators, that is, multiplying by $x^3 + x^2 - 6x = x(x+3)(x-2)$, we get

$$\begin{aligned} 2x + 6 &= A(x+3)(x-2) + Bx(x-2) + Cx(x+3) \\ &= A(x^2 + x - 6) + B(x^2 - 2x) + C(x^2 + 3x) \\ &= (A + B + C)x^2 + (A - 2B + 3C)x - 6A. \end{aligned}$$

Here we have collected together all the terms in x^2, all the terms in x, and the constant term. For the above equation to hold, the coefficients of x^2 on the left side and right side must match, that is, $0 = A + B + C$. Similarly, the coefficients of x must be the same: $2 = A - 2B + 3C$, and so must the constant terms: $6 = -6A$. This gives us three equations:

$$\begin{aligned} A + B + C &= 0 \qquad (x^2\text{-terms}), \\ A - 2B + 3C &= 2 \qquad (x\text{-terms}), \\ -6A &= 6 \qquad (\text{constant terms}). \end{aligned}$$

From the last equation we see that $A = -1$, so the first two equations become

$$\begin{aligned} B + C &= 1, \\ -2B + 3C &= 3. \end{aligned}$$

Adding the last equation to twice the previous one gives $5C = 5$, so $C = 1$ and $B = 1 - C = 0$.

Eliminating B, say, we find $C = 1$, $B = 0$. So

$$\frac{2x+6}{x^3+x^2-6x} = \frac{-1}{x} + \frac{0}{x+3} + \frac{1}{x-2},$$

The fact that $B = 0$ means that there is no term with denominator $x + 3$. In hindsight this is because we could have canceled $x + 3$ from the numerator and denominator of the original integrand. . .

and

$$\begin{aligned} \int \frac{2x+6}{x^3+x^2-6x}\, dx &= \int \left(\frac{-1}{x} + \frac{0}{x+3} + \frac{1}{x-2} \right) dx \\ &= -\int \frac{1}{x}\, dx + \int \frac{1}{x-2}\, dx = -\ln|x| + \ln|x-2| + c \\ &= \ln\left| \frac{x-2}{x} \right| + c. \end{aligned}$$

This example illustrates the principle behind the technique of partial fractions, but there are some complications. The first is that we have dealt with rational functions $\frac{p(x)}{q(x)}$ that are in **proper form**, that is, in which the numerator $p(x)$ is of lower degree than the denominator $q(x)$. If we encounter a situation in which $deg\, p(x) \geq deg\, q(x)$, we must first divide (by long division) to write $\frac{p(x)}{q(x)} = f(x) + \frac{r(x)}{q(x)}$, where $f(x)$ is a polynomial and $r(x)$ is the "remainder" in the division. In particular, $deg\, r(x) < deg\, q(x)$. Reducing a rational function to proper form should always be the first step.

EXAMPLE 3 $\int \frac{x^4+1}{x(x+1)(x-1)}$.

Solution The numerator has degree 4, and the denominator, $x(x+1)(x-1) = x^3 - x$, has degree 3, so the integrand is a rational function that is not proper. Performing a long division, we find $x^4 + 1 = (x^3 - x)x + x^2 + 1$, or $\frac{x^4+1}{x(x+1)(x-1)} = x + \frac{x^2+1}{x(x+1)(x-1)}$. We can concentrate on integrating the remainder term.

The previous discussion suggests trying to write

$$\frac{x^2+1}{x(x+1)(x-1)} = \frac{A}{x} + \frac{B}{x+1} + \frac{C}{x-1}.$$

We multiply by $x(x+1)(x-1)$ to clear the denominators and find

$$x^2 + 1 = A(x+1)(x-1) + Bx(x-1) + Cx(x+1).$$

We want to use this equation to find A, B, and C. We could use the approach of the last example, multiplying everything out, collecting terms according to the power of x, and solving the resulting equations, but there is another way to go about it.

Suppose we let $x = 0$ in the above equation. The last two terms will equal 0, since each of them has a factor of x, so the equation reads $0^2 + 1 = A(1)(-1) + 0 + 0$, or $1 = -A$, so $A = -1$. Similarly, letting $x = 1$ will make $x - 1 = 0$, so the terms containing a factor of $(x-1)$ will equal 0, and the equation will read $1^2 + 1 = 0 + 0 + C(1)(1+1)$, or $2 = 2C$, or $C = 1$. Finally, letting $x = -1$, we find that $(-1)^2 + 1 = 0 + B(-1)(-1-1) + 0$, or $2 = 2B$, or $B = 1$.

We can use this to evaluate the integral:

$$\begin{aligned}\int \frac{x^4+1}{x(x+1)(x-1)}\,dx &= \int \left(x + \frac{x^2+1}{x(x+1)(x-1)}\right)dx \\ &= \int \left(x - \frac{1}{x} + \frac{1}{x+1} + \frac{1}{x-1}\right)dx \\ &= \frac{1}{2}x^2 - \ln|x| + \ln|x+1| + \ln|x-1| + c \\ &= \frac{1}{2}x^2 + \ln\left|\frac{x^2-1}{x}\right| + c.\end{aligned}$$

The first new ingredient in this example was dividing to reduce the integrand to proper form. The second was an alternative way to find the constants A, B, and C. The idea was easy: After clearing the denominator we took the resulting equation and substituted cleverly chosen values for x. The values that were chosen were the ones that make one or other of the factors of the denominator equal 0. This had the effect of making each term on the right side except for one equal to 0, so it was easy to read off the value of the constant corresponding to that term. This technique is easier than the earlier method (so much so that you may be able to do it in your head).

The next complication we may encounter is that when we factor the denominator, we may find some factors repeated, that is, occurring with powers greater than 1. Suppose $(x-a)^m$ is the largest power of $(x-a)$ that divides the denominator. Then we have to try terms of the form

$$\frac{A_1}{x-a}+\frac{A_2}{(x-a)^2}+\frac{A_3}{(x-a)^3}+\ldots+\frac{A_m}{(x-a)^m}.$$

EXAMPLE 4 $\int \frac{9x}{(x+2)^2(x-1)}\,dx.$

Solution We try to write

$$\frac{9x}{(x+2)^2(x-1)}=\frac{A}{x+2}+\frac{B}{(x+2)^2}+\frac{C}{x-1}.$$

Clearing denominators, we find

$$9x = A(x+2)(x-1)+B(x-1)+C(x+2)^2.$$

Letting $x=1$ gives $9=0+0+C(1+2)^2=9C$, so $C=1$. Letting $x=-2$ gives $-18=0+B(-2-1)+0=-3B$, so $B=6$. There is no obvious substitution we can make for x that will tell us immediately what A is.

We could rewrite the equation using the values we have found for B and C. It becomes

$$\begin{aligned} 9x &= A(x^2+x-2)+6(x-1)+(x+2)^2,\\ A(x^2+x-2) &= 9x-6x+6-x^2-4x-4\\ &= -x^2-x+2. \end{aligned}$$

We find that A must equal -1. So

$$\begin{aligned} \int \frac{9x}{(x+2)^2(x-1)}\,dx &= \int\left(\frac{-1}{x+2}+\frac{6}{(x+2)^2}+\frac{1}{x-1}\right)dx\\ &= -\ln|x+2|-\frac{6}{x+2}+\ln|x-1|+c\\ &= \ln\left|\frac{x-1}{x+2}\right|-\frac{6}{x+2}+c. \end{aligned}$$

Here the trick we developed in Example 3 for finding the values of the constants did not give us a complete answer. It did allow us to find B and C easily, but to find A, we had to fall back on something more like what we did in Example 2.

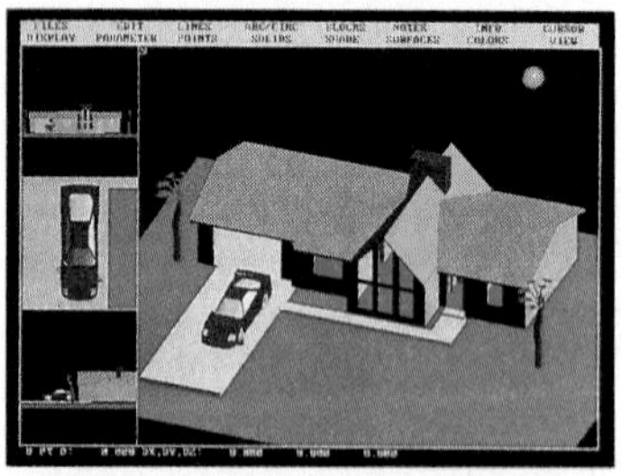

DesignCAD 3D *software was used to produce this image of a house, which permits architects to display their ideas in a form which is readily understood by clients. It is also easy to change the plan and see the effects of the changes.*

Of course, the calculations can get quite complicated if the polynomials are a little bigger.

EXAMPLE 5 $\int \frac{4x^5+2x^4-5x^3+x^2+8x-4}{x^2(x-1)^2(x+2)}\,dx.$

Solution The denominator is already factored for us; it has degree 5. Since the numerator has degree 5 too, the function is not in proper form, and we must divide. To do this, it is necessary to expand the denominator: $x^2(x-1)^2(x+2)=x^5-$

$3x^3 + 2x^2$. We find

$$\frac{4x^5 + 2x^4 - 5x^3 + x^2 + 8x - 4}{x^2(x-1)^2(x+2)} = \frac{4x^5 + 2x^4 - 5x^3 + x^2 + 8x - 4}{x^5 - 3x^3 + 2x^2}$$
$$= \frac{(4x^5 - 12x^3 + 8x^2) + (12x^3 - 8x^2) + 2x^4 - 5x^3 + x^2 + 8x - 4}{x^5 - 3x^3 + 2x^2}$$
$$= 4 + \frac{2x^4 + 7x^3 - 7x^2 + 8x - 4}{x^5 - 3x^3 + 2x^2}.$$

The next step is to try

$$\frac{2x^4 + 7x^3 - 7x^2 + 8x - 4}{x^5 - 3x^3 + 2x^2} = \frac{A}{x} + \frac{B}{x^2} + \frac{C}{x-1} + \frac{D}{(x-1)^2} + \frac{E}{x+2}.$$

Clearing denominators gives

$$2x^4 + 7x^3 - 7x^2 + 8x - 4 = Ax(x-1)^2(x+2) + B(x-1)^2(x+2)$$
$$+ Cx^2(x-1)(x+2) + Dx^2(x+2) + Ex^2(x-1)^2.$$

The trick of letting x equal certain well-chosen values will not work completely, but it will help. If we let $x = 0$, all the terms on the right side will be zero except for the one with B, and we find $-4 = 2B$, so $B = -2$. Letting $x = 1$ will leave nothing on the right side except $3D$, so $2 + 7 - 7 + 8 - 4 = 3D$, or $6 = 3D$ and $D = 2$. Finally, letting $x = -2$ gives $2(-2)^4 + 7(-2)^3 - 7(-2)^2 + 8(-2) - 4 = E(-2)^2(-2-1)^2$, or $32 - 56 - 28 - 16 - 4 = E(4)(9)$, or $-72 = 36E$ and $E = -2$. The equation becomes

$$2x^4 + 7x^3 - 7x^2 + 8x - 4 = Ax(x-1)^2(x+2) + (-2)(x-1)^2(x+2)$$
$$+ Cx^2(x-1)(x+2) + 2x^2(x+2) - 2x^2(x-1)^2$$
$$= Ax(x-1)^2(x+2) + Cx^2(x-1)(x+2)$$
$$+ (-2)(x-1)^2(x+2) + 2x^2(x+2) - 2x^2(x-1)^2$$
$$= Ax(x-1)^2(x+2) + Cx^2(x-1)(x+2)$$
$$- 2x^4 + 4x^3 + 2x^2 + 6x - 4,$$

so
$$4x^4 + 3x^3 - 9x^2 + 2x = Ax(x-1)^2(x+2) + Cx^2(x-1)(x+2)$$
$$= A(x^4 - 3x^2 + 2x) + C(x^4 + x^3 - 2x^2)$$
$$= (A + C)x^4 + Cx^3 - (3A + 2C)x^2 + 2Ax.$$

Equating the coefficients of each power of x, we find $A + C = 4$, $C = 3$, $3A + 2C = 9$, and $2A = 2$. From the last of these equations we see that $A = 1$. From the second equation we see that $C = 3$.

Now we can integrate:

$$\int \frac{4x^5 + 2x^4 - 5x^3 + x^2 + 8x - 4}{x^2(x-1)^2(x+2)}\,dx$$
$$= \int \left(4 + \frac{1}{x} - \frac{2}{x^2} + \frac{3}{x-1} + \frac{2}{(x-1)^2} - \frac{2}{x+2}\right)dx$$
$$= 4x + \ln|x| + \frac{2}{x} + 3\ln|x-1| - \frac{2}{x-1} - 2\ln|x+2| + c$$
$$= 4x + \frac{2}{x} - \frac{2}{x-1} + \ln\left|\frac{x(x-1)^3}{(x+2)^2}\right| + c.$$

Notice how we combine the ln terms, using $3\ln|x-1| = \ln|(x-1)^3|$ and $-2\ln|x+2| = \ln\frac{1}{|(x+2)^2|}$.

The third complication we may encounter is that it may not be possible to factor the denominator into linear factors. The quadratic polynomial $ax^2 + bx + c$ cannot be factored if $b^2 - 4ac < 0$ and can be factored into the product of two linear factors if $b^2 - 4ac \geq 0$ (cf. Theorem 1.3.4 and Equation 1.3.6). A quadratic polynomial that cannot be factored (i.e., for which $b^2 - 4ac < 0$) is called **irreducible** because it cannot be "reduced" into smaller pieces.

It is possible to prove that any polynomial of degree greater than 2 can be factored. This means that we can factor the denominator and then factor the factors and continue until we have written the denominator as a product of factors, each of which is linear or an irreducible quadratic. We have learned a few tricks to help us factor polynomials (cf. Theorem 1.3.7, p. 13, Example 1.3.10), but in practice it can be very difficult.

When we factor the denominator, we group together all the copies of an irreducible quadratic factor $ax^2 + bx + c$ as a power $(ax^2 + bx + c)^m$. We have to be careful to include factors that are a constant times $ax^2 + bx + c$. So, for instance, $(2x^2 + 1)(4x^2 + 2)$ should be written as $2(2x^2 + 1)^2$, not considered as two different factors.

If the denominator contains an irreducible quadratic factor $ax^2 + bx + c$, we must look for a term of the form $\frac{Ax+B}{ax^2+bx+c}$ in the expression of the rational function as a sum of simpler terms.

As before, if $b^2 - 4ac < 0$ and $(ax^2 + bx + c)^m$ is the largest power of $ax^2 + bx + c$ that divides the denominator, then we look for an expression of the form

$$\frac{A_1 x + B_1}{ax^2 + bx + c} + \frac{A_2 x + B_2}{(ax^2 + bx + c)^2} + \cdots + \frac{A_m x + B_m}{(ax^2 + bx + c)^m}.$$

EXAMPLE 6 $\int \frac{3x^4 - 9x^3 + 14x^2 - 9x + 2}{(x-1)(x^2-2x+2)^2}\,dx.$

Solution First note that the rational function is already proper, so division is unnecessary. Next note that the quadratic factor in the denominator is irreducible, since $b^2 - 4ac = (-2)^2 - 4(1)2 = 4 - 8 = -4 < 0$. Try

$$\frac{3x^4 - 9x^3 + 14x^2 - 9x + 2}{(x-1)(x^2 - 2x + 2)^2} = \frac{A}{x-1} + \frac{Bx + C}{x^2 - 2x + 2} + \frac{Dx + E}{(x^2 - 2x + 2)^2}.$$

Clearing denominators, we find

$$\begin{aligned} &3x^4 - 9x^3 + 14x^2 - 9x + 2 \\ &\quad = A(x^2 - 2x + 2)^2 + (Bx + C)(x - 1)(x^2 - 2x + 2) + (Dx + E)(x - 1). \end{aligned}$$

Letting $x = 1$, we find $1 = A(1)^2 = A$, so

$$\begin{aligned} &3x^4 - 9x^3 + 14x^2 - 9x + 2 - (x^2 - 2x + 2)^2 \\ &\quad = (Bx + C)(x^3 - 3x^2 + 4x - 2) + Dx^2 + Ex - Dx - E, \\ &3x^4 - 9x^3 + 14x^2 - 9x + 2 - (x^4 + 4x^2 + 4 - 4x^3 + 4x^2 - 8x) \\ &\quad = Bx^4 + Cx^3 - 3Bx^3 - 3Cx^2 + 4Bx^2 + 4Cx - 2Bx - 2C + Dx^2 + Ex \\ &\quad\quad - Dx - E \\ &2x^4 - 5x^3 + 6x^2 - x - 2 \\ &\quad = Bx^4 + (-3B + C)x^3 + (4B - 3C + D)x^2 + (-2B + 4C + E - D)x \\ &\quad\quad + (-2C - E). \end{aligned}$$

Equating the coefficients of each power of x, we find that we must solve

$$\begin{aligned} B &= 2, \\ -3B + C &= -5, \\ 4B - 3C + D &= 6, \end{aligned}$$

$$-2B + 4C - D + E = -1,$$
$$-2C - E = -2.$$

The first of these five equations tells us that $B = 2$; using this the second gives $C = 1$; using this the third gives $D = 1$ and the fifth gives $E = 0$.

This is easy: $B = 2$, $C = 1$, $D = 1$, and $E = 0$. So

$$\int \frac{3x^4 - 9x^3 + 14x^2 - 9x + 2}{(x-1)(x^2-2x+2)^2}\,dx$$
$$= \int \left(\frac{1}{x-1} + \frac{2x+1}{x^2-2x+2} + \frac{x}{(x^2-2x+2)^2} \right) dx.$$

At this point it helps to complete the square in the quadratic terms in the denominator: $x^2 - 2x + 2 = (x-1)^2 + 1$. Letting $u = x - 1$, we transform the in-

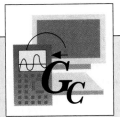

Some calculators can be used to solve systems of equations like the ones we found in Example 6. In the example there were five equations, but only four variables, B, C, D, E, so we only need to use four of the equations. Sometimes it matters which equations are chosen—certain choices may not determine the solution.

TI-85	SHARP EL9300
SIMULT 4 ENTER 1 ▼ 0 ▼ 0 ▼ 0 ▼ 2 F2(NEXT) (−) 3 ▼ 1 ▼ 0 ▼ 0 ▼ (−) 5 F2(NEXT) 4 ▼ (−) 3 ▼ 1 ▼ 0 ▼ 6 F2(NEXT) (−) 2 ▼ 4 ▼ (−) 1 ▼ 1 ▼ (−) 1 F5(SOLVE)	MENU 3(MATRIX) MENU *B*(EDIT) ▶ ENTER 4 ENTER 4 ENTER 1 ENTER −3 ENTER 4 ENTER −2 ENTER 0 ENTER 1 ENTER −3 ENTER 4 ENTER 0 ENTER 0 ENTER 1 ENTER −1 ENTER 0 ENTER 0 ENTER 0 ENTER 1 ENTER QUIT MENU ▼ ENTER 4 ENTER 1 ENTER 2 ENTER −5 ENTER 6 ENTER −1 ENTER QUIT (MAT *A* ÷ MAT *A* ÷ MAT *A*) × MAT *B* ENTER
CASIO f_x-7700*GB*	**HP 48SX**
MODE 0(MATRIX) F1(*A*) F6 F1(DIM) 4 EXE 4 EXE 1 EXE 0 EXE 0 EXE 0 EXE −3 EXE 1 EXE 0 EXE 0 EXE 4 EXE −3 EXE 1 EXE 0 EXE −2 EXE 4 EXE −1 EXE 1 EXE PRE F2(*B*) F6 F1(DIM) 4 EXE 1 EXE 2 EXE −5 EXE 6 EXE −1 EXE PRE F1(*A*) F4(A^{-1}) F1(*C* ▶ *A*) PRE F5(×)	↱ MATRIX 1 ENTER 0 ENTER 0 ENTER 0 ENTER ▼ 3 +/− ENTER 1 ENTER 0 ENTER 0 ENTER ▼ 4 ENTER 3 +/− ENTER 1 ENTER 0 ENTER ▼ 2 +/− ENTER 4 ENTER 1 +/− ENTER 1 ENTER ENTER 1/x ↱ MATRIX 2 ENTER ▼ 5 +/− ENTER ▼ 6 ENTER ▼ 1 +/− ENTER ENTER ×

```
[ 2]
[ 1]
[ 1]
[ 0]
```

Notice how we filled in the missing coefficients with 0; for example, the equation $B = 2$ becomes $1 \times B + 0 \times C + 0 \times D + 0 \times E = 2$, and so on.

tegral into

$$\int \left(\frac{1}{u} + \frac{2u+3}{u^2+1} + \frac{u+1}{(u^2+1)^2} \right) du = \int \frac{1}{u}\, du + \int \frac{2u}{u^2+1}\, du + \int \frac{3}{u^2+1}\, du$$
$$+ \int \frac{u}{(u^2+1)^2}\, du + \int \frac{1}{(u^2+1)^2}\, du.$$

We make the substitutions $v = u^2 + 1$, $dv = 2u\, du$ in the second and fourth integrals in the last expression, and $u = \tan\theta$, $du = \sec^2\theta\, d\theta$ in the fifth, giving

$$\int \frac{1}{u}\, du + \int \frac{1}{v}\, dv + 3\int \frac{1}{1+u^2}\, du + \frac{1}{2}\int \frac{1}{v^2}\, dv + \int \frac{\sec^2\theta}{(\sec^2\theta)^2}\, d\theta$$
$$= \ln|u| + \ln|v| + 3\arctan u - \frac{1}{2}\frac{1}{v} + \int \cos^2\theta\, d\theta$$
$$= \ln|u| + \ln|u^2+1| + 3\arctan u - \frac{1}{2}\frac{1}{1+u^2} + \frac{1}{2}\cos\theta\sin\theta + \frac{\theta}{2} + c.$$

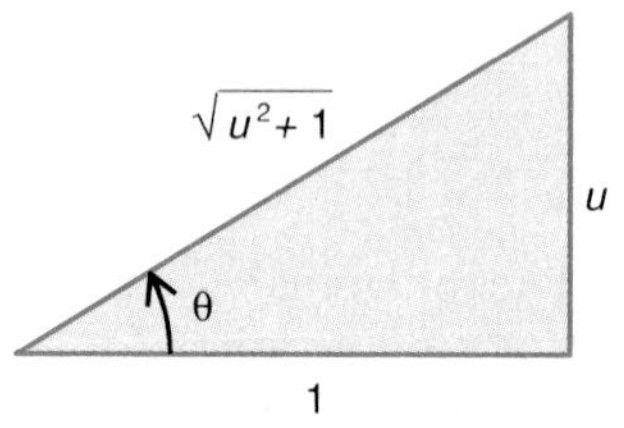

Figure 8.5.1

We use the triangle in Figure 8.5.1, in which $u = \tan\theta$, to find that $\cos\theta\sin\theta = \frac{u}{u^2+1}$. With this and the fact that $\theta = \arctan u$ the above expression becomes

$$\ln|u^3 + u| + 3\arctan u - \frac{1}{2}\frac{1}{1+u^2} + \frac{u}{2(u^2+1)} + \frac{1}{2}\arctan u + c$$
$$= \ln|u^3 + u| + \frac{7}{2}\arctan u + \frac{1}{2}\frac{u-1}{u^2+1} + c$$
$$= \ln|(x-1)^3 + x - 1| + \frac{7}{2}\arctan(x-1) + \frac{1}{2}\frac{x-2}{(x-1)^2+1} + c.$$

The individual steps in this solution are not too difficult, but the whole procedure is quite complicated, and of course there is the risk that a small error early in the calculation will go undetected and invalidate everything that follows.

Summary 8.5.1

Partial Fractions

This is a procedure for integrating any rational function, that is, a function that is the quotient of two polynomials.

1. If the rational function is not proper, perform a long division and express it as the sum of a polynomial and a proper rational function.
2. Factor the denominator as a product of linear factors and/or irreducible quadratic factors (i.e., factors of the form $q(x) = ax^2 + bx + c$ with $b^2 - 4ac < 0$). If the denominator has integer coefficients, use Theorem 1.3.7 to reduce the work required.
3. Express the proper rational function as a sum of terms of the following two types: If $(x-a)^m$ is the largest power of $(x-a)$ dividing the denominator, include terms of the form

$$\frac{A_1}{x-a} + \frac{A_2}{(x-a)^2} + \ldots + \frac{A_m}{(x-a)^m};$$

if $q(x) = ax^2 + bx + c$, with $b^2 - 4ac < 0$, and $q(x)^m$ is the largest power of $q(x)$

that divides the denominator, include terms of the form

$$\frac{B_1x + C_1}{q(x)} + \frac{B_2x + C_2}{q(x)^2} + \ldots + \frac{B_mx + C_m}{q(x)^m}.$$

4. Clear denominators and solve for all the unknown constants. You may want to simplify the work by first substituting for x the root of each linear factor after clearing denominators.
5. Integrate the resulting expression, one term at a time:

$$\int \frac{1}{x-a}\,dx = \ln|x-a| + c$$

$$\int \frac{1}{(x-a)^k}\,dx = -\frac{1}{k-1}\frac{1}{(x-a)^{k-1}} + c, \qquad \text{if } k > 1.$$

For the quadratic factors of the denominator, complete the square. The resulting integrals can be broken into the following forms:

$$\int \frac{u}{(u^2+a^2)^k}\,du = \frac{1}{2}\int \frac{dv}{v^k}, \qquad \text{with } v = u^2 + a^2,$$

$$= \begin{cases} -\frac{1}{2(k-1)}\frac{1}{(u^2+a^2)^{k-1}}, & \text{if } k > 1, \\ \frac{1}{2}\ln|u^2+a^2| + c, & \text{if } k = 1. \end{cases}$$

$$\int \frac{1}{(u^2+a^2)^k}\,du,$$

which can be integrated by letting $u = a\tan\theta$, transforming the integral into $a^{1-2k}\int \sec^{2-k}\theta\,d\theta$, which can be integrated by the methods of Theorem 8.1.3 (p. 565) or Theorem 8.2.5(ii) (p. 574), depending on the sign of $2-k$.

It can be shown that it is possible to express every proper rational function in the form outlined in Step 3 above. So in principle the technique of partial fractions will always work. In practice the stumbling block is often Step 2, factoring the denominator. Apart from that the technique is straightforward, though it can easily involve lengthy calculations, as the examples show. Sometimes we will be given the factorization of the denominator, and other times it will be possible to use Theorem 1.3.7 to help us guess an integer root. Long division will then enable us to begin the factorization.

Exercises 8.5

I

Integrate the following.

1. $\int \frac{1}{(x-1)(x+1)}\,dx$
2. $\int \frac{x}{(x-1)(x+1)}\,dx$
3. $\int \frac{x+1}{(x-1)(x+1)}\,dx$
4. $\int \frac{2x+3}{(x-1)(x+1)}\,dx$
5. $\int \frac{x+2}{(x-2)(x+3)}\,dx$
6. $\int \frac{3x-1}{x(x+4)}\,dx$
7. $\int \frac{x^2-4}{x(x+1)}\,dx$
8. $\int \frac{3x^2+x-1}{(x+2)(x+4)}\,dx$
9. $\int \frac{x^3+3x-1}{x^2(x+1)}\,dx$
10. $\int \frac{x^3+1}{(x+2)^2(x-1)}\,dx$

Evaluate the following integrals.

11. $\int \frac{x^2-x+1}{x(x^2+1)}\,dx$
12. $\int \frac{8-x}{x^3+4x}\,dx$

13. $\int \frac{1}{x^3+x}\,dx$

14. $\int \frac{x^2-2}{(x-1)(x^2+2x+2)}\,dx$

15. $\int \frac{x^3}{(x+2)(x^2-4x+8)}\,dx$

16. $\int \frac{3x^3+x-3}{2x^3+4x^2+4x}\,dx$

17. $\int \frac{4x^3-x^2+18x-4}{(x^2+4)^2}\,dx$

18. $\int \frac{6x^2-2x-4}{x^3-4x}\,dx$

19. $\int \frac{6x^2-4x-1}{(x+1)(2x^2-4x+3)}\,dx$

20. $\int \frac{x^4-2x^3}{x^5-4x^3}\,dx$

21. $\int 1/(x^4+2x^2+1)\,dx$

22. $\int (4x^3+2x^2+2x)/(x^4-1)\,dx$

23. $\int 1/(x^3-4x)\,dx$

24. $\int (2x^4+4x^2+2)/(x^2+1)^2 dx$

25. $\int (2x-6)/(x^3-9x)\,dx$

26. $\int (2x^2-x+1)/(x^3-x^2-x+1)\,dx$

27. $\int x^4/(x^3-x)\,dx$

28. $\int 3/(x^3-1)\,dx$

29. $\int (x^3+3x^2+2x)/(2x^2+6x+4)\,dx$

30. $\int x/\big((x-1)^3-4(x-1)\big)\,dx$

II

*31. Find $\int \frac{\cos\theta}{\sin^2\theta-2\sin\theta-8}\,d\theta$.

*32. Find $\int (1+e^t)/(1+e^{2t})\,dt$.

**33. Find $\int \frac{1}{x^4+1}\,dx$. (*Hint:* Factor $x^4+1=(x^2+rx+1)(x^2+sx+1)$.)

*34. Evaluate $\int \frac{x}{x^8+1}\,dx$.

*35. Evaluate $\int \frac{x^2+x+1}{(x^2+1)^2}\,dx$.

*36. Find an example to demonstrate that if there is a repeated factor in the denominator, then it may not be sufficient to consider the simple kind of partial fractions with a sum of constants divided by the linear and irreducible quadratic factors of the denominator.

Drill

37. $\int \frac{1}{x^2+4x+5}\,dx$

38. $\int \frac{1}{x^2+4x+3}\,dx$

39. $\int e^x/(e^x+3e^{x/2}+2)\,dx$

40. $\int \frac{\sin x}{\cos^2 x-1}\,dx$

41. $\int x\sin^3 x\,dx$

42. $\int \frac{\tan\theta}{2-\cos\theta}\,d\theta$

43. $\int x\arcsin x\,dx$

44. $\int e^{2x}\cos x\,dx$

45. $\int x^2\arctan x\,dx$

46. $\int x\arctan x\,dx$

47. $\int \ln(x^2+1)\,dx$

48. $\int \frac{1}{x^2}\arctan x\,dx$

49. $\int x/\big(1+\exp(x^2)\big)\,dx$

50. $\int (x^4+x^3+x^2+x+1)/(x^2+1)^2 dx$

51. $\int 1/(x^4+8x^2-9)\,dx$

52. $\int \cos(x)\cos(2x)\sin(2x)\,dx$

53. $\int e^x\sin e^{x/2}\,dx$

54. $\int \frac{1}{x}\sin\big(\ln(x^2)\big)\,dx$

55. $\int \sqrt{x}\ln x\,dx$

56. $\int \sqrt{x}e^{\sqrt{x}}\,dx$

III

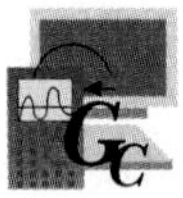

*57. Consider the integral $\int \frac{1}{x^3-2x^2-x+1}\,dx$. To apply partial fractions, it is necessary to factor the denominator, but using the Rational Root Theorem (Theorem 1.3.7, p. 13), we see that there are no rational roots.

(i) Use a graphics calculator or computer to find the roots of x^3-2x^2-x+1 with as much accuracy as possible. Use them to factor $x^3-2x^2-x+1=(x-a)(x-b)(x-c)$, and use partial fractions to integrate.

(ii) Use your answer to evaluate $\int_1^2 \frac{1}{x^3-2x^2-x+1}\,dx$.

(iii) Estimate this definite integral by Simpson's Rule S_n with $n=10$ and $n=20$.

Using Exercise 57 as a guide, evaluate the following.

58. $\int_{-1}^{0} \frac{1}{x^3-4x+2}\,dx$

59. $\int \frac{x+2}{x^3-3x+1}\,dx$

60. $\int \frac{1}{x^3+x+1}\,dx$

8.6 Rationalizing Substitutions

FRACTIONAL POWERS
$\tan(\theta/2)$ SUBSTITUTION

In Section 8.5 we learned how to integrate any rational function, so now it makes sense to try to convert other integrals to integrals of rational functions.

In this short section we learn two techniques for doing this in particular cases. Each of them can be used to convert a particular kind of integral into the integral of a rational function, which in principle we know how to evaluate. In both cases, however, the practical problem is that the rational function is likely to be quite complicated and difficult to integrate. Specifically, the denominator may have a fairly high degree and be difficult to factor.

EXAMPLE 1 Transform $\int \frac{x^{1/3}}{x^{3/4}-1}\,dx$ into the integral of a rational function.

Solution We need a substitution that will remove the fractional exponents. If $x = u^m$, then $x^{1/3} = u^{m/3}$, $x^{3/4} = u^{3m/4}$, and these powers of u will be integer powers if m is divisible by both 3 and 4. The smallest such m is $m = 12$; letting $x = u^{12}$, $dx = 12u^{11}\,du$, we find

$$\int \frac{x^{1/3}}{x^{3/4}-1}\,dx = 12\int \frac{u^4}{u^9-1}u^{11}\,du = 12\int \frac{u^{15}}{u^9-1}\,du.$$

This sort of technique will always transform integrands that are rational functions of fractional powers of x into rational functions of some new variable u. The idea is to let $x = u^m$, where m is divisible by the denominators of all the exponents in the original integrand.

However, the resulting integral may be difficult to evaluate. (Try to find the factorization of $u^9 - 1$ into linear and quadratic factors. . .).

Another type of integral that can be converted to the integral of a rational function is one whose integrand is a rational function of trigonometric functions, that is, the quotient of two "polynomials" in the *two* "variables" $\sin\theta$ and $\cos\theta$. Such integrals can often be simplified by other methods (substitution, integration by parts, etc.), but if all else fails, there is a substitution that always works. It is to let $x = \tan\frac{\theta}{2}$.

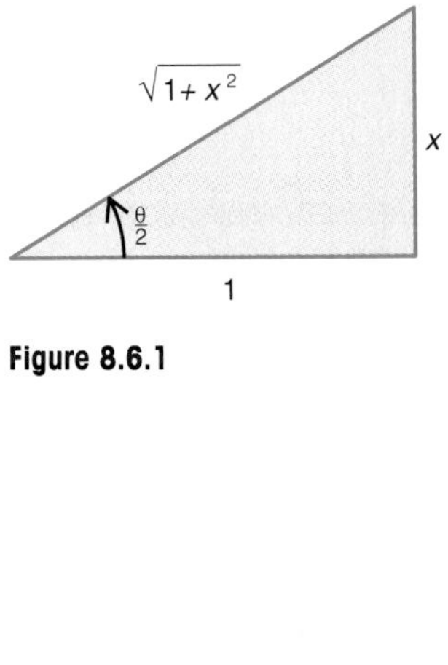

Figure 8.6.1

In this case, $\frac{\theta}{2} = \arctan x$, so $\theta = 2\arctan x$ and $d\theta = \frac{2}{1+x^2}\,dx$. From the triangle shown in Figure 8.6.1 we see that $\sin\frac{\theta}{2} = \frac{x}{\sqrt{1+x^2}}$, $\cos\frac{\theta}{2} = \frac{1}{\sqrt{1+x^2}}$, and from this we find

$$\sin\theta = 2\cos\frac{\theta}{2}\sin\frac{\theta}{2} = \frac{2x}{1+x^2},$$

$$\cos\theta = \cos^2\frac{\theta}{2} - \sin^2\frac{\theta}{2} = \frac{1-x^2}{1+x^2}.$$

EXAMPLE 2 Find $\int \frac{d\theta}{\cos\theta+\sin\theta}$.

Solution Letting $x = \tan\frac{\theta}{2}$, we have $d\theta = \frac{2}{1+x^2}\,dx$, so

$$\int \frac{d\theta}{\cos\theta+\sin\theta} = 2\int \frac{1}{\frac{1-x^2}{1+x^2}+\frac{2x}{1+x^2}}\frac{1}{1+x^2}\,dx$$
$$= 2\int \frac{1}{1+2x-x^2}\,dx.$$

To factor $x^2 - 2x - 1$, we find its roots using the Quadratic Formula (Theorem 1.3.4):

$$\frac{2 \pm \sqrt{4+4}}{2} = \frac{2 \pm 2\sqrt{2}}{2} = 1 \pm \sqrt{2},$$

so $x^2 - 2x - 1 = \left(x - (1+\sqrt{2})\right)\left(x - (1-\sqrt{2})\right)$.

We apply partial fractions:

$$\frac{1}{1+2x-x^2} = \frac{A}{x-(1+\sqrt{2})} + \frac{B}{x-(1-\sqrt{2})}.$$

Clearing denominators, being careful about signs, we get

$$\begin{aligned} -1 &= Ax - A(1-\sqrt{2}) + Bx - B(1+\sqrt{2}) \\ &= (A+B)x - (A+B) + (A-B)\sqrt{2}. \end{aligned}$$

So $A + B = 0$ and $(A+B) - (A-B)\sqrt{2} = 1$. We find that $A = -1/2\sqrt{2}$ and $B = 1/2\sqrt{2}$. From this we can integrate:

$$\begin{aligned} \int \frac{d\theta}{\cos\theta + \sin\theta} &= 2\int \frac{1}{1+2x-x^2}\,dx \\ &= -\frac{1}{\sqrt{2}}\int \frac{1}{x-(1+\sqrt{2})}\,dx + \frac{1}{\sqrt{2}}\int \frac{1}{x-(1-\sqrt{2})}\,dx \\ &= \frac{1}{\sqrt{2}}\left(-\ln\left|x-(1+\sqrt{2})\right| + \ln\left|x-(1-\sqrt{2})\right|\right) + c \\ &= \frac{1}{\sqrt{2}}\ln\left|\frac{x-1+\sqrt{2}}{x-1-\sqrt{2}}\right| + c \\ &= \frac{1}{\sqrt{2}}\ln\left|\frac{\tan\frac{\theta}{2}-1+\sqrt{2}}{\tan\frac{\theta}{2}-1-\sqrt{2}}\right| + c. \end{aligned}$$

Summary 8.6.1 **The tan $\frac{\theta}{2}$ Substitution**

$$x = \tan\tfrac{\theta}{2}, \qquad d\theta = \tfrac{2}{1+x^2}\,dx,$$

$$\sin\theta = \tfrac{2x}{1+x^2}, \qquad \cos\theta = \tfrac{1-x^2}{1+x^2}.$$

The substitution $x = \tan\frac{\theta}{2}$ always works in principle, but since the resulting rational function may be difficult to integrate, this approach should not be used before attempting to find an easier technique. It should be regarded as a last resort when everything else has been tried and rejected.

Exercises 8.6

I

Integrate each of the following.

1. $\int x^{1/3}/(x^{1/3}+1)\,dx$
2. $\int x^{1/4}/(x^{1/4}+4)\,dx$
3. $\int x^{1/2}/(x^{1/3}+1)\,dx$
4. $\int x^{1/3}/(x^{1/2}-1)\,dx$
5. $\int \frac{\sqrt{x}}{x+1}\,dx$
6. $\int \frac{\sqrt{x+1}}{x}\,dx$
7. $\int \frac{x-1}{\sqrt{x+1}}\,dx$
8. $\int \frac{1}{x+x^{3/4}}\,dx$
9. $\int x^{1/6}/(x^{2/3}+4x^{1/3})\,dx$
10. $\int \frac{1}{x}(x^2+1)^{1/3}dx$

Evaluate the following using the $\tan\frac{\theta}{2}$ substitution.

11. $\int \frac{1}{\sin\theta+1}\,d\theta$
12. $\int \frac{1}{(1-\cos\theta)^2}\,d\theta$
13. $\int \sec\theta\,d\theta$
14. $\int \frac{1}{3\cos\theta-4\sin\theta}\,d\theta$
15. $\int \frac{1}{1+\cos\theta}\,d\theta$
16. $\int \frac{\cos\theta}{1+\cos\theta}\,d\theta$
17. $\int \frac{\cos\theta}{\sin\theta(1+\cos\theta)}\,d\theta$
18. $\int \csc\theta\,d\theta$
19. $\int \frac{1}{1+\sin\theta+\cos\theta}\,d\theta$
20. $\int \frac{1}{4+\sin\theta+\cos\theta}\,d\theta$

II

*21. Suppose a and b are positive numbers with $a > b$. Find $\int \frac{1}{a+b\sin\theta}\,d\theta$.

*22. Suppose a and b are positive numbers. Find $\int \frac{1}{a\cos\theta+b\sin\theta}\,d\theta$.

**23. Find a substitution for integrating rational functions of the hyperbolic functions $\sinh t$ and $\cosh t$. Show that it converts any such integral into the integral of a rational function.

*24. Integrate $\int \frac{1}{1+\sinh t}\,dt$.

Drill

25. $\int \frac{x}{1+\sqrt{x}}\,dx$
26. $\int (x^{2/7}-3x^{-1/9})\,dx$
27. $\int \frac{1}{\sin(2\theta)}\,d\theta$
28. $\int \frac{1}{1+\sin(5\theta)}\,d\theta$
29. $\int \frac{\sin(2\theta)}{\sqrt{1+\sin\theta}}\,d\theta$
30. $\int x(x+8)^{-1/3}dx$
31. $\int \cos^2 x\sin x\sqrt{1+2\cos x}\,dx$
32. $\int x\sin(2x-3)\,dx$
33. $\int e^x(4+e^{x/2})^{1/3}dx$
34. $\int (1+\sin x)^{1/3}\cot x\,dx$

8.7 Additional Applications

AREA
ARCLENGTH
SURFACE AREA

In this chapter we have learned techniques for evaluating a number of types of integrals. Naturally, this extends the number of practical problems we can approach.

In fact, in Chapter 6 there were several instances in which interesting problems were left unsolved simply because we could not perform the necessary integrations. One place where this was particularly apparent was Section 6.9, where we developed the theory for finding the arclength of a curve but found that even for a simple parabola the resulting integral was beyond our capabilities.

The problem also arose in Section 6.1, where we had to use a complicated string of guesswork to evaluate the integral needed for finding the area of a circle, and in Section 6.7 there were regions whose centroids we could not find because the integrals involved were not ones we knew how to evaluate.

In this brief section we solve some of these problems. There is nothing really new here; we use techniques from the last chapter to solve some applied problem and techniques from this chapter to evaluate the necessary integrals.

EXAMPLE 1 **(cf. Example 6.1.3)** Find the area of the region inside a circle of radius r.

Solution We did this example in Chapter 6. At the time it was fairly tricky, but now it is easy.

The area we want is twice the area under the upper semicircle $y = \sqrt{r^2 - x^2}$, so it equals $2\int_{-r}^{r}\sqrt{r^2 - x^2}\,dx$. This integral cries out for a trigonometric substitution.

We let $x = r\sin\theta$, so $dx = r\cos\theta\,d\theta$, and

$$\begin{aligned}
2\int_{-r}^{r}\sqrt{r^2 - x^2}\,dx &= 2\int_{-\pi/2}^{\pi/2}\sqrt{r^2 - r^2\sin^2\theta}\,r\cos\theta\,d\theta \\
&= 2r^2\int_{-\pi/2}^{\pi/2}\sqrt{\cos^2\theta}\cos\theta\,d\theta = 2r^2\int_{-\pi/2}^{\pi/2}\cos^2\theta\,d\theta \\
&= 2r^2\frac{1}{2}(\theta + \cos\theta\sin\theta)\Big|_{-\pi/2}^{\pi/2} \\
&= r^2\left(\frac{\pi}{2} - \left(-\frac{\pi}{2}\right) + (0)(1) - (0)(-1)\right) \\
&= \pi r^2.
\end{aligned}$$

This agrees with the result we found in Example 6.1.3 by guessing the integral in Example 6.1.2. It is comforting to realize that now that we know about trigonometric substitution, there is a clear-cut method for approaching this problem.

EXAMPLE 2 Find the arclength of the parabola $y = x^2$ between $x = 0$ and $x = 1$.

Solution We know from Definition 6.9.1 that the arclength equals

$$\int ds = \int_0^1 \sqrt{1 + (y')^2}\,dx = \int_0^1 \sqrt{1 + 4x^2}\,dx.$$

Again a trigonometric substitution is called for; letting $2x = \tan\theta$, so $dx = \frac{1}{2}\sec^2\theta\,d\theta$, we find that the arclength is

$$\begin{aligned}
\int ds &= \int_0^1 \sqrt{1 + 4x^2}\,dx = \frac{1}{2}\int_0^{\arctan 2}\sqrt{1 + \tan^2\theta}\sec^2\theta\,d\theta \\
&= \frac{1}{2}\int_0^{\arctan 2}\sec^3\theta\,d\theta.
\end{aligned}$$

We found the new upper limit of integration by realizing that it had to be the value of θ for which $\tan\theta = 2$, since $\tan\theta = 2x$ and $2x = 2$ at the original upper limit of integration $x = 1$.

This integral can be evaluated by using the reduction formula of Theorem 8.2.5(ii) for integrals of powers of $\sec x$. We find that

$$\begin{aligned}
\int_0^{\arctan 2}\sec^3\theta\,d\theta &= \frac{1}{2}\sec\theta\tan\theta\Big|_0^{\arctan 2} + \frac{1}{2}\int_0^{\arctan 2}\sec\theta\,d\theta \\
&= \frac{1}{2}\big(\sec(\arctan 2)(2) - \sec(0)(0)\big) + \frac{1}{2}\ln|\sec\theta + \tan\theta|\Big|_0^{\arctan 2} \\
&= \sec(\arctan 2) + \frac{1}{2}\big(\ln|\sec(\arctan 2) + 2| - \ln|\sec(0) + \tan(0)|\big) \\
&= \sec(\arctan 2) + \frac{1}{2}\big(\ln|\sec(\arctan 2) + 2| - \ln(1)\big).
\end{aligned}$$

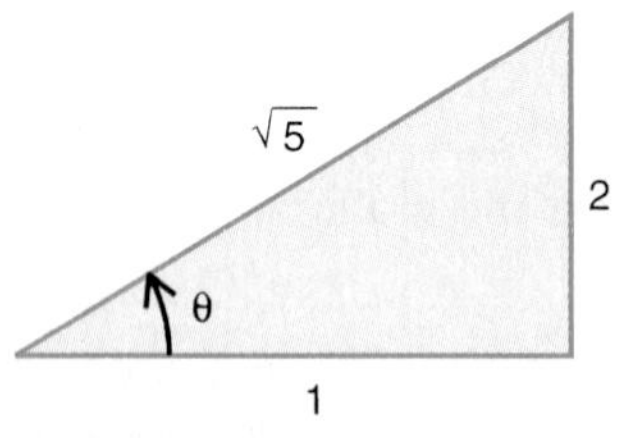

Figure 8.7.1

When $\theta = \arctan 2$, that is, $\tan\theta = 2$, we see from Figure 8.7.1 that $\sec^2\theta = 1 + \tan^2\theta = 5$, so $\sec\theta = \sqrt{5}$. The arclength equals

$$\int ds = \frac{1}{2}\Big(\sec(\arctan 2) + \frac{1}{2}\ln|\sec(\arctan 2) + 2|\Big) = \frac{\sqrt{5}}{2} + \frac{1}{4}\ln|2+\sqrt{5}|.$$

The numerical value of this expression is approximately 1.479. ◢

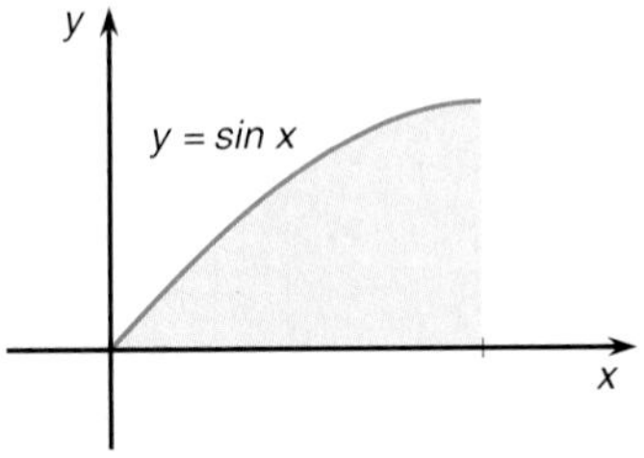

Figure 8.7.2

EXAMPLE 3 Find the centroid of the region enclosed between the x-axis and the curve $y = \sin x$ between the lines $x = 0$ and $x = \frac{\pi}{2}$, shown in Figure 8.7.2.

Solution Assuming that the density is $\rho = 1$, we find that the total weight is

$$M = \int_0^{\pi/2} \rho \sin x\, dx = -\cos x\Big|_0^{\pi/2} = 0 - (-1) = 1.$$

Using Theorem 6.7.9 (p. 472), we can find the moment about the x-axis:

$$\begin{aligned} M_x &= \frac{1}{2}\int_0^{\pi/2} \rho(\sin^2 x - 0)\, dx = \frac{1}{2}\int_0^{\pi/2} \sin^2 x\, dx \\ &= \frac{1}{4}(x - \cos x \sin x)\Big|_0^{\pi/2} = \frac{1}{4}\Big(\frac{\pi}{2} - 0 - (0)(1) + (1)(0)\Big) \\ &= \frac{\pi}{8}. \end{aligned}$$

The moment about the y-axis is

$$M_y = \int_0^{\pi/2} x\rho \sin x\, dx = \int_0^{\pi/2} x \sin x\, dx.$$

To evaluate this, we integrate by parts and find

$u = x,$	$dv = \sin x\, dx$
$du = dx,$	$v = -\cos x$

$$\begin{aligned} M_y &= \int_0^{\pi/2} x \sin x\, dx = -(x\cos x)\Big|_0^{\pi/2} + \int_0^{\pi/2} \cos x\, dx \\ &= -\Big(\frac{\pi}{2}(0) - (0)(1)\Big) + \sin x\Big|_0^{\pi/2} = 1 - 0 \\ &= 1. \end{aligned}$$

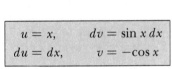

Putting everything together, we find that the centroid is

$$(\bar{x}, \bar{y}) = (M_y/M, M_x/M) = \Big(1, \frac{\pi}{8}\Big) \approx (1, 0.3927)$$

(see Figure 8.7.3). ◢

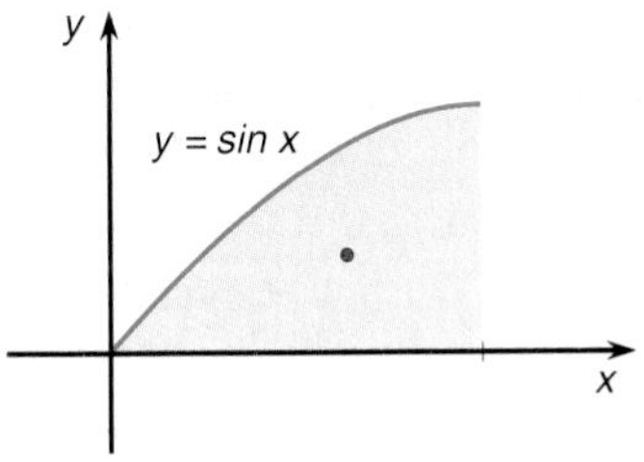

Figure 8.7.3

EXAMPLE 4 Find the arclength of the curve $y = e^x$ between $x = \ln(\sqrt{3})$ and $x = \ln(\sqrt{8})$.

Solution The arclength equals

$$\int ds = \int_{\ln(\sqrt{3})}^{\ln(\sqrt{8})} \sqrt{1+e^{2x}}\, dx = \int_{\ln(\sqrt{3})}^{\ln(\sqrt{8})} \sqrt{1+e^{2x}}\frac{1}{e^{2x}} e^{2x}\, dx$$

$u = 1 + e^{2x}$
$du = 2e^{2x}\, dx$

$$= \frac{1}{2}\int_4^9 \sqrt{u}\frac{1}{u-1}\, du$$

$$\begin{aligned}
\boxed{\begin{array}{l} u = v^2 \\ du = 2v\,dv \end{array}} \quad &= \int_2^3 v\frac{1}{v^2-1}v\,dv = \int_2^3 \frac{v^2}{v^2-1}\,dv \\
&= \int_2^3 \left(1 + \frac{1}{v^2-1}\right) dv \\
&= \int_2^3 \left(1 + \frac{1}{2}\left(\frac{1}{v-1} - \frac{1}{v+1}\right)\right) dv \\
&= \left(v + \frac{1}{2}\ln|v-1| - \frac{1}{2}\ln|v+1|\right)\Big|_2^3 \\
&= 1 + \frac{1}{2}(\ln 2 - \ln 4 - \ln 1 + \ln 3) \\
&= 1 + \frac{1}{2}\ln\frac{3}{2} \approx 1.2027.
\end{aligned}$$

$\ln 2 - \ln 4 - \ln 1 + \ln 3 = \ln\frac{(2)(3)}{(4)(1)} = \ln\frac{3}{2}$

In Example 4 we made two successive substitutions and then had to do a (simple) partial fractions calculation. To be able to make the first substitution $u = 1 + e^{2x}$, we multiplied and divided the integrand by e^{2x}, in order that the integrand would contain $e^{2x}dx$. This was needed, along with the constant $\frac{1}{2}$, to form du.

EXAMPLE 5 Find the surface area of the surface obtained by revolving about the x-axis the part of the curve $y = \cos x$ lying between $x = 0$ and $x = \frac{\pi}{2}$. See Figures 8.7.4(i) and (ii).

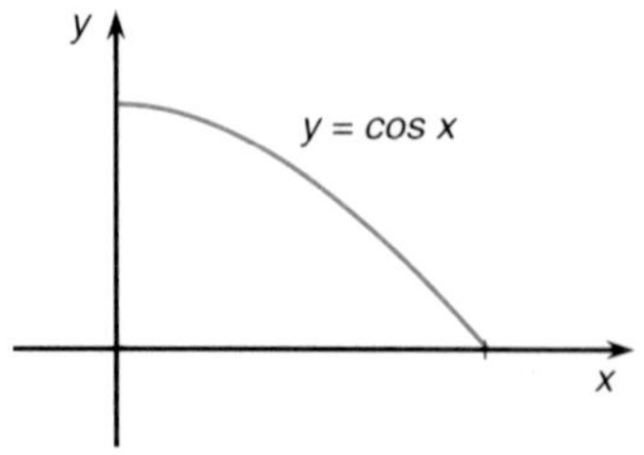

Figure 8.7.4(i)

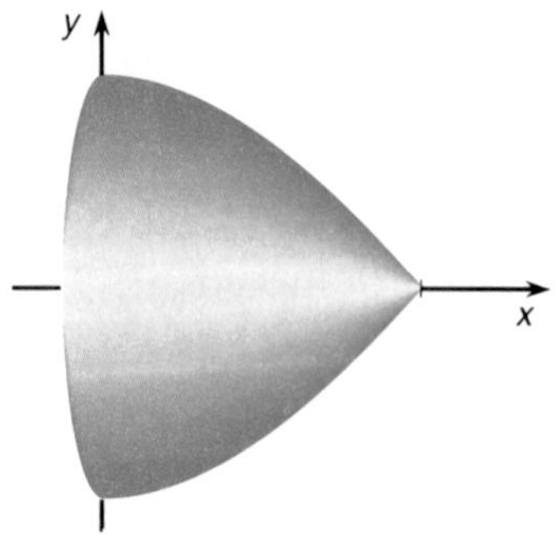

Figure 8.7.4(ii)

Solution The surface area is given by

$$\begin{aligned}
S = 2\pi \int \cos x\,ds &= 2\pi \int_0^{\pi/2} \cos x\sqrt{1 + (-\sin x)^2}\,dx \\
&= 2\pi \int_0^{\pi/2} \cos x\sqrt{1 + \sin^2 x}\,dx \\
\boxed{\begin{array}{l} u = \sin x \\ du = \cos x\,dx \end{array}} \quad &= 2\pi \int_0^1 \sqrt{1 + u^2}\,du \\
\boxed{\begin{array}{l} u = \tan\theta \\ du = \sec^2\theta\,d\theta \end{array}} \quad &= 2\pi \int_0^{\pi/4} \sqrt{1 + \tan^2\theta}\,\sec^2\theta\,d\theta \\
&= 2\pi \int_0^{\pi/4} \sec^3\theta\,d\theta \\
&= \pi\left(\sec\theta\tan\theta\Big|_0^{\pi/4} + \int_0^{\pi/4} \sec\theta\,d\theta\right) \\
&= \pi\left(\sqrt{2}(1) - (1)(0) + \ln|\sec\theta + \tan\theta|\Big|_0^{\pi/4}\right) \\
&= \pi\left(\sqrt{2} + \ln|\sqrt{2} + 1| - \ln|1 + 0|\right) \\
&= \pi\left(\sqrt{2} + \ln(1 + \sqrt{2})\right) \\
&\approx 7.2118.
\end{aligned}$$

Here we needed two substitutions, and then we had to integrate $\sec^3\theta$, which is beginning to be familiar, using the reduction formula of Theorem 8.2.5(ii).

Exercises 8.7

I

Find the area of the region lying between each of the following curves and the x-axis over the specified interval.

1. $y = x \sin x$, $x \in [0, \pi]$

2. $y = x \cos x$, $x \in \left[-\frac{\pi}{4}, \frac{\pi}{4}\right]$

3. $y = \tan^2 x \sec x$, $x \in \left[0, \frac{\pi}{4}\right]$

4. $y = \sec x \sin^2 x$, $x \in \left[-\frac{\pi}{4}, 0\right]$

5. $y = (x+1)/(x^2 - 4)$, $x \in [-1, 1]$

6. $y = e^x/(e^{2x} - 1)$, $x \in [-2, -1]$

7. $y = xe^x$, $x \in [-1, 1]$

8. $y = \cos^4 x$, $x \in [0, \pi]$

9. $y = \ln x$, $x \in [1, e]$

10. $y = 1/(3x^2 + 6x + 15)$, $x \in [3, 4]$

Find the arclength of each of the following curves.

11. $y = 3x^2$, $1 \le x \le 2$

12. $y = x^2 - 2x$, $0 \le x \le 1$

13. $y = \sqrt{2x}$, $2 \le x \le 8$

14. $y = \sqrt{x+1}$, $0 \le x \le 8$

15. $y = \sqrt{3x+1}$, $1 \le x \le 8$

16. $y = \ln(\sin x)$, $\frac{\pi}{6} \le x \le \frac{\pi}{3}$

17. $y = \ln(\cos x)$, $0 \le x \le \frac{\pi}{6}$

18. $y = \frac{1}{5}\ln\big(\sin(5x)\big)$, $\frac{\pi}{20} \le x \le \frac{\pi}{10}$

19. $y = \ln x$, $1 \le x \le 4$

20. $y = 3 + \ln(x^4)$, $1 \le x \le 2$

II

21. Find the arclength of the parabola $y = x^2$ between the points $(0, 0)$ and $(2, 4)$.

22. In Example 6.10.8 (p. 500) we used Simpson's Rule to estimate the integral obtained in figuring out the arclength of the curve $y = \frac{1}{2}x^2$ from $x = 0$ to $x = 2$. Evaluate this integral exactly, and compare with the value given in Example 6.10.8.

***23.** Find the arclength of the parabola $y = x^2$ from $x = 0$ to $x = a$, for any $a > 0$.

***24.** (i) Find the arclength of the parabola $y = x^2$ from $x = -a$ to $x = 0$, for any $a > 0$.

(ii) Find the arclength of the parabola $y = x^2$ from $x = a$ to $x = b$, for any $a < b$.

(iii) Use properties of trigonometric functions and the natural logarithm to show that the answer to part (i) equals the answer to Exercise 23, something that the geometrical situation shows us must be true.

***25.** Find the arclength of the parabola $y = kx^2$ from $x = a$ to $x = b$, for any $a < b$.

***26.** Consider the problem of finding the arclength of the parabola $y = 2x^2 + 8x + 10$ between $x = 0$ and $x = 1$.

(i) Solve this problem by evaluating the appropriate integral.

(ii) Solve the problem by using Exercise 25 without evaluating any integrals. (*Hint:* Draw a picture.)

***27.** Find the arclength of the curve $y = 1 + \frac{1}{2}\ln\big(\cos(2x)\big)$, for x between $x = \frac{\pi}{8}$ and $x = \frac{\pi}{6}$.

***28.** Find the arclength of the curve $y = e^{4x}$, for $0 \le x \le \ln 2$.

***29.** (i) Find the arclength of the curve $y = \ln x$, for x between $x = 1$ and $x = e$.

(ii) For the curve described in part (i), find the length of the line segment joining its endpoints, and compare with the arclength of the curve.

***30.** (i) Find the arclength of the curve $y = e^{3x}$, for x between $x = 0$ and $x = 1$.

(ii) Compare the length of the curve with the length of the straight line segment joining its endpoints.

***31.** Find the surface area of the surface of revolution obtained by revolving about the x-axis the curve $y = \sin x$, for $x \in \left[0, \frac{\pi}{3}\right]$.

***32.** Find the surface area of the surface of revolution obtained by revolving about the x-axis the curve $y = \sin\frac{x}{2}$, for $x \in [0, \pi]$.

***33.** A *torus* (doughnut shape) is obtained by revolving about the x-axis the circle of radius 1 with center at $(0, 2)$. See Figure 8.7.5. (The equation of the circle is $x^2 + (y-2)^2 = 1$.) Find the surface area of the surface. (*Hint:* Work separately with the upper and lower semicircles.)

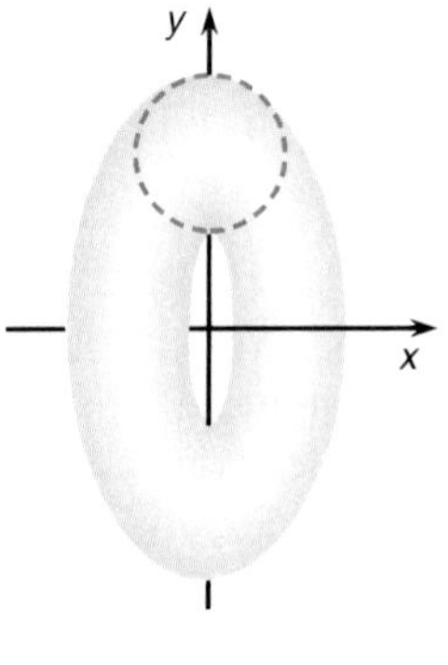

Figure 8.7.5

*34. Suppose $R > r > 0$. A torus is obtained by revolving about the x-axis the circle of radius r with center at $(0, R)$. (The equation of the circle is $x^2 + (y - R)^2 = r^2$.) Find the surface area of the surface.

35. (i) Find the centroid of the region between $y = \sin x$ and the x-axis and between $x = 0$ and $x = \frac{\pi}{4}$.
(ii) Compare this with the centroid of the triangular region whose vertices are the "corners" of the region in part (i).

*36. Find the centroid of the region between $y = \cos x$ and the x-axis and between $x = 0$ and $x = \frac{\pi}{3}$.

*37. Find the centroid of the region between $y = x \sin x$ and the x-axis and between $x = 0$ and $x = \frac{\pi}{2}$.

*38. Find the centroid of the region between $y = x^2 \sin x$ and the x-axis and between $x = 0$ and $x = \frac{\pi}{2}$.

8.8 Improper Integrals

DIVERGE
CONVERGE

In this section we consider what happens when we try to integrate a function that is not bounded. The integral of such a function does not exist, but under certain circumstances it is possible to think of it as a limit of integrals that do exist. In a similar way it is possible to think of integrals in which one or both limits of integration are infinite, expressing them as limits of ordinary integrals.

This topic is not really a technique of integration, so in a sense it does not belong in this chapter. However, it is convenient to study it at this point, now that we have become proficient at integrating.

EXAMPLE 1 Suppose we want to integrate $\int_0^\infty e^{-x}dx$. This formula does not make any sense as it stands, but as usual when we see the infinity sign, we should expect it to have something to do with limits.

Although we do not know what it means to have an integral with an infinite limit of integration, we are quite familiar with integrals like $\int_0^t e^{-x}dx$, which can be interpreted as the area under the graph of $y = e^{-x}$ for x between $x = 0$ and $x = t$. If we want to talk about $\int_0^\infty e^{-x}dx$, it ought to be the area under the curve $y = e^{-x}$ along the whole positive x-axis, as shown in Figure 8.8.1. One way to define this area would be to calculate the area under the curve between $x = 0$ and $x = t$, and then take a limit as $t \to \infty$ (see Figure 8.8.2).

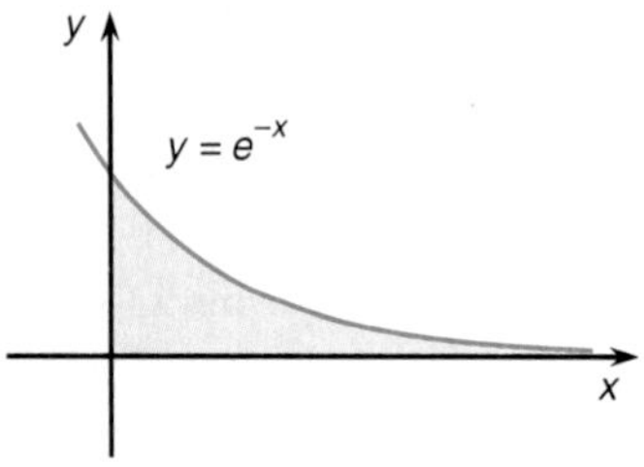

Figure 8.8.1

This suggests that we try letting

$$\int_0^\infty e^{-x}dx = \lim_{t\to\infty} \int_0^t e^{-x}dx.$$

It is easy to evaluate the integral on the right side: $\int_0^t e^{-x}dx = -e^{-x}\Big|_0^t = -e^{-t} - (-e^0) = 1 - e^{-t}$. If we let $t \to \infty$, we see that $e^{-t} = \frac{1}{e^t} \to 0$, so

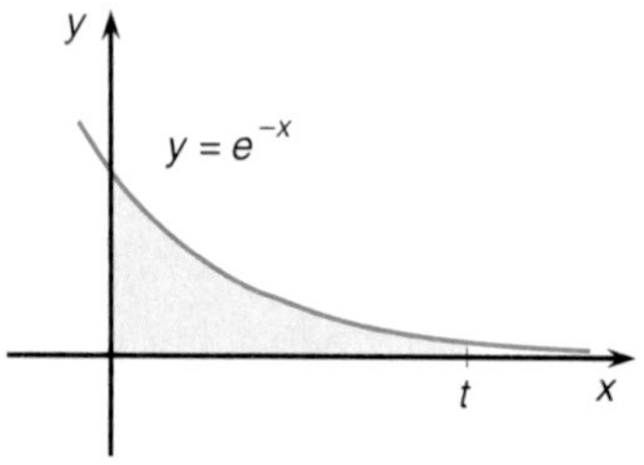

Figure 8.8.2

$$\int_0^\infty e^{-x}dx = \lim_{t\to\infty} \int_0^t e^{-x}dx = \lim_{t\to\infty} (1 - e^{-t})$$
$$= 1.$$

This example shows that we can define $\int_a^\infty f(x)\,dx$ to be $\lim_{t\to\infty} \int_a^t f(x)\,dx$, whenever the limit exists. Similarly, we define $\int_{-\infty}^a f(x)\,dx = \lim_{t\to-\infty} \int_t^a f(x)\,dx$, whenever the limit exists.

Finally, we should decide what is meant by $\int_{-\infty}^\infty f(x)\,dx$. This can be a little more tricky than it seems. A natural first choice might be to consider $\lim_{t\to\infty} \int_{-t}^t f(x)\,dx$. The difficulty with this is perhaps best explained by an example.

If we let $f(x) = x$, then for any $t > 0$, $\int_{-t}^{t} f(x)\,dx = \int_{-t}^{t} x\,dx = \frac{1}{2}x^2\big|_{-t}^{t} = \frac{1}{2}\left(t^2 - (-t)^2\right) = 0$. From this the limit is easily found: $\lim_{t\to\infty} \int_{-t}^{t} x\,dx = 0$.

The problem is that neither $\int_0^{\infty} x\,dx$ nor $\int_{-\infty}^{0} x\,dx$ exists. Indeed, we find $\lim_{t\to\infty} \int_0^t x\,dx = \lim_{t\to\infty} \frac{1}{2}(t^2 - 0^2) = \lim_{t\to\infty} \frac{t^2}{2}$, and this limit diverges to ∞. Practically the same calculation shows that $\lim_{t\to-\infty} \int_t^0 x\,dx = -\infty$.

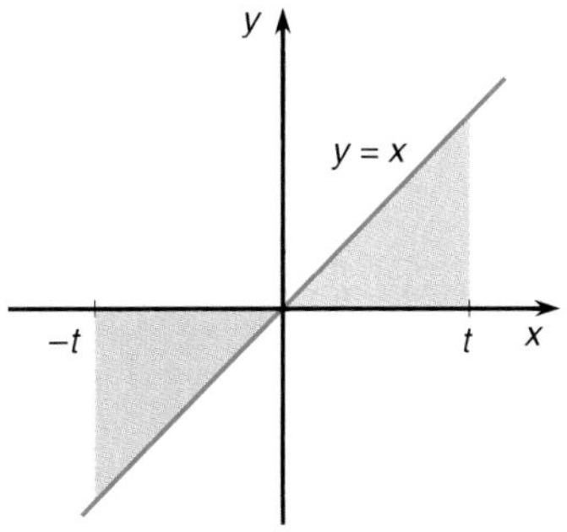

Figure 8.8.3

Neither of the integrals over half the real axis exists, and the fact that $\int_{-t}^{t} x\,dx$ tends to a limit as $t \to \infty$ is a result of the cancellation of the negative signed area to the left of the origin with the positive signed area to the right of the origin (see Figure 8.8.3).

It turns out that this kind of limit is not sufficient for many purposes, so it is preferable to define $\int_{-\infty}^{\infty} f(x)\,dx = \int_{-\infty}^{a} f(x)\,dx + \int_a^{\infty} f(x)\,dx$, *provided* both these integrals exist. It is necessary to take the limit at ∞ and the limit at $-\infty$ *separately* and require that both of them exist. It is not hard to show that the choice of a is not important. If both "one-sided" integrals exist for one choice of a, then the same is true for any other a, and moreover the sum of the two integrals will be the same no matter which a is used.

We summarize.

DEFINITION 8.8.1

If $f(x)$ is a function on $[a, \infty)$, we define the **improper integral**

$$\int_a^{\infty} f(x)\,dx = \lim_{t\to\infty} \int_a^t f(x)\,dx,$$

provided the limit exists.

Similarly, if $f(x)$ is a function on $(-\infty, a]$, we define the **improper integral**

$$\int_{-\infty}^{a} f(x)\,dx = \lim_{t\to-\infty} \int_t^a f(x)\,dx,$$

provided the limit exists.

Finally, if $f(x)$ is a function on $(-\infty, \infty)$, we define the **improper integral**

$$\int_{-\infty}^{\infty} f(x)\,dx = \lim_{t\to-\infty} \int_t^a f(x)\,dx + \lim_{t\to\infty} \int_a^t f(x)\,dx,$$

provided the limits exist. Any choice of a will give the same result.

If any of the limits does not exist, the corresponding improper integral is said to **diverge**. In this case we also say that the integral "does not exist." If the integral does exist, we sometimes say that the integral **converges** or **is convergent**.

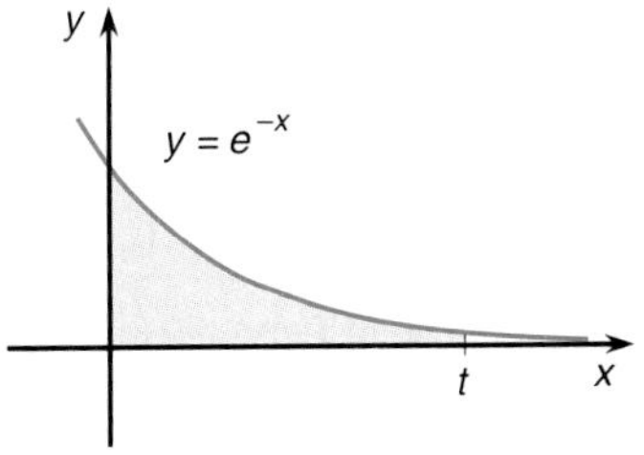

Figure 8.8.4(i)

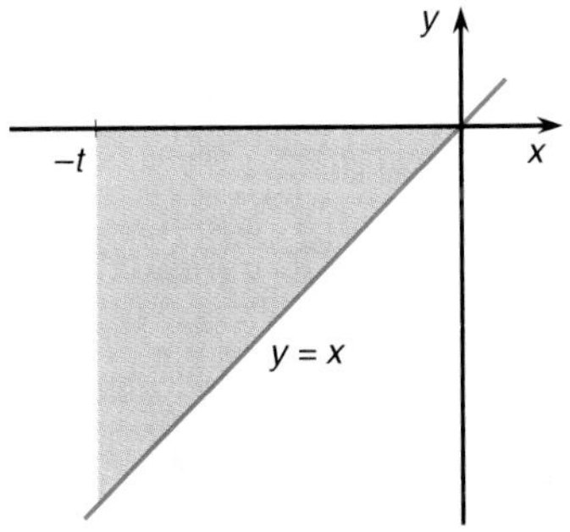

Figure 8.8.4(ii)

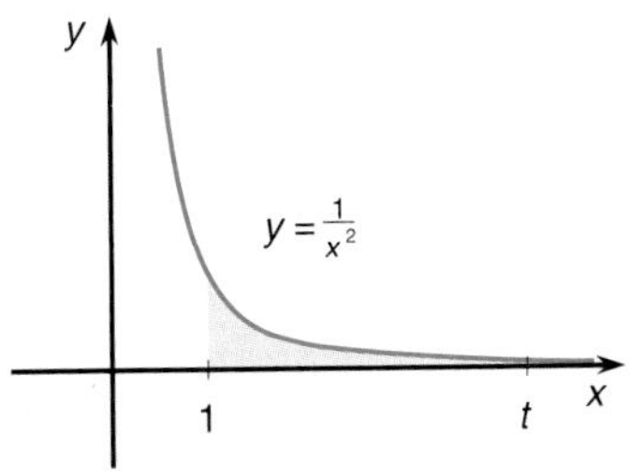

Figure 8.8.4(iii)

EXAMPLE 2

(i) We have already seen that $\int_0^{\infty} e^{-x}dx = \lim_{t\to\infty} (\int_0^t e^{-x}dx) = \lim_{t\to\infty} (-e^{-x})\big|_0^t = \lim_{t\to\infty} (1 - e^{-t}) = 1$. See Figure 8.8.4(i).

(ii) Consider $\int_{-\infty}^{0} x\,dx = \lim_{t\to-\infty} (\int_t^0 x\,dx) = \lim_{t\to-\infty} \frac{1}{2}(0^2 - t^2) = -\frac{1}{2} \lim_{t\to-\infty} t^2 = -\infty$. Since this limit diverges to $-\infty$, we say that the integral diverges. See Figure 8.8.4(ii).

(iii) Consider $\int_1^{\infty} \frac{1}{x^2}\,dx$, illustrated in Figure 8.8.4(iii). By definition it equals $\lim_{t\to\infty} \left(\int_1^t \frac{1}{x^2}\,dx\right) = \lim_{t\to\infty} \left(-\frac{1}{x}\right)\big|_1^t = \lim_{t\to\infty} \left(-\frac{1}{t} + \frac{1}{1}\right) = 1$.

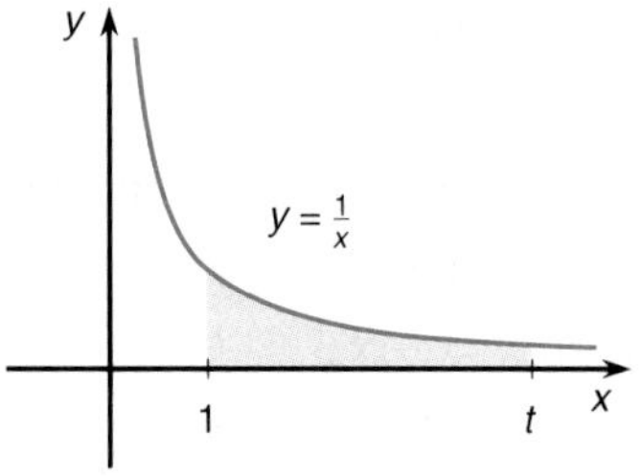

Figure 8.8.4(iv)

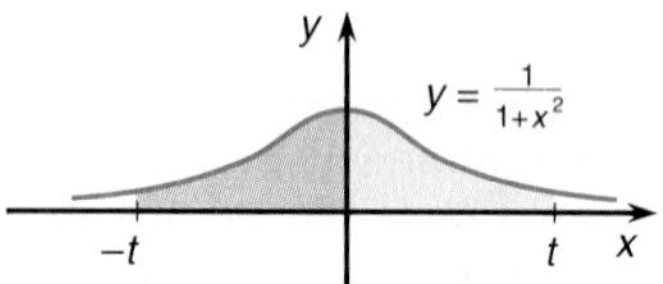

Figure 8.8.4(v)

(iv) Consider $\int_1^\infty \frac{1}{x}\,dx = \lim_{t\to\infty}\left(\int_1^t \frac{1}{x}\,dx\right) = \lim_{t\to\infty} \ln t = \infty$. The limit does not exist, so the integral diverges. See Figure 8.8.4(iv).

(v) The integral $\int_{-\infty}^{\infty} \frac{1}{1+x^2}\,dx$ must be broken up into two pieces:

$$\begin{aligned}\int_{-\infty}^{\infty} \frac{1}{1+x^2}\,dx &= \lim_{t\to-\infty}\left(\int_t^0 \frac{1}{1+x^2}\,dx\right) + \lim_{t\to\infty}\left(\int_0^t \frac{1}{1+x^2}\,dx\right)\\ &= \lim_{t\to-\infty}\left(\arctan x\Big|_t^0\right) + \lim_{t\to\infty}\left(\arctan x\Big|_0^t\right)\\ &= \lim_{t\to-\infty}(-\arctan t) + \lim_{t\to\infty}(\arctan t).\end{aligned}$$

Now as $t\to\infty$, we know that $\arctan t \to \frac{\pi}{2}$, and as $t\to-\infty$, $\arctan t \to -\frac{\pi}{2}$. Both limits exist, so

$$\int_{-\infty}^{\infty} \frac{1}{1+x^2}\,dx = -\left(-\frac{\pi}{2}\right) + \frac{\pi}{2} = \pi.$$

See Figure 8.8.4(v).

There is another way in which an integral can be "improper." It arises when the integrand tends to $\pm\infty$ at some point in the interval of integration.

EXAMPLE 3 Consider $\int_0^1 \frac{1}{\sqrt{x}}\,dx$. This integral does not make sense because the integrand tends to ∞ as $x \to 0^+$.

However, in the spirit of the previous definition we could let $t > 0$ and look at the integral $\int_t^1 \frac{1}{\sqrt{x}}\,dx$. This integral does exist for any positive t, and we could try taking the limit as $t \to 0^+$:

$$\begin{aligned}\lim_{t\to0^+}\left(\int_t^1 x^{-1/2}\,dx\right) &= \lim_{t\to0^+}\left(\frac{1}{\frac{1}{2}}x^{1/2}\Big|_t^1\right)\\ &= \lim_{t\to0^+} 2(\sqrt{1}-\sqrt{t}) = 2(1-0)\\ &= 2.\end{aligned}$$

This suggests a way to define the integral in cases in which the integrand tends to $\pm\infty$ at some point.

DEFINITION 8.8.2

Suppose $f(x)$ is a function on the interval $[a, b]$ and that $|f(x)| \to \infty$ as $x \to a^+$. Define the **improper integral** by

$$\int_a^b f(x)\,dx = \lim_{t\to a^+}\int_t^b f(x)\,dx,$$

provided the limit exists.

Similarly, if $\lim_{x\to b^-} |f(x)| = \infty$, define

$$\int_a^b f(x)\,dx = \lim_{t\to b^-}\int_a^t f(x)\,dx,$$

provided the limit exists.

In either case if the limit does not exist, the integral is said to **diverge**.

EXAMPLE 4

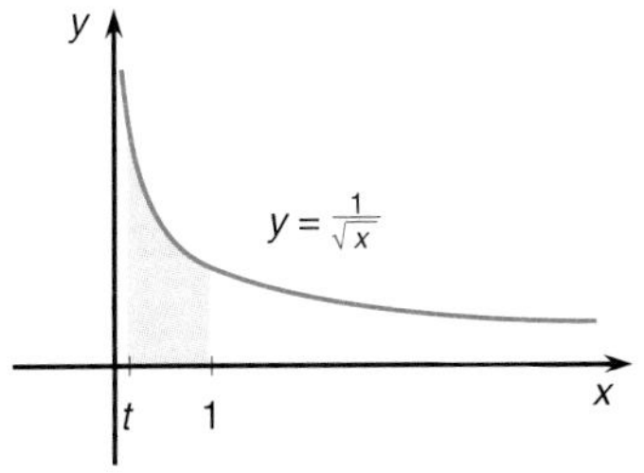

Figure 8.8.5(i)

(i) Example 3 above shows that

$$\int_0^1 \frac{1}{\sqrt{x}}\,dx = \lim_{t\to 0^+}\int_t^1 x^{-1/2}dx = \lim_{t\to 0^+} 2(\sqrt{1}-\sqrt{t}) = 2.$$

See Figure 8.8.5(i).

(ii) Consider $\int_0^2 \ln x\,dx$. The integrand tends to $-\infty$ as $x \to 0^+$, so we should consider the limit

$$\lim_{t\to 0^+}\int_t^2 \ln x\,dx = \lim_{t\to 0^+}(x\ln x - x)\Big|_t^2$$
$$= 2\ln 2 - 2 - \lim_{t\to 0^+}(t\ln t - t).$$

To find $\lim_{t\to 0^+}(t\ln t) = \lim_{t\to 0^+}\left(\frac{\ln t}{1/t}\right)$, we apply l'Hôpital's Rule:

$$\lim_{t\to 0^+}\frac{\ln t}{\frac{1}{t}} = \lim_{t\to 0^+}\frac{\frac{d}{dt}(\ln t)}{\frac{d}{dt}\left(\frac{1}{t}\right)} = \lim_{t\to 0^+}\frac{\frac{1}{t}}{-\frac{1}{t^2}}$$
$$= \lim_{t\to 0^+}(-t) = 0.$$

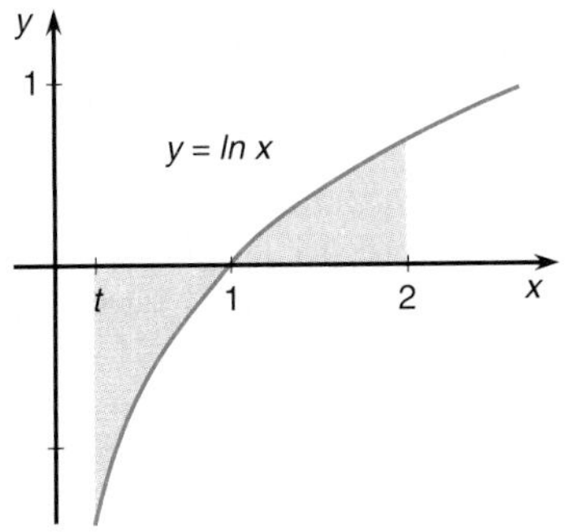

Figure 8.8.5(ii)

From this we find that

$$\int_0^2 \ln x\,dx = \lim_{t\to 0^+}\int_t^2 \ln x\,dx = \lim_{t\to 0^+}(x\ln x - x)\Big|_t^2$$
$$= \lim_{t\to 0^+}(2\ln 2 - 2 - t\ln t + t) = 2\ln 2 - 2.$$

The limit exists, so the improper integral exists. See Figure 8.8.5(ii).

(iii) Consider $\int_{-2}^2 \frac{1}{x-1}\,dx$. The difficulty here is that the integrand goes to $\pm\infty$ as x approaches 1 from either side, and 1 is inside the interval $[-2, 2]$ over which we are integrating.

The solution is to break the integral into integrals over $[-2, 1]$ and $[1, 2]$. Each of them is improper. Consider first

$$\int_{-2}^1 \frac{1}{x-1}\,dx = \lim_{t\to 1^-}\int_{-2}^t \frac{1}{x-1}\,dx$$
$$= \lim_{t\to 1^-}\ln|x-1|\Big|_{-2}^t = \lim_{t\to 1^-}(\ln|t-1| - \ln|-2-1|)$$
$$= \lim_{t\to 1^-}(\ln|t-1| - \ln 3).$$

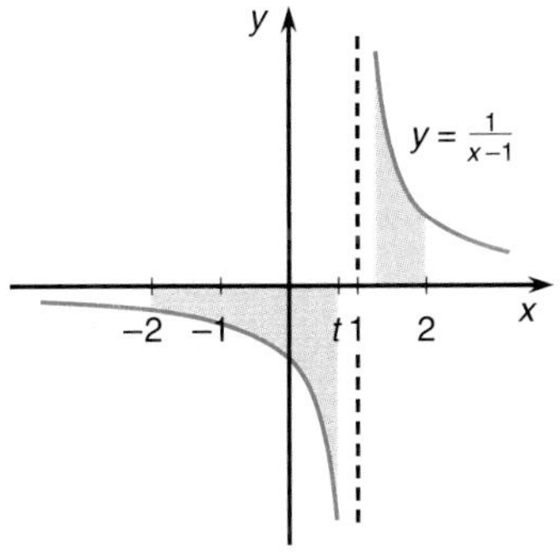

Figure 8.8.5(iii)

Now $\lim_{t\to 1^-}\ln|t-1| = \lim_{x\to 0^-}\ln|x| = \lim_{x\to 0^+}\ln x$, and this limit does not exist. (Recall the graph of $y = \ln x$, shown in Figure 8.8.5(ii), which tends to $-\infty$ as $x \to 0^+$.)

We see that the first integral does not exist, so without even looking at the second we conclude that the original integral $\int_{-2}^2 \frac{1}{x-1}\,dx$ does not exist. (In fact, the second one does not exist either, as can easily be checked.) See Figure 8.8.5(iii).

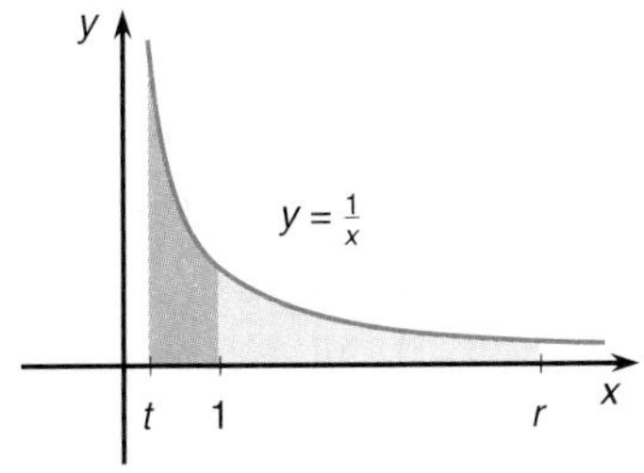

Figure 8.8.5(iv)

(iv) The integral $\int_0^\infty \frac{1}{x}\,dx$ is improper in two ways. First, it has an infinite limit of integration, and second, the integrand tends to ∞ as $x \to 0^+$. To deal with such a situation, we should break the integral into pieces, each of which contains only one of the difficulties. We consider $\int_0^1 \frac{1}{x}\,dx$ and $\int_1^\infty \frac{1}{x}\,dx$. See Figure 8.8.5(iv).

The first of these equals

$$\lim_{t\to 0+} \int_t^1 \frac{1}{x}\, dx = \lim_{t\to 0+} (\ln 1 - \ln t) = -\lim_{t\to 0+} \ln t = \infty.$$

In other words, the integral diverges.

The second is

$$\lim_{r\to\infty} \int_1^r \frac{1}{x}\, dx = \lim_{r\to\infty} \ln r = \infty.$$

We see that in fact both parts of the integral diverge. In terms of areas this means that the area under $y = \frac{1}{x}$ between $x = 0$ and $x = 1$ and the area under $y = \frac{1}{x}$ to the right of $x = 1$ are both *infinite.*

Notice that it is possible to have improper integrals that are improper at points *inside* the interval over which we are integrating, as in part (iii) of Example 4. In such a case it is necessary to break the integral into pieces, on each of which the integral is improper at one endpoint. Any convenient breaking point can be used.

It is also possible for an integral to be improper at more than one point, and possibly in different ways. In each case the same strategy is applied: Break the integral into simple pieces, each of which has only one endpoint at which the integral is improper.

In analyzing such an integral we check each piece to see whether it exists. If we find that one of the pieces diverges, then it is not necessary to check the others; we can say immediately that the original integral diverges.

Exercises 8.8

I

Decide whether or not each of the following improper integrals exists, and evaluate those that do.

1. $\int_1^\infty \frac{1}{x^3}\, dx$
2. $\int_{-\infty}^{-1} \frac{1}{x^5}\, dx$
3. $\int_0^1 x^{-1/3} dx$
4. $\int_1^\infty \ln x\, dx$
5. $\int_0^\infty \frac{1}{1+x^2}\, dx$
6. $\int_0^\infty \frac{x}{x^2+1}\, dx$
7. $\int_0^\infty \frac{x^2}{x^2+4}\, dx$
8. $\int_1^\infty e^{-3x} dx$
9. $\int_0^\infty xe^{-x^2} dx$
10. $\int_0^\infty \sin x\, dx$
11. $\int_0^4 \frac{1}{x-2}\, dx$
12. $\int_{-1}^6 \frac{1}{x^2-4}\, dx$
13. $\int_0^{\pi/2} \sec x\, dx$
14. $\int_0^\pi \cot x\, dx$
15. $\int_0^1 \frac{e^x}{e^x-1}\, dx$
16. $\int_{-1}^1 \frac{1}{e^x-e^{-x}}\, dx$
17. $\int_{-\infty}^\infty \frac{1}{x^2+9}\, dx$
18. $\int_0^{\pi/2} x \sec^2 x\, dx$
19. $\int_0^\infty 3\, dx$
20. $\int_{-1}^1 \ln|x|\, dx$

II

21. Let r be a real number and consider $\int_1^\infty x^r\, dx$.
    (i) For which values of r does the improper integral exist?
    (ii) Evaluate the integral when it exists.

22. Let r be a real number and consider $\int_0^1 x^r\, dx$.
    (i) For which values of r is the integral improper?
    (ii) For which values of r does the improper integral exist?
    (iii) Evaluate the integral when it exists.

*23. Suppose r is a real number and n is a positive integer. If $r \in [a, b]$, show that $\int_a^b \frac{1}{(x-r)^n}\, dx$ does not exist.

*24. Suppose $R(x) = \frac{p(x)}{q(x)}$ is a rational function and r is a real number for which $q(r) = 0$ and $p(r) \neq 0$. Show that if $r \in [a, b]$, then $\int_a^b R(x)\, dx$ does not exist. (*Hint:* See the previous exercise.)

25. In the discussion before Definition 8.8.1 (p. 607) we considered the improper integral $\int_{-\infty}^{\infty} x\,dx$. Show that $\lim_{t\to\infty} \int_{-t}^{t+1} x\,dx$ does not exist.

***26.** If $\int_{-\infty}^{\infty} f(x)\,dx$ exists, show that, unlike what happened in the previous exercise,

$$\lim_{t\to\infty}\left(\int_{-t}^{t+1} f(x)\,dx\right)$$

converges *and* equals $\int_{-\infty}^{\infty} f(x)\,dx$.

***27.** Give conditions on the real number A that determine whether or not the integral $\int_a^b \frac{e^x}{e^x - A}\,dx$ exists.

****28.** (i) Suppose $f(x)$ and $g(x)$ are continuous functions with $f(x) \geq g(x) \geq 0$ for all $x \in [a, \infty)$, and suppose the improper integral $\int_a^{\infty} f(x)\,dx$ exists. Show that the improper integral $\int_a^{\infty} g(x)\,dx$ exists.

(ii) Use the result of part (i) to show that $\int_1^{\infty} \frac{1}{x^4+3x^3+7x^2+11x+5}\,dx$ exists. (*Hint:* Let $f(x) = \frac{1}{x^4}$.)

29. In Example 2(v) (p. 608) we evaluated $\int_{-\infty}^{\infty} \frac{1}{1+x^2}\,dx$ by writing

$$\int_{-\infty}^{\infty} \frac{1}{1+x^2}\,dx = \int_{-\infty}^{0} \frac{1}{1+x^2}\,dx + \int_{0}^{\infty} \frac{1}{1+x^2}\,dx.$$

Confirm that it can also be done with $\int_{-\infty}^{\infty} \frac{1}{1+x^2}\,dx = \int_{-\infty}^{1} \frac{1}{1+x^2}\,dx + \int_1^{\infty} \frac{1}{1+x^2}\,dx$.

30. If $\int_{-\infty}^{\infty} f(x)\,dx$ converges, show that the sum $\int_{-\infty}^{a} f(x)\,dx + \int_a^{\infty} f(x)\,dx$ does not depend on the choice of a.

POINT TO PONDER

In Example 6.9.5 (p. 489), to find the surface area of a sphere we had to integrate

$$2\pi \int f(x)\sqrt{1+f'(x)^2}\,dx.$$

Since the function $f(x) = \sqrt{r^2 - x^2}$ is not differentiable at the endpoints $x = \pm r$, we had to integrate from $-a$ to a and then let $a \to r$ (see Figures 8.8.6(i) and (ii)).

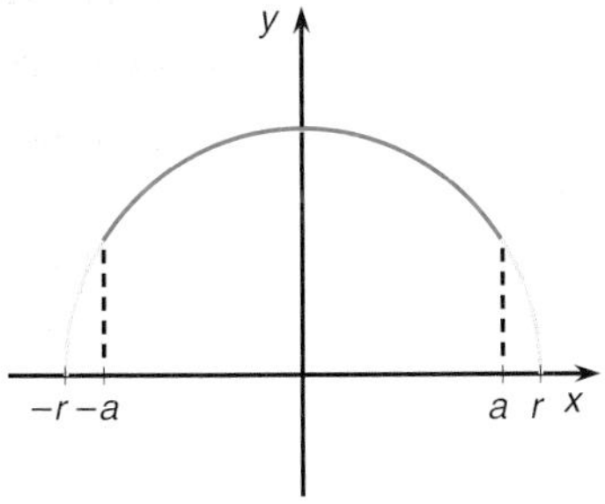

Figure 8.8.6(i)

This looks similar to what we do for improper integrals, but the reason is slightly different. The integrand does not tend to infinity; the difficulty is that it is not defined at the endpoints. Sometimes this is regarded as another type of improper integral.

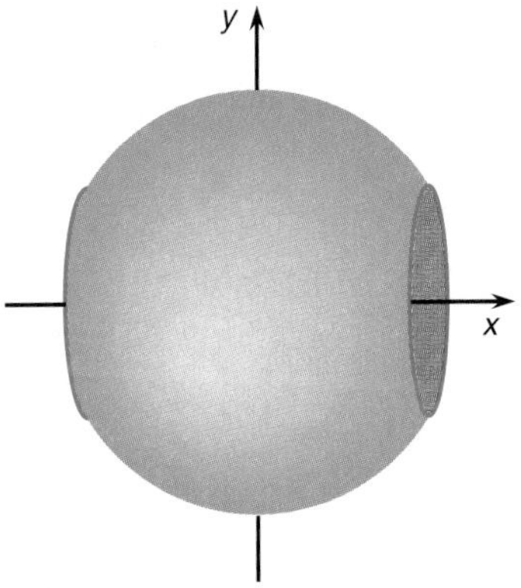

Figure 8.8.6(ii)

Chapter Summary

§8.1
integration by parts
reduction formula
definite integrals by parts
reduction formulas for $\cos^n x$, $\sin^n x$
(Theorem 8.1.3, p. 565)

§8.2
$\cos^2 x + \sin^2 x = 1$
$1 + \tan^2 x = \sec^2 x$
trigonometric integrals
integrating $\cos^m x \sin^n x$ (Summary 8.2.4, p. 571)
reduction formulas for $\tan^n x$, $\sec^n x$ (Theorem 8.2.5, p. 574)

integrating $\tan^m x \sec^n x$ (Summary 8.2.6, p. 576)

§8.3
trigonometric substitutions

§8.4
completing the square

§8.5
partial fractions
proper rational function
irreducible quadratic polynomial
$b^2 - 4ac < 0$

§8.6
fractional powers
$\tan\left(\frac{\theta}{2}\right)$ substitution (Summary 8.6.1, p. 600)

§8.7
additional applications
arclength

§8.8
improper integral (Definitions 8.8.1, p. 607 and 8.8.2, p. 608)
diverge
converge

Review Exercises

I

Evaluate the following integrals.

1. $\int (x+2)e^{3x}\,dx$
2. $\int (4e^{2x} + 2e^x)/(1 - e^{2x})\,dx$
3. $\int \frac{1}{x^2-6x+13}\,dx$
4. $\int \sin^3(5x)\,dx$
5. $\int \sqrt{x}\sin(\sqrt{x})\,dx$
6. $\int x^3\cos(x^2)\,dx$
7. $\int \frac{2x^4-10x^3+11x^2+4x-5}{x^3-6x^2+11x-6}\,dx$
8. $\int \frac{x^3+2}{x^2-1}\,dx$
9. $\int x\sin^2 x\cos x\,dx$
10. $\int x\cos^3 x\,dx$
11. $\int \sqrt{2x^2-12x+26}\,dx$
12. $\int \frac{1}{12+\sin x}\,dx$
13. $\int \frac{1}{x^2+6x+8}\,dx$
14. $\int \frac{2x-5}{(x+1)^3}\,dx$
15. $\int \frac{1}{1+\sqrt{x}}\,dx$
16. $\int \tan^2 x\,dx$
17. $\int \cos(\sqrt{x})\,dx$
18. $\int e^{\sqrt{x+7}}\,dx$
19. $\int \ln(1+x^2)\,dx$
20. $\int \frac{\cos(2x)}{1+\cos(2x)}\,dx$
21. $\int \frac{\sin(2x)}{1+\sin x}\,dx$
22. $\int \frac{\cos(2x)}{1+\cos x}\,dx$
23. $\int x\cos x\sin x\,dx$
24. $\int xe^{\sqrt{x}}\,dx$
25. $\int x(x+8)^{1/3}\,dx$
26. $\int \frac{1}{x}(x+1)^{1/3}\,dx$
27. $\int \sqrt{1+e^x}\,dx$
28. $\int x\sin x\sec^2 x\,dx$
29. $\int x\arctan x\,dx$
30. $\int x^2\arctan x\,dx$
31. $\int \csc^3 x\cot^3 x\,dx$
32. $\int x\sec^2(1+x^2)\,dx$
33. $\int x\sec^2 x\,dx$
34. $\int \frac{1}{x^2\sqrt{1-x^2}}\,dx$
35. $\int \frac{\sqrt{4-x^2}}{x^2}\,dx$
36. $\int \frac{2x^3+3x^2-8x-12}{x^4-16}\,dx$
37. $\int \frac{x+2}{2x^2-2x+1}\,dx$
38. $\int \frac{3x^{1/3}+5\sqrt{x}}{x^{2/3}+x}\,dx$
39. $\int \sqrt{x}\ln(\sqrt{x})\,dx$
40. $\int \ln(x^2-1)\,dx$
41. $\int_4^6 \frac{1}{x^3+3x^2+2x}\,dx$
42. $\int_{-\pi/4}^{\pi/4} (\sin^5 x\sec^4 x - 2\tan^3 x\sin^6 x - 3x^5)\,dx$
43. $\int \frac{\cos^2\theta}{\sin\theta}\,d\theta$
44. $\int_0^{\ln 5} \sqrt{4+e^x}\,dx$
45. $\int \sqrt{x}\ln x\,dx$
46. $\int \frac{x^3-x^2-3x}{x^3-1}\,dx$
47. $\int_0^1 \arcsin\left(\frac{x}{2}\right)dx$

48. $\int_0^{\pi/9} \sec(3x)\,dx$

49. $\int_1^4 \frac{1}{\sqrt{x}} \ln x\,dx$

50. $\int_2^4 \frac{(x+1)^2}{\sqrt{x^2-1}}\,dx$

Decide whether each of the following improper integrals exists. Evaluate those that do.

51. $\int_{-\infty}^{\sqrt{3}} \frac{1}{x^2+1}\,dx$

52. $\int_{-2}^{2} \frac{1}{x^2-1}\,dx$

53. $\int_4^{\infty} e^{3x}\,dx$

54. $\int_{-\infty}^{-1} e^{2x}\,dx$

55. $\int_1^{\infty} \frac{x}{1+x^2}\,dx$

56. $\int_0^{\infty} x^{-1/2}\,dx$

57. $\int_0^{\pi/2} \csc x\,dx$

58. $\int_{-\infty}^{0} \sin(2x)\,dx$

59. $\int_0^1 \frac{x^2-3x}{x^3+x}\,dx$

60. $\int_0^1 \frac{x^2-2x+1}{x^2-1}\,dx$

II

Suppose $a > 0$ and $b > 0$. Evaluate the following indefinite integrals.

61. $\int \frac{x}{\sqrt{ax+b}}\,dx$

62. $\int \frac{x^2}{\sqrt{ax+b}}\,dx$

63. $\int \sqrt{x^2 \pm a^2}\,dx$

64. $\int x^2\sqrt{x^2 \pm a^2}\,dx$

65. $\int \sqrt{a^2 - x^2}\,dx$

66. $\int x^2\sqrt{a^2 - x^2}\,dx$

*67. Find the centroid of the region lying between the x-axis and the curve $y = \sin 3x$ and between the lines $x = 0$ and $x = \frac{\pi}{9}$.

*68. Find the arclength of the curve $y = (x+3)^2$, from $x = 0$ to $x = 2$.

*69. Find the arclength of the curve $y = x^2 - 4x + 1$, from $x = 2$ to $x = 3$.

*70. Consider the integral $\int \frac{1}{x^2-1}\,dx$. Instead of using partial fractions, consider a trigonometric substitution. If $|x| < 1$, try $x = \sin\theta$. If $|x| > 1$, try $x = \sec\theta$. Work out the integral in both these cases, and then see whether you can find a single formula that works in both cases. Finally, reconcile your answer(s) with what you get using partial fractions.

71. (i) Evaluate $\int \frac{5x^3+14x^2-x+6}{x(x+3)(x^2+1)}\,dx$.
(ii) In writing the integrand in part (i) with partial fractions, $\frac{5x^3+14x^2-x+6}{x(x+3)(x^2+1)} = \frac{A}{x} + \frac{B}{x+3} + \frac{Cx+D}{x^2+1}$, we find that $B = 0$. What is the significance of this fact?

**72. Find the integral $\int (1+x^2)^{-3/2}\,dx$ using a hyperbolic substitution (cf. Exercise 47 in Section 8.3, p. 584). The answer will involve a hyperbolic function other than the one originally substituted; as with trigonometric substitutions, it is necessary to express this in terms of the function which was substituted.

For trigonometric substitutions this type of calculation was often made easier by using little triangles that showed the relationships among the various trigonometric functions. However, all the triangles really do is to remind us of the fundamental identity $\cos^2\theta + \sin^2\theta = 1$. In the hyperbolic situation we can use the corresponding identity $\cosh^2 t - \sinh^2 t = 1$.

*73. Compare the answers you get for $\int \frac{1}{1+\cos\theta}\,d\theta$ using (i) $x = \tan\frac{\theta}{2}$ or (ii) multiplying numerator and denominator by $1 - \cos\theta$.

Taylor's Formula

9

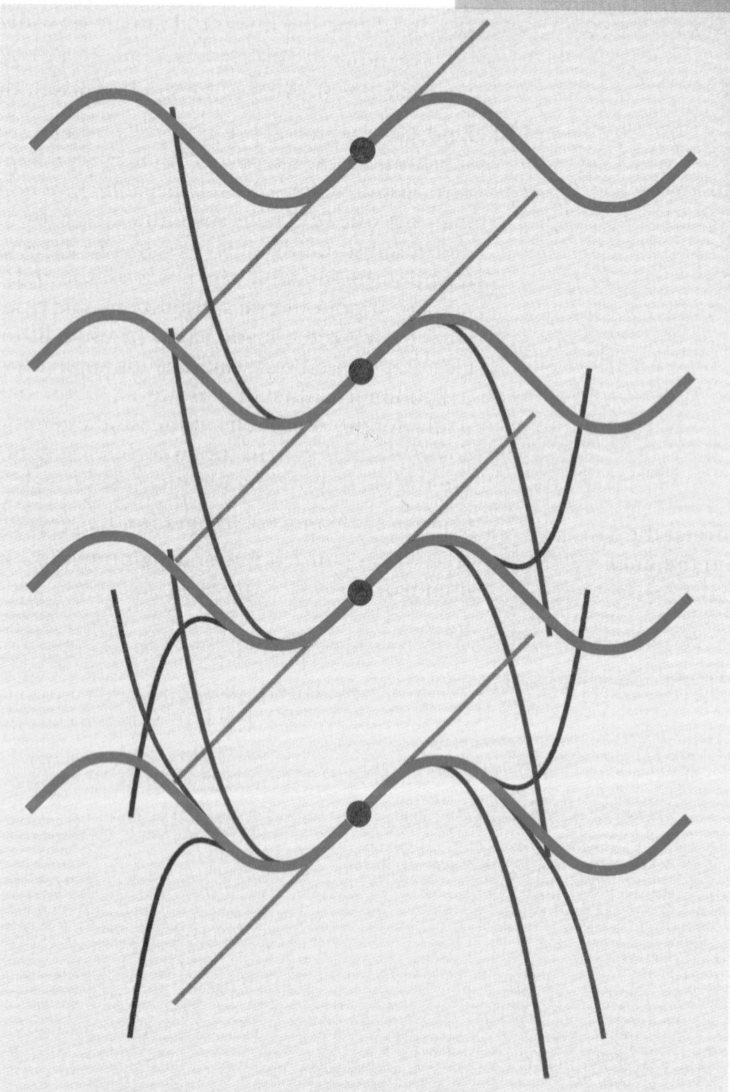

The straight line $y = x$ is fairly close to $y = \sin x$ near the origin, but $y = x - \frac{x^3}{6}$ is a better fit. The pictures show that $y = x - \frac{x^3}{6} + \frac{x^5}{120}$ is even better and $y = x - \frac{x^3}{6} + \frac{x^5}{120} - \frac{x^7}{5040}$ better yet.

In this chapter we learn how to find polynomials that approximate a complicated function like $y = \sin x$.

INTRODUCTION

In Section 4.10 we discussed the technique of linear approximation. The idea was that if we know the value of some function $f(x)$ at the point $x = a$, then we can estimate its value at a nearby point by assuming that the graph of $y = f(x)$ lies close to the tangent line at $x = a$. This works quite well for values of x near $x = a$ but, of course, is less accurate for points farther away.

The formula for the linear approximation is (cf. Equation 4.10.1)

$$f(a + h) \approx f(a) + f'(a)h.$$

This approximates the function $f(x)$ by its tangent line, which is the line having the same value as $f(x)$ at $x = a$ and whose derivative at $x = a$ is the same as $f'(a)$ (see Figure 9.1.1). If we write $y = \ell(x)$ for the tangent line, this means that $\ell(a) = f(a)$ and $\ell'(a) = f'(a)$.

A natural way to attempt to improve the accuracy of the approximation would be to use a parabola $y = p(x)$ instead of a straight line so that the curve of the parabola could fit more closely the curve of $y = f(x)$, at

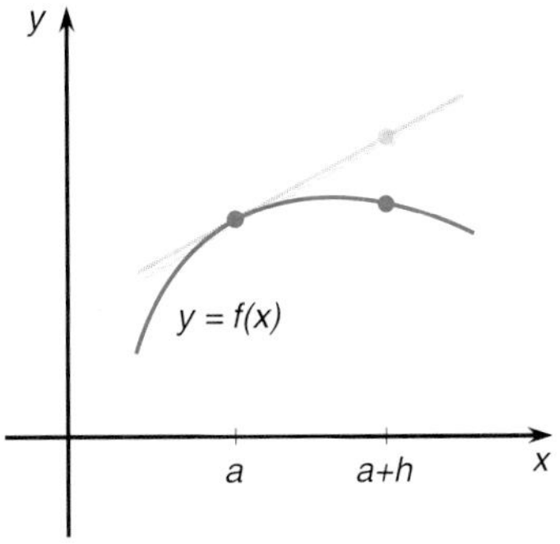

Figure 9.1.1

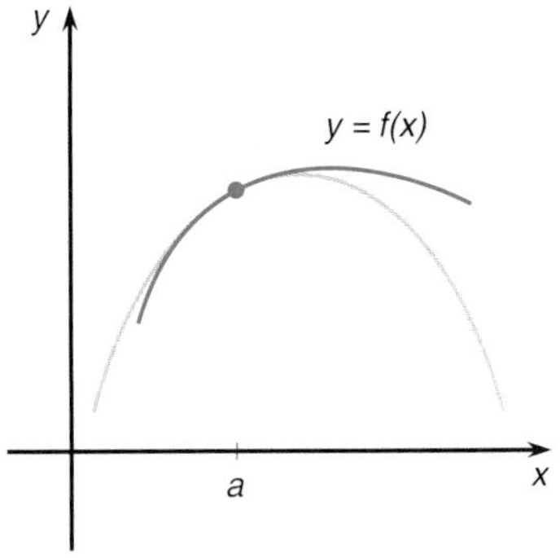

Figure 9.1.2

least near $x = a$. Exactly how to make the parabola fit the curve of $y = f(x)$ is not entirely clear, but a reasonable guess would be to make $p(a) = f(a)$, $p'(a) = f'(a)$, *and* $p''(a) = f''(a)$. If the parabola is chosen to have the same value, the same derivative and the same second derivative as $f(x)$ at the point $x = a$, then it seems likely that it will be close to $f(x)$, for values of x near a. See Figure 9.1.2.

Notice that it is not possible to do this with a straight line because the second derivative of a straight line function $y = mx + b$ is always zero at every point. There is no way to find a straight line with a specified second derivative unless that second derivative happens to be zero.

Of course, there is no particular reason to stop with the second derivative. If we want to approximate $y = f(x)$ near $x = a$, we could look for a polynomial $q(x)$ of degree n, chosen so that the values of $q(x)$ and its derivatives up to the nth derivative agree with the values of $f(x)$ and its derivatives at $x = a$, that is, $q(a) = f(a)$, $q'(a) = f'(a)$, $q''(a) = f''(a), \ldots, q^{(n)}(a) = f^{(n)}(a)$. It seems reasonable to expect that this is the best way to approximate $f(x)$ by a polynomial of degree n. It also seems reasonable to expect that using larger values of n may result in polynomials that give better approximations. (Notice that the $(n + 1)$th and higher derivatives of a polynomial of degree n are always zero, so there is no sense in asking that they agree with the corresponding derivatives of $f(x)$.)

It turns out that this is a very practical way to calculate values of complicated functions. It is relatively easy to evaluate polynomials, so if we want to evaluate a more complicated function, like e^x or $\sin x$, for example, we can approximate it with a polynomial and evaluate that. This is how computers evaluate functions like e^x, $\cos x$, and $\sin x$. Taylor's Formula, the subject of Section 9.1, tells us explicitly what the polynomial should be.

Any approximation is of limited value unless there is some way to assess the accuracy of the answer we calculate. Taylor's Formula with Remainder, discussed in Section 9.2, provides a way to measure the "remainder," the difference between the actual value and the approximation, in effect telling us how close the approximation is to the actual value. It can be used as follows. Suppose we want to know the value of $e^{1.345}$ correct to four decimal places. We can use Taylor's Formula with Remainder to tell us how large n, the degree of the approximating polynomial, will have to be in order to get an approximation that is at least this good.

Once we know that, it is reasonably straightforward to find and evaluate the polynomial. The material of this chapter will be extremely important in Chapter 10.

9.1 Taylor's Formula

TAYLOR POLYNOMIAL
FACTORIAL
MACLAURIN POLYNOMIAL

In this chapter we will write a polynomial $p(x)$ with the constant term first and highest term last:

$$p(x) = a_0 + a_1x + a_2x^2 + \ldots + a_{n-1}x^{n-1} + a_nx^n.$$

We are interested in the values of $p(x)$ and its derivatives at some point $x = a$, but for simplicity let us first consider the point $x = 0$. It is not difficult to express the values of $p(x)$ and its derivatives at $x = 0$ in terms of the coefficients a_i of the polynomial.

Certainly the value of $p(x)$ at $x = 0$ is just the constant term, $p(0) = a_0$. If we work out the derivative

$$p'(x) = a_1 + 2a_2x + \ldots + (n-1)a_{n-1}x^{n-2} + na_nx^{n-1},$$

we find that its value at $x = 0$ is just *its* constant term, that is, $p'(0) = a_1$. The idea is that when we substitute $x = 0$, any term with a positive power of x in it will become 0; only the constant term a_1 does not contain a positive power of x.

If we work out the second derivative

$$p''(x) = 2a_2 + (3)(2)a_3x + \ldots + (n-1)(n-2)a_{n-1}x^{n-3} + n(n-1)a_nx^{n-2},$$

we find that $p''(0) = 2a_2$. Notice that the terms involving a_0 and a_1 have become zero after two differentiations.

In the expression for the third derivative $p'''(x)$, the constant term is $(3)(2)a_3$, so $p'''(0) = (3)(2)a_3$.

The term a_3x^3 in the original polynomial $p(x)$ changes to $3a_3x^2$ after one differentiation, then to $(3)(2)a_3x$ after another, and finally to $(3)(2)a_3$ after the third. Terms that begin with a power of x lower than the third power will be zero after the third differentiation, and terms that begin with a power of x higher than the third power will still contain a positive power of x and will not contribute to $p'''(0)$.

What happens when we take the kth derivative and evaluate $p^{(k)}(0)$? As above, we have to look for the constant term. Each time we differentiate, each positive power of x is reduced by 1, so terms that begin with a power of x higher than the kth power will still contain a positive power of x after k differentiations. Terms that begin with a power of x lower than the kth power will have become zero, so the only nonzero term in $p^{(k)}(x)$ that does not contain a positive power of x is the term that comes from a_kx^k.

Differentiating a_kx^k gives ka_kx^{k-1}; differentiating again gives $k(k-1)a_kx^{k-2}$. Each differentiation has the effect of multiplying by the exponent and then lowering that exponent by 1. Doing this k times will give

$$\frac{d^k}{dx^k}(a_kx^k) = k(k-1)(k-2)\ldots 2a_k = k!a_k.$$

(Here we have used the abbreviation $k! = k(k-1)(k-2)\ldots(2)(1)$ for the product of the first k positive integers; the symbol $k!$ is read "k **factorial.**" See Definition 1.3.8, p. 14.)

Since only the kth term affects $p^{(k)}(0)$, we see that $p^{(k)}(0) = k!a_k$.

It is possible to turn this process around. Suppose we want to construct a polynomial of degree n, and suppose we know in advance the values we want the polynomial and its first n derivatives to have. Suppose $p(0) = b_0$, $p^{(1)}(0) = b_1$, $p^{(2)}(0) = b_2$, $p^{(3)}(0) = b_3, \ldots, p^{(n-1)}(0) = b_{n-1}$, $p^{(n)}(0) = b_n$ are the desired values for the polynomial and its first n derivatives at $x = 0$. Then dividing the above formula by $k!$, we see that the kth coefficient of the polynomial will have to be $a_k = \frac{p^{(k)}(0)}{k!} = \frac{b_k}{k!}$, for $k \geq 1$, and $a_0 = p(0) = b_0$.

If we allow ourselves to write $p^{(0)}(x)$ for $p(x)$ and agree that $0! = 1$, then the formula $a_k = \frac{p^{(k)}(0)}{k!} = \frac{b_k}{k!}$ is correct for $k = 0$ too. It may seem strange at first to let $0! = 1$, but in fact it is the right choice to make; it makes this formula particularly convenient, but there are also other reasons to do it this way. We encountered this convention when we discussed the Binomial Theorem (Theorem 1.3.10). One way to look at it is that we expect that $k! = k(k-1)!$, and for this to be true when $k = 1$ it is necessary that $0! = 1$.

EXAMPLE 1 A Polynomial with Specified Derivatives Find a polynomial $p(x)$ of degree 3 with $p(0) = 3$, $p^{(1)}(0) = 7$, $p^{(2)}(0) = -2$, and $p^{(3)}(0) = 12$.

Solution If we write $p(x) = a_0 + a_1x + a_2x^2 + a_3x^3$, then the above formula tells us that $a_3 = p^{(3)}(0)/3! = \frac{12}{(3)(2)(1)} = \frac{12}{6} = 2$. Also $a_2 = p^{(2)}(0)/2! = \frac{-2}{(2)(1)} = -1$, and $a_1 = p^{(1)}(0)/1! = \frac{7}{1} = 7$. Finally, $a_0 = p(0)/0! = \frac{3}{1} = 3$.

The polynomial is $p(x) = a_0 + a_1x + a_2x^2 + a_3x^3 = 3 + 7x - x^2 + 2x^3$. It is easy to check that it has the desired derivatives at $x = 0$.

The situation in which we are interested arises when we begin with a function $f(x)$ and want to construct a polynomial $p(x)$ of degree n with the same value and same first n derivatives at $x = 0$. In other words, we want $p(0) = f(0)$, $p^{(1)}(0) = f^{(1)}(0)$, $p^{(2)}(0) = f^{(2)}(0), \ldots, p^{(n-1)}(0) = f^{(n-1)}(0)$, $p^{(n)}(0) = f^{(n)}(0)$.

The formula we found above tells us that the coefficient a_k of x^k in $p(x)$ must be

$$a_k = \frac{f^{(k)}(0)}{k!}. \tag{9.1.1}$$

This is the first version of Taylor's Formula.

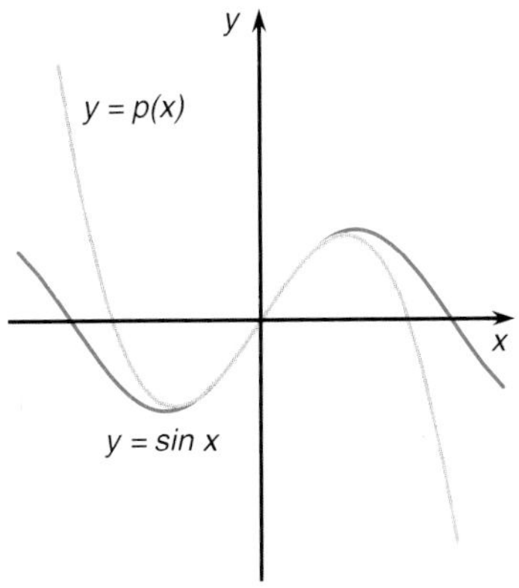

Figure 9.1.3

EXAMPLE 2 Find the polynomial of degree at most 4 that has the same value and same first four derivatives as $f(x) = \sin x$ at $x = 0$.

Solution We calculate the derivatives of $f(x)$, evaluate them at $x = 0$, and then use Equation 9.1.1 to find the coefficients a_i of the polynomial $p(x)$:

$f^{(0)}(x) = f(x) = \sin x,$	$f(0) = \sin 0 = 0,$	$a_0 = \frac{0}{1} = 0;$
$f^{(1)}(x) = f'(x) = \cos x,$	$f^{(1)}(0) = \cos 0 = 1,$	$a_1 = \frac{1}{1!} = 1;$
$f^{(2)}(x) = -\sin x,$	$f^{(2)}(0) = -\sin 0 = 0,$	$a_2 = \frac{0}{2!} = 0;$
$f^{(3)}(x) = -\cos x,$	$f^{(3)}(0) = -\cos 0 = -1,$	$a_3 = \frac{-1}{3!} = -\frac{1}{6};$
$f^{(4)}(x) = \sin x,$	$f^{(4)}(0) = \sin 0 = 0,$	$a_4 = \frac{0}{4!} = 0.$

Since $a_4 = 0$, the polynomial actually has degree 3, not 4 . . .

The polynomial is

$$p(x) = x - \frac{1}{6}x^3.$$

The graph of this polynomial is sketched in Figure 9.1.3; the part of the curve near $x = 0$ does look a lot like the corresponding part of the graph of $y = \sin x$. As we move farther from $x = 0$, the approximation is less good.

EXAMPLE 3 Find the polynomial of degree at most 5 that has the same value and same first five derivatives as $f(x) = \sin x$ at $x = 0$.

Solution We have already done most of the work in Example 2. What is left to do is find $f^{(5)}(x) = \frac{d}{dx}\left(f^{(4)}(x)\right) = \cos x$, so $f^{(5)}(0) = \cos 0 = 1$.

From this we find that the coefficient of x^5 in the polynomial will have to be $a_5 = \frac{1}{5!} = \frac{1}{120}$. The lower coefficients are the same as what we found above, so the polynomial we want is

The coefficient a_k of x^k is determined by the formula $a_k = \frac{f^{(k)}(0)}{k!}$, so the lower coefficients $a_0, a_1, \ldots, a_4$ are exactly the same as the ones in the previous example. . .

$$q(x) = x - \frac{1}{6}x^3 + \frac{1}{120}x^5.$$

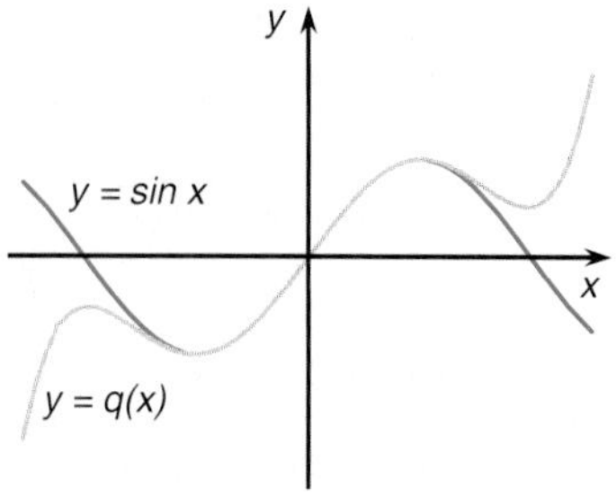

Figure 9.1.4

This polynomial is sketched in Figure 9.1.4, and it is a better approximation to $f(x) = \sin x$ than the polynomial $p(x)$ that we found in the last example, although even this one eventually goes off to infinity and becomes a bad approximation for x sufficiently far from $x = 0$.

The idea is that the higher the degree of the approximating polynomial, the more of its derivatives will agree with those of $f(x)$, so the more closely its behavior will agree with that of $f(x)$ and the better the approximation should be.

Now we would like to do the same thing at any fixed point $x = a$ rather than just the special point $x = 0$. Unfortunately, the derivative of $p(x)$ at $x = a$ is not as simple to evaluate at $x = a$, since it involves all the coefficients instead of only the constant term. The higher derivatives also involve all the coefficients, so it is much more complicated to say what the coefficients have to be in order that $p(x)$ and its derivatives will have specified values at $x = a$.

This difficulty arises because we express the polynomial in terms of powers of x, which are very convenient for evaluating at $x = 0$ but awkward anywhere else. If instead we express the polynomial in terms of powers of $(x - a)$, then the situation will improve considerably. If

$$p(x) = a_0 + a_1(x-a) + a_2(x-a)^2 + \ldots + a_{n-1}(x-a)^{n-1} + a_n(x-a)^n,$$

then $p(a) = a_0$, $p^{(1)}(a) = a_1$, $p^{(2)}(a) = 2a_2$, and so on.

The reason is really the same as it was in the case in which $a = 0$. The kth derivative of $a_k(x-a)^k$ is $k(k-1)\ldots(2)(1)a_k$; the kth derivative of any term containing a lower power of $(x - a)$ is zero, and the kth derivative of any term containing a power of $(x - a)$ higher than the kth will still contain a positive power of $(x - a)$ and so will be zero at $x = a$.

As in the case with $a = 0$, we see that with $p(x)$ written as above,

$$p^{(k)}(a) = k!a_k.$$

So if we want a polynomial of degree at most n that has the same value and the same first n derivatives at $x = a$ as $f(x)$, then we should use the polynomial $p(x) = a_0 + a_1(x-a) + a_2(x-a)^2 + \ldots + a_{n-1}(x-a)^{n-1} + a_n(x-a)^n$, with

$$a_k = \frac{f^{(k)}(a)}{k!}.$$

It is also convenient to write $p(x)$ using summation notation:

$$p(x) = \sum_{k=0}^{n} \frac{f^{(k)}(a)}{k!}(x-a)^k.$$

If we start with a function $f(x)$ that is n times differentiable, then we have just seen how to construct a polynomial of degree at most n so that its value and the values of its first n derivatives at $x = a$ are the same as those of the function $f(x)$. This polynomial can be used to approximate the original function $f(x)$, and it is extremely important. It is called the Taylor polynomial of degree n, after the English mathematician Brook Taylor (1685–1731). Notice that it depends on the choice of the point $x = a$ and on the degree n.

DEFINITION 9.1.2

Taylor Polynomials

Let $f(x)$ be a function defined on an interval containing $x = a$, and suppose its derivatives $f'(x) = f^{(1)}(x), f^{(2)}(x), \ldots, f^{(n)}(x)$ all exist at $x = a$. The corresponding **Taylor polynomial** of degree n at $x = a$ is defined to be the polynomial

$$P_n(x) = a_0 + a_1(x-a) + \ldots + a_{n-1}(x-a)^{n-1} + a_n(x-a)^n,$$

with

$$a_k = \frac{f^{(k)}(a)}{k!}.$$

It can also be written as

$$P_n(x) = \sum_{k=0}^{n} a_k(x-a)^k = \sum_{k=0}^{n} \frac{f^{(k)}(a)}{k!}(x-a)^k.$$

Notice that each $a_k = \frac{f^{(k)}(a)}{k!}$ is a *constant.* Only the powers $(x-a)^k$ depend on x.

Notice too that the polynomial depends on the choice of the starting point a. When we mention $P_n(x)$, it is necessary to make clear what a is being used. In the special case in which $a = 0$, Taylor polynomials are sometimes called **Maclaurin polynomials**, after the Scottish mathematician Colin Maclaurin (1698–1746).

Strictly speaking, the degree of $P_n(x)$ may be less than n (cf. Example 2).

The argument we have given above shows that $P_n(x)$ is determined by the property that it and its first n derivatives equal $f(x)$ and its first n derivatives at $x = a$.

EXAMPLE 4 **Taylor Polynomial of ln *x*** Find the Taylor polynomial $P_4(x)$ corresponding to the function $f(x) = \ln x$ at $a = 1$.

Solution The derivatives are easily calculated and evaluated:

$$\begin{aligned}
f^{(0)}(x) &= \ln x, & f^{(0)}(1) &= f(1) = \ln(1) = 0;\\
f^{(1)}(x) &= \tfrac{1}{x}, & f^{(1)}(1) &= \tfrac{1}{1} = 1;\\
f^{(2)}(x) &= -\tfrac{1}{x^2}, & f^{(2)}(1) &= -\tfrac{1}{1} = -1;\\
f^{(3)}(x) &= \tfrac{2}{x^3}, & f^{(3)}(1) &= \tfrac{2}{1} = 2;\\
f^{(4)}(x) &= -\tfrac{(3)(2)}{x^4}, & f^{(4)}(1) &= -\tfrac{6}{1} = -6.
\end{aligned}$$

From this we can easily write down the Taylor polynomial:

$$\begin{aligned}
P_4(x) &= 0 + \frac{1}{1!}(x-1) + \frac{-1}{2!}(x-1)^2 + \frac{2}{3!}(x-1)^3 + \frac{-6}{4!}(x-1)^4\\
&= (x-1) - \frac{1}{2}(x-1)^2 + \frac{1}{3}(x-1)^3 - \frac{1}{4}(x-1)^4.
\end{aligned}$$

At first it seems peculiar to write a polynomial in terms of powers of $(x-a)$ instead of powers of x, but in fact it is quite useful. After all, if we want to use $P_n(x)$ to approximate the value of $f(x)$ for some x near a, then we might write $x = a + h$, and $x - a$ is just h. It is actually easier to perform calculations with $x - a$ than it would be to use x.

With a graphics calculator we can plot the function $y = \ln x$ and the Taylor polynomial $P_4(x)$ of Example 4.

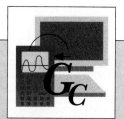

TI-85	SHARP EL9300
GRAPH F1 $(y(x) =)$ LN F1 (x) ▼ F1 (x) − 1 − (F1 (x) − 1) ^ 2 ÷ 2 + (F1 (x) − 1) ^ 3 ÷ 3 − (F1 (x) − 1) ^ 4 ÷ 4 M2(RANGE) 0 ▼ 2 ▼ ▼ (−) 2 ▼ 1 F5(GRAPH)	ln x/θ/T 2ndF ▼ x/θ/T − 1 − (x/θ/T − 1) x^2 ÷ 2 + (x/θ/T − 1) a^b 3 ▶ ÷ 3 − (x/θ/T − 1) a^b 4 ▶ ÷ 4 RANGE 0 ENTER 2 ENTER 1 ENTER −2 ENTER 1 ENTER 1 ENTER
CASIO f_x-6300G (f_x-7700GB) Range 0 EXE 2 EXE 1 EXE −2 EXE 1 EXE 1 EXE EXE EXE EXE Graph ln ■ EXE Graph ■ − 1 − (■ − 1) x^2 ÷ 2 + (■ − 1) x^y 3 ÷ 3 − (■ − 1) x^y 4 ÷ 4 EXE	HP 48SX ■ PLOT PLOTR ERASE ATTN ′ Y ■ LN X ↱ PLOT ■ DRAW ERASE 0 ENTER 2 XRNG 2 +/− ENTER 1 YRNG DRAW ATTN ′ Y ■ X − 1 − ■ X − 1 ▶ y^x 2 ÷ 2 + () X − 1 ▶ y^x 3 ÷ 3 − ■ X − 1 ▶ y^x 4 ÷ 4 ↱ PLOT ■ DRAW DRAW

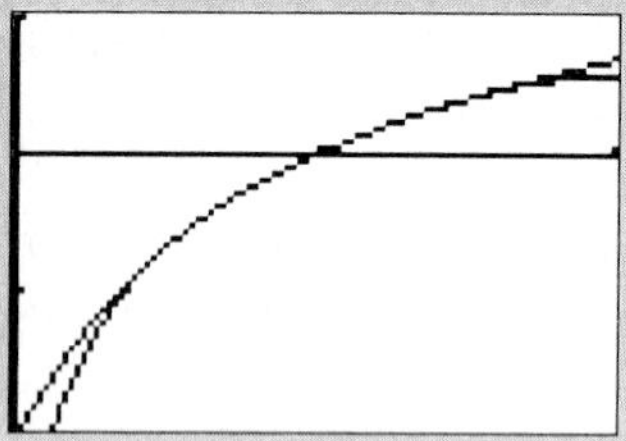

Identify which function is which, and notice how the approximation is better for values near $a = 1$.

EXAMPLE 5 **Approximating with the Taylor Polynomial** Use the Taylor polynomial $P_4(x)$ from Example 4 to approximate $\ln(1.1)$.

Solution The Taylor polynomial $P_4(x)$ was found with $a = 1$, so if $x = 1.1 = \frac{11}{10}$, then $x - a = x - 1 = \frac{1}{10} = 0.1$. Since

$$P_4(x) = (x-1) - \frac{1}{2}(x-1)^2 + \frac{1}{3}(x-1)^3 - \frac{1}{4}(x-1)^4,$$

we find

$$P_4(1.1) = 0.1 - \frac{1}{2}(0.1)^2 + \frac{1}{3}(0.1)^3 - \frac{1}{4}(0.1)^4 \approx 0.095\,3083.$$

This is a pretty good approximation to the actual value, which is 0.095 310, correct to six decimal places. Notice that it would actually be slightly more trouble to evaluate a fourth-degree polynomial written in terms of powers of x instead of powers of $(x - 1)$.

EXAMPLE 6 Find the Taylor polynomial $P_4(x)$ for $f(x) = \sin x$, with $a = \frac{\pi}{4}$.

Solution

$$\begin{aligned}
f(x) &= \sin x, & f\left(\tfrac{\pi}{4}\right) &= \sin\left(\tfrac{\pi}{4}\right) = \tfrac{1}{\sqrt{2}};\\
f^{(1)}(x) &= \cos x, & f^{(1)}\left(\tfrac{\pi}{4}\right) &= \cos\left(\tfrac{\pi}{4}\right) = \tfrac{1}{\sqrt{2}};\\
f^{(2)}(x) &= -\sin x, & f^{(2)}\left(\tfrac{\pi}{4}\right) &= -\sin\left(\tfrac{\pi}{4}\right) = -\tfrac{1}{\sqrt{2}};\\
f^{(3)}(x) &= -\cos x, & f^{(3)}\left(\tfrac{\pi}{4}\right) &= -\cos\left(\tfrac{\pi}{4}\right) = -\tfrac{1}{\sqrt{2}};\\
f^{(4)}(x) &= \sin x, & f^{(4)}\left(\tfrac{\pi}{4}\right) &= \sin\left(\tfrac{\pi}{4}\right) = \tfrac{1}{\sqrt{2}}.
\end{aligned}$$

From this it is possible to write down the Taylor polynomial:

$$\begin{aligned}
P_4(x) &= \frac{1}{0!\sqrt{2}} + \frac{1}{1!\sqrt{2}}\left(x - \frac{\pi}{4}\right) - \frac{1}{2!\sqrt{2}}\left(x - \frac{\pi}{4}\right)^2 - \frac{1}{3!\sqrt{2}}\left(x - \frac{\pi}{4}\right)^3\\
&\quad + \frac{1}{4!\sqrt{2}}\left(x - \frac{\pi}{4}\right)^4\\
&= \frac{1}{\sqrt{2}} + \frac{1}{\sqrt{2}}\left(x - \frac{\pi}{4}\right) - \frac{1}{2\sqrt{2}}\left(x - \frac{\pi}{4}\right)^2 - \frac{1}{6\sqrt{2}}\left(x - \frac{\pi}{4}\right)^3\\
&\quad + \frac{1}{24\sqrt{2}}\left(x - \frac{\pi}{4}\right)^4.
\end{aligned}$$

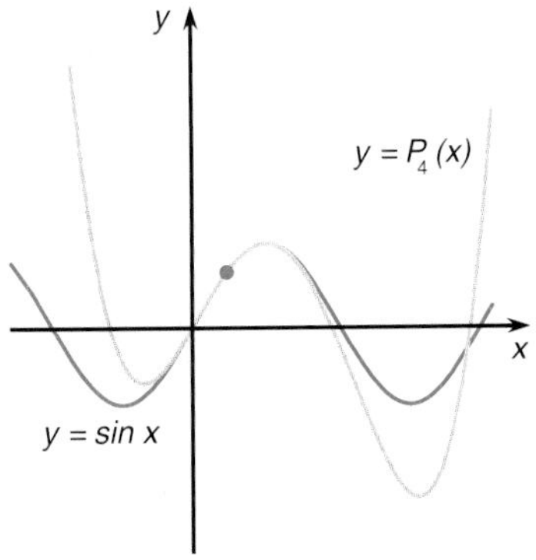

Figure 9.1.5

The interesting thing about this example is to compare it with Example 2. There we found the Taylor polynomial of degree 4 for $f(x) = \sin x$ but with $a = 0$. The result was a completely different polynomial.

This is not surprising when we realize that the polynomial in Example 2 approximates the function $f(x) = \sin x$, *starting from the point* $x = 0$, while the one we have just found approximates the same function, *starting from* $\frac{\pi}{4}$. It is graphed in Figure 9.1.5, where we see that it is close to $\sin x$ for x near $\frac{\pi}{4}$.

Exercises 9.1

I

For each function $f(x)$, find the indicated Taylor polynomial (i.e., $P_n(x)$, with the specified n and the specified value of the starting point a).

1. For $f(x) = \cos x$, $a = 0$, find $P_4(x)$.
2. For $f(x) = \cos x$, $a = \frac{\pi}{2}$, find $P_4(x)$.
3. For $f(x) = e^x$, $a = 0$, find $P_5(x)$.
4. For $f(x) = e^{-x}$, $a = 1$, find $P_4(x)$.
5. For $f(x) = \ln(x+1)$, $a = 0$, find $P_3(x)$.
6. For $f(x) = \cosh x$, $a = 0$, find $P_4(x)$.
7. For $f(x) = x^3$, $a = 1$, find $P_3(x)$.
8. For $f(x) = x^3$, $a = 2$, find $P_4(x)$.
9. For $f(x) = x^4$, $a = -1$, find $P_{12}(x)$.
10. For $f(x) = e^{\ln x}$, $a = 3$, find $P_6(x)$.

Use the Taylor polynomial of the specified degree and about the specified point a to estimate the following values. In each case, compare your answer with what you get using your calculator.

11. $\sin\left(\frac{1}{100}\right)$, $f(x) = \sin x$, $a = 0$, P_5
12. $\cos\left(\frac{1}{10}\right)$, $f(x) = \cos x$, $a = 0$, P_4
13. $e^{0.1}$, $f(x) = e^x$, $a = 0$, P_4
14. $e^{-0.03}$, $f(x) = e^x$, $a = 0$, P_4
15. $\sin\left(\pi - \frac{1}{5}\right)$, $f(x) = \sin x$, $a = \pi$, P_4

16. $\cos\left(\frac{1}{6}+\pi\right), f(x)=\cos x,\ a=\pi,\ P_4$

17. $\cosh(0.15), f(x)=\cosh x,\ a=0,\ P_4$

18. $\sinh\left(\frac{2}{15}\right), f(x)=\sinh x,\ a=0,\ P_4$

19. $\ln(0.9), f(x)=\ln x,\ a=1,\ P_5$

20. $f(17)$, with $f(x)=x^2-3x+1,\ a=2,\ P_3$

II

21. Find the Taylor polynomial of degree 5 for $f(x)=e^x$ about $a=0$, and use it to estimate $\sqrt{e}=e^{1/2}$.

22. Find the Taylor polynomial of degree 6 about the point $a=3\pi$ for the function $f(x)=\cos^2 x+\sin^2 x$.

23. (i) Estimate $\sin\left(\frac{\pi}{4}\right)$ using the Taylor polynomial of degree 4 with $a=0$.
(ii) Estimate the same value using $a=\frac{\pi}{2}$.
(iii) Which is the better approximation?

24. (i) Estimate $\ln(1.5)$ using the Taylor polynomial of degree 4 about $a=1$.
(ii) Make the same estimate with the Taylor polynomial of degree 4 about $a=2$.
(iii) Which is the better approximation?

***25.** If $f(x)$ and $g(x)$ are n times differentiable at $x=a$, let $P_n(x)$ and $Q_n(x)$ be their respective Taylor polynomials of degree n about the point $x=a$. Explain why the Taylor polynomial of degree n about $x=a$ of $h(x)=f(x)+g(x)$ is $S_n(x)=P_n(x)+Q_n(x)$.

***26.** (i) If $f(x)$ is a polynomial of degree n, what is its Taylor polynomial of degree n about $x=0$?
(ii) If $f(x)$ is a polynomial of degree n and $m\geq n$, what is its Taylor polynomial $P_m(x)$ about $x=a$?
(iii) Describe $P_m(x)$ if $m<n$.

***27.** Suppose $f(x)$ is a function and $p_n(x)$ is its Taylor polynomial of degree n about the point $x=a$. Let r be any nonzero number, let $F(x)=f(rx)$, and verify that the Taylor polynomial of degree n associated to $F(x)$ about $x=\frac{a}{r}$ is $P_n(x)=p_n(rx)$.

***28.** Compare the Taylor polynomial $P_4(x)$ for $f(x)=x^4-3x^2$ at $a=-1$ with the polynomial you get by writing $x=(x+1)-1$ and expanding $x^4=\left((x+1)-1\right)^4$ and $x^2=\left((x+1)-1\right)^2$ by the Binomial Theorem.

***29.** (i) Compare the Taylor polynomials of degree 4 about $a=0$ associated to $f(x)=\cos(2x)$ and $g(x)=\cos x$.
(ii) Compare the Taylor polynomials of degree 4 about $a=0$ associated to $F(x)=\cos(x^2)$ and $g(x)=\cos x$.

30. Using the Taylor polynomial $P_6(x)$ corresponding to the function $f(x)=\ln(x)$ about $a=1$, evaluate $P_6(2)$ to estimate $\ln 2$. How good is the estimate?

***31.** Sketch the graphs of the Taylor polynomials $P_1(x)$, $P_2(x)$, and $P_4(x)$ associated to the function $f(x)=\cos x$ at $a=0$.

***32.** Sketch the graphs of the Taylor polynomials $P_1(x)$, $P_3(x)$, and $P_5(x)$ associated to the function $f(x)=\sin x$ at $a=0$.

***33.** (i) Let $f(x)=\sin x$ and for each positive integer k evaluate $f^{(k)}(0)$. (*Hint:* Write $k=4m+i$, with $i=0,1,2,$ or 3.)
(ii) Write an expression for the kth coefficient of the nth Taylor polynomial $P_n(x)$ about $a=0$ (for $k\leq n$).

***34.** Write an expression for the kth coefficient of the nth Taylor polynomial $P_n(x)$ of $g(x)=\cos x$ about $a=0$ (for $k\leq n$).

III

***35.** Use a graphics calculator or computer to plot the graphs of several of the Taylor polynomials $P_1(x), P_2(x), P_3(x),\ldots$ for $f(x)=\sin x$ about the point $a=\frac{\pi}{2}$ (cf. the previous exercises).

***36.** (i) With a computer or programmable calculator, write a program to evaluate the Taylor polynomial $P_{10}(x)$ associated to the function $f(x)=e^x$ at $a=0$.
(ii) Try your program for different values of x, comparing the answers you get with what your calculator tells you to expect. You should find that if x is a reasonably small number, the results are remarkably close; on the other hand, there is clearly some trouble if x is very large (e.g., if $x=100$). The way your calculator finds e^x likely includes evaluating a Taylor polynomial, just like this, though it probably has a slightly more clever way of dealing with large numbers, but for small numbers this is exactly what it does.

***37.** (i) Write a program to evaluate the Taylor polynomial $P_{10}(x)$ for $f(x)=\ln(x)$ about $a=1$.
(ii) Evaluate $P_{10}(2), P_{10}(3), P_{10}(5)$, and compare with $\ln 2$, $\ln 3$, $\ln 5$.
(iii) Modify your program to evaluate $P_{100}(x)$ and repeat part (ii).

38. Plot $\sin x$ and the corresponding Taylor polynomials about $a=0$: $P_1(x), P_3(x), P_5(x), P_7(x), P_9(x)$.

39. Plot $\cos x$ and the corresponding Taylor polynomials about $a=0$: $P_2(x), P_4(x), P_6(x), P_8(x), P_{10}(x)$.

40. Plot $\sin x$ and the corresponding Taylor polynomials about $a=\frac{\pi}{4}$: $P_1(x), P_4(x), P_7(x), P_{10}(x)$.

9.2 Taylor's Formula with Remainder; Approximations

REMAINDER
LAGRANGE FORM
ESTIMATING THE REMAINDER

In Section 9.1 we saw how to construct the Taylor polynomials for a function $f(x)$ about a point $x = a$. The Taylor polynomial of degree n is the polynomial $P_n(x)$ of degree n (or less) that has the same value as $f(x)$ at $x = a$ and that has derivatives with the same values at $x = a$ up to and including the nth derivative. We expect that the polynomial $P_n(x)$ will behave much as $f(x)$ behaves, at least for x's near $x = a$.

We saw how it is possible to use Taylor polynomials to approximate the values of $f(x)$. This can be very useful in situations in which we have no way to evaluate $f(x)$ precisely. The examples that we did suggested that even with fairly small values of n (such as 4 or 5) it is possible to get fairly accurate approximations, especially if we are evaluating at a value x that is near a.

However, if we want to use this technique for calculations, it is essential to have some control over the accuracy of the approximation.

Suppose $f(x)$ is a function that is n times differentiable at $x = a$. Write $P_n(x)$ for the Taylor polynomial of degree n at $x = a$. The problem is to estimate the size of the difference $f(x) - P_n(x)$. This difference is called the **remainder**, and we write

$$R_n(x) = f(x) - P_n(x). \tag{9.2.1}$$

In fact there is a convenient formula that allows us to estimate the size of $R_n(x)$ (and with it the accuracy of the approximation).

THEOREM 9.2.2

Taylor's Formula with Remainder

Suppose the function $f(x)$ is defined on an interval that contains a and x, and suppose its $(n+1)$th derivative $f^{(n+1)}$ exists on this interval. Then the nth remainder is

$$R_n(x) = \frac{f^{(n+1)}(c)}{(n+1)!}(x-a)^{n+1},$$

for some number c between a and x. The number c depends on x.

This can also be written in the following way:

$$\begin{aligned} f(x) &= P_n(x) + R_n(x) \\ &= f(a) + \frac{f'(a)}{1!}(x-a) + \ldots + \frac{f^{(n)}(a)}{n!}(x-a)^n \\ &\quad + \frac{f^{(n+1)}(c)}{(n+1)!}(x-a)^{n+1}. \end{aligned}$$

In this form we see that the formula for the remainder is almost identical to what the next term would be if we continued the Taylor polynomial one more step (i.e., if we found $P_{n+1}(x)$), with the one exception that the $(n+1)$th derivative $f^{(n+1)}(c)$ is evaluated at c rather than at a. This makes it quite easy to remember.

This formula for the remainder was actually found by Joseph Louis Lagrange (1736–1813), a French-Italian mathematician, and is sometimes called the **Lagrange form of the remainder**. It is proved using Rolle's Theorem (Theorem 4.2.4), but the proof is tricky, so we postpone it until Example 7 at the end of the section. We prefer to consider some examples first in order to see the significance and usefulness of the theorem.

One apparent disadvantage of the Lagrange form of the remainder is that, like Rolle's Theorem itself, it does not give us any way to find the number c at which the $(n+1)$th derivative is evaluated. In practice this is not a terrible handicap; it is usually possible to estimate the size of that derivative without knowing the point exactly. We do know that c is between a and x, that is, that $a \le c \le x$ or $x \le c \le a$, so all that is required is a bound for the size of $f^{(n+1)}(u)$ for all $u \in [a, x]$ or $[x, a]$, as the case may be.

EXAMPLE 1 **Estimating *e*** If we find the Taylor polynomial $P_4(x)$ for $f(x) = e^x$ about the point $a = 0$ and use it to estimate the numerical value of e, how good will the approximation be?

Solution It is very easy to find the derivatives of $f(x) = e^x$; they are all the same: $f^{(k)}(x) = e^x$, so $f^{(k)}(0) = e^0 = 1$. The Taylor polynomial is

$$P_4(x) = 1 + \frac{x}{1!} + \frac{x^2}{2!} + \frac{x^3}{3!} + \frac{x^4}{4!}$$
$$= 1 + x + \frac{x^2}{2} + \frac{x^3}{6} + \frac{x^4}{24}.$$

The approximate value of $e = f(1)$ is $P_4(1) = 1 + 1 + \frac{1}{2} + \frac{1}{6} + \frac{1}{24} = 2\frac{17}{24} \approx 2.708\,333$.

Now Theorem 9.2.2 tells us that the remainder is $R_4(1) = \frac{1}{5!}f^{(5)}(c) = \frac{e^c}{120}$, for some $c \in [0, 1]$. What we need is to know how big this could be, which amounts to knowing how big e^c could be for $c \in [0, 1]$.

That is easy. After all, $f(x) = e^x$ is an increasing function (its derivative, e^x, is always positive), so its largest value on the interval $[0, 1]$ will be at the right end of the interval. We see that $e^c \le e^1 = e$, so $R_4(1) \le \frac{e}{120}$.

Knowing that $e < 3$, we see that the remainder is smaller than $\frac{3}{120} = \frac{1}{40} = 0.025$. We conclude that the estimate we found above for the value of e is correct to within an error of at most 0.025.

In fact, $e \approx 2.718\,282$, so our estimate is off by about 0.009 949, or less than half the error estimated by the formula for the remainder.

> This illustrates an important point. We do not expect to find *exactly* what the remainder is. After all, since $f(x) = P_n(x) + R_n(x)$, if we knew $R_n(x)$ exactly, then we could find $f(x)$ exactly too. That would be nice, but it is too much to expect. So the formula for the remainder allows us to give a bound—a limitation—on how large the remainder is, and that allows us to say with confidence how accurate the approximation is. In practice, the approximation will often be better than what we can guarantee using the estimate for the remainder.

The last example shows how it is possible to use the estimate of Theorem 9.2.2 for the remainder to assess the accuracy of a Taylor polynomial approximation. Most often, however, it is used the other way around. Instead of starting with the polynomial $P_n(x)$ and asking how accurate an approximation it gives for some particular x, we start with the function $f(x)$ and want to approximate its value at some point to within a specified accuracy. The problem we need to answer is: How large must n be in order that $P_n(x)$ has the desired accuracy?

EXAMPLE 2 Calculate e to within 0.000 001.

Solution We use $a = 0$ as the starting point, and we know that $f^{(k)}(0) = 1$, for every k, so it will be very easy to find the Taylor polynomial $P_n(x)$. The question is which n is the one to use.

The remainder for $P_n(1)$ is $R_n(1) = \frac{1}{(n+1)!}f^{(n+1)}(c)(x-a)^{n+1} = \frac{1}{(n+1)!}e^c(1-0)^{n+1} = \frac{1}{(n+1)!}e^c$. Since $0 \le c \le 1$, we see that $e^c \le e^1 = e$, and the remainder is less than or equal to $\frac{e}{(n+1)!}$. Since $e < 3$, we see that $R_n(1) < \frac{3}{(n+1)!}$.

All we need is to find a value of n for which this is less than 0.000 001. We try a few possibilities: $\frac{3}{5!} = \frac{3}{120} = 0.025$. This is much too big. Let us jump to $\frac{3}{8!} = \frac{3}{40{,}320} \approx 0.000\,0744$. This is still too big, but it is much closer. Evaluating $\frac{3}{9!} = \frac{3}{362{,}880} \approx 0.000\,008\,27$, we see that it is nearly what we want, but not quite. However, the next one is $\frac{3}{10!} = \frac{3}{3{,}628{,}800} \approx 0.000\,000\,827$, which is definitely smaller than 0.000 001.

We conclude that if $n + 1 = 10$, that is, $n = 9$, the remainder will be small enough, so we approximate $e = f(1)$ by $P_9(1)$:

$$\begin{aligned} P_9(1) &= 1 + \frac{1}{1!} + \frac{1^2}{2!} + \frac{1^3}{3!} + \frac{1^4}{4!} + \frac{1^5}{5!} + \frac{1^6}{6!} + \frac{1^7}{7!} + \frac{1^8}{8!} + \frac{1^9}{9!} \\ &= 1 + 1 + \frac{1}{2} + \frac{1}{6} + \frac{1}{24} + \frac{1}{120} + \frac{1}{720} + \frac{1}{5040} + \frac{1}{40{,}320} + \frac{1}{362{,}880} \\ &\approx 2.718\,281\,53. \end{aligned}$$

This certainly agrees to within the desired accuracy with $e \approx 2.718\,281\,83$.

EXAMPLE 3 **Estimating ln 2** If we want to estimate ln 2 by using a Taylor polynomial $P_n(x)$ associated to the function $f(x) = \ln x$ about the point $a = 1$, how large should n be in order to ensure that the approximation is correct to within 0.0001?

Solution The remainder associated to $P_n(x)$ is $R_n(x) = \frac{1}{(n+1)!}f^{(n+1)}(c)(x-a)^{n+1}$, for some c between a and x. In this case, $a = 1$ and the x we are interested in is 2.

As we see in the margin, the derivatives are not difficult to find; the pattern is fairly clear. The kth derivative contains the $-k$th power of x. Differentiating it changes the power to the $-(k+1)$th power and multiplies by $-k$. So the $(k+1)$th derivative is the product of $x^{-(k+1)}$ and the product of the integers $-1, -2, -3, \ldots, -k$, which is $(-1)^k k! x^{-(k+1)}$.

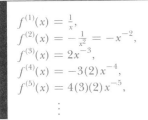

$f^{(1)}(x) = \frac{1}{x}$,
$f^{(2)}(x) = -\frac{1}{x^2} = -x^{-2}$,
$f^{(3)}(x) = 2x^{-3}$,
$f^{(4)}(x) = -3(2)x^{-4}$,
$f^{(5)}(x) = 4(3)(2)x^{-5}$,
$\vdots$

We see that $R_n(2) = \frac{1}{(n+1)!}\left((-1)^n n! c^{-(n+1)}\right)$, for some $c \in [1, 2]$. Since $x^{-(n+1)}$ is a *decreasing* function, its largest value will occur when $x = 1$ (at the left end of the interval). So

$$|R_n(2)| \le \frac{n!}{(n+1)!}1^{-(n+1)} = \frac{1}{n+1}.$$

To ensure the desired accuracy, we need to have $|R_n(2)| \le 0.0001$, and to do that, we need to have $\frac{1}{n+1} \le 0.0001$, or $n+1 \ge \frac{1}{0.0001} = 10{,}000$. This means that we need to take $n \ge 9999$ and evaluate $P_{9999}(x)$.

This is frustrating because it is not practical to evaluate so large a polynomial. If nothing else, round-off errors from a calculation this long are likely to make it unreliable anyway. It turns out that for some "nice" functions, like e^x and $\sin x$ and $\cos x$, for instance, the remainder gets quite small after only a moderate number of terms, which makes Taylor polynomials an efficient way of estimating these functions. However, the bad news is that there are other functions for which the remainder stays uncomfortably large unless n is itself extremely large, and as a practical tool Taylor polynomials have very limited use for these functions.

EXAMPLE 4 Find the value of $\sin(3^\circ)$ to within 0.0001.

Solution As usual, we will have to convert the angle 3° into radians; it is $3 \times \frac{\pi}{180} = \frac{\pi}{60}$ radians. If we let $f(x) = \sin x$, what we want to estimate is $f\left(\frac{\pi}{60}\right)$; we can do this by using Taylor polynomials about the point $a = 0$.

Notice that the first derivative is $f'(x) = \cos x$, the second derivative is $-\sin x$, the third derivative is $-\cos x$, and the fourth derivative is $\sin x$. After four derivatives we have come back to the original function $\sin x$. So the fifth, sixth, seventh, and eighth derivatives will be $\cos x$, $-\sin x$, $-\cos x$, and $\sin x$ again, and this pattern will keep repeating again and again. When we want to estimate the size of a remainder $R_n(x)$, we have to estimate the size of the $(n+1)$th derivative $f^{(n+1)}(x)$, and since every derivative is $\pm \sin x$ or $\pm \cos x$, we know that its absolute value is never bigger than 1.

The nth remainder is $R_n(x) = \frac{1}{(n+1)!} f^{(n+1)}(c)(x-a)^{n+1}$, for some $c \in [0, x]$. In our case, $a = 0$ and $x = \frac{\pi}{60}$, so the absolute value of the remainder satisfies $|R_n(x)| \le \frac{1}{(n+1)!}(1)|x-a|^{n+1} = \frac{1}{(n+1)!}\left(\frac{\pi}{60}\right)^{n+1}$.

The goal is to make this less than (or equal to) 0.0001, by choosing an appropriate n. Asking that $\frac{1}{(n+1)!}\left(\frac{\pi}{60}\right)^{n+1} \le 0.0001$ is the same as asking that $(n+1)!\left(\frac{60}{\pi}\right)^{n+1} \ge 10{,}000$. (We just inverted both sides of the inequality, remembering to reverse the direction of the inequality sign.)

With a calculator we find that $\frac{60}{\pi} \ge 19$, so it will be enough to find n so that $(n+1)!19^{n+1} \ge 10{,}000$ (since $(n+1)!\left(\frac{60}{\pi}\right)^{n+1}$ is even bigger ...). This we do by trial and error.

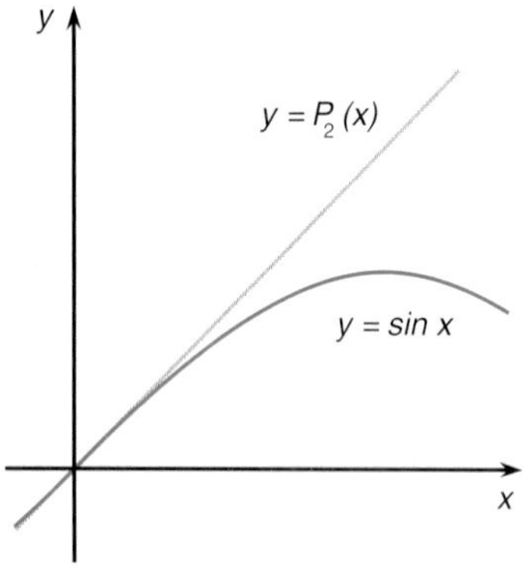

Figure 9.2.1

$f^{(0)}(0) = \sin(0) = 0,$
$f^{(1)}(0) = \cos(0) = 1,$
$f^{(2)}(0) = -\sin(0) = 0.$

First, with $n = 1$ we find that $(n+1)!19^{n+1} = 2!19^2 = 2 \times 19^2 = 722$, which is not big enough. With $n = 2$ we find that $(n+1)!19^{n+1} = 3! \times 19^3 = 6 \times 19^3 = 41{,}154$, which is more than sufficient.

In other words, it is only necessary to use the Taylor polynomial $P_2(x)$ in order to have the desired accuracy. Evaluating the derivatives, we find that $P_2(x) = 0 + \frac{1}{1!}(x-0) + 0 = x$. This Taylor polynomial is extraordinarily simple, but it works.

We find $P_2\left(\frac{\pi}{60}\right) = \frac{\pi}{60} \approx 0.052\,36$. (With a calculator, set to "degrees" for once, we find that $\sin(3^\circ) \approx 0.052\,336$, so our answer does agree to within the specified accuracy of 0.0001.)

Notice that the Taylor polynomial $P_2(x)$ that we used in this example is really just the tangent line at $x = 0$ (see Figure 9.2.1). We have used Lagrange's form of the remainder to check that $P_2(x)$ gives an approximation that is sufficiently accurate, for $x = \frac{\pi}{60}$. Then we calculated the value of the Taylor polynomial, which is in effect the same thing as the linear approximation (cf. Section 4.10).

Notice that because the second derivative $f^{(2)}(x)$ is zero at $x = 0$, the Taylor polynomial $P_2(x)$ is actually the same as the Taylor polynomial $P_1(x)$. When we estimated the remainder, the estimate we found for $R_1(x)$ was too big, so we had to use $R_2(x)$, for which we had a smaller estimate. Of course, since we know that $P_1(x)$ and $P_2(x)$ are the same, we see that the remainders are equal too. The point is that the *estimate* we made of the size of the remainder was smaller for $P_2(x)$ than for $P_1(x)$, and it was knowing the size of the estimate that enabled us to use $P_2(x)$ to find a value with confidence.

Since we are only *estimating* the size of the remainder, it may easily happen that the actual remainder is smaller than our estimate. It may also happen that we decide that it is necessary to use some $P_n(x)$ to get the desired accuracy when in fact a Taylor polynomial of some smaller degree would be adequate. This is not serious. At worst it means that we may have to calculate a few more terms of the polynomial than what is strictly necessary.

Another consideration is the choice of the "starting point" a about which we define the Taylor polynomials. So far we have not worried about which a to use, but in practice we are likely to be asked to approximate the value of some function at some particular point, and part of the solution will be to choose the starting point a.

The basic idea is to choose a to be as close as possible to the point at which the function is to be approximated. The catch is that it must be a point at which we know the values of the function and its derivatives so that we can construct the Taylor polynomials.

The formula for the remainder is

$$R_n(x) = \frac{f^{(n+1)}(c)}{(n+1)!}(x-a)^{n+1}.$$

If $|x-a|$ is small, then $(x-a)^{n+1}$ will approach 0 quickly as n increases. This will tend to make the remainder terms small without n having to be too large. If, on the other hand, the point x is far from a, then the remainder will contain increasing powers of the fairly large number $(x-a)$, which will make it less likely to become small for moderate values of n.

EXAMPLE 5 Estimate the value of $\arctan(-0.9)$ to within 0.0005.

Solution We could try constructing Taylor polynomials about the point $a = 0$, but that is actually fairly far from -0.9. A much better choice would be $a = -1$.

If $f(x) = \arctan x$, then $f^{(1)}(x) = \frac{1}{1+x^2} = (1+x^2)^{-1}$, and $f^{(2)}(x) = -(1+x^2)^{-2}(2x) = -2x(1+x^2)^{-2}$. We then find $f^{(3)}(x) = -2(1+x^2)^{-2} + (-2)(-2x)(1+x^2)^{-3}(2x) = -2(1+x^2)^{-2} + 8x^2(1+x^2)^{-3}$. Putting this over a common denominator, we find

$$f^{(3)}(x) = \frac{-2(1+x^2)+8x^2}{(1+x^2)^3} = \frac{6x^2-2}{(1+x^2)^3}.$$

Let us use this information to estimate the remainder $R_2(-0.9)$. We know that $R_2(-0.9) = \frac{1}{3!}f^{(3)}(c)\big(-0.9-(-1)\big)^3 = \frac{(0.1)^3}{6}f^{(3)}(c)$, for some c between -1 and -0.9.

We have to estimate the size of $f^{(3)}(c)$, for $c \in [-1, -0.9]$. It is probably easier to work separately with the numerator and the denominator. The numerator of $f^{(3)}(x)$ is $6x^2 - 2$. This is a function whose derivative $12x$ is always negative on the interval in question, so it is a decreasing function. Its largest value will occur at the left endpoint, where it equals $6(-1)^2 - 2 = 4$. So the numerator is always less than or equal to 4 on the interval.

For the denominator we have to be careful. What we want is to be able to say that $\frac{1}{(1+x^2)^3}$ is less than or equal to some number. Because of the way inequalities reverse when we take reciprocals, this means that we have to be able to say that $(1 + x^2)^3$ is *greater* than or equal to some number; we have to find the *minimum* possible value of the denominator. Certainly, $(1 + x^2)^3$ will have its smallest possible value at $x = -0.9$, so $\frac{1}{(1+x^2)^3} \le \frac{1}{(1+(-0.9)^2)^3}$.

From this we see that

$$f^{(3)}(c) = \frac{6c^2 - 2}{(1 + c^2)^3} \le \frac{4}{\left(1 + (-0.9)^2\right)^3} \approx 0.67457 \le 0.7.$$

Using this it is possible to estimate the remainder:

$$R_2(-0.9) = \frac{f^{(3)}(c)}{3!}\left(-0.9 - (-1)\right)^3 \le \frac{0.7}{6}(0.1)^3 \approx 0.000\,117.$$

With a graphics calculator we can plot the function $y = \arctan x$ and the Taylor polynomial $P_2(x)$ from Example 5. For comparison we also plot the Taylor polynomial of degree less than or equal to 2 around $a = 0$. Since $f(0) = f''(0) = 0$ and $f'(0) = 1$, we find that it is x (and its degree is actually 1).

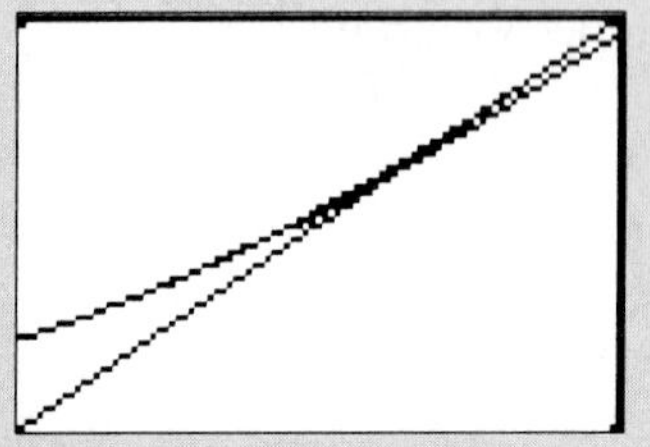

TI-85 GRAPH F1 ($y(x)$ =) TAN^{-1} F1 (x) ▼ (−) π ÷ 4 + (F1 (x) + 1) ÷ 2 + (F1 (x) + 1) ^ 2 ÷ 4 ▼ F1 (x) M2(RANGE) (−) 1 ▼ 0 ▼ ▼ (−) 1 ▼ 0 F5(GRAPH)	SHARP EL9300 [graph key] tan^{-1} x/θ/T 2ndF ▼ (−) π ÷ 4 + (x/θ/T + 1) ÷ 2 + (x/θ/T + 1) x^2 ÷ 4 2ndF ▼ x/θ/T RANGE −1 ENTER 0 ENTER 1 ENTER −1 ENTER 0 ENTER 1 ENTER [graph key]
CASIO f_x-6300G (f_x-7700GB) Range −1 EXE 0 EXE 1 EXE −1 EXE 0 EXE 1 EXE EXE EXE EXE Graph tan^{-1} X EXE Graph (−) π ÷ 4 + (X + 1) ÷ 2 + (X + 1) x^2 ÷ 4 EXE Graph X EXE	HP 48SX ↰ PLOT PLOTR ERASE ATTN ′ Y = ATAN X ↱ PLOT ↰ DRAW ERASE 1 +/− ENTER 0 XRNG 1 +/− ENTER 0 YRNG DRAW ATTN ′ Y = π ÷ 4 +/− + () X + 1 ► ÷ 2 + () X + 1 ► y^x 2 ÷ 4 ↱ PLOT ↰ DRAW DRAW ATTN ′ Y = X ↱ PLOT ↰ DRAW DRAW

The Taylor polynomial $P_2(x)$ is indistinguishable from arctan x near $x = -1$, but it gets farther away near $x = 0$. The Taylor polynomial around $a = 0$ is indistinguishable from arctan x near $x = 0$, but it gets farther away near $x = -1$.

$f(x) = \arctan x,$
$f(-1) = \arctan(-1)$
$= -\frac{\pi}{4};$
$f'(x) = \frac{1}{1+x^2},$
$f'(-1) = \frac{1}{1+(-1)^2} = \frac{1}{2};$
$f^{(2)}(x) = \frac{-2x}{(1+x^2)^2},$
$f^{(2)}(-1) = \frac{-2(-1)}{(1+(-1)^2)^2} = \frac{1}{2}.$

Using the Taylor polynomial $P_2(x)$ about the point $x = -1$ will certainly give us sufficient accuracy. To construct $P_2(x)$, we evaluate up to $f^{(2)}(-1)$. We find

$$P_2(x) = f(-1) + \frac{1}{1!}f'(-1)\left(x - (-1)\right) + \frac{1}{2!}f^{(2)}(-1)\left(x - (-1)\right)^2$$
$$= -\frac{\pi}{4} + \frac{x+1}{2} + \frac{(x+1)^2}{4}.$$

Evaluating at $x = -0.9$, we find the approximation $P_2(-0.9) = -\frac{\pi}{4} + \frac{0.1}{2} + \frac{(0.1)^2}{4} \approx -0.732\,898$. This is remarkably good; with a calculator we find that $\arctan(-0.9) \approx -0.732\,815$.

The importance of this example becomes clear when we consider what happens if we use $a = 0$ as the starting point. In estimating the size of the remainder $R_2(x)$ in that case we see that the smallest value of the denominator on $[-0.9, 0]$ is at 0, so on this interval $\frac{1}{(1+x^2)^3} \leq \frac{1}{1} = 1$. The numerator is more complicated in that it is sometimes positive and sometimes negative; it decreases from $6(-0.9)^2 - 2 = 2.86$ at $x = -0.9$ to $6(0)^2 - 2 = -2$ at $x = 0$. Its absolute value is never bigger than 2.86. The remainder has absolute value

$$|R_2(c)| = \left|\frac{f^{(3)}(c)}{3!}(-0.9 - 0)^3\right| \leq \left|\frac{1}{6}\frac{2.86}{1}(-0.729)\right| = 0.347\,49.$$

This is not nearly good enough to give us the accuracy we require. Moreover, to calculate the higher derivatives $f^{(4)}(x)$, $f^{(5)}(x)$, and so on becomes increasingly messy. It would be a major undertaking to find how large n would have to be in order that $P_n(x)$ constructed around $a = 0$ would be sufficiently accurate.

REMARK 9.2.3

1. To measure the accuracy of the approximation by $P_n(x)$, it is necessary to estimate the size of the remainder $R_n(x)$. Generally speaking, this means that we should estimate the *absolute value* of the remainder, which tells how far the estimate is from the actual value. It is sometimes possible to obtain additional information by using the sign of the remainder to say whether the actual value is above or below the approximation.
2. The Lagrange formula for the remainder $R_n(x)$ is the product of $\frac{1}{(n+1)!}$, $f^{(n+1)}(c)$, and $(x - a)^{n+1}$. The first and last are easy to evaluate exactly; it is the derivative that we must estimate, especially because the point c is not determined precisely. What is known is that c is between a and x. If $f^{(n+1)}$ is increasing between a and x (or decreasing), then it is easy to find the maximum possible value or, what is more important, the maximum absolute value. This can then be used in the estimate of the remainder.
3. It is frequently necessary to estimate the size of a quotient $\frac{p(x)}{q(x)}$. It often helps to write $\frac{p(x)}{q(x)} = p(x)\frac{1}{q(x)}$ and work separately with $p(x)$ and $\frac{1}{q(x)}$. To show that $\left|\frac{1}{q(x)}\right| \leq B$, for some constant B, it is necessary to show that $|q(x)| \geq \frac{1}{B}$. (It is very important to remember that the inequality is reversed when we take reciprocals.) So the idea is to show that $|p(x)| \leq A$, for some constant A, and that $|q(x)| \geq C$, for some constant $C > 0$. From this it is possible to conclude that the original quotient satisfies $\left|\frac{p(x)}{q(x)}\right| \leq \frac{A}{C}$.

4. From the formula for $R_n(x)$ we see that the two things that influence the size of the remainder are the size of the $(n+1)$th derivative of $f(x)$ and the size of $(x-a)$. The size of the various derivatives can vary tremendously. For some functions, such as e^x, $\cos x$, or $\sin x$, we have noticed that on a fixed interval the derivatives are all bounded by some constant that does not depend on n. Since the derivative is divided by $(n+1)!$, the remainder in such a case will decrease rapidly as n increases. The effect is that approximation by the Taylor polynomial $P_n(x)$ will likely be quite accurate with fairly small values of n. For other functions, such as $\ln x$, the absolute value of the $(n+1)$th derivative gets larger and larger as n increases, so the remainder does not decrease as rapidly. The practical effect may be that in order to have a sufficiently accurate approximation it will be necessary to use $P_n(x)$ with an unreasonably high n.

   The other consideration is the size of $x-a$, that is, how far away from the starting point a is the point at which we wish to estimate $f(x)$. If $x-a$ is very small, then the remainder $R_n(x)$ will likely become small for fairly small values of n even if the derivatives of the function increase in size. In this situation it will still be practical to use Taylor polynomials to approximate the value $f(x)$.

All of the examples we have done so far have estimated values that could be obtained almost instantly on a calculator, with better accuracy, so the importance of Taylor polynomials is not very clear. It is, of course, conceivable that we might want to calculate e to 300 decimal places, and in principle we now know how to set about doing that. There are instances in which calculator accuracy is insufficient, and it is reassuring to know that it is possible to have any desired degree of accuracy by using suitable Taylor polynomials $P_n(x)$. However, these uses are not encountered very often. Without necessarily realizing it, we rely on Taylor polynomials when we use our calculators, since calculators use Taylor polynomials to evaluate many built-in functions.

There are some other situations in which Taylor polynomials are very important and less easily circumvented. It is easy to write down examples of differential equations that we do not know how to solve. In this situation it is often easy to write down the Taylor polynomials of the solution and use them to find approximate values for the solution.

One of the simplest examples of this procedure is estimating an integral, which can be viewed as a differential equation because it asks for an indefinite integral, a function whose derivative equals the integrand.

EXAMPLE 6 Estimate to within 0.01 the following integral, which is very important in statistics:

$$\int_0^{1/2} e^{-x^2}\,dx.$$

Solution None of the techniques of integration that we have learned can help us to evaluate this integral. Instead, we consider the indefinite integral $F(x) = \int_0^x e^{-t^2}\,dt$, realizing that the value we want to estimate is $F\left(\frac{1}{2}\right)$. Let us approximate it by Taylor polynomials about the point $a=0$. By the Fundamental Theorem of Calculus we know that $F^{(1)}(x) = e^{-x^2}$, and from this we find that

$$
\begin{aligned}
F^{(2)}(x) &= -2xe^{-x^2}, & F^{(2)}(0) &= 0;\\
F^{(3)}(x) &= -2e^{-x^2} + 4x^2e^{-x^2}, & F^{(3)}(0) &= -2;\\
F^{(4)}(x) &= 4xe^{-x^2} + 8xe^{-x^2} - 8x^3e^{-x^2} = (12x - 8x^3)e^{-x^2}, & F^{(4)}(0) &= 0;\\
F^{(5)}(x) &= 12e^{-x^2} - 24x^2e^{-x^2} - 24x^2e^{-x^2} + 16x^4e^{-x^2} & &\\
&= (12 - 48x^2 + 16x^4)e^{-x^2}, & F^{(5)}(0) &= 12.
\end{aligned}
$$

Let us estimate $\left|R_3\left(\frac{1}{2}\right)\right|$. Since e^{-x^2} is a decreasing function of x for $x \geq 0$, we see that $e^{-x^2} \leq e^0 = 1$, for all $x \in \left[0, \frac{1}{2}\right]$. If $c \in \left[0, \frac{1}{2}\right]$, then $|c| \leq \frac{1}{2}$, and using the Triangle Inequality we find that

$$|F^{(4)}(c)| = |(12c - 8c^3)e^{-c^2}| \leq \left(12\left(\frac{1}{2}\right) + 8\left(\frac{1}{2}\right)^3\right) \times 1 = 7.$$

From this we find that $\left|R_3\left(\frac{1}{2}\right)\right| \leq \frac{1}{4!} \times 7 \times \left|\left(\frac{1}{2} - 0\right)^4\right| \approx 0.0182$.

This is not quite good enough, so we should try R_4. We find that $|F^{(5)}(c)| = |(12 - 48c^2 + 16c^4)e^{-c^2}| \leq \left|\left(12 + 48\left(\frac{1}{2}\right)^2 + 16\left(\frac{1}{2}\right)^4\right)(1)\right| = (12 + 12 + 1)(1) = 25$. From this we see that $\left|R_4\left(\frac{1}{2}\right)\right| \leq \frac{1}{5!}25\left(\frac{1}{2} - 0\right)^5 \approx 0.0065$.

This means that there will be sufficient accuracy if we calculate $P_4\left(\frac{1}{2}\right)$. To construct the polynomial, we note that $F(0) = 0, F^{(1)}(0) = 1, F^{(2)}(0) = 0, F^{(3)}(0) = -2$, and $F^{(4)}(0) = 0$. The Taylor polynomial is

$$
\begin{aligned}
P_4(x) &= \frac{0}{0!} + \frac{1}{1!}x + \frac{0}{2!}x^2 + \frac{-2}{3!}x^3 + \frac{0}{4!}x^4\\
&= x - \frac{1}{3}x^3.
\end{aligned}
$$

Evaluating at $x = \frac{1}{2}$, we find that $P_4\left(\frac{1}{2}\right) = \frac{1}{2} - \frac{1}{3}\frac{1}{8} = \frac{11}{24} \approx 0.4583$. From this we conclude that to within an accuracy of 0.01,

$$\int_0^{\frac{1}{2}} e^{-x^2}\, dx \approx 0.46.$$

It should be clear from Example 6 that estimating integrals using Taylor polynomials can be very useful. It is interesting because the answer is not something that can be checked instantly with a calculator. Although we will not deal with them here, similar remarks apply to approximating solutions of differential equations.

Here we present a proof of Taylor's Formula with Remainder.

EXAMPLE 7 Prove Theorem 9.2.2.

Solution To emphasize that the point at which the remainder is being evaluated is *fixed*, let us call it b (instead of x as in the theorem).

It is necessary to define a special function to which Rolle's Theorem can be applied. We let

$$D(t) = f(b) - \sum_{k=0}^{n} \left(\frac{f^k(t)}{k!}(b-t)^k \right) - R_n(b)\frac{(b-t)^{n+1}}{(b-a)^{n+1}}.$$

When $t = b$, the last quotient equals 0, and since $(b-t) = (b-b) = 0$, each term in the sum is zero except the one with $k = 0$. So $D(b) = f(b) - \frac{f^{(0)}(b)}{0!} = f(b) - \frac{f(b)}{1} = 0$.

When $t = a$ the last quotient equals 1, so

$$D(a) = f(b) - \sum_{k=0}^{n} \left(\frac{f^k(a)}{k!}(b-a)^k \right) - R_n(b) = f(b) - P_n(b) - R_n(b).$$

By the definition of the remainder R_n we see that $D(a) = 0$. Combining these two results, we have that

$$D(a) = D(b) = 0,$$

which means that Rolle's Theorem (Theorem 4.2.4) can be applied to $D(t)$. We conclude that there is a number c between a and b for which $D'(c) = 0$.

What remains is to calculate $D'(c)$.

Let us begin by differentiating the kth term of the sum. By the Product Rule, if $k \neq 0$,

$$\begin{aligned} \frac{d}{dt}\left(\frac{f^{(k)}(t)}{k!}(b-t)^k \right) &= \frac{f^{(k+1)}(t)}{k!}(b-t)^k + \frac{f^{(k)}(t)}{k!}k(b-t)^{k-1}(-1) \\ &= \frac{f^{(k+1)}(t)}{k!}(b-t)^k - \frac{f^{(k)}(t)}{(k-1)!}(b-t)^{k-1}. \end{aligned}$$

Notice that the two terms are very similar. When we add up these derivatives for every k, most of the parts will cancel one another. Using summation notation and applying Proposition 5.2.4 (about shifting the index of summation), we see that

$$\begin{aligned} \frac{d}{dt}\left(\sum_{k=0}^{n} \frac{f^{(k)}(t)}{k!}(b-t)^k \right) &= \sum_{k=0}^{n} \left(\frac{f^{(k+1)}(t)}{k!}(b-t)^k - \frac{f^{(k)}(t)}{k!}k(b-t)^{k-1} \right) \\ &= \sum_{k=0}^{n} \left(\frac{f^{(k+1)}(t)}{k!}(b-t)^k \right) - \sum_{k=1}^{n} \left(\frac{f^{(k)}(t)}{(k-1)!}(b-t)^{k-1} \right) \\ &= \sum_{k=0}^{n} \left(\frac{f^{(k+1)}(t)}{k!}(b-t)^k \right) - \sum_{\ell=0}^{n-1} \left(\frac{f^{(\ell+1)}(t)}{\ell!}(b-t)^{\ell} \right) \\ &= \frac{f^{(n+1)}(t)}{n!}(b-t)^n. \end{aligned}$$

This is the nth term in the first sum; all the other terms cancel.

The last term of $D(t)$ is easily differentiated with respect to t, and we find that

$$D'(t) = -\frac{f^{(n+1)}(t)}{n!}(b-t)^n - R_n(b)(n+1)\frac{(b-t)^n}{(b-a)^{n+1}}(-1).$$

To say that $D'(c) = 0$ means that

$$-\frac{f^{(n+1)}(c)}{n!}(b-c)^n + R_n(b)(n+1)\frac{(b-c)^n}{(b-a)^{n+1}} = 0.$$

Dividing by $(b-c)^n$, we find that

$$R_n(b)\frac{n+1}{(b-a)^{n+1}} = \frac{f^{(n+1)}(c)}{n!},$$

or

$$R_n(b) = \frac{f^{(n+1)}(c)}{(n+1)!}(b-a)^{n+1}.$$

If we return to writing x in place of b, this is the formula of Theorem 9.2.2, which we wanted to prove.

Exercises 9.2

I

In each of the following situations, estimate $f(x)$ to within the specified accuracy, using a Taylor polynomial $P_n(x)$ around the point a. (The first step is to find how large n must be in order to guarantee the desired accuracy.)

1. $f(x) = e^x$, $a = 0$, $x = \frac{1}{2}$, within 0.01
2. $f(x) = e^{2x}$, $a = 0$, $x = \frac{1}{4}$, within 0.01
3. $f(x) = \sin x$, $a = 0$, $x = 1$, within 0.0001
4. $f(x) = \cos x$, $a = 0$, $x = 2$, within 0.001
5. $f(x) = \ln x$, $a = 1$, $x = 1.2$, within 0.2
6. $f(x) = \tan x$, $a = 0$, $x = 0.1$, within 0.01
7. $f(x) = \ln x$, $a = 2$, $x = 1.9$, within 0.1
8. $f(x) = \sin x$, $a = \frac{\pi}{4}$, $x = 1$, within 0.01
9. $f(x) = x^{15}$, $a = 0$, $x = -\frac{1}{10}$, within 0.000 001
10. $f(x) = e^{x^2}$, $a = 0$, $x = \frac{1}{2}$, within 0.01

Estimate each of the following quantities to within the specified accuracy, using an appropriate Taylor polynomial $P_n(x)$ around an appropriate point a. (Specify which function $f(x)$ and which point a you are using, and calculate how large n must be in order to guarantee the desired accuracy.)

11. $\cos 4°$ to within 0.01
12. $\sin 6°$ to within 0.01
13. $\cos 43°$ to within 0.01
14. $\sin 66°$ to within 0.01
15. $\ln(1.2)$ to within 0.1
16. $\ln\left(\frac{5}{6}\right)$ to within 0.1
17. $\tan 3°$ to within 0.01
18. $\cot 46°$ to within 0.01
19. $\sinh(-0.12)$ to within 0.001
20. $\cosh(1.3)$ to within 0.01

II

21. Consider the Maclaurin polynomial $P_5(x)$ for $f(x) = \sin x$. What accuracy can you guarantee for approximations to $f(x)$ by $P_5(x)$ if $-0.5 \le x \le 0.5$?
22. What accuracy can you guarantee for the Taylor polynomial $P_6(x)$ associated to $f(x) = e^x$ about the point $a = 1$ if $0 < x < 2$? (You may want to use the inequalities $2.5 < e < 3$.)

*23. Suppose we want to approximate the value of $\sin^2(-2°)$ to within 0.0001. One way to do this would be to use the Taylor polynomials associated to $f(x) = \sin^2 x$ about the point $a = 0$. Unfortunately, the derivatives of $f(x)$ become fairly complicated as n increases. Another approach would be to estimate $\sin(-2°)$ by using Maclaurin polynomials for $g(x) = \sin x$ and squaring the result. This is probably easier, except that slightly more care is needed in determining which polynomial to use.
   (i) To within what accuracy must we estimate

$\sin(-2°)$ in order that its square will be accurate to within 0.0001? (*Hint:* Remember that $|\sin x| \leq |x|$, and assume that $|P_n(x)| \leq |x|$ too.)

(ii) Estimate $\sin^2(-2°)$ to within 0.0001.

Estimate each of the following to within the specified accuracy.

***24.** $\cos^2(132°)$ to within 0.001

***25.** $\sin^2(1°)\cos(1°)$ to within 0.0005

***26.** $\tan(-1°)$ to within 0.001

***27.** $e^{\sin(47°)}$ to within 0.0001

***28.** $\sec(43°)$ to within 0.001

***29.** $\int_0^{1/4} e^{-x^2}\,dx$ to within 0.01

***30.** $\int_{-1/8}^{1/8} e^{-x^2}\,dx$ to within 0.01

***31.** $\int_0^{1/5} \sin(x^2)\,dx$ to within 0.01

***32.** In Example 1 we had to estimate $\frac{e}{120}$, which we did by noting that $e < 3$. Strictly speaking, since the object of the procedure was to approximate e, we should not assume that we already know what e is. Show that $e < 3$ without using your calculator or looking up the value. (One way to proceed is to recall that e is determined by the fact that $\ln e = 1$, i.e., that $\int_1^e \frac{1}{x}\,dx = 1$. Since the derivative of $\ln x$ is positive, the function $\ln x$ is increasing, so it would be enough to show that $\ln 3 > 1$, i.e., that $\int_1^3 \frac{1}{x}\,dx > 1$. We could attempt to show this by noting that the integrand $\frac{1}{x}$ is greater than or equal to $\frac{1}{2}$ when $x \in [1,2]$ and that it is greater than or equal to $\frac{1}{3}$ when $x \in [2,3]$, so the integral is greater than or equal to $\int_1^2 \frac{1}{2}\,dx + \int_2^3 \frac{1}{3}\,dx = \frac{1}{2} + \frac{1}{3}$. Unfortunately, this is not greater than 1, so this attempt fails. However, you could use the same idea but divide the interval $[1,3]$ into more than two pieces . . .).

***33.** Consider the differential equation $f' = x^2 f$, and let $f(x)$ be a solution that satisfies $f(0) = 1$. (To say that it is a solution means that $f'(x) = x^2 f(x)$.)

(i) Using the equation, evaluate $f'(0)$. Then find a formula for $f''(x)$; your formula should involve $f'(x)$, but it is possible to substitute for $f'(x)$ using the differential equation and express $f''(x)$ in terms of $f(x)$. Do this and evaluate $f''(0)$.

(ii) Continue in this way and find $f^{(3)}(0)$, $f^{(4)}(0)$, and $f^{(5)}(0)$.

(iii) Write down the Taylor polynomial $P_5(x)$ associated to $f(x)$ at $a = 0$.

(iv) Use $P_5(x)$ to estimate $f(0.1)$.

(v) Recognizing that the differential equation is a separable equation, write $\frac{f'}{f} = x^2$ and integrate to find an explicit formula for $f(x)$. Use the fact that $f(0) = 1$.

(vi) Compare the answer to part (iv) with what your calculator gives you for $f(0.1)$ using the answer to part (v).

***34.** Write down the Maclaurin polynomial $P_5(x)$ for the function $f(x)$ that satisfies $f(0) = 4$ and $f'(x) = \cos(x^2)f(x)$.

***35.** Estimate to within 0.1 the integral $\int_0^1 e^{-x^2}\,dx$.

***36.** Estimate to within 0.01 the integral $\int_0^{1/4} \sin(x^2)\,dx$.

***37.** Estimate to within 0.01 the integral $\int_0^{1/10} \arctan(\sin x)\,dx$.

***38.** Estimate to within 0.05 the integral $\int_1^{3/2} \arctan(x^2 + 1)\,dx$.

III

39. The Maclaurin polynomial $P_6(x)$ for $f(x) = e^x$ gives a good approximation to the function. Use their graphs to estimate the smallest $x > 0$ for which $|f(x) - P_6(x)| \geq 0.0001$ and the smallest $x > 0$ for which $|f(x) - P_6(x)| \geq 0.01$.

40. Use the graphs of $f(x) = \sin x$ and its Maclaurin polynomial $P_5(x)$ to estimate the smallest $x > 0$ for which $|f(x) - P_5(x)| \geq 0.0001$ and the smallest $x > 0$ for which $|f(x) - P_5(x)| \geq 0.01$. (*Hint:* Graph $\sin x - P_5(x)$ and $y = 0.0001$, $y = -0.0001$, or $\sin x - P_5(x) - 0.0001$ and $\sin x - P_5(x) + 0.0001$.)

41. If you evaluate the Maclaurin polynomial $P_n(x)$ for e^x one term at a time, the answer on your calculator or computer will eventually stop changing, that is, additional terms no longer affect the digits you see. How large does n have to be for $P_n(1)$ to stop changing on your calculator? How about $P_n(2)$ or $P_n(10)$?

42. Use the Trapezoidal Rule T_n with $n = 10$ to estimate the integral in Example 6.

***43.** The function $f(x) = \cos x$ has a maximum at $x = 0$. The Maclaurin polynomial $P_2(x) = 1 - \frac{x^2}{2}$ is a downward-opening parabola. We expect that $P_4(x) = 1 - \frac{x^2}{2} + \frac{x^4}{4!}$ should be closer to $f(x)$, and since $\frac{x^4}{4!} \geq 0$, $P_2(x)$ lies below $P_4(x)$. This suggests that $P_2(x)$ is likely below the graph of $y = f(x)$, at least near $x = 0$. Sketch $y = f(x)$, $P_2(x)$, $P_4(x)$ on the same picture. This kind of argument is not entirely reliable, but it is a reasonable starting point.

***44.** Continuing from Exercise 43, consider the parabola whose vertex is at $(0, 1)$ and that crosses the x-axis at the first points where $y = \cos x$ does, that is, at $\left(\pm\frac{\pi}{2}, 0\right)$. (i) Do you expect this parabola to be above or below $y = \cos x$ near $x = 0$? (ii) Find the equation of the parabola. (iii) Sketch the parabola and $y = \cos x$, $P_2(x)$, and $P_4(x)$ on the same graph.

POINT TO PONDER

In Section 4.11 we discussed the use of Newton's Method to find approximate values of $\ln a$. The idea was that $\ln a$ is the value of x for which $f(x) = 0$, where $f(x) = e^x - a$, and Newton's Method is designed to find approximate values for the points at which some function is zero.

To be able to use this technique, it is necessary to be able to calculate the values of $f(x)$ and $f'(x)$ at each of the points $x = x_n$ generated by iterations of Newton's Method. When a computer finds $\ln a$, it uses this approach. When it comes to calculating $f(x_n)$ and $f'(x_n)$, it calculates approximate values, using Taylor polynomials.

The Taylor polynomials give very good approximations to e^x, so they are convenient for calculating $f(x_n)$ and $f'(x_n)$. These values can then be used to find the next approximation x_{n+1} using Newton's Method.

The Taylor polynomials for $\ln x$ do not work very efficiently, as we saw in Example 3, so it is not practical to use them to approximate $\ln a$. The combination of Taylor polynomials to approximate e^x and Newton's Method with these approximate values to find $\ln a$ is a remarkably efficient method for finding approximate values of natural logarithms.

You might like to try this approach to approximate $\ln 2$. By hand it is pretty tedious, but if you program a calculator or computer to evaluate the nth Taylor polynomial $P_n(x)$ for e^x, then it becomes more manageable. You can try using a fairly small n for the first few steps and then increasing the accuracy of the approximations of e^x as the iterations of Newton's Method bring you closer to the correct answer.

Chapter Summary

§9.1
factorial
$0! = 1$
Taylor polynomial
$P_n(x)$
Maclaurin polynomials

§9.2
remainder
Lagrange form of the remainder
estimating the remainder

Review Exercises

I

Write down the Taylor polynomial $P_n(x)$ of each of these functions $f(x)$ about the point a.

1. $f(x) = e^{-x}$, $n = 6$, $a = 0$
2. $f(x) = e^{3x}$, $n = 4$, $a = 0$
3. $f(x) = e^x$, $n = 6$, $a = 1$
4. $f(x) = e^{-2x}$, $n = 4$, $a = -1$
5. $f(x) = \sin(\pi x)$, $n = 5$, $a = \frac{1}{2}$
6. $f(x) = \cos\left(\frac{\pi}{2}x\right)$, $n = 4$, $a = -3$
7. $f(x) = \arctan x$, $n = 3$, $a = 1$
8. $f(x) = \cos(x + 5)$, $n = 4$, $a = -5$
9. $f(x) = x^{15}$, $n = 14$, $a = 0$
10. $f(x) = x^3$, $n = 2$, $a = 2$

Estimate each of the following to within the specified accuracy. State clearly which function $f(x)$ you are approximating and which a and n you are using.

11. $\frac{1}{e}$ within 0.0001
12. $\sqrt{e}$ within 0.0001
13. $\sqrt{1.1}$ within 0.005
14. $\sqrt{1.5}$ within 0.01
15. $\sec(1.1°)$ within 0.005
16. $\cos(-2°)$ within 0.005

17. $\sin(1.5^\circ)$ within 0.005

18. $\sin(10^\circ)$ within 0.01

19. $\tan(-0.1)$ within 0.005

20. $\cot(88^\circ)$ within 0.01

II

***21.** Suppose $f(x)$ is an even function (i.e., $f(-x) = f(x)$ for all x). Let $P_n(x)$ be the nth Maclaurin polynomial. Show that $P_n(x)$ is an even polynomial.

***22.** Suppose $f(x)$ is an odd function (i.e., $f(-x) = -f(x)$ for all x). Let $P_n(x)$ be the nth Maclaurin polynomial. Show that $P_n(x)$ is an odd polynomial.

***23.** Suppose $f(x)$ is n times differentiable at $x = a$. Let $g(x) = f(x + a)$. What is the relationship between the Taylor polynomials of $f(x)$ about the point a and the Maclaurin polynomials of $g(x)$?

Series

10

A hare trying to pass a tortoise must first catch up half the distance between them. Then he must catch up half the remaining distance. Then he must catch up half the remaining distance, and so on.
This argument, known as Zeno's Paradox, suggests that the hare can never overtake the tortoise.

INTRODUCTION

In this chapter we will study *infinite series,* which arise when we try to add up infinitely many numbers. For instance, if we consider

$$\begin{aligned}
&\tfrac{1}{2} &&= \tfrac{1}{2} = 0.5,\\
&\tfrac{1}{2}+\tfrac{1}{4} &&= \tfrac{3}{4} = 0.75,\\
&\tfrac{1}{2}+\tfrac{1}{4}+\tfrac{1}{8} &&= \tfrac{7}{8} = 0.875,\\
&\tfrac{1}{2}+\tfrac{1}{4}+\tfrac{1}{8}+\tfrac{1}{16} &&= \tfrac{15}{16} = 0.9375,\\
&\tfrac{1}{2}+\tfrac{1}{4}+\tfrac{1}{8}+\tfrac{1}{16}+\tfrac{1}{32} &&= \tfrac{31}{32} = 0.96875,
\end{aligned}$$

it certainly appears that these sums approach the value 1. Each time we add an additional term (equal to one-half the previous term), the total gets closer to 1. It is very tempting to think that if we keep going and add up *infinitely* many terms, the answer ought to be 1. We would like to be able to write

$$\frac{1}{2}+\frac{1}{4}+\frac{1}{8}+\frac{1}{16}+\frac{1}{32}+\ldots = 1.$$

It is to make sense of this perfectly reasonable idea that we introduce the concept of series. As the terminology we have just used suggests, the way to define a series will involve a kind of limit.

In Section 10.1 there is a discussion of *sequences*, which provide a way to discuss limits, and in Section 10.2 we consider a special type of series known as *geometric series*; the example just considered was of this type. In Section 10.3 we discuss the precise definition of series in terms of limits and consider the sometimes difficult question of whether the limit exists. A series is said to *converge* if the limit exists, and the next four sections deal with various criteria for deciding whether or not a particular series converges. Section 10.4 presents the *Integral Test*, Section 10.5 presents the *Comparison Test*, and Section 10.6 considers *alternating series* and the concept of *absolute convergence*. In Section 10.7 we discuss the *Ratio Test* and the *Root Test*. In these sections we will need to use l'Hôpital's Rule (p. 332).

In Chapter 9 we considered Taylor polynomials; for each positive n there is a Taylor polynomial $P_n(x)$ of degree n to approximate a function $f(x)$ near $x = a$. The interesting thing is that the terms in $P_n(x)$ are exactly the same as the corresponding terms in $P_{n+1}(x)$. To form $P_{n+1}(x)$, we add one additional term to $P_n(x)$, but the previous terms are left alone.

To get more accuracy, we keep adding more terms to the Taylor polynomial, so it seems reasonable to ask whether it is permissible to add up *all* the infinitely many terms as a series. This is the idea behind *Taylor series*, which are discussed in Section 10.8.

Section 10.9 is devoted to *power series*, a notion that is suggested by Taylor series in the previous section, and Section 10.10 considers how to differentiate and integrate power series.

10.1 Sequences

CONVERGENCE
DIVERGENCE
SQUEEZE THEOREM
SHIFTED SEQUENCE
PRECISE DEFINITION

Before going on to study infinite series in greater depth, we devote some time to the related topic of infinite *sequences*. A sequence just means an infinite list of numbers a_1, a_2, a_3, a_4, ... (a series, by contrast, means an infinite *sum* $\sum_{n=1}^{\infty} a_n$). The question we want to study about a sequence is whether or not it converges to a limit; for instance, the sequence $1, \frac{1}{2}, \frac{1}{3}, \frac{1}{4}, \frac{1}{5}, \frac{1}{6}, \ldots$, whose nth term is $\frac{1}{n}$, converges to 0, since $\frac{1}{n}$ approaches 0 as $n \to \infty$.

DEFINITION 10.1.1

> Suppose a_n is a real number, for each $n = 1, 2, 3, \ldots$. The family $\{a_n\}$ of all these numbers is called a **sequence**, and a_n is called its nth **term**.
>
> Another way of describing a sequence is to say that it is a function whose domain is the positive integers. The nth term a_n is the value of the function at the integer n.

Sequences sometimes begin with $n = 0$, $n = -1$, $n = 2$, or some other integer; in this case the sequence is a function whose domain is the set of all integers greater than or equal to 0 or -1 or 2, etc.

We have already observed that the sequence $a_n = \frac{1}{n}$ (i.e., 1, $\frac{1}{2}$, $\frac{1}{3}$, $\frac{1}{4}$, $\frac{1}{5}$, ...) converges to 0. See Figure 10.1.1(i).

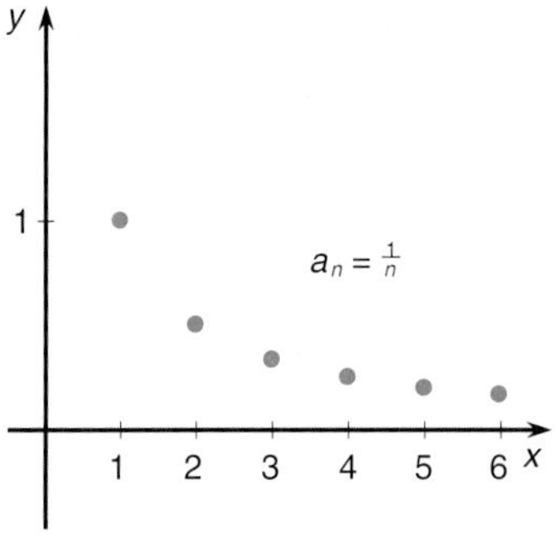

Figure 10.1.1(i)

The constant sequence −2, −2, −2, −2, −2, ... (i.e., $a_n = -2$, for every n) pretty clearly must converge to −2. See Figure 10.1.1(ii).

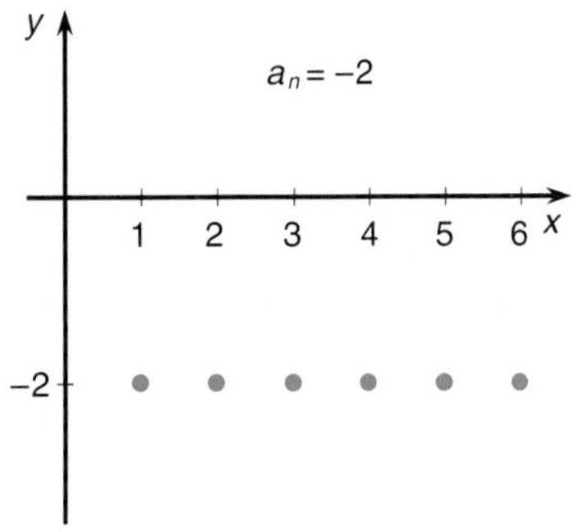

Figure 10.1.1(ii)

The sequence 1, −1, 1, −1, 1, −1, ... (i.e., $a_n = (-1)^{n+1}$) does not converge to any limit. See Figure 10.1.1(iii).

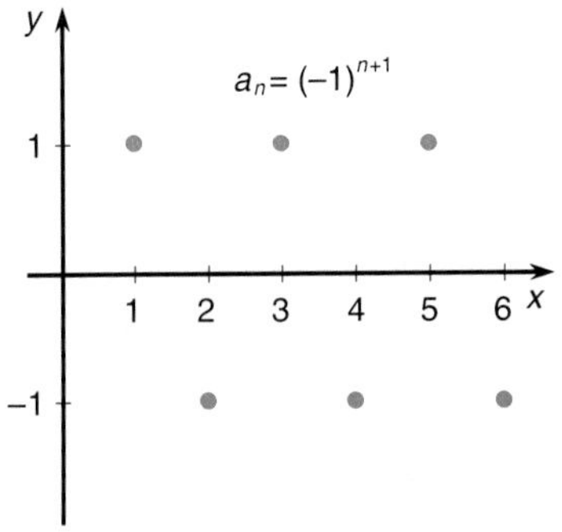

Figure 10.1.1(iii)

The idea of the limit of a sequence is a natural variant of the idea of the limit of a function. We will give the precise definition of this type of limit in Definition 10.1.7, but first we will explore how to use it.

DEFINITION 10.1.2

> The sequence $\{a_n\}$ **converges** to the limit L if its terms approach the number L as a limit as $n \to \infty$. In this case we write
>
> $$\lim_{n \to \infty} a_n = L.$$
>
> We say that the sequence $\{a_n\}$ **converges** (or is **convergent**) if it converges to some limit L and that it **diverges** (or is **divergent**) if it does not converge to any limit.

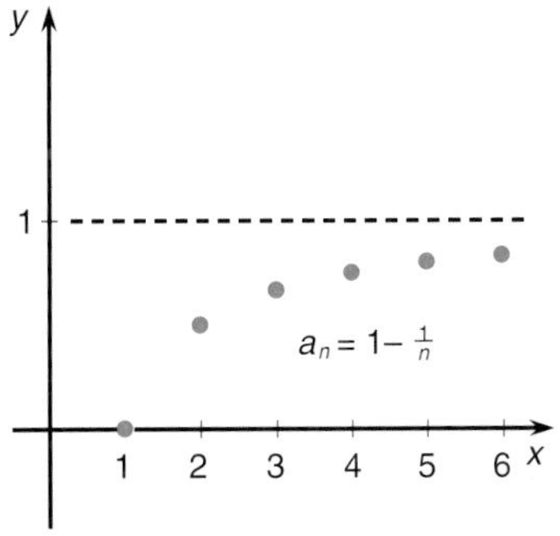

Figure 10.1.2(i)

As with limits of functions, if $\{a_n\}$ converges, then it converges to one and only one limit. (The terms cannot approach two different numbers ...).

EXAMPLE 1

(i) Consider the sequence $a_n = 1 - \frac{1}{n}$. Its nth term is slightly less than 1. The amount by which it is less than 1 is $\frac{1}{n}$, which approaches the limit 0, so it seems reasonable to say that $\lim_{n\to\infty} a_n = 1$. See Figure 10.1.2(i).

$b_1 = 3 - 1 = 2$
$b_2 = 3 + \frac{1}{4}$
$b_3 = 3 - \frac{1}{27}$
$b_4 = 3 + \frac{1}{256}$

(ii) Let $b_n = 3 + \left(-\frac{1}{n}\right)^n$. We evaluate a few terms at left. These terms certainly appear to be getting close to 3. The amount by which the nth term differs from 3 is $-\left(-\frac{1}{n}\right)^n$. This is sometimes negative and sometimes positive, which means that some of the terms are greater than 3 and some are less than 3, but the differences do get smaller and smaller. The sequence $\{(-\frac{1}{n})^n\}$ tends to 0 as $n \to \infty$, and $b_n \to 3$ (i.e., $\lim_{n\to\infty} b_n = 3$). See Figure 10.1.2(ii).

$c_1 = \sin\frac{\pi}{2} = 1$
$c_2 = \sin\pi = 0$
$c_3 = \sin\frac{3\pi}{2} = -1$
$c_4 = \sin 2\pi = 0.$

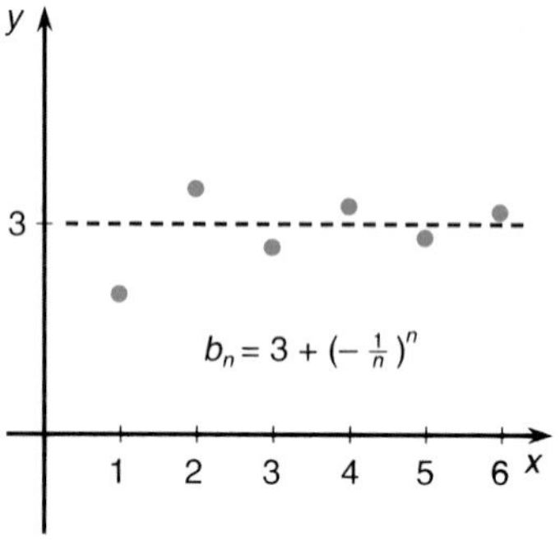

Figure 10.1.2(ii)

(iii) With $c_n = \sin\left(\frac{n\pi}{2}\right)$ we evaluate the first few terms at left to see what happens. After c_4 the sequence "repeats" (since the function $\sin x$ does). The sequence is $1,\ 0,\ -1,\ 0,\ 1, 0,\ -1,\ 0,\ 1,\ 0, -1, 0, \ldots$. It does not converge to any limit at all. See Figure 10.1.2(iii).

(iv) If $d_n = 3n^2 - 1$, then the terms get larger and larger and tend to infinity. Because of this they cannot converge to any (finite) limit L, so the sequence diverges. In this situation we sometimes say that the sequence $\{d_n\}$ **diverges to infinity**. See Figure 10.1.2(iv).

Parts (iii) and (iv) of Example 1 illustrate two different ways in which a sequence can diverge. The sequence $\{c_n\}$ in part (iii) diverges because its values "jump around." Some of them approach the limit 1, others -1, and others 0, but there is no one limit that they all approach. In part (iv), on the other hand, the sequence $\{d_n\}$ diverges to infinity because its terms tend to infinity, so they cannot approach any (finite) limit L.

There are other ways in which sequences can diverge, but these are two very common types of divergent sequence.

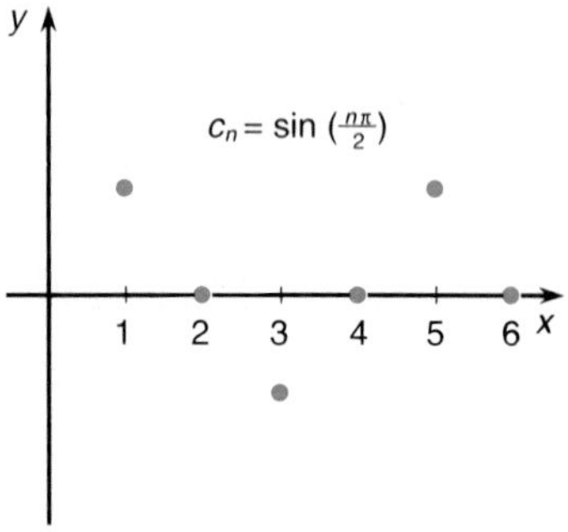

Figure 10.1.2(iii)

In discussing all these examples we have relied on the intuition we have about limits from our knowledge of limits of functions. In fact there is an extremely useful result that says that if a sequence $\{a_n\}$ is defined by $a_n = f(n)$ for some function $f(x)$ and if the function $f(x)$ has a limit as $x \to \infty$, then the sequence $\{a_n\}$ will converge to the same limit. Figure 10.1.3 shows the graph of $y = f(x)$, and the sequence $\{a_n\}$ is just its values at the integers.

THEOREM 10.1.3

Suppose $f(x)$ is a function for which we know that $\lim_{x\to\infty} f(x) = L$, for some number L. Define a sequence $\{a_n\}$ by $a_n = f(n)$. Then

$$\lim_{n\to\infty} a_n = L.$$

If $\lim_{x\to\infty} |f(x)| = \infty$, then $\{a_n\}$ diverges.

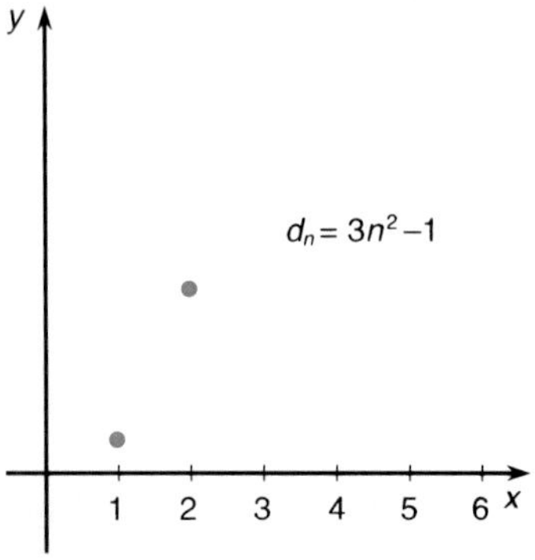

Figure 10.1.2(iv)

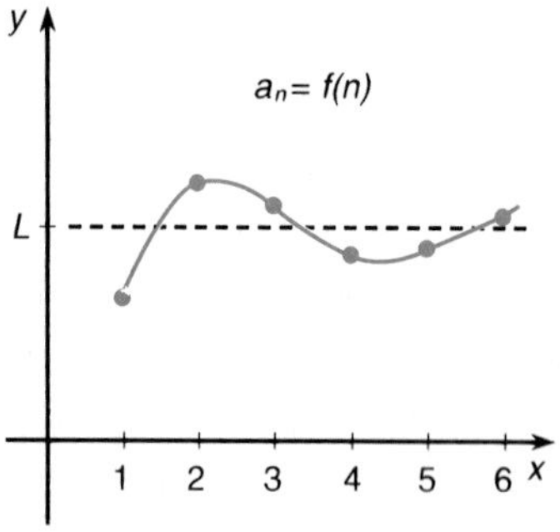

Figure 10.1.3

EXAMPLE 2

(i) We can use Theorem 10.1.3 to show that the sequences in parts (i) and (ii) of Example 1 do have the limits we think they should have. For $a_n = 1 - \frac{1}{n}$ we let $f(x) = 1 - \frac{1}{x}$, which certainly does tend to 1 as $x \to \infty$.

(ii) For $b_n = 3 + \left(-\frac{1}{n}\right)^n$ it is a little more difficult because it does not make sense to consider the function $3 + \left(-\frac{1}{x}\right)^x$. After all, the xth power of something is defined using the natural logarithm, which does not make sense for negative numbers.

It does make sense to consider the function $\left(\frac{1}{x}\right)^x$, but then it is necessary to multiply it by something that equals -1 when x equals an odd integer and is $+1$ when x equals an even integer. This is easier than it sounds. For instance, $\cos(\pi x)$ has that property. So we let $f(x) = 3 + \cos(\pi x)\left(\frac{1}{x}\right)^x$, and then $b_n = f(n)$.

The expected conclusion about b_n will follow if we can show that $\lim_{x\to\infty} f(x) = 3$, and this will follow by the Squeeze Theorem for limits at infinity if we can show that $\lim_{x\to\infty}\left(\frac{1}{x}\right)^x = 0$.

Now $\left(\frac{1}{x}\right)^x = e^{x\ln(1/x)} = e^{-x\ln x}$. For $x > e$, $\ln x > 1$, so $x\ln x > x$, and we conclude that $x\ln x \to \infty$, so

$$\lim_{x\to\infty}\left(\frac{1}{x}\right)^x = \lim_{x\to\infty} e^{-x\ln x} = \lim_{x\to\infty}\frac{1}{e^{x\ln x}} = 0,$$

using the version for limits at infinity of Theorem 2.3.2 about limits of composite functions.

From this we see that $\lim_{x\to\infty} f(x) = 3$, so the theorem applies, and we conclude that $\lim_{n\to\infty} b_n = 3$.

(iii) On the other hand, Theorem 10.1.3 does not help us with part (iii) of Example 1. There we have the sequence $c_n = \sin\left(\frac{n\pi}{2}\right)$ defined as the values $c_n = f(n)$, with $f(x) = \sin\frac{\pi x}{2}$. We also know that this function $f(x)$ does not have a limit as $x \to \infty$.

However, the theorem does not tell us anything in this situation. For instance, if we let $g(x) = \sin(\pi x)$, then $g(x)$ does not have a limit as $x \to \infty$. However, $g(n) = \sin(n\pi) = 0$ for *every* integer n. So if we define a sequence by $d_n = g(n)$, then the sequence $\{d_n\}$ *does* tend to a limit (0) as $n \to \infty$, even though the function $g(x)$ does not tend to any limit as $x \to \infty$.

The point is that even knowing that the values of $g(x)$ at the integers are a nice convergent sequence does not tell us anything about what $g(x)$ might do in between the integers (see Figure 10.1.4). It is because of this that we cannot try to use Theorem 10.1.3 "backwards." If the limit of $f(x)$ exists, then the limit of the corresponding sequence exists and is the same, but if $f(x)$ has no limit, then we cannot conclude that the same is true of the sequence $\{f(n)\}$.

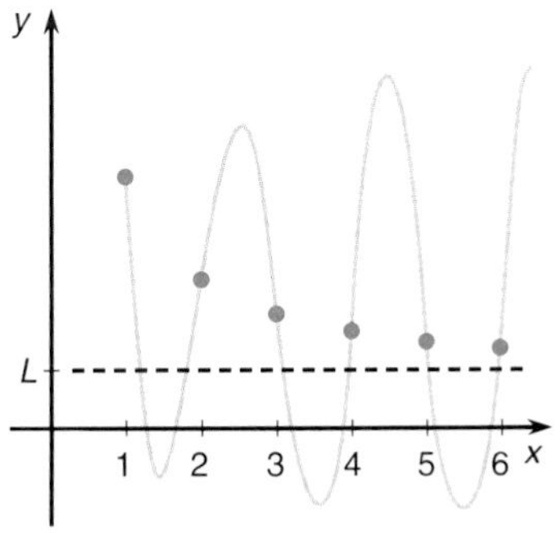

Figure 10.1.4

(iv) With $f(x) = 3x^2 - 1$ we find that $\lim_{x\to\infty} f(x) = \infty$, so if $d_n = 3n^2 - 1$, then $\{d_n\}$ diverges.

(v) Let $u > -1$ be a real number and let $k_n = n\ln\left(1 + \frac{u}{n}\right)$. To apply the theorem, we should consider the function $F(x) = x\ln\left(1 + \frac{u}{x}\right)$. We can write $F(x) = \frac{\ln(1+u/x)}{1/x}$ and observe that both numerator and denominator tend to 0 as $x \to \infty$.

This suggests applying l'Hôpital's Rule, from which we find that

$$\lim_{x\to\infty} F(x) = \lim_{x\to\infty} \frac{\ln\left(1+\frac{u}{x}\right)}{\frac{1}{x}} = \lim_{x\to\infty} \frac{\frac{1}{1+\frac{u}{x}}\left(-\frac{u}{x^2}\right)}{-\frac{1}{x^2}}$$

$$= \lim_{x\to\infty} \frac{1}{1+\frac{u}{x}} u = u\left(\lim_{x\to\infty} \frac{1}{1+\frac{u}{x}}\right)$$

$$= u.$$

Now using this result and applying Theorem 10.1.3, we see that the sequence $\{k_n\} = \{n\ln\left(1+\frac{u}{n}\right)\}$ converges to the limit u.

In Example 1(i) we considered $a_n = 1 - \frac{1}{n}$ and concluded that $\{a_n\}$ converges to 1 because $\frac{1}{n}$ converges to 0. In a way, what we have done here is to consider the constant sequence 1, 1, 1, 1, 1, . . ., which converges to 1, and the sequence $\{\frac{1}{n}\}$, which converges to 0, and looked at their difference $a_n = 1 - \frac{1}{n}$. We used this to see that $\{a_n\}$ converges to the difference $1 - 0 = 1$.

This is a familiar property of limits, and it encourages us to record a number of facts about limits of sequences that are very similar to analogous facts about limits of functions.

THEOREM 10.1.4

(i) The constant sequence $\{c\}$, whose nth term is c for every n, converges to c, that is,

$$\lim_{n\to\infty} c = c.$$

Now suppose $\{a_n\}$ and $\{b_n\}$ are convergent sequences so that $\lim_{n\to\infty} a_n = L$ and $\lim_{n\to\infty} b_n = M$. Then the sequence formed by multiplying $\{a_n\}$ by a number c and the sequences whose terms are the sum, the product, or the quotient of a_n and b_n all converge to the expected limits:

(ii) $\lim_{n\to\infty} ca_n = cL = c\lim_{n\to\infty} a_n$, for any $c \in \mathbb{R}$,

(iii) $\lim_{n\to\infty} (a_n + b_n) = L + M = \lim_{n\to\infty} a_n + \lim_{n\to\infty} b_n$,

(iv) $\lim_{n\to\infty} (a_n b_n) = LM = (\lim_{n\to\infty} a_n)(\lim_{n\to\infty} b_n)$,

(v) If $b_n \neq 0$ for each n and the limit M of $\{b_n\}$ is not zero, then

$$\lim_{n\to\infty} \frac{a_n}{b_n} = \frac{L}{M} = \frac{\lim_{n\to\infty} a_n}{\lim_{n\to\infty} b_n}.$$

(vi) Suppose $h(x)$ is a function that is defined at $x = a_n$ for every n and continuous at $x = L = \lim_{n\to\infty} a_n$, and define $d_n = h(a_n)$. Then the sequence $\{d_n\}$ is convergent, and $\lim_{n\to\infty} d_n = h(\lim_{n\to\infty} a_n) = h(L)$.

(vii) (**The Squeeze Theorem for Sequences**) Suppose $\{a_n\}$, $\{b_n\}$, and $\{c_n\}$ are sequences so that $\{a_n\}$ and $\{c_n\}$ both converge to the same limit L, and so that for each n, b_n lies between a_n and c_n, that is, either $a_n \leq b_n \leq c_n$ or $a_n \geq b_n \geq c_n$. Then the sequence $\{b_n\}$ is convergent, and $\lim_{n\to\infty} b_n = L$. In other words, if the sequence $\{b_n\}$ is "squeezed" between two sequences that both converge to L, then $\{b_n\}$ must also converge to L.

By using Theorems 10.1.3 and 10.1.4(i)–(vii) it is possible with many sequences to decide whether or not they converge and to evaluate the limit if they do.

EXAMPLE 3

See the Graphics Calculator Example on the next page.

(i) Let $u > -1$ be a real number and let $A_n = \left(1 + \frac{u}{n}\right)^n$. We observe that $\ln A_n = n\ln\left(1+\frac{u}{n}\right)$, which is the sequence $\{k_n\}$ that we considered in Example 2(v). There we saw that $\lim_{n\to\infty} k_n = u$. Since $A_n = \exp(k_n)$, we can apply part (vi) of Theorem 10.1.4 to see that $\{A_n\}$ converges and $\lim_{n\to\infty}\left(1+\frac{u}{n}\right)^n = \lim_{n\to\infty} A_n = \exp(\lim_{n\to\infty} k_n) = e^u$.

(ii) Consider the sequence $\{a_n\}$ defined by $a_n = \frac{n}{n+1}$. If we let $f(x) = \frac{x}{x+1}$, then $a_n = f(n)$, and we know that $\lim_{x\to\infty} f(x) = 1$, so Theorem 10.1.3 can be applied to conclude that $\{a_n\}$ converges to 1.

(iii) If $b_n = n + \frac{2}{n}$, we might consider separately the two parts of this sequence. The sequence $\{\frac{2}{n}\}$ converges to 0 (by Theorem 10.1.3, for instance).

The other part of the sequence is the sequence $\{n\}$, that is, 1, 2, 3, 4, . . ., which certainly does not converge. (It goes to infinity.)

The sequence $\{b_n\}$ is the sum of two sequences, one of which converges and the other of which diverges. This implies that $\{b_n\}$ itself must also diverge. For suppose it converged to some limit L. Since $\{-\frac{2}{n}\}$ converges to 0, Theorem 10.1.4(iii) would then tell us that $\{b_n + \left(-\frac{2}{n}\right)\}$ must converge to $L + 0 = L$. But $b_n + \left(-\frac{2}{n}\right) = n + \frac{2}{n} - \frac{2}{n} = n$, and we have already observed that the sequence $\{n\}$ does *not* converge. This contradiction shows that it was wrong to assume that $\{b_n\}$ converges, and we conclude that $\{b_n\}$ diverges.

(iv) If $c_n = (-1)^n$ and $d_n = (-1)^{n+1}$, that is,

$$\{c_n\} = -1,\ 1,\ -1,\ 1,\ -1,\ 1,\ldots,$$
$$\{d_n\} = 1,\ -1,\ 1,\ -1,\ 1,\ -1,\ldots,$$

then we know that neither of the sequences $\{c_n\}$ or $\{d_n\}$ converges.

However, their sum is the zero sequence, since $c_n + d_n = (-1)^n + (-1)^{n+1} = (-1)^n\left(1 + (-1)\right) = 0$.

This example shows that the sum of two divergent sequences may be convergent.

(v) Let $a_n = \frac{1}{2n+(-1)^{n+1}}$. We cannot simply change n to x to make a function $f(x)$ so that $a_n = f(n)$, because $(-1)^x$ makes sense only for certain values of x. However, observing that $(-1)^{n+1} \geq -1$, so $2n + (-1)^{n+1} \geq 2n - 1$ and $a_n = \frac{1}{2n+(-1)^{n+1}} \leq \frac{1}{2n-1}$, we find that $0 \leq a_n \leq \frac{1}{2n-1}$; since $\frac{1}{2n-1} \to 0$, the Squeeze Theorem shows us that $a_n \to 0$.

convergent + convergent = convergent
convergent + divergent = divergent
divergent + divergent = ?

Parts (iii) and (iv) of Example 3 illustrate an interesting phenomenon. The sum of two divergent sequences may be convergent (although it may also be divergent), but the sum of a divergent sequence and a convergent sequence can never be convergent.

Part (iv) of Example 3 also suggests another observation. In a way the sequences $\{c_n\}$ and $\{d_n\}$ defined there can be regarded as essentially the same sequence "shifted over." If $\{a_n\}$ is any sequence, then it is possible to consider the sequence whose nth term is a_{n-1}. This sequence starts with $n = 2$, and it has the same terms as $\{a_n\}$, but they are shifted over by one. For that matter we could consider the sequence $\{a_{n+1}\}$, which starts with $n = 0$, or the sequence that is obtained by shifting the sequence $\{a_n\}$ by k; this is the sequence $\{a_{n-k}\}$. It starts with $n = k + 1$.

With a graphics calculator we can plot the function $y = f(x) = \left(1 + \frac{1}{x}\right)^x$. This is related to the sequence $\{A_n\}$ of part (i) of Example 3, since $A_n = \left(1 + \frac{u}{n}\right)^n$: If $u = 1$, $A_n = f(n)$.

TI-85 GRAPH F1(y(x) =) (1 + 1 ÷ F1(x))^F1(x) M2(RANGE) 1 ▼ 20 ▼ ▼ 0 ▼ 3 F5(GRAPH)	SHARP EL9300 (1 + x/θ/T x^{-1}) a^b x/θ/T RANGE 1 ENTER 20 ENTER 5 ENTER 0 ENTER 3 ENTER 1 ENTER
CASIO f_x-6300G (f_x-7700GB) Range 1 EXE 20 EXE 5 EXE 0 EXE 3 EXE 1 EXE EXE EXE EXE Graph (1 + X x^{-1}) x^y X EXE	HP 48SX ↰ PLOT PLOTR ERASE ATTN ′ Y = ( ) 1 + 1/x X ► ► y^x X ↱ PLOT ↰ DRAW ERASE 1 ENTER 20 XRNG 0 ENTER 3 YRNG DRAW

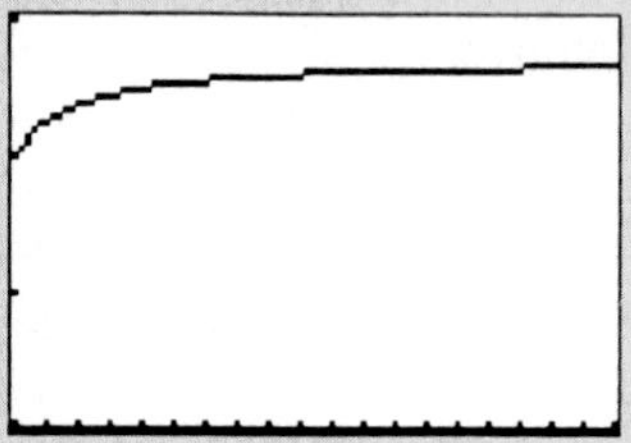

Notice the limiting behavior as x gets large. You might like to try it with different values of u, for example, $u = 2$, $u = -5$, $u = 20$.

Now if we look at one of these shifted sequences, it will tend to the same limit as the original sequence $\{a_n\}$, provided $\{a_n\}$ is convergent, and will diverge if $\{a_n\}$ diverges. Roughly speaking, the shifted sequence will display the same behavior as the original sequence but will do it "k terms later."

Another related question arises if we consider the sequence $0, 0, 3, \frac{1}{2}, \frac{1}{3}, \frac{1}{4}, \frac{1}{5}, \ldots$, whose nth term equals $\frac{1}{n-2}$ when $n \geq 4$ but that has some peculiar terms at the beginning. The important thing about a sequence of this kind is that its limiting behavior is not affected by the first few terms.

Since the sequence eventually equals $\frac{1}{n-2}$, which is the sequence $\frac{1}{n}$ shifted over by 2, and since that sequence converges to 0, we see that the peculiar sequence also converges to 0. We record these two important results as a theorem.

THEOREM 10.1.5

(i) If $\{a_n\}$ is a sequence, let $b_n = a_{n-k}$. The shifted sequence $\{b_n\}$ has the same properties of convergence as the original sequence $\{a_n\}$, that is, either they both diverge or they both converge to the same limit. (Strictly speaking, if the original sequence is defined for n beginning at 1, then $\{b_n\}$ is defined for n beginning at $k + 1$. If $\{a_n\}$ begins with $n = r$, then $\{b_n\}$ begins with $n = k + r$.)

(ii) Suppose $\{c_n\}$ and $\{d_n\}$ are two sequences that are the same after the first few terms. Then they have the same convergence properties, that is, either they both diverge or they both converge to the same limit.

EXAMPLE 4

(i) The first part of Theorem 10.1.5 shows immediately that $\{\frac{1}{n+k}\}$ converges to 0 for any integer k.

(ii) The second part of Theorem 10.1.5 can be used to show that the sequence $3,\ -2\pi,\ 4,\ -\sqrt{7},\ -e,\ 17,\ 0,\ 0, 0,\ 0,\ 0,\ 0,\ 0, \ldots$, which is all zeroes after the first six terms, converges to 0, even though there does not appear to be any reasonable pattern to the first six terms.

(iii) In Example 2(v) we considered the sequence $\{k_n\} = \{n \ln\left(1 + \frac{u}{n}\right)\}$, for $u > -1$. The restriction on u is needed so that $\ln\left(1 + \frac{u}{n}\right)$ makes sense for every n. For instance, if we let $u = -4$, then $k_n = n \ln\left(1 + \frac{u}{n}\right)$ is not defined for $n \leq 4$, but it does make sense for $n > 4$. Part (ii) of Theorem 10.1.5 says that if $u \leq -1$, we can ignore the first few missing terms, and the argument given in Example 2(v) will show that $\{k_n\} = \{n \ln\left(1 + \frac{u}{n}\right)\}$ converges to u for *any* real number u.

Similarly, in Example 3(i) we considered $A_n = \left(1 + \frac{u}{n}\right)^n$, for $u > -1$. In fact, A_n is defined for any real number u, and by the above remarks and Theorem 10.1.4(vi) we see that $\{\left(1 + \frac{u}{n}\right)^n\}$ converges, and

$$\lim_{n\to\infty}\left(1 + \frac{u}{n}\right)^n = \exp\left(\lim_{n\to\infty} n \ln\left(1 + \frac{u}{n}\right)\right) = e^u.$$

Part (iii) of Example 4 illustrates an important use of Theorem 10.1.5(ii). Frequently, a sequence is given by a formula $\{f(n)\}$ that does not make sense for a few small values of n. The theorem says that we can safely ignore this problem if $f(n)$ does make sense for all n past some fixed value.

EXAMPLE 5 **Powers of c** Suppose c is a positive number and consider the sequence of its powers: $a_n = c^n$. If $c > 1$, the powers c^n grow larger and larger and tend to infinity, so the sequence diverges. If $c = 1$, of course the sequence is a constant sequence and converges to 1.

So suppose $0 < c < 1$. Then $a_n = c^n = e^{n \ln c} = f(n)$, where $f(x) = e^{x \ln c}$. Since $0 < c < 1$, we know that $\ln c < 0$, and $x \ln c \to -\infty$ as $x \to \infty$, so $\lim_{x\to\infty} f(x) = \lim_{x\to\infty} e^{x \ln c} = 0$. By Theorem 10.1.3, $a_n = e^{n \ln c} \to 0$.

Now suppose $-1 < c < 0$ and consider the sequence $c^n = (-|c|)^n = (-1)^n|c|^n$. By the preceding argument we know that $\{|c|^n\}$ converges to 0, and c^n lies between $-|c|^n$ and $|c|^n$ (actually, it alternates between them). Both these sequences converge to 0, so by the Squeeze Theorem, Theorem 10.1.4(vii), $\{c^n\}$ converges to 0.

Finally, if $c = -1$, the sequence $\{(-1)^n\}$ diverges, and if $c < -1$, then $\{c^n\}$ diverges, since the absolute values of the terms tend to infinity.

We summarize:

$$\{c^n\} \quad \begin{cases} \text{diverges,} & \text{if } c > 1 \text{ or if } c \leq -1, \\ \text{converges to 1,} & \text{if } c = 1, \\ \text{converges to 0,} & \text{if } -1 < c < 1. \end{cases}$$

EXAMPLE 6 Let c be any positive number and consider the sequence $\{a_n\}$, where

$$a_n = \frac{c^n}{n!}.$$

It is always possible to find an integer N that is greater than c. If $n > N$, we can write $n! = (1)(2)(3)\dots(N)(N+1)(N+2)\dots n = N!(N+1)(N+2)\dots(n-1)n$, so

The inequality comes from making the denominator *smaller* by replacing each of $N+1, N+2, \dots, n$ by N.

$$\frac{c^n}{n!} = \frac{c^n}{N!(N+1)(N+2)\dots(n-1)n} \leq \frac{c^n}{N!(N)(N)\dots(N)}$$
$$= \frac{c^n}{N!N^{n-N}} = \frac{N^N}{N!}\left(\frac{c}{N}\right)^n.$$

Now $0 < \frac{c}{N} < 1$, so by Example 5 we see that $\{(\frac{c}{N})^n\}$ converges to 0, and the last line in the computation above also converges to zero as $n \to \infty$ because it is just $(\frac{c}{N})^n$ times the constant $\frac{N^N}{N!}$.

Since $\{\frac{c^n}{n!}\}$ lies between this sequence and the zero sequence, the Squeeze Theorem tells us that $\{\frac{c^n}{n!}\}$ converges to 0, that is,

$$\lim_{n\to\infty} \frac{c^n}{n!} = 0.$$

If $c < 0$, then $\frac{c^n}{n!} = (-1)^n\frac{|c|^n}{n!}$, which lies between the sequences $\{\frac{|c|^n}{n!}\}$ and $\{-\frac{|c|^n}{n!}\}$, both of which converge to 0. The Squeeze Theorem implies that $\{\frac{c^n}{n!}\}$ converges to 0.

We have shown that $\{\frac{c^n}{n!}\}$ converges to 0 if c is either positive or negative; if $c = 0$, the sequence is just the zero sequence, so it still converges to 0. So for *every* number c,

$$\lim_{n\to\infty} \frac{c^n}{n!} = 0.$$

Suppose a sequence $\{a_n\}$ is "increasing," that is, $a_{n+1} \geq a_n$ for every n. Then one possibility is that $\lim_{n\to\infty} a_n = \infty$, that is, the sequence diverges to infinity, and its terms eventually become larger than any number M.

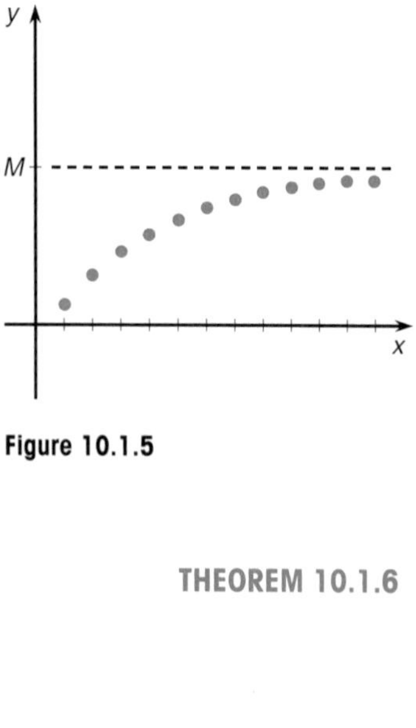

Figure 10.1.5

If, on the other hand, there is some number M so that $a_n \leq M$ for every n, then we say the sequence $\{a_n\}$ is "bounded." In this case the terms of the sequence are "trapped" between a_1 and M; they keep increasing but are never greater than M (see Figure 10.1.5). It seems reasonable to expect that in this situation they will have to converge to some limit, and this is in fact true.

THEOREM 10.1.6

Suppose $\{a_n\}$ is an increasing sequence, that is, $a_{n+1} \geq a_n$ for every n. If the sequence is **bounded**, that is, if there is some number M so that $|a_n| \leq M$ for every n, then it must converge. Otherwise, the sequence diverges to infinity, that is, $\lim_{n\to\infty} a_n = \infty$.

Notice that Theorem 10.1.6 does not say what the limit is, simply that there is one. This result is more useful than it sounds; we will use it in Section 10.4.

The Precise Definition of Limits for Sequences

As with limits of functions, we have discussed limits of sequences without giving a precise definition, concentrating instead on learning to work with the limits. Here we give the precise definition and show how it can be used to prove some of the results we have already discussed.

In fact the definition is very similar to the precise definition of limits for functions (Definition 2.7.1, p. 120).

DEFINITION 10.1.7

> The sequence $\{a_n\}$ **converges** to the limit L if and only if the following condition is satisfied.
>
> For every $\epsilon > 0$ it is possible to find an $N > 0$ so that $|a_n - L| < \epsilon$ whenever $n > N$.

In other words, up to any specified "degree of closeness" ϵ, the terms of the sequence must be within a distance of ϵ of the limit L if we go far enough out in the sequence.

EXAMPLE 7 We use this definition to prove the "obvious" fact that the sequence $\{\frac{1}{n}\}$ converges to 0.

Suppose $\epsilon > 0$ is fixed. We want to ensure that $\frac{1}{n} < \epsilon$. This is the same as $n > \frac{1}{\epsilon}$. (Notice the way the inequality is reversed when we take reciprocals.) So if we let N be any integer that is greater than $\frac{1}{\epsilon}$, then whenever $n > N$, we see that $n > N > \frac{1}{\epsilon}$, so $\frac{1}{n} < \epsilon$.

We have shown that given any $\epsilon > 0$, it is possible to find an N (e.g., any integer N that is greater than $\frac{1}{\epsilon}$) so that whenever $n > N$, we have that $\frac{1}{n} < \epsilon$. This is exactly what we need, since the limit is $L = 0$ and $|\frac{1}{n} - L| = |\frac{1}{n} - 0| = \frac{1}{n} < \epsilon$.

In fact it is not very difficult to use Definition 10.1.7 to prove all the Theorems 10.1.3, 10.1.4, 10.1.5 of this section.

Exercises 10.1

I

Say whether or not each of the following sequences $\{a_n\}$ converges, giving reasons, and for those that do converge, find the limit.

1. $a_n = \frac{1}{n} - 1$

2. $a_n = 2 + \frac{3}{n^2}$

3. $a_n = \frac{n-1}{4n}$

4. $a_n = \frac{\cos n\pi}{n}$

5. $a_n = \frac{n^2+n-2}{2n^2-3n}\cos(n\pi)$

6. $a_n = \frac{2n^2-3n+2}{n^3+4n^2-2}\sin\frac{n\pi}{2}$

7. $a_n = \left(\frac{2}{3}\right)^n$

8. $a_n = \left(-\frac{5}{4}\right)^n$

9. $a_n = \left(-\frac{1}{\sqrt{2}}\right)^n$

10. $a_n = \left(\frac{2}{\sqrt{3}}\right)^n$

11. $a_n = \frac{(n-1)!}{n!}$

12. $a_n = \frac{n!}{(n-2)!}$

13. $a_n = \frac{(n!)^2}{(n-1)!(n+1)!}$

14. $a_n = \frac{n!}{4^n}$

15. $a_n = \frac{3^n}{1-3^n}$

16. $a_n = \frac{1+5^n}{6^n}$

17. $a_n = 2\ln(3n)$

18. $a_n = \frac{\cos(n\pi)7^n}{n!}$

19. $a_n = n^{-1/2}$

20. $a_n = \sin(n)\frac{1}{3^n n!}$

II

***21.** We know that the function $f(x) = \frac{1}{x}$ approaches a limit as $x \to \infty$, namely, $\lim_{x\to\infty} f(x) = 0$. If we let $a_n = f(n)$, then Theorem 10.1.3 tells us that the sequence $\{a_n\}$ converges to 0. Find another function $g(x)$ so that $a_n = g(n)$ for each $n \geq 1$ (i.e., $g(n) = f(n)$ for each n) but so that $g(x)$ does not have a limit as $x \to \infty$.

***22.** Suppose $f(x)$ is a function satisfying $\lim_{x\to\infty} f(x) = L$. Define a sequence $\{a_n\}$ by $a_n = f(n)$. By Theorem 10.1.3, $\lim_{n\to\infty} a_n = L$. Find another function $g(x)$ that does not have a limit as $x \to \infty$ but so that $a_n = g(n)$ for each $n \geq 1$.

23. Show that $\{\frac{1}{n\ln n}\}$ converges and find its limit.

24. Show that $\left\{\left(\frac{n}{2}\right)^{-n/3}\right\}$ converges and find its limit.

25. Let $a_n = \frac{3^{2n}}{n!}$, for each $n \geq 1$, and decide whether or not $\{a_n\}$ converges, giving reasons and finding the limit if it does converge.

26. Let $b_n = (-1)^n \frac{3^{2n}}{n!}$, for each $n \geq 1$, and decide whether or not $\{b_n\}$ converges, giving reasons and finding the limit if it does converge.

27. Let $c_n = (-1)^{n+1} \frac{2^n (n-1)!}{n!}$, for each $n \geq 1$, and decide whether or not $\{c_n\}$ converges, giving reasons and finding the limit if it does converge.

*28. For what values of c does the sequence $\{n^c\}$ converge, if any? If there are any, find the corresponding limits.

*29. If $\{c_n\}$ is a bounded decreasing sequence (i.e., $c_{n+1} \leq c_n$ for every n), show that it is convergent.

*30. Find two examples of bounded divergent sequences.

10.2 Geometric Sums and Series and Applications

COMMON RATIO
INTEREST
GEOMETRIC SERIES
CONVERGENCE

In the introduction to this chapter we considered the following sums:

$$\begin{aligned}
&\tfrac{1}{2} &&= \tfrac{1}{2} = 0.5,\\
&\tfrac{1}{2} + \tfrac{1}{4} &&= \tfrac{3}{4} = 0.75,\\
&\tfrac{1}{2} + \tfrac{1}{4} + \tfrac{1}{8} &&= \tfrac{7}{8} = 0.875,\\
&\tfrac{1}{2} + \tfrac{1}{4} + \tfrac{1}{8} + \tfrac{1}{16} &&= \tfrac{15}{16} = 0.9375,\\
&\tfrac{1}{2} + \tfrac{1}{4} + \tfrac{1}{8} + \tfrac{1}{16} + \tfrac{1}{32} &&= \tfrac{31}{32} = 0.96875.
\end{aligned}$$

In all of these sums, each successive term is equal to one-half the previous term. The first term is $\frac{1}{2}$, the second term is half that, or $\frac{1}{4}$, the third term (if we continue that far) is $\frac{1}{8}$, and so on. The way to find the third term is to start with the first term $\frac{1}{2}$, multiply it by $\frac{1}{2}$ to get the second term, and multiply that by $\frac{1}{2}$ to get the third term, which is $\left(\frac{1}{2}\right)\left(\frac{1}{2}\right)\left(\frac{1}{2}\right) = \left(\frac{1}{2}\right)^3 = \frac{1}{8}$.

Similarly, the fourth term will be $\left(\frac{1}{2}\right)\left(\frac{1}{2}\right)\left(\frac{1}{2}\right)\left(\frac{1}{2}\right) = \left(\frac{1}{2}\right)^4 = \frac{1}{16}$, and the kth term will be the product of k factors of $\frac{1}{2}$, or $\left(\frac{1}{2}\right)^k$.

There is no particular reason for choosing $\frac{1}{2}$. We could just as easily have considered a sum like $\frac{1}{3} + \frac{1}{9} + \frac{1}{27} + \frac{1}{81} + \frac{1}{243} = \frac{1}{3} + \left(\frac{1}{3}\right)^2 + \left(\frac{1}{3}\right)^3 + \left(\frac{1}{3}\right)^4 + \left(\frac{1}{3}\right)^5$, or $-\frac{2}{3} + \left(-\frac{2}{3}\right)^2 + \left(-\frac{2}{3}\right)^3 + \left(-\frac{2}{3}\right)^4 + \left(-\frac{2}{3}\right)^5 + \left(-\frac{2}{3}\right)^6$.

It is also possible to look at $5 + 5^2 + 5^3 + 5^4 + 5^5 + 5^6 + 5^7$, but in this case the terms get larger and larger, so the sum is unlikely to have a (finite) limit as we take more and more terms.

Each of these sums is an example of what is called a *geometric sum*; the corresponding infinite series are called *geometric series*.

DEFINITION 10.2.1

Let a and r be real numbers, and let k be a positive integer. A sum of the form

$$\begin{aligned}
a + ar + ar^2 + ar^3 + \ldots + ar^k &= \sum_{n=0}^{k} ar^n \\
&= a(1 + r + r^2 + r^3 + \ldots + r^k)
\end{aligned}$$

is called a **geometric sum**. The number r is called the **common ratio**, since it is the ratio of each term to the previous one.

It is actually possible to find a simple formula for evaluating a geometric sum without adding up a lot of terms.

THEOREM 10.2.2

Geometric Sums

Suppose k is a positive integer and a, r are real numbers with $r \neq 1$. Then the geometric sum can be evaluated as follows:

$$a + ar + ar^2 + ar^3 + \ldots + ar^k = \sum_{n=0}^{k} ar^n = a(1 + r + r^2 + r^3 + \ldots + r^k)$$
$$= a\frac{1 - r^{k+1}}{1 - r} = \frac{a - ar^{k+1}}{1 - r}.$$

PROOF Let us first consider the case in which $a = 1$ and look at the sum $S = 1 + r + r^2 + r^3 + \ldots + r^{k-1} + r^k$. If we multiply the sum S by r we find

$$\begin{aligned} rS &= r(1 + r + r^2 + r^3 + \ldots + r^{k-1} + r^k) \\ &= r + r^2 + r^3 + r^4 + \ldots + r^k + r^{k+1}. \end{aligned}$$

Now what would happen if we were to subtract this from S? Both S and rS are sums of many terms, each of which is a power of r. In fact most of the terms they contain are the same, with the two exceptions that S begins with the term 1, which is not in rS, and rS ends with the term r^{k+1}, which is not in S. So if we subtract them, all the terms except these two will cancel, and we will be left with

$$\begin{aligned} S - rS &= (1 + \ r + r^2 + r^3 + \ldots + r^{k-1} + r^k) \\ &\quad - (r + r^2 + r^3 + r^4 + \ldots + r^k + r^{k+1}) \\ &= 1 - r^{k+1}, \end{aligned}$$

that is,

$$(1 - r)S = 1 - r^{k+1}.$$

Dividing this last equation by $1 - r$ (which is all right, since we assumed that $r \neq 1$) gives the result $S = 1 + r + r^2 + r^3 + r^4 + \ldots + r^{k-1} + r^k = \frac{1-r^{k+1}}{1-r}$.

In the general case, that is, when a is not necessarily equal to 1, the result follows by multiplying both sides of this equation by a. ◢

In the sum $a(1 + r + r^2 + r^3 + \ldots + r^{k-1} + r^k)$ the number k is the power of r in the last term. Since the first term contains $1 = r^0$, we see that there is a term for each power of r, starting with $r^0 = 1$ and ending with r^k. This means that there are $k + 1$ terms, not just k. We can find r^k by noting that it is the ratio of the last term to the first.

You may find it helpful to think of it this way: The number r is the common ratio of each term to the preceding one. The first term is a, and the last term is ar^k. If we added one more term, it would be ar^{k+1}. So for any particular geometric sum, find the first term a, the ratio r of any two adjacent terms, and the "next term after the last," ar^{k+1}. Then the sum can be evaluated by the formula $\frac{a-ar^{k+1}}{1-r}$.

Observe that if $r = 1$, then the sum is just $a + a + a + a + \ldots + a$, with a total of $k + 1$ terms, so it equals $(k + 1)a$. It was necessary to assume that $r \neq 1$ in order for the formula in the theorem to make sense, but this shows that when $r = 1$, if anything the sum is even easier to evaluate.

EXAMPLE 1 Evaluate each of the following: (i) $1 + \frac{1}{3} + \frac{1}{9}$, (ii) $\frac{1}{2} + \frac{1}{4} + \frac{1}{8} + \frac{1}{16}$, (iii) $\sum_{n=1}^{7} 5^n = 5 + 5^2 + 5^3 + 5^4 + 5^5 + 5^6 + 5^7$.

Solution

(i) The first term is $a = 1$, the common ratio is $\frac{1}{3}$, and the last term is $\frac{1}{9}$, so the "next term" would be $\frac{1}{27}$. (It is possible to check that this equals $\left(\frac{1}{3}\right)^{k+1}$ by noting that the number of terms ought to be $k + 1$, which is 3, which is correct.) The formula tells us that the sum equals

$$\frac{a - ar^{k+1}}{1 - r} = \frac{1 - \frac{1}{27}}{\frac{2}{3}} = \frac{26}{27}\frac{3}{2} = \frac{13}{9}.$$

This is easily checked by adding up the three terms.

With a programmable calculator it is not difficult to add up a geometric sum. For instance, we can do the sum $\sum_{n=1}^{7} 5^n$ from part (iii) of Example 1.

TI-85 PRGM F2(EDIT) GMSUM ENTER 0 STO N ENTER 0 STO S ENTER LbL B ENTER N + 1 STO N ENTER S + 5^N STO S ENTER If (N TEST F2(<) 7) ENTER Goto B ENTER S ENTER EXIT EXIT EXIT PRGM F1 (NAME) GMSUM ENTER	SHARP EL9300 ▼ ▼ (NEW) ENTER ENTER GSM ENTER N = 0 ENTER S = 0 ENTER COMMAND B 1 (Label) 1 ENTER N = N + 1 ENTER S = S + 5 a^b N ENTER COMMAND 3 (If) N COMMAND C 2 (<) 7 COMMAND B 2 (Goto) 1 ENTER COMMAND A 1 (Print) S ENTER COMMAND A 6 (End) ENTER (RUN) ENTER (GSM) ENTER
CASIO f_x-6300G (f_x-7700GB) MODE 2 EXE 0 → [redacted] : 0 → [redacted] : Lbl 1 : [redacted] + 1 → [redacted] : [redacted] + 5 x^y [redacted] → [redacted] : [redacted] < 7 ⇒ Goto 1 : [redacted] ◢ MODE 1 Prog 0 EXE (on the f_x-7700GB, : is PRGM F6, and Prog is PRGM F3.)	HP 48SX ATTN ↱ CLR [redacted] 0 SPC 1 SPC 7 SPC FOR N SPC 5 N y^x + NEXT ▼ ENTER ′ GSUM STO VAR GSUM ENTER [redacted]

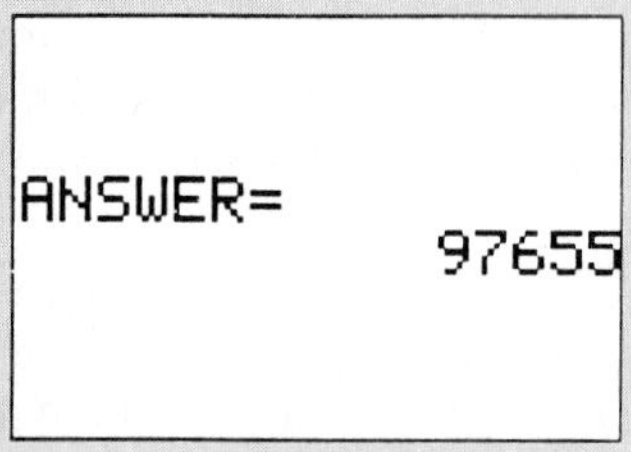

You might like to modify this program to evaluate other geometric sums.

(ii) This time we find that $a = \frac{1}{2}$, the common ratio is $r = \frac{1}{2}$, and the last term is $\frac{1}{16}$. The term after that would be $\frac{1}{32}$, so the sum equals

$$\frac{a - ar^{k+1}}{1 - r} = \frac{\frac{1}{2} - \frac{1}{32}}{1 - \frac{1}{2}} = \frac{\frac{16-1}{32}}{\frac{1}{2}} = \frac{15}{16}.$$

This confirms one of the sums evaluated in the introduction to this chapter and could in any event be checked quite easily.

(iii) For this example we find that $a = 5$, $r = 5$, the last term is 5^7, and the following term would be 5^8. The sum equals

$$\frac{a - ar^{k+1}}{1 - r} = \frac{5 - 5^8}{1 - 5} = \frac{5 - 390{,}625}{-4} = \frac{-390{,}620}{-4} = 97655.$$

This was probably a little easier than simply adding up the terms and less prone to errors.

Notice that in part (iii) of Example 1 the index of summation was n, and it ran from $n = 1$ to $n = 7$. The sum will still be a geometric sum no matter where the values of n begin and end, but we need to know the range of n in order to find the first and last terms.

Interest, Mortgages, and Annuities

Consider a bank account that earns interest at the effective rate of 10% per annum. This means that each year the amount in the account is increased by 10%, which is the same thing as multiplying it by 1.10. We say the balance increases by a **factor** of 1.10; the word "factor" reminds us to *multiply* by this constant.

Suppose the owner of the account pays \$1000 into it on the first day of each year. At the end of the first year the \$1000 in the account will be increased to $1.10 \times \$1000 = \$1100$. Then another \$1000 will be deposited, bringing the total to $\$1000 + 1100 = \$2100$ on the first day of the second year. At the end of the second year, interest will bring the total to $1.10 \times (1000 + 1.10 \times 1000)$. Then at the beginning of the third year another \$1000 will be added, and the total will be

$$1000 + 1.10 \times (1000 + 1.10 \times 1000) = 1000 + 1.10 \times 1000 + (1.10)^2 \times 1000.$$

A year later, this will have been multiplied by the factor 1.10 as interest was added, and an additional \$1000 will have been deposited. The total at the beginning of the fourth year will be

$$\begin{aligned} &1000 + 1.10 \times \big(1000 + 1.10 \times 1000 + (1.10)^2 \times 1000\big) \\ &= 1000 + 1.10 \times 1000 + (1.10)^2 \times 1000 + (1.10)^3 \times 1000 \\ &= 1000\big(1 + 1.10 + (1.10)^2 + (1.10)^3\big). \end{aligned}$$

With each passing year the total is multiplied by 1.10 and an additional \$1000 is added. We see that at the beginning of the nth year the balance in the account will be

$$\$1000\big(1 + 1.10 + (1.10)^2 + (1.10)^3 + \ldots + (1.10)^{n-2} + (1.10)^{n-1}\big).$$

At the end of the nth year the amount will have been increased by the factor 1.10, so it will be

$$\$1000\big(1.10 + (1.10)^2 + (1.10)^3 + \ldots + (1.10)^{n-1} + (1.10)^n\big).$$

(Notice that the highest powers occurring in these last two sums are not the same. Each sum consists of n terms, reflecting the fact that in each case there have been n deposits, but in the first the interest has been applied only $n-1$ times, and in the second the interest has been applied n times.)

Both these amounts are geometric sums, so it is possible to give formulas for evaluating them. For the balance at the beginning of the nth year the first term in the sum is $a = 1000$, the ratio is $r = 1.10$, and the last term is $1000(1.10)^{n-1}$. We see that the total amount is $\left(1000 - 1000(1.10)^n\right)/(1 - 1.10) = \frac{1000}{-0.10}\left(1 - (1.10)^n\right) = 10{,}000\left((1.10)^n - 1\right)$.

It is possible to evaluate the other sum in the same way, but it is easier to notice that it is just the first sum multiplied by 1.10, so it equals $11{,}000\left((1.10)^n - 1\right)$. For instance, at the end of the third year the amount in the account is $11{,}000\left((1.10)^3 - 1\right) = \$3641.00$. This can easily be checked by adding.

However, it would be slightly annoying to have to add up all the terms to find the amount at the end of 20 years. It is easier to calculate $11{,}000\left((1.10)^{20} - 1\right) = \$63{,}002.50$.

EXAMPLE 2 How much money will be in an account after five years if \$100 is deposited on the first of each month and 1% interest is paid at the end of each month?

Solution This situation is analogous to the discussion above. At the end of the first month, one deposit will have been made and interest will have been paid on it, so the balance will be 1.01×100. At the end of the second month another deposit and another interest payment will have been made; the balance will be $1.01\left(100 + 1.01(100)\right) = 100\left(1.01 + (1.01)^2\right)$.

After n months the balance will be $100\left(1.01 + (1.01)^2 + (1.01)^3 + \ldots + (1.01)^n\right)$. This geometric sum can be evaluated by using the formula. The first term is $100 \times 1.01 = 101$, the common ratio is 1.01, and the last term is $100(1.01)^n$, so the sum equals $\left(101 - 100(1.01)^{n+1}\right)/(1 - 1.01) = \frac{101}{-0.01}\left(1 - (1.01)^n\right) = 10{,}100\left((1.01)^n - 1\right)$.

In particular, in 5 years there are 60 months, so to find the balance after 5 years, we let $n = 60$ and find that the amount is $10{,}100\left((1.01)^{60} - 1\right) = \$8248.64$.

It is also possible to turn this process around. Consider a twenty-year mortgage for \$100,000. Somebody buying a house might have to borrow this money for the purchase, with the intention of paying it off in equal installments over a *term* of twenty years. For simplicity, let us assume that one payment is made each year (actual mortgages are often paid quarterly or monthly). Let us assume that the interest rate is 10% per year and that the annual payments are all the same size.

We write P for the principal, the original amount borrowed, which in this case is \$100,000. Let r be the factor by which an amount increases in one year due to interest; in this situation, with 10% annual interest, we have $r = 1.1$. Finally, write A for the amount of the annual payment.

At the end of the first year the amount owed will have increased from its initial value P to Pr, and then a payment of A will be made, reducing the balance owed to $Pr - A$. After another year this will have increased to $(Pr - A)r$, due to a year's interest, and the next payment will reduce it to $(Pr - A)r - A = Pr^2 - Ar - A$. After the third year the increase due to interest and the effect of the payment will leave $(Pr^2 - Ar - A)r - A = Pr^3 - Ar^2 - Ar - A$. At the end of the fourth year the balance owed will be $(Pr^3 - Ar^2 - Ar - A)r - A = Pr^4 - Ar^3 - Ar^2 - Ar - A$.

The pattern should be reasonably clear. After k years, that is, at the end of the kth year, the amount still owed will be

$$\begin{aligned} &Pr^k - Ar^{k-1} - Ar^{k-2} - \ldots - Ar^2 - Ar - A \\ &\quad = Pr^k - A(1 + r + r^2 + r^3 + \ldots + r^{k-2} + r^{k-1}). \end{aligned}$$

At the end of the term, that is, after twenty years, the amount owed should be zero, meaning that the mortgage has been paid off. This means that if n is the number of years in the term (in this case $n = 20$), then

$$Pr^n - A(1 + r + r^2 + r^3 + \ldots + r^{n-2} + r^{n-1}) = 0,$$

or

$$Pr^n = A(1 + r + r^2 + r^3 + \ldots + r^{n-2} + r^{n-1}).$$

The right side is a geometric sum that equals $A(1 - r^n)/(1 - r)$. Dividing both sides by the same amount, we can now solve to find a reasonably simple formula for A, namely,

$$A = \frac{Pr^n}{\frac{1-r^n}{1-r}} = \frac{(r-1)Pr^n}{r^n - 1}. \tag{10.2.3}$$

Returning to our example, in which $P = 100{,}000$, $n = 20$, and $r = 1.1$, we see that the annual payment will be $A = (r - 1)Pr^n/(r^n - 1) = (0.1)(100{,}000)(1.1)^{20} /\big((1.1)^{20} - 1\big) = \$11{,}745.96$. Notice that just to pay the 10% interest on \$100,000 would take \$10,000, so most of the payment is interest, and only \$1,745.96 of the first payment will go toward reducing the principal. In succeeding years, as the principal begins to diminish, the amount of interest paid will diminish, so a larger part of the payment of \$11,745.96 will go toward principal. See Figure 10.2.1.

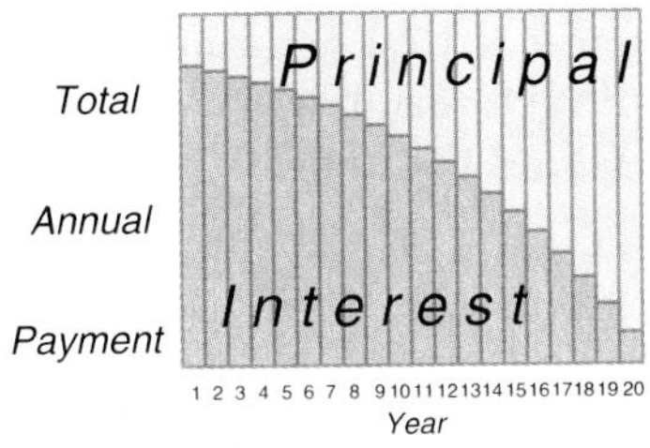

Figure 10.2.1

Formula 10.2.3 can be used to find the amount of the payment A, assuming that we know the principal P, the number of payments n, and the factor r by which interest increases the balance over the period between two payments. Notice that nothing in our calculation depended on letting the payments be once a year; this formula will work as long as the payments are evenly timed and for the same amount.

THEOREM 10.2.4

Mortgage Payments

Suppose equal payments of amount A are made to pay off a debt with principal P, and the interest increases the balance by the factor r between consecutive payments. Then the balance owing after k payments have been made is

$$Pr^k - A\frac{r^k - 1}{r - 1}.$$

If the debt is exactly paid off in n equal payments, then the amount of each payment is

$$A = \frac{(r-1)Pr^n}{r^n - 1}.$$

EXAMPLE 3 What will be the monthly payment for a car loan of \$20,000 paid over 3 years with 15% annual interest, compounded monthly?

Solution This situation is analogous to the mortgage discussed above. Here the principal is $P = 20{,}000$, the number of payments is the number of months in 3 years, which is $n = 36$, and the factor by which interest increases a balance in one month (the period between consecutive payments) is $r = 1 + \frac{15}{12}/100 = 1.0125$.

We divide 15% by 12 to get the (uncompounded) *monthly* interest.

From Theorem 10.2.4 we see that the monthly payment is $A = (r-1)Pr^n/(r^n - 1) = (0.0125)(20{,}000)(1.0125)^{36}/\big((1.0125)^{36} - 1\big) = \$693.31$.

Notice that since this loan is for a much shorter length of time than the mortgage discussed above, a smaller proportion of each payment will be for interest. In fact, one month's interest on $P = 20{,}000$ is $\frac{15}{12}\% = 1.25\%$ of P, or $0.0125 \times 20{,}000 = \$250$, which is much less than half of the monthly payment. See Figure 10.2.2.

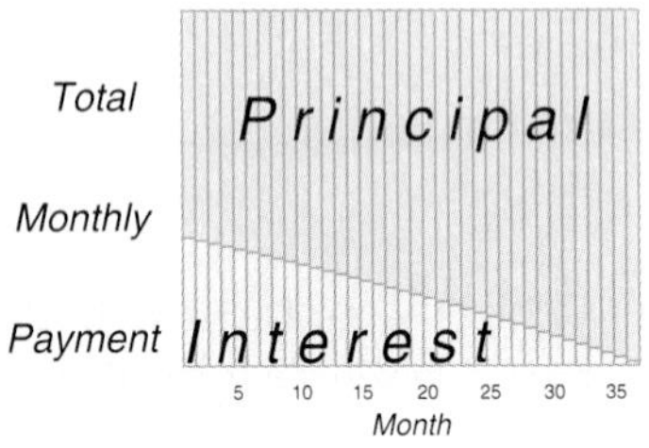

Figure 10.2.2

A person who is about to retire is likely to be interested in finding a way to provide a secure income for the rest of his or her life. One way of doing this is to purchase what is called an **annuity**. The idea is that the purchaser gives a large amount of money to a financial institution, which promises to pay a regular amount back to the purchaser. There are many possible variations; one type will pay a regular income for the rest of the purchaser's life, while another will last only for a fixed number of years. Some annuities are indexed to allow for inflation, and many other features are available.

However, the basic idea is approximately the same for any annuity.

EXAMPLE 4 Suppose a purchaser pays \$50,000 for an annuity that guarantees to pay \$7000 a year for 15 years. Assuming that interest is 10% per annum and that interest and disbursements are paid at the end of each year, how much profit will be left at the end of 15 years for the company selling the annuity (ignoring administration costs)?

Solution This situation is exactly like a mortgage except that the money moves in the opposite direction. If we write $P = 50{,}000$ for the amount invested, $r = 1.10$ for the factor by which the balance increases in one year due to interest, and $A = 7000$ for the annual payment, then the balance remaining after n years can be found by using Theorem 10.2.4. The amount remaining after n years is $Pr^n - A - Ar - Ar^2 - \ldots - Ar^{n-1} = Pr^n - A(r^n - 1)/(r-1)$.

In our example, with $n = 15$, we find that the amount left will be $50{,}000(1.1)^{15} - 7000\big((1.1)^{15} - 1\big)/(1.1 - 1) = -\$13{,}544.96$. The fact that the balance is negative means that the company had to pay out more money than there was left, which means that this annuity was badly planned and was a losing proposition for the company.

Geometric Series

In Theorem 10.2.2 we found that the geometric sum $a + ar + ar^2 + \ldots + ar^k$ is equal to $a\frac{1-r^{k+1}}{1-r}$. If we let k become larger and larger, which amounts to taking more and more terms in the sum, we can ask what happens to this value of the sum.

For instance, if $a = \frac{1}{2}$ and $r = \frac{1}{2}$, then the sum $\frac{1}{2} + \frac{1}{4} + \frac{1}{8} + \ldots + \frac{1}{2^{k+1}}$ equals $\left(\frac{1}{2} - \left(\frac{1}{2}\right)^{k+2}\right)/\left(1 - \frac{1}{2}\right) = 1 - \left(\frac{1}{2}\right)^{k+1}$. As $k \to \infty$, we know that $\left(\frac{1}{2}\right)^{k+1} \to 0$, so the value of the sum will tend to $1 - 0 = 1$.

We can write $\frac{1}{2} + \frac{1}{4} + \ldots + \left(\frac{1}{2}\right)^k + \ldots = 1$, where the "infinite sum" on the left side really means the limit of the finite sums as the number of terms becomes larger and larger.

What made this work was that $\lim_{k\to\infty} \left(\frac{1}{2}\right)^{k+1} = 0$. If we consider the sum $a + ar + ar^2 + \ldots + ar^k = \frac{a - ar^{k+1}}{1-r} = \frac{a}{1-r}(1 - r^{k+1})$, we see that the corresponding question concerns the behavior of r^{k+1} as $k \to \infty$.

If $|r| < 1$, then $\lim_{k\to\infty} r^{k+1} = 0$, by Example 10.1.5, and the value of the sum with k terms tends to $\frac{a}{1-r}(1-0) = \frac{a}{1-r}$ as $k \to \infty$. If $|r| > 1$, then this limit does not exist, and if $r = -1$, then r^{k+1} does not tend to a limit as $k \to \infty$ (since $(-1)^{k+1}$ is alternately 1 and -1). If $r = 1$, then Theorem 10.2.2 does not apply and we cannot use this argument, but in this case the sum with k terms is just $a + a + a + \ldots + a = ka$, and unless $a = 0$, this does not tend to a (finite) limit either.

DEFINITION 10.2.5

Suppose $a \neq 0$. The **geometric series**

$$a + ar + ar^2 + ar^3 + \ldots + ar^k + \ldots = \sum_{n=0}^{\infty} ar^n$$

means the limit as $k \to \infty$, if it exists, of the values of the geometric sum

$$a + ar + ar^2 + ar^3 + \ldots + ar^{k-1} = \sum_{n=0}^{k-1} ar^n = a\frac{1-r^k}{1-r}$$

with k terms in it.

The limit does exist precisely when $|r| < 1$, and in that case

$$a + ar + ar^2 + ar^3 + \ldots + ar^k + \ldots = \sum_{n=0}^{\infty} ar^n = \frac{a}{1-r}.$$

We say the geometric series **converges** if the limit exists and that it **diverges** if the limit does not exist.

EXAMPLE 5 Decide whether or not each of the following geometric series converges, and evaluate the ones that do: (i) $\sum_{n=0}^{\infty} 3^{-n}$, (ii) $\sum_{n=0}^{\infty} \frac{14}{5^{-n}}$, (iii) $\sum_{k=1}^{\infty} \frac{5}{(-4)^k}$.

Solution

(i) In this case the common ratio is the ratio of the nth term to the $(n-1)$th term, which is $\frac{3^{-n}}{3^{-(n-1)}} = 3^{-n+(n-1)} = 3^{-1} = \frac{1}{3}$. Since this is smaller than 1, the series converges. Its first term is $a = 3^{-0} = 1$, so Definition 10.2.5 tells us that $\sum_{n=0}^{\infty} 3^{-n} = \frac{a}{1-r} = 1/\left(1 - \frac{1}{3}\right) = 1/\left(\frac{2}{3}\right) = \frac{3}{2}$.

(ii) Here the ratio of the nth term to the $(n-1)$th term is $\frac{14}{5^{-n}} / \frac{14}{5^{-(n-1)}} = \frac{14}{5^{-n}} \frac{5^{-(n-1)}}{14} = \frac{5^{-(n-1)}}{5^{-n}} = 5^{-(n-1)-(-n)} = 5^1 = 5 > 1$. Since the ratio is greater than 1, the series diverges. It does not make sense to try to evaluate it.

(iii) The common ratio is $\left(\frac{5}{(-4)^k}\right) / \left(\frac{5}{(-4)^{k-1}}\right) = \frac{5}{(-4)^k} \frac{(-4)^{k-1}}{5} = \frac{(-4)^{k-1}}{(-4)^k} = (-4)^{k-1-k}$ $= (-4)^{-1} = -\frac{1}{4}$. Since the absolute value of this ratio is less than 1, the series converges.

Moreover, its value can easily be found. The first term in the series is the one with $k = 1$, which is $\frac{5}{(-4)^k} = \frac{5}{(-4)} = -\frac{5}{4}$. Since the common ratio is $r = -\frac{1}{4}$, we see that the value of the geometric series is $\frac{-5/4}{1-(-1/4)} = \frac{-5/4}{5/4} = -1$.

Notice that in part (iii) of Example 5, the sum was indexed by k running from $k = 1$ to $k = \infty$. Where the index runs does not affect the convergence of the series, since that depends only on the ratio of each term to the previous one. Of course, we do have to be careful because the range through which the index runs *does* determine the first term a, which we need to know in order to evaluate the series.

Exercises 10.2

I

Evaluate each of the following geometric sums.

1. $1 + \frac{1}{2} + \frac{1}{4} + \ldots + \frac{1}{64}$
2. $3 + 1 + \frac{1}{3} + \frac{1}{9} + \frac{1}{27}$
3. $1 - \frac{1}{2} + \frac{1}{4} - \ldots - \frac{1}{128}$
4. $\frac{1}{4} - \frac{1}{2} + 1 - 2 + 4 - \ldots + 64$
5. $\sum_{k=0}^{20} 2^{-k}$
6. $\sum_{n=0}^{10} 3^n$
7. $\sum_{\ell=1}^{8} 5(4^{-\ell})$
8. $\sum_{m=2}^{9} 10^{-m}$
9. $\sum_{n=-2}^{6} \frac{2}{(-6)^n}$
10. $\sum_{j=0}^{99} (-1)^j$

In each of the following situations, a loan is to be paid in equal installments. Find the size of the payment.

11. \$10,000, 12 annual payments, annual interest 10%.
12. \$10,000, 15 annual payments, annual interest 10%.
13. \$1000, monthly payments for 2 years, interest rate 12%, compounded monthly.
14. \$2000, monthly payments for 4 years, interest 18%, compounded monthly.
15. \$40,000, quarterly payments for 16 years, annual interest 10%, compounded quarterly.
16. \$60,000, quarterly payments for 20 years, annual interest 12%, compounded quarterly.
17. \$100,000, 15 annual payments, annual interest 12%.
18. \$100,000, monthly payments for 20 years, interest rate 12%, compounded monthly.
19. \$2000, daily payments for 1 year, interest 12%, compounded daily.
20. \$2000, weekly payments for 1 year, interest 12%, compounded weekly. (Assume that there are exactly 52 weeks in one year.)

Evaluate each of the following geometric series.

21. $1 + \frac{1}{2} + \frac{1}{4} + \ldots$
22. $49 + 7 + 1 + \frac{1}{7} + \frac{1}{49} + \ldots$
23. $9 - 3 + 1 - \frac{1}{3} + \frac{1}{9} - \ldots$
24. $\frac{3}{4} - \frac{9}{16} + \frac{27}{64} - \ldots$
25. $\sum_{m=0}^{\infty} 2^{-m}$
26. $\sum_{n=0}^{\infty} \left(\frac{3}{5}\right)^n$
27. $\sum_{k=1}^{\infty} 11(3^{-k})$
28. $\sum_{m=2}^{\infty} (\sqrt{2})^{-m}$
29. $\sum_{m=-4}^{\infty} \frac{5}{(-3)^m}$
30. $\sum_{j=0}^{\infty} (-0.9)^j$

II

31. What would be the monthly payment in the situation of Example 3 (p. 654) if interest rates went up to 18%?
32. Suppose a loan for \$15,000 is to be paid off in five equal annual installments. If the effective annual interest rate is 9%, what will be the size of the installments?

*33. With what size of annual payment would the company selling the annuity in Example 4 (p. 654) be able to break even?

*34. With what size of annual payment would the company selling the annuity in Example 4 be able to have a surplus of \$10,000 when the term of the annuity is over?

*35. (i) If the situation of Example 4 was repeated but interest rates rose to 11%, what effect would this have on the company's position?
(ii) What would happen if the interest rate was 12%?

*36. (i) A certain annuity, which costs \$30,000, promises to pay \$4402.90 per year for the rest of Grandfather's life. If the effective annual interest rate is 10%, how long will it be before all the money in the plan has been exhausted?

(ii) The best available actuarial predictions say that Grandfather's life expectancy is ten years. If Grandfather lives for exactly ten more years, will the company selling the annuity make or lose money (or break even)?

***37.** Suppose m and k are integers with $m < k$. Find a formula for evaluating the sum

$$\sum_{n=m}^{k} ar^n.$$

***38.** (i) What happens with the formula you found in the previous question if we allow m and k to be equal? How would you interpret this answer?

(ii) What happens with the formula you found in the previous question if $k = m - 1$? How would you interpret this answer?

***39.** Evaluate (i) $\sum_{k=0}^{12} 3^{2k}$, (ii) $\sum_{k=1}^{8} 2^{-3k}$.

***40.** Find a formula for evaluating the sum

$$\sum_{n=0}^{k} ar^{2n}.$$

***41.** (i) The sum $a + ar + ar^2 + \ldots + ar^n$ is a geometric sum with common ratio r. However, if we turn it around and write it the other way, $ar^n + ar^{n-1} + \ldots + ar^2 + ar + a$, it is a geometric sum with common ratio $\frac{1}{r}$. Find the formulas for evaluating each of these geometric sums.

(ii) Show that the two formulas in part (i) are actually equal.

42. (i) The repeating decimal $0.777\overline{7}\ldots$ can be written as a geometric series $\frac{7}{10} + \frac{7}{100} + \frac{7}{1000} + \ldots = 7\sum_{n=1}^{\infty}\left(\frac{1}{10}\right)^n$. Express $0.777\overline{7}\ldots$ as a rational number $\frac{r}{s}$.

(ii) Express the repeating decimal $2.41111\overline{1}\ldots$ as a rational number.

(iii) Express the repeating decimal $1.21212\overline{12}\ldots$ as a rational number.

43. Express the repeating decimal $2.112\,112\,\overline{112}\ldots$ as a rational number.

44. Express the repeating decimal $6.4223\,4223\,\overline{4223}\ldots$ as a rational number.

III

45. Verify the result of Example 3: With a calculator or computer, start with $P = 20{,}000$, add the monthly interest, and subtract the monthly payment $A = 693.31$. Add the next month's interest, subtract another payment, and so on. What happens after 3 years?

46. The calculation we did before Example 2 (p. 652) showed that $1000\left(1.10 + (1.10)^2 + \ldots + (1.10)^{20}\right) = 63{,}002.50$. Verify this by actually adding up the sum.

POINT TO PONDER

Suppose a rabbit is moving in a straight line, and consider how it moves one unit, that is, from 0 to 1. First it has to move from 0 to $\frac{1}{2}$, then from $\frac{1}{2}$ to $\frac{3}{4}$, then from $\frac{3}{4}$ to $\frac{7}{8}$, and so on. Since the rabbit has to go through all these steps, one after another, we conclude that it never reaches 1 (!).

This argument is known as **Zeno's Paradox**. It suggests that nothing can ever move anywhere (on p. 637 we consider another version, which shows that a fast hare can never overtake a slow tortoise).

Suppose for simplicity that the rabbit's speed is 1 unit per second. Then to get from 0 to $\frac{1}{2}$ takes it $\frac{1}{2}$ sec, to get from $\frac{1}{2}$ to $\frac{3}{4}$ takes it $\frac{1}{4}$ sec, to get from $\frac{3}{4}$ to $\frac{7}{8}$ takes it $\frac{1}{8}$ sec, and so on.

To go through all the infinitely many steps, from 0 to $\frac{1}{2}$, from $\frac{1}{2}$ to $\frac{3}{4}$, from $\frac{3}{4}$ to $\frac{7}{8}$, and so on takes a total time of

$$\frac{1}{2} + \frac{1}{4} + \frac{1}{8} + \ldots.$$

This is a geometric series, which we now know totals 1 sec. In 1 sec the rabbit actually goes through all the steps and does move 1 unit of distance.

The paradox can perhaps be explained by saying that the original argument tacitly assumed that each of the steps from 0 to $\frac{1}{2}$, from $\frac{1}{2}$ to $\frac{3}{4}$, from $\frac{3}{4}$ to $\frac{7}{8}$, and so on takes the same length of time. If the speed is constant, then they do not, and our knowledge of geometric series allows us to find the total time.

10.3 Convergence of Series

PARTIAL SUM
CONVERGENCE
DIVERGENCE
HARMONIC SERIES

In Section 10.2 we discussed geometric series and found that they *converge* when the absolute value of the common ratio is less than 1 and that there is a convenient formula for evaluating the series. In this section we discuss the notion of convergence for other series (i.e., infinite sums). Generally speaking, it can be quite difficult to decide whether or not a particular series converges, and it can be even more difficult to evaluate the series if it does converge.

The question we face is to try to make sense of an infinite sum

$$\sum_{n=1}^{\infty} a_n.$$

To say that this infinite sum equals the number S, say, ought to mean that if we add up enough terms of the sum, the total should approach the value S. In other words, we should consider the sum of the first N terms,

$$S_N = \sum_{n=1}^{N} a_n,$$

and see whether the *sequence* $\{S_N\}$ approaches the limit S.

DEFINITION 10.3.1

Consider the series $\sum_{n=1}^{\infty} a_n$. The sum

$$S_N = \sum_{n=1}^{N} a_n$$

of the first N terms is called the Nth **partial sum** of the series. The series $\sum_{n=1}^{\infty} a_n$ **converges** to a number S if the sequence of partial sums $\{S_N\}$ converges to S.

The series is **convergent** if it converges to some limit S, the *sum*, and **divergent** if it does not converge to any limit.

Let us emphasize again the distinction between sequences and series. A sequence $\{a_n\}$ is a *list* of numbers. A series $\sum_{n=1}^{\infty} b_n$ is like an infinite *sum*. Starting with a series, we can define the partial sums $S_N = \sum_{n=1}^{N} b_n$; each S_N is a number, and together they make a sequence $\{S_N\}$.

If you think back to the way we discussed geometric series, the way we considered convergence or divergence in Definition 10.2.5 was to evaluate the partial sums and see whether or not they tended to a limit. The thing that made that situation special was that there is an explicit formula for the partial sums. For many series no such formula is available.

EXAMPLE 1

(i) The easiest series to evaluate is the one whose terms are all zero, that is, $\sum_{n=1}^{\infty} 0$. If we add up the first N terms to find S_N, we are adding up N zeroes, so $S_N = 0$, for every N. The sequence of partial sums $\{S_N\}$ is the constant sequence $0, 0, 0, 0, \ldots$, which converges to 0. We have the completely unsurprising conclusion that $\sum_{n=1}^{\infty} 0 = 0$.

(ii) Consider the series $\sum_{n=1}^{\infty} a_n$, where the terms a_n are all the same, that is, $a_n = c$, for every n, where c is some constant. But now let us consider the case in which $c \neq 0$.

The partial sum S_N is obtained by adding up N terms, each of which equals c. So $S_N = \sum_{n=1}^{N} c = Nc$. The sequence of partial sums $\{S_N\} = \{Nc\}$ diverges (because the terms Nc tend to ∞ if $c > 0$ and to $-\infty$ if $c < 0$).

Combining parts (i) and (ii), we see that the series $\sum_{n=1}^{\infty} c$ converges to 0 if $c = 0$ and diverges otherwise.

$$\begin{aligned} S_1 &= b_1 = -1 \\ S_2 &= b_1 + b_2 = -1 + 1 = 0 \\ S_3 &= b_1 + b_2 + b_3 \\ &= -1 + 1 + (-1) = -1 \\ &\vdots \end{aligned}$$

(iii) Let $b_n = (-1)^n$ and consider $\sum_{n=1}^{\infty} b_n = \sum_{n=1}^{\infty} (-1)^n$. At left we find the first few partial sums. They are alternately -1 and 0; they do not converge to any limit, so the series diverges.

$$\begin{aligned} S_1 &= c_1 = \tfrac{1}{2} \\ S_2 &= c_1 + c_2 = \tfrac{1}{2} + \tfrac{1}{6} = \tfrac{2}{3} \\ S_3 &= c_1 + c_2 + c_3 = S_2 + c_3 \\ &= \tfrac{2}{3} + \tfrac{1}{12} = \tfrac{3}{4} \\ S_4 &= S_3 + c_4 = \tfrac{3}{4} + \tfrac{1}{20} = \tfrac{4}{5} \end{aligned}$$

(iv) Let $c_n = \frac{1}{n^2+n}$ and consider the series $\sum_{n=1}^{\infty} c_n$. At left we calculate some partial sums. From the ones we have, it is beginning to look as if $S_N = \frac{N}{N+1}$. This is in fact true. The reason is that if $S_N = \frac{N}{N+1}$, then S_{N+1} is obtained from it by adding $c_{N+1} = \frac{1}{(N+1)^2+(N+1)}$, so

$$\begin{aligned} S_{N+1} &= S_N + \frac{1}{(N+1)^2 + (N+1)} = \frac{N}{N+1} + \frac{1}{(N+1)(N+1+1)} \\ &= \frac{N(N+2)+1}{(N+1)(N+2)} = \frac{N^2+2N+1}{(N+1)(N+2)} = \frac{(N+1)^2}{(N+1)(N+2)} \\ &= \frac{N+1}{N+2}. \end{aligned}$$

We know that $S_N = \frac{N}{N+1}$ for $N = 1, 2, 3$, and 4, and we have just seen that it will always be true for the *next* value of N too. From this it is possible to conclude that it is true for *all* values of N. (This is an example of what is called a proof by *induction*; the concept is discussed at greater length in the appendix.)

Since the sequence of partial sums $\{S_N\} = \{\frac{N}{N+1}\}$ converges to 1, we see that the series $\sum_{n=1}^{\infty} c_n$ converges and $\sum_{n=1}^{\infty} c_n = \sum_{n=1}^{\infty} \frac{1}{n^2+n} = 1$.

In the last part of Example 1 we were able to find an exact formula for the partial sums of the series, which enabled us to find their limit and evaluate the series. Usually, however, this is quite difficult to do.

If the series $\sum_{n=1}^{\infty} a_n$ converges to L, say, then its partial sums must approach L. In particular, the partial sums S_N must eventually get close together (since they will all be close to L).

The difference between two consecutive partial sums is just one of the terms of the series: $S_N - S_{N-1} = a_N$. To say that these partial sums are close together is to say that a_N is small.

So we see that the terms of a convergent series must eventually get small. This is a very important observation.

In Example 1, the series in parts (i) and (iv) both converge. In the first case the terms are all zero, so the terms certainly converge to 0. The series in part (iv) is $\sum_{n=1}^{\infty} \frac{1}{n^2+n}$, and its terms also converge to zero, by Theorem 10.1.3.

The series in parts (ii) and (iii) of Example 1 diverge, and in part (ii) the terms converge to the nonzero number c, while in part (iii) the terms diverge.

THEOREM 10.3.2

The *n*th Term Test

Suppose the series $\sum_{n=1}^{\infty} a_n$ converges. Then its terms converge to 0, that is,

$$\lim_{n\to\infty} a_n = 0.$$

(Notice that this limit refers to the terms of the series as a *sequence.*)

Another way of saying exactly the same thing is this: If $\{a_n\}$ does *not* converge to 0 (i.e., if this sequence diverges or if it converges to something other than 0), then the corresponding series $\sum_{n=1}^{\infty} a_n$ diverges.

PROOF We have already observed that $a_n = S_n - S_{n-1}$. Suppose the series $\sum_{n=1}^{\infty} a_n$ converges to L. Then

$$\lim_{n\to\infty} a_n = \lim_{n\to\infty} (S_n - S_{n-1}) = \left(\lim_{n\to\infty} S_n\right) - \left(\lim_{n\to\infty} S_{n-1}\right) = L - L = 0.$$

(We have used Theorem 10.1.5 to show that the "shifted" sequence $\{S_{n-1}\}$ converges to the same limit as $\{S_n\}$, i.e., to L.)

We have already checked that this result is consistent with Example 1.

The way Theorem 10.3.2 is most often applied is to use the second form of it to prove that a series diverges by noticing that its terms do not converge to 0. It is in this sense that it is a test. We can test whether or not the nth term of a series $\sum_{n=1}^{\infty} a_n$ tends to zero, and if it does not, Theorem 10.3.2 allows us to conclude that the series diverges.

EXAMPLE 2

(i) Theorem 10.3.2 gives an almost instant argument that $\sum_{n=1}^{\infty} (-1)^n$ diverges because the sequence $-1, 1, -1, 1, \ldots$ of its terms does not converge to 0 (or to anything else).

(ii) The series $\sum_{n=1}^{\infty} \frac{n-1}{n}$ diverges because $\lim_{n\to\infty} \frac{n-1}{n} = 1$, by Theorem 10.1.3.

(iii) The series $\sum_{n=1}^{\infty} (3n-4)$ diverges because $\{(3n-4)\}$ diverges (it tends to ∞).

We see that Theorem 10.3.2 is extremely useful for proving that certain series diverge. It is important to realize that it does *not* allow us to conclude that if the terms $\{a_n\}$ of a series do tend to zero, then the series must converge. There is a very important example of a series that *diverges* even though its terms converge to zero.

The Harmonic Series

The series $\sum_{n=1}^{\infty} \frac{1}{n}$ whose nth term is $\frac{1}{n}$ is called the **harmonic series**. Its terms tend to 0, but the series diverges.

THEOREM 10.3.3

> The harmonic series diverges.

PROOF Instead of considering all the partial sums, let us concentrate on S_2, S_4, S_8, S_{16}, and so on, that is, on the partial sums S_N for which N is a power of 2.

For instance,

$$\begin{aligned} S_{16} &= 1 + \frac{1}{2} + \frac{1}{3} + \frac{1}{4} + \frac{1}{5} + \frac{1}{6} + \frac{1}{7} + \frac{1}{8} + \frac{1}{9} + \frac{1}{10} + \frac{1}{11} + \frac{1}{12} + \frac{1}{13} \\ &\quad + \frac{1}{14} + \frac{1}{15} + \frac{1}{16} \\ &= (1) + \left(\frac{1}{2}\right) + \left(\frac{1}{3} + \frac{1}{4}\right) + \left(\frac{1}{5} + \frac{1}{6} + \frac{1}{7} + \frac{1}{8}\right) \\ &\quad + \left(\frac{1}{9} + \frac{1}{10} + \frac{1}{11} + \frac{1}{12} + \frac{1}{13} + \frac{1}{14} + \frac{1}{15} + \frac{1}{16}\right) \end{aligned}$$

Now consider the expressions in the parentheses. Each of them is greater than or equal to $\frac{1}{2}$. The first one is 1; the second is $\frac{1}{2}$. The third is $\frac{1}{3} + \frac{1}{4}$. It consists of two terms, each of which is greater than or equal to $\frac{1}{4}$, so their sum must be greater than or equal to $\frac{1}{2}$. The next pair of parentheses enclose four terms, each greater than or equal to $\frac{1}{8}$, so it must be greater than or equal to $4\left(\frac{1}{8}\right) = \frac{1}{2}$. The last pair of parentheses contains eight terms, each greater than or equal to the last term, which is $\frac{1}{16}$, so the sum is greater than or equal to $8\left(\frac{1}{16}\right) = \frac{1}{2}$.

We could continue this way. If we consider all the terms that come after the 2^nth term up to the 2^{n+1}th term, then there are exactly 2^n such terms, and they are all greater than or equal to the last one, which is $\frac{1}{2^{n+1}} = 2^{-(n+1)}$. The sum of these terms is greater than or equal to $2^n \times 2^{-(n+1)} = \frac{1}{2}$.

So if we consider the partial sum S_{2^n}, we can group the terms in this way in $n + 1$ groups (up to 1, from there up to $\frac{1}{2}$, from there up to $\frac{1}{4}$, . . . , from $\frac{1}{2^{n-1}}$ up to $\frac{1}{2^n}$, . . .). Since the sum of each of these groups is greater than or equal to $\frac{1}{2}$, we see that the partial sum S_{2^n} is greater than or equal to $(n + 1)\left(\frac{1}{2}\right) = \frac{n+1}{2}$. (Since the first "group" is 1, we actually have $S_{2^n} \geq \frac{n+2}{2}$ . . .).

Since the terms of the sequence $\left\{\frac{n+1}{2}\right\}$ grow to infinity, we see that the partial sums of the harmonic series also diverge, so the harmonic series itself is divergent, as desired. ◢

See the Calculator Example on the next page.

This example is very important for many purposes, but the first thing it does is provide an example of a series whose terms converge to 0 but that is divergent. This shows that while Theorem 10.3.2 can be used to show that various series are divergent, there is no corresponding theorem in the other direction to show that they converge just because their terms approach zero.

We cannot show it yet, but later in the chapter we will learn that if we alter the harmonic series by putting in minus signs on every other term, then the resulting series converges.

With a programmable calculator we can test the idea of the proof of Theorem 10.3.3, for various values of n.

TI-85	SHARP EL9300
PRGM F2(EDIT) HARM ENTER F3 F1(Input) N EXIT ENTER 0 STO K ENTER 0 STO S ENTER Lbl B ENTER K + 1 STO K ENTER S + 1 ÷ K STO S ENTER F4 F1(If) (K TEST F2(<) 2^N) ENTER Goto B ENTER S ENTER EXIT EXIT EXIT PRGM F1(NAME) HARM ENTER 4 ENTER	▼ ▼ (NEW) ENTER ENTER HARM ENTER COMMAND A 3 (Input) *N* ENTER *K* = 0 ENTER *S* = 0 ENTER COMMAND B 1 (Label) 1 ENTER *K* = *K* + 1 ENTER *S* = *S* + (1 ÷ *K*) ENTER COMMAND 3 (If) *K* COMMAND C 2 (<) 2 a^b *N* COMMAND B 2 (Goto) 1 ENTER COMMAND A 1 (Print) *S* ENTER COMMAND A 6 (End) ENTER (RUN) ENTER (HARM) ENTER 4 ENTER
CASIO f_x-6300*G* (f_x-7700*GB*)	**HP 48SX**
MODE 2 EXE ? → [redacted] : 0 → [redacted] : 0 → [redacted] : Lbl 1 : [redacted] + 1 → [redacted] : [redacted] + (1 ÷ [redacted]) → [redacted] : [redacted] < 2x^y [redacted] ⇒ Goto 1: *S* ◢ MODE 1 Prog 0 EXE 4 EXE (on the f_x-7700*GB*, : is PRGM F6, and Prog is PRGM F3.)	ATTN ↱ CLR [redacted] "" KEY IN N ► SPC PROMPT SPC ′ *N* ► STO 0 SPC 1 SPC 2 SPC *N* y^x SPC FOR *K* SPC 1 *K* ÷ + NEXT ▼ ENTER ′ HARM STO VAR HARM 4 ENTER [redacted]

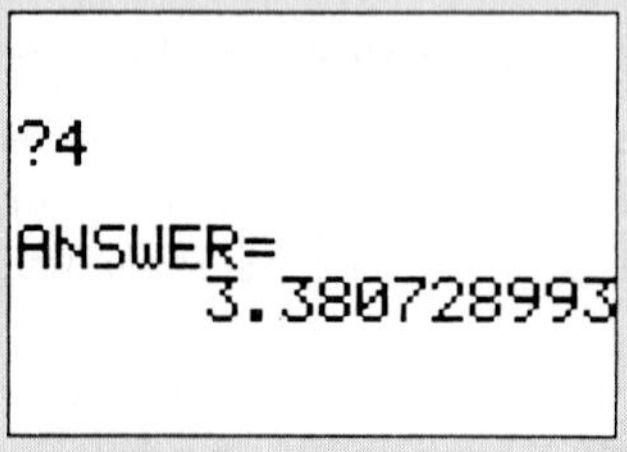

Now use the program with other values of n. With even moderately large values of n it can take quite a while to run. It is interesting to note that the sum up to $k = 2^n$ is always greater than $\frac{n+1}{2}$.

Generally speaking, it is quite delicate and difficult to decide whether or not series converge. The next three sections are devoted to developing some techniques for approaching problems of this sort.

We give one other example to demonstrate the sort of phenomena that can be encountered.

EXAMPLE 3 In Section 10.2 we observed that "shifting" a sequence does not change whether or not it converges, nor its limit if it does converge. The same is clearly true of a series.

For sequences we also observed that changing the first few terms does not affect the limit, but with series this is not true. After all, if we change some terms, the partial sums will change.

For instance, suppose we take a series $\sum_{n=1}^{\infty} a_n$ that converges to L. Define another series $\sum_{n=1}^{\infty} b_n$ by letting $b_1 = a_1 + 1$, $b_2 = a_2 - 2$, $b_3 = a_3 + 5$, and all other b_n's the same as the corresponding a_n's, that is, $b_n = a_n$, for $n > 3$.

Let us write S_N and T_N for the partial sums $S_N = \sum_{n=1}^{N} a_n$ and $T_N = \sum_{n=1}^{N} b_n$. Then

$$\begin{aligned}
T_1 &= b_1 = (a_1 + 1) = S_1 + 1, \\
T_2 &= b_1 + b_2 = (a_1 + 1) + (a_2 - 2) = a_1 + a_2 - 1 = S_2 - 1, \\
T_3 &= b_1 + b_2 + b_3 = (a_1 + 1) + (a_2 - 2) + (a_3 + 5) = a_1 + a_2 + a_3 + 4 \\
&= S_3 + 4, \\
T_4 &= b_1 + b_2 + b_3 + b_4 = (a_1 + 1) + (a_2 - 2) + (a_3 + 5) + a_4 \\
&= a_1 + a_2 + a_3 + a_4 + 4 = S_4 + 4.
\end{aligned}$$

In fact, for every $N > 3$,

$$\begin{aligned}
T_N &= \sum_{n=1}^{N} b_n = \sum_{n=1}^{3} b_n + \sum_{n=4}^{N} b_n = \big((a_1 + 1) + (a_2 - 2) + (a_3 + 5)\big) + \sum_{n=4}^{N} a_n \\
&= 4 + (a_1 + a_2 + a_3) + \sum_{n=4}^{N} a_n = 4 + \sum_{n=1}^{N} a_n \\
&= 4 + S_N.
\end{aligned}$$

We see that the first few partial sums change in different ways, but after we reach the stage at which the terms in the two series are the same, the partial sums always differ by 4. Since $\{S_N\}$ converges to L, we see that $\{T_N\}$ converges to $L + 4$, that is, $\sum_{n=1}^{\infty} b_n = \left(\sum_{n=1}^{\infty} a_n\right) + 4$.

Changing the first few terms may change the value of the infinite sum but does not change whether it converges or not.

THEOREM 10.3.4

(i) Consider two series $\sum_{n=1}^{\infty} a_n$ and $\sum_{n=1}^{\infty} b_n$. Suppose that the series are the same after the first few terms, that is, there is an integer K so that $a_n = b_n$ for every $n > K$.

Then either both series converge or else both series diverge. If they converge, then their sums differ in the obvious way owing to the changes in the first K terms:

$$\left(\sum_{n=1}^{\infty} a_n\right) - \left(\sum_{n=1}^{\infty} b_n\right) = \left(\sum_{n=1}^{K} a_n\right) - \left(\sum_{n=1}^{K} b_n\right) = \sum_{n=1}^{K} (a_n - b_n).$$

(ii) Suppose $\sum_{n=1}^{\infty} a_n$ converges. Then the shifted series $\sum_{n=k+1}^{\infty} a_{n-k}$ also converges, and the sums are the same. If either series diverges, then they both do.

(iii) Suppose $\sum_{n=1}^{\infty} a_n$ converges. Then for any $k > 1$ the series $\sum_{n=k}^{\infty} a_n$ that begins with a_k instead of a_1 also converges. If, on the other hand, either of these series diverges, then they both do.

PROOF

(i) Consider the partial sums $S_N = \sum_{n=1}^{N} a_n$ and $T_N = \sum_{n=1}^{N} b_n$. If $N > K$, then

$$\begin{aligned} S_N - T_N &= \left(\sum_{n=1}^{N} a_n\right) - \left(\sum_{n=1}^{N} b_n\right) = \sum_{n=1}^{N} (a_n - b_n) \\ &= \left(\sum_{n=1}^{K} (a_n - b_n)\right) + \left(\sum_{n=K+1}^{N} (a_n - b_n)\right) = \left(\sum_{n=1}^{K} (a_n - b_n)\right) + 0 \\ &= \left(\sum_{n=1}^{K} a_n\right) - \left(\sum_{n=1}^{K} b_n\right) = S_K - T_K. \end{aligned}$$

We see that for $N > K$, $S_N - T_N$ is a constant, independent of N. If either sequence converges, then so must the other, and

$$\begin{aligned} \left(\sum_{n=1}^{\infty} a_n\right) - \left(\sum_{n=1}^{\infty} b_n\right) &= \lim_{N\to\infty} S_N - \lim_{N\to\infty} T_N = \lim_{N\to\infty} (S_N - T_N) \\ &= \left(\sum_{n=1}^{K} a_n\right) - \left(\sum_{n=1}^{K} b_n\right), \end{aligned}$$

as needed.

(ii) Let $S_N = \sum_{n=1}^{N} a_n$, and let $T_M = \sum_{n=k+1}^{M} a_{n-k}$ be the partial sum of the shifted sequence. Then $T_{N+k} = S_N$ for every $N \geq 1$. By Theorem 10.1.5(i) these two sequences S_N and T_N must converge to the same limit (or must both diverge).

(iii) Define a new series by letting $b_n = 0$ if $n < k$ and $b_n = a_n$ if $n \geq k$. Then the series $\sum_{n=1}^{\infty} b_n$ is really the same thing as $\sum_{n=k}^{\infty} a_n$.

On the other hand, the terms a_n and b_n are equal after $n = k$, so by part (i) of this theorem, the two series are either both convergent or both divergent.

EXAMPLE 4

(i) The series $\frac{1}{5} + \frac{1}{6} + \frac{1}{7} + \frac{1}{8} + \frac{1}{9} + \ldots = \sum_{n=5}^{\infty} \frac{1}{n}$ is really just the harmonic series starting with its fifth term. By part (iii) of Theorem 10.3.4 it diverges.

(ii) Let $a_n = \frac{1}{3n+9}$ and consider the series $\sum_{n=1}^{\infty} a_n = \sum_{n=1}^{\infty} \frac{1}{3n+9}$. If we write it as $\sum_{n=1}^{\infty} \frac{1}{3n+9} = \frac{1}{3}\left(\sum_{n=1}^{\infty} \frac{1}{n+3}\right)$, then we see that this last sum is the same as the harmonic series shifted over by 3, starting from the fourth term.

Combining parts (i) and (ii) of Theorem 10.3.4, we see that the series diverges.

(iii) Consider the series with terms $3, 5, -4, 1, \frac{1}{2}, \frac{1}{4}, \frac{1}{8}, \ldots$, which is a geometric series after the first three terms. The first part of the theorem tells us that it converges and even tells us how to find the sum, but it is easier to write

$$\begin{aligned} 3 + 5 - 4 + 1 + \frac{1}{2} + \frac{1}{4} + \frac{1}{8} + \ldots &= 3 + 5 - 4 + \sum_{n=0}^{\infty} \left(\frac{1}{2}\right)^n \\ &= 4 + \frac{1}{1 - \frac{1}{2}} = 4 + 2 = 6. \end{aligned}$$

We evaluated this series by breaking off the first few anomalous terms and adding them up separately and then using the formula for geometric series to evaluate the other part of the series.

Exercises 10.3

I

Decide whether or not each of the following series converges, giving reasons, and evaluate the ones that do converge.

1. $\sum_{n=1}^{\infty} 2^{-n}$
2. $\sum_{n=1}^{\infty} 2^{n}$
3. $\sum_{n=1}^{\infty} (2^{-n} - 3)$
4. $\sum_{n=1}^{\infty} \left(n - \frac{1}{n}\right)$
5. $\sum_{n=1}^{\infty} 0$
6. $\sum_{n=1}^{3} (n^2 - n) + \sum_{n=4}^{\infty} 0$
7. $\sum_{n=1}^{7} (e^n + 2) + \sum_{n=8}^{\infty} (-5)$
8. $\sum_{n=4}^{\infty} \frac{2}{n}$
9. $\sum_{n=1}^{\infty} \left(\left(\frac{3}{2}\right)^n - (3^n - 2)/(2^n)\right)$
10. $\sum_{n=4}^{\infty} \left(\sin^n \frac{n\pi}{2}\right)$

Suppose $\sum_{n=1}^{\infty} a_n$ and $\sum_{n=1}^{\infty} b_n$ are convergent series. Suppose $\sum_{n=1}^{\infty} a_n = 3$ and $\sum_{n=1}^{\infty} b_n = -5$. Also suppose $a_1 = 3$, $a_2 = -5$, $a_3 = 0$, $a_4 = 1$, $b_1 = 2$, $b_2 = 2$, $b_3 = 3$, and $b_4 = 2$, and evaluate the following.

11. $\sum_{n=2}^{\infty} a_n$
12. $\sum_{n=3}^{\infty} (2a_n - 3b_n)$
13. $\sum_{n=1}^{\infty} (a_n - b_n)$
14. $\sum_{n=2}^{\infty} (a_n + 2b_n - 5^{-n})$
15. $\sum_{n=1}^{\infty} a_n - \sum_{n=3}^{\infty} a_n$
16. $\sum_{n=5}^{\infty} 4b_n$
17. $\sum_{n=4}^{\infty} b_n - \sum_{n=2}^{\infty} b_n$
18. $\sum_{n=2}^{\infty} b_{n-1}$
19. $\sum_{n=1}^{\infty} a_{n+3} - \sum_{n=4}^{\infty} a_n$
20. $\sum_{n=8}^{\infty} a_{n-7}$

II

*21. Find a series $\sum_{n=1}^{\infty} a_n$ whose Nth partial sum is $S_N = \frac{1}{N}$. Does your series converge, and if so, to what limit?

*22. Find a series $\sum_{n=1}^{\infty} a_n$ whose Nth partial sum is $S_N = \frac{1}{N+1}$. Does your series converge, and if so, to what limit?

*23. Find a series $\sum_{n=1}^{\infty} a_n$ whose Nth partial sum is $S_N = \frac{1}{N^2}$. Does your series converge, and if so, to what limit?

*24. Find a series $\sum_{n=1}^{\infty} a_n$ whose Nth partial sum is $S_N = 2 - \frac{1}{N^3}$. Does your series converge, and if so, to what limit?

*25. Consider the series $1 - 1 + \frac{1}{2} - \frac{1}{2} + \frac{1}{3} - \frac{1}{3} + \frac{1}{4} - \frac{1}{4} + \frac{1}{5} - \frac{1}{5} + \ldots + \frac{1}{n} - \frac{1}{n} + \ldots$, that is, the series whose even-numbered terms are $a_{2n} = -\frac{1}{n}$ and whose odd-numbered terms are $a_{2n-1} = \frac{1}{n}$. Find the partial sums and discuss its convergence.

III

26. Theorem 10.3.3 shows that the harmonic series diverges and in particular that $S_{2^n} = \sum_{k=1}^{2^n} \frac{1}{k} > \frac{n+1}{2}$. Use a programmable calculator or computer to evaluate S_4, S_8, S_{16}, S_{32}, S_{64}, S_{128}, S_{512}, S_{4096}, and verify this result. Although the partial sums S_N tend to ∞, they do it slowly.

POINT TO PONDER

In Example 1(iv) (p. 659) we let $c_n = \frac{1}{n^2+n}$ and considered the series

$$\sum_{n=1}^{\infty} c_n = \sum_{n=1}^{\infty} \frac{1}{n^2 + n}.$$

There is no particular reason to think about partial fractions in connection with series, but if we did, it might occur to us to write $\frac{1}{n^2+n} = \frac{1}{n(n+1)} = \frac{1}{n} - \frac{1}{n+1}$. The Nth partial sum is

$$S_N = c_1 + c_2 + c_3 + \ldots + c_{N-1} + c_N$$
$$= \left(1 - \frac{1}{2}\right) + \left(\frac{1}{2} - \frac{1}{3}\right) + \left(\frac{1}{3} - \frac{1}{4}\right) + \ldots$$
$$+ \left(\frac{1}{N-1} - \frac{1}{N}\right) + \left(\frac{1}{N} - \frac{1}{N+1}\right).$$

Each pair of parentheses encloses one term c_k. But we observe that the second part of each term cancels the first part of the next (e.g., the $-\frac{1}{2}$ cancels the $\frac{1}{2}$, the $-\frac{1}{3}$ cancels the $\frac{1}{3}$, etc.). In fact, almost everything cancels. The only things left are the 1 from the first term and the $-\frac{1}{N+1}$ from the last.

This is an example of what is called a "telescoping sum"; all the terms in the middle cancel, reminding us of the old-fashioned telescope shown in Figure 10.3.1, whose sections all collapse into one for easy carrying.

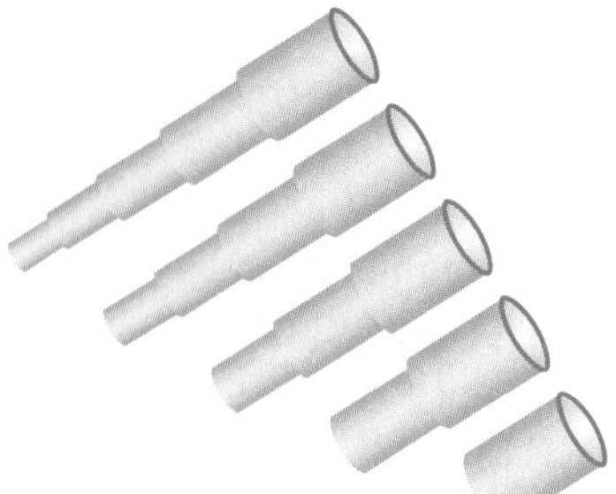

Figure 10.3.1

We find that $S_N = 1 - \frac{1}{N+1} = \frac{N+1-1}{N+1} = \frac{N}{N+1}$, exactly what we found in the example using a different method. It is usually difficult to find a formula for the partial sums of a series; here it was possible because of the special properties of telescoping sums.

10.4 The Integral Test

INCREASING SEQUENCE
INTEGRAL TEST
p-SERIES

We have learned about convergence and divergence of series and have learned in detail how to assess convergence and divergence for geometric series. Apart from that we have learned the nth Term Test (Theorem 10.3.2) for the convergence of a series; it is used to show that a series does *not* converge if its terms do not tend to 0, but it is no use for showing that any series *does* converge. We also saw that the harmonic series diverges, but apart from geometric series we have relatively few examples of series that we know are convergent. In this and the next two sections we will learn some additional tests for deciding whether various series converge or diverge. In this section and most of the next we will be considering series whose terms are all positive.

Suppose $\sum_{n=1}^{\infty} a_n$ is a series whose terms are all *positive*, that is, $a_n > 0$ for every n. As we form the partial sums, we see that $S_{N+1} = S_N + a_{N+1} > S_N$. This means that the partial sums are always **increasing**, since each one is obtained from the previous one by adding something positive.

Now from Theorem 10.1.6 we know that there are only two possibilities for an increasing sequence. Either it grows larger and larger and tends to infinity (i.e., "diverges to infinity") *or else* it is a **bounded sequence** and converges up to some limit.

Let us look at the series $\sum_{n=1}^{\infty} \frac{1}{n^2}$, whose terms are smaller than the corresponding terms of the harmonic series. It would be interesting to know whether or not it converges.

One obvious thing about this series is that its terms are found by evaluating the function $f(x) = \frac{1}{x^2}$ at the positive integers. We will find that it is possible to decide whether or not the series $\sum_{n=1}^{\infty} f(n)$ converges by looking at the improper integral $\int_1^{\infty} f(x)\,dx$.

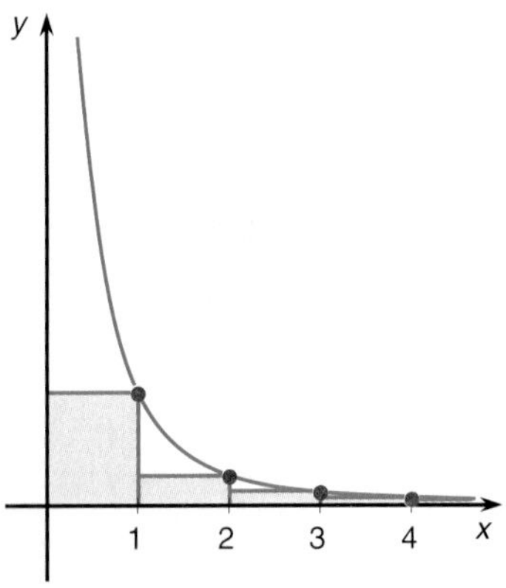

Figure 10.4.1

Let us try to think of the series $\sum_{n=1}^{\infty} \frac{1}{n^2}$ in terms of areas. In Figure 10.4.1 we see the graph of $y = f(x) = \frac{1}{x^2}$ with the first few positive integers marked on the

x-axis. The terms of the series are just the heights of the graph above the axis at the integer points $1, 2, 3, \ldots$.

Instead of thinking of them just as heights, we could also think of them as the areas of the rectangles with these heights and each of width equal to 1. These are the little shaded rectangles in Figure 10.4.1, each of which just reaches up to the curve at its right end (where its height is $f(n)$).

Apart from the first one, the rectangles lie below the graph $y = f(x)$ and to the right of the point $x = 1$. So the sum of the series, which equals the total area of the rectangles, should be less than the area of the first rectangle plus the area under the graph, that is, the area of the first rectangle plus the improper integral $\int_1^\infty f(x)\,dx$.

Strictly speaking, we should talk about the first N rectangles rather than all of them at once. What we see from Figure 10.4.1 is that if we could show that the area under the graph of $y = f(x)$ to the right of the point $x = 1$ is less than some number, M, for instance, then the total area of the first N rectangles must always be less than the area of the first rectangle plus M. This just means that the Nth partial sum $\sum_{n=1}^{N} a_n$ is always less than $a_1 + M$, for *every* N.

In particular, the partial sums are bounded, and we remarked at the beginning that an increasing sequence that is bounded must have a limit.

So if the whole area under the curve to the right of $x = 1$ is some (finite) number M, then the sequence of partial sums must be bounded by $a_1 + M$, and so it must converge. If we can show that the improper integral is *finite*, then the series must converge. This is the essence of the Integral Test.

In the particular case we were considering, the improper integral is easy to evaluate:

$$\int_1^\infty f(x)\,dx = \int_1^\infty \frac{dx}{x^2} = \lim_{r\to\infty}\left(\int_1^r \frac{dx}{x^2}\right) = \lim_{r\to\infty}\left(-\frac{1}{x}\right)\Bigg|_1^r = \lim_{r\to\infty}\left(1 - \frac{1}{r}\right)$$
$$= 1.$$

The integral is finite, so the Integral Test shows that the series converges. Notice that it does not tell us what the value of the series is, just that it is convergent. It is possible to conclude that the series is less than or equal to the sum of the first term and the improper integral, which in this case is 2, so in fact we have some (imprecise) information about the value of the series.

There are a number of subtleties that should be mentioned. One is that it seems awkward to leave out that first rectangle. It might make more sense to consider the integral $\int_0^\infty f(x)\,dx$ so that all the rectangles will lie under its graph.

The difficulty is that this integral may not be finite; in the particular case we have considered, with $f(x) = \frac{1}{x^2}$, the integral is improper at $x = 0$ too, and it diverges there, so the argument would not work.

In a sense, what we are doing is ignoring the first term of the series and showing that $\sum_{n=2}^\infty a_n$ is less than $\int_1^\infty f(x)\,dx$. Because of Theorem 10.3.4(iii), the convergence of this series is exactly equivalent to the convergence of the whole series, so this approach is perfectly reasonable, and it saves us from having to worry about the behavior of $f(x)$ near $x = 0$.

Another important point is that we have seen from Figure 10.4.1 that with $f(x) = \frac{1}{x^2}$ the graph of $y = f(x)$ is always above the rectangles. The reason is that if $x \leq n$, then $f(x) \geq f(n) = a_n$; the function $f(x)$ is *decreasing*, so it is always greater than its value at the next integer. To make the Integral Test work, we have to know that we are using a function $f(x)$ that is a decreasing function. In the example we know that $f(x) = \frac{1}{x^2}$ is decreasing because its derivative is $f'(x) = -\frac{2}{x^3}$, which is negative for all $x > 0$.

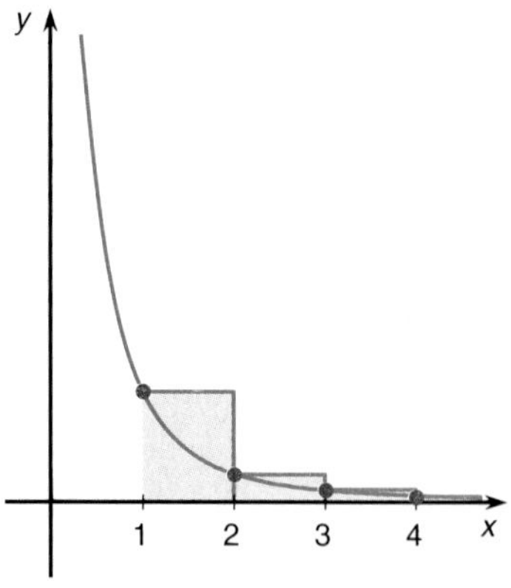

Figure 10.4.2

Finally we should observe that we could have drawn the rectangles a different way, as shown in Figure 10.4.2. Once again the areas of the rectangles equal the terms in the series, but this time the region under the graph of $y = f(x)$ is contained in the rectangles. This means that its area $\int_1^\infty f(x)\,dx$ must be less than the total area of the rectangles, that is, that the integral is less than the series.

These two facts combined tell us that the improper integral lies between the series without its first term and the whole series:

$$\sum_{n=2}^{\infty} a_n = \sum_{n=2}^{\infty} f(n) \le \int_1^\infty f(x)\,dx \le \sum_{n=1}^{\infty} a_n = \sum_{n=1}^{\infty} f(n).$$

If the integral is finite, then the series must converge, as we have already seen. If the series converges, then the integral must converge also. In other words, the convergence of the series is exactly equivalent to the convergence of the integral.

We also get some information on the value of the sum if it converges. Since $\sum_{n=2}^{\infty} a_n \le \int_1^\infty f(x)\,dx$, we add a_1 and find that $\sum_{n=1}^{\infty} a_n \le \int_1^\infty f(x)\,dx + a_1$, so $\int_1^\infty f(x)\,dx \le \sum_{n=1}^{\infty} a_n \le \int_1^\infty f(x)\,dx + a_1$.

THEOREM 10.4.1

The Integral Test

Suppose $f(x)$ is a decreasing function for $x \ge 1$ whose values are always positive, that is, $f(x) > 0$ for all $x \ge 1$. Let $a_n = f(n)$ for each integer $n \ge 1$, and consider the series $\sum_{n=1}^{\infty} a_n$. The series $\sum_{n=1}^{\infty} a_n$ converges if and only if the improper integral $\int_1^\infty f(x)\,dx$ converges, that is, $\int_1^\infty f(x)\,dx < \infty$. Moreover, if the series converges, then

$$\int_1^\infty f(x)\,dx \le \sum_{n=1}^{\infty} a_n \le \int_1^\infty f(x)\,dx + a_1.$$

The Integral Test is most often used to show that a series converges by showing that the corresponding integral converges.

EXAMPLE 1

(i) As the calculation done before Theorem 10.4.1 shows, $\int_1^\infty \frac{1}{x^2}\,dx = 1 < \infty$, so the corresponding series $\sum_{n=1}^{\infty} \frac{1}{n^2}$ converges.

(ii) Consider the series $\sum_{n=1}^{\infty} (2n)^{-3}$. Its terms can be found as the values at the positive integers of the function $f(x) = (2x)^{-3}$. This function is easily seen to be decreasing, since its derivative is $-6(2x)^{-4}$, which is negative for all $x \ge 1$. Accordingly, we should consider the integral of $f(x)$:

$$\int_1^\infty f(x)\,dx = \lim_{r\to\infty}\left(\int_1^r \frac{1}{(2x)^3}\,dx\right) = \lim_{r\to\infty}\left(\frac{1}{8}\int_1^r \frac{1}{x^3}\,dx\right)$$

$$= \frac{1}{8}\lim_{r\to\infty}\left(\frac{1}{-2}\frac{1}{x^2}\right)\Bigg|_1^r = \frac{1}{-16}\lim_{r\to\infty}\left(\frac{1}{r^2} - \frac{1}{1^2}\right) = \frac{1}{16}.$$

The improper integral converges, and the series must be convergent.

(iii) Consider the series $\sum_{n=2}^{\infty} \frac{1}{n\ln(n)}$. If we let $f(x) = \frac{1}{x\ln(x)}$, for $x \ge 2$, then the terms of the series are the values of $f(x)$ at the integers $n \ge 2$.

The derivative of the function f is $f'(x) = -\frac{1}{(x \ln x)^2}\left(\ln x + x\frac{1}{x}\right) = -\frac{1}{(x \ln x)^2}(\ln x + 1)$. For $x \geq 2$ this is always negative, so $f(x)$ is a decreasing function for $x \geq 2$.

To evaluate the integral $\int_2^\infty \frac{1}{x \ln x}\,dx$, we use the substitution $u = \ln x$, $du = \frac{1}{x}\,dx$, and find

$$\int_2^\infty f(x)\,dx = \lim_{r\to\infty}\left(\int_2^r \frac{1}{x\ln x}\,dx\right) = \lim_{r\to\infty}\left(\int_{\ln 2}^{\ln r} \frac{1}{u}\,du\right)$$

$$= \lim_{r\to\infty}(\ln u)\Big|_{\ln 2}^{\ln r} = \lim_{r\to\infty}\big(\ln(\ln r) - \ln(\ln 2)\big) = \infty,$$

because as $r \to \infty$, we know that $\ln r \to \infty$, so the natural logarithm of that must also tend to infinity, that is, $\ln(\ln r) \to \infty$.

We see that the integral diverges, so the series $\sum_{n=2}^\infty \frac{1}{n\ln(n)}$ diverges.

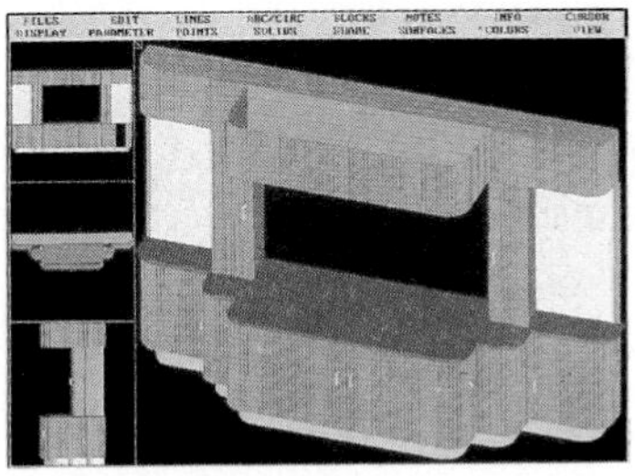

DesignCAD 3D *software was used to produce this image of a wall unit. The computer can generate pictures of the finished product and also plans for the carpenters who will build it.*

Notice the way we adjusted the Integral Test in part (iii) of Example 1 to allow for the fact that the series started with the $n = 2$ term. We considered the integral from $x = 2$ to ∞. For a series that begins with $n = k$, for instance, we consider the improper integral $\int_k^\infty f(x)\,dx$ and otherwise proceed in the same way.

In Section 10.3 we saw that the harmonic series $\sum_{n=1}^\infty \frac{1}{n}$ diverges, and in this section we have seen that $\sum_{n=1}^\infty \frac{1}{n^2}$ converges. We also considered an example with third powers in the denominator. It is natural to ask about the convergence of the series $\sum_{n=1}^\infty \frac{1}{n^p}$, where p is any positive number. A series of this form is sometimes called a p-series.

EXAMPLE 2 Convergence and Divergence of p-Series Let p be any fixed positive number, and let $f(x) = \frac{1}{x^p}$. Define $a_n = f(n) = \frac{1}{n^p}$, and consider the p-series $\sum_{n=1}^\infty \frac{1}{n^p}$.

To apply the Integral Test, we must first show that the function $f(x)$ is decreasing. Its derivative is $f'(x) = -p\frac{1}{x^{p+1}}$, which is negative for all $x > 0$, so $f(x)$ is decreasing and the integral test can be used. Since we have already considered the harmonic series, which is the p-series with $p = 1$, let us assume for the moment that $p \neq 1$.

We evaluate the improper integral:

$$\int_1^\infty f(x)\,dx = \lim_{r\to\infty}\left(\int_1^r f(x)\,dx\right) = \lim_{r\to\infty}\left(\int_1^r x^{-p}\,dx\right)$$

$$= \lim_{r\to\infty}\left(\frac{1}{1-p}x^{1-p}\right)\Big|_1^r = \frac{1}{1-p}\lim_{r\to\infty}(r^{1-p} - 1^{1-p}).$$

The important thing is what happens to r^{1-p} as $r \to \infty$, since nothing else depends on r. Of course, as $r \to \infty$, any positive power of it will tend to ∞, and any negative power of it will tend to 0.

We see that the integral converges if $1 - p < 0$, that is, if $p > 1$, and diverges if $1 - p > 0$, that is, if $p < 1$. We also know from Theorem 10.3.3 that the series diverges if $p = 1$.

In summary, the ***p*-series** $\sum_{n=1}^\infty \frac{1}{n^p}$ *converges* if $p > 1$ and *diverges* if $p \leq 1$.

EXAMPLE 3 (i) Does the series $\sum_{n=1}^{\infty} \frac{1}{\sqrt{n}}$ converge?

Solution This is a p-series with $p = \frac{1}{2}$, so it diverges.

(ii) Test the series $\sum_{n=4}^{\infty} (3n+6)^{-3}$ for convergence.

Solution This does not appear to be a p-series, but it can be rewritten as

$$\sum_{n=4}^{\infty} 3^{-3}(n+2)^{-3} = \frac{1}{27}\sum_{n=4}^{\infty}(n+2)^{-3} = \frac{1}{27}\sum_{m=6}^{\infty} m^{-3}.$$

The series in this last expression is nearly the p-series with $p = 3$; it is that series without its first five terms. The p-series with $p = 3$ converges, and by Theorem 10.3.4 we know that the same will be true of the series without its first five terms. The series converges.

(iii) Test the following series for convergence: $\sum_{n=1}^{\infty} \left(\frac{1}{n^2}\right)^{3/4}$.

Solution There is a temptation to look at the exponent $\frac{3}{4}$ and conclude that since it is less than 1, the series diverges, but upon closer inspection we realize that the terms of the series are $\left(\frac{1}{n^2}\right)^{3/4} = \left(n^{-2}\right)^{3/4} = n^{-2(3/4)} = n^{-3/2}$.

From this we see that the series is actually the p-series with $p = \frac{3}{2}$, so it converges.

Approximating the Value of a Series

The Integral Test also allows us to approximate the value of a series.

EXAMPLE 4 Corresponding to the p-series $\sum_{n=1}^{\infty} \frac{1}{n^p}$ we have evaluated the integral $\int_1^{\infty} \frac{1}{x^p}\,dx = \frac{1}{1-p}\lim_{r\to\infty}(r^{1-p} - 1^{1-p})$, and when $p > 1$, this equals $\frac{1}{p-1}$. Since the first term of a p-series is 1, Theorem 10.4.1 tells us that

$$\frac{1}{p-1} \le \sum_{n=1}^{\infty} \frac{1}{n^p} \le \frac{1}{p-1} + 1 = \frac{p}{p-1}.$$

For instance, with $p = \frac{11}{10}$, we see that

$$10 \le \sum_{n=1}^{\infty} \frac{1}{n^{11/10}} \le 11.$$

However, $\frac{1}{n^{1.1}}$ is close to $\frac{1}{n}$, and we know that while $\sum_{n=1}^{\infty} \frac{1}{n}$ diverges, it grows very slowly. From this we expect that $\sum_{n=1}^{\infty} n^{-11/10}$ will grow quite slowly. Using a calculator, we could add up the first few terms, and we find that

$$\sum_{n=1}^{10} n^{-11/10} \approx 2.680\,16, \qquad \sum_{n=1}^{50} n^{-11/10} \approx 3.828\,75, \qquad \sum_{n=1}^{100} n^{-11/10} \approx 4.278\,02.$$

To get anywhere near the actual value, for example, to get above 10, would take a very large number of terms. The estimate given by the integral is better than what we could find by adding up any reasonable number of terms.

With a programmable calculator we can evaluate $\sum_{n=1}^{m} n^{-11/10}$ for various values of m.

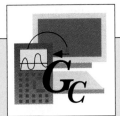

TI-85

PRGM F2(EDIT) PSER ENTER F3 F1(Input) M EXIT ENTER 0 STO N ENTER 0 STO S ENTER Lbl B ENTER N + 1 STO N ENTER S + N ^ ((−) 1.1) STO S ENTER F4 F1 (If) (N TEST F2(<) M) ENTER Goto B ENTER S ENTER EXIT EXIT EXIT PRGM F1(NAME) PSER ENTER 10 ENTER

SHARP EL9300

▼ ▼ (NEW) ENTER ENTER PSER ENTER COMMAND A 3 (Input) *M* ENTER *N* = 0 ENTER *S* = 0 ENTER COMMAND B 1 (Label) 1 ENTER *N* = *N* + 1 ENTER *S* = *S* + *N* a^b (−) 1.1 ENTER COMMAND 3 (If) *N* COMMAND C 2 (<) *M* COMMAND B 2 (Goto) 1 ENTER COMMAND A 1 (Print) *S* ENTER COMMAND A 6 (End) ENTER (RUN) ENTER (PSER) ENTER 10 ENTER

CASIO f_x-6300*G* (f_x-7700*GB*)

MODE 2 EXE ? → M : 0 → N : 0 → S : Lbl 1 : N + 1 → N : S + N x^y (−) 1.1 → S : N < M ⇒ Goto 1: S ◢ MODE 1 Prog 0 EXE 10 EXE

(on the f_x-7700*GB*, : is PRGM F6, and Prog is PRGM F3.)

HP 48SX

ATTN ↱ CLR « » "" KEY IN M ▶ SPC PROMPT SPC ′ *M* ▶ STO 0 SPC 1 SPC *M* SPC FOR *N* *N* SPC 1.1 +/− y^x + NEXT ▼ ENTER ′ PSER STO VAR PSER 10 ENTER ↰ CONT

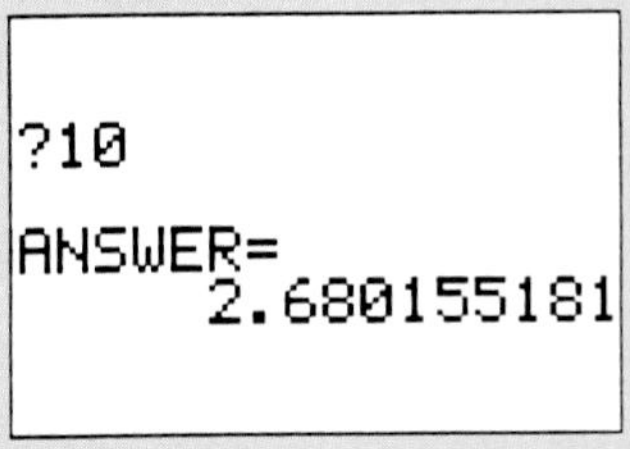

Now use the program with other values of n.

However, the estimate can still be improved considerably.

EXAMPLE 5 We continue with the p-series $\sum_{n=1}^{\infty} n^{-11/10}$. The error in the integral estimate comes from the size of a_1, since the answer lies between $\int_1^\infty f(x)\,dx$ and $\int_1^\infty f(x)\,dx + a_1$. Having figured out the sum $\sum_{n=1}^{10} n^{-11/10} \approx 2.680\,16$, we have to estimate what is left, that is, $\sum_{n=11}^{\infty} n^{-11/10}$. If we apply the Integral Test here, the error will depend on the first term $a_{11} = 11^{-11/10} \approx 0.071\,53$. We find that

$$\int_{11}^{\infty} x^{-11/10}\,dx \le \sum_{n=11}^{\infty} n^{-11/10} \le \int_{11}^{\infty} x^{-11/10}\,dx + a_{11}.$$

Now

$$\int_{11}^{\infty} x^{-11/10}\,dx = \lim_{r\to\infty}\int_{11}^{r} x^{-11/10}\,dx = \lim_{r\to\infty}\left(-10x^{-1/10}\right)\Big|_{11}^{r}$$
$$= 10(11)^{-1/10} \approx 7.867\,93.$$

So $\sum_{n=11}^{\infty} n^{-11/10}$ lies between 7.867 93 and $7.867\,93 + a_{11} \approx 7.939\,46$.

To get the value of the p-series, we add in the first ten terms: $\sum_{n=1}^{\infty} n^{-11/10} = \sum_{n=1}^{10} n^{-11/10} + \sum_{n=11}^{\infty} n^{-11/10}$. Since $\sum_{n=1}^{10} n^{-11/10} \approx 2.680\,16$, we find

$$10.54809 \leq \sum_{n=1}^{\infty} n^{-11/10} \leq 10.619\,62.$$

EXAMPLE 6 Estimate $\sum_{n=1}^{\infty} \frac{1}{n^2}$ to within 0.01.

Solution We want to add up $\sum_{n=1}^{K} \frac{1}{n^2}$ for some K and estimate the "tail" $\sum_{n=K+1}^{\infty} \frac{1}{n^2}$. The error that we find using the Integral Test depends on the first term in the tail, which is $\frac{1}{(K+1)^2}$. To make this ≤ 0.01, we need $(K+1)^2 \geq 100$, that is, $K \geq 9$. Adding up the first nine terms gives

$$\sum_{n=1}^{9} \frac{1}{n^2} \approx 1.539\,77.$$

The integral corresponding to the tail is

$$\int_{10}^{\infty} \frac{1}{x^2}\,dx = \lim_{r\to\infty}\left(\int_{10}^{r} \frac{1}{x^2}\,dx\right) = \lim_{r\to\infty}\left(-\frac{1}{x}\right)\Big|_{10}^{r} = \frac{1}{10}.$$

The Integral Test says that

$$\frac{1}{10} \leq \sum_{n=10}^{\infty} \frac{1}{n^2} \leq \frac{1}{10} + a_{10} = \frac{1}{10} + \frac{1}{100} = 0.11.$$

The original p-series $\sum_{n=1}^{\infty} \frac{1}{n^2} = \sum_{n=1}^{9} \frac{1}{n^2} + \sum_{n=10}^{\infty} \frac{1}{n^2}$ lies between $1.539\,77 + \frac{1}{10}$ and $1.539\,77 + 0.11$, that is,

$$1.639\,77 \leq \sum_{n=1}^{\infty} \frac{1}{n^2} \leq 1.649\,77.$$

In fact, though it is *not* obvious, $\sum_{n=1}^{\infty} \frac{1}{n^2} = \frac{\pi^2}{6}$.

We found that sufficient accuracy would be achieved by summing up to $K = 9$. This does *not* mean that $\sum_{n=1}^{9} \frac{1}{n^2}$ is the correct approximation, but rather that $\sum_{n=1}^{9} \frac{1}{n^2} + \int_{10}^{\infty} \frac{1}{x^2}\,dx$ is the correct approximation. It is accurate to within an error of a_{10}. In general, if $f(x)$ is a nonnegative decreasing function and $\sum_{n=1}^{\infty} a_n$ converges, where $a_n = f(n)$, then the value of the series is estimated by

$$\sum_{n=1}^{K} a_n + \int_{K+1}^{\infty} f(x)\,dx, \tag{10.4.2}$$

accurate to within a_{K+1}.

There is one final remark that should be made about the Integral Test. It is very important to check that the function $f(x)$ is decreasing, but it will sometimes happen that it increases for a little while and *then* decreases after some point $x = K$.

In this situation it is still possible to use the Integral Test because if we ignore the first few terms (the ones for $n < K$), then the rest of the series is obtained as the values at integer points of a decreasing function, and of course the convergence or divergence of a series is unaffected by changing the first few terms. What is important is that *after some point* K the function f must be decreasing.

EXAMPLE 7 Consider the series $\sum_{n=1}^{\infty} ne^{-n/12}$, and test it for convergence.

Solution If we let $f(x) = xe^{-x/12}$, then its derivative is $f'(x) = e^{-x/12} + xe^{-x/12} \times (-\frac{1}{12}) = e^{-x/12}(1 - \frac{x}{12})$. This derivative is negative for $x > 12$ but positive for $x < 12$.

We conclude that the function $f(x)$ is increasing for $x < 12$ and decreasing for $x > 12$. We can consider the series that begins with the twelfth term; it is constructed from the values at integer points of a decreasing function, so we can apply the Integral Test. The thing to do is to consider the improper integral $\int_{12}^{\infty} f(x)\,dx = \int_{12}^{\infty} xe^{-x/12}\,dx$, which can be integrated by parts. Letting $u = x$ and $dv = e^{-x/12}dx$, we find that $v = \int e^{-x/12}dx = -12e^{-x/12}$, so

$$\begin{aligned}
\int_{12}^{\infty} f(x)\,dx &= \lim_{r\to\infty}\left(\int_{12}^{r} xe^{-x/12}dx\right) = \lim_{r\to\infty}\left(\int u\,dv\right) \\
&= \lim_{r\to\infty}\left(uv - \int v\,du\right)\Big|_{12}^{r} = \lim_{r\to\infty}\left(-12xe^{-x/12} + 12\int e^{-x/12}dx\right)\Big|_{12}^{r} \\
&= \lim_{r\to\infty}\left(-12xe^{-x/12} + 12(-12)e^{-x/12}\right)\Big|_{12}^{r} \\
&= \lim_{r\to\infty}\left(-12re^{-r/12} - 144e^{-r/12} + 12(12)e^{-1} + 144e^{-1}\right).
\end{aligned}$$

We have to find what happens as $r \to \infty$, and the terms that depend on r are $-12re^{-r/12}$ and $-144e^{-r/12}$. The second of these is easy to deal with because as $r \to \infty$, we know that $-\frac{r}{12} \to -\infty$, so $e^{-r/12} \to 0$.

We write the other term as $-12re^{-r/12} = -12\frac{r}{e^{r/12}}$ and observe that as $r \to \infty$, both the numerator and the denominator of this quotient tend to ∞. This cries out for l'Hôpital's Rule. We find that

$$\lim_{r\to\infty} -12\frac{r}{e^{r/12}} = -12\lim_{r\to\infty}\frac{1}{\frac{1}{12}e^{r/12}} = -144\lim_{r\to\infty} e^{-r/12} = 0.$$

We find that the improper integral converges to a finite limit, which means that the series starting at the twelfth term converges. Consequently, the whole series converges.

Exercises 10.4

I

Decide whether or not each of the following series converges.

1. $\sum_{n=1}^{\infty} \frac{1}{n+5}$
2. $\sum_{n=1}^{\infty} \frac{1}{3n+5}$
3. $\sum_{n=1}^{\infty} \left(\frac{1}{n+7}\right)^4$
4. $\sum_{n=1}^{\infty} \frac{1}{(2n+3)^3}$
5. $\sum_{n=4}^{\infty} \frac{1}{\sqrt{n+7}}$
6. $\sum_{n=0}^{\infty} \frac{1}{\sqrt{(2n+3)^5}}$
7. $\sum_{n=1}^{\infty} \frac{1}{n^2+6n+9}$
8. $\sum_{n=1}^{\infty} \frac{1}{(3n-2)^3}$
9. $\sum_{n=4}^{\infty} ne^{-n^2}$
10. $\sum_{n=1}^{\infty} 4^{-n}$

Test the following series for convergence.

11. $\sum_{n=1}^{\infty} n^2e^{-n^3}$
12. $\sum_{n=1}^{\infty} \frac{1}{1+n^2}$
13. $\sum_{n=1}^{\infty} (n^2 + 2n + 2)^{-1}$
14. $\sum_{n=1}^{\infty} \frac{1}{n^2+4n+8}$

15. $\sum_{n=2}^{\infty} \ln(n)$

16. $\sum_{n=2}^{\infty} \left(n \ln^2(n)\right)^{-1}$

17. $\sum_{n=1}^{\infty} \frac{\ln(n)}{n^2}$

18. $\sum_{n=2}^{\infty} \frac{1}{n \ln(n^3)}$

19. $\sum_{n=1}^{\infty} \frac{1}{n^2+3}$

20. $\sum_{n=2}^{\infty} \left(n \ln^{5/6}(n)\right)^{-1}$

II

***21.** Find an example of a bounded sequence that does not converge.

***22.** Find an example of an increasing sequence that does not converge.

***23.** Test the series $\sum_{n=2}^{\infty} \frac{1}{n^2-1}$ for convergence. (*Hint:* After integrating, use properties of the natural logarithm.)

Does each of the following series converge?

***24.** $\sum_{n=3}^{\infty} \frac{1}{n^2-n-2}$

***25.** $\sum_{n=3}^{\infty} \frac{1}{n^2-3n+2}$

***26.** $\sum_{n=5}^{\infty} \frac{1}{n^2-n-12}$

***27.** $\sum_{n=3}^{\infty} \frac{1}{2n^2-6n+4}$

****28.** Suppose a, b, and c are any numbers with $a > 0$. Choose an integer K so that $ax^2 + bx + c$ is never zero for $x \geq K$. Show that the following series converges: $\sum_{n=K}^{\infty} \frac{1}{an^2+bn+c}$.

***29.** For what values of q does the following series converge: $\sum_{n=2}^{\infty} \frac{\ln(n)}{n^q}$?

***30.** Test the following series for convergence: $\sum_{n=1}^{\infty} \cosh^{-2}(n)$.

***31.** Test the following series for convergence: $\sum_{n=1}^{\infty} \cosh(n)$.

32. In Example 4 (p. 670) we mentioned that $\sum_{n=1}^{50} n^{-11/10} \approx 3.82875$. What estimate can you get from this for $\sum_{n=1}^{\infty} n^{-11/10}$?

33. In Example 4 we mentioned that $\sum_{n=1}^{100} n^{-11/10} \approx 4.27802$. What estimate can you get from this for $\sum_{n=1}^{\infty} n^{-11/10}$? Is your answer to this question consistent with what you found for Exercise 32?

34. For the series $\sum_{n=1}^{\infty} ne^{-n/12}$ of Example 7 (p. 673), find the value of K in Equation 10.4.2 for which the approximation is accurate within (i) 0.1, (ii) 0.001. (Do not evaluate.)

35. Estimate $\sum_{n=1}^{\infty} \frac{1}{n^4}$ to within 0.001.

36. How good is the approximation of $\sum_{n=1}^{\infty} \frac{1}{n^6}$ by Equation 10.4.2 with $K = 10$? Evaluate the approximation.

III

37. Use a programmable calculator or computer to confirm the values given in Example 4 for the partial sums $\sum_{n=1}^{N} n^{-11/10}$ with $N = 10, 50, 100$.

38. Estimate $\sum_{n=1}^{\infty} n^{-11/10}$ to within 0.001 (cf. Example 4) and Formula 10.4.2.

39. Estimate $\sum_{n=1}^{\infty} ne^{-n/12}$ to within 0.001 (cf. Example 7).

40. Estimate $\sum_{n=1}^{\infty} n^{-3/2}$ to within 0.001.

POINT TO PONDER

The Integral Test applies to a series $\sum_{n=1}^{\infty} a_n$ that can be expressed by $a_n = f(n)$, for some decreasing function $f(x)$. It is not always easy to find such a function, even apart from deciding whether it is decreasing. For instance, for the series $\sum_{n=1}^{\infty} (-1)^n \frac{1}{n^2}$ we cannot just change the n to x because $(-1)^x$ does not make sense unless x is an integer or a rational number with odd denominator. (For other x's the power involves ln, which is not defined for -1.) On the other hand, the function $\cos(\pi x)$ equals $(-1)^n$ when $x = n$ is an integer (why?), so we could let $f(x) = \cos(\pi x)\frac{1}{x^2}$. It is not decreasing; in fact the Integral Test cannot possibly apply because the a_n's are not all positive.

For the series $\sum_{n=1}^{\infty} \frac{1}{n!}$ we do not know a function $f(x)$ whose value at $x = n$ is $\frac{1}{n!}$ ($n!$ is only defined when n is an integer). There is no obvious way to use the Integral Test, even though the terms are positive and decrease to zero.

10.5 The Comparison Test

NONNEGATIVE SERIES
COMPARISON TEST
RATIONAL FUNCTIONS

In this section we learn a useful test for convergence of series whose terms are all nonnegative.

THEOREM 10.5.1

The Comparison Test

Suppose $\sum_{n=1}^{\infty} a_n$ and $\sum_{n=1}^{\infty} b_n$ are two series whose terms are all nonnegative. Suppose that $\sum_{n=1}^{\infty} a_n$ converges and that each term b_n is less than or equal to the corresponding term a_n, that is, $0 \le b_n \le a_n$. Then $\sum_{n=1}^{\infty} b_n$ converges.

PROOF To say that the series $\sum_{n=1}^{\infty} a_n$ converges means that its partial sums are bounded. On the other hand, each term b_n is less than or equal to the corresponding a_n, so the partial sums $T_N = \sum_{n=1}^{N} b_n$ must be less than or equal to the corresponding partial sums $S_N = \sum_{n=1}^{N} a_n$, that is, $T_N = \sum_{n=1}^{N} b_n \le \sum_{n=1}^{N} a_n = S_N$. In particular, if the S_N's are all bounded by some number M, then so are the T_N's.

This means that $\sum_{n=1}^{\infty} b_n$ converges, as required.

EXAMPLE 1

(i) Consider the series $\sum_{n=1}^{\infty} \frac{1}{n^4+1}$. Since $n^4 + 1 > n^4$, we can take reciprocals and find that $\frac{1}{n^4+1} < \frac{1}{n^4}$.

The series $\sum_{n=1}^{\infty} \frac{1}{n^4}$ is a p-series with $p = 4$, so it is convergent, and the Comparison Test shows that the "smaller" series $\sum_{n=1}^{\infty} \frac{1}{n^4+1}$ is also convergent. This is much easier than trying to apply the Integral Test.

(ii) Consider the series $\sum_{n=1}^{\infty} \frac{1}{e^n+2n+4}$. We note that

$$\frac{1}{e^n + 2n + 4} < \frac{1}{e^n},$$

for each $n \ge 1$. The series $\sum_{n=1}^{\infty} \frac{1}{e^n}$ is a geometric series with common ratio $\frac{1}{e}$. Since this ratio is less than 1, the geometric series converges, and the Comparison Test then tells us that the series $\sum_{n=1}^{\infty} \frac{1}{e^n+2n+4}$ converges.

(iii) Let us consider the series $\sum_{n=2}^{\infty} \frac{1}{\sqrt{n}-1}$. It appears reasonable to try to compare its terms with $\frac{1}{\sqrt{n}}$, and we find that $\frac{1}{\sqrt{n}-1} > \frac{1}{\sqrt{n}}$.

$\sqrt{n} - 1 < \sqrt{n}$, and taking reciprocals reverses the inequality.

We know that $\sum_{n=2}^{\infty} \frac{1}{\sqrt{n}}$ is a p-series with $p = \frac{1}{2}$, so it diverges. The series with which we began is greater than this divergent series, so it diverges too. After all, if it were *convergent*, then the Comparison Test would tell us that the smaller series $\sum_{n=2}^{\infty} \frac{1}{\sqrt{n}}$ would have to converge, and we know that it does not.

We conclude that the series $\sum_{n=2}^{\infty} \frac{1}{\sqrt{n}-1}$ diverges.

There are several things to notice in Example 1. The first is that while we might have managed to use the Integral Test for part (i), it would certainly have been more work. In part (ii) it seems that the Integral Test would be very difficult to use. Finally, in part (iii) we used the Comparison Test "backwards." We showed that the terms of the series in question are greater than the terms of a divergent series and concluded that it must also be divergent.

The examples we have considered so far have been constructed deliberately so that they work out well, but of course this will not always happen.

EXAMPLE 2 Suppose we look at the series $\sum_{n=2}^{\infty} \frac{1}{n^4-1}$. It is quite similar to part (i) of Example 1, but its terms are *not* less than or equal to the terms of the convergent series $\sum_{n=2}^{\infty} \frac{1}{n^4}$. They are all slightly *greater* than the corresponding terms.

The key is that word "slightly." The term $\frac{1}{n^4-1}$ is greater than $\frac{1}{n^4}$, but it is less than twice as much. Accepting this for the moment, we can use the Comparison Test because $\sum_{n=2}^{\infty} \frac{2}{n^4}$ is convergent (it is 2 times a convergent p-series without the first term). The terms of the series in which we are interested are less than the terms of this convergent series, so it converges.

What remains is the problem of actually showing that $\frac{1}{n^4-1} < \frac{2}{n^4}$, for $n \geq 2$. To work with an inequality like this, it is often easiest to work with the reciprocals, remembering that it is necessary to reverse the inequality.

So we want to show that $n^4 - 1 > \frac{n^4}{2}$, that is, $2(n^4 - 1) > n^4$. This amounts to showing that $2n^4 - 2 > n^4$, and subtracting n^4 from each side shows that this is the same as $n^4 - 2 > 0$ or $n^4 > 2$. This is certainly true if $n \geq 2$, so we have finished.

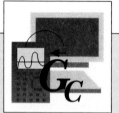

With a graphics calculator we can plot the graphs of the functions $\frac{1}{x^4-1}$, $\frac{1}{x^4}$, $\frac{2}{x^4}$, which are related to the terms of the series in Example 2.

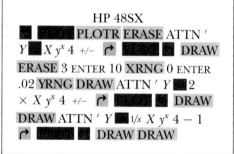

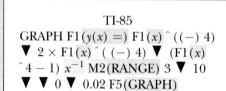

TI-85	SHARP EL9300
GRAPH F1 $(y(x) =)$ F1 (x) ^ ((−) 4) ▼ 2 × F1(x) ^ ((−) 4) ▼ (F1(x) ^4 − 1) x^{-1} M2(RANGE) 3 ▼ 10 ▼ ▼ 0 ▼ 0.02 F5(GRAPH)	x/θ/T a^b (−) 4 2ndF ▼ 2 × x/θ/T a^b (−) 4 2ndF ▼ (x/θ/T a^b 4 ► − 1) x^{-1} RANGE 3 ENTER 10 ENTER 1 ENTER 0 ENTER .02 ENTER .01 ENTER
CASIO f_x-7700GB/f_x-6300G	HP 48SX
Cls EXE Range 3 EXE 10 EXE 1 EXE 0 EXE .02 EXE .01 EXE EXE EXE EXE Graph [redacted] x^y (−) 4 EXE Graph 2 × [redacted] x^y (−) 4 EXE Graph ([redacted] x^y 4 − 1) x^{-1} EXE	[redacted] [redacted] PLOTR ERASE ATTN ′ Y [redacted] X y^x 4 +/− ↱ [redacted] [redacted] DRAW ERASE 3 ENTER 10 XRNG 0 ENTER .02 YRNG DRAW ATTN ′ Y [redacted] 2 × X y^x 4 +/− ↱ [redacted] [redacted] DRAW DRAW ATTN ′ Y [redacted] 1/x X y^x 4 − 1 ↱ [redacted] [redacted] DRAW DRAW

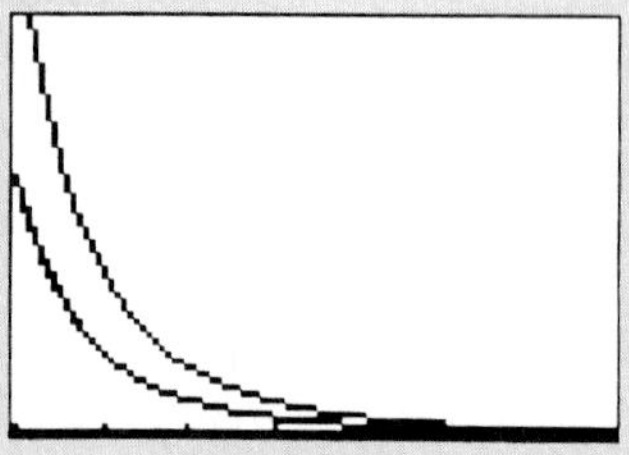

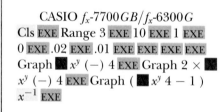

You may have to look closely to identify the three graphs. Check that $\frac{1}{n^4} < \frac{1}{n^4-1} < \frac{2}{n^4}$.

It might be helpful to write down this last argument in the reverse order, now that we have found what to do by working backwards. If $n \geq 2$, then certainly $n^4 > 2$. From this we see that $n^4 - 2 > 0$, and $2n^4 - 2 > n^4$. This can be written as $2(n^4 - 1) > n^4$, or $n^4 - 1 > \frac{n^4}{2}$. Taking reciprocals and reversing the inequality, we find that $\frac{1}{n^4-1} < \frac{2}{n^4}$, when $n \geq 2$.

Combined with the argument we gave above, this shows that $\sum_{n=2}^{\infty} \frac{1}{n^4-1}$ is convergent.

In this example we saw that it is not essential that the terms of the series be less than the terms of a convergent series in order for us to be able to apply the Comparison Test. It is just as good if they are less than some constant times the terms of the convergent series. This is an easy observation, but it is very useful.

A similar observation can be made based on the fact that the convergence of a series is not affected by changing the first few terms. So it will also be possible to use the Comparison Test in situations in which the terms of one series are less than a constant times the terms of another *after the first few terms.* This is also quite obvious, but it will often make the Comparison Test easier to use.

We record these minor modifications to the Comparison Test as a theorem.

THEOREM 10.5.2

The Improved Comparison Test

(i) If $\sum_{n=1}^{\infty} a_n$ is a convergent series and $C > 0$ is a constant so that $0 \leq b_n \leq Ca_n$ for every n greater than some number K, then $\sum_{n=1}^{\infty} b_n$ is convergent.

(ii) If $\sum_{n=1}^{\infty} a_n$ is a divergent series and $C > 0$ is a constant so that $0 \leq Ca_n \leq b_n$ for every n greater than some number K, then $\sum_{n=1}^{\infty} b_n$ is divergent.

Notice that Theorem 10.5.2 applies to series whose terms are all nonnegative.

EXAMPLE 3 Consider the series $\sum_{n=1}^{\infty} a_n$, where $a_n = \frac{n^4-3n^3+7n^2+1}{2n^5+3n^4-n^2+2n-5}$. We should ask what happens to a_n as $n \to \infty$, and from our work with limits of rational functions we know that the most important things are the terms of highest degree in the numerator and denominator. The idea is that as n becomes very large, the other terms will be so much smaller by comparison that they will not have much effect.

In this example this suggests that as $n \to \infty$, the terms of the series will be close to $\frac{n^4}{2n^5} = \frac{1}{2n}$. The series with these terms is divergent, so it looks as if the series we are considering is probably divergent too.

To make this idea work, we use a trick that we used with rational functions. We write

$$a_n = \frac{n^4 - 3n^3 + 7n^2 + 1}{2n^5 + 3n^4 - n^2 + 2n - 5} = \frac{n^4\left(1 - \frac{3}{n} + \frac{7}{n^2} + \frac{1}{n^4}\right)}{2n^5\left(1 + \frac{3}{2n} - \frac{1}{2n^3} + \frac{1}{n^4} - \frac{5}{2n^5}\right)}$$

$$= \frac{n^4}{2n^5} \frac{1 - \frac{3}{n} + \frac{7}{n^2} + \frac{1}{n^4}}{1 + \frac{3}{2n} - \frac{1}{2n^3} + \frac{1}{n^4} - \frac{5}{2n^5}}.$$

The last quotient tends to 1 as $n \to \infty$, and in particular it must eventually be greater than $\frac{1}{2}$. In other words, there is a number K so that that quotient is greater than $\frac{1}{2}$ whenever $n > K$.

This means that whenever $n > K$, $a_n > \frac{1}{2}\frac{1}{2n}$, and part (ii) of the improved version of the Comparison Test tells us that $\sum_{n=1}^{\infty} a_n$ diverges.

In a way, the idea in Example 3 is to compare the convergence or divergence of $\sum_{n=1}^{\infty} a_n$ with that of $\sum_{n=1}^{\infty} \frac{1}{2n}$, since the terms a_n are eventually close to $\frac{1}{2n}$. We could express this condition by considering the quotient

$$\frac{a_n}{\frac{1}{2n}} = \frac{n^4 - 3n^3 + 7n^2 + 1}{2n^5 + 3n^4 - n^2 + 2n - 5}\frac{2n}{1} = \frac{2n^5 - 6n^4 + 14n^3 + 2n}{2n^5 + 3n^4 - n^2 + 2n - 5}$$

and recognizing that as $n \to \infty$, the quotient tends to 1.

Even if the quotient tended to any other nonzero limit L, we should expect the convergence or divergence of the two series to be related.

THEOREM 10.5.3

The Limit Comparison Test

Suppose $\sum_{n=1}^{\infty} a_n$ and $\sum_{n=1}^{\infty} b_n$ are series with nonnegative terms. Suppose that

$$\lim_{n\to\infty} \frac{a_n}{b_n} = L,$$

for some nonzero number L. Then $\sum_{n=1}^{\infty} a_n$ and $\sum_{n=1}^{\infty} b_n$ *converge or diverge together*, that is, either they both converge or else they both diverge.

PROOF Suppose $\sum_{n=1}^{\infty} a_n$ converges. Since $\lim_{n\to\infty} \frac{a_n}{b_n} = L$, we know that there is some number K so that $\frac{a_n}{b_n}$ is greater than $\frac{L}{2}$ for every $n > K$. This amounts to $b_n < \frac{2}{L}a_n$, for $n > K$, and Theorem 10.5.2(i) shows that $\sum_{n=1}^{\infty} b_n$ is convergent.

Similarly, suppose $\sum_{n=1}^{\infty} a_n$ diverges. We can find K so that $\frac{a_n}{b_n} < 2L$ for every $n > K$. This says that $b_n > \frac{1}{2L}a_n$ for every $n > K$, and Theorem 10.5.2(ii) says that $\sum_{n=1}^{\infty} b_n$ diverges.

We can use Theorem 10.5.3 for any series whose terms are "a rational function of n," meaning that there is a rational function $R(x)$ so that the terms of the series are given by $a_n = R(n)$. To decide whether or not such a series converges, we look at the terms of highest degree in the numerator and denominator. Their quotient b_n will be a constant times some power of n, so the corresponding series $\sum_{n=1}^{\infty} b_n$ is a constant times a p-series, and we can immediately decide whether or not it converges.

An argument like the one in Example 3 shows that as $n \to \infty$, the quotient $\frac{a_n}{b_n}$ will tend to 1; the Limit Comparison Test then shows that both series converge or both diverge.

To decide whether or not the simpler series constructed from the highest-order terms is convergent or not, we observe that the p-series $\sum_{n=1}^{\infty} \frac{1}{n^d}$ will converge when $d > 1$. If d is an integer, this means that d must be 2 or greater. So if $R(x) = \frac{p(x)}{q(x)}$ is a rational function and $a_n = R(n)$, then the quotient of the highest-order terms of $p(n)$ and $q(n)$ is a constant times $\frac{1}{n^d}$, where d is the difference of the degrees of the highest-order terms of $q(x)$ and $p(x)$. The series converges when $d = \mathit{degree}\big(q(x)\big) - \mathit{degree}\big(p(x)\big) > 1$, that is, $d \geq 2$.

THEOREM 10.5.4

> If $R(x) = \frac{p(x)}{q(x)}$ is a **rational function** and $a_n = R(n)$, then the series $\sum_{n=k}^{\infty} a_n$ converges precisely when $degree(q(x)) > degree(p(x)) + 1$, that is, when the degree of the denominator is greater than the degree of the numerator by at least 2.

EXAMPLE 4

(i) The series $\sum_{n=2}^{\infty} \frac{n^2+3n-7}{n^5-2n^3+1}$ converges because the degree of the denominator is 5, which is 3 greater than the degree of the numerator.

(ii) The series $\sum_{n=3}^{\infty} \frac{5n-2}{n^2-4}$ diverges, since the degrees of its denominator and numerator differ by only 1. (Notice that this series starts at $n = 3$ because the denominator is zero when $n = 2$.)

> The Integral Test (Theorem 10.4.1) applies to series with decreasing positive terms, and the Comparison Test (Theorems 10.5.1, 10.5.2, 10.5.3) applies to series whose terms are nonnegative.

Exercises 10.5

Testing series for convergence or divergence is a skill that must be learned carefully. As with integration, it takes practice to develop a feel for what might be a suitable approach. It is extremely important to do a large number of exercises.

I

Give reasons why each of the following series is convergent or divergent.

1. $\sum_{n=3}^{\infty} \frac{1}{n+\sin(n\pi)}$
2. $\sum_{n=2}^{\infty} \frac{1}{n^2-2\cos(n\pi)}$
3. $\sum_{n=1}^{\infty} \frac{1}{4n^3+\ln n}$
4. $\sum_{n=1}^{\infty} \frac{1}{n^2\ln^3(n+1)}$
5. $\sum_{n=0}^{\infty} \left(e^n + \frac{n}{2}\right)$
6. $\sum_{n=2}^{\infty} \frac{1}{n^{3/2}-n}$
7. $\sum_{n=1}^{\infty} \frac{3}{2^n+4\ln(n)}$
8. $\sum_{n=-3}^{\infty} \frac{1}{7+\sqrt{n+3}}$
9. $\sum_{n=2}^{\infty} \frac{1}{2n^2\ln(n)}$
10. $\sum_{n=2}^{\infty} \frac{\ln(n)}{n^{3/2}}$
11. $\sum_{n=1}^{\infty} \frac{n^2-3n+4}{n^3-4n^2+n-12}$
12. $\sum_{n=2}^{\infty} \frac{3n^3+n^2-1}{2n^5-2}$
13. $\sum_{n=1}^{\infty} \frac{n^2+5n-2}{n^3\sqrt{n}+3n^3-2n+1}$
14. $\sum_{n=2}^{\infty} \frac{5n^3+2n^2-n+2}{n^5+2n^4-n^3-2}\sqrt{n}$
15. $\sum_{n=1}^{\infty} \frac{n^3-3n^2+n+3}{n^2-n+1}e^{-n/10}$
16. $\sum_{n=1}^{\infty} \frac{3n^3+n^2+4}{2n-3}$
17. $\sum_{n=2}^{\infty} \frac{n^2+2}{n^4-n^2+\cos(n\pi)}$
18. $\sum_{n=1}^{\infty} \frac{2n^3+4n^2+3n}{3n^4+3e^{-n}}$
19. $\sum_{n=1}^{\infty} \frac{4n+1}{n^3-2n^2+3}\left(2 - \sin(n\pi)\right)$
20. $\sum_{n=1}^{\infty} \frac{3n^2+6n+5e^n}{4n^5+3n^3+n-1}$

II

*21. Suppose $\sum_{n=1}^{\infty} a_n$ and $\sum_{n=1}^{\infty} b_n$ are series with nonnegative terms and $\sum_{n=1}^{\infty} a_n$ converges. Suppose too that $\lim_{n\to\infty} \frac{b_n}{a_n} = 0$. Show that $\sum_{n=1}^{\infty} b_n$ converges.

*22. Find an example in which $\lim_{n\to\infty} \frac{b_n}{a_n} = 0$, $\sum_{n=1}^{\infty} b_n$ converges and $\sum_{n=1}^{\infty} a_n$ diverges.

*23. Suppose $\sum_{n=1}^{\infty} a_n$ and $\sum_{n=1}^{\infty} b_n$ are series with nonnegative terms and $\sum_{n=1}^{\infty} a_n$ diverges. Suppose too that $\lim_{n\to\infty} \frac{b_n}{a_n} = \infty$. Show that $\sum_{n=1}^{\infty} b_n$ diverges.

***24.** Find an example in which $\lim_{n\to\infty} \frac{b_n}{a_n} = \infty$, $\sum_{n=1}^{\infty} b_n$ diverges and $\sum_{n=1}^{\infty} a_n$ converges.

Drill

Give reasons why each of the following series is convergent or divergent.

25. $\sum_{n=1}^{\infty} \frac{1}{2n^2+3n^{1/2}}$

26. $\sum_{n=2}^{\infty} \frac{7}{n\ln^3(n)}$

27. $\sum_{n=1}^{\infty} \frac{n}{e^{-3n}}$

28. $\sum_{n=1}^{\infty} \frac{1}{2n+3e^n}$

29. $\sum_{n=2}^{\infty} \frac{1}{\ln(n)+n\ln(n)}$

30. $\sum_{n=1}^{\infty} \csc(n^{-3})$

31. $\sum_{n=1}^{\infty} \frac{2}{ne^{6n}}$

32. $\sum_{n=1}^{\infty} \frac{5+n}{n^{5/2}\ln(n)+3}$

33. $\sum_{n=0}^{\infty} \left(\frac{1}{n+1} - \frac{1}{n+2}\right)$

34. $\sum_{n=0}^{\infty} \frac{\cos^2(n+\pi)}{n^2+e^{-2n}}$

III

35. (i) Use a programmable calculator or computer to evaluate $\sum_{n=2}^{100} \frac{1}{n^2\ln(n)}$. (ii) Estimate the accuracy of this approximation to $\sum_{n=2}^{\infty} \frac{1}{n^2\ln(n)}$ by considering the tail $\sum_{n=101}^{\infty} \frac{1}{n^2\ln(n)}$. Since $\frac{1}{n^2\ln(n)} < \frac{1}{n^2}$, argue that $\sum_{n=101}^{\infty} \frac{1}{n^2\ln(n)} < \sum_{n=101}^{\infty} \frac{1}{n^2}$ and estimate this last series, using the Integral Test.

36. Evaluate $\sum_{n=2}^{250} \frac{1}{n^2\ln(n)}$. How accurate is this approximation to $\sum_{n=2}^{\infty} \frac{1}{n^2\ln(n)}$?

Estimate each of the following series to within the specified accuracy.

37. $\sum_{n=1}^{\infty} \frac{1}{n^3\ln(n+1)}$ to within 0.0001

38. $\sum_{n=1}^{\infty} \frac{1}{n^2+5\ln(n)}$ to within 0.001

39. $\sum_{n=1}^{\infty} \frac{1}{n^2+e^n}$ to within 0.0001

40. $\sum_{n=1}^{\infty} \frac{1}{n^{5/2}+3}$ to within 0.0005

10.6 Alternating Series and Absolute Convergence

ALTERNATING SERIES
CONDITIONAL CONVERGENCE

In Sections 10.4 and 10.5 we considered series whose terms were all nonnegative. This was necessary for a result like the Comparison Test, as the following example shows. If we let $a_n = -\frac{1}{n}$, then $\sum_{n=1}^{\infty} a_n$ is the negative of the harmonic series, so it diverges. However, it is "less than" the convergent series $\sum_{n=1}^{\infty} \frac{1}{n^2}$ (in the sense that each term $a_n = -\frac{1}{n}$ is less than the corresponding term $\frac{1}{n^2}$). In fact it is less than *any* series whose terms are nonnegative.

Of course, this is no contradiction, since each a_n is negative and the Comparison Test does not apply. It also demonstrates that it is quite easy to deal with series whose terms are all negative; we simply consider the negative of such a series. The terms of *this* series will be all positive, and the Comparison and Integral Tests can be used for it. In this way we can often find whether or not the original negative series converges.

The situation is more complicated when some of the terms of a series are positive and some are negative. In this section we consider a very special case in which the terms of the series are *alternately* positive and negative.

$$S_1 = -1$$
$$S_2 = -1 + \frac{1}{2} = -\frac{1}{2}$$
$$S_3 = S_2 - \frac{1}{3} = -\frac{5}{6}$$
$$S_4 = S_3 + \frac{1}{4} = -\frac{7}{12}$$
$$S_5 = S_4 - \frac{1}{5} = -\frac{47}{60}$$
$$S_6 = S_5 + \frac{1}{6} = -\frac{37}{60}$$

EXAMPLE 1 Consider the series $\sum_{n=1}^{\infty} (-1)^n \frac{1}{n}$, which is obtained by inserting negative signs on the odd-numbered terms of the harmonic series.

We begin calculating partial sums at left. Since the terms of the series are alternately negative and positive, the partial sums will move alternately up and down. The amounts by which they rise and fall will be $\pm\frac{1}{n}$, and the important thing to notice is that these amounts get smaller and smaller, tending to zero.

The odd-numbered partial sums $S_1, S_3, S_5, \ldots$ are the "low" values, since the even-numbered sums are obtained from them by adding *positive* numbers. If we think about the odd-numbered sums by themselves for a while, we see that they are increasing. The reason is that to get, say, from S_1 to S_3, we add $a_2 + a_3 = \frac{1}{2} - \frac{1}{3}$, and the negative term $-\frac{1}{3}$ has a smaller absolute value than the positive term $\frac{1}{2}$, so the net effect is to add something positive (in this case $\frac{1}{2} - \frac{1}{3} = \frac{1}{6}$). Similarly, to get from S_3 to S_5, we add $a_4 + a_5 = \frac{1}{4} - \frac{1}{5}$, and once again the positive term $\frac{1}{4}$ outweighs the negative term $-\frac{1}{5}$. The net effect is to add a positive number to S_3.

In exactly the same way, to get from S_{2N+1} to the next odd-numbered partial sum S_{2N+3}, we add $\frac{1}{2N+2} - \frac{1}{2N+3} = \frac{(2N+3)-(2N+2)}{(2N+2)(2N+3)} = \frac{1}{(2N+2)(2N+3)}$, which is positive. This shows that the odd-numbered partial sums are increasing. See Figure 10.6.1.

Almost exactly the same argument shows that the even-numbered partial sums S_{2n} are *decreasing*, since to get from S_{2N} to S_{2N+2}, we add $a_{2N+1} + a_{2N+2} = -\frac{1}{2N+1} + \frac{1}{2N+2} = -\frac{1}{(2n+1)(2N+2)}$, which is negative.

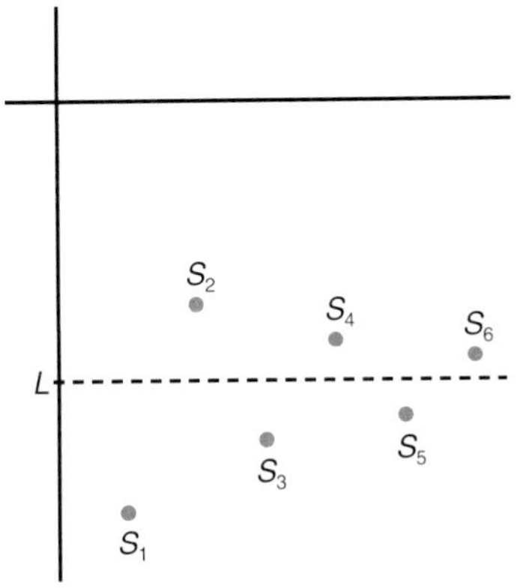

Figure 10.6.1

The sequence of odd-numbered partial sums is increasing, and in fact it is *bounded*. To see this, we notice that each odd-numbered partial sum S_{2N+1} is less than the preceding even-numbered sum S_{2N}, since $S_{2N+1} = S_{2N} - \frac{1}{2N+1}$. But the even-numbered sums are decreasing, so S_{2N} must always be less than S_2. Combining these two observations, we see that every odd-numbered partial sum S_{2N+1} is less than S_2.

This shows that the odd-numbered partial sums are a bounded increasing sequence, so they must converge to some limit L, shown as the dotted line in Figure 10.6.1.

Similarly, the even-numbered partial sums are a decreasing sequence. They approach closer and closer to the odd-numbered sums (after all, $S_{2N} = S_{2N-1} + \frac{1}{2N}$, and $\frac{1}{2N} \to 0$), so they must approach the same limit L.

This means that the sequence of all the partial sums $\{S_N\}$ converges to L. We have found the somewhat surprising result that if we put negative signs on alternate terms of the harmonic series, then the resulting series will be *convergent*.

In fact there was nothing so very special about the harmonic series. To make the argument work, all we really needed to know was that the terms were alternately positive and negative and that the absolute values of the terms decreased to zero.

DEFINITION 10.6.1

A series $\sum_{n=1}^{\infty} a_n$ is an **alternating series** if its terms are alternately positive and negative. Such a series can be written in one of the following two forms:

$$\sum_{n=1}^{\infty} (-1)^n |a_n| \quad \text{or} \quad \sum_{n=1}^{\infty} (-1)^{n+1} |a_n|,$$

depending on whether the first term is negative or positive. These can also be written as

$$\sum_{n=1}^{\infty} (-1)^n b_n \quad \text{or} \quad \sum_{n=1}^{\infty} (-1)^{n+1} b_n \qquad \text{with } b_n > 0.$$

THEOREM 10.6.2

Suppose $\sum_{n=1}^{\infty} a_n$ is an alternating series, and suppose the absolute values of its terms *decrease* to zero, that is, $|a_n| \geq |a_{n+1}|$ and $\lim_{n\to\infty} |a_n| = 0$. Then the series is convergent.

In addition, suppose we attempt to approximate the sum L with a partial sum S_N. Then the absolute value of the error in this approximation will be less than $|a_{N+1}|$ and will have the same sign as a_{N+1}, that is, $|L - S_N| < |a_{N+1}|$ and the sign of $L - S_N$ will be the same as the sign of a_{N+1}.

In other words, if we stop after N terms, the error can be estimated by the first "unused" term. The correct answer will be no farther from the approximation than the next term, and it will be in the same direction.

PROOF The proof is essentially the argument that was given in Example 1. Let us suppose that the first term is negative, so all the odd-numbered terms are negative. Then the odd-numbered partial sums are an *increasing* sequence because to get from S_{2N+1} to S_{2N+3}, we add $a_{2N+2} + a_{2N+3} = |a_{2N+2}| - |a_{2N+3}|$, which is positive, since the absolute values decrease.

On the other hand, the even-numbered partial sums are a decreasing sequence, so in particular they are all less than the first even-numbered sum S_2. Each odd-numbered sum is less than the immediately preceding even-numbered sum, since $S_{2N+1} = S_{2N} + a_{2N+1} = S_{2N} - |a_{2N+1}|$. In particular, each odd-numbered partial sum is less than S_2.

The odd-numbered partial sums are a bounded increasing sequence, so they converge to some limit L. The even-numbered sums approach closer and closer to the odd-numbered sums, since $S_{2N} - S_{2N-1} = a_{2N} \to 0$. This means that the sequence $\{S_N\}$ of all partial sums must converge to L.

The remark about the error is clear when we realize that L must be greater than each odd-numbered partial sum and less than each even-numbered partial sum. The proof is similar in the case in which the first term is positive.

EXAMPLE 2

(i) The series $\sum_{n=1}^{\infty} \frac{(-1)^n}{\ln(n+1)}$ is an alternating series. The absolute value of the nth term is $\frac{1}{\ln(n+1)}$, and since $\ln x$ is an increasing function that tends to ∞ as $x \to \infty$, we see that $\{\frac{1}{\ln(n+1)}\}$ is a sequence that is decreasing and which tends to 0.

Consequently, we can conclude that the series converges. (Notice that the series without the negative signs, $\sum_{n=1}^{\infty} \frac{1}{\ln(n+1)}$, diverges because for every $n \geq 1$, $n + 1 > \ln(n + 1)$, so $\frac{1}{\ln(n+1)} > \frac{1}{n+1}$, and the Comparison Test establishes the divergence.)

(ii) The series $\sum_{n=0}^{\infty} \frac{\cos(n\pi)}{\sqrt{n}}$ does not at first glance appear to be an alternating series, but of course $\cos(n\pi)$ equals 1 if n is even and -1 if n is odd, so the series could also be written in the exactly equivalent form $\sum_{n=0}^{\infty} (-1)^n \frac{1}{\sqrt{n}}$. Since $\{\sqrt{n}\}$ is an increasing sequence, we see that $\{\frac{1}{\sqrt{n}}\}$ is a decreasing sequence, and its limit is certainly zero. We conclude that the series converges.

(iii) Consider the series $\sum_{n=1}^{\infty} (-1)^{n+1} n e^{-n/8}$. It is an alternating series. To check the condition about the terms decreasing, we consider the function $f(x) = xe^{-x/8}$. Its derivative is $f'(x) = e^{-x/8} + xe^{-x/8}\left(-\frac{1}{8}\right) = \left(1 - \frac{x}{8}\right)e^{-x/8}$, which is negative for $x > 8$.

This means that the absolute values of the terms of the sequence decrease after the eighth term, but they increase up to the eighth term. The terms also tend to 0, as can easily be seen by applying l'Hôpital's Rule to the function $f(x) = \frac{x}{e^{x/8}}$.

As we have seen several times by now, the convergence of a series is not affected by the behavior of the first few terms. The fact that after the eighth term this series satisfies the conditions means that the conclusion of Theorem 10.6.2 will still be valid. The series converges.

(iv) Consider the series $\sum_{n=1}^{\infty}\left(1 + 2\cos(n\pi)\right)$. It is an alternating series. However, its terms do not tend to zero. (The even-numbered terms all equal 3, and the odd-numbered terms all equal -1.) A series whose terms do not tend to zero cannot possibly converge, by Theorem 10.3.2. The series diverges, even though it is an alternating series.

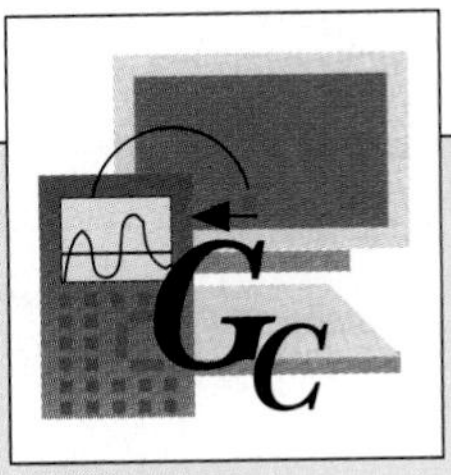

With a programmable calculator we can find partial sums and approximate a convergent series. For instance, with the series $\sum_{n=1}^{\infty}(-1)^{n+1}ne^{-n/8}$ of part (iii) of Example 2 we can evaluate the partial sum S_m.

TI-85

PRGM F2(EDIT) PSUM ENTER F3 F1(Input) M EXIT ENTER 0 STO N ENTER 0 STO S ENTER Lbl B ENTER N + 1 STO N ENTER S + N × e^x ((−) N ÷ 8) × ((−) 1)^(N + 1) STO S ENTER F4 F1(If) (N TEST F2(<) M) ENTER Goto B ENTER S ENTER EXIT EXIT EXIT PRGM F1(NAME) PSUM ENTER 50 ENTER

SHARP EL9300

▼ ▼ (NEW) ENTER ENTER PSUM ENTER COMMAND A 3 (Input) *M* ENTER *N* = 0 ENTER *S* = 0 ENTER COMMAND B 1 (Label) 1 ENTER *N* = *N* + 1 ENTER *S* = *S* + *N* × e^x ((−) *N* ÷ 8) × ((−) 1) a^b (*N* + 1) ENTER COMMAND 3 (If) *N* COMMAND C 2 (<) *M* COMMAND B 2 (Goto) 1 ENTER COMMAND A 1 (Print) *S* ENTER COMMAND A 6 (End) ENTER (RUN) ENTER (PSUM) ENTER 50 ENTER

CASIO f_x-6300*G* (f_x-7700*GB*)

MODE 2 EXE ? → [redacted] : 0 → [redacted] : 0 → [redacted] : Lbl 1 : [redacted] + 1 → [redacted] : [redacted] + [redacted] × e^x ((−) [redacted] ÷ 8) × ((−) 1) x^y ([redacted] + 1) → [redacted] : [redacted] < [redacted] ⇒ Goto 1: [redacted] ◢ MODE 1 Prog 0 EXE 50 EXE

(on the f_x-7700*GB*, : is PRGM F6, and Prog is PRGM F3.)

HP 48SX

ATTN ↱ CLR [redacted] "" KEY IN M ► SPC PROMPT SPC ′ *M* ► STO 0 SPC 1 SPC *M* SPC FOR *N* SPC *N* SPC *N* 8 +/− ÷ [redacted] 1 SPC +/− *N* SPC 1 + y^x × × + NEXT ▼ ENTER ′ PSUM STO VAR PSUM 50 ENTER [redacted]

```
?50
ANSWER=
 .2032963016
```

Since the 51st term of the series is $51 \times e^{-51/8} \approx 0.087$, we find that the approximation is correct to within this accuracy. Now use the program with other values of *m*. Which *m* will give accuracy to within 0.01?

Absolute Convergence

The example of the series $\sum_{n=1}^{\infty} (-1)^n \frac{1}{n}$ shows that it is possible for a series to converge even though the series obtained by replacing each term by its absolute value is divergent. (In this example the series with absolute values is the harmonic series.)

If we have a series $\sum_{n=1}^{\infty} a_n$, it may be interesting to consider the related series $\sum_{n=1}^{\infty} |a_n|$ and ask whether or not it converges. If it is convergent, the original series is called *absolutely convergent.*

DEFINITION 10.6.3

A series $\sum_{n=1}^{\infty} a_n$ is said to be **absolutely convergent** (or to **converge absolutely**) if the series of the absolute values of its terms is convergent, that is, if $\sum_{n=1}^{\infty} |a_n|$ is convergent. If $\sum_{n=1}^{\infty} a_n$ is convergent but *not* absolutely convergent, then it is said to be **conditionally convergent** (or to **converge conditionally**).

The importance of the concept of absolute convergence comes from the following theorem.

THEOREM 10.6.4

If $\sum_{n=1}^{\infty} a_n$ is absolutely convergent, then it converges, that is, if $\sum_{n=1}^{\infty} |a_n|$ converges, then $\sum_{n=1}^{\infty} a_n$ converges.

In words: An absolutely convergent series must be convergent.

PROOF Suppose $\sum_{n=1}^{\infty} |a_n|$ is convergent. Each term a_n satisfies $-|a_n| \le a_n \le |a_n|$, and adding $|a_n|$ to each part of this inequality, we find that

$$0 \le a_n + |a_n| \le 2|a_n|.$$

The series $\sum_{n=1}^{\infty} (a_n + |a_n|)$ has nonnegative terms, and its terms are less than those of the convergent series $\sum_{n=1}^{\infty} 2|a_n|$, so the Comparison Test shows that it converges.

We can write

$$\sum_{n=1}^{\infty} a_n = \sum_{n=1}^{\infty} (a_n + |a_n|) - \sum_{n=1}^{\infty} |a_n|,$$

which shows that the series $\sum_{n=1}^{\infty} a_n$ is the difference of two convergent series. This shows that it is convergent, as desired. ◢

So one way to prove that a series converges is to show that it converges absolutely. Since this involves the convergence of a series with nonnegative terms, it may be possible to use the Integral Test or the Comparison Test (which do not work with series whose terms may be negative).

However, we should always bear in mind that it is possible for a series to be conditionally convergent, that is, it may converge but not converge absolutely.

EXAMPLE 3

(i) The series $\sum_{n=1}^{\infty}(-1)^n\frac{1}{n}$ converges (by the condition for convergence of alternating series), but it is *not* absolutely convergent. It is conditionally convergent.

(ii) The series $\sum_{n=1}^{\infty}(-1)^n\frac{1}{n^{3/2}+2}$ converges absolutely (since $\sum_{n=1}^{\infty}\frac{1}{n^{3/2}+2}$ converges, by the Comparison Test, comparing it to the p-series $\sum_{n=1}^{\infty}\frac{1}{n^{3/2}}$).

In particular the original series is convergent. (This part could also be seen by noting that it is an alternating series.) Being absolutely convergent, it is *not* conditionally convergent.

(iii) The series $\sum_{n=2}^{\infty}\left(\cos\left(n\frac{\pi}{3}\right)\frac{2}{n\ln^2(n)}\right)$ is not an alternating series; notice that $\cos\left(2\frac{\pi}{3}\right)=-\frac{1}{2}$, $\cos\left(3\frac{\pi}{3}\right)=-1$, $\cos\left(4\frac{\pi}{3}\right)=-\frac{1}{2}$, $\cos\left(5\frac{\pi}{3}\right)=\frac{1}{2}$, $\cos\left(6\frac{\pi}{3}\right)=1$, $\cos\left(7\frac{\pi}{3}\right)=\frac{1}{2}$, and after that the "cos" part of the series repeats itself, with three positive terms followed by three negative terms and so on.

If we consider the corresponding positive series

$$\sum_{n=2}^{\infty}\left|\cos\left(n\frac{\pi}{3}\right)\frac{2}{n\ln^2(n)}\right|=\sum_{n=2}^{\infty}\left(\left|\cos\left(n\frac{\pi}{3}\right)\right|\frac{2}{n\ln^2(n)}\right),$$

it looks quite complicated. However, each of its terms is smaller than the corresponding term in the simpler series $\sum_{n=2}^{\infty}\frac{2}{n\ln^2(n)}$, and this looks like a series to which the Integral Test may apply. We consider the improper integral, making the substitution $u=\ln x$, $du=\frac{1}{x}\,dx$.

$$\begin{aligned}\int_2^{\infty}\frac{2}{x\ln^2(x)}\,dx&=2\lim_{r\to\infty}\left(\int_2^{r}\frac{1}{x\ln^2(x)}\,dx\right)=2\lim_{r\to\infty}\left(\int_{\ln 2}^{\ln r}\frac{1}{u^2}\,du\right)\\&=2\lim_{r\to\infty}\left(-\frac{1}{u}\right)\Big|_{\ln 2}^{\ln r}=2\lim_{r\to\infty}\left(-\frac{1}{\ln r}+\frac{1}{\ln 2}\right)\\&=\frac{2}{\ln 2}.\end{aligned}$$

The improper integral converges, so $\sum_{n=2}^{\infty}\frac{2}{n\ln^2(n)}$ converges, by the Integral Test. By the Comparison Test the series $\sum_{n=2}^{\infty}\left(\left|\cos\left(n\frac{\pi}{3}\right)\right|\frac{2}{n\ln^2(n)}\right)$ converges.

This means that the original series converges absolutely, which in particular means that it converges.

Notice that in part (iii) of Example 3 it was not possible to apply the test for alternating series but that it was possible to show that the series is absolutely convergent (and so convergent). The Integral Test and Comparison Test are only applicable to series with nonnegative terms, but now we see that it may be possible to use them with any series by considering the absolute values of the terms and showing that the series is absolutely convergent.

The concept of absolute convergence is important for another reason. When we are dealing with finite sums, we know that it makes no difference what order we use to add up the terms. With series, however, this is definitely *not* true. In fact, if we take the convergent series $\sum_{n=1}^{\infty}(-1)^n\frac{1}{n}$, it is possible to rearrange the terms so that the rearranged series diverges, and it is also possible to rearrange them so that the rearranged series converges to *any* limit L.

Roughly speaking, the way to find a rearrangement that converges to L is this: Start adding the positive terms $\frac{1}{2} + \frac{1}{4} + \frac{1}{6} + \ldots + \frac{1}{2N}$, and stop at the first N for which this sum is greater than L. Then add to this some negative terms $-1 - \frac{1}{3} - \frac{1}{5} - \ldots$, and stop after the first term with which the whole sum is less than L. Then add additional positive terms (starting with $\frac{1}{2N+2}$) until the whole sum exceeds L, then add additional negative terms until it is less than L, and so on.

Since $\frac{1}{n} \to 0$ as $n \to \infty$, these rearranged partial sums approach closer and closer to L; the rearranged series will converge to L. The "rearrangement" eventually uses all the terms of the original series.

This can be done for *any* value of L; the series can be rearranged so that it will converge to any limit. In fact, the same idea can be applied to any series that is *conditionally* convergent.

Because of this it is extremely important to be careful about rearranging series. On the other hand, for a series that is absolutely convergent there is no problem.

THEOREM 10.6.5

If $\sum_{n=1}^{\infty} a_n$ converges absolutely, then *any* rearrangement of this series also converges absolutely, and moreover the sums are the same. In other words, if $\sum_{n=1}^{\infty} b_n$ is a rearrangement of $\sum_{n=1}^{\infty} a_n$, then it converges absolutely, and

$$\sum_{n=1}^{\infty} b_n = \sum_{n=1}^{\infty} a_n.$$

(To say that $\sum_{n=1}^{\infty} b_n$ is a rearrangement of $\sum_{n=1}^{\infty} a_n$ means that it contains all the same terms but in a different order.)

This theorem is fairly difficult, so we omit the proof. It is another reason for the importance of the concept of absolute convergence and shows that in a sense a series that is absolutely convergent "converges very strongly" to its limit, so much so that even after being rearranged, it will still converge to the same limit.

Let us emphasize once again that "conditionally convergent" means convergent but *not* absolutely convergent.

Exercises 10.6

I

Decide whether or not each of the following series converges, and give your reasons.

1. $\sum_{n=1}^{\infty} (-1)^n \frac{1}{4n+3}$
2. $\sum_{n=1}^{\infty} (-1)^n \frac{1}{n^{4/3}}$
3. $\sum_{n=1}^{\infty} (-1)^{n+1} \frac{1}{n^{1/3}}$
4. $\sum_{n=1}^{\infty} (-1)^{n-1} \frac{1}{n^{-5/2}}$
5. $\sum_{n=3}^{\infty} (-1)^{n+7} \frac{n^3}{n^4-2n^2+1}$
6. $\sum_{n=2}^{\infty} (-1)^{n-3} \frac{1}{\ln(n)}$
7. $\sum_{n=2}^{\infty} (-1)^n \frac{\ln(n)}{n+4}$
8. $\sum_{n=1}^{\infty} (-1)^n \frac{\cos^2(n\pi)}{3n+2}$
9. $\sum_{n=2}^{\infty} \cos(n\pi) \frac{\ln(n)}{e^{-n}+1}$
10. $\sum_{n=1}^{\infty} (-1)^{2n} \frac{1}{n^{2/3}}$
11. $\sum_{n=1}^{\infty} \sin\left(n\frac{\pi}{5}\right) \frac{3n}{4n^3-2n+5}$
12. $\sum_{n=2}^{\infty} \sin\left(n\frac{\pi}{7}\right) \frac{1}{n \ln^3(n)}$
13. $\sum_{n=1}^{\infty} (-1)^n \frac{n+3}{3n^2+4n+2}$
14. $\sum_{n=1}^{\infty} \left(\frac{1}{n} - \frac{1}{n+1}\right)$

15. $\sum_{n=7}^{\infty} \cos(n\pi)\frac{\sqrt{n+3}}{4-\sqrt{n+10}}$
16. $\sum_{n=7}^{\infty} \cos(n\pi)\frac{1}{n\ln(n)}$
17. $\sum_{n=5}^{\infty} \sin(n^2-3)\frac{n^{4/3}}{1-n^{15/6}}$
18. $\sum_{n=1}^{\infty} \cos(e^{2n})\frac{1}{n^2+4}$
19. $\sum_{n=1}^{\infty} (-1)^{n+2}\cos\left(\frac{1}{n}\right)$
20. $\sum_{n=1}^{\infty} \cos(n\pi)\sin\left(\frac{1}{n}\right)$

Decide whether or not each of the following series converges absolutely, conditionally, or neither, and give your reasons.

21. $\sum_{n=1}^{\infty} (-1)^n\frac{1}{5n-3}$
22. $\sum_{n=2}^{\infty} (-1)^n\frac{1}{3n^2-4}$
23. $\sum_{n=1}^{\infty} \cos\left(n\frac{\pi}{5}\right)n^{-7/6}$
24. $\sum_{n=0}^{\infty} \cos^3(n\pi)\frac{1}{\ln(2n^2+4)}$
25. $\sum_{n=1}^{\infty} \cos\left(\pi(n^3-3n^2-n-1)\right)\sqrt{n}$
26. $\sum_{n=1}^{\infty} \sin(n+\pi)\frac{1+\exp(-n)}{1-\exp(n)}$
27. $\sum_{n=1}^{\infty} (-1)^n\tanh(n)n^{-1/7}$
28. $\sum_{n=1}^{\infty} (-1)^{n+1}\frac{\exp(n)}{\sinh(n)}$
29. $\sum_{n=1}^{\infty} \sin(3n^3+2n-\pi)\frac{\exp(n)}{\exp(2n)+1}$
30. $\sum_{n=2}^{\infty} \coth(3n^2-5)\frac{1}{\ln(n)}$

II

Show that each of the following series converges.

*31. $\sum_{n=1}^{\infty} \frac{\sin(n\pi/2)}{n}$

*32. $\sum_{n=1}^{\infty} \frac{\cos(n\pi/2)}{\ln(n+6)}$

33. $\sum_{n=1}^{\infty} \frac{\sin(n)}{n^{3/2}}$

34. $\sum_{n=4}^{\infty} \frac{\cos(n)}{e^{2n}+3n-1}$

*35. Suppose $a_n > 0$ for each n and $\sum_{n=1}^{\infty}(-1)^n a_n$ is conditionally convergent. Show that the series consisting of all the positive terms $\sum_{m=1}^{\infty} a_{2m}$ and the series $\sum_{m=1}^{\infty}(-a_{2m-1})$ consisting of all the negative terms must both be divergent.

*36. Explain how to find a rearrangement of $\sum_{n=1}^{\infty}(-1)^n\frac{1}{n}$ that diverges.

*37. Find the first twenty terms of a rearrangement of $\sum_{n=1}^{\infty}(-1)^n\frac{1}{n}$ that converges to $\frac{\pi}{2}$, following the suggestion in the text just before Theorem 10.6.5.

*38. Show that the series $\sum_{n=1}^{\infty}\sin\left(n\frac{\pi}{4}\right)\frac{1}{n}$ converges conditionally.

*39. Show that the series $\sum_{n=1}^{\infty}\cos(n^2\pi)\frac{1}{n}$ converges conditionally.

*40. Decide whether or not the series

$$\sum_{=1}^{\infty}(-1)^{n^2+n}\frac{1}{4n+5}$$

is convergent, and explain your reasons.

Drill

Decide whether or not each of the following series converges, and give your reasons.

41. $\sum_{n=1}^{\infty} \frac{n^2+\sqrt{n}}{2n^4-1}$
42. $\sum_{n=1}^{\infty} \frac{5n}{3n^2+2e^n}$
43. $\sum_{n=2}^{\infty} \frac{3^n}{2^n\ln(n)}$
44. $\sum_{n=1}^{\infty} e^{-2n^2}$
45. $\sum_{n=1}^{\infty} \frac{n+1}{(n^3+1)e^{-n}}$
46. $\sum_{n=1}^{\infty} \frac{ne^{-n}}{2+\cos(n+3)}$
47. $\sum_{n=1}^{\infty} (-1)^n\arctan\left(\frac{1}{n}\right)$
48. $\sum_{n=6}^{\infty} (-1)^n\frac{1}{2+\ln(\ln(n))}$
49. $\sum_{n=2}^{\infty} \frac{(n-2)!}{n!}$
50. $\sum_{n=2}^{\infty} \frac{1}{n^n}$

III

Use a programmable calculator or computer to estimate the following series with the specified accuracy.

51. $\sum_{n=1}^{\infty} (-1)^n\frac{1}{n}$, within 0.005
52. $\sum_{n=2}^{\infty} (-1)^n\frac{\ln(n)}{n}$, within 0.01
53. $\sum_{n=1}^{\infty} \frac{\cos(n)}{n^2}$, within 0.005
54. $\sum_{n=2}^{\infty} \frac{\sin(n^2)}{n^2\ln(n)}$, within 0.004

10.7 The Ratio Test and the *n*th Root Test

RATIO TEST
*n*TH ROOT TEST
TESTING CONVERGENCE OF SERIES

In this section we learn two convenient tests for convergence of series. For these tests it is not necessary to assume that the terms of the series are all nonnegative.

When we studied geometric series, we found that the convergence or divergence of a series depended on the *common ratio,* the ratio of each term to the previous one. If the absolute value of this ratio is less than 1, then the series converges; otherwise, it diverges.

For a series other than a geometric series the difficulty is that the ratio of each term to the preceding term will not be the same for different terms, so we cannot use the same idea. However, it sometimes happens that the ratios of the nth term to the preceding term will tend to a limit as $n \to \infty$. Suppose

$$\lim_{n\to\infty} \frac{a_n}{a_{n-1}} = \rho.$$

We use the Greek letter ρ ("rho," pronounced "row" as in "rowboat"), which is the Greek letter corresponding to our letter r, to remind us of the idea of the common ratio but to be sufficiently different that we will remember that it is not the ratio itself, but only a limit.

What this says is that for large values of n, each term is approximately ρ times the previous one. It suggests that the series behaves something like a geometric series whose common ratio is ρ. This is the idea behind the Ratio Test.

THEOREM 10.7.1

The Ratio Test

Consider the series $\sum_{n=1}^{\infty} a_n$, and suppose that $a_n \neq 0$, at least after the first few terms. Suppose the limit

$$\lim_{n\to\infty} \left| \frac{a_n}{a_{n-1}} \right|$$

exists and equals ρ. Then:

If $\rho < 1$, the series converges absolutely,

If $\rho > 1$, the series diverges,

If $\rho = 1$, the test is inconclusive.

PROOF Suppose the limit exists and equals ρ, with $\rho < 1$. Then we can find a number R between ρ and 1, that is, $\rho < R < 1$. Because $\left|\frac{a_n}{a_{n-1}}\right| \to \rho$, we know that there is an integer N so that $\left|\frac{a_n}{a_{n-1}}\right| < R$, for every $n \geq N$, i.e., $|a_n| < R|a_{n-1}|$, for $n \geq N$.

This means that past the Nth term, the absolute value of each term is less than R times the absolute value of the previous one. This means that the series $\sum_{n=N}^{\infty} |a_n|$ has terms that are less than or equal to the geometric series $|a_N|(1 + R + R^2 + R^3 + R^4 + \ldots)$.

This geometric series converges, since $R < 1$, and by the Comparison Test, $\sum_{n=N}^{\infty} |a_n|$ converges. From this we conclude that $\sum_{n=1}^{\infty} |a_n|$ converges, which means that $\sum_{n=1}^{\infty} a_n$ converges absolutely. This proves the first part of the theorem.

Suppose $\rho > 1$. As above, we see that there is an integer N so that $\left|\frac{a_n}{a_{n-1}}\right| > 1$ whenever $n \geq N$. This means that the absolute value of a_n is *greater* than the absolute value of a_{n-1}, for every $n \geq N$, which means that after the Nth term the absolute values of the terms are *increasing*.

In particular, this means that the terms cannot tend to zero, so the series must diverge. This proves the second part of the theorem.

Finally we give two examples to show that it is possible to have $\rho = 1$ for convergent or divergent series, which shows that when $\rho = 1$ it is impossible to draw any conclusions about convergence.

First consider the harmonic series $\sum_{n=1}^{\infty} \frac{1}{n}$. The quotient of its nth term over its $(n-1)$th term is $\left(\frac{1}{n}\right)/\left(\frac{1}{n-1}\right) = \frac{n-1}{n}$. As $n \to \infty$, this tends to 1, that is, $\lim_{n\to\infty} \frac{n-1}{n} = 1$. This is an example of a series that is divergent and for which the limit is $\rho = 1$.

Next consider the series $\sum_{n=1}^{\infty} \frac{1}{n^2}$. The quotient of its nth term over its $(n-1)$th term is $\left(\frac{1}{n^2}\right)/\left(\frac{1}{(n-1)^2}\right) = \frac{(n-1)^2}{n^2}$. As $n \to \infty$, this tends to 1, that is, $\lim_{n\to\infty} \frac{(n-1)^2}{n^2} = 1$. This is an example of a convergent series for which the limit is $\rho = 1$.

The existence of these two examples shows that the Ratio Test tells us nothing when $\rho = 1$. This concludes the proof of the theorem. ◢

> Notice that the limit in Theorem 10.7.1 is the limit of the *absolute values* of the ratios of consecutive terms rather than just the limit of the ratios themselves. This actually makes things easier.
>
> Also we should notice that unlike the situation with geometric series, in which a series with common ratio $r = 1$ must diverge, with the Ratio Test we are unable to draw any conclusion if the limit is $\rho = 1$. In this respect the situation is more delicate than it is for geometric series.

EXAMPLE 1

(i) Consider the series $\sum_{n=1}^{\infty} \frac{1}{n!}$. The ratio of its nth term by its $(n-1)$th term is $\left(\frac{1}{n!}\right)/\left(\frac{1}{(n-1)!}\right) = \frac{(n-1)!}{n!} = \frac{1}{n}$, which obviously tends to 0 as $n \to \infty$. This shows that the limit ratio is $\rho = 0$, and since $\rho < 1$, the Ratio Test tells us that the series converges absolutely.

(ii) Now we fix a number c and consider the series $\sum_{n=1}^{\infty} \frac{c^n}{n!}$. The ratio of its nth term over the preceding term is $\left(\frac{c^n}{n!}\right)/\left(\frac{c^{n-1}}{(n-1)!}\right) = \frac{c^n(n-1)!}{c^{n-1}n!} = \frac{c}{n}$. We let $n \to \infty$ and find that $\lim_{n\to\infty} \frac{c}{n} = 0$. The Ratio Test tells us the series converges absolutely (and so it converges).

Notice that this is true for any possible value of c. Also notice that since the terms of a convergent series must tend to 0, we have incidentally proved that $\lim_{n\to\infty} \frac{c^n}{n!} = 0$, something we did with much more work in Example 10.1.6.

(iii) Let $a_n = (-1)^n \frac{n^2}{2^n}$, and consider the series $\sum_{n=1}^{\infty} a_n$. To apply the Ratio Test, we find

$$\rho = \lim_{n\to\infty} \left|\frac{a_n}{a_{n-1}}\right| = \lim_{n\to\infty} \left|\frac{(-1)^n \frac{n^2}{2^n}}{(-1)^{n-1}\frac{(n-1)^2}{2^{n-1}}}\right| = \lim_{n\to\infty} \left|(-1)\frac{n^2}{(n-1)^2}\frac{2^{n-1}}{2^n}\right|$$

$$= \lim_{n\to\infty} \left(\frac{n^2}{(n-1)^2}\frac{1}{2}\right) = \frac{1}{2}\lim_{n\to\infty}\left(\frac{n^2}{n^2-2n+1}\right)$$

$$= \frac{1}{2}.$$

Since the limit ratio $\rho = \frac{1}{2}$ is less than 1, the Ratio Test tells us that the series converges absolutely.

(iv) For what values of c does the series $\sum_{n=1}^{\infty} \left((-1)^n(n^2-2)3^{n-2}c^n\right)$ converge?

Solution Consider the limit

$$\lim_{n\to\infty}\left|\frac{(-1)^n(n^2-2)3^{n-2}c^n}{(-1)^{n-1}\big((n-1)^2-2\big)3^{n-3}c^{n-1}}\right| = \lim_{n\to\infty}\left|(-1)\frac{n^2-2}{(n-1)^2-2}\frac{3^{n-2}}{3^{n-3}}\frac{c^n}{c^{n-1}}\right|$$
$$= \lim_{n\to\infty}\left(\frac{n^2-2}{n^2-2n-1}(3)|c|\right)$$
$$= 3|c|.$$

The Ratio Test says that the series will converge absolutely when this ratio is less than 1, that is, when $|c| < \frac{1}{3}$. It will diverge when $|c| > \frac{1}{3}$.

What remains is the more delicate case when $|c| = \frac{1}{3}$, that is, when $c = \pm\frac{1}{3}$. If $c = \frac{1}{3}$, then the nth term of the series is $(-1)^n(n^2-2)3^{n-2}\left(\frac{1}{3}\right)^n = (-1)^n(n^2-2)3^{n-2-n} = (-1)^n(n^2-2)3^{-2}$. As $n \to \infty$, this term does not tend to zero, so the series diverges. When $c = -\frac{1}{3}$, the result is similar; the only thing that changes is the sign of the nth term, which does not affect the fact that it does not tend to 0.

We see that the series converges when $-\frac{1}{3} < c < \frac{1}{3}$ and diverges otherwise.

> The Ratio Test tends to be particularly useful for series whose nth term involves factorials or nth powers.

The nth Root Test

The other test we should learn also tends to be applicable to series that involve nth powers in the nth term. The structure of this test is quite similar to that of the Ratio Test.

THEOREM 10.7.2

The nth Root Test

Consider the series $\sum_{n=1}^{\infty} a_n$. Suppose the limit

$$\lim_{n\to\infty}\sqrt[n]{|a_n|}$$

exists and equals ρ. Then:

If $\rho < 1$, the series converges absolutely,

If $\rho > 1$, the series diverges,

If $\rho = 1$, the test is inconclusive.

PROOF Suppose the limit exists and equals ρ, with $\rho < 1$. Let R be a number between ρ and 1, that is, $\rho < R < 1$. Then there is an integer N so that for every $n \geq N$ we have $\sqrt[n]{|a_n|} < R$. Another way of saying the same thing is that $|a_n| < R^n$, for every $n \geq N$.

Since $R < 1$, $\sum_{n=N}^{\infty} R^n$ is a convergent geometric series. By the Comparison Test, $\sum_{n=N}^{\infty} |a_n|$ is convergent, which means that $\sum_{n=1}^{\infty} |a_n|$ is convergent, which means that $\sum_{n=1}^{\infty} a_n$ is absolutely convergent. This proves the first part of the theorem.

If $\rho > 1$, then there is an integer N so that for every $n > N$ we have $\sqrt[n]{|a_n|} > 1$, which is the same as saying that $|a_n| > 1$ for every $n > N$. Since this means that $\{a_n\}$ cannot tend to 0, we know that the series diverges, and the second part of the theorem is established.

As before, we consider the harmonic series. The nth root of the nth term is $\sqrt[n]{\frac{1}{n}} = \left(\frac{1}{n}\right)^{1/n} = \exp\left(\frac{1}{n}\ln\left(\frac{1}{n}\right)\right)$. We know that as $n \to \infty$, $\frac{1}{n}\ln\left(\frac{1}{n}\right) \to 0$ (using l'Hôpital's Rule, for instance), so $\exp\left(\frac{1}{n}\ln\left(\frac{1}{n}\right)\right)$ tends to $e^0 = 1$.

The harmonic series is an example of a divergent series for which the nth root of the nth term tends to 1.

The nth root of $\frac{1}{n^2}$ is just the square of the nth root of $\frac{1}{n}$, so it must also approach the limit 1 as $n \to \infty$. This means that $\sum_{n=1}^{\infty} \frac{1}{n^2}$ is an example of a series for which the nth root of the nth term tends to 1 but that is convergent.

The existence of these two examples shows that the nth Root Test is inconclusive if $\rho = 1$. This completes the proof. ◢

EXAMPLE 2

(i) Consider the series $\sum_{n=1}^{\infty}\left(\frac{1}{n}\right)^n$. The nth Root Test suggests looking at $\left(\left(\frac{1}{n}\right)^n\right)^{1/n} = \frac{1}{n}$. Since this tends to $\rho = 0$ as $n \to \infty$, the nth Root Test says that the series converges.

(ii) Consider the series $\sum_{n=1}^{\infty}(3 - \cos(n\pi))\left(\frac{2}{3}\right)^n$. The nth root of its nth term is

$$\sqrt[n]{(3 - \cos(n\pi))\left(\frac{2}{3}\right)^n} = \frac{2}{3}\sqrt[n]{3 - \cos(n\pi)}.$$

See the Calculator Example on the next page.

Now $\cos(n\pi)$ is either 1 or -1, depending on whether n is even or odd, so $3 - \cos(n\pi)$ is either 4 or 2.

Now as $n \to \infty$, $\sqrt[n]{2} = 2^{1/n} = \exp\left(\frac{1}{n}\ln 2\right)$, and since $\frac{1}{n}\ln 2$ tends to 0, we see that $\sqrt[n]{2} \to \exp(0) = 1$. The same argument shows that $\sqrt[n]{4} \to 1$ as $n \to \infty$.

These two observations show that $\lim_{n\to\infty}\sqrt[n]{3 - \cos(n\pi)} = 1$, and from this we see that the nth root of the nth term of the series tends to $\frac{2}{3}\lim_{n\to\infty}\sqrt[n]{3 - \cos(n\pi)} = \frac{2}{3}$. The nth Root Test tells us that the series is convergent.

Notice that in this case the Ratio Test could not be used because when n is even, the ratio of the nth term to the $(n-1)$th term is

$$\frac{(3 - \cos(n\pi))\left(\frac{2}{3}\right)^n}{\left(3 - \cos\left((n-1)\pi\right)\right)\left(\frac{2}{3}\right)^{n-1}} = \frac{2\left(\frac{2}{3}\right)^n}{4\left(\frac{2}{3}\right)^{n-1}} = \frac{2}{4}\frac{2}{3} = \frac{1}{3},$$

but when n is odd, the ratio is

$$\frac{(3 - \cos(n\pi))\left(\frac{2}{3}\right)^n}{\left(3 - \cos\left((n-1)\pi\right)\right)\left(\frac{2}{3}\right)^{n-1}} = \frac{4\left(\frac{2}{3}\right)^n}{2\left(\frac{2}{3}\right)^{n-1}} = \frac{4}{2}\frac{2}{3} = \frac{4}{3}.$$

These ratios do not tend to a limit as $n \to \infty$, so it is not possible to use the Ratio Test. ◢

With a programmable calculator we can find partial sums and approximate the series $\sum_{n=1}^{\infty}(3 - \cos n\pi)\left(\frac{2}{3}\right)^n$ of part (ii) of Example 2. Notice that $\cos n\pi = (-1)^n$.

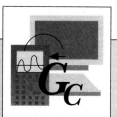

TI-85	SHARP EL9300
PRGM F2(EDIT) PTSUM ENTER F3 F1(Input) M EXIT ENTER 0 STO N ENTER 0 STO S ENTER Lbl B ENTER N +1 STO N ENTER S + (3 − ((−) 1) ^ N) × (2 ÷ 3) ^ N STO S ENTER F4 F1(If) (N TEST F2(<) M) ENTER Goto B ENTER S ENTER EXIT EXIT EXIT PRGM F1(NAME) PTSUM ENTER 20 ENTER	[program key] ▼ ▼ (NEW) ENTER ENTER PTSM ENTER COMMAND A 3 (Input) M ENTER N = 0 ENTER S = 0 ENTER COMMAND B 1 (Label) 1 ENTER N = N + 1 ENTER S = S + (3 − ((−) 1) a^b N) × (2 ÷ 3) a^b N ENTER COMMAND 3 (If) N COMMAND C 2 (<) M COMMAND B 2 (Goto) 1 ENTER COMMAND A 1 (Print) S ENTER COMMAND A 6 (End) ENTER [calc key] [program key] (RUN) ENTER (PTSM) ENTER 20 ENTER
CASIO f_x-6300G (f_x-7700GB)	**HP 48SX**
MODE 2 EXE ? → M : 0 → N : 0 → S : Lbl 1 : N + 1 → N : S + (3 − ((−) 1) x^y N) × (2 ÷ 3) x^y N → S : N < M ⇒ Goto 1 : S ► MODE 1 Prog 0 EXE 20 EXE (on the f_x-7700GB, : is PRGM F6, and Prog is PRGM F3.)	ATTN ↱ CLR ≪ ≫ "" KEYIN M ► SPC PROMPT SPC ′ M ► STO 0 SPC 1 SPC M SPC FOR N SPC 3 SPC 1 +/− N y^x − 2 SPC 3 ÷ N y^x × + NEXT ▼ ENTER ′ PSUM STO VAR PSUM 20 ENTER ↰ CONT

```
?20
ANSWER=
      6.398075337
```

Try to estimate the accuracy of this approximation. (*Hint:* If $n > 20$, then $\left(\frac{2}{3}\right)^n < \left(\frac{2}{3}\right)^{20} \ldots$).

Summary 10.7.3 Testing Convergence of Series

We have learned a number of tests for convergence and divergence of series. Given a series $\sum_{n=1}^{\infty} a_n$, it can be difficult to know which test to apply. In a sense there is no single answer, and we just try each test until we find one that works. However, the following observations may be helpful.

1. It is usually easy to apply the nth Term Test (Theorem 10.3.2), which may immediately show that the series diverges.
2. Next it is probably best to check whether the series is one of the special types we know how to deal with: (i) geometric series, (ii) alternating series (with terms

decreasing in absolute value), (iii) p-series (including the harmonic series), (iv) series whose terms are a rational function of n (cf. Theorem 10.5.4). It may also be possible to break the series into a sum of simpler series, to factor out constants, or to shift the index of summation to transform the series into one of the familiar types.

3. Next it may be possible to use the Comparison Test (Theorems 10.5.1, 10.5.2) or the Limit Comparison Test (Theorem 10.5.3). Here the difficulty is to find a suitable series with which to compare. Usually, this is done by modifying the original series in some way, often by simplifying or omitting part of the formula for a_n. It is vital to have series with nonnegative terms, but sometimes this means showing absolute convergence by working with $\sum_{n=1}^{\infty} |a_n|$.
4. The integral test may be useful, but only if the series can be expressed in terms of a function that we know how to integrate, and only if the terms are nonnegative. It too can be applied to $\sum_{n=1}^{\infty} |a_n|$.
5. The Ratio Test is most likely to be useful with series whose nth term involves factorials or exponentials or nth powers. The Root Test will have a better chance of working for series for which the nth term is an nth power. However, these are only guidelines, and both should be tried.
6. Finally, it may be necessary to combine tests. Often the Comparison Test is used to compare $\sum_{n=1}^{\infty} a_n$ with another series, which is then shown to converge or diverge by using the Integral, Ratio, or Root Test. Another common tactic is to break the series up as a sum and use different tests on each piece.

Some Familiar Convergent and Divergent Series

Geometric series	$a + ar + ar^2 + \dots = \sum_{n=0}^{\infty} ar^n = \frac{a}{1-r}$	Divergent when $\|r\| \geq 1$, (Absolutely) convergent when $\|r\| < 1$
Harmonic series	$1 + \frac{1}{2} + \frac{1}{3} + \frac{1}{4} + \dots = \sum_{n=1}^{\infty} \frac{1}{n}$	Divergent
p-series	$1 + \frac{1}{2^p} + \frac{1}{3^p} + \frac{1}{4^p} + \dots = \sum_{n=1}^{\infty} \frac{1}{n^p}$	Convergent when $p > 1$, divergent when $p \leq 1$
Alternating series	$\sum_{n=1}^{\infty} (-1)^n \|a_n\|$ or $\sum_{n=1}^{\infty} (-1)^{n+1} \|a_n\|$	Convergent if $\|a_n\|$ decreases to 0
Rational functions	$\sum_{n=1}^{\infty} \frac{p(n)}{q(n)}$ with $p(x)$, $q(x)$ polynomials	Converges precisely when $deg\ q(x) \geq deg\ p(x) + 2$

You might find it helpful to review the examples of this section in light of this summary. You could try looking at each example and trying to decide which approach to use *before* looking at the solution. Bear in mind that many examples can be done in more than one way.

Exercises 10.7

I

Test the following series for convergence, giving reasons for your conclusions.

1. $\sum_{n=1}^{\infty} \frac{4^n}{(2n)!}$
2. $\sum_{n=1}^{\infty} (-1)^n \frac{6^{n+1}}{(n-1)!}$
3. $\sum_{n=0}^{\infty} (-1)^{n+1} \frac{8^{2n+1}}{(2n+1)!}$
4. $\sum_{n=0}^{\infty} (-1)^n \frac{\pi^{2n}}{(2n)!}$
5. $\sum_{n=1}^{\infty} \frac{2^n}{n^2}$
6. $\sum_{n=1}^{\infty} \left(\frac{2}{3}\right)^n \ln(n+1)$
7. $\sum_{n=1}^{\infty} n^{-2n}$
8. $\sum_{n=1}^{\infty} \frac{n^n}{n!}$
9. $\sum_{n=1}^{\infty} n^{-n} n!$
10. $\sum_{n=1}^{\infty} \left(2 + \sin\left(\frac{n\pi}{2}\right)\right)\left(\frac{4}{5}\right)^n$
11. $\sum_{n=1}^{\infty} \frac{2^n}{\sqrt{n!}}$
12. $\sum_{n=1}^{\infty} \sin^n\left(\frac{1}{n}\right)$
13. $\sum_{n=1}^{\infty} ne^{-n}$
14. $\sum_{n=1}^{\infty} 2^{-1/n}$
15. $\sum_{n=0}^{\infty} 2^{-n} e^{n/2}$
16. $\sum_{n=1}^{\infty} \left(-\frac{1}{2}\right)^n \arctan(n^2)$
17. $\sum_{n=1}^{\infty} (n!)^{-n/3}$
18. $\sum_{n=1}^{\infty} \frac{\sin(n^2+2)}{n!}$
19. $\sum_{n=1}^{\infty} \sin\left(\frac{1}{n!}\right)$
20. $\sum_{n=1}^{\infty} \frac{n!(2n)!}{(3n)!}$

For what values of c is each of the following series convergent?

21. $\sum_{n=0}^{\infty} \frac{c^n}{(2n)!}$
22. $\sum_{n=0}^{\infty} (-1)^n \frac{c^{2n+1}}{(2n+1)!}$
23. $\sum_{n=0}^{\infty} (-1)^n \frac{c^{2n}}{(2n)!}$
24. $\sum_{n=1}^{\infty} \frac{c^n}{n^3}$
25. $\sum_{n=1}^{\infty} (-1)^n \frac{c^n}{n}$
26. $\sum_{n=1}^{\infty} (c-3)^n$
27. $\sum_{n=1}^{\infty} (-5)^n c^{-n}$
28. $\sum_{n=2}^{\infty} \frac{(4c)^n}{\ln(n)}$
29. $\sum_{n=1}^{\infty} (2n)! c^{-n}$
30. $\sum_{n=4}^{\infty} \ln(n)(3c)^{-n}$

II

31. Use the Ratio Test to do Example 2(i) (p. 691).

*32. In Example 2(ii) we showed that $\sum_{n=1}^{\infty} (3 - \cos(n\pi)) \left(\frac{2}{3}\right)^n$ is convergent. To do this, we used the nth root test and remarked that the Ratio Test would not work. Find a way to prove the convergence of this series without the nth Root Test, using the Comparison Theorem and the Ratio Test (or anything else you prefer).

33. For what values of c does the following series converge?

$$1 - \frac{c}{2} + \frac{c^2}{3} - \frac{c^3}{4} + \frac{c^4}{5} - \frac{c^5}{6} + \dots$$

*34. For what values of c does the following series converge?

$$\left(\frac{1}{2}c\right)^2 - \left(\frac{2}{3}c\right)^3 + \left(\frac{3}{4}c\right)^4 - \left(\frac{4}{5}c\right)^5 + \left(\frac{5}{6}c\right)^6 - \left(\frac{6}{7}c\right)^7 + \dots$$

*35. Suppose $p(x)$ is any polynomial, and consider the series $\sum_{n=1}^{\infty} p(n) r^n$. For what values of r is this series convergent?

*36. Suppose $R(x)$ is any rational function, and consider the series $\sum_{n=N}^{\infty} R(n) r^n$, where N is chosen so that $R(n)$ is defined for every $n \geq N$. For what values of r is this series convergent?

*37. What does the nth Root Test have to say about the convergence of the geometric series $\sum_{n=1}^{\infty} Cr^n$?

*38. Show that $\sum_{n=1}^{\infty} (e^{1/n^2} - 1)$ converges. (*Hint:* Estimate the nth term using the Mean Value Theorem.)

Drill

Test the following series for convergence, giving reasons for your conclusions.

39. $\sum_{n=1}^{\infty} \sin\left(\frac{1}{n^2}\right)$
40. $\sum_{n=1}^{\infty} \sin^2\left(\frac{1}{n}\right)$
41. $\sum_{n=1}^{\infty} \left(\frac{2n-1}{2n}\right)^n$
42. $\sum_{n=1}^{\infty} (-1)^n e^{-\sqrt{n}}$
43. $\sum_{n=1}^{\infty} n(n^2+1)^{-5/4}$
44. $\sum_{n=1}^{\infty} ne^{-3n}$
45. $\sum_{n=1}^{\infty} \sin\left(\frac{\pi}{n}\right)$
46. $\sum_{n=1}^{\infty} \cos\left(\frac{\pi}{2} + \frac{2}{n}\right)$
47. $\sum_{n=1}^{\infty} \frac{1}{\sqrt{n}} e^{-\sqrt{n}}$
48. $\sum_{n=1}^{\infty} \exp\left(\frac{n+1}{n^2+1}\right)$

49. $\sum_{n=1}^{\infty} \sqrt{\frac{n}{3^n}}$

50. $\sum_{n=1}^{\infty} \cos\left(\frac{3\ln(n)}{n^4}\right)$

51. $\sum_{n=1}^{\infty} \frac{(2n)!}{n^n}$

52. $\sum_{n=1}^{\infty} \frac{2^n n^{3/2}}{5^n}$

53. $\sum_{n=2}^{\infty} \frac{\arctan(n+1)}{n^2 \ln(n)}$

54. $\sum_{n=1}^{\infty} \left(\frac{n^3+1}{n^6+n^4+2}\right)^{1/5}$

55. $\sum_{n=1}^{\infty} \frac{1}{1+n+\cos(n)}$

56. $\sum_{n=1}^{\infty} \frac{n^n}{(3n)!}$

57. $\sum_{n=1}^{\infty} ne^{-n^3}$

58. $\sum_{n=1}^{\infty} n^{-\sqrt{n}}$

III

59. (i) Use a programmable calculator or computer to evaluate $\sum_{n=1}^{20} \frac{1}{n!}$. The hard part of this question is: (ii) Estimate how accurately this value approximates $\sum_{n=1}^{\infty} \frac{1}{n!}$.

60. Estimate $\sum_{n=1}^{\infty} \frac{5^n}{n!}$ to within 0.001.

10.8 Taylor Series

TAYLOR SERIES
MACLAURIN SERIES
REMAINDER

In Chapter 9 we studied Taylor polynomials. Given a function $f(x)$ and a point $x = a$, the Taylor polynomial of degree n for f about $x = a$ is a polynomial $P_n(x)$ of degree n (or less) that has the property that its derivatives at $x = a$ up to and including the derivative of order n are equal to the corresponding derivatives of f at $x = a$. The idea was that such a polynomial ought to be "close" to $f(x)$, at least for x near a.

In Definition 9.1.2 we found that

$$P_n(x) = \sum_{k=0}^{n} \frac{f^{(k)}(a)}{k!}(x-a)^k = \sum_{k=0}^{n} a_k(x-a)^k,$$

where $a_k = \frac{f^{(k)}(a)}{k!}$.

In this section we consider the possibility of continuing this finite sum and making it into an infinite series. To do this, we have to know that f can be differentiated infinitely often, that is, that $f^{(k)}(a)$ exists for *every* k. Second, we have to consider the question of whether or not the resulting series converges. Finally, at points at which it does converge, we have to consider the question of whether the limit to which the series converges equals the value of the original function.

Each of these can be a difficult problem, but surprisingly often everything works out very well. For many functions the series does converge to the value of the function, so it can be used to calculate, or at least estimate, the value of the function. It can also be used to study other aspects of the function's behavior.

DEFINITION 10.8.1

Taylor Series

Suppose $f(x)$ is a function that is *infinitely differentiable* at $x = a$, that is, so that $f^{(k)}(a)$ exists for each k.

The **Taylor series** for f about the point $x = a$ is the series

$$\sum_{k=0}^{\infty} \frac{f^{(k)}(a)}{k!}(x-a)^k = \sum_{k=0}^{\infty} a_k(x-a)^k,$$

where $a_k = \frac{f^{(k)}(a)}{k!}$.

EXAMPLE 1

(i) Consider the function $f(x) = e^x$. We know that for each k, $f^{(k)}(x) = \frac{d^k}{dx^k}(e^x) = e^x$, so for $a = 0$, for instance, we find that $a_k = \frac{e^0}{k!} = \frac{1}{k!}$. The Taylor series for e^x is

$$\sum_{k=0}^{\infty} a_k(x-a)^k = \sum_{k=0}^{\infty} \frac{1}{k!}x^k = \sum_{k=0}^{\infty} \frac{x^k}{k!}.$$

In fact we know from the Ratio Test (see Example 10.7.1(ii)), that this series converges for every value of x. Even more, we will soon see that it actually converges to the right value, that is, for every x, $e^x = \sum_{k=0}^{\infty} \frac{1}{k!}x^k$.

(ii) With the same function $f(x) = e^x$ and $a = 1$ we see that the coefficients of the series are now $a_k = \frac{f^{(k)}(1)}{k!} = \frac{e}{k!}$, and the series is

$$\sum_{k=0}^{\infty} a_k(x-a)^k = \sum_{k=0}^{\infty} \frac{e}{k!}(x-1)^k = e\sum_{k=0}^{\infty} \frac{(x-1)^k}{k!}.$$

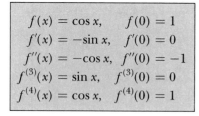

$$\begin{aligned} f(x) &= \cos x, & f(0) &= 1 \\ f'(x) &= -\sin x, & f'(0) &= 0 \\ f''(x) &= -\cos x, & f''(0) &= -1 \\ f^{(3)}(x) &= \sin x, & f^{(3)}(0) &= 0 \\ f^{(4)}(x) &= \cos x, & f^{(4)}(0) &= 1 \end{aligned}$$

(iii) Let $f(x) = \cos x$. We calculate its derivatives up to $f^{(4)}(x)$ and observe that after that, they repeat the same cycle of four over and over again. The odd-numbered derivatives are 0 at $x = 0$, and the even-numbered ones alternate between 1 and -1.

To find the coefficients of the Taylor series, we divide the kth derivative by $k!$, so the coefficients, starting with the $k = 0$ coefficient, are $1, 0, -\frac{1}{2!}, 0, \frac{1}{4!}, 0, -\frac{1}{6!}, 0, \frac{1}{8!}, 0, \ldots$. The Taylor series for $f(x) = \cos x$ about $a = 0$ is

$$1 - \frac{x^2}{2!} + \frac{x^4}{4!} - \frac{x^6}{6!} + \frac{x^8}{8!} - \frac{x^{10}}{10!} + \frac{x^{12}}{12!} - \ldots,$$

which can also be written as

$$\sum_{n=0}^{\infty} (-1)^n \frac{x^{2n}}{(2n)!}.$$

(Notice that in this last expression we use $2n$ rather than n to label the powers of x in the series, since we know that the odd powers of x do not appear.)

We can apply the Ratio Test to this sum; the ratio of the term containing x^{2n} over the previous nonzero term (the one containing x^{2n-2}) is

$$\frac{(-1)^n \frac{1}{(2n)!} x^{2n}}{(-1)^{n-1}\frac{1}{(2n-2)!}x^{2n-2}} = (-1)\frac{(2n-2)!}{(2n)!}x^2 = -\frac{x^2}{2n(2n-1)}.$$

The absolute value of this ratio tends to 0 as $n \to \infty$, for any fixed value of x, so by the Ratio Test the series converges absolutely for every x.

(iv) A similar calculation can be made for the function $f(x) = \sin x$. We find that its derivatives evaluated at $a = 0$, starting with $f^{(0)}(0) = f(0)$, are $0, 1, 0, -1, 0, 1, 0, -1, 0, \ldots$, and the Taylor series about $a = 0$ is

$$x - \frac{x^3}{3!} + \frac{x^5}{5!} - \frac{x^7}{7!} + \frac{x^9}{9!} - \ldots = \sum_{n=0}^{\infty} (-1)^n \frac{x^{2n+1}}{(2n+1)!}.$$

Again the Ratio Test can be used to show that the series converges for every value of x. Once again notice that the nth (nonzero) term of the series is not the one containing x^n but rather the one containing x^{2n-1}, because the even powers of x do not appear in the series.

$f^{(1)}(x) = \frac{1}{x}$
$f^{(2)}(x) = -\frac{1}{x^2}$
$f^{(3)}(x) = 2\frac{1}{x^3}$
$f^{(4)}(x) = (-2)(3)\frac{1}{x^4}$
$f^{(5)}(x) = (2)(3)(4)\frac{1}{x^5}$
$f^{(6)}(x) = (-2)(3)(4)(5)\frac{1}{x^6}$

(v) Let $f(x) = \ln x$. This function is defined only for $x > 0$, so let us try $a = 1$. We must find the derivatives of f and evaluate them at $x = 1$. We see that there is a pattern, and

$$f^{(k)}(x) = (-1)^{k+1}(2)(3)(4)(5)\ldots(k-1)\frac{1}{x^k} = (-1)^{k+1}(k-1)!\frac{1}{x^k}.$$

In particular, when we evaluate at $x = 1$, we find that $f^{(0)}(1) = f(1) = \ln(1) = 0$, and for $k \geq 1$, $f^{(k)}(1) = (-1)^{k+1}(k-1)!$. From this we find that for $k \geq 1$ the kth coefficient of the Taylor series is $a_k = \frac{f^{(k)}(1)}{k!} = \frac{(-1)^{k+1}(k-1)!}{k!} = (-1)^{k+1}\frac{1}{k}$.

The Taylor series for $f(x) = \ln x$ about the point $a = 1$ is

$$\sum_{k=1}^{\infty} a_k(x-a)^k = \sum_{k=1}^{\infty} (-1)^{k+1}\frac{1}{k}(x-1)^k = \sum_{k=1}^{\infty} (-1)^{k+1}\frac{(x-1)^k}{k}.$$

(Notice that the series begins with $k = 1$, since $a_0 = 0$.)

For this series it is possible to apply the Ratio Test. The ratio of the kth term to the $(k-1)$th term is $(-1)(x-1)\frac{k-1}{k}$, whose absolute value $|x-1|\frac{k-1}{k}$ tends to $|x-1|$.

From this we conclude that the series converges absolutely when $|x-1| < 1$ and diverges when $|x-1| > 1$. When $x - 1 = 1$, the series is $\sum_{k=1}^{\infty}(-1)^{k+1}\frac{1}{k}$, which converges because it is an alternating series, the absolute values of whose terms decrease to 0.

When $x - 1 = -1$, the series becomes

$$\sum_{k=1}^{\infty} (-1)^{k+1}\frac{(x-1)^k}{k} = \sum_{k=1}^{\infty} (-1)^{k+1}\frac{(-1)^k}{k} = \sum_{k=1}^{\infty} (-1)^{2k+1}\frac{1}{k} = -\sum_{k=1}^{\infty}\frac{1}{k}.$$

In this case we see that the Taylor series is the negative of the harmonic series, so it diverges.

Combining all this information, we see that the Taylor series converges absolutely when $0 < x < 2$, converges conditionally when $x = 2$, and diverges for $x \leq 0$ and for $x > 2$.

In a sense it is not too surprising that the series does not converge when $x \leq 0$, since the original function $f(x) = \ln x$ is not defined for these values of x, but it is disconcerting to find that for $x > 2$, values for which the function $\ln x$ is defined and perfectly reasonable, the series diverges. In fact, this sort of behavior is typical of Taylor series.

> Notice that in doing these calculations we had to find the value of the kth derivative of $f(x)$ evaluated at $x = a$. We found the derivative, as a function of x, but we did not evaluate it immediately. The point is that knowing $f^{(2)}(a)$, say, is not enough. We need to find the third derivative, and to do that, it is necessary to differentiate the second derivative, so we need to know the second derivative *as a function of* x. Only after we have found the derivatives $f^{(k)}(x)$ do we substitute $x = a$.

It is extremely important to bear in mind that the Taylor series $\sum_{n=0}^{\infty} a_n(x-a)^n$ is a *function* of x. It has different values for different x's, and as we have just seen, it may converge for some x's and diverge for others.

$$f'(x) = -\frac{1}{(1-x)^2}(-1)$$
$$= \frac{1}{(1-x)^2}$$
$$f''(x) = 2\frac{1}{(1-x)^3}$$
$$f^{(3)}(x) = (2)(3)\frac{1}{(1-x)^4}$$
$$f^{(4)}(x) = (2)(3)(4)\frac{1}{(1-x)^5}$$
$$f^{(5)}(x) = (2)(3)(4)(5)\frac{1}{(1-x)^6}$$
$$\vdots$$

EXAMPLE 2 Consider the function $f(x) = \frac{1}{1-x}$. Let us find its Taylor series about the point $a = 0$.

We find the first few derivatives; the pattern should be clear:

$$f^{(k)}(x) = (2)(3)\dots(k)\frac{1}{(1-x)^{k+1}} = k!\frac{1}{(1-x)^{k+1}}.$$

If we evaluate at $x = 0$, we find that $f^{(k)}(0) = k!$, and from this we can find that the kth coefficient of the Taylor series is $a_k = \frac{f^{(k)}(0)}{k!} = \frac{k!}{k!} = 1$. The Taylor series for $f(x) = \frac{1}{1-x}$ about the point $a = 0$ is

$$\sum_{k=0}^{\infty} a_k(x-a)^k = \sum_{k=0}^{\infty} x^k = 1 + x + x^2 + x^3 + x^4 + x^5 + \dots.$$

Having done this, we see that this series is a geometric series with common ratio x, and the formula for evaluating geometric series tells us that if $|x| < 1$, then the series equals $\frac{1}{1-x}$. This is just the function $f(x)$ that we began with.

We see that the Taylor series converges when $|x| < 1$ and that in this case it converges to $f(x)$. The Taylor series for $f(x)$ converges to $f(x)$, provided $|x| < 1$. ◢

As with Taylor polynomials, there is a special name for the series about $a = 0$.

DEFINITION 10.8.2

The Taylor series about the point $a = 0$ is sometimes called the **Maclaurin series**. The Maclaurin series of a function $f(x)$ is

$$\sum_{n=0}^{\infty} \frac{f^{(n)}(0)}{n!}x^n.$$

We refer to "Maclaurin series" or "Taylor series about 0" interchangeably.

We have just seen that the Maclaurin series for $f(x) = \frac{1}{1-x}$ is exactly what we would have expected from the formula for geometric series. This observation can save us a lot of trouble when we calculate series.

EXAMPLE 3

(i) Find the Maclaurin series for $F(x) = \frac{1}{1+x}$.

Solution We could try to find a formula for the kth derivative $F^{(k)}$, but it is easier to realize that $F(x) = \frac{1}{1+x} = \frac{1}{1-(-x)} = f(-x)$, where $f(x) = \frac{1}{1-x}$ is the function from Example 2.

This suggests that the series should be obtained by substituting $-x$ for x in the series for $f(x)$. The Maclaurin series for $F(x)$ is

$$\sum_{n=0}^{\infty} (-1)^n x^n = 1 - x + x^2 - x^3 + x^4 - x^5 + x^6 - \dots.$$

(ii) Find the Maclaurin series for $g(x) = \frac{1}{1+x^2}$.

Solution As above we realize that $g(x) = \frac{1}{1+x^2} = f(-x^2)$, so its series is

$$\sum_{n=0}^{\infty} (-x^2)^n = \sum_{n=0}^{\infty} (-1)^n x^{2n} = 1 - x^2 + x^4 - x^6 + x^8 - x^{10} + \dots.$$

By the Ratio Test we see that this series converges when $|x^2| < 1$, that is, when $|x| < 1$. From our knowledge of geometric series we know that for these values of x it converges to the "right" thing, $g(x)$.

(iii) Find the Maclaurin series for $h(x) = \arctan x$.

Solution We could begin finding the derivatives of $h(x) = \arctan x$. However, it is very tempting to remember that $\arctan x = \int \frac{1}{1+x^2}\,dx$ and just try integrating the series we have just found for $g(x) = \frac{1}{1+x^2}$.

There are several difficulties with this idea. One is that we will have to deal with the constant C that ought to be added to the indefinite integral; we could try instead to use an intelligently chosen definite integral. Another is that we do not really know anything about what it means to integrate a series. Finally, even if we perform the calculation, we will not know that the result is correct.

See the Calculator Example on the next page.

Despite all this, however, it does work, and if you find the first few derivatives of $\arctan x$, you will begin to see that that approach is quite messy, so the idea of integrating is quite appealing.

We know that $\arctan x = \int_0^x \frac{1}{1+t^2}\,dt$ and that for $|t| < 1$, $\frac{1}{1+t^2} = 1 - t^2 + t^4 - t^6 + \ldots = \sum_{n=0}^{\infty} (-1)^n t^{2n}$, so we can try to write

$$\arctan x = \int_0^x \left(\sum_{n=0}^{\infty} (-1)^n t^{2n}\right) dt = \sum_{n=0}^{\infty} \left(\int_0^x (-1)^n t^{2n}\,dt\right)$$

$$= \sum_{n=0}^{\infty} (-1)^n \frac{1}{2n+1} x^{2n+1}.$$

Since x was the upper limit of integration, we used t as the variable inside the integral, and for this reason we wrote out the series for $\frac{1}{1+t^2}$ in terms of t. This is necessary to avoid confusion.

Although we have no real justification for doing the calculation this way, the result does turn out to be correct. This is the Maclaurin series for $\arctan x$.

The calculation we did in the last part of Example 3 is extremely useful. It was much easier to do it this way than it would have been to find all the derivatives of the function $\arctan x$. We will return in the last section of the chapter to discuss this type of calculation, and there we will see that it really is permissible. *Roughly speaking*, operations like integrating and differentiating Taylor series are permissible when both the original series and the resulting series converge absolutely.

In principle, it is always possible to find Taylor series by finding the kth derivative of the function $f(x)$ at the point $x = a$, but in practice the calculations can be quite complicated. If it is possible to find the series by some trick like the one we have just used, then that may well be easier. There are times when we have no choice, but often we can avoid calculating derivatives.

EXAMPLE 4 (i) Find the Maclaurin series of the function $f(x) = \sin(x^3)$.

Solution We could start calculating derivatives, but instead we could just take the series for $\sin t$, which is

$$t - \frac{t^3}{3!} + \frac{t^5}{5!} - \frac{t^7}{7!} + \frac{t^9}{9!} - \ldots = \sum_{n=0}^{\infty} (-1)^n \frac{t^{2n+1}}{(2n+1)!},$$

and substitute into it $t = x^3$. Doing this gives us the Maclaurin series for $f(x) =$

$\sin(x^3)$; it is

$$x^3 - \frac{x^9}{3!} + \frac{x^{15}}{5!} - \frac{x^{21}}{7!} + \frac{x^{27}}{9!} - \dots = \sum_{n=0}^{\infty} (-1)^n \frac{x^{6n+3}}{(2n+1)!}.$$

(ii) Find the Maclaurin series for the function $g(x) = \ln(1+x)$.

Solution In Example 2(v) we found that the Taylor series for $\ln(t)$ about $a = 1$ is

$$\sum_{k=1}^{\infty} (-1)^{k+1} \frac{(t-1)^k}{k}.$$

To find the series for $g(x) = \ln(1+x)$, we try substituting $t = 1 + x$ in this series. We find that $t - 1 = (1 + x) - 1 = x$, and the series equals $\sum_{k=1}^{\infty} (-1)^{k+1} \frac{x^k}{k} = x - \frac{x^2}{2} + \frac{x^3}{3} - \frac{x^4}{4} + \dots$.

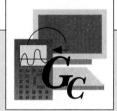

With a programmable calculator we can find partial sums and approximate the series for arctan x from part (iii) of Example 3.

TI-85	SHARP EL9300
PRGM F2 (EDIT) ARCT ENTER F3 F1 (Input) X ENTER F1 (Input) M EXIT ENTER (−) 1 STO N ENTER 0 STO S ENTER Lbl B ENTER N + 1 STO N ENTER S + ((−) 1) ^ N × X ^ (2 N + 1) ÷ (2 N + 1) STO S ENTER F4 F1 (If) (N TEST F2 (<) M) ENTER Goto B ENTER S ENTER EXIT EXIT EXIT PRGM F1 (NAME) ARCT ENTER 1 ENTER 20 ENTER	▼ ▼ (NEW) ENTER ENTER ARCT ENTER COMMAND A 3 (Input) X ENTER COMMAND A 3 (Input) M ENTER N = − 1 ENTER S = 0 ENTER COMMAND B 1 (Label) 1 ENTER N = N + 1 ENTER S = S + ((−) 1) a^b N × X a^b (2 N + 1) ÷ (2 N + 1) ENTER COMMAND 3 (If) N COMMAND C 2 (<) M COMMAND B 2 (Goto) 1 ENTER COMMAND A 1 (Print) S ENTER COMMAND A 6 (End) ENTER (RUN) ENTER (ARCT) ENTER 1 ENTER 20 ENTER
CASIO f_x-6300G (f_x-7700GB)	**HP 48SX**
MODE 2 EXE ? → X : ? → M : − 1 → N : 0 → S : LbL 1 : N + 1 → N : S + ((−) 1) x^y N × X x^y (2 N + 1) ÷ (2 N + 1) → S : N < M ⇒ Goto 1 : S ◢ MODE 1 Prog 0 EXE 1 EXE 20 EXE (on the f_x-7700GB, : is PRGM F6, and Prog is PRGM F3.)	ATTN ↱ CLR « » " " KEY IN X AND M ► SPC PROMPT SPC ′ M ► STO SPC ′ X ► STO SPC 0 SPC 0 SPC M SPC FOR SPC N SPC 1 +/− SPC N y^x X SPC 2 SPC N × 1 + y^x × 2 SPC N × 1 + ÷ + NEXT ▼ ENTER ′ ARCT STO VAR ARCT 1 ENTER 20 ENTER ↰ CONT

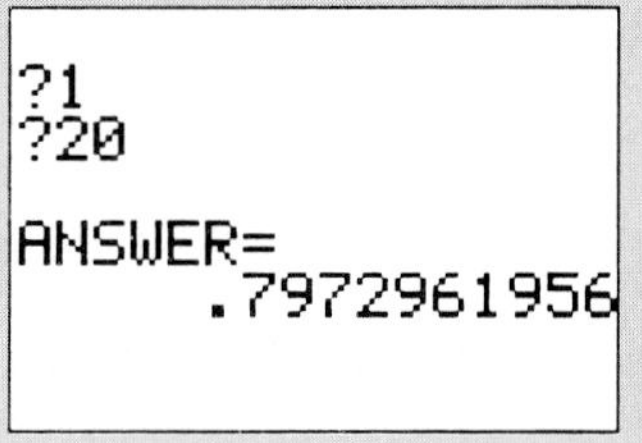

The value we get should be close to arctan $1 = \frac{\pi}{4} \dots$

To apply the Ratio Test to this series, we consider the ratio

$$\frac{(-1)^{k+1}\frac{x^k}{k}}{(-1)^k\frac{x^{k-1}}{k-1}} = (-1)(x)\frac{k-1}{k},$$

whose absolute value is $|x|\frac{k-1}{k}$, which tends to $|x|$ as $k \to \infty$. The Ratio Test tells us that the series converges absolutely for $|x| < 1$ and diverges for $|x| > 1$. For $x = 1$ it is an alternating series with terms whose absolute values decrease to 0, so it converges conditionally. If $x = -1$, it is the negative of the harmonic series, so it diverges.

The series converges for $x \in (-1, 1]$ and diverges elsewhere.

Convergence of Taylor Series to the Original Function

Now that we have learned how to find Taylor series and to test them for convergence, we should investigate the question of when they converge to the original function.

Let us begin with the Maclaurin series for $f(x) = e^x$, which is $1 + x + \frac{x^2}{2!} + \frac{x^3}{3!} + \ldots = \sum_{k=0}^{\infty} \frac{x^k}{k!}$. The partial sums of this series are just the Taylor polynomials $P_n(x) = \sum_{k=0}^{n} \frac{x^k}{k!}$.

To check whether or not the Taylor series converges to $f(x) = e^x$, we have to consider the difference $f(x) - P_n(x)$ and show that it tends to 0 as $n \to \infty$.

But this difference is exactly what we called the **remainder** $R_n(x) = f(x) - P_n(x)$ in Equation 9.2.1. In Theorem 9.2.2 we found a formula for the remainder:

$$R_n(x) = \frac{f^{(n+1)}(c)}{(n+1)!}(x-a)^{n+1},$$

for some number c between a and x.

In the case we are considering, $a = 0$ and $f(x) = e^x$, so the derivatives of $f(x)$ are all equal to e^x. Also the function e^x is an increasing function. If $x \geq 0$, then $0 \leq c \leq x$ and $e^c \leq e^x$. If, on the other hand, $x \leq 0$, then $x \leq c \leq 0$ and $e^c \leq e^0 = 1$. In either case, e^c can never be greater than the maximum of e^x and 1.

If we write M for that maximum, we see that

$$|R_n(x)| = \left|\frac{f^{(n+1)}(c)}{(n+1)!}x^{n+1}\right| = \left|\frac{e^c}{(n+1)!}x^{n+1}\right| \leq \frac{M}{(n+1)!}|x|^{n+1} = M\frac{|x|^{n+1}}{(n+1)!}.$$

As $n \to \infty$, the last expression tends to 0 (cf. Example 10.1.6). This means that $\lim_{n\to\infty} R_n(x) = 0$, and since $R_n(x) = f(x) - P_n(x)$, this means that $\lim_{n\to\infty} P_n(x) = f(x)$. Since $P_n(x)$ is the nth partial sum of the Taylor series of $f(x)$ about $a = 0$, we see that the series converges to $f(x)$, for each value of x.

This same approach can be used to show that many Taylor series converge to the appropriate function (for the values for which they converge).

EXAMPLE 5

(i) We have found that the Maclaurin series for $f(x) = \cos x$ is

$$\sum_{n=0}^{\infty} (-1)^n \frac{x^{2n}}{(2n)!}$$

and that it converges for every x. If we let $R_n(x) = f(x) - P_n(x)$ be the remainder for the Taylor polynomial of order n about $a = 0$, then we know that

$$R_n(x) = \frac{f^{(n+1)}(c)}{(n+1)!}(x-a)^{n+1},$$

for some c between $a = 0$ and x.

Now every derivative of $f(x) = \cos x$ is $\pm \sin x$ or $\pm \cos x$, and the absolute value of any one of them is always less than or equal to 1. Because of this,

$$\begin{aligned} |R_n(x)| &= \left| \frac{f^{(n+1)}(c)}{(n+1)!}(x-a)^{n+1} \right| \leq \left| \frac{1}{(n+1)!} x^{n+1} \right| \\ &= \left| \frac{x^{n+1}}{(n+1)!} \right|. \end{aligned}$$

We know that this last expression tends to zero as $n \to \infty$, so $\lim_{n\to\infty} |R_n(x)| = 0$. Since $R_n(x) = f(x) - P_n(x)$, this means that $\lim_{n\to\infty} P_n(x) = f(x)$. Since $P_n(x)$ is the nth partial sum of the Taylor series for $f(x)$ about $a = 0$, we see that the Taylor series converges to $f(x)$ for every x.

(ii) In the case of $F(x) = \frac{1}{1-x}$, whose Maclaurin series is $1 + x + x^2 + x^3 + \ldots = \sum_{n=0}^{\infty} x^n$, we saw that the series converges to $F(x)$ because it is a geometric series, and because we knew about geometric series, it was possible to show this convergence without using the formula for the remainder.

EXAMPLE 6 A Taylor Series That Converges to the Wrong Function In the examples so far the emphasis has been on finding a Taylor series, after which we have checked that it converges to the "right" value, which it always has. This is frequently what happens, but it is partly because we have chosen "nice" functions. It is possible to find Taylor series that converge, but to the wrong values.

For instance, let

$$f(x) = \begin{cases} e^{-1/x^2}, & \text{if } x \neq 0, \\ 0, & \text{if } x = 0. \end{cases}$$

Then for $x \neq 0$,

$$f'(x) = e^{-1/x^2}\left(\frac{2}{x^3}\right),$$

$$\begin{aligned} f''(x) &= e^{-1/x^2}\left(\frac{2}{x^3}\right)^2 - e^{-1/x^2}\left(\frac{6}{x^4}\right) \\ &= e^{-1/x^2}\left(\left(\frac{2}{x^3}\right)^2 - \left(\frac{6}{x^4}\right)\right) = e^{-1/x^2}\left(\frac{4-6x^2}{x^6}\right). \end{aligned}$$

If we continue taking higher derivatives $f^{(n)}(x)$, each of them will be e^{-1/x^2} times a rational function whose denominator is a power of x. By using l'Hôpital's Rule or Theorem 7.2.5 (p. 518) it is easy to see that $\lim_{x\to 0} e^{-1/x^2} x^{-k} = 0$, for any k, and we conclude that $\lim_{x\to 0} f^{(n)}(x) = 0$, for every n.

It takes some work to show that $f^{(n)}(x)$ exists at $x = 0$, for every n, but it does, and the argument just given suggests that $f^{(n)}(0) = 0$, for every $n \geq 0$. The Maclaurin series for $f(x)$ has every coefficient equal to 0, that is, it is $\sum_{n=0}^{\infty} 0(x^n)$. It converges for every x, and it converges to 0.

This gives an example of a function whose Maclaurin series converges for every x to the value 0, even though the function is nonzero for every x except $x = 0$.

Exercises 10.8

I

Find the Taylor series for each of the following functions about the specified point $x = a$.

1. $f(x) = e^{-2x}$ about $a = 0$
2. $f(x) = e^{1-x}$ about $a = 0$
3. $f(x) = \frac{1}{x}$ about $a = 1$
4. $f(x) = \frac{1}{-x}$ about $a = -1$
5. $f(x) = x^3 - 2x^2 - x + 5$ about $a = 0$
6. $f(x) = 2x^3 - 4x^2 + x - 3$ about $a = 2$
7. $f(x) = \cos x$ about $a = \frac{\pi}{2}$
8. $f(x) = \ln(x)$ about $a = 3$
9. $f(x) = \frac{1}{1+2x^2}$ about $a = 0$
10. $f(x) = xe^x$ about $a = 0$

Find the Taylor series for each of the following functions about the specified point $x = a$.

11. $f(x) = 2 - e^{3x}$ about $a = 1$
12. $f(x) = \ln(1 + x^2)$ about $a = 0$
13. $f(x) = \frac{1}{x-2}$ about $a = 0$
14. $f(x) = \arctan(x^2)$ about $a = 0$
15. $f(x) = x^2 + \cos x$ about $a = 0$
16. $f(x) = \cosh x$ about $a = 0$
17. $f(x) = e^{-2x}$ about $a = \frac{1}{2}$
18. $f(x) = \sqrt{x}$ about $a = 1$
19. $f(x) = \cos x + \sin x$ about $a = 0$
20. $f(x) = \cos^2 x + \sin^2 x$ about $a = \frac{\pi}{3}$

II

21. Find the Taylor series of $f(x) = \sqrt{1 + x}$ about the point $a = 0$.

22. (i) Find the Taylor series of the function $f(x) = \sin x$ about the point $a = \frac{\pi}{4}$.
    (ii) For what values of x does this series converge?

23. Find the first six derivatives of arctan x, and evaluate them at $x = 0$. Use these values to find the first seven terms of the Maclaurin series for arctan x, and confirm that they agree with what we found by integrating in Example 3(iii) (p. 699).

24. Find the first four derivatives of $\sin(x^3)$, and evaluate them at $x = 0$. Use these values to find the first five terms of the Maclaurin series for $\sin(x^3)$, and confirm that they agree with what we found in Example 4(i) (p. 699).

25. By differentiating the Taylor series about $a = 0$ for $\ln(1 + x^2)$, find the Taylor series about $a = 0$ for $\frac{x}{1+x^2}$. Confirm that it has the right value when $x = 0$.

26. Find the Maclaurin series for $\ln(1 + x)$ by integrating the geometric series for $\frac{1}{1+x}$.

27. Use the formula for the remainder to show that for every x the Maclaurin series for $f(x) = \sin x$ converges to $f(x)$.

28. Use the formula for the remainder to show that for every x the Taylor series for $f(x) = \cos x$ about the point $a = \frac{\pi}{4}$ converges to $f(x)$.

*29. Suppose $f(x)$ is an even function that is defined for every real number x, that is, $f(-x) = f(x)$. Show that the Maclaurin series for $f(x)$ does not contain any odd powers of x.

*30. Suppose $f(x)$ is an odd function that is defined for every real number x, that is, $f(-x) = -f(x)$. Show that the Maclaurin series for $f(x)$ does not contain any even powers of x.

*31. Use the formula for the remainder to show that the Taylor series for $f(x) = \ln x$ about the point $a = 1$ converges to $f(x)$ for every $x \in [1, 2]$.

*32. Let $p(x)$ be a polynomial, and show that the Maclaurin series for $p(x)$ is exactly equal to $p(x)$.

III

For each of the following functions $f(x)$, use a programmable calculator or computer to evaluate the partial sum of the corresponding Taylor series about the specified point $x = a$, up to and including the term with $(x - a)^N$, evaluated at the given x.

33. $f(x) = e^x$, $N = 22$, $a = 0$, $x = 3$

34. $f(x) = e^x$, $N = 18$, $a = 1$, $x = -1$

35. $f(x) = \sin(2x)$, $N = 25$, $a = 0$, $x = \pi$

36. $f(x) = \cos(x^2)$, $N = 20$, $a = 0$, $x = \pi$

37. $f(x) = \ln(1 + x^2)$, $N = 300$, $a = 0$, $x = \frac{1}{2}$

38. $f(x) = \sinh(x)$, $N = 39$, $a = 0$, $x = 2$

39. $f(x) = \int_0^x e^{-t^3}\, dt$, $N = 16$, $a = 0$, $x = 3$

40. $f(x) = \int_0^{x^2} e^{-t^2}\, dt$, $N = 22$, $a = 0$, $x = 1$

POINT TO PONDER

It is interesting to consider l'Hôpital's Rule in light of Taylor series. For instance, consider $\lim_{x\to 0} \frac{\sin x}{x}$. The numerator and denominator both tend to 0, so we can evaluate the limit using l'Hôpital's Rule.

If we write out the Maclaurin series for $\sin x$, which is

$$x - \frac{x^3}{3!} + \frac{x^5}{5!} - \frac{x^7}{7!} + \frac{x^9}{9!} - \cdots,$$

and divide by x, we obtain the series for $\frac{\sin x}{x}$,

$$1 - \frac{x^2}{3!} + \frac{x^4}{5!} - \frac{x^6}{7!} + \frac{x^8}{9!} - \cdots.$$

If we evaluate this series at $x = 0$, we find that its value is 1, which is the limit we want.

For another example, consider $\lim_{x\to 0} \frac{x\cos x}{1-e^x}$. Both numerator and denominator tend to 0. The series for $x \cos x$ can be obtained by multiplying the series for $\cos x$ by x; the result is $x - \frac{x^3}{2!} + \frac{x^5}{4!} - \frac{x^7}{6!} + \frac{x^9}{8!} - \cdots$. The series for $1 - e^x$ is $-x - \frac{x^2}{2!} - \frac{x^3}{3!} - \frac{x^4}{4!} - \frac{x^5}{5!} - \frac{x^6}{6!} - \cdots$.

Each of these series begins with x (i.e., not with a constant term). This reflects the fact that both functions are zero at $x = 0$. We can "factor out" x from each of them: The series for $x \cos x$ is $x\left(1 - \frac{x^2}{2!} + \frac{x^4}{4!} - \frac{x^6}{6!} + \frac{x^8}{8!} - \cdots\right)$. The series for $1 - e^x$ is $x\left(-1 - \frac{x}{2!} - \frac{x^2}{3!} - \frac{x^3}{4!} - \frac{x^4}{5!} - \frac{x^5}{6!} - \cdots\right)$.

The quotient is

$$\frac{x\left(1 - \frac{x^2}{2!} + \frac{x^4}{4!} - \frac{x^6}{6!} + \frac{x^8}{8!} - \cdots\right)}{x\left(-1 - \frac{x}{2!} - \frac{x^2}{3!} - \frac{x^3}{4!} - \frac{x^4}{5!} - \frac{x^5}{6!} - \cdots\right)} = \frac{1 - \frac{x^2}{2!} + \frac{x^4}{4!} - \frac{x^6}{6!} + \frac{x^8}{8!} - \cdots}{-1 - \frac{x}{2!} - \frac{x^2}{3!} - \frac{x^3}{4!} - \frac{x^4}{5!} - \frac{x^5}{6!} - \cdots}.$$

As $x \to 0$, the numerator tends to 1 and the denominator tends to -1, so the quotient tends to -1.

In effect we are using l'Hôpital's Rule, because we are really just calculating the quotient of the coefficients of x in the two series, which are equal to the derivatives of the numerator and denominator at $x = 0$.

The argument we have given is not fully justified, because we have not, for instance, defined dividing series or taking limits, but it can be made into a precise argument, and the underlying idea is correct.

10.9 Power Series

RADIUS OF CONVERGENCE
INTERVAL OF CONVERGENCE
CONVERGENCE OF FAMILIAR TAYLOR SERIES

In Section 10.8 we studied Taylor series, which are series of the form

$$\sum_{n=0}^{\infty} a_n (x - a)^n,$$

in which each term contains a *power* of $x - a$. The series obviously depends on the value of x, and it defines a function of x for those x's for which it converges.

In this section we study some properties of this type of series, and in particular we describe, in a fairly simple way, the set of x's for which the series converges.

In the next section we will continue our investigation of these series and consider the possibility of differentiating or integrating them by the simple technique of differentiating or integrating each term separately.

DEFINITION 10.9.1

A series of the form $\sum_{n=0}^{\infty} a_n(x-a)^n$ is called a **power series.**

EXAMPLE 1

(i) The obvious examples of power series are Taylor series. We know a number of examples of series of this type. They include the Maclaurin series for e^x, which is $\sum_{n=0}^{\infty} \frac{x^n}{n!}$. This series converges for every possible x. We also know the series for $\sin x$, which is $\sum_{n=0}^{\infty} (-1)^n \frac{x^{2n+1}}{(2n+1)!}$, and the series for $\cos x$, which is $\sum_{n=0}^{\infty} (-1)^n \frac{x^{2n}}{(2n)!}$. Both these series also converge for every x.

Another familiar series is the series for $\frac{1}{1-x}$, which is the geometric series $\sum_{n=0}^{\infty} x^n$; it converges absolutely when $|x| < 1$ and diverges if $|x| \geq 1$. We also found that the series for $\ln x$ about the point $a = 1$ is $\sum_{n=1}^{\infty} \frac{(-1)^{n+1}}{n}(x-1)^n$, which converges absolutely when $|x-1| < 1$, converges conditionally when $x = 2$, and diverges when $x \leq 0$ or when $x > 2$. Notice that the set of x's for which the series converges is the interval $(0, 2]$.

(ii) It is also possible to "make up" power series without obtaining them as Taylor series. For instance, $\sum_{n=0}^{\infty} 2^{-n}(3n^2-2n+1)(x-3)^n$ is a power series. Using the Ratio Test it is easy to see that it converges absolutely for $|x-3| < 2$ and diverges if $|x-3| > 2$. This means that the set of x's for which it converges must be some sort of interval whose endpoints are 1 and 5; we do not know whether either of the endpoints is included in the interval. In fact, if we substitute $x = 1$ into the series, it becomes $\sum_{n=0}^{\infty} 2^{-n}(3n^2-2n+1)(1-3)^n = \sum_{n=0}^{\infty} 2^{-n}(3n^2-2n+1)(-2)^n = \sum_{n=0}^{\infty} (3n^2-2n+1)(-1)^n$. The terms of this series do not approach 0, so the series diverges. Almost exactly the same thing happens with $x = 5$, and we see that the series converges for $x \in (1, 5)$ and diverges elsewhere.

(iii) Consider the series $\sum_{n=0}^{\infty} n!(x+4)^n$. To apply the Ratio Test, we consider the ratio

$$\frac{n!(x+4)^n}{(n-1)!(x+4)^{n-1}} = n(x+4).$$

If $x+4 > 0$, this ratio tends to ∞; if $x+4 < 0$, the ratio tends to $-\infty$; if $x+4 = 0$, the ratio tends to 0. We see by the Ratio Test that the series diverges unless $x = -4$.

These examples suggest that the Ratio Test is likely to be particularly useful for assessing where a power series converges. Suppose we consider the power series $\sum_{n=0}^{\infty} a_n x^n$. To use the Ratio Test, we have to take the ratio $\left|\frac{a_n x^n}{a_{n-1}x^{n-1}}\right| = \left|\frac{a_n}{a_{n-1}}x\right|$ and find its limit as $n \to \infty$.

Suppose the quotient $\left|\frac{a_n}{a_{n-1}}\right|$ of the absolute values of the nth coefficient and the preceding one tends to some limit ρ, that is,

$$\lim_{n\to\infty}\left|\frac{a_n}{a_{n-1}}\right| = \rho.$$

In this case the ratio we considered for the Ratio Test will tend to the limit

$$\lim_{n\to\infty}\left|\frac{a_n}{a_{n-1}}x\right| = |x| \lim_{n\to\infty}\left|\frac{a_n}{a_{n-1}}\right| = |x|\rho.$$

The Ratio Test tells us that the series converges absolutely if $|x|\rho < 1$ and that it diverges if $|x|\rho > 1$.

If $\rho = 0$, this means that the series will converge for every x. If $\rho = \infty$, it means that the series will only converge when $x = 0$. If ρ is a nonzero number, then the series will converge absolutely when $|x| < \frac{1}{\rho}$ and will diverge when $|x| > \frac{1}{\rho}$.

If we consider a power series $\sum_{n=0}^{\infty} a_n(x-a)^n$ about some point a (instead of just $a = 0$), then an identical argument shows that, provided $\lim_{n\to\infty}\left|\frac{a_n}{a_{n-1}}\right| = \rho$ exists, the series will converge absolutely when $|x-a|\rho < 1$ and diverge when $|x-a|\rho > 1$.

This suggests the following theorem.

THEOREM 10.9.2

Consider the power series $\sum_{n=0}^{\infty} a_n(x-a)^n$. Then there are three possibilities: Either

(i) the series converges for every x,

(ii) or the series diverges except for $x = a$,

(iii) or there is a positive number R so that the series converges absolutely when $|x-a| < R$ and diverges when $|x-a| > R$.

The number R is called the **radius of convergence** of the series; in case (i) we say that $R = \infty$, and in case (ii) we say that $R = 0$.

In case (iii) we see that the series converges absolutely when $x \in (a-R, a+R)$ and diverges when $x > a+R$ and when $x < a-R$. It is less clear what it will do for $x = a \pm R$. It may converge or diverge at either of them.

In any event the set of x's at which the series converges is one of the following intervals: $(a-R, a+R)$, $(a-R, a+R]$, $[a-R, a+R)$, or $[a-R, a+R]$. Whichever one it is, this interval is called the **interval of convergence**. In case (i) the interval of convergence is the whole line $(-\infty, \infty)$; in case (ii) the interval of convergence is $[a, a]$, which contains only one point.

Finally, if $\lim_{n\to\infty}\left|\frac{a_n}{a_{n-1}}\right| = \rho$ exists, then it equals the reciprocal of the radius of convergence: $\rho = \frac{1}{R}$. (This is true in a sense even in the case in which $\rho = \infty$, in which $R = 0$, and in the case in which $\rho = 0$, where $R = \infty$.)

In the discussion preceding the statement of the theorem, we have proved it under the assumption that the limit $\lim_{n\to\infty}\left|\frac{a_n}{a_{n-1}}\right|$ exists. Without this assumption the theorem is more difficult to prove.

EXAMPLE 2

See the Calculator Example on the next page.

(i) For the series $\sum_{n=0}^{\infty} \frac{x^n}{n!}$, $\sum_{n=0}^{\infty} (-1)^n \frac{x^{2n}}{(2n)!}$, and $\sum_{n=0}^{\infty} (-1)^n \frac{x^{2n+1}}{(2n+1)!}$, which are the Maclaurin series for e^x, $\cos x$, and $\sin x$, respectively, the radius of convergence is $R = \infty$.

If $a_n = \frac{1}{n!}$ (i.e., the coefficients of the series for e^x), then $\frac{a_n}{a_{n-1}} = \frac{1}{n} \to 0$ as $n \to \infty$. In this case, $\rho = 0$ and $R = \infty$.

For the series for $\cos x$ the odd-numbered coefficients are zero, so the ratio of consecutive coefficients will not make sense; for the series of $\sin x$ the same problem arises because the even-numbered coefficients are zero. (However, in each of these cases the quotient of consecutive *nonzero* coefficients makes sense, and this ratio tends to 0 in each case.)

(ii) Consider the series $\sum_{n=1}^{\infty} (-1)^n \frac{(x-1)^n}{n}$, which is the Taylor series about $a = 1$ of $f(x) = \ln x$. The absolute value of the ratio of consecutive coefficients is

$$\left| \frac{\frac{(-1)^n}{n}}{\frac{(-1)^{n-1}}{n-1}} \right| = \left| (-1)\frac{n-1}{n} \right| = \frac{n-1}{n},$$

which tends to 1 as $n \to \infty$. From the remark at the end of Theorem 10.9.2 we see that the radius of convergence is $R = 1$. From our study of this series earlier we know that the interval of convergence is $(0, 2]$.

(iii) For the series $\sum_{n=1}^{\infty} \frac{(-5)^n}{2n^3+1}(x+7)^{3n+1}$ we observe that the only nonzero coefficients are those corresponding to $(x+7)^4$, $(x+7)^7$, $(x+7)^{10}, \ldots$. The ratio of the coefficients of $(x+7)^m$ and $(x+7)^{m-1}$ is not a useful quantity to consider because at least one of these coefficients is always zero (so the ratio is either 0 or undefined).

However, suppose we rewrite the series as

$$\begin{aligned} \sum_{n=1}^{\infty} \frac{(-5)^n}{2n^3+1}(x+7)^{3n+1} &= (x+7)\sum_{n=1}^{\infty} \frac{(-5)^n}{2n^3+1}(x+7)^{3n} \\ &= (x+7)\sum_{n=1}^{\infty} \frac{(-5)^n}{2n^3+1}\left((x+7)^3\right)^n \\ &= (x+7)\sum_{n=1}^{\infty} \frac{(-5)^n}{2n^3+1}u^n, \end{aligned}$$

where $u = (x+7)^3$.

If we look now at this power series in the variable u, we see that it does make sense to consider the absolute value of the ratio of consecutive coefficients. It is

We could also consider the ratio of consecutive non-zero coefficients in the original series.

$$\left| \frac{\frac{(-5)^n}{2n^3+1}}{\frac{(-5)^{n-1}}{2(n-1)^3+1}} \right| = \left| (-5)\frac{2n^3 - 6n^2 + 6n - 1}{2n^3+1} \right| = 5\left| \frac{2n^3 - 6n^2 + 6n - 1}{2n^3+1} \right|,$$

which tends to 5 as $n \to \infty$.

So the power series in u has radius of convergence equal to $\frac{1}{5}$; this means that it converges absolutely when $|u| < \frac{1}{5}$ and diverges when $|u| > \frac{1}{5}$. Since $u = (x+7)^3$, we see that the series converges absolutely when $|x+7|^3 < \frac{1}{5}$, which is the same thing as saying $|x+7| < \sqrt[3]{\frac{1}{5}}$. By the same reasoning it diverges when $|x+7| > \sqrt[3]{\frac{1}{5}} = 5^{-1/3}$.

We see that the radius of convergence of the original series in powers of $x+7$ is $\sqrt[3]{\frac{1}{5}} = 5^{-1/3} \approx 0.5848$. It converges absolutely on the interval $(-7 - 5^{-1/3}, -7 + 5^{-1/3})$.

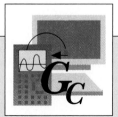

With a programmable calculator we can find partial sums for the Maclaurin series for e^x from part (i) of Example 2. We know that for every x the series converges to e^x. It is interesting to see how fast the partial sums converge for various values of x.

In programming with power series it is often more efficient to store the value of the nth term (our variable T) and use it to calculate the next term.

TI-85	SHARP EL9300
PRGM F2(EDIT) EXPML ENTER F3 F1(Input) X ENTER F1(Input) M EXIT ENTER 0 STO N ENTER 1 STO T ENTER 1 STO S ENTER Lbl B ENTER N + 1 STO N ENTER T × X ÷ N STO T ENTER S + T STO S ENTER F4 F1(If) (N TEST F2(<) M) ENTER Goto B ENTER S ENTER EXIT EXIT EXIT PRGM F1(NAME) EXPML ENTER 1 ENTER 10 ENTER	▼ ▼ (NEW) ENTER ENTER EXPML ENTER COMMAND A 3 (Input) X ENTER COMMAND A 3 (Input) M ENTER N = 0 ENTER T = 1 ENTER S = 1 ENTER COMMAND B 1 (Label) 1 ENTER N = N + 1 ENTER T = T × X ÷ N ENTER S = S + T ENTER COMMAND 3 (If) N COMMAND C 2 (<) M COMMAND B 2 (Goto) 1 ENTER COMMAND A 1 (Print) S ENTER COMMAND A 6 (End) ENTER (RUN) ENTER (EXPML) ENTER 1 ENTER 10 ENTER
CASIO f_x-6300G (f_x-7700GB)	HP 48SX
MODE 2 EXE ? → X : ? → M : 0 → N : 1 → T : 1 → S : LbL 1 : N + 1 → N : T × X ÷ N → T : S + T → S : N < M ⇒ Goto 1 : S ◢ MODE 1 Prog 0 EXE 1 EXE 10 EXE (on the f_x-7700GB, : is PRGM F6, and Prog is PRGM F3.)	ATTN ↱ CLR « » "" KEY IN X AND M ▶ SPC PROMPT SPC ′ M ▶ STO SPC ′ X ▶ STO SPC 1 SPC ′ T ▶ STO SPC 1 SPC 1 SPC M SPC FOR SPC N SPC T SPC X × N ÷ ′ T ▶ STO SPC T + NEXT ▼ ENTER ′ EXPL STO VAR EXPL 1 ENTER 10 ENTER ↰ CONT

```
?1
?10

ANSWER=
      2.718281801
```

Check how close the answer is to $e = e^1$. Now try the series with $x = 10$; see how close we get to e^{10} using the partial sum with $M = 10$ . . .

Summary 10.9.3 Convergence of Familiar Taylor Series

Function	Series	Radius of convergence	Interval of convergence
e^x	$\sum_{n=0}^{\infty} \frac{x^n}{n!} = 1 + x + \frac{x^2}{2} + \frac{x^3}{6} + \frac{x^4}{24} + \ldots$	∞	$(-\infty, \infty)$
$\cos x$	$\sum_{n=0}^{\infty} (-1)^n \frac{x^{2n}}{(2n)!} = 1 - \frac{x^2}{2} + \frac{x^4}{24} - \frac{x^6}{720} + \ldots$	∞	$(-\infty, \infty)$
$\sin x$	$\sum_{n=0}^{\infty} (-1)^n \frac{x^{2n+1}}{(2n+1)!} = x - \frac{x^3}{6} + \frac{x^5}{120} - \frac{x^7}{5040} + \ldots$	∞	$(-\infty, \infty)$
$\frac{1}{1-x}$	$\sum_{n=0}^{\infty} x^n = 1 + x + x^2 + x^3 + x^4 + \ldots$	1	$(-1, 1)$
$\frac{1}{1+x}$	$\sum_{n=0}^{\infty} (-1)^n x^n = 1 - x + x^2 - x^3 + x^4 - \ldots$	1	$(-1, 1)$
$\ln x$	$\sum_{n=1}^{\infty} (-1)^{n+1} \frac{(x-1)^n}{n}$ $= (x-1) - \frac{(x-1)^2}{2} + \frac{(x-1)^3}{3} - \frac{(x-1)^4}{4} + \ldots$	1	$(0, 2]$
$\ln(1+x)$	$\sum_{n=1}^{\infty} (-1)^{n+1} \frac{x^n}{n} = x - \frac{x^2}{2} + \frac{x^3}{3} - \frac{x^4}{4} + \ldots$	1	$(-1, 1]$
$\arctan x$	$\sum_{n=0}^{\infty} (-1)^n \frac{1}{2n+1} x^{2n+1} = x - \frac{x^3}{3} + \frac{x^5}{5} - \frac{x^7}{7} + \ldots$	1	$[-1, 1]$
$\cosh x$	$\sum_{n=0}^{\infty} \frac{x^{2n}}{(2n)!} = 1 + \frac{x^2}{2} + \frac{x^4}{24} + \frac{x^6}{720} + \ldots$	∞	$(-\infty, \infty)$
$\sinh x$	$\sum_{n=0}^{\infty} \frac{x^{2n+1}}{(2n+1)!} = x + \frac{x^3}{6} + \frac{x^5}{120} + \frac{x^7}{5040} + \ldots$	∞	$(-\infty, \infty)$

Exercises 10.9

I

Find the interval of convergence of each of the following power series.

1. $\sum_{n=1}^{\infty} nx^n$
2. $\sum_{n=1}^{\infty} \frac{1}{n}(x+2)^n$
3. $\sum_{n=0}^{\infty} \frac{1}{n^4+3}(x-3)^{n+1}$
4. $\sum_{n=1}^{\infty} \frac{1}{1+\ln(n)}(x-1)^n$
5. $\sum_{n=0}^{\infty} \frac{n!}{(2n)!}(x-\pi)^n$
6. $\sum_{n=1}^{\infty} \frac{n}{n^4+n^2+2}(2x-1)^n$
7. $\sum_{n=3}^{\infty} \frac{4}{(n-1)!}(\sqrt{2}-x)^n$
8. $\sum_{n=2}^{\infty} \left(1-\frac{1}{n}\right)^n (5x+2)^n$
9. $\sum_{n=2}^{\infty} \ln(4n)(x-12)^{n+3}$
10. $\sum_{n=1}^{\infty} 3^{n+2} 2^{-n/2}(2x+2)^n$
11. $\sum_{n=1}^{\infty} n^2 x^{2n+1}$
12. $\sum_{n=1}^{\infty} \frac{2n}{n+1}(2x-3)^{5n}$
13. $\sum_{n=2}^{\infty} 2^{-n} \frac{n}{n^3-1}(x-3)^{n+1}$
14. $\sum_{n=2}^{\infty} e^{-2n}(x+3)^n$
15. $\sum_{n=1}^{\infty} \frac{e^n}{n!}(x+5)^n$
16. $\sum_{n=1}^{\infty} \frac{\sqrt{n}}{n}(2-3x)^n$
17. $\sum_{n=1}^{\infty} \frac{n!}{n^5+n}(x-22)^n$
18. $\sum_{n=1}^{\infty} e^{n^2} x^n$
19. $\sum_{n=2}^{\infty} 2^{-3n} \frac{1}{n\ln(n)}(3x)^n$
20. $\sum_{n=1}^{\infty} 5^{2n+1} x^2 (3x-1)^n$

II

Find the radius of convergence of each of the following.

21. $\sum_{n=1}^{\infty} 2^{-n^2} x^n$
22. $\sum_{n=1}^{\infty} 2^{-n^2}(x-4)^n$
23. $\sum_{n=1}^{\infty} \frac{1}{n} x^{n^2}$

*24. Find the interval of convergence of $\sum_{n=1}^{\infty} 2^n \frac{1}{n} x^{n^2}$.

Find examples of power series whose intervals of convergence are each of the following.

*25. $(3, 5)$

*26. $(-1, 6]$

*27. $[-5, -2)$

*28. $[-4, 0]$

***29.** For the series in Example 2(iii) (p. 707), find the interval of convergence. Specifically, find what happens at the endpoints of the interval.

***30.** Suppose $\sum_{n=0}^{\infty} a_n x^n$ is a power series for which $\lim_{n\to\infty} \left|\frac{a_n}{a_{n-1}}\right|$ exists and equals ρ, where ρ could be any nonnegative number *or* ∞. Consider the related series $\sum_{n=0}^{\infty} n a_n x^n$, and show that its radius of convergence is the same as that of the original series.

***31.** Suppose $\sum_{n=0}^{\infty} a_n (x+3)^n$ is a power series for which $\lim_{n\to\infty} \left|\frac{a_n}{a_{n-1}}\right|$ exists and equals A, where A could be any nonnegative number *or* ∞. Consider the related series $\sum_{n=1}^{\infty} \frac{a_n}{n^2}(x+3)^n$, and show that its radius of convergence is the same as that of the original series.

***32.** (i) Find an example of a power series $\sum_{n=1}^{\infty} a_n x^n$ whose interval of convergence is $(-2, 2)$ but so that the interval of convergence of the related series $\sum_{n=1}^{\infty} \frac{a_n}{n^2} x^n$ is $[-2, 2]$.

(ii) Find an example of a power series $\sum_{n=1}^{\infty} a_n x^n$ whose interval of convergence is $[-4, 4]$ but so that the interval of convergence of the related series $\sum_{n=1}^{\infty} (n^3 + n^2 + 2) a_n x^n$ is $(-4, 4)$.

***33.** Suppose the radius of convergence of $\sum_{n=1}^{\infty} a_n (x-a)^n$ is R.

(i) What is the radius of convergence of $\sum_{n=1}^{\infty} 4^n a_n (x-a)^n$?

(ii) If r is any nonzero number, what is the radius of convergence of $\sum_{n=1}^{\infty} r^n a_n (x-a)^n$?

(iii) If $p(x)$ is any polynomial, what is the radius of convergence of $\sum_{n=1}^{\infty} p(n) a_n (x-a)^n$?

***34.** (i) If $\sum_{n=0}^{\infty} a_n x^n$ and $\sum_{n=0}^{\infty} b_n x^n$ are power series whose radii of convergence are R and r, respectively, show that the radius of convergence of $\sum_{n=0}^{\infty} (a_n + b_n) x^n$ is at least $\min(R, r)$.

(ii) Find an example in which the radius of convergence of the "sum" series is exactly equal to the minimum of the radii of convergence of the two original series.

(iii) Find an example in which the radius of convergence of the "sum" series is greater than the minimum of the radii of convergence of the two original series.

***35.** Suppose the series $\sum_{n=0}^{\infty} a_n (x+2)^n$ diverges with $x = 1$ and converges with $x = -5$. What is its radius of convergence?

POINT TO PONDER

In studying power series it is sometimes necessary to use *complex numbers*. Complex numbers contain the special element i, which satisfies $i^2 = -1$. Every complex number can be written as $z = x + iy$, for some real numbers x and y, and they can be added and multiplied. This allows us to make power series in which the terms are complex numbers. To get a picture, we can associate the complex number $z = x + iy$ to the point (x, y) in the plane; in this way the complex numbers are represented by all the points in the plane. If $y = 0$, then the number $x + iy$ is just the real number x; the complex numbers contain the real numbers, and in the picture of the plane the real numbers correspond to the horizontal axis.

Suppose a is a complex number and a_n is a complex number for each n. Then for any z we can consider the complex power series $\sum_{n=0}^{\infty} a_n (z-a)^n$. This series may or may not converge, but if it does, its limit will be a complex number.

It is not obvious, but it turns out that the set of complex numbers z in the plane for which the series converges is a disk around a. There will be a number R, the *radius of convergence*, so that inside the circle of radius R around a the series converges absolutely and outside the circle it diverges (two examples are shown in Figure 10.9.1). Its behavior on the circle may be complicated.

Now if a and a_n happen to be real numbers, then we can evaluate the series at the real numbers and get an

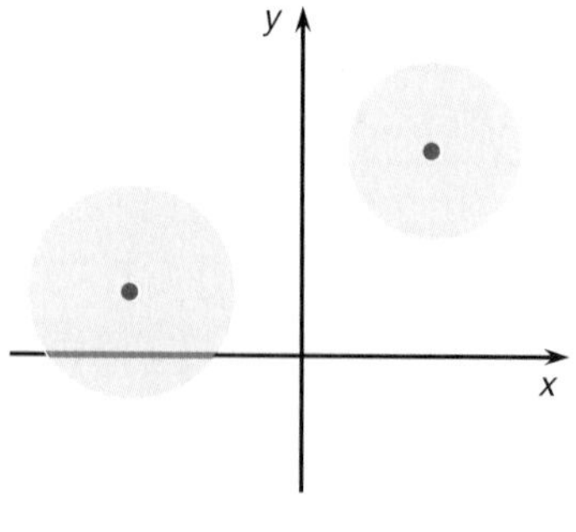

Figure 10.9.1

ordinary real power series. The real numbers at which the series converges will be the points on the real line that also happen to lie inside the disk, as shown in Figure 10.9.2. This set is an interval, and this is why we use the "radius of convergence" terminology for intervals on the line.

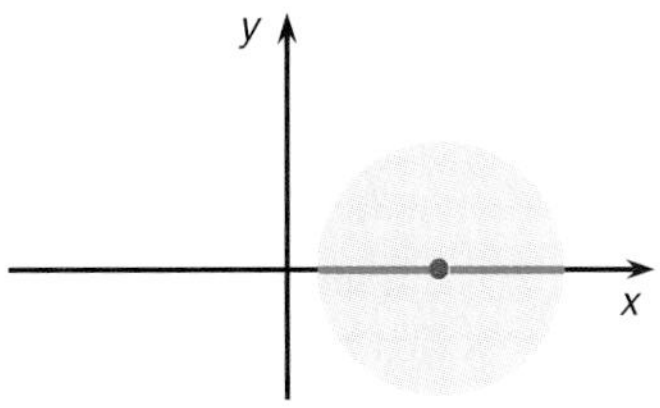

Figure 10.9.2

10.10 Derivatives and Integrals of Power Series

TERM-BY-TERM DIFFERENTIATION AND INTEGRATION

Our study of power series began with Taylor series, and in Section 10.9 we considered power series in their own right. The main result was Theorem 10.9.2, which describes the radius of convergence and the interval of convergence of a power series.

In this section we consider the question of differentiating or integrating a power series. As a preliminary example, consider the Taylor series of the exponential function e^x. It is

$$\sum_{n=0}^{\infty} \frac{x^n}{n!}.$$

Suppose we differentiate each term of this series; the derivative of the nth term is $\frac{d}{dx}\left(\frac{x^n}{n!}\right) = \frac{nx^{n-1}}{n!} = \frac{x^{n-1}}{(n-1)!}$, and we see that this is exactly the $(n-1)$th term. Actually, this is not true for the 0th term 1, whose derivative is 0, but the derivative of each of the other terms equals the preceding term. It is tempting to write

$$\frac{d}{dx}\left(\sum_{n=0}^{\infty} \frac{x^n}{n!}\right) = \sum_{n=0}^{\infty}\left(\frac{d}{dx}\left(\frac{x^n}{n!}\right)\right) = \sum_{n=1}^{\infty} \frac{x^{n-1}}{(n-1)!} = \sum_{m=0}^{\infty} \frac{x^m}{m!}.$$

Notice that after we differentiated, the series began with $n = 1$ (since the derivative of the term with $n = 0$ is zero), but since the nth term then involved the $(n-1)$th power of x, it was simpler to change the index of summation to m, where $m = n - 1$, so m runs from $m = 0$ to ∞.

What makes this calculation interesting is that it gives the right answer. We know from our study of Taylor series that $\sum_{n=0}^{\infty} \frac{x^n}{n!}$ converges for every x to the function e^x. So the derivative of the series ought to be the derivative of e^x, which is e^x, and the series we obtained *is* exactly the Maclaurin series of e^x.

In a similar vein, let us try to find the indefinite integral $\int_0^x \cos t\, dt$ by writing out the Maclaurin series $\sum_{n=0}^{\infty} (-1)^n \frac{t^{2n}}{(2n)!}$ for $\cos t$ and integrating. The calculation appears to be

$$\int_0^x \cos t\, dt = \int_0^x \left(\sum_{n=0}^{\infty} (-1)^n \frac{t^{2n}}{(2n)!}\right) dt = \sum_{n=0}^{\infty} (-1)^n \left(\int_0^x \frac{t^{2n}}{(2n)!}\, dt\right)$$

$$= \sum_{n=0}^{\infty} (-1)^n \left(\frac{1}{2n+1} \frac{t^{2n+1}}{(2n)!}\right)\Bigg|_0^x = \sum_{n=0}^{\infty} (-1)^n \frac{x^{2n+1}}{(2n+1)!}.$$

This is the Maclaurin series for sin x, which is exactly what we expect, since

$$\int_0^x \cos t\, dt = \sin t\big|_0^x = \sin(x) - \sin(0) = \sin x.$$

These examples suggest that it is possible to differentiate and integrate power series by dealing with each term separately. This is called differentiating and integrating *term-by-term.*

Unfortunately, it is not quite as simple as it appears. First there is the usual problem that the series might not converge. Clearly, we should only attempt this approach when the original series converges. Even so, we should then ask whether the series that results from term-by-term differentiation or integration will be convergent.

Now if both the original series and the one that results from the calculation are convergent, there is the question of whether the resulting series converges to the right value, that is, whether it converges to the derivative or the integral we want. We certainly hope it will, but since the series is defined by taking a limit, it is difficult to see what happens when we differentiate or integrate.

Another question arises if we attempt to find an *indefinite integral.* We can find the indefinite integrals of the terms, but we have no way to know what constant to add to each term. If we add peculiar constants to the terms, we can easily make the series divergent.

With all these potential difficulties and complications it is somewhat surprising that it all works.

THEOREM 10.10.1

Suppose $\sum_{n=0}^{\infty} a_n(x-a)^n$ is a power series whose radius of convergence is R, so the series is absolutely convergent on the interval $(a-R, a+R)$ (which means the entire real line if $R=\infty$).

For x in this interval, let f be the function defined by the series

$$f(x) = \sum_{n=0}^{\infty} a_n(x-a)^n.$$

Then

(i) f is differentiable on this interval, and its derivative can be found by differentiating the series **term-by-term**, that is,

$$f'(x) = \sum_{n=1}^{\infty} na_n(x-a)^{n-1},$$

for all $x \in (a-R, a+R)$.

Moreover, the series for the derivative has the same radius of convergence R as the original series.

(ii) The indefinite integral $\int_a^x f(t)\,dt$ can be found by integrating the series for f **term-by-term**, that is,

$$\int_a^x f(t)\,dt = \sum_{n=0}^{\infty} \frac{1}{n+1} a_n(x-a)^{n+1},$$

for each $x \in (a-R, a+R)$.

Moreover, this series for the indefinite integral has the same radius of convergence R as the original series.

(iii) In particular, if $c, d \in (a-R, a+R)$, then the definite integral can be found by **term-by-term** integration:

$$\int_c^d f(t)\,dt = \sum_{n=0}^{\infty}\left(\frac{a_n}{n+1}\left((d-a)^{n+1} - (c-a)^{n+1}\right)\right).$$

Moreover, this series for the definite integral converges absolutely whenever $c, d \in (a-R, a+R)$.

The proof of this theorem is difficult, and we omit it.

Notice that for the indefinite integral in part (ii) of Theorem 10.1.1 it is extremely important (for once) *not* to include a constant c added on to each term. Doing that might make the series diverge.

Notice too that the differentiated and integrated series have the same radius of convergence as the original series; this means that all three of them converge absolutely on the interval $(a-R, a+R)$ and diverge outside the corresponding closed interval $[a-R, a+R]$. The theorem does not say what happens at the endpoints; as the examples will show, it is possible that the three series may behave differently at the endpoints, but they may also all have the same convergence properties. Also remember that if $R = \infty$, then all three series converge absolutely for every possible value.

EXAMPLE 1

(i) For the series $\sum_{n=0}^{\infty} x^n$, the geometric series that is the Maclaurin series for $f(x) = \frac{1}{1-x}$, we know the radius of convergence is $R = 1$ and the interval of convergence is $(-1, 1)$.

The derivative of this function is $f'(x) = \frac{1}{(1-x)^2} = (1-x)^{-2}$, and Theorem 10.10.1 tells us that for $x \in (-1,1)$ it equals the series $\sum_{n=0}^{\infty} nx^{n-1} = \sum_{n=1}^{\infty} nx^{n-1}$. Notice the way we started the sum at $n = 1$, since the term with $n = 0$ is zero; to write it starting with $n = 0$ is not wrong, but it is slightly misleading, so it is preferable to make the change.

At the endpoints of the interval of convergence we see that with $x = \pm 1$ the nth term of the series does not tend to 0, so the differentiated series diverges at the endpoints, and its interval of convergence is $(-1, 1)$, the same as the interval of convergence for the original series for f.

(ii) Next we consider the indefinite integral $\int_0^x f(t)\,dt = \int_0^x \frac{1}{1-t}\,dt = -\ln(1-t)\big|_0^x = -\ln(1-x)$. By part (ii) of Theorem 10.10.1, for $x \in (-1,1)$ it equals $\sum_{n=0}^{\infty} \frac{1}{n+1}x^{n+1} = \sum_{n=1}^{\infty} \frac{1}{n}x^n$. This series converges absolutely for $x \in (-1, 1)$. If $x = 1$, it diverges (it is the harmonic series). If $x = -1$, it is an alternating series whose terms decrease to 0, so it converges. Its interval of convergence is $[-1, 1)$.

(iii) Suppose we integrate again and consider the indefinite integral $\int_0^x \big(-\ln(1-t)\big)\,dt = -\int_0^x \ln(1-t)\,dt$. While it is not terribly difficult, it might take us a moment or two to perform this integration, but writing down the corresponding series is easy. It is $\sum_{n=0}^{\infty} \frac{1}{n+1}\frac{1}{n+2}x^{n+2}$.

Notice that when $x = 1$, this series equals $\sum_{n=0}^{\infty} \frac{1}{(n+1)(n+2)}$, which is convergent. (The nth term is a rational function of n with the degree of the denominator two greater than the degree of the numerator.) The series also converges for $x = -1$, so the interval of convergence is $[-1, 1]$.

(iv) If $f(x) = \cosh x = \frac{1}{2}(e^x + e^{-x})$, then we could calculate its series by finding all its derivatives and evaluating them at $x = 0$, but of course it is much easier to combine the series for e^x and e^{-x}. What we find is

$$\frac{1}{2}\sum_{n=0}^{\infty}\left(\frac{x^n}{n!} + \frac{(-x)^n}{n!}\right) = \frac{1}{2}\sum_{n=0}^{\infty}\left(\frac{1}{n!} + \frac{(-1)^n}{n!}\right)x^n.$$

Since $1 + (-1)^n$ equals 2 when n is even and 0 when n is odd, this series can be written as

$$\frac{1}{2}\sum_{m=0}^{\infty}\frac{2}{(2m)!}x^{2m} = \sum_{m=0}^{\infty}\frac{x^{2m}}{(2m)!}.$$

(Notice the similarity between this series and the series for $\cos x$ . . .).

To find the series for the derivative of $\cosh x$, we simply differentiate. We see that the series for $\sinh x$ is

$$\sum_{m=0}^{\infty}\frac{2mx^{2m-1}}{(2m)!} = \sum_{m=1}^{\infty}\frac{x^{2m-1}}{(2m-1)!} = \sum_{n=0}^{\infty}\frac{x^{2n+1}}{(2n+1)!}.$$

Notice the similarity with the series for $\sin x$. Also notice the way we changed the index of summation to run from $m = 1$ to ∞ instead of starting at $m = 0$.

The Ratio Test shows that both the series for $\cosh x$ and the series for $\sinh x$ converge absolutely for every value of x.

In the last part of Example 1 we found the series for $\cosh x$ by combining the Maclaurin series for e^x and e^{-x}. This provides a series that converges for every value of x and that converges to $\cosh x$ for every x (since we know that the series for e^x and e^{-x} converge to these functions). However, we do not really know that what we have found is actually the Maclaurin series of $\cosh x$. On the face of it there is a possibility that there could be two distinct power series that converge to the same function $\cosh x$.

In fact this is not possible, and we can use Theorem 10.10.1 about term-by-term differentiation to see why.

THEOREM 10.10.2

Suppose $f(x)$ is a function. Suppose too that we have a power series $\sum_{n=0}^{\infty} a_n(x-a)^n$ whose radius of convergence is nonzero and that converges to $f(x)$ for each x in some interval about a, that is, for each $x \in (c, d)$, for some $c < a < d$. Then this series is the Taylor series for $f(x)$ about the point $x = a$.

PROOF Theorem 10.10.1 tells us that we can find not only the value of $f(x)$ at $x = a$ from the series, but also the values of all its derivatives $f^{(k)}(a)$. For instance, the derivative of $f(x)$ equals $\sum_{n=1}^{\infty} na_n(x-a)^{n-1}$, the second derivative equals $\sum_{n=2}^{\infty} n(n-1)a_n(x-a)^{n-2}$, the third derivative equals $\sum_{n=3}^{\infty} n(n-1)$

$(n-2)a_n(x-a)^{n-3}$, and so on. Notice that the series for $f^{(k)}(x)$ starts with the term having $n=k$, since the preceding terms are all zero.

The kth derivative $f^{(k)}(x)$ equals the series

$$\sum_{n=k}^{\infty} n(n-1)\dots(n-k+1)a_n(x-a)^{n-k}.$$

In particular, if we evaluate it at $x=a$, each term except the constant term is zero. The constant term is the term with $n=k$, so $f^{(k)}(a)$ equals the coefficient of this term, that is, $f^{(k)}(a)=k(k-1)\dots(k-k+1)a_k=k(k-1)\dots(1)a_k=k!a_k$.

Turned around, this says that $a_k=\frac{1}{k!}f^{(k)}(a)$, which is exactly the formula for the coefficients of the Taylor series for f. This proves that the series we started with must be the Taylor series, as desired.

This result is extremely useful because it enables us to use various tricks for manipulating power series without having to worry about whether the series that results will in fact be the Taylor series.

EXAMPLE 2

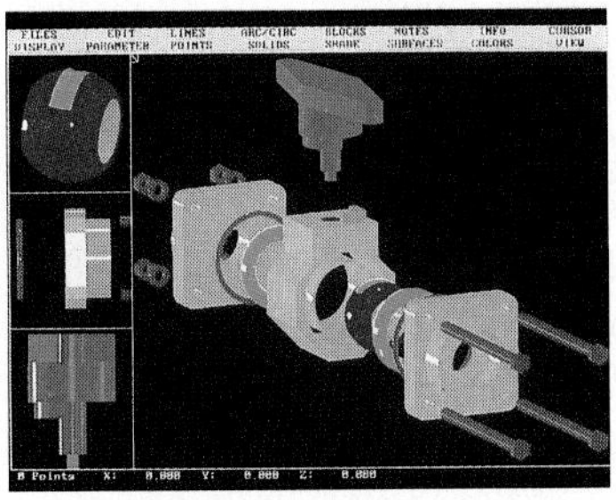

DesignCAD 3D *software was used to produce this exploded picture of valve assembly. The red ball stops fluid flow; the blue component at the top is the control.*

(i) To find the Maclaurin series for $f(x)=e^{-x^2}$, for instance, it is fairly complicated to find a formula for the kth derivative of f, but it is much easier to use the series for e^t and substitute $t=-x^2$. The resulting series is $\sum_{n=0}^{\infty}\frac{1}{n!}t^n=\sum_{n=0}^{\infty}\frac{1}{n!}(-x^2)^n=\sum_{n=0}^{\infty}(-1)^n\frac{1}{n!}x^{2n}$. Since the series for e^t converges to e^t for every value of t, we see that this series converges to e^{-x^2} for every value of x, so by Theorem 10.10.2 it must be the Maclaurin series.

(ii) From our study of geometric series we know that the series $1-x^2+x^4-x^6+x^8-\dots=\sum_{n=0}^{\infty}(-1)^n x^{2n}$ converges to $\frac{1}{1+x^2}$ for every $x\in(-1,1)$. From Theorem 10.10.2 we see that it must be the Maclaurin series of $\frac{1}{1+x^2}$.

By Theorem 10.10.1 we can integrate this series to find the series for arctan x. We find that for $x\in(-1,1)$,

$$\begin{aligned}\arctan x &= \int_0^x \frac{1}{1+t^2}\,dt=\int_0^x\left(\sum_{n=0}^{\infty}(-1)^n t^{2n}\right)dt\\ &=\sum_{n=0}^{\infty}\left(\int_0^x(-1)^n t^{2n}\,dt\right)=\sum_{n=0}^{\infty}\left((-1)^n\frac{1}{2n+1}x^{2n+1}\right).\end{aligned}$$

Theorem 10.10.1 assures us that for $x\in(-1,1)$ this series converges to arctan x, and Theorem 10.10.2 then tells us that it must be the Maclaurin series of arctan x.

(iii) Find the tenth and thirteenth derivatives of $f(x)=\arctan x$ at $x=0$, that is, $f^{(10)}(0)$ and $f^{(13)}(0)$.

Solution Instead of calculating derivatives, we look at the series we just found for arctan x and realize that since it is the Maclaurin series for arctan x, its coefficients are obtained by dividing the derivatives of f by factorials.

Specifically, the coefficient of x^{10} is $\frac{1}{10!}f^{(10)}(0)$, and the coefficient of x^{13} is $\frac{1}{13!}f^{(13)}(0)$. The coefficient of x^{10} is zero, so $f^{(10)}(0)=10!(0)=0$. The coefficient of x^{13} (i.e., x^{2n+1}, with $n=6$) is $(-1)^6\frac{1}{13}$. Since this equals $\frac{1}{13!}f^{(13)}(0)$, we see that $f^{(13)}(0)=13!(-1)^6\frac{1}{13}=12!=479{,}001{,}600$. This is easier than calculating directly.

(iv) For another example we consider the function $g(x) = \frac{\sin x}{x}$. We know the Maclaurin series for $\sin x$ is $\sum_{n=0}^{\infty} (-1)^n \frac{1}{(2n+1)!} x^{2n+1}$. To find the series for $g(x)$, the obvious thing to try is just dividing the series for $\sin x$ by x. This suggests the series $\sum_{n=0}^{\infty} (-1)^n \frac{1}{(2n+1)!} x^{2n}$. To be able to use the theorem, we must know that this series converges to $g(x)$ on some interval around $x = 0$.

The difficulty is that we were not very careful when we defined $g(x)$. Specifically, it is not defined when $x = 0$. Of course, we know that $\lim_{x \to 0} g(x) = 1$, so it makes sense to define a slightly different function

$$G(x) = \begin{cases} \frac{\sin x}{x}, & \text{if } x \neq 0, \\ 1, & \text{if } x = 0. \end{cases}$$

The series we found by dividing converges to $G(x)$ when $x = 0$ (since only the first term is nonzero) and also for any nonzero x (since in effect we have divided the series that converges to $\sin x$ by the "constant" x). The series converges to $G(x)$ for every x, so it is the Maclaurin series of $G(x)$.

Strictly speaking, we ought to check that it is possible to differentiate $G(x)$ infinitely often (i.e., so that it has a Maclaurin series in the first place). This is not easy, but in fact it can be derived as a consequence of the fact that there is a power series that converges to $G(x)$ for every point except $x = 0$. In a sense the power series has shown us how to "extend" the function $g(x) = \frac{\sin x}{x}$ (which is not defined at $x = 0$) into the function $G(x)$ (which is defined on the whole real line and can be differentiated infinitely many times).

(v) Evaluate the series $\sum_{n=1}^{\infty} nx^n$.

Solution We recognize that this series is obtained by taking the derivative of the Maclaurin series $\sum_{n=0}^{\infty} x^n$, which is $\sum_{n=1}^{\infty} nx^{n-1}$, and multiplying it by x. Since $\sum_{n=0}^{\infty} x^n$ is the Maclaurin series for $\frac{1}{1-x}$, we see that $\sum_{n=1}^{\infty} nx^n$ converges to $x\frac{d}{dx}\left(\frac{1}{1-x}\right) = x\frac{1}{(1-x)^2} = x(1-x)^{-2}$. The series converges for $x \in (-1, 1)$ to the function $x(1-x)^{-2}$. For instance, if $x = \frac{1}{2}$, we find that $\sum_{n=1}^{\infty} n2^{-n} = \frac{1}{2}\left(1 - \frac{1}{2}\right)^{-2} = \frac{1}{2}\left(\frac{1}{2}\right)^{-2} = \frac{1}{2}(2^2) = 2$.

We have seen that it is easy to add power series. In fact it's also possible to multiply and divide them. For example, suppose we want to find the series for $\frac{1}{(1+x)(1-2x)}$. One way to proceed would be to multiply the familiar series for $\frac{1}{1-x}$ and $\frac{1}{1-2x}$. We begin by writing down the two series, one above the other:

$$1 + x + x^2 + x^3 + x^4 + x^5 + x^6 + x^7 + x^8 + x^9 + \dots,$$
$$1 + 2x + (2x)^2 + (2x)^3 + (2x)^4 + (2x)^5 + (2x)^6 + (2x)^7 + (2x)^8 + (2x)^9 + \dots.$$

Then we multiply, beginning at the left. The idea is to work with one power of x at a time. The zeroth power of x is the constant term, and the only way to get a constant term in the product is to multiply together the constant terms of the two original series.

There are two ways to get terms containing $x = x^1$; we could multiply the constant term of the first series with the x-term of the second series or vice versa. Combining these, we find that the x-term in the product is $(1)(2x) + (x)(1) = 3x$.

For the x^2-term there are three possible combinations: the constant term from the first series times the x^2-term from the second, the product of the x-terms from both series, or the x^2-term from the first series multiplied by the constant term from the second series. Putting them together, we find that the x^2-term in the product series is $(1)(2x)^2 + (x)(2x) + (x^2)(1) = 7x^2$.

Similarly, for the x^3-term we have $(1)(2x)^3 + x(2x)^2 + x^2(2x) + x^3(1) = (8 + 4 + 2 + 1)x^3 = 15x^3$, and for the x^4-term we have $(1)(2x)^4 + x(2x)^3 + x^2(2x)^2 + x^3(2x) + x^4(1) = (16 + 8 + 4 + 2 + 1)x^4 = 31x^4$.

For the x^n-term we have $(1)(2x)^n + x(2x)^{n-1} + x^2(2x)^{n-2} + \ldots + x^{n-1}(2x) + x^n(1) = (2^n + 2^{n-1} + \ldots + 4 + 2 + 1)x^n$. The quantity in parentheses is a geometric sum with common ratio $\frac{1}{2}$, first term 2^n, and last term 1, so it equals $\frac{2^n - \frac{1}{2}}{1 - \frac{1}{2}} = 2^{n+1} - 1$.

This tells us the coefficient of the nth term of the product series. We see that the Maclaurin series for $\frac{1}{(1-x)(1-2x)}$ is

$$\sum_{n=0}^{\infty} (2^{n+1} - 1)x^n.$$

It may help to remember the process by which we multiply as follows. Write down the two series, one above the other. To find the term containing x^n, we must look for all products of one term from the first series with one from the second series whose powers of x add up to n. So for instance, to find the x^5-term, we multiply the $1 = x^0$-term from the first series and the x^5-term from the second, the $x = x^1$-term from the first and the x^4-term from the second, the x^2-term from the first and the x^3-term from the second, the x^3-term from the first and the x^2-term from the second, the x^4-term from the first and the $x = x^1$-term from the second, and finally the x^5-term from the first and the $1 = x^0$-term from the second.

We can illustrate this for the two series we just considered by joining each of these pairs of terms by a line segment, as shown below:

$$1 + x + x^2 + x^3 + x^4 + x^5 + x^6 + x^7 + x^8 + x^9 + \ldots,$$

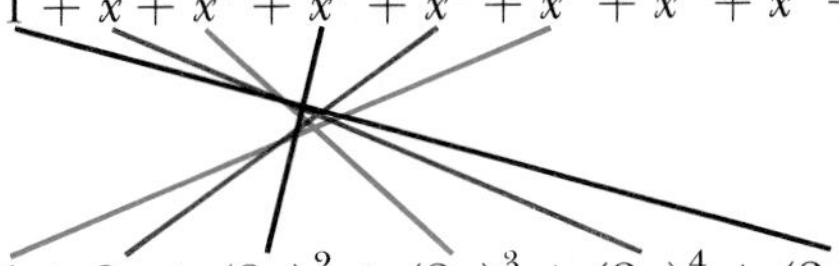

$$1 + 2x + (2x)^2 + (2x)^3 + (2x)^4 + (2x)^5 + (2x)^6 + (2x)^7 + (2x)^8 + (2x)^9 + \ldots.$$

Of course, it is easy to draw a similar picture for the x^7-term or for any other term in the product.

A related question would be the quotient of two power series. For instance, we might like to find the Maclaurin series for $\tan x$ by writing $\tan x = \frac{\sin x}{\cos x}$ and dividing the series for $\sin x$ and $\cos x$. We could attempt to do this by long division.

$$\begin{array}{r|l}
 & x + \frac{1}{3}x^3 + \frac{2}{15}x^5 + \ldots \\
\hline
\left(1 - \frac{x^2}{2} + \frac{x^4}{24}\ldots\right) & x - \frac{x^3}{6} + \frac{x^5}{120} - \ldots \\
 & x - \frac{x^3}{2} + \frac{x^5}{24} - \ldots \\
\hline
 & \frac{1}{3}x^3 - \frac{1}{30}x^5 + \ldots \\
 & \frac{1}{3}x^3 - \frac{1}{6}x^5 + \ldots \\
\hline
 & \frac{2}{15}x^5 + \ldots \\
 & \frac{2}{15}x^5 + \ldots \\
\hline
 & 0 + \ldots
\end{array}$$

Subtracting the fractions is a little tricky, but otherwise the division is fairly routine. We have found that the Maclaurin series for $\tan x$ begins as $x + \frac{1}{3}x^3 + \frac{2}{15}x^5 + \ldots$.

This kind of procedure will work for dividing any series, at least provided the constant term of the denominator is nonzero.

These techniques for multiplying and dividing series are extremely useful. Occasionally, as in the first example, they actually allow us to find a formula for the nth term so that it is possible to write down the entire series. But even when this is too difficult, they provide a technique for finding as many terms of the series as our patience will permit.

EXAMPLE 3 (i) Find the first three nonzero terms of the Maclaurin series for $\cos x \sin x$.

Solution As above, we multiply the familiar series for $\cos x$ and $\sin x$. We begin by writing down the two series, one above the other:

$$1 - \frac{x^2}{2} + \frac{x^4}{24} - \frac{x^6}{720} + \frac{x^8}{40{,}320} - \ldots,$$

$$x - \frac{x^3}{6} + \frac{x^5}{120} - \frac{x^7}{5{,}040} + \frac{x^9}{362{,}880} - \ldots.$$

Then we multiply, beginning at the left end. Since the second series has no constant term, there can be no constant term in the product. The lowest possible power of x is x itself, and the only way it occurs is by multiplying the 1 from the first series and the x from the second. The product will begin with x. There is no way to multiply a term from the first series and a term from the second to obtain a term containing x^2, but there are two x^3-terms. Adding them up, we get $(1)\frac{-x^3}{6} - \frac{x^2}{2}x = -\left(\frac{1}{6} + \frac{1}{2}\right)x^3 = -\frac{2}{3}x^3$.

There is no x^4-term, but for x^5 we find three: $(1)\frac{x^5}{120} - \frac{x^2}{2}\frac{-x^3}{6} + \frac{x^4}{24}x = \left(\frac{1}{120} + \frac{1}{12} + \frac{1}{24}\right)x^5 = \frac{16}{120}x^5 = \frac{2}{15}x^5$.

In principle it is possible to continue indefinitely in this fashion, but of course it becomes quite tedious. Still, we have found that the series for $\cos x \sin x$ begins with $x - \frac{2}{3}x^3 + \frac{2}{15}x^5 + \ldots$.

In fact, in this particular instance there is a better way to do it, which is to notice that $\cos x \sin x = \frac{1}{2}\sin(2x)$, then to take the series for $\sin t$, substitute $t = 2x$, and divide by 2. We find that the Maclaurin series for $\cos x \sin x$ is

$$\frac{1}{2}\sum_{n=0}^{\infty}(-1)^n \frac{2^{2n+1}}{(2n+1)!}x^{2n+1} = \sum_{n=0}^{\infty}(-1)^n \frac{2^{2n}}{(2n+1)!}x^{2n+1},$$

and it is easy to check that the first few terms agree with what we found by multiplying.

(ii) Find the first three nonzero terms in the Maclaurin series for $\sec x$.

Solution Noting that $\sec x = \frac{1}{\cos x}$, we realize that what is needed is to divide the constant function 1 by the function $\cos x$. The Maclaurin series for 1 is just 1 (i.e., a constant term and no other nonzero terms). We perform the division:

$$
\begin{array}{r|l}
 & 1 + \frac{1}{2}x^2 + \frac{5}{24}x^4 + \dots \\
\hline
\left(1 - \frac{x^2}{2} + \frac{x^4}{24}\dots\right) & 1 \\
 & 1 - \frac{x^2}{2} + \frac{x^4}{24} - \dots \\
\cline{2-2}
 & \frac{1}{2}x^2 - \frac{1}{24}x^4 + \dots \\
 & \frac{1}{2}x^2 - \frac{1}{4}x^4 + \dots \\
\cline{2-2}
 & \frac{5}{24}x^4 + \dots \\
 & \frac{5}{24}x^4 + \dots \\
\cline{2-2}
 & 0 + \dots
\end{array}
$$

We see that the series for $\sec x$ begins with $1 + \frac{1}{2}x^2 + \frac{5}{24}x^4 + \dots$.

(iii) Find $\frac{d^4}{dx^4}(\sec x)\big|_{x=0}$.

Solution Of course, it is possible to do this question by repeated differentiation, but on the other hand we know that the fourth derivative at $x = 0$ equals $4!$ times the coefficient of x^4 in the Maclaurin series. Having just found that that coefficient is $\frac{5}{24}$, we see that the value at $x = 0$ of the derivative is $4! \times \frac{5}{24} = 5$.

When we found the Taylor series for $\ln(x)$ about $x = 1$, we saw that it is

$$\sum_{n=1}^{\infty} (-1)^{n+1} \frac{1}{n} (x - 1)^n.$$

We also realized that the radius of convergence is 1 and that the interval of convergence is $(0, 2]$.

It is very tempting to ask what happens at the endpoint $x = 2$, where the series converges. If we knew that it converges to the "right" value, that is, to $\ln(2)$, we would have the extremely pretty result that $1 - \frac{1}{2} + \frac{1}{3} - \frac{1}{4} + \dots = \ln(2)$, that is, that the series obtained from the harmonic series by changing the signs of all the even-numbered terms converges to $\ln 2$.

This is in fact true, but it depends on the following theorem, which is very difficult. (This theorem contains the essence of a more technical theorem called Abel's Theorem.)

THEOREM 10.10.3

> Suppose the Taylor series for $f(x)$ about $x = a$ has radius of convergence R. Suppose the series converges to $f(x)$ for all $x \in (a - R, a + R)$. If the series converges at the endpoint $x = a + R$ and if the function f is continuous at $a + R$, then the series converges to $f(a + R)$ when $x = a + R$. The analogous result holds at $x = a - R$.

EXAMPLE 4

(i) As we remarked above, Theorem 10.10.3 implies that

$$\sum_{n=1}^{\infty} (-1)^{n+1} \frac{1}{n} = \ln(2).$$

(ii) In Example 2(ii) we found that the series for arctan x is $\sum_{n=0}^{\infty}(-1)^n \frac{1}{2n+1}x^{2n+1}$. Its radius of convergence is $R = 1$ (by the Ratio Test), and when $x = 1$, the series is an alternating series with terms whose absolute values decrease to 0. This means that the series converges when $x = 1$, and Theorem 10.10.3 tells us that it converges to the right value, namely, arctan $1 = \frac{\pi}{4}$. In symbols,

$$\sum_{n=0}^{\infty}\frac{(-1)^n}{2n+1} = 1 - \frac{1}{3} + \frac{1}{5} - \frac{1}{7} + \frac{1}{9} - \frac{1}{11} + \ldots = \frac{\pi}{4}.$$

This series can be used to estimate the value of π, though unfortunately it takes a large number of terms to achieve much accuracy.

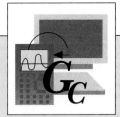

With a programmable calculator we can find partial sums and approximate the series for ln(2) from part (i) of Example 4.

TI-85 PRGM F2(EDIT) LOG2 ENTER F3 F1(Input) M EXIT ENTER 0 STO N ENTER 0 STO S ENTER Lbl B ENTER N + 1 STO N ENTER S + ((−) 1) ^ (N + 1) ÷ N STO S ENTER F4 F1(If) (N TEST F2(<) M) ENTER Goto B ENTER S ENTER EXIT EXIT EXIT PRGM F1(NAME) LOG2 ENTER 50 ENTER	SHARP EL9300 [key] ▼ ▼ (NEW) ENTER ENTER LOG2 ENTER COMMAND A 3 (Input) M ENTER N = 0 ENTER S = 0 ENTER COMMAND B 1 (Label) 1 ENTER N = N + 1 ENTER S = S + ((−) 1) a^b (N + 1) ÷ N ENTER COMMAND 3 (If) N COMMAND C 2 (<) M COMMAND B 2 (Goto) 1 ENTER COMMAND A 1 (Print) S ENTER COMMAND A 6 (End) ENTER [key] [key] (RUN) ENTER (LOG2) ENTER 50 ENTER
CASIO f_x-6300G (f_x-7700GB) MODE 2 EXE ? → M : 0 → N : 0 → S : Lbl 1 : N + 1 → N : S + ((−) 1) x^y (N + 1) ÷ N → S : N < M ⇒ Goto 1 : S ◢ MODE 1 Prog 0 EXE 50 EXE (on the f_x-7700GB, : is PRGM F6, and Prog is PRGM F3.)	HP 48SX ATTN ↱ CLR ≪ ≫ " " KEYIN M ▶ SPC PROMPT SPC ′ M ▶ STO SPC 0 SPC 1 SPC M SPC FOR SPC N SPC 1 +/− SPC N SPC 1 + y^x N ÷ + NEXT ▼ ENTER ′ LOG2 STO VAR LOG2 50 ENTER ↰ CONT

```
?50
ANSWER=
 .6832471606
```

Check how close the answer is to ln(2). Now try it with different values of M. Then try adapting the program to part (ii) of Example 4.

Exercises 10.10

I

Find the Maclaurin series for each of the following functions.

1. $\int_0^x e^{-t^2}\,dt$
2. $\int_{-x}^x e^{t^4}\,dt$
3. $\frac{d}{dx}\left(\frac{1}{1+x}\right)$
4. $\frac{d}{dx}\left(\frac{1}{1+x^3}\right)$
5. $\ln(2-x)$
6. $\arctan(4x)$
7. $\int_0^x \ln(1+2t)\,dt$
8. $\ln(1-x^2)$
9. $\frac{1}{(1-2x)(1-3x)}$
10. $\sin^2 x$

Find the Maclaurin series for each of the following functions.

11. $x^2 \sin x$
12. $(2x+1)\cos x$
13. $\sin\left(\frac{\pi}{2}-x\right)$
14. $(x^3-2x+3)\ln(1-x)$
15. $\frac{1-\cos x}{x^2}$
16. $\frac{\sin x - x}{x^3}$
17. $\frac{1}{2-x^2}$
18. $\cosh^2 x - \sinh^2 x$
19. $(1-x)^{-4}$
20. $e^{x^2}e^{-x^2}$

Find a formula for the coefficient of x^n in the Maclaurin series of each of the following functions. (For some of them it is likely more convenient to find a formula that is different in different cases, e.g., one formula for odd n and another for even n.)

21. $\frac{1}{1-x}e^x$
22. $\frac{1}{(1-2x)^2}$
23. $\cosh^2 x$
24. $\cos^2 x$
25. $\cos(\pi x)$
26. $\frac{1}{2-(1+x)}$
27. $\frac{x}{x-1}$
28. $\frac{x^2}{2-x}$
29. $\ln\left((1-x)^2\right)$
30. $\sqrt{e^{3x}}$

For each of the following functions, find the first three nonzero terms in its Maclaurin series.

31. $e^x\frac{1}{1-x}$
32. $\cos^3 x$
33. $e^{\sin x}$
34. $e^{\cos x}$
35. $\frac{x}{\sin x}$
36. $\frac{x}{\tan x}$
37. $\tanh x$
38. $x \coth x$
39. $\cot x$
40. $\frac{e^x}{\cos x}$

II

41. Find the indefinite integral in Example 1(iii) (p. 713).
42. Check that the series we found in Example 1(i) by differentiating term-by-term does in fact equal the Maclaurin series of the function $f'(x) = (x-1)^{-2}$, by finding a formula for the derivatives of $f'(x)$ and evaluating them at the point $x = 0$ in the usual way.

*43. Find the Maclaurin series for $\sinh^2 x$. (*Hint:* Square the function *before* you find the series.)

*44. Suppose the power series $\sum_{n=0}^{\infty} a_n x^n$ has radius of convergence $R = 1$. Then the series converges for each $x \in (-1, 1)$, and we can use it to define a function on this interval: For each $x \in (-1,1)$, define $f(x) = \sum_{n=0}^{\infty} a_n x^n$. Find an example of such a series that diverges at $x = 1$ but for which $\lim_{x\to 1^-} f(x)$ exists, that is, is a (finite) limit.

Evaluate each of the following series.

*45. $\sum_{n=1}^{\infty} n3^{-n}$

*46. $\sum_{n=0}^{\infty} (n+1)\left(-\frac{1}{2}\right)^n$

*47. $\sum_{n=2}^{\infty} n(n-1)\left(\frac{2}{3}\right)^n$

*48. Evaluate $\sum_{n=1}^{\infty} n^2 x^n$, stating clearly for what values of x the series converges.

*49. (i) Find the ninth derivative of $f(x) = \arctan x$ at $x = 0$, that is, $f^{(9)}(0)$.
(ii) Find the answer to part (i) in a different way. (If you used series, do it by differentiating, and vice versa.)

50. Find the first five nonzero terms in the Maclaurin series for $\tan x$.

51. Find $\frac{d^6}{dx^6}(\sec x)\Big|_{x=0}$.

*52. Find the Maclaurin series for $\frac{1}{(1-ax)(1-bx)}$.

53. Check the result of Example 3(iii) (p. 719) by differentiating, that is, find $\frac{d^4}{dx^4}(\sec x)\Big|_{x=0}$.

Chapter Summary

§10.1
sequence
term
convergence
divergence
diverges to infinity
bounded
Squeeze Theorem for sequences
bounded shifted sequences
precise definition of limits

§10.2
geometric sum
common ratio
interest
factor
mortgages
annuities
balance
monthly payment
geometric series
convergence
divergence
Zeno's Paradox

§10.3
partial sum
convergence
divergence
*n*th Term Test
harmonic series

§10.4
increasing sequence
bounded sequence
Integral Test
p-series

§10.5
Comparison Test
Limit Comparison Test
rational function

§10.6
alternating series
absolute convergence
conditional convergence

§10.7
Ratio Test
*n*th Root Test
testing convergence of series (Summary 10.7.3)

§10.8
Taylor series
Maclaurin series
remainder

§10.9
power series
radius of convergence
interval of convergence
convergence of familiar Taylor series (Summary 10.9.3)

§10.10
term-by-term differentiation
term-by-term integration

Review Exercises

I

Calculate each of the following, provided it exists.

1. $\sum_{n=1}^{\infty} 4(3^{-n})$

2. $\sum_{n=0}^{\infty} (-2)^{-3n}$

3. $\sum_{n=1}^{\infty} \frac{2^{n+1}}{3^n}$

4. $\sum_{n=0}^{\infty} \frac{3^{2n}}{4^n}$

5. The annual payment on a mortgage for \$20,000 at 12% compounded annually over a 25-year term.

6. The annual payment on a mortgage for \$16,000 at 10% compounded annually over a 20-year term.

7. The balance left owing after 10 years on a \$100,000 mortgage with a 20-year term at 12% annual interest.

8. The balance left owing after 15 years on a \$45,000 mortgage with a 20-year term at 14% annual interest.

9. The annual payment from an annuity that costs \$10,000 and that runs for 15 years with 10% interest.

10. The cost of an annuity that will pay \$12,000 per annum for 10 years, assuming interest of 9%.

For each of the following sequences, decide whether or not it converges; if it does, find its limit.

11. $\left\{\frac{n^2+1}{n^3}\right\}$

12. $\left\{\sin \frac{1}{n}\right\}$

13. $\{\cos(n\pi)\}$

14. $\{\sin(n\pi)\}$

15. $\left\{\frac{\sin(n)}{n}\right\}$

16. $\{\cos^4(n\pi)\}$

17. $\left\{\cos\left(\frac{1}{n}\right)\right\}$

18. $\left\{\cot\left(\frac{1}{n}\right)\right\}$

19. $\left\{\sin\left(\frac{\pi}{2} + \frac{1}{n}\right)\right\}$

20. $\left\{\exp\left(\sin \frac{\pi}{n}\right)\right\}$

Decide whether each of the following series is convergent, and explain your answer.

21. $\sum_{n=1}^{\infty} \frac{3n^4-n^2+4}{10n^3+7}$

22. $\sum_{n=1}^{\infty} \frac{2n^3+3n^2+1}{7n^4+2n^3-2}$

23. $\sum_{n=2}^{\infty} \frac{n^3+5n^2-2}{1-n^{3/2}}$

24. $\sum_{n=0}^{\infty} \frac{\sqrt{n+1}}{n^2+4}$

25. $\sum_{n=1}^{\infty} \frac{n^n}{n!}$

26. $\sum_{n=1}^{\infty} (-1)^{n+1} \sin \frac{1}{2n}$

27. $\sum_{n=1}^{\infty} (-1)^n \cos \frac{1}{n^2}$

28. $\sum_{n=1}^{\infty} (-1)^n \left(1 - \cos \frac{1}{n}\right)$

29. $\sum_{n=1}^{\infty} (-1)^n \sqrt{n} \sin \frac{1}{n}$

30. $\sum_{n=1}^{\infty} (-1)^{n+1} \sin\left((-1)^n \frac{1}{n}\right)$

31. $\sum_{n=1}^{\infty} \left(1 - \frac{n-1}{n}\right)$

32. $\sum_{n=1}^{\infty} 3^{n/3} \frac{1}{n!}$

33. $\sum_{n=1}^{\infty} 6^{4n+1} \frac{1}{n!}$

34. $\sum_{n=1}^{\infty} \cos(n^2 - 4) 3^{3-n/5}$

35. $\sum_{n=2}^{\infty} \ln \frac{1}{n}$

36. $\sum_{n=1}^{\infty} \frac{n^3}{n!}$

37. $\sum_{n=2}^{\infty} \ln(n) n^{-4/3}$

38. $\sum_{n=2}^{\infty} \ln(n) 2^{-n/4}$

39. $\sum_{n=1}^{\infty} (-1)^n \frac{1}{1+\ln(n)}$

40. $\sum_{n=1}^{\infty} n^{3/2} \frac{n!}{(n-3)!}$

For what values of c do each of the following series converge?

41. $\sum_{n=1}^{\infty} c^{-4n}$

42. $\sum_{n=1}^{\infty} nc^{-4n}$

43. $\sum_{n=1}^{\infty} n^{2c}$

44. $\sum_{n=1}^{\infty} \frac{1}{1+cn^2}$

45. $\sum_{n=1}^{\infty} (\sqrt{n}+1)^c$

46. $\sum_{n=1}^{\infty} c^n \frac{n!}{(2n)!}$

47. $\sum_{n=1}^{\infty} (n^3 - 2n + 12)^c$

48. $\sum_{n=1}^{\infty} c^n n^{-n}$

49. $\sum_{n=3}^{\infty} \left(1 - \frac{1}{n}\right)^{cn}$

50. $\sum_{n=1}^{\infty} \frac{n^{1-2c}+1}{n^c+2}$

Find the Taylor series of each of the following functions about the specified point.

51. $f(x) = \frac{1}{1-3x}$ about $a = 0$

52. $f(x) = \frac{1}{2-x}$ about $a = 0$

53. $f(x) = \cos x$ about $a = \pi$

54. $f(x) = \frac{1}{x}$ about $a = -1$

55. $f(x) = \ln x$ about $a = 4$

56. $f(x) = \ln(3x)$ about $a = \frac{1}{3}$

57. $f(x) = \ln |x|$ about $a = -1$

58. $f(x) = \ln x^2$ about $a = -2$

59. $f(x) = \arctan 2x$ about $a = 0$

60. $f(x) = x^{1/3}$ about $a = 1$

Find the radius of convergence of each of the following series.

61. $\sum_{n=1}^{\infty} n^3 (2x)^n$

62. $\sum_{n=1}^{\infty} n^{-4} \left(\frac{x}{7}\right)^n$

63. $\sum_{n=1}^{\infty} n! x^{3n}$

64. $\sum_{n=1}^{\infty} n^{n/2} x^n$

65. $\sum_{n=0}^{\infty} (n^2 - 4n + 6)^{1/3} x^{2n}$

66. $\sum_{n=1}^{\infty} \frac{n!}{(3n)!} x^n$

67. $\sum_{n=0}^{\infty} 2^n x^{2n}$

68. $\sum_{n=0}^{\infty} \frac{5^n}{n!} x^n$

69. $\sum_{n=1}^{\infty} \cos\left(\frac{1}{n}\right) x^n$

70. $\sum_{n=0}^{\infty} \sin\left(\frac{1}{n!}\right) x^n$

Find the first three nonzero terms in the Maclaurin series of each of the following functions.

71. $\cos^2 x \sin x$

72. $\cosh x \sinh x$

73. $\frac{1}{1-\ln(1-x)}$

74. $(1 - x)^{-3}$

75. $\exp(e^x)$

76. $\frac{x+1}{x-1}$

77. $\cos(e^x)$

78. $\sin(e^x)$

79. $\arctan(\sin x)$

80. $\ln(e^x)$

II

***81.** Without actually finding the series, show that the Taylor series of $f(x) = \sin x$ about *any* point $x = a$ converges to $f(x)$ at every x.

***82.** (i) Show that the Taylor series of $f(x) = e^x$ about *any* point $x = a$ converges to $f(x)$ at every x.

(ii) Find the Taylor series of $f(x) = e^x$ about the point $x = a$.

83. Suppose the Maclaurin series of $f(x)$ is $\sum_{n=0}^{\infty} a_n x^n$ and the Maclaurin series of $g(x)$ is $\sum_{n=0}^{\infty} b_n x^n$.

(i) What is the Maclaurin series of $f(x) - 3g(x)$?

(ii) Show that if the Maclaurin series for $f(x)$ converges at $x = u$ to $f(u)$, and if the Maclaurin series for $g(x)$ converges at $x = u$ to $g(u)$, then the

Maclaurin series for $f(x) + g(x)$ must converge at $x = u$ to $f(u) + g(u)$.

***84.** Find an example of two series $\sum_{n=1}^{\infty} a_n$ and $\sum_{n=1}^{\infty} b_n$ that are both divergent but whose sum $\sum_{n=1}^{\infty} (a_n + b_n)$ converges.

***85.** What is $f^{(7)}(0)$ if $f(x) = \frac{1}{1-2x}\frac{1}{1+2x}$?

86. Suppose the Maclaurin series of the function $f(x)$ is $\sum_{n=0}^{\infty} a_n x^n$, and suppose it converges for every x. Assuming that $a_0 \neq 0$, find the first four terms of the Maclaurin series of $\frac{1}{f(x)}$.

III

87. Graph the function

$$f(x) = \begin{cases} e^{-1/x^2}, & \text{if } x \neq 0, \\ 0, & \text{if } x = 0, \end{cases}$$

whose Maclaurin series we discussed in Example 10.8.6 (p. 702). On the same picture, graph $f'(x)$, $f''(x)$, and as many more derivatives as possible. Do they all tend to 0 as $x \to 0$?

Conics

11

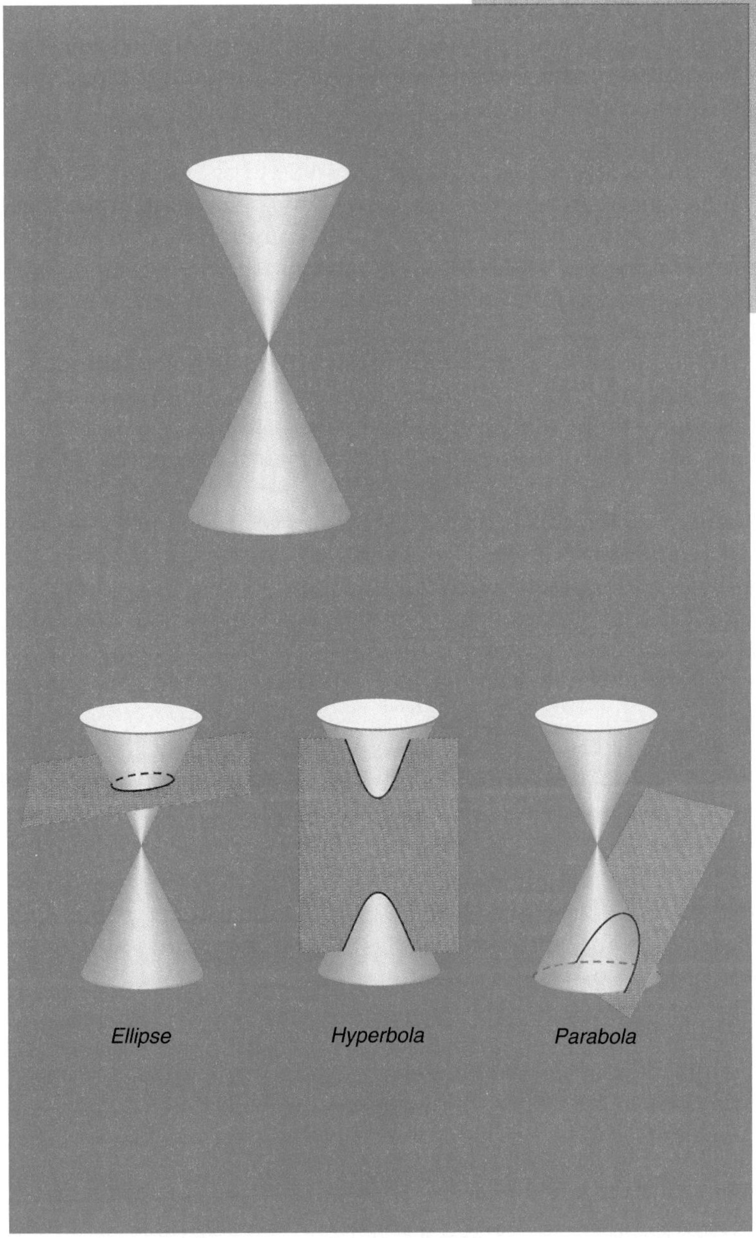

INTRODUCTION

In Section 1.6 we recalled some basic facts about conics, primarily with the goal of sketching their graphs. In this chapter we make a more systematic study, making use of some of the techniques we have learned.

One way of looking at conics is to take a two-ended cone, as shown at left, and intersect it with a plane. The points of intersection will usually be a conic. (There are some exceptions because it is possible to put the plane in special locations that result in *degenerate conics,* such as a single point or one or two straight lines.) This is the origin of the name "conic."

The illustration shows some of the possibilities.

However, in this chapter we will study conics by beginning with their equations and will not make much mention of the three-dimensional picture with the cone.

In Section 11.1 we recall briefly some of the material of Section 1.6, dealing with identifying and graphing conics from their equations. In Section 11.2 we consider conics from a more geometrical point of view, describing a way of constructing them and discussing some of their properties. Up to this point the conics under consideration all have equations without *cross-terms*, that is, terms involving xy. In Section 11.3 we will see that this amounts to insisting that the axes of the conic are parallel to the x- and y-axes. To study the general case, in which cross-terms are allowed, it is necessary to learn how to rotate the axes. The material of Section 11.3 is not used in later chapters.

11.1 Identifying and Graphing Conics

TRANSLATING AXES
COMPLETING SQUARES
CONICS IN STANDARD FORM

In this brief section we recall material that was covered in Section 1.6. If you are not familiar with conics, you should review that section first.

The equation of a conic is always a quadratic polynomial equation in the two variables x, y. For the present we will only consider equations without "cross-terms" xy, that is, equations of the form $ax^2 + by^2 + cx + dy + q = 0$ with a, b not both zero.

In Section 1.6 we found that it is often possible to simplify this type of equation by completing squares. Whenever a quadratic polynomial contains both an x^2-term and an x-term, we can complete the square and express the polynomial in terms of $(x - h)^2$, say. Letting $x' = x - h$, we rewrite the equation using the variable x'. The resulting polynomial will have an x'^2-term but no x'-term. A similar substitution can be used, if necessary, with y.

If the equation does not have an x^2-term, it is possible to combine the x-term and the constant term into x', resulting in an equation without a constant term. A similar procedure can be used to remove the constant term from an equation with no y^2-term. The process of replacing x with x' and y with y' is referred to as *translating coordinates.*

The value of doing this is that it gives a simpler equation that is more easily recognized as one of the standard types of conics. On the other hand, it does mean that we are considering the conic relative to a different set of coordinates (x', y'). The geometrical significance of changing coordinates is that the axes are *translated,* or shifted over; the conic is moved, and the origin changes. Being able to identify the conic is useful only if we keep track of the effect of changing coordinates.

Summary 11.1.1

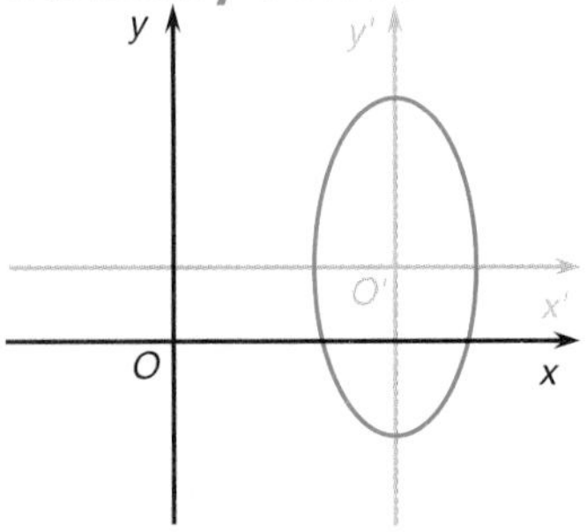

Figure 11.1.1

Translation of Axes

Suppose h and k are real numbers; let $O' = (h, k)$ and let $x' = x - h$, $y' = y - k$. Draw the x'-axis and the y'-axis through O'. To graph all the points (x, y) whose coordinates satisfy some equation $F(x, y) = 0$, it is equivalent to express $F(x, y)$ as a function of x' and y', that is,

$$F(x, y) = G(x', y') = G(x - h, y - k),$$

and then graph all points (x', y') that satisfy $G(x', y') = 0$; these points (x', y') must be plotted relative to the x'-axis and the y'-axis. See Figure 11.1.1.

The function G can be found by observing that $x = x' + h$ and $y = y' + k$, so

$$G(x', y') = F(x, y) = F(x' + h, y' + k).$$

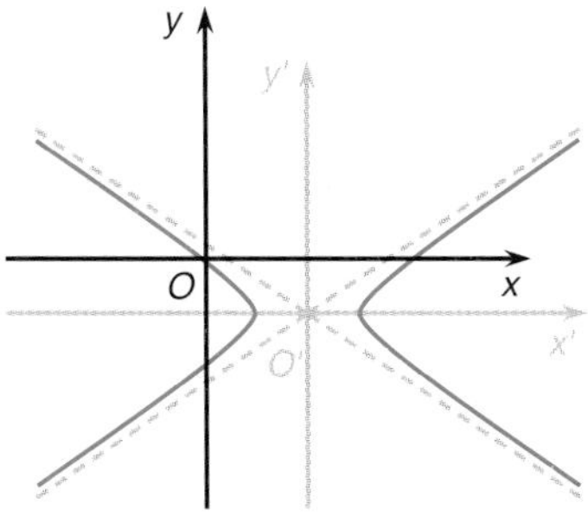

Figure 11.1.2

EXAMPLE 1 Consider the equation $4x^2 - 16x - 9y^2 - 18y + 3 = 0$. Completing the square both in x and y, we find

$$4(x^2 - 4x) - 9(y^2 + 2y) = -3 \quad \text{or} \quad 4(x-2)^2 - 16 - 9(y+1)^2 + 9 = -3$$

or $4x'^2 - 9y'^2 = 4$,

where $x' = x - 2$ and $y' = y + 1$.

We recognize this as the equation of a hyperbola whose transverse axis is horizontal. It is sketched in Figure 11.1.2, with the translated origin O' at $(x, y) = (2, -1)$ and the x'- and y'-axes shown.

For convenience we recall the standard forms of the various types of conics.

Table 11.1.2 Conics in Standard Form

$y = ax^2,\ a > 0$	Parabola	Opening upward
$y = ax^2,\ a < 0$	Parabola	Opening downward
$x = ay^2,\ a > 0$	Parabola	Opening to the right
$x = ay^2,\ a < 0$	Parabola	Opening to the left
$\frac{x^2}{a^2} + \frac{y^2}{b^2} = 1,\ a > b > 0,$	Ellipse	Intercepts $(\pm a, 0)$, $(0, \pm b)$; major axis horizontal
$\frac{x^2}{b^2} + \frac{y^2}{a^2} = 1,\ a > b > 0,$	Ellipse	Intercepts $(\pm b, 0)$, $(0, \pm a)$; major axis vertical
$\frac{x^2}{a^2} + \frac{y^2}{a^2} = 1,\ a > 0$	Circle	Radius a, center at $(0, 0)$
$\frac{x^2}{a^2} - \frac{y^2}{b^2} = 1$	Hyperbola	Intercepts $(\pm a, 0)$, transverse axis horizontal, asymptotes $y = \pm\frac{b}{a}x$
$\frac{y^2}{a^2} - \frac{x^2}{b^2} = 1$	Hyperbola	Intercepts $(0, \pm a)$, transverse axis vertical, asymptotes $y = \pm\frac{a}{b}x$

These nine types are illustrated in Figure 11.1.3.

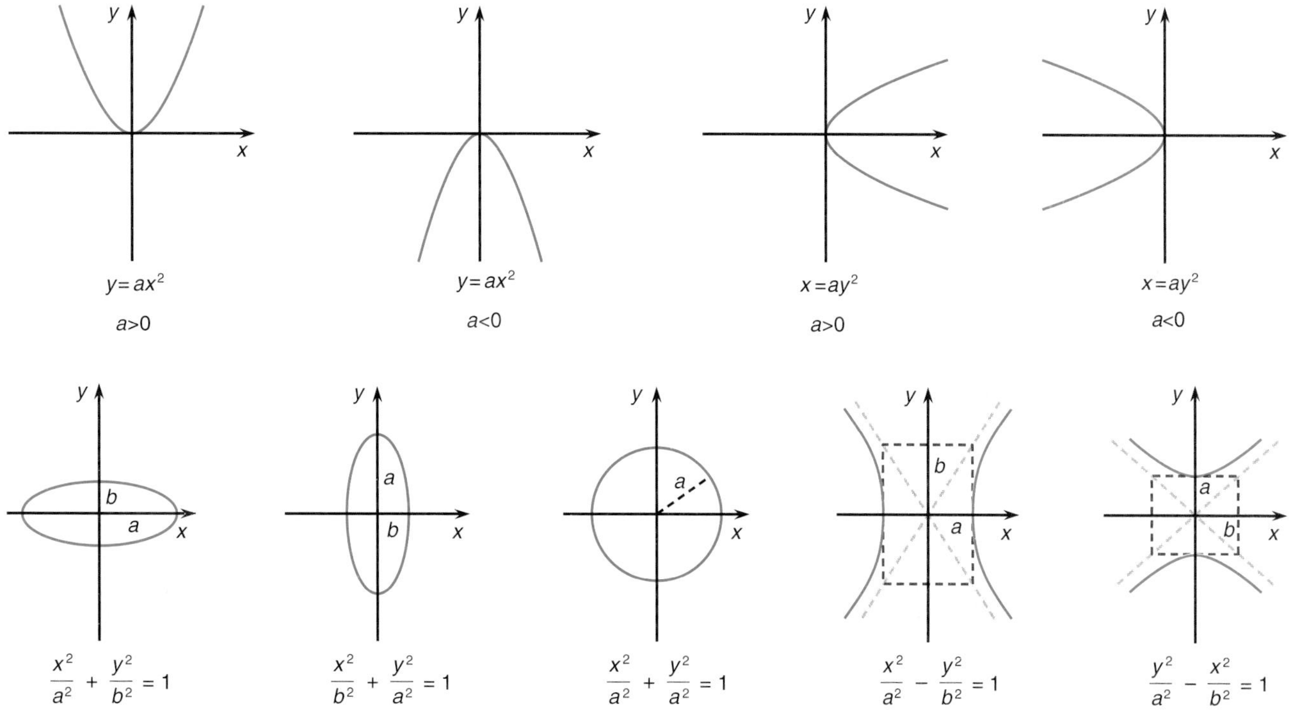

Figure 11.1.3

For the convenience of the reader we repeat the summary from Section 1.6 about how to sketch the graph of a conic.

Summary 11.1.2 Graphing Conics

Consider the quadratic equation $ax^2 + by^2 + cx + dy + q = 0$, which has no xy-term. To draw its graph, follow the following steps.

1. Complete squares if necessary; if the equation contains both an x^2-term and an x-term, that is, if $a \neq 0$ and $c \neq 0$, then complete the square in x. Similarly, if $b \neq 0$ and $d \neq 0$, complete the square in y.
2. Translate axes if necessary. Specifically, if you found it necessary to complete squares in Step 1, and the result of completing the square in x is a term involving $x - h$, then let $x' = x - h$. If completing the square in y gives a term involving $y - k$, let $y' = y - k$.

   If there is no x^2-term, then there must be a y^2-term. In this situation, combine the x-term and the constant term as $c(x - h)$, for some constants c and h, and let $x' = x - h$. Similarly, if there is no y^2-term, combine the y-term and the constant term as $d(y - k)$ and let $y' = y - k$.
3. If the axes have been translated, plot the "new origin" $O' = (h, k)$. (If you changed x to $x' = x - h$ but did not have to translate y, then $O' = (h, 0)$, that is, let $k = 0$. Similarly, if it was not necessary to translate x to x', then let $h = 0$.)
4. Express the equation in terms of x' and y' if translations were necessary. Inspect the equation to see whether it is the equation of a parabola, an ellipse, a circle, or a hyperbola. Find the intercepts.

   For a hyperbola $\frac{x^2}{a^2} - \frac{y^2}{b^2} = 1$ or $\frac{y^2}{a^2} - \frac{x^2}{b^2} = 1$, find the asymptotes and mark them on the picture. You may find it helpful to mark the intercepts (either $(\pm a, 0)$ or $(0, \pm a)$) and also the points on the other axis that are b units from the center, and use them to fill in a rectangle, as shown in Figure 11.1.5. The asymptotes pass through the corners of this box.

   Sketch the graph.
5. Occasionally, we encounter a quadratic equation that is not a parabola or an ellipse or a circle or a hyperbola. This kind of graph is called a **degenerate conic**. There are several possibilities, but we describe just two of them. In an equation like $\frac{x^2}{a^2} + \frac{y^2}{b^2} = -1$, the left side is the sum of two squares, so it is never negative, and we see that no point (x, y) can satisfy this equation. There are *no points* on the graph.

   Another possibility is an equation of the form $\frac{x^2}{a^2} - \frac{y^2}{b^2} = 0$. This is like a hyperbola except that the constant term is zero. Rearranging, we find that $\frac{x^2}{a^2} = \frac{y^2}{b^2}$, or $\frac{x}{a} = \pm\frac{y}{b}$. The graph is the two straight lines $y = \frac{b}{a}x$ and $y = -\frac{b}{a}x$. The name for this kind of graph comes from the observation that this is what results if a hyperbola "degenerates" into its two asymptotes.

In the standard forms for ellipses we always assume $a > b > 0$. In the standard forms for hyperbolas, the right side is always 1 (never -1), and a^2 is the denominator of the *positive* term on the left side. We may have $a > b > 0$ or $b > a > 0$ or $a = b > 0$.

EXAMPLE 2 (i) Sketch the graph of $2x - 4y^2 + 8y - 6 = 0$.

Solution We begin by completing the square in y, and this gives

$$0 = 2x - 4(y^2 - 2y) - 6 = 2x - 4(y - 1)^2 - 2 = 2(x - 1) - 4(y - 1)^2.$$

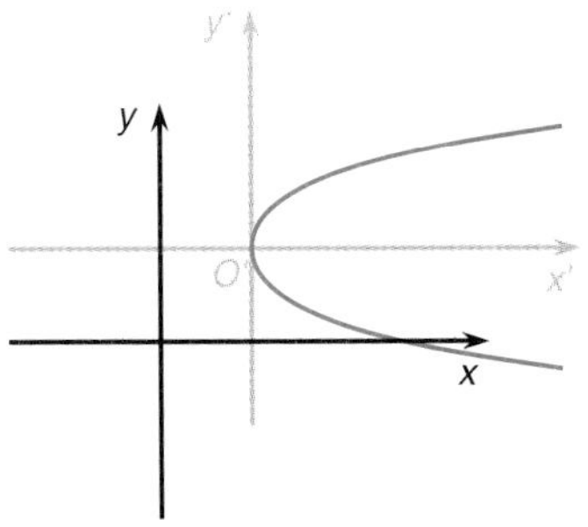

Figure 11.1.4

Letting $x' = x - 1$, $y' = y - 1$, we see that the conic is $2x' = 4y'^2$, which is the equation of a parabola whose vertex is at the "origin" O' and that opens to the right.

We plot the translated origin $O' = (1, 1)$ and the translated axes and then sketch the parabola using them. See Figure 11.1.4.

(ii) Sketch the graph of $9x^2 + 36x - 16y^2 + 180 = 0$.

Solution Completing the square in x, we write the equation as $9(x^2 + 4x) - 16y^2 = -180$, or $9(x + 2)^2 - 36 - 16y^2 = -180$. Letting $x' = x + 2$, we find that this is $9x'^2 - 16y^2 = -144$.

To put this into one of the standard forms of Table 11.1.2, we divide by -144 and find

$$\frac{y^2}{9} - \frac{x'^2}{16} = 1.$$

This is the equation of a hyperbola whose transverse axis is vertical and that, relative to the translated axes has its center at the origin O' and asymptotes $y = \pm\frac{3}{4}x'$.

In Figure 11.1.5 we sketch the translated origin $O' = (-2, 0)$ and the translated axes, the asymptotes, and the hyperbola itself.

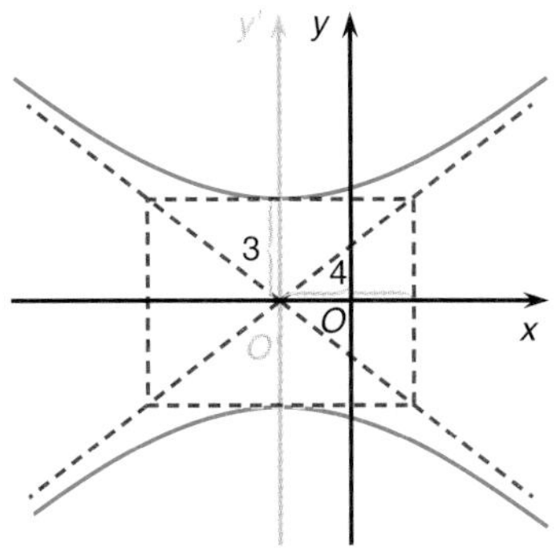

Figure 11.1.5

(iii) Sketch the graph of $7x^2 - 98x + 7y^2 + 28y + 371 = 0$.

Solution Completing both squares, we write the equation as $0 = 7(x^2 - 14x) + 7(y^2 + 4y) + 371 = 7(x - 7)^2 - 343 + 7(y + 2)^2 - 28 + 371$.

This looks like the equation of a circle until we let $x' = x - 7$, $y' = y + 2$ and make the final simplification, writing $7x'^2 + 7y'^2 = 0$, or just $x'^2 + y'^2 = 0$. The only way the left side of this equation can equal zero is for both x' and y' to equal zero, and this means that there is only one point that satisfies the equation. The graph of this equation is the degenerate conic that consists of the single point $(7, -2)$. See Figure 11.1.6.

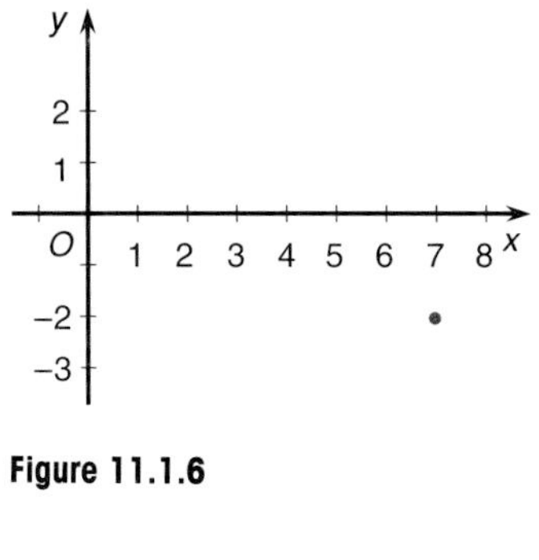

Figure 11.1.6

With a graphics calculator we can sketch the graphs of the conics in Example 2. We need to express y as a function of x; for the conic $9x^2 + 36x - 16y^2 + 180 = 0$ of part (ii), we write $16y^2 = 9x^2 + 36x + 180$, so $y = \pm\frac{1}{4}\sqrt{9x^2 + 36x + 180}$.

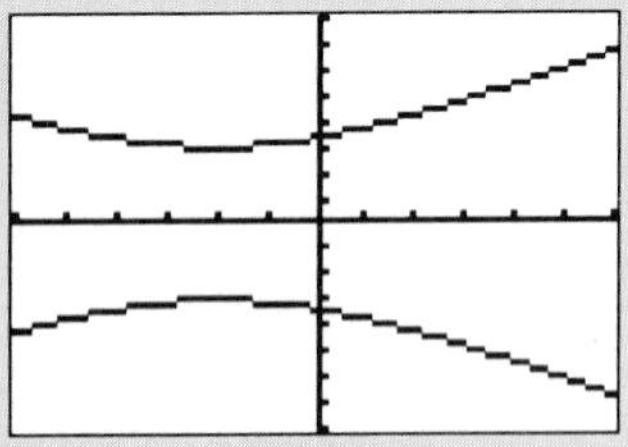

TI-85 GRAPH F1 (y(x) =) √ (9 F1 (x) x^2 + 36 F1 (x) + 180) ÷ 4 ▼ (−) √ (9 F1 (x) x^2 + 36 F1 (x) + 180) ÷ 4 M2(RANGE) (−) 6 ▼ 6 ▼ ▼ (−) 8 ▼ 8 F5(GRAPH)	SHARP EL9300 [graph key] √ 9 x/θ/T x^2 + 36 x/θ/T + 180 ► ÷ 4 2ndF ▼ (−) √ 9 x/θ/T x^2 + 36 x/θ/T + 180 ► ÷ 4 RANGE −6 ENTER 6 ENTER 1 ENTER −8 ENTER 8 ENTER 1 ENTER [graph key]
CASIO f_x-6300G (f_x-7700GB) Range −6 EXE 6 EXE 1 EXE −8 EXE 8 EXE 1 EXE EXE EXE EXE Graph √ (9 X x^2 + 36 X + 180) ÷ 4 EXE Graph (−) √ (9 X x^2 + 36 X + 180) ÷ 4 EXE	HP 48SX ↰ PLOT PLOTR ERASE ATTN ′ Y = √x () 9 × X y^x 2 + 36 × X + 180 ► ÷ 4 ↱ PLOT ↰ DRAW ERASE 6 +/− ENTER 6 XRNG 8 +/− ENTER 8 YRNG DRAW ATTN ′ Y = √x () 9 × X y^x 2 + 36 × X + 180 ► ÷ 4 +/− ↱ PLOT ↰ DRAW DRAW

To graph the parabola from part (i) of Example 2, it is necessary to use the quadratic formula to express y in terms of x, and again we find that there are two functions.

Exercises 11.1

I

Identify and sketch the graph of each of the following.

1. $x^2 - y^2 = -9$
2. $x^2 - 2x + 4y^2 - 16y + 16 = 0$
3. $x^2 + 6x + 2y = 1$
4. $4x^2 + y^2 - y = 0$
5. $2x^2 + y^2 + 4x + 1 = 0$
6. $2x^2 - 4y^2 + 3 = 0$
7. $3x^2 - 12x + y^2 - 4y = 0$
8. $\left(\frac{y}{x}\right)^2 + \frac{4}{x} + 3x^{-2} + 1 = 0$
9. $3y^2 - 12y + x - 7 = 0$
10. $x^4 + 2x^2y^2 + y^4 = 16$

Find the equation of each of the following. (Assume that there are no cross-terms in any of the equations.)

11. The parabola whose vertex is at $(-1, -2)$ that opens upward and contains the point $(-2, 2)$.
12. The ellipse whose center is $(5, -1)$ and whose horizontal and vertical axes have lengths 4 and 9, respectively.
13. The ellipse whose center is $(0, 3)$ and that contains $(0, 4)$ and $(2, 3)$.

14. The ellipse whose center is $(5, 2)$ and that contains $(1, 1)$ and $(3, 0)$.

15. The hyperbola with asymptotes $y = \pm 2(x - 1)$ that contains the point $(2, 0)$.

16. The hyperbola whose center is $(2, 1)$, whose transverse axis is vertical, and that contains $(5, -4)$ and $(2, 5)$.

17. The circle with center $(-3, 4)$ and radius 4.

18. The parabola that is symmetric about the y-axis and contains the points $(2, 0)$ and $(-3, -5)$.

19. The hyperbola that has center $(4, 0)$, that does not intersect the x-axis, and that contains the points $(4, 1)$ and $(3, 2)$.

20. The ellipse whose center is $(1, 2)$, whose minor axis has length 3, and that contains the point $(0, 5)$.

II

As usual, we assume for all the following questions that there are no cross-terms in any of the equations.

21. Sketch the graph of $4x^2 + 8x - 4y^2 - 8y = 0$.

22. There are two ellipses whose center is the origin, whose major axis is twice as long as the minor axis, and that pass through $(2, 1)$. Find the equation of the one whose major axis is longer. Draw a sketch of both ellipses.

23. Find the equations of both parabolas whose vertex is $(3, -1)$ and that pass through $(4, 3)$.

24. Find the equation of the ellipse whose center is $(4, -2)$ and that is tangent to the lines $y = 3$ and $x = 1$.

***25.** (i) Find an example of four points that do not lie on any ellipse.

(ii) Is it possible to find three points that do not lie on any ellipse?

***26.** The three points $(0, 0)$, $(1, 1)$, and $(2, 4)$ all lie on the parabola $y = x^2$. Find another parabola on which they all lie.

III

Use a graphics calculator or computer to sketch the graphs of the following conics.

27. $x^2 - 2x - \frac{y^2}{4} = 0$

28. $x^2 + 4x + 9y^2 - 18y + 9 = 0$

29. The ellipse with center $(2, -3)$, containing $(2, 4)$ and $(-1, -3)$.

POINT TO PONDER

When we listed the standard forms of the conics in Table 11.1.2, the idea was to use these forms to identify the type of any particular conic. For this approach to work, we have to know that after translating the axes, every conic can be put into one of these forms.

This is approximately true, but there are some subtleties. First, we have to make our blanket assumption that there are no cross-terms in the equation. Second, it is necessary to allow for the degenerate conics, which are not included in the list.

Some possible degenerate forms are (i) quadratic polynomials depending on only one of the two variables, (ii) equations like the equation of an ellipse or a circle except that the constant on the right side is 0 or -1, (iii) equations like the equation of a hyperbola except that the constant on the right side is 0.

(i) Try to make an argument to explain why every quadratic polynomial equation in x and y with no xy-term can be converted, by completing squares and translating coordinates if necessary into one of the standard forms or into one of these three degenerate forms.

(ii) What are the graphs of each of these types of degenerate conics? Which of them can you form by intersecting a plane with a two-ended cone?

(iii) Beginning with your argument that every (nondegenerate) conic can be expressed as one of the standard types, show that circles and ellipses are the only conics that are *bounded*, that is, contained inside some large circle.

11.2 Geometry of Conics

FOCUS
DIRECTRIX
ECCENTRICITY
REFLECTION PROPERTIES

So far we have concentrated on recognizing the type of a conic and sketching its graph. These are the things most often needed in the study of calculus. Now we discuss briefly some other aspects of conics.

Suppose we take two points in the plane, Q and R, and consider the set of all points P for which the sum of the distances $|PQ|$ and $|PR|$ equals some constant K. (Of course, K will have to be at least as large as the distance between Q and R, or there will be no such points P at all.)

One way to draw the graph of all these points P would be to take a piece of string that is K units long and (with thumbtacks) attach its ends to Q and R. Pulling the string tight with a pencil, we find that the point of the pencil is at a point P on the graph. If we move the pencil, keeping the string tight, it will draw the curve, which turns out to be an ellipse. See Figure 11.2.1.

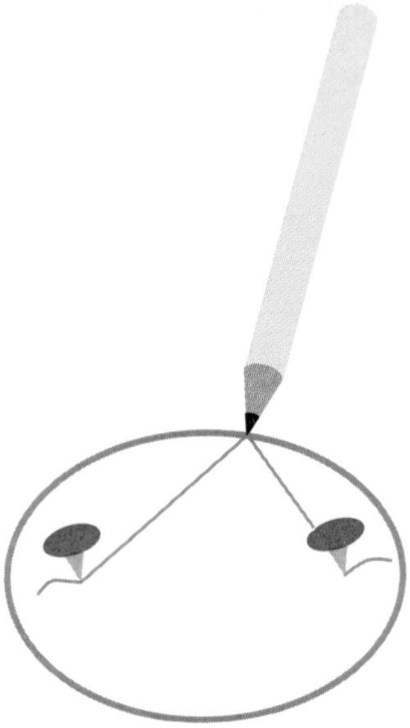

Figure 11.2.1

Each of the points Q and R is called a **focus** of the ellipse. (The plural of "focus" is "foci.")

Now suppose we draw the ellipse and mark on it the major axis (through the foci) and the minor axis (through the midpoint of QR). Consider the point P at one end of the minor axis. Its distance from the major axis is b, the semi-minor axis. Write c for the distance from either focus to the center of the ellipse. We see from Figure 11.2.2 that by the Pythagorean Theorem (Theorem 1.5.1) the distance $|QP|$ from P to a focus is $\sqrt{b^2 + c^2}$. The distance $|RP|$ to the other focus is the same. The constant K is the sum of these two distances, or $2\sqrt{b^2 + c^2}$.

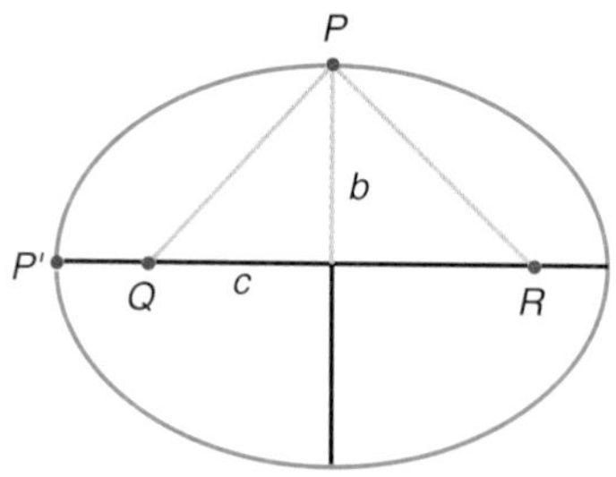

Figure 11.2.2

Now suppose P' is the point at one end of the major axis. Since the distance from P' to the center of the ellipse is the semi-major axis, a, we see that its distance from one focus is $|P'Q| = a - c$, and its distance from the other focus is $|P'R| = a + c$. The sum of these distances, which equals K, is $(a - c) + (a + c) = 2a$.

Since we also know that $K = 2\sqrt{b^2 + c^2}$, we see that $a = \sqrt{b^2 + c^2}$, or

$$a^2 = b^2 + c^2. \tag{11.2.1}$$

EXAMPLE 1 **Finding Foci** Consider the ellipse $\frac{x^2}{4} + \frac{y^2}{9} = 1$. We know that its semi-major axis is $a = \sqrt{9} = 3$ and its semi-minor axis is $b = \sqrt{4} = 2$. From Equation 11.2.1 we see that $c^2 = a^2 - b^2 = 9 - 4 = 5$, so $c = \sqrt{5}$. Since the foci lie on the major axis, which in this case is vertical, we see that the foci are $(0, -\sqrt{5})$ and $(0, \sqrt{5})$. See Figure 11.2.3.

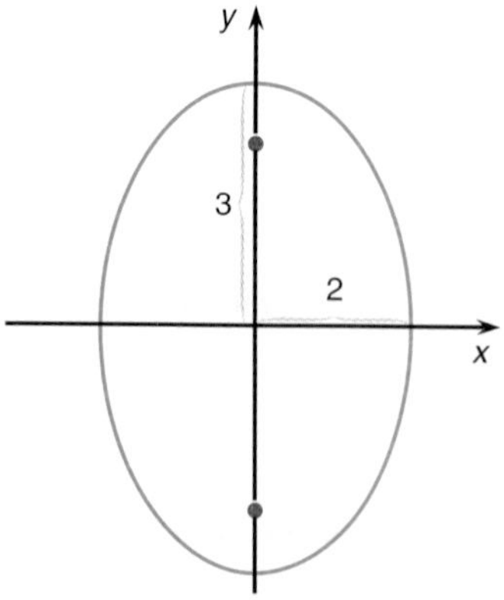

Figure 11.2.3

There is an interesting property of ellipses having to do with reflections. If an object starts at one focus and travels in a straight line until it hits the outside of the ellipse and then bounces off the edge (i.e., is "reflected"), then it will bounce so that it heads straight for the other focus.

For instance, if an ellipse was made of a reflective surface, the light from a flashbulb fired at one focus would all reflect back to the other focus. Moreover, since each ray of light would travel the same distance in going from one focus to the other (the distance K mentioned above), they would all arrive simultaneously (see Figure 11.2.4).

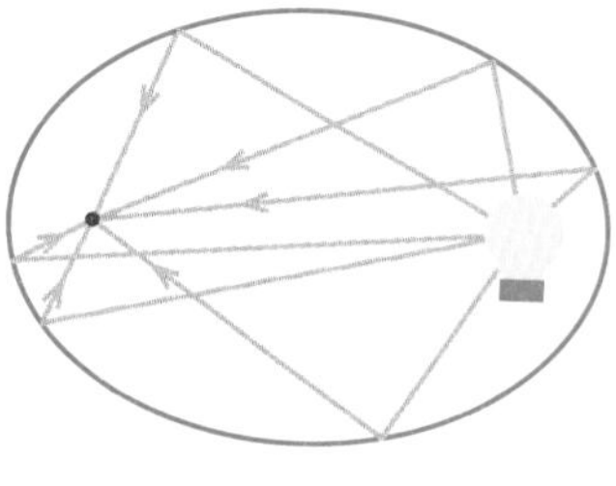

Figure 11.2.4

Some buildings are constructed with a room in the shape of a solid ellipse. Sound emitted at one focus will reflect off the walls and ceiling and bounce to the other focus. A person standing at one focus can hear quite clearly the sound of somebody whispering at the other focus, even though it would be inaudible to somebody standing much closer. Rooms of this design are sometimes called "whispering galleries"; two of the most famous are the one in the United States Capitol and the dome of St. Paul's Cathedral in London, England.

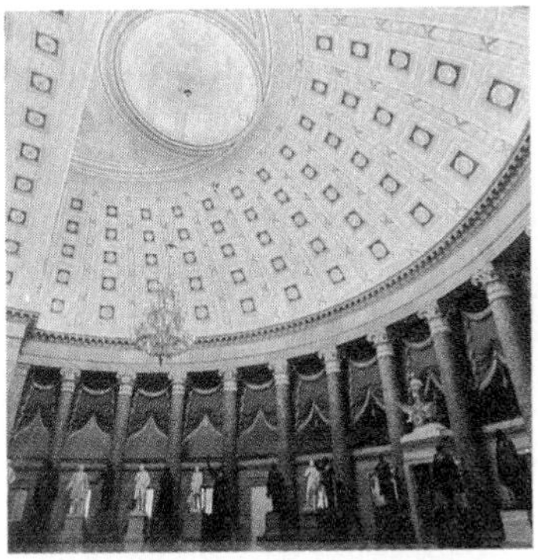

Statuary Hall (the Whispering Gallery), U.S. Capitol Building.

There is another way to describe conics geometrically that applies to parabolas and hyperbolas as well as to ellipses.

Fix a point F and a line d. Now consider all points P whose distance from the line d equals its distance from the point F. If we sketch a number of such points, as in Figure 11.2.5, we begin to suspect that the graph of all of them is a parabola, and this is correct.

We let the point be $F = (0, p)$, on the y-axis, and let the line d be $y = -p$. The advantage of this choice is that the origin lies exactly p units from both F and d, so it is on the graph.

Now consider a point $P = (x, y)$. Its distance from $F = (0, p)$ is $\sqrt{x^2 + (y - p)^2}$. Its distance from the line d is just its vertical distance from d, which is $|y + p|$. See Figure 11.2.6.

For P to lie on the graph, these two distances must be the same, that is, $\sqrt{x^2 + (y - p)^2} = |y + p|$. Squaring both sides, we find that $x^2 + y^2 - 2py + p^2 = y^2 + 2py + p^2$. Subtracting $y^2 - 2py + p^2$ from both sides results in $x^2 = 4py$, or $y = \frac{1}{4p}x^2$, which we recognize as the equation of a parabola with vertex at the origin.

In Figure 11.2.7 the illustration has been drawn with $p > 0$. If $p < 0$, the same calculations work, and in this case the parabola will open downward. If the line d is vertical, then the parabola opens to the left or to the right, depending on the relative positions of the focus and the directrix. Specifically, $y^2 = 4px$ is the equation

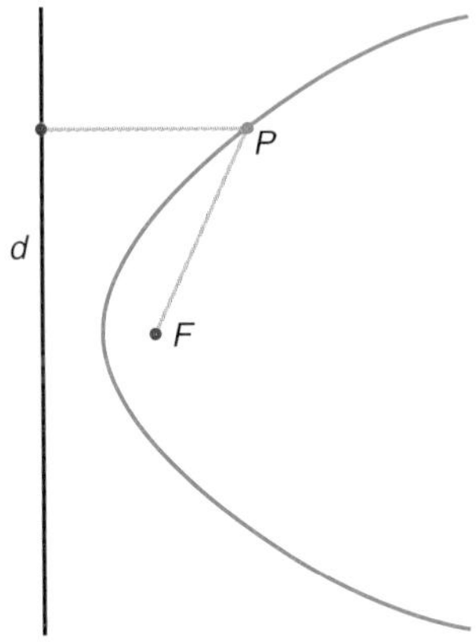

Figure 11.2.5

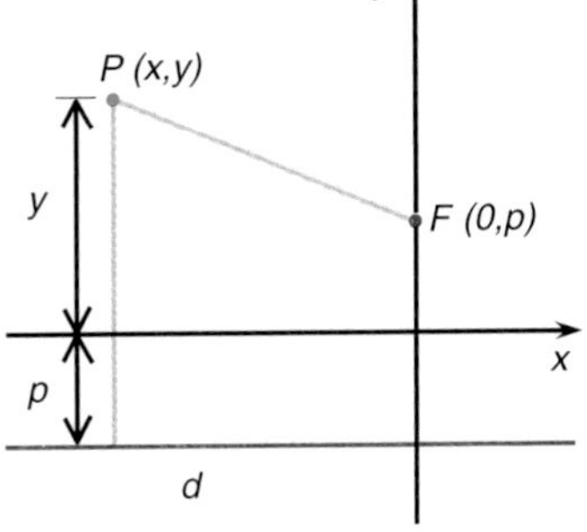

Figure 11.2.6

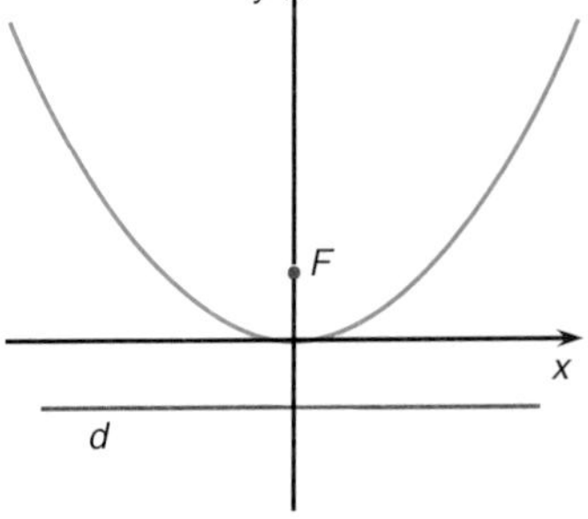

Figure 11.2.7

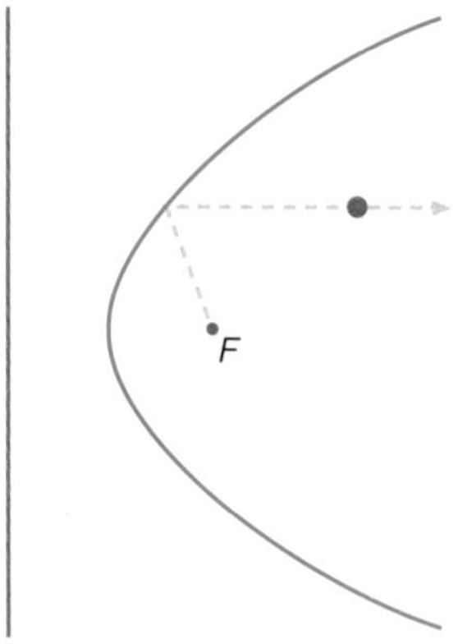

Figure 11.2.8

of a parabola with vertex at the origin, with $F = (p, 0)$ and with the line d equal to $x = -p$. It opens to the right if $p > 0$ and to the left if $p < 0$.

The point F is called the **focus** of the parabola, and the line d is the **directrix**. Every parabola can be described as the set of points that are equidistant from the focus and the directrix.

There is also a reflection property for parabolas. Suppose an object starts at the focus and moves in a straight line until it bounces off the parabola. Then after reflecting, it will be moving away from the directrix in a direction perpendicular to the directrix, as shown in Figure 11.2.8.

This property is the basic idea behind the parabolic reflector. If a reflecting material is formed in the shape of a parabola and a small light bulb is placed at its focus, then the rays of light moving away from the bulb will all reflect out in parallel lines, forming a narrow beam. Flashlights, automobile headlights, and huge searchlights are all built on this principle. See Figure 11.2.9.

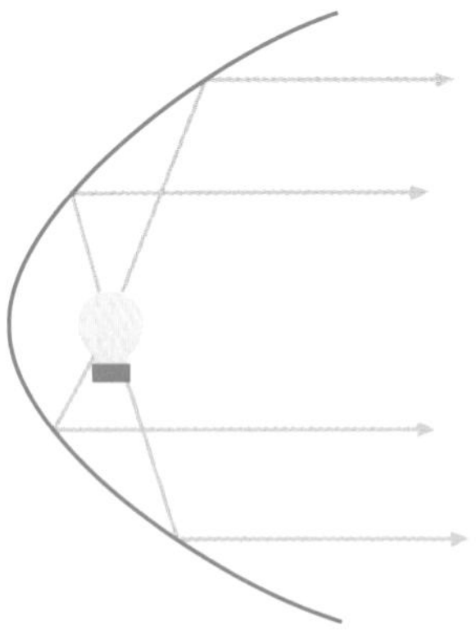

Figure 11.2.9

The same idea can also be used in reverse. If parallel rays of light enter a parabolic reflector, they will all be reflected and converge at the focus, as in Figure 11.2.10. This can be used to collect rays of the sun to heat water in a pipe, as in rooftop solar collectors, or even for cooking. It can also be used to collect sound waves by placing a microphone at the focus of a parabolic dish, which is the basis of some directional microphones. It can also be used to collect radio waves, as we see in the design of some radio telescopes and in satellite dishes (which collect radio signals

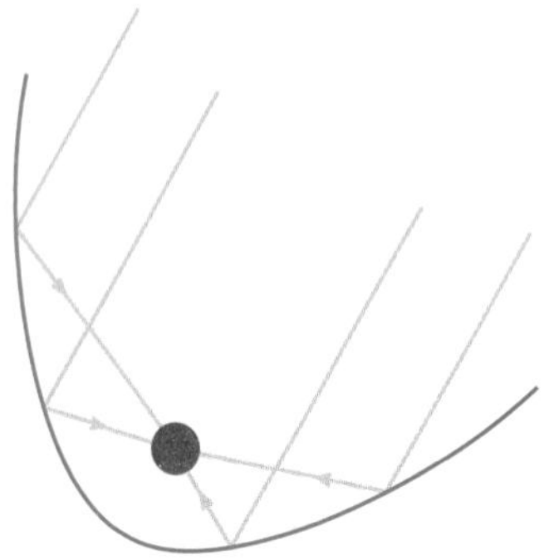

Figure 11.2.10

from a specific direction, i.e., the direction to the satellite, and concentrate them at the focus of the dish, where they can be picked up and connected to a television set).

Parabolic antenna for a VLA radio telescope in New Mexico.

EXAMPLE 2 Find the focus and directrix of the parabola $y^2 + 8x = 24$.

Solution There is no need to complete squares here, but writing the equation as $8(x - 3) = -y^2$ suggests letting $x' = x - 3$ and considering the translated origin $O' = (3, 0)$. The equation now reads $8x' = -y^2$. This is the equation of a parabola whose vertex is O' and that opens to the left (because of the minus sign). If we write it as $x' = -\frac{1}{8}y^2$, that is, $x' = \frac{1}{4p}y^2$, with $p = -2$, the formulas above suggest that the focus F is at the point where $x' = -2$, $y = 0$, and the directrix is the line $x' = -(-2) = 2$.

Converting this, we see that when $x' = -2$, $x = x' + 3 = 1$, and $x' = 2$ is the same as $x = 5$. The focus is $F = (1, 0)$, and the directrix is the line $x = 5$. The parabola is sketched in Figure 11.2.11.

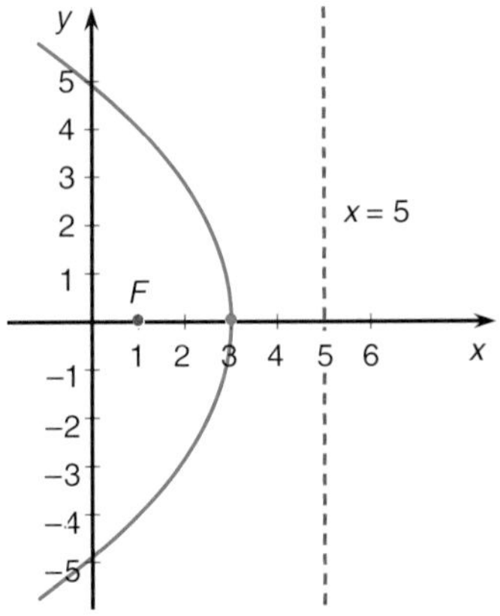

Figure 11.2.11

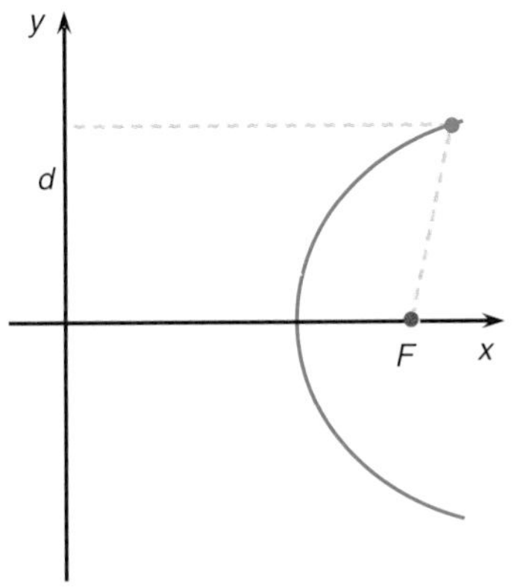

Figure 11.2.12

Now suppose we are given a point F and a line d. If $P = (x, y)$ is any point, we can consider the distance $|PF|$ and the distance from P to the line d. We have just seen that the set of all points P for which these distances are equal is a parabola. Suppose instead we consider the set of all points P for which the distance $|PF|$ is one-half the distance from P to d. Of course, there is no special reason to choose $\frac{1}{2}$; we could try $\frac{1}{3}$ or 17 just as easily. Let us suppose that e is some fixed positive number and consider the set of all $P = (x, y)$ for which the distance $|PF|$ is e times the distance from P to d. It is traditional to use the letter e for this purpose, but here it does *not* refer to the base of the natural logarithm. It is just any positive number.

To be specific, let d be the line $x = 0$ and let $F = (p, 0)$, with $p > 0$, as shown in Figure 11.2.12. The distance from $P = (x, y)$ to the line d is just $|x|$. Since $|PF| = \sqrt{(x-p)^2 + (y-0)^2} = \sqrt{(x-p)^2 + y^2}$, the condition on P amounts to

$$\sqrt{(x-p)^2 + y^2} = e|x|, \text{ or } (x-p)^2 + y^2 = e^2|x|^2$$

$$\text{or } x^2 - 2px + p^2 + y^2 = e^2x^2, \text{ or } (1-e^2)x^2 - 2px + p^2 + y^2 = 0,$$

$$\text{that is, } (1-e^2)\left(x^2 - \frac{2p}{1-e^2}x\right) + p^2 + y^2 = 0.$$

Completing the square, we have

$$(1-e^2)\left(x^2 - \frac{2p}{1-e^2}x + \frac{p^2}{(1-e^2)^2}\right) - (1-e^2)\frac{p^2}{(1-e^2)^2} + p^2 + y^2 = 0$$

$$(1-e^2)\left(x - \frac{p}{1-e^2}\right)^2 + p^2 - \frac{p^2}{(1-e^2)} + y^2 = 0$$

$$(1-e^2)x'^2 + y^2 = p^2\left(\frac{1}{1-e^2} - 1\right) = p^2\left(\frac{1-1+e^2}{1-e^2}\right) = \frac{e^2p^2}{1-e^2},$$

$$\text{or } \frac{(1-e^2)^2}{e^2p^2}x'^2 + \frac{1-e^2}{e^2p^2}y^2 = 1.$$

We have written $x' = x - \frac{p}{1-e^2}$. The last equation we recognize as the equation of a conic, with center at the point $O' = \left(\frac{p}{1-e^2}, 0\right)$. The line d is called the **directrix**; we chose it to be $x = 0$, that is, $x' = -\frac{p}{1-e^2}$.

This calculation was pretty complicated, but all we have really done is follow the procedures for completing squares and translating axes. Fortunately, the actual calculation is not terribly important for our purposes.

However, let us ask what kind of conic the equation represents. The important things are the signs of the coefficients of x'^2 and y^2. If $e < 1$, then $1 - e^2 > 0$ and both coefficients are positive. In this situation the conic is an ellipse. Since $1 - e^2 < 1$, we see that $(1-e^2)^2 < 1 - e^2$, so $\frac{1}{(1-e^2)^2} > \frac{1}{1-e^2}$. With $a^2 = \frac{e^2p^2}{(1-e^2)^2}$ and $b^2 = \frac{e^2p^2}{1-e^2}$, we see that $a^2 > b^2$ and the major axis of the ellipse is horizontal. Also $(1-e^2)a^2 = b^2$, so $a^2 - b^2 = a^2(1 - 1 + e^2) = e^2a^2$ and $e = \frac{\sqrt{a^2-b^2}}{a}$. Since $a^2 = \frac{e^2p^2}{(1-e^2)^2}$, $a = \frac{ep}{1-e^2} > 0$, and $p = \frac{a}{e}(1-e^2)$. The directrix is $x' = -\frac{p}{1-e^2} = -\frac{a}{e}$.

If $e > 1$, the coefficient of y^2 is negative, while the coefficient of x'^2, being a square, is positive. In this situation the conic is a hyperbola, $\frac{x'^2}{a^2} - \frac{y^2}{b^2} = 1$, with its transverse axis horizontal. We also find that $a^2(1-e^2) = -b^2$, so $a^2 + b^2 = a^2 - a^2(1-e^2) = e^2a^2$ and $e = \frac{\sqrt{a^2+b^2}}{a}$. Since $a^2 = \frac{e^2p^2}{(1-e^2)^2}$, $a = -\frac{ep}{1-e^2} > 0$, and $p = -\frac{a}{e}(1-e^2)$. The directrix is $x' = -\frac{p}{1-e^2} = \frac{a}{e}$.

If $e = 0$, we cannot write the last line of the calculation, which involved dividing by e^2, but in the previous line we see the equation becomes $x'^2 + y^2 = 0$, whose graph consists of the single point O', that is, $(x, y) = (p, 0)$.

If $e = 1$, then $1 - e^2 = 0$, so the above calculation, which required dividing by $1 - e^2$, is invalid. However, in that case we have already seen that the conic is a parabola.

Summary 11.2.2

The set of all points $P = (x, y)$ for which $|PF|$ equals e times the distance from P to d is a conic. The point F and the line d are the **focus** and the **directrix** of the conic. The number e is its **eccentricity**. We summarize the above discussion in a table:

Eccentricity	Conic
$e > 1$	Hyperbola
$e = 1$	Parabola
$0 < e < 1$	Ellipse
$e = 0$	Circle

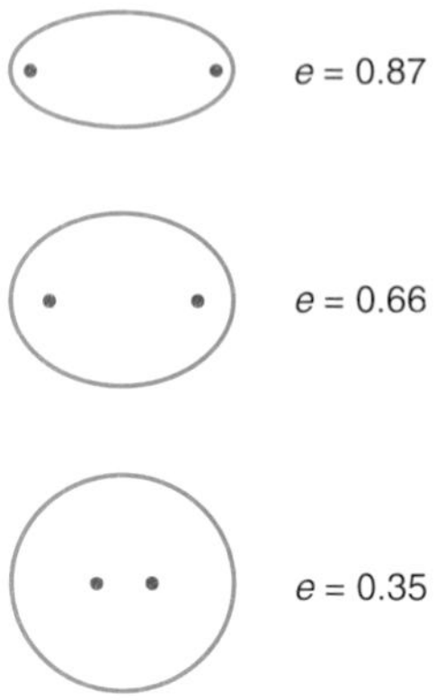

Figure 11.2.13

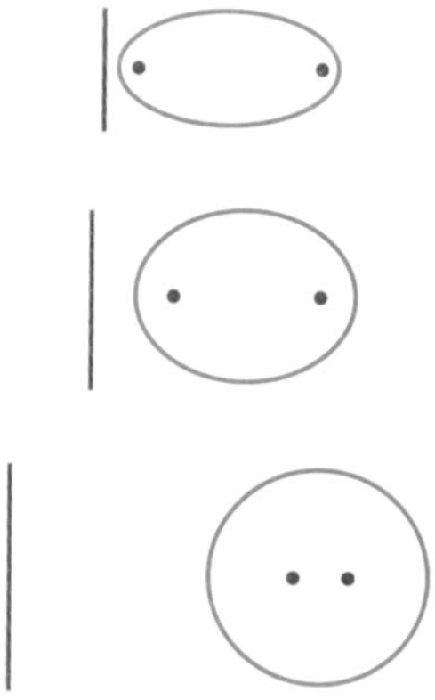

Figure 11.2.14

We have included the circle in this list (its presence gives some explanation of the word "eccentricity": A circle has zero eccentricity; the farther an ellipse is from being round, the greater is its eccentricity, and the more "eccentric" it is.) However, we ought to observe that it is not possible to define a circle by using a focus and directrix, as we did with the other conics. One way to think of it is to consider several ellipses with different eccentricities. The smaller the eccentricity, the closer the ellipse is to being round. See Figure 11.2.13.

The closer the ellipse is to being round, the closer together are its two foci. So if the eccentricity of an ellipse becomes smaller and smaller and approaches 0, then the foci of the ellipse move together. As the eccentricity actually becomes 0, the two foci merge into a single point, which is the center of the circle.

Meanwhile, we could ask what happens to the directrix. As the eccentricity approaches 0, the directrix moves farther and farther away from the ellipse. (After all, suppose the eccentricity e is very small. The distance from a point on the ellipse to a focus is e times the distance to the directrix, so any point on the ellipse must be much closer to the focus than it is to the directrix.) When the ellipse turns into a circle as e becomes 0, the directrix actually "moves off to infinity" (see Figure 11.2.14). This rather vague description is not at all precise, but it is intended to explain why no line will serve as the directrix of a circle.

For a hyperbola of the form $\frac{x^2}{a^2} - \frac{y^2}{b^2} = 1$ the transverse axis is horizontal, and the foci both lie on it. For a hyperbola with a vertical transverse axis the foci lie on the vertical axis. For this reason the transverse axis is sometimes called the **focal axis**.

We should also remark that just as an ellipse is the set of all points whose distances from Q and R add up to some constant K, a hyperbola can be expressed as the set of all points whose distances from the foci Q and R *differ* by some fixed number K, that is, the set of all P for which $|PQ| - |PR| = \pm K$. (We say $\pm K$ because we do not know which of $|PQ|$ and $|PR|$ is larger. In fact, if we just wrote K, we would get one branch of the hyperbola, and $-K$ would give the other branch.)

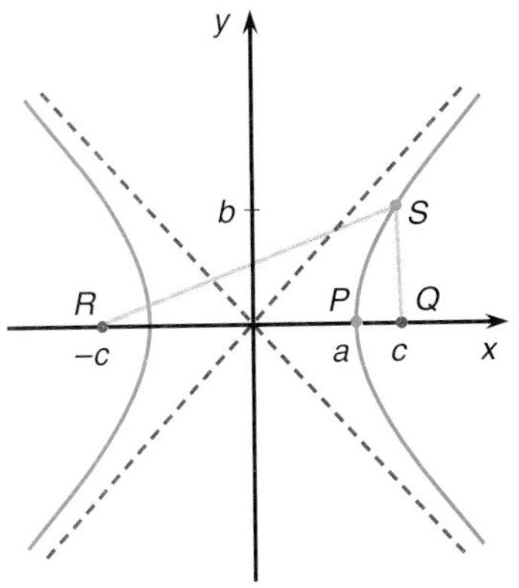

Figure 11.2.15

Finally, in analogy with the ellipse, for which we found that $a^2 = b^2 + c^2$, there are relations between the numbers a and b and the location of the foci of a hyperbola. Consider first the hyperbola $\frac{x^2}{a^2} - \frac{y^2}{b^2} = 1$, as illustrated in Figure 11.2.15. Let us write c for the distance from the origin to either focus, so the foci are the points $Q = (c, 0)$ and $R = (-c, 0)$.

Label the point $P = (a, 0)$ as shown. The distance from P to the left focus is $c + a$, and the distance from P to the right focus is $c - a$. The difference of these distances is $2a$, so the constant K referred to above is $K = 2a$.

Now consider the point S on the hyperbola whose y-coordinate is b and whose x-coordinate is positive. We can solve

$$\frac{x^2}{a^2} = 1 + \frac{y^2}{b^2} = 1 + \frac{b^2}{b^2} = 2,$$

so

$$x^2 = 2a^2$$

and

$$x = \sqrt{2}a.$$

So S is the point $(\sqrt{2}a, b)$. We know that $|SQ| - |SR|$ is $2a$ or $-2a$. Using Formula 1.5.2 for the distance between points, we can express this as

$$\pm\, 2a = \sqrt{(\sqrt{2}a - c)^2 + b^2} - \sqrt{(\sqrt{2}a + c)^2 + b^2}.$$

We rearrange and square both sides:

$$\pm 2a + \sqrt{(\sqrt{2}a + c)^2 + b^2} = \sqrt{(\sqrt{2}a - c)^2 + b^2}$$

$$4a^2 \pm 4a\sqrt{2a^2 + 2\sqrt{2}ac + c^2 + b^2}$$
$$+ 2a^2 + 2\sqrt{2}ac + c^2 + b^2 = 2a^2 - 2\sqrt{2}ac + c^2 + b^2$$
$$4a^2 \pm 4a\sqrt{2a^2 + 2\sqrt{2}ac + c^2 + b^2} = -4\sqrt{2}ac.$$

We rearrange so that everything is on the right side except the square root, then divide by $4a$ and square again:

$$\pm 4a\sqrt{2a^2 + 2\sqrt{2}ac + c^2 + b^2} = -4a^2 - 4\sqrt{2}ac$$
$$\pm\sqrt{2a^2 + 2\sqrt{2}ac + c^2 + b^2} = -a - \sqrt{2}c$$
$$2a^2 + 2\sqrt{2}ac + c^2 + b^2 = a^2 + 2\sqrt{2}ac + 2c^2$$
$$a^2 + b^2 = c^2.$$

We find that $c = \sqrt{a^2 + b^2}$.

Moreover, all these relations also hold for a hyperbola whose focal axis is vertical.

By applying the description before Summary 11.2.2 we see that $e = \frac{\sqrt{a^2+b^2}}{a}$, so $e = \frac{c}{a}$, and the directrix is $x = \frac{a}{e}$, that is, $x = \frac{a^2}{c}$.

Summary 11.2.3

	Hyperbola	Hyperbola	Ellipse	Ellipse
	$\frac{x^2}{a^2} - \frac{y^2}{b^2} = 1$	$\frac{y^2}{a^2} - \frac{x^2}{b^2} = 1$	$\frac{x^2}{a^2} + \frac{y^2}{b^2} = 1, a > b > 0$	$\frac{y^2}{a^2} + \frac{x^2}{b^2} = 1, a > b > 0$
Foci:	$(\pm c, 0)$, with $c = \sqrt{a^2 + b^2}$	$(0, \pm c)$, with $c = \sqrt{a^2 + b^2}$	$(\pm c, 0)$, with $c = \sqrt{a^2 - b^2}$	$(0, \pm c)$, with $c = \sqrt{a^2 - b^2}$
Eccentricity:	$e = \frac{c}{a} = \frac{1}{a}\sqrt{a^2 + b^2}$	$\frac{c}{a} = \frac{1}{a}\sqrt{a^2 + b^2}$	$\frac{c}{a} = \frac{1}{a}\sqrt{a^2 - b^2}$	$\frac{c}{a} = \frac{1}{a}\sqrt{a^2 - b^2}$
Directrix:	$x = \pm\frac{a}{e} = \pm\frac{a^2}{c}$	$y = \pm\frac{a}{e} = \pm\frac{a^2}{c}$	$x = \pm\frac{a}{e} = \pm\frac{a^2}{c}$	$y = \pm\frac{a}{e} = \pm\frac{a^2}{c}$

EXAMPLE 3 Find the equation of the hyperbola with center at the origin, eccentricity $e = 2$, and y-intercepts ± 4.

Solution Since the hyperbola has y-intercepts, its equation must be of the form $\frac{y^2}{a^2} - \frac{x^2}{b^2} = 1$, and since the intercepts are ± 4, we know that $a = 4$.

We also know that $2 = e = \frac{c}{a} = \frac{1}{4}\sqrt{4^2 + b^2}$. Squaring this equation, we see that $4 = \frac{1}{16}(16 + b^2)$, or $64 = 16 + b^2$, or $b^2 = 48$. The equation is $\frac{y^2}{16} - \frac{x^2}{48} = 1$.

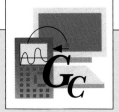

With a graphics calculator we can plot the hyperbola. We rewrite it as $\frac{y^2}{16} = 1 + \frac{x^2}{48}$, or $y = \pm 4\sqrt{1 + \frac{x^2}{48}}$.

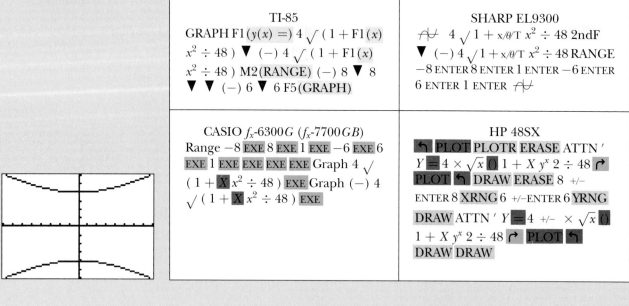

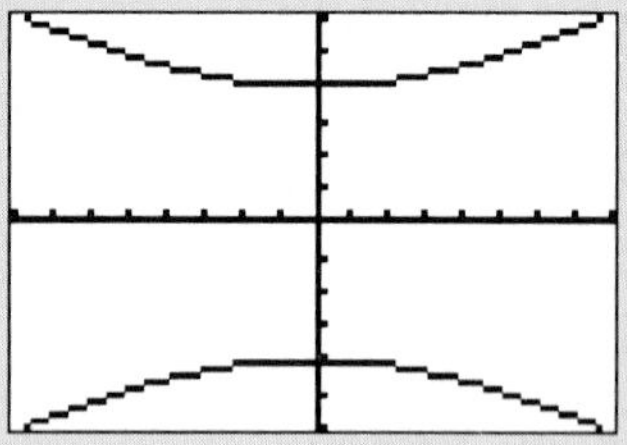

TI-85	SHARP EL9300
GRAPH F1 (y(x) =) 4 √ (1 + F1 (x) x^2 ÷ 48) ▼ (−) 4 √ (1 + F1 (x) x^2 ÷ 48) M2 (RANGE) (−) 8 ▼ 8 ▼ ▼ (−) 6 ▼ 6 F5 (GRAPH)	4 √ 1 + x/θ/T x^2 ÷ 48 2ndF ▼ (−) 4 √ 1 + x/θ/T x^2 ÷ 48 RANGE −8 ENTER 8 ENTER 1 ENTER −6 ENTER 6 ENTER 1 ENTER
CASIO f_x-6300G (f_x-7700GB)	**HP 48SX**
Range −8 EXE 8 EXE 1 EXE −6 EXE 6 EXE 1 EXE EXE EXE EXE Graph 4 √ (1 + X x^2 ÷ 48) EXE Graph (−) 4 √ (1 + X x^2 ÷ 48) EXE	↰ PLOT PLOTR ERASE ATTN ′ Y = 4 × √x () 1 + X y^x 2 ÷ 48 ↱ PLOT ↰ DRAW ERASE 8 +/− ENTER 8 XRNG 6 +/− ENTER 6 YRNG DRAW ATTN ′ Y = 4 +/− × √x () 1 + X y^x 2 ÷ 48 ↱ PLOT ↰ DRAW DRAW

Exercises 11.2

In these exercises, all conics are assumed to have their axes vertical and/or horizontal. In particular, this means that there are no xy-terms in their equations.

I

Find the foci and the directrices of the following conics.

1. $y = x^2 + 2x$
2. $y^2 = 2x + 4y - 8$
3. $\frac{x^2}{4} + y^2 = 1$
4. $2x^2 + 3y^2 = 1$
5. $x^2 - 4y^2 = 1$
6. $x^2 + 2x = y^2$
7. $x^2 + 4y^2 + 4x = 8y$
8. $\frac{x^2}{6} = \frac{2y^2}{3} - 6$
9. $9x^2 = 4y^2 + 18x + 8y + 11$
10. $100x^2 + y^2 = 20y$

Find the equation of each of the following conics.

11. The ellipse with center $(0, 0)$, eccentricity $e = \frac{1}{2}$, and foci $(\pm 3, 0)$.
12. The hyperbola with center $(0, 0)$ and eccentricity 3 and containing $(5, 0)$.
13. The parabola with directrix $x = 4$ and focus $(2, 2)$.
14. The hyperbola that opens sideways having transverse axis twice as long as its conjugate axis and center at $(2, -3)$ and that contains $(-2, -2)$.
15. The ellipse whose major axis is three times as long as its minor axis, whose center is $(4, 1)$, and that contains $(3, 3)$ but contains no point to the left of the y-axis.
16. The parabola containing $(1, 2)$, $(2, 4)$, and $(2, 1)$.
17. The hyperbola with center $(2, -1)$, eccentricity $e = 2$, and foci $(2, 1)$ and $(2, -3)$.
18. The conic whose eccentricity is $e = \frac{4}{3}$, whose center is $(0, 2)$, and that contains $(-6, 2)$.
19. The conic whose eccentricity is $e = 1$, that is symmetric about the line $x = 3$, and that contains the points $(3, 4)$ and $(2, 3)$.
20. The conic whose directrix is $x = 1$, that has the point $(4, 2)$ as a focus, and that contains the point $(3, 2)$.

Find each of the following.

21. The eccentricity of the ellipse $2x^2 + 8y^2 = 1$.
22. The eccentricity of $5x^2 - 8y^2 + 4y = 7$.
23. The foci of $3x^2 + 6x + 4y^2 = 12$.
24. The foci of $y^2 - x^2 = 4$.
25. The directrix of $x^2 + 4x - 8y = 12$.
26. The directrices of $x^2 + 4x - y^2 - 6y = 4$.
27. The eccentricity of $5x^2 + 3x - 7y = 12$.
28. The sum of the distances from any point on $4x^2 + 6y^2 = 24$ to the two foci.
29. The distance between the foci of $9x^2 - 4y^2 + 8y = 3$.
30. The possible distances from a focus to a directrix of $4x^2 + 9y^2 - 18y = 27$.

II

31. An architect wishes to fit a rectangular door inside a parabolic arch. If the door is 4 ft wide and 8 ft high and the arch is 6 ft wide at floor level, what is the height above the floor of the highest point of the arch?
32. Suppose a window is to be made in the shape of a horizontal ellipse with eccentricity $e = \frac{1}{2}$. There is to be a rectangular pane of clear glass measuring 1 ft by 2 ft in the center of the ellipse with its four corners just reaching the outer frame. The remaining space of the window will be filled with a design in stained glass. What is the long dimension of the ellipse?

*33. What is the relationship between the eccentricities of the ellipses
$$\frac{x^2}{a^2} + \frac{y^2}{b^2} = 1 \quad \text{and} \quad \frac{x^2}{a^2} + \frac{y^2}{b^2} = 4?$$

*34. What is the relationship between the eccentricities of the hyperbolas $\frac{x^2}{a^2} - \frac{y^2}{b^2} = k$ for different values of k?

*35. Find the equation of the hyperbola with center at the origin, having eccentricity $e = 4$, and containing the point $(2, 8)$.

*36. There are two ellipses with eccentricity $e = \frac{1}{2}$ with center at the origin that contain the point $(2, 1)$. Find their equations.

*37. Suppose the floor of an elliptically shaped room has its long dimension equal to 20 m and its short dimension equal to 15 m. How far out from the walls are the foci?

*38. How many directrices does each type of conic (ellipse, hyperbola, parabola, circle) have?

*39. Consider two circles C, D with different centers so that C is inside D. Consider all circles that lie inside D, tangent to D (at one point), and outside C, tangent to C (at one point). Show that the centers of all these circles lie on an ellipse. What are the foci of the ellipse?

****40.** Consider a line d and a point P that does not lie on d. Consider all circles through P that are tangent to d. Show that the centers of all such circles lie on a parabola.

****41.** Let $a > c > 0$, and let $F_1 = (c, 0)$, $F_2 = (-c, 0)$. Find the equation of the set of points $P = (x, y)$ determined by the condition $|PF_1| + |PF_2| = 2a$. You will probably want to let $b^2 = a^2 - c^2$.

****42.** The points $F_1 = (1, 1)$ and $F_2 = (-1, -1)$ do not lie on any horizontal or vertical line. Find the equation of the ellipse with F_1, F_2 as foci and semi-major axis of length 2.

****43.** How many ellipses are there with eccentricity $e = \frac{1}{2}$ and foci $(0, \pm 2)$?

****44.** How many hyperbolas are there with eccentricity $e = 2$ and asymptotes $y = \pm x$?

****45.** (Reflection property for parabolas) Consider a point P on the parabola $x = y^2$. In Figure 11.2.16 we draw the tangent line to the parabola at P and the segment PF from P to the focus F. We also show a ray of light coming horizontally from the right. To show that every such ray is reflected to the focus, we need to show that the angle of incidence, α, equals the angle of reflection, β (see Example 4.7.4, p. 295).

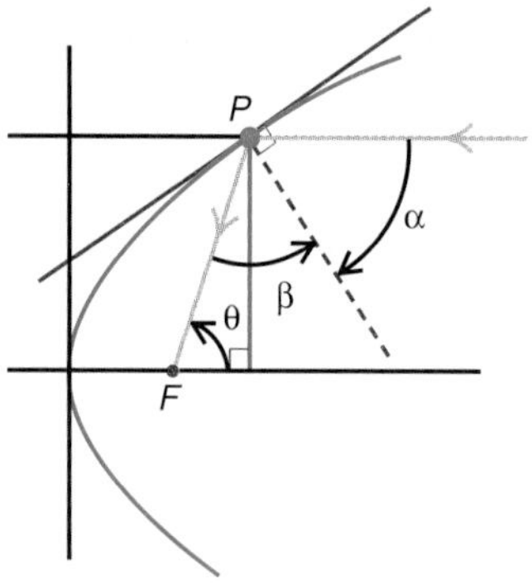

Figure 11.2.16

(i) If $P = (a^2, a)$, find a formula for the slope of the tangent at P; note that this equals $\cot \alpha$.

(ii) Find the coordinates of the focus F. Find a formula for $\tan \theta$ (θ is the angle between PF and the horizontal).

(iii) Use your knowledge of trigonometry to show that $\tan \beta = -\frac{\tan \theta + \tan \alpha}{1 - \tan \theta \tan \alpha}$.

(iv) Use part (iii) to conclude that $\tan \alpha = \tan \beta$, as required.

****46.** Prove the reflection property for the parabola $y^2 = 4px$. (Assume $p > 0$.)

***47.** Show that the equation of the tangent to the ellipse $\frac{x^2}{a^2} + \frac{y^2}{b^2} = 1$ at the point (u, v) is $\frac{ux}{a^2} + \frac{vy}{b^2} = 1$.

****48.** Use Exercise 47 to establish the reflection property for ellipses (following the general idea that we used for parabolas in Exercises 45 and 46).

***49.** An airplane starts at the origin and flies in a straight line northeast, that is, up and to the right along the line $y = x$. Transmitters at $(0, 0)$ and $(0, 150)$ (with distances measured in miles) send out synchronized radio signals, and delicate instruments in the plane measure the difference between their arrival times. If the speed of radio signals is known, it is possible to calculate that the plane is 50 miles farther from $(0, 0)$ than from $(0, 150)$. The set of all points at which this is true is one branch of a hyperbola. (i) Find the equation of the hyperbola, and identify the correct branch. (ii) Find the location of the airplane.

***50.** In the setting of Exercise 49, suppose winds have blown the aircraft off course, so it is no longer on the line $y = x$. Comparison of radio signals shows that it is 50 miles farther from $(0, 0)$ than from $(0, 150)$. Using signals from a third transmitter, it is possible to see that it is also 100 miles farther from $(0, 0)$ than from $(200, 0)$. Show that the location of the aircraft is determined by the points of intersection of two hyperbolas. (This principle can be used for navigation, but over long distances it is complicated by the fact that the surface of the earth is curved.)

III

GC

***51.** Use a graphics calculator or computer to sketch on one picture several ellipses with different eccentricities. Each should have its center at the origin, major axis horizontal, and semi-major axis equal to 1, and the eccentricities should be $\frac{1}{10}, \frac{1}{4}, \frac{1}{2}, \frac{3}{4}, \frac{9}{10}$. (You might begin by graphing the top half of each ellipse.)

***52.** Use a graphics calculator or computer to sketch several hyperbolas on the same picture. Each should have its center at the origin and transverse axis horizontal and should contain the point $(1, 0)$. The hyperbolas should have eccentricities $\frac{11}{10}, \frac{3}{2}, 2, 3, 5$.

11.3 Rotated Coordinates

ELIMINATING CROSS-TERMS
DISCRIMINANT
DEGENERATE CONICS

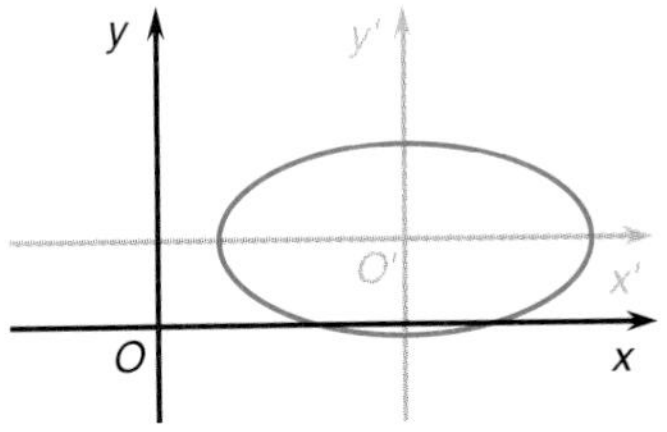

Figure 11.3.1

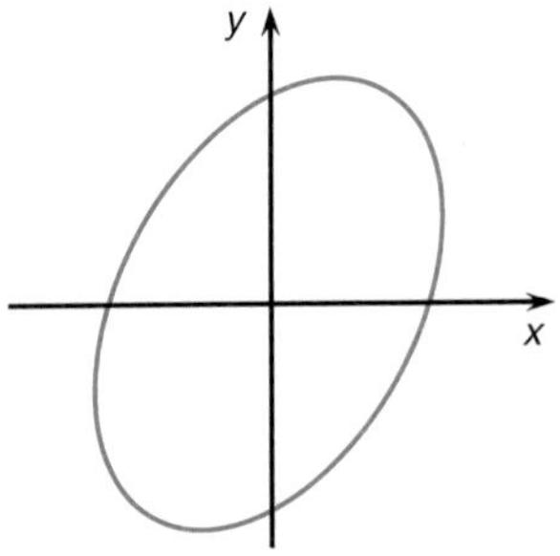

Figure 11.3.2(i)

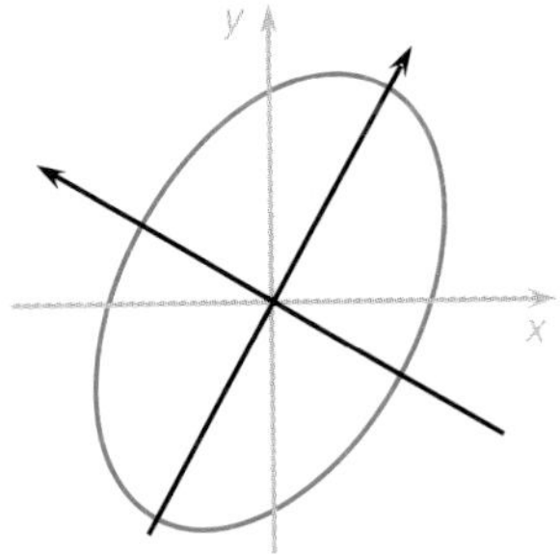

Figure 11.3.2(ii)

So far in our study of conics we have always assumed that the axes of the conic are horizontal or vertical, which is the same thing as assuming that the equation of the conic does not contain any cross-term. In this section we learn to deal with conics that do not satisfy this assumption.

Before proceeding, let us think briefly about the technique we learned in Sections 1.6 and 11.1 for *translating the axes.* Roughly speaking, if we consider a conic whose center is not the origin, then its equation will be complicated. It is possible to simplify the equation by considering a new set of axes that are "better placed" relative to the conic; the way to accomplish this is to place the translated origin at the center of the conic (see Figure 11.3.1).

Now suppose we consider an ellipse centered at the origin but placed so that its major and minor axes are not horizontal or vertical, as illustrated in Figure 11.3.2(i). Its equation is likely to be complicated and difficult to recognize, but taking a cue from the discussion above, we might try to find a new set of axes that are placed at an angle relative to the x- and y-axes, in such a way that they lie along the axes of the ellipse. This is illustrated in Figure 1.3.2(ii).

Relative to these rotated axes, the equation of the ellipse ought to be one of the familiar standard forms, and it should be easy both to identify what it is and also to sketch its graph. This is the idea that we explore in this section.

To make this approach work, it will be necessary to work with rotated axes. Specifically, suppose we know the coordinates (x', y') of a point P relative to a set of axes that are rotated at an angle θ relative to the x- and y-axes, as shown in Figure 11.3.3(i). Then we need to know how to find its coordinates (x, y) relative to the x- and y-axes.

It turns out that the most convenient way to approach this question is to use polar coordinates. Suppose the polar coordinates of P, *relative to the* x'- *and* y'*-axes,* are (r, ϕ), as shown in Figure 11.3.3(ii). In particular, this means that the rectangular

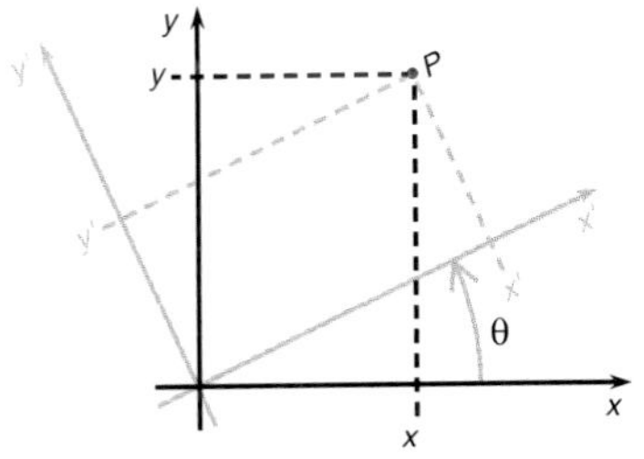

Figure 11.3.3(i)

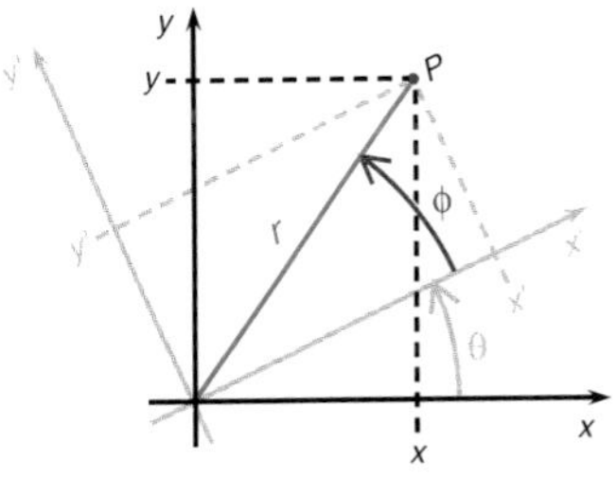

Figure 11.3.3(ii)

coordinates are

$$x' = r\cos\phi, \qquad y' = r\sin\phi$$

(see Equation 1.7.9, p. 50).

Now we consider the same point P relative to the x- and y-axes. To find its polar coordinates, we observe that the length r is the same, and the angle is $\theta + \phi$. Using

Equation 1.7.9 again, we see that the rectangular coordinates of P relative to the unrotated axes are

$$x = r\cos(\theta + \phi), \qquad y = r\sin(\theta + \phi).$$

Now we simplify, using the addition formulas for sin and cos (cf. Theorem 1.7.11, p. 51):

$$x = r\cos(\theta + \phi) = r\cos\phi\cos\theta - r\sin\phi\sin\theta = x'\cos\theta - y'\sin\theta,$$
$$y = r\sin(\theta + \phi) = r\cos\phi\sin\theta + r\sin\phi\cos\theta = x'\sin\theta + y'\cos\theta.$$

We summarize as a theorem.

THEOREM 11.3.1

Suppose the x'- and y'-axes are rotated through an angle θ relative to the x- and y-axes. If the coordinates relative to the rotated axes of a point P are (x', y'), then its coordinates relative to the (unrotated) x- and y-axes are

$$(x, y) = (x'\cos\theta - y'\sin\theta, x'\sin\theta + y'\cos\theta).$$

Conversely, if the coordinates of P are (x, y) relative to the x- and y-axes, then its coordinates relative to the x'- and y'-axes are

$$(x', y') = (x\cos\theta + y\sin\theta, -x\sin\theta + y\cos\theta).$$

PROOF The argument given before the statement of Theorem 11.3.1 proves the first formula. It is possible to obtain the second one in the same way, but it is easier to observe that the x- and y-axes are rotated through an angle of $-\theta$ relative to the x'- and y'-axes. Because of this, we can simply use the same formula, replacing θ with $-\theta$. Since $\cos(-\theta) = \cos\theta$ and $\sin(-\theta) = -\sin\theta$, the second formula drops out.

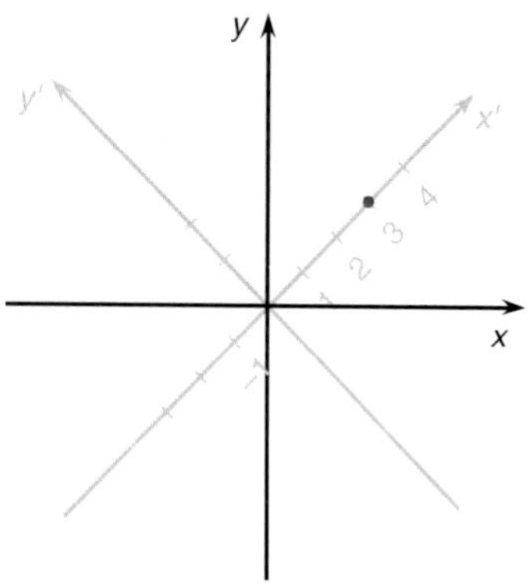

Figure 11.3.4

EXAMPLE 1 Consider the point P that lies on the line $y = x$ at a distance of 3 units above and to the right of the origin. It is not difficult to find its coordinates, but let us use Theorem 11.3.1.

We can imagine the x'-axis placed so that it lies along the line $y = x$, as shown in Figure 11.3.4. Then the point P lies on the positive x'-axis at a distance of 3 units from the origin, so its coordinates are $(3, 0)$ relative to the x'- and y'-axes.

Since these axes are rotated through an angle of 45° or $\frac{\pi}{4}$ relative to the usual axes, we can use Theorem 11.3.1 with $\theta = \frac{\pi}{4}$ to find that the coordinates of P relative to the usual axes are $x = x'\cos\theta - y'\sin\theta = 3\cos\frac{\pi}{4} - 0 = \frac{3}{\sqrt{2}}$ and $y = x'\sin\theta + y'\cos\theta = 3\sin\frac{\pi}{4} + 0 = \frac{3}{\sqrt{2}}$. The coordinates of P are $(x, y) = \left(\frac{3}{\sqrt{2}}, \frac{3}{\sqrt{2}}\right)$.

This result is not surprising, but it does demonstrate the use of Theorem 11.3.1 and confirm that it works.

We really want to use Theorem 11.3.1 to simplify the equation of a conic by removing the cross-term. We begin with an example.

EXAMPLE 2 Consider the conic whose equation is $4x^2 + 2y^2 - 2\sqrt{3}xy = 5$. Notice that it contains a cross-term, that is, a term containing xy.

Let us consider new axes that are rotated through an angle of $\frac{\pi}{3}$ relative to the usual axes. The idea is to express the equation in terms of the rotated coordinates (x', y'). From Theorem 11.3.1 we know that $x = x'\cos\frac{\pi}{3} - y'\sin\frac{\pi}{3} = \frac{x'}{2} - \frac{\sqrt{3}y'}{2}$ and $y = x'\sin\frac{\pi}{3} + y'\cos\frac{\pi}{3} = \frac{\sqrt{3}x'}{2} + \frac{y'}{2}$.

We intend to substitute these formulas into the equation. First we find that

$$x^2 = \frac{1}{4}x'^2 + \frac{3}{4}y'^2 - \frac{\sqrt{3}}{2}x'y',$$

$$y^2 = \frac{3}{4}x'^2 + \frac{1}{4}y'^2 + \frac{\sqrt{3}}{2}x'y',$$

$$xy = \frac{\sqrt{3}}{4}x'^2 - \frac{\sqrt{3}}{4}y'^2 + \left(\frac{1}{4} - \frac{3}{4}\right)x'y' = \frac{\sqrt{3}}{4}x'^2 - \frac{\sqrt{3}}{4}y'^2 - \frac{1}{2}x'y'.$$

Substituting into the original equation, we find that

$$\begin{aligned} 5 &= 4x^2 + 2y^2 - 2\sqrt{3}xy \\ &= (x'^2 + 3y'^2 - 2\sqrt{3}x'y') + \left(\frac{3}{2}x'^2 + \frac{1}{2}y'^2 + \sqrt{3}x'y'\right) - \left(\frac{3}{2}x'^2 - \frac{3}{2}y'^2 - \sqrt{3}x'y'\right), \end{aligned}$$

that is,

$$5 = x'^2 + 5y'^2.$$

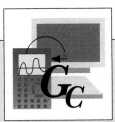

If we want to sketch this ellipse using a graphics calculator, we use the quadratic formula to solve for y. We write $4x^2 + 2y^2 - 2\sqrt{3}xy = 5$ as $2y^2 - (2\sqrt{3}x)y + (4x^2 - 5) = 0$, so $y = \frac{2\sqrt{3}x \pm \sqrt{12x^2 - 8(4x^2-5)}}{4} = \frac{2\sqrt{3}x \pm \sqrt{40 - 20x^2}}{4} = \frac{\sqrt{3}x \pm \sqrt{10 - 5x^2}}{2}$.

TI-85	SHARP EL9300
GRAPH F1 (y(x) =) (√(3) × F1 (x) + √ (10 − 5 F1 (x) x^2)) ÷ 2 ▼ (√(3) × F1 (x) − √ (10 − 5 F1 (x) x^2)) ÷ 2 M2(RANGE) (−) 2 ▼ 2 ▼ ▼ (−) 2 ▼ 2 F5(GRAPH)	[graph] (√ 3 ▶ × x/θ/T + √ 10 − 5 x/θ/T x^2 ▶) ÷ 2 2ndF ▼ (√ 3 ▶ × x/θ/T − √ 10 − 5 x/θ/T x^2 ▶) ÷ 2 RANGE −2 ENTER 2 ENTER 1 ENTER −2 ENTER 2 ENTER 1 ENTER [graph]
CASIO f_x-6300G (f_x-7700GB)	**HP 48SX**
Range −2 EXE 2 EXE 1 EXE −2 EXE 2 EXE 1 EXE EXE EXE EXE Graph (√ 3 × X + √ (10 − 5 X x^2)) ÷ 2 EXE Graph (√ 3 × X − √ (10 − 5 X x^2)) ÷ 2 EXE	↰ PLOT PLOTR ERASE ATTN ′ Y = () √x 3 × X + √x () 10 − 5 × X y^x 2 ▶ ▶ ÷ 2 ↱ PLOT ↰ DRAW ERASE 2 +/−ENTER 2 XRNG 2 +/−ENTER 2 YRNG DRAW ATTN ′ Y = () √x 3 × X − √x () 10 − 5 × X y^x 2 ▶ ▶ ÷ 2 ↱ PLOT ↰ DRAW DRAW

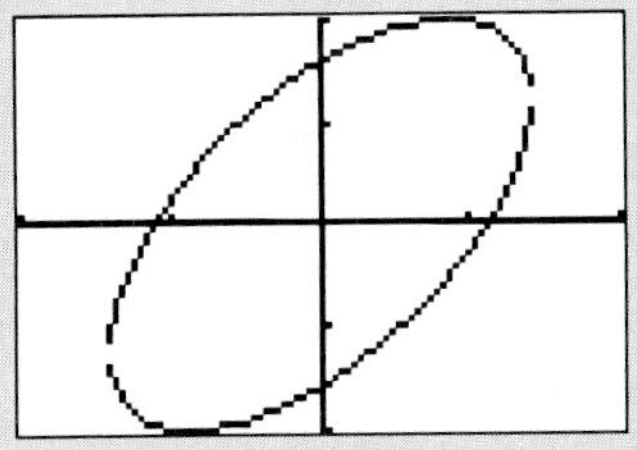

Distortion from unequal scales on the axes may make it difficult to guess the angle through which the conic was rotated relative to the standard position.

Of course, the interesting thing about this new equation is that it contains no cross-term. It is immediately recognized as the equation of an ellipse, and it is quite easy to sketch its graph. We draw in the rotated axes and then sketch the ellipse relative to them. See Figure 11.3.5.

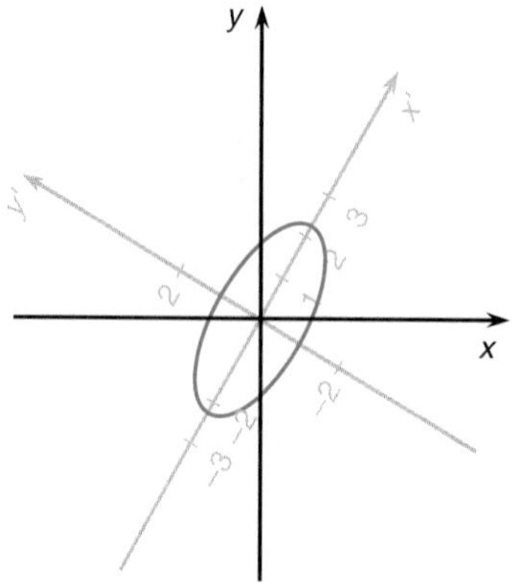

Figure 11.3.5

We were able to use the formulas from Theorem 11.3.1 that relate coordinates relative to the two sets of axes to convert the original equation into an equation in terms of the other set of coordinates. The resulting equation turned out to be easily recognized because it contained no cross-term.

Of course, in this example we were given the angle $\theta = \frac{\pi}{3}$, and in hindsight we see that this gives coordinate axes along the axes of the ellipse. On the other hand, if we simply guessed a value for θ, it is unlikely that it would correspond to the correct angle, and the converted equation would still have a cross-term and would still be difficult to identify and graph. The problem we have to consider now is this: Given an equation that contains a cross-term, how do we find a suitable angle θ through which to rotate the axes in order to remove the cross-term?

Consider the equation $Ax^2 + Bxy + Cy^2 + Dx + Ey = q$. Let us suppose that $B \neq 0$. If we rotate the axes through an angle θ and make the substitutions

$$x = x' \cos\theta - y' \sin\theta, \qquad y = x' \sin\theta + y' \cos\theta,$$

then we could actually do the calculation and work out the transformed equation exactly. But since we are principally interested in removing the cross-term, let us simply find the cross-term in the transformed equation and try to make it zero.

The coefficient of $x'y'$ in Ax^2 is $A(-2\cos\theta\sin\theta) = -2A\cos\theta\sin\theta$; the coefficient of $x'y'$ in Cy^2 is $2C\cos\theta\sin\theta$; and the coefficient of $x'y'$ in Bxy is $B(\cos^2\theta - \sin^2\theta)$. The remaining terms $Dx + Ey$ and k do not contain any $x'y'$-terms. The total $x'y'$-term is

$$\begin{aligned} &-2A\cos\theta\sin\theta + 2C\cos\theta\sin\theta + B(\cos^2\theta - \sin^2\theta) \\ &\quad = (C - A)\sin 2\theta + B\cos 2\theta. \end{aligned}$$

Our goal is to make this equal to zero, that is, $(A - C)\sin 2\theta = B\cos 2\theta$. This is the same as $\frac{A-C}{B} = \cot 2\theta$. So to find the angle θ of rotation that will remove the cross-term, we simply solve $\cot 2\theta = \frac{A-C}{B}$. Notice that we have assumed that $B \neq 0$, so there is no difficulty about dividing.

On the other hand, we are more familiar with tan than cot, so we might like to write the equation as $\frac{B}{A-C} = \tan 2\theta$, or $\theta = \frac{1}{2}\arctan\frac{B}{A-C}$. This might be preferable, especially for people whose calculators have an arctan button and not an arccot button, but it does have the disadvantage of dividing by 0 when $A = C$.

In Example 2 we rotated by $\theta = \frac{\pi}{3}$; if we want $-\frac{\pi}{4} < \theta \le \frac{\pi}{4}$, we could have used $\theta = -\frac{\pi}{6}$.

Let us consider what happens when $A = C$. Then $\cot 2\theta = 0$, which happens, for instance, when $2\theta = \frac{\pi}{2}$, that is, when $\theta = \frac{\pi}{4}$. It is convenient to remember this: When $A = C$, we choose $\theta = \frac{\pi}{4}$; otherwise, we use $\theta = \frac{1}{2}\arctan\frac{B}{A-C}$. As usual we can always assume that $-\frac{\pi}{2} < \arctan x < \frac{\pi}{2}$, so $-\frac{\pi}{4} < \theta \le \frac{\pi}{4}$. We have the following result.

THEOREM 11.3.2

Consider the equation $Ax^2 + Bxy + Cy^2 + Dx + Ey = q$, in which we assume that $B \neq 0$. The cross-term can be removed by rotating the axes through an angle θ, where

$$\theta = \begin{cases} \frac{\pi}{4}, & \text{if } A = C, \\ \frac{1}{2}\arctan\frac{B}{A-C}, & \text{otherwise.} \end{cases}$$

We can assume that $-\frac{\pi}{4} < \theta \le \frac{\pi}{4}$.

EXAMPLE 3 Identify and sketch the graph of $5x^2 + 26xy + 5y^2 = 72$.

Solution Noticing that the coefficients of x^2 and y^2 are equal, we realize that the angle of rotation should be $\theta = \frac{\pi}{4}$. We consider rotated axes and write

$$x = x'\cos\theta - y'\sin\theta = \frac{x'}{\sqrt{2}} - \frac{y'}{\sqrt{2}},$$

$$y = x'\sin\theta + y'\cos\theta = \frac{x'}{\sqrt{2}} + \frac{y'}{\sqrt{2}}.$$

From this we find that

$$x^2 = \frac{x'^2}{2} + \frac{y'^2}{2} - x'y', \qquad y^2 = \frac{x'^2}{2} + \frac{y'^2}{2} + x'y', \qquad xy = \frac{x'^2}{2} - \frac{y'^2}{2}.$$

Substituting these expressions into the original equation, we find that it becomes

$$\begin{aligned} 72 &= 5x^2 + 5y^2 + 26xy \\ &= 5\left(\frac{x'^2}{2} + \frac{y'^2}{2} - x'y'\right) + 5\left(\frac{x'^2}{2} + \frac{y'^2}{2} + x'y'\right) + 26\left(\frac{x'^2}{2} - \frac{y'^2}{2}\right) \\ &= 18x'^2 - 8y'^2, \end{aligned}$$

or

$$\frac{x'^2}{4} - \frac{y'^2}{9} = 1.$$

This is the equation of a hyperbola, which we sketch in Figure 11.3.6 (after marking the asymptotes).

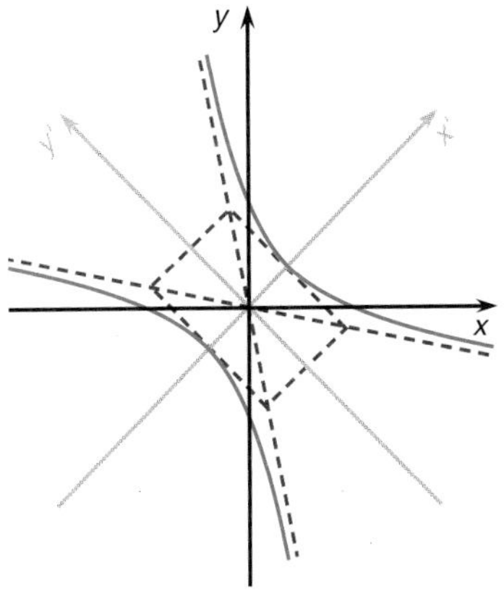

Figure 11.3.6

It is interesting to notice that although the coefficients of x^2 and y^2 in the original equation were both positive, the conic did not turn out to be an ellipse. It is dangerous to try to identify the conic from an equation containing a cross-term.

EXAMPLE 4 Identify and sketch the conic whose equation is $7x^2 + 24xy + 144 = 0$.

Solution From Theorem 11.3.2 we find that the correct angle is $\theta = \frac{1}{2}\arctan\frac{24}{7}$. Writing $2\theta = \arctan\frac{24}{7}$, that is, $\tan 2\theta = \frac{24}{7}$, we find

$$\frac{1}{\cos^2 2\theta} = \sec^2\theta = 1 + \tan^2 2\theta = 1 + \left(\frac{24}{7}\right)^2 = \frac{49+576}{49} = \frac{625}{49},$$

so

$$\cos 2\theta = \sqrt{\frac{49}{625}} = \frac{7}{25},$$

and

We use the half-angle formula (Theorem 1.7.13), noting that $\cos\theta > 0\ldots$

$$\cos\theta = \sqrt{\frac{\cos 2\theta + 1}{2}} = \sqrt{\frac{\frac{7}{25}+1}{2}} = \sqrt{\frac{\frac{32}{25}}{2}} = \sqrt{\frac{16}{25}} = \frac{4}{5}.$$

Since $\tan 2\theta > 0$, we see that $\theta > 0$ and $\sin\theta > 0$. So $\sin\theta = \sqrt{1-\cos^2\theta} = \sqrt{1-\frac{16}{25}} = \sqrt{\frac{9}{25}} = \frac{3}{5}$. We write

$$x = x'\cos\theta - y'\sin\theta = \frac{4}{5}x' - \frac{3}{5}y',$$

$$y = x'\sin\theta + y'\cos\theta = \frac{3}{5}x' + \frac{4}{5}y'.$$

Substituting, we find

$$\begin{aligned} 0 &= 7x^2 + 24xy + 144 \\ &= 7\left(\frac{16}{25}x'^2 + \frac{9}{25}y'^2 - \frac{24}{25}x'y'\right) + 24\left(\frac{12}{25}x'^2 - \frac{12}{25}y'^2 + \frac{7}{25}x'y'\right) + 144 \\ &= \frac{400}{25}x'^2 - \frac{225}{25}y'^2 + 144 = 16x'^2 - 9y'^2 + 144 \end{aligned}$$

or

$$\frac{y'^2}{16} - \frac{x'^2}{9} = 1.$$

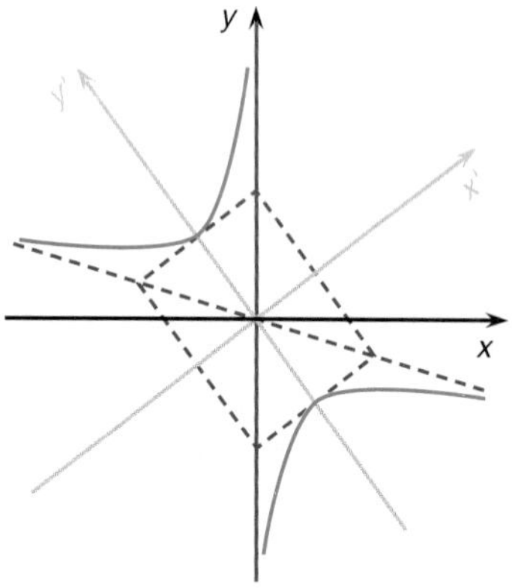

Figure 11.3.7

This is the equation of a hyperbola, which is easily sketched (see Figure 11.3.7). Observe that one of the asymptotes is the y-axis. Notice that once again it would not have been easy to guess from the original equation which type of conic it was.

EXAMPLE 5 Graph the conic $xy = 1$.

Solution This equation has a nonzero cross-term; in fact, that and the constant term are the only nonzero terms. The coefficients of x^2 and y^2 are both zero; in particular, they are the same, so we rotate the axes by $\theta = \frac{\pi}{4}$. Writing

$$x = \frac{x'}{\sqrt{2}} - \frac{y'}{\sqrt{2}}, \qquad y = \frac{x'}{\sqrt{2}} + \frac{y'}{\sqrt{2}},$$

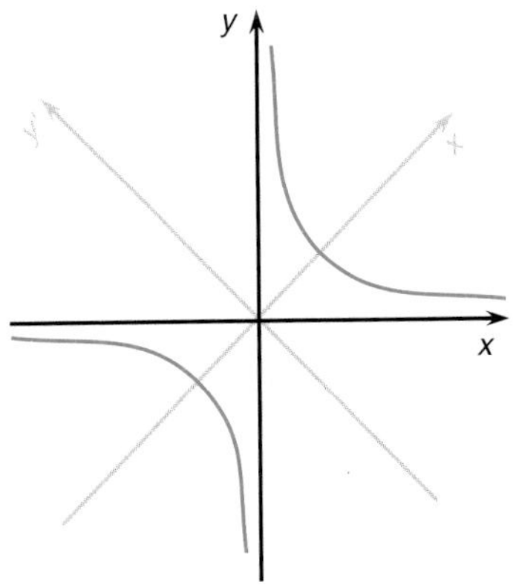

Figure 11.3.8

we find that the equation becomes

$$1 = xy = \frac{x'^2}{2} - \frac{y'^2}{2},$$

which is the equation of a hyperbola.

It is easily sketched, as shown in Figure 11.3.8.

In this particular case we might have rewritten the equation $xy = 1$ as $y = \frac{1}{x}$, and we would have seen how to graph it that way.

None of the examples we have considered so far has had an x-term or a y-term, but the way Theorem 11.3.2 was stated, it applies equally well to conics of that type. The complication is that if we start with an equation containing an x- and/or a y-term, we should expect that after rotating there will be an x'- and/or y'-term. To identify and graph the conic, it will likely be necessary to translate the coordinates *after* rotating them.

EXAMPLE 6 Sketch the graph of $7x^2 + 6\sqrt{3}xy + 13y^2 - (32 - 4\sqrt{3})x - (4 + 32\sqrt{3})y + 32 = 0$.

Solution We begin by finding the angle:

$$\theta = \frac{1}{2}\arctan\frac{B}{A-C} = \frac{1}{2}\arctan\frac{6\sqrt{3}}{-6} = \frac{1}{2}\arctan(-\sqrt{3}) = -\frac{1}{2}\frac{\pi}{3} = -\frac{\pi}{6}.$$

As usual, we write

$$x = x'\cos\theta - y'\sin\theta = \frac{\sqrt{3}}{2}x' + \frac{1}{2}y',$$

$$y = x'\sin\theta + y'\cos\theta = -\frac{1}{2}x' + \frac{\sqrt{3}}{2}y',$$

from which we find that

$$x^2 = \frac{3}{4}x'^2 + \frac{\sqrt{3}}{2}x'y' + \frac{1}{4}y'^2,$$

$$y^2 = \frac{1}{4}x'^2 - \frac{\sqrt{3}}{2}x'y' + \frac{3}{4}y'^2,$$

$$xy = -\frac{\sqrt{3}}{4}x'^2 + \frac{1}{2}x'y' + \frac{\sqrt{3}}{4}y'^2.$$

Substituting in the original equation, we find that it is

$$\begin{aligned}-32 &= 7x^2 + 13y^2 + 6\sqrt{3}xy - (32 - 4\sqrt{3})x - (4 + 32\sqrt{3})y\\ &= \frac{21}{4}x'^2 + \frac{7\sqrt{3}}{2}x'y' + \frac{7}{4}y'^2 + \frac{13}{4}x'^2 - \frac{13\sqrt{3}}{2}x'y' + \frac{39}{4}y'^2 - \frac{18}{4}x'^2\\ &\quad + \frac{6\sqrt{3}}{2}x'y' + \frac{18}{4}y'^2 - (16\sqrt{3} - 6)x' - (16 - 2\sqrt{3})y'\\ &\quad -(-2 - 16\sqrt{3})x' - (2\sqrt{3} + 48)y'\\ &= \frac{16}{4}x'^2 + \frac{64}{4}y'^2 + 8x' - 64y' = 4x'^2 + 16y'^2 + 8x' - 64y'.\end{aligned}$$

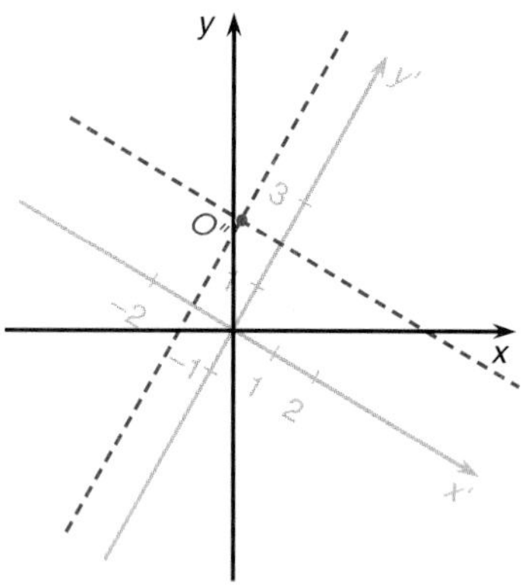

Figure 11.3.9(i)

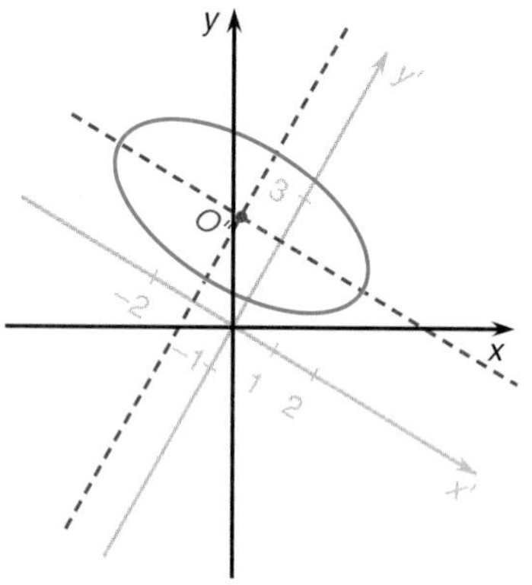

Figure 11.3.9(ii)

The equation has become $4x'^2 + 16y'^2 + 8x' - 64y' = -32$, or $x'^2 + 4y'^2 + 2x' - 16y' = -8$. We can complete the squares to find $(x' + 1)^2 - 1 + 4(y' - 2)^2 - 16 = -8$, or $x''^2 + 4y''^2 = 9$, where $x'' = x' + 1$ and $y'' = y' - 2$. This is

$$\frac{x''^2}{9} + \frac{4y''^2}{9} = 1,$$

which is the equation of an ellipse.

To graph it, we first plot the rotated x'- and y'-axes in Figure 11.3.9(i). Since the conic was translated relative to these axes, we draw the translated origin O'', whose coordinates relative to the x'- and y'-axes are $(x', y') = (-1, 2)$.

In Figure 11.3.9(ii) we draw the ellipse. The idea is to work relative to the rotated axes. The ellipse is drawn with its center at O'', the translated origin, and its axes parallel to the x'- and y'-axes. We could have labeled the axes through O'' as the x''- and y''-axes, but that is probably an unnecessary complication.

In doing this type of question there are two steps. First we rotated the axes to remove the xy-term, and then we translated the coordinates to remove the linear terms (i.e., the x'- and y'-terms). This second step was accomplished by completing the squares.

It is reasonable to ask whether it will work in the opposite order, that is, by completing the squares first. Curiously, it does not. The problem is that if we complete the squares and let $x' = x - k$, $y' = y - \ell$, say, then we can write $x = x' + k$ and $y = y' + \ell$. From this we find that $xy = x'y' + \ell x' + ky' + k\ell$, and translating the coordinates converts the xy-term into something that still contains x'- and y'-terms.

Because of this it makes sense to rotate the axes before trying to translate the axes to remove the linear terms.

Summary 11.3.3

Graphing Quadratic Polynomial Equations

Suppose we want to sketch the graph of the quadratic polynomial equation

$$Ax^2 + Bxy + Cy^2 + Dx + Ey = q.$$

1. If $B = 0$, we proceed according to the description in Summary 11.1.3 for graphing conics.
2. Suppose that $B \neq 0$. Then first it is necessary to rotate the axes to remove the cross-term Bxy. The necessary angle is $\theta = \frac{1}{2} \arctan \frac{B}{A-C}$, with the understanding that $\theta = \frac{\pi}{4}$ if $A = C$ and that in any case, $-\frac{\pi}{4} < \theta \leq \frac{\pi}{4}$.
   Use the rotated coordinates x' and y' defined by
   $$x = x' \cos\theta - y' \sin\theta,$$
   $$y = x' \sin\theta + y' \cos\theta.$$
   Substitute these formulas into the equation; the resulting equation should not have any cross-term.
3. Now the problem is one of graphing an equation that has no cross-term, so we proceed as in Summary 11.1.3. Specifically, we complete the squares and/or translate coordinates as needed to remove the x'-term if there is an x'^2-term and to remove the constant term if there is no x'^2-term (and similarly with the y'-terms).

4. Now it should be possible to express the equation in one of the standard forms and so identify its type.
5. To sketch the graph, the first step is to draw the rotated x'- and y'-axes. If the axes were translated, it is then necessary to plot the translated origin. It is important to plot this new origin *relative to the x'- and y'-axes.* After this is done, it should be straightforward to sketch the graph.

In doing the examples we have considered so far, we have noticed that it is difficult to predict the type of a conic from its equation if the equation contains a cross-term. In fact there is a way to do this, which we now describe.

THEOREM 11.3.4

Consider the equation

$$Ax^2 + Bxy + Cy^2 + Dx + Ey = q.$$

(i) If $B^2 - 4AC < 0$, then the graph is an ellipse or a circle if it is a conic, and if it is degenerate it can be a single point or no points at all.

(ii) If $B^2 - 4AC > 0$, then the graph is a hyperbola if it is a conic, and if it is degenerate it can be two intersecting lines.

(iii) If $B^2 - 4AC = 0$, then the graph is a parabola if it is a conic, and if it is degenerate it can be a line, a pair of parallel lines, or no points at all.

The quantity $B^2 - 4AC$ is called the **discriminant** of the equation $Ax^2 + Bxy + Cy^2 + Dx + Ey = q$. It is not accidental that it is the same formula that we encountered for the discriminant of a quadratic polynomial in one variable (cf. 1.3.4), but we will not pursue the connection here.

If you have trouble remembering which condition in Theorem 11.3.4 corresponds to which type of conic, it is easy to check it on the standard forms.

EXAMPLE 7 Discriminants of Some Conics in Standard Form

(i) For the equation $2x^2 + 7y^2 = 1$ the discriminant is $B^2 - 4AC = 0^2 - 4(2)(7) = -56$. The graph is an ellipse, corresponding to a negative discriminant. For the circle $x^2 + y^2 = 5$ the discriminant is -4.

It is also possible to try examples of degenerate conics. For instance, the equation $x^2 + y^2 = 0$ has only one solution, the point $(0, 0)$, and the equation $4x^2 + 9y^2 = -3$ has none at all. Both have negative discriminants.

(ii) The discriminant of $x^2 - y^2 = -1$ is $B^2 - 4AC = 0^2 - 4(1)(-1) = 4$. We see that this hyperbola corresponds to a positive discriminant. The related example $x^2 - y^2 = 0$, which has the same discriminant, can be written as $(x + y)(x - y) = 0$, so its graph is the two lines $x + y = 0$ and $x - y = 0$, which are the asymptotes of the hyperbola we just discussed.

(iii) The discriminant of $y = x^2$, that is, $x^2 - y = 0$, is $0^2 - 4(1)0 = 0$. This parabola has discriminant equal to zero.

It is also possible to find degenerate examples. For instance, the equation $x^2 = 0$ has as its graph the line $x = 0$. The equation $x^2 - x = 0$ can be written as $x(x - 1) = 0$, so its graph is the two (parallel) lines $x = 0$ and $x = 1$. The equation $y^2 = -7$ has no solutions. In each of these cases the discriminant is zero.

The value of the discriminant does not depend on the coefficients of x and y or the constant term. In fact, what makes the theorem work is that when we change an equation by rotating the axes, the discriminant does not change. (You can try this out on the examples we have already done.) Of course, this reflects the observation that rotating the axes does not change the nature of the graph, merely its orientation.

The proof of this fact is moderately messy; the proof of Theorem 11.3.4 is quite complicated. Strangely enough, it is easier to prove both of them by using the techniques of linear algebra. That is a subject outside the province of this book, but if you are familiar with linear algebra, there is a brief discussion in the "Point to Ponder" after the exercises.

The discriminant can be used to identify the type of a conic in a situation in which its exact graph may not be important and the calculations for the rotation are elaborate.

EXAMPLE 8 The equation $7x^2 - 5xy + 2y^2 - 12x + 8y = 13$ has discriminant $B^2 - 4AC = (-5)^2 - 4(7)2 = -31$. We can conclude without doing any more work that it is an ellipse, a circle, a point, or nothing at all. In fact it is an ellipse.

Exercises 11.3

I

For each of the following questions, the x'- and y'-axes are rotated through the specified angle θ relative to the x- and y-axes. The (x', y') or (x, y) coordinates of a point P are given; find its coordinates relative to the other set of axes.

1. $(x', y') = (1, 2), \theta = \frac{\pi}{4}$

2. $(x', y') = (-5, 5), \theta = \frac{3\pi}{4}$

3. $(x', y') = (0, 3), \theta = \frac{\pi}{3}$

4. $(x', y') = (2, -5), \theta = \pi$

5. $(x, y) = (1, 7), \theta = \frac{\pi}{6}$

6. $(x, y) = (1, -4), \theta = \frac{4\pi}{3}$

7. $(x, y) = (a, 0), \theta = \frac{5\pi}{6}$

8. $(x', y') = (11, -5), \theta = \frac{\pi}{2}$

9. $(x', y') = (0, 0), \theta = \frac{13\pi}{17}$

10. $(x, y) = (1, 0), \theta = \frac{2\pi}{3}$

Identify and sketch each of the following conics.

11. $13x^2 + 10xy + 13y^2 = 2$

12. $x^2 + \sqrt{3}xy = 2$

13. $x^2 - 2\sqrt{3}xy + 3y^2 - 2\sqrt{3}x - 2y = 0$

14. $x^2 + 6\sqrt{3}xy - 5y^2 = 8$

15. $52x^2 + 72xy + 73y^2 = 400$

16. $25x^2 + 120xy + 144y^2 - 156x + 65y = -507$ (*Hint:* Find $13\cos\theta$.)

17. $x^2 - 8xy + 7y^2 = 4$ (*Hint:* Find $\cos^2\theta$.)

18. $9x^2 - 6xy + 17y^2 = 72$ (*Hint:* Find $\cos^2\theta$.)

19. $(\sqrt{3} - 1)(x^2 + y^2) + 2(\sqrt{3} + 1)xy = 6$

20. $x^2 + xy = 2$

Identify and sketch each of the following conics.

21. $x^2 + 2xy + y^2 + 4x - 2y = 6$

22. $x^2 - 2\sqrt{3}xy - y^2 + x + \sqrt{3}y = -8$

23. $11x^2 + 24xy + 4y^2 - 18x - 16y + 5 = 0$

24. $x^2 - 4xy + 4y^2 + 2\sqrt{5}x - 9\sqrt{5}y = 0$

25. $31x^2 + 10\sqrt{3}xy + 21y^2 + (16 - 36\sqrt{3})x - (36 + 16\sqrt{3})y = 92$

26. $7x^2 - 2xy + 7y^2 + 18\sqrt{2}x - 30\sqrt{2}y + 54 = 0$

27. $x^2 - 6x - y^2 - 8y = 11$

28. $5x^2 + 6xy + 5y^2 - 4x - 12y = 0$

29. $x^2 + 4xy + 4y^2 + 50x = 0$

30. $2xy - 2\sqrt{2}x + 1 = 0$

II

***31.** Try sketching the graph of $5x^2 + 26xy + 5y^2 = 72$ by rotating the axes through an angle of $-\frac{\pi}{4}$. Compare your solution with what we did in Example 3 (p. 745) using $+\frac{\pi}{4}$. Is your equation the same? Is your picture the same?

***32.** Starting with the equation $x^2 + 3xy + y^2 + 2x - 4y = 12$, try completing the squares before you rotate the axes. Verify that this does not have the effect of removing the linear terms (i.e., the x- and y-terms).

***33.** For what values of q is $x^2 - xy + 2x = q$ the equation of a degenerate conic? Sketch the graph for several values of q, including degenerate and nondegenerate cases.

***34.** When we rotate a conic, we use the angle θ to know where to draw the rotated axes. On the other hand, for purposes of doing the calculations, all we really need to know is $\cos\theta$ and $\sin\theta$.

If we know that $\phi = \arctan \frac{B}{A-C}$, with $-\frac{\pi}{2} < \phi \le \frac{\pi}{2}$, then using Figure 11.3.10, we see that

$$\cos\phi = \frac{1}{\sqrt{1 + \frac{B^2}{(A-C)^2}}}, \qquad \sin\phi = \frac{\frac{B}{A-C}}{\sqrt{1 + \frac{B^2}{(A-C)^2}}}.$$

The angle of rotation is $\theta = \frac{1}{2}\phi = \frac{1}{2}\arctan\frac{B}{A-C}$.

(i) Use the above formulas and the half-angle formulas (Theorem 1.7.13, p. 52) to find expressions for $\cos\theta$ and $\sin\theta$.

(ii) Check your formula by applying it to Example 4 (p. 746).

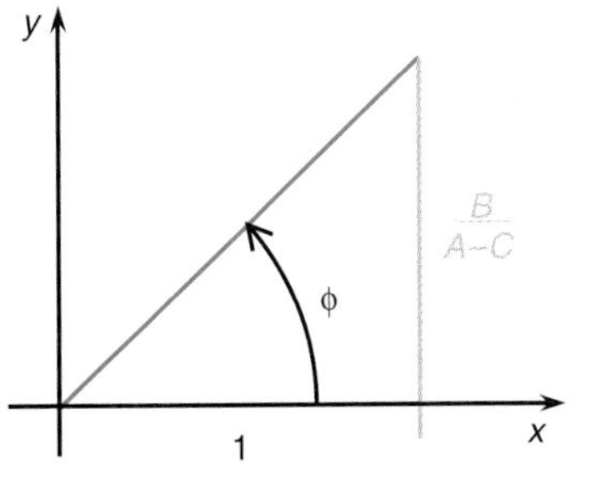

Figure 11.3.10

***35.** Prove the statement made in the text that the discriminant does not change if the axes are rotated. (*Hint:* Try it first for an equation with no linear terms, i.e., an equation of the form $Ax^2 + Bxy + Cy^2 = q$.)

***36.** Consider the equation $Ax^2 + Bxy + Cy^2 + Dx + Ey = q$, and suppose the coefficients A, B, and C are all integers, with $C \ne A$. Show that it is not possible to rotate the conic by $\theta = \frac{\pi}{6}$ so that the resulting conic has no cross-term.

****37.** (i) If $B \ne 0$ and $B^2 - 4AC = 0$, show that $Ax^2 + Bxy + Cy^2$ can be written in the form $Ax^2 + Bxy + Cy^2 = R(ax + by)^2$, for some numbers R, a, and b, with $R \ne 0$.

(ii) Then show that the equation $Ax^2 + Bxy + Cy^2 + Dx + Ey = q$ can be written as $R(ax + by)^2 + S(ax + by) + T(ax - by) = Q$, for some numbers S, T, and Q.

(iii) Use this to show that the graph is either a parabola, one or two straight lines, or the empty set.

***38.** (i) Consider the equation $x^2 + y^2 - 8x + 4y = 5$, and rotate it through an angle of $\theta = -\frac{\pi}{6}$. What is the coefficient of the cross-term in the resulting equation?

(ii) What is the coefficient of the cross-term after the original conic is rotated through an angle of $\theta = \frac{\pi}{8}$?

(iii) Find a formula for the coefficient of the cross-term in the equation that results from rotating the original equation through an angle θ.

39. Verify that the discriminant of the conic in Example 4 (p. 746) was not changed by rotating.

40. Verify that the discriminant of the conic in Example 6 (p. 747) was not changed by rotating.

Find the points on each of the following conics that are closest to the origin.

****41.** $x^2 + y^2 + 2xy + x - y = 3$

****42.** $2x^2 + 2\sqrt{3}xy + 1 = 0$

****43.** $15x^2 - 24xy + 8y^2 - 4 = 0$

****44.** Find the points on the graph of $x^2 + 2xy + 2y^2 = 0$ that are farthest from the origin.

****45.** Find the equation of the ellipse with foci $(2\sqrt{2}, -2\sqrt{2})$, $(-2\sqrt{2}, 2\sqrt{2})$ and semi-major axis 5.

****46.** Find the equation of the hyperbola with foci $(-1, 1)$, $(1, -1)$ and the coordinate axes as asymptotes.

47. Use a graphics calculator or computer to sketch the graph of the conic $7x^2 - 5xy + 2y^2 - 12x + 8y = 13$ from Example 8 (p. 750). What is the angle θ through which the axes are rotated?

POINT TO PONDER

At the end of this section it was remarked that the results about the discriminant could be proved by using linear algebra. This is the study of matrices and related topics, and it may be unfamiliar to you, but if you do happen to know about it, you may already have observed that the formula given in Theorem 11.3.1 for rotating coordinates can be written as a matrix multiplication:

$$\begin{pmatrix} x \\ y \end{pmatrix} = \begin{pmatrix} \cos(\theta)\,x' - \sin(\theta)\,y' \\ \sin(\theta)\,x' + \cos(\theta)\,y' \end{pmatrix} = \begin{pmatrix} \cos\theta & -\sin\theta \\ \sin\theta & \cos\theta \end{pmatrix} \begin{pmatrix} x' \\ y' \end{pmatrix}.$$

The matrix $\begin{pmatrix} \cos\theta & -\sin\theta \\ \sin\theta & \cos\theta \end{pmatrix}$ in this formula is what is often called a *change of basis matrix*.

Now we have observed that the discriminant $B^2 - 4AC$ does not depend on the linear or constant terms of an equation. If we consider just the quadratic terms, $Ax^2 + Bxy + Cy^2$, we can write them as a matrix product too:

$$Ax^2 + Bxy + Cy^2 = (x \quad y) \begin{pmatrix} A & \frac{B}{2} \\ \frac{B}{2} & C \end{pmatrix} \begin{pmatrix} x \\ y \end{pmatrix}.$$

This matrix product turns out to be a 1×1 matrix, which is just a number, and you can check that it is the right number, that is, $Ax^2 + Bxy + Cy^2$.

Next we notice that for the equation to have no cross-terms is exactly the same thing as for the big matrix in this formula to be diagonal. Removing the cross-terms from the equation is exactly the same problem as "diagonalizing" this matrix. This is one of the major problems that linear algebra enables us to solve.

In fact, if we use the first formula for $\begin{pmatrix} x \\ y \end{pmatrix}$ in terms of $\begin{pmatrix} x' \\ y' \end{pmatrix}$, this last formula can be written as

$$Ax^2 + Bxy + Cy^2 = (x \quad y) \begin{pmatrix} A & \frac{B}{2} \\ \frac{B}{2} & C \end{pmatrix} \begin{pmatrix} x \\ y \end{pmatrix}$$

$$= (x' \quad y') \begin{pmatrix} \cos\theta & \sin\theta \\ -\sin\theta & \cos\theta \end{pmatrix} \begin{pmatrix} A & \frac{B}{2} \\ \frac{B}{2} & C \end{pmatrix} \begin{pmatrix} \cos\theta & -\sin\theta \\ \sin\theta & \cos\theta \end{pmatrix} \begin{pmatrix} x' \\ y' \end{pmatrix}.$$

This is really just the *change of basis formula* for the matrix $\begin{pmatrix} A & B/2 \\ B/2 & C \end{pmatrix}$. In particular, we notice that the discriminant of the conic is just -4 times the determinant of the matrix, and in linear algebra we learn that the determinant is unchanged by changing bases. This explains why rotating does not change the discriminant.

Finally, we should remark that if $B = 0$, so the conic has no cross-term, then the matrix $\begin{pmatrix} A & B/2 \\ B/2 & C \end{pmatrix}$ is diagonal, and the coefficients A and C of x^2 and y^2 are precisely its *eigenvalues*. This was one of the first and is still one of the most important uses of eigenvalues.

If you do not know about linear algebra, then you have probably not managed to read this description, but you will be pleased to know that all the results of this section can be proved without any knowledge of linear algebra. However, some of the proofs are quite complicated; linear algebra actually provides a better framework within which to understand them.

None of this has much bearing on our use of the material in this section. You should *not* go out and try to learn all of linear algebra over the weekend.

Chapter Summary

§11.1
translation of axes
completing the square
conics in standard form (Table 11.1.2)
graphing conics (Summary 11.1.3)

§11.2
focus
reflection property of ellipse
directrix
reflection property of parabola

eccentricity
classification of conics by eccentricity
formulas for foci
focal axis

§11.3
rotated coordinates

$\theta = \frac{1}{2}\arctan\frac{B}{A-C}$
graphing quadratic equations with cross-terms (Summary 11.3.3)
degenerate conics
discriminant

Review Exercises

I

Identify and sketch each of the following conics.

1. $9x^2 + 16y^2 = 144$
2. $9y^2 - 4x^2 = 36$
3. $x - 4y^2 = 2$
4. $x^2 + y^2 = 2$
5. $x^2 + 2y^2 = 4$
6. $2x^2 - 3y^2 + 6 = 0$
7. $x^2 + 4y^2 - 4x = 0$
8. $4x^2 - 9y^2 + 8x + 54y = 113$
9. $x^2 + 4x - 2y = 3$
10. $3x^2 - 12x = 3y^2 + 6y$

Identify and sketch each of the following conics.

11. $41x^2 + 18xy + 41y^2 = 800$
12. $11x^2 - 10\sqrt{3}xy + y^2 = 48$
13. $9x^2 + 24xy + 16y^2 + 40x - 30y = 0$
14. $32x^2 - 52xy - 7y^2 + 180 = 0$
15. $4xy - 6\sqrt{2}x - 2\sqrt{2}y + 8 = 0$
16. $(x + 2y)^2 = x$
17. $93x^2 + 48xy + 107y^2 - 180x - 490y = 50$
18. $119x^2 - 240xy - 119y^2 + 312x - 130y = 0$
19. $31x^2 - 24xy - 14y^2 - 8\sqrt{17}x - 32\sqrt{17}y = 238$
20. $(x + y)^2 - (x - y)^2 = 4$

II

*21. For which value(s) of c is the following conic degenerate: $2x^2 + 3y^2 + 4x - 12y = c$? Identify the graph in each of the various cases.

*22. For which value(s) of c is the following conic degenerate: $x^2 - 4y^2 + cx = 4$? Identify the graph in each of the various cases.

*23. Sketch the graph of $x^2 + y^2 + 2x - 2y = c$ for the following values of c: (i) $c = 0$, (ii) $c = 2$, (iii) $c = -1$, (iv) $c = -2$, (v) $c = -3$.

*24. Sketch the graph of $4x^2 - 9y^2 + 16x + c = 0$ for the following values of c: (i) $c = 0$, (ii) $c = 16$, (iii) $c = -20$, (iv) $c = 52$.

*25. Sketch the graph of $x^2 + 2xy + y^2 + c(x - y) = 0$ for the following values of c: (i) $c = 1$, (ii) $c = \sqrt{2}$, (iii) $c = -1$, (iv) $c = 0$.

*26. Sketch the graph of $xy = c$ for the following values of c: (i) $c = 1$, (ii) $c = 2$, (iii) $c = -1$, (iv) $c = 0$.

*27. Consider the ellipse

$$\frac{x^2}{a^2} + \frac{y^2}{b^2} = 1.$$

(i) Show that for any value of t, the point $(x, y) = (a\cos t, b\sin t)$ lies on the ellipse.

(ii) This is analogous to the fact that $(x, y) = (r\cos\theta, r\sin\theta)$ lies on the circle $x^2 + y^2 = r^2$. As θ ranges from 0 to 2π, this point goes once around the circle, counterclockwise, starting and ending at the point $(r, 0)$. Describe what happens to the point $(a\cos t, b\sin t)$ as t ranges from 0 to 2π.

*28. Consider the hyperbola

$$\frac{x^2}{a^2} - \frac{y^2}{b^2} = 1.$$

(i) Show that for any value of t, the point $(x, y) = (a\cosh t, b\sinh t)$ lies on the hyperbola.

(ii) Describe what happens to the point $(a\cosh t, b\sinh t)$ as t ranges from $-\infty$ to ∞.

(iii) Describe a similar construction for points on the other branch of the hyperbola.

*29. Consider a hyperbola centered at the origin with no cross-terms in its equation. If we know that its eccentricity is $e > 1$, what are the possible equations of its asymptotes? (There are usually two possible pairs of asymptotes.)

*30. In Exercise 29 you found the possible asymptotes for a hyperbola with eccentricity e. For what value(s) of e will the two possible pairs of asymptotes actually be the same?

**31. We have considered transforming conics by rotating the axes, but it is possible to think of other ways to move them. One simple type of transformation is the *shear*, in which, for instance, the x-axis remains fixed and the y-axis is pulled sideways, as illustrated in the figure below.

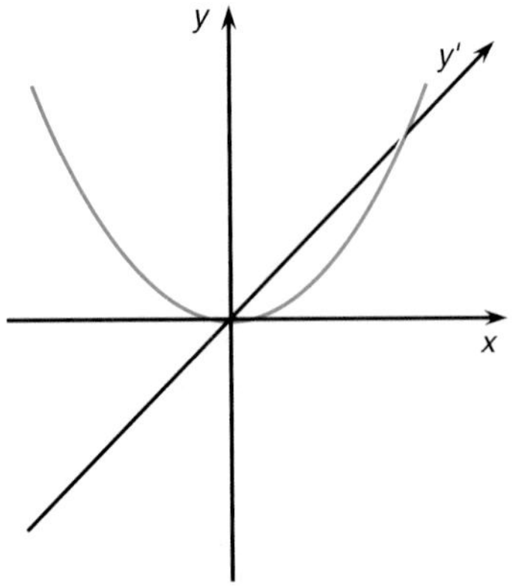

If we label the new axes x' and y', then the coordinates relative to them are

$$x' = x - y, \qquad y' = y.$$

(i) Consider the equation of the parabola $y = x^2$, and express it in terms of the sheared coordinates (x', y').

(ii) Identify the conic whose equation you found in part (i).

(iii) Sketch the graph of $y = x^2$, and then sketch the graph of the equation you found in part (i), but sketch it on a graph in which the x'-axis is horizontal and the y'-axis is vertical (i.e., in which they are not sheared).

It can be shown that the type of the conic does not change even when the axes undergo a shear.

**32. Repeat Exercise 31 for the parabola $x = y^2$.

**33. If the conic $Ax^2 + Bxy + Cy^2 + Dx + Ey = q$ is rotated through an angle θ to remove the cross-term, the slope of the rotated x'-axis is $\tan\theta$. Express this slope in terms of the coefficients A, B, C.

*34. Find the slope of the tangent to the hyperbola $\frac{x^2}{a^2} - \frac{y^2}{b^2} = 1$ at the point (u, v).

**35. Consider the part of the parabola $x = y^2$ with $0 \leq x \leq 1$. (i) Find the volume of the solid obtained by revolving it about the x-axis. (ii) Now consider the solid cone obtained by revolving the part of the line $y = x$ with $0 \leq x \leq 1$, and find its volume. (iii) Notice that the parabolic solid contains the cone and that they have the same cross-sections at $x = 0$, $x = 1$. What is the ratio of the two volumes?

**36. Compare the volumes of the solid obtained by revolving about the x-axis the parabola $x = y^2$, for $0 \leq x \leq a$, and the corresponding cone obtained by revolving the line segment that runs from $(0, 0)$ to $(a, \sqrt{a})$.

*37. (i) Show that the point on a parabola closest to the focus is the vertex.

(ii) A meteorite travels along a parabolic path, with the sun at the focus (see figure below). When

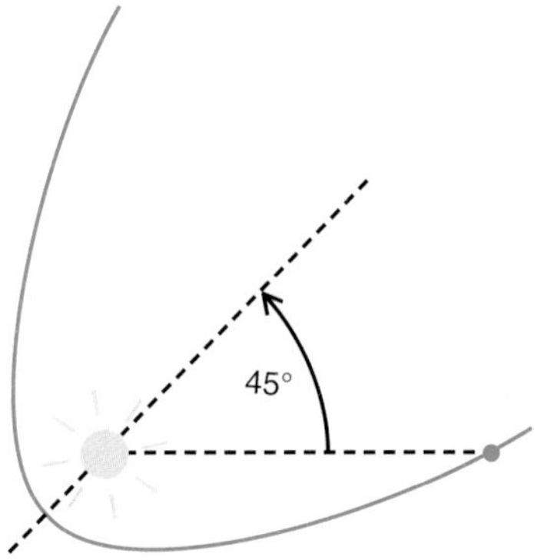

the meteorite is 50,000,000 km from the center of the sun, the line from the meteorite to the sun makes an angle of 45° with the axis of the parabola (the line through the vertex and the focus). How close will the meteorite come to the center of the sun?

Polar Coordinates

12

These curves can all be expressed quite simply in polar coordinates, reflecting the fact that they all have symmetries that involve angles. Some of them are surprisingly difficult to express in rectangular coordinates.

INTRODUCTION

In this chapter we consider polar coordinates and see how to use them with some of the material we have learned.

In Section 12.1 we recall the basic ideas and definitions. In Section 12.2 we discuss drawing graphs of equations that are given in polar coordinates. In Section 12.3 we consider how to use integration to find the area of the region enclosed inside the graph of an equation given in polar coordinates. In Section 12.4 we study conics from the point of view of polar coordinates and discuss the equations of various conics. We will see that the equations of different types of conics are nearly the same in polar coordinates, unlike the situation with rectangular coordinates, for which we had to learn different equations for each type. Finally, in Section 12.5 we discuss *parametric equations*, which are equations that describe the path of a moving point as a function of a parameter t, which usually represents the time. In this section we will usually write these equations in polar coordinates, but in Chapter 13 we will consider them in greater detail, writing them in rectangular coordinates or polar coordinates, depending on the situation.

12.1 Polar Coordinates

RADIUS
POLAR ANGLE
GRAPHING
POLAR/RECTANGULAR CONVERSIONS

In Section 1.7 we discussed **polar coordinates** briefly. The basic idea is quite simple: We identify a point P in the plane by specifying its distance r from the origin O and the angle from the positive horizontal axis to OP. This angle is labeled θ in Figure 12.1.1.

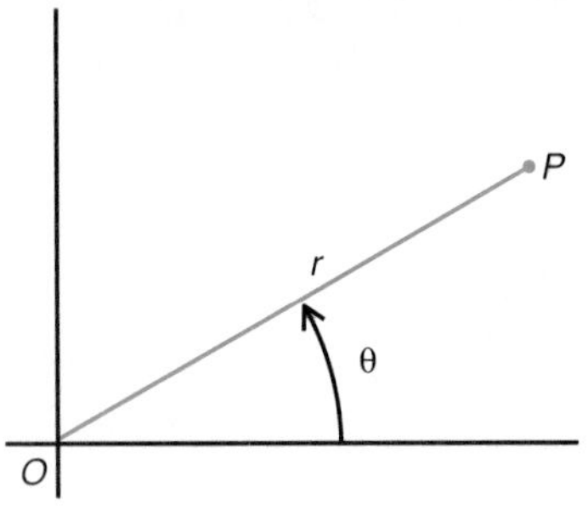

Figure 12.1.1

The main complication is that there are many values of θ that all correspond to the same point P. After all, if θ_0 is one such value, then $\theta_0 + 2\pi$ will do just as well, and so will $\theta_0 - 6\pi$ or $\theta_0 + 10\pi$. In fact the angles $\theta_0 + 2k\pi$ all correspond to the same direction in the plane, for any integer k (see Figure 12.1.2(i)).

A second problem is that it is convenient to allow r to be a negative number, in which case (r, θ) refers to a point on the opposite side of the origin. In Figure 12.1.2(ii), where $r < 0$, we see that the point $Q = (-r, \theta) = (|r|, \theta)$ has angle θ, while $P = (r, \theta)$ is at an angle of $\theta + \pi$. Note that $Q = -P$.

The third difficulty is that for the origin itself we have $r = 0$, but it is less clear how to specify the angle θ. We will say that with $r = 0$, *any* angle θ will specify the origin.

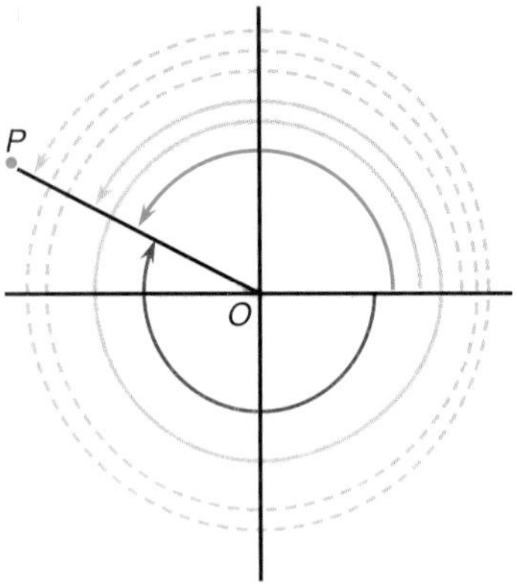

Figure 12.1.2(i)

EXAMPLE 1 If P is a point in the plane, then its polar coordinates are (r, θ), where $|r|$ is the distance from P to the origin. If $r > 0$, then θ is the angle from the positive horizontal axis to OP, measured in the counterclockwise direction. If $r < 0$, then θ is the angle from the positive horizontal axis to OQ, measured in the counterclockwise direction, where $Q = -P$ is the point opposite P.

If θ_0 is one such angle, then $\theta = \theta_0 + 2k\pi$ is another, for every integer k. If $P = O$ is the origin, then its polar coordinates are $(0, \theta)$, for *any* angle θ. Note that $(-r, \theta + \pi)$ refers to the same point as (r, θ).

The number r is sometimes called the **radius** or the **modulus** of the point P. The angle θ is called its **argument** or its **polar angle** or sometimes just "the angle."

It is easy to convert polar coordinates into rectangular or *Cartesian* coordinates. From Figure 12.1.3 we see that if (r, θ) are the polar coordinates of a point P and

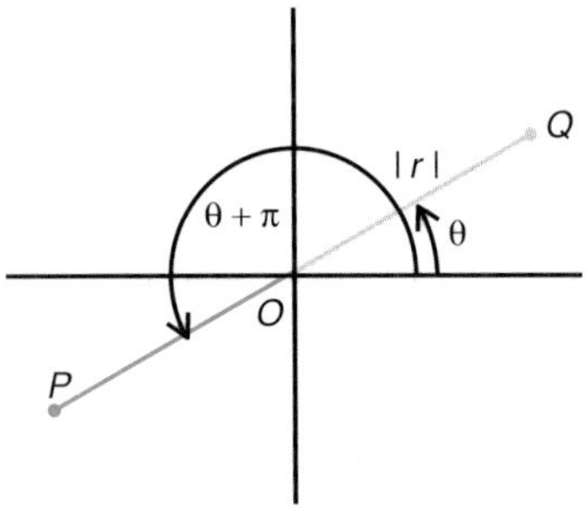

Figure 12.1.2(ii)

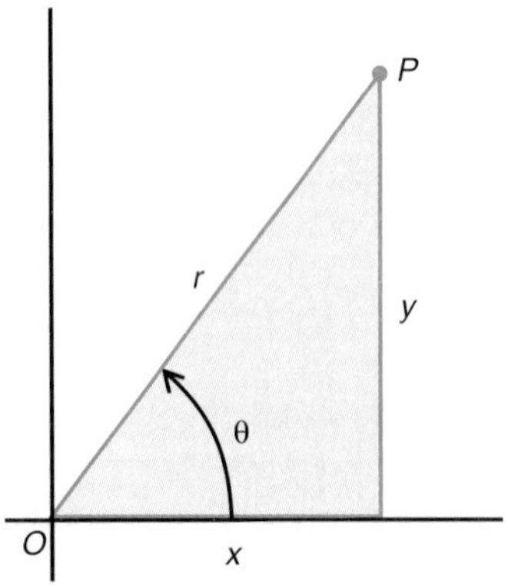

Figure 12.1.3

(x, y) are its rectangular coordinates, then they are related as follows:

$$x = r\cos\theta, \qquad y = r\sin\theta.$$

These formulas still work if $r < 0$; this is why we let θ be the angle of the *opposite* point when $r < 0$.

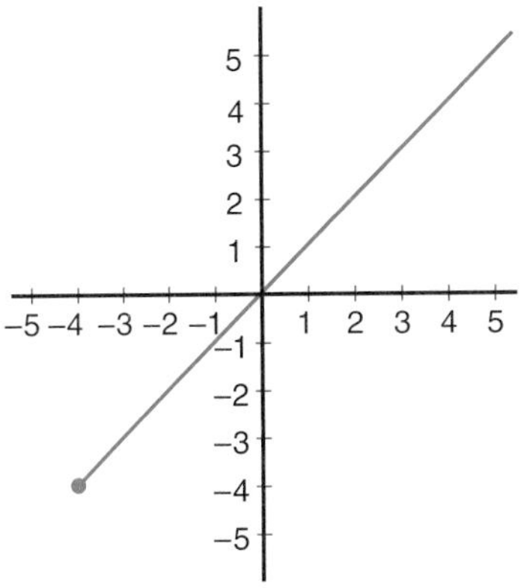

Figure 12.1.4

Going in the opposite direction, we let r be the distance from (x, y) to $(0, 0)$, that is, $r = \sqrt{x^2 + y^2}$. However, it is more complicated to express θ in terms of x and y. From Figure 12.1.3 we see that $\tan\theta = \frac{y}{x}$. The difficulties with this formula are two: First, it does not make sense if $x = 0$, and second, since $\tan(\theta + \pi) = \tan\theta$, the formula does not determine the angle θ completely.

It is necessary to use additional information from the picture to specify the angle. For instance, if we consider the point P whose rectangular coordinates are $(-4, -4)$, as illustrated in Figure 12.1.4, then we see that $r = \sqrt{16 + 16} = 4\sqrt{2}$. The argument θ satisfies $\tan\theta = \frac{y}{x} = \frac{-4}{-4} = 1$. We know that one angle whose tangent equals 1 is $\frac{\pi}{4}$, but as we have just remarked, another possibility is $\frac{\pi}{4} + \pi = \frac{5\pi}{4}$. We use the picture to see that $P = (-4, -4)$ must correspond to the angle $\frac{5\pi}{4}$ and not to $\frac{\pi}{4}$.

It is possible to check this by converting back to rectangular coordinates: $x = r\cos\theta = 4\sqrt{2}\cos\theta$. If $\theta = \frac{\pi}{4}$, then $\cos\theta = \frac{1}{\sqrt{2}}$ and $r\cos\theta = 4$. On the other hand, if $\theta = \frac{5\pi}{4}$, then $r\cos\theta = -4$. Since this is the correct value of x, we see that $\theta = \frac{5\pi}{4}$. Strictly speaking, we should write $\theta = \frac{5\pi}{4} + 2k\pi$, for any integer k.

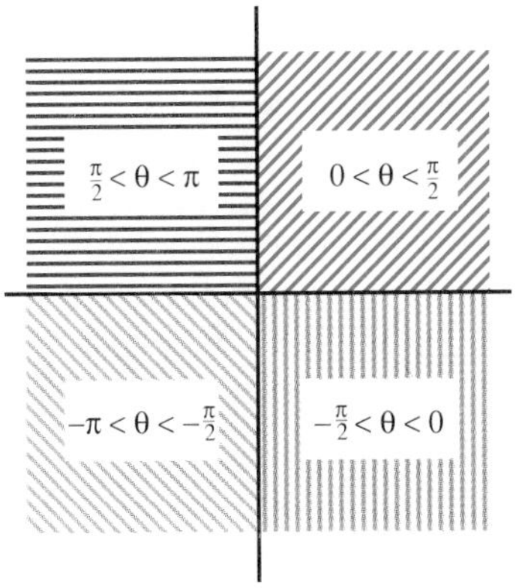

Figure 12.1.5

To figure out the correct value of the argument θ for a point P, we look at Figure 12.1.5. Observe that the argument is 0 for points along the positive x-axis and increases up to $\frac{\pi}{2}$ as P moves to the positive y-axis. In particular we see that we can always choose θ to satisfy $0 < \theta < \frac{\pi}{2}$ if P is above and to the right of the origin. For a point on the negative x-axis the argument is π, so if P moves to the left of the y-axis but stays above the x-axis, its argument will be between $\frac{\pi}{2}$ and π. Similarly, the argument will be between $-\frac{\pi}{2}$ and 0 if P is below and to the right of the origin and will be between $-\pi$ and $-\frac{\pi}{2}$ if P is below and to the left of the origin. These ranges are all indicated in the appropriate regions in Figure 12.1.5. This discussion assumes $r > 0$; if $r < 0$, we add π to the angle θ.

Of course, it is possible to add any integer multiple of 2π to any of these ranges. Notice that you do not have to memorize this picture; it is just a matter of figuring out the argument on each axis and then considering points in between.

It is usually easiest to decide the correct value for θ graphically, as we have just done, but it is also possible to write down a formula. We assume that $\arctan t$ is a number that always lies between $-\frac{\pi}{2}$ and $\frac{\pi}{2}$. Then using Figure 12.1.5, you can check that the value of the argument θ is given by the formula below.

THEOREM 12.1.1

Polar/Rectangular Conversions

Suppose the point P in the plane has polar coordinates (r, θ). Then its rectangular coordinates are (x, y), where

$$x = r\cos\theta, \qquad y = r\sin\theta.$$

Suppose the point P in the plane has rectangular coordinates (x, y). Then its polar coordinates are (r, θ), where $r = \sqrt{x^2 + y^2}$ and θ is determined by reference to Figure 12.1.5 or as follows.

If $r = 0$, then θ can be *any* angle. Otherwise, with the understanding that $-\frac{\pi}{2} < \arctan t < \frac{\pi}{2}$,

$$\theta = \begin{cases} \arctan \frac{y}{x} + 2k\pi, & \text{if } x > 0, \\ \arctan \frac{y}{x} + \pi + 2k\pi, & \text{if } x < 0, \\ \frac{\pi}{2} + 2k\pi, & \text{if } x = 0 \text{ and } y > 0, \\ -\frac{\pi}{2} + 2k\pi, & \text{if } x = 0 \text{ and } y < 0. \end{cases}$$

(In all these formulas, k is understood to be any integer.)

It is also possible to use $(-r, \theta + \pi)$ in place of (r, θ).

EXAMPLE 2 (i) Find the polar coordinates of the point P whose rectangular coordinates are $(x, y) = (0, -5)$.

We can use a graphics calculator to plot points in polar coordinates. For instance, we plot the points whose polar coordinates are $(r, \theta) = \left(3, \frac{3\pi}{4}\right)$, $\left(2, \frac{7\pi}{6}\right)$, and $(2, 1)$.

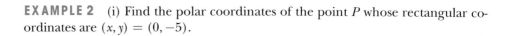

<table>
<tr><td>TI-85
MODE ▼ ▼ ▼ ▼ ENTER GRAPH F2(RANGE) (−) 3 ▼ 3 ▼ ▼ (−) 3 ▼ 3 ▼ F5(GRAPH) PtOn(3 COS (3 π ÷ 4) , 3 SIN (3 π ÷ 4)) ENTER PtOn(2 COS (7 π ÷ 6) , 2 SIN (7 π ÷ 6)) ENTER PtOn(2 COS 1 , 2 SIN 1) ENTER</td><td>SHARP EL9300
RANGE −3 ENTER 3 ENTER 1 ENTER −3 ENTER 3 ENTER 1 ENTER 2ndF PLOT ▼ ▶ ▼ ENTER 3 cos (3 π ÷ 4) ENTER 3 sin (3 π ÷ 4) ENTER 2ndF PLOT ▼ ▶ ▼ ENTER 2 cos (7 π ÷ 6) ENTER 2 sin (7 π ÷ 6) ENTER 2ndF PLOT ▼ ▶ ▼ ENTER 2 cos 1 ENTER 2 sin 1 ENTER</td></tr>
<tr><td>CASIO fx-6300G (fx-7700GB)
Range −3 EXE 3 EXE 1 EXE −3 EXE 3 EXE 1 EXE EXE EXE EXE Plot 3 cos (3 π ÷ 4) , 3 sin (3 π ÷ 4) EXE Plot 2 cos (7 π ÷ 6) , 2 sin (7 π ÷ 6) EXE Plot 2 cos 1 , 2 sin 1 EXE</td><td>HP 48SX
↰ PLOT PLOTR ERASE ATTN 3 ENTER π × 4 ÷ ENTER COS 3 × SWAP SIN 3 × PRG OBJ NXT R → C PRG DSPL NXT C → PX PIXON 7 ENTER π × 6 ÷ ENTER COS 2 × SWAP SIN 3 × PRG OBJ NXT R → C PRG DSPL NXT C → PX PIXON 1 ENTER COS 2 × 1 ENTER SIN 2 × PRG OBJ NXT R → C PRG DSPL NXT C → PX PIXON ↱ PLOT DRAW</td></tr>
</table>

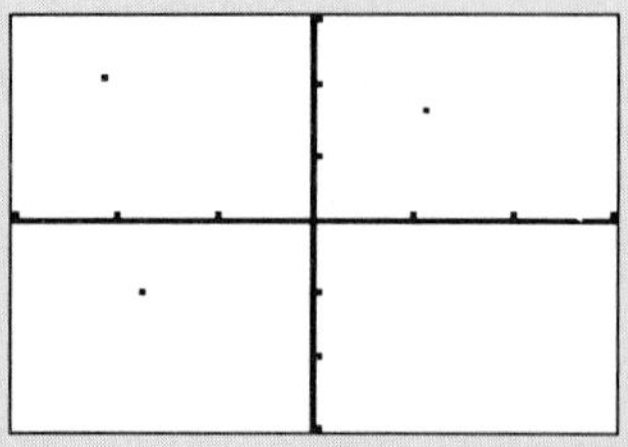

You can try plotting other points. Remember that the picture may be distorted because of the scales. Can you adjust it so that there is no distortion?

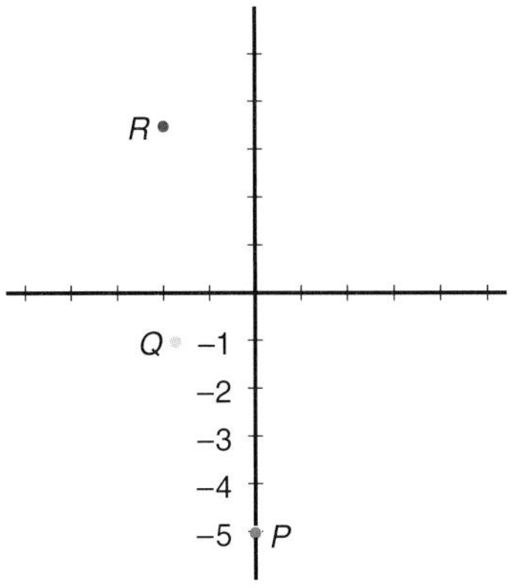

Figure 12.1.6

Solution We find immediately that $r = \sqrt{0^2 + (-5)^2} = 5$. We note that the point lies on the negative y-axis, directly below the origin, so its angle is $\theta = -\frac{\pi}{2}$, or, strictly speaking, $\theta = -\frac{\pi}{2} + 2k\pi$, for any integer k.

The point P is plotted in Figure 12.1.6.

(ii) Find the Cartesian coordinates of the point Q whose polar coordinates are $(r, \theta) = \left(2, \frac{7\pi}{6}\right)$.

Solution We know that $x = r\cos\theta = 2\cos\frac{7\pi}{6} = 2\left(-\frac{\sqrt{3}}{2}\right) = -\sqrt{3}$. Similarly, $y = r\sin\frac{7\pi}{6} = 2\left(-\frac{1}{2}\right) = -1$.

The point is $Q = (-\sqrt{3}, -1)$. It is also plotted in Figure 12.1.6.

(iii) Find the polar coordinates of the point R whose Cartesian coordinates are $R = (-2, 2\sqrt{3})$.

Solution We find that $r = \sqrt{4 + 12} = 4$. Since $\frac{y}{x} = -\sqrt{3}$, we consider $\arctan(-\sqrt{3}) = -\frac{\pi}{3}$. Since $-\frac{\pi}{3}$ corresponds to a point below and to the right of the origin, we realize that the correct angle is $\theta = -\frac{\pi}{3} + \pi + 2k\pi = \frac{2\pi}{3} + 2k\pi$.

The point R is also plotted in Figure 12.1.6.

Exercises 12.1

I

In each of the following questions, a point is specified by either its rectangular coordinates (x, y) or its polar coordinates (r, θ). Find the other kind of coordinates for the point.

1. $(r, \theta) = (2, 0)$

2. $(r, \theta) = \left(1, \frac{3\pi}{4}\right)$

3. $(x, y) = (6, -6)$

4. $(x, y) = \left(0, \frac{3}{2}\right)$

5. $(r, \theta) = \left(0, \frac{11\pi}{17}\right)$

6. $(r, \theta) = \left(\sqrt{3}, \frac{5\pi}{6}\right)$

7. $(x, y) = (-2\sqrt{3}, 2)$

8. $(x, y) = (4, -3)$

9. $(x, y) = (-9, 0)$

10. $(x, y) = (-4, -7)$

11. $(r, \theta) = (-1, 0)$

12. $(r, \theta) = \left(-2, -\frac{\pi}{2}\right)$

13. $(x, y) = (-5, 5)$

14. $(x, y) = (-7, -7)$

15. $(r, \theta) = (7, -23\pi)$

16. $(r, \theta) = \left(2, \frac{9\pi}{8}\right)$

17. $(x, y) = \left(-\frac{4}{\sqrt{3}}, -4\right)$

18. $(x, y) = (-12, 5)$

19. $(x, y) = \left(\frac{1}{\sqrt{2}}, -\frac{1}{\sqrt{2}}\right)$

20. $(x, y) = (0, -\pi)$

21. $(x, y) = (\sqrt{2}, 0)$

22. $(x, y) = (1, -\sqrt{3})$

23. $(r, \theta) = \left(-1, \frac{\pi}{2}\right)$

24. $(r, \theta) = \left(-2, \frac{3\pi}{4}\right)$

25. $(r, \theta) = \left(1, \frac{11\pi}{2}\right)$

26. $(r, \theta) = \left(-2, -\frac{7\pi}{4}\right)$

27. $(x, y) = (5, -12)$

28. $(x, y) = (-24, 7)$

29. $(r, \theta) = (2, \arctan 2)$

30. $(r, \theta) = \left(-4, \arctan\frac{7}{8}\right)$

II

***31.** (i) Consider the equation $x^2 + y^2 = 4$. By expressing x and y in terms of the polar coordinates r and θ, write the equation in terms of r and θ.

(ii) Express the equation $x^2 + 2x + y^2 - 4y = 4$ in polar coordinates. What is the graph of this equation?

***32.** Express the equation $3x - 4y = 12$ in polar coordinates. What is the graph of this equation?

33. Describe the region $\{(x, y) : x \geq 0, y \geq 0\}$ in terms of polar coordinates (r, θ).

34. Describe the region $\{(x, y) : x \leq y < 0\}$ in terms of polar coordinates (r, θ).

35. Describe the region $\{(x, y) : 1 < x^2 + y^2 < 4, 0 \leq -x \leq y\}$ in terms of polar coordinates (r, θ).

36. Describe the region $\{(x, y) : x^2 + y^2 \leq 9, |x| < y\}$ in terms of polar coordinates (r, θ).

37. Describe the region $\{(r, \theta) : r > 0, -\frac{\pi}{4} < \theta < \frac{\pi}{4}\}$ in terms of rectangular coordinates (x, y).

38. Describe the region $\{(r, \theta) : -1 \leq r \leq 1, \frac{\pi}{4} < \theta < \frac{3\pi}{4}\}$ in terms of rectangular coordinates (x, y).

***39.** Describe the region $\{(r, \theta) : r^2 - 2r \leq 0\}$ in terms of rectangular coordinates (x, y).

***40.** Describe the region $\{(r, \theta) : r > 0, \cos\theta < \sin\theta\}$ in terms of rectangular coordinates (x, y).

What is the graph of each of the following equations?

*41. $r = 10$

*42. $\theta = \frac{\pi}{6}$

*43. $\cos\theta = \frac{1}{2}$

*44. $\sin\theta = r\cos^2\theta$

III

If you have a machine that will plot in polar coordinates, use it to sketch the following curves.

45. $r = \sin(4\theta)$

46. $r = \cos^2\theta$

47. $r = 5 + \cos(6\theta)$

48. $r = 1 + 2\sin\theta$

12.2 Graphs in Polar Coordinates

CURVES
SYMMETRY
ROSES
CARDIOID
LEMNISCATE
LIMAÇON
TANGENTS

In this section we consider techniques for graphing equations that are expressed in polar coordinates. Let us begin with some easy examples.

EXAMPLE 1

(i) Consider the equation $r = 3$. The points (r, θ) that satisfy this equation, that is, for which $r = 3$, are the points on the circle of radius 3 around the origin, as sketched in Figure 12.2.1.

(ii) Consider the equation $\theta = \frac{4\pi}{3}$. The points (r, θ) that satisfy this equation, that is, for which $\theta = \frac{4\pi}{3}$, are exactly the points on the line through the origin that makes an angle $\frac{4\pi}{3}$ with the positive horizontal axis. The graph is sketched in Figure 12.2.2. Notice that the origin is included, because the origin can be written as $(r, \theta) = \left(0, \frac{4\pi}{3}\right)$, a point that satisfies the equation. The points below the horizontal axis have $r > 0$, while those above have $r < 0$.

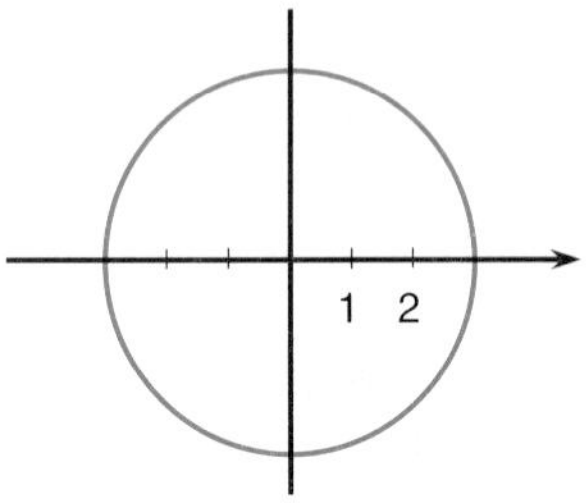

Figure 12.2.1

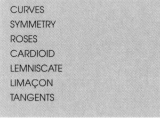

If we consider a point on the graph of $\theta = \frac{4\pi}{3}$, call it $\left(r, \frac{4\pi}{3}\right)$, for some r, then it is possible to write the same point with different coordinates, such as $\left(r, \frac{4\pi}{3} + 2\pi\right)$ or $\left(r, \frac{4\pi}{3} + 2k\pi\right)$ or $\left(-r, \frac{\pi}{3}\right)$. In any of these other forms, the point does *not* satisfy the equation. So when we draw a graph, we include a point P on it if any one of the possible sets of polar coordinates for P satisfy the equation, even if the others do not.

In this example we included the origin, which can be written as $(r, \theta) = (0, \theta)$ for *any* value of θ. Only one of these values satisfies the equation, but that is enough, and the origin is included.

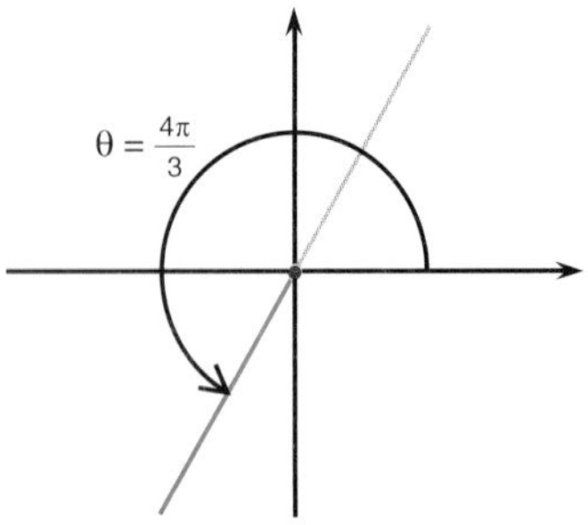

Figure 12.2.2

Next we consider some slightly more complicated examples.

EXAMPLE 2 **The Spiral of Archimedes** The equation $r = \theta, \theta \geq 0$ can be visualized as follows. We begin with $\theta = 0$, so $r = 0$, and we plot a point at the origin. Then suppose we let θ increase from 0 to 2π. This means that the points (r, θ) move around the origin counterclockwise. Since $r = \theta$, the distance from the origin also increases from 0 to 2π. The resulting points form a curve that begins at the origin and spirals out, rotating exactly once around the origin. It returns to the positive horizontal axis when $\theta = 2\pi$, so at that point r is also equal to 2π. This curve is illustrated in Figure 12.2.3(i).

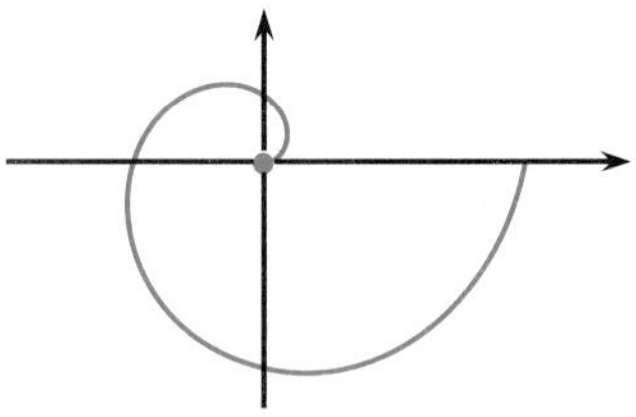

Figure 12.2.3(i)

However, there is no real reason to stop at $\theta = 2\pi$. We could let θ continue to increase, say, from 2π to 4π. This would correspond to having the curve spiral

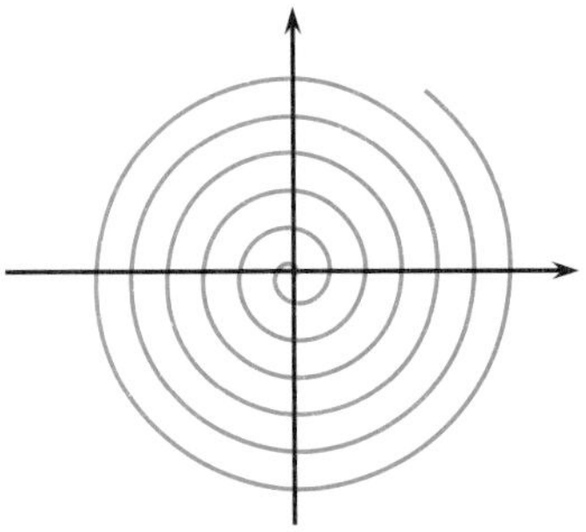

Figure 12.2.3(ii)

once more around the origin, this time spreading out from a radius of 2π at the beginning to 4π at the end. After that, θ could continue to increase, and the curve would continue to spiral outward. Each time it winds around the origin, it moves out so that its radius increases by 2π.

The resulting curve is illustrated in Figure 12.2.3(ii), which is drawn with a smaller scale than 12.2.3(i) in order to show more of the curve. This curve is sometimes referred to as the **Spiral of Archimedes**.

> In this example we see that different values of θ correspond to different points on the curve. Two values of θ that differ by 2π do *not* correspond to the same point.

EXAMPLE 3 Sketch the graph of $r = \cos\theta$.

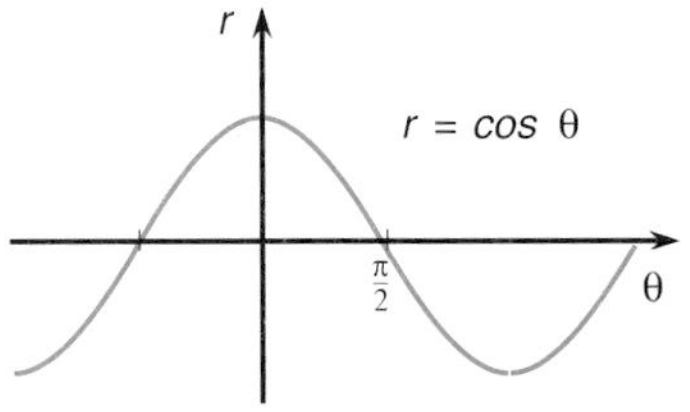

Figure 12.2.4

Solution It is helpful to begin by sketching the "ordinary" graph of $y = \cos\theta$, meaning the graph that is drawn by thinking of y and θ as rectangular coordinates. We make the θ-axis horizontal and the y-axis vertical, so the graph is just the familiar trigonometric curve, shown in Figure 12.2.4. It will be useful to refer to this graph when we draw the graph in polar coordinates.

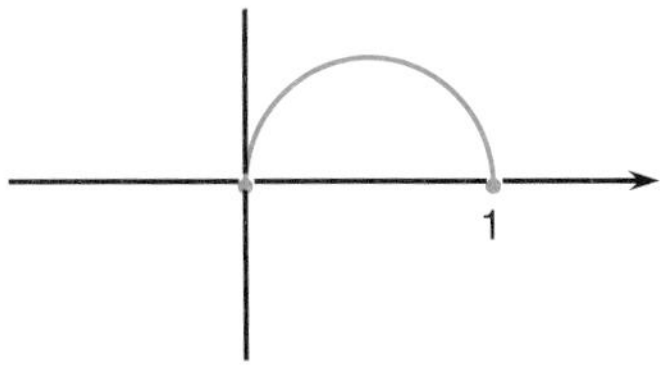

Figure 12.2.5(i)

As θ goes from 0 to $\frac{\pi}{2}$, we see that $r = \cos\theta$ starts at 1 and decreases to 0. The corresponding point on the polar graph starts at the point on the horizontal axis 1 unit to the right of the origin. It moves counterclockwise, getting closer and closer to the origin, finally reaching the origin when $\theta = \frac{\pi}{2}$. We expect a graph whose shape is that shown in Figure 12.2.5(i).

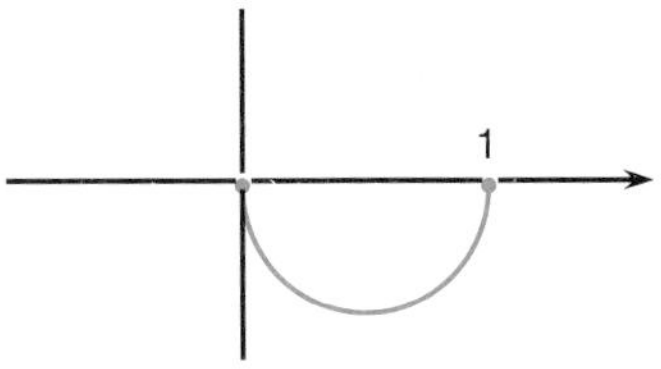

Figure 12.2.5(ii)

Similar remarks apply as θ starts at 0 and decreases to $-\frac{\pi}{2}$; we expect a curve like the one shown in Figure 12.2.5(ii).

The two parts together look a lot like a circle. If we multiply the equation $r = \cos\theta$ by r, it becomes $r^2 = r\cos\theta$. This is easily converted into Cartesian coordinates; it becomes $x^2 + y^2 = x$, or $x^2 - x + y^2 = 0$, or $\left(x - \frac{1}{2}\right)^2 + y^2 = \frac{1}{4}$. This is the equation of a circle of radius $\frac{1}{2}$ centered at $\left(\frac{1}{2}, 0\right)$. The circle is shown in Figure 12.2.5(iii).

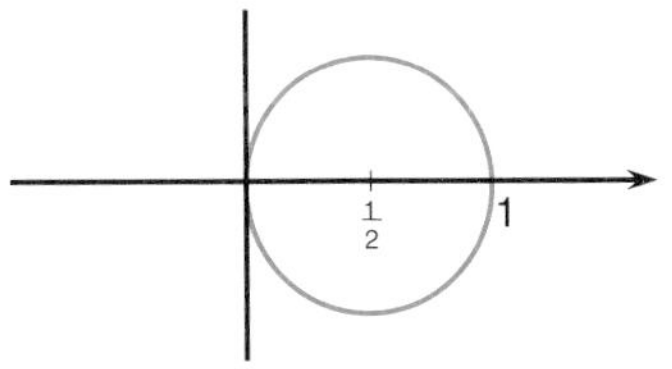

Figure 12.2.5(iii)

Suppose we let θ increase from $\frac{\pi}{2}$ to π. From Figure 12.2.4 we see that $\cos\theta$ starts at zero and decreases to -1. As the angle increases from $\frac{\pi}{2}$ to π, we expect the points on the curve to be above and to the left of the origin. However, because $r = \cos\theta$ is *negative*, the points are actually in the *opposite* direction, below and to the right of the origin.

The part of the curve with $\frac{\pi}{2} \leq \theta \leq \frac{3\pi}{2}$ retraces the circle we found when $-\frac{\pi}{2} \leq \theta \leq \frac{\pi}{2}$.

In Example 3 we were able to give a qualitative description of the shape of the curve, and then it was possible to make a lucky guess that it was a circle. Usually, we cannot expect to be so lucky. If the shape of the curve is unfamiliar or difficult to guess, we may have to resort to plotting some points in order to get a rough idea, exactly as we do when plotting in rectangular coordinates.

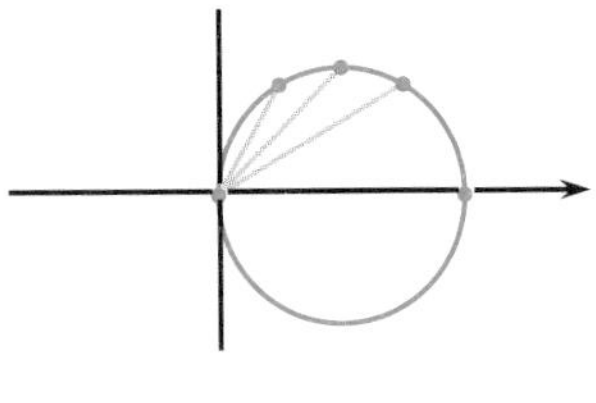

Figure 12.2.6

One way to approach this would be to draw some half-lines lying at some convenient angles, such as $\theta = 0, \frac{\pi}{6}, \frac{\pi}{4}, \frac{\pi}{3}, \frac{\pi}{2}$. For each angle we calculate the corresponding r (in the example it was $r = \cos\theta$). Then we mark a point on the half-line whose distance from the origin is r. This is done in Figure 12.2.6.

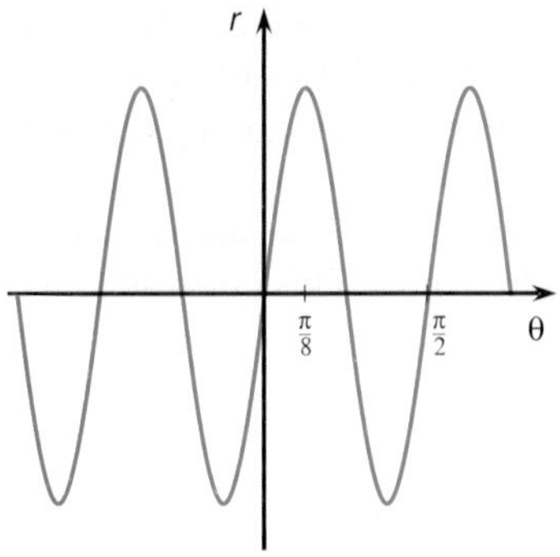

Figure 12.2.7

For this particular example it is not terribly difficult to proceed in this way, although it is not too easy to measure distances along half-lines unless they are horizontal or vertical. An alternative approach is to consider various points on the curve, such as $(\cos 0, 0)$, $\left(\cos\frac{\pi}{6}, \frac{\pi}{6}\right)$, $\left(\cos\frac{\pi}{4}, \frac{\pi}{4}\right)$, $\left(\cos\frac{\pi}{3}, \frac{\pi}{3}\right)$, $\left(\cos\frac{\pi}{2}, \frac{\pi}{2}\right)$. Instead of trying to plot them directly in polar coordinates, it is possible to convert each of them into rectangular coordinates $(x, y) = (r\cos\theta, r\sin\theta)$ and plot them that way. In many cases this approach is actually easier. You may also be able to obtain polar graph paper, which has angles and distances from the origin marked on it.

All that really matters is that if we cannot guess the shape of the curve, it is possible to plot as many points as necessary until we can see it.

EXAMPLE 4 Plot the graph of $r = \sin(4\theta)$.

Solution Once again we begin by plotting the ordinary graph $y = \sin(4\theta)$, shown in Figure 12.2.7.

If we start at $\theta = 0$ and let θ increase, we see that $r = \sin(4\theta)$ also starts at zero and increases up to $r = 1$ at $\theta = \frac{\pi}{8}$ and then decreases back down to $r = 0$ at $\theta = \frac{\pi}{4}$. This suggests that the curve will have a loop as shown in Figure 12.2.8(i), lying between the angles $\theta = 0$ and $\theta = \frac{\pi}{4}$ at which $r = 0$.

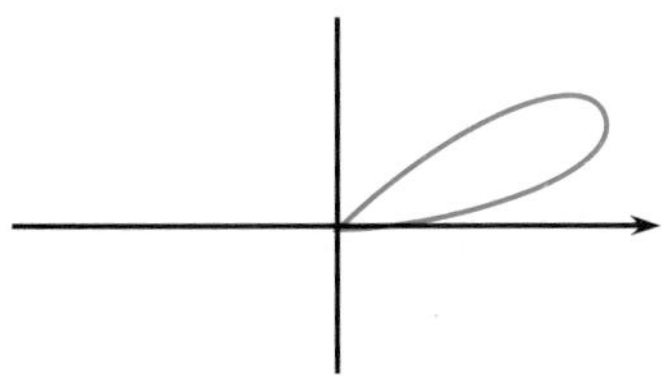

Figure 12.2.8(i)

Instead of letting θ increase further, let us look at the graph of $y = \sin(4\theta)$ in rectangular coordinates in Figure 12.2.7. There we see that y also increases from 0 up to 1 and then decreases back down to 0 when θ increases from $\frac{\pi}{2}$ to $\frac{3\pi}{4}$, and it does it again when θ increases from π to $\frac{5\pi}{4}$ and again between $\frac{3\pi}{2}$ and $\frac{7\pi}{4}$.

We expect three more loops as shown in Figure 12.2.8(ii).

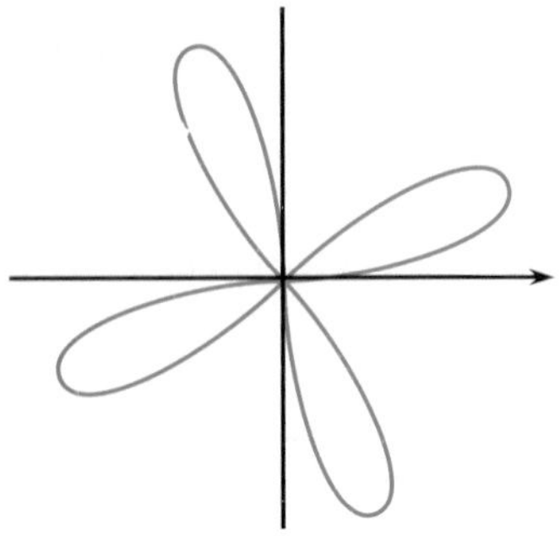

Figure 12.2.8(ii)

Now let us look at what happens when r is negative. This happens, for instance, when θ is between $\frac{\pi}{4}$ and $\frac{\pi}{2}$. The values of r decrease from 0 to -1 and then increase back up to zero again. Again we expect a loop, but since $r < 0$, we expect to find it on the opposite side of the origin. For instance, when $\theta = \frac{3\pi}{8}$, we have that $r = \sin(4\theta) = -1$. If we try to plot the corresponding point P in rectangular coordinates, we find it is

$$(x, y) = (r\cos\theta, r\sin\theta) = \left(-\cos\frac{3\pi}{8}, -\sin\frac{3\pi}{8}\right).$$

This point P is easily plotted; as we see in Figure 12.2.8(iii), it is exactly opposite the point $(x, y) = \left(\cos\frac{3\pi}{8}, \sin\frac{3\pi}{8}\right)$. We also plot the rest of the loop that occurs as θ increases from $\frac{\pi}{4}$ to $\frac{\pi}{2}$.

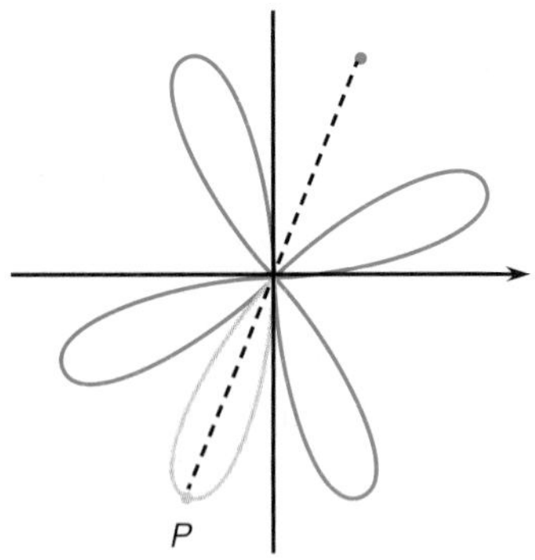

Figure 12.2.8(iii)

Finally, we notice that there will be three more analogous loops when $\theta \in \left[\frac{3\pi}{4}, \pi\right]$, $\left[\frac{5\pi}{4}, \frac{3\pi}{2}\right]$, $\left[\frac{7\pi}{4}, 2\pi\right]$. They are sketched in Figure 12.2.8(iv).

The graph of $r = \sin(4\theta)$, shown in Figure 12.2.8(iv), is an "eight-petaled rose." ◢

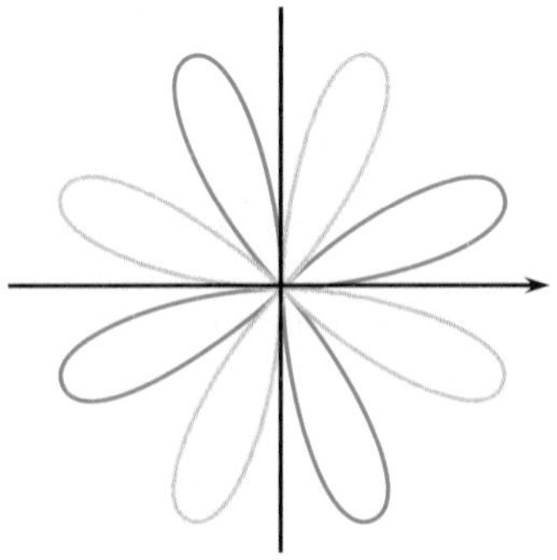

Figure 12.2.8(iv)

In doing this example we saw the usefulness of first sketching the equation in rectangular coordinates. We worked separately with the parts of the curve for which r was positive and negative. It was also helpful to locate the maxima and minima of $r = \sin(4\theta)$; these showed where we should expect loops. Loops begin and end at points where $r = 0$.

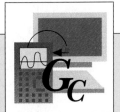

We can use a graphics calculator to sketch the graph of $r = \sin n\theta$, for various values of n.

TI-85	SHARP EL9300
MODE ▼ ▼ ▼ ▼ ▶ (Pol) ENTER EXIT 6 STO N ENTER GRAPH F1 $(r(\theta) =)$ SIN (N × F1(θ)) M2(RANGE) 0 ▼ 2 π ▼ .01 ▼ (−) 1 ▼ 1 ▼ ▼ (−) 1 ▼ 1 F5(GRAPH)	SET UP E 2 ⩘ sin 6 x/θ/T RANGE −1 ENTER 1 ENTER 1 ENTER −1 ENTER 1 ENTER 1 ENTER 0 ENTER 2 π ENTER .01 ENTER ⩘
CASIO f_x-7700GB	**HP 48SX**
MODE MODE − (POL) Range −1 EXE 1 EXE 1 EXE −1 EXE 1 EXE 1 EXE 0 EXE 2 π EXE .01 EXE Graph sin 6 θ EXE	↰ RAD ↰ PLOT ' R = SIN 6 × α ↱ F NEW P O L ENTER PTYPE POLAR PLOTR 1 +/− ENTER 1 XRNG 1 +/− ENTER 1 YRNG ' α ↱ F INDEP DRAW

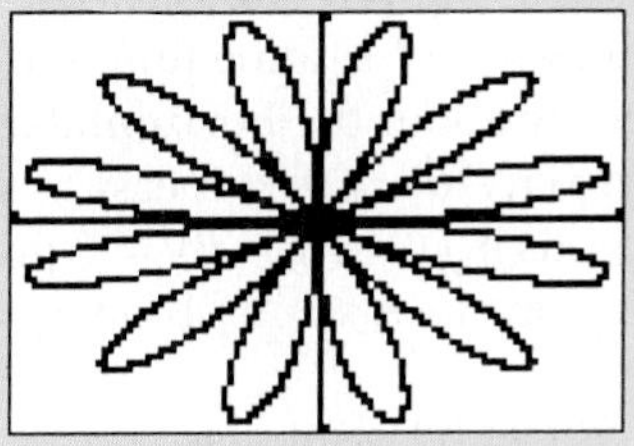

You can try plotting the curve with different values of n.

If we plot the graph of an equation of the form $r = \cos n\theta$ or $r = \sin n\theta$, we will get something like the "rose" we saw in the last example. The number of petals will depend on n. When we graphed $r = \cos\theta$ in Example 3, we found that there was only one petal. When we sketched the graph of $\cos\theta$ in rectangular coordinates, there were two loops as θ went from 0 to 2π, but the points coming from the loop in which r was negative turned out to be the same curve as the other loop, so there was only one petal. On the other hand, in Example 4, when we graphed $r = \sin(4\theta)$, we found that the "negative" loops resulted in new petals and there was a total of eight petals. This turns out to be part of a pattern.

Summary 12.2.1

An equation of the form

$$r = \cos(n\theta) \qquad \text{or} \qquad r = \sin(n\theta)$$

has a graph that is a **"rose."** The number of "petals" is n if n is odd and $2n$ if n is even.

A number of examples are sketched in Figure 12.2.9.

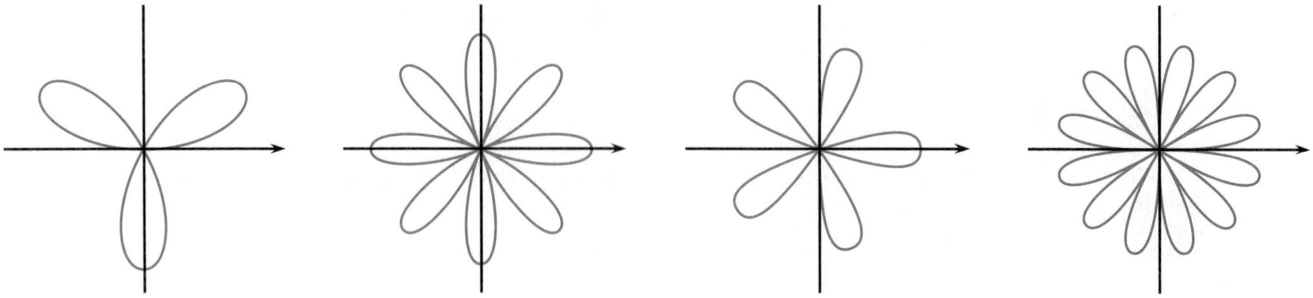

$r = \sin 3\theta$ $r = \cos 4\theta$ $r = \cos 5\theta$ $r = \sin 6\theta$

Figure 12.2.9

Before drawing any more graphs, let us remember that when we sketched graphs in rectangular coordinates, it was often possible to save a lot of work by noticing symmetries. For instance, in Example 3 we might have realized that the graph is symmetric about the horizontal axis (see Figure 12.2.5).

In terms of the equation $r = \cos\theta$, this happens because $\cos\theta = \cos(-\theta)$, so the values of r are the same for corresponding points above and below the axis.

If we consider an equation of the form $f(r, \theta) = 0$, where f is some function of two variables, then we see that its graph will be symmetric about the horizontal axis if the equation $f(r, \theta) = 0$ is the same as the equation $f(r, -\theta) = 0$. It is easy to see that it would also be symmetric about the horizontal axis if $f(-r, \pi - \theta) = 0$ gives the same equation as $f(r, \theta) = 0$. This is because the point $(-r, \pi - \theta)$ is obtained by reflecting the point (r, θ) in the horizontal axis (why?).

Similarly, the reflection of (r, θ) in the vertical axis can be written as $(r, \pi - \theta)$ or as $(-r, -\theta)$. From this we see that the graph is symmetric about the vertical axis if the equation is unchanged by substituting either of these points for (r, θ). (The fact that there are different possibilities comes from the different choices of polar coordinates for any particular point.)

Summary 12.2.2

Symmetry

Consider the graph of the equation $f(r, \theta) = 0$.

1. The graph is symmetric about the horizontal axis if the equation $f(r, \theta) = 0$ is unchanged if θ is changed to $-\theta$. It is also symmetric about the horizontal axis if the equation is unchanged when (r, θ) is replaced by $(-r, \pi - \theta)$.
2. The graph is symmetric about the vertical axis if the equation $f(r, \theta) = 0$ is unchanged if θ is changed to $\pi - \theta$. It is also symmetric about the vertical axis if the equation is unchanged when (r, θ) is replaced by $(-r, -\theta)$.
3. Because of the many different ways of writing a point, it is possible for a graph to be symmetric about either axis without the formula satisfying the conditions given here.

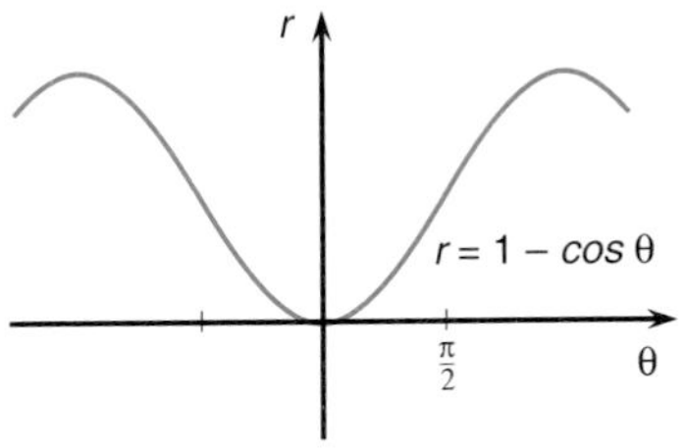

Figure 12.2.10

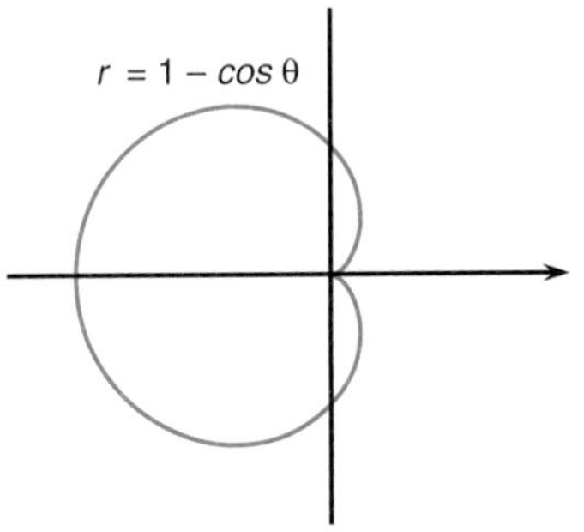

Figure 12.2.11

EXAMPLE 5 **Cardioid** Sketch the graph of the curve $r = 1 - \cos\theta$.

Solution As usual we begin with the plot in rectangular coordinates, shown in Figure 12.2.10. Observing that the equation remains the same if θ is changed to $-\theta$, we conclude that the graph in polar coordinates will be symmetric about the horizontal axis. We also see that the value of r increases from 0 at $\theta = 0$ to 2 at $\theta = \pi$. The graph is sketched in Figure 12.2.11. It is called a **cardioid**, because of its heartlike shape.

When we sketched the cardioid, we had to guess the exact shape of the curve at the origin. Notice that $r = 1 - \cos\theta$ is not zero except when $\cos\theta = 1$, that is, when $\theta = 2k\pi$. So if θ is near 0 but not equal to 0, we know that $r \neq 0$.

If we write the rectangular coordinates of the corresponding point P, they are $(x, y) = (r\cos\theta, r\sin\theta)$. If we consider the secant OP, as shown in Figure 12.2.12, its slope is $m = \frac{y-0}{x-0} = \frac{y}{x} = \frac{r\sin\theta}{r\cos\theta} = \tan\theta$. To do this calculation, we needed to know that $r \neq 0$.

Now as $\theta \to 0$, we expect that the slope $m = \tan\theta$ will approach $\tan 0 = 0$. This suggests that the tangent line will be horizontal. To put it another way, the cardioid approaches the origin with a horizontal tangent.

In fact, this argument depended on knowing that $r \neq 0$ for θ near 0, and with this assumption it is possible to draw the analogous conclusion for any curve in polar coordinates at points where it approaches the origin.

THEOREM 12.2.3

> Suppose the graph of $r = F(\theta)$ passes through the origin when $\theta = \theta_0$. Suppose that $\frac{dr}{d\theta}$ exists at $\theta = \theta_0$. Then the slope of the tangent line at the origin to the curve near $\theta = \theta_0$ is $m = \tan\theta_0$.

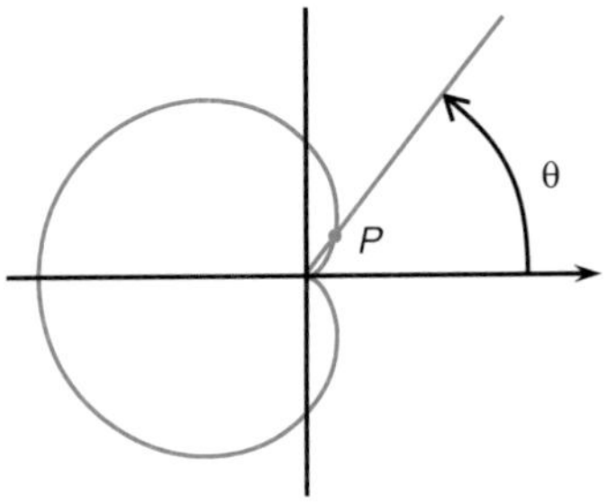

Figure 12.2.12

In other words the curve is tangent to the line $\theta = \theta_0$ at the origin.

EXAMPLE 6 Consider the curve $r = 2 - 2\sin\theta$. It passes through the origin when $\theta = \frac{\pi}{2} + 2k\pi$. The theorem says that the tangent line at this point has slope $m = \tan\frac{\pi}{2}$. This does not really make sense, but if we think of $\tan\theta$ as $\theta \to \frac{\pi}{2}^-$, we see that it tends to ∞. This means that secants through the origin and nearby points tend closer and closer to vertical lines, so we expect that the tangent line will be vertical.

The graph is shown in Figure 12.2.13; it is a cardioid. Notice how the tangent line at the origin is vertical. (One way to see immediately that it is a cardioid is to

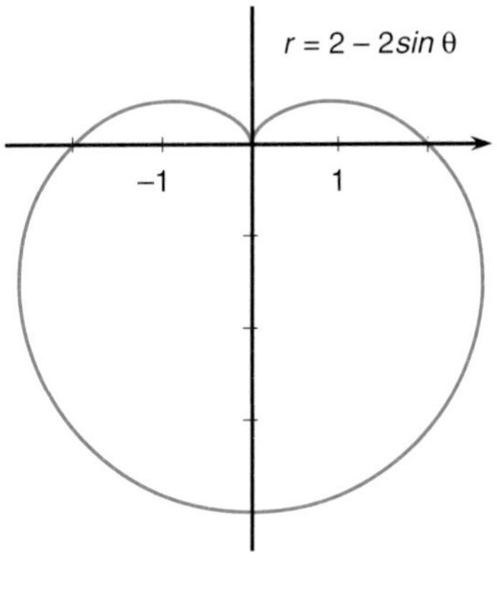

Figure 12.2.13

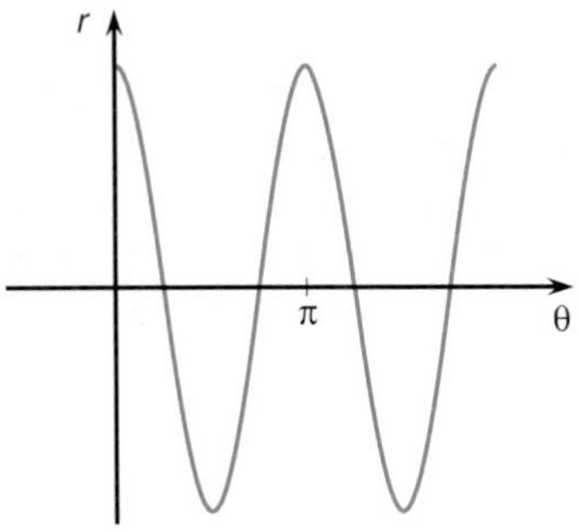

Figure 12.2.14(i)

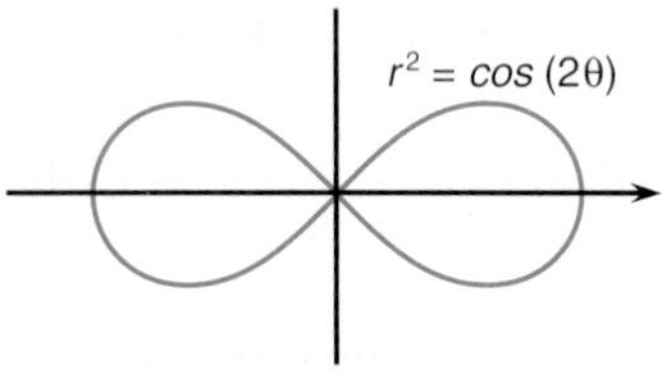

Figure 12.2.14(ii)

observe that $\sin\theta = \cos\left(\theta - \frac{\pi}{2}\right)$, so $r = 1 - \sin\theta$ is just the cardioid $r = 1 - \cos\theta$ rotated through an angle of $\frac{\pi}{2}$.)

EXAMPLE 7 Graph the equation $r^2 = \cos(2\theta)$.

Solution We begin by sketching the graph of $y = \cos(2\theta)$ in Figure 12.2.14(i). Notice that for it to equal r^2 we must have $\cos(2\theta) \geq 0$, that is, $\theta \in \left[-\frac{\pi}{4}, \frac{\pi}{4}\right]$ or $\theta \in \left[\frac{3\pi}{4}, \frac{5\pi}{4}\right]$ (plus the inevitable $2k\pi$).

The usual sort of analysis suggests that as θ moves from $-\frac{\pi}{4}$ to $\frac{\pi}{4}$, $r^2 = \cos(2\theta)$ must increase from 0 to 1 and then decrease back to 0. So $r = \sqrt{\cos(2\theta)}$ must also increase from 0 to 1 and then decrease back to 0. We expect a loop with tangents along the lines $\theta = \pm\frac{\pi}{4}$.

Similar remarks apply to $\theta \in \left[\frac{3\pi}{4}, \frac{5\pi}{4}\right]$, and we expect a similar loop there. The graph is sketched in Figure 12.2.14(ii). Notice the slopes of the tangents at the origin, and notice the symmetries. (Confirm them by using Summary 12.2.2.)

A curve of this shape is called a **lemniscate**. A similar shape in various positions will result from any of the following equations:

$$r^2 = K\cos(2\theta), \qquad r^2 = K\sin(2\theta).$$

The size of the curve depends on $|K|$; the location depends on whether the equation involves $\cos(2\theta)$ or $\sin(2\theta)$ and also on the sign of K. Some examples are sketched in Figure 12.2.15. You might like to check that it is easy to determine the position of the lemniscate by figuring out where r^2 has its maximum values.

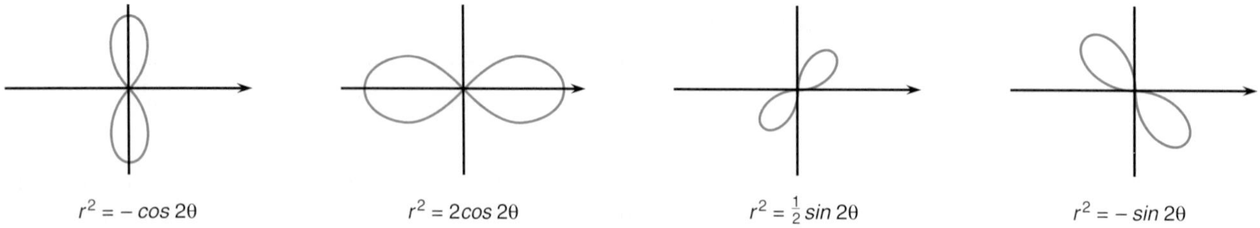

Figure 12.2.15

EXAMPLE 8 Find the points of intersection of the lemniscate $r^2 = 2\cos(2\theta)$ and the cardioid $r = 2 - 2\sin\theta$.

Solution If (r, θ) is on the cardioid, we know that $r^2 = (2 - 2\sin\theta)^2$, so to be on both curves, we need

$$2\cos(2\theta) = (2 - 2\sin\theta)^2 = 4 - 8\sin\theta + 4\sin^2\theta,$$

so

$$\begin{aligned} 2(\cos^2\theta - \sin^2\theta) &= 2(2 - 4\sin\theta + 2\sin^2\theta) \\ 1 - \sin^2\theta - \sin^2\theta &= 2 - 4\sin\theta + 2\sin^2\theta \\ 0 &= 1 - 4\sin\theta + 4\sin^2\theta \\ 0 &= (1 - 2\sin\theta)^2. \end{aligned}$$

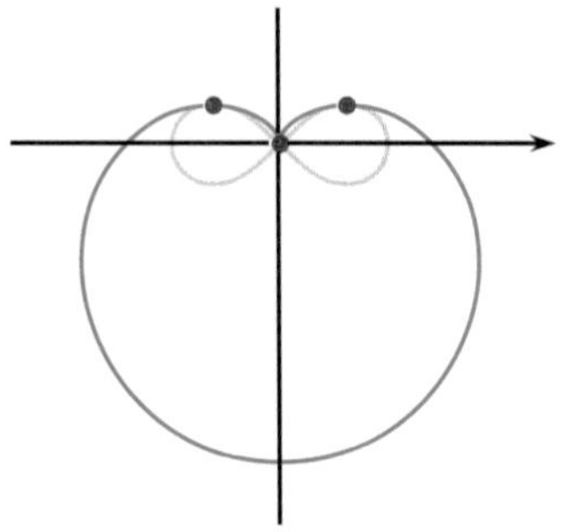
Figure 12.2.16

This means $\sin\theta = \frac{1}{2}$, so $\theta = \frac{\pi}{6}$ or $\frac{5\pi}{6}$. The corresponding points have $r = 2 - 2\sin\theta = 2 - 2\left(\frac{1}{2}\right) = 1$, so they are $(r, \theta) = \left(1, \frac{\pi}{6}\right)$ and $\left(1, \frac{5\pi}{6}\right)$. See Figure 12.2.16.

From the picture we see that the origin is on both curves. It is disconcerting that our calculation did not find it. The reason is that the origin occurs on the cardioid

$r = 2 - 2\sin\theta$ when $\theta = \frac{\pi}{2}$, but on the lemniscate it occurs when $\cos(2\theta) = 0$, that is, when $\theta = \frac{\pi}{4}, \frac{3\pi}{4}, \frac{5\pi}{4}, \frac{7\pi}{4}$. These angles do not coincide, so we do not find them from the formulas.

> In solving for points of intersection of curves in polar coordinates it is important to consider the possibility that a point (r, θ) can also be expressed in different ways: $(r, \theta + 2k\pi)$. The origin is $(0, \theta)$, for *any* θ, so it should be considered separately. Simply check whether the origin is on each curve. A sketch is often helpful.

We recall that the curve $r = 1 - \cos\theta$ is a cardioid. It seems reasonable to consider curves of the form

$$r = a + b\cos\theta \qquad \text{or} \qquad r = a + b\sin\theta.$$

The shape of any of these curves will depend on the signs and relative sizes of a and b. For instance, if $|a| > |b|$, then $|b\cos\theta| \le |b| < |a|$. Because of this, $r = a + b\cos\theta$ is never zero (see Figure 12.2.17(i)). In terms of the picture in polar coordinates this means that the curve will never pass through the origin. Such a curve is illustrated in Figure 12.2.17(ii).

On the other hand, if $|a| < |b|$, then $r = a + b\cos\theta$ will equal 0 for two values of θ, and r will be positive some of the time and negative some of the time. A curve of this type is shown in Figure 12.2.17(iii).

It turns out that there is one more distinction to be made. If $|\frac{a}{b}| \ge 2$, then the curve has a convex shape, whereas if $1 < |\frac{a}{b}| < 2$, the curve has a **dimple**; examples of these cases are shown in Figure 12.2.17(iv).

If $|a| = |b|$, the curve is a cardioid. Any of the other possibilities is called a **limaçon**, after the Latin word for snail. It is apparently pure coincidence that the "dimpled limaçon" of Figure 12.2.17(iv) looks like a lima bean, but this may help

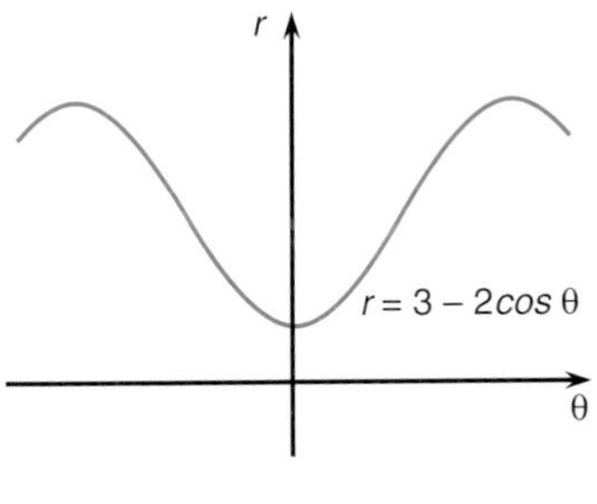

Figure 12.2.17(i)

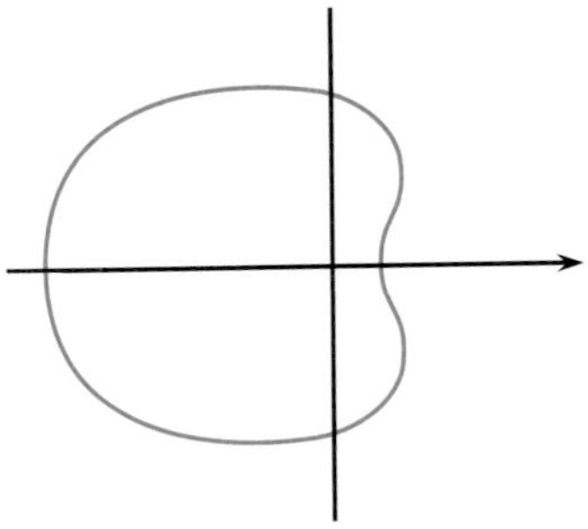

Figure 12.2.17(ii)

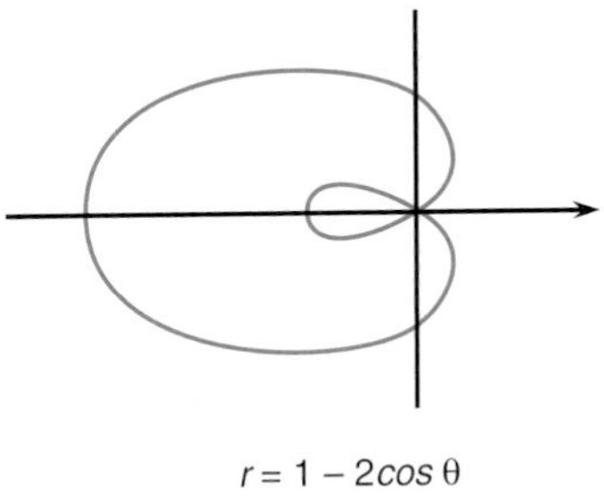

Figure 12.2.17(iii)

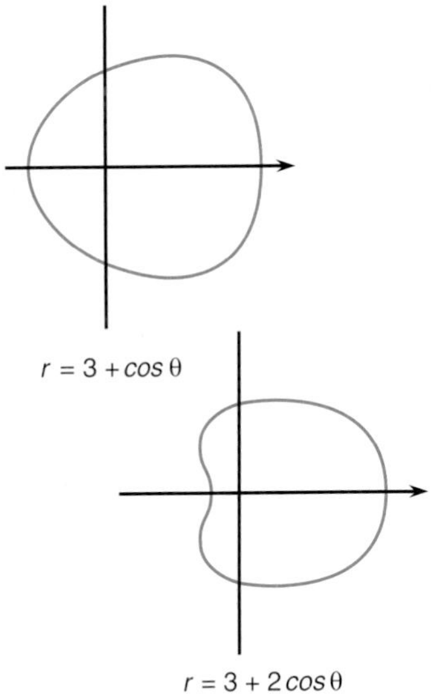

Figure 12.2.17(iv)

you to remember the name. The position of the limaçon is affected by the signs of a and b and the use of $\sin\theta$ or $\cos\theta$ in the equation.

Some of the possibilities for the curve $r = a + b\sin\theta$ are shown in Figure 12.2.18.

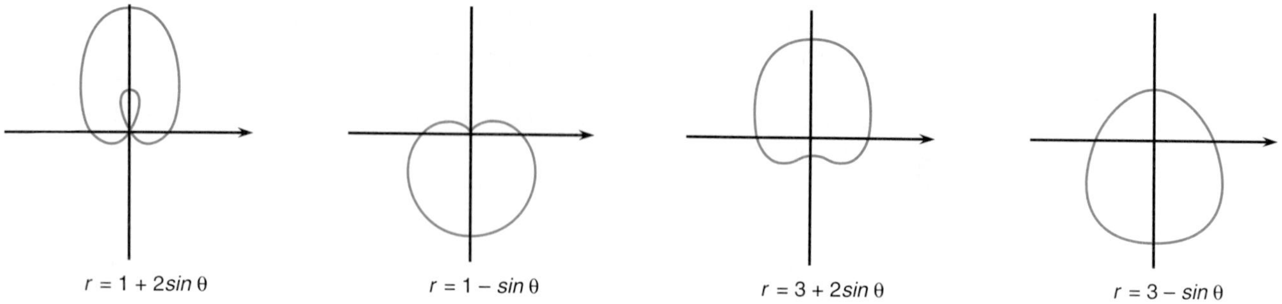

Figure 12.2.18

Summary 12.2.4 Curves in Polar Coordinates

$r = K$	Circle around origin, radius K
$\theta = K$	Line through origin
$\theta = K, r \geq 0$	Half-line from origin
$r = \theta, \theta \geq 0$	Spiral of Archimedes
$r = K\cos\theta$	Circle with radius $\frac{K}{2}$, center $\left(\frac{K}{2}, 0\right)$
$r = \cos(n\theta)$	Rose with n petals if n is odd, $2n$ petals if n is even
$r = 1 \pm \cos\theta$	Cardioid
$r^2 = \cos(2\theta)$	Lemniscate
$r = a + b\cos\theta$	Cardioid if $\lvert a\rvert = \lvert b\rvert$, otherwise limaçon; two loops if $\lvert a\rvert < \lvert b\rvert$, dimpled if $1 < \left\lvert\frac{a}{b}\right\rvert < 2$, convex if $\left\lvert\frac{a}{b}\right\rvert \geq 2$

Replacing cos by sin rotates any of these curves but does not change its type. Replacing $r = f(\theta)$ by $r = f(\theta + c)$ has the effect of rotating the curve *clockwise* by an angle c.

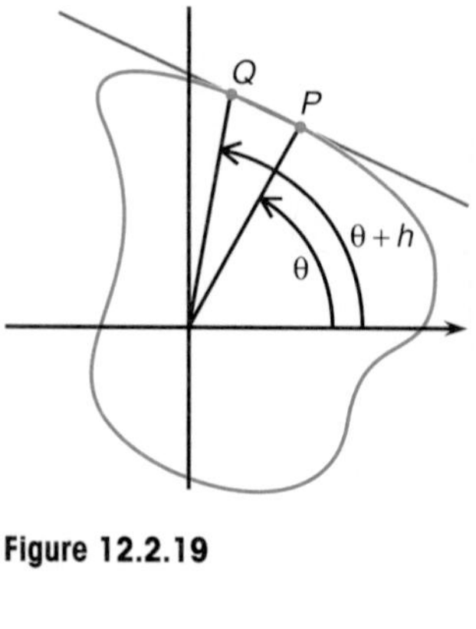

Figure 12.2.19

Slopes of Tangent Lines

Consider a curve $r = f(\theta)$, and let us try to find the slope of the tangent line at a point P on the graph. In Figure 12.2.19 we see the point P at $(r, \theta) = \big(f(\theta), \theta\big)$, and draw the nearby point Q at $\big(f(\theta + h), \theta + h\big)$. To find the slope of the secant PQ, we need to know the rectangular coordinates of P and Q.

Since $x = r\cos\theta$, $y = r\sin\theta$, we find that P is

$$(x_P, y_P) = \big(f(\theta)\cos\theta, f(\theta)\sin\theta\big)$$

and Q is

$$(x_Q, y_Q) = \big(f(\theta + h)\cos(\theta + h), f(\theta + h)\sin(\theta + h)\big).$$

The slope of the secant is

$$m = \frac{y_Q - y_P}{x_Q - x_P} = \frac{f(\theta + h)\sin(\theta + h) - f(\theta)\sin(\theta)}{f(\theta + h)\cos(\theta + h) - f(\theta)\cos(\theta)}.$$

To get the slope of the tangent, we should let Q approach P, that is, let $h \to 0$, and take a limit, assuming that the denominator of the difference quotient is not zero for $h \neq 0$.

Assuming that $f(\theta)$ is differentiable, we see that as $h \to 0$, the numerator and denominator both tend to zero. Applying l'Hôpital's Rule, we find that the slope of the tangent is

$$\begin{aligned} m &= \lim_{h\to 0} \frac{f(\theta + h)\sin(\theta + h) - f(\theta)\sin(\theta)}{f(\theta + h)\cos(\theta + h) - f(\theta)\cos(\theta)} \\ &= \lim_{h\to 0} \frac{f'(\theta + h)\sin(\theta + h) + f(\theta + h)\cos(\theta + h) - 0}{f'(\theta + h)\cos(\theta + h) - f(\theta + h)\sin(\theta + h) - 0} \\ &= \frac{f'(\theta)\sin\theta + f(\theta)\cos\theta}{f'(\theta)\cos\theta - f(\theta)\sin\theta}. \end{aligned}$$

We take derivatives with respect to h.

To do this, we have to know that the denominator $x_Q - x_P$ is nonzero for h near 0 but not equal to 0. One way to ensure this is to insist that $\frac{dx}{d\theta} \neq 0$, so the x-coordinate is an increasing (or decreasing) function of θ. Since $x = r\cos\theta = f(\theta)\cos\theta$, we find

$$\frac{dx}{d\theta} = \frac{d}{d\theta}\big(f(\theta)\cos\theta\big) = f'(\theta)\cos\theta - f(\theta)\sin\theta.$$

To say that this is not zero also conveniently ensures that the denominator is nonzero in our slope formula. We have just found the following result.

THEOREM 12.2.5

Suppose $f(\theta)$ is a differentiable function with $f'(\theta)$ continuous on some interval, and let C be the graph of $r = f(\theta)$. Suppose θ is a point in this interval at which

$$f'(\theta)\cos\theta - f(\theta)\sin\theta \neq 0.$$

Then the slope of the tangent to C at the corresponding point P is

$$m = \frac{f'(\theta)\sin\theta + f(\theta)\cos\theta}{f'(\theta)\cos\theta - f(\theta)\sin\theta}.$$

EXAMPLE 9 For the cardioid $r = 1 + \cos\theta$, the slope of the tangent is

$$\begin{aligned} m = \frac{f'(\theta)\sin\theta + f(\theta)\cos\theta}{f'(\theta)\cos\theta - f(\theta)\sin\theta} &= \frac{-\sin\theta\sin\theta + (1 + \cos\theta)\cos\theta}{-\sin\theta\cos\theta - (1 + \cos\theta)\sin\theta} \\ &= \frac{\cos^2\theta - \sin^2\theta + \cos\theta}{-2\cos\theta\sin\theta - \sin\theta}. \end{aligned}$$

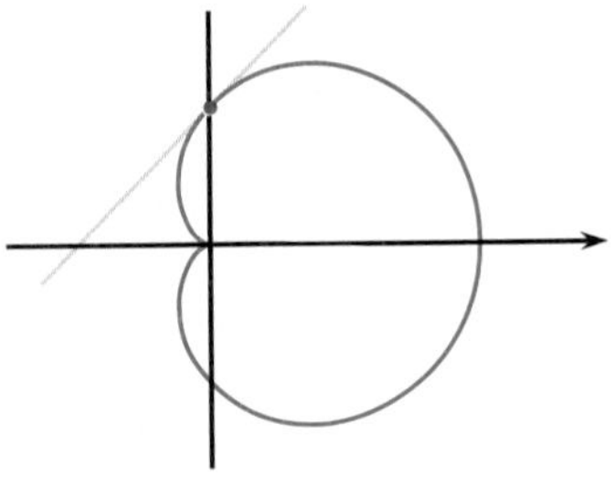

Figure 12.2.20(i)

In particular, when $\theta = \frac{\pi}{2}$, the slope is $m = \frac{0-1+0}{0-1} = 1$. The tangent at $\theta = \frac{\pi}{2}$ is shown in Figure 12.2.20(i).

If we want to find the highest point on the graph, we find where the slope of the tangent is zero. This means

$$\begin{aligned} 0 &= \cos^2\theta - \sin^2\theta + \cos\theta = \cos^2\theta - (1 - \cos^2\theta) + \cos\theta \\ &= 2\cos^2\theta + \cos\theta - 1. \end{aligned}$$

Writing $u = \cos\theta$ and solving $2u^2 + u - 1 = 0$, we find

$$\cos\theta = u = \frac{-1 \pm \sqrt{1+8}}{4} = \frac{-1 \pm 3}{4} = \frac{1}{2}, -1.$$

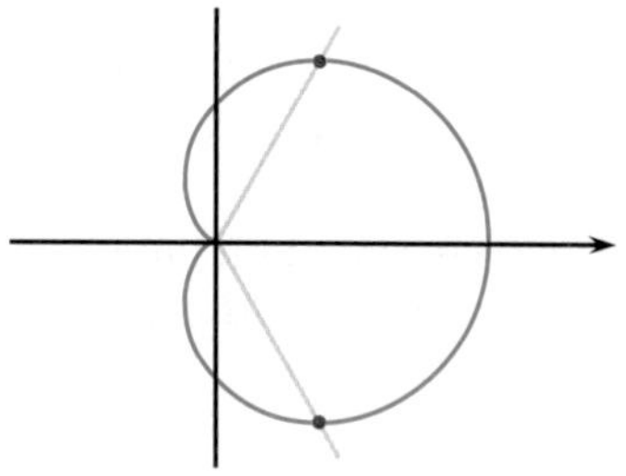

Figure 12.2.20(ii)

When $\cos\theta = -1$, we have $\sin\theta = 0$, and the denominator in the slope formula is zero, so the formula does not apply. The corresponding point on the graph is the cusp at the origin, where there is no tangent line.

The tangent line is horizontal when $\cos\theta = \frac{1}{2}$, that is, when $\theta = \frac{\pi}{3}$ (at the top point in Figure 12.2.20(ii)) and when $\theta = -\frac{\pi}{3}$ (at the bottom point).

The highest point on the graph occurs when $\theta = \frac{\pi}{3}$, so $r = f\left(\frac{\pi}{3}\right) = 1 + \cos\frac{\pi}{3} = \frac{3}{2}$. Its rectangular coordinates are

$$(x, y) = (r\cos\theta, r\sin\theta) = \left(\frac{3}{2}\left(\frac{1}{2}\right), \frac{3}{2}\left(\frac{\sqrt{3}}{2}\right)\right) = \left(\frac{3}{4}, \frac{3\sqrt{3}}{4}\right).$$

Exercises 12.2

I

Sketch the graph of each of the following curves.

1. $r = \sin\theta$

2. $r = 3\cos\theta$

3. $r = -2\cos\theta$

4. $r = 5$

5. $\sin\theta = \frac{1}{\sqrt{2}}, r \geq 0$

6. $r = \cos(5\theta)$

7. $r = 2\sin(2\theta)$

8. $r = \sin\left(3\theta + \frac{\pi}{2}\right)$

9. $r = -\theta, \theta > 0$

10. $r = \cos\theta + \sin\theta$

Sketch and identify the graph of each of the following curves.

11. $r = 1 + \sin\theta$

12. $r = 2 - \cos\theta$

13. $r = 2 - 3\cos\theta$

14. $r = 3 - 4\sin\left(\theta - \frac{\pi}{4}\right)$

15. $r = 1 + 10\cos\theta$

16. $r = \sqrt{\sin(2\theta)}$

17. $r^2 = \cos^2\theta - \sin^2\theta$

18. $r = 3 - 3\cos\theta$

19. $r^2 = \cos^2\theta + \sin^2\theta$

20. $r = 3\sin\theta\cos\theta$

II

Sketch the graph of each of the following curves.

***21.** $r = 5\cos\theta - 5\sin\theta$

***22.** $r = \cos 2\theta + \sin 2\theta$

***23.** $r = 3\cos\theta - 4\sin\theta$

***24.** $r = 1 + \cos\theta + \sin\theta$

Find the point(s) of intersection of each of the following pairs of curves. In each case draw a sketch.

***25.** The circle $r = \cos\theta$ and the line $r\cos\theta = \frac{1}{2}$.

***26.** The circle $r = -2\sin\theta$ and the line $r\sin\theta = -1$.

***27.** $r = 3 + 2\cos\theta$ and $r = 6 - 4\cos\theta$

***28.** $r = \cos\theta$ and $r^2 = \frac{3}{2}\cos(2\theta)$

***29.** $r = \cos(5\theta)$ and $r = \sin(5\theta)$

***30.** $r = \sin\theta$ and $r = \frac{2\theta}{\pi}$

****31.** $r = 1 + \cos\theta$ and $r = 3 + 5\cos\theta$

****32.** $r = \cos\theta - 2$ and $r = 2\cos\theta + 2$

***33.** Sketch the graph of $r = \cos^2\left(\frac{\theta}{2}\right)$.

34. In Example 3 (p. 761) it was stated that the part of the curve $r = \cos\theta$ with $\frac{\pi}{2} \leq \theta \leq \frac{3\pi}{2}$ is the same as the circle found when $-\frac{\pi}{2} \leq \theta \leq \frac{\pi}{2}$. Show that this is true by comparing the points whose polar coordinates are $(r, \theta) = (\cos\theta, \theta)$ and $\left(\cos(\theta + \pi), \theta + \pi\right)$.

35. Verify that if $r < 0$, then the point whose polar coordinates are (r, θ) is exactly opposite (i.e., the *negative*) of the point whose polar coordinates are $(|r|, \theta)$.

***36.** Verify the statement made in Summary 12.2.1 about the number of petals in a rose.

Sketch the graph of each of the following.

***37.** $r = |\sin(3\theta)|$

***38.** $r = |\cos(5\theta)|$

***39.** $r = |\cos\theta|$

***40.** $r = |\cos(4\theta)|$

***41.** Prove the remark made before Summary 12.2.2 that changing (r, θ) to $(-r, \pi - \theta)$ has the effect of reflecting the point in the horizontal axis.

***42.** Prove that replacing the point (r, θ) by $(r, \pi - \theta)$ or by $(-r, -\theta)$ has the effect of reflecting the point in the vertical axis.

***43.** Consider the equation $r^2 = \cos\theta$. Show that it is unchanged if we substitute $(-r, -\theta)$ for (r, θ) but that it changes if we substitute $(r, \pi - \theta)$. What can we conclude about its symmetry?

***44.** What are the polar coordinates of the points where the Spiral of Archimedes crosses the horizontal axis (cf. Example 2)?

***45.** What are the rectangular coordinates of the points where the Spiral of Archimedes crosses the horizontal or vertical axes (cf. Example 2)?

46. When we drew the Spiral of Archimedes in Example 2, we only considered values of θ that are nonnegative. Sketch the graph that results if we let $r = \theta$ and consider *all* possible values of θ.

For each of the following curves, find the rectangular coordinates of the highest and lowest points on the graph.

***47.** The right-hand petal of the rose $r = \cos 2\theta$.

***48.** The right-hand petal of the rose $r = \cos 3\theta$.

****49.** The lemniscate $r^2 = \cos 2\theta$.

***50.** The limaçon $r = 1 - 2\cos\theta$.

III

Use a graphics calculator or computer to find the points of intersection of the following pairs of curves. (You may find it easier to begin with the Cartesian graph of $y = f(\theta)$.)

51. $r^2 = 5\sin(2\theta), \quad r = \theta$

52. $r = 1 - 2\sin\theta, \quad r = \frac{\theta}{4}$

***53.** $r = \sin\frac{\theta}{2}, \quad r = \sin\theta$

***54.** $r^2 = 4\cos(4\theta), \quad r = 3 + 2\cos\theta$

POINT TO PONDER

When we consider the equation $\theta = \frac{4\pi}{3}$, we quite likely think of all points whose polar angle is $\frac{4\pi}{3}$, so it seems reasonable to draw the half-line consisting of all points (r, θ) with $\theta = \frac{4\pi}{3}$ and $r \geq 0$. On the other hand, for many other situations it is much more natural to allow r to be negative. Some books will write the entire line for $\theta = \frac{4\pi}{3}$, others the half-line. The best approach is probably to specify what we mean in any doubtful cases. For instance, for the spiral we drew in Example 2 we wrote $r = \theta$, $\theta \geq 0$. For the half-line we should write $\theta = \frac{4\pi}{3}$, $r \geq 0$.

In working with polar coordinates we often resort to converting to rectangular coordinates in order to draw something. In a way this may seem like cheating; if we are going to have to end up using Cartesian coordinates anyway, why not just use them all along?

In fact, polar coordinates are usually more awkward, especially for drawing graphs. On the other hand, for certain purposes they are very convenient. Obviously, they are suited to situations in which something goes around something else, and they are essential to one of the early successes of calculus, the study of the motion of the planets. The theory of planetary motion can best be expressed in terms of polar coordinates. It is so much more complicated in rectangular coordinates that it might never have been discovered if it had to be done in that form.

You might like to express the equation of a lemniscate in rectangular coordinates just for an example to show that something can be simpler in its polar form.

Polar coordinates are useful because they allow us to express certain problems in a convenient or suggestive form. We then use the techniques we know to try to solve the problems.

12.3 Area in Polar Coordinates

AREA FORMULA
LIMITS OF INTEGRATION
TRACING EXACTLY ONCE

For curves sketched in rectangular coordinates we were able to use integration to find the area contained under the graph of the curve. For curves in polar coordinates it is less clear what we might mean by the area "under" the graph, since the graph tends to wrap around the origin. Instead, it is reasonable to consider the region contained "inside" the curve if this makes sense.

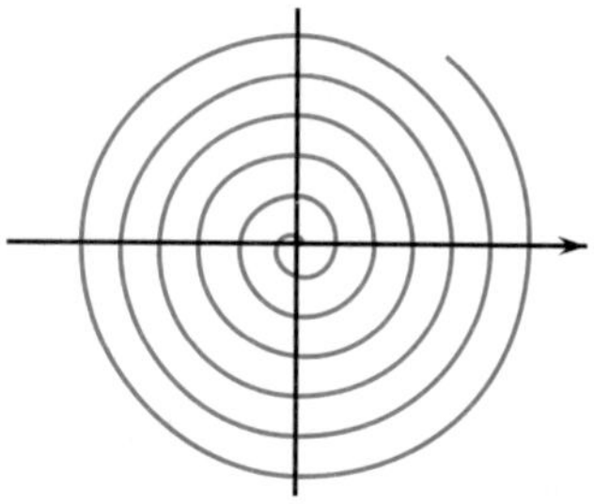

Figure 12.3.1

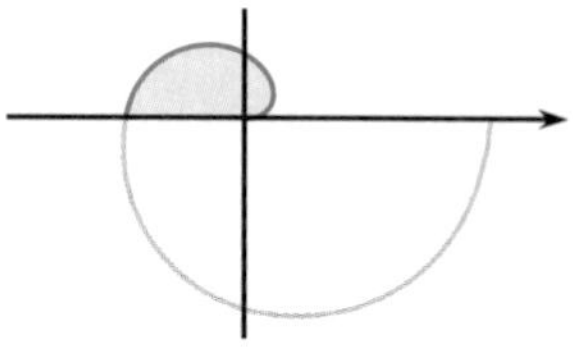

Figure 12.3.2

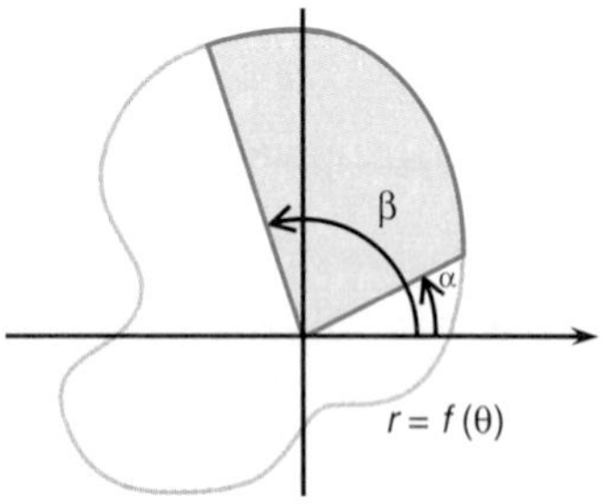

Figure 12.3.3

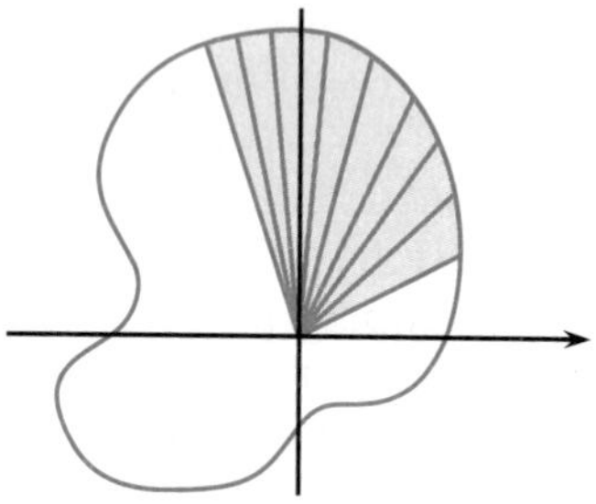

Figure 12.3.4

It certainly would make sense for some curves, like circles and lemniscates and limaçons and cardioids, but it is less clear what it might mean for the spiral shown in Figure 12.3.1.

The spiral does not "contain" anything. On the other hand, we might like to know the area of the shaded region in Figure 12.3.2, which is the region "inside" the spiral $r = \theta$, for $0 \leq \theta \leq \pi$.

For that matter, we might look at the graph of $r = f(\theta)$ and ask about the area of the region inside the graph for $\alpha \leq \theta \leq \beta$, for some angles α and β as shown in Figure 12.3.3. This should include the sort of region that is entirely enclosed inside a curve, but it will also cover other cases.

Our experience with integration suggests that we should break up the region into small sectors, as shown in Figure 12.3.4. A first guess might be to approximate the area of a little sector with the area of a little triangle, but in fact it is better to use a small circular sector.

Consider the small circular sector shown in Figure 12.3.5. Its radius is r and the central angle is $\Delta\theta$. We need to find its area.

The area of the whole circle of radius r is πr^2. The whole circle involves an angle of 2π at the center, so this little sector, with its angle $\Delta\theta$, is $\frac{\Delta\theta}{2\pi}$ of the whole circle, and its area is

$$\frac{\Delta\theta}{2\pi}\pi r^2 = \frac{1}{2}r^2\Delta\theta.$$

As usual, we consider a partition $\mathcal{P} = \{\theta_0, \theta_1, \ldots, \theta_n\}$ of $[\alpha, \beta]$ and break the region inside $r = f(\theta)$ into the little sectors with $\theta_{i-1} \leq \theta \leq \theta_i$. See Figure 12.3.6. The area of the ith sector is approximately equal to the area of the circular sector between θ_{i-1} and θ_i whose radius is $r = f(\theta^*)$ for any $\theta^* \in [\theta_{i-1}, \theta_i]$.

We just found that this area is $\frac{1}{2}r^2\Delta\theta = \frac{1}{2}f(\theta^*)^2(\theta_i - \theta_{i-1})$.

If we choose a point θ_i^* in $[\theta_{i-1}, \theta_i]$ for each i, then we can approximate the area of each little sector in this way and add them up to approximate the area of the whole region. The approximation to the area is

$$\sum_{i=1}^{n} \frac{1}{2}f(\theta_i^*)^2\Delta\theta_i = \sum_{i=1}^{n} \frac{1}{2}f(\theta_i^*)^2(\theta_i - \theta_{i-1}).$$

Of course, this is a Riemann sum.

Now if the mesh of $\mathcal{P}$ tends to 0 as we let $n \to \infty$, we see that the approximations to the area given by these Riemann sums will tend to a limit, which is the area.

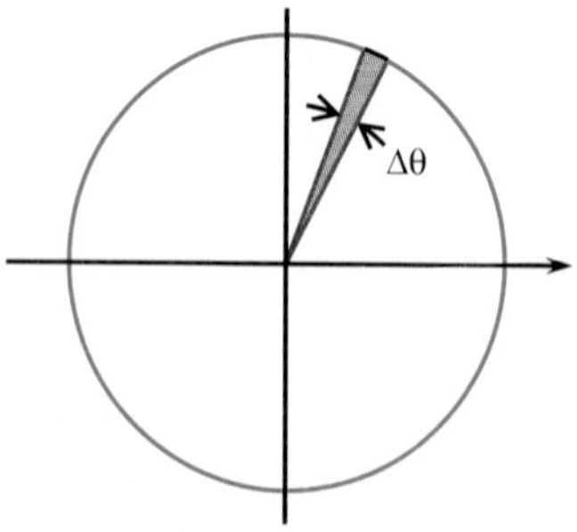

Figure 12.3.5

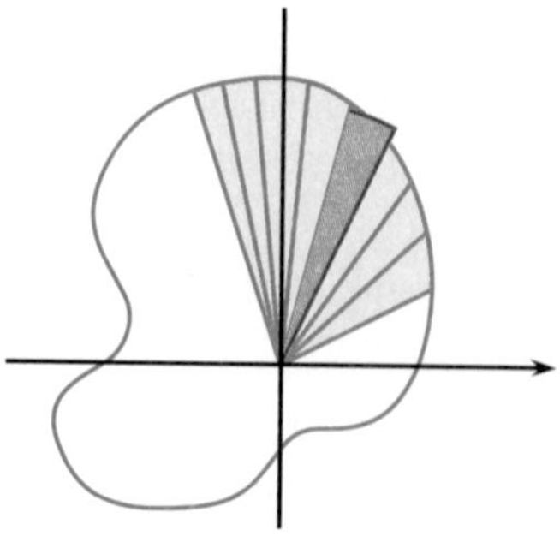

Figure 12.3.6

THEOREM 12.3.1

Consider the curve $r = f(\theta)$. The area of the region enclosed inside this curve as θ runs between the lines $\theta = \alpha$ and $\theta = \beta$ (i.e., for $\alpha \le \theta \le \beta$) is

$$Area = \frac{1}{2}\int_{\alpha}^{\beta} f(\theta)^2\, d\theta.$$

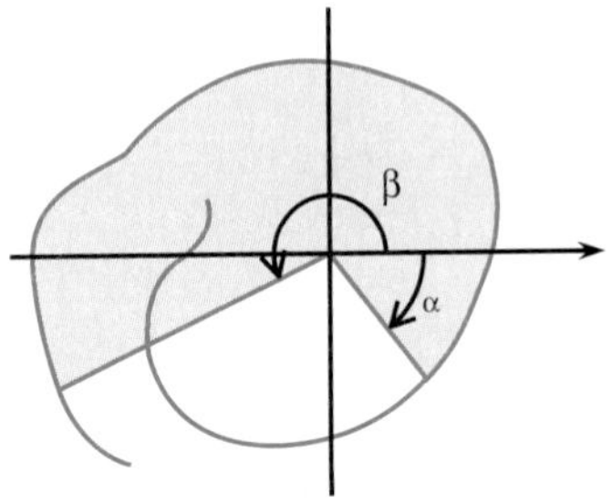

Figure 12.3.7

Such a region is illustrated in Figure 12.3.7.

Notice that the integrand $f(\theta)^2$ is always nonnegative even if the radius $r = f(\theta)$ is negative. So we always have a nonnegative integral and do not have to worry about "signed areas" the way we did for integrals in rectangular coordinates. However, we may have to worry about different parts of the region overlapping.

EXAMPLE 1 **Area of Cardioid** Find the area of the region inside the cardioid $r = 1 - \cos\theta$.

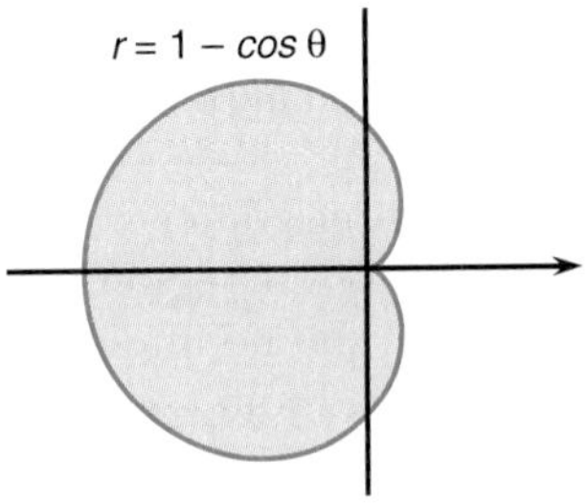

Figure 12.3.8

Solution In this case we want the entire region inside the curve, so we should let θ go from 0 to 2π (which takes us once around the origin). From the theorem we see that the area is

$$\frac{1}{2}\int_0^{2\pi} r^2\, d\theta = \frac{1}{2}\int_0^{2\pi} (1 - \cos\theta)^2\, d\theta = \frac{1}{2}\int_0^{2\pi} (1 - 2\cos\theta + \cos^2\theta)\, d\theta.$$

Using the familiar identity $\cos^2\theta = \frac{1}{2}(1 + \cos 2\theta)$ (cf. Theorem 1.7.13 and Example 5.7.5), we integrate and find that the area equals

$$\begin{aligned}\frac{1}{2}\int_0^{2\pi} (1 - 2\cos\theta + \cos^2\theta)\, d\theta &= \frac{1}{2}\int_0^{2\pi} 1\, d\theta \\ &\quad - \int_0^{2\pi} \cos\theta\, d\theta + \frac{1}{4}\int_0^{2\pi} (1 + \cos 2\theta)\, d\theta \\ &= \pi - (\sin\theta)\Big|_0^{2\pi} + \left(\frac{\theta}{4} + \frac{1}{8}(\sin 2\theta)\right)\Big|_0^{2\pi} \\ &= \frac{3\pi}{2}.\end{aligned}$$

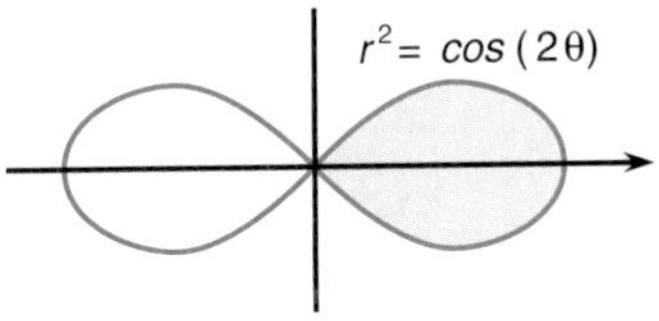

Figure 12.3.9(i)

The cardioid is illustrated in Figure 12.3.8. Its area is $\frac{3\pi}{2}$.

We have seen with integrals that finding the limits of integration can be a difficult part of a problem. This is frequently true with integrals in polar coordinates.

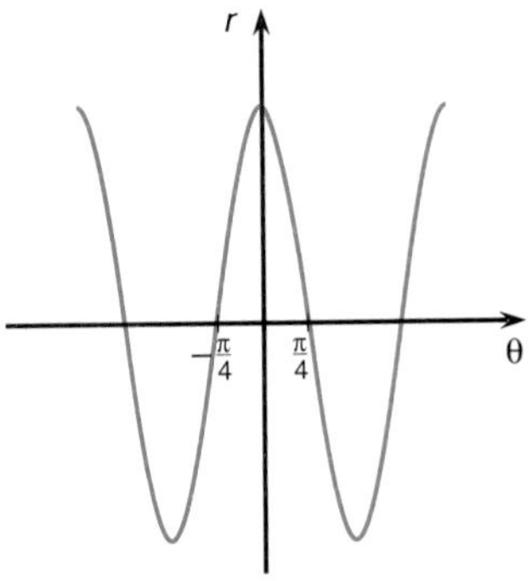

Figure 12.3.9(ii)

EXAMPLE 2 Find the area inside one half of the lemniscate $r^2 = \cos 2\theta$, which is illustrated in Figure 12.3.9(i).

Solution It may help to consider the graph of $y = \cos 2\theta$ in rectangular coordinates, which is shown in Figure 12.3.9(ii). There we see that $\cos 2\theta$ is positive for $-\frac{\pi}{4} < \theta < \frac{\pi}{4}$, and this is the range that corresponds to the right-hand loop of the lemniscate.

The area we want is given by

$$\frac{1}{2}\int_{-\pi/4}^{\pi/4} r^2\,d\theta = \frac{1}{2}\int_{-\pi/4}^{\pi/4} \cos(2\theta)\,d\theta$$
$$= \frac{1}{4}(\sin(2\theta))\Big|_{-\pi/4}^{\pi/4} = \frac{1}{4}\left(\sin\left(\frac{\pi}{2}\right) - \sin\left(-\frac{\pi}{2}\right)\right)$$
$$= \frac{1}{2}.$$

The required area is $\frac{1}{2}$ square unit.

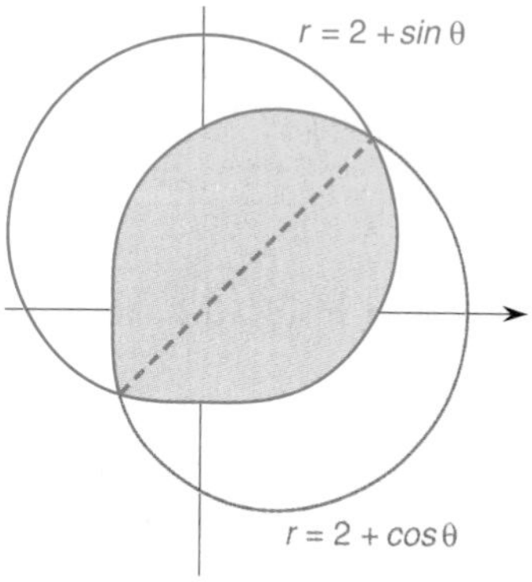

Figure 12.3.10

EXAMPLE 3 Find the area of the region consisting of all points inside the limaçon $r = 2 + \cos\theta$ and also inside the limaçon $r = 2 + \sin\theta$.

Solution We sketch the curves in Figure 12.3.10. The area we want to find is the shaded region in the picture. We begin by finding the points of intersection of the two curves.

For a point to be on both curves, we need $r = 2 + \cos\theta$ and $r = 2 + \sin\theta$, so $2 + \cos\theta = 2 + \sin\theta$, and $\cos\theta = \sin\theta$. This means $\tan\theta = 1$, which happens when $\theta = \frac{\pi}{4}$ and when $\theta = -\frac{3\pi}{4}$.

In the picture we see that the "upper left" part of the shaded region is determined by lying inside the curve $r = 2 + \cos\theta$, and the "lower right" part of the region lies inside $r = 2 + \sin\theta$. So to find the area of the upper left part, we should integrate $\frac{1}{2}(2 + \cos\theta)^2$. For limits of integration we should begin at $\alpha = \frac{\pi}{4}$. From there we should let the angle sweep counterclockwise until it reaches the other point of intersection of the two limaçons. This means that the angle will increase by π from $\frac{\pi}{4}$, so the upper limit of integration is $\beta = \frac{\pi}{4} + \pi = \frac{5\pi}{4}$. This angle is in the same position as $-\frac{3\pi}{4}$, but it is not the same angle.

To find the area of the lower right region, we integrate $\frac{1}{2}(2 + \sin\theta)^2$ from $\alpha = -\frac{3\pi}{4}$ to $\beta = \frac{\pi}{4}$. The total area is

$$\frac{1}{2}\int_{\pi/4}^{5\pi/4} (4 + 4\cos\theta + \cos^2\theta)\,d\theta + \frac{1}{2}\int_{-3\pi/4}^{\pi/4} (4 + 4\sin\theta + \sin^2\theta)\,d\theta$$
$$= \frac{1}{2}\left(4\theta + 4\sin\theta + \frac{1}{2}\left(\theta + \frac{1}{2}\sin 2\theta\right)\right)\Bigg|_{\pi/4}^{5\pi/4}$$
$$+ \frac{1}{2}\left(4\theta - 4\cos\theta + \frac{1}{2}\left(\theta - \frac{1}{2}\sin 2\theta\right)\right)\Bigg|_{-3\pi/4}^{\pi/4}$$
$$= \frac{1}{2}\left(4\pi + 4\left(-\frac{1}{\sqrt{2}} - \frac{1}{\sqrt{2}}\right) + \frac{\pi}{2} + \frac{1}{4} - \frac{1}{4}\right)$$
$$+ \frac{1}{2}\left(4\pi - 4\left(\frac{1}{\sqrt{2}} + \frac{1}{\sqrt{2}}\right) + \frac{\pi}{2} - \frac{1}{4} + \frac{1}{4}\right)$$
$$= \frac{9\pi}{2} - 4\sqrt{2}.$$

By symmetry the two parts of the region have the same area. We could evaluate one integral and double the result.

The area inside the two curves is $\frac{9\pi}{2} - 4\sqrt{2}$.

When we first found the angles θ of the points at which the two curves cross, we found $\frac{\pi}{4}$ and $-\frac{3\pi}{4}$. However, when we needed the limits of integration for the upper left part of the region, we began at $\frac{\pi}{4}$ but we ended at $\frac{5\pi}{4}$. It was important to look at the picture to realize that as θ increases from $\frac{\pi}{4}$, the angle at which it will reach the other intersection point is $\frac{5\pi}{4}$.

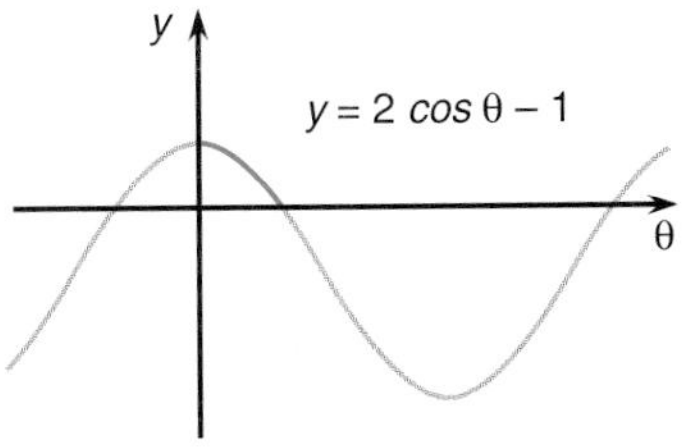

Figure 12.3.11(i)

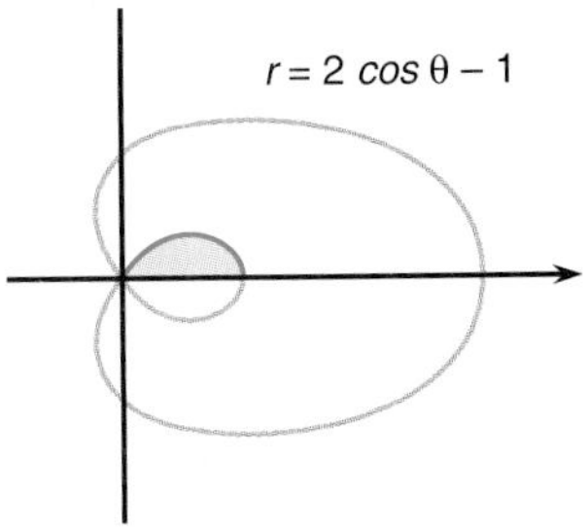

Figure 12.3.11(ii)

EXAMPLE 4 Find the area of the upper half of the inner loop of the limaçon $r = 2\cos\theta - 1$.

Solution We begin by sketching the graph of $y = 2\cos\theta - 1$ in Figure 12.3.11(i). In particular, it is positive between $\theta = -\frac{\pi}{3}$ and $\theta = \frac{\pi}{3}$, and negative for θ between $\frac{\pi}{3}$ and $\frac{5\pi}{3}$.

From this we can sketch the curve $r = 2\cos\theta - 1$ in Figure 12.3.11(ii). We observe that the inner loop is traced out as θ moves from $-\frac{\pi}{3}$ to $\frac{\pi}{3}$. The upper half of the inner loop will be traced out as θ moves from 0 to $\frac{\pi}{3}$.

Applying Theorem 12.3.1, we find that the area of the upper half of the inner loop is

$$\begin{aligned}\frac{1}{2}\int_0^{\pi/3}(2\cos\theta - 1)^2 d\theta &= \frac{1}{2}\int_0^{\pi/3}(4\cos^2\theta - 4\cos\theta + 1)\,d\theta \\ &= \frac{1}{2}(2\theta + \sin 2\theta - 4\sin\theta + \theta)\Big|_0^{\pi/3} \\ &= \frac{\pi}{3} - 0 + \frac{\sqrt{3}}{4} - 0 - \sqrt{3} + 0 + \frac{\pi}{6} - 0 \\ &= \frac{\pi}{2} - \frac{3\sqrt{3}}{4}.\end{aligned}$$

The area of the upper half of the inner loop is $\frac{\pi}{2} - \frac{3\sqrt{3}}{4}$.

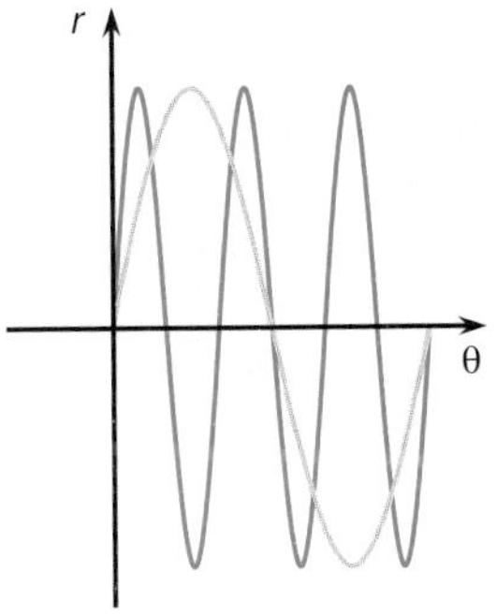

Figure 12.3.12(i)

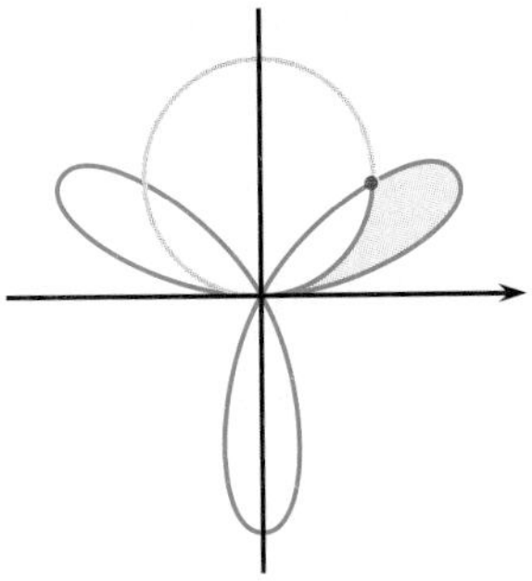

Figure 12.3.12(ii)

EXAMPLE 5 Find the area of the region consisting of all points that lie inside the upper right petal of the three-petaled rose $r = \sin 3\theta$ but outside the circle $r = \sin\theta$.

Solution We sketch the curves $y = \sin\theta$ and $y = \sin 3\theta$ in rectangular coordinates in Figure 12.3.12(i). Using this sketch, we draw the rose and the circle in Figure 12.3.12(ii), shading the specified region.

Before proceeding, we must know the coordinates of that point where the curves intersect. Since the curves are $r = \sin 3\theta$ and $r = \sin\theta$, we must solve $\sin\theta = \sin 3\theta$. Expanding $\sin 3\theta$ on the right side, we see that this amounts to

$$\sin\theta = \sin\theta\cos 2\theta + \cos\theta\sin 2\theta = \sin\theta\cos 2\theta + 2\cos^2\theta\sin\theta,$$

so $\sin\theta = 0,$ or $1 = \cos 2\theta + 2\cos^2\theta = \cos^2\theta - \sin^2\theta + 2\cos^2\theta,$

that is, $1 = 3\cos^2\theta - (1 - \cos^2\theta) = 4\cos^2\theta - 1.$

From this we see that $2 = 4\cos^2\theta$, so $\cos^2\theta = \frac{1}{2}$.

We consider the point $\theta = \frac{\pi}{4}$, where $\cos\theta = \frac{1}{\sqrt{2}}$, and it is easy to check that $\sin\frac{\pi}{4} = \frac{1}{\sqrt{2}} = \sin\frac{3\pi}{4}$. This is also consistent with Figure 12.3.12(i), where the two curves appear to cross at about $\theta = \frac{\pi}{4}$.

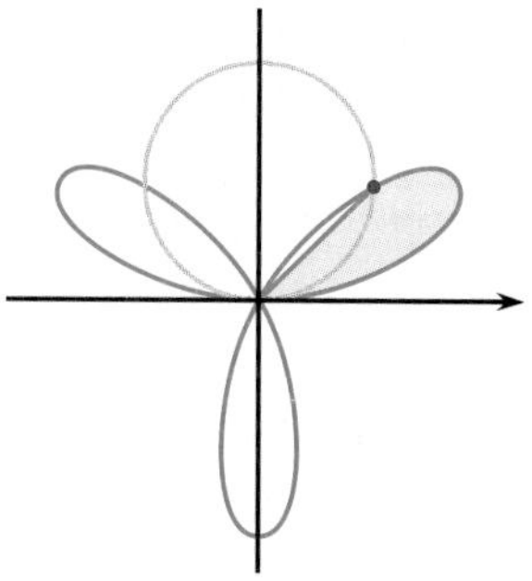

Figure 12.3.12(iii)

To find the desired area, we should find the area inside the rose for θ between 0 and $\frac{\pi}{4}$, which is the large shaded area in Figure 12.3.12(iii), and subtract the area of the part of this region that lies inside the circle. The area we want will be

$$\frac{1}{2}\int_0^{\pi/4} (\sin^2 3\theta - \sin^2\theta)\,d\theta = \frac{1}{4}\int_0^{\pi/4} (1 - \cos 6\theta - 1 + \cos 2\theta)\,d\theta$$

$$= \frac{1}{4}\left(-\frac{1}{6}\sin 6\theta + \frac{1}{2}\sin 2\theta\right)\Big|_0^{\pi/4}$$

$$= \frac{1}{4}\left(\frac{1}{6} + 0 + \frac{1}{2} - 0\right) = \frac{1}{6}.$$

The desired area is $\frac{1}{6}$.

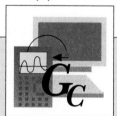

If we consider the similar question of finding the area of the region inside $r = \sin 3\theta$ but outside $r = \cos\theta + \sin\theta$, for $0 \le \theta \le \frac{\pi}{2}$, we can use a graphics calculator to sketch the graphs.

TI-85 MODE ▼ ▼ ▼ ▼ ▶ (Pol) ENTER EXIT GRAPH F1 ($r(\theta)$ =) 3 × SIN (3 × F1 (θ)) ▼ COS F1 (θ) + SIN F1 (θ) M2 (RANGE) 0 ▼ π ÷ 2 ▼ ▼ 0 ▼ 3 ▼ ▼ (−) .5 ▼ 3 F5 (GRAPH)	SHARP EL9300 SET UP E 2 [graph key] 3 sin 3 x/θ/T 2ndF ▼ cos x/θ/T + sin x/θ/T RANGE 0 ENTER 3 ENTER 1 ENTER 0 ENTER 3 ENTER 1 ENTER 0 ENTER π ÷ 2 ENTER .1 ENTER [graph key]
CASIO f_x-7700GB MODE MODE − (POL) Range 0 EXE 3 EXE 1 EXE 0 EXE 3 EXE 1 EXE 0 EXE π ÷ 2 EXE .01 EXE Graph 3 sin 3 θ EXE Graph cos θ + sin θ EXE	HP 48SX ↰ RAD ↰ PLOT ′ R = 3 × SIN 3 × α ↱ F NEW P O L ENTER PTYPE POLAR PLOTR 0 ENTER 3 XRNG 0 ENTER 3 YRNG ′ α ↱ F INDEP DRAW ATTN ′ R = COS α ↱ F ▶ + SIN α ↱ F ↱ PLOT ↰ DRAW DRAW

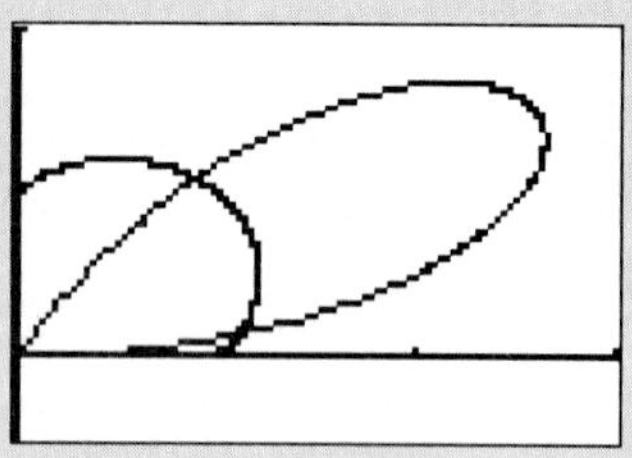

Unlike what happened in Example 5, it is not easy to solve for the intersection of these curves. But we can use the calculator to solve for where $\cos\theta + \sin\theta = 3\sin 3\theta$. We find two solutions $\theta = a, b \in \left[0, \frac{\pi}{2}\right]$, with $a < b$. Then we find that the required area is

$$\frac{1}{2}\int_a^b \left(9\sin^2 3\theta - (\cos\theta + \sin\theta)^2\right) d\theta$$

$$= \left(\frac{9}{4}\left(\theta - \frac{1}{6}\sin 6\theta\right) - \frac{\theta}{2} + \frac{1}{4}\cos 2\theta\right)\Big|_a^b.$$

With the help of the calculator we find that this is ≈ 1.603946.

Next we consider the following easy problem.

EXAMPLE 6 **Area of Rose** Find the area of the five-petaled rose $r = \sin 5\theta$.

Solution We sketch the function $y = \sin 5\theta$ in rectangular coordinates in Figure 12.3.13(i) and the rose in Figure 12.3.13(ii). It should be easy to find the area; all that is needed is to integrate $\frac{1}{2}r^2$.

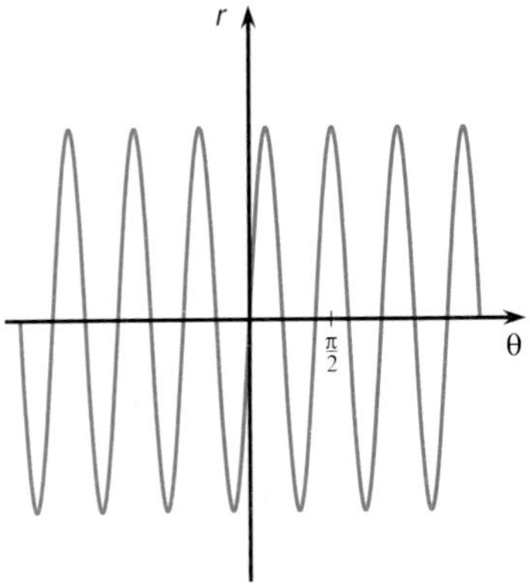

Figure 12.3.13(i)

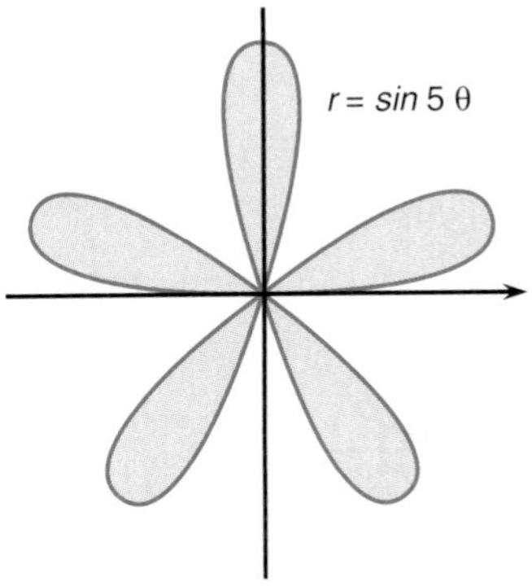

Figure 12.3.13(ii)

Since the rose goes all around the origin, we are tempted to integrate from $\theta = 0$ to $\theta = 2\pi$. This would be a blunder because, as we have already noted about roses with odd numbers of petals, the rose is traced out *twice* as θ goes from 0 to 2π. To go around the rose exactly once, it is necessary to let θ run from 0 to π. The area is

$$\frac{1}{2}\int_0^{\pi} \sin^2 5\theta\, d\theta = \frac{1}{4}\int_0^{\pi}(1 - \cos 10\theta)\, d\theta = \frac{1}{4}\left(\theta - \frac{1}{10}\sin 10\theta\right)\Big|_0^{\pi}$$

$$= \frac{\pi}{4} - 0 - \frac{1}{40}(0) + \frac{1}{40}(0) = \frac{\pi}{4}.$$

The area of the rose is $\frac{\pi}{4}$.

> Because we are allowing r to be positive or negative, it is possible that even though θ only runs from 0 to 2π, the moving point may cover part or all of a curve twice. If we want to find the area inside the curve, we have to be careful to choose the limits of integration in a way that traces out the appropriate part of the curve exactly once.

EXAMPLE 7 **Areas of Flowers** Two botanists are arguing about flowers. The colorful petals attract insects to help with pollination, so it is to the plant's advantage to have as large an area as possible exposed. One botanist argues that a bloom like a daisy with many petals arranged around its center will have a greater area than one with fewer petals of the same length. The other botanist points out that when there are more petals, each one is narrower, and concludes that the area will be the same or maybe even less. Which (if either) is right?

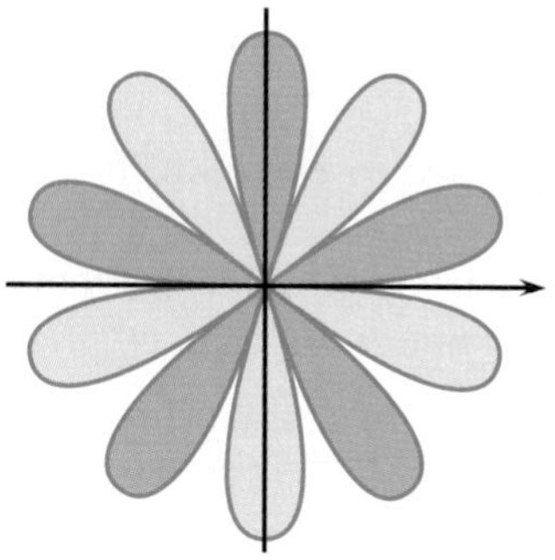

Figure 12.3.14

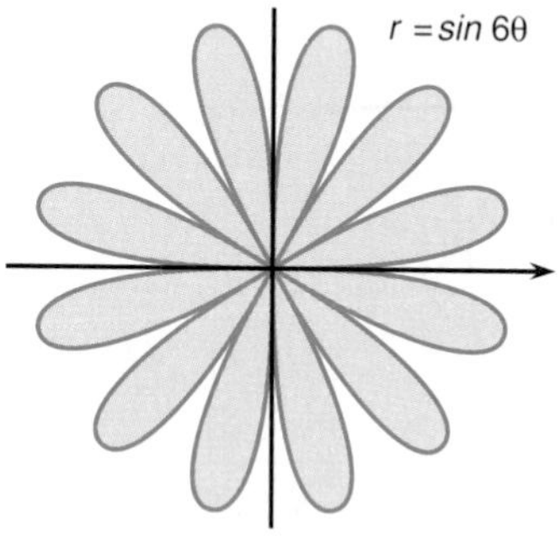

Figure 12.3.15

They decide to compare a hypothetical bloom with ten petals and one with twelve to see whether there is a difference, and they agree that the graphs we have been drawing for "roses" $r = \sin(n\theta)$ seem roughly representative of flower shapes.

In Example 6 we considered $r = \sin(5\theta)$, which has five petals, but we also know that $r = \sin(10\theta)$ has twenty. How do we get ten? The five-petaled rose $r = \sin(5\theta)$ has gaps in its petals: There is a petal with $0 \le \theta \le \frac{\pi}{5}$, but the next has $\frac{2\pi}{5} \le \theta \le \frac{3\pi}{5}$. Strictly speaking, there is a petal for $\frac{\pi}{5} \le \theta \le \frac{2\pi}{5}$, but it is "on the other side" (see Figure 12.3.13(ii)). To get a ten-petaled rose, we should fill in the gaps with another five-petaled rose, as shown in Figure 12.3.14.

The area of this bloom is twice the area we found in Example 6 for the five-petaled rose, so it is $\frac{\pi}{2}$.

In Figure 12.3.15 we see the twelve-petaled rose $r = \sin(6\theta)$. Its area is

$$\frac{1}{2}\int_0^{2\pi} \sin^2(6\theta)\,d\theta = \frac{1}{4}\int_0^{2\pi} \big(1 - \cos(12\theta)\big)\,d\theta = \left(\frac{\theta}{4} - \frac{1}{48}\sin(12\theta)\right)\Bigg|_0^{2\pi}$$

$$= \frac{2\pi}{4} - 0 - \frac{1}{48}(0) + \frac{1}{48}(0) = \frac{\pi}{2}.$$

The ten- and twelve-petaled flowers have the same area.

Exercises 12.3

I

In each of the following situations, find the area enclosed inside the given curve between the two angles specified.

1. $r = \theta;\ 0,\ \frac{\pi}{2}$
2. $r = \theta;\ \frac{\pi}{2},\ \pi$
3. $r = 6 - 2\cos\theta;\ 0,\ 2\pi$
4. $r = 4 - 4\sin\theta;\ 0,\ \pi$
5. $r = 7;\ -\frac{\pi}{4},\ \frac{\pi}{4}$
6. $r = 4\sin\theta;\ 0,\ \frac{\pi}{2}$
7. $r^2 = \sin 2\theta;\ \frac{\pi}{4},\ \frac{\pi}{2}$
8. $r = \sqrt{2 + \sin\theta};\ 0,\ \pi$
9. $r = \sin\theta;\ \pi,\ 2\pi$
10. $r = \cos^2\theta + \sin^2\theta;\ 0,\ 2\pi$

In each of the following situations, find the area enclosed inside the given curve.

11. $r = 2 - 2\sin\theta$
12. $r = 3 - 2\cos\theta$
13. $r = 4\cos\theta$
14. $r = -2\sin\theta$
15. $r = \cos\left(\theta + \frac{\pi}{4}\right)$
16. $r = 3\sin 4\theta$
17. $r^2 = \sin 2\theta$
18. $r^2 = 4\cos 2\theta$
19. $r = 2\sin\left(3\theta - \frac{\pi}{2}\right)$
20. $r^2 = \cos 3\theta$

In each of the following situations, find the area of the region enclosed in both the given curves.

21. $r = \sin\theta;\ r = \cos\theta$
22. $r = 1;\ r = 2\cos\theta$
23. $r = \cos\theta;\ r^2 = \frac{1}{2}\sin 2\theta$
24. $r = 1 + \cos\theta;\ r = 1$
25. $r = 1 + \cos\theta;\ r = 1 - \cos\theta$
26. $r = 1 + \cos\theta;\ r = \cos\theta - 1$
27. $r = 1 + \sin\theta;\ r = 1 + \cos\theta$
28. $r = 1 + \frac{1}{\sqrt{3}}\sin\theta;\ r = 1 + \cos\theta$
29. $r = 2\cos 4\theta;\ r = 1$
30. $r = 4 - \sin 5\theta;\ r = 1 + \sin 2\theta$

II

31. Find the area of the region lying inside the cardioid $r = 1 + \sin\theta$ and below and to the right of the line $y = x$.

32. Find the area of region lying inside the cardioid $r = 2 + 2\cos\theta$ and below and to the left of the line $x + y = 0$.

33. Find the area of the region contained in the "inner loop" of the limaçon $r = 1 - 2\sin\theta$.

****34.** (i) Find the area of the region contained in the "inner loop" of the limaçon $r = 1 + 5\sin\theta$.

(ii) Find the area of the region contained in the "inner loop" of the limaçon $r = 1 + a\sin\theta$, where $a > 1$.

***35.** Find the area of the region consisting of all points contained in both $r = \sin 3\theta$ and $r = \cos 3\theta$.

***36.** Find the area of the region consisting of all points contained in the circle $r = 2$ and above and to the right of the line $x + y = 2$.

***37.** (i) Find the area of the three-petaled rose $r = \cos 3\theta$.

(ii) Find the area of the seven-petaled rose $r = \cos 7\theta$.

(iii) Find the area of the n-petaled rose $r = \cos n\theta$ if n is an odd number.

***38.** (i) Find the area of the four-petaled rose $r = \cos 2\theta$.

(ii) Find the area of the eight-petaled rose $r = \cos 4\theta$.

(iii) Find the area of the $2n$-petaled rose $r = \cos n\theta$ if n is an even number.

***39.** Observe that on the rose in Example 6 the petals all have the same area. Calculate the area of one of the petals and multiply by 5 to get the total area.

***40.** The botanists from Example 7 realize that in actual flowers, adjacent petals may overlap. In light of this they propose a model that consists of the twelve-petaled rose $r = \sin(6\theta)$ superimposed on the other one $r = \cos(6\theta)$. The total surface area visible to an insect is not the sum of the two areas, since some petals are partly hidden behind others. Find the total visible area. By what percentage is it greater than the area of one twelve-petaled rose?

***41.** A leaf has the same shape as the limaçon $r = 3 + 2\cos\theta$, shown in Figure 12.3.16. If it measures 4 cm from the middle of the indentation to the opposite end, what is its surface area?

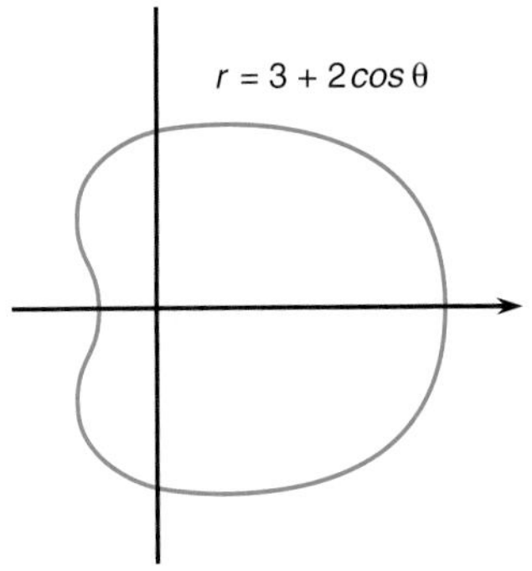

Figure 12.3.16

***42.** A leaf in the shape of a cardioid measures 2 in from the stem (at the indent) to the opposite end. Find its area.

III

Use a graphics calculator or computer to help you find the area of each of the following regions.

43. $r \le 3 + \cos\theta$, $r \le 2 - 2\sin\theta$

44. $r \le 4 + \cos\theta$, $r \le 3 - 2\cos\theta$

45. $r \le 1 - \cos\theta$, $r \le 2\sin\left(\theta + \frac{\pi}{4}\right)$

46. $r^2 \le 49\sin 2\theta$, $r \le 3 + 2\sin\theta$

12.4 Conics in Polar Coordinates

ECCENTRICITY
DIRECTRIX
FOCUS
ROTATIONS
STRAIGHT LINES

In Section 1.6 we reviewed conics from the point of view of sketching their graphs, and we returned to study them in more detail in Chapter 11. In both instances we wrote equations in the usual rectangular coordinates x and y, and we learned to recognize and classify conics from the appearance of their equations. We used various tricks to translate and/or rotate a conic until it is in one of the standard forms. In standard form a parabola has its vertex at the origin, and any other conic has its center at the origin.

In this section we consider what happens when we write the equations in polar coordinates r and θ. It turns out that to make the equation simple in polar coordinates, it is better to put a *focus* of the conic at the origin, rather than the center.

Let us consider a conic for which the origin is a focus. Let us also suppose that the directrix is the line $x = -p$, where p is some positive number. The situation is sketched in Figure 12.4.1, where we have marked a point P on the conic with polar coordinates (r, θ). We have also marked the point Q, which is the point on the directrix at the same height as P.

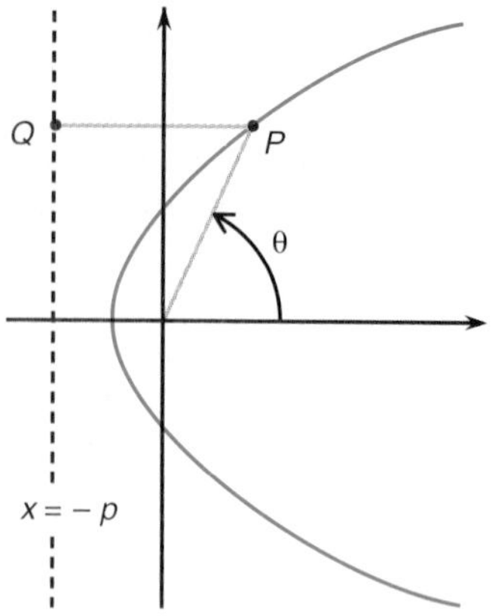

Figure 12.4.1

In Section 11.2 we saw that it is possible to describe a conic in terms of the distances from each of its points P to a **focus** and to a **directrix** (see Summary 11.2.2, p. 736). If the **eccentricity** of the conic is e, then the distance from P to the focus equals e times the distance from P to the directrix. In our situation the distance from P to the focus O is just r. The horizontal distance from the point P to the vertical axis is $r\cos\theta$, and the distance from that axis to the directrix $x = -p$ is p.

The equation of the conic is

$$r = e(p + r\cos\theta),$$

which can be simplified to

$$r = \frac{k}{1 - e\cos\theta}, \tag{12.4.1}$$

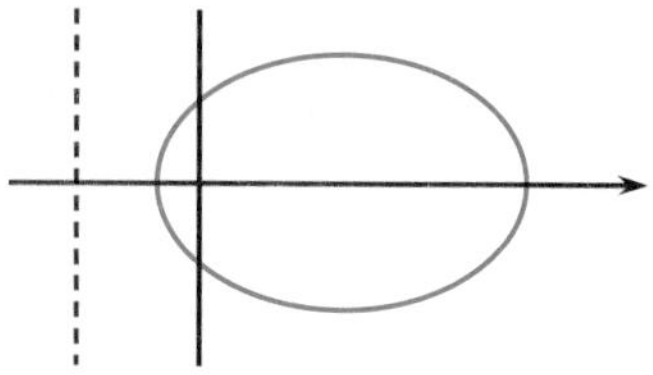

Figure 12.4.2

where k is some number (actually $k = ep$).

The interesting thing about this equation is that we do not have to know which type of conic we are dealing with. The form of the equation is the same for all types.

Observe that if $e < 1$, then $1 - e\cos\theta$ is never zero. In fact $1 - e\cos\theta \geq 1 - e$, so $r = \frac{k}{1-e\cos\theta} \leq \frac{k}{1-e}$. When $e < 1$, the conic is an ellipse, as shown in Figure 12.4.2; the curve is bounded, since r never gets larger than $\frac{k}{1-e}$.

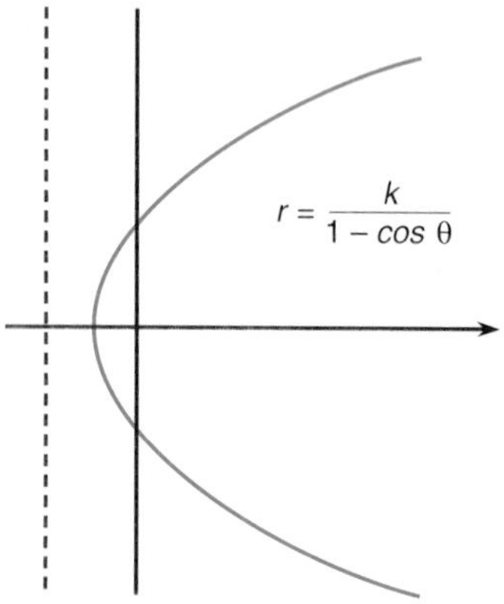

Figure 12.4.3

On the other hand, if $e = 1$, then the denominator $1 - e\cos\theta = 1 - \cos\theta$ tends to 0 as $\theta \to 0$, so $r \to \infty$. This is consistent with what we expect when $e = 1$, since the conic is a parabola (see Figure 12.4.3).

If $e > 1$, then there will be two values of θ at which the denominator is zero, and some of the time r will be positive and some of the time r will be negative. This can be seen in Figure 12.4.4; it is instructive to work out when r is positive and when it is negative, bearing in mind that r is measured from the origin, *not* from the center of the hyperbola. Notice that when the denominator tends to 0 from above or below the quotient $r = \frac{k}{1-e\cos\theta}$ tends to ∞ or $-\infty$, as we expect for points on a hyperbola.

Finally, if $e = 0$, the equation is $r = k$, which is just the equation of a circle of radius k around the origin.

EXAMPLE 1 **Rotated Ellipse** Consider the ellipse shown in Figure 12.4.5. It is rotated through an angle $\frac{\pi}{4}$ relative to the ellipse given by Equation 12.4.1, with some value of e that satisfies $0 < e < 1$. Find its equation in polar coordinates.

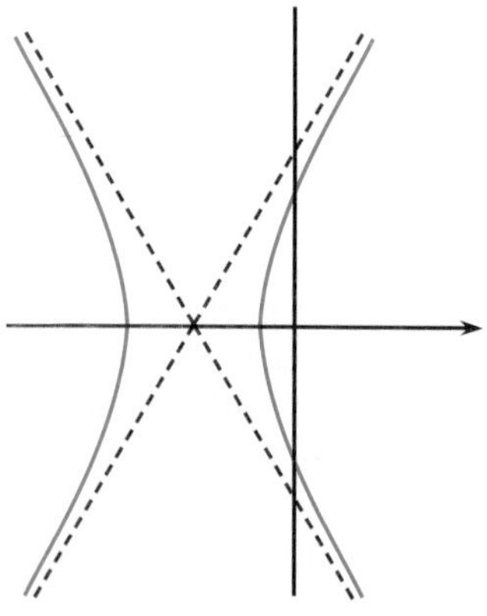

Figure 12.4.4

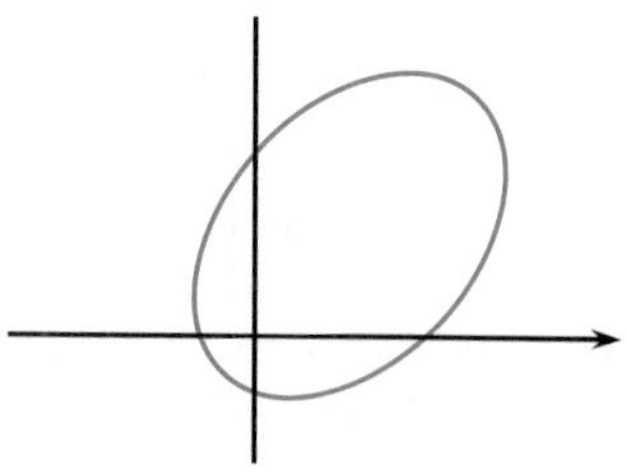

Figure 12.4.5

Solution Rather than go through the development we used to find Equation 12.4.1, it is much easier to notice that rotating a curve just amounts to "shifting" the angle θ. We should try to write the equation in the form

$$r = \frac{k}{1 - e\cos\theta'},$$

for some θ'.

By analogy with Figure 12.4.1, θ' should be the angle measured from the major axis of the ellipse (see Figure 12.4.6). In other words, θ' should be 0 when $\theta = \frac{\pi}{4}$, so $\theta' = \theta - \frac{\pi}{4}$. The equation of the ellipse is

$$r = \frac{k}{1 - e\cos(\theta - \frac{\pi}{4})}.$$

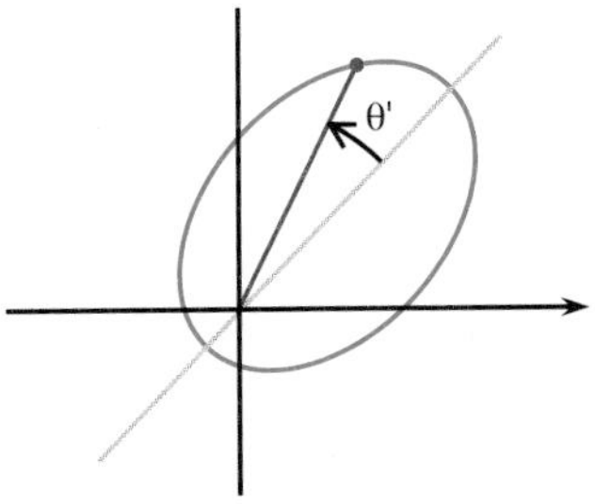

Figure 12.4.6

As Example 1 shows, the equation of a rotated conic is extremely simple, always supposing that the origin is a focus. If you have read Section 11.3, you know that this is in sharp contrast to the situation with rectangular coordinates. There the equations of rotated conics are much more complicated.

We summarize.

Summary 12.4.2

Conics in Polar Coordinates

The equation

$$r = \frac{k}{1 - e\cos\theta}$$

is the equation of a conic with eccentricity e and with the origin as a focus. If $e = 0$, it is a circle of radius k. Otherwise, we let $p = \frac{k}{e}$; the line $x = -p$ is a directrix.

The equation

$$r = \frac{k}{1 - e\cos(\theta - \theta_0)}$$

is the equation of a conic with eccentricity e and with the origin as a focus. This conic is obtained from the previous one by rotating counterclockwise through an angle of θ_0.

Several examples are sketched in Figure 12.4.7.

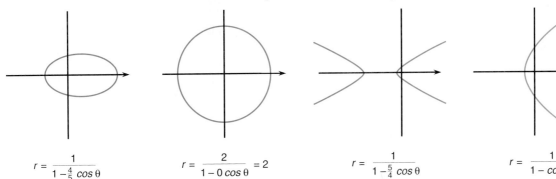

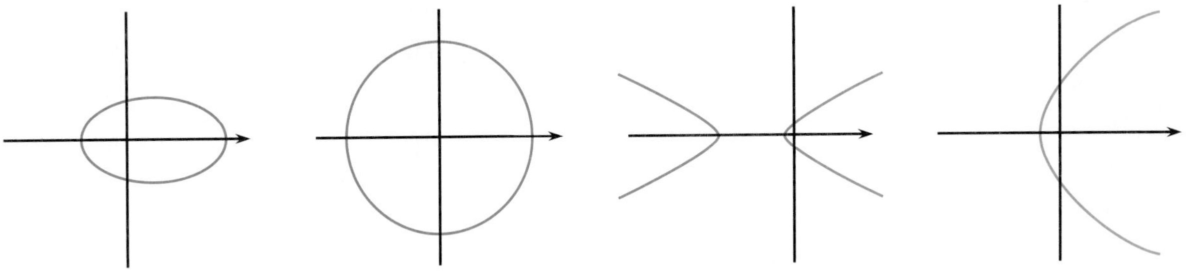

$r = \dfrac{1}{1 - \frac{4}{5}\cos\theta}$ $\quad r = \dfrac{2}{1 - 0\cos\theta} = 2$ $\quad r = \dfrac{1}{1 - \frac{5}{4}\cos\theta}$ $\quad r = \dfrac{1}{1 - \cos\theta}$

Figure 12.4.7

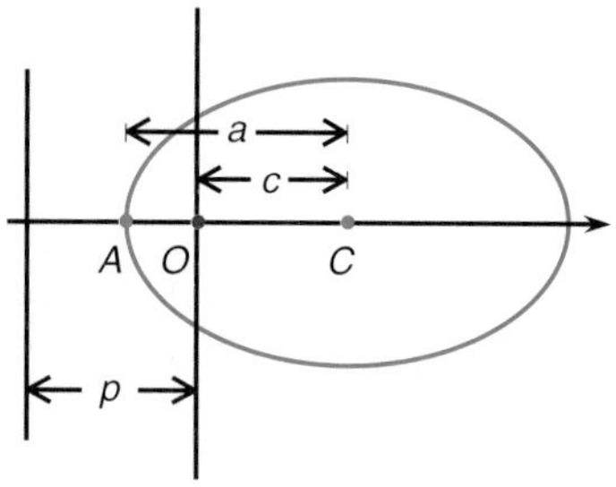

Figure 12.4.8

Notice that if the denominator is never zero, then the conic is either a circle or an ellipse. Circles are easily recognized because the denominator is constant. To distinguish between parabolas and hyperbolas, one approach is to see that for a parabola the denominator is zero at exactly one point, whereas for a hyperbola there are two points at which the denominator is zero (corresponding to the angles of the asymptotes).

EXAMPLE 2 Consider the ellipse whose eccentricity is e and whose semi-major axis is a. It is illustrated in Figure 12.4.8, where we have also plotted the focus at the origin and marked the point C which is the center of the ellipse.

We know (see Summary 11.2.3, p. 738) that the distance $c = |OC|$ between the focus and the center satisfies $c = ae$.

We can use a graphics calculator to sketch the conic $r = \frac{k}{1-e\cos\theta}$ for various values of k and e.

TI-85	SHARP EL9300
MODE ▼ ▼ ▼ ▼ ► (Pol) ENTER EXIT GRAPH F1 ($r(\theta)$ =) 1 ÷ (1 − .5 × COS F1 (θ)) ▼ 1 ÷ (1 − .8 × COS F1 (θ)) ▼ 1 ÷ (1 − .3 × COS F1 (θ)) ▼ 3 ÷ (1 − 2 × COS F1 (θ)) ▼ 5 ÷ (1 − 3 × COS F1 (θ)) M2 (RANGE) 0 ▼ 2 π ▼ ▼ (−) 5 5 ▼ ▼ (−) 5 ▼ 5 F5 (GRAPH)	SET UP E 2 [graph key] 1 ÷ (1 − .5 cos x/θ/T) 2ndF ▼ 1 ÷ (1 − .8 cos x/θ/T) 2ndF ▼ 1 ÷ (1 − .3 cos x/θ/T) 2ndF ▼ 3 ÷ (1 − 2 cos x/θ/T) 2ndF ▼ 5 ÷ (1 − 3 cos x/θ/T) RANGE −5 ENTER 5 ENTER 1 ENTER −5 ENTER 5 ENTER 1 ENTER 0 ENTER 2 π ENTER .02 ENTER [graph key]
CASIO f_x-7700GB	HP 48SX
MODE MODE − (POL) Range −5 EXE 5 EXE 1 EXE −5 EXE 5 EXE 1 EXE 0 EXE 2 π EXE .02 EXE Graph 1 ÷ (1 − .5 × cos θ) EXE Graph 1 ÷ (1 − .8 × cos θ) EXE Graph 1 ÷ (1 − .3 × cos θ) EXE Graph 3 ÷ (1 − 2 × cos θ) EXE Graph 5 ÷ (1 − 3 × cos θ) EXE	↰ RAD ↰ PLOT ′ R = 1 ÷ () 1 − .5 × COS α ↱ F NEW P O L ENTER PTYPE POLAR PLOTR 5 +/− ENTER 5 XRNG 5 +/− ENTER 5 YRNG ′ α ↱ F INDEP DRAW ATTN ′ R = 1 ÷ () 1 − .8 × COS α ↱ F ↱ PLOT ↰ DRAW DRAW ATTN ′ R = 1 ÷ () 1 − .3 × COS α ↱ F ↱ PLOT ↰ DRAW DRAW ATTN ′ R = 3 ÷ () 1 − 2 × COS α ↱ F ↱ PLOT ↰ DRAW DRAW ATTN ′ R = 5 ÷ () 1 − 3 × COS α ↱ F ↱ PLOT ↰ DRAW DRAW

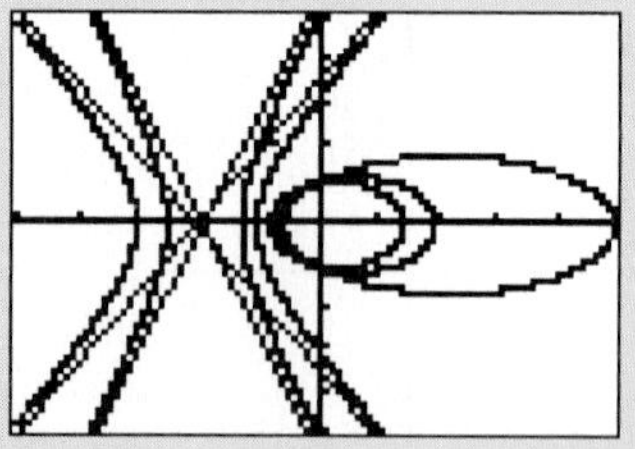

Identify each conic. Then try graphing the rotated conic $r = \frac{k}{1-e\cos(\theta-\theta_0)}$ for various values of k, e, and θ_0.

Let A be the point on the ellipse that is closest to the directrix, as shown in the illustration. From the picture we see that the distance $|AO|$ from A to the focus O is $|AO| = a - c = a - ae = a(1 - e)$. Writing p for the distance from O to the directrix, we use the defining property of conics to see that $|AO|$ is e times the distance from A to the directrix. Since this distance is $p - |AO|$, we have that $|AO| = e(p - |AO|)$, or $|AO|(1 + e) = ep$. Substituting $|AO| = a(1 - e)$ into this, we find that $a(1 - e^2) = ep$, so

$$p = a\frac{1 - e^2}{e}.$$

In writing Equation 12.4.1 we let $k = ep$, so in this case we see that $k = ep = a(1 - e^2)$. The equation of the ellipse is

$$r = \frac{a(1 - e^2)}{1 - e\cos\theta}.$$

It is also possible to use this idea in reverse.

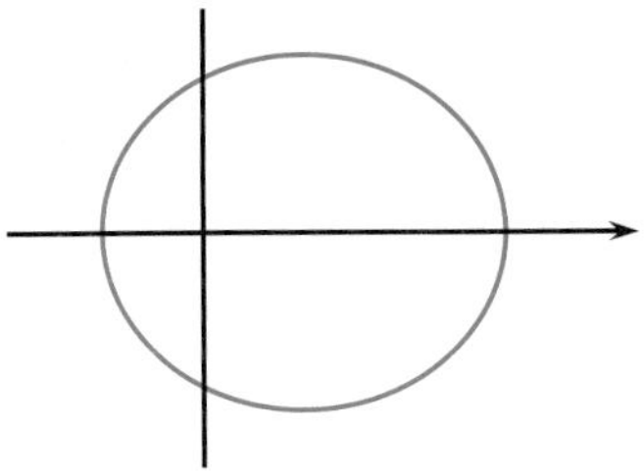

Figure 12.4.9

EXAMPLE 3 What are the semi-major and semi-minor axes of the ellipse

$$r = \frac{6}{2 - \cos\theta}?$$

Solution We begin by dividing numerator and denominator by 2 to write the equation in the form

$$r = \frac{3}{1 - \frac{1}{2}\cos\theta}.$$

From this we see that the eccentricity is $e = \frac{1}{2}$. As we saw in Example 2, $ep = a(1 - e^2)$, and in this case, knowing that $ep = 3$, we find that $a = 3/(1 - e^2) = 3/\left(1 - \frac{1}{4}\right) = 4$.

From Summary 11.2.3 we know that $c = ea = 2$. On the other hand, c also equals $\sqrt{a^2 - b^2}$, so $a^2 - b^2 = 2^2 = 4$. Since $a^2 = 16$, we find that $b^2 = 12$ and $b = 2\sqrt{3}$. The ellipse is illustrated in Figure 12.4.9.

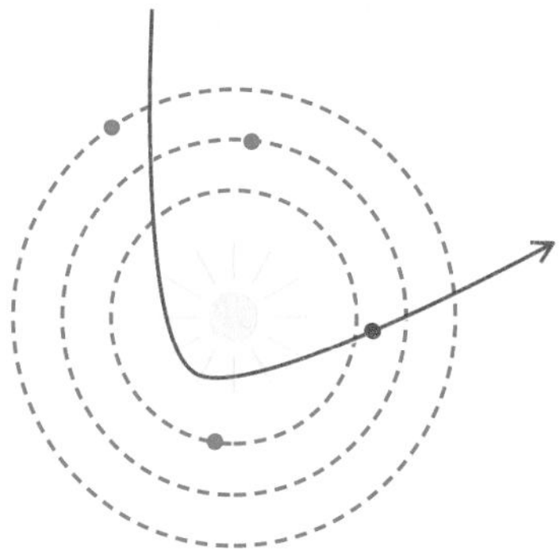

Figure 12.4.10

Much of the importance of describing conics in polar coordinates comes from the fact that the orbits of the planets are ellipses, with the sun at one of the foci. In fact, the orbit of a comet or any other object moving under the influence of the sun's gravity is also a conic, with the sun at a focus. If the orbit is an ellipse, the object is in a closed orbit, and it goes around and around the sun. If the orbit is a parabola or a hyperbola, the object will approach the sun, veer past, and fly off into space, never to return (see Figure 12.4.10).

Circles and Lines

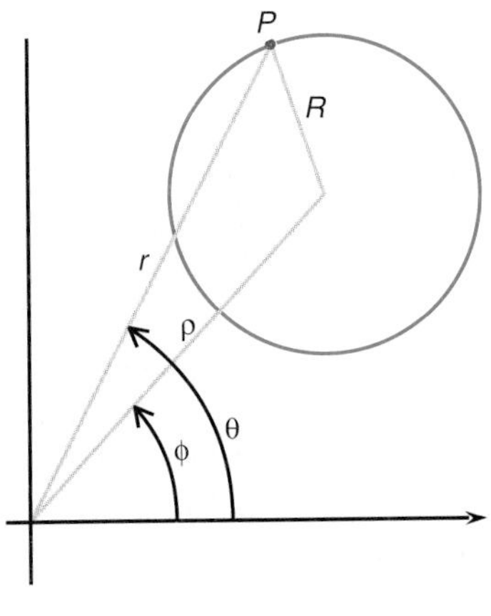

Figure 12.4.11

Finally, we consider the equations of lines and of circles whose centers are not necessarily at the origin.

Consider the circle shown in Figure 12.4.11. Its center is the point whose polar coordinates are (ρ, ϕ), and its radius is R. Also shown is a point P on the circle, whose polar coordinates are (r, θ). We apply the Law of Cosines (Theorem 1.7.10,

p. 50) to the triangle made by the origin, the point P, and the center of the circle. Noting that the angle in this triangle at the origin is $\theta - \phi$, we find that $R^2 = r^2 + \rho^2 - 2r\rho\cos(\theta - \phi)$, or

$$r^2 - 2r\rho\cos(\theta - \phi) + \rho^2 - R^2 = 0. \tag{12.4.3}$$

This is the equation of the circle. It is not very convenient to work with, however. For one thing, it does not express r as a function of θ. Instead, it gives a quadratic equation that r must satisfy. From the picture we see that for one value of θ there may be two different points on the circle, and they correspond to the two roots of the polynomial.

In special cases, of course, it will be simpler. For instance, if the center of the circle is at the origin, then $\rho = 0$ and the equation is just $r^2 = R^2$, which is just the ordinary circle $r = R$ of radius r about the origin.

Another special case occurs when $R = \rho$. This means that the origin is one of the points on the circle. The equation simplifies to $r^2 - 2rR\cos(\theta - \phi) = 0$, or

$$r = 2R\cos(\theta - \phi). \tag{12.4.4}$$

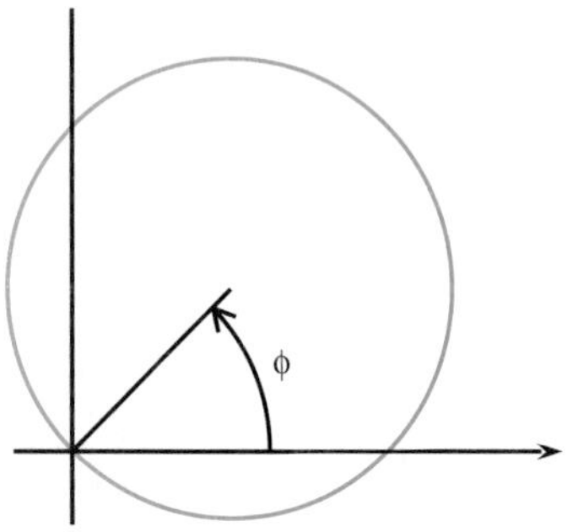

Figure 12.4.12

See Figure 12.4.12.

Next we consider straight lines.

We know that a line through the origin can be described by $\theta = \theta_0$, $r \in (-\infty, \infty)$, for some fixed θ_0.

The equation of the line $x = a$ is just $r\cos\theta = a$, or $r = a\sec\theta$.

Suppose we consider the line shown in Figure 12.4.13. The distance from the line to the origin is d. Suppose we draw the line segment from the origin that meets the line in a right angle, as shown, and suppose the polar coordinates of the point Q where this segment meets the line are (d, ϕ).

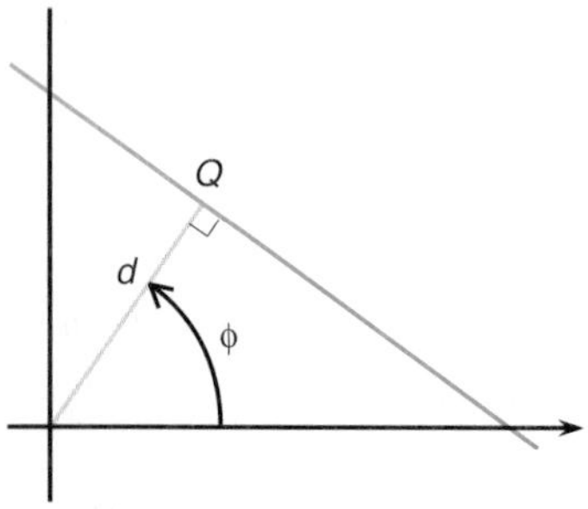

Figure 12.4.13

We could try to figure out the equation of the line, but it is much easier to notice that it can be obtained by rotating the line $x = d$ through an angle of ϕ. This means that the equation of the line is

$$r = d\sec(\theta - \phi). \tag{12.4.5}$$

EXAMPLE 4 Find the equation of the line $x + y = 2$ in polar coordinates.

Solution One approach is to write $x = r\cos\theta$, $y = r\sin\theta$, so the equation is $r(\cos\theta + \sin\theta) = 2$, that is,

$$r = \frac{2}{\cos\theta + \sin\theta}.$$

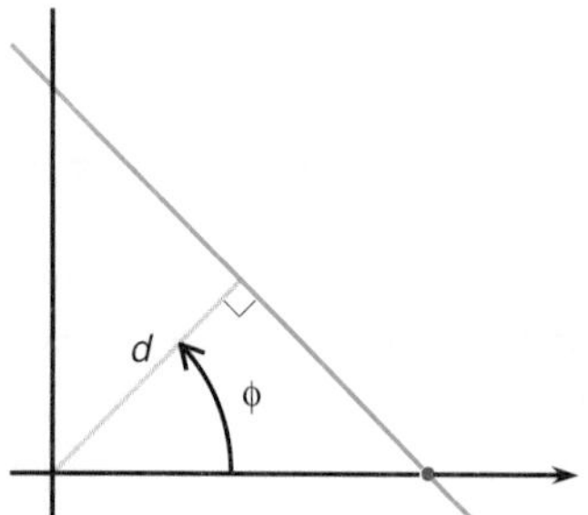

Figure 12.4.14

It is also possible to use the geometry of the situation. We know that the slope of this line is $m = -1$ (cf. Remark 1.5.10(ii)), so it makes an angle of $\frac{3\pi}{4}$ with the horizontal axis. The angle ϕ between the horizontal axis and the segment perpendicular to the line is $\frac{\pi}{4}$. See Figure 12.4.14.

From Equation 12.4.5 we know that the equation of the line is $r = d\sec\left(\theta - \frac{\pi}{4}\right)$, where d is the distance from the line to the origin. We could try to find the distance from the origin to the line. (Notice that what this means is the *smallest* distance from the origin to the line, so it requires finding the point on the line that is closest to the origin.) On the other hand, knowing that the equation is $r = d\sec\left(\theta - \frac{\pi}{4}\right)$, we could pick some convenient angle and solve.

Suppose we let $\theta = 0$. This means that we are considering points on the horizontal axis, that is, points (x, y) with $y = 0$. The only such point on the line $x + y = 2$ is $(x, y) = (2, 0)$. The polar coordinates of this point are $(r, \theta) = (2, 0)$, and they should satisfy $r = d\sec\left(\theta - \frac{\pi}{4}\right)$, that is, $2 = d\sec\left(-\frac{\pi}{4}\right) = d\sqrt{2}$. We conclude that $d = \sqrt{2}$.

The equation of the line is $r = \sqrt{2}\sec\left(\theta - \frac{\pi}{4}\right)$. It is not immediately obvious that this is the same as the other formula $r = \frac{2}{\cos\theta + \sin\theta} \cdots$

Exercises 12.4

I

Identify the type of each of the following curves and sketch its graph.

1. $r = 1/(1 - \cos\theta)$
2. $r = 1/\left(2 - 2\cos\left(\theta - \frac{\pi}{3}\right)\right)$
3. $r = 1/(2 - \cos\theta)$
4. $r = 4/(3 - \cos\theta)$
5. $r = 2/(2 + \cos\theta)$
6. $r = 5/(3 - \sin\theta)$
7. $r = 1/(1 - 2\cos\theta)$
8. $r = 2/(1 + \sqrt{2}\cos\theta)$
9. $r = 3/\left(1 + \sin\left(\theta + \frac{\pi}{2}\right)\right)$
10. $r = 1/\left(2 - \cos(\theta - \pi)\right)$
11. $r = \cos\theta$
12. $r = 3\sec\left(\theta + \frac{\pi}{3}\right)$
13. $r = 1/(1 - 2\sin\theta)$
14. $r(2 - 3\cos\theta) = 2$
15. $r = 4\csc\theta$
16. $r = -2\sec(\pi - \theta)$
17. $r = 6$
18. $r = 6\sin\left(\theta - \frac{\pi}{6}\right)$
19. $r\sin\left(\theta + \frac{\pi}{4}\right) = 2$
20. $r/\left(3\sin\left(\theta - \frac{\pi}{3}\right)\right) = 5$

II

*21. Find the equation of the conic whose eccentricity is e, that has a focus at the origin, and whose directrix is the line $x = p$, with $p > 0$.

*22. Find the equation of the conic whose eccentricity is e, that has a focus at the origin, and whose directrix is the line $y = -p$, with $p > 0$.

*23. Consider the hyperbola $x^2 - y^2 = 1$. Translate the axes until its right-hand focus is at the origin, and express the equation of the resulting conic in polar coordinates, in the form of Equation 12.4.1.

*24. Consider the hyperbola $y^2 - x^2 = 4$. Translate the axes until its upper focus is at the origin, and express the equation of the resulting conic in polar coordinates, in the form of the second equation in Summary 12.4.2.

*25. What is the (smaller) angle between the asymptotes of the hyperbola

$$r = \frac{1}{1 - 2\cos\theta}?$$

*26. If $e > 1$, what is the angle between the asymptotes of the hyperbola

$$r = \frac{1}{1 - e\cos\theta}?$$

*27. Find the equation of the line $x - y = 3$ in polar coordinates.

*28. Find the equation of the line $x - \sqrt{3}y = 4$ in polar coordinates.

*29. Identify and sketch the conic

$$r = \frac{1}{2 + \sin\theta + \cos\theta}.$$

*30. Consider the hyperbola whose eccentricity is $e > 1$, whose transverse axis is horizontal, and whose right-hand focus is at the origin. If the minimum distance between the two branches is $2a$, what is the equation of the hyperbola in polar coordinates?

*31. Find the *other* focus and the *other* directrix of the ellipse $r = \frac{1}{2 - \cos\theta}$.

**32. When we considered the equation of the line $x = a$, with $a \neq 0$, we wrote it in polar coordinates as $r\cos\theta = a$. Then we rewrote it as $r = a\sec\theta$. This involved dividing by $\cos\theta$, and we cannot do this when $\cos\theta = 0$. Explain why this does not matter, that is, that the set of points that satisfy $r\cos\theta = a$ is exactly the same as the set of points that satisfy $r = a\sec\theta$.

***33.** In Example 4 we remarked that it was not necessary to find the point on the line $x + y = 2$ that is closest to the origin in order to find the equation in polar coordinates. Nonetheless, it is possible to proceed this way. Find the point on the line $x + y = 2$ that is closest to the origin, and use it to find the distance from the origin to the line.

34. What point on the conic $r = \frac{2}{2-\cos(\theta+\pi/2)}$ is closest to the origin?

35. What are the points of intersection of the conics $r = \frac{2}{2+\cos\theta}$ and $r = \frac{2}{2+\sin\theta}$? Draw a picture.

36. What are the points of intersection of the conics $r = \frac{2}{2+3\cos\theta}$ and $r = \frac{1}{2+\cos\theta}$? Draw a picture.

***37.** Find the area of the ellipse $r = \frac{2}{2-\sin\theta}$. (It may be easier not to integrate.)

***38.** If $0 < e < 1$, find the area of the ellipse $r = \frac{k}{1+e\cos\theta}$.

***39.** (i) Find the slope of the tangent to $r = \frac{2}{2-\cos\theta}$ at the point (r, θ).

(ii) Find the rectangular coordinates of the highest point on the ellipse.

***40.** (i) If $0 < e < 1$, find the slope of the tangent to $r = \frac{k}{1+e\cos\theta}$ at the point (r, θ).

(ii) Find the rectangular coordinates of the highest point on the ellipse.

III

***41.** If you have a machine that will plot in polar coordinates, sketch $r = \frac{1}{1-e\cos\theta}$, with $e = 0, \frac{1}{4}, \frac{1}{2}, \frac{3}{4}, \frac{7}{8}, 1, 1.1, 1.5, 2, 4, 10$. Plot as many of these conics as possible on the same picture. In each case except $e = 0$ and $e = 1$, locate the second focus.

***42.** Sketch the graph of $y = \frac{1}{1-2\cos\theta}$ in rectangular coordinates, and then sketch the corresponding hyperbola $r = \frac{1}{1-2\cos\theta}$ in polar coordinates. Identify the parts of the polar graph where r is positive and negative.

POINT TO PONDER

Strictly speaking, since we used the directrix in deriving Equation 12.4.1, we have in effect assumed that the conic is not a circle, since a circle does not have a directrix (cf. the discussion after Summary 11.2.2). However, if we take Equation 12.4.1 and let $e = 0$, we obtain the equation $r = \frac{k}{1}$, or $r = k$. This is the equation of a circle of radius k around the origin. In Section 11.2 we remarked that it is possible to think of a circle as what happens when an ellipse "degenerates" by letting the two foci run together. In this sense, $r = k$ is the equation of a circle "with focus at the origin."

However, there is a fly in the ointment, which is that in deriving Equation 12.4.1 we let $k = ep$. This seems to suggest that when $e = 0$, we should have $k = 0$, and the equation becomes $r = 0$, the equation of a single point. Fortunately, as we saw in the discussion after Summary 11.2.2, if we think of an ellipse whose foci run together, then its directrix moves out to infinity. So although $e \to 0$, we also have that $p \to \infty$.

In Example 2 we saw that $k = ep = a(1 - e^2)$. So if we consider an ellipse with focus at the origin and semi-major axis a and let its eccentricity e tend to 0, then we

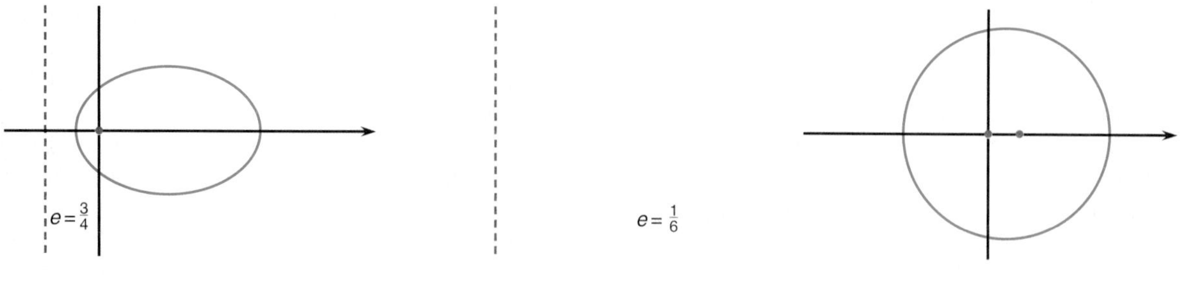

Figure 12.4.15

find that $k = a(1 - e^2) \to a$. Also the semi-minor axis b satisfies $a^2 - b^2 = c^2 = e^2a^2 \to 0$, which means that $b \to a$. From this we see that the two axes of the ellipse approach the same length a, and the ellipse approaches the round shape of a circle.

So as the ellipse degenerates, the graph deforms into a circle of radius a, and its equation $r = a(1 - e^2)/(1 - e\cos\theta)$ approaches the equation $r = \frac{a}{1}$, that is, $r = a$.

This argument shows that it really is legitimate to include the circle as one of the cases of Equation 12.4.1. See Figure 12.4.15.

12.5 Parametric Equations in Rectangular and Polar Coordinates

PARAMETER
MOVING POINT

In Section 12.4 we mentioned the fact that planets and comets move in the solar system along orbits that are conics. Most of the planets move on orbits that are nearly circular (for instance, the eccentricity of the earth's orbit is about 0.017, which is quite close to 0). The only orbits that have eccentricities greater than 0.1 are those of Mercury, whose orbit has eccentricity ≈ 0.206, and Pluto, whose orbit has eccentricity ≈ 0.248. There is a theory that Pluto may have had an orbit that was closer to circular but that it nearly collided with some large object and the gravitational field of the other object distorted its orbit.

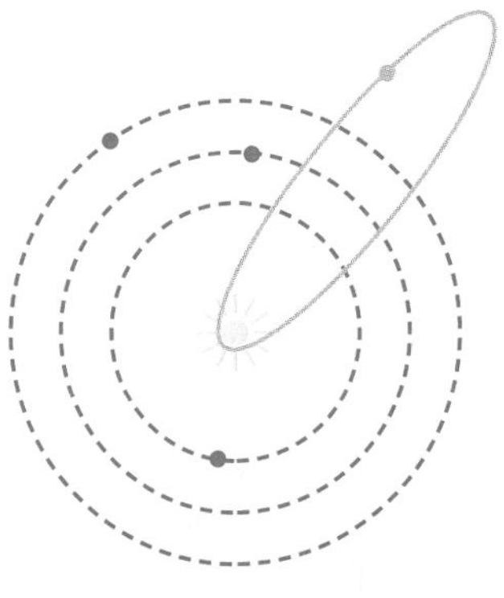

Figure 12.5.1

Comets often have orbits that are highly eccentric. For instance, the eccentricity of the orbit of Halley's Comet is about 0.97. We can imagine the comet as a point moving around its elliptical orbit (see Figure 12.5.1).

This suggests that its position can be determined at any time t, and we expect to be able to write its coordinates (x, y) as functions $x = x(t)$ and $y = y(t)$ of t. Evaluating these functions at any particular value of t will give the coordinates of the moving point at the time t. In this situation it is customary to refer to the variable t as a **parameter** and to $x = x(t)$ and $y = y(t)$ as **parametric equations** for the curve (or for the moving object).

Of course, in principle it should also be possible to express parametric equations in polar coordinates instead of rectangular coordinates. We should be able to write $\theta = \theta(t)$ and $r = r(t)$.

Naturally, this sort of idea does not apply only to astronomical motions. In Figure 12.5.2 we see a train moving along a straight track; we can certainly describe its position as a function of time.

Figure 12.5.2

In Figure 12.5.3 there is an illustration of the path of a ball that is thrown upward at an angle of 45° above the horizontal. It follows a parabolic arc until it lands, and we suppose that the maximum height it reaches is 16 ft. Here too it is possible to express the position of the moving object as a function of time.

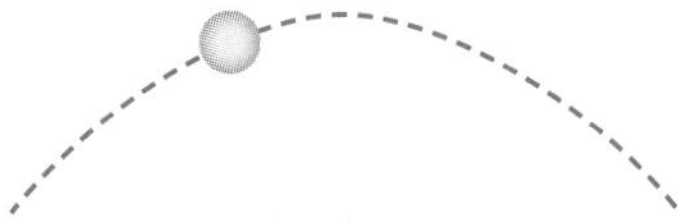

Figure 12.5.3

If a fast meteor approaches near the sun, the sun's gravity will pull it from the straight line it would otherwise follow, and its course will bend around the sun. It will fly off in another direction, which depends on how close it passes to the sun, and the

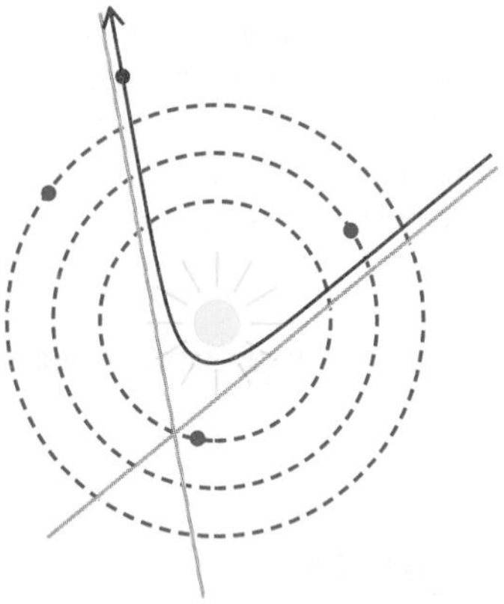

Figure 12.5.4

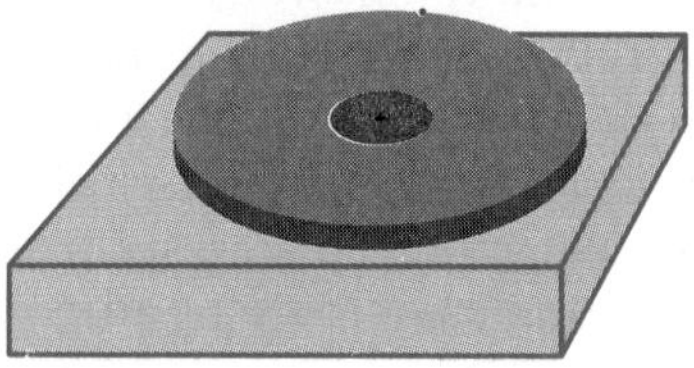

Figure 12.5.5

path it follows will be a hyperbola, as shown in Figure 12.5.4. The asymptotes of the hyperbola will be the straight line along the direction that the meteor would have followed if the sun had not been there to deflect it and the line along the direction toward which the meteor is deflected. The farther the meteor is from the sun, the less it will be affected by the sun's gravity, and this is why the parts of the hyperbola far away from the focus are very close to straight lines (the asymptotes). Here too we would like to express the position of the meteor as a function of time, giving parametric equations for the hyperbola.

Finally, we can imagine a speck of dust on the rim of a twelve-inch LP as it turns at $33\frac{1}{3}$ rpm. The speck of dust travels along a circle of radius 6, and it ought to be easy to express its position in terms of the time t (see Figure 12.5.5).

In fact it is fairly easy to find parametric equations that describe the position of a point moving along each of the curves mentioned above. The disconcerting thing is that there are many *choices* of formulas. Of course, this reflects the fact that an object may move along the same path in many different ways; even a train moving along a straight track can move at different speeds, can accelerate or decelerate, and for that matter can move in either of two directions. So to make practical use of formulas that express the position of a moving point as a function of t, it is essential to find the right formulas for a particular motion.

However, we begin by trying to find even one possible formula for each of the curves we have discussed.

EXAMPLE 1 **Moving Train** Suppose a train is moving along a straight track at a constant speed of 30 mph. For convenience let us think of the track as the real axis and let $t = 0$ at the moment when the train passes the origin O. See Figure 12.5.6.

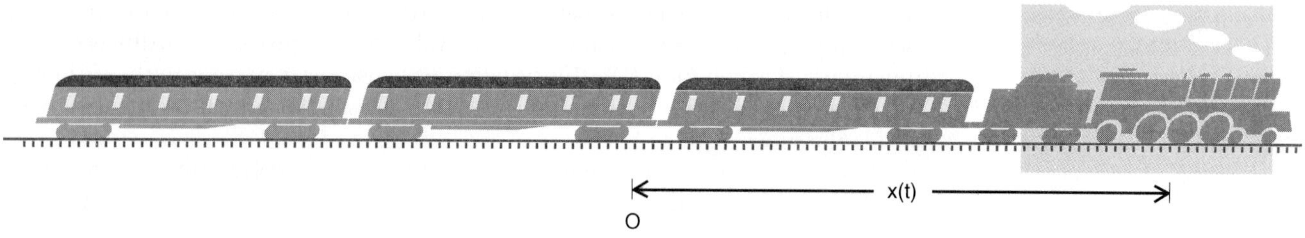

Figure 12.5.6

If we measure the time t in hours, and express the position of the train as $x = x(t)$, since the train moves at a constant speed of 30 mph, we find that $x'(t) = 30$. So $x(t)$ is an antiderivative of the constant function 30, so $x(t) = 30t + c$, for some constant c. Since the train is at the origin when $t = 0$, we find that $0 = 30(0) + c = c$, so $c = 0$ and $x(t) = 30t$.

The way we usually draw our axes, the train will be moving from left to right, and of course there is the other possibility that it moves from right to left. In that case the formula would be $x = x(t) = -30t$.

Notice that because this train is moving along a straight line, it is possible to express its position with a single coordinate x instead of the two needed for positions in the plane (either (x, y) or (r, θ)).

EXAMPLE 2 **Flying Ball** Next we consider the ball that has been thrown at an angle of 45° above the horizontal. Let us measure time t in seconds and assume that $t = 0$ at the moment when the ball is thrown. Let us also assume that its starting point is the origin in the plane.

In Remark 4.9.2 (p. 317) we observed that in this situation the ball will move so that its horizontal velocity is constant and its vertical height in feet will be given by $y = y(t) = -16t^2 + Bt + C$. Here B is the vertical speed with which the ball is thrown and C is the initial height, which we have assumed is 0, so $y(t) = -16t^2 + Bt$.

The maximum height will be reached when $y'(t) = 0$, that is, when $-32t + B = 0$, that is, when $t = \frac{B}{32}$. The height at this moment, which we know should equal 16, is $y\left(\frac{B}{32}\right) = -16\left(\frac{B}{32}\right)^2 + B\left(\frac{B}{32}\right) = -\frac{1}{64}B^2 + \frac{1}{32}B^2 = \frac{1}{64}B^2$. Solving to find when this equals 16, we see that $B = 32$. This tells us that the y-coordinate is given by $y = y(t) = -16t^2 + 32t$. In particular, the vertical velocity is $y'(t) = -32t + 32$, and at the moment when the ball is thrown, $y'(0) = 32$.

For the ball to move at a 45° angle above the horizontal, its horizontal and vertical velocities must be equal at $t = 0$, so its horizontal velocity at $t = 0$ must also equal 32 ft/sec. Since the horizontal velocity is constant, we know that it is 32 ft/sec at every time t, so the horizontal position is given by $x = x(t) = 32t$. See Figure 12.5.7.

The position of the ball is given by the parametric equations

$$x = x(t) = 32t \qquad \text{and} \qquad y = y(t) = -16t^2 + 32t.$$

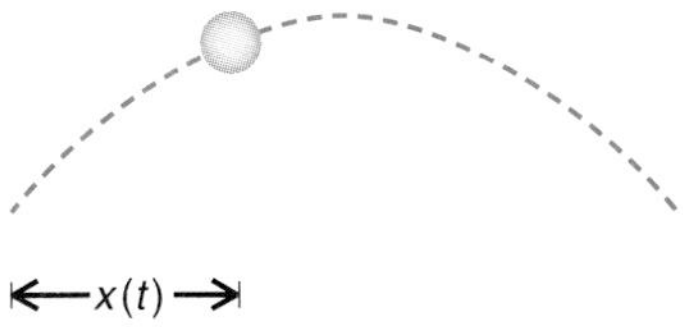

Figure 12.5.7

EXAMPLE 3 **Moving Speck on LP** Let us try to find parametric equations for a point moving around the circle $x^2 + y^2 = 36$. It is easy to check that the point $(x, y) = (6\cos u, 6\sin u)$ is on the circle, and this gives us parametric equations for the circle with the parameter u. We know that $(6\cos u, 6\sin u)$ is the point on the circle whose polar angle is $\theta = u$, as shown in Figure 12.5.8.

The complication comes when we want to make sure that the point moves around the circle at $33\frac{1}{3}$ rpm. Suppose we measure time t in minutes. We know that $(6\cos u, 6\sin u)$ goes around the circle exactly once each time u increases by 2π. We want the point to go around the circle $33\frac{1}{3}$ times each minute, so it should go around once in exactly $\frac{1}{33\frac{1}{3}}$ minute. This would be the case if u goes from 0 to 2π when t goes from 0 to $\frac{1}{33\frac{1}{3}}$, and we can arrange for this to happen by letting

$$u = 2\pi\left(33\frac{1}{3}t\right).$$

When $t = 0$, we see that $u = 0$, and when $t = \frac{1}{33\frac{1}{3}}$, we find that $u = 2\pi$.

If we let $x = 6\cos\left(2\pi\left(33\frac{1}{3}t\right)\right)$, $y = 6\sin\left(2\pi\left(33\frac{1}{3}t\right)\right)$, then the point (x, y) will move around the circle of radius 6 at a speed of $33\frac{1}{3}$ rpm.

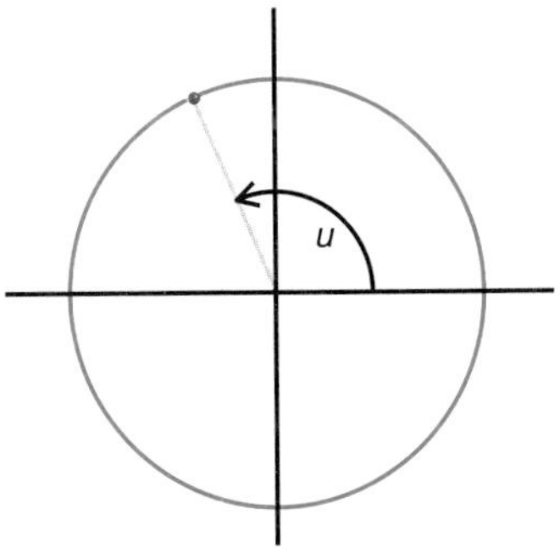

Figure 12.5.8

However, if we look at Figure 12.5.8 and imagine that it is the turntable seen from above, then as the angle u increases, the point moves counterclockwise. In fact, LPs turn clockwise, so we should adjust the parametric equations to make the point move in the opposite direction. The parametric equations we want for the point moving clockwise around the circle of radius 6 at $33\frac{1}{3}$ rpm are

$$x = 6\cos\left(-66\frac{2}{3}\pi t\right) = 6\cos\left(66\frac{2}{3}\pi t\right)$$

and

$$y = 6\sin\left(-66\frac{2}{3}\pi t\right) = -6\sin\left(66\frac{2}{3}\pi t\right).$$

EXAMPLE 4 Parametrizing an Ellipse If we wanted parametric equations for the circle $x^2 + y^2 = 4$ of radius 2 instead, we could let $x = 2\cos t$ and $y = 2\sin t$. This moving point would trace out the circle exactly once as the time t changes from 0 to 2π. It would move in the counterclockwise direction.

Now suppose we want to find parametric equations for a point moving around the ellipse $\frac{x^2}{a^2} + \frac{y^2}{b^2} = 1$. We can check that the point $(a\cos t, b\sin t)$ is on the ellipse. It is a little less clear exactly where it is. Unlike the case of the circle, it is not necessarily true that $(a\cos t, b\sin t)$ has polar angle t. When $t = 0$, for instance, we find that $(a\cos t, b\sin t) = (a, 0)$ does have polar angle 0. When $t = \frac{\pi}{2}$, then the point is $(0, b)$, which also has the expected angle, that is, $\frac{\pi}{2}$.

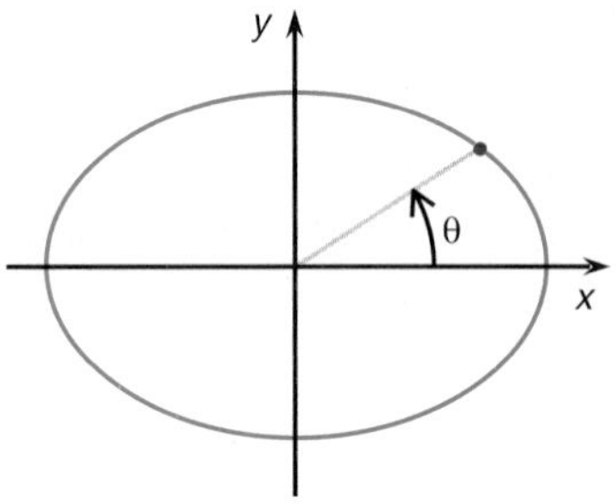

Figure 12.5.9

However, suppose we try $t = \frac{\pi}{4}$ and look at the corresponding point $\left(a\cos\frac{\pi}{4}, b\sin\frac{\pi}{4}\right) = \left(\frac{a}{\sqrt{2}}, \frac{b}{\sqrt{2}}\right)$, as shown in Figure 12.5.9. The polar angle θ cannot equal $\frac{\pi}{4}$ unless $a = b$.

On the other hand, $\tan\theta = \frac{b\sin t}{a\cos t} = \frac{b}{a}\tan t$ has the same sign as $\tan t$, and from this we conclude that as t moves from 0 to $\frac{\pi}{2}$, the angle θ also moves from 0 to $\frac{\pi}{2}$. Similar remarks can be made about the intervals $\left[\frac{\pi}{2}, \pi\right]$, $\left[\pi, \frac{3\pi}{2}\right]$, and $\left[\frac{3\pi}{2}, 2\pi\right]$. This means that the point $(a\cos t, b\sin t)$ moves around the ellipse in the counterclockwise direction, making one complete circuit as t ranges from 0 to 2π.

We find that $(x, y) = (a\cos t, b\sin t)$ are parametric equations for the ellipse $\frac{x^2}{a^2} + \frac{y^2}{b^2} = 1$, but the point does not move at a constant speed. In fact it moves more slowly out near the ends of the major axis than it does near the ends of the minor axis.

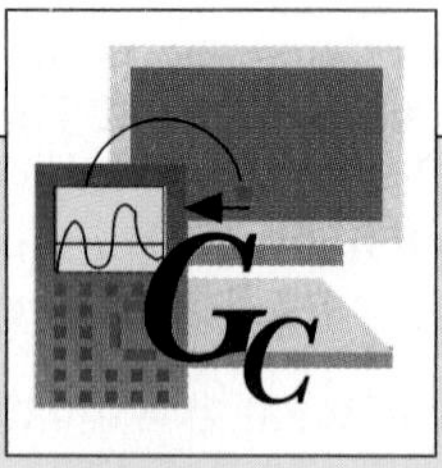

We can use a graphics calculator to sketch the ellipse $\frac{x^2}{9} + \frac{y^2}{4} = 1$, using the parametric equations $x = 3\cos t$ and $y = 2\sin t$. As t runs from 0 to 2π, the point moves around the ellipse counterclockwise.

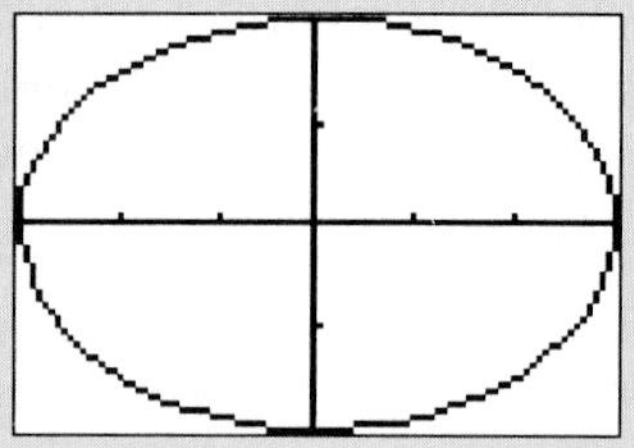

TI-85 MODE ▼ ▼ ▼ ▼ ▶ ▶ (Param) ENTER EXIT GRAPH F1($E(t)$ =) 3 COS F1(t) ▼ 2 SIN F1(t) M2(RANGE) 0 ▼ 2 π ▼ ▼ (−) 3 ▼ 3 ▼ ▼ (−) 2 ▼ 2 F5(GRAPH)	SHARP EL9300 SET UP E 3 ⩘ 3 cos x/θ/T ENTER 2 sin x/θ/T ENTER RANGE −3 ENTER 3 ENTER 1 ENTER −2 ENTER 2 ENTER 1 ENTER 0 ENTER 2 π ENTER .02 ENTER ⩘
CASIO f_x-7700GB MODE MODE × (PARAM) Range −3 EXE 3 EXE 1 EXE −2 EXE 2 EXE 1 EXE 0 EXE 2 π EXE .02 EXE Graph 3 cos T, 2 sin T) EXE	HP 48SX ↰ RAD ' 3 × COS T ▶ + α ↰ CST × 2 × SIN T ↰ PLOT NEW P A R ENTER PTYPE PARA PLOTR ↰ { } T SPC 0 SPC 6.3 ENTER INDEP AUTO

Try graphing $x^2 + 3y^2 = 3$ or $x^2 - y^2 = 1$ using parametric equations.

These parametric equations were relatively easy to write down and work with, but they turn out not to be the correct equations to describe the motion of a comet, for instance.

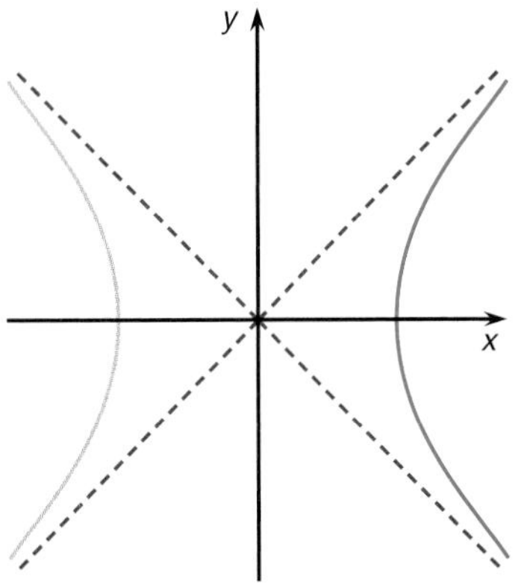

Figure 12.5.10

EXAMPLE 5 **Parametrizing a Hyperbola** In the same way we might look for parametric equations for a hyperbola. For convenience, consider the hyperbola $x^2 - y^2 = 1$. From looking at this equation we might be inspired to try letting $x = \cosh t$ and $y = \sinh t$. The standard identity $\cosh^2 t - \sinh^2 t = 1$ says that the point (x, y) lies on the hyperbola.

As t ranges from $-\infty$ to ∞, we know that $\sinh t$ ranges from $-\infty$ to ∞, but $\cosh t$ is always greater than or equal to 1. We conclude that as t moves along the real line, the point $(x, y) = (\cosh t, \sinh t)$ moves along the *right branch* of the hyperbola $x^2 - y^2 = 1$ (see Figure 12.5.10). In fact the point moves up from the bottom toward the top of the branch.

Similar considerations show that $(x, y) = (a\cosh t, b\sinh t)$ give parametric equations for the right branch of the hyperbola

$$\frac{x^2}{a^2} - \frac{y^2}{b^2} = 1.$$

All the examples up to now have given parametric equations for a moving point by expressing its rectangular coordinates (x, y) as functions of the parameter t. We think of t as the time, so $\big(x(t), y(t)\big)$ represents the position of the point at time t. But it may also be interesting to work with polar coordinates and write parametric equations for r and θ.

EXAMPLE 6 Consider the circle $r = 7$. We already know that the points on this circle are exactly the points whose polar coordinates are $(7, \theta)$, for any value of θ.

Another way of saying this is that as we let θ range over all possible values, the point $(7, \theta)$ will move around the circle. It will go around a full circuit each time θ increases by 2π, and it moves counterclockwise as θ increases.

This is slightly confusing because in a way what we are doing is using θ as the parameter. It might be more clear to write

$$r = 7, \qquad \theta = \theta(t) = t.$$

These are parametric equations; the value of r is the constant function 7, and we just let θ equal the parameter t.

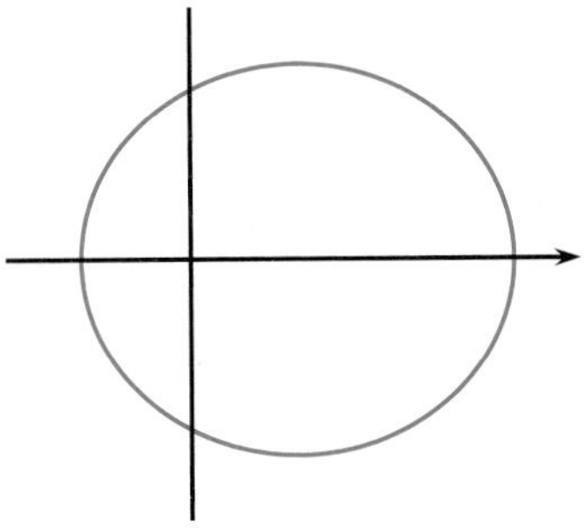
Figure 12.5.11

EXAMPLE 7 Consider the ellipse

$$r = \frac{1}{2 - \cos\theta},$$

as illustrated in Figure 12.5.11. In a way this equation can be used to give parametric equations; we let t be the parameter and let $\theta = t$, and then use the equation to write

$$r = \frac{1}{2 - \cos\theta} = \frac{1}{2 - \cos t}.$$

In this way we have expressed r and θ as functions of the parameter t. As t goes from 0 to 2π, the corresponding point will go around the ellipse exactly once.

The idea from Example 7 can be used for any curve that can be expressed by an equation of the form $r = f(\theta)$. We can use θ as the parameter, or, if we prefer, we can call the parameter t and let $\theta = t$. The equation can then be used to express r as a function of t.

Parametric equations obtained in this way have the property that the moving point "revolves at a constant rate" around the origin, in the sense that the angle θ increases at a constant rate (i.e., $\frac{d\theta}{dt} = \frac{dt}{dt} = 1$). Notice that the parametrization we found in Example 4 for the ellipse $\frac{x^2}{a^2} + \frac{y^2}{b^2} = 1$ does not have this property.

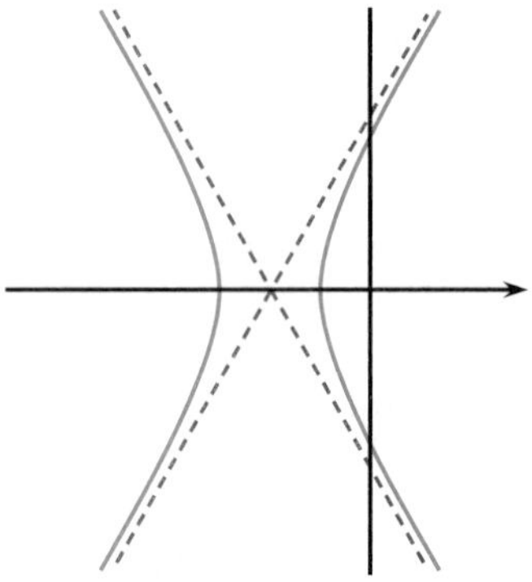

Figure 12.5.12

EXAMPLE 8 If we consider the hyperbola

$$r = \frac{1}{1 - 2\cos\theta},$$

as shown in Figure 12.5.12, then we can find parametric equations using the idea just discussed. With t as parameter they are

$$\theta = t, \qquad r = \frac{1}{1 - 2\cos t}.$$

The formula makes no sense if $t = \pm\frac{\pi}{3} + 2m\pi$, but otherwise the point (r, θ) always lies on the hyperbola. In fact as t goes from 0 to 2π, missing $\frac{\pi}{3}$ and $\frac{5\pi}{3}$, the point covers the whole hyperbola exactly once. If we want the point to move along just the left branch, we should let t range from $-\frac{\pi}{3}$ to $\frac{\pi}{3}$ or, to be more precise, over the open interval $\left(-\frac{\pi}{3}, \frac{\pi}{3}\right)$. The right branch will be traced out as t moves over $\left(\frac{\pi}{3}, \frac{5\pi}{3}\right)$.

From the picture we see that as t approaches $\pm\frac{\pi}{3}$ from either side, the moving point will move far out on one of the branches, approaching an asymptote. Even a very small change in t will result in a big change in r, and the point will move a large distance. This suggests that using this parametrization will make the point move faster and faster as it approaches the asymptotes.

We have seen that it is not terribly difficult to write down parametrizations for many familiar curves. Often the equation that defines the curve can be used to help express the coordinates in terms of some parameter.

For conics we have seen how to parametrize them in either rectangular or polar coordinates. Unfortunately, none of the parametrizations we have discussed gives the actual motion of a planet, comet, or meteor. One hint can be found in Example 8, in which we found that the speed gets larger and larger way out on the branches of a hyperbola. If a meteor moves on this path, we would expect that as it gets far away from the sun, it will experience very little gravitational force, so it will very nearly be coasting, and it seems logical to expect that its speed will approach a constant. There is certainly no reason to expect it to accelerate.

In order to solve this question we will need to be able to talk about the velocity of an object moving along a curve. Up until now we have been very careful to consider velocities of objects moving in straight lines or ones like the projectile of Example 2 whose motion can be seen as the combination of a motion in the horizontal direction and a motion in the vertical direction, each of which amounts to motion in a straight line.

Exercises 12.5

I

Find a parametrization of each of the following curves, and specify the range over which the parameter t must run in order to trace out the curve exactly once. Notice that there will be many possible answers.

1. $x^2 + y^2 = 25$
2. $4x^2 + 9y^2 = 36$
3. $r = \cos\theta$
4. $r = \sin 4\theta$
5. $3x - 4y = 7$
6. $y^2 = 2 - x^2$
7. $r = 3\sec\theta$
8. $r = 1 - 2\cos\theta$
9. $r^2 = \cos(2\theta)$
10. $x^2 + 2x + 4y^2 - 8y + 4 = 0$

II

*11. Using the fact that the eccentricity of the earth's orbit is about 0.017, calculate the ratio $\frac{b}{a}$ of its semi-minor axis to its semi-major axis.

*12. Using the fact that the eccentricity of Mercury's orbit is about 0.206, calculate the ratio $\frac{b}{a}$ of its semi-minor axis to its semi-major axis.

*13. Using the fact that the eccentricity of Halley's Comet's orbit is about 0.97, calculate the ratio $\frac{b}{a}$ of its semi-minor axis to its semi-major axis.

14. Find parametric equations for the motion of a rock that is thrown upward from ground level at an angle of 45° above the horizontal if the maximum height it reaches is 64 ft.

15. (i) Find parametric equations for a speck of dust moving around the rim of a twelve-inch LP that has mistakenly been put on a turntable revolving at 45 rpm.
    (ii) Find parametric equations for a speck of dust moving around on a 45 rpm disk at a distance of $1\frac{1}{2}$ in from the center.

16. Find parametric equations for the motion of a speck of dust on the rim of an LP revolving at $33\frac{1}{3}$ rpm so that the position of the speck at time $t = 0$ is the point $(0, 6)$.

17. Find parametric equations for the left-hand branch of the hyperbola $\frac{x^2}{a^2} - \frac{y^2}{b^2} = 1$.

18. Find parametric equations for the upper branch of the hyperbola $\frac{y^2}{a^2} - \frac{x^2}{b^2} = 1$.

*19. Find parametric equations for the upper semicircle of $x^2 + y^2 = 9$ using the parameter t, in such a way that the equation for x is $x = x(t) = t$. Be sure to specify the interval over which t runs.

*20. Find parametric equations for the right branch of $x^2 - y^2 = 4$ using the parameter t, in such a way that the equation for y is $y = y(t) = -3t$. Be sure to specify the interval over which t runs.

What is the equation of the curve that is parametrized by each of the following sets of equations?

*21. $x = t,\ y = t^2 - 1$

*22. $x = t - 3,\ y = t^3 + 2t - 1$

*23. $x = t^5 + 5,\ y = 4t^5 - 9$

*24. $x = t^2,\ y = 2t^2 + 5$

III

If you have a machine that can plot parametric equations, sketch the following curves.

25. $x = 2\sin t, \quad y = 3\cos t$
26. $x = 5\cosh t, \quad y = 4\sinh t$
27. $r = \frac{1}{2-\sin t}, \quad \theta = t$
28. $x = \cos^{1/3} t, \quad y = \sin^{1/3} t$
29. $x = \sin t, \quad y = \cos(3t)$
30. $x = \cos(3t), \quad y = \sin(5t)$

POINT TO PONDER

When we wrote the parametric equations $x = a\cos t$, $y = b\sin t$ for the ellipse $\frac{x^2}{a^2} + \frac{y^2}{b^2} = 1$, we observed that as t goes from 0 to 2π, the point (x, y) moves once around the ellipse, counterclockwise. On the other hand, we observed that t does not always equal the polar angle θ.

In fact it is possible to express θ in terms of t. After all, if $t \in \left[0, \frac{\pi}{2}\right)$, then $\tan\theta = \frac{b\sin t}{a\cos t}$, so $\theta = \arctan\left(\frac{b\sin t}{a\cos t}\right) = \arctan\left(\frac{b}{a}\tan t\right)$. This formula also holds if $t \in \left(-\frac{\pi}{2}, 0\right]$, but it must be modified slightly if $t \in \left(\frac{\pi}{2}, \frac{3\pi}{2}\right)$.

Still, in principle it is not difficult to express the polar angle θ in terms of the parameter t, and it is even easier to write $r = \sqrt{x^2 + y^2} = \sqrt{a^2 \cos^2 t + b^2 \sin^2 t}$.

If $a \neq b$, the polar angle θ is *not* the same as the parameter t. This reflects the fact that the point $(x, y) = (a \cos t, b \sin t)$ does not move in such a way that its angle θ increases at a constant rate.

From our formula for θ we find that

$$\frac{d\theta}{dt} = \frac{1}{1 + \frac{b^2}{a^2}\tan^2 t}\frac{b}{a}\sec^2 t = \frac{ab}{a^2 \cos^2 t + b^2 \sin^2 t}.$$

This formula for the derivative is valid for all values of t. In particular, notice that when $t = 0$, $\frac{d\theta}{dt} = \frac{ab}{a^2} = \frac{b}{a}$, and when $t = \frac{\pi}{2}$, $\frac{d\theta}{dt} = \frac{a}{b}$. This confirms the remark at the end of Example 4 that the point moves more slowly when $t = 0$ than when $t = \frac{\pi}{2}$ (assuming $a > b$).

Chapter Summary

§12.1
polar coordinates
radius
modulus
argument
polar angle
polar/rectangular conversions (Theorem 12.1.1)
graphing points in polar coordinates

§12.2
graphs in polar coordinates
Spiral of Archimedes
roses
symmetry in polar coordinates
cardioid
lemniscate
limaçon
dimple
slope of tangent line

§12.3
areas in polar coordinates
limits of integration
tracing curves exactly once

§12.4
conics in polar coordinates
focus
eccentricity
directrix
$r = \frac{k}{1 - e\cos\theta}$
circles
straight lines

§12.5
parameter
parametric equations

Review Exercises

I

Sketch each of the following curves.

1. $r = 2\csc\left(\theta - \frac{\pi}{4}\right)$
2. $r = 3 - 4\sin\theta$
3. $r = (3 - \sin\theta)^{-1}$
4. $r = (3\cos\theta - 2)^{-1}$
5. $r^2 = \sin\left(2\theta + \frac{\pi}{4}\right)$
6. $r = 2\cos\theta\sin\theta$
7. $r = 2 + 2\sin\theta$
8. $r = \sin\left(5\theta + \frac{\pi}{3}\right)$
9. $2\sin\theta = 1,\ r \geq 0$
10. $r + 3\theta = 0$

Find all points of intersection of each of the following pairs of curves. Make a sketch in each case.

11. $r = \theta, \theta \geq 0; \theta = 0, r \geq 0$
12. $r = \frac{1}{2}; r = \cos\theta$
13. $r = \sec\left(\theta + \frac{\pi}{4}\right); r = 1$
14. $r = \sec\theta; r = \sqrt{3}\csc\theta$
15. $r = \frac{1}{2}; r = \cos 2\theta$
16. $r = \csc\theta; r = 2\sin\theta$
17. $r^2 = \cos 2\theta; r = \frac{1}{\sqrt{2}}$
18. $r^2 = \cos^2\theta; r = \frac{1}{2}$
19. $r + \sin\theta = 0; \theta = \frac{\pi}{6}, r \geq 0$
20. $r^2 = \sin^2(2\theta) + \cos^2(2\theta); r = \sin(4\theta)$

In each of the following questions, find the area of the region enclosed inside both the curves.

21. $r = 1;\ r = -2\sin\theta$

22. $r = 1;\ r^2 = 2\sin\left(2\theta + \frac{\pi}{3}\right)$

23. $r = \sqrt{3}\cos\theta;\ r = \sin\theta$

24. $r = \cos(4\theta);\ r = \sin(4\theta)$

25. $r = 2 + \cos\theta;\ r = 2 - \sin\theta$

26. $r = 3 + \sin\theta;\ r = 3 - \cos\theta$

27. $r = \cos\theta;\ r = \sin\left(\theta - \frac{\pi}{2}\right)$

28. $r = 2 - 2\sin\theta;\ r = 2 + 2\cos\theta$

29. $r = 3 + \cos\theta;\ r = 1 - \cos\theta$

30. $r = 1 - \cos\theta;\ r^2 + 2r\cos\theta = 1$

Find a parametrization of each of the following curves, specifying the range over which the parameter must run in order to trace out the curve exactly once.

31. $y = 2x^2 - 3$

32. $2x + 3y^2 = 4$

33. $r = \cos 2\theta$

34. $r^2 = \cos 2\theta$

35. $r = 2\csc\theta$

36. $r = \sin 3\theta$

37. $5x + 4y + 3 = 0$

38. $y = 4$

39. $r + 3 = 3\sin\theta$

40. $r = (2 + 2\sin\theta)^{-1}$

II

Sketch each of the following curves.

41. $r = 2 + \cos\theta$

42. $r = 3 - \sin(4\theta)$

43. $r^2 = \sin 3\theta$

44. $r^2 = \sin^2(3\theta)$

45. $(r - \sec\theta)(r + \csc\theta) = 0$

46. $(r - \sin\theta)\big(r + \sin(2\theta)\big) = 0$

47. $r = \sqrt{\theta}$

48. $r = \theta^2$

49. $r = 2 + |\sin 3\theta|$

50. $r = 3 + \sin\frac{7\theta}{2}$

***51.** Consider the point whose polar coordinates are $(1, 0)$. Does it lie on the curve $r = \frac{\theta}{2\pi}$?

***52.** Consider the point $(r, \theta) = (1, \pi)$. Does it lie on the curve $r = \cos\theta - 2$?

***53.** When we write hyperbolas in rectangular coordinates, we can distinguish those whose transverse axis is horizontal or vertical by recognizing the two different standard forms in Summary 1.6.4 (p. 38):

$$\frac{x^2}{a^2} - \frac{y^2}{b^2} = 1 \quad \text{and} \quad \frac{y^2}{a^2} - \frac{x^2}{b^2} = 1.$$

If we consider hyperbolas with a focus at the origin and write their equations in polar coordinates, what will be the difference between those whose transverse axis is horizontal and those whose transverse axis is vertical?

***54.** Sketch the graph of $\theta = \sin r$, for $r \geq 0$.

***55.** Find the equations in polar coordinates of two ellipses with one focus at the origin, major axis horizontal, and eccentricity $e = \frac{1}{2}$ and containing the point $(r, \theta) = (1, 0)$.

***56.** Find the equations in polar coordinates of two ellipses with one focus at the origin, major axis vertical, and eccentricity $e = \frac{1}{4}$ and containing the point $(r, \theta) = (2, \pi)$.

***57.** Sketch the graphs of $r = \sec\theta$ and $r = 1 + \sec\theta$ on the same diagram.

***58.** Sketch the graphs of $r = 2\csc\theta$ and $r = 1 + 2\csc\theta$ on the same diagram.

Curves in the Plane

13

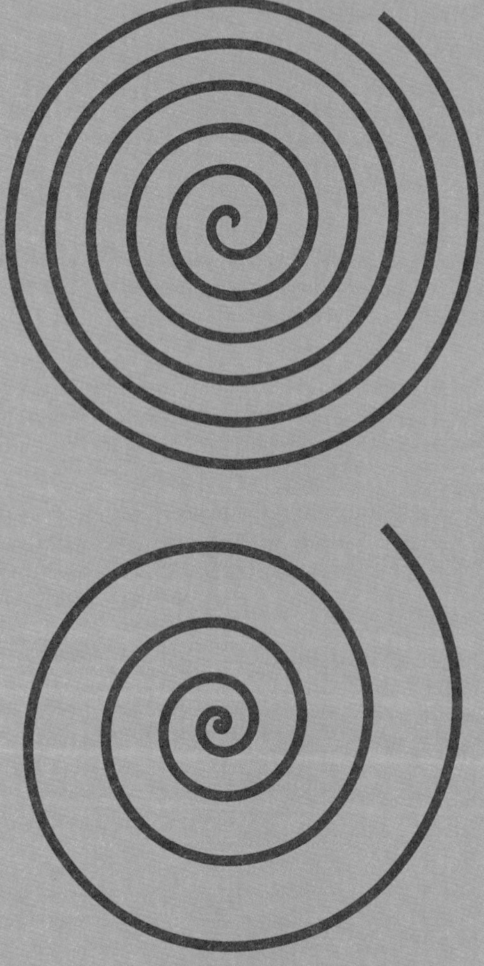

The Spiral of Archimedes (top), given by $r = \theta$, expands by the same amount each time it revolves once. It could be used, for instance, to describe the position of a coiled rope. The second curve is called a Lograithmic Spiral because its equation $r = \theta^k$ is a "straight-line" equation $\ln r = k \ln \theta$ after we take logarithms. It describes the shape of certain snail shells and the positions of leaves around the stems of certain plants, including roses.

INTRODUCTION

In the final section of Chapter 12 we considered parametric equations for various curves. The idea is to think of a point moving along some curve, a circle or a hyperbola, for instance, and to express its position as a function of time. We saw that it is possible to do this for many familiar curves, and it is possible to express the position using either rectangular or polar coordinates.

However, we began to realize that it is possible to find many different parametrizations for the same curve, just as it is possible to move along the curve in many different ways (with different speeds, different directions, different starting points, etc.). If we are to use this idea to study practical problems, it will be necessary to compare the motion described by various parametrizations, and for this we will need to discuss the idea of the velocity of an object moving along a curved path. That is the purpose of this short chapter.

In Section 13.1 we recall briefly the concept of parametric equations and consider some examples. In Section 13.2 we observe that the velocity of a moving point involves not only how fast it is moving (its speed) but also the direction of its motion. Because of this the velocity is best expressed as a *vector,* a quantity that has both a magnitude and a direction. In Section 13.3 we apply some of these ideas to the particular case of the motion of projectiles. Historically, this was one of the early uses of calculus, and it is still very important.

13.1 Parametric Equations

PARAMETER
CURVE
DOMAIN

Suppose an object is moving in the plane, as shown in Figure 13.1.1. At each particular time t it will be at some specific location (x, y), and we can think of the coordinates x and y as functions $x = x(t)$ and $y = y(t)$ of t. If we know these functions, then we know exactly how the object moves, and on the other hand if we know how an object moves, then in principle we ought to be able to find the functions that describe the motion.

It is often necessary to restrict the values that t can take. Usually, t will range over some interval (frequently the whole real line $\mathbb{R}$), but in some cases it will range over a union of intervals.

DEFINITION 13.1.1

Suppose I is an interval (open or closed, possibly infinite), and suppose

$$x = x(t), \qquad y = y(t)$$

are functions defined on I. Assume that they are continuous on I. Then they are called **parametric equations;** the variable t is called the **parameter.** The set of points

$$\{(x, y) = \big(x(t), y(t)\big) : t \in I\}$$

is called the **curve**, and it is sometimes also called the **image** of I. The interval I is called the **domain** of the parametric equations.

Sometimes several curves are combined to make one curve; in this situation it may be necessary to have parametric equations defined on several intervals.

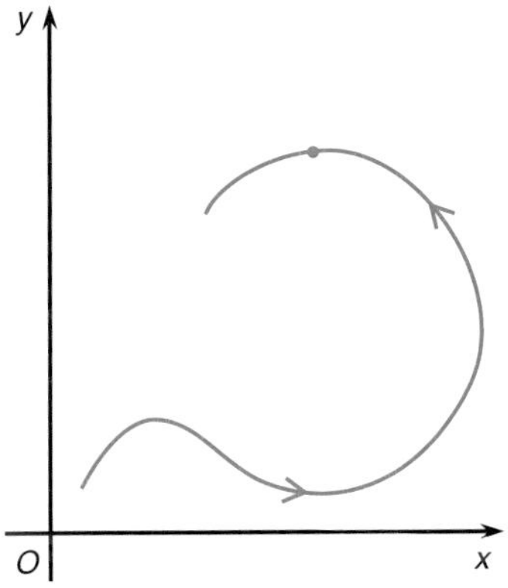

Figure 13.1.1

EXAMPLE 1

(i) Consider the ellipse $\frac{1}{4}x^2 + y^2 = 1$. The equations

$$x = 2\cos t, \qquad y = \sin t$$

are parametric equations. It is easy to check that the points $(x, y) = (2\cos t, \sin t)$ do all lie on the ellipse, and in fact we argued in Example 12.5.4 that as t ranges from 0 to 2π, the point $(x, y) = \big(x(t), y(t)\big)$ moves once around the ellipse counterclockwise, starting and finishing at the point $(x, y) = (2, 0)$. See Figure 13.1.2(i).

We can use the above equations to parametrize the ellipse, but strictly speaking, it is necessary to specify the interval I. One possibility is to let $I = [0, 2\pi)$. This would mean that the moving point would trace out the ellipse exactly once, returning to approach its starting point but never quite reaching it. Another possibility would be to use $I_2 = [0, 2\pi]$. The only difference is that

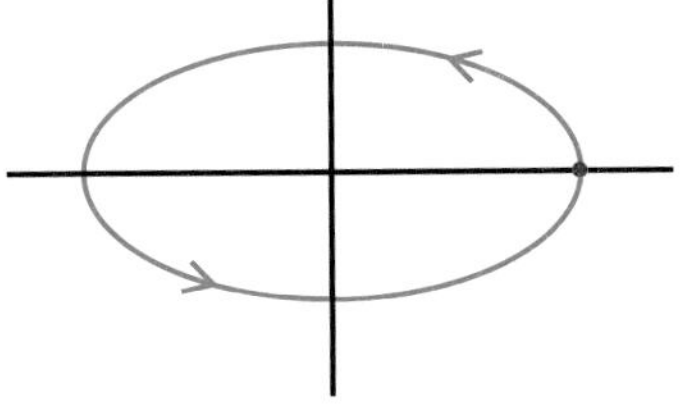
Figure 13.1.2(i)

this way the curve would actually return to its starting point. It would also be possible to use other intervals, such as $[-\pi, \pi)$ or $[-\pi, \pi]$.

Another possibility would be to let $I_3 = \mathbb{R}$. In this case the moving point would go around and around the ellipse infinitely many times. It is also possible to use the same equations but to let the domain be $I_4 = [0, \pi]$. This way the moving point will trace out only the upper half-ellipse. See Figure 13.1.2(ii).

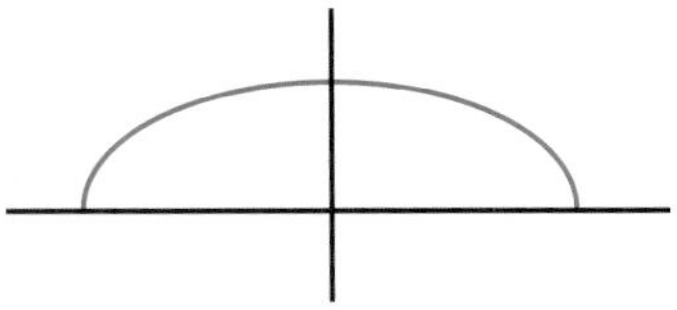

Figure 13.1.2(ii)

You can also check that the same ellipse can be parametrized by $x = -2\sin t$, $y = -\cos t$, for $t \in [0, 2\pi]$. In this situation the moving point moves once around the ellipse in the *clockwise* direction, beginning and ending at the point $(0, -1)$. See Figure 13.1.2(iii).

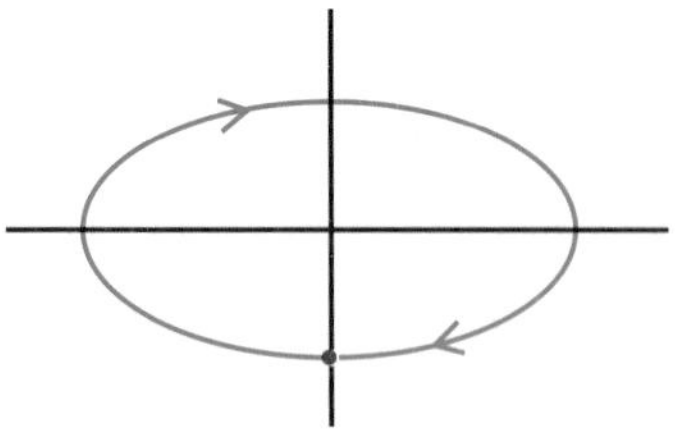

Figure 13.1.2(iii)

(ii) The straight line $2x - 3y = 2$, shown in Figure 13.1.3(i), can be parametrized in many different ways. To find one, we could try letting $x = t$ and then solving for y. Since $2x - 3y = 2$, we find that $3y = 2x - 2$, so $y = \frac{2x-2}{3}$. If $x = t$, then $y = \frac{2t-2}{3}$. These give parametric equations for the line, with domain $I = \mathbb{R}$.

It is easy to find other parametrizations. We could let $x = 2t$ and $y = \frac{4t-2}{3}$, for $t \in \mathbb{R}$. Another possibility is to let $x = 3t$, $y = 2t - \frac{2}{3}$. This has the slight advantage of involving fewer fractions.

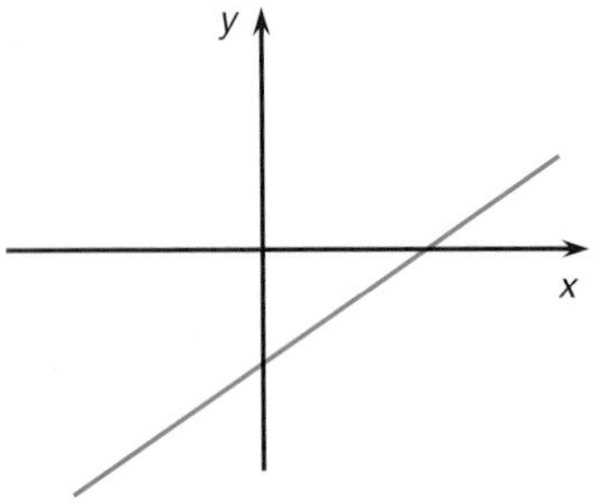

Figure 13.1.3(i)

If we let $y = t^2$, we can solve and find that $x = \frac{3}{2}t^2 + 1$. These points are always on the line $2x - 3y = 2$, but since $y = t^2$, we see that y is never negative. For that matter, $x = \frac{3}{2}t^2 + 1$ is never less than 1. We have found a parametrization of the part of the line above and to the right of the point $(1, 0)$; see Figure 13.1.3(ii). Moreover, since t and $-t$ correspond to the same point, the half-line is traced out twice as t runs over $\mathbb{R}$. If we let $I = [0, \infty)$, then the half-line is covered only once.

Finally, we observe that the whole line can also be parametrized by $x = \tan t$, $y = \frac{2}{3}\tan t - \frac{2}{3}$, for $t \in \left(-\frac{\pi}{2}, \frac{\pi}{2}\right)$. This example demonstrates that the domain can be a finite interval even for a curve that is unbounded, like a whole line. (Recall that for $t \in \left(-\frac{\pi}{2}, \frac{\pi}{2}\right)$ the range of the function $\tan t$ is all the real numbers.)

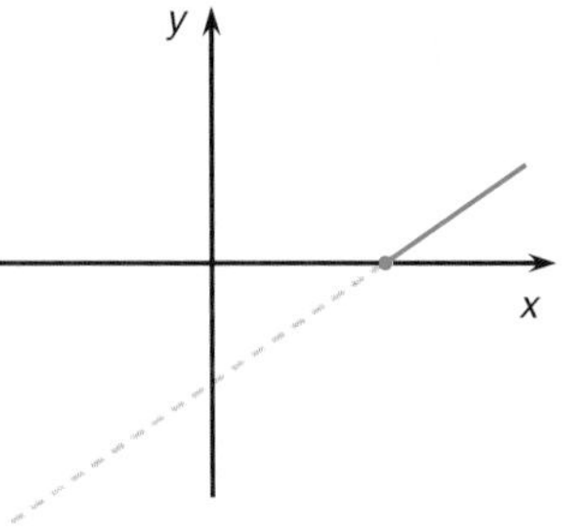

Figure 13.1.3(ii)

(iii) The graph of $y = x^5 - 10x + 1$ is a curve in the plane (see Figure 13.1.4). We think of a point moving along the graph from left to right. It can be parametrized by letting $x = t$, $t \in (-\infty, \infty)$, and then using the equation to solve for y. The parametric equations are $x = t$, $y = t^5 - 10t + 1$, for $t \in (-\infty, \infty)$.

(iv) Consider the rectangle whose corners are $(0, 0)$, $(3, 0)$, $(3, 2)$, and $(0, 2)$, as illustrated in Figure 13.1.5. If we want to parametrize it, the simplest thing to do would be to parametrize each of the four sides separately. For instance, to parametrize the bottom edge from $(0, 0)$ to $(3, 0)$, we could use the equations $x = t$, $y = 0$, for $t \in [0, 3]$.

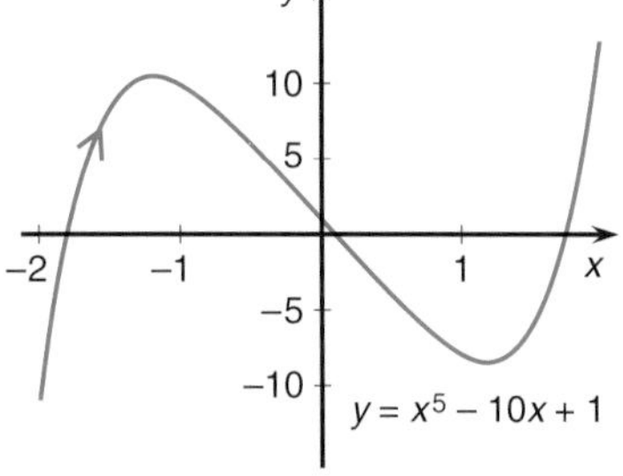

Figure 13.1.4

Next, to parametrize the right side, we need to let $x = 3$ and then to have y go from 0 to 2. Since we have already let t go from 0 to 3 in parametrizing the bottom edge, it might be confusing to use $[0, 2]$ as the domain for the parametrization of the right side. A better idea would be to let $y = t - 3$, for $t \in [3, 5]$.

Then, to move back along the top edge, we could let $x = 8 - t$ and $y = 2$, for $t \in [5, 8]$. Notice that as t moves from 5 to 8, the value of $x = 8 - t$ will go from 3 back to 0, so the corresponding point will move back from right to left along the top side of the rectangle.

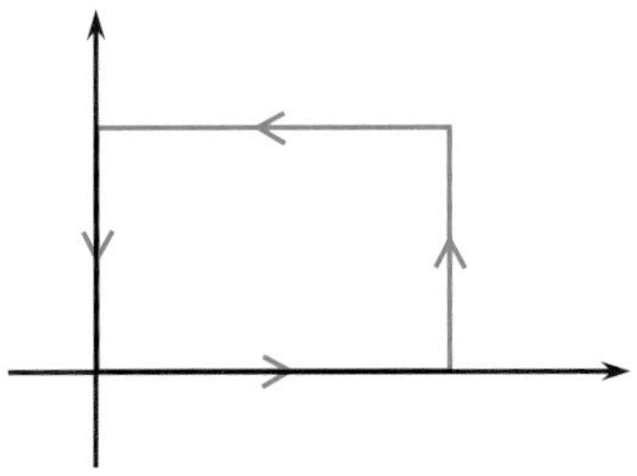

Figure 13.1.5

Finally, to move down the left side of the rectangle, we let $x = 0, y = 10 - t$, for $t \in [8, 10]$.

We have parametrized the rectangle by combining parametrizations on four different intervals:

$$(x, y) = \begin{cases} (t, 0), & \text{if } t \in [0, 3], \\ (3, t-3), & \text{if } t \in [3, 5], \\ (8-t, 2), & \text{if } t \in [5, 8], \\ (0, 10-t), & \text{if } t \in [8, 10]. \end{cases}$$

In fact we have fixed up this parametrization so that its domain is the interval $[0, 10]$. Strictly speaking, we began by working on each subinterval separately, but we arranged things so that they "join" at the corners of the rectangle. We could have expressed the parametrization as four separate parametrizations, but it is nice to be able to work with a single interval. There is a slight complication, which is that the functions $x(t)$ and $y(t)$ are not differentiable at $t = 3, 5, 8$, the points that correspond to the corners of the rectangle. To check this, you could begin by sketching the graphs of $x(t)$ and $y(t)$ for $t \in [0, 10]$.

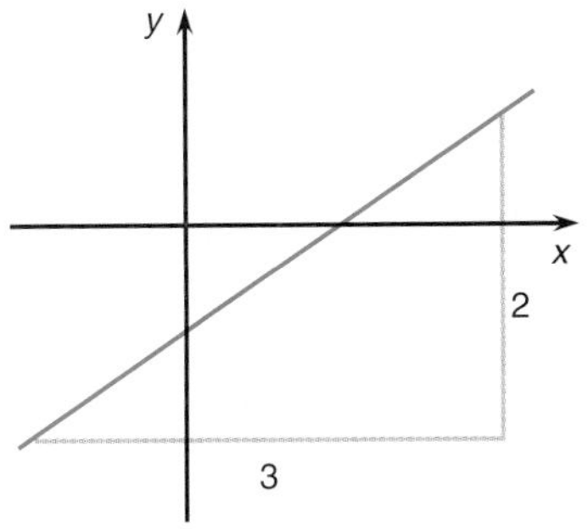

Figure 13.1.6

REMARK 13.1.2

In part (ii) of Example 1 we found the parametrization $x = 3t$, $y = 2t - \frac{2}{3}$ for the line $2x - 3y = 2$. It is interesting to make the following interpretation of these equations. Looking at the formulas $x = 3t$, $y = 2t - \frac{2}{3}$, we see that as t increases, x increases 3 times as fast as t does, and y increases twice as fast as t, which means that y increases $\frac{2}{3}$ as fast as x does.

As we move along the line, this suggests that we move more slowly in the vertical direction than in the horizontal direction, by a factor of $\frac{2}{3}$. This is exactly what we should expect, since the slope of the line is $\frac{2}{3}$; see Figure 13.1.6.

Summary 13.1.3

Parametrizations of Some Familiar Curves

(Many other parametrizations can be found; the ones given here are easy to describe.)

Line	$y = ax + b$;	$x = t$, $y = at + b$, $t \in \mathbb{R}$, if $a \neq 0$, can also use $y = t$, $x = \frac{t-b}{a}$, $t \in \mathbb{R}$,
Graph	$y = f(x)$;	$x = t$, $y = f(t)$, $t \in \text{domain}(f)$,
Circle	$x^2 + y^2 = r^2$;	$x = r\cos t$, $y = r\sin t$, $t \in \mathbb{R}$; counterclockwise, goes around once as t increases by 2π,
		$x = r\cos t$, $y = -r\sin t$, $t \in \mathbb{R}$; clockwise,
Ellipse	$\frac{x^2}{a^2} + \frac{y^2}{b^2} = 1$;	$x = a\cos t$, $y = b\sin t$, $t \in \mathbb{R}$; counterclockwise, goes around once as t increases by 2π,
Hyperbola	$\frac{x^2}{a^2} - \frac{y^2}{b^2} = 1$;	$x = a\cosh t$, $y = b\sinh t$, $t \in \mathbb{R}$; parametrizes right branch,
		$x = -a\cosh t$, $y = b\sinh t$, $t \in \mathbb{R}$; left branch,
Hyperbola	$\frac{y^2}{a^2} - \frac{x^2}{b^2} = 1$;	$x = b\sinh t$, $y = a\cosh t$, $t \in \mathbb{R}$; upper branch,
		$x = b\sinh t$, $y = -a\cosh t$, $t \in \mathbb{R}$; lower branch.

We can use a graphics calculator to sketch the hyperbola $\frac{x^2}{9} - \frac{y^2}{16} = 1$, using the parametric equations $x = 3\cosh t$, $y = 4\sinh t$ for the right branch and $x = -3\cosh t$, $y = 4\sinh t$ for the left branch.

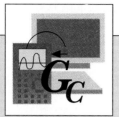

TI-85	SHARP EL9300
MODE ▼ ▼ ▼ ▼ ▶ ▶ (Param) ENTER EXIT GRAPH F1 ($E(t)$ =) 3 MATH F4F2 (cosh) M1 (t) ▼ 4 F1 (sinh) M1 (t) ▼ (−) 3 F2 (cosh) M1 (t) ▼ 4 F1 (sinh) M1 (t) EXIT M2 (RANGE) (−) 2 ▼ 2 ▼ ▼ (−) 10 ▼ 10 ▼ ▼ (−) 10 ▼ 10 F5 (GRAPH)	SET UP E 3 ⩘ 3 MATH ▼ ▶ ▼ ENTER x/θ/T ENTER 4 MATH ▲ ENTER x/θ/T 2ndF ▼ −3 MATH ▼ ENTER x/θ/T ENTER 4 MATH ▲ ENTER x/θ/T ENTER RANGE −10 ENTER 10 ENTER 1 ENTER −10 ENTER 10 ENTER 1 ENTER −2 ENTER 2 ENTER .02 ENTER ⩘
CASIO f_x-7700GB	**HP 48SX**
MODE MODE × (PARAM) Range −10 EXE 10 EXE 1 EXE −10 EXE 10 EXE 1 EXE −2 EXE 2 EXE .02 EXE Graph 3 MATH HYP F2 (csh) T , 4 F1 (snh) T) EXE Graph − 3 F2 (csh) T , 4 F1 (snh) T) EXE	↰ PLOT PLOTR 10 +/− ENTER 10 XRNG 10 +/− ENTER 10 YRNG ATTN ′ 3 × MTH HYP COSH T ▶ + α ↰ CST × 4 × SINH T ↰ PLOT NEW P A R ENTER PTYPE PARA PLOTR ↰ {} T SPC 2 +/− SPC 2 ENTER INDEP DRAW ATTN ′ 3 +/− × MTH HYP COSH T ▶ + α ↰ CST × 4 × SINH T ↱ PLOT ↰ DRAW DRAW

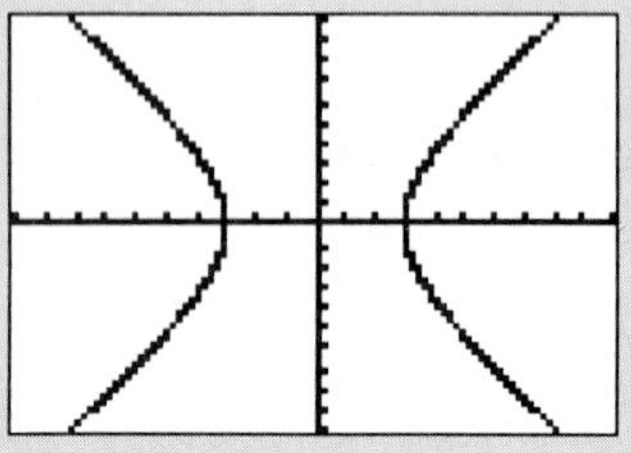

Try finding a different parametrization.

Suppose that we are given some parametric equations $x = x(t)$, $y = y(t)$, for $t \in I$. We know that these equations describe the position of a point moving along some curve in the plane, and it would be interesting to know what that curve is. In particular, we would like to be able to find a single equation that describes the curve. We have seen that sometimes just by looking at the parametric equations it is possible to get some idea of the shape of the curve.

In fact it is often possible to find an equation for the curve. The idea is to try to eliminate the parameter t from the two parametric equations. We illustrate with the following examples.

EXAMPLE 2

(i) Consider the parametric equations

$$x = 2 + 3t, \qquad y = 5t + 1.$$

We can solve for t in each equation and find that $t = \frac{x-2}{3}$, $t = \frac{y-1}{5}$. Equating these two expressions for t, we find that $\frac{x-2}{3} = \frac{y-1}{5}$.

This can be simplified to $5x - 10 = 3y - 3$, or $5x - 3y = 7$. We find that the curve we have parametrized is the straight line $5x - 3y = 7$.

(ii) Consider the parametric equations $x = t^3 - 1$, $y = 3t^6 - 2t^3 + 5$.

From the first of these equations we find that $t^3 = x + 1$, and substituting this into the second equation, we find that

$$y = 3(t^3)^2 - 2(t^3) + 5 = 3(x+1)^2 - 2(x+1) + 5 = 3x^2 + 4x + 6.$$

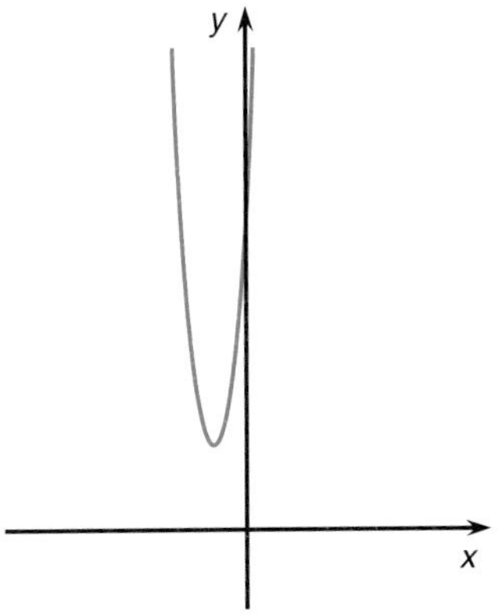

Figure 13.1.7

The curve is the parabola $y = 3x^2 + 4x + 6$, shown in Figure 13.1.7.

(iii) Consider the parametric equations

$$x = t, \qquad y = \sqrt{4 - t^2}.$$

We observe that for the square root to make sense we must have $4 - t^2 \geq 0$, which means $t \in [-2, 2]$.

Squaring the second equation, we find that $y^2 = 4 - t^2$, and since $x = t$, this amounts to $y^2 = 4 - x^2$. The parametric equations parametrize points that lie on the circle $x^2 + y^2 = 4$.

However, because $y = \sqrt{4 - t^2}$ is never negative, we see that they actually parametrize the upper half of the circle. As t runs from -2 to 2, the corresponding point (x, y) moves around the upper semicircle from $(-2, 0)$ to $(2, 0)$, as shown in Figure 13.1.8.

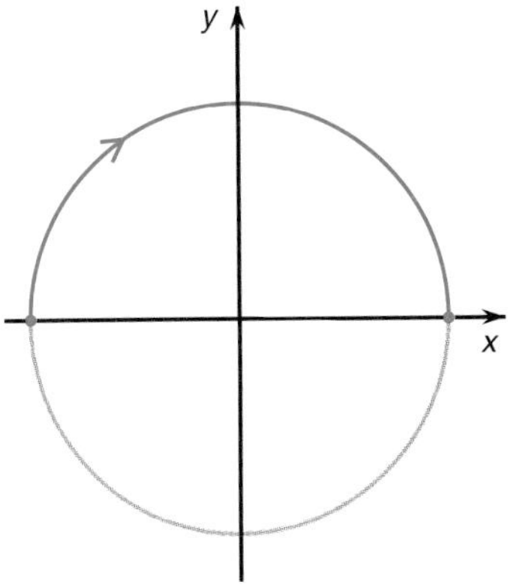

Figure 13.1.8

(iv) Consider the parametric equations

$$x = 2t, \qquad y = -\sqrt{1 + t^2}.$$

Here there is no problem about the square root, and t can be any real number.

From the first equation we can solve for t and find $t = \frac{x}{2}$. Squaring the second equation gives $y^2 = 1 + t^2 = 1 + \frac{x^2}{4}$. The point (x, y) is on the curve $y^2 - \frac{1}{4}x^2 = 1$, which is a hyperbola.

Since $y = -\sqrt{1 + t^2}$ is never positive, we see that the parametric equations trace out the bottom branch of the hyperbola, as shown in Figure 13.1.9.

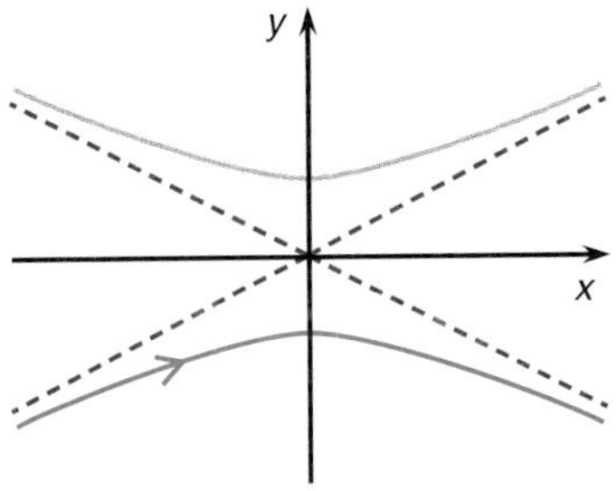

Figure 13.1.9

In Example 2 we remarked that certain parametrizations do not trace out the whole curve. In each such case we identified part of the curve on which the moving point lies, but we did not show that it passes through every point. You should convince yourself that the parametrizations actually do cover the stated parts of the curves.

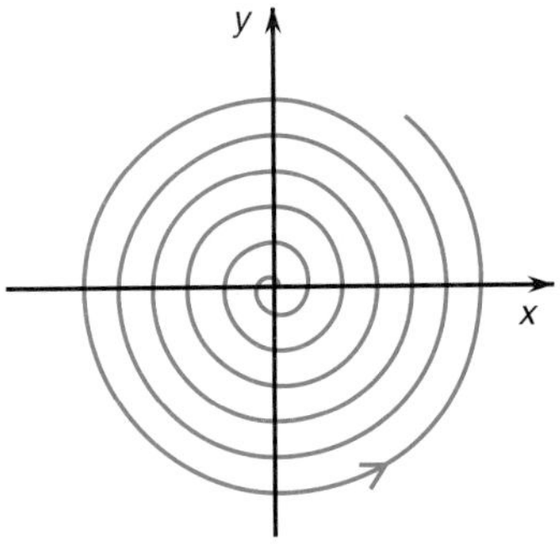

Figure 13.1.10

EXAMPLE 3 Parametrize the Spiral of Archimedes, given in polar coordinates by $r = \theta$, for $\theta \geq 0$. See Figure 13.1.10.

Solution We need to describe a point that goes around and around the origin. For instance, the point $(x, y) = (\cos t, \sin t)$ does this. The problem is that it always stays on the circle of radius 1.

We could also parametrize the spiral *in polar coordinates* by $r = t$, $\theta = t$, $t \in [0, \infty)$, and then find $x = r\cos\theta = t\cos t$, $y = r\sin\theta = t\sin t$.

We want a point that goes around and around but whose distance from the origin equals the angle. One way to achieve this is to let $(x, y) = (t\cos t, t\sin t)$. This point is at the same polar angle relative to the origin as $(\cos t, \sin t)$, so it goes around and around. However, its distance from the origin is $\sqrt{t^2\cos^2 t + t^2\sin^2 t} = |t|$. In particular, if $t \geq 0$, the distance is t, and this equals the polar angle. The spiral can be parametrized by the equations $(x, y) = (t\cos t, t\sin t)$, for $t \geq 0$.

Exercises 13.1

I

Find parametrizations for each of the following curves. Specify the domain and say how many times the curve is traced out. Notice that there will be many possible answers to these questions.

1. $x = 4y$
2. $3x + 4y = 5$
3. $x = y^2 + 2$
4. $2x - y^2 + 3y - 6 = 0$
5. $3x^2 + 4y^2 = 12$
6. $x^2 + 2x + y^2 = 3$
7. $y = 2$
8. $x^3 - 4y^2 - 2y + 8 = 0$
9. $y + 2\sin x = 3$
10. $y = \sec x$

Find an equation for the curve that is traced out by each of the following sets of parametric equations. Notice that there will be various correct answers, but try to put the equation into a standard sort of form from which the type of curve is easily recognized. Identify which portion of the curve is traced out by the moving point, how many times it is covered, and the direction in which the point moves over the curve.

11. $x = \sin t$, $y = \cos t$, $t \in [0, 2\pi]$
12. $x = 4\cos(2t)$, $y = 9\sin(2t)$, $t \in [-\pi, \pi]$
13. $x = 3t^4 - 5$, $y = 6 - 4t^4$, $t \in \mathbb{R}$
14. $x = 2t + t^2$, $y = t + 1$, $t \in (-\infty, 0)$
15. $x = 3 + e^t$, $y = 5 - 2e^t$, $t \in [0, \infty)$
16. $x = \frac{1}{1+t}$, $y = \frac{t}{1+t}$, $t \in [0, 1]$
17. $x = t^4$, $y = t^4 - 2t^2$, $t \in \mathbb{R}$
18. $x = 9\cosh t$, $y = 4\sinh t$, $t \in \mathbb{R}$
19. $x = \sinh t$, $y = 1 - \cosh t$, $t \geq 0$
20. $x = \sqrt{\arctan t}$, $y = 1 - \arctan t$, $t \in [0, \infty)$

II

21. Find a parametrization for the rectangle of Example 1(iv) that begins and ends at the point $(3, 2)$ and goes around the rectangle in the counterclockwise direction.
22. Find a parametrization for the rectangle of Example 1(iv) that begins and ends at the point $(0, 2)$ and goes around the rectangle in the clockwise direction.
23. Find a parametrization for the rectangle of Example 1(iv) that begins and ends at the point $(2, 0)$ and goes around the rectangle in the counterclockwise direction.
24. Consider the triangle whose vertices are $(1, 1)$, $(3, 1)$, and $(1, 3)$, and find a parametrization for it that begins and ends at $(1, 1)$ and moves counterclockwise.

*25. Consider the curve $y = \arctan x$ (i.e., the set of points in the plane that are the graph of this function). Show that it can be parametrized by either of the following pairs of functions:

$$x = t, \qquad y = \arctan t, \qquad \text{for } t \in \mathbb{R},$$

or

$$x = \tan t, \qquad y = t, \qquad \text{for } t \in \left(-\frac{\pi}{2}, \frac{\pi}{2}\right).$$

*26. At the end of Example 1(iv) it was observed that the functions given in that example for $x(t)$ and $y(t)$ are continuous on $[0, 10]$ but that they are not differentiable at 3, 5, and 8. Sketch the graph of each of these functions and verify these observations.

III

In each of the following exercises, use a graphics calculator or computer to plot the graphs of $x(t)$ and $y(t)$ as functions of t. Look at them and try to predict the approximate shape of the curve $(x(t), y(t))$ as t runs over the whole real line $\mathbb{R}$. If possible, graph the parametric curve and confirm your predictions.

27. $x(t) = t^3$, $y(t) = t^6$
28. $x(t) = t^4$, $y(t) = t^8$
29. $x(t) = \cos t$, $y(t) = \sin^2 t$
30. $x(t) = \sin t$, $y(t) = \arctan t$

POINT TO PONDER

In this section we have considered curves that lie in the plane. In practical situations we usually encounter paths lying in three-dimensional space, and it seems that for most purposes it will be necessary to work in that context.

Most of what we will do with curves in three-dimensional space is analogous to what happens with plane curves but a little more complicated. There are a few things that are genuinely different and have to be considered separately, but on the whole, plane curves are a good introduction to the general case.

It is also true that plane curves are more important in their own right than is apparent at first glance. If we think about the earth orbiting the sun in space, while it is true that it is moving through three-dimensional space, it is also true that the actual path of the earth lies in a flat plane. Since the sun also lies in this plane, everything that matters for studying the earth's motion does in fact lie in a plane.

We can study this motion as if it were just occurring in the ordinary xy-plane. One way to think of it is that if we choose the coordinates correctly, the sun and the earth's orbit will all lie in the horizontal plane, and

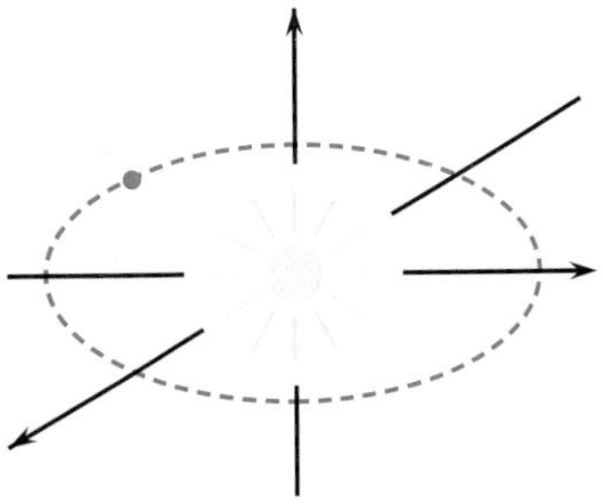

Figure 13.1.11

we can simply ignore the third coordinate; see Figure 13.1.11.

So while to some extent studying motions in the plane is an idealization of what actually happens in the real world, it is often quite accurate. We are all accustomed to looking at flat maps that represent the surface of the earth. Provided the region in question is not too large, there is no need to use a round globe even though it would theoretically be more accurate. In just the same way, a great many problems can be described by motions in a well-chosen plane. Sometimes this involves making approximations, and other times it is exact.

13.2 Vectors and Motion

DISPLACEMENT
VELOCITY
SPEED
TANGENT VECTOR
ARCLENGTH

To specify the velocity of an object moving along a straight line as shown in Figure 13.2.1, it is necessary to give its speed and the direction. For an object moving along a curved path, as in Figure 13.2.2, it is less clear what is meant by either of these concepts.

We begin by recalling what we did in the straight-line situation. If the position of the point on the line is given by $x = x(t)$, then to find its velocity at time $t =$

Figure 13.2.1

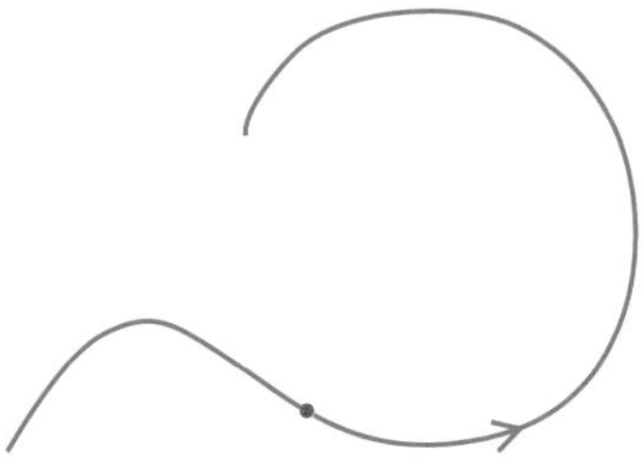

Figure 13.2.2

a, we considered the difference quotient $\frac{x(a+h)-x(a)}{h}$ and took the limit as $h \to 0$. The difference quotient represents the average velocity of the object over the short interval of time from a to $a+h$. Its numerator $x(a+h) - x(a)$ represents the change in the displacement of the point starting at $t = a$ and ending at $t = a + h$.

When we discussed velocity for a point moving along a straight line, we realized that it was necessary to use the idea of displacement. It is like distance along the line except that it is positive in one direction and negative in the other. On a line, there are only two directions. For motion in the plane it is not possible to specify a direction by a simple sign.

Vectors

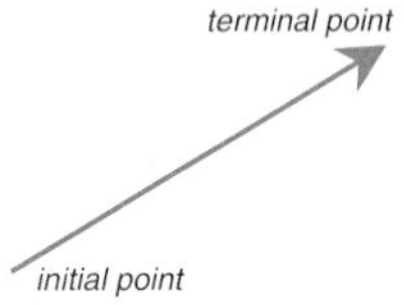

Figure 13.2.3

What is needed is a way of indicating a direction and a distance. This can be accomplished by using **vectors**. For this section we will think of a vector as an arrow in the plane (see Figure 13.2.3). It begins at some *initial point* and ends with the arrowhead at its *terminal point*. The arrow points in a particular direction, and its length is a nonnegative number that gives a distance. In this way vectors can be used to specify both a direction and a distance.

Frequently, we place the initial point of a vector **v** at the origin. In this case **v** is determined by the coordinates (x, y) of its terminal point. The numbers x and y are called the **components** of **v**; x is the **horizontal component**, and y is the **vertical component**. The arrow points x units horizontally and y units vertically. It is very useful to describe a vector **v** in this way, and we write $\mathbf{v} = (x, y)$.

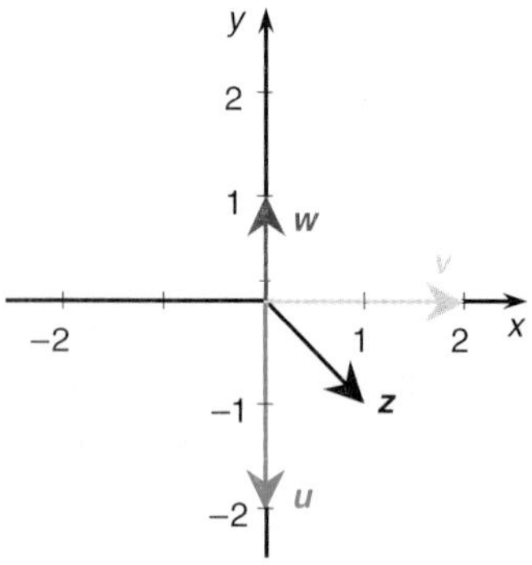

Figure 13.2.4

Recall that a vector has both a length and a direction. The vector $\mathbf{v} = (2, 0)$ points horizontally to the right and has length 2; the vector $\mathbf{u} = (0, -2)$ points straight down and has length 2; the vector $\mathbf{w} = (0, 1)$ points straight up and has length 1; the vector $\mathbf{z} = (1, -1)$ points down and to the right at an angle of $45°$ below the horizontal and has length $\sqrt{2}$. (To do this last one, we noticed that the vector $\mathbf{z} = (1, -1)$ is the arrow from the origin to the point $(1, -1)$ and that the distance from $(0, 0)$ to $(1, -1)$ is $\sqrt{2}$.) See Figure 13.2.4.

We will often use this idea, placing a vector with its initial point at the origin and then representing the vector by its components, the coordinates (a, b) of the terminal point. See Figure 13.2.5(i).

An important property of vectors is that they can be added. For instance, the vectors $\mathbf{u} = (a, b)$ and $\mathbf{v} = (c, d)$ can be added to form the vector $\mathbf{u} + \mathbf{v} = (a + c, b + d)$. Vectors are added by simply adding corresponding components.

To multiply a vector by a number, we just multiply each component by the number. For instance, with $\mathbf{u} = (a, b)$, $2\mathbf{u} = (2a, 2b)$. This is the vector that is twice as long as **u**, pointing in the same direction (see Figure 13.2.5(ii)). If the number by which we multiply is *negative*, then multiplying "turns the vector around," making it point in the opposite direction (see Figure 13.2.5(iii)).

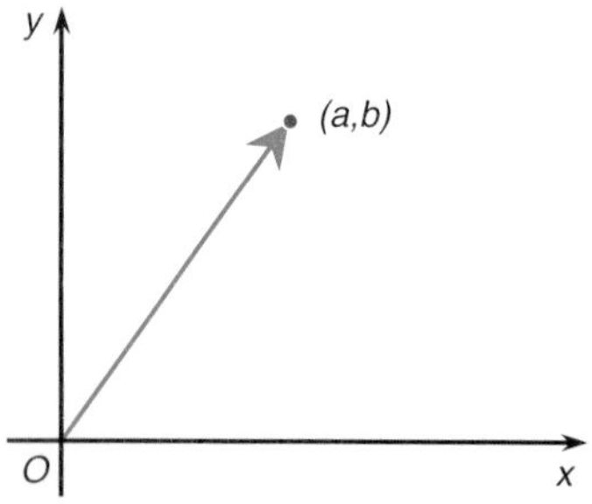

Figure 13.2.5(i)

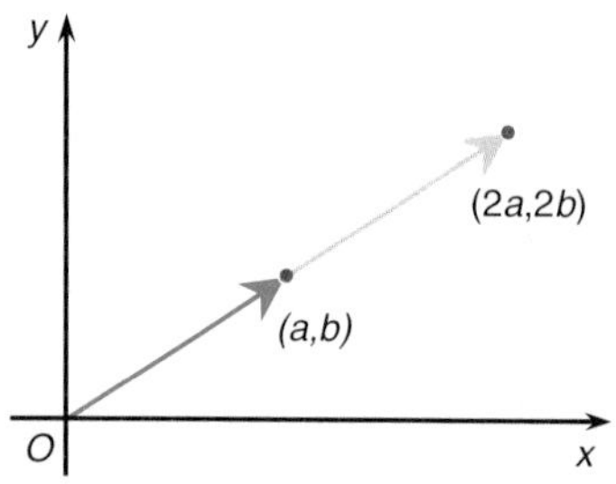

Figure 13.2.5(ii)

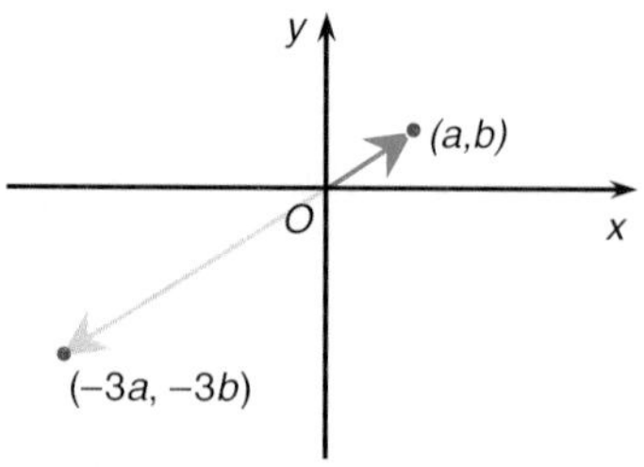

Figure 13.2.5(iii)

It also makes sense to consider the difference of two vectors; with $\mathbf{u} = (a, b)$ and $\mathbf{v} = (c, d)$, we find that $\mathbf{u} - \mathbf{v} = (a - c, b - d)$. For instance, $(3, -5) - (2, -1) = (1, -4)$ (see Figure 13.2.5(iv)).

The vectors **u** and **v** below are drawn with their initial points at the origin; **u** − **v** is not.

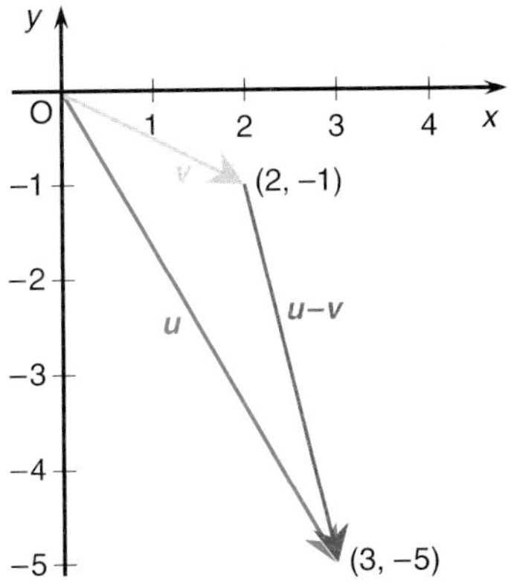

Figure 13.2.5(iv)

Notice that the length of the vector $\mathbf{u} - \mathbf{v}$ is $\sqrt{(a-c)^2 + (b-d)^2}$, which is just the distance from the point (a, b) to (c, d). The length of the difference vector is the distance between the terminal points of the two original vectors.

When we specify the position of a point in the plane by giving its coordinates (x, y), we can think of this as a vector that starts at the origin and ends at the point. We refer to it as the **displacement** of the point (meaning its displacement from the origin).

For the problem of a moving point, we want to consider the difference between the displacements of two nearby points. Suppose the position of the point is given by $(x, y) = \big(x(t), y(t)\big)$, so we should consider the difference $\big(x(a+h), y(a+h)\big) - \big(x(a), y(a)\big)$. To make sense of this formula, we think of the two points as *vectors* and write

$$\big(x(a+h), y(a+h)\big) - \big(x(a), y(a)\big) = \big(x(a+h) - x(a), y(a+h) - y(a)\big).$$

In other words, we just subtracted the two x-components and subtracted the two y-components. This is easy to understand graphically if we mark the two points $\big(x(a), y(a)\big)$ and $\big(x(a+h), y(a+h)\big)$ as shown in Figure 13.2.6. Their "difference" $\big(x(a+h), y(a+h)\big) - \big(x(a), y(a)\big) = \big(x(a+h) - x(a), y(a+h) - y(a)\big)$ can be visualized as the vector going from $\big(x(a), y(a)\big)$ to $\big(x(a+h), y(a+h)\big)$: If we start at $\big(x(a), y(a)\big)$ and move $x(a+h) - x(a)$ in the horizontal direction and $y(a+h) - y(a)$ in the vertical direction, then we will end up at the second point $\big(x(a+h), y(a+h)\big)$.

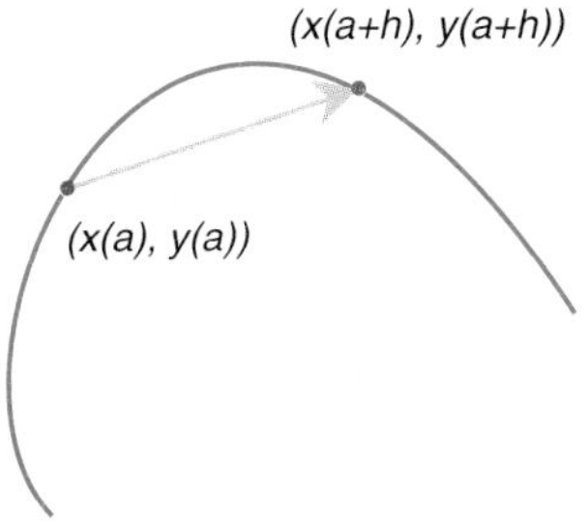

Figure 13.2.6

Returning to our moving point, we call the vector $\big(x(a+h) - x(a), y(a+h) - y(a)\big)$ the change in the **displacement** from $\big(x(a), y(a)\big)$ to $\big(x(a+h), y(a+h)\big)$. This change in the displacement includes both the idea of the distance between these points and the direction from one to the other.

DEFINITION 13.2.1

If (x, y) and (u, v) are two points in the plane, then the **change in the displacement** from (u, v) to (x, y) is the *vector*

$$(x, y) - (u, v) = (x - u, y - v),$$

which is the difference of the two vectors. We visualize it as an arrow that begins at the point (u, v) and ends at the point (x, y).

The length of the displacement vector is

$$\sqrt{(x-u)^2 + (y-v)^2},$$

which is the distance from (x, y) to (u, v).

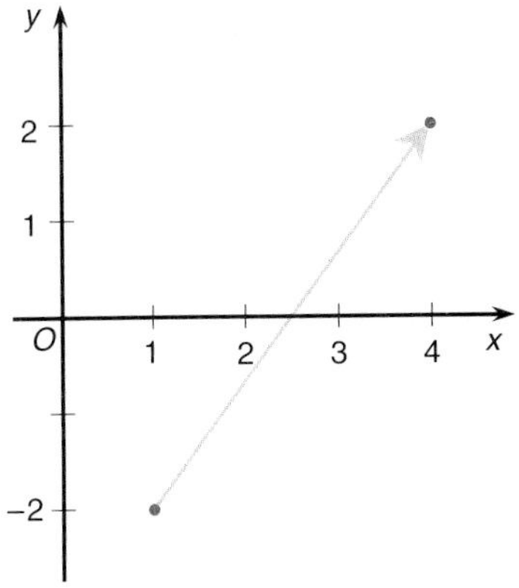

Figure 13.2.7

EXAMPLE 1

(i) The displacement from the point $(1, -2)$ to the point $(4, 2)$ is the vector $(4, 2) - (1, -2) = (3, 4)$. The distance between these points is the length of the displacement vector, which is $\sqrt{3^2 + 4^2} = 5$. See Figure 13.2.7.

(ii) The displacement from the origin $(0, 0)$ to a point (a, b) is the vector $(a, b) - (0, 0) = (a, b)$. This just means that to get from the origin to (a, b), it is necessary to go along the vector (a, b).

(iii) It is important to notice that the displacement from P to Q is not the same as the displacement from Q to P; in fact these two displacements are each other's negatives, meaning that they are vectors that have the same length but point in opposite directions.

Of course, this is obvious if we consider that moving from P to Q involves the same distance as moving from Q to P but in the opposite direction.

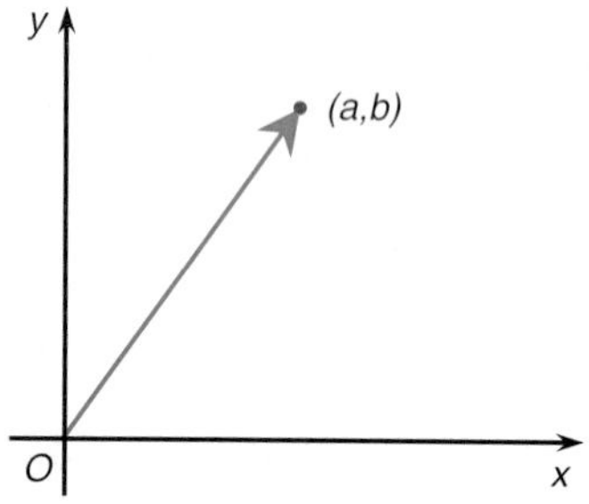

Figure 13.2.8

In part (ii) of Example 1 we wrote the expression (a, b) to mean two very different things. On the one hand, it represents the coordinates of a point in the plane. On the other, it represents a vector by its components. The vector is an arrow, and if we place its initial point at the origin, then its terminal point will be at the point (a, b). See Figure 13.2.8.

Because the vector and the point are closely related in this way, it is very convenient to use the same notation (a, b) for both of them, but in the backs of our minds we should remember that they are different things.

Velocity

Consider the moving point whose position at time t is the point $\big(x(t), y(t)\big)$. It is sometimes convenient to write

$$\mathbf{r} = \mathbf{r}(t) = \big(x(t), y(t)\big).$$

The letter $\mathbf{r}$ reminds us of "radius"; $\mathbf{r}(t)$ is the vector pointing from the origin out to the position of the moving point at time t.

Now as time t changes from $t = a$ to $t = a + h$, the moving point goes from $\mathbf{r}(a) = \big(x(a), y(a)\big)$ to $\mathbf{r}(a+h) = \big(x(a+h), y(a+h)\big)$. Its displacement changes by $\big(x(a+h) - x(a), y(a+h) - y(a)\big) = \mathbf{r}(a+h) - \mathbf{r}(a)$. By analogy with the straight line case we should try to find the average velocity over this short time. To do this, it is necessary to divide by h or, to put it another way, to multiply by $\frac{1}{h}$. It seems reasonable to expect that the average velocity over the interval from $t = a$ to $t = a + h$ is

To multiply a vector by $\frac{1}{h}$, we simply multiply each of its coordinates by $\frac{1}{h}$.

$$\frac{1}{h}\big(x(a+h) - x(a), y(a+h) - y(a)\big) = \left(\frac{x(a+h) - x(a)}{h}, \frac{y(a+h) - y(a)}{h}\right).$$

Notice that this formula for the average velocity is a *vector*, whose two components are difference quotients. Our original idea was to form the difference quotient for a moving point, allow h to tend to 0, and take the limit. We know that as $h \to 0$, the two difference quotients will both tend to limits: $\frac{x(a+h)-x(a)}{h} \to \frac{dx}{dt}(a) = x'(a)$ and $\frac{y(a+h)-y(a)}{h} \to \frac{dy}{dt}(a) = y'(a)$.

What we have found is that the *vector difference quotient*

$$\frac{1}{h}\big(\mathbf{r}(a+h) - \mathbf{r}(a)\big) = \frac{1}{h}\Big(\big(x(a+h), y(a+h)\big) - \big(x(a), y(a)\big)\Big)$$

$$= \left(\frac{x(a+h) - x(a)}{h}, \frac{y(a+h) - y(a)}{h}\right)$$

tends to the limit $\big(x'(a), y'(a)\big)$ as $h \to 0$. This vector represents the *instantaneous velocity* of the moving point at the time $t = a$; it points in the direction in which the point is moving.

DEFINITION 13.2.2

> Consider a point moving in the plane, and suppose that its position at time t is given by $\mathbf{r}(t) = \big(x(t), y(t)\big)$, where $x(t)$ and $y(t)$ are functions that are differentiable at $t = a$. The **velocity** of the moving point at time $t = a$ is the vector
>
> $$\mathbf{r}'(a) = \big(x'(a), y'(a)\big).$$

We found this formula by following the model of what we did for motion in a straight line, but it turns out to work very well.

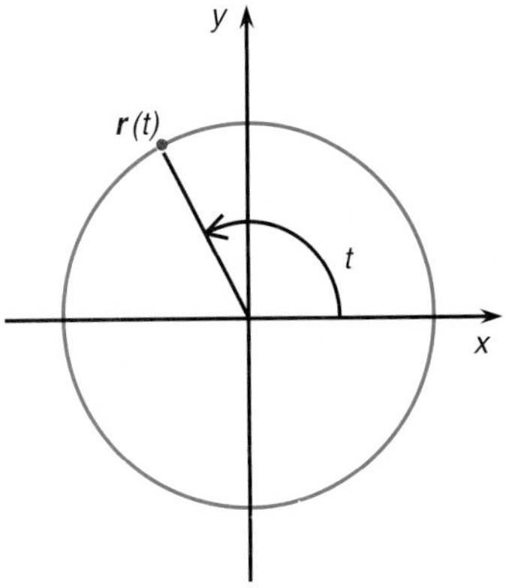

Figure 13.2.9(i)

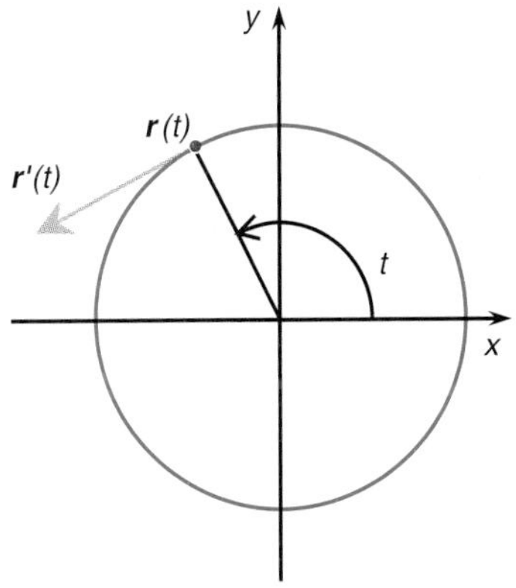

Figure 13.2.9(ii)

EXAMPLE 2 Consider the moving point

$$(x, y) = \mathbf{r}(t) = \big(x(t), y(t)\big) = (\cos t, \sin t).$$

We know that these are parametric equations for the circle of radius 1 around the origin. The point moves around the circle counterclockwise, making one complete revolution each time t increases by 2π.

The velocity vector is given by Definition 13.2.2. It is

$$\mathbf{r}'(t) = (-\sin t, \cos t).$$

Notice that it depends on t, meaning that the velocity changes and will be different at different times.

In Figure 13.2.9(i) we plot the point $(x, y) = \mathbf{r}(t) = (\cos t, \sin t)$. In Figure 13.2.9(ii) we also plot the corresponding velocity vector $\mathbf{r}'(t) = (-\sin t, \cos t)$, placing it so that its initial point is at the moving point $\mathbf{r}(t) = (\cos t, \sin t)$ on the circle. It turns out that the velocity vector is tangent to the circle, and its length is $\sqrt{\sin^2 t + \cos^2 t} = 1$.

From Example 2 we can see two very important features of velocity vectors. The first is that the direction in which the vector points is exactly the direction in which the moving point is moving. This direction may be different at different times, but it will be tangent to the curve. The second property is that the length of the velocity vector is the instantaneous *speed* at which the point is moving. In this example the length of the velocity vector is always 1. On the other hand, we know that the point $(\cos t, \sin t)$ goes exactly once around the circle in 2π units of time. Since it travels a distance of 2π, the circumference, we find that its speed is $\frac{2\pi}{2\pi} = 1$, equal to the length of the velocity vector.

We will use these observations to define the concepts of a tangent vector to a curve and of the speed of a moving point.

DEFINITION 13.2.3

> Consider a curve traced out by the moving point $\mathbf{r}(t) = \big(x(t), y(t)\big)$ for t in some interval I. Suppose $a \in I$ is a point at which both $x(t)$ and $y(t)$ are differentiable, and let $\mathbf{r}'(a) = \big(x'(a), y'(a)\big)$ be the velocity vector. Suppose that $\mathbf{r}'(a)$ is not the zero vector, that is, that $x'(a)$ and $y'(a)$ are not both zero.
>
> Then $\mathbf{r}'(a)$ is a **tangent vector** to the curve at the point $\mathbf{r}(a) = \big(x(a), y(a)\big)$, and its length $\|\mathbf{r}'(a)\| = \sqrt{x'(a)^2 + y'(a)^2}$ is the **speed** of the moving point at $t = a$.

EXAMPLE 3

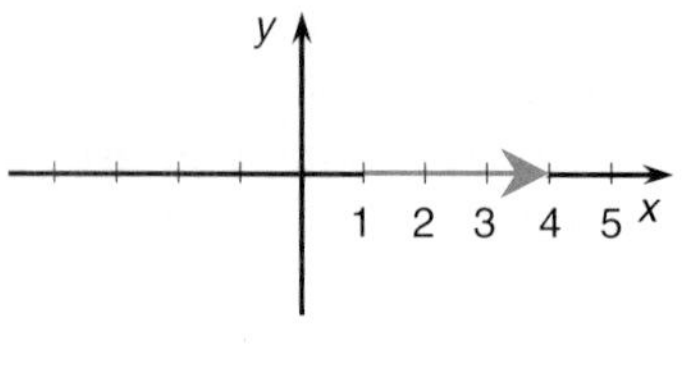

Figure 13.2.10

(i) Consider the parametric equations $x = 1 + 3t$, $y = 0$. They parametrize the horizontal axis. The velocity vector is $\mathbf{r}'(t) = (3, 0)$. Its horizontal component is 3, and its vertical component is 0.

Certainly this vector is tangent to the horizontal axis (see Figure 13.2.10). Moreover, the length of the velocity vector is 3, which means that the speed of the moving point is 3. This is confirmed by looking at the original equations. If t increases by 1 unit, then $x = 1 + 3t$ will increase by 3 units; this means that the point moves 3 units of distance in every 1 unit of time, which is exactly what it means to say its speed is 3.

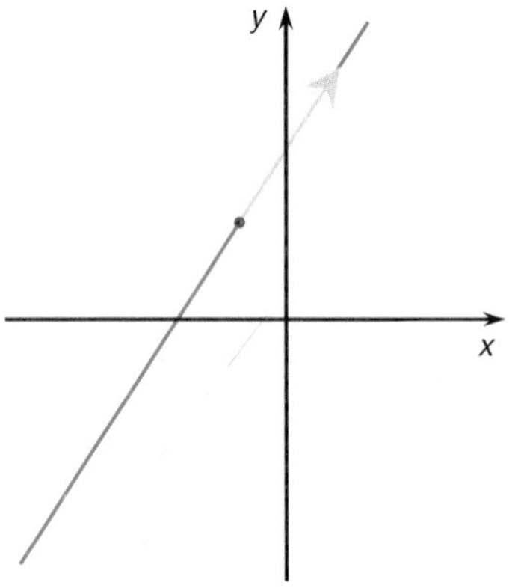

Figure 13.2.11

(ii) Consider the parametric equations $x = 2t - 1$, $y = 3t + 2$. Solving for t, we find that $t = \frac{x+1}{2}$ and $t = \frac{y-2}{3}$. Equating these expressions, we find that the curve is the straight line $\frac{x+1}{2} = \frac{y-2}{3}$, or $3x - 2y + 7 = 0$.

The velocity vector is $\mathbf{v} = \big(x'(t), y'(t)\big) = (2, 3)$. If we consider a time $t = a$ and draw the corresponding point on the line (see Figure 13.2.11), then we should place the velocity vector so that its initial point is at that point on the line. In the picture it seems that the vector is tangent to the line; to prove it, we can check that the slope of the line is $\frac{3}{2}$ (cf. Remark 1.5.10(ii), p. 27). The vector $(2, 3)$ points 2 units horizontally and 3 units vertically, so its slope is also $\frac{3}{2}$, and this is why it is tangent to the line.

Notice that the length of the velocity vector is $\sqrt{2^2 + 3^2} = \sqrt{13}$. The speed of the moving point is $\sqrt{13}$ units of distance per unit of time.

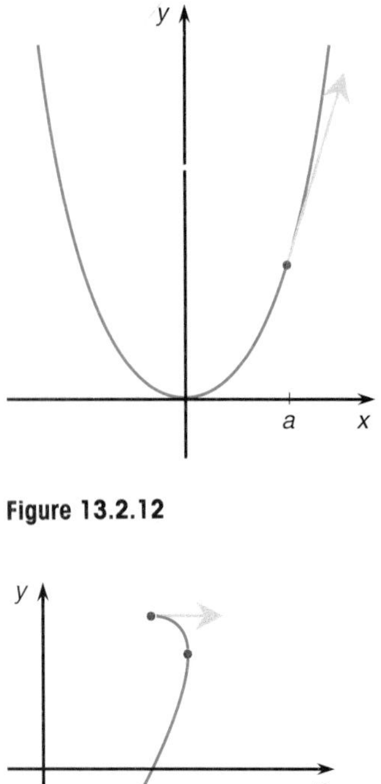

Figure 13.2.12

(iii) The parametric equations $x = t$, $y = t^2$ parametrize the parabola $y = x^2$. The velocity vector $\mathbf{r}'(t) = (1, 2t)$ changes as t varies. This is more complicated than the previous two examples, in both of which the velocity vector is constant.

Consider the time $t = a$ and the corresponding point (a, a^2) on the parabola, as shown in Figure 13.2.12. We know from our earlier work that the tangent to the parabola at this point has slope $2a$. On the other hand, the velocity vector we have found is $\mathbf{r}'(a) = (1, 2a)$, whose slope is also $\frac{2a}{1} = 2a$. This confirms that the velocity vector is in fact tangent to the curve at each point.

The speed of the moving point is the length of the velocity vector, which is $\sqrt{1^2 + (2a)^2} = \sqrt{1 + 4a^2}$. This depends on the time $t = a$. Notice that it is larger for large values of a; its smallest value will occur when $a = 0$. In terms of the illustration this means that the moving point moves most slowly at the vertex of the parabola and moves faster the farther it is from the origin.

(iv) Suppose a ship's course is parametrized by the moving point $\mathbf{r}(t) = (3 + 2t - t^2, 4 - t^2)$, for $t \in [0, 3]$. The axes are arranged so that the positive x-axis points due east and the positive y-axis points due north. At what moments is the ship sailing (a) east, (b) north, (c) northwest?

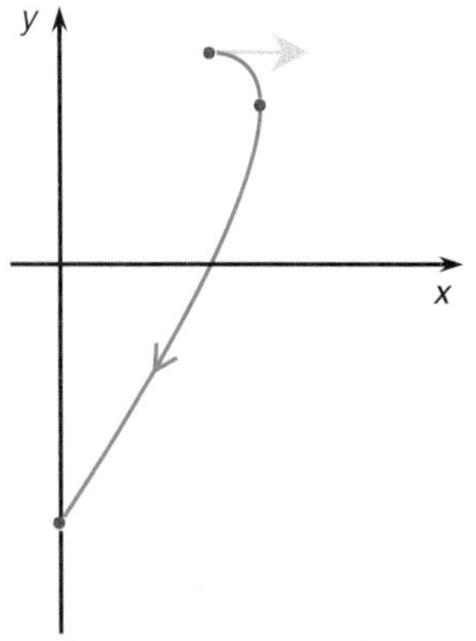

Figure 13.2.13

Solution

(a) The velocity vector is $\mathbf{r}'(t) = (2 - 2t, -2t)$. When the ship is sailing due east, the velocity vector points horizontally to the right, so it is $(k, 0)$, for some positive number k. For the vector $\mathbf{r}'(t) = (2 - 2t, -2t)$ to be of this form we must have $t = 0$. At $t = 0$ the velocity is $(2, 0)$, which does point due east. This means that the ship starts out sailing due east, as shown in Figure 13.2.13.

(b) If the ship is sailing due north, the velocity vector points straight up, that is, $\mathbf{r}' = (0, k)$, for some $k > 0$. In particular this means that $2 - 2t$ must equal zero, so $t = 1$. However, if we look at the velocity vector $\mathbf{r}'(1) = (0, -2)$, we realize that it points due *south*. There is no time at which the ship is sailing due north.

(c) If the ship is sailing northwest, its velocity vector must point up and to the left at a 45° angle above the horizontal. This means that the vector must be $(-k, k)$, for some $k > 0$. We solve to find when this is possible by setting $-(2 - 2t) = -2t$. Adding $2t$ to each side, we find that $4t - 2 = 0$, so $t = \frac{1}{2}$.

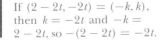

If $(2 - 2t, -2t) = (-k, k)$, then $k = -2t$ and $-k = 2 - 2t$, so $-(2 - 2t) = -2t$.

But when $t = \frac{1}{2}$, the velocity is $(1, -1)$, so the ship is sailing *southeast*. There is no time when the ship is sailing northwest.

Notice that it was possible to answer part (iv) of Example 3 without identifying the curve or finding its equation. The velocity vector is an extremely useful tool for looking at the motion of a point in the plane. It tells us the direction and the instantaneous speed of the motion.

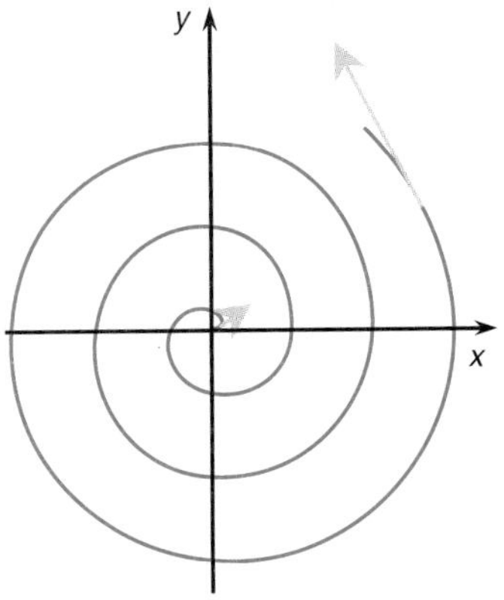

Figure 13.2.14

EXAMPLE 4 Recall the Spiral of Archimedes from Example 12.2.2, which we parametrized in Example 13.1.3 by the equations $(x, y) = \mathbf{r}(t) = (t\cos t, t\sin t)$. The velocity vector of the moving point at time t is $\mathbf{r}' = (\cos t - t\sin t, \sin t + t\cos t)$. When t is small, this is almost equal to the vector $(\cos t, \sin t)$, which points in the direction from the origin out to the point $(t\cos t, t\sin t)$. So at the beginning the point is mostly just moving away from the origin.

But when t is large, the corresponding velocity vector is nearly equal to $(-t\sin t, t\cos t)$ (i.e., the other part is much smaller). This is a vector that points in the direction tangent to a circle around the origin. For large t the point moves almost in a circle, with only a relatively small motion away from the origin. See Figure 13.2.14.

Parametric equations can also be used in situations having nothing to do with moving objects.

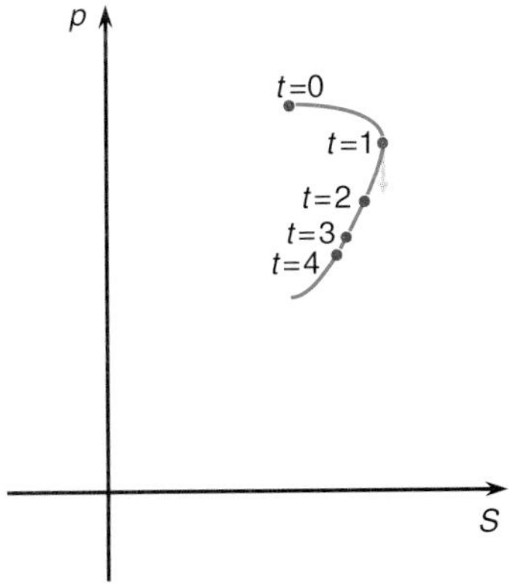

Figure 13.2.15

EXAMPLE 5 **Price Versus Sales** After a new consumer electronics product is introduced, sales rise quickly and the price gradually comes down. We measure time t in years, with $t = 0$ at the time of the first sales. Suppose the unit price at time t, in hundreds of dollars, is $p(t) = \frac{t^2+8}{t^2+4}$, and the monthly sales, in units of 100,000, are $S(t) = \frac{t^2+t+1}{t^2+1}$.

In Figure 13.2.15 we plot the curve $\mathbf{r}(t) = \big(S(t), p(t)\big)$. The dots represent the values at $t = 0, 1, 2, 3, 4$. From the shape of the curve we see that sales rise sharply at first and the price remains almost fixed. But as sales growth slows, competitive pressures force the price down.

A quick calculation shows that $S'(1) = 0$; this is the point at which the tangent vector points straight down. After 1 year, sales have peaked and the price is falling.

Arclength

A natural question to ask about a curve $\mathbf{r}(t) = \big(x(t), y(t)\big)$ is how to find its length, that is, its arclength. For instance, we are accustomed to finding that the length of the circle $\mathbf{r}(t) = (R\cos t, R\sin t)$ from $t = 0$ to $t = 2\pi$ is $2\pi R$.

Let us suppose the curve starts at $t = a$, and suppose the length of the curve from $\mathbf{r}(a)$ to $\mathbf{r}(t)$ is called $s = s(t)$. Further let us suppose that s is a differentiable function of t.

Then $s'(t)$ is the rate of change of s, that is, it is the rate at which the length of the curve is increasing at the point $\mathbf{r}(t)$. Another way to put it is that this is the rate of change of the distance traveled by the moving point since $t = a$.

But that is exactly what we call the *speed* of the moving point. We know that the speed is given by the length of the velocity vector $\mathbf{r}'(t) = \big(x'(t), y'(t)\big)$, so we find that

$$s'(t) = \sqrt{x'(t)^2 + y'(t)^2}.$$

Knowing the derivative of $s(t)$, we can find the arclength itself by integrating. The arclength of the curve between the points $\mathbf{r}(a)$ and $\mathbf{r}(b)$ is

$$s(b) = \int_a^b \sqrt{x'(t)^2 + y'(t)^2}\, dt.$$

To find this expression, we assumed that it makes sense to talk of arclength and also assumed that it is a differentiable function of the parameter t. Then we used our intuition about speed to find the formula. In a way this is not very satisfactory because we have never proved that what we call the "speed" of the moving point has anything to do with the rate of change of distance.

In fact these things are all correct, but instead of worrying about it, we will let the above argument explain where the formula for arclength comes from and why it seems reasonable. We will simply use the formula to "define" arclength.

We could also have approximated the curve by short straight segments; the sum of their lengths gives a Riemann sum. Letting the lengths of the segments tend to zero, we would find that the sum tends to this same integral, reassuring us that we have found the right definition.

DEFINITION 13.2.4

Suppose $x(t)$ and $y(t)$ are differentiable with continuous derivatives. If $a \le b$, the **arclength** of the curve $\mathbf{r}(t) = \big(x(t), y(t)\big)$ between the points corresponding to $t = a$ and $t = b$ is defined to be

$$s = \int_a^b \sqrt{x'(t)^2 + y'(t)^2}\, dt.$$

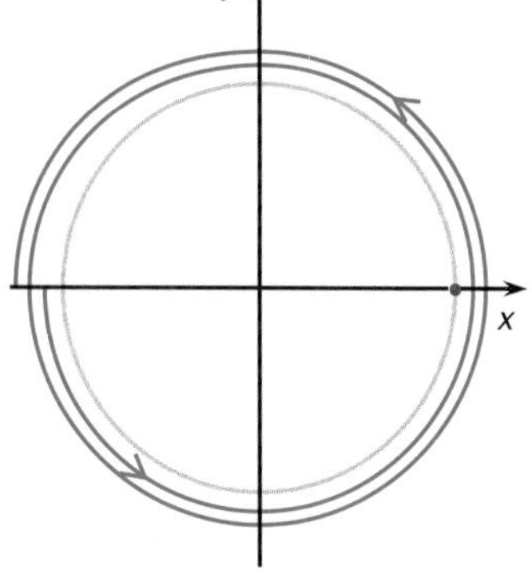

Figure 13.2.16

EXAMPLE 6 **Arclength of a Circle** Consider the circle $\mathbf{r}(t) = (R\cos t, R\sin t)$. Here the components are $x(t) = R\cos t$ and $y(t) = R\sin t$, so the arclength between $t = a$ and $t = b$ is

$$s = \int_a^b \sqrt{R^2\sin^2 t + R^2\cos^2 t}\, dt = \int_a^b R\, dt = R(b - a).$$

So, for instance, between $t = 0$ and $t = 2\pi$, which corresponds to going exactly once around the circle, the arclength is $R(2\pi - 0) = 2\pi R$. Of course, this is the circumference of the circle. See Figure 13.2.16.

Notice that the arclength between $t = -\pi$ and $t = 3\pi$ is $4\pi R$. As t runs from $-\pi$ to 3π, the point $\mathbf{r}(t)$ goes around the circle twice, and the distance it covers is twice the circumference.

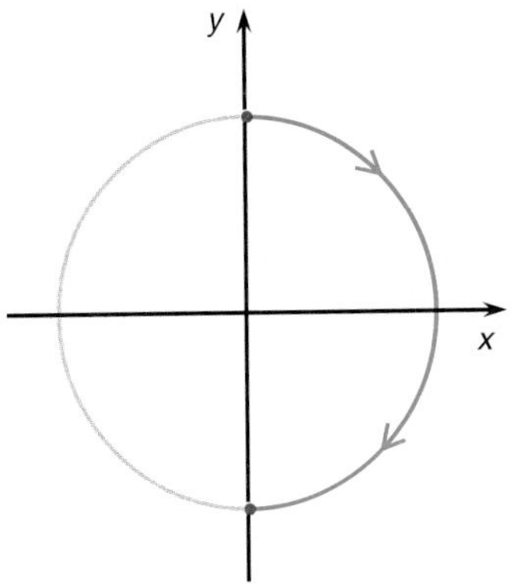

Figure 13.2.17

Of course, it is no great surprise that we can find the arclength of a circle, especially because the parametrization we chose is so convenient. However, the formula can be used for less well-chosen parametrizations.

EXAMPLE 7 Let $\mathbf{r}(t) = \big(\sin(\pi t^2), \cos(\pi t^2)\big)$, for $t \in [0, 1]$.

This point will move along the circle $x^2 + y^2 = 1$, starting at the point $\mathbf{r}(0) = (0, 1)$ and moving along the right half of the circle to $\mathbf{r}(1) = (0, -1)$ (see Figure 13.2.17). Its speed is not constant.

The arclength is given by

$$\begin{aligned} s &= \int_0^1 \sqrt{x'(t)^2 + y'(t)^2}\,dt = \int_0^1 \sqrt{(2\pi t)^2 \cos^2(\pi t^2) + (2\pi t)^2 \sin^2(\pi t^2)}\,dt \\ &= \int_0^1 2\pi t\sqrt{\cos^2(\pi t^2) + \sin^2(\pi t^2)}\,dt = 2\pi \int_0^1 t\,dt = \pi t^2\Big|_0^1 \\ &= \pi. \end{aligned}$$

It is reassuring to find that even when the point moves at a varying speed, we still get the correct arclength for the semicircle.

EXAMPLE 8 Find the arclength of the curve $\mathbf{r}(t) = (1 + t^2, 3 - 2t^3)$, for $t \in [0, 1]$.

Solution The arclength is

$$s = \int_0^1 \sqrt{(2t)^2 + (-6t^2)^2}\,dt = \int_0^1 \sqrt{4t^2 + 36t^4}\,dt = \int_0^1 2t\sqrt{1 + 9t^2}\,dt.$$

We substitute $u = 1 + 9t^2$, $du = 18t\,dt$, and get

$$\int_1^{10} \frac{1}{9}u^{1/2}du = \frac{1}{9}\frac{2}{3}u^{3/2}\Big|_1^{10} = \frac{2}{27}(10\sqrt{10} - 1).$$

The arclength is $\frac{2}{27}(10\sqrt{10} - 1) \approx 2.268$.

REMARK 13.2.5

In Definition 6.9.1 (p. 485) we found that the arclength for the curve formed by a graph $y = f(x)$ between $x = a$ and $x = b$ is given by

$$s = \int_a^b \sqrt{1 + f'(x)^2}\,dx.$$

If we want to represent this graph by parametric equations, the easiest thing to do is to write $\mathbf{r}(t) = \big(t, f(t)\big)$, for $t \in [a, b]$, that is, let $x(t) = t$ and $y(t) = f(t)$. Since $x'(t) = \frac{dt}{dt} = 1$, Formula 13.2.4 for arclength using parametric equations gives exactly the same formula as Definition 6.9.1 in this particular case. Of course, Definition 13.2.4 also applies to curves that are not represented as graphs, so it is more general than Definition 6.9.1.

Suppose we consider a curve given in polar coordinates by $r = f(\theta)$. It can be parametrized by $\big(x(\theta), y(\theta)\big) = (r\cos\theta, r\sin\theta) = \big(f(\theta)\cos\theta, f(\theta)\sin\theta\big)$, using θ as

We can use a graphics calculator to sketch the curve from Example 8. To set the ranges, we note that for $t \in [0, 1]$ the value of $x(t) = 1 + t^2$ is in $[1, 2]$, and the value of $y(t) = 3 - 2t^3$ is in $[1, 3]$.

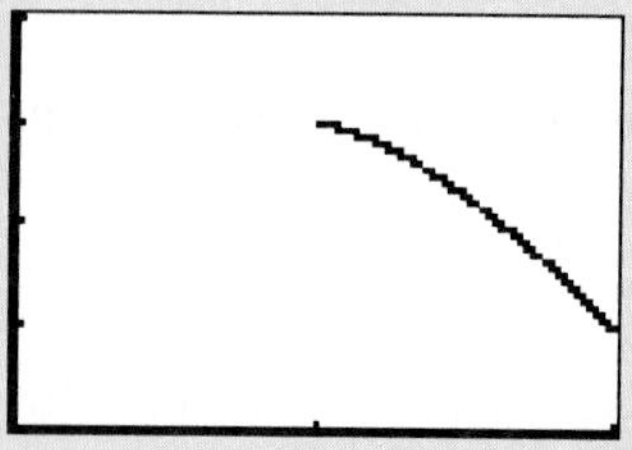

TI-85	SHARP EL9300
MODE ▼ ▼ ▼ ▼ ► ► (Param) ENTER EXIT GRAPH F1($E(t)$ =) 1 + F1(t) x^2 ▼ 3 − 2 F1(t) ^ 3 M2(RANGE) 0 ▼ 1 ▼ ▼ 0 ▼ 2 ▼ ▼ 0 ▼ 4 F5(GRAPH)	SET UP E 3 1 + x/θ/T x^2 ENTER 3 − 2 x/θ/T a^b 3 ENTER RANGE 0 ENTER 2 ENTER 1 ENTER 0 ENTER 4 ENTER 1 ENTER 0 ENTER 1 ENTER .01 ENTER
CASIO f_x-7700GB	**HP 48SX**
MODE MODE × (PARAM) Range 0 EXE 2 EXE 1 EXE 0 EXE 4 EXE 1 EXE 0 EXE 1 EXE .01 EXE Graph 1 + T x^2, 3 − 2 T x^y 3 EXE	′ 1 + T y^x 2 + α ↰ CST × () 3 − 2 × T y^x 3 ↰ PLOT NEW P A R ENTER PTYPE PARA PLOTR ↰ {} T SPC 0 SPC 1 ENTER INDEP AUTO

We did not need the picture to find the arclength, but it helps us to see that our answer is reasonable. For instance, how does our answer compare with the distance between the two endpoints of the curve?

the parameter. We find that

$$x'(\theta) = f'(\theta)\cos\theta - f(\theta)\sin\theta, \quad y'(\theta) = f'(\theta)\sin\theta + f(\theta)\cos\theta.$$

Since

$$\begin{aligned} x'(\theta)^2 + y'(\theta)^2 &= \big(f'(\theta)\cos\theta - f(\theta)\sin\theta\big)^2 + \big(f'(\theta)\sin\theta + f(\theta)\cos\theta\big)^2 \\ &= f'(\theta)^2\cos^2\theta - 2f'(\theta)f(\theta)\cos\theta\sin\theta + f(\theta)^2\sin^2\theta \\ &\quad + f'(\theta)^2\sin^2\theta + 2f'(\theta)f(\theta)\sin\theta\cos\theta + f(\theta)^2\cos^2\theta \\ &= f'(\theta)^2 + f(\theta)^2, \end{aligned}$$

we can apply Definition 13.2.4 to find the arclength.

THEOREM 13.2.6

Arclength for Curves in Polar Coordinates

If $f(\theta)$ is differentiable with continuous derivative, then the arclength of the part of the curve $r = f(\theta)$ between $\theta = \alpha$ and $\theta = \beta$ is

$$s = \int_{\alpha}^{\beta} \sqrt{f'(\theta)^2 + f(\theta)^2}\, d\theta.$$

EXAMPLE 9 **Arclength of a Cardioid** Find the arclength of the top half of the cardioid $r = 1 - \cos\theta$.

Solution Using the formula from Theorem 13.2.6 with $f(\theta) = 1 - \cos\theta, \theta \in [0, \pi]$, we find $f'(\theta) = \sin\theta$, so

$$\begin{aligned} f'(\theta)^2 + f(\theta)^2 &= \sin^2\theta + (1 - \cos\theta)^2 \\ &= \sin^2\theta + 1 - 2\cos\theta + \cos^2\theta \\ &= 2 - 2\cos\theta \\ &= 2(1 - \cos\theta). \end{aligned}$$

The arclength is $\sqrt{2}\int_0^\pi \sqrt{1 - \cos\theta}\, d\theta$. We substitute $u = 1 - \cos\theta$, so $\cos\theta = 1 - u$ and $\sin\theta = \sqrt{1 - \cos^2\theta} = \sqrt{2u - u^2}$. The arclength is

$u = 1 - \cos\theta$
$du = \sin\theta\, d\theta$

$v = 2 - u$
$dv = -du$

$$\begin{aligned} s &= \sqrt{2}\int_0^\pi \sqrt{1 - \cos\theta}\, d\theta \\ &= \sqrt{2}\int_0^2 \sqrt{u}\frac{1}{\sqrt{2u - u^2}}\, du = \sqrt{2}\int_0^2 \frac{1}{\sqrt{2 - u}}\, du \\ &= -\sqrt{2}\int_2^0 v^{-1/2}\, dv = -2\sqrt{2}v^{1/2}\Big|_2^0 \\ &= -2\sqrt{2}(0 - \sqrt{2}) = 4. \end{aligned}$$

The arclength of the top half of the cardioid is 4.

Exercises 13.2

I

Find the velocity vector $\mathbf{v}(t)$ for each of the following curves.

1. $x = 4t + 4, y = -3t - 3$
2. $x = t - 5, y = 6$
3. $x = 2t^2 - t + 3, y = 4t^2 - 2t - 6$
4. $x = t^2 - 1, y = 1 - t$
5. $x = \sqrt{1 + t^2}, y = t$
6. $x = 2t, y = \sqrt{1 - t^2}, t \in [-1, 1]$
7. $x = t^2, y = \arctan t$
8. $x = \frac{1}{1+t}, y = e^t, t \geq 0$
9. $x = 2\cos(3t), y = 2\sin(3t)$
10. $x = \sin^2 t + 3 + \cos^2 t, y = -5$

For each of the following curves, find all points at which the point is moving (i) straight up, (ii) horizontally to the left. Express the points by their (x, y)-coordinates and by the corresponding value(s) of the parameter t.

11. $x = \sin t, y = \cos t$
12. $x = t + 1, y = t^3 - 3t + 2$
13. $x = e^t, y = \sin(t^3 - t^2 + 3)$
14. $x = 3, y = t^2 - 4t + 1$
15. $x = 2t^2, y = t^3 - 3t^2 - 9t + 7$
16. $x = 1 - e^t, y = 1 - t^2$
17. $x = \sinh 2t, y = \cosh 2t$
18. $x = \arctan t, y = 1 - t^3$
19. $x = \exp(t^3 - 3t), y = |t|$
20. $x = \exp\big(1 - \arctan(1 - 2t^4) + 3\big)$,
    $y = 7 - 2\exp\big(1 - \arctan(1 - 2t^4) + 3\big)$

Find the arclength of the path followed by each of the following moving points for the specified values of the parameter.

21. $\mathbf{r}(t) = (3\sin t, 3\cos t), 0 \leq t \leq \pi$
22. $\mathbf{r}(t) = (\sin 3t, \cos 3t), 0 \leq t \leq \pi$
23. $\mathbf{r}(t) = (3t^2, 2t^3), 1 \leq t \leq 2$

24. $\mathbf{r}(t) = (t^2, t^3),\ 0 \leq t \leq 1$

25. $\mathbf{r}(t) = \big(4\cos(\pi t^2), -3\cos(\pi t^2)\big),\ 0 \leq t \leq 1$

26. $\mathbf{r}(t) = (e^{2t}, e^{3t}),\ -1 \leq t \leq 0$

27. $\mathbf{r}(t) = (\cosh t, t),\ 0 \leq t \leq 1$

28. $\mathbf{r}(t) = (t, \sqrt{4-t}),\ -4 \leq t \leq 4$

29. $\mathbf{r}(t) = (3, t^5),\ 0 \leq t \leq 2$

30. $\mathbf{r}(t) = (\cos t, \cos t),\ -\frac{\pi}{2} \leq t \leq \frac{\pi}{2}$

II

31. Suppose the course of a ship is given by $x = 5\cos \pi t$, $y = 5\pi t$, where the x-axis points due east, the y-axis points due north, distances are measured in kilometers, and time is measured in hours, with $0 \leq t \leq 1$.

(i) Find all times at which the ship is sailing northwest.

(ii) Find the speed of the ship at each time found in part (i).

(iii) What is the maximum speed of the ship for $0 \leq t \leq 1$?

***32.** Consider a ball thrown up into the air so that its path is parametrized by $x = 4t$, $y = 64t - 16t^2$, where t is measured in seconds and x and y are measured in feet.

(i) Find the velocity vector at time t.

(ii) When is the velocity vector exactly horizontal or exactly vertical?

(iii) At what point is the speed of the ball a minimum?

***33.** Sketch the curve that is parametrized by $x = \cos t$, $y = \sin 2t$, and find all values of t at which the moving point is going straight up or straight down.

***34.** Sketch the curve that is parametrized by $x = \cos t$, $y = \sin kt$, where k is a positive integer, and find all values of t at which the moving point is going straight up or straight down.

***35.** Sketch the curve that is parametrized by $x = \cos t$, $y = \sin \frac{t}{3}$, and find all values of t at which the moving point is going straight up or straight down.

***36.** Sketch the curve that is parametrized by $x = \cos 2t$, $y = \sin 3t$, and find all values of t at which the moving point is going straight up or straight down.

37. What is the average velocity of the point parametrized by $(x, y) = \mathbf{r}(t) = (t^2 + 1, t^4)$ over the interval from $t = 0$ to $t = 1$? (Remember that velocity is a vector.)

38. What is the average velocity of the point parametrized by $(x, y) = \mathbf{r}(t) = (t^2 + 1, t^4)$ over the interval from $t = -1$ to $t = 1$?

***39.** In Example 2 (p. 807) we observed that the velocity vector is always tangent to the circle. If we consider the velocity at time t, the initial point of the velocity vector is at the point $(\cos t, \sin t)$ on the circle, and to show that the velocity vector is tangent, it is enough to show that it is perpendicular to the line segment from the origin to $(\cos t, \sin t)$. Show that this is true.

***40.** For the ellipse parametrized by $x = 4\cos t$, $y = 9\sin t$, find all the values of t at which the tangent vector $\mathbf{r}'(t) = \big(x'(t), y'(t)\big)$ is perpendicular to the position vector $\mathbf{r}(t) = \big(x(t), y(t)\big)$, that is, to the segment running from the origin to the point $\mathbf{r}(t) = \big(x(t), y(t)\big)$.

***41.** Consider the functions

$$x(t) = \begin{cases} t^2, & \text{if } t < 0, \\ 0, & \text{if } 0 \leq t \leq 1, \\ (t-1)^2, & \text{if } t > 1; \end{cases}$$

$$y(t) = \begin{cases} -t^2, & \text{if } t < 0, \\ 0, & \text{if } 0 \leq t \leq 1, \\ (t-1)^2, & \text{if } t > 1. \end{cases}$$

(i) Sketch the graphs of each of these two functions of t, and show that they are both differentiable at every t.

(ii) Sketch the graph of the curve parametrized by the expressions $\mathbf{r}(t) = \big(x(t), y(t)\big)$.

(iii) Notice that the velocity vector is the zero vector for $t \in [0, 1]$ since both derivatives $x'(t)$ and $y'(t)$ are zero there. What is the tangent vector to the curve at the corresponding point?

***42.** Consider the curve parametrized by $x = t^2$, $y = t^3$.

(i) Find its velocity vector for every t.

(ii) The vector $\mathbf{r}'(t) = \big(x'(t), y'(t)\big)$ is zero when $t = 0$. In what direction is the point moving when t is slightly greater than zero and in what direction is it moving when t is slightly less than zero? What happens to these directions as t approaches 0 from above and from below?

(iii) Sketch the curve traced out by the moving point.

Exercises 41 and 42 have shown some of the types of behavior that are possible at points where both derivatives are zero. In particular they explain why it was necessary to assume that at least one derivative is nonzero in Definition 13.2.3.

43. Find the arclength of the circle $r = \sin\theta$.

44. Find the arclength of the right half of the circle $r = \cos\theta$, i.e., the right semicircle.

45. Find the arclength of the cardioid $r = 1 + \sin\theta$.

46. Find the arclength of the part of the cardioid $r = 1 + \cos\theta$ lying to the right of the vertical axis.

Use Simpson's Rule S_n with $n = 10$ to estimate the arclengths of the following curves.

47. $\mathbf{r}(t) = (\cos^2 t, \sin 3t),\ t \in [0, \pi]$

48. $\mathbf{r}(t) = (1 + t^2, \ln t),\ t \in [1, 3]$

49. $\mathbf{r}(t) = (t^3 - 2t + 2, t^4 + 2t^2 + 1),\ t \in [-1, 1]$

50. $\mathbf{r}(t) = (t^2 + t, t + \sin \pi t),\ t \in [0, 1]$

51. The piece of the hyperbola $x^2 - y^2 = 9$ between $(5, -4)$ and $(5, 4)$.

52. The short piece of the ellipse $4x^2 + 9y^2 = 36$ between $(3, 0)$ and $(0, 2)$.

POINT TO PONDER

Consider the point moving along the real axis whose position is given by $x(t) = t^2$. As t moves from -2 to 0, the point moves from $x = 4$ to $x = 0$. Its average velocity over this interval is $\frac{4-0}{-2-0} = -2$. Its average speed is 2.

As t goes from 0 to 2, the point moves from $x = 0$ to $x = 4$, so over this interval its average velocity is 2 and its average speed is 2.

However, as t goes from -2 to 2, the point starts at $x = 4$ and ends up at $x = 4$, so its net change of displacement is zero. The average velocity over this interval from $t = -2$ to $t = 2$ is 0. But since the total distance it has traveled is $4 + 4 = 8$ and the time it took is $2 + 2 = 4$, its average speed is $\frac{8}{4} = 2$.

So although the instantaneous speed is the absolute value of the velocity, the *average* speed (in this case 2) is not necessarily the absolute value of the *average* velocity (which in this case is 0).

The situation is similar for points moving in the plane. Consider the point $(x, y) = \mathbf{r}(t) = (\cos t, \sin t)$ that moves around the circle of radius 1. We saw that since it goes once around the circle when t increases from 0 to 2π, and the circumference of the circle is 2π, its average speed over the interval from $t = 0$ to $t = 2\pi$ is 1. However, since the point begins at the point $(x, y) = (1, 0)$ when $t = 0$ and returns to the same point when $t = 2\pi$, the net change in its displacement as t runs from 0 to 2π is zero, and its average velocity between $t = 0$ and $t = 2\pi$ is zero, that is, the zero vector $(0, 0)$.

Once again we see that the *average* speed may not equal the length of the *average* velocity vector.

13.3 Projectiles

HORIZONTAL AND VERTICAL COMPONENTS
MUZZLE VELOCITY
ANGLE OF ELEVATION

In this section we consider the particular case of the motion of a **projectile**, an object that has been thrown or "projected." The idea is that the object is made to move by some initial force but after that the only force acting on it is gravity.

This is the situation for a ball or a rock that has been thrown, an egg that has been dropped, or a cannonball that has been fired from a gun. It does not apply to an airplane or a rocket, which have engines that continue to exert a force on them, or to a falling feather, which encounters a considerable force due to air resistance. A hot air balloon, which is held up by the force of buoyancy in the air, or a glider, which experiences lift forces on its wings, does not qualify.

In these examples, we are in effect assuming that the force due to air resistance is so small that it can safely be ignored. This is approximately true for the sorts of examples we will consider.

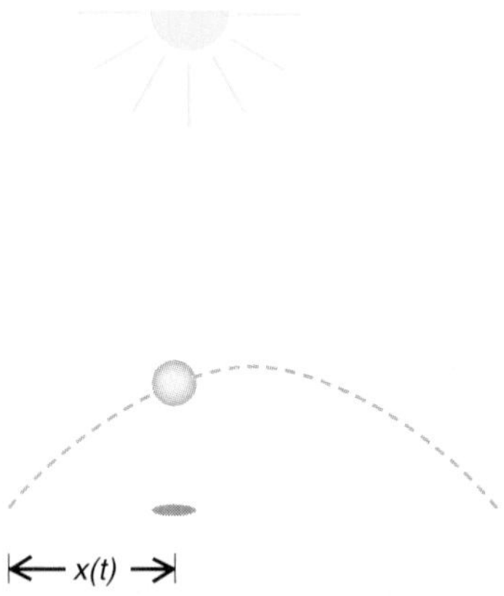

Figure 13.3.1

Consider a ball that has been thrown up into the air. Its position will change with time, and we can parametrize its motion by equations

$$x = x(t), \qquad y = y(t).$$

What these equations tell us, in effect, is that the horizontal position of the ball at time t is described by $x(t)$ and its height above the ground at time t is given by $y(t)$. If we imagine the sun shining down from directly overhead, then the ball casts a shadow on the ground, and $x(t)$ just measures the displacement of the shadow at time t. See Figure 13.3.1.

The remarkable thing is that these two "components" of the motion are independent. The vertical motion $y = y(t)$ behaves exactly as if there is no horizontal motion at all, and the horizontal motion is that of an object moving at a constant velocity (cf. Remark 4.9.2).

First we consider the vertical motion, as described by the height $y(t)$. As we saw in Section 4.9, the vertical motion undergoes a constant acceleration due to gravity. It is (approximately) 32 ft/sec^2 or 9.8 m/sec^2. This means that the second derivative is

$$y''(t) = -32,$$

where the minus sign reflects the fact that the acceleration is downward and the units of measurement are feet and seconds. (If we used meters and seconds, the condition would be $y''(t) = -9.8$.)

We conclude that $y'(t)$ must be given by

$$y'(t) = -32t + d,$$

for some constant d, and

$$y(t) = -16t^2 + dt + y_0. \tag{13.3.1}$$

Here y_0 is a constant; substituting $t = 0$ into Equation 13.3.1 shows that $y_0 = y(0)$, that is, that y_0 is the *initial height* or the height of the object at time $t = 0$.

Similarly, substituting $t = 0$ into the equation for y', we find that $d = y'(0)$ is the *initial velocity*, or, strictly speaking, the initial vertical velocity.

For the horizontal motion we are assuming that there is no force acting in a horizontal direction, which means that the horizontal acceleration is zero. So $x''(t) = 0$ and $x'(t) = c$, a constant. From this we find that

$$x(t) = ct + x_0, \tag{13.3.2}$$

where $x_0 = x(0)$ is the initial horizontal displacement and c is the initial horizontal velocity.

Combining Equations 13.3.1 and 13.3.2 we have the following theorem.

THEOREM 13.3.3

For an object moving under the influence of gravity near the earth's surface, experiencing no other forces, its path will be parametrized as follows:

$$x(t) = ct + x_0, \qquad y(t) = -16t^2 + dt + y_0,$$

where x_0 and y_0 are the initial horizontal and vertical displacements and c and d are the initial horizontal and vertical velocities, respectively. Measurements are in feet and seconds. If the units are meters and seconds, the equations become

$$x(t) = ct + x_0, \qquad y(t) = -4.9t^2 + dt + y_0.$$

One of the early applications of calculus was using these techniques to help aim artillery, and they can still be used to solve many practical problems.

EXAMPLE 1 A ball is thrown from ground level so that it reaches a maximum height of 64 ft and lands 128 ft from its starting point. At what speed does it hit the ground?

Solution Placing the origin at the starting point, we have $x_0 = y_0 = 0$, and the path is $x(t) = ct$, $y(t) = -16t^2 + dt$. The maximum value of $y(t)$ is 64, and it will occur at a point where $y'(t) = 0$, that is, $-32t + d = 0$ or $t = \frac{d}{32}$. This means that $64 = y\left(\frac{d}{32}\right) = -16\left(\frac{d}{32}\right)^2 + d\frac{d}{32} = -\frac{d^2}{64} + \frac{d^2}{32} = \frac{d^2}{64}$. We find that $d^2 = 64^2$ and the initial vertical velocity is $d = 64$. Incidentally, the maximum height is reached when $t = \frac{d}{32} = \frac{64}{32} = 2$, that is, 2 sec after starting.

To find when the ball hits the ground, we should solve $y(t) = 0$, that is, $-16t^2 + 64t = 0$, that is, $16t(-t + 4) = 0$, so $t = 0$ or $t = 4$.

We discard $t = 0$, the starting point, and find that the ball lands at $t = 4$. We are also told that the ball lands 128 ft from its starting position, which means that $x(4) = 128$. Since $x(t) = ct$, we find that $c = 32$, and the ball's path is parametrized by

$$x(t) = 32t, \qquad y(t) = -16t^2 + 64t, \qquad 0 \le t \le 4.$$

In particular, the velocity vector is

$$\mathbf{r}'(t) = (32, -32t + 64).$$

At the moment of impact, the velocity is $\mathbf{r}'(4) = (32, -64)$, and the speed equals $\sqrt{32^2 + (-64)^2} = 32\sqrt{5} \approx 71.6$ ft/sec.

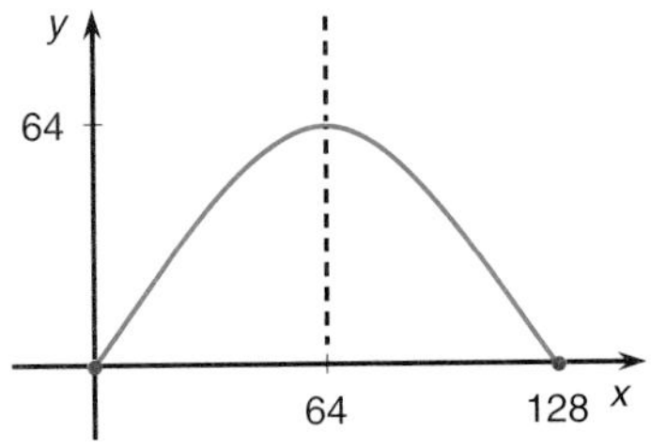

Figure 13.3.2

See the Graphics Calculator Example on the next page.

Notice that the path of the ball in Example 1, parametrized by $x = 32t$, $y = -16t^2 + 64t$, is the parabola $y = -\frac{1}{64}x^2 + 2x = -\frac{1}{64}(x - 64)^2 + 64$ (see Figure 13.3.2). It is symmetrical about the line $x = 64$ (which goes through the high point of the path). Because of this symmetry the speed of impact also equals the initial speed, that is, the speed at the starting point, which is the length of $\mathbf{r}'(0) = (32, 64)$.

The velocity vector is $\mathbf{r}'(t) = (32, -32t + 64)$. It tells us that the horizontal velocity is always 32 ft/sec and the vertical velocity starts at 64 ft/sec at $t = 0$, when the ball is rising fastest, drops to 0 at $t = 2$, when the ball is at its highest point, and falls to -64 ft/sec at $t = 4$.

The velocity vector is $(32, 0)$ at $t = 2$; this means the ball is moving horizontally at the top of its path at a speed of 32 ft/sec. At any other time the speed is $\sqrt{32^2 + (-32t + 64)^2}$ ft/sec, which is greater than 32 ft/sec.

The observation that the path is a parabola will be true for the motion of any projectile.

THEOREM 13.3.4

> If an object moves under the influence of gravity near the earth's surface, experiencing no other forces, then the path it follows will be a parabola or a vertical line segment.

PROOF From Theorem 13.3.3 we know that the path is parametrized by

$$x(t) = ct + x_0, \qquad y(t) = -16t^2 + dt + y_0.$$

We can use a graphics calculator to sketch the speed of the projectile from Example 1. We plot it as an ordinary function, not using parametric equations.

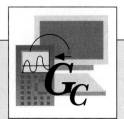

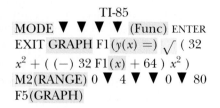

TI-85 MODE ▼ ▼ ▼ ▼ (Func) ENTER EXIT GRAPH F1 ($y(x)=$) $\sqrt{}$ (32 x^2 + ((−) 32 F1(x) + 64) x^2) M2(RANGE) 0 ▼ 4 ▼ ▼ 0 ▼ 80 F5(GRAPH)	SHARP EL9300 [key] $\sqrt{}$ 32 x^2 + (− 32 x/θ/T + 64) x^2 RANGE 0 ENTER 4 ENTER 1 ENTER 0 ENTER 80 ENTER 10 ENTER [key]
CASIO f_x-7700GB/f_x-6300G Cls EXE Range 0 EXE 4 EXE 1 EXE 0 EXE 80 EXE 10 EXE EXE EXE EXE Graph $\sqrt{}$ (32 x^2 + (− 32 X + 64) x^2) EXE	HP 48SX ↰ PLOT PLOTR ERASE ATTN ′ αY = $\sqrt{x}$ () 32 y^x 2 + () 32 +/− × αX + 64 ▶ y^x 2 ↱ PLOT ↰ DRAW ERASE 0 ENTER 4 XRNG 0 ENTER 80 YRNG DRAW

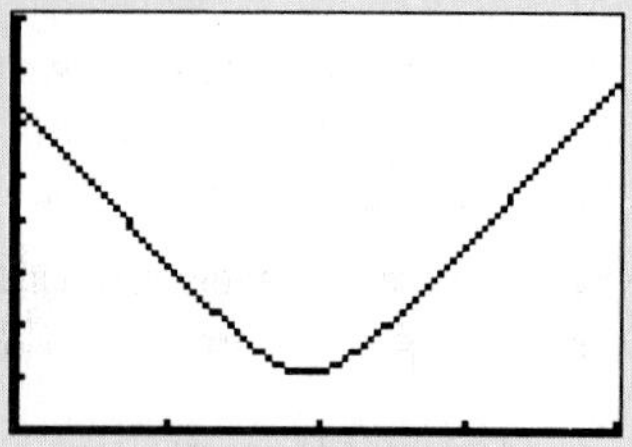

Notice how the speed has its minimum at the time corresponding to the *highest* point on the trajectory.

If $c = 0$, x is constant and we get a vertical line segment. Otherwise, letting $X = x - x_0 = ct$ and $Y = y - y_0 = -16t^2 + dt$, we see that $t = \frac{X}{c}$, so $Y = -16\left(\frac{X}{c}\right)^2 + d\frac{X}{c} = -\frac{16}{c^2}X^2 + \frac{d}{c}X$, the equation of a parabola, as required.

We can also write it as

$$y - y_0 = -\frac{16}{c^2}(x - x_0)^2 + \frac{d}{c}(x - x_0),$$

or

$$y = -\frac{16}{c^2}x^2 + \left(\frac{d}{c} + 32\frac{x_0}{c^2}\right)x + \left(y_0 - 16\left(\frac{x_0}{c}\right)^2 - \frac{dx_0}{c}\right).$$

EXAMPLE 2 If a ball is thrown at an angle of $\frac{\pi}{6}$ above the horizontal in such a way that it lands 50 ft from its starting point, what is the maximum height it will reach?

Solution We could do this question by finding the initial velocity, but it is easier to use the fact that the path of the ball is a parabola. Placing the origin at the starting point, we know that the path is a downward-opening parabola, so its equation is of the form $y = -Ax^2 + Bx + C$, and since the origin is on the path, we see that $C = 0$ and the path is $y = -Ax^2 + Bx$, with $A > 0$.

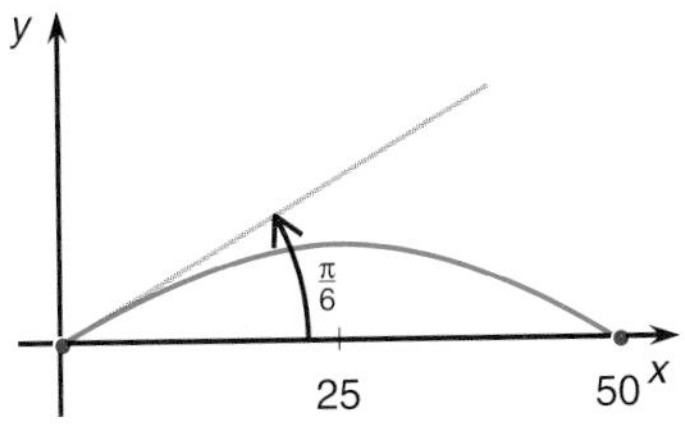

Figure 13.3.3

In Figure 13.3.3 we sketch the parabola. We know that the slope of its tangent at the origin is $\tan\frac{\pi}{6} = \frac{1}{\sqrt{3}}$. But this slope is also the derivative $\frac{dy}{dx}$ at $x = 0$, which is B.

The other thing we know is that the point $(50, 0)$ is on the parabola. Since the equation is $y = -Ax^2 + \frac{1}{\sqrt{3}}x$, we find that $0 = -A(2500) + \frac{1}{\sqrt{3}}(50)$, so $A = \frac{1}{50\sqrt{3}}$. The parabola is $y = -\frac{1}{50\sqrt{3}}x^2 + \frac{1}{\sqrt{3}}x$. It is easy to find that the highest point occurs when $x = 25$ (either by setting the derivative equal to 0 or by looking at the graph and using the symmetry). So the greatest value of y is

$$y = -\frac{1}{50\sqrt{3}}(25)^2 + \frac{1}{\sqrt{3}}(25) = \frac{25}{2\sqrt{3}} \approx 7.2.$$

The maximum height reached by the ball is about 7.2 ft. ◢

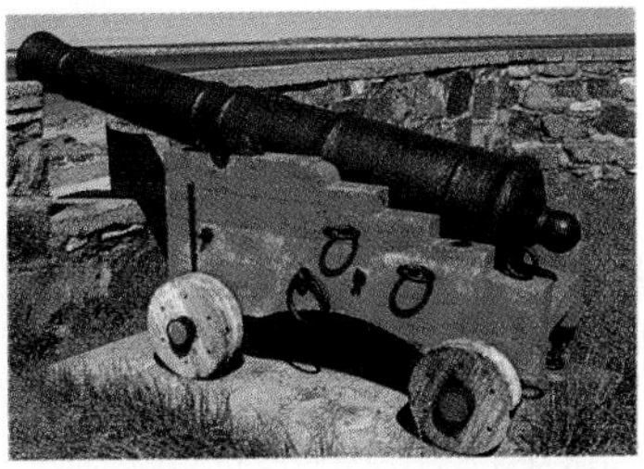

One of the antique cannons in the Cape Merry Battery in Churchill, Manitoba.

EXAMPLE 3 An old cannon shoots a lead ball with a "muzzle velocity" of 200 ft/sec. (i) At what angle of elevation should it be fired in order to reach a distance of 500 ft? (ii) What is its maximum range?

Solution

(i) Notice that the conventional term **muzzle velocity** actually means muzzle *speed*. We sketch the situation in Figure 13.3.4, where the angle of elevation is labeled θ.

As usual, it is convenient to place the origin at the starting point, so that the path of the cannonball is parametrized by $x = ct$, $y = -16t^2 + dt$. In Figure 13.3.5 we draw the initial velocity vector $\mathbf{r}'(0) = (c, d)$ and mark its horizontal and vertical components c and d, respectively. We are told that the initial speed is 200, that is, that $200 = \sqrt{c^2 + d^2}$. Since this is the length of the vector $\mathbf{r}'(0)$, we see from the diagram that $c = 200\cos\theta$ and $d = 200\sin\theta$.

As a result the parametric equations can be written as

$$x = 200\cos(\theta)\,t, \qquad y = -16t^2 + 200\sin(\theta)\,t.$$

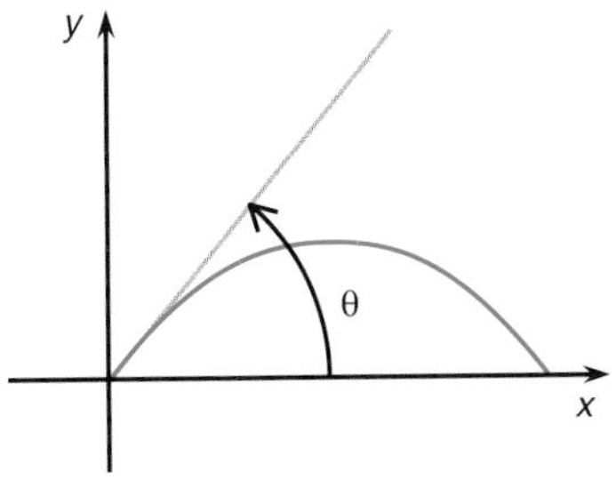

Figure 13.3.4

The cannonball will fall to the ground when $y(t) = 0$, that is, when $0 = -16t^2 + 200\sin(\theta)t = -8t(2t - 25\sin\theta)$, or when $t = \frac{25}{2}\sin\theta$ (we discard the starting point $t = 0$).

At this time the horizontal displacement will be $x\left(\frac{25}{2}\sin\theta\right) = 200\cos(\theta) \times \frac{25}{2}\sin(\theta) = 2500\cos\theta\sin\theta$. We want this to equal 500, that is, for the cannonball to land 500 ft from its starting point. This means that $\cos\theta\sin\theta = \frac{1}{5}$, that is, $\frac{1}{2}\sin(2\theta) = \frac{1}{5}$ or $\sin(2\theta) = \frac{2}{5}$.

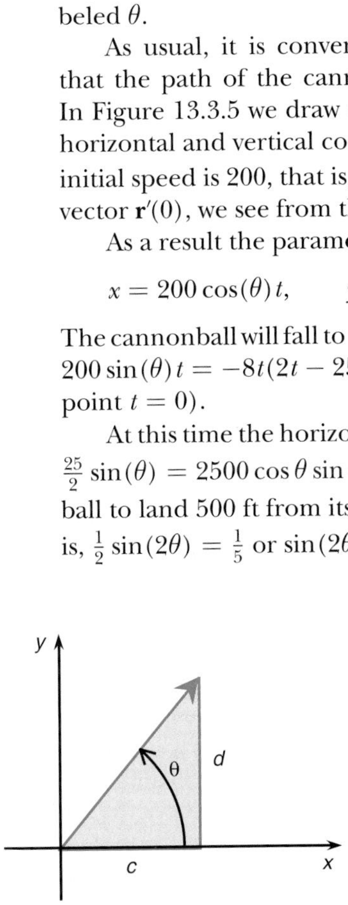

Figure 13.3.5

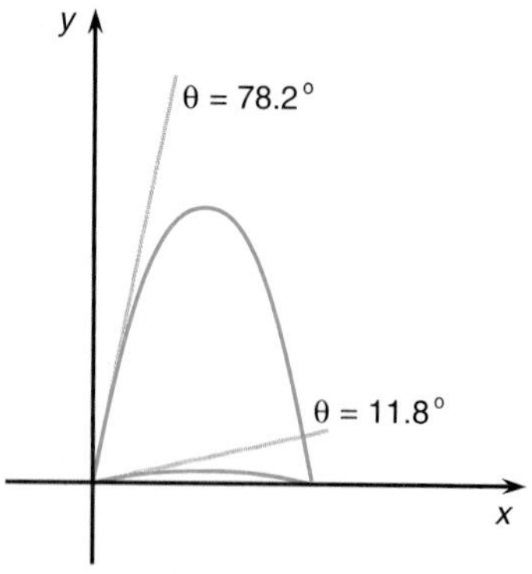

Figure 13.3.6

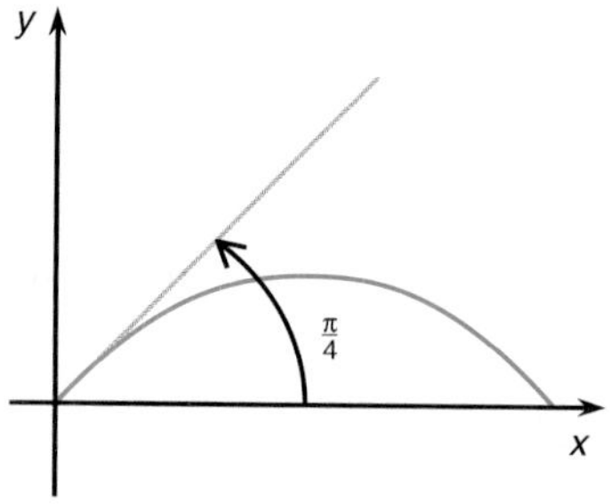

Figure 13.3.7

This means that $\theta = \frac{1}{2}\arcsin\frac{2}{5} \approx 0.206$ radians ($\approx 11.8°$). This is the required **angle of elevation** in order to shoot a distance of 500 ft.

There is another angle that will result in the same range, namely, $\frac{\pi}{2} - \frac{1}{2}\arcsin\frac{2}{5} \approx 1.365$ radians $\approx 78.2°$. See Figure 13.3.6.

(ii) We have already seen that if the cannon is elevated at an angle θ, then its range is $2500\cos\theta\sin\theta = 1250\sin(2\theta)$. Since the maximum value of $\sin 2\theta$ is 1, the maximum range is 1250 ft. Moreover, it occurs when $\sin 2\theta = 1$, which means $\theta = \frac{\pi}{4}$. See Figure 13.3.7.

Exercises 13.3

I

Find each of the following.

1. The speed of impact of a ball thrown so that its maximum height is 16 ft and its horizontal range is 32 ft.
2. The speed of impact of a ball thrown so that its maximum height is 40 ft and its horizontal range is 98 ft.
3. The speed of impact of a ball thrown so that its maximum height is 9 m and its horizontal range is 30 m.
4. The speed of impact of a ball thrown so that its maximum height is 5 m and its horizontal range is 5 m.
5. The horizontal distance covered by a ball thrown from ground level with initial velocity 50 ft/sec so that the maximum height it reaches is 30 ft.
6. The horizontal distance covered by a ball thrown from ground level with initial velocity 30 m/sec so that the maximum height it reaches is 10 m.
7. The initial speed of a ball thrown so that its horizontal range is 80 ft and its maximum height is 24 ft.
8. The maximum height reached by a ball thrown at 25 m/sec that lands after 4 sec.
9. The length of time in the air for a ball thrown at 64 ft/sec that lands 120 ft from its starting point.
10. The maximum height reached by a ball thrown with initial speed 32 ft/sec if it lands after 1 sec.
11. The angle of elevation for a cannonball to have a range of 800 ft if the muzzle velocity is 250 ft/sec.
12. The angle of elevation for a cannonball to have a range of 200 m if the muzzle velocity is 80 m/sec.
13. The range of a cannon with muzzle velocity 150 ft/sec with angle of elevation $\frac{\pi}{3}$.
14. The range of a cannon with muzzle velocity 150 ft/sec with angle of elevation $\frac{\pi}{6}$.
15. The maximum time the cannonball could be in the air if the cannon has muzzle velocity 100 ft/sec.
16. The muzzle velocity of a cannon whose maximum range is 400 m.
17. The maximum range of a cannon with a muzzle velocity of 100 m/sec.
18. The speed with which a ball is thrown if it reaches a maximum height of 40 ft when thrown at a 45° angle above the horizontal.
19. The speed with which a ball must be thrown at an angle of $\frac{\pi}{6}$ above the horizontal in order to land 75 ft from the thrower.

20. The two possible angles of elevation for a cannon with muzzle velocity 250 ft/sec if the range is to be 600 ft.

II

***21.** What is the maximum height reached by the cannonball in Example 3 (p. 819) when it is fired so as to land 500 ft from the cannon?

***22.** What is the maximum height reached by the cannonball in Example 3 when it is fired so as to have the maximum possible range?

***23.** At what elevation should the cannon of Example 3 be fired in order that the cannonball will reach the greatest possible height before falling to the ground? What is the maximum height it can reach?

***24.** If a cannon has a muzzle velocity V ft/sec, what is its greatest possible range, and what elevation should be used to achieve this range?

***25.** A manufacturer of toy cannons has discovered improvements that double the muzzle velocity. What will this do to the maximum range of the cannons?

***26.** A small boy has made a catapult that launches pebbles at a 45° angle above the horizontal; they land 60 feet away.

(i) How fast does his catapult launch the pebbles?

(ii) How fast should he launch a pebble if he wants it to go 80 ft?

***27.** A small girl is sitting 8 ft above ground level on a garage roof. She throws a ball at a 45° angle above the horizontal at a speed of 32 ft/sec. How much farther will it go than one thrown by her brother from ground level at the same speed and angle?

***28.** Suppose the sister and brother in Exercise 27 agree to throw balls at the same speed (32 ft/sec) but at an inclination of 30° above the horizontal. Before working it out, try to predict whether the difference between their two throws will be greater than, less than, or equal to what it was in Exercise 27 with inclinations of 45°. How much farther will the sister's ball go than her brother's?

III

***29.** Use a graphics calculator or computer to plot the trajectories of the cannonball of Example 3 corresponding to angles of elevation of $\theta = \frac{\pi}{10}, \frac{\pi}{8}, \frac{\pi}{6}, \frac{\pi}{5}, \frac{\pi}{4}, \frac{3\pi}{10}, \frac{\pi}{3}, \frac{3\pi}{8}, \frac{2\pi}{5}$. You should be able to do this question even if your equipment cannot plot parametric curves.

POINT TO PONDER

In our study of projectile motion we have assumed that the surface of the earth is perfectly flat. This is a reasonable assumption and quite accurate for motions over fairly short distances.

However, when an object is thrown near the earth's surface, what it actually does is go into an "orbit" around the center of the earth. As with planets and comets, this orbit is an ellipse with the center of the earth as one focus.

Of course, what happens is that the projectile hits the earth before it can follow much of the path, and all we ever see is a tiny piece at the end of a very elongated ellipse (see Figure 13.3.8).

The eccentricity of such a long ellipse is very close to 1, the eccentricity of a parabola. The little piece at the end of the ellipse is almost indistinguishable from a little piece of a parabola, and this is why for practical purposes we can work with parabolas.

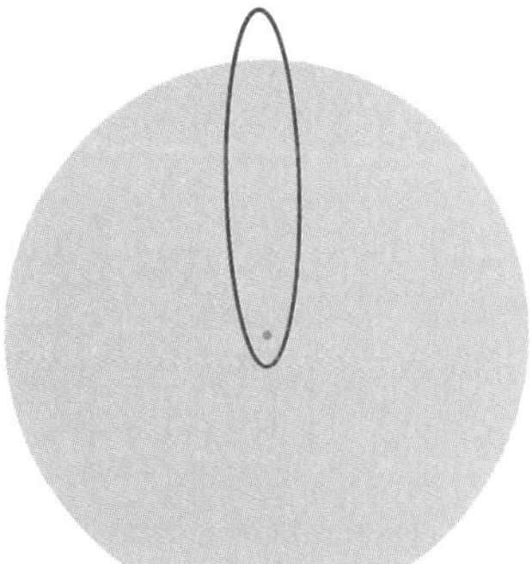

Figure 13.3.8

In effect, assuming that the earth is flat amounts to assuming that the center of the earth is "infinitely far away" (the farther away the center, the flatter the circle . . .). By the same token, if we take the long elliptical orbit and move the far focus "out to infinity," what happens is that the eccentricity approaches 1 and the ellipse approaches the shape of a parabola.

For motions involving reasonably small distances, the calculations with the parabola will give answers that are extremely close to the more correct ones obtained from the ellipse, so much so that there is no practical advantage in using the ellipse. And there is a practical advantage to using the parabola, that is, to assuming that a small piece of the earth's surface is flat: The calculations tend to be much easier.

Chapter Summary

§13.1
parametric equations
parameter
curve
domain
image
parametrization of lines, circles, ellipses, hyperbolas, graphs of functions

§13.2
vector
component
horizontal component
vertical component
displacement
change in the displacement
velocity vector
tangent vector
speed
arclength

§13.3
projectile
horizontal and vertical components
parametric equations for projectile motion
muzzle "velocity"
angle of elevation

Review Exercises

I

Parametrize each of the following curves.

1. $2x - 7y = 4$

2. $x = 3$

3. $y + 3 = 0$

4. $y = 2x^2 - 4x - 5$

5. $x + 3y^2 - 5y = 8$

6. $x = x^3 - 3y$

7. $2x^2 + 16y^2 = 8$

8. $x^2 + y^2 = 0$

9. $xy = 1$

10. $2xy^2 + 3xy = 12$

11. $x^2 - 2y^2 = 1$, right branch

12. $x^2 - y^2 + 1 = 0$, top branch

13. $5x^2 + 12y^2 = 30$, right half

14. $2x - 5y = 2, y \geq 0$

15. $4y^2 - 9x^2 = 36$, bottom branch

16. $y = \ln x$

17. $y = |x|$

18. $x^2 = y^2$, top half

19. $x = \sqrt{y}$

20. $e^x - 2 = \sin y$

Find an equation for each of the following curves, and identify the curve.

21. $x = 3t^2, y = t - 4$

22. $x = 3t^3, y = 4 - 9t^3$

23. $x = \cos t, y = 3 \sin t$

24. $x = \cos(5t), y = \sin(5t)$

25. $x = \cos^2 t - \sin^2 t, y = \cos t \sin t$

26. $x = \cos t, y = \sin\left(\frac{\pi}{2} - t\right)$

27. $x = -3 \sinh t, y = 2 \cosh t$

28. $x = \cos(t^2), y = 4 \sin(t^2)$

29. $x = \frac{1}{1+t^2}, y = \frac{t^2}{1+t^2}$

30. $x = 1 + \cos t, y = 2 - \sin t$

Find the velocity vector for each of the following parametrizations.

31. $x = 5t + 1, y = 2 - t$

32. $x = 3t + 2, y = 3t - 2$

33. $x = e^{2t}, y = 2t$

34. $x = \sin t, y = \cos(2t)$

35. $x = \sin t, y = \sin(3t)$

36. $x = t + 1, y = \ln t$

37. $x = \sin(t^2 + 2), y = 3 - \cos^2 t$

38. $x = \tan t, y = \sec t$

39. $x = \arctan(3t), y = \arcsin t$

40. $x = 2t + 3, y = x^2 - 1$

Find the length of the path traced out by each of the following moving points for the specified range of the parameter.

41. $\mathbf{r}(t) = (t^3, t^2), 0 \le t \le 1$

42. $\mathbf{r}(t) = (1 + t, 3 - 4t), 2 \le t \le 4$

43. $\mathbf{r}(t) = \left((1 - 2t)^{3/2}, \frac{3}{2}t^2\right), 0 \le t \le \frac{1}{2}$

44. $\mathbf{r}(t) = \left(2t - \frac{t^3}{3}, \sqrt{2}t^2\right), 0 \le t \le 1$

45. $\mathbf{r}(t) = \left(\cos(t + \pi), \sin(t + \pi)\right), 0 \le t \le \pi$

46. $\mathbf{r}(t) = \left(\cos(t + \frac{\pi}{2}), \cos t\right), 0 \le t \le \frac{\pi}{4}$

47. $\mathbf{r}(t) = (2e^{3t/2}, 3e^t), 0 \le t \le 2$

48. $\mathbf{r}(t) = \left(\frac{1}{2}\ln(1 + t^2), \arctan t\right), 0 \le t \le 1$

49. $\mathbf{r}(t) = (t^2, 2t^{5/2}), 0 \le t \le 1$

50. $\mathbf{r}(t) = \left(-4, \sin^2(-4t) + \cos^2(4t)\right), 0 \le t \le 2\pi$

Find each of the following.

51. All points where $x = 3t^3, y = \cos(\pi t)$ is moving horizontally.

52. The points where the tangent to $x = t^3 - 3t + 1, y = e^t$ is vertical.

53. The point where the moving point $x = 5t - 1, y = 2t^2 + 4t - 7$ has minimum speed.

54. The points at which $x = 9t^2, y = 2t^3 + 2$ has minimum speed.

55. The maximum height reached by a ball thrown at a speed of 48 ft/sec at an angle 30° above the horizontal.

56. The speed with which a ball must be thrown at an angle of 45° above the horizontal in order to reach a maximum height of 10 meters.

57. The length of time a ball is in the air if it reaches a maximum height of 16 ft and goes 84 ft horizontally.

58. The angle of elevation for a projectile that reaches a height of 20 m and covers a horizontal distance of 60 m.

59. The speed of impact of a projectile that was launched from ground level and flew 300 ft in 3 sec.

60. The greatest possible speed of impact for a projectile launched from ground level at a speed of 43 m/sec.

II

***61.** Parametrize the circle $x^2 + 2x + y^2 - 4y + 4 = 0$.

***62.** Parametrize the ellipse $4x^2 + 8x + 9y^2 = 32$.

***63.** Parametrize the upper branch of the hyperbola $y^2 - 4y - 9x^2 - 18x = 14$.

***64.** Parametrize the part of the line $2x - y = 0$ that lies above the parabola $y = x^2$.

***65.** Parametrize the part of the ellipse $4x^2 + 16y^2 = 16$ that lies below the line $y = x$.

***66.** Parametrize the part of the hyperbola $x^2 - y^2 = 1$ that lies between the y-axis and the line $x = 2$.

***67.** (i) Consider the curve $\mathbf{r}(t) = \left(x(t), y(t)\right)$, and suppose that $x(t)$ and $y(t)$ are differentiable at $t = a$. Assuming that $x'(a) \neq 0$, show that the slope of the tangent to the curve at the point $\left(x(a), y(a)\right)$ is

$$\frac{y'(a)}{x'(a)}.$$

(*Hint:* Consider the slope of the tangent vector.)

(ii) What happens at a point where $x'(a) = 0$?

***68.** Find the slope of the curve $\mathbf{r}(t) = \left(t\cos(t^2), e^{-3t}\right)$ at the point corresponding to $t = 0$.

***69.** Two friends each kick a ball from ground level. Both balls reach a maximum height of 12 ft, but one travels 30 ft horizontally and the other travels 72 ft horizontally. Compare the lengths of times the two balls are in the air.

***70.** If a ball is thrown horizontally at a speed of 20 ft/sec starting at a point 6 ft above ground level, how far will it travel in the horizontal direction before landing?

III

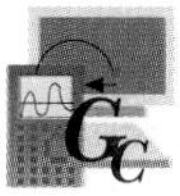

Use Simpson's Rule S_n with $n = 12$ to approximate the arclengths of the following curves.

71. $\mathbf{r}(t) = \left(\frac{1}{t}, \frac{1}{t^2}\right), t \in [1, 2]$

72. $\mathbf{r}(t) = (t^2, \sin \pi t), t \in [-1, 1]$

73. The top half of the ellipse $2x^2 + y^2 = 1$.

74. The part of the top branch of the hyperbola $4y^2 - x^2 = 4$ between $x = -2$ and $x = 2$.

APPENDIX

Mathematical Induction

Mathematical induction is a technique for proving certain types of theorems. It is most often applied to statements that are asserted for every positive integer n.

Induction proves the result for every n by showing it is true for $n = 1$ and then showing that knowing the result for n allows us to prove it for $n + 1$. If it is known for $n = 1$, then it must be true for the next integer, 2, and then for the next, 3, and so on for every n.

EXAMPLE **Sum of the first *n* integers** If n is a positive integer, show that the sum of the integers from 1 to n equals $\frac{n(n+1)}{2}$.

Solution First, with $n = 1$, the formula equals $\frac{1(1+1)}{2} = 1$, which *does* equal the sum of the integers from 1 to 1, that is, just 1.

Suppose we know it is true for some n, that is

$$1 + 2 + 3 + \cdots + n = \frac{n(n+1)}{2}.$$

Then to show it is true for $n + 1$ we consider the sum for $n + 1$ and find

$$\begin{aligned} 1 + 2 + 3 + \cdots + n + (n+1) &= (1 + 2 + 3 + \cdots + n) + (n+1) \\ &= \frac{n(n+1)}{2} + (n+1) = (n+1)\left(\frac{n}{2} + 1\right). \end{aligned}$$

(Here we used the formula for $1 + 2 + 3 + \cdots + n$ and then took out the common factor $n + 1$.) Now we find this equals

$$(n+1)\left(\frac{n+2}{2}\right) = \frac{(n+1)(n+2)}{2} = \frac{(n+1)((n+1)+1)}{2}.$$

This is exactly the required result (i.e., it is $\frac{n(n+1)}{2}$ with n replaced by $n + 1$.) This completes the proof.

Exercise: Use mathematical induction to prove the following formula for the sum of the squares of the first n positive integers:

$$1^2 + 2^2 + 3^2 + 4^2 + \cdots + n^2 = \frac{n(n+1)(2n+1)}{6}.$$

Answers to Odd-Numbered Exercises

Exercises 1.2 (p. 7)
1. $\frac{1}{84}$ **3.** $\frac{1}{84}$ **5.** $2 - \sqrt{3}$ **7.** $\sqrt{2} + \sqrt{3} - 3$ **9.** $x^2 + 3$
11. $[-3, 5]$ **13.** $(\frac{4}{3}, \frac{8}{3})$ **15.** $(-\infty, 0) \cup (1, \infty)$ **17.** $[-1, 0] \cup [2, 7]$
19. $(-\infty, -\sqrt{3}] \cup [\sqrt{3}, \infty)$

Exercises 1.3 (p. 16)
1. $(x - 3)(x + 4)$ **3.** $(x + 4)(x - 1)$ **5.** $2(x + 1)^2$
7. $(x + 2)(x - 2)$ **9.** $(x - 7 + \sqrt{7})(x - 7 - \sqrt{7})$ **11.** $\frac{-3 + \sqrt{17}}{2}$
13. $2 \pm \sqrt{5}$ **15.** no roots **17.** 0, 6 **19.** $\frac{4 \pm \sqrt{6}}{10}$ **21.** $(0, 2)$
23. $(-\infty, -3] \cup [-1, \infty)$ **25.** $(-\infty, -4] \cup [4, \infty)$
27. $(-\infty, 2 - \sqrt{2}] \cup [2 + \sqrt{2}, \infty)$ **29.** $(-\infty, \infty)$
31. $a = -1, q(x) = x^2 + 2x + 2$ **33.** $a = 2, q(x) = x^2 - x + 1$
35. $a = 1, q(x) = x^2 + x + 1$ **37.** $a = 3, q(x) = x^2 + 3x + 9$
39. $a = 0, q(x) = x^3 - 5x^2 + 3x - 11$ **41.** $x^3 + 12x^2 + 48x + 64$
43. $x^5 + 5x^4 + 10x^3 + 10x^2 + 5x + 1$ **45.** $x^6 - 6x^4 + 12x^2 - 8$
47. -1 **49.** $1 - 12x + 48x^2 - 64x^3$

Exercises 1.4 (p. 22)
1. $\mathbb{R}$ **3.** $\{x \in \mathbb{R} : x \neq -3\} = (-\infty, -3) \cup (-3, \infty)$ **5.** $[0, \infty)$
7. $(-1, \infty)$ **9.** $[0, 1) \cup (1, \infty)$ **11.** $[5, \infty)$ **13.** $[3, \infty)$ **15.** $[-\frac{1}{12}, \infty)$
17. $[0, \infty)$ **19.** $(0, 1]$

21.

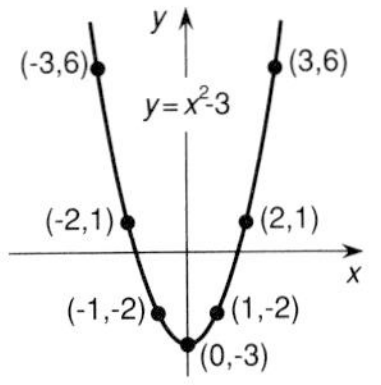

23.

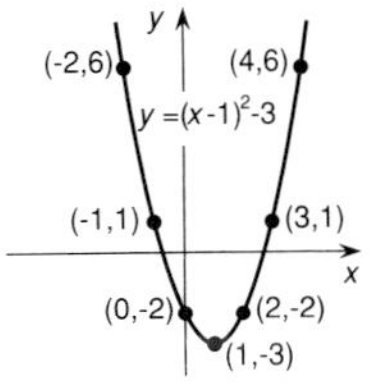

25.

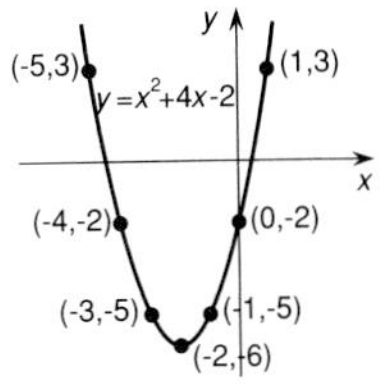

27.

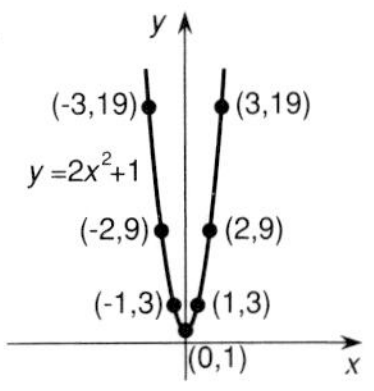

29.

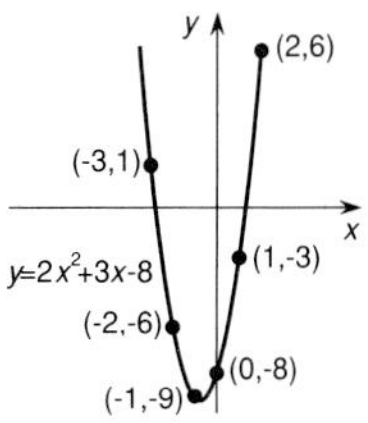

Exercises 1.5 (p. 32)
1. $\sqrt{580} \approx 24.083$ **3.** 5 **5.** 10 units **7.** 4 **9.** 4
11. $y - 2 = x - 1$ or $x - y + 1 = 0$ **13.** $x + 2y = 3$
15. $y + 6 = 2(x - 4)$ or $2x - y = 14$ **17.** $y = -x + 3$
19. $y - \frac{2}{3} = \frac{1}{3}x$ or $x - 3y + 2 = 0$ **21.** $-\frac{3}{2}$ **23.** $x + 2y = 11$
25. $\frac{7}{3}$ **27.** $\frac{7}{2}$ **29.** $(3, 2)$

Exercises 1.6 (p. 42)
1. ellipse, center $(0, 0)$, minor axis between x-intercepts $(\pm 3, 0)$, major axis between y-intercepts $(0, \pm 4)$ **3.** hyperbola, center $(0, 0)$, transverse axis between x-intercepts $(\pm 2, 0)$, conjugate axis between $(0, \pm 1)$, asymptotes $x \pm 2y = 0$ **5.** parabola, vertex $(2, 0)$, opening to left, x-intercept $(2, 0)$, y-intercepts $(0, \pm 2\sqrt{2})$ **7.** ellipse, center $(-2, 0)$, major axis between x-intercepts $(-5, 0)$, $(1, 0)$, minor axis from $(-2, -1)$ to $(-2, 1)$, y-intercepts $(0, \pm\sqrt{5}/3)$ **9.** parabola, vertex $(-1, -\frac{3}{2})$, opening upward, x-intercepts $(-1 \pm \sqrt{6}, 0)$, y-intercept $(0, -\frac{5}{4})$ **11.** ellipse, center $(0, 0)$, major axis between x-intercepts $(\pm\frac{1}{\sqrt{2}}, 0)$, minor axis between y-intercepts $(0, \pm\frac{1}{2\sqrt{2}})$ **13.** ellipse, center $(0, 0)$, minor axis between x-intercepts $(\pm\frac{3}{2}, 0)$, major axis between y-intercepts $(0, \pm 3)$ **15.** parabola, vertex $(\frac{1}{10}, 0)$, opening to the right, x-intercept $(\frac{1}{10}, 0)$, no y-intercepts **17.** ellipse, center $(2, 0)$, major axis between x-intercepts $(2 \pm \sqrt{5}, 0)$, minor axis between $(2, \pm\sqrt{\frac{5}{3}})$, y-intercepts $(0, \pm\frac{1}{\sqrt{3}})$ **19.** parabola, vertex $(\frac{5}{4}, 0)$, opening to left, x-intercept $(\frac{5}{4}, 0)$, y-intercepts $(0, \pm\frac{2}{\sqrt{5}})$
21. parabola, vertex $(0, \frac{1}{4})$, opening downward, x-intercepts $(\pm\frac{1}{4}, 0)$, y-intercept $(0, \frac{1}{4})$

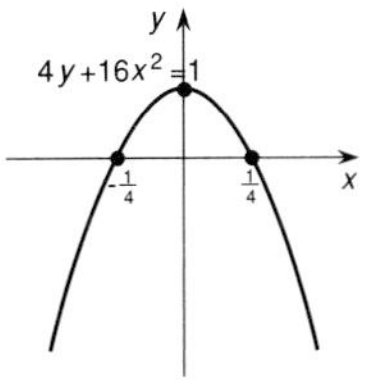

23. hyperbola, center (0, 0), transverse axis between y-intercepts $(0, \pm 1)$, no x-intercepts, conjugate axis between $(\pm 1, 0)$, asymptotes $y = \pm x$

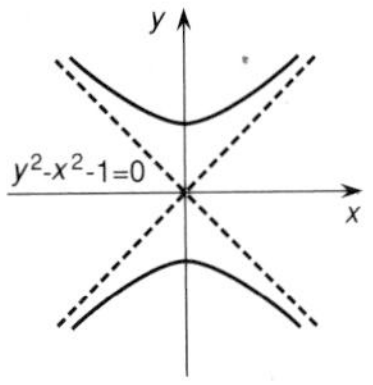

25. hyperbola, center (0, 0), no x-intercepts, transverse axis between y-intercepts $(0, \pm\frac{1}{\sqrt{2}})$, conjugate axis between $(\pm\sqrt{2}, 0)$, asymptotes $2y = \pm x$

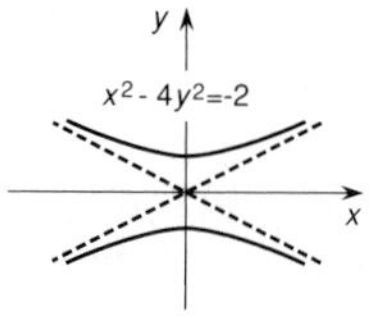

27. ellipse, center (0, 0), major axis between x-intercepts $(\pm\frac{\sqrt{7}}{2}, 0)$, minor axis between y-intercepts $(0, \pm\frac{\sqrt{7}}{3})$

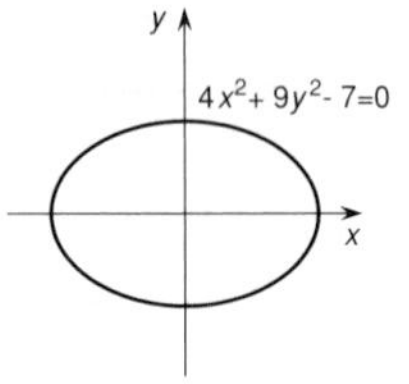

29. circle, center $(-\frac{1}{2}, -\frac{1}{2})$, radius $\sqrt{5}$

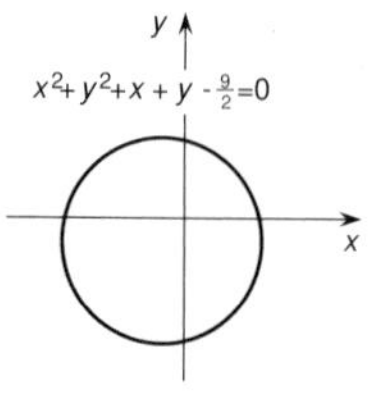

31. $(x + 1)^2 + (y - 3)^2 = 4$ or $x^2 + 2x + y^2 - 6y + 6 = 0$ **33.** $4x^2 - y^2 = 16$ **35.** $x^2 - 10x + 4y^2 - 16y + 37 = 0$ **37.** $x^2 + 4x + y^2 = 21$ **39.** $4x^2 - 16x + y^2 + 8y = 0$

Exercises 1.7 (p. 55)

1. -1 **3.** $-\frac{1}{\sqrt{2}}$ **5.** $\sqrt{3}$ **7.** $-\sqrt{2}$ **9.** $-\frac{1}{\sqrt{2}}$ **11.** $\sqrt{\frac{1}{2} - \frac{1}{2\sqrt{2}}} = \frac{1}{2}\sqrt{2 - \sqrt{2}}$ **13.** $\sqrt{\frac{1}{2} + \sqrt{3}/4} = \frac{1}{2}\sqrt{2 + \sqrt{3}}$ **15.** $-\frac{1}{2}\sqrt{2 + \sqrt{3}}$ **17.** $\frac{\sqrt{2 + \sqrt{2}}}{\sqrt{2 - \sqrt{2}}} = 1 + \sqrt{2}$ **19.** $-\frac{\sqrt{2 - \sqrt{2}}}{\sqrt{2 + \sqrt{2}}} = 1 - \sqrt{2}$ **21.** $(-3, 0)$ **23.** $(0, 2)$ **25.** $(\frac{1}{2}, -\frac{\sqrt{3}}{2})$ **27.** $(-2\sqrt{2}, 2\sqrt{2})$ **29.** $(-\sqrt{3}, 1)$ **31.** $(3, 2k\pi)$ **33.** $(2, -\frac{\pi}{2} + 2k\pi)$ **35.** $(2\sqrt{2}, -\frac{\pi}{4} + 2k\pi)$ **37.** $(4\sqrt{2}, \frac{5\pi}{4} + 2k\pi)$ **39.** $(\frac{1}{2\sqrt{3}}, -\frac{\pi}{6} + 2k\pi)$

Exercises 1.8 (p. 59)

1. $A + B$ **3.** $2A - B$ **5.** $\frac{1}{2}A$ **7.** $\frac{1}{3}A$ **9.** $10^B A$ **11.** 5 **13.** $\frac{1}{2}$ **15.** $-\frac{1}{3}$ **17.** -3 **19.** $\frac{1}{2}$

Chapter 1 Review Exercises (p. 60)

1. $x \in [-5, -3) \cup (-3, -1]$ **3.** $x \in (-1, -\frac{1}{4}) \cup (\frac{1}{4}, 1)$ **5.** $(-14, -7) \cup (-5, 2)$ **7.** $\{x: -\frac{1}{2} \le 1 - x^2 \le \frac{1}{2}\}$ or $[-\sqrt{3}/\sqrt{2}, -1/\sqrt{2}] \cup [1/\sqrt{2}, \sqrt{3}/\sqrt{2}]$ **9.** ϕ (the empty set) **11.** $(x + 12)(x - 5)$ **13.** $(x + \frac{1}{2} + \frac{\sqrt{5}}{2})(x + \frac{1}{2} - \frac{\sqrt{5}}{2})$ **15.** $x(x - 7)(x + 5)$ **17.** $(x^2 + 3)(x - \sqrt{5})(x + \sqrt{5})$ **19.** $x^3(x + 1)(x - 1)$ **21.** 12 **23.** 12 **25.** 0 **27.** 4 **29.** -32 **31.** $\{x : x \neq \pm\sqrt{3}\}$ **33.** $(-\infty, 1) = \{x : x < 1\}$ **35.** $\{y : y \ge -7\} = [-7, \infty)$ **37.** $5\sqrt{2} \approx 7.07107$ **39.** 4 **41.** $\sqrt{58} \approx 7.61577$ **43.** 3 **45.** 4 **47.** $x + y = 7$ **49.** $3x + 2y + 6 = 0$ **51.** $4x^2 + 9y^2 = 36$ **53.** $x^2 - 10x + y^2 - 8y + 32 = 0$ **55.** $y^2 - 4x^2 = 9$ **57.** $(-2, 1)$, $(-\frac{14}{5}, \frac{1}{5})$ **59.** $(-2, 1)$ **61.** parabola, vertex $(0, -2)$, opening upward **63.** hyperbola, center (0, 0), transverse axis between $(0, \pm 2)$, conjugate axis between $(\pm 2, 0)$, asymptotes $y = \pm x$ **65.** ellipse, center $(0, \frac{3}{2})$, major axis between $(\pm\sqrt{17}, \frac{3}{2})$, minor axis between $(0, \frac{1}{2}(3 \pm \sqrt{17}))$ **67.** hyperbola, center (0, 0), transverse axis between $(0, \pm\frac{1}{3})$, conjugate axis between $(\pm\frac{1}{4}, 0)$, asymptotes $3y = \pm 4x$ **69.** ellipse, center $(-2, 0)$, major axis between $(-3, 0)$ and $(-1, 0)$, minor axis between $(-2, \pm\frac{1}{\sqrt{3}})$ **71.** $-\frac{1}{\sqrt{2}}$ **73.** $-\sqrt{\frac{1}{2} - \frac{1}{2\sqrt{2}}} = -\frac{1}{2}\sqrt{2 - \sqrt{2}}$ **75.** $-\sqrt{3}$ **77.** $(\sqrt{3} - 1)/(2\sqrt{2})$ **79.** $\frac{1}{\sqrt{3}}$ **81.** $(r, \theta) = (4\sqrt{2}, \frac{5\pi}{4} + 2k\pi)$ **83.** $(r, \theta) = (3\sqrt{5}, -\arctan(2) + 2k\pi) \approx (6.708, -1.10715 + 2k\pi)$ **85.** $(r, \theta) = (8, -\frac{\pi}{2} + 2k\pi)$ **87.** $(x, y) = (-1, 0)$ **89.** $(x, y) = (-\frac{7}{\sqrt{2}}, -\frac{7}{\sqrt{2}})$ **91.** $\ln(10) + 2\ln a + \ln b$ **93.** $\ln(a)/\ln(10)$ **95.** $\ln a - \ln 2 - \ln b$ **97.** $1/(a^3)$ **99.** $\ln 2 + \ln a + b \ln 3$

Exercise 2.1 (p. 69)

1. 4 **3.** 2 **5.** 5 **7.** 12 **9.** 0 **11.** $2x - y = 3$ **13.** $x - y + 4 = 0$ **15.** $19x + y + 20 = 0$ **17.** $11x - y + 14 = 0$ **19.** $32x + y + 48 = 0$ **21.** $2ax - y = a^2$ **23.** 256 ft **25.** $32t$ ft/sec **27.** 8 ft **29.** **(iii)** $t = \frac{v}{32}$, $h = \frac{v^2}{64} + A$ **(iv)** after $(5 + \sqrt{41})/8$ sec **(v)** 12 ft/sec **33.** 25.1327 **37.** **(i)** 48, ≈ 6, ≈ 0.6

Exercises 2.2 (p. 79)

1. 1 **3.** -3 **5.** -1 **7.** 1 **9.** $11\sqrt{5} - 1$ **11.** 5 **13.** 1 **15.** 0 **17.** 0 **19.** 4 **21.** 2 **23.** 5 **25.** 0 **27.** -7 **29.** 0 **31.** -5 **33.** 0 **35.** does not exist **37.** $-\frac{3}{2}$ **39.** $\frac{1}{8}$ **41.** $a + 3$ **43.** no limit for $a = -1, -2$, otherwise $\frac{(a-1)(a-2)(a-5)}{(a+1)(a+2)}$. **45.** no limit for $a = 1$, otherwise $\frac{a(a-2)}{a-1}$. **47.** no limit for $a = 1$, limit is 1 if $a > 1$, -1 if $a < 1$ **49.** if $c \neq 0$, no limit for $a = c$, otherwise $\frac{a+c}{a-c}$; if $c = 0$, 1 for all a **53.** $-\frac{1}{2}$ at -1, does not exist at 1 **55.** does not exist **57.** 6 at -3

Exercises 2.3 (p. 91)

1. $\frac{1}{2}$ **3.** 3 **5.** $-\frac{1}{6}$ **7.** -32 **9.** $-2\sqrt{3}$ **11.** 3 **13.** -1 **15.** $(3 - \sqrt{2})^{5/2} \approx 3.1667$ **17.** $-\frac{1}{4}$ **19.** $\frac{5}{6}$ **21.** 0 **23.** 0 **25.** 0 **27.** 0 **29.** 0 **31.** does not exist **33.** $\frac{3}{2}$ **35.** does not exist **37.** $\frac{1}{108}$ **39.** does not exist **45.** **(i)** $\lim_{x \to a+c} f(x)$ **(ii)** $\lim_{x \to c-a} f(x)$ **53.** 2

Exercises 2.4 (p. 104)

1. ∞ **3.** 1 **5.** $\frac{1}{2}$ **7.** $-\infty$ **9.** ∞ **11.** no limit (limit from above is ∞, from below, $-\infty$). **13.** ∞ **15.** ∞ **17.** ∞ **19.** no limit, finite or infinite. **21.** 0 **23.** ∞ **27. (i)** 1 **(ii)** 6 **(iii)** -4 **(iv)** 0 **31.** 1 as $x \to 0^+$, -1 as $x \to 0^-$ **33.** ∞ from both sides **35.** ∞ from both sides **37.** $-\infty$ as $x \to 0$, ∞ as $x \to -1^-$, $-\infty$ as $x \to -1^+$, 0 as $x \to 1$ **39. (i)** $a \approx 2.325$

Exercises 2.5 (p. 112)

1. continuous everywhere **3.** (d) at 0, (a) elsewhere **5.** (c) at 0, (d) at $\pm\sqrt{3}$, (a) elsewhere **7.** (c) at 0, (a) elsewhere **9.** (b) at 0, (a) elsewhere **11.** bounded by $M = 8$ **13.** not bounded **15.** bounded by $M = 1$ **17.** bounded by $M = 1$ **19.** bounded by $M = 2$ **21.** min at $x = 1$, max at $x = 3, -1$ **23.** no max or min **25.** min at $t = 1$, max at $t = 0$ **27.** min at $u = 4$, max at $u = -2$ **29.** no max or min **33. (i)** $x \geq -4$ **(ii)** $\{t : t \neq -1\}$ **(iii)** $\{u : u \neq 0\}$ **43.** -2.7606 **45.** $-0.7748, 1, 1.9276$

Exercises 2.6 (p. 118)

1. 0 **3.** -2 **5.** 1 **7.** $\frac{1}{3}$ **9.** ∞ **11.** does not exist **13.** does not exist **15.** does not exist **17.** 0 **19.** 0 **21.** 0 **23.** 1 **25.** 0 **27.** $\frac{1}{2}$ **29.** $\frac{1}{2}$ **31. (i)** $a = \frac{\pi}{2} + k\pi, k = 0, \pm1, \pm2, \ldots$ **33.** $\cot x : a = k\pi$, $\tan^2 x : a = \frac{\pi}{2} + k\pi, k = 0, \pm1, \pm2, \ldots$

Exercises 2.7 (p. 124)

1. -5 **3.** 9 **5.** 1 **7.** 0 **9.** 6 **11.** does not exist **13.** does not exist **15.** 0 **17.** 2 **19.** does not exist **21.** does not exist **23.** does not exist **29.** 0.0001, 0.000 000 3, 0.000 000 000 3, 0.000 000 000 000 3

Chapter 2 Review Exercises (p. 126)

1. -1 **3.** $\frac{3}{4}$ **5.** 2 **7.** 0 **9.** $-\frac{5}{2}$ **11.** 2 **13.** $2\sqrt{5}$ **15.** $\frac{1}{12}$ **17.** $-\frac{9}{2}$ **19.** 2 **21.** 0 **23.** 1 **25.** π **27.** $\frac{\pi}{2}$ **29.** 3 **31.** 0 **33.** 0 **35.** does not exist (∞) **37.** $-\frac{1}{2}$ **39.** -2 **41.** $\frac{1}{2}$ **43.** $-\infty$ **45.** ∞ **47.** does not exist **49.** $-\infty$ **51.** $-2a$ **53.** $\frac{2}{3}$ **57.** removable at $x = -2$, infinite discontinuity at ±4 **61. (i)** $-2.115, 0.254, 1.861$ **63. (i)** $-2, \pm1$

Exercises 3.1 (p. 138)

1. 4 **3.** 1 **5.** 7 **7.** 2 **9.** 3 **11.** $4x - y = 3$ **13.** $x - y = 1$ **15.** $7x - y = 7$ **17.** $2x - y = 2$ **19.** $3x - y = 2$ **21.** $a = 1$ **23. (i)** $2x - y = 1, 2x + y = -1$ **(ii)** there are none **25. (i)** $32 - 32t$ ft/sec **(ii)** rising when $t < 1$, falling when $t > 1$ **(iii)** 22 ft **27. (i)** 6 qt-lb/sq in **(ii)** 0.2 qt **(iii)** bigger change when $P = 15$ **(iv)** ≈ 0.37333 **(v)** 0.375 **29. (i)** $2A + B$ **(ii)** $2aA + B$ **31.** $a = 0, f'(x) = 1$ if $x > 0, f'(x) = -1$ if $x < 0$ **37.** $2929 \leq N \leq 3071$, approx.

Exercises 3.2 (p. 152)

1. $3x^2 - 4x + 1$ **3.** $6x^2 + 6x + 6$ **5.** $18x^2 + 18x - 10$ **7.** $32x + 24$ **9.** $(-2x^2 - 2x + 2)/(x^2 + 1)^2$ **11.** $-1/(x - 1)^2$, except at $x = 1$ **13.** $-(8x)/(x^2 - 4)^2$, except at $x = \pm2$ **15.** $(2 - 6x)/(3x^2 - 2x)^2$, except at $x = 0, \frac{2}{3}$ **17.** $(8x^2 - 22x - 1)/(x^2 + 3x - 4)^2$, except at $x = 1, -4$ **19.** $(x^6 + 4x^3)/(1 + x^3)^2$, except at $x = -1$ **21. (i)** $12x^3 + 12x^2 - 2x - 2$ **(ii)** $72x^5 + 60x^4 - 9x^2 - 6x$ **(iii)** $-(3x^2 + 4x + 4)/(x^2 - x - 2)^2$, except at $x = 2, -1$ **23. (i)** $x - y + 1 = 0$ **(ii)** $4x + y + 4 = 0$ **25.** $12x - y + 16 = 0$ **31. (i)** falling when $0 < t < 2$, rising when $t > 2$ (or $t < 0$) **(ii)** after 2 sec, height is 2 ft **(iii)** no **33. (i)** $\frac{1}{2}g'(x)/f(x) = \frac{1}{2}g'(x)/\sqrt{g(x)}$ **(ii)** $x/\sqrt{1 + x^2}$ **(iii)** $(x + 1)(x^2 + 2x + 2)^{-1/2}$ **35. (i)** $4x^3 + 6x^2 - 2x$ **(ii)** $12x^2 + 12x - 2$ **(iii)** $24x + 12$ **43.** ≈ 2.7183 **45.** ≈ 0.5403

Exercises 3.3 (p. 164)

1. min 0 at $x = -1$, max 8 at $x = 1$ **3.** min -2 at $t = 0, 1$, max $-\frac{5}{4}$ at $t = \frac{1}{2}$ **5.** min -3 at $s = -\sqrt{2}$, max 1 at $s = 0, -2$ **7.** min 3 at $x = 0, 2$, max 12 at $x = 3$ **9.** min -226 at $u = -2$, max 230 at $u = 2$ **11.** min $\frac{1}{18}$ at $x = \pm4$, max $\frac{1}{2}$ at $x = 0$ **13.** min $-\frac{3}{2}$ at $x = 1$, max $-\frac{1}{4}$ at $x = -1$ **15.** min 0 at $x = 0$, max 4 at $x = -2$ **17.** min 0 at $x = 0$, max $\frac{1}{2}$ at $x = -1$ **19.** min -1 at $x = 0$, max 1 at $x = 2$ **21.** $21x^6 - 8x^3 + 1$ **23.** $\frac{2}{(x + 1)^2}$ **25.** $-\frac{2}{(2x + 3)^2}$ **27.** $-x^{-4}(4x + 3)$ **29.** $-(4x^3 + 4x)(x^4 + 2x^2 + 4)^{-2}$ **31.** 50 m **33.** ≈ 1056.3 cm^3 **35.** side of base $(200)^{1/3} \approx 5.848$ in, height $\frac{1}{2}(200)^{1/3} = 10(40)^{-1/3} \approx 2.924$ in **37. (i)** $(-\frac{2}{5}, \frac{1}{5})$ **(ii)** $(\frac{3}{2}, \frac{3}{2})$ **39.** $\frac{50}{3}$ cm by $\frac{25}{3}$ cm **41. (i)** square $\frac{4}{\pi + 4}$ m, circle $\frac{\pi}{\pi + 4}$ m **(ii)** all circle, no square **43. (i)** \$9 **(ii)** \$11.50 **45.** max $f(-1) = 4$, min ≈ 0.855 at $x \approx 0.311$ **47.** max $f(2) = 4$, min ≈ -10.4940 at $x \approx -1.5737$ **49.** max ≈ 1.8197 at $x \approx 2.0288$, min $f(0) = f(\pi) = 0$

Exercises 3.4 (p. 172)

1. $-2\cos x \sin x$ **3.** $-\sin(u)/(2 - \cos u)^2$ **5.** $2(w - 1)\sin(2w) + 2(w^2 - 2w - 1)\cos(2w)$ **7.** $\big(2\cos(2t)\sin(3t) - 3\sin(2t)\cos(3t)\big)/\sin^2(3t)$, not defined for $t = n\frac{\pi}{3}$, n an integer **9.** $3r^2 \tan r + r^3 \sec^2 r - \sec r \tan r$, except $r = \frac{\pi}{2} + n\pi$ **11.** $2\sec^2 x \tan x$, except for $x = (2n + 1)\frac{\pi}{2}$ **13.** 0, except for $x = n\pi$ **15.** $-\csc^2 x$, except $x = n\pi$ **17.** $-\csc^3 x - \csc x \cot^2 x = -\csc^3 x(1 + \cos^2 x)$, except $x = n\pi$ **19.** $\sec^2 x$, except $x = n\frac{\pi}{2}$ **21.** $x = (2n + 1)\frac{\pi}{2}$ **23.** $u = n\frac{\pi}{2}$ **25.** $v = (2n + 1)\frac{\pi}{4}$ **27.** $r = n\pi$ **29.** never zero **31.** $(1 + 2x^2)\cos x + 4x \sin x$ **33.** $-2\sin x \cos x/(1 + \sin^2 x)^2$ **35.** $[-(x^2 + 2x + 5)\sin x - (2x + 2)\cos x]/(x^2 + 2x + 5)^2$ **37.** $3x^2(\sin x + 2\cos x) + (x^3 - 8)(\cos x - 2\sin x)$ **39.** $2x + 2\tan x + 2x\sec^2 x + 2\tan x \sec^2 x$ **41. (i)** $\theta = -\frac{\pi}{6} + 2n\pi$ or $\frac{5\pi}{6} + 2n\pi$ **(ii)** none **43.** positive on $(-\frac{\pi}{2}, \frac{\pi}{2})$, negative on $[-\pi, -\frac{\pi}{2}) \cup (\frac{\pi}{2}, \pi]$, zero at $\pm\frac{\pi}{2}$ **45. (i)** $g'(x) = \sin x + x\cos x, g'(0) = 0$ **47.** $x = \frac{\pi}{4} + 2n\pi$ and $x = \frac{3\pi}{4} + 2n\pi$

Exercises 3.5 (p. 183)

1. $3t^2 \ln t + t^2$, for $t > 0$ **3.** $\ln(x^2) + 2 = 2\ln|x| + 2$, except $x = 0$ **5.** $e^r \ln r + \frac{1}{r}e^r, r > 0$ **7.** $(1 - 2x)e^{-2x}$ **9.** $1, u > 0$ **11.** $\frac{1}{x}$, for $x > 0$ **13.** $\frac{1}{t} + 1$, for $t > 0$ **15.** $-\frac{1}{t\ln^2 t}$, for $t > 0$ **17.** $\frac{2}{x}$, for $x \neq 0$ **19.** $(2x^2 + 2x)e^{2x+4}$, for $x > 0$ **21.** $\cosh^2 x + \sinh^2 x$ **23.** $\tanh(2x) + 2x\,\mathrm{sech}^2(2x)$ **25.** $e^x(\sinh x + \cosh x)$ **27.** 0 **29.** $3\cosh(3x) + 4\sinh(4x)$ **31.** $\big((2x^3 - 4)/x - 6x^2 \ln x\big)/(2x^3 - 4)^2$ **33.** $\big(2x(e^x + \sin x + 2) - (x^2 + 6)(e^x + \cos x)\big)/(e^x + \sin x + 2)^2$ **35.** $e^x\big((3 + x)\cos x - (2 + x)\sin x\big)$ **37.** $\big(2x(\tan^2 x + 2 + \sin x) - (x^2 + 3)(2\tan x \sec^2 x + \cos x)\big)/(\tan^2 x + 2 + \sin x)^2$ **39.** $3\sin^2 x \cos x$ **43.** $\frac{1}{x \ln 2}$ **45.** $a = 1$ **47. (i)** max $= \cosh c = \cosh(-c)$, min $= 1$ **(ii)** max $= 1$, min $= \mathrm{sech}\, c = \mathrm{sech}(-c)$ **(iii)** max $= \sinh c$, min $= \sinh(-c)$ **49. (i)** $\cosh(x + y) = \cosh x \cosh y + \sinh x \sinh y$ **(ii)** $\cosh(2x) = \cosh^2 x + \sinh^2 x$ **51.** $x\ln^2 x - 2x\ln x + 2x$ **53. (i)** 4×10^5 **(ii)** $100 \ln 2$ minutes **57.** $e \approx 2.7183$ **61.** ≈ 7.6009

Exercises 3.6 (p. 193)

1. $3(3x^4 - 2x^2 - 1)^2(12x^3 - 4x)$ **3.** $-2(x^2 + 4x + 5)^{-3}(2x + 4)$ **5.** $-2\sin(2x + 1)$ **7.** $2x/(x^2 + 4)$ **9.** $\sinh(\sin x)\cos x$ **11.** $(t - 1)/\sqrt{t^2 - 2t + 1}$, except $t = 1$ **13.** $\frac{1}{3}x^{-2/3}$, except $x = 0$ **15.** $\frac{-\sin r}{1 + \cos r}$, $r \neq (2n + 1)\pi$ **17.** $1, x > 0$ **19.** $\sinh(\sinh x)\cosh x$ **21.** $\frac{6x}{x^2 + 7}\ln^2(x^2 + 7)$ **23.** $4x^3\cos(\ln(x^4 + 2))/(x^4 + 2)$ **25.** $2x\ln(10)10^{x^2}$ **27.** $\frac{1}{2}\frac{\sinh x}{\sqrt{\cosh x}}$ **29.** 0 **31.** $-4x/(x^2 + 2)^3$ **33.** $\frac{1}{2}\sin x/\sqrt{4 - \cos x}$ **35.** $4(x + 3)^3(-x^2 - 6x + 1)/(x^2 + 1)^5$ **37.** $2x\cos(\exp(x^2 + 1))\exp(x^2 + 1)$ **39.** $\frac{5}{7}x^{-2/7}$ **45.** **(i)** max $\sqrt{2}$ at $x = \pm 1$, min 0 at $x = 0$ **(ii)** max $\frac{1}{\sqrt{2}}e^{-1/2}$ at $x = \frac{1}{\sqrt{2}}$, min $-\frac{1}{\sqrt{2}}e^{-1/2}$ at $x = -\frac{1}{\sqrt{2}}$ **47.** width $= \frac{2}{\sqrt{3}}R$, height $= \frac{2\sqrt{2}}{\sqrt{3}}R$ **49.** **(ii)** $\frac{1}{1 + t^2}$

Exercises 3.7 (p. 201)

1. $-\frac{2x + y}{x - 2y}$ **3.** $\frac{1 - y}{x + 3y^2} = \frac{y^3 - y}{(1 + x)(x + 3y^2)}$ **5.** $\frac{6x + y}{27y^2 - x}$ **7.** $\sec y$ **9.** $\tan x \tan y$ **11.** $\frac{2y - x}{2x + y}$ **13.** $\frac{x + 3y^2}{1 - y} = \frac{(x + 3y^2)(1 + x)}{y^3 - y}$ **15.** $\frac{27y^2 - x}{6x + y}$ **17.** $\cos y$ **19.** $\cot x \cot y$ **21.** $9x^8 + 8x^7 + 14x^6 - 42x^5 - 35x^4 - 64x^3 - 24x^2 - 16x$ **23.** $(x^3 + 4x)^{1/3} + \frac{x}{3}(3x^2 + 4)(x^3 + 4x)^{-2/3}$ **25.** $\frac{1}{2}\left(\sqrt{\frac{(x + 2)(x + 3)}{x + 1}} + \sqrt{\frac{(x + 1)(x + 3)}{x + 2}} + \sqrt{\frac{(x + 1)(x + 2)}{x + 3}}\right)$ **27.** $\frac{3}{2}\frac{x^2}{(x^2 + 2)^4\sqrt{x^3 + 1}} - \frac{8x\sqrt{x^3 + 1}}{(x^2 + 2)^3}$ **29.** $\left(\frac{(x + 1)(x - 2)}{(x + 3)(x^2 - 9)}\right)^{3/7}\left(\frac{3}{7}\left(\frac{1}{x + 1} + \frac{1}{x - 2} - \frac{1}{x + 3} - \frac{2x}{x^2 - 9}\right)\right)$ **35.** **(i)** $x + y = 2$ **(ii)** $6x - y = 12$ **37.** **(i)** $y = \pi$ **(ii)** $x + 12y = 3$ **39.** max $= \frac{18}{\sqrt{33}}$, min $= -\frac{18}{\sqrt{33}}$ **41.** max 4, min -12 **47.** ellipse **49.** ellipse

Exercises 3.8 (p. 217)

1. $2/(4x^2 + 16x + 17)$ **3.** $2x/(|x|(1 + x^2)\sqrt{x^2 + 2})$ **5.** $2x/((4 + x^2)\sqrt{x^4 + 8x^2 + 15})$ **7.** $-e^x/((5 + e^x)\sqrt{e^{2x} + 10e^x + 24})$ **9.** $-2\exp(\operatorname{arccot}(2x))/(1 + 4x^2)$ **11.** $2\sec^2 x\tan x/\sqrt{\sec^4 x - 1}$ **13.** $-\frac{2x + 2}{(x^2 + 2x + 3)^2 - 1}$ **15.** $1/(\operatorname{arccosh}(x)\sqrt{x^2 - 1})$ **17.** $\operatorname{sech} x$ **19.** $-\operatorname{csch} x$ **21.** $\frac{1}{5}$ **23.** $\frac{1}{4 - 4h^{-1}(u)} = \frac{1}{\sqrt{8 - 8u}}$ **25.** $(H^{-1})'(u) = 1/(3H^{-1}(u)^2 + 12H^{-1}(u))$ **31.** **(i)** $\sin x, x \in [-\frac{\pi}{2}, \frac{\pi}{2}]$ **(ii)** $\frac{1}{2}(\tan(u) - 1), u \in (-\frac{\pi}{2}, \frac{\pi}{2})$ **35.** $x + \pi, x \in (-\frac{3\pi}{2}, -\frac{\pi}{2})$ **37.** 104π mph ≈ 326.7 mph **39.** **(i)** $K(x) = \frac{x}{2.2}$, $P(x) = 2.2x$ **45.** ≈ 0.5236

Exercises 3.9 (p. 222)

1. $12x^2 - 18x + 4$ **3.** $(2 + 4x^2)\exp(x^2 + 1)$ **5.** $343e^{7x+1}$ **7.** $-\cos x$ **9.** 0 **11.** $2(x + 1)^{-3}$ **13.** $-162(3t + 2)^{-4}$ **15.** $-(e^{3x} - 4e^{2x} + e^x)(e^x + 1)^{-4}$ **17.** $x(1 - x^2)^{-3/2}$ **19.** $\exp(\sin x)(\cos^3 x - 3\cos x\sin x - \cos x)$ **21.** $\frac{n!}{(n - k)!}$, if $k \leq n$; 0, if $k > n$ **23.** $(-1)^{k-1}(k - 1)!\,x^{-k}$ **25.** **(i)** $0, \pm 1, \pm 2$ **27.** **(i)** 6 sec **(ii)** 144 ft **29.** $3\sqrt{58.8} \approx 23$ m/sec **33.** $\cos x$, if $n = 4k$; $-\sin x$, if $n = 4k + 1$; $-\cos x$, if $n = 4k + 2$; $\sin x$, if $n = 4k + 3$ **35.** $2^n e^{2x+1}$ **37.** $a^n\sin(ax + b)$, if $n = 4k$; $a^n\cos(ax + b)$, if $n = 4k + 1$; $-a^n\sin(ax + b)$, if $n = 4k + 2$; $-a^n\cos(ax + b)$, if $n = 4k + 3$ **39.** $\frac{625}{16}$ ft

Chapter 3 Review Exercises (p. 225)

1. $20x^4 - 6x$ **3.** $14x^6 - 16x^3$ **5.** $2 - \frac{1}{2\sqrt{x}}$, for $x > 0$ **7.** $\frac{3}{5}x^{-2/5} - \frac{1}{2}x^{-3/2}$, for $x > 0$ **9.** $\frac{10x}{(x^2 + 1)^2}$ **11.** $3\cos^2 x - 3\sin^2 x$ **13.** $6x\tan x + (1 + 3x^2)\sec^2 x, x \neq (2n + 1)\frac{\pi}{2}$ **15.** $2x^3/\sqrt{x^4 + 2}$ **17.** $-(2x^3 + x)(x^4 + x^2)^{-3/2}, x \neq 0$ **19.** $e^{2x-5}(2\cos x - \sin x)$ **21.** $(2x + 3)\cosh(x^2 + 3x - 2)$ **23.** $\tanh x$ **25.** $(12x^3 - 2)\operatorname{sech}^2(3x^4 - 2x)$ **27.** $\ln x + 1$, for $x > 0$ **29.** $3\sin^2 x\cos x\cosh^5 x + 5\sin^3 x\cosh^4 x\sinh x$ **31.** max $f(0) = -2$, min $f(-3) = -11$ **33.** max $f(0) = 4$, min $f(2) = -12$ **35.** max $f(0) = 1$, min $f(3) = \frac{1}{10}$ **37.** max $\ln 8$, min 0 **39.** max $f(0) = 0$, min $f(-\frac{\pi}{4}) = -1$ **41.** $\frac{2x + 3y}{2y - 3x}$ **43.** $-\frac{y}{x}$ **45.** 1 **47.** $(2x - \frac{y}{x})/\ln x$ **49.** $-(1 + (x + y)^2ye^{xy})/(1 + (x + y)^2xe^{xy})$ **51.** $\frac{1}{x^2 + 6x + 10}$ **53.** $-\frac{\cos x}{|\cos x|}$ **55.** $\frac{2x}{|x|\sqrt{x^2 + 1}}$ **57.** $\frac{2}{1 + 4x^2}\exp(\arctan(2x))$ **59.** $1/(\sqrt{1 - x^2}\arcsin x)$ **61.** **(i)** $\ln(u + \sqrt{u^2 - 1})$ **67.** **(ii)** $\frac{|\sin x|}{\sin x}\cos x = \cot x|\sin x|$

Exercises 4.1 (p. 234)

1. $36\pi \approx 113$ in/sec **3.** $\frac{30}{\sqrt{39}} \approx 4.8$ in/sec (downward), or $\frac{5}{2\sqrt{39}}$ ft/sec **5.** $\frac{5}{3}$ in^2/sec **7.** $\frac{103}{\sqrt{101}} \approx 10.25$ m/sec **9.** 3 ft^3/min **11.** 3 min **13.** $\frac{28}{\sqrt{365}} \approx 1.4656$ m/sec (upward) **15.** **(i)** 0.88 rad/sec ≈ 8.4 rpm **(ii)** 0.44 rad/sec ≈ 4.2 rpm **(iii)** $30\sqrt{2} \approx 42.4$ mph **(iv)** $\frac{200}{11}\pi \approx 57.1$ mph

Exercises 4.2 (p. 247)

1. abs max at $x = \frac{\pi}{2} + 2n\pi$, abs min at $x = \frac{3\pi}{2} + 2n\pi$ **3.** abs max at $x = (2n + 1)\pi$, abs min at $x = 2n\pi$ **5.** abs max at $x = 0$ **7.** abs min at $x = 0$ **9.** abs min at $x = 0$ **11.** $-\frac{1}{2}$ **13.** $\frac{1}{\sqrt{3}}$ **15.** 0 **17.** $\frac{\pi}{2}$ or $\frac{3\pi}{2}$ **19.** $\frac{1}{3}$ **21.** always increasing **23.** always increasing **25.** increasing if $x \geq 2$, decreasing if $x \leq 2$ **27.** increasing on $(-\frac{\pi}{2}, \frac{\pi}{2}), (\frac{\pi}{2}, \frac{3\pi}{2}), (-\frac{3\pi}{2}, -\frac{\pi}{2}), (\frac{3\pi}{2}, \frac{5\pi}{2}), \ldots$ **29.** always increasing **31.** $\frac{a + b}{2}$ **33.** 0 **35.** **(i)** $x^4 - 2x^2 - 3x$ **(ii)** $\frac{1}{5}x^5 - \frac{3}{4}x^4 + x^2 - x$ **37.** **(i)** $2x^4 - 2x^3 + 2x^2 - 3x$ **(ii)** $\sin(x) + 5$ **39.** $\frac{a}{4}$ **43.** not always **47.** **(ii)** above **49.** abs min $x \approx 2.141$, local min $x \approx -0.432$, local max $x \approx 0.541$ **51.** abs max $x \approx 2.455$, abs min $x \approx 5.233$ **53.** $\approx 0.7763, 1.8263$

Exercises 4.3 (p. 254)

1. ± 2 **3.** $n\pi, n = 0, \pm 1, \pm 2, \ldots$ **5.** none **7.** $(2n + 1)\frac{\pi}{2}$ **9.** $\pm 1, 0$ **11.** 0, not any kind of extremum **13.** 0, not any kind of extremum **15.** 0, local max, ± 2 abs min **17.** none **19.** 0 abs min **21.** -1 local max, 0 abs min, 2 abs max **23.** -3 abs min, 0 not any kind of extremum, 3 abs max **25.** -3 local min, -2 abs max, 2 abs min **27.** 0 abs min, $\frac{\pi}{2}$ abs max, $\frac{3\pi}{4}$ local min **29.** 0 abs min, 3 abs max **31.** $\frac{a + b}{2}$, abs min **35.** **(ii)** $x, x^3 + 3x$

Exercises 4.4 (p. 261)

1. abs min at 1 **3.** $-\sqrt{2}$ local max, $\sqrt{2}$ local min **5.** -1 abs max, 1 abs min **7.** $-\sqrt{2}$ abs max, $\sqrt{2}$ abs min **9.** ± 2 abs min, 0 local max **11.** $\frac{3}{2}$ abs min **13.** $(2n + 1)\frac{\pi}{2}$ abs max, $n\pi$ abs min **15.** $n\pi$ abs max, $(2n + 1)\frac{\pi}{2}$ abs min **17.** $-\frac{\pi}{2}$ abs min **19.** 0 abs min **21.** **(i)** every point abs max and abs min **(ii)** $t > 0$ abs max, $t < 0$ abs min **23.** **(i)** none: $a^2 - 3b < 0$, one: $a^2 - 3b = 0$, two: $a^2 - 3b > 0$ **25.** not differentiable: a, c, e, g, i, k, m; jumps: e, i, k; abs max: a, e, i, k; abs min c, g; local max f; $(k, m]$ local max and local min **27.** $x = 0$ abs min if k even, no extremum if k odd

Exercises 4.5 (p. 275)

1. $x = 1, y = x + 1$ **3.** $y = 1, x = \pm\sqrt{3}$ **5.** $y = x, x = \pm 2$ **7.** $y = x - 1, x = -1$ **9.** $x = 3, x = -2$

11.

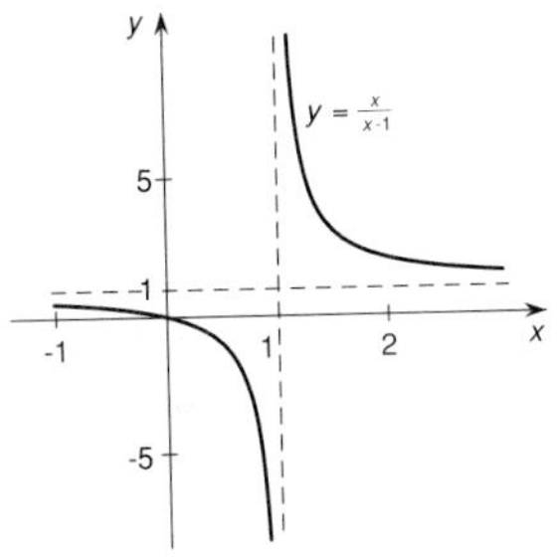

13.

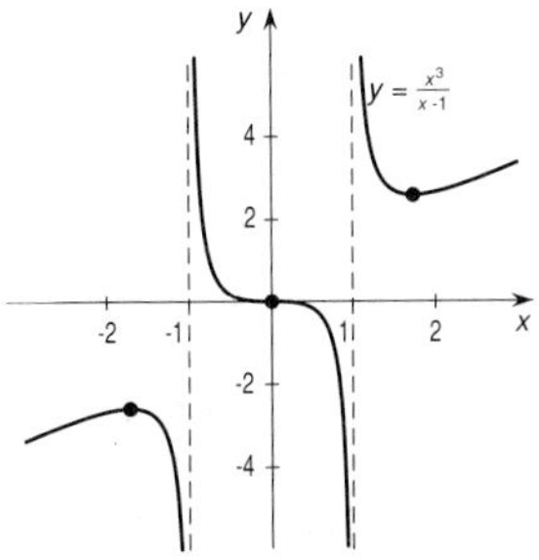

15.

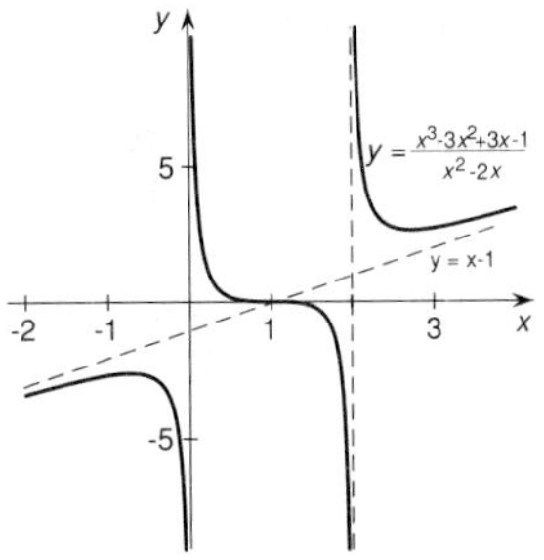

17.

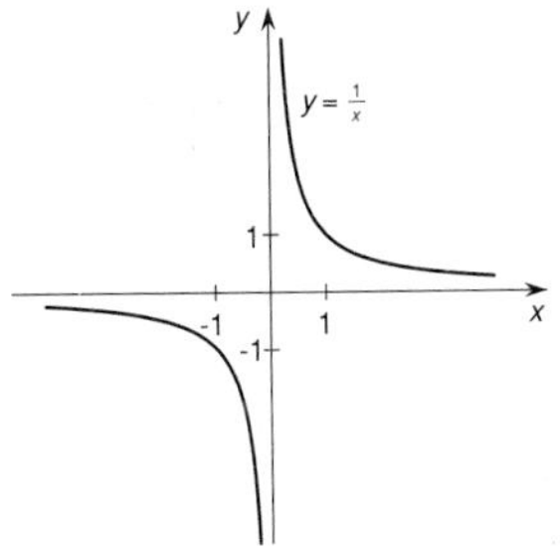

19.

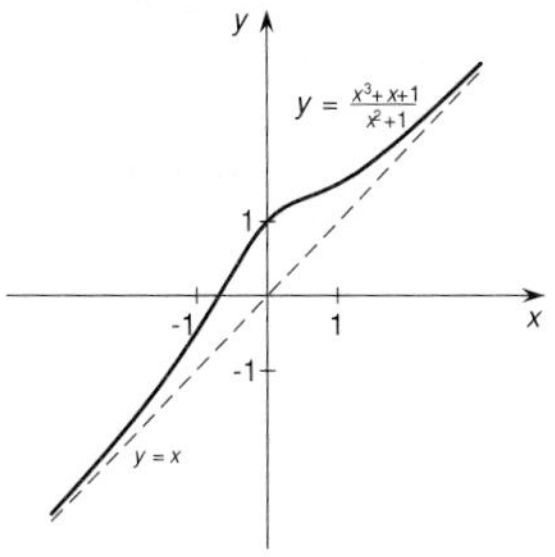

21.

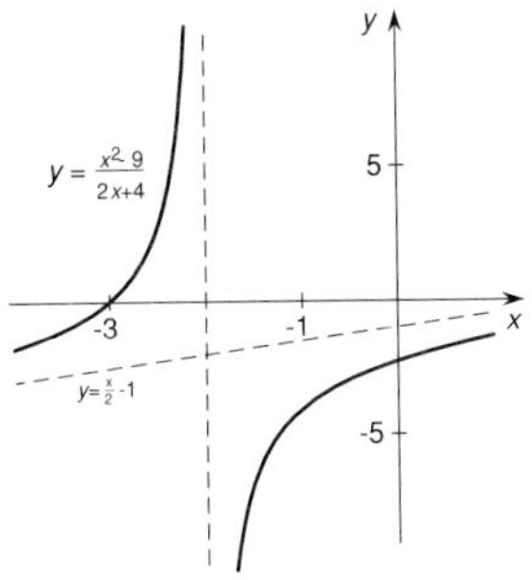

23. if $k = 0$, equal; $k > 0$ even, equal but infinite **25. (i)** $a_k = 0$, for every *odd* k **(ii)** $a_k = 0$, for every *even* k **27. (i)** $(3x^4 - x^2 - 1) + (-2x^3 + 7x)$

Exercises 4.6 (p. 289)

1. always concave up **3.** concave up for $x > 0$, down for $x < 0$ **5.** concave up $(-\infty, -\frac{2}{\sqrt{3}}) \cup (\frac{2}{\sqrt{3}}, \infty)$, down $(-\frac{2}{\sqrt{3}}, \frac{2}{\sqrt{3}})$ **7.** concave up $(\frac{\pi}{4} + n\pi, \frac{3\pi}{4} + n\pi)$, down $(-\frac{\pi}{4} + n\pi, \frac{\pi}{4} + n\pi)$ **9.** concave up $(-\infty, -\frac{1}{\sqrt{2}}) \cup (\frac{1}{\sqrt{2}}, \infty)$, down $(-\frac{1}{\sqrt{2}}, \frac{1}{\sqrt{2}})$ **11.** 0 **13.** 1 **15.** $(2n + 1)\frac{\pi}{2}$ **17.** $n\frac{\pi}{2}$ **19.** 0 **21.** $-\sqrt{2}$ local max, $\sqrt{2}$ local min, 0 inflection point

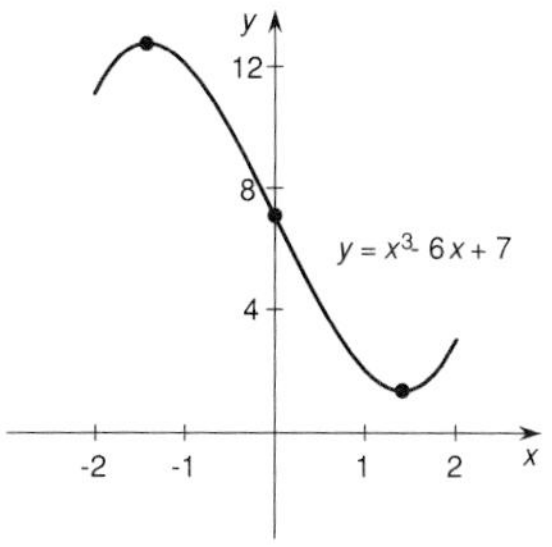

23. 0 infl pt.

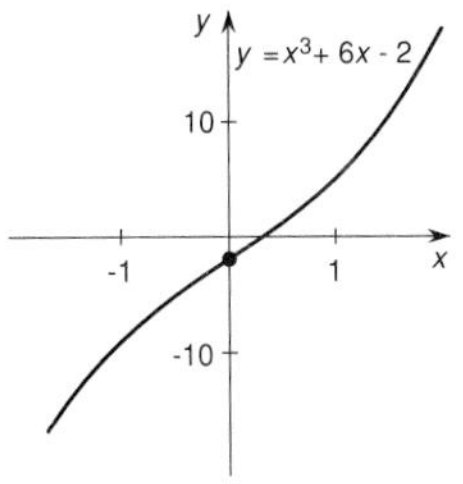

25. ± 1 abs min, 0 local max $\pm\frac{1}{\sqrt{3}}$ infl pts.

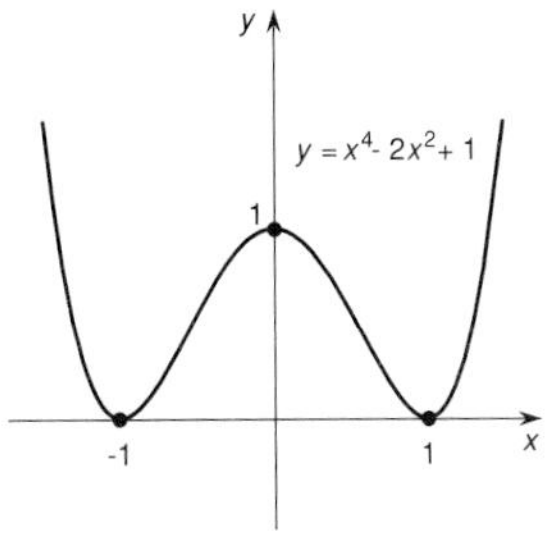

27. 0 local max, 2 local min

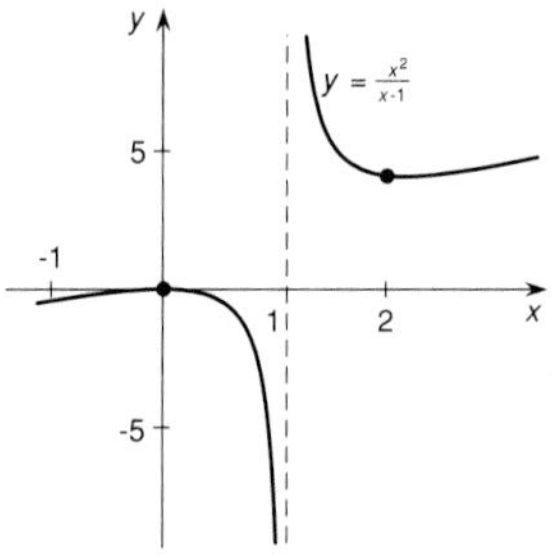

29. no extrema or infl pts.

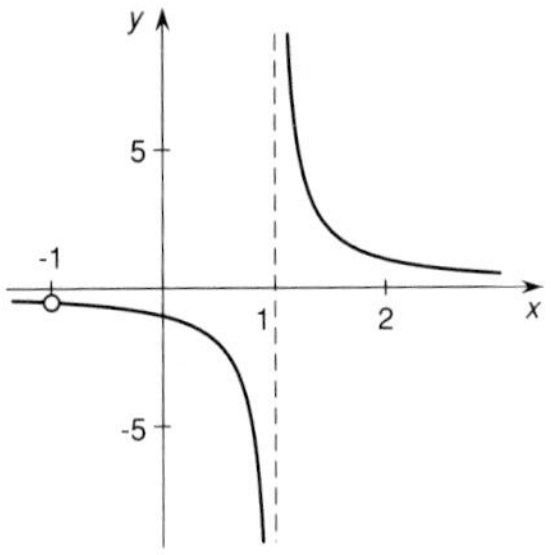

31. 0 max, $\pm\frac{1}{\sqrt{2}}$ infl pts.

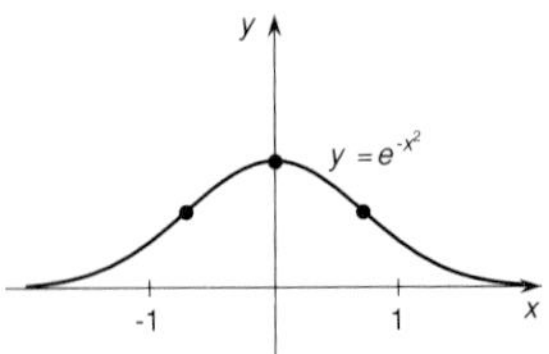

33. 0 infl pt.

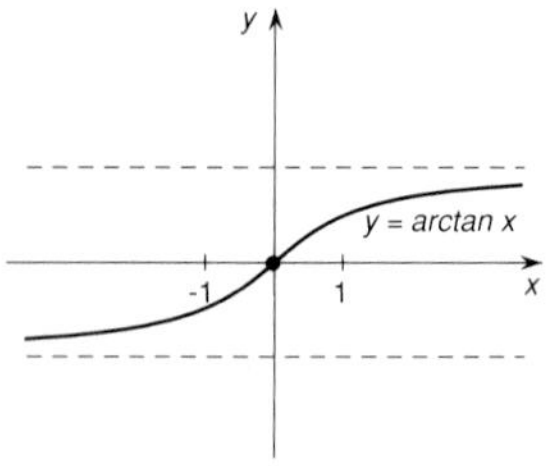

35. $\frac{1}{e}$ min

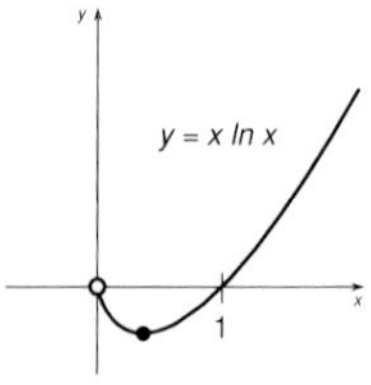

37. $k > 2$ odd **39.** **(i)** infl pt. **(ii)** infl pt. **(iii)** cusp **(iv)** neither **41.** -0.180 **43.** **(iii)** $(-\infty, 1.5)$

Exercises 4.7 (p. 300)

1. **(i)** $\frac{\sqrt{3}-1}{2}$ **(ii)** ≈ 1.196 rad $\approx 68.53°$ **3.** **(i)** $\frac{500{,}000}{116} \approx 4310$ ft **(ii)** $t = \frac{4}{\sqrt{3}} \approx 2.309$ min **5.** $(3 - \frac{1}{\sqrt{3}})$ km on land **7.** radius $(\frac{12}{\pi})^{1/3} \approx 1.563$ in, height $4(\frac{12}{\pi})^{1/3} \approx 6.253$ in **9.** **(ii)** radius $2(\frac{12}{\pi})^{1/3} \approx 3.126$ in, height $(\frac{12}{\pi})^{1/3} \approx 1.563$ in **11.** **(i)** one-third of the way from the point directly below A to the point directly below B **13.** **(i)** both $\frac{\pi}{4}$ (45°) **(ii)** $\theta = \arctan \frac{4}{3} \approx 0.927$ rad $\approx 53.1°$, $\phi = \frac{\pi}{2} - \theta \approx 0.644$ rad $\approx 36.9°$ **15.** $(1 + 3(12^{-1/3}))\sqrt{4 + 12^{2/3}} \approx 7.0235$ yd **17.** cost increases by \$350 **19.** vol $\frac{4}{27}\pi r^2 h$; radius $\frac{2}{3}r$, height $\frac{1}{3}h$ **21.** $100/(1 + 4^{1/3}) \approx 38.65$ m from weaker bulb **23.** $\frac{128{,}000\pi}{9\sqrt{3}} \approx 25{,}796$ **25.** **(i)** all hemisphere, no cylinder, radius $10\sqrt{\frac{10}{\pi}} \approx 17.84$ ft **(ii)** radius $10\sqrt{\frac{10}{3\pi}} \approx 10.30$ ft, height of cylinder $20\sqrt{\frac{10}{3\pi}} \approx 20.60$ ft **27.** **(i)** 4 **(ii)** 7 **(iii)** \$499.99 **29.** $\theta \approx 1.4701$

Exercises 4.8 (p. 312)

1. $2x^3 - 2x^2 + 3x + c$ **3.** $-\cos x + c$ **5.** $\arcsin x + c$ **7.** $\frac{2}{3}x^{3/2} + \frac{1}{5}x^5 + c$ **9.** $6\sqrt{x} + c$ **11.** $-\frac{1}{4}\cos(2x^2 - 3) + c$ **13.** $\frac{1}{2}\exp(x^2 + 2x - 5) + c$ **15.** $\frac{1}{2}\ln(x^2 + 1) + c$ **17.** $\frac{1}{6}(x^4 + 4)^{3/2} + c$ **19.** $\ln(e^x + 1) + c$ **21.** $\frac{1}{4}x^4 - x^2 + 3x + 5$ **23.** $x^3 - 2x^2 + x + 1$ **25.** $x + c$ for $x > 0$, $-x + d$ for $x < 0$ **27.** $\ln(x - 3) + c$ for $x > 3$, $\ln(3 - x) + d$ for $x < 3$ **29.** $\tan x + c_n$, for $x \in (n\pi - \frac{\pi}{2}, n\pi + \frac{\pi}{2})$, i.e., a different constant on each interval **31.** $-\ln|\cos x| + c_n$, for $x \in (n\pi - \frac{\pi}{2}, n\pi + \frac{\pi}{2})$, i.e., a different constant on each interval **33.** **(i)** $4000\sqrt{2} \approx 5657$ **(ii)** $\frac{\ln 5}{\ln 2}(40) \approx 92.88$ min **35.** $\tan x + c_n$, for $x \in (n\pi - \frac{\pi}{2}, n\pi + \frac{\pi}{2})$, with $c_0 = 4$ **37.** $c_1 \exp(\frac{1+\sqrt{5}}{2}t) + c_2 \exp(\frac{1-\sqrt{5}}{2}t)$ **39.** $\frac{1}{14}e^{4x} + c_1 e^{2x} + c_2 e^{-3x}$ **41.** **(ii)** $\frac{1}{2}\ln(1 + x^2)$ **43.** $t = \frac{1}{3}\ln 2 \approx 0.231$

Exercises 4.9 (p. 318)

1. $30t - 6$ **3.** $2t^3 - 2t^2 - 13t + 15$ **5.** 69 ft **7.** $\frac{20}{7}$ sec **9.** 3 sec **11.** **(i)** after $\frac{15}{4}$ sec **(ii)** $\frac{225}{4}$ ft **13.** $2\sqrt{(13.5)(4.9)} \approx 16.27$ m/sec **15.** **(i)** yes **(ii)** $\frac{2992}{15} \approx 199.47$ ft **17.** **(i)** 96 ft **(ii)** $48 + 24\sqrt{5} \approx 101.67$ ft **19.** $\frac{1}{16}v_0^2 \cos\theta \sin\theta$ **21.** 12,000 m **23.** $\approx 1.42737 \times 10^{16}$ ft^3/sec^2 $\approx 1.25673 \times 10^{12}$ mi^3/hr^2 **25.** **(i)** $t^5 - 4t^3$ **(ii)** $t \approx 2.1655$

Exercises 4.10 (p. 324)

1. $8\frac{1}{16}$ **3.** $8 - \frac{1}{16}(0.001) \approx 7.999\,937\,5$ **5.** 7.85 **7.** $\frac{1}{\sqrt{2}}(1 + \frac{\pi}{180}) \approx 0.7194$ **9.** $-1 - \frac{\pi}{90} \approx -1.0349$ **11.** 0.0099 **13.** $-\frac{1}{(x+1)^2}dx$ **15.** $\frac{\sin x - x\cos x}{\sin^2 x}dx$ **17.** $((y^2 - 2xy)/(x^2 - 2xy))\,dx$ **19.** $((\frac{x\sin x + \cos x}{x^2}y^2 - xy\cos x - y\sin x - 1)/(x\sin x + 2y\frac{\cos x}{x}))\,dx$ **21.** **(i)** 2.04 **23.** **(i)** 4800 in^3 **25.** $(\frac{6}{\pi})^{1/3}(1.01) \approx (\frac{6.18}{\pi})^{1/3} \approx 1.253$ in **27.** **(i)** -1.7 **(iii)** ≈ -1.6398

Exercises 4.11 (p. 329)

1. 4.3589 **3.** 7.2801 **5.** 0.328 268 856 **7.** 0.496 292 071 **9.** 1.764 542 338 **11.** 0.848 062 079 **13.** $-1.039\,072\,26$ **15.** $-0.547\,457\,039$, 2.118 253 365, etc. **17.** 1.412 743 866, $-1.166\,290\,826$ **19.** 8.062 257 748 **21.** 2.038 691 840 **23.** $x_n = \frac{1}{2n-1}x_1$ **25.** $\approx -0.152\,679\,653$

29. **(ii)** $\pm 2.028\ 757\ 838$ **(iv)** $\pm 1.076\ 873\ 986$
(v)

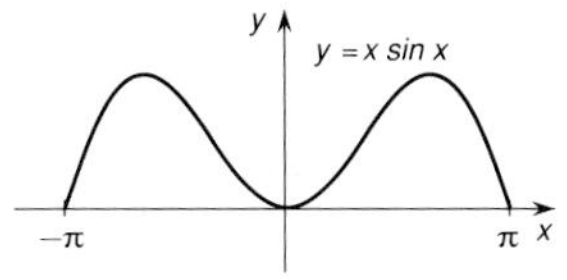

Exercises 4.12 (p. 338)

1. 2 **3.** $\frac{32}{7}$ **5.** $\frac{1}{2}$ **7.** 1 **9.** 1 **11.** ∞ **13.** $-\infty$ **15.** ∞ **17.** 0 **19.** 0 **21.** 0 **23.** 0 **25.** 0 **27.** $-\frac{2}{\pi}$ **29.** does not exist **31.** **(i)** $-\frac{1}{8}$ **(ii)** 1 **33.** 0

Exercises 4.13 (p. 341)

1. 100 **3.** motors \$100 profit, generators \$460 loss **5.** **(i)** 225 **(ii)** $\approx$ \$1.28 **(iii)** $\approx$ \$1.42 (at 179 units) **(iv)** at best they will lose \$212.50

Chapter 4 Review Exercises (p. 342)

1. $\frac{1}{8\pi} \approx 0.0398$ cm/sec **3.** 1.125 ft/sec **5.** 0.002 m^2/hr **7.** $\frac{27}{\sqrt{5}} \approx 12.075$ knots **9.** $\frac{27}{2}\pi \approx 42.41$ in^3/sec **11.** incrg. $[2, \infty)$, decrg. $(-\infty, 2]$ **13.** incrg. $(-\infty, -\sqrt{3}] \cup [\sqrt{3}, \infty)$, decrg. $[-\sqrt{3}, \sqrt{3}]$ **15.** decrg. $(-\infty, -1]$, incrg. $[-1, \infty)$ **17.** decrg. $[0, \infty)$, incrg. $(-\infty, 0]$ **19.** incrg. $[\frac{1}{e}, \infty)$, decrg. $(0, \frac{1}{e}]$ **21.** 0, no maxima/minima **23.** 0 abs min **25.** 3, -1 abs min, 1 local max **27.** no critical points **29.** 0 abs min **31.** no extrema **33.** -1 abs max, 1 abs min **35.** 0 abs min, $\frac{3}{2}$ abs max, 2 local min **37.** 1 abs max, abs min at every point in $[-3, 0]$ **39.** $-1 - \frac{1}{\sqrt{2}}$ local max, $-1 + \frac{1}{\sqrt{2}}$ abs min **41.** 2 abs max, -2 abs min; 0, $\pm 2\sqrt{3}$ infl pts., horizontal asymptote $y = 0$

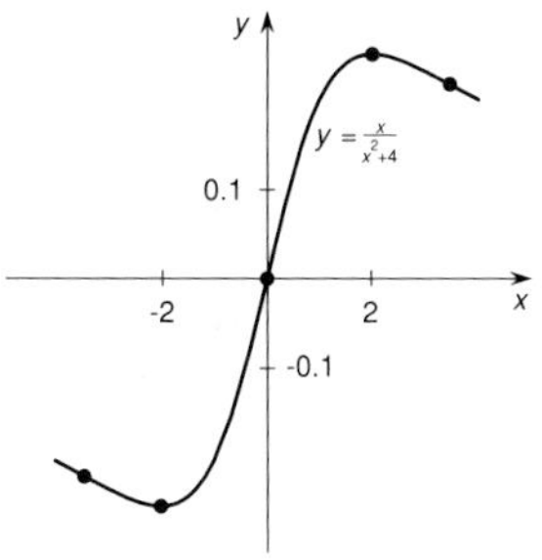

43. 0 local min, -6 local max, vert. asymptote at $x = -3$, asymptote $y = x - 3$ at $\pm\infty$

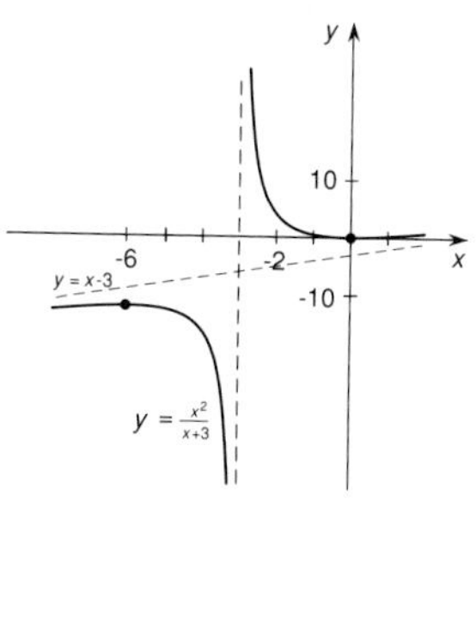

45. $n\pi$ abs max, $(2n + 1)\frac{\pi}{2}$ abs min, $(2n + 1)\frac{\pi}{4}$ infl pt.

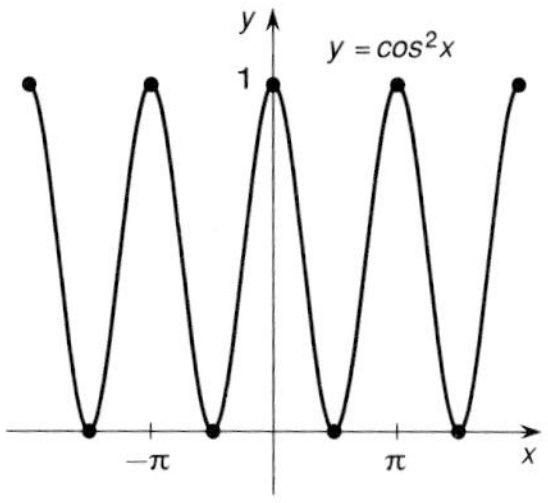

47. vert. asymptote $x = 0$, asymptote $y = x$ at $\pm\infty$

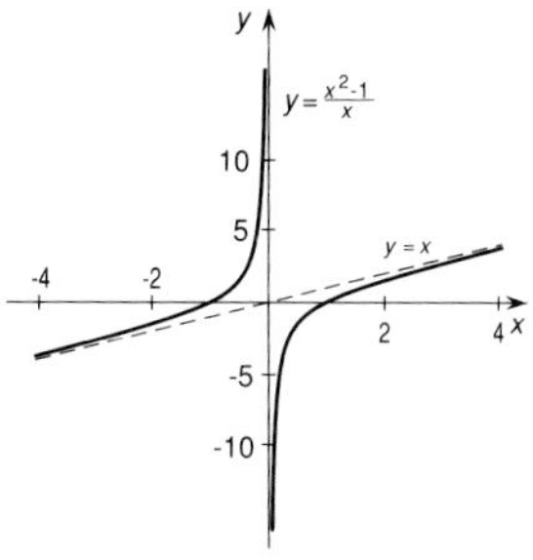

49. vert. asymptotes $x = \pm 3$, horiz. asymptote $y = 0$, infl pt. $x = 0$

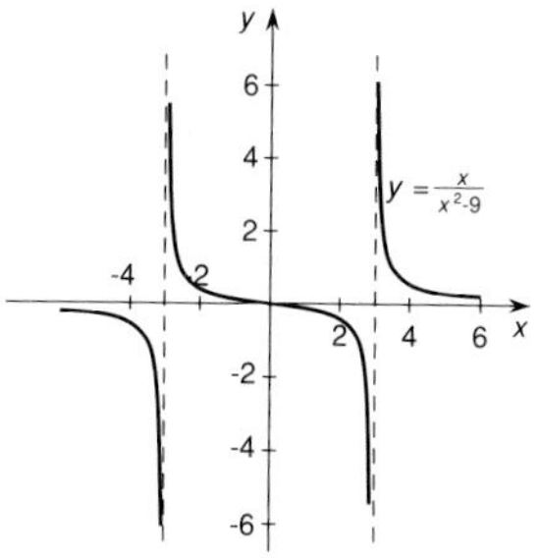

51. $\frac{1}{5}x^5 - x^3 + 2x^2 - x + c$ **53.** $\frac{12}{13}x^{13/4} + \frac{3}{2}x^2 - 4x + c$, for $x > 0$ **55.** $\frac{2}{3}(x + 7)^{3/2} + c$, for $x > -7$ **57.** $-\frac{1}{5}\cos(5x) + c$ **59.** $-\frac{1}{3}x^{-3} + x^2 + c$, if $x < 0$, $-\frac{1}{3}x^{-3} + x^2 + d$, if $x > 0$ **61.** 16 sec **63.** $16(2.6)^2 \approx 108$ ft **65.** $\frac{1210}{9} \approx 134$ ft **67.** $\frac{225}{19.6} \approx 11.48$ m **69.** $(154)^2/1350 \approx 17.57$ ft/sec^2 **71.** $9 - \frac{1}{36} \approx 8.9722$ **73.** $0.025 - \frac{1}{160{,}000} \approx 0.024\ 993\ 75$ **75.** $1024 - \frac{4}{150} \approx 1023.973\ 333$ **77.** $\frac{1}{2} + \frac{\pi\sqrt{3}}{1800} \approx 0.503\ 023$ **79.** $\sqrt{2} - \frac{\pi\sqrt{2}}{6000} \approx 1.413\ 473\ 1$ **81.** 6.403 124 237 **83.** 3.419 951 893 **85.** 1.373 400 767 **87.** 0.539 188 873 or 1.675 130 871 or $-2.214\ 319\ 743$ **89.** $\pm 0.765\ 366\ 865$ or $\pm 1.847\ 759\ 065$ **91.** 0 **93.** $-\frac{3}{4}$ **95.** 0 **97.** $-\infty$ **99.** 0 **101.** **(i)** $100\pi \approx 314.159$ ft/sec **(ii)** $164\pi \approx 515.22$ ft/sec

103. $\approx 66.15\ \text{in}^3$

105.

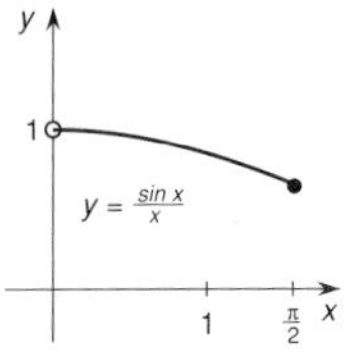

107. $\frac{3}{2}x^2 - 2x + c$ if $x > 0$, $\frac{3}{2}x^2 - 2x + c'$ if $x < 0$ **111.** (i) $dy = (\sin x + x\cos x)\,dx$ (ii) $\left((2xy^3 - 12x^2y)/(4x^3 - 3x^2y^2)\right)dx$ (iii) $\left(-1 - \tan(x+y)\right)dx$ **113.** $100 - \ln 10 \approx 97.697$ **115.** 800 **121.** $\frac{1}{18}$ rad/sec $= \frac{10}{\pi} \approx 3.183°$/sec ≈ 0.5305 rpm

123.

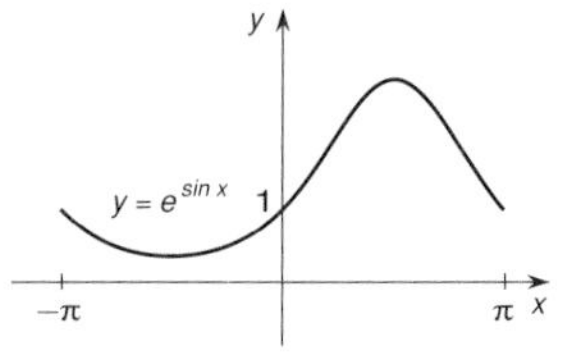

Exercises 5.1 (p. 351)

1. $\frac{14}{27} \approx 0.519$ **3.** $\frac{11}{25} = 0.44$ **5.** $\frac{33}{25} = 1.32$ **7.** $\frac{27}{2}$ **9.** $\frac{739}{216} \approx 3.421$ **11.** $\frac{\pi}{2}$ **13.** $\frac{\pi}{2}$ **15.** $3 + \sqrt{2} + \sqrt{3} \approx 6.146$ **17.** 0 **19.** 2π **21.** $\frac{1}{3}$ **23.** $\frac{7}{3}$ **25.** (i) 11 (iii) 9 **27.** less **29.** 5, $\frac{15}{4}, \frac{77}{25}, \frac{6767}{2500}$ **31.** $\approx$1.341, 1.095, 1.016, 1.004

Exercises 5.2 (p. 361)

1. 225 **3.** 70 **5.** 39 **7.** 121 **9.** 650 **11.** 7 **13.** $\frac{m(m-1)(2m-1)}{6}$ **15.** $\frac{(m+1)(m+2)(2m+3)}{6} - 1$ **17.** $\frac{1}{3}(2m+1)(m^2+m+3)$ **19.** 8 **21.** $\frac{n^2+n-m^2+m}{2}$ **23.** $\frac{n^2+n-m^2+m}{2}$ **25.** (i) −3 (ii) 4 **27.** 1 if n even; 0 if n odd **29.** $(a+b)^n = \sum_{k=0}^{n} \binom{n}{k} a^{n-k}b^k$ **31.** (i) 90 (ii) 90 **37.** 0.083 325 **39.** 1.017 120 611

Exercises 5.3 (p. 371)

1. $\frac{3}{8}$ **3.** $\frac{7}{32}$ **5.** 18 **7.** −10 **9.** $-\frac{3}{8}$ **11.** $\frac{\pi}{2}$ **13.** $\frac{4}{5}$ **15.** 0 **17.** $\frac{x^2}{2}$ **19.** $1 + e + e^2$ **25.** (i), (ii) all zero (iii) −1, −12, −3584 **27.** $\|\mathcal{P}_3\| = \frac{1}{4}$, $\|\mathcal{P}_4\| = \frac{1}{12}$ **31.** $\approx$0.9266 **33.** $\approx$0.373 **35.** (ii) $\frac{1}{m+1}$

Exercises 5.4 (p. 381)

1. $\frac{3}{2}$ **3.** 6 **5.** $25\frac{1}{2}$ **7.** $16\frac{1}{3}$ **9.** 8 **11.** −1 **13.** 4 **15.** −6 **17.** $18\frac{1}{3}$ **19.** −15 **29.** $c = 2, d = 3$ **31.** $b = 3, c = 2, d = -1$ **39.** $\frac{196}{81} \approx 2.4198$ **41.** 0.69011 **43.** 1.71828

Exercises 5.5 (p. 390)

1. $x^2 - 1$ **3.** $\sin x$ **5.** e^t **7.** $e^x - x^2$ **9.** 7 **11.** $\cos(x^2)(2x)$ **13.** e^{1-x} **15.** $\sin(1-u)$ **17.** $(\sin^2 r - 1)\cos r + (\cos^2 r - 1)\sin r = -(\cos^3 r + \sin^3 r)$ **19.** 0 **21.** incrg $[-2, 0] \cup [2, \infty)$, decrg $(-\infty, -2] \cup [0, 2]$ **23.** $-\frac{1}{2}$ local max, $\frac{1}{2}$ local min

Exercises 5.6 (p. 397)

1. $\frac{12}{5}$ **3.** 12 **5.** 2 **7.** 0 **9.** $e - 1$ **11.** $\frac{1}{2}(e^2 - 1)$ **13.** 0 **15.** $\frac{2}{\pi}$ **17.** $\frac{16}{3}$ **19.** $\frac{1}{36}(4^{18} - 3^{18}) = 1{,}898{,}112{,}673\frac{19}{36}$ **21.** $\frac{5}{2}$ **23.** $\frac{23}{3}$ **25.** $36\frac{1}{3}$ **27.** $11\frac{1}{12}$ **29.** $\frac{2}{3}$ **31.** (i) 0 (ii) $2(e^{1/2} - e^{-1/2})$ (iii) $-\frac{2}{\pi}$ **33.** (i) sinh(1) (ii) 0 (iii) tanh(2) **35.** (i) $10\frac{3}{4}$ (ii) $12\frac{1}{4}$ (iii) $8\frac{1}{4}$ **37.** (i) 1 (ii) $-\frac{5}{12}$ (iii) $2 + \sqrt{3}$ **43.** $-\frac{2}{3}$ **47.** $\approx$65.2864 **49.** $\approx$1.570 762 993

Exercises 5.7 (p. 408)

1. $\frac{1}{4}e^{4x-2} + c$ **3.** $\frac{1}{24}(t^3 - 2)^8 + c$ **5.** $\frac{33}{5}$ **7.** $-\frac{2}{\pi}$ **9.** $\frac{1}{3\pi}\sec^3(\pi x) + c$ **11.** $\frac{\pi}{8} - \frac{1}{4}$ **13.** π **15.** $\frac{\pi}{2} + \frac{1}{12}\sin(6x) + c$ **17.** $\frac{1}{2}$ **19.** $\frac{1}{3}\ln(\frac{26}{7})$ **21.** $\sqrt{1 + x^2} + c$ **23.** $\frac{1}{(\sqrt{x}+2)^2} + c$ **25.** $\frac{39}{14}2^{1/3}$ **27.** $\frac{1}{a}\sin(at) + c$ **29.** $\frac{1}{9}(2x^3 + 3x^2 + 2)^{3/2} + c$ **31.** $\frac{1}{3}\arctan(\frac{t}{3}) + c$ **35.** (i) 0 (ii) π **37.** (ii) $\frac{1}{6}\cos^6 x - \frac{1}{4}\cos^4 x + c$ (iii) $\frac{1}{3}\cos^3 x - \cos x + c$ (iv) $\frac{1}{7}\cos^7 x - \frac{1}{5}\cos^5 x + c$ (v) $\frac{1}{3}\sin^3 x - \frac{1}{5}\sin^5 x + c$ **41.** 0.636 567 **43.** 0.231 573

Chapter 5 Review Exercises (p. 410)

1. $\frac{15}{4}$ **3.** $\frac{81}{4}$ **5.** $\frac{37}{60}$ **7.** 49 **9.** 11 **11.** $\ln(6!) = \ln(720) \approx 6.579$ **13.** $\frac{31}{16}$ **15.** $36x$ **17.** $\frac{m(m-1)(2m-1)}{6}$ **19.** 0 **21.** 34 **23.** −20 **25.** 0 **27.** 0 **29.** $\frac{25}{12}$ **31.** $x^3 - 2x + 1$ **33.** −2 **35.** $\frac{4}{3}x^3 - 2x + \frac{2}{3}$ **37.** xe^x **39.** $-x$ **41.** $\frac{205}{4}$ **43.** $-\frac{7}{2}$ **45.** 2 **47.** 1 **49.** $\frac{216}{7}$ **51.** $\frac{1}{6}(1 + 2x^2)^{3/2} + c$ **53.** $\frac{191}{30}$ **55.** $\frac{2}{3}x^{3/2} - x + c$ **57.** $2\exp(\sqrt{t}) + c$ **59.** $x - 2\ln|x + 1| + c$ **61.** $\frac{n+1}{2}$ **63.** $\frac{1 - r^{k+1}}{1 - r}$ **65.** $\frac{\pi}{2}$ **67.** 0 local max, 1 local min, −2 (abs) min; $-\frac{1}{3} \pm \frac{\sqrt{7}}{3}$ infl pts **69.** $-e^2$ local max; $-e$ infl pt **73.** $\frac{\pi}{8} - \frac{1}{4}$ **75.** (i) $\frac{17}{3}$ (ii) 0 (iii) $\frac{17}{3}$ **79.** $-\sqrt{3}$

Exercises 6.1 (p. 423)

1. $\frac{2}{3}$ **3.** $\frac{17}{4}$ **5.** 1 **7.** 1 **9.** $\ln(2) \approx 0.69315$ **11.** $\frac{4}{3}$ **13.** $\frac{8}{3}\sqrt{2} \approx 3.77124$ **15.** 8 **17.** $\frac{256}{5}$ **19.** 9 **21.** 8 **23.** $\frac{2}{3}$ **25.** $\frac{4}{3}$ **27.** $\frac{9}{2}$ **29.** $\frac{9}{2}$ **31.** (i) 1 (ii) 1 **33.** $\frac{27}{4}$ **35.** 9 **37.** (i) $\frac{1}{4}t^4 - 2t^2 + 4$, if $t \in [-2, 0]$; $-\frac{1}{4}t^4 + 2t^2 + 4$, if $t \in [0, 2]$ (ii) 1 **39.** 8 g **43.** $\approx$1.895, area $\approx$ 0.421

Exercises 6.2 (p. 430)

1. $\frac{1}{3}$ **3.** $\frac{32}{3}$ **5.** 24 **7.** $\frac{1}{2}(e - 1)$ **9.** $\frac{1}{2}$ **11.** 23 **13.** −3 **15.** $-\frac{74}{3}$ **17.** $\frac{\pi}{2}$ **19.** $\frac{1}{2}(e^2 - e)$ **21.** $\frac{\pi}{2} + \frac{1}{3}$ **23.** $4\sqrt{3} - 4 - \frac{\pi}{3}$ **25.** (ii) $2(\frac{1}{n+1} - \frac{1}{m+1})$, if $m > n$ and $m - n$ is even; $(\frac{1}{n+1} - \frac{1}{m+1})$, if $m > n$ and $m - n$ is odd **27.** 3 **29.** $\frac{3}{2}$ **31.** $\frac{1}{2}|ad - bc|$ **33.** (i) $15 + \frac{24}{\pi}(1 + \frac{1}{\sqrt{2}}) \approx 28.0°\text{C}$ (ii) $15 + \frac{24}{\pi} - \frac{24}{\pi}\cos(\frac{17\pi}{12}) \approx 24.617°\text{C}$ **35.** 25 in^3 **37.** $23\frac{1}{3}$ m **39.** (i) $25e^{-2} \approx 3.3834$ ft^3 (ii) ln 25 $\approx$ 3.2189 min **41.** bh **43.** $\approx$5.010 **45.** $\approx$1.330 78 **47.** $\approx$2.114 216

Exercises 6.3 (p. 438)

1. 24π

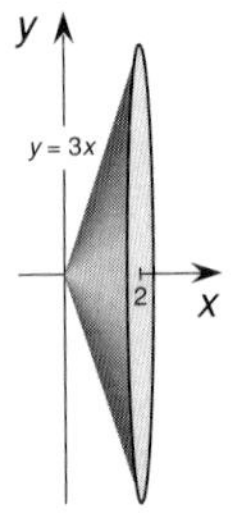

3. $\frac{8\pi}{3}$

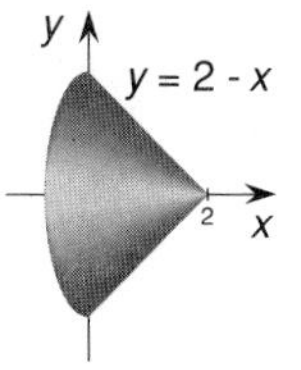

5. $\frac{\pi}{2}(e^2 - 1)$

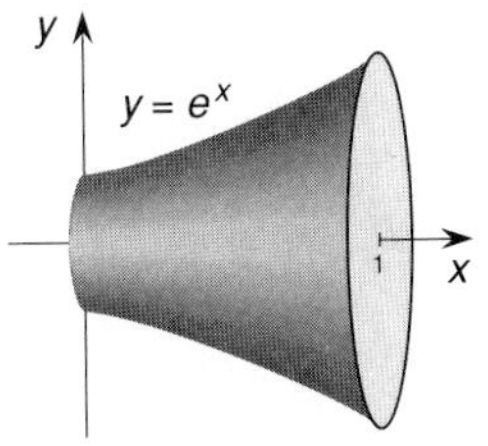

7. $\frac{\pi}{30}$

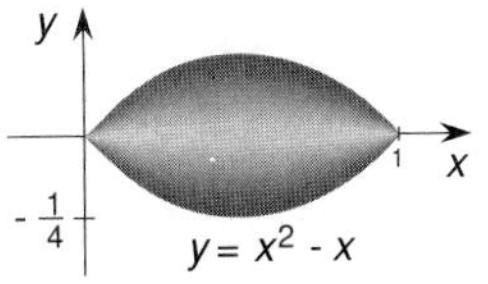

9. 12π

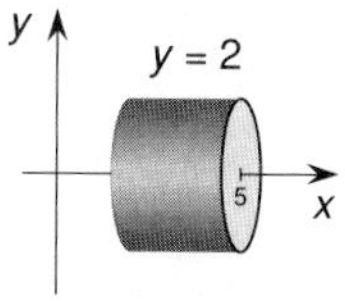

11. π

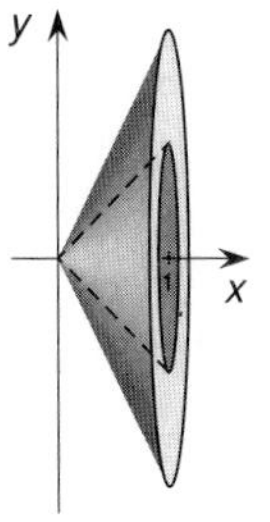

13. $\frac{1}{3}\pi^4 - \frac{1}{2}\pi^2$

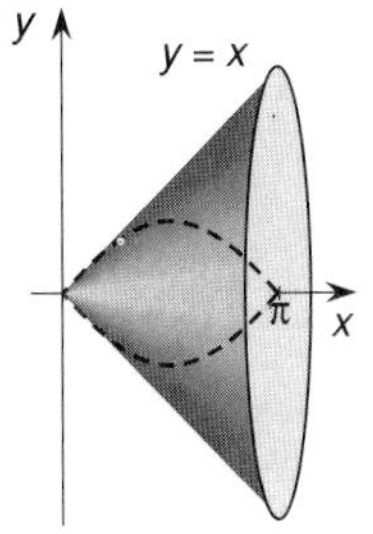

15. $\frac{56}{15}\pi$

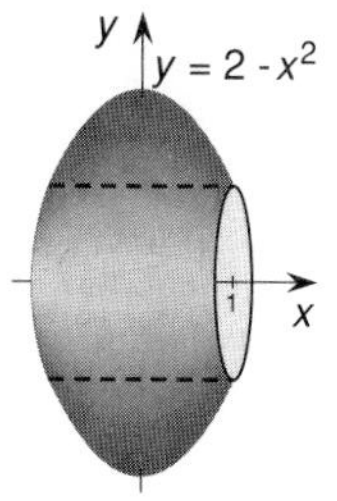

17. $3xh$

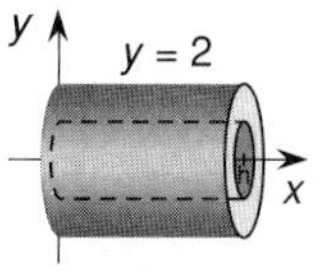

19. $\pi|M^2 - N^2|L$

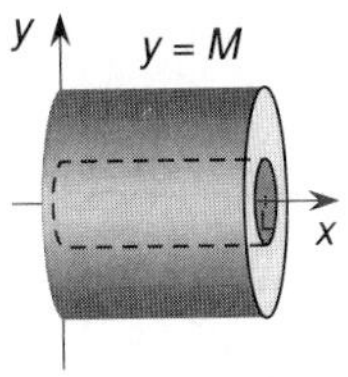

21. $\frac{2\pi}{35}$

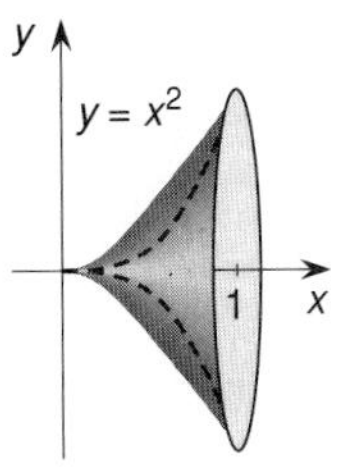

23. $\frac{72\pi}{5}\sqrt{3}$

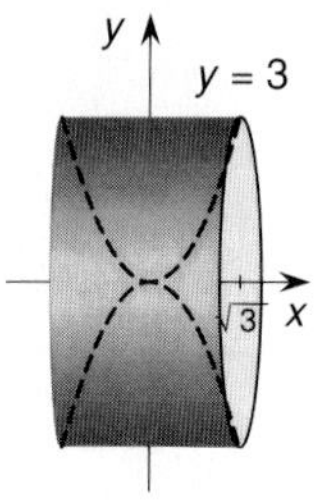

25. $\frac{8\pi}{117}$

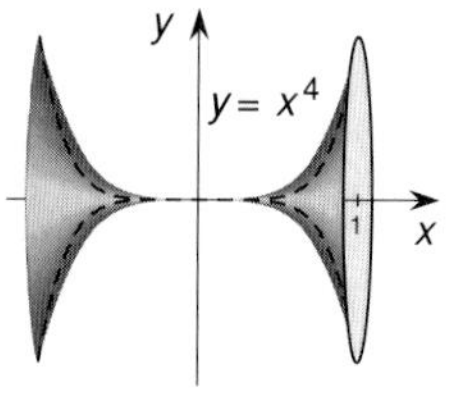

27. $\frac{8\pi}{3}$

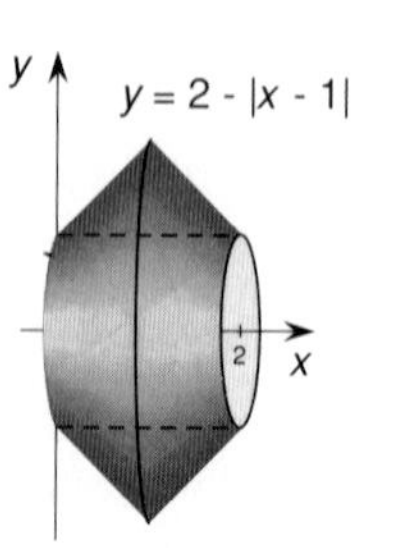

29. $\frac{4\pi}{3}$

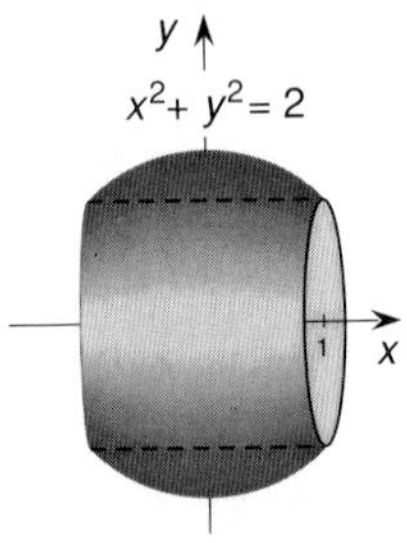

31. 16π

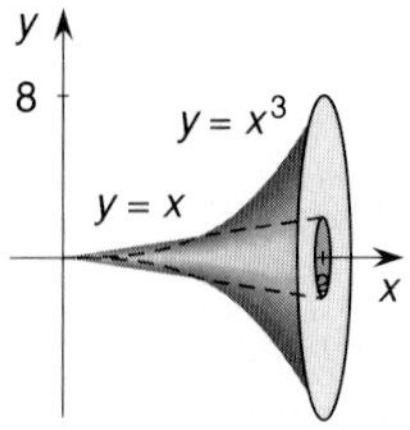

33. $\frac{8}{3}\pi^2 - 2\pi\sqrt{3}$ **35.** **(i)** doughnut ("solid torus") **(ii)** $4\pi^2$
37. $\frac{4}{3}\pi ab^2$
39.

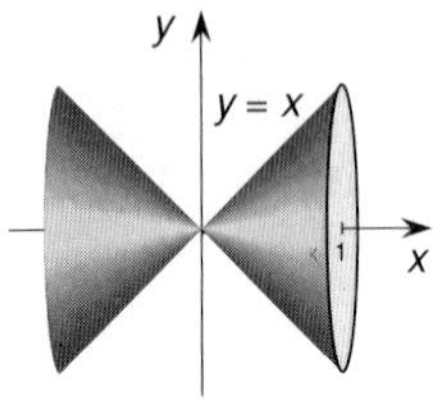

41. vol ≈ 183.511 **43.** 1.769

Exercises 6.4 (p. 445)
1. $\frac{29\pi}{21}$ **3.** $\frac{8\pi}{5}$ **5.** 8π **7.** $\frac{14\pi}{15}$ **9.** $\frac{332}{7}\pi$ **11.** $\frac{2809}{12}\pi$ **13.** $\frac{4\pi}{3}$ **15.** $\frac{5\pi}{24}$ **17.** $\pi(-\frac{13}{42} + \frac{9}{7}(3^{1/3}) + \frac{3}{2}(3^{2/5})) \approx 12.166$ **19.** $\frac{\pi}{3}(2 - \sqrt{2})$ **21.** $\frac{16\pi}{15}$ **23.** $\frac{2\pi}{3}$ **25.** $\frac{2\pi}{5}$ **27.** $\frac{256\pi}{5}$ **29.** π **31.** $\frac{4\pi}{15}$ **33.** $\frac{31\pi}{30}$ **35.** $\frac{128\pi}{15}$ **37.** 96π **39.** $\frac{7\pi}{6}$ **41.** $\frac{\pi}{3\sqrt{2}}$ **43.** $\frac{64}{3}\sqrt{2}\pi$ **45.** $\frac{\pi}{2m+1}a^{2m+1}$ **47.** $\frac{28\pi}{3} \approx 29.32$ oz **51.** **(i)** $\frac{4\pi}{15}$ in^3 **(ii)** ≈ 0.232 oz **53.** ≈ 2.598 **55.** ≈ 2.583

Exercises 6.5 (p. 454)
1. $\frac{4}{3}$ **3.** $e - 1$ **5.** $\frac{1}{4}\ln 5$ **7.** 0 **9.** $\frac{4}{\pi}$ **11.** $\frac{\pi}{4}$ **13.** 0 **15.** $\frac{2}{15\pi}$ **17.** 0 **19.** $\frac{2}{\pi}(e - 1)$ **21.** **(i)** 3 **(ii)** -7 **(iii)** C **23.** **(i)** $\frac{2}{5}(e - e^{-3/2}) \approx 0.998$ in **(ii)** $\frac{\pi}{5}(e^2 - e^{-3}) \approx 4.611$ in^2 **25.** **(i)** $\frac{\pi}{4}$ **(ii)** $\frac{2\pi}{3}$ **27.** **(i)** $\frac{5}{3}(e^{6/5} - 1) \approx 3.867$ g **(ii)** $\frac{20}{3}(e^{1/2} - e^{1/5}) \approx 2.849$ g **29.** **(i)** $\frac{1}{3}$ **(ii)** 1 **31.** $\frac{880}{21} \approx 41.90$ mph **35.** 0 **37.** **(i)** $\frac{A}{\sqrt{2}}$ **(ii)** $\frac{A}{\sqrt{2}}$ **(iii)** $A\sqrt{\frac{1}{2} - \frac{1}{\pi}}$ **39.** **(i)** $0.17(1 - e^{-10}) \approx 0.16999$ ppm **(ii)** $0.34(1 - e^{-5}) \approx 0.3377$ ppm **(iv)** $0.17(1 - e^{-10}) \times 1080\pi \times 10^{-6} \approx 0.000\,577$ in^3 **41.** 68° **43.** ≈ 0.3

Exercises 6.6 (p. 460)
1. 27 ft-lb **3.** $\frac{6}{\pi} \approx 1.9099$ ft-lb **5.** 0 **7.** 30 ft-lb **9.** -96 ft-lb **11.** 3 ft-lb **13.** $\frac{1}{4}$ joule **15.** 45 ft-lb **17.** $\frac{9}{1280}$ ft-lb **19.** $\frac{25}{6}$ ft-lb **21.** $52{,}000\pi$ ft-lb **23.** 1866.8π ft-lb **25.** $\frac{1}{\sqrt{2}} \approx 0.707$ ft ≈ 8.485 in **27.** **(i)** -16 lb **(ii)** 24 ft-lb **29.** **(i)** $h \times 2.112 \times 10^7/(h + 2.112 \times 10^7)$ ft-lb **(ii)** 2.112×10^7 ft-lb **31.** **(i)** $(\frac{4000}{4001})^2 \approx 0.9995$ lb **(ii)** $(\frac{4000 \times 5280}{1 + 4000 \times 5280})^2 \approx 0.999\,999\,905$ lb **33.** **(i)** 18 ft-lb **(ii)** $12\sqrt{2}$ ft/sec **35.** **(i)** $\frac{5}{4}$ ft-lb **(ii)** 10 ft-lb
37. $\frac{15}{2} - \frac{1}{14}\sqrt{105^2 - 4(1875)} \approx 3.259$ m
39. **(i)** $9800\pi(\frac{2Rd^3}{3} - \frac{d^4}{4})$J **(ii)** ≈ 1.979 m

Exercises 6.7 (p. 474)

1. 7 **3.** $x = 0$ **5.** $\frac{27}{16}$ **7.** $(2 - \frac{\pi}{2})/\ln(2) \approx 0.6192$ **9.** $\frac{7}{2}$ **11.** $(\frac{4}{3\pi}, 0)$ **13.** $(0, -\frac{4\sqrt{2}}{3\pi})$ **15.** $(-2, 2)$ **17.** $(0, \frac{12}{5})$ **19.** $(\frac{9}{20}, \frac{9}{20})$ **21.** 1 **23.** $\frac{3L}{4}$ from the light end, i.e., $\frac{3}{4}$ of the way along the rod **25.** **(ii)** $\frac{5}{3}$ **(iv)** $\frac{18}{11}, \frac{8}{5}$ **27.** $(\frac{2}{12 + 3\pi}, 0) \approx (0.0933, 0)$ **29.** $(0, \frac{28}{9\pi})$ **31.** $(0, \frac{212 + 192\pi}{72 + 27\pi}) \approx (0, 5.198)$ **33.** $(0, 0)$ **35.** $(2, 1)$ **37.** $(0, 0)$ **39.** at a distance of $\frac{3}{5}$ from center of larger circle, on the segment joining the two centers **41.** $\frac{2}{3}$ of the radius from the center of the heavy disk, on the segment joining the two centers **43.** $(\frac{9}{10}, \frac{9}{10})$ **45.** $(\frac{3}{2}, \frac{5}{12})$ **47.** $6\pi^2$ **49.** **(i)** $3\pi^2$ **(ii)** $2\pi^2 + \frac{4\pi}{3}$

Exercises 6.8 (p. 481)

1. $99(62.4) = 6177.6$ lb **3.** $\frac{25}{24}(62.4) = 65$ lb **5.** 27,300 lb **7.** $\frac{25}{128} \times 62.4 \times 59 \approx 719$ lb **9.** 5616π lb **11.** 7800π lb **13.** 1622.4π lb **15.** $16\sqrt{3} \times 62.4 \times (100 - \frac{4}{\sqrt{3}}) \approx 168{,}934$ lb **17.** $18.65 \times 62.4 \times 14 \approx 16{,}287$ lb **19.** 3369.6 lb **21.** $9(62.4)(10 - \frac{3}{2\sqrt{2}}) \approx 5020$ lb **23.** 5616π lb **25.** $62.4(72\sqrt{2}) \approx 6354$ lb **27.** $62.4\sqrt{3}(100 - \frac{1}{\sqrt{6}}) \approx 10{,}764$ lb **29.** $\frac{567}{16}(62.4)\sqrt{3} \approx 3830$ lb **31.** $16(62.4)(40 - 2\sqrt{2}) \approx 37{,}112$ lb **33.** 15,600 lb **37.** **(i)** $\frac{1500}{624} \approx 2.4$ ft **(ii)** $\frac{250}{39} \approx 6.41$ ft **(iii)** 199.68 tons **41.** **(i)** 208 lb **(ii)** 41.6 lb **43.** both 78,400 N

Exercises 6.9 (p. 492)

1. $\frac{8}{27}(10^{3/2} - (\frac{11}{2})^{3/2}) \approx 5.548$ **3.** $4\sqrt{2} \approx 5.657$ **5.** $\frac{8}{27}(10^{3/2} - (\frac{11}{2})^{3/2}) \approx 5.548$ **7.** $\frac{8}{27}(10^{3/2} - (\frac{13}{4})^{3/2}) \approx 7.634$ **9.** $\frac{8}{27}(10^{3/2} - (\frac{13}{4})^{3/2}) \approx 7.634$ **11.** $4\pi((\frac{29}{4})^{3/2} - (\frac{13}{4})^{3/2}) \approx 171.68$ **13.** $16\pi\sqrt{10} \approx 158.95$ **15.** $4\pi\sqrt{10} \approx 39.74$ **17.** 42π **19.** $2\pi rh$ **21.** $2\pi r$ **25.** $\pi r\sqrt{r^2 + h^2}(1 - \ell^2/h^2)$ **27.** 8π **29.** **(i)** $6\pi\sqrt{35}$ **(ii)** $7\pi\sqrt{35}$ **31.** $\frac{4\pi}{3}((b + \frac{1}{4})^{3/2} - (a + \frac{1}{4})^{3/2})$ **33.** $\pi(2 + \sqrt{17})$ **35.** elliptical solid has $\approx 7.7\%$ greater area **37.** ≈ 1.542 **39.** actual ≈ 2291.256; with $n = 10$, ≈ 2773 **41.** with $n = 10$, ≈ 25.039; with $n = 20$, ≈ 23.974 **43.** with $n = 10$, ≈ 14.277; with $n = 20$, ≈ 14.387

Exercises 6.10 (p. 504)

1. $T_4 = 8$, $S_4 = 8$, integral $= 8$ **3.** $T_6 \approx 4.661\,488\,4$, $S_6 \approx 4.666\,563\,1$, int $= 4\frac{2}{3}$ **5.** $T_4 \approx 1.896\,118\,9$, $S_4 \approx 2.004\,559\,8$, int $= 2$ **7.** $T_8 \approx 1.628\,968$, $S_8 \approx 1.610\,847$, int $= \ln 5 \approx 1.609\,438$ **9.** $T_4 \approx 0.780\,769\,2$, $S_4 \approx 0.807\,692\,3$, int $= \frac{1}{2}\ln 5 \approx 0.804\,719\,0$ **11.** $n = 20$, $S_{20} \approx 204.800\,853\,3$, int $= 204.8$ **13.** $n = 4$, $S_4 \approx 1.000\,134\,6$, int $= 1$ **15.** $n = 4$, $S_4 \approx -0.693\,25397$, int $= -\ln(2) \approx -0.693\,147\,18$ **17.** $n = 4$, $S_4 \approx 0.609\,418\,7$, int $= \frac{1}{3}(2\sqrt{2} - 1) \approx 0.609\,475\,7$ (might also use $n = 6$) **19.** $n = 12$, $S_{12} = 0$, int $= 0$ **21.** $n = 4$, $S_4 \approx 0.386\,260$ **23.** $n = 2$, $S_2 \approx 0.632\,334$ **25.** $n = 6$, $S_6 = 0$ **27.** $n = 2$, $S_2 \approx 4.639\,202$ **29.** $n = 4$, $S_4 \approx 0.301\,101$ **31.** $n = 4$, $S_4 \approx 1.147\,782\,3$ **33.** $n = 6$, $S_6 \approx 0.881\,374\,6$ **35.** $S_6 \approx 1.271\,270\,7$ **37.** $S_6 \approx 1.111\,431\,2$ **39.** $S_{12} \approx 0.835\,650\,9$ **41.** **(i)** $n = 10$ **(ii)** ≈ 1.479 **43.** $n = 58$ **45.** $S_{20} \approx 3.820$ **51.** $S_4 \approx 0.693\,25$ **53.** **(i)** 0.5026 **(ii)** $\approx 4.77208 \times 10^{-4}$ **55.** **(i)** $n = 8$ **(ii)** $n = 14$ **(iii)** 40 **(iv)** $S_{10} \approx 3.141\,592\,614$ **59.** $S_{20} \approx 0.693\,147$ **61.** $\approx 4.456\,65$ **63.** $\approx 12.001\,51$ **65.** $\approx 13.266\,03$

Chapter 6 Review Exercises (p. 506)

1. $\frac{8}{3}$ **3.** $\frac{20\sqrt{5}}{3}$ **5.** $\frac{1}{2}$ **7.** $2\pi - 4$ **9.** $\frac{\pi}{2} + \frac{1}{3}$ **11.** $\frac{\pi}{2}$ **13.** $12\sqrt{7} + 32(\arcsin(\frac{3}{4}) - \arcsin(\frac{\sqrt{7}}{4})) \approx 35.7595$ **15.** 0 **17.** $9\pi + 16 \arcsin \frac{\sqrt{5}}{2\sqrt{3}} - 18 \arcsin \frac{2\sqrt{5}}{3\sqrt{3}} \approx 20.843$ **19.** $3\sqrt{3} + \frac{4\pi}{3}$ **21.** $\frac{1}{4}\pi^2$

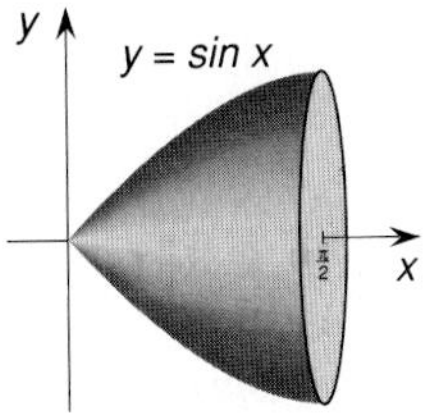

23. 24π

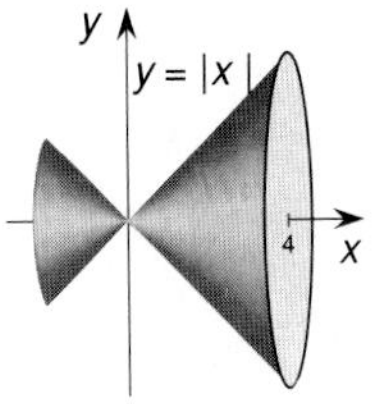

25. $\frac{1}{8}\pi^2$

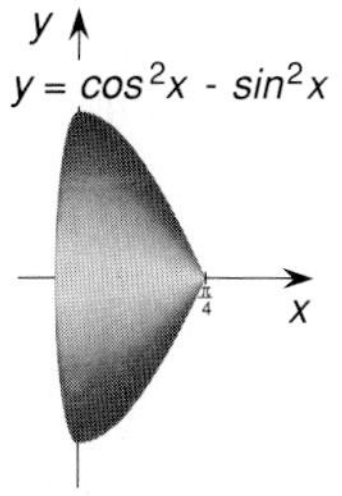

27. $\frac{1}{4}\pi^2$

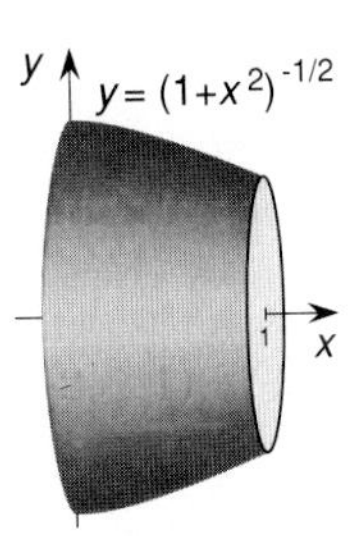

29. 8

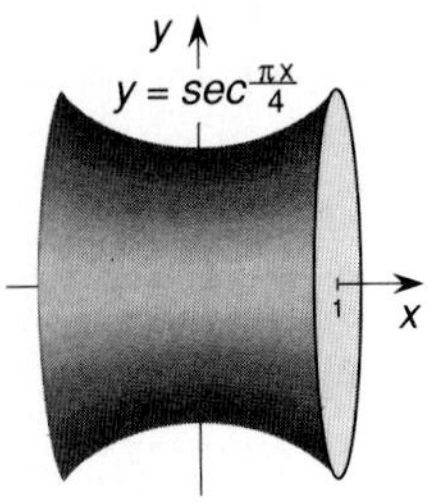

31. $\frac{13}{3}$ **33.** $\ln 2$ **35.** $-\frac{2}{\pi}$ **37.** 0 **39.** 1 **41.** 15 ft-lb **43.** 0.1764 J **45.** 18 ft-lb **47.** 30 ft-lb **49.** $(62.4)400\pi$ ft-lb **51.** $(0, 2)$ **53.** $(\frac{3}{2}, \frac{3}{2})$ **55.** $(\frac{8+3\pi}{6+3\pi}, \frac{8-3\pi}{6+3\pi})$ **57.** $(\frac{1}{2}, \frac{1}{2})$ **59.** $(\frac{2}{12+9\pi}, \frac{2}{12+9\pi})$ **61.** 7488 lb **63.** 4742.4 lb **65.** $(54{,}000 - 13{,}500\sqrt{2})\pi \approx 109{,}667$ kg $\approx 1{,}074{,}737$ N **67.** $2808\pi - 1123.2 \approx 7698$ lb **69.** $1497.6\sqrt{3} + 1331.2\pi \approx 6776$ lb **71.** $27\sqrt{2}\pi$ **73.** $\frac{\pi}{6}(17\sqrt{17} - 27)$ **75.** 6π **77.** $18\sqrt{2}\pi$ **79.** $\frac{\pi}{27}((36{,}865)^{3/2} - 10^{3/2}) \approx 823{,}579.5$ **81.** $S_8 \approx 0.500\,244$ **83.** $S_4 \approx 1.000\,135$ **85.** $S_4 = 0$ **87.** $S_8 \approx 1.856\,295$ **89.** $S_n = 0$ (all n) **91.** $\frac{28\pi}{3}$ **93.** $(\pi, 0)$ **103.** 3π **105.** sphere: $4\pi(\frac{3}{4})^{2/3} \approx 10.373$, other: $\frac{\pi}{3}(5^{3/2} - 1) \approx 10.661$ **107.** $S_{30} \approx 1.47894$, $S_{12} \approx 2.29559$, $S_{36} \approx 2.95789$

Exercises 7.1 (p. 513)

1. $a + b$ **3.** $a - 2b$ **5.** $ma + nb$ **7.** $-ma$ **9.** $\frac{a}{2}$ **11.** 0 **13.** 0 **15.** 0 **17.** 0 **19.** 1 **25.** ∞ **27.** **(i)** ≈ 11.06 **(ii)** ≈ 23.44 **(iii)** ≈ 29.40 **31.** $\approx 2.718\,281\,828$

Exercises 7.2 (p. 521)

1. 0 **3.** 0 **5.** 0 **7.** $\sqrt{5}x^{\sqrt{5}-1}$ **9.** 0 **11.** ∞ **13.** $\frac{1}{1+\sqrt{2}}x^{1+\sqrt{2}} + c$ **15.** $\frac{1}{\ln(2)}2^x + c$ **17.** $\frac{1}{2\ln 4}4^{x^2} + c$ **19.** $\frac{2}{\ln 10}10^{\sqrt{x}} + c$ **21.** 2 **23.** 2 **25.** $\frac{1}{x\ln(10)}$ **27.** $-\infty$ **29.** $\frac{\ln 3}{\ln 10}x + c$ **31.** e^3 **33.** $\frac{1}{e}$ **35.** e^{-4} **37.** e^{-2} **39.** 1 **41.** $2e^{2x}$ **43.** 2 **49.** 0 **51.** increasing everywhere **53.** **(i)** $(1 + \ln x)x^x$ **(ii)** 1 **(iii)** abs min at $\frac{1}{e}$
(iv)

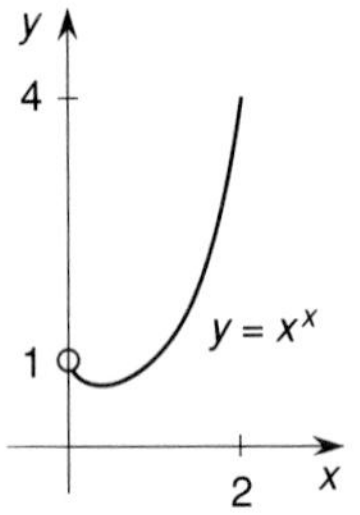

Exercises 7.3 (p. 528)

1. 12 **3.** 8 **5.** 128 **7.** $\frac{1}{9}$ **9.** $\frac{1}{32}$ **11.** 1 year **13.** $-3\frac{\ln 2}{\ln(0.75)} \approx 7.228$ years **15.** $\frac{1}{2}$ year **17.** $\frac{1}{4}$ year **19.** $-3\frac{\ln 2}{\ln(0.75)} \approx 7.228$ years **21.** $\frac{\ln(10)}{2\ln 2} \approx 1.661$ hr **23.** $-3\frac{\ln(0.9)}{\ln 2} \approx 0.456$ year **25.** $-30\frac{\ln 2}{\ln(0.4)} \approx 22.694$ years **27.** $(0.5)(7.4)^{-1/4} \approx 0.303$ g **29.** $\approx 7.9 \times 10^{29}$ g; 720 times the volume of earth

Exercises 7.4 (p. 542)

1. $-3\ln(2)/\ln(0.2) = 3\ln(2)/\ln(5) \approx 1.292$ hr **3.** $-\ln(2)/\ln(0.85) \approx 4.265$ days **5.** $-2\ln(2)/\ln(0.8) \approx 6.213$ hr **7.** $\frac{\ln 2}{2800} \approx 2.476 \times 10^{-4}$ **9.** $\frac{\ln 2}{34} \approx 0.02039$ **11.** **(i)** $\ln(\frac{5}{4})/\ln(\frac{161}{160}) \approx 35.8$ years after 3,200,000 **(ii)** $-\ln(\frac{15}{16})/\ln(\frac{161}{160}) \approx 10.36$ years **13.** $\frac{\ln(2)}{h}$ **15.** $Ie^{-t\ln(2)/h}$ **17.** **(ii)** $-\frac{140}{\ln(2)}\ln(0.01) \approx 930$ days **19.** $\approx 10.405\%$, $\approx 10.381\%$ **21.** $(e^{1/10} - 1) \times 100\% \approx 10.517\%$ **23.** $(e^r - 1) \times 100\%$ **25.** $\approx \$2246.99$ **27.** $\approx \$3245.86$ **29.** **(i)** $\approx 8.243\%$ **(ii)** $\approx 8.300\%$ **(iii)** $\approx 8.328\%$ **(iv)** $\approx 8.328\,667\%$ **(v)** $\approx 8.328\,707\%$ **31.** $\approx 6.183\,131\%$ **33.** **(i)** $\approx 7.250\,098\,3\%$, $7.250\,100\,3\%$ **(ii)** \$500,000 **35.** \$181,782.73 **37.** ≈ 16.1 years **39.** $100\ln(30) \approx 340$ days **41.** $\frac{1}{0.003946}\ln\frac{7}{2} \approx 317$ days **43.** **(i)** $1.3e^{5\ln(0.9/1.3)} \approx 0.207$ g **(ii)** $25\ln(\frac{0.5}{1.3})/\ln(\frac{0.9}{1.3}) \approx 65$ min **45.** $\approx 6.183\,632\,865\%$

Exercises 7.5 (p. 550)

1. $L = 4, P_0 = 2, c = \frac{1}{4}\ln 3$ **3.** $L = \frac{35}{3}, P_0 = 8, c = \frac{3}{35}\ln(\frac{11}{4})$ **5.** $L = 40{,}000, P_0 = 40{,}000/(1 + \exp(196\ln(\frac{5}{3}))), c = \frac{1}{400{,}000}\ln(\frac{5}{3})$ **7.** $L = \frac{85{,}096}{1891}, P_0 = \frac{85{,}096}{1891}(1 + \frac{83{,}205}{1891}\exp(10\ln(1935)))^{-1}, c = \frac{1891}{85{,}096}\ln(1935)$ **9.** $L = 10, P_0 = \frac{90}{13}, c = \frac{1}{10}\ln(\frac{3}{2})$ **11.** **(i)** 8 mg **(ii)** 7 days **17.** **(i)** $L = 14, P_0 = 5, c = \frac{1}{28}\ln(\frac{9}{5})$ **(ii)** $2\ln(\frac{695}{9})/\ln(\frac{9}{5}) \approx 14.8$ days **(iii)** $2\ln(50)/\ln(\frac{7}{5}) \approx 23.3$ days **(iv)** logistic: $2\ln(\frac{117}{5})/\ln(\frac{9}{5}) \approx 10.73$ days; exponential: $2\ln(\frac{13}{5})/\ln(\frac{7}{5}) \approx 5.68$ days **19.** **(i)** k **(ii)** $t = \frac{d}{a}$ **(iii)** $P_0 = k/(1 + e^d), c = \frac{a}{k}$ **21.** same a, d becomes $d + \ln(2 + e^{at_0 - d})$

Exercises 7.6 (p. 556)

1. $x = c\exp(\frac{t^3}{3})$ **3.** $x = \pm\sqrt{t^2 + c}$ **5.** $x = ct$ **7.** $x = 2 + c\exp(\frac{t^3}{3})$ **9.** $x = -\ln|c - e^t|$ **11.** $x = 2e^{t/2}$ **13.** $x = 3\exp((t^3 - 1)/3)$ **15.** $x = 1$ (constant function) **17.** $x = \arcsin(t + \frac{1}{\sqrt{2}})$ **19.** $x = \frac{1}{16}t^4 + \frac{(-4+\sqrt{5})}{2}t^2 + (-4 + \sqrt{5})^2$ **21.** $c_1e^{4t} + c_2e^{-3t}$ **23.** $c_1 + c_2e^{-t} + c_3e^{-2t}$ **25.** $c_1e^{-3t} + c_2te^{-3t}$ **27.** $c_1 + c_2e^{-t} + c_3te^{-t}$ **29.** $c_1 + c_2e^t + c_3te^t + c_4t^2e^t$ **31.** $f(t) = e^t$ **33.** $f(t) = 0$ **35.** $f(t) = t - 1$ **37.** 0 **39.** $-\frac{1}{15} + \frac{1}{40}e^{5t} + \frac{1}{24}e^{-3t}$

41.

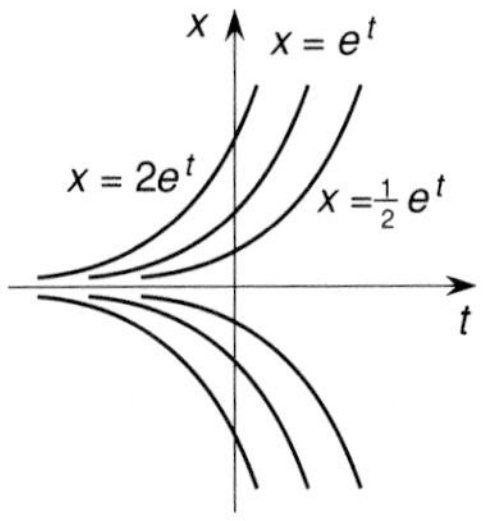

43.

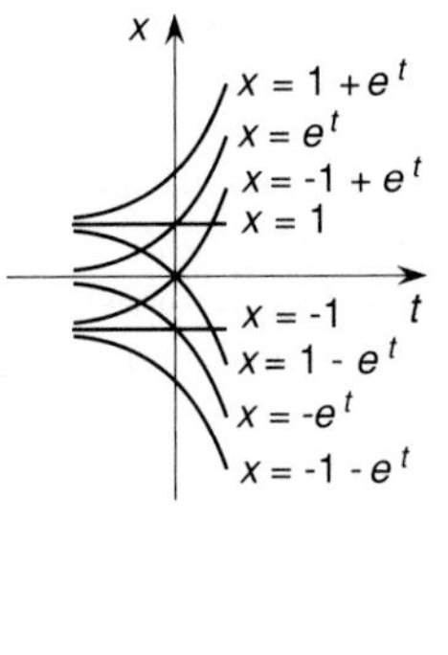

45.

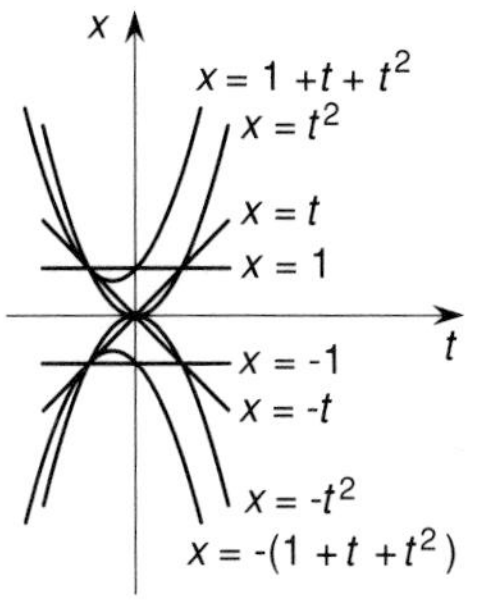

47. **(ii)** $f(t) = 0$ **51.** $(1 - e)e^t + (e - 1)te^t + ee^{-t}$

Chapter 7 Review Exercises (p. 558)

1. $(2 \ln x + 2)x^{2x}$ **3.** $(3 \ln x + 3 + \frac{5}{x})x^{3x+5}$ **5.** $\frac{2}{x} \log_2(x)x^{\log_2(x)}$ **7.** 0 **9.** 1 **11.** $10\frac{\ln(0.5)}{\ln(0.95)} \approx 135$ min **13.** $10{,}000e^{120 \ln(1.008)} \approx 26{,}017$ **15.** $20 + 60e^{10 \ln(5/6)} \approx 29.7$°C **17.** $\frac{1}{0.08} \ln(2) \approx 8.66$ years **19.** \$2949.11 **21.** $c_1 \exp(-(\frac{1}{2} - \frac{\sqrt{5}}{2})t) + c_2 \exp(-(\frac{1}{2} - \frac{\sqrt{5}}{2})t)$ **23.** $c_1 + c_2t + c_3e^{-3t}$ **25.** $-(3 \tan x + c)^{-1/3}$ **27.** $u = -\frac{1}{t+1} - \frac{1}{3t^3} + c$ **29.** $c \exp(-\frac{1}{2}x^2 - x)$ **31.** $(e^5 - 1)^{-1}(-(e^3 + 1)e^{2t} + (e^3 + e^5)e^{-3t})$ **33.** 0 **35.** e^{3t} **37.** 1 **39.** $\frac{\pi - 1}{2\pi}e^{\pi t} + \frac{\pi + 1}{2\pi}e^{-\pi t}$ **43.** $10^{(10^{10})}$ **45.** $\ln(3)/\ln(51/49) \approx 27.46$ years **47.** half: $\frac{\ln 99}{\ln(297/97)} \approx 4.1$ days; 90%: $\frac{\ln 981}{\ln(297/97)} \approx 6.1$ days

Exercises 8.1 (p. 568)

1. $-xe^{-x} - e^{-x} + c$ **3.** $(\frac{1}{3}x^2 - \frac{2}{9}x + \frac{2}{27})e^{3x} + c$ **5.** $-\frac{1}{10} \sin^4(2u) \cos(2u) - \frac{2}{15} \sin^2(2u) \cos(2u) - \frac{4}{15} \cos(2u) + c$ **7.** $x \tan x - \ln|\sec x| + c$ **9.** $(x^2 + x - 3)e^x + c$ **11.** $x \arcsin x + \sqrt{1 - x^2} + c$ **13.** $2 \ln(1 + x^2) + c$ **15.** $x^3 \ln|x| - \frac{1}{3}x^3 + c$ **17.** $\frac{1}{3}x^3 \ln|x| - \frac{1}{9}x^3 + c$ **19.** $e^{2x} (\frac{3}{2} - \frac{1}{10} \sin(4x) + \frac{1}{5} \cos(4x)) + c$ **21.** $\frac{1}{4}e^2 - \frac{5}{4}e^{-2}$ **23.** $-2(e^2 + 1)/(4 + \pi^2)$ **25.** 0 **27.** $-\frac{2}{\pi}$ **29.** $\frac{4}{3}\sqrt{2} \arctan(\sqrt{2}) - \frac{1}{3} + \frac{1}{3} \ln(\frac{3}{2}) - \frac{\pi}{6} \approx 1.079\,585$ **41.** **(i)** 0 **(ii)** π **43.** $\frac{x}{8} + \frac{1}{8} \cos x \sin^3 x - \frac{1}{8} \cos^3 x \sin x + c$ **45.** $-2x \cos(\frac{x}{2}) + 4 \sin(\frac{x}{2}) + c$ **47.** $-\frac{1}{3} \cos(x^3 + 4) + c$ **49.** $-\frac{1}{5}xe^{-5x} - \frac{1}{25}e^{-5x} + c$ **51.** $-\frac{2}{3} \cos^3 x + c$

Exercises 8.2 (p. 577)

1. $-\frac{1}{3} \sin^2 x \cos x - \frac{2}{3} \cos x + c$ **3.** $-\frac{1}{4} \sin^3 x \cos x + \frac{3}{8}x - \frac{3}{8} \cos x \sin x + c$ **5.** $\frac{1}{2} \sec x \tan x + \frac{1}{2} \ln|\sec x + \tan x| + c$ **7.** $\frac{1}{2} \tan^2 x + \ln|\cos x| + c$ **9.** $-\frac{1}{6} \sin^2(2x) \cos(2x) - \frac{1}{3} \cos(2x) + c$ **11.** $\frac{1}{4} \sin^4 x + c$ **13.** $\frac{1}{5} \cos^5 x - \frac{1}{3} \cos^3 x + c$ **15.** $\frac{1}{4} \tan^4 x + c$ **17.** $\frac{1}{3} \sec^3 x - \sec x + c$ **19.** $\frac{1}{8} \sin^4(2x) - \frac{1}{12} \sin^6(2x) + c$ **21.** $\frac{1}{8} \sin^8 x - \frac{1}{10} \sin^{10} x + c$ **23.** $\frac{1}{12} \cos(3x) \sin^3(3x) + \frac{x}{8} - \frac{1}{24} \cos(3x) \sin(3x) + c$ **25.** $\frac{1}{18} \tan^6 3x + c$ **27.** $-\frac{2}{3} \cos^3 x + c$ **29.** $-\cos x - 2 \sec x + \frac{1}{3} \sec^3 x + c$ **31.** $\frac{\pi}{2}$ **33.** $\frac{2}{15}$ **35.** 0 **37.** $\frac{1}{8}$ **39.** 0 **43.** $\frac{1}{12}$ **45.** **(ii)** 0 **47.** $\ln(1 + \sqrt{2})$ **49.** **(i)** $-\ln|\cot x + \csc x| + c$ **51.** **(i)** $-\frac{1}{3} \cot^3 x + c$ **(ii)** $\frac{1}{3} \csc^3 x - \frac{1}{5} \csc^5 x + c$ **53.** $\frac{1}{4} \sec^3 x \tan x + \frac{3}{8} \sec x \tan x + \frac{3}{8} \ln|\sec x + \tan x| + c$ **55.** $-\frac{1}{4} \cos^4 x + c$ **57.** $\frac{1}{3}(x^2 + 1)^{3/2} + c$ **59.** $\frac{1}{5}(x^2 + 1)^{5/2} + c$ **61.** $\frac{1}{2} \sin(x^2 - 3) + c$ **63.** $\tan x - \frac{3}{2}x + \frac{1}{2} \sin x \cos x + c$

Exercises 8.3 (p. 583)

1. $2 \arcsin(\frac{x}{2}) + \frac{x}{2}\sqrt{4 - x^2} + c$ **3.** $\operatorname{arcsec}(x) + c$ **5.** $-\frac{1}{x}\sqrt{1 - x^2} + c$ **7.** $\sqrt{x^2 - 4} - 2 \operatorname{arcsec}(\frac{x}{2}) + c$ **9.** $-\frac{1}{4x}\sqrt{4 + x^2} + c$ **11.** $\frac{x}{4}(1 - x^2)^{3/2} + \frac{3}{8} \arcsin x + \frac{3}{8}x\sqrt{1 - x^2} + c$ **13.** $\frac{9}{4} \arcsin(\frac{2x}{3}) + \frac{x}{2}\sqrt{9 - 4x^2} + c$ **15.** $\frac{x}{4}(1 + 5x^2)^{3/2} + \frac{3x}{8}\sqrt{1 + 5x^2} + \frac{3}{8\sqrt{5}} \ln|\sqrt{1 + 5x^2} + \sqrt{5}x| + c$ **17.** $\frac{1}{5}(8 + x^2)^{5/2} + c$ **19.** $\frac{1}{3}(x^2 - 1)^{3/2} + c$ **21.** $\frac{1}{2} \arctan x + \frac{1}{2} \frac{x}{1 + x^2} + c$ **23.** $\frac{x}{5}(5 - x^2)^{-1/2} + c$ **25.** $\frac{1}{2} \arcsin(x^2) + \frac{1}{2}x^2\sqrt{1 - x^4} + c$ **27.** $-\frac{1}{2}x^{-2}\sqrt{1 + x^4} + c$ **29.** $-\frac{1}{3}(4 - x^2)^{3/2} + 6 \arcsin(\frac{x}{2}) + \frac{3}{2}x\sqrt{4 - x^2} + c$ **31.** $\frac{\pi}{4}$ **33.** $\frac{14\sqrt{3}}{5}$ **35.** $\frac{1}{4} + \frac{1}{4\sqrt{2}} \ln(1 + \sqrt{2})$ **37.** $\frac{\pi}{8}$ **39.** $\frac{1}{2} \ln(10)$ **41.** $\frac{\sqrt{5}}{2} + \frac{1}{4} \ln(2 + \sqrt{5})$ **43.** $\frac{1}{2}(b\sqrt{1 + 4b^2} - a\sqrt{1 + 4a^2}) + \frac{1}{4} \ln(|2b + \sqrt{1 + 4b^2}|/|2a + \sqrt{1 + 4a^2}|)$ **45.** $-\frac{2}{3}(1 - x^2)^{3/2} + c$ **47.** $\operatorname{arcsinh} x + c$ **49.** $\arctan(e^x) + c$ **51.** $\frac{1}{3}(4 + x^2)^{3/2} + c$ **53.** $\frac{x}{4}(9 + x^2)^{3/2} - \frac{9x}{8}\sqrt{9 + x^2} - \frac{81}{8} \ln|x + \sqrt{9 + x^2}| + c$ **55.** $\frac{1}{2} \cosh^2 x + c$ **57.** $\frac{1}{3} \sin^3(x^2) + c$

Exercises 8.4 (p. 588)

1. $\arctan(x - 2) + c$ **3.** $\frac{1}{\sqrt{5}} \arctan(\frac{x + 3}{\sqrt{5}}) + c$ **5.** $\frac{1}{2} \ln(x^2 - 2x + 4) + \frac{2}{\sqrt{3}} \arctan(\frac{x - 1}{\sqrt{3}}) + c$ **7.** $x + \ln(x^2 - 2x + 2) + 3 \arctan(x - 1) + c$ **9.** $2 \arcsin(\frac{x + 1}{2}) + \frac{1}{2}(x + 1)\sqrt{3 - 2x - x^2} + c$ **11.** $\arcsin(\frac{x + 3}{5}) + c$ **13.** $-\frac{1}{2} \arctan(\frac{x - 2}{2}) + c$ **15.** $\frac{1}{2} \arctan(x - 1) + c$ **17.** $\sqrt{x^2 + 2x + 2} - \ln|x + 1 + \sqrt{x^2 + 2x + 2}| + c$ **19.** $(3x - 7)(x^2 - 4x + 5)^{-1/2} + c$ **21.** $\sqrt{3 + 2x - x^2} + 2 \arcsin(\frac{x - 1}{2}) + c$ **23.** $-\frac{1}{15}(x^2 + 2x + 17)^{-15/2} + c$ **25.** $\frac{1}{9}(x + 2)(5 - 4x - x^2)^{-1/2} + c$ **27.** $-\frac{1}{8}(x^2 + 4x + 13)^{-4} + c$ **29.** $\frac{1}{5}(1 - x^2)^{5/2} - \frac{1}{3}(1 - x^2)^{3/2} + c$ **31.** $\frac{x - 2}{4}(x^2 - 2x + 2)^{-2} + \frac{3}{8} \arctan(x - 1) + \frac{3}{8}(x - 1)(x^2 - 2x + 2)^{-1} + c$ **33.** $\frac{1}{2} \ln(x^2 + 2x + 10) + c$ **35.** $\frac{2}{5}(x - 1)^{5/2} + \frac{2}{3}(x - 1)^{3/2} + c$ **37.** $\frac{1}{5}(5 + 4x - x^2)^{5/2} - 3(5 + 4x - x^2)^{3/2} + c$ **39.** $e^{x/2}\sqrt{e^x - 4} - 4 \ln(e^{x/2} + \sqrt{e^x - 4}) + c$ **41.** $\frac{1}{8} \arcsin(2 \ln(x) - 1) + \frac{1}{4}(2 \ln(x) - 1)\sqrt{\ln x - \ln^2 x} + c$

Exercises 8.5 (p. 597)

1. $\frac{1}{2} \ln|\frac{x - 1}{x + 1}| + c$ **3.** $\ln|x - 1| + c$ **5.** $\frac{4}{5} \ln|x - 2| + \frac{1}{5} \ln|x + 3| + c$ **7.** $x + 3 \ln|x + 1| - 4 \ln|x| + c$ **9.** $x + \frac{1}{x} + 4 \ln|x| - 5 \ln|x + 1| + c$ **11.** $\ln|x| - \arctan x + c$ **13.** $\ln|x| - \frac{1}{2} \ln(x^2 + 1) + c$ **15.** $x - \frac{2}{5} \ln|x + 2| + \frac{6}{5} \ln(x^2 - 4x + 8) - \frac{4}{5} \arctan(\frac{x - 2}{2}) + c$ **17.** $2 \ln(x^2 + 4) - \frac{1}{2} \arctan(\frac{x}{2}) - (x^2 + 4)^{-1} + c$ **19.** $\ln|(x + 1)(2x^2 - 4x + 3)| + c$ **21.** $\frac{1}{2} \arctan x + \frac{1}{2}x/(1 + x^2) + c$ **23.** $\frac{1}{8} \ln|(x^2 - 4)/x^2| + c$ **25.** $\frac{2}{3} \ln|\frac{x}{x + 3}| + c$ **27.** $\frac{1}{2}(x^2 + \ln|x^2 - 1|) + c$ **29.** $\frac{1}{4}x^2 + c$ **31.** $\frac{1}{6} \ln(\frac{4 - \sin \theta}{\sin \theta + 2}) + c$ **33.** $\frac{1}{4\sqrt{2}} \ln((x^2 + \sqrt{2}x + 1)/(x^2 - \sqrt{2}x + 1)) + \frac{1}{2\sqrt{2}} \arctan(\sqrt{2}x + 1) + \frac{1}{2\sqrt{2}} \arctan(\sqrt{2}x - 1) + c$ **35.** $\arctan x - \frac{1}{2x^2 + 2} + c$ **37.** $\arctan(x + 2) + c$ **39.** $4 \ln(2 + e^{x/2}) - 2 \ln(1 + e^{x/2}) + c$ **41.** $-\frac{1}{3}x \sin^2 x \cos x - \frac{2}{3}x \cos x + \frac{1}{9} \sin^3 x + \frac{2}{3} \sin x + c$ **43.** $\frac{1}{2}x^2 \arcsin x - \frac{1}{4} \arcsin x + \frac{1}{4}x\sqrt{1 - x^2} + c$ **45.** $\frac{1}{3}x^3 \arctan x - \frac{1}{6}x^2 + \frac{1}{6} \ln(1 + x^2) + c$ **47.** $x \ln(x^2 + 1) - 2x + 2 \arctan x + c$ **49.** $-\frac{1}{2} \ln(1 + \exp(-x^2)) + c$ **51.** $\frac{1}{20} \ln|\frac{x - 1}{x + 1}| - \frac{1}{30} \arctan \frac{x}{3} + c$ **53.** $-2e^{x/2} \cos(e^{x/2}) + 2 \sin(e^{x/2}) + c$ **55.** $\frac{2}{3}x^{3/2} \ln x - \frac{4}{9}x^{3/2} + c$ **57.** **(ii)** $\approx -0.720\,128\,079$

Exercises 8.6 (p. 601)

1. $x - \frac{3}{2}x^{2/3} + 3x^{1/3} - 3 \ln|1 + x^{1/3}| + c$ **3.** $\frac{6}{7}x^{7/6} - \frac{6}{5}x^{5/6} + 2x^{1/2} - 6x^{1/6} + 6 \arctan(x^{1/6}) + c$ **5.** $2\sqrt{x} - 2 \arctan(\sqrt{x}) + c$ **7.** $\frac{2}{3}x^{3/2} - x + c$ **9.** $2\sqrt{x} - 24x^{1/6} + 48 \arctan(\frac{1}{2}x^{1/6}) + c$ **11.** $-2/(1 + \tan(\frac{\theta}{2})) + c$

13. $\ln(|1 + \tan\frac{\theta}{2}|/|1 - \tan\frac{\theta}{2}|) = \ln|\sec\theta + \tan\theta| + c$ **15.** $\tan(\frac{\theta}{2}) + c$
17. $\frac{1}{2}\ln|\tan(\frac{\theta}{2})| - \frac{1}{4}\tan^2(\frac{\theta}{2}) + c$ **19.** $\ln|1 + \tan(\frac{\theta}{2})| + c$
21. $\frac{2}{a}(1 - b^2/a^2)^{-1/2}\arctan\big([(\tan\frac{\theta}{2}) + b/a](1 - b^2/a^2)^{-1/2}\big) + c$
25. $\frac{2}{3}x^{3/2} - x + 2\sqrt{x} - 2\ln(1 + \sqrt{x}) + c$ **27.** $\frac{1}{2}\ln|\tan\theta| + c$
29. $\frac{4}{3}(1 + \sin\theta)^{3/2} - 4(1 + \sin\theta)^{1/2} + c$
31. $-\frac{1}{28}(1 + 2\cos x)^{7/2} + \frac{1}{10}(1 + 2\cos x)^{5/2} - \frac{1}{12}(1 + 2\cos x)^{3/2} + c$
33. $\frac{6}{7}(4 + e^{x/2})^{7/3} - 6(4 + e^{x/2})^{4/3} + c$

Exercises 8.7 (p. 605)

1. π **3.** $\frac{1}{\sqrt{2}} - \frac{1}{2}\ln(\sqrt{2} + 1) \approx 0.266$ **5.** $\frac{1}{2}\ln(3) \approx 0.549$
7. $2 - \frac{2}{e} \approx 1.264$ **9.** 1 **11.** $\sqrt{145} - \frac{1}{2}\sqrt{37} + \frac{1}{12}\ln(\frac{12 + \sqrt{145}}{6 + \sqrt{37}}) \approx 9.058$ **13.** $\sqrt{68} - \sqrt{5} + \frac{1}{2}\ln(\frac{4 + \sqrt{17}}{2 + \sqrt{5}})$
15. $\frac{5}{6}\sqrt{109} - \frac{5}{3} + \frac{3}{4}\ln\frac{10 + \sqrt{109}}{9} \approx 7.649$ **17.** $\frac{1}{2}\ln(3)$
19. $\sqrt{17} - \sqrt{2} + \frac{1}{2}\ln\big((\sqrt{17} - 1)(\sqrt{2} + 1)/[(\sqrt{17} + 1)(\sqrt{2} - 1)]\big)$
21. $\sqrt{17} + \frac{1}{4}\ln(4 + \sqrt{17})$ **23.** $\frac{a}{2}\sqrt{1 + 4a^2} + \frac{1}{4}\ln(2a + \sqrt{1 + 4a^2})$ **25.** $\frac{b}{2}\sqrt{1 + 4k^2b^2} + \frac{1}{4k}\ln(2kb + \sqrt{1 + 4k^2b^2}) - \frac{a}{2}\sqrt{1 + 4k^2a^2} - \frac{1}{4k}\ln(2ka + \sqrt{1 + 4k^2a^2})$
27. $\frac{1}{2}\ln(\frac{2 + \sqrt{3}}{1 + \sqrt{2}})$ **29.** **(i)** $\sqrt{1 + e^2} - \sqrt{2} + \frac{1}{2}\ln\big((\sqrt{1 + e^2} - 1)(\sqrt{2} + 1)/(\sqrt{1 + e^2} + 1)(\sqrt{2} - 1)\big) \approx 2.0035$
(ii) $\sqrt{e^2 - 2e + 2} \approx 1.988$ **31.** $\sqrt{2}\pi - \frac{\pi}{4}\sqrt{5} - \pi\ln(\frac{1 + \sqrt{5}}{2 + 2\sqrt{2}})$
33. $8\pi^2$ **35.** **(i)** $\bar{x} = \frac{1}{\sqrt{2}}(1 - \frac{\pi}{4})/(1 - \frac{1}{\sqrt{2}}) \approx 0.518\,095, \bar{y} = (\frac{\pi}{16} - \frac{1}{8})/(1 - \frac{1}{\sqrt{2}}) \approx 0.243\,603$ **(ii)** $(\frac{\pi}{6}, \frac{1}{3\sqrt{2}})$ **37.** $\bar{x} = \pi - 2, \bar{y} = \frac{1}{96}\pi^3 + \frac{1}{16}\pi$

Exercises 8.8 (p. 610)

1. $\frac{1}{2}$ **3.** $\frac{3}{2}$ **5.** $\frac{\pi}{2}$ **7.** diverges **9.** $\frac{1}{2}$ **11.** diverges **13.** diverges
15. diverges **17.** $\frac{\pi}{3}$ **19.** diverges **21.** **(i)** $r < -1$ **(ii)** $-\frac{1}{r+1}$
27. exists when $\ln A \notin [a, b]$

Chapter 8 Review Exercises (p. 612)

1. $\frac{x}{3}e^{3x} + \frac{5}{9}e^{3x} + c$ **3.** $\frac{1}{2}\arctan(\frac{x-3}{2}) + c$ **5.** $-2x\cos\sqrt{x} + 4\sqrt{x}\sin\sqrt{x} + 4\cos\sqrt{x} + c$ **7.** $x^2 + 2x + \ln|(x^2 - 3x + 2)/(x - 3)| + c$
9. $\frac{x}{3}\sin^3 x + \frac{1}{9}\sin^2 x\cos x + \frac{2}{9}\cos x + c$
11. $\frac{1}{\sqrt{2}}(x - 3)\sqrt{x^2 - 6x + 13} + 2\sqrt{2}\ln|x - 3 + \sqrt{x^2 - 6x + 13}| + c$ **13.** $\frac{1}{2}\ln\left|\frac{x+2}{x+4}\right| + c$ **15.** $2\sqrt{x} - 2\ln(1 + \sqrt{x}) + c$ **17.** $2\sqrt{x}\sin(\sqrt{x}) + 2\cos(\sqrt{x}) + c$ **19.** $x\ln(1 + x^2) - 2x + 2\arctan x + c$ **21.** $2\sin x - 2\ln(1 + \sin x) + c$
23. $\frac{x}{4}(\sin^2 x - \cos^2 x) + \frac{1}{4}\cos x\sin x + c$
25. $\frac{3}{7}(x + 8)^{7/3} - 6(x + 8)^{4/3} + c$
27. $2\sqrt{1 + e^x} + \ln\big((\sqrt{1 + e^x} - 1)/(\sqrt{1 + e^x} + 1)\big) + c$
29. $\frac{1}{2}(x^2 + 1)\arctan x - \frac{x}{2} + c$ **31.** $\frac{1}{3}\csc^3 x - \frac{1}{5}\csc^5 x + c$
33. $x\tan x + \ln|\cos x| + c$ **35.** $-\frac{1}{x}\sqrt{4 - x^2} - \arctan(\frac{x}{2}) + c$
37. $\frac{1}{4}\ln(2x^2 - 2x + 1) + \frac{5}{2}\arctan(2x - 1) + c$ **39.** $\frac{2}{3}x^{3/2}\ln\sqrt{x} - \frac{2}{9}x^{3/2} + c$ **41.** $\ln(\frac{5\sqrt{2}}{7})$ **43.** $\cos\theta + \frac{1}{2}\ln(\frac{1 - \cos\theta}{1 + \cos\theta}) + c$
45. $\frac{2}{3}x^{3/2}\ln x - \frac{4}{9}x^{3/2} + c$ **47.** $\frac{\pi}{6} + \sqrt{3} - 2$ **49.** $8\ln(2) - 4$
51. $\frac{5\pi}{6}$ **53.** diverges **55.** diverges **57.** diverges **59.** $\frac{1}{2}\ln(2) - \frac{3\pi}{4}$ **61.** $\frac{2}{3}a^{-2}(ax + b)^{3/2} - 2ba^{-2}(ax + b)^{1/2} + c$
63. $\frac{x}{2}\sqrt{x^2 \pm a^2} \pm \frac{a^2}{2}\ln|x + \sqrt{x^2 \pm a^2}| + c$ **65.** $\frac{a^2}{2}\arcsin(\frac{x}{a}) + \frac{x}{2}\sqrt{a^2 - x^2} + c$ **67.** $\bar{x} = \frac{1}{9}(3\sqrt{3} - \pi), \bar{y} = \frac{\pi}{6} - \frac{\sqrt{3}}{8}$
69. $\frac{\sqrt{5}}{2} + \frac{1}{4}\ln(2 + \sqrt{5})$ **71.** **(i)** $2\ln|x| + \frac{3}{2}\ln(x^2 + 1) - \arctan x + c$

Exercises 9.1 (p. 621)

1. $1 - \frac{x^2}{2} + \frac{x^4}{24}$ **3.** $1 + x + \frac{x^2}{2} + \frac{x^3}{6} + \frac{x^4}{24} + \frac{x^5}{120}$ **5.** $x - \frac{x^2}{2} + \frac{x^3}{3}$
7. $1 + 3(x - 1) + 3(x - 1)^2 + (x - 1)^3$
9. $1 - 4(x + 1) + 6(x + 1)^2 - 4(x + 1)^3 + (x + 1)^4$
11. 0.009 999 833 **13.** 1.105 170 833 **15.** 0.198 666 667
17. 1.011 271 094 **19.** −0.105 360 333
21. $1 + x + \frac{x^2}{2} + \frac{x^3}{6} + \frac{x^4}{24} + \frac{x^5}{120}$; $e^{1/2} \approx 1.648\,697\,917$
23. **(i)** 0.704 652 651 **(ii)** 0.707 429 207

31.

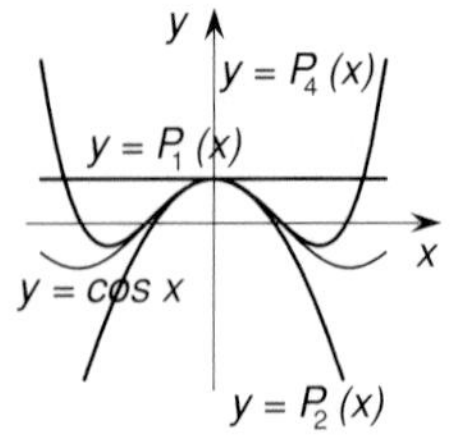

33. **(i)** $f^{(4m)}(0) = 0, f^{(4m+1)}(0) = 1, f^{(4m+2)}(0) = 0, f^{(4m+3)}(0) = -1$
(ii) $\frac{1}{k!}$, if $k = 4m + 1$; $-\frac{1}{k!}$, if $k = 4m + 3$; 0, if k is even

Exercises 9.2 (p. 633)

1. $P_3(\frac{1}{2}) \approx 1.646$ **3.** $P_7(1) \approx 0.841\,468$ **5.** $P_0(1.2) = 0$; $P_1(1.2) = 0.2$ **7.** $P_0(1.9) = \ln(2) \approx 0.693$; $P_1(1.9) \approx 0.643$
9. $P_0(-\frac{1}{10}) = 0$ **11.** $f(x) = \cos x, a = 0, P_1(\frac{4}{180}\pi) = 1$
13. $f(x) = \cos x, a = \frac{\pi}{4}, P_1(a - \frac{2}{180}\pi) \approx 0.7318$
15. $f(x) = \ln(x), a = 1, P_1(a + 0.2) = 0.2$ **17.** $f(x) = \tan x, a = 0, P_1(\frac{3}{180}\pi) = \frac{3}{180}\pi \approx 0.052\,360$
19. $f(x) = \sinh(x), a = 0, P_2(-0.12) = -0.12$
21. $\pm\frac{1}{720}\frac{1}{64} \approx \pm 2.17 \times 10^{-5}$ **23.** **(i)** ± 0.0014 is sufficient
(ii) 0.001218 (using P_1 for $\sin x$) **25.** using P_0 for $\cos x$ with $a = 0$, P_1 for $\sin x$ with $a = 0$, $\sin^2(1°)\cos(1°) \approx 3.046 \times 10^{-4}$
27. $f(x) = e^{\sin x}, a = \frac{\pi}{4}, P_2(a + \frac{2\pi}{180}) \approx 2.077\,919$
29. $f(t) = \int_0^t \exp(-x^2)\,dx, a = 0, P_2(\frac{1}{4}) = \frac{1}{4}$ **31.** $f(t) = \int_0^t \sin(x^2)\,dx, a = 0, P_1(\frac{1}{5}) = 0$ **33.** **(iii)** $P_5(x) = 1 + \frac{1}{3}x^3$ **(iv)** $P_5(0.1) = 1 + \frac{1}{3000}$
(v) $\exp(\frac{1}{3}x^3)$ **35.** ≈ 0.743, within 0.1 **37.** 0 or 0.005
39. $\approx 0.892, \approx 1.694$

Chapter 9 Review Exercises (p. 635)

1. $1 - x + \frac{x^2}{2} - \frac{x^3}{6} + \frac{x^4}{24} - \frac{x^5}{120} + \frac{x^6}{720}$
3. $e + e(x - 1) + \frac{e}{2}(x - 1)^2 + \frac{e}{6}(x - 1)^3 + \frac{e}{24}(x - 1)^4 + \frac{e}{120}(x - 1)^5 + \frac{e}{720}(x - 1)^6$ **5.** $1 - \frac{\pi^2}{2}(x - \frac{1}{2})^2 + \frac{\pi^4}{24}(x - \frac{1}{2})^4$
7. $\frac{\pi}{4} + \frac{1}{2}(x - 1) - \frac{1}{4}(x - 1)^2 + \frac{1}{12}(x - 1)^3$ **9.** 0
11. $f(x) = e^x, a = 0, P_7(-1) \approx 0.367\,857$
13. $f(x) = \sqrt{x}, a = 1, P_1(a + 0.1) = 1.05$
15. $f(x) = \sec x, a = 0, P_0(\frac{11\pi}{1800}) = 1$
17. $f(x) = \sin x, a = 0, P_1(1.5\frac{\pi}{180}) = \frac{\pi}{120} \approx 0.026\,180$
19. $f(x) = \tan x, a = 0, P_1(-0.1) = -0.1$ **23.** $P_n(x) = M_n(x - a)$

Exercises 10.1 (p. 647)

1. −1 **3.** $\frac{1}{4}$ **5.** diverges **7.** 0 **9.** 0 **11.** 0 **13.** 1 **15.** −1
17. diverges **19.** 0 **23.** 0 **25.** 0 **27.** diverges

Exercises 10.2 (p. 656)

1. $\frac{127}{64}$ **3.** $\frac{85}{128}$ **5.** $2 - 2^{-20}$ **7.** $\frac{5}{3}(1 - 4^{-8})$ **9.** $\frac{12}{7}(36 + 6^{-7})$
11. \$1467.63 **13.** \$47.07 **15.** \$1259.30 **17.** \$14,682.42

19. \$5.815 693 ≈ \$5.82 **21.** 2 **23.** $\frac{27}{4}$ **25.** 2 **27.** $\frac{11}{2}$ **29.** $\frac{1215}{4}$ **31.** \$723.05 **33.** \$6573.69 **35.** **(i)** loss of \$1608.04 at end **(ii)** profit of \$12,720.29 at end **37.** $a(r^m - r^{k+1})/(1 - r)$ **39.** **(i)** $\frac{1}{8}(9^{13} - 1)$ **(ii)** $\frac{1}{7}(1 - 8^{-8})$ **43.** $\frac{2110}{999}$

Exercises 10.3 (p. 665)

1. 1 **3.** diverges **5.** 0 **7.** diverges **9.** 2 **11.** 0 **13.** 8 **15.** −2 **17.** −5 **19.** 0 **21.** $a_1 = 1, a_n = -\frac{1}{n(n-1)}$, for $n > 1$ **23.** $a_1 = 1, a_n = \frac{1-2n}{n^2(n-1)^2}$, for $n > 1$ **25.** converges to 0

Exercises 10.4 (p. 673)

1. diverges **3.** converges **5.** diverges **7.** converges **9.** converges **11.** converges **13.** converges **15.** diverges **17.** converges **19.** converges **23.** converges **25.** converges **27.** converges **29.** $q > 1$ **31.** diverges **33.** ≈10.58132 **35.** 1.081 895 **39.** 143.903 745

Exercises 10.5 (p. 679)

1. diverges **3.** converges **5.** diverges **7.** converges **9.** converges **11.** diverges **13.** converges **15.** converges **17.** converges **19.** converges **25.** converges **27.** diverges **29.** diverges **31.** converges **33.** converges **35.** **(i)** ≈ 0.603 702 863 **(ii)** within 0.01 **37.** ≈ 1.604 61 **39.** 0.4147

Exercises 10.6 (p. 686)

1. converges **3.** converges **5.** converges **7.** converges **9.** diverges **11.** converges **13.** converges **15.** diverges **17.** converges **19.** diverges **21.** converges conditionally **23.** converges absolutely **25.** diverges **27.** converges conditionally **29.** converges absolutely **31.** converges conditionally **33.** converges absolutely **37.** $\frac{1}{2} + \frac{1}{4} + \cdots + \frac{1}{26} - 1 + \frac{1}{28} + \cdots + \frac{1}{38}$ **41.** converges **43.** diverges **45.** diverges **47.** converges **49.** converges **51.** ≈ 0.695 65 **53.** ≈ 0.3241

Exercises 10.7 (p. 694)

1. converges (Ratio Test) **3.** converges absolutely (Ratio Test) **5.** diverges (Ratio or *n*th Root Test) **7.** converges (*n*th Root Test) **9.** converges (Ratio Test) **11.** converges (Ratio Test) **13.** converges (Ratio or Integral Test) **15.** convergent geometric series **17.** converges (Ratio or Limit Comparison Test) **19.** converges absolutely (Ratio and Comparison Tests) **21.** all c **23.** all c **25.** $\in (-1, 1]$ **27.** $\{c : |c| > 5\}$ **29.** no nonzero c **33.** $c \in (-1, 1]$ **35.** $|r| < 1$, unless $p(n) = 0$, for all n, in which case all r **39.** converges (Limit Comparison Test) **41.** diverges (*n*th Term Test) **43.** converges (Comparison or Integral Test) **45.** diverges (Limit Comparison Test) **47.** converges (Integral Test) **49.** converges (Ratio Test) **51.** diverges (Ratio or *n*th Term Test) **53.** converges (Comparison Test) **55.** diverges (Comparison Test) **57.** converges (Ratio Test) **59.** **(i)** 1.718 281 828 459 045 **(ii)** 8.22×10^{-18}

Exercises 10.8 (p. 703)

1. $\sum_{n=0}^{\infty} \frac{1}{n!}(-2x)^n$ **3.** $\sum_{n=0}^{\infty} (-1)^n(x-1)^n$ **5.** $5 - x - 2x^2 + x^3$ **7.** $\sum_{n=0}^{\infty} (-1)^{n+1} \frac{1}{(2n+1)!}(x - \frac{\pi}{2})^{2n+1}$ **9.** $\sum_{n=0}^{\infty} (-1)^n 2^n x^{2n}$ **11.** $(2 - e^3) - e^3 \sum_{n=1}^{\infty} \frac{1}{n!}3^n(x-1)^n$ **13.** $-\frac{1}{2}\sum_{n=0}^{\infty} 2^{-n}x^n$ **15.** $1 + \frac{x^2}{2} + \sum_{n=2}^{\infty} (-1)^n \frac{1}{(2n)!}x^{2n}$ **17.** $\frac{1}{e}\sum_{n=0}^{\infty} (-1)^n 2^n \frac{1}{n!}(x - \frac{1}{2})^n$ **19.** $\sum_{n=0}^{\infty} \frac{s(n)}{n!}x^n$, where $s(n) = 1$ if $n = 4k$ or if $n = 4k + 1$, for some integer k, and $s(n) = -1$ if $n = 4k + 2$ or $n = 4k + 3$ for some integer k. **21.** $1 + \frac{x}{2} - \frac{1}{8}x^2 + \cdots + (-1)^{n+1}\frac{1(3)(5)\cdots(2n-3)}{n!2^n}x^n + \cdots = 1 + \frac{x}{2} + \sum_{n=2}^{\infty} (-1)^{n+1}\frac{1}{n!}\frac{(2n-3)!}{n!\,(n-2)!}2^{2-2n}x^n$ **25.** $\sum_{n=0}^{\infty} (-1)^n x^{2n+1}$ **33.** ≈ 20.085 536 923 **35.** $\approx 3.113 \times 10^{-7}$ **37.** ≈ 0.223 143 551 **39.** ≈ −18,155.343 286

Exercises 10.9 (p. 709)

1. $(-1, 1)$ **3.** $[2, 4]$ **5.** $(-\infty, \infty)$ **7.** $(-\infty, \infty)$ **9.** $(11, 13)$ **11.** $(-1, 1)$ **13.** $[1, 5]$ **15.** $(-\infty, \infty)$ **17.** $[22, 22]$ **19.** $[-\frac{8}{3}, \frac{8}{3})$ **21.** ∞ **23.** 1 **29.** $[-7 - 5^{-1/3}, -7 + 5^{-1/3}]$ **33.** **(i)** $\frac{R}{4}$ **(ii)** $\frac{R}{r}$ **(iii)** R **35.** 3

Exercises 10.10 (p. 721)

1. $\sum_{n=0}^{\infty} \frac{(-1)^n}{n!}\frac{x^{2n+1}}{2n+1}$ **3.** $\sum_{n=1}^{\infty} (-1)^n n x^{n-1}$ **5.** $\ln(2) - \sum_{n=1}^{\infty} \frac{2^{-n}}{n}x^n$ **7.** $\sum_{n=1}^{\infty} \frac{(-1)^{n+1}}{n(n+1)}2^n x^{n+1}$ **9.** $\sum_{n=0}^{\infty} (3^{n+1} - 2^{n+1})x^n$ **11.** $\sum_{n=0}^{\infty} \frac{(-1)^n}{(2n+1)!}x^{2n+3}$ **13.** $\sum_{n=0}^{\infty} \frac{(-1)^n}{(2n)!}x^{2n}$ **15.** $-\sum_{n=1}^{\infty} \frac{(-1)^n}{(2n)!}x^{2n-2} = \sum_{m=0}^{\infty} \frac{(-1)^m}{(2m+2)!}x^{2m}$ **17.** $\sum_{n=0}^{\infty} 2^{-n-1}x^{2n}$ **19.** $\frac{1}{6}\sum_{n=3}^{\infty} n(n-1)(n-2)x^{n-3} = \frac{1}{6}\sum_{m=0}^{\infty} (m+1)(m+2)(m+3)x^m$ **21.** $\frac{1}{0!} + \frac{1}{1!} + \frac{1}{2!} + \cdots + \frac{1}{n!}$ **23.** $1\frac{1}{(2n)!} + \frac{1}{2!}\frac{1}{(2n-2)!} + \frac{1}{4!}\frac{1}{(2n-4)!} + \cdots + \frac{1}{(2n)!}(1)$ **25.** $\frac{(-1)^n \pi^{2n}}{(2n)!}$ **27.** −1 for $n \geq 1$, and 0 for $n = 0$ **29.** $-\frac{2}{n}$ **31.** $1 + 2x + \frac{5}{2}x^2 + \cdots$ **33.** $1 + x + \frac{x^2}{2} + \cdots$ **35.** $1 + \frac{x^2}{6} + \frac{7}{360}x^4 + \cdots$ **37.** $x - \frac{x^3}{3} + \frac{2}{15}x^5 + \cdots$ **39.** $\frac{1}{x} - \frac{x}{3} - \frac{x^3}{45} + \cdots$ **41.** $x + (1 - x)\ln(1 - x)$ **43.** $-\frac{1}{2} + \frac{1}{2}\sum_{m=0}^{\infty} \frac{2^{2m}}{(2m)!}x^{2m}$ **45.** $\frac{3}{4}$ **47.** 24 **49.** **(i)** 40,320 **51.** 61 **53.** 5

Chapter 10 Review Exercises (p. 722)

1. 2 **3.** 4 **5.** \$2550.00 **7.** \$75,644.46 **9.** \$1314.74 **11.** 0 **13.** diverges **15.** 0 **17.** 1 **19.** 1 **21.** diverges (*n*th Term Test) **23.** diverges (*n*th Term Test) **25.** diverges (*n*th Term or Ratio Test) **27.** diverges (*n*th Term Test) **29.** convergent alternating series **31.** diverges (harmonic series) **33.** converges (Ratio Test) **35.** diverges (*n*th Term Test) **37.** converges (Integral Test) **39.** convergent alternating series **41.** $|c| > 1$ **43.** $c < -\frac{1}{2}$ **45.** $c < -2$ **47.** $c < -\frac{1}{3}$ **49.** no c **51.** $\sum_{n=0}^{\infty} 3^n x^n$ **53.** $\sum_{n=0}^{\infty} (-1)^{n+1}\frac{1}{(2n)!}(x - \pi)^{2n}$ **55.** $\ln 4 + \sum_{n=1}^{\infty} (-1)^{n+1}\frac{1}{n4^n}(x - 4)^n$ **57.** $-\sum_{n=1}^{\infty} \frac{1}{n}(x + 1)^n$ **59.** $\sum_{n=0}^{\infty} (-1)^n \frac{1}{2n+1}2^{2n+1}x^{2n+1}$ **61.** $\frac{1}{2}$ **63.** 0 **65.** 1 **67.** $\frac{1}{\sqrt{2}}$ **69.** 1 **71.** $x - \frac{7}{6}x^3 + \frac{61}{120}x^5 + \cdots$ **73.** $1 - x + \frac{x^2}{2} + \cdots$ **75.** $e + ex + ex^2 + \cdots$ **77.** $\cos(1) - \sin(1)x - \frac{1}{2}(\cos(1) + \sin(1))x^2 + \cdots$ **79.** $x - \frac{x^2}{2} + \frac{3}{8}x^5 + \cdots$ **83.** **(i)** $\sum_{n=0}^{\infty} (a_n - 3b_n)x^n$ **85.** 0

Exercises 11.1 (p. 730)

1. hyperbola, center (0, 0), asymptotes $y = \pm x$, transverse axis vertical

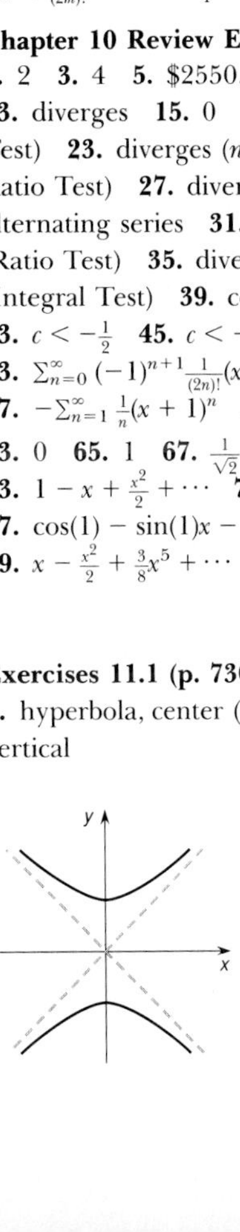

3. parabola, vertex $(-3, 5)$, opening downward

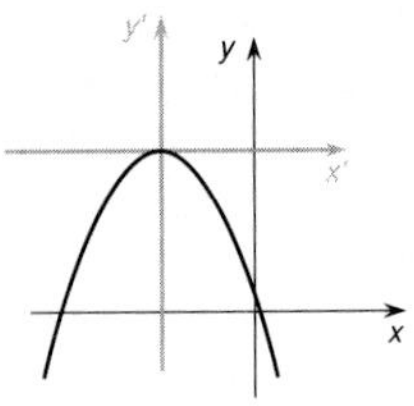

5. ellipse, center $(-1, 0)$, semi-major axis 1, vertical; semi-minor axis $\frac{1}{\sqrt{2}}$

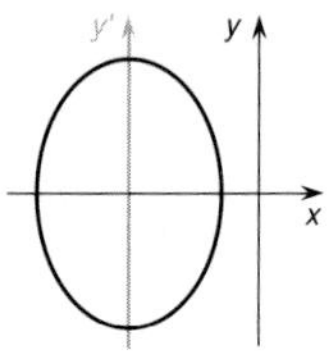

7. ellipse, center $(2, 2)$, semi-major axis 4, vertical; semi-minor axis $\frac{4}{\sqrt{3}}$

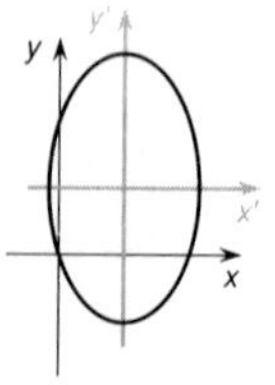

9. parabola, vertex $(19, 2)$, opening left

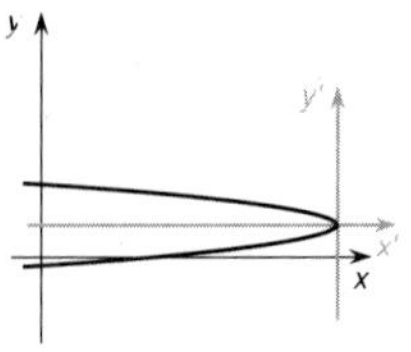

11. $4x^2 + 8x - y + 2 = 0$ **13.** $x^2 + 4y^2 - 24y + 32 = 0$ **15.** $4x^2 - 8x - y^2 = 0$ **17.** $x^2 + 6x + y^2 - 8y + 9 = 0$ **19.** $3x^2 - 24x - y^2 + 49 = 0$ **21.** two straight lines, slopes ± 1, through $(-1, -1)$

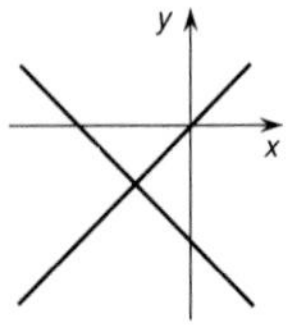

23. $4x^2 - 24x - y + 35 = 0$, $y^2 + 2y - 16x + 49 = 0$ **25. (i)** $(-1, 0)$, $(1, 0)$, $(0, 1)$, $(0, 2)$, for instance **(ii)** $(0, 0)$, $(1, 0)$, $(2, 0)$, for instance **27.** $y = \pm\sqrt{4x^2 - 8x}$ **29.** $y = -3 \pm 7\sqrt{1 - (x-2)^2/9}$

Exercises 11.2 (p. 739)

1. focus $(-1, -\frac{3}{4})$, directrix $y = -\frac{5}{4}$ **3.** foci $(\pm\sqrt{3}, 0)$, directrices $x = \pm\frac{4}{\sqrt{3}}$ **5.** foci $(\pm\frac{\sqrt{5}}{2}, 0)$, directrices $x = \pm\frac{2}{\sqrt{5}}$ **7.** foci $(\pm\sqrt{6} - 2, 1)$, directrices $x = -2 \pm \frac{8}{\sqrt{6}}$ **9.** foci $(1 \pm \frac{2\sqrt{13}}{3}, -1)$, directrices $x = 1 \pm \frac{8}{3\sqrt{13}}$ **11.** $\frac{x^2}{36} + \frac{y^2}{27} = 1$ **13.** $x = 3 - \frac{1}{4}(y-2)^2$, or $y^2 - 4y + 4x - 8 = 0$ **15.** $9(x-4)^2 + (y-1)^2 = 13$ **17.** $(y+1)^2 - \frac{1}{3}(x-2)^2 = 1$ **19.** $y = 4 - (x-3)^2$ **21.** $e = \frac{\sqrt{3}}{2}$ **23.** $(-1 \pm \frac{\sqrt{5}}{2}, 0)$ **25.** $y = -4$ **27.** $e = 1$ **29.** $2c = \frac{\sqrt{13}}{3}$ **31.** $\frac{72}{5} = 14.4$ ft **33.** equal **35.** $15y^2 - x^2 = 956$ **37.** $10 - \frac{5\sqrt{7}}{2} \approx 3.39$ m **39.** centers **41.** $\frac{x^2}{a^2} + \frac{y^2}{b^2} = 1$ **43.** one **49. (i)** $8(y-75)^2 - x^2 = 5000$, upper branch **(ii)** $(\frac{600 + 200\sqrt{2}}{7}, \frac{600 + 200\sqrt{2}}{7})$ **51.** plot $y = \pm\sqrt{1-e^2}\sqrt{1-x^2}$

Exercises 11.3 (p. 750)

1. $(\frac{1}{\sqrt{2}} - \sqrt{2}, \frac{1}{\sqrt{2}} + \sqrt{2})$ **3.** $(-\frac{3\sqrt{3}}{2}, \frac{3}{2})$ **5.** $(\frac{7+\sqrt{3}}{2}, \frac{7\sqrt{3}-1}{2})$ **7.** $(-\frac{a\sqrt{3}}{2}, -\frac{a}{2})$ **9.** $(0, 0)$ **11.** ellipse, center $(0, 0)$, semi-major axis $\frac{1}{2}$, along line $y = -x$; semi-minor axis $\frac{1}{3}$

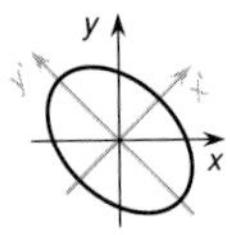

13. parabola, vertex $(0, 0)$, focus $(\frac{\sqrt{3}}{8}, \frac{1}{8})$

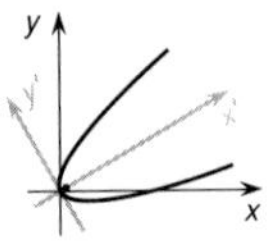

15. ellipse, center $(0, 0)$, semi-major axis 4, along line $3x + 4y = 0$; semi-minor axis 2

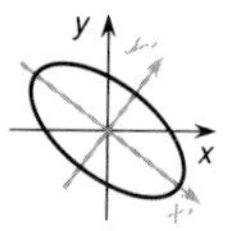

17. hyperbola, center $(0, 0)$, transverse axis along $2x + y = 0$, asymptotes $3y' = \pm x'$

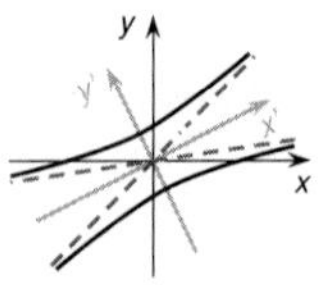

19. hyperbola, center (0, 0), transverse axis along $y = x$, asymptotes $y' = \pm 3^{1/4}x'$

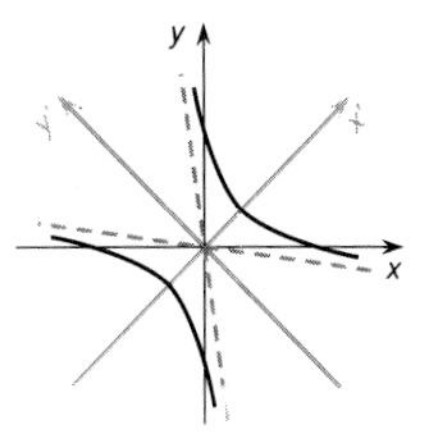

21. parabola, vertex $(\frac{19}{24}, -\frac{31}{24})$, focus $(\frac{5}{12}, -\frac{11}{12})$

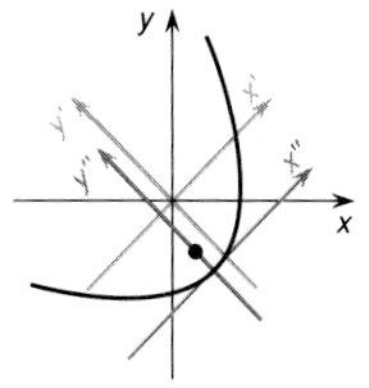

23. hyperbola, center $(\frac{3}{5}, \frac{1}{5})$, transverse axis along $3x - 4y - \frac{13}{5} = 0$

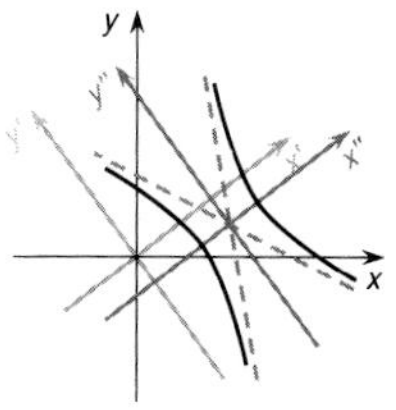

25. ellipse, center $(\frac{\sqrt{3}-1}{2}, \frac{\sqrt{3}+1}{2})$, semi-major axis 3, along $\sqrt{3}x + y = 2$; semi-minor axis 2

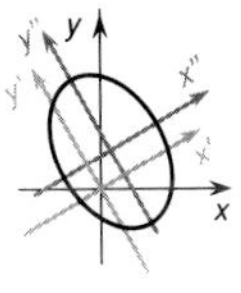

27. hyperbola, center $(3, -4)$, transverse axis horizontal, asymptotes $y = x - 7, y + x + 1 = 0$

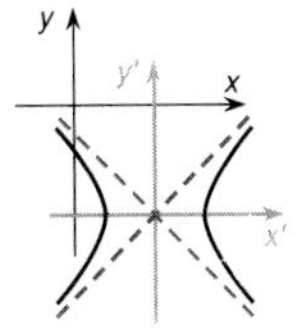

29. parabola, vertex $(-\frac{1}{2}, -\frac{9}{4})$, focus $(-\frac{5}{2}, -\frac{5}{4})$

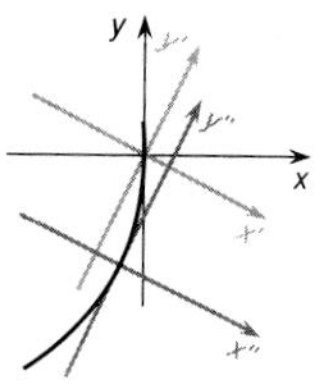

31. $y'^2/4 - x'^2/9 = 1$

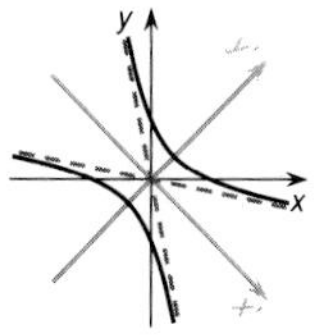

33. $q = 0$

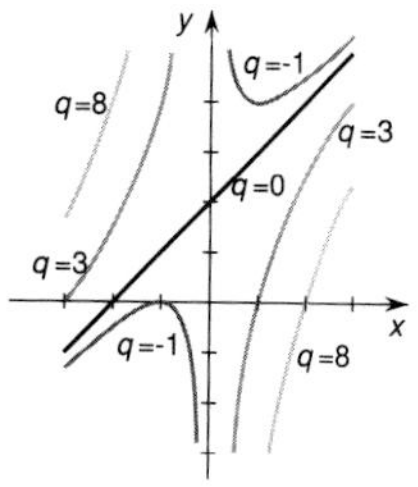

37. **(i)** $Ax^2 + Bxy + Cy^2 = A(x + \frac{By}{2A})^2$ **41.** $(\frac{3}{2}, -\frac{3}{2})$
43. $\pm(\frac{4}{5\sqrt{6}}, -\frac{3}{5\sqrt{6}})$ **45.** $17x^2 + 17y^2 + 16xy = 225$ **47.** $-\frac{\pi}{8}$

Chapter 11 Review Exercises (p. 753)

1. ellipse, semi-major axis 4, horizontal; semi-minor axis 3

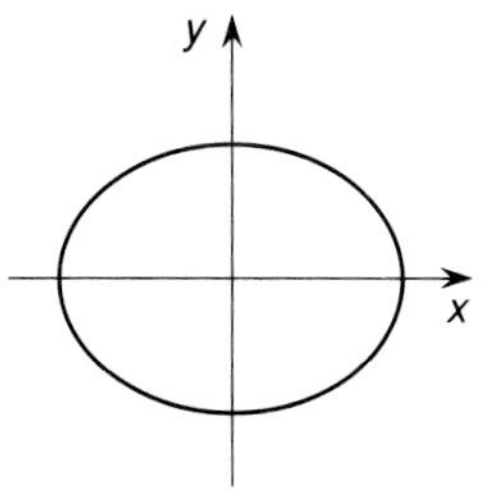

3. parabola, vertex (2, 0), focus $(\frac{33}{16}, 0)$

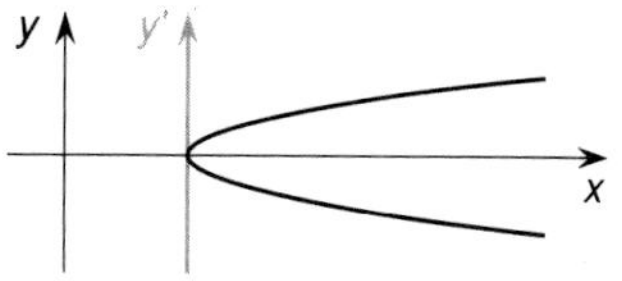

5. ellipse, semi-major axis 2, horizontal; semi-minor axis $\sqrt{2}$

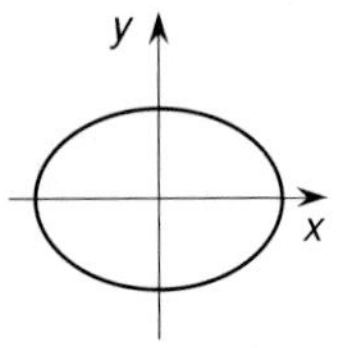

7. ellipse, center (2, 0), semi-major axis 2, horizontal; semi-minor axis 1

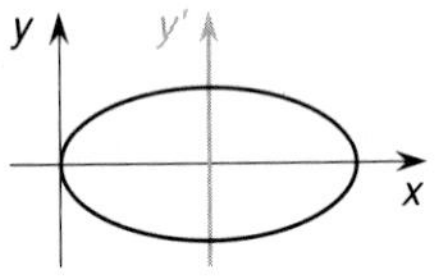

9. parabola, vertex $(-2, -\frac{7}{2})$, focus $(-2, -3)$

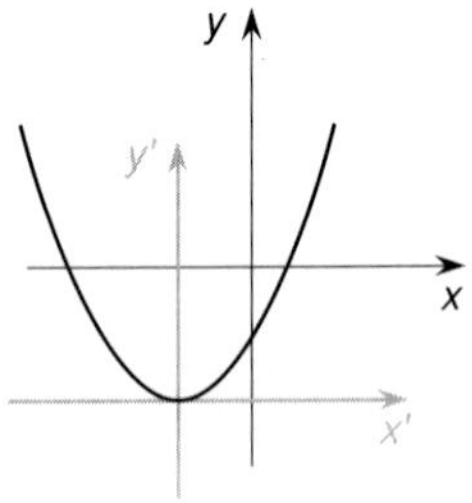

11. $\frac{x'^2}{16} + \frac{y'^2}{25} = 1$, ellipse, center (0, 0), semi-major axis 5, along $y = -x$; semi-minor axis 4

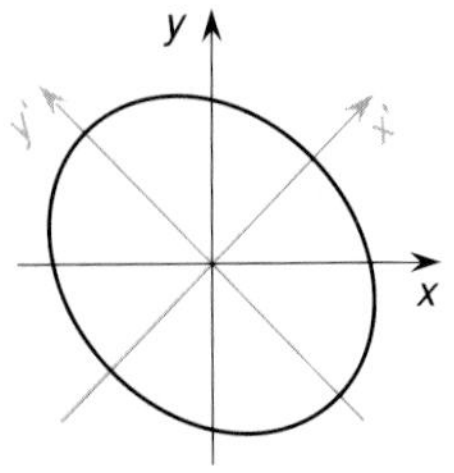

13. $y'^2 + 2x' = 0$, parabola, vertex (0, 0), focus $(-\frac{2}{5}, \frac{3}{10})$

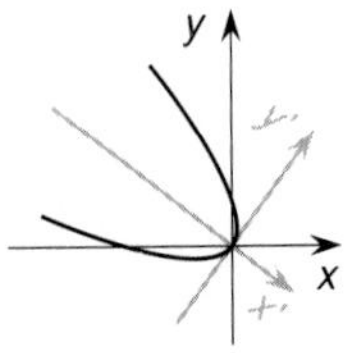

15. $y''^2 - x''^2 = 1$, hyperbola, center $(\frac{1}{\sqrt{2}}, \frac{3}{\sqrt{2}})$, asymptotes $x = \frac{1}{\sqrt{2}}$, $y = \frac{3}{\sqrt{2}}$, transverse axis along $x + y = 2\sqrt{2}$

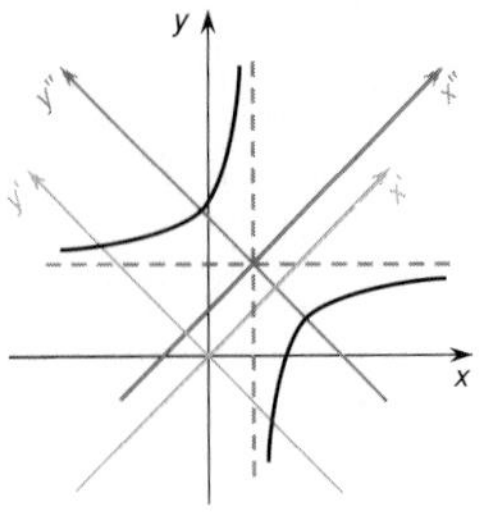

17. $\frac{3}{25}x''^2 + \frac{1}{5}y''^2 = 1$, ellipse, center $(\frac{2}{5}, \frac{11}{5})$, semi-major axis $\frac{5}{\sqrt{3}}$, along $3x + 4y = 10$; semi-minor axis $\sqrt{5}$

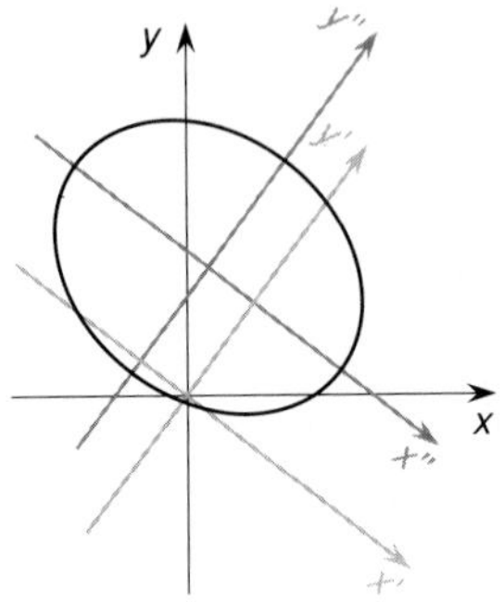

19. $y''^2 - 2x'^2 = 2$, hyperbola, center $(-\frac{4}{\sqrt{17}}, -\frac{16}{\sqrt{17}})$, transverse axis along $y = 4x$

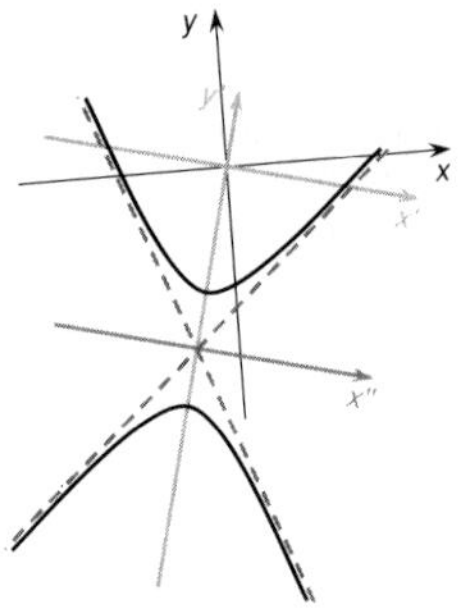

21. ellipse $c > -14$, point $c = -14$, empty set $c < -14$

23. **(i)** circle, radius $\sqrt{2}$, **(ii)** circle, radius 2, **(iii)** circle, radius 1, **(iv)** point $(-1, 1)$, **(v)** empty set

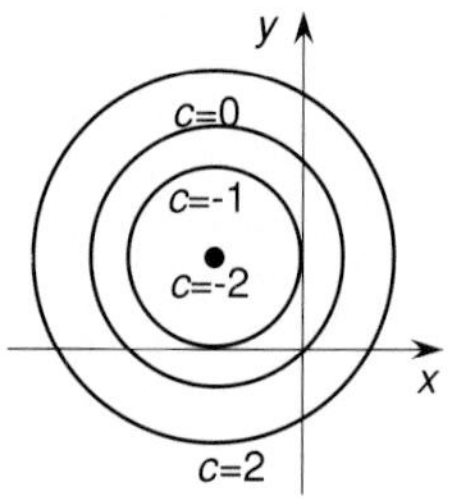

25. **(i)**, **(ii)**, **(iii)** parabolas, **(iv)** line $x + y = 0$

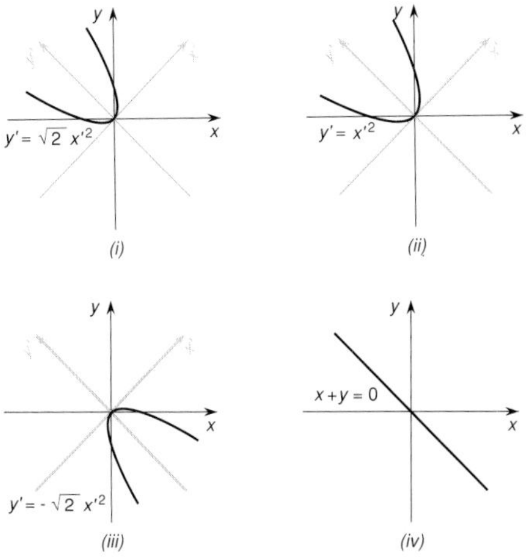

27. **(ii)** counterclockwise once around the ellipse, starting at $(a, 0)$ **29.** $y = \pm\sqrt{e^2 - 1}x$, $x = \pm\sqrt{e^2 - 1}y$
31. **(i)** $y'^2 - y' + 2x'y' + x'^2 = 0$ **(ii)** parabola
(iii)

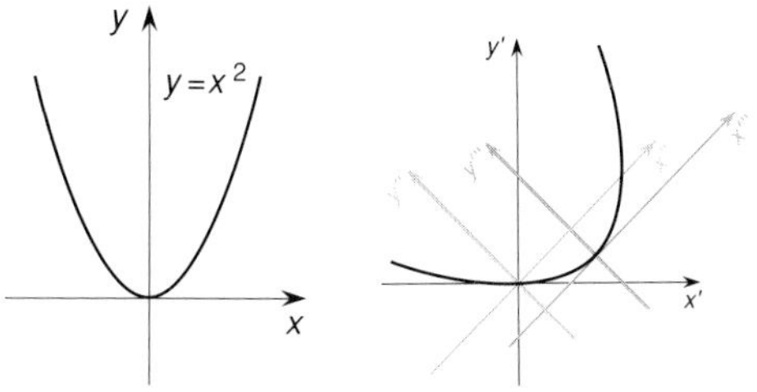

33. if $\frac{B}{A-C} > 0$,
$\sqrt{(\sqrt{(A-C)^2 + B^2} - |A-C|)/(\sqrt{(A-C)^2 + B^2} + |A-C|)}$; the negative of this if $\frac{B}{A-C} < 0$ **35.** **(i)** $\frac{\pi}{2}$ **(ii)** $\frac{\pi}{3}$ **(iii)** $\frac{3}{2}$
37. **(ii)** $\frac{50{,}000{,}000}{\sqrt{14 + 8\sqrt{2}}} \approx 9{,}937{,}843$ km

Exercises 12.1 (p. 759)
1. $(2, 0)$ **3.** $(6\sqrt{2}, -\frac{\pi}{4} + 2k\pi)$ **5.** $(0, 0)$ **7.** $(4, \frac{5\pi}{6} + 2k\pi)$
9. $(9, \pi + 2k\pi)$ **11.** $(-1, 0)$ **13.** $(5\sqrt{2}, \frac{3\pi}{4} + 2k\pi)$ **15.** $(-7, 0)$
17. $(\frac{8}{\sqrt{3}}, -\frac{2\pi}{3} + 2k\pi)$ **19.** $(1, -\frac{\pi}{4} + 2k\pi)$ **21.** $(\sqrt{2}, 2k\pi)$
23. $(0, -1)$ **25.** $(0, -1)$ **27.** $(13, \arctan(-\frac{12}{5}) + 2k\pi)$
29. $(\frac{2}{\sqrt{5}}, \frac{4}{\sqrt{5}})$ **31.** **(i)** $r = 2$ **(ii)** $r^2 + 2r(\cos\theta - 2\sin\theta) = 4$, circle, radius 3, center $(x, y) = (-1, 2)$ **33.** $r \geq 0$, $0 \leq \theta \leq \frac{\pi}{2}$
35. $1 < r < 2$, $\frac{\pi}{2} \leq \theta \leq \frac{3\pi}{4}$ **37.** $-x < y < x$ **39.** $x^2 + y^2 \leq 4$
41. circle around origin, radius 10 **43.** two half-lines, $y = \pm\sqrt{3}x$, $x \geq 0$ (i.e., $\theta = \pm\frac{\pi}{3}$); if we allow $r < 0$, get whole lines $y = \pm\sqrt{3}x$

Exercises 12.2 (p. 770)
1. circle, center $(x, y) = (0, \frac{1}{2})$, radius $\frac{1}{2}$

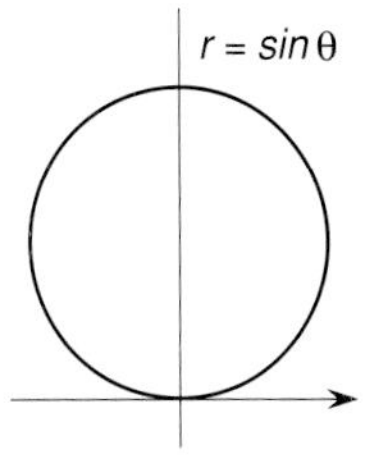

3. circle, center $(x, y) = (-1, 0)$, radius 1

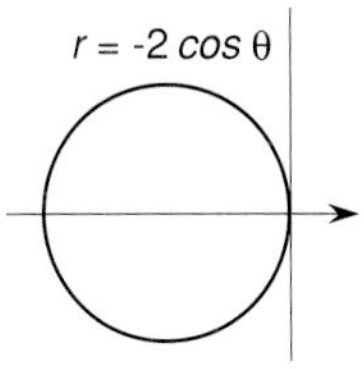

5. two half-lines $y = \pm x$, $y \geq 0$ (i.e., $\theta = \frac{\pi}{4}, \frac{3\pi}{4}$)

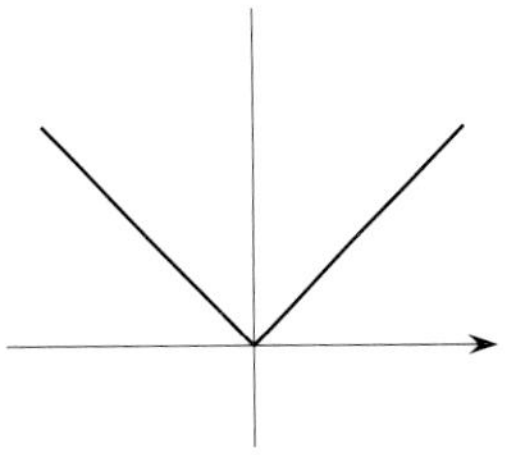

7. four-petaled rose (one petal in each quadrant)

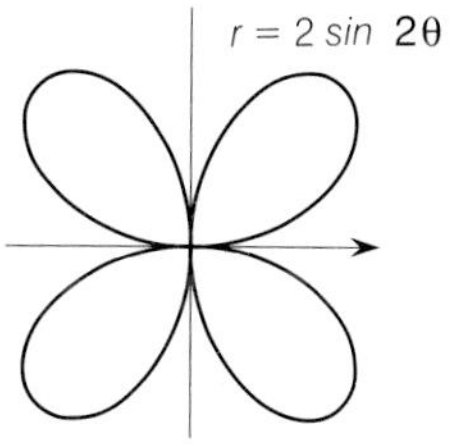

9. counterclockwise spiral, starting horizontally to the left from the origin

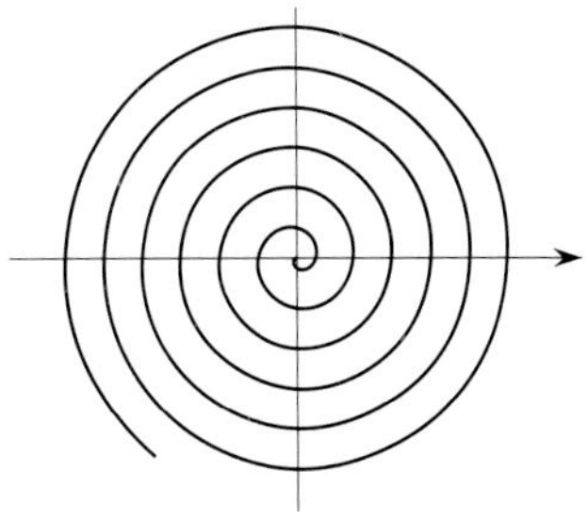

11. cardioid

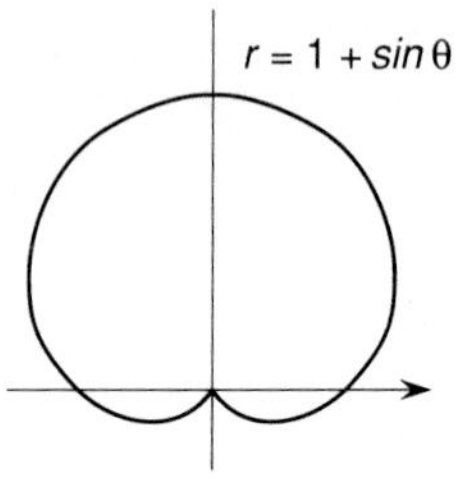

13. limaçon with inner loop

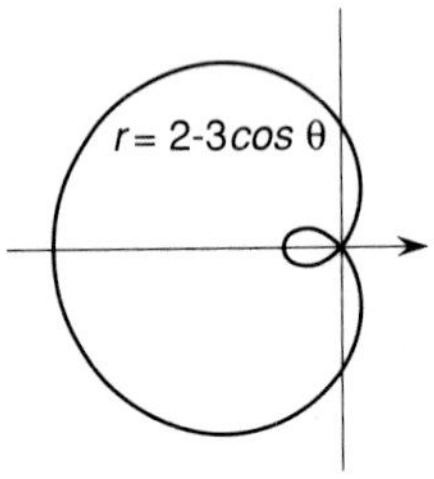

15. limaçon with large inner loop

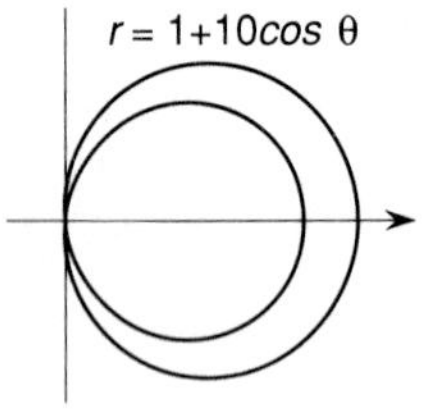

17. lemniscate

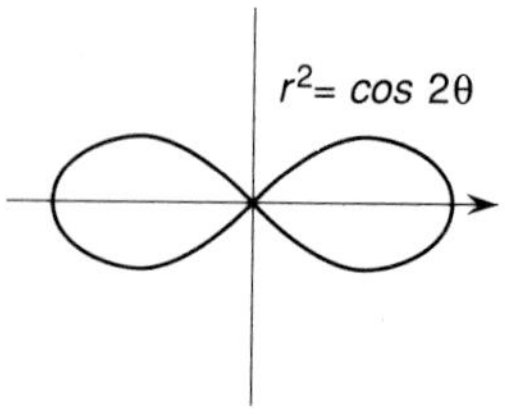

19. circle

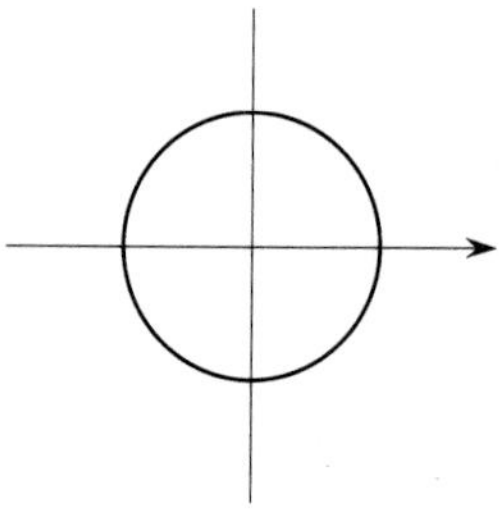

21. circle $r = 5\sqrt{2}\cos(\theta + \frac{\pi}{4})$

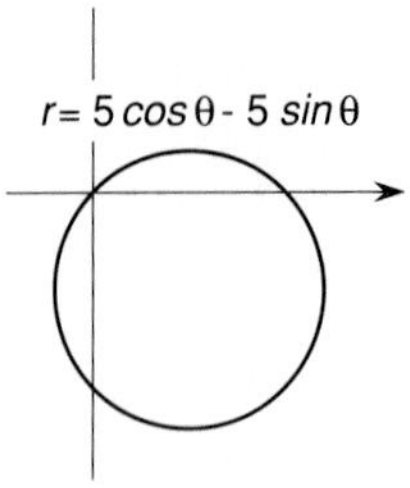

23. circle, radius $\frac{5}{2}$

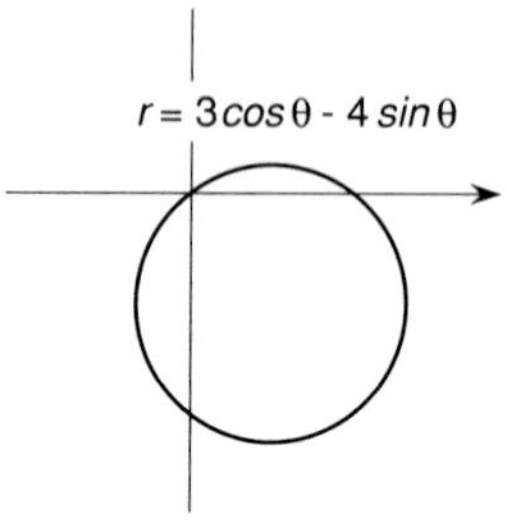

25. $(r, \theta) = (\frac{1}{\sqrt{2}}, \pm\frac{\pi}{4} + 2k\pi)$, i.e., $(x, y) = (\frac{1}{2}, \pm\frac{1}{2})$

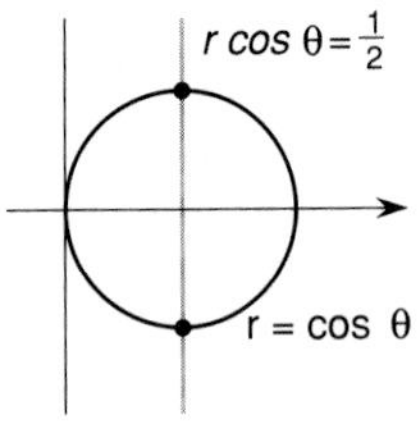

27. $(r, \theta) = (4, \pm\frac{\pi}{3})$

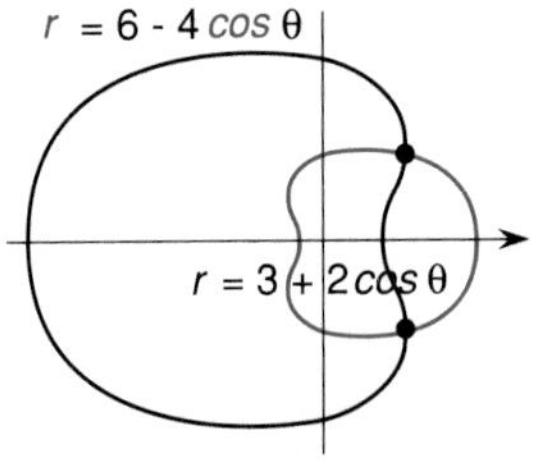

29. $r = \frac{1}{\sqrt{2}}$, $\theta = \frac{\pi}{20}, \frac{9\pi}{20}, \frac{17\pi}{20}, \frac{25\pi}{20}, \frac{33\pi}{20}$

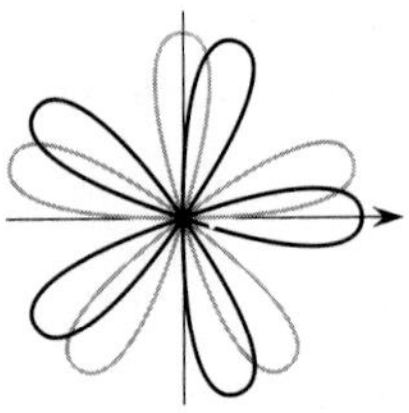

31. $(\frac{1}{2}, \pm\frac{2\pi}{3})$, $(2, 0)$, $(0, \pi)$

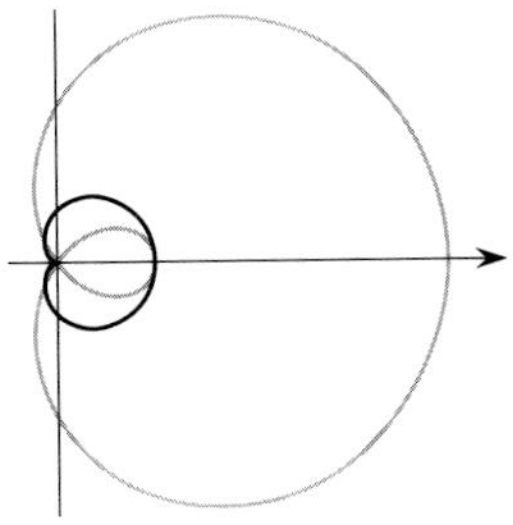

33. cardioid

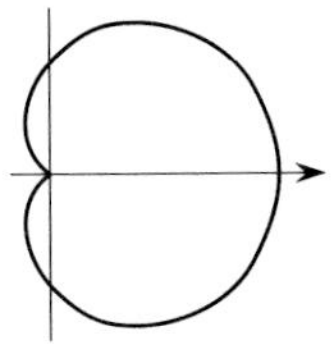

37. six-petaled rose

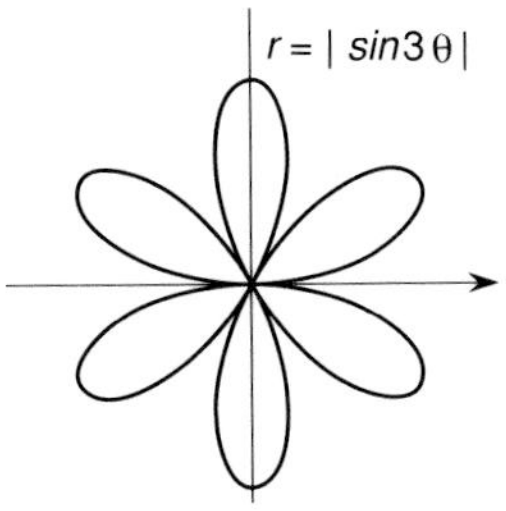

39. two circles

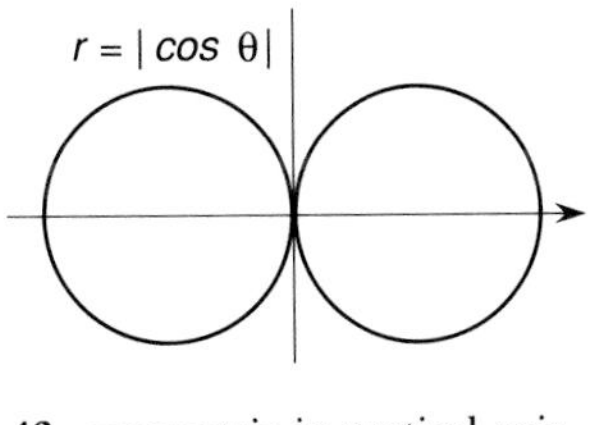

43. symmetric in vertical axis
45. horizontal axis: $((-1)^n n\pi, 0)$, $n = 0, 1, 2, \ldots$; vertical axis: $(0, (-1)^n(\frac{\pi}{2} + n\pi))$, $n = 0, 1, 2, \ldots$ **47.** $(\frac{2}{3}\sqrt{\frac{5}{6}}, \pm\frac{2}{3}\sqrt{\frac{1}{6}})$
49. highest: $(\pm\frac{\sqrt{3}}{2\sqrt{2}}, \frac{1}{2\sqrt{2}})$ lowest: $(\pm\frac{\sqrt{3}}{2\sqrt{2}}, -\frac{1}{2\sqrt{2}})$ **51.** two: $(r, \theta) = (0, 0)$, $\approx (1.376, 1.376)$ **53.** three: $(r, \theta) = (0, 0)$, $(\frac{\sqrt{3}}{2}, \frac{2\pi}{3})$, $(\frac{\sqrt{3}}{2}, \frac{\pi}{3})$

Exercises 12.3 (p. 778)
1. $\frac{x^3}{48}$ **3.** 38π **5.** $\frac{49\pi}{4}$ **7.** $\frac{1}{4}$ **9.** $\frac{\pi}{4}$ **11.** 6π **13.** 4π **15.** $\frac{\pi}{4}$
17. 1 **19.** π **21.** $\frac{\pi}{8} - \frac{1}{4}$ **23.** $\frac{\pi}{16}$ **25.** $\frac{3\pi}{2} - 4$ **27.** $\frac{3\pi}{2} - 2\sqrt{2}$
29. $\frac{4\pi}{3} - \sqrt{3}$ **31.** $\frac{3\pi}{4} - \sqrt{2}$ **33.** $\pi - \frac{3\sqrt{3}}{2}$ **35.** $\frac{\pi}{8} - \frac{1}{4}$ **37.** (i) $\frac{\pi}{4}$
41. $\frac{44\pi}{9}$ **43.** ≈ 13.3246 **45.** ≈ 3.5681

Exercises 12.4 (p. 785)
1. parabola, opening right

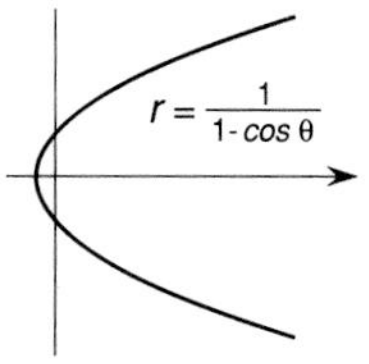

3. ellipse, $e = \frac{1}{2}$, major axis horizontal

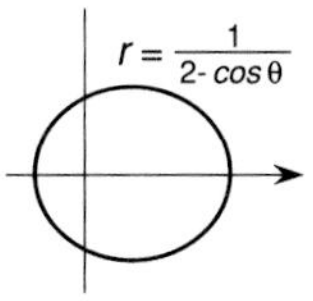

5. ellipse, $e = \frac{1}{2}$, major axis horizontal

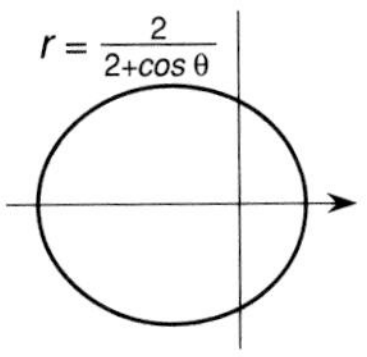

7. hyperbola, $e = 2$, transverse axis horizontal

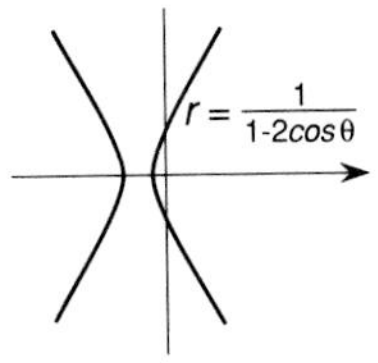

9. parabola, opening left

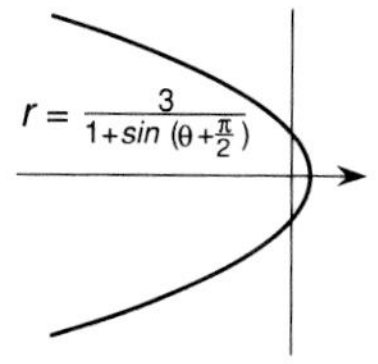

11. circle around $(x, y) = (\frac{1}{2}, 0)$, radius $\frac{1}{2}$

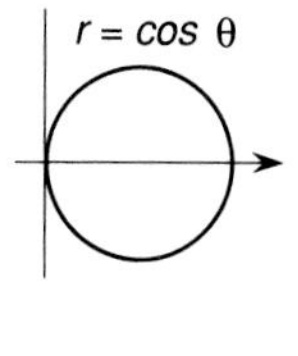

13. hyperbola, $e = 2$, transverse axis vertical

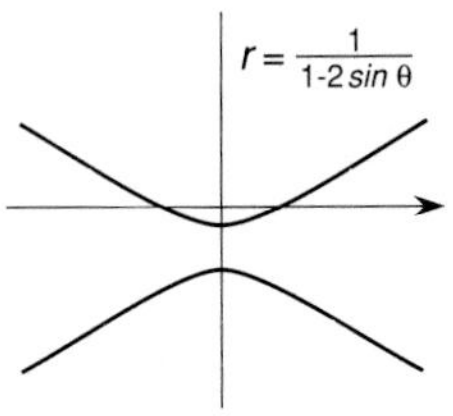

15. horizontal line $y = 4$

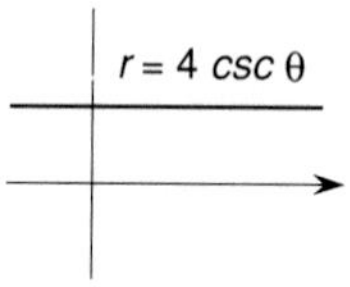

17. circle around origin, radius 6

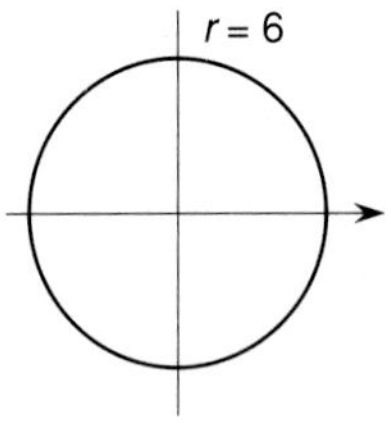

19. line $x + y = 2\sqrt{2}$

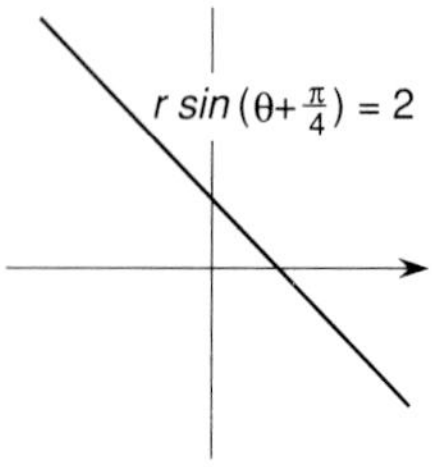

21. $r = \frac{pe}{1 + e\cos\theta}$ **23.** $r = \frac{1}{1 - \sqrt{2}\cos\theta}$ **25.** $\frac{\pi}{3}$ **27.** $r = \frac{3}{\sqrt{2}}\sec(\theta + \frac{\pi}{4})$
29. ellipse, $e = \frac{1}{\sqrt{2}}$, major axis along $\theta = \frac{5\pi}{4}$

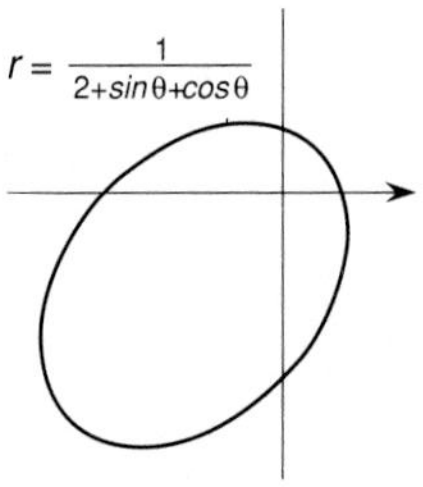

31. focus $(x, y) = (\frac{2}{3}, 0)$, directrix $x = \frac{5}{3}$ **33.** $(x, y) = (1, 1)$, distance $= \sqrt{2}$ **35.** $(r, \theta) = (\frac{2}{1 + 1/\sqrt{2}}, \frac{\pi}{4}), (\frac{2}{2 - 1/\sqrt{2}}, -\frac{3\pi}{4})$

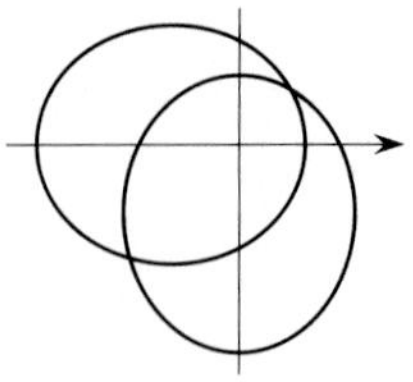

37. $\frac{8\pi}{3\sqrt{3}}$ **39.** **(i)** $\frac{1 - 2\cos\theta}{2\sin\theta}$ **(ii)** $(x, y) = (\frac{2}{3}, \frac{2\sqrt{3}}{3})$

Exercises 12.5 (p. 793)

1. $(x, y) = (5\cos t, 5\sin t)$; $0 \le t \quad 2\pi$ **3.** $(x, y) = (\cos^2 t, \cos t \sin t)$; $-\frac{\pi}{2} \le t < \frac{\pi}{2}$ **5.** $(x, y) = (4t, 3t - \frac{7}{4})$; $t \in (-\infty, \infty)$ **7.** $(x, y) = (3, t)$; $t \in (-\infty, \infty)$ **9.** $(r, \theta) = (\sqrt{\cos 2t}, t)$; $t \in [-\frac{\pi}{4}, \frac{\pi}{4}) \cup [\frac{3\pi}{4}, \frac{5\pi}{4})$, or $(x, y) = (\cos t\sqrt{\cos 2t}, \sin t\sqrt{\cos 2t})$; $t \in [-\frac{\pi}{4}, \frac{\pi}{4}) \cup [\frac{3\pi}{4}, \frac{5\pi}{4})$ **11.** ≈ 0.99986 **13.** 0.24 **15.** **(i)** one answer: $(x, y) = (6\cos(90\pi t), -6\sin(90\pi t))$, **(ii)** $(x, y) = (\frac{3}{2}\cos(90\pi t), -\frac{3}{2}\sin(90\pi t))$ **17.** $(x, y) = (-a\cosh t, b\sinh t)$; $t \in (-\infty, \infty)$, or $(x, y) = (-a\sqrt{1 + t^2/b^2}, t)$ $t \in (-\infty, \infty)$
19. $(x, y) = (t, \sqrt{9 - t^2})$; $t \in [-3, 3]$ **21.** $y = x^2 - 1$
23. $4x - y = 29$

Chapter 12 Review Exercises (p. 794)

1. straight line, slope = 1

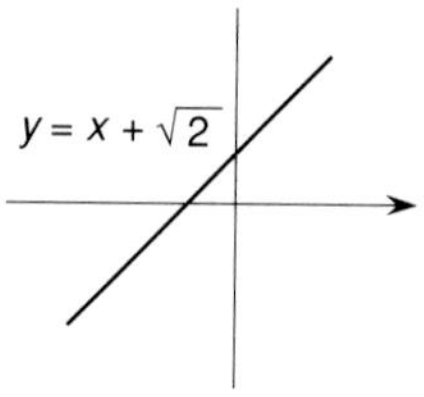

3. ellipse, $e = \frac{1}{3}$

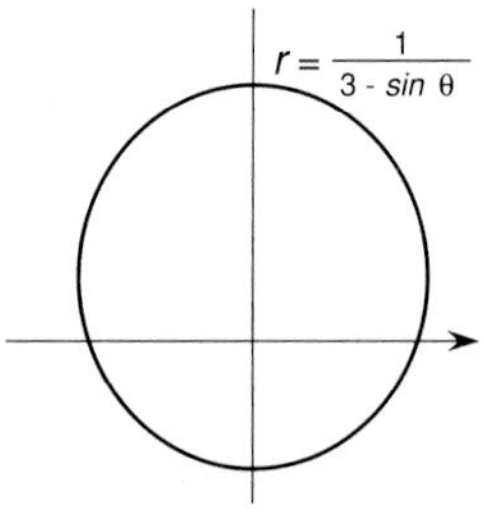

5. lemniscate, "transverse axis" along $\theta = \frac{\pi}{8}$

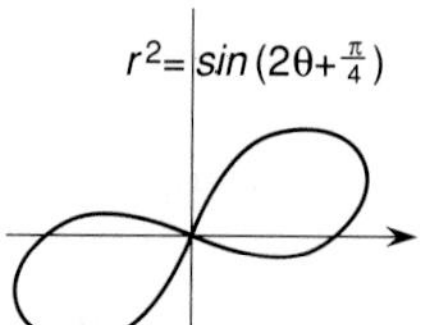

7. cardioid

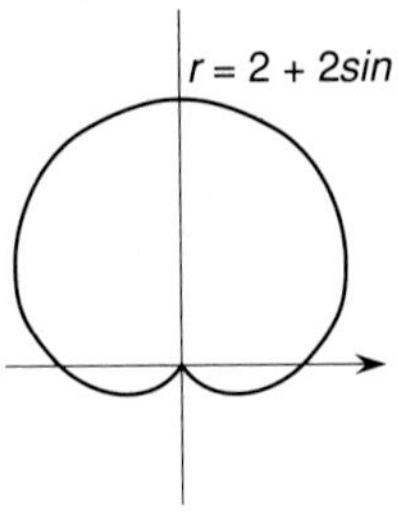

9. two half-lines, $\theta = \frac{\pi}{6}, \frac{5\pi}{6}$

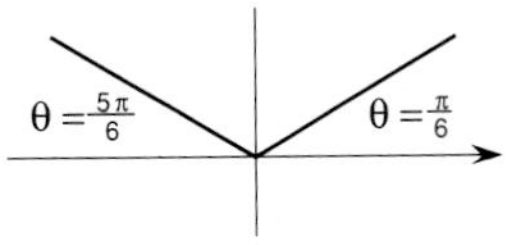

11. $(r, \theta) = (2k\pi, 2k\pi)$, $k = 0, 1, 2, \ldots$; i.e., $(x, y) = (2k\pi, 0)$

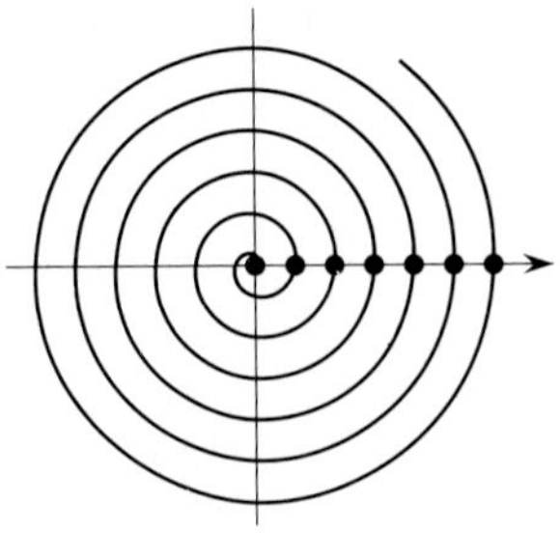

13. $(r, \theta) = (1, -\frac{\pi}{4})$, i.e., $(x, y) = (\frac{1}{\sqrt{2}}, -\frac{1}{\sqrt{2}})$

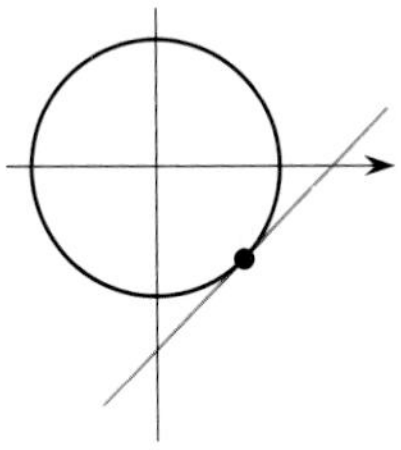

15. $(r, \theta) = (\frac{1}{2}, \theta)$, with $\theta = \pm\frac{\pi}{6}, \pm\frac{5\pi}{6}, \pm\frac{\pi}{3}, \pm\frac{2\pi}{3}$, i.e., $(x, y) = (\pm\frac{\sqrt{3}}{4}, \pm\frac{1}{4}), (\pm\frac{1}{4}, \pm\frac{\sqrt{3}}{4})$

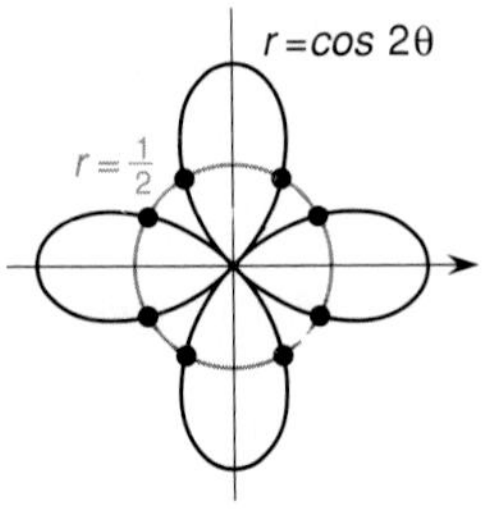

17. $(r, \theta) = (\frac{1}{\sqrt{2}}, \pm\frac{\pi}{6}), (\frac{1}{\sqrt{2}}, \pm\frac{5\pi}{6})$, i.e., $(x, y) = (\pm\frac{\sqrt{3}}{2\sqrt{2}}, \pm\frac{1}{2\sqrt{2}})$

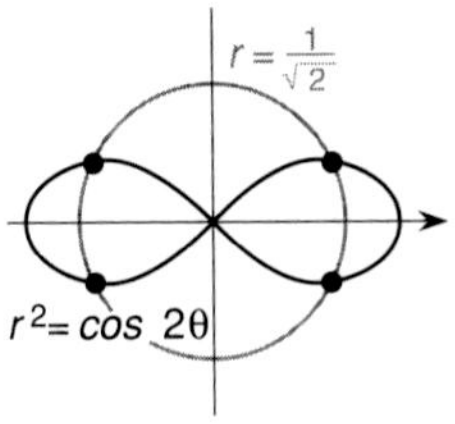

19. $(r, \theta) = (0, 0)$

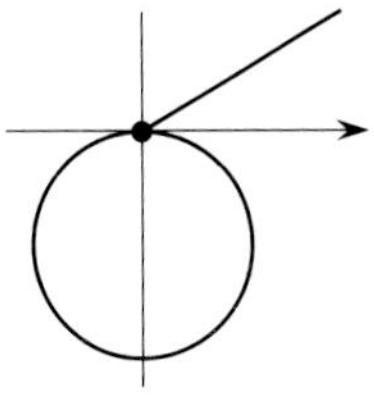

21. $\frac{2\pi}{3} - \frac{\sqrt{3}}{2}$ **23.** $\frac{5\pi}{24} - \frac{\sqrt{3}}{4}$ **25.** $\frac{9\pi}{2} - 4\sqrt{2}$ **27.** 0 **29.** $\frac{3\pi}{2}$
31. $(x, y) = (t, 2t^2 - 3)$; $t \in (-\infty, \infty)$ **33.** $(r, \theta) = (\cos 2t, t)$; $t \in [0, 2\pi)$, or $(x, y) = (\cos(2t)\cos t, \cos(2t)\sin t)$; $t \in [0, 2\pi)$
35. $(x, y) = (t, 2)$, $t \in (-\infty, \infty)$ **37.** $(x, y) = (4t, -5t - \frac{3}{4})$; $t \in (-\infty, \infty)$ **39.** $(r, \theta) = (-3 + 3\sin t, t)$; $t \in [0, 2\pi)$, or $(x, y) = (-3\cos t + 3\cos t \sin t, -3\sin t + 3\sin^2 t)$; $t \in [0, 2\pi)$
41. convex limaçon

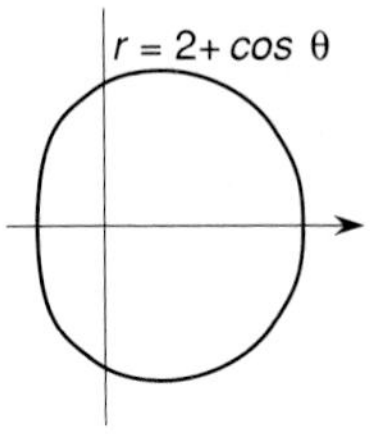

43. three-leaf "clover," six if we allow $r < 0$

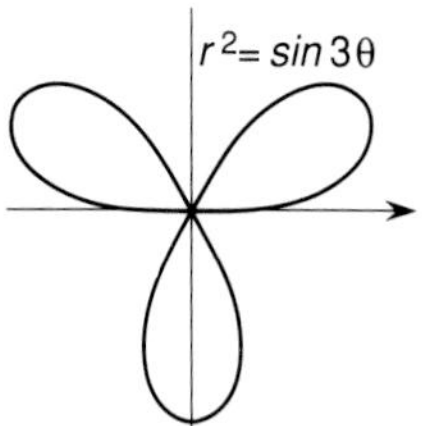

45. two lines $x = 1, y = -1$

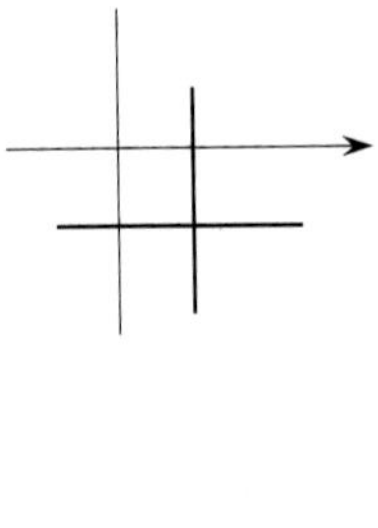

47. tightening spiral **49.** solid flower head with six lobes

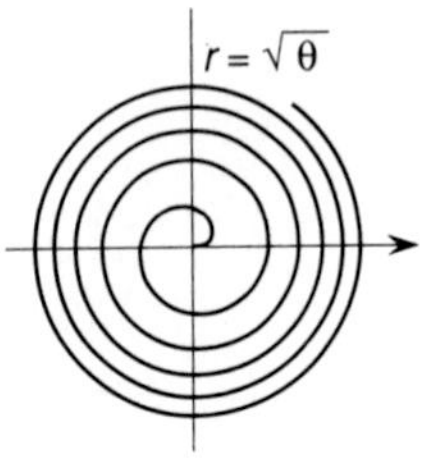

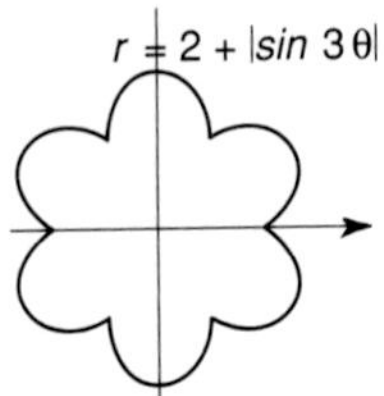

51. yes **53.** cos or sin in denominator
55. $r = \frac{1}{2 - \cos\theta}$, $r = \frac{3}{2 + \cos\theta}$
57.

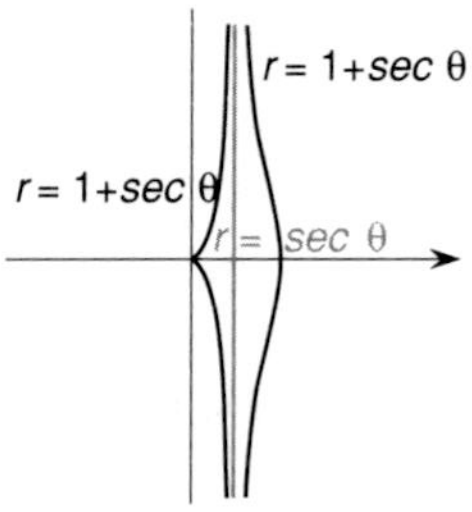

Exercises 13.1 (p. 802)
1. $(x, y) = (4t, t)$; $t \in (-\infty, \infty)$, traced once **3.** $(x, y) = (t^2 + 2, t)$; $t \in (-\infty, \infty)$, once **5.** $(x, y) = (2\cos t, \sqrt{3}\sin t)$; $t \in [0, 2\pi)$, once **7.** $(t, 2)$; $t \in (-\infty, \infty)$, once **9.** $(x, y) = (t, 3 - 2\sin t)$; $t \in (-\infty, \infty)$ once **11.** $x^2 + y^2 = 1$, circle, once around clockwise **13.** $4x + 3y + 2 = 0$, the half-line with $x \geq -5$; point moves in from the right, turns around, and goes out to the right, tracing the half-line twice **15.** $2x + y = 11$, the half-line with $x \geq 4$, traced once from left to right **17.** $x^2 - 4x - 2xy + y^2 = 0$, or $4x = (x - y)^2$; parabola, lower part is traced out twice, from right to left and back again **19.** $(y - 1)^2 - x^2 = 1$; hyperbola, right half of lower branch traced once from left to right **21.** $(x, y) = (3 - t, 2)$, for $t \in [0, 3)$; $(0, 5 - t)$ for $t \in [3, 5)$; $(t - 5, 0)$ for $t \in [5, 8)$; $(3, t - 8)$ for $t \in [8, 10]$ **23.** $(x, y) = (2 + t, 0)$, for $t \in [0, 1)$; $(3, t - 1)$ for $t \in [1, 3)$; $(6 - t, 2)$ for $t \in [3, 6)$; $(0, 8 - t)$ for $t \in [6, 8)$; $(t - 8, 0)$ for $t \in [8, 10]$ **27.** parabola **29.** parabola

Exercises 13.2 (p. 813)
1. $(4, -3)$ **3.** $(4t - 1, 8t - 2)$ **5.** $(t/\sqrt{1 + t^2}, 1)$ **7.** $(2t, \frac{1}{1+t^2})$ **9.** $(-6\sin(3t), 6\cos(3t))$ **11. (i)** $(x, y) = (-1, 0), t = \frac{3\pi}{2} + 2k\pi$ **(ii)** $(0, -1), t = \pi + 2k\pi$ **13. (i)** none **(ii)** none **15. (i)** none **(ii)** $(2, 12), t = -1$ **17. (i)** none **(ii)** none **19. (i)** $(e^{-2}, 1), t = 1$ **(ii)** none **21.** 3π **23.** $10\sqrt{5} - 4\sqrt{2}$ **25.** 10 **27.** $\sinh 1 = \frac{1}{2}(e - \frac{1}{e})$ **29.** 32 **31. (i)** $t = \frac{1}{2}$ **(ii)** $5\sqrt{2}\pi$ km/hr **(iii)** $5\sqrt{2}\pi$ km/hr **33.** "infinity sign"; $t = k\pi, k = 0, \pm 1, \pm 2, \ldots$

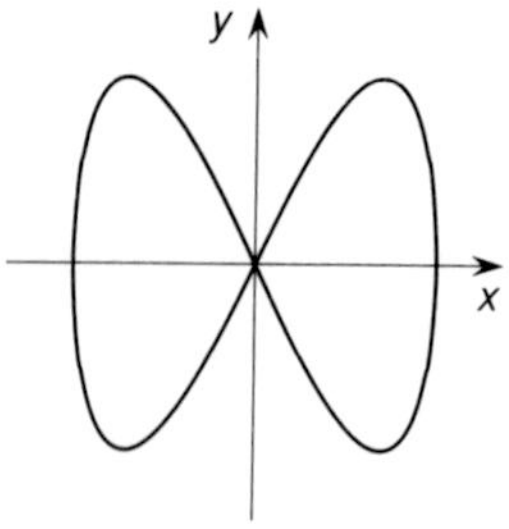

35. $t = k\pi, k = 0, \pm 1, \pm 2, \ldots$

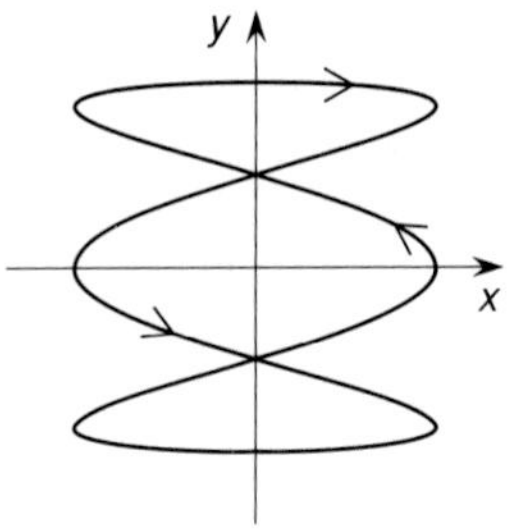

37. $(1, 1)$
41.

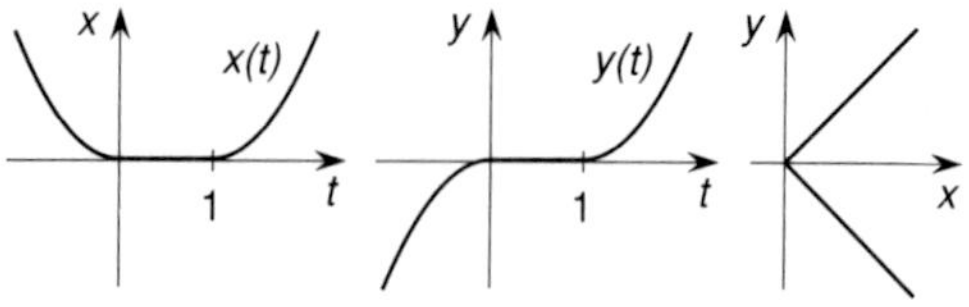

41. (iii) does not exist **43.** π **45.** 8 **47.** 6.5192 **49.** 7.1396 **51.** 9.1073

Exercises 13.3 (p. 820)
1. $16\sqrt{5} \approx 35.8$ ft/sec **3.** $\sqrt{298.9} \approx 17.3$ m/sec **5.** $5\sqrt{174} \approx 66$ ft **7.** $\sqrt{7808/3} \approx 51.0$ ft/sec **9.** $\frac{\sqrt{1335}}{16} \approx 2.28$ sec, or $\frac{\sqrt{2761}}{16} \approx 3.28$ sec **11.** $\approx 12.09° \approx 0.211$ rad, or $77.91°$ **13.** ≈ 609 ft **15.** 6.25 sec **17.** ≈ 1020 m **19.** ≈ 52.6 ft/sec **21.** ≈ 26.1 ft if $\theta \approx 11.8°$; ≈ 599 ft if $\theta \approx 78.2°$ (see Figure 13.3.6) **23.** $\theta = \frac{\pi}{2} = 90°$, i.e., straight up; height = 625 ft **25.** quadruple the range **27.** $16\sqrt{2} - 16 \approx 6.63$ ft farther **29.** plot $(x, y) = (200\cos(\theta)t, -16t^2 + 200\sin(\theta)t$

Chapter 13 Review Execises (p. 822)
1. $(x, y) = (7t, 2t - \frac{4}{7})$; $t \in (-\infty, \infty)$ **3.** $(x, y) = (t, -3)$; $t \in (-\infty, \infty)$ **5.** $(x, y) = (8 + 5t - 3t^2, t)$; $t \in (-\infty, \infty)$ **7.** $(x, y) = (2\cos t, \frac{1}{\sqrt{2}}\sin t)$; $t \in [0, 2\pi]$ **9.** $(x, y) = (t, \frac{1}{t})$; $t \in (-\infty, 0) \cup (0, \infty)$ **11.** $(x, y) = (\sqrt{1 + 2t^2}, t)$; $t \in (-\infty, \infty)$, or $(x, y) = (\cosh t, \frac{1}{\sqrt{2}}\sinh t)$; $t \in (-\infty, \infty)$ **13.** $(x, y) = (\sqrt{6}\cos\theta, \sqrt{\frac{5}{2}}\sin\theta)$; $\theta \in [-\frac{\pi}{2}, \frac{\pi}{2}]$ **15.** $(x, y) = (t, -\frac{3}{2}\sqrt{4 + x^2})$; $t \in (-\infty, \infty)$ **17.** $(x, y) = (t, |t|)$; $t \in (-\infty, \infty)$ **19.** $(x, y) = (\sqrt{t}, t)$; $t \in [0, \infty)$, or (t, t^2); $t \in [0, \infty)$ **21.** $x = 3(y + 4)^2$, parabola **23.** $9x^2 + y^2 = 9$, ellipse **25.** $x^2 + 4y^2 = 1$; ellipse **27.** $\frac{y^2}{4} - \frac{x^2}{9} = 1$, hyperbola **29.** $x + y = 1$; line segment from $(1, 0)$ to $(0, 1)$, omitting the endpoint $(0, 1)$ **31.** $(5, -1)$ **33.** $(2e^{2t}, 2)$ **35.** $(\cos t, 3\cos(3t))$ **37.** $(2t\cos(t^2 + 2), 2\cos t\sin t)$ **39.** $(\frac{3}{1 + 9t^2}, \frac{1}{\sqrt{1 - t^2}})$ **41.** $\frac{1}{27}(13\sqrt{13} - 8)$ **43.** $\frac{9}{8}$ **45.** π **47.** $2(1 + e^2)^{3/2} - 4\sqrt{2}$ **49.** $\frac{3886}{9375}\sqrt{29} + \frac{128}{9375}$ **51.** $t = \pm 1, \pm 2, \pm 3, \ldots$ The corresponding points are $(3n^3, \cos \pi n)$, $n = \pm 1, \pm 2, \pm 3, \ldots$ **53.** $t = -1$, $(x, y) = (-6, -9)$ **55.** 9 ft **57.** 2 sec **59.** $\sqrt{12304} \approx 111$ ft/sec **61.** $(x, y) = (-1 + \cos\theta, 2 + \sin\theta)$; $\theta \in [0, 2\pi]$ **63.** $(x, y) = (t, 2 + 3\sqrt{t^2 + 2t + 2})$; $t \in (-\infty, \infty)$, or $(-1 + \sinh t, 2 + 3\cosh t)$, $t \in (-\infty, \infty)$ **65.** $(x, y) = (2\cos t, \sin t)$; $t \in (\arctan(2) - \pi, \arctan(2))$ **67. (ii)** if $y'(a) \neq 0$, the tangent is vertical **69.** equal **71.** ≈ 0.905 **73.** $\approx 2.701\,29$

Index

Boldface page numbers indicate definitions.

Photo Credits

Page 19: The Bettmann Archive; Page 159: Stock•Boston/Daemmrich; Page 303: Aegma Infrared Systems/Science Photo Library; Page 311: Photo Researchers, Inc.; Page 317: Stock•Boston/John Elk; Page 317: Culver Pictures; Page 327: The Bettmann Archive; Page 331: Art Resource; Page 331: The Bettmann Archive; Page 359: The Bettmann Archive; Page 363: The Bettmann Archive; Page 453: Secchi-Lecaque/Roussel-UCLAF/CNRI/Science Photo Library; Page 513: Culver Pictures; Page 527: Runk/Schoenberger from Grant Heilman; Page 530: A.B. Dowsett/Science Photo Library; Page 534: Peter Arnold, Inc./Dennis Coleman; Page 542: The Picture Cube/Robert W. Ginn; Page 733: Architect of the Capitol; Page 734: The Picture Cube 1991/Ed Malitsky; Page 819: Stock•Boston/John Elk III 1985

All photographs of DESIGN CAD 3D are courtesy of American Small Business Computers.

Derivatives

$$\frac{d}{dx}(x^n) = nx^{x-1} \qquad \frac{d}{dx}(constant) = 0 \qquad \frac{d}{dx}(fg) = fg' + f'g$$

$$\frac{d}{dx}\left(\frac{f}{g}\right) = \frac{f'g - g'f}{g^2} \qquad \frac{d}{dx}(\sin x) = \cos x \qquad \frac{d}{dx}(\cos x) = -\sin x$$

$$\frac{d}{dx}(\tan x) = \sec^2 x \qquad \frac{d}{dx}(\cot x) = -\csc^2 x \qquad \frac{d}{dx}(\sec x) = \sec x \tan x$$

$$\frac{d}{dx}(\csc x) = -\csc x \cot x \qquad \frac{d}{dx}(\arctan x) = \frac{1}{1 + x^2} \qquad \frac{d}{dx}(\arcsin x) = \frac{1}{\sqrt{1 - x^2}}$$

$$\frac{d}{dx}(\text{arcsec } x) = \frac{1}{x\sqrt{1 - x^2}} \qquad \frac{d}{dx}(\ln x) = \frac{1}{x} \qquad \frac{d}{dx}(e^x) = e^x$$

$$\frac{d}{dx}(\cosh x) = \sinh x \qquad \frac{d}{dx}(\sinh x) = \cosh x \qquad \frac{d}{dx}(\tanh x) = \text{sech}^2 x$$

Integrals

$$\int x^n\,dx = \frac{1}{n+1}x^{n+1} + c, \quad n \neq -1 \qquad \int \frac{1}{x}\,dx = \ln|x| + c$$

$$\int \frac{1}{a^2 + x^2}\,dx = \frac{1}{a}\arctan\left(\frac{x}{a}\right) + c \qquad \int \frac{1}{\sqrt{a^2 - x^2}}\,dx = \arcsin\left(\frac{x}{|a|}\right) + c$$

$$\int \frac{1}{(a^2 + x^2)^2}\,dx = \frac{x}{2a^2(a^2 + x^2)} + \frac{1}{2a^3}\arctan\left(\frac{x}{a}\right) + c$$

$$\int \sqrt{a^2 - x^2}\,dx = \frac{x}{2}\sqrt{a^2 - x^2} + \frac{a^2}{2}\arcsin\left(\frac{x}{|a|}\right) + c$$

$$\int \frac{x}{a + bx}\,dx = \frac{x}{b} - \frac{a}{b^2}\ln|a + bx| + c \qquad \int \frac{dx}{\sqrt{x^2 \pm a^2}} = \ln\left|x + \sqrt{x^2 \pm a^2}\right| + c$$

$$\int \frac{1}{a^2 - x^2}\,dx = \frac{1}{2a}\ln\left|\frac{x + a}{x - a}\right| + c \qquad \int \frac{1}{x\sqrt{x^2 - a^2}}\,dx = \frac{1}{|a|}\text{arcsec}\left|\frac{x}{a}\right| + c$$

$$\int \ln x\,dx = x\ln x - x + c \qquad \int \sqrt{x^2 \pm a^2}\,dx = \frac{x}{2}\sqrt{x^2 \pm a^2} \pm \frac{a^2}{2}\ln\left|x + \sqrt{x^2 \pm a^2}\right| + c$$

If $b^2 - 4ac < 0$: $$\int \frac{dx}{ax^2 + bx + c} = \frac{2}{\sqrt{4ac - b^2}}\arctan\left(\frac{2ax + b}{\sqrt{4ac - b^2}}\right) + c$$

$$\int \sin x\,dx = -\cos x + c \qquad \int \cos x\,dx = \sin x + c$$

$$\int \tan x\,dx = \ln|\sec x| + c \qquad \int \sec^2 x\,dx = \tan x + c$$

$$\int \cos^2 x\,dx = \frac{x}{2} + \frac{1}{2}\cos x \sin x + c \qquad \int \sin^2 x\,dx = \frac{x}{2} - \frac{1}{2}\cos x \sin x + c$$

$$\int \cos^3 x\,dx = \frac{1}{3}\cos^2 x \sin x + \frac{2}{3}\sin x + c \qquad \int \sec x\,dx = \ln|\sec x + \tan x| + c$$